WATER ENCYCLOPEDIA

WATER QUALITY AND RESOURCE DEVELOPMENT

WATER ENCYCLOPEDIA

Editor-in-Chief
Jay Lehr, Ph.D.

Senior Editor
Jack Keeley

Associate Editor
Janet Lehr

Information Technology Director
Thomas B. Kingery III

Editorial Staff
Vice President, STM Books: **Janet Bailey**
Editorial Director, STM Encyclopedias: **Sean Pidgeon**
Executive Editor: **Bob Esposito**
Director, Book Production and Manufacturing: **Camille P. Carter**
Production Manager: **Shirley Thomas**
Senior Production Editor: **Kellsee Chu**
Illustration Manager: **Dean Gonzalez**
Editorial Program Coordinator: **Jonathan Rose**

WATER ENCYCLOPEDIA

WATER QUALITY AND RESOURCE DEVELOPMENT

Jay Lehr, Ph.D.
Editor-in-Chief

Jack Keeley
Senior Editor

Janet Lehr
Associate Editor

Thomas B. Kingery III
Information Technology Director

The *Water Encyclopedia* is available online at
http://www.mrw.interscience.wiley.com/eow/

A John Wiley & Sons, Inc., Publication

Copyright © 2005 by John Wiley & Sons, Inc. All rights reserved.

Published by John Wiley & Sons, Inc., Hoboken, New Jersey.
Published simultaneously in Canada.

No part of this publication may be reproduced, stored in a retrieval system, or transmitted in any form or by any means, electronic, mechanical, photocopying, recording, scanning, or otherwise, except as permitted under Section 107 or 108 of the 1976 United States Copyright Act, without either the prior written permission of the Publisher, or authorization through payment of the appropriate per-copy fee to the Copyright Clearance Center, Inc., 222 Rosewood Drive, Danvers, MA 01923, 978-750-8400, fax 978-646-8600, or on the web at www.copyright.com. Requests to the Publisher for permission should be addressed to the Permissions Department, John Wiley & Sons, Inc., 111 River Street, Hoboken, NJ 07030, (201) 748-6011, fax (201) 748-6008.

Limit of Liability/Disclaimer of Warranty: While the publisher and author have used their best efforts in preparing this book, they make no representations or warranties with respect to the accuracy or completeness of the contents of this book and specifically disclaim any implied warranties of merchantability or fitness for a particular purpose. No warranty may be created or extended by sales representatives or written sales materials. The advice and strategies contained herein may not be suitable for your situation. You should consult with a professional where appropriate. Neither the publisher nor author shall be liable for any loss of profit or any other commercial damages, including but not limited to special, incidental, consequential, or other damages.

For general information on our other products and services please contact our Customer Care Department within the U.S. at 877-762-2974, outside the U.S. at 317-572-3993 or fax 317-572-4002.

Wiley also publishes its books in a variety of electronic formats. Some content that appears in print, however, may not be available in electronic format.

Library of Congress Cataloging-in-Publication Data is available.

Lehr, Jay
　Water Encyclopedia: Water Quality and Resource Development

ISBN 0-471-73686-4
ISBN 0-471-44164-3 (Set)

Printed in the United States of America

10 9 8 7 6 5 4 3 2 1

CONTENTS

Preface ix
Contributors xi

Water Quality Control

Acid Mine Drainage—Extent and Character	1
The Control of Algal Populations in Eutrophic Water Bodies	2
Arsenic Compounds in Water	7
Arsenic Health Effects	15
Background Concentration of Pollutants	18
Waterborne Bacteria	20
Water Assessment and Criteria	24
Physiological Biomarkers and the Trondheim Biomonitoring System	28
Biomarkers, Bioindicators, and the Trondheim Biomonitoring System	29
Active Biomonitoring (ABM) by Translocation of Bivalve Molluscs	33
Biochemical Oxygen Demand and Other Organic Pollution Measures	37
Biodegradation	41
Bioluminescent Biosensors for Toxicity Testing	45
Biomanipulation	50
Genomic Technologies in Biomonitoring	58
Macrophytes as Biomonitors of Trace Metals	64
Biosorption of Toxic Metals	68
Bromide Influence on Trihalomethane and Haloacetic Acid Formation	74
Activated Carbon: Ion Exchange and Adsorption Properties	79
Activated Carbon—Powdered	86
Chlorination	88
Chlorination By-Products	91
Classification and Environmental Quality Assessment in Aquatic Environments	94
Coagulation and Flocculation in Practice	98
Colloids and Dissolved Organics: Role in Membrane and Depth Filtration	99
Column Experiments in Saturated Porous Media Studying Contaminant Transport	103
Cytochrome P450 Monooxygenase as an Indicator of PCB/Dioxin-Like Compounds in Fish	106
Water Related Diseases	111
Dishwashing Water Quality Properties	112
Disinfection By-Product Precursor Removal from Natural Waters	115
Alternative Disinfection Practices and Future Directions for Disinfection By-Product Minimization	118
Water Quality Aspects of Dredged Sediment Management	122
The Economics of Water Quality	127
Understanding *Escherichia Coli* O157:H7 and the Need for Rapid Detection in Water	136
Eutrophication and Organic Loading	142
Trace Element Contamination in Groundwater of District Hardwar, Uttaranchal, India	143
Iron Bacteria	149
Cartridge Filters for Iron Removal	152
Irrigation Water Quality in Areas Adjoining River Yamuna At Delhi, India	155
Water Sampling and Laboratory Safety	161
Municipal Solid Waste Landfills—Water Quality Issues	163
Land Use Effects on Water Quality	169
Monitoring Lipophilic Contaminants in the Aquatic Environment using the SPMD-TOX Paradigm	170
Use of Luminescent Bacteria and the *Lux* Genes For Determination of Water Quality	172
Water Quality Management	176
Water Quality Management and Nonpoint Source Control	184
Water Quality Management in an Urban Landscape	189
Water Quality Management in the U.S.: History of Water Regulation	193
Water Quality Management in a Forested Landscape	199
Trace Metal Speciation	202
Metal Ion Humic Colloid Interaction	205
Heavy Metal Uptake Rates Among Sediment Dwelling Organisms	211
Methemoglobinemia	219
Microbial Activities Management	223
Microbial Dynamics of Biofilms	228
Microbial Enzyme Assays for Detecting Heavy Metal Toxicity	233
Microbial Forms in Biofouling Events	239
Microbiological Quality Control in Distribution Systems	243
Water Quality Models for Developing Soil Management Practices	248
Water Quality Modeling—Case Studies	255
Field Sampling and Monitoring of Contaminants	263
Water Quality Models: Chemical Principles	269
Water Quality Models: Mathematical Framework	273
Environmental Applications with Submitochondrial Particles	278
Interest in the Use of an Electronic Nose for Field Monitoring of Odors in the Environment	281
Oil-Field Brine	284
Oil Pollution	290
Indicator Organisms	292
pH	294
Perchloroethylene (PCE) Removal	299
A Primer on Water Quality	301
Overview of Analytical Methods of Water Analyses With Specific Reference to EPA Methods for Priority Pollutant Analysis	304
Source-Water Protection	311
Protozoa in Water	313

Water Quality	314	California—Continually the Nation's Leader in Water Use	478
Water Quality	316	Lessons from the Rising Caspian	480
Emerging and Recalcitrant Compounds in Groundwater	316	Institutional Aspects of Water Management in China	484
Road Salt	319	Will Water Scarcity Limit China's Agricultural Potential?	488
Review of River Water Quality Modeling Software Tools	325	Water and Coastal Resources	489
River Water Quality Calibration	331	Water Use Conservation and Efficiency	489
Salmonella: Monitoring and Detection in Drinking Water	337	Conservation of Water	495
Lysimeter Soil Water Sampling	340	The Development of American Water Resources: Planners, Politicians, and Constitutional Interpretation	498
Regulatory and Security Requirements for Potable Water	343	Water Markets: Transaction Costs and Institutional Options	499
A Weight of Evidence Approach to Characterize Sediment Quality Using Laboratory and Field Assays: An Example For Spanish Coasts	350	Averting Water Disputes	501
Remediation and Bioremediation of Selenium-Contaminated Waters	355	Water Supply and Water Resources: Distribution System Research	509
Shellfish Growing Water Classification	360	Drought in the Dust Bowl Years	511
Sorptive Filtration	362	Drought Management Planning	514
Quality of Water in Storage	367	Drought and Water Supply Management	515
Quality of Water Supplies	370	Assessment of Ecological Effects in Water-Limited Environments	516
The Submitochondrial Particle Assay as a Biological Monitoring Tool	376	Reaching Out: Public Education and Community Involvement in Groundwater Protection	518
Microscale Test Relationships to Responses to Toxicants in Natural Systems	379	Integration of Environmental Impacts into Water Resources Planning	520
Toxicity Identification Evaluation	380	The Expansion of Federal Water Projects	522
Whole Effluent Toxicity Controls	382	Flood Control History in the Netherlands	524
Development and Application of Sediment Toxicity Tests for Regulatory Purposes	383	Food and Water in an Emergency	526
Algal Toxins in Water	387	Water Demand Forecasting	529
Ground Water Quality in Areas Adjoining River Yamuna at Delhi, India	392	Remote Sensing and GIS Application in Water Resources	531
Chlorine Residual	398	Globalization of Water	536
Source Water Quality Management	399	Water Science Glossary of Terms	541
Dose-Response of Mussels to Chlorine	401	Harvesting Rainwater	548
Metallothioneins as Indicators of Trace Metal Pollution	406	Urban Water Resource and Management in Asia: Ho Chi Minh City	552
Amphipod Sediment Toxicity Tests	408	Hydropower—Energy from Moving Water	554
Ciliated Protists as Test Organisms in Toxicity Assessment	413	Water Markets in India: Economic and Institutional Aspects	555
SOFIE: An Optimized Approach for Exposure Tests and Sediment Assays	418	Water Resources of India	559
		Water Infrastructure and Systems	567
Passive Treatment of Acid Mine Drainage (Wetlands)	423	Overview and Trends in the International Water Market	568
Biomarkers and Bioaccumulation: Two Lines of Evidence to Assess Sediment Quality	426	Best Management Practices for Water Resources	570
Lead and its Health Effects	432	Integrated Water Resources Management (IWRM)	574
Microbial Detection of Various Pollutants as an Early Warning System for Monitoring of Water Quality and Ecological Integrity of Natural Resources, in Russia	440	Management of Water Resources for Drought Conditions	576
		Water Resources Management	586
Luminescent Bacterial Biosensors for the Rapid Detection of Toxicants	453	NASA Helping to Understand Water Flow in the West	587
Development and Application of Sediment Toxicity Test for Regulatory Purposes	458	Transboundary Water Conflicts in the Nile Basin	590
E_h	464	Planning and Managing Water Infrastructure	594
		Application of the Precautionary Principle to Water Science	595

Water Resource Development and Management

		Water Pricing	603
Water Resources Challenges in the Arab World	470	Spot Prices, Option Prices, and Water Markets	606
Effluent Water Regulations in Arid Lands	475	Water Managed in the Public Trust	608

Water Recycling and Reuse: The Environmental Benefits	610	How We Use Water in These United States	650
State and Regional Water Supply	613	Valuing Water Resources	653
River Basin Decisions Support Systems	619	Water—Here, There, and Everywhere in Canada	656
Water Resource Sustainability: Concepts and Practices	624	Water Conservation—Every Drop Counts in Canada	660
The Provision of Drinking Water and Sanitation in Developing Countries	630	Ecoregions: A Spatial Framework for Environmental Management	667
Sustainable Management of Natural Resources	633	Flood of Portals on Water	668
Sustainable Water Management On Mediterranean Islands: Research and Education	638	Fuzzy Criteria for Water Resources Systems Performance Evaluation	674
Meeting Water Needs in Developing Countries with Tradable Rights	643	Participatory Multicriteria Flood Management	678
		Water Resources Systems Analysis	683
Water Use in the United States	645	Index	689

PREFACE

Cities, towns, states, and nations must manage their water resources wisely from both a quality and a quantitative perspective.

If we do otherwise and manage them with a narrow perspective, the public's needs will not be adequately met. In this volume of the *Water Encyclopedia*, authors from around the world have described a myriad of problems relating to individual water bodies as well as to geographic water resources and their management dilemmas.

Humans and other living creatures contribute to our water quality problems. Neither can be fully controlled. Even the nature of contaminant sources and programs for their elimination can be difficult to design. This volume contains the best and brightest ideas and case studies relating to the areas of water quality and resource management problems.

Quality problems deal with a diverse suite of subjects ranging widely from acid mine drainage to biosorption, colloids, eutrophication, protozoa, and recalcitrant compounds. Resource management features drought studies, flood control, river basin management, perennial overdraft, water banking, and a host of other subjects.

The perspective of scientists from nearly every continent of the world offers a truly catholic view of attitudes and biases harbored in different regions and how they affect scientific and regulatory outcomes.

The editors cannot imagine what has been left out, but we know of course that readers will at times come up short of finding an exact match to a problem they face. We hope they will contact us at our website and allow us the opportunity of adding additional subjects to our encyclopedia. At the same time, the reader will understand that many subjects in the area of water quality may have been addressed in our Surface Water category. It was often difficult to determine where an investigator would be more likely to look for a piece of information. (The complete index of all five volumes appears in the Ground Water volume as well as on our website.)

We trust all users of this encyclopedia will find it detailed, informative, and interesting. Not only are a wide range of subjects treated, but authors choose varying approaches to presenting their data to readers who may be professionals, students, researchers, as well as individuals simply satisfying their intellectual curiosity. We hope we are successfully serving all of these populations in some useful way.

Jay Lehr
Jack Keeley

CONTRIBUTORS

Absar Alum, *Arizona State University, Tempe, Arizona,* Water Quality Management in the U.S.: History of Water Regulation

Mohammad N. Almasri, *An-Najah National University, Nablus, Palestine,* Best Management Practices for Water Resources

Linda S. Andrews, *Mississippi State University, Biloxi, Mississippi,* Shellfish Growing Water Classification, Chlorine Residual

Hannah Aoyagi, *University of California, Irvine, California,* Cytochrome P450 Monooxygenase as an Indicator of PCB/Dioxin-Like Compounds in Fish

Robert Artinger, *TÜV Industrie Service GmbH, München, Germany,* Column Experiments in Saturated Porous Media Studying Contaminant Transport

Mukand Singh Babel, *Asian Institute of Technology, Pathumthani, Thailand,* Conservation of Water, Integrated Water Resources Management (IWRM)

Mark Bailey, *Centre for Ecology and Hydrology–Oxford, Oxford, United Kingdom,* Bioluminescent Biosensors for Toxicity Testing

Shimshon Balanson, *Cleveland State University, Cleveland, Ohio,* Macrophytes as Biomonitors of Trace Metals

Christine L. Bean, *University of New Hampshire, Durham, New Hampshire,* Protozoa in Water

Jennifer Bell, *Napier University, Edinburgh, United Kingdom,* Bioluminescent Biosensors for Toxicity Testing

Lieven Bervoets, *University of Antwerp, Antwerp, Belgium,* Active Biomonitoring (ABM) by Translocation of Bivalve Molluscs

J.M. Blasco, *Instituto de Ciencias Marinas de Andalucía, Cádiz, Spain,* A Weight of Evidence Approach to Characterize Sediment Quality Using Laboratory and Field Assays: An Example For Spanish Coasts

Ronny Blust, *University of Antwerp, Antwerp, Belgium,* Active Biomonitoring (ABM) by Translocation of Bivalve Molluscs

Marta Bryce, *CEPIS/PAHO, Delft, The Netherlands,* Flood of Portals on Water

Mario O. Buenfil-Rodriguez, *National University of Mexico, Cuernavaca, Morelos, Mexico,* Water Use Conservation and Efficiency

Jacques Buffle, *University of Geneva, Geneva, Switzerland,* Colloids and Dissolved Organics: Role in Membrane and Depth Filtration

Zia Bukhari, *American Water, Belleville, Illinois,* Understanding Escherichia Coli O157:H7 and the Need for Rapid Detection in Water

John Cairns, Jr., *Virginia Polytechnic Institute and State University, Blacksburg, Virginia,* Microscale Test Relationships to Responses to Toxicants in Natural Systems

Michael J. Carvan III, *University of Wisconsin–Milwaukee, Milwaukee, Wisconsin,* Genomic Technologies in Biomonitoring

M.C. Casado-Martínez, *Facultad de Ciencias del Mar y Ambientales, Cádiz, Spain,* A Weight of Evidence Approach to Characterize Sediment Quality Using Laboratory and Field Assays: An Example For Spanish Coasts

Tomás Ángel Del Valls Casillas, *Universidad de Cadiz, Cadiz, Spain,* Amphipod Sediment Toxicity Tests, Development and Application of Sediment Toxicity Test for Regulatory Purposes

Teresa A. Cassel, *University of California, Davis, California,* Remediation and Bioremediation of Selenium-Contaminated Waters

Augusto Cesar, *Universidad de Cadiz, Cadiz, Spain,* Amphipod Sediment Toxicity Tests

K.W. Chau, *The Hong Kong Polytechnic University, Hung Hom, Kowloon, Hong Kong,* Water Quality Models: Mathematical Framework

Paulo Chaves, *Water Resources Research Center, Kyoto University, Japan,* Quality of Water in Storage

Shankar Chellam, *University of Houston, Houston, Texas,* Bromide Influence on Trihalomethane and Haloacetic Acid Formation

X. Chris Le, *University of Alberta, Edmonton, Alberta, Canada,* Arsenic Compounds in Water

Russell N. Clayshulte, *Aurora, Colorado,* Water Quality Management in an Urban Landscape

Gail E. Cordy, *U.S. Geological Survey,* A Primer on Water Quality

Rupali Datta, *University of Texas, San Antonio, Texas,* Lead and its Health Effects

Joanna Davies, *Syngenta, Bracknell, Berkshire, United Kingdom,* The Control of Algal Populations in Eutrophic Water Bodies

Maria B. Davoren, *Dublin Institute of Technology, Dublin, Ireland,* Luminescent Bacterial Biosensors for the Rapid Detection of Toxicants

T.A. Delvalls, *Facultad de Ciencias del Mar y Ambientales, Cádiz, Spain,* Biomarkers and Bioaccumulation: Two Lines of Evidence to Assess Sediment Quality, A Weight of Evidence Approach to Characterize Sediment Quality Using Laboratory and Field Assays: An Example For Spanish Coasts

Nicolina Dias, *Centro de Engenharia Biológica, Braga, Portugal,* Ciliated Protists as Test Organisms in Toxicity Assessment

Galina Dimitrieva-Moats, *University of Idaho, Moscow, Idaho,* Microbial Detection of Various Pollutants as an Early Warning System for Monitoring of Water Quality and Ecological Integrity of Natural Resources, in Russia

Halanaik Diwakara, *University of South Australia, Adelaide, Australia,* Water Markets in India: Economic and Institutional Aspects

Francis G. Doherty, *AquaTox Research, Inc., Syracuse, New York,* The Submitochondrial Particle Assay as a Biological Monitoring Tool

Antonia A. Donta, *University of Münster, Centre for Environmental Research, Münster, Germany,* Sustainable Water Management On Mediterranean Islands: Research and Education

Timothy J. Downs, *Clark University, Worcester, Massachusetts,* Field Sampling and Monitoring of Contaminants, State and Regional Water Supply, Water Resource Sustainability: Concepts and Practices

Hiep N. Duc, *Environment Protection Authority, NSW, Bankstown, New South Wales Australia,* Urban Water Resource and Management in Asia: Ho Chi Minh City

Suzanne Du Vall Knorr, *Ventura County Environmental Health Division, Ventura, California,* Regulatory and Security Requirements for Potable Water

Sandra Dunbar, *Napier University, Edinburgh, United Kingdom,* Bioluminescent Biosensors for Toxicity Testing

Diane Dupont, *Brock University, St. Catharines, Ontario, Canada,* Valuing Water Resources

Michael P. Dziewatkoski, *Mettler-Toledo Process Analytical, Woburn, Massachusetts,* pH

Energy Information Administration—Department of Energy, Hydropower—Energy from Moving Water

Environment Canada, Water—Here, There, and Everywhere in Canada, Water Conservation—Every Drop Counts in Canada

Environmental Protection Agency, Water Recycling and Reuse: The Environmental Benefits

M. Eric Benbow, *Michigan State University, East Lansing, Michigan,* Road Salt

Teresa W.-M. Fan, *University of Louisville, Louisville, Kentucky,* Remediation and Bioremediation of Selenium-Contaminated Waters

Federal Emergency Management Agency, Food and Water in an Emergency

Huan Feng, *Montclair State University, Montclair, New Jersey,* Classification and Environmental Quality Assessment in Aquatic Environments

N. Buceta Fernández, *Centro de Estudios de Puertos y Costas, Madrid, Spain,* A Weight of Evidence Approach to Characterize Sediment Quality Using Laboratory and Field Assays: An Example For Spanish Coasts

Peter D. Franzmann, *CSIRO Land and Water, Floreat, Australia,* Microbial Activities Management

Christian D. Frazar, *Silver Spring, Maryland,* Biodegradation

Rajiv Gandhi Chair, *Jawaharlal Nehru University, New Delhi, India,* Oil Pollution

Suduan Gao, *USDA–ARS, Parlier, California,* E_h

Horst Geckeis, *Institut für Nukleare Entsorgung, Karlsruhe, Germany,* Metal Ion Humic Colloid Interaction

Robert Gensemer, *Parametrix, Corvallis, Oregon,* Effluent Water Regulations in Arid Lands

Mário Abel Gonçalves, *Faculdade de Ciências da Universidade de Lisboa, Lisboa, Portugal,* Background Concentration of Pollutants

Neil S. Grigg, *Colorado State University, Fort Collins, Colorado,* Planning and Managing Water Infrastructure, Drought and Water Supply

CONTRIBUTORS

Management, Drought Management Planning, Water Infrastructure and Systems, Water Resources Management

Håkan Håkanson, *University of Lund, Lund, Sweden,* Dishwashing Water Quality Properties

Carol J. Haley, *Virginia Water Resources Research Center,* Management of Water Resources for Drought Conditions

M.G.J Hartl, *Environmental Research Institute, University College Cork, Ireland,* Development and Application of Sediment Toxicity Tests for Regulatory Purposes

Roy C. Haught, *U.S. Environmental Protection Agency,* Water Supply and Water Resources: Distribution System Research

Joanne M. Hay, *Lincoln Ventures, Ltd., Lincoln, New Zealand,* Biochemical Oxygen Demand and Other Organic Pollution Measures

Richard M. Higashi, *University of California, Davis, California,* Remediation and Bioremediation of Selenium-Contaminated Waters

A.Y. Hoekstra, *UNESCO–IHE Institute for Water Education, Delft, The Netherlands,* Globalization of Water

Charles D.D. Howard, *Water Resources, Victoria, British Columbia, Canada,* River Basin Decisions Support Systems

Margaret S. Hrezo, *Radford University, Virginia,* Management of Water Resources for Drought Conditions

Enos C. Inniss, *University of Texas, San Antonio, Texas,* Perchloroethylene (PCE) Removal

James A. Jacobs, *Environmental Bio-Systems, Inc., Mill Valley, California,* Emerging and Recalcitrant Compounds in Groundwater

Chakresh K. Jain, *National Institute of Hydrology, Roorkee, India,* Water Quality Management, Trace Element Contamination in Groundwater of District Hardwar, Uttaranchal, India, Ground Water Quality in Areas Adjoining River Yamuna at Delhi, India, Irrigation Water Quality in Areas Adjoining River Yamuna At Delhi, India

Sanjay Kumar Jain, *National Institute of Hydrology, Roorkee, India,* Remote Sensing and GIS Application in Water Resources

Sharad K. Jain, *National Institute of Hydrology, Roorkee, Uttranchal, India,* Water Resources of India

H.A. Jenner, *KEMA Power Generation and Sustainables, Arnhem, The Netherlands,* Dose-Response of Mussels to Chlorine

Y. Jiang, *Hong Kong Baptist University, Kowloon, Hong Kong,* Algal Toxins in Water

B. Ji, *Hong Kong Baptist University, Kowloon, Hong Kong,* Algal Toxins in Water

N. Jiménez-Tenorio, *Facultad de Ciencias del Mar y Ambientales, Cádiz, Spain,* Biomarkers and Bioaccumulation: Two Lines of Evidence to Assess Sediment Quality

Zhen-Gang Ji, *Minerals Management Service, Herndon, Virginia,* Water Quality Modeling—Case Studies, Water Quality Models: Chemical Principles

Erik Johansson, *GS Development AB, Malmö, Sweden,* Dishwashing Water Quality Properties

B. Thomas Johnson, *USGS—Columbia Environmental Research Center, Columbia, Missouri,* Monitoring Lipophilic Contaminants in the Aquatic Environment using the SPMD-TOX Paradigm

Anne Jones-Lee, *G. Fred Lee & Associates, El Macero, California,* Water Quality Aspects of Dredged Sediment Management, Municipal Solid Waste Landfills—Water Quality Issues

Dick de Jong, *IRC International Water and Sanitation Centre, Delft, The Netherlands,* Flood of Portals on Water

Jagath J. Kaluarachchi, *Utah State University, Logan, Utah,* Best Management Practices for Water Resources

Atya Kapley, *National Environmental Engineering Research Institute, CSIR, Nehru Marg, Nagpur, India,* Salmonella: Monitoring and Detection in Drinking Water

I. Katsoyiannis, *Aristotle University of Thessaloniki, Thessaloniki, Greece,* Arsenic Health Effects

Absar A. Kazmi, *Nishihara Environment Technology, Tokyo, Japan,* Activated Carbon—Powdered, Chlorination

Keith O. Keplinger, *Texas Institute for Applied Environmental Research, Stephenville, Texas,* The Economics of Water Quality

Kusum W. Ketkar, *Jawaharlal Nehru University, New Delhi, India,* Oil Pollution

Ganesh B. Keremane, *University of South Australia, Adelaide, Australia,* Harvesting Rainwater

Rebecca D. Klaper, *University of Wisconsin–Milwaukee, Milwaukee, Wisconsin,* Genomic Technologies in Biomonitoring

Toshiharu Kojiri, *Water Resources Research Center, Kyoto University, Japan,* Quality of Water in Storage

Ken'ichirou Kosugi, *Kyoto University, Kyoto, Japan,* Lysimeter Soil Water Sampling

Manfred A. Lange, *University of Münster, Centre for Environmental Research, Münster, Germany,* Sustainable Water Management On Mediterranean Islands: Research and Education

Frédéric Lasserre, *Université Laval, Ste-Foy, Québec, Canada,* Water Use in the United States

N.K. Lazaridis, *Aristotle University, Thessaloniki, Greece,* Sorptive Filtration

Jamie R. Lead, *University of Birmingham, Birmingham, United Kingdom,* Trace Metal Speciation

G. Fred Lee, *G. Fred Lee & Associates, El Macero, California,* Water Quality Aspects of Dredged Sediment Management, Municipal Solid Waste Landfills—Water Quality Issues

Terence R. Lee, *Santiago, Chile,* Water Markets: Transaction Costs and Institutional Options, The Provision of Drinking Water and Sanitation in Developing Countries, Spot Prices, Option Prices, and Water Markets, Meeting Water Needs in Developing Countries with Tradable Rights

Markku J. Lehtola, *National Public Health Institute, Kuopio, Finland,* Microbiological Quality Control in Distribution Systems

Gary G. Leppard, *National Water Research Institute, Burlington, Ontario, Canada,* Colloids and Dissolved Organics: Role in Membrane and Depth Filtration

Mark LeChevallier, *American Water, Voorhees, New Jersey,* Understanding Escherichia Coli O157:H7 and the Need for Rapid Detection in Water

Nelson Lima, *Centro de Engenharia Biológica, Braga, Portugal,* Ciliated Protists as Test Organisms in Toxicity Assessment

Maria Giulia Lionetto, *Università di Lecce, Lecce, Italy,* Metallothioneins as Indicators of Trace Metal Pollution

Jody W. Lipford, *PERC, Bozeman, Montana, and Presbyterian College, Clinton, South Carolina,* Averting Water Disputes

Baikun Li, *Pennsylvania State University, Harrisburg, Pennsylvania,* Iron Bacteria, Microbial Dynamics of Biofilms, Microbial Forms in Biofouling Events

Rongchao Li, *Delft University of Technology, Delft, The Netherlands,* Transboundary Water Conflicts in the Nile Basin, Institutional Aspects of Water Management in China, Flood Control History in the Netherlands

Bryan Lohmar, *Economic Research Service, U.S. Department of Agriculture,* Will Water Scarcity Limit China's Agricultural Potential?

Inmaculada Riba López, *Universidad de Cadiz, Cadiz, Spain,* Amphipod Sediment Toxicity Tests

M.X. Loukidou, *Aristotle University of Thessaloniki, Thessaloniki, Greece,* Biosorption of Toxic Metals

Scott A. Lowe, *Manhattan College, Riverdale, New York,* Eutrophication and Organic Loading

G. Lyberatos, *University of Ioannina, Agrinio, Greece,* Cartridge Filters for Iron Removal

Kenneth M. Mackenthun, *Arlington, Virginia,* Water Quality

Tarun K. Mal, *Cleveland State University, Cleveland, Ohio,* Macrophytes as Biomonitors of Trace Metals

Philip J. Markle, *Whittier, California,* Toxicity Identification Evaluation, Whole Effluent Toxicity Controls

James T. Markweise, *Neptune and Company, Inc., Los Alamos, New Mexico,* Assessment of Ecological Effects in Water-Limited Environments, Effluent Water Regulations in Arid Lands

Pertti J. Martikainen, *University of Kuopio, Kuopio, Finland,* Microbiological Quality Control in Distribution Systems

M.L. Martín-Díaz, *Instituto de Ciencias Marinas de Andalucía, Cádiz, Spain,* A Weight of Evidence Approach to Characterize Sediment Quality Using Laboratory and Field Assays: An Example For Spanish Coasts, Biomarkers and Bioaccumulation: Two Lines of Evidence to Assess Sediment Quality

Maria del Carmen Casado Martínez, *Universidad de Cadiz, Cadiz, Spain,* Amphipod Sediment Toxicity Tests

K.A. Matis, *Aristotle University, Thessaloniki, Greece,* Sorptive Filtration

Lindsay Renick Mayer, *Goddard Space Flight Center, Greenbelt, Maryland,* NASA Helping to Understand Water Flow in the West

Mark C. Meckes, *U.S. Environmental Protection Agency,* Water Supply and Water Resources: Distribution System Research

Richard W. Merritt, *Michigan State University, East Lansing, Michigan,* Road Salt

Richard Meyerhoff, *CDM, Denver, Colorado,* Effluent Water Regulations in Arid Lands

J. Michael Wright, *Harvard School of Public Health, Boston, Massachusetts,* Chlorination By-Products

Cornelis J.H. Miermans, *Institute for Inland Water Management and Waste Water Treatment–RIZA, Lelystad, The Netherlands,* SOFIE: An Optimized Approach for Exposure Tests and Sediment Assays

Ilkka T. Miettinen, *National Public Health Institute, Kuopio, Finland,* Microbiological Quality Control in Distribution Systems

Dusan P. Miskovic, *Northwood University, West Palm Beach, Florida,* Oil-Field Brine

Diana Mitsova-Boneva, *University of Cincinnati, Cincinnati, Ohio,* Quality of Water Supplies

Tom Mohr, *Santa Clara Valley Water District, San Jose, California,* Emerging and Recalcitrant Compounds in Groundwater

M.C. Morales-Caselles, *Facultad de Ciencias del Mar y Ambientales, Cádiz, Spain,* Biomarkers and Bioaccumulation: Two Lines of Evidence to Assess Sediment Quality

National Drought Mitigation Center, Drought in the Dust Bowl Years

National Water-Quality Assessment (NAWQA) Program—U.S. Geological Survey, Source-Water Protection

Jennifer Nelson, *The Groundwater Foundation, Lincoln, Nebraska,* Reaching Out: Public Education and Community Involvement in Groundwater Protection

Anne Ng, *Swinburne University of Technology, Hawthorne, Victoria, Australia,* River Water Quality Calibration, Review of River Water Quality Modeling Software Tools

Jacques Nicolas, *University of Liege, Arlon, Belgium,* Interest in the Use of an Electronic Nose for Field Monitoring of Odors in the Environment

Ana Nicolau, *Centro de Engenharia Biológica, Braga, Portugal,* Ciliated Protists as Test Organisms in Toxicity Assessment

Diana J. Oakes, *University of Sydney, Lidcombe, Australia,* Environmental Applications with Submitochondrial Particles

Oladele A. Ogunseitan, *University of California, Irvine, California,* Microbial Enzyme Assays for Detecting Heavy Metal Toxicity, Cytochrome P450 Monooxygenase as an Indicator of PCB/Dioxin-Like Compounds in Fish

J. O'Halloran, *Environmental Research Institute, University College Cork, Ireland,* Development and Application of Sediment Toxicity Tests for Regulatory Purposes

Victor Onwueme, *Montclair State University, Montclair, New Jersey,* Classification and Environmental Quality Assessment in Aquatic Environments

Alper Ozkan, *Selcuk University, Konya, Turkey,* Coagulation and Flocculation in Practice

Neil F. Pasco, *Lincoln Ventures, Ltd., Lincoln, New Zealand,* Biochemical Oxygen Demand and Other Organic Pollution Measures

B.J.C. Perera, *Swinburne University of Technology, Hawthorne, Victoria, Australia,* River Water Quality Calibration, Review of River Water Quality Modeling Software Tools

Jim Philip, *Napier University, Edinburgh, United Kingdom,* Bioluminescent Biosensors for Toxicity Testing

Laurel Phoenix, *Green Bay, Wisconsin,* Source Water Quality Management, Water Managed in the Public Trust

Randy T. Piper, *Dillon, Montana,* Overview and Trends in the International Water Market

John K. Pollak, *University of Sydney, Lidcombe, Australia,* Environmental Applications with Submitochondrial Particles

Dörte Poszig, *University of Münster, Centre for Environmental Research, Münster, Germany,* Sustainable Water Management On Mediterranean Islands: Research and Education

Hemant J. Purohit, *National Environmental Engineering Research Institute, CSIR, Nehru Marg, Nagpur, India, Salmonella:* Monitoring and Detection in Drinking Water

Shahida Quazi, *University of Texas, San Antonio, Texas,* Lead and its Health Effects

S. Rajagopal, *Radboud University Nijmegen, Toernooiveld, Nijmegen, The Netherlands,* Dose-Response of Mussels to Chlorine

Krishna Ramanujan, *Goddard Space Flight Center, Greenbelt, Maryland,* NASA Helping to Understand Water Flow in the West

Lucas Reijnders, *University of Amsterdam, Amsterdam, The Netherlands,* Sustainable Management of Natural Resources

Steven J. Renzetti, *Brock University, St. Catharines, Ontario, Canada,* Water Demand Forecasting, Water Pricing, Valuing Water Resources

Martin Reuss, *Office of History Headquarters U.S. Army Corps of Engineers,* The Development of American Water Resources: Planners, Politicians, and Constitutional Interpretation, The Expansion of Federal Water Projects

I. Riba, *Facultad de Ciencias del Mar y Ambientales, Cádiz, Spain,* Biomarkers and Bioaccumulation: Two Lines of Evidence to Assess Sediment Quality, A Weight of Evidence Approach to Characterize Sediment Quality Using Laboratory and Field Assays: An Example For Spanish Coasts

Matthew L. Rise, *University of Wisconsin–Milwaukee, Milwaukee, Wisconsin,* Genomic Technologies in Biomonitoring

Arthur W. Rose, *Pennsylvania State University, University Park, Pennsylvania,* Acid Mine Drainage—Extent and Character, Passive Treatment of Acid Mine Drainage (Wetlands)

Barry H. Rosen, *US Fish & Wildlife Service, Vero Beach, Florida,* Waterborne Bacteria

Serge Rotteveel, *Institute for Inland Water Management and Waste Water Treatment–RIZA, Lelystad, The Netherlands,* SOFIE: An Optimized Approach for Exposure Tests and Sediment Assays

Timothy J. Ryan, *Ohio University, Athens, Ohio,* Water Sampling and Laboratory Safety

Randall T. Ryti, *Neptune and Company, Inc., Los Alamos, New Mexico,* Assessment of Ecological Effects in Water-Limited Environments

Masaki Sagehashi, *University of Tokyo, Tokyo, Japan,* Biomanipulation

Basu Saha, *Loughborough University, Loughborough, United Kingdom,* Activated Carbon: Ion Exchange and Adsorption Properties

Md. Salequzzaman, *Khulna University, Khulna, Bangladesh,* Ecoregions: A Spatial Framework for Environmental Management

Dibyendu Sarkar, *University of Texas, San Antonio, Texas,* Lead and its Health Effects

Peter M. Scarlett, *Winfrith Technology Centre, Dorchester, Dorset, United Kingdom,* The Control of Algal Populations in Eutrophic Water Bodies

Trifone Schettino, *Università di Lecce, Lecce, Italy,* Metallothioneins as Indicators of Trace Metal Pollution

Lewis Schneider, *North Jersey District Water Supply Commission, Wanaque, New Jersey,* Classification and Environmental Quality Assessment in Aquatic Environments

Wolfram Schuessler, *Institut für Nukleare Entsorgung, Karlsruhe, Germany,* Column Experiments in Saturated Porous Media Studying Contaminant Transport

K.D. Sharma, *National Institute of Hydrology, Roorkee, India,* Water Quality Management

Mukesh K. Sharma, *National Institute of Hydrology, Roorkee, India,* Ground Water Quality in Areas Adjoining River Yamuna at Delhi, India, Irrigation Water Quality in Areas Adjoining River Yamuna At Delhi, India

Daniel Shindler, *UMDNJ, New Brunswick, New Jersey,* Methemoglobinemia

Slobodan P. Simonovic, *The University of Western Ontario, London, Ontario, Canada,* Water Resources Systems Analysis, Fuzzy Criteria for Water Resources Systems Performance Evaluation, Participatory Multicriteria Flood Management

Shahnawaz Sinha, *Malcolm Pirnie Inc., Phoenix, Arizona,* Disinfection By-Product Precursor Removal from Natural Waters

Joseph P. Skorupa, *U.S. Fish and Wildlife Service,* Remediation and Bioremediation of Selenium-Contaminated Waters

Roel Smolders, *University of Antwerp, Antwerp, Belgium,* Active Biomonitoring (ABM) by Translocation of Bivalve Molluscs

Jinsik Sohn, *Kookmin University, Seoul, Korea,* Disinfection By-Product Precursor Removal from Natural Waters

Fiona Stainsby, *Napier University, Edinburgh, United Kingdom,* Bioluminescent Biosensors for Toxicity Testing

Ross A. Steenson, *Geomatrix, Oakland, California,* Land Use Effects on Water Quality

Leonard I. Sweet, *Engineering Labs Inc., Canton, Michigan,* Application of the Precautionary Principle to Water Science

Kenneth K. Tanji, *University of California, Davis, California,* E_h

Ralph J. Tella, *Lord Associates, Inc., Norwood, Massachusetts,* Overview of Analytical Methods of Water Analyses With Specific Reference to EPA Methods for Priority Pollutant Analysis

William E. Templin, *U.S. Geological Survey, Sacramento, California,* California—Continually the Nation's Leader in Water Use

Rita Triebskorn, *Steinbeis-Transfer Center for Ecotoxicology and Ecophysiology, Rottenburg, Germany,* Biomarkers, Bioindicators, and the Trondheim Biomonitoring System

Nirit Ulitzur, *Checklight Ltd., Tivon, Israel,* Use of Luminescent Bacteria and the *Lux* Genes For Determination of Water Quality

Shimon Ulitzur, *Technion Institute of Technology, Haifa, Israel,* Use of Luminescent Bacteria and the *Lux* Genes For Determination of Water Quality

U.S. Environmental Protection Agency, How We Use Water in These United States

U.S. Agency for International Development (USAID), Water and Coastal Resources

U.S. Geological Survey, Water Quality, Water Science Glossary of Terms

G. van der velde, *Radboud University Nijmegen, Toernooiveld, Nijmegen, The Netherlands,* Dose-Response of Mussels to Chlorine

F.N.A.M. van pelt, *Environmental Research Institute, University College Cork, Ireland,* Development and Application of Sediment Toxicity Tests for Regulatory Purposes

D.V. Vayenas, *University of Ioannina, Agrinio, Greece,* Cartridge Filters for Iron Removal

Raghuraman Venkatapathy, *Oak Ridge Institute for Science and Education, Cincinnati, Ohio,* Alternative Disinfection Practices and Future Directions for Disinfection By-Product Minimization, Chlorination By-Products

V.P. Venugopalan, *BARC Facilities, Kalpakkam, India,* Dose-Response of Mussels to Chlorine

Jos P.M. Vink, *Institute for Inland Water Management and Waste Water Treatment–RIZA, Lelystad, The Netherlands,* Heavy Metal Uptake Rates Among Sediment Dwelling Organisms, SOFIE: An Optimized Approach for Exposure Tests and Sediment Assays

Judith Voets, *University of Antwerp, Antwerp, Belgium,* Active Biomonitoring (ABM) by Translocation of Bivalve Molluscs

Mark J. Walker, *University of Nevada, Reno, Nevada,* Water Related Diseases

William R. Walker, *Virginia Water Resources Research Center,* Management of Water Resources for Drought Conditions

Xinhao Wang, *University of Cincinnati, Cincinnati, Ohio,* Quality of Water Supplies

Corinna Watt, *University of Alberta, Edmonton, Alberta, Canada,* Arsenic Compounds in Water

Janice Weihe, *American Water, Belleville, Illinois,* Understanding *Escherichia Coli* O157:H7 and the Need for Rapid Detection in Water

June M. Weintraub, *City and County of San Francisco Department of Public Health, San Francisco, California,* Chlorination By-Products, Alternative Disinfection Practices and Future Directions for Disinfection By-Product Minimization

Victor Wepener, *Rand Afrikaans University, Auckland Park, South Africa,* Active Biomonitoring (ABM) by Translocation of Bivalve Molluscs

Eva Ståhl Wernersson, *GS Development AB, Malmö, Sweden,* Dishwashing Water Quality Properties

Andrew Whiteley, *Centre for Ecology and Hydrology–Oxford, Oxford, United Kingdom,* Bioluminescent Biosensors for Toxicity Testing

Siouxsie Wiles, *Imperial College London, London, United Kingdom,* Bioluminescent Biosensors for Toxicity Testing

Thomas M. Williams, *Baruch Institute of Coastal Ecology and Forest Science, Georgetown, South Carolina,* Water Quality Management in a Forested Landscape

Parley V. Winger, *University of Georgia, Atlanta, Georgia,* Water Assessment and Criteria

M.H. Wong, *Hong Kong Baptist University, Kowloon, Hong Kong,* Algal Toxins in Water

R.N.S. Wong, *Hong Kong Baptist University, Kowloon, Hong Kong,* Algal Toxins in Water

J. Michael Wright, *Harvard School of Public Health, Boston, Massachusetts,* Alternative Disinfection Practices and Future Directions for Disinfection By-Product Minimization

Gary P. Yakub, *Kathleen Stadterman-Knauer Allegheny County Sanitary Authority, Pittsburgh, Pennsylvania,* Indicator Organisms

Yeomin Yoon, *Northwestern University, Evanston, Illinois,* Disinfection By-Product Precursor Removal from Natural Waters

M.E. Young, *Conwy, United Kingdom,* Water Resources Challenges in the Arab World

Mehmet Ali Yurdusev, *Celal Bayar University, Manisa, Turkey,* Integration of Environmental Impacts into Water Resources Planning

Karl Erik Zachariassen, *Norwegian University of Science and Technology, Trondheim, Norway,* Physiological Biomarkers and the Trondheim Biomonitoring System

Luke R. Zappia, *CSIRO Land and Water, Floreat, Australia,* Microbial Activities Management

Harry X. Zhang, *Parsons Corporation, Fairfax, Virginia,* Water Quality Management and Nonpoint Source Control, Water Quality Models for Developing Soil Management Practices

Igor S. Zonn, Lessons from the Rising Caspian

A.I. Zouboulis, *Aristotle University of Thessaloniki, Thessaloniki, Greece,* Biosorption of Toxic Metals, Arsenic Health Effects

WATER QUALITY CONTROL

ACID MINE DRAINAGE—EXTENT AND CHARACTER

ARTHUR W. ROSE
Pennsylvania State University
University Park, Pennsylvania

Acid mine drainage (AMD), also known as acid rock drainage (ARD), is an extensive environmental problem in areas of coal and metal mining. For example, the Appalachian Regional Commission (1) estimated that 5700 miles of streams in eight Appalachian states were seriously polluted by AMD. AMD is also serious near major metal mining districts such as Iron Mountain, CA and Summitville, CO (2,3). In streams affected by AMD, fish and stream biota are severely impacted and the waters are not usable for drinking or for many industrial purposes (4). In addition to deleterious effects of dissolved constituents (H^+, Fe, Al) on stream life, Fe and Al precipitates can cover the stream bed and inhibit stream life, and suspended precipitates can make the water unusable. In metal mining areas, heavy metals can add toxicity. General references on chemistry of AMD are Rose and Cravotta (5) and Nordstrom and Alpers (6).

CHEMISTRY OF FORMATION

AMD is formed by weathering of pyrite (FeS_2, iron sulfide) and other sulfide minerals, including marcasite (another form of FeS_2), pyrrhotite ($Fe_{1-x}S$), chalcopyrite ($CuFeS_2$), and arsenopyrite (FeAsS). The following reactions, involving oxygen as the oxidant, occur when pyrite is exposed to air and water:

$$FeS_2 + 3.5O_2 + H_2O = Fe^{2+} + 2SO_4^{2-} + 2H^+ \quad (1)$$

$$Fe^{2+} + 0.25O_2 + H^+ = Fe^{3+} + 0.5H_2O \quad (2)$$

$$Fe^{3+} + 3H_2O = Fe(OH)_3 + 3H^+ \quad (3)$$

The sum of these, representing complete oxidation and Fe precipitation, is

$$FeS_2 + 3.75O_2 + 3.5H_2O = Fe(OH)_3(s) + 2SO_4^{2-} + 4H^+ \quad (4)$$

In addition, the Fe^{3+} formed in Eq. 2 is a very effective oxidant of pyrite:

$$FeS_2 + 14Fe^{3+} + 8H_2O = 15Fe^{2+} + 2SO_4^{2-} + 16H^+ \quad (5)$$

These reactions also generate considerable heat, which tends to increase temperature and reaction rate.

In the above equations, the Fe precipitate is shown as $Fe(OH)_3$, but other ferric Fe phases can precipitate, depending on the conditions. Goethite (FeOOH), hematite (Fe_2O_3), ferrihydrite ($Fe_5HO_8 \cdot 4H_2O$), schwertmannite ($Fe_8O_8(OH)_6(SO_4)$), and jarosite ($KFe_3(SO_4)_2(OH)_6$) are among the products. The latter two products represent "stored acidity" that can react further to release additional acidity. Under evaporative conditions, $FeSO_4$ and other Fe sulfates can precipitate to form stored acidity.

Acid generation is dependent on a large number of factors, including the pH of the environment, temperature, the surface area of the pyrite or other source, the atomic structure of the pyrite, bacterial activities, and oxygen availability.

Oxidation of Fe^{2+} (Eq. 2) is relatively slow at pH below about 5. However, certain bacteria, such as Thiobacillus ferrooxidans, can catalyze the oxidation reaction under acid conditions. Bacterial action increases the reaction rate by a factor of about 10^6 (7). In addition, Fe^{3+}, the most effective oxidant via Eq. 4, has negligible solubility above about pH 3.5. As a result of these effects, severe AMD only develops in conditions where the water in contact with pyrite is highly acid and Fe-oxidizing bacteria are present (8). At higher pH, acid generation is relatively slow.

During natural weathering of pyrite-bearing rocks, the oxidation reactions happen slowly. In contrast, mining and other rock disturbances, such as road building, can result in greatly increased exposure of pyrite to oxidizing conditions, with resulting rapid acid generation. The water flowing from many underground mines is deficient in oxygen, and the above sequence proceeds only as far as reaction (1) [or perhaps reactions (1), (2) and (4)]. As a result, outflowing water contains elevated Fe^{2+} that oxidizes after it reaches the surface and generates additional acid owing to Fe precipitation after exposure to air. In such cases, pH can decrease downstream.

CHEMISTRY OF ACID MINE DRAINAGE

The H^+ generated by pyrite oxidation attacks various rock minerals, such as carbonates, silicates, and oxides, consuming some H^+ and releasing cations. For this reason, AMD commonly contains moderate to high levels of Ca, Mg, K, Al, Mn, and other cations balancing SO_4, the dominant anion. These reactions consume H^+ and increase the pH. If reaction with rock minerals is extensive, the resulting water may have a pH of 6 or even higher, and if oxidizing conditions exist, Fe may be relatively low. If carbonates are present in the affected rocks, the "AMD" may contain significant alkalinity as HCO_3. AMD is characterized by SO_4^{2-} as the dominant anion but can have a wide range of pH, Fe, and other cations (Table 1).

The pH of AMD typically ranges from about 2.5 to 7, but the frequency distribution of pH is bimodal, with most common values in the range 2.5 to 4 and 5.5 to 6.5 (5). Relatively fewer values are in the range 4 to 5.5. An extreme value of negative 3.6 is reported (2). Common ranges of other constituents are up to 100 mg/L Fe, up to 50 mg/L Al, up to 140 mg/L Mn, and up to 4000 mg/L SO_4.

A key variable characterizing AMD is "acidity." Acidity is commonly expressed as the quantity of $CaCO_3$ required to neutralize the sample to a pH of 8.3 by reaction (8). The acidity includes the generation of H^+ by reactions (1) and (3), as well as the effects of other cations that generate

Table 1. Analyses of Typical "Acid Mine Drainage" (5)

Constituent	Units	WMB A1	WMS 2K	LMS S2-15
pH		2.6	4.2	6.9
Acidity	mg/L CaCO$_3$	688	270	0
Alkalinity	mg/L CaCO$_3$	0	0	730
Fe	mg/L	174	34	5.7
Mn	mg/L	25.5	67	7.5
Al	mg/L	68.9	26	<0.14
Ca	mg/L	83	270	650
Mg	mg/L	74	280	230
SO$_4$	mg/L	913	1600	1900
Spec. Cond.	µS/cm	2120	1860	3890

acidity at pH<8.3, such as

$$Al^{3+} + 3H_2O = Al(OH)_3 + 3H^+ \qquad (6)$$

In common AMD, acidity includes contributions from H^+, Fe^{2+} (Eqs. 2 and 3), Fe^{3+} (Eq. 3), Al^{3+}, and Mn^{2+}. To accurately measure the contribution of these solutes, the acidity determination should include a step in which Fe and Mn are oxidized, commonly by addition of H_2O_2 and heating (9,10). In waters from metal mining areas, other heavy metals, such as Cu, may contribute to acidity. Acidity of AMD from coal mining areas is commonly up to 1000 mg/L as CaCO$_3$.

BIBLIOGRAPHY

1. Appalachian Regional Commission. (1969). *Acid Mine Drainage in Appalachia*. Appalachian Regional Commission, Washington, DC, p. 126.
2. Nordstrom, D.K., Alpers, C.N., Ptacek, C.J., and Blowes, D.W. (2000). Negative pH and extremely acidic mine waters from Iron Mountain, California. *Environ. Sci. Technol.* **34**: 254–258.
3. Plumley, G.S., Gray, J.E., Roeber, M.M., Coolbaugh, M., Flohr, M., and Whitney, G. (1995). The importance of geology in understanding and remediating environmental problems at Summitville. *Colorado Geological Survey* Special Publication **38**: 13–19.
4. Earle, J. and Callaghan, T. (1998). Impacts of mine drainage on aquatic life, water uses and man-made structures. In: *Coal Mine Drainage Prediction and Pollution Prevention in Pennsylvania*. PA Department of Environmental Protection, pp. 1-1–1-22. Available at: http://www.dep.state.pa.us/deputate/minres/districts/CMDP/main.htm.
5. Rose, A.W. and Cravotta, C.A. (1998). Geochemistry of coal mine drainage. In: *Coal Mine Drainage Prediction and Pollution Prevention in Pennsylvania*. PA Department of Environmental Protection, pp. 1-1–1-22. Available at: http://www.dep.state.pa.us/deputate/minres/districts/CMDP/main.htm.
6. Nordstrom, D.K. and Alpers, C.N. (1999). Geochemistry of acid mine waters. In: *Environmental Geochemistry of Mineral Deposits, Reviews in Economic Geology*. Vol. 6A. G.S. Plumley and M.J. Logsdon (Eds.). Society of Economic Geologists, Littleton, CO, pp. 133–160.
7. Singer, P.C. and Stumm, W. (1970). Acidic mine drainage—The rate-determining step. *Science* **167**: 1121–1123.
8. Kleinmann, R.L.P., Crerar, D.A., and Pacelli, R.R. (1981). Biogeochemistry of acid mine drainage and a method to control acid formation. *Mining Eng.* **33**: 300–313.
9. American Public Health Association. (1998). Acidity (2310)/titration method. In: L.S. Clesceri et al. (Eds.). *Standard Methods for the Examination of Water and Wastewater*, 20th Edn. American Public Health Association, Washington, DC, pp. 2.24–2.26.
10. U.S. Environmental Protection Agency. (1979). Method 305.1, Acidity (Titrimetric). In: *Methods for Chemical Analysis of Water and Wastes*. U.S. Environmental Protection Agency Report EPA/600/4-79-020. Available at: http://www.nemi.gov.

THE CONTROL OF ALGAL POPULATIONS IN EUTROPHIC WATER BODIES

JOANNA DAVIES
Syngenta
Bracknell, Berkshire, United Kingdom

PETER M. SCARLETT
Winfrith Technology Centre
Dorchester, Dorset, United Kingdom

INTRODUCTION

Eutrophication is a natural aging process occurring in lakes and reservoirs, which is characterized by increasing nutrient levels in the water column and increasing rates of sedimentation. For water bodies in urban and agricultural landscapes, this process is accelerated by increased nutrient inputs from agricultural fertilizers, sewage effluents, and industrial discharges. Eutrophication is accompanied by increased macrophyte and algal populations, which, without appropriate management, can develop to nuisance proportions.

Algae are microscopic plants that can reproduce rapidly in favorable conditions. Some species can form scum or mats near or at the water surface. The presence of excessive algae can disrupt the use of a water body by restricting navigational and recreational activities and disrupting domestic and industrial water supplies. In particular, algal blooms can have a severe impact on water quality, causing noxious odors, tastes, discoloration, and turbidity. Blue-green algal blooms are particularly undesirable due to their potential toxicity to humans, farm livestock, and wild animals. In addition, algae may block sluices and filters in water treatment plants and reduce water flow rates, which may in turn encourage mosquitoes and increase the risk of waterborne diseases such as malaria and bilharzia (schistosomiasis). Ultimately, the presence of excessive algal populations will limit light penetration through the water column, thus inhibiting macrophyte growth and leading to reduced biodiversity.

Methods for controlling algal populations can be divided into categories as follows:

1. Environmental methods involve limiting those factors, such as nutrients, that are essential for algal growth.
2. Chemical methods involve the application of herbicides, or other products such as barley straw, that have a direct toxic effect on algae.
3. Biomanipulation involves the control of zooplanktivorous fish to favor algal grazing by invertebrates and also describes the use of microbial products.

ENVIRONMENTAL METHODS

As algae are dependent on the availability of limiting nutrients required for growth, namely, phosphorus and occasionally nitrogen, the long-term control of algal populations requires measures to reduce the entry of nutrients from external sources and to reduce the internal release of nutrients from sediments.

Nutrients may enter water bodies from point sources, such as sewage and industrial effluents, and diffuse sources, such as runoff from agricultural land following the application of animal slurries or fertilizers. Methods for reducing external nutrient loading from point sources include diversion, effluent treatment, or installation of artificial wetlands. Methods for reducing loading from diffuse sources include buffer strips and adoption of good agricultural practice.

For many shallow water bodies, attempts to reduce external loading have been less successful due to the internal release of phosphate from the sediment, which delays the decline in the total phosphorus load. Under these circumstances, the successful management of algae requires simultaneous measures to reduce internal phosphate cycling. Techniques available for this purpose include dilution and flushing, hypolimnetic withdrawal, hypolimnetic aeration, artificial circulation, phosphorus inactivation, sediment oxidation, sediment sealing, and sediment removal.

A brief description of each method for reducing external and internal nutrient loading is provided below.

Diversion

Nutrient inputs can be reduced by the diversion of effluents away from vulnerable water bodies to water courses that have greater assimilative capacity. This option is only viable where there is an alternative sink within the vicinity of the affected water body as construction of pipe work to transport water over long distances is prohibitively expensive. Diversion also reduces the volume of water flushing through the water body and, therefore, may not be viable if such a reduction is predicted to significantly affect water body hydrology. Examples where diversion has successfully reduced external phosphate loading, leading to reductions in algal biomass, include Lake Washington in the United States (1). In this case, total phosphorus concentrations were reduced from 64 µg/L prior to diversion in 1967, to 25 µg/L in 1969. This reduction was accompanied by a fivefold decrease in the concentration of chlorophyll a over the same period.

Effluent Treatment

Reductions in external nutrient loading can be achieved by removing phosphate and/or nitrates from effluents, prior to their discharge. Phosphorus can be removed from raw sewage or, more commonly, final effluents, by the process of stripping, which involves precipitation by treatment with aluminum sulfate, calcium carbonate, or ferric chloride. The resulting sludge is spread onto land or transferred to a waste-tip. In contrast, removal of nitrates from effluents is more complex requiring the use of ion exchange resins or microbial denitrification. The process of phosphate stripping is a requirement of the Urban Waste Water Treatment Directive in some European countries including Switzerland and The Netherlands. Phosphate stripping has successfully reduced algal biomass in Lake Windermere in the United Kingdom. In this case, total internal phosphate concentrations were reduced from 30 µg/L in 1991 to 14 µg/L in 1997, while biomass of the filamentous algae, *Cladophora*, was reduced by 15-fold between 1993 and 1997 (2).

Constructed Wetlands

External nutrient loads may also be reduced by passing effluents through detention basins or constructed wetlands, which are areas of shallow water, planted with macrophytes, that are designed to retain and reduce nutrient concentrations by natural processes. Wetlands are particularly effective for reducing nitrate concentrations by denitrification and retaining phosphorus that is bound to particles. However, they will not permanently retain soluble phosphorus and may release phosphorus from sediments at certain times of year. The development of artificial wetland systems can also be prohibitively expensive as they require large areas of land and regular maintenance, including sediment dredging and macrophyte harvest, to remain efficient. An example where constructed wetlands have been developed as part of a lake restoration scheme is provided by Annadotter et al. (3).

Good Agricultural Practice and Buffer Strips

External nutrient loading from diffuse agricultural sources can be reduced by the adoption of good agricultural practices designed to minimize fertilizer use and reduce runoff into adjacent water courses. Strategies for reducing nutrient inputs include minimizing fertilizer applications, where possible, and using slow-release formulations. Measures to minimize opportunities for runoff include avoiding applications in wet weather, incorporating fertilizer into soil by ploughing after application, maintaining ground cover for as long as possible to minimize exposure of bare ground to rainfall, and, finally, ploughing along contours.

Nutrient loading from diffuse sources can also be reduced by the creation of buffer strips between cultivated land and vulnerable water courses, in which fertilizer applications are prohibited. The development of semi-natural vegetation within these strips will also serve to intercept and assimilate the nutrients in runoff, as described for constructed wetlands (4). Under current legislation in the European Union, buffer strips are only mandatory in situations where they are essential for protecting drinking water supplies from nitrate inputs and, in particular, in areas where drinking water abstractions have concentrations exceeding 50 mg nitrate per liter. As nitrate concentrations that contribute to eutrophication are much lower than 50 mg/L, this legislation is unlikely to assist in the control of algae. Various schemes exist at member state level to promote adoption of buffer strips as a normal farming practice although experience in the United Kingdom indicates that such schemes have not been widely adopted.

Dilution and Flushing

Dilution and flushing involve the influx of large volumes of low nutrient water into the affected water body, thus diluting and therefore reducing nutrient concentrations, and washing algal cells out of the water body. This approach is dependent on the availability of large volumes of low nutrient water and systems for its transport to the affected water body. Examples where dilution has been successfully used to control algae include Green Lake in Washington State (USA). In this case, phosphorus and chlorophyll *a* concentrations were reduced by 70% and 90%, respectively, within six years of initiation of dilution (5).

Phosphorus Precipitation from the Water Column

Surface water concentrations of phosphorus can be reduced by the application of aluminum salts to the water column. At water pH values between 6 and 8, these salts dissociate and undergo hydrolysis to form aluminum hydroxide, which is capable of binding inorganic phosphorus. The resulting floc settles to the bottom of the water body, where the precipitated phosphorus is retained. The precipitate also effectively seals the sediment, thus retarding the further release of phosphorus from the sediment. At pH values below 6 or above 8, the aluminum exists as soluble ions that do not bind phosphorus. As well as being ineffective for phosphorus precipitation, these forms of aluminum present a toxic hazard to fish and aquatic invertebrates. Therefore, the successful and safe use of aluminum salts requires careful calculation of the necessary dose based on water pH and may require the use of a buffer solution such as sodium aluminate. This technique is widely used in eutrophic water bodies and Welch and Cooke (6) report several case studies. Effects are typically rapid and have been reported to reduce phosphorus release from sediments for between ten and fifteen years, although the effectiveness and longevity of the treatment may be compromised by natural sedimentation and benthic invertebrate activity.

Alternatively, phosphorus can be precipitated from the water column by the application of iron or calcium salts. These salts do not pose such a toxicity hazard as aluminum but their successful use often requires additional management techniques such as aeration or artificial circulation in order to maintain the necessary water pH and redox conditions. Consequently, the use of iron and calcium salts is less widely reported (7,8).

Hypolimnetic Withdrawal, Aeration, and Artificial Circulation

The release of phosphorus from sediment can be inhibited by increasing the concentration of dissolved oxygen in hypolimnetic waters at the water–sediment interface. The hypolimnion is the layer of water directly above the sediment, which, in thermally stratified water bodies, is usually too deep to support photosynthesis. Continued respiration leads to depletion of dissolved oxygen and the concomitant release of phosphorus from iron complexes. In stratified lakes with low resistance to mixing, wind action and the resulting turbulence may cause temporary destratification and movement of phosphorus from the hypolimnion, through the metalimnion, and into the upper epilimnion layers. In susceptible water bodies, this natural process can be circumvented by implementation of hypolimnetic withdrawal, aeration, or artificial circulation processes. By default, these approaches have the advantage of extending the habitat available for colonization by fish and zooplankton.

Hypolimnetic withdrawal involves the removal of water directly from the hypolimnion through a pipe installed at the bottom of the water body. This process requires low capital investment and reduces the detention times of water in the hypolimnion, thus reducing the opportunity for the development of anaerobic conditions. The successful implementation of these systems depends on the availability of a suitable sink for discharge waters and measures to avoid thermal destratification, caused by epilimnetic waters being drawn downward, which would otherwise encourage the transport of hypolimnetic nutrients to surface waters. This risk can be reduced by careful control of the rate of withdrawal and redirection of inlet water to the metalimnion or hypolimnion. Examples where hypolimnetic withdrawal systems have been installed in lakes are reported by Nurnberg (9).

Aeration of the hypolimnion can be achieved by mechanical agitation, whereby hypolimnetic waters are pumped onshore where the water is aerated by agitation before being returned to the hypolimnion. More usually, the hypolimnion is aerated using airlift or injection systems, which use compressed air to force hypolimnetic waters to the surface where they are aerated on exposure to the atmosphere. Water is then returned to the hypolimnion with minimal increase in temperature. In contrast, artificial circulation involves mechanical mixing of anaerobic hypolimnetic water with the upper water body using pumps, jets, and bubbled air. By default, complete circulation causes destratification of the water body, with potential adverse consequences for cold water fish species, but may also serve to reduce the concentration of algal cells in the upper water body by increasing the mixing depth and relocating algal biomass to deeper water with reduced light availability (10). Detailed examples of these processes are provided by Cooke et al. (5).

Sediment Oxidation

The release of phosphorus from sediment can also be inhibited by oxidation of sediment by the "Riplox" process. This process involves the direct injection of calcium nitrate solutions into the sediment in order to stimulate microbial denitrification and thus restore an oxidized state, conducive to the binding of interstitial phosphate with ferric hydroxide. In some cases, where the inherent iron content of the sediment is inadequate, iron chloride is initially added to generate ferric iron. Similarly, calcium hydroxide may also be added in order to raise the sediment pH to levels required for optimum microbial activity (11). This process has been documented to reduce phosphate release by up to 90% under laboratory conditions and by 50–80% following the treatment of Lake Noon in the United States (5). However, sediment oxidation is only suitable for water bodies where phosphate binding is modulated by iron redox reactions. In shallow water bodies where phosphate release is predominantly influenced by fluctuating pH and temperature at the water–sediment interface, sediment oxidation may not significantly reduce phosphorus release.

Sediment Sealing

Sediment sealing involves the physical isolation of the sediment from the water column using plastic membranes or a layer of pulverized fly ash, a solid waste product from coal-burning power stations. These techniques may prove effective at reducing algal biomass in small enclosures used for recreational purposes or industrial water supplies but are unsuitable for conservation or restoration purposes, as fly-ash may contain high levels of undesirable heavy metals and open sediment is required to support aquatic plant growth (12).

Sediment Removal

The release of nutrients from the sediment can be averted by sediment removal either using conventional excavation equipment after drawdown or using a suction-dredger and pump. In both cases, the resulting wet sediment is then transferred to a suitable disposal site. Direct dredging is more commonly practiced but has the disadvantage that phosphates or other toxins adsorbed to sediment may be released into the water column when sediment is disturbed. Furthermore, benthic organisms will inevitably be disturbed and removed during dredging, although recolonization will mitigate any long-term effects. Examples where sediment removal projects have been implemented include Lake Finjasjon in Sweden and the Norfolk Broads in the United Kingdom (3,13).

CHEMICAL METHODS

Herbicides

The availability of aquatic herbicides has been limited by their small market value and the stringent toxicological and technical challenges presented by the aquatic environment. Aquatic herbicides, particularly those used to control algae, must be absorbed rapidly from dilute and often flowing aqueous solution. More recently, the number of products available for use in water has been reduced by the prohibitive cost of meeting the increasingly stringent requirements of pesticide registration. In 2004, the few products that remain on the market for use on a large scale are based on the active ingredients of copper, diquat, endothall, and terbutryn. Additional products based on other active ingredients are available for amateur use in enclosed garden ponds but are not registered for use in larger water bodies.

Use of herbicides to control algal blooms is often limited where water is required for domestic drinking water supply or for the irrigation of crops or livestock. Their use under these circumstances requires strict adherence to label recommendations and observation of recommended irrigation intervals. Despite these restrictions, chemical control may be preferable where immediate control is required or alternative, long-term measures are prohibited due to excessive cost.

Herbicide applications to water are generally made using hand-operated knapsack sprayers operated from bank or boat, or spray booms mounted to boats, tractors, helicopters, or planes. Much of this equipment is modified from conventional agricultural sprayers and nozzles, although the injection of herbicides into deep water or onto channel beds may require the use of weighted, trailing hoses fitted to boat-mounted spray booms.

Disadvantages associated with the use of herbicides include potential adverse effects on nontarget organisms including aquatic invertebrates and fish, development of herbicide-resistant algal strains, and excessive copper accumulation in treatment plant sludges, which may lead to disposal problems. The control of algae with herbicides can also create large quantities of decaying tissue, which cause deoxygenation of the water due to a high bacterial oxygen demand. This may lead to the death of fish and other aquatic organisms, particularly during summer months when deoxygenation is more rapid due to lower dissolved oxygen levels and increased rates of decomposition caused by higher water temperatures. Deoxygenation can largely be avoided by restricting applications to the early growing season. Where later treatments are essential, applications should be restricted to discrete localized areas of a water body, or slow-release formulations should be used to avoid a sudden buildup of decaying tissue.

Barley Straw

Research in the United Kingdom has led to the use of barley straw for the control and prevention of algal blooms in a range of water bodies including lakes, reservoirs, rivers, streams, and ponds (14). During decomposition, barley straw is known to release compounds, including active oxygen and hydrogen peroxide, which are toxic to algae (15). Although similar effects have been demonstrated with wheat, linseed, oil-seed rape, and lavender straw, barley straw provides the most effective and long-lasting control. For effective control, chopped straw is typically added to the water in loosely packed bales enclosed in mesh sacks, netting, or wire cages that are anchored to the bank or to the bottom of the water body. Depending on the type of water body and the severity of the algal bloom, typical application rates vary between 100 and 500 kg/ha. In order to promote the decomposition process and the release and distribution of the algicidal components, bales should be positioned at, or near, the water surface where water temperatures, movement, and aeration are the greatest. Depending on water temperature, the decomposition process may continue for 2–8 weeks after application before the straw releases sufficient quantities of the active components to cause an effect. These chemicals continue to be released until decomposition is complete, which may take up to six months. Therefore, for maximum efficacy, straw is typically applied in spring before algal blooms reach their peak and again in the autumn.

The use of straw has few, if any, adverse effects on nontarget organisms such as macrophytes, invertebrates, or fish. The main disadvantage associated with its use is the risk of deoxygenation caused by the high bacterial oxygen demand of those microorganisms responsible for straw decomposition. Deoxygenation may lead to the death of fish and other aquatic organisms, particularly during summer months when rates of decomposition are increased due to higher water temperatures. This risk can be reduced by early application during the spring and avoiding applications during prolonged periods of hot weather.

BIOMANIPULATION

Removal of Zooplanktivorous Fish

Algal populations may be controlled through the biomanipulation of foodwebs to favor algal grazing by invertebrates, by removing fish (16). Methods for the direct removal of fish include application of the piscicide, rotenone, to anesthetize fish prior to removal or to kill them outright. However, use of rotenone for this purpose is discouraged in some countries, including the United Kingdom, and requires special consent from the appropriate government authority. Alternative methods for fish removal include systematic electrofishing, seine netting, and fish traps. In smaller water bodies, drawdown may be used to concentrate fish into increasingly smaller water volumes to simplify their capture. Even where complete drawdown is possible, elimination of all fish is unlikely and additional measures may be necessary to prevent successful spawning of remaining fish. Spawning can be prevented by nets, placed in traditional spawning areas, to capture eggs, which are then removed.

The long-term success of fish removal schemes also requires measures to prevent recolonization from tributaries and floodwaters. In cases where water can be diverted and tributaries do not carry boat traffic, isolation of water bodies may be possible by the construction of dams. Where boat traffic requires access, the installation of electronic netting gates, which automatically lower to allow access, or engineered locks, which dose the water with rotenone on opening, may be required (12). Recolonization can also be avoided by creating fish-free enclosures, using fishproof barriers, within affected water bodies. Once restoration is complete within an enclosure, more enclosures may be built and eventually joined together until a large proportion of the water body is enclosed. This approach was used during the restoration of Hoveton Broad in the United Kingdom (12).

Alternative methods to control fish populations involve the introduction of piscivorous fish, such as pike or pikeperch in the United Kingdom or American largemouthed bass in the United States. The addition of piscivores can have immediate and dramatic impacts on invertebrate populations and has been practiced as part of restoration schemes in Lake Lyne in Denmark and Lakes Zwemlust and Breukeleveen in The Netherlands (12). However, the success of this approach is unlikely to be sustained without regular restocking as predation of the fish population will invariably lead to a decline in the piscivore population.

Microbial Products

The use of bacteria to control algal populations is a recent innovation adapted from the wastewater industry. As bacteria have a high surface area to volume ratio and a high uptake rate for nutrients relative to unicellular algae, they can out-compete algae for limiting nutrients, such as nitrogen and phosphorus, and have been demonstrated to suppress the growth of algal cultures under laboratory conditions (17). This observation has led to the commercial development of microbial products, containing bacteria and enzymes, that are designed to supplement natural microbial populations to the levels required to have a significant impact on algae. The number of available products has increased as the use of chemical algicides has become more restricted. However, few researchers report a significant reduction in algal growth following the use of microbial products under experimental conditions and their use on a large scale has yet to be widely documented (18).

CONCLUSIONS

While chemical control methods, such as the application of herbicides or barley straw, can provide rapid, short-term reductions in algal populations, the long-term and sustainable management of algae requires consideration of the cause and source of eutrophication and the implementation of techniques to reduce nutrient loading and to restore natural foodweb interactions. Evaluation of the suitability of the techniques discussed here, for use in individual cases, requires detailed assessments to determine the trophic status of the affected water body and, in particular, the relative contribution of point source, diffuse, and internal nutrient sources to total nutrient concentrations. Only when the causes of eutrophication are clearly identified can the symptom of excessive algal growth be efficiently managed.

BIBLIOGRAPHY

1. Edmondson, W.T. and Lehman, J.R. (1981). The effect of changes in the nutrient income on the condition of Lake Washington. *Limnol. Oceanography* **1**: 47–53.
2. Parker, J.E. and Maberley, S.C. (2000). Biological response to lake remediation by phosphate stripping: control of *Cladophora*. *Freshwater Biol.* **44**: 303–309.
3. Annadotter, H. et al. (1999). Multiple techniques for lake restoration. *Hydrobiologia* **395/396**: 77–85.

4. Abu-Zreig, M. et al. (2003). Phosphorous removal in vegetated filter strips. *J. Environ. Qual.* **32**(2): 613–619.
5. Cooke, G.D., Welch, E.B., Peterson, S.A., and Newroth, P.R. (1993). *Restoration and Management of Lakes and Reservoirs*, 2nd Edn. Lewis Publishers, Boca Raton, FL.
6. Welch, E.B. and Cooke, G.D. (1999). Effectiveness and longevity of phosphorous inactivation with alum. *J. Lake Reservoir Manage.* **15**: 5–27.
7. Randall, S., Harper, D., and Brierley, B. (1999). Ecological and ecophysiological impacts of ferric dosing in reservoirs. *Hydrobiologia* **395/396**: 355–364.
8. Prepas, E.E. et al. (2001). Long-term effects of successive $Ca(OH)_2$ and $CaCO_3$ treatments on the water quality of two eutrophic hardwater lakes. *Freshwater Biol.* **46**: 1089–1103.
9. Nurnberg, G.K. (1987). Hypolimnetic withdrawal as lake restoration technique. *J. Environ. Eng.* **113**: 1006–1016.
10. Brierly, B. and Harper, D. (1999). Ecological principles for management techniques in deeper reservoirs. *Hydrobiologia* **395/396**: 335–353.
11. Ripl, W. (1976). Biochemical oxidation of polluted lake sediment with nitrate—a new restoration method. *Ambio* **5**: 132.
12. Moss, B., Madgwick, J., and Phillips, G. (1997). *A Guide to the Restoration of Nutrient-Enriched Shallow Lakes*. Environment Agency, UK.
13. Phillips, G. et al. (1999). Practical application of 25 years research into the management of shallow lakes. *Hydrobiologia* **396**: 61–76.
14. Barrett, P.R.F and Newman, J.R. (1992). Algal growth inhibition by rotting barley straw. *Br. Phycol. J.* **27**: 83–84.
15. Everall, N.C. and Lees, D.R. (1997). The identification and significance of chemical released from decomposing barley straw during reservoir algal control. *Water Res.* **30**: 269–276.
16. Moss, B. (1992). The scope for biomanipulation in improving water quality. In: *Eutrophication: Research and Application to Water Supply*. D.W. Sutcliffe and J.G. Jones (Eds.). Freshwater Biological Association, Cumbria, UK, pp. 73–81.
17. Brett, M.T. et al. (1999). Nutrient control of bacterioplankton and phytoplankton dynamics. *Aquat. Ecol.* **33**(2): 135–145.
18. Duvall, R.J., Anderson, W.J., and Goldman, C.R. (2001). Pond enclosure evaluations of microbial products and chemical algicides used in lake management. *J. Aquat. Plant Manage.* **39**: 99–106.

ARSENIC COMPOUNDS IN WATER

Corinna Watt
X. Chris Le
University of Alberta
Edmonton, Alberta, Canada

Arsenic is the twentieth most abundant element in the earth's crust; it occurs naturally in the environment in both inorganic and organic forms. Arsenic is also released to the environment by anthropogenic activities such as pesticide use, wood preservation, mining, and smelting. Human exposure to arsenic by the general population occurs primarily from drinking water and food. In areas of endemic arsenic poisoning such as Bangladesh, India, Inner Mongolia, and Taiwan, the main exposure is through drinking water where inorganic arsenic levels can reach concentrations in the hundreds or thousands of micrograms per liter. The arsenic is released from natural mineral deposits into the groundwater in endemic areas. Groundwater is the primary drinking water source in these areas.

Arsenic is present in a variety of inorganic and organic chemical forms in water. This is a result of chemical and biological transformations in the aquatic environment. The specific arsenic compound present determines its toxicity, biogeochemical behavior, and environmental fate.

In natural waters, arsenic is typically found in the +5 and +3 oxidation states (1–6). The most common arsenic compounds detected in water are arsenite (As^{III}) and arsenate (As^V). Monomethylarsonic acid (MMA^V), monomethylarsonous acid (MMA^{III}), dimethylarsinic acid (DMA^V), dimethylarsinous acid (DMA^{III}), and trimethylarsine oxide (TMAO) have also been detected in water (Table 1). Several reviews provide important additional information regarding the cycling and speciation of arsenic in water and the environment (1,3,4,7–12).

Arsenic and arsenic compounds are classified as Group 1 carcinogens in humans by the International Agency for Research on Cancer (IARC). "The agent (mixture) is carcinogenic to humans" and "the exposure circumstance entails exposures that are carcinogenic to humans" (13). Chronic exposure to high levels of arsenic in drinking water has been linked to skin cancer, bladder cancer, and lung cancer, as well as to several noncancerous effects (8,11,14). Noncancerous effects from arsenic exposure include skin lesions, peripheral vascular disease (blackfoot disease), hypertension, diabetes, ischemic heart disease, anemia, and various neurological and respiratory effects (8,11,14). The reproductive and developmental effects of arsenic exposure have also been reported (14). The National Research Council and the World Health Organization have reviewed the most significant studies on arsenic exposure, toxicity, and metabolism (8,11,14). The association of arsenic with internal cancers has led to increased pressure for stricter guidelines for arsenic in drinking water.

Table 1. Arsenic Species Present in Water

Arsenic Species	Chemical Abbreviation	Chemical Formula
Arsenite, arsenous acid	As^{III}	H_3AsO_3
Arsenate, arsenic acid	As^V	H_3AsO_4
Monomethylarsonic acid	MMA^V	$CH_3AsO(OH)_2$
Monomethylarsonous acid	MMA^{III}	$CH_3As(OH)_2 [CH_3AsO]$
Dimethylarsinic acid	DMA^V	$(CH_3)_2AsO(OH)$
Dimethylarsinous acid	DMA^{III}	$(CH_3)_2AsOH$ $[((CH_3)_2As)_2O]$
Trimethylarsenic compounds	TMA	$(CH_3)_3As$ and precursors
Trimethylarsine oxide	TMAO	$(CH_3)_3AsO$
Oxythioarsenic acid		H_3AsO_3S

The World Health Organization (WHO) guideline for arsenic in drinking water is 10 µg/L. The Canadian guideline is 25 µg/L and is currently under review. The United States has recently lowered its maximum contaminant level (MCL) for arsenic in drinking water from 50 µg/L to 10 µg/L. The previous MCL was established as a Public Health Service standard in 1942 and was adopted as the interim standard by the U.S. Environmental Protection Agency (EPA) in 1975. The Safe Drinking Water Act (SDWA) Amendments of 1986 required the EPA to finalize its enforceable MCL by 1989. The EPA was not able to meet this request partly because of the scientific uncertainties and controversies associated with the chronic toxicity of arsenic. The SDWA Amendments of 1996 require that the EPA propose a standard for arsenic by January 2000 and promulgate a final standard by January 2001. In 1996, the EPA requested that the National Research Council review the available arsenic toxicity data and evaluate the EPA's 1988 risk assessment for arsenic in drinking water. The National Research Council Subcommittee on Arsenic in Drinking Water advised in its 1999 report that, based on available evidence, the MCL should be lowered from 50 µg/L (8). After much debate on what would be an appropriate MCL (3, 5, 10, or 20 µg/L), the EPA published its final rule of 10 µg/L in January 2001 under the Clinton administration. This was initially rescinded by the Bush administration. The reasons cited for this rejection were the high cost of compliance and incomplete scientific studies. In 2001, the National Research Council organized another subcommittee to review new research findings on arsenic health effects and to update its 1999 report on arsenic in drinking water (14). This second subcommittee concluded that "recent studies and analyses enhance the confidence in risk estimates that suggest chronic arsenic exposure is associated with an increased incidence of bladder and lung cancer at arsenic concentrations in drinking water that are below the MCL of 50 µg/L" (14). The MCL of 10 µg/L was approved later by the EPA and became effective in February 2002. The date for compliance has been set at January 23, 2006.

Most controversies over an appropriate MCL arise from a lack of clear understanding of the effects on health from exposure to low levels of arsenic. There are limited studies available that have examined the increased risk of cancer at low levels of arsenic exposure. Therefore, the National Research Council and the EPA had to base their assessments on epidemiological studies where the arsenic exposure is very high (8,11,14). Extrapolation of health effects from very high exposure data to the much lower exposure scenarios involves large uncertainties. Experimental animal studies were also consulted as well as the available human susceptibility information (8,14). Animal studies are of limited value when examining arsenic effects in humans because of differences in sensitivity and metabolism. Experimental animals are often exposed to very large doses of arsenic that are not representative of typical human exposure. In addition, the studies used to establish the new MCL do not take into account arsenic exposure through food intake and other beverages. Other confounding factors may include nutrition, metabolism, and predetermined genetic susceptibility (8,14).

The exact mechanism of arsenic's tumorigenicity is not clear. Therefore it is difficult to ascertain the risk of cancers from very low exposures to arsenic. Due to the limited information on the cancer risks posed by low-level arsenic exposure and to the unknown shape of the dose–response curve at low doses, the EPA used a default linear dose–response model to calculate the cancer risk. The linear extrapolation to zero assumes that there is no safe threshold of exposure at which health effects will not occur, whereas others argue that a safe threshold level or sublinear dose–response relationship may exist.

The cost of compliance with a lower MCL and the monitoring and treatment technology are other important considerations in setting the new MCL. According to the EPA, water systems that serve 13 million people and representing 5% of all systems in the United States would have to take corrective action at an MCL of 10 µg/L (15). However, the EPA believes that the new MCL "maximizes health risk reduction at a cost justified by the benefits" (15).

INORGANIC ARSENIC

As^{III} and As^V are the dominant arsenic species detected in most natural waters. As^{III} and As^V have been detected in all forms of natural water including groundwater, freshwater, and seawater (Table 2). In oxygen-rich waters of high redox potential, the As^V species H_3AsO_4, $H_2AsO_4^-$, $HAsO_4^{2-}$, and AsO_4^{3-} are stable (1,4). As^{III} species, which exist in reduced waters, may include H_3AsO_3, $H_2AsO_3^-$, and $HAsO_3^{2-}$ (1,4). However, in most natural waters, As^{III} is present as H_3AsO_3 (as its pK_a values are 9.23, 12.13, and 13.4) (8,16). The pK_a values of As^V are 2.22, 6.98, and 11.53 (8). At natural pH values, arsenate is usually present as $H_2AsO_4^-$ and $HAsO_4^{2-}$ (1,4,16).

The distribution of As^V and As^{III} throughout the water column varies with the season due to changes in variables such as temperature, biotic composition, pH, and redox potential (4,17–28). Biological activity also contributes to changes in speciation (1,3,4,6,20,22–24,29–39). The uptake of As^V by phytoplankton and marine animals results in the reduction of As^V to As^{III} and the formation of methylated arsenic compounds (4,19,23,24,29,32,34,37–42).

As^V is typically dominant in oxygen-rich conditions and positive redox potentials (4,6,33,43,44). As^V has been detected as the predominant species in most natural waters (Table 2). As^{III} is expected to dominate in anaerobic environments (4,7,10,25,33,45). As^{III} has been detected as the predominant species in groundwater (46–48) and is significant in the photic zone of seawater (43,45,49). As^{III} has also been associated with anoxic conditions in estuaries (31,50), seawater (33,43), and marine interstitial water (6,30). In lake interstitial water, 55% of the dissolved arsenic was present as arsenite and a few percent as DMA^V (51).

The As^{III}/As^V ratio typically does not reach thermodynamic equilibrium (1,4,6,20,24,33,50,52,53). This reflects biological mediation. The kinetics of the As^V–As^{III}

Table 2. Inorganic Arsenic Detected in Water[a]

Source	As^{III}	As^{V}	Reference
Groundwater			
Lead–zinc mine, Coeur d'Alene, Idaho, USA	0.1–42.3	N.A.	81
	0.1–1336.1[b]		81
Southwest United States	<5	21–120	44
Bowen Island, British Columbia, Canada	N.D.–220	N.D.–477.7	82
Six districts in West Bengal, India	10–2600	16–1100	83
Bangladesh (tubewell for drinking)	30–1200[b]		84
Bangladesh (deep tubewell for drinking)	50–80[b]		84
Bangladesh (deep tubewell for irrigation)	30–200[b]		84
Purbasthali (Burdwan), India	Present in 1 sample	N.D.–133.6	85
Taiwan (Pu-Tai) endemic area	285–683 (462 ± 129)	33–362 (177 ± 109)	48,86
Taiwan (I-Lan) comparison area	537–637 (572 ± 42)	24–67 (38 ± 18)	86
Taiwan (Hsin-Chiu) control area	<3	<12	86
Taiwan (Pu-Tai)	~20–720[b]		87
Taiwan (Fushing)	70.2 ± 2.6	870 ± 26	88
Taiwan (Chiuying)	51.6 ± 1.8	601 ± 22	88
Region Lagunera, northern Mexico	Trace–217	4–604	89
Fukuoka Prefecture, Japan	15–70	11–220	90
Kelheim, Lower Bavaria, Germany[c]	N.D.–176.3	N.D.–147.7	91
Rivers			
Hillsborough River, Tampa, Florida, USA	N.D.	0.25 ± 0.01	92
Withlacoochee River, Tampa, Florida, USA	N.D.	0.16 ± 0.01	92
Tamiami Canal, Florida, USA	0.47	0.07	49
Colorado River, Parker, Arizona, USA	0.11	1.95	49
Nine rivers in California, USA	0.02–1.3	0.07–42.5	49
Four rivers in California, USA	0.7–7.4[b]		20
Colorado River Slough, near Topock, California	0.08	2.25	93
North Saskatchewan River, Canada	0.21 ± 0.08	0.32 ± 0.01	94
Haya-kawa River, Japan	28[b]		95
River in Taiwan (Erhjin)	2.95 ± 3.52	4.59 ± 1.31	88
River in Taiwan (Tsengwen)	0.12 ± 0.06	2.57 ± 1.17	88
Lakes, Ponds, Reservoirs			
Mono Lake, California, USA (highly alkaline lake)	17250[b]		20
Pyramid Lake, California, USA	97.5[b]		20
Six lakes in California, USA	0.3–11.2[b]		20
Davis Creek Reservoir (filtered samples), California, USA	Trace–1.2	0.01–1.5	20
Davis Creek Reservoir (unfiltered samples), California, USA	0.8–2.6[b]		20
Elkhorn Slough, California, USA	1.1[b]		20
Donner Lake N. Shore, California, USA	0.06	0.07	49
Saddleback Lake S. End, California, USA	0.05	0.02	49
Squaw Lake (surface), California, USA	0.87	2.76	49
Squaw Lake (2-m depth), California, USA	0.53	2.96	49
Senator Wash Reservoir, California, USA	0.58	2.47	49
Lake Echols, Tampa, Florida, USA	2.74 ± 0.01	0.41 ± 0.01	92
Lake Magdalene, Tampa, Florida, USA	0.89 ± 0.01	0.49 ± 0.01	92
Remote pond, Withlacoochee Forest, Tampa, Florida, USA	N.D.	0.32 ± 0.01	92
University research pond, Tampa, Florida, USA	0.79 ± 0.01	0.96 ± 0.01	92
Coot Bay Pond, Everglades, Florida, USA	0.81	1.27	49
Paurotis Pond, Everglades, Florida, USA	0.04	0.02	49
Lake Washington, Seattle, USA	40% of TDA	60% of TDA	51
Lake Washington interstitial water, Seattle, USA (one sample)	55% of TDA	N.D.	51
Subarctic lakes[c] (pH 1), Yellowknife, NWT, Canada	N.A.	0.7–520	74
Subarctic lakes[c] (pH 6), Yellowknife, NWT, Canada	0–22	N.A.	74
Lake Biwa, Japan	<0.2	<1.9	64
Lake in Taiwan (Nanjen Hu)	0.05 ± 0.00	1.67 ± 0.07	88
Lake Pavin, France (measurements at different depths)	0–8.25	0.28–4.4	53

Table 2. (Continued)

Source	As^{III}	As^V	Reference
Seawater			
Pacific Ocean (<100 m)	1.1–1.6^b		39,96
Pacific Ocean, SE Taiwan	0.05 ± 0.02	1.28 ± 0.24	88
Western Atlantic Ocean (Stn.10)—surface	~0.3^b		45
Western Atlantic Ocean (Stn.10)—deep water (2200–3300 m)	1.5 ± 0.1^b		45
Western Atlantic Ocean (Stn.10)—bottom water (3950–4460 m)	1.5 ± 0.03^b		45
Western Atlantic Ocean (Stn.10)—80-m deep mixed layer (maximum)	0.03	N.D.	45
Western Atlantic Ocean (Stn.10)—700 m to bottom (maximum)	0.007	N.D.	45
Antarctic Ocean	0.003	1.0	54
East Indian Ocean	0.2	0.4	54
North Indian Ocean	0.2	0.5	54
Baltic Sea	N.D.–0.92	N.D.–1.01	43
Southern North Sea (over several months)	0.1–1.5^b		65
China Sea	0.2	0.3	54
Indonesian Archipelago	0.2	0.4	54
Estuaries and Coastal Waters			
Scripps Pier, La Jolla, California, USA	0.01–0.03	1.70–1.75	93
San Diego Trough, California, USA (surface to 100 m below surface)	0.01–0.06	1.32–1.67	93
Suisun Bay, California, USA	1.5^b		20
Southern California Bight (varying depths and locations)	N.D.–0.87	0.16–1.45	49
Tidal flat, Tampa, Florida, USA	0.6 ± 0.01	1.3 ± 0.01	92
Bay, Causeway, Tampa, Florida, USA	0.1 ± 0.01	1.4 ± 0.01	92
McKay Bay, Tampa, Florida, USA	0.06 ± 0.01	0.3 ± 0.01	92
Chesapeake Bay^c, Maryland, USA	N.D.–0.24	0.13–1.02	31
Patuxent River Estuary, Maryland, USA (over 2 years)	0.1–0.2	0.1–1.1	27
Saanich Inlet, Vancouver, British Columbia, Canada (oxic stations)	0.07–0.19	0.77–2.51	33
Saanich Inlet, Vancouver, British Columbia, Canada (anoxic stations)	0.07–1.91	N.D.–2.20	33
Hastings Arm^c porewaters, British Columbia, Canada	0.80–11.5	2.0–12.1	30
Alice Arm^c porewaters, British Columbia, Canada	2.9–28.8	2.3–28.8	30
Quatsino Sound^c porewaters (one station), British Columbia, Canada	N.A.	<3–10.5	30
Holberg Inlet^c porewaters (one station), British Columbia, Canada	N.A.	1.5–<13.5	30
Rupert Inlet^c porewaters (two stations), British Columbia, Canada	N.A.	<1.5–9.0	30
Surface waters of the Beaulieu Estuary, Hampshire, UK	<0.02–0.40	N.A.	17
	0.17–1.20^b		17
Deep waters of the Beaulieu Estuary, Hampshire, UK	0.43–1.07^b		17
Tamar Estuary^c, southwest England (primarily As^V)	~2.7^b		63
Tamar Estuary^c (porewaters), southwest England	~29.6^b		63
Tamar Estuary^c (porewaters), southwest England	5–60^b		97
Southampton water, UK	0.03–0.1	N.A.	26
Humbar Estuary^c, UK (over four seasons)	0.7–2.3^b		18
Thames Estuary and plume^c, UK (Feb.1989)	3.3 ± 0.9	N.A.	23
Thames Estuary and plume^c, UK (July 1990)	2.1 ± 1.0^b		23
Seine Estuary, France (over several months)	N.D.–1.2	0.4–2.1	22
Continental shelf off the Gironde Estuary, France	0.01–0.3	0.8–1.5	29
Tagus Estuary^c, Atlantic coast of Europe	2.8–14.7^b		67
Schelde Estuary^c, central Europe	0.06–0.4	1.8–4.8	50
Schelde Watershed^c, central Europe	0.003–8.2	0.9–27.7	50
Coastal water in Taiwan (Erhjin)	0.06 ± 0.04	0.88 ± 0.25	88
Coastal waters in Taiwan (Tsengwen)	0.03 ± 0.01	0.62 ± 0.05	88
Other			
Rainwater, Washington, USA	35% of TDA	65% of TDA	51
Rainwater, La Jolla, California, USA	<0.002	0.09–0.18	93
Hot Creek Gorge, Eastern Sierra Nevada, USA	0.61 ± 0.20		98
North drainage channel—Twin Butte Vista Hot Spring, Yellowstone National park, USA	610–1900	600–1900	99
Bottled water (carbonated)	1.6 ± 0.1	7.5 ± 0.3	100

N.A.: Not available (Not reported).
N.D.: not detected.
TDA: Total dissolved arsenic.
^{a}All concentrations in µg/L.
^{b}The sum of As^{III} + As^V.
^{c}Anthropogenic contamination.

transformation in natural waters is also known to be chemically slow (5,33,37,52,53). As a result, As^{III} has been observed in oxic waters, and As^V has been found in highly sulfidic water (25), contradictory to thermodynamic predictions. As^{III} was present in higher than expected amounts in the oxic epilimnion of the Davis Creek Reservoir (20). As^{III} has been detected in the oxic surface waters of lakes (24,53) and seawater (45,54). As^V has also been detected in anoxic zones (24,33,43,53).

The instability of arsenic species in water samples and the procedures used for sample handling and analysis may be the reason that As^V is commonly reported as the predominant species in water. Oxidation of As^{III} to As^V can occur during sample handling and storage. As^{III} may be present in higher concentrations in groundwater than previously reported (7). Many methods for preservation have been tested to prevent oxidation of As^{III} to As^V (55–60). On-site methods of analysis have been developed to avoid the need for preservation (61,62). As^{III} was determined as the predominant species in most groundwater samples measured from tanks and wells in Hanford, Michigan (47). In thirteen of sixteen wells and three of four tanks, As^{III} was present as $86 \pm 6\%$ of the total arsenic (47).

METHYLATED ARSENIC COMPOUNDS

Biological mediation is primarily responsible for the production and distribution of methylated arsenic species (1,3,4,6,19,22,29,31,34,35,37,42,43,45,49,63). MMA^V, MMA^{III}, DMA^V, DMA^{III}, and TMAO are the methylarsenicals present in natural waters (Table 3). It is estimated that methylarsenicals account for approximately 10% of the arsenic in the ocean (43). Methylarsenicals accounted for up to 59% of the total arsenic in lakes and estuaries in California (20). The pK_a values of MMA^V are 4.1 and 8.7, and the pK_a of DMA^V is 6.2 (8). At neutral pH, MMA^V occurs as $CH_3AsOOHO^-$ and DMA^V occurs as $(CH_3)_2AsOO^-$ (4). Trimethylarsenic (TMA) species have also been detected in some natural waters (Table 3). In the contaminated Tagus Estuary, trimethylarsenic was present at 0.010–0.042 µg/L (67). TMAO has been detected in marine interstitial waters (30) but not in surface waters (34). Methylated arsenic species are at a maximum in the euphotic zone of seawater (36,39,43). Similar to inorganic arsenic, the distribution of methylarsenicals is affected by seasonal changes in natural waters (19,20,22–24,27,32,64–66). Methylarsenicals usually occur predominantly in the pentavalent (+5) oxidation state (4).

Pentavalent arsenic species (As^V, MMA^V, DMA^V, and TMAO) are chemically or biologically reduced to trivalent arsenic species (As^{III}, MMA^{III}, and DMA^{III}), and biological methylation results in methylarsenicals. MMA^{III} and DMA^{III} are intermediates in the two-step methylation process. MMA^{III} and DMA^{III} appeared in minor fractions in Lake Biwa (eutrophic zone), Japan, and DMA^V was dominant in the summer (28,64).

PARTICULATE ARSENIC

In oxidizing conditions, As^V becomes associated with particulate material and may be released in reducing conditions. The amount of arsenic adsorbed to particulate can be substantial (62,68,69). The failure to account for arsenic in particulate results in an underestimation of total arsenic and inefficient treatment and removal of arsenic in water. The particulate matter in natural waters may occur as undissolved mineral (1,4,10) and organic species (70,71). As^V readily adsorbs and/or coprecipitates onto Fe^{III} oxyhydroxide particles (1,4,8,10,16,71). As^V and As^{III} can also react with sulfide ions to form insoluble arsenic sulfide precipitates (1,8,10,16,72,73). Under highly reducing conditions, the organic/sulfide fraction predominates (16,73).

In groundwater samples in the United States, particulate arsenic accounted for more than 50% of the total arsenic in 30% of the samples collected (69). In the Nile Delta Lakes, the arsenic budget consisted of 1.2–18.2 µg/L dissolved arsenic and between 1.2 and 8.7 µg/g of particulate arsenic (73). Significant amounts of particulate arsenic have also been detected in estuaries and coastal waters (17,23,70). In the drainage waters of an area impacted by mine wastes, particulate matter was greater than 220 times more concentrated than dissolved arsenic. Suspended matter in the deep water of Lake Washington contained up to 300 mg/g arsenic in the summer (51). In the Gironde Estuary, arsenic-containing suspended particulate matter varied with depth (5.1–26.8 µg/g) (29). The level of arsenic in phytoplankton was estimated at 6 µg/g compared to 20–30 µg/g in iron-rich and aluminum-rich terrigenous particles (29).

UNCHARACTERIZED ARSENIC SPECIES

Substantial amounts of arsenic species remain to be characterized. There are many reports of unidentified arsenic species in water (8,21,29,34,74–78). After UV irradiation of surface water from Uranouchi Inlet, Japan, the inorganic arsenic and dimethylarsenic concentrations detected by hydride generation increased rapidly (75). The UV-labile arsenic fractions represented 15–45% and 4–26% of the total dissolved arsenic in Uranouchi Inlet and Lake Biwa, respectively. In sediment porewater of Yellowknife, Canada, there was an increase of 18–420% in total dissolved arsenic concentration observed after irradiation (74). The difference between the sum of known arsenic species and total arsenic in the euphotic layer of an estuary was 13% (29).

In coastal waters, the concentration of total dissolved arsenic increased by approximately 25% following UV irradiation of the samples (78). In a National Research Council of Canada river water standard reference material, approximately 22% of the arsenic was unidentified (76). In estuarine waters, uncharacterized arsenic compounds corresponded to approximately 20% and 19% of the total arsenic content in summer and winter samples, respectively (77). Identification of these compounds is necessary to complete our understanding of the biogeochemical cycling of arsenic in the environment.

ARSENOTHIOLS

There is limited evidence that the precursors to methylarsine species detected by hydride generation may

Table 3. Organic Arsenic Species Detected in Water[a]

Source	MMA	DMA	TMA	Reference
Groundwater				
Taiwan (Pu-Tai)	~0.5–4.2	~2.0–6.9	~3.3–5.1	87
Region Lagunera, northern Mexico	<3	Trace–20		89
Rivers				
Withlacoochee River, Tampa, Florida, USA	0.06 ± 0.01	0.3 ± 0.01		92
Tamiami Canal, Florida, USA	<0.005	0.05		49
Colorado River, Parker, Arizona, USA	0.06	0.05		49
Nine rivers in California, USA	<0.007–0.56	<0.007–0.31		49
Colorado River Slough, near Topock, California	0.13	0.31		93
Haya-kawa River, Japan	N.D.	N.D.	2	95
Lakes, Ponds, Reservoirs				
Pyramid Lake, California, USA	N.D.	1.1		20
Six lakes in California, USA	N.D.–0.6	0.03–2.4		20
Davis Creek Reservoir (filtered samples), California, USA	0.18–0.3	0.03–0.9		20
Davis Creek Reservoir (unfiltered samples), California, USA	0.22–0.3	N.D.–0.8		20
Elkhorn Slough, California, USA	0.07	0.2		20
Donner Lake N. Shore, California, USA	<0.008	0.003		49
Saddleback Lake S. End, California, USA	<0.002	0.006		49
Squaw Lake (surface), California, USA	0.022	0.052		49
Squaw Lake (2-m depth), California, USA	0.020	0.11		49
Senator Wash Reservoir, California, USA	0.014	0.006		49
Lake Echols, Tampa, Florida, USA	0.11 ± 0.01	0.32 ± 0.01		92
Lake Magdalene, Tampa, Florida, USA	0.22 ± 0.01	0.15 ± 0.01		92
Remote Pond, Withlacoochee Forest, Tampa, Florida, USA	0.12 ± 0.01	0.62 ± 0.01		92
University research pond, Tampa, Florida, USA	0.05 ± 0.01	0.15 ± 0.01		92
Coot Bay Pond, Everglades, Florida, USA	<0.01	0.32		49
Paurotis Pond, Everglades, Florida, USA	<0.005	0.03		49
Lake Washington, Seattle, USA	N.D.	0.05		51
Lake Washington interstitial water, Seattle, USA (one sample)	N.D.	1		51
Subarctic lakes[c] (pH 1), Yellowknife, NWT, Canada	N.D.–0.5	N.D.–0.7	N.D.–0.2	74
Subarctic lakes[c] (pH 6), Yellowknife, NWT, Canada	N.D.–0.04	N.D.	N.D.	74
Lake Biwa, Japan	<0.05 <0.01 (MMA^{III})	<0.76 <0.01(DMA^{III})		64
Lake Pavin, France (measurements at different depths)	Not quantified	Not quantified		53
Seawater				
Pacific Ocean (<100 m)	0.009–0.02	0.02–0.2		39,96
Western Atlantic Ocean—upper 200 m	0.007	0.05		45
Antarctic Ocean	0.007	0.02		54
East Indian Ocean	0.03	0.05		54
North Indian Ocean	0.03	0.2		54
Baltic Sea	N.D.–0.03	N.D.–0.52		43
Southern North Sea (over several months)	0–0.1	0.05–0.3		65
China Sea	0.02	0.08		54
Indonesian Archipelago	0.03	0.09		54
Estuaries and Coastal Waters				
Scripps Pier, La Jolla, California, USA	0.01–0.02	0.12		93
San Diego Trough, California, USA (surface to 100 m below surface)	0.003–0.005	0.002–0.21		93
Suisun Bay, California, USA	0.1	0.07		20
Southern California Bight (varying depths and locations)	<0.002–0.031	0.01–0.26		49
Tidal flat, Tampa, Florida, USA	0.08	0.3		92
Bay, Causeway, Tampa, Florida, USA	N.D.	0.2		92
McKay Bay, Tampa, Florida, USA	0.07	1.0		92
Chesapeake Bay[c], Maryland, USA	N.D.–0.42	N.D.–0.34		31
Patuxent River Estuary, Maryland, USA (over 2 years)(max. values)	<0.3	0.2–0.6		27
Hastings Arm[c] porewaters, British Columbia, Canada	<0.03	0.09–0.23	<0.22 (TMAO)	30
Alice Arm[c] porewaters, British Columbia, Canada	<0.47	<0.19	<0.41 (TMAO)	30

Table 3. (*Continued*)

Source	MMA	DMA	TMA	Reference
Quatsino Sound[c] porewaters (one station), British Columbia, Canada	<0.18	<0.22	<0.70 (TMAO)	30
Holberg Inlet[c] porewaters (one station), British Columbia, Canada	<0.04	<0.04	<0.27 (TMAO)	30
Rupert Inlet[c] porewaters (two stations), British Columbia, Canada	<0.33	<0.24	<0.25 (TMAO)	30
Surface waters of the Beaulieu Estuary, Hampshire, UK	<0.02–0.32	<0.02–0.44		17
Deep waters of the Beaulieu Estuary, Hampshire, UK	0.03–0.26	0.19–0.40		17
Tamar Estuary[c], southwest England	0.2–0.5	0.02–1.3		63
Tamar Estuary[c] (porewaters), southwest England	0.4	0.5		63
Tamar Estuary[c] (porewaters), southwest England	0.04–0.7	0.1–0.5		97
Southampton Water, UK	0.03–0.08	0.05–0.5		26
Humbar Estuary[c], UK (over four seasons)	N.D.–0.02	N.D.–0.2		18
Thames Estuary and plume[c], UK (Feb.1989)	N.D.	N.D.		23
Thames Estuary and plume[c], UK (July 1990)	0.036 ± 0.016	0.138 ± 0.127		23
Seine Estuary, France (over several months)	N.D.–0.05	N.D.–0.3		22
Continental shelf off the Gironde Estuary, France	N.D.	0–0.1		29
Tagus Estuary[c], Atlantic coast of Europe	0.01–0.06	0.07–0.2	0.01–0.04	67
Other				
Rainwater, La Jolla, California, USA	<0.002	<0.002–0.024		93

N.D.: not detected.
TDA: Total dissolved arsenic.
[a]All concentrations in μg/L.
[b]The sum of $As^{III} + As^{V}$.
[c]Anthropogenic contamination.

be arsenic/sulfur compounds (40,74,79). Arsenic/sulfur compounds could dominate in reducing environments (4,34). The presence of arsenothiols in lake sediment porewater has been suggested (74). The presence of oxythioarsenate, $[H_3As^VO_3S]$, in water from an arsenic-rich, reducing environment has been demonstrated (80).

CONCLUDING REMARKS

Ingestion of arsenic from drinking water is a risk factor for several cancers and noncancerous health effects. Arsenic occurs naturally in the environment. Both inorganic and methylated arsenic species have been detected in aquatic systems. Inorganic As^V and As^{III} are predominant in most natural waters; methylated arsenic species occur at lower concentrations. Arsenothiols and uncharacterized arsenic species are also present in some aqueous environments. The relative abundance of various arsenic species in natural waters depends on both biological activity and chemical parameters, such as redox potential and pH.

BIBLIOGRAPHY

1. Ferguson, J.F. and Gavis, J. (1972). *Water Res.* **6**: 1259–1274.
2. Welch, A.H., Lico, M.S., and Hughes, J.L. (1988). *Ground Water* **26**: 333–347.
3. Tamaki, S. and Frankenberger, W.T. (1992). *Rev. Environ. Contam. Toxicol.* **124**: 79–110.
4. Cullen, W.R. and Reimer, K.J. (1989). *Chem. Rev.* **89**: 713–764.
5. Cherry, J.A., Shaikh, A.U., Tallman, D.E., and Nicholson, R.V. (1979). *J. Hydrol.* **43**: 373–392.
6. Andreae, M.O. (1979). *Limnol. Oceanogr.* **24**: 440–452.
7. Korte, N.E. and Fernando, L. (1991). *Crit. Rev. Environ. Control* **18**: 137–141.
8. National Research Council. (1999). *Arsenic in Drinking Water*. National Academy Press, Washington, DC.
9. Nriagu, J.O. (Ed.). (1994). *Arsenic in the Environment, Part I: Cycling and Characterization*. John Wiley & Sons, Hoboken, NJ.
10. Welch, A.H. (2000). *Ground Water* **38**: 589–604.
11. World Health Organization. (2001). *Environmental Health Criteria 224: Arsenic and Arsenic Compounds*, 2nd Edn. World Health Organization, Geneva, pp. 1–108.
12. Le, X.C. (2001). *Environmental Chemistry of Arsenic*. W.T. Frankenberger Jr. (Ed.). Marcel Dekker, New York, pp. 95–116.
13. IARC. (1987). International Agency for Research on Cancer, Suppl., Vol. 23. http://193.51.164.11/htdocs/Monographs/Suppl7/Arsenic.html.
14. National Research Council. (2001). *Arsenic in Drinking Water: 2001 Update*. National Academy Press, Washington, DC.
15. USEPA (U.S. Environmental Protection Agency) Federal Register. (2001). *National Primary Drinking Water Regulations. Arsenic and Clarifications to Compliance and New Source Contaminants Monitoring. Final Rule*, 40 CFR Parts 9, 141 and 142.
16. Mok, W.M. and Wai, C.M. (1994). *Arsenic in the Environment, Part I: Cycling and Characterization*. J.O. Nriagu (Ed.). John Wiley & Sons, Hoboken, NJ, pp. 99–117.
17. Howard, A.G., Arbab-Zavar, M.H., and Apte, S.C. (1984). *Estuar. Coast. Shelf Sci.* **19**: 493–504.
18. Kitts, H.J., Millward, G.E., Morris, A.W., and Ebdon, L. (1994). *Estuar. Coast. Shelf Sci.* **39**: 157–172.
19. Sanders, J.G., Riedel, G.F., and Osman, R.W. (1994). *Arsenic in the Environment, Part. I: Cycling and Characterization*. J.O. Nriagu (Ed.). John Wiley & Sons, Hoboken, NJ, pp. 289–308.
20. Anderson, L.C.D. and Bruland, K.W. (1991). *Environ. Sci. Technol.* **25**: 420–427.

21. Kuhn, A. and Sigg, L. (1993). *Limnol. Oceanogr.* **38**: 1052–1059.
22. Michel, P., Chiffoleau, J.F., Averty, B., Auger, D., and Chartier, E. (1999). *Cont. Shelf Res.* **19**: 2041–2061.
23. Millward, G.E., Kitts, H.J., Ebdon, L., Allen, J.I., and Morris, A.W. (1997). *Mar. Environ. Res.* **44**: 51–67.
24. Aurillo, A.C., Mason, R.P., and Hemond, H.F. (1994). *Environ. Sci. Technol.* **28**: 577–585.
25. Spliethoff, H.M., Mason, R.P., and Hemond, H.F. (1995). *Environ. Sci. Technol.* **29**: 2157–2161.
26. Howard, A.G. and Apte, S.C. (1989). *Appl. Organomet. Chem.* **3**: 499–507.
27. Riedel, G.F. (1993). *Estuaries* **16**: 533–540.
28. Sohrin, Y., Matsui, M., Kawashima, M., Hojo, M., and Hasegawa, H. (1997). *Environ. Sci. Technol.* **31**: 2712–2720.
29. Michel, P., Boutier, B., Herbland, A., Averty, B., Artigas, L.F., Auger, D., and Chartier, E. (1997). *Oceanol. Acta* **21**: 325–333.
30. Reimer, K.J. and Thompson, J.A.J. (1988). *Biogeochemistry* **6**: 211–237.
31. Sanders, J.G. (1985). *Mar. Chem.* **17**: 329–340.
32. Howard, A.G., Arbab-Zavar, M.H., and Apte, S.C. (1982). *Mar. Chem.* **11**: 493–498.
33. Peterson, M.L. and Carpenter, R. (1983). *Mar. Chem.* **12**: 295–321.
34. Francesconi, K.A. and Edmonds, J.S. (1997). *Adv. Inorg. Chem.* **44**: 147–189.
35. Maeda, S. (1994). *Arsenic in the Environment, Part I: Cycling and Characterization*. J.O. Nriagu (Ed.). John Wiley & Sons, Hoboken, NJ, pp. 155–187.
36. Francesconi, K.A. and Edmonds, J.S. (1994). *Arsenic in the Environment, Part I: Cycling and Characterization*. J.O. Nriagu (Ed.). John Wiley & Sons, Hoboken, NJ, pp. 221–260.
37. Sanders, J.G. (1980). *Mar. Environ. Res.* **3**: 257–266.
38. Johnson, D.L. and Burke, R.M. (1978). *Chemosphere* **8**: 645–648.
39. Santosa, S.J., Mokudai, H., Takahashi, M., and Tanaka, S. (1996). *Appl. Organomet. Chem.* **10**: 697–705.
40. Reimer, K.J. (1989). *Appl. Organomet. Chem.* **3**: 475–490.
41. Sanders, J.G. and Riedel, G.F. (1993). *Estuaries* **16**: 521–532.
42. Andreae, M.O. and Klumpp, D. (1979). *Environ. Sci. Technol.* **13**: 738–741.
43. Andreae, M.O. and Froelich, P.N. (1984). *Tellus* **36B**: 101–117.
44. Robertson, F.N. (1989). *Environ. Geochem. Health* **11**: 171–176.
45. Cutter, G.A., Cutter, L.S., Featherstone, A.M., and Lohrenz, S.E. (2001). *Deep Sea Res. II* **28**: 2895–2915.
46. Korte, N.E. (1991). *Environ. Geol. Water. Sci.* **18**: 137–141.
47. Hering, J.G. and Chiu, V.Q. (2000). *J. Environ. Eng.* 471–474.
48. Chen, S.L., Dzeng, S.R., Yang, M.H., Chiu, K.H., Shieh, G.M., and Wai, C.M. (1994). *Environ. Sci. Technol.* **28**: 877–881.
49. Andreae, M.O. (1978). *Deep Sea Res.* **25**: 391–402.
50. Andreae, M.O. and Andreae, T.W. (1989). *Estuar. Coast. Shelf Sci.* **29**: 421–433.
51. Crecelius, E.A. (1975). *Limnol. Oceanogr.* **20**: 441–451.
52. Cutter, G.A. (1992). *Mar. Chem.* **40**: 65–80.
53. Seyler, P. and Martin, J.M. (1989). *Environ. Sci. Technol.* **23**: 1258–1263.
54. Santosa, S.J., Wada, S., and Tanaka, S. (1994). *Appl. Organomet. Chem.* **8**: 273–283.
55. Raessler, M., Michalke, B., Schramel, P., Schulte-Hostede, S., and Kettrup, A. (1998). *Int. J. Environ. Anal. Chem.* **72**: 195–203.
56. Edwards, L., Patel, M., McNeill, L., Frey, M., Eaton, A.D., Antweller, R.C., and Taylor, H.E. (1998). *J. Am. Water Works Assoc.* **90**: 103–113.
57. Borho, M. and Wilderer, P. (1997). *SRT Aqua* **46**: 138–143.
58. Gallagher, P., Schwegel, C.A., Wei, X., and Creed, J.T. (2001). *J. Environ. Monitoring* **3**: 371–376.
59. Hall, G.E.M. and Pelchat, J.C. (1999). *J. Anal. At. Spectrom.* **14**: 205–213.
60. Lindemann, T., Prange, A., Dannecker, W., and Neidhart, B. (2000). *Fresenius J. Anal. Chem.* **368**: 214–220.
61. Kim, M.J. (2001). *Bull. Environ. Contam. Toxicol.* **67**: 46–51.
62. Le, X.C., Yalcin, S., and Ma, M. (2000). *Environ. Sci. Technol.* **24**: 2342–2347.
63. Howard, A.G., Apte, S.C., Comber, S.D.W., and Morris, R.J. (1988). *Estuar. Coast. Shelf Sci.* **27**: 427–433.
64. Hasegawa, H. (1997). *Appl. Organometal. Chem.* **11**: 305–311.
65. Millward, G.E., Kitts, H.J., Comber, S.D.W., Ebdon, L., and Howard, A.G. (1996). *Estuar. Coast. Shelf Sci.* **43**: 1–18.
66. Hasegawa, H. (1996). *Appl. Organometal. Chem.* **10**: 733–740.
67. de Bettencourt, A.M. (1988). *Neth. J. Sea Res.* **22**: 205–212.
68. Roussel, C., Bril, H., and Fernandez, A. (2000). *J. Environ. Qual.* **29**: 182–188.
69. Chen, H.W., Frey, M.M., Clifford, D., McNeill, L.S., and Edwards, M. (1999). *J. Am. Water Works Assoc.* **91**: 74–85.
70. Waslenchuk, D.G. and Windom, H.L. (1978). *Estuar. Coast. Shelf Sci.* **7**: 455–464.
71. Mucci, A., Richard, L.F., Lucotte, M., and Guignard, C. (2000). *Aqua. Geochem.* **6**: 293–324.
72. Kim, M.J., Nriagu, J.O., and Haack, S. (2000). *Environ. Sci. Technol.* **34**: 3094–3100.
73. Abdel-Moati, A.R. (1990). *Water Air Soil Pollut.* **51**: 117–132.
74. Bright, D.A., Dodd, M., and Reimer, K.J. (1996). *Sci. Total Environ.* **180**: 165–182.
75. Hasegawa, H., Matsui, M., Okamura, S., Hojo, M., Iwasaki, N., and Sohrin, Y. (1999). *Appl. Organomet. Chem.* **13**: 113–119.
76. Sturgeon, R.E., Siu, K.W.M., Willie, S.N., and Berman, S.S. (1989). *Analyst* **114**: 1393–1396.
77. de Bettencourt, A.M. and Andreae, M.O. (1991). *Appl. Organomet. Chem.* **4**: 111–116.
78. Howard, A.G. and Comber, S.D.W. (1989). *Appl. Organomet. Chem.* **3**: 509–514.
79. Bright, D.A., Brock, S., Cullen, W.R., Hewitt, G.M., Jafaar, J., and Reimer, K.J. (1994). *Appl. Organomet. Chem.* **8**: 415–422.
80. Schwedt, G. and Rieckhoff, M. (1996). *J. Chromatogr. A* **736**: 341–350.
81. Mok, W.M., Riley, J.A., and Wai, C.M. (1988). *Water Res.* **22**: 769–774.
82. Boyle, D.R., Turner, R.J.W., and Hall, G.E.M. (1998). *Environ. Geochem. Health* **20**: 199–212.

83. Chatterjee, A., Das, D., Mandal, B.K., Chowdhury, T.R., Samanta, G., and Chakraborti, D. (1995). *Analyst* **120**: 643–650.

84. Tanabe, K., Yokota, H., Hironaka, H., Tsushima, S., and Kubota, Y. (2001). *Appl. Organomet. Chem.* **15**: 241–251.

85. Nag, J.K., Balaram, V., Rubio, R., Alberti, J., and Das, A.K. (1996). *J. Trace Element Med. Biol.* **10**: 20–24.

86. Chen, S.L., Yeh, S.J., Yang, M.H., and Lin, T.H. (1995). *Biol. Trace Element Res.* **48**: 263–274.

87. Lin, T.H., Huang, Y.L., and Wang, M.Y. (1998). *J. Toxicol. Environ. Health* **53**: 85–93.

88. Hung, T.C. and Liao, S.M. (1996). *Toxicol. Environ. Chem.* **56**: 63–73.

89. Del Razo, L.M., Arellano, M.A., and Cebrian, M.E. (1990). *Environ. Pollut.* **64**: 143–153.

90. Kondo, H., Ishiguro, Y., Ohno, K., Nagase, M., Toba, M., and Takagi, M. (1999). *Water Res.* **33**: 1967–1972.

91. Raessler, M., Michalke, B., Schulte-Hostede, S., and Kettrup, A. (2000). *Sci. Total Environ.* **258**: 171–181.

92. Braman, R.S. and Foreback, C.C. (1973). *Science* **182**: 1247–1249.

93. Andreae, M.O. (1977). *Anal. Chem.* **49**: 820–823.

94. Yalcin, S. and Le, X.C. (2001). *J. Environ. Monitoring* **3**: 81–85.

95. Kaise, T.M. et al. (1997). *Appl. Organomet. Chem.* **11**: 297–304.

96. Santosa, S.J., Wada, S., Mokudai, H., and Tanaka, S. (1997). *Appl. Organomet. Chem.* **11**: 403–414.

97. Ebdon, L., Walton, A.P., Millward, G.E., and Whitfield, M. (1987). *Appl. Organomet. Chem.* **1**: 427–433.

98. Wilkie, J.A. and Hering, J.G. (1998). *Environ. Sci. Technol.* **32**: 657–662.

99. Gihring, T.M., Druschel, G.K., McCleskey, R.B., Hamers, R.J., and Banfield, J.F. (2001). *Environ. Sci. Technol.* 3857–3862.

100. Yalcin, S. and Le, X.C. (1998). *Talanta* **47**: 787–796.

ARSENIC HEALTH EFFECTS

I. Katsoyiannis
A.I Zouboulis
Aristotle University of Thessaloniki
Thessaloniki, Greece

INTRODUCTION

The toxicity of inorganic arsenic species depends mainly on their valence state (usually −3, +3, or +5) and also on the specific physical and chemical properties of the compounds in which it occurs. Trivalent (arsenite, As^{III}) compounds are generally more toxic than pentavalent (arsenate, As^V) compounds, whereas arsenic compounds that are more soluble in water usually are more toxic and more likely to present systemic effects, in comparison with less soluble compounds, which are more likely to cause chronic pulmonary effects, if inhaled. Among the most toxic inorganic arsenic compounds is arsine gas (AsH_3). Additionally, note that laboratory animals (commonly used for toxicity evaluation) are generally less sensitive than humans to the toxic effects of inorganic arsenic compounds.

The toxicity "scale" of arsenic compounds is as follows (in the order of decreased toxicity):

Arsine > Arsenites > Arsenates
> Organoarsenic compounds

Humans are usually exposed to arsenic compounds primarily through (contaminated) air, food, and/or water sources. The concentration of arsenic in air is usually in the range of a few ng As/m^3, although the relevant exposure may be higher in highly polluted areas. The consumption of contaminated water for drinking is also an important source of arsenic exposure. The concentration of arsenic is generally higher in groundwater than in surface water, especially when and where specific geochemical conditions favor dissolution of arsenic minerals.

Due to the widespread distribution of this element in major environmental compartments, it should not be surprising that most people experience a (more or less) measurable arsenic intake each day. It has been reported that exposures of the general population to inorganic arsenic are between 5 and 30 µg/day, which come from air, food, and water. In particular, it has been estimated that humans can get around 0.05 µg As/day from the air, 1–10 µg As/day from (drinking) water, and 5–20 µg As/day from food. Another 1–20 µg As/day can be absorbed, when someone is also a smoker. The U.S. Food and Drug Administration (U.S. FDA) has calculated the mean daily intake of inorganic arsenic at about 0.5 µg/kg body weight of an adult; this corresponds to a range of 30–38 µg/day for adults weighing 60–75 kg. The Food and Agriculture Organization (FAO) and the World Health Organization (WHO) have estimated the safe daily doses or tolerable daily intakes of inorganic arsenic. Provisional values for adults are 2.1 µg As/kg body weight, which equals about 126–160 µg As/day for adults weighing 60–75 kg.

From the previous information, it becomes obvious that the recent (up to 1998) maximum concentration limit (MCL) of 0.05 mg/L of arsenic in drinking water is not protective enough. Considering that humans consume about 2–3 liters of water daily, when the water contains about 50 µg/L of arsenic, this results in a daily arsenic intake of 100–150 µg/day, in addition to another 30–38 µg/day, which has been calculated by the U.S. FDA; therefore, the total arsenic intake of humans, who drink water containing 50 µg/L of arsenic, is up to 130–188 µg/day. This might be higher than the safe daily dose for arsenic, as proposed by the WHO. As a result, the concentration limit of arsenic in drinking water has recently been lowered. In Europe, the respective limit, according to the E.C. Directive 98/83, is now 10 µg/L. In the United States, after several discussions of this issue, the arsenic limit in drinking water remained 50 µg/L up to January 22, 2001. Then, the U.S. Environmental Protection Agency (1) published a relevant final rule, which includes the revised standard for arsenic in drinking water, of 10 µg/L.

METABOLISM AND DISPOSITION

Once arsenic compounds are ingested, the soluble forms of arsenic are readily absorbed from the gastrointestinal (GI) tract. Arsenate, whether in inorganic or in organic form, is better absorbed than arsenite because it is less reactive with the membranes of the GI tract.

The absorption of water-soluble inorganic arsenic compounds through the GI tract is very high. In humans, absorption rates of 96.5% for trivalent sodium arsenite and 94% for soluble sodium arsenate have been reported. In contrast, the GI absorption of less soluble trisulfide arsenate was reportedly only 20–30% in hamsters. The absorption of arsenic in the lungs depends on the size of the particles onto which arsenic compounds have been deposited, as well as on their respective solubility; respirable particles (i.e., 0.1–1 µm in diameter) can be carried deeper into the lungs, and therefore they are more likely to be absorbed.

Once absorbed, the blood transports arsenic to different body organs, such as the liver, kidney, lung, spleen, aorta, and skin. It is worth noting that, except for skin, clearance from these organs is relatively rapid. Arsenic also deposits extensively in the hair and nails (2). Arsenic compounds are then subject to metabolic transformation. Pentavalent arsenic compounds are reduced to trivalent forms, and then they are methylated in the liver to the less toxic methylarsinic acids. Typical levels in the blood of people who are not exposed to significant sources of arsenic pollution are in the range of 1–5 µg As/L (3). Finally, arsenic can be removed (cleared) from the body relatively rapidly, primarily through urine. Urinary excretion rates of 80% within 61 h, following oral doses, and 30–80% within 4–5 days following parenteral doses, have been measured in humans (4). Arsenic can also be lost from the body through the hair and nails, because they represent a nonbiologically available arsenic pool. In Fig. 1, the main routes of the fate, distribution, and

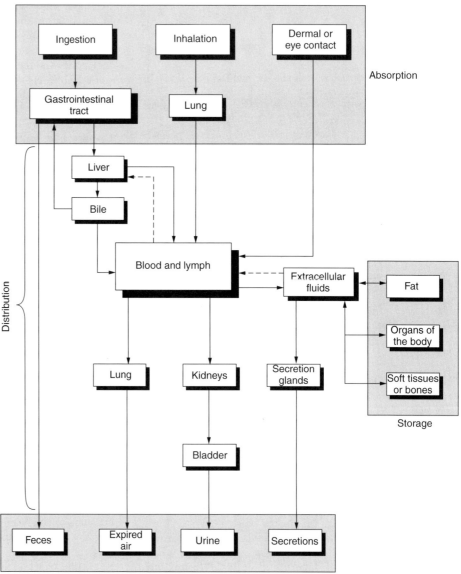

Figure 1. Fate, distribution, and excretion of toxic chemicals in the human body.

excretion of toxic chemicals, such as arsenic, for humans are presented.

MECHANISMS OF TOXICITY

The main inorganic arsenic species (i.e., As^V and As^{III}) have different mechanisms of action on which their toxicity depends. Arsenates behave similarly to phosphates. Consequently, they can substitute for phosphates in common cell reactions, whereas arsenites have high affinity for the thiol groups of proteins, causing inactivation of several enzymes.

In particular, the structural similarity of arsenates to phosphates allows them to substitute for phosphates in energy-producing reactions within the cell. First, arsenate can replace phosphate during glycolytic phosphorylation; if this occurs, glycolysis can continue, but ATP-producing reactions do not take place and, therefore, the cell produces less ATP. Second, arsenates can uncouple oxidative phosphorylation by substituting for phosphates in the ATP synthetase enzyme. As electrons are transferred to oxygen, ADP-arsenate (rather than ATP) is formed, which rapidly hydrolyzes. Hence, energy is wasted from the electron transport chain, because it cannot be stored appropriately. In addition, arsenate may also exert its toxic effects indirectly via reductive metabolism to arsenite.

The key to the toxicity of arsenite is its electrophilic nature; arsenite binds to electron-rich sulfydryl groups on proteins. Although such binding exerts adverse effects on structural proteins, such as the microfilaments and microtubules of the cytoskeleton in cultured cells, most of arsenite's toxicity is likely due to the inhibition of enzymes by binding to a thiol-containing active site. In particular, arsenite is known to inhibit enzymes of the mitochondrial citric acid cycle, and cellular ATP production decreases. Arsenite also uncouples oxidative phosphorylation. Arsenic's multipronged attack on the cell's energy production system can adversely affect cellular health and survival (5,6).

In contrast to inorganic arsenic, organoarsenic compounds, such as MMA or DMA, can bind strongly to biological molecules of humans, resulting in less toxic effects.

ARSENIC HEALTH EFFECTS

Acute arsenic exposure (i.e., high concentrations ingested during a short period of time) can cause a variety of adverse effects. The severity of effects depends strongly on the level of exposure. Acute high-dose oral exposure to arsenic typically leads to gastrointestinal irritation, accompanied by difficulty in swallowing, thirst, and abnormally low blood pressure. Death may also occur from cardiovascular collapse. The lethal dose to humans is estimated at 1–4 mg As/kg for an adult (3,7,8). Short-term exposure of humans to doses higher than 500 µg/kg/day can cause serious blood, nervous system, and gastrointestinal ill effects and also may lead to death.

Chronic arsenic uptake may have noncarcinogenic, as well as carcinogenic, effects on humans. Chronic exposure to low arsenic concentrations is of primary interest, when the health significance of arsenic in drinking

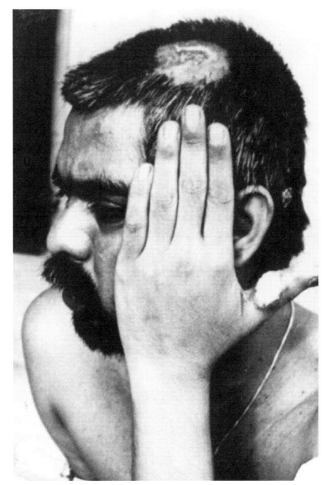

Figure 2. Skin cancer on the head caused by chronic arsenic intake.

water is evaluated. The most common signs of long-term, low-level arsenic exposure are dermal changes. These include variations in skin pigments, hyperkeratosis, and ulcerations. Vascular effects have also been associated with chronic arsenic exposure.

Chronic arsenic exposure can also lead to carcinogenesis in humans. Arsenic is classified as a human carcinogen, according to the U.S. EPA. This classification was based mainly on human data because the respective animal data were inadequate. Several epidemiological studies have documented an association between chronic exposure to arsenic in drinking water and skin cancer. A correlation between chronic arsenic exposure and cancer in the liver, bladder, kidney, lung, and prostate has also been documented. Arsenic contamination is of major concern in several countries, such as Bangladesh, Chile, and Taiwan. Figures 2–5, taken of people suffering from arsenicosis in these countries, show the significance of the arsenic problem, mainly due to the consumption of contaminated drinking water, and indicate clearly the urgent need for effective treatment of groundwater to remove arsenic (3,7,8).

As can be noticed in these photos, chronic exposure to arsenic can cause skin cancer in several parts of the human body: on the head, on the hand, and on the foot. In India,

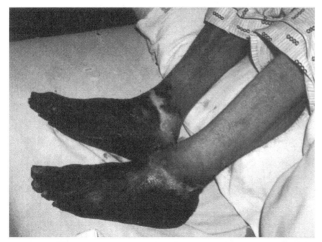

Figure 3. Skin cancer on the foot known as "black-foot" disease, caused by chronic arsenic intake.

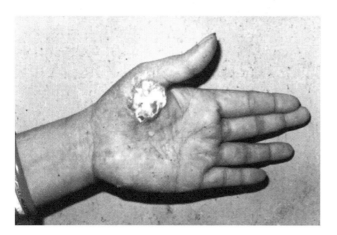

Figure 4. Skin cancer on the hand.

in the area of West Bengal, more than 40 million people suffer from different kinds of arsenicosis, and 37% of the water samples, which have been analyzed, contain arsenic concentrations higher than 50 µg/L. In Bangladesh, more than 70 million people suffer from several kinds of cancer due to chronic exposure to arsenic. (9). This problem also exists in several areas of the United States, as well as in Europe. In the United States, more than 3 million people drink water whose arsenic concentrations are higher than 50 µg/L, whereas in Europe, arsenic concentrations over the 10 µg/L concentration limit have often been reported in Germany, Hungary, Finland, and Greece (7).

BIBLIOGRAPHY

1. U.S. EPA. Arsenic in drinking water: health effects research. Available at: www.epa.gov/OGWDW/ars/ars10.html.
2. U.S. Environmental Protection Agency. (1984). *Health Assessment Document for Arsenic*. Office of Health and Environmental Assessment—Environmental Criteria and Assessment Office, EPA Report 600/8-32-021F.
3. Pontius, F.W., Brown, K.G., and Chen, C.-J. (1994). Health implications of arsenic in drinking water. *J. Am. Water Works Assoc.* **86**(9): 52–63.

Figure 5. Scientist measuring arsenic in contaminated groundwater wells of Bangladesh. (Courtesy Stevens Institute of Technology.)

4. Crecelius, E.A. (1977). Changes in the chemical speciation of arsenic following ingestion by man. *Environ. Health Perspect.* **19**: 147–150.
5. Desesso, J.M. et al. (1998). An assessment of the developmental toxicity of inorganic arsenic. *Reprod. Toxicol.* **12**(4): 385–433.
6. Abernathy, C.O., Lin, Y.P., Longfellow, D., Aposhian, H.V., Beck, B., Fowler, B., Goyer, R., Menler, R., Rossman, T., Thompson, C., and Waalkes, M. (1997). Proceedings of the Meeting on Arsenic: Health Effects, Mechanisms of Action and Research Issues, Hunt Valley, MD, Sept. 22–24. *Environ. Health Perspect.* **107**(7): 593–597.
7. World Health Organization. Toxic effects of arsenic in humans. Available at: www.who.int/peh-super/Oth-lec/Arsenic/Series4/002.htm
8. Saha, J.C., Dikshit, A.K., Barley, M.K., and Saha, K.C. (1999). A review of arsenic poisoning and its effects on human health. *Crit. Rev. Environ. Sci. Technol.* **29**(3): 281–313.
9. Karim, M.M. (1999). Arsenic in groundwater and health problems in Bangladesh. *Water Res.* **34**: 304–310.

BACKGROUND CONCENTRATION OF POLLUTANTS

MÁRIO ABEL GONÇALVES
Faculdade de Ciências da
Universidade de Lisoba
Lisoba, Portugal

The background concentration of an element or compound in a system may be considered as the averaged most common concentration defined from a set of representative samples. Traditionally, statistical techniques, such as moving averages and probability plots, are used to define these values quantitatively, which means that a statistically representative number of samples are required, and that samples represent the same general set of physical and chemical characteristics of the system. These approaches are most common to geochemical exploration surveys, through soil and rock analyses. Characterization of natural waters is based on the same principles, but the difficulties in achieving these principles are markedly different.

Defining background concentrations of pollutants in waters is of fundamental importance to establish water quality standards. Several pollutants occur in surface and groundwaters in a range of concentrations exclusively related to natural phenomena. However, anthropogenic activities induce large perturbations in most water systems. One major difficulty facing chemical characterization of surface and groundwater systems is their dynamic nature, which means that concentration patterns change with time. Once introduced in the water system, the fate and spatial distribution of a pollutant are determined by factors such as advection, diffusion, chemical reactivity, and biodegradation.

The large majority of the U.S. Environmental Protection Agency (EPA) list of priority pollutants refers to organic compounds that are solely due to human activities, and most of the inorganic pollutants occur in the natural waters at trace element concentrations (<1 mg/L). Thus, to define the background concentration of pollutants, one must distinguish those that occur naturally, with or without anthropogenic-induced disturbances, from those exclusively due to anthropogenic activities. Inorganic pollutants are clearly in the former group whereas organic compounds fall into the latter.

FACTORS DETERMINING WATER CHEMISTRY

Factors that influence the natural concentrations of major and minor elements in surface waters include the lithology, relief, climate, atmospheric dry and wet deposition, and human activities. Global river chemistry shows an enormous range in concentrations, ionic ratios, and proportion of ions in cation and anion sums spanning several orders of magnitude (1,2). It is unrealistic to define any kind of average of global water chemistry composition based on the large river basins. Besides, Maybeck (1) notes that river chemistry is very sensitive to alteration by human activities, and that, in the northern hemisphere, it is now difficult to find a medium-sized basin not significantly impacted by human activities. It must also be stressed that, between different chemical elements and species, different degrees of chemical reactivity exist, such as precipitation and incorporation into mineral phases, bonding to organic functional groups, and adsorption onto inorganic solid surfaces. These factors and the amount of colloids and suspended solid material in natural waters determine much of the element dispersion and concentration decay from source. Nriagu and Pacyna (3) estimated that human-induced mobilization of several trace metals far exceeds the natural fluxes, and enrichment factors in the range of 3–24 are reported for such elements as As, Cd, Pb, Se, and Hg.

Groundwaters are less sensitive to some of the external influences as the ones previously mentioned. However, understanding groundwater geochemistry is difficult because of the chemical heterogeneity of most aquifer systems. The residence time of groundwater is a factor that influences its final chemistry and depends on permeability and transmissivity and the amount of recharge of the aquifer. Groundwaters with long residence time in the reservoir become closer to chemical equilibrium with the minerals of the host rocks. In turn, low residence times of waters associated with the slow kinetics of silicate dissolution put severe constraints on groundwater chemistry. Nevertheless, cases where rapid ongoing dissolution and precipitation reactions are taking place, such as in some aquifers, result in relatively unaffected groundwater chemistry, irrespective of flow rate and direction. These are just some examples that illustrate the difficulty in defining acceptable background concentrations of chemical elements in several water systems.

BACKGROUND CONCENTRATION OF INORGANIC POLLUTANTS

Nitrate, ammonia, and trace metals are the most important inorganic pollutants present in waters. Several surface and groundwaters contain natural (background) concentrations of one or more of these and other chemical elements and species exceeding, for example, the U.S. EPA drinking water standards for reasons totally unrelated to human activities. Thus, identifying such natural sources is fundamental for regulatory decisions to avoid the assignment of unrealistic cleanup goals below such natural background concentrations. One of the most striking examples comes from groundwater arsenic contamination in Bangladesh (4), whose mechanisms, although attributable to iron oxide reductive dissolution, are still subject of intense debate. Of the millions of tube wells, about one-third have arsenic concentrations above the local drinking water standard (50 μg/L), and half of them do not meet the 10 μg/L guideline value of the World Health Organization (WHO) (4).

The approach to determine background concentrations of inorganic pollutants in water systems requires the sample of stream waters and/or groundwaters from boreholes in nearby areas in the same geological setting, where water chemistry has presumably not been affected by human activity. Regretfully, this approach is not always possible in several areas. In such cases, it might be reasonable to assume natural background concentrations equal to the measured concentrations in streams and groundwaters in the same general area and similar geologic environments, not forgetting that such factors as climate, relief, and aquifer recharge and hydraulic properties must also be evaluated. Sources of data on the "typical" chemical composition of groundwaters from different rock types also exist [see references in (5)]. However, the factors that influence water chemistry are varied and may differ between similar geological settings, and so the previous approaches are far from being reliable.

Use of cumulative probability plots of all the data for the area of interest can be a way to identify background concentrations, allowing the classification of samples into uncontaminated and contaminated groups (6). Some geostatistical techniques also aim to provide a separation between anomalous and background values in samples, especially factorial kriging (7). However, this does not avoid the need to have truly uncontaminated samples. Another shortcoming is that not all elements spread equally through a surface or groundwater system. Nitrates, for example, are highly mobile because they are not limited by solubility constraints. Adsorption onto solid phases is a mechanism that may rapidly remove trace metals from solution, such as Pb and Cd. Thus, what may be an uncontaminated water sample with respect to a given element may not hold in relation to other elements.

BACKGROUND CONCENTRATION OF ORGANIC POLLUTANTS

Organic pollutants are compounds that were mostly introduced in the environment by anthropogenic activities. Contrary to metals (the most important class of inorganic pollutants), several of these compounds are biodegradable, which, coupled to other mechanisms such as abiotic decomposition (including hydrolysis, oxidation reduction, and elimination), adsorption, dispersion, and dilution, contributes to what is called "natural attenuation." Their source is varied and includes spilling and leakage from underground storage tanks and landfills, and several commercial and industrial activities, to name but a few. Background concentration of these pollutants is solely the result of human impact on the water system. In this case, organic pollutants should not overcome the maximum concentration level for drinking water standard in agreement with the various national regulations. For water quality assessment, it is important to have a good knowledge of organic pollutant dispersion and persistence in water systems, and the same basic principles presented also apply to sampling in this case. As a result of the previously mentioned mechanisms, these compounds have a complex behavior in water systems. Chlorinated solvents, for example, sometimes behave as conservative solutes that are rapidly transported, but they also undergo several microbial degradation processes causing them to rapidly disappear and be replaced by the lightly chlorinated ethenes. Some case studies describe the development of a dynamic steady state in plumes of hydrocarbons that stopped spreading because the rate of input of soluble hydrocarbons was balanced by biodegradation mechanisms that consumed hydrocarbons in the plume. Thus, knowledge of these processes is fundamental to evaluate and prevent organic pollutant concentrations to build up above levels considered potentially harmful to human and ecological health.

BIBLIOGRAPHY

1. Meybeck, M. (2004). Global occurrence of major elements in rivers. In: *Treatise on Geochemistry—Surface and Ground Water, Weathering, and Soils*. J.I. Drever (Ed.). Vol. 5, pp. 207–223, Elsevier, B.V.
2. Gaillardet, J., Viers, J., and Dupré, B. (2004). Trace elements in river waters. In: *Treatise on Geochemistry—Surface and Ground Water, Weathering, and Soils*. J.I. Drever (Ed.). Vol. 5, pp. 225–272, Elsevier, Canada.
3. Nriagu, J.O. and Pacyna, J.M. (1988). Quantitative assessment of worldwide contamination of air, water and soils by trace-metals. *Nature* **333**: 134–139.
4. BGS and DPHE. (2001). *Arsenic Contamination of Groundwater in Bangladesh*, D.G. Kinniburgh and P.L. Smedley (Eds.). British Geological Survey, Keyworth, England.
5. Langmuir, D. (1996). *Aqueous Environmental Geochemistry*. Prentice-Hall, Upper Saddle River, NJ.
6. Hann (1977). *Statistical Methods in Hydrology*. Iowa University Press, Iowa City, IA.
7. Wackernagel, M. (1995). *Multivariate Geostatistics*. Springer-Verlag, Berlin.

WATERBORNE BACTERIA

BARRY H. ROSEN
US Fish & Wildlife Service
Vero Beach, Florida

CHARACTERISTICS OF BACTERIA

Bacteria are so widespread that only the most general statements can be made about their life history and ecology. They are everywhere on the earth, even in the most hostile environments, including extremes of temperature, such as freshwater hot springs. Many live in water; however, they are also an important part of soils and live in and on plants and animals; others are parasites of humans, animals, and plants. Only a handful of waterborne bacteria cause diseases, usually by producing toxins. Most bacteria have a beneficial role in decomposing dead material and releasing nutrients back into the environment.

Bacteria are a group of microorganisms that lack membrane-bound organelles and hence are considered simpler than plant and animal cells. They are separated into gram-positive and gram-negative forms based on the staining properties of the cell wall. Most bacteria are unicellular and are found in various shapes: spherical (coccus), rod-shaped (bacillus), comma-shaped (vibrio), spiral (spirillum), or corkscrew-shaped (spirochete). Generally, they range from 0.5 to 5.0 micrometers in size. Motile species (those that can move on their own) have one or more fine flagella arising from their surface. Many possess an outer, slimy capsule, and some have the ability to produce an encysted or resting form (endospore). Bacteria reproduce asexually by simple division of cells and rarely by conjugation.

Bacteria have a wide range of environmental and nutritional requirements. One way to classify them is based on their need for oxygen. *Aerobic bacteria* thrive in the presence of oxygen and require it for continued growth and existence. *Anaerobic bacteria* thrive in oxygen-free environments, as often found in lake or wetland sediments. *Facultative anaerobes* can survive in either environment, although they prefer oxygen. The way bacteria obtain energy provides another means of understanding and classifying organisms. *Chemosynthetic bacteria* are autotrophic and obtain energy from the oxidation of inorganic compounds such as ammonia, nitrite (to nitrate), or sulfur (to sulfate). *Photosynthetic bacteria* convert sunlight energy into carbohydrate energy. Bacteria such as the purple sulfur bacteria, purple nonsulfur bacteria, green sulfur bacteria, and green bacteria have a bacterial form of chlorophyll. Cyanobacteria, commonly called blue-green algae, are a separate group that contains chlorophyll a, a pigment that is common to eukaryotic algae and higher plants. *Heterotrophic* bacteria form a diverse group that obtains energy from other organisms, either while they are alive (parasites) or dead (saprophytes).

BACTERIAL IDENTIFICATION

Bacterial size and a lack of visual cues prevent the identification of organisms by traditional microscopic techniques. Typical bacterial identification involves isolating organisms by culturing them on a variety of media that reveal the organism's physiological or biochemical pathways. For example, purple nonsulfur photosynthetic bacteria can be isolated from lake sediments by inoculating a mineral salt–succinate broth and incubating in light at 30 °C in a bottle sealed to ensure anaerobic conditions.

INDICATORS OF BACTERIAL/FECAL CONTAMINATION

One emerging water quality issue is contamination of surface and groundwater by bacteria and other microorganisms that are defined as pathogens—organisms that cause disease in animals or plants. The difficulty in direct detection of bacteria in water has led to the use of fecal bacteria as an indicator of the presence of pathogens and the risk of disease. Rapid, inexpensive techniques have been standardized for determining if fecal material has contaminated water, although the sources of that contamination are numerous. Wildlife, pets and companion animals, agricultural animals, and humans are all possible sources (Fig. 1).

Although indicator bacteria are not pathogenic themselves, high numbers may indicate fecal contamination from leaky septic tanks, animal manure, or faulty wastewater treatment facilities. Some species also live in soil and on plants and are harmless. Total coliform is the broadest category (Table 1) of indicator bacteria, and it was originally believed that it indicates the presence of fecal pollution. Numerous nonfecal sources make this indicator too generic. Fecal coliform, a subgroup of total coliform, originates from the intestinal tract of warm-blooded animals. This subgroup is the most commonly used indicator of bacterial pollution in watersheds. *Escherichia coli* is a member of the fecal coliform subgroup. This subgroup is used because it correlates well with illness from swimming and can cause gastrointestinal problems. Fecal streptococci, also called fecal strep, are another grouping of bacteria, similar to the coliforms that are associated with feces from warm-blooded animals. Enterococci are a subgroup of fecal strep bacteria. This subgroup is used because it correlates well with human illness from recreational waterbodies. Enterococci and *E. coli* are considered to have a higher degree of association with outbreaks of gastrointestinal illness than fecal coliforms, as indicated by the U.S. Environmental Protection Agency. State and local government agencies commonly monitor for total or fecal coliform, and some monitor *E. coli* and enterococci. States and local health agencies may have more stringent standards than national guidelines (Table 1).

Escherichia coli O157:H7 is a potentially deadly fecal bacteria that can cause bloody diarrhea and dehydration in humans. The combination of letters and numbers in the name of this bacterium refers to the specific molecular markers found on its cell surface; they distinguish it from other types of *E. coli,* most of which are part of the normal bacterial flora in warm-blooded animal intestinal tracts. This organism does pose a threat to bathers or others from bodily contact in contaminated waters, although the majority of outbreaks are from contaminated food.

BACTERIAL SOURCE TRACKING

DNA fingerprinting, one tool for bacterial source tracking (BST), consists of a family of techniques that are under development to identify the sources of fecal contamination in various waterbodies. The goal is to identify the source, whether human, livestock, pets and companion animals, or wildlife. In theory, this allows targeting management activities for bacterial reduction on the appropriate sources.

The EPA periodically reports data submitted by states on impairments to rivers, lakes, and estuaries. Bacteria and pathogens, based on fecal indicator bacteria data collected by states, frequently exceed water quality criteria. An individual waterbody or segment of a river that exceeds these criteria is listed by the state [EPA's 303(d) list] and may trigger additional efforts to improve the waterbody. One mechanism for improving water quality that is currently being used by the EPA is the total maximum daily load (TMDL). Microbial source tracking techniques are currently the best technology for tracking sources of fecal contamination and can play an important role in TMDL development. For example, if fecal contamination is a major issue in a particular watershed, BST might show that the contamination originates from humans, not livestock. The TMDL should then reflect an appropriate proportion of reduction in bacterial loading from agriculture and from urban areas (leaky septic systems or ineffective sewage treatment plants).

Understanding the concept of how BST works, including its limitations, can help determine if these techniques can be used in a particular situation. Although DNA fingerprinting has received the greatest amount of attention recently, other BST methods are in use and can play an important part in fecal source tracking.

For most of these techniques, bacteria from known sources (humans, swine, raccoons, deer, cows, etc.) are collected directly from the animal, then isolated and grown

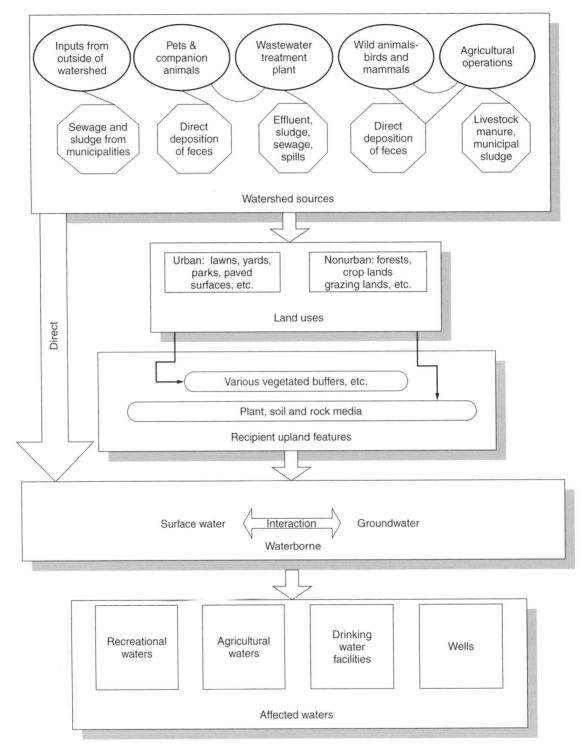

Figure 1. Potential pathways for pathogen movement into water.

in the laboratory. One species of bacteria, *E. coli*, has been extensively used, but others are more likely to be found in particular situations. For example, fecal streps (e.g., *Streptococcus feacalis*) are very numerous compared to *E. coli* in composted poultry litter or in high quality biosolids. These isolates grown in the laboratory are the basis of the library or database for subsequent comparison to unknown samples.

The current state of research normally involves isolating and culturing as many organisms as practical. Financial constraints usually limit this approach; several hundred may be needed from each major source to identify sufficiently the source in a water sample. In addition, regional differences in the genetics of isolates are just coming to light, potentially restricting the broader use of genetic libraries.

Table 1. Comparison of Fecal Bacteria Water Quality Indicators Commonly Used

Microbial Indicator	Properties	Federal Standard[a,b]
Total coliforms (TC)	• originally believed to indicate the presence of fecal pollution • widely distributed in nature: soils, water, flora, fauna • contains members of *Escherichia, Citrobacter, Klebsiella,* and *Enterobacter* identified by incubation at 35 °C	1000 CFU/100 mL
Fecal coliforms (FC)	• subgroup of TC • coliforms that originate specifically from intestinal tracts of warm-blooded animals • cultured by increasing the incubation temperature to 44.5 °C • remains the predominant indicator used to assess bacterial pollution in watersheds	200 CFU/100 mL
Escherichia coli	• member of the FC group • presence correlates with illness from swimming in both fresh and marine waters • has been shown epidemiologically to cause gastrointestinal symptoms • O157:H7 is a toxin-producing strain of this common bacterium	126 CFU/100 mL
Enterococci	• subgroup of the FC, including *Streptococcus faecalis, S. faecium,* and *S. avium* • commonly found in intestinal tracts of humans and other warm-blooded animals • presence correlates well with illness from both fresh and marine waters	33 CFU/100 mL

[a] Individual states may have higher standards but not lower; CFU = colony forming units.
[b] Primary contact water includes recreational use such as swimming and fishing.

CURRENT TECHNIQUES UNDER DEVELOPMENT

Molecular Methods (Genotype)

The overall intestinal environment in each type of animal is different enough to allow selective pressures on the microbial flora, resulting in populations of bacteria that have slightly different genetics from each species. The purpose of these molecular techniques is to find that difference among the genetics of the same type of bacteria isolated from these different animals. Commonly used methods, such as restriction endonuclease and polymerase chain reaction (PCR) used to amplify DNA, are sophisticated molecular techniques that rely on DNA extracted from isolates that have been cultured and purified before analysis.

Ribosomal RNA is the target of several molecular techniques because this portion of the bacterial genome is considered stable. The techniques associated with target ribosomal RNA are called "ribotyping." Other molecular methods target parts of the bacterial DNA and may use pulse field gel electrophoresis and randomly amplified polymorphic DNA—a technique that amplifies selected portions of the DNA by PCR. Highly qualified personnel trained on rather elaborate equipment must perform all of these molecular techniques, currently limiting the availability and cost of these procedures.

Biochemical Methods (Phenotype)

Biochemical methods have some advantages over molecular techniques in cost and efficiency. Molecular techniques are more precise, but more time-consuming, and are not suitable for analyzing large numbers of samples at this time. A combination of the two approaches may be best, the simpler, quicker, less costly biochemical approach followed by detailed molecular analysis of selected bacterial isolates to confirm the results.

One technique that falls under biochemical methods is antibiotic resistance analysis. The same principle that applied to the discussion of the intestinal tract selecting certain genetics in bacterial flora applies to microbes that acquire resistance to antibiotics or other biochemical traits. Each source is expected to have its own pattern of resistance; the general concept is that human fecal bacteria have developed the greatest diversity of resistance. Recent findings, however, have shown that domestic animals that receive antibiotics are hosts of antibiotic resistant bacteria and these bacteria can be found in waterbodies adjacent to farms.

Other biochemical methods that are being used include the F-Specific (F+ or FRNA) coliphage (although this is targeted at viral sources), sterols or fatty acid analysis from the cell walls and membranes of *E. coli*, and nutritional patterns.

STATUS OF THESE TECHNIQUES

The current set of techniques can distinguish between humans and animal sources. Separation of wildlife from domestic animals is also successful, although less accurate than human–animal separation. Distinguishing different types of domestic animals or different types of wildlife from one another is still under development but likely in the near future.

All of these techniques should be considered under development. Studies are in progress with more than one technique applied to the same set of conditions to determine the extent of similarity, effectiveness, cost, strengths, and weaknesses. Many site-specific studies have proved effective and valuable; however, regional variation needs to be addressed to move beyond these site-specific findings. Most likely, a combination of methods will be useful, based on the particular situation. Other nonbacterial techniques may also indicate the presence of wastewater, such as optical brighteners from laundry detergents and caffeine. Collectively, BST techniques will prove valuable for tracing fecal pollution from a variety of sources. State and local public health officials are increasingly acquiring this technology and may be good resources for information applicable on areawide or watershed scales.

WATER ASSESSMENT AND CRITERIA

Parley V. Winger
University of Georgia
Atlanta, Georgia

INTRODUCTION

Human activities have historically altered aquatic systems to some extent; however, with the Industrial Revolution and the shift from hunting and agricultural use of the land to amassing in population centers, pollution of adjacent waterways reached intolerable levels (1,2). Untreated municipal and industrial wastes were released directly into river, stream, and lake systems, and even with the development of sewage treatment in the mid-1800s, sewage outfalls continued to reduce the quality of downstream waters (1). Federal intervention into activities influencing water quality began in the late nineteenth century (e.g., River and Harbors Act of 1899) and continues today, primarily under the auspices of the Clean Water Act and its amendments (3,4). With the advent of legislative regulation for water quality, procedures for measuring and describing the status and quality of aquatic systems were needed for management and decision making. Developing from this need has been a progressive advancement of practices and procedures used to describe and categorize the quality of aquatic environments, culminating in the present-day regional approach to biological assessment and the development of biological criteria that can be used to evaluate ecosystem health and classify the biological integrity of aquatic systems. Biological assessment is defined as "an evaluation of the condition of a waterbody using biological surveys and other direct measurements of the resident biota in surface waters," and biological criteria are "numerical values or narrative expressions that describe the reference biological condition of aquatic communities inhabiting waters of a given designated aquatic life use and are used as benchmarks for water resources evaluation and management decision making" (5). A mandate of the Clean Water Act was to restore and maintain biological integrity of aquatic systems. Biological integrity encompasses all factors affecting an ecosystem and is defined as the "capability of supporting and maintaining a balanced, integrated, adaptive community of organisms having a species composition, diversity, and functional organization comparable to that of the natural habitat of the region" (6).

EARLY ASSESSMENTS

An extensive history and subsequent evolution of concepts, procedures, and understanding of processes associated with assessments of aquatic environments in relation to anthropogenic impacts exist (2). Basic to the development of appropriate and effective bioassessment procedures was the recognition that aquatic systems are more than just water and that the biological components represented by the myriad of microbes, plants, and animals are integral to these systems in carrying out the biological processes that ensure water quality. The diverse biological assemblages that reside in aquatic habitats, although generally unique for each system, are characteristic of the indigenous water quality. Early studies and measurements of water quality relied primarily on the chemical characteristics (pH, dissolved oxygen, etc.) of the water, but it was eventually recognized that these measurements provided little insight into the biological health of these bodies of water. Maintenance of clean, healthy, high-quality water is dependent on diverse and fully functional aquatic communities. Conversely, the biological components present in an aquatic system reflect the quality of the water and habitat.

DEVELOPMENT OF BIOASSESSMENT PROCEDURES

A number of factors and activities exist that can adversely impact aquatic systems. Aside from the general features that provide the underlying control of the structure and function of rivers and streams (climate, geology, soil types, etc.), anthropogenic activities can have marked influences on aquatic environments (7). Environmental factors (e.g., water quality, habitat structure, flow regime, energy source, and biotic interactions) influence the biological integrity of aquatic systems and anthropogenic activities generally influence one or more of these factors (5). Physical alterations of the habitat (e.g., agriculture, logging, channelization, flow alterations) can affect erosion, sedimentation, and hydrological characteristics of streams and rivers, and these, in turn, can drastically alter the biological communities. Inputs of contaminants and other pollutants from agricultural runoff and municipal and industrial effluents can also impact the biological components, but generally in a different manner from that elicited by physical disturbance. Similarly, different types of contaminants can adversely impact aquatic communities in unique ways (e.g., impacts from organic pollution are different from that shown by acid-mine drainage or an industrial chemical). Consequently, bioasssements need to be robust enough to be able to detect impairment of aquatic communities, regardless of the cause.

Although early assessments associated with organic pollution relied heavily on chemical analyses (basic water chemistry such as dissolved oxygen), the adverse effects of pollution on aquatic organisms were understood by the mid-1800s (4). However, most early documentation of the biological impacts from pollution in aquatic systems was qualitative. Results from biological measurements were generally lengthy lists of "indicator" organisms associated with zones of pollution or purification. For example, Kolkwitz and Marrson (8,9) identified the zones of purification/decomposition ("saprobia") downstream from sewage outfalls as poly-, meso-, and oligosaprobia, with each zone characterized by a unique assemblage of organisms.

The use of indicator organisms in categorizing water quality provided a framework for focusing on the biological components of aquatic systems and for the development of more robust bioassessment procedures (10). As "indicator organisms" were more descriptive of organic pollution (sewage) and the tolerances of organisms to the many types of pollution (sewage, metals, industrial contaminants, pesticides, etc.) were not well known, there was an impetus for development of bioassessment programs that incorporated community-based approaches (11,12). Patrick (13) pioneered the concept of using aquatic communities in stream assessments and compared histograms depicting the numbers of species of various taxonomic groups (blue-green algae, oligochaetes and snails, protozoans, diatoms and green algae, rotifers and clams, insects and crustaceans, and fish) comprising the communities in test streams with those from a clean or reference site. Patrick recognized that healthy streams had a great number of species representing the various taxonomic groups and no species were represented by a great number of individuals, whereas streams impacted by pollution had reduced numbers of species and those remaining were in great abundance.

A major drawback, however, of including most of the various biological components (algae, protozoa, benthic macroinvertbrates, and fishes) in these types of bioassessments was the amount of time required and the expertise needed to collect and identify all of the organisms comprising the aquatic communities at sites of concern. As a result of this criticism, aquatic scientists began to specialize or use specific taxonomic groups in bioassessments as representative components or surrogates of the biotic communities (14). Of course, each specialist considered their particular taxonomic group as the most appropriate for use in bioassessments, and for the most part, although each group had its advantages and disadvantages, they each provided meaningful information and generally reflected some level of impairment (5), albeit in much less time than that required for assessment of the full biological complement of taxa. A number of taxonomic groups have been touted as useful in assessing the quality of aquatic systems: diatoms (15), algae (16), protozoa (17,18), benthic macroinvertebrates (19–22), and fishes (23,24). The advantages of using periphyton (algae), benthic macroinvertbrates, and fish are outlined in Barbour et al. (25). The structure and function of specific biotic assemblages or taxonomic groups, such as benthic invertebrates or fish, integrate information about the past and present water quality conditions. Consequently, their inclusion in aquatic bioassessments strengthens the ability to categorize the prevailing water quality.

Refinements to aquatic bioassessment procedures resulted in more widespread use and acceptance, as well as a movement toward standardization of field and laboratory procedures (3,26,27). The inclusion of biotic indices (28–31) and diversity indices (32–34) in the analysis of biological data provided additional measures of community structure and function that aided in the interpretation and identification of impairment from pollution. These indices were also useful in reducing and synthesizing large amounts of data (data that reflected biotic responses to complex interactions and exposures to pollution) into comprehensible numeric values descriptive of water quality. However, bioassessment procedures were still deficient in their ability to fully and accurately describe the overall health of aquatic systems (4).

CURRENT STATE-OF-THE-ART BIOASSESSMENT PROCEDURES

The most useful measures of biotic integrity of aquatic systems reflect the biological responses of organisms and their populations to the prevailing (past and present) environmental conditions. The ability to classify the quality of streams and rivers (with indigenous biological assemblages) provides information that is integral for making management decisions and identifying anthropogenic impacts (35). Development of an "Index of Biotic Integrity" (IBI) using a combination of 12 community parameters (metrics) for fish assemblages provided direction into a new and innovative way of obtaining this type of information (24). The combination of these community attributes (multimetrics) into an IBI provided a classification of the quality of fish communities as a measure of water/habitat quality. The metrics used in the construction of the IBI encompassed several levels of fish community structure and function—species composition/richness and information on ecological factors (trophic structure and health). Successful application of multimetric indices depends on selecting the appropriate metrics used to describe and delineate the biological assemblages (36). The use of multimetrics provides a way of measuring the response of biological assemblages at several levels of organization (e.g., composition, condition, and function) to anthropogenic influences, and these measurable biological attributes can be tested, verified, and calibrated for use in the respective geographical areas of study (37–39). The multimetric concept was expanded on and modified to include benthic macroinvertebrates as well as fish in a set of procedures developed for the U.S Environmental Protection Agency called the Rapid Bioassessment Protocols (RBPs) for use in streams and rivers (40). The RBPs provided a cost-effective framework for bioassessment that could be used by federal, state, and local agencies for management purposes of screening, site ranking, and trend monitoring of the biotic integrity of streams. Modifications, improvements, and additions to these procedures were included in a more recent revision of the RBPs (25). The integrated assessments recommended in the RBPs include information (field collected) on the physical habitat and water quality as well as a number of measures (metrics) describing the biological assemblages of periphyton, benthic macroinvertebrate, and fish.

REFERENCE SITES AND ESTABLISHING BIOCRITERIA

Knowing the organisms that comprise or could comprise the aquatic assemblages in unimpacted streams provides the basis for application of current biological assessment procedures, such as the RBPs, and the development of biological criteria. Consequently, reference streams/sites provide the basis or benchmark for establishing suitable criteria that can be used for comparison with other streams/sites to detect impairment (5,41). Unfortunately, few streams and rivers are pristine anymore because of the widespread influence of anthropogenic activities. Consequently, streams/sites that are the least impacted (minimally impaired) are selected to establish the "attainable" condition or level that is used for comparison with the streams under study.

Reference sites must also be representative of the aquatic systems that are being evaluated. Two types of reference conditions, site specific and regional, are currently used for field bioassessments (25). However, multiple reference sites within a geographical region provide a realistic representation of the community assemblages and compensate for and include the inherent variability associated with individual sites (41). The selection of reference sites on an ecoregional or subregional basis provides an unbiased estimate of the least impacted (attainable) biological assemblages for that particular region of study (41,42). Using a suite (population) of reference streams/sites also allows the use of statistical analyses to establish the natural variability and ranges for each of the metrics and indices to be used for comparison with the study streams. The number of reference sites needed to describe the expected or attainable conditions may vary with the geographical area, but generally 3 to 20 sites are acceptable (41). Once the biological assessments are complete for the reference sites, the suitability and strengths of the metrics included in the analysis and the overall rating index used in detecting impairment can be evaluated using box-and-whisker plots (5) or other statistical analyses (43,44). The box-and-whisker-type plots for the metrics/indices determined for reference streams/sites depict the median, interquartile range (25th and 75th percentiles) and the range (minimum and maximum values). Metric and index values from the study sites that are above (low values for reference conditions) or below (high values for reference conditions) the interquartile range shown by the reference sites demonstrate impairment. The interquartile ranges for the various metrics and combined overall indices (combined scores for the individual metrics) establish the numeric biological criteria for that region.

The reference streams/sites must be representative of the resources at risk. They provide the benchmarks for the expectation or attainable level of biological integrity/water quality that the study areas will be compared with for classification of their level of impairment, and once defined, these biocriteria will describe the best attainable condition (5,37). The acceptable level of difference between the established criteria representing the reference sites and those of the study sites will vary depending on the designated aquatic life use (e.g., cold water fisheries, warm water fisheries, and endangered species) of the study sites (5). The narrative description based on the quantitative database or the numeric criteria (indices) calculated from the database from the reference sites can also be used as the biological criteria for identifying water quality and level of impairment.

An alternative approach to using multimetrics (45,46) in establishing reference conditions relies on multivariate analyses and predictive modeling (47–49). Models are developed that will predict the expected species composition of a "pristine" site given the physical and chemical characteristics (50). The species composition (and associated metrics) from this expected (predicted) community assemblage can be compared with those from the field study sites included in the bioassessments; assessment sites are compared with model-predicted reference conditions. These predictive models require extensive information on the physical and chemical characteristics of pristine streams and the biological organisms that would be associated with them (5). In the absence of field data (or as alternatives for defining reference conditions in the event that suitable sites no longer exist), historical data, simulation modeling, and expert consensus can be used to supplement the data needed for these models (5). Under the multivariate approach, the predicted species compositions (and associated metrics/indices) would establish the biological criteria for streams in that region.

BIOASSESSMENTS CRITICAL IN PRESERVING BIOTIC INTEGRITY

Current practices in biological assessment and the establishment and use of biological criteria based on community structure and function at reference sites (or predicted by models) are effective tools in describing biological integrity and classifying water quality (5,51). As these procedures are used, refinements will most likely be made (particularly at the regional level) as more information is generated and the selection/calibration of metrics is improved. Biological monitoring that includes bioassessment and the establishment of biocriteria on an ecoregional basis (46) is integral to meeting the demands for ecosystem health (52) and measurement of ecological integrity (53) of our nation's aquatic resources.

BIBLIOGRAPHY

1. Hynes, H.B.N. (1960). *The Biology of Polluted Water*. Liverpool University Press, Liverpool, UK, p. 202.
2. Warren, C.E. and Dourdoroff, P. (1971). *Biology and Water Pollution Control*. W.B. Saunders, Philadelphia, p. 434.
3. Mackenthun, K.M. and Ingram, W.M. (1967). *Biological Associated Problems in Freshwater Environments: Their Identification, Investigation and Control*. U.S. Department of

the Interior, Federal Water Pollution Control Administration, Washington, DC, p. 287.
4. Davis, W.S. (1995). Biological assessment and criteria: building on the past. In: *Biological Assessment and Criteria: Tools for Water Resource Planning and Decision Making*. W.S. Davis and T.P. Simon (Eds.). Lewis Publishers, Boca Raton, FL, pp. 1–29.
5. Gibson, G.R. Jr., Barbour, M.T., Stribling, J.B., Gerritesen, J., and Karr, J.R. (1996). *Biological Criteria, Technical Guidance for Streams and Small Rivers* (Rev. Edn.). U.S. Environmental Protection Agency, Office of Water, Washington, DC, EPA/822-B-96-001.
6. Karr, J.R. and Dudley, D.R. (1981). Ecological perspective on water quality goals. *Environ. Manage.* **5**: 55–68.
7. Hynes, H.B.N. (1970). *The Ecology of Running Water*. Liverpool University Press, Liverpool, UK, p. 555.
8. Kolkwitz, R. and Marrson, M. (1908). Ecology of plant saprobia. *Rep. German. Botan. Soc.* **26a**: 505–519.
9. Kolkwitz, R. and Marrson, M. (1909). Ecology of animal saprobia. *Int. Rev. Hydrobiol. Hydrogeogr.* **1**: 126–152.
10. Wilhm, J.L. (1975). Biological indicators of pollution. In: *River Ecology*. B.A. Whitton (Ed.). University of California Press, Berkeley, pp. 375–402.
11. Hynes, H.B.N. (1965). The significance of macroinvertebrates in the study of mild river pollution. In: *Biological Problems in Water Pollution, Third Seminar-1962*. C.M. Tarzwell (Ed.). U.S. Public Health Service, Robert A. Taft Sanitary Engineering Center, Cincinnati, Ohio, pp. 235–240.
12. Cairns J. Jr. (1974). Indicator species vs. the concept of community structure as an index of pollution. *Water Res. Bull.* **10**: 338–347.
13. Patrick, R. (1949). A proposed biological measure of stream conditions, based on a survey of the Conestoga Basin, Lancaster County, Pennsylvania. *Proc. Acad. Natural Sci. Phila.* **101**: 277–341.
14. Hart, C.W. Jr. and Fuller, S.L.H. (1974). *Pollution Ecology of Freshwater Invertebrates*. Academic Press, New York, p. 389.
15. Patrick, R. (1957). Diatoms as indicators of changes in environmental conditions. In: *Biological Problems in Water Pollution. 1956 Symposium*. C.M. Tarzwell (Ed.). U.S. Department of Health, Education and Welfare, Robert A. Taft Sanitary Engineering Center, Cincinnati, Ohio, pp. 71–83.
16. Palmer, C.M. (1957). Algae as biological indicators of pollution. In: *Biological Problems in Water Pollution*. C.M. Tarzwell (Ed.). U.S. Department of Health, Education and Welfare, Robert A. Taft Sanitary Engineering Center, Cincinnati, Ohio, pp. 60–69.
17. Cairns, J. Jr. (1974). Protozoans (protozoa). In: *Pollution Ecology of Freshwater Invertebrates*. C.W. Hart Jr. and S.L.H. Fuller (Eds.). Academic Press, New York, pp. 1–28.
18. Cairns, J. Jr. (1978). Zooperiphyton (especially protozoa) as indicators of water quality. *Trans. Am. Microscop. Soc.* **97**: 44–49.
19. Gaufin, A.R. and Tarzwell, C.M. (1956). Aquatic macroinvertebrate communities as indicators of organic pollution in Lytle Creek. *Sewage Industr. Wastes*. **28**: 906–924.
20. Goodnight, C.J. (1973). The use of aquatic macroinvertebrates as indicators of stream pollution. *Trans. Am. Microsop. Soc.* **92**: 1–13.
21. Roback, S.S. (1974). Insects (arthropoda: insecta). In: *Pollution Ecology of Freshwater Invertebrates*. C.W. Hart Jr. and S.L.H. Fuller (Eds.). Academic Press, New York, pp. 314–376.
22. Klemm, D.J., Lewis, P.A., Fulk, F., and Lazorchak, J.M. (1990). *Macroinvertebrate Field and Laboratory Methods for Evaluating the Biological Integrity of Surface Waters*. U.S. Environmental Protection Agency, Environmental Monitoring Systems Laboratory, Cincinnati, Ohio, EPA/600/4-90/030.
23. Doudoroff, P. and Warren, C.E. (1957). Biological indices of water pollution with special reference to fish populations. In: *Biological Problems in Water Pollution*. C.M. Tarzwell (Ed.). U.S. Public Health Service, Robert A. Taft Sanitary Engineering Center, Cincinnati, Ohio, pp. 144–163.
24. Karr, J.R. (1981). Assessment of biotic integrity using fish communities. *Fisheries* **6**: 21–27.
25. Barbour, M.T., Gerritsen, J., Snyder, B.D., and Stribling, J.G. (1999). *Rapid Bioassessment Protocols for Use in Streams and Wadeable Rivers: Periphyton, Benthic Macroinvertebrates and Fish*. U.S. Environmental Protection Agency, Office of Water, Washington, DC. EPA 841-B-99-002.
26. Slack, K.V., Averett, R.C., Greeson, P.E., and Lipscomb, R.G. (1973). *Methods for Collection and Analysis of Aquatic Biological and Microbiological Samples. Techniques of Water-Resource Investigations of the United States Geological Survey*. Book 5, Ch. 4, United States Department of the Interior, U.S. Government Printing Office, Washington, DC, p. 165.
27. C.I. Weber (Ed.). (1973). *Biological Field and Laboratory Methods for Measuring the Quality of Surface Waters and Effluents*. U.S. Environmental Protection Agency, Cincinnati, Ohio, EPA-670/4-73-001.
28. Beck, W.M. Jr. (1955). Suggested method for reporting biotic data. *Sewage Indust. Wastes* **27**: 1193–1197.
29. Chutter, F.M. (1972). An empirical biotic index of the quality of water in South African streams and rivers. *Water Res.* **6**: 19–30.
30. Hilsenhoff, W.L. (1982). *Using a Biotic Index to Evaluate Water Quality in Streams*. Technical Bulletin No. 132. Department of Natural Resources, Madison, WI, p. 22.
31. Hilsenhoff, W.L. (1987). An improved biotic index of organic stream pollution. *The Great Lakes Entomol.* **20**: 31–39.
32. Pielou, E.C. (1966). The measurement of diversity in different types of biological collections. *J. Theoretic Biol.* **13**: 131–144.
33. Wilhm, J.L. and Dorris, T.C. (1968). Biological parameters for water quality criteria. *BioScience* **18**: 477–481.
34. Washington, H.G. (1984). Diversity, biotic and similarity indices: a review with special relevance to aquatic ecosystems. *Water Res.* **18**: 653–694.
35. Hawkes, H.A. (1975). River zonation and classification. In: *River Ecology. Studies in Ecology*. Vol. 2. B.A. Whitton (Ed.). University of California Press, Berkeley, pp. 313–374.
36. Karr, J.R. (1999). Defining and measuring river health. *Freshwater Biol.* **41**: 221–234.
37. Barbour, M.T., Stribling, J.B., and Karr, J.R. (1994). Multimetric approach for establishing biocriteria and measuring biological condition. In: *Biological Assessment and Criteria, Tools for Water Resource Planning and Decision Making*. W.S. Davis and T.P. Simon (Eds.). Lewis Publishers, Boca Raton, FL, pp. 63–77.
38. Barbour, M.T. et al. (1996). A framework for biological criteria for Florida streams using benthic macroinvertebrates. *J. North Am. Benthologic Soc.* **15**: 185–211.
39. Fore, L.S., Karr, J.R., and Wisseman, R.W. (1996). Assessing invertebrate responses to human activities: evaluating alternative approaches. *J. North Am. Benthologic Soc.* **15**: 212–231.

40. Plafkin, J.L., Barbour, M.T., Porter, K.D., Gross, S.K., and Hughes, R.M. (1989). *Rapid Bioassessment Protocols for Use in Streams and Rivers. Benthic Macroinvertebrates and Fish*. U.S. Environmental Protection Agency, Office of Water Regulations and Standards, Washington, DC, EPA 440-4-89-001.
41. Hughes, R.M. (1995). Defining acceptable biological status by comparing with reference conditions. In: *Biological Assessment and Criteria, Tools for Water Resource Planning and Decision Making*. W.S. Davis and T.P. Simon (Eds.). Lewis Publishers, Boca Raton, FL, pp. 31–47.
42. Hughes, R.M., Larsen, D.P., and Omernik, J.M. (1986). Regional reference sites: a method for assessing stream potential. *Environ. Manage.* **10**: 629–635.
43. Hughes, R.M. et al. (1998). A process for developing and evaluating indices of fish assemblage integrity. *Can. J. Fish Aquat. Sci.* **55**: 1618–1631.
44. Kilgour, B.W., Somers, K.M., and Matthews, D.E. (1998). Using the normal range as a criterion for ecological significance in environmental monitoring and assessment. *Ecoscience* **5**: 542–550.
45. Karr, J.R. and Chu, E.W. (1999). *Restoring Life in Running Waters: Better Biological Monitoring*. Island Press, Washington, DC, p. 206.
46. Karr, J.R. and Chu, E.W. (2000). Sustaining living rivers. *Hydrobiologia* **422/423**: 1–14.
47. Moss, D., Furse, M.T., Wright, J.F., and Armitage, P.D. (1987). The prediction of the macro-invertebrate fauna of unpolluted running-water sites in Great Britain using environmental data. *Freshwater Biol.* **17**: 41–52.
48. Hawkins, C.P., Norris, R.H., Hogue, J.N., and Feminella, J.W. (2000). Development and evaluation of predictive models for measuring the biological integrity of streams. *Ecol. Applic.* **10**: 1456–1477.
49. Norris, R.H. and Hawkins, C.P. (2000). Monitoring river health. *Hydrobiologia* **435**: 5–17.
50. Clarke, R.T., Furse, M.T., Wright, J.F., and Moss, D. (1996). Derivation of a biological quality index for river sites: comparison of the observed with the expected fauna. *J. Appl. Stat.* **23**: 311–332.
51. U.S. Environmental Protection Agency. (2002). *Biological Assessment and Criteria: Crucial Components of Water Quality Programs*. Office of Water, Washington, DC, EPA 822-F-02-006.
52. Rapport, D.J. et al. (1999). Ecosystem health: the concept, the ISEH, and the important tasks ahead. *Ecosystem Health* **5**: 82–90.
53. Parrish, J.D., Braun, D.P., and Unnasch, R.S. (2003). Are we conserving what we say we are? Measuring ecological integrity within protected areas. *BioScience* **53**: 851–860.

PHYSIOLOGICAL BIOMARKERS AND THE TRONDHEIM BIOMONITORING SYSTEM

Karl Erik Zachariassen
Norwegian University of
Science and Technology
Trondheim, Norway

Poor water quality is frequently the result of the presence of potentially toxic substances. Detection of such substances by chemical analysis may be complicated, time consuming, and expensive, and it does not reveal much about possible toxic effects.

WHAT IS A BIOMARKER?

Biomonitoring of water quality takes advantage of the fact that organisms in the water may be moderately affected by the substances. A biomarker is a cellular, biochemical, or physiological parameter, which may be affected when bioactive chemicals are present in the water. One biomarker can respond to the presence of a wide variety of substances and respond before manifest toxic effects occur. Biomarkers are, therefore, particularly useful in early warning monitoring.

Physiological Biomarkers

However, biomarkers should be used by scientists with care, and an appropriate interpretation of their response requires insight into the integrated functional system they are part of, that is, by considering the biomarkers in a physiological context. In a healthy organism, optimal chemical and physical conditions are maintained by a variety of homeostatic regulatory mechanisms. Well-known examples of regulated parameters are body temperature, body fluid pH, and body fluid concentrations of ions, organic solutes, and hormones. Whenever such parameters, for some reason, are displaced beyond optimal range, regulatory mechanisms are activated to compensate for the disturbing stress.

The impact of a chemical substance on an organism can range from a trivial disturbance, which the organism can compensate for, to a serious and eventually lethal disturbance of a regulated parameter. When an organism is injured or killed by a toxic agent, it is because the agent changes a regulated parameter beyond the tolerated range. Regulated parameters as well as the activity of regulatory mechanisms can be physiological biomarkers to provide information about the severity of the impact of a toxic agent (1). As different toxic agents interfere with organisms in different ways, the combined effect pattern of several biomarkers may be more or less specific to the toxic agents, and the effect pattern may identify the agent in cases when the cause of a toxic impact in the environment is unknown or disputed.

THE TRONDHEIM BIOMONITORING SYSTEM

The Trondheim Biomonitoring System (TBS) was developed to extract the appropriate information out of the changes that occur in an organism affected by a toxic chemical. It is a so-called active biomonitoring system; that is, it uses organisms such as mussels placed individually in chambers where they can be exposed to the water that is to be monitored. It operates with three categories of biomarkers: biomarkers for exposure, biomarkers for evaluation of toxic effects, and fingerprint biomarkers to identify the toxic agent (Fig. 1).

Phase 1, Exposure Biomarkers

Exposure biomarkers are the first changes that occur in an affected organism, and they reflect the mechanisms

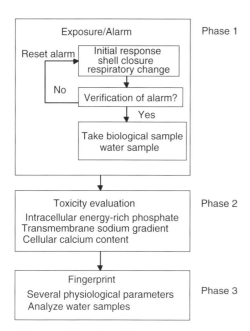

Figure 1. Trondheim Biomonitoring System.

that handle the impact. These changes may be increased activity of homeostatic regulatory mechanisms or the activation of detoxifying mechanisms such as cytochrome P_{450} for certain organic chemicals and metallothioneins (MTs) for certain toxic metals. It should be emphasized that the activation of detoxifying mechanisms such as P_{450} and MT production does not imply that an organism is seriously affected, only that it is compensating for a chemical impact. It should also be emphasized that these systems are activated even by substances of natural origin, and it may be difficult to distinguish natural from non-natural origins.

An ideal exposure biomarker should be sensitive to a wide variety of substances occurring at low concentrations, and it should be possible to monitor the biomarker continuously, for example, online on a computer. As moderate toxic impacts and the activation of many compensatory mechanisms are associated with changes in metabolic rate, the TBS has been using automatic measurement of oxygen consumption as an exposure biomarker. Water from chambers with and without animals is led to a separate chamber with an oxygen electrode connected to a computer, and the oxygen consumption is continuously determined from the difference in oxygen tension of water from chambers with and without animals.

Phase 2, Toxicity Biomarkers

As toxicity biomarkers, the TBS has so far been using tissue levels of energy-rich phosphates and cellular levels of sodium and calcium. The rationale behind the use of these parameters is that many toxic agents inhibit energy metabolism and ionic pumps or affect membrane permeability. Such impacts will affect cellular levels of ATP and active extrusion of sodium and calcium from cells, which will in turn increase the cellular levels of these ions. A substantially reduced transmembrane energy gradient for sodium or high cellular levels of calcium are inevitably lethal, and moderate changes in these parameters indicate that the cells are under serious stress and may become injured if the stress persists (2).

Phase 3, Fingerprint Biomarkers

To obtain pollutant-specific fingerprint biomarkers, a combination of exposure and toxicity stress biomarkers has been used. In addition, intracellular and extracellular concentrations of free amino acids have been included. The specificity of the response of the latter is enhanced because they include more than 20 different substances and they are influenced by a variety of stress factors such as a drop in the transmembrane sodium gradient, anaerobiosis, and cell volume changes. In principle, any measurable parameter may be used, but the number of parameters should be kept at the minimum required to obtain a pollutant-specific effect pattern. The development of an operative system of fingerprint biomarkers is complicated because different pollutants sometimes act together and the changes associated with a single pollutant may depend on its concentration (hormesis).

Toxic agents may also affect a variety of other processes, such as nervous transmission, hormonal control systems and immune responses, osmoregulation, respiration, and heart activity, and effects may be acute or caused by chronic, long-term exposure. Depending on the mechanisms of action of the various toxic chemicals, appropriate biomarkers may be selected and included in the biomarker categories of the TBS.

BIBLIOGRAPHY

1. Aunaas, T. and Zachariassen, K.E. (1993). Physiological biomarkers and the Trondheim biomonitoring system. In: *Biomonitoring of Coastal Waters and Estuaries*. K.J.M. Kramer (Ed.). CRC Press, Boca Raton, FL, pp. 107–130.
2. Børseth, J.F. et al. (1995). Transmembrane sodium gradient and calcium content in the adductor muscle of *Mytlius edulis* L. in relation to the toxicity of oil and organic chemicals. *Aquatic Toxicol.* **31**: 263–276.

BIOMARKERS, BIOINDICATORS, AND THE TRONDHEIM BIOMONITORING SYSTEM

Rita Triebskorn
Steinbeis-Transfer Center for
Ecotoxicology and
Ecophysiology
Rottenburg, Germany

Environmental effects resulting from chemical exposure can be traced at different levels of biological organization. In this context, the term "biomarker" is used mainly for functional measures expressed at the organism to suborganism level of biological organization, while the term "bioindicator" refers to effect endpoints at higher biological levels including population and community-level attributes (1).

During the last two decades, interest in using biomarkers or bioindicators as monitoring tools to assess environmental pollution has steadily increased. This is mainly due to the fact that chemical analyses, which are well established in ecotoxicological monitoring programs, can exclusively give information on chemicals present in the environment or accumulated in biota. However, they cannot provide information on exposure-related effects, which possibly occur in exposed organisms. In addition, chemical analyses can often only provide information on a defined spectrum of chemicals (being searched for) available and are mainly "snapshots" of a defined exposure scenario (depending on the time of sampling). Furthermore, low dose effects have received increasing attention during the last few years, especially in the context of endocrine disruption. Therefore, the detection of effects that may be attributed to chemicals present in the environment in concentrations at or beyond the detection limits has become more and more important. On the other hand, it has also become evident that the presence of a chemical in an organism or an environment does not necessarily imply that the organism or the environment is affected or threatened by that chemical.

Biomarkers are suitable not only for delivering information on the health status of exposed organisms (biomarkers of effect), but also on the quality and/or quantity of the exposure situation (biomarkers of exposure). Thus, they can be used both as measures of toxic effects and as fingerprints for chemical exposure (2–4). Depending on the character of biomarkers selected for risk assessment, the evaluator will receive more data on either adverse effects in organisms or on the exposure situation itself (Fig. 1).

Whereas established ecological surveys (e.g., the water quality assessment system) are often based on the disappearance of species under stressed conditions (thus using mortality as the endpoint of toxic effect), the biomarker approach uses responses of organisms (still) living under unfavorable conditions as early warnings before populations or ecosystems are severely damaged (5–7). Since biomarkers respond quickly to environmental stressors before ecological effects become apparent, they serve to detect effects at an early stage. They are highly sensitive and, in addition, integrate effects of animals' exposure to contaminants and confounding factors over time (8–10).

To date, most biomarker research has been conducted in aquatic ecosystems (summarized in Reference 1). In the last decade, however, the field of biomarker application in terrestrial environments has also expanded and several techniques have been adapted for soil organisms (11,12).

A concept to integrate physiological biomarkers into a monitoring system was elaborated by a group of scientists from Trondheim, Norway, ten years ago (2). This "Trondheim Biomonitoring System" (TBS) is a practically operative integrated system for biomonitoring of marine environments using the mussel *Mytilus edulis* as the monitoring organism. The technical setup of this system consists of ten exposure chambers each containing one mussel. The exposure chambers, through which seawater flows, are mounted into a submersible exposure unit and are connected to a respirometer. Thus, data for oxygen consumption can be recorded continuously for each mussel and displayed on a computer screen. In addition, samples (e.g., tissue, hemolymph) for other purposes can be taken from the exposed organisms. The aims of this biomonitoring system are as follows:

1. To detect the presence of environmental pollution at the earliest possible stage in order to enact possible countermeasures before ecological damage becomes obvious.
2. To evaluate the stress status of organisms or ecosystems exposed to environmental pollutants.
3. To identify the pollutants responsible for the detected stress status.

To achieve these aims, the TBS uses physiological biomarkers in three phases. In the first phase, "the

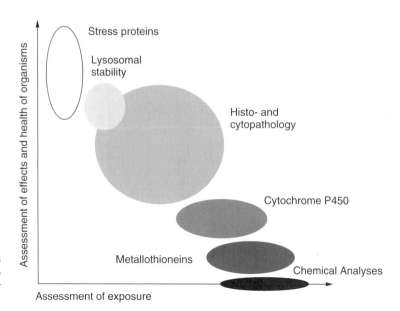

Figure 1. Suitability of different biomarkers to assess effects and health of organisms on the one hand and the exposure situation on the other. Chemical analyses only provide information on the exposure situation.

alarm phase," very sensitive biological parameters (e.g., shell closure or oxygen consumption of mussels) are continuously recorded as "alarm parameters." In the event the alarm is verified as not due to changes in natural environmental parameters, mussel samples are taken for phase 2.

In the second phase, "the toxicity evaluation phase," biomarkers indicating toxicity stress or strain (e.g., intracellular energy-rich phosphates, strombine formation, transmembrane sodium gradient, cellular calcium content) are used to assess the health status of the exposed organisms.

In the third phase, "the fingerprint phase," a set of chemical-specific biomarkers (e.g., transmembrane distribution of free amino acids) are applied in parallel with chemical analyses of samples automatically taken in phase 1. These investigations provide information on the nature of the chemicals responsible for the effects observed in phases 1 and 2.

For freshwater monitoring of pollution, the Multispecies Freshwater Biomonitor (MFB) has been established by LIMCO Int. (13). Similar to the TBS, the MFB uses flow-through exposure chambers for the exposure of limnic organisms including gammarids, chironomids, or fish. As biomarkers/bioindicators, breathing activity and locomotion of animals can be recorded continuously in this system. In case of an incidence, samples can be taken for different analytical purposes.

In order to evaluate biomarker responses with respect to their toxicological and ecological relevance (i.e., their chemical specificity and importance for effects at higher biological levels including populations and ecosystems), an extended monitoring project called VALIMAR (*vali*dation of bio*mar*kers for the assessment of small stream pollution) was conducted at two small streams in southern Germany between 1995 and 2000 using fish as monitoring organisms (10). The project was based on a "middle-up and middle-down" approach using flow-through aquaria in the field as intermediates between controlled lab and uncontrolled field situations.

According to the results of VALIMAR, for stream systems mainly influenced by wastewater release and agricultural activities, a combination of cellular (e.g., histopathology of kidney or liver, ultrastructure of gills or liver) with biochemical biomarkers (e.g., EROD activity, stress protein expression, catalase or esterase activity) can be recommended as useful tools for assessing risk in fish populations. Biomarkers proved to be sensitive early warning sentinels, which indicate the general stress status and the health of fish in streams influenced by different sources of pollution, but also in streams influenced by temporary or locally restricted spills (6,10). In addition, biomarkers are very useful in assessing the recovery of aquatic ecosystems (14). Although in the VALIMAR project the ultrastructure of the gills could clearly be identified as the best discriminator between the two streams in both brown trout and stone loach (15), the use of single biomarkers to assess the holistic effects of pollution, either structural or biochemical, is not recommended. Whereas biochemical biomarkers are sensitive and respond rapidly to stressors, their disadvantage is that they are also highly variable and are quick to recover following removal of stress. Structural responses, on the other hand, are generally less variable, but also less sensitive in response to short-term events. Since structural parameters require more time to recover than biochemical responses, structural responses are useful in understanding past water quality conditions for some days or weeks after a stress event. For the experimental situation of VALIMAR, a combination of biomarkers to discriminate between the two test streams in trout was the gill ultrastructure, esterase, and cytochrome P450 enzymes. A suitable combination of biomarkers in stone loach was gill ultrastructure, acid phosphatase, and alanine aminotransferase (15).

As cellular biomarkers, histological and ultrastructural changes in monitoring organs (e.g., liver or gills of fish, or hepatopancreas or kidney of invertebrates) were shown to be suitable for pollution assessment in terrestrial and aquatic ecosystems (16–18). The *in situ* evaluation of cellular injury allows a direct diagnosis of adverse effects qualitatively but also quantitatively by means of morphometry or different semiquantitative assessment methods (19). Since histology and cytology not only reflect the impact of xenobiotics but also the general metabolic status of an organism, it is essential for an investigator of cellular biomarkers to be aware of the diversity and plasticity of cellular reactions occurring in an organ under unpolluted conditions. Many studies using light and electron microscope techniques as tools of prospective as well as retrospective risk assessment have been conducted. The majority of this work has been done with fish, but experience is also available for molluscs, annelids, mites, diplopods, isopods, and collembolans in response to heavy metals and organic pollutants (reviewed in References 11 and 20).

At the subcellular/organelle level, the lysosomal system has been identified as a very sensitive target for xenobiotics (21). Lysosomes are membrane-bound vesicles containing acid hydrolases. In intact lysosomes, these hydrolases are prevented from reacting with cytoplasmic compounds by an intact membrane. In response to stress, the lysosomal membrane stability decreases and its permeability increases. As a result, hydrolases are released into the cytoplasm. One of the easiest techniques to determine lysosome membrane integrity is the neutral-red retention assay, which is based on the uptake of the dye neutral-red by transmembrane diffusion. The capacity of healthy lysosomes to take up and maintain the dye depends on the efficiency of membrane-bound proton pumps (22). In stressed cells, neutral-red gradually leaks from the lysosomes into the surrounding cytoplasm. The time of dye retention in the lysosomes is the measure for the intensity of stress. In the past, most experience with the neutral-red retention assay exists in the marine field (21); however, a few studies have also been conducted with soil invertebrates exposed to environmental pollutants (23,24).

At the physiological, biochemical, and molecular levels, several biomarkers are well established, which differ with respect to their capability to indicate exposure and/or effects. The most prominent biomarkers are biotransformation enzymes (e.g., cytochrome P450 and glutathione-S-transferase), metallothioneins, and stress proteins. In the TBS, respiratory rates, intracellular energy-rich phosphates, strombine formation, the transmembrane sodium gradient, the cellular calcium content, the transmembrane distribution of free amino acids, and other physiological parameters are recommended for evaluation of chemical stress in marine ecosystems. Physiological, biochemical, and molecular biomarkers are very sensitive and respond quickly to environmental stressors. Therefore, they are also useful tools to detect influences on organisms in minor contaminated environments. In contrast to cellular biomarkers, however, which show saturation response kinetics with increasing intensity of the stressor, response kinetics of biochemical and molecular biomarkers often follow an optimum curve (Fig. 2). Thus, low values for molecular biomarker responses are difficult to interpret since they may occur in healthy as well as in moribund organisms, in the latter having already surpassed the optimal biomarker response threshold. Whether these low values result from exposure to very low or very high stressor intensities can only be decided when biochemical markers are combined with markers that are characterized by optimum response kinetics, for example, histopathological diagnostics (25). In addition, the dependence of physiological, biochemical, and molecular marker responses on exogenous and/or endogenous stress factors other than xenobiotics (e.g., temperature or reproduction status) has to be taken into account by the investigator.

The question of whether biomarker responses are relevant for effects at higher biological levels (i.e., populations or ecosystems) can be answered either experimentally (as in Reference 26) or by using a weight of evidence approach (27). Köhler and colleagues (26) studied the relevance of stress protein responses in the slug *Deroceras reticulatum* exposed to metals for different population parameters. They showed that the hsp70 levels in slugs after 2–3 weeks of exposure to the respective metals reflected effects in fecundity, reproduction, and mortality of lifetime exposure. Thus, in this case, biomarker measurements were proved to be early warning tools to predict long range consequences of xenobiotic exposure at the level of life cycle parameters.

Since long generation times of many organisms often do not permit such direct evidence of causal relationship between effects at low and high biological levels, a weight of evidence approach commonly used in human medicine is suitable to relate causes and effects on the basis of epidemiological data. By means of Hill's causality criteria (28), for example, the criteria of biological plausibility, strength of association, consistency of association, or experimental evidence, causality can be established even when direct experimental proof of causality is lacking. Based on this, in the VALIMAR project (10), a simultaneous collection of data for responses at different levels of biological organization in combination with a detailed characterization of the exposure situation made it possible to establish causality between different levels of biological organization as well as between exposure and effects (27).

BIBLIOGRAPHY

1. Adams, S.M. (2002). Biological indicators of aquatic ecosystem stress: introduction and overview. In: *Biological Indicators of Aquatic Ecosystem Stress*. S.M. Adams (Ed.). American Fisheries Society, Bethesda, MD, pp. 1–11.

2. Aunaas, T. and Zachariassen, K.E. (1994). Physiological biomarkers and the Trondheim biomonitoring system. In: *Biomonitoring of Coastal Waters and Estuaries*. K.J.M. Kramer (Ed.). CRC Press, Boca Raton, FL, pp. 107–130.

3. Depledge, M.H. and Fossi, M.C. (1994). The role of biomarkers in environmental assessment (2). Invertebrates. *Ecotoxicology* **3**: 161–172.

4. Peakall, D.B. and Walker, C.H. (1994). The role of biomarkers in environmental assessment (3). Vertebrates. *Ecotoxicology* **3**: 173–179.

5. Ham, K.D., Adams, S.M., and Peterson, M.J. (1997). Application of multiple bioindicators to differentiate spatial and temporal variability from the effects of contaminant exposure on fish. *Ecotoxicol. Environ. Safety* **37**: 53–61.

6. Adams, S.M., Bevelhimer, M.S., Greeley, M.S., Levine, D.A., and Teh, S.J. (1999). Ecological risk assessment in a large river-reservoir: 6: bioindicators of fish population health. *Environ. Toxicol. Chem.* **18**(4): 628–640.

7. Adams, S.M. (2000). Assessing sources of stress to aquatic ecosystems using integrated biomarkers. *Biological Resource Management Connecting Science and Policy*, pp. 17–29.

8. Van Gestel, C.A.M. and van Brummelen, T.C. (1996). Incorporation of the biomarker concept in ecotoxicology calls for a redefinition of terms. *Ecotoxicology* **5**: 217–225.

9. Triebskorn, R. et al. (1997). Induction of heat shock proteins, changes in liver ultrastructure, and alterations of fish behavior: are these biomarkers related and are they useful to reflect the state of pollution in the field? *J. Aquat. Ecosys. Stress Recov.* **6**: 57–73.

10. Triebskorn, R. et al. (2001). The project VALIMAR (VALIdation of bioMARkers for the assessment of small stream pollution): objectives, experimental design, summary of results, and recommendations for the application of biomarkers in risk assessment. *J. Aquat. Ecosyst. Stress Recov.* **8** (3/4): 161–178.

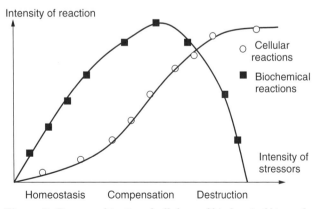

Figure 2. Response kinetics of cellular and biochemical biomarkers.

11. Kammenga, J.E. et al. (2000). Biomarkers in terrestrial invertebrates for ecotoxicological soil risk assessment. *Rev. Environ. Contam. Toxicol.* **164**: 93–147.
12. Köhler, H.R. and Triebskorn, R. (2004). Stress im Boden: Früherkennung ökotoxikologischer Effekte durch Biomarker. *BIUZ* **4**: 240–248.
13. Gerhardt, A. (2001). A new Multispecies Freshwater Biomonitor for ecological relevant control of surface waters. In: *Biomonitors and Biomarkers as Indicators of Environmental Change*, Vol. II, F. Butterworth et al. (Eds.). Kluwer Plenum Press, New York, pp. 301–317.
14. Munkittrick, K.R., Servos, M.R., Carey, J.H., and van der Kraak, G.J. (1997). Environmental impacts of the pulp and paper wastewater: evidence for a reduction in environmental effects at North American pulp mills since 1992. *Water Sci. Technol.* **35**(2/3): 329–338.
15. Dietze, U. et al. (2001). Chemometric discrimination between streams based on chemical, limnological and biological data taken from freshwater fishes and their interrelationships. *J. Aquat. Ecosyst. Stress Recov.* **8**(3/4): 319–336.
16. Triebskorn, R. and Köhler, H.-R. (1996). The impact of heavy metals on the grey garden slug *Deroceras reticulatum* (Müller): metal storage, cellular effects and semi-quantitative evaluation of metal toxicity. *Environ. Pollut.* **93**: 327–343.
17. Gernhöfer, M., Pawert, M., Schramm, M., Müller, E., and Triebskorn, R. (2001). Ultrastructural biomarkers as tools to characterize the health status of fish in contaminated streams. *J. Aquat. Ecosyst. Stress Recov.* **8**(3/4): 241–260.
18. Triebskorn, R. et al. (2004). Toxic effects of the non-steroidal anti-inflammatory drug diclofenac. Part II: cytological effects in liver, kidney, gills and intestine of rainbow trout (*Oncorhynchus mykiss*). *Aquat. Toxicol.* **68**(2): 151–166.
19. Schramm, M., Müller, E., and Triebskorn, R. (1998). Brown trout *(Salmo trutta)* f. *fario* liver ultrastructure as biomarkers of small stream pollution. *Biomarkers* **3**(2): 93–108.
20. Triebskorn, R. et al. (1991). Invertebrate cells as targets for hazardous substances. *Z. Angew. Zool.* **78**(3): 277–287.
21. Moore, M.N. (1990). Lysosomal cytochemistry in marine environmental monitoring. *Histochemistry* **22**: 187–191.
22. Seglen, P.P. (1983). Inhibitors of lysosomal function. *Methods Enzymol.* **96**: 737–765.
23. Weeks, J.M. and Svendsen, C. (1996). Neutral red retention by lysosomes from earthworm (*Lumbricus rubellus*) coelomocytes: a simple biomarker of exposure to soil copper. *Environ. Toxicol. Chem.* **15**(10): 1801–1805.
24. Harreus, D., Köhler, H.-R., and Weeks, J. (1997). Combined non-invasive cell isolation and neutral-red retention assay for measuring the effects of copper on the lumbricid *Aporrectodea rosea* (Savigny). *Bull. Environ. Contam. Toxicol.* **59**: 44–49.
25. Triebskorn, R. and Köhler, H.-R. (2003). Cellular and molecular stress indicators as tools to assess effects and side-effects of chemicals in slugs. *BCPC Symp. Proc.* **80**: 69–76.
26. Köhler, H.-R. et al. (1998). Validation of hsp70 stress gene expression as a marker of metal effects in *Deroceras reticulatum* (pulmonata): correlation with demographic parameters. *Environ. Toxicol. Chem.* **17**: 2246–2253.
27. Triebskorn, R. et al. (2003). Establishing causality between pollution and effects at different levels of biological organization: the VALIMAR project. *Hum. Ecol. Risk Assessment* **9**(1): 171–194.
28. Bradford-Hill, A. (1965). The environment and disease. Association or causation? *Proc. R. Soc. Med.* **9**: 295–300.

ACTIVE BIOMONITORING (ABM) BY TRANSLOCATION OF BIVALVE MOLLUSCS

Roel Smolders
Judith Voets
Lieven Bervoets
Ronny Blust
University of Antwerp
Antwerp, Belgium

Victor Wepener
Rand Afrikaans University
Auckland Park, South Africa

DEFINITION

To start with, it is necessary to clearly define what is understood by active biomonitoring (ABM) using bivalve molluscs in this context. Here, the definition that will be used is: "the translocation of bivalve molluscs from one place to another for the purpose of environmental quality monitoring" (1–3).

The use of bivalves obviously indicates that the focus will be on the aquatic environment. However, ABM has also been performed in the terrestrial environment, using, for example, snails (4,5), mosses (6,7), lichens (8), or grasses (9).

Thus, an ABM approach employs organisms that are collected from an (generally) unstressed, unpolluted population and that, afterwards, are translocated to potentially polluted sites. The chemical and biological consequences of this translocation, which usually involves the caging of the organisms, can then be followed in space and time to estimate the effects of exposure on selected endpoints (3,10).

A second approach in which organisms are used to reflect environmental pollution is passive biomonitoring (PBM), which comprises "the collection of organisms from their habitats at sites where a natural population exists" (1–3). Although PBM has also frequently been used, ABM has a number of properties that advocate its use in environmental quality monitoring.

(DIS) ADVANTAGES OF ABM

The advantages of ABM for biomonitoring purposes are:

- Experiments can be performed for a **known exposure period**.
- All organisms have a **similar life history** at the start of the exposure.
- It is easy to **compare different sites**, even if the organisms are not normally present at the exposure locations.
- A **comparison between transplanted and indigenous organisms** can indicate to what extent the indigenous organisms are adapted to the location or can give information about uptake kinetics of pollutants. Indigenous species can be "genetically

protected," and thus the use of transplanted organisms can circumvent this adaptation process, making the comparison among different sites more precise and the use of nonadapted species more sensitive.
- Compared with laboratory-bred organisms, transplanted organisms are **better acclimated** to changing environmental conditions, and thus the impact of field exposure through, for example, climatological shock, will be less pronounced.

Although these advantages advocate the use of ABM under field conditions, a number of drawbacks also exist that have to be kept in mind when evaluating and interpreting the results of ABM studies:

- **Food availability** may skew the results by overestimating or underestimating the instream toxicity. As mussels depend mainly on algae as a food source, exposure in nutrient-poor but clean water may indicate stressful situations not related to pollutants but to a decreased quality of food resources (11,12). On the other hand, high amounts of food in eutrophic streams may mitigate the impact of pollutant exposure on aquatic organisms. Several authors have demonstrated the positive impact of increased food availability on growth and reproduction of pollutant-exposed organisms (13–15).
- The **loss of cages** containing the transplanted animals can make interpretation of the results much more difficult or even impossible.
- Compared with indigenous organisms, ABM might give **an overestimation of effects** because prolonged exposure to low levels of contamination eventually could lead to genetic adaptation of indigenous organisms. ABM does not take this kind of evolutionary processes into account.

WHY USE BIVALVE MOLLUSCS?

ABM studies have been performed with a wide variety of organisms, but bivalve molluscs have a number of important advantages compared with other aquatic organisms, which advocate their usefulness in biomonitoring.

- Although subsequent handling of organisms had a demonstrably negative effect on bivalve growth (16,17), **handling stress** and crowding are less important in bivalves compared with other organisms.
- Many bivalve molluscs are **relatively resistant to pollution**, although this feature is obviously species-dependent. The most commonly used test species for both marine and freshwater environments have a relatively high pollution tolerance, although this does not necessarily mean that they are insensitive to it.
- As filter feeders, mussels are **exposed to water, food, and suspended matter** as potential sources of contamination, which provides an integrated measure of pollution exposure, which is often not the case with other test species.
- Mussels have a **very high bioaccumulation** and a low biotransformation potential for both organic and inorganic contaminants (18,19), which means that there will be no or low transformation of, for example, organic micropollutants, and mussels can thus be used as long-term bioaccumulators.
- Mussels are **sedentary organisms** and are often easily collected. Mostly, they are attached to rocky or woody substrates in high numbers, which makes it easy to collect a large number of organisms from one location.
- Bivalves often have a **commercial value**, which makes them target organisms for environmental protection and legislation efforts.

However, a number of negative features connected to the use of bivalve molluscs as a biomonitoring tool also exist.

- **Reproduction**, an ecotoxicological endpoint with very high ecological relevance, is relatively difficult to measure and is highly dependent on seasonality (20).
- **No clear reference populations** exist, whereas other aquatic organisms (e.g., daphnids or zebrafish) have clearly defined responses under "no stress" or laboratory situations, which might complicate the comparison between different studies when organisms for transplantation are taken from geographically different collection sites.
- Mussels have no or a relatively **low biotransformation potential** (also a benefit, see above), which limits the use of biotransformation enzymes as biomarkers of exposure (18).

THE USE OF BIVALVE MOLLUSCS FOR BIOMONITORING

Biomonitoring is regarded as the regular systematic use of organisms to evaluate changes in the environment or water quality (21). It may take on many forms, from the measurement of chemical residues in tissues of living organisms (i.e., chemical monitoring) to the quantification of biological endpoints (i.e., biological effect monitoring or biomarkers) including changes to various biochemical, physiological, morphological, and behavioral factors (22).

The most popular bioindicator species used in the marine bioaccumulation and biomarker ABM are species from the family Mytilidae (e.g., *Mytilus edulis* and *M. galloprovincialis*) and *Perna* spp. from the family Pernidae. With the subsequent expansion of ABM to freshwater environments, the Dreissenidae (e.g., *Dreissena polymorpha*) and species from the family Unionidae (e.g., *Anodonta cygnea, Elliptio complanata*) are the most used bioindicator organisms. These four families make up over 70% of the bioindicator organisms used in ABM studies worldwide. An evaluation of three databases revealed over 350 references related to the implementation of transplanted bivalves in biomonitoring programs. These studies that were carried out between the early 1980s and 2004 showed two very definite trends in the application of bivalves in ABM.

Chemical Monitoring with Bivalve Molluscs

Physical and chemical measurements of dissolved and particulate bound compounds reveal valuable information concerning the presence and the dispersion of pollutants in aquatic environments. However, important shortcomings exist in predicting the impact of pollution in the environment. Due to changes in water composition and environmental conditions, representative samples cannot easily be obtained (23,24). Furthermore, measurements of pollutants in the water and/or sediments do not necessarily represent their bioavailability. Therefore, measurements of accumulated micropollutants in the biota can help to determine the presence, bioavailability, and impact of pollution in the environment.

During the last three decades, bivalves have widely been employed as biomonitors to evaluate the bioavailability of micropollutants. Marine mussels, such as *M. edulis*, *M. galloprovincialis*, and *P. viridis*, have been used since 1975 in monitoring programs to evaluate the bioavailability and effects of pollutants present in estuarine and coastal ecosystems (25–29). In several areas of the world, this so-called "Mussel Watch Program" has been implemented (30). Following these marine biomonitoring programs, freshwater mussels have been used since the late 1980s to monitor the quality of freshwater ecosystems. Freshwater bivalves that have been used are, among others, *E. complanata*, the Asiatic clam *Corbicula fluminea*, and *Anodonta* sp. (31,32). In Europe, the most frequently used bivalve species is the zebra mussel, *D. polymorpha* (33–36). One matter that is of special interest when molluscs are used for chemical monitoring involves the question whether concentrations of micropollutants in resident mussels might vary from transplanted mussels. Therefore, comparison of the accumulation of contaminants between transplanted and indigenous mussels is essential to evaluate and validate the applicability of caged organisms as a chemical monitoring tool. Among others, an extensive study by Bervoets et al. (37) showed that transplanted zebra mussels, exposed for a period of 6 weeks in summer, accumulated a broad range of micropollutants up to levels comparable with levels measured in resident mussels.

Table 1 lists a number of selected studies that have used ABM with bivalve molluscs to monitor the presence of pollutants in freshwater, estuarine, and marine environments.

Biological Monitoring with Bivalve Molluscs

The realization, however, that it is virtually impossible to monitor all contaminants present in the environment (both of anthropogenic and natural origin), together with the increased attention that the relationship between chemical body burdens and toxic effects was receiving, led to the application of biomarkers in ABM (59,60). Biomarkers involve the assessment of the overall quality of the aquatic environment by examining biochemical, physiological, behavioral, or population responses that reflect the potential of contaminants to impair biological processes in exposed organisms (61). From the early 1990s, biomarkers became frequently employed in ABM studies as indicators of biological responses, which were related to exposure and toxic effects of environmental chemicals. Various biochemical parameters in bivalves have been tested for their responses to toxic substances and their potential use as biomarkers of exposure or effect (22). The biomarkers that have been investigated the most extensively are related to those biochemical parameters that provide short-term indication of long-term effects, that is, responses at the lower level of biological organization (e.g., antioxidant enzymes, metallothioneins, acetylcholine esterase). However, the technical demands and lack of ecological relevance have been identified as a

Table 1. A Brief Overview of Chemical Monitoring Studies with Bivalve Molluscs

Area	Species	Common Name	Pollutant	Reference
Marine waters	*Mytilus edulis*	Common mussel	Heavy metals	39
	Mytilus edulis	Common mussel	PAHs	40
	Mytilus edulis	Common mussel	PCBs and hydrocarbons	41
	Mytilus galloprovincialis	Bay mussel	Heavy metals	42
	Mytilus galloprovincialis	Bay mussel	PAHs	43
	Mytilus californianus	California mussel	Heavy metals	44
	Macoma balthica	Baltic tellin	Heavy metals	39, 45
	Perna viridis	Green mussel	Heavy metals	46, 47
Estuarine	*Perna perna*	Brown mussel	Hg, MeHg	48
	Ragnia cuneata	Wedge clam	Heavy metals	49
	Ragnia cuneata	Wedge clam	Dioxins and furans	50
	Potamocorbula amurensis	Amur River clam	Hydrocarbons	51
Freshwater	*Dreissena polymorpha*	Zebra mussel	Organotins	52, 53
	Dreissena polymorpha	Zebra mussel	Heavy metals	37, 38, 54
	Dreissena polymorpha	Zebra mussel	PCBs, PBDEs, DDE, HCB	37
	Corbicula fluminea	Asian clam	Heavy metals	31
	Pyganodon grandis	Giant floater	MeHg	55
	Elliptio complanata	Eastern elliptio	PAH and PCB,	56, 57
	Hyridella depressa		Heavy metals	58

concern in the application of these biomarkers in ABM. In this regard, energy budgets at different levels of biological organization can potentially link these different levels and provide a holistic and integrative indication of the effects observed within exposed organisms. This concept has been applied from as early as the beginning of the 1980s to demonstrate the scope for growth changes in bivalves along a pollution gradient. In recent years, energetic responses are increasingly used as biomarkers in ABM and have been statistically correlated to negative effects on growth and condition in bivalves transplanted along a pollution gradient and other invertebrate community-related responses (2,3,62,63).

CONCLUSIONS

Bivalve molluscs have a number of properties that advocate their use in both chemical and biological monitoring studies. Active biomonitoring, being the translocation of organisms from one (usually unpolluted) location to another one (usually polluted), can provide valuable information about the state of the environment, the presence and biological availability of pollutants, and the effects of these pollutants on organisms and ecosystems.

BIBLIOGRAPHY

1. De Kock, W.C. and Kramer, K.J.M. (1994). In: *Biomonitoring of Coastal Waters and Estuaries*. K.J.M. Kramer (Ed.). CRC Press, Boca Raton, FL.
2. Bervoets, L., Smolders, R., Voets, J., and Blust, R. (2003). In: *Biological Evaluation and Monitoring of the Quality of Surface Waters*. J.J. Symoens and K. Wouters (Eds.). Cercle Hydrobiologique de Bruxelles, p. 129.
3. Smolders, R., Bervoets, L., Wepener, V., and Blust, R. (2003). *Hum. Ecol. Risk Assess.* **9**: 741–760.
4. Pihan, F. and de Vaufleury, A. (2000). *Ecotoxicol. Environ. Safety* **46**: 137–147.
5. Scheifler, R., Ben Brahim, M., Gomot-de Vaufleury, A., Carnus, J.-M., and Badot, P.-M. (2003). *Environ. Pollut.* **122**: 343–350.
6. Couto, J.A., Aboal, J.R., Fernández, J.A., and Carballeira, A. (2004). *Chemosphere* **57**: 303–308.
7. Fernández, J.A., Aboal, J.R., and Carballeira, A. (2004). *Ecotoxicol. Environ. Safety* **59**: 76–83.
8. Jeran, Z., Byrne, A.R., and Batic, F. (1995). *The Lichenologist* **27**: 375–385.
9. Dietl, C., Reifenhäuser, W., and Peichl, L. (1997). *Sci. Total Environ.* **205**: 235–244.
10. Salazar, M.H. and Salazar, S.M. (1999). *Standard Guide for Conducting in-situ Field Bioassays with Caged Bivalves*. ASTM International Guideline E2122-02.
11. Borcherding, J. (1995). *Malacologia* **36**: 15–27.
12. Madon, S.P., Schneider, D.W., Stoeckel, J.A., and Sparks, R.E. (1998). *Can. J. Fish. Aquat. Sci.* **55**: 401–413.
13. Bridges, T.S., Farrar, J.D., and Duke, B.M. (1997). *Environ. Toxicol. Chem.* **16**: 1659–1665.
14. Emery, V.L., Moore, D.W., Gray, B.R., Duke, B.M., Gibson, A.B., Wright, R.B., and Farrar, J.D. (1997). *Environ. Toxicol. Chem.* **16**: 1912–1920.
15. Stuijfzand, S.C., Helms, M., Kraak, M.H.S., and Admiraal, W. (2000). *Ecotoxicol. Environ. Safety* **46**: 351–356.
16. Salazar, M.H. and Salazar, S.M. (1996). In: M.A. Champ and P.F. Seligman (Eds.). *Organotin*. Chapman & Hall, London.
17. Monroe, E.M. and Newton, T.J. (2001). *J. Shellfish Res.* **20**: 1167–1171.
18. Daubenschmidt, C., Dietrich, D.R., and Schlatter, C. (1997). *Aquat. Toxicol.* **37**: 295–305.
19. Nasci, C. et al. (1999). *Environ. Pollut.* **39**: 255–260.
20. Seed, R. and Suchanek, T.H. (1992). In: *The Mussel Mytilus: Ecology, Physiology, Genetics and Culture*. E.M. Gosling (Ed.). Elsevier, Amsterdam.
21. De Zwart, D. (1995). *Monitoring Water Quality in the Future*. Vol. 3. *Biomonitoring*. National Institute of Public Health and Environmental Protection (RIVM), Bilthoven, The Netherlands.
22. Van der Oost, R., Beyer, J., and Vermeulen, N.P.E. (2003). *Environ. Toxicol. Pharmacol.* **13**: 57–149.
23. Luoma, S.N. (1983). *Sci. Total Environ.* **28**: 1–22.
24. van Griethuysen, C., Meijboom, E.W., and Koelmans, A.A. (2003). *Environ. Toxicol. Chem.* **22**: 457–465.
25. Goldberg, E.D. et al. (1978). *Environ. Conserv.* **5**: 101–125.
26. Farrington, J.W. et al. (1983). *Environ. Sci. Technol.* **14**: 490–496.
27. Walsh, A.R. and O'Halloran, J. (1998). *Environ. Toxicol. Chem.* **17**: 1429–1438.
28. Chase, M.E. et al. (2001). *Mar. Pollut. Bull.* **42**: 491–505.
29. de Astudillo, L.R. et al. (2002). *Arch. Environ. Contamin. Toxicol.* **42**: 410–415.
30. Blackmore, G. and Wang, W.X. (2003). *Environ. Toxicol. Chem.* **22**: 388–395.
31. Andres, S. et al. (1999). *Environ. Toxicol. Chem.* **18**: 2462–2471.
32. Gewurtz, S.B., Lazar, R., and Haffner, G.D. (2000). *Environ. Toxicol. Chem.* **19**: 2943–2950.
33. Regoli, L., Chan, H.M., de Lafontaine, Y., and Mikaelian, I. (1994). *Aquat. Toxicol.* **53**: 115–126.
34. Mersch, J., Wagner, P., and Pihan, J.C. (1996). *Environ. Toxicol. Chem.* **15**: 886–893.
35. Hendriks, A.J., Pieters, H., and de Boer, J. (1998). *Environ. Toxicol. Chem.* **17**: 1885–1898.
36. Zimmermann, S. et al. (2002). *Environ. Toxicol. Chem.* **21**: 2713–2718.
37. Bervoets, L. et al. (2004). *Environ. Toxicol. Chem.* **23**: 1973–1983.
38. Kraak, M.H.S., Scholten, M.C.T., Peeters, W.M.H., and Dekock, W.C. (1991). *Environ. Pollut.* **74**: 101–114.
39. Rainbow, P.S. (1995). *Mar. Pollut. Bull.* **31**: 183–192.
40. Baumard, P. et al. (1999). *Mar. Environ. Res.* **47**: 17–47.
41. Porte, C. and Albaiges, J. (1994). *Arch. Environ. Contamin. Toxicol.* **26**: 273–281.
42. Conti, M.E. and Cecchetti, G. (2003). *Environ. Res.* **93**: 99–112.
43. Baumard, P. et al. (1998). *Estuarine Coastal Shelf Sci.* **47**: 77–90.
44. Gutierrez-Galindo, E.A. and Munoz-Barbosa, A. (2003). *Cien. Mars.* **29**: 21–34.
45. Regoli, F. et al. (1998). *Arch. Environ. Contamin. Toxicol.* **35**: 594–601.
46. Yap, C.K., Ismail, A., Tan, S.G., and Omar, H. (2002). *Environ. Int.* **28**: 117–126.

47. Avelar, W.E.P. et al. (2000). *Water Air Soil Pollut.* **118**: 65–72.
48. Kehrig, H.D.A., Costa, M., Moreira, I., and Malm, O. (2001). *Environ. Sci. Pollut. Res.* **8**: 275–279.
49. Mcconnell, M.A. and Harrel, R.C. (1995). *Am. Malacologic Bull.* **11**: 191–201.
50. Harrell, R.C. and McConnell, M.A. (1995). *Estuaries* **18**: 264–270.
51. Prereira, W.E., Hostettler, F.D., and Rapp, J.B. (1992). *Mar. Pollut. Bull.* **24**: 103–109.
52. Stab, J.A. et al. (1995). *Environ. Toxicol. Chem.* **14**: 2023–2032.
53. Regoli, L., Chan, H.M., de Lafontaine, Y., and Mikaelian, I. (2001). *Aquat. Toxicol.* **53**: 115–126.
54. Camusso, M., Balestrini, R., and Binelli, A. (2001). *Chemosphere* **44**: 263–270.
55. Malley, D.F., Stewart, A.R., and Hall, B.D. (1996). *Environ. Toxicol. Chem.* **15**: 928–936.
56. Gewurtz, S.B., Drouillard, K.G., Lazar, R., and Haffner, G.D. (2002). *Arch. Environ. Contamin. Toxicol.* **43**: 497–504.
57. Gewurtz, S.B., Lazar, R., and Haffner, G.D. (2003). *J Great Lakes Res.* **29**: 242–255.
58. Vesk, P.A. and Byrne, M. (1999). *Sci. Total Environ.* **225**: 219–229.
59. De Maagd, P.G.J. (2000). *Environ. Toxicol. Chem.* **19**: 25–35.
60. Chapman, P.M. (1997). *Mar. Pollut. Bull.* **34**: 282–283.
61. McCarty, J.F. and Shugart, L.R. (1990). In: J.F. McCarty and R.L. Shugart (Eds.). *Biomarkers of Environmental Contamination.* Lewis Publishers, Boca Raton, FL, pp. 3–16.
62. Widdows, J., Phelps, D.K., and Galloway, W. (1981). *Mar. Environ. Res.* **4**: 181–194.
63. Smolders, R., De Coen, W., and Blust, R. (2004). *Environ. Pollut.* **132**: 245–263.

BIOCHEMICAL OXYGEN DEMAND AND OTHER ORGANIC POLLUTION MEASURES

Neil F. Pasco
Joanne M. Hay
Lincoln Ventures, Ltd.
Lincoln, New Zealand

Biochemical (also called biological) oxygen demand (BOD) is an empirical test indicator of biological activity and follows a specified laboratory procedure to determine the oxygen requirements of wastewaters, effluents, and polluted waters (1,2). The BOD test measures the amount of oxygen consumed in breaking down organic matter using both aerobic biological and chemical degradation processes, giving an indication of the organic strength of a wastewater. The test has its widest application in measuring waste loadings to treatment plants and in evaluating the BOD-removal efficiency of such treatment systems.

BOD is determined by incubating a water sample in a sealed container for a specified time and measuring the loss of oxygen that occurs between the beginning and end of the test. Samples must often be diluted prior to incubation.

WHAT DOES IT MEASURE?

Biodegradable organic matter is a composite substrate and is difficult to quantify directly. The BOD test estimates the amount of dissolved oxygen used during the degradation of organic waste as a surrogate for measuring the waste itself.

The amount of molecular oxygen measured during the incubation period can accrue from three sources:

1. The oxygen used for the biochemical degradation of organic material (carbonaceous biochemical oxygen demand, CBOD).
2. The oxygen used to oxidize inorganic materials such as sulfides and ferrous iron.
3. The oxygen used to oxidize reduced forms of nitrogen (nitrogenous demand), although a nitrifying inhibitor is often used to suppress the nitrogenous demand.

The earth's surface is mildly oxidizing and a tendency exists for reduced compounds, such as carbohydrates, to react with oxygen to form the more stable endproducts, carbon dioxide and water. BOD is the manifestation of this tendency in an aqueous environment, and two of the three sources listed above, (1) and (3), require microorganisms as the catalyst, whereas in (2), oxygen is capable of directly oxidizing some classes of inorganic materials.

Primarily, BOD values indicate the amount of carbonaceous organic material. The oxygen demand made by the oxidation of ammonia and organic nitrogen can be inhibited using 2-chloro-6-(trichloromethyl) pyridine or allyl thiourea in order to remove the nitrogenous demand and isolate the CBOD value. In the remainder of this article, BOD refers to CBOD only.

HISTORICAL CONTEXT

The complete biochemical oxidation of organic matter in a water sample can take up to 90 days, an impractical duration for an analytical test. The current international regulatory standard method for determining biodegradable organic compounds in wastewaters has been set as a five-day BOD assay (BOD_5). BOD_5 quantifies the amount of dissolved oxygen required for the microbial oxidation of carbonaceous organic material in five days under specified conditions (1). The five-day duration has no theoretical grounding but is based on an historical convention reported in 1912, as follows (3):

In a report prepared by the Royal Commission on Sewage Disposal in the United Kingdom at the beginning of last century, it was recommended that a five-day, 18.3 °C, BOD value be used as a reference in Great Britain. These values were selected because British rivers do not have a flow time to the open sea greater than five days and average long-term summer temperatures do not exceed 18.3 °C. The temperature has been rounded upward to 20 °C, but the five-day time period has become the universal scientific and legal reference.

The standardized test was first published in the *American Public Health Association (APHA) Standard Methods* in 1917.

UNITS OF MEASURE

BOD is expressed either as milligrams of oxygen per liter (mg O_2/L) or parts per million (ppm).

WHAT IS IT USED FOR?

Despite its dubious ancestry, the BOD_5 test is still the most important and widely used environmental index for monitoring organic pollutants in wastewaters, mainly as proof of compliance with relevant legislation. Most developed countries strictly regulate the permissible BOD loadings of water as it is discharged from a treatment plant into receiving waters (e.g., sea or river), which is typically required to be less than 20 mg O_2/L. The BOD_5 value for potable water is close to zero; domestic wastewater effluent entering a treatment plant typically has a BOD_5 of around 300 mg O_2/L; and industrial effluents can have BOD_5 values greater than 20,000 mg O_2/L.

Figure 1. Standard 300 mL BOD bottle.

HOW DOES IT WORK?

Heterotrophic microorganisms require a metabolizable source of organic carbon, such as sugars, amino acids, nucleotides, organic acids, and fats, in order to grow. Depending on their physiology and the prevailing environmental conditions, microorganisms can harness energy from these food sources using either respiratory or fermentative biochemical pathways. Respiration is defined as the oxidation of an energy source (such as organic carbon) in which electrons are removed from the carbon source and donated to an inorganic terminal electron acceptor within the cell. Aerobic respiration uses oxygen as the terminal electron acceptor. Hence, through a complexity of genetics, physiology, and biochemistry, the breakdown of organic carbon is coupled to oxygen reduction and is facilitated by microorganisms.

Natural organic detritus plus organic waste from wastewater treatment plants, agricultural, and urban runoff provides food sources for waterborne microorganisms. The lab-based BOD_5 test is unique in that it simulates the microbial decomposition of organic wastes in the environment, thereby approximating the response of natural ecosystems (4), which helps explain the enduring relevance of the BOD test over so many years.

MEASUREMENT PROTOCOL

The BOD_5 test consists of a sample of wastewater held in a full, airtight bottle (usually 300 mL) (Fig. 1) for five days at 20 °C in the absence of light. The amount of dissolved oxygen present in the sample is measured at the beginning of the test period and at the end of five days using a dissolved oxygen probe (Fig. 2). The difference in dissolved oxygen between the day 0 and day 5 measurements represents the biochemical oxygen demand. The test requires the presence of three

Figure 2. Dissolved oxygen probe.

key ingredients: dissolved oxygen, organic substrate, and microorganisms, as prescribed by the standard test protocol. For complete specifications on how to set up and calculate a BOD_5 analysis, refer to Reference 1.

COMPONENTS OF THE BOD$_5$ TEST

Dilution Water

The dilution water (saturated with dissolved oxygen) must be free of organic matter and bioinhibitory substances and has a twofold function:

- It ensures that aerobic conditions prevail throughout the test. Water samples with high BOD loadings contain insufficient dissolved oxygen for the complete oxidation of the compounds present in the sample. Without dilution, the sample will become anaerobic at some point during the five-day incubation period, causing the microorganisms to switch from a respiratory to a fermentative mode. Diluting these samples ensures that sufficient oxygen is maintained to oxidize all the organic matter present, with some residual oxygen left at the end of the test.
- It supplies the inorganic nutrients necessary for microbial growth (magnesium sulfate, calcium chloride, ferric chloride) and a buffering capacity in order to maintain a pH suitable for bacterial growth.

Seed (Microorganisms)

The seed refers to the microorganisms responsible for the biodegradation of organic matter in the sample. Although microorganisms occur naturally in the environment, a diluted water sample may require seeding with additional microbes to ensure a sufficient number of microbes are present to degrade the organic matter in the sample. The preferred seed is effluent from a wastewater treatment plant. The seed material has a BOD of its own, which must be measured separately in a control (seed in dilution water with no added substrate) and subtracted from the BOD of the samples. In a 300 mL BOD bottle, the volume of seed is typically 2 mL.

Wastewater Sample

The wastewater is the source of the organic matter to be measured. Given that biodegradable organic matter is a composite substrate, the BOD measurement is an aggregate of the wide range of substances present. It is also dependent on the microbial species present and the relative biodegradability of the substrates present. Depending on the expected BOD strength of the wastewater sample, a known volume is added to the BOD bottle and made up to 300 mL with dilution water.

STANDARDS

Glucose–Glutamic Acid (GGA)

The primary standard for calibrating BOD$_5$ tests is a glucose and glutamic acid solution (150 mg glucose/L, 150 mg glutamic acid/L). The GGA solution is intended to be a reference point for evaluating the dilution water quality, seed effectiveness, and analytical technique.

Organization for Economic Cooperation and Development (OECD)

The GGA solution has increasingly been questioned as a valid standard, particularly for the developing rapid BOD tests (see later). Many microorganisms preferentially use simple sugars like glucose and will actively repress the metabolism of other carbon sources when glucose is present, which has led to an underestimation of BOD values of samples containing compounds that are less readily biodegradable. A more appropriate standard solution with similar biodegradative properties to the sample being analyzed may be warranted, especially for rapid BOD tests. The OECD-defined wastewater standard solution (5) has been proposed as an alternative and more appropriate choice of standard for analysis of treated wastewater (6,7). The OECD solution, however, is not currently recognized as a regulatory standard.

QUALITY CONTROLS

No absolute BOD value of a sample exists, and BOD results are test defined, meaning BOD values are based on the parameters of the test method, not on any "true" BOD value. Some specific requirements exist that must be met for a BOD$_5$ analysis to be valid and which ensure that the methodology used is correct. Controls to be included each time a BOD$_5$ test is performed are a dilution water blank, a seed (the microorganisms) sample, and a GGA standard. If any of these quality control checks fall outside the recommended ranges (specified below), the results of the BOD$_5$ test should be rejected.

- *BOD blank* (bottle containing only dilution water and no seed) must not show a dissolved oxygen depletion of more than 0.2 mg O_2/L after five days.
- *Seed blank* (microorganisms added to water sample) should contribute between 0.6 and 1.0 mg O_2/L uptake per BOD bottle.
- *The glucose–glutamic acid standard* should report a BOD of 198 ± 31 mg O_2/L after a five-day incubation (i.e., the oxidation of the glucose–glutamic acid solution consumes 198 mg O_2/L over five days).

Sample dilutions should report a 50% decrease in dissolved oxygen over the five-day incubation period. At minimum, there should be an oxygen depletion of at least 2 mg O_2/L between the initial and final reading, plus a residual dissolved oxygen reading of at least 1 mg O_2/L.

HOW MUCH IS MEASURED?

Wastewater contains both biodegradable and non-biodegradable material, the latter not being represented in the BOD measurement. However, in general, the BOD$_5$ test does not represent the total biodegradable fraction of organic material present. Depending on the type of wastewater, the rates at which the biodegradable material in a given sample can be oxidized over five days can be between 50% and 90%.

Where the composition of the organic substrate is chemically defined, as is the case for the primary GGA standard, the amount of oxygen required for its complete oxidation can be calculated from a stoichiometric equation.

For example, the stoichiometric equation for the oxidation of glucose ($C_6H_{12}O_6$) is

$$C_6H_{12}O_6 + 6O_2 \rightarrow 6CO_2 + 6H_2O$$

For the complete oxidation of 1 mole of glucose, 6 moles of oxygen is required. A 150 mg glucose/L is reported to have a BOD_5 of 99 mg/L (1). On a molar basis, 150 mg glucose/L corresponds to 0.83 mmol/L, and the stoichiometric oxygen requirement is 5 mmol/L. The BOD_5 glucose requirement of 99 mg/L converts to 3.094 mmol O_2/L, and therefore the biological oxidation occurring after five days equates to 61.9% conversion. A similar exercise with glutamic acid, excluding the oxidation of nitrogen, equates to 58.4% conversion; thus the mean molar conversion of the standard glucose and glutamic acid mix in a one-hour BOD_5 test is 60.5% (8).

PROBLEMS

Although the BOD_5 test sounds simple, practical difficulties associated with it exist, and it has significant limitations. Although the test has been refined over the years, the basic approach of using a dilution technique has remained unchanged. Unfortunately, these dilutions reduce the concentration of substrates and microorganisms in the samples, thereby decreasing the overall kinetics. This decrease, along with the arbitrary time period of five days, means the actual BOD of the sample may not be reflected. The presence of toxic substances, such as copper, mercury, or chromium, even at low concentrations, in the wastewater can affect, or kill, the bacteria. A decrease in BOD_5 values may be because of a decrease in organic load or may be because of the presence of a toxin in the sample. The test is also insensitive and imprecise at low concentrations; the method requires experience and has an accepted margin of error of 15–20%, possibly because of the heterogeneity of the microbial populations in wastewater treatment systems and their widely differing responses to substrates.

However, the major drawback of the traditional BOD test continues to be the five-day measurement time. Such a delay is unacceptable for any form of active intervention for environmental monitoring and/or process control. It is suggested that the low solubility of oxygen in water is rate-limiting in the BOD_5 test; an O_2 saturated solution at 20 °C contains 9.07 mg O_2/L. The limited oxygen storage capacity of water is responsible for aqueous environments being so susceptible to nutrient deoxygenation and is why wastewater samples are often required to be substantially diluted to keep the oxygen depletion within the 1–9 mg O_2/L working range of the BOD_5 test.

ALTERNATIVE MEASUREMENT TECHNOLOGIES

In an attempt to overcome the difficulties in achieving rapid, reproducible results, a number of analytical wet chemical techniques have been developed to monitor organic pollution, as alternatives to BOD_5 measurements (9).

Chemical Oxygen Demand (COD)

The COD test measures the chemical oxidation of wastewater by using a strong oxidizing agent under acid conditions. As for BOD, it measures the amount of oxygen required as a surrogate for measuring the organic waste component directly. However, the COD test does not differentiate between biologically available and inert organic matter and measures the total quantity of oxygen required to oxidize all organic matter into carbon dioxide and water. COD values are always greater than BOD values, but the results can be obtained within two hours and the presence of toxic compounds in the sample do not affect COD measurements. In many cases, the BOD/COD ratio is used to provide an indication of the biodegradability of the wastewater. However, the chemicals used, such as acid, chromium, silver, and mercury, produce liquid hazardous waste that requires careful handling and disposal. Finally, the precision and accuracy of the COD test are questionable at low BOD values.

Total Organic Carbon (TOC)

This test measures carbon, where the carbon is oxidized by catalytic combustion (either at low temperatures by ultraviolet light and the addition of persulfate reagent, or by using a catalyst at high temperatures) to carbon dioxide, which is then measured. However, good correlations between BOD and TOC measurements have not been demonstrated, and again, the test is unable to differentiate between biodegradable and nonbiodegradable matter. The main advantage of the TOC test is its speed; determinations can be made within minutes, facilitating a greater number of measurements compared with either BOD or COD tests.

RAPID BOD TECHNOLOGIES

Oxygen-Based Sensors

In response to the restrictive five-day time lag inherent in the traditional BOD_5 test, a number of rapid BOD techniques have been devised. They have evolved down two development paths, both respiratory-based: microbial biosensors and respirometers.

Microbial BOD biosensors consist of a microbial biofilm immobilized onto a dissolved oxygen probe. Since the first BOD biosensor of this sort was developed (10), various sensor designs have been reported, and their response time ranges from 3 to 30 minutes. In all reported cases, the decrease in dissolved oxygen forms the basis of the analytical signal. One consequence of oxygen limitation for these sensors is that the amount of organic material biodegraded in a short-time assay is a very small fraction (∼1%) of the total biodegradable organic content, and they can only respond to the easily assimilated compounds in the wastewater.

Commercially available respirometers, such as RODTOX®, BIOX®, BIOMONITOR®, and RACOD®, are designed to replicate the operation of a wastewater treatment plant, rather than the methodology of the

BOD test. They consist of one or more activated-sludge bioreactors maintained at steady-state conditions by continuous stirring. The inclusion of organic material increases microbial respiration, in turn increasing the rate of oxygen uptake, which is recorded by an oxygen sensor. Oxygen consumption is then correlated with BOD; these devices also report a short-term BOD estimation in 3–30 minutes.

Nonoxygen-Based Sensors

A second generation of rapid BOD sensors have been reported based on the measurement of charge transfer, either to an alternative electron acceptor, such as a synthetic redox-active mediator (8,11), or directly to an electrode (12), rather than the depletion of oxygen.

Mediated Sensors

The oxidation (loss of electrons and the concurrent transfer of charge) of organic substrates can be coupled to the reduction of a redox mediator, such as potassium hexacyanoferrate. These mediators readily accept or donate (i.e., shuttle) electrons in biochemical reactions. In a BOD application, the redox mediator substitutes for oxygen as the terminal electron acceptor during aerobic oxidation and transfers the electrons to an electrode, generating an analytical signal that can be converted to a BOD value. By replacing oxygen with a more soluble synthetic mediator, higher numbers of microorganisms can be incorporated into the assay, significantly reducing the incubation time to one or two hours (8).

Direct Electron Transfer

Direct electron transfer to an electrode is exhibited by some microorganisms, such as *Shewanella putrefaciens*, thereby eliminating the need for a redox mediator. The direct reducing action of the electrochemically active microorganisms can be exploited in a microbial fuel cell device, and the current produced has been shown to be dependent on the concentration of organics in the wastewater water sample (12).

CONCLUSION

For all its shortcomings and the availability of a number of related measurement techniques, the BOD_5 assay remains the preferred test for reporting the oxygen requirements of wastewaters, effluents, and polluted waters, which is, in part, because of its international regulatory status, but also because it uniquely characterizes the impact of oxygen depletion developing from biological activity in breaking down organic matter on receiving waters.

Concern over water quality continues to increase and has led to the implementation of stricter environmental pollution control legislation worldwide. Legislation, such as the European Urban Wastewater Treatment directive, is framed in terms of compliance with BOD_5 and will entrench the test even further. Although the need for a more rapid test is well recognized, the success of these alternative tests will remain limited until they can demonstrate robust comparability with the BOD_5 measurement.

BIBLIOGRAPHY

1. APHA (American Public Health Association). (1995). *Standard Methods for the Examination of Water and Wastewater*, 19th Edn. A.D. Eaton, L.S. Clesceri, and A.E. Greenberg (Eds.). APHA, Washington, DC, pp. 5.1–5.9.
2. JIS. (1993). *Japanese Industrial Standard Committee Testing Methods for Industrial Wastewater*. Tokyo, Japan.
3. Tchobanoglous, G. and Schroeder, E. (1987). *Water Quality*. Addison-Wesley, Reading, MA.
4. Clark, D.W. (1992). BOD: the modern alchemy. *Public Works* **50–51**: 86–89.
5. OECD. (1981). *OECD Guideline 209—Activated Sludge, Respiration Inhibition Test*. OECD Guideline for testing of chemicals. OECD, Paris, France.
6. Tanaka, H., Nakamura, E., Minamiyama, Y., and Toyoda, T. (1994). BOD biosensor for secondary effluent from wastewater treatment plants. *Water Sci. Technol.* **30**: 215–227.
7. Liu, J. and Mattiasson, B. (2002). Microbial BOD sensors for wastewater analysis. *Water Res.* **11**(15): 3786–3802.
8. Pasco, N.F., Baronian, K.H., Jeffries, C., and Hay, J. (2000). Biochemical mediator demand—a novel rapid alternative for measuring biochemical oxygen demand. *Appl. Microbiol. Biotechnol.* **53**(5): 613–618.
9. Bourgeois, W., Burgess, J.E., and Stuetz, R.M. (2001). On-line monitoring of wastewater quality: a review. *J. Chem. Technol. Biotechnol.* **76**(4): 337–348.
10. Karube, I., Matsunaga, T., Mitsuda, S., and Suzuki, S. (1977). Microbial electrode BOD sensors. *Biotechnol. Bioeng.* **19**: 1535–1547.
11. Yoshida, N., Yano, K., Morita, T., McNiven, S.J., Nakamura, H., and Karube, I. (2000). A mediator-type biosensor as a new approach to biochemical oxygen demand estimation. *Analyst* **125**(12): 2280–2284.
12. Kim, M. et al. (2003). Practical field application of a novel BOD monitoring system. *J. Environ. Monitor.* **5**(4): 640–643.

BIODEGRADATION

Christian D. Frazar
Silver Spring, Maryland

The beginnings of life on this planet arose during a time of extreme conditions. For life to arise in an inhospitable world, the first organisms needed to be simple but hearty. The life forms of today considered most similar to those early ancestors of life are bacteria and archaea—simple, primitive organisms. Despite their primitive nature, these organisms have a diverse and powerful metabolic potential.

As life and other processes transformed the earth's air, land, and water, early organisms were in a new and changing environment. A wider variety of ecological niches became available. For five-sixths of the earth's existence, it was inhabited exclusively by microorganisms (1). Given their short generation time, microorganisms have had unfathomable generations to adapt to earth's many chemicals. As a result it is quite possible that for every natural compound that exists, there is a microorganism that has evolved to metabolize it and use it for growth.

Microorganisms are at the base of the food chain and play a key role in both secondary production and in decomposition. The ability of microorganisms to decompose or break down compounds is known as biodegradation. This term is often used to refer to the degradation of environmental pollutants by bacteria or fungi. This discussion will focus on the degradation process itself and the factors that influence and limit biodegradation.

MICROBIAL DIVERSITY AND METABOLIC POTENTIAL

The use of biodegradation to remove a contaminant from the environment is known as bioremediation. Understanding microbial diversity and metabolic potential is essential to understanding how to use biodegradation for remediation projects. During the twentieth century, a vast number of synthetic and natural compounds were released into the environment. A complex community of microorganisms often has the mechanisms to use or detoxify many of the compounds that humans have released.

Some organisms have a very narrow range of metabolic potential, and others have a wider capacity. For example, it has been demonstrated that the organism *Pseudomonas putida,* strain F1, degrades approximately 105 compounds, and it has been suggested that it can degrade many more (2). Other compounds, many of which are synthetic, are resistant to microbial degradation. Resistant compounds are known as recalcitrant compounds; many factors contribute to a compound's recalcitrance, which will be discussed later.

Many synthetic compounds are considered xenobiotics. The term xenobiotic is often used to refer to compounds that are foreign to biological systems. The term originates in mammalian systems, not microbial systems. It does not always describe a compound accurately in microbial terms, because microbial diversity is so vast. What is foreign to one group of bacteria may be readily metabolized by a different group.

Microorganisms benefit from biodegradation by metabolizing a compound to yield carbon and energy. Energy is derived from the transfer of electrons and the conservation of that energy in the form of molecules such as adenosine triphosphate (ATP) and nicotinamide-adenine dinucleotide (NAD^+). ATP conserves this energy through high energy phosphate bonds and is used to drive energy-requiring reactions in the cell. NAD^+ is an electron carrier involved in oxidation–reduction reactions. If metabolism results in complete breakdown of a compound to inorganic products such as CO_2 and H_2O, this is referred to as mineralization. As the compound of concern disappears from the environment, the number of organisms that use this compound as a substrate for metabolism increases. This is known as growth-linked biodegradation, and mineralization is a common result.

ENRICHMENT CULTURES AND BIODEGRADATION RESEARCH

When research scientists attempt to identify an organism capable of degrading a particular compound, they collect microorganisms from the environment and inoculate them into a culture flask containing minimal medium, and the compound of interest provides the sole carbon source. The minimal medium provides other nutrients, such as nitrogen and phosphorus, that are necessary for microbial growth. Organisms unable to use this compound will not be able to grow, whereas those that can metabolize the compound will flourish. The organisms for this enrichment culture are often collected from a site where the compound of interest is found naturally or as a contaminant. For instance, if a researcher were attempting to isolate a xylene-degrading microorganism, a sample might be collected from an area of a gasoline spill because xylene is a component of gasoline.

Scientists need to be cautious of optimism because one organism may seem to degrade a chemical efficiently under laboratory conditions, but the actual environment where the degradation is to take place may have many inhibiting factors that cannot be controlled.

It is important for researchers to identify degradation products because some products are more toxic than the parent compound. For example, the hydrolysis of the herbicide 2,4-D is catalyzed by soil microorganisms, which yields a form of the compound toxic to plants (3).

FACTORS LIMITING BIODEGRADATION

The mere presence of a compound and a capable microorganism does not guarantee the compound's degradation. Microbial growth requires water, nitrogen, phosphorus, trace elements, carbon, energy, temperature, and suitable pH conditions. Biodegradation commonly supplies the carbon or energy or both. But growth may be inhibited by other toxins or limited by other essential elements or conditions. Under environmental conditions, predation can also be a factor that limits biodegradation. Protozoa consume bacteria; bacteriophages are viruses that can infect and lyse bacteria.

If no organisms can use the compound as the sole carbon or energy source, a compound may still be cometabolized. Cometabolism involves the degradation of a compound by an organism without providing carbon, energy, or any other factors required for growth. Unlike mineralization, cometabolism is not growth-linked. Even though the compound is not functioning as the carbon or energy source, it is possible for degradation to be quite rapid (4). Often other organic molecules in the environment provide the carbon or energy source, which leads to an increase in the microbial population that is responsible for the increased rate of degradation.

ENZYMATIC MECHANISMS

The proper enzyme must exist for metabolism or transformation of a compound. Over geologic time, microorganisms have adapted to compounds and structures to which they are exposed regularly. It is often believed that microbial enzymes have not evolved rapidly enough to cope with the onslaught of synthetic compounds that have been released into the environment in the last century (2).

Biodegradation may occur inside or outside the cell. Intracellular enzymes require the organism to have some means of transporting the compound inside the organism for degradation. Extracellular enzymes are released by the organism and degrade the compound before the products are transported into the cell, where the final degradation occurs. This is a common microbial tactic when the carbon or energy source is too large to internalize. Extracellular enzymes are manufactured in comparatively high concentrations compared to intracellular enzymes.

Degradative enzymes are often highly specific for a particular reaction. The enzyme must be able to fit the correct molecular structure into its active site. The active site of an enzyme is responsible for the cleavage or molecular rearrangement of a portion of the compound. The three-dimensional structure of the functional group fits neatly into the active site of the enzyme. Functional groups are often the targets of these enzymes (Table 1). Compounds can have a combination of functional groups, and the transformation of these compounds will depend on the spatial arrangement of the functional groups and the enzyme's ability to catalyze the degradation products. If it is known that an organism can degrade a structurally similar compound, then it may be possible to enrich for an organism capable of degrading the compound of concern.

ACCLIMATION PHASE

Initially, the organisms in the environment that can degrade a particular compound will often be insufficient to cause a noticeable change in the quantity of compound present. This initial period, referred to as the acclimation phase, can vary in length among compounds, environments, and organisms. Bioavailability and compound concentration can influence the length of the acclimation phase. If the compound is present in the environment at a concentration that is too high, then it can be toxic to organisms that would break it down (5).

Research has shown that organisms that have been exposed to a particular compound can respond more quickly to a second exposure to the same compound (6). The first exposure selects for a community of organisms capable of degrading the compound. Upon reintroduction of that same compound, the community is already capable of degrading the compound, and the acclimation phase is shortened. The shortened acclimation phase can be attributed either to a more metabolically active community or a more rapid increase in the number of organisms capable of degrading the compound (6). This response has been identified among organisms capable of degrading several different pesticides (6).

RECALCITRANCE

Recalcitrant molecules are either resistant to degradation or are degraded at a very slow rate. There are several reasons that this occurs. As suggested previously, it is possible that the necessary enzyme does not exist. It is also possible that the correct enzyme may exist, but it cannot reach the active site and catalyze the necessary reaction. Finally, the enzyme necessary for the degradation of the compound may never be turned on by the organism. In this case, the enzyme is encoded on the organism's DNA, but no signal induces enzyme production. Some compounds, such as 2,4-D, are readily degraded under aerobic conditions but not under anaerobic conditions (4). Other compounds, such as the organophosphate herbicide, fenitrothion, are preferentially degraded under anaerobic conditions (4). Thus, the environment can play a role in a compound's recalcitrance.

BIOAVAILABILITY

A compound must be bioavailable before degradation will occur. The organism and its enzymes must be capable of coming into contact with the compound. To understand fully the limitation that bioavailability presents, we need to examine the microscopic scale on which microorganisms exist. Most bacteria are no longer than 10 μm, and the distance between one soil particle and another may be a journey that a bacteria never makes. Most bacteria in soil or sediment are attached to particles, but some are motile. These motile organisms often simply pass through pore spaces and do not really contribute to the microenvironment. If examined in great detail, it would be possible to see that the surfaces of soil and sediment particles consist of depressions and valleys, which provide habitats for native microorganisms. Some microorganisms will never leave these tiny spaces. These organisms rely on the movement of compounds and nutrients for survival. Some pores are too small for microorganisms

Table 1. Common Functional Groups in Contaminants

Name	Structure	Example(s)
Halogen	R–X where X=Cl, Br, or F	Metolachlor
Aromatic ring	(benzene ring)	Benzene, toluene, ethylbenzene, xylenes (BTEX), polyaromatic hydrocarbons (PAHs)
Carboxylic acid	$R_1-\overset{\overset{O}{\|\|}}{C}-OH$	Picloram
Alkene	$R_1-C=C-R_2$	Aldrin, dieldrin, and heptachlor
Alcohol (hydroxyl group)	$HO-R_1$	2,4-D
Ester	$R_1-\overset{\overset{O}{\|\|}}{O-C}-R_2$	Malathion
Ether	R_1-O-R_2	2,4-D, MTBE

but are large enough to provide a protective area for the compound.

Over time, the bioavailability of a compound tends to decrease due to sequestration of the compound. When the compound is plentiful, microorganisms will use the molecules that are easiest to obtain. Available compound is used, whereas the excess becomes bound to soil and sediment particles. Compound bioavailability then limits biodegradation. Researchers have attempted to model the diffusion and sorption of a compound to soil or sediment (7). One model is the diffusion–sorption bioavailability model that attempts to describe the rate of biodegradation in the presence of soil or sediment aggregates (7). Many biodegradation studies are done in flasks and do not take into account environmental factors such as diffusion and sorption.

NONAQUEOUS PHASE LIQUIDS AND COMPOUNDS OF LOW WATER SOLUBILITY

Nonaqueous phase liquids (NAPLs) generally have low water solubility. Dense nonaqueous phase liquids can be found below an aquifer, whereas light nonaqueous phase liquids are found on the surface of the water. A very small fraction of a NAPL is soluble in water, and this quantity is available to microorganisms for biodegradation. The vast majority of NAPLs are not in the aqueous phase, so they are often considered unavailable to the organisms capable of degrading or metabolizing them, thus presenting a factor limiting biodegradation. NAPLs that have polluted the environment include hydrocarbons (mainly from oil spills) and industrial solvents.

Microbial growth on compounds of low water solubility is generally slow; however; in some cases, exponential growth has been observed (8). There are several ways by which a compound that has low water solubility can be degraded. The small fraction of the compound that enters the aqueous phase is readily accessible to the organism. The organism also can come in direct contact with the compound at the water–compound interface (9). Finally, some organisms can make the compound more accessible through a process known as pseudosolubilization (9). Pseudosolubilization occurs via compounds released by the organism, which act as carrier molecules (10). The compound becomes enclosed in a micelle, which may diffuse through the aqueous phase (10). Microorganisms in the aqueous phase can then use the carbon source within the micelle. Surface active compounds can also be released by the organism. These compounds are composed of various lipids and interact with low solubility compounds such as hydrocarbons to facilitate their uptake by the organism (11,12).

GROUNDWATER BIOREMEDIATION

Groundwater contamination is of high priority because so many people rely on groundwater as a source of drinking water. Porosity, the empty space between soil and sediment grains, determines how much water an aquifer can hold. Water moves slowly through underground aquifers and that movement is a function of the type of soil or rock that composes the aquifer. Darcy's law allows scientists to approximate the discharge of water from an aquifer and takes the material that makes up the aquifer and the hydraulic gradient into consideration. Once the plume is located and the concentration of compound has been determined, removing it can be difficult and costly.

For *in situ* biodegradation to occur, it is often necessary to supply oxygen and nutrients to the microorganisms. The absence or presence of oxygen can make a substantial difference to the active microbial community and its ability to degrade the contaminant. The contaminated groundwater can also be pumped from the ground into a reactor, where microorganisms degrade the contaminant (for a review, see Reference 13). These reactors function similarly to sewage treatment plants. Contaminated groundwater and nutrients are pumped into a reactor where microorganisms are grown so that they are attached to the reactor. As the contaminated groundwater passes the microorganisms, they degrade the compound. The water is then disposed of or returned to the aquifer. Bioreactors are often limited by their high costs. Pumping of NAPLs is often difficult due to their low solubility in water.

MARINE OIL SPILLS

Marine oil spills can be remediated by bioremediation. Dispersants are used to spread the nonaqueous phase contaminants across the water's surface to create more favorable biodegradation conditions. The dispersants themselves are biodegraded in the process. It is estimated that 50% of the oil from the *Exxon Valdez* spill was biodegraded on the water surface, on the shore, or in the water column (14). Fertilizers are often used to enhance the rate of oil degradation and to supply otherwise limiting nutrients.

ALGAL BIOREMEDIATION

Biodegradation is not limited only to bacteria. Fungi play an important role in soil systems; in many aquatic systems, algae can play an important role. Several algal species are capable of degrading some pesticides in freshwater systems (15). Algae can be used to clean industrial pulp mill effluent before the effluent is discharged. They are effective in removing both the dark color of the effluent and organic halides (16). Some algae can bioaccumulate some polycyclic aromatic hydrocarbons in marine systems (17). Bioaccumulation is the sequestration of a compound by an organism, where the compound remains unchanged and the organism is unharmed by its presence. Over time, significant quantities of compound can be accumulated by the organism.

The metabolic diversity of microorganisms is a powerful tool for degrading unwanted compounds. Numerous sources are available to provide additional information regarding biodegradation processes. The journal *Biodegradation*, published quarterly, contains scientific advancements in biodegradation. Biodegradation databases are available that provide information on biodegradable compounds and the enzymes and microbial species that degrade them. A useful database is the University of Minnesota Biocatalysis/Biodegradation Database, found on the web at: www.umbbd.ahc.umn.edu/index.html.

BIBLIOGRAPHY

1. Madigan, M., Martinko, J., and Parker, J. (2000). *Brock: Biology of Microorganisms*, 9th Edn. Prentice-Hall, Upper Saddle River, NJ.
2. Wackett, L. and Hershberger, C. (2001). *Biocatalysis and Biodegradation: Microbial Transformation of Organic Compounds*. ASM Press, Washington, DC.
3. Vlitos, A. and King, L. (1953). Fate of sodium 2,4-dichlorophenoxy-ethyl-sulphate in the soil. *Nature* **171**: 523.
4. Liu, D. et al. (2000). Factors affecting chemical biodegradation. *Environ. Toxicol.* **15**: 476–483.
5. Saxena, A., Zhang, R., and Bollag, J. (1987). Microorganisms capable of metabolizing the herbicide metolachlor. *Appl. Environ. Microbiol.* **53**: 390–396.
6. Robertson, B.K. and Alexander, M. (1994). Growth-linked and cometabolic biodegradation: possible reason for occurrence of absence of accelerated pesticide biodegradation. *Pest. Sci.* **41**: 311–318.
7. Scow, K. and Hutson, J. (1992). Effect of diffusion and sorption kinetics of biodegradation: theoretical considerations. *Soil Sci. Soc. Am. J.* **56**: 119–127.
8. Stucki, G. and Alexander, M. (1987). Role of dissolution rate and solubility in biodegradation of aromatic compounds. *Appl. Environ. Microbiol.* **53**: 292–297.
9. Goswami, P.C., Singh, H.D., Bhagat, S.D., and Baruah, J.N. (1983). Mode of uptake of insoluble solid substrates by microorganisms. I: sterol uptake by arthrobacter species. *Biotechnol. Bioeng.* **25**: 2929–2943.
10. Velanker, S.K., Barnett, S.M., Houston, C.W., and Thompson, A.R. (1975). Microbial growth on hydrocarbons—some experimental results. *Biotechnol. Bioeng.* **17**: 241–251.
11. Geurra-Santos, L., Käppeli, O., and Fiechter, A. (1984). *Pseudomonas aeruginosa* biosurfactant production in continuous culture with glucose as a carbon source. *Appl. Environ. Microbiol.* **48**: 301–305.
12. MacDonald, C.R., Cooper, D.G., and Zajic, J.E. (1981). Surface-active lipids from *Nocardia erthropolis* grown on hydrocarbons. *Appl. Environ. Microbiol.* **41**: 117–123.
13. Langwaldt, J.H. and Puhakka, J.A. (2000). On-site biological remediation of contaminated groundwater: a review. *Environ. Pollut.* **107**: 187–197.
14. Sugai, S., Lindstrom, J., and Braddock, J. (1997). Environmental influences on the microbial degradation of *Exxon Valdez* oil on the shorelines of Prince William Sound, Alaska. *Environ. Sci. Technol.* **31**: 1564–1572.
15. U.S. EPA. (1976). *Degradation of Pesticides by Algae*. Office of Research and Development, Environmental Research Laboratory, Athens, GA.
16. Dilek, F.B., Taplamacioglu, H.M., and Tarlan, E. (1999). Colour and AOX removal from pulping effluents by algae. *Appl. Microbiol. Biotechnol.* **52**: 585–591.
17. Kirso, U. and Irha, N. (1998). Role of algae in fate of carcinogenic polycyclic aromatic hydrocarbons in the aquatic environment. *Ecotoxicol. Environ. Safety* **41**: 83–89.

BIOLUMINESCENT BIOSENSORS FOR TOXICITY TESTING

JIM PHILIP
JENNIFER BELL
FIONA STAINSBY
SANDRA DUNBAR
Napier University
Edinburgh, United Kingdom

SIOUXSIE WILES
Imperial College London
London, United Kingdom

ANDREW WHITELEY
MARK BAILEY
Centre for Ecology and
Hydrology–Oxford
Oxford, United Kingdom

BACKGROUND

This article summarizes recent developments in whole cell bioluminescent biosensors for estimating toxicity at wastewater treatment plants (WWTPs). This started with the use of natural marine bioluminescent bacteria, and there is now at least a 20-year history of toxicity testing with these organisms. Molecular biology has given rise to many new bioluminescent strains for toxicity testing to overcome some specific limitations of *Photobacterium (Vibrio) fischeri*. Luminescence decay toxicity tests have also been developed in higher organisms, up to *Caenorhabditis elegans* (1), to try to improve the relevance of these tests. Custom-made biosensors for protecting wastewater treatment plants are already in development. These developments coincide with the pattern of environmental legislation in Europe.

LEGISLATIVE DRIVERS

The Urban Wastewater Treatment Directive and the Water Framework Directive

In the European Union, under the Urban Wastewater Treatment Directive (2), the functioning of wastewater treatment works is considered important, and reaching nontoxic levels is of primary concern (3). In common with most legislation in the area, the Directive is directed toward regulating discharges to the environment, rather than to the WWTP itself. The more recent EU Water Framework Directive (4) is concerned mostly with protecting receiving waters from pollution and toxic discharges.

IPPC

The Integrated Pollution Prevention and Control Directive, adopted in 1996 (5), is one of the cornerstones of the EU's environmental legislation. It is likely that fines based on toxicity levels, in addition to the current physicochemical parameters, will be assessed on companies that discharge industrial effluents at concentrations above guidelines (6).

ENVIRONMENTAL QUALITY STANDARDS

Industrial effluents tend to contain more toxic substances than domestic wastewater, and numerical limits are set for discharges to ensure compliance with Environmental Quality Standards (EQSs). Compliance is then monitored by chemical analysis. There are several problems in this approach:

- There are many substances for which there is no EQS (over 99%).
- There are no ecotoxicological data for thousands of chemicals, and most EQSs are based on limited data.
- Difficulties are experienced in predicting the interaction of chemicals with each other and the subsequent effect on the environment.
- There are analytical difficulties for many chemicals, and great cost in separating and identifying all constituents.

TOXICITY CONSENTS AND DIRECT TOXICITY ASSESSMENT

In 1993, the predecessor to the U.K. Environment Agency, the National Rivers Authority (NRA), began research into setting discharge consents based on toxicity. This became known as Direct Toxicity Assessment (DTA). DTA provides a direct measure of acute toxicity, and it is not necessary to identify the substances causing effluent toxicity to treat or reduce it (7); a property can be measured and treated directly. If toxicity is detected but considered acceptable, a toxicity consent can be derived and applied. If the toxicity of the effluent is unacceptably high, however, remedial action may be needed, and then a toxicity consent can be derived (Fig. 1). In the United States, a similar process is applied to discharges to receiving waters; the process is called Whole Effluent Toxicity (WET).

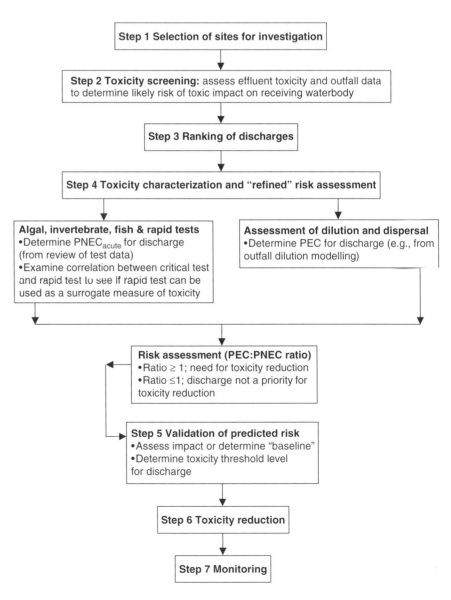

Figure 1. The Direct Toxicity Assessment process.

Two Types of "Legislative" Toxicity

From the above, it can be recognized that there is a need to measure toxicity in two different senses. On the one hand, there is the need to protect the biological WWTP, especially the activated sludge plant, from toxic influents that can kill the treating biomass, leading to loss of treatment and concomitant financial and regulatory implications. On the other hand, there is the need to measure the toxicity of the treated effluent from the WWTP for environmental protection.

Toxicity, An Imperfect Analyte

Toxicity is not a well-defined analyte for a biosensor. Acute toxicity refers normally to lethality as a result of exposure to a high concentration of a toxicant for a short period. Chronic toxicity may be lethal or sublethal but refers to exposure to a low dose for a long period. Although chronic toxicity does not necessarily cause death, it can result in long-term debilitation of some form.

Nevertheless, to make meaningful comparisons, it is necessary to put numbers to toxicity. These are some of the most commonly encountered numbers:

- LC_{50}: the concentration that produces 50% mortality in test organisms;
- LD_{50}: the dose that produces 50% mortality in test organisms;
- IC_{50}: the concentration at which 50% of growth or activity is inhibited;
- EC_{50}: concentration at which 50% of the predicted effect is observed;
- LOEC (the lowest observable effect concentration): the concentration at which the lowest effect is seen;
- NOEC (the no observable effect concentration): the maximum concentration at which no effect is observed.

In bioluminescence tests, the figures quoted are usually EC_{50} or IC_{50}. Their direct correlation with lethality (e.g., LC_{50}) in higher organisms is fraught with difficulty. It should be expected that the EC_{50} will be lower than the LC_{50} because impairment of function should precede total inhibition (8).

THE RELATIONSHIP BETWEEN BACTERIAL BIOLUMINESCENCE AND TOXICITY

The bacterial bioluminescence reaction involves oxidation of a long-chain aldehyde (RCHO) and reduced flavin mononucleotide ($FMNH_2$), resulting in the production of oxidized flavin (FMN) and a long-chain fatty acid (RCOOH), along with the emission of light (Fig. 2).

$FMNH_2$ production depends on functional electron transport, so only viable cells produce light. This relationship between cellular viability and light endows bioluminescence with the ability to report bacterial injury and recovery. It can be established readily that any substance that impairs the biochemistry will lead to a reduction in light output. Of crucial importance to the

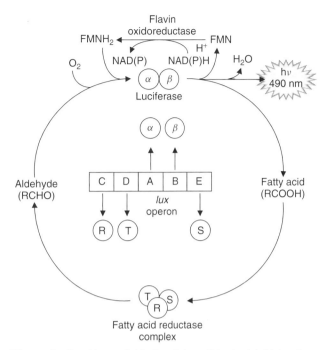

Figure 2. Genetics and biochemistry of bacterial bioluminescence. The rectangles represent genes and the circles represent enzymes or enzyme subunits.

commercial development of the bacterial bioluminescence toxicity tests was the fact that the change in light output is proportional to the concentration of the toxicant. Thus, it was possible to develop quantitative tests and compare the results to toxicity tests with higher organisms (Fig. 3).

BIOLUMINESCENT TOXICITY BIOSENSORS: POTENTIAL FOR USE IN DTA

What is required of these sensors is not necessarily greater sensitivity, resulting in lower EC_{50} values. The real value to DTA of treated effluents discharged to receiving water bodies would be relevance (good correlation) to the results of toxicity tests with higher organisms. The tests based on naturally occurring marine bioluminescent bacteria as used in the Microtox® test or the ToxAlert® test are in widespread use, and they correlate inhibition of luminescence with toxicity (9–12).

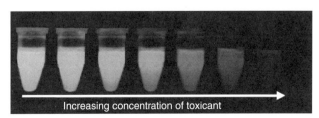

Figure 3. A series of tubes containing bioluminescent bacteria responding to increasing concentrations of toxicant by producing less light. (Courtesy of Remedios Ltd, http://www.remedios.uk.com/.)

The tests are rapid, and reliable toxicity testing has been demonstrated (13–19). Their use has resulted in extensive databases on the toxicity of pure chemicals (18). The potential for using the Microtox test as an alternative to animal testing has been demonstrated (19). It was found more sensitive and less prone to false positives than two common toxicity tests, the L-929 Minimal Eagle's Medium (MEM) elution cytotoxicity test and the Draize rabbit eye irritancy test. It offers the often quoted advantages of low cost and convenience, and it is far less controversial than these latter tests.

Despite these advantages, application to environmental samples has not always been successful (20,21), and correlation with other toxicity assays may require much more intrusive modification. In particular, the marine origin of these natural bacteria may act as a barrier to their use in DTA. For example, they are sensitive to pH and osmotic conditions. The revolution in genetic engineering, however, has made it routine to transfer genes for bioluminescence to other bacteria. This has opened up new possibilities for biosensor design.

BIOLUMINESCENT TOXICITY BIOSENSORS: MONITORING THE PERFORMANCE OF WWTPs

The limitations of marine bioluminescent bacteria become much more apparent when the biosensor is to be used for predicting toxicity in WWTPs. Then, it is not the treated, biodegraded effluent that is to be monitored, but the toxic influent before treatment; so the concentrations of pollutants are much higher, depending on the particular industrial wastewater. The issues for the plant operator are somewhat similar to those of DTA, regulatory and financial if a fine is levied for noncompliance. However, the situation for the operator is more serious if the influent to a biological wastewater treatment plant is toxic to the microbial community that performs the purification.

LOSS OF TREATMENT SPELLS REGULATORY AND FINANCIAL TROUBLE

Partial or complete loss of treatment through toxicity to mean liquor suspended solids (MLSS) will cause out-of-compliance and can also result in serious operational problems (such as sludge bulking and foaming) and the attendant need for untreated disposal or storage. In the event of total loss of viability, restoring the biological plant can take weeks to months, with very serious financial and regulatory consequences. Until recently, total loss of viability has been considered anecdotal, but real instances have been recorded. Much more frequent occurrences are instances of partial or chronic toxic events. For example, up to 50% inhibition of nitrification has been attributed to toxicity from industrial wastewater (22).

MARINE BIOLUMINESCENCE TOXICITY TESTS ARE NOT FIT FOR THE CHALLENGE

The irrelevance of marine bioluminescent biosensors can be demonstrated with phenol as an example. During coal coking, an aqueous effluent is produced containing several hundred mg/L of phenols. Biological treatment of these wastewaters is an accepted full-scale practice; high efficiencies of removal are achievable, but occasionally treatment fails through toxicity to the treating biomass. However, the 5-minute EC_{50} for phenol to the Microtox strain is 18 mg/L (23). Used as a toxicity biosensor upstream of such an activated sludge plant, this strain would cause continuous false alarms while the biodegrading community functioned normally.

ENTER GENETIC ENGINEERING

Thus, there is a need for biosensor strains that are more relevant to the wastewater treatment system under consideration. The best correlations between toxicity tests and the actual behavior of an activated sludge plant are obtained when microorganisms from the activated sludge plant are used in toxicity tests (24). In this manner, Kelly et al. (20) developed a bioluminescent toxicity biosensor by inserting the *luxCDABE* genes from *V. fischeri* into a *Pseudomonas* strain from an activated sludge plant. By similar means, a wide range of alternative luminous bacteria has been developed for use in toxicity testing of contaminated waters (25–30).

MEASURING LIGHT PRODUCTION AND DECAY

One reason for the popularity of bioluminescence biosensors is the ease of detecting and quantifying light. Light does not accumulate or diffuse and facilitates real-time and *in situ* measurements. It can be measured linearly over several magnitude orders, an extremely important consideration. As a result, different techniques are available for detection. The most widely used is luminometry.

Luminometry is rapid and sensitive, and its sensitivity is increasing with time due to technical innovations. Laboratory-based luminometers generally rely on high sensitivity photomultiplier tubes and have a range of advanced features. For field deployment of toxicity biosensors, a more portable instrument is required. The choice is normally silicon photodiode technology, which has advanced rapidly in the last 10 years, so that very inexpensive, compact, and sensitive photodiodes are now available, making them suited to field use. The technology has advanced to the stage where an instrument is available with 24 individual photodiodes so that a 24-well microtiter plate can be read in an instrument that is small enough to be handheld (Fig. 4).

Example Data

A generic example is presented here, where a handheld instrument was used to measure bioluminescent decay from a genetically modified biosensor designed to respond to phenol (30). The data in Fig. 5a show the response of the sensor to different concentrations of a toxicant (in this case trichlorophenol) with time. The instrument measured light every second, and each decay curve represents a concentration of toxicant. Note that an increasing concentration of toxicant depresses the end point of the

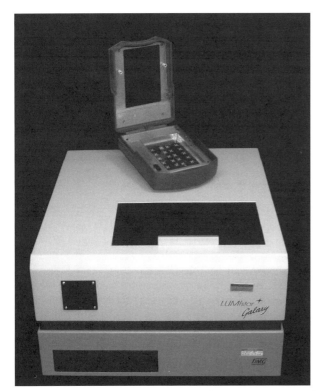

Figure 4. Two types of luminometer. On the bottom is a research laboratory instrument based on photomultiplier technology. On the top is a silicon photodiode instrument containing multiple photodiodes for microtitre plate reading (courtesy of Cybersense Biosystems, http://www.cybersensebiosystems.com/).

reaction with reference to a nontoxic control. These data were then manipulated to give a mean endpoint figure, which was then plotted against concentration (Fig. 5b).

BIBLIOGRAPHY

1. Lagido, C., Pettitt, J., Porter, A.J.R., Paton, G.I., and Glover, L.A. (2001). *FEBS Lett.* **493**: 36.
2. European Commission Council Directive. (1991). Urban Waste Water Treatment Directive 91/271/EEC. EEC, Brussels.
3. Farré, M. and Barceló, D. (2003). *Trends Anal. Chem.* **22**: 299.
4. European Commission Parliament and Council Directive. (2000). *Water Framework Directive* 2000/60/EC. EEC, Brussels.
5. European Commission Council Directive. (1996). *Integrated Pollution Prevention and Control Directive* 96/61/EC. EC, Brussels.
6. dos Santos, L.F., Defrenne, L., and Krebs-Brown, A. (2002). *Anal. Chim. Acta* **456**: 41.
7. Scottish Environmental Protection Agency. (2003). *Direct Toxicity Assessment*. Guidance no. 03-DLM-COPA-EA3.
8. Connell, D.W., Lam, P., Richardson, B., and Wu, R. (1999). *Introduction to Ecotoxicology*. Blackwell Science, Oxford.
9. Bulich, A.A. and Isenberg, D.L. (1981). *ISA Trans.* **20**: 29.
10. Isenberg, D.L. (1993). The Microtox™ toxicity test a developer's commentary (I). In: *Ecotoxicology Monitoring*. M. Richardson (Ed.). VCH Publishing, New York, p. 3.
11. Merck. (1997). *ToxAlert™ 10 Operating Manual*.
12. Microbics Corp. (1992). *Microtox System Operating Manual*.
13. Chang, J.C., Taylor, P.B., and Leach, F.R. (1981). *Bull. Environ. Contam. Toxicol.* **26**: 150.
14. Dutka, B.J., Kwan, K.K., Rao, S.S., Jurkovic, A., and Lui, D. (1991). *Environ. Monitor. Assess.* **16**: 287.
15. Kaiser, K.L.E., Ribo, J.M., and Kwasniewska, K. (1988). *Water Pollut. Res. J. Can.* **23**: 356.
16. Steinberg, S.M., Poziomek, E.J., Engelmann, W.H., and Rogers, K.R. (1995). *Chemosphere* **30**: 2155.
17. Yates, I. and Porter, J. (1982). *Appl. Environ. Microbiol.* **44**: 107.

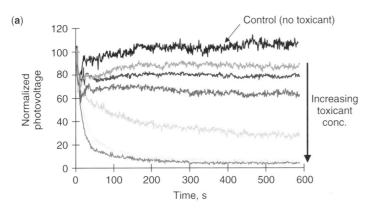

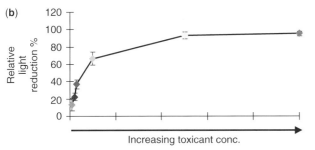

Figure 5. Sample data measuring the response of a genetically modified bioluminescent bacterium to the presence of trichlorophenol. (**a**) The output from an instrument recording light with time. (**b**) the percentage light reduction at a suitable endpoint has been measured against the nontoxic control [the black trace in (a)], and the relationship to concentration has been plotted. The shaded data points on (**b**) correspond to the shades of the decay curves on (**a**).

18. Kaiser, K.L.E. and Palabrica, V.S. (1991). *Water Pollut. Res. J. Can.* **26**: 361.
19. Bulich, A.A., Tung, K.K., and Scheibner, G. (1990). *J. Biolumin. Chemilumin.* **5**: 71.
20. Kelly, C.J., Lajoie, C.A., Layton, A.C., and Sayler, G.S. (1999). *Water Environ. Res.* **71**: 31.
21. Novaha, M., Vogel, W.R., and Gaugitsch, H. (1995). *Environ. Int.* **21**: 33.
22. Jönsson, K., Grunditz, C., Dalhammar, G., and Jansen, J.L.C. (2000). *Water Res.* **34**: 2455.
23. Blum, D.J.W. and Speece, R.E. (1991). *Res. J. WPCF* **63**: 198.
24. Gernaey, K., Petersen, B., Ottoy, J.-P., and Vanrolleghem, P. (1999). *Water Qual. Int.* June: 16.
25. Boyd, E.M., Killham, K., Wright, J., Rumford, S., Hetheridge, M., Cumming, R., and Meharg, A.A. (1997). *Chemosphere* **35**: 1967.
26. Brown, J.S., Rattray, E.A.S., Paton, G.I., Reid, G., Caffoor, I., and Killham, K. (1996). *Chemosphere* **32**: 1553.
27. Layton, A.C., Gregory, B., Schultz, T.W., and Sayler, G.S. (1999). *Ecotoxicol. Environ. Safety* **43**: 222.
28. Paton, G.I., Palmer, G., Burton, M., Rattray, E.A.S., McGrath, S.P., Glover, L.A., and Killham, K. (1997). *Lett. Appl. Microbiol.* **24**: 296.
29. Sinclair, G.M., Paton, G.I., Meharg, A.A., and Killham, K. (1999). *FEMS Microbiol. Lett.* **174**: 273.
30. Wiles, S., Whiteley, A.S., Philp, J.C., and Bailey, M.J. (2003). *J. Microbiol. Methods* **55**: 667.

BIOMANIPULATION

MASAKI SAGEHASHI
University of Tokyo
Tokyo, Japan

Biomanipulation is a currently used lake purification method, which modifies a lake ecosystem to an ideal state. Prior to discussing biomanipulation, lake purification methods are reviewed, and the principles and current status of biomanipulation are described afterward.

CLASSIFICATION OF LAKE RESTORATION METHODS

Some measures have been designed and implemented to prevent the eutrophication of lakes. These methods can be divided into two types, the "temporary type" and the "sustainable type." The separation of artificially aggregated algae (1–4) and aeration (5–7) are temporary types of purification methods. These methods are always immediately and strongly effective, and special knowledge of the methods is not required because the mechanisms are relatively simple. These methods, however, do not sustain their effects, and whenever algal blooms appear, we should use these methods. Therefore, this type of method is not essentially a lake purification method. On the other hand, external nutrient loading reductions (8), flow control (9), dredging of sediments (10), and "biomanipulation" are the sustainable type. There are many unclear points as to the effects of these types of methods; however, they restore a lake ecosystem, and long-term stable effects can be expected in contrast to the temporary types of methods. External nutrient loading reductions are the most effective methods; however, it is difficult to prevent completely external nutrient loading due to the considerable amount of so-called nonpoint nutrient loading sources. Furthermore, internal nutrient elution from sediments is always abundant. In other words, it is difficult to prevent on algal bloom with only external nutrient loading or flow control. On the other hand, dredging of sediments is costly. That is why interest in "biomanipulation" has been growing.

PRINCIPLES OF BIOMANIPULATION

The theoretical possibility that ecosystems have more than one equilibrium has been discussed (11). Lakes make complicated ecosystems (Fig. 1) (12) and also have alternative equilibria, a clear one and a turbid one (13, 14). This idea is based on the observation that the restoration of turbid eutrophic lakes by nutrient reduction seems to be often prevented by ecological feedback mechanisms; moreover, the clear state also possesses a number of stabilizing feedback mechanisms (15). The negative effect of the vegetation of aquatic macrophytes on turbidity has been one of the reasons to expect alternative stable states in freshwater ecosystems. Scheffer (15) showed a schematic representation of the effect of vegetation on the nutrient–turbidity relationship and consequences for system equilibria if the vegetation completely disappears from a lake at a critical turbidity (Fig. 2). The equilibrium curve is also affected by the depth of the lake (Fig. 3). Scheffer summarized the relationship using his model and proposed the "marble-in-a-cup" theory (Fig. 4). The minima of the resulting curves represent stable equilibria, whereas the hilltops are unstable breakpoints. The hysteresis behavior of lake ecosystems (13) is thus derived from this relationship. At a low nutrient level, the system has one globally stable equilibrium, a clear water state. Any increase in the nutrient level gradually changes the shape of the stability landscape and gives rise to an alternative turbid equilibrium. However, no major disturbances occur, and

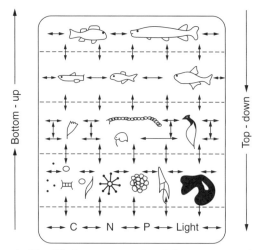

Figure 1. Simplified structure of a lake ecosystem (arrows: downward = predation, grazing, impact on the resource basis; upward = use of resources; horizontal = competition) (12).

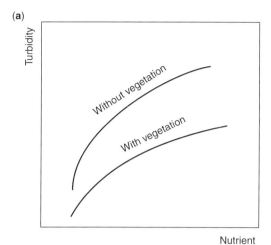

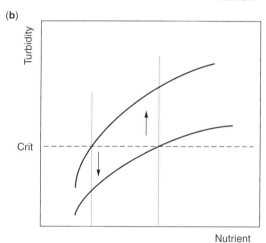

Figure 2. Schematic representation of (**a**) the effect of vegetation on the nutrient–turbidity relationship and (**b**) consequences for systems equilibria if vegetation completely disappears from a lake at a critical turbidity (crit) (15).

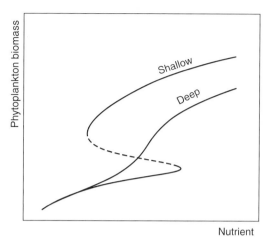

Figure 3. Equilibrium density of phytoplankton as a function of nutrient level computed from Scheffer's model for a shallow lake of homogeneous depth ("shallow") and for a deep lake with gradually declining slopes ("deep"). The dashed part represents unstable "breakpoint" equilibria (15).

the system will stay in its current state, responding only slightly to enrichment. If the nutrient level is further raised, the stability of the clear state decreases, and slight perturbations are enough to cause a switch in the turbid equilibrium. At still higher nutrient levels, the clear water equilibrium disappears, inevitably causing an irreversible jump to a turbid state. Efforts to restore the system by decreasing the nutrient level change the stability landscape again, but even if the nutrient level at which the system was formerly clear is realized, there will hardly be a response to the measures. An alternative clear equilibrium may be present, but the locally stable turbid state is sustained. Only a severe reduction in nutrient level results in a switch to the clear state.

The reduction of the nutrient level as a measure to restore a lake ecosystem, which has the alternative equilibria mentioned before, gives poor results, but "biomanipulation" (16) as an additional measure can have significant effects, provided that the nutrient level has been sufficiently reduced to allow the existence of a stable alternative clear water equilibrium. A schematic diagram of the restoration of water quality by biomanipulation is shown in Fig. 5. Biomanipulation may be defined as the restructuring of a biological community to achieve a favorable response, usually a reduction in the algal biomass, attainment of clear water, and promotion of a diverse biological community (17). The theory of biomanipulation is based on the prediction that increased piscivore abundance results in decreased planktivore abundance and that increased zooplankton grazing pressure leads to reductions in phytoplankton abundance and improved water clarity (18).

REVIEW OF TRIALS OF BIOMANIPULATION AND ITS PERSPECTIVE

A number of fish manipulations in lakes and ponds have been done in various water ecosystems around the world. Some researchers have reviewed them (17,18,20). Based on these reviews, the biomanipulation trials so far attempted are summarized in Table 1. The manipulations, lake size, and outcomes are mentioned in this table. Table 1 shows that there has been a considerable spectrum of fish manipulation occurring throughout the world. Note that not every manipulation was aimed at algal growth suppression. Some of them were an unexpected change in fish biomass. In early stages, such as Hrbácek et al. described (21), little attention was given to reducing the algal biomass by manipulating the fish biomass.

These trials can be divided into several groups based on their results. The following categories are used for classifying the trials:

A: biomass of phytoplankton decreased ("successful")
B: dominated by larger zooplankton species or an increase in the biomass of zooplankton/obvious change in phytoplankton community was not observed ("insufficient")
C: no obvious changes in both the phytoplankton and zooplankton communities ("failure")
D: harmful algal dominance was observed ("risky")
E: others ("others")

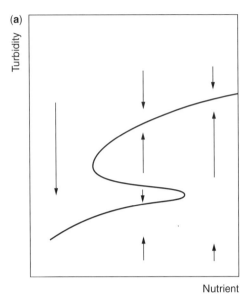

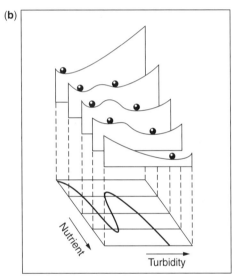

Figure 4. (a) Equilibrium turbidity as a function of nutrient level; direction and speed of change are indicated by arrows (example computed from vegetation–phytoplankton model). (b) "Marble-in-a-cup" representation of stability at five different nutrient levels. The minima correspond to stable equilibria; the tops to unstable breakpoints (15).

There are 35 trials listed in Table 1. Twenty cases of "success" (57.1%), five cases of "insufficient" (14.3%), three cases of "failure" (8.6%), four cases of "risky" (11.4%), and three cases of "others" (8.6%) are categorized. Categories A and B can be considered "positive cases"; C and D are the "negative cases." Based on this consideration, there are 71.4% positive and 20.0% negative results.

There are various manipulations in the "positive" cases. A-1, A-2, and B-3 are the cases that are naturally caused. Case A-2 was also affected by summer stratification. A-6–8, and B-4 are enclosure experiments, which examined the effect of planktivorous fish (A-6–8: bluegill, bream, roach, and crucian carp) or piscivores (B-4: walleye). Fish were killed by fish toxin (mainly rotenone) in A-4, 9, 10, 13, and 14, and B-1 and B-2. Selective catchment, which is more theoretical than poisoning, was found in cases A-12 and A-16–20. The stocking of piscivores was tried in cases A-3, 5, 11, 13, 15, 16, and 19 and B-1, 2, and 5. The piscivores stocked were pike, pikeperch, rudd, salmonine, trout, bass, and walleye. Case A-15 is one of the most perfect biomanipulation trials because the lake was emptied by pumping, and daphnids and macrophytes were stocked as well as piscivores.

Some reasons could be postulated for the "no effect" cases. In general, the removal of fish was not enough (53). This means that the calculation of sufficient biomass removal of fish to improve water quality is essential for successful biomanipulation.

The transition to an ecosystem dominated by harmful algae, which is the most undesirable result, was observed in cases D-1–4. In these cases, large size blue-green algae, which could not be predated by zooplankton, dominated the phytoplankton community. This is one of the contradictions in the effect of biomanipulation. Increased zooplankton predate small, edible phytoplankton. Large, inedible algae such as filamentous blue-green algae then occupy the phytoplankton's niche. Hosper and Meijer (48) pointed out that filamentous blue-green algae and the possible development of invertebrate predators (*Neomysis, Leptodora*) on *Daphnia* are uncertain factors for successful biomanipulation. Therefore, it should be determined whether or not inedible phytoplankton will increase in a waterbody prior to the biomanipulation.

The long-term stability of biomanipulation still needs to be investigated because these trials were performed over relatively short periods. Hosper and Meijer (48)

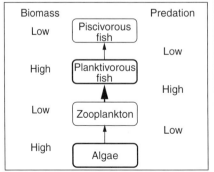

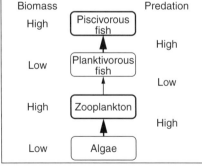

Figure 5. The paradigm of biomanipulation (19).

Table 1. Biomanipulation Trials Reported in the Literature

Type #	Manipulation	Waterbody	Appearances Premanipulation[a]	Appearances Postmanipulation[a]
A-1	**Natural case: 1960s** decrease in *Neomysis mercedis* (planktivores) perhaps due to increase in long-fin smelt (22)	**Lake Washington**, Washington, USA	<1950s> **F:** large amount of *Neomysis* **P:** dominance of *Oscillatoria*	<1970–1980> **F:** abruptly scarce of *Neomysis* **Z:** daphnid dominance **P:** decrease in *Oscillatoria*
A-2	**Die-off (winterkill): in winter, 1964–65** DO depletion due to ice cover (23)	**Lake Severson**, Minnesota, USA area = 10.4 ha z_{max} = 5.3 m; z_{ave} = 2.2 m	<1955–1965> **Z:** small species **P:** 300–600 mg chl.a·m^{-3} water bloom of blue-greens **O:** summer stratification	<1965> **Z:** large species **P:** <30 mg chl.a·m^{-3}; small greens **F:** disappearance of planktivores **M:** increased **O:** disappearance of stratification
A-3	**Stocking (piscivores): 1972** ca. 200 ind.·ha^{-1} of trout (24)	**Hubenov Reservoir**, the Czech Republic area = 92.5 ha z_{max} = 31.5 m z_{ave} = 8.93 m control; Klicana; Vrchlice res.	<1976> (control reservoirs) **F:** dense population: roach, perch, other small cyprinids **Z:** small cladocerans **P:** 10.5(annual);11.4 (summer) mg chl.a·m^{-3} **S:** 2.1 (annual); 2.5 (summer) m **fit well to the Dillon–Rigler regression**	<1976> (Hubenov Reservoir) **Z:** large daphnids **P:** 2.4(annual);2.3 (summer) mg chl.a·m^{-3} **S:** 4.9 (annual); 5.0 (summer) m **half or even less than half the value expected from the Dillon–Rigler regression**
A-4	**Poisoning (rotenone): Nov. 1973** (25,26)	**Lake Lilla Stocke-lidsvatten** Sweden area = 1 ha z_{max} = 8 m;	<1973> **Z:** small cladocerans (*Bosmina*) **P:** 400 mm^3·m^{-2} (in August) **S:** 1.2 m (in August) **PP:** 18 mg C·m^{-3}·h^{-1} (average, surface)	<1975,76> **Z:** large copepods (*Eudiaptomus*) **P:** 25 mm^3·m^{-2} (in August) **O:** Secci: 5.8 m (in August) **PP:** 2 mg C·m^{-3}·h^{-1} (average, surface)
A-5	**Piscivore stocking: 1975–** An aggressive salmonine stocking program (27)	**Lake Michigan**, Wisconsin, USA area = 58,188 km^2 z_{max} = 285 m z_{ave} = 84 m	<1975> **F:** Alewife = 110,000 t total **Z:** calanoid copepods dominant **P:** 0.7–3.5 mg chl.a·m^{-3} **S:** 3–8 m	<1983,84> **F:** alewife = 18,000–20,000 t total **Z:** large daphnids dominant **P:** 0.2–2.5 mg chl.a·m^{-3} **S:** 5–15 m
A-6	**Enclosure exp. (fish stocking): June 15–25, 1975** bluegill (0–4 ind.·encl.$^{-1}$) (28,29)	**Enclosures in Pleasant Pond**, Minnesota, USA area = 0.25 ha z_{max} = 2.5 m 1 m in diamter; 1.8 m in depth; closed bottom	[with fish] **Z:** small cladocerans **P:** increased with the biomass of fish (20—80 mg chl.a·m^{-3})	[without fish] **Z:** large daphnids **P:** low (<5 mg chl.a·m^{-3})
A-7	**Enclosure exp. (fish stocking): 1976** without fish and with fish (90 g m^{-2} of bream and roach) from June to October (30,31)	**Enclosures in Lake Trummen**, Sweden; enclosure: 3 m in diameter 2.2 m in depth open for sediments	[with fish] **Z:** rotifers, cyclopoid copepodites; **P:** 440 mg chl.a·m^{-3} (in August) *Microcystis* dominant (in summer; 83 g fresh·m^{-3}) [lake] soft-water, dense fish population (bream, roach, and pike)	[without fish] **Z:** large daphnids **P:** 20.4 mg chl.a m^{-3} (in August) a mixed population of cryptomonads, small blue-greens, and diatoms

(*continued overleaf*)

Table 1. (Continued)

Type #	Manipulation	Waterbody	Appearances Premanipulation[a]	Postmanipulation[a]
A-8	**Enclosure exp. (fish stocking): 1976** without fish and with fish (220 g m^{-2} of crucian carp) June to October (30,31)	**Enclosures in Lake Bysjön,** Sweden Enclosure: 3 m in diameter 2.2 m in depth open for sediments	[with fish] **Z:** rotifers (*Keratella*) **P:** 286 mg chl.*a* m^{-3} *Microcystis, Aphanizomenon, Pediastrum, Scenedesmus, Synedra, Stephanodiscus, Melosira, Cryptomonas* [lake] hard water, low fish population (crucian carp and rainbow trout)	[without fish] **Z:** large daphnids **P:** <5 mg chl.*a* m^{-3} cryptomonads, *Aphanizomenon*
A-9	**Poisoning (rotenone): Sept. 22, 1980** 0.5 ppm; 165 kg·ha^{-1} of whitefish died in 2 days (32)	**Lake Haugatjern,** Norway area = 9.1 ha z_{ave} = 7.6 m; z_{max} = 15.5 m	<1980> **F:** whitefish (700 kg·ha^{-1}) **P:** diatoms (in June); *Anabaena* (in summer); *Staurastrum* (in autumn) summer biomass = 16 g ww/m^3	<1981,1982> **P:** reduced in biomass (1.5 g ww/m^3) growth rate: increase (five times) *Rhodomonas* (1981, 1982); *Cryptomonas* (1982)
A-10	**Poisoning (rotenone): late 1980** (33)	**Lake Haugatjern,** Norway area = 9.1 ha z_{max} = 15.5 m; z_{ave} = 7.6 m	**Z:** rotifers **P:** large-size *Anabaena flos-aquae* and *Staurastrum luetkemuelleri*	**Z:** daphnids; increased in size (1.3–1.8 mm) **P:** fast growing species, gelatinous greens, 78% decrease in biomass
A-11	**Stocking: May 1981** pikeperch of 20 specimens (2–4 kg ind.) (34,35)	**Lake Gjersjøen,** Norway area = 250 ha z_{max} = 63.5 m	<1980> **F:** juvenile roach = 12,000–15,000 fish·ha^{-1} (pelagic areas) roach dominance (>95%) (littoral areas) **P:** blue-greens (*Oscillatoria*) = 60%	<–1991> **F:** juvenile roach = 250 fish·ha^{-1} (pelagic areas) (1988–1991) perch dominance (>50%) (littoral areas) first spawning of pikeperch (1982) **P:** the biomass was reduced by half blue-greens (*Oscillatoria*) = 20%
A-12	**Removal: 1986–87** roach and bream: 160 kg·ha^{-1} (34,36–38)	**Lake Væng,** Denmark area = 15 ha z_{max} = 2 m; z_{ave} = 1.2 m	<1986> **Z:** rotifer dominance; 0.4 g DW·m^{-3} **P:** *Anabaena* and *Stephanodiscus* dominance; 5.9 g DW·m^{-3}	<1987,1988> **Z:** large cladoceran dominance; 1.3–2.7 g DW·m^{-3} **P:** cryptophytes and periodically larger species; 0.8–3.0 g DW·m^{-3}
A-13	**Poisoning (rotenone): Oct. 1987** 3 ppm rotenone **Stocking: 1988–1990** largemouth bass, walleye (39)	**Lake Christina,** Minnesota, USA area = 1619 ha z_{max} = 4.0 m; z_{ave} = 1.5 m	<1985–1987> **F:** planktivores and bentivores **Z:** small cladoceran (<100 ind L^{-1}) **P:** 40–60 mg-chl.*a*·m^{-3} (summer ave.); diatom, nonfilamentous blue-green dominance **S:** <0.3 m (midsummer)	<1988–1990> **Z:** large daphnids (>100 ind. L^{-1}) **P:** 1988,89;30–50 mg chl.*a*·m^{-3} (summer); filamentous blue-green dominance 1990; <20 mg chl.*a*·m^{-3} (summer); cryptophyceae dominance **S:** 0.3–1 m (midsummer)
A-14	**Poisoning (rotenone): Sept. 1987** 0.5 ppm all the fish (mainly whitefish: 4600 kg) were killed (40)	**Lake Mosvatn,** Norway area = 46 ha z_{max} = 3.2 m; z_{ave} = 2.1 m	<1978–1987> **Z:** rotifer dominance **P:** 23 mg chl.*a*·m^{-3} (in summer) **S:** 1.7 m (in summer)	<1988> **Z:** daphnid dominance; individually doubled in weight **P:** 7 mg chl.*a*·m^{-3} (in summer) **S:** >2.3 m (in summer)

Table 1. (*Continued*)

Type #	Manipulation	Waterbody	Appearances Premanipulation[a]	Postmanipulation[a]
A-15	**Removal: Mar. 1987** all fish; emptied by pumping **Stocking: 1987** juvenile pike (1600); rud(140); *Daphnia*; macrophytes (37,41–44)	**Lake Zwemlust**, The Netherlands area = 1.5 ha z_{max} = 2.5 m; z_{ave} = 1.5 m	<–1986> **P:** *Microcystis* bloom; 250 mg chl.a·m^{-3} **M:** 5% of the lake bottom (6g DW m^{-2}) **F:** 800 kg·ha^{-1}; bream dominance (75%) **S:** 0.3 m	<1987–1990> **F:** rud = 395(1990), 300(1991) kg·ha^{-1} pike = 44 kg·ha^{-1} (1990–91) **P:** *Microcystis* bloom disappeared; 5–13 mg chl.a·m^{-3} **M:** 70% (87 g DW m^{-2}; 1988) and almost 100% (200 g DW m^{-2}; 1989) of the lake bottom **S:** often reached the lake bottom
A-16	**Removal: in Apr. 1987** 85% (2000 kg) of fish (breams of 1200 kg; carp of 550 kg) **Addition: May–July 1987** pike perch of 800 ind. (3.0 cm long) (37,45)	**Galgje site, Lake Bleiswijkse Zoom**, The Netherlands area = 3.1 ha z_{ave} = 1.1 m control: Zeeltje site area = 11.3 ha; z_{ave} = 1.1 m	(control area) in 1987 **F:** breams (430 kg·ha^{-1}); carp (240 kg·ha^{-1}) **Z:** 100 g·m^{-3}, large cladocera **P:** 130 mg chl.a·m^{-3}, greens and blue-greens	(treated area) in 1987 **F:** breams (45 kg·ha^{-1}), carp (59 kg·ha^{-1}) **Z:** 300 g·m^{-3}, large cladocera **P:** 10 mg·m^{-3}, greens and blue-greens
A-17	**Removal: 1988** bream: 100 kg·ha^{-1} (17% reduction) (36)	**Lake Søbygård** Denmark area = 40 ha z_{max} = 2 m; z_{ave} = 1.0 m	<1984–1985> **Z:** 0.02 g DW·m^{-3}; rotifers dominance (in summer) **P:** 52 g DW·m^{-3}; *Scenedesmus* dominance	<1986–88> **Z:** 0.8–1.2 g DW·m^{-3}; cladoceran dominance (in summer) **P:** 14 g DW·m^{-3}; diverse community
A-18	**Removal: 1988–1990** major part of grass carp (46)	**Senec gravel-pit lake**, West Slovakia area = 116 ha uneven bottom z_{max} = 6.5–8.5 m	<1986–1988> **Z:** rotifers and Gastrortica **P:** 3.9–30.6 mg chl.a·m^{-3} (annual max.); *Microcystis* bloom (in 1987) **M:** considerably reduced	<1989–90> **Z:** rotifers (1989), copepods (1990) **P:** 10.7–16.1 mg chl.a·m^{-3} (annual max) **M:** gradually restored
A-19	**Removal: Nov. 1990–June 1991** 75% of fish (425,000 kg); mainly bream and roach **Stocking: May 1990** 600,000 young pikes (3–4 cm) (47,48)	**Lake Wolderwijd**, The Netherlands area = 2700 ha z_{max} = 2.5 m; z_{ave} = 1.5 m	<1990, 1991> **F:** 200 kg·ha^{-1}: bream(95), roach(40), ruffe(30) **Z:** *Daphnia* (in spring); hardly present (in summer) **P:** filamentous blue-green dominance 20–40 (in spring); 100–135 mg chl.a·m^{-3} **S:** 0.25–0.5 m **O:** mysid shrimp = 30 ind·m^{-2}	<1997> **F:** 45 (June); 115 (Sept.) kg·ha^{-1} mainly ruffe 0+ of cyprinids = 10–15 kg·ha^{-1} 92% of introduced pikes died from predatory fish = 5–10 kg·ha^{-1} **Z:** daphnid dominance (300 ind.·L^{-1}) (May); collapsed in July **P:** various species including filamentous blue-greens; <50 mg chl.a·m^{-3} **S:** 1.8 m max; 1.1 m average
A-20	**Removal: Nov. 1992–June 1993** fish (bream and roach) of 3582 kg (119.4 kg·ha^{-1}; 80%) (49)	**Lake Duinigermeer**, The Netherlands area = 28 ha z_{ave} = 1 m	<1992> **Z:** *Daphnia*; peak at the end of May (100 ind. L^{-1}) **P:** 60 mg chl.a·m^{-3} (in summer); cyanobacteria (*Oscillatoria*) **M:** disappeared **S:** 0.3–0.6 m (in summer)	<1993> **Z:** *Daphnia*; same biomass as 1992, appeared earlier, individuals were larger than in 1992 **P:** 20 mg chl.a·m^{-3} (in summer); cyanobacteria (*Oscillatoria*) **M:** covered 50% of the lake bottom **S:** 0.7 m–bottom

(*continued overleaf*)

Table 1. (*Continued*)

Type #	Manipulation	Waterbody	Appearances Premanipulation[a]	Appearances Postmanipulation[a]
B-1	**Poisoning (rotenone): 1951** **Stocking (rainbow trout): every year** 15–18 cm; 370–740 ha^{-1} (50)	**Paul Lake**, Michigan, USA area = 1.21 ha z_{max} = 12.0 m	<–1951> **F:** yellow perch, largemouth bass dominance **Z:** daphnids (70–90%); low biomass	<1951–> **F:** brown trout dominance **Z:** increased daphnid biomass
B-2	**Poisoning (Fishtox): 1955, 1956** 68730 (907 kg) fish [346 (18.6 kg) pike] were killed **Stocking: 1957** largemouth bass (150 ind. ha^{-1}) (21)	**Poltruba backwater**, Czech Republic area = 0.18 ha z_{max} = 5.56 m; z_{ave} = 2.77 m	<1955> **Z:** small species (*Bosmina* etc.) **P:** diatoms dominant	<1957> **Z:** large species (*Daphnia* sp.) **P:** cryptophycae dominant
B-3	**Dieoff (disease): spring, 1967** a massive dieoff of alewives (51)	**Lake Michigan**, Wisconsin, USA area = 58,188 km^2 z_{max} = 285 m; z_{ave} = 84 m southeastern part	<until 1966> **F:** alewives: uncommon (1954); dominant (1966) **Z:** large sp. (1954); small sp. (1966)	<1968> **Z:** large species
B-4	**Partitioning exp. (fish stocking): May–Oct. 1976** northern part: walleye fry (10,000 ind.) introduced southern part: control (28)	**Pleasant Pond** Minnesota, USA area = 0.25 ha z_{max} = 2.5 m	[the southern part: control] **F:** minnow **Z:** rotifers **P:** *Chlamidomonas*; *Cryptomonas*; *Aphanizomenon* (two peaks); *Oscillatoria*; *Dactylococcopsis*; *Anabaenopsis*; *Kirchneriella*	[the northern part: fish stocked] **F:** 306 ind. (7.7 kg) of walleye (Oct.) **Z:** large cladocerans **P:** *Chlamidomonas*; *Cryptomonas*; *Aphanizomenon*; a variety of algal species
B-5	**Stocking: Mar. and Apr. 1981** predatory fish (perch and trout) 5150 kg (82 ind.) (52)	**A flooded quarry**, Dresden, Germany area = 0.044 ha z_{ave} = 7.6 m; z_{max} = 15.5 m	<1979, 1980> **F:** large population of *Leucaspius*, a few tehches and crucian carps (230 kg·ha^{-1} total) **Z:** rotifers, small cladocerans, cyclopoid copepods, Chaoborus (ca. 8–12 g ww·m^{-2} in summer) **P:** *Peridinium* and *Cryptomonas* (winter); various types (summer)	<1981> **F:** decreased rapidly **Z:** large daphnids (ca. 16–24 g ww·m^{-2} in summer) **P:** biomass was not changed drastically *Oocystis* and *Chlamydomonas* became dominant *Peridinium* vanished completely
C-1	**Removal: 1986** roach, bream, and crucian carp: 275 kg·ha^{-1} (78% reduction) **Stocking: 1986** perch: 15 kg·ha^{-1} (15%) (36)	**Frederiksborg Castle Lake**, Denmark area = 21 ha z_{max} = 8 m; z_{ave} = 3.1 m	<1986> **Z:** *Daphnia* and *Eudiaptomus* dominance; 0.8–1.4 g DW·m^{-3} **P:** 12–20 g DW·m^{-3}; *Microcystis* dominance	<1987> **Z:** *Daphnia* and *Eudiaptomus* dominance; 1.2–1.3 g DW·m^{-3} **P:** 12–13 g DW·m^{-3}; *Aphanizomenon* dominance **only minor changes**
C-2	**Removal: 1986–1989** 19.7 t (141 kg·ha^{-1}) of bream and roach **Stocking: 1987–1989** asp (predatory cyprinid) (53)	**Lake Rusutjärvi**, Finland area = 140 ha z_{ave} = 2.0 m	<–1985> a eutrophic lake	<–1990> **Water quality has not improved**
C-3	**Removal: 1990–91** Vendace of 34 t (34)	**Lake Froylandsvatn**, Norway a shallow lake	<–1990> **F:** vendace and small trout **Z:** small biomass **P:** large biomass	<–1994> **A possible effect has not been monitored**

Table 1. (*Continued*)

Type #	Manipulation	Waterbody	Appearances Premanipulation[a]	Appearances Postmanipulation[a]
D-1	**Stocking: 1977–** 20,000–80,000 of zander parrs were stocked every year pike fry and fingerlings (1985–) adult pike, eel, and catfish (1991–) **Restrictions catch: 1979–** piscivores (52,54)	**Bautzen Reservoir**, Dresden, Germany area = 533 ha $z_{ave} = 7.4$ retention time = 193 d first impounded in 1974	<1974–1980> **F:** large pike and perch population (1974) (it was opened for fishing in May 1976); large stock of small perch and roach (1976) zooplankton **Z:** total 6.52 ± 2.85 (g ww·m^{-3}) daphnids: 2.30 ± 1.42 cyclopoid copepods: 1.07 ± 0.85 individual body weight: 0.011 ± 0.004 (mg ww) **P:** 7.75 ± 3.44 (g ww·m^{-3}) **S:** 1.71 ± 0.55 (m)	<1981–> **Z:** total 7.47 ± 1.57 (g ww·m^{-3}) daphnids: 4.71 ± 1.54 cyclopoid copepods: 0.63 ± 0.40 individual body weight: 0.024 ± 0.007 (mg ww) daphnid dominance **P:** 22.31 ± 17.44 (g ww·m^{-3}) lower (in 1981), higher (in 1982–) blue-green species increased **S:** 1.89 ± 0.53 (m)
D-2	**Poisoning (rotenone): Sept. 23, 1980** **Stocking (fish): Oct. 1980** largemouth bass; walleye; planktivorous bluegills channel catfish (55,56)	**Round Lake**, Minnesota, USA area = 12.6 ha $z_{max} = 10.5$ m; $z_{ave} = 2.9$ m	< –1980> **F:** largemouth bass (1960–1970) bluegills, and *Pomoxis nigromaculatus* (1976–1980) **Z:** small cladocerans **P:** *Cryptomonas*; various greens and filamentous blue-greens	<1981,1982> **Z:** large daphnids (1981) large daphnids -> small cladocerans (1982) **P:** *Cryptomonas*; filamentous blue-greens, filamentous blue-green algae were responsible for algal peak
D-3	**Removal: 1985–1992** selective catchment of planktivore **Stocking: 1988–1992** pikeperch (57)	**Lake Feldberger Haussee**, Germany area = 13 ha $z_{max} = 12$ m; $z_{ave} = 6.3$ m	< –1985> **Z:** rotifer dominance **P:** green dominance; rare filamentous blue-greens (2.2 g ww·m^{-3})	<1987– > **Z:** cladocerans **P:** filamentous blue-green dominance (21.5 g ww·m^{-3}) significant relationship between filamentous blue-green and *Eudiaptomus*
E-1	**Planktivore addition: 1955?** inadvertently introduced into an upstream of glut herring (planktivore) (58)	**Crystal Lake**, Connecticut, USA	<1942> **Z:** large species dominant	<1964> **Z:** small species dominant
E-2	**Poisoning(rotenone): June 4, 1960** 0.5 ppm rotenone was added (59)	**Fern Lake**, Washington, USA area = 21.2 acres $z_{max} = 25$ feet	<until 1959> **Z:** appearance at all sights of the lake	<1960> **Z:** open water, shore edge:absent dense weed patches:resist
E-3	**Aeration: summer, 1975, 1976, 1978, 1979** to supply O$_2$ for low temperature hypolimnion (60)	**Tory Lake**, area = 1.23 ha $z_{max} = 10$ m $z_{ave} = 4.5$ m	<1973,1974> **Z:** localization at 0–2 m in depth (1974) **O:** anoxic situation was found at 0.3 m (May 1973) and 5 m (summer 1974)	<1975–1979> **F:** fat head minnow **Z:** uniform distribution at 0–8 m in depth (in 1975 and 1976) cladocerans: localized at surface; rotifers:also in hypolimnion where DO<1 ppm; small species dominance (1978 and 1979)

[a] F: fish; Z: zooplankton; P: phytoplankton; PP: primary productivity; M: macrophytes; S: Secci depth; O: others.

evaluated the long-term studies of biomanipulation in The Netherlands. They surveyed nine trials (Lakes Zwelmust, Bleiswijkse Zoom, Noorddiep, Breukeleveense plas, Klein Vogelenzang, IJzeren Man, Sondelerleijen, Wolderwijd, and Duinigermeer). Their survey indicated that six cases (Lakes Zwelmust, Bleiswijkse Zoom, Noorddiep, IJzeren Man, Wolderwijd, and Duinigermeer) showed positive results; however, only three lakes (Noorddiep, IJzeren Man, and Wolderwijd) showed long-term (>5 years) stability of clear water. Thus, long-term stability resulting from biomanipulation must be examined. In other words, the ecological sustainability of treated lakes determines their long-term stability. One of the essential methods for elucidating long-term stability is to quantify the material flow in a lake ecosystem as a means of numerical modeling. For this purpose, some numerical models that describe the lake ecosystem and predict the effect of biomanipulation are available in the literature (19,61).

BIBLIOGRAPHY

1. Bernhardt, H. et al. (1982). *Proc. Japanese-German Workshop Waste Water Sludge Treatment*, pp. 583–637.
2. Edward, J.K. and Wingler, B.J. (1990). *Aqua* **39**: 24–35.
3. Longhurst, S.J. and Graman, N.J.D. (1987). *Public Health Eng.* **14**(6): 71–76.
4. Zabel, T. (1985). *J. AWWA* **77**(5): 42–46.
5. Lindenschmidt, K.E. and Hamblin, P.F. (1997). *Water Res.* **31**(7): 1619–1628.
6. Little, J.C. (1995). *Water Res.* **29**(11): 2475–2482.
7. Soltero, R.A. et al. (1994). *Water Res.* **28**(11): 2297–2308.
8. Daley, R.D. and Pick, F.R. (1990). *Verh. Int. Verein. Limnol.* **24**: 314–318.
9. Johnson, M.G. et al. (1965). *Can. Fish. Cult.* **35**: 59–66.
10. Recknagel, F. et al. (1995). *Water Res.* **29**(7): 1767–1779.
11. May, R.M. (1977). *Nature* **269**: 471–477.
12. Benndorf, J. (1995). *Int. Rev. Gesamten Hydrobiol.* **80**(4): 519–534.
13. Hosper, S.H. (1989). *Hydrobiol. Bull.* **23**: 5–11.
14. Timms, R.M. and Moss, B. (1984). *Limnol. Oceanogr.* **29**: 472–486.
15. Scheffer, M. (1990). *Hydrobiologia* **200/201**: 475–486.
16. Shapiro, J. et al. (1975). *Proc. Symp. Water Qual. Manage. Biological Control.* University of Florida, pp. 85–96.
17. Perrow, M.R. et al. (1997). *Hydrobiologia* **342/343**: 355–365.
18. DeMelo, R. et al. (1992). *Limnol. Oceanogr.* **37**(1): 192–207.
19. Sagehashi, M. et al. (2000). *Water Res.* **34**(16): 4014–4028.
20. Shapiro, J. (1990). *Hydrobiologia* **200/201**: 13–27.
21. Hrbácek, J. et al. (1961). *Verh. Int. Verein. Limnol.* **14**: 192–195.
22. Edmondson, W.T. and Litt, A.H. (1982). *Limnol. Oceanogr.* **31**: 1022–1038.
23. Schindler, D.W. and Comita, G.W. (1972). *Arch. Hydrobiol.* **69**: 413–451.
24. Hrbácek, J. et al. (1978). *Verh. Int. Verein. Limnol.* **20**: 1624–1628.
25. Stenson, J.A.E. et al. (1978). *Verh. Int. Verein. Limnol.* **20**: 794–801.
26. Henrikson, L. et al. (1980). *Hydrobiologia* **68**: 257–263.
27. Scavia, D. et al. (1986). *Can. J. Fish. Aquat. Sci.* **43**: 435–443.
28. Lynch, M. (1979). *Limnol. Oceanogr.* **24**: 253–272.
29. Lynch, M. and Shapiro, J. (1982). *Limnol. Oceanogr.* **26**: 86–102.
30. Andersson, G. et al. (1978). *Hydrobiologia* **59**: 9–15.
31. Bjork, S. (1978). *Proc. MAB Project 5 Regional Workshop Land use Impacts on Lake and Reservoir Ecosystems*, Warsaw, May 26–June 2, 1978, pp. 196–219.
32. Reinertsen, H. and Olsen, Y. (1984). *Verh. Int. Verein. Limnol.* **22**: 649–657.
33. Langeland, A. (1990). *Hydrobiologia* **200/201**: 535–540.
34. Bratli, J.L. (1994). *Mar. Pollut. Bull.* **29**(6–12): 435–438.
35. Brabrand, A. et al. (1993). *Oecologia (Heidenberg)* **95**(1): 38–46.
36. Jeppesen, E. et al. (1990). *Hydrobiologia* **200/201**: 205–218.
37. Meijer, M.L. et al. (1994). *Hydrobiologia* **275–276**: 457–466.
38. Nielsen, S.N. (1994). *Ecol. Modelling* **73**: 13–30.
39. Hanson, M.A. and Butler, M.G. (1994). *Hydrobiologia* **279/280**: 457–466.
40. Sanni, S. et al. (1990). *Hydrobiologia* **200/201**: 263–274.
41. Janse, J.H. et al. (1995). *J. Aquat. Ecol.* **29**(1): 67–79.
42. Kornijow, R. et al. (1992). *Arch. Hydrobiol.* **123**(3): 337–347.
43. Kornijow, R. et al. (1992). *Arch. Hydrobiol.* **123**(3): 349–359.
44. Van Donk, E. et al. (1990b). *Hydrobiologia* **200/201**: 275–289.
45. Jayaweera, M. and Asaeda, T. (1996). *Ecol. Modelling* **85**: 113–127.
46. Horecka, M. et al. (1994). *Biologia, Bratislava* **49**: 141–146.
47. Meijer, M.L. et al. (1994). *Hydrobiologia* **275/276**: 31–42.
48. Hosper, S.H. and Meijer, M.L. (1999). *J. Jpn. Soc. Water Environ.* **22**(1): 18–24.
49. Van Berkum, J.A. et al. (1995). *Neth. J. Aquat. Ecol.* **29**(1): 81–90.
50. Kitchell, J.A. and Kitchell, J.F. (1980). *Limnol. Oceanogr.* **25**: 389–402.
51. Wells, L. (1970). *Limnol. Oceanogr.* **15**: 556–565.
52. Benndorf, J. et al. (1984). *Int. Rev. Gesamten Hydrobiol.* **69**: 407–428.
53. Pekkarinen, M. (1990). *Aqua Fennica* **20**(1): 13–25.
54. Benndorf, J. and Schweiz, Z. (1988). *Hydrobiologia* **49**(2): 237–248.
55. Shapiro, J. and Wright, D.I. (1984). *Freshwater Biol.* **14**: 371–383.
56. Wright, D.I. and Shapiro, J. (1984). *Verh. Int. Verein. Limnol.* **22**: 518–524.
57. Kasprzak P. et al. (1993). *Arch. Hydrobiol.* **128**(2): 149–168.
58. Brooks, J.L. and Dodson, S.I. (1965). *Science* **150**: 28–35.
59. Kiser, R.W. et al. (1963). *Trans. Am. Fish. Soc.* **92**: 17–24.
60. Taggart, C.T. (1984). *Can. J. Fish. Aquat. Sci.* **41**: 191–198.
61. Sagehashi, M. et al. (2001). *Water Res.* **35**(7): 1675–1686.

GENOMIC TECHNOLOGIES IN BIOMONITORING

MICHAEL J. CARVAN III
MATTHEW L. RISE
REBECCA D. KLAPER
University of
Wisconsin–Milwaukee
Milwaukee, Wisconsin

Biological monitoring, or biomonitoring, is defined as using measurements of the resident biota in surface waters to evaluate the condition of the aquatic environment (1). In part, biological monitoring in the United States is designed to satisfy the goals of the Clean Water Act, which aims to protect wildlife in and on the water and eliminate discharge in toxic amounts into navigable water. Biomonitoring has inherent advantages over simply measuring the presence or concentration of a particular stressor or toxin due to the fact that organisms represent an integrated response to factors in their environment over time. Biomonitoring in aquatic habitats traditionally involved measuring the diversity within the community of macroinvertebrates, fish, or periphyton using species diversity, composition with respect to functional groups, or genetic diversity (2). Community indices have inherent problems. For example, streams can include a large spatial variance that is not correlated with problems in habitat. In addition, sampling effort can affect diversity measurements, resulting in an increase in the number of species sampled with longer or more frequent sampling times (3). Identification of species is time consuming and can increase the costs associated with these types of measurements due to personnel-hours required (4).

As one alternative, measurement of the concentration of chemicals has been used as a surrogate for biological criteria. These chemical water quality criteria (TMDL or total maximum daily load) have been developed by determining concentrations of compounds that cause acute toxicity to standard test organisms. However, simply measuring the concentration of a chemical present in a system in one snapshot of time does not indicate the actual exposure of any particular organism to that chemical, neither to the organism as a whole nor to a target of action. Variation in exposure within a single species may be influenced by habitat utilization in the environment, feeding behaviors, and metabolism and physiology. To provide this information, chemical data can be used in conjunction with biomarkers, which are enzymes, proteins, or other products that can indicate an organism has been exposed to a toxin or that a toxic response has occurred. This involves sampling one or several species in a community to collect tissues, or whole organisms, for detection of the biomarker. A problem with this method in the past has been that biomarkers, for the most part, indicate a general response that can occur with exposure to several different compounds and are not specific enough to indicate a particular exposure or toxic response.

Genomic tools could potentially solve some of the problems with each of these techniques, while at the same time complementing traditional methods of monitoring. Examining the genes that are being expressed within an organism, or across several organisms, in the community provides hundreds to thousands of potential biomarkers for exposure and effects (5,6). As the number of biomarkers is several orders of magnitude greater, there is also the capability of differentiating exposures and effects to provide a more specific indicator, even from a complex mixture. Here, we provide an overview of the technologies of genomics (and other -omics) and provide a few examples of the development of their use for environmental monitoring.

HIGH THROUGHPUT TECHNOLOGIES AND THEIR APPLICATION TO BIOMONITORING

Genomics

There are a number of high throughput genomic technologies with application to environmental biomonitoring. Quantitative reverse transcription polymerase chain reaction (qRT-PCR) is one of the most sensitive techniques for detection of low abundance mRNA. The reverse transcription process creates a complementary DNA molecule (cDNA) from an mRNA template. The PCR process amplifies the gene-specific cDNA target in a series of thermal cycles that include DNA denaturation, target-specific primer binding, and primer extension using a thermostable polymerase. Utilization of fluorescent labeling allows the quantitative detection of cDNA molecules following each round of amplification. Gene-specific fluorescent labeling of targets can be accomplished using TaqMan® (Applera Corporation) or molecular beacon (7) probes that are designed to anneal to an internal sequence of the PCR product. This approach allows multiple target sequences to be analyzed using multiple fluorescent labels in the same reaction (multiplex qRT-PCR). Intercalating agents like SYBR green are used as nonspecific probes that bind DNA and track the accumulation of PCR product. These assays are less expensive and very flexible because SYBR green will bind to any target. SYBR green can also be used to detect multiple targets if used in conjunction with melting point analysis of the PCR products (8).

Relative quantitation of the PCR products compares fluorescent intensity and the number of cycles required for the fluorescent signal to reach a fixed threshold (generally multiple standard deviations above background fluorescence) generally in the logarithmic portion of the amplification cycle. The parameter Ct (cycle threshold) is defined as the cycle number at which the fluorescence emission exceeds the fixed threshold. Ct is proportional to the number of target molecules in the sample and is used to derive mRNA abundance estimates (Fig. 1). The use of a normalizer gene, expressed at approximately the same level in the experimental and control samples used in microarray and qRT-PCR studies, corrects for technical inaccuracies (i.e., slight differences in amounts of RNA template due to pipetting error, or variability in efficiency of reverse transcriptase reactions). While there are no perfect normalizer genes, a number of transcripts are commonly reported in the literature, including glyceraldehyde-3-phosphate dehydrogenase, beta actin, major histocompatibility complex I, cyclophilin, various ribosomal proteins, and ribosomal RNA. Comparative quantitation uses Ct data to determine the relative abundance of specific transcripts in different samples. A new approach for the relative quantitation of targets in qRT-PCR (9) takes into account primer-dependent differences in the efficiency of amplification resulting in more accurate relative estimates of target abundance. Absolute quantitation of mRNA abundance in a sample requires comparison of Ct for the target gene to that of samples with known quantities of target. A standard curve can be generated using a dilution series of known targets and can then be used to estimate the number of target molecules in a particular sample. Some real-time PCR systems (e.g., Stratagene Mx3000P) include user-friendly software for performing data analyses such as conversion of Ct values to fold differences using comparative or absolute quantitation methods corrected with normalizer gene data.

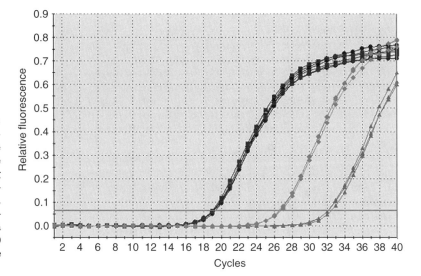

Figure 1. qRT-PCR amplification plots. Curves show amplification from template A (control, non-exposed tissue) and template B (exposed tissue). The horizontal line is the threshold. The Ct values for the amplification plots are virtually identical, indicating that the normalizer gene is expressed at approximately the same level in control and exposed tissues. Analysis of these triplicate data using Mx3000P software (Stratagene) and the relative quantification method (9) revealed that the gene of interest (GOI) was 44-fold downregulated in the exposed sample relative to the control sample.

DNA arrays (also known as Gene Chips®, DNA chips, or gene arrays) have revolutionized the way scientists think about biological problems. DNA arrays allow the simultaneous detection and quantitation of hundreds to thousands of DNA targets within a single complex mixture. The unlabeled probes for DNA arrays are printed (spotted) using high speed robots and immobilized on a solid substrate (generally nylon membranes, glass slides, or quartz wafers). Macroarrays are low density DNA arrays that are usually printed on nylon membranes or glass slides with spot diameters generally greater than 300 μm. Microarrays are high density DNA arrays that are usually printed on glass slides or quartz wafers with spot diameters generally less than 200 μm. The highest density microarrays utilize tens of thousands of oligonucleotide probes applied to, or synthesized directly on, quartz wafers. For DNA array analyses, the targets in a single sample are generally labeled with a fluorescent (or radioactive) tag and hybridized to the immobilized probes (spots on the DNA array). Hybridization of specific targets is quantified relative to standard genes on a typical one-color (or radioactive) array. Two- (or multi-) color arrays utilize multiple fluorescent labels, one for each sample to be compared, and allow the quantitation of the relative abundance of each of the specific targets in each sample. The most common experimental approach is to use the fluorescent dyes Cy3 and Cy5 to label cDNA from the control and experimental samples (or vice versa), which are then hybridized to the DNA array. This process is diagrammed in Fig. 2. Through competitive hybridization, the abundance and proportion of bound target in the two samples should reflect the abundance and proportion of corresponding transcripts in the original samples. Images of the hybridized microarray are generally acquired by laser confocal scanning using dye-specific lasers and/or filters. Quantitation of the data at each spot requires sophisticated image analysis and statistical software and the final output depicts the ratio of cDNA abundance between the two samples for each probe by scatter plot, dendrogram, heat map, or Venn diagram. The differential gene expression data from microarray experiments are usually confirmed by qRT-PCR.

Microarray analysis is an extremely sensitive technique for the analysis of mRNA abundance, and small variations in the experimental conditions may lead to dramatically different results and conclusions. As a result, the Microarray Gene Expression Data (MGED) Society was formed as an international organization of biologists, computer scientists, and data analysts that aims to facilitate the sharing of microarray data generated by functional genomics and proteomics experiments. The MGED has established a set of guidelines called the Minimum Information About a Microarray Experiment (MIAME) with the goal that such information will allow one to "interpret unambiguously and potentially reproduce and verify an array-based gene expression monitoring experiment" (www.mged.org) (10). Adherence to the MIAME guidelines and submission of all microarray images and data to a public data repository such as Gene Expression Omnibus (http://www.ncbi.nlm.nih.gov/geo/) are required by most of the top scientific journals in the United States and Europe.

Serial analysis of gene expression (SAGE) is a powerful tool that allows the analysis of overall gene expression patterns (11). SAGE can be used to identify and quantitate genes from model organisms with a sequenced genome (12) and from those with few known genes (13). The SAGE technique is based on the concept that short sequence tags (10–14 bp) can contain sufficient information to uniquely identify a transcript. Sequence tags are linked together in concatamers, which are then sequenced. Quantitation of the number of times a particular tag is observed provides the expression level of the corresponding transcript. SAGE has contributed to our understanding of numerous biological processes including differentiation, development, carcinogenesis, and responses to environmental chemicals and stressors (14).

Massively parallel signature sequencing (MPSS) technology generates sequence information from millions of DNA fragments without the need for individual sequencing reactions and the physical separation of DNA

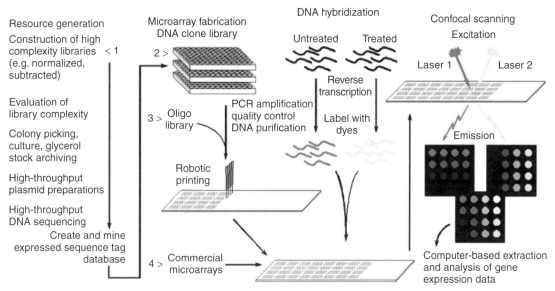

Figure 2. Microarray fabrication and utilization for transcriptional profiling. There are a number of entry points into the microarray fabrication process that are indicated by the arrows (>). The utilization of transcriptional profiling and other functional genomic technologies will require the generation of genomic resources for the vast majority of ecologically important species. This process begins, at step <1, with construction of high complexity libraries and expressed sequence tag (EST) databases. Investigations that utilize organisms for which DNA clone or oligonucleotide (oligo) libraries are available (entry points 2> and 3>, respectively) can derive fragments from these libraries for spotting on prepared glass slides. Investigations using human, mouse, *Drosophila*, yeast, or sequenced microorganisms can purchase commercial slides and enter at step 4>.

fragments (15,16). Amplified DNA molecules are bound to Megaclone™ microbeads, which are arrayed in a flow cell, and the reagents required for sequencing flow over the microbeads. The proprietary protocol elicits sequence-dependent fluorescent responses from the microbeads, which are recorded by a CCD camera after each cycle, to construct a short 17–20 bp signature sequence. The sequences are then grouped to determine abundance and compared to public and private databases to identify the gene represented by the individual signature sequences and to determine putative function of the gene product (17,18).

Proteomics

High throughput protein expression analysis is primarily performed using three techniques: two-dimensional (2D) gel electrophoresis, mass spectrometry, and protein arrays. Two-dimensional gel electrophoresis separates labeled proteins from a biological sample by their isoelectric point and electrophoretic mobility (size), producing a 2D fingerprint of protein content. Changes in protein expression can be determined by analyzing 2D gels from different samples and isolation of the elements that vary with experimental conditions. N-terminal protein sequencing can be used to identify the protein represented by a single element of a 2D gel (19). Alternatively, the isolated element can be analyzed by mass spectrometry to determine peptide mass and/or sequence, leading to its identification. The primary types of mass spectrometry used for protein expression analysis are matrix-assisted laser desorption ionization time of flight (MALDI-TOF) and Electrospray ionization MS quadrupole (ESI-MS). MALDI-TOF allows peptide fingerprinting of a protein sample following tryptic digestion. The peptide fingerprint is then compared to a database with data from known proteins or computationally derived data for theoretical peptides (20). ESI-MS provides peptide sequence information in addition to any post-translational modifications, for example, acylation, phosphorylation, or glycosylation (21).

Protein arrays are assay systems using immobilized probes on a solid support, which include glass, membranes, microtiter wells, and mass spectrometer plates. Protein arrays are used to assay a biological sample for a number of interactions including ligand binding, protein–protein, protein–DNA, protein–drug, receptor–ligand, and enzyme–substrate (22–24). Depending on the design of the array, proteins of interest can be either immobilized probes or labeled targets for binding to immobilized high affinity probes (antibodies, engineered scaffolds, single protein domains, peptides, or nucleic acids). Since this field is very new and rapidly evolving, a variety of methods are under development for utilization of protein from a number of sources, application to different formats and surfaces, immobilization of probes, fabrication of arrays, and detection of signal. Many of the techniques available today rely on adaptation of DNA array technologies and the bioinformatics tools developed for data analysis. We now know that the number of proteins produced by a given organism is, on the average, an order of magnitude greater than the number of genes in the genome. Many of the protein array technologies are focused on specific pathways and do not try to analyze the entire proteome.

Metabolomics

The foundation of metabolomics is the ability to detect changes in metabolic processes that are initiated in response to cellular stressors and enzyme system inhibitors. A number of analytical techniques can be used to generate metabolomic databases, including liquid chromatography in conjunction with mass spectrometry, high resolution nuclear magnetic resonance spectroscopy, high pressure liquid chromatography, and optical spectroscopic techniques (25–27). These techniques produce a comprehensive profile of small molecule signals. When samples derived from stressed tissues (diseased, toxicant exposed, etc.) are compared with appropriate control tissues, relatively small stresses can result in dramatic changes in metabolite profiles (28).

DEVELOPING TOOLS FOR ECOTOXICOGENOMIC RESEARCH AND BIOMONITORING

Genomics is defined as the comprehensive study of an organism's genes and their functions. Toxicogenomic research is aimed at understanding how changes in gene expression, elicited by chemical exposures, contribute to toxicological effects. DNA microarrays, important tools in toxicogenomic research, allow the relative expression of thousands of genes to be evaluated simultaneously. A primary goal of ecotoxicological research is to shed light on the mechanisms by which toxins exert their effects on various wildlife species, and ecotoxicogenomics involves the use of genomic resources and methods in ecotoxicological research.

Expressed sequence tag (EST) databases and DNA microarrays are important tools in ecotoxicogenomic research. EST databases, generally arising from high complexity (e.g., subtracted, normalized) cDNA libraries, allow researchers to identify the genes expressed in a given tissue under select conditions and at a particular point in time. For example, a suppression subtractive hybridization (SSH) cDNA library construction kit (e.g., from Clontech) could be used with mRNA from two tissues (e.g., toxin-exposed liver and control nonexposed liver) to generate reciprocal SSH libraries. One of these libraries would be enriched for genes upregulated in toxin-exposed liver relative to nonexposed liver, and the other library would be enriched for genes suppressed by the exposure. Characterization of these libraries, involving colony (clone) picking and overnight culture, plasmid DNA preparation, and automated sequencing of clone inserts (generally from one end, hence the term "tag" in EST) is generally performed in 96- or 384-well format. EST databases are designed by bioinformaticians to perform quality control such as trimming of poly A tails, low quality sequence, and vector. In addition, EST databases are generally designed to perform assemblies to identify ESTs contributing to contiguous sequences (putative transcripts). Assembled ESTs and their putative translations are aligned (BLASTN and BLASTX, respectively) with public nucleic acid or amino acid sequence databases (e.g., NCBI nt or nr, respectively) for identification. The top (most negative E-value) BLAST hits (gene names, regions of sequence overlap, percent identity/similarity, and significance of hit) and functional annotations associated with these genes (gleaned from public databases such as Gene Ontology Consortium or Swiss-Prot) may be linked to sequences within an EST database (4,29). With steadily decreasing costs associated with DNA sequencing, EST projects for ecologically important (e.g., indicator) species are now relatively affordable.

An EST project usually precedes design and construction of DNA microarrays (see Fig. 2). For model organisms (e.g., *D. rerio, X. laevis, C. elegans, Drosophila*, rat, and mouse) and some organisms of agricultural importance (e.g., rice, barley, cow, Atlantic salmon), genomic resources (e.g., EST databases and DNA microarrays) have been developed and are available. However, ecotoxicogenomic researchers generally wish to study wildlife species for which little or no genomic data previously exist. If a researcher has access to a fully equipped molecular biology laboratory, it is possible to develop genomic reagents for an ecologically important species for under $50,000, not including personnel costs. Reciprocal SSH libraries can be constructed for about $5000; 10,000 sequences from these high complexity libraries (likely representing approximately 5000 different transcripts) can be obtained for under $20,000; 1000–2000 selected transcripts (e.g., relevant to xenobiotic metabolism) can be PCR amplified, cleaned, and spotted onto glass slides for under $20,000; reagents for conducting preliminary microarray experiments (50 chips) would cost approximately $5000. An example of such a project might be to study the effects of atrazine exposure on gene expression in a larval amphibian. Reciprocal SSH libraries could be constructed from atrazine-exposed versus nonexposed larval tissues. Transcripts of interest could be selected from the resulting EST database for inclusion on a microarray. It is now affordable for a researcher to partially characterize a nonmodel organism's transcriptome, generate genomic tools (EST database and DNA microarray), and acquire meaningful data on how the sentinel organism responds transcriptomically to a particular insult. The cost of constructing custom DNA microarrays will decrease as the number of array facilities increases. Methods for construction depend on whether the research desires a cDNA or oligonucleotide microarray.

A typical, simple toxicogenomic experimental objective might be to evaluate the impact of a single dose of toxin (e.g., atrazine) on global gene expression in a specific organ (e.g., liver) at various time points (Fig. 3). An alternative experimental approach would use microarrays to evaluate gene expression changes in a tissue associated with various toxin doses at a single time point (e.g., 24-h exposure). For all genes represented on the DNA microarrays used, expression can be compared in the exposed versus unexposed tissues. Microarray experiments are generally run on pooled samples (e.g., pooled $n = 5$ exposed liver versus

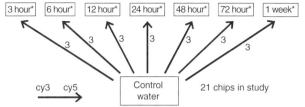

Figure 3. Microarray experimental design for determining expression of candidate molecular biomarkers of toxin exposure over time. All RNA samples are from a single toxin (e.g., atrazine) challenge at a single concentration (e.g., 10 ppb).

pooled $n = 5$ control liver), followed by real-time qRT-PCR for select transcripts of interest on individual fish samples to validate microarray results and reveal biological variability in expression of informative genes. Validated toxin-responsive genes can be functionally annotated using public sequence databases such as Swiss-Prot (http://us.expasy.org/sprot/). The example microarray experiment (Fig. 3) would be relatively easy to run, and statistical analyses of resulting data would be straightforward (e.g., K mean clustering of background-corrected, Lowess-normalized Cy5/Cy3 ratio values using software such as GeneSpring by Silicon Genetics, Fig. 4). It is important to note, however, that organisms in nature generally do not encounter a single toxin in a single dose for an extended period of time. Actual toxin exposures in the wild (e.g., exposure of larval amphibians to agricultural runoff) are continuums with fluctuating toxin concentrations and dynamic ambient conditions (e.g., temperature, pH). Furthermore, ecologically important organisms often encounter mixtures of toxins, such as effluents or automobile exhaust. Realistic modeling of this type of complexity will require extremely well-designed ecotoxicogenomic experiments supported by collaborations with mathematicians, statisticians, and computer scientists.

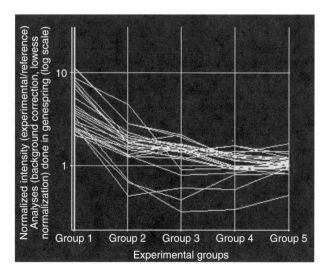

Figure 4. K means clustering identifies suites of genes having similar expression profiles.

RECENT APPLICATION OF GENOMIC TOOLS IN BIOMONITORING

Only very recently have genomic tools been used to understand environmental processes or to determine environmental quality, and we will present a few examples. Random shotgun sequencing of DNA from an environmental sample (e.g., a natural biofilm) can be used to determine species composition and provide evidence of survival strategies in these harsh environments (30). The relatively small number of dominant bacterial species in many harsh environments allows for extensive analyses without culturing organisms in the laboratory or having a previously sequenced genome to aid analyses (30).

Stressor-specific signatures in gene expression profiles can be used to diagnose which stressors are impacting populations in the field (31). Consensus PCR primers can be designed using sequence information from several animal sequences. It may then be possible to amplify some of these well-characterized stress-induced genes from organisms of interest in ecotoxicology. cDNAs representing stress-induced genes from a number of species can be arrayed on a single glass slide for rapidly screening environmental samples (31).

A few cDNA microarrays are available for fish species important in the aquaculture industry. These arrays are valuable for expression profiling following exposure to specific toxicants (32) or bacterial infection (28). While these studies are laboratory based, they demonstrate the potential for cDNA microarrays to provide insight into organism health as part of a larger biomonitoring effort. Currently available microarray resources can also be used to analyze samples from many closely related species (33,34). Thus, biomonitoring projects may be able to take advantage of genomic resources developed for model organisms in other fields and may not need to generate a species-specific microarray for every organism of interest. Targeted low density macroarrays from a few environmentally relevant species can detect alterations in gene expression following exposure to "environmental estrogens" (35,36). This technology is useful for identifying endocrine disruption; however, it may have limited utility in the analysis of the multiple stressors found in complex environments.

It is important to note that genomic information must be tied to traditional measures of toxicity to be useful. Gene expression profiling provides a picture in a snapshot of time where toxic endpoints are an accumulation of exposure and effect (5,37). To include microarray and other types of genomic data in risk assessments will involve measurements of gene expression over time and over several tissues and organisms.

BIBLIOGRAPHY

1. Gibson, G.A., Barbour, M.T., Stribling, J.B., Gerritsen, J., and Karr, J.R. (1996). *Biological Criteria: Technical Guidance for Streams and Rivers*. EPA Office of Science and Technology, Washington, DC.

2. Karr, J.R. and Chu, E.W. (1999). *Restoring. Life in Running Waters: Better Biological Monitoring*. Inland Press, Washington, DC.

3. Li, J. et al. (2001). *Freshwater Biol.* **46**(1): 87–97.
4. Resh, V.H. (1993). Rapid assessment approaches to biomonitoring using benthic macroinvertebrates. In: *Freshwater Biomonitoring and Benthic Macroinvertebrates*. D.M. Rosenburg and V.H. Resh (Eds.). Chapman and Hall, New York, pp. 195–233.
5. Klaper, R. and Thomas, M.A. (2003). *BioScience* **54**: 403–412.
6. Thomas, M.A. and Klaper, R. (2004). *Trends Ecol. Evol.* **19**: 440–445.
7. Tyagi, S. and Kramer, F.R. (1996). *Nat. Biotechnol.* **14**(3): 303–308.
8. Siraj, A.K. et al. (2002). *Clin. Cancer Res.* **8**(12): 3832–3840.
9. Pfaffl, M.W. (2001). *Nucleic Acids Res.* **29**(9): e45.
10. Brazma, A. et al. (2001). *Nat. Genet.* **29**(4): 365–371.
11. Velculescu, V.E., Zhang, L., Vogelstein, B., and Kinzler, K.W. (1995). *Science* **270**(5235): 484–487.
12. Fujii, S. and Amrein, H. (2002). *EMBO J.* **21**(20): 5353–5363.
13. Lorenz, W.W. and Dean, J.F. (2002). *Tree Physiol.* **22**(5): 301–310.
14. Tuteja, R. and Tuteja, N. (2004). *Bioessays* **26**(8): 916–922.
15. Brenner, S. et al. (2000). *Nat. Biotechnol.* **18**(6): 630–634.
16. Brenner, S. et al. (2000). *Proc. Natl. Acad. Sci. U.S.A.* **97**(4): 1665–1670.
17. Brandenberger, R. et al. (2004). *BMC. Dev. Biol.* **4**(1): 10.
18. Reinartz, J. et al. (2002). *Brief. Funct. Genomic Proteomic* **1**(1): 95–104.
19. Hu, Y., Huang, X., Chen, G.Y., and Yao, S.Q. (2004). *Mol. Biotechnol.* **28**(1): 63–76.
20. Sheffield, L.G. and Gavinski, J.J. (2003). *J. Anim. Sci.* **81**(Suppl 3): 48–57.
21. Eymann, C. et al. (2004). *Proteomics* **4**(10): 2849–2876.
22. Espina, V. et al. (2004). *J. Immunol. Methods* **290**(1–2): 121–133.
23. Gerstein, M. and Echols, N. (2004). *Curr. Opin. Chem. Biol.* **8**(1): 14–19.
24. Predki, P.F. (2004). *Curr. Opin. Chem. Biol.* **8**(1): 8–13.
25. Frederich, M. et al. (2004). *Phytochemistry* **65**(13): 1993–2001.
26. Griffin, J.L. (2004). *Philos. Trans. R. Soc. London B* **359**(1446): 857–871.
27. Purohit, P.V., Rocke, D.M., Viant, M.R., and Woodruff, D.L. (2004). *OMICS.* **8**(2): 118–130.
28. Kell, D.B. (2004). *Curr. Opin. Microbiol.* **7**(3): 296–307.
29. Rise, M.L. et al. (2004). *Physiol. Genomics* **20**: 21–35.
30. Tyson, G.W. et al. (2004). *Nature* **428**(6978): 37–43.
31. Snell, T.W., Brogdon, S.E., and Morgan, M.B. (2003). *Ecotoxicology* **12**(6): 475–483.
32. Koskinen, H. et al. (2004). *Biochem. Biophys. Res. Commun.* **320**(3): 745–753.
33. Renn, S.C., Aubin-Horth, N., and Hofmann, H.A. (2004). *BMC. Genomics* **5**(1): 42.
34. Rise, M.L. et al. (2004). *Genome Res.* **14**(5): 478–490.
35. Larkin, P., Sabo-Attwood, T., Kelso, J., and Denslow, N.D. (2003). *Ecotoxicology* **12**(6): 463–468.
36. Larkin, P. et al. (2003). *EHP Toxicogenomics* **111**(1T): 29–36.
37. Neumann, N.F. and Galvez, F. (2002). *Biotechnol. Adv.* **20**(5–6): 391–419.

MACROPHYTES AS BIOMONITORS OF TRACE METALS

Shimshon Balanson
Tarun K. Mal
Cleveland State University
Cleveland, Ohio

INTRODUCTION

Aquatic macrophytes are macroscopic aquatic plants that play an important role in the health of aquatic ecosystems. Not only do macrophytes provide vital structural support in aquatic ecosystems (1), but they also contribute toward food production, nutrient regeneration, and habitat for a wide range of organisms. Because of their roles in aquatic ecosystem health, aquatic macrophytes can be utilized as indicators of stream health (2). In fact, aquatic macrophytes may be indicative of a variety of ecosystem stressors including nutrient runoff, changes in hydrologic regime, and exotic species invasion (3,4). Therefore, species richness of aquatic macrophytes, their composition, and density have all been incorporated into different macrophyte indicators (5,6). As such, bioindicators estimate the pollutants affecting the biota in an ecosystem and are purely qualitative in nature.

BIOINDICATORS VERSUS BIOMONITORS

Of course, the use of aquatic macrophytes as *qualitative* bioindicators of stream health does not reveal the amount of pollutants in the water column and sediments. Therefore, another technique must be utilized when attempting to *quantify* the direct impact of in-stream pollutants on aquatic macrophytes. The use of biota as monitors of ecosystem health relies on changes in the morphology, biochemical pathways, and life cycle of the organisms being tested as well as direct measurement of pollutant levels in the tissues (2,7,8). Biomonitors have been preferred to abiotic monitors in many instances. Abiotic monitors such as water and sediments do detail the amount of a pollutant in the ecosystem but do not reveal the quantities that are actually *available* to organisms living in an ecosystem (9).

In some ways, however, bioindicators and biomonitors share many of the same purposes. Like bioindicators, biomonitors can be utilized to identify changes in plants over the course of time, especially morphological and biochemical changes (8,10–12). Changes can be quantified at contaminated sites and patterns compared to those in plants unaffected by pollutants. Experiments can also be conducted to explore what pollutants and at what level are causing the changes in the organism. Furthermore, small discrete changes can be indicative of low contamination levels and can serve as early warning signs (12).

TRACE METAL POLLUTION IN THE AQUATIC ENVIRONMENT

Pollution is a global crisis and affects both terrestrial and aquatic environs. In aquatic ecosystems, pollution can enter through various vectors. Agricultural and industrial pollutants enter aquatic ecosystems through point sources and nonpoint sources. Atmospheric deposition, runoff, groundwater seepage, precipitation, and direct discharge are the pathways by which pollutants enter aquatic ecosystems. Trace metals are important pollutants that infiltrate aquatic ecosystems. Because many of them are essential nutrients for biota at low concentrations in certain forms and deadly at high concentrations in other forms (13), trace metals can affect aquatic ecosystems in a way akin to that of phosphorus and nitrogen.

Trace metals, in excessive quantities, have serious health ramifications for humans and other organisms. In many areas of the world, where industrial pollution is less regulated and sophisticated water treatment facilities are not in place, trace metal pollution can significantly impact drinking water and food supplies (14,15). The effects of certain trace metal species or high concentrations of essential trace metals on human health include and are not limited to: increased rates of cancer, allergic reactions, behavioral problems, lower cognitive function, lowered immunity, reduced fertility, and central nervous system dysfunction (13,16–20). Therefore, it is imperative that water resources utilized for recreation, irrigation, and drinking water supply be investigated for potential impacts of trace metal contamination on human health and ecosystem function.

AQUATIC MACROPHYTES AS BIOMONITORS

Aquatic macrophytes are prime candidates as biomonitors of trace metals in aquatic ecosystems. Their ability to accumulate trace metals many times the background levels in the environment is a significant instrument for assessing the availability of trace metals to biota (14,21–23). Aquatic macrophytes can amass trace metals by accumulation of metal ions in tissues through the roots, shoots, and leaves or through adsorption of metals to the surface of plants (21,22,24–27). The ability to bioaccumulate in roots, shoots, and leaves, termed hyperaccumulation, allows aquatic plants to be utilized as biomonitors in three distinct ways.

The tissues of aquatic macrophytes can be analyzed for trace metal concentration. Through a rather elaborate sample preparation process, with special attention to avoid introducing outside contaminants to the sample, tissue can then be tested for the concentrations of various metals generally using atomic absorption spectrophotometry (8,22,25,27–29). The direct testing of metal concentration in macrophyte tissues delivers a quantitative value regarding the accumulation potential of a given plant and can help identify the potential impact of trace metals to biota in the ecosystem (30,31). This is of primary importance when monitoring aquatic ecosystems utilized for agricultural and drinking water purposes (14). Direct testing also allows for quantification of accumulated trace metals in plant parts. This information is especially useful for plants that may be consumed. Active *in situ* experiments are preferred over passive ones by allowing temporal impacts and controlling background concentration (2).

However, that is not to say that direct tissue testing is not without its difficulties. The cost of directly testing plant tissues can be extraordinarily high and the results may not be indicative of current ecosystem conditions (32). Time-dependent accumulation and adsorption must also be taken into account (24,32). Background metal concentration also impacts uptake potential for aquatic plants. As such, *in situ* experiments must monitor sediment and water column metal concentrations and pay special attention to local environmental conditions. Finally, metal concentration is influenced by pH and other factors relating to sediment/soil composition (33).

Beyond direct testing for metal concentration in plant tissues, there exists the potential to indirectly monitor the concentration of metal through its impact on plants. Trace metals can be indirectly monitored by analyzing biochemical changes in aquatic macrophytes. Various trace metals, cadmium (Cd), lead (Pb), copper (Cu), and chromium (Cr), have been shown to reduce carbohydrate, protein, and chlorophyll production in aquatic macrophytes (see Fig. 1) (10,34). Reduction in essential nutrient uptake has also been observed in plants affected by trace metals (35), which may account for the decrease in production of other compounds in plants. Additionally, biomarkers of metal stress can indicate the presence of metal, and through research a stress-response pattern can be elucidated (36). Toxicity studies are quite similar in approach and may also help estimate levels of trace metal contamination that significantly reduce survivorship in aquatic macrophytes. Laboratory studies are far more reliable sources of information regarding indirect effects since *in situ* experiments may be confounded by synergistic effects of various compounds in the water column or other unidentifiable factors (32).

The third mechanism for testing the effects of trace metals on aquatic macrophytes is through studies of morphological changes. Changes in plant height, biomass, shoot length, and structural symmetry may all be indicative of metal stress (7,11,35,37,38). The goal in biomonitoring is to relate the degree of morphological changes in organisms to metal concentrations in the environment and available to biota. Thus, the long-term goal of utilizing morphological changes is linked to direct testing. Morphological monitoring is significantly limited,

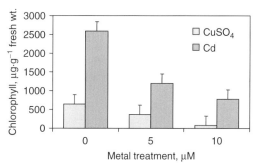

Figure 1. Changes in chlorophyll concentration as a function of trace metal treatment after eight days (CuSO$_4$, copper sulfate; Cd, cadmium). (Adapted from Refs. 10 and 34.)

much like indirect testing, to laboratory media where metal-specific responses can be controlled.

HYPERACCUMULATION POTENTIAL OF VARIOUS AQUATIC MACROPHYTES

Laboratory and *in situ* experiments on the hyperaccumulation capability of various aquatic macrophytes have yielded a wide range of results. Figure 2 exhibits the maximum observed accumulation potential of various metals by aquatic macrophytes. While aquatic macrophytes as a whole may accumulate significant levels of trace metals from the water column and sediments, no single species accumulates all metals well enough to be utilized as a single biomonitor for aquatic ecosystems (39). Figure 2 also indicates the maximum accumulation potential of various trace metals by different aquatic macrophytes. Please note that the greatest concentration of different metals has been observed in different aquatic macrophytes. However, the concentrations of metal in the tissue are not adjusted for background concentration.

While different aquatic macrophytes accumulate metals in different concentrations, it also has been found that different tissues vary in metal accumulation. In general, roots hyperaccumulate the greatest concentration of trace metals (28,30). Additional studies have shown metal concentrations in inflorescence and fruiting bodies of plants to be at startling levels especially considering their importance as food for both humans and other organisms (Fig. 3) (14,40).

Since roots are shown to be the greatest accumulators of heavy metals, it would seem obvious that rooted aquatic

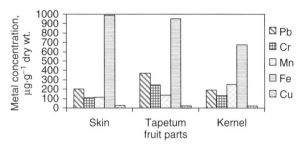

Figure 3. Accumulation of various metals in the edible plant *Trapa natans* (Pb, lead; Cr, chromium; Mn, manganese; Fe, iron; Cu, copper). (Adapted from Ref. 14.)

macrophytes would be preferred to floating macrophytes. However, this is not true in all cases. Floating macrophytes of the genus *Lemna* are utilized frequently in phytotoxicity studies (34,41). Ease of use and consistency with previous studies often influence the species of macrophytes to be used, especially for laboratory studies. *In situ* studies generally rely on the available local species, although submerged rooted species of the genus *Potamogeton* almost always have been investigated.

MACROPHYTES AS BIOMONITORS OF INDUSTRIAL POLLUTANTS

Aquatic macrophytes are already being utilized as keystone monitors of industrial pollutants in various bodies of water (15,21,22,42). For instance, aquatic macrophytes have been used to monitor pollutants in discharge of water from mines and ore processing facilities. Aquatic macrophytes, living in such highly polluted places, have adapted to extreme conditions and therefore are more likely to survive than any new species introduced as a monitor.

Aquatic macrophytes are also utilized as biomonitors of drinking water in parts of India (15). In instances where the indigenous population relies on streams and lakes for drinking water, industrial pollution can significantly harm the human population. In such cases, the concentration of metal in aquatic macrophytes can indicate to resource managers and decision-makers that something must be done to protect the residents. Direct testing of aquatic macrophytes can also indicate to farmers that irrigating crops from a specific body of water may have ramifications on human health down the line. Additionally, aquatic macrophytes may be a vital link in the foodweb dynamics of a region and significantly impact the health of animals as well as humans (22).

FUTURE RESEARCH OPPORTUNITIES

Before aquatic macrophytes can be fully employed as biomonitors of trace metals, various avenues of research must be undertaken. Direct correlations between biochemical changes and metal concentration in both plant tissues and background environments must be fully realized through intensive research. Synergistic effects of various metals should also be assessed. Studies of

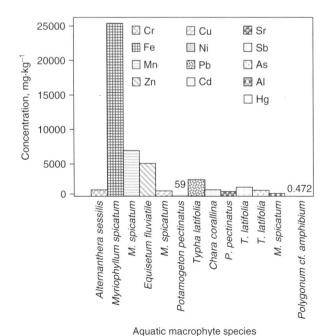

Figure 2. Maximum accumulation of trace metals by aquatic macrophytes (Cr, chromium; Fe, iron; Mn, manganese; Zn, zinc; Cu, copper; Ni, nickel; Pb, lead; Sr, strontium; Sb, antimony; As, arsenic; Al, aluminum; Hg, mercury). (Adapted from Refs. 15, 21, 22, and 25.)

morphological changes due to heavy metal accumulation can be undertaken in both laboratory and *in situ* settings. Additionally, time-dependent mechanisms relating to hyperaccumulation of trace metals and their impacts on morphology and biochemical processes should be investigated. Those utilizing biomonitoring for practical purposes must know how long macrophytes retain metals in their tissues and if they are re-released into the water column in any significant way. Studies should also take into account the implication of introducing new aquatic macrophyte species as biomonitors and their potential impact on the aquatic ecosystem in general.

IMPORTANCE OF BIOMONITORING TO OTHER AREAS OF SCIENTIFIC RESEARCH

Biomonitoring of aquatic macrophytes can provide additional insight into other areas of science. The information on hyperaccumulation of heavy metals has a profound impact on the use of plants in bioremediation (8,15,42). Plants may be of vital use in extracting pollutants from a given site. Additional research on the re-release of trace metals back into the water column and sediment will have tremendous value for developing bioremediation programs as well.

Furthermore, understanding biomonitoring potential for aquatic macrophytes can also provide insight into the mechanisms and factors that influence the metal ion accumulation in aquatic plants (43). Metal accumulation by macrophytes is also a significant component of metal cycling in aquatic ecosystems (44). Thus, the scientific knowledge gained through research on aquatic macrophytes as biomonitors may have far-reaching impacts on other important realms of scientific exploration.

BIBLIOGRAPHY

1. Small, A.M. et al. (1996). A macrophyte-based rapid biosurvey of stream water quality: restoration at the watershed scale. *Restoration Ecol.* **4**: 124–145.
2. Lovett Doust, J., Schmidt, M., and Lovett Doust, L. (1994). Biological assessment of aquatic pollution: a review, with emphasis on plants as biomonitors. *Biol. Rev.* **69**: 147–186.
3. Haury, J. (1996). Assessing functional typology involving water quality, physical features and macrophytes in a Normandy river. *Hydrobiologia* **340**: 43–49.
4. King, S.A. and Buckney, R.T. (2000). Urbanization and exotic plants in northern Sydney streams. *Austral. Ecol.* **25**: 455–461.
5. Thiebaut, G., Guerold, F., and Muller, S. (2002). Are trophic and diversity indices based on macrophyte communities pertinent to monitor water quality? *Water Res.* **36**: 3602–3610.
6. Newbold, C. and Holmes, N.T.H. (1987). Nature conservation: water quality criteria and plants as water quality monitors. *Water Pollut. Control* **86**: 345–364.
7. Mal, T.K., Adorjan, P., and Corbett, A.L. (2002). Effect of copper on growth of an aquatic macrophyte, *Elodea Canadensis. Environ. Pollut.* **120**: 307–311.
8. Sinha, S., Saxena, R., and Singh, S. (2002). Comparative studies on accumulation of Cr from metal solution and tannery effluent under repeated metal exposure by aquatic plants: its toxic effects. *Environ. Monitor. Assess.* **80**: 17–31.
9. Phillips, D.J. (1978). Use of biological indicator organism to quantitate organochlorine pollutants in aquatic environments—a review. *Environ. Pollut.* **16**: 167–229.
10. Babu, T.S. et al. (2003). Similar stress responses are elicited by copper and ultraviolet radiation in the aquatic plant *Lemna gibba*: implication of reactive oxygen species as common signals. *Plant Cell Physiol.* **44**: 1320–1329.
11. Mal, T.K., Uveges, J.L., and Turk, K.W. (2002). Fluctuating asymmetry as an ecological indicator of heavy metal stress in *Lythrum salicaria. Ecol. Indicators* **1**: 189–195.
12. Lovett Doust, L., Lovett Doust, J., and Schmidt, M. (1993). Forum: in praise of plants as biomonitors—send in the clones. *Funct. Ecol.* **7**: 754–458.
13. NCEH (National Center for Environmental Health). (2002). *Toxicity of Heavy Metals and Radionuclides*. Centers for Disease Control, Atlanta. Available at http://www.cdc.gov/nceh/radiation/savannah/SRSHES_Toxicity_jan02.htm. Accessed on 10/04/2004.
14. Rai, U.N. and Sinha, S. (2001). Distribution of metals in aquatic edible plants: *Trapa natans* (Roxb.) Makino and *Ipomoea aquatica* Forsk. *Environ. Monitor. Assess.* **70**: 241–252.
15. Rai, U.N., Sinha, S., and Chandra, P. (1996). Metal biomonitoring in water resources of Eastern Ghats, Koraput (Orissa), India by aquatic plants. *Environ. Monitor. Assess.* **43**: 125–137.
16. Sta, F. et al. (1997). Changes in natural killer cell subpopulations in lead workers. *Int. Arch. Occup. Environ. Health* **69**: 306–310.
17. Tong, S. et al. (1996). Lifetime exposure to environmental lead and children's intelligence at 11–13 years: the PortPirie cohort study. *Br. Med. J.* **312**: 1569–1578.
18. Undeger, U., Basaran, N., Canpinar, H., and Karsu, E. (1996). Immune alterations in lead exposed workers. *Toxicology* **109**: 167–172.
19. Sokol, R.Z., Okuda, H., Nagler, H.M., and Berman, N. (1994). Lead exposure *in vivo* alters the fertility potential of sperm *in vitro. Toxicol. Appl. Pharmacol.* **124**: 310–316.
20. Assennato, G. et al. (1986). Sperm count suppression without endocrine dysfunction in lead exposed men. *Arch. Environ. Health* **41**: 387–390.
21. Samecka-Cymerman, A. and Kempers, A.J. (2004). Toxic metals in aquatic plants surviving in surface water polluted by copper mining industry. *Ecotoxicol. Environ. Safety* **59**: 64–69.
22. Hozhina, E.I., Khramov, A.A., Gerasimov, P.A., and Kumarkov, A.A. (2001). Uptake of heavy metals, arsenic and antimony by aquatic plants in the vicinity of ore mining processing industries. *J. Geochem. Explor.* **74**: 153–162.
23. Zurayk, R., Sukkariyah, B., and Baalbaki, R. (2001). Common hydrophytes as bioindicators of nickel, chromium and cadmium pollution. *Water Air Soil Pollut.* **127**: 373–388.
24. Keskinkan, O., Goksu, M.Z.L., Basibuyuk, M., and Forster, C.F. (2004). Heavy metal adsorption properties of a submerged aquatic plant (*Certaophyllum demersum*). *Bioresour. Technol.* **92**: 197–200.
25. Zakova, Z. and Kockova, E. (1999). Biomonitoring and assessment of heavy metal contamination of streams and reservoirs in the Dyje/Thaya River basin, Czech Republic. *Water Sci. Technol.* **39**: 225–232.

26. Siebert, A. et al. (1996). The use of the aquatic moss *Fontinalis anitpyretica* L. ex. Hedw. as a bioindicator for heavy metals. *Sci. Total Environ.* **177**: 137–144.
27. Lewander, M., Greger, M., Kaustky, L., and Szarek, E. (1996). Macrophytes as indicators of bioavailable Cd, Pb, and Zn flow in the river Przemsza, Katowice Region. *Appl. Geochem.* **11**: 169–173.
28. Cardwell, A.J., Hawker, D.W., and Greenway, M. (2002). Metal accumulation in aquatic macrophytes from southeast Queensland, Australia. *Chemosphere* **48**: 653–663.
29. Markett, B. (1995). Sample preparation (cleaning, drying, homogenization) for trace element analysis in plant matrices. *Sci. Total Environ.* **176**: 45–61.
30. Baldantoni, D. et al. (2004). Assessment of macro and microelement accumulation capability of two aquatic plants. *Environ. Pollut.* **130**: 149–156.
31. Stankovic, Z., Pajevic, S., Vuckovic, M., and Stojanovic, S. (2000). Concentrations of trace metals in dominant aquatic plants of the Lake Provala (Vojvodina, Yugoslavia). *Biol. Plantarium* **43**: 583–585.
32. Seaward, M.R.D. (1995). Use and abuse of heavy metal bioassays in environmental monitoring. *Sci. Total Environ.* **176**: 129–134.
33. Sparling, D.W. and Lowe, T.P. (1998). Metal concentrations in aquatic macrophytes as influenced by soil and acidification. *Water Air Soil Pollut.* **108**: 203–221.
34. Mohan, B.S. and Hosetti, B.B. (1997). Potential phytotoxicity of lead and cadmium to *Lemna minor* grown in sewage stabilization ponds. *Environ. Pollut.* **98**: 233–238.
35. Singh, R.P. et al. (1997). Response of higher plants to lead contaminated environment. *Chemosphere* **34**: 2467–2493.
36. Keltjens, W.G. and van Beusichem, M.L. (1998). Phytochelatins as biomarkers for heavy metal stress in maize (*Zea mays* L.) and wheat (*Triticum aestivum* L.): combined effects of copper and cadmium. *Plant Soil* **203**: 119–126.
37. Uveges, J.L., Corbett, A.L., and Mal, T.K. (2002). Effects of lead contamination on the growth of *Lythrum salicaria* (purple loosestrife). *Environ. Pollut.* **120**: 319–323.
38. Nedelkoska, T.V. and Doran, P.M. (2000). Characteristics of heavy metal uptake by plant species with potential for phytoremediation and phytomining. *Minerals Eng.* **13**: 549–561.
39. Schlacher-Hoenlinger, M.A. and Schlacher, T.A. (1998). Differential accumulation patterns of heavy metals among the dominant macrophytes of a Mediterranean seagrass meadow. *Chemosphere* **37**: 1511–1519.
40. Villar, C. et al. (1999). Trace metal concentrations in coastal marshes of the Lower Parana River and the Rio de la Plata Estuary. *Hydrobiologia* **397**: 187–195.
41. Blinova, I. (2004). *Use of Freshwater Algae and Duckweeds for Phytotoxicity Testing.* Wiley-Interscience, Hoboken, NJ. Available at http://www.interscience.wiley.com. Accessed 09/14/2004.
42. Karathanasis, A.D. and Johnson, C.M. (2003). Metal removal potential by three aquatic plants in an acid mine drainage wetland. *Mine Water Environ.* **22**: 22–30.
43. Nigam, K.D.P., Srivastav, R.K., Gupta, S.K., and Vasudevan, P. (1998). A mathematical model for metal ion uptake by aquatic plants for waste water treatment. *Environ. Model. Assess.* **3**: 249–258.
44. Jackson, L.J. (1998). Paradigms of metal accumulation in rooted aquatic vascular plants. *Sci. Total Environ.* **219**: 223–231.

BIOSORPTION OF TOXIC METALS

M.X. LOUKIDOU
A.I. ZOUBOULIS
Aristotle University of Thessaloniki
Thessaloniki, Greece

GENERAL ASPECTS

Metals may be disseminated in the environment "naturally" by geochemical and biological activities, but anthropogenic actions today produce a greater input. Several industrial activities, such as the mining, electroplating, and metal finishing industries, have intensified environmental pollution problems, as well as the discharge of heavy metals in the environment. It is well known that metals are not biodegradable and can be transformed from one chemical form to another. Growing attention is being given to the potential hazard shown by their increased concentrations in the environment because their accumulation in living tissues throughout the food chain may pose a serious human health problem. Therefore, metal-laden waste sources should be effectively treated (1).

Different factors, such as the degree of environmental risk assessment, the probable future increase in market price, and the depletion rate of reserves are usually applied to set them into different categories—high, medium, or low priority. Under public and media pressure, the European Union and the U.S. EPA have introduced and progressively enforced stricter regulations on metal discharges, particularly for effluents from industrial operations. Currently available techniques for treating metal-bearing effluents, including mainly precipitation, evaporation, adsorption, ion exchange, membrane processing, and solvent extraction, are not effective enough, or rather expensive and occasionally inadequate, considering the large wastewater quantities produced. The main performance characteristics of common heavy metal wastewater treatment technologies are summarized in Table 1 (2).

The method that consists of oxidation/reduction, chemical precipitation, and solid/liquid separation has been traditionally the most commonly applied. Depending on the volumes involved, this treatment can be operated either in batch or continuous mode. It is also essential to control carefully the addition of chemical agents, as well as the pH. One of the main problems of this method is the production of toxic sludge; it transforms an aquatic pollution problem into another of subsequent safe solid waste disposal.

As a consequence, research efforts are intensive to apply more efficient and cost-effective treatment technologies. Biosorption promises to fulfill these requirements because it is a competitive, effective, and rather cheap method.

Generally, biosorption is a treatment process that uses inexpensive biomass, usually by-product, to sequester toxic metals from contaminated effluents. Biosorbents can

Table 1. Performance Characteristics of Common Heavy Metal Removal Technologies to Treat Industrial Wastewater

Method	Disadvantages	Advantages
Chemical precipitation	• Difficult separation • Disposal of resulting toxic sludge • Not very effective	• Simple • Relatively cheap
Electrochemical treatment	• Applicable for high metal concentrations • Sensitive to specific conditions, such as the presence of certain interfering compounds	• Metal recovery
Reverse osmosis	• Application of high pressures • Membrane scaling/fouling • Expensive	• Pure effluent/permeate (available for recycling)
Ion exchange	• Sensitive to the presence of particles • Expensive resins	• Effective • Possible metal recovery
Adsorption	• Not very effective for certain metals	• Conventional sorbents

be prepared from natural abundant biomass sources, such as seaweeds, or from waste biomass of algae, fungi, or bacteria (Fig. 1) (3–6).

The unique capabilities of certain types of biomass to concentrate and immobilize particularly heavy metals can also offer some degree of selectivity. This property depends on the following parameters:

- the specific type of biomass,
- the mixing conditions of the suspension,
- the type of biomass preparation, and
- the specific physicochemical conditions.

Broad-range biosorbents can collect most of the heavy metals from an aqueous solution (wastewater), presenting rather a smaller degree of selectivity among them. A specific metal can be separated either during sorption uptake, by manipulating the properties of a biosorbent, or upon desorption during the regeneration cycle of the biosorbent.

Metal uptake by living or dead cells can consist of two different modes (5). The first, independent of metabolic activity, is referred to as *biosorption* or *passive uptake*. The second mode of metal uptake into the cell (and across the cell membrane) depends on cell metabolism and is referred to as *bioaccumulation, active uptake*, or *intracellular uptake*. Both living and dead cells can take up metals. The use of dead cells seems to be the preferred alternative for the majority of studies. The wider acceptability of dead cells is devoted to the absence of toxicity limitations, as well as of requirements for growth media and nutrient supplements in the solutions to be treated. The biosorbed metals can also be easily desorbed, creating a more concentrated solution, from which metals can be recovered for subsequent reuse by using convenient technologies, such as electrolysis; simultaneously, the regenerated biomass can also be

Figure 1. SEM microphotograph of **(a)** *Penicillium chrysogenum* fungi and **(b)** *Saccharomyces cerevisiae* yeast, commonly used as biosorbents.

Table 2. The Main Characteristics of Metal Biosorption and Bioaccumulation

Feature	Biosorption	Bioaccumulation
Metal affinity	High under favorable conditions	Toxicity will affect metal uptake
Rate of metal uptake	Rapid	Slower than biosorption
Selectivity	Poor	Better than biosorption
Temperature tolerance	Within a modest range	Inhibited by relatively low or high temperatures
Versatility	pH dependent, affected by anions or other molecules	Requires an energy source; dependent on plasma membrane

reused. The main characteristics of metal biosorption and bioaccumulation processes are presented in Table 2.

Biological processes to remove toxic metals can be more effectively used to treat metal solutions of low concentration (in the range of few mg/L), as a final polishing step to remove any residual metal. Therefore, the appropriate pretreatment of metal-laden wastewater containing elevated concentrations (in the range of several mg/L) is considered necessary and is usually done by common methods, such as chemical precipitation.

Although the use of living organisms is often successful for treating even highly toxic organic contaminants, living microorganisms in conventional biological treatment systems have not been found particularly useful for treating wastewater containing heavy metal ions. When the metal concentration becomes high enough, then the metabolism of the microorganism is disrupted, causing its death. The defense mechanisms of living microorganisms are responsible for the observed decrease of metal sorption. This disadvantage is not operative when nonliving microorganisms, or biological materials derived from microorganisms, are used to remove metal ions from dilute aqueous solutions.

DESORPTION

Whenever the biosorption process is used as an alternative treatment method for metal-laden wastewater, regeneration of loaded biosorbent is an important issue, to keep the process costs low and to establish the possibility of recovering the metal extracted from the liquid phase (4,7). The desorption process could result in a more concentrated metal solution, keeping the metal uptake high upon subsequent reuse of the biomass and when appropriately performed, without causing physical changes in or damaging the biosorbent. The use of dilute acids was found quite effective, and careful control of pH may also provide a simple means to increase the selectivity of metal recovery. Carbonates and/or bicarbonates were also suggested as desorptive agents; they appear to have the most commercial potential, though carbonates may sometimes affect the biomass structure.

The selection among possible metal recovery methods, subsequently applied to the resulting concentrated metal solutions, depends on the ease of metal removal from the biomass and on the commercial (or environmental) value of the metal or the biomass, as a reusable entity. When the waste biomass is relatively cheap or when it is used to accumulate valuable metals, destructive recovery by incineration or strong acid/alkali dissolution of the biomass may be considered economically feasible. The recovery method also depends on the mechanism of metal accumulation. Biosorption is often reversible, whereas metabolism-dependent intracellular accumulation is usually irreversible and, hence, necessitates the use of destructive treatment to recover metals.

MECHANISMS OF METAL BIOSORPTION

The complexity of a microorganism's structure implies that there are many different ways for the metal to be captured by a (living or dead) cell. Therefore, biosorption mechanisms can be varied, and in some cases, they are still not very well understood. They may be divided into several categories, according to the dependence on cell metabolism, metabolism-dependent and nonmetabolism-dependent, as well as on the location, where the metal will be removed from the solution, through extracellular accumulation/precipitation, cell surface sorption/precipitation, or intracellular accumulation, as shown in Fig. 2 (3).

Transport of the metal across the cell membrane yields intracellular accumulation, which depends on cell metabolism; this implies that this kind of biosorption may take place only with the viable cells. It is often associated with the active defense system of microorganisms, which react in a toxic environment (in this case, the metal).

In physicochemical interaction between a metal and the functional groups on a cell surface that takes place during physical adsorption, ion exchange, and/or surface complexation mechanisms, sorption onto a cell surface would not depend on metabolism. Cell walls of microbial biomass, composed mainly of polysaccharides, proteins, and lipids, can offer several metal-binding functional groups, such as carboxylate, hydroxyl, sulfate, phosphate, and amino groups. Biosorption of metals through physicochemical interaction occurs relatively rapidly and can be reversible. In this case, biomass has the major characteristics of an ion exchange resin, or of an activated carbon, implying certain advantages, such as lower cost, in the industrial application of biosorption.

Precipitation of metals may depend on cellular metabolism or may be independent of it. Metal removal from solution is often associated with the active defense system of microorganisms. Metal removal by precipitation, which does not depend on cellular metabolism, may be a consequence of the chemical or physical interaction

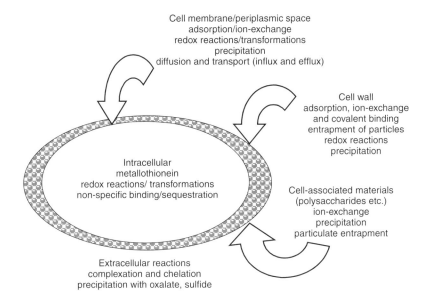

Figure 2. Several processes that contribute to microbial uptake and detoxification of aqueous solutions containing toxic metals.

between the metal and the cell surface. It becomes evident that the biosorption mechanisms are varied, and they can also take place simultaneously.

Generally, biosorption by living microorganisms consists of two basic steps. First, an independent metabolism binding step to cell walls and second, metabolism-dependent intracellular uptake, whereby metal ions are transported across a cell membrane into the cell (8). Cadmium, zinc, copper, and cobalt biosorption by the dead biomass of algae, fungi, and yeasts takes place mainly through electrostatic interactions between the metal ions in solution and cell walls. Cell walls of microorganisms contain polysaccharides as basic building blocks. The ion exchange properties of natural polysaccharides have been studied in detail, and it is well established that bivalent metal ions may exchange with the counterions of polysaccharides. Alginates of marine algae usually occur as natural salts of K^+, Na^+, Ca^{2+}, and/or Mg^{2+}. These metallic ions can be exchanged with counterions, such as Co^{2+}, Cu^{2+}, Cd^{2+}, and Zn^{2+}, resulting in biosorptive uptake of the metals. Knowledge of metal biosorption mechanisms could lead to the development of economically attractive sorbent materials.

MODELING

Mathematical modeling of biosorption offers an extremely powerful tool for a number of tasks on different levels; it is essential for process design and optimization. The biosorptive capacity and required contact time are two of the most important parameters. Sorptive equilibrium is established when the concentration of sorbate in the bulk solution is in dynamic balance with that on the interface. An equilibrium analysis is the most important information required to evaluate the affinity or capacity of a biosorbent. However, thermodynamic data can predict only the final state of a system from an initial nonequilibrium mode. It is important to determine how sorption rates depend on the concentrations of solution and how these rates are affected by sorptive capacity. From kinetic analysis, the solute uptake rate, which determines the residence time required to complete the sorption reaction, may be established (9). Equilibrium and kinetic analyses allow estimation of sorptive capacities and rates and also lead to suitable rate expressions that are characteristic of possible reaction mechanisms.

A wide range of equilibrium-based models has been used to describe the sorption of metals onto solid particles. They include the common isotherm equations, such as those developed by Langmuir and Freundlich. The Langmuir equation models monolayer sorption onto surfaces that contain a finite number of accessible sites:

$$q_e = \frac{q_{\max} b C_e}{1 + b C_e} \quad (1)$$

where $q_{\max}$ is the maximum quantity of metal ions per unit weight of biomass to form a complete monolayer on the surface (mg/g) and b is a constant related to the affinity of binding sites for the metal ions (sorbate, L/mg). Note that $q_{\max}$ represents a practical limiting adsorptive capacity corresponding to the surface of a sorbent fully covered by metal ions. This quantity is particularly useful for assessing adsorption performance, especially when the sorbent does not reach its full saturation, because it enables an indirect comparison between different sorbents (10).

The empirical Freundlich equation accounts macroscopically for sorption onto heterogeneous surfaces:

$$q_e = K_F C_e^{1/n} \quad (2)$$

where K_F is an indicator of adsorptive capacity (L/g) and n is adsorptive intensity (dimensionless) (10).

Table 3 lists several types of biomass examined for their ability to remove Cd(II), Ni(II), Cu(II), and Cr(VI), which are metals considered of high environmental priority. The values of specific metal uptake q (mg of metal biosorbed per gram of biomass) are quite high in several cases

Table 3. Selected Literature Survey: Comparison of Different Biosorbents for Cd(II), Cu(II), Ni(II), and Cr(VI) Removal Capacity

Metal	Biosorbent	q_{max}, mg/g	pH	T, °C	Biomass Concentration, g/L	Reference
Cd(II)	B. licheniformis	142.7	7	20	1	11
	B. laterosporus	159.5	1	25	1	11
	Sphaerotilus natans	115	6	25	0–1	3
	Sargassum fluitans	101.4	7	25	—	3
	Rhizopus arrhizus	26.8	7	26	3	11
	Saccharomyces cerevisiae	0.56	5	25	2	8
Cu(II)	Chlorella vulgaris	37.6	4	25	0.75	16
	Spirulina platensis	10	6	25	1	3
	Synechocystis sp.	23.4	4.5	25	1	3
Cr(VI)	Chlorella vulgaris	24	2	25	1	11
	Zooglera ramigera	3	2	25	—	3
	B. laterosporus	72.6	1	25	1	11
	Rhizopus arrhizus	62	2	25	1	13
	Rhizopus nigricans	123.45	2	25	1	13
Ni(II)	Ascophyllum nodosum	70	6	25	1	16
	Chlorella vulgaris	42.3	5	25	1	16
	Synechocystis sp.	15.8	5	25	1	16

and showed that selected biosorbents can be effective for removing heavy metals.

Sorption kinetics may be controlled by several independent processes that can act in series or in parallel, such as (i) bulk diffusion, (ii) external mass transfer (film diffusion), (iii) chemical reaction (chemisorption), and (iv) intraparticle diffusion (9). Any of these four steps, or any combination among these steps may be the rate-controlling factor. Transport in the solution is sometimes the rate-determining parameter in large-scale field processes. At sufficiently high agitation speed in the reaction vessel, the bulk diffusion step can be safely ignored because then sorption onto sorbent particles is decoupled from mass transfer in the bulk mixture.

IMMOBILIZATION

In the industrial application of biosorption, an immobilization procedure for biosorbent materials is often necessary to improve subsequent solid/liquid separation. The immobilization of biomass within appropriate solid structures creates a new sorptive material that has the appropriate size, mechanical strength, rigidity, and porosity necessary for use in conventional unit operations, such as fixed beds, adsorption columns, or mixing tanks. The principal techniques for the application of biosorption are based on entrapment in a strong, but rather permeable matrix, such as encapsulation within a membrane-like structure and bonding of smaller particles. It has been recognized that immobilizing biomass in a polymer matrix may improve the biomass performance and capacity, and, simultaneously, there is no need for the solid–liquid separation of a metal-laden biomass from the treated solution (11). Many polymeric matrices have been used for biomass immobilization, including polysulfone, polyurethane, alginate, polyacrylamide, and polyethyleneimine (PEI) (3). Several modifications of biomass immobilization have been used for metal removal, such as Chlorella homosphaera immobilized by sodium alginate; Sargassum fluitans and Ascophyllum nodosum cross-linked with formaldehyde, glutaraldehyde, or embedded in polyethylene imine; sphagnum peat moss immobilized in porous polysulfonate beads; and Rhizopus arrhizus immobilized in reticulated polyurethane foam (5). Modified polyacrylonitrile (denoted as PAN) was also examined as an efficient binder of fungal biomass (dead cells of P. chrysogenum, a waste by-product during antibiotic production) (12).

BIOSORPTION OF HEAVY METAL ANIONS

Several studies were carried out and highlighted that biosorptive removal of metal cations from aqueous solution depends mainly on the interactions between metal ions and specific groups, which are associated with the biosorbent surface. The uptake of anions by biosorption has become a growing concern, because they cannot be easily removed as insoluble hydroxides or salts by the common application of chemical precipitation.

Several chemically modified sorbents were examined for their enhanced ability to bind anions (3). The selective modification of Rhizopus nigricans using cetyl trimethylammonium bromide was also examined; polyethyleneimine improved the biosorptive efficiency to high levels (13). The modification of P. chrysogenum with common cationic surfactants (amines) or with a cationic polyelectrolyte overcame the low capacity of biomass for treating arsenic-contaminated wastewater (14). Extensive studies were carried out and reported on the biosorption of molybdate or (hexavalent) chromium anions (15). The removal of these anions can be related to the various functional groups of biomass surface, such as amino groups or proteins, which can provide positive surface charges, therefore, causing the electrostatic attraction of anions.

BIOSORPTION TREATMENT PLANTS

Following equilibrium and dynamic sorption studies, the quantitative basis for the sorption process can be established, including process performance models. Biosorption process feasibility is assessed for well-selected cases. It is necessary to realize that there are mainly two types of pilot plants, which eventually can be run side by side (Fig. 3) (4):

- biomass processing pilot plant
- biosorption pilot plant

The process of biosorption for removing metals can show performance comparable to its closest commercially used competitor, ion exchange treatment. Effluent qualities of the order of only a few ppb (µg/L) of residual metal can be achieved. Commercial ion exchange resins are rather costly, and the price tag of biosorbents can be an order of magnitude cheaper, only 10% of the ion exchange resin cost. The main attraction of biosorption seems to be the cost-effectiveness of this process. Biosorption is in its early developmental stages, and further improvements in both performance and costs can be expected in the near future.

EXISTING TECHNOLOGIES

Several biosorption processes have been developed and introduced for removing metal contaminants from surface water and groundwater. The BIOCLAIM® process employs microorganisms, principally bacteria of the genus *Bacillus*, which have been treated with strong caustic solution to enhance the accumulation of metal, washed with water to remove residual caustic, and immobilized in extruded beads using polyethyleneimine (PEI) and glutaraldehyde (3). The AlgaSORB™ process is a proprietary family of products that consists of several types of nonliving algae in different immobilization matrices. The BIO-FIX® process includes several types of biomass, including sphagnum peat moss, algae, yeast, bacteria, and/or aquatic flora, which are immobilized in polysulfone. Immobilized *Rhizopus arrhizus* has also been evaluated for recovering uranium from an ore bioleach solution. The application of flotation for the subsequent separation of metal-laden biomass, when applied in suspension, was also suggested (17).

FUTURE ASPECTS

It should be evident from this brief survey that the possibilities for microbial metal removal and recovery are abundant. Biosorption is a multidisciplinary scientific area extending into several sciences, such as biology, chemistry, and engineering. The biosorption process could be used, even with a relatively low degree of understanding of its metal-binding mechanisms, but better understanding is necessary to increase the effectiveness and to optimize the relevant applications. Regarding technological development, scientific investigation should result in new families of products, as well as of biosorbents. The accompanying services should involve:

- assessing the type of problem to be remedied,
- assessing the applicability of biosorption,
- developing a customized biosorption process,
- designing and building the plant, and
- recovering the removed metals for subsequent reuse (4).

BIBLIOGRAPHY

1. Wase, J. and Forster, C. (1997). *Biosorbents for Metal Ions*. Taylor & Francis, London.
2. Eccles, H. (1999). Treatment of metal-contaminated wastes: Why select a biological process. *Trends Biotechnol.* **17**: 462–465.
3. Vegliò, F. and Beolchini, F. (1997). Removal of metals by biosorption: A review. *Hydrometallurgy* **44**: 301–316.
4. Volesky, B. (2001). Detoxification of metal-bearing effluents: Biosorption for the next century. *Hydrometallurgy* **59**: 203–216.
5. Volesky, B. (1990). *Biosorption of Heavy Metals*. CRC Press, Boca Raton, FL, pp. 84–301.
6. Kefala, M.I., Matis, K.A., and Zouboulis, A.I. (1999). Biosorption of cadmium ions by Actinomycetes and separation by flotation. *Environ. Pollut.* **104**: 283–293.
7. Gadd, G.M. (1990). Biosorption. *Chem. Ind.* **2**: 421–426.
8. Volesky, B. and Holan, Z.R. (1995). Biosorption of heavy metals—review. *Biotech. Prog.* **11**: 235–250.
9. Ho, Y.S., Ng, J.C.Y., and McKay, G. (2000). Kinetics of pollutant sorption by biosorbents: Review. *Sep. Purif. Methods* **29**: 189–232.
10. Yiacoumi, S. and Tien, C. (1995). *Kinetics of Metal Ion Adsorption from Aqueous Solutions*. Kluwer Academic, Boston.
11. Zouboulis, A.I., Loukidou, M.X., and Matis, K.A. (2004). Biosorption of toxic metals from aqueous solutions by bacteria strains isolated from metal-polluted soil. *Process Biochem.* **39**: 909–916.
12. Zouboulis, A.I., Matis, K.A., Loukidou, M., and Sebesta F. (2003). Metal biosorption by PAN-immobilized fungal biomass in simulated wastewaters. *Colloids Surf. A: Physicochem. Eng. Aspects* **212**: 185–195.

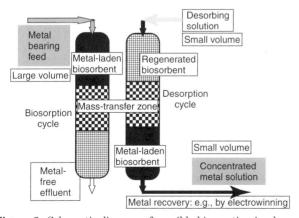

Figure 3. Schematic diagram of possible biosorption implementation, using packed bed columns for biosorption and desorption (4).

13. Bai, R.S. and Abraham, T.E. (2001). Biosorption of Cr(VI) from aqueous solution by *Rhizopus nigricans*. *Bioresource Technol.* **79**: 73–81.
14. Loukidou, M.X., Matis, K.A., Zouboulis, A.I., and Liakopoulou-Kyriakidou, M. (2003). Removal of As(V) from wastewaters by chemically modified fungal biomass. *Water Res.* **37**: 4544–4552.
15. Kratochvil, D. and Volesky, B. (1998). Advances in the biosorption of heavy metals. *Trends Biotechnol.* **16**: 291–300.
16. Donmez, G.C., Aksu, Z., Ozturk, A., and Kutsal, T. (1999). A comparative study on heavy metal biosorption characteristics of some algae. *Process Biochem.* **34**: 885–892.
17. Zouboulis, A.I., Lazaridis, N.K., and Matis, K.A. (2002). Removal of toxic metal ions from aqueous systems by biosorptive flotation. *J. Chem. Technol. Biotechnol.* **77**: 958–964.

BROMIDE INFLUENCE ON TRIHALOMETHANE AND HALOACETIC ACID FORMATION

SHANKAR CHELLAM
University of Houston
Houston, Texas

INTRODUCTION

General Background

Chemical disinfectants are added to drinking water primarily to kill or inactivate microorganisms, as discussed in other sections of this encyclopedia. However, an undesired and unintentional consequence of their use is that they react with natural organic matter (NOM) to form disinfection by-products (DBPs) (1,2). DBPs have been studied extensively since the first reports of trihalomethanes (THMs) and haloacetic acids (HAAs) in drinking water by Rook (3) and Christman et al. (4), respectively. THMs and HAAs constitute the two most important classes of DBPs measured on a weight basis in chlorinated drinking water (5). Because of length limitations, this section can only succinctly describe the important influence of the bromide ion on THM and HAA formation and speciation. The reader is directed to the references cited to obtain more detailed information. Additionally, only DBPs formed because of the application of free chlorine will be considered herein because most municipalities around the world use it for disinfection.

Simplistically, the formation of organic DBPs including THMs and HAAs can be written as follows:

$$\text{NOM} + \text{chlorine} + \text{bromide} \rightarrow \text{THMs} + \text{HAAs} + \text{other halogenated DBPs} \quad (1)$$

In the absence of the bromide ion, the only THM that can be formed is $CHCl_3$ (chloroform) whereas three HAAs, namely, chloroacetic acid (ClAA), dichloroacetic acid (Cl_2AA), and trichloroacetic acid (Cl_3AA), can be formed. However, bromide can be present in drinking water supplies due to natural geological occurrence, salt water intrusion, or human activities such as coal mine drainage or re-injection of oil field brine (6). When Br^- is present in water along with NOM, addition of chlorine can also result in formation of three additional THMs, namely, dichlorobromo methane, chlorodibromo methane ($CHClBr_2$), and bromoform ($CHBr_3$), and six additional HAAs, namely, bromo acetic acid (BrAA), dibromo acetic acid (Br_2AA), tribromo acetic acid (Br_3AA), bromochloro acetic acid (BrClAA), dichlorobromo acetic acid (Cl_2BrAA), and chlorodibromo acetic acid ($ClBr_2$AA). Hence, a total of four THMs and nine HAAs contain bromine and chlorine. As can be inferred from Equation 1, changes in the concentration of the reactants (including Br^-) will influence both the formation and relative speciation of the four THMs and nine HAAs.

Health Effects and Regulations

Epidemiological studies suggest a slightly increased risk of bladder, colon, and rectal cancers after long-term exposure to chlorinated water because of the presence of DBPs (7), warranting their designation as "primary" drinking water contaminants. However, partially because of limited data available for the occurrence, formation, and toxicology of the mixed chlorobromo acetic acids and Br_3AA, current federal regulations (8) simplistically address the sum of mass concentrations of THMs and only five HAAs (HAA5, viz, ClAA, Cl_2AA, Cl_3AA, BrAA, and Br_2AA) expressed in μg/L.

Literature reports indicate that increasing concentrations of the bromide ion cause a general shift from chlorinated toward brominated THMs (9–14), HAAs (15–18), as well as other DBPs classes (19). These studies have unequivocally established the formation of the unregulated mixed bromochloro HAA species (ClBrAA, Cl_2BrAA, and $ClBr_2$AA) and Br_3AA when chlorine is added to waters containing bromide. Importantly, recent evidence points to the increased cytotoxicity, mutagenicity, and genotoxicity of brominated acetic acids and nitromethanes compared with their chlorinated counterparts (20,21). Hence, an expressed need exists to delineate the role of bromide ion in controlling THM and HAA9 speciation, with the ultimate aim of reducing health risks associated with DBPs in drinking water.

CHEMISTRY OF BROMINATED THM AND HAA FORMATION

The relative proportion of individual THM and HAA species depends on several water chemistry factors including the concentrations of Br^- and NOM (quantified often as dissolved organic carbon, DOC), Cl_2 dosage, temperature, pH, and contact time (22). Additionally, NOM characteristics, such as polarity and molecular weight distribution, also influence THM and HAA speciation (23,24). Hence, seasonal variations in NOM characteristics, temperature, and bromide levels can be expected to induce changes in THM and HAA formation with season.

The oxidation of bromide by free chlorine (hypochlorous acid) to hypobromous acid is a very fast second-order reaction:

$$\text{HOCl} + \text{Br}^- \rightarrow \text{HOBr} + \text{Cl}^- \quad (2)$$

Note that this reaction also demonstrates that Br⁻ could exert a chlorine demand in natural waters. The simultaneous and parallel reactions of HOBr and HOCl with NOM results in the formation of halogenated DBPs. A higher fraction of HOBr ($pK_a = 8.8$) will remain undissociated at pH values commonly encountered in drinking water treatment than of HOCl ($pK_a = 7.5$), which allows increased formation of halogenated DBPs in bromide-containing water supplies. Starting from the early report of THM formation (3), bromine has been understood to be more reactive than chlorine in forming THMs and HAAs (13,14,16).

Trihalomethanes

Several investigations have shown that, for fixed NOM concentrations and chlorine doses, increasing bromide substantially increases total THM formation on a mass basis (9,12,13). This phenomenon is corroborated by the positive exponents for bromide ion in empirical THM models (25–29). Hence, Br⁻ is a THM precursor in addition to NOM. The increase in total THM levels is smaller when calculated on a molar basis because bromine has a higher atomic weight (79.9 g/mole) than does chlorine (35.5 g/mole).

Increasing Br⁻/DOC ratio also induces changes in speciation under most circumstances, wherein $CHCl_3$ mole fraction monotonically decreases and $CHBr_3$ mole fraction monotonically increases. In contrast, $CHCl_2Br$ and $CHClBr_2$ mole fractions first increase and then decrease. The chlorine dose has to be increased to adjust for increasing chlorine demand for higher DOC waters in order to obtain a constant residual at the end of the predetermined incubation time when performing tests under simulated distribution system (30) or uniform formation conditions (31). These changes in Br⁻/Cl₂ ratios also influence THM formation and speciation. Simultaneous changes in Br⁻/DOC and Br⁻/Cl₂ ratios may confound analysis of THM speciation and formation, even though the above-mentioned trends are generally observed (9,10,12–14). In general, high pH favors the formation of THMs (2,32) possibly because they are the end products of hydrolysis of several DBPs (33).

The degree of bromine substitution in THMs can be quantified for the bromine incorporation factor n (34):

$$n = \frac{\sum_{k=0}^{3} k \times [CHCl_{3-k}Br_k]}{\sum_{k=0}^{3} [CHCl_{3-k}Br_k]} \quad (3)$$

where each THM species is expressed in μM. If $CHCl_3$ is exclusively formed $n = 0$, whereas $n = 3$ corresponds to a situation when only $CHBr_3$ is formed. In most water treatment scenarios, $0 < n < 3$, which corresponds to the simultaneous formation of all or several THM species. As can be anticipated, increasing Br⁻, decreasing DOC, and decreasing Cl₂ dose generally result in increasing THM bromine incorporation factor (9,10,12–14).

Haloacetic Acids

In contrast to the numerous THM studies available in the literature, investigations on the role of bromide ion on HAA9 formation and speciation had been limited until relatively recently by the nonavailability of commercial standards for the mixed chlorobromo and tribromo species. Early HAA9 studies employing instrumental standards for several mixed chlorobromo species synthesized in academic laboratories revealed a gradual shift from the chlorinated to the mixed bromochloro and then to the brominated species with increasing Br⁻ concentration (15,16). However, unlike the THMs, the molar yield of HAA9 does not appear to strongly depend on Br⁻ (15,16), which indicates that Br⁻ may not necessarily be a precursor for HAAs. Studies that only monitored HAA5 or HAA6 (HAA5 and BrClAA) have indicated a decrease in the regulated HAAs with increasing Br⁻ (10,35), which should be interpreted with caution because shifts toward the unmonitored and more brominated species would cause this phenomenon rather than a decrease in the total molar concentration of all nine HAAs. These patterns are consistent with the negative exponent for Br⁻ in empirical power law functions derived to model HAA6 formation (26) and purely chlorinated HAAs (27) and positive exponent for BrAA and Br_2AA (27).

Monohalogenated HAAs (ClAA and BrAA) are typically formed at very low levels after chlorination of drinking water supplies, whereas the di- (Cl_2AA, ClBrAA, Br_2AA) and trihalogenated (Cl_3AA, Br_3AA, Cl_2BrAA, and $ClBr_2AA$) HAAs are the dominant species. Their relative dominance may depend on specific ultraviolet absorbance at 254 nm and pH (16,36). As a general rule, HAAs are preferentially formed at low pH values partially because trihalogenated HAAs undergo base catalyzed hydrolysis at high pH values (33).

Equation 3 can be extended to define a HAA9 bromine incorporation factor n':

$$n' = \frac{\begin{array}{c}[BrAA] + [BrClAA] + [BrCl_2AA]\\ +2 \times [Br_2ClAA] + 2 \times [Br_2AA] + 3 \times [Br_3AA]\end{array}}{\begin{array}{c}[ClAA] + [Cl_2AA] + [Cl_3AA] + [BrAA] + [BrClAA]\\ +[BrCl_2AA] + [Br_2ClAA] + [Br_2AA] + [Br_3AA]\end{array}} \quad (4)$$

Similar to the THM bromine incorporation factor, $0 \leq n' \leq 3$ and n' increases with the degree of bromine substitution.

EFFECTS OF TREATMENT PROCESSES ON SPECIATION

During drinking water purification, changes in the relative amounts of Br⁻ and DOC are most often brought about by decreases in NOM rather than by increases in Br⁻. In other words, the Br⁻/DOC ratio is increased by decreasing the denominator rather than by increasing the numerator, as in most previous investigations of bromide effects on THM and HAA formation and speciation (9,12,13,15,16,35). This result is because processes such as coagulation, precipitative softening, nanofiltration (NF), biological filtration, advanced oxidation, and granular activated carbon (GAC) treatment are capable of DBP precursor (or NOM) removal, whereas Br⁻ passes almost conservatively through them (37). These changes in Br⁻/DOC inherent to drinking water treatment processes will induce changes in THM and HAA formation and speciation as discussed next. Because of length limitations, only NF and GAC will be considered herein. Additional information on other treatment processes can be found elsewhere (1,33,38).

Effects of Granular Activated Carbon Treatment

Granular activated carbon treatment has been designated as one of the best available technologies for DBP precursor removal by the U.S. Environmental Protection Agency and is widely employed for NOM removal. Several excellent reviews of GAC use in drinking water treatment are available including its use for DBP control (39,40). However, changes in DBP speciation have not been emphasized in these reviews. Hence, this section focuses primarily on speciation issues and the reader is directed elsewhere for a more holistic understanding of GAC use for DBP control.

Figure 1 depicts several aspects of THM and HAA precursor removal and changes in speciation after GAC treatment for a pretreated surface water. The breakthrough curves corresponding to a gradual increase in effluent DOC and UV_{254} values are shown in Fig. 1a. Consequently, the effluent Br^-/DOC ratio decreases (Fig. 1e) because GAC treatment has very little effect on bromide ion. Figure 1e also depicts decreasing Br^-/Cl_2 decrease with progressive breakthrough of NOM, which is caused by increasing disinfectant demand, which necessitates higher Cl_2 doses to maintain a ~ 1 mg/L residual at the end of the 24 h incubation time (SDS testing conditions). These changes result in an increase in the formation of brominated THMs and HAAs in the effluent with operating time (Fig. 1b and 1c). Note that after ~ 3200 bed volumes, $CHBr_3$ concentrations were greater in the effluent than in the influent. Similar changes in THM speciation have been discussed earlier (41,42). Importantly, it should be noted that even though each individual THM and HAA increased on mass basis in the effluent (Fig. 1b and 1c), only the mole fraction of chloroform increased substantially (0.10 to 0.27). This result was accompanied by a significant decrease in $CHBr_3$ mole fraction (0.33 to 0.13). Mole fractions of $CHCl_2Br$ remained approximately constant, whereas that of $CHClBr_2$ increased slightly from 0.18 to 0.23. Of the unregulated HAAs, two mixed chlorobromo species (ClBrAA and Cl_2BrAA) were detected in both the GAC influent and effluent, whereas Br_3AA and $ClBr_2AA$ were never formed (probably because of the relatively low influent levels of $Br^- = 72$ µg/L). In the feed water ($Br^-/DOC = 5.8$ µM/mM), the speciation of the detectable HAAs was in the order: $Cl_2AA > BrClAA > Cl_3AA \sim Br_2AA > Cl_2BrAA$. After 4800 bed volumes ($Br^-/DOC = 13.4$ µM/mM), the speciation changed to $Cl_2AA \sim Br_2AA > BrClAA > Cl_2BrAA \sim Cl_3AA$ because the DOC was reduced to 0.8 mg/L with Br^- remaining at 72 µg/L. This change in Br^-/DOC ratio increased the mole fraction of Br_2AA from 0.13 in the influent water to 0.21 in the effluent. It should be noted that different speciation patterns should be expected if Br^- levels in the influent were higher.

Decreasing Br^-/DOC with GAC operation (increasing bed volumes) reduced both THM and HAA bromine incorporation factors as shown in Fig. 1(d). Another method of analyzing halogen incorporation is to calculate DBP-X/DBP, which refers to the molar concentration of Cl or Br incorporated in THMs or HAAs divided by

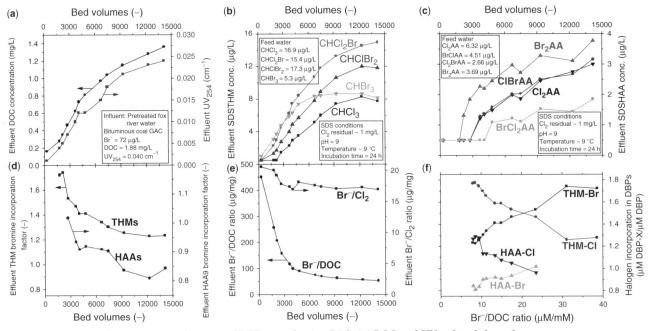

Figure 1. Several aspects of DBP control using GAC. (**a**) DOC and UV_{254} breakthrough curves; (**b**) and (**c**) formation of individual THMs and HAAs in the effluent, respectively; decreasing THM and HAA bromine incorporation factors (**d**) are caused by decreases in Br^-/DOC and Br^-/Cl_2 ratios (**e**). Increasing Br incorporation and decreasing Cl incorporation in GAC effluent for both THMs and HAAs with Br^-/DOC is shown in (**f**).

the total molar concentration of THM or HAA (Fig. 1f). Bromine incorporation into THMs and HAAs formed in GAC effluent was the highest at early times (high Br^-/DOC ratio) and decreased progressively with bed operation. As may be anticipated, the opposite trend was observed for chlorine incorporation. Bromine and chlorine incorporation in both THMs and HAAs were found to be equal at Br^-/DOC of ~20 µM/mM, corresponding to a Br^-/Cl_2 molar ratio of 15.5, which confirms that HOBr is more reactive than HOCl is in forming THMs and HAAs.

Effect of Nanofiltration

The early work by Taylor and co-workers simply documented that NF membranes were capable of very high NOM removal and capable of allowing municipalities to easily meet or exceed federal DBP regulations even if chlorine is employed for both primary and secondary disinfection (43). As demonstrated in more recent investigations (44–46), DBP precursor rejection decreases with increasing feed water recovery (Fig. 2a), which indicates that diffusion is the controlling mechanism of NOM and DBP precursor removal by NF membranes. Hence, increasing recovery reduces the Br^-/DOC ratio (Fig. 2f), which consequently increases the formation of the more chlorinated THMs ($CHCl_3$ and $CHCl_2Br$) while decreasing the more brominated ones ($CHClBr_2$ and $CHBr_3$) as depicted in Fig. 2c. Importantly, decreasing HAA9 and THM4 formation as recovery is lowered (increasing Br^-/DOC) indicates that NF waters are precursor limited.

Because NF achieved >90% removal of DOC (Fig. 2a) while allowing the almost complete passage of Br^-, a substantial increase in more brominated THM and HAA species was observed in the permeate compared with the feed water at a fixed operating recovery (70%) (Fig. 2b and 2e). Such speciation changes also decrease n and n' values with increasing recovery (Fig. 2d). Note that the mixed chlorobromo species were all detected in the permeate water. Similar to Fig. 1e, the Br^-/Cl_2 ratio also decreased with increasing DOC because greater Cl_2 doses were necessary to maintain a ~2 mg/L residual at the end of the 24 h holding time. Formation and speciation trends in Fig. 1 are qualitatively similar to those reported earlier (44,47).

Importantly, chlorination conditions (pH, temperature, incubation time, and dose) have been shown to influence THM and HAA speciation more than the membrane material and molecular weight cutoff or source water location (35,47).

Figure 3 depicts individual THM species formed in the permeate water normalized by the corresponding value in the feed water for three different NF membranes. Under the SDS conditions employed, NF preferentially reduced $CHCl_3$ and $CHCl_2Br$ concentrations, whereas NF increased concentrations of the more brominated species ($CHClBr_2$ and $CHBr_3$). Cumulatively, even though ~90% of THM precursors were removed by the same 400 Da molecular weight cutoff membrane shown in Fig. 1, it

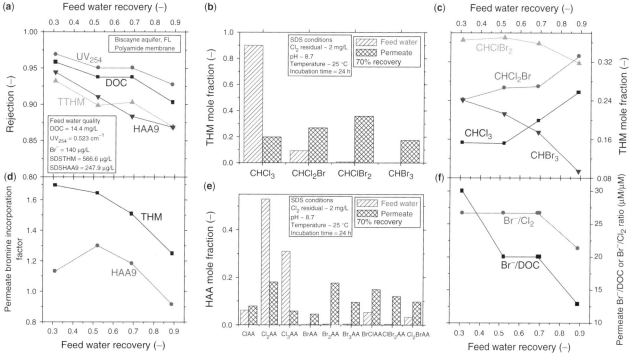

Figure 2. Several aspects of DBP control using NF technology. (**a**) Decreasing rejection of DOC, UV_{254}, THM, and HAA precursors with increasing recovery; (**b**) and (**e**) a general increase in the brominated THMs and HAAs in the permeate compared with the feed water; increasing DOC with recovery decreases both Br^-/DOC and Br^-/Cl_2 ratios (**f**), causing changes in individual THM formation (**c**). Decreases in Br^-/DOC and Br^-/Cl_2 ratios shown in (**f**) also cause a general decrease in THM and HAA bromine incorporation factors.

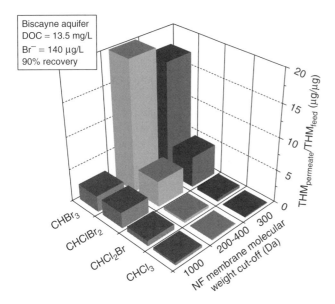

Figure 3. Effects of nanofiltration and membrane molecular weight cutoff on THM speciation for a naturally brominated groundwater measured under simulated distribution system conditions mentioned in Fig. 2.

also induced a fourfold increase in $CHClBr_2$ and a 20-fold increase in $CHBr_3$. As reported earlier (37), the shift toward the more brominated THMs was generally consistent with a decrease in MWCO, which increases Br^-/DOC ratio.

In summary, Figs. 1–3 as well as other investigations (44,47) unequivocally demonstrate the formation of the unregulated HAAs after advanced treatment. In fact, one study showed that the mole fraction of unregulated HAAs could be as high as $\sim$50% following NF (47). Such substantial changes in speciation are not captured in existing DBP regulations, which only limit the sums of mass concentrations of four THMs and five HAAs. This important pitfall should be considered in formulating more equitable DBP regulations designed to protect human health.

Acknowledgments

This work has been funded by a grant from the National Science Foundation CAREER program (BES-0134301). The contents do not necessarily reflect the views and policies of the sponsors nor does the mention of trade names or commercial products constitute endorsement or recommendation for use. The assistance of Dr. Ying Wei from the City of Houston and Ramesh Sharma in measuring THMs and HAAs is appreciated.

BIBLIOGRAPHY

1. Singer, P.C. (Ed.). (1999). *Formation and Control of Disinfection By-Products in Drinking Water*. American Water Works Association, Denver, CO.
2. Singer, P.C. (1994). *J. Environ. Eng.* **120**: 727–744.
3. Rook, J.J. (1974). *Wat. Treat. Exam.* **23**: 234–243.
4. Christman, R.F. et al. (1983). *Environ. Sci. Technol.* **17**: 625–628.
5. Krasner, S.W. et al. (1989). *J. Am. Water Works Assoc.* **81**: 41–53.
6. Nawrocki, J. and Bilozor, S. (1997). *J. Water SRT-Aqua* **46**: 304–323.
7. Zavaleta, J.O. et al. *Formation and Control of Disinfection By-Products in Drinking Water*. P.C. Singer (Ed.). American Water Works Association, Denver, CO, pp. 95–117.
8. USEPA. (1998). *National Primary Drinking Water Regulations: Disinfectants and Disinfection By-Products*. Final Rule.
9. Minear, R.A. and Bird, J.C. (1980). *Water Chlorination: Environmental Impact and Health Effects*. R.L. Jolley, W.A. Brungs, and R.B. Cumming (Eds.). Vol. 3, Ann Arbor Science, Ann Arbor, MI, pp. 151–160.
10. Krasner, S.W. et al. (1996). *Disinfection By-Products in Water Treatment: The Chemistry of Their Formation and Control*. R.A. Minear and G.L. Amy (Eds.). CRC Press, Boca Raton, FL, pp. 59–90.
11. Krasner, S.W. et al. (1994). *J. Am. Water Works Assoc.* **86**: 34–47.
12. Oliver, B.G. (1980). *Water Chlorination: Environmental Impact and Health Effects*, Vol. 3, R.L. Jolley, W.A. Brungs, and R.B. Cumming (Eds.). Ann Arbor Science, Ann Arbor, MI, pp. 141–149.
13. Symons, J.M. et al. (1993). *J. Am. Water Works Assoc.* **85**: 51–62.
14. Symons, J.M. et al. (1996). *Disinfection By-Products in Water Treatment: The Chemistry of their Formation and Control*. R. Minear and G.L. Amy (Eds.). Lewis Publishers, Boca Raton, FL, pp. 91–130.
15. Pourmoghaddas, H. et al. (1993). *J. Am. Water Works Assoc.* **85**: 82–87.
16. Cowman, G.A. and Singer, P.C. (1996). *Environ. Sci. Technol.* **30**: 16–24.
17. Wu, W.W. and Chadik, P.A. (1998). *J. Environ. Eng.* **124**: 932–938.
18. Roberts, M.G. et al. (2002). *J. Am. Water Works Assoc.* **94**: 103–114.
19. Richardson, S.D. et al. (1999). *Environ. Sci. Technol.* **33**: 3378–3383.
20. Kargalioglu, Y. et al. (2002). *Teratogen. Carcinogen. Mutagen.* **22**: 113–128.
21. Plewa, M.J. et al. (2004). *Environ. Sci. Technol.* **38**: 62–68.
22. Singer, P.C. and Reckhow, D.A. (1999). *Water Quality and Treatment*, 5th Edn. R.D. Letterman (Ed.). McGraw Hill, New York, pp. 12.11–12.51.
23. Schnoor, J.L. et al. (1979). *Environ. Sci. Technol.* **13**: 1134–1138.
24. Liang, L. and Singer, P.C. (2003). *Environ. Sci. Technol.* **37**: 2920–2928.
25. Solarik, G. et al. (2000). *Natural Organic Matter and Disinfection By-Products: Characterization and Control in Drinking Water*. S.E. Barrett, S.W. Krasner, and G.L. Amy (Eds.). *ACS Symposium Series* Vol. 761. American Chemical Society, Washington, DC, pp. 47–66.
26. Sohn, J. et al. (2004). *Water Res.* **38**: 2461–2478.
27. Chowdhury, Z.K. and Amy, G.L. (1999). *Formation and Control of Disinfection By-Products in Drinking Water*. P.C. Singer (Ed.). American Water Works Association, Denver, CO, pp. 53–64.
28. Westerhoff, P. et al. (2000). *J. Am. Water Works Assoc.* **92**: 89–102.
29. Amy, G.L. et al. (1987). *J. Am. Water Works Assoc.* **79**: 89–97.
30. Koch, B. et al. (1991). *J. Am. Water Works Assoc.* **83**: 62–70.
31. Summers, R.S. et al. (1996). *J. Am. Water Works Assoc.* **88**: 80–93.

32. Reckhow, D.A. et al. (1990). *Environ. Sci. Technol.* **24**: 1655–1664.
33. Xie, Y.F. (2000). *Disinfection Byproducts in Drinking Water*. Lewis Publishers, Boca Raton, FL.
34. Gould, J.P. et al. (1983). *Water Chlorination: Environmental Impact and Health Effects*, Vol. 4, R.L. Jolley, W. Brungs, J. Cotruva, J. Mattice, and V. Jacobs (Eds.). Ann Arbor Science Publishers, Ann Arbor, MI, pp. 297–310.
35. Laine, J.-M. et al. (1993). *J. Am. Water Works Assoc.* **85**: 87–99.
36. Hwang, C.J. et al. (2000). *Natural Organic Matter and Disinfection By-Products: Characterization and Control in Drinking Water*. S.E. Barrett, S.W. Krasner, and G.L. Amy (Eds.). *Vol. ACS Symposium Series* 761, American Chemical Society, Washington, D.C., pp. 173–187.
37. Jacangelo, J.G. et al. (1995). *J. Am. Water Works Assoc.* **87**: 64–77.
38. Letterman, R.D. (Ed.) (1999). *Water Quality and Treatment*, 5th Edn. McGraw Hill, New York.
39. Snoeyink, V.L. et al. (1999). *Formation and Control of Disinfection By-Products in Drinking Water*. P.C. Singer (Ed.). American Water Works Association, Denver, CO, pp. 259–284.
40. Snoeyink, V.L. and Summers, R.S. (1999). *Water Quality and Treatment*, 5th Edn. R.D. Letterman (Ed.). McGraw Hill, New York, pp. 13.11–13.83.
41. Amy, G.L. et al. (1991). *Water Res.* **25**: 191–202.
42. Summers, R.S. et al. (1993). *J. Am. Water Works Assoc.* **85**: 88–95.
43. Taylor, J.S. and Jacobs, E.P. (1996). *Water Treatment Membrane Processes*. J. Mallevialle, P.E. Odendaal, and M.R. Wiesner (Eds.). McGraw Hill, New York, pp. 9.1–9.70.
44. Chellam, S. (2000). *Environ. Sci. Technol.* **34**: 1813–1820.
45. Allgeier, S.C. and Summers, R.S. (1995). *J. Am. Water Works Assoc.* **87**: 87–99.
46. Chellam, S. and Taylor, J.S. (2001). *Water Res.* **35**: 2460–2474.
47. Chellam, S. and Krasner, S.W. (2001). *Environ. Sci. Technol.* **35**: 3988–3999.

ACTIVATED CARBON: ION EXCHANGE AND ADSORPTION PROPERTIES

BASU SAHA
Loughborough University
Loughborough, United
Kingdom

INTRODUCTION

The importance of environmental pollution control has increased significantly in recent years. Environmentalists are primarily concerned with the presence of heavy metals, pesticides, herbicides, chlorinated hydrocarbons, and radionuclides in groundwater, surface water, drinking water, and aqueous effluents because of their high toxicity and impact on human and aquatic life.

Several techniques have been developed and used to remove and/or recover a wide range of micropollutants from water and a variety of industrial effluents. Adsorption using activated carbon is well established for the removal of organic molecules from aqueous solution, but to a much lesser extent for the removal of toxic heavy metals.

Here, the preparation, properties, and metal sorption performance of a range of as-received and oxidized samples of granular activated carbon that were either prepared or modified in the author's laboratory are discussed. Samples were evaluated for the removal of trace toxic metal ions from aqueous solutions. Batch and column experiments have been performed to elucidate the relationship between sorptive performance and the physical and chemical structure of these materials. A detailed discussion of adsorption and ion-exchange properties of engineered activated carbons and carbonaceous materials can be found in a recent publication by Streat et al. (1).

ACTIVATED CARBONS

Activated carbon is the generic name given to amorphous carbonaceous materials with an extensively developed internal pore structure. The characteristics of activated carbons depend on the carbonaceous precursor and the technique of activation used in their production. Activated carbon is obtained by thermal decomposition, combustion, and partial combustion of different carbonaceous substances. The carbonaceous substances are usually cheap vegetable material high in carbon content and low in ash content. The raw material can be wood, peat, fruit stones, nutshells, or sawdust. (2). Nonvegetable sources include blood, bones, leather waste, fish, petroleum products, and by-products (3). Currently, a bulk of the activated carbon comes from natural coal, which is because of its low cost and availability.

Activated carbons have proved versatile adsorbents mainly because of their large surface area and surface reactivity. They also possess favorable pore size, which makes the internal surface accessible, enhances the adsorption rate, and enhances mechanical strength (2). According to the International Union of Pure and Applied Chemistry (IUPAC) system, the pore structures are divided into three categories, namely, micropores (<2 nm in diameter), mesopores (between 2 and 50 nm in diameter), and macropores (>50 nm). The pores exist in different shapes, which can be V-shaped pores, tapered pores, capillary shaped with one end closed or open at both ends, contracted entrance pores, and so on.

Micropores constitute the bulk of the area, and their sizes are usually comparable with the adsorbate molecules. The mechanism of adsorption in micropores is micropore filling, and it comes as a result of the adsorption force field that encompasses the whole volume of micropores. This type of mechanism leads to higher heat of adsorption energy as compared with the adsorption energy on surface or mesopore regions. Mesopores have a dual function that involves adsorption and channels for the adsorbate to access the interior of the adsorbent. The mechanism of adsorption is via capillary condensation. Macropores serve as passages for the adsorbate to access the interior of the adsorbent, and they do not contribute significantly to the adsorption capacity.

Activated carbon is a microcrystalline, nongraphitic form of carbon and can be produced in the form of powders, granules, and shaped products satisfying any modern engineering requirements. Its exceptional adsorptive properties are attributed to the controlled size and distribution of pores within the carbon matrix. The strong market position of the activated carbons relates to their unique properties and lower cost compared with that of other possible competitive adsorbents.

Through the choice of precursor, the method of activation, and control of processing conditions, the adsorptive properties of activated carbons can be specially designed for many diverse applications such as the purification of potable water and adsorption of various emissions from motor vehicles. In drinking water treatment, the carbon is used to remove dissolved organics with molecular weights in excess of 45 g/mol (4).

HISTORICAL BACKGROUND

The use of activated carbon as an adsorbent dates back to the Egyptians in about 1500 B.C. (2). It was used in the form of carbonized wood mainly for medical purposes as well as a purifying agent. However, the basis of its industrial production and use was established in the late eighteenth century and its uses have broadened since then. The first industrially manufactured activated carbons were Eponit decolorizing carbons. They have been produced since 1909 by heating wood charcoal with steam and carbon dioxide in a furnace specially designed for this purpose. Activated carbons are now produced industrially by several different methods, as many new production technologies have appeared.

Granular activated carbon found uses for purification of gases and extraction of vapors from gas streams, recovery of liquid petroleum from natural gas, extracting benzene from manufactured coal gas, and recovering volatile solvents from various industrial applications. Activated carbon has also been applied to the removal of color, odor, taste, and many chemical undesirable species from aqueous solutions either by adsorption or ion exchange (1). It is now extensively used for potable, domestic, and industrial waste water purification, sugar refining, and metal recovery (2,3,5,6).

Over the last two decades, an interest has developed in activated carbon, and applications continue to evolve in response to growing demands for environmental protection and emerging technologies.

PRECURSORS

The materials used for the production of activated carbon are of high carbon content and contain a low concentration of inorganic substances. Precursor materials include wood, animal bones, nutshells, coconut shell, fruit stones, brown coal, bituminous coal, charcoal, peat, lignin, lignite, pitch, mineral oil products, and petroleum and lubricant wastes (3,7). However, only wood, bones, coal, lignite, coconut shell, and peat have been used on an industrial scale because they are currently the only economically viable raw materials (3,4).

As already mentioned, many potential raw materials exist for the production of activated carbon, but not all of them produce activated carbon of high quality. The reason behind this lies in the high amount of inorganics that exist in the structure of the raw materials as well as in their scarcity. Poor quality can also result from poor storage conditions and difficult workability of the raw materials.

The principal properties of manufactured active carbons depend on the type and properties of the raw material used. The properties of the final product will also depend on the nature of the raw material but mostly on the nature of the activating agent and the conditions of the activation process, which makes the carbonization and activation processes the most important factors in producing a good quality activated carbon. The next section gives a brief review on how activated carbon is produced and discusses how the activation of the carbon can result in a material suitable for specialized applications.

CARBONIZATION

The first stage of production of the activated carbons is the carbonization process, where the properties of the activated carbon can be modified. The carbonization process involves thermal decomposition and elimination of the noncarbon species and can introduce a simple pore structure that depends principally on the physical properties of the precursor and can be controlled by the rate of heating.

Generally, carbonization is conducted by heating the initial material up to 973 K in the absence of oxidizing gases. During this process, the raw material is dehydrated, that is, oxygen- and hydrogen-containing compounds are decomposed and eliminated from the precursor to produce a skeleton possessing a latent pore structure. If the temperature of carbonization is low, impurities such as tar may be left in the pores without decomposition. The presence of impurities, which block the pores, may result in a low surface area and limited surface activity. The mechanical and adsorptive properties of an activated carbon can also be introduced and controlled by various additives introduced during the stabilization and carbonization process.

ACTIVATION

The activation process is the second stage of production, where any disorganized carbon from the surface and the carbon from the aromatic sheets under the surface are burned off by contacting the carbonized material either with gases or with chemical reagents, which results in wider pores and new porosity with an oxidized surface.

Generally, two main activation processes are used by industry to manufacture activated carbon: physical (or thermal) and chemical activation. The main difference between the two is that in physical activation, the process is done at high temperature and the yield of activated carbon is low, whereas in chemical activation, the process is carried out at a lower temperature and results in a much higher yield.

Physical activation involves carbonization of the raw material and the activation of the carbonized product. Carbonization is generally performed at temperatures around 873–923 K in the absence of oxidizing gases (8). The resulting material is enhanced by activation via partial gasification with mild oxidizing agents such as steam, carbon dioxide, or a mixture of both at 1073–1273 K (9). In the case of activation with air or oxygen, excessive burning of the carbon exists and the reaction is difficult to control, which is the main reason that these oxidizing agents are not used for activation purposes. Generally, microporosity is introduced when activation is done using carbon dioxide, whereas meso- or macroporosity develops if steam is used for the activation.

Chemical activation, on the other hand, involves the reaction of a carbon precursor with a reagent. The contact of the raw material with the reagent takes place while the temperature increases from 623 K to 1173 K. The commonly used reagents for chemical activation in industry until 1970 were zinc chloride, sulfuric acid, and phosphoric acid. The preferred precursors exposed to this type of activation are lignocellulosic materials, wood being the most common. Today, only nonmineral precursors can be applied successfully, with wood and olive stones the most commercially used. Jankowska and Swiatkowski (10) reported that this activation process produces a carbon matrix that contains a considerable amount of heteroatoms (including residual of phosphates, sulfates, and zinc).

PHYSICAL AND CHEMICAL PROPERTIES OF ACTIVATED CARBONS

The structure of activated carbon is best described as a twisted network of defective carbon layer planes, cross-linked by aliphatic bridging groups (11). The lattice is composed of pores sizes whose size distribution can be controlled by the choice of carbon feedstock and mode of preparation. The surface of activated carbons is not uniform in either structural or in energetic respect. Atoms at the carbon surface are in a different electronic state than those in a pore phase, especially on the edges of the carbon layers; where defects, dislocations, and discontinuities of the layer planes are present. Such sites, called the "active" sites, are associated with high concentration of unpaired electron spin centers and are therefore expected to play a significant role in chemical and adsorptive interaction of the activated carbon with different compounds (12).

Today, it is well established that almost all types of activated carbons are covered with oxygen complexes, unless special care is taken to eliminate them. Therefore, the adsorptive properties of an activated carbon are not only determined by its porous structure but also by its chemical composition. These complexes are often the source of the properties, which make an activated carbon useful or effective in certain applications. Two main types of surface functional groups exist: acidic and basic surface groups, which are discussed in the following section.

ACID SURFACE FUNCTIONAL GROUPS

Several types of oxides having an acidic character have been identified to exist on the surface of activated carbons by different types of techniques: specific organic reactions, potentiometry, Fourier transform infrared (FTIR), and X-ray photoelectron spectroscopy (XPS). The structures presented in Fig. 1 are the oxygen containing groups that are found at the edges of the graphene layers. The most important ones are the carboxyl (a), carbonyl (f), and single hydroxyl (phenolic) (e) groups.

Other groups can then be generated from a combination of these three because the simultaneous presence of various types of organic groups may lead to more complex surface structures. Thus, carboxylic anhydrides (b) can be produced when carboxyl groups are close together, lactone groups (c) when hydroxyl and carboxyl groups are joint, and lactoles (d) when carboxyl and carbonyl groups come close enough (Fig. 2). Carbonyl groups can also be present on the surface of the carbon as conjugated structures called quinones (g). Lastly, ether type structures (h), which are stable even at high temperatures, can also be found on the activated carbon surface.

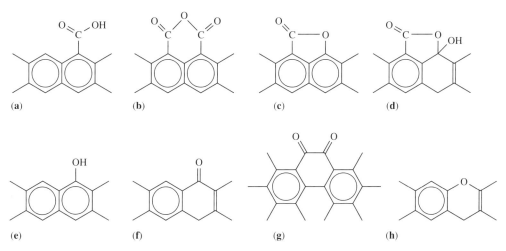

Figure 1. Possible structures of surface oxygen groups (redrawn from Ref. 13).

Figure 2. Possible interaction of oxygen-containing groups (redrawn from Ref. 13).

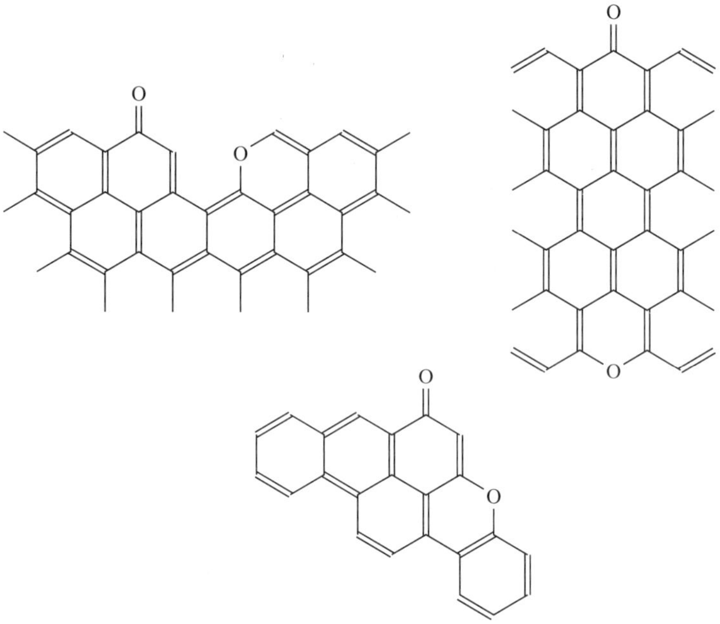

Figure 3. Possible structures of basic surface sites on a graphene layer (redrawn from Ref. 16).

BASIC SURFACE FUNCTIONAL GROUPS

Very scant information is available about basic surface functional groups. They are formed when a carbon is heat treated above 973 K, cooled in an inert atmosphere, and re-exposed to oxygen at room temperature. In order to explain the uptake of acids, Garten et al. (14) and Garten and Weiss (15) proposed a chromene structure, but g-pyrone-like structures are more likely (16). These structures are schematically shown in Fig. 3, and they characteristically possess two oxygen groups situated on different rings of the graphitic layer.

Evidence also exists that protons can be adsorbed on basal planes either by reaction with pyrones or by donor–acceptor interactions. The end result is the formation of basic surface structures, which can be represented by pyrone-like structures with oxygen atoms, in general, located in two different rings of a graphite layer; the positive charge being stabilized by the resonance (Fig. 4).

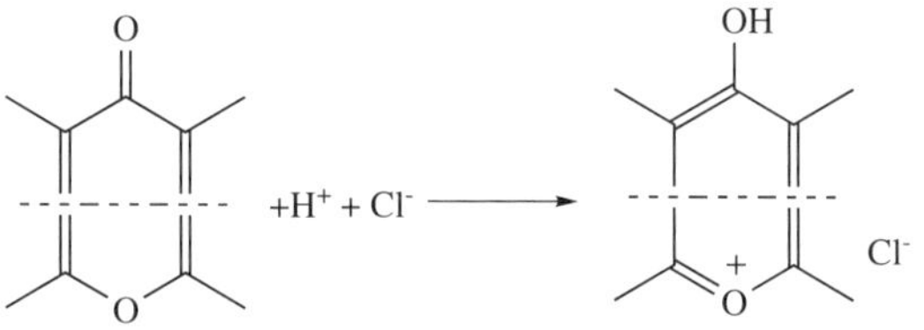

PREPARATION AND PROPERTIES OF ACTIVATED CARBON

The synthesis, structure, and adsorption properties of engineered activated carbons with reference to the body of work carried out in the author's research laboratory are reviewed. The main emphasis is granular activated carbon (GAC). The majority of engineered carbons discussed here have graphitic or disordered graphitic microstructures. Also, for most engineered carbon materials, the originating precursor is organic and the materials develop from heat treatment of the precursor in inert atmospheres (carbonization). A selection of technically important carbons developing from solid, liquid, and gaseous organic precursors are presented in Table 1.

Figure 4. Reaction of proton with pyrone-like structure (redrawn from Ref. 16).

The role of pore structure and surface area of engineered carbons and their impact on adsorbent performance are discussed elsewhere (1,17,18). The surface chemical structure affects interaction with polar and nonpolar molecules because of the presence of chemically reactive functional groups (see Fig. 1). The adsorption is a complex interplay between the chemical and porous surface structure of the carbon. Important factors including the nature and relative amounts of surface functional groups, the surface area, and pore size distribution, as well as the characteristics of the adsorbate molecule (e.g., size and

Table 1. Starting Precursors for Carbons (Adapted from Ref. 1)

Primary	Secondary	Example of Carbons
Hydrocarbon gases		Pyrocarbons, carbon blacks, vapor grown carbon fibers, matrix carbon[a]
Petroleum derived		Delayed coke, calcined coke
	Petroleum pitch	Needle coke, carbon fibers, binder and matrix carbon[a]
	Mesophase pitch	Mesocarbon microbeads, carbon fibers
Coals		Semicoke, calcined coke
	Coal chars	Activated carbons
	Coal tar pitch	Premium cokes, carbon fibers, binder and matrix carbons[a]
	Mesophase pitch	Mesocarbon microbeads, carbon fibers
Polymers	Polyaddition and polycondensation type resins	PAN-based carbon fibers
		Glassy carbons, binder and matrix carbons, carbon beads, graphite films and monoliths
Biomass	Coconut shells, apricot, olive and peach stones, pits	Activated carbons

[a]Precursor as a binder in granular carbons and graphites and in carbon–carbon composites.

nature of the cation/anion, polarity and chemical structure of the molecule, molecular/ionic dimensions) are discussed by Streat et al. (1). The importance of precise characterization of the surface and structure of engineered carbons in order to tailor adsorbents for specific applications is outlined by Saha et al. (17–19).

Granular oxidized carbon samples showed similar ion-exchange properties to active carbon fibers with respect to influence of solution pH on ion-exchange behavior. The pH dependence may be attributed to acidic surface groups that dissociate as a function of solution pH. The point-of-zero charge (PZC) plays an important role in the ion-exchange behavior of the carbon adsorbents in removing metal ions from solution (17).

SORPTION OF TRACE METALS ONTO ACTIVATED CARBON

Previous publications have discussed the adsorption of heavy metal ions from aqueous solutions by carbonaceous materials (20–24). It is widely accepted that surface acidic functional groups are responsible for metal ion binding (20). Chemical oxidation is commonly used to introduce these functional groups onto the surface of carbons.

The results of detailed minicolumn experiments to determine breakthrough characteristics and regeneration performance of selected oxidized activated carbons are outlined in this section. Samples of BGP (a wood-based activated carbon) were tested using solutions containing target metals to determine overall performance criteria, which consisted of testing samples of BGP against solutions containing copper, nickel, zinc, or cadmium. All breakthrough capacities were calculated based on the total amount of metal removed before 5% or 50% of the feed concentration of metals was detected in the outlet, which was to give an indication of uptake performance of each material for the four metals used. The results of the minicolumn experiments are summarized in Table 2.

Table 2. Uptake Capacity of Carbons from 1 mM Metal Solutions (pH 4.7)

			Breakthrough Capacity (mmol/g)	
Sample[a]	Metal	pH	5%	50%
BGP OxII	Cu	4.7	0.35	0.42
BGP AOx	Cu	4.7	0.04	0.05
BGP Unoxidized	Cu	4.7	—	0.004
BGP OxII	Zn	4.7	0.13	0.16
BGP AOx	Zn	4.7	—	0.13
BGP OxII	Ni	4.7	0.14	0.20
BGP AOx	Ni	4.7	0.03	0.06
BGP OxII	Cd	4.7	0.16	0.21
BGP AOx	Cd	4.7	—	0.02

[a]AOx, air oxidized; OxII, alkali washed, acid oxidized for 9 h.

An unoxidized sample of BGP was used to generate a breakthrough curve to compare copper sorption capacity with oxidized samples. The copper breakthrough curves can be seen in Fig. 5.

Results show that significant improvement occurred in copper sorption capacity on oxidation of the original sample. The copper sorption capacities increased by a factor of up to 100 by oxidizing the original material. This change can be attributed to oxygen surface groups introduced onto the surface of Ceca BGP. The 5% copper breakthrough capacities of BGP OxII and BGP AOx were 0.35 mmol/g and 0.04 mmol/g, respectively. Over 100 bed volumes of 1 mM solution were passed before any copper could be detected leaving the column using the acid oxidized sample, BGP OxII.

BGP AOx was also used to generate breakthrough curves for removal of nickel, zinc, and cadmium. Results for these experiments can be seen in Fig. 6. It shows that the metal sorption capacity of air-oxidized samples was much less than acid-oxidized samples for a feed concentration of 1 mM and pH of 4.7. Breakthrough capacities at 5% were 0.02 mmol/g for nickel, and for other samples, the 5% breakthrough was instantaneous. The low capacity of BGP

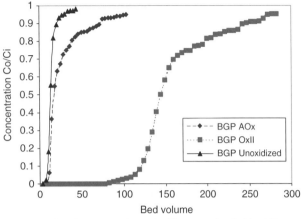

Figure 5. Copper breakthrough curves for samples of BGP.

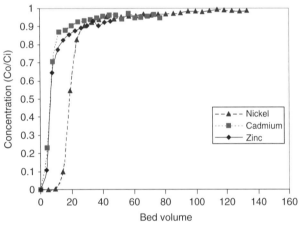

Figure 6. Breakthrough curves for nickel, zinc, and cadmium using BGP AOx sample.

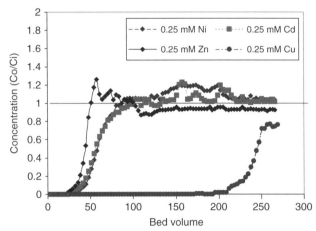

Figure 7. Breakthrough curve using a feed solution containing copper, nickel, zinc, and cadmium with a sample of BGP OxII.

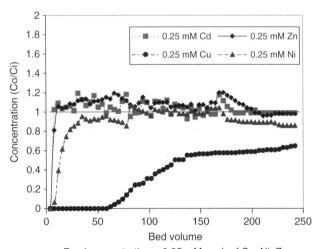

Figure 8. Breakthrough curve using a feed solution containing copper, nickel, zinc, and cadmium with a sample of BGP AOx.

AOx sample can be partially attributed to the high reaction temperature during the oxidation rising above 723 K, which would have resulted in desorption of some oxygen surface groups. Very poor breakthrough performance was obtained for the BGP AOx sample with the exception of the nickel experiment when 10 bed volumes were treated before any metal was detected in the outlet.

A combined feed containing all four metals, copper, zinc, nickel, and cadmium, was passed through the column containing 1 g of BGP OxII (see Fig. 7). The feed concentration of all metals was 0.25 mM and the feed pH was 4.7, which indicate the following selectivity series: $Cu^{2+} > Ni^{2+}, Cd^{2+} > Zn^{2+}$. The increase in the concentration above the feed was because of chromatographic elution. A solution containing copper, zinc, nickel, and cadmium was used to obtain the selectivity of BGP AOx. The results shown in Fig. 8 indicate a selectivity order of $Cu^{2+} > Ni^{2+} > Zn^{2+}, Cd^{2+}$. These studies confirm that both materials are highly selective toward copper in the presence of other metal ions.

The BGP OxII elution results are shown in Fig. 9. Most of the metals were eluted with 0.1M HCl in the first 10 bed volumes except for cadmium. All the copper was recovered, but only 86% of nickel, 90% of zinc, and 85% of cadmium were eluted. Elution efficiency of BGP AOx was 100% for all metals. The results show that most of the metals were recovered before 10 bed volumes of eluant had been passed through the columns.

The sorptive capacity and selectivity of oxidized carbons vary for different metal ions, and higher valency metal ions are usually preferred to those of lower valency (20,21). It has also been observed that selectivity differs even within a series of metals with the same valency. Among divalent metals, Cu^{2+} is generally the most preferred ion. However, the reason for the higher affinity of oxidized carbons toward this particular metal ion has not been clearly identified. The formation of metal surface complexes on the oxidized carbon involving cooperative action seems quite likely (see Fig. 10).

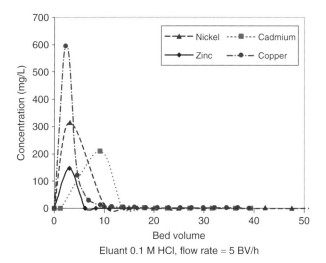

Figure 9. Elution curves of BGP AOx.

It can be concluded that air and acid oxidation significantly enhance the sorption capacity of Ceca BGP. BET and Langmuir surface areas were relatively unchanged by air oxidation, but some surface area was lost during acid oxidation. Acid-oxidized samples have a much higher copper uptake capacity than the as-received material. Air-oxidized samples show lower metal uptake capacity compared with acid-oxidized samples. Acid-oxidized samples are heat treated and alkali washed to remove organic by-products that are formed during the oxidation. At every stage, some material is lost, reducing the yield of material based on weight of as-received material compared with the final product. The sorption capacity of the materials was reduced by these treatments, indicating the metal binding capacity of these organic by-products formed during oxidation. Equilibrium capacity of modified samples decreased with a reduction in the pH because of the weakly acidic nature of the sorbents. Kinetic experiments indicate that about 80% of the metals are removed in the first twenty minutes for all samples. Minicolumn experiments show that modified carbons are efficient sorbents for the removal of trace metal ions from solution. After initial breakthrough, complete loading of carbons occurs slowly, which is most likely because of the presence of micropores and small mesopores within the structure of the materials. All samples are regenerated using 0.1 M HCl solution. Both modified samples demonstrated good regeneration efficiencies with 100% of the copper recovered during elution cycles compared with the amount of metal removed during the sorption experiments.

CONCLUSIONS

Engineered activated carbons can be manufactured in a range of physical forms and from a wide variety of starting precursor materials. Pyrolysis and activation combined with further surface pretreatment results in a range of adsorbent materials with widely varying physicochemical properties. It is argued that liquid-phase sorption of solutes on carbon surfaces involves a complex interplay between the delocalized π-electron system developing from the aromatic backbone of the graphite microstructure and the electron withdrawing surface functional groups (e.g., oxygen functional groups). This interplay not only lowers the point-of-zero charge of the surface but also reduces the dispersive adsorption potential by decreasing the π-electron density in the graphene layers (2). Optimization of the physical properties of carbonaceous adsorbents, that is, surface area and a well-developed pore structure, are essential for the application of the adsorbents in conventional water treatment technologies (e.g., using fixed-bed adsorbers). In addition, it is vital that the chemical properties of the adsorbent surface must be tailored to suit the individual application. Carbon surface functional groups influence the selectivity of the adsorbent for metal ions in solution. Oxidation treatment of commercially available carbons dramatically enhances their ion-exchange properties and results in reversible adsorption of metal ions from solution. Design of adsorption systems using engineered carbons requires careful optimization of process parameters with due attention to the speciation chemistry of metal ions and the solution pH that markedly influences the interaction of the solute species with the adsorbent surface. We are now in a position to reverse-engineer tailored structured carbonaceous adsorbents with optimized physicochemical properties suited for selective separation of solutes from aqueous streams.

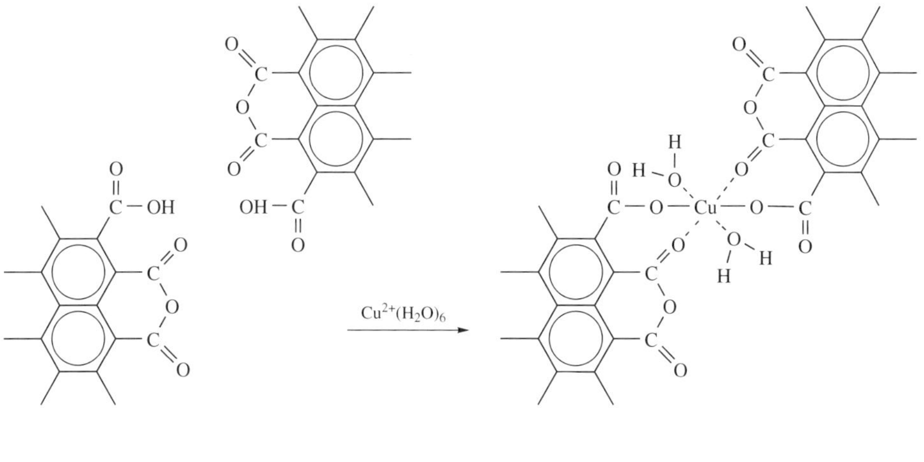

Figure 10. Postulated complexation reaction between copper (II) and oxidized carbon surface.

Acknowledgments

I would like to express my sincere gratitude to Professor Michael Streat for his support, continuous encouragement, and advice for conducting the research in the area of adsorption and ion exchange and development of tailored sorbent materials for environmental remediation. I am indebted to Dr. Hadi Tai, Purazen Chingombe, and Eleni Karounou for their painstaking experimental work. Funding from European Community, EPSRC, Severn Trent Water Ltd., and Commonwealth Scholarship Commission is gratefully acknowledged.

BIBLIOGRAPHY

1. Streat, M., Malik, D.J., and Saha, B. (2004). In: *Ion Exchange and Solvent Extraction*. Vol. 16. A.K. SenGupta and Y. Marcus (Eds.). Marcel Dekker, New York, pp. 1–84.
2. Bansal, C.R., Donnet, J.B., and Stoeckli, F. (1988). *Active Carbon*. Marcel Dekker, New York, pp. 1–17.
3. Hassler, J.W. (1967). Activated carbon. *Chemical and Process Engineering Series*. pp. 1–171.
4. Noll, K.E. (1991). *Adsorption Technology for Air and Water Pollution Control*, Lewis Publishers, Chelsea, MI, pp. 1–211.
5. Donnet, J.B. and Bansal, R.C. (1984). *Carbon Fibres*. Marcel Dekker, New York, pp. 1–40.
6. Roy, G.M. (1995). *Activated Carbon Applications in Food and Pharmaceutical Industries*. Technomic, pp. 1–15.
7. Mantell, C.L. (1968). *Carbon and Graphite Handbook*. John Wiley & Sons, Hoboken, NJ, pp. 6–34.
8. Smíšek, M. and Cerný, S. (1967). *Active Carbon: Manufacture Properties and Applications*. Elsevier, Amsterdam.
9. Freeman, J.J., Gimblett, F.G.R., Roberts, R.A., and Sing, K.S.W. (1987). *Carbon* **25**(4): 559–563.
10. Jankowska, H. and Swiatkowski, A. (1991). *Active Carbon*. Ellis Horwood Limited, pp. 9–74.
11. McEnaney, B. and Mays, T.J. (1989). *Introduction to Carbon Science*. Butterworths, London, pp. 153–156.
12. Puri, B.R. (1970). *Chemistry and Physics of Carbon*. Vol. 6. Marcel Dekker, New York, pp. 1–191.
13. Boehm, H.P. (1994). *Carbon* **32**(5): 759–769.
14. Garten, V.A., Weiss, D.E., and Willis, J.B. (1957). *Austral J. Chem.* **10**: 295–308.
15. Garten, V.A. and Weiss, D.E. (1957). *Rev. Pure. Appl. Chem.* **7**: 69–122.
16. Voll, M. and Boehm, H.P. (1971). *Carbon* **9**: 481–488.
17. Saha, B., Tai, M.H., and Streat, M. (2001). *Trans. IChemE.* **79**(Part B): 211–217.
18. Saha, B., Tai, M.H., and Streat, M. (2003). *Chem. Eng. Res. Des.* **81**: 1343–1353.
19. Saha, B., Tai, M.H., and Streat, M. (2001). *Trans. IChemE.* **79**(Part B): 345–351.
20. McEnaney, B. (1999). In: *Carbon Materials for Advanced Technologies*. T.D. Burchell (Ed.). Pergamon, New York, pp. 1–33.
21. Kadirvelu, K., Faur-Brasquet, C., and Le Cloirec, P. (2000). *Langmuir*. **16**: 8404–8409.
22. Shim, J.-W., Park, S.-J., and Ryu, S.-K. (2001). *Carbon* **39**: 1635–1642.
23. Corapcioglu, M.O. and Huang, C.P. (1987). *Water Res.* **21**: 1031–1044.
24. Budinova, T.K., Gergova, K.M., Petrov, N.V., and Minkova, V.N. (1994). *J. Chem. Technol. Biotechnol.* **60**: 177–182.

ACTIVATED CARBON—POWDERED

Absar A. Kazmi
Nishihara Environment
Technology
Tokyo, Japan

INTRODUCTION

The most commonly used adsorbent in water and wastewater treatment is activated carbon. Activated carbon is manufactured from carbonaceous material such as wood, coal, or petroleum residues. A char is made by burning the material in the absence of air. The char is oxidized at higher temperatures to create a very porous structure. The "activation" steps provide irregular channels and pores in the solid mass, resulting in a very large surface-area-per-mass ratio. Surface areas ranging from 500 to 1500 m^2/g have been reported, and all but a small surface area is in the pores. Once formed, activated carbon is pulverized to a very fine powder (Fig. 1). The size is predominantly less than 0.075 mm (200 sieve). Dissolved organic material adsorbs to both exterior and interior surfaces of the carbon. The characteristics of powdered activated carbon are summarized in Table 1.

APPLICATION IN WATER TREATMENT

Seasonal application of powdered activated carbon (PAC) at the raw water intake or rapid mix unit is used by some plants to correct short-term raw water quality problems such as algal blooms. PAC is used to correct taste and

Figure 1. Powdered activated carbon.

Table 1. Characteristics of Powdered Activated Carbon

Parameter	Unit	Values
Total surface area	m^2/g	800–1800
Bulk density	kg/m^3	360–740
Particle density, wetted in water	kg/L	1.3–1.4
Particle size range	mm	5–50
Effective size	mm	—
Uniformity coefficient	UC	—
Mean pore radius	Å	20–40
Iodine number	—	800–1200
Abrasion number	Minimum	70–80
Ash	%	≤6
Moisture as packed	%	3–10

odor problems, which are primarily aesthetic qualities of water. Contact time is needed to allow adsorption to occur. Powdered activated carbon is contacted with water in open vessels where it is maintained in suspension for the necessary contact time and then removed by conventional solids removal processes (Fig. 2). The general feeling among water treatment plant operators is that, when everything else fails, it is time to use PAC. It is quite reliable, but its effectiveness depends on the type of carbon, the dosage, and the point of application. Usual carbon dosage ranges from 0.5 to 5.0 mg/L.

Normally, the PAC dose is based on operator experience, although laboratory tests may provide more precise dosages. In these tests, a series of carbon dosages is added to odorous water and the residual odor levels are determined. When a single odorous chemical is present, the initial and residual concentrations can be measured by using advanced laboratory identification techniques. Where such equipment is not available, the odor source is unknown, or numerous compounds are involved, the threshold odor number (TON) method may suffice. These data then can be plotted in one of several ways to clarify the removal response, to produce a predictive trend, and to determine the dosages to achieve acceptable water quality.

A simple plot of carbon dose versus TON (or the residual concentration) might be sufficient to highlight the relationship; however, the preferred method is to identify constants in an equation for removal and apply these to dosage calculations. The Freundlich and Langmuir isotherm equations are commonly used. The primary difference is that the Freundlich isotherm equation is strictly an empirical approach, accepted as standard; the Langmuir isotherm has a theoretical basis and has wider applicability.

The Freundlich isotherm is expressed by Eq. 1 and its corollary by Eq. 2

$$\frac{X}{M} = kC^{1/n} \tag{1}$$

$$\ln\left(\frac{X}{M}\right) = \ln(k) + \frac{1}{n}\ln C \tag{2}$$

X/M = mass of adsorbate per unit mass of adsorbent, mg adsorbate/g activated carbon
C = final or equilibrium concentration of the adsorbate, mg/L
k = empirical constant (y intercept)
n = slope inverse constant

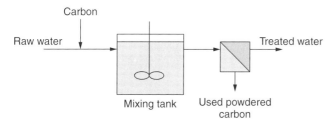

Figure 2. Powdered activated carbon for water treatment.

Similarly, the Langmuir isotherm is defined by Eqs. 3, 4, and 5:

$$\frac{X}{M} = \frac{abC}{1+bC} \tag{3}$$

$$\frac{C}{XM} = \frac{1}{ab} + \left(\frac{1}{a}\right)C \tag{4}$$

$$\left(\frac{1}{X/M}\right) = \left(\frac{1}{ab}\right)\left(\frac{1}{C}\right) + \frac{1}{a} \tag{5}$$

where a = maximum number of moles adsorbed per mass of adsorbent at monolayer saturation
b = empirical constant, L/mg

The Langmuir isotherm can be plotted in terms of $C/(X/M)$ versus C or $1/(X/M)$ versus $1/C$. The plot that produces a clear trend provides more accurate constants.

ANALYSIS AND DESIGN OF POWDERED ACTIVATED CARBON CONTACTOR

For a powdered activated carbon (PAC) application, isotherm adsorption data can be used in conjunction with a material mass balance to obtain an approximate estimate of the amount of carbon that must be added, as illustrated below. Here again, because of the many unknown factors involved, column and bench tests are recommended to develop the necessary design data. If the mass balance is written around the contactor (i.e., batch reactor) after equilibrium has been reached, the resulting expression is

Amount absorbed = initial amount of absorbate present
− final amount of absorbent present

$$q_e m = VC_o - VC_e, \tag{6}$$

where q_e = adsorbate phase concentration after equilibrium, mg adsorbate/g adsorbent
m = mass of adsorbent, g
V = volume of liquid in the reactor, L
C_o = initial concentration of the adsorbate, mg/L
C_e = final equilibrium concentration of the adsorbate after absorption

If Eq. 6 is solved for q_e, the following expression is obtained:

$$q_e = \frac{V(C_o - C_e)}{m} \tag{7}$$

The equation can be rewritten as

$$\frac{V}{m} = \frac{q_e}{C_o - C_e} \tag{8}$$

The term V/m, defined as the specific volume, represents the volume of liquid that can be treated with a given amount of carbon. The reciprocal of the specific volume corresponds to the dose of adsorbent that must be used.

APPLICATION IN WASTEWATER TREATMENT

A proprietary process, "PACT," combines the use of powdered activated carbon with the activated-sludge process. In this process, when the activated carbon is added directly to the aeration tank, biological oxidation and physical adsorption occur simultaneously. A feature of this process is that it can be integrated into an existing activated sludge system at nominal capital cost. The addition of powdered activated carbon has several advantages, including (1) system stability during shock loads, (2) reduction of refractory priority pollutants, (3) color and ammonia removal, and (4) improved sludge settleability. In some industrial waste applications, where nitrification is inhibited by toxic organics, the application of PAC may reduce or limit this inhibition.

The use of powdered activated carbon in secondary wastewater systems results in an inseparable mixture of biological solids and carbon. Thermal regeneration of the carbon also results in destruction of biomass, eliminating the need for other sludge processing and disposal techniques, but increases the size of the carbon regeneration system.

The dosage of powdered activated carbon and the mixed liquor–powdered activated carbon suspended solids concentration are related to solids retention time as follows:

$$X_p = \frac{X_i \text{SRT}}{\tau} \tag{9}$$

where X_p = equilibrium powdered activated carbon–MLSS content, mg/L
  X_i = powdered activated carbon dosage, mg/L
  SRT = solids retention time, days
  τ = hydraulic retention time, days

Carbon dosages typically range from 20 to 200 mg/L. At higher SRT values, the organic removal per unit of carbon is enhanced, thereby improving process efficiency. Reasons cited for this phenomenon include (1) additional biodegradation due to decreased toxicity, (2) degradation of normally degradable substances due to increased exposure time to the biomass through adsorption on the carbon, and (3) replacement of low molecular weight compounds with high molecular weight compounds, resulting in improved adsorption efficiency and lower toxicity (Fig. 3).

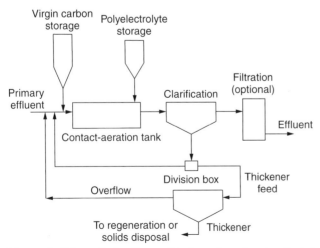

Figure 3. Schematic for the application of powdered activated carbon.

READING LIST

Metcalf & Eddy, Inc. (2003). *Wastewater Engineering: Treatment, Disposal, Reuse*, 4th Edn. McGraw-Hill, New York, pp. 1138–1162.

Peavy, H.S., Rowe, D.R., and Tchbanoglous, G. (1985). *Environmental Engineering*, International ed. McGraw-Hill, New York, pp. 195–197.

Qasim, S.R, Motley, E.M., and Zhu, G. (2000). *Water Works Engineering, Planning, Design & Operation*. Prentice-Hall, Upper Saddle River, NJ, pp. 437–463.

CHLORINATION

Absar A. Kazmi
Nishihara Environment
Technology
Tokyo, Japan

INTRODUCTION

Chlorine is added to water to kill disease-causing bacteria, parasites, and other organisms. At low concentrations, chlorine probably kills microorganisms by penetrating the cell and reacting with the enzymes and protoplasm. At higher concentrations, oxidation of the cell wall destroys the organism. Chlorination also removes soluble iron, manganese, and hydrogen sulfide from water.

CHLORINE COMPOUNDS

The principal chlorine compounds used in water treatment are chlorine (Cl_2), sodium hypochlorite (NaOCl), and calcium hypochlorite $Ca(OCl)_2$. Chlorine gas can be liquefied by compression and shipped to the site in compact containers. Because it can be regasified easily and has a solubility of approximately 700 mg/L in water at the pH and temperatures generally found in water treatment plants, this form of chlorine is usually preferred. Hypochlorites tend to raise the pH, thus driving the reaction more toward the less effective OCl^-. Commercially available calcium hypochlorite contains approximately 70–80% available chlorine, whereas NaOCl contains only 3–15% available chlorine. Some practical difficulty is involved in dissolving $Ca(OCl)_2$, and both hypochlorites are more expensive on an equivalence basis than liquefied Cl_2. There are other considerations, however, that sometimes dictate using hypochlorites.

CHLORINATION CHEMISTRY

Chlorine may be applied to water in gaseous form (Cl_2) or as an ionized product of solids [$Ca(OCl)_2$, NaOCl]. The reactions in water are as follows:

$$Cl_2 + H_2O \longrightarrow H^+ + HOCl \tag{1}$$

$$Ca(OCl)_2 \longrightarrow Ca^{2+} + 2OCl^- \tag{2}$$

$$NaOCl \longrightarrow Na + OCl^- \tag{3}$$

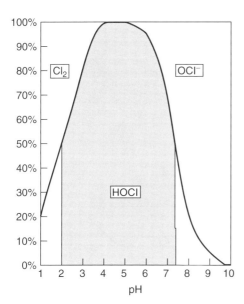

Figure 1.

The hypochlorite (HOCl) and the hypochlorite ion (OCl$^-$) in these equations are further related by

$$\text{HOCl} \rightleftharpoons \text{H}^+ + \text{OCl}^- \tag{4}$$

A relationship governed primarily by pH is shown in Fig. 1.

The sum of HOCl and OCl$^-$ is called the free chlorine residual and is the primary disinfectant used. HOCl is a more effective disinfectant. As indicated in the previous equations, HOCl is produced on a one-to-one basis by the addition of Cl$_2$ gas, along with the reduction of pH, which limits the conversion of OCl$^-$. Cell membranes are rapidly penetrated by HOCl because lipid structures accept nonpolar materials. Passage of OCl$^-$ is much slower, and this mitigates its killing. Although a low pH is best for disinfection by chlorine, the pH is often adjusted upward to reduce the chlorine-like odor and because swimmers suffer eye irritation when the concentration of HOCl is high.

AFFECTING FACTORS

Factors affecting the process are:

1. *Form of Chlorine.* Hypochlorous acid is more effective than the hypochlorite ion by approximately two orders of magnitude.
2. *pH.* Because the free-chlorine species are related to pH, one would expect a relationship between efficiency and pH. Empirically, it has been found that chlorine dosages must be increased to compensate for higher pH.
3. *Concentration, Contact Time.* The chlorine concentration and contact time relationship is often expressed by

$$C^n t_p = k, \tag{5}$$

C = concentration of chlorine, mg/L

t_p = time required for given percent kill, min

n, k = experimentally derived constants for a given system

4. *Temperature.* The effects of temperature variations can be modeled by the following equation derived from the van't Hoff–Arrhenius equation:

$$\ln \frac{t_1}{t_2} = \frac{E'(T_2 - T_1)}{R_{T_1, T_2}} \tag{6}$$

where t_1, t_2 = time required for given kills
T_1, T_2 = temperature corresponding to t_1 and t_2, K
R = gas constant, 1.0 cal/K·mol
E' = activation energy related to pH

CHLORINATION PROCESS

Chlorine is a strong oxidant and reacts with almost any material that is in a reduced state. In water, this usually consists of Fe^{2+}, Mn^{2+}, H$_2$S, and organics. Ammonia (NH$_3$) is sometimes present in small quantities or may be added for purposes to be presently discussed. These oxidizable materials consume chlorine before it has a chance to act as a disinfectant. The amount of chlorine required for this purpose must be determined experimentally because the exact nature and the quantity of oxidizable material in water are seldom known. A typical titration curve is shown in Fig. 2.

The products of organics oxidized by chlorine are undesirable. Organic acids (humic, fulvic) form chlorinated hydrocarbons or trihalomethanes (THMs) that are suspected of being carcinogenic. Because THMs are very seldom associated with groundwater, they are a primary concern where surface water is used. Lifetime consumption of water with THMs at a level greater than 0.10 mg/L is considered a potential cause of cancer by the Environmental Protection Agency. Minute quantities of phenolic compounds react with chlorine to form severe taste and odor problems. The original organics must be removed before chlorination, and undesirable compounds must be removed after chlorination, or the compounds must be prevented from forming. The compounds can be removed by adsorption onto activated carbon, or their formation can be prevented by the substitution of chloramines, which do not react with organics or phenols, for free chlorine. Chloramines can be formed by adding a small quantity of ammonia to the water. The reactions of chlorine with ammonia are as follows:

$$\text{NH}_3 + \text{HOCl} \longrightarrow \text{NH}_2\text{Cl (monochloramine)} + \text{H}_2\text{O} \tag{7}$$

$$\text{NH}_2\text{Cl} + \text{HOCl} \longrightarrow \text{NHCl}_2 \text{ (dichloramine)} + \text{H}_2\text{O} \tag{8}$$

$$\text{NHCl}_2 + \text{HOCl} \longrightarrow \text{NCl}_3 \text{ (nitrogen dichloride)} + \text{H}_2\text{O} \tag{9}$$

These reactions are dependent on several factors; the most important are pH, temperature, and reactant quantities. At a pH greater than 6.5, monochloramine is the predominant species. Combined residuals

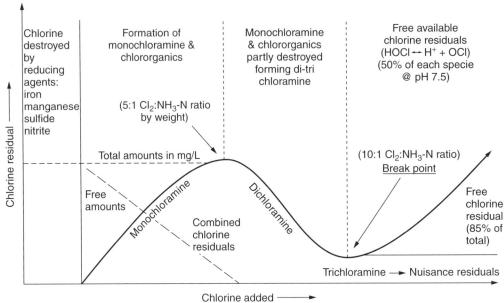
Figure 2.

are less effective as a disinfectant, so a concentration of 2–3 mg/L and contact time in excess of 30 min are often required. Chloramines are persistent and provide continued protection against regrowth in the distribution system.

APPLICATION AND DESIGN CONSIDERATIONS

Provision may be made for applying chlorine at several points within the water treatment processes. When treating raw water of good quality, no early application may be necessary, yet it is advisable to design a plant to allow for easy addition of early applications later, should future conditions require them.

Chlorine may be added to the incoming flow (prechlorination) to assist in oxidizing inorganics to arrest biological action that may produce undesirable gases in the sludge at the bottom of clarifiers. Chlorine is frequently added prior to filtration to keep algae from growing at the medium surface and to prevent large populations of bacteria from developing within the filter medium.

Mixing is one of the most important aspects of the chlorination process. The velocity gradient must be sufficient to ensure a uniform concentration of chlorine throughout the water and to break up any remaining flocculant material that might shield microorganisms from contact with chlorine. A contact chamber must be provided to ensure adequate kill time. In water treatment plant operations, mixing and contact operations may be accomplished by sectioning off part of the clear well.

CHLORINE GAS SAFETY CONSIDERATIONS

Chlorine gas is a very strong oxidant that is toxic to humans. It is heavier than air, so it spreads slowly at ground level. Therefore, extreme care must be exercised in its manufacture, shipping, and use. Accounts of evacuations of populated areas because of rail or barge accidents involving chlorine gas are common news items. The use of hypochlorite is often mandated when large quantities of chlorine are needed in treatment plants in highly populated areas.

Safe and effective application of chlorine requires specialized equipment and considerable care and skill on the part of the plant operator. Liquefied chlorine is delivered to water treatment plants in tanks containing from 75 to 1000 kg. Large plants may be designed to allow using chlorine directly from a tank car. In such cases, designers should be aware of the Interstate Commerce Commission (ICC) and Occupational Health and Safety Agency (OSHA) regulations for shipping and handling chlorine.

Safety considerations mandate storing chlorine tanks in a separate room. Storage and operating rooms should not be directly connected to other enclosed areas of the treatment plant. All doors to these facilities should be open to the outside, and windows should be provided for visual inspection from the outside. Safety equipment, including masks with air tanks, chlorine detection devices, and emergency repair equipment, should be provided in strategic locations.

READING LIST

Metcalf & Eddy, Inc. (2003). *Wastewater Engineering: Treatment, Disposal, Reuse*, 4th Edn. McGraw-Hill, New York, pp. 1218–1286.

Peavy, H.S, Rowe, D.R., and Tchbanoglous, G. (1985). *Environmental Engineering*, International ed. McGraw-Hill, New York, pp. 182–190.

CHLORINATION BY-PRODUCTS

J. Michael Wright
Harvard School of Public Health
Boston, Massachusetts

Raghuraman Venkatapathy
Oak Ridge Institute for Science
and Education
Cincinnati, Ohio

June M. Weintraub
City and County of San
Francisco Department of Public
Health
San Francisco, California

BACKGROUND

Disinfection of drinking water is considered one of the most successful public health interventions, effectively eliminating many waterborne diseases (e.g., cholera) from many parts of the world. In 1908, chlorine was the first disinfectant to be used on a continuous basis to treat water in the United States (1). Chlorine can be used for primary or secondary disinfection via gaseous chlorine, sodium hypochlorite, or calcium hypochlorite. According to an American Water Works Association (AWWA) survey of medium (10,000–100,000 people served) and large (>100,000 people served) water systems in the United States, chlorine gas was used for disinfection by 84% of the surveyed utilities (2). Chlorination was also the predominant choice for small systems (<10,000 people served) surveyed by the AWWA, with all of the groundwater users relying on chlorine gas or sodium/calcium hypochlorite for disinfection (3). Chlorine is the primary disinfectant used for water treatment worldwide, but chlorine dioxide and ozone are also common in many European countries (see Alternative Disinfectants chapter).

Chlorination is a well-defined and fairly cost-effective treatment process. Chlorine provides adequate residual for treatment throughout the distribution system to combat bacteria, viruses, and some protozoa (e.g., *Giardia*) but may not efficiently inactivate resistant pathogens, such as *Cryptosporidium* (4). Gaseous chlorine is explosive and can present safety concerns, so many U.S. systems are switching to less hazardous types of application (5). An additional disadvantage of chlorine disinfection is the formation of potentially hazardous disinfection by-products (DBPs).

DBP FORMATION AND OCCURRENCE

In 1974, the trihalomethanes (THMs) were the first DBPs to be identified (6,7). DBPs are formed when organic or inorganic matter present in water combines with disinfectants such as chlorine. DBP formation can vary seasonally and is influenced by numerous water quality and treatment parameters, including contact time, pH, temperature, natural organic matter, chlorine dose and residual, and bromide levels (8). As shown in Table 1, different water treatment processes will influence the type and amount of DBPs formed. Recent analytical improvements have led to the identification of over 600 DBPs in laboratory scale studies and another 250 DBPs in distribution system samples (17). While numerous halogenated DBPs have been identified, they account for less than half of the total organic halide concentrations detected in chlorinated water (18).

The main chlorination DBPs include halogenated species such as the THMs, haloacetic acids (HAAs), haloacetonitriles (HANs), haloacetaldehydes (HAs), haloketones (HKs), and halonitromethanes (HNMs). The

Table 1. By-products Formed During Disinfection of Drinking Water

Disinfectant	Organohalogenated DBPs[a]	Inorganic DBPs	Nonhalogenated DBPs
Chlorine	THMs,[b] HAAs,[b] HANs, HAs, HNMs, HKs, NDMA, bromohydrins, chlorophenols, MX and brominated MX analogues	Chlorate, iodate	Aldehydes, cyanoalkanoic acids, alkanoic acids, carboxylic acids, benzene
Chloramines	THMs[b], HAAs,[b] HAs, NDMA, HKs, MX, cyanogen chloride, organic chloramines	Chlorate, nitrate, nitrite, hydrazine	Aldehydes, ketones
Chlorine dioxide	HAAs, HANs, HAs, HKs, MX	Chlorate, chlorite	Carboxylic acids
Ozone	Bromoform, monobromoacetic acid, dibromoacetic acid, dibromacetone, cyanogen bromide	Bromate, chlorate, iodate, ozonates, hydrogen peroxide, hypobromous acid, epoxides	Aldehydes, ketones, ketoacids, carboxylic acids, iodic acid, aldo acids

[a]Abbreviations: DBPs, disinfection by-products; THMs, trihalomethanes; HAAs, haloacetic acids; HAs, haloacetaldehydes; HANs, haloacetonitriles; HNMs, halonitromethanes; HKs, haloketones; NDMA, *N*-nitrosodimethylamine; MX, 3-chloro-4-(dichloromethyl)-5-hydroxy-2(5*H*)-furanone.
[b]Includes the iodinated DBPs.
Source: References 9–16.

two most predominant groups of DBPs include the THMs and the HAAs (19). Chloroform (TCM) is typically the most prevalent THM found in chlorinated water, although the brominated THMs can predominate in bromine-rich water. Dichloroacetic acid (DCAA) and trichloroacetic acid (TCAA) are the most prevalent of the nine HAAs typically measured in drinking water.

THM and HAA concentrations found in U.S. finished water are usually less than 0.1 mg/L, while the HANs, HKs, HAs, and HNMs rarely exceed 0.01 mg/L (15,20,21). 3-Chloro-4-(dichloromethyl)-5-hydroxy-2(5H)-furanone (MX) is typically found at an order of magnitude less than most other halogenated DBPs (22,23), but recent findings in the United States and Russia have detected MX levels exceeding 300 ng/L (15,24). Other chlorinated DBPs such as N-nitrosodimethylamine (NDMA) are found in chlorinated and chloraminated water at concentrations usually below 10 ng/L (25,26).

DISINFECTION AND DBP REGULATIONS

As part of the Stage 1 Disinfection By-products Rule (DBPR), the U.S. EPA (27) established maximum residual disinfectant levels in water distribution systems for chlorine (4 mg/L as Cl_2), chloramines (4 mg/L as Cl_2), and chlorine dioxide (0.8 mg/L). Stage 1 also established maximum contaminant levels (MCLs) for bromate (0.010 mg/L), chlorite (1.0 mg/L), and HAA5 [sum of monochloroacetic acid (MCAA), DCAA, TCAA, monobromoacetic acid, and dibromoacetic acid] (0.060 mg/L) and reduced the THM4 [sum of chloroform (TCM), bromodichloromethane (BDCM), dibromochloromethane (DBCM), and bromoform (TBM)] MCL to 0.080 mg/L. The Stage 2 DBPR proposes to change the monitoring requirement for DBPs from a running annual average to a locational running annual average (28).

Members of the European Union are required to regulate total THMs at 0.1 mg/L and bromate at 0.25 mg/L, based on the Council of the European Union directive from 1998. As shown in Table 2, member countries have the discretion to establish other standards or set lower maximum acceptable concentrations (MACs) for individual DBPs. Canada has a similar MAC for THM4 and has recently proposed a MAC of 0.016 mg/L for BDCM (31). Australia has nonenforceable national guidelines for several DBPs, including 0.25 mg/L for THM4 (32). The World Health Organization (16) drinking water guidelines include values for TCM (0.20 mg/L), BDCM (0.06 mg/L), DBCM (0.10 mg/L), TBM (0.10 mg/L), dibromoacetonitrile (0.07 mg/L), MCAA (0.02 mg/L), TCAA (0.20 mg/L), cyanogen chloride (0.07 mg/L), formaldehyde (0.90 mg/L), and trichlorophenol-2,4,6 (0.20 mg/L). WHO provisional guideline values include DCAA (0.05 mg/L), bromate (0.01 mg/L), chloral hydrate (0.01 mg/L), dichloroacetonitrile (0.02 mg/L), chlorite (0.70 mg/L), and chlorate (0.70 mg/L).

HEALTH EFFECTS OF CHLORINE AND CHLORINATION BY-PRODUCTS

Existing toxicologic data on chlorine carcinogenicity is equivocal. No dose–response relationships have been

Table 2. Standards for Disinfection By-products in Drinking Water

Country	Disinfection By-product Standards,[a] mg/L				
	THM4	BDCM	HAA5	Bromate	Chlorite
United States	0.08		0.06	0.01	1.00
Canada	0.10	0.016[b]			
United Kingdom	0.10			0.01	
Germany	0.01				0.20
Italy	0.03			0.25	0.80
Australia[c]	0.25				
WHO[c]		0.06		0.01	0.70[d]

[a]Abbreviations: THM4, sum of bromodichloromethane, dibromochloromethane, bromoform, chloroform; BDCM, bromodichloromethane; HAA5, sum of monochloroacetic acid, dichloroacetic acid, trichloroacetic acid, monobromoacetic acid, dibromoacetic acid; WHO, World Health Organization.
[b]Proposed guideline.
[c]Nonenforceable guidelines.
[d]Provisional guideline.
Source: References 16, 28–33.

observed, but positive findings for lymphoma and leukemia were detected among female rats (34,35). Chloroform is the most studied DBP and was the first to be recognized as a carcinogen in 1976. Liver and kidney cancer in mice and rats and central nervous system depression in mice have been reported following administration of high doses of chloroform (36,37). Similar effects have been noted in animals following exposure to BDCM (37–40). Other DBPs reported to be carcinogenic in animals include DBCM, TBM, MX, NDMA, the HANs, and the HAAs (41–45). Despite exposure assessment limitations, epidemiologic studies have also reported associations between DBP exposure and bladder cancer (46,47).

Much of the ongoing DBP research is focused on potential reproductive and developmental toxicity. Previous studies showed teratogenic effects in animals following administration of HAAs and HANs (48–53). Full litter resorptions were reported in rats following dosing with BDCM and bromoform, and sperm abnormalities have also been reported in male rats following ingestion of BDCM in water (54,55). Klinefelter and colleagues (56) and Tyl (57) have reviewed most of the existing reproductive and developmental toxicity data. Associations between DBPs and a variety of reproductive and developmental outcomes reported in the epidemiologic literature have also been reviewed in detail elsewhere (58–61). Overall, the weight of evidence for reproductive and developmental effects of DBPs is mixed for many endpoints. Toxicological and epidemiological studies examining various effects are ongoing and may further elucidate the relationship between DBP exposures and potential health effects.

ROUTES OF EXPOSURE

The main route of exposure for nonvolatile chlorinated DBPs (e.g., HAAs) is from ingestion of water or water-based beverages. Exposure to volatile DBPs, such as the THMs, can occur through inhalation, dermal absorption, or ingestion. Since inhalation and dermal exposures to volatile DBPs may be greater than from ingestion, activities such as bathing, showering, and swimming need to be considered when determining overall exposure levels (62). Point-of-use filtration devices can be used to minimize DBP ingestion exposures (24). Boiling water prior to consumption is also effective at removing many DBPs, especially the volatile compounds (63,64). In addition to point-of-use treatment, DBPs can be reduced from a variety of changes in water treatment processes or from source water protection efforts (see Alternative Disinfectants chapter). Changes in water treatment implemented to reduce DBPs need to be carefully balanced with protection against microbial pathogens, so that drinking water supplies are not compromised.

BIBLIOGRAPHY

1. Wigle, D.T. (1998). Safe drinking water: a public health challenge. *Chronic Dis. Can.* **19**(3): 103–107.
2. American Water Works Association (2000). Committee report: disinfection at large and medium-sized systems. *J. Am. Water Works Assoc.* **92**(5): 32–43.
3. American Water Works Association (2000). Committee report: disinfection at small systems. *J. Am. Water Works Assoc.* **92**(5): 24–31.
4. Korich, D.G. et al. (1990). Effects of ozone, chlorine dioxide, chlorine, and monochloramine on *Cryptosporidium parvum* oocyst viability. *Appl. Environ. Microbiol.* **56**(5): 1423–1428.
5. Weinberg, H.S., Unnam, V.P., and Delcomyn, C.D. (2003). Bromate in chlorinated drinking waters: occurrence and implications for future regulation. *Environ. Sci. Technol.* **37**(14): 3104–3110.
6. Rook, J.J. (1974). Formation of haloforms during chlorination of natural waters. *Water Treat. Exam.* **23**(2): 234–243.
7. Bellar, T.A., Lichtenberg, J.J., and Kroner, R.C. (1974). The occurrence of organohalides in chlorinated drinking water. *J. Am. Water Works Assoc.* **66**(22): 703–706.
8. Singer, P.C. (1994). Control of disinfection by-products in drinking water. *J. Environ. Eng.* **4**: 727–744.
9. Andrews, R.C. et al. (1990). Occurrence of the mutagenic compound MX in drinking water and its removal by activated carbon. *Environ. Technol.* **11**: 685–694.
10. Bichsel, Y. and von Gunten, U. (1999). Oxidation of iodide and hypoiodous acid in the disinfection of natural waters. *Environ. Sci. Technol.* **33**: 4040–4045.
11. Plewa, M.J. et al. (2004). Chemical and biological characterization of newly discovered iodoacid drinking water disinfection byproducts. *Environ. Sci. Technol.* **38**(18): 4713–4722.
12. Richardson, S.D. et al. (1994). Multispectral identification of chlorine dioxide disinfection byproducts in drinking water. *Environ. Sci. Technol.* **28**(4): 592–599.
13. Richardson, S.D. et al. (1999). Identification of new ozone disinfection by-products in drinking water. *Environ. Sci. Technol.* **33**(19): 3368–3377.
14. Richardson, S.D. et al. (2003). *Environ. Sci. Technol.* **37**(17): 3782–3793.
15. Weinberg, H.S., Krasner, S.W., Richardson, S.D., and Thruston, A.D. Jr. (2002). The occurrence of disinfection by-products (DBPs) of health concern in drinking water: results of a nationwide DBP occurrence study, EPA/600/R02/068, U.S. Environmental Protection Agency, National Exposure Research Laboratory, Athens, GA, 2002. http://www.epa.gov/athens/publications/reports/EPA_600_R02_068.pdf.
16. WHO (2004). *Guidelines for Drinking-Water Quality*, 3rd Edn. World Health Organization, Geneva. http://www.who.int/water_sanitation_health/dwq/gdwq3/en/.
17. Sadiq, R. and Rodriguez, M.J. (2004). Fuzzy synthetic evaluation of disinfection by-products—a risk-based indexing system. *J. Environ. Manage.* **73**(1): 1–13.
18. Stevens, A.A., Moore, L.A., and Miltner, R.J. (1989). Formation and control of non-trihalomethane disinfection by-products. *J. Am. Water Works Assoc.* **81**(8): 54–60.
19. Singer, P.C. (2002). Occurrence of haloacetic acids in chlorinated drinking water. *Water Supply* **2**(5–6): 487–492.
20. Krasner, S.W. et al. (1989). The occurrence of disinfection byproducts in US drinking water. *J. Am. Water Works Assoc.* **81**: 41–53.
21. Roberts, M.G., Singer, P.C., and Obolensky, A. (2002). Comparing total HAA and total THM concentrations using ICR data. *J. Am. Water Works Assoc.* **94**: 103–114.
22. Wright, J.M. et al. (2002). 3-Chloro-4-(dichloromethyl)-5-hydroxy-2(5H)-furanone (MX) and mutagenic activity in Massachusetts drinking water. *Environ. Health Perspect.* **110**(2): 157–164.
23. Meier, J.R. et al. (1987). Studies on the potent bacterial mutagen 3-chloro-4-(dichloromethyl)-5-hydroxy-2(5H)-furanone: aqueous stability, XAD recovery and analytical determination in drinking water and in chlorinated humic acid solutions. *Mutat. Res.* **189**: 363–373.
24. Egorov, A. et al. (2003). Exposures to drinking water chlorination by-products in a Russian City. *Int. J. Hyg. Environ. Health* **206**: 1–13.
25. Najm, I. and Trussell, R.R. (2001). NDMA formation in water and wastewater. *J. Am. Water Works Assoc.* **93**(2): 92–99.
26. California Department of Health Services (2002). Studies on the occurrence of NDMA in drinking water. http://www.dhs.ca.gov/ps/ddwem/chemicals/NDMA/studies.htm.
27. U.S. Environmental Protection Agency (2001). Office of Water (4606M), EPA 816-F-01-010, May. Stage 1 Disinfectants and Disinfection Byproducts Rule: A Quick Reference Guide. http://www.epa.gov/safewater/mdbp/qrg_st1.pdf.
28. U.S. Environmental Protection Agency (2003). Office of Water (4606M), EPA 816-D-03-002, November. The Stage 2 Disinfectants and Disinfection Byproducts Rule (Stage 2 DBPR) Implementation Guidance-Draft for Comment. http://www.epa.gov/safewater/stage2/pdfs/st2_dbpr_ig_11-11-03.pdf.
29. Hydes, O. (1999). European regulations on residual disinfection. *J. Am. Water Works Assoc.* **91**(1): 70–74.
30. Collivignarelli, C. and Sorlini, S. (2004). Trihalomethane, chlorite and bromate formation in drinking water oxidation of Italian surface waters. *J. Water SRT Aqua* **53**: 159–168.
31. Health Canada (2004). Federal–Provincial–Territorial Committee on Drinking Water. Trihalomethanes in Drinking Water. August. http://www.hc-sc.gc.ca/hecs-sesc/water/pdf/trihalomethanes_drinking_water.pdf.
32. Simpson, K.L. and Hayes, K.P. (1998). Drinking water disinfection by-products: an Australian perspective. *Water Res.* **32**(5): 1522–1528.
33. Drinking Water Inspectorate (2000). Water, England and Wales: The Water Supply (Water Quality) Regulations, No. 3184. http://www.dwi.gov.uk/regs/si3184/3184.htm#.
34. Soffritti, M., Belpoggi, F., Lenzi, A., and Maltoni, C. (1997). Results of long-term carcinogenicity studies of chlorine in rats. *Ann. N.Y. Acad. Sci.* **837**: 189–208.

35. National Toxicology Program (1992). NTP toxicology and carcinogenesis studies of chlorinated water (CAS Nos. 7782-50-5 and 7681-52-9) and chloraminated water (CAS No. 10599-90-3) (deionized and charcoal-filtered) in F344/N rats and B6C3F1 mice (drinking water studies). *Natl. Toxicol. Program Tech. Rep. Ser.* **392**: 1–466.
36. National Cancer Institute (1976). Report on the carcinogenesis bioassay of chloroform. NTIS PB-264-018. National Cancer Institute, Bethesda, MD.
37. Balster, R.L. and Borzelleca, J.F. (1982). Behavioral toxicity of trihalomethane contaminants of drinking water in mice. *Environ. Health Perspect.* **46**: 127–136.
38. Tumasonis, C.F., McMartin, D.N., and Bush, B. (1985). Lifetime toxicity of chloroform and bromodichloromethane when administered over a lifetime in rats. *Ecotoxicol. Environ. Safety* **9**: 233–240.
39. Dunnick, J.K., Eustis, S.L., and Lilja, H.S. (1987). Bromodichloromethane, a trihalomethane that produces neoplasms in rodents. *Cancer Res.* **47**: 5189–5193.
40. National Toxicology Program (1986). Toxicology and carcinogenesis of bromodichloromethane in F344/N rats and B6C3F1 mice. Technical Report Series No. 321. NIH Publication No. 88–2537, CAS No. 75-27-4. U.S. Department of Public Health and Human Services, Washington, DC.
41. Bull, R.J. et al. (1985). Mutagenic and carcinogenic properties of brominated and chlorinated acetonitriles: by-products of chlorination. *Fund. Appl. Toxicol.* **5**: 1065–1074.
42. Bull, R.J. et al. (1990). Liver tumor induction in B6C3F$_1$ mice by dichloroacetate and trichloroacetate. *Toxicology* **63**: 341–359.
43. Komulainen, H. et al. (1997). Carcinogenicity of the drinking water mutagen 3-chloro-4-(dichloromethyl)-5-hydroxy-2(5*H*)-furanone in the rat. *J. Natl. Cancer Inst.* **89**(12): 848–856.
44. National Toxicology Program (1985). Toxicology and carcinogenesis studies of chlorodibromomethane in F344/N rats and B6C3F$_1$ mice. TR 282. DHHS Publ. No. (NIH) 85–2538. National Institutes of Health, Bethesda, MD.
45. National Toxicology Program (1989). Toxicology and carcinogenesis of tribromomethane (bromoform) in F344/N rats and B6C3F1 mice. Technical Report Series No. 350. NIH Publication No. 89–2805, CAS No. 75-25-2. U.S. Department of Public Health and Human Services, Washington, DC.
46. Villanueva, C.M. et al. (2004). Disinfection byproducts and bladder cancer: a pooled analysis. *Epidemiology* **15**: 357–367.
47. Morris, R.D. et al. (1992). Chlorination, chlorination by-products, and cancer: a meta-analysis. *Am. J. Public Health* **82**: 955–963.
48. Epstein, D.L. et al. (1992). Cardiopathic effects of dichloroacetic acid in Long-Evans rat. *Teratology* **46**: 225–235.
49. Hunter, E.S. III, Roger, E.H., and Schmid, R.A. (1996). Comparative effects of haloacetic acids in whole embryo culture. *Teratology* **54**: 57–64.
50. Smith, M.K. et al. (1988). Teratogenic effects of trichloroacetonitrile in the Long-Evans rat. *Teratology* **38**: 113–120.
51. Smith, M.K., Randall, J.L., Strober, J.A., and Read, E.J. (1989). Developmental toxicity of dichloroacetonitrile: a by-product of drinking water disinfection. *Toxicology* **12**: 765–772.
52. Smith, M.K., Randall, J.L., Read, E.J., and Strober, J.A. (1989). Teratogenic activity of trichloroacetic acid. *Teratology* **10**: 445–451.
53. Smith, M.K., Randall, J.L., Read, E.J., and Strober, J.A. (1992). Developmental toxicity of dichloroacetic acid in Long-Evans rat. *Teratology* **46**: 217–223.
54. Klinefelter, G.R., Suarez, J.D., Roberts, N.L., and DeAngelo, A.B. (1995). Preliminary screening for the potential of drinking water disinfection byproducts to alter male reproduction. *Reprod. Toxicol.* **9**: 571–578.
55. Bielmeier, S.R., Best, D.S., Guidici, D.L., and Narotsky, M.G. (2001). Pregnancy loss in the rat caused by bromodichloromethane. *Toxicol. Sci.* **59**: 309–315.
56. Klinefelter, G.R., Hunter, E.S., and Narotsky, M. (2001). Reproductive and developmental toxicity associated with disinfection by-products of drinking water. In *Microbial Pathogens and Disinfection By-Products in Drinking Water: Health Effects and Management of Risks*, G.F. Craun, F.S. Hauchman, and D.E. Robinson (Eds.). ILSI Press, Washington, DC, pp. 309–323.
57. Tyl, R.W. (2000). Review of animal studies for reproductive and developmental toxicity assessment of drinking water contaminants: disinfection by-products (DBPs). RTI Project No. 07639. Research Triangle Institute, Research Triangle Park, NC.
58. Nieuwenhuijsen, M.J. et al. (2000). Chlorination disinfection by-products in water and their association with adverse reproductive outcomes: a review. *Occup. Environ. Med.* **57**(2): 73–85.
59. Graves, C.G., Matanoski, G.M., and Tardiff, R.G. (2001). Weight of evidence for an association between adverse reproductive and developmental effects and exposure to disinfection by-products: a critical review. *Regul. Toxicol. Pharmacol.* **34**: 103–124.
60. Bove, F., Shim, Y., and Zeitz, P. (2002). Drinking water contaminants and adverse pregnancy outcomes: a review. *Environ. Health Perspect.* **110**(Suppl. 1): 61–74.
61. Hwang, B.F. and Jaakkola, J.J. (2003). Water chlorination and birth defects: a systematic review and meta-analysis. *Arch. Environ. Health* **58**(2): 83–91.
62. Backer, L.C. et al. (2000). Household exposures to drinking water disinfection by-products: whole blood trihalomethane levels. *J. Expo. Anal. Environ. Epidemiol.* **10**(4): 321–326.
63. Wu, W.W., Benjamin, M.M., and Korshin, G.V. (2001). Effects of thermal treatment on halogenated disinfection by-products in drinking water. *Water Res.* **35**(15): 3545–3550.
64. Krasner, S.W. and Wright, J.M. (2005). The effect of boiling water on disinfection by-product exposure. (In review).

CLASSIFICATION AND ENVIRONMENTAL QUALITY ASSESSMENT IN AQUATIC ENVIRONMENTS

HUAN FENG
VICTOR ONWUEME
Montclair State University
Montclair, New Jersey

LEWIS SCHNEIDER
North Jersey District Water Supply Commission
Wanaque, New Jersey

Water plays an important role in our daily life. Natural water in each compartment on Earth is balanced through a hydrological cycle. With the population increase, industrialization, and rapid urbanization, water quality has been affected and an increased demand for water use has occurred. Anthropogenic substances are discharged into our aquatic environment. As a consequence, water contamination has been found in many rivers, streams, and lakes globally. Facing our current water environmental problems, water resources management and aquatic environmental protection and restoration are important issues to us. In the United States, water quality has been improved significantly since 1972 when the Clean Water Act was enacted by the U.S. Environmental Protection Agency. In this article, we introduce the classification of waters based on the reservoir they reside in and discuss the water quality, water pollution, and aquatic environmental legislations. This article gives a general overview of water classification, quality assessment, and legislations. Greater efforts are still needed to protect the aquatic environment and improve water quality.

INTRODUCTION

Water is essential to human life. A water molecule is made up of two positively charged atoms of hydrogen and one negatively charged atom of oxygen ($2H^+ + O^{2-} = H_2O$). Each hydrogen atom of the water molecule is linked to the oxygen atom by a covalent bond. It is a polar molecule. Because of the distribution of the charges, the molecules interact with each other by forming hydrogen bonds between the molecules. Therefore, the water molecule is stable. The properties of water are determined by the structure of the water molecule. It has a high specific heat (1 calorie $°C^{-1}$ g^{-1}), a high latent heat of fusion (80 calories), and evaporation (540 calories). Light can only travel a maximum of a few hundred meters through water, sound can travel thousands of kilometers through water, and water is essential to life. Pure water freezes at $0\,°C$ and boils at $100\,°C$ at 1 atmosphere pressure. The maximum density of freshwater is at $4\,°C$ (for seawater it is at its freezing point $-1.9\,°C$) at 1 atmosphere pressure. The pH ($pH = -\log_{10}[H^+]$) value of pure water is $pH = 7$, rainwater $pH = 5.0–5.6$, river water $pH = 6–7$ (6.8 on average), and seawater $pH = 7.5–8.5$ (7.8 on average).

CLASSIFICATION OF WATER AND HYDROLOGIC CYCLE

Based on their physical and chemical properties and their storage reservoirs, the waters on Earth are classified as follows (1–3):

Rainwater—Precipitation from the atmosphere when water vapor in the air condenses and water droplets grow sufficiently large.

River (Stream) Water—Precipitation that does not evaporate or infiltrate into the ground runs off the surface and flows back toward the ocean. Best measure of water volume carried by a river is discharge that is defined as the amount of water that passes a fixed point in a given amount of time. Usually, it is expressed as cubic meter per second.

Lake and Pond—Lakes are inland depressions that hold standing freshwater year-round. Ponds are generally considered as small bodies of water.

Groundwater—The second largest reservoir of freshwater. The source of groundwater is caused by infiltration, a process of water percolating through the soil and into fractures and permeable rocks. Presently, groundwater is being removed faster than it can be replenished in many areas.

Seawater—Oceans contain more than 97% of all liquid water in the world. In open ocean, seawater has approximately 35 g of salts in 1000 g of seawater; thus, salinity = 35‰.

The hydrological cycle keeps the water balance on Earth. It describes the circulation of water when it evaporates from land, water, and organisms or transpires from plants, enters the atmosphere and condenses, and then precipitates back to the Earth's surface. Water can move underground by infiltration or overland by runoff into streams, rivers, lakes, and oceans. Renewable water supplies are made up of surface runoff and infiltration into accessible freshwater aquifers. It is estimated that 1.5 billion people lack access to an adequate supply of drinking water and nearly 3 billion lack acceptable sanitation (3). Regarding the water use in each sector, on the worldwide average, agriculture consumes about 69% of total water withdrawal and industry accounts for about 25% of all water use (3).

ENVIRONMENTAL QUALITY ASSESSMENT

Water Pollution

Since industrialization, many inorganic and synthetic organic chemicals have been used to make pesticides, plastics, pharmaceuticals, pigments, and so on. They are highly persistent and toxic and tend to bioaccumulate in food chains. These characteristics are also true of many heavy metals such as mercury, lead, cadmium, and nickel. The two most important sources of toxic, inorganic and organic chemicals in water are improper disposal of industrial and household wastes and runoff of pesticides from high-use areas (such as fields, roadsides, and golf courses). It was reported that more than 25% of the mass of all materials dumped into the ocean is dredged materials from ports and waterways (2,4). Industrial waste is one major waste dumped at sea. Between 1890 and 1971, for example, 1.4×10^6 m^3 solid waste has been dumped into the water of the New York Bight (5,6). In 1972 the U.S. Marine Protection, Research, and Sanctuaries Act (the Ocean Dumping Act) prohibited dumping of all types of materials into ocean waters without a permit from the EPA. In fact, the United States has stopped its ocean dumping except for uncontaminated dredged materials with a permit from the U.S. Corps of Engineers.

Any physical, biological, or chemical change in water quality that adversely affects living organisms can be considered pollution. Pollutants can be from point sources (i.e., discharge pollutants from specific locations such as factories) and non-point sources (i.e., no specific location of discharge, such as agricultural fields and feedlots). At present, at least 2.5 billion people in less developed countries lack adequate sanitation, and about half lack access to clean drinking water (7). Sewage treatment plants and pollution-control devices can greatly reduce pathogens (8,9). To remove intestinal bacteria, drinking water is generally disinfected via chlorination.

Most urban sewer systems in the United States were built in the late 1800s and early 1900s. In 1972, the Federal Clean Water Act, in an effort to make U.S. water "fishable and swimmable," mandated the upgrade of sewage treatment to the secondary treatment level by 1977 (8,9). Although the discharge flow has been increasing over the years because of a population growth and an increasing discharge of waste by industry, total suspended solids, 5-day biochemical oxygen demand (BOD), and oil and grease have decreased in the waste stream. In most areas, heavy metal concentrations in the waste stream have been reduced as a result of enhanced sewage treatment and more strict federal and state policies and criteria that prevent these contaminants from entering the waste stream (10,11). The amount of dichlorodiphenyltrichloroethane (DDT) shows a dramatic decrease with time, although it is being gradually flushed from the land. Because of their toxic effect, polychlorinated biphenyls (PCBs) are also no longer used in a large quantity at present. Nutrient concentrations in coastal waters have been changed with land practice change, and the increasing supply of nutrients results in deoxygenation in some coastal waters as the surplus plant material decays (12–14).

Water Legislation

Water legislation in the United States was designed primarily to protect the nation's water from pollution, which included the Clean Water Act (CWA) (1972), a major revision of the prior Federal Water Pollution Control Act; and the Safe Drinking Water Act (1974) to protect populations consuming water and prevent groundwater contamination; as well as other laws such as the Resource Conservation and Recovery Act (RCRA) and the Comprehensive Environmental Response, Compensation and Liability Act (CERCLA) to prevent and mitigate contaminated groundwater. "The objective of this act (CWA) is to restore and maintain the chemical, physical and biological integrity of the Nation's waters." A major concern of the CWA was to prevent discharges to navigable water bodies of the United States. The CWA did this by making it illegal to discharge to a navigable waterway without a permit. The National Pollutant Discharge Elimination System (NPDES) defines this permitting process and implements the requirements for a point source discharger to monitor conventional, toxic, and nonconventional pollutants. Reduction (conventional pollutants) or limiting (toxicants) of these effluent discharges is achieved by the discharger through pollution control technologies. Pretreatment standards for indirect dischargers are also required in the CWA. Industries discharging indirectly to a publicly owned water treatment work (POTW) through a wastewater collection system must also comply with pollutant limits. Section 404 of the CWA includes permits oversighted by the U.S. Army Corp of Engineers to discharge dredged spoils to waters of the United States. Section 404 has been interpreted to protect adjacent and nonadjacent waterbodies that covers wetlands. Water quality standards are imposed to maintain the goal of "fishable/swimable waters," which are adopted by states with prior approval from the EPA. The water quality standards include designated use(s) of the waterbody, an antidegradation policy, and numerical and narrative criteria. Non-point source pollution is a key concern for the waterways of the United States and is addressed in the 1972 amendments to the CWA, which include state primacy in land use planning and regulations to control non-point source pollution. Storm water discharges may be considered point source discharges when conveyed through storm sewers. In 1987, the CWA amendments required discharges from municipal storm sewers to meet criteria under the act.

The primary goals of the SDWA were to ensure potable water at the consumer's tap and protect groundwater from contamination. The act applies to public water systems (PWSs), those servicing over 15 service connections or at least 25 persons. PWSs are in turn divided into community and noncommunity systems. The differentiation of community versus noncommunity is that community systems are used year-round, whereas use of noncommunity systems may vary from at least 60 days to more the 6 months. To protect the public health, the SDWA establishes primary and secondary standards. The primary standards are based on specific feasible maximum contaminant levels (MCLs) and maximum contaminant level goals (MCLGs) for the future. The primary standards include microbiological monitoring, inorganics (primarily metals and nitrate/nitrite), volatile organic compounds, synthetic organic compounds and pesticides, asbestos, and radionuclides. The SDWA establishes monitoring frequencies and analytical reporting, public notification, recordkeeping, and remediation requirements and procedures for these primary standards. Secondary contaminants are based on aesthetics, which include taste, odor, and appearance of the water, and adversely affect public welfare. Recommended ranges and limits are given for secondary standards that include physical, chemical, and biological characteristics. Similar to the primary standards, secondary standards establish monitoring requirements, public notification, and remediation. State underground injection wells are also regulated by the SDWA. Sole source aquifers are protected as well as wellhead areas and defined watersheds for surface water sources. Amendments to the SDWA, particularly in 1996, included grant and loan funding, a consumer confidence report designating water quality for annual customer distribution, and the use of risk assessment and cost–benefit analysis for adding new contaminant standards.

Water Quality Today

Significant progress in water protection and quality improvement has been made since the early 1970s. Most progress is from municipal sewage treatment facilities. In the United States, the Environmental Protection Agency has focused more on watershed-level monitoring and protection since 1998. States are required to identify waters not meeting water quality goals and develop total maximum daily loads for each pollutant and each listed waterbody. However, some problems remain. Greatest impediments to achieving national goals in water quality are sediment, nutrients, and pathogens, especially from non-point source discharges. Globally, according to Cunningham et al. (3), sewage treatment in wealthier countries of Europe generally equals or surpasses the United States. In Russia, only about half of the tap water supply is safe to drink. In urban areas of South America, Africa, and Asia, 95% of all sewage is discharged untreated into rivers. Two-thirds of India's surface waters are contaminated sufficiently to be considered dangerous to human health.

Pollution control has been advocated and practiced for many years, which includes source reduction and wastewater treatment. The public has recognized that the most effective way to reduce pollution is not producing it or not releasing it into our environment. To reduce the water pollution in our living environment, the wastewater treatment technologies have been developed, which include primary treatment (i.e., physical separation of large solids from the waste stream), secondary treatment (i.e., biological degradation of dissolved organic compounds), and tertiary treatment (i.e., removal of plant nutrients from secondary effluent) (8,9). The present water remediation technologies include oxidation, reduction, neutralization, or precipitation of polluted water. Bioremediation technologies mainly use living organisms to effectively break down contaminants in polluted waters (15).

CONCLUSION

Natural water is essential to our human life and living plants and organisms on Earth. The global hydrological cycle keeps the water balance in all compartments on Earth. Water resources management is important in regulating water use, discharge, and protecting the environment. With human population growth and increasing urbanization, demand on the water supplies is becoming increasingly high. Because of industrialization and economic development, water quality in some regions is becoming a serious environmental problem. Some rivers and streams are polluted by various anthropogenic contaminants. Since the 1970s, water pollution has been realized as a serious concern and water protection has received more and more attention. Since then, water quality has been improved to varying degrees in different countries. Overall, great efforts are still needed to conservatively use our water resource and protect the aquatic environment.

BIBLIOGRAPHY

1. Drever, J.I. (1997). *The Geochemistry of Natural Waters*, 3rd Edn. Prentice-Hall, Upper Saddle River, NJ.
2. Sverdrup, K., Duxbury, A.C., and Duxbury, A.B. (2003). *An Introduction to the World's Oceans*, 7th Edn. McGraw-Hill, New York.
3. Cunningham, W., Cunningham, M.A., and Saigo, B. (2005). *Environmental Science: A Global Concern*, 8th Edn. McGraw-Hill, New York.
4. Sverdrup, K., Duxbury, A.C., and Duxbury, A.B. (2003). *An Introduction to the World's Oceans*, 7th Edn. McGraw Hill, New York.
5. Freeland, G.L. and Swift, D.J.P. (1978). *Surficial Sediments: Marine EcoSystems Analysis (MESA) Program*, MESA New York Bight Atlas Monograph 10, New York Sea Grant Institute, Albany, NY.
6. Duedall, I.W., O'Connors, H.B., Wilson, R.E., and Parker, J.H. (1979). *The Lower Bay Complex*. MESA New York Bight Atlas Monograph, 29, New York Sea Grant Institute, Albany, NY.
7. WHO/UNICEF (World Health Organization/United Nations Children's Fund). (2000). *GlobalWater Supply and Sanitation Assessment 2000 Report*. New York.
8. LaGrega, M.D., Buckingham, P.L., and Evans, J.C. (2003). *Hazardous Waste Management*, 2nd Edn. McGraw-Hill, New York.
9. Metcalf and Eddy. (2003). *Wastewater Engineering*, 4th Edn. McGraw-Hill, New York.
10. Al-Sayed, M., Al-Mershed, A., and Al-Askar, I. (1999). Evaluation and classification of water use rate policy using the Points Credit Index. *Desalination* **123**(2–3): 233–239.
11. Guerbet, M. and Jouany, J.M. (2002). Value of the SIRIS method for the classification of a series of 90 chemicals according to risk for the aquatic environment. *Environ. Impact Assessment Rev.* **22**(4): 377–391.
12. Aller, R.C. et al. (1985). Early chemical disgenesis, sediment–water solute exchange, and storage of reactive organic matter near the mouth of the Changjian, East China Sea. *Continental Shelf Res.* **4**: 227–251.
13. Gu, D., Iricanin, N., and Trefry, J.H. (1987) The geochemistry of interstitial water for a sediment core from the Indian River Lagoon, Florida. *Florida Scientist* **50**: 99–110.
14. Ha., H. and Stenstrom, M.K. (2003). Identification of land use with water quality data in stormwater using a neural network. *Water Res.* **37**(17): 4222–4230.
15. Rittmann, B.E. and McCarty, P.L. (2001). *Environmental Biotechnology: Principles and Applications*. McGraw-Hill, New York.

READING LIST

NJDEP. (2002). *N.J.A.C 7:10A Licensing of Water Supply and Wastewater Treatment System Operators*.

NJDEP. (2004). *NJSDWA N.J.A.C. 7:10*.

USEPA. (1999). *US. 40 CFR 141 National Primary Drinking Water Regulations*.

Powell, F.M. (1998). *Law and the Environment*. West Educational Publications, Indiana.

Tarazona, J.V. et al. (2000). Assessing the potential hazard of chemical substances for the terrestrial environment. Development of hazard classification criteria and quantitative environmental indicators. *Sci. Total Environ.* **247**(2–3): 151–164

Federal Water Pollution Control Act. 33 U.S.C. 2002. USEPA. pp. 1251–1387.

Calamari, D. (2002). Assessment of persistent and bioaccumulating chemicals in the aquatic environment. *Toxicology* **181/182**: 183–186.

El-Manharawy, S. and Hafez, A. (2001). Water type and guidelines for RO system design. *Desalination* **139**(1–3): 97–113.

Mukherjee, A.B., Zevenhoven, R., Brodersen, J., Hylander, L.D., and Bhattacharya, P. (2004). Mercury in waste in the European Union: Sources, disposal methods and risks. *Resources, Conservation and Recycling* **42**(2): 155–182.

COAGULATION AND FLOCCULATION IN PRACTICE

Alper Ozkan
Selcuk University
Konya, Turkey

INTRODUCTION

All waters, particularly surface waters and process wastewaters, contain both dissolved and suspended particles. Very fine particles only a few microns in diameter settle extremely slowly by gravity alone. Therefore, several methods are used to increase the size of solid particles in solid/liquid separation. The processes of coagulation and flocculation are the major techniques used to aggregate particles. Solid particles can be flocculated into relatively large lumps, called flocs, that settle out more rapidly and thus particles are separated more easily from water (1,2). Although the terms coagulation and flocculation are often used interchangeably, or the single term "flocculation" is used to describe both, they are two different processes. Fine particles can be aggregated by neutralizing the electrical charge of the interacting particles, which is "coagulation," and bridging the particles with polymolecules is called "flocculation" (3,4).

COAGULATION

The major interactive forces between suspended particles are of two kinds, attractive and repulsive. The stability of particle suspensions depends on the properties of these forces. Attractive forces arise from van der Waals forces, which are effective only at very close range, whereas repulsive forces are due to the overlap of similarly charged electrical double layers of particles. Aggregation between particles can occur if the magnitude of the interparticle repulsive force does not exceed the corresponding attractive force. Electrostatic forces are responsible mostly for repulsion, so if suspension properties are adjusted to minimize such forces, particles can be coagulated (4,5).

Most solid particles suspended in water have a negative charge; since they have the same type of surface charge, they repel each other when they come close together. Repulsive forces prevent coagulation of particles and also retard their settlement. The specific surface charge of a particle and its magnitude, measured as zeta potential, are indicative of the strength of the repulsive forces due to charge. The charge on particle surfaces can be controlled by suspension pH. At some pH, called point-of-zero charge (PZC), the particles have no net charge and are unstable with respect to coagulation. pH control is a simple method for destabilizing suspensions. However, it may be impractical if the PZC occurs in an inconvenient pH range. Therefore, coagulants are often used to eliminate the negative charge on particle surfaces (4,6). Coagulants have a charge opposite to that of the particle surfaces; this causes charge neutralization that allows particles to come into contact and start to associate in bunches. Inorganic salts, such as $CaCl_2$, $MgCl_2$, $FeCl_3$, $AlCl_3$, $FeSO_4$, and $Al_2(SO_4)_3$, are used as coagulants. However, aluminum and iron salts are the most commonly used in wastewater treatment. The ability of most inorganic salt ions to compress the electrical double layer or the magnitude of the zeta potential is related to the valence of the ionic charge itself. In addition, hydroxy complexes of these are very active; thus they adsorb strongly on solid surfaces and can reverse the sign of the zeta potential. When hydrolyzable metal ions are used in the pH range and at the concentration level where the metal hydroxide is precipitated, sweep coagulation of suspension occurs by precipitating hydroxides. This effect leads to overall destabilization of a particle suspension (4–7).

FLOCCULATION

Polymeric flocculants are used effectively to destabilize fine particle suspensions. The advantage of polymeric flocculants is their ability to produce large, stronger flocs compared to those obtained by coagulation. The flocculating action of polymeric flocculants proceeds via either "charge patch attraction" or "polymer bridging." Charge patch attraction occurs when the particle surface is negatively charged and the polymer is positively charged. On the other hand, bridging is considered a consequence of the adsorption of segments of flocculant macromolecules onto the surfaces of more than one particle. Such bridging links the particles into loose flocs, and incomplete surface coverage ensures that there is sufficient unoccupied surface on each particle for adsorption during the collisions of chain segments attached to the particles. Most flocculants are synthetic polymers based on repeating units of acrylamide and its derivatives, which may contain either a cationic or anionic charge. Both anionic and nonionic polymers can be manufactured at very high molecular weights and thus can form large, rapid-settling, good-compacting flocs. They are widely used in industrial applications (2,4,8).

The kind of initial adsorption of polymers onto a particle varies, depending on the respective charges of both polymer and particle. It may be purely electrostatic if these charges are opposite in sign. If not, then other physicochemical reactions may take place. A cationic polymer, positively charged, can adsorb on the surface of negatively charged particles via electrostatic bonding. The most likely mechanism of adsorption for nonionic polymers is through hydrogen bonding. Anionic polymers and negatively charged suspensions may also adsorb via hydrogen bonding. In suspensions to which a cation has been added, polymer adsorption often also occurs through cation bridging; that is, the cations can form an electrostatic bridge between the negatively charged particle surface and the negatively charged polymer (4,5).

Theoretically, flocculants may be used either after destabilization of the suspension by coagulation, that is, predestabilization, or without destabilization. Polymeric flocculants are more effective in destabilized fine particle suspensions and so the first option that involves predestabilization by coagulation is always better. Destabilization prior to flocculant addition can usually be accomplished by eliminating the charge barrier by pH control or by adding an inorganic salt. This permits small flocs to form; the bridging action of the polymer links them into substantially larger units (2,6,9,10).

BIBLIOGRAPHY

1. Wills, B.A. (1985). *Mineral Processing Technology*, 3rd Edn. Pergamon Press, New York.
2. Laskowski, J.S. (2000). *Aggregation of fine particles in mineral processing circuits. Proc. 8th Int. Miner. Process. Symp.*, Antalya, Turkey, Oct. 16–18, 2000, Balkema, Rotterdam, pp. 139–147.
3. Laskowski, J.S. (1992). In: *Colloid Chemistry in Mineral Processing*. J.S. Laskowski and J. Ralston (Eds.). Elsevier, New York, Chap. 7, pp. 225–241.
4. Klimpel, R.R. (1997). *Introduction to Chemicals Used in Particle Systems*. ERC Particle Science & Technology, Gainesville, FL.
5. Somasundaran, P. (1980). In: *Fine Particle Processing*. P. Somasundaran (Ed.). AIME, New York, Vol. 2, pp. 947–975.
6. Hogg, R. (1999). Flocculation and dewatering. *Int. J. Miner. Process.* **58**: 223–236.
7. Letterman, R.D., Amirtharajah, A., and O'Melia, C.R. (1999). In: *Water Quality and Treatment*. R.D. Letterman (Ed.). McGraw-Hill, New York, Chap. 6, pp. 6.1–6.66.
8. Hogg, R. (1984). Collision efficiency factors for polymer flocculation. *J. Colloid Interface Sci.* **102**: 232–236.
9. Kitchener, J.A. (1972). Principles of action of polymeric flocculants. *Br. Polym. J.* **4**: 217–229.
10. Hogg, R. (1992). *Process design for flocculation, clarification and thickening of fine-particle suspensions. 18th Int. Miner. Process. Congr.*, Sydney, Australia, May 23–28, The Australasian Institute of Mining and Metallurgy, Sydney, pp. 1315–1319.

COLLOIDS AND DISSOLVED ORGANICS: ROLE IN MEMBRANE AND DEPTH FILTRATION

GARY G. LEPPARD
National Water Research Institute
Burlington, Ontario, Canada

JACQUES BUFFLE
University of Geneva
Geneva, Switzerland

INTRODUCTION

A colloid is operationally defined as any particulate with a least dimension in the range of 1.0–0.001 µm; the range includes macromolecules, but not conventional dissolved substances, which tend to be smaller (1,2). For environmental scientists, aquatic colloids are being redefined as any particulate that provides a molecular milieu into and onto which chemicals can escape from bulk water and whose movement is not significantly affected by gravitational settling (3). In practice, colloids and dissolved substances are often defined by a filter operation (4,5). Living microbes, when sufficiently small, can also be considered colloids, albeit colloids capable of very complex behavior (2).

Filtration for water scientists is the separation of particulates from a liquid, achieved by passing a suspension through a porous solid. Through the application of membrane filters and/or depth filters, filtration protocols are commonly used (1) to concentrate or remove particulate matter and (2) to separate suspended particulates into size classes for subsequent biological, chemical, or physical analyses. In water treatment, one discriminates between membrane filters, which are porous membranes that concentrate particulates at or close to the upper filter surface, and depth filters. The latter are relatively inexpensive filters with high collection efficiencies and flow rates that entrap particulates within the filter interior.

Filters for use in the analysis of aquatic samples may be separated into three categories:

1. filters made of a random assembly of fine fibers;
2. filters made of an impermeable material, which has discrete cylindrical holes traversing it and which acts as a true sieve; and
3. "depth filters," which have a spongy structure.

These three categories of filters and their relationships to applicable size ranges and to standard sieves and dialysis membranes are shown in Fig. 1. The various filters in use are complementary in pore size ranges, pore size distributions, and chemical nature, thus providing versatility in minimizing sample perturbation by filters.

PRACTICAL ADVANTAGE OF BETTER UNDERSTANDING

Understanding better the roles of colloids and dissolved organics in a filtration process is particularly useful when considering the following goals:

1. understanding the processes of fouling and biofouling as they relate to water treatment filters (6,7);
2. optimizing the design of filters for improved performance in specific situations (5,6), focused on filters designed to be cleaned *in situ*; and
3. improving the quality of size fractionations for limnologists and oceanographers focused on capturing and analyzing natural microbiological consortia with minimal artifact (7).

Some specific concepts and advances are noteworthy. Biofouling of filters is an unwanted progressive deposition of microorganisms and their consequent formation of biofilms, which clog filter pores and eventually establish an additional (but different) filter at the upper surface

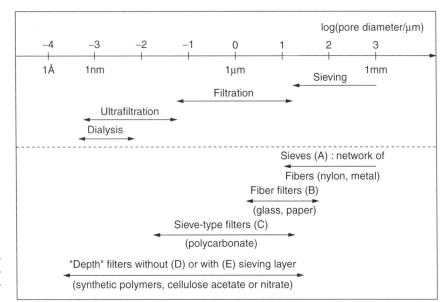

Figure 1. Size ranges for sieving and filtration, including depth filters, and for application of the most important filter types. Reproduced with permission from Buffle et al. (5).

of a membrane filter. The economic consequences of this process are considerable (6). New technology is leading to the identification of specific nanoscale organic entities that microorganisms produce to facilitate their attachment to filter surfaces (8). Such research may lead to better methods for cleaning filters *in situ*. In general, fouling may also result from purely physicochemical interactions between organic and inorganic colloids and a membrane surface, leading to colloid coagulation (5). In such cases, controls over the flow rate and physicochemical conditions are essential.

As realization increases of the many biogeochemical processes dominated by certain abundant living colloids in the oceans, the need to analyze isolated important biological species grows in importance (7). To isolate these microbes and separate them according to size (and then possibly according to dominant species) for subsequent characterization and assessment of their biogeochemical roles, there is a requirement for well-controlled filtration procedures. An additional and valuable consideration is filter selection according to filter compatibility with various techniques of analytical optical microscopy for studying fresh isolates and their minimally perturbed natural consortia while they sit on filters.

ABUNDANT COLLOIDS AND DISSOLVED ORGANICS

In natural waters, the most commonly encountered colloids are fractal aggregates of fulvic and humic acids, living cells of colloidal size, viruses, fibrils (linear aggregates of biopolymers rich in polysaccharide), refractory skeletal fragments of organisms, clay minerals, and oxyhydroxides of iron and manganese (9). Three major groups enable one to describe colloid coagulation properties: inorganic (compact) colloids, fibrillar (rather rigid) large biopolymers, and small (nanometers) fulvic acids (1). Fulvic and humic acids are at the borderline between dissolved and colloidal organics, whereas organic metabolites emanating from leaky cells, lysing cells, or excreta are usually dissolved. In engineered aquatic ecosystems, such as water treatment facilities, the same colloids and dissolved organics tend to be present but in different relative amounts. These colloids and dissolved organics can have an impact on the biological activities and clarity of both natural and engineered waters through their influences on flocculation (10,11). The impact on filtration processes of those most abundant can be considerable, leading to defective filter performance and misleading information (5).

KINDS OF FILTERS AND THEIR APPLICATION: GENERAL CONSIDERATIONS

An in-depth review of the various kinds of filters, used currently for water treatment and for the size fractionation of aquatic particles and colloids, is beyond the scope of this article. Water treatment plants and their apparatus are evolving toward greater cost-effectiveness through design changes, which are adaptations to the specific properties of the water being received. Thus, different sites are likely to have different kinds of membrane filters and use different flow rates, and changes are made continually. General considerations are found in Buffle et al. (5) and in Droppo (4). General research needs are outlined well in a 1998 Committee Report of the AWWA (12). For water treatment, there is a strong research effort to develop plastic filters with well-controlled properties, an ongoing work that is progressing well. Coupled to this are considerable interests in developing filters that can be cleaned *in situ* and improved filters for removing waterborne pathogens. Filtration can be used in conjunction with chemical aids, such as flocculants, to minimize fouling. This is a research topic that is likely to expand quickly.

BIOPHYSICOCHEMICAL INTERACTIONS BETWEEN FILTERS AND COLLOIDS/DISSOLVED ORGANICS

As stated earlier and revealed by Figs. 1 and 2, the various types of filters are complementary in properties. The nature of filtration problems, however, varies from one filter type to another, in particular because flow rates through the various membrane types differ.

Filters with a large pore size (>0.1 μm) are used for sieving and filtration (Figs. 1 and 2). A large solution flow rate can be achieved, even though a rather low pressure is applied to the filter. At a large flow rate, the concentration of colloids at the filter surface may become orders of magnitude larger than in the bulk solution (the so-called concentration polarization effect), leading to instantaneous coagulation (Fig. 3) and consequent clogging of the filter (5). Under such conditions, the nature of the membrane is not very relevant. The way to minimize clogging is to use minimum flow rates. Although very low flow rates are usually required to avoid surface coagulation, they cannot be so low as to permit bulk coagulation (Fig. 3).

Filtration membranes with pore sizes in the range 1–100 nm are used for ultrafiltration. High pressure (1–4 atm) must be applied to the membrane because of the small pore size, but flow rates are never large. When the effective hole size is of the same order of magnitude as the molecule, dialysis or reverse osmosis occurs, whereby the solvent does not move through the membrane. With such membranes and techniques, clogging due to surface coagulation is low or negligible, and it decreases with the membrane pore size. On the other hand, when molecules and pore sizes become more and more similar, interactions between the solute (the organics especially) and the membrane body become increasingly important (Fig. 3). In

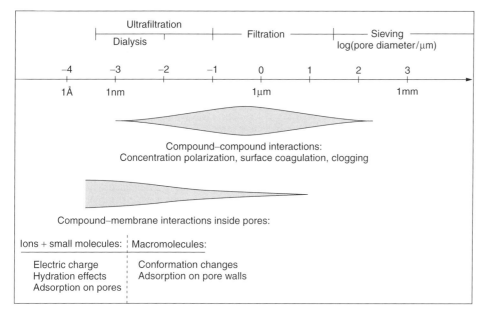

Figure 2. Schematic of size ranges where important secondary filtration effects are expected to play a significant role. The thickness of hatched zones reflects the relative importance of the corresponding factors. The word "compound" designates any component different from water, be it particulate or colloidal or dissolved. Reproduced with permission from Buffle et al. (5).

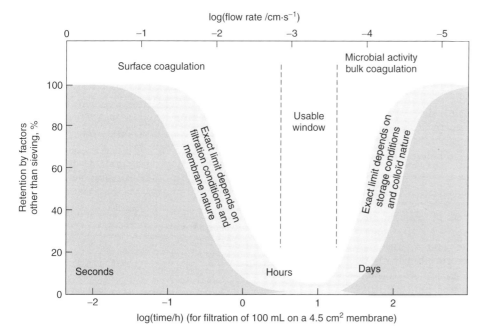

Figure 3. Semiquantitative representation of the change in retention with flow rate and with aggregation in the filtration cell, revealing the flow rate window of minimal artifact. Reproduced with permission from Leppard and Buffle (2).

such cases, the chemical nature of the membrane plays an important role. Thus it must be realized that the problems encountered in sieving and filtration are rather different from those in ultrafiltration, dialysis, and reverse osmosis. Detailed descriptions of the various types of interactions are given in Buffle et al. (5). The concept of the optimal flow rate window is illustrated in Fig. 3.

In addition to the aforementioned physicochemical interactions between membranes, suspended colloids, and dissolved organics, biological effects may play an important role. In particular, once microorganisms have been collected on the membrane filter, some of them secrete adhesive organic macromolecules (as cell attachment devices) directly onto membrane surfaces, leading to the development of a biofilm (an attached consortium of individual colloidal cells enmeshed in a porous matrix of their extracellular biopolymers). Attempts to identify specific important bioadhesives are receiving an impetus from transmission electron microscopy examinations of colloid/filter associations, using fouled membrane filters (8).

SIZE FRACTIONATION

Various sequential size fractionation procedures have been subjected to comparative analysis. In analytical applications, one can use either washing (also called diafiltration) or concentration techniques and, for each case, either sequential (also called cascade) or parallel filtration. In the concentration technique, the solution is pushed through the membrane by applying pressure. In the washing technique, a constant volume is maintained over the filtration membrane by compensating with synthetic solution. It has been found that the washing technique is more reliable because it avoids concentrating the particles in the retentate, thus minimizing coagulation problems. In a sequential filtration procedure, the same solution is filtered successfully through a series of membranes of decreasing pore size; the filtrate of one cell is filtered on the following membrane. The proportion of colloids in each size fraction is obtained by analyzing the various filtrates. In the parallel procedure, aliquots of the same sample are filtered through several membranes of different pore size. The proportions of colloids in the size fractions are obtained, based on the analysis of the filtrates and calculations. Because the reproducibility of filtration is not better than 5–10% error, the accumulation of errors becomes exceedingly large over five filtration steps in all cases, but sequential filtration is preferred because it minimizes aggregation problems. The rate of coagulation increases with particle concentration and heterogeneity.

OPTIMIZATION

The factors that affect the retention of colloids are summarized in Fig. 3. At large flow rates ($>3 \times 10^{-3}$ cm/s), large concentration polarization occurs and retention is due mostly to surface coagulation, which often leads to clogging. At very low flow rates, such surface coagulation is not important, but coagulation in the bulk solution becomes an important factor (for flow rates $<10^{-4}$ cm/s) because filtration times can be exceedingly long (days). For this reason, in chemical analysis, it is also preferable to avoid dialysis for which very long equilibration times are required. At intermediate flow rates ($10^{-4} - 3 \times 10^{-3}$ cm/s), the separation of colloids is effectively based on sieving and depends on membrane pore size.

The aforementioned limiting values of flow rates are valid for stirred conditions with a diffusion layer thickness (δ) of 10 µm. However, surface and bulk coagulation strongly depend on hydrodynamic conditions, which influence the value of δ (and thus surface coagulation), and on the bulk coagulation rate. Thus, both surface and bulk coagulation depend on the filtration mode (batch filtration with stirring, cross-flow filtration, pulsed filtration, filtration on rotary cylinders or disks, hollow fiber filtration). Decreasing δ below a few micrometers, to minimize surface coagulation, is very difficult; thus the schematic curves of Fig. 3 can be seen as the upper limits for both surface and bulk coagulation. It is interesting to note that the actual pore size of a membrane filter might not play a direct role in particle retention; the actual pore size might influence only the flow rate and thus the coagulation efficiency.

NEW VISTAS

In the last few years, novel chemical treatments have been applied to both porous and nonporous surfaces to create modified surfaces that kill bacteria on contact (13). The treatments are especially advantageous because they are not expected to exacerbate the problem of growing bacterial resistance to antibiotics. In practice, an antibacterial chemical (which disrupts drastically the integrity of the bacterial surface to kill bacteria before they can establish a biofilm) is bonded to the surface of interest so that it can neither be washed away nor modified upon interaction with oncoming bacteria. The treatments are currently applicable to glass surfaces, polymer surfaces, and carbohydrate-based porous materials. Can their use with membrane and depth filters be far behind? Considering the explanations given previously, this membrane treatment should be combined with a physicochemical means to minimize "cake" formation at the surface of filters by surface coagulation.

An interesting complementary idea is the creation of living filters in which a biofilm is purposely formed at the surface of the pores of a filter. Even though the concept needs considerable development, it can be envisioned that such a filter could be used (1) selectively to extract certain valuable substances from water for recovery or (2) selectively to degrade organic contaminants that become bound within the biofilm matrix (and thus are subject to enzymatic activities of nearby microorganisms).

Acknowledgments
The authors thank the Fonds National Suisse and the Natural Sciences and Engineering Research Council of Canada for generous long-term support.

BIBLIOGRAPHY

1. Buffle, J. et al. (1998). A generalized description of aquatic colloidal interactions: the three-colloidal component approach. *Environ. Sci. Technol.* **32**: 2887–2899.

2. Leppard, G.G. and Buffle, J. (1998). Aquatic colloids and macromolecules: effects on analysis. In: *The Encyclopedia of Environmental Analysis and Remediation*. R.A. Meyers (Ed.). John Wiley & Sons, Hoboken, NJ, Vol. 2, pp. 349–377.
3. Gustaffson, O. and Gschwend, P.M. (1997). Aquatic colloids: concepts, definitions, and current challenges. *Limnol. Oceanogr.* **42**: 519–528.
4. Droppo, I.G. (2000). Filtration in particle size analysis. In: *Encyclopedia of Analytical Chemistry*. R.A. Meyers (Ed.). John Wiley & Sons, Chichester, pp. 5397–5413.
5. Buffle, J., Perret, D., and Newman, M. (1992). The use of filtration and ultrafiltration for size fractionation of aquatic particles, colloids, and macromolecules. In: *Environmental Particles*. J. Buffle and H.P. van Leeuwen (Eds.). Lewis, Chelsea, MI, Vol. 1, pp. 171–230.
6. Flemming, H.-C., Griebe, T., and Schaule, G. (1996). Antifouling strategies in technical systems—a short review. *Water Sci. Technol.* **34**(5–6): 517–524.
7. Cole, J.J. (1999). Aquatic microbiology for ecosystem scientists: new and recycled paradigms in ecological microbiology. *Ecosystems* **2**: 215–225.
8. Liao, B.Q. et al. (2004). A review of biofouling and its control in membrane separation bioreactors. *Water Environ. Res.* (in press).
9. Leppard, G.G. (1992). Evaluation of electron microscope techniques for the description of aquatic colloids. In: *Environmental Particles*. J. Buffle and H.P. van Leeuwen (Eds.). Lewis, Chelsea, MI, Vol. 1, pp. 231–289.
10. Ferretti, R., Stoll, S., Zhang, J., and Buffle, J. (2003). Flocculation of hematite particles by a comparatively large rigid polysaccharide: schizophyllan. *J. Colloid Interface Sci.* **266**: 328–338.
11. Leppard, G.G. and Droppo, I.G. (2004). Overview of flocculation processes in freshwater ecosystems. In: *Flocculation in Natural and Engineered Environmental Systems*. I.G. Droppo, G.G. Leppard, T.G. Milligan, and S.N. Liss (Eds.). Lewis/CRC Press, Boca Raton, FL, pp. 25–46.
12. AWWA. (1998). American Water Works Association Membrane Technology Research Committee, Committee Report: Membrane Processes. *J. Am. Water Works. Assoc.* **90**(6): 91–105.
13. Borman, S. (2002). Surfaces designed to kill bacteria. *C&EN* **80**(23): 36–38.

COLUMN EXPERIMENTS IN SATURATED POROUS MEDIA STUDYING CONTAMINANT TRANSPORT

Wolfram Schuessler
Institut für Nukleare Entsorgung
Karlsruhe, Germany

Robert Artinger
TÜV Industrie Service GmbH
München, Germany

Column experiments are one key technique to investigate transport phenomena of contaminants in water-bearing porous media. The main applications of column experiments are the determination of (1) sorption coefficients for contaminants in different porous media (e.g., aquifers), (2) transport phenomena of dissolved matter (e.g., filtration of aquatic colloids), and (3) reaction mechanisms of dissolved matter (e.g., dissociation kinetics of aquatic complexes).

The typical setup of a column experiment is given in Fig. 1. From a reservoir a fluid (e.g., groundwater or contaminated water) is pumped through a column and the effluent of the column is analyzed online or by collecting individual samples. To control the flow-through either a flowmeter or a balance is used. A second reservoir can be used to switch from one feeding solution to another [e.g., from a groundwater solution to a contaminated (spiked) groundwater]. In addition to the analysis of the effluent, the distribution of sorbing tracers or filtrated colloids in the column can be analyzed, either online or postmortem. Depending on the samples under investigation the whole experiment can be performed in a glove box, for example, for radiation protection reasons or to control the ambient air composition.

Most of the column experiments found in the literature fall into three categories: experiments with (1) intact natural porous media to the greatest possible extent, (2) artificially packed natural porous media, and (3) artificial porous media (e.g., ion exchange resins).

For the experiments in category 1 the most difficult task is to get an undisturbed (or least disturbed) column from the field site into the lab and to maintain the natural boundary conditions.

Category 2 experiments somehow represent a compromise between categories 1 and 3. The natural material can be well characterized before using it in the column experiment. During the process of column packing the natural steady state between the solid and the liquid phase is disturbed. Therefore, prior to a column experiment under "near natural" conditions the whole system, porous medium and feed solution, has to be equilibrated until a steady state is reached. This new steady state is not necessarily the one found in nature. To come as close as possible to natural conditions, parameters such as surrounding gas atmosphere or redox state have to be monitored and, if possible, controlled.

Categories 1 and 2 aim at the quantitative determination of parameters "directly" applicable to the natural systems under investigation. However, even being as close as possible to the natural conditions, one has to be aware of possible artifacts and the problems of upscaling, relevance, and transferability (1).

Category 3 experiments are used to investigate properties of dissolved matter like aquatic complexes of contaminants and aquatic colloids. Here the main goal is to identify and quantify processes under well defined conditions, which normally is not possible in a complex natural system. In these experiments well defined standard porous media (e.g., quartz sand) or porous media with special characteristics (e.g., ion exchange resins) are used.

COLUMN PREPARATION

The best preparation method strongly depends on the properties of the system under investigation. Therefore,

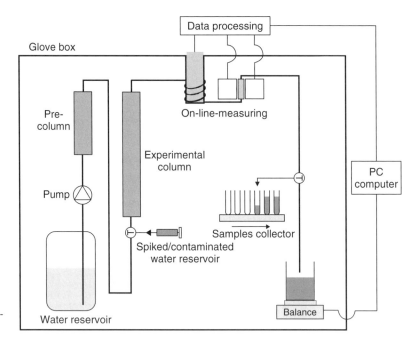

Figure 1. Experimental setup for column experiments.

only a few main hints, resulting from the long experience with column experiments of D. Klotz (2) and our own experiments with radionuclides, are given here.

The dimensions of the column are determined by the porous medium under investigation. It turned out that the column diameter should be at least 30 mm and larger than 25 mean grain diameter; the length should be larger than four times the column diameter. This setup enables evaluation of the experiments assuming an almost homogeneous distributed porous medium (not necessarily meaning homogeneous grain size) and minimizes the influence of boundary effects at the inlet and the outlet of the column.

Suitable materials for the columns are stainless steel or acrylic glass. The use of acrylic glass enables eye inspection of the readily prepared column. The disadvantage in case of experiments with natural samples is the growth of algae if the experiment is not run in the dark and at limited pressure. If the column is kept in the dark except for short periods of light (e.g., for sample taking), the growth of algae is prevented. At the inlet and the outlet of the column a frit is used in order to keep the porous medium inside the column and to distribute or sample the water over the whole cross section of the column.

As examples, two preparation methods (2) for columns are given here. In preparing a column with a cohesive medium, a slurry of one part porous medium and one part water is made. The slurry is filled into the vertical adjusted column up to a height of 1 cm. Then the water is extracted from the bottom of the column with a slow flow rate (1 cm³/h or less). A second batch of slurry is put on top of the condensed porous medium in the column and the water is again extracted. This procedure is applied until the column is filled to the top. In the case of a sandy porous medium, the vertical adjusted column is filled with 1 cm of water and then 1 cm of the sandy medium is put into the column. The sandy medium is compacted by stamping with a bead-molding in the column and knocking with a rubber mallet from outside the column. Next, the supernatant is discarded and the process is continued unless the column is filled. This procedure guarantees a tightly packed reproducible bed inside the column.

EXPERIMENTAL PROCEDURE

In general, the flow through the column should be directed from the bottom to the top. Remaining gas bubbles originating from the packing procedure in the column are driven by the water flow and the buoyant force toward the outlet and are thus removed from the column during the equilibration process. If the column is operated from the top to the bottom, the buoyant force is always directed toward the column inlet and thus always some bubbles remain in the column, influencing the flow characteristics in an uncontrollable way.

In the case of sandy media, a pump is located at the inlet of the column and the effluent is recovered in an open sample reservoir at the outlet. For cohesive media, at the column outlet pressure is applied to extract the effluent from the column. At the outlet a sample collector can be used to allow solution analysis of individual fractions of the effluent. Flow-through can easily be monitored using a balance and a data acquisition program. In addition, online detection methods can be applied, for example, radiometric or spectroscopic methods or ion-sensitive electrodes.

Initially, the column is manipulated until a steady state in chemical composition of the effluent and hydraulic properties of the column is reached. The hydraulic properties are determined and monitored by using conservative tracers. When the steady state is reached, either a short pulse injection or a continuous step injection may be applied for a contaminated water. Which method is used depends on the system under investigation. Experiments with radioactive contaminants often use the pulse injection method to keep the activity introduced into the column as low as possible (3). In contrast, experiments with water contaminated by heavy metals frequently use the step injection method. However, the corresponding breakthrough curves of the tracers in both experiments contain the same information although they look very different at first glance (4).

A column may be used several times. However, one has to make sure that the subsequent experiments are not influenced by the preceding ones. This has to be kept in mind for the analysis of the breakthrough curves as well as for the postmortem analysis of the tracer distribution in the column.

EXAMPLES

Sorption Coefficient of Cadmium onto Sea Sand

The sorption coefficient of cadmium (Cd) onto sea sand (5) in a synthetic groundwater is determined using a plexiglas column (length 500 mm, diameter 50 mm) (Fig. 2). The column is equilibrated with the water for a period of 3 weeks. Bromide is applied as a conservative tracer for the determination of the hydraulic properties of the column. Cd breakthrough curves are determined from pulse injection (300–800 mL) experiments with initial Cd concentrations of 8.9×10^{-6} mol/L. The Cd concentration in the eluate is determined online as a function of time, that is, the Cd breakthrough curve, by means of a Cd-sensitive electrode and after collection of individual fractions by ICP-MS.

In the case where all chemical reactions involved in the experiment are in equilibrium, the sorption coefficient can be determined from the breakthrough curve according to

$$R_f = 1 + \gamma K_d V/V_p$$

where γ is the density of the packed column (here 1.7 g·cm^{-3}), V is volume of the column (here 1000 cm^3), V_p is pore volume of the column (here 400 cm^3), and the retardation factor $R_f = v_D/v_t$ with v_D being the Darcy velocity and v_t the pore water flow velocity. Here $R_f =$ $V_{max}/V_p = 8000/400 = 20$ and therefore $K_d = 4.5$ g·cm^{-3}. V_{max} is the solution volume eluted from the column corresponding to maximum tracer concentration in the effluent (maximum of the breakthrough curve).

Transport Phenomena of Dissolved Matter, Size Exclusion of Humic Colloids

Column experiments (6) were carried out to investigate the influence of humic colloids on subsurface uranium migration (Fig. 3). The columns were packed with well characterized aeolian quartz sand and equilibrated with groundwater rich in humic colloids (DOC: 30 mg·dm^{-3}). U migration was studied under an Ar/1% CO$_2$ gas atmosphere. Precolumns (250 mm long, and 50 mm in diameter) were used as filters to remove possibly generated larger particles and to pre-equilibrate the groundwater. The whole system was equilibrated with groundwater over a period of at least three months. Tritiated water (HTO) was used as a conservative tracer to determine the hydraulic properties of the column. The breakthrough curve of HTO was measured using a flow-through monitor for β-radiation; U was determined using single fraction analysis. The uranium-spiked groundwater was prepared by adding aliquots of an acidic U(VI) stock solution to the groundwater. The U concentration in the spiked groundwater was 10^{-6} mol·dm^{-3}.

The breakthrough curves (Fig. 3) indicate that U was transported as a humic colloid-borne species with a velocity up to 5% faster than the mean groundwater flow (monitored by HTO). This could be attributed to size exclusion of the humic colloids from small water-bearing pores within the column.

Dissociation Kinetics of Am-Humic Colloids

The column experiments (3) were performed using the same groundwater/sand system as described for the U experiment (Fig. 4). The breakthrough curves of HTO and Am were measured using flow-through monitors for β and γ radiation and additionally by single fraction

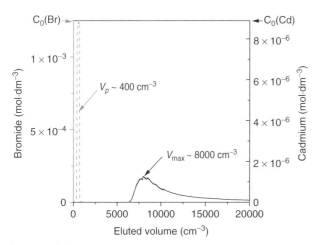

Figure 2. Column experiment to determine the sorption coefficient of Cd onto sea sand.

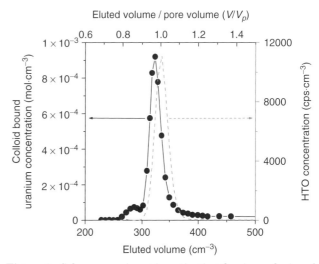

Figure 3. Column experiment investigating the size exclusion of humic colloids in a sandy aquifer.

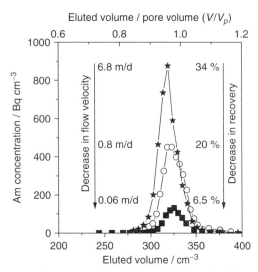

Figure 4. Column experiment investigating the dissociation kinetics of Am-humic colloids.

analysis. The ^{241}Am concentrations in eluate fractions and batch experiments were further determined by liquid scintillation counting. The Am-spiked groundwater was prepared in the same manner as described for U. The Am concentration in the spiked groundwater was 2×10^{-8} mol $\cdot$ dm^{-3}. The breakthrough curves again indicate that Am was transported as a humic colloid-borne species with a velocity up to 5% faster than the mean groundwater flow (monitored by HTO).

By varying the groundwater flow velocity about two orders of magnitude (0.06–7 m/d), an increase of the Am recovery (ΣAm injected into the column/ΣAm recovered from the column) is obtained ranging from 6% to 34% (Fig. 4). That means the longer the residence time of humic colloid-borne Am in a column, the higher is the amount of Am sorbed onto sand. The significant dependence of the Am recovery on the groundwater flow velocity is due to time-dependent turnover of Am from humic colloids onto the sand surface. These types of experiments were further used to investigate the actinide humic colloid interaction kinetics (7–10).

BIBLIOGRAPHY

1. Schüßler, W. et al. Humic colloid-borne contaminant migration: applicability of laboratory results to natural systems. *Nucl. Sci. Eng.* (in press).
2. Klotz, D. (1991). *Experiences with column experiments to determine contaminant migration (in German)*. GSF-Bericht 7/91.
3. Artinger, R., Kienzler, B., Schuessler, W., and Kim, J.I. (1998). Effects of humic substances on the ^{241}Am migration in a sandy aquifer: batch and column experiments with Gorleben groundwater/sediment systems. *J. Contam. Hydrol.* **35**: 261–275.
4. Bürgisser, C.S., Cernik, M., Borkovec, M., and Sticher, H. (1993). Determination of nonlinear adsorption isotherms from column experiments: an alternative to batch studies. *Environ. Sci. Technol.* **27**: 943–948.
5. Schüssler, W., Artinger, R., and Kienzler, B. (1998). Geochemical modelling of transport processes: surface complexation (in German). *Wiss. Ber. FZKA* **6051**: 57–76.
6. Artinger, R. et al. (2002). Humic colloid-borne migration of uranium in sand columns. *J. Contam. Hydrol.* **58**: 1–12.
7. Artinger, R. et al. (2000). Humic colloid-borne Np migration: influence of the oxidation state. *Radiochim. Acta* **88**(9–11): 609–612.
8. Artinger, R., Schüßler, W., Schäfer, T., and Kim, J.I. (2002). A kinetic study of Am(III)/humic colloid interactions. *Environ. Sci. Technol.* **36**: 4358–4363.
9. Artinger, R. et al. (2002). ^{241}Am migration in a sandy aquifer studied by long-term column experiments. *Environ. Sci. Technol.* **36**: 4818–4823.
10. Schäfer, T. et al. (2003). Colloid-borne americium migration in Gorleben groundwater: significance of iron secondary phase transformation. *Environ. Sci. Technol.* **37**: 1528–1534.

CYTOCHROME P450 MONOOXYGENASE AS AN INDICATOR OF PCB/DIOXIN-LIKE COMPOUNDS IN FISH

Hannah Aoyagi
Oladele A. Ogunseitan
University of California
Irvine, California

POLYCHLORINATED BIPHENYL AND DIOXIN-LIKE COMPOUNDS AS ENVIRONMENTAL TOXICANTS

Polychlorinated biphenyls, dioxins, and furans are ubiquitous and persistent organic pollutants (POPs). These three categories of chemicals are among the 12 priority POPs that have been scheduled for global phase-out according to the Stockholm Convention on Persistent Organic Pollutants (SCOPOP). SCOPOP was inaugurated in May 2001, and it has been ratified by more than 100 countries under the auspices of the United Nations Environment Program (1). Despite the stringent regulatory restrictions now placed on the manufacture and distribution of polychlorinated biphenyls and dioxin-like compounds, their recalcitrance in polluted aquatic environments necessitates the continuation of monitoring programs that have been implemented to reduce negative impacts on biotic and abiotic components of vulnerable ecosystems.

Polychlorinated biphenyls (PCBs) include 209 congeners consisting of two to ten chlorine atoms attached to a biphenyl structure. PCBs are highly soluble, clear, oily liquids at room temperature. They are chemically inert and thermally stable. These properties make them useful as insulators and coolants for capacitors and transformers in a variety of industries. Monsanto began production of PCBs in 1930, marketing congener mixtures under the brand name Aroclor. PCBs were also sold under the names Clophen (Germany), Kanechlor (Japan), Fenchlor (Italy), and Phenochlor (France). PCB production was banned in the United States in 1977 due to potential carcinogenicity (2,3).

PCBs tend to deposit in sediments and soils where they persist and bioaccumulate; their half-lives are in the range of months to decades. Due to their high affinity for lipids, PCBs have been detected in the adipose tissues of animals worldwide, including humans (3). Consumption of contaminated fish is the primary route of exposure for adults, whereas infants can be exposed to high concentrations of PCBs in breast milk (4).

The adverse health effects associated with PCB exposure include neurological impairment, developmental effects, carcinogenicity, hepatotoxicity, thyroid problems, dermal and ocular abnormalities, and changes in immune function (3). Occupational studies provided the first evidence of chloracne (skin eruptions) and neurological impairment, but detailed acute and chronic health effects of PCB exposure in the general population were provided by the Yusho and Yu-cheng poisoning incidents in Taiwan where 1843 adults, children, and pregnant women consumed PCB-contaminated rice oil (5,6).

The increasing recognition of human health effects due to PCB exposure has led to suspicions of their ecotoxicological impacts on various organisms inhabiting polluted ecosystems (7). Aquatic animals and their predators are at risk of cancer, neurotoxicity, reproductive toxicity, immune impairment, and endocrine effects. Due to the bioaccumulative nature of PCBs, health risks are highest for the top predators in the food chain (7). No threshold level of exposure has been identified for immune effects in animals. Moreover, it has been found that reproductive impairment can endure long after exposure in some organisms, indicating that PCBs may pose a serious threat to the fundamental survival of many species (2).

Chlorinated Dibenzofurans

CDFs are a class of chemicals composed of 135 congeners; each consists of a furan ring surrounded by two benzene rings, and one to eight chlorine atoms are attached to the benzene structures (Fig. 1). They are colorless solids, soluble in organic solvents, and do not interact with acids or alkalis. However, they are less stable than PCBs and degrade at a temperature of 700 °C (4). CDFs are not manufactured for commercial use, but they are by-products of PCB and chlorinated herbicide manufacturing, solid waste incineration, and bleaching processes. Currently there are no specific regulations pertaining to CDFs (8). They have been detected in the atmosphere and in surface waters but tend to adhere to sediments where they can bioaccumulate in fish and other aquatic organisms (4). Human exposure routes include eating contaminated fish and drinking milk from cows that have ingested feed contaminated by CDFs from atmospheric deposition. There are no data on the human health effects of CDFs alone because exposures have typically occurred simultaneously with PCBs and other similar compounds. Animal studies indicate that exposure to CDFs leads to adverse health outcomes similar to those of PCB exposure, including hepatotoxicity, thyroid alterations, and reproductive effects (8).

Chlorinated Dibenzo-p-dioxins

CDDs comprise a group of 75 congeners; each consists of two benzene rings connected by two oxygen bridges, and one to eight chlorine atoms surround the structure

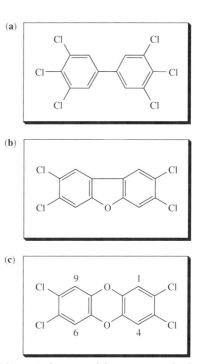

Figure 1. Schematic diagram of the structure of polychlorinated biphenyls (**a**), chlorinated dibenzofurans (**b**), and chlorinated dibenzo-p-dioxins (**c**) compounds.

(Fig. 1). CDDs are colorless solids or crystals of low solubility. They readily partition in air, water, and soil (4). CDDs were originally produced as herbicides, such as Agent Orange. In 1983, the Environmental Protection Agency (EPA) banned the use of chlorophenoxy herbicides on agricultural lands. CDDs are still emitted from incineration and chlorine bleaching processes in paper and pulp mills, as well as from the combustion of fossil fuels, wood, and cigarettes. Recent regulations have attempted to reduce paper and pulp mill effluents, as well as reducing dioxin emissions from incineration (9). CDDs are found worldwide in small concentrations, in the atmosphere (incinerator emissions), surface water (paper and pulp mill effluents), and in soils and sediments (herbicide spraying and atmospheric deposition). CDDs in soil and sediments can bioaccumulate through the food chain and reach humans through fish, meat, and milk. Occupational exposures can occur during the manufacture of pesticides and herbicides. Human and animal health effects are similar to PCBs, including cancer, immunosupression, reproductive and endocrine effects, chloracne, and neurotoxicity (9).

CYTOCHROME P450 ENZYMES

The cytochrome P450 superfamily (P450s) of microsomal hemoproteins is a set of oxygen activating terminal oxidases that catalyze several biosynthetic and metabolic processes. The monooxygenase activity of P450 enzymes is expressed through a series of electron transfer reactions involving the heme moiety (Fig. 2). The overall reaction can be summarized by the schematic equation

$$RH + O_2 + 2H^+ + 2e^- \longrightarrow ROH + H_2O$$

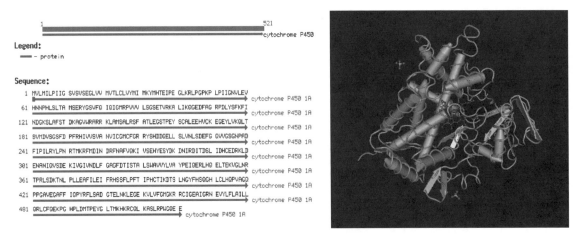

Figure 2. Amino acid sequence of salmon P4501A, and the three-dimensional structure of eukaryotic cytochrome P450 monooxygenase. The amino acid sequence was recovered from the National Center for Biotechnology Information database (http://www.ncbi.nlm.nih.gov/) (12). The image was generated from another P450 sequence imported into the Cn3D4.1 software program. The string and barrel structure shows the position of the essential iron atom.

The RH substrate represents a wide variety of compounds, including chlorinated aromatic pollutants. More than a hundred different P450 enzymes have been identified in animals, plants, fungi, and microorganisms, although they appear to play the most complex role in animals (10).

P450s perform a variety of functions, including the capacity to hydroxylate different structural types of organic molecules. They were first identified in liver tissue and later in tissues such as the adrenal gland and lymphocytes (10). In the liver, P450s catalyze the first step in metabolizing environmental toxicants as well as endogenous steroid hormones and drugs. These reactions have important implications for the use of cytochromes P450 as biomarkers of environmental toxicant exposures (12).

In the typical metabolic reaction, cytochrome P450 binds with the substrate, oxygenates it, and transforms it into a reactive state. In the second step of metabolism, the substrate undergoes a conjugation reaction (glucuronidation) and eventually elimination. In the reactive state, the substrate has the potential to form DNA adducts, thereby initiating carcinogenesis. However, ecotoxicologists are more interested in the mechanism of P450 induction and how changes in enzyme synthesis can be used as an indicator of exposure to specific environmental toxins. Cytochrome P450 assays are useful because genetic regulation of the enzyme can be induced by certain substrates (12).

There are two types of P450 induction; one requires phenobarbitone as a substrate, and the other requires carcinogenic polycyclic hydrocarbons such as PCBs, CDDs, and CDFs as a substrate. This second type of induction occurs with cytochrome P450 1A1 (CYP1A1), which is believed to have evolved at least 800 million years ago as a way for animals to metabolize plant chemicals (12). Cytochrome P450 1A1 can be induced in animals by planar molecules such as planar PCBs and CDDs; 2,3,7,8-tetrachlorodibenzo-p-dioxin (TCDD) is the most potent inducer known (12). These specific planar molecules bind readily to the cytosolic aryl hydrocarbon receptor (AhR) due to a similarity shared in the number of chlorine atoms present at the para and meta positions of the chemical structure. Although the main function of the AhR is unclear, it allows planar molecules to be translocated into the nucleus and induce CYP1A1 by suppressing normal genomic control of transcription (12).

Cytochrome P450 1A1 induction is a potent biomarker of ecosystem-wide exposure to PCBs and dioxin-like compounds due to its occurrence in most animal species. CYP1A1 induction is quantifiable, it indicates specific types of exposure, and its induction is manifested as a reversible subclinical effect (14). Rather than assessing the toxicological endpoints of PCB and dioxin exposure, CYP1A1 provides a measurement of alterations that affect the bioavailability and metabolism of carcinogens and other toxins (14). Sustained CYP1A1 induction can lead to an increase in the rate of metabolism and activation of carcinogens and mutagens, leading to increased formation of oxidative DNA adducts (12). Activated metabolites can also be transported to distant tissues where they cause oxidative damage and other adverse effects (15).

ECOLOGICAL IMPACTS OF PCBS AND DIOXIN-LIKE COMPOUNDS ON FISH

PCBs and dioxin-like compounds have impacted aquatic ecosystems through paper mill effluents, other industrial discharges, urban runoff, and atmospheric deposition (4). These contaminants persist in sediments and bioaccumulate, causing chronic exposure in many fish species (Table 1). Of all fish studied to date, salmonids are the most sensitive to chlorinated hydrocarbons, especially in regard to reproductive effects. They are thus useful as test organisms for the many possible physiological effects of exposure to PCBs and dioxin-like contaminants (16).

Chronic exposure to TCDD increases adult mortality in female rainbow trout (*Oncorhynchus mykiss*) in a dose-dependent fashion after a relatively long exposure of at

Table 1. Chemical Properties of PCBs, CDFs, and CDDs

Compounds	Commercial Uses and Sources into the Environment	Estimated Environmental Half-Life[a]
Polychlorinated biphenyl	Insulators, coolants in capacitors and transformers	Vapor phase: 12.9 days for monochlorobiphenyl to 1.31 years for heptachlorobiphenyl Freshwater: 6.8–9 years
Chlorinated dibenzofuran	By-product of heating PCBs, incineration, and chlorine bleaching	Surface water: 7 days–4 weeks Soil: 7 days–4 weeks Aerobic aquatic: 7 days–4 weeks Anaerobic aquatic: 28 days–16 weeks
Chlorinated dibenzo-p-dioxin	Herbicides, by-product of incineration and bleaching	Surface water: 1.15–1.62 years Soil: 1.15–1.62 years Aerobic aquatic: 1.15–1.62 years Anaerobic aquatic: 4.58–6.45 years

[a]References 4, 11.

least 26 weeks (16). TCDD exposure, it is also believed affects behavior in rainbow trout, making them less active and responsive (17). Reproductive outcomes are manifested in the decreased survival of young and a large number of deformities, although egg characteristics such as weight, lipid content, and total number of eggs produced do not differ from those of unexposed populations. TCDD-exposed rainbow trout young experience an unusually high incidence of yolk sac edema, a disease rarely seen in unexposed fish (17).

Chlorinated hydrocarbon exposure can also lead to AhR-mediated induction of CYP1A1 in rainbow trout and in many other species of fish, making it one of the most consistently observed outcomes in fish toxicology. CYP1A1 levels naturally fluctuate in female rainbow trout and decline as they approach their reproductive phase. It has been observed that the fluctuations are associated with the health of the fish and its environmental conditions. Although it is unclear exactly what role the enzyme plays in the life cycles of fish, it may be involved in mitigating yolk sac edema and a number of other adverse health effects, both directly and indirectly (16).

Several factors affect the intensity of cytochrome P450 induction and PCB/dioxin toxicity in fish. The potency of the compound, or mixture of compounds, determines the extent of induction and subsequent toxicity. In addition to being the most potent CYP1A1 inducer, 2,3,7,8-TCDD is considered the chlorinated polycyclic hydrocarbon most toxic to all life and thus has a toxic equivalency factor (TEF) of 1. In rainbow trout and carp, TCDD is followed in toxicity by CDFs, whose TEFs range from 0.359 to 0.028, and nonortho PCBs, whose TEFs range from 0.00016 to 0.000041 (17,18).

Lipid levels also play a role in CYP1A1 induction, toxicant bioavailability, and baseline health conditions of fish. Linoleic acid (18:2) controls hepatic cytochrome P450 concentrations, high levels of 18:2 decrease enzymatic activity (19). Chlorinated polycyclic hydrocarbons have high affinity for lipids and can be readily stored in the adipose tissues of fish. This type of storage is believed beneficial by reducing chronic exposure, although it can be harmful if the tissue is mobilized.

METHODS FOR ASSESSING CYP1A INDUCTION AND THE RATIONALE FOR INCLUDING IT IN ECOTOXICOLOGICAL ASSESSMENTS

CYP1A1 induction is an appropriate biomarker of PCB/dioxin exposure in fish because it is easily and rapidly quantified, specific, and has a temporary, reversible effect. There are various methods for assaying cytochrome P450 induction, using well-established criteria to characterize accurately the extent of induction in the liver and other tissues of interest. The level of induction may be used as a measure of potential adverse health effects in fish, without the manifestation of disease end points. Ideally, a biomarker should quickly signal adverse effects before they become serious and irreversible (14). CYP1A1 induction fits this criterion; it is not a disease end point itself, but an indicator of potential DNA damage through carcinogen activation, and may be reversed once exposure ends (15).

Induction can be examined in terms of mRNA production, protein synthesis, or enzymatic activity, and there are methods for assessing cytochrome P450 induction at each of these stages: mRNA quantification, immunologic protein determination, and assays of enzymatic activity (20). mRNA quantification is a relatively simple and rapid method compared to immunodetection and enzyme assays and may be a more useful tool for measuring aquatic pollution through cytochrome P450 induction. One process involves isolating CYP1A RNA from tissues of interest, synthesizing cDNA, designing primers to amplify specific regions of CYP1A cDNA, and quantifying using reverse transcription-polymerase chain reaction (RT-PCR) and gel electrophoresis techniques (20,21). Other studies have used slot blot analysis and RNA spectrophotometry to quantify CYP1A induction (22).

Immunodetection methods, such as Western blotting or immunohistochemistry, require identifying antibodies for specific cytochromes P450; monoclonal antibody 1-12-3 is specific to CYP1A1 and can detect CYP1A in all fishes studied thus far (23). The antibody and developer are applied to tissue samples and incubated until staining has developed. The resulting staining pattern indicates which cells have expressed CYP1A1 and to what degree. The intensity of CYP1A staining is measured on a scale of 0 to 15, where zero indicates no induction of the enzyme (23). The advantages of immunodetection methods are that they have the sensitivity to detect very small quantities of cytochrome P450 in all types of cells (15). The drawbacks are that these methods require a longer induction time than mRNA quantification, and it is uncertain whether appropriate antibodies can be found for all species of fish (20).

Enzyme assays measure the biotransformational activities of induced CYP1A on specific substrates. One type of assay examines the deethylation of 7-ethoxyresorufin (EROD) to resorufin, which can be compared to a resorufin reference (12). The dealkylation of 7-pentoxyresorufin (PROD) and UDP-glucuronosyltransferase activity (UDP-GT) are also used, and all three types of biotransformation may be quantified by using fluorescence spectrophotometry (24). Enzyme assays provide a good quantitative method for determining CYP1A induction, but the induced enzyme is subject to denaturation prior to EROD, potentially underrepresenting the extent of induction. Furthermore, certain pollutants may inhibit EROD, masking their effect on CYP1A1 induction (20).

CASE STUDIES USING P450 AS AN INDICATOR OF POLLUTANTS IN FISH

The Atlantic Tomcod in the Hudson River Estuary of the United States

The Hudson River contains aromatic hydrocarbon contaminants released from industry, municipal wastewater treatment plants, and sewage outflows. During a span of 30 years, hundreds of thousands of pounds of PCBs were released from two major sites impacting the Hudson River estuary (22). CDDs and CDFs were released during a 20-year period by a company manufacturing the herbicide 2,3,4-T, and from municipal waste incineration. Therefore, the estuarine sediments contain high levels of contaminants that have bioaccumulated in fish, including the Atlantic tomcod (22).

Yuan and co-workers (22) used Atlantic tomcod to compare hepatic CYP1A1 mRNA expression in relationship to sediment levels and hepatic burdens of chlorinated aromatic hydrocarbons and to evaluate CYP1A1 as a biomarker for mapping sediment contaminant concentrations in the Hudson River estuary. Previous studies had linked tomcod body burdens of PCBs, CDDs, and CDFs to regions of high sediment contamination and linked body burden of chlorinated aromatic hydrocarbons to CYP1A1 induction. The authors observed significant differences in CDD and CDF body burdens that varied across the estuary and sediment contamination levels. They also found that CYP1A1 induction varied significantly across sediment contamination levels. However, CYP1A1 induction did not correspond to the type of chlorinated aromatic hydrocarbon or to the TEF. The authors hypothesized that polyaromatic hydrocarbons (PAHs)—also present in the Hudson River estuary but not included in the study—were partially responsible for the observed CYP1A1 induction. Additionally, chronic exposure to PCBs, CDFs, and CDDs, it was shown, depletes AhR cells, thereby lowering the level of CYP1A1 inducibility and restricting the ability to predict exposure in fish based on enzyme levels.

Carp Fish in Hikiji River, Kanagawa Prefecture, Japan

Between 1992 and 2000, dioxin-contaminated raw sewage was released into the Hikiji River, Japan from an industrial waste incinerator (25). The toxic equivalent (TEQ) of 3.0 g of TCDD was released, and water concentrations ranged from 0.93–8.7 pg-TEQ/L in the year 2000. Agricultural runoff and municipal incinerators contributed a variety of other CDDs as well as CDFs. Sakamoto and co-workers examined the level of CYP1A induction in carp from the Hikiji River, both upstream and downstream from the effluent discharge, in conjunction with indicators of their reproductive health (25).

Muscle tissue analysis of carp from the contaminated downstream region showed higher concentrations of several different CDDs, CDFs, and coplanar PCBs. Liver microsomal CYP1A content was significantly higher in the exposed male and female carp, although EROD activity was increased only in females. Serum estrogen levels, an important factor in the initial stages of oocyte development, were low in the exposed fish and significantly correlated with CYP1A content. Reproductive health markers were also correlated to estrogen levels; exposed females produced fewer oocytes containing yolk protein, although the difference was not statistically significant. The investigators noted that estrogen poses a problem in interpreting CYP1A induction due to its interactions with the enzyme. It is known that, of all endogenous compounds, estrogen has the largest impact on CYP1A activity. High estrogen levels during fish reproductive phases are associated with low CYP1A levels. However, the study found a consistent pattern of increased CYP1A induction in exposed carp across all serum estrogen levels.

PROSPECTS FOR USING CYTOCHROME P450 AS A BIOINDICATOR

Though its mechanisms are not fully understood, CYP1A induction, it has been shown, is a very effective biomarker for fish exposure to PCBs and dioxin-like compounds. As Yuan et al. demonstrated, it is difficult to distinguish the effects of PCBs from those of CDDs, CDFs, and PAHs, and thus it is not yet possible to use CYP1A1 induction as a direct measure of one particular aquatic contaminant or class of chemicals. Nor are there adequate data on the level of response for all fish species residing in contaminated waterways. Future research must better quantify the dose–response relationship between exposure and induction across species and elucidate the various modifications of CYP1A induction by endogenous agents, fish health status, and other environmental conditions. Toward this end, the development and breeding of genetically engineered indicator fish with reporter genes such as bioluminescence or green fluorescent protein linked to the inducer locus of cytochrome P450 genes may ultimately prove indispensable.

Acknowledgments

This work was supported in part by the Program in Industrial Ecology at the University of California, Irvine, and by the UC Toxic Substances Research and Teaching Program.

BIBLIOGRAPHY

1. United Nations Environment Programme. (2003). *Stockholm Convention on Persistent Organic Pollutants*. ONLINE. Available: http://www.pops.int/ [10 July 2003].
2. Agency for Toxic Substances and Disease Registry. (2000). *Toxicological Profile for Polychlorinated Biphenyls*. ONLINE. Available: http://www.atsdr.cdc.gov/toxprofiles/tp17.html [18 June 2003].
3. U.S. Environmental Protection Agency. (2002). *Health Effects of PCBs*. ONLINE. Available: http://www.epa.gov/opptintr/pcb/effects.html [18 June 2003].
4. Food and Agricultural Organization. (2000). *Fact Sheets on Chemical Compounds*. ONLINE. Available: http://www.fao.org/DOCREP/003/X2570E/X2570E08.htm [10 July 2003].
5. Kashimoto, T. et al. (1985). PCBs, PCQs and PCDFs in blood of yusho and yu-cheng patients. *Environ. Health Perspect.* **59**(2): 73–78.
6. Hsu, S.T. et al. (1985). Discovery and epidemiology of PCB poisoning in Taiwan: a four-year follow-up. *Environ. Health Perspect.* **59**(2): 5–10.
7. Chiu, A. et al. (2000). Effects and mechanisms of PCB ecotoxicity in food chains: algae, fish, seal, polar bear. *Environ. Sci. Health* **18**: 127–152.
8. Agency for Toxic Substances and Disease Registry. (1994). *Toxicological Profile for Chlorodibenzofurans*. ONLINE. Available: http://www.atsdr.cdc.gov/toxprofiles/tp32.html [18 June 2003].
9. Agency for Toxic Substances and Disease Registry. (1998). *Toxicological Profile for Chlorinated Dibenzo-p-dioxins*. ONLINE. Available: http://www.atsdr.cdc.gov/toxprofiles/tp104.html [18 June 2003].
10. Omura, T., Ishimura, Y., and Fuji-Kuriyama, Y. (1993). *Cytochrome P-450*, 2nd Edn. Kodansha, Tokyo, pp. 6–7.
11. Howard, P.H. et al. (1991). *Handbook of Environmental Degradation Rates*. Lewis, Chelsea, MI, pp. 510, 657.
12. Ruckpaul, K. and Rein, H. (1990). *Frontiers in Biotransformation: Principles, Mechanisms and Biological Consequences of Induction*. Vol. 2, Taylor and Francis, Philadelphia, pp. 5, 6–10, 13–16, 39.
13. Arukwe, A. (2002). Complementary DNA cloning, sequence analysis and differential organ expression of beta-napthoflavone-inducible cytochrome P4501A in Atlantic salmon (*Salmo salar*). *Comp. Biochem. Physiol.* **133**: 613–624.
14. Gil, F. and Pla, A. (2001). Biomarkers as biological indicators of xenobiotic exposure. *J. Appl. Toxicol.* **21**: 245–255.
15. Pelkonen, O. and Raunio, H. (1997). Metabolic activation of toxins: tissue-specific expression and metabolism in target organs. *Environ. Health Perspect.* **105**(4): 767–774.
16. Giesy, J.P. et al. (2002). Effects of chronic dietary exposure to environmentally relevant concentrations of 2,3,7,8-tetrachlorodibenzo-p-dioxin on survival, growth, reproduction and biochemical responses of female rainbow trout (*Oncorhynchus mykiss*). *Aquatic Toxicol.* **59**(1–2): 35–53.
17. Hornung, M.W., Zabel, E.W., and Peterson, R.E. (1996). Toxic equivalency factors of polybrominated dibenzo-p-dioxin, dibenzofuran, biphenyl, and polyhalogenated diphenyl ether congeners based on rainbow trout early life stage mortality. *Toxicol. Appl. Pharmacol.* **140**(2): 227–234.
18. Loganathan, B.G. et al. (1995). Isomer-specific determination and toxic evaluation of polychlorinated biphenyls (PCBs), polychlorinated/brominated dibenzo-p-dioxins and dibenzofurans, polybrominated biphenyl ethers, and extractable organic halogen in carp from the Buffalo River, New York. *Environ. Sci. Technol.* **29**(7): 1832–1838.
19. Landis, W.G. and Yu, M.H. (1999). *Introduction to Environmental Toxicology: Impacts of Chemicals upon Ecological Systems*, 2nd Edn. Lewis, Boca Raton, FL, pp. 144–145.
20. Cousinou, M. et al. (2000). New methods to use fish cytochrome P4501A to assess marine organic pollutants. *Sci. Total Environ.* **247**(2–3): 213–225.
21. Rees, C.B. et al. (2003). Quantitative PCR analysis of CYP1A induction in Atlantic salmon (*Salmo salar*). *Aquatic Toxicol.* **62**: 63–78.
22. Yuan, Z. et al. (2001). Is hepatic cytochrome P4501A1 expression predictive of hepatic burdens of dioxins, furans, and PCBs in Atlantic tomcod from the Hudson River estuary? *Aquatic Toxicol.* **54**(3–4): 217–230.
23. Moore, M.J. et al. (2003). Cytochrome P4501A expression, chemical contaminants and histopathology in roach, goby and sturgeon and chemical contaminants in sediments from the Caspian Sea, Lake Balkhash and the Ily River Delta, Kazakhstan. *Mar. Pollut. Bull.* **46**: 107–119.
24. Huuskonen, S. and Lindström-Seppä, P. (1995). Hepatic cytochrome P4501A and other biotransformation activities in perch (*Perca fluviatilis*): The effects of unbleached pulp mill effluents. *Aquatic Toxicol.* **31**(1): 27–41.
25. Sakamoto, K.Q. et al. (2003). Cytochrome P450 induction and gonadal status alteration in common carp (*Cyprinus carpio*) associated with the discharge of dioxin contaminated effluent to the Hikiji River, Kanagawa Prefecture, Japan. *Chemosphere* **51**(6): 491–500.

WATER-BASED DISEASE. See WATER RELATED DISEASES

WATERBORNE DISEASE. See WATER RELATED DISEASES

WATER RELATED DISEASES

MARK J. WALKER
University of Nevada
Reno, Nevada

Water related diseases can be placed into four categories: (1) water-based disease, (2) waterborne disease, (3) water vectored disease, and (4) water washed disease. Infectious microbes that cause disease within these categories include bacteria, enteric viruses, parasitic protozoans, helminthes, and algae. The health effects of infection vary considerably, from symptomless to temporary mild illness to prolonged illness to death. Each category represents a distinct pathway and mechanism for infection by a microbial pathogen. Symptoms of infection, such as gastrointestinal upset, may be common across categories.

The categories represent differences in strategies that can be applied to control the prevalence of infection.

Water-based diseases are caused by microorganisms whose reproductive cycles involve passage through an aquatic animal. If the aquatic animal is not present, the microorganism cannot reproduce and, consequently, cannot infect humans. As an example, trematodes, *Shistosoma* spp. (blood flukes), cause the disease schistosomiasis. This disease occurs in China, Japan, the Philippines, Indonesia (infection with spp. *japonica*), throughout Africa (spp. *haematobium*), in Southeast Asia (spp. *mekongi*), and in western and central Africa (spp. *intercalatum*). Shistosomes must pass through a snail in the aquatic environment to develop to a form that infects people (the cercacia). The cercacia enters the bloodstream by passing through skin when people come in contact with contaminated water. The parasite is reintroduced into water as an egg contained in human feces. Control programs to prevent water-based diseases are often designed to eliminate the intermediate aquatic host, using molluscicides, snails that compete for the same habitat but do not act as intermediate hosts, or snail eating fish or birds. The most effective control programs include education about the life cycle, including proper disposal of human wastes.

Waterborne diseases are part of the fecal–oral cycle of infection. This type of disease may be transmitted by many routes, including water contaminated by feces from an infected host. Microorganisms, in a life cycle stage that may infect a new host, are excreted and enter water that is then consumed. Common waterborne microbial pathogens include protozoa (such as *Cryptosporidium* and *Giardia*), viruses (such as *hepatitis*), bacteria (such as *Salmonella* and *Shigella*), helminthes (such as *Ascaris*) and algae (such as *Schizothrix*). The likelihood of infection upon consumption depends upon the virulence of the strain present, survival in the aquatic environment, the number of organisms ingested, the estimated infectious dose, and the susceptibility of the host. As an example, cryptosporidiosis is a gastrointestinal upset caused by ingestion of the environmentally resistant stage of *Cryptosporidium parvum,* the oocyst. This parasitic protozoan has a wide range of hosts, including domesticated animals and wildlife. Oocysts of *Cryptosporidium* spp. are found in surface waters throughout the world. The disease is usually self-limiting, unless the infected subject is immunocompromised.

Immunocompetent hosts develop an immune response to the parasite which dramatically diminishes the likelihood of illness after initial exposure. Control measures for waterborne diseases include water treatment, using flocculants, settling and filtration processes, followed by application of disinfectants such as sodium hypochlorite. Education is an important component of strategies for reducing the prevalence of waterborne diseases, especially education about the fecal–oral route of infection and the role of proper human waste disposal and personal hygiene in preventing disease transmission.

Water vectored diseases are caused by microorganisms that have intermediate hosts in insects that reproduce in aquatic environments and transmit pathogenic microorganisms as part of feeding on human blood. If the aquatic environment needed for the insect to reproduce is not present, the prevalence of the disease is reduced. Malaria, yellow fever, and onchocerciasis are water vectored diseases associated with mosquitos and black flies, respectively. The parasitic microorganisms that initiate disease in humans undergo a portion of their life cycles in the insect host and are transmitted during the bite directly into the human bloodstream. In malaria, the protozoan species of genus *Plasmodium* must develop in an anopheline mosquito. When the female mosquito bites an infected person, it draws blood containing forms of the parasite that develop further within the mosquito, eventually entering the salivary glands. The infected mosquito then injects the developed form into the next human host in saliva secreted during the bite. The anopheline mosquito requires standing, fresh waters for reproduction. Common habitats for breeding include ponds and marshes. Control programs to prevent water vectored diseases focus on eliminating aquatic habitat for insect hosts and reducing the prevalence of disease in human hosts. Activities for this include draining perennially wet areas to remove standing water and application of pesticides or oils designed to kill developing mosquito larvae.

Water washed diseases are caused by contact with water contaminated with fecal material and microorganisms that colonize and reproduce in skin lesions or vulnerable tissues. Common microbial pathogens that cause this type of infection include *Pseudomonas* spp. and *Staphylococcus* spp. This type of water related illness is most associated with lack of adequate supplies, which leads to inadequate personal hygiene. Typical water washed diseases include skin irritations and ulcers, conjunctivitis, and trachoma. Increased supply, coupled with improved sanitation and education, often reduces the prevalence of water washed disease.

WATER VECTORED DISEASE. See WATER RELATED DISEASES

WATER WASHED DISEASE. See WATER RELATED DISEASES

DISHWASHING WATER QUALITY PROPERTIES

Eva Ståhl Wernersson
Erik Johansson
GS Development AB
Malmö, Sweden

Håkan Håkanson
University of Lund
Lund, Sweden

INTRODUCTION

Water is used as a rinse medium in dishwashing to remove soil from the dishes. An important requirement of the process is that the consumption of fresh water should be moderate, because the supply of fresh water in many countries is limited. Furthermore, the dishwater should reduce the microbiological activity (even when a dishwasher is not used) to diminish the risk of cross-contamination from the dishwater to the dishware.

In dishwashing, it is not only the dishwater that contributes to the cleaning of the dishware, but also the dishwashing process and the detergents. The dishware covers a range from glasses to pots and pans, which consist of a variety of materials such as glass, china, ceramics, thermostable plastics, stainless steel, and aluminum. The dishwashing processes use techniques from hand dishwashing or machine dishwashing to granule-assisted pot washing. For hand dishwashing, the dishware is soaked in water with hand dishwashing detergents (pH 7 to 8) before cleaning. The contact time between the dishware and the dishwater is in the range of 20 to 60 minutes, whereas in dishwashers, the contact time between dishwater and dishware is on the order of 1 to 10 minutes. The dishwashers use detergents with a pH from 12 to 14 that in the dishwater solution becomes pH 9 to 11. In all dishwashing processes, the importance of rinsing the dishware with fresh water is unquestionable (1). In the international standards (2,3), either hot fresh water at a temperature of 85 °C or cold fresh water with additives of disinfecting chemicals such as chlorine should be used to achieve hygienically clean dishware.

The microbiological status of the dishwater is of great importance, because this could cause cross-contamination and thereby the spread of infections (4,5). Therefore, this subject will be addressed more thoroughly.

CHARACTERIZING DISHWATER QUALITY

Dishwater consists of fresh water, detergent, and food residues. Depending on the hardness of the fresh water supply, detergents with more or less complex-acting properties are used. In hand dishwashing, the concentration of the washing-up detergent is around 0.05% (6), and in dishwashers, it is normally about 0.2% (7). The soil in the dishwater can be anything ranging from lipstick residues on glasses to carbohydrates like starch or vegetable fibers, proteins, animal or vegetable fats, and microorganisms from food residues.

There are several parameters to consider when characterizing dishwater, such as the conductivity, the pH, the temperature, the chemical oxygen demand (COD), the general microbiological level expressed as aerobic plate count (APC), and the characterization of the microbiological content. The amount of solid substances in dishwater is low; it varies from 5 mg/L (dishwasher for glasses and cutlery) to 5000 mg/L (pots and pans machine) according to Harpel (8) and our own measurements.

The conductivity is commonly used to monitor the concentration of the detergent used in dishwashers. However, the conductivity can be affected by the ions of the soil dispersed in the dishwater, and this could cause a malfunction of the dosage of the detergent.

The pH is a result of the amount and type of detergent used and the soil on the dishware. The organic soil in

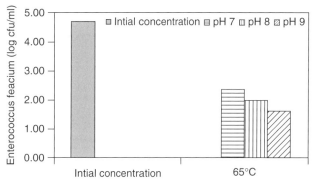

Figure 1. The reduction, 30 minutes after injection of *E. faecium* in dishwater, as a function of pH at 65 °C dishwashing temperature.

the dishwater is hydrolyzed by the high pH into smaller components. The influence of the pH on the microbiological growth in dishwater is shown in Fig. 1. From the figure, it can be seen that a change of the pH from 7 to 9 at 65 °C has a relatively small influence on the reduction of the concentration of the investigated bacteria, *Enterococcus feacium*. The samples were taken and analyzed 30 minutes after the addition of the bacteria, where the initial concentration was measured to be 50000 cfu/mL (^{10}log 4.7). The detergent used in the experiment was Gigant (Diskteknik, Sweden), an ordinary alkaline dishwasher detergent used in professional kitchens. The chosen test microorganism in this investigation is a rather thermostable bacteria, used in international standards (3) and in literature (8).

The temperature of the dishwater is determined by the process used. For hand dishwashing, the temperature is normally 40–45 °C (9), and for dishwashers, the temperature is 55–65 °C (7). The influence of the temperature is vital for the reduction of the concentration of bacteria in dishwater (see Fig. 2). For this microorganism, the reduction after 30 minutes is negligible for temperatures lower than 55 °C. However, at 65 °C there is a noticeable reduction of the concentration of *E. feacium*. For comparison, in Fig. 2, the reduction of the concentration of the bacteria for the conditions in

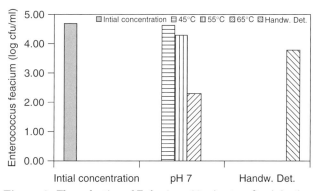

Figure 2. The reduction of *E. faecium*, 30 minutes after injection in dishwater, as a function of temperature of the dishwaters. The pH was kept to 7. For comparison, the reduction of *E. faecium* for the same time, using hand dishwashing detergent at 45 °C dishwashing temperature.

hand dishwashing is included. The decrease in microbiological activity depends solely on the antibacteriological properties of the detergent (Dizzy, Diskteknik, Sweden). Even with longer exposure times of microorganisms in dishwaters containing hand dishwashing detergents, the reduction rate will be low (6,10). Because of safety regulations in many countries, hand dishwashing cannot be performed at temperatures above 55 °C.

Another important factor of the temperature is the emulsion of fats into dishwater, which is considerably more effective at higher temperatures.

The COD depends mainly on the type and the amount of soil on the dishware. Also, the detergent adds to the value of the COD for dishwater. Values ranging from 500 to 5700 mg/L are reported in literature (8) and confirmed by our own measurements (see Fig. 3). Dishwater samples were taken in a professional kitchen from a pot and pan machine. The first sample was taken before the changing of dishwater. The second sample was taken after draining and filling the dishwater tank with fresh water and detergent, and the following ones after each dishwashing cycle. In this experiment, the increase of the COD is about 270 mg/L for each cycle. The manufacturers of one-tank machines recommend changing the dishwater after about 20 dishwashing cycles.

The microbiological content of the dishwater is more difficult to predict. There are several types of active bacteria in dishwater. In the studies of Mattick et al. (9) and Mospuye and van Holy (11), it is obvious that the distribution of the species found in the food residues are transferred to the dishwater. The most frequently reported microorganisms are *Bacillus cereus*, *Bacillus* spp., *Campylobacter* spp., *Clostridium prefringens*, *E.coli*, Enterobacteriaceae, *Salmonella* spp., and *Staphylococcus* spp. (9,11–13). Some of these are foodborne pathogens that pose a considerable risk of the spread of infection, if cross-contamination occurs (4).

Reported values of the APC in dishwater show that hand dishwashing exhibits higher values of the APC than does dishwater from dishwashers (see Table 1) (9,12,13). The content of Enterobacteriaceae and *Bacillus cereus* in the dishwaters from the studies can also be seen in the same table. The level of the APC ranges from ^{10}log 2.6 ± 0.8 to ^{10}log 4.0 ± 1.2. The observations confirm that in hand dishwashing conditions, with a lower temperature and less aggressive detergents, the microbiological activity is higher than in dishwater from dishwashers.

ENVIRONMENTAL ASPECTS

An important factor from an environmental point of view is the consumption of water. The largest consumption of water is for hand dishwashing. In professional dishwashers, the dishwater is reused up to 20–30 cycles, and only diluted with fresh water from each rinse cycle with about 10% of the tank volume. The tank volume ranges from 50 to 250 L in dishwashers. In dishwashers, where the cleaning process is assisted by plastic granules, the granules are reused for about 2000 dishwashing cycles.

To reduce the microbiological activity in dishwater, several methods can be used, for instance, ozone, high temperature and high pH, and disinfecting chemicals such as chlorine (14). From an environmental point of view, high temperature is the most agreeable and efficient parameter to use (see Fig. 2). In many countries, detergents containing aggressive chemicals, like chlorine, are not recommended, which is consistent with an environmental approach. Even the use of ozone is questionable because its general reactivity is well documented.

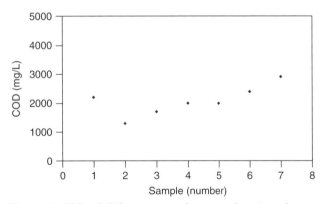

Figure 3. COD of dishwater samples as a function of sample numbers.

Table 1. The APC, Entereobacteriaceae, and *Bacillus cereus* Found in Different Dishwaters (cfu/mL) from Professional and Domestic Kitchens

Type of Establishment	Domestic Homes (9)	Hospices (9)	Street-Vended Food (11)	Professional Kitchens (13)
Dishwashing process	Hand dishwashing	Hand dishwashing	Hand dishwashing	Dishwasher
Number of samples	52	10	18	16
APC				
Mean	$1.3 \cdot 10^5$	$2.3 \cdot 10^4$	$1.0 \cdot 10^4$	$2.7 \cdot 10^2$
Range	10^1 to $>3 \cdot 10^5$	$3 \cdot 10^3$ to $1 \cdot 10^5$	$3 \cdot 10^3$ to $3 \cdot 10^6$	<10 to 10^4
Enterobacteriaceae				
Mean	$7.1 \cdot 10^4$	$7.2 \cdot 10^2$	ND	1.2
Range	<1 to $>3 \cdot 10^5$	<1 to $>6 \cdot 10^3$	ND	<1 to 10^2
Bacillus cereus				
Mean	ND	ND	46% of 65 isolates were determined	$2.7 \cdot 10^2$
Range				<10 to 10^4

ND: not determined.

CONCLUSIONS

The temperature is the most important factor in maintaining a low level of microbiological activity in the dishwater. This parameter is also from an environmental point of view less harmful than ozone or detergents containing other oxidizing chemicals. According to the international standard, the heat load on dishware should exceed 3600 HUE (Heat Unit Equivalence) to achieve hygienically clean dishware.

To shorten the dishwashing cycle time, granule-assisted dishwashing could decrease the energy and detergent consumption in the cleaning process. The rinsing procedure with fresh water must follow the above-mentioned standard.

Hand dishwashing consumes more fresh water than do dishwashers. It also uses a lower temperature, which has a considerably lower effect on the reduction of the microbiological activity of the dishwater. It is important to keep an acceptable quality level of the reused dishwater, because otherwise the dishwashing process and the rinse process will not be able to obtain clean dishware. To improve the dishwater quality, some manufacturers have integrated cyclones or filters into their dishwashers.

BIBLIOGRAPHY

1. Cogan, T.A., Sladder, J., Bloomfield, S.F., and Humphrey, T.J., (2002). Achieving hygiene in the domestic kitchen: the effectiveness of commonly used cleaning procedures. *J. Appl. Microbiol.* **92**: 885–892.
2. American National Standard/NSF International Standard 3. (2001). Ann Arbor, MI.
3. Deutsche Norm DIN 10512. (2001). *Food Hygiene—Commercial Dishwashing with Onetank-Dishwashers—Hygiene Requirements*, type testing, Gewerbliches Geschirrspülen mit Eintank-Geschirrspülmaschinen Dezember, Beuth Verlag GmbH, Berlin, Germany.
4. Ståhl Wernersson, E., Johansson, E., and Håkanson, H. (2004). Cross-contamination in dishwashers. *J. Hospital Infection* **56**: 312–317.
5. Cowan, M.E., Allen, J.P., and Pilkington, F. (1995). Small dishwashers for hospital ward kitchens. *J. Hospital Infection* **29**: 227–231.
6. Charnock, C. (2003). The antibacterial efficacy of Norwegian hand dishwashing detergents. *Food Protection Trends* **23**: 10, 790–796.
7. Ståhl Wernersson, E., Johansson, E., and Håkanson, H. (2004). Granule-assisted dishwashing improves hygiene. *Food Service Technol.* **4**: 129–137.
8. Harpel, S., Tilkes, F., and Krüger, S. (1994). Forderungen an die maschinelle Dekontamination vor Geschirr unter hygienischen Gesichtspunkten sowie an eine geringe Belastung der Abwässer—ein Widerspruch? (Requirements to the decontamination in dish-washing machines under hygienic and environmental aspects—a contradiction in terms?) *Zbl. Hyg.* **195**: 377–383.
9. Mattick, K., Durham, K., Hendrix, M., Slader, J., Griffith, C., Sen, M., and Humphrey, T. (2003). The microbiological quality of washing-up water and the environment in domestic and commercial kitchens. *J. Appl. Microbiol.*, **94**: 842–848.
10. Kusumaningrum, H.D., van Putten, M.M., Rombouts, F.M., and Beumer, R.R. (2002). Effects of antibacterial dishwashing liquid on foodborne pathogens and competitive microorganisms in kitchen sponges. *J. Food Protection* **65**: 61–65.
11. Mosupye, F.M., and von Holy, A. (1999). Microbiological quality and safety of ready-to-eat street vended foods in Johannesburg, South Africa. *J. Food Protection* **62**: 1278–1284.
12. Mosupye, F.M. and von Holy, A. (2000). Microbiological hazard identification and exposure assessment of street food vending in Johannesburg. *Int. J. Food Microbiol.* **61**: 137–145.
13. Ståhl Wernersson, E., Håkanson, H., Lindvall, I., and Trägårdh, C. (2003). Hygiene in warewashers utilizing blasting granules that foodservice establishments use. *Food Protection Trends* **23**(10): 797–807.
14. von Krüger, S. and Zschaler, R. (1996). Überprüfung der Dekontaminationswirkung in Mehrtank-Spühlmaschinen. *Arch. Lebensmittelshygiene* **47**: 105–128.

DISINFECTION BY-PRODUCT PRECURSOR REMOVAL FROM NATURAL WATERS

Yeomin Yoon
Northwestern University
Evanston, Illinois

Shahnawaz Sinha
Malcolm Pirnie Inc.
Phoenix, Arizona

Jinsik Sohn
Kookmin University
Seoul, Korea

BACKGROUND

Water is one of Earth's most important natural resources, and its role in our everyday life cannot be ignored. Civilizations began and were centered within regions of abundant water supplies. However, water present in our natural environment is not always potable for human consumption. Many lives have been lost as recently as the early part of last century because of waterborne diseases such as cholera and typhoid fever. Tragically, this battle still continues elsewhere in developing countries where infant mortality is often the result of waterborne diseases (1). To be "potable," water needs to be disinfected with a method such as chlorination, which has been successfully practiced in the United States for over a century.

However, chlorinated disinfection by-products (DBPs) have been a concern for many years because of their adverse effects on human health. A great deal of research has been conducted on these chlorinated by-products. Several chlorinated DBPs have already been identified since the discovery of trihalomethanes (THMs) in 1974 (2). Other chlorinated disinfection by-products include haloacetic acids (HAAs), chloral hydrate, haloketone, and haloacetonitriles. One way to identify the total amount of chlorinated DBPs is to measure the total organic halide (TOX) concentration. However, if all identified chlorinated DBPs were considered, they would only account for about 50% of the TOX (3). The unidentified

DBPs may be too large (nonvolatile) or too small (polar) in molecular weight to be detected by currently accepted analytical methods such as gas chromatography.

In recent years, other alternative disinfectants have been discovered and used by scientists, including chloramines, chlorine dioxide, ozone, potassium permanganate, and ultraviolet irradiation. However, because of proven reliability, cost, and availability, chlorine has remained the chemical of choice. As a result, of all the municipal water supplies that are being chemically disinfected, at least 70% use chlorine (4). Unfortunately, in the process of attacking these unwanted microorganisms, the chlorine reacts with other impurities such as natural organic matter (NOM) and bromide, and it can lead to the formation of carcinogenic DBPs (5). Alternative disinfectants may help to minimize these DBPs. However, these oxidants/disinfectants may produce other DBPs, which may also be harmful (4).

Health-related studies on these carcinogenic by-products suggest that the presence of as low as 6 µg/L of chloroform (chlorinated DBP) can have a lifetime cancer risk of 10^{-6} (6). Another recent DBP-related risk study (7) showed that as the bromide concentration increases from 50 µg/L to 200 µg/L, it substantially increases overall DBP-related cancer risk. To reduce the public health risk from these toxic compounds, the U.S. Environmental Protection Agency (EPA) has regulated THMs since 1979, with a maximum contaminant level (MCL) of 100 µg/L. In view of recent studies, the EPA has proposed a more stringent disinfection/disinfection by-products (D/DBPs) rule, which will be implemented in two stages. Stage 1 (effective from 1998) of the D/DBP rule would reduce the current MCL for THMs from 100 µg/L to 80 µg/L, with a new MCL restriction of 60 µg/L for HAA_5 (sum of five HAA species). Stage 2 (effective from 2002) would further reduce both of these MCLs for THMs and HAAs (as HAA_6) by an additional 50% (8). However, the success of water utilities in meeting these stringent and challenging MCLs will require a better understanding of the DBP precursor material, NOM, and its reactivity with chlorine in forming DBPs.

The most effective strategy to decrease the formation of DBPs would be to reduce the amount of NOM, which is the main precursor of DBPs, before chlorination. The processes that are available to remove this precursor material include membranes, granular activated carbon adsorption, biofiltration processes, and chemical treatment by coagulation. Coagulation has remained the most widely accepted method by utilities to remove NOM from natural waters.

PRECURSOR REMOVAL BY COAGULATION

Coagulation

Coagulation is an age-old practice in water treatment and has not changed much over the years. Until recently, coagulation has primarily removed color (e.g., NOM) and turbidity from water. More attention has been given to NOM removal from water. In the early 1990s, the EPA proposed the enhanced coagulation (EC) rule to further reduce health-related risks from these DBPs and compelled utilities to practice EC to further remove NOM. In enhanced coagulation, NOM removal is based on raw water total organic carbon (TOC) and alkalinity content (Table 1). EC has already been evaluated by several researchers and has been found to be effective in both NOM removal and DBP minimization. Other benefits include reduction in chlorine demand in finished water (9). The main objective of EC is to achieve 15% to 50% removal of organic matter before disinfection. However, some of the disadvantages of EC also reported from these studies include increased chemical use, which translates into higher operational cost; higher sludge production; and additional chemical costs associated with pH adjustments. Thus, more research is still needed to overcome these disadvantages.

Table 1. Percent of Organic Removal for Enhanced Coagulation Requirements Proposed for the DBP Rule

TOC, mg/L	Alkalinity, mg/L as $CaCO_3$		
	<60	60–120	>120
2–4	35%	25%	15%
4–8	45%	35%	25%
>8	50%	40%	30%

Coagulants Used for NOM Removal

Various types of metal coagulants are available to remove NOM. However, two of the most common conventional metal coagulants in water treatment are aluminum sulfate ($Al_2(SO_4)_3 \cdot 14H_2O$, commonly referred to as alum) and ferric chloride ($FeCl_3 \cdot 6H_2O$, commonly referred to as iron salt). Alum is currently the most widely used. However, over the past years, the application of iron salt as a coagulant has increased dramatically (10). Conventional coagulants of other types include aluminum-based sodium aluminate, aluminum chloride, and ferric sulfate (11). However, the application of these coagulants in potable water has been limited. Recently, various other types of (alternative) coagulants have been introduced. These specialized coagulants are more common in Europe and Japan than in the United States (12). These coagulants (patented products) are claimed to be more effective than the conventional coagulants. Conventional coagulants are basically salts of a strong acid (e.g., HCl or H_2SO_4) and a weak base ($Al(OH)_3$ or $Fe(OH)_3$). Thus, it is a mixture of cations (from bases) and anions (from acids). However, alternative coagulants are specially prepared and may include additional anions as additives (i.e., anions such as sulfate).

Effectiveness of Coagulation in Precursor Minimization

Numerous studies have focused on removal of the "humic" fraction of NOM that is believed to be the precursor for THMs. Thus, most of the research has focused on the removal of this fraction by coagulation and other processes. Oliver and Lawrence (13) studied raw water samples collected from treatment plants in the Great Lakes Basin to investigate the effect of alum coagulation in haloform reduction and reported a dramatic drop in haloform formation by alum coagulation. Chadik and Amy

(14) studied natural waters in six different geographical locations to evaluate the effectiveness of metal coagulants (alum and ferric chloride) in removing THM precursors. THM formation potential (THMFP) reduction by alum was in a range between 36% and 66%, whereas the reduction by iron was in a range between 33% and 77%. Alum was found to be more effective than iron in reducing TOC and THMFP. Jodellah and Weber (15) investigated THM precursor removal by alum coagulation, lime softening, and activated carbon adsorption, and they found that precursor removal was variable and concluded that greater removal of organic matter does not necessarily indicate greater reduction of THMFP. No correlation was found between TOC removal by these treatments and overall THMFP reduction. It can be concluded that the reduction of THMFP is dependent on the qualitative nature of the organic matter, rather than on the quantitative amount of organic matter. Thus, a careful assessment of the physical and chemical composition remaining after different treatments of NOM mixtures is required before a mechanistic explanation can be provided.

Hubel and Edzwald (16) studied the THM precursor removal by alum, high-molecular-weight polymers, cationic polymers (usually used as coagulant aids), and various combinations of these coagulants. In this study, it was concluded that cationic polymers were less effective in precursor removal as compared with alum when used as a sole coagulant, where alum removed 50% to 60% more organic matter than did cationic polymers. This study also indicated that the type of polymers used (polyacrylamide versus polyquaternary amine) (i.e., charge density and molecular weight) had slight effects on precursor removal. Cationic polymers as sole coagulants were found not to be effective in THM precursor removal, but they were effective in turbidity removal by forming denser and better settling flocs than alum alone. Knocke et al. (17) studied the effect of seasonal temperature variation in precursor removal in natural water by alum and ferric sulfate. The results showed that lower temperature (coagulation at pH 5.0) significantly reduced the rate of THM formation (2 °C versus 22 °C) for equivalent organic content.

SUMMARY

This study provides a general overview on coagulation and its effectiveness in removing NOM from natural waters for DBP minimization with various metal coagulants. Although other technologies such as activated carbon adsorption and membrane filtration are available to reduce DBP precursors, these alternative technologies are comparatively expensive processes (18). For the time being, coagulation may remain the technology of choice for DBP minimization. In addition, identification of DBP precursors based on detailed NOM characterization and fractionation techniques can provide mechanisms for how these precursor materials could effectively be controlled in natural waters. Once identified, these potential precursor materials could then be effectively removed by in-depth knowledge of coagulants and coagulation techniques.

BIBLIOGRAPHY

1. Carol, H.T. and Fox, K. (1990). *Health and Aesthetic Aspect of Water Quality in Water Quality and Treatment—a Handbook of Community Water Supplies*, 4th Edn. American Water Works Association/McGraw Hill, New York.
2. Rook, J.J. (1974). Formation of haloforms during chlorination of natural water. *Water Treatment Exam* **23**: 234.
3. Krasner, S.W. et al. (1989). The occurrence of disinfection by products in US drinking water. *J. Am. Water Works Assoc.* **81**: 41–53.
4. Oxenford, J. (1996). *Disinfection By-products: Current Practices and Future Direction in Disinfection By-products in Water Treatment*. CRC/Lewis Publishers, Boca Raton, FL.
5. Rook, J.J. (1976). Haloforms in drinking water. *J. Am. Water Works Assoc.* **68**: 168–172.
6. Bull, R.J. and Kopfler, F.C. (1991). Health effect of disinfectants and disinfection by-products. Proceedings of American Water Works Association Annual Conference, Denver, CO.
7. Sohn, J., Sinha, S., Amy, G.L., and Zhu, H.W. (1997). The role of coagulation in DBP related risk reduction. Proceedings of American Water Works Association Water Quality & Technology Conference, Denver, CO.
8. Pontius, F.W. (1996). An update of federal regs. *J. Am. Water Works Assoc.* **88**: 36–46.
9. Sinha, S. (1996). *Disinfection by-products precursors removal by enhanced coagulation*, Masters Thesis, The University of Kansas-Lawrence.
10. Baltpurvins, K.A., Burns, R.C., and Lawrance, G. (1996). Effect of pH and anion type on the aging of freshly precipitated iron (III) hydroxide sludge. *Environ. Sci. Technol.* **30**: 939–944.
11. Van Benschoten, J.E. and Edzwald, J.K., (1990). Measuring aluminum during water treatments. *J. Am. Water Works Assoc.* **82**: 71–78.
12. Amirtharajah, A. and O'Melia, C.R. (1990). *Coagulation Process: Destabilization, Mixing and Flocculation, in Water Quality and Treatment*, 4th Edn. American Water Works Association/McGraw Hill, New York.
13. Oliver, B.G. and Lawrence, J. (1979). Haloforms in drinking water: a study of precursors and precursor removal. *J. Am. Water Works Assoc.* **71**: 161–163.
14. Chadik, P.A. and Amy, G.L. (1983). Removing trihalomethane precursors from various natural waters by metal coagulants. *J. Am. Water Works Assoc.* **75**: 532–536.
15. Jodellah, A.M. and Weber, W.J. (1985). Controlling trihalomethane formation potential by chemical treatment and adsorption. *J. Am. Water Works Assoc.* **77**: 95–100.
16. Hubel, R.E. and Edzwald, J.K. (1987). Removing THM precursor by coagulation, *J. Am. Water Works Assoc.* **79**: 98–106.
17. Knocke, W.R., West, S., and Hoehn, R.C. (1986). Effect of low temperature on the removal of trihalomethane precursors by coagulation. *J. Am. Water Works Assoc.* **78**: 189–195.
18. Singer, P.C. (1994). Control of disinfection by products in drinking water. *J. Environ. Eng.-ASCE* **120**: 727–744.

ALTERNATIVE DISINFECTION PRACTICES AND FUTURE DIRECTIONS FOR DISINFECTION BY-PRODUCT MINIMIZATION

June M. Weintraub
City and County of San Francisco Department of Public Health
San Francisco, California

J. Michael Wright
Harvard School of Public Health
Boston, Massachusetts

Raghuraman Venkatapathy
Oak Ridge Institute for Science and Education
Cincinnati, Ohio

INTRODUCTION

Disinfection of public water supplies is required in many countries to inactivate microbial pathogens and protect public health. Chlorine is the predominant disinfectant used worldwide for this purpose, and while chlorine provides significant benefit, it can create by-products that may cause adverse health effects. Disinfection by-products (DBPs) have been linked to bladder and rectal cancers and to adverse reproductive outcomes (1,2). The Safe Drinking Water Act amendments of 1996 required the United States Environmental Protection Agency (U.S. EPA) to develop regulations that balanced the conflicting goals of pathogen control with DBP minimization. The Stage 1 Disinfection By-products Rule established standards for trihalomethanes (THMs), haloacetic acids (HAAs), bromate, and chlorite in U.S. drinking water (3).

The use of alternative disinfectants in the United States has received increasing attention in light of DBP regulatory changes. The formation of DBPs depends both on the type of disinfectant used and the presence of precursors. Thus, there are essentially three approaches to reduce the occurrence of any particular DBP: (1) alternative disinfection, (2) other treatment options (e.g., that reduce either the presence of DBP precursors or the amount of disinfectant needed), and (3) approaches aimed at source water, such as watershed protection to limit the addition of DBP precursors to waterways, or the use of source water with naturally low precursor levels. The advantages and disadvantages of these options are briefly outlined below, including usage information, operational concerns, disinfectant properties, and potential impact on the occurrence of specific DBPs in drinking water supplies.

ALTERNATIVE DISINFECTANTS

A number of disinfectants are appropriate to treat drinking water on a large scale. Chlorine, chloramine, chlorine dioxide, ozone, and ultraviolet radiation are among the most widely used disinfectants. Ozone and chlorine dioxide are more common in some European countries, but most water systems use chlorine for either primary (predistribution) or secondary (residual) disinfection (4). Of all the choices, only chlorine and chloramine are suited for secondary disinfection to maintain residual protection in the distribution system.

Chloramine

Chloramines were first used to treat water in the United States and Canada in 1917 (5). Chloramination is increasing in the United States, with about 30% of medium and large utilities reporting its use in 1998, compared to 20% in 1989. Chloramination involves the addition of ammonia to chlorinated water; aqueous chlorine reacts with ammonia to form inorganic chloramines (6). A principal operational consideration is that the chlorine/ammonia dosage ratio must be monitored carefully to prevent the growth of nitrifying bacteria in distribution systems and storage reservoirs (7).

Chloramines are weaker disinfectants compared to chlorine, ozone, or chlorine dioxide. Chloramine CT (product of disinfectant concentration and contact time) values to control *Giardia*, *Cryptosporidium*, or viruses are up to four orders of magnitude higher than for the other disinfectants, making chloramines an impractical choice for primary disinfection (Table 1). Chloramines are therefore typically used only for secondary disinfection; the maximum residual disinfectant level (MRDL)

Table 1. CT (Concentration × Time) Values[a] for Alternative Disinfectants

Disinfectant	Units	Virus			Giardia			Cryptosporidium
		2 log	3 log	4 log[b]	1 log	2 log	3 log[b]	2 log[b]
Chlorine	mg·min/L	3	4	6	35	69	104	>4,000
Chloramine	mg·min/L	643	1067	1491	615	1230	1850	>25,000
Chlorine dioxide	mg·min/L	4	13	25	8	15	23	40[c]
Ozone	mg·min/L	<1	<1	1	<1	<1	1	3.9–10
UV radiation	mJ/cm²	40	N/A[d]	N/A	N/A	N/A	N/A	12

[a]All values are at 10 °C, pH 6–9.
[b]Required removal for large public drinking water systems in the United States.
[c]Chauret et al. (8) report variable CT values for 2 log inactivation of *Cryptosporidium* at 21 °C, ranging from 75 to 1000 mg·min/L.
[d]N/A, not available.
Source: References 6, 8–10.

for chloramines is 4 mg/L. Compared to chlorine, chloramines are more stable secondary disinfectants and may reduce taste and odor complaints. Chloramination may also be more effective than chlorine at inhibiting biofilm and bacterial regrowth in distribution systems (11).

The use of chloramines reduces the formation of two of the major regulated DBP groups—trihalomethanes and haloacetic acids. However, some DBPs, including dihalogenated acids and NDMA (N-nitrosodimethylamine), may increase in chloraminated systems (12,13). Source water chemistry and ammonia application timing influence the formation of chloramination by-products. For example, in one chloraminated system, high levels of iodoacids were attributed to the simultaneous application of chlorine and ammonia to source water high in iodide. When ammonia application is applied after significant free chlorine contact time, iodoacid formation is minimized (14).

Chlorine Dioxide

During the 1940s, the Niagara Falls, New York water treatment plant was the first North American utility to use chlorine dioxide for water treatment (15). Chlorine dioxide was used by 8% of medium and large utilities in the United States in 1998, more than triple the proportion that utilized it in 1989 (7). Chlorine dioxide must be generated on site, which requires significant operator skill; thus, it is relatively expensive in comparison to chlorine. Chlorine dioxide is flammable and explosive, and its use therefore requires significant safety precautions.

Compared to chlorine, chlorine dioxide is more effective at inactivation of *Cryptosporidium* and *Giardia* but less effective for *E. coli* and rotaviruses (see Table 1). Although chlorine dioxide CT values of 40 mg · min/L at 10 °C are reported to accomplish 2 log removal, (i.e., 99%) of *Cryptosporidium* (6), highly variable CT values have been reported at other temperatures (8). An advantage of chlorine dioxide is that disinfection efficiency is not dependent on pH or the presence of ammonia (16). However, the half-life decreases at pH > 9 and with increasing chlorine dioxide concentrations (17). The MRDL for chlorine dioxide is 0.8 mg/L (6).

Chlorine dioxide is effective at controlling DBP formation, since it can destroy precursors to trihalomethanes and haloacetic acids and does not oxidize bromide ion to bromate. The predominant DBPs formed by chlorine dioxide are chlorate and chlorite, which may lead to adverse human health effects (see the chapter entitled Health Effects of Commonly Occurring Disinfection By-products in the Municipal Water Supply). Chlorine dioxide can produce other DBPs such as dihalogenated HAAs and haloacetaldehydes (18). Due to the inability of chlorine dioxide to remove hydroxyfuranone precursors, high levels of MX and the brominated MX analogues have been found in waters with elevated total organic carbon and bromide (19).

Ozone

Ozonation was first used in The Netherlands in 1893 and remains a popular disinfection method in Europe (6). While ozonation was used for disinfection by less than 10% of large water utilities in 1998, its use is increasing in the United States (7). In the proposed Stage 2 Disinfectants and Disinfection By-products Rule, the U.S. EPA reported that in 2000 there were 332 plants of various sizes using ozone and 58 plants that were planning to install ozonation (20). Ozone is one of the strongest oxidants used to disinfect water and is the predominant choice of the bottled water industry. Although ozonation is a fairly expensive and complex treatment process, it has several operational advantages. These advantages include short contact time, taste and odor control, and improved coagulation and filtration efficiency. Ozone is very effective at microbial inactivation and is effective against viruses, *Giardia*, and *Cryptosporidium* (see Table 1). Because ozone can also degrade more complex organic matter that may promote microbial regrowth in distribution systems (21), and because ozone does not provide a protective residual, it must be used in combination with chlorine or chloramine in systems where residual disinfection is necessary.

Use of ozone limits the formation of chlorinated DBPs, but it can combine with bromide in water to form bromate, brominated organics, and other brominated by-products (22,23). Ozonation can also produce organic DBPs such as the aldehydes, ketones, ketoaldehydes, carboxylic acids, aldo acids, keto acids, hydroxy acids, alcohols, esters, and alkanes (23). Furthermore, if ozonation is followed by chlorination, the formation of aldehydes and ketones may allow subsequent formation of their halogenated derivatives (e.g., chloropicrin and other halonitromethanes) (19,21,24).

Ultraviolet Radiation

Ultraviolet (UV) radiation is not widely used to treat drinking water in the United States, but has been used extensively to treat wastewater (20). UV radiation can inactivate viruses, bacteria, and protozoans (9). The use of UV radiation is somewhat limited by intensive operational and maintenance requirements, and it is most effective in low turbidity waters. Advantages of UV radiation include a very short detention time requirement, so dedicated contact chambers are not necessary (25). However, a slow flow rate is required, and this may pose operational constraints. UV disinfection does not provide residual protection against bacterial regrowth in the distribution system. UV radiation alone forms few DBPs but DBP formation can still occur when UV disinfection is combined with secondary disinfectants (26).

OTHER TREATMENT OPTIONS

In addition to alternative disinfection, many treatment options are available to water utilities to reduce DBP formation. These include pH adjustment, disinfectant timing or treatment process changes, and physical or oxidative removal of DBP precursors. Changes to these complex treatment processes can have different effects on the various types of DBPs.

The impact of pH on DBP concentrations is reported by a number of researchers (27–32). In general, the concentrations of THMs increase while those of HAAs, HANs, and haloketones decrease with increasing pH. Many of the halogenated DBPs are hydrolyzed above pH 8, reducing their formation (33).

Treatment plant operators may be able to limit DBP formation by reducing contact time or the initial disinfectant dose. Changing the order of disinfectant application within the treatment chain should also be considered. This may include moving the primary disinfectant step to follow filtration, or changing the application point for the secondary disinfectant application (34). Distribution system maintenance may decrease residual disinfectant demand, allowing operators to use smaller amounts of disinfectant.

The major precursors to DBPs of concern are organic substances (e.g., humic and fulvic acids), bromine, and iodine. If treatment processes or primary disinfection methods that reduce DBP precursors are used, then secondary disinfection with chlorine or chloramine will produce lower levels of their associated by-products. Precursor removal can be accomplished by coagulation, filtration, or oxidative processes.

Coagulation is an important step in converting soluble precursors to particulate, which can then be removed by filtration or settling (35). Enhanced coagulation and enhanced softening processes are those that improve removal of DBP precursors over conventional treatment. Coagulants specified by the U.S. EPA as appropriate for precursor removal to achieve compliance with the Stage 1 Disinfection By-Products Rule include aluminum or iron salts, polyaluminum chloride, and cationic polymers, as well as anionic and nonionic polymers (36).

Physical removal of precursors includes processes such as microfiltration and biofiltration, which are used primarily in larger plants. Biofiltration through the use of sand filters or granulated activated carbon can reduce formation of halogenated DBPs (37). Passage of ozonated and chlorinated water samples through a rapid sand filter reduced the concentration of aldehydes by 26–62%, due to biological activity in the sand filters (38,39). However, removal of organic contaminants through biofiltration does not affect bromide concentrations in the treated water and results in greater formation of brominated DBPs (39). Other physical precursor removal techniques like electroporation (40) show promise but are in the early stages of development for practical use.

Oxidative precursor removal is another effective means of reducing DBPs. Ozone is effective at destroying precursors to MX and trihalogenated HAAs (19). Potassium permanganate is very effective at removing DBP precursors, both by oxidation and by enhancing coagulation (6). Some advanced oxidation processes (AOPs) that combine oxidation techniques show promise for utilities (10). Typical AOPs include combinations of ozone, hydrogen peroxide, UV radiation, solid catalysts, or sodium hypochlorite. Titanium dioxide photocatalysis, a combination of titanium dioxide and UV radiation (26), may be applicable for smaller utilities.

WATER QUALITY PROTECTION MEASURES

In the third edition of its guidelines for drinking water quality published in 2004, the World Health Organization (WHO) emphasized source water quality protection as a critical element to providing safe drinking water (41). Source water quality is highly variable, with some regions having high quality source waters and naturally low levels of DBP precursors. For example, some waters contain low bromide levels and low levels of total organic carbon. Such high quality waters are less likely to form certain by-products upon disinfection and may require minimal treatment to remove DBP precursors.

Source water protection, such as covered reservoirs, may help reduce DBP levels by limiting the amount of runoff of organic matter into surface water supplies. Other watershed management practices can reduce the amount of dissolved organic carbon in a water source and concomitant production of DBPs (42). Highly protected source waters, such as naturally filtered groundwater, may require minimal disinfection to remove microbial pathogens.

Providing safe, palatable water necessitates the use of a multiple barrier approach in which watershed protection, water filtration and disinfection, and maintenance of the water distribution system are addressed (43). Future trends in drinking water treatment in developed countries may shift the focus from water treatment to pollution prevention (44) and concentrate resources on maintaining outdated distribution systems and protecting the quality of source water.

BIBLIOGRAPHY

1. Nieuwenhuijsen, M.J., Toledano, M.B., Eaton, N.E., Fawell, J., and Elliott, P. (2000). Chlorination DBPs in water and their association with adverse reproductive outcomes: a review. *Occup. Environ. Med.* **57**(2): 73–85.
2. Villanueva, C.M. et al. (2004). DBPs and bladder cancer: a pooled analysis. *Epidemiology* **15**(3): 357–367.
3. U.S. Environmental Protection Agency (2001). Office of Water (4606M), EPA 816-F-01-010, May. *Stage 1 Disinfectants and Disinfection Byproducts Rule: A Quick Reference Guide.* http://www.epa.gov/safewater/mdbp/qrg_st1.pdf.
4. Connell, G.F. (1998). European water disinfection practices parallel U.S. treatment methods. Chlorine Chemistry Council. *Drinking Water Health Q.* **4**(3).
5. Hall, E.L. and Dietrich, A.M. (2000). A brief history of drinking water. *Opflow* **26**(6).
6. U.S. Environmental Protection Agency (1999). Office of Water (4607), EPA 815-R-99-014, April. *Alternative Disinfectants and Oxidants Guidance Manual.* http://www.epa.gov/safewater/mdbp/pdf/alter/cover_al.pdf.
7. AWWA Water Quality Division Disinfection Systems Committee (2000). Committee report: disinfection at large and medium size systems. *J. Am. Water Works Assoc.* **92**(5): 32–43.
8. Chauret, C.P., Radziminski, C.Z., Lepuil, M., Creason, R., and Andrews, R.C. (2001). Chlorine dioxide inactivation of *Cryptosporidium parvum* oocysts and bacterial spore indicators. *Appl. Environ. Microbiol.* **67**(7): 2993–3001.

9. Driedger, A.M., Rennecker, J.L., and Marinas, B.J. (2001). Inactivation of *Cryptosporidium parvum* oocysts with ozone and monochloramine at low temperature. *Water Res.* **35**(1): 41–48.

10. Mofidi, A.A. et al. (2002). *Advanced Oxidation Processes and UV Photolysis for Treatment of Drinking Water*. Metropolitan Water District of Southern California, La Verne, California.

11. Momba, M.N.B., Cloete, T.E., Venter, S.N., and Kfir, R. (1998). Evaluation of the impact of disinfection processes on the formation of biofilms in potable surface water distribution systems. *Water Sci. Technol.* **38**(8–9): 283–289.

12. Diehl, A.C. et al. (2000). DBP formation during chloramination. *J. Am. Water Works Assoc.* **92**(6): 76–90.

13. California Department of Health Services (2002). *Studies on the Occurrence of NDMA in Drinking Water*. http://www.dhs.ca.gov/ps/ddwem/chemicals/NDMA/studies.htm.

14. Plewa, M.J. et al. (2004). Chemical and biological characterization of newly discovered iodoacid drinking water DBPs. *Environ. Sci. Technol.* **38**(18): 4713–4722.

15. Aieta, E.M. and Berg, J.D. (1986). A review of chlorine dioxide in drinking water treatment. *J. Am. Water Works Assoc.* **78**(6): 62–72.

16. Hoehn, R.C. and Gates, D.J. (1999). Control of disinfection by-product formation using chlorine dioxide. In: *Formation and Control of Disinfection By-Products*. P.C. Singer (Ed.). American Water Works Association, Denver, pp. 205–221.

17. Gordon, G., Kieffer, R.G., and Rosenblatt, D.H. (1972). The chemistry of chlorine dioxide. *Prog. Inorg. Chem.* **15**: 201–286.

18. Krasner, S.W. et al. (2001). The occurrence of a new generation of DBPs (beyond the ICR). In: *Proceedings of the American Water Works Association Water Quality Technology Conference*. American Water Works Association, Denver.

19. Weinberg, H.S., Krasner, S.W., Richardson, S.D., and Thruston, A.D. Jr. (2002). *The occurrence of disinfection by-products (DBPs) of health concern in drinking water: results of a nationwide DBP occurrence study*, EPA/600/R02/068, U.S. Environmental Protection Agency, National Exposure Research Laboratory, Athens, GA.

20. U.S. Environmental Protection Agency (2003). *40 CFR Parts 141,142 and 143*. National Primary Drinking Water Regulations: Stage 2 Disinfectants and Disinfection Byproducts Rule (Proposed Rules). 68 *Fed. Reg.* 49548 (August 18).

21. Reckhow, D.A. (1999). Control of disinfection by-product formation using ozone. In: *Formation and Control of Disinfection By-Products*. P.C. Singer (Ed.). American Water Works Association, Denver, pp. 179–204.

22. von Gunten, U. (2003). Ozonation of drinking water: Part II. Disinfection and by-product formation in presence of bromide, iodide or chlorine. *Water Res.* **37**(7): 1469–1487.

23. Richardson, S.D. et al. (1999). Identification of new ozone disinfection by-products in drinking water. *Environ. Sci. Technol.* **33**: 3368–3377.

24. Hoigné, J. and Bader, H. (1988). The formation of trichloronitromethane (chloropicrin) and chloroform in a combined ozonation/chlorination treatment of drinking water. *Water Res.* **22**(3): 313.

25. Malley, J.P. (1999). Control of disinfection by-product formation using ultraviolet light. In: *Formation and Control of Disinfection By-Products*. Singer, P.C. (Ed.). American Water Works Association, Denver, pp. 223–235.

26. Richardson, S.D. et al. (1996). Identification of TiO_2/UV disinfection byproducts in drinking water. *Environ. Sci. Technol.* **30**(11): 3327–3334.

27. Stevens, A.A., Slocum, C.J., Seeger, D.R., and Robeck, G.G. (1976). Chlorination of organics in drinking water. *J. Am. Water Works Assoc.* **68**(11): 615–620.

28. Lange, A.L. and Kawczynski, E. (1978). Controlling organics: the Contra Costa County Water District experience. *J. Am. Water Works Assoc.* **70**(11): 653–660.

29. Trussell, R.R. and Umphres, M.D. (1978). The formation of trihalomethanes. *J. Am. Water Works Assoc.* **70**(11): 604–612.

30. Miller, W.J. and Uden, P.C. (1983). Characterization of nonvolatile aqueous chlorination products of humic substances. *Environ. Sci. Technol.* **17**(3): 150–152.

31. Reckhow, D.A. and Singer, P.C. (1985). Mechanisms of organic halide formation during fulvic acid chlorination and implications with respect to preozonation. In: *Water Chlorination: Chemistry, Environmental Impact and Health Effects. Vol. 5*, R.L. Jolley, R.J. Bull, and W.P. Davis (Eds.). Lewis Publishers, Chelsea, MI, pp. 1229–1257.

32. Nieminski, E.C., Chaudhuri, S., and Lamoreaux, T. (1993). The occurrence of DBPs in Utah drinking waters. *J. Am. Water Works Assoc.* **85**(9): 98–105.

33. Singer, P.C. (1994). Control of disinfection by-products in drinking water. *J. Environ. Eng.* **4**: 727–744.

34. Carlson, M. and Hardy, D. (1998). Controlling DBPs with monochloramine. *J. Am. Water Works Assoc.* **90**(2): 95–106.

35. Huehmer, R.P. (2001). *Strategies for Disinfection Byproduct Mitigation in Low-Pressure Membrane Systems*. Presentation at Annual Conference of the Association of State Drinking Water Administrators.

36. U.S. Environmental Protection Agency (1999). Office of Water (4607), EPA 815-R-99-012, May. *Enhanced Coagulation and Enhanced Precipitative Softening Guidance Manual*. http://www.epa.gov/safewater/mdbp/coaguide.pdf.

37. Lykins, B.W. and Clark, R.M. (1988). *GAC for Removing THMs* (EPA 600/9-88/004). U.S. Environmental Protection Agency, Washington, DC.

38. Van Hoof, F., Janssens, J.G., and van Dijck, H. (1985). Formation of mutagenic activity during surface water preozonation and its removal in drinking water treatment. *Chemosphere* **14**(5): 501–509.

39. Sketchell, J., Peterson, H.G., and Christofi, N. (1995). Disinfection by-product formation after biologically assisted GAC treatment of water supplies with different bromide and DOC content. *Water Res.* **29**(12): 2635–2642.

40. Schlager, K.J. (2001). *Final Report: An Electroporation Drinking Water Disinfection System*. http://cfpub.epa.gov/ncer_abstracts/index.cfm/fuseaction/display.abstractDetail/abstract/1525/report/F.

41. World Health Organization (2004). *WHO Guidelines for Drinking-Water Quality*, 3rd Edn. WHO, Geneva. http://www.who.int/water_sanitation_health/dwq/en/gdwq3_contents.pdf.

42. Fleck, J.A., Bossio, D.A., and Fujii, R. (2004). Dissolved organic carbon and disinfection by-product precursor release from managed peat soils. *J. Environ. Qual.* **33**(2): 465–475.

43. Ford, T.E. (1999). Microbiological safety of drinking water: United States and global perspectives. *Environ. Health Perspect.* **107**(S1): 191–206.

44. U.S. Environmental Protection Agency (2000). Office of Water (4606M), EPA 816-F-00-002, *Drinking Water: Past, Present, and Future*. February. http://www.epa.gov/safewater/consumer/dwppf.pdf.

WATER QUALITY ASPECTS OF DREDGED SEDIMENT MANAGEMENT

Anne Jones-Lee
G. Fred Lee
G. Fred Lee & Associates
El Macero, California

INTRODUCTION

The United States has 25,000 miles of navigable waterways, which include coastal waterways, such as the Intercoastal Waterway along the Eastern and Gulf Coasts; the major inland rivers, such as the Mississippi and Ohio; and major ports and harbors, such as New York, Galveston, Houston, Los Angeles, San Diego Bay, San Francisco Bay, Puget Sound, and those of the United States–Canadian Great Lakes. The U.S. Army Corps of Engineers (COE) is charged by the U.S. Congress with maintaining the navigation depth of these waterways as they fill with sediment. Maintaining navigation depth is considered to be of national importance to minimize transportation costs of bulk materials. To accomplish this, the COE dredges, or issues permits for contractors to dredge, about 500 million yd^3/yr of sediment from major waterways as well as smaller harbors, marinas, ports, and channels.

In the 1960s, it was beginning to be recognized that the sediments of some U.S. waterways were highly contaminated with chemicals that have the potential to be pollutants—that is, to impair the beneficial uses of waterbodies as domestic water supply sources or for aquatic life/fisheries. As the conventional method of disposal of dredged sediment was to dump it in deeper open water, concern developed about the solubilization and desorption of chemical constituents that were associated with sediment particles when sediments were dredged and then introduced into a watercolumn at a disposal site. In response to water quality impact concerns, the U.S. Army Corps of Engineers developed a $30 million five-year Dredged Material Research Program (DMRP) to investigate the water quality significance of potential pollutants in dredged sediments. As part of that program, the authors conducted extensive laboratory and field studies to investigate the release of sediment-associated contaminants and their impacts associated with open water disposal of dredged sediment (1–3), results of which are summarized or referenced herein.

IMPACT OF TYPE OF DREDGING/DISPOSAL OPERATION

The development of an approach to evaluate and regulate potential water quality impacts of dredging and dredged sediment disposal requires an understanding of what transpires during such operations. In his *Handbook of Dredging Engineering*, Herbich (4) described the various methods of dredging and dredged sediment disposal. Dredging is typically accomplished by either mechanical or hydraulic means.

Mechanical Dredging/Barge Disposal

Mechanical dredging typically employs a clamshell or dragline bucket to remove sediments, which are then deposited onto a barge. Water introduced during dredging drains off the barge. After the barge is towed to the designated disposal site, the sediment is released and drops as a fairly cohesive mass to the bottom. As limited mixing occurs of the sediments being dredged with water, mechanical dredging results in limited opportunity for release of constituents from the sediments to the water, which limits the potential for sediment-associated contaminants to adversely impact water quality.

Hydraulic Dredging

Hydraulic dredging involves sucking up the sediment as a slurry (Fig. 1) and pumping it to an open water or confined on land disposal area or into the holding hoppers of a hopper dredge for transport to an open water disposal site. The sediment:water ratio of the slurry is characteristically 1:4 by volume. The slurrying results in the mixing of the sediments with water, which tends to promote the release of constituents associated with the sediment and interstitial water into the slurry water. If hopper dredge is used, the excess water is typically allowed to drain off at the dredging site or during transport to the disposal site to optimize efficiency of sediment removal.

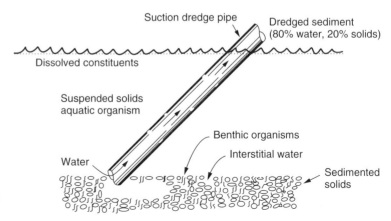

Figure 1. Diagrammatic representation of hydraulic dredging.

Hopper-Dredge Disposal

Figure 2 illustrates the watercolumn during disposal of dredged sediment from a hopper dredge. When the hopper dredge is at the designated disposal location, the hopper doors are opened and the sediment is released. Most of the sediment rapidly descends to the bottom as a fairly cohesive mass. As indicated in Fig. 2, the descent of the dredged sediment from the hopper to the bottom of the watercolumn forms a turbid cloud in the waters of the region.

In their evaluation of the nature and water quality impacts of dredging and dredged sediment disposal, Lee et al. (1) and Jones and Lee (2) monitored the watercolumn during more than 10 open water disposal operations by measuring more than 30 physical and chemical parameters at various depths in the watercolumn before, during, and after the passage of the turbid plume for each. Figure 3 shows the characteristic pattern of turbidity marking the passage of the turbid plume during open water disposal of dredged sediment. As indicated in Fig. 3, near the surface (2-meter depth) the turbidity persisted at a location a few tens of meters downcurrent from the dump for about 2 minutes. Near the bottom, at 14 meters, the turbid plume turbidity persisted for about 7 minutes. The plume typically dispersed to indistinguishable from ambient turbidity of the disposal area in about one hour.

Lee et al. (1) found that, in some instances, some increase occurred in concentration of some chemical constituents in this turbid cloud as well as some decrease in the dissolved oxygen (DO) concentration, as is shown in Fig. 4.

In the studies conducted by Lee et al. (1), the only constituent of potential concern that was released was ammonia. The potential impact of ammonia on aquatic organisms at an open water disposal site would have to be reviewed on a site-by-site basis, because it would be controlled largely by the sensitivity of the organisms there and the rate of dilution.

The turbid plume associated with open water dumping of mechanically dredged sediment is even less pronounced than with hopper-dredge disposal, because the

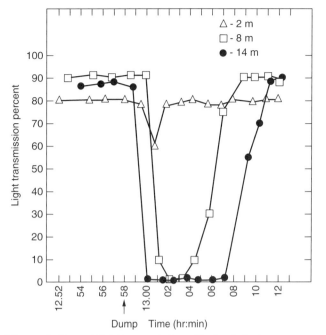

Figure 3. Percent light transmission with depth during passage of turbid plume from open water disposal of hopper-dredged sediment.

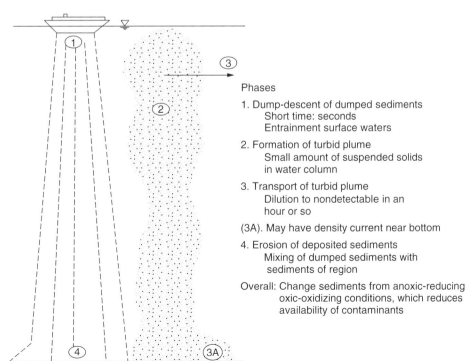

Figure 2. Hopper-dredge disposal of dredged sediment.

mechanically dredged sediment is a more cohesive mass. It was concluded by Lee et al. (1) that open water disposal of even contaminated sediments would not be expected to cause water quality problems in the disposal site water-column because of the limited releases and the short exposures that aquatic organisms could experience from such releases.

Pipeline Disposal

The open water pipeline disposal of hydraulically dredged sediments, illustrated in Fig. 5, presents a situation significantly different from that of dumping of mechanically or hopper-dredged sediments in open waters. In this process, the dredged sediment slurry is transported via pipeline to the discharge location where it is discharged. The slurry sinks to the bottom and moves downcurrent as an approximately one-meter-thick density current along the bottom. This density current has been found to persist for thousands of meters from the point of discharge during disposal operations. As this dredged sediment density current is typically characterized by low DO and/or the release of constituents such as ammonia, its presence at a location for extended periods of time (many hours to a day or so) could represent a significant adverse impact on aquatic organisms residing on the bottom in the path of the current. Ordinarily, however, such density currents do not persist in one location for extended periods because of the intermittent nature of pipeline disposal operations owing to frequent mechanical adjustments and the movement of the dredge. Further, the point of discharge is frequently moved because of the accumulations of sediments near the point of discharge.

Upland-"Confined" Disposal of Dredged Sediments

It has been commonly assumed that "confined" disposal of dredged sediment on land or behind a dike to form an island in water is less environmentally damaging than open water disposal. This assumption misjudges the adverse impacts of open water disposal and misjudges the water quality protection afforded by "confined" disposal. It is now more widely recognized that upland disposal of contaminated dredged sediments has a greater potential for causing adverse environmental impact than open water disposal of dredged sediments.

As generally practiced, "confined" disposal does not truly confine deposited sediment or eliminate adverse impact. Rather, it provides a settling area where the larger, denser particles settle and the supernatant water and fine materials from the hydraulic dredging overflow the confinement and enter the nearshore areas of the nearby watercourse. Ironically, it is this overflow water that would represent a threat to water quality. It is the fine materials that do not readily settle in the disposal area that contain chemicals that could become available to adversely affect water quality. The area where the confined disposal overflow occurs (i.e., nearshore) is generally the

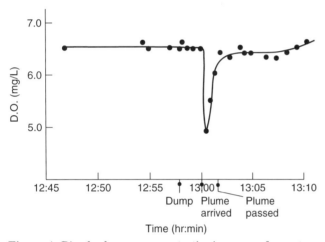

Figure 4. Dissolved oxygen concentration in near-surface waters during passage of turbid plume from hopper-dredge disposal of dredged sediment.

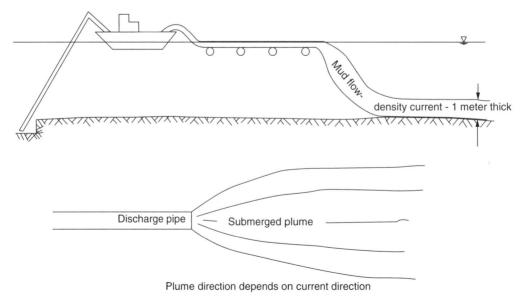

Figure 5. Hydraulic dredging with open water pipeline disposal.

most ecologically sensitive area of the waterbody. Thus, rather than being more protective of ecological/water quality, "confined" disposal operations pose a potentially greater risk than open water disposal.

Potential adverse impacts associated with the materials that remain in the confined area also exist. Studies conducted at the COE Waterways Experiment Station (5) have shown that when the sediments in a confined disposal area dry, they tend to release constituents such as heavy metals, which appears to be related to the oxidation of the amorphous sulfides present in the sediments, which keep the heavy metals in particulate form while the sediment is wet, and the development of acidic conditions. It may also be related to the aging of the ferric hydroxide precipitate (hydrous oxides). Although freshly precipitated ferric hydroxide has a substantial holding power for constituents, aged precipitates, especially those that dry out, lose some of this holding power. It is therefore not surprising to find that dredged sediments that dry in a confined disposal area release appreciable heavy metals when the area is subjected to precipitation or receives additional dredged sediment. The contaminants released then drain from the confinement area to the nearby nearshore waters. The potential for chemicals released from the confined dredged sediments to pollute area groundwater also exists. It is important, therefore, that those who advocate upland disposal of contaminated dredged sediments conduct a proper, critical review and evaluation of the potential adverse impacts of the constituents associated with the sediments that leave the confined disposal area during overflow during dredging operations or via drainage or seepage from the disposal site.

Contaminant Release from Redeposited Sediment

In addition to concern existing about the potential watercolumn impacts of contaminants in dredged sediment, a need exists to evaluate the potential for chemicals in the redeposited dredged material to be released to the watercolumn at the disposal site at sufficient rate to be adverse to water quality. In order to assess whether measured or unmeasured chemicals could have an adverse impact on watercolumn organisms or organisms that might colonize the redeposited dredged sediments shortly after deposition, Lee et al. (1) developed a dredged sediment elutriate screening toxicity test. The dredged sediment toxicity screening test involves introduction of standard test organisms into the settled elutriate from a standard oxic elutriate test procedure (oxic mixing of 1 volume sediment with 4 volumes water and oxic settling). Grass shrimp (*P. pugio*) were used for marine conditions and daphnids were used for freshwater. Lee et al. (1) and Jones and Lee (2) found that sediments collected near urban and industrial areas showed toxicity to aquatic life under the conditions of the laboratory test. Typically, in tests in which toxicity was found, from 10% to 50% of the test organisms were killed in the 96-hour test period. Toxicity was substantially less than may have been expected based on the elevated concentrations of the myriad measured and unknown contaminants in the sediments. Thus, although many of the sediments tested had high concentrations of heavy metals and chlorinated hydrocarbons of various types, those constituents were present in the sediments largely in unavailable, nontoxic forms. In the case of contaminated New York Harbor sediments, Jones and Lee (6) and Jones-Lee and Lee (7) found that the toxicity discovered in the toxicity testing was because of ammonia. Other investigators are also finding that ammonia is one of the principal causes of sediment toxicity to aquatic life. In general, however, it has been found that disposal sites are of sufficiently high energy and the rates of release of pollutants sufficiently slow that water quality problems in the watercolumn do not occur from redeposited sediments.

However, the bioaccumulation or buildup of chemical contaminants in higher trophic-level aquatic organisms that recolonize the disposal site is of concern. A few constituents exist, including methylmercury, PCBs, and PAHs, that have the potential for trophic magnification in aquatic environments. The bioaccumulation of constituents within organisms occurs from both direct uptake of constituents from the water and the consumption of particles that have constituents associated with them as well as the consumption of other organisms. The bioaccumulation and subsequent impact of chemical contaminants cannot be predicted based on their concentration in the sediment. Evaluation of this potential impact needs to be made on a site-specific basis using organism bioaccumulation tests of the type described by the U.S. EPA and COE (8,9).

REGULATING DREDGED SEDIMENT DISPOSAL

In the 1972 amendments to the Federal Water Pollution Control Act (PL 92–500), Congress specified in Section 404 that the disposal of dredged sediments in U.S. waters may take place, provided that an avoidance of "unacceptable effects" exists. It further stated that the disposal of dredged sediments should not result in violation of applicable water quality standards after consideration of dispersion and dilution, toxic effluent standards, or marine sanctuary requirements, and should not jeopardize the existence of endangered species. In order to implement these regulations, the U.S. EPA and Corps of Engineers developed two dredged sediment testing manuals, *Evaluation of Dredged Material Proposed for Ocean Disposal - Ocean Testing Manual* (8), and *Evaluation of Dredged Material Proposed for Discharge in Waters of the U.S. - Inland Testing Manual* (9). The Ocean Testing Manual was developed to implement Section 103 of Public Law 92–532 (the Marine Protection, Research, and Sanctuaries Act of 1972). These manuals prescribe testing procedures to assess biological effects such as toxicity and bioaccumulation of hazardous chemicals from dredged sediments.

The reliable evaluation of the potential adverse effects that may occur from dredging and dredged sediment disposal in a particular manner requires effort and understanding of the biological, chemical, and physical processes involved. It has been long recognized that potential adverse impacts cannot be determined based on the bulk chemical composition of the sediment; this understanding had, in fact, led to the DMRP studies of the 1970s. However, in an attempt to simplify the process and make determinations more administratively expedient, the U.S. EPA and other regulatory agencies have again attempted to regulate dredged sediment disposal based on chemical concentration-based "sediment quality criteria." One such approach that has been advanced is based on the "co-occurrence" in a sediment of a total concentration of a chemical and a biological response not even necessarily caused by the chemical of concern. Lee and Jones-Lee (10) reported on their reviews of co-occurrence-based so-called sediment quality guidelines and discussed why they are technically invalid and cannot provide a reliable evaluation for any purpose. Lee and Jones-Lee (11) recommended that a "best professional judgment" triad, weight-of-evidence approach be used to evaluate the water quality significance of chemicals in aquatic sediments. This approach is based on the integrated use of aquatic toxicity, bioaccumulation, organism assemblage information, and physical and chemical information focused on identifying the cause of aquatic life toxicity. Additional information on the water quality impacts of chemicals in sediments is available from Jones-Lee and Lee (7), DOER (12), and in the U.S. Army Corps of Engineers bulletin, "Dredging Research," available on request from www.wes.army.mil/el/dots/drieb.html.

DREDGING FOR SEDIMENT CLEANUP

In addition to dredging for maintaining navigation depth of waterways, dredging is being undertaken to remove (remediate) contaminant-laden sediments from Superfund and other sites. Evison of the U.S. EPA Office of Emergency and Remedial Response summarized the magnitude of the problem of contaminated sediments at Superfund sites in a presentation entitled, "Contaminated Sediment at Superfund Sites: What We Know So Far" at a U.S. EPA and U.S. Army Corps of Engineers national workshop (13). At the same workshop, Ellis (14), Sediments Team Leader with the U.S. EPA Office of Emergency and Remedial Response, presented a discussion entitled "Superfund Cleanup Issues at Contaminated Sediment Sites." The dredging of contaminated "Superfund" sediments and their disposal require attention to the same water quality/environmental quality, and contaminant availability, transport, and impact issues that are faced in dredging of contaminated sediment for channel depth maintenance.

An example of the use of dredging for sediment remediation is the current cleanup of PCB-contaminated sediments in the Hudson River in New York State. The Hudson River sediments from Hudson Falls to New York City (a distance of 200 miles) were contaminated by PCBs discharged for decades in wastewaters by General Electric Company. U.S. EPA (15) provides information on the past and proposed dredging of the PCB-contaminated sediments in that area. Another example of Superfund dredging is the dredging of PCB-contaminated sediments from the Wisconsin Fox River (16). There, an issue of considerable concern is the appropriateness of placing PCB-contaminated sediments in municipal landfills.

BIBLIOGRAPHY

1. Lee, G.F., Jones, R.A., Saleh, F.Y., Mariani, G. M., Homer, D.H., Butler, J. S., and Bandyopadhyay, P. (1978). Evaluation of the Elutriate Test as a Method of Predicting Contaminant Release during Open Water Disposal of Dredged Sediment and Environmental Impact of Open Water Dredged Materials Disposal, Vol. II: Data Report. Technical Report D-78-45, U.S. Army Engineer Waterways Experiment Station, Vicksburg, MS.

2. Jones, R.A. and Lee, G.F. (1978). Evaluation of the Elutriate Test as a Method of Predicting Contaminant Release during Open Water Disposal of Dredged Sediment and Environmental Impact of Open Water Dredged Material Disposal, Vol. I: Discussion. Technical Report D-78-45, U.S. Army Engineer Waterways Experiment Station, Vicksburg, MS.

3. Lee, G.F. and Jones-Lee, A. (2000). Water quality aspects of dredging and dredged sediment disposal. In: *Handbook of Dredging Engineering*, 2nd Edn. J.B. Herbich (Ed.). McGraw Hill, New York, pp. 14.1–14.42. Available at http://www.gfredlee.com/dredging.html.

4. Herbich, J.B. (Ed.). (2000). *Handbook of Dredging Engineering*, 2nd Edn. McGraw Hill, New York.

5. Palermo, M.R. (1986). Development of a Modified Elutriate Test for Estimating the Quality of Effluent from Confined Dredged Material Disposal Areas. Technical Report D-86-4, Department of the Army, U.S. Army Corps of Engineers, Washington, DC.

6. Jones, R.A. and Lee, G.F. (1988). Toxicity of U.S. waterway sediments with particular reference to the New York Harbor area. In: *Chemical and Biological Characterization of Sludges, Sediments, Dredge Spoils and Drilling Muds*. ASTM STP 976 American Society for Testing Materials, Philadelphia, PA, pp. 403–417.

7. Jones-Lee, A. and Lee, G.F. (1993). Potential significance of ammonia as a toxicant in aquatic sediments. In: *Proc. First International Specialized Conference on Contaminated Aquatic Sediments: Historical Records, Environmental Impact, and Remediation*. International Assoc. Water Quality, Milwaukee, WI, pp. 223–232, Available at http://www.gfredlee.com/psedqual2.htm.

8. U.S. EPA and COE (U.S. Army Corps of Engineers). (1991). *Evaluation of Dredged Material Proposed for Ocean Disposal—Ocean Testing Manual*. U.S. EPA Office of Water (WH-556F), EPA-503-8-91/001, Washington, DC.

9. U.S. EPA and COE (U.S. Army Corps of Engineers). (1998). *Evaluation of Dredged Material Proposed for Discharge in Waters of the U.S.—Inland Testing Manual*. Office of Water, EPA-823-B-98-004, Washington, DC.

10. Lee, G.F. and Jones-Lee, A. (2004). Unreliability of Co-Occurrence-Based Sediment Quality 'Guidelines' for Evaluating the Water Quality Impacts of Sediment-Associated Chemicals. Report of G. Fred Lee & Associates, El Macero, CA. Available at http://www.gfredlee.com/psedqual2.htm.

11. Lee, G.F. and Jones-Lee, A. (2004). Appropriate Incorporation of Chemical Information in a Best Professional Judgment 'Triad' Weight of Evidence Evaluation of Sediment Quality, Presented at the 2002 Fifth International Symposium on Sediment Quality Assessment (SQA5), In: *Aquatic Ecosystem Health and Management* M. Munawar (Ed.). **7**(3): 351–356. Available at http://www.gfredlee.com/BPJWOEpaper-pdf.

12. DOER (U.S. Army Corps of Engineers Dredging Operations and Environmental Research). (2002). Program Products of U.S. Army Corps of Engineers Operation and Maintenance

Navigation Program, Volume II (FY 1997–2002), CD available from U.S. Army Engineer Research and Development Center (ERDC) Waterways Experiment Station, Vicksburg, MS. Available at http://www.wes.army.mil/el/dots/doer.

13. Evison, L. (2003). Contaminated sediment at superfund sites: what we know so far. Presented at U.S. EPA and U.S. Army Corps of Engineers national workshop *Environmental Stability of Chemicals in Sediments*. San Diego, CA.

14. Ellis, S. (2003). Superfund cleanup issues at contaminated sediment sites. Presented at U.S. EPA and U.S. Army Corps of Engineers national workshop *Environmental Stability of Chemicals in Sediments*. San Diego, CA. Available at http://www.sediments.org/sedstab/agenda.pdf.

15. U.S. EPA. (2004). Hudson River PCBs. U.S. Environmental Protection Agency Region 2 Superfund, New York. Available at http://www.epa.gov/region02/superfund/npl/0202229c.htm.

16. U.S. EPA. (2004). Lower Fox River and Green Bay Area of Concern. U.S. Environmental Protection Agency Region 5, Chicago, IL. Available at http://www.epa.gov/glnpo/aoc/greenbay.html.

THE ECONOMICS OF WATER QUALITY

KEITH O. KEPLINGER
Texas Institute for Applied
Environmental Research
Stephenville, Texas

The role of economics in water quality decision-making is a topic receiving increasing attention today. A number of factors have spurred this interest. On the policy front, the total maximum daily load program has emerged as a premier water quality program in the United States, engendering fresh opportunities for economic analysis, because the means of attaining ambient targets are not prescribed. The increased relevance of economic analysis in the policy arena has been accompanied by a blossoming of the discipline of environmental economics and greater public and governmental acceptance of pursuing cost-effective and market-based solutions to environmental problems.

The theory of environmental economics has been extensively developed, but the utility and form of policy instruments are highly influenced by the institutional landscape. Consequently, this report begins with a brief assessment of water quality policy in the United States, focusing on the landmark legislation that constitutes the lion's share of current water quality policy, the 1972 Clean Water Act (CWA). Next, the meaning of costs and benefits in a water quality context are discussed, and the merits of two commonly used economic instruments in water quality assessments are explored: cost–benefit analysis and cost-effectiveness analysis. Finally, equity issues and the economic implications of water quality policy alternatives are discussed, with particular emphasis on the historical development and potential for incentive-based mechanisms.

WATER QUALITY POLICY IN THE UNITED STATES

Federal water quality policy in the United States can be succinctly characterized as initially state-based and ambient-based, then switching to a federal system dominated by technology standards with passage of the 1972 CWA. More recently, federal water quality policy has seen a return to the ambient-based approach for impaired waters as the total maximum daily load (TMDL) program has emerged. An ambient-based water quality approach targets the attainment of a specified level of ambient (in-stream) water quality as its objective. Loads or concentrations of pollutants are typically used as water quality measures. Ambient-based water quality approaches contrast with technology-based approaches, which specify the application of controls (e.g., best available technology), based on industrial classification, without explicit regard for the ambient conditions of receiving waters.

A legal scholar has noted that the breathtakingly ambitious goals of the CWA suggest a virtually cost-blind determination to control water pollution (1). It might also be added that the CWA takes a benefits-blind approach as well. Although not completely absent, references to costs or benefits in the CWA are few. The major regulatory mechanism of the CWA, the National Pollutant Discharge Elimination System (NPDES), focuses on technical feasibility as the overriding factor in establishing pollution controls. Considering the political and historical context surrounding the passage of the CWA, this largely prescriptive approach to restoring water quality can be considered neither surprising nor irrational. After years of inaction under prior national clean water legislation in the face of deteriorating water quality and mounting political pressure, the Senate was keen to pass legislation that forced action (2).

Portney (3) opines that

> When the U.S. Congress drafted the environmental legislation of the early 1970's, these and other problems were thought to be so obvious and so serious—and, in fact, they were—that members simply could not envision directing EPA to address them only if it was affordable to do so. As a consequence, most of the early laws effectively directed the agency's administrator to disregard cost altogether in setting ambient air quality standards and some water quality standards.

Houck (2) attributes the CWA's shift in focus to technology standards to "a new ethical premise, that water should simply be clean."

Unlike the former state administered ambient water quality approach, the prescriptive NPDES approach largely dispensed with the need for scientific endeavor and did not require the monitoring of water quality, estimation of the environmental impacts of control actions, or the estimation of economic costs and benefits.

During debate of the 1972 amendments, classic economic arguments were put forth in favor of continuing ambient-based federal water quality legislation. For example, in House testimony, the Chairman of the Council of Economic Advisors (CEA) stated:

CEA is in agreement with the use of water quality targets appropriate to the conditions and expected uses of water in particular areas of the country. That is basic to the concept of relating the costs of programs to the benefits received from them. To abandon that concept for a nationally legislated standard which focuses on the level of pollutants removed and is unrelated to water quality uses and standards is economically unwise because it means a necessary misallocation of our inevitably scarce economic resources (4).

Countering this line of argumentation was a consensus, which ultimately prevailed, that the existing state-based water quality approach, which relied on ambient standards, had been given ample time to succeed, but had overwhelmingly failed. "The Senate found the standards weak, late, widely disparate, scientifically doubtful, largely unenforced, and probably unenforceable" (2). Thus, whatever the economic and scientific demerits of the prescriptive NPDES approach, it was at least viewed as an approach that could be effectively implemented and would actually produce cleaner water—the benefits of which were, presumably, greater than the administrative and social costs of the program.

Of the two approaches, it can be argued that the NPDES approach was preferable only because it actually produced quantifiable water quality improvements, whereas the former federal legislation was largely effete. Tietenburg (5) concludes, "the wrong inference was drawn from the early lack of legislative success [in achieving water quality]," suggesting that equal or greater water quality gains could have been made at a lesser cost if correctly structured and adequately enforced incentive-based programs had been implemented. It can also be argued, however, that prescriptive regulation may be preferable when the problem is dire, the benefits great, and when scientific data are lacking, conditions which appear to have largely existed when NPDES was drafted. The National Research Council (6), however, concludes, "the data and science have progressed sufficiently over the past 35 years to support the nation's return to ambient-based water quality management." The initial implementation of NPDES and subsequent return to an ambient water quality approach can also be seen within a historical context. Ackerman and Stewart (7) suggest that the prescriptive technology approach "made some sense as a crude first-generation strategy." Portney (3) attributes the growing role of economics in environmental decision-making to often much higher costs and fewer benefits of additional pollution reduction, the impact of global integration, and to the loss (since the 1970s) of moral imperative status for pollution control.

It is also likely that Congress and the American public believed that effluent limitations based on best available technology applied to point sources would largely solve the nation's water quality problems. Nonetheless, Section 303(d) of the CWA, the legislation establishing the TMDL program, was included as a backup measure to achieve state water quality criteria if technology measures failed. Reminiscent of and in fact modeled after the 1965 Clean Water Act, Section 303(d) enumerates a number of steps designed to achieve ambient water quality targets consistent with designated uses. The TMDL program, thus, exemplifies an ambient-based approach to water quality.

Largely ignored for the first two decades after passage of the 1972 CWA, a series of lawsuits starting in the 1980s forced both the states and the U.S. Environmental Protection Agency (EPA) to gear up and prepare for the onslaught of TMDL activity to come. The trickle of TMDLs seen in the 1980s has swelled to the point where TMDLs have become a prominent (and controversial) component of CWA compliance. Recent state lists indicate in excess of 21,000 impaired water bodies throughout the nation requiring in excess of 40,000 TMDLs (6). These figures suggest that the ambient-based TMDL program will continue as a major component of water quality control for decades to come.

Although Section 303(d) does not explicitly mention cost considerations, the nature of the legislation facilitates an important role for economic analysis. The goal of attaining quantitative water quality targets is paramount, but the means of attainment are unaddressed, thereby permitting a potentially broad array of attainment strategies; each has its own unique set of control measures and associated economic costs. Consequently, an analysis of viable alternatives, including their economic consequences, can play an important role within the TMDL paradigm.

AN ECONOMIC CHARACTERIZATION OF WATER QUALITY

Water pollution is a classic externality produced by various types of economic activity. An industrial facility, for example, may discharge toxic effluent into a river causing fish kills and environmental impairment further downstream. Because the fisherman and recreationists that suffer harm are not parties to the plant's decision to discharge, that harm is external to the decision process. Another key feature of water pollution is that, in almost all cases, it harms many users and nonusers simultaneously; hence, in economic lingo, it is a *public* bad. From the reverse perspective, clean water is a public good that produces benefits to often-large groups of users but requires costs to attain it. The externality and the public good nature of water quality/water pollution often lead to market failure—a situation where market forces alone do not result in socially desirable outcomes. The major causes for market failure in water quality are undefined water quality rights and control obligations, and the practical obstacles that discourage numerous individuals who receive relatively small water quality benefits from coordinating their efforts and directly contracting with discharges. Societies have sought to correct market failures through various forms of social intervention. Although not exhaustive, alternative externality correction devices include market emergence, merger, economic incentives, regulation, prohibition, pseudomarkets, and moral suasion.

COMPONENTS OF ECONOMIC ANALYSIS

The two fundamental components of applied microeconomic analyses are costs and benefits. Water pollution involves sacrifices for some groups (costs) and economically benefits others. The benefits of attaining clean water are equivalent to the costs of water pollution, whereas the cost of attaining clean water is equivalent to the benefits that accrue to the production of water pollution. To avoid confusion, the following sections discuss costs and benefits from a water quality perspective.

Costs

Water quality is a public good requiring costs for its attainment. Due to the constraints imposed by nature and current technology, producing more of one good, including water quality, is typically accomplished only at the expense of producing less of other goods. Occasionally, analyses suggest that management or structural measures designed to improve water quality may also result in net benefits to dischargers. Such "win–win" solutions are the exception but might occur for one of two reasons: (1) current technology and resources are not being efficiently utilized, or (2) recent technology has opened up new opportunities for improving water quality while simultaneously producing benefits in other areas. Numerous water pollution control measures have been identified for municipalities, industry, and agriculture. Economic and engineering studies often estimate the costs of implementing specified measures and typically express those costs in monetary units, for example, dollars, which represent the foregone opportunity of producing and consuming other goods.

Benefits

The benefits of mitigating water pollution can be classified as environmental and economic. Environmental benefits are measured by a large number of environmental indexes, such as biological diversity and increased fish populations. Economic benefits are the monetary values society is willing to pay for the associated environmental benefits. Numerous studies have quantified the economic benefits of water quality (8,9). Following Freeman's taxonomy, benefits include improvements in (1) recreation (swimming, fishing, boating, water fowl hunting), (2) nonuser benefits (amenity, aesthetic, and ecological benefits not directly associated with activities on or near a waterbody, but for which households may be willing to pay), (3) diversionary uses (reducing risk to human health and decreased costs for municipal water supply treatment), and (4) commercial fisheries. Many of the benefits listed above are nonmarket; that is, they are not purchased or sold in ordinary markets and therefore have no observable price. Lack of organized markets for water quality complicates the estimation of water quality benefits (10), though a number of indirect methods have been developed.

ECONOMIC INSTRUMENTS FOR INFORMING WATER QUALITY POLICY

Cost–Benefit Analysis

Applied to water quality, cost–benefit analyses generally estimate and compare both the economic costs and benefits of a water quality project. Such analyses are nominally prescriptive; benefits that exceed costs may represent a sufficient condition for proceeding with a project. A more exacting and universally accepted welfare economic criterion is the Pareto criterion, which states that everyone should be made at least as well off while nobody is made worse off. A policy producing net benefits (total benefits exceeding total costs) is considered a *potential* Pareto improvement because, in theory, the gains of the policy action can be redistributed such that everyone is made at least as well off as before. In practice, a strict Pareto improvement is difficult to achieve and seldom attempted. Thus, a potential Pareto improvement, that is, the generation of positive net benefits, is often used as a decision rule.

In theory, the optimal level of ambient water quality is such that the benefits produced by one additional unit of water quality just equals the costs to attain it. Thus, to fully assess the merits of a water quality project, detailed information on both its benefits and costs is required. Significant difficulties, limitations, and complications inherent in environmental and economic benefit estimation techniques, as described in succeeding sections, suggest one reason why cost–benefit analysis has not factored more prominently in water quality policy.

It is noteworthy that authors of environmental economics textbooks, for the reasons cited, typically exhort students to be cautious in their interpretation of cost–benefit analyses. For example, "Although cost-benefit analysis, with its reliance on numbers and economic theory, may seem as if it might be a precise science, this is not the case" (11). On somewhat of a counterpoint, Hartwick and Olewiler (12) argue, "Cost benefit analysis has been maligned by those who say that it is useless because the numbers are so bad. This criticism misses the point of the exercise. ... The estimate of the benefits of environmental control is crucial if we are to evaluate the tremendous costs of reducing pollution and make informed judgments about the social value of improving environmental quality."

When benefits cannot be adequately quantified because of uncertainty or other technical hurdles, benefits can only be described. By contrast, estimating the costs of structural or managerial pollution control measures is relatively straightforward and generally involves much less uncertainty. Comparing quantified costs to descriptive, though potentially large, benefits can unduly discount benefits. Although they broadly support the use of cost–benefit analysis, a group of highly respected economists also highlighted some of its limitations in a draft set of principles (13). One of many caveats was that "care should be taken to assure that quantitative factors do not dominate important qualitative factors in decision-making" (14).

Another factor hindering the use of formal cost–benefit analyses as a decision-making tool is the perception that such analyses are prone to manipulation, especially when gaps in knowledge and sources of uncertainty are substantial. Recounting his experience as the senior economist for environmental and resource policy at the Council of

Economic Advisors, Jason Shogren (15) comments: "But now I understand that some people see our orthodoxy as not just simply confining but as downright prehistoric. ...Because after all, cost–benefit analysis is just naked self-interest dressed up in banker's pajamas, isn't it?" Many studies, for instance, have indicated that the nonuse or existence value of environmental amenities may far outweigh their commercial and use value, while other studies omit the estimation of these values altogether.

A final factor discouraging the more widespread use of cost–benefit analysis for water quality is that the estimation of benefits has no practical relevance in either technology-based or ambient-based water quality policy. NPDES technology-based controls, for instance, are mandated on the basis of industrial classification and available technology, while TMDL ambient-based controls are designed to achieve ambient criteria, regardless of the economic benefits produced.

Estimating the Economic Benefits of Water Quality. Three linkages must be established and quantified to estimate water quality benefits: (1) the link between a pollution mitigation strategy and water quality indicators, (2) the link between water quality indicators and environmental benefits, and (3) the translation of environmental benefits into economic benefits. The first linkage is often estimated by simulating a pollution mitigation strategy by using a water quality model. Biophysical models may also be employed to estimate environmental benefits such as increased fish populations. Estimating economic benefits requires knowledge of both environmental benefits and how society values those benefits.

When water pollution causes direct damage to a market commodity resulting in either a decrease in production (e.g., in fisheries) or an increase in expense (e.g., increased treatment cost for municipal drinking water), calculating the value of water quality improvements is relatively straightforward. Because of the public good nature of many water quality benefits, however, many benefits are not directly reflected in markets. Economists have developed a number of indirect methods for estimating nonmarket values of water quality improvement. The hedonistic technique is often used to reveal a lower bound on amenity or aesthetic value. This technique infers the value of a nonmarket commodity by analyzing the value of a commodity whose value is influenced by the nonmarket commodity, for example, comparing waterfront property values across regions experiencing different levels of water pollution. Differences in human behavior, such as increased recreation, due to water quality improvements, can also be used in conjunction with travel cost models to estimate the value of recreational water quality benefits. The travel cost model assumes that the willingness to pay for a recreational trip is at least equal to the cost incurred in traveling to the recreational site. Hedonic and travel cost models can be classified as revealed preference methods because their estimation techniques depend on observable (or revealed) behavior.

If values for water quality are not intimately related to use, for example, the desire to provide clean water environments to current and future generations, they are not necessarily revealed through either changes in market prices or other aspects of human behavior, such as travel to a recreational site or home location. Observational methods, thus, cannot reveal nonuse valuations. Researchers have circumvented this hurdle by describing hypothetical markets to respondents who are asked to indicate (or state) their willingness to pay for a nonmarket commodity (16). This relatively recent benefits estimation methodology has come to be known as the contingent valuation (CV) method, because willingness to pay values are contingent upon the particular hypothetical market described. Because the situation is hypothetical and an actual transaction or observable behavioral pattern has not been effected, value estimates based on surveys are controversial. Participants may answer strategically, for example, may indicate too high a value because they think it will improve the chance that a water quality improvement will be made, or indicate too low a value if they think that their reported willingness to pay might result in increased taxes. If participants are convinced that their responses will have no effect on actual outcomes, the incentive to reveal an inaccurate valuation is reduced (8).

The public good nature of water quality improvements also complicates benefits estimation and adds to data collection expense (10). The value of water quality improvements is typically spread across a wide population of users, and researchers must determine the relevant population for a particular improvement. Another obstacle in estimating nonmarket benefits is that most people do not have well-formed values for the vast array of environmental resources and therefore may resort to heuristics and simple protocols to construct dollar value estimates. These protocols, however, vary widely among individuals, and many irrelevant issues, such as the appropriateness of the payment vehicle or who should pay, often influence responses (17). A few experiments suggest that stated willingness to pay for existence values may be two to ten times higher than actual contributions or payments (18).

Despite the theoretical and methodological difficulties, quantifying the benefits of achieving clean water has often changed the character of debate (18). It is typical that the costs required to achieve water quality are highly concentrated but that benefits are widely distributed. Because benefits to any one user may be quite small, only bearers of costs would have a great incentive to change the outcome of a debate. Given the public good nature of many water quality improvements, however, a CV analysis may show that benefits, though widely dispersed, far outweigh costs.

In sum, certain benefits of water quality are reflected in markets, but many are not. The nonmarket benefits of water quality are important and potentially very large. Measuring these benefits, however, presents unique challenges and methodological difficulties, which have been addressed by a number of ingenious nonmarket benefits estimation techniques; CV is among the most promising. The accuracy, completeness, and potential bias of such techniques, especially CV, however, remain controversial issues.

Cost-Effectiveness Analysis

Cost-effectiveness analysis circumvents the difficulty and controversy of estimating economic benefits by focusing on the costs of achieving a quantified noneconomic objective. Once an objective has been chosen, costs of alternative strategies that reach the objective, for example, a water quality indicator, can be compared. Because the estimation of water quality indicators is generally considered more reliable than estimating the economic benefits of different levels of those indicators, cost-effectiveness analyses generally engender less controversy than cost–benefit analyses.

Cost-effectiveness analyses, in a water quality context, estimate the costs and environmental effectiveness of defined control measures or combinations of measures. Effectiveness is often measured as a reduction in the ambient load or concentration of a pollutant. Cost-effectiveness ratios can be calculated by dividing the cost of a control measure by its effectiveness. This ratio quantifies the cost of achieving a one-unit improvement in the effectiveness measure. Cost-effectiveness ratios can be easily compared across control measures to determine the most cost-effective measures—those measures that have the lowest ratios. To achieve a water quality target when any one control measure does not single-handedly achieve the target, the most cost-effective solution is typically found by applying the most cost-effective measure first, followed by the second most cost-effective measure, and so on, until a specified water quality target is achieved.

Cost-effectiveness analysis is particularly adapted to TMDL deliberations for the following reasons. First, numerical water quality targets are an essential component of TMDL plans. Second, estimating the environmental effectiveness of control measures is also central to TMDL stakeholder deliberations. Descriptions of control measures and their effectiveness are required elements of TMDL implementation plans (19). Third, the means of attaining water quality targets are not prescribed, allowing the comparison of alternative strategies. Fourth, although not a requirement, there is no prohibition on considering the costs of alternative strategies and individual control measures in determining which set of measures to implement. The EPA has characterized the TMDL process as a "cost-effective framework" for achieving water quality and has endorsed cost-effectiveness assessments (20). Fifth, from a welfare economic perspective, reducing or minimizing the cost of achieving a given water quality target increases social welfare. Sixth, compared to estimation of economic benefits, the calculation of pollution mitigation costs is relatively straightforward and less controversial. TMDL assessments, thus, are natural candidates for the application of cost-effectiveness analyses. Despite the suitability of cost-effectiveness analysis within the TMDL framework, rigorous analyses have been pursued for relatively few TMDLs, and there remains a large potential for improving the economic performance of TMDL (and other ambient-based water quality program) allocations.

Cost Minimization. Cost-effectiveness ratio analysis, as outlined above, minimizes costs within the limitations imposed by ratio analysis, for example, linearity and additivity. Under more complex (and realistic) assumptions, cost-effectiveness analysis requires more completely defined relationships among costs, water quality indicators, and control measure adoption. When costs and control measure are not additive, combinations of control measures should be assessed to capture interactive effects. Control measure costs are not necessarily linear. The same control measure at different locations or implemented at different levels can involve different unit environmental responses and entail different unit costs. More sophisticated forms of analyses are needed to analyze partial and site-specific implementation of control measures. Nonlinear and nonadditive relationships among varying levels and combinations of control measures can be addressed by employing a generalized cost minimization framework capable of solving for cost minimizing levels of control measures.

EQUITY CONSIDERATIONS

The scope of equity considerations within technology-based programs, such as NPDES, is rather limited because the procedures for prescribing water quality remedies are highly delineated within the law and regulation. Equity considerations, however, can become highly important within the ambient-based TMDL approach because particular remedies are not prescribed. In many cases, numerous entities that often belong to two or more industries are responsible for exceeding ambient standards. Loads can be reduced to TMDL targets by any combination of control responsibilities as long as the load reductions sum to the total reduction needed to achieve the TMDL target. Cost minimization can be used as the sole criterion for assigning loads, but such assignments may be rejected because they are not perceived as equitable. Hence, the flexibility of the TMDL approach creates challenges to promulgating implementation plans that are deemed equitable. Solutions deemed equitable, however, may compromise economic efficiency, that is, implementing the most cost-effective solution.

Control costs are generally concentrated in particular industrial sectors or entities. The fairness of assigning control costs is almost always at issue in water quality deliberations. Subsidizing water quality improvements by transferring mitigation costs from a small group of dischargers to broader publics is sometimes viewed as an equitable means of distributing control costs. State and federal funds are often appropriated to defray additional expenses of new control obligations. Such transfers can promote the acceptance of new controls. When such funding is not available, the legality of mandating new control obligations, especially for nonpoint sources, is often challenged.

Equity criteria can be analyzed by rigorous analytical methods, such as the attainment of a Pareto optimal. More generally, however, equity issues are dealt with subjectively; nonetheless, concern for fairness is an important driver that influences many water quality decisions. This is particularly true within the TMDL program, where attaining ambient standards often necessitates additional efforts by industry and agriculture. Most states assemble advisory boards of watershed stakeholders, typically composed of public servants, environmental interests, and industry representatives, to recommend key TMDL decisions, including load allocations. Concern for fairness is an important factor influencing load assignments. Recommended load assignments are a product of the internal dynamics of advisory board deliberations, but appeals to equity or fairness are repeatedly raised in these deliberations and undoubtedly play a large role in determining load assignments. It may be deemed equitable for each identified source, or group of sources, to reduce its water quality impact by roughly the same percentage or absolute amount. There are many potential equity criteria, however, and often no concensus on which one is best. Imposing pollution reduction obligations on small businesses and family farms, especially when economic viability is jeopardized, may be deemed unfair (as well as politically infeasible). Even if concensus if reached rearding equitable control assignments, implementation strategies deemed equitable may significantly depart from cost-effective solutions, although a correctly structured incentive-based mechanism can potentially overcome this limitation.

WATER QUALITY POLICY INSTRUMENTS

Poor water quality is symptomatic of market failure and provides a rationale for social intervention. This intervention can take the form of direct regulation, prohibition, or one of a variety of incentive-based mechanisms. Complete prohibition of discharge is sometimes resorted to for extremely toxic substances, when even a very small amount might cause harm. Regulatory approaches, often referred to as "command and control" regulation, typically prescribe control remedies that allow little flexibility in the means for achieving goals. By contrast, incentive-based mechanisms encourage behavior that reduces pollution through market forces and signals and allow at least some degree of flexibility in the means for achieving goals.

The NPDES permit system for point sources, which bases acceptable discharges on best available technology, exemplifies the regulatory approach. The regulatory approach, and NPDES in particular, have been criticized in recent years for reasons that relate mostly to questions of economic efficiency: that it is overly prescriptive and lacks flexibility, that it seeks to regulate inputs rather than outputs (water quality), that it does not adequately consider costs, and even that it has reduced real incomes (7,21). Whether or not a particular market- or incentive-based policy is, more efficient in practice, than a particular "command and control" policy is, however, an empirical question. Alternative policy instruments possess unique biophysical and economic implications, and the efficacy of any given policy depends on a number of site-specific behavioral, environmental, and economic factors that may be difficult to establish a priori. Nonetheless, incentive-based mechanisms for achieving water quality have attracted considerable academic interest because of their theoretical efficiency characteristics, and a number of incentive-based programs have been initiated.

Incentive-Based Mechanisms for Improving Water Quality

Incentive-based mechanisms are advocated primarily as a means of reducing the financial burden of control costs while maintaining or making further progress on water quality. The defining hallmark of incentive-based systems is that they are designed to be incentive compatible, that is, by pursuing self-interests, such as profitability, environmental goals are simultaneously achieved. Unlike prescriptive regulation, incentive-based mechanisms also promote the development and implementation of new pollution control technology. Though recommended by environmental economists for the past several decades, environmental organizations have more recently promoted incentive-based mechanisms (particularly effluent trading) for achieving water quality (22–24).

While generally viewed as a cost saving measure, making pollution control more affordable can also enhance its political acceptability and, hence, promote additional pollution control. Tietenburg (25), for instance, maintains that "with the inclusion of a tradable permits program for sulfur in the [acid rain bill], the compliance cost was reduced sufficiently to make passage politically possible." Incentive-based mechanisms for improving environmental quality include emission fees or taxes, tradable permits, deposit-refund systems, subsidies for pollution control, removal of subsidies with negative environmental impacts, reductions in market barriers, and performance standards (26–28). We focus here on the two most prominent incentive-based mechanisms, pollution taxes and tradable permits.

Pollution Taxes. Economists have long recognized the allocational inefficiencies of environmental externalities. In the early part of the twentieth century, the noted economist Arthur Cecil Pigou (29) recognized that "when the divergences between private and social costs are pervasive throughout the economy, market prices cannot be used as measurements of consumer satisfaction." Pigou proposed an externality tax on pollution to internalize the damages caused by pollution. During his time, Pigou's pollution tax was regarded as an academic exercise and did not gain practical significance. The environmental consciousness of the 1970s, however, spurred a revival of interest in pollution taxes as a solution to environmental externalities, and several European countries and Japan adopted pollution taxes. The consensus among economists is that these fees have typically not had noticeable effects on pollution because they have not been set at levels that affect behavior (27).

Simple models of pollution abatement costs and benefits suggest that pollution taxes yield results identical to those of tradable permits within a cap if the cap or tax is appropriately set. This conclusion can be drawn from a simple graph where pollution is represented on the horizontal axis, control costs are represented on the

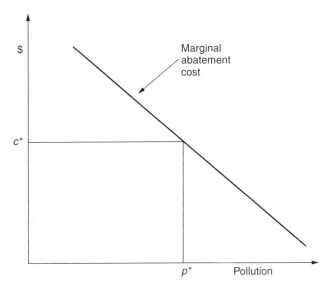

Figure 1. Theoretical equivalence of a pollution tax (c^*) and a pollution cap (p^*).

vertical axis, and declining marginal abatement costs are plotted within the graph (Fig. 1). Any point on the marginal abatement cost function can be mapped to a level of pollution (p^*) and a control cost (c^*). A pollution cap is represented as a vertical line and the equivalent pollution tax by a horizontal line, each intersecting the marginal abatement cost function at the same point. For a given pollution tax, the firm would have an incentive to reduce pollution up to the point where control costs exceed the tax, which could also be achieved by specifying a pollution cap at this level of pollution. Despite the theoretical equivalence of pollution taxes and "cap and trade" programs for the simple model portrayed in Fig. 1, a number of practical advantages are often cited in favor of tradable permits, especially when attaining ambient standards is the overriding consideration.

Tradable Permits. The theoretical underpinnings of a tradable permits system has been attributed to Coase (30), who, in his classic work, *The Problem of Social Cost*, forcefully argued the importance of clear liability rules and bargaining between parties to achieve socially desirable levels of harmful effects (31). Practical applications of Coase's basic premise were extended to air (32) and water pollution (33). Whereas Coase envisioned the bargaining of social harms between the producer and recipient of those harms (regardless of who was liable), Dales set out the parameters of the type of "cap and trade" program in widespread use today. The basic components of the program involve setting a cap on total waste, issuing emission permits, the aggregate of which equals the cap, and allowing the sale and purchase of permits among polluters (33). Marketable credits are formed when dischargers who have low mitigation costs reduce loads below permitted levels. These credits can in turn be sold to entities that have high mitigation costs so that both parties gain, thereby providing an incentive to trade.

Historical precedent for tradable permit systems for water quality is found in the allocation of natural resources, in water supply and fisheries (34). Tietenberg (25) traces the acceptance of tradable permits in the United States to the introduction of "pollution offsets" by the U.S. Congress in 1977, which opened the door to emissions trading in the air arena. In the years since, tradable emission permit systems have been successfully implemented at the federal level for the lead phaseout program, to reduce ozone-depleting chemicals under the Montreal Protocol, and for the sulfur allowance program (25). A number of programs have also been developed at the state level. Emissions trading in the air arena has generally been deemed quite successful. Burtraw and Mansur (35), for instance, estimate that trading of sulfur dioxide emissions resulted in a 13% reduction in compliance costs and produced health related benefits of nearly $570 million in 1995. The actual use of tradable permits for water quality control is of recent origin. In 1992, Willey wrote, "no water emissions markets exist yet," whereas several functioning effluent trading programs were operating by the late 1990s (26,36). Of the 37 trading and offset programs summarized by Environomics (36), 11 were considered well along in being implemented; trades were under way or completed; 5 had specific trading mechanisms approved and were very near implementation; 6 had completed the development and program approval process, but no specific trades had yet been identified; 12 were in various stages short of program approval, including study, discussion, planning, and/or development; 1 was exclusively a study; and 2 were inactive or discontinued.

The advantages of "cap and trade" programs relative to pollution taxes are, first and foremost, that a new pollution tax on industry may be politically unacceptable whereas the issuing of permits produces a potentially valuable property right. A revenue-neutral tax would reduce or eliminate this advantage, although such programs are rarely considered. Second, under conditions of uncertainty, a permit system would guarantee the attainment of the specified pollution level, whereas a large degree of guesswork would be involved in setting the appropriate tax level. Several adjustments of the tax might be required to find the level that would produce the sought for level of pollution control. Third, the amount of pollution reduction achieved by a given pollution tax is also unstable over time. Other things being equal, economic growth would foster increased emissions, as more pollution producing products and services were demanded and produced, because the pollution tax is a variable component of production costs. On the other hand, given expected advances in technology, levels of pollution might decrease if more cost-effective pollution technology were implemented over time. In summation, pollution taxes and tradable permits have relative advantages and disadvantages, but tradable permits systems are considered preferable when achieving quantitative pollution targets is the paramount consideration. Consequently, tradable effluent permits are conceptually well suited to achieving ambient water quality standards.

Because of numerous factors, TMDL and other ambient-based control allocations typically do not exhibit the greatest degree of cost-effectiveness. A tradable permit system holds the promise of reallocating any initial set of load assignments so that a least cost allocation may be attained. Whether or not a tradable permit system, in practice, is more efficient than its absence depends on the level of transactions costs relative to the gains from trade. In general, a trading system is deemed advisable only if the benefits attributable to trading exceed associated transactions costs. Transaction costs include the costs to industry of quantifying their mitigation costs, finding trading partners, initiating a trade, and bargaining, as well as the administrative costs (often borne by the public) of setting up and maintaining a market for pollution credits. If load allocations and implementation strategies are initially made in a relatively efficient manner, there is little potential for further reducing compliance costs that might be realized by a tradable permit system.

Trading has the most potential when there are large differences between marginal pollution control costs and potentially large amounts of pollution credits that can be traded. For these reasons, some researchers see the greatest potential in trades between point and nonpoint sources because the cost of nonpoint mitigation is often believed much less than that of point source mitigation (37,38). Point/nonpoint source trading, however, entails an additional set of challenges, including monitoring and measuring nonpoint pollution and uncertainty regarding the effectiveness of nonpoint controls (38,39).

Within a water quality context, effluent trading is most attractive for waterbodies where pollutants are well mixed and where there are at least several large dischargers and large differences in control costs. Spatial characteristics of water pollution can pose challenges to the development of viable markets. If pollutants are not well mixed, likely in a stream or river, the waterbody may need to be broken into several markets, or conditions to trading would need to be applied so that improvements in one area would not lead to unacceptable degradation in other areas. For this reason, it is also likely that sale of pollution credits downstream would be more likely than upstream sales for streams and rivers. Trading programs for water pollution control reflect the fact that pollution in waterbodies is generally more localized and spatially isolated than in air sheds. The potential for viable water pollution trading programs depends on various site-specific factors, many of them dictated by nature. Under the right conditions, however, a properly structured trading program can be a useful tool for achieving cost-effective water pollution control, either within or outside the TMDL process.

Having achieved some notable successes in the air arena, economists and policy makers have investigated the potential of applying the tradable permit concept to water pollution control and several programs have been developed (26,38). In 1996, the EPA issued a *Draft Framework for Watershed-Based Trading* designed to promote, encourage, and facilitate trading wherever possible, provided that equal or greater water pollution control could be attained at an equal or lower cost (40). Despite considerable effort by the EPA and the states to implement watershed trading, programs have been developed at relatively few locations, and trading within these programs has been limited, probably less than anticipated. It remains to be seen whether efforts by EPA to vigorously implement the TMDL program will greatly increase the use of watershed trading, as suggested by EPA (26), or if spatial and administrative challenges will continue to limit its adoption to a small percentage of impaired or potentially impaired waters.

SUMMARY

Like all environmental issues, the maintenance and attainment of water quality involves intrinsic economic dimensions. For a long time, however, the economic dimensions of water quality have been subordinated by prescriptive water quality policy. The emergence of an ambient-based water quality approach in the United States (in the form of the TMDL program), however, has opened the door to a more flexible approach to water quality attainment. While both benefits and costs are important elements of water quality, benefit estimation is fraught with inherent complexity and controversy. Neither technology-based nor ambient-based policy relies on cost–benefit criteria. Control costs and their estimation, however, are often important elements in TMDL (and other ambient-based program) deliberations. Hence, cost-effectiveness analyses are particularly well adapted to ambient-based water quality approaches.

Equity considerations can increase the costs of achieving water quality targets. A correctly structured incentive-based mechanism, however, can achieve both equity and economic efficiency. Pollution taxes and tradable permits are two incentive-based mechanisms that can, in principle, achieve water quality targets at lower cost. Because maintaining a quantified level of water quality (often measured by loads) is a paramount feature of ambient-based approaches, a tradable permit (cap and trade) system is better suited to TMDLs and other ambient-based programs than pollution taxes. Whether a "cap and trade" program is more efficient in practice than its absence is highly dependent on site-specific factors and the magnitude of transaction costs. The potential for viable tradable permit systems for water quality and the extent to which they may eventually be adopted remain to be seen. What is quite apparent, however, is the increasing use of economic analysis in the planning and implementation of individual water quality projects and the more prominent role that economic assessments have played in forming and evaluating water quality policy.

BIBLIOGRAPHY

1. Percival, R.V., Miller, A.S., Schroeder, C.H., and Leape, J.P. (1992). *Environmental Regulation: Law, Science, and Policy*. Little, Brown, Boston, Toronto, London.
2. Houck, O.A. (1997). TMDLs: the resurrection of water quality standards-based regulation under the clean water act. *Environ. Law Rep. Environmental Law Institute* **27**: 10329–10344.
3. Portney, P.R. (1998). Counting the cost: the growing role of economics in environmental decisionmaking. *Environment* March.

4. *Water Pollution Control Legislation—1972: Hearings on H.R. 11896, H.R. 11895 Before the Comm. On Public Works, 92d Cong. 483 (House Hearings)*. 1971. At 202–03 (statement of a panel composed of Russell E. Train, Council on Environmental Quality, and Paul V. McCracken, Council of Economic Advisors).
5. Tietenberg, T.H. (1992). *Environmental and Natural Resource Economics*, 3rd Edn. HarperCollins, New York.
6. National Research Council. (2001). *Assessing the TMDL Approach to Water Quality Management*. National Academy Press, Washington, DC.
7. Ackerman, B.A. and Stewart, R.B. (1988). Reforming environmental law: the democratic case for market incentives. *Columbia J. Environ. Law* **13**: 171–199.
8. Freeman, A.M. III (1982). *Air and Water Pollution Control: A Benefit-Cost Assessment*. John Wiley & Sons, Hoboken, NJ.
9. Ribaudo, M.O. (1989). *Water Quality Benefits from the Conservation Reserve Program*. AER-606. U.S. Dept. Agric., Econ. Res. Serv., Feb.
10. Ribaudo, M.O. and Hellerstein, D. (1992). *Estimating Water Quality Benefits: Theoretical and Methodological Issues*. TB-1808. U.S. Dept. of Agric., Econ. Res. Serv.
11. Kahn, J.R. (1998). *The Economic Approach to Environmental and Natural Resources*, 2nd Edn. Harcourt Brace College Publishers, Fort Worth.
12. Hartwick, J.M. and Olewiler, N.D. (1986). *The Economics of Natural Resource Use*. Harper & Row, New York.
13. *Inside EPA's Risk Policy Report*. (1996). *Leading Economist Warn There Are Limits to Benefit-Cost Analysis*. An Inside Washington Publication. Jan. 31.
14. Arrow, K.J., Cropper, M.L., Eads, G.C., Hahn, R.W., Lave, L.B., Noll, R.G., Portney, P.R., Russell, M., Schmalensee, R., Kerry Smith, V., and Stavins, R.N. (1996). *Benefit-Cost Analysis in Environmental, Health, and Safety Regulation: A Statement of Principles*. American Enterprise Institute, The Annapolis Center, and Resources for the Future. AEI Press.
15. Shogren, J.F. (1998). Do All the Resource Problems in the West Begin in the East? *J. Agric. Resour. Econ.* **23**(2): 309–318.
16. Mitchell, R.C. and Carson, R.T. (1989). *Using Surveys to Value Public Goods: The Contingent Valuation Method*. Resources for the Future, Washington, DC.
17. Schkade, D.A. (1995). *Issues in the Valuation of Environmental Resources: A Perspective from the Psychology of Decision Making*. In: *Review of Monetary and Nonmonetary Valuation of Environmental Investments*, T.D. Feather, C.S. Russell, K.W. Harrington, and D.T. Capan (Eds.). (Appendix B). U.S. Army Corps of Engineers. IWR Report 95-R-2.
18. Loomis, J.B. (1997). Use of non-market valuation studies in water resource management assessments. *Water Resour. Update* **109**: 5–9.
19. U.S. EPA 841-D-99-001. (1999). *Draft Guidance for Water Quality-based Decisions: The TMDL Process (Second Edition)*. Office of Water, Washington, DC.
20. Fox, C.J. (2000). *Letter to The Honorable Bud Shuster, U.S. House of Representatives*. April 5.
21. Meiners, R.E. and Yandle, B. (1993). Clean water legislation: reauthorize or repeal? In: *Taking the Environment Seriously*. R.E. Meiners and B. Yandle (Eds.). Rowman & Littlefield, Lanham, MD.
22. Oates, W.E. (1999). Forty years in an emerging field: economics and environmental policy in retrospect. *Resources* **137**: 8–11.
23. Faeth, P. (1999). *Market-Based Incentives and Water Quality. Drawn from a forthcoming report by WRI: Trading as an Option: Market Based Incentives and Water Quality by Paul Faeth*. World Resources Institute. <http://www.econet.apc.org/wri/wri/incentives/faeth.html>.
24. Environmental Defense. (2000). *From Obstacle to Opportunity: How Acid Rain Emissions Trading Is Delivering Cleaner Air*.
25. Tietenberg, T. (1999). Tradable permit approaches to pollution control: faustian bargain or paradise regained? In: *Property Rights, Economics, and the Environment*. M.D. Kaplowitz (Ed.). JAI Press, Stamford, CT.
26. Environmental Protection Agency (EPA). (2001). *The United States Experience with Economic Incentives for Protecting the Environment*. EPA-240-R-01-001. Office of the Administrator, Washington, DC.
27. Hahn, R.W. (2000). The impact of economics on environmental policy. *J. Environ. Econ. Manage.* **39**: 375–399.
28. Stavins, R.N. (1998). *Market Based Environmental Policies*. Discussion Paper 98–26. Resources for the Future, Washington, DC.
29. Pigou, A.C. (1920). *The Economics of Welfare*. Macmillan, London.
30. Tietenberg, T. *Editors Introduction to the Evolution of Emissions Trading: Theoretical Foundations and Design Considerations*. Introduction to the first volume of a two-volume collection of previously published articles. Environmental Economics Reference Series, Ashgate Press, in press.
31. Coase, R. (1960). The problem of social cost. *J. Law Econ.* October.
32. Crocker, T.D. (1966). The structuring of atmospheric pollution control systems. *The Economics of Air Pollution*. H. Wolozin (Ed.). W. W. Norton, New York, pp. 61–86.
33. Dales, J.H. (1968). Land, Water and Ownership. *Can. J. Econ.* **1**: 791–804.
34. Colby, B.G. (2000). Cap and trade challenges: a tale of three markets. *Land Econ.* **76**(4): 638–658.
35. Burtraw, D. and Mansur, E. (1999). *The Effects of Trading and Banking in the SO_2 Allowance Market*. Discussion Paper 99–25. Resources for the Future, Washington, DC.
36. Environomics. (1999). *A Summary of U.S. Effluent Trading and Offset Projects*. Prepared for: Dr. Mahesh Podar, USEPA, Office of Water. <http://www.epa.gov/owow/watershed/hotlink.htm>.
37. Letson, D., Crutchfield, S., and Malik, A. (1993). *Point-Nonpoint Source Trading for Managing Agricultural Pollutant Loadings: Prospects for Coastal Watershed*. USDA, ERS, AER-674.
38. Letson, D. (1992). Point/nonpoint source pollution reduction trading: an interpretive survey. *Nat. Resour. J.* **32**(spring): 219–232.
39. Stephenson, K. and Norris, P. (1998). Watershed-based effluent trading: the nonpoint source challenge. *Contemporary Econ. Policy* **16**(4): 412–421.
40. Environmental Protection Agency (1996). *Draft Framework for Watershed-Based Trading*. EPA 800-R-96-001. Office of Water, Washington, DC. <http://www.epa.gov/OW/watershed>

UNDERSTANDING *ESCHERICHIA COLI* O157:H7 AND THE NEED FOR RAPID DETECTION IN WATER

Zia Bukhari
Janice Weihe
American Water
Belleville, Illinois

Mark LeChevallier
American Water
Voorhees, New Jersey

INTRODUCTION

Escherichia coli is a facultative anaerobe that commonly inhabits the gastrointestinal tracts of humans within hours of birth. This gram-negative bacillus of the family Enterobactericeae is normally a harmless commensal that helps the host by suppressing pathogenic bacteria (due to competition) and by producing vitamin K and vitamin B complex (1). Individuals who suffer severe immunosuppression (both acquired and due to chemotherapy) or a breached gastrointestinal barrier can experience infection by nonpathogenic *E. coli*. More likely, however, are instances of infection by pathogenic *E. coli* that can include urinary tract infections, sepsis/meningitis, or diarrhea. Within the pathogenic *E. coli* responsible for diarrheagenic disease, differentiation of the strains has led to development of five classes (virotypes): enteropathogenic *E. coli* (EPEC), enterohemorrhagic *E. coli* (EHEC), enterotoxigenic *E. coli* (ETEC), enteroinvasive *E. coli* (EIEC), and enteroaggregative *E. coli* (EAEC) (2). Differentiation of pathogenic strains of *E. coli* is based on the organism's surface antigen profiles of O (somatic), H (flagellar), and K (capsular). This serological classification was originally proposed by Kauffman in 1944 and a modified form is still in use today. Although this serotyping of *E. coli* is a reproducible mechanism for identifying *E. coli* associated with particular clinical disease, the serotypes themselves are not responsible for virulence but are only identifiable chromosomal markers for virulence factors (2).

This article focuses on *E. coli* O157:H7, a member of the EHEC class of diarrheagenic *E. coli*, and specifically discusses transmission routes, particularly focused on the waterborne route and the methods that may be employed for detecting contamination.

HISTORY

Escherichia coli O157:H7 was first identified in 1975 by the Centers for Disease Control (CDC) but did not gain widespread attention until 1982 from the investigation of two outbreaks of gastrointestinal disease. Individuals were experiencing severe cramping, abdominal pains, watery diarrhea leading to bloody diarrhea, and low-grade fever. The illness, hemorrhagic colitis (HC), was associated with the consumption of undercooked hamburgers from a fast food restaurant, and examination of stool cultures from affected individuals revealed the presence of a rare *E. coli* serotype, O157:H7 (3). In addition, sporadic cases of hemolytic uremic syndrome (HUS) were reported by Karmali et al. (4), who found fecal cytotoxins and cytotoxin-producing *E. coli* in stool samples. Karmali et al. (5) also reported that verocytotoxin/Shiga-like toxins were the common virulence factor between HC and HUS. This led to the recognition of *E. coli* O157:H7 as an emerging enteric pathogen that was responsible for both intestinal and renal disease. To determine whether *E. coli* O157:H7 was really an emerging pathogen (rather than one that had not been detected previously), retrospective studies were conducted by the CDC and Public Health Laboratory Services (PHLS) in the United Kingdom, which indicated the presence of *E. coli* O157:H7 in only 1 of 3000 samples from 1973 to 1983 and in 1 of 15,000 samples from 1978 to 1982, respectively (2). It has been suggested that the reason for this pathogenic emergence was the acquisition of virulence factors such as the Shiga toxin gene by horizontal gene transfer and recombination (6). *Escherichia coli* O157:H7 produces Shiga-like toxins that allow categorizing it among other Shiga toxin-producing *E. coli* (STEC), alternatively called verocytotoxin-producing *E. coli* (VTEC). These terms are synonymous and are seen in the literature interchangeably. Various strains of *E. coli* O157:H7 can produce Shiga toxin 1 (stx1) and/or Shiga toxin 2 (stx2). Stx1 is virtually identical to the toxin produced by *Shigella dysenteriae* whereas stx2 is approximately 56% homologous to stx1 (6).

CLINICAL FEATURES AND TREATMENT

Infection by *E. coli* O157:H7 brings a wide range of clinical manifestations ranging from asymptomatic carriage to HC and HUS. The incubation period between exposure and disease onset ranges from 1 to 8 days. The illness typically presents with abdominal cramps and nonbloody diarrhea that progresses to bloody diarrhea after 1–2 days in 70% of cases. Blood content in stools can vary from low levels to almost entirely blood (7). Fever occurs in 30% of patients and vomiting in 30–60% of infected individuals (6). Most cases of HC resolve themselves within 1–2 weeks, and the treatment focuses on supportive care, emphasizing nutrition and hydration. It is not generally recommended to treat with antimotility agents, as this could delay O157 from being voided from the body, thus potentially increasing toxin exposure (8).

From 3% to 20% of the more severe cases of HC progress to HUS. Factors that may contribute to this progression include fever, age related susceptibility (i.e., children and the elderly are more susceptible); elevated serum leukocyte count; and treatment with antimotility agents (6). The role of antibiotic treatment in the progression to HUS remains controversial. Conflicting studies have shown both increased and decreased levels of HUS among patients who receive antibiotic therapy for HC. Death can occur in 3–5% of those infected with HUS, whereas 12–30% can experience various long-term complications, including permanent renal damage (2). There is no therapy for abatement of HUS, but dialysis and transfusions are supportive treatment, along with careful management of fluids, electrolytes, and blood pressure.

One highly touted treatment drug was Synsorb-Pk, which consists of a synthetic analog of the receptor for the Stx toxins (Gb3) linked to diatomaceous earth. This compound binds stx1 and stx2 from the intestines of affected patients, and by preventing additional toxin damage, it had the potential to limit the progression from HC to HUS. Initial clinical trials showed promise (2) but final phase III trials showed no effect of stx-binding Synsorb-Pk on the severity of HUS (9).

In some HUS patients, particularly the elderly, the disease also progresses to thrombotic thrombocytopenic purpura (TTP). TTP involves a greater incidence of fever and neurological deficits and involves organs other than the kidneys, specifically the heart and brain (7). There is also a distinct skin darkening that occurs with TTP in 90% of cases. Patients suffering from TTP typically have reduced platelet and red blood cell counts compared with HUS patients. Neurological findings in TTP range from confusion and headaches to coma and seizures. Cardiac involvement ranges from simple rhythmic disturbances to sudden cardiac arrest. Both cardiac and neurological problems are likely to be due to the occurrence of a classic TTP clinical finding: hyaline thrombi (clots composed of platelets and fibrin) that occlude the small arterioles and capillaries. Treatment of TTP consists largely of plasmapheresis (plasma volume exchange) or plasma infusions that continue until the platelet count stabilizes and hemolysis ceases. The increased use of two preventive measures has assisted in reducing mortality rates to 30%; however, some patients (11–28%) may experience a recurrent episode (10). Due to the lack of effective therapies for O157 infections, using vaccines to reduce colonization of the human intestine has been attempted by targeting proteins involved in attachment and effacement or by specifically targeting stx or the O157 lipopolysaccharide. Usually, children are most highly affected by O157 infections and implementation of vaccines in these populations can be difficult. Nonetheless, vaccination of cattle, which can be a major primary source of contamination in food and water, could reduce environmental burdens from this organism. The large numbers of cattle needing vaccination as well as the rapid turnover of these cattle may make such large-scale vaccination programs challenging (11).

PATHOGENESIS OF *E. COLI* O157:H7

Escherichia coli O157:H7 is highly acid resistant, survives the stomach environment, and passes into the gut (8). There, it adheres to the large intestine and disrupts the brush border, causing the characteristic attachment and effacing lesions of O157 infections. *Escherichia coli* O157:H7 strains can produce stx1 and/or stx2; the latter is more damaging to human renal cells (12) and leads to a greater number of HUS cases (13). The Shiga toxins are bacteriophage encoded and consist of five B subunits and one A subunit. The B subunits bind to globotriaosylceramide (Gb3), a glycolipid found in high levels in the human kidney (14), especially in infants and children (13). Upon binding to the cell membrane, the A subunit is internalized by receptor-mediated endocytosis.

Within the cell, the A subunit is disrupted and produces a fragment that has RNA N-glycosidase activity, which then acts on the 28S rRNA of the 60S ribosomal subunit. This prevents the peptide chain elongation step of protein synthesis and results in cell death (8). It is believed that this process is the primary cause of renal cell damage in HUS (7).

Although the Shiga toxins produced by O157 are the most frequently discussed virulence factors, other factors can also be responsible for host pathogenesis. The presence of a 60-MDa plasmid that encodes for an enterohemolysin may cause lysis of erythrocytes as seen in the bloody diarrhea of HC. The heme and hemoglobin released from lyzed erythrocytes could be a source of iron for O157 (2,6). A second virulence factor important in O157 pathogenicity is the locus of enterocyte effacement (LEE). The LEE contains the *eaeA* and *tir* genes. The *eae* gene encodes for intimin, a bacterial cell surface protein, which along with a translocated intimin receptor, encoded by *tir*, allows O157 to adhere to intestinal mucosa. This produces the attachment and effacing lesions seen in *E. coli* O157 infections (7,15).

EPIDEMIOLOGY

The CDC estimated that the annual disease burden due to *E. coli* O157:H7 is around 20,000; up to 250 deaths occur at a cost of $250–350 million (16). However, the true incidence may be underestimated due to the failure of laboratories to isolate or detect this pathogen successfully. Some researchers estimate that the incidence is as high as 73,000 annually; 11,000 of those are from waterborne transmission (17). The disease is of greater importance in the developed countries of the Northern Hemisphere, where 80% of all STEC infections are caused by *E. coli* O157:H7 (13). Within the United States, more cases are reported in the northern states, specifically the Northeast, than in southern states. There is also an association with seasonality; most cases are reported in the summer months when peak shedding of O157 in animals has been noted (6).

Stx-producing *E. coli*, specifically O157:H7, are found in the flora of various animals, including cattle, sheep, goats, deer, pigs, cats, dogs, chickens, gulls, and even flies. Cattle are recognized as the important animal species responsible for harboring *E. coli* O157:H7; infection rates typically range from 2% to 16%, however, incidence rates up to 87% have been identified (18). Farm animals are likely reservoirs of human infection because the organism can survive for extended periods in animal feces (19), soil (20,21), and water (22,23).

In human infections, there is a wide variation in the shedding patterns of *E. coli* O157:H7; 66% of stool samples are negative for *E. coli* O157:H7 7 days after diarrhea, despite a lack of antibiotic therapy (24). A later study noted a median shedding period of 17 days (range 2–62 days) (25). Others found a median shedding period of 21 days (range 5–124 days, during which most of the patients were asymptomatic). In addition, changes in the *E. coli* profile of two patients were noticed over 2 weeks of shedding, specifically a loss in the Shiga toxin (Stx) gene, as noted by pulsed field gel electrophoresis (PFGE). This occurrence has important implications for epidemiological studies based on molecular techniques (26).

TRANSMISSION ROUTES

Enterohemorrhagic E. coli (EHEC) can be transmitted by any route where fecally contaminated material is ingested by a susceptible host. Epidemiological investigations during outbreaks have found that the infectious dose for EHEC is low (i.e., 10–200 organisms) (27). This low infectious dose combined with the possibility of asymptomatic carriage of E. coli O157:H7 allows direct person-to-person transmission. This is especially likely in institutional settings such as day-care facilities and hospitals (25). Incidences of infection from both recreational and occupational exposure to farm animals have been reported (28) as well as occupational exposure of healthcare and laboratory personnel (2,29). A documented outbreak of E. coli O157:H7 infection was linked to exposure from a contaminated building. Infected individuals had visited a county fair building, in which E. coli O157:H7 was isolated from the dust within the rafters. No other links were established, so this suggested airborne dispersion of the organism (30).

A common route of transmission is food, and a number of outbreaks have been reported from consumption of undercooked ground beef either in restaurants or within individual homes. This suggests important implications for modern food processing practices, where beef from various sources is ground together and distributed statewide or nationwide. In addition to meat, other foods were implicated, including mayonnaise, unpasteurized apple juice, fermented hard salami, lettuce, radish sprouts, and alfalfa sprouts. Indirect contamination of these foods may have occurred due to improper hygiene among food handlers or by contact between uncooked foods and raw meats contaminated with E. coli O157:H7. There is also the possibility of direct contamination of produce in the fields by manure from infected animals (6). Foodborne outbreaks have also occurred following consumption of yogurt and milk (31,32).

WATERBORNE TRANSMISSION

A number of E. coli O157:H7 outbreaks have been linked to recreational waters. Outside Portland, Oregon, 59 individuals were infected after swimming in a recreational lake in 1991 (33). Similarly, water from a children's paddling pool was responsible for transmitting infection to two other children in Scotland, UK, in 1992 (34). Epidemiological investigations did not reveal conclusive information; however, it was postulated that a child with diarrhea contaminated the pool with E. coli O157:H7, and as the pool water was not changed or disinfected, it served as the vehicle of infection for the other two children, who in turn infected others by person-to-person contact. A water park outbreak in Georgia in 1998 resulted in dozens of infections and the death of one child and was linked to low chlorine levels in the pool water (35).

The first municipal waterborne E. coli O157:H7 outbreak occurred in Missouri in 1989 (36); more than 240 people were infected and four individuals died. The outbreak, it was suspected, occurred from backflow during two water main breaks, leading to intrusion of the pathogen. In Wyoming, 157 people were infected by E. coli O157:H7 during the summer of 1998 as a result of a community water system supplied by a spring and two wells under the influence of surface water. The water in this system was not chlorinated, and the outbreak may have occurred from fecal contamination by wildlife near the spring (37). Another large waterborne outbreak occurred in New York in 1999 where an untreated well at a county fair was contaminated by an adjacent septic system. A total of 127 confirmed cases of E. coli O157:H7 infection occurred that led to 14 cases of HUS and 2 deaths (38).

The most publicized waterborne outbreak of E. coli O157:H7 occurred in May 2000 in Walkerton, Ontario. Approximately 1350 cases of gastroenteritis were reported in individuals exposed to Walkerton municipal water: overall, it was estimated that the number of cases associated with Walkerton exceeded 2300 individuals. Stool samples confirmed 167 cases of E. coli O157:H7 infection; 65 individuals were hospitalized, and 27 individuals developed hemolytic uremic syndrome. Four deaths were directly due to the E. coli outbreak in Walkerton, and E. coli O157:H7 was a contributing factor that led to three additional deaths. During the outbreak, a supply source well was contaminated with coliforms and E. coli O157:H7. This well was prone to surface contamination, especially following flooding conditions, which is what preceded the Walkerton incident. Furthermore, environmental testing of livestock farms, especially a farm adjacent to this well, indicated the presence of E. coli O157:H7 infections in the livestock. Had appropriate chlorine residuals been maintained during the flooding and well contamination, it would have prevented the widespread outbreak (39,40).

SUSCEPTIBILITY OF E. COLI O157:H7 TO CHLORINATION

The susceptibility of various strains of E. coli O157:H7 to chlorine disinfection has been examined in a number of studies, and, except for an occasional resistant strain (41), the organism can be rapidly inactivated by low levels of disinfectant. For example, in the study reported by Zhao et al. (41), free chlorine concentrations of 0.25, 0.5, 1.0, and 2.0 mg/L at 23 °C and sampling times of 0, 0.5, 1.0, and 2.0 min resulted in >7 log reduction. It was determined that a 0.25 mg/L free chlorine concentration required only 1 min to achieve these levels of inactivation, suggesting that this organism had no unusual tolerance to chlorine. In another study, Rice et al. (42) compared the chlorine susceptibility of E. coli O157:H7, isolated from two waterborne outbreaks, with four wild-type E. coli isolates and reported approximately 4 log inactivation using a chlorine residual of 1.1 mg/L and exposure times up to 120 s. The biocidal activity of chlorine is temperature dependent, and the temperatures used in the Rice et al. (42) study (5 °C) were intended to represent worst case conditions for both groundwater or winter surface water. This was probably a factor leading to lower inactivation levels in this study compared to those reported by Zhao et al. (41).

According to a Water Quality Disinfection Committee survey (1992), it was determined that United States water utilities maintain a median chlorine residual of 1.1 mg/L and a median exposure time of 45 minutes before the point of first use in the distribution system. This suggested that *E. coli* O157:H7 is unlikely to survive conventional water treatment in the United States at these levels of chlorination. In nondisinfected drinking water, *E. coli* O157:H7 has environmental survival or chlorine susceptibility similar to wild-type *E. coli* (43), which suggests that wild-type *E. coli* can be an adequate indicator organism of fecal contamination of water.

DIAGNOSTIC PROCEDURES

Biochemical

Unlike the majority of other *E. coli* isolates, most strains of *E. coli* O157:H7 do not ferment sorbitol within 24 h (44). This allows easy identification of O157 on sorbitol-containing media, such as sorbitol-MacConkey (sMAC) agar. Colonies of O157 appear colorless whereas sorbitol fermenters are pink or mauve. The use of sMAC agar is one of the primary screening methods used in the clinical laboratory to analyze patient specimens for the presence of serotype O157:H7 (45). It has been reported that sMAC agar has a detection sensitivity of 50–60% (14) and the use of sMAC agar is most successful early in the disease; the detection rate declines further (down to 33%) as the infection progresses. The use of cefixime-tellurite sorbitol-MacConkey agar (46) increases sensitivity and inhibits other nonsorbitol fermenting organisms. *Escherichia coli* O157:H7 also does not produce β-glucuronidase and therefore is negative by the methylumbelliferyl glucuronide (MUG) assay (47). A variety of selective media have been formulated to use these two biochemical characteristics by adding fluorogenic or chromogenic compounds (2). Although many of these selective media are useful, isolating and identifying *E. coli* O157:H7 by these methods has limitations. One of the greatest limitations involves the time-consuming procedures. Current USDA methods involve four enrichment and culturing steps followed by 10 biochemical confirmation tests that require a minimum of 3 days (17). In addition, there may be misidentification of *E. coli* O157:H7 because other intestinal bacteria (i.e., *E. hermanii* and *Hafnia* spp.) share similar phenotypes and resemble serotype O157:H7 on a sorbitol-containing medium. In addition, there are O157 that are not the H7 serotype or pathogenic and also cannot ferment sorbitol (48). There is also evidence to indicate that serotype O157:H7 in sorbitol-containing foods can mutate to become a sorbitol-fermenting phenotype (49). Isolation of sorbitol-fermenting *E. coli* O157 from HUS patients has also been reported (50) and these strains were also MUG assay positive. These findings were initially considered atypical, but an increasing prevalence of sorbitol-fermenting *E. coli* O157:H7 from HUS patients has been reported (51). There is also the issue of other flora within an environmental or clinical sample that mask low levels of O157:H7. In addition, it has been shown that *E. coli* O157:H7 can enter a viable but nonculturable state, especially when present for a long time in cold water. In such a situation, these culture techniques would not detect the organism isolated from water samples (22).

Immunologic

Currently, a number of tests are available commercially for detecting *E. coli* O157:H7; the majority of these assays are antibody based, using polyclonal, preadsorbed, affinity purified or monoclonal antibodies that target the surface-expressed antigens of O157 and H7, although some tests are specific for stx toxins. These antibody based tests have been used primarily in an enzyme-linked immunosorbent assay (ELISA); the most common assay is the "sandwich" type ELISA, where the same antibody is used both to capture and to detect the specific antigens. Such antibodies have also been effective for use in agglutination or latex agglutination tests and enable rapid screening or serological confirmation of isolates. Anti-O157 antibodies have also been coupled to paramagnetic beads (Dynal, Inc., Lake Success, NY) to isolate this pathogen selectively from the sample matrix by immunomagnetic separation (IMS) (52). IMS procedures have also been combined with ELISA, using colorimetric or fluorometric detection procedures. Simpler procedures have used lateral flow immunoassays, where the antibody–antigen complex is allowed to migrate along a membrane and then captured by a second antibody that is immobilized in a specific region of the membrane. The lateral flow format, which is similar to pregnancy test kits, produces a visual signal (with the aid of colloidal gold or dye encapsulating liposomes) when samples are positive and can be conducted without specialized skills or equipment. Obviously, using antibody based isolation and detection procedures can be prone to cross-reactivity with other organisms (i.e., *Citrobacter freundii*, *E. hermanii*, *Yersinia enterocolitica*) or other non-H7 serotypes; however, these can be reduced either by preadsorption procedures or by selecting specific target antigens (53).

Molecular

Molecular assays typically do not have the limitations of immunologic and biochemical methods. Molecular methods may include pulsed field gel electrophoresis (PFGE), ribotyping, and polymerase chain reaction (PCR). A number of PCR conditions have been developed for identifying *E. coli* O157:H7 based on virulence factors such as the stx, enterohemolysin, and intimin genes (2,14,15). Some multiplex PCRs have been developed that are more specific, namely, a PCR based assay that amplifies an allele in the *uidA* gene that is unique to serotype O157:H7, including phenotypic variants of serotype O157:H, that are sorbitol and methyl-umbelliferyl glucuronide positive (53). Coupled with primers specific for *stx* genes, this multiplex PCR assay simultaneously identifies isolates of serotype O157:H7 and the type of Shiga toxin they encode (54).

The advantages of molecular methods include specificity, sensitivity, and the ability to detect phenotypic variants of serotype O157:H7. However, the complexity and expense of these assays may limit their application in routine water microbiological analysis. Fortunately, some molecular detection procedures, such as the BAX® system by Dupont-Qualicon, do offer a user-friendly molecular assay. In this procedure, primers target sequences derived through random amplified polymorphic DNA (RAPD) analysis and all the PCR components are provided in a tablet format. Simple rehydration of the tablet with the lyzed test sample, followed by sample incubation in a thermal cycler (under preset thermocycling conditions) enables specific detection of *E. coli* O157. Although this assay provides a user-friendly format for specific detection of *E. coli* O157:H7, a drawback is that it uses proprietary equipment and software that necessitate an initial capital outlay of approximately $25,000.

CURRENT STATUS OF *E. COLI* O157:H7 MONITORING IN WATER

Currently, monitoring for microbial contamination of public water systems is regulated under a number of different United States Environmental Protection Agency (U.S. EPA) regulations such as the Safe Drinking Water Act (SDWA) of 1974, and amendments introduced in 1986 and 1996. A list published every 5 years contains contaminants that are known or anticipated in public water systems and could possibly require regulation. Microbial contaminants within this list are regulated by the Surface Water Treatment Rule and Total Coliform Rule of 1989. According to U.S. EPA regulations, a system that operates at least 60 days per year and serves 25 people or more or has 15 or more service connections is regulated as a public water system. Under the Total Coliform Rule, routine monitoring of tap water is required for all public water systems; the frequency and number of samples vary with the size of the system and the number of people that are served. The presence of total coliforms indicates that the system is contaminated or vulnerable to fecal contamination. For any size system, two consecutive total coliform positive samples from one site and one fecal coliform or *E. coli* positive sample result in an acute violation of the Maximum Contaminant Level (MCL) and lead to mandatory immediate notification of the public. Various procedures have been documented in *Standard Methods for the Examination of Water and Wastewater*, 20th ed. (55) to determine the presence of total coliforms, fecal coliform, and *E. coli*. However, *E. coli* O157:H7 has biochemical characteristics that do not allow detection by the techniques used for these broader groups.

The occurrence of waterborne or water-associated outbreaks has demonstrated the need for rapid user-friendly methods for isolating and detecting *E. coli* O157:H7 from source and finished water samples. The absence of such methods has hampered the accumulation of epidemiological information, the effectiveness of water treatment plants for removal or inactivation, and limited timely responsiveness during outbreak investigations, thus delaying employment of adequate control strategies to prevent further disease transmission. In a recent study, Bukhari et al. (56) evaluated four lateral diffusion (pregnancy type test kits) immunologic assays (Reveal®, Eclipse™, ImmunoCard STAT!, and Mionix) and one molecular assay (BAX). They determined that the Reveal test was the most promising and had a sensitivity level of 1×10^4 CFU. Using a molecular detection procedure (BAX® *E. coli* O157:H7 assay), a sensitivity level of 10 CFU was ascertained. Both the Reveal and BAX *E. coli* O157:H7 detection procedures, combined with an overall method, consisting of 200 mL water sample concentration (membrane filtration), enrichment in tryptic soy broth, and simultaneous capture by immunomagnetic separation, enabled rapid detection (less than 8 h) of 0.03–0.12 CFU/mL. The sensitivity and reproducibility of this overall procedure were determined by using spiked environmental water samples in multilaboratory round-robin trials, which indicated that the Reveal assay was more sensitive and reproducible when using spiked finished water samples; however, the BAX method had greater specificity, an important factor when evaluating raw water samples. It is anticipated that user-friendly procedures such as these will allow water utilities to employ these methods routinely to confirm the presence of *E. coli* O157:H7 rapidly in all total or fecal coliform positive samples. Alternatively, water providers could use these methods directly to gather epidemiological information, assess environmental survival, develop treatment strategies, or investigate risks during outbreak situations.

CONCLUSION

In conclusion, numerous waterborne outbreaks of human *E. coli* O157:H7 infection have occurred in the past; however, employing routine monitoring programs and coupling them with adequate disinfection would help reduce the risk of waterborne disease transmission. Historically, routine monitoring was not feasible for water utilities because the methods available were tedious and time-consuming; however, it has become viable with the methods recently described by Bukhari et al. (56), which would allow water utilities to rapidly arm themselves with the relevant information to use appropriate contingency measures to ensure continued safety from this significant human pathogen.

BIBLIOGRAPHY

1. Weir, E. (2000). *Escherichia coli* O157:H7. *CMAJ* **163**(2): 205.
2. Nataro, J.P. and Kaper, J.B. (1998). Diarrheagenic *Escherichia coli*. *Clin. Micobiol. Rev.* **11**: 142–201.
3. Riley, L.W., Remis, R.S., Helgerson, S.D., McGee, H.B., Wells, J.G., Davis, B.R., Hebert, R.J., Olcott, E.S., Johnson, L.M., Hargrett, N.T., Blake, P.A., and Cohen, M.L. (1983). Hemorrhagic colitis associated with a rare *Escherichia coli* serotype. *N. Engl. J. Med.* **308**: 681–685.
4. Karmali, M.A., Petric, M., Lim, C., Fleming, P.C., and Steele, B.T. (1983). *Escherichia coli* cytotoxin, haemolytic-uraemic syndrome, and haemorrhagic colitis. *Lancet* **ii**: 1299–1300.

5. Karmali, M.A., Steele, B.T., Petric, M., and Lim, C. (1983). Sporadic cases of haemolytic uremic syndrome associated with faecal cytotoxin and cytotoxin producing *Escherichia coli* in stools. *Lancet* **i**: 619–620.
6. Mead, P. and Griffin, P. (1998). *Escherichia coli* O157:H7. *Lancet* **352**: 1207–1212.
7. Coia, J. (1998). Clinical, microbiological and epidemiological aspects of *Escherichia coli* O157 infection. *FEMS Immunol. Medical Microbiol.* **20**: 1–9.
8. Paton, J.C. and Paton, A.W. (1998). Pathogenesis and diagnosis of Shiga toxin-producing *Escherichia coli* infections. *Clin. Microbiol. Rev.* **11**: 450–479.
9. Trachtman, H., Cnaan, A., Christen, E., Gibbs, K., Zhao, S., Acheson, D.W., Weiss, R., Kaskel, F.J., Spitzer, A., and Hirshman, G.H. (2003). Effect of an oral Shiga toxin-binding agent on diarrhea-associated hemolytic uremic syndrome in children: a randomized control study. *JAMA* **290**(10): 1337–1344.
10. Rust, R.S. (2002). Thrombotic thrombocytopenic purpura and hemolytic uremic syndrome. Available: www.emedicine.com/neuro/topic499.html (accessed November 2003).
11. Horne, C., Vallance, B.A., Deng, W., and Finlay, B.B. (2002). Current progress in enteropathogenic and enterohemorrhagic *Escherichia coli* vaccines. *Expert Rev. Vaccines* **1**(2): 89–99.
12. Li, W. and Drake, M. (2001). Development of a quantitative competitive PCR assay for detection and quantification of *Escherichia coli* O157:H7 cells. *Appl. Environ. Microbiol.* **67**: 3291–3294.
13. Ochoa, T.J., and Cleary, T.G. (2003). Epidemiology and spectrum of disease of *Escherichia coli* O157. *Curr. Opin. Infect. Dis.* **16**: 259–263.
14. Kehl, S.C. (2002). Role of the laboratory in the diagnosis of enterohemorrhagic *Escherichia coli* infections. *J. Clin. Microbiol.* **40**: 2711–2715.
15. Ibekwe, M.A., Watt, P.M., Grieve, C.M., Sharma, V.K., and Lyons, S.R. (2002). Multiplex flurogenic real-time PCR for detection and quantification of *Escherichia coli* O157:H7 in dairy wastewater wetlands. *Appl. Environ. Microbiol.* **68**: 4853–4862.
16. Boyce, T.G., Swerdlow, D.L., and Griffin, P.M. (1995). Current concepts: *Escherichia coli* O157:H7 and the hemolytic-uremic syndrome. *N. Engl. J. Med.* **333**: 364–368.
17. Shelton, D.R. and Karns, J.S. (2001). Quantitative detection of *Escherichia coli* O157 in surface waters by using immunomagnetic electrochemiluminescence. *Appl. Environ. Microbiol.* **67**: 2908–2915.
18. Orr., R. 2000. The prevalence of *E. coli* O157:H7 in cattle. Available at: www.foodsafetynetwork.ca/animal/prevalence-O157-cattle.htm (accessed January 2004).
19. Wang, G., Zhao, T., and Doyle, M.P. (1996). Fate of enterohemorrhagic *Escherichia coli* O157:H7 in bovine faeces. *Appl. Environ. Microbiol.* **62**: 2657–2570.
20. Maule, A. (2000). Survival of verocytotoxigenic *Escherichia coli* O157 in soil, water and on surfaces. *Symp. Ser. Soc. Appl. Microbiol.* **29**: 71S–78S.
21. Ogden, I.D., Hepburn, N.F., MacRae, M., Strachan, N.J., Fenlon, D.R., Rusbridge, S.M., and Pennington, T.H. (2002). Long-term survival of *Escherichia coli* O157 on pasture following an outbreak associated with sheep at a scout camp. *Lett. Appl. Microbiol.* **34**: 100–104.
22. Wang, G. and Doyle, M.P. (1998). Survival of enterohemorrhagic *Escherichia coli* O157:H7 in water. *J. Food Prot.* **61**: 662–667.
23. Chalmers, R.M., Aird, H., and Bolton, F.J. (2000). Waterborne *Escherichia coli* O157. *J. Appl. Microbiol. Symp. Suppl.* **88**: 124S–132S.
24. Tarr, P.I., Neill, M.A., Clausen, C.R., Watkins, S.L., Christie, D.L., and Hickman, R.O. (1990). *Escherichia coli* O157:H7 and the hemolytic uremic syndrome: Importance of early cultures in establishing the etiology. *J. Infect. Dis.* **162**: 553–556.
25. Belongia, E.A., Osterholm, M.T., Soler, J.T., Ammend, D.A., Braun, J.E., and Macdonald, K.L. (1993). Transmission of *Escherichia coli* O157:H7 infection in Minnesota child daycare facilities. *JAMA* **269**: 883–888.
26. Karch, H., Russmann, J., Schmidt, H., Schwarzkopf, A., and Heesemann, J. (1995). Long-term shedding and clonal turnover of enterohemorrhagic *Escherichia coli* O157 in diarrheal disease. *J. Clin. Microbiol.* **33**: 1602–1605.
27. Willshaw, G.A., Thirwell, J., Jones, A.P., Parry, S., Salmon, R.L., and Hickey, M. (1994). Verocytotoxin-producing *Escherichia coli* O157 in beefburgers linked to an outbreak of diarrhea, haemorrhagic colitis and haemolytic uraemic syndrome in Britain. *Lett. Appl. Microbiol.* **19**: 304–307.
28. Coia, J., Sharp, J.C.M., Campbell, D.M., Curnow, J., and Ramsay, C.N. (1998). Environmental risk factors for sporadic *Escherichia coli* O157 infection in Scotland: results of a descriptive epidemiology study. *J. Infect.* **36**(3): 317–321.
29. Griffin, P.M. (1995). *Escherichia coli* O157:H7 and other enterohemorrhagic *Escherichia coli*. In: M.J. Blaser, P.D. Smith, J.I. Ravdin, H.B. Greenberg, and R.L. Guerrent (Eds.), *Infections of the Gastrointestinal Tract*, Raven Press, New York.
30. Varma, J.K., Greene, K.D., Reller, M.E., DeLong, S.M., Trottier, J., Nowicki, S.F., DiOrio, M., Koch, E.M., Bannerman, T.L., York, S.T., Lambert-Fair, M., Wells, J.G., and Mead, P.S. (2003). An outbreak of *Escherichia coli* O157 infection following exposure to a contaminated building. *JAMA* **290**: 2709–2712.
31. Morgan, D., Newman, C.P., Hutchinson, D.N., Walker, A.M., Rowe, B., and Majid, F. (1993). Verotoxin producing *Escherichia coli* O157 infections associated with the consumption of yogurt. *Epidemiol. Infect.* **111**: 181–187.
32. Upton, P. and Coia, J.E. (1994). Outbreak of *Escherichia coli* infection associated with pasteurized milk. *Lancet* **344**(8928): 1015.
33. Keene, W.E., McAnulty, J.M., Hoesly, F.C., Williams, L.P., Hedberg, K., Oxman, G.L., Barrett, T.J., Pfaller, M.A., and Fleming, D.W. (1994). A swimming-associated outbreak of hemorrhagic colitis caused by *Escherichia coli* O157:H7 and *Shigella sonnei*. *N. Engl. J. Med.* **331**: 579–584.
34. Brewster, D.H., Browne, M.I., Robertson, D., Houghton, G.L., Bimson, J., and Sharp, J.C.M. (1994). An outbreak of *Escherichia coli* O157 associated with a children's paddling pool. *Epidemiol. Infect.* **112**: 441–447.
35. Gilbert, L. and Blake, P. (1998). Outbreak of *Escherichia coli* O157:H7 infections associated with a waterpark. *Georgia Epidemiol. Rep.* **14**: 1–2.
36. Swerdlow, D.L., Woodruff, B.A., Brady, R.C., Griffin, P.M., Tippen, S., Donnell, H.D., Geldreich, E., Payne, B.J., Meyer, A., and Wells, J.G. (1992). A waterborne outbreak in Missouri of *Escherichia coli* O157:H7 associated with bloody diarrhea and death. *Ann. Intern. Med.* **117**: 812–819.
37. Olsen, S.J., Miller, G., Breuer, T., Kennedy, M., Higgins, C., Walford, J., McKee, G., Fox, K., Bibb, W., and Mead, P. (2002). A waterborne outbreak of *Escherichia coli* O157:H7 infections and hemolytic uremic syndrome: Implications for rural water systems. *Emerg. Infect. Dis.* **8**: 370–375.

38. New York Department of Heath. (2000). Health commissioner releases *E. coli* outbreak report. Available: www.health.state.ny.us/nysdoh/commish/2000/ecoli.html (accessed November 2003).
39. Hrudey, S.E., Payment, P., Huck, P.M., Gillham, R.W., and Hrudey, E.J. (2003). A fatal waterborne disease epidemic in Walkerton, Ontario: Comparison with other waterborne outbreaks in the developed world. *Water Sci. Technol.* **47**: 7–14.
40. O'Conner, D.R. (2002). Report of Walkerton Inquiry: Part 1—The events of May 2000 and related issues. Available: http://web.utk.edu/~hydro/Geol685/Walkerton_Summary.pdf (accessed November 2003).
41. Zhao, T., Doyle, M.P., and Zhao, P. (2001). Chlorine inactivation of *Escherichia coli* O157:H7 in water. *J. Food Prot.* **64**: 1607–1609.
42. Rice, E.W., Clark, R.M., and Johnson, C.H. (1999). Chlorine inactivation of *Escherichia coli* O157:H7. *Emerg. Infect. Dis.* **5**: 461–463.
43. Rice, E.W., Johnson, C.H., Wild, D.K., and Reasoner, D.J. (1992). Survival of *Escherichia coli* O157:H7 in drinking water associated with a waterborne disease outbreak of hemorrhagic colitis. *Lett. Appl. Microbiol.* **15**: 38–40.
44. Farmer, J.J. III and Davis, B.R. (1985). H7 antiserum-sorbitol fermentation medium: A single-tube screening method for detecting *Escherichia coli* O157:H7 associated with hemorrhagic colitis. *J. Clin. Microbiol.* **22**: 620–625.
45. March, S.B., and Ratnam, S. (1986). Sorbitol-MacConkey medium for detection of *Escherichia coli* O157:H7 associated with hemorrhagic colitis. *J. Clin. Microbiol.* **23**: 869–872.
46. Zadik, P.M., Chapman, P.A., and Siddons, C.A. (1993). Use of tellurite for the selection of vero/cytotoxigenic *Escherichia coli* O157:H7. *J. Med. Microbiol.* **39**: 155–158.
47. Thompson, A., Hodge, D., and Borczyk, A. (1990). Rapid biochemical test to identify verocytotoxin-positive strains of *Escherichia coli* serotype O157. *J. Clin. Microbiol.* **28**: 2165–2168.
48. Willshaw, G., Smith, H., Roberts, D., Thirwell, J., Cheasty, T., and Rowe, B. (1993). Examination of raw beef products for the presence of verocytotoxin producing *Escherichia coli*, particularly those of serogroup O157. *J. Appl. Bacteriol.* **75**: 420–426.
49. Fratamico, P.M., Buchanan, R.L., and Cooke, P.H. (1993). Virulence of an *Escherichia coli* O157:H7 sorbitol-positive mutant. *Appl. Environ. Microbiol.* **59**: 4245–4252.
50. Gunzer, F., Bohm, H., Russman, H., Bitzan, M., Aleksic, S., and Karch, H. (1992). Molecular detection of sorbitol-fermenting *Escherichia coli* O157 in patients with hemolytic-uremic syndrome. *J. Clin. Microbiol.* **30**: 1807–1810.
51. Bitzen, M., Ludwig, K., Klemt, M., Konig, H., Buren, J., and Muller-Wiefel, D.E. (1993). The role of *Escherichia coli* O157 infections in the classical (enteropathic) haemolytic urameic syndrome: Results of a Central European, multicenter study. *Epidemiol. Infect.* **110**: 183–196.
52. Wright, D.J., Chapman, P.A., and Siddons, C.A. (1994). Immunomagnetic separation as a sensitive method for isolating *Escherichia coli* O157 from food samples. *Epidemiol. Infect.* **113**: 31–39.
53. Feng, P. (1993). Identification of *Escherichia coli* O157:H7 by DNA probe specific for an allele of uidA gene. *Mol. Cell Probes* **7**: 151–154.
54. Cebula, T.A., Payne, W.L., and Feng, P. (1995). Simultaneous identification of strains of *Escherichia coli* serotype O157:H7 and their Shiga-like toxin type by mismatch amplification mutation assay-multiplex PCR. *J. Clin. Microbiol.* **33**: 248–250.
55. *Standard Methods for the Examination of Water and Wastewater*. 20th ed. (1998). American Public Health Association, American Water Works Association, Water Environment Federation, joint publishers.
56. Bukhari, Z., Weihe, J., and LeChevallier, M. (2005). Detection methods for *Escherichia coli* O157:H7 American Water Works Association Research Foundation Report.

EUTROPHICATION AND ORGANIC LOADING

Scott A. Lowe
Manhattan College
Riverdale, New York

The connection between eutrophication and organic loading was first realized in the 1870s by engineers in London. They concluded that when sewage was discharged into local rivers, the level of dissolved oxygen (DO) dropped, and the amount of algae increased. In other words, the rivers were eutrophied.

Raw sewage contains organic matter (including pathogens that make it so harmful when mixed in drinking water). The typical composition by weight of raw sewage is about 42% carbon (C), 7% nitrogen (N), and 1% phosphorus (P). The remaining 50% is water. This is also our human stoichiometry.

When discharged into a receiving water such as a lake, river, or estuary, the sewage begins to break down into its components. The sewage carbon, for example, oxidizes to CO_2:

$$C \to CO_2 \tag{1}$$

Stoichiometrically, it can be seen that $2 \times 16\,\text{g}\,O$ is required to oxidize $12\,\text{g}$ of carbon, or expressed as a ratio, $2.67\,\text{g}\,O/\text{g}\,C$. Similarly, for sewage nitrogen, which is mainly bound up as ammonia, NH_3:

$$NH_3 \to NO_3^- + H_2O + H^+ \tag{2}$$

The stoichiometry is $4 \times 16\,\text{g}\,O$ for every $14\,\text{g}\,N$; as a ratio, $4.57\,\text{g}\,O/\text{g}\,N$.

Phosphorus in raw sewage is bound up mainly as orthophosphate, PO_4, which is already in an oxidized state and so remains unchanged.

The oxygen that is consumed in this degradation process is supplied by the receiving water DO—hence the drop in DO that was observed in those London rivers. The other observed consequence of high levels of organic loading is excessive algal activity. To understand this phenomenon, one must consider what algae, typically phytoplankton, require to grow.

Phytoplankton have an internal stoichiometry that is similar in proportion to that described above for sewage, that is, 42% C, 7% N, and 1% P. The rest is water. To grow, they need to consume C, N, and P in this proportion. Being plants, they obtain their carbon from dissolved CO_2 in water, in the well-known process of photosynthesis. This also means that phytoplankton require sufficient light to grow.

The N and P that phytoplankton require come from dissolved N and P in the water. This N and P results from the breakdown of organic matter, as described before. If all four items (light, C, N, and P) are supplied in excess, then extreme algal growth occurs. This creates obvious visual effects such as "pea soup" in a pond or lake or "red tides" in the ocean.

To control this phenomenon, C is not important because there is always excess dissolved CO_2 in the water compared to the other nutrients, N and P. Light is also considered uncontrollable in natural systems. Therefore, control efforts center around N and P. The ratio at which algae need N and P is called the nitrogen to phosphorus ratio, simply N/P. It is approximately 7; algae need approximately seven times as much N as P, per their internal composition (7%N and 1%P). Therefore, if N and P in the water are measured (N as dissolved NH_3 and NO_3 and P as dissolved PO_4), then the ratio of actual N/P can be compared to the algal N/P ratio of 7.

The implication of this comparison is that if the measured N/P ratio is greater than, say, 10, then there is excess nitrogen in the water and the lack of phosphorus is limiting algal growth. Conversely, a low N/P ratio would imply that a lack of nitrogen is limiting algal growth. This is known as the limiting nutrient concept. Once the limiting nutrient (either N or P) for a system is identified, then efforts at eutrophication control will focus on further restricting inputs of this nutrient to the system. If the system is relatively balanced in both N and P, then whichever of the two is easier and more cost-effective to control is focused on.

In general, it has been found that freshwater systems are P limited, whereas marine systems, including estuaries, are N limited. Some real world examples of attempts to control eutrophication include:

- the ban on phosphorus in detergents and fertilizers in the watersheds that drain to the Great Lakes, which are P-limited;
- the upgrading of wastewater treatment plants to remove N that discharge to Long Island Sound, which is N-limited;
- the installation of tertiary treatment to remove P in wastewater treatment plants on watersheds that drain into New York City's drinking water reservoirs, which are P-limited; and
- the upgrading of wastewater treatment plants to remove N that discharge to Chesapeake Bay, which is N-limited.

Two final points—first, organic loads can come from raw sewage, and also from a myriad of other sources such as

- agricultural runoff
- industrial discharges
- forests
- urban runoff
- stormwater
- wastewater treatment plants

Second, sediment often contains large amounts of organic matter as a result of organic loading. One consequence of this is that even when the external organic load is reduced, for example, by treating raw sewage in a wastewater treatment plant, there still remains an internal source of nutrients for algal growth. Additionally, sediments with high organic content also exert an oxygen demand on the water, known as SOD—sediment oxygen demand.

TRACE ELEMENT CONTAMINATION IN GROUNDWATER OF DISTRICT HARDWAR, UTTARANCHAL, INDIA

CHAKRESH K. JAIN
National Institute of Hydrology
Roorkee, India

The trace element contamination in the groundwater of District Hardwar in the state of Uttaranchal (India) has been assessed to find the suitability of groundwater for drinking purposes. Forty-eight groundwater samples from shallow and deep aquifers were collected during both pre- and postmonsoon seasons in 2002 and analyzed for Fe, Mn, Cu, Ni, Cr, Pb, Cd, and Zn. The presence of trace elements in groundwater was recorded in all samples, but these were not significantly higher. Water quality standards were violated for iron and manganese at certain locations. The concentration of iron varies from 210 to 7004 µg/L during the premonsoon season and from 163 to 6555 µg/L during the postmonsoon season against the maximum permissible limit of 1000 µg/L; that of manganese varies from 7.5 to 839 µg/L during the premonsoon and from 9 to 859 µg/L during the postmonsoon season against the limit of 300 µg/L. The concentration of copper, chromium, lead, cadmium, and zinc were well within the permissible limits in all samples from District Hardwar.

INTRODUCTION

The contamination of groundwater by trace elements has received great attention in recent years due to their toxicity and accumulative behavior. These elements, in contrast to most pollutants, are not biodegradable and undergo a global ecobiological cycle in which natural waters are the main pathways. The major sources of trace elements in groundwater include weathering of rock minerals, discharge of sewage and other waste effluents on land, and runoff water. The water used for drinking should be free from any toxic elements, living and nonliving organisms, and excessive amounts of minerals that may be hazardous to health. Some trace elements are essential for humans, for example, cobalt and copper, but large quantities of them may cause physiological disorders. Cadmium, chromium, and lead are highly toxic to humans even in low concentrations.

The creation of the new state of Uttaranchal has posed many challenges for planners and policy makers. Problems such as drinking water, transportation, power sector, housing and construction, and safety from natural hazards are very serious and require immediate attention. For sustainable development of a society, it is essential that the natural resources are used judiciously for the benefit of the existing population and also to meet the needs and aspirations of future generations. Drinking water is one such precious commodity for which a planned strategy is needed for immediate demands and for sustainability for future needs also. A large part of the state of Uttaranchal lies in the hills, where distribution of the drinking water supply and its quality are major problems needing immediate attention. The physicochemical and bacteriological characteristics of the groundwater of District Hardwar in the state of Uttaranchal have been described in an earlier report (1). In this article, the groundwater quality is examined with reference to trace elements.

STUDY AREA

District Hardwar is part of the Indo-Gangetic plains that lie between latitude 29°30′ and 30°20′ N and between longitude 77°40′ and 78°25′ E in the state of Uttaranchal (Fig. 1). It is the largest district (populationwise) of Uttaranchal State and occupies an area of about 2360 km². Per the 2001 census, the population of the District Hardwar is 14,44,213 and the population density is 612 per km².

Physiographically, the area is generally flat except for the Siwalik Hills in the north and northeast. The area is devoid of relief features of any prominence except for deep gorges cut by ravines and rivers flowing through the area. The area is bounded by River Yamuna in the west and River Ganga in the east. The climate of the area is characterized by a moderate type of subtropical monsoon. The average annual rainfall in the region is about 1000 mm; the major part is received during the monsoon period. The major land use is agriculture, and there is no effective forest cover. The soils of the area are loam to silty loam and are free from carbonates. The most common groundwater use is by hand pumps and tube wells. Based on the lithologic logs and water table fluctuation data, two types of aquifers have been delineated in the area. The upper is a shallow unconfined aquifer that generally extends to depths around 25 m. The deeper is confined to semiconfined and located about 25 to 150 m below ground level separated by three to four aquifers at average depths of 25 to 55, 65 to 90, and 120 to 150 m. Water table contours in the area indicate the southward trend of groundwater flow in both unconfined and confined aquifers.

EXPERIMENTAL METHODOLOGY

Forty-eight groundwater samples from District Hardwar were collected during both pre- (June 2002) and postmonsoon (October 2002) seasons from various abstraction sources at various depths covering extensively populated area, commercial, industrial, agricultural, and residential colonies, so as to obtain a good areal and vertical representation; they were preserved by adding ultrapure nitric acid to a pH of 2.0 (2,3). The hand pumps and tube wells were continuously pumped prior to sampling to ensure that groundwater sampled was representative of the groundwater aquifer. The details of sampling locations and source and depthwise distribution are given in Tables 1 and 2, respectively.

Metal ion concentrations were determined by atomic absorption spectrometry using a Perkin-Elmer Atomic Absorption Spectrometer (Model 3110) using an air–acetylene flame. Operational conditions were adjusted in accordance with the manufacturer's guidelines to yield optimal determination. Quantification of metals was based upon calibration curves of standard solutions of respective metals. These calibration curves were determined several times during the period of analysis. The detection limits for iron, manganese, copper, nickel, chromium, lead, cadmium and zinc are 0.003, 0.001, 0.001, 0.004, 0.002, 0.01, 0.0005, and 0.0008 mg/L, respectively.

RESULTS AND DISCUSSION

The Bureau of Indian Standards (BIS), earlier known as the Indian Standards Institution (ISI), laid down standard specifications for drinking water during 1983; they have been revised and updated from time to time. To enable the users to exercise their discretion toward water quality criteria, the maximum permissible limit has been prescribed especially where no alternate source is available. The national water quality standards describe essential and desirable characteristics that must be evaluated to assess the suitability of water for drinking (4). The trace element data for the two sets of samples collected during pre- and postmonsoon seasons are presented in Figs. 2–9. The toxic effects of these elements and the

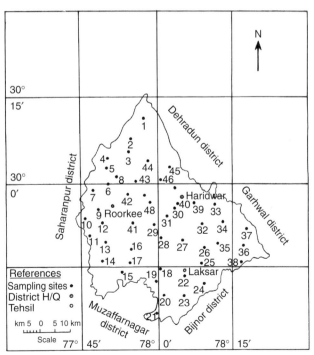

Figure 1. Study area showing location of sampling sites (District Hardwar).

Table 1. Description of Groundwater Sampling Locations in District Hardwar

Sample No.	Location	Source	Depth, m
1	Mohand	OW	10
2	Banjarewala	OW	10
3	Buggawala	HP	30
4	Kheri	HP	38
5	Dadapatti	HP	38
6	Bahbalpur	HP	15
7	Sikandarpur	HP	25
8	Bhagwanpur	HP	30
9	Chudiala	HP	40
10	Balswa Ganj	HP	25
11	Manakpur	HP	40
12	Iqbalpur	HP	30
13	Jharera	HP	30
14	Sherpur	HP	30
15	Narsen	HP	30
16	Manglour	HP	35
17	Libarheri	HP	35
18	Mahesari	HP	20
19	Sahipur	HP	35
20	Khanpur	HP	15
21	Chandpuri Kalan	HP	15
22	Laksar	HP	15
23	Kalsiya	HP	10
24	Niranjanpur	HP	15
25	Sultanpur	HP	15
26	Shahpur	HP	25
27	Pathri	HP	15
28	Subashgarh	HP	20
29	Marghubpur	HP	30
30	Bahadarabad	HP	20
31	Alipur	HP	20
32	Katarpur	HP	10
33	Kankhal	HP	25
34	Shyampur	HP	30
35	Rasiya Garh	HP	40
36	Gandikhatta	HP	10
37	Laldhang	HP	90
38	Kottawali	HP	40
39	Hardwar	HP	30
40	Jwalapur	HP	30
41	Roorkee	HP	10
42	Gumanwala	HP	30
43	Manubas	HP	10
44	Bandarjud	HP	15
45	Beriwala	HP	5
46	Hazara	HP	10
47	Aurangbad	HP	10
48	Daulatpur	HP	35

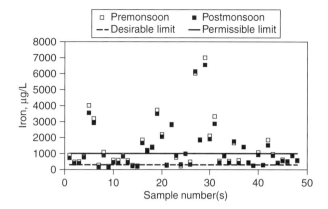

Figure 2. Distribution of iron at different sampling locations.

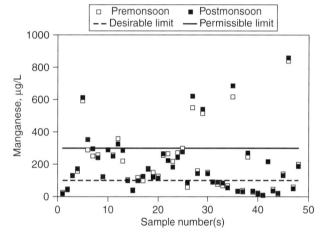

Figure 3. Distribution of manganese at different sampling locations.

extent of their contamination in the groundwater are discussed in the following sections.

Iron (Fe)

The concentration of iron in the groundwater of District Hardwar ranges from 210 to 7004 µg/L during the premonsoon season and from 163 to 6555 µg/L during the postmonsoon season; its distribution at different sites is shown in Fig. 2. The Bureau of Indian Standards has recommended 300 µg/L as the desirable limit and 1000 µg/L as the maximum permissible limit for drinking water (4). It is

Table 2. Source and Depthwise Distribution of Sampling Sites in District Hardwar

Source Structure	Depth Range			Total number
	<0–20 m	20–40 m	>40 m	
Hand pumps	6, 18, 20, 21, 22, 23, 24, 25, 27, 28, 30, 31, 32, 36, 41, 43, 44, 45, 46, 47	3, 4, 5, 7, 8, 9, 10, 11, 12, 13, 14, 15, 16, 17, 19, 26, 29, 33, 34, 35, 38, 39, 40, 42, 48	37	46
Tube wells	—	—	—	—
Open wells	1, 2	—	—	2
Total	22	25	1	48

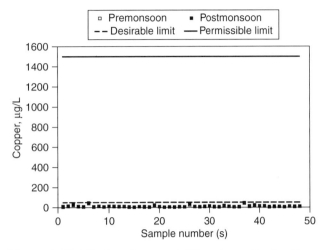

Figure 4. Distribution of copper at different sampling locations.

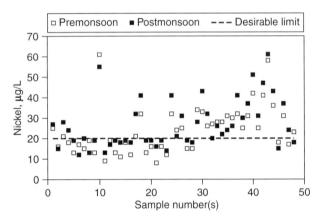

Figure 5. Distribution of nickel at different sampling locations.

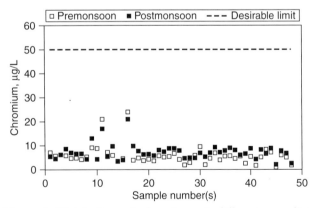

Figure 6. Distribution of chromium at different sampling locations.

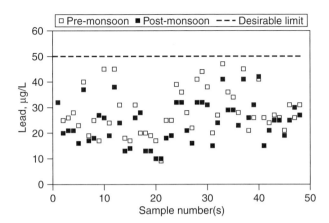

Figure 7. Distribution of lead at different sampling locations.

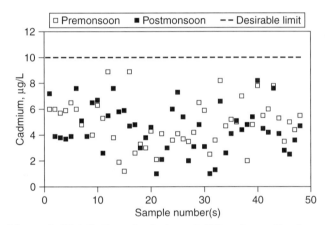

Figure 8. Distribution of cadmium at different sampling locations.

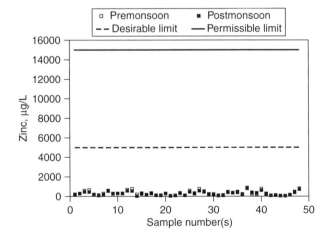

Figure 9. Distribution of zinc at different sampling locations.

evident from the results that only about 15% of the samples fall within the desirable limit of 300 µg/L and about 35% of the samples even cross the maximum permissible limit of 1000 µg/L. High concentrations of iron generally cause an inky flavor and a bitter, astringent taste. It can also discolor clothes and plumbing fixtures and cause scaling, which encrusts pipes. Excessive concentration may promote bacterial activities in pipe and service mains, causing objectionable odors and red-rot disease in water. Well water containing soluble iron remains clear while pumped out, but exposure to air causes precipitation of iron due to oxidation and a resulting rusty color and turbidity.

The increase in iron concentrations during the postmonsoon season may be due to mixing phenomena of recharge water that has more oxygen to react with iron ore, which may be available in clay lenses in the aquifer. The limits of iron in water supplies for potable use have not been laid down from health considerations but are due to the fact that iron in water supplies may cause discoloration of clothes, plumbing fixtures, and porcelain ware. The "red-rot" disease of water, caused by bacterial precipitation of hydrated oxides of ferric iron that give an unaesthetic appearance to water, clog pipes, pit pipes, and cause foul smells, is due to the presence of relatively high iron in water. The concentration of iron in natural water is controlled by both physicochemical and microbiological factors. The weathering of rock and discharge of waste effluents on land are generally considered the main sources of iron in groundwater. Iron migrates adsorbed to suspended matter, as insoluble hydrated iron compounds complexed to inorganic and organic ligands, and also as hydrated ions. Dissolved carbon dioxide, pH, and E_h of water affect the nature of aqueous iron species in the water. In groundwater, iron generally occurs in two oxidation states, ferrous (Fe^{2+}) and ferric (Fe^{3+}) forms.

Manganese (Mn)

The concentration of manganese was recorded at a maximum level of 839 µg/L during the premonsoon season and 859 µg/L during the postmonsoon season; its distribution at different sites is shown in Fig. 3. Manganese is an essential element, which does not occur naturally as a metal but is found in various salts and minerals frequently in association with iron compounds. In general, the concentration of manganese in groundwater is low due to geochemical control. A concentration of 100 µg/L has been recommended as a desirable limit and 300 µg/L as the permissible limit for drinking water (4). The WHO has prescribed 500 µg/L as a provisional guideline for drinking water (5). The presence of manganese above the permissible limit in drinking water often imparts an alien taste to water. It also has adverse effects on domestic uses and water supply structures.

It is evident from the results that about 35% of the samples from District Hardwar fall within the desirable limit of 100 µg/L, 50% of the samples cross the desirable limit but are within the permissible limits, and about 10–15% of the samples exceed the permissible limit of 300 µg/L. The high concentration of manganese at few locations may be attributed to the reducing conditions of the water and dissolution of manganese bearing minerals from the soil strata. Manganese may gain entry into the body by inhalation, consumption of food, and through drinking water.

Copper (Cu)

The concentration of copper was recorded at a maximum level of 47 µg/L during the premonsoon season and 41 µg/L during the postmonsoon season. The distribution of copper at different sites during the pre- and postmonsoon seasons is shown in Fig. 4. The Bureau of Indian Standards has recommended 50 µg/L as the desirable limit and 1500 µg/L as the permissible limit in the absence of alternate sources (4). Beyond 50 µg/L, the water has an astringent taste and causes discoloration and corrosion of pipes, fittings, and utensils. The World Health Organization has recommended 2000 µg/L as the provisional guideline for drinking water (5).

All samples from District Hardwar fall below the desirable limit of 50 µg/L. As such, the groundwater of District Hardwar can be safely used as a source of drinking water supplies. In general, the principal sources of copper in water supplies are corrosion of brass and copper pipe and addition of copper salts during water treatment for algae control. The toxicity of copper to aquatic life depends on the alkalinity of the water. At lower alkalinity, copper is generally more toxic to aquatic life. Copper, if present in excess amount in public water supplies, enhances corrosion of aluminum and zinc utensils and fittings. High intake of copper may result in damage to the liver. The industrial sources of copper that enhance the concentration in groundwater include industrial effluents from electroplating units, textiles, paints, and pesticides.

Nickel (Ni)

The concentration of nickel in the study area was recorded at a maximum level of 61 µg/L during the pre- and postmonsoon seasons; its distribution at different sites is shown in Fig. 5. The World Health Organization has recommended 20 µg/L as the guideline for drinking water (5).

More than 50% of the samples from District Hardwar exceed the WHO limit of 20 µg/L during pre- as well as postmonsoon seasons. The violation of the BIS limit could not be ascertained as no permissible limit of nickel has been prescribed in BIS drinking water specifications. Nickel at trace levels is essential to human nutrition, and no systemic poisoning from nickel is known in this range. The level of nickel usually found in food and water is not considered a serious health hazard. Some of the important nickel minerals include garnierite, nickeliferous limonite, and pentiandite. Certain nickel compounds have carcinogenic effects on animals; however, soluble compounds are not currently regarded as human or animal carcinogens.

Chromium (Cr)

The concentration of chromium in the study area was recorded at a maximum level of 24 µg/L during the premonsoon season and 21 µg/L during the postmonsoon season; in this range, it is not harmful to humans. The distribution of chromium at different sites during pre- and postmonsoon seasons is shown in Fig. 6. A concentration of 50 µg/L has been recommended as a desirable limit for drinking water (4). The WHO has also prescribed 50 µg/L as the guideline for drinking water (5). All samples from the study area fall well within the desirable limit for drinking water.

The two important oxidation states of chromium in natural waters are +3 and +6. In well-oxygenated waters, Cr(+6) is the thermodynamically stable species. However, Cr(+3) is kinetically stable and could persist bound to naturally occurring solids. Interconversions of Cr(+3) and Cr(+6) occur in conditions similar to those in natural waters. Municipal wastewater releases a considerable amount of chromium into the environment. Chromium is not acutely toxic to humans due to the high stability of natural chromium complexes in abiotic matrices. In addition, the hard acid nature of chromium imparts a strong affinity for oxygen donors rather than sulfur donors present in biomolecules. However, Cr(+6) is more toxic than Cr(+3) because of its high rate of adsorption through intestinal tracts. In the natural environment, Cr(+6) is likely to be reduced to Cr(+3), thereby reducing the toxic impact of chromium discharges.

Lead (Pb)

The concentration of lead in the study area was recorded at a maximum level of 47 μg/L during the premonsoon season and at 42 μg/L during the postmonsoon season. The distribution of lead at different sites during the pre- and postmonsoon seasons is shown in Fig. 7. The Bureau of Indian Standards has prescribed 50 μg/L lead as the desirable limit for drinking water (4). Beyond this limit, water becomes toxic. WHO has also prescribed the same guideline for drinking water (5).

All samples from District Hardwar fall within the permissible limit for drinking water during the pre- as well as the postmonsoon seasons, and therefore, the groundwater in the study area can be safely used as a source of drinking water supplies. The major source of lead contamination is the combustion of fossil fuel. Lead is removed from the atmosphere by rain, falls back on the earth's surface, and seeps into the ground. Lead passes from the soil to water and to plants and finally into the food chain. In drinking water, it occurs primarily due to corrosion of lead pipes and solder, especially in areas of soft water. Dissolution of lead requires extended contact time, so lead is most likely to be present in tap water after being in the service connection piping and plumbing overnight.

Cadmium (Cd)

Cadmium is a nonessential, nonbeneficial element known to have a high toxic potential. The cadmium content in the study area varies from 1.2 to 9.0 μg/L during the premonsoon season and from 1.0 to 8.2 μg/L during the postmonsoon season. The distribution of cadmium at different sites during the pre- and postmonsoon seasons is shown in Fig. 8. The Bureau of Indian Standards has prescribed 10 μg/L cadmium as the desirable limit for drinking water (4). Beyond this limit, water becomes toxic. WHO has prescribed 3 μg/L cadmium as the guideline for drinking water (5).

All samples from District Hardwar were within the desirable limit of 10 μg/L, as prescribed by BIS. It is obvious, therefore, that the groundwater of District Hardwar does not present any cadmium hazards to humans. The levels of cadmium in public water supplies are normally very low; generally, only small amounts exist in raw water, and many conventional water treatment processes remove much of the cadmium. Drinking water that contains more than 10 μg/L of cadmium can cause bronchitis, emphysema, anemia, and renal stone formation in animals. Cadmium can also enter the environment from a variety of industrial sources, including mining and smelting, electroplating, and pigment and plasticizer production. Drinking water is generally contaminated by galvanized iron pipe and plated plumbing fittings of the water distribution system. The U.S. EPA has classified cadmium as a probable human carcinogen based on positive carcinogenicity testing. However, no such health hazard is expected in the groundwater of District Hardwar.

Zinc

The concentration of zinc in the study area ranges from 44 to 878 μg/L during the premonsoon season and from 15 to 787 μg/L during the postmonsoon season. The distribution of zinc at different sites during the pre- and postmonsoon seasons is shown in Fig. 9. The Bureau of Indian Standards has prescribed 5000 μg/L zinc as the desirable limit and 15,000 μg/L as the permissible limit for drinking water (4). The WHO has prescribed 3000 μg/L as the guideline for drinking water (5). In the study area, all samples analyzed were within the desirable limits prescribed by the BIS (4) and the WHO (5).

CONCLUSIONS AND RECOMMENDATIONS

The groundwater quality in District Hardwar varies from place to place and with the depth of the water table. The water drawn for domestic applications should be tested and analyzed to ensure the suitability of groundwater for human consumption. The heavy metals except iron and manganese, which are present in appreciable concentration in groundwater, have been below the prescribed maximum permissible limits. The concentration of iron varies from 210 to 7004 μg/L during the premonsoon season and from 163 to 6555 μg/L during the postmonsoon season compared to the maximum permissible limit of 1000 μg/L; that of manganese varies from 7.5 to 839 μg/L during premonsoon and from 9 to 859 μg/L during the postmonsoon season compared to the 300 μg/L limit. The concentration of copper, chromium, lead, cadmium, and zinc were well within the permissible limits in all samples from the study area.

BIBLIOGRAPHY

1. Jain, C.K. (2003). *Ground water quality in District Hardwar, Uttaranchal*. Technical Report, National Institute of Hydrology, Roorkee, India.
2. Jain, C.K. and Bhatia, K.K.S. (1988). *Physico-Chemical Analysis of Water and Wastewater, User's Manual, UM-26*. National Institute of Hydrology, Roorkee, India.
3. APHA. (1992). *Standard Methods for the Examination of Water and Waste Waters*, 18th Edn. American Public Health Association, Washington, DC.
4. BIS. (1991). *Specifications for Drinking Water, IS:10500:1991*. Bureau of Indian Standards, New Delhi.
5. WHO. (1996). *Guidelines for Drinking Water, Recommendations*, Vol. 2. World Health Organization, Geneva.

IRON BACTERIA

BAIKUN LI
Pennsylvania State University
Harrisburg, Pennsylvania

GENERAL INTRODUCTION

Iron is one of the most abundant elements found in nature, accounting for at least 5% of Earth's crust. On the surface of Earth, iron exists naturally in two oxidation states, ferrous (Fe^{2+}) and ferric (Fe^{3+}). Dissolved or ferrous iron is usually found only in groundwater supplies that have not been exposed to oxygen. Water containing ferrous iron is clear and colorless. At contact with the air, the dissolved ferrous iron reacts with oxygen and forms insoluble ferric oxide, also known as iron oxide or "red rust." Water containing ferric iron has undesirable odors and tastes. Iron staining has caused problems in well systems and water distribution pipelines. With the presence of iron bacteria, iron may cause the corrosion of steel piping components.

Iron bacteria are found in all parts of the world, with the exception of the southern polar region. The "true" iron bacteria are microorganisms in which the oxidation of iron is an important source for their metabolic energy. This group is most often associated with filamentous and stalked forms that are encrusted with iron *(Thiobacillus, Leptothrix, Clonothrix, Gallionella, Sphaerotilus)*. During the oxidation, soluble ferrous iron (Fe^{2+}) is transferred to insoluble ferric iron (Fe^{3+}), which then precipitates out of solution. The overall reaction of spontaneous ferrous iron oxidation is

$$Fe^{2+} + 0.25\ O_2 + 2.5\ H_2O \rightarrow Fe(OH)_3 + 2H^+$$

The energy released from this reaction promotes the growth of thread-like slimes, which, with the ferric iron, form a voluminous mass that has caused clogging in the water supply system.

Thiobacullus ferrooxidans (Fig. 1) and *Leptothrix ferrooxidans* (Fig. 2) are the best-known iron-oxidizing bacteria. Both of them can grow autotrophically with ferrous iron (Fe^{2+}) as the electron donor. They are very common in acid-polluted environments such as coal-mining dumps. Also, acidophilic iron bacteria exist, such as *Ferrobacillus ferrooxidans* and *Ferroplasma acidophilum* (archaeon), which can oxidize iron in low pH environments (even pH = 0) at high temperature (50 °C), and are most commonly found associated with acid mine waste. The acidophilic bacteria are not commonly found in drinking water supplies.

Iron bacteria also cause the majority of problems in water pumps and pipes, blocking them with a brown, slimy, clay-like material (Fig. 3). Sometimes a bad odor comes from the water, like rotten eggs. Iron bacteria are not harmful to humans, but they are harmful to pipelines and pumps. If left untreated, they will cause the eventual blockage of the bore and failure of the pump.

Normally, iron bacteria refer to those in which the oxidation of iron is an important source for their

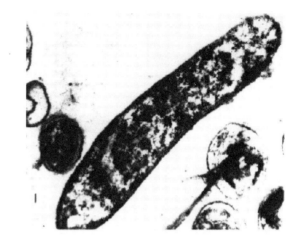

Figure 1. *T. ferrooxidans* cell suspension viewed by an electron microscope (magnified 30,000 times) (1).

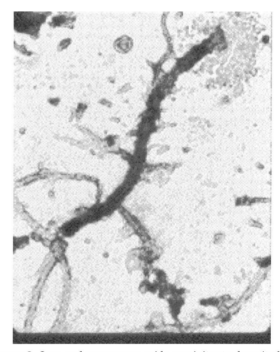

Figure 2. Iron and manganese oxide precipitants deposited on filaments of *Leptothrix ferrooxidans* (2).

metabolic energy, but another type of iron bacteria exists: iron-reducing bacteria, which get metabolic energy by reducing ferric iron to ferrous iron under anoxic or anaerobic condition with the presence of organic matters or other electron donors. Important ferric reducers include *Shewanella, Geobacter, Geospirillum,* and *Geovibrio*. Iron-reducing bacteria are important at natural cleanup and energy production.

IRON-OXIDIZING BACTERIA

Most iron-oxidizing bacteria grow and multiply in water and use dissolved iron as part of their metabolism. They oxidize ferrous (Fe^{2+}) iron into its insoluble ferric state (Fe^{3+}) and deposit it in the slimy gelatinous material that

Figure 3. The cut section of a cast-iron pipe shows a severe iron bacteria buildup on the pipe walls.

surrounds their cells. These filamentous bacteria grow in stringy clumps and are found in most iron-bearing surface waters. They have been known to proliferate in waters containing iron as low as 0.1 mg/L. To carry out the oxidation function, these aerobic bacteria need at least 0.3 ppm of dissolved oxygen in the water.

Iron-Oxidizing Bacteria in Groundwater and Surface Water System

Although iron bacteria do not cause health problems in people, they may have the following unpleasant and possibly expensive effects: cause unpleasant odors and tastes in water, corrode plumbing equipment, reduce well yields (clog screens and pipes), and increase chances of sulfur bacteria infestation for pipeline corrosion. Iron bacteria can build up in a stagnant section of the water supply system, such as the quick disconnect fittings of recoil hoses that are not regularly flushed. If the iron bacteria buildup becomes thick enough, anaerobic conditions may develop at the wall of the fitting or pipe that can cause corrosion of stainless steel or growth of sulfate-reducing bacteria.

The growth of iron bacteria also causes biofouling and blockage in groundwater bores. Although most groundwater contains some iron, iron bacteria do not affect all bores in the absence of oxygen. When iron bacteria infect a bore, the resulting growth may be suspended in water or deposited onto the bore hole equipment. Iron oxide (because of the oxidation by iron bacteria) suspended in the bore water is usually exhibited as biofouling in the irrigation pipes, which will deteriorate the pipeline surface and cause corrosion and shorten the pipeline lifetime. Bacterial growth deposits impact the bore hole equipment as well as the irrigation plumbing.

Normally, the presence of iron and oxygen in water triggers the growth of iron bacteira. Although iron is present in water, it is seldom found at concentrations greater than 10 mg/L or 10 parts per million (ppm). The U.S. Environmental Protection Agency (EPA) did not include iron in the National Primary (health-related) Drinking Water Regulations, but they did set a Secondary Regulation limit of 0.3 mg/L based on aesthetic and taste considerations. Surface water supplies usually have concentrations below the standard level (0.05–0.20 mg/L), although higher levels may be present because of corrosion of pipes. Higher iron levels (1.0–10 mg/L) are most common in groundwater supplies.

Iron-Oxidizing Bacteria in Acid Mining Drainage (AMD)

Mine drainage is metal-rich water formed from chemical reaction between water and rocks containing sulfur-bearing minerals. The runoff formed is usually acidic and frequently comes from areas where ore- or coal-mining activities have exposed rocks containing pyrite—a sulfur-bearing mineral with the major component of FeS. When pyrite is exposed to air and reacts with oxygen and water, sulfuric acid and dissolved iron (Fe^{2+}) are produced (Reaction 1). Ferrous iron is oxidized to ferric (Fe^{3+}) when reacting with oxygen (Reaction 2) and forms ferric hydroxide ($Fe(OH)_3$, Reaction 3). Ferric hydroxide precipitates to form the red, orange, or yellow sediments in the bottom of streams containing mine drainage. The acid runoff further dissolves heavy metals such as copper, lead, and mercury into ground or surface water. In the natural environment, the formation of acid mine drainage would take a long time to develop. But the rate and degree by which acid mine drainage proceeds can be dramatically sped up by the action of certain bacteria (e.g., *Thiobacillus ferrooxidans*).

$$4FeS_{2(s)} + 14O_{2(g)} + 4H_2O_{(l)} \rightarrow$$
$$4Fe^{2+}_{(aq)} + 8SO_4^{2-}_{(aq)} + 8H^+_{(aq)} \quad (1)$$

$$4Fe^{2+}_{(aq)} + O_{2(g)} + 4H^+_{(aq)} \rightarrow 4Fe^{3+}_{(aq)} + 2H_2O_{(l)} \quad (2)$$

$$4Fe^{3+}_{(aq)} + 12H_2O_{(l)} \rightarrow 4Fe(OH)_{3(s)} + 12H^+_{(aq)} \quad (3)$$

The overall reaction for pyrite oxidation and precipitation (Reaction 4) is as follows:

$$4FeS_{2(s)} + 15O_{2(g)} + 14H_2O_{(l)} \rightarrow$$
$$4Fe(OH)_{3(s)} + 8SO_4^{2-}_{(aq)} + 16H^+_{(aq)} \quad (4)$$

In acid creeks in mining regions like Contrary Creek, VA (Fig. 4), iron-rich precipitates often contain sulfur and are yellow. Many iron-oxidizing bacteria such as *T. thiooxidans* and *T. ferrooxidans* produce this "yellow boy," a yellowish-orange precipitate that turns the acidic runoff in the streams to an orange or red color and covers the stream bed with a slimy coating.

Although iron-oxidizing bacteria play a nuisance role in acid mine drainage, they are the central role for microbial mining, in which the leaching of copper from low-grade ore can recover copper from the drainage water of mines and improve the copper content and ore grade. Many ores that are sources of metals (copper) are sulfide (HS^-), which are highly insoluble minerals. If the concentration of metal in the ore is low, it may not be economically feasible to concentrate the mineral by conventional chemical means. On the other hand, the oxidized status of sulfide–sulfate (SO_4^{2-}) forms a water-soluble mineral, which can improve

Figure 4. Acid creeks in mining regions at Contrary Creek, VA (picture from USGS website) (3).

the copper content in ores. The natural oxidization from sulfide to sulfate is slow, but it could be catalyzed with the presence of iron-oxidizing bacteria (e.g., *T. ferrooxidans*), thus aiding in the solubilization of the metal (Reactions 5 and 6). Through these processes, the soluble copper sulfate can be recovered from ore (4).

$$Fe^{2+} + \tfrac{1}{4}O_2 + H^+ \rightarrow Fe^{3+} + \tfrac{1}{2}H_2O$$
$$\text{(by } T.\ ferrooxidans,\ \text{or } L.\ ferrooxidans) \quad (5)$$

$$CuS(\text{insoluble copper}) + 8Fe^{3+} + 4H_2O \rightarrow$$
$$Cu^{2+}(\text{soluble copper}) + 8Fe^{2+} + SO_4^{2-} + 8H^+ \quad (6)$$

IRON-REDUCING BACTERIA IN BIOREMEDIATION AND ENERGY PRODUCTION

Besides aerobic iron-oxidizing bacteria (normally referred to as iron bacteria), another type of bacteria exists: anaerobic iron-reducing bacteria, which use ferric iron as an electron acceptor for energy metabolism. A wide variety of both chemoorganotrophic and chemolithotrophic bacteria can have this feature. Ferric reduction can be coupled to the oxidation of several organic and inorganic electron donors. Various organic compounds, including aromatic compounds, can be oxidized anaerobically by iron-reducers. Much research on the energetics of ferric iron reduction has been done in the gram-negative bacterium: *Shewanella putrefaciens*, in which ferric-dependent anaerobic growth occurs with various organic electron donors. Other important ferric reducers include *Geobacter*, *Geospirillum*, and *Geovibrio*.

Geobacter metallireducens has been a model for study of the physiology of Fe^{3+} reduction. This organism can oxidize acetate with Fe^{3+} as an acceptor as follows:

$$\text{Acetate} + 8Fe^{3+} + 4H_2O \rightarrow 2HCO_3^- + 8Fe^{2+} + 9H^+$$

Geobacter can also use H_2 or other organic electron donors including aromatic hydrocarbon toluene as the electron donor, which may be environmentally significant because toluene from accidental spills or leakage from hydrocarbon storage tanks often contaminates ferric-rich aquifers, and it has been suggested that organisms like *Geobacter* may be natural cleanup agents in such environments (4).

Another distinct feature of the *Geobacter* species is that it has the ability to directly transfer electrons onto the surface of electrodes, which has made it possible to design novel microbial fuel cells that can efficiently convert waste organic matter to electricity. It has been found that dissimilatory metal-reducing microorganisms, such as *Geobacter* and *Rhodoferax* species (5,6), have the novel ability to directly transfer electrons throughout the cell membrane onto the surface of electrodes, which has led to the construction of microbial fuel cells that are more efficient than are traditional microbial fuel cells in that previously described microbial fuel cells depended on the toxic electron shuttling mediator (e.g., thionine, benzylviologen, phenazines, and phenoxazines) (7,8) to transfer electrons from cell membrane to electrode. As *Geobacter* can transfer electrons onto the electrodes by themselves, no electron shuttling mediator is needed, which can reduce the toxicity to bacterium and make microbial fell cells more cost-effective (Fig. 5). In addition, because *Geobacter* can biodegrade many contaminants, it is possible to harvest electricity from many types of waste organic matter or renewable biomass (5).

SOLUTION FOR BIOFOULING AND WATER COLOR CAUSED BY IRON BACTERIA

Because of the expensive cost incurred by biofouling, corrosion, and red water in the water supply system, efficient approaches should be employed to prevent and control the growth of iron bacteria in water supply and groundwater systems. Currently, chemical and mechanical methods for treating iron bacteria problems are available.

Chemical Treatment

Because it is difficult to get rid of iron bacteria once they exist in well systems, prevention is the best safeguard against accompanying problems. For well drillers, prevention means disinfecting everything that goes into the ground with a strong chlorine solution (250 ppm). As iron bacteria are nourished with the presence of carbon and other organics, it is essential

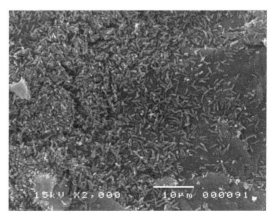

Figure 5. *Geobacter* colonizing a graphite electrode surface in a microbial fuel cell (9).

that these are not introduced into any part of the well system during the drilling process. Also, temperature is normally low in groundwater drilling systems, which leads to the much slower chemical reactions in wells. Therefore, bacterial cells need a long exposure to the chemical for the treatment to be effective.

Mechanical Treatment

In addition to the chemical treatment, mechanical approaches are available to control the iron bacteria in community water systems. Stagnant water conditions can be avoided by looping dead-end plumbing lines and periodically flushing low-flow lines to reduce bacteria. Forcing hot water or steam into a well to disperse the slime and kill the bacteria has also worked well. In addition, flushing large quantities of heated water into the aquifer has been found successful in field tests.

BIBLIOGRAPHY

1. Ehrlich, H.L. (1990). *Geomicrobiology*, 2nd Edn. Marcel Dekker, New York.
2. http://soils1.cses.vt.edu/ch/biol_4684/Microbes/Leptothrix.html.
3. http://pubs.usgs.gov/publications/text/Norrieviewing.html.
4. Madican, M.T., Martinko, J.M., and Parker, J. (2003). *Brock Biology of Microorganisms*. Person Education, Upper Saddle River, NJ, pp. 588 and 671.
5. Bond, D.R., Holmes, D.E., Tender, L.M., and Lovley, D.R. (2002). Electrode-reducing microorganisms that harvest energy from marine sediments, *Science* **295**: 483–485.
6. Chaudhuri, S.K. and Lovley, D.R. (2003). Electricity generation by direct oxidation of glucose in mediatorless microbial fuel cells. *Nature Biotechnol.* **21**(10): 1229–1232.
7. Roller, S.D. et al. (1984). *J. Chem. Technol. Biotechnol.* **34B**: 3.
8. Kim N., Choi, Y., Jung, S., and Kim, S. (2000). *Biotechnol. Bioeng.* **70**: 109.
9. Bond, D.R. and Lovley, D.R. (2003). Electricity production by *Geobacter sulfurreducens* attached to electrodes. *Appl. Environ. Microbiol.* **69**(3): 1548–1555.

CARTRIDGE FILTERS FOR IRON REMOVAL

D.V. Vayenas
G. Lyberatos
University of Ioannina
Agrinio, Greece

INTRODUCTION

The presence of iron in a water supply at concentrations that exceed the secondary drinking water standards of 0.2 mg/L (1) is undesirable for the following reasons (2): (1) iron precipitates give water a reddish color when exposed to air, (2) iron gives water an undesirable metallic taste, (3) deposition of iron in the distribution system can reduce the effective pipe diameter and eventually clog a pipe, and (4) iron is a substrate for bacterial growth in water mains. When iron bacteria die and slough off, bad odors and unpleasant tastes may result. Consequently, it is very important to remove iron from surface water before it may be used. Iron may be removed from water through a physical-chemical method, a biological method, or a combination of the two.

The oxidation state of iron may be either +2 (ferrous) or +3 (ferric). Surface water iron is generally found in its precipitated ferric form and does not require any further treatment to oxidize it. Groundwater iron is in its ferrous state in the deeper layers of some water reserves that lack oxygen, or in groundwater, in which case it is in a reduced dissolved form (Fe^{2+}) and often chelated. The oxidation state of iron in water depends on the pH and the redox potential. A dissolved form of iron (Fe^{2+} or $FeOH^-$) can be transformed to a precipitated form [$FeCO_3$, $Fe(OH)_2$, or $Fe(OH)_3$] by increasing either the oxidation potential, the pH, or both (3).

Aeration followed by solid–liquid separation is the most commonly used physical-chemical treatment method for iron removal. Aeration is generally recommended for oxidizing ferrous iron in water with high iron concentrations (>5 mg/L), so that costs for chemicals can be avoided (4). Sedimentation and/or filtration (sand or dual-media filtration) are used for the solid–liquid separation. If necessary, other treatments may be added, such as pH correction (lime softening), chemical oxidation by adding an oxidizing agent (such as chlorine, potassium permanganate, or ozone), ion oxidation, settling, and biological oxidation. Unfortunately, several problems are associated with physical-chemical treatments for iron removal that lead to relatively low iron removal rates. Substances such as humic acids, silicates, phosphates, and polyphosphates inhibit oxidation, precipitation, and filtration of ferric hydroxide. Thus, adding reagents is often necessary for pH correction and flocculation. The iron is not retained in a compact form, thus leading to low removal capacity, low filtration rates, large amounts of wash water, and difficult sludge treatment. The effective size of the filter medium can be between 0.5 and 1 mm and filtration rates between 5 and 15 m/h.

The rates of ferrous iron oxidation by air increase with pH; about 90% conversion may be achieved in 10–20 min at pH 7 (5). Iron oxidation is substantially slower at a pH below 6 and reduced forms may persist for some time in aerated waters (6). The rates of ferrous iron oxidation may be increased by the action of microorganisms.

Several genera of bacteria oxidize dissolved iron by different mechanisms. *Gallionella* sp., *Leptothrix ochracea*, and *Crenothrix polyspora* cause intracellular oxidation by enzymatic action; extracellular oxidation is caused by the catalytic action of polymers excreted by the filaments comprising the stalks produced by *Gallionella* sp.; and the sheaths from *Leptothrix* sp., *Crenothrix* sp., *Clonothrix* sp., and *Sphaerotilus* sp. Secondary extracellular oxidation is also caused by the catalytic action of polymers, excreted by the microscopic filaments of capillary appearance that are abandoned by the gliding trichomes of *Toxothrix thrichogenes* and extracellular polymers that are excreted by various Siderocapsaceae (7).

Biological iron removal systems have the following advantages compared with physicochemical iron removal systems: they require smaller installations because higher filtration rates may be used and/or because aeration and filtration can take place simultaneously in the same vessel; they allow longer lasting filtration runs because the retention time of iron in the filter increases due to the formation of denser precipitates and the use of coarser media; they generate denser backwash sludge that is easier to thicken and dewater; they lead to higher net production of clarified water because less water is required for backwashing, though it is possible to use raw water for backwash; they require no added chemicals; they do not deteriorate water quality over time; and they require lower capital and operating outlays (8).

The metabolic activities of iron bacteria are not fully understood, but it is believed that the same oxidation of iron is carried out by some variation in the physical-chemical reaction,

$$4Fe^{2+} + O_2 + 10H_2O \rightarrow 4Fe(OH)_3 + 8H^+ + energy$$

Whatever the metabolic pathway for the iron oxidation reactions, the biological process is catalytic and causes rapid oxidation. The red insoluble precipitates are all slightly hydrated iron oxides that, beneficially, are forms more compact than the precipitates formed when using physical-chemical processes. This feature partially explains the greater iron retention capacity between backwashes of biological filters compared to physical-chemical treatment filters.

It is important to recognize that iron bacteria catalyze oxidation reactions under conditions of pH and E_h that are intermediate between those of natural groundwater and those required for conventional treatment. The range of conditions allowing biological iron oxidation thus straddles the theoretical boundaries between the fields of Fe^{2+} and Fe^{3+} stability expected strictly by chemical thermodynamics. This can be visualized in a stability diagram, using pH and E_h as ordinates, as shown in Fig. 1.

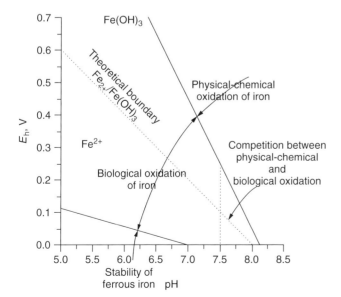

Figure 1. Stability diagram for biological iron removal.

By aerating the raw water and raising the dissolved oxygen concentration, the raw water's rH can be increased into the pH and E_h range where biological oxidation can take place. By controlling the quantity of supplied air, the rH can be adjusted to a level that is ideal for iron removal.

Once a biological iron removal plant is constructed, the system must be given time to "seed" with bacteria naturally present in the water source. This seeded biomass is naturally and continuously regenerated during the life of the plant and is periodically removed partially by backwashing. For iron removal plants, the seeding period is from 1 day to 1 week. At the end of the seeding period, the metal concentration in the effluent falls to near detection levels. The seeding period is affected by the temperature of the water source and is generally longer for cold water sources.

Iron bacteria are generally robust, and because of the variety of species involved, one type or another can thrive under most environmental conditions. Given the correct pH, from 6–8, and E_h, the bacteria can normally oxidize iron at temperatures ranging from 5 to 50 °C. Inhibition of the biological process can, however, be caused by H_2S, chlorine, NH_4^+, and some heavy metals normally not present in water sources.

AMMONIA AND MANGANESE INHIBITION OF BIOLOGICAL IRON REMOVAL

Water sources that contain iron often also contain ammonia and manganese. Ammonia, iron, and manganese may be removed chemically or biologically from a water supply. Biological removal of these pollutants is preferable because there is no need to add extra chemicals and the volume of sludge generated is appreciably smaller and hence easier to handle (9). Simultaneous biological removal of the three elements is very difficult mainly due to the different redox potential values needed to oxidize them. Mouchet (7) summarized the methods for iron and manganese removal from potable water and concluded that due to the different values required for dissolved oxygen concentration, redox potential, and pH, the basic design of plants for iron and manganese removal is substantially different. He also concluded that it is generally not possible to achieve simultaneous removal of Fe^{2+} and Mn^{2+} in a single filter, except in some cases of extremely low filtration rates. Gouzinis et al. (10) studied the simultaneous removal of ammonia, iron, and manganese from potable water using a pilot-scale trickling filter and the influences these pollutants have on filter performance and efficiency. They concluded that iron has a strong negative effect on ammonia removal and ammonia has a very low impact on iron removal. They also concluded that iron has a strong effect on Mn removal and manganese seems to have a much lower effect on iron removal. The main conclusion of that work was that iron should be removed before ammonia and manganese oxidation. For low iron concentration, oxidation by extended aeration could reduce iron at very low levels, and subsequent biological ammonia and manganese removal could be carried out without any problems. For higher iron concentration, an initial biological iron oxidation stage should be necessary.

FILTERS FOR IRON REMOVAL

A key feature of biological iron removal lies in the high rate of iron oxidation. Thus, the engineering systems that have been developed for biological iron removal can be classified by the type of water flow and by aeration into pressurized filters, gravity-flow units (9), and trickling filters.

Pressurized Filters

In a pressurized unit, the main components of the plant (Fig. 2) are the reactor, which is a filter specially designed to operate at high rates and contains sand somewhat coarser than normal for conventional filters; a calibrated raw water aeration system where oxygen is supplied either by an air injector fitted with a rotameter (to adjust the air flow rate, thereby ensuring that E_h and dissolved oxygen concentrations in the preaerated raw water are conducive to optimum growth of iron bacteria) or by recirculating a portion of the treated water, aerated near saturation (this configuration is selected when the dissolved oxygen content at the filtration inlet must be very low and tightly controlled); and an optional aeration system for filter effluent, followed by an optional nonchlorinated washwater storage tank and a storage tank for chlorinated water (7).

Filters can be backwashed by using raw water (in which case, the need for a washwater storage tank is eliminated) or treated water, always beginning with a simultaneous air scour. Aeration systems are simple baffle mixers. Filtration rates can in certain cases attain 40 or even 50 m/h.

Gravity-Flow Units

Gravity-flow units contain an atmospheric pressure aeration system followed by an open or closed filter (Fig. 3). Filtration can also be carried out by gravity flow in these units. The atmospheric pressure aeration system may be cascade aeration (designed to supply water that features optimum oxidation conditions for biological iron removal, as determined by the raw water pH) or spray aeration. These types of open aeration systems are particularly recommended when the raw water contains H_2S. Gravity filtration is usually designed for plants of high capacity.

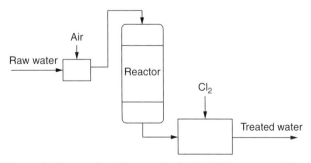

Figure 2. Process flow diagram for biological iron removal in a pressurized filter.

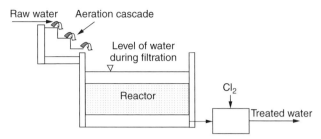

Figure 3. Process flow diagram for biological iron removal in a gravity plant.

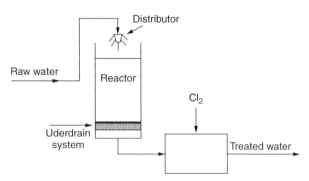

Figure 4. Process flow diagram for biological iron removal in a trickling filter.

Filtration rates can in certain cases attain 20–23 m/h. The filter medium is usually sand of an effective size between 1.0 and 1.5 mm (7).

Trickling Filters

The modern trickling filter consists of a bed of a highly permeable medium to which microorganisms are attached and through which water is percolated or trickled (Fig. 4). The filter media usually consist of either rock or a variety of plastic packing. Filters are constructed with an underdrain system for collecting the treated water and any biological solids that have become detached from the media. This underdrain system is important both as a collection unit and as a porous structure through which air can circulate. The collected liquid is passed to a settling tank where the solids are separated from the treated water (11).

An adequate flow of air is of fundamental importance to the successful operation of a trickling filter. The principal factors responsible for air flow in an open top filter are natural draft and wind forces. In natural draft, the driving force for airflow is the temperature difference between the ambient air and the air inside the pores. If the water is colder than the ambient air, the pore air will be cold and the direction of flow will be downward. If the ambient air is colder than the water, the flow will be upward. The latter is less desirable for mass transfer because the partial pressure of oxygen (and thus the oxygen-transfer rate) is lowest in the region of highest oxygen demand. Thus, a trickling filter has the advantage that it does not require an external air supply or an aeration system. If, in addition, the mean diameter of the filter media is sufficiently small (up to 5 mm), then complete aeration and very good filtration may be effected at the same time (12). Filtration rates are intermediate between those of pressurized filters and gravity-flow plants.

ADVANTAGES AND DISADVANTAGES OF FILTERS FOR IRON REMOVAL

Pressurized Filters

These filters are specially designed to operate at high rates and use particles of small mean diameter as support material, thus increasing the specific surface area available for ferrous iron oxidation. As a result, this type of filters exhibits high iron removal rates. On the other hand, the small particles lead to small pores, which may be clogged quite easily due to bacterial growth and iron precipitate filtration. Consequently, frequent backwashing is necessary. Finally, an external air supply is necessary, leading to increased operating cost.

Gravity Units

These filters are designed for plants where large quantities of drinking water have to be treated. The flow rates are quite low compared to the flow rates in pressurized systems. The support material is usually fine sand, which enables very good filtration but also needs frequent backwashing. The use of cascade aeration keeps the operating cost low and makes the system ideal for H_2S removal from drinking water. The capital cost of these systems is relatively high.

Trickling Filters

Trickling filters are quite simple constructions and do not require an external air supply or an aeration system, leading thus to low capital and operating costs. The support material is larger than that for pressurized systems to secure adequate air circulation. Filtration also takes place simultaneously with oxidation but a separate settling tank may be needed to ensure complete removal of iron precipitates. The rates in these filters lie between the high rates of pressurized systems and the low rates of gravity plants. Figure 5 presents typical iron concentration profiles along the depth of a pilot-scale trickling filter.

Technology offers various solutions for iron removal from drinking water. Cartridge filters combine physical-chemical and biological iron removal. The type of filter that is used depends on the physicochemical characteristics of the water, iron concentration, volumetric flow rate, and, of course, the costs. Thus, a feasibility study is necessary before deciding on the type of cartidge filter that will be used.

BIBLIOGRAPHY

1. *EEC-Official Journal of the European Communities*, No 80/779, August 30, 1980.
2. Kontari, N. (1988). Groundwater, iron and manganese: an unwelcome trio. *Water/Eng. Manage.* 25–26.
3. Hem, J.D. (1961). Stability field diagrams as aids in iron chemistry studies. *J. Am. Water Works Assoc.* **53**(2): 211.
4. Wong, J.M. (1984). Chlorination-filtration for iron and manganese removal. *J. Am. Water Works Assoc.* **76**: 76–79.

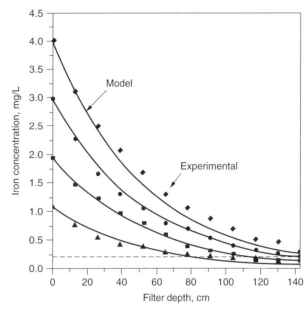

Figure 5. Typical iron concentrations profiles along the depth of a pilot-scale trickling filter (13).

5. Sundstrom and Klei (1979). *Wastewater Treatment*. Prentice-Hall, Englewood Cliffs, NJ.
6. Sawyer, C.N. and McCarty, P.L. (1978). *Chemistry for Environmental Engineering*, 3rd Edn. McGraw-Hill, Singapore.
7. Mouchet, P. (1992). From conventional to biological removal of iron and manganese in France. *J. Am. Water Works Assoc.* **84**(4): 158–166.
8. Gage, B., O'Dowd, D.H., and Williams, P. (2001). Biological iron and manganese removal, pilot and full scale applications. *3rd Water Works Assoc. Conf.*, Ontario.
9. Degremont, (1991). *Water Treatment Handbook*, 6th Edn. Lavoisier, Paris.
10. Gouzinis, A., Kosmidis, N., Vayenas, D.V., and Lyberatos, G. (1998). Removal of Mn and simultaneous removal of NH_3, Fe and Mn from potable water using a trickling filter. *Water Res.* **32**(8): 2442–2450.
11. Metcalf & Eddy. (1991). *Wastewater Engineering, Treatment, Disposal, Reuse*, 3rd Edn. McGraw-Hill, New York.
12. Dimitrakos Michalakos, G., Martinez Nieva, J., Vayenas, D.V., and Lyberatos, G. (1997). Removal of iron from potable water using a trickling filter. *Water Res.* **31**(5): 991–996.
13. Tekerlekopoulou, A.G., Vasiliadou, I., Vakianis, I., and Vayenas, D.V. (2004). Physicochemical and biological iron removal from potable water. *Water Res.* (submitted).

IRRIGATION WATER QUALITY IN AREAS ADJOINING RIVER YAMUNA AT DELHI, INDIA

CHAKRESH K. JAIN
MUKESH K. SHARMA
National Institute of Hydrology
Roorkee, India

The groundwater quality of areas adjoining the River Yamuna in Delhi (India) has been assessed for the suitability of groundwater for irrigation. Thirty-eight groundwater samples from shallow and deep aquifers were collected each during pre- and postmonsoon seasons in the year 2000 and analyzed for various parameters. The suitability of groundwater for irrigation has been evaluated based on salinity, sodium adsorption ratio (SAR), residual sodium carbonate (RSC), and boron content. According to the U.S. Salinity Laboratory classification of irrigation water, more than 50% of the samples fall in water type C3-S1 (high salinity and low SAR). Such water cannot be used on soils that have restricted drainage. Even with adequate drainage, special management for salinity control may be required and plants with good tolerance should be selected. About 30% of the samples fall in water type C3-S2 (high salinity and medium SAR). Such water will induce an appreciable sodium hazard in finely textured soils that have good cation exchange capacity, especially under low leaching conditions.

INTRODUCTION

Groundwater plays an important role in agriculture for both watering crops and for irrigating dry season crops. It is estimated that about 45% of the irrigation water required is from groundwater sources. Unfortunately, the availability of groundwater is not unlimited, nor is the groundwater protected from deterioration. In most instances, the extraction of excessive quantities of groundwater has resulted in drying up of wells, damaged ecosystems, land subsidence, saltwater intrusion, and depletion of the resource. Groundwater quality is being increasingly threatened by agricultural, urban, and industrial wastes. It has been estimated that once pollution enters the subsurface environment, it may remain concealed for many years, disperses over wide areas of the groundwater aquifer, and renders groundwater unsuitable for consumption and other uses. The rate of depletion of groundwater levels and the deterioration of groundwater quality are of immediate concern in major cities and towns of the country.

The National Capital Territory (NCT) of Delhi is facing severe problems in managing groundwater quality and quantity (1). Surface waterbodies play a significant role in groundwater flow. The hydraulic gradient plays a significant role in the lateral and vertical migration of contaminants in groundwater aquifers. Therefore, the present study has been carried out to assess the suitability of groundwater for irrigation in areas adjoining the River Yamuna at Delhi.

STUDY AREA

Delhi generates about 1900 MLD of sewage against installed capacity of 1270 MLD of sewage treatment. The balance of untreated sewage along with a significant quantity of partially treated sewage is discharged into the River Yamuna every day. The river receives sewage and industrial wastes through sixteen drains, which join River Yamuna between Wazirabad and Okhla. Thus Delhi is the largest contributor of pollution to the River Yamuna, which receives almost 80% of its pollution load through these drains.

The climate of Delhi is influenced mainly by its inland position and the prevalence of the continental type air during the major part of the year. Extreme dryness with an intensely hot summer and cold winter are the characteristics of the climate. During the monsoon months, air of oceanic origin penetrates to this area and causes increased humidity, cloudiness, and precipitation. The normal annual rainfall in the National Capital Territory of Delhi is 611.8 mm. The rainfall increases from southwest to northeast; about 81% of the annual rainfall is received during the three monsoon months of July, August, and September; the balance of the annual rainfall is received as winter rains and as thunderstorm rain during pre- and postmonsoon months.

METHODOLOGY

Groundwater samples were collected in polyethylene bottles during the pre- and postmonsoon seasons during the year 2000 from areas adjoining the River Yamuna at Delhi and were preserved by adding an appropriate reagent (2,3). All samples were stored in sampling kits maintained at 4 °C and brought to the laboratory for detailed physicochemical analysis. The details of sampling locations are given in Table 1. The physicochemical analyses were performed following standard methods (2,3).

RESULTS AND DISCUSSION

The hydrochemical data for the two sets of samples collected from the areas adjoining the River Yamuna at Delhi (Fig. 1) during the pre- and postmonsoon seasons are given in Table 2. The pH of the groundwater of areas adjoining the River Yamuna are mostly within the range of 6.7 to 8.3 during the premonsoon season and 6.6 to 8.2 during the postmonsoon season; most of the samples point toward the alkaline range in both seasons. The total dissolved solids (TDSs) in the groundwater vary from 402 to 2266 mg/L during the premonsoon season and from 397 to 2080 mg/L during the postmonsoon season. Higher values of TDSs in areas nearby the River Yamuna indicate high mineralization of the groundwater.

Alkalinity varies from 116 to 380 mg/L during the premonsoon season and from 106 to 310 mg/L during the postmonsoon season. From the view point of hardness, only one sample exceeds the maximum permissible limit of 600 mg/L. The concentration of chloride was within the permissible limit. The sulfate content limit in groundwater is exceeded at two locations (Himmat Puri and Golf Course). The nitrate content in drinking water is considered important because of its adverse health effects. The nitrate content in the groundwater of areas adjoining the River Yamuna varies from 1 to 286 mg/L during the premonsoon season. The higher level of nitrate at certain locations may be attributed to surface disposal of domestic sewage and runoff from agricultural fields. It has also been observed that the groundwater samples collected from hand pumps at various depths have a high nitrate

Table 1. Description of Groundwater Sampling Locations

Sample No.	Label*	Location	Type of Well (Depth in ft)[a]
1	1	Bhagwanpur Khera	HP (25)
2	2	Loni Road	BW (20)
3	3	Kabul Nagar	OW (35)
4	4	Naveen Shahdara	HP (20)
5	5	Seelampur	HP (30)
6	6	Shastri Park	HP (25)
7	7	Lakshmi Nagar	HP (100)
8	8	Prit Vihar	TW (40)
9	9	Shankar Vihar	HP (200)
10	A	Pratap Nagar	TW (100)
11	B	Himmatpuri	HP (80)
12	C	Civil Lines	TW (250)
13	D	Rajpur road	OW (25)
14	E	Malka Gunj	HP (20)
15	F	Tripolia	BW (100)
16	G	Gulabi Bagh	BW (40)
17	H	Gulabi Bagh	BW (160)
18	I	Shastri Nagar	BW (100)
19	J	Shastri Nagar	BW (60)
20	K	Lekhu Nagar	HP (40)
21	L	Ram Pura	HP (40)
22	M	Punjabi Bagh West	HP (40)
23	N	Rajghat	BW (50)
24	P	JLN Marg	HP (60)
25	Q	GB Pant Hospital	BW (20)
26	R	Panchkuin Marg	TW (450)
27	S	Panchkuin Marg	HP (60)
28	T	Rajendra Nagar	BW (100)
29	U	Rajendra Nagar	BW (300)
30	V	Shankar Road	OW (20)
31	W	IARI, Pusa	TW (120)
32	X	Zoological Park	HP (60)
33	Y	Golf Course	BW (80)
34	Z	Rabindra Nagar	TW (150)
35	a	Teen Murti Chowk	HP (60)
36	b	Malcha Marg	BW (100)
37	c	Sardar Patel Road	TW (150)
38	d	Janpath	BW (250)

[a]HP: hand pump; OW: open well; BW: bore well; TW: tube well.

content, which may be attributed to wellhead pollution. The fluoride content was within the maximum permissible limit of 1.5 mg/L in all samples.

WATER QUALITY EVALUATION FOR IRRIGATION

Irrigation water quality refers to its suitability for agricultural use. The concentration and composition of dissolved constituents in water determine its quality for irrigation. The quality of water is an important consideration in any appraisal of salinity or alkali conditions in an irrigated area. Good quality water can result in maximum yield under good soil and water management practices. The most important characteristics of water that determine the suitability of groundwater for irrigation are as follows:

1. Salinity
2. Relative proportion of sodium to other cations
3. Residual sodium carbonate
4. Boron content

The safe limits of electrical conductivity for crops of different degrees of salt tolerance under varying soil textures and drainage conditions are given in Table 3. The quality of water is commonly expressed by classes of relative suitability for irrigation with reference to salinity levels. The recommended classification with respect to electrical conductivity, sodium content, sodium absorption ratio (SAR), and residual sodium carbonate (RSC) are given in Table 4. The values of SAR, % Na, and RSC for groundwater samples collected from areas adjoining the River Yamuna are given in Table 5.

Salinity

Salinity is broadly related to total dissolved solids (TDSs) and electrical conductivity (EC). A high concentration of TDSs and electrical conductivity in irrigation water may increase the soil salinity, which affect the salt intake of the plant. The salts in water affect plant growth directly and also affect the soil structure, permeability, and aeration, which indirectly affect plant growth. Soil water passes into the plant through the root zone by osmotic pressure. As the dissolved solid content of the soil water in the root zone increases, it is difficult for the plant to overcome the osmotic pressure and the plant root membranes are unable to assimilate water and nutrients. Thus, the dissolved solids content of the residual water in the root zone also has to be maintained within limits by proper leaching. These effects are visible in plants by stunted growth, low yield, discoloration, and even leaf burns at the margin or top.

Relative Proportion of Sodium to Other Cations

A high salt concentration in water leads to the formation of saline soil, and high sodium leads to the development of an alkaline soil. The sodium or alkali hazard in the use of water for irrigation is determined by the absolute and relative concentration of cations and is expressed in terms of the sodium adsorption ratio (SAR). If the proportion of sodium is high, the alkali hazard is high; conversely, if calcium and magnesium predominate, the hazard is lower. There is a significant relationship between SAR values of irrigation water and the extent to which sodium is absorbed by soil. If water used for irrigation is high in sodium and low in calcium, the cation-exchange complex may become saturated with sodium. This can destroy the soil structure owing to dispersion of clay particles. A simple method of evaluating the danger of high-sodium water is the sodium adsorption ratio (SAR) (4):

$$\text{SAR} = \frac{\text{Na}^+}{\sqrt{(\text{Ca}^{2+} + \text{Mg}^{2+})/2}} \quad (1)$$

The percentage sodium is calculated as

$$\%\text{Na} = \frac{\text{Na}^+ + \text{K}^+}{\text{Ca}^{2+} + \text{Mg}^{2+} + \text{Na}^+ + \text{K}^+} \times 100 \quad (2)$$

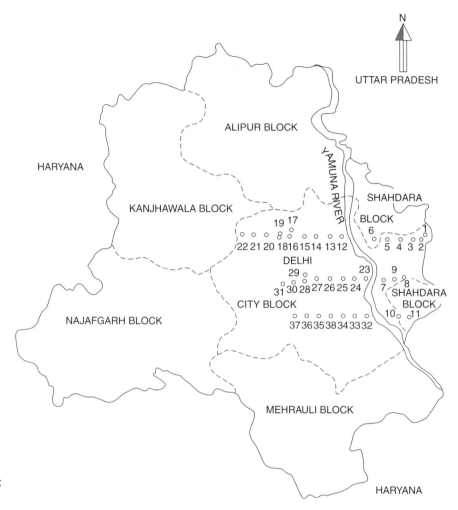

Figure 1. Study area showing sampling locations.

where all ionic concentrations are expressed in milliequivalents per liter.

The values of SAR and percentage sodium of the groundwater samples collected from the area adjoining the River Yamuna are given in Table 5. Calculation of SAR for a given water provides a useful index of the sodium hazard of that water for soils and crops. A low SAR (2 to 10) indicates little danger from sodium; medium hazards are between 7 and 18, high hazards between 11 and 26, and very high hazards above that. The lower the ionic strength of the solution, the greater the sodium hazards for a given SAR (4).

Table 2. Hydrochemical Data on Groundwater Samples from Delhi (Pre- and Postmonsoon, 2000)

Parameter	Minimum[a]		Maximum[a]		Mean[a]	
pH	6.7	(6.6)	8.3	(8.2)	7.1	(7.1)
Conductivity, μS/cm	628	(620)	3540	(3250)	1463	(1370)
TDSs, mg/L	402	(397)	2266	(2080)	936	(879)
Alkalinity, mg/L	116	(106)	380	(310)	201	(194)
Hardness, mg/L	116	(114)	841	(792)	235	(225)
Chloride	17	(17)	400	(390)	86	(86)
Sulfate, mg/L	43	(48)	690	(680)	180	(177)
Nitrate, mg/L	1.0	(ND)	286	(287)	78	(68)
Phosphate, mg/L	0.07	(0.03)	0.28	(0.21)	0.16	(0.08)
Fluoride, mg/L	0.33	(ND)	1.31	(0.76)	0.80	(0.41)
Sodium, mg/L	55	(51)	340	(322)	160	(157)
Potassium, mg/L	4.2	(2.2)	121	(103)	26	(20)
Calcium, mg/L	25	(21)	240	(227)	59	(59)
Magnesium, mg/L	9.0	(7.0)	64	(56)	21	(19)

[a]Values in parentheses represent postmonsoon data.

Table 3. Safe Limits of Electrical Conductivity for Irrigation Water

Sample No.	Nature of Soil	Crop Grown	Upper Permissible Safe Limit of EC in Water, µS/cm
1	Deep black soil and alluvial soils having clay content of more than 30% that are fairly to moderately well drained	Semitolerant	1500
		Tolerant	2000
2	Textured soils having clay content of 20–30% that are well drained internally and have good surface drainage	Semitolerant	2000
		Tolerant	4000
3	Medium textured soils having clay of 10–20%, internally very well drained, and having good surface drainage	Semitolerant	4000
		Tolerant	6000
4	Light textured soils having clay of less than 10% that have excellent internal and surface drainage	Semitolerant	6000
		Tolerant	8000

The SAR values in the groundwater of the study area ranged from 1.70 to 12.06 during the premonsoon season and from 1.65 to 11.24 during postmonsoon. As evident from the SAR values, the groundwater of the study area falls in the category of low sodium hazard except for two samples (Civil Lines and Sardar Patel Road), which reveals that the groundwater in the study area is free from any sodium hazard. The percentage sodium in the study area varied from 30.4 to 83.7 during the premonsoon season and from 34.5 to 82.9 during postmonsoon. More than 50% of the samples in the study area exceeded the recommended value of 60% during both pre- and postmonsoon seasons.

Residual Sodium Carbonate

In addition to total dissolved solids, the relative abundance of sodium with respect to alkaline earths and boron and the quantity of bicarbonate and carbonate in excess of alkaline earths also influence the suitability of water for irrigation. This excess is denoted by residual sodium carbonate (RSC) and is determined by the following formula:

$$RSC = (HCO_3^- + CO_3^-) - (Ca^{2+} + Mg^{2+}) \quad (3)$$

where all ionic concentrations are expressed in equivalents per mole (epm). Groundwater containing high concentrations of carbonate and bicarbonate ions tends to precipitate calcium and magnesium as carbonate. As a result, the relative proportion of sodium increases and is fixed in the soil, thereby decreasing soil permeability. If the RSC exceeds 2.5 epm, the water is generally unsuitable for irrigation. Excessive RSC causes the soil structure to deteriorate because it restricts water and air movement through soil. If the value is between 1.25 and 2.5, the water is of marginal quality; values less than 1.25 epm indicate that the water is safe for irrigation. During the present study, the RSC values were mostly negative. The RSC values clearly indicate that the groundwater in the study area does not have a residual sodium carbonate hazard.

Boron

Boron is essential to the normal growth of all plants, but the concentration required is very small and if exceeded may cause injury. Plant species vary both in boron requirement and in tolerance to excess boron, so that concentrations necessary for the growth of plants having high boron requirements may be toxic for plants sensitive to boron. Though boron is an essential nutrient for plant growth, generally it becomes toxic beyond 2 ppm in irrigation water for most field crops. It does not affect the physical and chemical properties of the soil, but at high concentrations, it affects the metabolic activities of plants.

U.S. Salinity Laboratory Classification

The sodium concentration is an important criterion in irrigation water classification because sodium reacts with soil to create sodium hazards by replacing other cations. The extent of this replacement is estimated by the sodium adsorption ratio (SAR). A diagram for studying the suitability of groundwater for irrigation is based on the SAR and the electrical conductivity of water expressed in µS/cm.

The chemical analysis data of groundwater samples collected from the area adjoining the River Yamuna in Delhi has been processed per the U.S. Salinity Laboratory classification (Fig. 2), and the results have been summarized in Table 6. It is evident from the results that during the premonsoon season, only three samples (7.8%) fall into water type C2-S1 (medium salinity and low SAR). Such water can be used if a moderate amount of leaching occurs, and plants with moderate salt tolerance can be grown in most cases without special practices for salinity control. More than 50% of the

Table 4. Guidelines for Evaluating Irrigation Water Quality

Water Class	Sodium (Na), %	Electrical Conductivity, µS/cm	SAR	RSC
Excellent	<20	<250	<10	<1.25
Good	20–40	250–750	10–18	1.25–2.0
Medium	40–60	750–2250	18–26	2.0–2.5
Bad	60–80	2250–4000	>26	2.5–3.0
Very bad	>80	>4000	>26	>3.0

Table 5. Values of Sodium Adsorption Ratio (SAR), Percentage Sodium (%Na), and Residual Sodium Carbonate (RSC)

Sample No.	Sample Location	Type (Depth)	Premonsoon 2000			Postmonsoon 2000		
			SAR	%Na	RSC	SAR	%Na	RSC
1	Bhagwanpur Khera	HP (25)	4.93	65.1	0.24	5.35	69.8	1.52
2	Loni Road	BW (20)	3.81	56.4	0.27	4.10	58.4	−0.31
3	Kabul Nagar	OW (35)	2.99	55.6	−0.04	2.35	54.1	0.18
4	Naveen Shahdara	HP (20)	2.74	51.9	−0.30	2.39	49.2	−0.43
5	Seelampur	HP (30)	1.70	41.7	0.11	1.65	41.0	0.17
6	Shastri Park	HP (25)	3.20	67.9	1.44	2.87	66.9	1.38
7	Lakshmi Nagar	HP (100)	2.79	55.1	0.58	2.89	54.8	0.41
8	Prit Vihar	TW (40)	4.84	64.1	1.76	4.86	65.7	1.75
9	Shankar Vihar	HP (200)	5.49	71.2	1.31	5.93	72.7	1.92
10	Pratap Nagar	TW (100)	5.88	71.7	1.05	6.31	74.1	0.98
11	Himmatpuri	HP (80)	7.40	68.3	−2.75	7.00	68.6	−2.22
12	Civil Lines	TW (250)	12.06	83.7	1.30	11.24	82.9	1.24
13	Rajpur road	OW (25)	2.26	48.7	0.39	2.30	50.7	0.57
14	Malka Gunj	HP (20)	3.01	59.6	0.22	4.83	69.0	0.31
15	Tripolia	BW (100)	6.15	67.4	−1.74	6.05	67.1	−1.86
16	Gulabi Bagh	BW (40)	3.31	44.9	−6.06	3.99	48.9	−6.01
17	Gulabi Bagh	BW (160)	1.94	30.4	−6.71	2.33	34.5	−5.46
18	Shastri Nagar	BW (100)	3.58	64.0	0.40	3.37	61.0	−0.10
19	Shastri Nagar	BW (60)	2.48	47.4	−2.31	2.64	47.7	−1.79
20	Lekhu Nagar	HP (40)	2.99	59.0	0.82	2.64	50.3	0.00
21	Ram Pura	HP (40)	2.39	52.2	0.60	2.48	52.4	0.53
22	Punjabi Bagh West	HP (40)	4.31	59.0	−0.69	4.94	60.4	−1.46
23	Rajghat	BW (50)	4.42	61.2	−1.07	4.25	60.8	−1.22
24	JLN Marg	HP (60)	8.61	82.0	2.04	8.15	81.8	2.12
25	GB Pant Hospital	BW (20)	4.28	65.9	−0.08	4.11	64.3	−0.23
26	Panchkuin Marg	TW (450)	8.00	74.3	0.75	6.74	68.4	−0.91
27	Panchkuin Marg	HP (60)	4.15	62.7	0.08	3.81	58.9	−0.43
28	Rajendra Nagar	BW (100)	2.22	54.8	0.01	2.05	49.1	−0.59
29	Rajendra Nagar	BW (300)	2.13	49.4	−0.32	2.47	48.9	−1.16
30	Shankar Road	OW (20)	2.70	49.9	−1.66	4.32	60.5	−0.05
31	IARI, Pusa	TW (120)	5.44	61.1	−1.89	5.28	59.7	−1.63
32	Zoological Park	HP (60)	3.49	54.5	−1.34	3.25	57.1	−0.62
33	Golf Course	BW (80)	4.33	44.3	−12.9	4.32	45.2	−12.06
34	Rabindra Nagar	TW (150)	5.84	65.1	−1.54	5.89	65.0	−1.27
35	Teen Murti Chowk	HP (60)	7.23	74.1	0.52	6.98	72.5	0.82
36	Malcha Marg	BW (100)	9.23	75.3	0.34	9.16	76.0	0.58
37	Sardar Patel Road	TW (150)	10.71	79.6	1.59	10.99	81.3	1.75
38	Janpath	BW (250)	5.63	67.6	−0.39	5.94	70.2	−0.16

samples fall into water type C3-S1 (high salinity and low SAR). Such water cannot be used on soils with restricted drainage. Even with adequate drainage, special management for salinity control may be required, and plants with good salt tolerance should be selected. About 30% of the samples fall into water type C3-S2 (high salinity and medium SAR). Such water cannot be used on soils with restricted drainage. Even with adequate drainage, special management for salinity control may be required, and plants with good salt tolerance should be selected. The water will also present an appreciable sodium hazard in finely textured soils that have good cation exchange capacity, especially under low leaching conditions. Three samples (7.8%) fall into water type C3-S3 (high salinity and high SAR). Such water cannot be used on soils with restricted drainage. Even with adequate drainage, special management for salinity control may be required, and plants with good salt tolerance should be selected. The water will also produce harmful levels of exchangeable sodium in most soils and will require special soil management, good drainage, high leaching, and organic matter additions. One sample falls into water type C4-S2 (very high salinity and medium SAR). Such water is not suitable for irrigation. High salinity is harmful to plant growth and changes soil structure, permeability, and aeration, which in turn affect plant growth and yield considerably. An almost similar trend was observed during the postmonsoon season.

CONCLUSIONS

The quality of groundwater varies from place to place with the depth of the water table. It also shows significant variation from one season to another. The U.S. Salinity Laboratory classification for irrigation water indicates that more than 50% of the samples fall into water type C3-S1 (high salinity and low SAR). Such water cannot be used on soils with restricted drainage. Even with adequate drainage, special management for salinity control may be

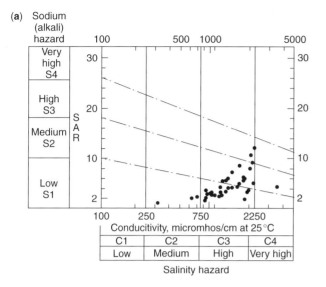

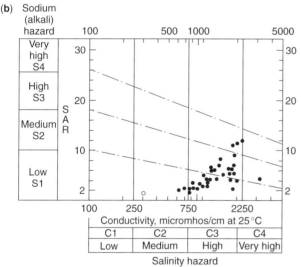

Figure 2. (a) U.S. Salinity Laboratory classification (premonsoon 2000). (b) U.S. Salinity Laboratory classification (postmonsoon 2000).

required, and plants with good salt tolerance should be selected. About 30% of the samples fall into water type C3-S2 (high salinity and medium SAR). Such water will also induce an appreciable sodium hazard in finely textured soils that have good cation exchange capacity, especially under low leaching conditions.

Table 6. Results of U.S. Salinity Laboratory Classification

	Sample Numbers	
Classification/Type	Premonsoon 2000	Postmonsoon 2000
C2-S1	21,28,29	3,21,28
C3-S1	2,3,4,5,6,7,13,14,15,16,17,18, 19,20,22,23,25, 27,30,32	2,4,5,6,7,8,13,14,16,17,18, 19,20,23,25,27,29,30,32
C3-S2	1,8,9,10,11,24,26,31,34, 35,38	1,9,10,11,15,22,24,26,31, 34,35,36,38
C3-S3	12,36,37	12,37
C4-S2	33	33

BIBLIOGRAPHY

1. CGWB and CPCB. (2000). *Status of Ground Water Quality and Pollution Aspects in NCT-Delhi*, January 2000.
2. APHA. (1992). *Standard Methods for the Examination of Water and Waste Waters*, 18th Edn. American Public Health Association, Washington, DC.
3. Jain, C.K. and Bhatia, K.K.S. (1988). *Physico-chemical Analysis of Water and Wastewater, User's Manual, UM-26*. National Institute of Hydrology, Roorkee.
4. Richards, L.A. (Ed.). (1954). *Diagnosis and Improvement of Saline and Alkali Soils, Agricultural Handbook 60*. U.S. Department of Agriculture, Washington, DC.

WATER SAMPLING AND LABORATORY SAFETY

TIMOTHY J. RYAN
Ohio University
Athens, Ohio

Water sampling is done for a wide variety of reasons, including water quality, thermal effluent assessments, point and nonpoint industrial source impacts, microbial contamination from human wastes, and accidental releases. For example, samples can be taken from rivers, lakes, streams, and oceans to determine the level of a specific chemical component, such as nitrates or sulfates. Or tests may be done to investigate the biological oxygen demand (BOD) of the water, as an indicator of the overall health of the water body. Water is routinely sampled for the presence of harmful pathogens, which potentially pose a risk not only to the end user but to the sampling personnel as well. Because sampling may place the person in unsafe locations or even perilous environments, and because sample collection represents a significant allocation of time and resources, safeguards for personnel and samples alike are important.

The first decision in water sampling methodology is the categorical type of sample to be collected: general survey or site-specific. For general survey sampling, locations and analyses should be selected that are representative of the water source, treatment plant, storage facilities, distribution network, or points of use to be characterized. At a minimum, duplicate samples should be collected at each site. To accurately assess large or widely distributed bodies of water, consideration should be given to the use of a statistically sound sampling algorithm.

A site-specific sampling regimen is inherently simpler in that the sampling location and analysis parameters are known, easily anticipated, or even proscribed by a regulatory agency.

Actual procedures for sampling and analysis frequently depend on the type of analysis to be conducted. As a general truism, the following items are essential to conduct an effective sampling trip: labeled bottles appropriate for the analysis to be run, field equipment for commonly determined parameters (pH, water temperature, etc.), and sampling collection tools. Additional useful and typical equipment might include preservatives, ice packs and coolers, personal gear suitable for all weather conditions and intense insect exposure, a camera or video equipment (if needed to document the actual sampling operation), and laboratory chain-of-custody and analysis requisition forms. All sampling equipment and material should be checked before departing on the collection campaign, and real-time instruments, such as dissolved oxygen spectrophotometers should be calibrated, have their batteries fully charged, and be packed securely to avoid damage. Of all items taken on a sampling trip, the field log book is perhaps the most important. Although it will slow the sampling mechanics, all information pertinent to the sampling trip should be fastidiously recorded in it. Special bound books with water-resilient paper and bindings are available for such use, and they are well worth their added cost because of the increased information security they provide.

Immediately before sample collection, the sampling site should be adequately described in the log book. The use of global positioning satellite (GPS) transponders for precise location identification is always a sound practice and is in fact critical if repeated sampling in the same location on different dates is a possibility. Still-frame or video photographic documentation may be helpful in this regard in that readily identifiable landmarks can easily be recorded and communicated to third parties at a later date. The log book should also document the specific sampling collection information, such as time of day, prevailing wind direction, and weather conditions. Water-related aspects that may be important to the interpretation of specific sampling results include ambient temperature, water temperature, dissolved oxygen, pH, and salinity. Any unusual aspects about the site or water being sampled, such as oil slicks, excessive algae growth, or unusual color or odor, should be noted in the log, along with both typical or atypical human activities in the sampling vicinity. Such activities might include logging, commercial fishing with nets, dredging, industrial permitted point sources, construction work, aerial agricultural spraying, and utility company line clearing or defoliant operations.

The physical security and integrity of water samples is essential to ensuring the accuracy and legal defensibility of their analysis results. Toward this end, there are several things that one can to do assure and maintain sample quality in the field. First and foremost, one must take all necessary precautions to protect samples from contamination and deterioration. Maintaining the sterilized state of the bottle in which biological samples are to be collected is obviously important. Any specific sampling instructions from the analysis laboratory for handling sample containers must be closely followed. After samples are collected, they are normally kept refrigerated and care must be exercised that this does not result in the sample actually freezing. Unless specifically instructed otherwise, all samples should be place in a dark place away from other contaminant sources while awaiting transport or analysis. The person conducting the sampling should have clean hands and abstain from eating, gum chewing, or smoking while working with sampling media or other equipment that could be cross contaminated by such habits. To prevent sample or site confusion, bottles for particular sample types (e.g., volatile organics, chloride, total organic halides) can be color coded for more confident identification.

Sample collection can take place from a variety of locations, including from shore, a boat, access from a bridge, adjacent to the effluent stream, and from receiving waters. Different procedures and safety precautions should be taken with each situation. In general, it is best to take samples from mid-stream rather than near the shore to reduce the possibility of local shore effects contaminating the sample. It bears reiteration to note in the sampling log book the exact spot of sampling, so that if further sampling is needed (e.g., in response to a contested sample result), the identical location can be found for further rounds of testing. Water samples can also be taken in the home or business, such as from a water faucet, well head, or septic system. During these samplings, the same precautions need to be taken concerning the physical security and integrity of the sample. As for outdoor samples, notes should be kept in a log book about pertinent aspects of the water such as location in the building where the sample was collected, temperature of the water, time of day, and overall appearance of the water (e.g., turbid, milky, clear).

After field sample collection, precautions must also be taken by the individual handling water samples in the laboratory. As in the field, normal prohibitions against eating and smoking should be followed, both for the safety of personnel as well as for the integrity of the sample because contamination can result from such sources. Ensuring a hygienic laboratory environment when testing biological samples is a high priority, primarily as a safeguard against cross contamination but also to safeguard the health of the laboratory staff. Because water samples are most often collected in specially cleaned glass bottles (e.g., a standard 300 milliliter BOD bottle), containers should routinely be transported in shock-absorbent carriers to protect them from breakage, resultant sample loss, and possible injury. Note that even when using plastic bottles, precautions should be taken because such containers can break, leak, be punctured, or fail under pressure or heat.

In approximately 25 states and territories, laboratories handling water samples may be subject to federal Occupational Safety and Health Administration (OSHA) standards; laboratories in other states must comply with similar regulations promulgated under state authority. Regardless of their origin, laboratory safety regulations and hazard communication rules are to be followed by persons handling, processing, and otherwise analyzing water samples. Such work can involve exposure to highly corrosive agents (e.g., potassium chromate for chloride analyses), highly reactive catalysts (e.g., metallic sodium for total chlorine), or moderately toxic reagents (e.g., mercuric nitrate for total chlorine). Accordingly, basic minimal protective clothing should be mandatory for handling water samples. This personal protective equipment should include a laboratory coat or chemically resistant apron, gloves impermeable to the reagents handled, and an appropriately selected and worn respirator (where airborne hazards exist, and fumehoods are not available). Perhaps of most importance, protection of the eyes from chemical splashes is required, preferably in the form of chemical goggles or as safety glasses in conjunction with a face shield.

READING LIST

Title 29, Labor. *Code of Federal Regulations, Part 1910.* Standard 1910.1450, Occupational exposure to hazardous chemicals in laboratories.

Test Methods for Evaluating Solid Waste, Physical/Chemical Methods. EPA publication SW-846. Available: http://www.epa.gov/epaoswer/hazwaste/test/sw846.htm (accessed September 22, 2004).

WHO—Water Sampling and Analysis. Available: http://www.who.int/water_sanitation_health/dwq/en/2edvol3d.pdf (accessed September 11, 2004).

MUNICIPAL SOLID WASTE LANDFILLS—WATER QUALITY ISSUES

G. Fred Lee
Anne Jones-Lee
G. Fred Lee & Associates
El Macero, California

INTRODUCTION

Here we provide information on significant shortcomings inherent in the management of municipal solid waste (MSW) and industrial "nonhazardous" waste by the Subtitle D "dry tomb" landfilling practiced today. Focus is on why following the prescriptive measures outlined for such landfills cannot be relied on to provide protection of public health, groundwater resources, or the interests of nearby property owners.

Historically, the goal of municipal solid wastes and industrial "nonhazardous" waste management has been to get the wastes out of sight in the least costly manner. Solid wastes from urban areas were deposited on nearby low-value lands, frequently wetlands, to create waste dumps. This approach evolved into the deposition of wastes into areas excavated for greater disposal capacity. Wastes in the dumps were often burned to reduce volume. Beginning in some areas in the 1950s, wastes placed in dumps were covered daily with a layer of soil to reduce odors and access to the wastes by vermin, flies, birds, and so on, which was the beginning of the "sanitary landfill." Although the design of sanitary landfills diminished some environmental and public health concerns, it did not address issues of the potential for the wastes to cause groundwater pollution, for the gas generated in the landfill to cause explosions, or for the waste to cause other public health or environmental problems. The pollution of groundwater by landfill leachate from conventional sanitary landfills was recognized in the 1950s (2), but it was not until the 1990s that national regulations existed directed at control of groundwater pollution by landfills.

U.S. EPA SUBTITLE D DRY TOMB LANDFILLING

Recognizing the pollution of groundwaters by sanitary landfills, Congress passed the Resource Conservation and Recovery Act (RCRA) that directed the U.S. EPA to address the issue. As this was not accomplished in a timely manner, the U.S. EPA was sued; in response to that suit, the U.S. EPA developed the dry tomb approach to landfilling of MSW. Despite its acknowledgment of technical shortcomings in the approach, but under the pressure of the lawsuit, the U.S. EPA delineated the prescriptive standards for dry tomb landfills in Subtitle D of RCRA. These prescriptive standards have been adopted by some state regulatory agencies as minimum design standards for MSW landfills.

The concept of dry tomb landfilling was built on the premise that because water in contact with the wastes led to the formation of leachate that traveled to groundwater, isolation of the wastes in dry tombs would prevent groundwater pollution by landfill leachate. A dry tomb landfill, as implemented by the U.S. EPA, is illustrated in Fig. 1. It relies on a liner and cap to keep the wastes dry, leachate collection and removal systems to keep the wastes from polluting groundwater, and gas collection and removal systems. It also relies on groundwater monitoring to signal incipient pollution of groundwater by landfill leachate and specifies a "postclosure" period of 30 years, during which time the facility is to be maintained and monitored.

These provisions notwithstanding, the U.S. EPA realized in its development of these requirements that such a system would not, in itself, ensure the protection of groundwaters forever. As part of adopting the RCRA Subtitle D regulations, the U.S. EPA stated in the draft regulations (3),

> First, even the best liner and leachate collection system will ultimately fail due to natural deterioration, and recent improvements in MSWLF (municipal solid waste landfill) containment technologies suggest that releases may be delayed by many decades at some landfills.

Further, the U.S. EPA (4) Criteria for Municipal Solid Waste Landfills states,

> Once the unit is closed, the bottom layer of the landfill will deteriorate over time and, consequently, will not prevent leachate transport out of the unit.

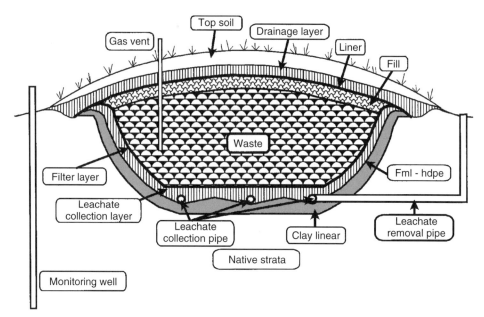

Figure 1. Single-composite liner landfill containment system.

Thus, even if the dry tomb system could be installed without flaw, it will eventually fail to isolate the wastes, which is significant because many of the contaminants in MSW landfills do not decompose to inert forms (e.g., heavy metals) and are hence a threat forever. Furthermore, although many organics can eventually be bacterially decomposed, such decomposition requires water, which is excluded from the landfill under ideal dry tomb landfill conditions. It is therefore important to consider the characteristics and shortcomings of the components of a dry tomb landfill to provide protection of groundwater quality for as long as the wastes in the landfill remain a threat, and how they can be improved to provide more reliable protection.

LANDFILL COVER

Allowed closure of Subtitle D landfills is begun by covering the waste with soil that is placed and shaped to serve as the base for a low-permeability plastic sheeting layer. The plastic sheeting is overlaid by a one- to two-foot deep drainage layer. The drainage layer is overlaid by a few inches to a foot or so of topsoil that serves as a vegetative layer designed to promote the growth of vegetation to reduce erosion of the landfill cover. In principle, water that enters the vegetative layer would either be taken up by the vegetation or penetrate through the root zone to the porous (drainage) layer. When the moisture reaches the low-permeability plastic sheeting layer of the cover, it is supposed to move laterally to the outside of the landfill (see Fig. 1).

However, in addition to deficiencies introduced during the installation of the plastic sheeting, landfill covers are subject to a variety of factors that can breach their integrity, including burrowing animals, differential settling of the wastes beneath the cover, and deterioration of the plastic sheeting. The typical approach to monitoring landfill covers for integrity in keeping moisture out of the landfill, that is advocated by landfill owners and operators and allowed by regulatory agencies, involves visual inspection of the surface of the vegetative soil layer of the landfill cover. If cracks or depressions in the surface are found, they are filled with soil. Such an approach, however, will not detect or remediate cracks in or deterioration of the plastic sheeting layer, which is the basis for the moisture removal system for the cover. As a result, moisture that enters the drainage layer and comes in contact with the plastic sheeting layer will penetrate into the wastes rather than be directed off the landfill. If this occurs during the postclosure care period, the increased leachate generation could be detected. However, it could also readily occur in year 31 after closure or thereafter, when there could well be no one monitoring leachate generation.

LEACHATE COLLECTION AND REMOVAL SYSTEM

The key to preventing groundwater pollution by a dry tomb landfill is collecting all leachate that is generated in the landfill, in the leachate collection and removal system. As shown in Fig. 1, a Subtitle D system prescribes a relatively thin (0.06 inch) plastic sheeting layer (high-density polyethylene—HDPE) and a 2-ft thick compacted soil/clay layer, which together, in intimate contact, form a "composite" liner beneath the wastes. Atop the liner is the leachate collection system consisting of gravel or some other porous medium, which is intended to allow leachate to flow rapidly to the top of the HDPE liner for removal. This porous layer is overlaid by a filter layer, which is supposed to keep the solid waste from migrating into the leachate collection system. Thus, in principle, leachate that is generated in the solid waste passes through the filter layer to the leachate collection layer beneath. Once it reaches the sloped HDPE liner, it is supposed to flow across the top of the liner to a collection pipe, where it would be transported to a sump from which it could be pumped from the landfill. According to regulations, the maximum elevation of leachate ("head") in the sump is to be no more than 1 ft.

However, leachate collection and removal systems, as currently designed, are subject to many problems. Biological growth, chemical precipitates, and "fines" derived from the wastes all tend to cause the leachate collection system to plug. This, in turn, increases the head of the leachate above the liner upstream of the area that is blocked, which further stresses the integrity of the system. Although the potential exists to back-flush some of these systems, this back-flushing will not eliminate the problem.

One of the most significant problems with leachate collection systems' functioning as designed is that the HDPE liner, which is the base of the leachate collection system, develops cracks, holes, rips, tears, punctures, and points of deterioration. Some of these are caused at the time of installation, and HDPE integrity deteriorates over time. When the leachate that is passing over the liner reaches one of these points, it starts to pass through the liner into the underlying clay layer. If the clay layer is in intimate contact with the HDPE liner, the rate of initial leakage through the clay is small. If, however, problems existed achieving or maintaining intimate contact between the clay and HDPE liner, such as the development of a fold in the liner, the leakage through the HDPE liner hole can be quite rapid. Under those conditions, the leachate can spread out over the clay layer and can leak at a substantial rate through the clay. As noted above, even in the establishment of the Subtitle D regulations, the U.S. EPA recognized that such a liner will deteriorate over time.

GROUNDWATER MONITORING

U.S. EPA Office of Solid Waste Emergency Response senior staff members have indicated that the fact that HDPE liners will fail to prevent leachate from passing into the underlying groundwaters does not mean that the Subtitle D regulations are fundamentally flawed because of the groundwater monitoring system requirement. They claim that when leachate-polluted groundwaters first reach the point of compliance for groundwater monitoring, they are detected by the groundwater monitoring system with sufficient reliability so that a remediation program can be initiated. The point of compliance for groundwater monitoring at Subtitle D landfills is specified as being no more than 150 meters from the downgradient edge of the waste deposition area and must be on the landfill owner's property. However, serious technical deficiencies exist in that position.

It was pointed out by Cherry (5) that initial leakage through HDPE-lined landfills will be through areas where there are holes, rips, tears, or points of deterioration of the HDPE liner, which, as illustrated in Fig. 2, can lead to relatively narrow plumes of polluted groundwaters at the point of compliance for groundwater monitoring. The typical groundwater pollution plume in a sand, gravel, or silt aquifer system will likely be on the order of 10 to 20 ft wide at the point of compliance. In order to detect incipient leakage from a Subtitle D landfill, therefore, these narrow

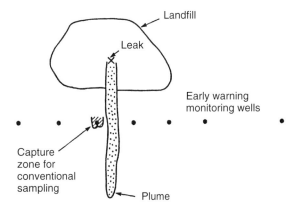

Figure 2. Leakage from HDPE-lined landfill (adapted from Ref. 5).

plumes would have to be detected by the groundwater monitoring well array at the point of compliance. Typically, federal and state regulatory agencies allow monitoring wells to be placed 100 or more feet apart at the point of compliance; each monitoring well has a zone of capture of about 1 ft. Thus, if the wells were 200 ft apart, 198 ft would exist between wells where a 10- to 20-ft wide plume of leachate-polluted groundwater could pass undetected. It is virtually impossible to reliably monitor groundwater for leachate contamination in fractured rock or cavernous limestone areas. There, leachate can travel great distances and in unexpected directions via cracks and caverns.

It is because of the unreliability of groundwater monitoring systems that are based on vertical monitoring wells at the point of compliance that some states (such as Michigan) require that a double-composite liner be used at municipal solid waste landfills. In that system, a leak detection system is in place between the composite liners to detect leakage through the upper composite before the lower liner is breached. Although this approach is not foolproof in its ability to detect when both liner systems fail, it has a much greater probability of detecting when the upper composite liner fails. It is for this reason that Lee and Jones-Lee (6) recommended that all Subtitle D landfills incorporate a double-composite liner with a leak detection system between the two liners.

LANDFILL GAS

Some organics in municipal solid wastes can serve as a source of food for bacteria that will, in a moist landfill environment, produce methane and CO_2 (landfill gas). MSW landfills also release a number of other volatile chemicals, including highly hazardous volatile organic compounds (VOCs) and odorous compounds, which are a threat to the health and welfare of those within the sphere of influence of the landfill. This sphere can extend for several miles, depending on the topography of the area and the tendency for atmospheric inversions to occur.

Although landfill advocates claim that the practice of daily cover of the wastes will reduce the gaseous (odorous) releases from landfills, even diligent covering does not prevent them. Further, when landfill owners/operators become sloppy in operations, greater-than-normal landfill gas emissions occur, which are typically detected through landfill odors. Although some attempt to minimize the significance of smelling landfill gas on adjacent properties as only being an aesthetic problem, in fact, as discussed by Shusterman (7), it is now known that noxious odors can cause illness in people. Odors should be controlled so that they do not trespass across the landfill adjacent property owner's property line.

The only way at present to reliably ensure the protection of adjacent and nearby property owners/users from landfill gas is to provide a sufficient landfill-owner-owned buffer land about the landfill. If an adjacent or nearby property owner/user can smell the landfill, then inadequate buffer land exists between the landfill and adjacent properties, which should make it necessary for the landfill owner/operator to either acquire adjacent buffer land or to use more than the minimum approach for controlling gaseous releases from the landfill. It is also important to control future land use within the landfill area so that releases from the landfill would not be adverse to the land use. For example, agriculture in these areas should be restricted, because releases from the landfill could contaminate some crops. As long as landfill owners are allowed to use adjacent properties for their waste disposal buffer zones, justified NIMBY ("not in my back yard") issues by adjacent property owners will exist.

A common misconception held by landfill applicants and some regulators is that landfill gas production will cease after a comparatively short time after closure of a dry tomb landfill. Estimates of landfill gas production rate and duration are typically based on a model assuming unbagged, homogeneous wastes that are allowed to interact with moisture, which is not the situation that is found in Subtitle D dry tomb MSW landfills. The key to landfill gas production is the presence of sufficient moisture to allow bacteria to metabolize certain of the organic materials in the landfill to methane and CO_2 (landfill gas). Thus, the rate of moisture penetration through the cover and the mixing of that moisture with the waste components control the rate and duration of landfill gas production. Once the landfill is closed, Subtitle D landfills aim to keep moisture out of contact with the buried waste; the dryer the waste is kept, the slower the rate of landfill gas production. Furthermore, because much of the municipal waste that is placed in Subtitle D landfills is contained within plastic bags, and because those plastic bags are only crushed and not shredded, the crushed bags will "hide" fermentable components of the waste. Thus, once moisture does breach the cover and is allowed to contact with the wastes, the period of landfill gas production will be prolonged further, until after the plastic bags decompose and the bagged wastes allowed to interact with the moisture, which can extend gas production many decades, to a hundred or more years.

POSTCLOSURE MONITORING AND MAINTENANCE

The Subtitle D relies on the 30-year postclosure period of monitoring and maintenance to continue the protection of groundwater. The mistaken idea existed that 30 years after closure, the waste in a dry tomb landfill would no longer be a threat. From the characteristics of wastes and their ability to form leachate, as well as the processes than can occur in a landfill, it is clear that 30 years is an infinitesimally small part of the time that waste components in a landfill, especially a dry tomb landfill, would be a threat to cause groundwater pollution through leachate formation. A critical review of the processes that can take place in a MSW landfill that can generate leachate shows that the containment system of a dry tomb landfill, for which there is at least an initial effort to reduce the moisture entering the wastes, will eventually fail to prevent groundwater pollution for as long as the wastes are a threat. The municipal solid wastes in a classic sanitary landfill where no attempt to prevent moisture from entering the wastes exists have been found to generate leachate for thousands of years. Dry tomb landfilling delays and then prolongs leachate generation.

Subtitle D regulations also require that a small amount of assured funding be available for 30 years of postclosure monitoring and maintenance. Some regulatory agencies will allow landfill companies to be self-insured or insured through an insurance company that is backed by a landfill company. Such approaches should not be allowed, because landfill companies are amassing large liabilities for the ultimate failure of the landfill liner systems and the groundwater pollution that will occur as a result of those failures. It is well understood that, ultimately, private landfill companies will not likely be able to comply with Subtitle D regulations for funding of remediation. The amount of postclosure monitoring and maintenance funding that is currently required is grossly inadequate compared with the funding levels that could be necessary during the 30-year mandatory postclosure period, much less the extended period over which the wastes in the landfill will be a threat. Although Congress, through Subtitle D, required that the regulations include provisions to potentially require additional funding at the expiration of the 30-year postclosure care period, the likelihood of obtaining this funding from private landfill companies, even if they still exist 30 years after a landfill has been closed, or from a public agency that developed or owns landfills, is remote.

Lee (8) discussed the importance of solid waste management regulatory agencies requiring that landfill owners, whether public or private, prepare for the inevitable failure of the landfill containment system and provide for funding to address this failure. Designation of a 30-year postclosure assured funding period in RCRA, which is implemented not as a minimum but rather as a definitive period, leaves the public to pay the significant balance of the cost of landfilling; the public is left to deal with the long-term impacts of MSW landfills on public health and the environment, and eventual remediation of the landfill. This significant deficiency in Subtitle D regulations is recognized not only in the technical community but also by various other groups and individuals who have reviewed this issue. For example, in the executive summary of its report *Funding of Postclosure Liabilities Remains Uncertain,* in a section labeled, "Funding Mechanisms Questionable," the U.S. Congress General Accounting Office (GAO) (9) concluded,

> Owners/operators are liable for any postclosure costs that may occur. However, few funding assurances exist for postclosure liabilities. EPA only requires funding assurances for maintenance and monitoring costs for 30 years after closure and corrective action costs once a problem is identified. No financial assurances exist for potential but unknown corrective actions, off-site damages, or other liabilities that may occur after the established postclosure period.

Further, the U.S. EPA Inspector General (10), in a report entitled "RCRA Financial Assurance for Closure and Postclosure," came to similar conclusions:

> There is insufficient assurance that funds will be available in all cases to cover the full period of landfill postclosure monitoring and maintenance. Regulations require postclosure activities and financial assurance for 30 years after landfill closure, and a state agency may require additional years of care if needed. We were told by several state officials that many landfills may need more than 30 years of postclosure care. However, most of the state agencies in our sample had not developed a policy and process to determine whether postclosure care should be extended beyond 30 years, and there is no EPA guidance on determining the appropriate length of postclosure care. Some facilities have submitted cost estimates that were too low, and state officials have expressed concerns that the cost estimates are difficult to review.

LANDFILL SITING ISSUES

In the development of Subtitle D landfill regulations, the U.S. EPA failed to address one of the most important issues that should be addressed in developing a minimum Subtitle D landfill, namely, the need to site landfills at geologically suitable sites for a landfill of this type. Although the EPA does require that minimum Subtitle D landfills not be sited too close to airports, where major bird problems could exist for aircraft, too near an earthquake fault, or within a flood plain, the EPA does not address the issue of siting minimum Subtitle D landfills where the underlying geological strata do not provide natural protection of the groundwaters from pollution by landfill leachate when the landfill liner systems eventually fail. In accordance with current regulations, minimum Subtitle D landfills can be sited over highly important aquifers that serve as domestic water supply sources. They can also be sited in fractured rock and cavernous limestone areas, where it is impossible to reliably monitor for the pollution of groundwaters by landfill leachate using vertical monitoring wells.

Subtitle D landfill regulations also fail to address one of the most important causes for people to object to landfills; the deposition of wastes is allowed very near the landfill property owner's property line. Under these conditions, the landfill gases, blowing paper, birds, rodents, vermin, and so on associated with the landfill can impinge on adjacent properties. For example, it is well established that landfill gas can readily travel a mile or more from a landfill, and thereby be adverse to the adjacent property owners' use of their properties. It is recommended that at least a mile, and preferably two miles, of landfill-owned buffer lands exist between the area where wastes are deposited and adjacent property owners' property lines, which would provide for dissipation of releases from the landfill on the landfill owner's property. Such an approach would greatly reduce the trespass of waste-derived materials from the landfill onto adjacent properties.

ROLE OF 3 Rs IN MSW LANDFILLING

Considerable efforts are being made in many parts of the country to increase reuse, recycling, and reduction (the "3Rs") of MSW as part of conserving natural resources and landfill space. As discussed by Lee and Jones-Lee (11), the 3 Rs should be applied to the maximum extent practicable to reduce the number of new and expanded Subtitle D landfills that will eventually pollute groundwaters.

JUSTIFIED NIMBY

NIMBY is an acronym for "not in my backyard," a commonly dismissed plea presumed to be made by those who, without justification, simply oppose having a landfill nearby. However, Lee and Jones-Lee (12) and Lee et al. (13) discussed many of the technical issues that, in fact, justify a NIMBY position for Subtitle D landfills. Table 1 presents typical, real, adverse impacts of landfills on nearby property owners/users.

IMPROVING LANDFILLING

The degree of protection of public health and the environment from adverse impacts of MSW disposed of in dry tomb landfills can be improved to address some of the deficiencies of the approaches and specifications currently commonly accepted.

Siting

Landfills should be sited so that they provide, to the maximum extent possible, natural protection of groundwaters when the liner systems fail. Siting landfills above geological strata that have groundwater whose flow

Table 1. Adverse Impacts of Dry Tomb Landfills on Adjacent/Nearby Property Owners/Users[a]

- Public health, economic and aesthetic impacts on groundwater and surface water quality
- Methane and VOC migration—public health hazards, explosions, and toxicity to plants
- Illegal roadside dumping and litter near landfill
- Truck traffic
- Noise
- Dust and wind-blown litter
- Odors
- Vectors, insects, rodents, birds
- Condemnation of adjacent property for future land uses
- Decrease in property values
- Impaired view

Source: Lee et al. (13).

paths are not readily amenable to monitoring for leachate-polluted groundwaters should be avoided. Of particular concern are fractured rock and cavernous limestone areas, as well as areas with sandy lenses.

Design

MSW landfills should incorporate double-composite liners with a leak detection system between the two liners.

Closure

MSW landfills should incorporate leak-detectable covers that will indicate when the low-permeability layer of the landfill cover first fails to prevent moisture from entering the landfill.

Monitoring

The primary monitoring of liner leakage should be associated with the leak detection system between the two composite liners. If vertical monitoring wells are used, the spacing between the vertical monitoring wells at the point of compliance should be such that a leak in the HDPE liner caused by a 2-ft-wide rip, tear, or point of deterioration at any location in the landfill would be detected based on the plume that is generated at the point of compliance with a 95% reliability.

Landfill Gas Collection

For those landfills that contain wastes that can produce landfill gas, a landfill gas collection system should be designed, installed, and maintained for as long as the wastes in the dry tomb landfill have the potential to generate landfill gas, giving proper consideration to how and for how long gas will be generated in the system. The landfill gas collection system should be designed to have at least a 95% probability of collecting all landfill gas generated at the landfill.

Maintenance

The maintenance of the landfill cover, monitoring system, gas collection system, and so on. should be conducted for as long as the wastes in the landfill will be a threat, with a high degree of certainty of detecting landfill containment system and monitoring system failure, which will extend well beyond the 30-year period typically established, especially with improved provisions for excluding moisture.

Funding

The funding for closure, postclosure monitoring, maintenance, and groundwater remediation should be established at the time the landfill is established, in a dedicated trust fund of sufficient magnitude to address plausible worst-case scenario failures for as long as the wastes in the landfill will be a threat. Unless appropriately demonstrated otherwise, it should be assumed that the period of time for which postclosure care funding will be needed will be infinite.

Buffer Land

At least several miles of landfill-owner-owned buffer lands should exist between where wastes are deposited and adjacent properties.

Adoption of these approaches (or as many of them as possible) will improve the protection of groundwater quality, public health, and environmental quality from adverse impacts of the dry tomb landfilling of MSW.

Further information on the topics discussed is provided by Jones-Lee and Lee (14), Lee and Jones-Lee (6,12,15,16,18), and Lee (17).

BIBLIOGRAPHY

1. U.S. EPA. (1991). Solid waste disposal facility criteria; final rule, part II. *Federal Register.* 40 CFR Parts 257 and 258, U.S. Environmental Protection Agency, Washington, DC.

2. ASCE. (1959). *Sanitary Landfill.* Report Committee on Sanitary Landfill Practice of the Sanitary Engineering Division of the American Society of Civil Engineers, New York.

3. U.S. EPA. (1988). Solid waste disposal facility criteria: proposed rule. *Federal Register* **53**(168): 33,314–33,422–. 40 CFR Parts 257 and 258, U.S. Environmental Protection Agency, Washington, DC.

4. U.S. EPA. (1988). *Criteria for Municipal Solid Waste Landfills.* U.S. Environmental Protection Agency, Washington, DC.

5. Cherry, J.A. (1990). Groundwater monitoring: some deficiencies and opportunities. In: *Hazardous Waste Site Investigations: Towards Better Decisions* B.A. Berven and R.B. Gammage (Eds.). Proceedings of the 10th ORNL Life Sciences Symposium, Gatlinburg, TN, Available from gfredlee@aol.com as "LF019."

6. Lee, G.F. and Jones-Lee, A. (1998). *Assessing the Potential of Minimum Subtitle D Lined Landfills to Pollute: Alternative Landfilling Approaches.* Proc. Air and Waste Management Assoc. 91st Annual Meeting, San Diego, CA. Available on CD ROM as paper 98-WA71.04(A46), p. 40 and http://www.gfredlee.com/alternative_lf.html.

7. Shusterman, D. (1992). Critical review: the health significance of environmental odor pollution. *Arch. Environ. Health.* **47**(1): 76–87.

8. Lee, G.F. (2003). *Workshop on Landfill Postclosure and Financial Assurance.* Comments submitted to Mike Paparian, California Integrated Waste Management Board, by G. Fred Lee & Associates, El Macero, CA. Available at http://www.gfredlee.com/paparian10-30-03T.pdf.

9. GAO. (1990). *Funding of Postclosure Liabilities Remains Uncertain.* GAO/RCED-90-64, U.S. General Accounting Office Report to the Congress, Washington, DC.

10. U.S. EPA. (2001). *RCRA Financial Assurance for Closure and Postclosure.* Audit Report, 2001-P-007, U.S. Environmental Protection Agency, Office of Inspector General, Washington, DC.

11. Lee, G.F. and Jones-Lee, A. (2000). *Three Rs Managed Garbage Protects Groundwater Quality.* Proc. AWMA 93rd annual conference, Salt Lake City, UT, paper 00-454 CD ROM, Air & Waste Management Association, Pittsburgh, PA. Available at http://www.gfredlee.com/3rpap_sli.pdf.

12. Lee, G.F. and Jones-Lee, A. (1994). Addressing Justifiable NIMBY: A Prescription for MSW Management.

13. Lee, G.F., Jones-Lee, A., and Martin, F. (1994). *Landfill NIMBY and Systems Engineering: A Paradigm for Urban Planning.* In: Proc. National Council on Systems Engineering Fourth Annual International Symposium, Vol. 1, pp. 991–998. Available at http://www.members.aol.com/duklee2307/NIMBY-NCO2.pdf.
14. Jones-Lee, A. and Lee, G.F. (1993). Groundwater pollution by municipal landfills: leachate composition, detection and water quality significance. *Proc. Sardinia '93 IV International Landfill Symposium.* Sardinia, Italy, pp. 1093–1103. Available at http://www.gfredlee.com/lf-conta.htm.
15. Lee, G.F. and Jones-Lee, A. (1995). *Overview of Landfill Post Closure Issues.* Presented at American Society of Civil Engineers Convention session devoted to "Landfill Closures - Environmental Protection and Land Recovery." San Diego, CA. Available at http://www.gfredlee.com/plandfil2.htm#postclosure.
16. Lee, G.F. and Jones-Lee, A. (1998). *Deficiencies in Subtitle D Landfill Liner Failure and Groundwater Pollution Monitoring.* Presented at the NWQMC National Conference "Monitoring: Critical Foundations to Protect Our Waters. U.S. Environmental Protection Agency, Washington, DC. Available at http://www.gfredlee.com/nwqmcl.html.
17. Lee, G.F. (2002). Solid Waste Management: USA Lined Landfilling Reliability. An invited submission for publication in *Natural Resources Forum*, a United Nations Journal, New York. Available at http://www.gfredlee.com/UNpaper-landfills.pdf.
18. Lee, G.F. and Jones-Lee, A. (2004). *Flawed Technology of Subtitle D Landfilling of Municipal Solid Waste.* Report of G. Fred Lee & Associates, El Macero, CA. Available upon request from gfredlee@aol.com.

LAND USE EFFECTS ON WATER QUALITY

Ross A. Steenson
Geomatrix
Oakland, California

Factors that determine water quality at a site include climate, geology, biology, and land use (anthropogenic practices). Water quality varies naturally as the result of differing natural environments, but anthropogenic practices can override these natural variations. During the last several decades, research and regulatory efforts in the United States have largely focused on understanding and limiting water quality effects from specific or localized anthropogenic sources, which are termed point sources.

Point sources include farms; industrial facilities where chemicals are manufactured, used, or stored; landfills; sewage treatment plants; hazardous waste sites; mines; and construction sites.

There are numerous routes through which sources can affect water quality (e.g., downward migration of source chemicals, wet weather runoff, and direct discharge). The water quality effects observed from sources include increased concentrations of dissolved chemicals (e.g., fertilizers/nutrients, pesticides/herbicides, industrial chemicals, trace elements, and minerals), loss of dissolved gases (e.g., oxygen, carbon dioxide), and change in physical characteristics (e.g., sediment load, temperature). Waters potentially susceptible to these effects include nearby surface waterbodies (e.g., streams and rivers) and underlying shallow groundwater or groundwater that is connected hydraulically to affected surface waterbodies.

More recently, studies and regulatory efforts have shifted beyond water quality effects at specific sites to understanding the broader effects. These efforts at understanding the broader water quality effects of land use have included studies of larger geographic regions and studies of the effects of increased human populations. Regional studies have combined available water quality data with measures of local land use and/or census information and identified general correlations between areal water quality, land use, and population density (1). As a result of the broadened view of water quality effects, the concept of nonpoint sources has evolved. Nonpoint sources can include multiple point sources or more areal sources (or practices).

Nonpoint sources include agricultural activities and agricultural and urban runoff. These diffuse sources have a greater potential to affect water quality and are much harder to address than point sources. An example of a common contaminant that is largely the result of nonpoint sources is nitrate, which is considered the most widespread contaminant in groundwater, particularly in agricultural areas (2). Runoff from urban areas is a potential source of many chemicals (e.g., spilled oils), debris, and sediment that can be rapidly discharged to surface waterbodies following wet weather. Thus, urbanization of watersheds has been viewed increasingly as ever more significantly affecting water quality (3).

In the last decade, concern has increased regarding the indirect effects on water quality from human populations and the pharmaceuticals and personal care products used by individuals. The number of these products is rapidly increasing, and it is believed that many of these chemicals find their way primarily into surface waterbodies, but also into groundwater, through incomplete treatment at sewage treatment plants or disposal in landfills (4).

BIBLIOGRAPHY

1. Eckhardt, D.A.V. and Stackelberg, P.E. (1995). Relation of ground-water quality to land use on Long Island, New York. *Ground Water* **33**(6, Nov.–Dec.): 1019–1033.
2. Hallberg, G.R. and Keeney, D.R. (1993). Nitrate. In: *Regional Ground-Water Quality*, W.M. Alley (Ed.). Van Nostrand Reinhold, New York, pp. 297–322.
3. Horner, R.R., Skupien, J.J., Livingston, E.H., and Shaver, H.E. (1994). *Fundamentals of Urban Runoff Management: Technical and Institutional Issues.* Terrene Institute, Washington, DC, in cooperation with the U.S. Environmental Protection Agency.
4. Daughton, C.G. and Ternes, T.A. (1999). Pharmaceuticals and personal care products in the environment: agents of subtle change? *Environ. Health Perspect.* **107**(Suppl. 6, Dec.).

MONITORING LIPOPHILIC CONTAMINANTS IN THE AQUATIC ENVIRONMENT USING THE SPMD-TOX PARADIGM

B. Thomas Johnson
USGS—Columbia
Environmental Research Center
Columbia, Missouri

SPMD-TOX PARADIGM

SPMD-TOX (Fig. 1) is a monitoring tool that uses microscale techniques to assess the environmental impact of civilization on an aquatic resource. To determine a relevant profile, the sample must reflect the chemical contaminant's effect, not just its presence. To collect and concentrate bioavailable lipophilic contaminants, a semipermeable membrane device is used instead of live animals. To make risk assessments that are both effective and efficient, microscale toxicity tests (TOX) are employed. (Use of specific products by USGS and its laboratories does not constitute an endorsement.)

AQUATIC ENVIRONMENTAL SAMPLING

Water by its very physical and chemical nature is a natural repository and carrier of domestic, industrial, and agricultural products. In the last 50 years, analytical chemists have made significant strides in collecting, separating, and identifying these environmental waterborne contaminants (1). However, to determine the effect of these contaminants on an aquatic ecosystem, additional questions must be answered: Is the contaminant simply a chemical out of place, or is it a pollutant, a chemical toxin injurious to the health of the ecosystem? Where is the contaminant? Does the contaminant move? Is the contaminant a pollutant that is bioavailable and toxic to the ecosystem?

Bioavailability

Bioavailability is a critical concept in ecotoxicology because a harmful effect occurs only when biota are exposed to bioavailable pollutants (2). In an aquatic ecosystem, a chemical tends to exist in one of three forms: (1) dissolved in water, (2) passively sorbed directly into the sediment and on biotic or abiotic components suspended in the water column, or (3) actively integrated in some biota and bioaccumulated through respiratory or food ingestion pathways. It is evident that water-soluble contaminants are dissolved and bioavailable to all biota in the water column, but lipophilic contaminants are fat-soluble and, as a result, present a different challenge to identify their toxic effects on biota.

Samples

Samples from an aquatic environment come with some general knowledge. A site may be industrial, rural, urban, recreational, or commercial. The area of concern may be a result of stormwater, a chemical spill, a fish kill, a careless citizen, or a business error. Based on the Paracelsus adage, toxicologists know that all things are toxic and the dosage makes the "toxicant." Paracelsus (1493–1541) is often considered the father of modern toxicology. He brought empirical evidence into toxicology with his writings *"What is there that is not a poison? All things are poison and nothing without poison. Solely the dose determines that a thing is not a poison."* In this context, a toxicant must be defined both qualitatively (identified) and quantitatively (how much); toxicity is clearly dose-responsive. Simply stated, a waterborne chemical in the environment may be a contaminant at one concentration and a toxicant at another concentration; dosage makes the difference. Thus, the modern-day toxicity bioassay is predicated on the dose–response experimental design.

SPMDs

To collect and concentrate a lipophilic sample, a semipermeable membrane device (SPMD) that mimics the bioaccumulation of waterborne organic contaminants by fish is

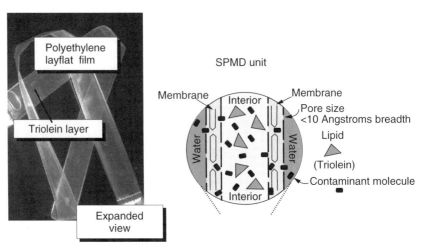

Figure 1. SPMD unit.

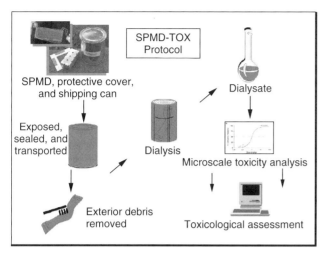

Figure 2. SPMD-TOX protocol.

used (SPMDs are sold exclusively by Environmental Sampling Technology in St. Joseph, MO); the SPMD permits estimates of tissue residues from the exposure site. The SPMD is a small low-density polyethylene lay-flat tube that contains neutral lipid triolein with a membrane surface area to lipid ratio of about 400 cm^2/mL (Fig. 2). The SPMD is mounted in a protective stainless steel container and submerged generally 2 to 4 feet below the surface for about 30 days. Waterborne organic molecules diffuse through transitory openings (so called transport corridors) in the polyethylene membrane and sorb to the triolean, thus simulating the passive diffusion of a chemical across a fish gill's membrane. The SPMD monitors the lipophilic contaminant's bioavailability by a log K_{ow} value of less than one, a molecular weight of 600 daltons or less, and a size less than 10 Å. The entire unit is returned to the laboratory and the SPMD is chemically dialyzed using an organic solvent to recover sequestered chemicals.

The SPMDs as environmental contaminant-concentrating tools have many advantages. (1) They are abiotic, which means SPMDs do not metabolize sequestered products but provide a true reflection of bioavailable contaminants. (2) They are time integrative samplers, which means SPMDs will collect all lipophilic materials introduced into the water column during the exposure period, even if the introductions are episodic. (3) SPMDs can survive in heavily polluted, toxic environments where living organisms may not survive. (4) They are not temperature specific; SPMDs can be used in both cold and warm water. (5) SPMDs are easily transported to sites of interest for sampling and to laboratories for processing. (6) SPMD dialysates are analyzed by traditional GC/MS to identify the sequestered contaminants; algorithms and sampling rates are used to back-calculate concentrations in the water column. (7) SPMDs are cost-effective and, as a result, can be used in large monitoring programs.

TOXICITY ASSAYS

Microtox (3) determines the acute toxicity of SPMD dialysates by measuring the changes in light produced when samples are exposed to bioluminescent bacteria under standard conditions. Microtox bacteria *Vibrio fischeri* are clonal cultures, which diminish possible genetic differences and ensure quality control of the tester strain with greater assay sensitivity and precision. The freeze-dried bacteria are available on demand and all test media and glassware are prepackaged, standardized, and disposable. The test requires minimal laboratory space and limited dedicated equipment. The test is well defined, computer assisted, and user friendly.

Microtox is microscale. Dialysates are prepared in the basic dose–response design and placed in an SDI Model 500 Analyzer for incubation. [Columbia Environmental Research Center (CERC) uses Microtox materials and equipment sold by Strategic Diagnostics Inc. (SDI) in Newark, DE, to preserve the Microtox protocol. SDI provides comprehensive instructive guides, manuals and computer software to operate the Microtox test at their Web site (www.azurenv.com). The Microtox protocol described here is a standard USGS SOP.] The amount of light remaining in the sample is used to determine the sample's relative toxicity, which can then be compared to the standard reference's toxicity. As the toxicant's concentration increases, bacterial light emissions decrease in a dose-dependent manner. Supporting computer software is used to determine a 50% loss of light in the test bacteria, that is, the effective concentration (EC_{50}) value. Tests are completed and data are available in <30 min. The lower the EC_{50} value, the greater the acute toxicity of the sample.

Mutatox (4) determines the DNA-damaging substances in SPMD dialysates by measuring the increase in light produced when samples are exposed to dark mutant stains of *V. fischeri*. (Mutatox is prepared and sold exclusively by SDI. The Mutatox protocol described here is a standard USGS SOP.) Similar to Microtox, the test bacteria are stored freeze-dried under vacuum and require no preculturing; aseptic techniques are minimal; all test media and glassware are prepackaged, standardized, and disposable. The test requires minimal laboratory space and limited dedicated equipment. Because most bacterial cells such as these dark mutant strains fail to duplicate vertebrate biotransformation of progenotoxins into active genotoxins (DNA-damaging substances), a vertebrate activation system, typically rat or fish hepatic enzymes, is incorporated into the test.

Mutatox is microscale; all tests are conducted in microvolumes with microcuvettes. A single reaction cuvette contains the dark mutant bacteria, the diluent, the dialysate, a metabolic activation system, and a carrier solvent such as DMSO. Test samples are serially prepared over a 100-fold dose range, preincubated at 37 °C (metabolic activation) for 15 to 30 min, and then grown in the dark at 27 °C for 18 to 24 h. Benzopyrene, a commonly found PAH, is used as a positive control. A genotoxic response is defined as an increased light value of at least three times the light intensity of the solvent control blank. Mutatox is a qualitative test that provides a yes-no assessment of potentially DNA-damaging substances.

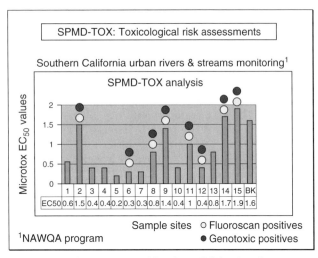

Figure 3. Risk assessment of Southern California urban streams with SPMD-TOX.

Case Study: A Simulated Oil Spill (5)

SPMD-TOX monitored a simulated oil spill during an 8-week period along the St. Lawrence River in Canada. An SPMD was placed weekly in each of the 20 oiled plots and removed 1 week later. The dialysates from the first 2 weeks fluoresced, indicating the presence of PAHs, and Mutatox analyses confirmed these findings with a yes response. Microtox analyses initially showed evidence of acute toxicity followed by a rapid decline and by week 4 no detectable acute toxicity. Samples from weeks 3 through 8 showed no evidence of the oil spill in the water column. SPMD-TOX analyses of these samples during the 8-week exposure period presented a clear picture of the occurrence and rapid recovery of the simulated oil spill site.

Case Study: Urban River Sites (6)

During a 30-day exposure period, SPMD-TOX was used to monitor 15 urban sites at rivers and streams in Southern California for the presence of bioavailable waterborne lipophilic ecotoxins. Fluoroscan, Microtox, and Mutatox data suggested the probable presence of PAHs and the initial identity of both acute toxic and genotoxic substances that were bioavailable in the water column (Fig. 3).

SUMMARY

SPMD-TOX is a sensitive, technically simple, and cost-effective assessment tool for monitoring waterways for bioavailable lipophilic chemical contaminants that may adversely affect aquatic communities and water quality. The SPMD-TOX paradigm can rapidly identify chemical incursions that may impact the quality of the water and provide the ecotoxologist and resource manager with valuable risk assessments of areas of concern.

BIBLIOGRAPHY

1. Manahan, S.E. (1989). *Toxicological Chemistry*. Lewis, Chelsea, MI.
2. Rand, G.M., Wells, P.G., and McCarty L.S. (1995). Introduction to aquatic toxicology. In: *Fundamentals of Aquatic Toxicology*, G.M. Rand, (Eds.). Taylor & Francis, Washington, DC. pp. 3–66.
3. Microbics Corporation. (1992). *Microtox Manual*. Microbics Corporation, Carlsbad, CA., Vols. I to V.
4. Johnson, B.T., Petty, J.D., and Huckins, J.N. (2000). Collection and detection of lipophilic chemical contaminants in water, sediment, soil and air—SPMD-TOX. *Environ. Toxicol.* **15**: 248–252.
5. Johnson, B.T. et al. (2004). Hazard assessment of a simulated oil spill on intertidal areas of the St. Lawrence River with SPMD-TOX. *Environ. Toxicol.* **19**: 329–335.
6. Johnson, B.T., Petty, J.D., Huckins, J.N., Goodbred, S., and Burton, C. (2002). Assessment of aquatic habitat quality in urban streams in the Western United States using SPMD-TOX. 23rd *Annu. Meet. Soc. Environ. Chem. Toxicol.*, November 15–20, 2002, Salt Lake City, Utah. Abstract.

Johnson, B.T. (1997). Microtox® Toxicity Test System—new developments and applications. In: *Microscale Testing in Aquatic Toxicology*, P.G. Wells, K. Lee, and C. Blaise (Eds.). CRC Press, Washington, DC, pp. 201–218.

Johnson, B.T. (2000). Revised: Activated Mutatox assay for detection of genotoxic substances. Technical Methods Section. *Environ. Toxicol.* **15**(3): 253–259.

USE OF LUMINESCENT BACTERIA AND THE *LUX* GENES FOR DETERMINATION OF WATER QUALITY

Nirit Ulitzur
Checklight Ltd.
Tivon, Israel

Shimon Ulitzur
Technion Institute of Technology
Haifa, Israel

INTRODUCTION

Water quality monitoring tests have expanded over the years to include the use of bioassays for routine testing and screening of water quality in an attempt to overcome several drawbacks of using the physicochemical tests. The identification of chemicals by the sophisticated chemical and instrumental methods are conditioned by the presence of a reference sample. Such methods are unable to estimate possible biological synergistic or antagonistic interactions between chemicals, and their routine use is limited due to their high cost and the requirement for skilled personnel. The routine use of aquatic bioassays has been limited by the time required (usually days) to perform such tests as well as the cost and skill level. In order to overcome these limitations, numerous short-term bioassays have been developed in the past fifteen years, but the most commonly used and thoroughly validated rapid tests are the bioluminescence-based bioassays based on the use of luminescent bacteria. The following sections describe the various approaches and technologies that utilize luminescent bacteria for determination of water quality.

USE OF NATURAL LUMINESCENT BACTERIA FOR ON-SITE DETERMINATION OF WATER QUALITY

The generation of light by luminescent bacteria involves the activity of the enzyme luciferase that simultaneously catalyzes the oxidation of $FMNH_2$ and long chain aldehyde with the aid of molecular oxygen to yield FMN, long chain fatty acids, water, and light. Seven genes (*luxRICDABE*) are involved in the regulation and synthesis of the *Vibrio fischeri lux* proteins. The intricate molecular aspects of the bacterial *lux* system have been reviewed extensively (1–4).

The use of intact luminescent bacteria for analytical purposes has some clear advantages: the light of a single bacterium may reach about 30,000 quanta/s/cell; thus, one can use a simple luminometer to detect light generated by a few hundred cells/mL. Chemophysical and biological factors that affect cell respiration, the rate of protein or lipid synthesis, promptly alter the level of luminescence. In 1979, Bulich first showed that luminescent bacteria used for toxicity testing could be freeze-dried and provided to testing laboratories in a convenient reagent configuration (5). Freeze-dried (lyophilized) luminescent bacteria reagent can be stored for years at $-18\,°C$. Upon hydration, freeze-dried cultures can yield 95–98% of their original viability, as well as their original level of *in vivo* luminescence. Thus, the use of these preparations for analytical purposes is not different, in practice, from other biochemical tests. Numerous studies described the use of natural luminescent bacteria or recombinant *lux*-carrying bacteria for rapid assessment of toxic components in aqueous environments. The next sections describe in more detail the bioluminescent tests for acute and chronic water toxicants.

MICROTOX TEST

The use of the bacterial luminescence system for the rapid assessment of aquatic toxicity was first suggested by Bulich (5) and later commercially marketed as Microtox. The bioassay is based on a preparation of lyophilized cultures of *Vibrio fischeri* (NRRL B11177). The lyophilized culture is hydrated with high purity water from which 10 μL aliquots each containing 10^6 cells are added to serial dilutions of the tested water supplemented with NaCl. Light measurements from the diluted water sample and clean water control are obtained after 5–15 min of incubation at $15\,°C$. The minimal concentration of tested water that results in 50% inhibition of light relative to the light level of the control is defined as IC_{50} (or EC_{50}) of the said water sample, or of the tested chemical spiked into it. Figure 1 shows typical results and determination of the IC_{50} value for a toxic agent spiked into a clean water sample. The Microtox test has been successfully applied worldwide for quality testing of industrial water, sewage, effluents, and contaminated sediments (6,7) and qualified as a standard for certain types of samples by various standards organizations in the United States and Europe. For many chemicals and toxic water samples, the sensitivity of the tested water in Microtox agreed well with the results obtained with daphnia or fish-based bioassays (8). In order to meet the need for the

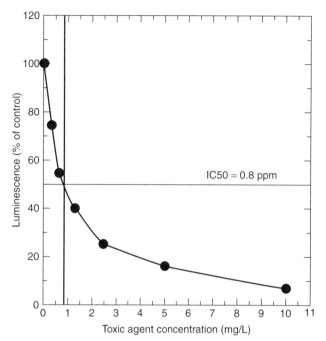

Figure 1. Typical response of luminescent bacteria to a toxic agent (see text for details).

higher sensitivity required for drinking water analysis, we developed for AZUR Environmental a 22-h chronic test that involves bacterial growth and *de novo* synthesis of the *lux* system in *V. fischeri* (9,10). The test has been successfully used to screen drinking water samples for the presence of unexpected chemical contamination. The Microtox Test System is commercially available from Strategic Diagnostics, Inc.

TOXSCREEN TEST

A suitable test for drinking water toxicants should be rapid and sensitive enough to detect very low concentrations of diverse groups of toxic agents. Other requirements from such a test include simple operation, minimal false positive or negative responses, affordability, and the capability to run the test outside the laboratory. A novel bioassay (ToxScreen) that demonstrates those requirements has been developed by CheckLight (11). ToxScreen utilizes a marine luminescent bacterium (*Photobacterium leiognathi*) and applies unique assay buffers that markedly increase the sensitivity of the test and also enable the preliminary discrimination between organic and cationic metal toxicants. Cationic heavy metals can be detected at sub-milligram/liter levels; many of them were detected at 10–50 μg/L range. Similarly, pesticides, PAHs, and chlorinated hydrocarbons were detected in a separate test, all within 20–45 min. For most of the tested toxic agents, ToxScreen was found to be markedly (over 10–100-fold) more sensitive than Microtox (11). A practical feature of the new bioassay is the option of running the test at ambient temperatures ($18–35\,°C$). In addition, the stability of the freeze-dried bacterial reagent preparation precludes the need for refrigeration or freezing during shipment. These features, together with its simple operation procedure, suggest that it has potential applications as a cost-effective prescreening tool to appraise chemical toxicity in various water sources. The higher sensitivity and simplicity of ToxScreen has recently been confirmed by the U.S. EPA's Environmental Technology Verification (ETV) program of rapid toxicity technologies (http://www.epa.gov/etv/verifications/vcenter1-12.html).

USE OF LUMINESCENT BACTERIA FOR ONLINE MONITORING OF WATER TOXICITY

The increasing concern for drinking water quality, together with the growing terrorism awareness, has emphasized the need for real-time monitoring of drinking water quality. Two kinds of bioluminescence-based online monitors for water quality have been developed so far. An online monitor that utilizes lyophilized cultures of luminescent bacteria was first developed by AZUR Environmental and Siemens in the mid-1990s. The device (Microtox-OS) was designed to automate the standard Microtox test and operate on-site, unattended for two weeks. The Microtox-OS Test System was equipped with a statistical process control software to continuously monitor data and was user-configurable to identify test results that were statistically different from preceding values. Due to its high operation cost, however, only a few units were sold.

The TOXcontrol-BioMonitor (now available through Microlan, The Netherlands) is based on freshly cultivated light-emitting *V. fischeri* cells that are continuously mixed with clean water and tested water. Light is detected using an online dual monitoring luminometer as it passes through twin flow cells. Any significant difference in light detected from the reference and monitoring streams results in a number of programmable events: an alarm can be programmed or a valve can be switched to collect samples for further analysis. No information has been published on the sensitivity of this automatic bioluminescent test. In general, luminescent bacteria growing in liquid medium in the presence of nutrients are not expected to show high sensitivity toward toxic agents. The system is mainly used for river biomonitoring and wastewater monitoring in Europe.

The integration of the ToxScreen test with an automatic online monitor was recently introduced by CheckLight Ltd. and Systea Srl and named MicroMac-ToxScreen. MicroMac-ToxScreen is an innovative automated online water quality monitoring system that utilizes the ToxScreen reagents to detect microgram/liter concentrations of toxic organic and inorganic chemical pollutants in surface or groundwater, as well as raw and treated drinking water. Freeze-dried luminescent bacteria are hydrated and kept in the device at 4 °C to maintain a stable luminescent culture. The instrument is resupplied with newly hydrated luminescent bacteria and a fresh inventory of liquid assay every two weeks. Automatic safeguards have been engineered into the system to assure reagent and data quality and appropriate instrument functioning. The instrument is also equipped with autocalibration features to assure reliable instrument performance; microprocessor-based system controls provide for data storage, data downloading, real-time communication with a remote PC, and user adjustable alarm levels.

Cho et al. (12) have recently adapted a 384-multiwell plate monitor for determination of water toxicity. Lyophilized *luxAB* carrying *Janthinobacterium lividum* cells are exposed to the water in question for 15 min followed by luminescence recording after addition of aldehyde. Similar to other terrestrial *lux*-carrying systems, this test is more sensitive to metals than the Microtox test.

LUX REPORTER GENES FOR DETERMINATION OF GENERAL STRESS-INDUCING AGENTS OR SPECIFIC TOXIC CHEMICALS

With the development of recombinant DNA technology, various recombinant bioluminescent bacteria have been engineered by fusing the *lux* gene with a stress-responding inducible promoter or with a specific promoter for a given toxicant (13). Exposing these recombinant cells to active toxic agents resulted in dramatic increase in bacterial luminescence. Such bacteria can be extremely useful for bioremediation studies; they can be utilized to determine the presence and concentrations of specific pollutants such as Hg(II) (14), or detection of linear alkanes (15) and naphthalene (15), or general responses to environmental stress (16,17).

A novel approach for identification of water toxicants by using *Escherichia coli* carrying *lux* genes fused with stress promoters was suggested by Ben Israel et al. (18). Twenty-three out of 25 tested chemicals were identified by a novel testing matrix using a 3-h test procedure. This biological recognition strategy also provides a tool for evaluating the degree of similarity between the modes of action of different toxic agents. Recently, Kim and Gu (19) described a high throughput toxicity monitoring and classification biosensor system using four immobilized bioluminescent *E. coli* strains that have plasmids bearing a fusion of a specific promoter to the *lux*CDABE operon. These strains are immobilized in a single 96-well plate using an LB-agar matrix. The bioluminescence increases when the water sample contains chemicals that cause either oxidative damage, membrane damage, DNA damage, or protein damage.

USE OF LUMINESCENT BACTERIA FOR DETERMINATION OF MUTAGENIC AND CARCINOGENIC AGENTS

Ulitzur et al. (20,21) and Weiser et al. (22) were the first to use luminescent bacteria for determination of mutagenic and genotoxic agents. The first test, later commercialized as Mutatox (23,24), was based on a spontaneous dark variant (M169) of the luminescent bacterium *Photobacterium leiognathi*. Light production was restored when cells were incubated in the presence of subacute concentrations of genotoxic agents, including base substitution, frameshift, DNA synthesis inhibitors, DNA damaging agents, and DNA intercalating agents (25). Mutatox test showed very good correlation with the Ames test (26). The primary genetic lesion of M169 was discovered later to be due to the presence of mutation in the *rpoS* gene (4).

Vitotox® is another example of a bioluminescent test for genotoxic agents (27,28). In this test, two recombinant *Salmonella typhimurium* TA104 strains are used to determine both genotoxicity and cytotoxicity of the tested sample. Under normal growth conditions, the luciferase reporter gene is repressed; in the presence of a DNA damaging agent, the SOS response is stimulated and leads to derepression of the strong promoter controlling the expression of the luciferase gene.

A new mutagenicity assay, in which a series of genetically modified strains of marine bacterium *Vibrio harveyi* was used for detection of mutagenic pollution of marine environments, has recently been described by Czyz et al. (29).

USE OF LUMINESCENT BACTERIA FOR DETERMINATION OF NUTRIENTS IN WATER

AOC (Assimilable Organic Carbon)

Regrowth potential of heterotrophic bacteria in potable water depends mainly on the presence of an assimilable organic carbon source. Many bacteria are capable of dividing in water containing as low as 5–10 µg/L of various carbon sources. Most of the standard or proposed AOC methodologies are hindered by their long duration and require between 5 and 14 days for completion. A bioassay that utilizes a nutrient-deprived culture of autoinducer-requiring mutant of *V. fischeri* has been developed and is commercially available (CheckLight Ltd). Luminescence in this strain is directly proportional to the concentration of utilizable organic material in the water sample. This 2-h test enables the detection of as little as 5 µg/L organic carbon and the test has shown high correlation with the standard 7-d long AOC test.

Haddix et al. (30) have recently described a test system in which *Pseudomonas fluorescens* P-17 and *Spirillum* species strain NOX bacteria were engineered with a *luxCDABE* operon fusion and inducible transposons that were selected on minimal medium. Independent mutants were screened for high luminescence activity and predicted AOC assay sensitivity. All mutants tested were able to grow in tap water under AOC assay conditions. Peak bioluminescence and plate count AOC tests were linearly related. Bioluminescence results that were obtained 2–3 d postinoculation were comparable with the 5-d ATP pool luminescence AOC assay and 8-d plate count assay.

BOD (Biochemical Oxygen Demand)

The biochemical oxygen demand (BOD) test (BOD5) is a crucial environmental index for monitoring organic pollutants in wastewater but is practically limited by the 5-d requirement for completing the test. A rapid BOD test based on an autoinducer-requiring mutant of *V. fischeri* is now commercially available (CheckLight Ltd.). The test procedure requires the water in question to be boiled for 30 min in the presence of 1 N HCl followed by addition of 1 N NaOH to neutralize the sample. This brief hydrolysis step is sufficient to break down proteins and complex carbohydrates into oligomers or monomers that can be assimilated by the bacteria. A good correlation was found between the 2-h bioluminescent test and the 5-d standard BOD test.

A BOD sensing system based on bacterial luminescence using recombinant *E. coli* containing *lux A-E* genes from *V. fischeri* has recently been described by Sakaguchi et al. (31). Frozen cells of luminescent recombinants of *E. coli* were applied to measure and detect organic pollution due to biodegradable substances in various wastewaters. The data from this study showed a similar correlation with that of the conventional method for BOD determination (BOD5).

Other Test Systems

Due to the intense use of synthetic nitrogen fertilizers and livestock manure in current agriculture practices, food (particularly vegetables) and drinking water may contain unacceptable concentrations of nitrate. A plasmid-borne transcriptional fusion between the *E. coli* nitrate reductase (*nar*G) promoter and the *Photorhabdus luminescens lux* operon provided *E. coli* with a highly bioluminescent phenotype when in the presence of nitrate concentrations as low as 5×10^{-3} mol/L (0.3 ppm) (32).

Recently, Gillor et al. (33) have described a reporter system for nitrogen availability. The promoter of the glutamine synthetase-encoding gene, P *gln*A, from *Synechococcus* sp. strain PCC7942 was fused to the *luxAB* luciferase-encoding genes of the bioluminescent bacterium *Vibrio harveyi*. The resulting construct was introduced into a neutral site on the *Synechococcus* chromosome to yield the reporter strain GSL. Light emission by this strain was dependent on ambient nitrogen concentrations. The linear response range of the emitted luminescence was 1 mM to 1 µM for the inorganic nitrogen species tested (ammonium, nitrate, and nitrite) and 10–50-fold lower for glutamine and urea. When water samples collected from a depth profile in Lake Kinneret (Israel) were exposed to the reporter strain, the bioluminescence of the reporter strain mirrored the total dissolved nitrogen concentrations chemically determined for the same samples. This test strain was shown to be a sensitive indicator of bioavailable nitrogen.

CONCLUSIONS

The bioluminescence tests for detecting water toxicants as well as for measuring elementary nutrients have been shown to be simple, rapid, and relatively inexpensive tools for monitoring water quality. Still, the relatively high cost ($3000–25,000) of commercially available luminometers restricts the routine applications of these technologies. The use of the *lux* genes in recombinant microorganisms has not been commercially applied yet mainly due to the restriction on using recombinant DNA in commercial products. Wider application of bioluminescent tests for screening water quality will ensure a better supply of adequate quality and will minimize the danger of water contamination by terrorist acts.

BIBLIOGRAPHY

1. Wilson, T. and Hastings, J.W. (1998). Bioluminescence. *Annu. Rev. Cell Biol.* **14**: 197–230.

2. Meighen, E.A. and Dunlap, P.V. (1993). Physiology, biochemical and genetic control of bacterial bioluminescence. *Adv. Microbol. Physiol.* **34**: 1–67.
3. Ulitzur, S. and Dunlap, P. (1995). The role of LuxR protein in regulation of the lux system of *Vibrio fischeri*. *Photochem. Photobiol.* **62**: 625–632.
4. Ulitzur, S., Matin, A., Fraley, C., and Meighen, E. (1997). H-NS protein represses transcription of the lux systems of *Vibrio fischeri* and other luminous bacteria cloned into *Escherichia coli*. *Curr. Microbiol.* **35**(6): 336–342.
5. Bulich, A.A. (1979). Use of luminescent bacteria for determining toxicity in aquatic environments. In: *Aquatic Toxicology, ASTM STP 667*. L.L. Markings and R.A. Kimerle (Eds.). American Society of Testing and Materials, Washington, DC.
6. Bulich, A.A. and Isenberg, D.L. (1981). Use of the luminescent bacterial system for the rapid assessment of aquatic toxicity. *ISA Trans.* **20**: 20–33.
7. Bulich, A.A. and Bailey, G. (1995). Environmental toxicity assessment using luminescent bacteria. In: *Environmental Toxicology Assessment*. M. Richardson (Ed.). Taylor and Francis, London, pp. 29–40.
8. Kaiser, K.L.E. (1993). Qualitative and quantitative relationships of Microtox data with toxicity data for other aquatic species. In: M. Richardson (Ed.). *Ecotoxicology Monitoring*, VCH, Weinheim.
9. Bulich, A.A. and Huynh, H. (1995). Measuring chronic toxicity using luminescent bacteria. *Can. Tech. Rep. Fish Aquat Sci.* **2050**: 23.
10. Backhaus, T., Froehner, K., Altenburger, R., and Grimme, L.H. (1997). Toxicity testing with *Vibrio fischeri*: a comparison between the long term (24 h) and the short term (30 min) bioassay. *Chemosphere* **35**: 2925–2938.
11. Ulitzur, S., Lahav, T., and Ulitzur, N. (2002). A novel and sensitive test for rapid determination of water toxicity. *Environ. Toxicol.* **17**: 291–296.
12. Cho, J.C. et al. (2004). A novel continuous toxicity test system using a modified luminescent freshwater bacterium. *Biosens. Bioelectron.* **20** (2): 338–344.
13. Ross, P. (1993). The use of bacterial luminescence systems in aquatic toxicity testing. In: M.L. Richardson (Ed.). *Ecotoxicology Monitoring*, VCH Verlagsgesellschaft, New York, p. 185–195.
14. Selifonova, O., Burlage, R., and Barkay, T. (1993). Bioluminescent sensors for detection of bioavailable Hg(II) in the environment. *Appl. Environ. Microbiol.* **59**(9): 3083–3090.
15. Werlen, C., Jaspers, M.C.M., and van der Meer, J.R. (2004). Measurement of biologically available naphthalene in gas and aqueous phases by use of a *Pseudomonas putida* biosensor. *Appl. Environ. Microbiol.* **70**: 43–51.
16. Belkin, S., Vollmer, A.C., van Dyk, T.K., Smulski, D.R., Reed, T.R., and LaRossa, R.A. (1994). Oxidative and DNA-damaging agents induce luminescence in *E. coli* harboring lux fusions to stress promoters. In: *Bioluminescence & Chemiluminescence: Fundamentals & Applied Aspects*. A.K. Campbell, L.J. Krica, and P.E. Stanely (Eds.). John Wiley, Chichester, UK, pp. 509–512.
17. DuBow, M.S. (1998). The detection and characterization of genetically programmed responses to environmental stress. *Ann. NYAS Online* **851**: 286–291
18. Ben-Israel, O., Ben-Israel, H., and Ulitzur, S. (1998). Identification and quantification of toxic chemicals by use of *Escherichia coli* carrying lux genes fused to stress promoters. *Appl. Environ. Microbiol.* **64**: 4346–4352.
19. Kim, B.C. and Gu, M.B. (2003). A bioluminescent sensor for high throughput toxicity classification. *Biosens. Bioelectron.* **18**(8): 1015–1021.
20. Ulitzur, S., Weiser I., and Yannai S. (1980). A new, sensitive, fast and simple bioluminescence assay for mutagenic compounds. *Mutat. Res.* **74**: 113–124.
21. Ulitzur, S. and Weiser, I. (1981). Acridine dyes and other DNA-intercalating agents induce the luminescence system of luminescent bacteria and their dark variants. *Proc. Nat. Acad. Sci. U.S.A.* **78**: 3338–3342.
22. Weiser, I., Ulitzur, S., and Yannai, S. (1981). DNA damaging agents and DNA synthesis inhibitors induce the luminescence in dark variants of luminescent bacteria. *Mutat. Res.* **91**: 443–450.
23. Kwan, K.K., Dutka, B.J., Rao, S.S., and Liu, D. (1990). Mutatox test: a new test for monitoring environmental genotoxic agents. *Environ. Pollut.* **65**(4): 323–332.
24. Bulich, A. (1992). Mutatox: a genotoxicity assay using luminescent bacteria. *Schriftenr Ver Wasser Boden Lufthyg* **89**: 763–770.
25. Ulitzur, S. (1986). Bioluminescence test for genotoxic agents. *Methods Enzymol.* **133**: 264–274.
26. Sun, T.S. and Stahr, H.M. (1993). Evaluation and application of a bioluminescent bacterial genotoxicity test. *J. AOAC Int.* **76**: 893–898.
27. van der Leile, D. et al. (1997). The VITOTOX test, and SOS bioluminescence *Salmonella typhimurium* test to measure genotoxicity kinetics. *Mutat. Res.* **389**: 279–290.
28. Verschaeve, L. et al. (1999). Vitotox® bacterial genotoxicity and toxicity test for the rapid screening of chemicals. *Environ. Mol. Mutagen.* **33**: 240–248.
29. Czyz, A. et al. (2002). Comparison of the Ames test and a newly developed assay for detection of mutagenic pollution of marine environments. *Mutat. Res.* **519** (1–2): 67–74.
30. Haddix, P.L., Shaw, N.J., and LeChevallier, M.W. (2004). Characterization of bioluminescent derivatives of assimilable organic carbon test bacteria. *Appl. Environ. Microbiol.* **70**(2): 850–854.
31. Sakaguchi, T. et al. (2003). A rapid BOD sensing system using luminescent recombinants of *Escherichia coli*. *Biosens. Bioelectron.* **19**(2): 115–121.
32. Prest, A.G., Winson, M.K., Hammond, J.R., and Stewart, G.S. (1997). The construction and application of a lux-based nitrate biosensor. *Lett. Appl. Microbiol.* **24**(5): 355–360.
33. Gillor, O. et al. (2003). *Synechococcus* PglnA::luxAB fusion for estimation of nitrogen bioavailability to freshwater cyanobacteria. *Appl. Environ. Microbiol.* **69**(3): 1465–1474.

WATER QUALITY MANAGEMENT

CHAKRESH K. JAIN
K.D. SHARMA
National Institute of Hydrology
Roorkee, India

INTRODUCTION

Water is one of the basic necessities for the survival of human beings and the prosperity of civilization. The ever increasing growth of population, industrialization, the steady rise in irrigation, urbanization, and the high level of living standards exert tremendous pressure on the available water which is highly uneven in its spatial and temporal distribution in quantity and quality. Water is put to different uses such as irrigation, industry, power generation, drinking, bathing, recreation, fisheries, wild life propagation, and pollution abatement. For each of the uses, water is required in appropriate quantity and quality. Nowadays, there is an increasing awareness of water quality, especially in urban areas. Due to the rapid increase in population and the growth of industrialization in the country, the pollution of natural water by municipal and industrial wastes has increased tremendously.

The term "water quality" is a widely used expression, which has an extremely broad spectrum of meanings. Each individual has vested interests in water for a particular use. The term water quality, therefore, must be considered relative to the proposed use of water. From the user's point of view, the term "water quality" is defined as the physical, chemical, and biological characteristics of water by which the user evaluates the acceptability of water. For example, for the sake of human health, we require a pure, wholesome, and potable water supply.

In recent years, rivers are being used indiscriminately for disposal of municipal, industrial, and agricultural wastes, thereby polluting the river water beyond permissible limits. Due to this, river waters are gradually becoming unfit even for irrigation in some places. Thus it has become essential to evaluate the environmental impacts of water resources to minimize the progressive deterioration in water quality. Therefore, a detailed study of water quality in all vulnerable rivers is imperative for better management and use of water.

WATER QUALITY MONITORING

Water quality problems stem from two factors, the natural hydrology of a river basin and the development and use of land and water resources by human beings. Depending on the interrelation of these two factors, a wide variety of quality problems can result. Each river basin, therefore, is unique, and it must be subjected to individual and intensive water quality assessment to provide a proper basis for judicious management of land and water.

The following points should be considered for proper planning of a water quality monitoring program:

What are the objectives of the program?
From where are samples to be taken?
Which determinants are of interest?
When and how often are samples to be taken?
What is to be done with the results?

These questions will provide a framework for defining measurement programs. The results obtained from such a program should be regularly reviewed to decide if any changes (e.g., determinants or sampling frequency) in the monitoring program are necessary.

Objectives of the Monitoring Program

It is obvious that the objectives of a program should be clearly and precisely formulated by the user. If it is not done, inappropriate analytical data may well be provided, and/or the user is likely to call for needlessly large or unduly small numbers of results. Furthermore, if the objectives are not precisely expressed, it will be difficult or impossible to decide the extent to which they are achieved.

It is suggested that analytical information should be requested regularly only when the user knows beforehand that the results will be used in a precisely known fashion to answer one or more defined questions on quality. Requests for analysis based on the thought that the results may ultimately prove useful should be avoided, particularly when the analytical sampling effort is limited. There is an almost infinite number of analyses that might be useful in most situations, but it is completely impracticable to attempt any such comprehensive coverage. Therefore, selection from all possible objectives and determinants is essential for proper planning of a monitoring program.

As a further means of optimizing measurement programs, users should formulate their information needs as quantitatively as possible. As an extreme and perhaps rather artificial example of a badly defined requirement, consider the statement, "to obtain information on the quality of the river." Such a statement is almost completely useless as a basis for designing a monitoring program for the following reasons:

1. The determinants are not specified, so that the analyses required are not known.
2. The particular river and the locations on the river are not defined, so that inappropriate sampling positions may be chosen.
3. No indication is given of the timescale or sampling frequency, so that too few or too many samples may be collected and analyzed.
4. As a result of all of the above deficiencies, there is no indication of the amount of data that needs to be processed and the nature of the data treatment, so that appropriate data handling techniques cannot be defined.

Therefore, users of analytical results must seek to avoid uncertainties such as those in the preceding example by careful and quantitative definition of every aspect of their requirements. Thus, an objective such as the above would be better expressed by a statement of the form, "to estimate each year the annual average concentration of ammonia ($NH_3 + NH_4^+$) at all river sites used for the production of potable water." Appropriate statements of this type for other determinants of interest then provide a set of quantitative targets essential in optimizing the choice of sampling, analytical, and data handling techniques.

Objectives of GEMS/WATER

The fundamental objectives of the water quality monitoring system within the GEMS/WATER program are to:

1. Assess the impacts of human activities upon the quality of the water and its suitability for required uses.
2. Determine the quality of water, in its natural state, which might be available to meet future needs.

3. Keep under observation the source and pathways of specified hazardous substances.
4. Determine the trend of water quality at representative stations.

The first objective is met by the establishment of impact stations, the second by baseline stations, the third by either impact or baseline stations depending upon whether the hazardous substance is of artificial or natural origin, and the fourth by trend stations.

Baseline stations are located where no direct diffuse or point sources of pollutants are likely to be found. They are used to establish the natural background level of variables, to check if no synthetic compounds are found in remote areas (e.g., DDT), and to assess the long-term trends of surface water quality resulting from global atmospheric pollution. Impact stations are situated in waterbodies where there is at least one major use of the water or which are greatly affected by human activities. Four type of impact stations can be identified according to different uses of water:

1. *Drinking water*: at the raw water intake before treatment for drinking water.
2. *Irrigation*: at the water intake before distribution for irrigation.
3. *Aquatic life*: river and lake stations representative of the general quality of the waterbody.
4. *Multiple impacts*: several water uses at the station and/or of the waterbody.

Trend stations are set up specially to assess the trends in water quality. They must be representative of a large area with various types of human activities. These stations should be more frequently sampled to increase the statistical significance of the average concentrations and to validate the trends.

Sampling Locations and Points

Sampling location and point mean the general position within a waterbody and the exact position at which samples are obtained. The objectives of a program sometimes immediately define the sampling locations. For example, when the concern is to measure the efficiency of a chemical plant for purifying water, sampling locations will be required before and after the plant. Similarly, when the effect of effluent discharge on the water quality of a receiving river is of interest, samples will be required from locations upstream and downstream of the discharge. For larger scale water bodies (e.g., a river basin, a large estuary, a large urban drinking water distribution system), however, the objectives may be defined in terms that do not indicate sampling locations. For example, objectives such as "to measure river quality within a river basin" or "to measure the quality of water in a distribution system" give no indication of which of the large number of possible sampling locations are of interest. Such broadly expressed objectives are completely inadequate for detailed planning of efficient sampling programs and should always be sharpened, so that they do indicate the positions of sampling locations. A commonly useful device for helping in this respect is to consider the intended use of the water because this aids in indicating these positions in a waterbody where quality is of key importance.

Figure 1 illustrates the hypothetical case of a river system along with criteria for choosing different sampling sites. Figures 2 and 3 illustrate the location of sampling sites for lakes and groundwater with corresponding criteria for choosing of different sampling sites.

Determinants of Interest

The particular determinants appropriate for a program depend critically on the type of water and the objectives of the study. Depending on the intended use of a receiving water, the parameters listed in Table 1 are significant for water quality characterization. They are guidelines for analysis of wastewater quality for treatment and control. The parameters listed in Table 2 are frequently used to identify various types of pollution from industrial wastewater.

Some of the most important and most frequently used tests for analyzing water are the nonspecific tests listed in Table 3. These tests often measure a property of a group of substances. For example, alkalinity indicates the capacity of water to neutralize hydrogen ions. Many of these tests are used to determine the suitability of natural water for industrial or municipal use and to determine the type and degree of treatment required. Table 4 lists some of the more frequently measured parameters in pollution studies.

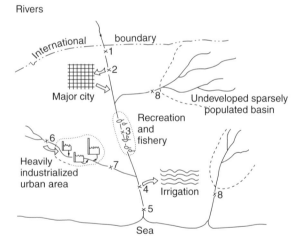

Station	Type	Criteria
1	Impact	Immediately downstream of an international (or baseline) boundary
2	Impact	Abstraction for public supply of large town
3	Impact	Important fishing, recreation, and amenity zone
4	Impact	Abstraction for large-scale agricultural area
5	Trend	Freshwater tidal limit of major river
6	Impact	Abstraction for large industrial activity
7	Impact	Downstream of industrial effluent discharge and important tributary influencing main river
8	Baseline	Station where water is in a natural state (no direct or indirect pollution, no water use)

Figure 1. Monitoring site selection—rivers.

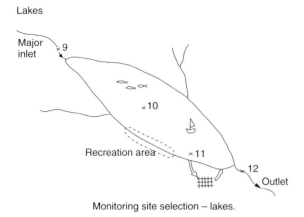

Station	Type	Criteria
9	Impact (or baseline)	Principal feeder tributary
10	Impact (or baseline)	General water quality of lake
11	Impact	Water supply of major city
12	Impact (or baseline)	Water leaving lake

Figure 2. Monitoring site selection—lakes.

Frequency and Time of Sampling

The quality of water in various waterbodies is rarely constant in time. There may be some relationship between the role of change in different variables, but others alter independently. In measuring the mean, maximum, and minimum values of variables over a period of time, the closeness of the monitored values to the true values depend upon the variability of the variables and the number of samples taken. The larger the number of samples from which the mean is derived, the narrower are the limits of the probable difference between the observed and true means. These confidence limits are not directly proportional to the number of samples but to the square of the number. Therefore, to double the reliability of a mean value, the number of samples must be increased fourfold.

Variations in water quality are caused by changes (increase or decrease) in the quantity of any of the inputs to a waterbody. Such changes may be natural or manmade and either cyclic or random. Water quality variation may therefore be similarly cyclic or random.

Variability differs among rivers, lakes, and underground waters. It is most pronounced in rivers, and the ranges are greater, the nearer the sampling point is to the source or sources of variability. As the distance from the source increases, longitudinal mixing smooths out irregularities and fewer samples are needed to meet given confidence limits. However, as the distance between the source of variability and the sampling point increases, there is a reduction in the range of variation, and there is also dilution. Some variables are reduced by self-purification,

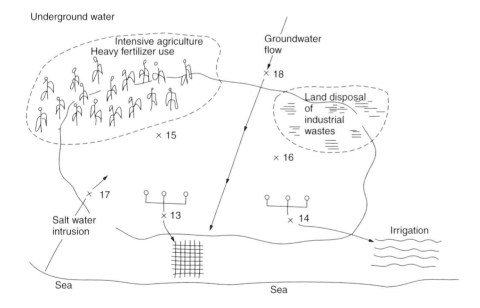

Station	Type	Criteria
13	Impact	Water supply to large town threatened by fertilizer residues and saline intrusion
14	Impact	Water for large-scale irrigation threatened by leachate from waste tips
15 } 16 } 17 }	Impact	Water supply of major city
18	Baseline	No human activities in the groundwater recharge area

Figure 3. Monitoring site selection—underground water.

Table 1. Parameters for Water Quality Characterization

Water Use	Quality Parameters
Domestic water supply	Color, odor, taste
	Organic content: chlorine demand, COD, BOD, TOC, phenols
	Carcinogens and toxic compounds: insecticides, pesticides, detergents
	Turbidity, salinity
	Alkalinity, pH
	Total hardness, Ca, Mg, Fe, Si, etc.
	Pathogenic organisms, total bacterial count (37 °C), *E. coli*, plankton count
Fish, shellfish, wildlife, and recreation	Color, odor
	Toxic compounds
	Turbidity, floating matter, sludge deposits, salinity
	Temperature
	Dissolved oxygen, BOD
	Alkalinity, pH, pathogenic organisms, plankton count, nitrogen, phosphorus (inorganic nutrients that support algal blooms and other undesirable aquatic growth)
Agricultural irrigation	Salinity and Na-Ca content
	Alkalinity, pH
	Pesticides, growth regulators
	Persistent synthetic chemicals (e.g., polyethylene derivatives, asphalt sprays.)
	Pathogenic organisms
Watering livestock	Salinity
	Toxic compounds
	Pathogenic organisms
	Plankton count

Table 2. Significance of Parametric Measurements

Test(s)	Significance
Dissolved solids	Soluble salts may affect aquatic life or future use of water for domestic or agricultural purposes
Ammonia, nitrites	Degree of stabilization (oxidation) or nitrates and total organic nitrogenous matter, organic nitrogen
Metals	Toxic pollution
Cyanide	Toxic pollution
Phenols	Toxic pollution, odor, taste
Sulfides	Toxic pollution, odor
Sulfates	May affect corrosion of concrete, possible biochemical reduction to sulfides
Calcium and magnesium	Hardness
Synthetic detergents	Toxic pollution

deposition, and adsorption. These effects must be considered if a sampling station used for quality control is located some distance from the point of use.

In lakes, the mass of water and good lateral mixing provide inertia against any rapid changes from modifications in input and output. Many lakes exhibit marked seasonal variations due to thermal stratification, overturn, and biological activity. Depending on the type of lake, it may be sampled with a seasonal bias related to the natural cycles of the lake.

Underground water has a lower variability than that of either rivers or lakes. The rate of quality change depends on the depth of sampling, the volume of the aquifer, and hydraulic conductivity. The time elapsing between changes in land use and in surface recharge water and their effect upon the underground water depends on the time of percolation. Variations are often, but not invariably, seasonal with a time lag according to the rate of percolation. Direct injection into boreholes or saline intrusion from subterranean sources may take effect more rapidly.

The time of sampling is also of main concern when the quality of the water shows more or less regular variations, for example, diurnal variations in the concentration of dissolved oxygen in rivers and variations of quality on start-up and shutdown of industrial plants.

It is seldom that the quality of water at only one instant is of interest. Normally, information is required for a period during which quality may vary. The problem arises, therefore, of deciding the time at which to collect samples, so that they will adequately represent the quality during the period of interest.

The best technical solution to this problem is often to use an automatic, on-line instrument that provides continuous analysis of the water of interest. This approach can be of great value in that, in principle, a continuous record of quality is obtained and the problems of selecting particular times for sampling do not arise.

Table 3. Nonspecific Water Quality Parameters

Physical	Chemical	Physiological
Filterable residues	Hardness	Taste
Salinity	Alkalinity and acidity	Odor
Density	BOD	Color
Electrical conductance	COD	Suspended solids
	Total carbon	Turbidity
	Chlorine demand	

Table 4. Tests Used for Measuring the Pollution of Natural Waters

Nutrient Demand	Specific Nutrients	Nuisance	Toxic
DO	Nitrogen	Sulfide	Cyanide
BOD	Ammonia	Sulfite	Heavy metals
COD	Nitrate	Oil and grease	Pesticides
Total carbon	Nitrite	Detergents	
	Organic nitrogen	Phenols	
	Phosphorus		
	Orthophosphate		
	Polyphosphate		
	Organic phosphorus		

MAJOR WATER QUALITY ISSUES

Water Scarcity

Due to uneven distribution of rainfall in time and space and ever-increasing demand of water for agricultural, industrial, and domestic activities, water resources are overexploited. This results in shrinking or even drying up of many waterbodies for a considerable period in a year.

Maintaining targeted water quality in such waterbodies is very difficult. If the objectives of the Water Act of 1974 are to be fulfilled, maintaining a minimum level of water needs to be specified, for example, restrictions on water abstraction from different waterbodies (rivers, lakes, or ground water). The Central Ground Water Authority (CGWA) formed under the Environment (Protection) Act of 1986 has already taken such initiatives. The restrictions need to be extended also to rivers and lakes.

Targets for conservation include reducing demands by optimum use, minimization of waste, efforts to reduce percolation and evaporation losses, conservation efforts in domestic uses, groundwater recharging, rainwater harvesting, afforestation, recycling, and reuse.

There are a number of cases where salinity is increasing in both surface water and groundwater. The increase in groundwater salinity is mainly due to increased irrigation or seawater intrusion in coastal areas. The salinity in surface water is increasing mainly due to discharges of industrial wastewater or agricultural return water. Salinity impairs the fitness of water for drinking or irrigation. It may also affect the ecosystem in surface waters.

Oxygen Depletion

A large portion of wastewater mostly from domestic sources is discharged into watercourses without any treatment. Such wastewater contains high amounts of organic matter. Industries also discharge effluents containing high levels of organic matter, for example, agro industries. When oxidized in water by microbes, this organic matter consumes dissolved oxygen. Water has limited availability of oxygen, and when consumption exceeds the availability, oxygen depletion results. Thus, the survival of aquatic life becomes difficult.

In many waterbodies, a massive input of organic matter sets off a progressive series of chemical and biological events in downstream water. The stretch is characterized by high bacterial population, cloudy appearance, high BOD, and strong disagreeable odor—all indicating general depletion of oxygen. Masses of gaseous sludge rising from the bottom are often noticed floating near the surface of the water. During the monsoon, the sludge deposited in such stretches is flushed and stays in suspension, causing a rise in oxygen uptake downstream. Due to such sudden oxygen depletion, heavy fish mortality occurs every year during the first flushing after the onset of the monsoon.

Pollution Due to Urbanization

Urbanization has encouraged people to migrate from villages to urban areas. This has given rise to a number of environmental problems, such as water supply and wastewater generation, collection, treatment, and disposal. In urban areas, water is tapped for domestic and industrial use from rivers, streams, wells, and lakes. Approximately 80% of the water supplied for domestic use passes out as wastewater. In most cases, wastewater is let out untreated, and either it percolates into the ground and in turn contaminates the groundwater or is discharged into the natural drainage system causing pollution in downstream areas.

Nonpoint Source Pollution

Nonpoint sources of water pollution have been recognized as of greater importance than point sources. This is due in part to the continuing efforts to reduce pollution from point sources over the past few decades, as well as recognition that nonpoint sources, such as stormwater, may contain harmful contaminants. Pesticides and nutrients, particularly, are of major concern because of high toxicity and eutrophication. This has increased the need to identify and quantify major sources of nutrients and pesticides deposited in river systems.

For proper understanding of the nature and the magnitude of diffuse water pollution under different circumstances, it is necessary to consider some source related characteristics, briefly discussed below, for diffuse pollution from some more common sources.

In rural areas, most people use open fields for defecation, a few use pit latrines or septic tanks. Much of the bathing and washing (clothes, utensils, etc.) is in or near the waterbody, reducing abstraction and transport of water but causing *in situ* diffuse pollution.

Eutrophication

The discharge of domestic wastewater, agricultural return water or runoff water, and many industrial effluents contributes nutrients such as phosphates and nitrates. These nutrients promote excess growth of algae in waterbodies. This is not desirable for a balanced aquatic ecosystem.

Salinity

Salinity is increasing in many waterbodies, especially groundwater, due to leaching of salt buildup in agricultural areas under intense irrigation. A number of industrial activities discharge wastewater with high dissolved solids that increase the salinity of water.

Due to the discharge of toxic effluents from many industries and increased use of chemicals in agriculture and their subsequent contribution to waterbodies, many waterbodies in the country are polluted by toxic substances. Toxic substances impair water quality and make it unfit for human consumption, aquatic life, and irrigation.

Natural Contaminants

By far the greatest water quality problem in developing countries is the prevalence of waterborne diseases. Yet, in addition to human induced pollution problems, water supplies also have specific natural quality problems, which are mainly related to local geology. Some of the specific problems of natural origin include fluoride and arsenic.

Pathogenic Pollution

Waterborne diseases are the most important water quality issues in India. This is mainly due to inadequate arrangements for transport and treatment of wastewaters. A major portion of the wastewater generated from human settlements is not properly transported and treated before discharge into natural waters. This results in contamination of both surface water and groundwater. Moreover, the contribution of pathogens from diffuse sources is also quite significant. Thus, most of the surface waterbodies and many groundwater sources are contaminated.

A large population in the country still uses water directly for drinking or contact use without any treatment and thus are exposed to waterborne diseases. This is the single major cause of mortality due to water pollution.

Ecological Health

A large number of areas in our aquatic environment support rare species and are ecologically sensitive. They need special protection. The Water Act of 1974 provides for maintaining and restoring the wholesomeness of aquatic resources, which is directly related to the ecological health of waterbodies, so it is important that the ecological health of waterbodies is given first priority in water quality goals.

GROUNDWATER QUALITY

In recent years, groundwater pollution due to human activity has become an increasing threat. The adverse effects on groundwater quality are the result of human activity on the ground, unintentionally from agriculture, domestic and industrial effluents, and unexpectedly from subsurface or surface disposal of sewage and industrial wastes.

A major problem in urbanized areas is the collection and disposal of domestic wastewater. Because a large volume of sewage is generated in a small area, the waste cannot be adequately disposed of by conventional septic tanks and cesspools. Therefore, special disposal sites may be required to collect and dispose of such wastes in densely populated areas.

The quality of groundwater is of great importance in determining the suitability of a particular groundwater for a certain use (public water supply, irrigation, industrial applications, power generation, etc.). The quality of groundwater is the result of all the processes and reactions that have acted on the water from the moment it condensed in the atmosphere to the time it is discharged by a well. Therefore, the quality of groundwater varies from place to place, with the depth of water table, and from season to season, and is primarily governed by the extent and composition of dissolved solids in it.

The wide range of contamination sources is one of the many factors contributing to the complexity of groundwater assessment. It is important to know the geochemistry of chemical–soil–groundwater interactions to assess the fate and impact of pollutants discharged onto the ground. Pollutants move through several different hydrologic zones as they migrate through the soil to the water table. The serious implications of this problem necessitate an integrated approach to undertake groundwater pollution monitoring and abatement programs.

The supply of groundwater is not unlimited, nor is it always available in good quality. In many cases, the abstraction of excessive quantities of groundwater has resulted in drying up of wells, saltwater intrusion, and drying up of rivers that receive their flows in dry seasons from groundwater. Groundwater quality is increasingly threatened by agricultural, urban, and industrial wastes, which leach or are injected into underlying aquifers. Once pollution has entered the subsurface environment, it may remain concealed for many years, becoming dispersed over wide areas, and rendering groundwater supplies unsuitable for human use.

A vast majority of groundwater quality problems are caused by contamination, overexploitation, or combination of the two. Most groundwater quality problems are difficult to detect and hard to resolve. The solutions are usually very expensive, time-consuming, and not always effective. Groundwater quality is slowly but surely declining everywhere. Groundwater pollution is intrinsically difficult to detect; it may well be concealed below the surface, and monitoring is costly, time-consuming, and somewhat hit-or-miss by nature.

Groundwater in several areas, where sewage is discharged without proper treatment, has been adversely affected by contaminants from sewage. Due to indiscriminate use of fertilizers, very high concentrations of potassium and nitrate have been found in groundwater at several places in the states of Punjab, Haryana, and Uttar Pradesh. Excessive concentrations of nitrate in groundwater that may originate from fertilizers or sewage, or both, have been reported to the extent of several hundred, mg/L in various parts of the country. Groundwater pollution from some industrial sources has reached alarming levels. High levels of hexavelant chromium at Ludhiana and Faridabad, lead near Khetri in Rajasthan, nickel in Coimbatore, and cadmium in Kanpur and parts of Delhi are some of the manifestations of heavy metal pollution. Arsenic concentrations in groundwater have been found in West Bengal in excess of the permissible limit of 0.05 mg/L. The population in these areas is suffering from arsenic dermatosis from drinking arsenic-rich groundwater.

The problem of groundwater pollution in several parts of the country has become so acute that unless urgent steps for detailed identification and abatement are taken, extensive groundwater resources may be damaged. All activities carried out on land have the potential to contaminate groundwater, whether from urban, industrial, or agricultural activities. Large-scale, concentrated sources of pollution, such as industrial discharges, landfills, and subsurface injection of chemicals and hazardous wastes, are obvious sources of groundwater pollution. These concentrated sources can be easily detected and regulated, but the more difficult problem is from diffuse sources of pollution, such as leaching of agrochemicals and animal wastes, subsurface discharges from latrines and septic tanks, and infiltration of polluted urban runoff and sewage where sewerage does not exist or is defunct. Diffuse sources can affect entire aquifers, which are difficult to control and treat. The only solution to diffuse sources of pollution is to integrate land use with water management. Table 5 presents land-use activities and their potential threat to groundwater quality.

Common Groundwater Contaminants

- *Nitrates*: Dissolved nitrate is the most common contaminant in groundwater. High levels can cause blue baby disease (methemoglobinemia) in children, may form carcinogens, and can accelerate eutrophication in surface waters. Sources of nitrates include sewage, fertilizers, air pollution, landfills, and industries.
- *Pathogens*: Bacteria and viruses that cause waterborne diseases such as typhoid, cholera, dysentery, polio, and hepatitis. Sources include sewage, landfills, septic tanks, and livestock.
- *Trace metals*: Include lead, mercury, cadmium, copper, chromium, and nickel. These metals can be toxic and carcinogenic. Sources include industrial and mine discharges and fly ash from thermal power plants due to either fallout or disposal in ash ponds.
- *Organic compounds*: Include volatile and semi-volatile organic compounds such as petroleum derivatives, PCBs, and pesticides. Sources include agricultural activities, street drainage, sewage landfills, industrial discharges, spills, and vehicular emission fallout.

Table 5. Land-Use Activities and Their Potential Threat to Groundwater Quality

Land Use	Activities Potential to Groundwater Pollution
Residential	Unsewered sanitation
	Land and stream discharge of sewage
	Sewage oxidation ponds
	Sewer leakage, solid waste disposal, landfill
	Road and urban runoff, aerial fallout
Industrial and commercial	Process water, effluent lagoon
	Land and stream discharge of effluent
	Tank and pipeline leakage and accidental spills
	Well disposal of effluent
	Aerial fallout
	Landfill disposal and solid wastes and hazardous wastes
	Poor housekeeping
	Spillage and leakage during handling of material
Mining	Mine drainage discharge
	Process water, sludge lagoons
	Solid mine tailings
	Oilfield spillage at group gathering stations
Rural	Cultivation with agrochemicals
	Irrigation with wastewater
	Soil salinization
	Livestock rearing
Coastal areas	Saltwater intrusion

LEGAL CONSIDERATIONS

The basic objective of the Water (Prevention and Control Pollution) Act of 1974 is to maintain and restore the wholesomeness of national aquatic resources. Due to the large variation in type, size, shape, quality, and quantity of water available from our aquatic resources, each one has a very specific waste load receiving capacity. This implies the need for prescribing different effluent standards based on the assimilative capacity of recent systems. Notwithstanding its merits, it is difficult to administer compliance with widely varying standards for specific types of effluents. To reduce the administrative difficulties of relating effluent quality to ambient water quality, the concept of Minimum National Standards (MINAS) was developed by the Central Pollution Control Board, whereby minimum effluent limits are prescribed for each category of discharge, regardless of receiving water requirements. Where water quality standards cannot be reached by imposing a standard level of treatment alone, then and only then will the conditions of the receiving waters dictate more stringent controls (State Boards can make the MINAS more stringent). The conventional methods of treatment cannot cope with these specific situations, and the polluters are required to introduce new methods for specific requirements. The number of such situations in our country is gradually rising due to water scarcity in many waterbodies and some times due to the nature of effluents.

APPROACH TO WATER QUALITY MANAGEMENT

The Water (Prevention and Control of Pollution) Act of 1974 reflects the national concern for water quality management. The basic objective of the Act is to maintain and restore the wholesomeness of water through prevention and control of pollution. The Act does not define the level of wholesomeness to be maintained or restored in different waterbodies of the country. To define these levels for different waterbodies, the Central Pollution Control Board (CPCB) had initially taken the use of water as a basis for identifying water quality objectives for different waterbodies. Hence, it has classified the national aquatic resources according to their uses. CPCB has also identified primary water quality criteria for different uses of water as a yardstick for preparing different pollution control programs. The use-based classification system is given in Table 6.

The "designated best use" yardstick, as explained above, is based on a limited number of parameters called primary criteria. While implementing such criteria, there are many practical difficulties, which need scientific support. The criteria should be such that they can support the designated uses and also the ecological sustainability of the waterbody (wholesomeness), which is the prime objective of the Water Act of 1974. Hence, the CPCB, with the help of an expert group, has revised the approach.

Table 6. Use-Based Classification of Surface Waters in India

Designated Best Use	Quality Class	Primary Quality Criteria
Drinking water source without conventional treatment, but with chlorination	A	6.5 to 8.5 (1); 6 or more (2); 2 or less (3); 50.5–200%, and 20–50% (4); NIL (5–8)
Outdoor bathing (organized)	B	6.5 to 8.5 (1); 5 or more (2); 3 or less (3); 500, 5%-2000, and 20%-500 (4); NIL (5–8)
Drinking water source with conventional treatment	C	6.5 to 8.5 (1); 4 or more (2); 3 or less (3); 5000, 5%-20000, and 20%-5000 (4); NIL (5–8)
Propagation of wildlife and fisheries	D	6.5 to 8.5 (1); 4 or more (2); NIL (3–4); 1.2 (5); NIL (6–8)
Irrigation, industrial cooling, and controlled waste disposal	E	60 to 8.5 (1); NIL (2–5); 2250 (6); 26 (7); 2 (8)

(1) pH.
(2) Dissolved oxygen, mg/L.
(3) BOD, (20 °C) mg/L.
(4) Total coliform (MPN/100 mL).
(5) Free ammonia, mg/L.
(6) Electrical conductivity, µmho/cm.
(7) Sodium adsorption ratio.
(8) Boron, mg/L.

The Revised Approach

Human interests cannot be ignored or downgraded in importance, but it is now widely accepted that the long-term interests of human beings themselves lie in maintaining the environment and ecosystems in an overall healthy condition. A system totally oriented to the objective of protecting direct beneficial uses by humans and of classifying water quality on that basis may not be adequate. In many cases (such as for small waterbodies or those with no significant current water use), the objective may have to be the protection of basic environmental qualities for sustaining ecosystems. Even where current or potential beneficial uses can be clearly defined and a "designated best-use," the first priority in water quality assessment and management would be maintaining and restoring the desirable level of general environmental quality or "wholesomeness." With this approach, protection of designated best use is not being abandoned, but the overall health of ecology is given its rightful place, ahead of any direct current beneficial uses by humans. Such ecological considerations shall ensure the biological integrity of a waterbody, including structural as well as functional integrity.

For effective management of water quality, it is necessary to involve local administrative bodies at the district, block, municipal, and village panchayat levels, NGOs, and the citizens themselves in conducting at least a crude, rapid, and overall assessment of water quality.

Water quality requirements for specific large-scale organized uses may continue to be prescribed and even strengthened, but the stakeholders could share responsibility for water quality monitoring and management in such cases. It would lead to a two-pronged approach; defining water-quality requirements and classification with the objective of (1) maintaining and restoring wholesomeness of water for the health of ecosystems and the environment in general (basic water quality requirement) and (2) protecting designated organized uses of water by humans and expecting stakeholders to share responsibility in monitoring and managing such quality by associating with them local administrations, NGOs, and citizens in monitoring and managing the desired quality to the extent possible.

All waterbodies whose existing water quality is below acceptable levels should be identified on a priority basis to take remedial measures to restore their quality to acceptable levels within a stipulated period.

For waterbodies whose water is of acceptable quality, but below desirable quality, the limits of the relevant parameters should be identified and a quality restoration program should be implemented within a reasonable period.

WATER QUALITY MANAGEMENT AND NONPOINT SOURCE CONTROL

HARRY X. ZHANG
Parsons Corporation
Fairfax, Virginia

INTRODUCTION OF NONPOINT SOURCE CONTROL

Nonpoint source (NPS) pollution, unlike pollution from industrial and sewage treatment plants, comes from many diffuse sources. NPS pollution is caused by rainfall or snowmelt moving over and through the ground. As the runoff moves, it picks up and carries away natural and human-made pollutants, finally depositing them into lakes, rivers, wetlands, coastal waters, and even our underground sources of drinking water.

The United States has made tremendous advances in the past 25 years to clean up the aquatic environment by controlling pollution from industries and sewage treatment plants. Despite those successes, approximately 39% of assessed rivers and streams and 39% of assessed lakes are not safe for fish consumption, and 37% of its surveyed estuarine area is not safe for basic uses such as swimming or fishing. The estimates for nonattainment of swimming in surveyed rivers and lakes were 32% and 30%. Today, nonpoint source pollution is the leading remaining cause of water quality problems. It is the main reason that approximately 40% of our surveyed rivers, lakes, and estuaries are not clean enough for basic uses such as fishing or swimming (1,2).

Nonpoint source pollutants include (1) excess fertilizers, herbicides, and insecticides from agricultural lands and residential areas; (2) oil, grease, and toxic chemicals from urban runoff and energy production; (3) sediment from improperly managed construction sites, crop and forest lands, and eroding stream banks; (4) salt from irrigation practices and acid drainage from abandoned mines; (5) bacteria and nutrients from livestock, pet wastes, and faulty septic systems; and (6) atmospheric deposition and hydromodification (3). NPS pollution is widespread because it can occur any time activities disturb the land or water. Agriculture, forestry, grazing, septic systems, recreational boating, urban runoff, construction, physical changes to stream channels, and habitat degradation are potential sources of NPS pollution. Careless or uninformed household management also contributes to NPS pollution problems. The most common NPS pollutants are sediment and nutrients. These wash into waterbodies from agricultural land, small and medium-sized animal feeding operations, construction sites, and other areas of disturbance.

EXISTING REGULATIONS RELATED TO NONPOINT SOURCE CONTROL

During the last 10 years, the United States has made significant progress in addressing NPS pollution. At the federal level, recent NPS control programs include the Nonpoint Source Management Program established under Section 319 of the 1987 Clean Water Act (CWA) Amendments, and the Coastal Nonpoint Pollution Program established by the 1990 Coastal Zone Act Reauthorization Amendments (CZARA).

The most prominent of these was the Nonpoint Source Management Program set up by the 1987 amendments (4). Under this program, states were to identify waterways that would not be expected to meet ambient standards because of nonpoint sources. For these areas, states were to develop NPS programs that identified best management practices for the categories of nonpoint sources involved. Because EPA had little or no authority to regulate many nonpoint sources, it relied upon state and local governments to implement controls. The program provides grants to states, territories, and tribes to implement NPS pollution controls described in approved NPS pollution management programs. In the 1987 amendments to the Clean Water Act, toxic waste and pollution from NPS received more attention than they had previously, and stormwater discharges from urban areas and industrial facilities were included in the NPDES permit program.

Under Section 303(d) of the 1972 Clean Water Act, states, territories, and authorized tribes are required to develop lists of impaired waters. The law requires that these jurisdictions establish priority rankings for waters on the lists and develop total maximum daily loads (TMDLs) for these waters. In 1991, EPA issued the Guidance for Water Quality-Based Decisions: The TMDL Process (5). In 1992, the EPA clarified the goals of the Nonpoint Source Management Program and moved toward a watershed protection approach, which tailors NPS pollution control strategies to fit conditions in particular watersheds and gives state and local governments flexibility. In addition to engineering controls on nonpoint source pollution, management strategies might include outreach and education programs as well as government action to change land use patterns and activities that contribute to NPS pollution.

The 1987 amendments also created the National Estuary Program (NEP), which provides a basis for managing NPS pollution. The NEP relied on existing statutory authority at the state and municipal levels to accomplish program goals. Each National Estuary Program is charged to create and implement a Comprehensive Conservation and Management Plan (CCMP) that addresses all aspects of environmental protection for the estuary, including issues such as water quality, habitat, living resources, and land use.

Another regulation on nonpoint source pollution is the Coastal Zone Act Reauthorization Amendments Section 6217, which addresses coastal NPS pollution problems in coastal waters. This program is administered jointly with the National Oceanic and Atmospheric Administration (NOAA). In its program, a state or territory describes how it will implement nonpoint source pollution controls, known as management measures, that conform with those described in Guidance Specifying Management Measures for Sources of Nonpoint Pollution in Coastal Waters (6). The plans must also specify abatement measures that are "economically achievable" and yet "reflect the greatest degree of pollutant reduction achievable by applying the best available nonpoint source control practices, technologies, processes, siting criteria, operating methods, or other alternatives.

RELATED FEDERAL PROGRAMS FOR NONPOINT SOURCE CONTROL

There are several federal programs available to address nonpoint source control.

U.S. Environmental Protection Agency (EPA)

The EPA administers Section 319 of the Clean Water Act, also known as the Nonpoint Source Management Program. Under Section 319, states, territories, and tribes apply for and receive grants from the EPA to implement NPS pollution controls. The EPA administers other sections of the CWA to help states, territories, and tribes plan for and implement water pollution control programs, which can include measures for NPS control. These include Section 104(b)(3), Water Quality Cooperative Agreements; Section 104(g), Small Community Outreach; Section 106, Grants for Pollution Control Programs; Section 314, Clean Lakes Program; Section 320, National Estuary Program; and Section 604(b), Water Quality Management Planning. Together with the NOAA, EPA helps administer Section 6217 of the 1990 Coastal Zone Act Reauthorization Amendments, a program that tackles nonpoint source pollution affecting coastal waters (7).

National Oceanic and Atmospheric Administration (NOAA)

NOAA administers Section 306 of the Coastal Zone Management Act that provides funds for water pollution control projects, including NPS management activities, in states that have coastal zones. Together with the EPA, NOAA also helps administer Section 6217 of the CZARA.

U.S. Department of Agriculture (USDA)

The USDA administers incentive-based conservation programs through the Consolidated Farm Services Agency, the Natural Resources Conservation Service (NRCS), and the U.S. Forest Service to help control NPS pollution from agriculture and forestry.

U.S. Department of Transportation (DOT)/Federal Highway Administration (FHWA)

Under the Intermodal Surface Transportation Efficiency Act of 1991, the Federal Highway Administration developed erosion control guidelines for federally funded construction projects on roads, highways, and bridges.

U.S. Department of the Interior (DOI)

Within the U.S. Department of the Interior, the Bureau of Reclamation (BLR), the Bureau of Land Management (BLM), and the Fish and Wildlife Service (FWS) administer several programs to help states manage NPS pollution by providing technical assistance and financial support.

NONPOINT SOURCE CONTROL BY SOURCE CATEGORIES

NPS Pollution from Agriculture

The latest National Water Quality Inventory indicates that agriculture is the leading contributor to water quality impairments; it degraded 60% of the impaired river miles and half of the impaired lake acreage surveyed by states, territories, and tribes (1). Agricultural activities that cause NPS pollution include confined animal facilities, grazing, plowing, pesticide spraying, irrigation, fertilizing, planting, and harvesting. The major agricultural NPS pollutants that result from these activities are sediment, nutrients, pathogens, pesticides, and salts. Agricultural activities can also damage habitat and stream channels. Agricultural impacts on surface water and groundwater can be minimized by properly managing activities that can cause NPS pollution.

Managing Sedimentation. Excessive sedimentation clouds water, which reduces the amount of sunlight reaching aquatic plants; covers fish-spawning areas and food supplies; and clogs the gills of fish. In addition, other pollutants such as phosphorus, pathogens, and heavy metals are often attached to soil particles and wind up in the waterbodies with the sediment. Farmers and ranchers can reduce erosion and sedimentation by 20–90% by applying management measures to control the volume and flow rate of runoff water, keep the soil in place, and reduce soil transport.

Managing Nutrients. When nutrients such as phosphorus, nitrogen, and potassium in the form of fertilizers or manure are applied in excess of plant needs, they can wash into aquatic ecosystems where they can cause excessive plant growth, which reduces swimming and boating opportunities, creates a foul taste and odor in drinking water, and kills fish. In drinking water, high concentrations of nitrate can cause methemoglobinemia, a potentially fatal disease in infants also known as blue baby syndrome. Farmers can implement nutrient management plans that help maintain high yields and save money on the use of fertilizers while reducing NPS pollution.

Managing Confined Animal Facilities and Livestock Grazing. Confined animal and livestock areas are major sources of animal waste. Runoff from poorly managed facilities can carry pathogens (bacteria and viruses), nutrients, and oxygen-demanding substances that contaminate shellfishing areas and cause other major water quality problems. Groundwater can also be contaminated by seepage. Discharges can be limited by storing and managing facility wastewater and runoff using an appropriate waste management system.

Overgrazing exposes soils, increases erosion, encourages invasion by undesirable plants, destroys fish habitat, and reduces the filtration of sediment necessary for building streambanks, wet meadows, and floodplains. To reduce the impacts of grazing on water quality, farmers and ranchers can adjust grazing intensity, keep livestock out of sensitive areas, provide alternative sources of water and shade, and revegetate rangeland and pastureland.

Managing Irrigation. Irrigation water is applied to supplement natural precipitation or to protect crops from freezing or wilting. Inefficient irrigation can cause water quality problems. In arid areas, for example, where rainwater does not carry residues deep into the soil, excessive irrigation can concentrate pesticides, nutrients, disease-carrying microorganisms, and salts—all of which impact water quality—in the top layer of soil. Farmers can reduce NPS pollution from irrigation by increasing water use efficiency.

Managing Pesticides. Pesticides, herbicides, and fungicides are used to kill pests and control the growth of weeds and fungus. These chemicals can enter and contaminate water through direct application, runoff, wind transport, and atmospheric deposition. To reduce NPS contamination from pesticides, people can apply integrated pest management (IPM) techniques based on the specific soils, climate, pest history, and crop for a particular field.

NPS Pollution from Urban Runoff

The most recent National Water Quality Inventory reports that runoff from urban areas is the leading source of impairments to surveyed estuaries and the third largest source of water quality impairments to surveyed lakes. Runoff from rapidly growing urban areas will continue to degrade coastal waters.

Increased Runoff. Nonporous urban landscapes such as roads, bridges, parking lots, and buildings do not let runoff slowly percolate into the ground. Receiving waters are often adversely affected by urban runoff due to alterations in hydraulic characteristics of streams receiving runoff such as higher peak flow rates, increased frequency and duration of high flows and possible downstream flooding, and reduced baseflow levels.

Increased Pollutant Loads. Urbanization also increases the variety and amount of pollutants transported to receiving waters. Sediment from development and new construction; oil, grease, and toxic chemicals from automobiles; nutrients and pesticides from turf management and gardening; viruses and bacteria from failing septic systems; road salts; and heavy metals are examples of pollutants generated in urban areas (8). Sediments and solids constitute the largest volume of pollutant loads to receiving waters in urban areas.

Measures to Manage Urban Runoff

To protect surface water and groundwater quality, urban development and household activities must be guided by plans that limit runoff and reduce pollutant loadings.

Managing Urban Runoff from New Development. New developments should attempt to maintain the volume of runoff at predevelopment levels by using structural BMP controls and pollution prevention strategies. Plans for managing runoff, sediment, toxics, and nutrients can establish guidelines to help achieve both goals. Management plans are designed to protect sensitive ecological areas, minimize land disturbances, and retain natural drainage and vegetation.

Managing Urban Runoff from Existing Development. Existing urban areas can target their urban runoff control projects to make them more economical. Runoff management plans for existing areas can first identify priority pollutant reduction opportunities, then protect natural areas that help control runoff, and finally begin ecological restoration and retrofit activities to clean up degraded waterbodies.

Managing Runoff for On-Site Disposal Systems. The control of nutrient and pathogen loadings to surface waters can begin with the proper design, installation, and operation of on-site disposal systems. These septic systems should be situated away from open waters and sensitive resources such as wetlands and floodplains and should also be inspected, pumped out, and repaired at regular time intervals.

NPS Pollution from Forestry

Sources of NPS pollution from forestry activities include removal of streamside vegetation, road construction and use, timber harvesting, and mechanical preparation for planting trees. Road construction and road use are the primary sources of NPS pollution on forested lands; they contribute up to 90% of the total sediment from forestry operations. Harvesting trees in the area beside a stream can affect water quality by reducing the streambank shading that regulates water temperature and by removing vegetation that stabilizes the streambanks. These changes can harm aquatic life by limiting sources of food, shade, and shelter (7).

Establishing Streamside Management Areas. Plans for buffer strips or riparian zones often restrict forestry activities in vegetated areas near streams, thereby establishing special Streamside Management Areas (SMAs). Vegetation in the SMA stabilizes streambanks, reduces runoff and nutrient levels in runoff, and traps sediment generated from upslope activities before it reaches surface waters. SMA vegetation moderates water temperature by shading surface water and provides habitat for aquatic life.

Managing Road Construction and Maintenance. Good road location and design can greatly reduce the transport of sediment to waterbodies. Roads should follow the natural contours of the land and be located away from steep gradients, landslide-prone areas, and areas of poor drainage. Proper road maintenance and closure of unneeded roads can help reduce NPS impacts from erosion over the long term.

Managing Timber Harvesting. Most detrimental effects of harvesting are related to access and movement of vehicles and machinery and dragging and loading trees or logs. These effects include soil disturbance, soil compaction, and direct disturbance of stream channels. Poor harvesting and transport techniques can increase sediment production ten to twenty times and disturb as much as 40% of the soil surface.

Other NPS sources include those induced by hydromodification/habitat alteration, marinas/boating, and roads/highways/bridges, which will not be discussed in detail here.

EMPLOYING BEST MANAGEMENT PRACTICES

Public and private groups have developed and used pollution prevention and pollution reduction initiatives and NPS pollution controls, known as management measures, to clean up our water efficiently. The use of best management practices (BMPs) in addressing runoff problems was frequently identified as a NPS control measure, although additional research and field studies on the performance of BMPs were also needed. A wide variety of BMPs, both structural and nonstructural, are available to address urban storm water runoff and discharges (9–12). In summary, many BMPs are used primarily for water quantity control (i.e., to prevent flooding), although they may provide ancillary water quality benefits. Some BMP types have been analyzed for performance in terms of site-specific pollutant removal, but not extensively enough to allow for generalizations. It is still difficult to develop comparisons of cost-effectiveness for various BMP types. Other BMP types, particularly nonstructural and those that do not have discrete inflow or outflow points, are difficult to monitor. In addition, only limited cost studies have been conducted for stormwater BMPs. Therefore, the pollutant removal performance of certain BMP types is essentially undocumented. The role of chemical pollutant monitoring versus receiving stream biological monitoring in evaluating BMP performance on a watershed scale is not well documented or needs further research.

One particular new design that is becoming popular and is widely used in recent years is low impact development (LID). LID is an innovative stormwater management approach using a basic principle that is modeled after nature; managing rainfall at the source using uniformly distributed decentralized microscale controls. Its goal is to mimic a site's predevelopment hydrology by using design techniques that infiltrate, filter, store, evaporate, and detain runoff close to its source (13). Another important concept in controlling urban NPS pollution is smart growth. Although it recognizes the many benefits of growth, smart growth practice invests time, attention, and resources in restoring community and vitality to existing cities and older suburbs. The purposes of smart growth and low impact development practices are to promote a number of practices that can lessen the environmental impacts of development, to preserve ecosystem functions as well as specific critical areas, and to design developments to maintain predevelopment conditions (14).

WATER QUALITY MANAGEMENT AND TMDL PROGRAMS

Water quality management has become increasingly more complicated. Problems such as toxic contaminants, sediments, nutrients, and habitat alteration result from a variety of point and nonpoint sources. The Water Quality Planning and Management Regulation (40 CFR 130) links a number of Clean Water Act sections, including Section 303(d), to form the water quality approach to protecting and cleaning up the nation's waters. Previous practices for implementing 303(d) have focused primarily on point sources and wasteload allocations (WLA). In recent years, nonpoint source contributions to water quality problems have become better understood, and it is now clear that EPA and state implementation of 303(d) must encompass nonpoint source pollution problems and seek to address problems occurring over large geographic areas (5).

The water quality approach emphasizes the overall quality of water within a waterbody and provides a mechanism through which the amount of pollution entering a waterbody is controlled by the intrinsic conditions of that body of water and the standards set to protect it. Section 303(d) of the Clean Water Act addresses those impaired waters that are not "fishable, swimmable" by requiring states to identify the waters and to develop a total maximum daily load (TMDL). EPA's regulations establish a TMDL as the sum of point source load allocation (WLA) plus nonpoint source allocation (LA) plus a margin of safety (MOS). TMDLs can be expressed in terms of either mass per time, toxicity, or other appropriate measures that relate to a state's water quality standard. If BMPs or other nonpoint source pollution control actions make more stringent load allocations practicable, then WLAs can be made less stringent. Thus, the TMDL process provides for nonpoint source control trade-offs.

The TMDL process is viewed as an effective approach to watershed protection and a water quality management tool for both point and nonpoint source pollution control on a watershed scale. A watershed protection approach is a strategy for effectively protecting and restoring aquatic ecosystems and protecting human health. This strategy has as its premise that many water quality and ecosystem problems are best solved at the watershed level rather than at the individual waterbody or discharger level (15). The current regulations that address water quality problems have tended to focus on particular sources, pollutants, or water uses and have not resulted in an integrated environmental management approach. Consequently, significant gaps exist in our efforts to protect watersheds from the cumulative impacts of a multitude of activities. Existing water pollution prevention and control programs are, however, excellent foundations on which to build a watershed approach. Increasingly, state and tribal water resource professionals are turning to watershed management to achieve greater results from their programs. Major features of a watershed protection approach include targeting priority problems, promoting a high level of stakeholder involvement, integrated solutions that use the expertise and authority of multiple agencies, and measuring success through monitoring and other data gathering.

TMDL IMPLEMENTATION AND NPS CONTROL

NPDES permits are usually sought for point sources, but nonpoint source controls may be established by implementing BMPs, so that surface water quality objectives are met. These controls should be based on load allocations developed by using the TMDL process. When establishing permits for point sources on the watershed scale, the record should show that in the case of any credit for future nonpoint source reductions, there is reasonable assurance that nonpoint source controls will be implemented and maintained or that nonpoint source reductions are demonstrated through an effective monitoring program.

However, it is difficult to ensure, *a priori*, that implementing nonpoint source controls will achieve expected load reductions. Nonpoint source control measures may fail to achieve projected pollution or chemical load reductions due to inadequate selection of BMPs, inadequate design or implementation, or lack of full participation by all contributing sources of nonpoint pollution. States should describe nonpoint source load reductions and establish a procedure for reviewing and revising BMPs in TMDL documentation. The TMDL implementation plan must contain a description of best management practices or other management measures. The implementation plan may deal with sources on a watershed basis, as long as the scale of the implementation plan is consistent with the geographic scale for which the TMDL pollutant load allocations are established. This makes water quality monitoring an important component during TMDL implementation.

The achievement of nonpoint source load reductions is a complex challenge. The variability in nonpoint source loadings due to hydrologic variability can often make it difficult to discern short-term trends. Therefore, achievement of water quality standards is tracked using selected long-term milestones and measures. Tracking the implementation of management actions over time will provide valuable information. The tracking of implementation will assist in determining the success of the load allocation, the adequacy of funding and resources, the potential for water quality improvement, and the need for corrective actions. For areas that have predominantly nonpoint source controls, the use of tracking information can support demonstration of progress in the absence of clear benefit through water quality monitoring (16).

SUMMARY AND CONCLUSION

Today, nonpoint source pollution remains the largest source of water quality problems in the United States. It is the main reason that approximately 40% of our surveyed rivers, lakes, and estuaries are not clean enough for basic uses such as fishing or swimming (1,17). This remaining problem impacting water quality is not easily remedied because NPS pollution comes from pipes and also from diffuse sources such as agricultural and forestry operations, urban areas, and construction sites.

Best management practices or other innovative management measures such as low impact development are frequently identified to address the NPS problem. However, additional research and field studies on the long-term performance of BMPs, especially on the watershed scale, are much needed.

The past decade has seen a shift toward an emphasis from "end-of-pipe" treatment for point source pollution to the watershed approach, which encourages a holistic take on identifying problems and implementing the integrated solutions that are needed to overcome multiple causes of water quality impairment, including NPS. In addition to the watershed approach, the water quality based TMDL program has evolved as an effective tool for addressing more complicated water quality problems (e.g., for waters that remain polluted after the application of technology-based standards). TMDL implementation and NPS control on a watershed basis are two interconnected components for successful water quality management.

BIBLIOGRAPHY

1. U.S. EPA. (1996). *Nonpoint Source Pollution: The Nation's Largest Water Quality Problem*. EPA841-F-96-004A.
2. U.S. EPA. (2002). *National Water Quality Inventory: 2000 Report*. EPA-841-R-02-001.
3. U.S. EPA. (1994). *What is Nonpoint Source (NPS) Pollution? Questions and Answers*. Report EPA-841-F-94-005, http://www.epa.gov/owow/nps/qa.html
4. Ortolano, L. (1997). *Environmental Regulation and Impact Assessment*. John Wiley & Sons, Hoboken, NJ.
5. U.S. EPA. (1991). *Guidance for Water Quality-Based Decisions: The TMDL Process*. EPA 440/4-91-001, Washington, DC.
6. U.S. EPA. (1993). *Guidance Specifying Management Measures for Sources of Nonpoint Pollution in Coastal Waters*. Report EPA-840-B-93-001c, Washington, DC.
7. U.S. EPA. (1996). *The Nonpoint Source Management Program*. EPA841-F-96-004D.
8. U.S. EPA. (1996). *Managing Urban Runoff*. EPA841-F-96-004G.
9. American Society of Civil Engineers (ASCE). (2001). National Stormwater Best Management Practices (BMPs) Database, http://www.bmpdatabase.org/.
10. American Society of Civil Engineers (ASCE) and Water Environment Federation (WEF). (1998). *Urban Runoff Quality Management*. ASCE, Alexandria, VA and WEF, Reston, VA.
11. U.S. EPA. (1999). *Preliminary Data Summary of Urban Stormwater Best Management Practices*. Report EPA-821-R-99-012.
12. U.S. EPA. (2002). *Urban Stormwater BMP Performance Monitoring*. Report EPA-821-B-02-001.
13. Low Impact Development Center, Inc. (LID). (2002). <http://www.lid-stormwater.net/>.
14. U.S. EPA. (2002c). Environmental Protection and Smart Growth, <http://www.epa.gov/smartgrowth/>.
15. U.S. EPA. (1996d). *Why Watershed Approach*. Report EPA800-F-96-001.
16. U.S. EPA. (1999). *Draft Guidance for Water Quality-based Decisions: The TMDL Process (Second Edition)*. Report EPA-841-D-99-001.
17. U.S. EPA. (2001a). *National Management Measures to Control Nonpoint Source Pollution from Agriculture*. Washington, DC.

READING LIST

Novotny, V. and Olem, H. (1994). *Water Quality: Prevention, Identification, and Management of Diffuse Pollution*. Van Nostrand Reinhold, New York.

U.S. EPA. (2001b). *National Management Measures to Control Nonpoint Source Pollution from Forestry (Draft)*. Washington, DC.

U.S. EPA. (2002d). Total Maximum Daily Loads, <http://www.epa.gov/owow/tmdl>.

U.S. EPA. (2002e). *National Management Measures to Control Nonpoint Source Pollution from Urban Areas*. Report EPA-842-B-02-003, Washington, DC.

Wanielista, M., Kersten, R., and Eaglin, R. (1997). *Hydrology: Water Quantity and Quality Control*. John Wiley & Sons, Hoboken, NJ.

WATER QUALITY MANAGEMENT IN AN URBAN LANDSCAPE

RUSSELL N. CLAYSHULTE
Aurora, Colorado

Urban landscape broadly defines a built environment influenced by different combinations of land uses. Residential and commercial development, industrial areas, parks, open space, and roadways are examples of urban land uses. The construction of wastewater, water, and stormwater pipelines, and power lines, generally called utility infrastructure, also change the natural environment and help shape the urban landscape. Land use patterns strongly influence surface water quality and to a lesser extent groundwater. Land use choices are interactive parts of water quality management, restoration, or enhancement programs. Urban land use decisions must consider water quality management strategies and goals. Conversely, water quality must be considered in zoning and plotting processes used by local governments. A goal in shaping urban areas is to produce a sustainable urban landscape that incorporates reasonable and effective water quality strategies to restore, protect, or enhance the urban environment.

Scale affects how water quality is managed and often who is responsible for this management. Water quality management targets overall health and the quality of water and environmental resources. Management also provides direction in fixing specific water quality problems in targeted waterbodies. Consequently, water quality management in urban settings occurs on two basic levels. Macroscale landscapes include all combined urban land uses for a single city or a combination of cities and towns. Water quality management on the macroscale is often linked to watershed restoration or protection efforts. Microscale landscapes can include smaller urban watersheds, drainages, or other limited geographic areas where site-specific water quality management is necessary to control targeted urban-caused pollution. Macroscale watersheds are often measured in square miles (square kilometers); microscale landscapes are identified in acres (hectares).

Water quality management programs in urban landscapes are divided into three general source categories: point sources, nonpoint sources, and stormwater runoff. Point sources are discrete discharges that go from a pipe source (e.g., wastewater treatment plant discharge) back into the environment. Point sources can be permitted to meet very specific discharge limits under local, state, and federal regulations. Nonpoint sources are a type of diffuse pollution without a single point of origin or not introduced into the environment from a single pipe. Common nonpoint sources are agriculture, forestry, urban, mining, construction, hydromodification (e.g., dams, channels, and ditches), land disposal, and saltwater intrusion. Stormwater runoff is a type of nonpoint source runoff usually associated with urban landscapes, which is subject to permitting requirements for medium and large cities. Stormwater and other nonpoint sources in urban landscapes are collectively called urban runoff.

Urban runoff can flow across residential and urban streets, roofs, lawns, open space, hard (impervious) surfaces, and other areas. Urban runoff occurs under both wet and dry weather conditions. Dry-weather runoff from irrigation and leaky pipes contributes various amounts of polluted runoff. Urban runoff carries many different types of chemicals, sediments, debris, and pathogens. An excessive amount of urban pollution impairs water quality in wetlands, streams, rivers, lakes, reservoirs, or groundwater (receiving waters). Construction or other development activities introduce large amounts of sediments and other pollutants into receiving waters. Sediment and solids contribute the largest volume of pollutant load to urban receiving waters. Increased loads can harm fish and wildlife populations, kill native vegetation, reduce or limit recreational uses, and contaminate drinking water.

The terms pollution and pollutants are different. This difference is very important from a regulatory perspective. The 1972 federal Clean Water Act (Public Law 92-500) defines the term "pollution" to mean the "man-made or man-induced alteration of the chemical, physical, biological, and radiological integrity of water." Generally, pollution is the presence of a substance in the environment that because of its chemical composition or quantity prevents the functioning of natural processes and produces undesirable environmental and/or health effects. The act defines the term "pollutant" to mean any substance introduced into the environment that adversely affects the usefulness of a resource or the health of humans, animals, or ecosystems. Common types of pollutants include dredged spoil, solid waste, incinerator residue, sewage, garbage, sewage sludge, munitions, chemical wastes, biological materials, radioactive materials, heat, wrecked or discarded equipment, rock, sand, cellar dirt, and industrial, municipal, and agricultural waste discharged into water. Specific pollutants can be targeted and regulated through a number of local, state, and federal processes. Harmful impacts from pollutants and pollution are often linked to risk studies with measurable impact levels.

Pollution management in urban landscapes is a relatively new requirement for cities and towns. Water quality management, as a national or even global problem, became a higher priority public concern in the late 1940s. The first legislation to control water pollution was the Water Pollution Control Act of 1948 (Public Law 845, 80th Congress). In 1972, the United States Congress enacted the first comprehensive national clean water legislation in response to growing public concern for serious and widespread water pollution. The federal Clean Water Act (Public Law 92-500; as amended in 1977, Public Law 95–217 and as amended in 1987, Public Law 100-4) is the primary federal law in the United States that protects national waters, including wetlands, lakes, rivers, aquifers, and coastal areas. The goal of the federal Clean Water Act is to make waters of the United States "fishable and swimmable, where applicable." Meeting the goals of the act requires an extensive and integrated approach to water pollution abatement.

Federal regulations require all states to maintain, improve, and in some cases restore the quality of their water resources. This federal law significantly altered water resource management in urban settings. Federal regulations provide a clear path to clean water and an effective national water program. In 1972, only a third of U.S. waters were safe for fishing and swimming. Wetlands losses were estimated in excess of 460,000 acres annually with an alarming loss of urban wetlands. The United States Environmental Protection Agency reported almost 100% wetland loss for many larger cities. Agricultural and construction runoff caused erosion of over 2.25 billion tons of soil per year. The deposition of large amounts of phosphorus and nitrogen, referred to as nutrients, into receiving waters was attributed, in a large part, to urban centers. This nutrient enrichment caused many ponds, lakes, and reservoirs to experience excessive algal growth, resulting lost fisheries, and water supply contamination. Sewage treatment plants or point sources served about 85 million people. A large amount of human wastes were flowing into receiving waters with little or no treatment.

Water quality in the United States was characterized in a 1998 report to the American Congress as part of a

Table 1. Leading Sources of Water Quality Impairment Related to Urban Landscapes for Rivers, Lakes, and Estuaries[a]

	Rivers and Streams	Lakes, Ponds, Rivers	Estuaries
Pollutants	Siltation (38%)[b]	Nutrients (44%)	Pathogens (47%)
	Pathogens (36%)	Metals (27%)	Organic enrichment (42%)
	Nutrients (28%)	Siltation (15%)	Metals (27%)
Sources[c]	Hydromodification (20%)	Hydromodification (20%)	Municipal point sources (28%)
	Urban runoff/storm sewers (12%)	Urban runoff/storm sewers (12%)	Atmospheric deposition (23%)
			Urban runoff/storm sewers (23%)

[a] Adapted from Reference 1.
[b] Values in parentheses represent the percentage of surveyed river miles, lake acres, or estuary square miles that are classified as impaired.
[c] Excluding unknown, natural, and "other" sources.

national water quality inventory (1). Table 1 lists leading sources of water quality impairment in urban areas. This report showed that about 40% of receiving waters still did not meet the fishing and swimming goals of the federal Clean Water Act. In fact more than 291,000 miles of assessed rivers and streams did not meet state adopted water quality standards. Sediment and siltation is one of the most common pollutants from urban development and construction. Other major pollutants are pathogens, nutrients, and metals. Runoff from agricultural lands and urban areas is a major source of these pollutants. Across all types of waterbodies, states, territories, tribes, and other jurisdictions have shown that poor water quality affects drinking water, aquatic life, fish consumption, swimming, and other recreational uses.

However, the national water quality picture has improved since the passage of the Clean Water Act. By 2001, two-thirds of the nation's waters were safe for fishing and swimming. The rate of annual wetlands losses is reduced to about 70,000–90,000 acres. The amount of soil lost due to agricultural runoff and urban construction activities has been cut by over 1 billion tons annually. Phosphorus and nitrogen levels in receiving waters are lower. Modern wastewater treatment facilities serve over 173 million people. Aggressive water quality management programs in urban areas have helped make these differences.

States adopt use classifications and numerical water quality standards for receiving waters in urban watersheds. Within urban watersheds, waters are divided into individual stream segments for classification and standard setting. Site-specific water quality classifications are used to protect any existing or intended use of the target water. Typical use classification categories include recreation, agriculture, aquatic life, domestic water supply, or wetlands. These use classification categories are referred to as "beneficial uses." Beneficial use classification types vary from state to state and, in some cases, even within a state. A water quality standard consists of four basic elements: (1) *designated uses* of the water body (e.g., recreation, water supply, aquatic life, agriculture), (2) *water quality criteria* to protect designated uses (numerical pollutant concentrations and narrative requirements), (3) an *antidegradation policy* to maintain and protect existing uses and high quality waters, and (4) *general policies* addressing implementation issues (e.g., low flows, variances, mixing zones).

Regulatory and enforcement programs provide the basic framework for water quality management in an urban environment. These programs generally reside at the state level. Although the federal Clean Water Act and other federal laws provide guidance for regulatory programs, states have the flexibility to establish programs to fit conditions within the state. Often standards and classifications of surface and groundwater drive management options. Urban watershed efforts ensure that waters within an urban landscape meet underlying standards and beneficial uses. Enforcement actions and lawsuits that target the amount of allowable pollutant load discharged into receiving waters (wasteload litigation) have triggered more water quality management programs in urban centers.

An urban watershed protection approach is an integrated, holistic strategy to protect or attain the desired beneficial uses of waters within an urban area. The approach is more effective than isolated efforts under existing programs that do not consider the watershed as a whole. Nonpoint source or stormwater runoff control on a watershed level can have a significant impact on the protection of beneficial uses. A watershed protection approach addresses point source discharges along with nonpoint source and stormwater pollutant loads. An urban watershed approach also considers other human activities that may affect the uses and quality of water resources.

Development patterns change the natural landscape and replace it with hard or impervious surfaces (i.e., concrete, asphalt, or hard-packed dirt), which in turn change the hydrologic or flow patterns from surfaces and into receiving waters. Urbanization causes increased stormwater runoff, which is more intense, has a higher volume, and has less residual runoff than natural runoff conditions. A typical city block characterized by pavement and rooftops generates over nine times more runoff than a woodland area of equal size. Less water gets into underlying groundwater and more water enters receiving waters as runoff flow. Aquifers under urban areas are called "starved" because water recharge is minimized. Natural runoff from an area without any kind of disturbance is called background or sometimes baseflow. Urbanization that alters runoff patterns can decrease baseflow and increase runoff flow. In most urban landscapes, lawn irrigation or other forms of dry-weather discharge can greatly increase runoff flow.

Urban hydrologic changes or major features like roads and parking lots cause many types of water quality problems such as nutrient enrichment, chemical pollutants, turbidity or sedimentation, biological pathogen introduction, toxic conditions, or increased temperature. Transportation systems have the highest pollutant concentrations compared with other urban land uses. Common types of transportation-related pollutants include gasoline, exhaust, motor oil and grease, antifreeze, undercoating, brake linings, rubber, asphalt, concrete, and diesel fuel. Landscaping in urban areas can contribute large amounts of nutrients from fertilizer application and pesticide or herbicide residues. Grass clippings and leaves in stormwater systems are oxygen-demanding substances, which can cause low oxygen conditions in receiving waters and stress aquatic life.

Widespread water quality data from urban areas across the country show that water quality problems are directly caused by increased watershed imperviousness. Ecological stress becomes apparent in an urban watershed when the amount of impervious cover reaches 10–20%. Typically, residential areas have impervious cover ranging from 25% to 60%; commercial areas can reach 100%. Increased pollutant loading is directly linked to the amount of impervious surface. Urban landscapes with greater than 50% impervious surface have highly stressed ecosystems. Low impact development is a new urban management concept designed to reduce the amount of impervious surface and promote more natural runoff patterns and processes.

Changes in flow regimes also cause increased frequency of flooding and peak flow volumes in receiving waters. These floods cause loss of aquatic or riparian habitat and changes in stream physical characteristics (channel width and depth). Stormwater management requires urban areas to develop stormwater drainage systems to avoid flood damage. These systems accumulate runoff from storms and send these flows away from the urban area. Flood control structures and modifications to natural drainage systems decrease natural stormwater purification functions. Runoff quality has not been well addressed in the design criteria of flood control structures. In some cases, stormwater systems are used as a convenient method of waste disposal.

Water quality monitoring for urban runoff facilitates the critical element of relating the physical, chemical, and biological characteristics of receiving waters to land use characteristics. Without current information on water quality and pollutant sources, the effects of land-based activities on water quality cannot be fully assessed, effective management and remediation programs cannot be implemented, and program success cannot be evaluated. Monitoring programs in the past have been the responsibility of state and federal agencies; recent local watershed management has shifted monitoring to the watershed or microscale environment. Some common types of water and environmental quality indicator parameters monitored in urban landscapes are shown in Table 2.

Not all urban runoff is polluted. The severity of the water quality problem in receiving water is directly related to the beneficial uses assigned to the waterbody by the regulating agency. Urban runoff quality is assessed against established standards and classifications. Many stream segments in urban areas have site-specific water quality standards and classifications. Consequently, streams or rivers flowing through urban centers with multiple segments can have differing sets of standards and classifications. Water quality parameters identified as pollutants in one segment may not be pollutants in the next contiguous downstream segment. This factor makes pollutant characterization, management, and control very difficult in urban areas.

There are many implementation methods or tools used in watershed planning and management. The method used for a specific watershed program depends on the problem and available resources. Best management practices (BMPs) are water quality management tools. BMPs used in urban water quality programs fall into two categories: (1) erosion control that prevents discharge

Table 2. Common Indicator Parameters of Urban Runoff and Landscape Alteration

	Common Indicator Parameters
Urban Runoff Pollutants	Nutrients (total phosphorus, o-phosphorus, total nitrogen, nitrite, nitrate, ammonia); chemical oxygen demand; total organic carbon; salts (deicer products like sodium chlorides, calcium chlorides); biochemical oxygen demand (caused by grass clippings, animal wastes, leaves, or other carbon compounds); pathogens (total coliform bacteria, fecal coliform bacteria, *E. coli* bacteria, viruses, or protozoa); solids (total suspended solids, total dissolved solids, settleable solids); pH; temperature; aluminum, arsenic, boron, barium, cadmium, chromium, chromium(VI), chlorine, copper, cyanide, iron, lead, manganese, magnesium, mercury, nickel, radium, selenium, silver, sulfate, zinc, uranium; phenols, BTEX (benzene and xylene), organic (oil, gasoline or grease), and inorganic compounds; and herbicides and other pesticides
Urban Hydrologic or Habitat Alteration	Changes in nitrogen and phosphorus concentrations (nutrient enrichment); increased chlorophyll concentrations in urban ponds, lakes, and reservoirs (measure of algal production); bioaccumulation of toxins in plant and animals; changes in species biodiversity; deposition of sediment (sand, soil, or silt); decreases in dissolved oxygen; increased runoff; increased flow velocity; increased or decreased temperatures in riparian corridors; increased sediment and bed-load transport; streambank and coastal erosion and deposition; hydromodification (dams and channels); increased impervious or paved surfaces; and loss of riparian habitat

of pollutants or provides improved water quality in runoff from construction sites or development areas and (2) urban stormwater practices intended to reduce loads in the built urban environment. Similar best management practices are applicable to both stormwater runoff in urban areas and construction site runoff. Many urban watershed programs rely on BMPs to help solve pollution and/or pollutant problems. BMPs include structural and nonstructural methods, measures, or practices that help prevent, reduce, or mitigate adverse water quality problems caused by urbanization.

Structural BMPs are constructed to treat runoff passively before it enters receiving waters. Such BMPs used on a construction or development site can be either temporary or permanent and are designed to reduce sediment load or other runoff waste products for the life of the project. Additionally, these practices should protect aquatic or riparian environments. Nonstructural BMPs include prevention and source control that minimize or eliminate a problem before it occurs. Source control BMPs are sometimes referred to as "good housekeeping" measures because a clean site produces fewer pollutants than a dirty site. Site planning and design of BMPs may, in and of itself, be considered a nonstructural BMP.

Even as treatment facilities have improved, water quality goals have become more difficult and costly to meet. The physical, biological, and ecological characterization of urban water resources intensified after the passage of the 1972 Clean Water Act, and substantial efforts are needed to resolve problems and find workable solutions. A locally defined *balanced ecological community* achievable through water quality protection and water resource management is often not a community goal because of resource limitations. Achieving urban water quality goals requires a balance between the natural environment and designated resource uses. Effective and cost-efficient water quality management and supply require an integrated resource management program.

Maintenance, improvement, and restoration of water resources at a specific, watershed or regional scale are issues of concern to local governments, special districts, state agencies, and federal agencies. Institutional responsibilities vary among these entities in the water quality management system. Many states have approached water quality planning and management through state, regional, or locally linked programs using management and operating agencies. In most states, local management agencies are responsible for implementing state water quality management plans. Local agencies can decide on the need for and specific characteristics of a wastewater treatment process and then implement activities to meet specified state and federal goals.

The protection of surface and groundwater resources from urban growth and development can best be managed by plans that limit runoff and reduce pollutant loads. Local communities can use management plans to address water quality problems on a site-specific (e.g., stream segment or drainage basin) and a larger watershed level. Management planning in the United States and in many places around the world has shifted from a governmental process to a local effort. Consequently, hundreds of new water quality management agencies, particularly in urban areas, have begun grassroots efforts to protect and enhance local water resources.

Local management agencies may be individual municipal governments, watershed associations and authorities, general-purpose governments that hold a state or federal discharge permit, or other special districts responsible for planning and approving permitted facilities. Local governments or affiliated agencies can enter into agreements to form watershed associations or authorities with a single management agency. A municipality can be defined as any regional commission, county, metropolitan district offering sanitation service, sanitation district, water and sanitation district, water conservancy district, metropolitan sewage disposal district, service authority, city and county, city, town, watershed association or authority, Indian tribe or authorized Indian tribal organization, or any two or more of them that act jointly in a domestic wastewater treatment works. Water quality management in urban landscapes is an evolving and dynamic effort of pollution and pollutant abatement involving a growing number of groups, agencies, and governments. Land use decision makers must consider growth and development impacts on urban water resources and identify offsetting mitigative practices.

BIBLIOGRAPHY

1. United States Environmental Protection Agency. (2000). *Water Quality Conditions in the U.S.—A profile from the 1998 National Water Quality Inventory Report to Congress*, EPA 841-F-00-006. Environmental Protection Agency, Office of Water.

WATER QUALITY MANAGEMENT IN THE U.S.: HISTORY OF WATER REGULATION

ABSAR ALUM
Arizona State University
Tempe, Arizona

Water is the true lifeline of our planet because of the fundamental role it plays in all kinds of biological activities and its catalytic contribution/functions in major chemical and geologic processes. These fundamental roles make water the single most important factor determining the course of ecological, social, and economic changes on the earth. The sustained growth of any human society is intricately associated with the quality of freshwater resources. The Industrial Revolution in the early years of the last century resulted in much ecological pollution. Biological and chemical contamination of freshwater resources is the greatest threat to sustainability. Water quality management strategies used to prevent progression in water pollution are based on a decision-making process. Mandatory laws have traditionally been proved to be the most effective means to achieve the goals of any society. Historically, human societies have relied on some types of cultural norms, social standards, and ethical, cultural, or religious guidelines to maintain the integrity of water resources.

The Industrial Revolution of the early twentieth century and the environmental pollution quagmire resulting as its aftermath have acted as catalysts of change in water quality management. As a result, mandatory regulations have become a preferred option as opposed to voluntary guidelines to ensure proper water quality.

One of the hallmarks of the last century is increased public awareness of water quality, which has been augmented by plummeting water quality and steadily increasing population. Though water has historically been considered a commodity free for all, in the recent past, water quality and quantity have emerged as the bases of major conflicts on local, regional, national, and international levels. Under the new paradigm, water is perceived as a marketable commodity. Like any marketable commodity, the quality of water is dictated by the type of usage and is valued based on market laws of supply and demand. In the United States, it is estimated that one acre-foot of water used in agriculture and industry would generate revenue of $400 and $400,000, respectively. In the United States, the water quality management program has evolved over several decades. Based on the area of use and type of water, the history of water quality management can be divided into drinking water and wastewater regulations. Early legislation focused on ensuring the appropriate quality of drinking water, whereas the emergence of wastewater management is relatively new. Developing and implementing water quality standards and criteria achieve the appropriate quality of finished drinking water and reclaimed wastewater.

HISTORY OF DRINKING WATER REGULATIONS

The plight of public health during the nineteenth century acted as a driving force to understand waterborne disease outbreaks. By late nineteenth century, scientists established a link between gastroenteritis and drinking water contaminants, which resulted in realization of the need for water quality management. Therefore, the drinking water industry became the first sector with elaborate water quality assurance and management plans; however, there was no regulatory control over this industry. During the early 1900s, most drinking water plants built in the United States were driven by aesthetic need to reduce turbidity that resulted in an incidental reduction in disease causing agents. Slow sand filtration was commonly used by cities (such as Philadelphia) to reduce turbidity. In 1908, Jersey City (New Jersey) had the first water system to use chlorine as primary disinfectant. Other disinfectants (such as ozone) were not employed in the United States until several decades later.

The history of federal regulations for water quality goes back to the late nineteenth century. The River and Harbor Act of 1886 was the first federal regulation to address water quality. In 1893, the Interstate Quarantine Act, enacted by Congress, empowered the United States Public Health Service (PHS) to establish drinking water standards. The objective of the Interstate Quarantine Act of 1893 was "to prevent the introduction, transmission and spread of communicable diseases." The River and Harbor Act of 1886, recodified in 1899, empowered the Corps of Engineers to issue discharge permits.

In 1914, the U.S. Public Health Service (PHS) introduced the first drinking water quality standards. The standards for bacteriological (only contagious) quality of drinking water were applied only to interstate carriers like ships and trains. These standards were revised in 1925. The early 1940s was the beginning of the "Chemical Revolution" with the rapid introduction of synthetic and natural chemicals in the industrial, agricultural, and social sectors. Soon, it was realized that these chemicals were contaminating freshwater sources. Drinking water quality standards were revised again in 1946, and some new chemical standards were introduced.

The Federal Water Pollution Control Act of 1948 (FWPCA) was the first legislation to address the issue comprehensively. Principles of limited federal enforcement and limited federal financial assistance were adopted to develop state and federal cooperative programs. These principles provided the framework for the Federal Water Pollution Control Act of 1956 and the Water Quality Act of 1965. In 1962, the PHS introduced its last revision of drinking water standards. Although the emphasis was on microbial diseases, 28 chemicals were also regulated. These revisions included about 20 standards and, for the first time, recommended the use of qualified personnel. Some of the important recommendations were use of water from protected sources, control of pollution sources, and chlorination of water drawn from unprotected sources. Under the Water Quality Act of 1965, states were required to develop and establish water quality standards; by the early 1970s, all states had developed their standards.

Passage of the Water Quality Act of 1965 (October 2, 1965) also resulted in the creation of the Federal Water Pollution Control Authority (FWPCA) as a separate entity in the Department of Health, Education, and Welfare (HEW); however, FWPCA did not reside there long. In 1966, it was transferred to the Department of the Interior (Reorganization Plan No. 2 of 1966). Under the Water Quality Improvement Act of 1970 (84 Stat. 113, April 3, 1970), FWPCA was abolished and its functions were folded into the new United States Environmental Protection Agency (U.S. EPA).

The sporadic amendments of the FWPCA of 1948 and repeated restructuring and reorganization of the responsible agency resulted in inefficient implementation of these laws and regulations. During the early 1970s, increased awareness of environmental and health issues and emergence of environmental activism along with complacent environmental regulations resulted in an exigent need to overhaul the system and consolidate authority. To resolve these issues, FWPCA was amended in 1972 and Congress passed the Clean Water Act (CWA) of 1972. Under this act (the Clean Water Act of 1972), the U.S. EPA was given the mandate to develop regulations to restore and maintain the chemical, physical, and biological integrity of the nation's water. This act manifests a basic change in strategy, and the focus of enforcement was changed from implementing water quality standards to effluent regulations. The National Pollution Discharge

Elimination System (NPDES) was established under which each point source is required to obtain a discharge permit. Under the Clean Water Act, the U.S. EPA was given a mandate to establish technologically based effluent limitations.

The Clean Water Act of 1972 mandated that all publicly owned sewage treatment works use the "best practical control technology" for industrial wastes and secondary treatment by July 1, 1977, "best available treatment economically achievable" by July 1, 1983, and the elimination of all discharges of pollution by 1985. However, these deadlines were relaxed on the recommendations of the National Commission on Water Quality in 1977, and the zero discharge goals were eliminated. The commission recommended a comprehensive recycling and reuse program using alternative and innovative technologies to achieve the goals of the Clean Water Act.

Over the course of years, the CWA evolved into a multifaceted statute ensuring the achievement of its objectives through the construction of local wastewater treatment plants (construction grant program) and the creation of a major national wastewater discharge permitting system.

Local public agencies were required to submit proposals for each proposed construction project. These projects were funded through the federal grants allocated to the states. During the period 1972–1990, more than $60 billion in federal grant money was invested in municipal sewage treatment works, along with more than $20 billion from state and local governments.

Under the CWA amendments of 1987, the construction grant program was changed, and beginning in 1990, the method of municipal financial assistance was shifted from grants to loans provided by state revolving funds. Since 1990, all the states have created "revolving loan funds," which serve as low-cost financing sources for water quality infrastructural projects.

The Safe Drinking Water Act (SDWA), passed in 1974, established stringent regulatory standards for the quality of drinking water for all public water supply systems except the very smallest. Currently, the Safe Drinking Water Act regulates more than 175,000 public water systems in the United States.

The Surface Water Treatment Rule was passed in 1989. This rule covers all public water supply systems using surface water or groundwater under the direct influence of surface water and intends to protect against exposure to pathogens such as *Giardia*, viruses, and *Legionella*. It requires 4 and 3 log removal of enteric viruses and *Giardia* during water treatment. It also requires a 0.2 mg/L of residual disinfectant at the point of water entry and maintaining a detectable level of residual disinfectant throughout the distribution system. The Interim Enhanced Surface Water Treatment Rule (IESWTR) was promulgated in 1989. It covers all public water supply systems that serve 10,000 people or more and use surface water or groundwater under the direct influence of surface water. It requires 2 log removal of *Cryptosporidium* from filtered water systems.

The Stage I Disinfectants and Disinfection By-products (D-DBP) Rule of 1998 was aimed at providing public health protection to an additional 20 million households by reducing the levels of disinfectants and disinfection by-products in drinking water. This rule established the maximum residual disinfectant level goals (MRDLGs) for chlorine, chloramines, and chlorine dioxide; maximum contaminant level goals (MCLGs) for four trihalomethanes, two haloacetic acids, bromate, and chlorite; and National Primary Drinking Water Regulations (NPDWRs) for three disinfectants (chlorine, chloramines, and chlorine dioxide), two groups of organic disinfection by-products, and two inorganic disinfection by-products (chlorite and bromate). The NPDWRs consist of maximum residual disinfectant levels (MRDLs) or maximum contaminant levels (MCLs) or treatment techniques for these disinfectants and their by-products. The NPDWRs also include monitoring, reporting, and public notification requirements for these compounds.

In 2000, the U.S. EPA proposed a groundwater rule (GWR). It uses a multiple barrier strategy to reduce waterborne health risks. The following five components are the basis of this multiple barrier strategy:

- periodic sanitary survey of groundwater systems
- hydrogeologic assessment to identify wells sensitive to fecal contamination
- source water monitoring for systems drawing from sensitive wells without treatment or with other indications of risk
- correction of significant and fecal contamination
- compliance monitoring to ensure that disinfection treatment is reliably operated

The original Federal Water Pollution Control Act has had six major amendments in the intervening years. The passage of Federal Water Pollution Control Act Amendment of 1972, which is also known as the Clean Water Act (CWA), divides the history of water pollution control (including wastewater treatment) in the United States into two eras: the pre- and post-Clean Water Act periods. Before 1972, states had the regulatory authority for water pollution control, whereas federal involvement was limited to cases involving interstate waters. Due to unique socioeconomic and political circumstances, states have historically shown different levels of motivation and willingness to address water pollution control issues.

The history of water quality management and policy formulation is complicated (Table 1). The three most important lessons learned are institutional reform, improved processes for conflict resolution, and increased use of modern planning and decision-making procedures. All these lessons learned through the evolution of U.S. water quality management policy may serve as a guideline for other countries to avoid repeating mistakes and to ensure efficient use of water resources.

Table 1. Time Line of Water Quality Regulations in the United States

Year	Events	Comments
1890	Harbors and Rivers Act	Prohibited discharge of any refuse in to interstate waters
1893	Interstate Quarantine act	Intended to control spread of communicable diseases from one area to another
1899	Harbors and Rivers Act Amendment	Exempted the "refuse flowing from streets and sewers and passing therefrom in a liquid form"
1912	The Public Health Service Act	First water related regulation that prohibited use of a common cup on interstate carriers
1914	"Treasury Standards" or USPHS standards	First microbial standards applicable to water systems supplying water to interstate carriers. HPC count should be less than 100 per mL and not more than one of the five 10 mL portions of a sample can be positive for *E. coli*
1924	Oil and Pollution Control Act	To control discharge of oil damaging coastal waters
1925	Revised USPHS Standards	1 coliform/100 mL; first physical-chemical standards for Pb, Cu, Zn, and soluble mineral salts
1943	Revised USPHS Standards	Standards revised to include more contaminants
1946	Revised USPHS Standards	Standards revised to include more contaminants
1948	Federal Water Pollution Control Act	Public Law 8-845: first law to address comprehensively the issue based on the principle of limited federal enforcement and limited federal financial assistance
1948	Establishment of Division of Water Supply and Pollution Control (DWSPC),	DWSPC established in PHS, FSA, to administer the Water Pollution Control Act, June 30, 1948
1956	Federal Water Pollution Control Act Amendment	Authorized construction grants for municipal treatment works
1961	Water Pollution Control Act	Congress advocated 85% removal of pollutants using comprehensive programs and plans for water pollution abatement and control
1962	Revised USPHS Standards	Included 28 constituents
1965	Water Quality Act October 2	States are required to have water quality standards to receive grants
1965	Federal Water Pollution Control Administration (FWPCA)	FWPCA was established in the Department of Health, Education, and Welfare (HEW)
1969	Community Water Supply Study (CWSS) initiated	Initiated a survey of 969 public water systems (representing only 5% of nation's water systems) for compliance with 1962 standards
1970	CWSS report released	41% of public water systems surveyed for compliance with 1962 standards
1970	Water Quality Improvement Act	No new provisions regarding standards. The act continued the authority of states to set water quality standards and the authority of FWPCA to approve such standards. Questioned the Environmental Protection Agency's authority to require uniform treatment limitation for water discharge
1972	Louisiana (Mississippi River) Water Quality Report Published	36 organic compounds found in drinking water and source waters.
1972	"Clean Water Act" (CWA) or FWPCA amendment of 1972	Set discharge limits and required waste treatment prior to discharge and established National Pollution Discharge Elimination System (NPDES)
1973	GAO report	Out of 446 community water systems investigated, only 60 were found to comply with USPHS standards
1974	Consumer Reports (three part series) published	Organic contaminants in New Orleans
1974	Environmental Defense Fund	Organic contaminants in New Orleans
1974	Trihalomethanes (THMs) identified	A major public health concern
1974	National Organic Reconnaissance Survey	Survey of THM occurrence nationwide
1974	Safe Drinking Water Act (SDWA)	Established stringent regulatory standards for the quality of drinking water for all public water supply systems except the very smallest. Signed on Dec.16, 1974, as Public Law 93-523
1975	National interim primary drinking water regulations (NIPDWRs)	Based on 1962 USPHS standards, interim rules for 18 organic, inorganic, and microbial contaminants
1976	1st NIPDWRs amendment	Standards for four radionuclides
1976	Toxic Substances Control Act (TSCA)	Regulation of all compounds that have or may possess toxic properties
1977	Best practical control technology currently available (BPCTCSA)	Defined best available waste treatment for different industries
1979	2nd NIPDWRs amendment	Total THMs
1980	Comprehensive Environmental Restoration, Compensation, And Liability Act (CERCLA)	"Superfund" bill that established a detailed program to describe priority pollutants and first national database of pollutants
1980	3rd NIPDWRs amendment	Monitoring for corrosion and sodium
1983	4th NIPDWRs amendment	Identifies the best available mean to comply with THMs rule

Table 1. (Continued)

Year	Events	Comments
1986	SDWA amendment	Major revisions introduced in SDWA. Set standards for pollutants in drinking water; authorizes contamination warning and alert system
1987	Volatile organic contaminants standards	Standards for eight contaminants
1988	SDWA amendment	Public Law 100-572: Lead Contamination Control Act
1989	Surface Water Treatment Rule	The Surface Water Treatment Rule regulates public water systems using a lake, stream, or pond water as a source of drinking water. It was intended to ensure the safety of drinking water
1989	Total Coliform Rule (TCR)	Establishes microbiological standards and related monitoring requirements for all community and noncommunity water systems
1991	National Primary Drinking Water Regulation for Radon-222	EPA proposed to regulate radon at 300 pCi/L. Public water supplies (systems serving over 25 individuals or with greater than 15 connections) are required to monitor, report, and notify requirements for radon (July 18, 1991) (56 FR 33050)
1991	Phase II SOCs and IOCs also lead and copper	Standards for 27 new contaminants
1992	Phase V SOCs and IOCs	Standards for 22 new contaminants
1996	SDWA amendment	Public Law 104-182 requires U.S. EPA to publish an MCLG and promulgate NPDWR for contaminants of public health significance
1996	Information Collection Rule	Monitoring only to provide information on the occurrence of disinfection by-products (DBPs) and pathogens, including *Cryptosporidium* (May 14, 1996)
	Analytical Methods for Radionuclides	Approval of additional (66) analytical methods to monitor gross alpha, gross beta, tritium, uranium, radium-226, radium-228, gamma emitters, and radioactive cesium, iodine, and strontium in drinking water (Mar. 5, 1997)
1997	Withdrawal of 1991 proposed rule on radon-222	Withdrawal of drinking water regulations proposed for radon-222 (Aug. 6, 1997)
1997	Small System Compliance Technology List for the Surface Water Treatment Rule	The list of technologies that small systems can use to comply with the Surface Water Treatment Rule (SWTR) (Aug. 6, 1997)
1998	Revisions to State Primacy Requirements	The regulations requiring states to obtain and retain primary enforcement authority (primacy) for the Public Water System Supervision (PWSS) program under Section 1413 of the Safe Drinking Water Act (SDWA) are amended (Apr. 28, 1998)
1998	Drinking Water Contaminant Candidate List	The list of contaminants that are known or anticipated in public water systems and may require regulations under the SDWA (Mar. 2, 1998)
1998	Variances and Exemptions Rule	The rule includes procedures and conditions under which a primacy state/tribe or the EPA Administrator may issue small system variances to public water systems serving less than 10,000 persons (Aug. 14, 1998)
1998	Consumer Confidence Reports	Community water systems are required to prepare annual consumer confidence reports on the quality of the water delivered by the systems and provide these reports to their customers (Aug. 14, 1998)
1998	Stage I D-DBP Rule	Standards for 11 new contaminants (Dec. 16, 1998)
1998	Interim Enhanced Surface Water Treatment Rule	Improve control of microbial pathogens, including specifically the protozoan *Cryptosporidium*, in drinking water; and address risk trade-offs for disinfection by-products (Dec. 16, 1998)
1999	Suspension of Unregulated Contaminant Monitoring Requirements for small public water systems	This rule exempts small and medium public water systems from monitoring unregulated contaminants during 1 year every 5 years, as required under UCMR (Jan. 8, 1999)
1999	Revisions to the Unregulated Contaminant Monitoring Rule.	This final rule includes a list of contaminants to be monitored. The data in the database will be used to identify contaminants on the Drinking Water Contaminant Candidate List (CCL) (Sept. 17, 1999)
1999	Analytical methods for chemical and microbiological contaminants and revisions to laboratory certification requirements	EPA has approved the updated versions of 25 American Society for Testing and Materials (ASTM), 54 Standard Methods for Examination of Water and Wastewater (Standard Methods or SM), and 13 Environmental Protection Agency (EPA) analytical methods for compliance determinations of chemical contaminants in drinking water (Dec. 1, 1999)
1999	Underground Injection Control Regulations for Class V Injection Wells	Class V Underground Injection Control (UIC) regulations are revised and new requirements are added for two categories of endangering Class V wells to ensure protection of underground sources of drinking water (Dec. 7, 1999)
1999	Lead and Copper Rule minor revisions	These minor revisions are intended to streamline requirements, promote consistent national implementation, and, in many cases, reduce the burden for water systems. The LCRMR do not change the action levels or MCLGs for lead and copper, established by the Lead and Copper Rule (Dec. 20, 1999)

(continued overleaf)

Table 1. (Continued)

Year	Events	Comments
2000	Analytical Methods for Perchlorate and acetochlor	This rule specifies the approved analytical methods for measuring of perchlorate and acetochlor in drinking water and also includes minor technical changes to correct or clarify the rule published on Sept. 17, 1999 (Mar. 2, 2000)
2000	Public Notification Rule	Owners and operators of public water systems are required to notify persons served when they fail to comply with the requirements of the National Primary Drinking Water Regulations (NPDWR), have a variance or exemption from the drinking water regulations, or are facing other situations posing a risk to public health (May 4, 2000)
2000	Removal of the MCLG for chloroform	In accordance with the order of the U.S. Court of Appeals for the District of Columbia Circuit, the zero MCLG for chloroform is removed from National Primary Drinking Water Regulations (NPDWRs) (May 30, 2000)
2000	Drinking Water State Revolving Fund Rule	This interim final rule gives each state considerable flexibility to determine the design of its DWSRF program and to direct funding toward its most pressing compliance and public health problem. It also explains the ways to receive a capitalization grant, how to use capitalization grant funds intended for infrastructure projects, and also the roles of both the states and EPA in managing and administering the program
2000	Radionuclides Rule	This rule establishes the MCLGs, MCLs, monitoring, reporting, and public notification requirements for radionuclides. Uranium is added to the list, and the monitoring requirements for combined radium-226 and radium-228, gross alpha particle radioactivity, and beta particle and photon radioactivity have been revised (Dec 7, 2000)
2001	Unregulated Contaminant Monitoring List 2 Rule	The analytical methods for thirteen chemical contaminants on List 2 are approved and also the schedule for *Aeromonas* monitoring is described (Jan. 11, 2001)
2001	Arsenic Rule	A health-based, nonenforceable MCLG for arsenic of zero and an enforceable MCL for arsenic of 0.01 mg/L (10 µg/L) is established for nontransient noncommunity water systems (Jan. 22, 2001)
2001	Filter Backwash Recycling Rule	The purpose of the FBRR is to further protect public health by requiring public water systems (PWSs), where needed, to institute changes to the return of recycle flows to a plant's treatment process that may otherwise compromise microbial control (June 8, 2001)
2002	Long Term 1 Enhanced Surface Water Treatment Rule (Jan. 14, 2002)	The rule requires systems to meet strengthened filtration requirements as well as to calculate levels of microbial inactivation
2002	Unregulated Contaminant Monitoring Regulation for Public Water Systems	Establishment of Reporting Date (Aug. 9, 2002) for water systems (serving more than 10,000 persons) to report all contaminant monitoring results they receive before May 13, 2002 for the Unregulated Contaminant Monitoring Regulation (UCMR) monitoring program; Direct Final Rule (Mar. 12, 2002)
2002	Guidelines establishing test procedures for the analysis of pollutants under the Clean Water Act; National Primary and Secondary Drinking Water Regulations	The test procedures (i.e., analytical methods) for the determination of chemical, radiological, and microbiological pollutants and contaminants in wastewater and drinking water are revised and updated; Final Rule (Oct. 23, 2002)
2002	Unregulated Contaminant Monitoring Regulation: Analytical methods for chemical and microbiological contaminants	Approval of the analytical method and an associated minimum reporting level (MRL) to support Unregulated Contaminant Monitoring Regulation's (UCMR) (for chemical and microbiological contaminants) List 2 *Aeromonas* monitoring; Final Rule (Oct. 29, 2002)
2003	Propose MCLG and National Primary Drinking Water Regulations	Regulations for any contaminant selected from contaminant candidate list (Aug. 2003)
2004	National Primary and Secondary Drinking Water Regulations	Approval of additional method (Colitag™) for detecting coliforms and *E. coli* in drinking water; Final Rule (Feb. 13, 2004)

READING LIST

U.S. National Archives and Records Administration. (2004). Records of the Federal Water Pollution Control Administration (FWPCA) (Record Group 382, Years 1948–68) www.archives.gov

U.S. EPA (1997). *The Safe Drinking Water Act—One Year Later—Success in Advancing Public Health Protection*. EPA 810-F-97-002.

U.S. EPA (1999). *25 Years of the Safe Drinking Water Act: History and Trends*, EPA 816-R-99-007.

U.S. EPA, Office of Water. http://www.epa.gov/water/.

WATER QUALITY MANAGEMENT IN A FORESTED LANDSCAPE

THOMAS M. WILLIAMS
Baruch Institute of Coastal
Ecology and Forest Science
Georgetown, South Carolina

In 1955, Trousdell and Hoover published a paper in the *Journal of Forestry* that documented a rise in the water table following clear-cutting of a stand of loblolly pine on poorly drained soil. They warned that this rise may lead to saturated soils that inhibit regeneration and preclude using heavy equipment on the site. In the first century A.D., Pliny the Elder stated, "Often after woods have been cut down, springs on which the trees used to feed, emerge" (cited in Reference 1). Modern scientific methods can document truth that has been observed and debated for millennia. The influence of forests on water is such a subject. During two millennia of observation and nearly a century and a half of scientific investigation, the role of forests in water has been empirically demonstrated, but it is still not understood at a fundamental level.

Forested land produces the highest quality water of any land use (2), and forests have been used to protect drinking water supplies around the world. Forested streams also support food webs essential to the health of fish and other aquatic organisms. Rocky streambeds in forests are essential to aquatic macroinvertebrates that are used by some of our most desirable game fish. Protection of water was one of the objectives of the Organic Act of 1897, which formed the National Forest system, and the Weeks Act of 1911, which added 21 million acres of National Forests in the eastern United States (3).

In addition to protecting water, forests are also used by humans for harvest of timber products, harvest of game and fish, and outdoor recreation. Forests are often managed to enhance their usefulness for these purposes. Humans have used fire to alter ecosystems for hunting for at least 10,000 years, and exploitive logging has been practiced throughout the historical period. Management of forests to produce timber and fiber products has been widely practiced since the late nineteenth century. Much of our knowledge of the influence of forests on water has been gained by our efforts to evaluate the impact of forest management on water.

Modern forest hydrology research began in France in the nineteenth century. Widespread deforestation following the French revolution was blamed for a series of floods and landslides that plagued the country (1). During that debate, proponents of forests argued that reforestation would increase stream flow, regulate floods, and prevent erosion and landslides. Most of these claims were supported by observations of several romantic era writers and references to ancient Greek scholars. To resolve these disputes, the influence of forests became an area of active scientific inquiry. Surell (4) was one of the early researchers to collect numerical data on mountain watersheds and landslides. Observation of stream flow and rainfall began by 1850 and examination of watersheds with different land uses began by 1860. The controversy over the value of forests in regulating streams continued in the United States by the early 1900s. Pinchot (5) lamented that "the writing and talking on this branch of forestry has had little definite fact or trustworthy observation behind it." At that time, the U.S. Forest Service began efforts to collect trustworthy observations.

Studies at Wagon Wheel Gap, Colorado, introduced the most widely used technique in examining water in forested regions (6). The use of paired experimental watersheds (7) has produced statistically valid tests of many ideas that had been debated for centuries. This technique involves measuring the process of interest for several years in two closely matched watersheds, applying a treatment to alter the process in one watershed, and then continuing measures in both watersheds for several years. From a strong correlation between the data from the two watersheds prior to treatment, changes in the treated watershed can be determined with high confidence. Over the last century, paired watersheds have been used to study most aspects of forest management. A solid empirical basis is available for evaluating forest management impacts on water quality. We have a relatively clear understanding how forests produce clean water and which aspects of forest management have the potential to degrade water quality.

Water from undisturbed forests has very little suspended sediment. Estimates of total sediment loss from forested watersheds across the United States are generally less than 0.5 Mg/ha/yr (8). Very low rates of erosion are due to the mechanism of peak flow generation in forested watersheds. Between rains, flow in most streams is derived from groundwater, which has little associated sediment. Sediment is added to streams as rainwater travels across the soil surface and suspends soil particles. Horton (9) explained the generation of surface flow as rain that fell more quickly than water could enter the soil. If water could enter the soil at 3 mm/h and it rained 10 mm/h, then 7 mm of rain must flow across the surface and into streams. However, in temperate forests, measured infiltration rates may approach 1000 mm/h and any incoming rain will enter the soil (10). However, we know that forested streams rise quickly after rain begins. A number of different mechanisms have been found that deliver water quickly to streams in forested watersheds (11). These mechanisms do not include water moving very far along the surface of the ground. In forested areas, most of the sediment found in streams has been eroded from stream banks.

High rates of infiltration on forest soils result from two mechanisms. The tree canopy intercepts raindrops and, once it is wetted, allows drops to fall a short distance to the forest floor, thus protecting the forest floor from high energy impacts during intense rain. Litter covering the soil and organic matter incorporated into the soil create large pores and allow them to stay open throughout the storm. As long as these soil pores remain open, surface water does not erode the soil and add sediment to streams. Protection of the soil pores by litter and organics is the more important mechanism. High infiltration can occur if the litter and organic matter are not disturbed when the canopy is removed.

Disturbance of forest soil by harvest and replanting of trees in commercial forestry is the most likely way that forest management can degrade water quality. Removing forest products from the site requires cutting trees, skidding products to a landing, loading them on a truck, and hauling them to a mill. Each aspect can damage the forest floor. Sediment losses from skid trails and landings may be as high as 7000 Mg/ha and from roads can be as high as 1000 Mg/ha (12). On a watershed basis, sediment losses of 10 Mg/ha/yr have been found following logging and preparation of the watershed for replanting (13,14). This soil loss rate was similar to that considered acceptable to maintain agricultural productivity and occurs only for 3–5 years of a 25-year rotation. Since those studies were concluded, forestry has been challenged to improve water quality associated with harvesting and regeneration. Throughout the United States, foresters have examined techniques to minimize exposure of bare soil and to build roads and landings in a way that maintains a buffer of undisturbed forest soil for infiltration before runoff reaches a stream bank. These techniques, known collectively as best management practices (BMPs), are now being used across the United States (15). A number of studies have shown that BMPs reduce sediments by roughly 90% (16).

Cutting the forest does increase the amount of water that becomes stream flow and will cause an unavoidable increase in sediment in alluvial streams. More water is returned to the atmosphere by evaporation of intercepted rain and transpiration by trees than any other land cover. Examining runoff data from 94 paired watershed experiments, Bosch and Hewlett (17) concluded that there was a generally linear increase in steam flow with increasing percentage of forest removed. They also concluded that cutting conifers resulted in a greater increase than cutting deciduous hardwoods. Sahin and Hall (18) looked at 145 such experiments and estimated that stream flow increased 20–25 mm for each 10% reduction in the cover of conifers and 17–19 mm for each 10% reduction in hardwoods. Increases in stream flow tend to increase the size of alluvial channels, causing an unavoidable increase in suspended sediment. Van Lear et al. (19) found an increase from 0.040 Mg/ha/yr to 0.15 Mg/ha/yr in sediment loss from a watershed where the only source of sediment was a stream channel. However, the stream flow returns to preharvest values as the new forest grows. In some rapidly growing forests, water yield may actually decline below the pretreatment level when the growth rate is at a maximum (20).

An early experiment to determine if forest regeneration was responsible for the decline in stream flow revealed the importance of nitrogen cycling in forest harvesting. At Hubbard Brook, New Hampshire, all trees on a watershed were cut and all vegetation killed for three successive years to examine stream flow (21). Stream flow increased by more than 300 mm during the first year. This increase was maintained as long as herbicide was applied to the watershed and declined when vegetation was allowed to grow. In addition to an increase in flow, the concentration of nitrate, calcium, and magnesium increased greatly. Nitrate concentrations exceeded drinking water standards. The large pool of nitrogen in the vegetation and the forest floor was mineralized. Nitrification of this mineralized nitrogen yielded nitrate and hydrogen ions; nitrate was leached from the system, and hydrogen ions displaced cations, which were then lost.

Due to the concerns raised by the Hubbard Brook experiment, cycling and retention of nitrogen is now better understood than any other biogeochemical process in forested ecosystems. Nitrogen is tightly held in forested ecosystems and export of nitrate to streams is highly limited. Ten separate chemical mechanisms for retention of mineralized organic nitrogen have been identified in forest soils (22,23). The most common are ammonium uptake by bacteria that decompose high C:N material, ammonium or nitrate uptake by plants, and denitrification of nitrate to nitrous oxide or nitrogen gas.

When a forest is harvested, a great deal of organic nitrogen is deposited in the form of logging slash, and moisture and temperature increase in the forest floor. The logging slash is also high in carbon and decomposition bacteria outcompete nitrifiers for mineralized ammonium. As carbon is lost as carbon dioxide, the ratio of carbon to nitrogen declines and decomposition bacteria are less able to compete with nitrifiers. Unless regrowth is inhibited, plants become competitors for ammonium and also can use nitrate produced by nitrifiers. Finally, if nitrate is leached from the forest slope, it moves to the riparian zone where soils are often saturated. In these saturated zones, bacteria deplete oxygen from the soil and then use nitrate as an electron receptor in anaerobic respiration. Nitrate is reduced to nitrous oxide or to neutral nitrogen and lost from the system as a gas. Unless all these mechanisms are limited, normal forest harvesting does not result in nitrate concentrations of more than 1 mg/L. These mechanisms are most effective in moist, warm, temperate forests.

In addition to the original Hubbard Brook experiment, nitrate leaching has been occasionally found in other places, although stream concentrations have generally been found below 10 mg/L. Clear-cutting other watersheds in the mountains of New England also showed elevations of nitrate near 10 mg/L (24). Use of herbicide, especially hexazinone (25), in repeated treatments (26) may increase stream concentrations to the 2–5 mg/L range. In northwestern Europe, nitrogen input from air pollution appears to be increasing stream nitrate concentrations. Nitrate concentrations between 0.5 and 1 mg/L are found in undisturbed forested watersheds in Scandinavia (27). Lepistö (28) found evidence that a watershed in Finland had become saturated with nitrogen and showed nitrate leaching even during the growing season. This occurred after atmospheric nitrogen input had increased from 2.9 kg/ha/yr in 1969–1979, to 6.2 kg/ha/yr in 1971–1979, to 8.5 kg/ha/yr in 1980–1990.

A number of other anions and cations are found in low concentrations in streams draining forested landscapes. The concentrations of these ions and ratios of one ion to another are related to location and subsurface geology. The ratios of various ions have been used widely to interpret pathways of water within the watershed. Mulholland (29) used ratios of sulfate and calcium to differentiate among vadose zone, shallow groundwater, and deep groundwater sources of stream flow in a watershed dominated by dolomite. In this case, atmospheric input of sulfate resulted in high concentrations in throughfall, but adsorption by iron and aluminum in the soil B horizon resulted in low sulfate in shallow groundwater. Likewise, calcium was very low in throughfall but very high in groundwater that had been in contact with the dolomite bedrock. Eisenbeer and Lack (30) used a similar technique with silica and potassium to examine surface runoff and pipe flow in an Amazonian watershed. Intense weathering of the soil and bedrock resulted in leaching of silica into the water passing through the soil. Intense weathering also isolated potassium into the living biomass; it was leached from leaves in throughfall and surface flow. The ratio of potassium to silica could be used in the same way that sulfate and calcium were used in the other geologic setting.

Hydrogen ions are generally not of great concern in forest streams with pH slightly acid, unless the geology is dominated by calcareous materials. Nitrification produces an H^+ ion for each NO_3^- ion and can be a source of acidification in stream water. Generally, that is not the case because H^+ is held more tightly in the soil cation exchange complex, and potassium, calcium, or magnesium is displaced to be carried to the stream. Afforestation acidified streams in the British Isles on former moorland (31).

As forest management strives to increase the growth of forest products, pesticides and fertilizer have been used on more forest plantations. Although these two activities are not closely related, contamination from both can be avoided in the same way. If rain dissolves fertilizer and pesticide from the soil surface, they will be quickly immobilized within the soil. Contamination from either can be eliminated by avoiding application in streams or near stream areas that may saturate in a large rain. Buffering both perennial and ephemeral streams when fertilizer or pesticide is applied will prevent contamination. Both materials are often aerially applied, so a number of management practices relating to atmospheric conditions are followed in development of best management practices. Most problems with these materials have been caused when aerial sprays are inadvertently applied to surface waters.

Fire is a tool that has been used to manage forest ecosystems since early man first learned to use it. Fire can have a profound impact on water quality. The impact of fire is closely related to the fate of the forest floor. If the forest floor is damp and fire intensity is low, there is little impact on water quality. As fire intensity increases, the impact on water also increases. If trees are killed and the forest floor remains intact, the result is similar to logging with an increase in flow but small changes in sediment or chemistry. However, fire effects can be much larger than those from logging if the forest floor is dry or the fire is intense enough to dry and consume the forest floor. As more bare soil is exposed, infiltration declines and water runs across the surface carrying sediment and cations from the ash. If the fire is highly intense, volatilized organic compounds move from the forest floor into the soil. As these compounds cool, soil particles become hydrophobic and prevent water from entering the soil (32). In this case, all rainwater runs across the surface, there is severe erosion, and streams are filled with sediment. Thus, fire may result in sediment losses from 1–2 Mg/ha to 500 Mg/ha, depending on fire severity and land slopes (33). In areas of high air pollution, severe fire may also result in nitrate concentrations that exceed drinking water standards (34). The goal of fire management is to prevent such intense fires by using prescribed, less intense fires, mechanical fuel reduction, and fire suppression.

Undisturbed forests produce the highest quality water of any land use. However, forested watersheds also lose more water to the atmosphere though evaporation and transpiration than other land uses. Human use of the forest can degrade the quality of water produced if care is not taken to protect the forest floor. Careless logging, especially poorly designed roads, can contribute to stream sedimentation and degrades water for both human use and as fish habitat. Most of these problems can be avoided if best management practices are used. There are few problems with chemical water quality, except for nitrate, which may become a problem if forest regrowth is prevented or in areas with prolonged high atmospheric nitrogen deposition. Such problems have been seen in northwestern Europe and after fire in the Los Angeles basin.

BIBLIOGRAPHY

1. Andréassian, V. (2004). Waters and forests: From historical controversy to scientific debate. *J. Hydrol.* **291**: 1–27.
2. Brooks, K. and Achouri, M. (2003). Sustainable use and management of freshwater resources: The role of forests. In: State of the Worlds Forests 2003. Food and Agricultural Organization, United Nations, Rome, Italy, pp. 74–85.
3. Sedell, J., Sharpe, M., Dravnieks Apple, D., Copenhagen, M., and Furniss, M. (2000). Water and the Forest Service. USDA, Forest Service, Washington Office, Washington, DC, p. 2000.
4. Surell, A. (2004). *Etude sur les torrents des Hautes Alps*. Carilian-Goeury et Victor Dalmont, Paris, p. 1841 (cited in Reference 1).
5. Pinchot, G. (1905). A primer of forestry. Bureau of Forestry,. U.S. Dept. of Agriculture, Government Printing Office, Washington, DC (cited in Reference 1).
6. Bates, C.G. and Henry, A.J. (1928). Forest and streamflow experiment at Wagon Wheel Gap, Colorado. *Mon. Weather Rev. Suppl.* No. 3. USDA Weather Bureau, Washington, DC.
7. Hewlett, J.D. (1982). *Principles of Forest Hydrology*. University of Georgia Press, Athens, GA, pp. 86–88.
8. Binkley, D. and Brown, T. (1993). Forest practices as non-point sources of pollution in North America. *Water Resour. Bull.* **29**: 729–740.
9. Horton, R.E. (1933). The role of infiltration in the hydrologic cycle. *Trans. Am. Geophys.* **14**: 446–460.
10. Hursh, C.R. and Brater, E.F. (1941). Separating storm-hydrographs form small drainage areas into surface and subsurface flow. *Trans. Am. Geophys. Union.* **Part III**: 863–871.
11. Bonnel, M. (1993). Progress in understanding of runoff generation in forests. *J. Hydrol.* **150**: 217–275.
12. Elliot, W.J. (2000). Roads and other corridors. Drinking water from forests and grasslands: a synthesis of the scientific literature. G.A. Dissmeyer (Ed.). USDA Forest Service,

13. Hewlett, J.D. (1979). Forest water quality: an experiment in harvesting and regenerating Piedmont forests. School of Forest Resources, University of Georgia, Athens, GA.
14. Pye, J. and Vitousek, P.M. (1985). Soil and nutrient removals and windrowing on a Southeastern US Piedmont site. *Forest. Ecol. Manage.* **11**: 145–155.
15. Shepard, J. (2002). *For. Best Manage. Pract. Symp.*, Gainesville, FL. National Council on Air and Stream Improvement, CD-ROM.
16. Ice, G.G., Megahan, W.F., McBroom, W.M., and Williams, T.M. (2003). Opportunities to assess past, current and future impacts of forest management. In: *Total Maximum Daily Load (TMDL) Environmental Regulations II.* American Society of Agricultural Engineers, St, Joseph, MI, pp. 243–248.
17. Bosch, J.M. and Hewlett, J.D. (1982). A review of catchment experiments to determine the effect of vegetation changes on water yield. *J. Hydrol.* **55**: 3–23.
18. Sahin, V. and Hall, M.J. (1996). The effects of afforestation and deforestation on water yields. *J. Hydrol.* **178**: 293–309.
19. Van Lear, D.H., Douglass, J.E., Cox, S.K., and Augspurger, M.K. (1985). Sediment and nutrient export in runoff from burned and harvested pine watersheds in the South Carolina. Piedmont. *J. Environ. Qual.* **14**: 169–174.
20. Cornish, P.M. and Vertessy, R.A. (2001). Forest age induced changes in evapotranspiration and water yield in a eucalypt forest. *J. Hydrol.* **242**: 43–63.
21. Likens, G.E. et al. (1970). Effects of forest cutting and herbicide treatment on nutrient budgets in the Hubbard Brook watershed-ecosystem. *Ecological Monogr.* **40**: 23–47.
22. Vitousek, P.M. and Melillo, J.M. (1979). Nitrate losses from disturbed forests: patterns and mechanisms. *Forest Sci.* **25**: 605–619.
23. Nadelhoffer K.J., Aber, J., and Melillo, J.M. (1984). Seasonal patterns of ammonium and nitrate uptake in nine temperate forest ecosystems. *Plant and Soil* **80**: 321–335.
24. Johnson, C.E., Johnson A.H., and Siccama, T.G. (1991). Whole-tree clear-cutting effects on exchangeable cations and soil acidity. *Soil Sci. Soc. Am. J.* **55**: 502–508.
25. Neary, D.G., Bush, P.B., and Grant, M.A. (1986). Water quality of ephemeral forest streams after site preparation with the herbicide hexazinone. *Forest Ecol. Manage.* **14**: 23–40.
26. Munson, A.D., Margolis, H.A., and Brand, D.G. (1993). Intensive silviculture treatment: impacts on soil fertility and planted conifer response. *Soil Sci. Soc. Am. J.* **57**: 246–255.
27. Arheimer, B., Andeson, L., and Lepistö, A. (1996). Variation of nitrogen concentration in forests streams—influences of flow, seasonality and catchment characteristics. *J. Hydrol.* **179**: 281–304.
28. Lepistö, A. (1995). Increased leaching of nitrate at two forested catchments in Finland over a period of 25 years. *J. Hydrol.* **172**: 103–123.
29. Mulholland, P.J. (1993). Hydrometric and stream chemistry evidence of three storm flowpaths in Walker Branch Watershed. *J. Hydrol.* **151**: 291–310.
30. Eisenbeer, H. and Lack, A. (1996). Hydrometric and hydrochemical evidence for fast flowpaths at La Cuenca, Western Amazonia. *J. Hydrol.* **180**: 237–250.
31. Neal, C., Smith, C.J., and Hill, S. (1992). Forestry impacts on upland water quality. Institute of Hydrology Report Series No. 30. Institute of Hydrology, Wallingford, UK.
32. Debano, L.F. (2000). The role of fire in soil heating and water repellency in wildland environments: a review. *J. Hydrol.* **231–232**: 195–206.
33. Landsburg, J.D. and Tiedemann, A.R. (2000). Fire management. In: Drinking water from forests and grasslands: a synthesis of the scientific literature. USDA Forest Service. Asheville, NC, General Technical Report SRS-39, Chap. 12, pp. 124–138.
34. Riggan, P.J. et al. (1994). Effects of fire severity on nitrate mobilization in watersheds subject to chronic atmospheric deposition. *Environ. Sci. Technol.* **28**: 369–376.

TRACE METAL SPECIATION

JAMIE R. LEAD
University of Birmingham
Birmingham,
United Kingdom

INTRODUCTION

Trace metals may be defined as those metals whose concentration is low enough so that their presence exerts a negligible influence on the overall chemistry of a water body. They can be contrasted with major ions that are present in sufficient amounts to affect the chemistry of the system. Major metal ions include the alkali and alkaline earth metals (Na, K, Ca, Mg) and perhaps Al, Fe, and Mn. Although there is no strict definition of a "trace metal," we usually consider the transition metals, certain metalloids, and lanthanides. Radioactive elements are often considered separately. The potential effects on ecological and human health created by a certain number of trace metals mean that these metals are of particular importance and are the subject of intensive study. The effects of these metals may often be deleterious, but many metals are also micronutrients. In no order, some of the most important and most investigated trace metals are chromium (Cr), nickel (Ni), copper (Cu), zinc (Zn), arsenic (As), cadmium (Cd), tin (Sn), mercury (Hg), and lead (Pb).

Trace metals in natural waters can be present at different concentrations but can also be distributed between different physicochemical forms or species. This distribution of a metal between different forms is termed its speciation. Important examples of speciation in the environment are: (1) the oxidation state, for example, Fe(II) and Fe(III); (2) the nature of the ligand in a solution phase metal complex, for example, Cd^{2+} and $CdCl_4^{2-}$; and (3) the distribution of metal between dissolved, colloidal, and particulate phases. The exact form in which the trace metal exists substantially mediates the bioavailability and toxicity of the metal and its transport in the environment and has important implications for biomagnification (i.e., increased concentration of pollutant as it passes through the food chain), human and ecological health, and biogeochemical cycling. Speciation is thus intrinsically important and also relevant to understanding wider environmental processes. We can examine the three examples above and discuss some of the possible environmental implications of speciation.

1. *The Oxidation State.* [Fe(II)/Fe(III)]—Fe(II) occurs in reducing (oxygen-deficient) conditions, such as porewaters in sediments and bottom waters of stagnant and/or biologically active lakes, particularly in the summer months. Conversely, Fe(III) occurs in oxidizing (oxygen-rich) conditions, such as surface river, lake, or marine waters. Additionally, Fe(II) tends to be present in the dissolved phase, Fe^{2+} (aq), whereas Fe(III) tends to hydrolyze (react with water) very strongly and form solid colloidal or particulate oxide or hydroxide phases, $Fe(OH)_3$ (s), where the (aq) denotes the aqueous phase and (s) denotes solid. This has important implications for the behavior of Fe, which are discussed below in example 3. In addition, trace metals may bind to the surface of solid Fe(III) particles. Interconversion of Fe(II) and Fe(III) at interfaces between oxidizing and reducing conditions may have an impact on the environmental fate and behavior of any associated metals. See Davison and de Vitre (1) for an in-depth discussion of iron chemistry in aquatic systems.

2. *The Nature of Solution Phase Ligand (Cd^{2+} and $CdCl_4^{2-}$).* The salinity of water (the concentration of the chloride ligand) determines the extent to which metals such as cadmium are present as the positively charged hydrated Cd^{2+} or as the negatively charged chloro complex $CdCl_4^{2-}$. Intermediate numbers of chloride ions in the complex are also possible. This situation is most commonly encountered in estuarine systems, where freshwater meets seawater and the chloride concentration of the water gradually increases. The change in charge from the positive Cd to the negative $CdCl_4$ affects properties of the metal such as permeability through biological membranes and sorption/desorption to clay, organic, and oxide particles. These changes have implications for metal bioavailability and transport. In addition to chloride, other important ligands exist such as sulfate, amino acids, and EDTA.

3. *Dissolved, Colloidal, and Particulate Distribution.* The distribution of metals between these three phases has long been recognized as essential to the fate and behavior of trace metals. Particular prominence is given to the free (dissolved) metal ion, as it is predicted that the biological response is a function of the dissolved concentration in bulk water (2). The relationship between speciation and bioavailability may, however, be less straightforward; complex and poorly understood physicochemical and biological interactions mediate biological uptake of metal (3,4). Transport through the environment is also affected by the partitioning of metal (5); microbiological activity, aggregation of colloids, and sedimentation of particles are among the dominant processes. Somewhat simplistically, dissolved and colloid-bound metals tend to remain in the water column, and particle-bound metals tend to settle out to the sediments. This geochemical fractionation results in different transportation behavior and alters the ultimate fate of the metal.

Table 1. Approximate Concentrations of Trace Metals in Seawater and Freshwater[a]

Metal	Seawater, molar	Freshwater, molar
Mn	10^{-9}	$10^{-5} - 10^{-9}$
Fe	10^{-9}	$10^{-4} - 10^{-7}$
Cu	$10^{-8} - 10^{-9}$	$10^{-6} - 10^{-9}$
Zn	$10^{-8} - 10^{-10}$	$10^{-5} - 10^{-8}$
Ni	10^{-8}	$10^{-6} - 10^{-10}$
Cd	$<10^{-9}$	$10^{-7} - 10^{-10}$
Pb	$<10^{-10}$	$10^{-6} - 10^{-9}$
Hg	10^{-11}	$10^{-7} - 10^{-10}$

[a] Data from Reference 6.

As indicated briefly in these examples, the question of trace metal speciation is complex. Our understanding of these difficult areas is in part due to the physicochemical and biological complexity of natural waters. Natural waters contain all or almost all elements of the periodic table. Trace metals can be present at extremely low levels, often at or below analytical limits of detection; concentrations are as low as 10^{-12} M in some cases. See Table 1 for illustrative examples of element concentrations, although it must be kept in mind that a wide range of concentrations exists. Additionally, a range of dissolved, colloidal, and particulate organic matter and suspended particles exist that have complex and unknown structures and effects on metal speciation. These include a range of chemistries (silicates, oxides, carbonates, humic substances, etc.) and sizes (1 nm upward). Microorganisms also play an important (and again poorly understood) role in metal speciation, through biochemical processes, sorption to cell walls and membranes, and the production and excretion of complexing ligands into the environment.

Our understanding of speciation is further hindered by the variability of natural waters spatially (e.g., surface waters in lakes may have a higher oxygen content than bottom waters and have significant effects on metal speciation) and temporally (e.g., daily and seasonal cycles in chemistry, often mediated by changes in biological activity) often over very small scales.

CHEMICAL REACTIONS THAT AFFECT METAL SPECIATION

We can distinguish between two broad types of chemical reactions that may affect speciation (7). First is redox reactions where the oxidation state of the atoms is altered. An example of this is the Fe(II)/Fe(III) interconversions in the examples above. Second are reactions where changes in the "coordinative relationships" of participating atoms occur. An example of this is the change from Cd^{2+}, where the coordinative partner is water (i.e., Cd and water are chemically bound to each other) to $CdCl_4^{2-}$, where the coordinative partner is the chloride ions (Cd and Cl are chemically bound to each other). Both acid–base, precipitation and complexation reactions fit this category. Reactions occurring in environmental systems are often impossible to separate conceptually because of the system complexity (8). For instance, example 1 in the Introduction

[Fe(II) to Fe(III) interconversions] involves all of these types of reactions. Detailed discussion of these reactions can be found in chemistry textbooks and in relation to environmental systems in Stumm and Morgan (7) and Buffle and Stumm (6). In this review, we concentrate on complexation reactions.

A metal complex is formed by the donation of electrons from a ligand to a metal; this is a stable association of metal and ligand whose characteristics are different from the individual components. The simplest example is that of the hydrated metal, sometimes termed "free" or "dissolved". (Note: It is important to distinguish between the term dissolved in this sense and the operationally used term dissolved, which is used to denote metals passing through a standard membrane during filtration. In the latter definition, a fraction of colloid-bound metals is included.) The hydrated metal ion is often written simply by excluding the coordinated water (as in example 2 above with Cd). Other ligands will always be indicated, as when the Cd is bound to chloride ligands.

Typical ligand groups include

1. oxygen groups: water; various types of organic groups (carboxylic acids, ketones, etc.), which may be in solution or more importantly as part of colloids and surfaces; various types of inorganic groups (silicates, oxides) mostly occurring as solid-phase material;
2. sulfur groups: thiols, thioethers, and other compounds, which may either be simple solution compounds or as part of larger compounds such as humic substances; and
3. nitrogen groups: for instance, amino acids and humic substances.

Solid phases containing ligand groups capable of binding metals in a comparable way to solution phase ligands play a vital role in determining metal speciation. Often this role is dominant as solid material contains a much higher concentration of ligand groups and their associated metal ions compared with the solution phase.

Metal speciation (e.g., the distribution of metal between solid and solution phase) depends on both the solution conditions and the nature of the solid phase. For instance, pH (related to concentration of hydrogen ions, H^+) has a major impact on speciation because it competes directly with metal ions for ligand sites. Generally, as the pH is lowered (higher concentration of hydrogen ions), metal ions tend to occur in the solution rather than in the solid phase. Other factors, which influence metal speciation through complex mechanisms, include ionic strength and the concentration of dissolved organic matter. The role of solid-phase material is also essential in determining the solid:solution distribution; the roles of oxides and organic matter are particularly important.

MEASUREMENT OF METAL SPECIATION

Despite the importance of metal speciation and the major advances in the last few years, speciation measurement remains complicated because of the difficulty of unambiguously interpreting an analytical signal in terms of a chemical species. Ion-selective electrodes can specifically measure the free metal ion activity (concentration) of metals such as copper, but generally they cannot measure the low levels of metal in unpolluted waters (9). Other techniques capable of measuring labile metal complexes include voltammetric methods and various competition methods using ion exchange resins, solid phases or solvent extraction (10). In addition, there are a number of physical separation methods in use such as filtration, centrifugation, and dialysis. When combined with a method for measuring total metal concentration, these give size-based metal speciation data (5). However, the data extracted are, at least to some extent, operational.

MODELING METAL SPECIATION

Similarly to the measurement of metal speciation, modeling has made immense advances during the last 10–15 years. However, the complexity of the systems studied and the difficulties of speciation analysis mean that many fundamental difficulties have not been solved. For instance, speciation modeling assumes that the aquatic system modeled is at chemical equilibrium (11). For many environmental systems, this is not the case. For instance, redox reactions and many sorption reactions of metals to organic surfaces may not be at equilibrium. The difficulties in fully and rigorously modeling these systems can be understood if we consider some of the components present in natural waters. These include

- solution phase inorganic ligands (e.g., chloride, carbonate, sulfate)
- solution phase organic ligands (e.g., amino acids, EDTA)
- inorganic colloids and surfaces (e.g., oxides, silicates, carbonates)
- organic colloids and surfaces (e.g., humic substances, microbial exudates)
- mixtures of ligands and metals
- variety of solution conditions

A number of model types have been used successfully to interpret metal interactions in laboratory systems. Generally, solution phase speciation is relatively well understood. Additionally, much progress has been made in modeling metal interactions to oxide surfaces and humic substances (7,11), and development and validation continue (12). Individual models have been brought together in easy-to-use computer packages called speciation codes. There are now dozens of speciation codes that have been developed from the 1960s onward, often for very different purposes and with their own strengths and weaknesses. Despite this, the use of a valid, predictive speciation model for use in natural waters remains an elusive goal, toward which much effort is focused.

SUMMARY

Trace metal speciation is defined as the distribution of the metal among different physicochemical forms. Speciation has important implications for the bioavailability, toxicity, and environmental transport of metals. For instance, free and colloidal forms of the metal may tend to remain in solution; larger particle-bound metal tends to sediment out and is removed from the water column. This difference in behavior results in geochemical fractionation of different metal forms and results in differences in both the fate and behavior of the metal. Additionally, the size of a metal complex may be an important parameter in understanding metal uptake by organisms, although in more complex ways because of biological processes. However, despite major advances in our understanding, methods of measuring and modeling metal speciation require significant development before they can confidently provide rigorous understanding in this important area.

BIBLIOGRAPHY

1. Davison, W. and de Vitre, R.R. (1992). Iron particles in freshwaters. In: *Environmental Particles*. J. Buffle and H.P. van Leeuwen (Eds.). Lewis, Boca Raton, FL.
2. Campbell, P.G.C. (1995). Interactions between trace metals and aquatic organisms: a critique of the free-ion activity model. In: *Metal Speciation and Bioavailability in Aquatic Systems*. John Wiley & Sons, Chichester, UK.
3. Mirimanoff, N. and Wilkinson, K.J. (2000). Regulation of Zn accumulation by a freshwater gram-positive bacterium (*Rhodococcus opacus*). *Environ. Sci. Technol.* **34**: 616–622.
4. Tessier, A., Buffle, J., and Campbell, P.G.C. (1994). Uptake of trace metals by aquatic organisms. In: *Chemical and Biological Regulation of Aquatic Systems*. J. Buffle and R.R. de Vitre (Eds.). Lewis, Boca Raton, FL.
5. Lead, J.R., Davison, W., Hamilton-Taylor, J., and Buffle, J. (1997). Characterising colloidal material in natural waters. *Aquat. Geochem.* **3**: 219–232.
6. Buffle, J. and Stumm, W. (1994). General chemistry of aquatic systems. In: *Chemical and Biological Regulation of Aquatic Systems*. J. Buffle and R.R. de Vitre (Eds.). Lewis, Boca Raton, FL.
7. Stumm, W. and Morgan, J.J. (1996). *Aquatic Chemistry: Chemical Equilibria and Rates in Natural Waters*. John Wiley & Sons, Hoboken, NJ.
8. Tessier, A. (1992). Sorption of trace element on natural particles in toxic environments.
9. Mota, A.M. and Correira dos Santos, M.M. (1995). Trace metal speciation of labile chemical species in natural waters and sediments: electrochemical approaches. In: *Metal Speciation and Bioavailability in Aquatic Systems*. John Wiley & Sons, Chichester, UK.
10. Apte, S. and Batley, G.E. (1995). Trace metal speciation of labile chemical species in natural waters and sediments: non-electrochemical approaches. In: *Metal Speciation and Bioavailability in Aquatic Systems*. John Wiley & Sons, Chichester, UK.
11. Turner, D. (1995). Problems in trace metal speciation modelling. In: *Metal Speciation and Bioavailability in Aquatic Systems*. John Wiley & Sons, Chichester, UK.
12. Vulkan, R. et al. (2000). Copper speciation and impacts on bacterial responses in the pore water of copper contaminated soils. *Environ. Sci. Technol.* 5115–5121.

METAL ION HUMIC COLLOID INTERACTION

Horst Geckeis
Institut für Nukleare Entsorgung
Karlsruhe, Germany

INTRODUCTION

Humic and fulvic acids in natural water are well known to have significant impact on metal ion bioavailability and mobility in aquifer systems. In natural water, toxic heavy metals, rare earth elements, and radionuclides such as the actinide ions are frequently found to be attached to natural organic matter such as humic and fulvic acids. These complexes are known to be of high kinetic and thermodynamic stability under given geochemical conditions and therefore may control trace metal transfer to bioorganisms and their transport in soil and groundwater systems. For those reasons, the interaction of metal ions with such humic matter is a major issue in a number of research fields such as environmental monitoring, risk assessment of contaminated sites, and nuclear waste disposal. In order to understand and to assess the impact of humic and fulvic acids on the behavior of trace metals, we need to know the interaction mechanisms, complexation kinetics, and finally thermodynamic data that can be implemented into geochemical speciation codes. A large spectrum of experimental methods is used for obtaining (1) insight into the humic acid structure and its properties as a complexing ligand, (2) quantitative information on metal ion speciation in humic containing water, and (3) molecular insight into the metal–humic interaction mechanism.

The ongoing development of understanding and experimental methods still makes the research on humic matter quite a dynamic area. Besides this short overview of the current state of the art in humic–metal ion interaction research, more comprehensive reviews on various aspects of metal–humic interaction are available (e.g., see Reference 1).

BASIC PROPERTIES OF HUMIC ACID

Humic/fulvic acid is a class of organic substances in the natural carbon cycle with a high stability and comparably high residence time (2,3). It is found in different sources, such as dissolved in natural water and as constituent, especially in soil, sediments, and brown coal. Humic acid is used both as a collective term for humic substances with a sufficiently high content of hydrophilic groups that dissolve in pH neutral range and as a specific term for the fraction that is dissolved in pH neutral range and flocculates in the acidic range with the complementary fraction of fulvic acid, soluble also under acidic conditions. In the following, humic acid (HA) is used in the general sense (humic and fulvic acids).

The composition, mass distribution, and functional group content vary within limits, reflecting different origins and histories. Disregarding sulfur and nitrogen (present in varying low concentrations generally on the order of up to around 1% by weight), the atomic composition is dominated by carbon, oxygen, and hydrogen. The contributions of these substances vary around $CO_{0.5}H$; that is, the formal oxidation state of carbon is around zero.

It should be borne in mind that description of the metal ion–humic colloid (or humic acid and fulvic acid as humic colloid constituents) interaction has been hampered in the past by large uncertainty concerning the structure of humic and fulvic acids. For a very long time, the molecular mass was given by number from around 10,000 to 100,000 or higher (2). Recent application of advanced mass spectroscopic methods [matrix assisted laser desorption ionization-time-of-flight mass spectrometry (MALDI-TOF), electrospray ionization mass spectrometry (ESI-MS), and TOF–secondary ion mass spectrometry (TOF-SIMS)], Overhauser nuclear magnetic resonance (NMR), and the use of size/diffusion velocity-based methods [size exclusion chromatography (SEC) and flow-field flow fractionation (FFF)] show that the mass distribution maxima are considerably lower, centering at masses below 1000 daltons and ending at around 3000 mass units (4,5). This means that several assumptions made in the past, concerning, for example, assumed pronounced macromolecular polyelectrolyte character, have to be reassessed today. In dissolved form, the molecules are highly hydrated and possibly form associates. Therefore, the size distribution is generally determined to lie within a few nanometers in diameter. The proton exchange capacity can reach up to 14 meq/g (6), where protonation/deprotonation takes place from about pH 10 down into the acidic range (below pH 3 a considerable number of the groups are still ionized). Carboxylic types of groups are normally quantified with about one-third of the total capacity. The rest are mainly ascribed to phenolic-type groups. The carboxylic groups with pK_a values around 4 account for the humic negative charge at even low pH, while the phenolic sites tend to deprotonate in the high pH region only (pK_a around 10).

MODELING APPROACHES

Transformation of species distribution data to complexation constants is not trivial for humic acid. Today there is still no general agreement. Major challenges are the difficulty to take the inherent heterogeneity of humic acid functional group characteristics into account, the appropriate selection of an electrostatic model, and finally the applicability to complex multicomponent real systems with a broad variability in water chemistries. Different model approaches have been developed.

A relatively simple nonelectrostatic approach has been applied to estimate metal ion speciation in a humic-containing groundwater (7). The authors found the accessibility of functional groups for interaction with metal ions to vary with water chemistry (pH, ionic strength) and the origin of the humic acid under consideration. Taking such variability into account by an experimentally determined value for the capacity of metal ion loading under given conditions allowed the determination of complexation constants for a variety of metal ions that are invariant with metal ion concentration, ionic strength, and pH. The hitherto most widely applied humic acid complexation models (8) are the more sophisticated WHAM (Windermere humic aqueous model) (9) and NICA (nonideal competitive adsorption)–Donnan model (10). Both approaches take metal ion interaction with carboxylate and phenolic groups into account. Heterogeneity of functional groups is considered in WHAM by a discrete binding model postulating four different subsites for each functional group type, that is, carboxylic and phenolic sites. In the NICA–Donnan approach, a continuous site distribution is implemented. Both models use Donnan-type expressions for taking electrostatic effects into account. The model assumption of a Donnan gel phase varying with ionic strength is certainly debatable for the very small humic acid molecules but appears to be able to mimic the surrounding diffuse double layer (11). A comprehensive data set on generic values on deprotonation and metal ion complexation constants for a variety of metal ions is available for both codes (12,13) and at least allows for a first estimation of metal ion speciation in complex systems containing humic matter even though parameter adjustment turns out necessary in many specific cases.

Figure 1 shows trends for complexation constants derived from different model approaches for several metal ions. Complexation constants taking an experimentally obtained loading capacity into account are shown in Fig. 1a (14). The bars reflect the uncertainty of the data obtained by applying a multitude of different speciation methods ranging from ultrafiltration to fluorescence spectroscopy. Log ß* values describing the complexation reaction $Me^{z+} + HA(z) \rightleftharpoons MeHA(z)$ increase—as one may expect—more or less with increasing charge of the metal cations and their tendency to hydrolysis in solution. $HA(z)$ corresponds to the accessible site concentration. The fact that the derived complexation constants do not exactly follow the trend in hydrolysis constants suggests the possible influence of steric hindrances. A similar picture can be obtained by plotting generic data obtained from reanalysis of literature data by using the NICA–Donnan approach (Fig. 1b) (12). Those constants K_1 and K_2 represent median affinities of metal ions to HA proton exchanging groups, which are modeled by a continuous site distribution with two maxima.

SPECIATION METHODS

Experimental approaches to differentiate humic-bound metal ions from other species and to characterize the metal–humic compound can be roughly subdivided in two groups:

1. Physical or chemical separation of bound humic from molecular metal ions (invasive speciation).
2. Spectroscopic methods (noninvasive speciation).

Different experimental methods are used for obtaining the different types of information. For the purpose of determining interaction constants, quantification of the noncomplexed and complexed ions is required.

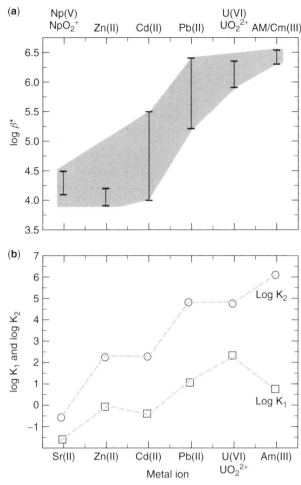

Figure 1. HA complexation constants for various metal ions obtained by using different model approaches. Data in part (**a**) are taken from Reference 14; those in part (**b**) correspond to the NICA–Donnan model approach (12). For explanation see text.

Experimental methods frequently used for determination of the species distribution in the metal ion–humic acid system are given in Table 1. Species distribution can be obtained by separation of the metal ion and the humic acid (complex) by size, density, or charge. Application of such methods is associated with uncertainty with respect to loss of substance, incomplete or uncertain species separation, possible kinetic rearrangements during separation, and problems related to changes in chemical composition. For this reason, methods are preferred where equilibrium is maintained. Frequently applied speciation tools for a number of trace metal ions such as Cu, Cd, Zn, and Pb are the electrochemical methods. Notably, anodic stripping voltammetry (ASV) achieves very low detection limits but intervenes in chemical equilibria and in many cases provides some kind of "operational" speciation. It is possible to differentiate "labile" metal ion species (i.e., free ions and readily dissociating complexed species) from stable or inert complexes.

Spectroscopic methods shown in Table 1 can give the species distribution without disturbing the equilibrium conditions. In general, absorption spectroscopy requires relatively high metal ion concentrations exceeding the ranges relevant to natural conditions. Fluorescence spectroscopy using appropriate fluorescent metal ions as probes [e.g., U(VI), Eu(III), Cm(III)] usually provide higher sensitivity for both excitation and emission. The humic acid itself is also fluorescent. The humic acid fluorescence intensity is quenched by metal ion complexation and thus the degree of fluorescence quenching can be used to study the metal ion–HA interaction and to quantify the complexation reaction. Probing the coordination environment for insight into the structure of the complex is specifically possible by X-ray absorption spectroscopy methods such as EXAFS and XANES as well as multinuclide nuclear magnetic resonance (NMR) spectroscopy. Examples for all these methods are given below.

METAL ION–HA INTERACTION MECHANISMS

Interaction of metal ions with humic/fulvic acids in principle may proceed via different pathways (see Table 2). Those individual mechanisms cannot be differentiated just by physical and/or chemical separation. In many cases spectroscopy can help to identify the binding mode and provide even details of the metal ion coordination environment. The following section is certainly not complete and only aims to give a rough overview into recent investigations on metal binding to humic/fulvic matter.

Complexation to Humic/Fulvic Acid

Various spectroscopic methods have been applied to gain insight into the coordination of the metal–humic/fulvic acid binding. Cu(II) is found to interact with four oxygen or two oxygen and two nitrogen atoms in a tetragonal coordination environment as shown by ESR (27). Later studies combining EXAFS and XANES spectroscopic experiments seem to verify the contribution of N ligands to the Cu(II) coordination (28). Studies on Cd(II) complexation with humic acid by EXAFS resulted in coordination via six oxygen atoms (distances 2.297 Å) (29). It seems to be clear that only carboxylate groups coordinate with Cd(II) as suggested by ^{113}Cd-NMR (30). Similar findings are recorded for the Ni(II) coordination with soil fulvic acid by EXAFS (31). Different to those d-transition metal ions, organomercury ions such as $Hg(CH_3)^+$ mainly bind to organic sulfur groups, for example, the thiol or disulfane groups. After saturation of those groups, coordination to oxygen atoms is found by EXAFS (32). Inorganic Hg(II) ions appear to form bidentate complexes with one thiol and one oxygen containing functional group (33). Al(III)–FA interaction studied by ^{27}Al-NMR suggests a coordination by carboxylic acid groups very similar to that observed in the oxalate complex (34). Formation of binary humate and ternary hydroxo and carbonato humate complexes are found for the interaction of trivalent actinides such as Am(III) and Cm(III) by UV/Vis absorption and notably the time-resolved laser fluorescence spectroscopy (TRLFS) (35,36). Such ternary species have rarely been taken into account by most model approaches up to now, but they appear to be even dominant

Table 1. Examples of Experimental Methods Used for Determination of Species Distribution (Complexed and Noncomplexed Metal Ions) in the Metal Ion–Humic Acid System and Methods Providing Analytical Information on the Nature of Such Complexes

Methods	Principle
Invasive	
Ultrafiltration, dialysis, ultracentrifugation, flow-field flow fractionation (FFF), size exclusion chromatography (SEC)	Size- or density-based separation (15–18,48)
Ion exchange, electrophoresis, Donnan membrane	Charge-based separation (19–22)
Ion selective electrodes (potentiometry), anodic stripping voltammetry (ASV)	Separate determination of noncomplexed or "labile" metal ion species (23,24)
Noninvasive (or Minimum Invasive)	
Fluorescence quenching, time-resolved laser fluorescence spectroscopy (TRLFS), UV/Vis, IR absorption	Detection and quantification of the complexation reaction via spectral shifts, emission intensity variations, fluorescence lifetime variations (provides to some extent information on coordination environment) (25,26) (see section below)
X-Ray absorption spectroscopy (EXAFS, XANES)	Absorption spectrum fine structure provides information on coordination environment (see section below)
Nuclear magnetic resonance spectroscopy (NMR)	Chemical shift provides chemical information on coordination environment (see section below)

in the near-neutral range in typical groundwater. It is noteworthy that TRLFS studies can be carried out at metal ion trace concentrations. For Cm(III) concentrations as low as 10^{-8} mol/L are sufficient. Such low metal ion concentrations are not accessible by most other spectroscopic methods. The tetravalent actinide Th(IV) studied by XPS and EXAFS turned out to mainly interact with carboxylic groups and was found to be coordinated by 9–10 oxygens at a distance of 2.43 Å (37), similar to trivalent ions such as Tb(III) (38). The coordination of trivalent metal ions like Eu(III) via carboxylate groups has been unequivocally verified by studying the C1s NEXAFS spectra of the humic acid (39). The $\pi^*_{C=O}$ transition of the carboxylate groups is clearly depressed upon complexation with Eu(III) ions. Hexavalent UO_2^{2+} ions are also found to interact only with carboxylate groups as shown by EXAFS study. Phenolic groups appear to be of minor importance as a coordinating ligand (40).

Metal Ion Binding to Inorganic–Humic Agglomerates

Metal ion–HA interaction in natural systems tends to be complicated by the existence of both mineral surfaces and nanosized mineral particles, both partly coated with HA. Simple additive approaches—that is, assuming the metal complexation to be accomplished by interaction with the mineral surface and independently with sorbed and dissolved humic acid—often tend to underestimate the sorbed metal ion fraction (41,42). One possibility for the discrepancy may be the formation of ternary complexes such as ≡S–O–Me–HA (where ≡S–O stands for a deprotonated mineral surface hydroxyl group and Me for the metal ion). A recent study claims, on the evidence of fluorescence spectroscopic experiments, that such ternary complexes indeed exist in the γ-Al_2O_3/HA/Cm(III) system (43).

Formation of Inorganic Colloids Stabilized by Humic/Fulvic Acid

In many cases, redox reactions are involved in the metal ion–humic interaction. Such processes can also be followed by EXAFS and XPS as has been shown for the reactions of Np(V), Pu(VI), and Tc(VII) (44–46). Humic bound Fe(II) and/or quinone groups are thought to act as electron donors. In all those cases it turned out that precipitating oxyhydroxides of the reduced tetravalent species to some extent are kept in solution as colloidal species stabilized by humic acid coatings. The differentiation of Me–HA complexes and colloid–HA agglomerates becomes challenging in such systems because pronounced long-term kinetic effects are involved in such redox reactions. Those studies were performed to investigate the possible behavior of radionuclides tentatively released from a nuclear waste repository. But inorganic colloid–humic aggregates have also been found to dominate the transport of trace elements like iron (47). The binding state of Fe(III) in a river colloidal matter was scrutinized by combined electron paramagnetic resonance (EPR), transmission electron microscopy (TEM), and FFF experiments. It turned out that the total Fe(III) was distributed between humic acid complexes, FeOOH colloids, and structurally

Table 2. Possible Binding Modes of Metal Ion Interaction with Humic Acid (HA)

Interaction Mechanism	Dissociation Rates[a]
Formation of inner-sphere complexes with functional groups of the HA macromolecules or Interaction with more than one HA entity. In both cases the formation of ternary complexes with HA and other dissolved ligands are possible.	Increasing ↑
Surface sorption to inorganic colloids (e.g., aquatic clay colloids), which in turn are stabilized by humic/fulvic acid coating.	
Inclusion of the metal ion into inorganic colloids stabilized by humic/fulvic coating due to formation of oxyhydroxide colloids as a consequence of redox reactions or formation of solid solutions.	

[a] In a qualitative way, the dissociation rates of metal ions from colloidal species increase from bottom to top

incorporated in clay colloids, the last two agglomerated with humic matter. Similar observations have also been made for trace elements such as the naturally abundant lanthanides, uranium, and thorium in a humic-rich groundwater (48).

It is obvious that the metal ion interaction mode with humic matter is of importance for the assessment of HA influence on the geochemical cycle of trace elements. Notably, the kinetics of the metal ion desorption from the individual HA compounds varies considerably depending on the interaction mechanism (see Table 2). Even though there is still uncertainty concerning the exact coordination environment of HA bound metal ions, the coordination of most metal ions with HA carboxylate groups is now verified by a number of spectroscopic studies.

KINETICS

The metal ion interaction shows a sequence of reaction/exchange velocities. Primary association is rapid, followed by rearrangement to kinetically more stable binding. Dissociation is considerably slower and rates are decreasing with increasing aging time. Dissociation shows different kinetic modes, varying from seconds to weeks (49,50). The metal ion is believed to either migrate to stronger binding sites (51) or move to inner sites of the humic macromolecular structure (52). In view of the now generally accepted relatively small size of the humic acid molecules, the existence of a significant inner space becomes somewhat implausible. It is more likely that intermolecular bridging by polyvalent cations leading to the interaction of metal ions with more than one humic acid entity (see Table 2) become responsible for the observed time-dependent decrease of dissociation rates. The existence of such metal-induced humic agglomeration even at trace metal concentrations has been visualized, for example, by flow-field flow fractionation studies (48) or by using fluorescence probes (53). The metal-induced aggregation of natural organic matter is furthermore well known from studies in the marine environment (54).

Investigating humic matter in a natural groundwater revealed that part of the highly charged metal ion inventory of the humic acid shows a very slow dissociation, basically irreversible for time scales that can be studied under laboratory conditions. These metal ions show a different behavior from that observed for metal ion–humate complexes synthesized under laboratory conditions (48). The exact nature of binding for this inventory is not yet understood. One possibility is association with nanosized inorganic mineral presumably as solid solution. Various support for that assumption is found in analyzing various groundwater humic colloid size fractions: while humic acid is found mainly at sizes <10 nm, a major fraction of natural colloid-borne polyvalent elements such as thorium and rare earth elements are found in larger colloids >30 nm. Since metal ion dissociation from such types of colloids is very slow, their colloid-mediated transport becomes very relevant. Such processes are of interest in areas such as colloid-facilitated radionuclide transport in nuclear waste disposal research. A final conclusion also with regard to the relevance of such mechanisms to the toxic metal transport will require further investigations.

OUTLOOK

Metal ion interaction with humic acid is still a challenging field. A lot of new insight into the metal ion coordination could be obtained by the application of sensitive spectroscopic and other analytical methods. Knowledge about the specific binding state at least allows the estimation of humic-bound metal behavior in a semiquantitative way. There are chemical models available that—with some drawbacks—are able to predict metal ion speciation to some extent in humic-containing media. Open questions still remain concerning the "real" structure of humic/fulvic substances under *in situ* conditions and regarding the "real" configuration of metal ion–humic complexes. Application of newly emerging advanced analytical methods will certainly drive the research field and improve both the insight into the "real" metal–humic interaction mechanisms and the accuracy of model predictions.

BIBLIOGRAPHY

1. Clapp, C.E., Hayes, M.H.B., Senesi, N., Bloom, P.R., and Jardine, P.M. (Eds.). (2001). *Humic Substances and Chemical Contaminants*. Soil Science Society of America, Madison, WI, Section 3, p. 303.
2. Stevenson, F.J. (1994). *Humus Chemistry*, 2nd Edn. John Wiley & sons, Hoboken, NJ.
3. Buckau, G. et al. (2000). Origin and mobility of humic colloids in the Gorleben aquifer system. *Appl. Geochem.* **15**: 171–179.
4. Simpson, A.J. et al. (2002). Molecular structures and associations of humic substances in the terrestrial environment. *Naturwissenschaften* **89**: 84–88.
5. Leenheer, J.A. and Croue, J.-P. (2003). Characterizing aquatic dissolved organic matter. *Environ. Sci. Technol.* **37**: 18A–26A.
6. Elkins, K.M. and Nelson, D.J. (2002). Spectroscopic approaches to the study of the interaction of aluminum with humic substances. *Coord. Chem. Rev.* **228**: 205–225.
7. Kim, J.I. and Czerwinski, K.R. (1996). Complexation of metal ions with humic acid: metal ion charge neutralization model. *Radiochim. Acta* **73**: 5–10.
8. Dudal, Y. and Gerard, F. (2004). Accounting for natural organic matter in aqueous chemical equilibrium models: a review of the theories and applications. *Earth Sci. Rev.* **66**: 199–216.
9. Tipping, E. and Hurley, M.A. (1992). A unifying model of cation binding by humic substances. *Geochim. Cosmochim. Acta* **56**: 3627–3641.
10. Benedetti, M.F., Milne, C.J., Kinniburgh, D.G., van Riemsdijk, W.H., and Koopal, L.K. (1995). Metal ion binding to humic substances: application of the non-ideal competitive adsorption model. *Environ. Sci. Technol.* **29**: 446–457.
11. Benedetti, M.F., van Riemsdijk, W.H., and Koopal, L.K. (1996). Humic substances considered as a heterogeneous Donnan gel. *Environ. Sci. Technol.* **30**: 1805–1813.
12. Milne, C.J., Kinniburgh, D.G., van Riemsdijk, W.H., and Tipping, E. (2003). Generic NICA-Donnan model parameters for metal-ion binding by humic substances. *Environ. Sci. Technol.* **37**: 958–971.
13. Tipping, E. (1998). Humic ion-binding model VI: an improved description of the interactions of protons and metal ions with humic substances. *Aquatic Geochem.* **4**: 3–48.
14. Marquardt, C.M. (Ed.). (2000). Influence of humic acids on the migration behaviour of radioactive and non-radioactive substances under conditions close to nature. *Forsch. Wiss. Ber. FZKA* **6557**: 14.
15. Kim, J.I., Rhee, D.S., and Buckau, G. (1991). Complexation of Am(III) with humic acids. *Radiochim. Acta* **52/53**: 49–55.
16. Czerwinski, K.R., Buckau, G., Scherbaum, F., and Kim, J.I. (1994). Complexation of the uranyl ion with aquatic humic acid. *Radiochim. Acta* **65**: 111.
17. Glaus, M.A., Hummel, W., and van Loon, L.R. (1995). Equilibrium dialysis–ligand exchange: adaptation of the method for determination of conditional stability constants of radionuclide–fulvic acid complexes. *Anal. Chim. Acta* **303**: 321.
18. Hasselöv, M., Lyven, B., Haraldsson, C., and Sirinawin, W. (1999). Determination of continuous size and trace element distribution of colloidal material in natural water by on-line coupling of flow field-flow fractionation with ICPMS. *Anal. Chem.* **71**: 3497–3502.
19. Burba, P., Rocha, J., and Klockow, D. (1994). Labile complexes of trace metals in aquatic humic substances: investigation by means of an ion exchange-based flow procedure. *Fresenius J. Anal. Chem.* **349**: 800–807.
20. Sonke, J.E. and Salters, V.J.M. (2004). Determination of neodymium–fulvic acid binding constants by capillary electrophoresis inductively coupled plasma mass spectrometry (CE-ICP-MS). *J. Anal. At. Spectrom.* **19**: 235–240.
21. Seibert, A. et al. (2001). Complexation behaviour of neptunium with humic acid. *Radiochim. Acta* **89**: 505–510.
22. Temminghoff, E.J.M., Plette, A.C.C., Van Eck, R., and van Riemsdijk, W.H. (2000). Determination of the chemical speciation of trace metals in aqueous systems by the Wageningen Donnan membrane technique. *Anal. Chim. Acta* **417**: 149–157.
23. Florence, T.M. (1982). The speciation of trace elements in waters. *Talanta Rev.* **29**: 345–364.
24. Filella, M., Buffle, J., and van Leeuwen, H.P. (1990). Effect of physico-chemical heterogeneity of natural complexes—Part I: Voltammetry of labile metal–fulvic complexes. *Anal. Chim. Acta* **232**: 209.
25. Cook, R.L. and Cooper, H.L. (1995). Metal ion quenching of fulvic acid fluorescence intensities and lifetimes: nonlinearities a possible three-component model. *Anal. Chem.* **67**: 174.
26. Kim, J.I., Rhee, D.S., Wimmer, H., Buckau, G., and Klenze, R. (1993). Complexation of trivalent actinide ions (Am^{3+}, Cm^{3+}) with humic acid: a comparison of different experimental methods. *Radiochim. Acta* **62**: 35–43.
27. Senesi, N., Bocian, D.F., and Sposito, G. (1985). Electron spin resonance investigation of copper(II) complexation by soil fulvic acid. *Soil Sci. Soc. Am. J.* **49**: 114–115.
28. Frenkel, A.I., Korshin, G.V., and Ankudinov, A.L. (2000). XANES study of Cu^{2+}-binding sites in aquatic humic substances. *Environ. Sci. Technol.* **34**: 2138–2142.
29. Liu, C., Frenkel, A.I., Vairavamurthy, A., and Huang, P.M. (2001). Sorption of cadmium on humic acid: mechanistic and kinetic studies with atomic force microscopy and X-ray absorption fine structure spectroscopy. *Can. J. Soil Sci.* **81**: 337–348.

30. Otto, W.H., Carper, W.R., and Larive, C.K. (2001). Measurement of cadmium(II) and calcium(II) by fulvic acids using ^{113}Cd-NMR. *Environ. Sci. Technol.* **35**: 1463–1468.
31. Strathmann, T.J. and Myneni, S.C.B. (2004). Speciation of aqueous Ni(II)-carboxylate and Ni(II)-fulvic acid solutions: combined ATR-FTIR and XAFS analysis. *Geochim. Cosmochim. Acta* **68**: 3441–3458.
32. Qian, J. et al. (2002). Bonding of methyl mercury to reduced sulfur groups in soil and stream organic matter as determined by X-ray absorption spectroscopy and binding affinity studies. *Geochim. Cosmochim. Acta* **66**: 3873–3885.
33. Haitzer, M., Aiken, G.R., and Ryan, J.N. (2003). Binding of mercury(II) to aquatic humic substances: influence of pH and source of humic substances. *Environ. Sci. Technol.* **37**: 2436–2441.
34. Lee, N.C.Y. and Ryan, D.K. (2004). Study of fulvic–aluminum(III) ion complexes by ^{27}Al solution NMR. In: *Humic Substances—Nature's Most Versatile Materials*. E. Ghabbour and G. Davies (Eds.). Taylor and Francis, New York, pp. 219–228.
35. Buckau, G., Kim, J.I., Klenze, R., Rhee, D.S., and Wimmer, H. (1992). A comparative spectroscopic study of the fulvate complexation of trivalent transuranium ions. *Radiochim. Acta* **57**: 105–111.
36. Panak, P., Klenze, R., and Kim, J.I. (1996). A study of ternary complexes of Cm(III) with humic acid and hydroxide or carbonate in neutral pH range by time-resolved laser fluorescence spectroscopy. *Radiochim. Acta* **74**: 141–146.
37. Denecke, M.A., Bublitz, D., Kim, J.I., Moll, H., and Farkes, I. (1999). EXAFS investigation of the interaction of hafnium and thorium with humic acid and Bio-Rex70. *J. Synchrotron Rad.* **6**: 394–396.
38. Monsallier, J.M. et al. (2003). Spectroscopic study (TRLFS and EXAFS) of the kinetics of An(III)/Ln(III) humate interaction. *Radiochim. Acta* **91**: 567–574.
39. Plaschke, M., Rothe, J., Denecke, M.A., and Fanghänel, Th. (2004). Soft X-ray spectromicroscopy of humic acid europium(III) complexation by comparison to model substances. *J. Electron Spectrosc. Relat. Phenom.* **135**: 53–62.
40. Denecke, M.A. et al. (1998). *Radiochim. Acta* **82**: 103–108.
41. Vermeer, A.W.P., McCulloch, J.K., van Riemsdijk, W.H., and Koopal, L.K. (1999). Metal ion adsorption to complexes of humic acid and metal oxides: deviations from the additivity rule. *Environ. Sci. Technol.* **33**: 3892–3897.
42. Rabung, T., Geckeis, H., Kim, J.I., and Beck, H.P. (1998). The influence of anionic ligands on the sorption behavior of Eu(III) on natural hematite. *Radiochim. Acta* **82**: 243–248.
43. Wang, X.K. et al. (2004). Effect of humic acid on the sorption of Cm(III) onto γ-Al$_2$O$_3$ studied by the time-resolved laser fluorescence spectroscopy. *Radiochim. Acta*, **92**: 691–734.
44. Zeh, P., Kim, J.I., Marquardt, C.M., and Artinger, R. (1999). The reduction of Np(V) in groundwater rich in humic substances. *Radiochim. Acta* **87**: 23–28.
45. Marquardt, C.M. et al. (2004). The redox behaviour of plutonium in humic rich groundwater. *Radiochim. Acta* **92**: 619–662.
46. Maes, A. et al. (2004). Evidence for the interaction of technetium colloids with humic substances by X-ray absorption spectroscopy. *Environ. Sci. Technol.* **38**: 2044–2441.
47. Benedetti, M.F., Ranville, J.F., Allard, T., Bednar, A.J., and Menguy, N. (2003). The iron status in colloidal matter from the Rio Negro, Brasil. *Coll. Surf. A* **217**: 1–9.
48. Geckeis, H., Rabung, Th., Ngo Manh, T., Kim, J.I., and Beck, H.P. (2002). Humic colloid-borne natural polyvalent metal ions: dissociation experiment. *Environ. Sci. Technol.* **36**: 2946–2952.
49. Langford, C.H. and Gutzman, D.W. (1992). Kinetic studies of metal ion speciation. *Anal. Chim. Acta* **256**: 183–201.
50. Schüssler, W., Artinger, R., Kienzler, B., and Kim, J.I. (2000). Conceptual modeling of the humic colloid-borne americium(III) migration by a kinetic approach. *Environ. Sci. Technol.* **34**: 2608–2611.
51. Rate, W., McLaren, R.G., and Swift, R.S. (1993). Response of copper(II)–humic acid dissociation kinetics to factors influencing complex stability and macromolecular conformation. *Environ. Sci. Technol.* **27**: 1408–1414.
52. Choppin, G.R. and Clark, S.B. (1991). The kinetic interactions of metal-ions with humic acids. *Mar. Chem.* **36**: 27–38.
53. Engebretson, R. and Von Wandruszka, R. (1998). Kinetic aspects of cation enhanced aggregation in aqueous humic acid. *Environ. Sci. Technol.* **32**: 488–493.
54. Chin, W.C., Orellana, M.V., and Verdugo, P. (1998). Spontaneous assembly of marine dissolved organic matter into polymer gels. *Nature* **391**: 568–572.

HEAVY METAL UPTAKE RATES AMONG SEDIMENT DWELLING ORGANISMS

Jos P.M. Vink
Institute for Inland Water
Management and Waste Water
Treatment–RIZA
Lelystad, The Netherlands

INTRODUCTION

Sediments of rivers, river banks, and floodplains are commonly characterized by elevated levels of heavy metals such as Cd, Cu, Cr, Ni, Pb, Zn, and As (1). Organisms that occur in these sediments are known to accumulate these metals (2,3), which, in turn, are passed on through the foodweb. However, the development of well-defined quality criteria for (aquatic) sediments is seriously hindered mainly because of three reasons:

1. Laboratory exposure tests that are conducted with organisms (bioassays) do not consider the geochemical composition of the sediment *in situ* and do not link temporal changes in chemical speciation with toxicological effects.
2. It is usually unclear which routes of exposure contribute to accumulation in organisms and dictate the adverse effects.
3. There is a general uncertainty about the relative contributions of the various metal species that take part in biological uptake.

The general concept of (bio)availability of metals becomes clear when body concentrations of either aquatic or terrestrial organisms collected from the field are compared with those performed in exposure tests using disturbed, oxygenated, and spiked sediments or waters. In the latter case, concentrations are mostly significantly higher, both in the close environment and in the organism itself. This phenomenon is explained by chemical speciation, which is based on the lability of metal species. The most labile form is the free dissolved ion, followed by dissolved inorganic metal–ion pairs (e.g., M–OH, M–CO_3, M–Cl) and organically complexed forms (M–DOM).

This concept of lability, however, has in recent years been more often theorized than actually shown by empirical data. This is mainly due to the fact that only very few techniques have become available to actually measure free ion activities, let alone to measure these periodically during exposure of organisms. The concept of the free ion activity model (FIAM) assumes that free aqueous metal ion concentrations, rather than the total or dissolved concentration, largely determine the toxicological or biological effect that is observed in organisms that are exposed to water or sediment containing heavy metals.

In a critical literature review, Campbell (4) concluded that studies that are suitable for testing the FIAM concept in the presence of natural dissolved organic matter (DOM) are extremely scarce. There are numerous studies that report on the effects of DOM on metal bioavailability, but virtually all of these are qualitative in nature (speciation is underdefined). The few quantitative studies (5–9,11) are, however, not in agreement, and the applicability of FIAM to natural waters (with DOM) remains to be demonstrated. It is concluded that "future progress would be greatly aided by the development of methods capable of measuring the free ion concentration of metals in the presence of DOM."

Recently, a new technique was introduced aimed at tackling the difficulties described above. This technique was introduced as Sediment Or Fauna Incubation Experiment, SOFIE in short (12). With this technique, it is possible to quantify very labile metal fractions, including free metal ions, in pore water, at low oxygen and reduced conditions, repeatedly, and in a nondestructive manner to the sample. Bioassays are conducted simultaneously in the same setting. Sediment samples are obtained in an undisturbed manner, including the overlying surface water.

Here, we deal with the gradients and speciation of metal species, including free ion activities, over water–sediment interfaces. The effect of metal uptake by sediment dwelling organisms on metal speciation, and vice versa, is discussed.

METHODS

Materials

This study was performed with a novel experimental technique (EU patent 1018200/02077121.8, October 2001, J. Vink, Rijkswaterstaat), which was introduced as Sediment Or Fauna Incubation Experiment, or SOFIE®. The method was described in detail by Vink (12). In short, this "cell" consists of a core, 190 mm radius, 200 mm height, which is used as a sampling device to obtain undisturbed sediment, and the corresponding surface water. After sampling, this water–sediment system becomes part of the cell. Specific pore water probes, constructed from permeable PES polymers, are attached at 5-mm layer increments. Probes are directly connected to preconditioned ion-exchange microcolumns (MIC). A MIC contains a metal chelating polymer (styrene divinylbenzene copolymers, RI-718, containing paired iminodiacetate ions that act as chelating groups), which is highly selective to metals. Retained metals are liberated from the polymer with an acid-extraction procedure. In this setup, ion exchange between pore water and exchange polymer is practically instantaneous and is therefore executed at the reigning geochemical status, including redox potential, of the probed sediment layer.

With these cells, water–sediment interfaces from six different river systems in The Netherlands were sampled (Table 1). Sites were selected for their degree of contamination, physical properties, or their general representativeness for sediments commonly found in the river. Aqueous steady-state conditions were determined by probing the water/sediment on three different occasions. Location MB was probed every 5 mm, but the other locations were probed in intervals that coincided with the layering of the sediment—which could be observed through the transparent cell wall. After the introduction of test organisms, chemical speciation was monitored in time and depth.

Exposure Tests with Benthic Organisms

Species of the oligochaete *Limnodrilus* (family Tubificidae) and the mosquito larvae *Chironomus riparius* were chosen for accumulation test because of their wide abundance in aquatic systems and their importance for the ecological food chain. Both species are sediment dwellers, which means that they live in close contact with the sediment. Populations were bred in the laboratory under contaminant-free conditions. Oligocheate populations consisted mainly of *L. claparedeanus* and *L. hoffmeisteri* in equal amounts. Individuals with corresponding life stages (L2) were introduced in the cell in field-realistic densities (approximately 1250 oligochaetes and 220 chironomids, respectively). For locations MB and SW, only exposure tests with oligochaetes were performed. Body concentrations were measured periodically by sampling individuals from the population on seven occasions at 2–3 day intervals. They were allowed to void their gut for 48 h, which, based on dissection, was an adequate period. The organisms were freeze dried for 48 h and the dry weight was determined. Digestion was done in 500 μL of 14.9 M HNO_3 (Ultrex) at 180 °C. Metal concentrations in the digests were determined using inductively coupled plasma mass spectrometry (ICP-MS, Perkin Elmer Elan 6000). Dolt-2 (certified by the Community Bureau of Reference, BCR, Brussels, Belgium) was used as biological reference material.

Chemical Analysis

Aqueous concentrations of NO_3^-, NO_2^-, SO_4^{2-}, PO_4^{3-}, and Cl^- were measured with segmented ion chromatography (Dionex AS50 chromatography compartment, ED_{50} electrochemical detector). Limits of detection (LOD) ranged from 0.025 mg·L^{-1} for NO_2^- to 0.05 mg·L^{-1} for PO_4^{3-}. As, Al, Ca, Cd, Co, Cu, Cr, Fe, Mg, Mn, Ni, Pb, and Zn analyses were carried out with ICP-MS (LOD min/max: 0.003 μg·L^{-1} for Cd to 6.0 μg·L^{-1} for Zn). NH_4^+ analyses were performed by Skalar segmented flow analyzer SAN+

6250 matrix fotometer (LOD: 0.022 mg·L^{-1}). Dissolved organic carbon was determined with a Shimadzu 5000 TOC analyzer (LOD: 0.45 mg·L^{-1}).

RESULTS AND DISCUSSION

Table 1 shows the solid phase characteristics of the six sites. Steady-state concentrations in pore water and overlying surface water are shown in Fig. 1.

Typically, steep redox gradients occur over the water–sediment interface, more or less pronounced for the various sites. Redox-sensitive compounds follow these gradients: denitrification precedes sulfate reduction, and anaerobic conditions were reached in most cases within 15 mm from the water–sediment interface. The increasing reduction of ferric(III)oxyhydroxides to Fe(II), and manganese(V)oxides to Mn(II) results in increasing concentrations with depth of aqueous Fe^{2+} and Mn^{2+}. Pore water concentrations of Cd, Cu, Ni, Pb, and Zn in most cases tend to decrease with depth and coincide with sulfate reduction patterns. These metals most probably associate with sulfides and therefore become insoluble. Heavy metals, however, compete for binding opportunities between the reactive sulfide phase and relatively large amounts of dissolved organic matter, which keeps them in the aqueous phase. There is a pH range between the sites of almost 2 units.

Time-Varying Exposure Concentrations

Immediately after introduction of the test organisms to the water–sediment system, steady-state concentrations shifted. Oligochaetes and chironomids are sediment dwellers. They live in burrows and irrigate these with oxygen-rich surface water. Consequently, their immediate environment is oxidized, which in turn directly affects the chemical reactivity and bioavailability of metals that remain in the sediment. In Fig. 2, an example is shown of how aqueous concentrations of copper species change with time and depth during the exposure test (site MB).

Dissolved metal concentrations change with time, and so do specific metal species such as M^{2+}. These time-varying changes may follow first-order reaction kinetics:

$$dC/dt = kC$$

so the external concentration, or exposure concentration, at a given time $C(t)$ is written as

$$C(t) = C_i e^{k_0 t} \quad (1)$$

in which C_i is the initial concentration in (pore) water, and k_0 is a rate term that describes the increase or decrease rate of the initial steady-state concentration. In analogy, the free ion activity concentration at a given time, M^{2+} (t), may be written as

$$[M^{2+}](t) = [M_i^{2+}] e^{k_{0,\text{act}} t} \quad (2)$$

The rates at which these concentration shifts occurred proved to highly affect bioaccumulation patterns, which is shown later.

Table 1. Sediment Properties

	Location					
	MB	GA	KV	HD	AS	SW
River System	Rhine	Waal	IJssel South	Hollands Diep	Meuse	IJssel North
Depth (m)	3.5	0.3	1	5.2	0.4	0.4
Dry weight (%)	55	80	73	34	45	43
<63 μm (%)	72.6	4.1	15.7	21.8	54.2	65.5
<16 μm (%)	59.3	3.1	9.7	17.8	23.3	49
<2 μm (%)	35.2	1.8	5.65	10.4	14	29.55
Organic C (%)	2.4	0.2	0.55	1.14	2.25	4.15
Inorganic C (%)	1.1	0.6	1.3	0.87	0.55	0.56
pH	6.9	8.1	7.4	7.6	7.2	7.6
E_h (mV)	−448	−163	−149	23	−147	−171
Al (g/kg)	23.4	5.6	3.5	17.3	9.2	11
As (mg/kg)	nd	5.6	4	6.5	8.5	22
Ca (g/kg)	24.1	10.7	19.6	35.1	23.2	32
Cd (mg/kg)	7.49	0.37	0.14	2.59	1.57	2.35
Co (mg/kg)	15.3	7.8	3.3	7.7	8.6	12
Cr (mg/kg)	102.1	32.4	16.8	45.9	27.7	57.5
Cu (mg/kg)	93.1	11.4	4.9	26.5	25.7	48
Fe (g/kg)	29.3	7.6	8.0	18.2	22.2	24.5
Hg (mg/kg)	1.5	0.2	0.055	0.31	0.205	1.20
Mg (g/kg)	7.79	2	3.06	5.24	3.48	8
Mn (mg/kg)	654	241.7	196.5	430.5	524.5	530
Ni (mg/kg)	35.6	6.6	8.35	19.4	20.8	31.5
Pb (mg/kg)	703.6	28	11.5	39.8	79	91.5
Zn (mg/kg)	1091	133	33	231	320.5	365

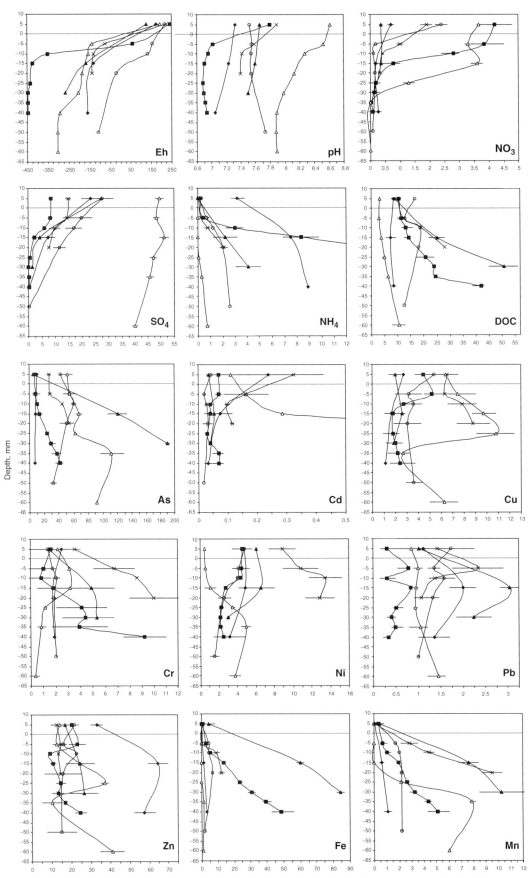

Figure 1. Aqueous steady-state characteristics for the six sites. MB (■); GA (△); KV (×); HD (○); AS (♦); SW (▲). E_h in mV; Fe, Mn, NO_3, SO_4, NH_4, DOC in mg/L; As, Cd, Cu, Cr, Ni, Pb, Zn in µg/L.

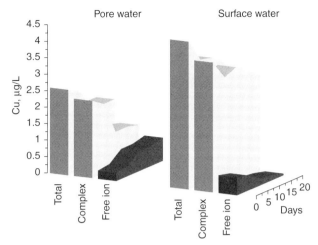

Figure 2. Example of time-dependent speciation shifts in overlying surface water and pore water (20 mm below water–sediment interface) during an exposure test. The free ion concentration in pore water increases, through oxidation, and is exhausted from the overlying water due to uptake by biota.

Uptake and Elimination by *Limnodrilus* and *Chironomus*

Metal uptake and excretion rates were approximated by a time/concentration-dynamic, two-compartment model. It is assumed that body concentrations vary with time and depend on uptake from the water phase with a certain rate. Uptake is at the same time accompanied by elimination (excretion) from the organism's body. Therefore,

$$dQ/dt = k_1 C(t) - k_2 Q(t)$$

with Q being the internal body concentration ($\mu g \cdot g^{-1}$ dw), k_1 and k_2 being uptake and elimination rate constants, respectively (d^{-1}), and t being time. This yields, for $Q(0) = 0$,

$$Q(t) = \frac{k_1 C_0}{k_2 - k_0}(e^{-k_0 t} - e^{-k_2 t}) \quad (3)$$

where C and k_0 are derived from Equations 1 and 2.

In analogy with the free ion activity,

$$Q(t) = \frac{k_1 [M^{2+}]}{k_2 - k_{0,\text{act}}}(e^{-k_{0,\text{act}} t} - e^{-k_2 t}) \quad (4)$$

With the introduction of k_0, time-dependent concentrations determine the overall exposure of organisms, and thus $Q(t)$. Note that elimination may partly be attributed to growth of the organisms during the exposure period. This separate loss term is not considered in the model.

To determine the consequence of *not* considering time-dependent concentrations, a one-compartment model was included for comparison. With this model, a constant uptake rate, a, is assumed, as opposed to concentration-dependent uptake. Body concentrations are then described as

$$dQ/dt = a - kQ$$

which yields

$$Q(t) = C_0 e^{-k \cdot t} + \frac{a}{k}(1 - e^{-k \cdot t}) \quad (5)$$

Figure 3 shows bioaccumulated amounts of some priority elements and heavy metals by oligochaetes and chironomids (site GA). Typically, metals are taken up at a certain rate (k_1), leading to increased body concentrations. After some time, elimination (k_2) exceeds uptake, and concentrations decline to reach a steady state. In general, bioaccumulated amounts can be ranked according to Ca > Fe > Mg > Zn > Al > Mn > Cu > Ti ~ Pb > Cr > As ~ Cd ~ Ni. Of both species, chironomids appear the most sensitive since they tend to take up these elements faster than oligochaetes and accumulate larger amounts (normalized to dry weight).

Examples of accumulation patterns for oligochaetes and chironomids for all locations are shown in Fig. 4. To test whether these sediment dwellers actually show different patterns at different locations (i.e., to determine location specificity), the accumulation data of oligochaetes and chironomids were statistically compared (two-tailed F-test, Mann–Whitney, $p < 0.05$). Results are summarized in Table 2 and show that locations do indeed matter. When distinguishing locations, both test species show the lowest sensitivity for chromium and nickel, which seem nondiscriminating. On the other hand, the combinations cadmium–chironomids and zinc–oligochaetes show a high discriminating potential. The fact that accumulation patterns on various locations are indeed different—also on locations with comparable dissolved concentrations—indicates that there is a difference in the actual bioavailability of metals.

Relation Between Time-Dynamic Speciation and Body Concentrations

To analyze the performance of the uptake and elimination models with varying exposure concentrations and speciation, the overall data set was divided into four subsets: two "environment compartments" (overlying surface water, and sediment pore water), each with two chemical "species" (total dissolved metal concentration, and free ion activities). Performance of the uptake–elimination models (Eqs. 3, 4, and 5) were tested with each subset. Results of measurements and corresponding model parameters C_i and k_0 were subsequently subjected to statistical analyses. As a descriptor for the goodness of prediction for the uptake–elimination model, the standard error of each model parameter was used rather than the correlation coefficient of the model with the measurements. The standard error is a measurement of the variability of a data point Y around the predicted value Y_p. It represents information about the goodness of fit in the same manner as the standard deviation does about the spread around the mean. The standard error is written as

$$sy \cdot x = \sqrt{\frac{\Sigma(Y - Y_p)^2}{n}}$$

By minimizing the sum of squares, values for k_1 and k_2 and corresponding standard errors were derived. Considering only Cu, Cd, Ni, Pb, Zn, and As, the data set consisted of over 1800 chemical and 608 biological measurements.

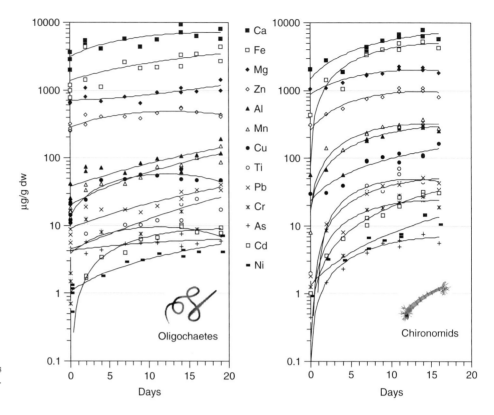

Figure 3. Bioaccumulation patterns of priority elements and heavy metals in two benthic species.

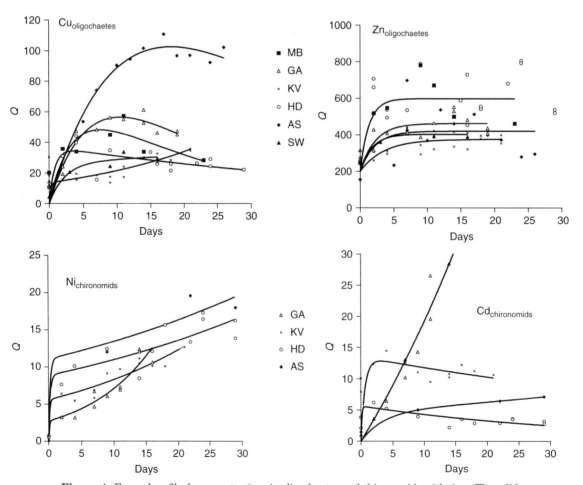

Figure 4. Examples of body concentrations in oligochaetes and chironomids with time. The solid line is the model prediction based on Equation 4 (Cd, Cu, Ni) and Equation 5 (Zn).

Table 2. Variance Between Locations: Accumulation Patterns in Oligochaetes[a] and Chironomids

Cd	MB	GA	KV	HD	AS	SW	Cu	MB	GA	KV	HD	AS	SW
MB		D	ns	ns	ns	ns	MB		ns	D	ns	D	ns
GA			D	D	D	D	GA			D	ns	D	D
KV	D			ns	ns	ns	KV	D			ns	D	ns
HD	D	D			ns	ns	HD	D	ns			D	ns
AS	D	D	ns			ns	AS	ns	ns	ns			D
SW							SW						

Ni	MB	GA	KV	HD	AS	SW	Pb	MB	GA	KV	HD	AS	SW
MB		ns	ns	ns	D	ns	MB		D	D	D	D	D
GA			ns	ns	ns	ns	GA			D	ns	ns	D
KV	ns			ns	D	ns	KV	D			D	D	ns
HD	ns	D			D	ns	HD	ns	D			ns	ns
AS	ns	ns	ns			D	AS	ns	ns	ns			ns
SW							SW						

Zn	MB	GA	KV	HD	AS	SW	As	MB	GA	KV	HD	AS	SW
MB		D	D	ns	ns	D	MB		D	D	D	D	ns
GA			D	D	ns	D	GA			ns	D	D	ns
KV	ns			D	ns	D	KV	D			D	ns	ns
HD	D	D			D	D	HD	D	ns			ns	ns
AS	ns	ns	ns			D	AS						ns
SW							SW						

Cr	MB	GA	KV	HD	AS	SW
MB		ns	ns	ns	D	ns
GA			ns	ns	D	ns
KV	ns			ns	ns	ns
HD	ns	D			D	ns
AS	ns	ns	ns			D
SW						

[a]Data for oligochaetes shaded gray; D, significantly different; ns, not significantly different.

Only those combinations with standard errors that were of the same magnitude or smaller than the values of k_1 and k_2 themselves were considered significant. This does not mean that other relationships between compartments, chemical species, and accumulation do not exist; this statistical criterion is necessary to identify the best model predictor, and therefore the most significant source of metals to biota.

The outcome of the modeling exercise and the statistical comparisons are summarized in Table 3. Chemical speciation and concentration shifts for the two compartments are shown on the left side. These were used as input for the biotic uptake models, for which results are shown on the right side. Only combinations where the statistical criterion was met ($sy \cdot x < k_1, k_2$) are shown. Obviously, the other combinations are thefore of less significance.

Negative k_0 values indicate a decrease in concentrations, positive values indicate an increase. Combinations of measured C_i and k_0 values were used as initial constants in the iteration procedure. The advantage of this approach is that variables that have a relatively large certainty or reliability do not participate in the iteration process as such. They do not need to be included, since $C(t)$, M^{2+}, and k_0 were actually measured with time. This is a large advantage, since the optimization procedure is now focused on the accurate estimation of the remaining variables k_1 and k_2, that is, the uptake and elimination rates.

The most important conclusion from Table 3 is that both benthic species appear to obtain their metals primarily from the overlying surface water and not from the sediment pore water compartment. Total concentrations in the sediment (mg/kg) are not indicative and do not reflect the magnitude of total body concentrations. Both test species accumulated the largest amounts of cadmium at site GA (chironomids even faced lethal concentrations), although sediment contents are second-lowest. In all cases, and therefore conclusively, the most important source for cadmium and nickel for oligochaetes is the free ion activity in the overlying surface water. In 100% of the cases, the measured C_{act} and k_0 combination for free ions in the surface water compartment yielded the best descriptors for accumulation patterns based on the two-compartment model. For copper, this seems valid in half of the cases. On some occasions, free ion concentrations dropped to a point where the contribution as a relevant source probably becomes negligible (see also Fig. 2). Potential sources of free Cu ions, such as humic complexes, have mostly slow dissociation kinetics that cannot make up for Cu-uptake rates by biota. Chemical dissociation is therefore rate-limiting.

Uptake from the sediment pore water was observed only occasionally, and solely, for lead. The source of Pb appears more variable, which is in agreement with the fact that the physiological mechanisms of Pb uptake are still unclear. It is occasionally suggested that uptake and elimination of Pb obey different mechanisms than in the case of the other heavy metals (i.e., an interaction between the free metal ion and a channel and/or carrier transport system in the external and/or internal epithelium), but this is yet to be demonstrated.

For zinc and arsenic, the performance of the one-compartment model was in 80% of the cases better than the model that used two compartments. External concentrations (i.e., outside the organism) obviously never became rate-limiting for uptake. In river sediments, Zn is mostly present in fair amounts, and the fraction of labile species, among them free ions, is mostly the largest of all metals. Most organisms regulate their Zn body concentrations very efficiently, and steady state is often quickly reached (see also Fig. 4). Dissolved arsenic does not yield significant free ion concentrations, being an oxyanion; the predominant species are arsenate or arsenite, and these are negatively charged.

It is also shown that uptake rates are not "universal constants." The variation of k_1 values over the locations may be significant. Kinetic processes therefore do indeed matter. The fact that the overlying water plays such a significant role in metal uptake should actually not be very surprising. Benthic organisms are often considered to be in close contact with sediment pore water, but species that live in burrows are actually in closer contact with overlying water because (1) the burrow inhibits direct contact between the organism and pore water, and (2) organisms exchange the water in their burrows with overlying water.

These presented findings provide a significant contribution to our understanding of chemical speciation of metals in water–sediment systems and subsequent metal

Table 3. Chemical Speciation in and Uptake from Different Compartments[a]

		Speciation in Surface Water				Speciation in Pore Water					Oligochaetes									Chironomids								
											Uptake from Surface Water				Uptake from Pore Water					Uptake from Surface Water					Uptake from Pore Water			
Metal, Site	Total, mg/kg	C_{tot}	k_0	C_{act}	k_0	% act	C_{tot}	k_0	C_{act}	k_0	% act	Q_{max}	k_1-C	sy·x	k_1-act	sy·x	k_2	r^2	Q_{max}	k_1-C	sy·x	k_1-act	sy·x	k_2	r^2		
Cd																											
MB	7.49	0.08	0.037	0.025	−0.277	32.1	0.05	0.033	0.01	0.157	9.6	2.2			20.16	0.01	0.03	0.47	nd								
GA	0.37	0.20	−0.003	0.050	0.360	25.0	0.30	0.038	0.07	0.010	23.3	10.0			18.45	6.65	0.03	0.94	29.0	8.20	0.01			−0.03	0.95		
KV	0.14	0.23	0.005	0.060	−0.005	26.7	0.09	0.001	0.03	−0.001	34.8	0.7			0.31	0.11	0.05	0.58	11.0	401.0	4.65			8.37	0.13		
HD	2.59	0.05	0.002	0.014	0.0003	30.0	0.03	0.001	0.01	0.001	28.8	2.0			48.07	0.01	0.20	0.71	6.3	208.0	26.30			1.34	0.37		
AS	1.57	0.39	−0.004	0.053	0.0003	13.6	0.12	0.005	0.06	−0.001	53.4	0.6			53.50	0.05	8.18	0.99	7.4	2.42	1.17			0.13	0.88		
SW	2.35	0.04	−0.001	0.003	0.0001	7.3	0.04	0.001	0.02	−0.001	51.2	0.9			10.27	3.15	3.47	0.85	nd								
Cu																											
MB	93.1	4.48	−0.180	0.45	−0.380	10.0	2.51	−0.095	0.33	0.012	13.1	57.5			45.96	0.01	0.09	0.85	nd								
GA	11.4	6.00	0.094	0.70	0.088	11.7	5.38	0.095	2.00	0.096	37.2	55.0			20.01	5.34	0.09	0.84	140.0	3.25	2.27			−0.05	0.82		
KV	4.9	6.56	−0.043	0.57	−0.036	8.7	10.17	−0.043	0.79	−0.043	35.0	35.0	37.25	0.01			16.76	0.41	51.0	2.85	0.92			0.18	0.31		
HD	26.5	5.41	−0.100	1.23	−0.041	22.7	3.07	−0.070	0.12	−0.012	3.8	39.7	6.06	1.61			0.76	0.71	48.0	3.40	1.82			0.55	0.71		
AS	25.7	2.74	−0.040	0.07	−0.003	2.6	1.90	−0.066	0.12	−0.006	6.3	103.0	21.25	17.71			1.24	0.55	152.0	6.65	4.82			0.33	0.78		
SW	48.0	2.55	−0.082	0.49	−0.027	19.2	2.87	−0.063	0.70	−0.041	24.3	32.0			31.08	24.9	0.68	0.88	nd								
Ni																											
MB	35.6	4.57	0.020	1.02	−0.042	22.3	2.69	0.039	0.50	−0.602	18.6	12.0			1.77	2.19	0.15	0.87	nd					−0.03	0.81		
GA	6.60	1.00	0.081	0.05	0.024	5.0	3.00	0.200	0.05	0.021	1.7	5.1			7.34	2.44	−0.01	0.75	14.7			14.96	4.71	4.31	0.94		
KV	8.35	10.27	−0.066	1.16	−0.012	11.3	16.29	−0.086	5.86	−0.086	36.0	11.0			0.44	0.36	0.44	0.54	13.0			11.01	5.68				
HD	19.4	4.82	0.016	0.73	−0.031	15.1	2.65	0.053	0.73	−0.012	27.4	7.6			6.41	2.56	0.42	0.51	17.3			8.60	3.05	0.57	0.94		
AS	31.5	5.44	0.094	0.58	0.009	10.7	5.70	−0.048	1.97	−0.028	34.6	3.9			0.47	0.24	0.07	0.52	21.0	9.32	0.01			4.36	0.90		
SW	6.17	6.17	0.027	1.27	0.055	20.6	5.16	−0.031	0.30	0.018	5.7	9.0			1.28	0.29	0.09	0.91	nd								
Pb																											
MB	703.6	0.30	0.024	0.00	0.047	1.0	0.83	−0.015	0.14	0.080	16.8	135.6			27.20	0.01	0.21	0.10	nd								
GA	28.0	2.00	−0.040	0.10	−0.001	5.0	1.00	−0.028	0.05	0.016	5.0	35.0			10.68	5.88	0.54	0.71	51.0	0.75	0.18			0.05	0.92		
KV	11.5	1.93	0.100	0.64	−0.007	33.2	1.39	−0.007	0.25	−0.005	17.8	21.0	0.53	0.18			−0.06	0.66	13.5					0.09	0.94		
HD	39.8	1.74	−0.080	0.20	−0.009	11.8	1.33	−0.073	0.20	−0.019	14.7	26.4			74.82	18.75	0.44	0.68	32.7			38.70	7.25	0.29	0.96		
AS	79.0	2.60	−0.081	0.17	0.000	6.5	2.51	−0.004	0.11	0.014	4.3	26.1	1.21	0.29			0.05	0.81	97.0			138.8	7.96	0.12	0.99		
SW	91.5	3.70	0.135	0.56	−0.030	15.2	4.68	−0.155	0.44	−0.062	9.4	20.3					18.52	3.61	0.05	0.86	nd						
Zn																											
MB	1091	20.60	−0.020	7.83	−0.034	38.0	10.96	0.114	2.78	0.050	25.4	781			30.57	0.01	0.39	0.82	nd								
GA	133.0	14.17	−0.144	3.01	−1.117	21.2	37.00	−0.262	19.00	−0.261	51.4	480	0.26	0.15					0.72	1100	8.69	1.36			0.87		
KV	33.0	11.27	−0.006	4.47	−0.006	39.7	13.28	−0.006	4.00	−0.006	30.1	395	0.19	0.12					0.56	710					0.70		
HD	231.0	21.70	−0.125	4.12	−0.043	19.0	15.96	0.048	4.51	−0.157	28.3	730	29.23	0.01					0.67	512	5.08	7.77			0.35		
AS	320.5	33.40	0.091	4.80	0.018	14.4	64.70	−1.526	11.30	−0.396	17.5	530	0.17	0.03					0.95	765			17.83	4.78	0.06	0.95		
SW	365	23.96	0.539	0.14	−0.017	0.6	18.81	0.216	0.43	0.377	2.3	429	7.27	2.52					0.72	nd							
As																											
MB	nd	8.61	4.865				23.95	7.262				23.0	0.10	0.06					0.90	nd							
GA	5.6	44.27	0.949				71.57	0.518				7.4	0.05	0.12					0.54	7.5	0.06	0.05			0.87		
KV	4.0	23.95	1.633				46.0	−0.072				10.0	−0.09	0.05					0.78	3.3	0.67	0.15			0.96		
HD	6.5	42.44	−0.319				50.76	−0.378				10.1	42.40	0.01					0.12	4.0	0.73	0.55			12.42	0.91		
AS	8.5	12.70	−0.323				5.90	0.599				14.6	0.01	0.01					0.96	nd							
SW	22.0	4.60	0.215				107.8	0.589				23.1	−0.25	0.02					0.99	nd							

[a] C, concentration (μg/L); tot, total dissolved; act, free ion activity; k_0, k_1, k_2 are rate constants (d^{-1}); Q_{max}, maximum C in biota (μg/g dw); r^2, correlation coefficient. Values printed in italic are derived with the one-compartment uptake model.

uptake by organisms. It is demonstrated that the presence of benthic organisms in a water–sediment system has a significant effect on chemical speciation, and this has to be considered when performing bioassays and chemical tests. Body concentrations are regulated by temporal variations in metal concentrations and chemical speciation. The measurements illustrate that biological kinetics (uptake) are for most metals faster than chemical kinetics (dissociation of less labile or complexed metals). Exposure can therefore only be estimated and modeled when time-related changes in metal concentrations are taken into account. The construction of generic models that are capable of transferring chemical information to a biological response has undoubtedly a high priority. Knowledge of the free metal ion activity is a significant step forward. Still, it may in cases be insufficient to predict the biological response, since one must also consider potential interactive effects of hydrogen ions (4,10) or effects of other macrochemicals (13). In any case, it is concluded that an integrated cell such as SOFIE provides the necessary experimental tool to support, in a mechanistic way, environmental risk assessments of contaminants. This knowledge may in time be incorporated in the derivation of better-founded quality criteria for sediments.

BIBLIOGRAPHY

1. Chapman, P.M. and Wang, F. (2001). *Environ. Toxicol. Chem.* **1**: 3–22.
2. DiToro, D.M. et al. (1992). *Environ. Sci. Technol.* **26**: 96–101.
3. Hare, L., Tessier, A., and Warren, L. (2001). *Environ. Toxicol. Chem.* **4**: 880–889.
4. Campbell, P.C.G. (1995). *Metal Speciation and Bioavailability in Aquatic Systems*. John Wiley & Sons, Hoboken, NJ.
5. Sunda, W.G. and Lewis, J.A.M. (1978). *Limnol. Oceanogr.* **23**: 870–877.
6. Giesy, J.P., Newell, A., and Leversee, G.J. (1983). *Sci. Total Environ.* **28**: 23–28.
7. Borgmann, U. and Charlton, C.C.J. (1984). *J. Great Lakes Res.* **10**: 393–401.
8. Daly, H.R., Jones, M.J., Hart, B.T., and Campbell, I.C. (1990). *Environ. Toxicol. Chem.* **9**: 1013–1018.
9. Meador, J.P. (1991). *Aquat. Toxicol.* **19**: 13–32.
10. Brown, P.L. and Markich, S.J. (2000). *Aquat. Toxicol.* **51**: 177–194.
11. Davison, W. and Zhang, H. (1994). *Nature* **367**: 546–548.
12. Vink, J.P.M. (2002). *Environ. Sci. Technol.* **36**: 5130–5138.
13. DeSchamphelaere, K.A.C. and Janssen, C.R. (2002). *Environ. Sci. Technol.* **36**: 48–54.

METHEMOGLOBINEMIA

Daniel Shindler
UMDNJ
New Brunswick, New Jersey

INTRODUCTION

Methemoglobinemia is a disorder of the oxygen-carrying hemoglobin in red blood cells. One cause of methemoglobinemia is contaminated well water. There are also many non-water-related causes. Individuals are particularly susceptible to the development of methemoglobinemia during early infancy and during pregnancy. Exposure to certain foods, drugs, and chemicals may cause methemoglobinemia. Methemoglobinemia can also occur in the absence of environmental exposure as part of several inherited disorders.

BLUE BABY SYNDROME AND WELL WATER

In 1945, Dr. Hunter Comly from Iowa City, Iowa described the blue baby syndrome. Two infants developed a bluish-gray skin discoloration (medically referred to as cyanosis) as a result of being fed infant formula diluted with well water.

The water from these wells was found to have high concentrations of nitrate. Once nitrate contamination was recognized as a cause of methemoglobinemia in infants, an additional 278 cases in 14 different states were reported by Walton in 1951. Yet, despite awareness and well water surveillance for nitrate and nitrite content, cases of methemoglobinemia continue to be reported.

High concentrations of nitrate in drinking water can cause high levels of methemoglobin in the body. Nitrite (converted from the ingested nitrate) oxidizes iron in the hemoglobin of red blood cells to form methemoglobin, which lacks hemoglobin's oxygen-carrying ability. Oxidation occurs when oxygen-carrying ferrous (Fe^{2+}) iron loses an electron and becomes ferric (Fe^{3+}). If unrecognized and untreated, rising levels of methemoglobin progressively decrease the ability of red blood cells to carry oxygen.

SUSCEPTIBLE INDIVIDUALS

Infants are particularly prone to develop methemoglobinemia from nitrate in drinking water until about 6 months of age. In early life, their ratio of fluid intake to body weight is greater than in adults. The stomach bacteria of infants are more capable of changing ingested nitrate to methemoglobin-begetting nitrite. Gastrointestinal problems that frequently occur in infants, such as diarrhea and vomiting, seem to further increase their predisposition to develop clinically significant methemoglobinemia.

Pregnant women also have an increased predisposition to develop methemoglobinemia. Increased methemoglobin levels during pregnancy may cause spontaneous abortion. During March 1993, the LaGrange County (Indiana) Health Department identified three women who reported a total of six spontaneous abortions during 1991–1993 and who resided in proximity to each other; each had obtained drinking water from nitrate-contaminated private wells. A fourth woman from another part of the county had two spontaneous abortions after she had moved into a new home with a nitrate-contaminated private well.

NITRATES IN DRINKING WATER

Nitrate contamination of well water may come from easily identifiable sources such as sewage disposal systems or nearby livestock facilities. Other times the cause of water contamination is less obvious, originating from distant landfills, fertilized cropland, parks, golf courses, lawns, and gardens. Episodes of methemoglobinemia continue to occur in rural farm settings especially with shallow (less than 100 feet deep) domestic wells in a setting where nitrogen-based fertilizers are heavily used. These wells may not be tested frequently enough for fluctuating concentrations of nitrate, which (according to the Environmental Protection Agency) should be no greater than 10 parts per million (ppm) at all times. Runoff of contaminated surface water from cultivated fields into a well may happen after a flood. This mode of well water contamination may be especially common when the well is too shallow, poorly located, or badly constructed.

OTHER WATER-RELATED CAUSES

Contaminated well water is not the only water-related cause of methemoglobinemia. Nor is the danger of undiagnosed methemoglobinemia limited to infants. Nitrite containing boiler fluid additives contaminating the faucet water can be responsible. An outbreak in New Jersey grammar school children in October 1992 was caused by dilution of canned soup with hot tap water. The water boiler had been serviced 5 months earlier with commercial conditioner fluid containing nitrite, and it had not been started until the morning of the incident. In another incident in 1996, six adult office workers developed methemoglobinemia after drinking coffee prepared with water from a hot water faucet connected to a boiler with a defective backflow-prevention valve. The conditioning fluid in this boiler also contained nitrites. Boiling water before using it does not decrease the level of existing nitrate or nitrite in the water. Prolonged boiling may actually increase the concentration of existing nitrate and nitrite in the water.

As water is almost invariably involved in food preparation, it is also important to recognize dietary causes of methemoglobinemia. Although vegetables are washed and cooked in water, nitrates may not be originating from the water used for rinsing, cooking, or serving the vegetables, but rather from the vegetables themselves. Broccoli, carrots, cauliflower, collard greens, onions, and spinach all contain some nitrate. Fortunately, the amount of vegetable nitrate in the normal diet is insufficient to cause significant methemoglobinemia in adults. Yet, ingestion of concentrated vegetables (carrot juice, carrot soup, spinach water) has resulted in methemoglobinemia in susceptible infants. Methemoglobinemia caused by excessive vegetable ingestion is not confined to humans. A herd of cows consumed approximately 20 kg of onions per cow per day for 6 weeks, which caused methemoglobinemia in the cows. Five cows died, and two had spontaneous abortions.

Diaper washing and reuse has also been implicated in methemoglobinemia. Anilines are dyes that are used in laundry ink. Aniline labeled diapers have caused infant methemoglobinemia by direct absorption through the skin and possibly by spreading to other poorly rinsed diapers in the same laundry water. Children's crayons no longer contain methemoglobin-producing dyes, but commercial markers such as red wax crayons may still contain them. Skin contact with dyed blankets and freshly dyed shoes can also potentially cause methemoglobinemia.

NON-WATER-RELATED CAUSES

Nitrates are found in the normal diet, not just in vegetables but also in meat and fish preservatives, and as meat color enhancers. Sodium and potassium nitrate have both caused life-threatening methemoglobinemia when they were mistaken for salt or used inappropriately in meat preparation.

Certain medications contain nitrates. Silver nitrate is used on the skin to treat burns. Nitroglycerin is administered orally, intravenously, under the tongue (sublingual), and on the skin in patients with heart disease. Amyl nitrite is an inhaled medication that is abused as a recreational drug. Amyl nitrite and isobutyl nitrite may be sold as "room deodorizers" to avoid drug laws. Nitrite can also be inhaled accidentally. Firefighters responding to a fire where isobutyl nitrite was stored developed methemoglobinemia because of exposure to the products of combustion.

Many non-nitrate medications have caused methemoglobinemia. Several antibiotics have caused methemoglobinemia during routine administration. Topical anesthetics are common during medical and dental procedures. There are many reports of methemoglobinemia after their routine administration at all ages. They may be given to teething infants and consequently get absorbed into the bloodstream through the mucous membranes of the mouth. Topical anesthetic ointments can also cause methemoglobinemia if they are absorbed through the skin. Anesthetic ointments given for diaper rash should be discontinued if the treated infant develops methemoglobinemia. Infants have also developed methemoglobinemia by accidentally ingesting a small number of tablets of a urinary tract analgesic called phenazopyridine.

Some industrial chemicals can cause methemoglobinemia. They may be ingested accidentally by infants and children when found in the household. Examples include naphthalene in mothballs, nitrobenzene in solvents such as gun cleaners, and copper sulfate in plant fungicides.

HEREDITARY CAUSES

Methemoglobinemia can also occur because of an inherited lack of an enzyme in red blood cells. The enzyme is called NADH diaphorase, or cytochrome b_5 reductase. Several mutations of the enzyme have been identified. The mode of transmission is autosomal recessive, which means that the gene abnormality has to be present in both parents to cause methemoglobinemia. When both parents are carriers of the trait, there is a 25% chance of a child inheriting abnormal genes from both parents, and therefore of developing methemoglobinemia. There is a 50% chance of a child inheriting one abnormal gene and becoming a carrier without actually developing methemoglobinemia. Many reports of such cases describe siblings derived from inbred populations.

Also, rare abnormalities of hemoglobin are inherited in an autosomal dominant fashion—a 50–50 chance of passing the disorder to the children. These inherited abnormal forms of hemoglobin are unstable and, as a result, are more readily transformed to methemoglobin.

DIAGNOSIS

Methemoglobinemia is potentially lethal if it is left undiagnosed. Transient blue or grayish skin discoloration of an infant may be mistakenly dismissed as being caused by changes in temperature. Familiarity with methemoglobinemia and a high index of suspicion by the physician are potentially lifesaving. Untreated infants can become irritable and lethargic. Some may develop seizures. The heart rate may become slow. The blood pressure may drop. They may eventually lapse into a coma and die.

Isolated cases of methemoglobinemia are rare. The diagnosis of this potentially life-threatening disorder can only be made if a physician is familiar with the clinical clues and manifestations. A patient who is visibly cyanotic will initially be suspected to have a primary cardiac or a respiratory problem. Consequently, oxygen is routinely administered to patients with cyanosis. The lack of a response to the oxygen provides a valuable clue to the diagnosis. Patients with methemoglobinemia will remain blue or ashen even after oxygen administration.

Pulse oximetry will be inaccurate and unreliable in these patients. The pulse oximeter is a simple finger-clip device containing a light sensor. It was designed to measure the percentage of oxygen saturation of hemoglobin by measuring the amount of light passing through the finger or earlobe. Methemoglobin absorbs light at wavelengths similar to normal hemoglobin, and as a result, oximetry cannot reliably distinguish the two.

The blood from a patient with methemoglobinemia will be dark red when drawn and remain dark even when exposed to room air. Dark blood drawn from a vein of a patient with heart or lung disease will turn bright red on exposure to the oxygen in room air or after bubbling pure oxygen through the blood.

Methemoglobinemia should also immediately come to mind if multiple cases of cyanosis are discovered. An environmental exposure may then be a likely cause.

As soon as the diagnosis of methemoglobinemia is suspected, a blood methemoglobin level should be measured to confirm the diagnosis. Patients will appear cyanotic when the methemoglobin level is approximately 10% or greater.

TREATMENT

Normal methemoglobin levels in the blood range from 1% to less than 3%. If blood methemoglobin levels are only mildly elevated (less than 20%), patients are observed and given oxygen to saturate the remaining normal hemoglobin.

In more serious cases, where methemoglobin levels are greater than 30%, patients are given intravenous methylene blue. Patients with underlying conditions such as heart problems or severe anemia may get treated with methylene blue even with a lower methemoglobin level.

Although at first glance giving blue dye to a blue patient seems counterintuitive, the treatment is simple and can be life saving. Methylene blue acts as a nonenzymatic electron carrier. It accelerates the rate with which red blood cells convert methemoglobin back to oxygen-carrying hemoglobin. Large doses of methylene blue can indeed cause methemoglobinemia, so if there is no response to the first dose, it is usually only repeated once.

In some patients, however, even a single dose of methylene blue may pose a danger. Patients with a deficient glucose-6-phosphate dehydrogenase enzyme system will not respond to the methylene blue, and their red cells may get destroyed (hemolyzed). In such cases, other more specialized treatments such as hyperbaric oxygen, or packed red blood cell transfusions, may be necessary.

In cases of environmental exposure, such as contaminated well water, it is equally important to identify the excess nitrate source and to prevent repeat exposure to it by the susceptible individual.

READING LIST

Well Water

Comly, H. (1945). Cyanosis in infants caused by nitrates in well water. *JAMA* **129**: 112–116.

Reprinted as a landmark article in *JAMA* **257**: 2788–2792 (1987).

Walton, G. (1951). Survey of literature relating to infant methemoglobinemia due to nitrate-contaminated water. *Am. J. Public Health* **41**: 987–995.

Bucklin, R. and Myint, M.K. (1960). Fatal methemoglobinemia due to well water nitrates. *Ann. Intern. Med.* **52**: 703–705.

Carlson, D.J. and Shapiro, F.L. (1970). Methemoglobinemia from well water nitrates: a complication of home dialysis. *Ann. Intern. Med.* **73**: 757–759.

Gastrointestinal Problems

Yano, S.S., Danish, E.H., and Hsia, Y.E. (1982). Transient methemoglobinemia with acidosis in infants. *J. Pediatr.* **100**: 415–417.

Lanir, A., Borochowitz, Z., and Kaplan, T. (1986). Transient methemoglobinemia with acidosis and hyperchloremia in infants. *Am. J. Pediatr. Hematol. Oncol.* **8**: 353–355.

Smith, M.A., Shah, N.R., Lobel, J.S. et al. (1988). Methemoglobinemia and hemolytic anemia associated with *Campylobacter jejuni* enteritis. *Am. J. Pediatr. Hematol. Oncol.* **10**: 35–38.

Kay, M.A., O'Brien, W., Kessler, B. et al. (1990). Transient organic aciduria and methemoglobinemia with acute gastroenteritis. *Pediatrics* **85**: 589–592.

Murray, K.F. and Christie, D.L. (1993). Dietary protein intolerance in infants with transient methemoglobinemia and diarrhea. *Pediatrics* **122**: 90–92.

Hanukoglu, A. and Danon, P.N. (1996). Endogenous methemoglobinemia associated with diarrheal disease in infancy. *J. Pediatr. Gastroenterol. Nutr.* **23**: 1–7.

Pregnancy

MMWR (1996). Spontaneous abortions possibly related to ingestion of nitrate-contaminated well water—LaGrange County, Indiana, 1991–1994. *Morb. Mortal. Wkly. Rep.* **45**: 569–572.

Boiler Fluid

Askew, G.L., Finelli, L., Genese, C.A. et al. (1994). An outbreak of methemoglobinemia in New Jersey in 1992. *Pediatrics* **94**: 381–384.

Nitrate in Vegetables

Smeenk, C. and van Oudheusden, A.P. (1968). Hemoglobinemia caused by spinach water. *Ned. Tijdschr. Geneeskd.* **112**: 1378–1379.

Keating, J.P. et al. (1973). Infantile methemoglobinemia caused by carrot juice. *N. Engl. J. Med.* **288**: 824–826.

Chan, T.Y. (1996). Food-borne nitrates and nitrites as a cause of methemoglobinemia. *Southeast Asian J. Trop. Med. Public Health* **27**: 189.

Sanchez-Echaniz, J., Benito-Femandez, J., and Mintegui-Raso, S. (2001). Methemoglobinemia and consumption of vegetables in infants. *Pediatrics* **107**: 1024–1028.

Bryk, T., Zalzstein, E., and Lifshitz, M. (2003). Methemoglobinemia induced by refrigerated vegetable puree in conjunction with supraventricular tachycardia. *Acta Paediatr.* **92**: 1214–1215.

Rae, H.A. (1999). Onion toxicosis in a herd of beef cows. *Can. Vet. J.* **40**(1): 55–57.

Dyes and Crayons

Graubarth, J., Bloom, C.J., Coleman, F.C., and Solomon, H.N. (1945). Dye poisoning in the nursery: a review of seventeen cases. *JAMA* **128**: 1155.

Howarth, B.E. (1951). Epidemic of aniline methaemoglobinaemia in newborn babies. *Lancet* **1**: 934–935.

Rieders, F. and Brieger, H. (1953). Mechanism of poisoning from wax crayons. *JAMA* **151**: 1490.

Kearney, T.E., Manoguerra, A.S., and Dunford, J.V. Jr. (1983). Chemically induced methemoglobinemia from aniline poisoning. *West. J. Med.* **140**: 282.

Poisoning from Food Preparation

Roueche, B. (1947). *Eleven Blue Men*. Little Brown & Co., Boston, MA.

Orgeron, J.D. et al. (1957). Methemoglobinemia from eating meat with high nitrite content. *Public Health Rep.* **72**(3): 189–193.

Singley, T.L. III. (1963). Secondary methemoglobinemia due to the adulteration of fish with sodium nitrite. *Ann. Intern. Med.* **57**: 800–803.

Silver Nitrate for Burns

Cushing, A.H. and Smith, S. (1967). Methemoglobinemia with silver nitrate therapy of a burn. *J. Pediatr.* **74**: 613–615.

Inhalation

Guss, D.A., Normann, S.A., and Manoguerra, A.S. (1985). Clinically significant methemoglobinemia from inhalation of isobutyl nitrite. *Am. J. Emerg. Med.* **3**: 46–47.

Hovenga, S., Kcenders, M.E., van der Werf, T.S. et al. (1996). Methemoglobinemia after inhalation of nitric oxide for treatment of hydrochlorothiazide-induced pulmonary edema. *Lancet* **348**: 1035–1036.

Nitroglycerin and Amyl Nitrite

Gibson, G.R. et al. (1982). Methemoglobinemia produced by high-dose intravenous nitroglycerin. *Ann. Intern. Med.* **96**: 615.

Paris, P.M., Kaplan, R.M., Stewart, R.D., and Weiss, L.D. (1986). Methemoglobin levels following sublingual nitroglycerin in human volunteers. *Ann. Emerg. Med.* **15**: 171.

Forsyth, R.J., Moulden, A. (1991). Methaemoglobinaemia after ingestion of amyl nitrite. *Arch. Dis. Child* **66**: 152.

Antibiotics

Damergis, J.A., Stoker, J.M., and Abadie, J.L. (1983). Methemoglobinemia after sulfamethoxazole and trimethoprim. *JAMA* **249**: 590.

Koirala, J. (2004). Trimethoprim-sulfamethoxazole—induced methemoglobinemia in an HIV-infected patient. *Mayo. Clin. Proc.* **79**: 829–830.

Moreira, V.D. et al. (1998). Methemoglobinemia secondary to clofazimine treatment for chronic graft-versus-host disease. *Blood* **92**: 4872.

Reiter, W.M. and Cimoch, P.J. (1987). Dapsone-induced methemoglobinemia in a patient with P. carinii pneumonia and AIDS. *N. Engl. J. Med.* **317**: 1740.

Coleman, M.D., Rhodes, L.E., Scott, A.K. et al. (1993). The use of cimetidine to reduce dapsone-dependent methaemoglobinaemia in dermatitis herpetiformis patients. *Br. J. Clin. Pharmacol.* **34**: 244.

Wagner, A., Marosi, C., Binder, M. et al. (1995). Fatal poisoning due to dapsone in a patient with grossly elevated methaemoglobin levels. *Br. J. Dermatol.* **133**: 816.

Sin, D.D. and Shafran, S.D. (1996). Dapsone- and primaquine-induced methemoglobinemia in HIV infected individuals. *J. Acquir. Immune. Defic. Syndr. Hum. Retrovirol.* **12**: 477–481.

Ward, K.E. and McCarthy, M.W. (1998). Dapsone-induced methemoglobinemia. *Ann. Pharmacother.* **32**: 549–553.

Mandrell, B.N. and McCormick, J.N. (2001). Dapsone-induced methemoglobinemia in pediatric oncology patients. *J. Pediatr. Oncol. Nurs.* **18**: 224–228.

Brabin, B.J., Eggelte, T.A., Parise, M., and Verhoeff, F. (2004). Dapsone therapy for malaria during pregnancy: maternal and fetal outcomes. *Drug. Saf.* **27**: 633–648.

Topical Anesthetics

Duncan, P.G. and Kobrinsky, N. (1983). Prilocaine-induced methemoglobinemia in a newborn infant. *Anesthesiology* **59**: 75.

Collins, J.F. (1990). Methemoglobinemia as a complication of 20% benzocaine spray for endoscopy. *Gastroenterology* **98**: 211.

Postiglione, K.F. and Herold, D.A. (1992). Benzocaine-adulterated street cocaine in association with methemoglobinemia. *Clin. Chem.* **38**: 596.

Brown, C.M., Levy, S.A., and Sussan, P.W. (1994). Methemoglobinemia: life-threatening complication of endoscopy premedication. *Am. J. Gastroenterol.* **89**: 1108–1109.

Guerriero, S.E. (1997). Methemoglobinemia caused by topical benzocaine. *Pharmacotherapy* **17**: 1038.

Cooper, H.A. (1997). Methemoglobinemia caused by benzocaine topical spray. *South Med. J.* **90**: 946.

Brisman, M. et al. (1998). Methaemoglobin formation after the use of EMLA cream in term neonates. *Acta Paediatr.* **87**: 1191.

Balicer, R.D. and Kitai, E. (2004). Methemoglobinemia caused by topical teething preparation: a case report. *Sci. World J.* **4**: 517–520.

Moore, T.J., Walsh, C.S., and Cohen, M.R. (2004). Reported adverse event cases of methemoglobinemia associated with benzocaine products. *Arch. Intern. Med.* **164**: 1192–1196.

Phenazopyridine Overdose

Nathan, D.M., Siegel, A.J., and Bunn, H.E. (1977). Acute methemoglobinemia and hemolytic anemia with phenazopyridine: possible relation to acute renal failure. *Arch. Intern. Med.* **137**: 1636–1638.

Gavish, D. et al. (1986). Methemoglobinemia, muscle damage and renal failure complicating phenazopyridine overdose. *Isr. J. Med. Sci.* **22**: 45.

Christensen, C.M., Farrar, H.C., and Kearns, G.L. (1996). Protracted methemoglobinemia after phenazopyridine overdose in an infant. *J. Clin. Pharmacol.* **36**: 112.

Gold, N.A. and Bithoney, W.G. (2003). Methemoglobinemia due to ingestion of at most three pills of pyridium in a 2-year-old: case report and review. *J. Emerg. Med.* **25**: 143–148.

Other Causes of Methemoglobinemia

Banning, A., Harley, J.D., O'Gorman Hughes, D.W. et al. (1965). Prolonged methaemoglobinaemia after eating a toy-shaped deodorant. *Med. J. Aust.* **2**: 922–925.

Chilcote, R.R., Williams, B., Wolff, L.J. et al. (1977). Sudden death in an infant from methemoglobinemia after administration of "sweet spirits of nitre." *Pediatrics* **59**: 280–282.

Sherman, J.M. and Smith, K. (1979). Methemoglobinemia owing to rectal-probe lubrication. *Am. J. Dis. Child* **133**: 439–440.

Shessler, R., Dixon, D., Allen, Y. et al. (1980). Fatal methemoglobinemia from butyl nitrite ingestion. *Ann. Intern. Med.* **92**: 131–132.

Ellis, M., Hiss, Y., and Shenkman, L. (1992). Fatal methemoglobinemia caused by inadvertent contamination of a laxative solution with sodium nitrite. *Isr. J. Med. Sci.* **28**: 289.

Mallouh, A.A. and Sarette, W.O. (1993). Methemoglobinemia induced by topical hair oil. *Ann. Saudi. Med.* **13**: 78–88.

Osterhoudt, K.C., Wiley, C.C., Dudley, R. et al. (1995). Rebound severe methemoglobinemia from ingestion of a nitroethane artificial fingernail remover. *J. Pediatr.* **126**: 819–821.

Methylene Blue Treatment

Clifton, J. II and Leikin, J.B. (2003). Methylene blue. *Am. J. Ther.* **10**: 289–291.

Goluboff, N. and Wheaton, R. (1961). Methylene blue induced cyanosis and acute hemolytic anemia complicating the treatment of methemoglobinemia. *J. Pediatr.* **58**: 86–89.

Beutler, E. and Baluda, M.C. (1963). Methemoglobin reduction. Studies of the interaction between cell populations and of the role of methylene blue. *Blood* **22**: 323.

Ng, L.L., Naik, R.B., and Polak, A. (1982). Paraquat ingestion with methaemoglobinaemia treated with methylene blue. *BMJ* **284**: 1445.

Harvey, J.W. and Keitt, A.S. (1983). Studies of the efficacy and potential hazards of methylene blue therapy in aniline-induced methaemoglobinaemia. *Br. J. Haematol.* **154**: 29.

Fitzsimons, M.G., Gaudette, R.R., and Hurford, W.E. (2004). Critical rebound methemoglobinemia after methylene blue treatment: case report. *Pharmacotherapy* **24**: 538–540.

Khanduri, U., Sharma, A., and Kumar, P. (2004). Unique red cell abnormality in a case of fatal methylene blue induced haemolysis. *Hematology* **9**: 239–241.

Rosen, P.J., Johnson, C., McGehee, W.G., and Beutler, E. (1971). Failure of methylene blue treatment in toxic methemoglobinemia. Association with glucose-6-phosphate dehydrogenase deficiency. *Ann. Intern. Med.* **75**: 83–86.

Hibbard, B.Z. Jr. et al. (1978). Severe methemoglobinemia in an infant with glucose-6-phosphate dehydrogenase deficiency. *J. Pediatr.* **93**: 816–818.

Red Blood Cell Exchange

Golden, P.J. and Weinstein, R. (1998). Treatment of high-risk, refractory acquired methemoglobinemia with automated red blood cell exchange. *J. Clin. Apheresis.* **13**: 28–31.

MICROBIAL ACTIVITIES MANAGEMENT

Peter D. Franzmann
Luke R. Zappia
CSIRO Land and Water
Floreat, Australia

INTRODUCTION

Microbial activity can be either detrimental or beneficial to the quality of water in water treatment and distribution systems. The growth of microbial community biofilms on the pipe walls in potable water distribution systems, and in water supply reservoirs, can significantly reduce the aesthetic, microbiological, and chemical quality of the water. Microbial processes can also be beneficial in processes such as biofiltration, to prevent reactions that occur in the distribution system that are otherwise detrimental to the water quality.

Normally more than 99.9% of microbial cells in water distribution systems occur in biofilms rather than in the planktonic form. Pipe wall biofilm communities can act as reservoirs of opportunistic pathogenic microorganisms as bacteria embedded in biofilm matrices are protected from water disinfectants (1). Biofilm communities may also impact the chemical and aesthetic quality of distributed waters by concentrating, oxidizing, and releasing metals that can result in "dirty water" problems (2), producing taste and odor compounds (3), enhancing corrosion (4), and reducing disinfectant residuals in chlorinated (5) and chloraminated distribution systems (6).

GENERAL CONTROL OF BIOFILM GROWTH AND ACTIVITY

Many water treatment processes are focused on the management of biofilm growth and activity in distribution systems generally, and more specifically on the control of biofilm processes such as nitrification and manganese oxidation. The parameters that can manage biofilm growth and activity exert their control collectively in management strategies, but their contributions to management of microbial activity are best understood by examination of their individual effects. In a comparative study on the control of microbial activity in four different distribution systems in the United States and France, the authors concluded that it was always difficult to make definitive conclusions from full-scale systems where no operating parameter was truly controlled (7); however they emphasized that maintenance of a disinfectant residual was the strongest controller of bacterial growth in their systems. On the other hand, other studies have suggested that with careful management and appropriate treatments, a disinfectant residual is not required (8).

Control of Biodegradable Organic Carbon in Distributed Water

Control of microbial activity in distribution systems is required so that the water quality meets community health and aesthetic expectations. It can be achieved first by distribution of "biologically stable water," that is, water that will not promote microbial growth and activity in the distribution system, and second by avoiding the use of pipe materials that leach substrates that can support microbial growth (9). Biologically stable water can be produced without the use of a disinfectant by reducing the concentration of biodegradable organic carbon (10) and other growth-promoting compounds such as methane, ammonium, or reduced sulfur compounds, in the water that leaves the treatment plant.

Types of Organic Carbon in Drinking Water. For the distribution of biologically stable water, water treatment plants aim to reduce dissolved organic carbon (DOC), specifically the fractions of the DOC that are consumed by microbes. In practice, two types of DOC bioassay are used at treatment plants to determine if biostable water is being produced, and each assay measures different but related fractions of the bioreactive DOC.

The first of these fractions, the biodegradable dissolved organic carbon (BDOC), is measured by the decrease of DOC over time, usually up to 10 days, after the water has been inoculated with washed biomass from a sand filter (11) or a microbial population that has been grown on glass beads. BDOC is the measure of the change in DOC because of its consumption by a mixed microbial community until that consumption effectively stops. The remaining DOC is considered refractory to biodegradation in reasonable time. Methods for the measurement of BDOC have been detailed by Joret and Lévy (11) and Servais et al. (12), and rapid flow-through methods for the determination of BDOC have been proposed (13).

The other bioreactive fraction of the DOC is the assimilable organic carbon (AOC) and is the fraction used by cells to build biomass. Most water treatment plants measure this parameter by the method developed by van der Kooij (8). In this method, AOC is measured by the increase in viable numbers of two strains of bacteria, *Pseudomonas fluorescens* strain P17 and *Spirillum* sp. strain NOX, over time in water that was pasteurized at 60 °C for 30 min before inoculation. The amount of growth is compared with a standard curve of growth of both strains in AOC-free water to which standard additions of a carbon source, usually acetate, are added. In most water samples, growth reaches the stationary phase in about 10 days. The growth of each strain is taken as the difference between the number of viable cells in the water at completion of growth and the number of cells in the water just after inoculation. AOC is mainly composed of organic matter with a molecular weight of <1000 (14) and is expressed in the units µg (acetate-C equivalents)·L^{-1}. Other methods for AOC measurement have been described, using increase in turbidity or ATP as indicators of biomass production (13). As the water industry is increasingly demanding standardized methods so that parameters from different plants can be compared, most AOC determinations now use the van der Kooij method, which is the method detailed in the entry STANDARD METHODS FOR THE EXAMINATION OF WATER AND WASTEWATER.

AOC and BDOC can be measured in chlorine-containing waters only if the chlorine is reduced with thiosulfate by the application of about 2 moles of thiosulfate per mole of chlorine. The ratio of BDOC or AOC to DOC varies in untreated surface and aerated groundwaters, but BDOC usually represents about 10% to 30% of the DOC, and AOC usually accounts for 1% to 3% of the DOC (15). In treated waters, the correlation of AOC or BDOC to DOC concentration is poor (16). van der Kooij (8) showed that distribution of water with an AOC concentration below 10 µg·L^{-1} without a disinfectant limited the mean heterotrophic plate count to below 100 colony-forming units·mL^{-1}, a guideline value considered by many regulators to indicate water of good microbiological quality. Biologically stable water does not promote growth or activity of microorganisms to a significant extent. Various parameter values have defined biological stability of distributed waters, and some of these are compared in Table 1.

For measures of biostability, Escobar and Randall (19) argued that both AOC and BDOC measures should be used. BDOC has a detection limit of about 100 µg·L^{-1}, whereas for AOC, the detection limit is about 10 µg·L^{-1}.

Table 1. Various Parameter Values that Define Biologically Stable Water

Attribute	Value	Comment	Reference
AOC	10 µg·L^{-1}	AOC uptake by biofilm limited at <10 µg·L^{-1}	9
AOC	<50 µg·L^{-1}	For AOC in a system with a chlorine residual	17
BDOC	200 µg·L^{-1}	In water with no disinfectant residual	18
BDOC	500 µg·L^{-1}	In water <15 °C with no disinfectant residual	18
BDOC	<500 µg·L^{-1}	In water with a chlorine residual of 0.1 mgd·L^{-1}	18
BDOC	≤150 µg·L^{-1}	Water without a disinfectant	10

Some treatments like nanofiltration can reduce BDOC almost completely, but considerable AOC can remain in the permeate. Measurement of AOC alone does not give information on the pool of hydrolyzable BDOC than can form AOC in the distribution system; thus, measurement of AOC alone can give an overestimate of the biostability of distributed water (19).

The Effect of Treatment Processes on AOC and BDOC. As the quantity and type of dissolved organic carbon in water is the major factor that promotes bacterial growth (and hence activity) in distribution systems, modern water treatment plants usually investigate the effect of different water treatment methods on the quantity and type of DOC in their treated water. Some treatments, such as the use of oxidants for destruction of pesticides in treatment plant influent, or chlorine for disinfection, can partially oxidize the DOC to form biologically available AOC or BDOC. Other processes, notably those that use methods of biological treatment such as slow sand filtration and biofiltration using a granulated activated carbon support (GAC), can produce water with low AOC and BDOC.

A range of treatments and their effects on AOC and BDOC concentrations are shown in Table 2. The data in Table 2 oversimplify the outcomes of the different treatment options, as combinations of treatment options can have different outcomes, individual treatment options can be modified in many ways, and the success of each treatment type is dependent on the nature of the preceding treatments that were used. For example, in one study, there was 57% less AOC in finished waters from GAC filters than for anthracite sand filters when using polymer as the coagulant, whereas when using ferric ions as the coagulant, that improvement was only 13% [derived from the data of Table 4 of Volk and LeChevallier (20)]. The data in Table 2 provides a list of site-specific examples of the effect of different treatments on AOC and BDOC.

Choice of a treatment option to obtain a desired water quality will require testing by each treatment plant using its receiving water, and its implementation will require some optimization of the process parameters that can be manipulated. For substantial removal of AOC and BDOC, some form of biological treatment is generally required (21), although initial results with a magnetic ion exchange treatment for DOC removal, MIEX (WesTech Engineering, Salt Lake City, Utah), has shown exceptionally good removal of BDOC, and AOC when subsequently filtered (Table 2), in addition to removal of 60–70% of the DOC (22).

AOC concentrations in the distribution system are dependent not only on the water type and treatment system but also on the choice of materials used. Materials in contact with the water can leach considerable amounts of AOC and affect the biofilm growth on pipe surfaces. For example, nonplastic polyvinyl chloride supports an order of magnitude less biofilm than does high-density polyethylene (9).

Although the concentration of the biologically available DOC fraction is arguably the most important determinant of microbial activity in distribution systems, it does not act in isolation, and its effects are modulated by other parameters that control microbial activity in the distribution system, such as disinfectant type and concentration, temperature, residence time, and concentrations of nonorganic oxidizable microbial substrates such as ammonium or manganese. Mathematical models that attempt to explain those interactions have been developed (e.g., Ref. 28), as have relatively simple, descriptive, decision-support criteria (16).

Disinfection to Control Microbial Activity

The maintenance of a disinfectant residual is probably the strongest inhibitor of microbial activity through a distribution system (7). High concentrations of disinfectant, even in the presence of relatively high concentrations of dissolved organic carbon or AOC, can limit microbial activity such as biofilm growth in distribution systems, e.g., 160 $\mu g \cdot L^{-1}$ AOC with a chloramine residual of >2 $mg \cdot L^{-1}$, or 107 $\mu g \cdot L^{-1}$ AOC with >0.8 $mg \cdot L^{-1}$ free chlorine [in Pine Hills Distribution System; (7)]. However, as high concentrations of disinfectant are the most common cause of customer taste or odor complaints (29), and as there are increasing concerns about disinfection by-products (30), the goal of water treatment utilities is often to deliver water with a minimal amount of disinfectant residual, often with reduced DOC, to control microbial activity and maintain microbiologically safe water.

As with treatment methods for removal of AOC and BDOC, many different concentrations of disinfectant have been shown to prevent biofilm growth, or specific microbial activities depending on the water quality, temperature, residence time, or type of disinfectant used. As already discussed, if the water quality is high (i.e., biologically stable), in some cases, no disinfectant is required to deliver safe and aesthetically acceptable water to the consumer (9).

Table 2. Some Examples of the Effects of Water Treatments on DOC, BDOC, and AOC Concentrations. The Examples are Site Specific, as the Outcome from the Use of any Treatment is Dependent on Many Water Parameters, Especially the Nature of the Natural Organic Matter in the Receiving Water, and on the Pretreatment Processes Employed

Treatment	Effect	Reference
Ozonation	↑AOC 560%	23
Enhanced coagulation	↓29% DOC, ↓38% BDOC, no change AOC	24
Ultrafiltration	↓85% DOC, ↓76% to ↑8.7% AOC[a]	24
Alum coagulation	↓16% AOC, ↓49% BDOC	25
Two-stage GAC biofilter	↓51–72% AOC for ozonated water	26
Two-stage GAC biofilter	↓37–70% AOC for non-ozonated water	26
Biofiltration	↓80 to 90% AOC depending on concentration	27
Slow sand filtration	↓48% AOC	21
Chlorination	↑80% AOC	21
MIEX treatment	↓>88% BDOC, ↓24% AOC, ↓>95% when filtered	Our data

↑= increase, ↓= decrease.
[a] Dependent on water source.

Table 3. Disinfectant Residual Concentrations and Their Effects on Microbial Activity in Pilot-Scale and Actual Water Distribution Systems

Treatment: Effect	Reference
Free chlorine 0.05 mg·L^{-1}: no biofilm growth initiation in pipes	33
Free chlorine > 0.5 mg·L^{-1} or chloramine > 1.0 mg·L^{-1}: required to decrease probability of coliform occurrence	16
Chloramine 2.1 mg·L^{-1}: inhibited bacterial activity	
Total chlorine 0.3 mg·L^{-1}: prevents establishment of biofilms	Our data
Total chlorine < 0.02 mg·L^{-1}: allows establishment of biofilms	Our data
Chlorine 1.0–3.0 mg·L^{-1}: disrupts established biofilms	34
Free chlorine residual 3 mg·L^{-1}: no effect on biofilms in iron pipes	17
Chlorine or chloramine 1 mg·L^{-1}: inactivates bacteria on copper PVC or galvanized pipes	17
Free chlorine 0.2 mg·L^{-1}: reduced biofilm concentration	35

Although site and water specific, some examples of recommended disinfectant levels to prevent biofilm growth or microbial activity in distribution systems are given in Table 3. In reality, the target level of a disinfectant residual should be decided only after appropriate testing. Once established, biofilms can exert a chlorine demand that increases linearly with the BDOC concentration in the water and with the ratio of the surface area to volume of the biomass, whereas the chlorine demand of detached biomass is related to the amount of biomass (5). The effectiveness of a disinfectant residual in controlling microbial activity is also markedly affected by pipe material, retention time of the water in the distribution system, and water chemistry (17).

Chloramine has been shown to provide better control of microbial activity than has chlorine (31) which may be related to slower decay and greater penetration into biofilms (32).

Temperature and Microbial Activity

Temperature has a marked effect on microbial activity. Generally, microbial growth and microbial processes such as nitrification in distribution systems are limited at temperatures of less than 15 °C, and deterioration of water quality in colder climates can be temperature limited (16).

Control of Nitrification in Distribution Systems

Nitrification, the oxidation of ammonium through nitrite to nitrate, is a microbial activity that is problematic for many water utilities that distribute water with chloramine as the disinfectant. The activity depletes the residual chloramine and can make conditions more conducive to growth of heterotrophic bacteria in the distribution system (36). Ammonium in source waters can lead to biological instability in distribution systems if not treated. For example, ammonium concentrations throughout Illinois averaged 0.62 mg N·L^{-1}, which equates to an oxygen demand of 2.8 mg O$_2$·L^{-1} (15).

For distribution systems that do not use chloramine as the disinfectant, ammonium can be removed economically from source waters by nitrifying bacteria. Attached biofilm processes such as GAC-biofiltration, fluidized bed filtration, or rapid sand filtration of nonchlorinated water are generally used because of the slow growth rates of nitrifying bacteria (15). In this way, microbial activity that would otherwise occur in the distribution system is used in a treatment process at the plant to convert ammonium to nitrate. Larger support media (1 mm diameter) should be used in nitrifying biofilters to lessen sheer stresses on slow-growing populations of nitrifying bacteria (15).

In chloraminated systems, nitrification needs to be controlled to prevent the loss of the disinfectant with subsequent increase in biofilm activity. In cooler climates, nitrification may be controlled for much of the year without intervention, as nitrification usually occurs when water temperatures are above 15 °C (6). Active methods of control of nitrification in distribution systems include maintenance of a chloramine residual of 2.5 mg·L^{-1} with a chlorine-to-ammonia-nitrogen ratio of 4:1, which limits biofilm development in the distribution system by reduction of the organic carbon of the water, active cleaning of the distribution system by flushing or pigging, reducing the detention time of the water in the system, eliminating "dead ends" in the mains (36), and limiting reactive surfaces on pipes by coating with carbonate or polymer additives (37).

Control of Microbial Activity Involved in "Dirty Water" Formation

"Dirty water" caused by iron and manganese precipitates is a common problem for water supply utilities (38). Dirty water precipitates result from abiotic or biological oxidation of reduced iron or manganese, or organic materials with which they may be complexed. Iron and manganese deposits and sludges can limit the penetration of disinfectants into biofilms. Microbial manganese oxidation in a distribution system has been controlled by maintaining a chlorine residual of >0.2 mg·L^{-1} and reducing manganese concentrations to below 0.02 mg·L^{-1} (2).

Removal of manganese and iron before distribution usually involves chemical oxidation to their insoluble forms Fe^{3+} and Mn^{4+} by aeration, chlorine, chlorine dioxide, potassium permanganate, or ozone, with subsequent removal by flocculation, sedimentation or filtration (38), or combined chemical and biological oxidation and removal by biofiltration (15,39). Biofilters are usually slow to develop and generally require the development of MnO$_2$(s) precipitates to help catalyze the oxidation of soluble Mn^{2+} (15). Biofilters can be used for the simultaneous removal of AOC, BDOC, ammonium, iron, and manganese (40).

CONCLUSIONS

Microbial activity in water distribution systems affects the health compliance and aesthetics of the distributed water. Control of microbial activity in distribution systems is never complete, but it can be sufficiently achieved by limiting microbial energy and food sources (i.e., by removing as much AOC, BDOC, DOC, ammonia, reduced metals, and reduced sulfur compounds as is possible), in most cases by maintaining a reasonable residual disinfectant in the distributed water, and by careful selection and maintenance of distribution infrastructure, so that minimal microbiological growth and chemical reactions can occur on surfaces.

Water utilities have many treatment and disinfection choices for the control of microbial activity in distribution systems, and those choices will be guided by the quality of the waters they receive, their geography, the retention time of water in the system, and the expectations of the community. The most appropriate treatment options for control of microbial activity by any one utility must be guided by analysis of the performance of treatment methods in other treatment plants, and extensive testing and optimization of treatment methods alone, and in combination with the full treatment system.

BIBLIOGRAPHY

1. LeChevallier, M.W., Cawthon, C.D., and Lee, R.G. (1988). Factors promoting survival of bacteria in chlorinated water supplies. *App. Environ. Microbiol.* **54**: 649–654.

2. Sly, L.I., Hodgkinson, M.C., and Arunpairojana, V. (1990). Deposition of manganese in a drinking water distribution system. *Appl. Environ. Microbiol.* **56**: 628–639.

3. Franzmann, P.D., Heitz, A., Zappia, L.R., Wajon, J.E., and Xanthis, K. (2001). The formation of malodorous dimethyl oligosulphides in treated groundwater: the role of biofilms and potential precursors. *Water Res.* **35**: 1730–1738.

4. Wagner, D. and Chamberlain, A.H.L. (1997). Microbiologically influenced copper corrosion in potable water with emphasis on practical relevance. *Biodegradation* **8**: 177–187.

5. Lu, W., Kiéné, L., and Lévi, Y. (1999). Chlorine demand of biofilms in water distribution systems. *Water Res.* **33**: 827–835.

6. Wolf, R.L., Lieu, N.I., Izaguirre, G., and Means, E.G. (1990). Ammonia-oxidising bacteria in a chloraminated distribution system: seasonal occurrence, distribution, and disinfection resistance. *Appl. Environ. Microbiol.* **56**: 451–462.

7. Najm, I. et al. (2000). *Case Studies of the Impacts of Treatment Changes of Biostability in Full Scale Distribution Systems*. AWWA Research Foundation, Denver, CO.

8. van der Kooij, D. (1992). Assimilable organic carbon as an indicator of bacterial regrowth. *J. Am. Water Wastewater Assoc.* **84**: 57–65.

9. van der Kooij, D., van Lieverloo, J.H.M, Schellart, J., and Heimstra, P. (1999). Maintaining quality without a disinfectant residual. *J. Am. Water Wastewater Assoc.* **91**: 55–64.

10. Servais, P., Laurent, P., and Randon, G. (1993). Impact of biodegradable dissolved organic carbon (BDOC) on bacterial dynamics in distribution systems. *Proceedings AWWA WQTC Conference*, Miami, FL, 7–10 November.

11. Joret, J.C. and Lévi,Y. (1986). Méthode rapide d'évaluation du carbone éliminable des eaux par voie biologique. *Trib. Cebedau* **39**: 3–9.

12. Servais, P., Billen, G., and Hascoët, M.C. (1987). Determination of the biodegradable fraction of dissolved organic matter in waters. *Water Res.* **21**: 445–450.

13. Frías, J., Ribas, F., and Lucena, F. (1992) Comparison of methods for the measurement of biodegradable organic carbon and assimilable organic carbon in water. *Water Res.* **29**: 2785–2788.

14. Hem, L.J. and Efraimsen, H. (2001). Assimilable organic carbon in molecular weight fractions of natural organic matter. *Water Res.* **35**: 1106–1110.

15. Rittmann, B.E. and Snoeyink, V.L. (1984). Achieving biologically stable drinking water. *J. Am. Water Wastewater Assoc.* **76**: 106–114.

16. Volk, C.J. and LeChevallier, M.W. (2000). Assessing biodegradable organic matter. *J. Am. Water Wastewater Assoc.* **92**: 64–76.

17. LeChevallier, M.W., Schulz, W., and Lee, R.G. (1991). Bacterial nutrients in drinking water. *Appl. Environ. Microbiol.* **57**: 857–862.

18. Piriou, P., Dukan, S., and Kiene, L. (1998). Modelling bacteriological water quality in drinking water distribution systems. *Water Sci. Technol.* **38**: 299–307.

19. Escobar, I.C. and Randall, A.A. (2001). Assimilable organic carbon (AOC) and biodegradable dissolved organic carbon (BDOC): complementary measurements. *Water Res.* **18**: 4444–4454.

20. Volk, C.J. and LeChevallier, M.W. (2002). Effects of conventional treatment on AOC and BDOC levels. *J. Am. Water Wastewater Assoc.* **94**: 112–123.

21. Easton, J. (1993). *Measurement and Significance of Assimilable Organic Carbon*. Foundation of Water Research FR0373, Marlow, UK.

22. Slunjski, M. et al. (2002). MIEX—good research commercialised. *Water* **29**: 42–47.

23. Huck, P.M. (2000). Measurement of biodegradable organic matter and bacterial growth potential in drinking water. *J. Am. Water Wastewater Assoc.* **82**: 78–86.

24. Volk, C.J. et al. (2000). Impact of enhanced and optimized coagulation on removal of organic matter and its biodegradable fraction in drinking water. *Water Res.* **34**: 3247–3257.

25. Owen, D.M., Amy, G.L., and Chowdhury, Z.K. (1993) *Characterization of Natural Organic Matter and Its Relationship to Treatability*. AWWA Research Foundation, Denver, CO.

26. Vahala, R., Ala-Peijari, T., Rintala, J., and Laukkanen, R. (1998). Evaluating ozone dose for AOC removal in two-step GAC filters. *Water Sci. Technol.* **37**: 113–120.

27. Hacker, P.A., Paszkokolva, C., Stewart, M.H., Wolfe, R.L., and Means, E.G. (1994). Production and removal of assimilable organic-carbon under pilot-plant conditions through the use of ozone and peroxone. *Ozone-Sci. Eng.* **16**: 197–212.

28. Gagon, G.A., Ollos, P.J., and Huck, P.M. (1997). Modelling BOM utilisation and biofilm growth in distribution systems: review and identification of research needs. *J. Water Supply Res. Technol. Aqua* **46**: 165–180.

29. Suffet, I.H., Corado, A., Chou, D., Butterworth, S., and McGuire, M.J. (1994). AWWA Taste and Odour Survey. *Proceedings of the 1993 Water Quality Technology Conference*, Miami, FL November 7–11.

30. Richardson, S.D. (2003). Disinfection by-products and other emerging contaminants in drinking water. *Trends Anal. Chem.* **22**: 666–684.

31. Neden, D.G., Jones, R.J., Smith, J.R., Kirmeyer, G.J., and Frost, G.W. (1992). Comparing chlorination and chloramination for controlling bacterial regrowth. *J. Am. Water Wastewater Assoc.* **84**: 80–88.

32. Momba, M.N.B., Kfir, R., Venter, S.N., and Cloete, T.E. (2000). An overview of biofilm formation in distribution systems and its impact on the deterioration of water quality. *Water South Africa* **26**: 59–66.
33. Lund, V. and Ormerod, K. (1995). The influence of disinfection processes on biofilm formation in water distribution systems. *Water Res.* **29**: 1013–1021.
34. Armon, R., Arbel, T., and Green, M.. (1998). A quantitative and qualitative study of biofilm disinfection on glass, metal and PVC surfaces by chlorine, bromine, and bromochloro-5,5-dimethylhydantoin (BCDMH). *Water Sci. Technol.* **38**: 175–179.
35. Hallam, N.B., West, J.R., Forster, C.F., and Simms, J. (2001). The potential for biofilm growth in water distribution systems. *Water Res.* **35**: 4063–4071.
36. Kirmeyer, G.J., Odell, L.H., Jacangelo, J., Wilczak, A., and Wolfe, R. (1995). *Nitrification Occurrence and Control in Chloraminated Systems*. AWWA Research Foundation and American Water Works Association, Denver, CO.
37. Woolschlager, J.E., Rittmann, B.E., Piriou, P., and Schwartz, B. (2001). Developing a effective strategy to control nitrifier growth using the comprehensive disinfection and water quality model (CDWQ). In: *Bridging the Gap: Meeting the World's Water and Environmental Resources Challenges: "World Water Congress CD-ROM". Proceedings of the World Water and Environmental Resources Congress*, Orlando, FL, May 20–24.
38. Larson, A.L. (1993). Case pilot study: Organic fouling of a coated media pressure filter. *Proceedings AWWAWQTC Conference*, Miami, FL, 7–10 November.
39. Sly, L.I., Arunpairojana, V., and Dixon, D.R. (1993). Bioloigcal removal of manganese from water by immobilized manganese-oxidising bacteria. *Water* **20**: 38–40.
40. Korth, A., Bendinger, B., Czekalla, C., and Wichmann, K. (2001). Biodegradation of NOM in rapid sand filters for removing iron and manganese. *Acta Hydrochim. Hydrobiol.* **29**: 289–295.

MICROBIAL DYNAMICS OF BIOFILMS

BAIKUN LI
Pennsylvania State University
Harrisburg, Pennsylvania

BIOFILM FORMATION

The solid–liquid interface between a surface and an aqueous medium (e.g., water, blood) provides an ideal environment for the attachment and growth of microorganisms. Bacteria have a marked tendency to attach to surfaces exposed to an aqueous environment and initiate biofilm formation, characterized as a matrix of cells and cellular products attached to a solid substratum. The biofilm systems that result can be beneficial, as exemplified by fixed-film wastewater treatment processes (e.g., trickling filters and rotating biological contactors). However, this biological matrix can increase operational costs and/or decrease product quality in a variety of industries, including oil and paper production, semiconductor manufacture, and drinking water distribution. In water distribution systems and heat transfer equipment, for example, biofilms can cause substantial energy losses resulting from increased fluid frictional resistance and increased heat transfer resistance. Moreover, the pathogens trapped in microbial biofilm in a water supply pipeline have caused public health concerns.

Microorganisms grow on surfaces enclosed in biofilm and form microcolonies of bacterial cells, which are encased in adhesive polysaccharides excreted by the cells. Biofilm can provide advantages for microorganisms' survival in natural environment, especially under oligotrophic condition, because biofilms trap nutrients for growth of the microbial population and help prevent detachment of cells on surfaces present in flowing systems. In drinking water systems (oligotrophic environment), the percentage (70%) of positive (viable) cells was significantly higher at surface-adhered biofilm cells than that (40%) in the planktonic cells (1). It has been postulated that bacteria attached to surfaces are more active because of enrichment of nutrients at the surface. Therefore, the growth of cells at the surface is predominant over deposition and has advantage over planktonic (free swimming) in an oligotrophic environment.

Biofilms may form on a wide variety of surfaces, including living tissues, indwelling medical devices, industrial or potable water system piping, or natural aquatic systems. The variable nature of biofilms has been observed from an industrial water system and a medical device (Figs. 1 and 2). The water system biofilm is highly complex, containing corrosion products, clay material, freshwater diatoms, and filamentous bacteria. The biofilm on the medical device, on the other hand, seems to be composed of a single, coccoid organism and the associated extracellular polymeric substance (EPS) matrix.

It has been found that microbial motility may play a role in structure formation in biofilms (2). After the initial attachment to the substratum, *Pseudomonas aeruginosa* evidently moves on the substratum by means of twitching motility, which suggests that the initial microcolonies are formed by aggregation of bacteria. Moreover, microcolonies of bacterial cells in biofilms are the basic structural units in sessile communities, different from that of planktonic cells

Figure 1. Scanning electron micrograph of a native biofilm that developed on a mild steel surface in an 8-week period in an industrial water system. Rodney Donlan and Donald Gibbon, authors. Licensed for use, American Society for Microbiology Microbe Library. Available from: http://www.microbelibrary.org/.

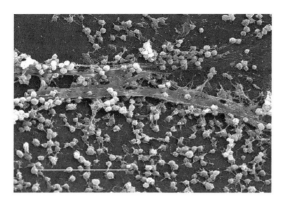

Figure 2. Scanning electron micrograph of a staphylococcal biofilm on the inner surface of an indwelling medical device. Bar, 20 μm. Used with permission of Lippincott Williams & Wilkins. Photograph by Janice Carr, CDC (http://www.cdc.gov/ncidod/eid/vol8no9/02-0063.htm).

and there is different gene expression in sessile bacteria. A large fraction of the biofilm bacteria (*P. aeruginosa*, *Discosoma* sp.) occasionally swim from one microcolony to another. In the initial phase of biofilm formation, bacteria are nonmotile and sessile inside the microcolonies. After several days of biofilm formation, biofilm bacteria start to swim rapidly in circles, the compact microcolonies are dissolved, and loose structures containing bacteria from different microcolonies are formed, along with flow channels. These results suggest that biofilms contain both sessile populations and planktonic populations.

BIOFILM STRUCTURE HETEROGENEITY

Biofilm Physical Structure

It had long been assumed that biofilm had a homogeneous structure, with microorganisms uniformly distributing throughout the entire biofilm at both structure and dynamics levels (3). Recently, biofilm was discovered heterogeneous in structure and function by microelectrode and confocal laser scanning microscope (CLSM). Numerous microbiology and microbial ecology of microbial ecosystem studies revealed that biofilms typically contain many layers with microbial ecology variation in each layer. Biofilms are not composed of bacterial cells randomly distributed in a homogeneous matrix, but they are composed of microcolonies of matrix-enclosed cells with intervening open-water channels. This complex and fascinating organization of biofilm communities was seen in flow-cell biofilms of several *Pseudomonas* species and in mixed-species natural biofilms formed on rock surfaces (4).

Microorganisms tend to accumulate in the bottom layer inside biofilm toward substratum, which leads to higher biomass density in the depth of biofilm. With the higher biomass density, the effective diffusivity distribution in heterogeneous biofilm decreased toward the bottom of a biofilm. It has been measured that the ratio of effective diffusivity in biofilm to the bulk solution diffusivity (Df/D) decreased from 0.68–0.81 in the top layer to 0.38–0.45 in the bottom layer (5). The highest biomass density, 120 g/L, was near the bottom of the biofilm, whereas the lowest density was measured near the top layers of biofilm, 20–30 g/L (6). It is anticipated that in the bottom layer of biofilm, metabolite and other by-products accumulate and have difficulties to diffuse out of biofilm, which lead to high mass.

Besides the mass density and diffusivity, other physiological features also vary along the depth of biofilm, which offer the heterogeneous structure in biofilm property (7). The solid densities (with units of mg-TS (total solid)/cm^3 total biofilm) in the bottom layers were 5–10 times higher than were those in the top layers; the ratio of living cells to total biomass decreased from 72–91% in the top layers to 31–39% in the bottom layers; the porosities of biofilms changed from 89–93% in the top layers to 58–67% in the bottom layers. In contrast, the mean pore radius of biofilms decreased from approximately 1.7–2.7 μm in the top layers to 0.3–0.4 μm in the bottom layers. Thus, bacterial cells living in a dense multilayer matrix face the problem of limited transport of substrate and nutrients into the film inner layers as well as export of waste products out of these deep regions.

Extracellular Polymeric Substances (EPSs)

A biofilm is an assemblage of microbial cells that is irreversibly associated with a surface and enclosed in a matrix of primarily polysaccharide material. The microcolonies in a biofilm are surrounded by large amounts of EPS. It has been found that EPSs play an important role in gene transformation and cell–cell contact in biofilm. Although it is not clear that EPSs have a role in shaping the spatial structure in biofilms, evidence has been provided that EPSs can facilitate storage of nutrients in biofilms for subsequent intake during periods of carbon limitation (8–10). The EPSs play several important roles for the biofilm, including (1) protection from naturally occurring or manmade biocides and antibiotics, (2) stabilization of bacterial adhesion to the substratum and to other cells in the biofilm, and (3) entrapment and concentration of soluble nutrients in proximity to biofilm members.

EPSs may account for 50–90% of the total organic carbon of biofilms (11) and can be considered the primary matrix material of the biofilm. EPSs synthesized by microbial cells may vary greatly in their composition, but the major component is polysaccharide. Some are neutral macromolecules, but most are polyanionic because of the presence of either uronic acids (D-glucuronic acid, D-galacturonic, D-mannuronic acids) or ketal-linked pyruvate. Inorganic residues, such as phosphate or rarely sulfate, may also confer polyanionic status (10). A few EPSs may even be polycationic, as exemplified by the adhesive polymer obtained from strains of *Staphylococcus epidermidis* associated with biofilms (12).

The amount of EPS synthesis within the biofilm is known to depend greatly on the availability of carbon substrates (both inside and outside the cell) and on the balance between carbon and other limiting nutrients (e.g., nitrogen, calcium) (10). Moreover, different organisms produce differing amounts of EPSs, and the amount of EPSs increases with age of the biofilm. Because EPSs are highly hydrated, they prevents desiccation in some natural biofilms. EPSs may also contribute to the antimicrobial resistance properties of biofilms by impeding the mass transport of antibiotics through the biofilm, probably by binding directly to these agents (13). This process can lead to the high antibiotics resistance of pathogens growing in biofilm.

Channel Structures in Biofilm

Biofilms are highly hydrated open structures containing a high fraction of EPSs and large void spaces between microcolonies. Most of the bacteria in the nutrient-sufficient parts of the biosphere grow in distinct microniches in highly structured multispecies communities. Bacterial growth zones form as discrete microcolony structures separated by channel boundaries, and these structures are maintained over time (Fig. 3). Also, open spaces extend from the top of the biofilms to the deep inner regions, which forms a channel-like network. Mushroom-shaped structures, separated by channels and voids, were observed in P. fluorescens biofilm (14).

It has been found that when biofilm grew from 20 to 300 μm in thickness, many vertical and horizontal channels permeated the film (15). These channels extended from the surfaces into at least one-half of the depth of the film. Moreover, on anaerobic fixed-bed reactor communities, integrative networks of channel-like structures were found throughout the microbial granules (15), which has been interpreted as playing a role in transport of cooperative substrates, nutrients, and gases (16). This porous architecture of biofilms may represent a more optimal arrangement for the influx of substrate and nutrients and transfer of wastes.

The biofilm heterogeneous structure of microbial clusters of cells and EPSs was also verified by microelectrode measurement (17). The microbial clusters of cells and EPSs were separated by interstitial voids. The cell clusters were 300 μm wide, and the voids were 100 μm wide. Biofilms are not flat homogeneous layers of cells, in which transport is exclusively diffusional. The transport of metabolites involved both convection and diffusion, which offers a relative importance depending on the hydraulic regime. The velocity of the bulk liquid had a strong influence on the oxygen concentration and mass distribution in the biofilm, with high flow velocity increasing oxygen transferring inside biofilm and concentration differences and thus increasing convectional transport.

BIOFILM METABOLIC HETEROGENEITY

Metabolic heterogeneity may account for the phenomenal success of microbial biofilms in nature and in device-related and other chronic bacterial infections, and it may be a major factor in the inherent resistance of medical biofilms to antibiotics.

Heterogeneity within a population (differing activity of single cells associated with increasing depth in a biofilm) might provide useful insight into aspects of substrate and mass transport limitation. The established biofilm population was composed of cells having a variety of growth rates. Such differences might reflect different positions of cells within the biofilm matrix; for example, cells located near the surface of the biofilm may not be as limited by mass transport of, or competition by surrounding cells for, nutrients (18).

Cell ribosome content changes at different growth phases, with higher ribosome content in new cells and lower ribosome content in old cells. The variation in ribosome content can provide a useful tool for evaluating the status of a great variety of microorganism growing at different generation times inside biofilm (19). Quantifying the fluorescent light emitted from in situ rRNA hybridized cells indicated that sulfate-reducing bacteria (SRB) in a young multispecies biofilm were more active than were the sulfate-reducing bacteria in an established biofilm (20). When microcolonies of a P. putida strain growing in a biofilm reaches a critical size, the light emitted by the cells decreased in the center of the microcolonies and eventually throughout the microcolonies, which indicates that the cells displayed different levels of growth activity correlating with their location in the biofilm and with the biomass of biofilm (18).

MICROBIAL STRUCTURE AND DYNAMIC HETEROGENEITY IN BIOFILM IN WATER/WASTEWATER TREATMENT

The heterogeneity of microbial structure and dynamics is very important in biological wastewater treatment. Trickling filter, which depends on a variety of microorganisms growing in different layers in biofilm, is a major process in nutrient removal. Wastewater biofilms are complex multispecies biofilms, displaying considerable heterogeneity with respect to both the microorganism present and their physicochemical microenvironment. Successive vertical zonations of predominant respiratory processes occurring simultaneously in close proximity have been found in aerobic wastewater biofilms with a typical thickness of only a few millimeters (21). Numerous environmental microbiology and molecular biology studies have revealed the correlation between the microbial structure and dynamic heterogeneity and the availability of oxygen and substrate in two biofilm systems: sulfur-reducing biofilm and nitrification/denitrification biofilm.

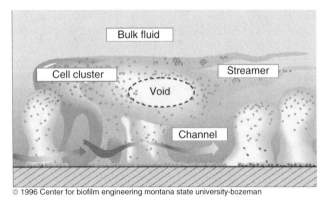

Figure 3. The diagram of cluster-channel structure in biofilm (Image from http://www.erc.montana.edu/Res-Lib99-SW/Image_Library/Structure-Function/default.htm.)

Sulfate-Reducing Biofilm

The reduction of sulfate (SO_4^{2-}) or sulfite (SO_3^{2-}) to sulfide (S^{2-}) occurs in an anaerobic zone deeper inside biofilm with organic carbon as the electron donor. It is carried out by anaerobic SRB. The presence of oxygen and nitrate in the deep layer of biofilm has been found to inhibit the anaerobic SRB growth, because oxygen and nitrate compete with sulfate or sulfite to be the electron acceptor in anaerobic biofilm. The correlation of the vertical distribution of SRB with the gradients of oxygen and hydrogen sulfide (the product of the reducing sulfate by SRB) in a trickling filter revealed that hydrogen sulfide was predominant in the anoxic zone of the biofilm, and most of the SRB were found in the anoxic layer as dense clusters of assemblies (22). Oxygen respiration occurred in the upper 0.2–0.4 mm of the biofilm, whereas sulfate reduction occurred in deeper, anoxic parts of the biofilm (below 1.5 mm inside biofilm) (23). Moreover, the presence of sulfate in biofilm significantly affects the stratification of microbial community inside biofilm, because sulfate is the main electron acceptor for SRB in anoxic reaction. Total sulfate reduction rates (μmol S/cm^3/h) increased when the availability of SO_4^{2-} or organic substrate increased as a result of deepening the sulfate reduction zone or an increase in the sulfate reduction intensity.

SRB can even survive in the aerobic zone in biofilm and conduct good sulfur-reducing ability. In one study, SRB (approximately 10^9–10^{10} cell per cm^3 of biofilm) was evenly distributed throughout the biofilm, even in the oxic surface (21). Oxygen penetrated only about 100 μm from the surface in a biofilm approximately 1000 μm thick. Sulfide was produced from SRB in a narrow zone of 150–300 μm below the surface at a maximum specific rate of 21 μmol H$_2$S /cm^3/h.

Nitrification/Denitrification Biofilm

Nitrogen is removed in biological wastewater treatment by two types of bacteria: aerobic autotrophic nitrifier (oxidize NH_4^+ to NO_3^-) and anoxic heterotrophic denitrifer (reduce NO_3^- to N_2). Two major groups of nitrifiers exist: ammonia oxidizing bacteria (AOB, oxidize NH_4^+ to NO_2^-) and nitrite oxidizing bacteria (NOB, oxidize NO_2^- to NO_3^-). In biofilm, aerobic nitrifier and anoxic denitrifer dominate in the different zones because of the transfusion of oxygen and organic and inorganic substrates. The nitrifying and denitrifying bacterial populations and their activities in biofilms have been investigated by the application of fluorescent *in situ* hybridization (FISH) and microsensor technology. A dense layer of ball-shaped clusters of *Nitrosomonas* cells (ammonium oxidizers) in close physical association with surrounding smaller clusters of *Nitrobacter* cells (nitrite oxidizers) was found in the aerobic zone of a nitrifying biofilm from the trickling filter of an aquaculture (17). The nitrifying zone (measured with microelectrode) was restricted to the outer 100–150 μm of the biofilm, and the nitrifiers were found in this zone as a dense layer of clusters. The central part of the biofilm was found anoxic and inactive for nitrification, and significantly fewer nitrifiers were found there.

The dynamic studies of nitrifers have revealed their stratification and activity inside biofilm. The active ammonia-oxidizing zone was located in the outer part of a biofilm, whereas the active nitrite-oxidizing zone was located just below the ammonia-oxidizing zone (24). The number of nitrite-oxidizing bacteria (NOB) is low compared with the number of ammonia-oxidizing bacteria (AOB). Oxygen penetration depth was about 1200 μm in the wastewater biofilm because of the loose structure and low microbial activity. Nitrification occurred in the top 200 μm, whereas denitrification occurred in the deeper anoxic layers (below a depth of about 1200 μm). The nitrification layer is 150–175 μm, AOB activity is 0.13–0.08 μmol NH_4^+/cm^2/h, and average specific nitrifying rate is 8.7–4.6 μmol NH_4^+/cm^3/h. NOB activity is 0.22–0.3 μmol NO_2^-/cm^2/h, and the specific rate is 14.7–17.1 μmol NO_2^-/cm^3/h. AOB (*Nitrosomonas*) were detected throughout the biofilm, whereas NOB (*Nitrospira*-like bacteria) were found mainly in the inner parts of biofilm and always found in vicinity of AOB clusters, which may have reflected the synthophic association between AOB and NOB (25).

Besides the microbial stratification and activity in biofilm, some parameters were examined to indicate the different microbial processes inside biofilm. The stratification of microbial processes and the associated redox potential (ORP) changes in biofilms were studied by microelectrodes (26). In aerobic/sulfate-reducing biofilm, a clearly stratified structure with depth was observed: aerobic oxidization took place only in a shallow layer near the surface, and sulfate reduction occurred in the deeper anoxic zone. The boundary between these two processes was well defined by redox potential (ORP increased from −100 mV to +100 mV). In the aerobic/nitrifying biofilm, a stratified structure with depth was also observed, but it was less well defined (ORP increased from +300 mV to +330 mV). In this biofilm, aerobic oxidation took place throughout the biofilm depth, and more nitrification occurred in the deeper section of the biofilm.

The competition occurs between anoxic denitrifer and SRB in biofilm (27). In SRB biofilm, oxygen was depleted in the first 0.2 mm of the biofilm (1.5 mm thickness). SRB occurred in the anoxic zones inside biofilm. However, when nitrate is added in the bulking solution, the SRB zone withdrew deeper into the biofilm. Nitrate penetrated into 0.4 mm depth inside biofilm, which indicates that denitrification occurred. Nitrate decreased the sulfide content either by oxidation of produced sulfide or by inhibition of sulfate reduction.

Oxygen Limitation in Biofilm

Growth rate limitation occurs in biofilms presumably as a result of restriction for a particular nutrient (such as oxygen and substrate) that fails to fully penetrate the biofilm. The expression of alkaline phosphatase (APase) could be a physiological indicator. It was found that the band of APase (to indicate enzyme activities) on the *P. aeruginosa* biofilm surface decreased from 46 μm thick with biofilms grown in pure oxygen to 2 μm thick with biofilms grown in pure nitrogen (28). The expression of APase on the biofilm surface revealed marked spatial physiological heterogeneity within biofilms in which active protein synthesis was restricted by oxygen availability inside the biofilm. This heterogeneity has implications for microbial ecology and for understanding the reduced susceptibilities of biofilms to antibiotics agents under limited nutrients in the establishment of physiological gradients.

The direct observation of living biofilms has found many different microniches occupied by individual cells within these complex communities. Cells near the surface of the microcolony would obtain high concentrations of the oxygen delivered by the water in the channels while oxygen is rapidly depleted below the microcolony surface, and the centers of the microcolonies that comprise biofilms growing in air are virtually anaerobic (29). The steep oxygen gradient is paralleled by gradients of differing steepness for every ion and molecule needed or produced by cells in biofilms, and the interaction of these gradients produces a different microniche for every individual cell in the complex heterogeneous community.

BIOFILM DETACHMENT

Biofilm development is a balance between new biofilm growth and old biofilm detachment. Biofilm detachment is the removal of cells from the existing biofilm to the bulk fluid. It is a significant factor that balances cell growth in the biofilm and brings the system to steady state. Biofilm cells may be dispersed either by shedding of daughter cells from actively growing cells, detachment as a result of nutrient levels, or shearing of biofilm aggregates (continuous removal of small portions of the biofilm) because of flow effects.

The mechanisms underlying the process of shedding by actively growing cells in a biofilm are not well understood. It has been recognized that several physical forces are involved in biofilm detachment (30): erosion (continuous removal of small particles of biofilm as a result of shear forces), sloughing (random removal of large portions of biofilm, often occurring in old and thick biofilms or when exposed to dramatic environmental changes), abrasion (removal caused by collisions of solid particles with biofilms particularly in fluidized bed reactors), and predator grazing (consumption of biofilm by larger organisms such as protozoa).

Besides these physical forces, the microbial metabolic dynamic inside biofilm also contributes to biofilm detachment. As biofilm matures, oxygen will be depleted in the upper layer of biofilm and anoxic/anaerobic zones will dominate in the deeper layers of biofilm. The anoxic/anaerobic reactions produce a variety of biogas (N_2, H_2S, CO_2, CH_4, H_2, etc.) and might increase up to 20% in the established biofilm (31). The biogas contributes to the biofilm detachment. Moreover, the amount of EPSs in biofilm also changes during biofilm formation and maturation. The content of EPSs decreased within the biofilm in the course of biofilm maturation. As a consequence, the adhesion strength of biofilm adjacent to substratum becomes more weakened, which leads to biofilm detachment (32).

BIBLIOGRAPHY

1. Manz, W. et al. (1993). *In situ* identification of bacteria in drinking water and adjoining biofilming by hybridization with 16S and 23S rRNA-directed fluorescent oligonucleotide probes. *Appl. Environ. Microbiol.* **59**(7): 2293–2298.
2. Nielson, T.T. et al. (2000). Development and dynamics of *Pseudomonas* sp. biofilm. *J. Bacteriol. Microbiol.* **192**(22): 6482–6489.
3. Dalsgaard, T. and Revsbech, N.P. (1992). Regulation factors of denitrification in trickling filter biofilms as measured with the oxygen/nitrous oxide microsensor. *FEMS Microbiol. Ecol.* **101**: 151–164.
4. Costerton, J.W. et al. (1995). *Microbial Biofilms Annu. Rev. Microbiol.* **49**: 711–745.
5. Beyenal, H., Tanyolaç, A., and Lewandowski, Z. (1998). Measurement of local effective diffusivity in heterogeneous biofilms. *Water Sci. Technol.* **38**(8–9): 171–178.
6. Zhang, T.C. and Bishop, P.L. (1994). Evaluation of tortuosity factors and effective diffusivities in biofilms. *Water Res.* **28**(11): 2279–2287.
7. Zhang, T.C. and Bishop, P.L. (1994). Density, porosity and pore structure of biofilms. *Water Res.* **28**(11): 2267–2277.
8. Hussain, M., Wilcox, M.H., and White, P.J. (1993). The slime of coagulase-negative staphylococci: biochemistry and relation to adherence. *FEMS Microbiol. Rev.* **104**: 191–208.
9. Leriche, V., Sibille, P., and Carpentier, B. (2000). Use of an enzyme-linked lectinsorbent assay to monitor the shift in polysaccharide composition in bacterial biofilms. *Appl. Environ. Microbiol.* **66**: 1851–1856.
10. Sutherland, I.W. (2001). Biofilm exopolysaccharides: a strong and sticky framework. *Microbiology* **147**: 3–9.
11. Flemming, H-C., Wingender, J., Griegbe, and Mayer, C. (2000). Physico-chemical properties of biofilms. In: *Biofilms: Recent Advances in Their Study and Control*. L.V. Evans (Ed.). Harwood Academic Publishers, Amsterdam, pp. 19–34.
12. Mack, D. et al. (1996). The intercellular adhesin involved in biofilm accumulation of *Staphylococcus epidermidis* is a linear ß-1,6-linked glucosaminoglycan: purification and structural analysis. *J. Bacteriol.* **178**: 175–183.
13. Donlan R.M. (2000). Role of biofilms in antimicrobial resistance. *ASAIO J.* **46**: 47–52.
14. Korber, D.R., Lawrence, J.R., and Caldwell, D.E. (1994). Effect of motility on surface colonization and reproductive success of *Pseudomonas fluorescens* in dual-dilution continuous culture and batch culture systems. *Appl. Environ. Microbiol.* **60**: 1421–1429.
15. Massol-deya, A.A., Whallon, J., Hickey, R., and Tiedjie, J.M. (1995). Channel structure in aerobic biofilms of fixed-film reactors treating contaminated groundwater. *Appl. Environ. Microbiol.* **61**(2): 769–777.
16. MacLeod, F.A., Giot, S.R., and Costerton, J.W. (1990). Layered structure of bacterial aggregates produced in an upflow anaerobic sludge bed and filter reactor. *Appl. Environ. Microbiol.* **56**: 1598–1607.
17. De Beer, D. and Stoodley, P. (1995). Relation between the structure of anaerobic biofilm and transport phenomena. *Water Sci. Technol.* **32**(8): 11–18.
18. Tolker-Nielsen, T. et al. (2000). Development and dynamics of *Pseudomonas* sp. biofilms. *J. Bacteriol.* **182**: 6482–6489.
19. Poulsen, L.K., Ballard, G., and Stahl, D.A. (1993). Use of rRNA fluorescence *in situ* hybridization for measuring the activity of single cells in young and established biofilms. *Appl. Environ. Microbiol.* **59**(5): 1354–1360.
20. Ito, T., Okabe, S., Satoh, H., and Watanabe, Y. (2002). Successional development of sulfate-reducing bacterial populations and their activities in a wastewater biofilm growing under

21. Okabe, S., Itoh, T., Satoh, H., and Watanabe, Y. (1995). Analyses of spatial distributions of sulfate-reducing bacteria and their activity in aerobic wastewater biofilms. *Appl. Environ. Microbiol.* **65**(11): 5107–5116.
22. Ramsing, N.B., Juhl, M., and Jorgensen, B.B. (1993). Distribution of sulfate-reducing bacteria, O2, and H2S in photosynthetic biofilms determined by oligonucleotide probes and microelectrodes. *Appl. Environ. Microbiol.* **59**(11): 3840–3849.
23. Kuhl, M. and Jorgensen, B.B. (1992). Microsensor measurements of sulfate reduction and sulfide oxidation in compact microbial communities of aerobic biofilms. *Appl. Environ. Microbiol.* **58**(4): 1164–1174.
24. Okabe, S., Satoh, H., and Watanabe, Y. (1999). In situ analysis of nitrifying biofilms as determined by *in situ* hybridization and the use of microelectrodes. *Appl. Environ. Microbiol.* **65**(7): 3182–3191.
25. Gieseke, A. et al. (2001). Community structure and activity dynamics of nitrifying bacteria in a phosphate-removing biofilm. *Appl. Environ. Microbiol.* **67**(3): 1351–1362.
26. Bishop, P.L. and Yu, T. (1999). A microelectrode study of redox potential change in biofilms. *Water Sci. Technol.* **39**(7): 179–185.
27. Santegoeds, C.M., Muyzer, G., and de Beer, D. (1998). Biofilm dynamics studied with microsensors and molecular techniques. *Water Sci. Technol.* **37**(4–5): 125–129.
28. Xu, K.D. et al. (1998). Spatial physiological heterogeneity in *Pseudomonas aeruginosa* biofilm is determined by oxygen availability. *Appl. Environ. Microbiol.* **64**(10): 4035–4039.
29. Lewandowski, Z. (2000). Structure and function of biofilms. In: *Biofilms: Recent Advances in Their Study and Control.* L.V. Evans (Ed.). Harwood Academic Publishers, Amsterdam, pp. 1–17.
30. Peyton, B.M. and Characklis, W.G. (1993). A statistical analysis of the effect of substrate utilization and shear stress on the kinetics of biofilm detachment. *Biotechnol. Bioeng.* **41**(7): 728–735.
31. Ohashi, A. and Harada, H. (1994). Characterization of detachment mode of biofilm developed in an attached-growth reactor. *Water Sci. Technol.* **30**(11): 35–45.
32. Horan, N.J. and Eccels, C.R. (1986). Purification and characterization of extracellular polysaccharide from activated sludgs. *Water Res.* **20**: 1427–1432.

MICROBIAL ENZYME ASSAYS FOR DETECTING HEAVY METAL TOXICITY

Oladele A. Ogunseitan
University of California
Irvine, California

THE PERSISTENT PROBLEM OF TOXIC METALS IN THE WATER ENVIRONMENT

Throughout the world, industrial development has been associated with increased mining, refining, and use of potentially toxic metals in large-scale manufacturing processes. Industrialization is also tightly linked to increases in the combustion of fossil fuels such as coal and petroleum, which produce considerable amounts of toxic metals such as mercury and lead, respectively (1). Inevitably, many industrial activities generate environmental pollution that changes the natural biogeochemical cycling of elements by modifying the size and flux of metal reservoirs. The primary consequence of metal pollution in natural water systems is toxic effects on aquatic organisms, but ultimately human exposure occurs from consumption of contaminated water, seafood, and plant crops that are cultivated in land irrigated with polluted water.

The United States Environmental Protection Agency (EPA) has established maximum contaminant levels (MCL) for eleven metals that dominate the inorganic contaminants most commonly encountered in drinking water and groundwater sources (Table 1). The symptoms associated with these toxic metals range from skin disease from chromium (MCL = 0.1 mg/L) to more serious conditions such as kidney disease from cadmium (MCL = 0.005 mg/L) and mercury (MCL = 0.002 mg/L). In addition, nonenforceable standards have been established for secondary pollutants such as Al, Cu, Fe, Mn, Ag, and Zn because they are regarded as nuisance contaminants with the assumption of negligible health consequences at the established concentrations permitted in water (Table 2).

Metals are also ranked third among the top 100 contaminant categories most commonly found in violation of established water quality standards such as the total maximum daily load (TMDL; see Table 3). Metals represent the largest proportion (1977 or 20.6%) of the total number of TMDLs approved for contaminant categories in water systems (Table 4). Mercury, copper, and lead are the three metals found most frequently in impaired water systems. These issues make metal pollution one of the most pressing water quality problems. In addition, there is a lot of uncertainty about metal doses that can produce disease because of the variability of individual genetic and behavioral susceptibility factors. Therefore, new techniques are continuously sought to produce consistent and reliable measures of ecological and human health risks from metal exposure. Microorganisms are involved in three of the multifaceted strategies designed to deal with metal contamination of water resources: first, in the control of acid mine drainage; second, in the bioremediation of contaminated water systems; and third, as indicators of metal bioavailability in toxicity testing. This chapter focuses on the third aspect of microbial interactions with metals in water, the use of microbial enzymes to indicate bioavailability and toxicity. A case study of microbial enzyme response to Pb is presented as an example.

MICROBIAL METALLOENZYMES

Several enzymes used by microorganisms to catalyze key reactions of the major biogeochemical cycles are metalloenzymes; to function properly, they require metal ions positioned at or near the active site. At least a third of all proteins in the biosphere are metalloproteins, and several metals are commonly found in physiological systems in small concentrations. For example Ca, K, Na, Fe, Mo, Mn, Mg, Cu, Zn, Co, Ni, W, and Al are commonly

Table 1. Inorganic Chemicals Category of Water Contaminants with Established Maximum Contaminant Levels (MCLs)

Contaminant	MCLG[b], mg/L[c]	MCL or TT[b], mg/L[c]	Potential Health Effects from Ingestion of Water	Sources of Contaminant in Drinking Water
Antimony	0.006	0.006	Increase in blood cholesterol; decrease in blood sugar	Discharge from petroleum refineries; fire retardants; ceramics; electronics; solder
Arsenic	0[d]	0.010 as of 01/23/06	Skin damage or problems with circulatory systems and may have increased risk of cancer	Erosion of natural deposits; runoff from orchards; runoff from glass and electronics production wastes
Asbestos (fiber >10 μm)	7 million fibers per liter	7 million fibers per liter	Increased risk of developing benign intestinal polyps	Decay of asbestos cement in water mains; erosion of natural deposits
Barium	2	2	Increase in blood pressure	Discharge of drilling wastes; discharge from metal refineries; erosion of natural deposits
Beryllium	0.004	0.004	Intestinal lesions	Discharge from metal refineries and coal-burning factories; discharge from electrical, aerospace, and defense industries
Cadmium	0.005	0.005	Kidney damage	Corrosion of galvanized pipes; erosion of natural deposits; discharge from metal refineries; runoff from waste batteries and paints
Chromium (total)	0.1	0.1	Allergic dermatitis	Discharge from steel and pulp mills; erosion of natural deposits
Copper	1.3	TT[e]; action level = 1.3	Short term exposure: gastrointestinal distress. Long term exposure: liver or kidney damage. People with Wilson's disease should consult their personal doctors if the amount of copper in their water exceeds the action level	Corrosion of household plumbing systems; erosion of natural deposits
Cyanide (as free cyanide)	0.2	0.2	Nerve damage or thyroid problems	Discharge from steel/metal factories; discharge from plastic and fertilizer factories
Fluoride	4.0	4.0	Bone disease (pain and tenderness of the bones); children may get mottled teeth	Water additive that promotes strong teeth; erosion of natural deposits; discharge from fertilizer and aluminum factories
Lead	zero	TT[e]; action level = 0.015	Infants and children: delays in physical or mental development; children could show slight deficits in attention span and learning abilities. Adults: kidney problems; high blood pressure	Corrosion of household plumbing systems; erosion of natural deposits
Mercury (inorganic)	0.002	0.002	Kidney damage	Erosion of natural deposits; discharge from refineries and factories; runoff from landfills and croplands
Nitrate (measured as nitrogen)	10	10	Infants below the age of six months who drink water containing nitrate in excess of the MCL could become seriously ill and, if untreated, may die. Symptoms include shortness of breath and blue-baby syndrome.	Runoff from fertilizer use; leaching from septic tanks, sewage; erosion of natural deposits

Table 1. (Continued)

Contaminant	MCLG[a], mg/L[b]	MCL or TT[a], mg/L[b]	Potential Health Effects from Ingestion of Water	Sources of Contaminant in Drinking Water
Nitrite (measured as nitrogen)	1	1	Infants below the age of six months who drink water containing nitrite in excess of the MCL could become seriously ill and, if untreated, may die. Symptoms include shortness of breath and blue-baby syndrome.	Runoff from fertilizer use; leaching from septic tanks, sewage; erosion of natural deposits
Selenium	0.05	0.05	Hair or fingernail loss; numbness in fingers or toes; circulatory problems	Discharge from petroleum refineries; erosion of natural deposits; discharge from mines
Thallium	0.0005	0.002	Hair loss; changes in blood; kidney, intestine, or liver problems	Leaching from ore-processing sites; discharge from electronics, glass, and drug factories

[a] Asbestos, cyanide, nitrate, and nitrite are the few nonmetallic chemicals in this category. They are included here for comparison of standards and also because their presence in water may affect the biological availability of metals and the nature of microbial interaction with available metal ions.

[b] Definitions: **Maximum contaminant level (MCL)**: The highest level of a contaminant that is allowed in drinking water. MCLs are set as close to MCLGs as feasible using the best available treatment technology and taking cost into consideration. MCLs are enforceable standards.
Maximum contaminant level goal (MCLG): The level of a contaminant in drinking water below which there is no known or expected risk to health. MCLGs allow for a margin of safety and are nonenforceable public health goals.
Treatment Technique (TT): A required process intended to reduce the level of a contaminant in drinking water.

[c] Units are in milligrams per liter (mg/L) unless otherwise noted. Milligrams per liter are equivalent to parts per million.

[d] MCLGs were not established before the 1986 Amendments to the Safe Drinking Water Act. Therefore, there is no MCLG for this contaminant.

[e] Lead and copper are regulated by a treatment technique that requires systems to control the corrosiveness of their water. If more than 10% of tap water samples exceed the action level, water systems must take additional steps. For copper, the action level is 1.3 mg/L, and for lead it is 0.015 mg/L.

Source: U.S. EPA (http://www.epa.gov/safewater/mcl.html#mcls).

associated with diverse proteins in microorganisms (see Table 5). For each of these "essential" metals, there are at least two other metals that can compete for their interaction with proteins, but these competing "nonessential" metals are likely to be very toxic because the proteins to which they bind do not function properly (2). To survive in environments with fluctuating metal concentrations and parameters of metal bioavailability, many microorganisms have evolved specific uptake systems for essential metals, whereas specific detoxification systems are evolved to resist the detrimental effects of nonessential metals. For example, the uptake of iron is facilitated by the production of pyoverdins in fluorescent *Pseudomonas*, whereas many members of this genus also produce enzymes such as mercuric reductase which has the only known function of mercury detoxification (3). Therefore, measurement of metalloenzyme kinetics in the presence and absence of toxic metal ions can provide a direct assay for the biological availability of metallic contaminants in water systems.

Metal bioavailability is influenced by many factors, including pH, particulate matter that can adsorb metals, and inorganic elements and ions such as sulfur, chloride, and carbonate that can affect metal speciation (4,5). Therefore, monitoring metal concentrations in water based solely on chemical extraction and physical analysis typically generates "total concentration" data, but the biologically available concentration remains unknown. This is the main reason for investigating enzyme bioassays to estimate metal toxicity. When cell-based assays are used, enzyme assays reflect the quantity of metals which must have entered the cell to provoke physiological responses. Ogunseitan (3) and Ogunseitan and colleagues (6,7) have used such responses to monitor the biological availability of Hg and Pb, respectively. The following section describes the use of a microbial enzyme to assess the toxicity of biologically available Pb in water systems.

Table 2. Permissible Concentrations of Six Metals and Some Nonmetal Components[a]

Contaminant	Secondary Standard
Aluminum	0.05 to 0.2 mg/L
Chloride	250 mg/L
Color	15 (color units)
Copper	1.0 mg/L
Corrosivity	Noncorrosive
Fluoride	2.0 mg/L
Foaming agents	0.5 mg/L
Iron	0.3 mg/L
Manganese	0.05 mg/L
Odor	3 threshold odor number
pH	6.5–8.5
Silver	0.10 mg/L
Sulfate	250 mg/L
Total dissolved solids	500 mg/L
Zinc	5 mg/L

[a] The concentrations of the metals are monitored according to National Secondary Drinking Water Regulations (NSDWRs or Secondary Standards), which are nonenforceable guidelines regulating contaminants that may cause cosmetic effects (such as skin or tooth discoloration) or aesthetic effects (such as taste, odor, or color) in drinking water. EPA recommends secondary standards to water systems but does not require systems to comply. However, states may choose to adopt them as enforceable standards. The co-occurrence in water of nonmetallic components of this list may affect the biological availability of metal ions, which is a major factor in toxicity assessments based on microbial enzymes.

Table 3. Metals Are Ranked Third Among the Top 100 Impairments from Water Pollution; Hg, Cu, and Pb are the Most Frequently Cited Pollutants

Impairment	Number of Impairments
Chromium VI	1
Contaminated sediments (zinc)	1
Zinc	351
Contaminated sediments (alum)	1
Contaminated sediments (lead)	2
Molybdenum	1
Thallium	4
Arsenic	177
Contaminated sediments (cadmium)	1
Contaminated sediments (copper)	1
Contaminated sediments (silver)	1
Heavy metals	3
Iron	154
Lead	442
Manganese	52
Aluminum	48
Cadmium	227
Contaminated sediments (gold)	1
Silver	48
Trace elements	2
Boron	23
Chromium	75
Contaminated sediments (chromium)	1
Copper	528
Metals	2597
Metals—mercury	3
Contaminated sediments (mercury)	2
Mercury	697
Nickel	44
Selenium	173

Source: U.S. EPA (http://oaspub.epa.gov/pls/tmdl/waters_list.impairments?p_impid=13).

Table 4. Approved TMDLs by Metallic Pollutants Since January 1, 1996

Pollutant	Number of TMDLs
Aluminum	415
Arsenic	13
Boron	9
Cadmium	69
Chromium	12
Chromium VI	3
Contaminated sediments (copper)	1
Contaminated sediments (mercury)	2
Contaminated sediments (nickel)	1
Copper	79
Iron	439
Lead	95
Manganese	409
Mercury	146
Metals	49
Nickel	29
Selenium	65
Silver	29
Thallium	1
Zinc	104

Source: U.S. EPA (http://oaspub.epa.gov/pls/tmdl/waters_list.tmdl_pollutants?p_polid=13).

MICROBIAL AMINOLEVULINATE DEHYDRATASE AS A BIOSENSOR FOR Pb BIOAVAILABILITY

delta-Aminolevulinate dehydratase (ALAD; porphobilinogen synthase; EC: 4.2.1.24) is a 128-kDa metalloenzyme that catalyzes the first common step in the biosynthetic pathway for all tetrapyrroles, including oxygen carriers (e.g., heme), light gathering arrays (e.g., chlorophyll), vitamin B_{12}, and cofactor F430. The enzyme catalyzes the reaction between two molecules of aminolevulinic acid (ALA), which engage in a Knorr-type condensation reaction to produce porphobilinogen, the universal precursor of all pyrroles (Fig. 1) (7). ALAD activity has been demonstrated in all microorganisms that have been investigated across several phylogenetic groups. Microbial ALAD requires either zinc, magnesium, or both for optimum functioning, although these metals can be substituted by heavy metals such as Pb and Hg (Fig. 2).

In the proximity of the enzyme active site is the metal-binding domain, which for zinc consists of three cysteines at positions 133, 135, and 143, and a solvent molecule, presumably a hydroxide ion. Asparagine at positions 135 and 143, it has been postulated, is essential for Mg-dependent ALAD (8). Few microbial ALAD systems have been described, and the metallic component of ALAD differs among species. For example, Zn is required for ALAD activity in yeasts, whereas Mg is required in Bradyrhizobium japonicum and Pseudomonas aeruginosa, and both Zn and Mg are required by the Escherichia coli ALAD, although the requirement for Mg in E. coli is not stringent (8). Moreover, the sensitivity of ALAD to toxic metals depends on the identity of the metallic cofactor (8). ALAD activity in aquatic organisms has long been used as a biomarker of lead exposure (9,10), and a polymorphism in the ALAD locus has been described in humans that has implications for susceptibility to lead poisoning (11).

ALAD activity in an environmental strain of Pseudomonas putida ATCC 700097 originally isolated from an urban wastewater stream was sensitive to Pb and other toxic metals (12). There is a statistically significant dose–response relationship between ALAD activity in cells of this organism and Pb (Pearson correlation coefficient $= -0.985$; $r^2 = 0.97$; and $p < .001$). The highest level of inhibition of ALAD activity was approximately 74% of the normal level when cells were incubated with $Pb^{2+} > 500$ μM. The relationship between Pb and ALAD activity was statistically described by $\log_{10} [Pb] = 3 : 68 - 1.41$ [ALAD activity]. ALAD activity is measured in μmol of porphobilinogen synthesized per mg protein per minute. The direct exposure of protein extracted from P. putida showed a stronger inhibitory effect on ALAD activity, represented by a reduction of up to 85% in response to 500 μM of Pb. A higher concentration of Pb is needed to produce a comparable level of ALAD inhibition in P. putida cells seeded into natural freshwater; this suggests that Pb is not completely biologically available in freshwater. Therefore, an immobilized enzyme biosensor can be expected to be an even more sensitive tool for assessing biologically reactive forms of Pb in water, although this measurement precludes the barrier represented by cellular integrity (7).

The reduced bioavailability of Pb in water systems can be attributed to the presence of ligand chemicals, which

Table 5. Examples of Metals Required by Key Enzyme Functions[a]

Metal	Enzyme Examples	Relevant Organisms and/or Ecological Function
Calcium	Collagenase Calpain protease	*Clostridium histolyticum* Pathogenesis
Cobalt	Halomethane methyltransferase	Facultative methylotrophs Chloromethane degradation
Copper	Copper amine oxidases	*Arthrobacter globoformis* Deamination of primary amines to corresponding aldehydes
Iron	Cytochrome P450 Soluble methane Monooxygenase	*Pseudomonas putida*; P450 cam; polyaromatic hydrocarbon biodegradation
Magnesium	Magnesium chelatase	*Rhodobacter sphaeroides* Chlorophyll synthesis
Manganese	Manganese-dependent peroxidases	*Phanerochaete chrysosporium* Carbon cycling–lignin degradation
Molybdenum	Xanthine dehydrogenase Nitrate reductase; nitrogenase	*Pseudomonas putida* *Rhizobium* Purine (caffeine) degradation Nitrogen fixation
Selenium	Selenocysteine (UGA codon) enzymes; glutathione peroxidase; hydrogenases	*Escherichia coli* *Methanococcus voltae* Nitrogen respiration
Tungsten	"True W-enzymes" Aldehyde ferredoxin oxidoreductase	*Thermophilic archaea; Pyrococcus furiosus* Molybdenum antagonist
Zinc	Aminolevulinate dehydratase (inhibited by Pb)	Several bacteria and nearly all Archaea Porphyrin synthesis

[a]The biological availability of metal ions affects enzyme function, potentially constituting the basis for the development of protein-based biosensors.

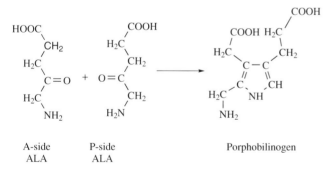

Figure 1. The metalloenzyme, delta-aminolevulinate dehydrate, catalyzes the condensation of two molecules of aminolevulinic acid to produce porphobilinogen, the universal precursor of tetrapyrroles.

are anions or molecules that form coordination compounds or complexes with metals, and commonly occur in natural waters (5). The most important inorganic ligands that form complexes with metals in natural waters are F^-, Cl^-, SO_4^{2-}, OH^-, HCO_3^-, CO_3^{2-}, HPO_4^{2-}, and NH_3; HS^- and S^{2-} are influential in anoxic waters. The covalent bonding tendency ($\Delta\beta$) and the ionic bonding tendency (Z^2/r) of metals determine the types of ligands that could reduce bioavailability (5). Metals such as Pb are classified as borderline (for Pb, $2 > \Delta\beta > 0$; and $2.5 < Z^2/r < 7.0$).

As such, not all of these ligands are relevant to the bioavailability of Pb, but clearly many commonly detected regulated and unregulated inorganic ions in water systems affect Pb bioavailability. This limitation of bioavailability is likely to exist for other toxic heavy metals.

In conclusion, the pervasiveness of metal contamination of water systems indicates a great need to develop sensitive measures of biologically available concentrations of toxic heavy metals in water. The use of microbial enzymes assays for detecting heavy metal toxicity is an emerging technology with abundant resources represented by the physiological diversity inherent in microbial communities. However, further research is required to produce easily interpreted information on enzyme responses to environmental metals in a manner that can supplement the monitoring of other water quality parameters.

Acknowledgment

The Program in Industrial Ecology at the University of California, Irvine, supported this work.

BIBLIOGRAPHY

1. Pang, D.J. (1995). Lead in the environment. In: *Handbook of Ecotoxicology*. D.J. Hoffman, B.A. Rattner, G.A. Burton, and J.C. Cairns Jr. (Eds.). Lewis, Boca Raton, FL, pp. 356–391.

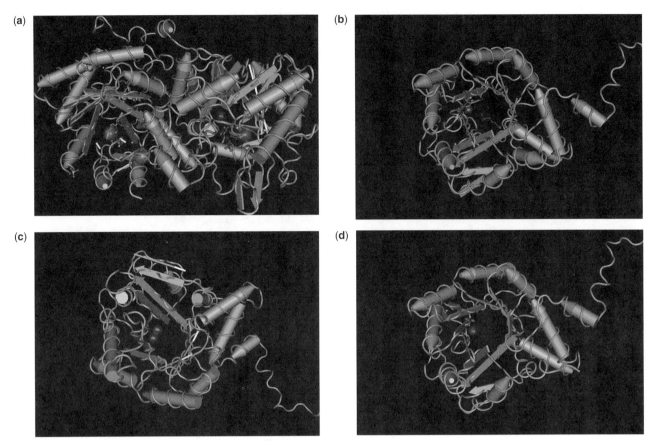

Figure 2. The three-dimensional structure of aminolevulinate dehydratase (ALAD) protein showing the metal-binding domains is presented in (**a**) and (**b**). (**a**) The Mg-dependent ALAD of *P. aeruginosa* with the position of Mg indicated by the arrow. The N-terminal arms of the *P. aeruginosa* ALAD do not project out of the octamer, and the dimers are asymmetric, unlike in Zn-dependent ALAD found in *Archaea* microbes ((**b**) Arrow indicates metal-binding domain with zinc. Phylogenetic analysis of the amino acid sequence of the metal-binding domain of ALAD indicates that the occurrence of various metal ions in ecosystems inhabited by microorganisms exerts a strong selective pressure on the evolution of preference for specific metal ions. For example, the metal-binding domain of ALAD accepts Pb (**c**) and Hg (**d**). Protein structures were created and manipulated by the Cn3D software, version 4.1.

2. Ogunseitan, O.A. (1999). Microbial proteins as biomarkers for ecosystem health. In: *Integrated Assessment of Ecosystem Health* K. Scow, G. Fogg, D. Hinton, and M. Johnson (Eds.). Lewis, Boca Raton, FL, pp. 207–222.

3. Ogunseitan, O.A. (1998). Protein method for investigating mercuric reductase gene expression in aquatic environments. *Appl. Environ. Microbiol.* **64**: 191–204.

4. Hsieh, K.M., Murgel, G.A., Lion, L.W., and Shuler, M.L. (1994). Interactions of microbial biofilms with toxic trace metals: 2. Prediction and verification of an integrated computer model of lead (II) distribution in the presence of microbial activity. *Biotechnol. Bioeng.* **44**: 232–239.

5. Newman, M.C. and Jagoe, C.H. (1994). Ligands and the bioavailability of metals in aquatic environments. In: *Bioavailability: Physical, Chemical, and Biological Interactions* J.L. Hamelink, P.F. Landrum, H.L. Bergman, and W.H. Benson (Eds.). Lewis, Boca Raton, FL, pp. 39–62.

6. Ogunseitan, O.A., Yang, S., and Scheinbach, E. (1999). The delta-aminolevulinate dehydratase activity of *Vibrio alginolyticus* is resistant to lead (Pb). *Biol. Bull.* **197**: 283–284.

7. Ogunseitan, O.A. Yang, S. and Ericson, J. (2000). Microbial delta-aminolevulinate dehydratase as a biosensor of lead (Pb) bioavailability in contaminated environments. *Soil Biol. Biochem.* **32**: 1899–1906.

8. Chauhan, S. and O'Brian, M.R. (1995). A mutant *Bradyrhizobium japonicum* delta-aminolevulinic acid dehydratase with an altered metal requirement functions *in situ* for tetrapyrrole synthesis in soybean root nodules. *J. Biol. Chem.* **270**: 19, 819–823, 827.

9. Conner, E.A. and Fowler, B.A. (1994). Biochemical immunological properties of hepatic delta-aminolevulinate dehydratase in channel catfish (*Ictalurus punctatus*). *Aquatic Toxicol.* **28**: 37–52.

10. Overman, S.R. and Krajicek, J.J. (1995). Snapping turtles (*Chelydra serpentina*) as biomonitors of lead contamination of the big river in Missouri's old lead belt. *Environ. Toxicol. Chem.* **14**: 689–695.

11. Claudio, L., Lee, T., Wolff, M.S., and Wetmur, J.G. (1997). A murine model of genetic susceptibility to lead bioaccumulation. *Fundam. Appl. Toxicol.* **35**: 84–90.

12. Ogunseitan, O.A. (1996). Removal of caffeine in sewage by *Pseudomonas putida*: implications for water pollution index. *World J. Microbiol. Biotechnol.* **12**: 251–256.

MICROBIAL FORMS IN BIOFOULING EVENTS

BAIKUN LI
Pennsylvania State University
Harrisburg, Pennsylvania

BIOFOULING FORMATION: NUTRIENT ADSORPTION

The term *biofouling* refers to organic debris as a result of the accumulation of inorganic and organic particulate matter along with the growth of various organisms on a surface. The development of a biofouling layer depends on the initial attachment of bacteria to this surface, their eventual increase in number, and the development of a protective slime layer, or glycocalyx.

Biofouling leads to many undesirable problems in industry, such as decreased heat transfer in cooling towers, regrowth in drinking water distribution systems (Fig. 1), and deterioration of materials. In the medical arena, biofilms are responsible for dental plaque and persistent infections on medical implants. It has long been observed that biofilms are much less susceptible to antimicrobial agents than are their planktonic (free-swimming) counterparts.

Most microorganisms, especially heterotrophic bacteria, can easily access nutrients previously deposited on a surface. The attachment of microorganisms to surfaces has a survival function, because the surface in oligotrophic conditions served as sites of nutrients for microorganisms' growth even though they cannot multiply in the bulk aqueous phase. One possible reason for the rapid bacterial growth at surfaces in aquatic environments was the evolvement of microbially produced extracellular enzymes (1). These enzymes can metabolize the macromolecular or particulate substrates. In the bulk aqueous phase, the contact between enzyme and substrate was rare, and products of enzymatic hydrolysis would diffuse away and be unavailable to the bacterium. At the surface, on the other hand, the enzyme and substrate would be in close contact, and provided the enzyme remained active in the adsorbed state, the bacterium should benefit from the end products of the reaction.

The adhesion of microorganisms to a solid surface is a strategy for access to nutrients. If the bulk water had very few dissolved organic nutrients in it, the growth and reproduction of bacteria would be a very slow process. However, the attachment to a solid surface can substantially accelerate bacterial growth. The bacterial attachment to surfaces shows a great advantage: attached bacteria, known as *sessile or adherent bacteria*, become exposed to a plentiful supply of fresh water containing fresh nutrients that they can extract continuously. Another advantage that sessile bacteria have is their ability to adsorb dissolved organic molecules (food sources for many bacteria) from the water layer. This attachment or adsorption facilitates the growth of microorganisms, especially in oligotrophic environment.

Because of the easy access to nutrients on the surface, the attached bacteria normally have a bigger size than do free-swimming bacteria. When acetate was a substrate, the attached bacteria were larger (average vol. = $0.271\ \mu m^3$) than were planktonic bacteria (average vol. = $0.118\ \mu m^3$), had a higher frequency of dividing cells, and were responsible for most of the acetate uptake (2). These larger attached cells suggest that many of the planktonic (free-swimming) cells were starving and, hence, remained small, whereas the attached cells were growing on substrate available at the surface (3).

MICROORGANISM FORMS IN BIOFOULING FORMATION

One advantage of attachment to a surface is that microorganisms can easily get access to nutrients on the surface and survive in the adverse environment. Starved cells of the marine *Vibrio* DW1 are more adhesive than are unstarved cells, which leads to the suggestion that adhesion may be a tactic in the survival strategy of these organisms under starvation conditions (4). It has been widely accepted that microorganisms attach to available surfaces and initiate the formation of monolayer covering on surfaces; then they develop into thick layers of extensive biofilms, enveloped within extracellular polymeric matrices (5). Generally, there are five diversities in bacterial growth patterns adhered onto surfaces (Fig. 2). These diversities are as follows:

1. *Mother–Daughter or Shedding Cells*. After bacteria adhere to a surface, the small starved cells (<0.5 μm diameter) grow to normal size (>1.0 μm diameter), presumably with substrate accumulating on the surface. The growing cells (now the mother cell) attach to the surface in a perpendicular orientation, and as each gowing cell begins to divide, the daughter cells exhibit a rapid spinning motion before release into the aqueous phase. The mother cell, which remains on the surface, repeats the processes of cellular growth and reproduction. Many daughter cells return to colonize the surface further to get nutrients, and the extensive secondary colonization of the surface results in nutrient uptake exceeding

Figure 1. The cut section of a cast-iron pipe shows a severe iron bacteria buildup on the pipe walls.

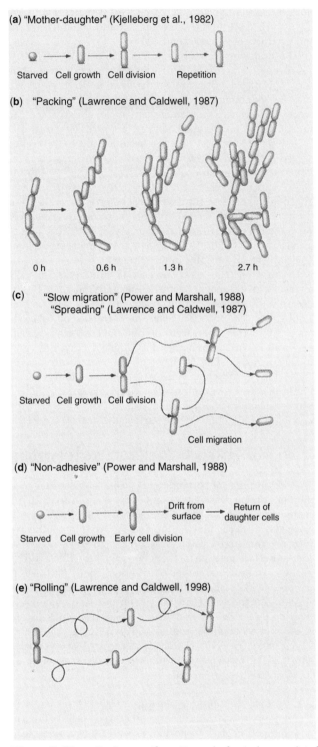

Figure 2. Diversity in growth patterns in bacteria associated with surface. (**a**) Mother–daughter or shedding cells, (**b**) packing cells, (**c**) slow migratory or spreading cells, (**d**) reversibly adhering cells, and (**e**) rolling cells. (Courtesy of Ref. 6.)

nutrient input to the surface, and hence, the cells cease dividing and revert to typical small starvation forms (6).

2. *Packing Cell*. Bacteria divide in a manner that results in the daughter cells aligning adjacent to each other, which eventually leads to the formation of a contiguous monolayer (7).

3. *Slow Migratory or Spreading Cells*. When *Pseudomonas* JD8 cells are in a medium lacking an energy source, the adhering, small, and starved cells grow to normal size and begin to divide, and then the daughter cells migrate very slowly (0.15 μm/min) and at random for variable distances (5–50 times their cell length) across the surface before the next division occurs (8).

4. *Reversibly Adhering or Rolling Cells*. *Vibrio* MH3 is capable of metabolizing surface-bound static acid, the starved cells grow to normal size, and cellular division begins near the surface, but the dividing cells return to the aqueous phase for the final production of daughter cells. That cell division proceeds to completion in the aqueous phase was also described by Lawrence and Caldwell (7) as rolling; that is, cells loosely attached and somersaulting across the surface as they grow and divide similarly may develop as a result of the bacteria reversibly adhering to the surface.

5. *Gliding Cells*. Many cytophage-like gliding bacteria can use complex polysaccharide substrata as an energy substrate. They have a unique form of adhesion that allows them to remain attached to a solid substrate in aqueous condition, and they can glide (translocate) across the surface (9). After the initial attachment to the substratum, *P. aeruginosa* evidently moves on the substratum by means of twitching motility, which suggests that the initial microcolonies are formed by aggregation of bacteria. A large fraction of the biofilm bacteria (*P. aeruginosa*, *Discosoma* sp.) occasionally swim from one microcolony into another (10). This characteristic certainly makes for efficient use of the solid substrates they metabolize.

THE RELATIONSHIP OF MICROORGANISMS INSIDE BIOFOULING LAYERS

In the biofouling events, different types of microorganisms form either competition or syntrophic correlation. In the mixed biofilms formed by fungi and bacteria, the interactions vary with different microbial species and forms. In the fungal biofouling, *P. aeruginosa* (bacteria) cells readily adhere to *Candida albicans* (a pathogenic fungi) hyphae, but not to the yeast form of *Candida*. Over a 24–48 hour time course, bacteria attach to the *C. albicans* hyphae and form biofilms consisting of densely packed cells surrounded by extracellular matrix (ECM). These findings suggest that the bacteria in biofilm extract nutrients from the underlying *Candida* hyphae, eventually killing the hyphal form of the yeast, and several virulence genes of *P. aeruginosa* that are involved in the killing of *Candida* hyphae in mixed biofilms (11).

Although *P. aeruginosa* is harmful to *Candida* in the biofouling formation, *Streptococcus mutans* can make *C. albicans* strongly adhere to the surface. It was found that *C. albicans* is detached very easily when growing alone, even by a slight movement, when it adheres to plastic rods in a medium containing sucrose (12). However, in mixed cultures of *C. albicans* and *S. mutans*, *Candida* tend to grow superficially on the surface of a polysaccharide layer formed by the cocci that have colonized the rod surface and are strongly adhered. These findings are confirmed in a subsequent study showing that firm adhesion of *C. albicans* to acrylic resin occurs when the yeasts are incubated simultaneously with *S. mutans* in the presence of glucose. (13). Other studies also found that *S. mutans* grow under and above the yeast and hyphal layers and are clearly adherent to both morphological forms of *C. albicans* (14). This association in the mixed species biofilm makes the biofilm develop very strongly and the fouling is not easy to remove.

Moreover, in a two-species bacterial biofouling event, different metabolic interactions are formed in chemostats (bulk water) and biofouling layers (15). When *P. putida* R1 and *Acinetobacter* C6 strains were grown together in chemostats with limiting concentrations of benzyl alcohol as substrate, they competed for the primary carbon source, which gave *Acinetobacter* C6 a growth advantage. *Acinetobacter* strain C6 outnumbered *P. putida* R1 (cell number ratio: 500:1). However, under similar growth conditions in biofilms, *P. putida* R1 was present in higher numbers than was *Acinetobacter* strain C6 (cell number ratio: 5:1). In addition, these two organisms compete or display commensal interactions depending on their relative physical positioning in the biofilm. In the initial phase of biofilm development, the growth activity of *P. putida* R1 was shown to be higher near microcolonies of *Acinetobacter* strain C6. After a few days, *Acinetobacter* strain C6 colonies were overgrown by *P. putida* R1 cells and new structures developed, in which microcolonies of *Acinetobacter* strain C6 cells were established in the upper layer of the biofilm. In this way, the two organisms developed structural relationships allowing *Acinetobacter* strain C6 to be close to the bulk liquid with high concentrations of benzyl alcohol and allowing *P. putida* R1 to benefit from the benzoate leaking from *Acinetobacter* strain C6.

In sticky biofouling layers, a variety of microorganisms (e.g., hetetrotrophic, autotrophic, aerobic, and anoxic) attach to each other and form a stable and complex microecological community, so the slime layer increases in size and the bacteria become firmly attached to the surface. With the biofilm becoming thicker, oxygen depletion (anaerobiosis) occurs in the lower layers of the biofouling layers, caused by the mass transfer resistance inside the biofilm and the metabolic activity of an aerobic microorganism on the surface layer of the biofilm. Under these conditions, bacteria that require oxygen-free conditions to grow can start to develop. These bacteria include those that can grow in both oxygenated and deoxygenated conditions (known as facultative anaerobic bacteria). Examples are the coliform bacteria, such as *E. coli*, and certain species of *Pseudomonas*. Other bacteria are strictly anaerobic, growing well only in the absence of oxygen. These bacteria include certain species of sulfate reducing bacteria (SRB) and methanogenic bacteria, which are predominant in the deeper layer of biofilm near substratum. The biogas (N_2, H_2S, CO_2, CH_4, etc.) produced by these anoxic and anaerobic bacteria may cause the detachment of the thick biofilm from the surface.

BIOFOULING IN A COOLING TOWER

Biofouling in a cooling tower has been a long-standing problem in industry (Fig. 3). The inorganic and organic contaminants in a cooling tower may precipitate from the flowing water and lead to fouling problems (16). Several major types of microorganisms are involved in cooling tower biofouling, for example, iron bacteria, sulfur reducing/oxidizing bacteria, algae, and fungi.

Major aerobic bacteria that may cause problems in cooling systems are iron-oxidizing bacteria. This group of bacteria are commonly found in well waters that contain soluble iron and large amounts of carbon dioxide. The iron bacteria oxidize the soluble ferrous (Fe^{2+}) to insoluble ferric (Fe^{3+}) as a source of energy, and they use carbon dioxide as a source of carbon for cell growth. They generally produce slimy red-brown flocs of hydrated ferric oxide that can cause plugging of filters and heat exchangers.

Another group of aerobic bacteria in cooling systems is sulfur-oxidizing bacteria, which produce yellow stringy slimes. These bacteria oxidize and reduce sulfur compounds (H_2S) to sulfuric acid (SO_4^{2-}) in the aerobic environment, which reduces the pH of the water to even lower than 2. The sulfuric acid produced by the sulfur bacteria may cause slime production and corrosion of concrete basins. This group of bacteria is mainly found in refinery cooling towers contaminated with reduced sulfur compounds.

Many types of algae contain the green pigment chlorophyll, which traps the energy from sunlight. Algae synthesize carbon from carbon dioxide from air and require energy from sunlight. Growth of algae primarily occurs in the tower deck area because of ample sunlight. Having been introduced from the atmosphere as spores to water, algae readily become established in areas where plenty of sunlight exists, such as the top of the deck and at the sides of the tower. Algae produce mainly green or blue–green gelatinous slimes, which may be long and filamentous or

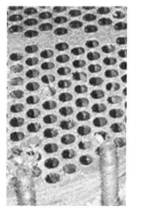
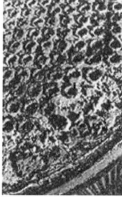

Figure 3. Cooling tower corrosion caused by biofouling. Left: cooling tower wall without biofouling. Right: cooling tower wall with biofouling.

loose and flocculent. The slime species most commonly found is the green algae, the cyanobacteria (originally the blue–green algae) and the diatoms. Prolific algal growth may produce green or blue–green scums in some cooling tower sumps. Algae slimes can plug distribution nozzles and troughs in the cooling tower deck, which causes poor water distribution across the tower and hence reduces cooling efficiency. In the biofouling development, the growth of algae may provide a food source that encourages the growth of other organisms, such as bacteria and fungi.

Fungi are another major group of microorganism that causes problems in the cooling tower. The most insidious problem caused by the growth of fungi is the attack on cooling tower wood, which can go unnoticed for many years. Molds and yeasts can break down the cellulose in the wood, which leads to structural weakening.

BIOFOULING IN MEMBRANE WATER TREATMENT SYSTEM

Biological contamination, known as biofouling, occurs most often during nanofiltration and reverse osmosis in water treatment systems. The attachment of inorganic/organic matters and microorganisms has damaging, often irreversible effects on nanofiltration and reverse osmosis systems (Fig. 4). The major impact of biofouling is the substantial reduction of the flux through the membrane systems, which results in a higher pressure, that causes higher operation cost (18). In some cases, membrane materials are suitable environments for microorganisms to grow, which will cause the membrane to be completely destroyed in a short period of time.

The types of microorganisms, their growth factors, and their concentration in a membrane system greatly depend on critical factors, such as temperature, the presence of sunlight, pH, dissolved oxygen concentrations, and the presence of organic and inorganic nutrients. Both aerobic (oxygen-dependent) and anaerobic (oxygen-independent) bacteria can attach onto the membrane surface and develop biofilm. Aerobic bacteria usually live in an environment of warm, shallow, and sunlit water, with a high dissolved oxygen content, a pH of 6.5 to 8.5, and an abundance of organic and inorganic nutrients. Anaerobic bacteria, on the other hand, are usually present in closed systems with little to no dissolved oxygen and become active when a sufficient amount of nutrients (e.g., organic matter or the debris of dead microorganisms) is present.

Another major type of microorganism in membrane biofouling is algae, which can grow on membrane parts exposed to sunlight. The amount of sunlight determines the amount of oxygen that is produced by algae. Aerobic bacteria, which are oxygen-dependent, need the oxygen produced by algae when the dissolved oxygen content in the feed water is insufficient for metabolism. Therefore, the growth of algae boosts the aerobic bacteria growth and further causes biofouling. While the algae die off, they become a food source for bacteria, because they release organic nutrients that bacteria need for growth in a membrane system, especially in an oligotrophic water environment.

METHODS OF BIOFOULING CONTROL

The most efficient approach to solve biofouling is to kill living attached microorganisms by using biocides, or to inhibit the life processes of living organisms without actually killing them by using biostat. It is possible for a chemical to be a biostat to certain organisms at a low concentration and a biocide to the same organisms at a higher concentration. Both biocides and biostats solve the slime problem by affecting various parts of the microbial cell.

Two types of biocides exist: organic biocides, classified as nonoxidizing biocides because they do not oxidize other chemicals or oxidize them only very slightly, and inorganic biocides, classified as oxidizing biocides (e.g., chlorine) because they have a very strong oxidation capacity for many organic and inorganic matters. Nonoxidizing (organic) molecules react with various parts of the microbial cell, generally at a specific site. The most common groups of nonoxidizing biocides are aldehydes, organo-sulfur biocides, chlorinated isocyanurates, organonitrogen biocides, amine salts, and heavy metals.

Examples of useful oxidizing biocides are halogens such as chlorine (Cl_2), chlorine dioxide (ClO_2) (19), hydrogen peroxide (H_2O_2), ozone, and chlorine, bromine donors such as chlorinated isocyanurates and hydantoins. The main drawback with the oxidizing biocides is their high level of reactivity to reduce compounds such as iron and H_2S and organic matter during the penetration through biofilm (little residual is left to control microorganisms). So they do not effectively penetrate and disperse microbial slimes. Moreover, some oxidizing biocides (e.g., chlorine) are not effective at solving clogging on the membrane surface. At low chlorine levels, biofilm bacteria can produce extracellular material faster than chlorine can diffuse through biofilm, so they are shielded in slime and make the slime problem worse (20). Therefore, chlorine is usually supplemented by the addition of nonoxidizing biocides in many membrane systems.

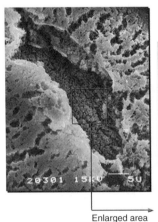

Figure 4. Scanning electron micrographs of mature biofilms formed on cellulose acetate (CA) reverse osmosis (RO) membranes. The RO membranes were fed with a pretreated municipal effluent at Water Factory 21 in Orange County, California. (Courtesy of Ref. 17.)

BIBLIOGRAPHY

1. ZoBell, C.E. and Anderson, D.Q. (1943). The effect of solid surfaces upon bacterial activity. *J. Bacteriol* **46**: 39–56.
2. Pedros-Alio, C. and Brock, T.D. (1983). The importance of attachment to particles for planktonic bacteria. *Arch. Hydrobiol.* **98**: 354–379.
3. Morita, R.Y. (1982). Starvation-survival of heterotrophs in the marine environment. *Adv. Microbiol. Ecol.* **6**: 171–198.
4. Dawson, M.P., Humphrey, B.A., and Marshall, K.C. (1981). Adhesion: a tactic in the survival strategy of a marine *Vibrio* during starvation. *Curr. Microbiol.* **6**: 195–1999.
5. Costerton, J.W. et al. (1987). Bacterial biofilms in nature and disease. *Annu. Rev. Microbiol.* **41**: 435.
6. Fletcher, M. (1996). *Bacterial-Adhesion-Molecular and Ecological Diversity*. Wiley-Liss, New York, p. 71.
7. Lawrence, J.R. and Caldwell, D.E. (1987). Behavior of bacterial stream populations within the hydrodynamic boundary layers of surface microenvironments. *Microbiol. Ecol.* **14**: 15–27.
8. Power, K. and Marshall, K.C. (1988). Cellular growth and reproduction of marine bacteria on surface-bound substrate. *Biofouling* **1**: 163–174.
9. Humphrey, B.A., Dickson, M.R., and Marshall, K.C. (1979). Physicochemical and *in situ* observations on the adhesion of gliding bacteria to surface. *Arch. Microbiol.* **120**: 231–238.
10. Nielson, T.T. et al. (2000). Development and dynamics of *Pseudomonas* sp. biofilm. *J. Bacteriol. Microbiol.* **192**(22): 6482–6489.
11. Hogan, D.A. and Kolter, R. (2002). *Pseudomonas–Candida* interactions: an ecological role for virulence factors. *Science* **296**: 2229–2232.
12. Arendorf, T.M. and Walker, D.M. (1980). *Candida albicans*: its association with dentures, plaque and the oral mucosa. *J. Dent. Assoc. South Africa.* **35**: 563–569.
13. Branting, C., Sund, M.L., and Linder, L.E. (1989). The influence of *Streptococcus mutans* on adhesion of *Candida albicans* to acrylic surfaces *in vitro*. *Arch Oral Biol.* **34**: 347–353.
14. Adams, B., Baillie, G.S., and Douglas, L.J. (2002). Mixed species biofilms of *Candida albicans* and *Staphylococcus epidermidis*. *J. Med. Microbil.* **51**: 344–349.
15. Christensen, B.B., Haagensen, J.A.J., Heydorn, A., and Molin, S. (2002). Metabolic commensalism and competition in a two-species microbial consortium. *Appl. Environ. Microbiol.* **68**(5): 2495–2502.
16. Rogers, J. et al. (1994). Influence of temperature and plumbing material selection on biofilm formation and growth of *Legionella pneumophila* in a model potable water system containing complex microbial flora. *Appl. Environ. Microbiol.* **60**(5): 1585–1592.
17. http://www.desertwildlands.com/AquaMem/AquaMem/biofouling/biofilm_links.html.
18. Vrouwenvelder, J.S. et al. (2003). Tools for fouling diagnosis of NF and RO membranes and assessment of the fouling potential of feed water. *Desalination* **157**(1–3): 361–365.
19. Walker, J.T. et al. (1995). Control of *Legionella pneumophila* in a hospital water system by chlorine dioxide. *J. Ind. Microbiol.* **15**: 384–390.
20. Saeed, M.O. (2002). Effect of dechlorination point location and residual chlorine on biofouling in a seawater reverse osmosis plant. *Desalination* **143**(3): 229–235.

MICROBIOLOGICAL QUALITY CONTROL IN DISTRIBUTION SYSTEMS

MARKKU J. LEHTOLA
ILKKA T. MIETTINEN
National Public Health Institute
Kuopio, Finland

PERTTI J. MARTIKAINEN
University of Kuopio
Kuopio, Finland

MICROBES IN DRINKING WATER

Good drinking water should not be harmful for health or materials of the distribution system. Undesired health effects can be caused by chemicals or biological agents, including viruses, bacteria, fungi, and protozoa (1–7).

Most of bacteria found in drinking water cannot cause disease in the general population (8). Usually waterborne epidemics are caused by accidental contamination of drinking water, for example, by flooding, surface runoff, or leakage of wastewater pipelines (6). Pathogenic fecal microbes like enteric viruses, protozoan parasites, *Campylobacter*, enterohemorragic *Escherichia coli* (EHEC), *Yersinia enterocolitica*, *Microsporidia*, *Helicobacter pylori*, *Salmonella*, and *Shigella* from contaminated raw water sources may also enter the distribution system as a result of inadequate water treatment (2–6,9,10). These microbes probably do not multiply in the drinking water environment, but they may survive there, especially in biofilms (9,11).

The growth of native nonpathogenic microbes in the distribution system can have several undesired effects on the water quality: the microbial growth can complicate bacteriological water quality monitoring, iron bacteria can precipitate iron-producing iron flocs, the growth of actinomycetes and fungi gives an unpleasant taste and odor to the water, and microbes can destroy the piping materials resulting from their biocorrosion (9,12,13). There are some pathogens or opportunistic pathogens like *Legionella*, *Pseudomonas aeruginosa*, *Mycobacteria*, and *Aeromonas*, which can grow in drinking water distribution systems (9,10,14,15). Bacteria and fungi also promote the occurrence of protozoa and higher animals by serving as food for them (13,16).

Microbial cells are attached to almost any surface in aquatic environments (17). Bacteria in biofilms represent the most important part of the bacterial biomass in drinking water distribution networks (18,19), and detachment of bacteria from biofilms accounts for most planktonic cells present in water (20). Biofilms promote the survival and growth of pathogenic bacteria and survival of viruses and parasites (4,9,11,21–23), and they increase the disinfection resistance of bacteria (9,21,24,25).

MICROBIAL QUALITY CONTROL

The growth of microbes in a drinking water distribution system is affected by factors like disinfection, availability of nutrients, temperature, and water residence time.

Waterworks normally disinfect drinking waters by chlorination, ozonation, UV-radiation, or some combination of them. Nutrients, temperature, and residence time are important in microbial growth control when water is distributed without disinfection residuals (ozonation, UV-disinfection) or when disinfection residual is low (chlorine in peripheral parts of distribution system). Also, the pipe material and water hydraulics affect microbial growth in biofilms.

Disinfection

Chlorine is added to water as free chlorine, chlorine dioxide, or chloramines. Free chlorine destroys bacteria by reactions with bacterial enzymes (26), whereas chloramine reacts with nucleic acids, tryptophan, and sulfur-containing amino acids (27). The following factors affect the disinfection efficiency: (1) nature of disinfectant, (2) concentration of disinfectant, (3) length of the contact time, (4) temperature, (5) type and concentration of organisms, and (6) pH (26). One major disadvantage of using chlorine as a disinfectant is the formation of harmful by-products in reactions with organic matter (28,29). Chlorine also degrades organic matter and increases the content of easily available organic carbon for bacteria (AOC) (13,30,31).

There are many advantages to using chloramines instead of free chlorine. Chloramines have a good residual effect and the ability to destroy microbes in biofilms (27). Also, the formation of unwanted chlorinated organic by-products is low with chloramines (32,33). However, the addition of chloramines may enhance growth of ammonia oxidizing bacteria and thus accumulation of nitrite in drinking water (34,35).

Ozonation is a commonly used technique for removing pathogenic microbes, taste, and odor from water (36,37). Ozone is an unstable gas, which has to be generated onsite in the waterworks. The disinfection mechanism is based on the reaction with the double bonds in fatty acids of bacterial cell walls and cell membranes and the protein capsid of viruses (36). In water, ozone is decomposed rapidly back to oxygen and has therefore no residual effects in the distribution system. As a strong oxidant, ozone effectively degrades natural organic matter (30,38). Reactions with natural organic matter increase the content of easily available organic carbon (AOC) and microbially available phosphorus (MAP) in water and thus microbial growth in the distribution network (30,39–41). Ozonation can also produce ozonation by-products; of particular concern is carcinogenic bromate, which is produced when water containing bromide is ozonated (42,43).

UV-radiation ($\lambda = 253.7$ nm) is effective in destroying bacteria and viruses in water (44–47). The destructive effect of UV-radiation is based on DNA and RNA damage such as thymine dimer formation, hydrate formation in the DNA, denaturation of the DNA double strand, and polymerization between nucleic acids and proteins. The absorption maximum of DNA and RNA 255–265 nm is near the wavelengths emitted from mercury low-pressure lamps (253.7 nm). Most microorganisms are inactivated by relatively low UV_{254} dosages—usually in the range of 2–6 mWs/cm^2, although viruses tend to be more resistant to UV_{254} radiation than bacteria (45). There is no evidence on the formation of undesirable by-products or increase in the contents of microbial nutrients like biodegradable organic carbon or microbially available phosphorus by UV-disinfection (45,48,49). Disadvantage of UV-disinfection is similar to ozonation: the lack of residual activity in the distribution system (45).

Biofilms. Some countries use phosphate-based anticorrosion (50–52) chemicals in their distribution systems. The efficiency of phosphates to control biofilm formation is based on neutralization of positively charged corrosion products, such as goethite (α-FeOOH), to negatively charged $FePO_4$, which lowers the adhesion of bacteria to pipe surfaces (50,52). Some studies have shown that these chemicals have not increased microbial growth (50,51,53,54). However, there are geographical regions where use of phosphate against corrosion should be considered carefully because there availability of phosphorus is a limiting factor for microbial growth in drinking waters (see later).

It is difficult to prevent formation of biofilms by chlorine disinfection because residual concentration >1–2 mg/L is needed (20,21,55). Inactivation of fixed bacteria in biofilms needs even a higher concentration of chlorine (>3 mg/L) (25). The inactivation efficiency depends on the composition of the pipe material, the disinfection agent (chloramine vs. chlorine) (27,55), temperature, and water velocity (17).

Microbial Nutrients

Organic Carbon. Except photoautotrophs and chemoautotrophs, organisms need organic compounds for their carbon and energy sources. Organic compounds are partially assimilated into the cell material and partially oxidized to provide energy (56). Bacteria can use a wide range of substrates, but some substrates like amino acids, carboxylic acids, and carbohydrates are more readily usable than others (57).

Aromatic humus substances in drinking water are difficult to use as microbial substrates; usually only a small part of total organic carbon is easily used by microbes (39). There are two commonly used approaches to analyze the microbial usability of aquatic natural organic matter: determination of assimilable organic carbon (AOC) or biodegradable organic carbon (BDOC). AOC is that part of organic matter that can be converted to cell mass. It is expressed as a carbon concentration by means of a conversion factor (39). BDOC is the part of organic carbon that can be mineralized by heterotrophic microbes (58,59).

The first method for AOC assay was published in 1982 by Van der Kooij et al. (39). The test is based on the growth of certain bacterial strains in a water sample, and AOC is calculated using empirical yield values for these bacteria. The bacterial strains used in the test are *Pseudomonas fluorescens* strain P-17 and *Spirillum* strain NOX (60). In regions with high content of organic matter and low amount of phosphorus in water, a modification of the test is required. Miettinen et al. (61) modified the test by adding inorganic nutrients to the water sample to ensure that only organic carbon limits the bacterial growth in the water and suggested the term $AOC_{potential}$. They used the term AOC_{native} for the assimilable carbon determined without addition of inorganic nutrients.

In the BDOC method, organic carbon is mineralized by the natural microbial community in water, and BDOC is measured as a difference in content of dissolved organic carbon (DOC) before and after (>10 days) incubation of the inoculated water sample (58). There also is a biofilm reactor application for the method. In this application, a water sample is filtered through a column where microbes are attached in biofilm (e.g., glass/sand matrix), and the difference in DOC between inlet and outlet is analyzed (62,63).

Phosphorus. Phosphorus, as a macronutrient, is an essential nutrient for microbes. Bacteria need phosphorus for biosynthesis of nucleic acids, lipopolysaccharides, and phospholipids (56). Phosphate is a vital component of the intracellular energy-transferring ATP system (56). The optimum C:N:P ratio (carbon:nitrogen:phosphorus) for bacterial growth is 100:10:1.

Phosphorus occurs in nature only in the form of chemical compounds, either as inorganic orthophosphate (HPO_4^{2-}, $H_2PO_4^-$) or in organic compounds. In a humus-rich environment, phosphorus is associated with high-molecular-weight humic materials, especially in the presence of iron or manganese (64,65).

Soluble inorganic phosphate is considered to be entirely biologically available (66). All living organisms possess the enzyme, alkaline phosphatase, to convert organic phosphorus to inorganic phosphorus, but only microbes and fungi can excrete the enzyme outside of their cells (exoenzymes) to mineralize and dissolve organic phosphates (66).

Chemical phosphorus analysis has two steps: (1) conversion of phosphorus compounds to dissolved orthophosphate, and (2) colorimetric determination of the dissolved orthophosphate. With these colorimetric standard methods, the detection limit for phosphorus in water is 2–10 µg P/L (60).

Another approach for phosphorus analysis is to measure microbially available phosphorus (MAP) with a bioassay (67). There inorganic nutrients (except phosphorus) and organic carbon are added to the water samples to ensure that inorganic nutrients or organic carbon do not restrict microbial growth. Samples are inoculated with *Pseudomonas fluorescens* bacteria, and the maximum growth of *Pseudomonas fluorescens* is transformed to the amount of microbially available phosphorus with a conversion factor taken from the standardization of the method. The MAP bioassay is very sensitive, and the detection limit is 0.08 µg P/L. MAP represents the part of total phosphorus that is easily available for bacteria. It is important that MAP does not correlate to the content of total phosphorus in water (67).

Nutrient Limitations of Bacterial Growth in Drinking Water. Microbial growth in drinking water is generally limited by AOC or BDOC (39,68–71). In many cases, organic carbon limits microbial growth also in biofilms (18,53–55,72).

The importance of AOC over phosphorus is based on the high ratio of carbon to phosphorus. For example, in an American study, the C:P ratio in drinking water was in the range 100:250 to 100:43, whereas the typical ratio for optimal microbial activity is 100:1 (73). LeChevallier et al. (74) proposed that to limit the growth of coliform bacteria in drinking water, AOC levels should be below 50 µg/L. According to Van der Kooij (13), the AOC concentration should be less than 10 µg/L to limit the microbial aftergrowth. For BDOC, the guideline value of 0.15 mg BDOC/L for biologically stable water has been proposed (69).

In some regions, the correlation between AOC and heterotrophic growth response is weak (73,75) or there is no correlation at all (76,77). Kerneis et al. (78) found no correlation between BDOC and growth of heterotrophic microbes in a drinking water distribution network. It was first found in Finland that microbial growth in drinking water is limited by the availability of phosphorus (79). Subsequently, phosphorus limitation has been reported for drinking waters in Japan (80,81).

In regions where microbial growth in drinking water is limited by phosphorus, a major increase in microbial growth has been achieved with additions of only 1–5 µg/L of PO_4-P (80,82). Sathasivan and Ohgaki (81) reported that phosphorus could become a limiting nutrient at concentrations of 1–3 µg/L. In phosphorus-limited drinking water, the formation of biofilms was also affected by the availability of phosphorus (83).

The reasons for the phosphorus limitation in treated drinking waters are high organic carbon concentration in raw waters and more effective phosphorus removal than that of AOC (31). In fact, in many cases, the AOC content even increases during the drinking water purification process, especially if the waterworks uses ozonation.

BIBLIOGRAPHY

1. Gerba, C.P., Wallis, C., and Melnick, J.L. (1975). Viruses in water: The problem, some solutions. *Environ. Sci. Technol.* **9**(13): 1122–1126.

2. Lippy, E.C. and Waltrip, S.C. (1984). Waterborne disease outbreaks—1946-1980: a thirty-five-year perspective. *J. AWWA* **77**(2): 60–67.

3. Lahti, K. and Hiisvirta, L. (1995). Causes of waterborne outbreaks in community water systems in Finland: 1980–1992. *Water Sci. Technol.* **31**(5–6): 33–36.

4. Ford, T.E. (1999). Microbiological safety of drinking water: United States and global perspectives. *Environ. Health Perspect.* **107**(Supplment I): 191–206.

5. Kukkula, M., Maunula, L., Silvennoinen, E., and von Bonsdorf, C-H. (1999). Outbreak of viral gastroenteritis due to drinking water contaminated by norwalk-like viruses. *J. Infect. Dis.* **180**: 1771–1776.

6. Miettinen, I.T., Zacheus, O., von Bonsdorff, C-H., and Vartiainen, T. (2001). Waterborne epidemics in Finland in 1998–1999. *Water Sci. Technol.* **43**(12): 67–71.

7. Kramer, M.H., Quade, G., Hartemann, P., and Exner, M. (2001). Waterborne diseases in Europe-1986-96. *J. AWWA* **1**: 48–53.

8. Hunter, P. (2002). Epidemiological evidence of disease linked to HPC bacteria. In: *Heterotrophic Plate Counts and Drinking-water Safety*, J. Bartram et al. (Eds.), WHO/IWA Publishing, London, UK, pp. 119–136.
9. Percival, S.L. and Walker, J.T. (1999). Potable water and biofilms: A review of the public health implications. *Biofouling* **42**(2): 99–115.
10. Szewzyk, U., Szewzyk, R., Manz, W., and Schleifer, K.-H. (2000). Microbiological safety of drinking water. *Annu. Rev. Microbiol.* **54**: 81–127.
11. Storey, M.V. and Ashbolt, N.J. (2001). Persistence of two model enteric viruses (B40-8 and MS-2 bacteriophages) in water distribution pipe biofilms. *Water Sci. Technol.* **43**(12): 133–138.
12. LeChevallier, M.W. and McFeters, G.A. (1985). Interactions between heterotrophic plate count bacteria and coliform organisms. *Appl. Environ. Microbiol.* **49**(5): 1338–1341.
13. Van der Kooij, D. (1990). Assimilable organic carbon (AOC) in drinking water. In: *Drinking Water Microbiology, Progress and Recent Developments*, G.A. McFeters (Ed.). Springer-Verlag, Michigan, pp. 57–87.
14. Rusin, P.A., Rose, J.B., Haas, C.N., and Gerba, C.P. (1997). Risk assessment of opportunistic bacterial pathogens in drinking water. *Rev. Environ. Contam. Toxicol.* **152**: 37–38.
15. Norton, C.D., LeChevallier, M.W., and Falkinham, J.O. III (2004). Survival of *Mycobacterium avium* in a model distribution system. *Water Res.* **38**(6): 1457–1466.
16. Sibille, I., Sime-Ngando, T., Mathieyu, L., and Block, J-C. (1998). Protozoan bacteriovory and *Escherichia coli* survival in drinking water distribution systems. *Appl. Environ. Microbiol.* **64**(1): 197–202.
17. Characklis, W.G. and Marshall, K. (1990). Biofilms: A basis for an interdisciplinary approach. In: *Biofilms*, W.G. Characklis and K.C. Marshall (Eds.). John Wiley & Sons, Hoboken, NJ.
18. Laurent, P., Servais, P., and Randon, G. (1993). Bacterial development in distribution networks. study and modelling. *Water Supply* **11**(3/4): 387–398.
19. Zacheus, O.M., Lehtola, M.J., Korhonen, L.K., and Martikainen, P.J. (2001). Soft deposits the key site for microbial growth in drinking water distribution networks. *Water Res.* **35**(7): 1757–1765.
20. Van der Wende, E., Characklis, W.G., and Smith, D.B. (1989). Biofilms and bacterial drinking water quality. *Water Res.* **23**(10): 1313–1322.
21. LeChevallier, M.W., Babcock, T.M., and Lee, R.G. (1987). Examination and characterization of distribution systems biofilms. *Appl. Environ. Microbiol.* **53**(12): 2714–2724.
22. Keevil, C.W. (1989). Chemostat models of human and aquatic corrosive biofilms. In: *Proc. Recent Advances in Microbial Ecology, ISME 5*, T. Hattori et al. (Eds.). Japan Scientific Societies Press, Tokyo.
23. Buswell, C.M. et al. (1998). Extended survival and persistence of *Campylobacter* spp. in water and aquatic biofilms and their detection by immunofluorescent-antibody and—rRNA staining. *Appl. Environ. Microbiol.* **64**(2): 733–741.
24. LeChevallier, M.W., Lowry, C.D., and Lee, R.G. (1990). Disinfecting biofilms in a model distribution system. *J. AWWA* **82**(7): 87–99.
25. Gilbert, P., Allison, D.G., and McBain, A.J. (2002). Biofilms *in vitro* and *in vivo*: do singular mechanisms imply cross-resistance? *J. Appl. Microbiol. Symp. Suppl.* **92**: 98S–110S.
26. White, G.C. (1986). *The Handbook of Chlorination*, 2nd Edn., Van Nostrand Reinhold Company, New York.
27. LeChevallier, M.W., Cawthon, C.D., and Lee, R.G. (1988). Inactivation of biofilm bacteria. *Appl. Environ. Microbiol.* **54**: 2492–2499.
28. Rook, J.J. (1974). Formation of haloforms during chlorination of natural waters. *Water Treatm. Exam.* **23**: 234–243.
29. Boorman, G.A. et al. (1999). Drinking water disinfection byproducts: Review and approach of toxicity evaluation. *Environ. Health Perspect.* **107**(Supplement I): 207–217.
30. Miettinen, I.T., Vartiainen, T., and Martikainen, P.J. (1998). Microbial growth in drinking waters treated with ozone, ozone/hydrogen peroxide or chlorine. *Ozone Sci. Eng.* **20**: 303–315.
31. Lehtola, M.J., Miettinen, I.T., Vartiainen, T., and Martikainen, P.J. (2002). Changes in content of microbially available phosphorus, assimilable organic carbon and microbial growth potential during drinking water treatment processes. *Water Res.* **36**(15): 3681–3690.
32. Trussel, R.R. (1998). An overview of disinfectant residuals in drinking water distribution systems. In: *Preprint of Specialized Conference on Drinking Water Distribution With of Without Disinfectant Residual*, M. Gerlach and R. Gimbel (Eds.). IWW Rheinisch-Westfälisches Institut fur Wassenforschung, Mulheim an der Ruhr, Germany.
33. Nissinen, T.K., Miettinen, I.T., Martikainen, P.J., and Vartiainen, T. (2002). Disinfection by-products in Finnish drinking waters. *Chemosphere* **48**: 9–20.
34. Odell, L.H. et al. (1996). Controlling nitrification in chloraminated systems. *J. AWWA* **88**(7): 86–98.
35. Wilczak, A. et al. (1996). Occurrence of nitrification in chloraminated distribution systems. *J. AWWA* **88**(7): 74–85.
36. Anselme, C., Suffet, I.H., and Mallevialle, J. (1988). Effects of ozonation on tastes and odor. *J. AWWA* **October**: 45–51.
37. Langlais, B., Reckhow, D.A., and Brink, D.R. (Eds.). (1991). *Ozone in Water Treatment, Application and Engineering*. CRC Press/Lewis Publishers, Boca Raton, FL.
38. Glaze, W.H. (1987). Drinking water treatment with ozone. *Environ. Sci. Technol.* **21**(3): 224–230.
39. Van der Kooij, D., Visser, A., and Hijnen, W.A.M. (1982). Determination the concentration of easily assimilable organic carbon in drinking water. *J. AWWA* **74**: 540–545.
40. Van der Kooij, D. and Hijnen, W.A.M. (1984). Substrate utilization by an oxalate-consuming *Spirillum* species in relation to its growth in ozonated water. *Appl. Environ. Microbiol.* **47**: 551–559.
41. Lehtola, M.J., Miettinen, I.T., Vartiainen, T., Myllykangas, T., and Martikainen, P.J. (2001). Microbially available organic carbon, phosphorus and microbial growth in ozonated drinking water. *Water Res.* **35**(7): 1635–1640.
42. Fielding, M. and Hutchison, J. (1995). Formation of bromate and other ozonation by-products in water treatment. *Water Supply* **13**(1): 71–84.
43. Myllykangas, T., Nissinen, T., and Vartiainen, T. (2000). Bromate formation during ozonation of bromide containing drinking water—a pilot scale study. *Ozone Sci. Eng.* **22**: 487–499.
44. Anghern, M. (1984). Ultraviolet disinfection of water. *Aqua* **2**: 109–115.
45. Wolfe, R.L. (1990). Ultraviolet disinfection of potable water. *Environ. Sci. Technol.* **24**(6): 768–773.
46. Oppenheimer, J.A., Jacangelo, J.G., Laine, J-M., and Hoagland, J.E. (1997). Testing the equivalency of ultraviolet

light and chlorine for disinfection of wastewater to reclamation standards. *Water Environ. Res.* **69**(1): 14–24.

47. Parrotta, M.J. and Bekdash, F. (1998). UV-disinfection of small groundwater supplies. *J. AWWA* **90**(2): 71–81.

48. Shaw, J.P., Malley, J.P. Jr., and Willoughby, S.A. (2000). Effects of UV irradiation on organic matter. *J. AWWA* **92**(4): 157–167.

49. Lehtola, M.J. et al. (2003). Impact of UV-disinfection on microbially available phosphorus, organic carbon, and microbial growth in drinking water. *Water Res.* **37**(5): 1064–1070.

50. Abernathy, C.G. and Camper, A.K. (1998). The effect of phosphorus based corrosion inhibitors and low disinfection residuals on distribution biofilms. *Proc. Water Quality Technology Conference.* AWWA, CD-Rom. San Diego, CA.

51. Rompre, A., Prevost, M., Coallier, J., Brisebois, P., and Lavole, J. (2000). Impacts of implementing a corrosion control strategy on biofilm growth. *Water Sci. Technol.* **41**(4–5): 287–294.

52. Appenzeller, B.M., Duval, Y.B., Thomas, F., and Block, J-C. (2002). Influence of phosphate on bacterial adhesion onto iron oxyhydroxide in drinking water. *Environ. Sci. Technol.* **36**(4): 646–652.

53. Appenzeller, B.M.R. et al. (2001). Effect of adding phosphate to drinking water on bacterial growth in slightly and highly corroded pipes. *Water Res.* **35**(4): 1100–1105.

54. Chandy, J.P. and Angles, M.L. (2001). Determination of nutrients limiting biofilm formation and the subsequent impact on disinfectant decay. *Water Res.* **35**(11): 2677–2682.

55. Block, J-C. (1992). Biofilms in drinking water distribution systems. In: *Biofilms, Science and Technology*, L.F. Melo et al. (Eds.), Kluwer Academic Publishers, The Netherlands.

56. Schlegel, H.G. (1997). *General Microbiology*, 7th Edn., Cambridge University Press, Cambridge, UK.

57. Van der Kooij, D., Visser, A., and Oranje, J.P. (1982). Multiplication of *fluorescent pseudomonads* at low substrate concentrations in tap water. *Antonie van Leeuwenhoek* **48**: 229–243.

58. Servais, P., Billen, G., and Hascoet, M-C. (1987). Determination of the biodegradable fraction of dissolved organic matter in waters. *Water Res.* **21**(4): 445–450.

59. Huck, P.M. (1990). Measurement of biodegradable organic matter and bacterial growth potential in drinking water. *J. AWWA* **82**(7): 78–86.

60. Clescerl, L.S., Greenberg, A.E., and Eaton, A.D. (Eds.). (1999). *Standard Methods for Examination of Water and Wastewater*, 20th Edn. American Public Health Association, Washington, DC.

61. Miettinen, I.T., Vartiainen, T., and Martikainen, P.J. (1999). Determination of assimilable organic carbon in humus-rich drinking waters. *Water Res.* **33**(10): 2277–2282.

62. Lucena, F., Frias, J., and Ribas, F. (1990). A New dynamic approach to the determination of biodegradable dissolved organic carbon in water. *Environ. Technol.* **12**: 343–347.

63. Frias, J., Ribas, F., and Lucena, F. (1992). A method for the measurement of biodegradable organic carbon in waters. *Water Res.* **26**(2): 255–258.

64. Jones, R.I., Salonen, K., and De Haan, H. (1988). Phosphorus transformations in the epilimnion of humic lakes: abiotic interactions between dissolved humic materials and phosphate. *Freshwater Biol.* **19**: 357–369.

65. Shaw, P.J., Jones, R.I., and De Haan, H. (2000). The influence of humic substances on the molecular weight distributions of phosphate in epilimnic lake waters. *Freshwater Biol.* **45**(4): 383–393.

66. Jones, R.D. (1997). Phosphorus cycling. In: *Manual of Environmental Microbiology*, J.C. Hurst, G.R. Knudsen, M.J. McInerney, L.D. Stetzenbach, and M.V. Walter (Eds.). ASM Press, Washington, DC, pp. 343–348.

67. Lehtola, M.J., Miettinen, I.T., Vartiainen, T., and Martikainen, P.J. (1999). A new sensitive bioassay for determination of microbially available phosphorus in water. *Appl. Environ. Microbiol.* **65**: 2032–2034.

68. Joret, J.C., Levi, Y., and Volk, C. (1991). Biodegradable dissolved organic carbon (BDOC) content of drinking water and potential regrowth of bacteria. *Water Sci. Technol.* **24**(2): 95–101.

69. Servais, P., Laurent, P., and Randon, G. (1995). Comparison of the bacterial dynamics in various French distribution systems. *J Water SRT-Aqua* **44**(1): 10–17.

70. Prévost, M. et al. (1998). Suspended bacterial biomass and activity in full-scale drinking water distribution systems: impact of water treatment. *Water Res.* **32**(5): 1393–1406.

71. Niquette, P., Servais, P., and Savoir, R. (2001). Bacterial dynamics in the drinking water distribution system of Brussels. *Water Res.* **35**(3): 675–682.

72. Van der Kooij, D., Veenendaal, H.R., Baars-Lorist, C., Van der Klift, D.W., and Drost, Y.C. (1995). Biofilm formation on surfaces of glass and teflon exposed to treated water. *Water Res.* **29**(7): 1655–1662.

73. Zhang, W. and DiGiano, F.A. (2002). Comparison of bacterial regrowth in distribution systems using free chlorine and chloramine: a statistical study of causative factors. *Water Res.* **36**: 1469–1482.

74. LeChevallier, M.W., Schultz, W., and Lee, R.G. (1991). Bacterial nutrients in drinking water. *Appl. Environ. Microbiol.* **57**(3): 857–862.

75. Noble, P.A., Clark, D.L., and Olson, B.H. (1996). Biological stability of ground water. *J. AWWA* **88**(5): 87–96.

76. Gibbs, R.A., Scutt, J.E., and Croll, B.T. (1993). Assimilable organic carbon concentrations and bacterial numbers in a water distribution system. *Water Sci. Technol.* **27**(3–4): 159–166.

77. Miettinen, I.T., Vartiainen, T., and Martikainen, P.J. (1997). Microbial growth and assimilable organic carbon in Finnish drinking water. *Water Sci. Technol.* **35**(11–12): 301–306.

78. Kerneis, A., Nakache, F., Deguin, A., and Feinberg, M. (1995). The effects of water residence time on the biological quality in a distribution network. *Water Res.* **29**(7): 1719–1727.

79. Miettinen, I.T., Vartiainen, T., and Martikainen, P.J. (1996). Contamination of drinking water. *Nature* **381**: 654–655.

80. Sathasivan, A., Ohgaki, S., Yamamoto, K., and Kamiko, N. (1997). Role of inorganic phosphorus in controlling regrowth in water distribution system. *Water Sci. Technol.* **35**(8): 37–44.

81. Sathasivan, A. and Ohgaki, S. (1999). Application of new bacterial regrowth potential method for water distribution system—a clear evidence of phosphorus limitation. *Water Res.* **33**(1): 137–144.

82. Miettinen, I.T., Vartiainen, T., and Martikainen, P.J. (1997). Phosphorus and bacterial growth in drinking water. *Appl. Environ. Microbiol.* **63**: 3242–3245.

83. Lehtola, M.J., Miettinen, I.T., and Martikainen, P.J. (2002). Biofilm formation in drinking water affected by low concentrations of phosphorus. *Can. J. Microbiol.* **48**(6): 494–499.

WATER QUALITY MODELS FOR DEVELOPING SOIL MANAGEMENT PRACTICES

Harry X. Zhang
Parsons Corporation
Fairfax, Virginia

INTRODUCTION

Water quality models are used extensively in water quality planning and pollution control. Models are applied to answer a variety of questions, support watershed planning and analysis, and develop total maximum daily loads (TMDLs). Watershed loading models include techniques designed primarily to predict pollutant movement from the land surface to waterbodies. TMDLs result in allocating the quantity of pollution that can be discharged from point sources (wasteload allocation) and nonpoint sources (load allocation) to ensure that water quality standards are achieved within a specified margin of safety (1). Management measures can be applied to achieve all of the pollution control needed by pollutant sources to achieve the load allocation. Best management practice (BMP) is generally defined as a practice or combination of practices that are the most effective and practicable (including technological, economic, and institutional considerations) means of controlling point and nonpoint pollutants at levels compatible with economic and environmental quality goals (2).

It is important to estimate the collective impacts of all management activities in a watershed to evaluate whether water quality goals will be achieved. In watersheds that have easily characterized problems, it may be easy to project that water quality benefits will be achieved by implementing management measures such as erosion and sediment control and runoff minimization. However, in a watershed that has multiple land uses where agriculture is considered to contribute groups of pollutants, it is more complicated to estimate the combined impacts of a variety of management measures and practices on a fairly large number of diverse farming operations. In this type of situation, computer modeling may be needed. A variety of models exist to help assess the relative benefits of implementing practices at the field and watershed level. Managers can use models to develop goals and objectives for watershed and waterbody use and protection, to develop and to evaluate the effectiveness of various management and mitigation alternatives to improve existing water quality, and to determine the response time of a waterbody after implementing a management alternative. In particular, models can help understanding key "cause-and-effect" processes within the natural environment, developing waste load allocations and TMDL analysis, and evaluating the effectiveness of various contaminant reduction scenarios.

BRIEF OVERVIEW OF NONPOINT SOURCE WATER QUALITY MODELS

Watershed loading models range from simple loading rate assessments in which loads are a function of land use type only, to complex simulation techniques that more explicitly describe the processes of rainfall, runoff, sediment detachment, and transport to receiving waters. Some loading models operate on a watershed scale, integrating all loads within a watershed, and some allow subdividing the watershed into contributing subbasins. Given the complex nature of the processes and interactions that determine the fate and transport of a pollutant or chemical in nature, mathematical models can attempt to represent only the principal components of the environment that influence a given water quality variable. It is virtually impossible to incorporate fully all of the mechanisms and processes that occur in nature. Therefore, there are varying degrees of simplicity or complexity embodied in watershed and water quality modeling.

Many computer models are available to assess soil management practices. Based on their complexity, operation, time step, and simulation technique, watershed-loading models can be grouped into three categories: simple methods, midrange models, and detailed models.

Simple Models

The major advantage of simple methods is that they can provide a rapid means of identifying critical areas with minimal effort and data requirements. Simple methods are typically derived from empirical relationships between physiographic characteristics of the watershed and pollutant export. Simple methods are often used when data limitations and budget and time constraints preclude using complex models. They are used to diagnose nonpoint source pollution problems, when information is relatively limited. They can be used to support an assessment of the relative significance of different sources, guide decisions for management plans, and focus continuing monitoring efforts.

Typically, simple methods rely on a large-scale aggregation of characteristics and neglect features of small patches of land. They rely on generalized sources of information and therefore have low to medium requirements for site-specific data. Default values provided for these methods are derived from empirical relationships that are based on regional or site-specific data. The estimates are usually expressed as mean annual values. Simple methods provide only rough estimates of sediment and pollutant loadings and have limited predictive capability.

In addition, field-scale models, which have traditionally specialized in agricultural systems, are loading models that are designed to operate on a smaller, more localized scale. Field-scale models have often been employed to aid in selecting management measures and practices.

Midrange Models

The advantage of midrange watershed loading models is that they evaluate pollution sources and impacts over broad geographic scales and, therefore, can assist in defining target areas for pollution mitigation programs in a watershed. The midrange model compromises between the empiricism of simple methods and the complexity of detailed mechanistic models (e.g., relating pollutant loading to hydrologic and erosion processes). Mechanistic models usually include detailed input–output features.

Detailed Models

Detailed models best represent the current understanding of watershed processes that affect pollution generation. Detailed models are best able to identify causes of problems rather than describe overall conditions. If properly applied and calibrated, detailed models can provide relatively accurate predictions of variable flows and water quality at any point in a watershed. The additional precision they provide, however, comes at the expense of considerable time and resources for data collection and model application.

Model Comparison

Detailed models incorporate the manner in which watershed processes change over time continuously rather than relying on simplified terms for rates of change. Algorithms in detailed models more closely simulate the physical processes of infiltration, runoff, pollutant accumulation, in-stream effects, and groundwater–surface water interaction. The input and output of detailed models also have greater spatial and temporal resolution. Moreover, the manner in which physical characteristics and processes differ over space is incorporated within the governing equations. However, input data file preparation and calibration of detailed models require professional training and adequate resources. Their added accuracy might not always justify the amount of effort and resources required.

In comparison, simple methods use large simulation time steps to provide longer term averages or annual estimates. Their accuracy may decrease in estimating seasonal or storm loading because they cannot capture the large fluctuations of pollutant loading or concentration usually observed when smaller time steps are employed. Neither the simple nor the midrange models consider degradation and transformation processes, and few incorporate detailed representations of pollutant transport within and from the watershed. Although model applications might be limited to relative comparisons, they can often provide useful information to water quality managers for watershed-level planning decisions. If adequate site-specific data are not available, values can be estimated and assumptions can be made, but these estimates will increase the uncertainty in the model predictions. It will be very difficult to validate whether simple or complex models provide better estimates. Therefore, if data availability does not reach the level that a detailed model requires, it is recommended that a simpler model be employed.

REVIEW OF WATER QUALITY MODELS FOR DEVELOPING SOIL MANAGEMENT PRACTICES

Hydrologic Simulation Program–FORTRAN (HSPF) and WinHSPF (in BASINS 3.0)

Developed and maintained by the EPA (http://www.epa.gov/ost/basins/) and the USGS (http://water.usgs.gov/software/hspf.html), the Hydrological Simulation Program–FORTRAN (HSPF) is one of the most comprehensive modeling systems for simulating watershed hydrology, point and nonpoint loading, and receiving water quality for both conventional pollutants and toxicants.

The HSPF model was developed in the early 1960s as the Stanford Watershed Model. In the 1970s, water quality processes were added. The development of a Fortran version incorporating several related models using software engineering design and development concepts was funded by the EPA Research Laboratory (Athens, Ga.) in the late 1970s. In the 1980s, the USGS and EPA jointly developed preprocessing and postprocessing software, algorithm enhancements, and use of the USGS Watershed Data Management (WDM) system.

HSPF simulates hydrologic and water quality processes on land surfaces, streams, and impoundments. HSPF is generally used for a watershed-based analysis of the effects of land use, point and nonpoint source treatment alternatives (e.g., soil management practices), and flow diversions. Time series of the runoff flow rate, sediment yield, and user-specified pollutant concentrations can be generated at any point in the watershed. Compared with many other watershed loading models, HSPF has several advantages. First, HSPF is both a watershed model and a receiving water model; it can simulate both the pollutant loads on land surfaces and in-stream water quality in a complex watershed. Second, HSPF can be run under both continuous and storm situations. Data requirements for HSPF are extensive, and calibration and verification are strongly recommended. Because of its comprehensive nature, the HSPF model requires highly trained personnel. A team effort is recommended to apply it to real case studies.

WinHSPF is a new interface to HSPF Version 12 that is replacing Nonpoint Source Model (NPSM) from Better Assessment Science Integrating Point and Nonpoint Sources (BASINS) Version 2.0. Within BASINS V3.0 system, GenScn is a model postprocessing and scenario analysis tool that is used to analyze output from HSPF and SWAT. WDMUtil is used to manage and create the WDM files that contain the meteorologic data and other time series data used by HSPF. BASINS 3.0 includes several new data and functions including 1° DEM grids, an automatic delineation tool that creates watershed boundaries based on DEM grids, and a new watershed report function for land use, topography, and hydrologic response units.

Applications. HSPF and BASINS are widely used to for develop TMDLs for bacteria and many other wet weather pollutants. HSPF and BASINS produce pollutant concentration time series output that make it easy to determine the frequency and duration of water quality standard exceedances. Control scenario alternatives (e.g., soil management practices) can be evaluated for effectiveness in addressing water quality standard exceedances and the extent to which they would bring excessive costs to a community. The water quality impact of land use changes can also be conveniently evaluated. In addition, HSPF and BASINS can support source water assessments to examine the potential pollution risks to public water supplies.

The latest BASINS V3.0 provides a range of watershed models from which watershed managers can select the one that best meets their needs for an appropriate level of model complexity, control scenario detail, uncertainty, and analysis cost. The three watershed models in BASINS 3.0 are HSPF v12, Soil and Water Assessment Tool (SWAT), and the simple PLOAD model. SWAT and PLOAD models will be further discussed.

Watershed Modeling System (WMS)

The Watershed Modeling System (WMS) (http://www.emrl.byu.edu/wms.htm) is a comprehensive simulation model for hydrologic analysis. It was developed by the Environmental Modeling Research Laboratory of Brigham Young University in cooperation with the U.S. Army Corps of Engineers Waterways Experiment Station (USACE-WES). WMS merges information obtained from terrain models and GIS with industry-standard, lumped-parameter, hydrologic analysis models such as HEC-1, TR-20, and HSPF. WMS V6.1 provides tools for all phases of watershed modeling, including automated watershed and subbasin delineation, geometric parameter computation, hydrologic parameter computation, and result visualization. WMS is a powerful tool for analyzing and visualizing watersheds, which includes interfaces to several industry-standard hydrologic models. Each model can be set up and run within WMS. Model schematics are created and maintained automatically by WMS as you delineate a watershed from Map, Triangulated Irregular Networks (TINs), or Digital Elevation Models (DEMs) data. Terrain models can obtain geometric attributes such as area, slope, and runoff distances. WMS contains several calculators to aid in hydraulic/hydrologic design and analysis. These include an interface to HY-8 for culvert design/analysis, an open channel calculator, a detention basin calculator, a curb and gutter calculator, and a weir calculator. In addition, a calculator for Lag Time and/or Time of Concentration is included in WMS to aid in hydrologic model input. This calculator allows using any basin or flow path equation to compute travel times in the watershed and evaluate the hydrologic impact of management practices.

Soil and Water Assessment Tool (SWAT)

The Soil and Water Assessment Tool (SWAT) (http://www.brc.tamus.edu/swat/) is a watershed-scale model developed by Dr. Jeff Arnold for the USDA Agricultural Research Service (ARS). SWAT simulates the effect of agricultural management practices such as crop rotation, conservation tillage, residue, nutrient, pesticide management, and improved animal waste application methods on water quality. SWAT incorporates features of several ARS models and is a direct outgrowth of the SWRRB model (Simulator for Water Resources in Rural Basins) (3,4). The SWRRB model has been used on several watersheds to assess management practices and to test its validity (5,6). Specific models that contributed significantly to the development of SWAT were CREAMS (Chemicals, Runoff, and Erosion from Agricultural Management Systems) (7), GLEAMS (Groundwater Loading Effects on Agricultural Management Systems) (8), and EPIC (Erosion-Productivity Impact Calculator) (9).

SWAT is a continuous time model, that is, a long-term yield model. The model is not designed to simulate detailed, single flood routing. Instead, SWAT was developed to predict the impact of land management practices on water, sediment, and agricultural chemical yields in large complex watersheds where soils, land use, and management conditions vary over long periods of time. To satisfy this objective, the model is physically based. Rather than incorporating regression equations to describe the relationship between input and output variables, SWAT requires specific information about weather, soil properties, topography, vegetation, and land management practices in the watershed. The physical processes associated with water movement, sediment movement, crop growth, and nutrient cycling, are directly modeled by SWAT using this input data. The benefits of this approach are (1) watersheds without monitoring data (e.g., stream gauge data) can be modeled, (2) the relative impact of alternative input data (e.g., changes in management practices, climate, vegetation) on water quality or other variables of interest can be quantified, and (3) uses readily available inputs. Furthermore, SWAT is computationally efficient. Very large basins or a variety of management strategies can be simulated without excessive investment of time or money. SWAT enables users to study long-term impacts.

A new watershed model, the SWAT 2002 has been added to the EPA BASINS V3.0 system. SWAT is a watershed-scale model developed to predict the impact of land management practices on water, sediment, and agricultural chemical yields in large complex watersheds where soils, land use, and management conditions vary over long periods of time.

Applications. SWAT is ideally suited for simulating agricultural watersheds where fertilizer and pesticide application rates are known and where altering agricultural management practices is viewed as a viable and necessary way to meet water quality standards. SWAT is ideally suited for nitrogen, phosphorous, and pesticide TMDLs due to its use of prepackaged nutrient and pesticide application loads in standard forms (e.g., by type of manure or fertilizer and by the common name of the pesticide) as well as other agriculture specific inputs including tillage practices and crop rotation. SWAT requires a considerable amount of information on agricultural management practices but readily houses and internally processes this information within the interface. Management options are already included, so there is a direct translation to soil conservation practices and BMP-based controls. SWAT can be used to determine the likely reduction in sediment loads from installing small dams in critical areas in watersheds or how employing other land-based best management practices, such as terraces, vegetative buffer strips, or other cropland BMPs, would cut silt runoff/sediment loads to lakes and reservoirs.

CREAMS and GLEAMS

CREAMS, first developed in 1970s by the USDA, is a field-scale model designed to simulate the impact of land management on water, sediment, nutrients, and pesticides leaving the edge of a field. A number of other USDA models such as GLEAMS, SWRRB, AGNPS and EPIC, trace their origins to the CREAMS model.

GLEAMS is a nonpoint source model that focuses on pesticide and nutrient groundwater loadings. GLEAMS (10) simulates the effects of management practices and irrigation options on edge of field surface runoff, sediment, and dissolved and sediment-attached nitrogen, phosphorus, and pesticides. The model considers the effects of crop planting date, irrigation, drainage, crop rotation, tillage, residue, commercial nitrogen and phosphorus applications, animal waste applications, and pesticides on pollutant movement. The model has been used to predict the movement of pesticides (11), nutrients, and sediment from various combinations of land uses and management (12,13).

Erosion–Productivity Impact Calculator (EPIC)

The Erosion–Productivity Impact Calculator (EPIC) (http://www.brc.tamus.edu/epic/index.html) was originally developed to simulate the impact of erosion on crop productivity and has now evolved into a comprehensive agricultural management, field-scale, nonpoint source loading model. In the early 1980s, teams of USDA Agricultural Research Service (ARS), Soil Conservation Service (SCS) (now Natural Resources Conservation Service, or NRCS), and Economic Research Service (ERS) scientists developed EPIC to quantify the costs of soil erosion and the benefits of soil erosion research and control in the United States. Developed in late 1980s, the model uses a daily time step to simulate weather, hydrology, soil temperature, erosion–sedimentation, nutrient cycling, tillage, crop management and growth, pesticide and nutrient movement with water and sediment, and field-scale costs and returns. EPIC (14) simulates the effect of management strategies on edge of field water quality and nitrate nitrogen and pesticide leaching to the bottom of the soil profile. The model considers the effect of crop type, planting date, irrigation, drainage, rotations, tillage, residue, commercial fertilizer, animal waste, and pesticides on surface and shallow groundwater quality. The EPIC model has been used to evaluate various cropland management practices (15,16).

AGNPS

Agricultural Nonpoint Source Pollution Model 2001 (AGNPS 2001) (http://www.sedlab.olemiss.edu/agnps.html) is a joint USDA–ARS and NRCS system of computer models developed to predict nonpoint source pollutant loadings within agricultural watersheds. AGNPS is a midrange model. It contains a continuous simulation, surface runoff model designed to assist in determining BMPs, the setting of TMDLs, and for risk and cost/benefit analyses.

AGNPS 2001 is a tool for evaluating the effect of management decisions that impact a watershed system. The AGNPS 2001 system is a direct update of the AGNPS 98 system of modules containing many enhancements. These enhancements have been included to improve the capability of the program and to automate many of the input data preparation steps needed for use with large watershed systems. The capabilities of RUSLE, used by USDA-NRCS to evaluate the degree of erosion on agricultural fields and to guide development of conservation plans to control erosion, have been incorporated into AnnAGNPS. The AnnAGNPS model has been applied to many field and watershed size areas to estimate pollutant runoff from various land uses and management practices (6,15,17–19).

Revised Universal Soil Loss Equation (RUSLE)

The Revised Universal Soil Loss Equation (RUSLE) (http://www.sedlab.olemiss.edu/rusle/) is an easily and widely used computer program that estimates rates of soil erosion caused by rainfall and associated overland flow. The most current version of RUSLE is Version 1.06b released on January 2001 (20). In the United States, the USDA–NRCS is the principal user of RUSLE and has implemented this approach in most of its local field offices. The NRCS is the major source for data needed to apply RUSLE and is the leading authority on field application of RUSLE. RUSLE is used by numerous government agencies, private organizations, and individuals to assess the degree of erosion, identify situations where erosion is serious, and guide development of conservation plans to control erosion. RUSLE has been applied to cropland, rangeland, disturbed forest lands, landfills, construction sites, mining sites, reclaimed lands, military training lands, parks, land disposal of waste, and other land uses where mineral soil material is exposed to the erosive forces of raindrop impact and overland flow. RUSLE estimates average annual soil loss, expressed as mass per unit area per year, which is defined as the amount of sediment delivered from the slope length assumed in the RUSLE computation.

Water Erosion Prediction Project (WEPP)

The Water Erosion Prediction Project (WEPP) model (http://topsoil.nserl.purdue.edu/nserlweb/weppmain/) is a process-based, distributed parameter and continuous simulation model for erosion prediction. WEPP is expected to become a major component of the "conservation tool kit." It is a computer program designed to be employed by the same personnel currently using USLE (Universal Soil Loss Equation) or RUSLE. By analyzing how farming and land use affect soil erosion and sediment delivery, WEPP promises better conservation planning, better project planning, and optimum resource inventory and assessment. It will help users select the best erosion control practices, aid in choosing optimum locations for future project sites, and evaluate erosion and sedimentation over specified areas. It is applicable to small watersheds (field-sized) and can simulate small profiles up to large fields. It mimics the natural processes that are important in soil erosion.

WEPP can be applied on the field scale to simulate single hillslope erosion or more complex watershed-scale erosion: Hillslope Applications and Watershed Applications. WEPP (21) simulates water runoff, erosion, and sediment delivery from fields or small watersheds. Management practices, including crop rotation, planting and harvest date, tillage, compaction, strip-cropping, row arrangement, terraces, field borders, and windbreaks, can be simulated. The WEPP model has been applied to various land use and management conditions (22,23).

Nitrate Leaching and Economic Analysis Package (NLEAP)

The Nitrate Leaching and Economic Analysis Package (NLEAP) (http://gpsr.ars.usda.gov/products/nleap/nleap.htm) is a field-scale computer model developed to provide a rapid and efficient method for determining potential nitrate leaching from agricultural practices. It uses basic information concerning on-farm management practices, soils, and climate to project nitrogen budgets and nitrate leaching indexes. NLEAP calculates potential nitrate leaching below the root zone and to groundwater supplies. NLEAP has three levels of analysis to determine leaching potential: an annual screening, a monthly screening, and an event-by-event analysis.

The NLEAP model was designed to answer questions regarding potential leaching of nitrate. The processes modeled include movement of water and nitrate, crop uptake, denitrification, ammonia volatilization, mineralization of soil organic matter, nitrification, and mineralization–immobilization of crop residue, manure, and other organic wastes. The limitations of NLEAP include modeling soil nitrogen processes in organic soils that is not currently available. NLEAP does not predict yield reductions caused by pests or nutrient deficiencies. However, the user should consider the effects of these problems when estimating crop yield. The model should not be used where rapid water infiltration, leaching, denitrification, and ammonium volatilization require time steps smaller than one day. NLEAP (24) evaluates the potential of nitrate nitrogen leaching due to land use and management practices. The NLEAP model has been used to predict the potential for nitrogen leaching under various management options (25,26). Most recently, it has been applied to compare the impact of TMDL allocation scenarios on nitrate leaching (27).

Root Zone Water Quality Model (RZWQM)

The Root Zone Water Quality Model (RZWQM) (http://gpsr.ars.usda.gov/products/rzwqm.htm) was developed during the past 10 years by a team of USDA–ARS scientists. A majority of the team members are part of the present Great Plains Systems Research Unit in Fort Collins, Colorado. RZWQM is a one-dimensional process-based model that simulates the growth of a plant and the movement of water, nutrients, and agrochemicals over, within, and below the crop root zone of a unit area of an agricultural cropping system under a range of common management practices. The primary use of RZWQM will be as a tool for assessing the environmental impact of alternative agricultural management strategies on the subsurface environment. These alternatives may include conservation plans on a field-by-field basis; tillage and residue practices; crop rotations; planting date and density; and irrigation, fertilizer, and pesticide scheduling (method of application, amounts, and timing). The model predicts the effects of these management practices on the movement of nitrate and pesticides to runoff and deep percolation below the root zone.

Pesticide Root Zone Model (PRZM)

The EPA's Pesticide Root Zone Model (PRZM) (http://www.epa.gov/ceampubl/gwater/przm3/index.htm) is a one-dimensional, dynamic, compartmental model that can be used to simulate chemical movement in unsaturated soil systems within and immediately below the plant root zone. It has two major components: hydrology/hydraulics and chemical transport. The hydrologic component for calculating runoff and erosion is based on the Soil Conservation Service curve number technique and USLE. The chemical transport component can simulate pesticides or organic and inorganic nitrogen species.

PRZM can simulate multiple zones. This allows combining different root zone and vadose zone characteristics into a single simulation. Zones can be visualized as multiple land segments joined together horizontally. PRZM (28) simulates the movement of pesticides in unsaturated soils within and immediately below the root zone. Several different field crops can be simulated and up to three pesticides are modeled simultaneously as separate parent compounds or metabolites. The PRZM model has been used under various conditions to assess pesticide leaching under fields (11,13).

Generalized Watershed Loading Function (GWLF)

The Generalized Watershed Loading Function (GWLF) model (http://www.avgwlf.psu.edu/) was developed by Haith and Shoemaker (29). The GWLF model can simulate runoff, sediment, and nutrient loadings from a watershed, given variable-size source areas (i.e., agricultural, forested, and developed land). ANSWERS (30), which has been used in GWLF, is a spatially distributed watershed model. The model is primarily a runoff and sediment model because soil nutrient processes are not simulated. The ANSWERS model has been applied to several small field-sized areas using various management practices (6,31).

During the last 5–10 years, the Pennsylvania Department of Environmental Protection (PADEP) has recognized the indispensability of GIS technology and endeavored to integrate it into all of the agency's internal program areas. Pennsylvania State University has been assisting PADEP in developing and implementing various GIS-based watershed assessment tools, including one called AVGWLF that facilitates using the GWLF model via a GIS software (ArcView) interface. The comprehensive modeling approach was based on this GWLF/ArcView interface that enables accurate prediction of nutrient and sediment loads in watersheds throughout the state of Pennsylvania, particularly for those watersheds for which historical stream monitoring data do not exist. This methodology relies on using statewide data sets to derive reasonably good estimates of various critical model parameters that exhibit significant spatial variability within the state. GWLF is a continuous simulation model that uses daily time steps for weather data and water balance calculations. Monthly calculations are made for sediment and nutrient loads, based on the daily water balance accumulated to monthly values.

Source Loading and Management Model (SLAMM)

The Source Loading and Management Model (SLAMM) (http://wi.water.usgs.gov/slamm/) was originally developed in the late 1970s to understand better the relationships between sources of urban runoff pollutants and runoff quality. Runoff is calculated by a method developed by the University of Alabama at Birmingham for small-storm hydrology. SLAMM is strongly based on actual field observations, and relies minimally on theoretical processes that have not been adequately documented or confirmed in the field. SLAMM is used mostly as a planning tool to understand better sources of urban runoff pollutants and their control. SLAMM incorporates unique process descriptions to predict more accurately the sources of runoff pollutants and flows for the storms of most interest in water quality analyses and quantification of the impacts of management practices.

Regression Methods and Simple Spreadsheet Models

USGS Regression Method. The USGS has developed equations for determining pollutant loading rates based on regression analyses of data from sites throughout the country (76 gauging stations in 20 states). The regression approach is based on a statistical description of historic records of storm runoff responses on a watershed level (32,33). This method may be used for rough preliminary calculations of annual pollutant loads when data and time are limited. Input required for this level of modeling includes drainage data, percent imperviousness, mean annual rainfall, general land use pattern, and mean minimum monthly temperature. Application of this method provides mean planning loads and corresponding confidence intervals for storms.

Simple Method by MWCOG. The "simple method," as its name implies, is an easy-to-use empirical equation for estimating pollutant loadings of an urban watershed by the Metropolitan Washington Council of Governments (MWCOG). The method is applicable to watersheds less than 1 square mile in area, and can be used for analyzing a smaller watershed or site planning. The method was developed using the database generated during a Nationwide Urban Runoff Program (NURP) study (34) in the Washington, DC, area and the national NURP data analysis. The equations, however, may be applied anywhere in the country. Some precision is lost as a result of the effort to make the equation general and simple. The method is adequate for decision-making at the site-planning level.

FHWA Model. The Federal Highway Administration (FHWA) has developed a simple statistical spreadsheet procedure to estimate pollutant loading and impacts to stream and lakes that receive highway stormwater runoff (35). The FHWA model uses a set of default values for pollutant event mean concentrations that depends on traffic volume and whether the setting of the highway's pathway is rural or urban. The FHWA uses this method to identify and quantify the constituents of highway runoff and their potential effects on receiving waters and to identify areas that might require controls. The FHWA model is well suited for screening application and can evaluate lake and stream impacts of highway stormwater discharges. However, it assesses seasonal variability in a limited manner and does not consider the soluble fraction of pollutants or precipitation and settling of pollutants in lakes.

PLOAD Model. PLOAD is a simplified GIS-based watershed loading model. It can model combined point and nonpoint source loads in either small urban areas or in rural watersheds of any size. The user may calculate the nonpoint source loads using either the export coefficient or the EPA's simple method approach. Best management practices (BMPs), which reduce both point and nonpoint source loads, may also be included in computing total watershed loads. PLOAD broadly addresses pollutant loading by land use categories and subwatersheds but does not get at individual nonpoint sources or at actual pollutant fate and transport processes.

PLOAD was designed to be generic, so that it can be applied as a screening tool in a wide range of applications, including NPDES stormwater permitting, watershed management, and reservoir protection projects. The uncertainty of export coefficients is high, but the PLOAD model provides a good fit to the regulatory concept of the phased TMDL. Uncertainty can be reduced by calibrating to local conditions based on multiyear datasets and can be addressed in later phases of the TMDL as more detailed source identification and seasonal data become available (36).

SUMMARY AND CONCLUSION

Water quality models of different complexity level (ranging from simple, midrange, to detailed) for developing soil management practices have been reviewed. In general, complex models potentially have greater accuracy in their predictions than simple models because of the comprehensive simulation of environmental processes. However, more site-specific data are required to use these complex models fully. Application of complex models is usually more expensive than using simple models for developing management practice. Therefore, the trade-off among these models should be one of the key considerations during model selection.

Although guidance is still needed for determining nonpoint source loadings from urban and rural areas (37), the common approaches used currently for estimating wet weather nonpoint pollution loads are (38) (1) unit load concepts (long-term statistical average loads are related to land uses), (2) event-oriented models (water and pollutant loads are calculated for a single storm), and (3) continuous models (water and pollutant loads are calculated using daily or shorter time intervals and complex, mostly deterministic hydrologic models).

The realization that management of nonpoint sources of pollutants is as important an issue as management of the more readily controllable point sources has led to the concept of watershed management and planning. Increased computational power, as well as a desire for watershed planning, has resulted in the advancement of mathematical modeling tools. One of the trends is developing integrated modeling systems that link the models, data, and user interface within a single system. New developments in modeling systems have increasingly relied on a geographic information system (GIS) and a database management system to support modeling and analysis (1). U.S. EPA's BASINS offers a comprehensive modeling system that brings key data and analytical components together "under one roof" (2). USDA's benchmark watershed model SWAT has been incorporated into the BASINS V3 system. The Watershed Modeling System (WMS) is another useful tool for watershed modeling and management practice development. These multipurpose environmental analysis systems are expected to be extensively used by federal, regional, state, and local agencies in watershed and water quality studies in the future.

BIBLIOGRAPHY

1. U.S. EPA. (1997). *Compendium of Tools for Watershed Assessment and TMDL Development*. EPA 841-B-97-006, Washington, DC.
2. U.S. EPA. (2000). *National Management Measures to Control Nonpoint Source Pollution from Agriculture (Draft)*. Washington, DC.
3. Williams, J.R., Nicks, A.D., and Arnold, J.G. (1985). Simulator for water resources in rural basins. *J. Hydraul. Eng.* **111**(6): 970–986.
4. Arnold, J.G., Williams, J.R., Nicks, A.D., and Sammons, N.B. (1990). *SWRRB: A Basin Scale Simulation Model for Soil and Water Resources Management*. Texas A&M University Press, College Station, TX.
5. Arnold, J.G. and Williams, J.R. (1987). Validation of SWRRB—simulator for water resources in rural basins. *J. Water Resour. Plann. Manage.* **113**: 243–256.
6. Bingner, R.L., Murphree, C.E., and Mutchler, C.K. (1987). *Comparison of Sediment Yield Models on Various Watersheds in Mississippi*. American Society of Agricultural Engineers Paper 87–2008, St. Joseph, MI.
7. Knisel, W.G. (1980). *CREAMS, a Field Scale Model for Chemicals, Runoff and Erosion from Agricultural Management Systems*. USDA Conservation Research Report. No. 26.
8. Leonard, R.A., Knisel, W.G., and Still, D.A. (1987). GLEAMS: groundwater loading effect on agricultural management systems. *Trans. ASAE* **30**(5): 1403–1428.
9. Williams, J.R., Jones, C.A., and Dyke, P.T. (1984). A modeling approach to determining the relationship between erosion and soil productivity. *Trans. ASAE* **27**(1): 129–144.
10. Knisel, W.G., Leonard, R.A., and Davis, F.M. (1991). Water balance components in the georgia coastal plain: a GLEAMS model validation and simulation. *J. Soil Water Conserv.* **46**(6): 450–460.
11. Zacharias, S., Heatwole, C.D. Dillaha, T., and Mostighimi, S. (1992). *Evaluation of GLEAMS and PRZM for Predicting Pesticide Leaching Under Field Conditions*. American Society of Agricultural Engineers Paper 92–2541, St. Joseph, MI.
12. Knisel, W.G. and Leonard, R.A. (1989). Irrigation impact on groundwater: model study in humid region. *J. Irrig. Drainage Eng.* **15**(5): 823–837.
13. Smith, M.C., Bottcher, A.B., Campbell, K.L., and Thomas, D.L. (1991). Field testing and comparison of the PRZM and GLEAMS models. *Trans. Am. Soc. Agric. Eng.* **34**(3): 838–847.
14. Sharply, A.N. and Williams, J.R. (1990). *EPIC—Erosion/Productivity Impact Calculator: 1. Model Development*. USDA Bulletin No. 1768, p. 235.
15. Sugiharto, T., McIntosh, T.H., Uhrig, R.C., and Lavdinois, (1994). Modeling alternatives to reduce dairy farm and watershed nonpoint source pollution. *J. Environ. Qual.* **23**: 18–24.
16. Edwards, D.R. and Daniel, T.C. (1994). Quality of runoff from fescue grass plots treated with poultry litter and inorganic fertilizer. *J. Environ. Qual.* **23**: 579–584.
17. Bingner, R.L. and Theurer. F.D. (2001). AnnAGNPS: estimating sediment yield by particle size for sheet & rill erosion. In: *Proceedings of the Sediment: Monitoring, Modeling, and Managing, 7th Federal Interagency Sedimentation Conference*. Reno NV, 25-29 March 2001. pp. I-1–I-7.
18. Line, D.E., Coffey, S.W., and Osmond, D.L. (1997). WATERSHEDSS—GRASS AGNPS modeling tool. *Trans. Am. Soc. Agric. Eng.* **40**(4): 971–975.
19. Young, R.A., Onstad, C.A., Bosch, D.D., and Anderson, W.P. (1994). *Agricultural Nonpoint Source Pollution Model, Version 4.03 AGNPS User's Guide*. USDA Agricultural Research Service, Morris, MN.
20. USDA. (2001). Revised Universal Soil Loss Equation, http://www.sedlab.olemiss.edu/rusle.
21. Flanagan, D.C. and Nearing, M.A. (1995). *USDA-Water Erosion Prediction Project: Hillslope Profile and Watershed Model Documentation*. NSERL Report No.10, West Lafayette, IN.
22. Tiscareno-Lopez, M., Lopes, V.L., Stone, J.J., and Lane, L.J. (1993). Sensitivity analysis of the WEPP watershed model for rangeland applications: I. Hillslope processes. *Trans. ASAE* **36**: 1659–1672.
23. Liu, B.Y., Nearing, M.A., Baffault, C., and Ascough, J.C. (1997). The WEPP watershed model: three comparisons to measured data from small watersheds. *Trans. Am. Soc. Agric. Eng.* **40**: 945–952.
24. Follet, R.F., Kenney, D.R., and Cruse, R.M. (1991). *NLEAP (Nitrogen Leaching and Economic Analysis Package)*. Soil Science of America, Inc., Madison, WI.
25. Wylie, B.K. et al. (1994). Predicting spatial distributions of nitrate leaching in northeastern colorado. *J. Soil Water Conserv.* **49**: 288–293.
26. Wylie, B.K., Shaffer, M.J., and Hall, M.D. (1995). Regional assessment of NLEAP NO_3-N leaching indices. *Water Resour. Bull.* **31**: 399–408.
27. Culver, T.B. et al. (2002). Case study of impact of total maximum daily load allocations on nitrate leaching. *ASCE J. Water Resour. Plann. Manage.* **128**(4): 262–270.
28. Mullens, J.A., Carsel, R.F., Scarbrough, J.E., and Ivery, I.A. (1993). *PRZM-2, A Model for Predicting Pesticide Fate in the Crop Root and Unsaturated Soil Zones: User Manual for Release 2.0*. Office of Research and Development, U.S. EPA, Athens, GA, EPA Report 600/R-93/046.
29. Haith, D.A. and Shoemaker, L.L. (1987). Generalized watershed loading functions for stream flow nutrients. *Water Resour. Bull.* **23**(3): 471–478.

30. Beasley, D.B., Huggins, L.F., and Monke, E.J. (1980). ANSWERS: a model for watershed planning. *Trans. Am. Soc. Agric. Eng.* **23**: 938–944.
31. Griffin, M.L., Beasley, D.B., Fletcher, J.I., and Foster, G.R. (1988). Estimating soil loss on topographically non-uniform field and farm units. *J. Soil Water Conserv.* **43**: 326–331.
32. Tasker, G.D. and Driver, N.E. (1988). Nationwide regression models for predicting urban runoff water quality at unmonitored sites. *Water Resour. Bull.* **24**(5): 1091–1101.
33. Driver, N.E. and Troutman, B.M. (1989). Regression models for estimating urban storm-runoff quality and quantity in the United States. *J. Hydrology* **109**: 221–236.
34. U.S. EPA. (1983). *Results of the Nationwide Urban Runoff Program*. Vol. Final Report, Washington, DC.
35. Federal Highway Administration (FHWA). (1996). *Evaluation and Management of Highway Runoff Water Quality*. Publication No. FHWA-PD-96-032.
36. Cocca, P.A. (2001). *BASINS 3.0: New Tools for Improved Watershed Management*. Proc. ASCE World Water Congr. 2001, Orlando, FL.
37. U.S. EPA. (1998). *Report of the Federal Advisory Committee on the Total Maximum Daily Load (TMDL) Program*. EPA-100-R-98-006, Washington, DC.
38. Novotny, V. (1999). Integrating diffuse/nonpoint pollution control and water body restoration into watershed management. *J. Am. Water Resour. Assoc.* **35**(4): 717–727.

READING LIST

American Society of Civil Engineers (ASCE). (2001). National Stormwater Best Management Practices (BMPs) Database, http://www.bmpdatabase.org/.

Arnold, J.G., Williams, J.R. and Maidment, D.R. (1995). Continuous-time water and sediment-routing model for large basins. *J. Hydraul. Eng.* **121**(2): 171–183.

Bicknell, B.R., Imhoff, J.C., Kittle, J.L Jr., and Donigian, A.S. Jr. (1996). *Hydrological Simulation Program—FORTRAN, User's Manual for Release 11*. Environmental Research Laboratory, U.S. EPA, Athens, GA.

Carrubba, L. (2000). Hydrologic modeling at the watershed scale using NPSM. *J. Am. Water Resour. Assoc.* **36**(6): 1237–1246.

Cooper, A.B., Smith, C.M., and Bottcher, A.B. (1992). Predicting runoff of water, sediment, and nutrients from a New Zealand grazed pasture using CREAMS. *Trans. Am. Soc. Agric. Eng.* **35**(1): 105–112.

Deliman, P.N., Glick, R.H., and Ruiz, C.E. (1999). *Review of Watershed Water Quality Models*. U.S. Army Corps of Engineers, Technical Report W-99-1.

Donigian, A.S. and Huber, W.C. (1991). *Modeling of Nonpoint Source Water Quality in Urban and Non-urban Areas*. EPA 600/3-91/039, U.S. EPA, Athens, GA.

Evans, B.M., White, R.A., Petersen, G.W., Hamlett, J.M., Baumer, G.M. and McDonnell, A.J. (1994). *Land Use and Non-point Pollution Study of the Delaware River Basin*. Environmental Resources Research Institute, Penn State University, Publication No. ER94-06, p. 76.

Haith, D.R., Mandel, R., and Wu, R.S. (1992). *GWLF: Generalized Watershed Loading Functions User's Manual, Version 2.0*. Cornell University, Ithaca, NY.

Novotny, V. and Olem, H. (1994). *Water Quality: Prevention, Identification, and Management of Diffuse Pollution*. Van Nostrand Reinhold, New York.

Water Environment Research Foundation (WERF). (2001). *Water Quality Models: A Survey and Assessment*, Final Report on Project 99-WSM-5.

Whittemore, R.C. and Beebe, J. (2000). EPA's BASINS model: good science or serendipitous modeling. *J. Am. Water Resour. Assoc.* **36**(3): 493–499.

Yoon, K.S., Yoo, K.H., Wood, C.W., and Hall, B.M. (1994). Application of GLEAMS to predict nutrient losses from land application of poultry litter. *Trans. Am. Soc. Agric. Eng.* **37**(2): 453–460.

Young, R.A., Onstad, C.A., Bosch, D.D., and Anderson, W.P. (1989). AGNPS: a nonpoint source pollution model for evaluating agriculture watersheds. *J. Soil Water Conserv.* **44**: 168–173.

WATER QUALITY MODELING—CASE STUDIES

ZHEN-GANG JI
Minerals Management Service
Herndon, Virginia

This entry focuses on case studies of water quality modeling. Information on water quality models is provided in WATER QUALITY MODELS.

MODEL SELECTION

Although a variety of models are available for studying water quality processes, it is still a complex task to select a model (or models) that best meets all (or most) of the study objectives. Model selection should be balanced among competing demands. Time and resources are always limited, to one degree or another, so the goal should be to identify the model(s) that addresses all of the important processes that affect the waterbody. Selecting a model that is too simple can result in lack of accuracy and certainty that are needed in decision making; selecting an overly complex model may result in misdirected resources, study delays, and increased cost.

It is helpful to ask the following questions during model selection:

1. What are the key hydrodynamic processes?
2. What are the water quality concerns?
3. What spatial and temporal scales are adequate for resolving these processes?
4. How will the model be used in supporting management decision making?

It might be desirable to select the simplest model that can meet the study needs. However, in real practice, comprehensive models are often preferred to simple models for a variety of reasons. Typical features of comprehensive models include:

1. three-dimensional (3-D) in space and time dependent
2. turbulent scheme for vertical mixing
3. hydrodynamic, thermal, sediment, and eutrophication processes.

Generally speaking, comprehensive models should (but not always) have better mathematical, physical, chemical, and biological representations of natural waters than simple models. It is often advantageous to adopt a more detailed model to address various scientific and engineering applications than to switch models continuously from one phase of a project to another or from one project to another (1). As discussed later and in the entry WATER QUALITY MODELS: CHEMICAL PRINCIPLES, the Environmental Fluid Dynamics Code (EFDC) model is a comprehensive model (2).

In many applications, comprehensive models have advantages compared to simple models for the following:

1. *The Evolving Understanding of the Waterbody Studied.* A waterbody is understood gradually as the modeling effort progresses. Detailed understanding of the system is often achieved after the modeling study is finished. At the beginning of a study, it is sometimes difficult to know which simple model should be able to describe the system adequately. Therefore, it is helpful to have a model that has the most capability (and flexibility for further model enhancement) to describe important processes (known and to be known) in the system.
2. *The Management Needs.* A primary goal of a modeling study is often to support management decision making, which might change with time and cannot always be foreseeable. It is desirable to choose a model that can meet these "expanding" needs. It is generally more cost-effective to choose a comprehensive model that can address current and future needs than to use a simple model that is later found inadequate and has to be replaced by one more advanced.
3. *The Falling Costs of Computing.* The progress in computer technology has dramatically reduced the costs of computing and makes comprehensive models more affordable.
4. *The Modelers.* A modeler often needs to study hydrodynamic, sediment, toxic, and/or water quality processes in a variety of waterbodies and, therefore, needs models and tools that are versatile enough for these applications. It is common for a modeler to have several modeling studies during any given period and to complete tens of modeling projects over the years. Rather than learning and using a variety of models for different applications, it is advantageous to stick with one (or a few) comprehensive model, know it well, and apply it to modeling rivers, lakes, and estuaries.

Comprehensive models can be applied appropriately only by well-trained and experienced modeling professionals and often need extensive measured data for model calibration and verification. The progress in computer technology, the enrichment of measured data, and the enhancement of water quality models are making comprehensive models powerful tools for water quality management.

Two cases studies are introduced here: the first is on sediment and toxic metals in a shallow river, and the second is on eutrophication in a deep reservoir.

CASE STUDY I: BLACKSTONE RIVER

Many waterbodies are polluted by sediments and toxic metals. This case study illustrates the modeling of hydrodynamics, sediment transport, and toxic metals in the Blackstone River, Massachusetts (3).

Background

The Blackstone River Basin (Fig. 1) consists of approximately 1657 square kilometers encompassing 30 cities and towns. The Blackstone River flows from Worcester, Massachusetts, to Pawtucket, Rhode Island. The river is 77 km long and has 133 meters of total fall. The distances shown in Fig. 1 are river kilometers from Slaters Mill Dam. There are presently 14 dams and impoundments on the mainstem of the Blackstone that are significant to the hydrodynamic and water quality processes in the river. The Blackstone has been the largest source of pollutants discharging into Narragansett Bay, principally from industrial discharge of metals and resuspension of contaminated sediments behind the low head dams in the river. Sediment and metals in certain sections of the riverbed can be traced back 200 years to the American Industrial Revolution.

This study is intended to provide the United Sates Environmental Protection Agency (EPA) with a relatively simplified example of how comprehensive models can be used to evaluate the distributions of heavy metals in a shallow, narrow, urban, industrial river.

Hydrodynamic, Sediment, and Metals Simulations

The Blackstone River model is developed within the framework of the EFDC model (2). The EFDC model is a public-domain modeling package used for simulating three-dimensional flow, transport, and biogeochemical processes in surface water systems, including rivers, lakes, estuaries, reservoirs, wetlands, and coastal regions. The EFDC model has been extensively tested and documented in more than 80 modeling studies and is currently supported by the EPA. The model is used by a number of organizations, including universities, governmental agencies, and environmental consulting firms. Representative applications of the EFDC model include modeling of sediment and metals transport in the Blackstone River (3), wetting and drying simulation of Morro Bay (4,5), and simulating hydrodynamic, thermal, and sediment processes in Lake Okeechobee (6–8). More information on the EFDC model is also provided in the entry WATER QUALITY MODELS: CHEMICAL PRINCIPLES.

The Blackstone River model includes a hydrodynamic submodel, a sediment submodel, and a toxicant submodel. They are coupled together and are executed simultaneously. The hydrodynamic model simulates velocity, water elevation, and turbulent mixing for sediment modeling. The output of the hydrodynamic and sediment models are linked to the toxicant model to simulate the five metals in the river: cadmium (Cd), chromium (Cr), copper (Cu), nickel (Ni), and lead (Pb).

The typical width of the Blackstone River is around 25 meters and varies from less than 10 m upstream to

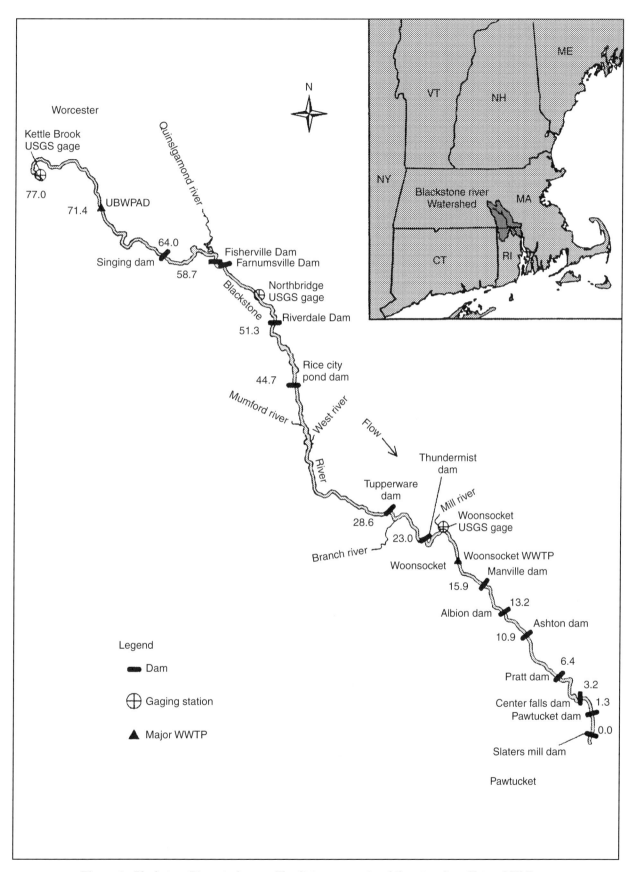

Figure 1. Blackstone River study area. The distances are river kilometers from Slaters Mill Dam.

more than 35 m downstream. It has an average drop of 1.73 m/km, so the Blackstone is a gravity-driven river. The grid of the Blackstone River model has one cell across the river and one layer in the vertical. Along the river, there are 256 grid cells of varying cell widths and a uniform cell length of 300 m. A time step of 30 seconds is used throughout the simulation.

Figure 2 presents the time series of the modeled and measured concentrations of total suspended solids (TSS) along the Blackstone River during a storm (November 2–6, 1992). The horizontal axis is in days from November 2, 1992, and the vertical axis is in milligram/liter. The closed circles represent the measured TSS, and the solid line represents the model results. The river kilometers are also shown in the plots. The model realistically simulates the sediment resuspension processes at dams, including Singing Dam (km = 64.0), Fisherville Dam (km = 58.4), Riverdale Dam (km = 51.3), and Rice City Pond Dam (km = 44.7). The high sediment concentration at km = 37.3 is caused by flows from upstream and from tributaries. Figure 3 is similar to Fig. 2, except for Pb concentrations and shows that the model simulates Pb processes well.

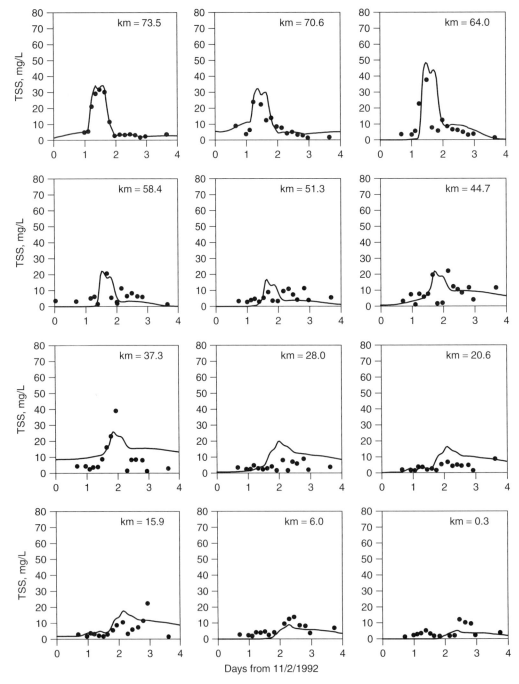

Figure 2. Measured and modeled total suspended solids (TSS) concentration along the Blackstone River from November 2–6, 1992.

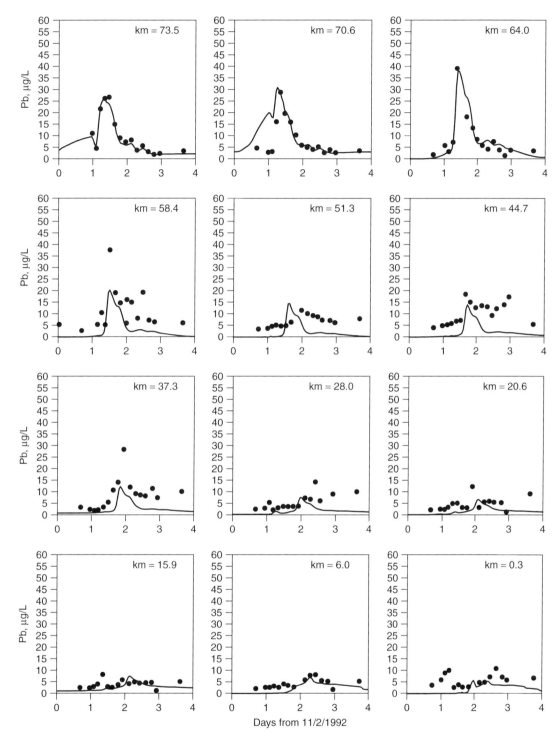

Figure 3. Measured and modeled lead (Pb) concentration along the Blackstone River from November 2–6, 1992.

To conduct a statistical analysis, the model results are also saved at the same locations and times at which the data were collected. Table 1 summarizes the error analyses of seven variables: river flow rate (Q), TSS, Cd, Cr, Cu, Ni, and Pb. The table presents the values of observed mean, modeled mean, absolute mean error, root mean square (rms) error, observed change, and the percentage of relative rms error. The relative rms error is defined as the rms error divided by the observed change. These analyses show that the simulations agree well with the observed data. The relative rms errors range from 6.17% (Q) to 17.27% (TSS), and the relative rms errors for the five metals are no more than 15.03%.

Conclusions

It is a challenge to apply coupled hydrodynamic, sediment, and contaminant fate and transport models to studies of surface water systems. Contaminant transport depends

Table 1. Error Analysis of Observed and Modeled River Flow Rate (Q), Total Suspended Sediment (TSS), Cadmium (Cd), Chromium (Cr), Copper (Cu), Nickel (Ni), and Lead (Pb)

Variable	Obs. Mean	Modeled Mean	Mean abs. err.	rms err.	Obs. Change	Relative rms err. (%)
Q (m³/s)	9.53	10.95	1.86	3.42	55.39	6.17
TSS (mg/L)	6.80	8.20	4.54	6.53	37.80	17.27
Cd (μg/L)	0.88	1.22	0.52	0.71	5.12	13.95
Cr (μg/L)	4.08	1.99	2.27	3.17	27.08	11.71
Cu (μg/L)	12.27	12.09	5.77	8.69	57.80	15.03
Ni (μg/L)	7.92	7.73	2.31	3.52	25.70	13.68
Pb (μg/L)	7.52	4.57	4.00	5.43	37.80	14.38

on hydrodynamic and sediment conditions. Heavy metals and toxic chemicals can preferentially adsorb and desorb with solids in the water column and sediment bed. High flow events, such as storms, increase solid loadings from the watershed, increase river flow velocity, reintroduce previously deposited chemicals back into the water column via resuspension, and transport the resuspended contaminants further downstream until they settle out in deposition zones.

So far, there are few published modeling studies on sediment and metal transport in rivers that simulated storms on an hourly basis and used comprehensive data sets for model input and model calibration. This study can simulate the sediment and metal transport processes in detail. Statistical analysis and graphic presentation indicate that the model results are in very good agreement with the data. For shallow and narrow rivers like the Blackstone, the EFDC model with a 1-D grid can represent the hydrodynamic, sediment, and metals processes reasonably well.

CASE STUDY II: LAKE TENKILLER

Water quality management requires scientifically credible numerical models for evaluating pollution control options. This case study summarizes the eutrophication and water quality modeling of Lake Tenkiller, Oklahoma (9).

Lake Tenkiller (Fig. 4) is a reservoir located in northeastern Oklahoma. The lake is 48 km long, up to 3 km wide, and 70 km² in area. Its depth varies from more

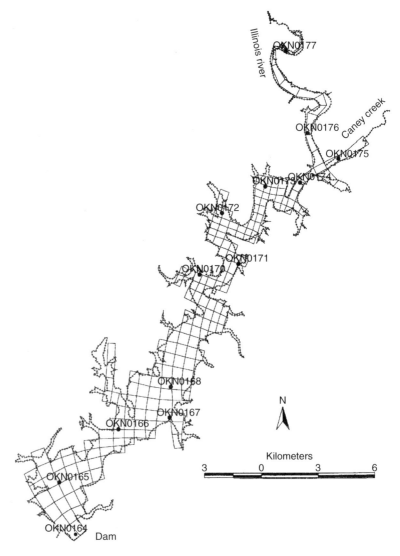

Figure 4. Lake Tenkiller study area, model grid, and data stations.

than 45 m near the dam to less than 6 m in the northern section. The lake has a retention time of 1.76 years. Major water quality issues include nutrient enrichment, eutrophication, and hypolimnetic dissolved oxygen (DO) depletion. Because of the 3-D variability of the lake, it is critical to simulate the hydrodynamic and water quality processes using a 3-D model, so that water quality parameters in the lake can be described in detail and cost-effective water management approaches can be proposed and evaluated.

Water Quality Simulations

The Lake Tenkiller model is also developed within the framework of the EFDC model (2) and includes the hydrodynamic, sediment, and water quality submodels. The water quality parameters considered, as presented in the entry WATER QUALITY MODELS: CHEMICAL PRINCIPLES, include algae, DO, organic nitrogen, ammonia, nitrite and nitrate, organic phosphorus, and phosphates.

The study area is divided into a grid of discrete cells. Figure 4 shows the model grid overlaying an outline of

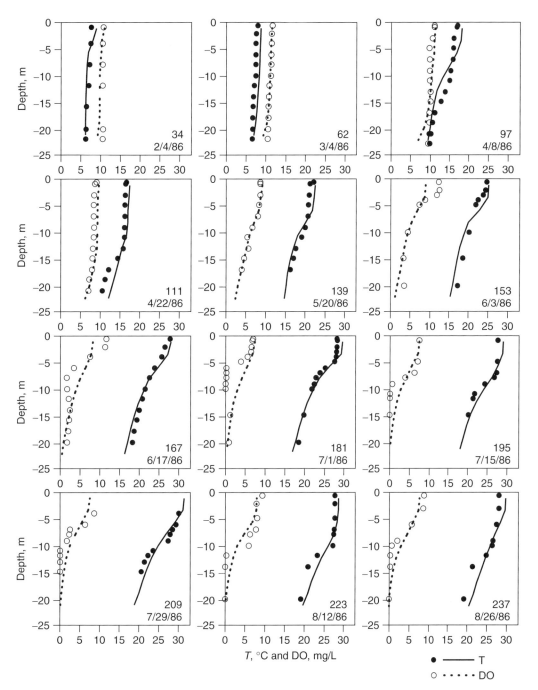

Figure 5. Vertical profiles of temperature and dissolved oxygen (DO) at Station OKN0166. Solid line = modeled temperature, dashed line = modeled DO, closed circles = measured temperature, and open circles = measured DO.

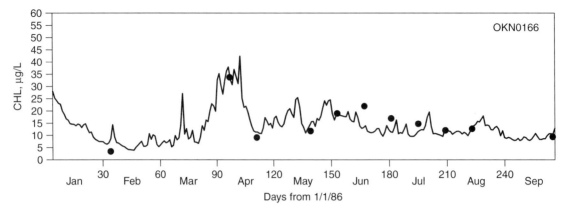

Figure 6. Model-data comparison of chlorophyll-a (CHL) at Station OKN0166. Solid line = modeled surface CHL, and closed circles = measured surface CHL.

Lake Tenkiller. The numerical grid consisted of 198 cells in the horizontal plane and 10 vertical sigma layers. The 10 vertical layers are necessary and important for resolving the vertical temperature and DO profiles in the lake. The time period for model calibration is 262 days from January 5, 1986 to September 24, 1986. Solutions of the model are obtained using a 50-second time step. Using a 2.4-Ghz Pentium IV PC, approximately 6 CPU hours are required for a 262-day simulation.

The 12 plots in Fig. 5 show the vertical profiles of modeled and measured water temperature (T) and DO at Station OKN0166. The horizontal axis represents T in °C and DO in mg/L. The vertical axis represents water depth in meters. The solid curve is the modeled T, and the closed circles are the measured T. The dashed curve is the modeled DO, and the open circles are the measured DO. Overall, the modeled T and DO are consistent with the measured data. Both the model and the data indicate that DO is vertically mixed at values around 10 mg/L in the winter and is very stratified at 8 mg/L or more on the surface and almost 0 mg/L at the bottom in the summer. The lake started to become stratified in April and exhibited strong thermal stratification in the summer; the surface-bottom T difference was up to 15 °C, and the DO difference up to 10 mg/L. It also appears that the model might need more vertical layers to resolve DO stratifications better in the summer.

Algae levels are often measured by chlorophyll-a concentrations (CHL) in water quality models. Figure 6 gives the modeled CHL (solid line) and the measured CHL (closed circle) at OKN0166, the same station whose T and DO are displayed in Fig. 5. Figure 6 reveals that algae in the lake vary from less than 10 µg/L in the winter to around 25 µg/L in the summer, both the model and the data indicate a strong algae bloom in the spring.

Conclusions

The dynamic simulation of eutrophication in a deep water system is a very complicated and computationally intensive endeavor because a large number of chemical, biological, and biochemical processes interact and the reaction rates and external inputs vary with time. Despite the progress in 3-D hydrodynamic, water quality, and sediment diagenesis models and their successful applications in estuaries and bays (10), few similar 3-D modeling studies on lakes and reservoirs have been published.

The primary purpose for developing the Lake Tenkiller model is to use the model as a tool for proposing and testing load-management strategies aimed at limiting eutrophication processes in the lake. The Lake Tenkiller model is calibrated and represents existing hydrodynamic and water quality processes in the lake satisfactorily and is a useful tool for managing eutrophication in the lake (11).

BIBLIOGRAPHY

1. Nix, J.S. (1990). Mathematical modeling of the combined sewer system. In: *Control and Treatment of Combined Sewer Overflows*. P.E. Moffa (Ed.). Van Nostrand Reinhold, New York, pp. 23–78.

2. Hamrick, J.M. (1992). *A Three-Dimensional Environmental Fluid Dynamics Computer Code: Theoretical and Computational Aspects*. The College of William and Mary, Virginia Institute of Marine Science, Special Report 317.

3. Ji, Z.-G., Hamrick, J.H., and Pagenkopf, J. (2002). Sediment and metals modeling in shallow river. *J. Environ. Eng.* **128**: 105–119.

4. Ji, Z.-G., Morton, M.R., and Hamrick, J.M. (2000). Modeling hydrodynamic and sediment processes in Morro Bay. In: *Estuarine and Coastal Modeling: Proc. 6th Int. Conf.* M.L. Spaulding and H.L. Butler (Eds.). ASCE, pp. 1035–1054.

5. Ji, Z.-G., Morton, M.R., and Hamrick, J.M. (2001). Wetting and drying simulation of estuarine processes. *Estuarine Coastal Shelf Sci.* **53**: 683–700.

6. Jin, K.R. and Ji, Z.G. (2003). Modeling of sediment transport processes and wind-wave impact in a large shallow lake. *J. Hydraul. Eng.* ASCE (tentatively accepted).

7. Jin, K.R. and Ji, Z.G. (2003). Application and validation of a 3-D model in a shallow lake. *J. Waterway Port Coastal, Ocean Eng.* (accepted).

8. Jin, K.R., Ji, Z.G., and Hamrick, J.M. (2002). Modeling winter circulation in Lake Okeechobee, Florida. *J. Waterway Port Coastal Ocean Eng.* **128**: 114–125.

9. Ji, Z.-G., Morton, M.R., and Hamrick, J.M. (2003). Three-dimensional hydrodynamic and water quality modeling in a reservoir. In: *Estuarine and Coastal Modeling: Proc. 8th*

Int. Conf. M.L. Spaulding and H.L. Butler (Eds.). ASCE (accepted).
10. Cerco, C.F. (1999). Eutrophication models of the future. *J. Environ. Eng.* **125**(3): 209–210.
11. Tetra Tech. (2000). *Water Quality Modeling Analysis in Support of TMDL Development for Tenkiller Ferry Lake and the Illinois River Watershed in Oklahoma*. Technical report to United States Environmental Protection Agency, Region 6, and Department of Environmental Quality, State of Oklahoma. Tetra Tech, Inc., Fairfax, VA.

FIELD SAMPLING AND MONITORING OF CONTAMINANTS

TIMOTHY J. DOWNS
Clark University
Worcester, Massachusetts

INTRODUCTION

Field sampling and monitoring of water/wastewater contaminants are strategically important for more sustainable integrated water resources management (IWRM). One of the vital first steps in IWRM is the collection of an inventory that characterizes existing watershed conditions. Relevant features include geographic and environmental aspects (e.g., land use, topography, wetlands), infrastructure (e.g., supply, sewerage and drainage), municipal data (e.g., numbers of people, growth rate, regulations, by laws), pollution sources/discharges (point and nonpoint sources, e.g., landfills, underground storage tanks), and the characteristics of receiving surface waters and groundwaters (indicators of flow rates and physical, chemical, biological, and microbiological quality).

Water quality problems that require characterization through sampling and monitoring are a combination of the effects of human use and degradation of water in domestic, industrial, commercial, and agricultural activities (and its discharge as wastewater), and the impact of human activities that change land use/land cover on the water quality of runoff. Wastewater discharge can be handled by treatment technologies, but during rainstorms, combined sewer overflows (CSOs) and sanitary sewer overflows (SSOs) can degrade water quality. In urban areas, domestic wastewaters contain sanitary sewage composed of oxygen-demanding wastes, nutrients, and pathogens. Industrial wastewaters contain metals and organic pollutants. Urban runoff from paved surfaces contains hydrocarbons, runoff from eroded land surfaces produces sediment-laden flows to receiving waters.

Chief among inventory data are contaminant-related data (1):

- pollutants of concern and effects on water resources;
- a base map for locating pollution sources and controls;
- areas of concern where pollutant loadings pose high public health and ecological health risks and where control efforts should focus;
- areas of high water quality where protection should focus; and
- information for developing water quality models, if needed.

The inventory of existing data reveals gaps to be filled by sampling and monitoring. Chambers Dictionary defines sampling as: the "taking of, testing of a sample; the examination and analysis of data obtained from a random group in order to deduce information about the population as a whole"; to monitor is to "watch, check, supervise" (2). The inventory also provides information resources for setting clear concise IWRM goals. The goals and data gaps determine sampling and monitoring regimes.

Six basic practical considerations of environmental sampling/monitoring are as follows

- Which environmental media to sample (surface water, groundwater, air, soil, sediment, biota), and which standardized sampling method to use to obtain the sample?
- Where in the contaminant-environment cycle to take samples (physical location)?
- Which indicators/parameters of environmental quality to measure and which laboratory analysis methods to apply to detect them?
- When to take samples (times of the year, e.g., wet and dry seasons, cold and hot seasons)?
- How often to sample (frequency of sampling, subhourly, hourly, daily, weekly, monthly)?
- How to report contamination results (implied risks, impacts, data and modeling limitations and assumptions)?

Figure 1 shows stages in sampling/monitoring as an adaptive process and its input data.

Knowing the following greatly informs water–contaminant system sampling and characterization:

- physicochemical characteristics of the contaminant of interest especially its degree of water phase affinity (hydrophilicity) or water phase aversiveness (hydrophobicity),
- dominant removal processes—decay by chemical reaction (e.g., hydrolysis, aqueous oxidation or reduction), sunlight degradation (photolysis), and/or microbial (bio-) degradation;
- dominant transport processes: sediment transportation in runoff, soil erosion and washing into rivers and lakes, leaching into aquifers, dry deposition, or wet deposition from air into water;
- contaminant toxicity and risk: human cancer and noncancer health effects from sufficient exposure by ingestion, inhalation, and/or dermal/ocular contact; and nonhuman health effects (ecotoxicology).

SPATIAL CONSIDERATIONS—WHERE TO SAMPLE AND MONITOR

Conservation of mass (see Eq. 1) always applies, so scientific sampling and monitoring of water/wastewater quality

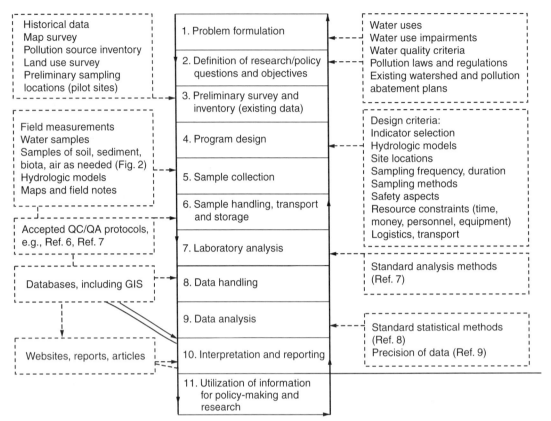

Figure 1. Stages of water quality monitoring. Shows determining factors, sources of data, and standard methods (expanded from Ref. 5). Stages 1–11 can be thought of as an adaptive feedback loop, responding to changing conditions, interests, capabilities, and priorities over time. In this way, the sampling/monitoring becomes sustainable. A key dimension to the relative sustainability is the strengthening of capacities/resources (human, informational, policy-making, financial, and equipment).

seek to take measurements that provide information about the following aspects of the contaminant source stream and the receiving water stream (4):

- transport processes of contaminants from sources to receiving waters (advection, diffusion, volatilization, atmospheric deposition)
- contaminant transformation processes from pollution source to receiving waters
- the timescale on which dominant processes operate
- the spatial scale on which dominant processes operate
- the factors governing these time and space rates
- the existing data providing information about the above
- the missing data needed to be collected
- the models that can be used to represent the processes and to explore contaminant control and water management change scenarios.

$$\begin{aligned}\text{Rate of contaminant mass increase in water}\\ \text{(or any bounded space)} = \text{rate of mass input}\\ - \text{rate of mass output} + \text{rate of mass created}\\ \text{internally} - \text{rate of mass lost internally}\end{aligned} \quad (1)$$

Under different hydrologic conditions, the relative contribution to pollution loading by nonpoint versus point sources can change dramatically. For example, agricultural sources of nutrients and urban runoff sources of hydrocarbons may become dominant during storms compared to factory effluents and sewage treatment effluents that dominate pollution loading in dry conditions. Figure 2 provides a useful schematic framework for sampling/monitoring water and wastewater flows.

TEMPORAL CONSIDERATIONS—WHEN AND HOW OFTEN TO SAMPLE

Watershed assessments can be simple or detailed; the latter require more data collected over an extended period of time. Simple assessments (taking weeks) can be used to design more detailed ones (taking months or years) or are used when there are extreme constraints on resources that can be applied (the case in many poor countries). The following conditions require a detailed assessment (4):

- high variability in water quality and quantity in time and/or space;
- receiving waters for pollution are complex (e.g., large lakes, rivers, nonuniform aquifers), requiring two- or three-dimensional sampling over time;

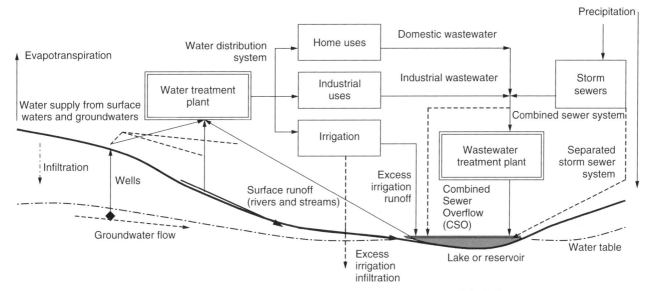

Figure 2. Water and wastewater flows to and from receiving waters. A full hydrologic picture is needed to assess where sampling of water quality and quantity should occur for baseline (preaction) characterization and where sampling should be done to monitor the performance of contaminant controls and water management actions (expanded from Ref. 3).

- existing data are insufficient to model the impacts of management alternatives on the water resources; and/or
- complex supply and sanitation systems need to be modeled to optimize their performance under varying conditions (e.g., treatment plants that have multiple unit process stages between influent and effluent).

Contaminants are sampled and monitored to determine the nature and magnitude of contamination and the sources of it. Simple assessments rely largely on existing data; detailed water quality assessments involve considerable field sampling to estimate the hydrograph (flow vs. time) and its associated pollutograph (pollution levels vs. time).

Sampling design is a statistical question. A hypothesis is posed, for example, "Groundwater contamination by arsenic (a highly toxic, naturally occurring substance that has potential cancer and noncancer health effects) represents a potential health risk (if enough water is consumed) because levels exceed health standards on the average." To test such hypotheses requires taking samples that allow us to estimate the 'real' variability of the values. Samples allow us to determine the sample mean (x_m) and sample variance (s^2), that are used to estimate the population/complete system's mean (μ) and variance (σ^2). The challenge, as in any sampling, is to characterize population variability using a representative sample. Sample statistics and a knowledge of the way values are distributed by probability (normally, log normally, or otherwise), allows us to accept or reject a hypothesis at a given confidence level. Detailed sampling methods for specific situations are given in Reference (10).

The minimal water quality sample requirements are three samples, representative of low, moderate, and high contaminant concentrations. But to gauge "representativeness" requires knowledge of variability. The goal of a good sampling program is to provide information to scientists and policy-makers on the following (4):

- probable maximum and minimum values;
- variability on different timescales (hours, days, months, years, decades as required);
- correlations with associated variables (hydrological, meteorological, pollutant production rates); and
- evidence about sources of contamination, environmental transport, and degradative/transformative processes.

Point-pollution sources, such as factories and sewage treatment plants, can be highly variable in time for the level (concentration as mass of contaminant per unit volume) and type (microbial, organic chemical, inorganic chemical, conservative, or nonconservative) of contaminants. Adding natural variability in environmental conditions and receiving waters (flow rates, temperatures, pH, other contaminant levels), it is evident that data quality and reliability are highly sensitive to the sampling regime selected.

Samples are either discrete or composite. If sampling seeks merely to estimate average levels of contaminants in water or wastewater, composite samples can save time by combining subsamples (aliquots) of discrete samples together. However, information about variability (mean, median, mode, variance, ranges, 90–95% confidence intervals, probability distributions) is lost, and it is these data that are most relevant to risk assessments and management policy. There are important trade-offs between sampling/monitoring costs and information value. Clear sampling/monitoring objectives must guide the optimization of this trade-off.

Rapidly changing conditions of water quantity and quality over time often call for flow- or quality-sensitive sampling: the higher the variability, the higher the sampling frequency. Under such conditions, it is desirable to monitor the qualities and quantities of contaminant sources (wastewater flows) and receiving waters simultaneously.

WATER QUALITY INDICATORS—CRITERIA FOR SAMPLING/MONITORING PARAMETERS

It is helpful to classify water contaminants into eight groups: pathogens, oxygen-demanding wastes, nutrients, salts, thermal pollution, heavy metals, pesticides, and volatile organic compounds (11). Each water use has quality requirements monitored by these indicators.

Pathogens (Disease-Causing Organisms)

These belong to four main groups:

1. bacteria—causing diseases like cholera, bacillary dysentery (shigellosis), typhoid, and paratyphoid fever;
2. viruses—diseases including infectious hepatitis, poliomyelitis;
3. protozoa—causing amoebic dysentery, giardiasis, and cryptosporidiosis;
4. helminths (parasitic worms)—causing schistosomiasis, and dracunculiasis (guinea worm)

There are four main types of water pollution diseases: waterborne (by ingestion); water-washed (lack of clean water for washing); water-based (involve water contact, no ingestion); and water-related (pathogen host relies on water habitat; humans need no water contact).

Oxygen-Demanding Wastes (ODWs)

Dissolved oxygen (DO) is a primary water quality indicator (8 mg/L needed for coldwater fish; 5 mg/L often used for others) and a saturation value 8–15 mg/L (depending on temperature, salinity). ODWs oxidize in receiving waters as bacteria use oxygen to degrade them. As DO drops, aquatic life is threatened, and odors, tastes, and color become undesirable. ODWs are mainly biodegradable wastes from households and livestock and effluents from food processing and paper production. The wastewater indicators are chemical oxygen demand (COD)—the amount of oxygen needed to oxidize wastes chemically—and biochemical oxygen demand (BOD)—the amount needed by microbes. The BOD of a diluted wastewater sample after 5 days (BOD_5) is used to measure the performance of wastewater treatment plants (WWTPs).

Nutrients

The chemicals essential to life—nitrogen (N), phosphorus (P), carbon (C), calcium (Ca), potassium (K), iron (Fe), manganese (Mn), boron (B), and cobalt (Co)—are viewed as contaminants when their levels allow excessive algal growth and compromise drinking uses, recreation, and habitat for other species. Nutrient enrichment occurs by *eutrophication*, important in lakes. C, N, and P are the most important nutrients. The limiting nutrient controls growth, and it is usually N in seawaters and P in freshwaters. Major N sources are WWTPs, animal feedlot runoff, chemical fertilizers, and N deposits from polluted air. NO_3 (nitrate) is the most common form in water. P is limited in nature, but human actions contribute it: runoff of fertilizers and domestic sewage (feces and detergents).

Salts

Dissolved solids, cations, anions—total dissolved solids (TDS)—is the indicator used: TDS <1500 mg/L is freshwater, 1500–5000 mg/L is brackish water, and saline water is >5000 mg/L. Seawater has 30–34,000 mg/L TDS. Drinking water levels must be <500 mg/L for humans but can be higher for livestock. Irrigation waters with TDS >500 mg/L and poor drainage can lead to salt buildup in soils that affect crops. Water is unsuitable for irrigation above 2100 mg/L.

Thermal Pollution

Cooling waters from power plants change water temperature, affecting aquatic species. Warming benefits some species over others, but power outages create sudden drops in temperature. Increases in temperature increase organism metabolic rates so oxygen demand goes up, waste degradation rates also go up, and the amount of DO that water can hold drops. All these factors mean that DO is drawn down.

Heavy Metals

These elements give up one or more electrons in aqueous solution to form cations (X^{n+}). "Heavy" means that the specific gravity is >4–5. Those most important for environmental impact are mercury (Hg), lead (Pb), cadmium (Cd), and arsenic (As). Metals are nondegradable and become toxic above trace levels.

Pesticides

Insecticides, herbicides, rodenticides, and fungicides are of three main chemical groups: organochlorines (OCPs or chlorinated hydrocarbons, e.g., DDT, Kepone®, aldrin, dieldrin), organophosphates (OPPs, e.g., parathion, malathion, diazinon), and carbamates (e.g., aldicarb). OCPs are persistent and tend to bioaccumulate in fatty tissue. OPPs are not persistent, but they are more acutely toxic to humans than OCPs. Carbamates are also short-lived and of high acute toxicity to humans.

Volatile Organic Compounds (VOCs)

Their high volatility means we find low levels in surface waters (a few μg/L), but when released as solvent waste into soil, they migrate to groundwater, and levels can be 100–1000 times surface water levels. Five VOCs are very toxic: vinyl chloride (VC), tetrachloroethylene (PCE), trichloroethylene (TCE), 1,2-dichloroethane (metal degreaser), and carbon tetrachloride (household cleaner, grain fumigant, solvent). VC is a known human carcinogen that is used in the production of PVC resins. TCE, a solvent cleaner used on engines, electronics, even septic tanks, is the most common groundwater contaminant in the United States.

WATER QUALITY STANDARDS—U.S., EUROPEAN UNION, AND WORLD HEALTH ORGANIZATION

Water quality (and treatment) standards depend on the usage of the water—drinking, recreational, industrial, agricultural, hydroelectric, or ecological.

World Health Organization (WHO) Standards

WHO standards are used as the reference for all national standards (12,13). There are five types of sampling/monitoring parameters:

1. Bacteriological: *E. coli* or thermotolerant coliform bacteria and total coliform bacteria must not be detectable in drinking water anywhere in the distribution system (once leaving the treatment plant).
2. Inorganic chemical (levels in mg/L): 19 inorganics.
3. Organic chemical (levels in µg/L): 11 classes of organic substances: (i) chlorinated alkanes (five parameters, e.g., dichloromethane); (ii) chlorinated ethenes (five, e.g., vinyl chloride); (iii) aromatic hydrocarbons (six, e.g., benzene); (iv) chlorinated benzenes (five, e.g., 1,2-dichlorobenzene); (v) miscellaneous organics (nine, e.g., acrylamide); (vi) pesticides (30, e.g., lindane); chlorophenoxy herbicides (six, e.g., 2,4-DB); (vii) disinfectants (five, e.g., chlorine); (viii) disinfectant by-products (12, e.g., bromoform); (ix) chlorinated acetic acids (five, e.g., dichloroacetic acid); (x) haloacetonitriles (six parameters, e.g., chloropicrin).
4. Radionuclide (Bq/L): gross alpha-particle and gross beta-particle activity screening levels.
5. Secondary physicochemical "nuisance" parameters: 18 inorganic/physical parameters (e.g., iron, copper, color, turbidity); and 13 organics.

Standards for carcinogens, such as the pesticide alachlor, are given as levels for a given acceptable risk, based on the precautionary assumption that any exposure to a carcinogen has a finite risk.

European Union Standards

Domestic Supply. In the European Union (EU), raw surface waters intended for domestic supply (the most stringent quality) are placed into three categories—A1, A2, A3—based on the sampling of 46 parameters from pH, odor, through metals, organics, nitrate, nitrogen, ammonia, to total and fecal coliforms, fecal streptococci and *Salmonella*. Each category has a guide limit and a (higher) mandatory limit, and each must undergo different water treatment, ranging from simple physical treatment and disinfection (A1), through normal physicochemical treatment and disinfection (A2), to intensive physicochemical treatment and disinfection (A3) (14).

Once the waters have been treated, drinking water must be sampled and monitored to meet the new (1998) EU drinking water standard grouped as follows (15):

- Microbiological monitoring: Tap water—*E. coli* and enterococci (both 0/100 mL); water for sale in bottles and containers—*E. coli* and enterococci (both 0/250 mL), bacterial colony counts at 22 °C (100/mL), 37 °C (20/mL), *Pseudomonas aeruginosa* (0/250 mL).
- Chemical monitoring: 26 parameters—acrylamide, antimony, arsenic, benzene, benzo(a)pyrene, boron, bromate, cadmium, chromium, copper, cyanide, 1,2-dichloroethane, epichlorohydrin, fluoride, lead, mercury, nickel, nitrate, nitrite, specified pesticides, total specified pesticides, specified polyaromatic hydrocarbons (PAHs), selenium, tetrachloroethene plus trichloroethane, total specified trihalomethanes, and vinyl chloride.
- Indicator monitoring parameters: aluminum, ammonium, chloride, *Clostridium perfringens*, color, conductivity, pH, iron, manganese, odor, sulfate, sodium, taste, bacterial colony count, coliform bacteria, total organic carbon, turbidity, tritium (radioactivity), and total indicative radiation dose.

Sampling is done exclusively at consumers' taps under the EU Directive (98/83/EEC).

Surface Waters Not Used Primarily for Supply. The EU Freshwater Fish Directive specifies 14 physicochemical parameters to protect salmonid and cyprinid waters: temperature (where there is a thermal discharge), dissolved oxygen, pH, suspended solids, biochemical oxygen demand, total phosphorus, nitrite, phenolic compounds, petroleum hydrocarbons, nonionized ammonia, total ammonium, total residual chlorine, total zinc, and dissolved copper. Guide and mandatory values are given. The EU also has a Bathing Water Directive (76/110/EEC) for inland and coastal waters. Coastal and estuarine waters are monitored using these physicochemical and microbial water quality parameters, most sampled every two weeks or more frequently when water quality is compromised (15):

- Microbiological: total coliforms, fecal coliforms, fecal streptococci, *Salmonella*, entero viruses;
- Physicochemical: pH, color, mineral oils, surfactants that react with methylene blue, phenols, transparency, dissolved oxygen (% saturation), tarry residues and floating debris, ammonia, nitrogen (Kjeldahl), plus any other substances chosen as appropriate indicators of pollution (site specific).

Wastewaters. The EU monitoring of wastewaters is broken into toxic, persistent, and potentially bioaccumulating substances (black list) and less toxic substances or those whose effects are confined to a limited area depending on the nature of the receiving water (gray list). The Dangerous Substances Directive provides for monitoring the following parameters in wastewater streams:

- List 1 ("black" list): organohalogen compounds (or substances that may form them in aquatic environments), organophosphorus compounds, organotin compounds, carcinogens in the aquatic environment (from gray list), mercury and its compounds, cadmium and its compounds, persistent mineral oils and petroleum hydrocarbons, and persistent synthetic substances.
- List 2 ("gray" list): metalloids/metals and their compounds, biocides and their derivatives, substances that adversely affect taste and odor, toxic/persistent silicon compounds (or substances that may form them in aquatic environments), inorganic compounds of phosphorus and elemental phosphorus, nonpersistent mineral oils and petroleum hydrocarbons, cyanides, fluorides, and oxygen-depleting substances such as ammonia and nitrites.

Groundwaters. EU List 1 substances must be prevented from entering groundwaters, and pollution by List 2 substances must be strictly regulated and monitored. Strict hydrologic and environmental impact must be assessed before any listed substance can be licensed for disposal to groundwater in the EU.

U.S. Standards

U.S. Chemical Standards. In the United States, drinking water standards are of two types: (1) primary standards, specified as maximum contaminant levels (MCLs) derived from health concerns; and (2) unenforceable secondary standards based on taste, odor, color, hardness, and corrosivity. Maximum contaminant level goals (MCLGs) are levels that represent no anticipated health effects, including a margin of safety, regardless of attainment cost and feasibility. Inorganic substances that must be monitored (and for which there are MCLs in the U.S.) include toxic heavy metals (e.g. arsenic, cadmium, lead, mercury); nitrites (NO_2) and nitrates (NO_3); fluoride; and asbestos fibers. Organic chemical contaminants that have MCLs in the United States can be put in three groups (16):

1. Synthetic organic chemicals (SOCs)—used in the manufacture of agricultural and industrial products, such as insecticides and herbicides.
2. Volatile organic chemicals (VOCs)—synthetic substances that vaporize at room temperature. Examples include degreasing agents, paint thinners, glues, dyes. Specific examples include benzene, trichloroethylene (TCE), and vinyl chloride.
3. Trihalomethanes (THMs)—by-products of water chlorination, including chloroform, dibromochloromethane, and bromoform.

The levels are measured in milligrams per liter (mg/L). For the MCLs of primary drinking water contaminants regulated in the United States see (17,18).

U.S. Microbiological Standards. Instead of testing routinely for single pathogenic organisms, simpler tests are used that measure indicator organisms, principally coliform bacteria (e.g., *E. coli*), that indicate the contamination by fecal material. Levels of coliform bacteria in feces and untreated domestic wastewater are very high. In the United States, microbial standards require that for 'large' water supplies (serving more than 1000 people), no more than 5% of test samples can show any coliforms in smaller systems, testing less than 40 samples/month, no more than one sample can test positive (16). This method works when pathogen survival rates in water (outside their host) are less than those of coliforms. But nonbacterial pathogens—especially viruses, *Cryptosporidium* and *Giardia* cysts—often outlive coliforms and can be present when no coliform are detected. Treatment techniques for these contaminants are specified instead of MCLs in the United States. Coliform sampling is also used to monitor water-contact recreational water quality; typical standards are 1000 coliforms per 100 milliliters.

U.S. Radionuclide Standards. Naturally occurring radionuclide contaminants include radon and radium-226, often found in groundwater. Others, such as strontium-90 and tritium contaminate surface water after nuclear fallout. The levels are measured in picocuries per liter (pCi/L). Dissolved radon gas is the most prominent groundwater contaminant of this type. Colorless, odorless, and tasteless, the risk derives not from ingestion but from inhalation as it outgases from heated shower water.

BIBLIOGRAPHY

1. Bingham, D.R. (1998). Watershed management in U.S. urban areas. In *Watershed Management—Practice, Policies and Coordination*. R.J. Reimold (Ed.). McGraw-Hill, New York, pp. 181–189.
2. Kirkpatrick, E.M. (Ed.). (1983). *Chambers 20th Century Dictionary*. W & R Chambers Ltd., Edinburgh.
3. Masters, G.M. (1996). *Introduction to Environmental Engineering and Science*, 2nd Edn. Prentice-Hall, Englewood Cliffs, NJ.
4. Heathcote, I.W. (1998). *Integrated Watershed Management—Principles and Practice*. John Wiley & Sons, Hoboken, NJ, pp. 196–206.
5. Gray, N.F. (1999). *Water Technology—An Introduction for Environmental Scientists and Engineers*. Arnold, London, p. 135.
6. ISO (1990). *Water Quality Sampling—International Standard 5667*. International Organization for Standardization, Geneva.
7. APHA (1989). *Standard Methods for the Examination of Water and Wastewater*. American Public Health Association, Washington, DC.
8. Demayo, A. and Steel, A. (1996). Data Handling and Presentation. In: *Water Quality Assessments*. D. Chapman (Ed.). E and FN Spon, London, pp. 511–612.
9. Caulcutt, R. and Boddy, R. (1983). *Statistics for Analytical Chemists*. Chapman and Hall, London.
10. Ostler, N.K. and Holley, P.K. (Eds.) (1997). *Sampling and Analysis, Environmental Technology Series*, Vol. 4. Prentice-Hall, Upper Saddle River, NJ.

11. Masters, G.M. (1996). *Introduction to Environmental Engineering and Science*, 2nd Edn. Prentice-Hall, Englewood Cliffs, NJ, pp. 175–184.
12. WHO (1993). *Revision of WHO Guidelines for Drinking Water Quality*. World Health Organization, Geneva.
13. WHO (1993). *Guidelines for Drinking Water Quality*, Vol. 2, Health Criteria and Other Supporting Information. World Health Organization, Geneva.
14. Gray, N.F. (1999). *Water Technology—An Introduction for Environmental Scientists and Engineers*. Arnold, London, pp. 1–5.
15. Gray, N.F. (1999). *Water Technology—An Introduction for Environmental Scientists and Engineers*. Arnold, London, pp. 16–18.
16. Masters, G.M. (1996). *Introduction to Environmental Engineering and Science*, 2nd Edn. Prentice-Hall, Englewood Cliffs, NJ, pp. 269–274.
17. Hammer, M.J. and Hammer, M.J. Jr. (1996). *Water and Wastewater Technology*, 3rd Edn. Prentice-Hall, Englewood Cliffs, NJ.
18. USEPA (1995). *National Primary Drinking Water Standards*, U.S. Environmental Protection Agency, Washington, DC.

WATER QUALITY MODELS: CHEMICAL PRINCIPLES

Zhen-Gang Ji
Minerals Management Service
Herndon, Virginia

Water quality represents the physical, chemical, and biological characteristics of water resources and is used to measure the ability of a waterbody to support beneficial uses. Eutrophication is a process of nutrient overenrichment of a waterbody, resulting in accelerated biological productivity (growth of algae and weeds). Symptoms of eutrophication include algal blooms, reduced water clarity, and oxygen depletion. In modeling studies, water quality and eutrophication are sometimes used interchangeably to represent the processes of waterbody enrichment by nutrients. Key state variables in water quality models include algae, nutrients, and dissolved oxygen (DO).

This entry focuses on water quality models. Information on applications of water quality models is provided in Water Quality Modeling—Case Studies.

ALGAE AND PHYTOPLANKTON

Algae are a group of aquatic microscopic plants that contain chlorophyll and grow by photosynthesis. Algae takeup nutrients, including phosphate, ammonium, nitrate, silica, and carbon dioxide, from water or benthic sediments and release oxygen to the water. Algae may be free-floating or rooted on the bottom of a waterbody. An overabundance of algae in a waterbody is known as eutrophication.

Phytoplankton are a group of tiny, free-floating plants transported by currents. In water quality modeling studies, phytoplankton are frequently referred to as free-floating algae and are often more important than rooted aquatic vegetation in the basic food production of an ecosystem. They are the most biologically active plants in aquatic ecosystems and generally have greater influences on water quality than other plants. Phytoplankton is also the most important element and the first stage in the food chain: it provides food for zooplankton, fish, and small aquatic animals.

NUTRIENTS

Nutrients are chemical elements or compounds necessary for algal growth. The essential nutrients include nitrogen, phosphorus, carbon dioxide, and silica. A host of other micronutrients, such as iron, manganese, potassium, sodium, copper, zinc, and molybdenum, are also required for algal growth. However, these minor nutrients are generally not considered in water quality modeling because they are needed only in trace amounts, and they are usually present in quantities sufficient for algal growth.

Excessive nutrient levels can be harmful to ecosystems by causing eutrophication and unwanted algal blooms. Major processes affecting nutrient concentrations include

1. algal uptake
2. mineralization and decomposition of dissolved organic nutrients
3. hydrolysis converting particulate organic substances into dissolved organic form
4. chemical transformations of nutrients
5. sediment sorption and desorption
6. settling of particulate matter
7. nutrient fluxes from benthic sediments
8. external nutrient loadings

The two nutrients of greatest concern are nitrogen and phosphorus. Nitrogen, one of the most abundant elements on the earth, makes up 78% of the earth's atmosphere as a gas. Nitrogen exists in these different forms:

1. organic nitrogen (ON)
2. ammonia (NH_3)
3. nitrite and nitrate ($NO_2 + NO_3$)
4. nitrogen gas (N_2)
5. algae
6. zooplankton and aquatic animals

Phosphorus is one of the vital nutrients for algal growth. However, too much phosphorus can cause excessive algal growth and can lead to eutrophication. Compared to nitrate (NO_3), phosphates dissolve less readily and tend to attach to sediment particles. Phosphorus exists in different forms, including

1. organic phosphorus
2. phosphates (largely orthophosphates—PO_4)
3. algae
4. zooplankton and aquatic animals

In modeling water quality processes, these are the major differences between nitrogen and phosphorus:

1. Unlike nitrogen, phosphorus does not have a gas phase, which can be a major nitrogen loss mechanism in certain aquatic systems.
2. Forms of inorganic nitrogen are easily dissolved in water; phosphates are often strongly sorbed to sediment. Therefore, phosphates can settle with sediment solids to the bottom of a waterbody, such as a lake or a reservoir, and become a phosphorous source to the waterbody. In many waterbodies, bottom sediments contain enough phosphorus to accelerate eutrophication, even after external sources have been terminated.

Figure 1 illustrates processes that affect nutrients in an aquatic system:

1. Physical transport: Nutrients are advected and dispersed within the water column and are transported into the system by inflow and out of the system by outflow.
2. Exchange with the atmosphere: Atmospheric deposition adds nutrients to a waterbody, and volatilization removes gaseous nutrients from the waterbody. Reaeration adds DO to the waterbody.
3. Sorption and desorption: Exchanges between particulate nutrients and dissolved nutrients are affected by the total suspended solid concentrations and the partition coefficients.
4. Reaction and algal uptake: Chemical reactions and algal uptake transform nutrients and reduce the concentrations of dissolved nutrients.
5. Exchange on the bed–water interface: Dissolved nutrient is exchanged between the sediment bed and the water column via diffusion, and the particulate nutrient can be settled onto or be resuspended from the bed, depending on flow conditions.
6. Sediment diagenesis: In the sediment bed, sediment diagenesis can be a significant factor in nutrient cycling and oxygen balance in the water column.

Note that not all processes shown in Fig. 1 are essential to every nutrient. For instance, volatilization is insignificant to phosphorus cycling, and sorption and desorption are not essential to nitrogen transformation.

DISSOLVED OXYGEN

Dissolved oxygen, the amount of oxygen that is dissolved in water, is one of the most important parameters of water quality. Most fish and aquatic insects need DO to survive. Fish, especially larvae, die when DO levels get too low. Low DO is a sign of possible pollution in a waterbody. As DO levels in water drop below 5.0 milligrams/liter (mg/L), aquatic life is put under stress. The lower the concentration, the greater the stress. Low DO levels for a prolonged period can result in large fish kills.

Oxygen enters water by reaeration from the atmosphere and by plant photosynthesis. Oxygen concentrations in the water column fluctuate under natural conditions, but severe oxygen depletion usually results from human pollution. Figure 2 shows the measured DO for 20 months from November 1992 to June 1994 in Lake Wister, Oklahoma (1). The lake exhibits strong seasonal DO variations. During the winter months (November, December, and January), DO in the lake is high; values are around 10 mg/L, and it is well mixed vertically. In the summer, DO is very stratified; values are less than 1.0 mg/L in the lower portion of the lake (water depth from 9 to 18 m), whereas the surface DO can still be 8 mg/L or higher. This large DO gradient in the vertical is caused primarily by temperature stratification of the lake, biochemical processes in the water column, and benthic oxygen demand at the lake bottom.

WATER QUALITY MODELS

Variations in water quality variables, such as algae, nutrients, and dissolved oxygen, are often described by

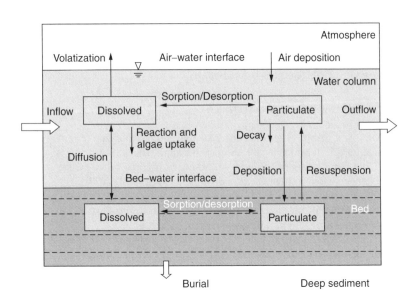

Figure 1. Processes of nutrient transformation.

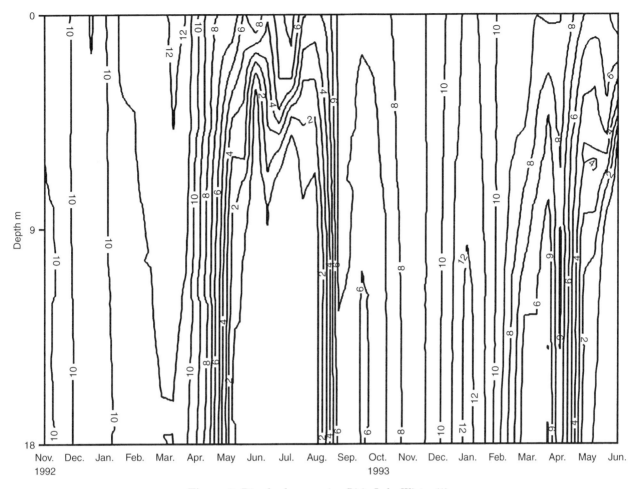

Figure 2. Dissolved oxygen (mg/L) in Lake Wister (1).

using a set of mass conservation equations in a water quality model. The conservation of mass accounts for all material entering or leaving a waterbody; transport of material within the waterbody; and physical, chemical, and biological transformations of the material. Therefore, all governing equations for water quality processes have a form similar to

$$\frac{\partial C}{\partial t} + \frac{\partial (uC)}{\partial x} + \frac{\partial (vC)}{\partial y} + \frac{\partial (wC)}{\partial z} = \frac{\partial}{\partial x}\left(K_x \frac{\partial C}{\partial x}\right)$$
$$+ \frac{\partial}{\partial y}\left(K_y \frac{\partial C}{\partial y}\right) + \frac{\partial}{\partial z}\left(K_z \frac{\partial C}{\partial z}\right) + S_C \quad (1)$$

where
C = concentration of a water quality state variable
u, v, w = velocity components in the x, y, and z directions, respectively
K_x, K_y, K_z = turbulent diffusivities in the x, y, and z directions, respectively
S_C = internal and external sources and sinks per unit volume

Equation 1 incorporates transport due to flow advection and dispersion, external pollutant inputs, and kinetic interaction among water quality variables. The last three terms on the left-hand side of Equation 1 account for advection transport. The first three terms on the right-hand side of Equation 1 account for diffusion transport. The term S_C represents kinetic processes and external loads.

In water quality models, kinetic processes are often decoupled from physical transport processes. By linearizing S_C with respect to C, the equation for kinetic processes and external loadings, called the kinetic equation, is

$$\frac{\partial C}{\partial t} = S_C = k \times C + R \quad (2)$$

where k = kinetic rate (time^{-1})
R = source/sink term due to external loadings and/or internal reactions (mass volume^{-1} time^{-1}).

The governing equations, Equations 1 and 2, are widely used in water quality models. Water quality and eutrophication processes are very complicated. Even though all kinetic equations are based on mass-balance equations, empirical formulations are often used as approximations for specifying model parameters, such as k and R. Therefore, for the same water quality process, there might be a variety of ways to describe it mathematically. The major differences among water quality models are

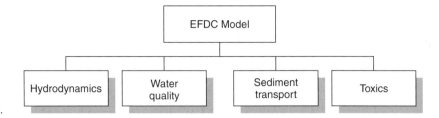

Figure 3. Primary modules of the EFDC model.

primarily in the way the kinetic equations are specified, how many nutrients are considered, and how many state variables are used to describe each nutrient cycle. Mathematical expressions of these terms can vary from model to model (such as References 2 and 3).

A water quality model typically has the following water quality variable groups:

1. algae
2. organic carbon
3. phosphorus
4. nitrogen
5. silica
6. other water quality variables

Each group may consist of several state variables, representing different components of the group. For example, a nutrient can be dissolved, particulate, refractory, or labile.

A water quality model is a discretized version of a set of differential equations, such as Equations 1 and 2, that describe processes in a waterbody. The discretized set of equations is converted into computer code, so that numerical solutions of the model can be derived by computer.

Numerical models span a wide variety of approaches and levels of sophistication and are typically categorized in terms of their representation of space and time:

1. steady-state or time-dependent (dynamic)
2. zero-, one-, two, or three-dimensional

The temporal characteristics indicate whether the model is steady-state (inputs and outputs constant over time) or time-dependent (dynamic). A steady-state model uses constant values of input variables to produce time-independent results, in which the state variables are independent of time. In contrast, a time-dependent model describes the temporal variability of a waterbody.

Since the 1990s, three-dimensional hydrodynamic and water quality modeling has matured from a research subject to a practical analytical technology. The rapid development of the computer industry provides more and more powerful computing ability for numerical simulation. Computational requirements for realistic three-dimensional modeling have changed from supercomputers and high-end workstations to desktop personal computers. Water quality models are more and more widely used in water resource management (4,5).

A variety of water quality models have been developed for modeling rivers, lakes, estuaries, and coastal areas. They can be categorized as

1. One-dimensional and steady-state models. They are one-dimensional in space and do not vary with time, such as the QUAL2E model (2).
2. Two-dimensional and time-dependent models. They are two-dimensional in space and time-dependent, such as the laterally averaged W2 model (6).
3. Comprehensive models. They are (1) three-dimensional in space and time-dependent, (2) with a hydrodynamic submodel to provide physical transport to the water quality submodel, and (3) coupled with a sediment diagenesis submodel to calculate sediment fluxes from the benthic sediments. The Environmental Fluid Dynamics Code (EFDC) (7,8), supported by the U.S. Environmental Protection Agency, is in this category. The primary modules of the EFDC model are shown in Fig. 3, which indicates submodels for hydrodynamics, water quality process, sediment transport, and toxic processes.

BIBLIOGRAPHY

1. OWRB. (1996). *Final Report for Cooperative "Clean-Lakes" Project, Phase I: Diagnostic and Feasibility Study on Wister Lake*. Oklahoma Water Resources Board, United States Army Corps of Engineers, and Oklahoma State University, Stillwater, OK.

2. Brown, L.C. and Barnwell, T.O. (1987). *The Enhanced Stream Water Quality Models QUAL2E and QUAL2E-UNCAS: Documentation and User Manual*. U.S. Environmental Protection Agency, Athens, GA. EPA/600/3-87-007.

3. Cerco, C.F. and Cole, T.M. (1994). *Three-Dimensional Eutrophication Model of Chesapeake Bay: Volume 1, Main Report*. Technical Report EL-94-4, U.S. Army Engineer Waterways Experiment Station, Vicksburg, MS.

4. Ji, Z.-G., Hamrick, J.H., and Pagenkopf, J. (2002). Sediment and metals modeling in a shallow river. *J. Environ. Eng.* **128**: 105–119.

5. Ji, Z.-G., Morton, M.R., and Hamrick, J.M. (2004). Three-Dimensional Hydrodynamic and Water Quality Modeling in a Reservoir. In: *Estuarine and Coastal Modeling: Proc. 8th Int. Conf.* M.L. Spaulding and H.L. Butler (Eds.). ASCE (accepted).

6. Cole, T.M. and Buchak, E.M. (1995). *CE-QUAL-W2: A Two-Dimensional, Laterally Averaged, Hydrodynamic and Water Quality Model, Version 2.0*. Instruction Report EL-95, U.S. Army Engineer Waterways Experiment Station, Vicksburg, MS.

7. Hamrick, J.M. (1992). *A Three-Dimensional Environmental Fluid Dynamics Computer Code: Theoretical and Computational Aspects*. The College of William and Mary, Virginia Institute of Marine Science, Special Report 317.
8. Park, K., Kuo, A.Y., Shen, J., and Hamrick, J.M. (1995). *A Three-Dimensional Hydrodynamic–Eutrophication Model (HEM3D): Description of Water Quality and Sediment Processes Submodels*. The College of William and Mary, Virginia Institute of Marine Science, Special Report 327.

WATER QUALITY MODELS: MATHEMATICAL FRAMEWORK

K.W. CHAU
The Hong Kong Polytechnic University
Hung Hom, Kowloon, Hong Kong

INTRODUCTION

In recent decades, accompanying agricultural and industrial development and population growth, it has often been observed that an increasing tendency exists for some water bodies, for example, the North Sea (1), along the coast of Brittany to marine waters (2), some rivers in western North America (3), the Chesapeake Bay in the United States (4), and the coastal waters in Hong Kong (5–7), to exhibit increases in the severity and frequency of algae blooms and growth of aquatic weeds apparently as a result of elevated levels of nutrients, including nitrogen and phosphorus. Many of these problems, including bottom-water anoxia, decline in fisheries, and loss of submerged aquatic vegetation, are associated with the deteriorated water quality caused by eutrophication.

Water quality models are employed to describe the balance of mass and energy within an aquatic ecosystem. The problem can be defined as the input of polluting materials including organic and inorganic nutrients into a water body that stimulate the growth of algae or rooted aquatic plants resulting in the interference with desirable water uses of aesthetics, recreation, fish maintenance, and water supply. This topic has been discussed extensively, and many studies, including water quality control and modeling, have been conducted (4,8). Initially, water quality modeling focused on seasonal steady-state conditions (9,10) and, later, time-varied modeling for an entire year was explored with reasonable success (11). Nowadays, the forefronts of water quality modeling are the integration of hydrodynamic and water quality models and the development and incorporation of sediment layers to investigate the long-term recovery of an aquatic ecosystem (4,12,13).

TYPICAL MODEL FRAMEWORK

Typically, in an ecosystem under an aquatic environment, several physical, chemical, biochemical, and biological processes can affect the transport and interaction among the nutrients, dissolved oxygen (DO), phytoplankton, zooplankton, and carbonaceous material (10). The mass and energy transformations are regulated by processes such as growth, respiration, mortality, and decomposition; these in turn are governed by environmental quality parameters such as temperature, toxicity, and nutrient concentrations. The principal variables of importance then in the analysis of eutrophication are as follows:

1. Solar radiation at the surface and with depth, water temperature, and chloride
2. Geometry of water body
3. Flow, velocity, diffusion, and dispersion
4. Nutrients: phosphorus, nitrogen, silica, and so on
5. Phytoplankton, zooplankton, and so on

As shown in Fig. 1, the system is highly coupled, and energy and mass balance for individual constituents are invariably linked to several others. A system of state variables, including organic parameters (carbon, nitrogen and phosphorus), inorganic parameters (DO, ammonia nitrogen, nitrite + nitrate nitrogen, and orthophosphate), and biological constituents (phytoplankton and zooplankton), is used to develop an unsteady multilayered two-dimensional or even three-dimensional water quality model. It should be recognized that the nutrients above are present in several forms in the water body, and not all forms are readily available for uptake by phytoplankton nor are limited nutrients for growth of phytoplankton. In the system, ammonia, nitrite + nitrate, and orthophosphate are taken as the available nutrients uptaken by phytoplankton (13). Moreover, available historical field data of solar radiation intensity, layer-averaged water temperature, and salinity are often taken as the known variables that will mainly affect some parameters and the saturated DO.

GOVERNING EQUATIONS

Because of the difficulty in obtaining regular, high-frequency spatial and temporal field data with reasonable accuracy, it is almost improbable to select a unique correct formulation for a specific process in an ecosystem. Such difficulties can partly be overcome by examination of the actual processes in the laboratory, but this is only, as indicated, partly a solution, because the processes in the ecosystem can never be isolated from their environments and it is impossible to take into account all natural coupling mechanisms. Consequently, each process has several alternative mathematical formulations that, at this stage of ecological modeling, are equally valid to a certain extent.

A water quality model is established to solve the partial differential equations describing conservation of mass and momentum of incompressible fluid and pollutants over the depth of each layer. The layer-averaged water quality variables and transport equations of pollutants solved in the model are defined and derived in a similar way to those of the hydrodynamic model. The hydrodynamic and water quality models are run simultaneously. The water quality algorithms are integrated directly into the hydrodynamic

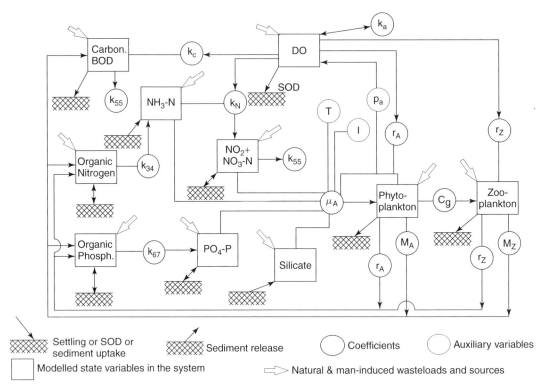

Figure 1. State variable interactions in a water quality model.

model, so that they share the same computational grid and time steps. The water body can be very irregular on the lateral topography and includes some small side cove. To adapt the complicated boundaries of the water body, a boundary-fitted orthogonal curvilinear grid system can be employed. It can be generated numerically by solving two elliptical equations (14). A grid "block" technique (15,16) has been introduced to overcome the computational difficulty caused by the unsteady fluctuation of the water surface level.

In the following sections, subscript "u" and "l" of all variables mean the value in the upper and the lower layer, respectively, and subscript "0", "s", and "b" denote the value at the layer interface, at the free water surface, and at the bed, respectively. Subscript k is the index of layer: $k = u$ or l. Notation $X_{y,20}$ designates the value of X_y at temperature 20 °C, and θ is the corresponding temperature correction coefficient. In this way, the general differential transport equation for the layer-averaged concentration of each state variable in the boundary-fitted orthogonal curvilinear coordinate system is expressed as follows:

$$\frac{\partial (h_k \varphi_k)}{\partial t} + \frac{1}{g_{11}g_{22}} \left\{ \frac{\partial}{\partial \xi} \left[g_{22} h_k \left(u_k^* \varphi_k - \frac{\Gamma_{\xi,k}}{g_{11}} \frac{\varphi_k}{\partial \xi} \right) \right] \right.$$
$$\left. + \frac{\partial}{\partial \eta} \left[g_{11} h_k \left(v_k^* \varphi_k - \frac{\Gamma_{\eta,k}}{g_{22}} \frac{\varphi_k}{\partial \eta} \right) \right] \right\} = \Phi_{\varphi,k} \quad (1)$$

$$\Phi_{\varphi,k} = w_0 \varphi_0 + \text{Flux}_{s,\varphi} - \text{Flux}_{0,\varphi} + S_{\varphi,k} \quad (2)$$

where u_k^* and v_k^* = kth-layer-averaged velocity components in $\xi - \eta$ orthogonal curvilinear coordinates; h_k = depth of kth layer; φ_k = kth-layer-averaged concentration of a state variable, and it may represent phytoplankton, DO, and so on; $\Gamma_{\xi,\kappa}$ and $\Gamma_{\eta,\kappa}$ = longitudinal and transverse dispersive coefficients, respectively, in the kth layer (17); g_{11} and g_{22} = coordinate transformation coefficients between Cartesian and orthogonal curvilinear coordinates (ξ, η); $S_{\varphi,k}$ represents reaction kinetics, settling, sediment release, and external sources and/or sinks in the kth layer, and its expression for each water quality variable is listed in Table 1; w_0 = vertical velocity at the layer interface (positive upward); and φ_0 = concentration of φ at the layer interface.

In this context, $\text{Flux}_{s,\varphi}$, $\text{Flux}_{b,\varphi}$, and $\text{Flux}_{0,\varphi}$ are the vertical diffusive fluxes of φ across the water surface, the bed, and the layer interface, respectively. $\text{Flux}_{s,\varphi}$ is assumed to be zero. Density stratification affects the mixing across the layer interface, and $\text{Flux}_{0,\varphi}$ is derived as follows:

$$\text{Flux}_{0,\varphi} = \left(\Gamma_m \frac{\partial \varphi}{\partial Z} \right)_{z=Z_0} \approx \Gamma_m \frac{\varphi_u - \varphi_l}{\delta},$$
$$\Gamma_m = \Gamma_{m0}(1 + \beta_C \text{Ri})^{\alpha_C}, \Gamma_{m0} = \frac{\varepsilon_{m0}}{\sigma_t}, \quad (3)$$

in which Γ_m = vertical flux diffusive coefficient, Γ_{m0} and $\varepsilon_{m0} (\approx \kappa h_l(1 - h_l/h)u_*)$ = values of Γ_m and turbulent viscosity coefficient for the case without density stratification, respectively; h = total water depth; u_* = friction velocity; δ = mixing layer thickness at the interface and is approximated by $\delta \approx \kappa h_l (1 - h_l/h)^{1/2}$. κ (≈ 0.4) and σ_t (≈ 0.9) are the von Kármán constant and turbulent Schmidt number, respectively; β_C ($= 10/3$) and α_C ($= -1.5$) are the empirical coefficients (18). Ri is a gradient Richardson number

Table 1. List of Notations and Source Terms for Water Quality Variables

Description	Notation	Sources and/or Sinks (S_φ)
(1) Phytoplankton (chlorophyll-a)	A (μg Chl-a/L)	$h(\mu_A - r_A - M_A)A - w_{SA}A - hC_g AZ + S_A$
(2) Zooplankton (organic carbon equivalence)	Z (mg ZooplC/L)	$h(\alpha_2 \dfrac{K_Z}{K_Z + A} C_g \alpha_{12} A - r_Z - M_Z)Z - w_{SZ}Z + S_Z$
(3) Organic nitrogen	N (μg N/L)	$h[f_{ON}\alpha_{13}(r_A + M_A)A + f_{ON}\alpha_{23}(r_Z + M_Z)Z - k_{34}N] - w_{SN}(1 - f_{DON})N + \text{SEDI}_N + S_N$
(4) Ammonia nitrogen	NH_3 (μg N/L)	$h\{(1 - f_{ON})[\alpha_{13}(r_A + M_A)A + \alpha_{23}(r_Z + M_Z)Z] + k_{34}N - f_{pref}\alpha_{14}\mu_A A - k_N NH_3\} + \text{SEDI}_{NH_3} + S_{NH_3}$
(5) Nitrite + nitrate nitrogen	NO_{23} (μg N/L)	$h[k_N NH_3 - (1 - f_{pref})\alpha_{15}\mu_A A - k_{55} NO_{23}] + \text{SEDI}_{NO_{23}} + S_{NO_{23}}$
(6) Organic phosphorus	P (μg P/L)	$h\{f_{OP}[\alpha_{16}(r_A + M_A)A + \alpha_{26}(r_Z + M_Z)Z] - k_{67}P\} - w_{SP}(1 - f_{DOP})P + \text{SEDI}_P + S_P$
(7) Orthophosphate phosphorus	PO_4 (μg P/L)	$h\{(1 - f_{OP})[\alpha_{16}(r_A + M_A)A + \alpha_{26}(r_Z + M_Z)Z] - \alpha_{17}\mu_A A + k_{67}P\} - w_{SPO_4}(1 - f_{DOP_4})PO_4 + \text{SEDI}_{PO_4} + S_{PO_4}$
(8) Carbonaceous BOD	BOD (mg O_2/L)	$h[2.67(\alpha_{18}M_A A + \alpha_{28}M_Z Z) - (k_C \text{BOD} + \dfrac{5}{4}\alpha_{ON}k_{55}NO_{23})] - w_{SBOD}\text{BOD} + S_{BOD}$
(9) Dissolved oxygen	DO (mg O_2/L)	$h\{k_a(\text{DO}^s - \text{DO}) - \alpha_{19}r_A A - \alpha_{29}r_Z Z + \mu_A[\alpha_{19} + \dfrac{48}{14}\alpha_{13}(1 - f_{pref})]A - (k_C \text{BOD} + \alpha_{49}k_N NH_3)\} - \text{SOD} + S_{DO}$

Note: A in the equations for zooplankton, carbonaceous BOD, and DO is in mg Chl-a/L; Z in the equations for organic nitrogen, ammonia, organic phosphorus, and orthophosphate is in μg ZooplC/L; and NH_3 in the equation for DO and NO_{23} in the equation for CBOD are in mg N/L.

at the layer interface and is calculated by

$$\text{Ri} = -\dfrac{g}{\rho}\dfrac{\dfrac{\partial \rho}{\partial z}}{\left(\dfrac{\partial u}{\partial z}\right)^2} \approx \dfrac{\rho_l - \rho_u}{\rho}\dfrac{g\delta}{[(u_u^* - u_l^*)^2 + (v_u^* - v_l^*)^2]} \quad (4)$$

in which ρ_u and ρ_l = density of water in the surface and the bottom layer, respectively; ρ = averaged density of water column; and $g(= 9.81$ m/s^2) = acceleration due to gravity. Temporal water temperature and salinity are estimated from the field data.

The quantities of water quality variables decrease because of the corresponding settling—w_{SA}, w_{SZ}, w_{SN}, w_{SP}, w_{SPO_4}, and w_{SBOD} (m/day). The algal growth relates to many factors, including available nutrients, water temperature, solar radiation, zooplankton grazing, and tidal flushing. Organic nitrogen undergoes a bacterial decomposition whose end product is ammonia (NH_4-N). Organic phosphorus converts to PO_4-P by mineralization. Ammonia nitrogen, in the presence of nitrifying bacteria and oxygen, is converted to NO_3-N (k_N). Denitrification (k_{55}) occurs under anaerobic conditions and CBOD decreases because of stabilization (k_C). Both NH_4 and $NO_2 + NO_3$ are available for algae uptake, although the preferred form is NH_4-N for physiological reasons. The ammonia preference is described by a factor f_{pref} (9):

$$f_{pref,k} = \dfrac{NH_{4,k}}{K_{mNN} + NO_{23,k}} \times \left(\dfrac{NO_{23,k}}{K_{mNN} + NH_{4,k}} + \dfrac{K_{mNN}}{NH_{4,k} + NO_{23,k}}\right) \quad (5)$$

A by-product of photosynthetic carbon fixation is the production of DO. An additional source of oxygen from algal growth occurs when the available ammonia nutrient source is exhausted and the phytoplankton begins to use the available nitrate. For nitrate uptake, the initial step is a reduction to ammonia, which produces oxygen. DO also increases by atmospheric re-aeration (k_a, day^{-1}) because of the deviation from the saturation concentration DO^s (mg O_2/L) (19). The re-aeration rate in natural waters depends on internal mixing and turbulence caused by velocity gradients and fluctuation, temperature, wind mixing, waterfalls, dams, rapids, surface films, and so on. It is expressed in the following forms:

$$k_a = k_{a,20}\theta^{T-20}, \theta = 1.005 \sim 1.030 \approx 1.024 \quad (6)$$

$k_{a,20}$ represents re-aeration as a function of velocity and depth, with the following formula (19):

$$k_{a,20} = \begin{cases} \dfrac{3.9\,V^{0.5}}{h^{1.5}} & \text{when } W < 6.0 \text{ m/s} \\ \dfrac{3.9\,V^{0.5}}{h^{1.5}} + \dfrac{0.728W^{0.5} - 0.317W + 0.0372W^2}{h} & \text{when } W \geq 6.0 \text{ m/s} \end{cases} \quad (7)$$

$$\ln \text{DO}^s = -139.3441 + \dfrac{1.5757 \times 10^5}{T + 273.15} - \dfrac{6.6423 \times 10^7}{(T + 273.15)^2}$$

$$+ \dfrac{1.2438 \times 10^{10}}{(T + 273.15)^3} - \dfrac{8.6219 \times 10^{11}}{(T + 273.15)^4} \quad (8)$$

$$- S\left[1.7674 \times 10^{-2} - \dfrac{1.0754 \times 10}{T + 273.15} + \dfrac{2.1407 \times 10^3}{(T + 273.15)^2}\right]$$

where V is velocity of flow in meters/second; W is wind speed in meters/second at 10 m above water surface (it will be taken into account when wind speed $W \geq 6$ m/s); and T (°C) and S (ppt) are the relevant water temperature and salinity. A minimum value of 1.6/h is imposed on $k_{a,20}$ (9).

Moreover, an adsorption–desorption interaction occurs between dissolved inorganic phosphorus and suspended particulate matter in the water column. The subsequent settling of the suspended solids together with the sorbed inorganic phosphorus can act as a significant loss mechanism in the water column and is a source of phosphorus to the sediment.

Table 2 lists typical kinetic parameters used in water quality models from the literature (9,19,20). The growth and proliferation of phytoplankton is a result of the utilization and conversion of inorganic nutrients into organic plant material through photosynthesis. μ_A depends on three principal components: (1) temperature—T, (2) solar radiation—I, and (3) nutrients—N_{nutri}. Multiplicative effects are assumed, that is, $\mu_A = f(T, I, N_{\text{nutri}}) = g(T)g(I)g(N_{\text{nutri}})$.

The relationship with temperature is as follows:

$$g(T)_k = \mu_{A,k}^{\max} = \mu_{A,20}^{\max} \theta^{T_k - 20} \quad (9)$$

where $\mu_{A,20}^{\max}$ = maximum growth rate of phytoplankton at 20 °C under optimal light and nutrient conditions in the kth layer. The light-limitation function, represented by $g(I)$ over a given water depth of layer (h_k), is approximately integrated as follows (19):

$$g(I)_k = \frac{1}{h_k} \int_{h_u}^{h_u + h_k} \frac{I}{I_s} e^{(1 - I/I_s)} dz = \frac{2.718}{\gamma_k h_k}[e^{-\alpha_2} - e^{-\alpha_1}]$$

$$\alpha_2 = \alpha_1 e^{-\gamma_k h_k}, \alpha_0 = \frac{I_0}{I_s}, \alpha_1 = \alpha_0 e^{-\gamma_u h_u} \quad (10)$$

$$\frac{dI}{dz} = -\gamma_k I, \quad I = I_0 @ z = 0$$

and z-axis is upward vertically

where I, I_0, and I_s = light intensity at depth z, incoming solar radiation intensity just below the surface, and saturating light intensity (the optimum light intensity at which the relative photosynthesis is a maximum), respectively; and γ_k = overall extinction coefficient in m^{-1}. Phytoplankton may adjust its chlorophyll composition to adapt to the changes in solar radiation. Therefore, I_s is determined to be the weighted average of the light intensity for the previous 3 days as follows: $I_s = 0.7I_1 + 0.2I_2 + 0.1I_3$, where $I_i = 0.5 \times$ [daily average visible light intensity beneath the surface] i days earlier (20). The annual average of daily solar radiation intensity is used in the modeling.

The minimum value of nutrient limitations computed by Michaelis–Menten-type expression is chosen for

Table 2. Typical Kinetic Parameters Used in Water Quality Models

Parameters	Description	Value (and θ)
$\mu_A^{\max}$	Maximum phytoplankton growth rate	2.10 (1.066)
K_{mNN}	Half-saturation constant for nitrogen uptake	15.0
K_{mNP}	Half-saturation constant for phosphorus uptake	1.50
r_A	Endogenous respiration rate of phytoplankton	0.05 (1.08)
M_A	Nonpredatory mortality rate of phytoplankton	0.10
C_g	Grazing (filtering) rate of zooplankton	0.30 (1.066)
r_Z	Endogenous respiration rate of zooplankton	0.02 (1.045)
M_Z	Mortality rate of zooplankton	0.10
α_{12}	Assimilated carbon per unit algae mass ingested (average CCHL)	112.5
α_2	Zooplankton assimilation efficiency	0.6
K_Z	Half-maximum-efficiency food level for zooplankton filtering	12.0
k_{34}	Conversion efficiency of organic nitrogen to ammonia nitrogen	0.05 (1.08)
α_{13}	Stoichiometric ratio of cell nitrogen to algae chlorophyll-a	10.0
α_{23}	Stoichiometric ratio of cell nitrogen to zooplankton carbon	α_{13}/CCHL
k_N	Nitrification rate of ammonia to nitrate via nitrite nitrogen	0.05 (1.08)
k_{55}	Denitrification rate	0.09 (1.045)
k_{67}	Conversion efficiency of organic phosphorus to inorganic one	0.03 (1.08)
α_{16}	Stoichiometric ratio of phosphorus to algae chlorophyll-a	1.00
α_{26}	Stoichiometric ratio of phosphorus to zooplankton carbon	α_{16}/CCHL
k_C	Deoxygenation or decay rate of carbonaceous BOD	0.23 (1.047)
BOD$_{u-5}$	Ratio of the ultimate to 5-day carbonaceous BOD	1.54
α_{18}	Stoichiometric ratio of phytoplankton to organic carbon	CCHL
α_{28}	Stoichiometric ratio of zooplankton to organic carbon	1.00
α_{ON}	Oxygen to nitrogen ratio	2.28
α_{19}	Stoichiometric ratio of phytoplankton to oxygen	2.67CCHL
K_{mpc}	Half saturation constant for phytoplankton affects mineralization	10.0
K_{NIT}	Half saturation DO constant for oxygen limitation of nitrification	2.0
K_{NO}	Half-max. DO constant for oxygen limitation of denitrification	0.1
K_{BOD}	Half saturation DO constant for CBOD deoxygenation.	0.5
w_{SA}	Settling rate of phytoplankton	1.10
w_{SN}	Settling rate of particulate organic nitrogen	0.30

$g(N_{\text{nutri}})$. Thus, the full expression for layer-averaged phytoplankton growth rate is expressed as follows (19):

$$\mu_{A,k} = \mu_{A,k}^{\max} g(I)_k \min \times \left\{ \frac{\text{NH}_{4,k} + \text{NO}_{23,k}}{K_{\text{mNN}} + \text{NH}_{4,k} + \text{NO}_{23,k}}, \frac{\text{PO}_{4,k} \cdot f_{\text{DPO}_4}}{K_{\text{mNP}} + \text{PO}_{4,k} \cdot f_{\text{DPO}_4}} \right\} \quad (11)$$

in which $\text{NH}_{4,k}$ (μg N/L), $\text{NO}_{23,k}$ (μg N/L), and $\text{PO}_{4,k}$ (μg P/L) = concentrations of NH_4-N, $\text{NO}_2 + \text{NO}_3$-N, and PO_4-P in kth layer, respectively; and K_{mNN} (μg N/L) and K_{mNP} (μg P/L) = Michaelis constants for nitrogen and phosphorus uptake by algae, respectively.

Depending on the history of the algal cells, the CCHL ratio (mg C/mg Chl-a) is affected by light intensity, temperature, and nutrient availability as follows (21):

$$\text{CCHL} = \frac{\alpha I_s}{2.718 \mu_{A,k}^{\max}} \quad (12)$$

where α can be obtained from laboratory results.

A saturating recycle equation is used for hydrolysis and bacterial decomposition of organic nitrogen to ammonia (k_{34}) and the mineralization of organic phosphorus to inorganic phosphorus (k_{67}):

$$k_{34} = k_{34,20} \theta^{T_k - 20} \frac{A_k}{K_{\text{mpc}} + A_k}$$

$$k_{67} = k_{67,20} \theta^{T_k - 20} \frac{A_k}{K_{\text{mpc}} + A_k} \quad (13)$$

This expression is a compromise between the conventional first-order temperature-corrected mechanism and a second-order recycle mechanism, with the recycle rate being directly proportional to the amount of phytoplankton biomass. It approaches second-order dependency at low phytoplankton concentrations ($A \ll K_{\text{mpc}}$), where K_{mpc} is the half-saturation constant for recycle, and it approaches first-order recycle when the phytoplankton greatly exceeds the half-saturation constant. This mechanism slows the recycle rate if the algal population is small, but it does not permit the rate to increase continuously as phytoplankton increases.

The following processes, namely, nitrification (k_N) of NH_4 to NO_3 via NO_2-N, denitrification (k_{55}) of NO_3-N, and deoxygenation (k_C) of CBOD, are considered temperature and oxygen dependent:

$$k_N = k_{N,20} \theta^{T_k - 20} \frac{\text{DO}_k}{K_{\text{NIT}} + \text{DO}_k}$$

$$k_{55} = k_{55,20} \theta^{T_k - 20} \frac{K_{\text{NO}}}{K_{\text{NO}} + \text{DO}_k}$$

$$k_C = k_{C,20} \theta^{T_k - 20} \frac{\text{DO}_k}{K_{\text{BOD}} + \text{DO}_k} \quad (14)$$

where DO_k (mg O_2/L) = concentration of DO; K_{NIT} and K_{BOD} (mg O_2/L) = half-saturation DO constants for oxygen limitation of nitrification and of organic carbon stabilization; and K_{NO} (mg O_2/L) = half-maximum DO constant for denitrification.

Primary production by phytoplankton in surface waters is a major source of labile organic carbon to coastal sediment. Particles from the euphotic zone sink to the sediment–water interface, where benthic organisms rapidly degrade the labile organic compounds present in the settled materials (22). Sediment algal carbon is expressed as follows (23):

$$\frac{dC_{\text{sedi}}}{dt} = \alpha_{\text{sediCA}} w_{\text{SA}} A_l - \frac{v_{\text{sedi}}}{h_{\text{sedi}}} C_{\text{sedi}} - k_{\text{Csedi}} C_{\text{sedi}} \quad (15)$$

where C_{sedi} (g C/m^2) = sediment algal carbon; α_{sediCA} (g C/g Chl-a) = sediment algal carbon per unit algal mass settled; v_{sedi} (m/day) = sediment accumulation rate; h_{sedi} (m) = thickness of the sediment layer; k_{Csedi} (day^{-1}) = oxidation rate of sediment algal carbon.

SUMMARY

Temporal simulation of phytoplankton growth in a coastal water system is complicated, resource demanding, and computationally intensive because several chemical, biological, and biochemical processes interact, and the reaction rates and external inputs vary with time. Furthermore, the flow and associated circulation are also functions of time, with time scales ranging from minutes to months, even years. Flow and pollutant transport in a natural water body commonly interact with density stratification that resulted from salt water intrusion and solar radiation, which may be described as a layered system. Nowadays, with the advances in computing resources, water quality models are capable of addressing this problem efficiently and flexibly because of their grid system and good numerical performance. In the near future, more work should be undertaken on seeking more accurate pollution source data, including the point/nonpoint sources and atmospheric loads carried directly to the water surface by rain and wind, and for more information on the sediment release of nutrients to obtain much better simulations by the present model.

BIBLIOGRAPHY

1. van Eden, A. (1993). Marine pollution and the environment of the North Sea. *Eur. Water Pollut. Contr.* **3**(5): 16–24.
2. Fera, P. (1993). Marine eutrophication along the Brittany coasts—origin and evolution. *Eur. Water Pollut. Contr.* **3**(4): 26–32.
3. Bothwell, M.L. (1992). Eutrophication of rivers by nutrients in treated kraft pulp mill effluent. *Water Pollut. Res. J. Canada* **27**(3): 447–472.
4. Cerco, C.F. and Cole, T. (1993). Three-dimensional eutrophication model of Chesapeake Bay. *J. Environ. Eng. ASCE* **119**(6): 1006–1025.
5. Chau, K.W. and Jiang, Y.W. (2002). Three-dimensional pollutant transport model for the Pearl River estuary. *Water Res.* **36**(8): 2029–2039.
6. Chau, K.W. and Sin, Y.S. (1992). Correlation of water quality parameters in Tolo Harbour, Hong Kong. *Water Sci. Technol.* **26**(9-11): 2555–2558.
7. Sin, Y.S. and Chau, K.W. (1992). Eutrophication studies on Tolo Harbour, Hong Kong. *Water Sci. Technol.* **26**(9–11): 2551–2554.

8. Lung, W.S., Martin, J.L., and McCutcheon, S.C. (1993). Eutrophication analysis of embayments in Prince William Sound, Alaska. *J. Environ. Eng. ASCE* **119**(5): 811–824.
9. Ambrose, R.B., Wool, T.A., Connolly, J.P., and Schanz, R.W. (1988). *WASP4, a hydrodynamic and water quality model–model theory, user's manual, and programmer's guide.* U.S. Environmental Protection Agency Report, EPA/600/3-87/039.
10. Orlob, G.T. (1982). *Mathematical Modeling of Water Quality: Streams, Lakes, and Reservoirs.* John Wiley & Sons, New York.
11. Lung, W.S. (1993). *Water Quality Modeling, Vol.III: Application to Estuaries.* CRC Press, Boca Raton, FL.
12. Chau, K.W. and Jiang, Y.W. (2003). Simulation of transboundary pollutant transport action in the Pearl River delta. *Chemosphere* **52**(9): 1615–1621.
13. Chau, K.W. and Jin, H.S. (1998). Eutrophication model for a coastal bay in Hong Kong. *J. Environ. Eng. ASCE* **124**(7): 628–638.
14. Thompson, J.F., Warsi, Z.U.A., and Mastin, C.W. (1985). *Numerical Grid Generation.* Elsevier Science Publishing, New York.
15. Chau, K.W. and Jin, H.S. (1995). Numerical solution of two-layer, two-dimensional tidal flow in a boundary fitted orthogonal curvilinear coordinate system. *Int. J. Numer. Methods Fluids* **21**(11): 1087–1107.
16. Chau, K.W. and Jiang, Y.W. (2001). 3d numerical model for Pearl River estuary. *J. Hydraul. Eng. ASCE* **127**(1): 72–82.
17. Fischer, H.B., List, E.J., Koh, R.C.Y., Imberger, J., and Brooks, N.H. (1979). *Mixing in Inland and Coastal Waters.* Academic Press, New York.
18. Rodi, W. (1980). *Turbulence Models and Their Application in Hydraulics, State-of-the-Art.* IAHR Publications, Delft, The Netherlands.
19. Thomann, R.V. and Mueller, J.A. (1987). *Principles of Surface Water Quality Modeling and Control.* Harper & Row Publishers, New York.
20. Lee, J.H.W., Wu, R.S.S., Cheung, Y.K., and Wong, P.P.S. (1991). Forecasting of dissolved oxygen in marine fish culture zone *J. Environ. Eng. ASCE* **117**(6): 816–833.
21. Jin, H.S., Egashira, S., and Chau, K.W. (1998). Carbon to chlorophyll-*a* ratio in modeling long-term eutrophication phenomena. *Water Sci. Technol.* **38**(11): 227–235.
22. Di Toro, D.M., Paquin, P.R., Subburamu, K., and Gruber, D.A. (1990). Sediment oxygen demand model: methane and ammonia oxidation. *J. Environ. Eng. ASCE* **116**(5): 945–986.
23. Chau, K.W. (2002). Field measurements of SOD and sediment nutrient fluxes in a land-locked embayment in Hong Kong. *Adv. Environ. Res.* **6**(2): 135–142.

ENVIRONMENTAL APPLICATIONS WITH SUBMITOCHONDRIAL PARTICLES

DIANA J. OAKES
JOHN K. POLLAK
University of Sydney
Lidcombe, Australia

It is now recognized that the supply of sufficient clean and safe water for drinking and growing food as well as for supplying adequate volumes of water for the maintenance of ecosystems will be one of the most significant problems facing the population of the twenty-first century. Shortages of clean water have already a direct effect on human health, causing many deaths as well as severely damaging plant and animal life and thus affecting the ecology as well as the production of agricultural crops. Hence, it is important that the quality of water is routinely evaluated.

The classic toxicological methods that have been developed and used by scientists over many years for risk assessments often evaluate the outcomes of exposures to single chemicals. However, in the present environment, in which exposures to mixtures of chemicals have been significantly increased, it is inappropriate to only evaluate the toxic effects of individual chemicals or toxins. It is now widely recognized that the toxic effects caused by exposures to chemical mixtures may even be caused by extremely low concentrations of individual chemicals because of the additive or even synergistic actions of other chemicals in a water sample. Thus, it was demonstrated in 105 extracts of sediments collected from aquatic environments that many of the extracts with low or barely detectable levels of priority pollutants were highly toxic when evaluated with the Microtox bioassay (1). Jacobs et al. (1) suggested that "... chemical data alone provides no direct indication of the potential effects that contaminant mixtures may have on the aquatic biota. Toxicity bio-assays furnish reasonable estimates concerning the biological impacts of sediment contaminants; both chemical and biological analyses should be performed." Although some of these toxic samples did not contain any significant concentrations of priority pollutants, it is likely that these extracts contained unidentified and possibly unregulated pollutants at low concentrations.

Even nontoxic chemicals have the potential to increase the toxicity of other chemicals in a mixture. An example of such an effect is the increased toxicity of phenytoin by the presence of lactose in the mixture (2). In this context, it is questionable whether the guidelines for permissible concentrations of individual chemicals (such as ADI—acceptable daily intake; MRL—minimal risk level; NOAEL—no observable adverse effect level) provide meaningful safety levels either for the ecology or for human health, because these parameters focus only on individual chemicals and do not consider the potential effects of other chemicals on the overall toxicity of a chemical mixture. It is, therefore, important that we use bioassays for the evaluation of the toxic effects of aquatic samples.

For the past 24 years, the Organization for Economic Co-operation and Development (OECD) has provided standardized test protocols to evaluate aquatic toxicity with laboratory-based representative single-species toxicity tests, such as fish, Daphnia spp., and algae (3). These bioassays provide useful data of the overall toxicities of chemical mixtures that are present in aqueous samples.

An example of a subcellular bioassay is the submitochondrial particle (SMP) test. With this test, we use the cellular organelle, the mitochondrion, which is present in all eukaryotic (nucleated) cells. Mitochondria are effectively the powerhouses of the cell, providing over 90% of all energy produced in all eukaryotic organisms. Mitochondria catalyze reactions in which molecules with a low energy content are oxidized (lose electrons) to give rise to "energy-rich" molecules, such as adenosine-triphosphate (ATP). ATP is required for synthetic metabolic processes or for mechanical or heat energy. The metabolic activities of mitochondria are so fundamental to aerobic life that their functional organization, their proteins and enzymes, are very similar in all aerobic organisms (from amoeba, plants, or even fungi to humans). Hence, the evaluation of cellular toxicities by mitochondria from one species is likely to apply to other species. Stable SMPs are produced by the disruption of the internal membranes of isolated bovine heart mitochondria (4,5). Once prepared, the SMPs are stable for at least 2 years when kept at $-80\,°C$. SMPs are now commercially available in the United States from Harvard Bioscience, Inc., Madison, WI. SMPs consist of inverted, vesicular portions of the inner mitochondrial membranes that can perform the integrated enzymatic functions of electron transport and oxidative phosphorylation (6).

The test measures inhibitory effects that chemicals present in contaminated water may exert on SMP enzyme activity. Two different assay procedures have been devised for SMPs (7,8) as illustrated in Fig. 1. Figure 1a illustrates the complexes of the inner mitochondrial membrane that are involved in the so-called forward electron transfer reaction (ETR), whereas Fig. 1b illustrates the reverse transfer reaction (RET) in which access to complex III and complex IV is blocked with the inhibitor antimycin A.

The ETR assay measures the flow of electrons from NADH via complexes I, II, III, and IV to molecular oxygen. Measurements of the oxidation are carried out in six 1-mL cuvettes. Five cuvettes receive different concentrations of the samples that are tested, whereas one cuvette serves as the control. The extent of the inhibition of the oxidation of NADH provides an indication of the toxic effects of the aqueous sample that is being evaluated. The assay buffer consisted of 0.25 M sucrose, 20 mM Hepes buffer (pH 7.4), 6 mM $MgCl_2$, and 200 nM NADH (25 µL). The decrease in the absorption at 340 nm measures the oxidation of NADH to NAD with a Beckman DU650 spectrophotometer (Fullerton, CA), with the automatic six-cell holder that is maintained at $30\,°C$.

In the RET assay, succinate is added as a substrate and because access to complexes III and IV is blocked (see Fig. 1b), the NAD that has been added is reduced to NADH by the SMPs. Hence, the RET assay measures the reduction of NAD to NADH leading to the increased absorption at 340 nm. The assay buffer consisted of 0.18 M sucrose, 50 mM Tris buffer (pH 7.5), 0.1 M sodium succinate, 0.7 µg/mL antimycin A, 30 mM $MgCl_2$, and 1 mM NAD. The NAD reduction was measured at 340 nm with a Beckman DU650 spectrophotometer, with an automatic six-cell holder maintained at $30\,°C$. This test evaluates the inhibitory effect exerted by chemicals in the aqueous samples on the oxidation of succinate that leads to the reduction of NAD to NADH. For further details of the SMP assay protocols see Oakes and Pollak (9).

The SMP test has evaluated the toxicities of 113 different chemicals (10,11). SMP test procedures have also been validated in field trials through the participation in several ongoing U.S. Environmental Protection Agency

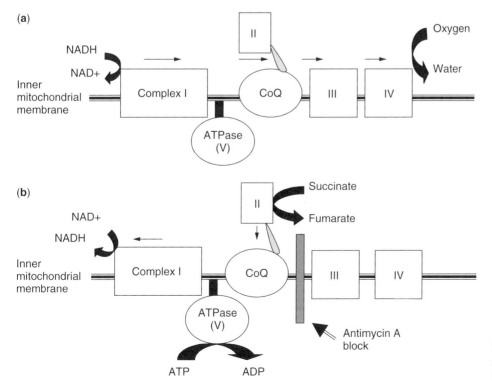

Figure 1. (a) Diagram of the ETR adapted from Ref. 7. (b) Diagram of the RET adapted from Ref. 7.

Geological Surveys and by the Wisconsin Department of Natural Resources Bio-Monitoring Projects, as well as in studies in the Illinois and Minnesota River Basins (11). Other studies have shown that the results obtained by the SMP test correlate well with the results obtained with *in vitro* and *in vivo* bioassays, such as Microtox, mammalian cell cultures, nematode, Daphnia spp., and fish assays or sea urchin fertilization and larval development tests (6,10,12–15). A further example of the useful information that the SMP test can provide was established when the toxicity of the herbicide formulation Tordon 75D was investigated (9). The herbicide Tordon 75D consists of a mixture of the triisopropanolamine salts of 2,4-dichlorophenoxy acetic acid (2,4-D) and 4-amino-3,5,6-trichloropicolinic acid (Picloram) as well as other components, such as the solvent triisopropanolamine and diethyleneglycol monoethyl ether, a silicone defoamer, and a proprietary surfactant (polyglycol 26-2). Using both the ETR and the RET assays and testing the complete formulation as well as the individual components, researchers have shown that although the proprietary surfactant by itself severely inhibited the oxidative functions of the SMPs, the mixture of only the active components had no significant effects on the oxidative functions of the SMPs. These results suggested that the surfactant damaged the inner mitochondrial membrane and was the primary component of Tordon 75D that inhibited the oxidative functions of the SMPs. A later study assessed the potential toxicity of three related herbicide formulations containing the ester derivatives of 2,4,5-T and 2,4-D. It was shown that in these formulations, the "other components" such as diesel oil and surfactants contributed significantly (approximately 50%) to the overall toxicity in an additive manner (16). More recently, the SMP test evaluated the toxicities of leachates that emanated from the Sydney Olympic Site. The leachates were also evaluated with several other bioassays as well as by chemical analyses. The SMP test, although less sensitive than other bioassays, such as the sea urchin fertilization and development tests, provided useful information as a rapid prescreening tool (15). Of the two SMP assays, only the RET was suited for evaluation of the leachate toxicity because it is known that samples containing significant levels of divalent cations can activate the SMP activity when measured with the ETR assay, effectively masking any inhibitory effect (6,15).

Overall, these studies have demonstrated that the SMP test is a useful prescreening tool for evaluating the toxic effects of aqueous media or chemical formulations. The SMP test is less sensitive than the Microtox assay to changes in pH and to the presence of solvents; hence, it can evaluate a greater range of aqueous samples. The major limitation of the SMP test as well as other cellular and subcellular *in vitro* assays is that they cannot account for pharmacokinetic and pharmacodynamic processes that may affect the actions of chemicals *in vivo*, nor can the SMP test evaluate the action of chemicals on specific receptors or enzymes (e.g., acetyl choline esterase) other than mitochondrial enzymes.

BIBLIOGRAPHY

1. Jacobs, M.W. et al. (1993). Comparison of sediment extract Microtox toxicity with semi-volatile organic priority pollutant concentrations. *Arch. Environ. Contam. Toxicol.* **24**: 461–468.
2. Tyrer, J.H., Eadie, M.J., Sutherland, J.M., and Hooper, W.D. (1970). Outbreak of anticonvulsant intoxication in an Australian city. *Br. Med. J.* **4**: 271–273.
3. Organization for Economic Co-operation and Development (OECD). (1981). *Guidelines for Testing of Chemicals*. OECD, Paris.
4. Blondin, G.A., Knobeloch, L.M., Read, H.W., and Harkin, J.M. (1987). Mammalian mitochondria as *in vitro* monitors of water quality. *Bull. Environ. Contam. Toxicol.* **38**: 467–474.
5. Knobeloch, L.M., Blondin, G.A., and Harkin, J.M. (1990). Use of submitochondrial particles for prediction of chemical toxicity in man. *Bull. Environ. Contam. Toxicol.* **44**: 661–668.
6. Read, H.W., Harkin, J.M., and Gustavson, K.E. (1997). Environmental applications with sub-mitochondrial particles. In: Wells, P.G., Lee, K., and Blaise C. (Eds.). *Microscale Testing in Aquatic Toxicology—Advances, Techniques and Practice*. CRC Press, Boca Raton, FL.
7. Knobeloch, L.M., Blondin, G.A., and Harkin, J.M. (1994). A rapid bio-assay for toxicity assessment of chemicals: forward electron transport assay. *Environ. Toxicol. Water Quality* **9**: 79–82.
8. Knobeloch, L.M., Blondin, G.A., and Harkin, J.M. (1994). A rapid bio-assay for toxicity assessment of chemicals: reverse electron transport assay. *Environ. Toxicol. Water Quality* **9**: 231–234.
9. Oakes, D.J. and Pollak, J.K. (1999). Effects of a herbicide formulation, Tordon 75D, and its individual components on the oxidative functions of mitochondria. *Toxicology* **136**: 41–52.
10. Knobeloch, L.M., Blondin, G.A., Read, H.W., and Harkin, J.M. (1990). Assessment of chemical toxicity using mammalian mitochondrial electron transport particles. *Arch. Environ. Contam. Toxicol.* **19**: 828–835.
11. Harkin, G.M., Blondin, G.A., Knobeloch, L.M., and Read, H.W. (1991). Mitochondrial bio-assay for toxic substances in water. *Gov. Rep. Announce. Index (U.S.)* **91**(16). Abstr. No. 143,481 (only seen in abstract form in *Chem. Abstr.* **116**: 417).
12. Blondin, G.A., Knobeloch, L.M., Read, H.W., and Harkin, J.M. (1989). An *in vitro* Submitochondrial Bio-assay for predicting acute toxicity in fish. In: *Aquatic Toxicology and Environmental Fate*. G.W. Suter and M.A. Lewis (Eds.). American Society for Testing and Materials, Vol. 11, pp. 551–563.
13. Degli-Esposti, D.M., Ngo, A., and Myers, M.A. (1996). Inhibition of mitochondrial complexI may account for IDDM induced by intoxication with the rodenticide Vacor. *Diabetes* **45**: 1531–1534.
14. Argese, E. et al. (1998). Comparison of *in vitro* submitochondrial particle and Microtox assays for determining the toxicity of organotin compounds. *Environ. Toxicol. Chem.* **17**: 1005–1012.
15. Byrne, M., Pollak, J., Oakes, D., and Laginestra, E. (2003). Comparison of the submitochondrial particle test, Microtox and sea urchin fertilization and development tests: parallel assays with leachates. *Aust. J. Ecotox.* **9**: 19–28.
16. Oakes, D.J. and Pollak, J.K. (2000). The *in vitro* evaluation of the toxicities of three related herbicide formulations containing ester derivatives of 2,4,5-T and 2,4-D using submitochondrial particles. *Toxicology* **151**: 1–9.

INTEREST IN THE USE OF AN ELECTRONIC NOSE FOR FIELD MONITORING OF ODORS IN THE ENVIRONMENT

JACQUES NICOLAS
University of Liege
Arlon, Belgium

INTRODUCTION

The increasing number of complaints related to malodors generated by agricultural, landfill, and wastewater treatment facilities give rise to a growing interest in the measurement of olfactive annoyance. Instrument manufacturers have sought to provide suitable environmental monitoring solutions. Because the need for such devices is generally driven by legislation, compliance with the standardized validation of techniques is often mandatory. So, current standard methods remain the references to measure odor annoyance.

Two procedures are generally proposed: either olfactometry, based on odor assessment by human panels, or chemical analysis by chromatographic techniques. However, neither meets the requirements of on-line monitoring for environmental applications.

New emerging technologies such as electronic noses, based on nonspecific sensor arrays, may offer objective and on-line instruments for assessing environment odors. But, despite the recent appearance of commercial devices that can be implemented in the field, a number of limitations can be highlighted, both from the technology itself and its application to environmental studies.

BASIS OF THE CHIEF ODOR MEASUREMENT METHODS

Olfactometry is the scientific measurement of odor concentration using a system of sampling and a methodology regulated to European (EN 13 725) (1) and American (ASTM E679-91) (2) standards. The odor concentration of a gaseous sample is determined by presentation to a panel of observers, who have known acuity to odor, in varying dilutions. The procedure aims at determining the dilution at which only half the panel can detect the odor. The odor concentration is then the number of dilutions required for the sample to reach this threshold. It is expressed in multiples of one odor unit per m^3 (ou/m^3). This technique provides directly comparable data for different odor types. Although this technique gives the right human sense evaluation and is now based on standard methodology, it remains strongly influenced by subjectivity and is time-consuming, labor-intensive, and expensive. Olfactometry laboratories are often remote from the odor source: it is obviously not appropriate for real-time and continuous operation on-site.

Alternatively, chemical analysis of the odorous mixture by gas chromatographic techniques (such as GC-MS) provide an accurate concentration of specific compounds in the sample (3). Some instruments can be used on-site and for continuous assessment, but analytic methods never provide the global olfactive perception, as most environmental odors are complex mixtures of compounds. Moreover, they do not take into account the interactions between different odorants and interfering background substances, which may lead to synergistic or antagonistic effects.

The electronic nose is a more recent technique based upon a fixed array of nonspecific gas sensors, each of which responds differently to an odorant and the global pattern of responses that is characteristic of that odorant (4).

The heart of the electronic nose is an array of chemical sensors (metal oxide semiconductors, conductive polymers, quartz microbalances). When exposed to a gaseous atmosphere, these individual sensors provide individual signals. Subsequently, a signal pattern is deduced by analyzing these signals by suitable statistical and mathematical methods. The diagnosis is thus based on the "fingerprint" of the gas mixture classified by pattern recognition techniques (artificial neural networks, discriminant function analysis). Before applying such an instrument on a real atmosphere for on-line recognition of odors in the environment, it must be trained with typical gas mixtures for this environment. After the training, the system can identify one of the learned odors by establishing similarities between the actually observed pattern of signals and those previously observed. Hence, the electronic nose is an analytical instrument that can characterize an odor without reference to its chemical composition. The signal provided is the odor, considered as a whole. That makes this method particularly attractive and shows its great potential.

STATE OF THE ART

Reported developments in complex odor analysis are focused chiefly on quality assessment in the food, drink, and perfume industries. So far, few attempts have been made to characterize and monitor complex odor in the environment. Although original work carried out with laboratory-based systems and more recent field investigations have shown promising results, the number of trials carried out under realistic environmental conditions is relatively limited.

Some papers review the current status of sensor array technology and discuss its potential application to the assessment of olfactive annoyance in the environment by referring to exhaustive literature surveys (5–8).

Early investigations include the analysis of single substances, but some works deal with more complex odors. In a review of odor measurement techniques for sewage treatment works, Gostelow et al. (9) listed some examples of electronic nose application to environmental odor problems. However, the use of electronic noses to monitor water quality in the field remains a virtually unexplored domain with only few applications reported so far.

TYPICAL USER NEEDS

In the field of environmental monitoring and more particularly of water quality, the typical needs of the final user of electronic nose require both qualitative (classification) and quantitative (regression) approaches.

From a qualitative point of view, different annoyance odors must be discriminated, and the identity of unknown samples must be predicted using a previously calibrated learning model. A range of data processing techniques may be used to analyze sensor array data, but due to the large number of variables (i.e., number of sensors) and samples, pattern recognition techniques (such as multivariate statistics and artificial neural networks) are employed to reduce the dimensionality of the sensor array data. The relationships between the samples can then be compared and correlated using simple scatter plots. The choice of analysis technique depends on the amount and nature of information available and the type of information required from the analysis.

Figure 1 could refer to a typical result of a discriminant function analysis (DFA) of observations made at a landfill site by an electronic nose. Three types of odor tonalities can be perceived: the fresh refuse, the odor of landfill gas, and one of the leachates. The scatter plot of the observations in the plane of the two calculated roots shows clustering into the three expected groups. The model so calibrated by DFA can further be used by the landfill manager to classify a new observation. For example, the emergence of an odor identified as "leachates" can let us suspect a problem in the leachate treatment system.

This ability of an electronic nose to discriminate rapidly between slight variations in complex mixtures makes the technique ideal for on-line process diagnostics and screening across a wide range of applications.

Quantitative approaches imply regression procedures by correlating sensor responses with some quantitative parameter obtained either by another instrument or by human perception. In some cases, reasonable fits are obtained when the sensor responses are compared with odor concentration resulting from olfactometry measurements. However, this approach is impeded by the subjectivity of the olfactive evaluation, the nonlinear relation between the perceived intensity of an odor and its concentration, as well as by adaptation phenomena (perception) as a result of extended exposure to that odor (8). This complicates the elaboration of quantitative models. Moreover, the model so calibrated is often site specific, and its validity over long periods of time must be demonstrated before sensor array systems can be used for quantitative analysis.

In the field of water quality, the overall electronic nose output is often compared with classical water characterization variables, such as volatile suspended solids (10) or global organic parameters (biochemical oxygen demand and total organic carbon) (11). Studies generally conclude that a number of different wastewater quality relationships could be formulated from the electronic nose analysis of a sewage liquid. In all cases, the principal advantages put forward by the authors to justify an electronic nose with respect to traditional analytical methods are the rapidity of the response and the noninvasive nature of the measurement.

The model obtained by regression could be used as an on-line monitoring tool for a process. For example, Fig. 2 highlights a "stress" episode in a composting process, caused by the absence of aeration of the compost pile. A specific indicator is constructed by relating the sensor signals to the four families of compounds emitted during that "stress" phase (nitrogen compounds, carboxylic acids, ammonia, and chlorinated compounds). It is constructed by a canonical correlation analysis. The figure shows the temporal evolution of that combination of the sensor signals applied to the whole data set obtained by continuous signal monitoring for 11 days. The root value exhibits a peak during the "stress" period. Of course, such a global odor indicator could be calculated for other environmental processes, such as wastewater treatment, and it could be monitored as a global process control variable.

More generally, the purposes of continuous, *in situ* monitoring of odorous emissions by an electronic nose could be

— to predict the rise of malodor in the background before it becomes an annoyance for the surroundings;
— to use the odor as a process variable, aiming at a better understanding of the odor release and

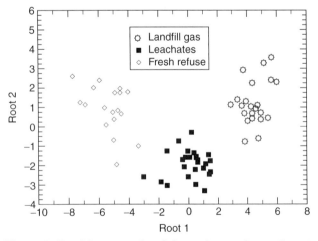

Figure 1. Possible scatter plot of electronic nose observations at a landfill site in the plane of the two first roots of a discriminant function analysis.

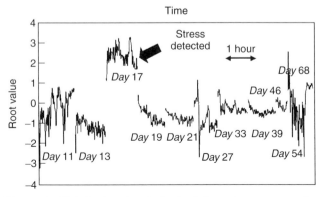

Figure 2. Global odor signal obtained from the response of a sensor array to detect a stress episode in a composting process.

relating this emission to the process phase, or the problem that caused the emission;
— to control odor abatement techniques in real time, such as atomization of neutralizing agents.

Simple sensory early warning devices that can detect sudden changes in a process stream obviously present great interest.

APPLICATIONS RELATED TO WATER QUALITY

The use of electronic noses to monitor water quality in the field may involve the agricultural and industrial effluents allowed to seep into groundwater or to flow into streams or rivers (6). The electronic nose can be used in these applications by collecting samples of the effluent. Boreholes can also be employed to collect samples to test groundwater contamination.

A very promising area is the detection of hydrophobic and highly volatile organic compounds (VOCs) in water systems (8). The recent developments in on-line detection of oils and petroleum hydrocarbons (12) have arisen from the recent advances in sensor technology as well as from progress in computing and pattern recognition. Studies aim also at real-time control of wastewater treatment plants (13) and at the quality tests of potable water (14).

A specific obstacle when dealing with liquid samples is to generate or collect a headspace gas that can be reliably and safely measured by the sensor array. Because parameters such as flow, temperature, or suspended solids constantly change, the major difficulty lies in drawing a headspace sample that is representative of the liquid phase (8). But several applications involve also monitoring odors emitted in the air from wastewater (9). Odor abatement and control are major issues facing municipal sewage treatment facilities.

SPECIFIC PROBLEMS FOR ENVIRONMENTAL APPLICATIONS

Much of the original work on environmental applications has used sensor arrays in laboratory-based conditions. However, to understand the effects of environmental parameters on localized odor pollution, it is necessary to translate these laboratory-based experiences into formats that can be applied to measurements under variable conditions (15). Although new sensor materials and designs are continuously being reported, the major limitation of currently available sensors remains their sensitivity to changes in temperature, humidity, and flow rate.

The solution to this obstacle is generally to work under fixed experimental conditions, that is, to incorporate the effects of variations in ambient temperature and humidity into the design of instruments by using sample pretreatment systems. That was demonstrated for headspace pretreatment in on-line wastewater monitoring (13). However, when considering the practical application of a sensor array, this can make the overall instrument more complex and expensive and can also affect its portability or limit sample throughput.

A second approach is to measure these parameters and calibrate the sensors under varying humidity levels to compensate for changes in subsequent data analysis (16). Yet, the method has still to be validated for real-life environmental applications where pollutants must be analyzed rapidly under ever changing background conditions and in the presence of interfering compounds.

Alternatively, the development of a classification model under a wide range of operating and meteorologic conditions makes those conditions "neutral" for odor recognition (17). It is an effective but sometimes fastidious solution.

Sensor drift is another serious impairment of chemical sensors. They alter over time and so have poor repeatability because they produce different responses for the same odor. That is particularly troublesome for electronic noses for which the sensor signals can drift during the learning phase, requiring repeated exposure to the same type of odor.

The consequences of sensor drift on classification results exclude the possibility of further odor identification (18). Though drift correction algorithms could be applied for each individual sensor, the usual way to counteract the global drift of a sensor array is to correct the whole pattern, using multivariate methods. First, the main direction of the drift is determined in the first component space of a multivariate method, such as principal component analysis (PCA). The drift component can then be removed from the sample gas data, thus correcting the final score plot of PCA (19).

Moreover, the combination of multiple sensors, cross-selectivity, and pattern recognition analysis makes the usual techniques of instrument calibration difficult to apply to electronic noses. As a result of this issue there are, to date, no international standards that directly refer to electronic nose measurement.

An additional problem occurs when the concentration of volatiles is low, which is often the case in the environment: the limit of detection of the sensors and the limits of recognition of the electronic nose are reached. A possible solution to this problem should be to improve the sample uptake, for example, by preconcentrating the analytes prior to investigation (20).

Other requirements of field monitoring include low cost, low maintenance, small size, low response time, and low power consumption where autonomous, portable instruments are concerned.

TRENDS AND FUTURE WORK

Electronic nose technology exhibits several possibilities for environmental monitoring. There is a true potential for sensor array in this area, provided that some conditions are respected. With the technology used so far, it is unrealistic to envisage a universal electronic nose to cope with any odor type. Specific data processing and sometimes specific instruments must be designed for each application. In the field of water quality, sewage odor profiles are specific for individual treatment works and for different unit processes within a work. These different sewage odor compositions induce a scatter of observations, which must be removed before trying to find a relationship with odor concentration determined by other methods (21).

Results to date have been based mainly on assessing environmental odors measured near emission sources. This is due to the constraints of using commercial chemical sensors that exhibit a limit of detection often higher than the threshold of smell (generally on the order of 1 ppmv). So, environmental monitoring by an electronic nose must be envisaged only in the surroundings of the emission.

Potential applications in odor assessment by electronic noses are numerous, but, before these specific applications can become a reality, a number of challenges still need to be overcome. Programs of experimental work should be undertaken to assess various characteristics of electronic nose performance, including drift, comparability of sensors before and after replacement, environmental influence, sensitivity, and signal-to-noise ratio.

There are not, so far, any possible comparisons between different electronic nose results. Even an instrument whose sensors were replaced is no longer the same as the original: the learning phase must be started again. In the frame of environmental applications, that particularly excludes the possibility of using the electronic nose to test compliance with standards.

So, it is necessary to develop calibration procedures and to realize suitable calibration artifacts. Any new "calibration standards" will need to consist of generic mixtures that reflect the gas components typically found in the environment and have a significant effect on the various detector responses. It is also essential to validate the quantitative assessment of sensor array responses against other methods, such as olfactometry measurements, to confirm comparisons with human perception.

There is a need to develop internationally acceptable methodologies for harmonizing electronic nose characterization and performance testing. This requirement can be best met by developing standardized procedures and protocols that enable making quantitative and objective comparisons between different types of instruments. These procedures could then form the basis of future international standards in this area.

BIBLIOGRAPHY

1. European Standard EN13725. (2003). *Air Quality-Determination of Odour Concentration by Dynamic Olfactometry*.
2. ASTM - E679-04. (2004). *Standard Practice for Determination of Odor and Taste Thresholds By a Forced-Choice Ascending Concentration Series Method of Limits. ASTM International.*
3. Ramel, M. and Nomine, M. (2000). Physicochemical characterisation of odours. *Analusis* **28**: 171–179.
4. Gardner, J.W. and Bartlett, P.N. (1999). *Electronic Noses: Principles and Applications*. Oxford University Press, Oxford.
5. Pearce, T.C., Schiffman, S.S., Nagle, H.T., and Gardner, J.W. (Eds.). (2002). *Handbook of Machine Olfaction: Electronic Nose Technology*. Wiley-VCH, Weinheim.
6. Nagle, H.T., Gutierrez-Osuna, R., Kermani, B.G., and Schiffman, S.S. (2002). Chapter 22: Environmental monitoring. In: Ref. 5, pp. 419–444.
7. Stuetz, R.M. and Nicolas, J. (2001). Sensor arrays: an inspired idea or an objective measurement of environmental odours? *Water Sci. Technol.* **44**: 53–58.
8. Bourgeois, W., Romain, A.C., Nicolas, J., and Stuetz, R.M. (2003). The use of sensor arrays for environmental monitoring: Interests and limitations. *J. Environ. Monitoring* **5**: 852–860.
9. Gostelow, P., Parsons, S.A., and Stuetz, R.M. (2001). Odour measurements for sewage treatment works. *Water Res.* **35**: 579–597.
10. Dewettinck, T., Van Hege, K., and Verstraete, W. (2001). The electronic nose as a rapid sensor for volatile compounds in treated domestic wastewater. *Water Res.* **35**: 2475–2483.
11. Stuetz, R.M., Georges, S., Fenner, R.A., and Hall, S.J. (1999). Monitoring wastewater BOD using a non-specific sensor array. *J. Chem. Technol. Biotechnol.* **74**: 1069–1074.
12. Ueyama, S., Hijikata, K., and Hirotsuji, J. (2001). *Water monitoring system for oil contamination using polymer-coated quartz crystal microbalance chemical sensor. Instrum. Control Automation*, Malmo, Sweden, pp. 287–292.
13. Bourgeois, W. and Stuetz, R.M. (2000). Measuring wastewater quality using a sensor array: prospects for real-time monitoring. *Water Sci. Technol.* **41**: 107–112.
14. Gardner, J.W., Shin, H.W., Hines, E.L., and Dow, C.S. (2000). An electronic nose system for monitoring the quality of potable water. *Sens. Actuators B* **69**: 336–341.
15. Flint, T.A., Persaud, K.C., and Sneath, R.W. (2000). Automated indirect method of ammonia flux measurement for agriculture: Effect of incident wind angle on airflow measurements. *Sens. Actuators B* **69**: 389–396.
16. Orts, J., Llobet, E., Vilanova, X., Brezmes J., and Correig, X. (1999). Selective methane detection under varying moisture conditions using static and dynamic sensor signals. *Sens. Actuators B* **60**: 106–117.
17. Nicolas, J., Romain, A-C., and Maternova, J. (2001). Chemometrics methods for the identification and monitoring of an odour in the environment with an electronic nose. In: M.T. Ramirez (Ed.), *Sensors and Chemometrics*. Research Signpost, Kerala, pp. 75–90.
18. Romain, A.C., Nicolas, J., and Andre, Ph. (2002). Three years experiment with the same tin oxide sensor arrays for the identification of malodorous sources in the environment. *Sens. Actuators B* **84**: 271–277.
19. Artursson, T. et al. (2000). Drift correction for gas sensors using multivariate methods. *J. Chemometrics* **14**: 711–723.
20. Nakamoto, T. (2002). Chapter 3: Odor handling and delivery system. In: Ref. 5.
21. Stuetz, R.M., Fenner, R.A., and G. Engin, G. (1999). Assessment of odours from sewage treatment works by an electronic nose, H_2S analysis and olfactometry. *Water Res.* **33**: 452–461.

OIL-FIELD BRINE

Dusan P. Miskovic
Northwood University
West Palm Beach, Florida

INTRODUCTION

From the beginnings of oil production, oil producers started to separate the oil field brine (OFB) mixture of oil and salt water from crude oil. The basic saline constituents of an OFB are as follows: chlorides, sulfates, hydrocarbonates, and sulfides of Na, Ca, Mg, ferrous Fe, and Sr. Much smaller concentrations of K, Li, B, Zn, Cu, bromides, and iodides can be attributed to total salinity as well. The actual salt content (qualitative and quantitative) may vary in a broad range of appearance and concentrations (1,2). Very often, the total concentration of salts in OFB can exceed several times the salinity of sea water. The separation of oil and OFB is a logical consequence of the oil production process due to the fact that much more OFB than oil is more often produced, especially in older oil wells. However, until recently neither the oil drillers nor any governmental agency paid much attention to disposal of OFB as a by-product of crude oil production. A simple procedure for OFB was diluting it in streams, rivers, and natural fresh and marine waterbodies. That practice took place until interest in crude oil production and consequently consumption of derivatives dramatically increased. Widespread oil production worldwide is constantly increasing, but the pollution aspects came to a focal point of environmental concern. To keep a high profit margin in the industry and satisfy strict environmental quality regulations, more sophisticated technologies for OFB treatment are becoming viable alternatives to the conventional OFB management practice.

HISTORICAL REVIEW OF OFB CONTROL PRACTICE

A chronology of OFB control was presented recently in the context of a concurrent increase in oil production efficiency and environmental quality self-regulation (3). It was pointed out through ethical consideration that technically improved efficiency of oil production did not successfully resolve pollution problems caused by OFB.

Generally, since the beginning of commercial oil production in the 1920s, a relatively small amount of OFB was lifted from formations. In most cases, oil producers did not realize that they had a disposal problem until someone objected. There are few records about early OFB control practices worldwide. In most inland oil fields, OFB was dumped into shallow evaporation pond (pits).

Since the first lawsuits were filed, simple OFB control practices were used. During the 1930s in California, for example, up to 200,000 barrels/day were simply discharged into the ocean (3). A common way to "control" OFB was in the context of early wastewater control practice; "dilution can be a solution for pollution." Saline water from oil fields would be released into dry channels when rain turned these into fast moving streams. Other practical modes to cope with the lack of systematic regulation in a particular industry were an oil producer's acceptance of the costs associated with storage reservoirs, evaporation ponds, and pipes to the ocean. These were ways to prevent damage suits.

The period of OFB management, as a method of pollution prevention, started in the 1930s. In Texas, for example, limited production from most wells was introduced. A more technical approach to the problem was addressed to prevent the field's hydrostatic pressure from dropping too quickly. Reducing and stopping production when the water-to-oil ratio exceeded a certain value was recommended. Also, it was concluded that evaporation ponds that created more concentrated brine did not resolve the disposal problem. A new alternative for OFB disposal in the form of brine injection wells became more suitable.

The brine injection technology seemed to be very promising as a way of disposing of OFB and also as a way to boost oil production. It was their own technology, which was being improved until now (see next section). In the beginning, an "environmentally friendly" method had been put to additional good use. Not long after the practice was widely accepted, some abandoned injecting wells came to life either as artesian salt springs or as steady salty streams polluting surface freshwater bodies.

Summarizing this brief historical review of OFB pollution control practice, it is possible to point out that a positive shift from increasing industrial efficiency to environmental quality improvement occurred after long-term experience in monitoring. Finally, a new approach to OFB control in the form of wastewater treatment technology emerged.

OFB USE FOR IMPROVING OIL PRODUCTION

Crude oil produced by natural reservoir pressure is referred to as primary production. Any additional production of oil from the introduction of artificial energy into a reservoir is considered enhanced recovery. As mentioned before, OFB injection technology was introduced primarily to recover as much as 50–70% of the original remaining petroleum. A comprehensive historical development emphasizing the environmental issues was presented (3), and a concise review of oil field water injection technology is given in Reference 2. In addition, there is much information scattered in oil production literature (4–6). Specific data focused on improving the quality of OFB before reinsertion is found elsewhere (7–10).

A whole industry supporting well injection was established. It deals with the conditioning of the physico-chemical properties of OFB, to remove remaining oil, as well as to prevent scaling (11). It is of great importance to establish injection water (OFB) quality criteria to prevent interruption or a decrease in oil production. There are three categories of substances affecting reinsertion of OFB: inorganic (dissolved gases, TDS, and TSS), organic (residual oil), and microorganisms. More important qualitative features of OFB and their pollution consequences after well injection will be discussed later.

To understand the many sources of wastewater originating in oil production and some of the polluting characteristics most relevant for OFB management as its major component, the schematic presented in Fig. 1 will be discussed.

Assuming that up to 90% of all fluid lifted from an oil well is practically OFB, the main source of wastewater in oil production is OFB itself. The initial volume of OFB is somehow decreased after two steps of crude oil dehydration, free water knock out (FWKO)

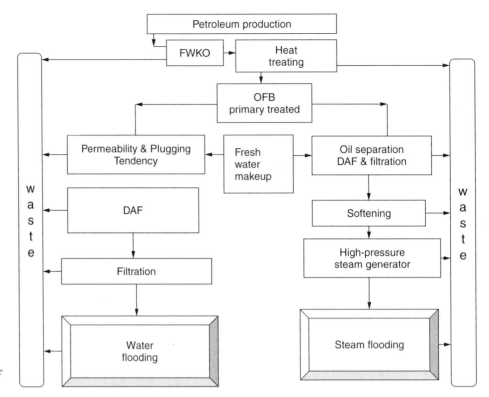

Figure 1. Schematic of OFB use for improving oil production.

and heat treating. Following the removal of oil and gas from the fluid produced, primary treated OFB is now prepared for additional treatment in the course of enhanced recovery. Water and steam flooding are the most common processes for secondary recovery of crude oil. Several steps of additional preparation/purification of the produced water (OFB) have to be conducted, regardless of the type of process applied. In flooding, specific tests have to be done to evaluate permeability and the plugging tendency of a core subjected to a particular floodwater (3). The produced water has to be clarified after mixing with supplemental fresh makeup water. A combination of dissolved air flotation (DAF) and filtration methods can be applied to produce brine physicochemically compatible with the formation rock (12). Also, a more sophisticated method such as radially entrained vapor extraction (REVEX) was considered for removing oil and solids from brine (13). This technology uses well-accepted principles of gravity and centrifugal sedimentation, as well as flotation.

On the other hand, a more complex train of consecutive stepwise processes have to be performed to prepare OFB for steam flooding. Here are three steps of OFB treatment:

— physicochemical removal of oil by gravity separation, primary/secondary DAF, and filtration;
— two-stage softening to provide water suitable for a high-pressure boiler;
— high-pressure steam generation based on previously deoiled and clarified OFB.

The serious disadvantage of steam flooding is condensed water, which contains all solids (salts) concentrated several times compared to the feed water. Practically, very concentrated brine as a disposal stream must be reinserted. An improvement in the existing process was achieved recently in German and Dutch oil fields (14). This new process allows the production of high quality boiler feed water by using mechanical vapor compression, without salt content for steam generation. It was accomplished by ensuring a closed water cycle, thus preventing the excessive volume of injected steam from brine, by continuously adding fresh water instead.

It is possible to conclude that no matter which process is chosen to improve oil production, a far from ideal elimination of excess OFB by reinsertion is a reality. Generally speaking, the OFB used in crude oil production by reinserting it into wells still remains a major source of groundwater pollution (see next section).

POLLUTION ASPECTS OF OFB

The extensive use of injection wells for disposal of a variety of chemical wastes, including OFB, started in the late 1960s. The first concrete action to protect drinking water supplies from possible contamination by injection wells came from the Safe Drinking Water ACT of 1974 (3). The same practice continued, and during the 1990s, there were approximately 170,000 active and more than 2 million abandoned injection wells in the United States. Generally, to estimate the total quantitative or qualitative budget of OFB even on a single oil field requires a very systematic analysis of a long-term record. That was also the problem in pollution control treatment (15).

For example, the concentration range of TDS contained in an OFB varies from 10 to more than 350 g/L, of which sodium chloride contributes about 80% or more. In contrast to the extreme high levels of TDS, TSS values are at incomparably lower levels, in most cases around 100 mg/L, even though, three categories of constituents affect OFB well injection: inorganic, organic, and microorganisms. The crucial impact on freshwater pollution is due to TDS content.

A concise review of the impact of OFB liquidation on water quality in reverse chronological order is presented in Table 1. Based on the available data, it is possible to distinguish two types of research activity: the water management approach and specific characterization of OFB in selected regions with different implications for water quality. In most cases, two categories of water sources were investigated. Because deep well injection is the most common OFB liquidation technique, particular attention was paid to groundwater quality change (16–21). After many years in use, OFB pits and injection wells have started to impact the quality of streams and shallow aquifers too (22–30). Some cases applied an integrated approach to a combined quality control management of groundwater and surface water affected by OFB disposal (31).

In addition to the two main categories of water sources affected by OFB contamination, there is a third group of more specific and very diversified implications of OFB management practice. Some illustrative examples are briefly described below.

An interesting result was obtained from a simulating study of the use of OFB as a deicer on a gravel roadbed (16). Constituents of brine were detected in water samples from groundwater monitoring wells. The map of an OFB plume was generated based on groundwater quality and surface geophysical methods.

A recently published article illustrates research in progress that quantifies human health risks from naturally occurring radioactive materials (NORM) related to OFB streams, pipe scale, and sludge (33). In another risk-based approach, the status of the water quality of a former OFB disposal pit, whose operation was discontinued in the mid-1940s, was evaluated (34). After a thorough water analysis, it was found that a long-term

Table 1. Impact of OFB on Water Quality

Year	Research Status (Region)	Impact/Implication	Ref.
	Management / Modeling	*Groundwater*	
1992	Use of a OFB as a deicer on roads (OH)	OFB plume in groundwater	16
1987	Groundwater management model for OFB plume containment (KS)	The potential for improving groundwater quality affected by OFB	17
1985	An inventory of long-term (1930s–1950s) disposal of OFB by ponds, wells, and pits (KS)	The progressive deterioration of groundwater quality during the time	18, 19
1985	As simplified model of differentiating and mixing of groundwater; OFB and natural brine (OK)	Potential for contaminating a freshwater aquifer	20
1984	General study on OFB contamination (NM)	OFB pit and deep groundwater quality (10,000 ft)	21
	Monitoring	*Surface Water, Shallow Aquifers*	
2001	Sodium adsorption ratios and salinity levels in aquifers (TX)	Agriculture and irrigation	22
1997	Nitrate, chloride, and bromide concentrations in OFB increased as well depth decreased (TX)	Cropland, pasture, and groundwater quality	23
1989	OFB pits, oil-exploration holes, and watershed (TX)	The salinization of aquifers and hydrocarbon concentration increase in artesian wells	24
1980	Colorado River water quality (TX)	Salinization of a river flow	25
1976	OFB recharge and groundwater quality fluctuation (OH)	Shallow groundwater aquifer contamination	26
1976	OFB and river water quality (KS)	An extremely high chloride concentration in river water	27
1975	Water supply and OFB pits and injection wells (OH)	Salinization up to 28 g/L of surface water	28
1975	Safe future development of water supply (OK)	Salinization of shallow alluvial aquifers	29
1975	The estimate of the quantities of a waste by deep well injection; general and OFB	Potential impact on water quality in the context of the Water Pollution Control Act of 1972	30
	Integrated Management	*Surface / Groundwater*	
1985	Study on the interrelationship of surface and groundwater quality management (IL)	OFB as a quality impact parameter	31
	Specific Monitoring	*Diversified*	
2000	Isolation and characterization of bacteria; nitrate-reducing; sulfide-oxidizing (Canada)	Characterization of OFB biogeochemical cycle	32
1999	Naturally occurring radioactive materials (NORM) in OFB (TX, LA, OK, MS, CA)	Human health risk	33
1997	Remediation of water quality of a former disposal pit (TX)	Surface water and aquatic life risk assessment for recreation	34
1994	Isolation of anaerobic bacteria (OK)	New specific anaerobic bacteria identified in OFB	35
1981	Chemical characterization of brines in salt mines (LA)	Potential impact of OFB on salt mines	36

natural remediation process gradually turned brine into freshwater that sustained aquatic plants. The pond was categorized for potential recreational use.

Specific microbiological investigation has proven that OFB can be a good substrate for the growth of strictly anaerobic bacteria (35). In another microbiologically focused study, two novel nitrate-reducing and sulfide-oxidizing strains of bacteria were isolated from production fluid (32).

An interesting study was performed on the impact of OFB on salt mines (36). Based on isotopic studies, it was hypothesized that active leaks, stalagmites, and stalactites in salt mines, which contained high levels of Ca, K, and bromides, arose from OFB formation water.

Generally, it is possible to conclude that a very broad spectrum of monitoring research data clearly indicates the significant impact of OFB management practice on the deteriorating quality of different water resources.

OFB CONTROL BY WASTEWATER TREATMENT METHODS

The previously considered aspects of OFB management such as environmental monitoring, socioeconomic aspects and improvement of crude oil production by using OFB were based on a broad pool of literature sources. In contrast, reports on the control of OFB by wastewater treatment methods are rather limited and sporadic.

It is worthwhile mentioning that there are some attempts to treat oil field effluents in a different engineering context. Cationic polyelectrolytes were developed to partially treat effluents containing a high percentage of residual free oil, as well as stable emulsions (37). Also, there are reports of treating OFB by using configurations similar to those employed for oil refinery wastewater (38–41). Two case studies of systematic approaches will be presented below. In both studies the OFB was considered wastewater that needed to be purified and discharged or reused.

Table 2. Alternatives of Microbiological[a] and Advanced Treatment[b]

No.	Description
Alternative I	[a]Activated sludge
	[b]Advanced treatment with GBAC
Alternative II	[a]Activated sludge + 25% dilution with freshwater
	[b]Advanced treatment with GBAC
Alternative III	[a]Activated sludge + PAC
	[b]Advanced treatment with GBAC
Alternative IV	[a]Activated Sludge + 25% dilution with freshwater + PAC
	[b]Advanced treatment with GBAC
Alternative V	[a]Activated sludge + 50% dilution with freshwater + PAC
	[b]Advanced treatment with GBAC

Various aspects of an innovative technology for OFB treatment were investigated on a laboratory and pilot scale (15,42–44). Both free and dispersed oily matter were separated by gravitation and sedimentation. Apart from physicochemical oil removal processes, special attention was paid to different variants of improved microbiological treatment (see Table 2). In that regard, two approaches were applied; dilution with freshwater and powdered activated carbon (PAC). The aim was to intensify the microbiological process and neutralize its inhibitors by adsorption on PAC. Advanced treatment was carried out on granular biological activated carbon (GBAC). It was found that the GBAC removed up to three times more organic matter than its adsorption capacity, and thus its adsorption power was not exhausted. This was a clear indication that the microorganisms present in the biofilm on GBAC oxidize the adsorbed pollutants and regenerate it at the same time. A technological scheme for complete treatment was proposed (see Fig. 2). It was concluded that the complete treatment employed here was very efficient. The organic matter content in the effluent did not exceed

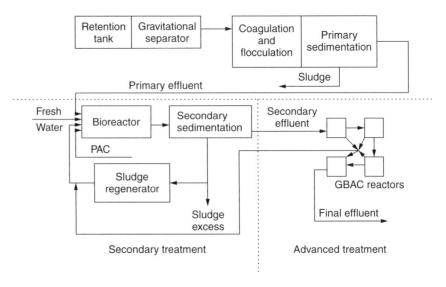

Figure 2. Block schematic of a complete OFB treatment.

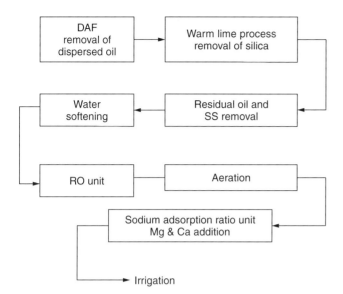

Figure 3. Block schematic of RO pilot unit for treating OFB.

2.5 ppm of BOD. Also, a final decrease of salinity was achieved by mixing effluent in an adequate proportion with the recipient water (44). Thus the treated OFB can be discharged into recipients of low self-purification capacity. Based on a preliminary cost analysis, it was estimated that a complete OFB treatment would cost 10% more than a conventional PAC/GBAC treatment of municipal wastewater.

The second case study of complete OFB treatment represents a state-of-the-art process based on reverse osmosis (RO). The configuration applied converts an oil field effluent into freshwater (45). The simplified block schematic of the process is presented in the Fig. 3. The input to the process contained a relatively low level of salinity (about 7 g/L).

The concentration of a soluble oil was also at a moderate level, below 200 ppm. Still, due to the passage of oil through the filters, RO membrane fouling was observed. Despite that, the overall results showed clearly successful conversion of OFB into freshwater of irrigation quality. The operating cost was less then 10¢/bbl.

Concluding this section, it is possible to draw out some positive outcomes. Crude oil still represents the most important energy source, so it is crucial to continue further improvement of OFB treatment technologies. The case studies presented clearly show the feasibility and the cost-effectiveness of the complete OFB treatment. It is quite obvious that OFB pollution control by wastewater treatment methods is superior to the liquidation–reinsertion practice of the past.

BIBLIOGRAPHY

1. Stoessell, R.K. (1997). *Ground Water* **35**(3): 409–418.
2. Kemmer, F.N. (Ed.). (1988). *The Nalco Water Handbook*, 2nd Edn. McGraw-Hill, New York, pp. 43.1–43.23.
3. Gorman, H.S. (1999). *Business History Review* **73**(4): 601–640.
4. American Petroleum Institute (1983). *Primer of Oil and Gas Production*.
5. Collins, A.G. (1975). *Geochemistry of Oilfield Waters*. Elsevier, Amsterdam.
6. American Petroleum Institute (1974). *Treating Oil Field Emulsions*, 3rd Edn.
7. Wright, C.C., Cross, D., Ostroff, A.G., and Stanford, J.R. (1977). *Oil Field Subsurface Injection of Water*. ASTM, Philadelphia.
8. Muvrin, B. and Salopek, B. (1992). *Nafta* **43**(7): 349–362.
9. Ariman, T. and Veziroglu, T.N. (Eds.). (1987). *Particulate and Multiphase Processes*. Hemisphere. Vol. 2, pp. 535–550.
10. Hoke, S.H. and Collins, A.G. (1981). *Proceedings of the Second ASTM Symposium on Water for Subsurface Injection*, Ft. Lauderdale, FL, pp. 34–48.
11. M-I Drilling Fluids Co. (1992). 1992–93 General Catalog, Houston.
12. Miskovic, D. and Sears, B. (1994). *Private Communication*. Lenox Institute of Water Technology, MA.
13. Advanced Separation Technologies, Inc. (1994). *REVEX Technology*, Houston.
14. Koren, A. and Nadov, N. (2002). *Mechanical Vapor Compression to Treat Oil Field Produced Water*. I.D.E. Technologies Ltd., Ra'anana, Israel.
15. Miskovic, D. et al. (1989). *Water Sci Technol.* **21**: 1849–1852.
16. Chapman, M.J. and Bair, E.S. (1992). *Ground Water Monitoring Rev.* **12**(3): 203–209.
17. Heidari, M. et al. (1987). *Water Resour. Bull. WARBAQ* **23**(2): 325–335.
18. Whittemore, D.O. et al. (1985). *Proceedings of a Symposium on Groundwater Contamination and Reclamation*, Tucson, AZ, pp. 109–116.
19. Sophocleus, M.A. (1984). *J. Hydrology* **69**(1–4): 197–222.
20. Mast, V.A. (1985). *Ground Water Monitoring Rev.* **5**(3): 65–69.
21. Stephens, D.B. and Spalding, C.P. (1984). *Proc. Ogallala Aquifer. Symposium II*, Lubbock, TX, pp. 440–450.
22. Hudak, P.F. (2001). *Int. J. Environ. Stud.* **58**(3): 331–341.
23. Hudak, P.F. and Blanchard, S. (1997). *Environ. Int.* **23**(4): 507–517.
24. Dutton, A.R. et al. (1989). *Ground Water GRWAAR* **27**(3): 375–383.
25. Rawson, J. (1980). *Geological Survey Austin TX*, 80–224, p. 66.
26. Pettyjohn, W.A. (1976). *Ground Water* **14**(6): 472–480.
27. Leonard, R.B. and Kleinschmidt, M.K. (1976). *Kansas Geological Survey*. Lawrence Chemical Quality Series 3, September. p. 24.
28. Pettyjohn, W.A. (1975). *Ground Water* **13**(4): 332–339.
29. U.S. EPA-660/3-74-033, April (1975).
30. Conrad, E.T. and Hopson, N.E. (1975). *Water Resour. Bull.* **11**(2): 370–378.
31. Crandall, D.A. (1984). *Proc. 20th Annu Conf. Ame Water Resour Assoc. Symp.*, Washington, DC, pp. 193–199.
32. Gevertz, D. et al. (2000). *Appl Environ Microbiol.* **66**(6): 2491–2501.
33. Anonymous (1999). *Oil Gas J.* **97**(1): 29–32.
34. Alvillar, S.B. and Hahn, R.W. (1997). *Proc. Air Waste Manag Assoc 90*. Annual Meeting, Toronto, 8–13.
35. Bhupathiraju, V.K. et al. (1994). *Int. J. Systematic Bacteriol.* **44**(3): 565–572.

36. Kumar, M.B. and Martinez, J.D. (1981). *J. Hydrology* **54**(1–3): 107–140.
37. Mahanta, D. (1994). *Chem. Ind. Dig.* **7**(2): 125–127.
38. Roev, G.A. (1981). *Ochistnye Sooruzheniya Gazonefte-Perekachivayushchih Stancii I Neftebaz*. Nedra, Moskva.
39. Jones, H.R. (1973). *Pollution Control in the Petroleum Industry*. Noyes Data Corp., Park Ridge, NJ, pp. 175–202.
40. Barnes, D. et al. (1984). *Surveys in Industrial Wastewater Treatment 2-Petroleum and Organic Chemicals Industries*. Pitman, London, pp. 29–52.
41. Miskovic, D. et al. (1986). *Water Sci. Technol.* **18**(9): 105–114.
42. Dalmacija, B. et al. (1995). *Water Res.* **30**(2): 295–298.
43. Dalmacija, B. et al. (1996). *Water Res.* **30**(5): 1065–1068.
44. Miskovic, D. *Private communications with local oil company*. Novi Sad, 1983–1986.
45. Tao, F.T. (1993). *Oil Gas J.* **91**(38): 88–91.

OIL POLLUTION

KUSUM W. KETKAR
RAJIV GANDHI CHAIR
Jawaharlal Nehru University
New Delhi, India

The word oil refers to hydrocarbons—crude oil, gasoline, heating oil, fuel oil, diesel, jet fuel, and other refined oil-based products. Oil pollution refers to input of hydrocarbons into a water body, either on the surface of the earth or underground. Oil may directly enter surface water bodies like ponds, rivers, streams, and oceans, and it may leach into underground water. Recognizing that moving water cleans itself, oil pollution occurs when more oil enters water than natural, physical, and chemical processes can absorb it.

Accidental oil spills by oil tankers are only one of many causes of oil pollution, but they attract a lot of media attention. Media attention is justified because accidents spill a lot of oil in a short period of time. For example, oil tanker *Exxon Valdez* spilled 11 million gallons of crude oil in Prince William Sound, Alaska, in 1989. In addition to accidental oil spills, oil may enter marine environment through natural seepage, off-shore oil drilling, illegal disposal of waste water containing oil from tankers, tank barges and other vessels, land-based petrochemical industrial complexes, runoff containing oil from roads, oil-contaminated stormwater, and sludge from municipalities and numerous other land-based activities. Initially, oil-contaminated water may enter local streams, rivers, lakes, or ponds, but ultimately it is dumped into the oceans. According to Oceana, 80% of oil dumped into the sea is from inland operators, of which 44% is the result of direct dumping or coastal drainage, 33% is transported through the atmosphere, and the remaining 20% comes from accidental or deliberate spills from vessels and marine facilities (see Fig. 1) (1).

Worldwide, accidental oil spills account for about 12% of the total oil input into our oceans, whereas deliberate dumping from vessels and illegal dumping from tankers on the high seas account for 33%. Every year, over 6000

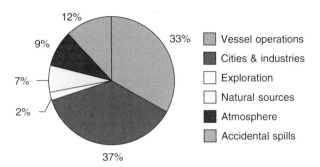

Figure 1. Causes of oil pollution. *Source*: Euoropa Oceana.

oil tankers use oceans as highways to transport crude oil. In 1983, 1.21 billion tons of crude oil was transported, and by 2002, that quantity had risen to 2 billion tons of crude. Table 1 gives estimated world maritime operational and accidental sources of oil entering marine environment.

On average, 25 major accidents and many smaller ones spill over 700 tons of oil into the seas every year (see Table 2). Overall, 2 to 10 million tons of oil leaks into the oceans annually.

Table 3 lists oil spills over a million gallons (3367 tons) in U.S. waters. After 1990, no spill in U.S. waters has been over a million gallons, and overall, there has been a sharp decline in large oil spills since the passage of the Oil Pollution Act of 1990. Most oil spill incidents in the United States occur along Alaska and the western and southern coasts. Alaska, Texas, and Louisiana are the major oil-producing states in the United States.

Since World War II, unprecedented increases in the world demand for energy and the rapid economic growth of the world economy have gone hand in hand. Oil is an indispensable source of energy in the production of goods and services in general and particularly in transportation. Most countries are poorly endowed with this valuable product also known as "black gold." The result is that oil-poor countries have to import it from oil-rich countries. Petroleum and petroleum products are transported in tank ships or tank barges. According to Oceana, over 6000 tankers traverse world oceans carrying crude oil and oil products. The United States is the world's largest consumer of oil, and it imports over 50% of the oil that it needs. Oil comes from both short-hauls like Mexico, Venezuela, and Nigeria and long hauls like Saudi Arabia,

Table 1. World Maritime Operational and Accidental Sources of Oil Entering Water Bodies (Million Tons Annually)

Source	1990	1981–85	1973–75
Bilge and fuel oil	0.25	0.31	—
Tanker operational losses	0.16	0.71	1.08
Accidental spillage			
Tanker accidents	0.11	0.41	0.20
Nontanker accidents	0.01		0.10
Marine terminal operations	0.03	0.04	0.50
Dry-docking	0.01	—	—
Scrapping of ships			

Source: U.S. Coast Guard, OPA-90, Double Hull Tanker Study, 1991.

Table 2. World's 25 Biggest Oil Spills Since 1970

Number	Quantity (10³ Tons)	Location	Year	Accident
1	1,000	Arab Gulf	1991	Gulf War
2	816	Mexico	1979	Platform Ixtoc I
3	476	South Africa	1983	Castillo de Bellver
4	267	France	1978	Amoco Cadiz
5	234	Italy	1991	Haven
6	158	Canada	1988	Odessey
7	146	Trinidad	1979	A. Empress & A. Captain
8	145	Barbados	1979	Atlantic Empress
9	141	Greece	1980	Irene Seranade
10	135	UK	1967	Torrey Canyon
11	130	Oman	1972	Sea Star
12	115	North Sea	1971	Texaco
13	110	North Pacific	1977	Hawaiian Patriot
14	103	Kuwait	1981	Tank
15	100	Spain	1976	Urquiola
16	95	Turkey	1999	Independentza
17	92	Portugal	1969	Julius Schindler
18	85	Portugal	1975	Jacob Maersk
19	84	Nigeria	1979	Tank
20	83	UK	1993	Brear
21	81	Spain	1992	Aegean Sea
22	80	Morocco	1989	Kharg S
23	77	Spain	2002	Prestige
24	72	UK	1996	Sea Empress
25	72	Mozambique	1992	Kalina

Source: Reference 1.

Russia, the North Sea, and Indonesia. Over 4000 tankers make port calls every year, and the ports experiencing the highest tanker traffic are the ports of New York-New Jersey, Houston, Los Angeles/Long Beach, and San Francisco Bay. As the New York seaboard is shallow and has a soft bottom, only smaller tankers make port calls there. Crude from the Middle East and other long-haul countries is transported in very large crude carriers (VLCCs) and ultra-large crude carriers (ULCCs) to New York and other shallow ports. Lightering operation is used offshore to bring crude in smaller tankers to the shallow ports. These long-haul tankers (VLCCs) that bring crude to the United States have a capacity of 200,000 tons or more. They make calls to ports on the West Coast and the Gulf of Mexico because the coastal waters there are deep with rocky bottoms. Also, tank vessels use U.S. coastal waters, rivers, and lakes to distribute over 90% of petroleum and petroleum products, which results in over 20,000 port calls annually. As the vessel traffic increases, the risk of casualties and oil spills increases. Navigational risk in coastal waters is significantly influenced by vessel traffic, width of shipping channels, and the number of course changes. But the risk rises less than proportionately as vessel traffic increases (2). In addition to transportation of oil and refined products, there is offshore drilling and the crude is transported via pipelines to the mainland refineries. Alaskan crude is transported in tankers to the refineries. Thus, oil drilling, transfer, processing, and distribution make intensive use of the country's marine resources. With this intensive use comes the risk of oil input into the marine environment. The consumption of crude-based products generates waste, which releases oil into the natural environment. Hydrocarbons that are disposed of on land ultimately enter oceans.

The question is as follows: What is the impact of input of hydrocarbons on the marine environment? Scientists seem to agree that, in the long run, little adverse impact occurs on the marine environment. Take the Persian Gulf, for instance. Oil-producing Middle Eastern countries use ports on the Persian Gulf coast to ship crude oil and refined products. But biologists have observed that the Persian Gulf has experienced sharp increases in the concentration of oil-eating bacteria as oil input in the Gulf has increased (3). The environmental impact of oil input depends on the spill amount, its type—heavy or light crude or refined oil—the time of the year, location, previous exposure, and current length of exposure (4). The 1989 *Exxon Valdez* spill of 37,000 tons of crude in winter in Prince William Sound did more damage to the marine environment and the beaches than did the spill of 83,000 tons of crude by *Brear* on the high seas off the coast of Scotland. Temperate climates tend to have fewer oil-eating bacteria than do the tropics, and natural degradation takes much longer in the former climates than in the latter.

Generally, industrial, urban, and heavily populated areas are located near waterways and ports. Inputs of stormwater, sewerage, and industrial waste containing oil and oil-based products and input of oil from operational and accidental spills causes severe damage to the marine environment of the coastal waterways. For instance,

Table 3. Oil Spills in U.S. Coastal Waters

Number	Quantity (10³ Tons)	Location	Year	Accident
1	37.00	Prince William Sound, Alaska	1989	Exxon Valdez
2	25.30	Nantucket, Massachusetts	1976	Argo Merchant
3	13.00	Gulf of Mexico, Texas	1990	Mega Borg
4	12.60	Pilot Town, Louisiana	1981	Apex Houston
5	9.30	Off Cameron, Louisiana	1984	Alvenus
6	6.90	South of Semidi Islands, Arkansas	1988	UMTB 283
7	4.24		1985	Exxon Barge No. 283
8	4.20	San Francisco, California	1984	Puerto Rican
9	3.54	Mississippi River Montx, Louisiana	1982	Arkas
10	3.53	Donaldsville, Louisiana	1983	Barge No. 218

Source: U.S. Coast Guard.

coastal waters around New Orleans, Houston, Galveston, New York, and many other ports are dead in the sense that they can support no marine life. These waters continue to violate the mandate of the Clean Water Act of 1977. Coastal waters around New York–New Jersey ports are neither suitable for fishing nor swimming. To go fishing or swimming in New Jersey, one has to drive about 50 miles south from New York City. Similarly, the Hudson and East Rivers around New City have not been able to support any marine life until recently. The ecological impact of oil trapped in environmentally sensitive areas like marshes, wetlands, estuarine and freshwater communities, and intertidal and subtidal zones is the heaviest, although these effects appear to last from a few months to 15–20 years. But there can be exceptions. For instance, long-term and concentrated exposure of oil has killed colonies of coral reefs in Indonesian coastal waters and is suspected to have adversely affected invertebrates and plants in the coral reef community. The number 2 fuel oil that oil barge *Bouchard* 65 spilled in the west entrance of the Cape Cod canal in Buzzards Bay, Massachusetts, in 1974 entered salt marshes of Winsor Cove; and these mashes had not fully recovered by 2000 (5). The National Research Council observed that the most damaging effects of petroleum are oiling and tarring of beaches through the modification of benthic communities along polluted coastlines and the endangering of seabird species (6). Neither human health nor the food chain seems to be endangered by oil spills. The images of blackened beaches and dead birds are eye-catching, and television and the print media have effectively used them to win ratings and earn advertising revenues (7).

Large oil tankers crisscrossing the world's oceans and coastal waters pose a major threat for accidents leading to the spillage of large quantities of oil in a short period of time. Oil spreads quickly and evaporates. Heavy oils take longer to evaporate than do lighter and refined oils. If the spill is not quickly contained, it can trap sea birds and choke fish and other sea animals to death. If the spill is near a coastal area, it can kill life and damage property onshore and adversely affect commercial and subsistence fishing. The oil spill from the *Exxon Valdez* spread over hundreds of miles into numerous and remote coves, inlets, and bays from Prince William Sound through the Southern Kenai Peninsula and further south to Kodiak Island. Public outcry over the spoilage of pristine, rugged Alaskan environment cost Exxon-Mobile Corporation several billion dollars in cleanup costs. Organisms under rocks where hot water washing was used are taking longer to return to normal levels than those under those rocks that were left for nature to clean. In the aftermath of the accidental oil spill by the *Exxon Valdez*, the U.S. Congress passed the Oil Pollution Act in 1990 with the goal of preventing future oil spills and minimizing damage from accidental oil spills from operations and transportation of crude or refined petroleum products. The European Union followed the American example and passed a similar act after oil spills off the Spanish and British coasts. Among the preventive measures, the most controversial measures are the replacement of single hull tankers by double hull

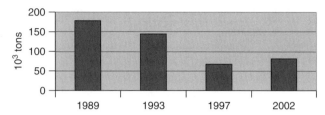

Figure 2. The world maritime operational oil spills. *Source*: International Tanker Owners Pollution Federation.

tankers and tanker owners' liability in case of a spill. As a result of various regulatory measures and increased vigilance by tanker ship owners, the number of large oil spills world wide has declined.

According to INTERTANKO, the quantity of oil spilled due to maritime operations has declined worldwide since 1989 because of the Oil Pollution Act of 1990 and the adoption of International Convention MARPOL 73/78 by the International Maritime Organization. The European Union followed the U.S. example and passed a law to prevent future oil spills after accidental spills by *Brear* off the coast of Scotland and by the *Aegean Sea* off the coast of Spain in the early 1990s. Figure 2 gives the quantity of oil spills from maritime operations.

BIBLIOGRAPHY

1. OCEANA. Available: http://www.euoropa.oceana.org.
2. Ketkar, K.W. (2003). The Oil Pollution Act of 1990—a decade later. *Spill Sci. Technol.* **7** (1–2).
3. Holloway, M. (1991). Soiled shores. *Sci. Am.* **October**.
4. Mielke, J.E. (1990). *Oil in the Ocean: The Short - and Long-Term Impacts of Spills*. Congressional Research Service, the Library of Congress, Washington, DC.
5. Hampson, G. Destruction and recovery of the Winsor Cove, Cataumet, MA Salt Marsh from a #2 fuel oil spill: a 25 year history. *Environment Cape Cod* **3** (1).
6. National Research Council. (1985). *Oil in the Sea: Inputs, Fates and Effects*. National Academy Press, Washington DC.
7. Ketkar, K.W. (1995). The US Oil Pollution Act of 1990 and the future of the maritime industry. *Marine Policy* **15**(5).

INDICATOR ORGANISMS

GARY P. YAKUB
Kathleen Stadterman-Knauer
Allegheny County Sanitary
Authority
Pittsburgh, Pennsylvania

INTRODUCTION

Assessing biological contamination has always been an important part of any water quality study. Sources of fecal contamination include natural or agricultural sources and infrastructure problems that result in the release of untreated sewage into natural waters. This contamination can result in releasing dangerous human pathogens into recreational waters and source waters used to supply potable drinking water. People who contact contaminated water through bodily contact or ingestion are at risk for a variety of infections.

Since 1920, 1884 waterborne disease outbreaks, 882,144 cases of illness, and 1169 deaths in the United States have been from contaminated drinking and recreational waters (1). The causative agents responsible for these diseases could be bacteria, fungi, viruses, helminthes (worms), protozoa, or other pathogenic organisms. An attempt to characterize the biological quality of a water sample by investigating all of the known causative agents of waterborne disease would be impractical due to the large number of possible agents and the time-consuming methodology required. In addition, some agents can cause disease even at very low concentrations, and other causative agents can be difficult, if not impossible, to culture in the laboratory. These factors point to the need for a system to monitor biological water quality quickly, easily, and cost-effectively.

Indicator bacteria are the most common way of assessing biological water quality. Indicator bacteria are bacteria that can predict the presence of pathogenic organisms. To be a successful indicator, the following criteria must be met:

1. Indicator bacteria must be easy to cultivate in the laboratory using simple, cost-effective procedures.
2. Indicator bacteria must be present in high enough concentrations to be readily detectable in a manageable sample volume.
3. Indicator bacteria must effectively predict the risk of illness from exposure to the waterborne pathogens present in contaminated water.

Indicator bacteria are used in epidemiological studies to establish a correlation between bacteria levels and the rate of disease in a given population. From this information, water quality standards can be developed for a particular indicator bacteria based on acceptable risk limits. These water quality standards can then be used for quantitative evaluation of biological water quality.

Indicator bacteria are currently used to assess water quality for a variety of applications. Wastewater treatment facilities use fecal coliforms or *Escherichia coli* as indicators of proper disinfection. Drinking water plants monitor total coliforms as an indicator of water quality in their distribution systems. Recreational water quality standards are based upon fecal coliforms, *E. coli*, or enterococcus bacteria levels depending upon water type and usage.

THE COLIFORM GROUP

The coliform group is the group of bacteria that has received the most attention for application as an indicator of fecal pollution. Coliform bacteria are found in the intestinal tract of humans and other animals; therefore, their presence in ambient water indicates fecal pollution and the potential presence of pathogens (2).

During the late 1940s and early 1950s, the United States Public Health Service began a series of studies based on coliform bacteria levels (3). Its goal was to determine if coliform concentrations in recreational water were correlated with the incidence of gastroenteritis in swimmers. From these studies, total coliform was used as the indicator bacteria for recreational waters. In the mid 1960s, total coliforms were replaced by fecal coliforms, which showed a greater specificity for mammalian fecal pollution and less variation due to environmental conditions. Fecal coliforms made up about 18% of total coliforms (3). In 1976, the U.S. EPA recommended fecal coliforms to the regulatory community as indicator organisms for the presence of pathogens in recreational waters (2). In 1986, based on new research data, U.S. EPA recommended new monitoring guidelines, including the use of *E. coli* as an indicator to replace fecal coliforms. The presence of *E. coli*, it was found, correlated with swimming-associated gastrointestinal illness better than fecal coliform levels (2).

THE ENTEROCOCCUS GROUP

In its 1986 Bacterial Water Quality Criteria report, the U.S. EPA also concluded that members of the enterococcus group of bacteria were acceptable as an indicator of fecal pollution due to the high correlation between enterococcus bacteria levels and swimming-related gastrointestinal illness. The strong correlation was noted in both freshwater and saltwater, and enterococci have been proposed as an indicator of water quality for both environments (3).

Enterococcus bacteria are a subgroup of the fecal streptococci that includes *S. faecalis, S. faecium, S. gallinarum,* and *S.avium*. Enterococci are differentiated from other streptococci by their ability to grow in 6.5% sodium chloride, at pH 9.6, and at 10 °C and 45 °C.

There has been some attempt to use the fecal coliform to fecal strep ratio to differentiate the source of fecal pollution. A ratio higher than four was proposed to suggest human sources, and a ratio of less than 0.7 was proposed to suggest nonhuman sources. However, factors such as a difference in die-off rates among the enterococci and a variability to chlorine disinfection resulted in a recommendation that the fecal coliform to fecal strep ratio should not be used for source differentiation (4).

THE FUTURE OF INDICATOR ORGANISMS

There are many organisms that do not correlate with the indicators used to determine fecal contamination in surface waters. The ecology, prevalence, and resistance to stress of indicator organisms may differ from many of the pathogenic microorganisms for which they serve as proxies. For example, fecal coliform does not indicate many disease-causing viruses, such as hepatitis A or E, Coxsackie viruses, adenoviruses, and Norwalk viruses or indigenous bacteria such as *Helicobacter* or *Legionella*. *Giardia* and *Cryptosporidium*, two parasitic protozoa, also do not correlate with fecal coliform levels (5). A

2001 American Academy of Microbiology report advocates the advancement of molecular techniques for detecting waterborne pathogens in the context of a risk-based analysis (5). Molecular tools could allow rapid detection of an organism, and also determine its virulence, prevalence, and source of contamination. However, many of these techniques are in development and may not be available to utilities for many years. Therefore, indicator organisms will continue to play a role in water quality analysis until these new techniques become available and cost-effective.

BIBLIOGRAPHY

1. American Water Works Association. (1999). *Waterborne Pathogens*. Manual of Water Supply Practice M48, Denver, CO.
2. U.S. EPA. (1999). *Action Plan for Beaches and Recreational Waters*. EPA/600/R-98/079. Washington, DC.
3. U.S. EPA. (1986). *Ambient Water Quality Criteria for Bacteria—1986*. EPA/440/5-84-002. Washington, DC.
4. American Public Health Association, American Water Works Association, and Water Environment Federation. (1992). *Standard Methods for the Examination of Water and Wastewater*. 18th Edn. Washington, DC.
5. Rose, J.B. and Grimes, D.J. (2001). *Reevaluation of Microbial Water Quality: Powerful New Tools for Detection and Risk Assessment*. A Report from American Academy of Microbiology, Washington, DC.

pH

MICHAEL P. DZIEWATKOSKI
Mettler-Toledo Process
Analytical
Woburn, Massachusetts

pH THEORY

Some chemical compounds, when dissolved in water, dissociate to yield hydrogen ions (H^+, e.g., protons), as in Equation 1, and are called acids. Other compounds called bases yield hydroxide ions (OH^-), when dissolved in water, as in Equation 2.

$$HA = H^+ + A^- \tag{1}$$

$$MOH = M^+ + OH^- \tag{2}$$

The chemical species in Equations 1 and 2 represented by A^- and M^+ are anions (negatively charged ions) and cations (positively charged ions), respectively. Acids and bases can be organic (containing two or more of the following atoms: carbon, hydrogen, oxygen, and nitrogen), such as acetic acid ($HC_3H_2O_2$), or inorganic (consisting of a metal and nonmetal), such as the base sodium hydroxide (NaOH). In addition to classification as organic or inorganic, acids and bases can also be ranked according to their strengths. Some common acids and bases are listed in Table 1.

The relative strength of an acid or base is related to the degree of dissociation that the compound undergoes according to either Equation 1 or 2. A strong acid is one that dissociates completely to yield a relatively large number of hydrogen ions, whereas a strong base is a compound that yields a relatively large number of hydroxide ions.

For strong acids and strong bases, the reactions described by Equations 1 and 2 proceed from left to right to form products and the reverse reactions are negligible. The converse is true for weak acids and weak bases; only a relatively small number of either hydrogen ions or hydroxide ions are produced. In the case of a weak acid or base, the reverse reactions described by Equations 1 and 2 are significant. Weak acids and bases are characterized by an acid dissociation constant (K_a) or a base dissociation constant (K_b) (see Table 1). The larger the value of either K_a or K_b, the greater is the degree of dissociation of the acid or base.

The hydrogen ion concentration in a solution of 1 mol/L hydrochloric acid would be approximately 1 mol/L. However, for a 1 mol/L sodium hydroxide solution (a base), the hydrogen ion concentration is approximately

Table 1. Common Strong and Weak Acids and Bases

Strong Acids		Strong Bases	
HCl, Hydrochloric acid		LiOH, Lithium hydroxide	
H_2SO_4, Sulfuric acid		NaOH, Sodium hydroxide	
HNO_3, Nitric acid		KOH, Potassium hydroxide	
$HClO_4$, Perchloric acid		$Ca(OH)_2$, Calcium hydroxide	
HBr, Hydrobromic acid		$Sr(OH)_2$, Strontium hydroxide	
HI, Hydroiodic acid		$Ba(OH)_2$, Barium hydroxide	
Weak Acids	K_a@25 °C[a]	Weak Bases	K_b@25 °C[b]
HF, Hydrofluoric acid	6.8×10^{-4}	$(CH_3)_2 NH$, Dimethylamine	5.1×10^{-4}
$HCHO_2$, Formic acid	1.7×10^{-4}	NH_3, Ammonia	1.8×10^{-5}
$HC_7H_5O_2$, Benzoic acid	6.3×10^{-5}	N_2H_4, Hydrazine	1.7×10^{-6}
$HC_2H_3O_2$, Acetic acid	1.7×10^{-5}	NH_2OH, Hydroxylamine	1.1×10^{-8}
HClO, Hypochlorous acid	3.5×10^{-8}	$C_6H_5 NH_2$, Aniline	4.2×10^{-10}
H_3BO_3, Boric acid	5.9×10^{-10}	NH_2CONH_2, Urea	1.5×10^{-14}

[a] Reference 1, p. 672.
[b] Reference 1, p. 680.

10^{-14} mol/L. In pure water, the hydrogen ion concentration is 10^{-7} mol/L. In the preceding examples, the hydrogen ion concentration varies by a factor of 10^{14} in going from acid to base. Therefore, it is more convenient to express the relative acidity of a solution by using a logarithmic scale based on hydrogen ion concentration and referred to as pH.

pH is expressed mathematically as

$$\mathrm{pH} = -\log[\mathrm{H}^+] \quad (3)$$

where $[\mathrm{H}^+]$ is the hydrogen ion concentration in the solution expressed in mol/L or molarity (M). Using this definition, the pH values for 1 M hydrochloric acid, water, and 1 M sodium hydroxide are 0, 7, and 14, respectively. Although the pH scale is technically defined from 0 to 14, pH values below 0 and above 14 are possible.

To contrast, consider the pH of a 1 M HCl solution and a 1 M acetic acid solution. From the previous example, we know that the pH of the 1 M HCl solution is 0, whereas the 1 M acetic acid solution will have a pH of 2.4. The difference in pH between the two solutions lies in the fact that acetic acid is a "weak" acid, and hence it does not completely dissociate in aqueous solution. For this reason, a simple relationship between pH and weak acid or weak base concentration does not hold true.

In our previous example, we stated that the pH of a 1 M NaOH is 14. To arrive at this result, one must first calculate the pOH:

$$\mathrm{pOH} = -\log[\mathrm{OH}^-] \quad (4)$$

and then calculate the pH:

$$\mathrm{pH} = 14 - \mathrm{pOH} \quad (5)$$

pH BUFFERS

The reaction shown in Equation 1 will proceed from left to right to a limited extent for a weak acid. In this case, the reverse reaction (formation of HA) can also occur to a significant extent. The same is true for weak bases (Equation 2). As an example, consider the weak acid acetic acid (Equation 6) and the weak base ammonia (Equation 7):

$$\mathrm{HC_2H_3O_2} = \mathrm{H}^+ + \mathrm{C_2H_3O_2}^- \quad (6)$$

$$\mathrm{NH_3} + \mathrm{H_2O} = \mathrm{NH_4}^+ + \mathrm{OH}^- \quad (7)$$

Acetic acid dissociates to form hydrogen ion and acetate ion ($\mathrm{C_2H_3O_2}^-$). Acetate ion is referred to as the conjugate base to acetic acid because it can acquire a hydrogen ion and reform acetic acid. Ammonia, on the other hand, acquires a hydrogen ion from a water molecule to form ammonium ion ($\mathrm{NH_4}^+$) and hydroxide ion. Ammonium ion is referred to as the conjugate acid of ammonia because it can lose a hydrogen ion to reform ammonia.

Solutions that can resist pH changes upon the addition of limited quantities of an acid or base are called buffers. Buffer solutions may contain either an acid and conjugate base or a base and conjugate acid. The presence of both species (acid/conjugate base or base/conjugate acid) permits the unique properties of these solutions. Buffers can be acidic, neutral, or basic.

Buffers can be prepared in the laboratory by combining the acid or base species with a salt consisting of the corresponding conjugate acid or conjugate base. In the case of acetate buffer, acetic acid is combined with an acetate salt such as sodium acetate. The quantity of each component used can be obtained from reference literature or calculated (1). Standard buffer "recipes" are available for numerous buffer compositions and pH ranges (2).

Alternatively, buffers can be purchased commercially. These are typically available in pH values from 1 to 14. The accuracy with which a buffer is prepared can depend upon the manufacturer and composition of the buffer. The National Institute of Standards and Technology (NIST) sells salts for preparing buffer standards with an uncertainty of ±0.005 pH units. Other manufacturers typically sell buffers having uncertainties ranging from 0.01 to 0.02 pH units. Most buffers will state that they are traceable to NIST standards. This means that during the manufacture of the buffer, its value is standardized using NIST buffers.

The pH values of buffer solutions depend on temperature. For example, the pH value for the standard NIST borax buffer at 25°C is 9.180. At 50°C, the pH decreases to 9.011 (3). Commercially available buffers usually state the pH value at 25°C. pH values for other temperatures can be obtained from the buffer manufacturer.

pH MEASUREMENT

There are two well-accepted methods for determining solution pH: colorimetry and sensor-based technology. Each method has advantages and disadvantages, and each has utility based primarily on the application and desired accuracy of the determination.

Colorimetry

The colorimetric determination of pH employs dye compounds, called indicators, which change color depending upon solution pH. The dyes used exhibit colors in the visible spectrum and can undergo one or more color changes as a function of pH. The indicator can be immobilized on a paper substrate (pH paper) or can be prepared as a solution using water or water combined with an organic solvent. The simplest indicators are those that undergo only a single color change such as litmus paper or phenolphthalein solution. For example, acidic solutions cause the color of litmus paper to change from blue to red whereas basic solutions cause phenolphthalein solution to change from colorless to red (or pink). Table 2 lists several acid–base indicators along with their color changes as a function of pH.

The pH paper method offers greater utility because the immobilized indicators can undergo several different color changes across the full pH scale. Some papers are designed to indicate only a single unit pH change; others can indicate changes of 0.1 or 0.2 pH units.

Sensor-Based Measurement

These technologies employ sensors that either generate a potential (glass pH sensors and solid-state sensors) or use

Table 2. Compounds Used to Indicate Solution pH

Indicator	Transition pH	Acid Color	Base Color
Methyl violet	0.0–1.6	Yellow	Blue
Cresol purple	1.2–2.8	Red	Yellow
Methyl orange	3.1–4.4	Red	Orange
Congo red	3.0–5.0	Violet	Red
Methyl red	4.8–6.0	Red	Yellow
Litmus	5.0–8.0	Red	Blue
Phenolphthalein	8.0–9.6	Colorless	Red
Thymolphthalein	8.3–10.5	Colorless	Blue
Alizarin yellow	10.1–12.0	Yellow	Orange red

optical principles to determine the pH. Signal conditioning electronics are required in addition to the sensor for measurement, display, and data transmission. Both types of sensor technology can be used in the laboratory or in the field.

Glass pH Sensor. A glass pH electrode consists of two electrode elements; a silica glass membrane that selectively responds to hydrogen ions and a reference element. The two elements can be separate or combined as one into a single sensor, as shown in Fig. 1. When the elements are combined, the unit is referred to as a combination pH electrode. The combined electrode and the separate pH and reference elements are available commercially.

pH Glass Element. A pH glass element consists of a short section (<1 cm) of special glass that produces a voltage in the presence of hydrogen ions. This section of glass is attached to a shaft of lead glass that can be 10 or more centimeters long. The bulk of the membrane glass matrix usually consists of silicon dioxide. Other compounds, such as oxides of lithium, calcium, lanthanum, cesium, and uranium, are added to provide unique characteristics such as low resistance, high selectivity to hydrogen ions, and high strength. When immersed in aqueous solution, the glass matrix forms a hydrated layer, often called a gel layer, which is approximately 5 to 20 nm thick (4). The processes responsible for the sensing function of the glass can be viewed as an equilibrium binding of hydrogen ions within the gel layer (Fig. 2). The inside of the pH glass element is filled with a buffer saturated with silver chloride. Electrical contact is completed through a silver chloride coated silver wire immersed in the buffer. Usually the buffer is designed to have a pH of 7. Therefore, when the electrode is immersed in a pH 7 buffer, the potential difference across the glass membrane will be 0 volts versus the reference element. The shaft glass plays no role in the function of the electrode, aside from providing a substrate that is a good electrical insulator to which the membrane glass can be attached.

Reference Element. The reference element provides a reference potential against which the potential of the pH element is measured. The pH element and reference element form a complete electrical circuit when they are immersed together in the same solution. The reference is characterized by its ability to maintain a constant potential regardless of solution conditions. Additionally, the potential of the reference must not drift with time and should exhibit little or no hysteresis when the element undergoes a change in temperature. The reference element is simply a tube filled with an electrolyte into which a porous material called a frit is imbedded (Fig. 1). The frit allows the solution being tested to come in contact with the reference electrolyte. Usually, references are designed so that a small amount of the reference electrolyte will flow through the frit into the test solution. Ideally, there is no bulk flow of test solution into the reference

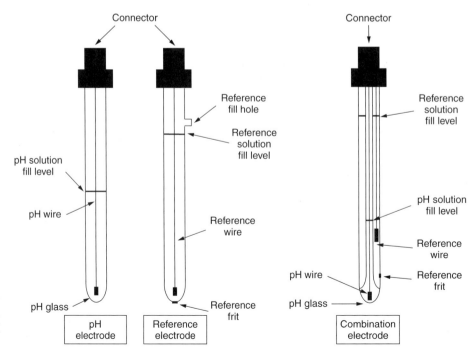

Figure 1. Separate pH and reference electrodes and the combination pH electrode.

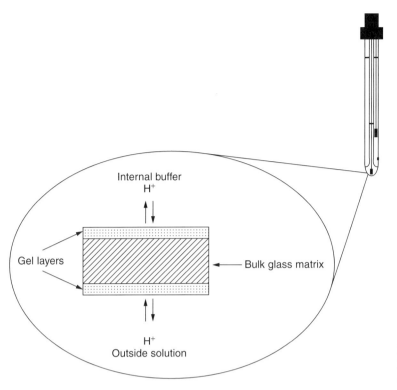

Figure 2. Representation of a cross section of a pH glass membrane showing gel layers and bulk glass membrane.

element. However, some test solution components may diffuse across the frit and contaminate the reference electrolyte after exposure for an extended period of time. The most common reference is based on a silver–silver chloride electrochemical cell. The system consists of a silver chloride coated silver wire immersed in a solution containing a high concentration of potassium chloride, usually either 3.5 M or saturated potassium chloride.

In use, the pH and reference elements are electrically connected to a high input impedance measuring device called a meter. Both the pH and reference elements are immersed to a depth great enough to ensure that the test solution completely covers the pH glass and the reference frit. The potential measured and displayed by the meter (E_{meter}) is the difference in potential between the two elements:

$$E_{meter} = E_{pH} - E_{ref} + E_J \quad (8)$$

where E_{pH} and E_{ref} are the potentials of the pH and reference elements, respectively. The other potential in the equation, E_J, is the reference junction potential. This potential results from the charge separation occurring due to differences in the migration rates of ions across the reference electrolyte/test solution interface. Reference electrodes are designed to keep this term small and constant, regardless of solution conditions.

The potential E_{meter} as a function of solution pH is given by the following expression;

$$E_{meter} = \left(\frac{2.303\,RT}{F}\right)(7 - \mathrm{pH}) \quad (9)$$

where R is the ideal gas law constant (8.314 J/mol·K), T is the temperature in K, and F is Faraday's constant (96485 C/mol). At 25 °C, the slope term in ( ) has a value of 59.2 mV. This is the operational form of the Nernst equation applied to a glass pH electrode (see Reference 4 for a discussion of the Nernst equation). Equation 9 yields a plot that has a negative slope, as shown in Fig. 3.

When the pH is measured at temperatures different from the calibration temperature, a measurement error can result. This is due to the change in electrode response as a function of temperature. For this reason, practically all commercially available pH meters provide automatic temperature compensation (ATC). This feature corrects only for the effect of temperature on electrode response, not for the true temperature coefficient of a solution resulting from chemical equilibria shifts with temperature. For ATC, the temperature at calibration and measurement are either entered into the meter manually, or a temperature sensor sends this information to the meter automatically. The temperature sensor can be built into the pH probe, or it may be separate.

Calibration. The probe and meter are calibrated as a pair. During calibration, the meter relates the potential produced by the electrode immersed in a buffer to the pH value of the buffer. Calibration is performed either using standard buffer solutions or a sample solution of known pH value. In laboratory use, the calibration is generally done either daily or just before the system is used for a series of measurements. The time interval between calibrations in an industrial process depends upon how quickly solution conditions degrade electrode performance. The interval can be as often as every few days or as little as a month or more. Detailed procedures for the proper use and calibration of glass pH electrodes can be found in reference sources (5,6).

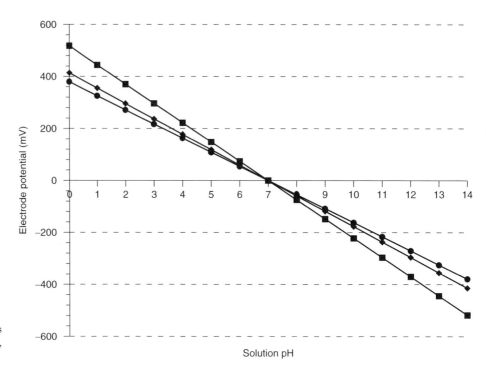

Figure 3. pH electrode potential as a function of solution pH at 0 °C (●), 25 °C (♦), and 100 °C (■).

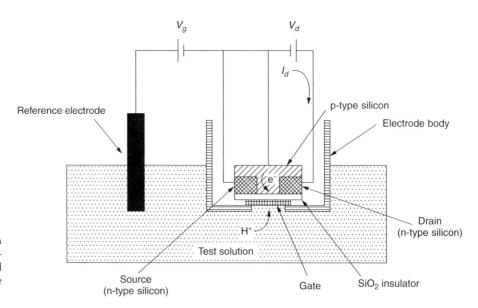

Figure 4. Typical arrangement of an ISFET sensor used for pH measurement. V_g and V_d are the gate and drain voltages, respectively. I_d is the drain current.

Solid State Sensors. More recently, pH sensitive electrodes have been developed based on metal-oxide semiconductor field effect transistor (MOSFET) technology. The sensors are similar in construction and operation to a traditional MOSFET, except that the gate is modified by an ion selective coating. The sensor is then referred to as an ion-selective field effect transistor (ISFET) (Fig. 4) (7–9).

The ISFET is enclosed in the electrode body so that only the gate is exposed to the solution. The sensors can be made to respond selectively to many different ions. For pH measurement, the gate is coated with either an oxide of silicon, aluminum, or tantalum or a nitride of silicon. In operation, a bias voltage is applied to the gate and the drain. Protons adsorbed to the gate alter the conductance of the p-type silicon above the gate. This results in current flow in the n-p-n region between the source and drain. Under these conditions, the drain current (I_d) is approximated by the following equation (10):

$$I_d = AV_d[V_c + V_g - E_N(\text{pH})] \qquad (10)$$

where A is a constant, V_d is the drain potential, V_c is a combination of the threshold potential of the p-n junction and phase boundary potentials, V_g is the potential applied to the gate, and E_N is the Nernst potential (or slope, equal to 59.2 mV at 25 °C). Either the voltage or current generated at the ISFET can be monitored by the meter. If the gate voltage is monitored, then the drain current is kept constant, or conversely, if the drain current is monitored, the gate voltage is kept constant.

The ISFET sensor uses the same type of reference element as the glass electrode. Attempts have been made to modify the ISFET to act as a reference electrode (11). Thus far, work has not produced significant improvement over a conventional reference system.

Optical pH Sensors. The concept of pH determination using colorimetry and absorbance spectrophotometry has been integrated into a sensor that is the size of a conventional laboratory glass pH electrode (12). The measurement is based on using indicator dyes immobilized in a polymer film. Several different indicator films are available, depending upon the pH range desired for the measurement.

BIBLIOGRAPHY

1. Ebbing, D.D. (1993). *General Chemistry*, 4th Edn. Houghton Mifflin Company, Boston, pp. 695–696.
2. Perrin, D.D. and Dempsey, B. (1974). *Buffers for pH and Metal Ion Control*. Chapman and Hall, London.
3. Bates, R.G. (1964). *Determination of pH: Theory and Practice*. John Wiley & Sons, Hoboken, NJ, p. 76.
4. Galster, H. (1991). *pH Measurement: Fundamentals, Methods, Applications and Instrumentation*. VCH, Weinheim, p. 117.
5. ASTM Standard E70-97. (1998). *Standard Test Method for pH of Aqueous Solutions with the Glass Electrode*.
6. Eaton, A.D., Clesceri, L.S. and Greenberg, A.E. (Eds.). (1995). *Standard Methods for the Examination of Water and Wastewater*, 19th Edn. American Public Health Association, Washington, DC, pp. 4-65–4-69.
7. Ref. 4, pp. 189–195.
8. Zemel, J.N. (1975). *Anal. Chem.* **47**: 255A.
9. Wohltjen, H. (1984). *Anal. Chem.* **56**: 87A.
10. Ref. 4, p. 192.
11. Ref. 4, pp. 193–194.
12. Ocean Optics, Inc. (2002). Product literature. Aviable: www.oceanoptics.com.

PERCHLOROETHYLENE (PCE) REMOVAL

Enos C. Inniss
University of Texas
San Antonio, Texas

INTRODUCTION

Perchloroethylene (C_2Cl_4) or PCE is also known by other names, tetrachloroethylene and perc. Often in chemistry, one might see the -ethylene replaced with -ethene. PCE is a chemically synthesized compound that has been used as a solvent since the 1940s because of its nonflammable nature. This characteristic eliminated many accidental fires in the workplace, which resulted in increased employee safety and reduced cost of operation. Since it is a volatile organic compound (VOC), PCE was considered "clean" because it quickly volatilizes when exposed to air. Appropriate applications of this chlorinated solvent include dry cleaning, metal degreasing, brake cleaning (replacing 1,1,1-trichloroethane), and production of HFC134a (a hydrofluorocarbon alternative to the chlorofluorocarbons or CFCs phased out as refrigerants in the early 1990s) (1).

By the early 1980s widespread use of PCE coupled with historically poor handling and disposal practices led to its regulation under the Safe Drinking Water Act Amendments of 1986 (SDWAA–1986) (2). Now a common groundwater contaminant and a priority pollutant, PCE has been observed to degrade both biologically and chemically.

The biological transformation process is reductive dechlorination, an anaerobic process where microorganisms use the chlorinated ethylenes as electron acceptors (3). Once reduced, the reduction products may also be further transformed by cometabolic oxidation, a gratuitous metabolic reaction under aerobic conditions where the microorganisms are growing on another substance and gain no energy from the cometabolized substance (4).

The chemical transformations may also be either reductions or oxidations. The reductions are achieved through interactions with zero valent metals (ZVMs). Oxidative processes include potassium permanganate or photochemical oxidation.

REDUCTIVE DECHLORINATION

Microbial metabolism requires that energy and carbon sources be readily available. Most sugars, a common hydrocarbon, have the ability to give electrons by simply breaking the carbon–hydrogen bond and are therefore referred to as electron donors. The highly oxidized nature of PCE (with all of the chlorine substitutions) makes it nearly impossible to further oxidize. This quality promotes its recalcitrance or persistence in the environment. Therefore, any transformations of this compound require that a reduction (with the chlorinated ethylene serving as the electron acceptor) first take place. In the presence of microorganisms, this process is referred to as reductive dechlorination (Fig. 1). Under anaerobic conditions a number of identified microorganisms with the proper dehalogenase enzyme are expected to be able to sequentially reduce PCE to trichloroethylene (TCE), then to several isomers of dichloroethylene (DCE), then to vinyl chloride (VC), and finally to ethylene (ETH), where it could be taken up by microorganisms and plants as part of regular metabolic pathways (7).

Because of the significant energy required to push this reaction pathway to the innocuous ethylene product, the process is more likely to halt at either DCE or VC production. Note that the higher the potential ($E^{\circ\prime}$) the more likely the compound will be reduced. All of the chlorinated ethylenes (PCE, TCE, DCE, and VC) are considered priority pollutants under SDWAA–1986. So an additional transformation step is usually required to achieve removal of all contaminants of concern.

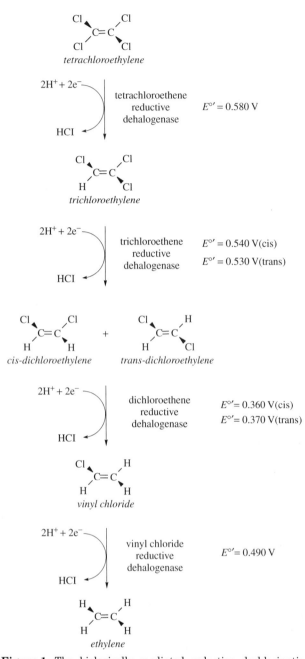

Figure 1. The biologically mediated reductive dechlorination process from perchloroethylene all the way to (the nonchlorinated) ethylene (5) with the corresponding standard reduction potentials, $E^{\circ\prime}$, for each half-reaction (6).

COMETABOLIC OXIDATION

Concerns for the incomplete reduction of the chlorinated ethylenes led to consideration of additional transformation pathways/mechanisms. Biologically, the reductive dechlorination may be forced to try to get beyond the VC production or a different pathway may be followed. Once PCE is initially reduced to TCE, then this and the remaining reduction products may more readily be cometabolically oxidized. The pathway shown for TCE (Fig. 2) oxidation is assumed to be similar for DCE and VC oxidation (8). The energetics of this process also suggest that the lesser chlorinated products are more easily oxidized by microorganisms possessing the appropriate oxygenase enzymes.

The commonality of the enzymes is that they do chemistry on the carbon nucleus of the compound, weakening the carbon–carbon double bond. These transformations are further considered cometabolic because the enzymes are produced to do chemistry on compounds with similar chemical structures to that of the chlorinated ethylenes (Fig. 3). Sustainability of this process therefore requires the addition of both the similar target compound (e.g., phenol) and reducing equivalents [e.g., NAD(P)H] as cometabolites (9).

The resulting products from this cometabolic oxidation are innocuous compounds that can readily be incorporated into typical metabolic function of microorganisms.

METAL REDUCTION

Similar to the biologically mediated reductive dechlorination, zero valent metals (ZVMs, typically granular iron) when they corrode either produce hydrogen or serve as an electron donor directly (10). ZVMs are used in permeable reactive barriers (PRBs) as well as in canister filters to allow for on-site treatment.

CHEMICAL OXIDATION

Photocatalytic oxidation uses a titanium dioxide catalyst, which is excited by the UVA spectrum (315–400 nm) of ultraviolet light to produce hydroxyl radicals. The process is used with volatilized contaminants at ambient temperature and pressure. Therefore, photocatalytic oxidation is an appropriate treatment option to use in conjunction with either air stripping or soil vapor extraction (11).

Potassium permanganate is another stronger oxidizer that rapidly degrades chlorinated ethylenes either in solution or in soils. This compound, similar to the biological oxygenase enzymes, attacks and weakens the carbon–carbon double bond (12).

CONCLUSION

There are both chemical and biological treatment options available for removal of PCE. Depending on cost, availability, and preference of options, one can choose from biologically mediated reductive dechlorination of PCE and its reduction products. Recognizing the likely limitation in achieving complete transformation to innocuous compounds, one could switch to biologically mediated cometabolic oxidation. Or depending on the form the chlorinated ethylenes are in, solubilized versus volatilized, a chemical reduction (ZVMs) or oxidation process (photocatalytic or potassium permanganate) may be chosen.

BIBLIOGRAPHY

1. HSIA (Halogenated Solvents Industry Alliance, Inc.) (1999). Perchloroethylene white paper. Available at http://www.hsia.org/white_papers/perc.htm.

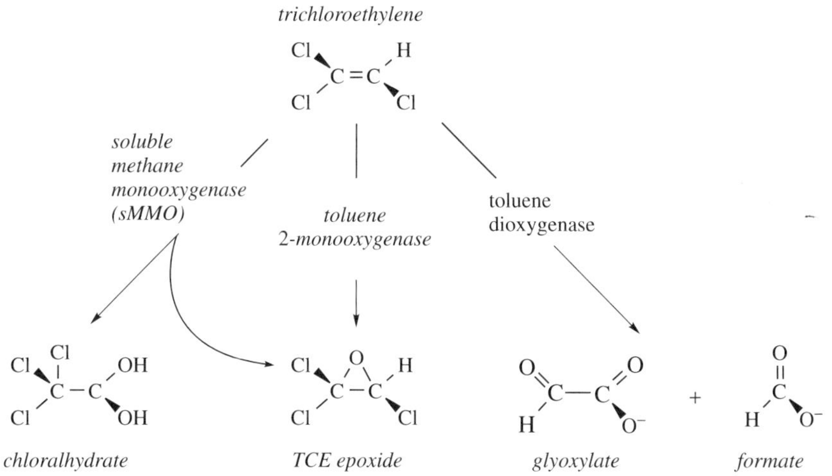

Figure 2. The first step of biologically mediated chlorinated ethylene oxidation (8).

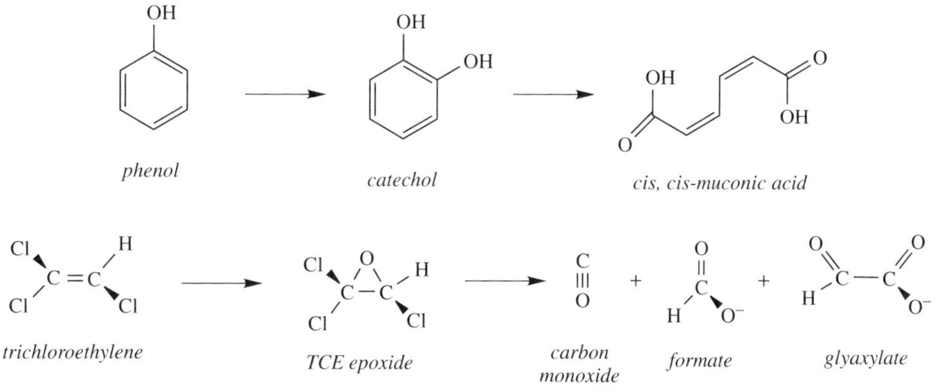

Figure 3. Illustration of the cometabolic nature of chlorinated ethylene oxidation where chemistry similar to that performed on the target compound (phenol) is also done on the chlorinated compound.

2. Safe Drinking Water Act Amendments (1986). Conference report (to accompany S.124), United States Congress (99th House, 2nd Session:1986), U.S. Government Printing Office, Washington, DC.
3. McCarty, P.L. (1994). An overview of anaerobic transformation of clorinated solvents. In: *Symposium on Intrinsic Bioremediation of Ground Water*, U.S. Environmental Protection Agency, Washington, DC, EPA 540/R-94/515, pp. 135–142.
4. Atlas, R.M. and Bartha, R. (1993). *Microbial Ecology: Fundamentals and Applications*, 3rd Edn. Benjamin/Cummings Publishing, Redwood City, CA.
5. Ellis, L. (updated April 2000). Anaerobic tetrachloroethylene graphic pathway map, UMBBD (The University of Minnesota Biocatalysis/Biodegradation Database). Available at http://www.labmed.umn.edu/umbbd/tce2/tce2_image_map.html.
6. Vogel, T.M., Criddle, C.S., and McCarty, P.L. (1987). Transformations of halogenated aliphatic compounds. *Environ. Sci. Technol. Crit. Rev.*, 722–736.
7. Ellis, L. (updated April 2000). Tetrachloroethylene pathway map (anaerobic), UMBBD (The University of Minnesota Biocatalysis/Biodegradation Database). Available at http://www.labmed.umn.edu/umbbd/tce2/tce2_map.html.
8. Jung Oh, D. (updated April 2000). TCE graphical pathway map, UMBBD (The University of Minnesota Biocatalysis/Biodegradation Database). Available at http://www.labmed.umn.edu/umbbd/tce/tce_image_map.html.
9. Whittaker, M., Monroe, D., and Jung Oh, D. (updated April 2000). Trichloroethylene pathway map, UMBBD (The University of Minnesota Biocatalysis/Biodegradation Database). Available at http://www.labmed.umn.edu/umbbd/tce/tce_map.html.
10. Tratnyek, L. (updated September 2002). Chemistry of contaminant reduction with zero-valent iron metal, CGR (Center for Groundwater Research). Available at http://cgr.ese.ogi.edu/iron/.
11. Hager, S., Bauer, R., and Kudielka, G. (2000). Photocatalytic oxidation of gaseous chlorinated organics over titanium dioxide. *Chemosphere* **41**: 1219–1225.
12. Huang, K.-C. et al. (2001). Oxidation of chlorinated ethenes by potassium permanganate: a kinetics study. *J. Hazardous Mater.* **B87**: 155–169.

A PRIMER ON WATER QUALITY

GAIL E. CORDY
U.S. Geological Survey

WHAT IS IN THE WATER?

Is it safe for drinking? Can fish and other aquatic life thrive in streams and lakes that are affected by human

---

This article is a US Government work and, as such, is in the public domain in the United States of America.

A PRIMER ON WATER QUALITY

activities? What is the water quality? To answer these questions, it is helpful to understand what "water quality" means, how it is determined, and the natural processes and human activities that affect water quality.

WHAT DO WE MEAN BY "WATER QUALITY"?

Water quality can be thought of as a measure of the suitability of water for a particular use based on selected physical, chemical, and biological characteristics. To determine water quality, scientists first measure and analyze characteristics of the water such as temperature, dissolved mineral content, and number of bacteria. Selected characteristics are then compared to numeric standards and guidelines to decide if the water is suitable for a particular use.

HOW IS WATER QUALITY MEASURED?

Some aspects of water quality can be determined right in the stream or at the well. These include temperature, acidity (pH), dissolved oxygen, and electrical conductance (an indirect indicator of dissolved minerals in the water). Analyses of individual chemicals generally are done at a laboratory.

WHY DO WE HAVE WATER-QUALITY STANDARDS AND GUIDELINES?

Standards and guidelines are established to protect water for designated uses such as drinking, recreation, agricultural irrigation, or protection and maintenance of aquatic life. Standards for drinking-water quality ensure that public drinking-water supplies are as safe as possible. The U.S. Environmental Protection Agency (USEPA) and the States are responsible for establishing the standards for constituents in water that have been shown to pose a risk to human health. Other standards protect aquatic life, including fish, and fish-eating wildlife such as birds.

HOW DO NATURAL PROCESSES AFFECT WATER QUALITY?

Natural water quality varies from place to place, with the seasons, with climate, and with the types of soils and rocks

through which water moves. When water from rain or snow moves over the land and through the ground, the water may dissolve minerals in rocks and soil, percolate through organic material such as roots and leaves, and react with algae, bacteria, and other microscopic organisms. Water may also carry plant debris and sand, silt, and clay to rivers and streams making the water appear "muddy" or turbid. When water evaporates from lakes and streams, dissolved minerals are more concentrated in the water that remains. Each of these natural processes changes the water quality and potentially the water use.

WHAT IS NATURALLY IN THE WATER?

The most common dissolved substances in water are minerals or salts that, as a group, are referred to as dissolved solids. Dissolved solids include common constituents such as calcium, sodium, bicarbonate, and chloride; plant nutrients such as nitrogen and phosphorus; and trace elements such as selenium, chromium, and arsenic.

In general, the common constituents are not considered harmful to human health, although some constituents can affect the taste, smell, or clarity of water. Plant nutrients and trace elements in water can be harmful to human health and aquatic life if they exceed standards or guidelines.

Dissolved gases such as oxygen and radon are common in natural waters. Adequate oxygen levels in water are a necessity for fish and other aquatic life. Radon gas can be a threat to human health when it exceeds drinking-water standards.

HOW DO HUMAN ACTIVITIES AFFECT WATER QUALITY?

Urban and industrial development, farming, mining, combustion of fossil fuels, stream-channel alteration, animal-feeding operations, and other human activities can change the quality of natural waters. As an example of the effects of human activities on water quality, consider nitrogen and phosphorus fertilizers that are applied to crops and lawns. These plant nutrients can be dissolved easily in rainwater or snowmelt runoff. Excess nutrients carried to streams and lakes encourage abundant growth of algae, which leads to low oxygen in the water and the possibility of fish kills.

Chemicals such as pharmaceutical drugs, dry-cleaning solvents, and gasoline that are used in urban and industrial activities have been found in streams and ground water. After decades of use, pesticides are now widespread in streams and ground water, though they rarely exceed the existing standards and guidelines established to protect human health. Some pesticides have not been used for 20 to 30 years, but they are still detected in fish and streambed sediment at levels that pose a potential risk to human health, aquatic life, and fish-eating wildlife. There are so many chemicals in use today that determining the risk to human health and aquatic life is a complex task. In addition, mixtures of chemicals typically are found in water, but health-based standards and guidelines have not been established for chemical mixtures.

WHAT ABOUT BACTERIA, VIRUSES, AND OTHER PATHOGENS IN WATER?

The quality of water for drinking cannot be assured by chemical analyses alone. The presence of bacteria in water, which are normally found in the intestinal tracts of humans and animals, signal that disease-causing pathogens may be present. Giardia and cryptosporidium are pathogens that have been found occasionally in public-water supplies and have caused illness in a large number of people in a few locations. Pathogens can enter our water from leaking septic tanks, wastewater-treatment discharge, and animal wastes.

HOW CAN I FIND OUT MORE ABOUT MY WATER QUALITY?

Contact your local water supplier and ask for information on the water quality in your area. The USEPA requires public-water suppliers to provide water-quality data to the public on an annual basis in an understandable format. State agencies that deal with health, environmental quality, or water resources also can provide information on the quality of your water. Additional resources can be found on the Internet at:

http://water.usgs.gov/nawqa
http://www.epa.gov/safewater

OVERVIEW OF ANALYTICAL METHODS OF WATER ANALYSES WITH SPECIFIC REFERENCE TO EPA METHODS FOR PRIORITY POLLUTANT ANALYSIS

Ralph J. Tella
Lord Associates, Inc.
Norwood, Massachusetts

INTRODUCTION

This article is an overview of the analytical chemistry behind the analyses of common environmental contaminants. Specific reference is made to the U.S. Environmental Protection Agency (EPA) "priority pollutants." The list of priority pollutants was first published in the *Federal Register* in 1980 as the list of toxic pollutants for which categorically defined industries were required to analyze their effluent streams to obtain wastewater discharge permits under the National Pollution Discharge Elimination System (NPDES).

To require industry to analyze for pollutants, the EPA was compelled to develop standardized test procedures. As scientists and engineers became familiar with these procedures, the priority pollutant list developed into a commonly requested set of analyses used to characterize and evaluate the toxicity of various environmental media including soils, groundwater, and air.

The priority pollutant list is comprised of several classes of compounds or "fractions," grouped according to similar chemical and physical characteristics. A complete list of the priority pollutants is provided in Table 1. These fractions include 111 organic compounds, and 15 inorganic elements and compounds, a total of 126 parameters.

Note that these 126 parameters are not the only toxic pollutants in the world. Rather, they are only a convenient list of common pollutants for which there are standardized test methods. Many more pollutants can be analyzed for, some use the same analytical techniques. However, to gain this information from the laboratory, you must specifically ask for it. Many commercial laboratories have developed economical techniques, which "screen" for these nonpriority pollutants, such as "library searches," or "extended runs." Many of these techniques are, however, only semiquantitative and qualitative.

Because petroleum products in the aqueous environment comprise a large focus of attention in the field of environmental restoration, this article also describes various methods of analyses for petroleum products. Note that only some components of petroleum are regulated as EPA "priority pollutants."

GAS CHROMATOGRAPHY

Gas chromatography (GC) is a powerful and highly versatile instrumental method of analysis that was developed in the early 1950s. The number of technological advances in the technique since then has been so tremendous that a complete and detailed discussion of the technique is beyond the scope of this article. Rather, this discussion is limited to those techniques that are commonly used in environmental analyses.

Theory

This is the basic process by which a GC works. A sample is introduced into the instrument in a vapor state, and the various gaseous components are separated so that they can be individually identified and quantitatively measured. A schematic diagram of this process is provided in Fig. 1. The basic components are a gas cylinder with a reducing valve, a constant pressure regulator, a port for injecting the sample, a chromatographic column, a detector, an exit line, and a data collection device such as a strip chart recorder or computer.

The gas cylinder contains a carrier gas such as hydrogen, helium, or nitrogen, which is continuously swept through the chromatographic column at a constant temperature and flow rate. A small sample for analysis is injected, usually with a syringe, into the sample port, where it is flash-evaporated to convert its components into a gas. The constantly flowing stream of carrier gas carries the gaseous constituents through the chromatographic column. The gases travel through at different rates so that they emerge from the column at different times. Their presence in the emerging carrier gas is detected by chemical or physical means, and the response of the detector is fed to the data collection device. A typical chromatogram produced by this process is shown in Fig. 2. Each peak represents a specific chemical compound or mixture of compounds that has the same rate of movement through the column. The time for each compound to emerge from the column is a characteristic of the compound known as its retention time. The area under the peak is proportional to the concentration of the compound in the sample.

Standard packed chromatographic columns are generally glass or metal tubes varying from about 1–10 m in length and about 3–6 mm in diameter. The tube is packed with a inert solid impregnated with a nonvolatile liquid, such as silicone oil or polyethylene glycol. As a gaseous sample passes through the column, its components are partitioned between the stationary liquid phase of the column and the moving gas phase. Components that are relatively soluble in the liquid phase move through the column at slower rates than components that are not so soluble. Most gaseous materials can be separated by selecting suitable stationary phases and column lengths. More recently, capillary columns with diameters of 0.2–0.4 mm and lengths of 20–30 m have replaced standard packed columns. These columns contain the solid phase on the inside wall of the capillary tube. The gas containing the organic components passes through the center of the capillary tube, and the organics partition themselves between the gas and the stationary phase, emerging from the column at different times.

Table 1. 126 EPA Priority Pollutants

A. Chlorinated benzenes
Chlorobenzene
1,2-Dichlorobenzene
1,3-Dichlorobenzene
1,4-Dichlorobenzene
1,2,4-Trichlorobenzene
Hexachlorobenzene

B. Chlorinated ethanes
Chloroethane
1,1-Dichloroethane
1,2-Dichloroethane
1,1,1-Trichloroethane
1,1,2-Trichloroethane
1,1,2,2-Tetrachloroethane
Hexachloroethane

C. Chlorinated phenols
2-Chlorophenol
2,4-Dichlorophenol
2,4,6-Trichlorophenol
Parametachlorocresol
(4-chloro-3-methyl phenol)

D. Other chlorinated organics
Chloroform (trichloromethane)
Carbon tetrachloride
(tetrachloromethane)
Bis(2-chloroethoxy)methane
Bis(2-chloroethyl)ether
2-Chloroethyl vinyl ether (mixed)
2-Chloronaphthalene
3,3-Dichlorobenzidine
1,1-Dichloroethylene
1,2-trans-Dichloroethylene
1,2-Dichloropropane
1,2-Dichloropropylene
(1,3-dichloropropene)
Tetrachloroethylene
Trichloroethylene
Vinyl chloride (chloroethylene)
Hexachlorobutadiene
Hexachlorocyclopentadiene
2,3,7,8-Tetrachloro-dibenzo-p-dioxin
(TCDD)

E. Haloethers
4-Chlorophenyl phenyl ether
2-Bromophenyl phenyl ether
Bis(2-chloroisopropyl) ether

F. Halomethanes
Methylene chloride (dichloromethane)
Methyl chloride (chloromethane)
Methyl bromide (bromomethane)
Bromoform (tribromomethane)

Dichlorobromomethane
Chlorodibromomethane

G. Nitrosamines
N-Nitrosodimethylamine
N-Nitrosodiphenylamine
N-Nitrosodi-n-propylamine

H. Phenols (other than chlorinated)
2-Nitrophenol
4-Nitrophenol
2,4-Dinitrophenol
4,6-Dinitro-o-cresol
(4,6-dinitro-2-methylphenol)
Pentachlorophenol
Phenol
2,4-Dimethylphenol

I. Phthalate esters
Bis(2-ethylhexyl)phthalate
Butyl benzyl phthalate
Di-n-butyl phthalate
Di-n-octyl phthalate
Diethyl phthalate
Dimethyl phthalate

J. Polnuclear aromatic hydrocarbons
(PAHS)
Acenaphthene
1,2-Benzanthracene (benzo(a)
anthracene)
Benzo(a)pyrene (3,4-benzo-pyrene)
3,4-Benzofluoranthene (benzo(b)
fluoranthene)
11,12-Benzofluoranthene (benzo(k)
fluoranthene)
Chrysene
Acenaphthalene
Anthracene
1,12-Benzoperylene (benzo(ghi)
perylene)
Fluorene
Fluoranthene
Phenanthrene
1,2,5,6-Bibenzanthracene (dibenzo(ah)
anthracene)
Indeno (1,2,3-cd) pyrene
(2,3-o-phenylene pyrene)
Pyrene

K. Pesticides and metabolites
Aldrin
Dieldrin
Chlordane (technical mixture and
metabolites)
Alpha-endosulfan

Beta-endosulfan
Endosulfan sulfate
Endrin
Endrin aldehyde
Heptachlor
Heptachlor epoxide
(BHC-hexachlorocyclohexane)
Alpha-BHC
Beta-BHC
Gamma-BHC (Lindane)
Delta-BHC Toxaphene

L. DDT and metabolites 4,4-DDT
4,4-DDE (p,p-DDX)
4,4-DDD (p,p-TDE)

M. Polychlorinated biphenyls (PCBS)
PCB-1242 (Arochlor 1242)
PCB-1254 (Arochlor 1254)
PCB-1221 (Arochlor 1221)
PCB-1232 (Arochlor 1232)
PCB-1248 (Arochlor 1248)
PCB-1260 (Arochlor 1260)
PCB-1016 (Arochlor 1016)

N. Other organics
Acrolein
Acrylonitrile
Benzene
Benzidine
2,4-Dinitrotoluene
2,6-Dinitrotoluene
1,2-Diphenylhydrazine
Ethylbenzene
Isophorone
Naphthalene
Nitrobenzene toluene

O. Inorganics
Antimony
Arsenic
Asbestos
Beryllium
Cadmium
Chromium
Copper
Cyanide
Total lead
Mercury
Nickel
Selenium
Silver
Thallium
Zinc

The absence of packing minimizes the dispersion of the contaminants so that excellent resolution of the different compounds is obtained, even when there are 20 or more in the sample. This makes it possible to identify a greater number of compounds in highly complex mixtures.

The temperature of the sample and column must be held sufficiently high during movement through the column and subsequent detection to maintain them in the gaseous state. Only materials that can be volatilized or converted to volatile compounds can be detected by gas chromatography.

Sample Introduction Techniques

Samples may be introduced into the GC system in only two phases: liquid and gaseous. For successful liquid phase introduction, the injection port and column temperature

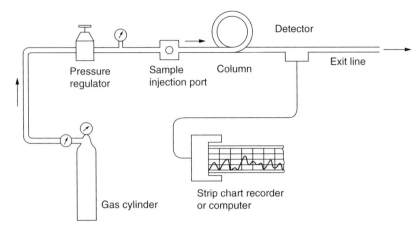

Figure 1. Schematic of a gas chromatograph system.

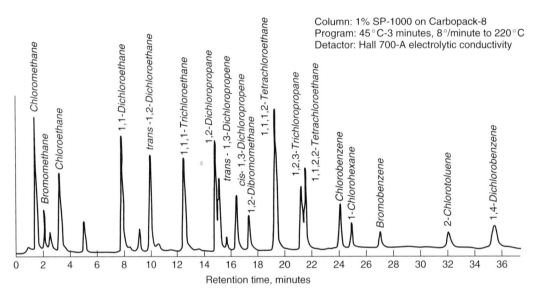

Figure 2. Gas chromatogram of purgeable halocarbons.

must be sufficiently high to permit complete volatilization of the material to be analyzed. A solid sample may be introduced by dissolving the sample in a low boiling solvent, thereby treating the sample as a liquid. Gas samples may be either directly injected via a gastight syringe or by using a gas sampling valve that uses the principle of having the carrier gas purge, under pressure, a calibrated volume loop within which the gas sample was trapped.

A major consideration in sample analysis is the required level of sensitivity or limit of detection. A common way of increasing a procedure's detection limit is to concentrate the substance being analyzed (the analyte). Three common procedures are used in gas chromatography: purge and trap, headspace, and solvent extraction.

Purge and Trap. The purge and trap procedure is commonly used in environmental analysis to determine volatile organic compounds or "purgeables," such as in EPA Methods 624, 601, and 602. In each of these methods, the purge and trap apparatus must be similar (although trap materials differ). The purge and trap device consists of three separate pieces of equipment: the sample purger, the trap, and the desorber.

An inert gas is bubbled through the sample in a specially designed purging device at ambient temperature. The purgeables are efficiently transferred from the aqueous phase to the vapor phase. The vapor is swept through a sorbent column where the purgeables are trapped. After purging is completed, the sorbent column is heated and backflushed with the inert gas to desorb the purgeables into a GC column.

Headspace. The headspace technique is commonly used for analyses of solid samples, although it can as easily be used for liquids. In this technique, EPA Method 5020, a known volume or weight of sample is collected in a sealed glass container and allowed to equilibrate at 90 °C for 1 hour. A sample of the headspace gas is then withdrawn for analysis. The advantages of this technique are its speed of sample preparation and its ability to leave potentially interfering semivolatile compounds behind.

Solvent Extraction. Solvent extraction is any procedure by which a solvent is used to extract an analyte into a liquid phase for direct injection (with or without preconcentration) into a GC. Solvent extraction techniques include both liquid/liquid and solid/liquid procedures. Although these procedures work best for semivolatile and nonvolatile compounds, the EPA has established similar procedures for analyzing volatiles in solid materials.

A popular liquid/liquid technique is used in EPA Method 625 (or 8270) for acid/base/neutral extractable organics. In this method, a liquid sample is separately extracted under both acidic and base/neutral conditions. A variety of glassware and apparatus has been developed for use in this method, although the separatory funnel is probably the simplest and most commonly used.

Solid/liquid techniques typically employ either Soxhlet extraction glassware or ultrasound. A Soxhlet extractor is a device that ensures intimate contact of the sample matrix with the extraction solvent by continuously deluging the sample with solvent. Alternatively, the EPA has approved ultrasound (Method 3550) for analyzing certain solid matrices.

Air Sampling Techniques. A variety of techniques are used to extract and concentrate air pollutants. The two most common procedures use either solid or liquid trap materials. Liquid traps, called "impingers," are used when the analyte can be easily absorbed into a stable liquid phase, such as cyanide into sodium hydroxide.

Solid trap materials may range from simple paper filters to exotic polymers. The difficulty in using solid trap materials is getting the analyte off of the trap back into a gaseous phase. This can be accomplished by either eluting the trap material with a solvent or by thermal adsorption.

If the analysis does not require concentration, air samples may be collected in any inert gastight device of known volume such as gastight syringes, evacuated glass bulbs, inert bags, or canisters. The methods most commonly used in the environmental industry are the EPA "TO" series of methods or others adapted from NIOSH approved procedures.

Detectors

The photoionization detector (PID) has become well known as a useful tool in the environmental industry. It is a nondestructive detector that lends itself to series operation with other detectors; thereby it enhances the selectivity and sensitivity of the GC.

The PID is the required detector for EPA Method 602 (or 8020/8021) for purgeable aromatics. Because it does not require the use of a fuel gas, it has become a popular detector for handheld portable organic vapor analyzers.

The PID operates on the principle of photoionization. The fact that certain molecules can be ionized by light is related to their characteristic molecular structure. The PID consists of a high-energy ultraviolet lamp located adjacent to an ionization chamber that contains a pair of electrodes. When a positive potential is applied to one electrode, the field created collects any ions formed by the absorption of UV light at the other electrode. The measured current is proportional to the concentration.

A PID responds to all compounds whose ionization potentials are less than or equal to the energy of the UV light source. Typical UV lamp energies in use are 10.2 eV and 11.7 eV. The magnitude of the ionization potential does not by itself dictate the sensitivity of the PID response. The degree of ionization depends on molecular structure. The number, type, and structure of the carbon bond, therefore, dictate the sensitivity of the detector to the compound. In general, the PID shows the greatest response to condensed-ring aromatic compounds (i.e., polyaromatic hydrocarbons), followed by aromatic compounds. These are other general selectivity rules:

- aromatics > alkanes > alkenes
- ketones > aldehydes > esters > alcohols > alkanes
- cyclic compounds > noncyclic
- branched compounds > nonbranched
- iodine-substituted > bromine-substituted > chlorine-substituted > fluorine-substituted

Because of the PID's response to aromatic compounds, it is the detector of choice for EPA Method 602 (or 8020/8021). The aromatic compounds in petroleum-based products such as benzene, toluene, and xylene are easily detected by this procedure. The aliphatic compounds in petroleum-based products such as methane, pentane and octane are suppressed. The flame ionization detector responds well to these compounds.

The Flame Ionization Detector. The flame ionization detector (FID) is probably the oldest and most generally used GC detector. The FID responds to most organic compounds, which can be both a benefit and liability in its application. As a benefit, the FID can be nearly universally used in a number of applications. This property lends itself to "fingerprinting" analysis of complex mixtures such as petroleum products. As a liability, the FID may respond to so many compounds that individual compound separation in a complex mixture, may become difficult, if not impossible.

The FID operates on the principle of flame ionization. As gases are eluted from a column, they are burned in a tiny hydrogen flame. A device continuously monitors the electrical conductivity of this flame. As organic compounds ionize in the flame, this conductivity is altered, proportional to the concentration, and measured. Most FIDs have sufficient energy to ionize any organic species with an ionization potential (IP) of 15.4 eV or less.

The use of FIDs in portable instrumentation has been hampered by the problem of maintaining a stable flame and fuel gas requirements. However, now there are many popular field instruments that have overcome these obstacles and are frequently used where gases such as methane are of interest.

The Electron Capture Detector. The electron capture detector (ECD) is used to detect compounds that have halogen atoms or polar functional groups such as pesticides, herbicides, and PCBs. EPA Method 608 (or 8080) uses an ECD for organochlorine pesticides and PCBs. The ECD uses a small radioactive source such as nickel 63 or tritium to emit beta particles. The ECD measures the ability of compounds to capture these particles from the ion current. The reduced current is measured and converted into an electrical signal for detection.

Mass Spectroscopy. Mass spectroscopy is the most rapidly developing analytical technology today. When coupled with a gas chromatograph (GC/MS), it is a powerful tool for positive identification and quantification of a large number of organic compounds. Recent advances in computer technology have dramatically expanded these systems' capabilities, speed of analyses, and cost efficiency. EPA Methods 624 or 8260 (purgeables) and 625 or 8270 (acid/base/neutral extractables) use this technique.

The mass spectrometer sorts charged gas molecules or ions according to their masses. The substance to be analyzed is vaporized in the GC and converted to positive ions or ion fragments by bombardment with rapidly moving electrons. These ions or ion fragments are pulled from the gas stream by an electrical field, accelerated, and separated by their mass-to-charge ratio. A computer then records the fragmentation pattern of the molecule, which is characteristic of a particular compound. The relative retention time and abundances of key ions and secondary ions are used to identify the compound qualitatively.

Most commercial labs have access to spectral libraries with information on more than 50,000 compounds. Compounds identified in this manner are referred to as "tentatively identified compounds" or "TICs." As the name implies, the data collected in this manner should be used with caution because an actual analytical reference standard has not been used. The quanitation of TICs may be orders of magnitude in error.

Other Detectors

Microcoulometric Detectors. Microcoulometric detectors such as the "Hall" detector are used for halide specific analyses such as EPA Method 601 (or 8010) for purgeable halocarbons. In this detector, halide-containing compounds are catalytically reduced using hydrogen, and the resulting "HX" is measured by a highly sensitive conductivity meter.

Thermal Conductivity. Thermal conductivity detectors use a heated wire that is placed in the stream of eluting gases. Each gas has a different thermal conductivity or ability to carry away heat from the wire and thus changes the temperature of the wire. These changes alter the resistivity of the wire from which a signal is generated.

Thermal conductivity detectors respond to a wide range of compounds but are relatively insensitive to low concentrations. For these reasons, they are usually used in the environmental industry for permanent gas analysis such as automobile and stack gas emissions. Many portable combustible gas meters also operate on the principle of thermal conductivity, such as units known as "explosimeters."

Flame Photometric Detectors. The flame photometric detector uses the optical emission of a gas burned in a flame to detect pollutants, typically sulfur and phosphorus containing compounds such as H_2S and phosphorus-based pesticides. The EPA has several interim pesticide and herbicide procedures that use this detection system.

Dry Electrolytic Conductivity Detectors. The dry electrolytic conductivity detectors (DELCD) are halogen specific. Often paired with other detectors such as FIDs, the DELCD can be used for highly selective identification of halogen containing compounds in gas streams.

HIGH PERFORMANCE LIQUID CHROMATOGRAPHY

High performance liquid chromatography (HPLC) is another analytical tool used for compounds not readily accessible by GC. As the name implies, HPLC uses a liquid carrier instead of a gas to sweep the sample through the column for separation. Compounds that are high in molecular weight (such as polynuclear aromatic hydrocarbons) or are thermally liable (i.e., decompose when heated) are good candidates for this technique. Many detectors are available to analyze both inorganic and organic species.

METALS ANALYSES

There are thirteen priority pollutant metals (see Table 1), and two EPA accepted methodologies used to analyze for them, atomic absorption spectroscopy (AA) and inductively coupled plasma (ICP). The former method uses the technique of absorption spectroscopy and the latter emission spectroscopy. Each technique has its advantages and disadvantages relative to selectivity, sensitivity, interferences, cost, and speed of analysis.

Atomic Absorption Spectroscopy

Atomic absorption spectroscopy (AA) is the most commonly used technique to analyze for metals. This technique has been around since the 1950s and was the first instrumental method of metal analysis accepted to replace what were previously colorimetric or gravimetric methods of analyses.

Because instrumental operating conditions vary from metal to metal, the EPA has published specific methods for each metal. Collectively, these methods are known as the "200 series" in water or the "7000" series in solid materials.

Theory. Atomic absorption is based on the principle that all atoms absorb light at particular wavelengths corresponding to the energy requirements of a particular atom. For example, copper atoms absorb light strongly at 324.7 nm because this particular wavelength has the right energy to alter the atom's electronic state from a ground state to an excited state. Other elements have their own specific wavelengths. Because each atom may have several different electronic states, a spectrum of these characteristic wavelengths can be created. The wavelength lines that are used in AAS are those that originate in the ground state, called resonance lines.

The basic AA system consists of a stable light source of particular wavelength, a flame or other excitation device, a

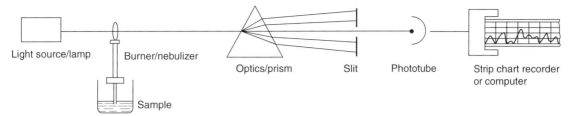

Figure 3. Schematic diagram of atomic absorption spectrometer.

prism to disperse and isolate emission lines, and a detector with appropriate amplifiers (see Fig. 3). The light source is directed into the flame or other excitation device, and its intensity is measured with and without the introduction of a sample. The decrease in intensity or absorption is proportional to the concentration. Two different excitation devices are commonly used. Flame AA is used to detect concentrations in the approximate parts per million range and flameless AA or a heated graphite furnace is used to detect in the parts per billion range. Flameless AA or a heated graphite furnace uses an electrically heated graphite tube to concentrate the atoms, instead of a flame.

Because mercury is unique in that it can be liberated in its ground state at room temperature a special cold vapor technique, EPA Method 245.1 should be used to measure low (ppb) concentrations. This technique uses a chemical reaction to liberate mercury atoms instead of a heated excitation source as in flame or graphite furnace techniques.

Inductively Coupled Plasma

Inductively coupled plasma (ICP) is another commonly used technique for metals analyses. The main attraction of ICP is that it can analyze many elements at once, and for certain refractory elements such as beryllium, it is much more sensitive.

However, AA must still be used to detect low levels of mercury and other metals. For these reasons, many laboratories that have ICP systems use both techniques to analyze for the 13 priority pollutant metals. Because of its multielement capabilities, ICP is also very useful in screening unknown samples. The EPA method for multielement ICP analysis is Method 6010.

Theory. The technique by which ICP operates is emission spectroscopy. Using a radio-frequency generated, argon sustained plasma, atoms are excited to a much higher energy state that can be achieved in normal AA systems. As the electrons fall back to their original state, they emit radiation at element specific or characteristic wavelengths. This radiation is in turn sensed by a detector; the signal is amplified and recorded. Prior to recent computer advances, this signal processing was often extremely time-consuming and subject to interferences.

Metal Sample Preparation

Nearly all samples require some preparation prior to metals analysis. This is required to isolate selectively the element you are analyzing for (the "analyte") from the rest of the sample (its "matrix") and to put it into a more readily analyzable state. This preparation usually begins in the field with the addition of nitric acid as a preservative. Nearly all aqueous samples for metals analysis should be preserved with nitric acid if they cannot be delivered to the laboratory on the same day.

The exception to this rule is discriminating between valence states of certain metals such as chromium. Samples of hexavalent chromium should not be preserved with acid. Although the chemistry of hexavalent chromium is not completely understood, it is believed that acid may promote the reduction of the hexavalent to the trivalent state.

Acidification of the sample helps to (1) promote digestion of the sample matrix and dissolution of the metal and (2) inhibits its precipitation or "plating" out on the walls of the sample container. Preserving metals samples with acid is not intended to inhibit microbial activity. It does not matter if microbes ingest or otherwise alter the metallic compounds because the sample will be digested and analyzed for total metals, regardless of valence state. For this same reason, samples for total metals analyses need not be refrigerated.

Samples submitted for dissolved metals should not be acidified until they have been filtered. Acidification would promote dissolution of the larger particles and bias the results. Note that "dissolved" metals is an empirically defined term. Specifically, "dissolved" metals mean all metals that are filterable (pass through) a 0.45-micron membrane filter.

"Total" metals mean the entire element irrespective of valence or physical state. To enhance the analysis and reduce the chance of matrix interferences, most metals samples undergo rigorous acid digestion. Through this procedure, solids and semisolid samples may be broken down into a dissolved aqueous state that allows the nebulizer of the instrument to aspirate or pass the sample into the flame.

Metals samples may also be treated or prepared through various solvent extraction or preconcentration techniques to enhance sensitivity and/or eliminate matrix effects. However, most of these techniques are not required by the modern instrumentation manufactured today.

CYANIDE AND TOTAL PHENOLICS

Cyanide and total phenolics are typically analyzed by colorimetry. These techniques involve lengthy wet chemical digestions and distillations to eliminate interference.

Once these steps are taken, the analyte is reacted chemically to produce a color, whose intensity is proportional to concentration. Various procedures exist that cover the variety of instrumentation manufactured to help automate these analyses. However, the most commonly used methods for these analyses are EPA Method 335.1 for cyanide and EPA Method 420.1 for total phenolics.

Sample Preparation

Cyanide. It is extremely important to preserve all aqueous samples for cyanide analysis with sodium hydroxide at a pH greater than 12 to ensure that the sample is not volatilized to hydrogen cyanide, an extremely toxic gas. In addition, the sample should be kept refrigerated.

Total Phenolics. Approved preservatives for total phenolics have varied in recent years. However, the generally agreed upon preservative is either phosphoric or sulfuric acid. Although not specifically required, copper sulfate may be added to inhibit microbial growth and initiate the analysis.

ASBESTOS

Although asbestos is a priority pollutant, it is rarely analyzed in water during routine or initial site investigations because its occurrence in groundwater is unusual. When it does occur, it is typically the result of localized dumping of asbestos insulation. Exceptions to this occur where asbestos fibers are locally mined, are part of an industrial process effluent, or when an asbestos cement pipe has deteriorated.

Because the focus of concern with asbestos is its previously common use as an insulator, it has been largely considered an industrial hygiene or indoor air pollution problem. As such, it has been the responsibility of the National Institute for Occupational Safety and Health (NIOSH) to promulgate standardized testing procedures for the material.

Generally, asbestos-containing material (ACM) contains chrysotile or amosite asbestos in a matrix of other fibers or particles such as glasswool or gypsum. This fibrous matrix makes positive identification of asbestos difficult, and it is best left to the experts. State laboratory certification for asbestos analysis is generally not available. Therefore, one should ascertain if the laboratory satisfactorily passed the latest EPA "round-robin" asbestos proficiency exam.

Bulk asbestos can be efficiently identified by phase-contrast microscopy. Because its effective range is for fibers approximately 5 microns or more in length, more sensitive microscopic methods are often employed for water analyses or other surface (filter) counting techniques. Transmission electron microscopy (TEM) is often specified for full-spectrum fiber size evaluations or comparisons of fiber abundances. Scanning electron microscopy (SEM) is also a powerful tool and can provide semiqualitative and quantitative analysis when coupled with energy dispersive X-ray instrumentation.

NONPRIORITY POLLUTANT PETROLEUM HYDROCARBON ANALYSES

Petroleum hydrocarbon products such as gasoline, diesel, fuel oil, and asphalt are complex mixtures of hundreds of individual compounds. They are manufactured according to performance-based standards, not chemical composition per se; therefore, their composition is not fully known and also varies based on the crude source. These petroleum compounds are primarily alkanes, although depending upon the distillation fraction (as well as the original parent crude), various amounts of unsaturated hydrocarbons (aromatics), nitrogen and sulfur compounds may also be present. In addition to these "base" compounds, various manufacturers add different proprietary additives such as tetraethyl lead, methyl *tert*-butyl ether, ethyl *tert*-butyl ether, ethanol, detergents, lubricants, and antifreezes.

Complicating matters for the environmental scientist, as soon as many of these compounds enter the environment, they begin transforming partitioning into the atmosphere, water, and biosphere. This all happens at highly variable rates depending on temperatures, pressures, redox potentials, available bacteria, and nutrients. No single analytical method can be used to quantify or qualify all petroleum hydrocarbons. The scientist must instead carefully decide which compounds are of most significance and whether or not an empirically derived method or surrogate parameter will suffice to describe conditions.

The lighter molecular weight volatile aromatic compounds, benzene, toluene, ethylbenzene, and xylenes (BTEX) can be resolved in any of the EPA purgeable VOC methods, such as 624 (or 8260), 602 (or 8021). The PID detector for a 602 analysis is highly sensitive to aromatics and where only petroleum hydrocarbons are a concern, is the cost-effective method of choice. There are also specialized petroleum analysis labs that routinely analyze for additives, other petroleum hydrocarbon background components, and can match ("fingerprint") and date petroleum types.

Heavier molecular weight polynuclear aromatic compounds associated with petroleum compounds may be detected via EPA Method 625 (or 8270). These compounds include benzo(*a*)pyrene, anthracene, and pyrene. As previously described, the PNA fraction may also be resolved by HPLC techniques.

Due to the difficulty in analyzing petroleum products as a "whole," it has become common to use various surrogate parameters for the different petroleum hydrocarbon fractions within different product types. Some of these surrogate parameters include total ionizable compounds (TIC); total oil and grease; total petroleum hydrocarbon oil and grease (TPH); total petroleum hydrocarbons by FID (TPH); volatile petroleum hydrocarbons (VPH); extractable petroleum hydrocarbons (EPH); total organic compounds (TOC); total volatile solids (TVS); and except for aromatics, chemical oxygen demand (COD). A discussion of each of these procedures is beyond the scope of this article, but because national regulatory guidelines have been established for total petroleum hydrocarbon oil and grease, a discussion of these procedures is in order.

Total Oil and Grease

Total oil and grease by EPA Method 418.1 or Standard Method 503 includes any compound that is extractable by Freon® and may be quantified either gravimetrically (503 A) or by infrared spectroscopy (503 B). More recently, Freon has been replaced by hexane due to Freon's ozone depletion properties. Because the solvent is not specific (no solvent is), a variety of "interfering" compounds which may not be of particular importance to the scientist, such as sulfuric compounds, humic substances, and chlorophyll, may also be extracted. To help discriminate between oil and grease of petroleum origin and that of animal or vegetable origin, one should request total petroleum hydrocarbon oil and grease (TPH), Method 503 E. This method takes the additional step of passing the Freon extract through silica gel. Silica gel preferentially adsorbs polar compounds such as fatty acids. The material that passes through the silica gel is defined as TPH. This separation is by no means absolute. Many complex aromatic hydrocarbons, sulfur and nitrogen compounds may also be adsorbed by the silica gel.

The scientist must carefully choose between the procedure and the relative benefits and drawbacks of each before requesting laboratory services. In general, gravimetric procedures should be used only when a higher detection limit is adequate (approximately 5 ppm for water) and the "oil and grease" does not volatilize below 700 °C. This would include most components of gasoline and No. 2 fuel oil. When lower detection limits (approximately 0.1 ppm for water) are required and the "oil and grease" is a lighter fraction, the IR procedure is the method of choice.

Gas chromatography largely replaced infrared spectroscopy in the analysis of TPH in the 1990s. A modification of EPA Method 8100 lends itself well to producing a FID chromatogram of the various petroleum hydrocarbons from C_6 through higher than C_{38}. By simply altering the temperature program of the GC, gasoline and diesel range organics can be selectively identified. Moreover, by combining various detector arrangements with selective resins to presort and separate alkane fractions from aromatic fractions, more selective groupings of various hydrocarbon fractions can be made. These selective groupings have been useful as surrogate parameters for assessing the toxicity of petroleum in environmental risk assessments.

SOURCE-WATER PROTECTION

National Water-Quality Assessment (NAWQA) Program—U.S. Geological Survey

Not all water resources are equally vulnerable to contamination. Even areas having similar land uses and sources of contamination can have different degrees of

vulnerability and, therefore, different response rates to protection and management strategies. NAWQA findings clearly demonstrate that natural features—such as geology, soils, and hydrology—and land-management practices—such as tile drainage and irrigation—can affect the movement of chemicals over land or to aquifers. Effective management of nonpoint source pollution may, therefore, require targeted strategies based on different degrees of vulnerability rather than uniform treatment of contaminant sources. Linking knowledge on natural features with the use, occurrence, and transport of chemicals through the watershed makes it easier to set priorities in streams and aquifers most vulnerable to contamination and increase the cost-effectiveness of strategies designed to protect water resources in diverse settings.

State of Washington. The Washington State Department of Health, in concert with USGS, assessed the vulnerability of public water-supply wells to pesticide contamination based on geology, well characteristics, land-use activities, and low levels of detection. NAWQA information on pesticide contamination enabled the health department to identify wells with low vulnerability to contamination and obtain waivers for quarterly monitoring required under the Federal Amendments to the Safe Drinking Water Act, 1996. By using the information to meet USEPA requirements for safe drinking water, Washington State was able to protect their drinking-water source while saving at least $6 million in costly additional monitoring. This is an annual savings of as much as $70 per household on small public supply systems that were granted full monitoring waivers.

This article is a US Government work and, as such, is in the public domain in the United States of America.

State of New Jersey. NAWQA data on organic compounds are used heavily in New Jersey's source-water assessment. USGS and the New Jersey Department of Environmental Protection are developing models to assess the vulnerability of public water supplies (including surface-water intakes and ground-water community and non-community wells) in the State to contamination by regulated compounds.

I coordinate the Upper Mississippi River Source Water Protection Initiative, an effort that will lead to the development of source-water protection plans for public water suppliers within the upper Mississippi River basin. The water quality data that have been generated and documented through NAWQA will figure prominently in the preparation of these plans. The information on certain contaminant levels in various settings within the basin, and the information describing the sources of contaminants provide documentation and solid rationale for identifying source water protection strategies, priorities, and protection measures for public water suppliers. In my opinion, more than any single information source, the Upper Mississippi River NAWQA provides an extremely valuable substantive basis for source water protection in the Upper Mississippi River basin (Mr. David Brostrom, Coordinator, Upper Mississippi River Source Water Protection Initiative, March 2001).

In addition to the examples cited above, more than 30 other states use USGS information to develop source-water protection plans for drinking-water sources. The collaborative projects in these states address nearly 40 percent of the nation's public water supply, serving more than 90 million people.

USGS information is also used widely by states to develop management plans for constituents, such as pesticides, nutrients, and MTBE. Specifically, state environmental and natural resource agencies prioritize streams and ground-water areas for assessment of these constituents on the basis of vulnerability concepts, contaminant occurrence data, and quality-assurance protocols of the NAWQA Program.

- *State of Kansas.* NAWQA findings on elevated concentrations of atrazine (frequently approaching or exceeding the USEPA drinking-water standard) in water-supply reservoirs in the Lower Kansas River Basin were used by the Kansas State Board of Agriculture as the basis for establishing a pesticide management area in northern Kansas (Delaware River Basin). Within this management area, the State of Kansas called for both voluntary and mandatory restrictions on pesticide usage on cropland to improve water quality. The management area was the first in the nation to focus on reducing atrazine in runoff to streams and reservoirs.
- *State of Washington.* The Washington State Department of Ecology created a Ground Water Management Area to protect ground water from nitrate contamination. The management area covers Grant, Franklin, and Adams counties, located in an intensive agricultural region of the Central Columbia Plateau. NAWQA information and communication of those findings in the USGS publication "Nitrate Concentrations in Ground Water of the Central Columbia Plateau" provided the scientific basis for implementing the management area. As follow-up to the NAWQA findings, USGS works with the Department of Ecology to (1) identify areas with lower nitrate concentrations, which could potentially serve as sources of future drinking-water supplies, (2) statistically correlate nitrate concentrations with natural features and human activities to better assess vulnerability; and, (3) design a long-term monitoring strategy for assessment of changes in nitrate concentrations over time.
- *State of California.* USGS works with the California State Water Resources Control Board and Department of Health Services to assess the vulnerability of public supply wells to contamination. The State uses USGS ground-water-age-dating analyses as one indicator of vulnerability. In addition, on the basis of NAWQA findings on the occurrence of industry-related and petroleum-based chemicals in ground water, the State has included the collection and analysis of VOCs in their vulnerability assessment. More than 200 wells have been sampled in southern California, and these efforts will be extended to northern California and the Central Valley.
- *State of Idaho.* NAWQA information formed the framework for predictive models and maps showing the vulnerability of ground water to contamination by the widely used herbicide atrazine in Idaho. The maps are used by the Idaho State Department of Agriculture to develop its State Pesticide Management Plan. Atrazine data from the NAWQA study in the Upper Snake River Basin were used to calibrate and verify the predictive models, which showed that significant factors associated with elevated atrazine concentrations in ground water were atrazine use, land use, precipitation, soil type, and depth to ground water. These modeling tools aid in the design of cost-effective programs for monitoring and protecting ground-water resources throughout the State.
- *State of Pennsylvania.* The Pennsylvania Department of Environmental Protection works with USGS as a follow-up to NAWQA findings on the prevalence of MTBE in ground water and its potential to contaminate public drinking-water supplies. Through the partnership, consistent and quality-assured data will be compiled, and a qualitative vulnerability rating for MTBE will be developed for different hydrogeologic settings throughout the State of Pennsylvania. The State will use the results to prioritize areas where MTBE should be assessed and where public-supply wells should be tested, and to target inspections of gasoline storage tanks.
- *National scale.* At the request of the USEPA's Office of Ground Water and Drinking Water, the NAWQA Program published a national map that shows the patterns of risk for nitrate contamination of shallow ground water (available in *Nitrate in Ground Waters*

of the United States—Assessing the Risk, USGS FS-092-96). By targeting regions with the highest risk of nitrate contamination, resources can be directed to areas most likely to benefit from pollution-prevention programs and long-term monitoring. Use of risk guidelines to locate areas for prevention of contamination also can result in cost-effective management. Once ground water is contaminated, it is expensive and, in many cases, virtually impossible to clean up.

VULNERABILITY CAN CHANGE OVER TIME

NAWQA findings show that the vulnerability to contamination of streams and ground water can differ seasonally in nearly every basin. For example, in streams that drain agricultural areas in many parts of the nation, the highest levels of nutrients and pesticides occur during spring and summer when recently applied chemicals are washed away by spring rains, snowmelt, and irrigation. Excessive amounts of contaminants can also enter streams during storm events. For example, sampling of nutrients and pesticides through a large storm event on the Potomac River in 1996 showed that concentrations and total amounts of nutrients and atrazine can increase during localized large storms, sometimes with overwhelming effects on receiving waters, such as the Chesapeake Bay. In this case, concentrations of individual compounds exceeded USEPA drinking-water standards during and following the extreme storm events. Such information helps water suppliers better understand the role of the short-term and seasonal events, and raises considerations related to timing of withdrawals, mixing, and storage to most effectively deliver high quality water at a minimum cost.

PROTOZOA IN WATER

Christine L. Bean
University of New Hampshire
Durham, New Hampshire

Protozoa are eukaryotic unicellular animals that possess membrane-bound genetic material and other cellular organelles that carry out various life functions (1). They are colorless and lack cell walls. These motile Eukarya obtain food by ingesting other organisms or organic particles. The uptake of these macromolecules occurs by a process called pinocytosis (2), whereby fluid is sucked into a channel formed by cell membrane invagination and enclosing the fluid within a membrane-bound vacuole. Particulate matter such as bacteria may also be ingested by phagocytosis.

Protozoa may be found in a variety of freshwater and marine habitats and in soil. Pathogenic protozoa include *Giardia lamblia, Entamoeba histolytica*, and *Cryptosporidium parvum. Giardia* and *Cryptosporidium* have been frequent causes of waterborne outbreaks in the United States. Many protozoa are parasitic in animals as well as humans. Reservoirs of protozoa include wild and domestic animals and humans. *Giardia* may be found in cats, dogs, sheep, pigs, goats, and cattle. *Cryptosporidium* has long been known as a pathogen of calves.

Morphologic differences in groups of protozoa can identify and differentiate them microscopically. They move and reproduce in various ways that can categorize them into groups or phyla such as amoeba (move by means of pseudopodia), flagellates, and ciliates. Most protozoa that can be found in water multiply by binary fission and are transmitted to humans through the ingestion of the infective stage of the parasite.

Free living protozoa such as flagellates, ciliates, and thecamoebae may feed on bacterial communities and actually multiply in water distribution systems (3).

FLAGELLATES

The taxonomic group of protozoa that use flagella are called Mastigophora. Motility is performed with the flagellum by either pushing or pulling the organism. Flagella can occur singly, in pairs, or in large numbers. Nearly one-half of the known species of protozoa are flagellates (4). Many of these are free-living, but several are parasitic for both humans and animals. Free-living protozoa include the green flagellate *Euglena*, which may be found in freshwater ponds (4). *Giardia lamblia* is an example of a parasitic freshwater flagellate that has a cyst and a trophozoite form. The cyst is the resting, environmentally resistant, infective form, and the trophozoite is the motile, feeding stage.

AMOEBAS

The taxonomic group of protozoa that use pseudopods for motility are called Sarcodina. Pseudpods may capture algae, bacteria, and other protozoa in the process of phagocytosis (4). Amoebic dysentery is caused by *Entamoeba histolytica*, which results in ulcerations of the intestinal tract and bloody diarrhea. Exposure may be from fecally contaminated freshwater or food supplies as well as person-to-person transmission of the cyst form. *Acanthamoeba* spp. is free-living amoebas that normally live in soil or in fresh water. They are capable of causing human infection in the brain, skin, and eye after exposure takes place through the respiratory tract, skin, or mucosal ulceration or open wounds (1).

CILIATES

The taxonomic group of protozoa that move by beating of cilia are called Ciliophora. Cilia are hair-like structures arranged in longitudinal or spiral rows that beat and result in an organized motion that moves the organism in one direction. Ciliates have two kinds of nuclei: the micronucleus and the macronucleus. The micronucleus is concerned only with inheritance, and the macronucleus is involved in the production of mRNA for cell growth and function (2). The genus *Paramecium* is a common protozoan in water. These are huge ciliated cells measuring about 50 μm in diameter. *Balantidium coli* is an example of a pathogenic ciliate. It is primarily a parasite of domestic animals such as pigs, but it has been shown to parasitize humans also.

SPOROZOANS

The taxonomic group of protozoa once believed to be spores is called Apicomplexa or Sporozoa. These protozoa have complex life cycles in which sexual and asexual reproductive phases occur sometimes in different hosts (4). Sporozoa do not have cilia, flagella, or pseudopods. Microsporidia have recently been recognized as a waterborne pathogen. The term "Microsporidia" describes protozoans belonging to the phylum *Microspora* (5). These protozoa are very small (<2 μm in diameter), and detection has been problematic. *Toxoplasma gondii* is another sporozoan highly associated with immunocompromised patients. The oocyst is shed in cat feces and is highly resistant to environmental degradation.

BIBLIOGRAPHY

1. AWWA (American Water Works Association). (1999). Waterborne Pathogens, Manual of Water Supply Practices. AWWA M48 177–182.
2. Brock, T., Madigan, M., Martinko, K., and Parker, J. (1994). *Biology of Microorganisms*, 7th Edn. Prentice-Hall, Englewood Cliffs, NJ, pp. 845–856.
3. Sibille, I., Sime-Ngando, T., Mathiew, L., and Block, J. (1998). Protozoan bacterivory and *Escherichia coli* survival in drinking water distribution systems. Appl. Environ. Microbiol. **64**(1): 197–202.
4. Pommerville, J.C. (2004). *Fundamentals of Microbiology*, 7th Edn. Jones and Bartlett Publishing, Sudbury, MA, pp. 594–630.
5. Dowd, S., Gerba, C., and Pepper, I. (1998). Confirmation of the human-pathogenic Microsporidia Enterocytozoon bieneusi, Encephalitozoon intestinalis, and Vittaforma corneae in water. Appl. Environ. Microbiol. **64**(9): 3332–3335.

WATER QUALITY

KENNETH M. MACKENTHUN
Arlington, Virginia

To most people, water quality means water whose characteristics represent a high degree of purity for the use to which the water is put. For drinking, this means cool, clear water, free from harmful contaminants and conditions that produce off-flavors. Other uses require a quality whose characteristics best support the particular use.

The Water Quality Act of 1965 first designated broad water uses when requiring the establishment of water quality standards for "public water supplies, propagation of fish and wildlife, recreational purposes, and agricultural, industrial, and other legitimate uses." The current Clean Water Act maintains those uses in Section 303(c)(2)(A) and adds language to take into consideration the use and value for navigation.

CHRONOLOGY OF EVENTS

In the decades of the 1950s and 1960s, many waterways in the United States showed signs of gross pollution. The Potomac River near the nation's capital was too dirty for swimming; the Cuyahoga River near Cleveland burned because of ignited oil on its surface; the Androscoggin River in Maine carried logs to pulp and paper mills and was treated with sodium nitrate to enhance dissolved oxygen through nitrification; the city of St. Louis collected its garbage, ground it, and barged it to the center of the Mississippi River for dumping; and the Flambeau River in Wisconsin carried so much black liquor from sulfite paper pulping that its bottom was covered with wood fibers, the river would rumble from escaping gases of decomposition, and boils of wood fibers and slime would rise to the river surface 6 inches thick and 10 feet in diameter. Many lakes of that era were overcome with algal and other vegetation growths from excessive nitrogen and phosphorus nutrients. These lakes included Lake Sebasticook in Maine, Lake Okeechobee in Florida, the Madison Lakes in Wisconsin, Detroit Lakes in Minnesota, and Lake Washington in Washington. Boston Harbor emitted sufficient hydrogen sulfide to blacken white-painted shoreside houses because of decomposing marine algae stimulated by nitrogen and phosphorus from the discharge of inadequately treated sewage.

The decade of the 1960s awakened the nation to the fact that water pollution must be better controlled. Prior to the 1970s, water quality went through a period of field investigation, investigative equipment development, and understanding of the relationship among aquatic organisms, chemistry, and physical conditions. Lake and stream sampling equipment was developed by Ekman (1), Petersen (2), and Surber (3). The writings of Kolkwitz and Marsson (4), Forbes and Richardson (5), Birge and Juday (6), Bartsch (7), and Hutchinson (8) influenced relationship understanding. The North American Benthological Society, formed in 1953, had its first meeting in Havana, Illinois, to share investigative information among the state biologists and coordinate taxonomic difficulties associated with stream benthic organisms.

In the decade of the 1970s, federal regulatory controls were instituted. Initial laws and regulations were refined in later years, increased emphasis was placed on identifying and controlling toxic pollutants, and efforts were made to manage nonpoint source pollutants and stormwater runoff. Water quality standards and point source discharge permits were refined and enhanced. The North American Lake Management Society held its first meeting in 1979.

LAWS AND REGULATIONS

The basic law controlling freshwater quality, with implementing regulations, dates to the River and Harbor Act of 1899 (33 USC 401). There have been many amendments along the way, and the present law was rewritten in the Federal Water Pollution Control Act Amendments of 1972 (33 USC 1251), which was significantly amended by the Clean Water Act of 1977 to emphasize toxic pollutant control. The latter laws now are commonly referred to as the Clean Water Act. The basic laws controlling marine water quality include the Marine Protection, Research and Sanctuaries Act (33 USC 1401), the Coastal Zone Management Act (16 USC 1451), the Act to Prevent Pollution from Ships (33 USC 1901), and the Magnuson–Stevens Fishery Conservation and Management Act (16 USC 1801). Other federal laws that impact actions that may affect water quality include the Endangered Species Act (16 USC 1531), the Fish and Wildlife Conservation Act (16 USC 2901), the Marine Mammal Protection Act (16 USC 1361), the National Environmental Policy Act (42 USC 4321), the National Invasive Species Act (16 USC 4701), and the Oil Pollution Act (33 USC 2701).

PROGRAMS

Defining water quality has a long record of accomplishments. Ellis (9) reviewed the literature for 114 substances and, in a 72-page document for the U.S. Bureau of Fisheries, listed lethal concentrations found by the authors. California published a 512-page book on water quality criteria that contained 1369 references. This publication was revised and enhanced the following year to include 3827 references. In 1968, the Secretary of the Interior appointed nationally recognized scientists to a national technical advisory committee to develop water quality criteria for the five recognized uses of water. This resulted in the Green Book, which constituted the most comprehensive documentation of its time on water quality requirements for defined water uses. The Blue Book followed in 1974 as the result of a contract with the National Academy of Sciences and the National Academy of Engineering. Section 304 of the Clean Water Act mandates that the Environmental Protection Agency (EPA) develop and issue Federal water quality criteria, which resulted in the Red Book, "Quality Criteria for Water," in 1976. The EPA published 65 separate water criteria documents for the 65 toxic pollutants in 1980. The Gold Book summaries followed in 1986. Presently, water quality criteria are defined in 63 *Federal Register* 68354, 10 December 1998.

Section 303 of the Clean Water Act requires that water quality standards be developed by the states and approved by EPA for all waters of the United States. Water quality standards along with national pollutant discharge elimination system permits are the hallmarks of water quality controls in the Clean Water Act. All permits required by the Act must be written in a manner that does not violate water quality standards, and Section 401 of the Act provides for state certification to that fact.

The discharge of any pollutant to the waters of the United States is controlled in Section 402 of the Clean Water Act by a national pollutant discharge elimination system (NPDES) permit. Section 301 (b)(1)(C) of the Act requires that such permits meet water quality standards. The elements of a NPDES permit include identification, description, and a unique number for each discharge; effluent limits based upon effluent guidelines, new source performance standards, toxic effluent standards, applicable water quality standards, and the best professional judgment of the permit writer; requirements for monitoring and reporting; delineation of best management practices; and a compliance schedule.

A Clean Lakes Program was established by the EPA in 1972, pursuant to Section 314 of the Clean Water Act, to provide financial and technical assistance to states in restoring publicly owned lakes. The program has funded approximately $145 million in grant activities, but there have been no appropriations since 1994.

The storage and retrieval of water quality data are vested in the EPA's STORET system. Begun in 1962, STORET contains more than 206 million analyses of chemical and physical water quality constituents at more than 761,000 locations across the United States, including coastal and international waters. Biological data are also included.

States are required by Section 305(b) of the Clean Water Act to submit a biennial report to the EPA that describes the water quality of the state. The EPA, in turn, prepares a National Water Quality Inventory, which is submitted to the Congress biennially. According to these reports, the United States has 3.5 million miles of rivers and streams; 41 million acres of lakes, ponds, and reservoirs; 34,000 square miles of estuaries; 58,000 miles of ocean shoreline; and 277 million acres of wetlands such as marshes, swamps, bogs, and fens. Rivers are polluted by bacteria, siltation, nutrients, oxygen-depleting substances, metals, habitat alterations, and suspended solids. Lakes are polluted by nutrients, siltation, oxygen-depleting substances, metals, suspended solids, pesticides, and organic toxic pollutants.

WATER QUALITY ETHIC

Nationally, water quality is significantly improved over the conditions of the 1950s and 1960s. Fish now inhabit waterways where once they were restricted from doing so because of unsuitable water quality. There is evidence, also, that an environmental ethic is gaining ground among the general population. An environmental ethic means exercising respect and consideration for the integrity, stability, and beauty of the water and its inhabitants. Such an ethic needs to be nurtured and expanded in its application. Regulations are required to keep societal goals focused on future attainment and to protect and preserve the environment for humans. A dedicated environmental ethic, coupled with viable regulations allowing flexibility for adjustment to a particular circumstance, will foster the Clean Water Act's Section 101 national goal that, wherever attainable, water quality shall provide for the protection and propagation of fish, shellfish, and wildlife and provide for recreation in and on the water.

BIBLIOGRAPHY

1. Ekman, S. (1911). Neue apparate zur qualitativen und quantitativen erforschung der bordenfauna der seen. *Int. Rev. Hydrobiol.* **7**: 164–204.

2. Petersen, C.G.J. (1911). Valuation of the sea. *Danish Biol. Sta. Report I*, **20**: 1–76.

3. Surber, E.W. (1936). Rainbow trout and bottom fauna production in one mile of stream. *Trans. Am. Fish. Soc.* **66**: 193–202.
4. Kolkwitz, R. and Marsson, M. (1909). Oekologie der tierischen saprobien. *Int. Rev. Gesamten Hydrobiol.* **2**: 126–152.
5. Forbes, S.A. and Richardson, R.E.. (1913). Studies on the biology of the Upper Illinois River. *Ill. Nat. Hist. Surv. Bull.* **9**(10): 481–574.
6. Birge, E.A. and Juday, C. (1922). *The Inland Lakes of Wisconsin. The Plankton. I. Its Quantity and Chemical Composition. Wis. Geol. and Natural History Bull. No. 64, Scientific Series No. 13.* 1–222.
7. Bartsch, A.F. (1948). Biological aspects of stream pollution. *Sewage Works J.* **20**(2): 292–302.
8. Hutchinson, G.E. (1957). *A Treatise on Limnology*. John Wiley & Sons, New York.
9. Ellis, M.M. (1937). *Detection and Measurement of Stream Pollution. U.S. Bureau of Fisheries Bull. No. 48*. 365–437.

WATER QUALITY

U.S. Geological Survey

Water quality is a term used to describe the chemical, physical, and biological characteristics of water, usually in respect to its suitability for a particular purpose. Although scientific measurements are used to define a water's quality, it's not a simple thing to say that "this water is good," or "this water is bad." After all, water that is perfectly good to wash a car with may not be good enough to serve as drinking water at a dinner party for the President! When the average person asks about water quality, they probably want to know if the water is good enough to use at home, to play in, to serve in a restaurant, etc., or if the quality of our natural waters are suitable for aquatic plants and animals.

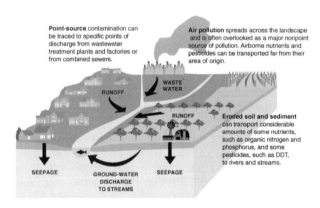

As the diagram above shows, assessment of the occurrence of chemicals that can harm water quality, such as nutrients and pesticides in water resources, requires recognition of complicated interconnections among surface water and ground water, atmospheric contributions,

natural landscape features, human activities, and aquatic health. The vulnerability of surface water and ground water to degradation depends on a combination of natural landscape features, such as geology, topography, and soils; climate and atmospheric contributions; and human activities related to different land uses and land-management practices.

More and more nowadays we are hearing about situations where the quality of our water is not good enough for normal uses. Bacteria and microorganisms have gotten into drinking-water supplies, sometimes causing severe illness in a town; chemical pollutants have been detected in streams, endangering plant and animal life; sewage spills have occurred, forcing people to boil their drinking water; pesticides and other chemicals have seeped into the ground and have harmed the water in aquifers; and runoff containing pollutants from roads and parking lots have affected the water quality of urban streams.

Yes, water quality has become a very big issue today, partly because of the tremendous growth of the Nation's population and urban expansion and development. Rural areas can also contribute to water-quality problems. Medium- to large-scale agricultural operations can generate in animal feed, purchased fertilizer, and manure, more nitrogen and phosphorus than can be used by crops or animals. These excess nutrients have the potential to degrade water quality if incorporated into runoff from farms into streams and lakes All this growth puts great stress on the natural water resources, and, if we are not diligent, the quality of our waters will suffer.

EMERGING AND RECALCITRANT COMPOUNDS IN GROUNDWATER

TOM MOHR
Santa Clara Valley Water District
San Jose, California

JAMES A. JACOBS
Environmental Bio-Systems, Inc.
Mill Valley, California

During the last two decades, the dominant groundwater contamination issues have focused on petroleum hydrocarbons (including gasoline, diesel, motor oil, lubricants, and jet fuel), chlorinated solvents, metals, and pesticides. Recently, attention has shifted to a variety of newly discovered compounds in groundwater. Some of these compounds were additives to fortify or stabilize the previously mentioned products. As additives, these compounds were present at low ratios, ranging from 10% to less than 1% of the total volume of the product. Though widely used in industry for decades, these emerging compounds were, until recently, not analyzed or regulated. Emerging and recalcitrant compounds include, but are not

---

This article is a US Government work and, as such, is in the public domain in the United States of America.

limited to, methyl tertiary-butyl ether (MtBE), a gasoline oxygenate additive; perchlorate, the major component of solid rocket motors and a minor ingredient of highway safety flares, and various stabilizers of chlorinated solvents such as 1,4-dioxane. Many of these recalcitrant compounds are also suspected to be carcinogenic or to cause other adverse health effects. Consequently, advisory drinking water action levels have been adopted for many emerging contaminants; regulatory agencies have added these compounds to their list of Compounds of Interest at remediation sites.

In the past, when confronted by new and emerging contaminants, consultants in the environmental industry generally did not have all the required remedial information because it was not immediately available. Consequently, many of these contaminants, when first encountered, are described in the literature as recalcitrant. Some compounds that were thought to be recalcitrant years ago, such as PCBs and TNT, have since been cleaned up using a variety of biological or chemical methods, following years of laboratory testing.

In approximately 1979, MtBE was added to gasoline as a fuel oxygenate to reduce air emissions and improve performance. MtBE was originally thought recalcitrant by many investigators. As an ether compound, MtBE was recognized as a compound significantly different from other petroleum hydrocarbon compounds and was not thought to be easily treated using existing hydrocarbon remedial technologies. As remedial solutions emerged, the cost of MtBE remediation was considered prohibitive. The state of the art of MtBE remediation has continued to improve in effectiveness and affordability, following years of research in the laboratory, as well as failures and successes in the field.

MtBE is highly soluble in groundwater and has low affinity for organic carbon in the soil. As a result of these physical and chemical characteristics, MtBE retardation by organic matter in soil is minimal. The solubility of MtBE in water is 43,000 mg/L, compared with the 1780 mg/L solubility of benzene. Consequently, MtBE is often found at the leading edge of gasoline plumes, followed by benzene (1).

Over the years, it has been determined that MtBE does break down using chemical oxidation (ozone, Fenton's reagent) as well as aerobic and anaerobic *ex situ* bioremediation methods. The chemical characteristics of MtBE require prompt source control of the fuel leak to minimize groundwater impacts and remediation costs (2). Successful management of MtBE contamination at fuel release sites requires adequate site characterization, selection of the appropriate remedial technology, sound engineering remedial design plans and implementation, proper operation and maintenance of the remedial system, and a cooperative approach with the regulatory agencies. Although MtBE moves quickly in groundwater, when detected soon after a spill, MtBE can be cleaned up at a cost equal to or slightly higher than that for other hydrocarbons, such as benzene. However, at sites where fuel has been leaking for some time, the MtBE plume can be significantly larger than the corresponding benzene plume, and the MtBE cleanup can cost much more.

The widespread release of chlorinated solvents in industrial and commercial cleaning and degreasing operations allowed solvents to impact groundwater throughout the nation from the 1960s through the 1980s. One place where solvent groundwater plumes are abundant is in Silicon Valley, south of San Francisco, California, where printed circuit board and semiconductor manufacturers using various solvents in degreasing operations created numerous large (1 mile or longer) chlorinated solvent plumes in shallow groundwater.

There has been innovation in restoring aquifers impacted with chlorinated solvents, and progress has been made at many of Silicon Valley's solvent release sites. The additives to solvents sometimes have the potential to be as problematic as the solvents themselves. Numerous additives are routinely included with most industrial solvents to ensure that the solvents perform as needed in their intended degreasing application. These additives are collectively known as *solvent stabilizers*, or inhibitors. They mitigate or prevent reactions in the degreaser with water, acids, and metals and inhibit degradation from heat, light, and oxygen. Stabilizers are generally added to TCE and PCE at volumetrically small proportions, generally totaling 1% or less. A few stabilizers, however, were added in the percent range. For example, 1,1,1-trichloroethane (TCA) was stabilized with 1,4-dioxane at 2–5% by volume, and some citations list as much as 8%. 1,4-Dioxane is a cyclic ether that inhibits reactions with metals, particularly aluminum salts.

The stabilizers most commonly associated with the four main solvents, TCA, trichloroethylene (TCE), tetrachloroethylene (PCE), and dichloromethane (DCM), are listed in Table 1.

Once released to the subsurface, the relative rates of migration of TCA and 1,4-dioxane are markedly different, governed by their physicochemical properties, as contrasted in Table 2. 1,4-Dioxane is resistant to both abiotic degradation and biotransformation, and owing to its infinite solubility and low affinity for sorption to soil organic matter, moves through the subsurface relatively unimpeded. Among 123 organic compounds ranked for their subsurface mobility, 1,4-dioxane is ranked first; it is deemed the most mobile among the compounds ranked (3).

Column experiments in the literature and modeling exercises have shown that 1,4-dioxane is extremely mobile in the subsurface. Accordingly, site investigation and remedial designs may require modification to account

Table 1. Solvents and Their Stabilizers

TCA	1,4-Dioxane	1,3-Dioxalane
	Nitromethane	1,2-Butylene oxide
TCE	1,2-Butylene oxide	Cresol
	Tetrahydrofuran	Epichlorohydrin
	Diisopropylamine	Alkyl pyrroles
DCM	Cyclohexane	Cyclohexene
PCE	Cyclohexene	Amines
	Butoxymethyloxirane	Phenols

Table 2. Physicochemical Properties of TCA and 1,4-dioxane

Property	1,4-Dioxane CASRN 123-91-1 $C_4H_8O_2$	TCA CASRN 71-55-6 $C_2H_3Cl_3$
Molecular weight	88.10	133.4
H_2O solubility mg/L at 20 °C	Miscible	1,360
Boiling point at 760 mmHg	101.1 °C	74.1 °C
Vapor pressure mmHg at 20 °C	37 mmHg at 25 °C	96 mmHg at 20 °C
Vapor density	3.03	5.45
Henry's constant, atm·m^3/mol	3×10^{-6}	1.5×10^{-2}
Log K_{OW}	−0.42	2.49
Log K_{OC}	0.54	2.85
Specific gravity	1.03 at 20 °C	1.34 at 20 °C

for the presence of 1,4-dioxane. Where 1,4-dioxane has been discovered at solvent release sites, regulators found it necessary to require that monitoring well networks be expanded, due to the larger footprint of 1,4-dioxane occurrence, and capture zones increased. Similarly, it has been necessary to add remediation equipment capable of removing 1,4-dioxane.

Existing groundwater treatment systems designed to remove chlorinated solvents are generally ineffective for remediating 1,4-dioxane, due to its K_{OC} and low Henry's law constant. In El Monte, California, a liquid granular activated carbon treatment system consisting of two 20,000-pound carbon vessels and treating 500 gallons per minute of solvent-contaminated groundwater was ineffective in reducing influent 1,4-dioxane concentrations at 14 µg/L to the treatment target of 3 µg/L (Fig. 1). In the City of Industry, California, an air stripper designed to remove 1.2 mg/L chlorinated solvents at 70 gallons per minute reduced 610 µg/L influent 1,4-dioxane to 430 µg/L in effluent. Conventional activated sludge and other common municipal wastewater treatment technologies have also proved ineffective in removing 1,4-dioxane.

The remedial technologies most commonly employed in removing 1,4-dioxane from groundwater *ex situ* are advanced oxidation processes (AOP), often in combination with ultraviolet light. AOP processes include ultraviolet light with ozone, hydrogen peroxide with ultraviolet light, ozone and hydrogen peroxide in combination, and Fenton's reagent (hydrogen peroxide and ferrous iron). Ultraviolet light causes release of hydroxyl radicals from hydrogen peroxide added to influent contaminated water. The hydroxyl radicals can react with 1,4-dioxane to oxidize the molecule to harmless reaction products (water, carbon dioxide, and residual chloride).

Biodegradation of 1,4-dioxane *in situ* is not presently considered a viable remediation option. The ether bond is a highly stable linkage and not readily biodegraded under ambient conditions. Recent research has established that there is promise for engineered bioreactors treating 1,4-dioxane *ex situ* and for *in situ* bioremediation using propane as a substrate. These methods of enhanced biodegradation show promise, whereas the more passive approach of monitored natural attenuation remains a poor candidate for effective management of 1,4-dioxane in groundwater.

SUMMARY

As many sites near closure or switch over to monitored natural attenuation, the specter of finding new and emerging compounds that are very mobile and recalcitrant is real. Discoveries of previously unknown contaminants may delay completion of site cleanup and require expensive revisions to remedial infrastructure to address their presence. Laboratory methods should be employed to obtain a full appreciation of the compounds likely to be present. Many of these emerging recalcitrant compounds are not presently subject to legal standards such as a maximum contaminant level, so there is often confusion regarding how the detection of these compounds affects progress at cleanup sites. Better regulatory guidance for the full gamut of compounds likely to be present at degreasing, fuel release, defense, and other sites is needed to ensure that time, money, and effort are well spent to complete the remedial investigation/feasibility study (RIFS) process once instead of repeating it upon the discovery of each new contaminant.

A more detailed treatment of this topic, with complete reference listings, may be found in the revised *Solvent Stabilizers White Paper* (4). The presentation binder from the Groundwater Resources Association symposium titled *Characterization and Remediation of Emerging and Recalcitrant Contaminants* in San Jose on June 14–15, 2001, and the December 10, 2003 symposium on *1,4-Dioxane and Other Solvent Stabilizers in the Environment* also contains additional information and articles on recalcitrant compounds (5).

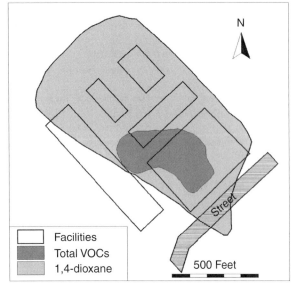

Figure 1. Silicon Valley solvent recycling facility plumes (10 µg/L contour).

BIBLIOGRAPHY

1. Morrison, R., Beers, R., and Hartman, B. (1998). *Petroleum Hydrocarbon Contamination: Legal and Technical Considerations*. Argent Communications Group, Foresthill, CA.
2. Wilson, J. (2003). Fate and Transport of MtBE and Other Gasoline Compounds. In: *MtBE Remediation Handbook*. E. Moyer and P. Kostecki (Eds.). Amherst Scientific, Amherst, MA.
3. Roy, W.R. and Griffin, R.A. (1985). Mobility of organic solvents in water-saturated soil materials. *Environ. Geol. Water Sci.* **7**: 241–247.
4. Mohr, T.K.G. (2001). *Solvent Stablizers White Paper*. Presented to *Groundwater Resour. Assoc. 2nd Symp. Groundwater Contaminants, Characterization and Remediation of Recalcitrant and Emerging Contaminants*, San Jose, CA June 14–15, 2001. Available for download at: http://www.valleywater.org/Water/Water_Quality/Protecting_your_water/_Lustop/_Publications_and_Documents/_PDFs_Etc/SolventStabilizers.pdf
5. Groundwater Resources Association. (2001). *Characterization and Remediation of Emerging and Recalcitrant Contaminants*. 2nd Symp. GRA Contaminant Ser., June 14–15, 2001, San Jose, CA.

ROAD SALT

M. Eric Benbow
Richard W. Merritt
Michigan State University
East Lansing, Michigan

INTRODUCTION

Maintenance of safe winter roadways and walkways is frequently accomplished by the application of chemical deicers and anti-icing compounds before or during snowfall and ice cover (Fig. 1). Road salt has been used since the 1940s and continues to be extensively applied throughout much of North America and Europe. It is estimated that 10 million metric tons of road salt are spread on American roadways each year, while another 3 million metric tons are used in Canada (1,2) (Fig. 2). The total annual cost of applying road salt can be greater than $62 billion in the United States because of damaged roads, bridges, and motor vehicles (1). By far, the most regularly used deicing agent is sodium chloride (NaCl), although several other inorganic salts are available (3). For instance, in the winter of 1997–1998, approximately 5 million tons of NaCl were applied to Canadian roadways compared with 120,000 tons of calcium chloride ($CaCl_2$) (4).

The widespread and increasing use of road salt has raised concerns about its effects on natural resources and drinking water. During roadway application, the components of road salt are left in the environment, typically on the soils adjacent to treated roads or into aquatic ecosystems, such as streams, rivers, wetlands, and lakes. Some salt enters groundwater during its movement. This large-scale input of salt into aquatic ecosystems is thought to have an impact on biological communities. However, the magnitude of these effects on specific ecosystem components is relatively unknown

Figure 1. A road salt storage facility open to the elements that can increase runoff during heavy rain effects. This facility is on a parking lot near Lansing, MI (photo by M.E. Benbow).

Figure 2. Residual road salt on the pavement of a parking lot near Lansing, MI (photo by M.E. Benbow).

compared with other pollutants, and recent research is beginning to provide important information. In an extensive review by Environment Canada (4), it was found that: "Based upon the available data, it is considered that road salts that contain inorganic salts with or without ferrocyanide salts are entering the environment in a quantity or concentration or under conditions that have or may have an immediate or long-term harmful effect on the environment on which life depends. Therefore, it was concluded that road salts... are 'toxic' as defined in Section 64 of the Canadian Environmental Protection Act, 1999." Concerns exist about the effects of road salt on the natural environment that are balanced by transportation safety and economic issues.

For this topic on road salt, we will focus on the most commonly used form, NaCl. We briefly discuss its basic physical and chemical properties, ice control alternatives to NaCl, and the potential impacts on surface water biological communities.

DEFINITION AND PROPERTIES

Sodium chloride is the most common road salt used for melting snow and ice on pavements (3,4). Road salt can also be defined as calcium chloride, magnesium chloride ($MgCl_2$), and potassium chloride (KCl_2). In addition, an anticaking agent is often added to road salt, which is usually sodium ferrocyanide ($Na_4Fe[CN]_6 \cdot H_2O$), but is sometimes ferric ferrocyanide ($Fe_4[Fe[CN]_6]_3$), a component that has been listed as a toxic substance in the Clean Water Act (5). These anticaking agents are typically added in amounts ranging from 20 to 100 ppm (6).

The common physical and chemical properties of NaCl are given in Table 1. Pure sodium chloride is composed of 60.7% elemental chlorine and 39.3% sodium that form cubic crystals. Rock salt NaCl usually has other trace elements, including metals such as phosphorus, sulfur, nitrogen, zinc, and copper (4). By lowering the eutectic freezing point of water, the crystalline structure of pure NaCl provides anti-icing properties that results in melting when salt crystals combine with water (H_2O) molecules. The eutectic freezing point is the lowest freezing temperature for water after combining with a salt. Thus, pure NaCl melts snow and ice down to a temperature of $-21.12\,°C$ (Table 1). However, NaCl eliminates freezing when temperatures are above $-11\,°C$, and as the difference between ambient temperature and the eutectic temperature decreases, the melting rate of ice and snow slows down (4). It has been reported that the lowest effective deicing temperature for NaCl is from $0\,°C$ to $-15\,°C$ (4). When temperatures go below this range, other inorganic road salts are sometimes mixed with NaCl for more effective treatment on roadways, or used in place of NaCl. In addition, abrasives such as sand are mixed with rock salt during application when temperatures drop below the optimal effective temperatures. They also are added to help the salt stick to the pavement surface. Sand and other abrasives can have negative effects on roadside communities (4,7,8).

ALTERNATIVES TO NACL

Historically, NaCl has been the preferred deicing agent because of its economic cost, easy storage, quick melting properties, and ease of handling and application. However, with recent concerns about its impact on the natural environment (see below), several alternatives have been proposed and tested (1). Table 2 gives a summary of the most common deicing and anti-icing alternatives to NaCl by comparing their eutectic temperatures and relative cost differences. The other inorganic salts ($CaCl_2$, $MgCl_2$, KCl) are more expensive than NaCl and are usually more corrosive to vehicle and roadway structures such as bridges and overpasses (1,6). They are often used in mixtures for snow and ice melting at higher rates and lower temperatures than using NaCl alone, but they also are more expensive (1). Calcium magnesium acetate (a mixture of calcium and magnesium acetates) is even more expensive, but can work well at temperatures (e.g., $-9.44\,°C$) lower than those most effective for NaCl, and has been reported to be the least toxic of the alternatives (1). Other alternatives have been tested, such as methanol and urea. However, methanol is considerably more volatile, toxic, and flammable than NaCl, and urea is a nutrient fertilizer that can quickly contribute to eutrophication of surface waterbodies. The overwhelming argument by government agencies for the continued use of NaCl is based on economic cost–benefit comparisons to the alternatives, despite the potential effects on the environment.

IMPACTS ON AQUATIC ENVIRONMENTS

Road salt enters the air, soil, surface, and groundwater through vehicle splashing and direct runoff into adjacent areas of roadways. During windy conditions, splash and spray from melted road salt slurries can be transported

Table 1. Physical and Chemical Properties of Sodium Chloride

Property	Value or Description
Chemical formula	NaCl
Molecular weight	58.45
Atomic weight of Na	22.9898
Atomic weight of Cl	35.4527
Appearance	Clear or white solid
Solubility	35.7 g/100g water at $0\,°C$
Density	2.2×10^3 kg/m^3
Melting point	$801\,°C$
Boiling point	$1465\,°C$
Eutectic freezing point	$-21.12\,°C$

Table 2. Economic Cost and Melting Temperature Comparisons Among Chemical Alternatives to NaCl for Controlling Ice and Snow

Chemical	Relative Economic Comparison	Eutectic Temperature, °F	Eutectic Temperature, °C	Eutectic Concentration, % w/w
Sodium chloride (NaCl)	1	−6	−21	23.3
Calcium chloride ($CaCl_2$)	7	−67	−55	29.8
Magnesium chloride ($MgCl_2$)	7	−28	−33	21.6
Potassium chloride (KCl)	4	13	−11	19.5
Calcium acetate	35	5	−15	44.0
Magnesium acetate	35	−22	−30	31.0
Methanol	10	−144	−98	100.0
Urea	7	11	−12	32.5

Source: Data were taken and modified from Table 1 in the Salt Institute (6).

several meters from the side of the road and can cover roadside vegetation (4). As NaCl quickly dissociates into its component ions Na^+ and Cl^- when in solution, it is the concentrations of these ions that can affect communities adjacent to treated roadways. Although Na^+ ions readily bond to negatively charged soil particles and are typically consumed in biological processes in the sediments, Cl^- ions are more mobile and tend to be transported in surface and groundwater, where concentrations can accumulate and persist if the geological conditions are favorable (e.g., clay-lined wetlands). However, if substantial ground–surface water exchange exists, Cl^- will move with the direction of water flow in the system.

ROAD SALT LOADING AND CHLORIDE LEVELS

Studies have shown that deicing salts accumulate in rivers, streams, wetlands, and lakes during application, during snow removal and disposal, and from losses during handling and storage (4,9–11). For instance, it was estimated that road salt input into the Great Lakes was nearly two million tons by 1960: about 700,000 tons in Lake Erie; 500,000 in Lake Ontario; 400,000 in Lake Michigan; 190,000 in Lake Huron; and 90,000 in Lake Superior (12). Environment Canada (4) estimated, from a report by Morin and Perchanok (13), that the total chloride road salt loading for Canada between 1997 and 1998 was nearly 5 million tons. In 1992, 17 states in the United States reported that road salt application was a significant source of groundwater contamination, and four reported degraded wetland integrity because of salinity (14).

Increases have been documented in sodium and chloride concentrations in surface and groundwater over the past 50 years (10,15–17). In New York, it was reported that chloride transport rates increased by 145% from the 1950s to the 1970s, and that road salt application accounted for 69% of that increase (16). Other studies have shown similar trends in other parts of the United States and Canada (18–20). However, Blasius and Merritt (7) reported rather low chloride concentrations from two Michigan streams during winter months, but explained that increased discharge during spring thaw acted to dilute the chloride additions from winter road salt application. This inverse relationship of discharge and chloride levels also has been reported in other studies (21,22).

Environment Canada (4) summarized various chloride studies of urban lakes and retention ponds and reported chloride levels from 200 to 5910 mg/L. In the interstitial peat mat waters of an Indiana wetland, Wilcox (23) reported chloride levels up to 1215 mg/L. Benbow and Merritt (24) measured chloride concentrations from 43 water bodies in 37 counties of Michigan and reported a range from 18 to 2700 mg/L. From a wetland adjacent to a sand-salt storage facility, chloride was reported in another study to be 12,463 mg/L (25).

INDIRECT EFFECTS OF NACL

Residual road salt runoff is known to directly impact aquatic flora and fauna from both elevated chloride levels, and potentially from the anticaking chemicals routinely mixed for application (26,27). Sodium ferrocyanide is one such agent reported to release cyanide ions via photolysis (28); however, little information exists on the amount of ferrocyanide that might be entering impacted waterbodies during road salt application. Another indirect effect is the potential release of heavy metals already present in benthic (stream, wetland, or lake bottom) sediments through ion exchange mechanisms (29–31). For example, Doner (29) found that sodium chloride increased the mobility of Ni(II), Cu(II), and Cd(II) in soil leach columns. Furthermore, the release of mercury-adsorbed sediment particles (exchanged with Na^+) has been documented to increase with chloride ion concentrations in freshwater sediments (32). In laboratory column experiments, nearly 90% of mercury sorbed to sediments was mobilized when salt was introduced (33). Road salt also has been suggested as an agent of lake and stream acidification via ion exchange of Na^+ for H^+ (34), and is suspected to reduce soil fertility through accelerated leaching of calcium and magnesium (31).

Road salt accumulation affects small lake mixing through the establishment of short- and long-term chemoclines that prevent mixing and leading to anoxic conditions (19,35,36). Dense saline layers accumulate at lake bottoms, reducing the likelihood of mixing and prolonging anaerobic conditions of the lower layers, thus resulting in fish kills and destruction of the benthic fauna on which fish feed. Road salts entering surface waters may also trigger excessive blue-green algae growth, stimulating nuisance algal blooms (37). The mechanism of this growth involves facilitated phosphate absorption by algae in the presence of inorganic ions (38). Seasonal influxes of road salt runoff have been reported to maintain high chloride concentrations of lakes so that salt outflow is abnormally high throughout the year (39), thereby creating long-term water chemistry changes downstream.

These studies provide evidence that road salt can have indirect effects on aquatic biological communities beyond that of direct toxicity; however, direct mortality of aquatic animals can have cascading effects on the rest of the ecosystem. As macroinvertebrates commonly serve as biological indicators of aquatic ecosystems and are a dominant food source for higher consumers (40,41), we have provided a more thorough review of road salt toxicity on these organisms. Only a brief summary for zooplankton and fish is given. We refer readers to Environment Canada (4) for more information on these organisms and protozoans, bacteria, and fungi.

DIRECT EFFECTS ON AQUATIC MACROINVERTEBRATES

Before direct toxicity effects can be evaluated, the natural salinity tolerance for invertebrates should be reviewed; and this has been done in several studies. Williams et al. (42) found relationships between macroinvertebrate community structure and chloride levels of Canadian springs. In that study, they attributed high chloride levels to groundwater contamination by road salt (42). Several authors have reported benthic macroinvertebrate salinity tolerances along natural salt gradients in estuaries (43,44), salt marshes (45), and more recently in hyporheic habitats along a salinity gradient from a freshwater stream into an estuary (46). Overall, these studies have shown that distinct invertebrate communities are often associated with salinity ranges. But what about macroinvertebrates typically found in freshwater that are exposed to elevated salt concentrations from residual road salt?

Road salt toxicity studies on aquatic macroinvertebrates have shown variable results (4,7,24). Early studies found compromised osmoregulatory processes for certain aquatic insects (47–49), and species-specific tolerance variation is common in the literature (Table 3). Most acute (24–96 h) and chronic toxicity (>7 d) studies report LC_{50} values that vary among species and test conditions. The LC_{50} is the lethal concentration at which 50% of the test

Table 3. Review of Field and Laboratory Studies of NaCl Effects on Aquatic Macroinvertebrates

Endpoint Effect	NaCl, Cl$^-$, or Road Salt[a] Concentrations, mg/L	Reference
LC_{50} for *Culex* sp. (Culicidae)	10,254	Dowden and Bennett (50)
100% Mortality for 12-h exposure for *Chironomus attenatus* (Diptera: Chironomidae)	9995	Thornton and Sauer (47)
100% Mortality for 48-h exposure for *Cricotopus trifascia* (Diptera: Chironomidae)	8865	Hamilton et al. (51)
100% Mortality for 48-h exposure for *Hydroptila angusta* (Trichoptera: Hydroptilidae)	10,136	Hamilton et al. (51)
100% Mortality for 48-h exposure for *Nais variabilis* (Oligochaeta: Naididae)	3735	Hamilton et al. (51)
No effect on drift or mortality for *Gammarus pseudolimnaeus* (Amphipoda: Gammaridae)	800	Crowther and Hynes (27)
No effect on drift or mortality for *Hydropsyche betteni* and *Cheumatopsyche analis* (Trichoptera: Hydropsychidae)	1650	Crowther and Hynes (27)
Some drift activity of macroinvertebrates observed in field experiments	>1000	Crowther and Hynes (27)
No effect on composition or diversity of Ephemeroptera, Plecoptera, Trichoptera and Coleoptera in the field	50–67	Molles (8)
Mortality of *Hydropsyche betteni*, *H. bronta*, and *H. slossonae* (Trichoptera: Hydropsychidae) unaffected after 10 d	800	Kersey (52)
80% Mortality for 6-day exposure for *H. betteni*	6000	Kersey (52)
Similar growth rates among treatments for *Hexagenia limbata* (Ephemeroptera: Ephemeridae)	0, 2000, 4000, 8000	Chadwick (53)
96-h LC_{50} at 28 °C and 18 °C, respectively, for *H. limbata*	2400 and 6300	Chadwick (53)
96-h LC_{50} for *Tricorythus* sp. (Ephemeroptera: Tricorythidae)	2200–4500	Goetsch and Palmer (54)
Approximate LC_{50} for *Hydropsyche betteni* (Trichoptera: Hydropsychidae); however, no dose-response was tested	13,308	Kundman (55)
No significant drift observed for *H. betteni* in lab	2000–8000	Kundman (55)
96-h LC_{50} for *Lepidostoma* sp. (Trichoptera: Lepidostomatidae)	6000	Williams et al. (10)
70% mortality 96-h exposure for *Nemoura trispinosa* (Plecoptera: Nemouridae)	6000	Williams et al. (10)
100% mortality 96-h exposure *Gammarus pseudolimnaeus*	6000	Williams et al. (10)
No effect on macroinvertebrate functional feeding group assemblages in two rivers	8–16 (Cl$^-$)	Blasius and Merritt (7)
Drift effects on *Gammarus pseudolimnaeus* (Amphipoda): from 55%—68% in laboratory experiments over 24 h	2500–10,000	Blasius and Merritt (7)
Variable and no significant drift of Perlidae, Limnephilidae, Heptageniidae, Tipulidae in laboratory experiments over 24 h	1000–10,000	Blasius and Merritt (7)
Significantly lower leaf pack decomposition, but due to sand not salt	—	Blasius and Merritt (7)
96 h LC_{50} for *Gammarus* sp. (Amphipoda)	7700	Blasius and Merritt (7)
96 h LC_{50} two limnephilid caddisflies (Trichoptera)	3526	Blasius and Merritt (7)
No significant mortality for two Perlidae (Plecoptera) and one Tipulidae (Diptera) species in lab	>10,000	Blasius and Merritt (7)
100% mortality after 7 days for *Hexagenia limbata* (Ephemeroptera)	12,000	Chadwick et al. (56)
Loss of osmoregulation and increased mortality	>8000	Chadwick et al. (56)
96 h LC_{50} estimated for *Callibaetis fluctuans* (Ephemeroptera) and *Physella integra* (Gastropoda) under different test conditions (lab and field)	5000–10,000 (road salt)	Benbow and Merritt (24)
96 h LC_{50} estimated for *Hyallela azteca* (Amphipoda) and *Chaoborus americanus* (Diptera: Chaoboridae) under different test conditions (lab and field)	>10,000 (road salt)	Benbow and Merritt (24)

[a] When Cl$^-$ or road salt was used instead of NaCl in a study, it is indicated in parentheses.
Source: The table is modified and expanded from Blasius and Merritt (7).

organisms die. In sublethal tests, the EC_{50} is reported, which is defined as the effective concentration where the tested variable (e.g., number of eggs or growth rate) is reduced by 50%. We provide Table 3 as a summary of tolerance levels of macroinvertebrates reported in the literature. This table was modified and expanded from Blasius and Merritt (7).

From the literature in Table 3, it is apparent that salt tolerance for aquatic macroinvertebrates is relatively high, with highest acute toxicity reported from mayflies (Ephemeroptera: *Hexagenia limbata*). Acute toxicity studies reported LC_{50} values (or greater mortality) from 2400 to >13,000 mg/L NaCl (or road salt). Variable mortality was evident among taxa with *Culex* mosquitoes (Culicidae) having an LC_{50} of 10,254 mg/L NaCl and caddisfly (Trichoptera) taxa with values from 3526 to 13,308. Nonlethal effects also were variable. Drift effects for various taxa (amphipods, caddisflies, mayflies, stoneflies, and craneflies) ranged from 2500 to 10,000 mg/L NaCl, with some drift effects found at concentrations >1000 mg/L NaCl. Growth rates for a mayfly (*H. limbata*) did not differ among treatments from 0–8000 mg/L NaCl; however, osmoregulation was lost at higher concentrations. This variation in lethal and nonlethal road salt tolerance may be a product of artificial testing conditions that do not represent natural environmental changes in temperature and other variables.

In addition, many taxonomic groups have not been tested, indicating that unstudied, yet sensitive species may exist. Most studies did not evaluate the background chloride levels of waterbodies typically occupied by the invertebrates tested, and when background chloride levels were measured, they were often below the invertebrate toxicity levels. In several studies, it was noted that chloride levels were sometimes diluted with snow melt runoff at the time of the highest expected impact (early spring). Nonlethal effects (e.g., drift, fecundity, production) have been understudied and may be important to invertebrates impacted by residual road salt runoff. Additional data and studies can be found in the review by Environment Canada (4).

EFFECTS ON ZOOPLANKTON AND FISH

Road salt tolerance variability has been reported for several species of zooplankton, with *Daphnia pulex*, *D. magna*, and *Ceriodaphnia dubia* being the most commonly tested taxa (4). Results from short-term acute toxicity studies (≤24 to 96 h) reported LC_{50} values ranging from 2308 to 7754 mg/L NaCl (4). In studies lasting 7–10 d, LC_{50} values ranged from about 2000 to 6000 mg/L NaCl, whereas EC_{50} for zooplankton mean brood size, number of broods, or total progeny ranged from about 1400 to 6000 mg/L NaCl (4).

Fish road salt toxicity was low during studies <24 h, with LC_{40} and LC_{50} values ranging from 20,000 to 50,000 mg/L NaCl (4). In 24 h studies, road salt toxicity was higher, but LC_{50} values were >7000 mg/L NaCl for bluegill (*Lepomis macrochirus*), indian carp fry (*Cirrhinius mrigalo*, *Labeo rohoto*, and *Catla catla*), brook trout (*Salvelinus fontinalis*), and rainbow trout (*Oncorhynchus mykiss*) (4). Longer acute toxicity studies reported LC_{50} values for bluegill, rainbow trout, and indian carp fry to be >9500, 11,112, and 4980 mg/L NaCl, respectively (4). The American eel (*Anguilla rostrata*) was reported to have LC_{50} values from about 18,000 to 22,000 mg/L NaCl in one study, and fathead minnow (*Pimephales promelas*) values were from >7500 to 10,831 mg/L NaCl (4). In seven to ten day acute toxicity studies on fathead minnow and frog (*Xenopus laevis*) embryos, eggs, and larvae, LC_{50} values ranged from 1440 to 5490 mg/L NaCl depending on life stage (4).

CONCLUSIONS

The sustained and increasing use of road salt for winter roadway maintenance is becoming a popular issue as a result of raised awareness of potential ecological effects on natural waterways. Balancing human transportation safety with potential ecological repercussions drives this issue, and it is clear from the literature that the answer will be complicated, if ever resolved. The economics associated with using deicer alternatives suggest that, at present, road salt is the best practical agent for winter snow and ice removal. Chlorides are known to have effects on organisms in nature, and in many reports, it is evident that chloride concentrations are rising in waterbodies associated with roadways that are heavily treated with road salt. However, short-term toxicity studies have shown that mortality among aquatic zooplankton, macroinvertebrates, and fish is highly variable, species-dependent, and related to test conditions. Many toxicity studies are done under laboratory conditions that are far removed from the natural variation of climatic and biological components of natural waterbodies, which may influence true mortality of certain taxa. Apart from toxicity studies, larger scale studies that evaluate complex community and ecosystem level responses are needed for a better understanding of road salt effects in nature.

BIBLIOGRAPHY

1. D'Itri, F.M. (1992). *Chemical Deicers and the Environment*. Lewis Publishers, Chelsea, MI.
2. Salt Institute. (2000). Deicing salt facts: a quick reference. *SI Pub.* **2**(77): 83.
3. Field, R. and O'Shea, M.L. (1992). The USEPA research program on the environmental impacts and control of highway deicing salt pollution. In: *Chemical Deicers and the Environment*. F.M. D'Itri (Ed.). Lewis Publishers, Chelsea, MI, pp. 117–133.
4. Environment Canada. (2001). *Priority Substances List Assessment Report—Road Salts*.
5. Environmental Protection Agency. (2003). *Administrative Determination: EPA Clarifies that Ferric Ferrocyanide is one of the "Cyanides" in the Clean Water Act's List of Toxic Pollutants*. Environmental Protection Agency, Washington, DC, EPA-821-F-03-012.
6. Salt Institute. (2004). *Highway Salt and our Environment*. Salt Institute, Alexandria, VA.
7. Blasius, B.J. and Merritt, R.W. (2002). Field and laboratory investigations on the effects of road salt (NaCl) on

stream macroinvertebrate communities. *Environ. Pollut.* **120**: 219–231.

8. Molles, M.C. Jr. (1980). *Effects of Road Salting on Aquatic Invertebrate Communities*. U.S. Department of Agriculture, Forest Service, Rocky Mountain Forest and Range Experiment Station, Fort Collins, CO, p. 10.
9. Scott, W.S. (1976). The effect of road deicing salts on sodium concentration in an urban water-course. *Environ. Pollut.* **10**: 141–153.
10. Williams, D.D., Williams, N.E., and Cao, Y. (2000). Road salt contamination of groundwater in a major metropolitan area and development of a biological index to monitor its impact. *Water Res.* **34**(1): 127–138.
11. Scott, W.S. (1980). Road salt movement into two Toronto streams. *J. Environ. Eng. Div.*
12. Struzeski, E. (1971). *Environmental Impact of Highway Deicing*. United States Environmental Protection Agency, Washington, DC.
13. Morin, D. and Perchanok, M. (2000). *Road Salt Loadings in Canada. Supporting Document for the Road Salt PSL Assessment*. Commercial Chemicals and Evaluation Branch, Environment Canada, Hull, Quebec.
14. United States Environmental Protection Agency. (1994). *National Water Quality Inventory: 1992*. U.S. EPA, Washington DC.
15. Bubeck, R.C. et al. (1971). Runoff of deicing salt: effects on Irondequoit Bay, Rochester, New York. *Science* **172**(3988): 1128–1132.
16. Peters, N.E. and Turk, J.T. (1981). Increases in sodium and chloride in the Mohawk River, New York, from the 1950s to the 1970s attributed to road salt. *Water Res. Bull.* **17**(4): 586–598.
17. Mayer, T., Snodgrass, W.J., and Morin, D. (1999). Spatial characterization of the occurrence of road salts and their environmental concentrations as chlorides in Canadian surface waters and benthic sediments. *Water Qual. Res.* **34**(4): 545–574.
18. Environment Canada (2001). *Priority Substances List Assessment Report—Road Salts*. Environment Canada, Hull, Quebec.
19. Judd, J.H. (1970). Lake stratification caused by runoff from street deicing. *Water Res.* **4**: 521–532.
20. Lathrop, R.C. (1975). *A Limnological Study of Two Neighboring Lakes: First Sister Lake and Second Sister Lake, Washtenaw County, Michigan* [MS]. Natural Resources, The University of Michigan, Ann Arbor, MI.
21. Emmett, W.W. (1975). *The Channels and Waters of Upper Salmon River, Idaho*. U.S. Department of Interior, Washington, DC.
22. Drever, J.I. (1982). *The Geochemistry of Natural Waters*. Prentice Hall, Englewood Cliffs, NJ.
23. Wilcox, D.A. (1986). The effects of deicing salts on water chemistry in Pinhook Bog, Indiana. *Water Res. Bull.* **22**(1): 57–65.
24. Benbow, M.E. and Merritt, R.W. (2004). Road-salt toxicity of select Michigan wetland macroinvertebrates under different testing conditions. *Wetlands* **24**(1): 68–76.
25. Ohno, T. (1990). Levels of cyanide and NaCl in surface waters adjacent to road salt facilities. *Environ. Pollut.* **67**: 123–132.
26. Demers, C.L. (1997). Effects of road deicing salt on aquatic invertebrates in four Adirondack streams. In: *Chemical Deicers and the Environment*. F.M. D'Itri (Ed.). Lewis Publishers, Ann Arbor, MI, pp. 245–251.
27. Crowther, R.A. and Hynes, H.B.N. (1977). The effect of road deicing salt on the drift of stream benthos. *Environ. Pollut.* **14**: 113–126.
28. Hsu, M. (1984). *Anticaking Agents in Deicing Salt*. Materials and Research Division, Maine Department of Transportation, p. 84–4.
29. Doner, H.E. (1978). Chloride as a factor in mobilities of Ni(II), Cu(II), and Cd(II) in soil. *Soil Sci. Soc. Am. J.* **42**: 882–885.
30. Jones, P.H., Jeffrey, B.A., Watler, P.K., and Hutchon, H. (1992). Environmental impact of road salting. In: *Chemical Deicers and the Environment*. F.M. D'Itri (Ed.). Lewis Publishers, Chelsea, MI, pp. 1–116.
31. Shanley, J.B. (1994). Effects of ion exchange on stream solute fluxes in a basin receiving highway deicing salts. *J. Environ. Qual.* **23**: 977–986.
32. Wang, J.S., Huang, P.M., Liaw, W.K., and Hammer, U.T. (1991). Kinetics of the desorption of mercury from selected freshwater sediments as influenced by chloride. *Water Air Soil Poll.* **56**: 533–542.
33. Macleod, C.L., Borcsik, M.P., and Jaffe, P.R. (1996). Effect of infiltrating solutions on the desorption of mercury from aquifer sediments. *Environ. Technol.* **17**: 465–475.
34. Norton, S.A., Henriksen, A., Wright, R.F., and Brakke, D.F. (1987). Episodes of natural acidification of surface waters by natural sea salts. In: *Geomon, Int. Workshop on Geochemistry and Monitoring in Representative BasinsóExtended Abstracts*. B. Moldan and T. Paces (Eds.). Geological Survey, Prague, Czechoslovakia, pp. 148–150.
35. Hoffman, R.W., Goldman, G.R., Paulson, S., and Winters, G.R. (1981). Aquatic impacts of deicing salts in the Central Sierra Nevada Mountains, California. *Water Res. Bull.* **17**(2): 280–285.
36. Kjensmo, J. (1997). The influence of road salts on the salinity and the meromictic stability of Lake Svinsjoen, southeastern Norway. *Hydrobiologia*. **347**: 151–158.
37. Sharp, R.W. (1971). *Road Salt as a Polluting Element*. State University College of Forestry at Syracuse University, Syracuse, NY.
38. Shapiro, J., Chamberlain, W., and Barrett, J. (1969). Factors influencing phosphate use by algae. Paper presented at *Proc. 4th Int. Conf. Wat. Pollut. Res.* Prague.
39. Cherkauer, D.S. and Ostenso, N.A. (1976). The effect of salt on small, artificial lakes. *Water Res. Bull.* **12**(6): 1259–1267.
40. Rosenberg, D.M. and Resh, V.H. (1993). *Freshwater Biomonitoring and Benthic Macroinvertebrates*. Chapman and Hall, New York.
41. Merritt, R.W. and Cummins, K.W. (Eds.). (1996). *An Introduction to the Aquatic Insects of North America*, 3rd Edn. Kendall/Hunt, Dubuque, IA.
42. Williams, D.D., Williams, N.E., and Cao, Y. (1997). Spatial differences in macroinvertebrate community structure in springs in southeastern Ontario in relation to their chemical and physical environments. *Can. J. Zool.* **75**: 1404–1414.
43. Williams, D.D. and Williams, N.E. (1998). Seasonal variation, export dynamics and consumption of freshwater invertebrates in an estuarine environment. *Estuarine, Coastal Shelf Sci.* **46**: 393–410.
44. Williams, D.D. and Williams, N.E. (1998). Aquatic insects in an estuarine environment: densities, distribution and salinity tolerance. *Freshwater Biol.* **39**: 411–421.
45. Giberson, D.J., Bilyj, B., and Burgess, N. (2001). Species diversity and emergence patterns of Nematocerous flies (Insecta: Diptera) from three coastal salt marshes in Prince Edward Island, Canada. *Estuaries* **24**: 862–874.

46. Williams, D.D. (2003). The brackishwater hyporheic zone: invertebrate community structure across a novel ecotone. *Hydrobiologia* **510**: 153–173.
47. Thornton, K.W. and Sauer, J.R. (1972). Physiological effects of NaCl on *Chironomus attenuatus* (Diptera: Chironomidae). *Ann. Entomol. Soc. Am.* **65**(4): 872–875.
48. Sutcliffe, D.W. (1961). Studies on salt and water balance in caddis larvae (Trichoptera): I. Osmotic and ionic regulation of body fluids in *Limnephilus affinis* Curtis. *J. Exp. Biol.* **38**: 501–519.
49. Sutcliffe, D.W. (1961). Studies on salt and water balance in caddis larvae (Trichoptera): II. Osmotic and ionic regulation of body fluids in *Limnephilus stigma* Curtis and *Anabolia nervosa* Leach. *J. Exp. Biol.* **38**: 521–530.
50. Dowden, B.F. and Bennett, H.J. (1965). Toxicity of selected chemicals to certain animals. *J. Water Pollut. Control Fed.* **37**: 1308–1316.
51. Hamilton, R.W., Buttner, J.K., and Brunetti, R.G. (1975). Lethal levels of sodium chloride and potassium chloride for an oligochaete, a chironomid midge, and a caddisfly of Lake Michigan. *Environ. Entomol.* **4**(6): 1003–1006.
52. Kersey, K. (1981). *Laboratory and Field Studies on the Effects of Road De-icing Salt on Stream Invertebrates*. University of Toronto, Institute for Environmental Studies, Snow and Ice Control Working Group, Toronto, Canada, SIC-9.
53. Chadwick, M.A.J. (1997). *Influences of Seasonal Salinity and Temperature on Hexagenia limbata (Serville) (Ephemeroptera: Ephemeridae) in the Mobile River*. Masters Thesis. Auburn University, Auburn, AL.
54. Goetsch, P.-A. and Palmer, C.G. (1997). Salinity tolerances of selected macroinvertebrates of the Sabie River, Kruger National Park, South Africa. *Arch. Environ. Contam. Toxicol.* **32**: 32–41.
55. Kundman, J.M. (1998). *The Effects of Road Salt Runoff (NaCl) on Caddisfly (Hydropsyche betteni) Drift in Mill Run, Meadville, Pennsylvania*. Senior Thesis. Department of Environmental Science, Allegheny College, Meadville, PA.
56. Chadwick, M.A., Hunter, H., Feminella, J.W., and Henry, R.P. (2002). Salt and water balance in *Hexagenia limbata* (Ephemeroptera: Ephemeridae) when exposed to brackish water. *Florida Entomol.* **85**(4): 650–651.

REVIEW OF RIVER WATER QUALITY MODELING SOFTWARE TOOLS

ANNE NG
B.J.C. PERERA
Swinburne University of Technology
Hawthorne, Victoria, Australia

INTRODUCTION

Rivers supply valuable water resources for humans, and many aquatic ecosystems. However, due to population increase and its adverse effects on the rivers, and other adverse activities, the water quality in rivers has generally declined. Therefore, appropriate river water quality management strategies aimed at controlling and improving water quality should be seriously considered. To manage river water quality in the most effective and efficient way, the cause and effect relationships of the river system must first be investigated. River water quality modeling tools are extensively used in water quality management to identify these cause and effect relationships.

River water quality modeling software is designed to model the water quality in a river system. Many generic water quality modeling software tools are widely available, and most of them are in the public domain and available at no cost. The applicability of these software tools depends on the study objectives. Therefore, it is necessary to review available water quality software modeling tools, so that the most appropriate software tool can be selected for the specific application.

Here, we mainly concentrate on reviewing available public domain river water quality modeling software tools, although a brief review of catchment water quality modeling software tools is also presented. A case study is also included to identify and select the most suitable software for modeling of water quality in the Yarra River, Australia.

WATER QUALITY MODELING SOFTWARE TOOLS

In general, the water quality modeling software tools can be categorized into three broad groups, namely, catchment, river, and integrated software tools. Under these groups, the available public domain software is shown in Fig. 1. Since the focus here is on river water quality software tools, they are further divided into two groups, namely, steady and unsteady state modeling software tools.

Catchment Water Quality Modeling Softwares

Catchment water quality modeling software tools are used to estimate the amount of pollutant loadings generated from different land surfaces in catchments, which affects the water quality in streams and rivers. A listing of commonly used catchment water quality modeling software is shown in Fig. 1.

The most commonly used catchment water quality modeling software is the Agricultural Nonpoint Source Pollution Software (AGNPS) (1), which was developed by the United States Agricultural Research Services. AGNPS can be used in both event and continuous simulation modes, to estimate sediment and nutrient loads from agricultural areas. The catchment is divided into a number of cells to determine pollutant loadings. Tim et al. (2) used AGNPS (linked with a GIS system) to examine the effect of varying widths of vegetated buffer strips on sediment yield of the bluegrass catchment in Iowa, USA. AGNPS was also used to predict pollutants generated from site-specific catchment characteristics in Missouri, USA (3)

Storm Water Management Model (SWMM) is an urban stormwater quantity and quality software tool that was developed by the United States Environmental Protection Agency (U.S. EPA) (4), which can be used in both event and continuous simulation modes. Data required by SWMM are relatively intensive. It has the most versatile hydrological and hydraulic simulation modules, while the water quality simulation is relatively weak

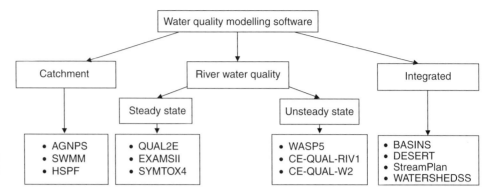

Figure 1. Summary of available public domain water quality software.

in representation of the true physical, biological, and chemical processes (5).

The Hydrological Simulation Program–FORTRAN (HSPF) is one of the comprehensive software tools, which simulates the catchment runoff processes together with a minor component on river water quality (6). This software tool can be used to simulate nonpoint source runoff. Major conventional water quality constituents such as dissolved oxygen (DO), biochemical oxygen demand (BOD), and all forms of nitrogen and phosphorus can be simulated. As discussed in References 5 and 7, HSPF is a highly complex software tool, which requires extensive resources and data. Moore et al. (8) used HSPF on the North Reelfoot Creek catchment, located in the northwest corner of Tennessee, USA to examine several best management practices in reducing erosion and sedimentation. Runoff and sediment load in the Upper Changjiang River Basin in China was simulated using HSPF (9). They found that HSPF underestimated the suspended solid concentration by up to 72% of the actual. The error in the model may have resulted from lack of data to produce a well-calibrated model.

River Water Quality Modeling Software Tools

River water modeling software tools are used to simulate the effects of pollutants generated from catchment and point sources (stormwater/treated sewage discharge outfall), on river and stream water quality. They do not estimate the nonpoint source pollutant load from the catchments. This category is reviewed in detail, since the focus is on review of river water quality modeling software tools. Two types of software tools exist, namely, steady and unsteady state.

Steady-State Software Tools. The steady-state river water quality modeling software assumes that the magnitude of flow and pollutant entering the stream do not vary with time (10). Therefore, in these software, the "average" inputs of flows and pollutants are considered for the flow event, giving "average" values for output water quality concentrations. Although these software tools cannot assess the water quality for time varying conditions, they can be useful in determining the critical water quality concentrations under design conditions. The results obtained from steady-state software are always more conservative than the results obtained with the unsteady-state software (11). The steady-state software tools are commonly used because they are less complex, easy to use, and require less input data. Below is a review of some commonly used public domain steady-state river water quality modeling software tools and their applications.

The Enhanced Stream Water Quality Model (QUAL2E) is a one-dimensional steady-state river water quality simulation software tool, which was developed and is supported by U.S. EPA (12). In 1995, QUAL2E was upgraded with a Windows interface to enhance its user-friendliness. Although QUAL2E is a steady-state flow software tool, it can also account for diurnal variation (difference in temperature between the warmest and coolest parts of the day) in temperature or algal photosynthesis and respiration. QUAL2E can model all conventional water quality constituents, as well as three other user-defined water quality constituents. QUAL2E can be applied to different waterbody types and allows modeling of multiple waste discharges and diversions. It also includes components that allow implementation of uncertainty analysis of model parameters using first-order error analysis, one-at-a-time, and Monte Carlo simulation (see UNCERTAINTY ANALYSIS IN WATERSHED MODELING in this encyclopedia).

QUAL2E has been extensively used in many applications. The majority of these applications used this software tool to simulate major conventional water quality constituents in rivers. Some recent applications on modeling conventional water quality constituents in rivers include the work of Ghosh and McBean (13), Cvitanic and Kompare (14), and Ning et al. (15). In comparison, limited applications were cited in the literature using QUAL2E to simulate microbial water quality (16).

Ghosh and McBean (13) used QUAL2E to develop a water quality model of the Kali River in India. Although they experienced data limitations in their application, they commented that provided adequate data were available, QUAL2E can be effectively used to model water quality in rivers that have water pollution problems. This is consistent with the findings of Barnwell et al. (17), who stated that it is important to have site-specific data representing the properties of the actual system to model river water quality successfully.

Cvitanic and Kompare (14) applied QUAL2E to simulate and predict the possible changes in water quality in the River Sava in Croatia with the construction of impoundments. However, they found that QUAL2E was not suitable for their application, because the model prediction could not be validated during the validation stage. They concluded that a two-dimensional model is more suitable to predict water quality in impoundments, since large variations of river water quality exist in the river (laterally and vertically) during summer periods. They could have simply avoided this problem by selecting the most suitable software tool for their study, since they had a good prior knowledge of lateral and vertical water quality variations of the river.

Ning et al. (15) developed and calibrated a model for the Kao-Ping River Basin in Taiwan using QUAL2E. They successfully used the model as a simulation tool to assess the water quality standard requirements downstream by hypothetically eliminating pig farming activities and constructing a sewer system in the upstream areas. Although QUAL2E has been successfully used as a simulation tool in this study, they commented that an economic instrument for controlling and reducing the wasteload allocations would also be needed in the long term.

Steynberg et al. (16) used QUAL2E in an effort to simulate the effect of various management strategies, which can result in the desired level of fecal coliform in the Rietspruit catchment in South Africa. It was found that the model was unsuccessful in predicting the measured fecal coliform level. Due to the time varying discharge of effluent and different levels of wastewater treatment, the use of the steady-state model QUAL2E was inadequate for their study.

Reviewing the river water quality modeling software, Shanahan et al. (18) reported that QUAL2E has become the standard modeling software tool and has shown to be most applicable in situations where point source pollutants are dominant. Therefore, QUAL2E has been integrated and linked into a number of other modeling software tools. For example, QUAL2E has been integrated into decision support systems (DSSs), as in BASIN (5). Mulligan and Brown (19) and Ng (20) linked QUAL2E with genetic algorithm optimization software, GENESIS (21). As stated in Yang et al. (22), the use of remotely sensed water quality data with QUAL2E in the Te-Chi Reservoir in Taiwan can accurately interpret the spatial variation in water quality and monitor the water quality. De Azevedo et al. (23) linked QUAL2E with a water quantity network flow allocation model MODSIM of Labadie (24) to assess and evaluate six management alternatives for strategic river basin planning.

Exposure Analysis Modelling System (EXAMSII) (25) can be used in modeling of streams, rivers, and reservoirs in one-, two-, and three-dimensional modes. It accounts for many water quality transformation processes, such as photolysis, hydrolysis, oxidation, and sorption with sediments and biota (5).

SYMTOX4, the Simplified Method Program–Variable Complexity Stream Toxics Model (26), is a one-dimensional software tool that can be used to simulate the water column and benthic toxicity caused by point sources discharged into rivers. Major conventional water quality constituents can be simulated using SMYTOX4. This software is Windows based and has the capability to perform uncertainty and sensitivity analyses. One example was cited in Reference 5 using SMPTOX 4 on the Flint River, Michigan, in the United States.

Unsteady-State Software Tools. Unsteady-state (or dynamic) water quality software tools can be used to simulate water quality response in rivers whose flow and water quality characteristics change with time. All natural rivers and streams have unsteady-state flow characteristics, especially during high flow period, and therefore unsteady-state modeling software tools are more realistic. However, they require more data inputs compared to steady-state software tools. Below is a review of the available public domain unsteady-state water quality software tools (which are also listed in Fig. 1), together with their applications.

Water Quality Analysis Simulation Program (WASP5) is a well-known unsteady-state water quality simulation software tool supported by U.S. EPA (27). This software has flexible compartments such as hydrodynamics, eutrophication (DO/nutrients/algal/carbon), and toxins. The user can use these compartments selectively or all compartments simultaneously. This software tool can be used to model rivers and streams in one-, two-, and three-dimensional modes.

Lung and Larson (28) successfully used WASP5 to predict the impact of eutrophication under steady state in the Upper Mississippi River and Lake Pepin in the United States. They recognized that the unsteady-state mode should be used to study algal growth; however, relevant data inputs for algal growth dynamics were not available to them for unsteady-state flow modeling. They justified the use of steady-state mode after ascertaining that the phytoplankton population did not vary greatly from hour to hour.

Suárez et al. (29) developed an unsteady-state water quality model of the Nalón River in Spain using WASP5. This model was used to assess the impact of combined sewer flow (with daily fluctuations) on river quality and its effect on the aquatic system. The majority of the water quality related inputs required in WASP5 (e.g., reaeration rate methods, decay rates) were obtained from preliminary water quality modeling using QUAL2E. The WASP5 model was successfully calibrated, which adequately simulated the water quality and activities occurring in the river accounting for time variation. However, the decay rates were required to be the same for all reaches of the river in WASP5, which they found to be one of the main deficiencies.

A eutrophication model was developed for the Tolo Harbour in Hong Kong using WASP5 by Lee and Arega (30). This model accounts for sediment water interaction together with time and spatial variation in water quality. They successfully simulated DO and chlorophyll-a concentrations and matched them with observations. The model was developed to study the long-term trends of eutrophication in the harbor.

As reported by the World Bank (31), WASP5 is not appropriate for basins with large catchment areas, since it is complex and time consuming to calibrate and simulate water quality conditions of rivers and streams associated with these large basins.

The Hydrodynamic and Water Quality Model for Streams (CE-QUAL-RIV1) is a one-dimensional software tool developed by The U.S. Army Experiment Waterways Experiment Station (32). It has two separate compartments: hydrodynamics and water quality. The results obtained from the hydrodynamics compartment are used as input to the water quality compartment. Many conventional water quality constituents can be modeled, including the effects of algae and macrophytes. One advantage of using this model is that it allows modeling of river structures such as dams. This model is less widely used compared to QUAL2E and WASP5 (33). CE-QUAL-RIV1 has been applied to the Cumberland River, the Chattahoochee River, and the lower Ohio River in the United States (5).

CE-QUAL-W2, also developed by The U.S. Army Waterways Experiment Station (32), is a two-dimensional experiment, laterally averaged, hydrodynamic and water quality model. It contains one module, which models both hydrodynamics and water quality. It can model DO, nutrients, and algae interactions. Since this software accounts for variations in longitudinal and vertical directions (not in lateral direction), this software is best used in situations where large variations in lateral velocities and water quality concentrations do not occur (5). Martin (34) used CE-QUAL-W2 for DeGray Lake in Arkansas (USA) and demonstrated its usefulness.

Integrated Water Quality Modeling Software Tools. The integrated software tools consist of several stand-alone tools in one package. For example, catchment and river modeling software tools can be integrated into one package to analyze both flow and water quality in rivers and associated catchments. When some form of decision support is available in integrated software, they are called decision support systems (DSSs). There is an increased use of DSSs in river water quality management in recent times. The purpose of a DSS is to effectively allow decision-makers to simulate the whole process of decision-making, related to the particular application (e.g., improving river water quality), to investigate and simulate alternative decision management scenarios, and to improve the effectiveness of decision-making. Below are descriptions of four public domain integrated water quality software tools (or DSSs) found in the literature.

Better Assessment Science Integrating Point and Nonpoint Sources (BASINS), developed by the U.S. EPA Office of Water (35), consists of a catchment water quality modeling software tool (NPSM) and a river water quality modeling software tool (QUAL2E). The Nonpoint Source Model (NPSM) is a Windows interface that works with the catchment model HSPF (6). The graphical system in BASINS uses Arc-View GIS software. One disadvantage of this system is that the data management module is less useful to countries other than the United States, since all relevant information and data are only applicable for basins in the United States, which are updated annually.

Decision Support System for Evaluating River Basin Strategies (DESERT) is a flexible, Microsoft Windows-based tool for decision support for water quality management at the catchment scale. DESERT was developed by two organizations jointly: International Institute for Applied Systems Analysis in Austria (IIASA) and the Institute for Water and Environmental Problems in Russia (36). This software tool provides a powerful instrument for developing least-cost river catchment policies, and for assessing these policies under conditions that are deviating from the design scenario (37). Fan (38) used DESERT to identify the most efficient water quality management strategy in terms of wastewater treatment alternatives for the Veszprémi-Séd River in Budapest, Hungary.

StreamPlan (Spreadsheet Tool for River Environmental Assessment Management and Planning) was developed at IIASA in 1996 (39). It is a DSS that allows decision-makers to evaluate river and catchment water quality policies considering local and regional water quality goals, effluent standards, costs, financing, economic instruments, municipal water management issues, and generation of wastewater treatment plant alternatives. StreamPlan was developed for use on a Microsoft Excel platform, which is familiar to most model users. Jolma et al. (40) discussed the use of StreamPlan in three degraded river catchments in Central and Eastern Europe: Narew (Poland), Morava (The Czech Republic), and Nitra (Slovak Republic).

Water, Soil and Hydro-Environmental Decision Support System (WATERSHEDSS) is similar to BASINS, except WATERSHEDSS is more focused on nonpoint source pollution (41). The U.S. EPA Office of Research developed this system in 1994, with the cooperation of North Carolina State University water quality group and the Department of Biological Engineering of Pennsylvania State University (42).

EVALUATION AND SELECTION OF WATER QUALITY SOFTWARE FOR THE YARRA RIVER, AUSTRALIA

To manage river quality in the most effective and efficient way, the cause and effect relationships of the river system must first be identified. A river water quality modeling tool is required for the Yarra River, Australia, for this purpose mainly to identify the cause and effect relationships in the river from different settings of effluent license limits from sewage treatment plants (STPs). Furthermore, this modeling should be able to simulate and study the effects of various "what-if" management strategies prior to implementation.

The catchment water quality software tools mainly deal with the generation and transport of overland pollution and do not directly consider river water quality. Therefore, they are not suitable for modeling Yarra River water quality, although quantifying overland pollutant runoff into streams is an important component in the modeling of river water quality. However, modeling of overland pollutant runoff is outside the scope of modeling in-stream water quality in the Yarra River. The river water quality software tools deal with river water quality and therefore are relevant to modeling water quality in the Yarra River. Although the integrated water quality modeling software (or DSS) tools are very efficient simulation and management systems as complete decision-making tools, they require extensive data, which were not available for the Yarra River at the time of selection of the appropriate modeling tools and, therefore, were not considered. Thus, the river water quality software tools were further investigated for modeling the Yarra River, its tributaries, and associated STPs.

Four criteria, as listed below, were used in selecting the river water quality software for use in the Yarra River, from the computer software tools listed in Fig. 1.

- Data availability for use in the modeling software and the purpose of using the model
- Ability to simulate major conventional water quality constituents such as DO, BOD, and nutrients
- Ability to produce longitudinal profiles (upstream/ downstream) of water quality concentrations
- Wider successful usage of the software

These broader criteria will certainly reduce the number of river water quality software tools that can be used for the Yarra River. QUAL2E and WASP5 were the only two software tools that fitted the above criteria and therefore can be used for development of the Yarra River water quality model. Both software tools can simulate conventional water quality constituents, can produce longitudinal water quality profiles, and have been used successfully on many applications.

As stated in References 5 and 18, both QUAL2E and WASP5 software tools are well known and credible, with extensive capabilities and wide usage. These software tools were further evaluated for modeling the Yarra River. A summary of the evaluation results is given in Table 1. Three main categories, namely, fundamentals, water quality, and others, were considered in classifying the attributes of these software tools. Some of these attributes were considered in a comparative study by Ambrose et al. (27).

The first category consists of attributes, which forms the basic structural framework of the software tools. Most attributes are self-explanatory. Hydrodynamics is an extremely important attribute in water quality modeling, because the movement of water affects the fate of the water quality constituents. WASP5 has an independent compartment for simulating hydrology of the water body system, whereas QUAL2E requires hydrology (or flows)

Table 1. Evaluation Summary of QUAL2E and WASP5 Attributes

Fundamentals		QUAL2E	WASP5
Operational requirements	Documentation	Y	Y
	Support	Y	Y
	Credibility	Y	Y
Water body type	River, stream	Y	Y
	Estuary	N	N
	Lake	N	Y
	Reservoir	N	Y
Dimension	one	Y	Y
	Two	N	Y
	Three	N	Y
Transport	Advection	Y	Y
	Dispersion	Y	Y
Hydrodynamics	Input	Y	Y
	Simulated	N	Y
Steady/unsteady	Steady state	Y	Y
	Unsteady state	N	Y
Discretization	—	Y	Y
Hydraulic structures	—	Y	N
Water quality constituents			
DO	Reaeration/(built-in equations)	Y/Y	Y/Y
	CBOD	Y	Y
	NH_3	Y	Y
	NO_2	Y	Y
	SOD	Y	Y
	Algae	Y	Y
Nitrogen forms	Org-N	Y	Y
	NH_3	Y	Y
	NO_2	Y	N
	NO_3	Y	Y
	Algae	N	Y
Phosphorus forms	Org-P	Y	Y
	Diss–P	Y	Y
	Algae	Y	Y
Temperature		Y	N
Settling/benthos		Y/Y	Y/Y
Toxicity		N	Y
Others			
Uncertainty and sensitivity analysis		Y	N

as input. Both software can be operated in a steady-state environment, which is the most common water quality modeling application, although WASP5 can also be used in an unsteady-state environment. A one-dimensional longitudinal process can be modeled with both software tools and is considered as the dominant transport process in most river systems, since the water quality process in the river is considered well mixed in both lateral and vertical directions (10). WASP5 can simulate river water quality in two and three dimensions. The possible increase in DO concentration by water quality structures such as dams and weirs can only be considered in QUAL2E.

The second category is related to water quality constituents. Both software tools account for most sinks on DO processes. Built-in reaeration formulas are available in both QUAL2E and WASP5. QUAL2E accounts for the four forms of nitrogen in the nitrogen cycle: Org-N, NH_3, NO_2, and NO_3 (10). WASP5 combines NO_2 and NO_3, in the overall nitrification process from NH_3 to NO_3. However, as stated in Reference 7, lumping of NO_2 and NO_3 does not cause any significant effect in the overall result, since the transformation of NO_2 to NO_3 is rapid. All phosphorus processes including the algae cycle can be accounted for in both software tools. Both software tools have the ability to simulate both settling and benthic activity, which are important for streams with low velocity. QUAL2E is the only software tool that can simulate temperature using the atmospheric heat balance equation.

The third category deals with the additional attributes. QUAL2E provides a built-in uncertainty and sensitivity analysis module, which is useful in determining the sensitivity of input parameters to output water quality. The uncertainty and sensitivity analysis of model parameters is a major component in the overall model development.

Based on the above evaluation, WASP5 is considered to be "over qualified" for the development of the Yarra River water quality model. Steady-state simulation is considered sufficient for this study, because it can be used to determine the critical water quality concentrations under design conditions. This is necessary when the model is used to simulate and study the effect of different effluent license limits on river water quality. Furthermore, only grab sample water quality data of the river and at sewage treatment plants are available for the Yarra River. This data limitation can at best be considered as steady state and only suitable to model water quality in one dimension. QUAL2E is also less complex and provides all the essential elements that are required for modeling Yarra River water quality. These elements include modeling of interaction of conventional water quality constituents and the built-in uncertainty and sensitivity analysis. The use of QUAL2E is also supported by a large number of applications in river water quality modeling (15,19).

CONCLUSION

Management of river water quality has become increasingly important due to a decline in water quality caused by human activities. Successful implementation of efficient management strategies requires the development of river water quality models. These water quality models can be used to simulate and assess the cause and effect relationships of river water quality and then to study various management strategies to improve water quality, before their implementation.

Many water quality software tools are available in the public domain, which can be used to develop river water quality models. They were reviewed here. After a detailed evaluation of the river water quality modeling tools, QUAL2E was considered as the most suitable tool for use in the Yarra River model development. The major purpose of the development of the Yarra River water quality model is to assess the effect of various sewage treatment plants (STPs) effluent license limits on river water quality and this can be adequately done with QUAL2E software. The major evaluation criteria were the appropriateness of available data for use in the modeling software and the purpose of using the model.

BIBLIOGRAPHY

1. Bingner, R., Theurer, F., Cronshey, R., and Darden, R. (2001). AGNPS 2001 Web Site. Available at http://www.sedlab.olemiss.edu/agnps.html.
2. Tim, U., Jolly, R., and Liao, H. (1995). Impact of landscape feature and feature placement on agricultural non-point source pollution control. *ASCE J. Water Resour. Plan Manage.* **121**(6): 463–470.
3. Trauth, K. and Adams, S. (2004). Watershed-based modeling with AGNPS for storm water management. *ASCE J. Water Resour. Plan. Manage.* **130**(3): 206–214.
4. Huber, W.C. and Dickinson, R. (1988). *Storm Water Management Model User's Manual, Version 4*, EPA/600/3-88/001a (NTIS PB88-236641/AS). Environmental Protection Agency, Athens, GA. Available at http://ccee.oregonstate.edu/swmm/.
5. U.S. EPA (1997). *Compendium of Tools for Watershed Assessment and TMDL Development*, Report No. EPA/841/B/97/006. Office of Water, Washington, DC.
6. Bicknell, B., Imhoff, J., Kittle, J., Donigian, A., and Johnson, R. (1997). *Hydrological Simulation Program–FORTRAN (HSPF): User's Manual for Release 11.0*, Report No. EPA 600SR97080. National Service Center for Environmental Publication, Cincinnati, OH. Available at http://www.epa.gov/ceampubl/swater/hspf/index.htm.
7. U.S. EPA (1997). *Technical Guidance Manual for Developing Total Maximum Daily Loads—Book 2: Streams and Rivers*, Report No. EPA/823/B/97/002. Office of Water, Washington, DC.
8. Moore, L., Chew, C., Smith, R., and Sahoo, S. (1992). Modeling of best management practices on North Reelfoot Creek, Tennessee. *Water Environ. Res.* **64**(3): 241–247.
9. Hayashi, S., Murakami, S., Watanabe, M., and Hua, X. (2004). HSPF simulation of runoff and sediment loads in the Upper Changjiang River Basin, China. *ASCE J. Environ. Eng.* **130**(7): 801–815.
10. Thomann, R. and Mueller, J. (1987). *Principles of Surface Water Quality Modeling and Control*. HarperCollins Publishers, New York.
11. U.S. EPA (1987). *Selection Criteria for Mathematical Models Used in Exposure Assessments*, Report No. EPA/600/8/87/042). Office of Health and Environmental Assessment, Washington, DC.

12. Brown, L.C. and Barnwell, J.O. (1987). *The Enhanced Stream Water Quality Models QUAL2E and QUAL2E - UNCAS: Documentation and User Manual*, Report No. EPA/600/3-87/OC. U.S. EPA, Athens, GA. Available at http://www.epa.gov/docs/QUAL2E_WINDOWS/index.html.
13. Ghosh, N.C. and McBean, E. (1998). Water quality modeling of the Kali River, India. *Water, Air Soil Pollut.* **102**: 91–103.
14. Cvitanic, I. and Kompare, B. (1999). Lessons learned by modelling impounded river water quality with QUAL2E. In: *Water Pollution V—Modelling, Measuring and Prediction*, WIT Press, Greece.
15. Ning, S., Chang, N., Yang, L., Chen, H., and Hsu, H. (2001). Assessing pollution prevention program by QUAL2E simulation analysis for the Kao-Ping River Basin, Taiwan. *J. Environ. Manage.* **61**, 61–76.
16. Steynberg, M.C. et al. (1995). Management of microbial water quality: new perspectives for developing areas. *Water Sci. Technol.* **32**(5–6): 183–191.
17. Barnwell, T., Brown, L., and Whittemore, C. (2004). Importance of field data in stream water quality modeling using QUAL2E-UNCAS. *ASCE J. Environ. Eng.* **130**(6): 643–647.
18. Shanahan, P., Henze, M., Koncsos, L., Rauch, W., Reichert, P., Somlyódy, L., and Vanrolleghem, P. (1998). River water quality modelling: problems of the art. In: *IAWQ 19th Biennial International Conference*, 21–26 June 1998, Vancouver, Canada, pp. 253–260.
19. Mulligan, A.E. and Brown, L.C. (1998). Genetic algorithm for calibrating water quality models. *ASCE J. Environ. Eng.* **124**(3): 202–211.
20. Ng, A.W.M. (2001). *Parameter Optimisation of River Water Quality Models Using Genetic Algorithms [Ph.D. thesis]*. Victoria University, Melbourne, Australia.
21. Grefenstette, J. (1995). *A User's Guide to GENESIS (Version 5.0)*. Available at http://www.cs.purdue.edu/coast/archive/clife/Welcome.html.
22. Yang, M., Merry, C., and Sykes, R. (1999). Integration of water quality modeling, remote sensing, and GIS. *J. Am. Water Resour. Assoc.* **35**(2): 253–263.
23. De Azevedo, L. et al. (2000). Integration of water quantity and quality in strategic river basin planning. *ASCE J. Water Resour. Plan. Manage.* **126**(2): 85–97.
24. Labadie, J. (1994). MODSIM-DSS: generalized river basin decision support system and network flow model. Department of Civil Engineering, Colorado State University. Available at http://modsim.engr.colostate.edu/.
25. Burns, L. (2004). *Exposure Analysis Modelling System: User's Guide For EXAMSII Ver 2.98*, Report No. EPA/600/R-00/081. U.S. EPA, Athens, GA. Available at http://www.epa.gov/ceampubl/swater/exams/exams29804.htm.
26. CEAM (1993). SMPTOX3 Simplified Method Program—Variable Complexity Stream Toxics Model Version 2.01. Available at http://www.epa.gov/ceampubl/swater/smptox3/index.htm.
27. Ambrose, R., Wool, T., and Martin, J. (1988). *The Water Quality Analysis Simulation Program, WASP5; Part A: Manual Documentation*, Report No. EPA/600/3-87/039. Environmental Research Laboratory, U.S. EPA, Athens, GA. Available at http://www.epa.gov/ceampubl/swater/wasp/index.htm.
28. Lung, W. and Larson, C. (1995). Water quality modeling of Upper Mississippi River and Lake Pepin. *ASCE J. Environ. Eng.* **121**(10): 691–699.
29. Suárez, J., Ascorbe, A., Liaño, A., Sáinz, J., Temprano, J., and Tejero, I. (1995). Dynamic simulation of water quality in rivers. WASP5 application to the River Nalón (Spain). In: Wrobel, L. and Latinopoulos, P. (Eds.), *Water Pollution III: Modelling, Measuring and Prediction*, Computational Mechanics Publications, UK, pp. 178–188.
30. Lee, J. and Arega, F. (1999). Eutrophication dynamics of Tolo Harbor, Hong Kong. *Mar. Pollut. Bull.* **39**(1–12): 187–192.
31. World Bank. (1997). *Pollution Prevention and Abatement Handbook, Part II, Water Quality Models*. Available at http://www.esd.worldbank.org/pph/home.html.
32. USACE (1990). *CE-QUAL-RIV1: A Dynamic One-Dimensional (Longitudinal) Water Quality Model for Streams*, Instruction Report E-90-1. U.S. Army Engineer Waterway Experiment Station, Vicksburg, MI. Available at http://www.wes.army.mil/el/elmodels/index.html#wqmodels.
33. Gore, J.A. and Petts, G.E. (1989). *Alternatives in Regulated River Management*. CRC Press, Boca Raton, FL.
34. Martin, J. (1988). Application of two-dimensional water quality model. *ASCE J. Environ. Eng.* **114**(2): 317–336.
35. U.S. EPA (2001). Better Assessment Science Integrating point and Nonpoint Sources (BASINS). Available at http://www.epa.gov/waterscience/basins/basinsv3.htm.
36. Ivanov, P., Masliev, I., De Marchi, C., Kularathna, M., and Somlyódy, L. (1996). *DESERT User's Manual: Decision Support System for Evaluating River Basin Strategies*. International Institute for Applied System Analysis, Austria. Available at http://www.iiasa.ac.at/Research/WAT/docs/getit.html.
37. Perera, B.J.C. and Somlyódy, L. (1998). Decision support systems for river water quality management. In: *Second International Conference on Environmental Management*, February 10–13, 1998, Wollongong, Australia, pp. 677–684.
38. Fan, D. (1996). *Water Quality Management of the Veszprémi-Séd River and Malom and Nádor Channel System in Hungry*. International Institute for Applied System Analysis, Austria.
39. De Marchi, C., Jolma, A., Masliev, I., Perera, B.J.C., Smith, M.G., and Somlyódy, L. (1996). *StreamPlan User's Manual: Spreadsheet Tool for River Environment Assessment, Management and Planning*. International Institute for Applied System Analysis, Austria. Available at http://www.water.hut.fi/wr/research/StreamPlan/.
40. Jolma, A. et al. (1997). StreamPlan: a support system for water–quality management on a river basin scale. *Environ. Software Model.* **12**(4): 275–284.
41. Osmond, D. et al. (1997). WATERSHEDSS: a decision support system for watershed-scale nonpoint source water quality problems. *J. Am. Water Resour. Assoc.* **33**(2): 327–341.
42. NCSU (1994). WATERSHEDSS—a decision support system for nonpoint source pollution control. Available at http://www.water.ncsu.edu/watershedss/.

RIVER WATER QUALITY CALIBRATION

ANNE NG
B.J.C. PERERA
Swinburne University of Technology
Hawthorne, Victoria, Australia

CALIBRATION OF RIVER WATER QUALITY MODELS

Introduction

River water quality models play an important role in water quality management. Through these models, the cause and effect relationships and the river assimilative capacity can be determined so that the appropriate strategies can be implemented in managing river water quality.

The process of the model development involves data collection and model selection, model assembly, calibration, validation, and uncertainty and sensitivity analysis. Once the river water quality model is developed using the appropriate river water quality modeling software (refer to REVIEW OF RIVER WATER QUALITY MODELING SOFTWARE TOOLS in this encyclopedia) and data, it is necessary to calibrate the model before it can be confidently used as a decision-making tool. Model calibration is necessary when the model parameters cannot be measured physically. One such parameter (which cannot be physically measured) is the decay rate of biochemical oxygen demand (BOD), and there are other such parameters in river water quality models. Model calibration is frequently referred to as parameter estimation (1), because the calibration yields the model parameters.

Here, methods and techniques used in model calibration (specifically in water related applications) are reviewed first. Then the overall methodology used in calibrating the river water quality model developed for the Yarra River, Australia, using genetic algorithms is presented.

Model Calibration Methods

Model calibration techniques can broadly be divided into two categories: manual and automatic, as shown in Fig. 1. The manual method is a trial and error method, where the model output due to different parameter sets is compared with observations visually and the parameter set that best matches the model output with observations is selected as the optimum parameter set (2,3). This method is subjective and time consuming. It can also miss the optimum parameter set. It may even lead to unrealistic parameter sets (4,5).

The automatic calibration method provides some measure of objectivity to parameter estimation and generally is conducted through optimization. Parameter optimization is achieved in the automatic calibration method through an objective function, which minimizes or maximizes a user-defined function. Several objective functions have been used in the past to assess modeled and observed responses in mathematical models in determining the optimum parameter set. The minimization of the sum of the squared difference of modeled output (due to different model parameters) and observations residual sum of squares between the actual and modeled values has been commonly used as the objective function in many hydrological studies (5–9). The use of this objective function is known as the least squares method. There are two variations from the least squares method, and they are the simple least squares method (5) and the weighted least squares method (8,9). The difference between these two methods is that the weighted least squares method requires different weights to be attached to each data point in the objective function, whereas in the simple least squares method, the weights are assumed to be equal. As stated by Sorooshian and Gupta (4), the selection of the objective function can be subjective and can produce different results for different objective functions. For example, the optimum model parameters obtained from a catchment model using two different objective functions—one considering peak flows and the other considering runoff volumes—can produce different results.

In parameter optimization, an optimization technique is used to determine the optimum parameter set within a prescribed parameter space. As Fig. 1 shows, the optimization techniques can be divided into two broad categories: deterministic and stochastic. The deterministic techniques (also defined as local search methods) determine the optimum parameter set through a systematic search. They are designed to locate the optimum parameter set when the response surface defined by the user-defined function is unimodal (i.e., a single peak/trough). However, if the response surface is multimodal, the parameter set obtained from deterministic methods may not produce the global optimum, since the solution can be trapped at a local optimum point. Starting the optimization with different "seeds" (i.e., starting with different parameter sets for the optimization) may alleviate this problem to a certain extent. Sorooshian and Gupta (4) stated that in calibration of hydrologic models, the optimum parameter set is very rarely found through deterministic methods, since the hydrological problems contain multimodal response surfaces. Duan et al. (10) showed that there were more than hundreds of local optimum solutions in their rainfall and runoff model.

Direct and indirect search methods are two deterministic optimization methods. The direct method seeks the optimum by "hopping" around the search space of a predefined grid, where each grid point defines a parameter set, and assessing the objective function at each of these grid points. Basically, these methods use the objective functions of previous two points to determine the next point to be considered in the optimization. Sorooshian and Gupta (4) listed the most common direct search methods used in hydrologic models: Rosenbrock (11), pattern search (12), and Nelder–Mead downhill simplex methods (13). The indirect method (also known as the gradient search method) seeks the optimum solution by defining the next search point, considering both the objective function value and its gradient. Steepest decent, Hessian matrix, and Newton method are examples of indirect methods (4). These three methods have the common feature that they start from a user-defined starting point, but they differ

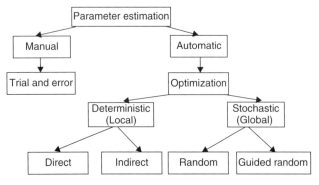

Figure 1. Broad methods in model parameter estimation.

from each other in the direction of moves and the lengths of moves.

As compared to deterministic methods, the stochastic methods are more efficient in locating the optimum parameter set, when the response surface is multimodal. However, they can also be used when the response surface is unimodal. Stochastic methods are also known as global search methods, since they are designed to produce the global optimum parameter set. They can be subdivided into two main categories, namely, random and guided random. The random search method generates the parameter sets randomly from the parameter range and optimizes the parameter sets. Generally, the random search method generates a parameter set from a uniform distribution of parameters. It does not consider the history of previous solutions in terms of optimality to determine the next parameter set, and hence the method can be inefficient. On the other hand, the guided random search method provides guided information for the next search based on the history of previously considered points (14).

Several guided random search methods exist, such as simulated annealing, adaptive random search, shuffled complex algorithms, and evolutionary algorithms (EAs) (10,14). Of these four methods, EAs have recently been used by many researchers successfully and have attracted wide attention from diverse fields, such as different areas of engineering, computer science, operations research, mathematics, and political science. The growing number of these applications is due to their ease of interfacing, simplicity, and extensibility (15).

Evolutionary Algorithms (EAs)

Evolutionary algorithms (EAs) are stochastic optimization methods that utilize the natural process of evolution (16). It has been demonstrated that EAs are robust search techniques that outperform the traditional optimization methods in many applications, in particular, when the response surface is discontinuous, noisy, nondifferentiable, and multimodal (9,17,18). Bäck and Schwefel (19) stated that EAs have become common and successful techniques in model parameter optimization and have been successfully used in model parameter optimization by Mulligan and Brown (9) and Seibert et al. (20).

There are three main forms of EAs: evolutionary programming (21), evolutionary strategies (22), and genetic algorithms (GAs) (17). Apart from these three forms of EAs, there are two other forms that were originally derived from GAs: classifier system (17) and genetic programming (23). These five forms share the common conceptual principle of EAs. That is, they repeatedly apply sequential evolutionary operators that simulate the evolution of parameter sets from the search space. These evolutionary operators are parameter representation, parameter initialization, selection, crossover, and mutation to yield offsprings (or new parameter sets) for the next generation. Depending on the type of EAs, these evolutionary operators are applied simultaneously or selectively. Of these five forms, GAs have proved to provide robust search in complex parameter search spaces. It is also the only form of EAs that utilizes all above five evolutionary operators (i.e., parameter representation, population initialization, selection, crossover, and mutation) (24).

Genetic Algorithms (GAs)

Genetic algorithms (GAs) are the most prominent and powerful optimization techniques that have been applied successfully recently in many disciplines (25). They are robust search techniques that are based on concepts of natural selection and genetics. For this reason, the terminology used in GAs is borrowed from natural genetics. The reader is referred to Goldberg (17) for details of GAs. The overall GA process as applicable to parameter optimization of river water quality models (or any mathematical model) is described below.

Mathematical models have their own model parameters. According to the genetics terminology, each model parameter is a gene, while a complete set of model parameters is a chromosome. The process of a GA begins with an initial population of a number of model parameter sets, which are chosen at random or are heuristic (requires some prior knowledge of the likely "optimum" parameter set) within a specified parameter range. This is the first generation of a number of generations (generally with a constant population size for all generations) in a GA run. Each model parameter set is then evaluated via an objective function to yield its fitness value.

The second and subsequent populations (or generations) are generated by combining model parameter sets with high fitness values from the previous population (i.e., parent) through selection, crossover, and mutation operations to produce successively fitter model parameter sets (i.e., offspring). The selection GA operator favors those parent parameter sets with high fitness value over those of lower fitness value in producing offspring. The crossover operator exchanges model parameter values from two selected model parameter sets. The mutation operator adds variability to randomly selected model parameter sets by altering some of the values randomly. Several generations are considered in one GA run, until no further improvement (within a certain tolerance) is achieved in the objective function.

The proper selection of GA operators is important for efficient optimization of model parameters. Franchini and Geleati (26) and Ng and Perera (27) investigated the effect of these GA operators on model parameters and found that they were insignificant in achieving the optimum parameter set provided that the widely used GA operators methods and values are used. Wardlaw and Sharif (28), on the other hand, found that different GA operators could produce different optimal solutions.

CALIBRATION OF RIVER WATER QUALITY MODEL USING GENETIC ALGORITHMS

River Model Development

The procedure of calibrating a river water quality model is discussed in this section. The case study discussed is of Yarra River in Australia. No attempt is made to discuss technical results; rather, the overall methodology of calibration using GAs is presented to give readers an overview of the calibration procedure. GAs were selected because they have proved to provide a robust search in complex parameter search space. Interested readers can refer to Ng (29) for complete details of the calibration and the results of the case study.

The QUAL2E (30) river water quality modeling software tool was selected and used to develop the Yarra River Water Quality Model (YRWQM) (refer to REVIEW OF RIVER WATER QUALITY MODELING SOFTWARE TOOLS in this encyclopedia). A precalibration uncertainty/sensitivity analysis was first undertaken to identify the sensitive and insensitive model parameters. This step was necessary because more effort could then be given to calibration of sensitive parameters during the model calibration. Monte Carlo simulation (MCS) was used for precalibration uncertainty/sensitivity analysis, since it has the ability to consider many different input parameter sets sampled from their distributions and to analyze the output response probability distributions. Interested readers can refer to PARAMETER UNCERTAINTY AND SENSITIVITY ANALYSIS IN RIVER WATER QUALITY MODELS (in this encyclopedia) for a review of different methods used for uncertainty/sensitivity analysis of model parameters. The precalibration uncertainty/sensitivity analysis of YRWQM model parameters and results were discussed in detail in Ng and Perera (31).

Linking of YRWQM and GA Software

A public domain GA software, namely, GENESIS (32), was used in calibration of YRWQM. GENESIS was selected for this study, since it has been used successfully for different applications (9,33,34) in the past. However, it was necessary to link YRWQM and GENESIS in order to perform the model calibration, since the calibration requires several simulation runs of YRWQM with model parameters generated from GENESIS. Linking was done through input and output files of YRWQM and GENESIS for each calibration event.

Procedure in River Water Quality Calibration

Several flow events were available for modeling of Yarra River water quality. Each event had information on flow in the river and tributaries, emissions from sewage treatment plants (STPs), and water quality measurements. Three flow events were used in calibration (35), and a further three events were used in validation.

Eleven decay rates, which were responsible for nitrogen, phosphorus, and dissolved oxygen, were considered in this calibration (Table 1). This table also shows the influence and relationships of these decay rates on water quality output responses. To use GAs, an appropriate decay rate range is required. These ranges for BOD_d, NH_3, NO_2, $Org-P_d$, $Org-P_d$, BOD_s, $Org-N_s$, and $Org-P_s$ reaction rates were derived by applying the standard first-order reaction equation (36) using field data (35). The decay rate range for SOD, NH_3 benthos, and Diss-P benthos could not be estimated, because of lack of data. Therefore, the range of decay rates for the three parameters were obtained from Bowie et al. (37), who had compiled results from previous studies.

Four water quality output responses, total kjeldahl nitrogen (TKN), total nitrogen (TN), total phosphorus

Table 1. Water Quality Decay Rates Considered in Calibration

Decay Rates	Symbols Used in Text	Influence on Output Responses
Org-N decay	$Org-N_d$	TKN
Org-N settling	$Org-N_s$	TKN
NH_3 decay	NH_{3-d}	TKN
NH_3 benthos	NH_3ben	TKN
NO_2 decay	NO_{2-d}	TN
Org-P decay	$Org-P_d$	TP
Org-P settling	$Org-P_s$	TP
Diss-P benthos	Diss-Pben	TP
CBOD decay	$CBOD_d$	DO
CBOD settling	$CBOD_s$	DO
SOD (sediment oxygen demand)	SOD	DO

(TP), and dissolved oxygen (DO), were used in calibration and compared with respective observations at six water quality monitoring stations located along the Yarra River. These four output responses require the estimation of eleven water quality decay rates (Table 1). There is some interaction between these output responses (i.e., TKN affects TN, and both TN and TP affect DO) as shown in Fig. 2.

The procedure for estimating decay rates was done in a systematic way as shown in Fig. 2 and also stated by McCutheon (38), Wesolowski (39), and U.S. EPA (40). First, the parameters of the water quality constituents that were not affected by other water quality constituents were estimated. Then, these parameters were kept constant, and the parameters of other water quality constituents were estimated as previously. In this study, the objective function based on simple least squares was used, since no information was available on the weights. The simple least squares objective function is given in Equation 1. This equation considers the minimization of squared difference between observed and modeled water quality concentrations at the water quality stations. This squared difference in Equation 1 is known as the fitness in GA. The lower the value, the fitter is the parameter set.

$$\text{Min} \sum_{i=6} (\text{OBS}_i - \text{MOD}_i)^2 \qquad (1)$$

where OBS_i is the observed water quality concentration at water quality station i, and MOD_i is the modeled water quality concentration at water quality station i.

The first set of parameters considered was $Org-N_d$, $Org-N_s$, NH_{3-d}, and NH_3 benthos, which are responsible for TKN. Then, these parameters were kept constant as discussed previously and the second parameter set, NO_{2-d}, was optimized considering the output response of TN. The third set of parameters includes $Org-P_d$, $Org-P_s$, and Diss-P benthos and was optimized using the output response of TP. Optimization of decay rates for phosphorus can be done in parallel with TKN and/or TN, since TKN and TN are not influenced by TP and vice versa. The last set of parameters of $CBOD_d$, $CBOD_s$, and SOD was "optimized" using DO, keeping all other parameters at their optimized values.

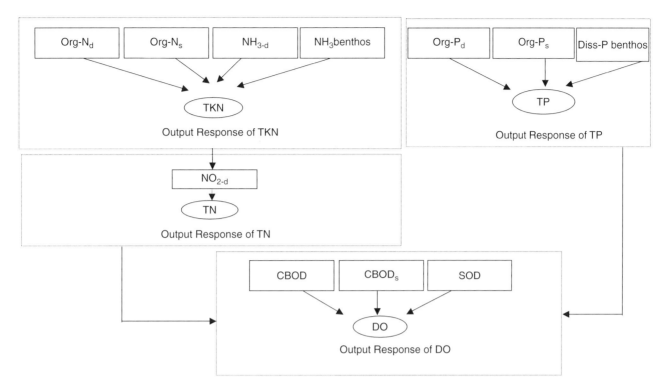

Figure 2. Systematic process used in calibration of YRWQM.

Selection of a Single Optimized Parameter Set

Three flow events used in calibration produced three sets of optimum model parameter sets. Selection of a single optimum parameter set from these sets for use in YRWQM requires subjective judgment. An attempt was made to select the single "optimum" parameter set, which models the observed water quality with reasonable accuracy for all three events. The standard student t-test (41) showed the three optimal decay rate sets (obtained from three events) produced equally good water quality predictions that match the observations at 95% significant level. In other words, the use of the three respective optimal decay sets in YRWQM has produced output water quality prediction that does not show any significant difference between modeled and observed water quality. Therefore, another statistical test, known as cumulative absolute relative error (CARE) cited in Reference 40, was used to quantitatively assess which parameter set had produced the overall lowest error with respect to all three events. The CARE value was determined using the following function:

$$\text{Min} \sum_{i=1}^{3} \sum_{j=1}^{4} \sum_{k=1}^{6} \left(\frac{\text{OBS}_{i,j,k} - \text{MOD}_{i,j,k}}{\text{MOD}_{i,j,k}} \right) \quad (2)$$

where OBS is observed water quality, MOD is modeled water quality, i is event used in the calibration (three events), j is output water quality constituents of TKN, TN, TP, and DO (four constituents), and k is water quality sampling stations (six stations).

Once the single optimum parameter set was obtained, this set was then used in the validation process. Validation is a process that assesses the predictability of the model once it has been calibrated. This was done using three independent events, which were not used in the calibration. The model validation will enhance the confidence in using YRWQM for analysis for various management schemes in improving water quality. The student t-test results also showed that the observed and modeled water quality concentrations were not significantly different from each other at 95% significance level for all three validation events.

CONCLUSION

Simulation models are used to assess various management scenarios in improving river water quality. In order to use these simulation models confidently, the models must be well calibrated. Model calibration (or often referred to as parameter estimation) yields a set of model parameters that best estimate conditions that match with the observations. The calibrated model can then be used to simulate various management scenarios so that the implementation of water quality policy can be done in the most efficient way.

Model calibration can be done using manual and automatic methods. The manual methods use trial and error approaches, which are time consuming and require subjective judgment in defining the optimum parameter set. They can often miss the optimum parameter set. The automatic calibration methods provide some measure of objectivity in calibrating the mathematical models and obtaining the optimum model parameters. Genetic algorithms (GAs) are widely used stochastic search methods, based on the concepts of natural selection and genetics, which have proved to be successful and efficient in identifying the optimum parameter set in many water-related applications.

Due to the proven success of GAs as a calibration method, this method was used here to demonstrate the overall methodology in calibrating a river water quality model using the case study of the Yarra River in Australia. The methodology, including model development and linkage with GAs, the process of water quality calibration, and validation, was discussed.

BIBLIOGRAPHY

1. Beck, M. (1987). Water quality modeling: a review of the analysis of uncertainty. *Water Resour. Res.* **23**(8): 1393–1442.
2. Tsihrintzis, V., Fuentes, H., and Rodriguez, L. (1995). Calibration and verification of QUAL2E water quality model in sub-tropical canals. In: *First International Conference*, Aug. 14–18, 1995, Texas, USA, pp. 214–217.
3. Janssen, P. and Heuberger, P. (1999). Calibration of process-oriented models. *Ecol. Model.* **83**: 55–66.
4. Sorooshian, V. and Gupta, K. (1995). Model calibration. In: Singh, V.P. (Ed.), *Computer Models of Watershed Hydrology*, Water Resources Publications, Littleton, CO, pp. 26–68.
5. Mohan, S. (1997). Parameter estimation of non-linear Muskingum models using genetic algorithm. *ASCE J. Hydraul. Eng.* **123**(2): 137–142.
6. Johnston, P. and Pilgrim, D. (1976). Parameter optimisation for watershed models. *Water Resour. Res.* **12**(3): 477–485.
7. Wang, Q. (1991). The genetic algorithm and its application to calibrating conceptual rainfall-runoff models. *Water Resour. Res.* **27**(9): 2467–2471.
8. Little, K. and Williams, R. (1992). Least squares calibration of QUAL2E. *Water Environ. Res.* **64**(2): 179–185.
9. Mulligan, A.E. and Brown, L.C. (1998). Genetic algorithm for calibrating water quality models. *ASCE J. Environ. Eng.* **124**(3): 202–211.
10. Duan, Q., Sorooshian, S., and Gupta, V. (1992). Effective and efficient global optimization for conceptual rainfall-runoff models. *Water Resour. Res.* **28**(4): 1015–1031.
11. Rosenbrock, H. (1960). An automatic method for finding the greatest or least value of a function. *Comput. J.* **3**: 175–184.
12. Hooke, R. and Jeeves, T. (1961). Direct search solutions of numerical and statistical problems. *J. Assoc. Comput. Machine* **8**(2): 212–229.
13. Nelder, J. and Mead, R. (1965). A simplex method for function minimisation. *Comput. J.* **7**: 308–313.
14. Filho, J., Alippi, C., and Treleaven, P. (1994). Genetic algorithm programming environments. *IEEE Comput. J.* **27**(6): 28–63.
15. Dasgupta, D. and Michalewicz, Z. (1997). Evolutionary algorithms—an overview. In: Dasgupta, D. and Michalewicz, Z. (Eds.), *Evolutionary Algorithms in Engineering Applications*, Springer, New York, pp. 3–28.
16. De Jong, K., Fogel, D., and Schwefel, H. (1997). A history of computation. In: Bäck, T., Fogel, D., and Michalewicz, Z. (Eds.), *Handbook on Evolutionary Computation* (Vol. Part A, Sec. 2.3), Institute of Physics Publishing, Oxford University Press, New York, pp. 1–12.
17. Goldberg, D.E. (1989). *Genetic Algorithms in Search, Optimisation and Machine Learning*. Addison-Wesley Publishing, Reading, MA.
18. Schwefel, H. (1997). Advantages and disadvantages of evolutionary computation over other approaches. In: Bäck, T., Fogel, D., and Michalewicz, Z. (Eds.), *Handbook on Evolutionary Computation* (Vol. Part A, Sec. 1.3), Institute of Physics Publishing, Oxford University Press, New York, pp. 1–2.
19. Bäck, T. and Schwefel, H. (1993). An overview of evolutionary algorithms for parameter optimisation. *Evol. Comput.* **1**(1): 1–23.
20. Seibert, J., Uhlenbrook, S., Leibundgut, C., and Halldin, S. (2000). Multiscale calibration and validation of a conceptual rainfall-runoff Model. *Phys. Chem. Earth* **25**(1): 59–64.
21. Fogel, D.B. (1993). Evolving behaviors in the iterated prisoner's dilemma. *Evol. Comput.* **1**: 77–97.
22. Schwefel, H. (1981). *Numerical Optimisation of Computer Models*. John Wiley & Sons, Chichester, UK.
23. Kinnear, J., Smith, R., and Michalewicz, Z. (1997). Derivative methods. In: Bäck, T., Fogel, D., and Michalewicz, Z. (Eds.), *Handbook on Evolutionary Computation* (Vol. Part B, Sec. 1.5), Institute of Physics Publishing, Oxford University Press, New York, pp. 1–15.
24. Eshelman, L. (1997). Genetic algorithms. In: Bäck, T., Fogel, D., and Michalewicz, Z. (Eds.), *Handbook on Evolutionary Computation* (Vol. Part B, Sec. 1.2), Institute of Physics Publishing, Oxford University Press, New York, pp. 1–11.
25. Paz, E. (1998). A survey of parallel genetic algorithms. Available at http://www.illigal.ge.uiuc.edu.
26. Franchini, M. and Galeati, G. (1997). Comparing several genetic algorithm schemes for the calibration of conceptual rainfall-runoff models. *Hydrol. Sci.* **42**(3): 357–379.
27. Ng, A.W.M. and Perera, B.J.C. (2003). Selection of genetic algorithm operators for water quality model calibration. *Eng. Appl. Artificial Intell. Elsevier* **16**(5–6): 529–541.
28. Wardlaw, R. and Sharif, M. (1999). Evaluation of genetic algorithms for optimal reservoir system operation. *ASCE J. Water Resour. Plan. Manage.* **125**(1): 25–33.
29. Ng, A.W.M. (2001). *Parameter optimisation of river water quality models using genetic algorithms* [Ph.D. thesis]. Victoria University, Melbourne, Australia.
30. Brown, L.C. and Barnwell, J.O. (1987). *The Enhanced Stream Water Quality Models QUAL2E and QUAL2E -UNCAS: Documentation and User Manual*, Report No. EPA/600/3-87/OC. U.S. EPA, Athens, GA.
31. Ng, A.W.M. and Perera, B.J.C. (2001). Uncertainty and sensitivity analysis of river water quality model parameters. In: *First International Conference on Water Resources Management*, Halkidiki, Greece, 24–26 Sep., pp. 175–184.
32. Grefenstette, J. (1995). *A User's Guide to GENESIS (Version 5.0)*. Available at http://www. cs.purdue.edu/coast/archive/clife/Welcome.html.
33. Whitley, D. (1989). The genitor algorithm and selection pressure: why rank-based allocation of reproductive trials is best. In: *Proceedings of the Third Conference on Genetic Algorithms*, June 4–7, 1989, George Mason University, Virginia, pp. 116–121.
34. Liong, S.Y., Chan, W.T., and ShreeRam, J. (1995). Peak-flow forecasting with genetic algorithm and SWMM. *ASCE J. Hydraul. Eng.* **121**(8): 613–617.
35. Ng, A.W.M. and Perera, B.J.C. (2000). *Data analysis and preliminary estimates of parameters for river water quality model development*. In: *Proceedings of the 26th Hydrology and Water Resources Symposium*, Perth, Australia, pp. 773–778.
36. Thomann, R. and Mueller, J. (1987). *Principles of Surface Water Quality Modeling and Control*. HarperCollins Publishers, New York.

37. Bowie, G., Mills, W., Porcella, D., Campbell, C., Pagenkopf, J., Rupp, G., Johnson, K., Chan, P., Gherini, S., and Chamberlin, C. (1985). *Rates, Constants and Kinetic Formulations in Surface Water Quality Modeling*, Report No. EPA/600/3-85/040. U.S. EPA, Athens, GA.

38. McCutcheon, S.C. (1989). *Water Quality Modeling, Volume I: Transport and Surface Exchange in Rivers*. CRC Press, Boca Raton, FL.

39. Wesolowski, E. (1994). *Calibration, Verification, and Use of a Water Quality Model to Simulate Effects of Discharging Treated Wastewater to the Red River of the North at Fargo, North Dakota*, Report No. 94-4058. U.S. Geological Survey, North Dakota.

40. U.S. EPA (1997). *Technical Guidance Manual for Developing Total Maximum Daily Loads—Book 2: Streams and Rivers*, Report No. EPA/823/B/97/002. Office of Water, Washington, DC.

41. Haldar, A. and Mahadevan, S. (2000). *Probability, Reliability and Statistical Methods in Engineering Design*. John Wiley & Sons, Hoboken, NJ, pp. 97–99.

SALMONELLA: MONITORING AND DETECTION IN DRINKING WATER

HEMANT J. PUROHIT
ATYA KAPLEY
National Environmental
Engineering Research Institute,
CSIR
Nehru Marg, Nagpur, India

INTRODUCTION

Providing safe drinking water is one of the challenges faced by most countries due to increasing population and industrialization. The Water Quality Guidelines stated that the most simple waterborne disease risk management involves, among other factors, identifying potential sources of contamination (1). The World Health Organization (WHO) report of March 2001 states that more than three million people die annually from waterborne related diseases. Unsafe drinking water and inadequate sanitation are two of the reasons for this problem. The WHO, conducting a workshop on hazard characterization of pathogens in food and water, launched a program of work with the objective of providing expert advice on risk assessment of microbiological hazards. *Salmonella* has been identified as one of the key pathogens for risk analysis.

Salmonellosis and typhoid fever are the main diseases caused by infection of *Salmonella*. Salmonellosis is a major public health problem because of its large and varied animal reservoir, the existence of human and animal carrier states, and the lack of a concerted nationwide program to control *Salmonella*. However, the epidemiology of typhoid fever and other enteric fevers primarily involves person-to-person spread because these organisms lack a significant animal reservoir. Contamination with human feces is the major mode of spread, and the usual vehicle is contaminated water. Typhoid fever is a public health problem of which there are an estimated 33 million cases, resulting in 500,000 deaths each year worldwide (2).

Different culture media and enrichment methods have been proposed for the isolation of *Salmonella* species from environmental samples (3,4). Conventional methods for testing *Salmonella* include the most probable number (MPN) technique (5). Additionally, selective growth media like tetrathionate broth (TB), tetrathionate broth with brilliant-green (TTBG), bismuth sulfide (BS) agar, *Salmonella*-Shigela (SS) agar, and xylose lysine deoxycholate (XLD) agar are used for enumeration of *Salmonella* (3). However, these organisms are sometimes difficult to detect or enumerate from the natural ecosystems (6). This is due to *Salmonella* often entering into the viable and nonculturable state, when exposed to environmental stresses. The routine surveillance of drinking water relies on the growth and biochemical properties of the microorganisms, which is time consuming (3,4). These drawbacks are overcome with the use of molecular tools that can detect even the viable but nonculturable forms with sensitivity and specificity.

Molecular tools involve the use of gene probes, enzyme linked assays, and the polymerase chain reaction (PCR). Use of molecular tools for detection of *Salmonella* has mainly been demonstrated in food samples. There are not many reports that describe detection methods in water samples. Purohit and Kapley (7) have proposed the use of PCR as an option in microbial quality control of drinking water. It is estimated that there are over 40,000 references to PCR that describe its use in various applications (8). The bottleneck is in getting the PCR as a diagnostic tool from the laboratory to the field, but this needs to be done as a case-specific solution.

METHODOLOGY FOR DETECTION USING MOLECULAR TOOLS

Sampling and Generation of DNA

Methods recommended by the United States Food and Drug Administration require the culture of a sample prior to testing. An aliquot of the overnight grown sample is to be inoculated onto selective media for *Salmonella*. The suspected colonies are identified by a set of biochemical tests and, if required, further confirmed by conventional serology for *Salmonella* detection (9). Use of DNA probes and immunodetection systems also require culturing or pre-enrichment of the test sample for a few hours on selective media prior to testing with antibody or DNA probe. Immunofluorescence methods (10) and enzyme immunosorbent assays (11,12) have been successfully used to enrich *Salmonella* in primary enrichment broths. An electrochemical enzyme-linked immunosorbent assay (ELISA) coupled with flow injection analysis (ELISA-FIA) was used for detecting *Salmonella* in meat after only 5 h of incubation of pre-enrichment broth (13). Meckes and MacDonald (14) demonstrated the use of a commercially available molecular probe system to isolate and enumerate *Salmonella* spp. in sludge in less time than cultural techniques with biochemical confirmation. Nucleic acid sequence-based amplification (NASBA) results demonstrated the detection of *Salmonella* in food samples after 18 h of pre-enrichment at initial inoculum levels of 10^2 and 10^1 CFU per 25 g food sample. The primer and probe set were based on mRNA sequences of the *dnaK* gene of *Salmonella* (15).

Direct detection of *Salmonella* spp. in water samples has also been demonstrated without the use of any pre-enrichment steps using molecular methods (6,16,17). These methods have the advantage of being able to detect *Salmonella* that are in viable but nonculturable form. Bacteria enter into the nonculturable state on prolonged exposure to river or seawater, due to environmental stresses and limiting nutrient conditions (16,18,19). Such bacteria can still work as etiological agents when they come across the suitable host. Knight et al. (6) demonstrated the use of gene probe for detection from 500–1500 mL seawater samples. Kapley et al. (16) have shown the use of multiplex PCR to detect *Salmonella* from river water. In this case 5-liter samples were collected and filtered on-site through a 0.2 μm filter. The filter paper with residue was suspended in phosphate buffered saline and brought to the lab on ice within 2 h for analysis. The filter paper with buffer was thoroughly mixed and the cells were harvested from buffer. The total DNA was extracted using proteinase K treatment. The lysed cell preparation was used directly as template for PCR for detection of *Salmonella*.

PCR Based Monitoring

The polymerase chain reaction (PCR) is an established molecular technique, which reliably identifies a segment of DNA; and it uses a set of specific subsequences to amplify target segment DNA from a mixed template population. There are some loci reported for amplification target *Salmonella* along with other members of the Enterobacteriaceae family to detect the main waterborne pathogens but some are highly specific to *Salmonella*. Way et al. (20) have used multiplex PCR to detect *Salmonella* and other coliform bacteria. *PhoP* primers, specific to the phoP/phoQ loci of coliform pathogenic bacteria such as *Salmonella*, *Shigella*, *Escherichia coli*, and *Citrobacter* species served as presumptive indicators of enteric bacteria. In addition to the *phoP* primers, the *Hin* primers, 5′CTAGTGCAAATTGTGACCGCA 3′ and 5′ CCCCATCGCGCTACTGGTATC 3′, targeting a 236-bp region of hin/H2 and the *H-1i* primers, 5′ AGCCTCGGC-TACTGGTCTTC 3′ and 5′ CCGCAGCAAGAGTCACCTCA 3′, amplifying a 173-bp region of the H-1i flagellin gene, were used. Both *Hin* and *H-1i* primers are specific to motile *Salmonella* species and are not present in *Shigella*, *E. coli*, or *Citrobacter* species. Cohen et al. (21) evaluated the suitability of the *fimA* gene amplification by PCR as a specific method for detection of *Salmonella* strains. *Salmonella typhimurium* and other pathogenic members of the Enterobacteriaceae family produce morphologically and antigenically related, thin, aggregative, type 1 fimbriae. Waage et al. (22) demonstrated the detection of low numbers of *Salmonella* in environmental water by a nested polymerase chain reaction assay. The target loci selected were the conserved sequences within a 2.3 kb randomly cloned DNA fragment from the *Salmonella typhimurium* chromosome. The nested PCR assay correctly identified 128 of a total of 129 *Salmonella* strains belonging to subspecies I, II, IIIb, and IV with a sensitivity of the assay was 2 CFU. No PCR products were obtained from any of the 31 non-*Salmonella* strains examined. Riyaz-Ul-Hassan et al. (23) also describe a PCR reaction using *Salmonella* enterotoxin gene (*stn*) as target loci. The protocol describes the detection of less than 10 cells of *Salmonella* in 250 mL of blood and approxmately 1 cell in 1 mL of water without any enrichment.

Duplex or multiplex reactions further improve over the limitation of the locus-specific detection in PCR or gene probes. A duplex PCR described by Kapley and co-workers demonstrates the use of *phoE* primers to detect *Salmonella* in drinking water where *Vibrio* specific primers were also used in the same reaction (24). Multiplex PCR to detect *Salmonella* along with different pathogens in a single reaction has also been demonstrated (17,25). While Kapley and co-workers have demonstrated the use of the gene locus that has been shown to be essential for the invasion of *Salmonella* into epithelial cells of the host's intestine (inv), Kong and co-workers have used the invasion plasmid antigen B (*ipaB*) gene of *Salmonella typhimurium*. The primers for *ipaB*, forward primer 5′ GGACTTTTTAAAAGCGGCGG 3′ and reverse primer 5′ GCCTCTCCCAGAGCCGTCTGG 3′, amplified a 314 bp region. A multiplex PCR that amplifies five different loci in a single reaction has been reported for detection of *Salmonella* from river water samples (16). The gene markers used were *invA*, *SpvA*, *Spv*, and *phoE* for pathogenic *Salmonella*, and 16S rRNA specific primers to assess the total eubacterial load in the water samples. The *inv* locus was chosen as its expression has been shown to be essential for the invasion of *Salmonella* into epithelial cells of the host's intestine; causing the gastrointestinal infections (26). The *invA* gene codes for a protein, which is necessary for virulence of the bacterium. The primers used, upper 5′-CCTGATCGCACTGAATATCGTACTG 3′ and lower 5′GACCATCACCAATGGTCAGCAGG 3′, amplified a 598 bp fragment. The other pathogenic determinants in *Salmonella* are the plasmid coded virulence *Spv* genes; *SpvA*, *SpvB*, and *SpvC* (27). More than 80% of all *Salmonella* isolated from food and clinical specimens contain the virulence factors coded through plasmid (28). The primers reported for *spvA* upper 5′ TGTATGTTGATAC-TAAATCC 3′ and lower 5′ CTGTCATGCAGTAACCAG 3′, amplify a 470 bp product, while the primers reported for *spvB*, upper 5′ ATGAATATGAAT CAGACCACC 3′, and lower 5′ GGCGTATAGTCG GCGGTTTTC 3′ amplify a 669 bp product (16). The *phoE* gene encodes for a phosphate limitation-inducible outer membrane pore protein and has been shown to be *Salmonella* specific. The primers, upper 5′ AGCGCCGCGGTACGGGCGATAAA 3′ and lower 5′ ATCATCGTCATTAATGCCTAA CGT 3′, from the *phoE* locus have already been tested for 133 different *Salmonella* strains (29). The key to this detection tool has been provided through the thermocycling steps used in the M-PCR. The protocol has used gradient temperature steps, which has provided the selective annealing of primers. The developed multistep program could coamplify the target gene markers, used in the study. The specificity of the program has been proved with the DNA template derived from even the river water samples, which represents the heterogeneous microbial population.

The most widely used selective primer set for detection of *Salmonella* has been reported, which are designed from *invA* locus (30). An extended determination of selectivity by using 364 strains showed that the inclusivity was 99.6% and the exclusivity was 100% for the *invA* primer set (31).

FUTURE OPTIONS IN DETECTION OF *SALMONELLA* IN WATER SAMPLES

With the increasing incidences of spread of waterborne diseases, there is a great need to be able to detect waterborne pathogens like *Salmonella* (31). The key to this problem is to monitor subtle impacts that may have long-term effects that are hazardous. Genomic tools provide an option for analyzing basic resources at the nucleic acid level of an organism or population of microorganisms. Purohit et al. (32) have reviewed the use of genomic tools in monitoring and assessment processes that can help evaluate the environment impact. These tools can be extrapolated for assessing the mixing of raw sewage waters with water resources. A shortened PCR time with quantification and assured selectivity could be achieved by real time PCR. Based on this concept, an advanced nucleic acid analyzer (ANNA) has been reported that can detect bacteria within 7 min (33). This battery-operated device can be used in the field and has software that can be used by first timers. Inclusion of fluorescent dyes to the DNA during amplification has led to fresh advances in detection methods. Use of molecular beacons, oligonucleotide probes that become fluorescent upon hybridization, have been used in real time PCR assay to detect as few as 2 colony forming units (CFU) per reaction of *Salmonella* species. This method uses a 122 bp section of the *himA* as the amplification target. This method could also discriminate between amplicons obtained from similar species such as *E. coli* and *C. freundii*. The assay could be carried out entirely in sealed PCR tubes, enabling fast and direct detection of *Salmonella* in a semiautomated format (34). Detection of 1 colony forming unit/mL in food products was also demonstrated by real time PCR using *SipB* and *SipC* as target loci (35). Beyond the real time PCR, protocols need to be developed that can directly detect the pathogen using a biosensor. Kramer and Lim (36) have demonstrated a rapid and automated fiberoptic biosensor assay for the detection of *Salmonella* in sprout rinse water. Alfalfa seeds contaminated with various concentrations of *Salmonella typhimurium* were sprouted. The spent irrigation water was assayed 67 h after alfalfa seed germination with the RAPTOR (Research International, Monroe, WA), an automated fiberoptic-based detector. *Salmonella typhimurium* was identified in spent irrigation water when seeds were contaminated with 50 CFU/g. Viable *Salmonella typhimurium* cells were also recovered from the waveguides after the assay. This biosensor assay system has the potential to be directly connected to water lines and can assist in process control to identify contaminated water.

BIBLIOGRAPHY

1. Deere, D., Stevens, M., Davison, A., Helm, G., and Dufour, A. (2001). Management strategies. In: *Water Quality—Guidelines, Standards and Health: Assessment of Risk and Risk Management for Water-Related Infectious Disease*. World Health Organization, Geneva, Switzerland, World Health Organization, TJ International (Ltd), Padstow, Cornwall, UK.

2. Garmory, H.S., Brown, K.A., and Titball, R.W. (2002). *Salmonella* vaccines for use in humans: present and future. *FEMS Microbiol. Rev.* **26**: 339–353.

3. Hussong, D, Enkiri, N.K., Burge, W.D. (1984). Modified agar medium for detecting environmental *Salmonella* by the most-probable number method. *Appl. Environ. Microbiol.* **48**: 1026–1030.

4. Morinigo, M.A., Borrego, J.J., and Romero, P. (1986). Comparative study of different methods for detection and enumeration of *Salmonella* spp. in natural waters. *J. Appl. Bacteriol.* **61**: 169–176.

5. Russ, C.F. and Yanko, W.A. (1981). Factors affecting salmonellae repopulation in composted sludges. *Appl. Environ. Microbiol.* **41**: 597–602.

6. Knight, I.T., Shults, S., Kaspar, C.W., and Colwell, R.R. (1990). Direct detection of *Salmonella* spp. in estuaries by using a DNA probe. *Appl. Environ. Microbiol.* **56**: 1059–1066.

7. Purohit, H.J. and Kapley, A. (2002). PCR as an emerging option in the microbial quality control of drinking water. *Trends Biotechnol.* **20**: 325–326.

8. White, T.J. (1996). The future of PCR technology: diversification of technologies and applications. *Trends Biotechnol.* **14**: 478–483.

9. Durango, J., Arrieta, G., and Mattar, S. (2004). Presence of *Salmonella* as a risk to public health in the Caribbean zone of Colombia. *Biomedica* **24**: 89–96.

10. Thomason, B.M. (1981). Current status of immunofluorescent methodology for salmonellae. *J. Food Prot.* **44**: 381–384.

11. Anderson J.M. and Hartman P.A. (1985). Direct immunoassay for detection of salmonellae in foods and feeds. *Appl. Environ. Microbiol.* **49**: 1124–1127.

12. Mattingly, J.A. and Gehle, W.D. (1984). An improved enzyme immunoassay for the detection of *Salmonella*. *J. Food Sci.* **49**: 807–809.

13. Croci, L., Delibato, E., Volpe, G., De Medici, D., and Palleschi G. (2004). Comparison of PCR, electrochemical enzyme-linked immunosorbent assays, and the standard culture method for detecting *Salmonella* in meat products *Appl. Environ. Microbiol.* **70**: 1393–1396.

14. Meckes, M.C. and MacDonald, J.A. (2003). Evaluation of a DNA probe test kit for detection of salmonellae in biosolids. *J. Appl. Microbiol.* **94**: 382–386.

15. D'Souza, D.H. and Jaykus L.-A. (2003). Nucleic acid sequence based amplification for the rapid and sensitive detection of *Salmonella enterica* from foods. *J. Appl. Microbiol.* **95**: 1343–1350.

16. Kapley, A., Lampel, K., and Purohit, H.J. (2001). Rapid detection of *Salmonella* in water samples by multiplex PCR. *Water Environ. Res.* **73**: 461–465.

17. Kong, R.Y.C., Lee, S.K.Y., Law, T.W.F., Law, S.H.W., and Wu, R.S.S. (2002). Rapid detection of six types of bacterial pathogens in marine waters by multiplex PCR. *Water Res.* **36**: 2802–2812.

18. Roszak, D.B., Grimes, D.J., and Colwell, R.R. (1984). Viable but non recoverable stage of *Salmonella enteritidis* in aquatic systems. *Can. J. Microbiol.* **30**: 334–338.
19. Roszak, D.B. and Colwell, R.R. (1987). Survival strategies of bacteria in the natural environment. *Microbiol. Rev.* **51**: 365–379.
20. Way, J.S. et al. (1993). Specific detection of *Salmonella* spp. by multiplex polymerase chain reaction. *Appl. Environ. Microbiol.* **59**: 1473–1479.
21. Cohen, H.J., Mechanda, S.M., and Lin, W. (1996). PCR amplification of the *fim*A gene sequence of *Salmonella typhimurium*, a specific method for detection of *Salmonella* spp. *Appl. Environ. Microbiol.* **62**: 4303–4308 (1996).
22. Waage, A.S., Vardund, T., Lund, V., and Kapperud, G. (1999). Detection of low numbers of *Salmonella* in environmental water, sewage and food samples by a nested polymerase chain reaction assay. *J. Appl. Microbiol.* **87**: 418–428.
23. Riyaz-Ul-Hassan, S., Verma, V., and Qazi, G.N. (2004). Rapid detection of *Salmonella* by polymerase chain reaction. *Mol. Cell Probes* **18**: 333–339.
24. Kapley, A., Lampel, K., and Purohit, H.J. (2000). Development of Duplex PCR for *Salmonella* and *Vibrio*. *World J Microbiol. Biotechnol.* **16**: 457–458.
25. Kapley, A., Lampel, K., and Purohit, H.J. (2000). Thermocycling steps and optimization of multiplex PCR. *Biotechnol. Lett.* **22**: 1913–1918.
26. Clark, C.G., MacDonald, L.A., Ginocchio, C.C., Galan, J.E., and Johnson, R.P. (1996). *Salmonella typhimurium* invA expression probed with a monoclonal antibody to the C-terminal peptide of InvA. *FEMS Microbiol. Lett.* **136**: 263–268.
27. Suzuki, S. et al. (1994). Virulence region of plasmid pNL2001 of *Salmonella enteritidis*. *Microbiology* **140**: 1307–1318.
28. Williamson, C.M., Baird, G.D., and Manning, E.J. (1998). A common virulence region on plasmids from eleven serotypes of *Salmonella*. *J. Gen. Microbiol.* **134**: 975–982.
29. Spierings, G., Elders, R., vanLith, B.M., Hofstra, H., and Tommassen, J. (1992). Characterization of the *Salmonella typhimurium* phoE gene and development of *Salmonella* specific DNA probes. *Gene* **122**: 45–52.
30. Rahn, K. et al. (1992). Amplification of an invA gene sequence of *Salmonella typhimurium* by polymerase chain reaction as a specific method of detection of *Salmonella*. *Mol. Cell Probes* **6**: 271–279.
31. Malorny, B., Hoorfar, J., Bunge, C., and Helmuth R. (2003). Multicenter validation of the analytical accuracy of *Salmonella* PCR: towards an international standard. *Appl. Environ. Microbiol.* **69**: 290–296.
32. Purohit, H.J., Raje, D.V., Kapley, A., Padmanabhan, P., and Singh, R.N. (2003). Genomics tools in environmental impact assessment. *Environ. Sci. Technol.* **37**: 356A–363A.
33. Belgrader, P. et al. (1999). PCR detection of bacteria in seven minutes. *Science* **284**: 449–450.
34. Chen, W., Martinez, G., and Mulchandani, A. (2000). Molecular beacons: a real-time polymerase chain reaction Assay for detecting salmonella. *Anal. Biochem.* **280**: 166–172.
35. Ellingson, J.L.E., Anderson, J.L.A., Carlson, S.A., and Sharma, V.K. (2004). Twelve hour real-time PCR technique for the sensitive and specific detection of *Salmonella* in raw and ready-to-eat meat products. *Mol. Cell. Probes* **18**: 51–57.
36. Kramer, M.F. and Lim D.V. (2004). A rapid and automated fiber optic-based biosensor assay for the detection of *Salmonella* in spent irrigation water used in the sprouting of sprout seeds. *J. Food Prot.* **67**: 46–52.

LYSIMETER SOIL WATER SAMPLING

KEN'ICHIROU KOSUGI
Kyoto University
Kyoto, Japan

INTRODUCTION

The solute concentration of soil water provides important information regarding spatial and temporal distribution of plant nutrients, salinity, and trace elements (1); biological and chemical reactions in soil; and soil and groundwater contamination by industrial wastes and pesticides. For quantifying convective chemical transport in soil, both the water flux and the solute concentration of flowing water should be measured (2). In such a case, a sampling methodology for soil water needs to provide quantitative information on soil water flow as well as to maintain the physical soil environment of the water-sampling profile in a state similar to that of the natural soil profile, as the soil environment affects biological and chemical reactions.

TENSION-FREE LYSIMETER

One of the traditional techniques for sampling soil water is the tension-free lysimeter, which is also referred to as the zero-tension lysimeter. In this method, a horizontally buried pan intercepts infiltrating water, forming a temporary saturated zone above the pan, and the water drains into a sampling bottle. The tension-free lysimeter collects water only when the soil immediately above the pan has a positive pressure (3). Therefore, the soil in the sampling profile above the lysimeter is wetter than the soil surrounding the lysimeter. With this matric pressure gradient, water will tend to flow from the lysimeter into the surrounding dry-soil region, resulting in an underestimation of the natural water flux (e.g., 4,5).

CAPILLARY LYSIMETER

In the capillary lysimeter, a wick made of glass or nylon fibers is attached to the base of the water-collecting pan (Fig. 1) in order to establish drier conditions above the lysimeter and to lessen the problem of bypass flow around the lysimeter (6,7). Thus, this device is also referred to as the wick lysimeter. In order to make the water-sampling rate the same as the natural water flux, the length, number, and material of the wicks of the capillary lysimeter must be designed to match the local soil hydraulic properties as well as to respond to the range of fluxes to be encountered (8). However, Gee et al. (3) pointed out that capillary lysimeters tend to undersample when the soil water flux is less than 1000 mm/yr, even if the wick length is optimized for a given soil type. In a laboratory test, Kosugi (9) showed that the capillary lysimeter resulted in wetter conditions in the sampling soil profile than in a natural soil profile, given heavy irrigation of 18.6 mm/h. Recently, Gee et al. (3) proposed extending an impermeable pipe, of about the same diameter as the water-collecting pan, from the base of the pan to a height of

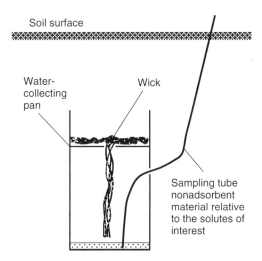

Figure 1. Schematic diagram of the capillary lysimeter.

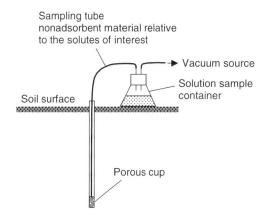

Figure 2. Schematic diagram of the tension lysimeter.

15 cm or more, in order to control water divergence around the lysimeter. Although they succeeded in reducing the flux divergence to less than 10% for coarse sand by using a 15-cm-long pipe, they concluded that finer-textured soils might require a pipe that was at least 60 cm long. Such a long pipe might disturb the water-sampling profile and might reduce root water uptake by cutting root systems. Detailed descriptions of the capillary lysimeter are found in Selker (10).

TENSION LYSIMETER

A generalized version of the capillary lysimeter is the tension lysimeter, which is one of the most frequently used water-sampling techniques. In this method, water is collected through a porous cup or plate by applying suction with a vacuum source instead of a wick. In most cases, porous cup samplers (also referred to as suction cups) are used because of their ease of installation, simplicity of design, and low cost (Fig. 2). However, owing to their small sphere of measurement, the suction cups provide only "point information," which does not adequately integrate spatial variability (11,12).

With the tension lysimeter method, the suction for water extraction is fixed at an empirically decided value between about 20 and 85 kPa (1,13). As a result, the soil moisture condition in the water-sampling profile can be altered depending on the suction applied. In addition, as the volume of soil that is sampled by the tension lysimeter depends on the soil moisture condition at the time of sampling, the soil hydraulic properties, and the flow properties of the lysimeter (12), the water-sampling rate is not necessarily the same as the natural water flux. Moreover, it is reported that the solute concentrations in the sampled water depend on the duration and degree of the sampler vacuum because water extracted from large pores at low suctions may have a different composition from that extracted from micropores at high suctions (14).

Corwin (1) provides more information on the tension lysimeter with regard to equipment preparation, installation procedure, and problems.

CONTROLLED-TENSION LYSIMETER

In contrast to the above-mentioned techniques, some recent studies have proposed controlling the suction for extracting soil water by referring to matric pressure observations made in the surrounding natural soil profile. The suction control is intended to make the rate of water extraction by the lysimeter the same as the unsaturated water flux in the natural soil profile. Such a "controlled-tension lysimeter" seems to be the most accurate alternative to the methods traditionally used to measure water and convective chemical fluxes in soil.

In the controlled-tension lysimeter, the suction is manually or automatically adjusted to a target value determined from tensiometer observations in the natural soil profile (2,15–17), or the suction control is set so that the soil matric pressure immediately above the porous plate should be similar to the matric pressure at the same depth in the natural soil profile (9,18–20). When the two matric pressures are the same, the rate of water extraction by the lysimeter is expected to be similar to the unsaturated water flux in the natural soil profile, because the water-sampling profile has the same upper (i.e., the flux boundary condition defined by rain and irrigation rate) and lower (i.e., the hydraulic head boundary condition) boundary conditions as the natural profile. Thus, the controlled-tension lysimeter method is soundly in accordance with unsaturated flow theory.

Figure 3 shows a schematic of a type of controlled-tension lysimeter proposed by Kosugi (9) and Kosugi and Katsuyama (20). The equipment consists of two tensiometers (TE) and a ceramic porous plate (PP) connected to a suction system by a sampling tube. The porous plate is buried horizontally in the water-sampling profile. One tensiometer monitors the soil matric pressure immediately above the plate, ψ_a. The other tensiometer monitors the matric pressure, ψ_b, at the same depth in a natural soil profile, adjacent to the sampling profile. In the sampling profile, infiltrating water is extracted through the porous plate by applying suction so that $\psi_a = \psi_b$. The suction system consists of a water collection container (WC) connected to a vacuum pump (PU), a valve for releasing the suction (RV), and a pressure transducer (PT) for monitoring the air pressure, p_c, in the water collection

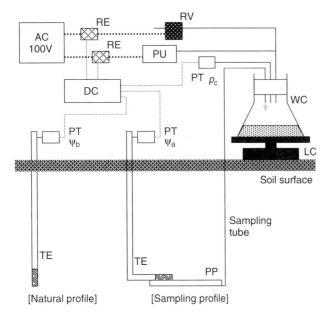

Figure 3. Schematic diagram of the controlled-tension lysimeter (modified from Ref. 9).

DC: Datalogger and controller
LC: Load cell
PP: Porous plate
PT: Pressure transducer
PU: Vacuum pump
RE: Relay
RV: Release valve
TE: Tensiometer
WC: Water container

container. The values of ψ_a, ψ_b, and p_c are continuously monitored at 3-second intervals. When $\psi_a > \psi_b$ (i.e., the sampling profile is wetter than the natural profile), the vacuum pump is turned on in order to extract the water above the porous plate. When $\psi_a < \psi_b$ (i.e., the sampling profile is dryer than the natural profile), the pump is turned off and the valve is opened in order to immediately stop water extraction by releasing the suction in the water collection container. During a field test for more than 400 d, the lysimeter maintained the soil moisture condition in the sampling profile similar to that in the natural soil profile and extracted a reasonable amount of water (20).

COMMENTS

Problems common to the capillary, tension, and controlled-tension lysimeters include sample biases caused by contaminations and adsorption as the solution passes through wicks and porous cups or plates. Sample contamination can be significantly reduced by pretreatment of the sampler with an acid wash (1 M HCl solution) and repeated rinsing with a salt solution similar to the soil solution that will be sampled (21). Ceramic, Teflon, and stainless steel are commercially available as materials for porous cups and plates. From among these, ceramic is the greatest in trace metal adsorption (22). Various studies have detailed the adsorption of NH_3, P, and K by ceramic cups. Nonetheless, ceramic is most commonly used for porous cups and plates because it has the lowest cost. Although Teflon is the least reactive, it is hydrophobic and has poor flow characteristics. Hence, Teflon is not necessarily a suitable material for porous cups and plates (1).

Another complicating issue is the change in pH that results from CO_2 degassing (23). To reduce CO_2 degassing, it is effective to minimize the gas/liquid ratio in the sample container. Collected samples should be tightly capped with the smallest airspace possible remaining in the sample bottle.

BIBLIOGRAPHY

1. Corwin, D.L. (2002). Suction cups. In: *Methods of Soil Analysis, Part 4—Physical Methods*. J.H. Dane and G.C. Topp (Eds.). Soil Science Society of America, Madison, WI, pp. 1261–1266.

2. Brye, K.R., Norman, J.M., Bundy, L.G., and Gower, S.T. (1999). An equilibrium tension lysimeter for measuring drainage through soil. *Soil Sci. Soc. Am. J.* **63**: 536–543.

3. Gee, G.W., Ward, A.L., Caldwell, T.G., and Ritter, J.C. (2002). A vadose zone water fluxmeter with divergence control. *Water Resour. Res.* **38**: 10.1029/2001WR000816.

4. Jamison, J.M. and Fox, R.H. (1992). Estimation of zero-tension pan lysimeter collection efficiency. *Soil Sci.* **153**: 85–94.

5. Chiu, T.F. and Shackelford, C.D. (2000). Laboratory evaluation of sand under-drains. *J. Geotech. Geoenviron. Eng.* **126**: 990–1001.

6. Holder, M. et al. (1991). Capillary-wick unsaturated zone soil pore water sampler. *Soil Sci. Soc. Am. J.* **55**: 1195–1202.

7. Maeda, M., Liyanage, B.C., and Ozaki, Y. (1999). Water collection efficiency of wick samplers under steady state flow conditions. *Soil Sci. Plant Nutr.* **45**: 485–492.

8. Knutson, J.H. and Selker, J.S. (1994). Unsaturated hydraulic conductivities of fiberglass wicks and designing capillary pore-water samplers. *Soil Sci. Soc. Am. J.* **58**: 721–729.

9. Kosugi, K. (2000). A new sampling method of vertical infiltration water in unsaturated soil without disturbing soil moisture condition. *J. Jpn. Soc. Hydrol. & Water Resour.* **13**: 462–471 (in Japanese with English summary).

10. Selker, J.S. (2002). Passive capillary samplers. In: *Methods of Soil Analysis, Part 4—Physical Methods*. J.H. Dane and G.C. Topp (Eds.). Soil Science Society of America, Madison, WI, pp. 1266–1269.

11. Biggar, J.W. and Nielsen, D.R. (1976). Spatial variability of the leaching characteristics of a field soil. *Water Resour. Res.* **12**: 78–84.

12. Hart, G.L. and Lowery, B. (1997). Axial-radial influence of porous cup soil solution samplers in a sandy soil. *Soil Sci. Soc. Am. J.* **61**: 1765–1773.

13. Tokuchi, N. (1999). Sampling method of solute in soil. In: *Methods of Forest Environment Measurement*. Hakuyusha, Tokyo, Japan, pp. 186–188 (in Japanese)*.

14. Rhoades, J.D. and Oster, J.D. (1986). Solute content. In: *Methods of Soil Analysis, Part 1: Physical and Mineralogy Methods*. A. Klute (Ed.). Soil Science Society America, Madison, WI, pp. 985–1006.

15. Ozaki, Y. (1999). Bury-type lysimeter to collect infiltrated soil water. In: *Proc. Symp., Study on practical problems on monitoring of NO_3-N leaching from crop field*. Tsukuba, Japan. Nov. 4–5. National Agriculture Research Center, Tsukuba, Japan, pp. 9–19 (in Japanese)*.

*The title is tentative translation from the original Japanese title by the authors of this paper.

16. Ciglasch, H., Aoungsawad, K., Amelung, W., Pansombat, K., Sombuttun, T., Tarn, C., Totrakool, S., and Kaupenjohann, M. (2002). Fluxes of agrochemicals in an upland soil in Northern Thailand as investigated with novel techniques. Paper 1566. In: Abst., 17th World Congress of Soil Science. Bangkok, Thailand, Aug. 14–21.
17. Lentz, R.D. and Kincaid, D.C. (2003). An automated vacuum extraction control system for soil water percolation samplers. *Soil Sci. Soc. Am. J.* **67**: 100–106.
18. Duke, H.R. and Haise, H.R. (1973). Vacuum extractors to assess deep percolation losses and chemical constituents of soil water. *Soil Sci. Soc. Am. Proc.* **37**: 963–964.
19. van Grinsven, J.J.M. et al. (1988). Automated *in situ* measurement of unsaturated soil water flux. *Soil Sci. Soc. Am. J.* **52**: 1215–1218.
20. Kosugi, K. and Katsuyama, M. (2004). Controlled-suction period lysimeter for measuring vertical water flux and convective chemical fluxes. *Soil Sci. Soc. Am. J.* **68**: 371–382.
21. Creasey, C.L. and Dreiss, S.J. (1988). Porous cup samplers: cleaning procedures and potential sample bias from trace element contamination. *Soil Sci.* **145**: 93–101.
22. McGuire, P.E., Lowery, B., and Helmke, P.A. (1992). Potential sampling errors: trace metal adsorption on vacuum porous cup samplers. *Soil Sci. Soc. Am. J.* **56**: 74–82.
23. Suarez, D.L. (1986). A soil water extractor that minimizes CO_2 degassing and pH errors. *Water Resour. Res.* **22**: 876–880.

REGULATORY AND SECURITY REQUIREMENTS FOR POTABLE WATER

SUZANNE DU VALL KNORR
Ventura County Environmental
Health Division
Ventura, California

INTRODUCTION

Water is essential for the nutrients, metabolic processes, and cellular activity that all living bodies need to survive and thrive. One adult human requires approximately 2.5 liters of potable water from food or liquid sources every day (1). Potable water is water that is fit to drink, free of pathogenic microorganisms or chemicals that are harmful to human health, and free of offensive taste and odors. Drinking water sources include surface waters such as lakes and rivers, and groundwater accessed by wells. Desalination can produce potable water from ocean water, though cost often prohibits widespread use of this method. Public water systems include community, noncommunity, and transient water systems. A water system includes the source water, treatment steps, storage, and the distribution network. Contamination can occur at any point in the water system.

Standard treatment methods typically involve several filtration steps to remove larger particles and objects from the source water and sedimentation to allow smaller particles to settle out. Coagulation and flocculation follow sedimentation to remove unsettleable particles. Disinfection to inactivate pathogens commonly involves chlorine, bromine, or ozone, and fine filtration to remove protozoa. The treated water may also be pH-stabilized. These treatment methods control most pathogens and inorganic contaminants, such as arsenic and lead, and the secondary drinking water standards of taste, odor, color, and balance. The most common types of industrial chemicals, volatile organic compounds (VOCs), such as benzene (a known carcinogen), toluene, and methyl-tertiary-butyl ether, have been routinely found in groundwater. Removing VOCs requires more extensive treatment methods.

Known waterborne disease pathogens can infect healthy adults as well as sensitive populations (young children, pregnant women, elderly adults, and immunocompromised individuals). The Centers for Disease Control (CDC) estimated that over 900,000 incidents of waterborne intestinal disease occur annually, resulting in 900 deaths per year (2). More recently, 39 waterborne disease outbreaks occurred between 1999 and 2000, affecting over 2000 people and resulting in two deaths. Twenty-eight of the 39 were from drinking water that used groundwater sources.

Treatment and protection of drinking water sources will not prevent contamination after delivery (point-of-use). Illness resulting from point-of-use contamination is not considered a waterborne disease incident by the CDC. Household point-of-use treatment methods typically do not remove many organic chemicals (3). Drinking water requirements change as the federal Environmental Protection Agency (EPA) and the CDC further discover the causes of waterborne disease. This article briefly reviews selected contaminants and current regulatory and security requirements that seek to ensure safe and reliable sources of drinking water.

SELECTED BIOLOGICAL CONTAMINANTS

Cryptosporidium

Cryptosporidium parvum oocysts (cysts) are the infective form of a protozoan smaller than *Giardia* that is highly resistant to chlorine disinfection and can cause symptoms of acute watery diarrhea (cryptospiridiosis) from as few as 10 cysts. The surface water treatment rule in effect in 1993 did not require the fine filtration needed to remove *Cryptosporidium* cysts. It is estimated that over 80% of surface water sources are contaminated with *Cryptosporidium* or *Giardia* (4). *Cryptosporidium* can spread in fully chlorinated water, even in the absence of detectable coliforms (5), and medical therapy to treat cryptospiridiosis is limited.

The most notable outbreak of waterborne disease caused by *Cryptosporidium parvum* happened in Milwaukee, Wisconsin, in 1993. This outbreak affected an estimated 403,000 people, resulting in 4400 hospitalizations and 104 deaths (5). It cost city residents an estimated $54 million in medical costs and lost wages (2). Two weeks passed before the waterborne nature of the outbreak was recognized and boil water notices were issued to the public (6). Widespread absenteeism, increased emergency room visits for diarrhea, and a citywide shortage of over-the-counter antidiarrheal medicines led to the detection of the cryptospiridiosis outbreak (5).

Investigation of the outbreak revealed that the public water supply came from Lake Michigan and was treated and chlorinated before distribution. Decreased raw source water quality and decreased coagulation and flocculation effectiveness led to increased turbidity in the treated water. The turbidity standards in effect in 1993 allowed daily turbidity measurements in a month not to exceed 1.0 NTU, though periodic spikes above 1.0 NTU were allowed. The highest turbidity measurement was 1.7 NTU, and the treated water had met all state and federal standards at that time (6). The investigation did not find the original source of contamination, but it was determined that the water quality standards "were not adequate to detect the outbreak" (6). This outbreak led to the Interim Enhanced Surface Water Rule in 1998 and later adoption of a primary regulatory standard requiring 99% cyst removal by filtration. *Cryptosporidium* contaminated water can be disinfected with ozone (4). Point-of-use treatment includes boiling water for 1 minute or filtration with a pore size of 1 micron or smaller (4), which is capable of 99% removal.

Escherichia coli O157:H7

Escherichia coli is a fecal coliform, found in the intestines of both humans and animals, that may contaminate water through contact with sewage discharges, leaking septic systems, or water runoff from animal feedlots. *Escherichia coli* O157:H7 is an enterohemorrhagic strain of one of the four classes of virulent pathogenic *E. coli* that cause gastroenteritis in humans. It produces a potent verotoxin that damages the intestinal lining, often accompanied by severe bloody diarrhea, particularly in sensitive populations. Severe cases of *E. coli* O157:H7 can cause renal failure and loss of kidney function (hemolytic uremic syndrome). Unlike other fecal coliforms, *E. coli* O157:H7 can cause illness from ingesting as few as 10 organisms. One study of a waterborne *E. coli* O157:H7 outbreak involved 243 patients who used an unchlorinated groundwater supply (32 hospitalizations, 4 deaths) in Burdine township, Missouri, between 12/15/89 and 1/20/90 (7). The case study determined that the largest number of cases of bloody diarrhea was in the municipal water supply service area (7). More recently, four waterborne outbreaks of *E. coli* O157:H7 occurred between 1999 and 2000, including one in Albany, New York, in 1999. The suspected source in Albany, New York, was contaminated well water consumed at a fair that caused 127 cases of illness, 71 hospitalizations, including at least 10 children with bloody diarrhea; 14 cases of hemolytic uremic syndrome; and two deaths (8,9).

The detection of coliforms during routine sampling leads to detecting the presence of fecal coliforms, including *E. coli*. Standard treatment methods inactivate or remove *E. coli* O157:H7 or other fecal coliforms. These and other waterborne disease pathogens are included in Table 1.

SELECTED CHEMICAL CONTAMINANTS

Methyl-*tertiary*-butyl Ether

Methyl-*t*-butyl ether (MTBE) was introduced in the 1970s as a substitute for lead in gasoline, to reduce emissions and maintain the oxygen level required by the Clean Air Act. MTBE is water-soluble and moves rapidly through soil into groundwater. It is the second most frequently detected chemical in monitoring wells, according to the U.S. Geological Survey (4). Contamination can come from leaking underground fuel

Table 1. Selected Waterborne Pathogens and Parasites[a]

Pathogens	Disease	Incubation Period	Estimated Infective Dose, # Organisms	Survival in Water, Days
Hepatitis A virus	Type A infectious hepatitis	Up to 50 days	$10-50^b$	Up to 140 days in groundwater
Salmonella typhi	Typhoid fever	7–21 days	10,000–100,000	Unknown[c]
E. coli O157:H7	Hemorrhagic colitis; hemolytic uremic syndrome	2–8 days	As few as 10^d	Up to 20 days in wastewater[e]
E. coli (all others)	Enteropathogenic diarrhea	12–72 hours	10^8	Up to 45 days in groundwater
Vibrio cholerae	Cholera	1–5 days	10^6-10^9	5–16 days in surface water
Shigella spp.	Bacillary dysentery	1–7 days	10–100	35 days in groundwater; 24 months in surface water
Cryptosporidium spp.	Cryptosporidiosis	2–21 days	10–100 cysts	Up to 6 months in a moist environment (ova)
Giardia lamblia	Giardiasis	6–22 days	5–100 cysts	Up to 4 months in surface water (ova)
Leptospira spp.	Leptospirosis, hemorrhagic jaundice	4–19 days	Unknown	3–9 days in surface water
Entamoeba histolytica	Amebic dysentery	2–4 weeks	10–20 cysts	Up to 1 month in surface water (ova)
Ascaris lumbricoides	Ascariasis	2 months	1 cyst, egg or larvae	Up to 7 years in soil

[a] Information compiled from Reference 10, unless otherwise indicated.
[b] Based on contaminated food consumption (11).
[c] 240 days survival in ice (12).
[d] The infective dose is unknown, but the FDA estimates that the dose may be similar to that of *Shigella* spp., based on outbreak data and the organism's ability to transmit person-to-person (13).
[e] Based on information in agricultural research (14).

tanks, leaking gasoline pipelines, and surface spills. It was detected in groundwater in 55% (3180 out of 5738 sites) of underground leaking gasoline tank sites under investigation in 1998 (15).

The oral LD_{50} for MTBE is approximately 4000 mg/kg, and MTBE was tentatively classified as a possible carcinogen (16). MTBE use has been banned in 11 states. In 2000, the California Senate approved an executive order requesting the ban of MTBE and a proposed phase-out by December 31, 2003 (17). This proposed phase-out has not yet occurred. California has set both primary and secondary standards for MTBE. At the federal level, MTBE is an unregulated contaminant, subject to the Unregulated Contaminant Monitoring Rule (UCMR), and is on the EPA drinking water contaminant candidate list (18). The UCMR requires that all large and selected small public water systems monitor for MTBE. MTBE can be removed in varying amounts through granular activated carbon, air stripping, or advanced oxidation.

Perchlorate

Perchlorate is an ion with four chlorine attachments, most commonly used in industrial applications. Ammonium perchlorate is used as an energetics booster in explosives, pyrotechnics, rocket fuel, and highway safety flares (19). It has been detected in groundwater in several states, including the Colorado River which supplies water to millions of people (20), and was found most recently in milk and lettuce (21,22). The concern is that perchlorate interferes with iodide uptake by the thyroid gland, leading to a disruption of thyroid hormones that regulate metabolism and growth; continuous thyroid disruptions could cause a hormone imbalance, particularly in pregnant women, developing fetuses, and infants. Several studies show conflicting results, though two studies indicated adverse health effects in susceptible populations at levels as low as 0.01 µg/kg (20,23). Perchlorate can be removed by treatment with ion exchange, ultraviolet light and peroxide, or peroxide and ozone (24).

A maximum contaminant level (MCL) has not been established by EPA. The Perchlorate Community Right to Know Act of 2003 mandated an enforceable national perchlorate contaminant standard by July 1, 2004; however, the projected research completion date set by the EPA, necessary for setting a standard, extends beyond this deadline (20,25). The EPA set a reference dose (RfD) of 1 ppb (or 2.1 µg/kg for a 70 kg adult) (26) as part of the draft Health Assessment on Perchlorate in 2002, currently under review by the National Academy of Sciences (NAS) (18). NAS estimates that it will complete the review by December 2004 (13). The RfD is an estimate of the daily dose below which health risks would be considered negligible based on lifetime exposure.

The California Department of Health Services (CDHS) was required by law to establish a maximum contaminant level (MCL) for perchlorate by January 1, 2004, although one has not yet been established (27). CDHS set an action level of 6 ppb (28), equivalent to the public health goal (PHG) set by the California Environmental Protection Agency (Cal-EPA) Office of Environmental Health Hazard Assessment (OEHHA). The studies used to establish the PHG were criticized by environmental organizations (20). The action level and the PHG are advisory levels, not regulatory requirements. If an action level is exceeded in groundwater, the water system must notify the local governing body (city council or board of supervisors), and CDHS recommends public notification. If an action level is exceeded by ten times, CDHS recommends discontinuing consumption until treated. Eight other states have set advisory levels for perchlorate, ranging from 1 to 18 ppb (29).

REGULATORY DRINKING WATER STANDARDS

Drinking water must meet the requirements of the Safe Drinking Water Act (SDWA), with monitoring and reporting as specified in the regulation. Individual states may either adopt the federal standard (SDWA) or stricter standards. California adopted stricter standards through the Calderon-Sher Safe Drinking Water Act. Other local drinking water requirements can be found at http://www.epa.gov/safewater/dwinfo.htm. Drinking water sources that include surface waters must be disinfected prior to distribution to the consumer. Current laws and regulations do not require routine treatment of groundwater, unless blended with surface water, although EPA has proposed the Ground Water Rule (GWR) which would require disinfection (8,18). Primary drinking water standards are legally enforceable standards applicable to public drinking water systems, defined as water systems with 15 or more service connections, serving 25 or more people, or operating at least 60 days per year (30). CDHS regulates water systems with five or more service connections. These standards address both biological and chemical contamination and are summarized in Table 2. The maximum contaminant levels (MCL), or specific treatment techniques, such as filtration, used in place of the MCL, as required by the SDWA, are established through evaluation and research to protect public health. Ensuring a safe potable water supply requires vigilance by wholesale and retail water purveyors and water treatment facilities.

SDWA requires testing for coliforms, a group of aerobic or facultative anaerobic bacteria commonly found in soil, vegetation, and the intestinal tracts of warm-blooded animals. The presence of coliforms indicates potential fecal coliform contamination and the probability that other pathogens or parasites are also present. Some fecal coliforms survive in water as long as 10 weeks (32). The maximum contaminant level for coliforms is less than or equal to 5% of the water samples that are positive for coliforms when 40 or more samples are collected per month, or one positive sample when less than 40 samples are collected per month (33). The largest public water systems, serving millions of people, must take at least 480 samples per month (34). Smaller systems must take at least five samples per month, unless the state conducted a sanitary survey during the previous 5 years (34). Under certain circumstances, the smallest systems, those serving less than 1000 people, could take one sample per month, not counting repeat sampling (34). Any samples positive for coliforms must also be tested for fecal coliforms. In California, analytical results must be

Table 2. Current Primary Regulatory Standards for Selected Contaminants[a]

Contaminant	Potential Health Effect	EPA, MCL in mg/L[b]	CDHS, MCL in mg/L[b]
Cryptosporidium spp.	Gastrointestinal illness	99% removal	99% removal
Giardia lamblia	Gastrointestinal illness	99.9% removal	99.9% removal
Total coliforms	Pathogen indicator organism	5.0%[j]	5.0%[j]
Viruses, enteric	Gastrointestinal illness	99.99% removal or inactivation	99.99% removal or inactivation
Arsenic	Circulatory problems, skin damage, increased risk of cancer	0.010	0.05[k]
Asbestos	Intestinal polyps from ingestion	7 MFL	7 MFL
Benzene	Anemia, reduced blood platelets, increased risk of cancer	0.05	0.001
Chromium[e]	Variable	0.1	0.05
Copper[c]	Gastrointestinal illness, liver and kidney damage	1.3	1.3
Free cyanide	Nervous system and thyroid problems	0.2	0.2
Ethylene dibromide (EDB)	Liver, stomach, kidney, and reproductive problems; increased risk of cancer	0.00005	0.00005
Fluoride	Bone disease; dental discoloration and pitting	4.0[f,g]	2.0
Lead[c]	Developmental effects in children	0.015	0.015
Inorganic mercury	Kidney damage	0.002	0.002
MTBE	Tentatively classed as possible carcinogen	None	0.013[g]
Nitrate (as nitrogen)	Blue baby syndrome	10	10
Nitrite (as nitrogen)	Blue baby syndrome	1	1
Polychlorinated biphenyls (PCBs)	Immune, thymus, reproductive and nervous system damage, increased risk of cancer	0.0005	0.0005
Styrene (vinyl benzene)	Liver and kidney damage; circulatory problems	0.1	0.1
Toluene (methyl benzene)	Nervous system; liver and kidney damage	1.0	0.15
Trichloroethylene (TCE)[h]	Liver damage; increased risk of cancer	0.005	0.005
1,1,1-Trichloroethane (1,1,1-TCH)	Liver, nervous system, and circulatory system problems	0.2	0.2
Trihalomethanes (TTHMs)	Disinfection by-product; liver, kidney, central nervous system effects; increased risk of cancer	0.08[f]	0.10
Uranium	Kidney toxicity, increased risk of cancer from radionuclides	30 µg/L	20 pCi/L
Vinyl chloride	Increased risk of cancer	0.002	0.0005
Xylenes (total)	Nervous system damage	10	1.750[d]

[a] Information from U.S. EPA and California Department of Health Services websites, unless otherwise indicated.
[b] MFL = million fibers greater than 10 µm in length per fluid liter; pCi/L = picocuries per liter; 1 pCi = 37 becquerels (Bq) per cubic meter. Food intake is generally measured in Bq/kg or Bq/L. EPA uses 0.9 pCi/µg as a conversion factor (11).
[c] U.S. EPA has a maximum contaminant level goal (MCLG) of 1.3 mg/L for copper and 0.015 mg/L for lead based on the number of samples taken at point-of-use. Additional monitoring, corrosion control, and treatment are required if the regulatory action level is exceeded in more than 10% of samples. Public education is required if the regulatory action level is exceeded for lead.
[d] As a single isomer or the sum of all isomers.
[e] All chromium levels. Hexavalent chromium causes severe diverse health affects, including cancer. It was the groundwater contaminant affecting public health in Hinkley, California, portrayed in the movie "Erin Brockovich."
[f] EPA standard effective January 1, 2004 for all water systems.
[g] Also has a secondary standard.
[h] TCE was the groundwater contaminant involved in a large leukemia cluster (28 cases in 20 years, four times the national average) in Woburn, Massachusetts, portrayed in the movie, "A Civil Action" (31).
[j] Based on 40 or more samples per month, or one positive sample when less than 40 samples are collected per month.
[k] California must adopt the federal standard, or stricter, by 2006. EPA handles enforcement until the new standard is adopted.

reported electronically by the tenth day of the following month. The water supplier must retain bacteriological analysis records for at least 5 years, and chemical analysis records for at least 10 years. Wholesale and retail water suppliers are required to provide an annual water quality report (AWQR) to their customers. In California, the Consumer Confidence Report (CCR) acts as the AWQR, but must adhere to CDHS regulations, which requires contaminant reporting in easy to read table format (35).

A violation of a primary drinking water standard requires that the water system take corrective action and notify its customers. The notification must include a description of the violation, what it means using appropriate health effects language, when it occurred, what action they should take, and who to contact for more information. Violations are classified by tiers under the Public Notification Rule (36,37). A Tier 1 violation requires public notification within 24 hours by radio, television, hand delivery, or other approved means, and concurrent consultation with the appropriate implementing agency (CDHS or EPA where it directly implements the program). Examples of Tier 1 violations are exceeding the MCL for nitrates or nitrites, failure to test for fecal coliforms or *E. coli*, a repeat fecal coliform positive sample, a waterborne disease outbreak, or other emergency. A Tier 2 violation requires public notification within 30 days. An example of a Tier 2 violation is exceeding the MCL for total coliforms or turbidity. Water system customers must receive notification of a Tier 3 violation within 12 months of the violation. Examples of Tier 3 violations are testing procedure violations or nonfecal coliform monitoring violations. The implementing agency must receive a copy of the public notification within 10 days of initiation for Tier 2 and 3 violations. The water system must send out repeat notifications every 3 months for Tier 2 violations, or every 12 months for Tier 3 violations, for as long as the violation exists. Copies of public notices for Tier 1, 2, and 3 violations must be retained for at least 3 years. The water system must also conduct public notification within 12 months of available monitoring results for unregulated contaminants (36).

The implementing agency has formal and informal enforcement actions available to address a failure on the part of the water system to take corrective action or to notify the public or the implementing agency properly. Formal enforcement includes administrative orders, penalties, and civil and criminal actions. States may also refer a recalcitrant water system to the EPA for further enforcement action.

Secondary drinking water standards are nonenforceable guidelines for secondary contaminants that may cause cosmetic effects, such as tooth discoloration from high fluoride levels, or aesthetic effects such as a disagreeable taste, odor, or color. Secondary contaminants are defined as any contaminant that adversely affects taste, odor, or appearance, or that may cause a substantial number of people to discontinue use, or that may otherwise adversely affect public welfare. Selected secondary drinking water standards are summarized in Table 3. The water purveyor must notify the public when a secondary drinking water standard is exceeded. Treatment is at the option of the purveyor or the customer at point-of-use.

ENHANCED SECURITY REQUIREMENTS

The danger of a contaminated water supply to large populations became one focus of the Department of Homeland Security as part of a national directive to protect critical infrastructures. An intentionally introduced biological or chemical contaminant may have a limited effect unless introduced in sufficient quantity. However, the 1993 Milwaukee *Cryptosporidium* outbreak demonstrated that sometimes a small quantity is all that is necessary to affect a significant portion of the population adversely.

Table 3. Current Secondary Drinking Water Standards for Selected Contaminants[a]

Constituent	U.S. EPA[b]	CDHS[b]
Color	15 color units	15 color units
Chloride	250 mg/L	250 mg/L
Fluoride	2.0 mg/L	None[c]
Iron	0.3 mg/L	0.3 mg/L
Odor	3 threshold odor units	3 threshold odor units
pH	6.5–8.5	6.5–8.5
Turbidity	0.5 NTU	0.5 NTU
Sulfates	250 mg/L[d]	250 mg/L[d]
Total dissolved solids	500 mg/L	500 mg/L
MTBE	None	0.005 mg/L

[a] Information from U.S. EPA and California Department of Health Services websites, unless otherwise indicated.
[b] NTU = nephelometry turbidity units. 0.5 NTU is required for 95% of the daily measurements taken in a monthly period, with no spikes above 1.0 NTU.
[c] Primary standard in California.
[d] 250 mg/L is the lower limit; 500 mg/L is the upper limit.

Two federal laws involving two agencies were passed in 2002: the Homeland Security Act and the Public Health Security and Bioterrorism Preparedness and Response Act (Bioterrorism Act). Both laws pertain to public water systems. The Homeland Security Act requires reporting knowledge of a potential threat to a critical infrastructure, including public water systems, to the Department of Homeland Security. The Bioterrorism Act amended the SDWA to address terrorist and intentional or malevolent acts (threats) against public drinking water supplies. It requires that each public water system serving more than 3300 people conduct a vulnerability assessment (VA), to submit the VA to EPA by deadlines mandated by system size, to prepare an emergency response plan (ERP), and to implement any necessary security enhancements. Interaction between both laws and both agencies must be coordinated within a legal framework.

The purpose of the VA is to evaluate the susceptibility of the water system to potential threats and the risk to the community served by the water system and to plan for reducing risks. The VA must include six basic elements (38):

1. characterization of the water system, its mission, and objectives
2. identification and prioritization of adverse consequences to avoid
3. determination of critical assets that may be subject to a threat and potential undesirable consequences
4. assessment of the probability of different types of threats
5. evaluation of existing countermeasures
6. analysis of current identified risks and development of a prioritized plan to reduce these risks

The first element involves prioritizing the utility services and the facilities necessary to provide those services to the community it serves. Characterizing the water system may include a review of operating procedures and management practices for each element within the water system. The systems' objectives also depend on other utilities, such as power, and the need to evaluate the risk to the system if a threat affects another critical infrastructure.

The next two elements involve a review of each component within the water system in relation to any possible threat, including but not limited to physical barriers, water collection, treatment and pretreatment, storage facilities, distribution (including all piping, conveyances, and valves), automated systems, computer systems, and chemical use and storage areas. Preparing these two elements must consider how a threat could affect each element and the potential consequences, such as service disruption, illness, death, or economic impacts. For example, the introduction of a threat contaminant (TC) could involve concentration changes at every stage in the water system as it mixes with clean water or involve reactions with water or piping. Even a weak solution could adversely affect public health, either as an acute exposure to TC dangerous at low doses, or as a chronic health threat to low TC doses over long exposure periods. Each potential disruption in a system's ability to provide a safe and reliable water supply requires prioritization based on potential magnitude.

The fourth element involves identifying different types of threats the system has experienced in the past or that could occur in the future and the probability of each identified possibility. Specific types of threats such as a fire, vandalism, or cyberattacks, might already have specific countermeasures in place. The fifth element involves evaluating the capabilities and limitations of existing risk reduction methods. Examples of these include intrusion detection systems, delay mechanisms (locks, fencing, vehicle access points), and water quality monitoring alarms. The sixth element requires reviewing the vulnerabilities and countermeasures identified in the first five elements and providing recommendations for improvement.

VAs were to be completed and sent directly to EPA no later than 3/31/03 for water systems serving 100,000 or more people, no later than 12/31/03 for systems serving 50,000 to 99,000 people, and no later than 6/30/04 for water systems serving less than 50,000 people (39). Compliance with this requirement was voluntary for water systems serving less than 3300 people and updates to the VA are also voluntary (39). The EPA will determine system size based on data in the Safe Drinking Water Information System, submitted by each state in 2002 (30).

Some state agencies may require duplicate submission; neither CDHS nor Cal-EPA required a copy of the VA. The EPA plans to review each VA (38).

The next requirement of the Bioterrorism Act was for each water system to prepare an Emergency Response Plan (ERP), incorporating the results of the VA. Emergency response planning is essential for the water system to continue to provide a safe and reliable water supply in the event of an emergency. The ERP must include all procedures and equipment that will significantly reduce the impact of a threat, including basic information such as owner, address, emergency contacts, system components, population served, and number of service connections. The ERP must also identify alternative equipment and water sources, access control, emergency response, and incident specific procedures. Examples of incident specific procedures include policies for handling bomb threats, contamination incidents, or workplace violence.

The ERP is to be maintained securely on site. The water system is required to certify to EPA that the ERP was completed within 6 months of VA submission. The water system has the responsibility for ensuring and upgrading security procedures and equipment.

CONCLUSION

Drinking water laws and regulations were enacted to protect public health. Both the EPA and CDHS also provide information for individuals on private wells so they can protect their water quality. Some waterborne disease outbreaks demonstrate a need to evaluate water quality monitoring methods, and, as in the case of the 1993 Milwaukee cryptospiridiosis outbreak, could lead to a change in regulatory standards. Concerns over the security of critical infrastructures and evaluation of existing systems may also lead to changes that will ultimately enhance protecting public health. As science and technology advances and as more is known about the health affects of waterborne chemical and biological contaminants, drinking water laws and regulations will continue to change. The goal of both the regulatory agencies and the water purveyors should always be to ensure a safe and reliable drinking water supply.

BIBLIOGRAPHY

1. Whitney, E. and Boyle, M. (1984). *Understanding Nutrition*, 3rd Edn. West, pp. 7, 368, 369.
2. Lee, G.F. and Jones-Lee, A. (1993). "Public Health Significance of Waterborne Pathogens," Report to Cal-EPA Comparative Risk Project, December 1993, http://www.gfredlee.com/phealthsig_080801.pdf, accessed September 21, 2004.
3. Santa Clara Valley Water District (SCVWD). (2003). Frequently asked questions, perchlorate. May 20, 2003. http://www.valleywater.org/Water/Water_Quality/Protecting_your_water/_Perchlorate_Information/_pdf/Perchlorate FAQ 5-21-03.pdf −436.0 KB, accessed July 28, 2004.
4. Los Angeles Department of Water and Power (LADWP). (2003). MtBE, General Information. http://www.ladwp.com/ladwp/cms/ladwp001420.jsp, accessed July 26, 2004.
5. MacKenzie et al. (1994). A massive outbreak in Milwaukee of *Cryptospirodium* infection transmitted through public water supply. *N. Engl. J. Med.* **331**(3): 161–167.
6. Centers for Disease Control (CDC). (1996). Surveillance for Waterborne-Disease Outbreaks—U.S., 1993–1994, April 12, 1996. http://www.cdc.gov/epo/mmwr/preview/mmwrhtml/00040818.htm, accessed July 23, 2004.
7. Swerdlow et al. (1993). A waterborne outbreak in Missouri of *Escherichia coli* O157:H7, associated with bloody diarrhea and death. Centers for Disease Control, Enteric Diseases Branch. *Annu. Intern. Med.* **119**(3): 249–250.

8. Centers for Disease Control (CDC). (2002). Surveillance for Waterborne-Disease Outbreaks—U.S., 1999–2000, Surveillance Summaries. November 22, 2002, http://www.cdc.gov/epo/mmwr/preview/mmwrhtml/ss5108a1.htm, accessed September 21, 2004.
9. New York Department of Health (DOH). (2000). Health commissioner releases *E. coli* outbreak report. March 31, 2000. http://www.health.ny.us/nysdoh/commish/2000/ecoli.html, accessed September 27, 2004.
10. Salvato, J.A. (1992). *Environmental Engineering and Sanitation*, 4th Edn. John Wiley & Sons, Hoboken, NJ.
11. Food and Drug Administration (FDA). (2001). Center for Food Safety and Applied Nutrition. Outbreaks associated with fresh and fresh-cut produce, incidence, growth, and survival of pathogens in fresh and fresh-cut produce. September 30, 2001, http://vm.cfsan.fda.gov/~comm/1ft3-4a.html, accessed September 20, 2004.
12. Health Canada. (2001). Office of Laboratory Security, "Material Safety Data Sheet-Infectious Substances, *Salmonella typhi*. May 15, 2001. http://www.hc-sc.gc.ca/pphb-dgspsp/msds-ftss/msds134e.html, accessed September 10, 2004.
13. Food and Drug Administration (FDA). (2003). Center for Food Safety & Applied Nutrition, Foodborne Pathogenic Microorganisms and Natural Toxins Handbook, Bad Bug Book, *Escherichia coli* O157:H7, updated January 7, 2003, http://vm.cfsan.fda.gov/~mow/chap15.html, accessed September 10, 2004.
14. McCaskey, T.A. et al. (1998). Constructed wetlands controlling *E. coli* O157:H7 and *Salmonella* on the farm. *Highlights Agric. Res.* **45**(1): http://www.ag.auburn.edu/aaes/communications/highlights/spring98/ecoli.html, accessed September 20, 2004.
15. Fogg, G.E. et al. (1998). Hydrologic Sciences, Department of Land, Air, and Water Resources, University of California Davis, Impacts of MTBE on California Groundwater. *Health and Environmental Assessment of MTBE*, Volume 4, Report to the Governor and Legislature of the State of California as Sponsored by SB 521, November 12, 1998. http://www.tsrtp.ucdavis.edu/mtberpt/vol4_1.pdf, accessed September 10, 2004.
16. Office of Environmental Health Hazard Assessment (OEHHA). (1999). California Environmental Protection Agency (Cal-EPA). MTBE in drinking water, California public health goal", March 12, 1999, p. 33. http://www.oehha.org/water/phg/pdf/mtbe_f.pdf, accessed September 27, 2004.
17. Office of the Governor. (2002). Governor Davis allows more time for ethanol solution. Press Release PR02:139, March 15, 2002. http://www.energy.ca.gov/releases/2002_releases/2002-03-15_governor_mtbe.html, accessed September 27, 2004.
18. Environmental Protection Agency (EPA). (2002). National Academy of Sciences' Review of EPA's Draft Perchlorate Environmental Contamination: Toxicological Review and Risk Characterization. http://cfpub2.epa.gov/ncea/cfm/recordisplay.cfm?deid=72117, accessed September 10, 2004.
19. Calgon Carbon Corporation. (2001). Groundwater Remediation Technologies Analysis Center, About Perchlorate. http://www.perchloratenews.com/about-perchlorate.html, accessed July 28, 2004.
20. Environmental Working Group (EWG). (2004). Rocket fuel in drinking water: Perchlorate pollution spreading nationwide. http://www.ewg.org/reports/rocketwater/, accessed July 28, 2004.
21. Lee, M. (2004). Perchlorate threat looms for farmers. *The Sacramento Bee*, July 26. http://www.sacbee.com/content/news/v-print/story/10140109p-11060888c.html, accessed July 26, 2004.
22. Bustillo, M. (2003). Tests link lettuce to toxin in water. *The Mercury News*, April 28. http://www.mercurynews.com/mld/mercurynews/news/5734263.htm, accessed July 28, 2004.
23. Silva, A. (2002). Perchlorate goes from obscurity to point of controversy. *San Bernardino County Sun*, December 14. http://www1800lawinfo.com/practice/printnews.htm?story_id=4009, accessed July 28, 2004.
24. Calgon Carbon Corporation. (2001). Groundwater Remediation Technologies Analysis Center, Case study, Calgon Carbon Corporation—ISEP (R) Continuous Ion Exchange. http://www.perchloratenews.com/case-study1.html, accessed July 28, 2004.
25. Spillman, B. (2003). Boxer's perchlorate bill doesn't give EPA enough time. *The Desert Sun*, March 10. http://www.thedesertsun.com/, accessed July 28, 2004.
26. Food and Drug Administration (FDA). (2004). Center for Food Safety and Applied Nutrition, Office of Plant and Dairy Foods. Perchlorate questions and answers. June 22, 2004. http://www.cfsan.fda.gov/~dms/clo4qa.html, accessed July 28, 2004.
27. Thompson, D. (2004). State a year tardy in water standards for rocket fuel. *Contra Costa Times*, January 29, 2004. http://www.contracostatimes.com/mld/cctimes/news/7823750.htm, accessed July 28, 2004, SB 1822 Statutes of 2002, available for review at http://www.leginfo.ca.gov.
28. California Department of Health Services (CDHS). (2004). Perchlorate in drinking water: action level July 1, 2004, http://www.dhs.ca.gov/ps/ddwem/chemicals/perchl/actionlevel.htm, accessed July 28, 2004.
29. Office of Environmental Health Hazard Assessment (OEHHA). (2003). California Environmental Protection Agency (Cal-EPA). Frequently asked questions about the public health goal for perchlorate. http://wwwoehha.org/public_info/facts/perchloratefacts.html, accessed July 28, 2004.
30. Environmental Protection Agency (EPA). (2003). Large water systems emergency response plan outline: Guidance to assist community water systems in complying with the public health security and bioterrorism preparedness and response act of 2002. Document EPA 810-F-03-007, July 2003. http://www.epa.gov/safewater/security, accessed September 21, 2004.
31. Durant, J., Chen, J., Hemond, H., and Thilly, W., (1995). Elevated incidence of childhood leukemia in Woburn, Massachusetts. *Environ. Health Perspect. Suppl.* **103**(56): September.
32. Kerr, M. et al. (1999). Survival of *Escherichia coli* O157:H7 in bottled natural mineral water. *J. Appl. Microbiol.* **87**(6): 833–849.
33. California Health and Safety Code, Section 111145, 111165, 116455 http://www.leginfo.ca.gov/accessed August 4, 2004.
34. Environmental Protection Agency (EPA). (2002). *E. Coli* O157:H7 in drinking water. http://www.epa.gov/safewater/ecoli.html, November 26, 2002, accessed September 10, 2004.
35. California Department of Health Services (CDHS). (2004). Consumer Confidence Reports February 13, 2004, http://www.dhs.ca.gov/ps/ddwem/publications/ccr/ccrguideca-02-13-04.pdf, accessed September 28, 2004.
36. California Department of Health Services (CDHS). (2002). Public notification requirements for drinking water regulation violations—proposed regulations. Document R-59-01, pp. 49–52, 58, November 1, 2002.

37. Environmental Protection Agency (EPA). (2002). Final state implementation guidance for the public notification (PN) rule. http://www.epa.gov/safewater/pws/impguide.pdf, accessed September 28, 2004.
38. Ware, P. (2002). EPA will check drinking water assessments for completeness but will not grade them. December 27, 2002. http://www.amsa-cleanwater.org/advocacy/security/articles02.cfm, accessed September 10, 2004.
39. Safe Drinking Water Act, 2002, Title IV—Drinking Water Safety and Security, Sections 1433–1435.

READING LIST

Environmental Protection Agency (EPA). (2002). MTBE in drinking water. http://www.epa.gov/safewater/mtbe.html, accessed September 14, 2004.

Environmental Protection Agency (EPA). (2002). Proposed ground water rule. EPA Document 815-F-00-003, April 2002. http://www.epa.gov/safewater/gwr.html, accessed September 27, 2004.

Environmental Protection Agency (EPA). (2002). Radionuclides in drinking water. November 26, 2002. http://www.epa.gov/safewater/radionuc.html, accessed September 27, 2004.

Los Angeles Department of Water and Power (LADWP). (2003). Cryptosporidium, General Information. http://www.ladwp.com/ladwp/cms/ladwp001423.jsp, accessed July 26, 2004.

A WEIGHT OF EVIDENCE APPROACH TO CHARACTERIZE SEDIMENT QUALITY USING LABORATORY AND FIELD ASSAYS: AN EXAMPLE FOR SPANISH COASTS

T.A. DelValls
M.C. Casado-Martínez
I. Riba
Facultad de Ciencias del Mar y Ambientales
Cádiz, Spain

M.L. Martín-Díaz
J.M. Blasco
Instituto de Ciencias Marinas de Andalucía
Cádiz, Spain

N. Buceta Fernández
Centro de Estudios de Puertos y Costas
Madrid, Spain

THE WEIGHT OF EVIDENCE APPROACH—DEFINITIONS AND LINES OF EVIDENCE

During recent years, different initiatives have been carried out to use multiple lines of evidence with the aim to assess sediment quality in aquatic ecosystems. One of the most useful and widely used methods was the design and application of integrative assessments to establish sediment quality (1). These methods comprise the synoptic use of different methodologies such as chemical analysis in sediments to establish contamination, in sediment toxicity to address the biological effects under laboratory conditions, chemical residue analyses in organism tissues to determine the bioavailability of the contaminants, and the structure of the benthic community or the histopathological lesions in resident organisms to determine the biological effects under field conditions (1,2). One of these integrative assessments widely used in the studies of sediment quality is the Sediment Quality Triad that was used synoptically for the first time in Spain and in Europe in the early 1990s (3,4). This method also permits using convenient statistical tools to derive sediment quality guidelines (SQGs), which can be used to derive concentrations associated and not associated with the biological effect in the overlapping area of the different lines of evidences (Fig. 1).

Recently, these integrative assessments, and especially the SQT, have suffered some modifications in their interpretation and the representation of their results. In this sense, new methodologies have been incorporated to establish the chronic effects under laboratory conditions, some new parameters were incorporated to avoid the influence of natural casuistry, and especially some new format or representation of the data have been proposed to avoid loss of information, which lacks association with the classic representations (1,6).

The main change suffered by the integrative assessment is the change in the names of these methods and they are more precisely known during recent years as weight of evidence approaches (WOE) and defined as concepts instead of as methods. The WOE approach is defined by Burton et al. (7) as "...a process used in environmental assessment to evaluate multiple lines-of-evidence concerning ecological conditions... and include

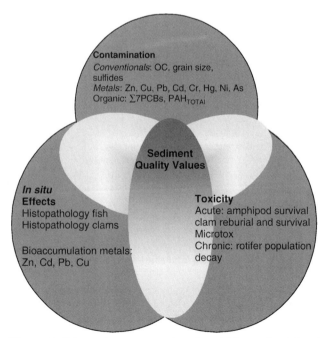

Figure 1. Schematic representation of a widely used weight of evidence approach such as the Sediment Quality Triad (SQT). Adapted from Riba et al. (5).

assessment of impairment, prioritization of site contamination, and decision-making on management actions." This definition goes farther than the classic definition related to the integrative assessment and the Sediment Quality Triad that were considered methods or concepts instead of processes, including the decision-making process.

In this sense, we have reviewed here the evolution of the different lines of evidence that have been growing during recent years associated with the use of this kind of WOE approach in Spain from the classic application of the Triad in 1990s and with emphasis on the potential new application for the decision-making processes in the management of sediment quality, including dredged material characterization. Basically, it describes the application of the different lines of evidence in new environmental problems in Spain that have been and are the marine accidental wastes, the management of dredged material, and the assessment of sediment quality in aquatic ecosystems. It describes the use of a conveniently designed battery of sediment toxicity tests to be incorporated in the tier testing approach for the management of the dredged material in Spain, the use of caged animals to help in the relationship between biological effects measured in laboratory and field conditions, the use of bioaccumulation studies either at field or at laboratory conditions to determine the bioavailability of contaminants, and finally, the use of histopathological methods as part of the weight of evidence approach and its use in combination with bioaccumulation of contaminants in different tissues to derive tissue quality values (TQVs) (5,8,9). Finally, some recommendations of the correct process to integrate these results under a weight of evidence approach using the sediment quality triad is proposed for further sediment quality assessment in Spain.

DESIGN OF ACUTE SEDIMENT TOXICITY TESTS FOR THE CHARACTERIZATION OF DREDGED MATERIAL IN SPAIN

The incorporation of the toxicity test in the management of dredged material disposal is part of a tier testing that is under final development by CEDEX and uses a battery of sediment toxicity tests as a complementary tool to the classic physicochemical analysis using a modification of the sediment quality guidelines (10). This battery of toxicity tests includes the commercial screening test Microtox® in the same samples in which chemical analyses are necessary. If doubts about hazardousness of these samples are found in the other two tests, the bioassay using amphipods and the bioassay using sea urchin larvae are conducted classifying the samples as hazardous if any of the tests show positive responses compared with the biological guidelines proposed.

THE USE OF CHRONIC, BIOACCUMULATION, AND SUBLETHAL BIOASSAYS TO CHARACTERIZE SEDIMENT QUALITY IN DREDGED MATERIAL

One of the recent improvements proposed by different agencies is the possibility to use chronic bioassays having endpoints different from mortality. The main objective is to compare the chronic and acute responses using bioassays to characterize the quality of sediments or dredged material.

The toxicity tests involve the comparison of the three proposed acute sediment toxicity tests versus different chronic bioassays using fish (*Solea senegalensis*) and clams (*Scrobicularia plana*) and using different sublethal endpoints such as the modification of biomarkers of exposure (methallotioneins, EROD, enzymatic activities of the oxidant stress, etc.) and biomarkers of effect (vitellogenin/vitellin; histopathology). Besides, the complete cycle of an amphipod species (*Corophium volutator*) is used to address the sediment quality of this material. Results suggest that the biomarkers are a powerful tool to identify the effect in the gray areas in which acute toxicity tests do not show positive responses. In this sense, the use of histopathological studies has permitted the identification of moderate toxicity associated with PCBs that showed absence of effect using the battery of acute toxicity tests (11).

The complexities at issue are even greater in the case of evaluating concerns for fish and wildlife at a management site. Assessing the potential for impacts of concern on such receptors requires information about the bioaccumulation potential of contaminants from the sediments or dredged material (e.g., results from a bioaccumulation test), how receptors use the management site, and how specific contaminants will move through the local food web. The proposed chronic tests can be used in the same design as bioaccumulation tests and different chemicals could be analyzed in the same tissues in which biomarker, including histopathological, analysis was performed. These studies will permit one to address the mobility of the different contaminants through the food chain and, specifically in some contaminants, to identify the hazardousness associated with the dredged material, such as in the case of PCBs that can suffer biomagnification processes.

The link between the chemical residues and the analysis of biomarker, especially including histopathological diseases in the same tissues by means of bioaccumulation-chronic tests, could permit one to derive tissue quality values (TQVs) in a similar approach to those used to derive sediment quality values or guidelines (SQGs) as reported by Riba et al. (5,9). These TQVs could be used to prevent the risk associated with the contaminants in human food of commercial species and contaminants that cannot be identified by the standardized acute toxicity tests.

USING *in situ* APPROACHES TO CHARACTERIZE DREDGED MATERIAL IN SPAIN—CAGING ANIMALS

Sediment toxicity tests provide information about toxicity or hazard to sediment-dwelling organisms. However, exposure conditions for benthos at a management site may differ significantly from those occurring in the laboratory. Understanding this difference will require information about the management site, including hydrodynamics (is the site dispersive or depositional?), the area extent of coverage once the material is deposited, how organisms at the disposal site interact with the sediment and overlying water, and so on. In this sense, different initiatives have been carried out in Spain to address the biological effects under field conditions funded by the Ministry of Science and Technology (DREDGED REN2002_01699/TECNO; TRIADA; VEM2003_−20563/INTER).

One of the studies designed to address part of the deficiencies associated with laboratory assays is the use of caged animals at the dredged or disposal site to the measurement of different endpoints, either lethal (12,13) or sublethal (8,13). Some designs of benthic chambers are used with this aim; an example is shown in (Fig. 2). The organisms, basically benthic, are confined in the chamber for ideally the same period of exposure as that used in the bioaccumulation-chronic laboratory assays. The tests need the anchorage of either field-collected or laboratory-cultured animals under field conditions exposed to the sediment of concern. The advantages compared with laboratory-exposed devices are that the exposure conditions are representatives of actual conditions at the site. The disadvantages of these field assays are related to the complexities of field-based experimentation, including the difficulties to discriminate the endpoints from noncontaminant variables.

In caged experiments, the same parameters as those registered in the laboratory assays are included in the bioassays under field conditions: biomarkers of exposure and effects, chemical concentration in the biological tissues, mortality, and other sublethal endpoints. Another advantage associated with these studies is the potential complementary use with the laboratory tests that permits one to determine the biological effect under field conditions and, by comparing with the laboratory responses, to discriminate some natural causality that is associated with other kinds of approaches to establish biological effect under field conditions, such as the macrobenthic community studies (10).

Figure 2. Schematic representation of the benthic chambers used to expose caged organisms under field conditions. The animals are transported from the laboratory in cold poliespan boxes to the boat. The chambers (50 cm by 25 cm by 15 cm) were immersed and anchored in each sampling station from every port by scuba divers.

FUTURE DEVELOPMENTS

The complexity of the matrix defined by sediments or dredged material usually located in highly dynamic ecosystems, such as the littoral areas and especially the estuaries, ensures that successfully using any analytical tool requires having an appreciation for the uncertainties associated with its use. No single test or evaluative tool is able to provide a complete and accurate picture of the complexities inherent in evaluating contaminated sediments. In Table 1, some of the sources of uncertainty associated with applying biological tests to manage dredged material are shown.

Notwithstanding the uncertainties included in Table 1, biological tests provide the most direct and certain means for assessing the potential for toxicity and contaminant bioavailability in sediments.

Other contaminants exist that are not usually considered in the studies of sediment quality, such as the presence of pathogens in sediments. Microbial pathogens in aquatic systems can contaminate drinking water supplies and/or shellfish and are responsible for hundreds of beach closure events annually. The proximity of sediments to be dredged to sources of pathogens (e.g., sewage outfalls and agricultural runoff) and the presence of conditions favorable for the long-term viability of microbes makes pathogens in sediments or dredged material a matter of growing concern. Unfortunately, standard microbial methods in use worldwide fail to provide adequate information on which to base management decisions. Some pathogens of concern are not associated with the widely used fecal coliform indicator. In addition, some pathogens cannot be cultured from environmental samples using standard media and conditions. Molecular methods are currently being developed to address the deficiencies in current approaches. The molecular approaches under development are not based on the need to culture the pathogens of concern but on extracting and analyzing their DNA. The use of gene-probe technology has the potential to provide a reliable and highly specific method for detecting a pathogen as well as its virulence potential, that is, its ability to cause disease. When such methods are refined and available for widespread use, more credible assessments of pathogens in sediments will be possible.

Another point of future research is related to the effects of two key environmental variables, such as pH and salinity in aquatic ecosystems. These two variables control most processes occurring in these environments, and they are especially of interest in the border of aquatic systems, such as those defined by estuaries between marine and freshwater environments. In this kind of ecosystem, these variables, and especially salinity, can play a significant role not only in the partitioning of the contaminants but in the bioavailability and thus in the toxic effects associated with the chemicals. Recently, Riba et al. (14,15) reported the high influence of these variables in the bioavailability of metals from a mining spill that occurred in Spain in April 1998 that affected the Guadalquivir estuary. Their results showed that the same sample tested at different salinity or pH values using a truly estuarine species (*Ruditapes phillipinarum*) showed significantly higher

Table 1. Summarized List of Most Common Uncertainties Associated with Biological Testing of Dredged Material. The Uncertainty Description and the Palliative Action to Minimize the Effects on the Final Results Are Included

Topic	Uncertainty	Palliative-Minimizing
Handling and sampling	How the sample is taken and handled prior to testing, and the conditions the sample is subjected to during testing, can affect the accuracy and precision of test results	Careful and standardized procedures conducted during sediment assessments are necessary to minimize the source of uncertainty in sediment assessment
Extrapolation and number of species in the battery	The battery of tests will be composed of a relatively small number of taxa in comparison with the number residing at the site of concern	Testing with multiple species that are closely associated with the sediment and have a demonstrated sensitivity to the contaminants of concern
Dynamism of the ecosystem	Provide only a snapshot of the processes affecting contaminant exposure and effects in sediments	Development of more significant tests including chronic tests (at least one cycle of life of the organism)
Space scale is limited	The laboratory tests are conducted under controlled conditions of space and may overestimate the extent of exposure and effects given that the movement of the organism are restricted to the material being tested, and water movement (i.e., flow) is minimal	Conduct bioassays under field conditions increasing the space scale in which they are performed to avoid the lack in the extent of exposure and effects associated with the mobility of the organisms
Extrapolation to ecological scales	Biological tests most commonly measure effects on individual organisms, whereas the ecological impacts of concern occur at the level of populations and communities	Selection of the appropriate battery of ecologically relevant organisms
Experimental variables	Test organisms will respond positively and negatively to different experimental variables that are unrelated to the degree of contamination present in a sediment (grain size, food frequency, etc.)	Developing a thorough experimental understanding of the biology and ecology of the test organisms used in biological tests
Results analysis (statistical)	Over or underestimate of the hypothesis used in the different tests	Ensuring that adequate replication is prescribed to detect a desired magnitude of effect

toxicity at salinity values of 10 or lower or at pH values of 6 or lower compared with the toxicity tested at higher salinity or pH values. They relate the effect based on the higher mobility of some metals from sediments to water and also in the speciation of them that changed when the pH and salinity values changed. It informs one of the necessity to incorporate this kind of environmental effect in the final design of the weight of evidence approach, taking into account the potential influence of the pH and salinity values, especially under estuarine conditions.

Finally, we recommend to integrative assessment or the more recently cited weight of evidence approaches to avoid or at least minimize most of the deficiencies and uncertainties associated with laboratory and field bioassays. Furthermore, specifically applied WOEs for sediments or dredged material quality assessment is the use of the sediment quality triad incorporating some modifications and improvements to assess the sediment quality in both the disposal and the dredged site. One premise of the method is the design of a tier testing approach based on the definition of the Triad (Fig. 3). In this sense, the incorporation of new chemicals of concern specifically detected in the area to be dredged should be taken into account in the first tiers together with some biological tests for screening of the biological effects. The next steps will lead one to conduct laboratory sediment toxicity and bioaccumulation (if contaminants of concern are present) tests. In this sense, design specific and low-cost chronic tests, including sublethal endpoints such as histopathology, to avoid false negatives of toxicity such as those associated with specific contaminants. In the last tiers, the use of in situ surveys or those assays under field conditions using caged animals to establish the overall pollution status of the sediment to be dredged are included. The integration of these measurements will permit one to address the environmental quality of the systems and, furthermore, to establish the correct decision-making when selecting the beneficial use and/or the disposal options of the dredged material.

These recommended methods are not unique and a more extensive list of potential lines of evidence (7) that can be incorporated into the final tier testing using the weight of evidence approaches, such as the Triad, are included in Table 2. Each of these lines of evidence (LOEs) has their own advantages and limitations, but what is clear is that only one LOE will not be sufficient to address the sediment quality to determine the different management options for dredged material, and the use of several of these LOEs under the final design of a tier testing is recommended using a weight of evidence approach, such as that shown in Fig. 3.

Acknowledgments

This work has been partially funded by grants funded by the Spanish Ministry of Development (BOE 13-12-02) and of Education and Science (REN2002_01699/TECNO and VEM2003_−20563/INTER). The design of the acute toxicity tests for dredged material characterization was conducted under joint contract between CEDEX and the University of Cádiz (2001 and 2003).

Table 2. Description of Different Lines of Evidence, Including Their Advantages and Limitations When Used to Assess Sediment Quality and Dredged Material Characterization (7)

Line of Evidence	Advantage	Limitation
Sediment chemistry	Relatively simple and standardized. Widely used and utility proved. Cause information	Lack in the effect of information. Assumed feasibility to measure all existing chemicals
Toxicity tests (laboratory)	Relatively simple and standardized. Utility proved and widely used. Effect information	Lack in the determination of cause. Difficulty in field extrapolation. Natural stressors not assessed
Tissue chemistry	Determination of bioavailability and measure of exposure. If biomagnifications proved: useful in food chain models (human risk)	Lack in the effect measurements. Difficult to eliminate influence from essential and nonessential chemicals, food chain measurements, acclimation/adaptation
In situ alteration (field studies)	Measurements under real environment conditions	No cause is identified. No standard method. Difficult to discriminate natural effects
Biomarkers of exposure (EROD, methallotioneins, etc.)	Indicators of exposure under either field or laboratory conditions, including natural variables	No clear identification of cause, although relative information. Relative relationship to effects. Influenced by natural variables
Biomarkers of effects (histology, etc.)	Indicators of effects under either field or laboratory conditions, including natural variables	Lack in the cause identification. Nonspecific to contaminants and affected by natural variables
Benthic community structure	Standard methods, *in situ* conditions. Long-term measure of effects, including natural variables	Highly affected by natural variables. Not predictive. Confounding factors different from stressors. Absence of organisms in highly disturbed (natural or not) areas
Laboratory TIE	Partitioning of chemicals under laboratory conditions	Not widely proved. Insensitive. Not available for all chemicals
Field TIE	Partitioning of chemicals under field conditions	Highly complex design on the device used. Problems with interpretation of the results. Not possible for all the chemicals. Not standardized
Bethic fluxes of nutrients and/or contaminants	Functional capability of the ecosystems	High complexity in the devised use. No standard methods. Not widely used for sediment quality but for biogeochemical cycles
Caging animals	Toxicity and chemical residues under field conditions. Direct exposure to sediments	Difficult to clearly extrapolate to real conditions. Complex to devise and casualty inherent
Tissue quality values (TQVs)	Relationship between sublethal and chemical residues in organisms tissues	Not standardized. Relatively new. Large and temporal data base. Not easily available for all chemicals

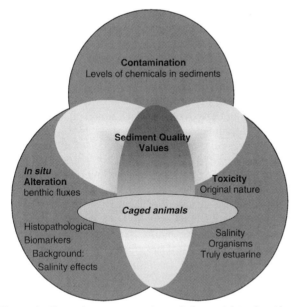

Figure 3. Synoptic representation of the weight of evidence approach recommended in the design of a convenient tier testing method either to assess sediment quality in littoral ecosystems or to manage dredged material. Each circle represents different areas including different lines of evidence (Table 2) to be integrated in the overall integrated approach. The overlapping area permits derivation of sediment quality values and tissue quality values for those overlapping areas, including histopathological and bioaccumulation studies under either field or laboratory conditions.

BIBLIOGRAPHY

1. Chapman, P.M. (2000). *Int. J. Environ. Pollut.* **13**: 1.
2. DelValls, T.A., Riba, I., and Martín-Díaz, M.L. (2004). *Using the triad to characterize dredged material in estuaries: lacks and improvements needed*, 4th SedNet Workshop "Harmonization of Impact Assessment Tools for Sediment and Dredged Materials." June 10–11, San Sebastian, Spain.
3. DelValls, T.A., Forja, J.M., and Gómez-Parra, A. (1998). *Environ. Toxicol. Chem.* **17**: 1073.
4. DelValls, T.A. and Chapman, P.M. (1998). *Cienc. Mar.* **24**: 336.
5. Riba, I. et al. (2004). *Chemosphere* (in press).
6. Riba, I., Forja, J.M., Gómez-Parra, A., and DelValls, T.A. (2004). *Environ. Pollut.* in press.
7. Burton, G.A. et al. (2002). *Hum. Ecol. Risk. Assess.* **8**: 1675.
8. Martín-Díaz, M.L. (2004). *Determinación de la calidad ambiental de sistemas litorales y de estuario de la Península Ibérica utilizando ensayos de campo y laboratorio*, Tesis Doctoral, Universidad de Cádiz.
9. Riba, I., Blasco, J., Jiménez-Tenorio, N., and DelValls, T.A. (2004). *Chemosphere* in press.
10. DelValls, T.A., Lubián, L.M., Forja, J.M., and Gómez-Parra, A. (1998). *Environ. Toxicol. Chem.* **16**: 2323.
11. DelValls, T.A. et al. (2005). *Track. Trend. Anal. Chem.* (submitted).
12. Riba, I., Casado-Martínez, M.C., Forja, J.M., and DelValls, T.A. (2004). *Environ. Toxicol. Chem.* **23**: 271.

13. Martín Díaz, M.L., Blasco, J., Sales, D., and Del Valls, T.A. (2003). *International Conference on Remediation of Contaminated Sediments*. Batelle Press.
14. Martín Díaz, M.L., Jiménez-Tenorio, N., Sales, D., and DelValls, T.A. (2005). *Mar. Pollut. Bull.* (submitted).
15. Riba, I., García-Luque, E., Blasco, J., and DelValls, T.A. (2003). *Chem. Speciation Bioavailab.* **15**: 101.

REMEDIATION AND BIOREMEDIATION OF SELENIUM-CONTAMINATED WATERS

RICHARD M. HIGASHI
TERESA A. CASSEL
University of California
Davis, California

JOSEPH P. SKORUPA
U.S. Fish and Wildlife Service

TERESA W.-M. FAN
University of Louisville
Louisville, Kentucky

GENERAL AND HISTORICAL BACKGROUND

The trace element selenium (atomic symbol Se) occurs naturally in soils, water, and biota, including food. It is nutritionally required due to various Se-bearing proteins that incorporate selenocysteine, now recognized as the 21st essential amino acid. In addition to being a required nutrient, Se in excess is toxic to biota (1). The first recorded cases of Se toxicity were penned by Marco Polo during his travels in western China, based on symptoms he observed in livestock (2). In the 1930s, the element Se was attributed to these symptoms, by then known variously as "alkali disease" and "blind staggers," which afflicted livestock in the western United States (3). However, in certain cases, such symptoms could have been due to other factors (4).

For Se, the margin between nutritional requirement and toxicity is unusually narrow and depends on the individual species and circumstance (3). This fact creates a quandary when setting environmentally protective Se criteria. Health benefits of consuming vegetables high in Se are widely touted (5), as cases of human toxicity and crop damage from Se are rare (6). Livestock face both deficiency and toxicity (7), while toxicity is the primary concern with fish, birds, and in particular various aquatic-associated wildlife (8). The latter issue roared to the news headlines as the aquatic bird disaster at California's Kesterson Reservoir surfaced in the early 1980s (9)—a harbinger of widespread Se problems such as fish mortality and deformities in Belews Lake (10) and numerous other cases internationally (8).

A closer look at the affected biota and circumstances surrounding Se toxicity reveals *water* as the transmitting medium of highest concern (8). Natural water processes (precipitation) and human activities relocate the naturally occurring Se in soils, rocks, and groundwater into collected water bodies that harbor and attract wildlife (8), which might also be used for growing forage and watering livestock (6). Human mobilization of Se mainly consists of irrigation, mine drainage, and surface discharge of groundwater; the latter includes petroleum processing, agriculture, and urban uses. We focus here on the factors important in Se-contaminated water; for the reader interested in the physiology and biochemistry of Se wildlife toxicity, there is an excellent recent review on this complex topic by Spallholz and Hoffman (11).

UNDERSTANDING THE PROBLEM: FUNDAMENTALS OF SELENIUM BIOGEOCHEMISTRY

In order to solve a problem, it is essential to first understand the nature of the problem. In the case of Se toxicosis of wildlife, exposure to Se primarily occurs through the diet, not by direct exposure to water (12). Therefore, it is vital to understand the "biogeochemistry" of Se: how it moves from contaminated water and sediment to work its way up the foodweb. Figure 1 illustrates the various paths comprising the biogeochemistry of Se. This figure illustrates the first four fundamental facts about wildlife exposure to Se:

1. The paths of Se are numerous and interrelated.
2. Most of the paths eventually lead up the foodweb.
3. Se changes chemical form from an inorganic salt to various organic forms.
4. Some of the organic forms lead to volatilization, a natural path for Se to leave an aquatic system altogether.

A fifth important fact, not evident in Fig. 1, is that Se "bioaccumulates" as it heads up the foodweb, generally becoming more concentrated in tissues. This concept may be familiar to many readers, as this phenomenon is discussed widely in the general press and health advisories regarding foodweb bioaccumulation of mercury, pesticides, polychlorinated biphenyls (PCBs), and many other contaminants. Table 1 (13) is an example from the scientific literature that documents this process for Se. Note the very large variation in bioaccumulation factors (BCFs), illustrating the effects of site-specific biogeochemistry in different aquatic systems. A comprehensive study illustrating the complex pathways of bioaccumulation has recently been published (14).

SETTING THE REMEDIATION TARGET: THE AQUATIC LIFE CRITERION

For remediation to occur, there must first be measurable targets to achieve. The complexity of Se biogeochemistry makes setting such a target extremely difficult to accomplish or even define. An earlier United States Environmental Protection Agency (U.S. EPA) Water Quality Criterion for Se, which set a maximum of 5 µg/L total Se, provided a clear target (15).

Figure 1. Biogeochemical cycling of Se in aquatic ecosystem. This scheme is modified from Reference 13. Arrows indicate processes that can lead to risk from foodweb accumulation of Se ("ecotoxic" risk). Other arrows trace the Se volatilization process by which Se can be lost from the aquatic system.

a, Uptake and transformation of Se oxyanions by aquatic primary and secondary producers; much of the biotransformation pathway is yet to be defined.
b, Release of selenonium and other organic Se metabolites by aquatic producers.
c, Uptake of organic Se compounds by aquatic producers.
d, Abiotic oxidation of organic Se compounds to Se oxyanions.
e, Release of alkylselenides from selenonium or other alkylated Se precursors through abiotic reaction.
f, Release of alkylselenides from selenonium or other alkylated Se precursors through aquatic producers.
g, Volatilization of alkyselenides into the atmosphere.
h, Oxidation of alkylselenides to Se oxyanions.
i, Formation of red amorphous Se element by aquatic and sediment producers.
j, Detrital formation from aquatic producers.
k, Se bioaccumulation into the foodchain with potential ecotoxic consequences; the toxic form(s) are yet to be defined.
l, Assimilation of waterborne selenium oxyanions into sediment biota.
m, Oxidation of sediment Se(0) to oxyanions.
n, Reduction of sediment Se(0) to Se(−II) or vice versa.
o, Assimilation of sediment Se(−II) into sediment biota.
p, Oxidation of sediment Se(−II) to selenite.

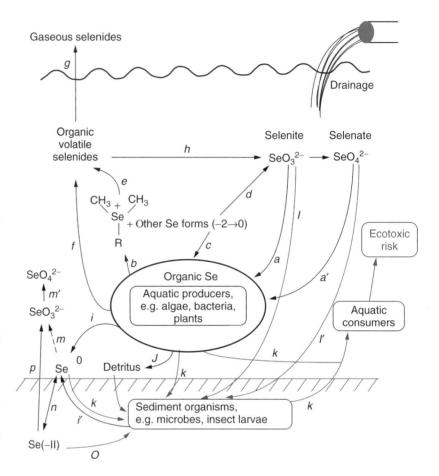

The setting of this temporary value spawned numerous attempts to apply the three traditional strategies of remediation, which are containment, removal, and treatment (16).

However, such a criterion for remediation remains very much a "moving target" due to the complexity of the biogeochemistry. In particular, there are currently attempts to develop a more scientifically sound U.S. EPA Aquatic Life Criterion for Se (17), for which there is now a draft (18). What follows is a very abbreviated description of the issues surrounding the setting of an Aquatic Life Criterion. For a more complete understanding, the reader should refer to the cited literature.

Several independent studies submitted to the U.S. EPA estimated that the 5-μg/L Se toxicity threshold was either too high or too low; interestingly, arguments for higher or lower thresholds appear to be closely associated with whether researchers were affiliated with the corporate-service (arguing that 5 μg/L is too low) or public-service (arguing that 5 μg/L is too high) scientific communities. A slightly different formulation of this observation has also been presented elsewhere (19). Additionally, the proposed California Toxics Rule of 5 μg/L Se in water (20) was judged by the U.S. Fish and Wildlife Service to be too high (21), jeopardizing 15 species protected by the Endangered Species Act (22), the majority of which were aquatic-dependent wildlife that do not actually live *in* the water. The lesson here is that, because of the biogeochemistry of Se, it is important to include consideration of the entire aquatic-based foodweb, not just aquatic organisms.

Recently, the U.S. EPA contracted the Great Lakes Environmental Center to derive chronic Se criteria on a fish-tissue basis, rather than the traditional water concentration basis. The results of this analysis were considered to have limited applicability because it produced a threshold (7.9 μg Se/g tissue) based on an LC_{20} (lethal concentration at which 20% mortality is expected), which is not adequately protective of the species of concern (22). Several other flaws were found in the study, which essentially compiled experimental data from 17 published studies (18). One of these (23) is qualitatively distinct because it incorporated a "winter-stress" design accounting for the increase in toxicity of dietary Se to birds, fish, and mammals under low temperatures. Adjustment of these data for the

Table 1. Excerpted Data from Reference 13 Illustrating Bioaccumulation of Se and the Great Variance in the Bioconcentration Factors of Se

	Salinity, ppth	Waterborne Se, μg/L	Body Burden, μg/g	BCF
Algae	76	13.5	11.8	873
Algae	14	7.9	14.8	1877
Algae	41	7.0	16.0	2281
Algae	136	13.1	9.7	745
Algae	80	13.3	22.6	1695
Algae	10	6.0	1.5	247
Algae	47	5.2	11.7	2273
Algae	66	5.1	9.9	1954
Algae	54	8.7	8.6	978
Average	58.2	8.8	11.8	1436
S.D.	38.1	3.5	5.8	738
Midge larvae	12	5.2	18.1	3489
Benthic composite	76	5.8	7.4	1288
Water column composite	76	5.8	12.3	2130
Corixid	39	7.5	8.9	1187
Corixid	90	8.8	6.8	778
Artemia + corixid	126	11.5	11.7	1012
Artemia	108	12.2	9.0	740
Corixid	108	12.2	10.1	834
Artemia	47	5.6	9.8	1749
Artemia	66	4.9	16.6	3372
Artemia	54	9.5	10.9	1145
Midge larvae	8.3	2.5	42.3	16886
Artemia	67	506.0	19.6	39
Average	67.5	46.0	14.1	2665
S.D.	35.8	138.3	9.4	4389

chronic-level protection needed (<5% mortality) would result in a limit of less than 5.8 μg/g on a whole-body fish-tissue basis.

Situations under which fish tissues would be contaminated to that level, however, may still be unacceptable to birds eating lower on the trophic level (e.g., invertebrate eaters) (24). Bioconcentration factors relating the transfer of Se from aquatic invertebrates to whole-body tissue of invertebrate-eating fish have been documented in the field to vary from 0.67 to 1.36, when whole-body fish tissue contains 5–10 μg/g Se. Applying those bioconcentration factors to the recently proposed fish-tissue criterion for Se of 7.9 μg/g (18) suggests that the corresponding Se content of aquatic invertebrates would range from 5.8 to 11.8 μg/g, which would result in an EC_{20}–EC_{85} range (effects concentrations, a range for which 20–85% of a population would show toxic effects) in mallard ducks (and other birds) (23) that also utilize the invertebrates as food (24).

STRATEGIES FOR REMEDIATION

From the preceding section, it is evident that there is no universally accepted water or tissue threshold concentrations for the protection of wildlife, and the issues are likely to remain unsettled for some time. Therefore, there is no fixed "target concentration" for remediation to achieve. This section on remediation strategies is written with this situation in mind.

Most of the past remediation attempts are mentioned only briefly here, as they were based on the obsolete 5-μg/L water criterion and do not incorporate the vital concepts of Se biogeochemistry. The following description attempts to discuss Se remediation efforts on the basis of the three strategies of remediation, which are containment, removal, and treatment (16). However, as the reader will see for Se, sometimes none of these three textbook categories are appropriate.

Containment

The first strategy of remediation, containment, remains difficult to achieve in many cases, as much Se contamination occurs in open lentic systems such as reservoirs, in lotic systems such as streams and rivers, and in receiving waters such as deltas and estuaries. To this end, the U.S. EPA has embarked on regulation based on the total maximum daily load (TMDL) for lotic and receiving waters, under Section 303(d) of the Clean Water Act (25). More recently, the best waterfowl protective estimates (based on empirical data from multiple environments) argue for reduction of the old 5-μg/L limits (8), forcing TMDL limits downward. This is an illustration of the "moving target" frustrating remediation efforts. Clearly, any remediation strategy must be compatible with changing regulations, but despite this clarion call, no traditional remediation strategies have met the challenge. For further description of TMDLs in general, the reader is referred to Reference 25.

The TMDL for Se basically collapses the complexity of Se biogeochemistry into a single box and a single total Se value, which may be protective of a given receiving water. However, this approach does not engender remediation solutions, since it does not account for Se biogeochemistry, provides little guidance for lotic system remediation, and provides no guidance at all for remediation of lentic systems.

In the United States, the lotic Se containment strategy is already implemented under TMDL regulations in California and elsewhere (25). For example, selected agricultural drainage from California's San Joaquin Valley is metered into the San Joaquin River, the allowed volume of which is based on quick turnaround time of Se analysis of the discharged water. For descriptions and issues surrounding the implementation of Se TMDL, the reader is referred further to Reference 25.

Simple containment of lotic Se into terminal water evaporation basins can avoid restrictive TMDL regulations since there is no discharge of water into lotic systems, and this strategy therefore has been suggested as worthy of further investigation (26). However, those authors also recognized that this simply obligates the discharger to use another remediation strategy, since the real problem is foodweb accumulation from either lotic or lentic systems. Therefore, while such total containment might "remediate" the downstream lotic systems, there is still a need to remediate the contaminated lentic system that has been created in the process.

Removal

The strategy of removal of Se is highly vulnerable to changing threshold regulations, as discussed earlier. Moreover, simple removal of Se is not necessarily "remediation" because the focus is put on waterborne concentrations of Se. In spite of this, several removal strategies have been tested in the past and tests of several more are currently underway. For the few that are in the peer-reviewed published literature, the reader is referred to a recent compilation (27). Removal of Se has been categorically difficult because of its typically low starting concentrations in the parts-per-billion range, its chemical similarity to sulfur (which can be present at more than a millionfold higher concentrations), the very high volumes of water for agricultural drainage, and especially because of the ever-lowering threshold target, as discussed previously.

Treatment

The third textbook strategy, treatment of Se, has much overlap with removal approaches, and again the focus here has been on reducing waterborne concentrations of Se, which does not equate to remediation (28).

An instructive example of a combined removal–treatment strategy is the algal–bacterial process (29) using the dissimilatory reduction capability of bacteria to reduce selenate into elemental Se, which has been developed to the point of large-scale trials. Although it can remove approximately 80% of the total Se from agricultural drainage water, the microbial and algal action on Se increased the levels of the more bioavailable selenite and organic Se, resulting in 2–4 times greater Se concentrations in the test invertebrates (30).

The lesson here is that waterborne Se is an inappropriate focus for remediation, one of the key reasons for drafting a new U.S. EPA Aquatic Life Criterion for Se based on tissue concentrations (18). The algal–bacterial process is also an example of how TMDL regulations, while written to be protective of wildlife in receiving waters, can inadvertently misguide major remediation efforts because it is based on waterborne Se.

Management for Mitigation

For Se remediation, there is a fourth strategy, that of management. This strategy is embodied by, for example, a California Waste Discharge Requirement under the Porter-Cologne Water Quality Act, which specified "a program of management actions to reduce, avoid, and mitigate for adverse environmental impacts to wildlife" (26). Management with this goal is a bona fide remediation strategy.

Along this vein, there has been much effort, especially in the western United States, to manage impacts on wildlife using a watershed-based approach, some of which are briefly outlined in Reference 6. In that paper, Engberg and co-workers have stated that "of the five management options presented, mitigation (option 5) may be the only long-term solution to managing selenium." A key tool in such mitigation is the construction of "alternative" and "compensation" habitats to build wetland habitats with clean water for impacted species (6,26). Note that this remediation strategy does not draw upon any of the textbook "remediation" strategies of containment, removal, or treatment.

The mitigation approach also has its downsides. In the western United States, water supply issues can complicate the Se remediation in water drainage. For example, the cost of land and water is a major obstacle to building and maintaining compensation and alternative habitats. In general, the management approach may require extraordinary effort: it has been stated that "managers must translate the complexities of selenium chemistry and biochemistry into cogent management and regulatory approaches, all the while understanding and blending financial, economic, and social constraints into the solutions" (6). Furthermore, since Se is a semi metal (or metalloid) with a wide variety of forms in many types of samples, it is exceptionally challenging to analyze.

Natural Bioremediation of Se

As mentioned above, the containment of waterborne Se into terminal water basins eliminates downstream impacts but creates a contaminated lentic system. Use of the mitigation approach is probably needed (26), but combinations of water treatment and management of the terminal basins have also been investigated. Only a few of these take advantage of the natural biogeochemistry to focus on removal of Se from the natural foodweb accumulation. Such an approach differs from the above strategies, which have an underlying assumption that the contaminated system presents only a problem that must be "fixed" somehow (containment, removal, treatment) or bypassed (management–mitigation). Rarely is consideration given that the aquatic system may be mostly functioning "right," and only a few particular processes that are awry need to be addressed.

"Biovolatilization" of Se is one such approach, since it takes advantage of the natural biogeochemistry to remove Se. The problem with biovolatilization of any type, as Fig. 1 outlines, is that the process also draws Se into the biota, and consequently up the foodweb. This tendency has been particularly troublesome in attempts to utilize aquatic vascular plants to volatilize Se. For example, vascular plants volatilize a relatively small amount of Se while sequestering it in bioavailable foodweb materials such as the shoots and roots. Although the shoots could be harvested and disposed, the Se is mostly contained in the below-ground portions of the plants (31), which are not practical for harvesting.

Fan and Higashi (28,32) have described natural algal Se volatilization as part of an alternative remediation process in terminal basins. The overall remediation process, outlined in Fig. 2, combines volatilization of Se with interrupting the foodweb accumulation of Se (33). In this strategy, photosynthetic algae function to volatilize Se while serving as food to macroinvertebrates (brine shrimp). The brine shrimp graze the algae, ideally preventing algal accumulation and participation in the detrital cycle. In turn, the brine shrimp are harvested as a marketable product, thereby intercepting the foodweb-accumulated Se before it can impact fish and birds. Both volatilization and harvesting of brine shrimp result in a net removal of Se from the aquatic system. The resulting blockages to the foodweb accumulation of Se are shown by the X's in Fig. 2.

A full-scale evaporation basin system has been monitored for over three years in this regard (33). The terminal evaporation basins are brine, which allows the flourishing of brine shrimp that has commercial value—in fact, at California's Tulare Lake Drainage District, there has been successful commercial harvesting of brine shrimp for over five years (34). The efficacy of the approach is clear: (1) the algae that volatilize Se and feed the brine shrimp grow naturally in these basins; (2) the brine shrimp also grow naturally in these basins to a high density; (3) much of the scheme utilizes water management that is familiar to drainage operators; and (4) costs of encouraging and managing brine shrimp growth is offset by marketing harvested materials.

Comparisons with analogous systems where brine shrimp harvesting is not implemented has demonstrated that, under active brine shrimp harvesting (33):

- Waterborne Se concentrations are not increased.
- Se volatilization—a removal process of Se—appears to be enhanced.
- Concentrations of Se in algae and macroinvertebrates—a risk indicator for fish and bird toxicity—are not increased.
- Benthic macroinvertebrate biomass—a key foodweb accumulation indicator—is greatly decreased.
- Total Se as well as organic Se deposited to the sediment (available for biogeochemical refluxing as in Fig. 2) is decreased >90%.

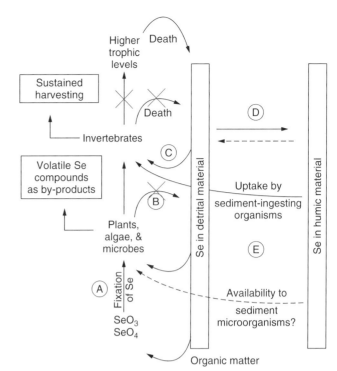

Figure 2. Bioremediation in drainage basins by reducing Se ecotoxic risk through invertebrate harvest and Se volatilization. In this "biogeochemical reflux" scheme, the drainage inorganic Se forms are initially biologically fixed by aquatic algae and microbes Ⓐ. The fixed Se does not directly head up the foodchain in the water column, as is often portrayed. Instead, a major fraction enters into organic matter, taking a detour through detritus (recently dead organic matter) and sediment Ⓑ, then reentering the foodchain at several trophic levels Ⓒ. Over longer periods, part of the detrital material is converted to recalcitrant humic material Ⓓ, locking up the Se until sediment-ingesting organisms reintroduce them to the foodchain Ⓔ. Through sustained harvesting (upper left box) of water-column invertebrates that consume algae and microbes, the bioavailable Se is removed from water, plus detrital formation resulting from the death of water-column organisms is also blocked. Both types of blockages are shown by the three X's. In turn, this would help minimize the sediment–detritus foodchain pathway for Se. In the meanwhile, additional Se can be removed by manipulating the algae/microbe community for optimal Se volatilization (lower left box).

Research is continuing to determine how this strategy can be applied to other systems and to achieve long-term sustainability. At present, the combined strategy of impounding Se-contaminated water in basins, applying the algal volatilization–foodweb interruption process, and constructing compensation/alternative habitats illustrates the advantages of an integrated management approach. In fact, such watershed-scale management appears to be required in order to remediate a contaminant with complex biogeochemistry such as Se.

BIBLIOGRAPHY

1. Cooper, W.C. and Glover, J.R. (1974). The toxicity of selenium and its compounds. In: *Selenium*. R. Zingaro and W.C. Cooper (Eds.). Van Nostrand Reinhold, New York.

2. Marsden, X. (1962). *The Travels of Marco Polo (1271–1285)*. The Heritage Press, Norwich, CT.
3. Combs, G.F. Jr., Levander, O.A., Spallholz, J.E., and Oldfield, J.E. (Eds.). (1987). *Selenium in Biology and Medicine*. Van Nostrand Reinhold, New York.
4. O'Toole, D. and Raisbeck, M.F. (1998). Magic numbers, elusive lesions: comparative pathology and toxicology of Selenosis in waterfowl and mammalian species. In: *Environmental Chemistry of Selenium*. W.T. Frankenberger and R.A. Engberg (Eds.). Marcel Dekker, New York, pp. 355–395.
5. Keck, A.-S. and Finley, J.W. (2004). Cruciferous vegetables: cancer protective mechanisms of glucosinolate hydrolysis products and selenium. *Integrative Cancer Therapies* 3(1): 5–12.
6. Engberg, R.A., Westcot, D.W., Delamore, M., and Holz, D.D. (1998). Federal and state perspectives on regulation and remediation of irrigation-induced selenium problems. In: *Environmental Chemistry of Selenium*. W.T. Frankenberger, R.A. Engberg (Eds.). Marcel Dekker, New York, pp. 1–25.
7. Maas, J. (1998). Selenium metabolism in grazing ruminants: deficiency, supplementation, and environmental implications. In: *Environmental Chemistry of Selenium*. W.T. Frankenberger and R.A. Engberg (Eds.). Marcel Dekker, New York, pp. 113–128.
8. Skorupa, J.P. (1998). Selenium poisoning of fish and wildlife in nature: lessons from twelve real-world examples. In: *Environmental Chemistry of Selenium*. W.T. Frankenberger and R.A. Engberg (Eds.). Marcel Dekker, New York, pp. 315–354.
9. Ohlendorf, H.M. (2002). The birds of Kesterson Reservoir: a historical perspective. *Aquat. Toxicol.* **57**: 1–10.
10. Lemly, A.D. (2002). Symptoms and implications of selenium toxicity in fish: the Belews Lake case example. *Aquat. Toxicol.* **57**: 39–49.
11. Spallholz, J.E. and Hoffman, D.J. (2002). Selenium toxicity: cause and effects in aquatic birds. *Aquat. Toxicol.* **57**: 27–37.
12. Presser, T.S. and Ohlendorf, H.M. (1987). Biogeochemical cycling of selenium in the San Joaquin Valley, California, USA. *Environ. Manage.* **11**: 805–821.
13. Fan, T.W.-M., Teh, S.J., Hinton, D.E., and Higashi, R.M. (2002). Selenium biotransformations into proteinaceous forms by foodweb organisms of selenium-laden drainage waters in California. *Aquat. Toxicol.* **57**: 65–84.
14. Stewart, A.R., Luoma, S.N., Schlekat, C.E., Doblin, M.A., and Hieb, K.A. (2004). Food web pathway determines how selenium affects aquatic ecosystems: a San Francisco Bay case study. *Environ. Sci. Technol.* **38**: 4519–4526.
15. U.S. Environmental Protection Agency (1987). *Ambient Water Quality Criteria for Selenium*. EPA, Washington, DC, EPA440/5-87/083.
16. Brusseau, M.L. and Miller, R.M. (1996). Remediation. In: *Pollution Science*. I.L. Pepper, C.P. Gerba, and M.L. Brusseau (Eds.). Academic Press, San Diego, Chap. 11, p. 154.
17. Sappington, K.G. (2002). Development of aquatic life criteria for selenium: a regulatory perspective on critical issue and research needs. *Aquat. Toxicol.* **57**: 101–113.
18. Aquatic Life Water Quality Criteria for Selenium (2002). *Draft Document Prepared for the United States Environmental Protection Agency*, Washington, DC, by the Great Lakes Environmental Center, Traverse City, MI, p. 67.
19. Hamilton, S.J. (2004). Review of selenium toxicity in the aquatic food chain. *Sci. Total Environ.* **326**: 1–31.
20. EPA (1997). Water quality standards; establishment of numeric criteria for priority toxic pollutants for the State of California. *Fed. Reg.* **62**(150): 42159–42208.
21. *Final Biological Opinion on the Effects of the U.S. Environmental Protection Agency's "Final Rule for the Promulgation of Water Quality Standards: Establishment of Numeric Criteria for Priority Toxic Pollutants for the State of California"* (2000). United States Department of Interior: Fish and Wildlife Service and United States Department of Commerce: National Marine Fisheries Service, p. 323
22. United States Code [U.S.C.] (2000). *The Endangered Species Act of 1973. 16 U.S.C.*, Sections 1531–1544.
23. Lemly, A.D. (1993). Metabolic stress during winter increases the toxicity of selenium to fish. *Aquat. Toxicol.* **27**: 133–158.
24. Ohlendorf, H.M. (2003). Ecotoxicology of selenium. In: *Handbook of Ecotoxicology*, 2nd Edn. D.J. Hoffman, B.A. Rattner, G.A. Burton Jr., and J. Cairns Jr. (Eds.). Lewis Publishers, Boca Raton, FL, pp. 465–500.
25. The TMDL regulations and literature are very extensive—for more information, access the website http://www.epa.gov/owow/tmdl/.
26. Tanji, K. et al. (2003). Evaporation ponds as a drainwater disposal management option. *Irrigation Drainage Syst.* **16**: 279–295.
27. Frankenberger, W.T. and Engberg, R.A. (Eds.). (1998). *Environmental Chemistry of Selenium*. Marcel Dekker, New York.
28. Fan, T.W.-M. and Higashi, R.M. (1998). Biochemical fate of selenium in microphytes: natural bioremediation by volatilization and sedimentation in aquatic environments. In: *Environmental Chemistry of Selenium*. W.T. Frankenberger and R.A. Engberg (Eds.). Marcel Dekker, New York, pp. 545–563.
29. Gerhardt, M.B. et al. (1991). Removal of selenium using a novel algal-bacterial process. *Res. J. Water Pollut. Control Fed.* **63**: 799–805.
30. Amweg, E.L., Stuart, D.L., and Weston, D.P. (2003). Comparative bioavailability of selenium to aquatic organisms after biological treatment of agricultural drainage water. *Aquat. Toxicol.* **63**: 13–25.
31. Terry, N. and Zayed, A. (1998). Phytoremediation of selenium. In: *Environmental Chemistry of Selenium*. W.T. Frankenberger and R.A. Engberg (Eds.). Marcel Dekker, New York, pp. 633–655.
32. Fan, T.W.-M. and Higashi, R.M. (2000). Microphyte-mediated selenium biogeochemistry and its role in *in situ* selenium bioremediation. In: *Phytoremediation of Contaminated Soil and Water*. N. Terry and G.S. Banuelos (Eds.). CRC Press, Boca Raton, FL, pp. 283–302.
33. Higashi, R.M., Rejmankova, E.J., Gao, S., and Fan, T.W.-M. (2003). Mitigating selenium ecotoxic risk by combining foodchain breakage with natural remediation. *University of California Salinity/Drainage Program Annual Report*.
34. Rofen, R. (2001). Agricultural evaporation ponds—source of marketable brine shrimp. *Curr. Newlett. UC Center for Water Resources* **2**(2): 9–10.

SHELLFISH GROWING WATER CLASSIFICATION

LINDA S. ANDREWS
Mississippi State University
Biloxi, Mississippi

Seafood-related illnesses in the United States are primarily associated with the consumption of bivalve molluscan shellfish (1,2). As these animals are water filter feeders and are often consumed raw, the water that they are grown in affects their safety and quality.

Bivalve molluscs feed by filtering microscopically sized particles out of the water. Via their inhalant siphon, they pump water through their gills. The gills sort out particles of the correct size and "feel" and direct them to their gut (3). The large volume of water that the bivalves pump through their siphons causes concern among public health officials. Consequently, it is imperative that water authorities, usually shellfish producer states, monitor the quality of water in and around the shellfish beds or reefs. For commercial markets, harvesting is only permitted from approved growing areas. The microbial acceptability of growing areas is classified on the basis of a sanitary survey of the shoreline to detect potential pollution sources and bacteriological analysis of water samples taken from the area (4).

Shellfish sanitation programs originated at the national level in the United States in 1925 after a series of typhoid epidemics spread by the consumption of raw shellfish threatened the collapse of the oyster industry (3). Today, through the National Shellfish Sanitation Program, NSSP (5), State and federal public health regulators have established guidelines for water quality in shellfish growing areas to decrease the risk of illness associated with these seafood products.

Sanitation surveys are normally performed by the state regulatory authority and include a written report of a shoreline survey, bacterial quality of the water, and an evaluation of the effect of any meteorological, hydrodynamic, and geographic characteristics of the growing area. Each surveying state maintains records of the surveys. For example, for Mississippi, survey results are filed with the Mississippi Department of Marine Resources (http://www.mississippiwebsite.com/deptmarinerec.htm). An analysis of the data from this survey determines the appropriate growing area classification. During shoreline surveys, the state authority identifies and evaluates all actual and potential sources of pollution to the growing area. If a pollution source exists, the distance from pollution to growing area is determined with an assessment of the potential impact. The effectiveness of sewage treatment and other wastewater treatment systems is evaluated for removal of the microbial and/or chemical contaminants.

Comprehensive sanitation surveys for each growing area must be performed at least once every 12 years with triennial reevaluation based on water quality analysis and potential for any new sources of pollution. If this triennial evaluation determines that conditions have changed from the 12-year survey, the area in question may then be reclassified. On an annual basis, the sanitary survey shall be updated to reflect changes in the conditions of the growing area based on "drive through" surveys, observations made during routine sample collection, or information from other sources.

In general, each growing area is classified as approved, conditionally approved, restricted, conditionally restricted, or prohibited based on the survey results (see Table 1). Status of a growing area is separate from its classification and may be either considered open or closed for

Table 1. Shellfish Harvest Area Classification Categories (5)

Classification	
Approved	Growing areas are classified as approved when the sanitation survey finds the area safe for the direct marketing of shellfish or is not subject to human or animal fecal pollution.
Conditional	Growing areas are classified as conditional when the area in question is in the open status for a reasonable period of time and when pollution factors are predictable. There may be direct potential for distribution of pollutants based on unusual conditions or specific times of the year when bacterial numbers are increased by heavy water runoff that affects wastewater treatment plant function.
Restricted	Growing areas are classified as restricted when the sanitary survey indicates a limited degree of pollution and when levels of fecal pollution, human pathogens, or poisonous or deleterious pollutants are at such levels that shellstock can be made safe through either relaying or depuration.
Prohibited	Growing areas are classified as prohibited when no current sanitary survey exists or when the survey determines that the growing area is adjacent to a sewage treatment plant outfall or other point source with public health significance or when the water is polluted because of previous or current sources of contamination.

harvesting of shellstock. All correctly classified growing areas are normally open for harvesting unless they are classified as prohibited. Closures may occur temporarily because of adverse weather conditions, the presence of biotoxins in concentrations of public health significance, or when the state authority fails to complete the written sanitary survey or triennial review evaluation report. A growing area may be placed into a remote status if the sanitation survey determines that the area has no human habitation and is not impacted by human pollution sources.

The NSSP allows for a growing area to be classified using either a total or fecal coliform standard. Two sample collection strategies are used: adverse pollution condition and systematic random sampling. Each state authority may choose one of these sampling plans or use both depending on the location of the shellfish bed. Each state authority determines the number and location of sampling stations based on potential sources of pollution contamination as determined by the shoreline survey. Except for prohibited areas, the original or new classification of a shellfish growing area that has the potential to be impacted by a pollution source requires a minimum of 30 samples collected under various environmental conditions. For an area not impacted by a pollution source, 15 samples are required for initial classification. When using fecal coliforms (FC) as the indicator bacteria, the water quality must meet the following standards: FC median or geometric mean most-probable-number (MPN) may not exceed 14 per 100 mL, with not more than 10% of samples exceeding 43 MPN per 100 mL for 5 tube decimal dilution or 49 MPN per 100 mL for 3 tube decimal dilution. When classifying point source growing sites, the bacterial quality of every station in the growing area must meet the fecal coliform standards as described above. Sample stations must be located adjacent to actual or potential sources of pollution. Refer to the NSSP Model Ordinance Chapter IV for details on sampling to achieve statistically reliable results (5).

Separate standards are used for shellstock that will be processed by depuration posterior to harvesting. For the restricted classification of growing areas that are affected by point sources or nonpoint sources, FC median or geometric mean MPN of the water samples shall not exceed 88 per 100 mL and not more than 10% of the samples shall exceed an MPN of 260 per 100 mL for a 5 tube decimal dilution test or 300 MPN per 100 mL for a 3 tube decimal dilution test.

Numerous factors are involved in interpreting FC data. Fecal coliforms are common as indicators of human fecal pollution and potential human fecal pathogens. However, pathogenic microorganisms typically do not occur in direct proportion to the number of FC present (3). This is especially true with viruses that move through tidal water at a much faster rate than heavier coliforms. Also, FC may die off at a rate different from other pathogenic organisms, thus giving a false sense of security that the shellfish are safe for human consumption. Another problem with using FC as an indicator of potential human pathogens is that FC may originate from sources other than humans. In rural areas, wild animals are likely to be the primary source of FC found in runoff. Major human pathogens of concern in shellfish waters usually originate on land, but exceptions include sewage discharges from boats and sewage treatment facilities. Runoff from land varies with rainfall, soil type, and degree of saturation. Consequently, one or even a few samples collected periodically may indicate very little about the pollution potential. Numerous samples collected under all weather conditions (heavy rain and high tides) are the only way to properly evaluate and classify a particular growing site.

The restrictive rules for classification of shellfish growing waters have been established to provide the safest shellfish possible. All shellfish for human consumption are marked with a harvesters tag that indicates that they were harvested only from approved harvest areas. Today, illness outbreaks caused by the consumption of human pathogens that are present in raw molluscan shellfish are rare because of these rules for classification.

BIBLIOGRAPHY

1. Anonyomous. (1988). Seafood Safety: Seriousness of Problems and Efforts to Protect Consumers. GAP/RCED-88-135, United States General Accounting Office, Washington, DC.
2. CDC. (1999). *Morbidity and Mortality Reports*. Center for Disease Control, Atlanta, GA. Available: www.cdc.gov.
3. Croonenberghs, R.E. (2000). Contamination in shellfish-growing areas. Section VI, Seafood safety. In: *Marine and Freshwater Products Handbook*, pp. 665–696.
4. Cook, D.W. (1991). Microbiology of bivalve molluscan shellfish. In: *Microbiology of Marine Food Products*, pp. 19–39.
5. NSSP. (2002). Model Ordinance, Guide for the Control of Molluscan Shellfish, Chap. IV. U.S. Department of Health and Human Services, Public Health Service, Food and Drug Administration. Available: http://www.ISSC.org.

SORPTIVE FILTRATION

K.A. Matis
N.K. Lazaridis
Aristotle University
Thessaloniki, Greece

INTRODUCTION

Although filtration is one of the principal unit operations in the treatment of potable water, the filtration of effluents is less practiced; usual examples are supplemental removals of suspended solids from wastewaters of biological and chemical treatment processes and also the removal of chemically precipitated phosphorus (1). The available references dealing with filtration are quite voluminous—see, among others, the articles on filter presses, deep bed, cartridge, batch, variable volume, and continuous filters (2). Hence, the information provided in this section serves only as an introduction to the subject; for additional details, the literature should be consulted. In this article, no mention will be made of chromatographic applications (3).

The seepage of rainfall and runoff into porous solids and rocks and the storage and movement of groundwaters in open-textured geologic formations are important elements in the resource and quality management of water and wastewater. Although fine-textured granular materials remove pollutants, the water drawn from them is acceptable only when natural filtration together with a time lapse between pollution and use bar the transport of pollutants to springs, wells, and infiltration structures (4).

SORPTION

A typical recent example of sorption is the removal of copper by suspended particulate matter from river water (5). Iron, silicon, and aluminum oxides were used as adsorbents, serving as models of naturally occurring suspended particles. On the other hand, sorption upstream of filtration constitutes a solution; the combined process has been also termed adsorptive filtration (6). In this module (offered by Separation Technologies), fundamental aspects of the process are illustratively given, such as inertial impaction and diffusional interception; however, the latter is effective only in gas filtration. Sorption, in general, is defined as a surface process irrespective of mechanism, adsorption or precipitation (7).

Batch and column sorption studies of zinc, cadmium, and chromium were conducted on calcite to determine its retention capacity for these elements common in industrial effluents and to explore its behavior as a purification filter in a continuous flow system (8). Fibrous carbon materials were examined for heavy oil sorption (9); oil spill accidents have caused serious problems in the environment, including disasters in living systems. Elsewhere (10), biologically active carbon (followed by ultrafiltration) was used for simultaneous sorption and biodegradation of organic constituents. Naturally occurring diatomaceous earth was tested as a potential sorbent for lead ions; the intrinsic exchange properties were further improved by modification with manganese oxides (11). Today, effective low-cost adsorbents for toxic metals are available; for instance, the case study to remove (by Metsorb™ adsorptive filtration media) depleted uranium from contaminated water at a U.S. Army testing site (12).

In recent years, contamination of ground and surface waters with heavy metals has become a major concern. The knowledge of the oxidation state of pollutant ions is often a prerequisite for the application of efficient treatment methods, as in the case of arsenic. The inhibition of conventional metal precipitation due to the presence of chelating or complexing compounds, such as acetate, citrate, and tartrate ions, ammonia, and EDTA, which may be present in most real wastewater streams, is another problem for examination. Thermodynamic equilibrium diagrams and software packages (such as Mineql+) have been employed to construct aqueous speciation diagrams for the metals under investigation and, then, to interpret the removal mechanism involved. Sorption isotherm equations for equilibrium uptake, such as that of Freundlich and Langmuir, have been often used to fit the experimental data, and the activation energy of the process was calculated (13).

Figure 1a presents some typical kinetic results for hexavalent chromium ions in batch sorptive removal under different conditions by (uncalcined) hydrotalcite as the appropriate sorbent, noting that desorption, particularly in successive stages, has received much less attention in the literature. The data were effectively fitted by a second-order kinetics equation.

Another suitable (for metal cations) inorganic sorbent material is the group of natural or synthetic zeolites (Fig. 1b); the nonselectivity of the process is also apparent, sorbing zinc or calcium. Stress was given to electrokinetic measurements (expressed as zeta-potential) of the system for its surface charge under the applied conditions to predict its behavior (14). The solution pH is important and influences metal speciation. So, for example, at pH lower than 6.8, corresponding to the pristine point of zero charge of α-FeOOH, the surface of these particles is positively charged. Therefore, in the pH range examined, the adsorption of anionic metal species was more pronounced (15).

Fixed-Bed Sorption

Ferric hydroxides, such as goethite, are another good example extensively studied; fixed-bed operation is the appropriate configuration for large-scale applications such as wastewater treatment (see Fig. 2a). The adsorbent material was granulated by crystallization through controlled freezing. Column adsorbers, due to pressure drop, require a suitable shape and size of the bonding material used. The experimental results are presented, expressed typically by the breakthrough curve concept, service time versus breakthrough (column outlet concentration related to the initial metal concentration—$C_{outlet}/C_{initial}$), as a percentage. In the inplots, the isoremoval lines of treatment time versus the respective quantity of sorbent are also presented at various breakthrough points.

The major aim when sizing adsorptive columns is the ability to predict the service time until the column effluent exceeds a predefined solute (pollutant) concentration. The bed depth–service time model (abbreviated as BDST) relates the service time of a fixed-bed to the height of adsorbent in the bed, hence to its amount, because quantity is directly proportional to bed height. The measurement of sorbent quantity is more precise than the determination of the respective volume, especially for granules. Therefore, the sorbent quantity is preferably used instead of the bed height (15,16).

Surface complexation models were used to describe metal cation adsorption on adsorbent materials; models, such as constant capacitance, diffuse layer, and triple layer were applied, showing that they could simulate experimental results. The fundamental concepts on which all these models are based remain, more or less, the same (17).

The computer program PHREEQC (of the U.S. Geological Survey) permits a wide variety of aqueous geochemical calculations; surface-complexation modeling, applying mainly the diffuse double layer (DDL) model,

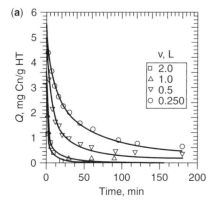
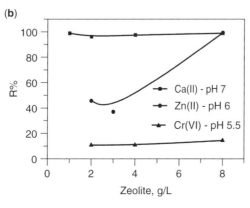

Figure 1. (a) Time variation of chromium(VI) loading on hydrotalcite (1 g/L) at various eluant volumes. Reprinted with permission from Reference 13; copyright (2004) American Chemical Society. (b) Removal of calcium, zinc, or chromate ions (50 mg/L) by zeolites as a function of their concentration, at different pH values. Reprinted with permission from Reference 14; copyright (2004) Elsevier.

is included in the program. The program uses a modification of the usually applied Newton–Raphson method iteratively to revise the values of the variables until a solution of the equation system is accepted within specified tolerances. A limitation of the program may be that uncertainties occur in determining the number of sorption sites, the surface area, the composition of sorbed species, and the appropriate surface-complexation constants. It was demonstrated that some of the surface complexation model (SCM) parameters might be incorrect due to an experimental artifact, though they give good simulations of data. This may be due to the success of least-squares fitting programs, such as FITEQL, in finding model parameters to describe experimental data sets.

The results of the model, according to surface-complexation reactions and constants (see Table 1), are presented in Fig. 2b, in comparison with the experimental data for arsenic(III) oxyanions; they show good agreement, although the diffusion effects (external, or liquid film, and intraparticle transport resistance) were not taken into consideration (18). Similar conclusions were obtained with As(V). The arsenic problem is of particular concern

Table 1. As(III) Species Distribution and Surface-Complexation Reactions and Their Reaction Constants[a]

Species Distribution	Reaction Constants, log K
$AsO_3^{3-} + 4H^+ = H_4AsO_3^+$	34.44
$AsO_3^{3-} + 3H^+ = H_3AsO_3$	34.74
$AsO_3^{3-} + 2H^+ = H_2AsO_3^-$	25.52
$AsO_3^{3-} + H^+ = HAsO_3^{2-}$	13.41
Surface-Complexation Reaction	
$Goeth\_OH + H_3AsO_3 = Goeth\_H_2AsO_3 + H_2O$	4.51

[a]Reference 18.

for small communities in rural areas around the world (a characteristic case is that of Bangladesh), where groundwater comprises the main drinking water source. Arsenite is favored under reducing (i.e., anaerobic) conditions. In this *Encyclopedia*, refer to another article on SORPTION KINETICS in the chapter on Physics and Chemistry of Water.

For a number of ion exchangers of ultrafine particle size (for efficient use in packed column beds), a convenient binding polymer, such as modified polyacrylonitrile, has been used. The resulting sorbents were tested for separation and preconcentration of different contaminants, including radioactive wastes. The sorption characteristics of these composite materials were not affected by the binding polymer, whereas their physicochemical properties (hydrophilicity, mechanical strength, etc.) can be modified by the degree of cross-linking of the polymer, the use of suitable copolymers, or by changing the composition and temperature of polymerization (19). Certainly, an alternative to facilitate their solid/liquid separation downstream, particularly if fine adsorbents are to be dispersed, is to apply flocculation (20); note that many low-cost synthetic sorbents are produced as powders.

THE COMBINED PROCESS

Liquid-phase carbon adsorption is a full-scale technology in which groundwater is pumped through one or more vessels containing activated carbon to which dissolved organic contaminants adsorb (21). When the concentration of contaminants in the effluent from the bed exceeds a certain level, the carbon can be regenerated in place; removed and regenerated at an off-site facility; or removed and disposed of. Carbon used for explosive- or metal-contaminated groundwater probably cannot be regenerated and should be removed and properly disposed of. Adsorption by activated carbon has a long history of use in treating municipal, industrial, and hazardous wastes. Figure 3a illustrates a sorptive filtration system.

Another innovation has been proposed by introducing a two-stage process in a compact microfiltration (MF) hybrid cell (Fig. 3b). A large number of techniques have been used to limit membrane fouling; among them is air bubbling that also constitutes the transport medium in flotation, as applied in wastewater treatment; flotation is

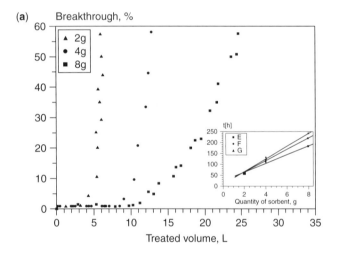

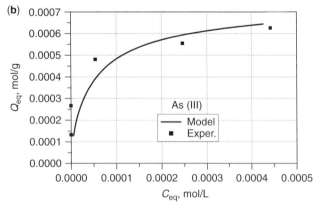

Figure 2. (**a**) Breakthrough curves for goethite (at different quantities) and chromates; as an inset, isoremoval lines for 20%, 35%, and 50% breakthrough. Reprinted with permission from Reference 15; copyright (2001) Elsevier. (**b**) Sorption of trivalent arsenic on goethite mineral: experimental vs. theoretical (DDL modeling) values. Reprinted with permission from Reference 18; copyright (1999) Kluwer.

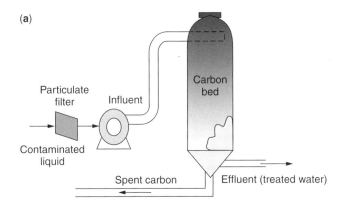

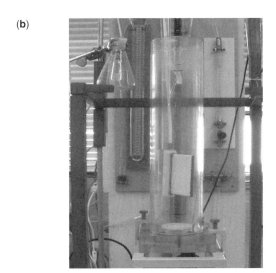

Figure 3. (a) Typical fixed-bed carbon adsorption system. From Reference 21. (b) A close-up photograph of the hybrid MF laboratory cell based on flotation (22).

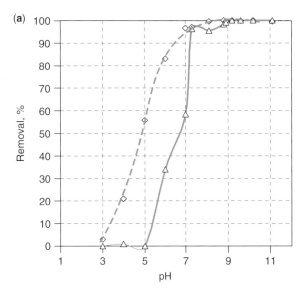

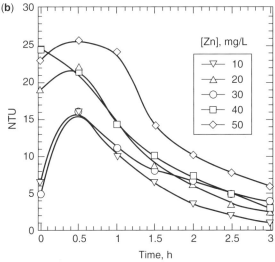

Figure 4. (a) Removal of copper by biosorption (rhombic symbols) or simply, precipitation (triangles): A comparison of the two processes. Modified from experimental results published by Zouboulis et al. (24). (b) Influence of linear velocity on residual concentration expressed as turbidity (pH 9.5). Reprinted with permission from Reference 26; copyright (2002) Dekker.

suitable as a pretreatment stage for microfiltration. The former accounted for around 90% solids removal.

Ceramic flat-sheet membrane modules of multichannel geometry were used in this cell. The objective was to apply microfiltration by submerged membranes for efficient separation of metal-loaded zeolites. By this hybrid process, low residual metal concentration in treated water at maximum water yield and a higher concentration of metal bonding agent in the concentrates can be achieved (22). During the experimental series of continuous-flow separations by the membranes, all samples collected showed 100% zeolite recovery due to membrane filtration, hence, efficient solid/liquid separation. The metal (zinc) removal depended only on the amount of zeolite sorbent used; the remaining process parameters, examined in detail, had absolutely no influence on the former.

One of the early notable investigations of adsorptive filtration is that by the EPA (23) at Superfund sites, those sites classified as uncontrolled and abandoned places where hazardous waste is located, possibly affecting the local ecosystem or people. In this application, toxic metals were removed by attachment to a thin layer of iron oxides that were immobilized on the surface of an appropriate filter medium, such as sand grains.

An alternative sorbent that has been tried quite extensively is microbial biomass (24). The reason is that thousands of tons of residual biomass are produced each year from the fermentation, pharmaceutical, and chemical (i.e., citric acid biosynthesis) industries and also from biological wastewater treatment plants. The bacterial cell surfaces are known to be anionic, due to ionized groups in the various cell wall polymers. From a comparison of biosorption and other metal separation methods, such as filtration and centrifugation, it was found that the former favors the removal efficiency and applicability at lower (acidic) pH. In these conventional processes, metal cation removal is mainly due to their precipitation as hydroxides from pH alteration; this is presented in Fig. 4a showing a pH front/edge moving to the right toward alkaline values; precipitation was followed by flotation with dodecylamine as a surfactant. Of course, in sorption, metals are bonded on dead biomass (*Streptomyces rimosus*). Oxidation processes leading to insoluble products can also be mediated by specific bacteria, for example, for Mn(II) and Fe(II) (25); in these papers, from our lab, upflow filtration units were tested. One of the objectives was to obtain low sludge volumes, which often depends upon the method of precipitation.

Upflow filtration has a theoretical advantage because coarse-to-medium filtration can be achieved by a single medium (such as sand) with almost perfect gradation of pore space and grain size from coarse to fine in the direction of flow. The bed is backwashed in the same direction but at higher flowrates, so the desired relative positions of fine media are maintained or reestablished by each run (1,4,26). The shortcomings of upflow filtration were overcome by using floating filter media. This type of filter can be backwashed with minimum water consumption and at lower rates. It also appeared from the laboratory experiments that metal removal was not limited by the solubility of the respective zinc hydroxide but by the effectiveness of the subsequent solid/liquid separation method (Fig. 4b). This semibatch unit, consisting of two columns in series and external recirculation of the filtered wastewater from the unit to the feed tank, was suitable for small- to medium-scale applications.

In conclusion, sorptive filtration presents a promising technique. Today, there is a tendency for combined and compact processes, offering both the capacity for effluent treatment for environmental reasons, plus the recovery of metals that otherwise would be lost and, more importantly, water reuse.

Acknowledgments

Thanks are due to Dr. A.I. Zouboulis (AUTh) and Prof. A. Grohmann (WABOLU, Berlin) for their help.

BIBLIOGRAPHY

1. Metcalf & Eddy Inc. (1991). *Wastewater Engineering: Treatment, Disposal, Reuse*. Revised. McGraw-Hill, Singapore, p. 248.
2. Purchas, D.B. (Ed.) (1977). *Solid/Liquid Separation Equipment Scale-Up*. Uplands Press, Croydon, U.K., pp. 289, 319, 365, 445.
3. Wankat, P.C. (1986). *Large-Scale Adsorption and Chromatography*. CRC Press, Boca Raton, FL.
4. Fair, G.M., Geyer, J.G., and Okun, D.A. (1968). *Water and Wastewater Engineering*. John Wiley & Sons, New York, Vol. 2, p. 27-1.
5. Grassi M.T., Shi B., Allen H.E. 1997. Sorption of copper by suspended particulate matter, *Coll. Surf. A* **120**: 199–203.
6. Univ. Pretoria, accessed 2004. Training module, Principles of Filtration, site: http://www.up.ac.za/academic/chemeng/Facilities/Tribology_presentations/05aSepTech.pdf.
7. Sposito, G.A. (1984). *The Surface Chemistry of Soils*. Oxford University Press, Oxford.
8. García-Sánchez, A. and Álvarez-Ayuso, E. (2002). Sorption of Zn, Cd, and Cr on calcite. Application to purification of industrial wastewaters. *Miner. Eng.* **15**: 539–547.
9. Inagaki, M., Kawahara, A., Nishi, Y., and Iwashita, N. (2002). Heavy oil sorption and recovery by using carbon fiber felts. *Carbon* **40**: 1487–1492.
10. Pirbazari, M., Ravindran, V., Badriyha, B.N., and Kim, S.-H. (1996). Hybrid membrane filtration process for leachate treatment. *Water Res.* **30**: 2691–2706.
11. Al-Degs, Y., Khraisheh, M.A.M., and Tutunji, M.F. (2001). Sorption of lead ions on diatomite and manganese oxide modified diatomite. *Water Res.* **35**: 3724–3728.
12. HydroGlobe, Inc., accessed 2004. Site www.hydrogobe.com
13. Lazaridis, N.K., Pandi, T.A., and Matis, K.A. (2004). Chromium(VI) removal from aqueous solutions by Mg-Al-CO_3 hydrotalcite: Sorption–desorption kinetic and equilibrium studies. *Ind. Eng. Chem. Res.* **43**: 2209–2215.
14. Zamboulis, D., Pataroudi, S.I., Zouboulis, A.I., and Matis, K.A. (2004). The application of sorptive flotation for the removal of metal ions. *Desalination* **162**: 159–168.
15. Lehmann, M., Zouboulis, A.I., and Matis, K.A. (2001). Modelling the sorption of metals from aqueous solutions on goethite fixed-beds. *Environ. Pollut.* **113**: 121–128.
16. McKay, G. (1996). *Use of Adsorbents for the Removal of Pollutants from Wastewaters*. CRC Press, Boca Raton, FL.
17. Dzombak, D.A. and Morel, F.M.M. (1990). *Surface Complexation Modelling: Hydrous Ferric Oxide*. John Wiley & Sons, New York.
18. Matis, K.A., Lehmann, M., and Zouboulis, A.I. (1999). Modelling sorption of metals from aqueous solution onto mineral particles: The case of arsenic ions and goethite ore. In: *Natural Microporous Materials in Environmental Technology*. P. Misaelides, F. Macašek, T.J. Pinnavaia, and C. Colella (Eds.). Kluwer Academic, The Netherlands, pp. 463–472.
19. Zouboulis, A.I., Matis, K.A., Loukidou, M., and Šebesta, F. (2003). Metal biosorption by PAN-immobilized fungal biomass in simulated wastewaters. *Coll. Surf. A* **212**: 185–195.
20. Matis, K.A., Zouboulis, A.I., Mandjiny, S., and Zamboulis D. (1997). Removal of cadmium from dilute solutions by hydroxyapatite. Part III. Flocculation studies, *Sep. Sci. Tech.* **32**: 2107–2128.
21. Federal Remediation Technology Roundtable. Ex Situ Physical/Chemical Treatment (assuming pumping), site: http://www.frtr.gov.
22. Lazaridis, N.K., Blöcher, C., Dorda, J., and Matis, K.A. (2004). A hybrid MF process based on flotation. *J. Membrane Sci.* **228**: 83–88; (b) Matis, K.A., Blöcher, C., Mavrov, V., Chmiel, H., and Lazaridis, N.K., Verfahren und vorrichtung zur membranunterstützten flotation, Patents DE 10214457 C1 & EP 1354632 A1.
23. EPA (1993). *Emerging technology summary: Metals treatment at Superfund sites by adsorptive filtration*. EPA/540/SR-93/515, Cincinnati, OH.
24. Zouboulis, A.I., Rousou, E.G., Matis, K.A., and Hancock I.C. (1999). Removal of toxic metals from aqueous mixtures: Part 1. Biosorption. *J. Chem. Tech. Biotechnol.* **74**: 429–436.
25. Katsoyiannis, I.A. and Zouboulis, A.I.: (a)(2002). Removal of arsenic from contaminated water sources by sorption onto iron-oxide-coated polymeric materials; (b) (2004). Biological treatment of Mn(II) and Fe(II) containing groundwater: Kinetic considerations and product characterization. *Water Res.* **36**: 5141–5155; **38**: 1922–1932.
26. Zouboulis, A.I., Lazaridis, N.K., and Grohmann, A. (2002). Toxic metals removal from waste waters by upflow filtration with floating filter medium. I. The case of zinc. *Sep. Sci. Tech.* **37**: 403–416.

QUALITY OF WATER IN STORAGE

Paulo Chaves
Toshiharu Kojiri
Water Resources Research Center
Kyoto University, Japan

Water is our most precious natural resource. It is not only indispensable for the survival of all living beings but also has social and economic importance. It is no coincidence that most early human settlements were located near sources of freshwater. However, available water resources are increasingly under pressure from human population growth, increased standard of living, urbanization, and development of extensive areas for agriculture. As a result, available water may, at times, become insufficient to fulfill all the needs of human living, particularly during dry periods, which are caused by the natural randomness of the hydrological cycle.

Therefore, it becomes necessary to find and to develop new sources of freshwater in order to secure a constant and appropriate supply of water. Certainly, many methods to maintain the stable supply of water exist, such as groundwater exploitation and water deviation from wet to dry areas. Undoubtedly, however, the most popular way worldwide to stabilize water supply is through the operation of stored water, particularly in lakes and manmade reservoirs. This concept was appreciated long ago, and reservoir construction has a long history, with the first dam believed to be built in ancient Egypt around 2800 B.C. (1).

In the last few decades, most of these water storage systems have suffered from poor water quality, mainly caused by human activities within their watersheds and inefficient operation of these reservoirs. Although a variety of techniques to improve water quality in storage exist, most are either very costly or not sustainable. Therefore, efforts to find new methods to achieve better water quality are extremely important. One way is by taking advantage of the close relationship between quality (limnological) and quantity (hydrological and climatological characteristics) of water to better manage stored volumes to yield optimal benefits for water use while at the same time improving its quality.

In the following sections, the main issues that impact on the quality of water in storage are presented. First, the important relationship between quantity and quality of water, and the potential improvement of water quality through the proper operation of storage volumes are highlighted. Then, the main issues related to evaluation of water quality are explained, showing the importance of proper water quality for different uses. Finally, all these topics are combined within the storage management and operation framework when water quality is considered.

WATER QUALITY AND STORAGE

Manmade reservoirs and lakes used for water storage, like other water bodies, are vulnerable to increasingly frequent problems of water quality. Some of the most common problems concerning the quality of water in storage are organic pollution, anoxia (deficiency in oxygen), eutrophication (increase of phytoplankton because of high nutrient concentrations), toxicity, turbidity, siltation, and waterborne diseases. In the same way, a variety of different possible causes of these problems exist, such as industrialization, increase of population, overexploitation of agricultural land, bad farming practices, poverty, lack of proper sanitation systems, runoff from cities, deforestation, lack of proper enforcement of the regulations for pollution control, and overuse of water resources with decrease of water volumes and levels.

Probably, the most effective method to control pollution and improve quality of water in storage is through comprehensive watershed management, which includes reduction and control of nonpoint and point sources of pollution, such as nutrient and organic matter loads. On the other hand, other techniques exist that may be applied directly to the water bodies themselves, such as artificial mixing, aeration, sediment removal, chemical manipulation, algaecides, and biomanipulation of species like fishes and aquatic plants. However, most of these techniques are costly and difficult to implement.

Basically, three main components of a reservoir system exist: inflow, storage, and release. Inflows may be defined as natural events, whereas storage and release may be considered under human control. Nevertheless, both natural variability and human decision can greatly influence the water quality of a reservoir. Reservoirs tend to present water of different quality during wet and dry hydrologic periods. With the increased runoff during periods of rain, higher nutrient loads would be expected in the reservoir, and these nutrients stimulate algae blooms and, therefore, may lower the water quality. Moreover, higher levels of inflow may also increase the inputs of organic matters resulting in dissolved oxygen (DO) depletion, which is defined as the decrease of dissolved oxygen concentrations, vital for sustaining aquatic life. On the other hand, higher inflows increase flushing rates and enhance mixing characteristics, decreasing stratification and potentially improving water quality. Hydrological factors, however, are not the only influences on water quality. Climatologic characteristics also play an important role, because during warm periods a higher probability of reservoir thermal stratification exists, because of a difference in temperature between water layers, leading to a difference in water density. The phenomenon of stratification, defined as the difference in density through the water column, which is commonly caused by variations of temperature, restricts the mixing of water between different layers. The isolated lower layer, called hypolimnion, can suffer from DO depletion by anaerobic processes and an increase in nutrients and other materials from bottom sediments.

Water quality is closely tied to storage volumes. Clearly an increase of water volume results in increased dilution of pollutants, but, in addition, other physical, chemical, and biological processes that influence the quality of water may be closely related to controllable hydrological factors, such as the variation of release discharge, storage volumes, and water levels. For example, an increased retention time (defined as the ratio between reservoir volume and discharge rate) can influence water quality in many ways, such as decreasing mixing proprieties, increasing phosphorus retention, increasing the difference between surface-bottom temperatures (stratification), and decreasing turbidity that can enhance chances of algae bloom. Moreover, fluctuation of water levels may also enormously affect water quality, such as increased erosion from the shores that may result in higher turbidity, increasing washout of margins, and decreasing reproduction of some fish species (that lay eggs on coastal vegetation), consequently changing the phytoplankton (algae) population of the aquatic system. The quality of stored water may also be affected by the release of water with poor quality from different levels, such as bottom-spillway discharge.

Zalewski et al. (2) emphasizes the importance of ecohydrology to use ecosystem properties as management tools for sustainability of water resources, such as through adequate management of quantities to improve water quality. Moreover, his paper presents an example of the relationship between storage levels and the eutrophication process for the Sulejow Reservoir. Straskraba and Tundisi (3) discuss, in a general way, many aspects related to the management of reservoir water quality, such as ecotechnologies, relation between storage and water quality, and the possible impacts of different volumes and quality of discharged water from reservoirs.

Therefore, the proper operation of release discharges and storage volumes considering single or multiple outlets and offtakes can also be considered an efficient method to improve the quality of stored water, by discharging water of undesirable quality or by increasing the storage volume with additional water, hence diluting water of poor quality (4,5). Although some general idea of the relationships between hydrologic and climatologic variables and water quality condition could be described, it is not possible to make a definitive statement about such relations. Specific studies for each reservoir system are needed to truly understand its limnology characteristics. Such studies need to include data collection and analysis, application of mathematical models, biological and hydrologic investigations, and, lastly, an understanding of the socioeconomic characteristics of the reservoir basin.

Mathematical models can play an important role in the decision-making process, as these tools may support the decision makers in making the "best" decisions. The outcome of different decisions may be simulated and explored virtually with relatively low costs and in a short period of time. For more detailed explanation behind the mathematics of physical, chemical, and biological concepts of the quality models, the readers are referred to Jorgensen (6), Orlob (7), and Thomann and Mueller (8).

WATER QUALITY EVALUATION

Water quality directly affects virtually all water uses. Fish survival, diversity, and growth; recreational activities; corrosion of turbines of hydropower plants; municipal, industrial, and domestic water supplies; agricultural uses like irrigation and livestock; waste disposal; and general aesthetics are all affected by the physical, chemical, biological, and microbiological conditions that exist in water bodies. Some water uses can affect water quality, which can influence other water uses. For instance, navigation may cause increased bank erosion leading to accelerated loss of storage volumes (reservoir sedimentation) and higher water turbidity. Water bodies used for wastewater assimilation may suffer from depletion of DO, which is required to sustain aquatic life, and an increase in total dissolved solids (TDSs), which is highly relevant to water supply, as higher TDSs increase treatment costs.

The minimum acceptable quality of water depends very much on the water use itself. Water for irrigation, for example, should be low in dissolved salts, but water intended for livestock should be low in bacteria. Water used in industrial processes should usually be of a much higher quality than water used for industrial cooling. As for municipal supply, water must not only be safe to drink, but ideally should contain low concentrations of materials such as calcium and iron, as they may cause costly infrastructure damage or add unpleasant characteristics to the water even after treatment (9).

A water quality index may be an efficient way to objectively measure water quality using a vast number of existing water quality parameters, helping decisions to be taken in a more efficient and less subjective way. Many of the water quality indices tend to be developed based on local characteristics and local expertise (10). Water quality indices may also be developed to express the conformity or net benefits of a water body with regard to a particular type of water use. In this scenario, different indices may be developed for the evaluation for a specific use or of the natural state of the system, such as for domestic drinking water supply, recreational use, wildlife preservation, trophic state situation (measure of eutrophication), and toxicity levels of water. Indices can be based on a single quality parameter or can even be presented as a combination of many quality parameters, such as the well-known Water Quality Index developed by the U.S. Environmental Protection Agency (EPA). Therefore, their use in the management and operation framework has to be closely studied and discussed among all stakeholders involved in the decision-making process. As a result of the vagueness involved in defining these indices, some new attempts have been made to evaluate water quality using fuzzy theory, such as in works by Norwick and Turksen (11), Dombi (12), and Chang et al. (13).

MANAGEMENT OF STORAGE WATER QUALITY

Water volume has always been a prime consideration in the design and construction of storage reservoirs, but quality has not. In the same way that quantity is regulated, water can be stored and released for the improvement of water quality and net benefits. As described in the previous sections, assessment, evaluation, and modeling of water quality are very complex tasks for water resource managers. Some of the reasons why water quality is still being left out from the decision-making framework process include:

- Multiple stakeholders and multiple objectives are present in most cases.
- A large number of physical, chemical, and biological water quality parameters exist to be considered.
- Quality conditions present great spatial and temporal variability.
- It is difficult to economically evaluate water and environmental quality.
- Quality may be highly influenced by human activities and natural conditions.
- Complete and representative spatial and temporal water quality data are seldom available.
- Water quality simulation is a complex task because of the high levels of uncertainty involved in this process.

The uncertainties when analyzing water quality are much greater than when dealing with water quantity. Uncertainty may strongly affect water quality management and the decision-making process of storage operation. However, from where does uncertainty come? Basically, uncertainty is inherent to many processes, such as measurement, natural randomness, and human perception. In a more concrete way, some sources of uncertainties assessing water quality include:

- Construction problems of treatment facilities
- Structures and parameters of a simulation model
- Uncertainty of mismeasurement and incomplete data sets
- Future and changing conditions (e.g., new socioeconomic activities in the watershed, global warming)
- Vagueness of objectives (e.g., "improvement" of water quality)
- Situations of accidents and disaster (e.g., traffic accidents or chemical leakage from factories)
- Natural randomness of climate and hydrological variability (e.g., extreme events such as floods or long drought periods)

Proper identification of the system uncertainties can support a more flexible and effective management framework, where components identified with greater uncertainty are considered for further investigation and the ones with less uncertainty may become more important decision tools, as the consequences can be more accurately predicted. Moreover, it may increase the public acceptance of a certain decision regarding reservoir management. Results after uncertainty analysis can be given in different ways, such as probability density function and confidence intervals, giving a range of possible outcomes avoiding the common "unexpected" results.

Moreover, information on cost is frequently needed to improve the water quality management prior to the water quality impact assessment. Such costs may be related to water treatment; penalties for excessive pollution; socioeconomic losses, such as associated with tourism activities; and decrease of benefits of certain objectives when a multipurpose operation is considered.

The operation of single and multiple outlet reservoir systems can present different objectives, such as improvement of water quality in stored volumes, released discharges, and intake water. Moreover, the management horizon can be specified to deal with different problems, such as related to the long-term planning, real-time operation, and emergency situations.

Management of water quality can be considered together with the various types of reservoirs that may present different water use, such as reservoirs for water supply, irrigation, hydropower generation, flood control, recreation, and, of course, multipurpose systems. As explained previously, for each type of reservoir, a different evaluation method may be necessary.

Besides the improvement of water quality condition itself, many advantages to considering water quality in the operation process exist, such as wider acceptance of reservoirs among the general public, decrease in maintenance costs, increase in net benefits, preservation of aquatic life, and achievement of sustainable policies.

Straskraba and Tundisi (3) present many management models that combine quality simulation with management tools. Research in this field is in constant development, and some of the new models are imbedded with complex components, such as remote sensing information, geographic information system (GIS), management and optimization techniques, artificial intelligence (AI), and decision support systems (DSSs).

It seems that much still exists to be accomplished and developed in the field of water quality in storage. The focus should be on integrated quality management, which must consider quantity and quality along with the three main components of a storage reservoir system: inflow, storage, and release. Moreover, other topics should be considered such as simultaneous combination of stochastic optimization and water quality simulation models, including uncertainty analysis, multiple water quality parameters in a multiobjective optimization framework, and multipurpose reservoirs considering the different stakeholders and water uses. The potential success in solving water quality related problems will depend on the ability of different professionals, including researchers, hydrologists, limnologists, and decision-makers, to work together in an interdisciplinary and holistic manner.

BIBLIOGRAPHY

1. Biswas, A.K. (1970). *A History of Hydrology*. North Holland Publishing, Amsterdam.
2. Zalewski, M. et al. (1997). *Ecohydrology: A New Paradigm for Sustainable Use of Aquatic Resources*. UNESCO, Paris, France.
3. Straskraba, M. and Tundisi, J.G. (1999). *Guidelines of Lake Management—Reservoir Water Quality Management*. Vol. 9. ILEC, Japan.
4. Welch, E.B. and Patmont, C.R. (1980). Lake restoration by dilution, Moses Lakes, Washington. *Water Res.* **14**: 1317–1325.

5. Chaves, P., Kojiri, T., and Yamashiki, Y. (2003). Optimization of storage reservoir considering water quantity and quality. *Hydrol. Processes* **17**: 2769–2793.
6. Jorgensen, S.E. (1983). *Application of Ecological Modeling in Environment Management*. Elsevier, Amsterdam.
7. Orlob, G.T. (1983). *Mathematical Modeling of Water Quality: Streams, Lakes and Reservoirs*. John Wiley & Sons, Chichester, UK.
8. Thomann, R.V. and Mueller, J.A. (1990). *Principles of Surface Water Quality Modeling and Control*. HarperCollins Publishers, New York.
9. Kiely, G. (1997). *Environment Engineering*. McGraw-Hill, Berkshire.
10. Thanh, N.C. and Biswas, A. (1990). *Environment—Sound Water Management*. Oxford University Press, Oxford, UK.
11. Norwick, A. and Turksen, I. (1984). A model for the measurement of membership and consequences of its empirical implementation. *Fuzzy Sets Syst.* **12**(1): 1–25.
12. Dombi, J. (1990). Membership function as an evaluation. *Fuzzy Sets Syst.* **35**: 1–21.
13. Chang, N.B., Chew, H.W., and Ning, S.K. (2001). Identification of river water quality using the fuzzy synthetic evaluation approach. *J. Environ. Manag.* **63**: 293–305.

QUALITY OF WATER SUPPLIES

XINHAO WANG
DIANA MITSOVA-BONEVA
University of Cincinnati
Cincinnati, Ohio

Adequate quantity and quality of water are necessary for human living and activities. A broad definition of water supply includes any quantity of available water. A more restricted concept of water supply refers to water collected and distributed to the general public or to other public or private utilities for residential, commercial, and industrial use. An even narrower definition of water supply only includes the water collected and conveyed for use in a community or a region (1). In the discussion here, water supply refers to water from the process of collection and distribution to meet human demand.

WATER RESOURCES

Water resources are renewable but finite and scarce. As shown in Fig. 1, 97.22% of earth's water is captured in oceans. Its total volume is enormous—1.3 billion cubic kilometers—but inadequate for human consumption because of the salt content. Approximately 2% of water is locked up in polar icecaps and glaciers. Water found in land, including surface water and groundwater, makes up less than 1% of the earth's water resources (2). Groundwater is found in aquifers, moisture-laden strata where water fills the spaces between rock particles. Although only about half of groundwater can be economically withdrawn for human use, it represents more than 97% of the usable freshwater resources. In addition, groundwater is a major source of replenishment for surface water.

WATER CONSUMPTION

Human water consumption consists of three major categories—agricultural, industrial, and domestic water use (3). According to the Food and Agriculture Organization of the United Nations (FAO), agricultural use accounts for 70% of global water withdrawals, while industrial and municipal use account for another 20% and 10%, respectively (4). Farming withdraws large amounts of water from rivers, lakes, reservoirs, and wells in order to irrigate crop fields and sustain plant growth in agricultural and horticultural practices. Almost 60% of the world's freshwater withdrawals are used for irrigation purposes. In Asia, irrigated agriculture averages approximately 82% of all water consumption. The proportion in the United States is 41% and about 30% in Europe (5). Irrigation water is applied through irrigation systems for the purposes of field preparation, preirrigation, application of chemicals, weed elimination, dust reduction, plants cooling, frost protection, leaching salts from roots, and harvesting. Irrigation water is also used by golf courses, parks, nurseries, turf farms, and cemeteries.

Industry consumes the second largest amount of water. In heavily industrialized regions, the percentage could

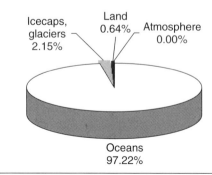
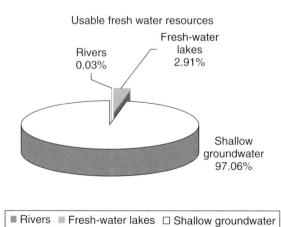

Figure 1. Earth's water resources.

be much higher. In Europe, industry accounts for 52% of the total water consumption and in North America, 44% (6). Almost one-half of the industrial withdrawals are utilized for generating thermoelectric power with steam-driven turbine generators. Much of the industrial water is used for purposes such as cooling and is returned to its source. Large amounts of water are required also by food processing and paper producing industries, oil refineries, and chemical and metallurgical plants.

Domestic water usage accounts for the last 10% of the total water consumption. This category refers to domestic consumption of water for drinking, hygiene, and other household uses, as well as commercial and institutional uses. The consumed domestic water is collected and treated in wastewater treatment plants before being discharged back to a natural water body.

TYPES OF WATER SUPPLY

Water supply systems refer to private (self-supplied, such as private wells) or public (may be operated by public agencies or private companies) withdrawals from ground and surface water designated to serve specific users. The most common is the public or municipal water supply system, which provides water to domestic, public, and commercial users and, occasionally, to industries, hydroelectric and thermoelectric plants, or for irrigation (7). The majority of the population throughout the world, especially urban population, is served by public water supply systems. Figure 2 illustrates the average percentage of the large city population served by each type of water supply service, by region. Household connection or yard tap and public tap are public water supplies. In the more developed regions, such as Europe and North America, more than 96% of the urban population has household connection to public water supply services. In Africa and Asia, considerable numbers of people are not served by public water supply systems.

QUALITY OF WATER SUPPLY

Water is an excellent solvent that can dissolve many elements. The amount of each element affects water quality, reflected in the chemical, physical, or biological condition of water. Therefore, water quality is often measured with what is contained in water, such as dissolved oxygen, suspended solids, pesticides, metals, oils, minerals, and nutrients. Pathogenic bacteria might affect human and animal health and so they are often measured. Physical conditions are another set of water quality measurement, such as water temperature, water color, and turbidity. Water quality also can be assessed with biological indicators, which measure the type and diversity of fish, macroinvertebrate species, and plants (9).

Some uses of water for industrial activities, such as mining, cooling water supply, washing, fire protection, or oil well repressurization, may not depend on water quality. Different industries may have different requirements for water quality, such as the turbidity or temperature. Normally, industrial water use does not require filtration and chlorine treatments.

Quality criteria for irrigation water include measurements of its salinity, alkalization, toxicity, and electrical conductivity. Salinity indicates the presence of high concentrations of chloride, sulfate, and bicarbonate compounds of sodium, calcium, and magnesium. In regions with significant rainfalls, those soluble salts are flushed out from the crop root zone on a regular basis. In arid and semiarid areas, water percolating through the soil is not sufficient to leach those salts. The increasing concentrations may adversely affect crops with low level salt tolerance thresholds. Therefore, irrigation waters must be controlled for their content of dissolved salts in order to prevent further deterioration of the soil quality and reduction of crop yields. There is a direct positive relationship between salt concentrations and the electrical conductivity of irrigation water. Alkalization occurs when the soil is more saturated with sodium than calcium and magnesium. Excess sodium results in high pH, reduced oxygen content, soil erosion, and less nutrient absorption by plants. Some constituents of the irrigation waters such as boron, chloride, and sodium are toxic to crops. Irrigation waters containing such chemicals are not recommended for use in watering soybeans, citrus, or grape plants (10).

Water quality of domestic water supplies has a direct effect on human health. Microbial pathogens that are found to cause enteric diseases in humans include bacteria (such as *Vibrio cholerae, Shigella, Salmonella, Campylobacter*, and *Escherichia coli*), viruses (such as hepatitis A virus and rotavirus), and parasites (such a *Giardia lamblia* and *Cryptosporidium*). Bacterial waterborne infections include cholera, dysentery, salmonellosis, typhoid fever, "traveler's disease," and *E. coli* outbreaks. The most common viral disease contracted through contaminated water is hepatitis A. Sewage-tainted drinking water can also cause certain parasitic protozoan diseases such as amebic dysentery and giardiasis. Those diseases are usually contracted by drinking water or eating food contaminated with pathogens of fecal origin. They are characterized by severe diarrhea, vomiting, and cramps, and in some cases, by high fever, headache, and muscle pains. The most common cause of death associated with those ailments is rapid dehydration and subsequent collapse of the vascular system (11).

Domestic water supply for drinking purposes is the most critical of all. The World Health Organization (WHO) estimates that each year approximately 2.5 million children all over the globe, ages 0–12, die of waterborne diarrhea illnesses. Waterborne diseases are the leading cause of death in most developing nations in Africa, Asia, and Latin America. Despite efforts to render safe drinking water to an increasing number of people, as a result of the unprecedented population growth, approximately 1.1 billion people still use unsafe drinking water (12). Water disinfection in the last century has largely reduced the incidence of infectious waterborne diseases. Health officials today and the general public, however, are more and more concerned about chronic health effects potentially caused by the presence of toxic chemicals in drinking water. Ingestion of even small amounts of some inorganic pollutants such as arsenic, lead, cadmium, chromium, mercury, asbestos, cyanide, or nitrates can

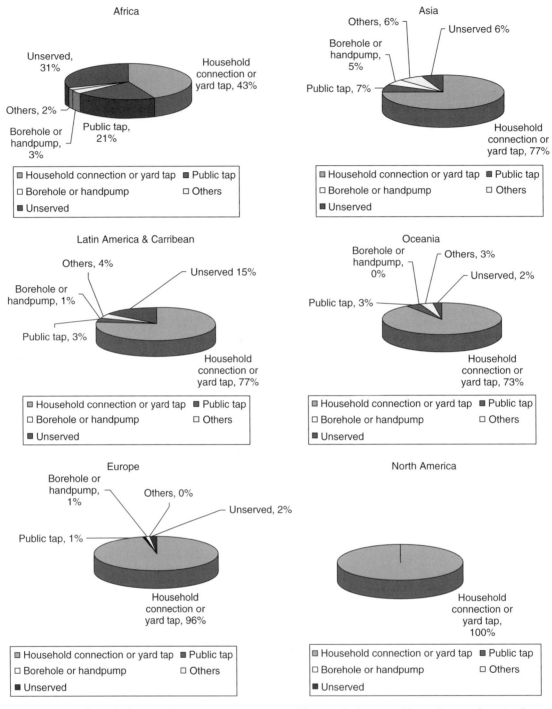

Figure 2. Water supply in the largest cities: average percentage of the population served by each type of service, by region (8).

cause a wide range of health problems. Especially vulnerable to lead poisoning are young children. Ingested lead can result in anemia, neurological disorders, kidney problems, sterility, and birth defects. Evidence suggests that exposure to even very low levels of asbestos in drinking water can result in cancers of the gastrointestinal tract. Although conclusive evidence about cause-and-effect link between chemical contaminants and cancers, birth defects, miscarriages, nervous and reproductive system disorders, and organ damage has not yet been established, there is little doubt that prolonged exposure to some organic and inorganic substances can result in chronic health conditions. Nitrates are another inorganic chemical with serious health effects. The intake of nitrate-containing water by infants under the age of six months may cause methemoglobinemia, known also as the "blue baby syndrome" (13).

The organic compounds detected in drinking water include a wide variety of herbicides and pesticides, disinfection by-products, PCBs, solvents, oils, and so on.

They are volatile and nonvolatile in nature. Many of them are suspected carcinogens (14). *Radionuclides* are also known to be present in some drinking water supplies.

The quality of the drinking water supplies in the United States is one of the highest on earth. With the adoption of the Safe Drinking Water Act in 1974, subsequently amended in 1986 and 1996, the U.S. Congress empowered the U.S. Environmental Protection Agency (U.S. EPA) to enforce regulations aimed at protecting the quality of drinking water supplies. The U.S. EPA has established two sets of standards for water quality: *primary* and *secondary*. The objective of the primary standards is to protect human health. They are based on toxicological, epidemiological, and clinical data. The *maximum contaminant level* (MCL) is the primary standard enforced by the U.S. EPA. Secondary drinking water standards refer to contaminants that create unpleasant odor, taste, and color or may be corrosive or staining. They can cause minor skin irritations or damage clothing and appliances (15). Table 1 displays the major contaminant categories, their potential health effects, and sources of contaminants in drinking water.

FACTORS AFFECTING THE QUALITY OF WATER SUPPLY

The quality of the water source is critical for the quality of water supply. Water pollution refers to the deteriorating quality of water due to the presence of pollutants in the water. Effluent discharges from industries and municipal sewage plants are considered point source pollution. They are relatively easy to identify, regulate, control, and monitor since they are usually located near factories, wastewater treatment plants, septic tanks, or stormwater outfalls. In the United States, industries discharging directly into rivers and lakes are required to apply for permits in accordance with the National Pollutant Discharge Elimination System (NPDES) program. The permit specifies the volume, type, and permissible concentrations of pollutants allowed to be released in the waterways. The other option for controlling industrial discharges is to reuse the treated wastewater instead of discharging it into receiving waters.

Nonpoint source pollutants consist of suspended solids, nutrients, petroleum, inorganic, and radioactive compounds. A major contributor of sediment to the waterways is the runoff from agricultural fields and construction sites. Silt and soil increase the turbidity of water, which impairs the productivity of photosynthetic plants, reduces water depth, endangers bottom-dwelling animals, and suffocates fish. Nutrients enter waterways in the form of fertilizer and sewage runoff, leaves and grass, or as runoff from livestock farms and pastures. Many hazardous chemicals such as asbestos, hydrocarbons, PCBs, pesticides, mercury, and lead enter the waterways

Table 1. The U.S. EPA's National Primary Drinking Water Standards

Contaminant	Potential Health Effects from Ingestion of Water	Sources of Contaminant in Drinking Water
Microorganisms	Gastrointestinal illness, legionnaire's disease	Human and fecal animal waste
Disinfection by-products	Increased risk of cancer, anemia; in infants and young children—central nervous system effects, liver and kidney problems; increased risk of cancer	Human and fecal animal waste
Inorganic chemicals	Increase in blood cholesterol; problems with circulatory system; increased risk of developing benign intestinal polyps; intestinal lesions; kidney damage; allergic dermatitis; nerve damage or thyroid problems; bone disease (pain and tenderness of the bones); delays in physical or mental development of infants and children	Discharge from petroleum refineries; ceramics; erosion of natural deposits; runoff from glass manufacturing sites; decay of asbestos cement in water mains; erosion of natural deposits; discharge from metal refineries; corrosion of galvanized pipes; discharge from steel and pulp mills, plastic and fertilizer factories; water additive that promotes strong teeth; corrosion of household plumbing systems; erosion of natural deposits
Organic chemicals	Nervous system or blood problems; increased risk of cancer; eye, liver, kidney, adrenal gland, or spleen problems; anemia; cardiovascular system or reproductive problems; circulatory system problems; thymus gland problems; immune deficiencies	Added to water during sewage/wastewater treatment; runoff from herbicide used on row crops; leaching from gas storage tanks and landfills; leaching from linings of water storage tanks and distribution lines; leaching of soil fumigant used on rice and alfalfa; discharge from chemical plants, agricultural chemical factories, and other industrial activities; residue of banned termiticide; runoff from herbicide used on rights of way; runoff/leaching from soil fumigant used on soybeans, cotton, pineapples, and orchards; emissions from waste incineration and other combustion; discharge from wood preserving factories; discharge from rubber and plastic factories; leaching from landfills; discharge from factories and dry cleaners; runoff/leaching from insecticide used on cotton and cattle; discharge from textile finishing factories; breakdown of heptachlor; residue of banned insecticide

through airborne deposition. Acid rainfall is a typical example of airborne pollution that affects the waterways.

In addition to water quality at sources, the contamination of distribution pipelines may be a major concern for the quality of water supply. One cause of contamination is cross-connection, where there is a physical connection between a drinking water supply line and a source of contamination, such as viruses, bacteria, and nitrate. The materials of pipes could also be the source of contamination, such as lead dissolved from pipes and plumbing fixtures.

TREATMENT OF WATER SUPPLIES

A water supply system normally consists of facilities for water withdrawal, storage, treatment, and distribution. After treatment, water is usually resent into the distribution system or to water storage reservoirs (16). Water suppliers apply different types of treatment according to the type of water use and the quality of their source water. Some groundwater supplies can meet the standard requirements without any additional treatment. In most cases, however, and especially when it comes to surface water supplies, a combination of treatment methods is applied to remove contaminants from raw water.

The first stage of the water treatment process is usually to settle the water in order to separate the large solid debris. During the process of *chemical coagulation* the synthetic organic polymers (cationic and anionic) are applied to coagulate colloidal material. During *flocculation*, often preceded by *rapid mixing*, tiny particles combine into larger and heavier "flocs." *Sedimentation* is used to reduce sediment loads in raw water both before and after flocculation. It allows for the removal of sand, silt, gravel, and alum floc. Some hard waters require an additional chemical treatment before filtration. Lime-soda softening is applied to raw waters rich in calcium and magnesium. As in flocculation, softening converts hardness-creating ions into insoluble precipitates, which are then removed through settling and filtration. Some raw water supplies containing iron, manganese, and volatile organic compounds need to be treated by *aeration*. This method allows for the oxidation of chemicals, formation of precipitates, and their subsequent removal by filtration. *Filtration* is mostly a physical process in which the water flows through beds of gravel, sand, anthracite, or diatomaceous earth. However, in order to increase its effectiveness, coagulants are also applied throughout the filtration process. *Adsorption* involves the application of activated carbon, which adsorbs nonvolatile organics and some taste- and odor-causing compounds. Ion exchange is used to remove inorganic chemicals (such as arsenic, chromium, or excess fluoride) as well as nitrates and radionuclides (17).

Chlorination, ozonation, and ultraviolet radiation are methods used during the proccess of *disinfection*. Chlorination is very effective in killing bacteria but less effective in eliminating viruses and protozoans. However, it has the advantage to continue the process of disinfection in the distribution pipes. Ozonation and UV radiation have powerful germicidal effect but both have disadvantages. Ozone has a very short half-life and must be generated electrically at the point and at the moment of use. UV disinfection is applied through UV lamps, which allows for contamination (18). Figure 3 illustrates the process in a domestic water treatment plant in Cincinnati, Ohio (19).

Desalination is not a typical water treatment method. However, in arid coastal areas or in regions with mounting pressure on existing freshwater resources, desalination is routinely used to transform seawater into potable water. Saudi Arabia is the leading producer of desalted water in the world. Two methods of desalination are usually applied. Smaller plants rely mainly on reverse osmosis, while larger producers use multistage flash distillation processes (20).

PROTECTION OF WATER RESOURCES

Rapid industrialization since the nineteenth century has been accompanied by reckless waste management practices, which had turned many surface water bodies into sewers or algae-coated cesspits containing microorganisms, toxic chemicals, heavy metals, organic pollutants, pesticides, fertilizers, and their by-products. In addition, these contaminants penetrate the soil, where they can

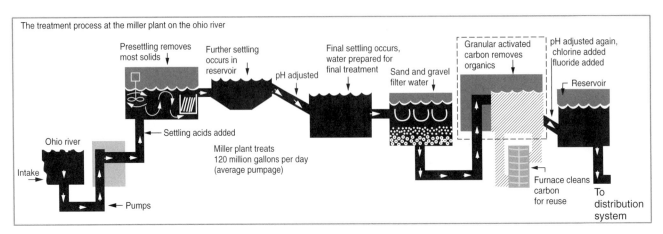

Figure 3. A domestic water treatment plant.

reach the water table and deteriorate the quality of groundwater. Since most of the freshwater supplies rely on groundwater and the slow flow of groundwater, it is critical to keep groundwater from being contaminated. Both the United States and the European Commission have come to the conclusion that setting chemical quality standards for groundwater is not appropriate, as it creates the impression of an allowed level of pollution.

The World Bank estimates that during the past century water consumption has risen sixfold, which is twice as fast as the rate of population growth. By the year 2025, four billion people, particularly in Africa, the Middle East, and South Asia, are expected to live under conditions of severe water stress, if appropriate actions are not taken (21). The total volume of water consumption also affects the quality of water supply. In order to keep the demand within the safe yield of the system, many cities are implementing water-saving policies and restrictions (22).

Three major irrigation methods—sprinkler, microirrigation, and surface (flood)—are applied (23). In general, only a small portion of the irrigation water is actually reaching its intended target. The majority is transformed into vapor through the processes of evaporation and transpiration or is lost in transit (24). Many conventional methods should be supported to increase field water use efficiency and the crop productivity of unit water consumption. Through more efficient irrigating practices such as the automated drip irrigation, alternate irrigation in controlled root zones, and water-deficit-regulated irrigation can substantially reduce water demand in agriculture without reducing crop yield (25). Research has shown that comprehensive water-saving farming technology can increase irrigation productivity 10–15% (26).

Water conservation approaches should also be applied in manufacturing and mining industries and in urban areas to establish a water-saving society. To do so, industrial wastewater should be treated and reused to increase the utilization ratio. Sustainable water resources will not only efficiently support the existing socioeconomic systems but will also provide the required quantity and quality of water at reasonable cost in the future (27).

BIBLIOGRAPHY

1. *The American Heritage Dictionary of the English language*, 4th Edn. (2000). Houghton Mifflin, Boston.
2. U.S Geological Survey (2003). Water science homepage. Available at http://ga.water.usgs.gov/edu/earthwherewater.html.
3. World Resources Institute Database (2000). Available at http://earthtrends.wri.org/searchable_db/index.cfm?theme=2.
4. Food and Agriculture Organization of the United Nations (FAO) (2003). Water Resources, Development and Management Service. *AQUASTAT Information System on Water and Agriculture: Review of World Water Resources by Country*. Available online at http://www.fao.org/waicent/faoinfo/agricult/agl/aglw/aquastat/water_res/index.htm.
5. Nadakavukaren, A. (2000). *Our Global Environment*, Waveland Press, Prospects Heights, NY.
6. World Resources Institute Database (2000). Available at http://earthtrends.wri.org/searchable_db/index.cfm?theme=2.
7. *USGS National Handbook of Recommended Methods for Water Data Acquisition*, Chap. 11. Available at http://water.usgs.gov/pubs/chapter11/chapter11C.html. Accessed 2004.
8. WHO Global Water Supply and Sanitation Assessment 2000 Report. Available at http://www.who.int/docstore/water_sanitation_health/Globassessment/Global4.3.htm. Accessed 2004.
9. U.S. EPA (2004). *Monitoring and Assessing Water Quality Guidelines*. Available at http://www.epa.gov/owow/monitoring/elements/elements.html#d. Accessed 2004
10. Hoffman, G.J. (2004). Water quality criteria for irrigation, EC97-782, University of Nebraska, Institute of Agriculture and Natural Resources, issued in cooperation with the U.S. Department of Agriculture. Available at http://ianrpubs.unl.edu/irrigation/ec782.htm#sum. Accessed 2004.
11. WHO drinking water quality and infectious diseases transmission and prevention. Available at http://www.who.int/water_sanitation_health/dwq/infectdis/en/print.html. Accessed 2004.
12. WHO Joint Monitoring Program Report (2004). Meeting the MDG drinking-water and sanitation target: a mid-term assessment of progress. Available at http://www.who.int/water_sanitation_health/monitoring/jmp2004/en/. 2004.
13. National Sanitation Foundation (1991). *Standard NSF 61-1991 Drinking water system components—health effects*.
14. National Sanitation Foundation (1991). *Standard NSF 61-1991 Drinking water treatment chemicals—health effects*.
15. U.S. Environmental Protection Agency (2002). National Primary Drinking Water Regulations. Available at http://www.epa.gov/safewater/mcl.html. EPA 816-F-02-013.
16. Litke, D.W. and Kauffman, L.F. (1993). Analysis of residential use of water in the Denver metropolitan area, Colorado, 1980–87. *U.S. Geol. Surv. Water Resour. Invest. Rep.* **92–4030**, p. 69.
17. Logsdon, G., Hess, A., and Horsley, M. (1999). Guide to selection of water treatment processes. In: *Water Quality and Treatment—A Handbook of Community Water Supplies*, 5th Edn. R.D. Letterman (Ed.). McGraw-Hill, New York, pp. 3.1–3.26.
18. Haas, C.N. (1999). Disinfection. In: *Water Quality and Treatment - A Handbook of Community Water Supplies*, 5th Edn. R.D. Letterman (Ed.). McGraw-Hill, New York, pp. 14.1–14.60.
19. Greater Cincinnati Water Works. Available at http://www.cincinnati-oh.gov/water/pages/-3283-/.
20. *Encyclopedia of Desalination and Water Resources* (1997). EOLSS Publishers Co. Ltd., UK.
21. World Bank Water Resources Strategy (2003).
22. Jenkins, M.W. et al. (2004). Optimization of California's water supply system: results and insights. *J. Water Resour. Plan. Manage.* **130**(4): 271–275.
23. U.S. Department of Agriculture, National Agricultural Statistics Service. Drought and Fire Survey 2000, Montana's 2000 Drought/Fire Survey Results: press release. Available at http://www.nass.usda.gov/mt/pressrls/misc/firesurv.htm. Accessed June 25, 2003.
24. Hutson, S.S., Barber, N.L., Kenny, J.F., Linsey, K.S., Lumia, D.S., and Maupin, M.A. (2004). *U.S. Geological Survey Estimated Use of Water in the United States in 2000*. USGS Circular 1268, released March 2004, revised May 2004.
25. Kang, S. and Li, Y. (1997). The water economic trends and countermeasures for the 21st century. *Acta Agric. Technol.* **18**: 32–36.

26. Feng, P.G. (1997). Study of the principles of support for water resource sustainability and water economical utilization. *Agric. Res. Arid Area* **15**: 1–7.
27. Loucks, D.P. (2000). Sustainable water resources management. *Water Int.* **25**(1): 2.

THE SUBMITOCHONDRIAL PARTICLE ASSAY AS A BIOLOGICAL MONITORING TOOL

Francis G. Doherty
AquaTox Research, Inc.
Syracuse, New York

BACKGROUND

While studies assessing the toxicity of environmental contaminants to aquatic organisms date back to the early 1900s, the use of toxicity testing as a tool for monitoring the environmental acceptability of discharges from wastewater treatment and manufacturing facilities to surface waters (i.e., rivers, lakes, and oceans) did not begin to evolve until the 1970s. One of the primary motivating factors behind the development of biological monitoring, or biomonitoring, was the manufacture of exponentially increasing numbers of new chemicals for which analytical procedures were generally lacking. Biomonitoring was viewed as an approach to detect the presence of potentially toxic chemicals in facility discharges without necessarily having to identify the contaminants at that stage of the assessment. Additional advantages included the reduced expense associated with biologically testing significantly greater numbers of samples in comparison with analytical procedures. This approach also focused on only biologically available forms of the chemicals in solution reflecting any interactive effects (synergistic or antagonistic) that might be occurring among contaminants within a complex sample.

Initial research efforts in the biomonitoring field tended to focus on identifying a "most sensitive species" that satisfied a variety of other criteria. These criteria included, but were not necessarily limited to, widespread geographic distribution and ease of culture in the laboratory so as to provide a steady supply of consistently healthy test organisms. A significant amount of effort was also invested in the development of standardized protocols that focused on the control of test variables to provide guidance to laboratories conducting aquatic toxicity tests, thereby permitting direct comparisons of toxicity data from multiple sources.

This preliminary round of research tended to use common endemic fish species (e.g., salmonids and cyprinids) and planktonic invertebrates (e.g., Cladocerans) that served as a food source for larval fish stages as the primary test species in whole organism assays. Lethality and growth or reproduction served as test endpoints. These efforts ultimately led to the development of acute ("short-duration") and chronic ("long-duration") standardized protocols most often employing the fathead minnow, *Pimephales promelas*, and the Cladoceran *Ceriodaphnia dubia* for discharges to freshwater systems (1,2). Companion protocols were also developed for discharges to brackish and marine environments employing saltwater organisms.

This initial focus, however, quickly and almost simultaneously expanded to include other less common fish and invertebrate species along with surrogate microorganisms and *in vitro* biochemical assays in a continuing attempt to identify that elusive "most sensitive species or assay." Alternately, these activities addressed refining or enhancing existing protocols with the primary, preferred species. Other objectives included in this broad research effort included accelerating the pace at which analyses were conducted by primarily reducing the exposure time for standardized tests, which also resulted in a reduction in the cost of testing.

SUBMITOCHONDRIAL PARTICLE ASSAY

One such surrogate assay developed during the early to mid-1980s was the *in vitro* submitochondrial particle (SMP) assay (3–6). This assay exposes processed mammalian mitochondria to samples of interest and spectrophotometrically monitors the concentration of NADH to assess the level of "toxicity" exerted by contaminants present in the test solution. SMPs are prepared from bovine heart mitochondria according to the procedure detailed in Hansen and Smith (7). Briefly, the patented SMP production process involves sonically disrupting isolated mitochondria and separating the inner and outer membranes by differential centrifugation. The inner-membrane isolates retain more than 60 fully functional interacting enzymes associated with cellular electron transport and oxidative phosphorylation. Since the test system is in direct contact with test material, there are no physical barriers interfering with the interaction between the test material and receptors as there are in whole organism tests, thereby creating a potentially ultrasensitive assay.

There are two distinct and functional complementary assays utilizing SMPs. The electron transfer (ETr) assay assesses forward electron transfer through the entire electron transport chain. The test endpoint is the loss of NADH monitored as a decrease in absorbance at 340 nm in a standard spectrophotometer using 1-cm path length disposable methyl acrylate cuvettes. The reverse electron transfer (RET) assay assesses effects on electron transfer and energy coupling processes. In this test, energy derived from the hydrolysis of ATP drives electrons in the reverse direction, up the electron transport chain. The test endpoint is the production of NADH also monitored at 340 nm on a standard spectrophotometer. Sample preparation and absorbance measurement patterns for the RET assay parallel that described for the ETr assay.

Processed SMP and associated reagents for conducting tests are commercially available. Reagents for conducting the assays include SMP diluent for diluting particles to 1.5 mg/mL; RET and ETr concentrated reaction mixtures (CRM) containing buffer, electron donors (RET CRM only), and cofactors required by the SMP, adenosine-5-triphosphate (ATP) for activating the RET reaction, and ß-nicotinamide adenine dinucleotide sodium salt, reduced

Table 1. Final Concentrations of Particles, Activating Agents, and Concentrated Reaction Mixture (CRM) Components for the Reverse Electron Transport (RET) and Electron Transport (ETr) Submitochondrial Particle (SMP) Assays

SMP	RET	ETr
Particles	0.05 mg/mL	0.0167 mg/mL
Activating agents	3.33 mM ATP	0.137 mM NADH
CRM components	50 mM HEPES (pH 7.5)	50 mM HEPES (pH 7.5)
	6 mM Mg^{2+}	6 mM Mg^{2+}
	5 mM K^+ succinate	
	1 mM NAD^+	

form ($NADH\ 2Na^+$) for activating the ETr reaction. Unpublished data generated in the author's laboratory has revealed that particles held at $-80\,°C$ for up to six months remain fully functional. In contrast, some vials of particles held at $-20\,°C$ for that same period of time exhibit a reduction to complete loss of activity that is not directly related to storage duration. The final concentrations of particles and reagents in the two assays are provided in Table 1.

The objective of an individual test is to determine the median effective concentration (EC_{50}): the concentration at which the rate of NADH loss (ETr) or production (RET) is twofold the rate observed in the control treatment. Individual rates of NADH loss or production (as expressed by the slopes of the lines-of-best-fit for each treatment) are determined by simple linear regression analysis for the relationship between time and the mean NADH concentration between replicates within a treatment. Individual slopes are then used to calculate percent inhibition values relative to the control treatment for each toxicant treatment. The EC_{50} for an individual test is determined from the linear portion of the regression of individual treatment percent inhibition values versus concentration. Determination of linearity (i.e., which points to include in the generation of the line-of-best-fit) is achieved through both qualitative and quantitative criteria. Plots are visually assessed to identify a linear portion to the curve and supplemented with the generation of coefficient of determination (R^2) values to confirm the most consistent relationship among points to include in the lines-of-best-fit. The statistical analyses can be conveniently performed by routine spreadsheet software.

EXISTING LITERATURE REVIEW

Published studies assessing the sensitivity of SMPs to test materials can be segregated into two broad classes—those dealing with pure chemicals and those dealing with environmental samples. Studies focused on assessing the toxicity of pure chemicals did not directly relate to the use of SMPs for environmental monitoring. Rather, these studies for the most part simply attempted to demonstrate the potential for use of SMPs for biomonitoring by invoking the limitations associated with existing methods and comparing the sensitivity of the SMP assay with other more established tests. For example, Bragadin and Dell'Antone [8] cited the cost and duration of acute fish tests as motivation for the development of alternatives such as the SMP assay. They reported the EC_{50} values for a few dozen toxic compounds using a modified approach to the assay described above and concluded that the results from their assay compared favorably with published values of toxicity for other species that included rainbow trout and the commercial bacterial luminescent assay, Microtox®.

Similar approaches were adopted using the RET variant by Argese et al. [9] with linear alkylbenzene sulfonates, nonylphenol polyethoxylates, and their derivatives; by Argese et al. [10] with chlorophenols; by Argese et al. [11] with heavy metals; and by Argese et al. [12] with chloroanilines. Each study compared their in-house generated toxicity values from the RET variant of the SMP assay with published literature values for other species, and in general concluded there were significant levels of correlation with many of the existing assays. Researchers associated with the development of the SMP assays also published their own results with a substantial set of diverse chemicals ($n = 162$) tested with one of the SMP assay variants while comparing toxicity values with published endpoints for other species [13]. Their conclusions mirrored those of other researchers in this field while they also summarized other current issues related to the state of the SMP assays.

A second group of studies also reported relationships between toxicities of test materials (both pure chemicals and environmental samples) generated through the SMP assays and other species but relied on data generated within their own laboratories rather than from published literature values. This approach tends to be more reliable since each investigator has control over test variables associated with each assay conducted within their lab. Data drawn from multiple published studies containing data generated by a variety of individuals is inherently more variable due to inconsistencies in test protocols and investigator techniques from one laboratory to another.

Miana et al. [14] compared the toxicity of tributyltin to the green alga *Selenastrum capricornutum* and the Cladoceran *Daphnia magna* to submitochondrial particles and concluded that SMP assay was fast, reproducible, and easily handled. Bettermann et al. [15] assessed the toxicity of elutriate samples from sediments previously demonstrated to be toxic to the amphipod *Hyalella azteca*. They reported a significant correlation between ETr EC_{50} values and amphipod survival for two of three watershed sites. No such relationship was observed between results from the RET assay and sediment elutriates from any of the watersheds. Dutka et al. [16] used a multispecies test battery to assess water and sediment toxicity from two river basins in Chile. The SMP RET assay was among those suggestive of toxicity in several of the samples collected during this study.

Weideborg et al. [17] compared the RET variant with six other routine toxicity tests to assess the toxicity of 82 soluble and non-water-soluble chemicals. They reported similar EC_{50} values for both the RET variant and Microtox tests for low toxicity chemicals, with Microtox being the more sensitive of the two. However, the relationships among all of the tests for a diverse set of test substances was variable and may have been a function of test design rather than test sensitivity. Sherry et al. (18) assessed the toxicity of refinery effluents with Microtox, a variety of algae, invertebrate, and fish tests, and the RET variant of the SMP assay. They reported that acute toxicity was detected only by the Microtox and SMP tests. Argese et al. (19) compared the RET assay with Microtox for their sensitivity to 14 organotin compounds. They reported good agreement between the two assays with comparable sensitivity for organotins with an R^2 of 0.92. Sheesley et al. (20) used the RET variant, the invertebrate *Ceriodaphnia dubia*, and the green alga *S. capricornutum* to assess the toxicity of aqueous and organic solvent extracts of atmospheric particulate matter from the Lake Michigan airshed. Based on a direct comparison of LC_{50} and EC_{50} values, the RET SMP assay was the most sensitive of the three for aqueous extracts of particulate matter from three different sites.

ENVIRONMENTAL MONITORING WITH SMPs

As outlined above, studies assessing the potential for SMPs as a routine biomonitor generally attempted to compare the sensitivity of one or more of the SMP assay variants with that of an established assay. Investigators either ranked assays by comparing each assay's sensitivity to an individual chemical or environmental sample or generated regression analyses for pairs of assays over a range of toxicities from groups of test substances. However, while sensitivity should be a factor in the selection of surrogate assays, it is not the only parameter that needs to be considered.

Biological monitoring, as taken in the current context, was instituted for protection of natural aquatic systems, not wildlife or human health protection. To that end, the initial species selected for that task were considered representative of the aquatic environment. Consequently, the most appropriate surrogate species or assay for that purpose would be one that is equally, but not overly, sensitive to a broad range of chemicals as the primary biomonitoring species while incorporating advantages not provided by the primary species. Such advantages might include reduced test duration time or reduced expense associated with the conductance of those tests. Adopting a surrogate assay that exhibits increased sensitivity relative to the primary test species may prompt expensive remedial actions that are not necessary for the protection of the system under study. SMP assays have exhibited the potential to be among the most sensitive assays conducted with a group of chemicals or environmental samples.

One of the confounding issues related to sensitivity rankings is the change in rank among assays for different types of contaminants. The SMP assay variants are not immune to this type of variability as demonstrated by some of the studies cited above. The solution to this dilemma involves the development of multispecies test batteries with the components of those batteries exhibiting alternating sensitivities to varying chemical classes. Positive (toxic) responses by one or more components within the test battery should then be taken as an indication of contamination. SMP assays have exhibited the potential for being a constructive component of a multispecies or multiassay test battery.

As noted above, any assay adopted for biomonitoring purposes needs to be standardized so that results from different groups of investigators are compatible and comparable. In this regard, the SMP assays are only in a rudimentary stage of development. Manual test protocols only include an approach for generating EC_{50} values involving serial dilution of test solutions. There are no formalized screening procedures for whole environmental samples or order-of-magnitude dilutions for chemical samples. Additionally, the data analysis approach currently in use involves a subjective selection of data points used in the generation of an EC_{50}. This aspect of the assay will need to be evaluated so as to minimize differences in the test endpoints not necessarily related to the toxicity of the test material itself.

Included within the standardization process is the reproducibility of an assay. There have been no studies addressing between-laboratory reproducibility for either of the SMP assay variants. The only published study related to this point focused on the within-laboratory repeatability of the ETr and RET assay variants (21). They reported coefficients-of-variation that were comparable to more traditional assays using three different reference toxicants. Since the level of variability exhibited by an assay increases in between-laboratory round robin studies in comparison with in-laboratory results, such a study is necessary to further consider SMP assays as surrogates for traditional aquatic toxicity tests.

Lastly, SMPs are a manufactured test system in contrast with existing traditional assays that use whole organisms. A great deal of research has been invested in developing culture methods for standard test species so that tests are conducted with consistently healthy test organisms. There are no published studies available assessing variabilities associated with the manufacturing process aside from a preliminary study conducted within the author's laboratory.

CONCLUSION

The SMP assay variants have exhibited the potential to serve as surrogate biomonitoring assays in freshwater systems. Test results have demonstrated that the one or both variants are as sensitive or more sensitive than traditional assays. However, SMP assays are only in a rudimentary stage of development with respect to biomonitoring applications. Both the manufacturing and testing protocols need to be standardized and evaluated prior to serious consideration of SMP assays as a legitimate surrogate biomonitoring option.

BIBLIOGRAPHY

1. Weber, C.I. (1991). *Methods for Measuring the Acute Toxicity of Effluents and Receiving Waters to Freshwater and Marine Organisms*, 4th Edn. Environmental Monitoring Systems Laboratory, U.S. Environmental Protection Agency, Cincinnati, OH, EPA/600/4-90/027.

2. Lewis, P.A., Klemm, D.J., Lazorchak, J.M., Nordberg-King, T.J., Peltier, W.H., and Heber, M.A. (1994). *Short-Term Methods for Estimating the Chronic Toxicity of Effluents and Receiving Waters to Freshwater Organisms*, 3rd Edn. Environmental Monitoring Systems Laboratory, U.S. Environmental Protection Agency, Cincinnati, OH, EPA/600/4-91/002.
3. Blondin, G.A., Knobeloch, L.M., Read, H.W., and Harkin, J.M. (1987). Mammalian mitochondria as *in vitro* monitors of water quality. *Bull. Environ. Contam. Toxicol.* **38**: 467–474.
4. Blondin, G.A., Knobeloch, L.M., Read, H.W., and Harkin, J.M. (1989). An *in vitro* submitochondrial bioassay for predicting acute toxicity in fish. In: *Aquatic Toxicology and Environmental Fate: Eleventh Volume, ASTM STP 1007.* G.W. Suter, II and M.A. Lewis (Eds.). American Society for Testing and Materials, Philadelphia, pp. 551–563.
5. Knobeloch, L.M., Blondin, G.A., Lyford, S.B., and Harkin, J.M. (1990). A rapid bioassay for chemicals that induce pro-oxidant states. *J. Appl. Toxicol.* **10**: 1–5.
6. Knobeloch, L.M., Blondin, G.A., Read, H.W., and Harkin, J.M. (1990). Assessment of chemical toxicity using mammalian mitochondrial electron transport particles. *Arch. Environ. Contam. Toxicol.* **19**: 828–835.
7. Hansen, M. and Smith, A.L. (1964). Studies on the mechanism of oxidative phosphorylation. VII. Preparation of a submitochondrial (ETP_H), which is capable of fully, coupled oxidative phosphorylation. *Biochim. Biophys. Acta* **81**: 214–222.
8. Bragadin, M. and Dell'Antone, P. (1994). A new *in vitro* toxicity test based on the response to toxic substances in solutions of mitochondria from beef heart. *Arch. Environ. Contam. Toxicol.* **27**: 410–414.
9. Argese, E. et al. (1994). Submitochondrial particle response to linear alkylbenzene sulfonates, nonylphenol polyethoxylates and their biodegradation derivatives. *Environ. Toxicol. Chem.* **13**: 737–742.
10. Argese, E. et al. (1995). Submitochondrial particles as toxicity biosensors of chlorophenols. *Environ. Toxicol. Chem.* **14**: 363–368.
11. Argese, E. et al. (1996). Submitochondrial particles as *in vitro* biosensors of heavy metal toxicity. *J. Aquat. Ecosys. Health* **5**: 125–134.
12. Argese, E. et al. (2001). Assessment of chloroaniline toxicity by the submitochondrial particle assay. *Environ. Toxicol. Chem.* **20**: 826–832.
13. Read, H.W., Harkin, J.M., and Gustavson, K.E. (1998). Environmental applications with submitochondrial particles. In: *Microscale Testing in Aquatic Toxicology.* P.G. Wells, K. Lee, and C. Blaise (Eds.). CRC Press, Boca Raton, FL, pp. 31–52.
14. Miana, P., Scotto, S., Perin, G., and Argese, E. (1993). Sensitivity of *Selenastrum capricornutum*, *Daphnia magna* and submitochondrial particles to tributyltin. *Environ. Technol.* **14**: 175–181.
15. Bettermann, A.D., Lazorchak, J.M., and Dorofi, J.C. (1996). Profile of toxic response to sediments using whole-animal and *in vitro* submitochondrial particle (SMP) assays. *Environ. Toxicol. Chem.* **15**: 319–324.
16. Dutka, B.J. et al. (1996). Water and sediment ecotoxicity studies in Temuco and Rapel River Basin, Chile. *Environ. Toxicol. Water Qual.* **11**: 237–247.
17. Weideborg, M., Arctander Vik, E., Daae øfjord, G., and Kjønnø, O. (1997). Comparison of three marine screening tests and four Oslo and Paris Commission procedures to evaluate toxicity of offshore chemicals. *Environ. Toxicol. Chem.* **16**: 384–389.
18. Sherry, J., Scott, B., and Dutka, B. (1997). Use of various acute, sublethal and early life-stage tests to evaluate the toxicity of refinery effluents. *Environ. Toxicol. Chem.* **16**: 2249–2257.
19. Argese, E. et al. (1998). Comparison of *in vitro* submitochondrial particle and Microtox® assays for determining the toxicity of organotin compounds. *Environ. Toxicol. Chem.* **17**: 1005–1012.
20. Sheesley, R.J. et al. (2004). Toxicity of ambient atmospheric particulate matter from the Lake Michigan (USA) airshed to aquatic organisms. *Environ. Toxicol. Chem.* **23**: 133–140.
21. Doherty, F.G. and Gustavson, K.E. (2002). Repeatability of the submitochondrial particle assay. *Ecotox. Environ. Safety* **53**: 122–128.

MICROSCALE TEST RELATIONSHIPS TO RESPONSES TO TOXICANTS IN NATURAL SYSTEMS

John Cairns, Jr.
Virginia Polytechnic Institute
and State University
Blacksburg, Virginia

A primary purpose of ecotoxicology is to estimate toxicity thresholds below which no observable deleterious effects can be detected upon ecosystems. Since uncertainty is unacceptably high when extrapolations are made from single species (1), tests at higher levels of biological organization are much more effective (2). Moreover, such levels of testing permit increased opportunities of testing for alteration of functional attributes (3). Large-scale tests are both risky and expensive; consequently, microscale testing has been developed.

Microscale testing can be carried out with standardized communities, which are superb in furnishing a validated set of results. However, standardized communities do not closely resemble communities from natural systems. Communities from any two natural systems rarely contain the same species. However, any naturally derived community will contain species that span a range of tolerance to stress. This range of tolerance can be generalized from one natural community to another. This extrapolation provides predictive capabilities, which are the ultimate goal of ecotoxicological testing. Furthermore, culturing test organisms for use in microscale testing customarily consumes as much or more time than the testing itself.

Microscale toxicity testing with communities obtained from natural systems has many benefits, in addition to reducing culture problems typical of laboratory-assembled communities. Methods are even available for carrying out microscale toxicity testing in natural systems (4). Communities of organisms that have colonized artificial substrates can be collected. These collections are not artificial communities, although convenience sometimes dictates naming these communities "artificial" in order to distinguish them from collections made directly from natural substrates. Artificial substrates (5) have major advantages in microscale testing.

1. The age of the community is known.
2. A determination can be made of when the community has reached a dynamic equilibrium comparable to that of the system that furnished the colonizing organisms.
3. Natural systems are not disturbed by sampling efforts.
4. Communities can be moved to the microscale test system with minimal disturbance to the community that is the subject of the test.
5. Communities in the natural system can be monitored to determine if changes in either structure or function have occurred that were evident in the microscale system as a result of toxic stress.
6. Since microbial species have a cosmopolitan distribution, results from different geographic regions can be compared.
7. Since microorganisms and macroinvertebrates are much more abundant than larger organisms higher in the food chain, their use in toxicity testing is more likely to be accepted by animal rights groups.
8. Microscale tests allow a setup of many more concentrations in the same space that would be necessary for far fewer larger scale tests.
9. Automation is possible for the counting of numbers of individuals per species and total number of some species used in microscale toxicity tests (6).

Microscale testing units are not miniature ecosystems. Properly designed, they can mimic selected cause–effect pathways (e.g., nutrient cycling and prey relationships) (7) not observable at lower levels of biological complexity. However, they do not have all the cause–effect pathways found in natural ecosystems and landscapes. Some relevant discussions of complex system testing (8) and examples of microscale testing (9,10) are available in the literature. Suitability of the microscale test design can be confirmed by comparing measurements made in it to similar measurements in natural systems (11). Methods are available for the statistical analysis of this unique evidence (12–14). One of the most difficult attributes to include in microscale testing is the invasive pressure of organisms that are attempting to colonize (12). This invasion pressure is one of the major factors in succession of species in natural systems.

The major disadvantage to microscale toxicity testing is that sometimes a modest degree of taxonomic ability is essential. Taxonomy is often not taught in many American institutions of higher learning. However, staff members with no taxonomic ability can be instructed in just a few months in levels of identification necessary in some microscale tests, although years of experience are necessary for the more complex microscale toxicity tests.

BIBLIOGRAPHY

1. Cairns, J. Jr. (1983). Are single species toxicity tests alone adequate for estimating environmental hazard? *Hydrobiologia* **100**: 47–57.
2. Cairns, J. Jr. (Ed.) (1986). *Community Toxicity Testing*, STP920. American Society for Testing and Materials, Philadelphia.
3. Cairns, J. Jr. and Pratt, J.R. (Eds.) (1989). *Functional Testing of Aquatic Biota for Estimating Hazards of Chemicals*, STP988. American Society for Testing and Materials, Philadelphia.
4. Arnegard, M.E., McCormick, P.V., and Cairns, J. Jr. (1998). Effects of copper on periphyton communities assessed *in situ* using chemical-diffusing substrates. *Hydrobiologia* **385**: 163–170.
5. Cairns, J. Jr. (1982). *Artificial Substrates*. Ann Arbor Science Publishers, Ann Arbor, MI.
6. Cairns, J. Jr. and Niederlehner, B.R. (Eds.) (1995). *Ecological Toxicity Testing: Scale, Complexity and Relevance*. Lewis Publishers, Chelsea, MI.
7. Clements, W.H., Cherry, D.S., and Cairns, J. Jr. (1989). The influence of copper exposure on predator–prey interactions in aquatic insect communities. *Freshwater Biol.* **21**: 483–488.
8. Hoffman, D.J., Rattner, B.A., Burton, G.A. Jr., and Cairns, J. Jr. (2003). *Handbook of Ecotoxicology*, 2nd Edn. Lewis Publishers/CRC Press, Boca Raton, FL.
9. McCormick, P.V., Belanger, S.E., and Cairns, J. Jr. (1997). Evaluating the hazard of dodecyl alkyl sulphate to natural ecosystems using indigenous protistan communities. *Ecotoxicology* **6**: 67–85.
10. Orvos, D.R., Scanferlato, V.S., Lacy, G.H., and Cairns, J. Jr. (1989). Fate and effects of genetically-engineered *Erwinia carotovora* in terrestrial and aquatic microcosms. *89th Annual Meeting of the American Society for Microbiology*, New Orleans, Louisiana, p. 355.
11. Niederlehner, B.R. and Cairns, J. Jr. (1994). Consistency and sensitivity of community level endpoints in microcosm tests. *J. Aquat. Ecosys. Health* **3**: 93–99.
12. Pontasch, K.W. and Cairns, J. Jr. (1989). Establishing and maintaining laboratory-based microcosms of riffle insect communities: their potential for multispecies toxicity tests. *Hydrobiologia* **175**: 49–60.
13. Pontasch, K.W., Smith, E.P., and Cairns, J. Jr. (1989). Diversity indices, community comparison indices and canonical discriminant analysis: interpreting the results of multispecies toxicity tests. *Water Res.* **23**: 1229–1238.
14. Smith, E.P. and Cairns, J. Jr. (1993). Extrapolation methods for setting ecological standards for water quality: statistical and ecological concerns. *Ecotoxicology* **2**: 203–219.

TOXICITY IDENTIFICATION EVALUATION

Philip J. Markle
Whittier, California

Bioassays for aquatic chronic and acute toxicity testing using standardized methods have been used to measure toxic effects and in some cases can even estimate the magnitude of toxicity for many years. Their use has been incorporated into regulatory requirements for point and nonpoint discharges to both marine and freshwater environments. However, these standard methods provide virtually no information on which constituent(s) are responsible for the observed effect. Reliance on only chemical-specific analyses to determine the causative agent(s) in a water sample exhibiting toxicity or suspected of being toxic to aquatic organisms can be misleading. Initially, the list of potential toxicants that would need to be quantified is nearly limitless, not to mention costly.

Additionally, even GC/MS analyses may not be sensitive enough to detect low concentrations of some toxic substances. Furthermore, even if elevated levels of a particular compound are identified, the compound may or may not be bioavailable, and receiving water characteristics (i.e., total organic carbon, hardness, salinity, pH) may confound the observed toxicological sensitivity. Chemical quantification will also not identify any additive, synergistic, or antagonistic properties of many toxicant combinations. It was specifically because of these concerns that aquatic toxicity testing was developed and incorporated into many water pollution control plans.

The U.S. EPA has developed and published several manuals detailing specific methods for identifying toxic constituents (1–4). These methods are collectively referred to as "toxicity identification evaluations" (TIEs) and are often incorporated as part of a larger site-specific study identified as a "toxicity reduction evaluation" (TRE). The ultimate goal of a TRE is to identify toxic constituents, identify their sources, evaluate toxicity control options, and eventually eliminate the toxicity. The TIE is typically used to identify the toxic constituents and to evaluate the effectiveness of the toxicity control mechanisms. During the initial stages of a TIE (phase 1), toxicity is characterized through a series of sample manipulations followed by toxicity testing. These manipulations typically include pH adjustments, chemical additions (EDTA and sodium thiosulfate), chemical extractions (solid phase extraction), and physical manipulations, (filtration and aeration) followed by toxicity testing to assess any increases or decreases resulting from the manipulation.

The relative increases or decreases in observed toxicity in the subsequent toxicity tests conducted on the manipulated sample often characterize the properties of the chemical constituent(s). For example, a decrease in toxicity observed in the EDTA addition tests would indicate possible metal toxicity. A list of commonly applied phase 1 TIE manipulations and their interpretation is contained in Table 1. Other manipulations and result interpretations are specifically detailed in the U.S. EPA manuals. Once characterized in phase 1, the investigator would proceed to phase 2 to identify the individual constituent(s). The phase 2 TIE procedures incorporate additional toxicity testing and/or analytical methods. The testing and/or analytical measurements vary, depending on the phase 1 characterization interpretations. The last phase of the TIE procedure (phase 3) consists of a series of final confirmation steps designed to provide a "weight of evidence" approach in confirming the causative agent(s). Through the establishment of causal links and correlations, the approach is designed to ensure that suspected toxicants are conclusively identified and consistent from sample to sample.

Identification of what is "toxic" in aquatic toxicity tests is most often determined quantitatively through the regulatory process as permit limits or objectives and no definitive standards have been defined or are

Table 1. Summary of Commonly Applied Phase 1 TIE Manipulations and Associated Interpretation

TIE Manipulation	Purpose/Interpretation
Baseline—no manipulation	Conducted to determine if toxicity is present
EDTA addition	Controls most metal (cationic) toxicity
Sodium thiosulfate addition	Reduces easily oxidizable compounds such as chlorine or some metals
Piperonyl butoxide addition	Controls toxicity due to organophosphate pesticides
	An increase in toxicity would indicate possible pyrethroid pesticide toxicity
pH control at 6.5	Controls ammonia/metal toxicity
	A decrease in toxicity may indicate ammonia/metal toxicity
pH control at 8.5	Controls some metal toxicity
	An increase in toxicity would indicate ammonia and/or some metal toxicity
Aeration	Removes volatile compounds and some surfactants
Filtration	Removes solids and some surfactants
Solid phase extraction (SPE)	Removes nonpolar organic compounds and some surfactants
	Recovery of toxicity from the SPE column indicates nonpolar organic toxicity
	Failure to recover toxicity from the SPE column indicates surfactant toxicity

universally used. Although no universal definitions exist, several measures of toxicity are more commonly employed and seem to have achieved greater acceptance. These include the use of the "no observed effect concentration" (NOEC) or "no observed effect level" (NOEL). The NOEC and NOEL use hypothesis testing to identify statistically significant differences relative to a nontoxic control. The NOEC or NOEL is defined as the highest concentration not significantly different from the control. Another commonly used analysis involves various point estimation techniques. These quantitative analyses use dose–response data to estimate "safe" or "nontoxic" concentrations or effects using metrics such as the LC_{50} (lethal concentration causing a 50% effect) or EC/IC_{25} (effective or inhibition concentration causing a 25% effect). Due to the large amount of toxicity testing required in the TIE process, reductions in testing replication and exposure volume and duration are often made to save on sample and staffing resources. Results are typically represented qualitatively as opposed to quantitatively that is, relative increases or decreases in toxicity compared to the no manipulation treatment or appropriately manipulated control were observed. Although fairly straightforward, several things can significantly complicate the TIE process. To identify the causative agent, a sample must first exhibit toxicity. For this reason, the persistence of toxicity is often a confounding factor. Multiple samples eventually need to be collected and tested during the course of the analysis. If toxicity is episodic or sporadic, collection and subsequent testing of a suitably toxic sample will also be more unlikely. Most TIEs are conducted during the course of 1 to 2 weeks requiring extended holding of samples. Constituents that rapidly degrade during this extended holding time can often confound identification due to a loss of toxicity. Furthermore, the numerous manipulations and increased holding time necessary in conducting a TIE may also alter basic water quality characteristics that may artificially affect the observed toxicity. For these reasons, successful completion of a TIE is much more likely in samples exhibiting a large amount of toxicity. Another common complication involves multiple constituents. If a sample contains two or more toxic constituents, the investigator will need to identify at least one of the constituents and selectively remove or control its toxicity without impacting the others prior to attempting to identify the other constituent(s).

Although much more work conducted using freshwater organisms has resulted in greater acceptance and standardization in the freshwater methods, the described TIE procedures have been successfully used to identify and ultimately control toxic compounds in effluents and freshwater and marine receiving waters. It would be impractical to detail all of the possible TIE manipulations and their interpretation in this overview. The referenced manuals provide specific details on all aspects of the conduct and interpretation of TIE results. Copies of the referenced manuals may be obtained directly from the U.S. EPA (http://www.epa.gov/waterscience/WET/).

BIBLIOGRAPHY

1. EPA/600/6-91/005F. (1992). *Toxicity Identification: Characterization of Chronically Toxic Effluents, Phase 1*. U.S. EPA Office of Research and Development, Washington, DC.
2. EPA/600/6-91/003. (1991). *Methods for Aquatic Toxicity Identification Evaluations-Phase 1 Toxicity Characterization Procedures*, 2nd Edn. U.S. EPA Office of Research and Development, Washington, DC.
3. EPA/600R-92/080. (1993). *Methods for Aquatic Toxicity Identification Evaluations-Phase 2 Toxicity Identification Procedures for Samples Exhibiting Acute and Chronic Toxicity*. U.S. EPA Office of Research and Development, Washington, DC.
4. EPA/600/3-88/036. (1989). *Methods for Aquatic Toxicity Identification Evaluations-Phase III Toxicity Confirmation Procedures*. U.S. EPA Environmental Research Laboratory, Duluth, MN.

WHOLE EFFLUENT TOXICITY CONTROLS

PHILIP J. MARKLE
Whittier, California

The selection of an appropriate control sample/water for whole effluent toxicity testing is a critical experimental design component. Whole effluent toxicity (WET) testing has been an important tool for toxicity compliance and research testing alike to identify acute and chronic toxicity in effluents and receiving waters. The U.S. EPA and others have developed several manuals detailing specific WET methods; many have been incorporated into the National Pollutant Discharge Elimination System (NPDES) permit program for compliance determination. The primary objective of these aquatic toxicity tests is to estimate a maximum "safe" concentration of toxic substances in a discharge or receiving water that will allow for normal growth and propagation of aquatic organisms. A typical WET test experimental design consists of exposing organisms under controlled laboratory conditions to a series of effluent, receiving water, and/or toxicant dosed samples and recording biological observations such as reproduction, growth, and survival.

A set of exposures is concurrently conducted using control water. The biological results obtained from the effluent/receiving water exposures are then qualitatively or quantitatively compared to the results obtained in the control water exposure to identify relative reductions in biological response as an indication of toxicity. Effects relative to the control are used to estimate toxicity in a WET test, so control water selection can be critical. Control water that stimulates an unusually large response or enhancement when compared to a more normal response in the effluent/receiving water is more likely to identify falsely that a sample is "toxic," when it was not. Likewise, control water that exhibits a suppressed response when compared to an effluent/receiving water that exhibits toxicity is more likely to be identified as nontoxic when in fact it was toxic. Nontoxicity related factors associated with control waters can elicit a stimulatory or inhibitory response in test organisms. These factors need to be identified by the investigator and carefully controlled to minimize any misidentification of toxicity or nontoxicity in a WET test.

In multiconcentration tests, the same control water is also used to dilute the effluent or receiving water to the appropriate concentration. For this reason, control water is often referred to as *dilution water*. Although the amount of sample increases as concentration increases, the amount of dilution water likewise decreases as concentration increases. The U.S.EPA whole effluent toxicity testing protocols define acceptable dilution water as water that is appropriate for the test objectives, supports adequate test organism response measured in the test, is consistent in quality, and is free of contaminants that could cause toxicity (1–3). Although test organisms may have a wide range of tolerance to many commonly measured water characteristics (pH, hardness, conductivity, salinity, etc.), it is likely that the organisms have a much narrower "optimal" range. Differences in these basic water characteristics between control water and effluent samples may account for an apparent observed difference in toxicological response in a WET test when in reality, the sample was actually nontoxic. Furthermore, control water/effluent sample interactions in multiple concentration tests could increase or decrease apparent toxicity and result in improper identification of toxicity.

Selection of an appropriate control water will depend in some part on the objectives of the experiment. If the primary objective of the test is to quantify the relative toxicity of a sample over time or location (i.e., is sample "A" more or less toxic than sample

"B," or is discharge "A" more toxic this month than it was last month), synthetic control water would be most appropriate. Dilution water characteristics such as hardness, pH, magnesium/calcium ratios, and many others can drastically effect the toxicological response in a test. It is impossible to understand or predict all of the dilution water/sample interactions that may effect the toxicological response, so the incorporation of standard synthetic control water allows the investigator to eliminate these concerns when the objective of toxicity testing is to measure relative toxicity. Synthetic dilution water is prepared by dissolving fixed amounts of reagent grade salts into high purity de-ionized water. For synthetic freshwater, $NaHCO_3$, $CaSO_4$, $MgSO_4$, and KCl are the most commonly incorporated reagent salts. Commercially available seawater reagents are available for testing with estuarine and marine species. The referenced U.S. EPA method protocols provide detailed preparatory procedures for many of these waters. The advantages of synthetic waters are that they are very consistent from batch to batch, would not be expected to contain toxic constituents, and are easily prepared. The disadvantages of synthetic waters are that they are typically nutrient poor and although adequate, may not be "optimal" water for the test organisms.

If the objective of the test is to estimate or predict ecological impacts on the receiving water, control water with similar receiving water characteristics should be used. The more similar the control water characteristics are to the receiving water, the more predictive the toxicity test will be. Characteristics most commonly simulated include pH, hardness, alkalinity, and conductivity. However, differences between receiving water and control water in other less understood characteristics such as total organic carbon and micronutrient concentration, to name only a few, may significantly compromise the predictive ability of the toxicity test and may require further investigation. An immediate upstream receiving water not influenced by the discharge may be a suitable source. The advantages to using receiving water is that sample/control water interactions are accounted for in the toxicity test design with little effort and surface waters tend to be higher in micronutrient content than synthetic waters. Disadvantages of using receiving water include the possibility that they may contain toxic constituents, they may contain pathogens or predatory organisms, water quality characteristics are not typically consistent over time, and upstream receiving water may not be available.

Additionally, any manipulations of a sample prior to or during a toxicity test should also be incorporated into the control/dilution water. For instance, if a sample requires aeration to maintain appropriate dissolved oxygen levels, the control should be similarly aerated to compensate for any "artifactual" increase or decrease in toxicological response that may occur from the manipulation. This becomes increasingly more important when conducting toxicity identification evaluations (TIEs). In a TIE, toxic effluent or receiving waters are manipulated to isolate or remove specific compounds or classes of compounds. Many of these manipulations involve chemical extractions and/or the addition of substances that could affect the response.

Appropriately manipulated control water aliquots must also be tested and compared to identify any potentially stimulatory or inhibitory effects.

BIBLIOGRAPHY

1. U.S. Environmental Protection Agency. (2002). *Short-term Methods for Estimating the Chronic Toxicity of Effluents and Receiving Waters to Freshwater Organisms*, 4th Edn. EPA/821/R-02/013. Office of Water, Washington, DC.
2. U.S. Environmental Protection Agency. (2002). *Short-Term Methods for Estimating the Chronic Toxicity of Effluents and Receiving Waters to Marine and Estuarine Organisms*. 3rd Edn. EPA/821/R-02/014. Office of Water, Washington, DC.
3. U.S. Environmental Protection Agency. (2002). *Methods for Measuring the Acute Toxicity of Effluents and Receiving Waters to Freshwater and Marine Organisms*, 5th Edn. EPA/821/R-02/012. Office of Water, Washington, DC.

DEVELOPMENT AND APPLICATION OF SEDIMENT TOXICITY TESTS FOR REGULATORY PURPOSES

M.G.J HARTL
F.N.A.M. VAN PELT
J. O'HALLORAN
Environmental Research Institute
University College Cork,
Ireland

INTRODUCTION

Traditional approaches to sediment toxicity assessment have employed chemical analysis to identify and quantify pollutants present. This approach, however, will only provide information on chemical classes that are analyzed and, when used alone, is of little value in ecotoxicological assessment, because toxicity cannot be determined on the basis of chemistry alone (1). Toxicity is ultimately defined as a measurable biological response to a particular substance or mixture of substances (2,3). Toxicity testing provides a more direct means of assessing the potential adverse effects of contaminants. In a complementary way, ecotoxicological assessment should provide a measure of the combined effects of the compounds in a complex sample, thereby taking into account any additive, antagonistic, or synergistic effects and include a degree of biological relevance.

An ecotoxicological sediment assessment necessitates a tiered approach using different endpoints and several test species representing different trophic levels, because the effect of pollutants may differ between species. Thus, a battery of bioassays rather than single species assays should be employed. Battery style approaches in the evaluation of sediment toxicity (both freshwater and marine) have been described (2,4–8). For example, the SED-TOX Index recommends the integration of multitrophic and multiexposure route tests (different sediment phases) in toxicity assessment of sediments (8). The ecotoxicological triad explores a similar concept (9). In this article, we define sediments, consider their ecotoxicological significance, and summarize some of the key sediment toxicity test systems in use and/or development.

WHAT ARE AQUATIC SEDIMENTS?

Sediments represent an open, dynamic, and heterogeneous biogeochemical system (10) that is formed by an accumulation of particulate matter derived from continental runoff, coastal erosion, or atmospheric deposition, which precipitates to the bottom of a water body. Typically, sediments are an accumulation of particulate mineral matter, inorganic matter of biogenic origin, organic matter in various stages of synthesis or decomposition, and water (11,12). Sediments normally consist of an inorganic matrix coated with organic matter (13), giving rise to a wide variety of physical, chemical, and biological characteristics. Control sediments can be formulated from particulate matter of known origin and characteristics for use in toxicity testing (12,14,15).

The ecotoxicological significance lies in the tendency of many pollutants, especially the less polar organic contaminants and trace elements, to show a strong affinity to suspended particulate matter (16,17). They are sequestered from the water column and incorporated into the sediment. Redox conditions influence the chemical speciation, sorption behavior, and partition coefficients of incorporated compounds and trace elements. Undisturbed sediments tend to accumulate many chemical compounds, and so act as sinks. The retention capacity of sediments for many pollutants is dependent on salinity, pH, E_h, and/or mechanical disturbance. Changes in these conditions can result in release of the contaminants, and therefore sediments may act not only as sinks but also as secondary sources, directing often highly concentrated pulses of pollutants at benthic organisms (i.e., organisms intimately associated with sediments). Fine-grained, organically rich sediments play a major role in the biogeochemical fate of chemicals, both of natural and anthropogenic origin, and, along with water quality, have increasingly become the focus of attention in assessing the state of the aquatic environment.

In situ sediment toxicity assessments are rarely performed because of logistics, the difficulties of identifying reference and control sites, and controlling or correcting for confounding environmental variables. Thus field-collected sediment samples are used in laboratory-based toxicity test models. Sediments vary on both a spatial and temporal scale and are structured systems of oxic and anoxic zones (18). These two zones display very different chemical conditions (19,20). Accordingly, during the collection of sediment samples, one must ensure that these zones are not mixed, as this may result in differences in redox status, which will affect the bioavailability of contaminants. The oxic layer of the sediment is preferably sampled and used for toxicity testing especially because this layer interfaces with the water column *in situ* (21,22).

Sediments are not homogeneous but are composed of the following phases: whole sediment, sediment–water interface, pore water, and elutriates. Examination of any single sediment phase may be insufficient to give an accurate ecotoxicological assessment (23,24).

Recent investigations using field-collected sediment samples have demonstrated that the whole sediment phase can be used in toxicity assessment under controlled laboratory conditions (see Tables 1–4).

TEST SYSTEMS

A comprehensive assessment of potential sediment toxicity requires the consideration of multiple exposure phases and multiple test models representing different trophic levels and sediment related habitats.

Primary criteria for test species selection for assessing sediment contamination and toxicity include the species' ecological and/or economical importance and its relative sensitivity to sediment contamination, predictable and consistent response of control organisms, ease of culture and maintenance, short duration, replicable, relatively inexpensive, comparable, and ecologically relevant (6,25,26). In addition to species selection, the endpoints in sediment toxicity tests depend on the question being addressed in the environmental risk assessment (2) and may include acute and long-term toxicity, endocrine, reproductive, and genotoxic effects.

The majority of test systems in use for regulatory purposes are commercial test kits assessing acute general effects on microorganisms using sediment extracts (Table 1). Sediment pore water extracts (also known as interstitial water) are defined as the water occupying the space between sediment particles (27). Contaminants in pore water represent the water-soluble, bioavailable fraction and, as a result, may be a major route of exposure to infaunal species (28–30). The use of elutriate extracts, as opposed to pore water extracts, provides information on the leaching potential of sediment-associated contaminants and may therefore yield important data on the potential adverse effects to benthic organisms, following disturbance of the underlying sediment (6,31). Methods applicable to whole sediment, sediment suspension, sediment elutriate, pore-water extracts, and/or sediment extracts from the marine and freshwater environment have been previously reviewed (32).

These test systems are well established, validated and reproducible, fast, cheap, and require little specialized training. In addition, *in vitro* models, using cells derived from a range of taxa, are currently being developed and validated (22). All these tests are suitable for screening purposes and initial hazard identification. However, they have limited ecological relevance because they are often restricted to nonspecific endpoints or a single trophic level. Therefore, the use and development of a multiple test system or test battery, using various endpoints for both general toxicity (see Tables 1 and 2) and specific toxicities (e.g., genotoxic and reproductive effects; see Tables 3 and 4) in sediment-associated organisms from several taxa representing different trophic levels, is desirable. These models may have a higher ecological relevance than the microbial test systems. In some cases, multiple endpoints for different toxic effects (acute, long term, and specific)

Table 1. Examples of Test Systems Used for Sediment Toxicity Assessment—Acute Toxicity Tests

Endpoint	Taxa	Chemicals Identified[a]	In vivo	In vitro	Test Phase Sediment	Extracts[b]	Reference
Enzyme inhibition/ bioluminescence	**Bacteria**						
	Microtox©	PAHs, POPs, metals		+	+	p,e	6,34–38
	ToxiChromoPad©	PAHs, POPs		+	+		38
	LUMIStox©	unspecified		+	+	e	39
	BioTox©	Pesticides	+			p,e,o	3,40
	MetPAD©	Metals		+		e	41
	Toxi-Chromotest©	PAHs, POPs		+	+	e	6,38
Behavior							
Motility	**FW diatom**		+			e	42
	Invertebrates						
	(Daphtoxkit©)	Pesticides	+			p,e,o	3,40
Reburial		Resin-acid	+		+		43

[a] Major chemical classes identified in the sediments.
[b] p: porewater extracts; e: elutriates; o: organic solvent extracts; PAHs: polycyclic aromatic hydrocarbons; POPs: persistent organic pollutants.

Table 2. Examples of Test Systems Used for Sediment Toxicity Assessment—Subchronic Toxicity Tests

Endpoint	Taxa	Chemicals Identified[a]	In vivo	In vitro	Test Phase Sediment	Extracts[b]	Reference
Survival							
	Invertebrates	PAHs, POPs, metals	+		+	e, p	32,44,45
	Vertebrates	PAHs, POPs, metals	+			e, o, p	32,46
Growth inhibition							
	FW Algaltoxkit©	PAHs, PCBs		+		o	38
	FW microalgae		+		+		47
	Marine microalgae	PAHs, PCBs	+		+		6
Behavior							
	Invertebrates	Organotins, metals	+		+		48–50
	Vertebrates	Metals	+		+		51,52
Enzyme induction							
EROD	**Invertebrates**	PAHs, PCBs	+		+		53
	Vertebrates	PAHs, OCPs, metals	+	+	+	o	54–67

[a] Major chemical classes identified in the sediments.
[b] p: porewater extracts; e: elutriates; o: organic solvent extracts; PAHs: polycyclic aromatic hydrocarbons; POPs: persistent organic pollutants.

or single endpoints in multiple organ systems have been used in the same test species in order to observe potential toxicity on various levels of biological organization (33). Unfortunately, this type of approach is currently not accepted for regulatory purposes, because many of the bioassays involved are generally less well validated.

Therefore, we recommend that sediment toxicity assessments should be further developed and evaluated using a tiered approached, consisting of screening using short-term general toxicity tests (Tier 1); hazard identification applying more specific (multiple) endpoints in multiorganism experiments, representing different trophic levels and habitats associated with sediments, as well as different modes of bioavailability by using both sediment extracts and whole sediments (Tier 2); and *in situ* ecosystem function, for example, lifetime reproductive success, and components of biodiversity (Tier 3).

BIBLIOGRAPHY

1. Heida, H. and van der Oost, R. (1996). *Water Sci. Technol.* **34**: 109–116.
2. Chapman, P.M. et al. (2002). *Mar. Pollut. Bull.* **44**: 271–278.
3. Fernandez-Alba, A.R., Guil, M.D.H., Lopez, G.D., and Chisti, Y. (2002). *Anal. Chim. Acta* **451**: 195–202.
4. Dutka, B.J., Tuominen, T., Churchland, L., and Kwan, K.K. (1989). *Hydrobiologia* **188**: 301–315.
5. Giesy, J.P. and Hoke, R.A.J. (1989). *Gt. Lakes Res.* **15**: 539–569.

Table 3. **Examples of Test Systems Used for Sediment Toxicity Assessment—Genotoxicity Tests**

Endpoint	Taxa	Chemicals Identified[a]	In vivo	In vitro	Test Phase Sediment	Extracts[b]	Reference
Mutation							
	Bacteria						
	AMES test	PAHs, POPs		+		o	68
	Mutatox©	PAHs, POPs		+	+	p,e,o	69
Micronucleus							
	Vertebrates						
		PAHs, POPs	+		+		70
SSBs							
	Invertebrates						
		POPs	+		+		71
		PAHs, metals	+		+		33
	Vertebrates						
		PAHs, POPs	+		+		65,70–74
DNA adducts							
	Vertebrates						
		PAHs, PCBs	+		+		59,75

[a] Major chemical classes identified in the sediments.
[b] p: porewater extracts; e: elutriates; o: organic solvent extracts; PAHs: polycyclic aromatic hydrocarbons; POPs: persistent organic pollutants.

Table 4. **Examples of Test Systems Used for Sediment Toxicity Assessment—Endocrine and Reproduction Tests**

Endpoint	Taxa	Chemicals Identified[a]	In vivo	In vitro	Test Phase Sediment	Extracts[b]	Reference
	Invertebrates						
Imposex		Organotins	+		+		76
Larval development		PAHs, metals	+			e,o,p	32,77
Spermiotoxicity		PAHS, POPs		+		p	44,78
Emergence		Unspecified	+		+		32
	Vertebrates						
Estrogen-like activity		PAHS, POPs	+			o	79
Fertility		PAHS, POPs		+	+		80
Vitellogenin		Unspecified	+		+		81,82

[a] Major chemical classes identified in the sediments.
[b] p: porewater extracts; e: elutriates; o: organic solvent extracts; PAHs: polycyclic aromatic hydrocarbons; POPs: persistent organic pollutants.

6. Cheung, Y.H. et al. (1997). *Arch. Environ. Contam. Toxicol.* **32**: 260–267.
7. Matthiessen, P. et al. (1998). *Mar. Environ. Res.* **45**: 1–15.
8. Bombardier, M. and Bermingham, N. (1999). *Environ. Toxicol. Chem.* **18**: 685–698.
9. Chapman, P.M. (2000). *Int. J. Environ. Pollut.* **13**: 351.
10. Chapman, P., Bishay, F., Power, E., Hall, K., Harding, L., McLeay, D., Nassichuk, M., and Knapp, W. (Eds.). (1991). 17th Annual Aquatic Toxicology Workshop. In: *Canadian Technological Report in Fisheries and Aquatic Science*. Vol. 1. Vancouver, BC, pp 323–330.
11. Knezovich, J.P., Harrison, F.L., and Wilhelm, R.G. (1987). *Water Air Soil Pollut.* **32**: 233.
12. ASTM. (1994). *American Society for Testing and Materials: Terminology*. ASTM, Provo, UT.
13. Rand, G.M., Wells, P.G., and McCarty, L.S. (1997). *Fundamentals of Aquatic Toxicology—Effects, Environmenatl Fate and Risk Assessment*, 2nd Edn. G.M. Rand (Ed.). Taylor and Francis, London, pp. 3–67.
14. Suedel, B.C. and Rodgers, J.H. (1994). *Environ. Toxicol. Chem.* **13**: 1163.
15. Quevauviller, P. and Ariese, F. (2001). Trac-Trends. *Anal. Chem.* **20**: 207–218.
16. IEEE. (1987). *Organotin Symposium Oceans'86.* Vol. 4. IEEE, Washington, DC, pp. 1370–1374.
17. Ragnarsdottir, K.V.J. (2000). *Geol. Soc.* **157**: 859–876.
18. Fenchel, T. (1969). *Ophelia* **6**: 1–182.
19. Machan, R. and Ott, J.A. (1972). *Limnol. Oceanogr.* **17**: 622–626.
20. Aller, R.C. (1978). *Am. J. Sci.* **278**: 1185–1234.
21. ASTM. (1990). *Standard guide for collection, storage, characterization and manipulation of sediments for toxicity testing*. ASTM, Provo, UT.
22. Ni Shuilleabhain, S., Davoren, M., Hartl, M.G.J., and O'Halloran, J. (2003). *In vitro Methods in Aquatic Toxicology*. C. Mothersill, B. Austin (Eds.). Springer Verlag, Berlin, pp. 327–374.
23. Burton, G.A. (1991). *Environ. Toxicol. Chem.* **10**: 1585–1627.
24. Bettermann, A.D., Lazorchak, J.M., and Dorofi, J.C. (1996). *Environ. Toxicol. Chem.* **15**: 319–324.
25. Traunspurger, W. and Drews, C. (1996). *Hydrobiologia* **328**: 215–261.

26. Doherty, F.G. (2001). *Water Qual. Res. J. Can.* **36**: 475–518.
27. Ingersoll, C.G. (1995). In: *Fundamentals of Aquatic Toxicology; Effects, Environmental Fate, and Risk Assessment.* G.M. Rand (Ed.). Taylor and Francis, New York.
28. Carr, R.S., Williams, J.W., and Fragata, C.T.B. (1989). *Environ. Toxcol. Chem.* **8**: 533–543.
29. Carr, R.S. (1998). In: *Microscale Testing in Aquatic Toxicology. Advances, Techniques and Practice.* Wells, P.G., Lee, K., and Blaise, C. (Eds.). CRC Press, Boca Raton, FL, pp. 523–538.
30. Nipper, M. et al. (2002). *Mar. Pollut. Bull.* **44**: 789–806.
31. Wong, C.K.C., Cheung, R.Y.H., and Wong, M.H. (1999). *Environ. Pollut.* **105**: 175–183.
32. Nendza, M. (2002). *Chemosphere* **48**: 865–883.
33. Coughlan, B.M. et al. (2002). *Mar. Pollut. Bull.* **44**: 1359–1365.
34. Kemble, N.E. et al. (2000). *Arch. Environ. Contam. Toxicol.* **39**: 452–461.
35. Cook, N.H. and Wells, P.G. (1996). *Water Qual. Res. J. Can.* **31**: 673–708.
36. Grant, A. and Briggs, A.D. (2002). *Mar. Environ. Res.* **53**: 95–116.
37. Ankley, G.T., Hoke, R.A., Giesy, J.P., and Winger, P.V. (1989). *Chemosphere* **18**: 2069–2075.
38. Côte, C. et al. (1998). *Environ. Toxicol. Water Qual.* **13**: 93–110.
39. Guzzella, L. (1998). *Chemosphere* **37**: 2895–2909.
40. Kahru, A., Pollumaa, L., Reiman, R., and Ratsep, A. (1999). *ATLA-Altern. Lab. Anim.* **27**: 359–366.
41. Bitton, G., Campbell, M., and Koopman, B. (1992). *Environ. Toxicol. Water Qual.* **7**: 323–328.
42. Cohn, S.A. and McGuire, J.R. (2000). *Diatom Res.* **15**: 19–29.
43. Hickey, C.W. and Martin, M.L. (1995). *Environ. Toxicol. Chem.* **14**: 1401–1409.
44. Ho, K.T. et al. (1997). *Environ. Toxicol. Chem.* **16**: 551–558.
45. Williams, T.D. (1992). *Mar. Ecol.-Prog. Ser.* **91**: 221–228.
46. U.S. EPA. (2001). Methods for Assessing the Chronic Toxicity of Marine and Estuarine Sediment-associated Contaminants with the Amphipod *Leptocheirus plumulosus*, 1st Edn. United States Environmental Protection Agency, Washington, DC.
47. Keddy, C.J., Greene, J.C., and Bonnell, M.A. (1995). *Ecotox. Environ. Safety* **30**: 221–251.
48. Stronkhorst, J., van Hattum, B., and Bowmer, T. (1999). *Environ. Toxicol. Chem.* **18**: 2343.
49. Phelps, H.J., Hardy, J.T., Pearson, W.H., and Apts, C.W. (1983). *Mar. Pollut. Bull.* **14**: 452–455.
50. Byrne, P.A. and O'Halloran, J. (2000). *Environ. Toxcol.* **15**: 456–468.
51. Byrne, P.A. and O'Halloran, J. (1999). *Mar. Pollut. Bull.* **39**: 97–105.
52. Weis, J.S. et al. (2001). *Can. J. Fish. Aquat. Sci.* **58**: 1442–1452.
53. Michel, X. et al. (2001). *Polycycl. Aromat. Compd.* **18**: 307–324.
54. Gagne, F., Blaise, C., and Bermingham, N. (1996). *Toxicol. Lett.* **87**: 85–92.
55. Gunther, A.J. et al. (1997). *Mar. Environ. Res.* **44**: 41–49.
56. Escartin, E. and Porte, C. (1999). *Mar. Pollut. Bull.* **38**: 1200–1206.
57. Myers, M.S. et al. (1998). *Mar. Pollut. Bull.* **37**: 92–113.
58. Willett, K.L. et al. (1997). *Environ. Toxicol. Chem.* **16**: 1472–1479.
59. Rice, C.A. et al. (2000). *Mar. Environ. Res.* **50**: 527–533.
60. Johnson, L. et al. (1993). *Mar. Environ. Res.* **35**: 165–170.
61. Besselink, H.T. et al. (1998). *Aquat. Toxicol.* **42**: 271–285.
62. Beyer, J.U. et al. (1996). *Aquat. Toxicol.* **36**: 75–98.
63. Mondon, J.A., Duda, S., and Nowak, B.F. (2001). *Aquat. Toxicol.* **54**: 231–247.
64. Wong, C.K.C., Yeung, H.Y., Woo, P.S., and Wong, M.H. (2001). *Aquat. Toxicol.* **54**: 69–80.
65. Digiulio, R.T. et al. (1989). *Environ. Toxicol. Chem.* **8**: 1103–1123.
66. Gadagbui, B.K.M. and Goksoyr, A. (1996). *Biomarkers* **1**: 252–261.
67. Zapata-Perez, O. et al. (2000). *Mar. Environ. Res.* **50**: 385–391.
68. Bombardier, M., Bermingham, N., Legault, R., and Fouquet, A. (2001). *Chemosphere* **42**: 931–944.
69. Johnson, B.T. and Long, E.R. (1998). *Environ. Toxicol. Chem.* **17**: 1099–1106.
70. Bombail, V., Aw, D., Gordon, E., and Batty, J. (2001). *Chemosphere* **44**: 383–392.
71. Everaarts, J.M. (1995). *Mar. Pollut. Bull.* **31**: 431–438.
72. Theodorakis, C.W, Dsurney, S.J., and Shugart, L.R. (1994). *Environ. Toxicol. Chem.* **13**: 1023–1031.
73. Pandrangi, R., Petras, M., Ralph, S., and Vrzoc, M. (1995). *Environ. Mol. Mutagen.* **26**: 345–356.
74. Kilemade, M.F. et al. (2004). *Environ. Mol. Mutagen.* **44**: 56–64.
75. Stein, J.E. et al. (1992). *Environ. Toxicol. Chem.* **11**: 701–714.
76. Diez, S., Abalos, M., and Bayona, J.M. (2002). *Water Res.* **36**: 905–918.
77. Del Valls, T.A., Forja, J.M., and Gomez-Parra, A. (1998). *Environ. Toxicol. Chem.* **17**: 1073–1084.
78. Carr, R.S. et al. (1996). *Environ. Toxicol. Chem.* **15**: 1218–1231.
79. Khim, J.S. et al. (2001). *Arch. Environ. Contam. Toxicol.* **40**: 151–160.
80. Nagler, J.J. and Cyr, D.G. (1997). *Environ. Toxicol. Chem.* **16**: 1733–1738.
81. Lye, C.M., Frid, C.L.J., and Gill, M.E. (1998). *Mar. Ecol. Prog. Ser.* **170**: 249–260.
82. McArdle, M.E., McElroy, A.E., and Elskus, A.A. (2004). *Environ. Toxicol. Chem.* **23**: 953–959.

ALGAL TOXINS IN WATER

B. Ji
M.H. Wong
R.N.S. Wong
Y. Jiang
Hong Kong Baptist University
Kowloon, Hong Kong

INTRODUCTION

Algal toxins have various adverse implications on the ecosystem directly and indirectly. Cyanotoxins, a group of algal toxins produced by toxic cyanobacteria, are of great concern. The understanding of their occurrence, properties, detection, and removal from water bodies is of importance for monitoring and management of cyanotoxins in freshwater bodies.

ECOLOGY OF TOXIC CYANOBACTERIA IN WATER

Over 40 genera of cyanobacteria are toxic, and the toxic species with particular public health concern generally inhabit freshwater bodies. They are classified into three groups according to the toxin produced: hepatotoxin-producing species (e.g., *Anabaena, Anabaenopsis, Microcystis, Nodularia, Oscillatoria,* and *Cylindrospermopsis*); neurotoxin-producing species (e.g., *Anabaena, Aphanizomenon,* and *Lyngbya*); and dermatotoxin-producing species (e.g., *Lyngbya*; and *Schizothrix*) (1,2).

The toxic cyanobacterial bloom usually occurs in the surface, near-shore area of slow-flowing freshwater bodies. The formation of cyanobacterial bloom and the associated occurrence of cyanotoxins are affected by a number of environmental parameters such as light intensity, temperature, nutrients, trace metals, and organic matters (3). The progressive eutrophication has been reported to be the major cause for the formation of cyanobacterial blooms (4).

CYANOTOXINS AND RELATED HEALTH PROBLEMS

Cyanotoxins are basically classified into three types according to their toxicological targets: hepatotoxins (e.g., microcystins, nodularins, cylindrospermopsins), neurotoxins (e.g., anatoxins, saxitoxins), and dermatotoxins (e.g., aplysiatoxin, lyngbyatoxin A) (5).

Hepatotoxins

The hepatotoxins mainly target the liver. The three major hepatotoxins are microcystins, nodularins, and cylindrospermopsins. Microcystins are the most implicated cyanotoxins in animal and human poisonings and can be produced by several genera, including *Microcystis, Anabaena, Oscillatoria, Nostoc,* and *Anabaenopsis*. Microcystins are cyclic peptides that are made up of five nonprotein amino acids and two protein amino acids (Fig. 1). The toxicity of microcystins is associated with the presence of Adda (3-amino-9-methoxy-2,6,8-trimethyl-10-phenyldeca-4,6-dienoic acid), which has a characteristic violet absorbance at 238 nm (6). Structural variants of microcystins differ in the presence of two protein amino acids (X and Z), two groups (R1 and R2), and two demethylated positions (3 and 7). The molecular weight of microcystins ranges from 800 to 1100 daltons (7).

Microcystins can inhibit hepatocyte protein phosphatases 1 and 2A (8,9). The contamination of microcystins in drinking water is associated with a high incidence of primary liver cancer. The LD_{50} of the extreme toxic microcystin, microcystin-LR, is 50 $\mu g\,kg^{-1}$ body weight in mice by intraperitoneal (i.p.) injection (10).

Nodularins are cyclic pentapeptides and structurally similar to microcystins (Fig. 2). Until now, only four variants of nodularin have been identified in *Nodularia spumigena*. Nodularin can also inhibit the activities of protein phosphatases 1 and 2A and potently promote tumor formation (11,12). The LD_{50} of nodularin in mice is 30 $\mu g\,kg^{-1}$ body weight by i.p. injection (10). In addition, long-term exposure to a low dose of nodularins or microcystins may cause progressive changes in liver tissue, leading to chronic inflammation and focal degeneration of hepatocytes (13).

Cylindrospermopsins are tricyclic alkaloids (Fig. 3) and are produced by *Cylindrospermopsis raciborskii, Aphanizomenon ovalisporum, Umezakia natans,* and *Raphidiopsis curvata*. Cylindrospermopsins are potent inhibitors of protein synthesis and their main target organ is the liver. Unlike the other hepatotoxins, cylindrospermopsins can also affect other organs (e.g., kidney, thymus, and heart). The LD_{50} of cylindrospermopsin in mice is 2100 $\mu g\,kg^{-1}$ body weight at 24 hours by i.p. injection (14).

Neurotoxins

The cyanobacterial neurotoxins are relatively unstable and less common when compared with hepatotoxins. They are, however, diverse in terms of the chemical structures and mammalian toxicities. The major neurotoxins are anatoxins and saxitoxins.

Anatoxins are produced mainly by *Anabaena* species. The two common anatoxins are anatoxin-a and anatoxin-a(s) (Fig. 4). Anatoxin-a is a cholinergic agonist that binds to nicotinic acetylcholine receptors in nerves and neuromuscular junctions. Anatoxin-a(s) mainly inhibits acetylcholinesterase acitivity. The values of LD_{50} of anatoxin-a

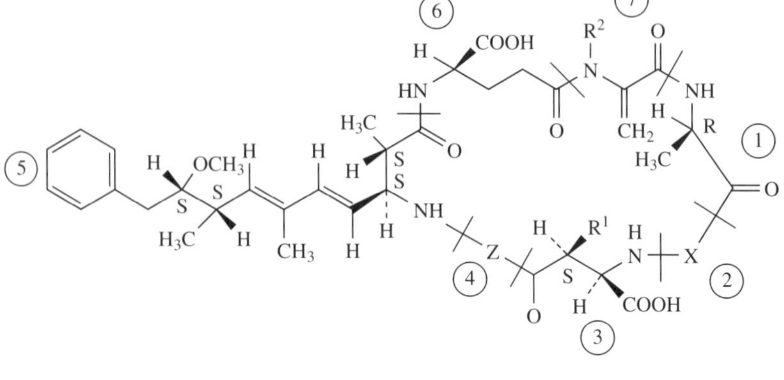

Figure 1. General structure of microcystins [cyclo-D-Ala^1-X^2-D-methyl-isoaspartic acid (MeAsp)^3-Z^4-Adda^5-D-isoglutamic acid (Glu)^6-N-methyldehydroalanine (Mdha)^7, R=H or CH$_3$].

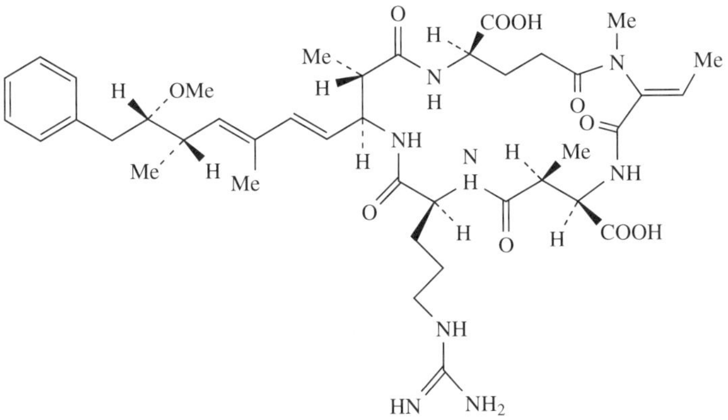

Figure 2. Structure of nodularin.

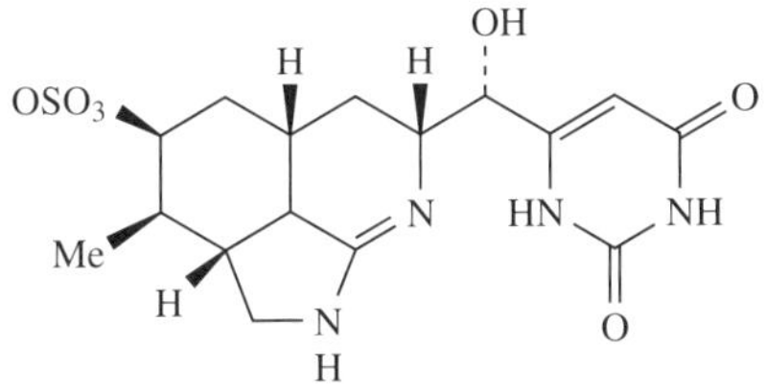

Dermatotoxins

Dermatotoxins are usually produced by benthic marine cyanobacteria (e.g., *Lyngbya majuscule*). Lyngbyatoxin A and aplysiatoxin are two specific dermatotoxins and are related to acute dermatitis and animal death. Long-term exposure to these toxins may lead to skin tumors (16).

DETECTION OF CYANOTOXINS IN WATER

The general methods developed for detection of cyanotoxins include chemical analyses, for example, high-performance liquid chromatography (HPLC), mass spectroscopy (MS), and nuclear magnetic resonance (NMR); biological assays, for example, mouse, invertebrate, and bacterial bioassays; and biochemical assays, for example, protein phosphatase inhibition assay and immunoassay (ELISA).

In chemical analyses, HPLC is widely used for the detection of most cyanotoxins (17). LC/MS is the best method for the detection of saxitoxins and its variants (18). To identify and differentiate the variants of microcystins, fast atom bombardment mass spectroscopy (FABMS) and FABMS/MS methods have been developed (19).

Mouse bioassay is the only method for the detection of all types of cyanotoxins compared with other methods (17). In invertebrate bioassay, brine shrimp, mosquito, fruitfly, house fly, and locust are commonly used target organisms (20,21). The invertebrate bioassay, however, is limited in the detection of some specific cyanotoxins. For example, house fly and locust are only sensitive to saxitoxins, whereas brine shrimp can be used only for detection of microcystins. Bacterial bioassay is seldom used because of its weakness on the reflection of the correlation between the actual concentrations of known cyanotoxin and the responses of the testing bacterium (17).

Protein phosphatase inhibition assay is the most sensitive biochemical method for the analysis of microcystins and nodularins. It uses ^{32}P-labeled glycogen phosphorylase and the colorimetric protein phosphatase inhibition assay. This assay has proved to be an efficient screening method for water samples because of its simple operation. The protein phosphatase inhibition assay is currently adopted for the analysis of microcystins in drinking water.

Figure 3. Structure of cylindrospermopsin.

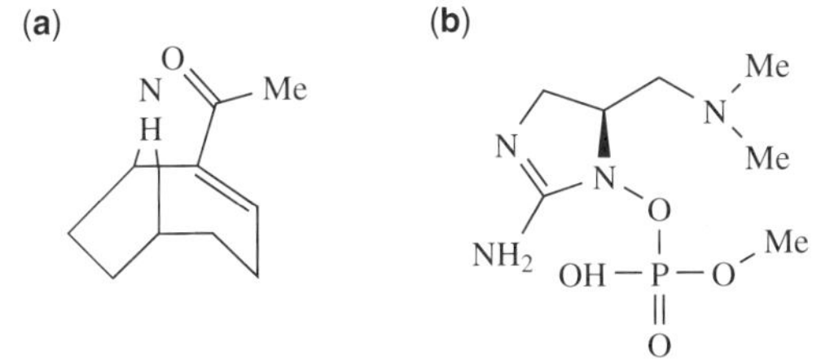

Figure 4. Structures of Anatoxin-a (a) and Anatoxin-a(s) (b).

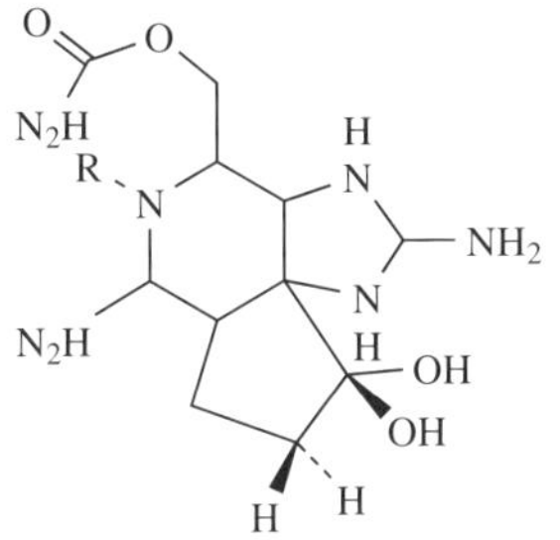

Figure 5. Structure of saxitoxin.

and anatoxin-a(s) are 200 µg kg^{-1} and 20 µg kg^{-1} body weight in mice by i.p. injection, respectively (15).

Saxitoxins (Fig. 5), known as marine paralytic shellfish poisons (PSPs), are recently found in freshwater cyanobacteria including *Anabaena circinalis*, *Lyngbya wollei*, *Cylindrospermopsis raciborskii*, and *Planktothrix* sp. The toxicity of saxitoxins in mice is different depending on their variants. Saxitoxin is the most potent PSP with LD$_{50}$ of 10 µg kg^{-1} body weight in mice by i.p. injection (5).

The detection limits for microcystin-LR in raw and finished drinking water are $0.87\ \mu g\,L^{-1}$ and $0.09\ \mu g\,L^{-1}$, respectively (22). The protein phosphatase inhibition assay, however, may be affected by various other noncyanobacterial toxins and metabolites (e.g., okadaic acid and tautomycin). Consequently, additional confirmation should be made to validate the presence of specific cyanobacterial hepatotoxins (23).

ELISA is a sensitive and specific method for the detection of cyanotoxins using polyclonal or monoclonal antibodies (24). It has been successfully employed for the quantitative detection of cyanobacterial hepatotoxins including microcystin-LR, -RR, -YR, and nodularins in domestic water supplies with the detection limits of $0.05\ \mu g\,L^{-1}$ for microcystin-LR in water samples, for instance (25–27).

STABILITY OF CYANOTOXINS IN WATERS

Cyanotoxins are normally present inside cyanobacterial cells, which enter into the water bodies during cell senescence and lysis. The release of cyanotoxins is influenced by several factors, including light intensity, application of algicide, chlorination, and the unfavorable conditions that accelerate cell death (28). Hence, the ratio of dissolved cyanotoxins to intracellular ones is different at different growth phases of cyanobacteria. In field study, the breakdown of a cyanobacterial bloom may lead to an increase of dissolved microcystins in water (29), although the concentrations detected are relatively low compared with cyanotoxins inside the cells.

The existence of cyanotoxins in the aquatic environment is also affected by their stability. Factors such as temperature, pH, light intensity, and the existence of other microorganisms and oxidants with respect to their effect on cyanotoxin stability have been investigated. Anatoxin-a, for example, is stable at pH 4 for at least 21 days in reservoir water, whereas less than 5% of the original concentration was detected after 14 days at pH 8–10 (30). Microcystins and cylindrospermopsins can be oxidized by strong oxidants, for example, ozone and sodium hypochlorite (NaOCl), and their degradation under intense UV light can be accelerated by the addition of titanium dioxide (TiO_2) (31–33). Similarly, anatoxin-a and saxitoxins are also unstable when exposed to ozone. Photolysis of microcystin may occur in the presence of humic substances and pigments of cyanobacterial cells (31,34). In addition, microcystin-LR and -RR can be degraded *in vitro* in bacterium *Sphingomonas* sp. by microcystinase (35). The degradation of nodularin can be performed with the presence of the extract of *Nodularia spumigena* (36). Anatoxin-a can be degraded micobiologically by *Pseudomonas* sp. in pure culture (37). On the other hand, microcystins and anatoxins can also adsorb onto sediments, suspended solids, and dissolved organic matters, which may correlate to the accumulation and persistence of these toxins in the environment.

REMOVAL OF CYANOTOXINS FROM WATER

Cyanotoxins may impose fatal, acute, or chronic effects on animals and humans and present a health problem to the public through contamination of drinking water, recreationally used water, and fish or shellfish. The extent of cyanotoxin poisoning can be decreased by reducing human exposure to cyanotoxins, which can be achieved through preventing toxic cyanobacterial bloom formation; monitoring of the number of cyanobacteria or the concentration of cyanotoxins in water; notifying the public of the possible hazards in the water; and providing technical and scientific advice for removal of cyanotoxins in water.

The World Health Organization (WHO) has suggested the value of $1\ \mu g\,L^{-1}$ for microcystin-LR in drinking water resources and provides guidelines on monitoring the number of cyanobacterial cells and the concentration of chlorophyll a in drinking and recreationally used waters, which indicates that the concentrations of microcystins can be forecasted from the densities of cyanobacterial cells if microcysin-producing cyanobacteria are dominant (38,39). Cyanobacterial blooms usually can be controlled in several ways, for example, prevention of eutrophication by reduction of external nutrient loading, control of cyanobacteria by raw water abstraction, and application of algicide. The conventional methods adopted in water treatment, including coagulation, flocculation, sand filtration, ultrafiltration, and microfiltration, are effective for removal of cyanobacterial cells from drinking water. However, the cyanotoxins may enter into water because of the lysis of cyanobacterial cells during the water treatment process. The efficiency of these methods to remove dissolved cyanotoxins in water is low (40).

The more effective and practical chemical and physical methods to eliminate the contamination of dissolved cyanotoxins in water are developed based on the adsorption and destruction of the toxins. Ozonation and potassium permanganate, for example, are effective for the destruction of dissolved microcystins. The treatments by chemical oxidants, however, require optimal doses, temperature, pH, and the optimal concentrations of dissolved organic carbon (41). Although the effect of chlorination on destruction of hepatotoxins has been investigated subsequently, the efficiency of conventional chlorination in water treatment on removal of microcystins is low, which might be because of the insufficient chlorine available for microcystin oxidation. The other pertinent techniques for removal of cyanotoxins from drinking water, for example, reverse osmosis, nanofiltration, powdered activated carbon adsorption, granular activated carbon adsorption, and biodegradation, have also been investigated (42–44). However, the efficiencies of these methods for removal of different cyanotoxins are variable. No sole method regarding the removal of all types of cyanotoxins from water bodies efficiently can be adopted. The selection of the appropriate treatment method depends on the target toxin. Further investigation is required to assess the optimization of these treatment methods.

BIBLIOGRAPHY

1. Carmichael, W.W. (1997). The cyanotoxins. In: *Advances in Botanical Research*. P. Callow (Ed.). Academic Press, London, pp. 211–256.

2. Whitton, B.A. (1992). Diversity, ecology and taxonomy of the cyanobacteria. In: *Photosynthetic Prokaryotes*. N.H. Mann and N.G. Carr (Eds.). Plenum Press, New York, pp. 1–51.
3. Kaebernick, M. and Neilan, B.A. (2001). Ecological and molecular investigations of caynotoxins production. *FEMS Microbiol. Ecol.* **35**: 1–9.
4. Reynolds, C.S. (1987). Cyanobacterial waterblooms. In: *Advances in Botanical Research*. P. Callow (Ed.). Academic Press, London, pp. 17–143.
5. Briand, J.F., Jacquet, S., Bernard, C., and Humbert, J.F. (2003). Health hazards for terrestrial vertebrates from toxic cyanobacteria in surface water ecosystems. *Vet. Res.* **34**: 361–377.
6. Rinehart, K.L. et al. (1988). Nodularin, microcystin and the configuration of Adda. *J. Am. Chem. Soc.* **110**: 8557–8558.
7. Carmichael, W.W. (1992). Cyanobacterial secondary metabolites—the cyanotoxins. *J. Appl. Bacteriol.* **72**: 445–459.
8. Runnegar, M. et al. (1995). In vivo and in vitro binding of microcystin to protein phosphatases 1 and 2A. *Biochem. Biophys. Res. Commun.* **216**: 162–169.
9. Fischer, W.J. et al. (2000). Microcystin-LR toxicodynamics, induced pathology, and immunohistochemical localization in livers of blue-green algae exposed rainbow trout (*Oncorhynchus mykiss*). *Toxicol. Sci.* **54**: 365–373.
10. Rinehart, K.L., Namikoshi, M., and Choi, B.W. (1994). Structure and biosynthesis of toxins from blue-green algae (cyanobacteria). *J. Appl. Phycol.* **6**: 159–176.
11. Ohta, T. et al. (1994). Nodularin, a potent inhibitor of protein phosphatases 1 and 2A, is a new environmental carcinogen in male F344 rat liver. *Cancer Res.* **54**: 6402–6406.
12. Carmichael, W.W., Eschedor, J.T., Patterson, G.M.L., and Moore, R.E. (1988). Toxicity and partial structure of a hepatotoxic peptide produced by the cyanobacterium *Nodularia spumigena* Mertens emend. L 575 from New Zealand. *Appl. Environ. Microbiol.* **54**: 2257–2263.
13. Elleman, T.C., Falconer, I.R., Jackson, A.R.B., and Runnegar, M.T. (1978). Isolation, characterization and pathology of the toxin from a *Microcystis aeruginosa* (*Anacystis cyanea*) bloom. *Aust. J. Biol. Sci.* **31**: 209–218.
14. Hawkins, P.R. et al. (1997). Isolation and toxicity of *Cylindrospermopsis raciborskii* from an ornamental lake. *Toxicon.* **35**: 341–346.
15. Falconer, I.R. (1998). Algal toxins and human health. In: *The Handbook of Environmental Chemistry, Part C: Quality and Treatment of Drinking Water II*. J. Hrubec (Ed.). Springer-Verlag, Berlin, pp. 53–82.
16. Osborne, N.J., Webb, P.M., and Shaw, G.R. (2001). The toxins of *Lyngbya majuscula* and their human and ecological health effects. *Environ. Int.* **27**: 381–392.
17. Lawton, L.A., Beattie, K.A., Hawser, S.P., Campbell, D.L., and Codd, G.A. (1994). Evaluation of assay methods for the determination of cyanobacterial hepatotoxicity. In: *Detection Methods for Cyanobacterial Toxins. Proceedings of the First International Symposium on Detection Methods for Cyanobacterial Toxins*. G.A. Codd, T.M. Jeffries, C.W. Keevil, and E. Potter (Eds.). The Royal Society of Chemistry, Cambridge, UK, pp. 111–116.
18. Lawrence, J.F., Menard, C., and Cleroux, C. (1995). Evaluation of prechromatographic oxidation for liquid chromatographic determination of paralytic shellfish poisons in shellfish. *J. AOAC Int.* **78**: 514–520.
19. Harada, K.I. (1996). Chemistry and detection of microcystins. In: *Toxic Microcystis*. M.F. Watanabe, K.I. Harada, W.W. Carmichael, and H. Fujiki (Eds.). CRC, Boca Raton, FL, pp. 103–148.
20. Falconer, I.R. (1993). Measurement of toxins from blue-green algae in water and foodstuffs. In: *Algal Toxins in Seafood and Drinking Water*. I.R. Falconer (Eds.). Academic Press, London, pp. 165–175.
21. McElhiney, J., Lawton, L.A., Edwards, C., and Gallacher, S. (1998). Development of a bioassay employing the desert locust (*Schistocerca gregaria*) for the detection of saxitoxin and related compounds in cyanobacteria and shellfish. *Toxicon.* **36**: 417–420.
22. Lambert, T.W., Boland, M.P., Holmes, C.F.B., and Hrudey, S.E. (1994). Quantitation of the microcystin hepatotoxins in water at environmentally relevant concentrations with the protein phosphatase bioassay. *Environ. Sci. Technol.* **28**: 753–755.
23. Metcalf, J.S., Bell, S.G., and Codd, G.A. (2001). Colorimetric immuno-protein phosphatase inhibition assay for specific detection of microcystins and nodularins of *Cyanobacteria*. *Appl. Environ. Microbiol.* **67**: 904–909.
24. Nagata, S. et al. (1995). Novel monoclonal antibodies against microcystin and their protective activity for hepatotoxicity. *Nat. Toxins.* **3**: 78–86.
25. An, J. and Carmichael, W.W. (1994). Use of a colorimetric protein phosphatase inhibition assay and enzyme linked immunosorbent assay for the study of microcystins and nodularins. *Toxicon.* **32**: 1495–1507.
26. Chu, F.S., Huang, X., and Wei, R.D. (1990). Enzyme-linked immunosorbent assay for microcystins in blue-green algal blooms. *J. Assoc. Anal. Chem.* **73**: 451–456.
27. Chen, G., Chen, C.W., Yu, S.Z., and Ueno, Y. (1998). Evaluation of different water treatment to remove microcystins by using a highly sensitive ELISA. In: *Mycotoxins and Phycotoxins—Developments in Chemistry, Toxicology and Food Safety*. M. Miraglia, H. van Egmond, C. Bera, and J. Gibert (Eds.). Alaken Inc, Fort Collins, CO.
28. Welker, M., Steinberg, C., and Jones, G.J. (2001). Release and persistance of microcystins in natural waters. In: *Cyanotoxins—Occurrence, Causes and Consequences*. I. Chorus (Ed.). Springer-Verlag Telos, Berlin, pp. 85–103.
29. Jones, G.J., Bourne, D.G., Blakeley, R.L., and Doelle, H. (1994). Degradation of the cyanobacterial hepatotoxin microcystin by aquatic bacteria. *Nat. Toxins* **2**: 228–235.
30. Stevens, D.K. and Krieger, R.I. (1991). Stability studies on the cyanobacterial nicotinic alkaloid anatoxin-a. *Toxicon.* **29**: 167–179.
31. Tsuji, K. et al. (1995). Stability of microcystins from cyanobacteria II: effect of UV light on decomposition and isomerization. *Toxicon.* **33**: 1619–1631.
32. Chiswell, R.K. et al. (1999). Stability of cylindrospermopsin, the toxin from the cyanobacterium, *Cylindrospermopsis raciborskii*: effect of pH, temperature, and sunlight on decomposition. *Environ. Toxicol.* **14**: 155–161.
33. Senogles, P.J., Scott, J.A., Shaw, G., and Stratton, H. (2001). Photocatalytic degradation of the cyanotoxin cylindrospermopsin using titanium dioxide and UV irradiation. *Water Res.* **35**: 1245–1255.
34. Welker, M. and Steinberg, C. (2000). Rates of humic substance photosensitized degradation of microcystin-LR in natural waters. *Environ. Sci. Technol.* **34**: 3415–3419.
35. Bourne, D.G. et al. (1996). Enzymatic pathway for the bacterial degradation of the cyanobacterial cyclic peptide toxin microcystin-LR. *Appl. Environ. Microbiol.* **62**: 4086–4094.

36. Twist, H. and Codd, G.A. (1997). Degradation of the cyanobacterial hepatotoxin, nodularin, under light and dark conditions. *FEMS Microbiol. Lett.* **151**: 83–88.
37. Kiviranta, J., Sivonen, K., Lahti, K., and Niemelasi, R. (1991). Production and biodegradation of cyanobacterial toxins—a laboratory study. *Arch. Fur. Hydrobiol.* **121**: 281–294.
38. WHO. (1998). *Guidelines for Drinking Water Quality.* World Health Organization, Geneva.
39. WHO. (1999). *Toxic Cyanobacteria in Water: A Guide to their Public Health Consequences, Monitoring and Management.* World Health Organization, Geneva.
40. Falconer, I.R. (1999). An overview of problems caused by toxic blue-green algae (cyanobacteria) in drinking and recreational water. *Environ. Toxicol.* **14**: 5–12.
41. Newcombe, G. and Nicholson, B. (2004). Water treatment options for dissolved cyanotoxins. *J. Water Sup. Res. Technol.—Aqua.* **53**: 227–239.
42. Newcombe, G. et al. (2003). Treatment options for microcystin toxins: similarities and differences between variants. *Environ. Technol.* **24**: 299–308.
43. Falconer, I.R. et al. (1989). Using activated carbon to remove toxicity from drinking water containing cyanobacterial blooms. *J. Am. Water Works Assoc.* **81**: 102–105.
44. Svrcek, C. and Smith, D.W. (2004). Cyanobacteria toxins and the current state of knowledge on water treatment options: a review. *J. Environ. Eng. Sci.* **3**: 155–184.

GROUND WATER QUALITY IN AREAS ADJOINING RIVER YAMUNA AT DELHI, INDIA

CHAKRESH K. JAIN
MUKESH K. SHARMA
National Institute of Hydrology
Roorkee, India

The groundwater quality in areas adjoining the River Yamuna at Delhi (India) has been assessed to determine the suitability of groundwater for domestic use. Thirty-eight groundwater samples from shallow and deep aquifers were collected each during pre- and postmonsoon seasons in the year 2000 and analyzed for various physicochemical and bacteriological parameters and trace elements. The study indicated concentrations of total dissolved solids, nitrate, sulfate, and sodium higher than water quality standards. The presence of total coliforms indicates bacterial contamination in the groundwater. The grouping of samples according to their hydrochemical facies indicated that the majority of the samples fall into Na-K-Cl-SO$_4$ followed by Na-K-HCO$_3$ and Ca-Mg-Cl-SO$_4$ hydrochemical facies. The qualitative analysis of data depicted higher concentrations of various physicochemical and bacteriological parameters on the western side of River Yamuna, even in deep aquifers.

INTRODUCTION

Water is an essential and vital component of our life support system. In tropical regions, groundwater plays an important role within the context of fluctuating and increasing contamination of surface water resources. Groundwater has unique features (excellent natural quality, usually free from pathogens, color, and turbidity), which render it particularly suitable for a public water supply. Groundwater also plays an important role in agriculture for watering of crops and for irrigation of dry season crops. It is estimated that about 45% of the irrigation water requirement is met from groundwater sources. Unfortunately, the availability of groundwater is not unlimited nor is it protected from deterioration. In most instances, the extraction of excessive quantities of groundwater has resulted in drying up of wells, damaged ecosystems, land subsidence, saltwater intrusion, and depletion of the resource. Groundwater quality is being increasingly threatened by agricultural, urban, and industrial wastes. It has been estimated that once pollution enters the subsurface environment, it may remain concealed for many years, disperses over wide areas of the groundwater aquifer, and renders groundwater supplies unsuitable for consumption and other uses. The rate of depletion of groundwater levels and the deterioration of groundwater quality are of immediate concern in major cities and towns of the country.

The National Capital Territory (NCT) of Delhi is facing severe problems in managing groundwater quality and quantity. Surface water bodies play a significant role in groundwater flow. The hydraulic gradient has a significant role in lateral and vertical migration of contaminants in groundwater aquifers. Therefore, the present study has been carried out to assess the suitability of groundwater for domestic uses in areas adjoining the River Yamuna at Delhi and to examine the likely impact of Yamuna River water quality on groundwater. The suitability of each well for drinking has been reported in an earlier report (1).

STUDY AREA

Delhi generates about 1900 MLD of sewage against installed capacity of 1270 MLD of sewage treatment. The balance of untreated sewage along with a significant quantity of partially treated sewage is discharged into the River Yamuna every day. The river receives sewage and industrial wastes through various drains, which join the River Yamuna between Wazirabad and Okhla. Thus Delhi is the largest contributor of pollution to the River Yamuna, which receives almost 80% of its pollution load through these drains.

The climate of Delhi is influenced mainly by its inland position and the prevalence of continental type air during the major part of the year. Extreme dryness with an intensely hot summer and cold winter are the characteristics of the climate. Only during the monsoon months does air of oceanic origin penetrate to this area and cause increased humidity, cloudiness, and precipitation. The normal annual rainfall in the National Capital Territory of Delhi is 611.8 mm. The rainfall increases from southwest to northeast; about 81% of the annual rainfall is received during the three monsoon months of July, August, and September. The balance of annual rainfall is received as winter rains and as thunderstorm rain during pre- and postmonsoon months.

Thirty-eight groundwater samples from shallow and deep aquifers were collected from both sides of the River Yamuna at Delhi (Fig. 1). The details of sampling locations are given in Table 1.

METHODOLOGY

The groundwater samples were collected in polyethylene bottles during pre- and postmonsoon seasons in the year 2000 from open wells, hand pumps, and tube wells and were preserved by adding an appropriate reagent (2,3). The water samples for trace element analysis were collected in acid-leached polyethylene bottles and preserved by adding ultrapure nitric acid (5 mL/L). Samples for bacteriological analysis were collected in sterilized high-density polypropylene bottles. All samples were stored in sampling kits maintained at 4 °C and brought to the laboratory for detailed chemical and bacteriological analysis. The details of sampling locations are given in Table 1. The physicochemical and bacteriological analyses were performed following standard methods (2,3).

RESULTS AND DISCUSSION

The National Capital Territory of Delhi has the peculiar feature of infiltration of surface water to groundwater from the River Yamuna and from various drains in addition to customary recharge from rainfall. Groundwater recharge also occurs through stagnant water pools in low-lying areas, where surface runoff water collects. The quartzite ridge, which is the prolongation of the Aravalli mountain range, forms the principal watershed in the south, southeast, and southwest parts of Delhi. Because of it, the eastern surface runoff and drainage join the River Yamuna, whereas the runoff from the western part of Delhi goes into the Najafgarh drain. The ever increasing discharge of domestic and industrial wastes into improperly lined sewage drains in Delhi leads to a high risk of contaminating the groundwater. The excessive groundwater uplift also has an adverse impact on water quality in groundwater aquifers of limited thickness.

General Characteristics

The hydrochemical data for the two sets of samples collected from the areas adjoining the River Yamuna at Delhi are given in Table 2. The pH in the groundwater

Figure 1. Study area showing sampling locations.

Table 1. Description of Groundwater Sampling Locations

Sample No.	Location	Type of Well (depth in ft.)[a]
1	Bhagwanpur Khera	HP (25)
2	Loni Road	BW (20)
3	Kabul Nagar	OW (35)
4	Naveen Shahdara	HP (20)
5	Seelampur	HP (30)
6	Shastri Park	HP (25)
7	Lakshmi Nagar	HP (100)
8	Prit Vihar	TW (40)
9	Shankar Vihar	HP (200)
10	Pratap Nagar	TW (100)
11	Himmatpuri	HP (80)
12	Civil Lines	TW (250)
13	Rajpur road	OW (25)
14	Malka Gunj	HP (20)
15	Tripolia	BW (100)
16	Gulabi Bagh	BW (40)
17	Gulabi Bagh	BW (160)
18	Shastri Nagar	BW (100)
19	Shastri Nagar	BW (60)
20	Lekhu Nagar	HP (40)
21	Ram Pura	HP (40)
22	Punjabi Bagh West	HP (40)
23	Rajghat	BW (50)
24	JLN Marg	HP (60)
25	GB Pant Hospital	BW (20)
26	Panchkuin Marg	TW (450)
27	Panchkuin Marg	HP (60)
28	Rajendra Nagar	BW (100)
29	Rajendra Nagar	BW (300)
30	Shankar Road	OW (20)
31	IARI, Pusa	TW (120)
32	Zoological Park	HP (60)
33	Golf Course	BW (80)
34	Rabindra Nagar	TW (150)
35	Teen Murti Chowk	HP (60)
36	Malcha Marg	BW (100)
37	Sardar Patel Road	TW (150)
38	Janpath	BW (250)

[a] HP: Hand pump; OW: open well; BW: bore well; TW: tube well.

of areas adjoining the River Yamuna is mostly within the range 6.7 to 8.3 during the premonsoon season and 6.6 to 8.2 during the postmonsoon season; most of the samples point toward the alkaline range in both seasons. The pH of all the samples is within the limits prescribed by BIS (4) and WHO (5) for various uses of water, including drinking and other domestic supplies. The conductivity varies from 628 to 3540 µS/cm during the premonsoon season and from 620 to 3250 µS/cm during postmonsoon; the conductivity of more than 70% of the samples was above 1000 µS/cm during both pre- and postmonsoon seasons. The maximum conductivities of 3540 and 3250 µS/cm were observed at Golf Course during the pre- and postmonsoon seasons, respectively. Higher values of conductivity in the areas nearby the River Yamuna indicate high mineralization of the groundwater.

The total dissolved solids (TDSs) in the groundwater vary from 402 to 2266 mg/L during the premonsoon season and from 397 to 2080 mg/L during the postmonsoon season. The TDSs of only about 10% of the samples analyzed were within the desirable limit of 500 mg/L. The TDSs of more than 80% of the samples were above the desirable limit of 500 mg/L. An almost similar trend was observed during the postmonsoon season. Water containing more than 500 mg/L of TDSs is not considered desirable for drinking water, though more highly mineralized water is also used where better water is not available. For this reason, 500 mg/L as the desirable limit and 2000 mg/L as the maximum permissible limit has been suggested for drinking water (4). Water containing TDSs of more than 500 mg/L causes gastrointestinal irritation (4). One sample at Golf Course (BW, 80) even crosses the maximum permissible limit of 2000 mg/L.

Carbonates, bicarbonates, and hydroxides are the main cause of alkalinity in natural waters. Bicarbonates represent the major form because they are formed in considerable amount from the action of carbonates on basic materials in the soil. The alkalinity value in the groundwater varies from 116 to 380 mg/L during the premonsoon season and from 106 to 310 mg/L during postmonsoon. About 60% of the samples from the study area fall within the desirable limit of 200 mg/L, and about 40% of the samples cross the desirable limit but are within the maximum permissible limit of 600 mg/L. No sample of the study area crosses the maximum permissible limit of 600 mg/L. The high alkalinity in the study area may be attributed to the action of carbonates upon basic materials in the soil. Such water has an unpleasant taste.

Calcium and magnesium along with their carbonates, sulfates, and chlorides make the water hard, both temporarily and permanently. A limit of 300 mg/L has been recommended as the desirable limit for drinking water (4). The total hardness in the study area ranges from 116 to 871 mg/L during the premonsoon season and from 114 to 792 mg/L during postmonsoon. More than 80% of the samples were found well within the desirable limits for domestic use during both pre- and postmonsoon seasons. However, one sample from Golf Course even crosses the maximum permissible limit of 600 mg/L during both seasons.

The desirable limits for calcium and magnesium in drinking water are 75 and 30 mg/L, respectively (4). In groundwater in the study area, the levels of calcium and magnesium vary from 25 to 240 and from 9 to 64 mg/L, respectively, during the premonsoon season. Slightly lower levels of calcium and magnesium were observed during the postmonsoon season. In groundwater, the calcium content generally exceeds the magnesium content in accordance with their relative abundance in rocks. The increase in magnesium is proportionate to calcium in both seasons.

The concentration of sodium in the study area varies from 55 to 340 mg/L during the premonsoon season and from 51 to 322 mg/L during postmonsoon. The sodium concentration of more than 50 mg/L makes the water unsuitable for domestic use. The sodium concentration was higher at all sites in the study area. The high sodium values may be attributed to base-exchange phenomena. Groundwater with high sodium is unsuitable for irrigation due to the sodium sensitivity of crops/plants.

Table 2. Hydrochemical Data of Groundwater Samples of Delhi (Pre- and Postmonsoon, 2000)[a]

Parameter	Minimum		Maximum		Mean	
pH	6.7	(6.6)	8.3	(8.2)	7.1	(7.1)
Conductivity, μS/cm	628	(620)	3540	(3250)	1463	(1370)
TDSs, mg/L	402	(397)	2266	(2080)	936	(879)
Alkalinity, mg/L	116	(106)	380	(310)	201	(194)
Hardness, mg/L	116	(114)	841	(792)	235	(225)
Chloride	17	(17)	400	(390)	86	(86)
Sulfate, mg/L	43	(48)	690	(680)	180	(177)
Nitrate, mg/L	1.0	(ND)	286	(287)	78	(68)
Phosphate, mg/L	0.07	(0.03)	0.28	(0.21)	0.16	(0.08)
Fluoride, mg/L	0.33	(ND)	1.31	(0.76)	0.80	(0.41)
Sodium, mg/L	55	(51)	340	(322)	160	(157)
Potassium, mg/L	4.2	(2.2)	121	(103)	26	(20)
Calcium, mg/L	25	(21)	240	(227)	59	(59)
Magnesium, mg/L	9.0	(7.0)	64	(56)	21	(19)
MPN Coliform, per 100 mL	Nil	(Nil)	2400	(150)	—	(—)
Total count, per 100 mL	10	(12)	1850	(540)	—	(—)
Copper, mg/L	0.006	(0.003)	0.178	(0.085)	0.023	(0.012)
Iron, mg/L	0.390	(0.128)	5.740	(5.842)	1.960	(1.216)
Manganese, mg/L	0.009	(0.008)	0.944	(0.837)	0.162	(0.183)
Cobalt, mg/L	0.005	(Nil)	0.034	(0.037)	0.016	(0.011)
Nickel, mg/L	0.005	(0.009)	0.043	(0.113)	0.028	(0.028)
Chromium, mg/L	0.006	(0.003)	0.033	(0.078)	0.013	(0.011)
Lead, mg/L	0.010	(0.012)	0.064	(0.098)	0.033	(0.037)
Cadmium, mg/L	0.003	(0.005)	0.010	(0.021)	0.006	(0.011)
Zinc, mg/L	0.021	(0.012)	1.110	(0.732)	0.320	(0.190)

[a]Values in parentheses represent postmonsoon data.

The concentration of potassium in groundwater varies from 4.2 to 121 mg/L during the premonsoon season and from 2.2 to 103 mg/L during postmonsoon. Potassium is an essential element for humans, plants, and animals and is derived in the food chain mainly from vegetation and soil. The main sources of potassium in groundwater include rainwater, weathering of potash silicate minerals, potash fertilizers, and surface water used for irrigation. It is more abundant in sedimentary rocks and commonly present in feldspar, mica, and other clay minerals. The Bureau of Indian Standards has not included potassium in drinking water standards. However, the European Economic Community has prescribed a guideline level for potassium at 10 mg/L in drinking water. Though potassium is found extensively in some igneous and sedimentary rocks, its concentration in natural waters is usually quite low because potassium minerals are resistant to weathering and dissolution. A higher potassium content in groundwater is indicative of groundwater pollution.

The concentration of chloride varies from 17 to 400 mg/L during the premonsoon season. Almost the same trend was observed during the postmonsoon season. The maximum chloride content in groundwater was recorded at Golf Course. No sample in the study area crosses the maximum permissible limit of 1000 mg/L. The limits of chloride have been laid down primarily from taste considerations. A limit of 250 mg/L chloride has been recommended for drinking water supplies (4,5). However, no adverse health effects on humans have been reported from intake of waters containing a higher chloride content.

The concentration of sulfate in the study area varies from 43 to 690 mg/L during the premonsoon season and from 48 to 680 mg/L during postmonsoon. Most of the samples fall within the permissible limit (400 mg/L) for drinking water supplies. Only two samples from Himmat Puri and Golf Course exceeded the maximum permissible limit. In groundwater, sulfate generally occurs as soluble salts of calcium, magnesium, and sodium. The sulfate content of water may change significantly with time during infiltration of rainfall and groundwater recharge, which takes place mostly from stagnant water pools, puddles, and surface runoff water collected in low-lying areas.

The nitrate content of drinking water is considered important for its adverse health effects. The occurrence of high levels of nitrate in groundwater is a prominent problem in NCT-Delhi. The nitrate content in the groundwater of areas adjoining the River Yamuna varies from 1 to 286 mg/L during the premonsoon season. Almost the same trend was observed during postmonsoon. Of the 38 samples analyzed, 20 samples (52.5%) had nitrate content less than 45 mg/L, whereas in about 20% of the groundwater samples, the nitrate content exceeded even the maximum permissible limit of 100 mg/L. The higher level of nitrate at certain locations may be attributed to the surface disposal of domestic sewage and runoff from agricultural fields. It has also been observed that the groundwater samples collected from hand pumps at various depths have a high nitrate content, which may be attributed to wellhead pollution.

Nitrate is an effective plant nutrient and is moderately toxic. A limit of 45 mg/L has been prescribed by WHO (5) and BIS (4) for drinking water supplies. A concentration above 45 mg/L may be detrimental to human health. In higher concentrations, nitrate may produce a disease known as methemoglobinemia (blue baby syndrome), which generally affects bottle-fed infants. Repeated heavy doses of nitrates by ingestion may also cause carcinogenic diseases.

Fluoride is present in soil strata due to the presence of geological formations such as fluorspar and fluorapatite and amphiboles such as hornblende, tremolite, and mica. Weathering of alkali, silicate, igneous, and sedimentary rocks, especially shales, contributes a major portion of fluorides to groundwaters. In addition to natural sources, considerable amounts of fluoride may be contributed by human activities. Fluoride salts are commonly used in the steel, aluminum, brick, and tile industries. Fluoride containing insecticides and herbicides may be contributed by agricultural runoff. Phosphatic fertilizers, which are extensively used, often contain fluorides as an impurity, and they may increase the levels of fluoride in soil. The accumulation of fluoride in soil eventually results in leaching by percolating water, thus increasing the fluoride concentration in groundwater. However, in the groundwater of the area adjoining the River Yamuna, the fluoride content was within the maximum permissible limit of 1.5 mg/L in all samples.

Bacteriological Parameters

The presence of coliforms in water is an indicator of contamination by human or animal excrement. The presence of fecal coliforms in groundwater is a potential public health problem, because fecal matter is a source of pathogenic bacteria and viruses. Groundwater contamination from fecal coliform bacteria is generally caused by percolation from contamination sources (domestic sewage and septic tank) into the aquifers and also by poor sanitation. Shallow wells are particularly susceptible to such contamination. The bacteriological contamination of groundwater in Delhi is mostly attributable to indiscriminate dumping of waste and garbage without any precautions or scientific disposal practices. In Delhi, most of the hand pumps withdraw groundwater from upper strata, which is most susceptible to contamination from polluted surface water.

The groundwater samples collected from the area adjoining the River Yamuna in Delhi have significantly high total coliform. Bacteriological analysis indicates the presence of coliforms in more than 75% of samples during both pre- and postmonsoon seasons. The presence of coliforms was reported mostly from hand pumps. More than 30% of the samples have MPN coliforms >100 per 100 mL during the premonsoon season. Inadequate maintenance of hand pumps and unhygienic conditions around the structures may be responsible for bacterial contamination. Indiscriminate land disposal of domestic waste on the surface, improper disposal of solid waste, and leaching of wastewater from landfills further heighten the chances of bacterial contamination.

Trace Elements

Most of the trace metals are of immediate concern because of their toxicity and nonbiodegradable nature. Cadmium, chromium, and lead are highly toxic to humans even in low concentrations. The concentrations of heavy metals in groundwater except iron, which is present in appreciable concentration, were below the prescribed maximum permissible limits in most of the samples. The concentration of iron varies from 0.39 to 5.74 mg/L during the premonsoon season and from 0.128 to 5.842 mg/L during postmonsoon. The concentration ranges of copper and zinc were well below maximum permissible limits.

The study clearly indicated higher concentration of total dissolved solids, electrical conductivity, nitrate, sulfate, and sodium. The presence of total coliforms indicates bacterial contamination in groundwater. The presence of heavy metals in groundwater was recorded in many samples, but the levels were not significant. Water quality standards have been violated for TDSs, nitrate, sulfate, and sodium at a few places.

Chadha's Diagram for Hydrochemical Classification

Chadha's diagram (6) is a somewhat modified version of the Piper trilinear diagram (7). In this diagram, the difference in milliequivalent percentage between alkaline earths (calcium plus magnesium) and alkali metals (sodium plus potassium), expressed as percentage reacting values, is plotted on the x axis and the difference in milliequivalent percentage between weak acidic anions (carbonate plus bicarbonate) and strong acidic anions (chloride plus sulfate) is plotted on the y axis. The resulting field of study is a square or rectangle depending on the size of the scales chosen for x and y coordinates. The milliequivalent percentage differences between alkaline earth and alkali metals and between weak acidic anions and strong acidic anions would plot in one of the four possible subfields of the diagram. The main advantage of this diagram is that it can be made simply on most spreadsheet software packages.

The square or rectangular field describes the overall character of the water. The diagram has all the advantages of the diamond-shaped field of the Piper trilinear diagram and can be used to study various hydrochemical processes, such as base cation exchange, cement pollution, mixing of natural waters, sulfate reduction, saline water (end product water), and other related hydrochemical problems.

The chemical analysis data of all the samples collected from the area adjoining the River Yamuna in Delhi have been plotted on Chadha's diagram (Fig. 2) and results have been summarized in Table 3. It is evident that during the premonsoon season, most of the samples fall in Group 7 (Na-K-Cl-SO_4) followed by Group 8 (Na-K-HCO_3 type) and Group 6 (Ca-Mg-Cl-SO_4). An almost similar trend was observed during the postmonsoon season.

Impact of River Water Quality on Groundwater

Surface waterbodies play an important role in groundwater flow. The infiltration of surface water to groundwater usually occurs in a recharge geographical area; base flow from groundwater to surface waterbodies may occur in a discharge geographical area. In a discharge area, the hydraulic head increases with depth, and net saturated flow is upward toward the water table, but in a recharge area, the water table lies at a considerable depth beneath a thick unsaturated zone. The relationship of surface water

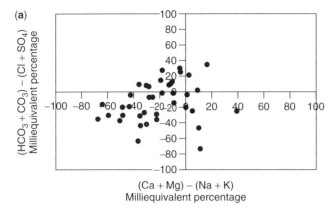

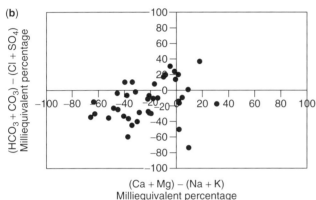

Figure 2. (a) Chadha's diagram for hydrochemical classification (premonsoon 2000); (b) Chadha's diagram for hydrochemical classification (postmonsoon 2000).

to groundwater and its recharge/discharge characteristics may change seasonally or once during a longer time span. In deep groundwater aquifers, the movement of water from a recharge to a discharge area may take place during several years, but in shallow aquifers, recharge and discharge may be much closer and even adjacent to each other. The hydraulic gradient plays a significant role in lateral and vertical migration of contaminants in groundwater aquifers.

During the present study, 38 groundwater samples were collected within 10 km of the eastern and western banks of the River Yamuna to ascertain the impact of river water on groundwater. The qualitative analysis of data showed higher concentrations of various physicochemical and bacteriological parameters on the western side of River Yamuna, even in deep aquifers. However, due to a paucity of hydrochemical, geologic, and water level data, no specific inferences could be drawn. Further studies are being planned to investigate the impact of Yamuna River water quality on the groundwater system.

CONCLUSIONS

The suitability of groundwater in the area adjoining the River Yamuna in Delhi has been examined per BIS and WHO standards. The quality of the groundwater varies from place to place with the depth of the water table. It also shows significant variation from one season to another. Only about 10% of the total samples analyzed were within the desirable limit of 500 mg/L for TDSs, and more than 80% of the samples were above the desirable limit but within the maximum permissible limit of 2000 mg/L. From the viewpoint of hardness, more than 80% of the samples were well within the desirable limits for domestic use during both pre- and postmonsoon seasons. More than 50% of the samples had a nitrate content of less than 45 mg/L, whereas in about 20% of the groundwater samples, the nitrate content even exceeded the maximum permissible limit of 100 mg/L. The higher level of nitrate at certain locations may be attributed to the surface disposal of domestic sewage and runoff from agricultural fields. The grouping of samples according to their hydrochemical facies clearly indicates that the majority of the samples fall in Na-K-Cl-SO_4 followed by Na-K-HCO_3 and Ca-Mg-Cl-SO_4 hydrochemical facies. The qualitative analysis of data showed higher concentrations of various physicochemical and bacteriological parameters on the western side of River Yamuna, even in deep aquifers. However, due to the paucity of hydrochemical, geologic, and water level data, no specific inferences could be drawn. More detailed studies including contaminant transport modeling studies are needed to understand better the impact of Yamuna River water quality on the groundwater aquifer.

BIBLIOGRAPHY

1. Jain, C.K. and Sharma, M.K. (2001). *Groundwater Quality in Adjoining Areas of River Yamuna at Delhi*. Technical Report, National Institute of Hydrology, Roorkee, India.
2. APHA. (1992). *Standard Methods for the Examination of Water and Waste Waters*, 18th Edn. American Public Health Association, Washington, DC.
3. Jain, C.K. and Bhatia, K.K.S. (1988). *Physico-chemical Analysis of Water and Wastewater*, User's Manual, UM-26. National Institute of Hydrology, Roorkee, India.
4. BIS. (1991). *Specifications for Drinking Water*, IS:10500:1991, Bureau of Indian Standards, New Delhi, India.

Table 3. Summarized Results of Water Classification

	Sample Numbers	
Classification/Type	Premonsoon 2000	Postmonsoon 2000
Na-K-HCO_3	1,2,3,4,6,7,8,14,20,21	1,2,3,6,7,13,20,21
Ca-Mg-HCO_3	5,13	4,5
Ca-Mg-Cl-SO_4	16,17,19,29,30,33	16,17,19,28,29,33
Na-K-Cl-SO_4	9,10,11,12,15,18,22,23,24,25,26,27,28, 31,32,34,35,36,37,38	8,9,10,11,12,14,15,18,22,23,24,25,26, 27,30,31,32,34,35,36,37,38

5. WHO. (1996). *Guidelines for Drinking Water, Recommendations*, Vol. 2, World Health Organization, Geneva.
6. Chadha, D.K. (1999). A proposed new diagram for geochemical classification of natural waters and interpretation of chemical data. *Hydrogeol. J.* **7**(5): 431–439.
7. Piper, A.M. (1944). A graphical procedure in the geochemical interpretation of water analysis. *Trans. Am. Geophys. Union.* **25**: 914–923.

CHLORINE RESIDUAL

Linda S. Andrews
Mississippi State University
Biloxi, Mississippi

Chlorine is a disinfectant added to drinking water to control microbial contamination. Both commercial water bottlers and public water systems use chlorine for this purpose. Levels added are calculated to maintain a disinfectant (chlorine) residual throughout the distribution system from the treatment facility to the end user.

Historically, the disinfection and sanitation of water, food products, and food processing equipment have used halogen-containing disinfectants such as chlorine (1). The worldwide use of chlorine has improved our quality of life and prevented many disease outbreaks caused by waterborne pathogens. Chlorine was first used as a disinfectant in 1897 to treat polluted water mains after an outbreak of typhoid fever in England, and later, in 1912, it was used after a typhoid fever outbreak in Niagara Falls, New York (2). Chlorine was introduced to the food processing industries in 1935. Today, disinfection is considered to be the primary mechanism for the inactivation/destruction of many pathogenic organisms to prevent the spread of waterborne diseases to downstream users and the environment (3).

Three common methods of disinfection exist in the United States and Canada: chlorination, ozonation, and ultraviolet (UV) disinfection. Chlorine, the most widely used disinfectant for municipal wastewater, destroys target organisms by oxidation of the cellular material (3). Chlorine may be applied as chlorine gas, hypochlorite solutions, or other chlorine compounds in solid or liquid form. Hypochlorous acid present in aqueous chlorine solutions is the biocidal "active chlorine" (4).

Advantages for use of chlorination include the following (3):

- Chlorination is an established method for disinfection.
- Chlorination is inexpensive compared with other methods (UV and ozone).
- Chlorine residuals remaining in the wastewater effluent will enhance the long-term effects of disinfection after the initial treatment and can be measured to determine the effectiveness.
- Chlorine is very effective against a wide variety of microorganisms, especially enteric pathogens.
- Chlorine has the ability to oxidize certain undesirable organic and inorganic compounds.
- Levels of chlorine can be adjusted according to biological load.
- Oxidative capacities of chlorine can eliminate off odors during disinfection.

In recent years, some health concerns have developed over the discovery of potentially carcinogenic by-products generated during chlorination in the presence of organic material. These by-products, termed "disinfection by-products," include such compounds as trihalomethanes, chloroform, and chlorophenols. Tests of chlorine-treated water have identified more than 250 compounds, approximately 10% of which have been classified as potentially mutagenic and/or carcinogenic (5). Both disinfectants and by-products can have adverse health effects (6).

Other disadvantages for use of chlorination include the following (3):

- Chlorine residuals, even in low concentrations, can be toxic to aquatic life and thus may require a dechlorination process.
- Shipping, storage, and handling of chlorine chemicals can pose a safety risk to workers and to the environment.
- Levels of dissolved solids are increased in chlorine-treated effluent.
- Chlorine residuals are unstable in the presence of heavy biological load to the chlorination system, requiring increased dose applications.
- Certain parasitic species known to cause problems in water supplies are relatively resistant to chlorine oxidation treatments. These species include the oocysts of *Cryptosporidium parvum* and the cysts of *Entamoeba histolytica* and *Giardia lamblia* as well as the eggs of some parasitic worms.
- Other long-term effects of chlorination and the presence of chlorine residuals are yet unknown.

Continuous chlorination is a necessity for surface water supplies from lakes, springs, ponds, or cisterns. The effectiveness of chlorine as a disinfectant is a function of contact time, the chlorine solution used, the temperature of the water and environment, and the level of residual chlorine from first introduction to the water system and continuing to the end user. Continuous chlorination typically uses a chlorine residual of 3–5 ppm. Municipalities, however, use lower levels of 0.2–0.5 ppm because they represent larger distribution systems that provide a longer contact time. Also, the higher levels of chlorine residual may cause an objectionable flavor and odor (7).

Several types of chlorine residuals can be measured in treated water. Free chlorine residual is the measure of the disinfectant safety margin and is in the form of hypochlorous acid. The combined chlorine residual includes chlorine as chloramines and chlororganics that are produced when chlorine reacts with ammonia products that may be present because of biological load. The total chlorine residual is the sum of free and combined residuals (7). A growing number of water systems have naturally occurring ammonia and organic contaminants in their water supplies. These contaminants react with their chlorine treatment to form the combined chlorine residuals (8). These combined residuals require up to 100 times the contact time or 25 times the chlorine concentration to be effective disinfectants in water (9).

It is estimated that 3–4% of the total chlorine produced in the United States is used in potable water treatment, wastewater treatment, swimming pools, and cooling water biocide applications. Many health risks are associated with water that has not been disinfected, and health risks are associated with total residual chlorine. Residual chlorines are produced from water treatment facilities and in processing plants. Large power producing plants that use chlorinated water to prevent biofouling of heat exchangers in once-through cooling towers are also a major source of residual chlorine. Approximately 90% of such plants chlorinate cooling water on a periodic basis as compared with a continuous basis for water and wastewater treatment. Typically, chlorine is dosed at 1–2 mg/L for 20–30 minutes two to three times daily. To minimize chlorine residuals in once-through cooling systems, chlorinated effluent from one group of condensers can be blended with nonchlorinated cooling water, or targeted chlorination can be practiced (10).

DECHLORINATION

Chlorination has been used widely for disinfection of wastewater before discharge. Before the passage of the 1972 Federal Water Pollution Control Act and for several years after, significant levels of residual chlorine were routinely discharged into the environment. As the impact of the toxic by-products became known, dechlorination was instituted to remove residual chlorine from wastewater before discharge into sensitive aquatic waters (11). Dechlorination effectively removes free or total combined chlorine residuals remaining after chlorination. A common method of dechlorination is to expose the effluent to sulfur dioxide or sulfite salts. Although expensive, carbon absorption has also been used when total dechlorination is desired. State regulatory agencies set the policy for allowable levels of chlorine residuals, which effectively makes dechlorination essential for certain industries. Dechlorination protects aquatic life from the toxic effects of residual chlorine and prevents formation of harmful chlorinated compounds in drinking water. However, chemical dechlorination can be difficult to control with the potential for significant overdosing with sulfite, which suppresses the dissolved oxygen content and lowers the pH (11).

BIBLIOGRAPHY

1. Biocide Int., Inc. (1994). *ECO-Benefits of Oxine-A Chlorine Dioxide Based Antimicrobial*. Biocide Intl., Norman, OK.
2. Sconce, J.S. (1962). *Chlorine-Its Manufacture, Properties and Uses*. R.E. Dreiger Publ., Malabar, FL, pp. 28–37; 514–537.
3. Solomon, C., Casey, P., Mackne, C., and Lake, A. (1998). *Chlorine Disinfection*, a Technical Overview. Environmental Technology Initiative, U.S. EPA.
4. Andrews, L.S., Key, A., Martin, R., Grodner, R., and Park, D. (2002). Chlorine dioxide wash of shrimp and crawfish an alternative to aqueous chlorine. *Food Micro.* **19**: 261–267.
5. Sen, A.C., Owuso-Yaw, J., Wheeler, W., and Wei, C. (1989). Reactions of aqueous chlorine and chlorine dioxide with tryptophan, n-methyltryptophan and 3-indolelactic acid: kinetic and mutagenicity studies. *J. Food Sci.* **54**: 1057–1060.
6. *Federal Register* 66FR 16858. March 28, 2001.
7. Quality Water Solutions, Chlorination. (2004). Available at: http://www.qh20.com/chlorination.
8. Spon, R. (2002). *Do You Really Have a Free Chlorine Residual: How to Find Out and What You Can Do About It*. RR Spon & Associates, Roscoe, IL.
9. Disinfection with Chlorine. (2004). Available at: http://www.geocities.com.
10. Lew, C.C., Mills, W., and Loh, J. (1999). Power plant discharges of total residual chlorine and trihalomethanes into rivers: Potential for human health and ecological risks. *Hybrid Methods Eng.* **1**: 19–36.
11. USEPA (2000). *Wastewater Technology Fact Sheet, Dechlorination*. EPA 832-F-00-022.

SOURCE WATER QUALITY MANAGEMENT

LAUREL PHOENIX
Green Bay, Wisconsin

Water utilities serving water authorities or municipalities can have many levels of control over the quality of their source waters. The more control they have over protecting the water quality of their sources, the less money they have to spend on water treatment, and the lower their risk when dealing with unforeseen emergencies. When they have no control of changing water quantities from drought, flood, or industrial pollution, they must spend more money to access backup supplies. Here, the phrase "water utilities" will be used interchangeably with water authorities or municipalities to refer to any local government entity providing public drinking water.

HIGH CONTROL OF SURFACE WATER SOURCES

Water authorities with the highest control own the entire watershed of a surface water source and protect the water quality by keeping people and development out. Others may rely on a river surrounded on all sides by a pristine forest with little or no human activity (e.g., hiking, camping) and would try to extract their water at a point just below this pristine area rather than farther downstream where other land uses could degrade the water. For utilities that cannot practically deny access to the entire watershed or the waterbody itself, strict guidelines for land use in that area can be locally imposed instead. These guidelines can include such things as restricting the amount of impervious surfaces that can be added through development (e.g., buildings, driveways, roads), forbidding dumping of hazardous materials or use of pesticides, or restricting recreational activities on land and water to prevent animal waste, sediment, or fossil fuels from watercraft from contaminating the water supply. Some states grant water authorities extraterritorial land use control over land in a different municipality in order to create and enforce restrictions that will protect the water supply.

HIGH CONTROL OF GROUNDWATER SOURCES

Groundwater sources can be highly protected if the utility has total control over land use in the aquifer's recharge zone. This type of control occurs when the utility owns all the land in the recharge zone, or where state or local laws require recharge zones to be fenced with signs directing people to stay out. Some states or local governments require this absolute protection of recharge zones, no matter who owns the land or what municipality has political jurisdiction. When total denial of human access or development is impossible, guidelines like the ones aforementioned can be mandated for any land lying over the recharge zone. The distance from the land surface to the water table and the substrate geology can then guide how strict the guidelines are; more restrictions being placed on land parcels overlying more vulnerable aquifers.

LITTLE CONTROL OF WATER SOURCES

Few water utilities have the high degree of control previously mentioned. This is because protecting surface or groundwater supplies can only be done through land use controls, and land use decisions have traditionally been the right of each local government. Therefore, most utilities may have to use a variety of strategies to deal with issues over which they have no control. This section discusses four common problems: (1) when source water is degraded, (2) when supply is uncertain, (3) when drinking water regulations are tightened, or (4) when emergency situations arise.

Degraded Source Water Quality

Surface waters can easily be polluted from increased drainage area development, farm runoff, or heavy rains. Development in the watershed increases erosion and causes sediment loading to surface waters. Farm runoff includes pesticides, sediment, fertilizer or manure, and silage leachate (nitrogen-rich liquid that drips from stored grains in silos). Heavy rains mobilize nonpoint source pollutants from urban landscapes, mines, farms, or silvicultural areas into water bodies. In fact, this "first flush" of runoff into water bodies carries high concentrations of pollutants. This multitude of diverse pollutants—sediment, chemicals, metals, pet waste, and so on—causes different kinds of problems for water treatment plants.

Groundwater can be polluted for the same reasons just stated, because, except for sediment, these nonpoint source pollutants can pollute groundwater by percolating down to the water table. Groundwater can also be degraded as the water table drops, because the quality of the water from that level may be higher in total dissolved solids (TDSs), be briny, or have arsenic or radon. Groundwater can also become unfit for use if compounds such as MTBE, perchlorate, solvents, or methane and diesel fuel from hydraulic fracturing migrate into the aquifer.

Even desalting ocean water does not protect utilities from pollution problems. As coastal waters become more polluted, the costs of the reverse osmosis plants increase. Therefore, source water pollution from poor land use practices, type of geological substrate, or industrial pollutants often leaves municipalities vulnerable to forces they cannot control.

Variability in Surface and Groundwater Supplies—Quantity

Surface water supply quantities can vary for several reasons. First, precipitation patterns can vary from drought conditions that reduce the amount of water available to very stormy conditions that cause flooding and subsequently degrade those floodwaters with everything they come in contact with. In states with riparian water law, the drought conditions will typically cause all utilities using the same water source to equally share the burden of insufficient water. Second, well pumping next to a river or lake can reduce surface water levels for anyone next to or downstream of the pumping. If state water law has no conjunctive use rules to prevent well owners from unfairly "stealing" water from surface supplies, then a utility may receive less water than it expects. Third, a utility may lose access to a water source if it was leased and will not be renewed when the lease is up. And finally, in states with prior appropriation water law, a utility must consider the problems that come along with junior water rights on a stream or lake. In times of drought, a very junior right on a stream may mean it gets little or no water from that stream. If they have water rights on several streams, the different adjudication dates for each stream and the amount of water in each stream will help it calculate which streams will be able to give them how much "wet" water that year. In prior appropriation law, "first in time is first in right." The year the right is obtained determines the order of use, so owners of earlier (and older) water rights will get their turn to use their full amount of allotted water before someone with a more junior right may divert their water.

Groundwater can become inaccessible the moment other well-pumpers over that aquifer draw down the water table to a level below the water utility's well.

Consequently, the uncertainty of water supplies due to the weather, leasing versus owning a water right, or state water law rules means that utilities best serve their public by having numerous water sources available to them. In this way, diverting water from an alternate or several alternate sources protects them from inadequate water available in drought times or highly degraded water resulting from flood. If they do as Phoenix does and purchase a "water ranch" (a local ranch with attached groundwater rights), then they are buying not only a water source, but a clean and nonevaporative storage site as well.

New Regulation

If new federal drinking water regulations are promulgated with lower allowable concentrations of some pollutant, then a utility's water supply becomes automatically degraded with respect to the new regulations. Arsenic is a good example of a pollutant that recently had its maximum allowable concentrations in drinking water reduced. Utilities either must stop using that water, spend enormous amounts to clean that pollutant out of the water, or mix that water with water from another source to "dilute" it down to the new, acceptable drinking water standard.

Planning for Emergency Situations

Preparing for emergencies by planning strategies is always good to do whenever possible. Water source emergencies can be a pollutant spill (e.g., fuel truck falls off highway into a river, accidental release of industrial materials into a river) or a pipeline break. These types of emergencies can only be efficiently dealt with when a utility has several sources from which to draw or sources designated as emergency backup supplies.

THE FUTURE OF SOURCE WATER QUALITY MANAGEMENT

Blending different source waters to upgrade quality, having multiple sources to give a utility options for accessing better quality or larger amounts, or instituting conservation to "create" new water are only a few strategies that utilities must employ to prepare for the future and attenuate rising costs.

Future strategies to manage source waters will need a foundation of federal protection first. Not only is a significant strengthening of the Water Quality Act and Clean Water Act necessary to reduce discharges into water sources, but also much higher, and therefore effective, fines for industrial spills or NPDES permit violations need to be mandated and enforced. After this foundation is laid, there are a few additional strategies that utilities can use. Cooperative interlocal agreements can result in larger, regional water treatment systems that are more efficient. A community downstream of its surface water source or having its groundwater recharge area under a neighboring community can craft agreements to protect the supplies. For example, one community can pay another to reduce development over a recharge area by buying some of their land outright or convincing them to only use conservation subdivision designs with reduced impervious surfaces and installation of water gardens. An upstream community can be paid to improve its stormwater pollution controls or reduce land use practices that create nonpoint source pollution. For source water quality management to improve in the future, it needs not only more cooperative interlocal strategies recognizing land use regulations as the key problem to address, but also effective water pollution controls at the federal level.

DOSE-RESPONSE OF MUSSELS TO CHLORINE

S. Rajagopal
G. van der Velde
Radboud University Nijmegen
Toernooiveld, Nijmegen, The Netherlands

V.P. Venugopalan
BARC Facilities
Kalpakkam, India

H.A. Jenner
KEMA Power Generation and Sustainables
Arnhem, The Netherlands

Chlorine, like any other biocide, produces a dose-dependent response in organisms. However, the response is dependent on a number of parameters. Chlorination of cooling water leaves residuals and by-products that are potential pollutants in the receiving water body and can impact nontarget organisms. Therefore, it is imperative to generate data on the chlorine dose–response relationships of important fouling organisms, such as various mussel species, for efficient but environmentally acceptable biofouling control. Data are presented to show the effects of mussel size (shell length), season of sample collection (spawning vs. nonspawning season), nutritional status (fed vs. nonfed), and acclimation temperature (5–30 °C) on the mortality pattern of different mussel species under continuous chlorination (0.5–5 mg L^{-1}). From the data, it can be concluded that although various factors can influence the dose–response relationships of mussels, generalization is not possible because of species specificity. Among the various parameters, mussel species, mussel size, status of byssal attachment, spawning season, and acclimation temperature have significant effect on chlorine tolerance of mussels, whereas nutritional status shows very little effect.

INTRODUCTION

Chlorination is the most commonly used disinfectant and biocide for treatment of drinking and industrial cooling water (1,2). Depending on the organisms involved and the end result desired, chlorine may be dosed either intermittently or continuously (3,4). A third important criterion that often has a bearing on the applied dose is the concentration of the chlorine (and its reaction products) that reaches the consumer (in the case of drinking water chlorination) or the environment (in the case of cooling water chlorination) (2,5). A fourth factor is chlorine demand, which exhibits variations that are both seasonal and geographical (1). Accordingly, in an actual chlorination program, the chlorine dose is modulated in such a way that the measured level of chlorine residuals is sufficient to bring about the desired result, whether it be microbiological control in drinking water or biofouling control in cooling water (6,7).

For any given organism, one can generate a dose–response curve on the basis of how the organism responds to a given dose. However, in the case of chlorine, it is appropriate to represent this in the form of a concentration–response curve, because dosing of a certain amount of chlorine results in a variable quantity of chlorine that is available for biocidal action. In other words, the available quantity of chlorine may vary depending on the chlorine demand of the water (8).

FACTORS INFLUENCING THE CHLORINE TOXICITY

The response of an organism to chlorine would vary depending on a number of factors. Among the various factors, the type of organism is an important criterion. Shelled organisms (such as mussels) can withstand relatively long-term exposure to chlorinated water (9,10), when compared with soft-bodied organisms, such as hydroids or ascidians (11). A review of literature clearly indicates all other fouling organisms succumb faster than mussels. Therefore, chlorine regime targeted against mussels would also eliminate other fouling organisms. In the present review, chlorine dose–response relationships of mussels are discussed in relation with various factors, including mussel species, mussel size, spawning season, byssus attachment, nutritional status, and acclimation temperature, which can influence the chlorine tolerance of mussels.

MUSSEL SPECIES

Efficacy of chlorine as an antifoulant depends on various parameters, most importantly residual levels of chlorine and contact time (6,9). A survey of existing literature shows that at residual levels commonly employed (1 mg L^{-1}) in power station cooling circuits, mortality takes several days. For example, at 1 mg L^{-1} continuous chlorination, *Mytilus edulis* (blue mussels) takes about 480 h for 100% mortality (12). On the other hand, in tropical marine mussels, *Perna viridis* (green mussels) takes about 816 h for 100% mortality when 1 mg L^{-1} residual chlorine is applied continuously (13). In *Dreissena polymorpha* (zebra mussel), 95% mortality is observed after about 552 h exposure to 1 mg L^{-1} residual chlorine (14). In comparison, *Mytilopsis leucophaeata* (dark false mussel) takes about 1104 h to achieve 100% mortality at 1 mg L^{-1} residual chlorine (Fig. 1). The exposure time required for 100% mortality of *M. leucophaeata* at different chlorine concentrations are much higher than that required for *D. polymorpha* (588 h) and *Mytilus edulis* (966 h).

MUSSEL SIZE

In the case of common fouling organisms such as mussels and barnacles, it is often seen that the size (or age) of the organism is an important factor that influences its sensitivity to chlorine. It has been shown that, for several organisms, a size-dependent variation in the response exists, with larger organisms showing increased tolerance (Fig. 1). However, such size-dependent nature of toxicity is not universal and there are organisms which exhibit uniform sensitivity to chlorine, irrespective of size. An example for mussel size influencing tolerance to chlorine is the mussel *M. leucophaeata* (9). Here, the tolerance is maximum in medium-size mussels (about 10 mm), whereas smaller (2 mm) and larger (20 mm) mussels show greater sensitivity (Fig. 1). This pattern is in contrast with the results reported for *Mytilus edulis*, where the tolerance linearly increases with shell size (15). In the case of *D. polymorpha*, shell size has no effect on the sensitivity of the organism to chlorine (10,16). Therefore, the relationship between mussel size and chlorine toxicity is not similar among different mussel species, and generalizations regarding the size effect should be made after careful observation.

BYSSUS ATTACHMENT

Mussels use byssus threads to attach themselves to hard substrata. The status of attachment is also an important criterion that determines the response of mussels to chlorine (15,17). It has been experimentally shown that mussels, which normally are attached with the help of their byssus threads, become more sensitive to chlorine when they are exposed to it under unattached condition (Fig. 2). Once detached from a substratum, the mussel tries to reattach itself by producing new byssus threads, for which it has to open its bivalve shell and extend its "foot" outside. This kind of enhanced byssogenic activity increases the exposure of the soft tissues of the mussel to chlorine, thereby increasing the toxic effect. On the other hand, attached mussels are byssogenically less active, and in a chlorinated environment their shells remain mostly closed, thereby protecting their soft body from chlorine (12,18).

SPAWNING SEASON

Physiological status of the organism is also an important factor that influences chlorine toxicity. Research using a number of organisms has shown that chlorine toxicity is significantly higher during breeding seasons than during nonbreeding seasons. Mussels collected during the spawning season and those collected during nonspawning season behave quite differently with respect to their sensitivity to chlorine (Fig. 3). Mussel species collected during their spawning season were less tolerant to chlorine, whereas those collected during the nonspawning season were more tolerant. The difference in tolerance between the two groups was nearly 29% (9). Kilgour and Baker (16) and Jenner et al. (1), who reported similar results for *D. polymorpha*, attribute the greater tolerance of mussels during the nonspawning season to low metabolic rates and reduced filtration rates, which would result in reduced exposure to the toxicant. Lower energetic demands during nonbreeding seasons may be the reason for reduced toxicant uptake. On the other hand, mussels tend to be weaker after spawning when they have little energy reserves in the body (19), with the result that

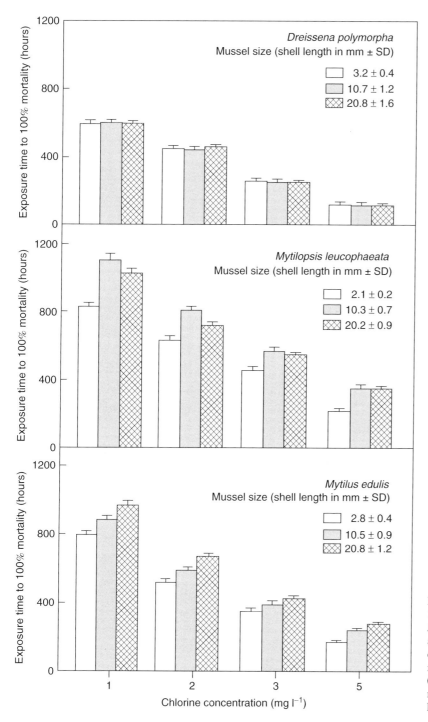

Figure 1. Comparison of exposure times to reach 100% mortality of different size groups of *Mytilopsis leucophaeata*, *Dreissena polymorpha*, and *Mytilus edulis* at different chlorine concentrations. Mortality data are expressed as mean ±SD ($n = 80$) of four replicate experiments ($n = 20$ individuals in each experiment). Test methods and mortality determinations were similar in all toxicity studies of species.

they are less tolerant to biocide. The data point to the importance of judicious sampling while carrying out toxicity experiments using seasonal breeders.

FED VS. NONFED MUSSELS

Status of feeding may have an effect on the toxicity of chlorine to organisms. Kilgour and Baker (16) showed that mussels *D. polymorpha*, when maintained on a diet of *Chlorella*, were consistently more sensitive to hypochlorite than starved mussels. The effect was attributed to an increased tendency of the fed mussels to filter water, which incidentally increases the exposure of their body parts to chlorine. Mussels that are fed with microalgae are likely to filter more water than those that are unfed. On the other hand, Rajagopal et al. (13) showed that, in the case of the mussel *Perna viridis*, fed and starved individuals showed similar mortality rates when exposed to chlorine. It must be kept in mind that chlorine may act as a strong suppressant of filtration activity in bivalve mussels, which has been shown by Rajagopal et al. (18) using Mussel-Monitor®, an automated instrument with which one can

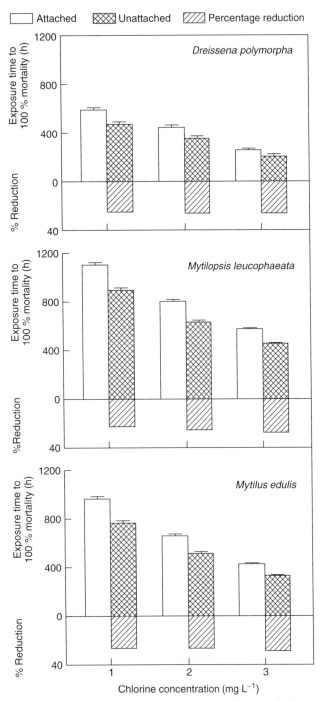

Figure 2. Time difference (%) between byssally attached and unattached *Dreissena polymorpha* (17), *Mytilopsis leucophaeata*, and *Mytilus edulis* (15) for 100% mortality at different chlorine concentrations.

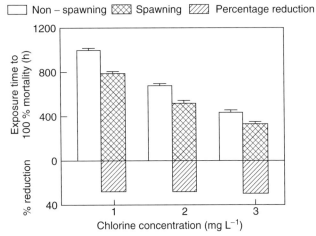

Figure 3. Cumulative mortality (%) of spawning and nonspawning *Mytilopsis leucophaeata* at different chlorine concentrations (TRC = total residual chlorine). Eighty mussels were used at each chlorine dose.

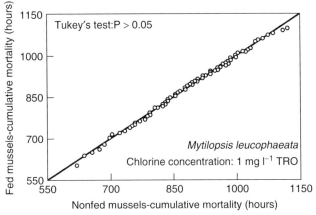

Figure 4. Cumulative mortality (%) of fed and nonfed *Mytilopsis leucophaeata* at different chlorine concentrations (TRO = total residual oxidant).

monitor the opening and closing of mussel shells (1,4). Shell valve movement of *M. leucophaeata* tested with unfiltered brackish water from the Noordzeekanaal in The Netherlands showed little or no filtration in presence of 1 mg L^{-1} residual chlorine. Obviously, presence of microalgae would have no significant effect on *M. leucophaeata* at a residual chlorine concentration of 1 mg L^{-1} (Fig. 4). Rajagopal et al. (20) have also shown that filtration activity in *D. polymorpha* stops almost completely at residual chlorine level of 0.5 mg L^{-1} and higher.

ACCLIMATION TEMPERATURE

Temperature is yet another important factor that influences the sensitivity of organisms to chlorine. Mussels acclimated to different temperatures show significantly different tolerances to chlorine (Fig. 5). Decrease in acclimation temperature from 30 °C to 5 °C increases chlorine tolerance (0.5 mg L^{-1} residual chlorine) of *M. leucophaeata* by 52 days. Such increase in chlorine tolerance at lower acclimation temperatures has also been reported for other mussel species, such as *D. polymorpha* (11,14) and *Mytilus edulis* (1,12). However, at acclimation temperatures above 35 °C, temperature has overriding effects when compared with chlorine. Harrington et al. (21) showed that at 36 °C, combined use of temperature and chlorine resulted in mortality

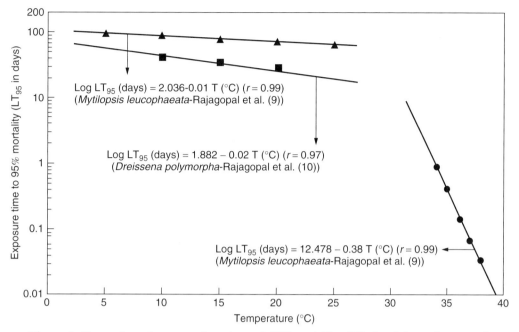

Figure 5. Comparison of exposure times to reach 95% mortality of *Mytilopsis leucophaeata* and *Dreissena polymorpha* at 0.5 mg L^{-1} residual chlorine depending on the acclimation temperature. Triangles and circles: data of Rajagopal et al. (9); Rectangles: data of Rajagopal et al. (10). Lines are linear regressions.

of *D. polymorpha* at rates similar to that obtained with heat alone.

CONCLUSION

Data available in the literature show that various factors can influence the sensitivity of organisms estimated using chlorine bioassays. Among the parameters, mussel species, mussel size, spawning season, acclimation temperature, and status of attachment seem to have significant influence on chlorine tolerance. Therefore, chlorine bioassays using mussels or such organisms need to be carried out after taking the above factors into consideration.

Acknowledgments

We thank M.G. Versteeg and M. Van der Gaag for assistance with field studies. KEMA Power Generation and Sustainables, Arnhem and Schure-Beijerinck-Popping Fonds, Amsterdam, The Netherlands financially supported this research.

BIBLIOGRAPHY

1. Jenner, H.A., Whitehouse, J.W., Taylor, C.J.L., and Khalanski, M. (1998). Cooling water management in European power stations: biology and control. *Hydroecol. Appl.* **1–2**: 1–225.

2. White, G.C. (1999). *Handbook of chlorination and alternative disinfectants*. John Wiley & Sons, Hoboken, NJ, p. 1569.

3. Claudi, R. and Mackie, G.L. (1994). *Practical Manual for Zebra Mussel Monitoring and Control*. Lewis Publishers, London, p. 227.

4. Rajagopal, S., Van der Velde, G., Van der Gaag, M., and Jenner, H.A., (2003). How effective is intermittent chlorination to control adult mussel fouling in cooling water systems? *Water Res.* **37**: 329–338.

5. Allonier, A.S., Khalanski, M., Camel, V., and Bermond, A. (1999). Characterization of chlorination by-products in cooling effluents of coastal nuclear power stations. *Mar. Pollut. Bull.* **38**: 1232–1241.

6. Mattice, J.S. and Zittel, H.E. (1976). Site-specific evaluation of power plant chlorination. *J. Water Pollut. Cont. Fed.* **48**: 2284–2308.

7. Rajagopal, S. (1997). *The ecology of tropical marine mussels and their control in industrial cooling water systems*. Ph.D. thesis, University of Nijmegen, The Netherlands, p. 184.

8. Rajagopal, S., Azariah, J., Nair, K.V.K., Van der Velde, G., and Jenner, H.A. (1996). Chlorination and mussel control in the cooling conduits of a tropical coastal power station. *Mar. Environ. Res.* **41**: 201–221.

9. Rajagopal, S., Van der Gaag, M., Van der Velde, G., and Jenner, H.A. (2002). Control of brackish water fouling mussel, *Mytilopsis leucophaeata* (Conrad) with sodium hypochlorite. *Arch. Environ. Contam. Toxicol.* **43**: 296–300.

10. Rajagopal, S., Van der Velde, G., Van der Gaag, M., and Jenner, H.A. (2002). Laboratory evaluation of the toxicity of chlorine to the fouling hydroid *Cordylophora caspia*. *Biofouling* **18**: 57–64.

11. Rajagopal, S., Van der Velde, G., and Jenner, H.A. (2002). Effects of low-level chlorination on zebra mussel, *Dreissena polymorpha*. *Water Res.* **36**: 3029–3034.

12. Lewis, B.G. (1985). *Mussel control and chlorination*. Report No TPRD/L/2810/R85, Central Electricity Research Laboratories, Leatherhead, Surrey, England, p. 33.

13. Rajagopal, S., Venugopalan, V.P., Van der Velde, G., and Jenner, H.A. (2003). Tolerance of five species of tropical marine mussels to continuous chlorination. *Mar. Environ. Res.* **55**: 277–291.

14. Van Benschoten, J.E., Jensen, J.N., Harrington, D.K., and DeGirolamo, D. (1995). Zebra mussel mortality with chlorine. *J. Am. Water Works Assoc.* **87**: 101–108.
15. Rajagopal, S., Van der Velde, G., Van der Gaag, M., and Jenner, H.A. (2004). Byssal detachment underestimates tolerance of mussels to toxic compounds. *Mar. Pollut. Bull.* (in press).
16. Kilgour, B.W. and Baker, M.A. (1994). Effects of season, stock, and laboratory protocols on survival of zebra mussels (*Dreissena polymorpha*) in bioassays. *Arch. Environ. Contam. Toxicol.* **27**: 29–35.
17. Rajagopal, S., Van der Velde, G., and Jenner, H.A. (2002). Does status of attachment influence survival time of zebra mussel, *Dreissena polymorpha* exposed to chlorination? *Environ. Toxicol. Chem.* **21**: 342–346.
18. Rajagopal, S., Van der Velde, G., and Jenner, H.A. (1997). Shell valve movement response of dark false mussel, *Mytilopsis leucophaeta*, to chlorination. *Water Res.* **31**: 3187–3190.
19. Bayne, B.L., Thompson, R.J., and Widdows, J. (1976). Physiology. In: *Marine Mussels: Their Ecology and Physiology.* B.L. Bayne (Ed.). Cambridge University Press, Cambridge, pp. 121–206.
20. Rajagopal, S., Van der Velde, G., Van der Gaag, M., and Jenner, H.A., (2002). Sublethal responses of zebra mussel, *Dreissena polymorpha* to low-level chlorination: an experimental study. *Biofouling* **18**: 95–104.
21. Harrington, D.K., Van Benschoten, J.E., Jensen, J.N., Lewis, D.P., and Neuhauser, E.F. (1997). Combined use of heat and oxidants for controlling adult zebra mussels. *Water Res.* **31**: 2783–2791.

METALLOTHIONEINS AS INDICATORS OF TRACE METAL POLLUTION

Trifone Schettino
Maria Giulia Lionetto
Università di Lecce
Lecce, Italy

INTRODUCTION

Metallothioneins (MTs) are low molecular weight, cystein-rich metal-binding proteins that are involved in detoxification and homeostasis of heavy metals. Since their discovery in the horse kidney by Margoshes and Vallee (1), metallothioneins have been further identified in most living organisms, vertebrates, invertebrates, algae, fungi, and plants. Today, the term "metallothionein" is used to designate a series of well-known molecules showing a large degree of structural and functional similarities to those first described for horse kidney metallothionein. A nomenclature system for metallothionein was adopted in 1978 (2) and then extended by introducing a subdivision of all MTs into three classes (3). By this convention Class I includes metallothioneins with locations of cysteines closely related to those in the horse kidney metallothionein; Class II comprises metallothionein with locations of cysteines only distantly related to those in horse kidney metallothionein; and Class III subsumes low molecular weight metalloisopolypeptides containing gammaglutamyl-cysteinyl units resembling in their features mammalian MTs. They are called phytochelatin and occur predominantly in plants and fungi. As the number of known MT sequences has grown, this subdivision has become inadequate. In order to better differentiate the several known MTs, in 1999 (4,5) a new classification system was proposed based on sequence similarities and phylogenetic relationships. This system subdivides the MT superfamily into families, subfamilies, subgroups, and isolated isoforms and alleles. As before, the metallothionein superfamily is defined phenomenologically as comprising all polypeptides that resemble equine renal metallothionein in several of their features (2,3). Such general features are low molecular weight, high metal content, characteristic amino acid composition (low content of aromatic amino acid residues, high cystein content, which accounts for MT heavy metal affinity and binding capacity), unique amino acid sequence with characteristic distribution of Cys (i.e., Cys–X–Cys, where X stands for an amino acid residue other than cysteine), and spectroscopic manifestations characteristic of metal thiolate clusters, which provide the protein with a highly stable tertiary structure. The metal affinity for the binding sites follows the general order found for inorganic thiolates: $Hg(II) > Ag(I) > Cu(I) > Cd(II) > Zn(II)$. The MT superfamily is subdivided into several families, each of them including MTs that share a particular set of sequence specific characters. Members of a family can belong to only one family and are thought to be evolutionarily related. The inclusion of a MT in a family presupposes that its amino acid sequence is alignable with that of all members. A common and exclusive sequence pattern, a profile, and a phylogenetic tree can therefore be connected with each family. Each family is identified by its number and its taxonomic range. To date, 15 MT families are known: Family 1, vertebrate MTs; Family 2, mollusc MTs; Family 3, crustacean MTs; Family 4, echinodermata MTs; Family 5, diptera MTs; Family 6, nematoda MTs; Family 7, ciliata MTs; Family 8, fungi-I MTs; Family 9, fungi-II MTs; Family 10, fungi-III MTs; Family 11, fungi-IV MTs; Family 12, fungi-V MTs; Family 13, fungi-VI MTs; Family 14, prokaryota MTs; and Family 15, planta MTs.

BIOLOGICAL FUNCTION OF METALLOTHIONEINS

The ubiquitous distribution of MTs in virtually all types of organisms studied to date attests to the conserved nature of MTs and their function. The biological function of MTs is likely related to the physiologically relevant metals that these proteins bind. In mammals, MT is found to bind zinc and copper under normal physiological conditions. Both zinc and copper are trace metals that are essential for life. Recent studies have produced strong evidence to support the idea that MT functions as a metal chaperone for the regulation of gene expression and for synthesis and functional activity of proteins, such as metalloproteins and metal-dependent transcription factors (6–10). MT could thus serve as a reservoir of essential metals.

MTs are inducible proteins. Exposure of the organisms to high levels of heavy metals (e.g., Zn, Cu, Cd, and Hg) and the following increase of heavy metal cations in the cells stimulates metalloprotein neosynthesis by enhancing MT gene transcription. Cis-actin sequences, termed metal response elements (MREs), located in multiple copies along the promoter, allow heavy metal ion induction of MT transcripts (11,12). The MT mRNA is translated by cytosolic free ribosomes, leading to an increase of apometallothioneins that will rapidly react with free metal cations, sequestering them and protecting cell structures from nonspecific interaction with heavy metal cations. This process circumvents cellular damage preventing metal toxicity under overload conditions. MT induction can be measured as the concentration or rates of formation of the responsible mRNA, MT, and levels of MT-bound metals. Each of them provides different information on the inductive process and may display differential dynamics.

METALLOTHIONEINS AS BIOMARKERS

The importance of metallothionein (MT) in toxicologic responses to heavy metals was early recognized for potential application as a "biomarker" of organism exposure to heavy metals in aquatic environments. A biomarker is a pollutant-induced variation in cellular or biochemical components or processes, structures, or functions that is measurable in a biological system or sample (13). The biomarker approach in environmental monitoring has been increasingly used in the last 20 years for the ecotoxicological assessment of aquatic ecosystems. Since the harmful effects of pollutants are typically manifested at lower levels of biological organization before disturbances are realized at the population, community, or ecosystem levels (14), the use of biomarkers measured at the cellular level has been proposed as sensitive "early warning" tools for biological effect measurement in environmental quality assessment (15).

One aspect of environmental degradation in aquatic environments is pollution from heavy metals, which are persistent and accumulable by aquatic organisms. A great number of metal pollution events have been reported in fresh waters, coastal waters, and groundwaters, where abnormal metal levels occurred as the consequence of natural and, especially, of manufactured sources of pollution. A crucial aspect of heavy metal pollution, determining the actual ecological risk, is the bioavailability of heavy metals in aquatic environments. Bioavailable heavy metals represent that portion of the total environmental metal load that is of direct ecotoxicological relevance (16). Apart from chemical analytical techniques, some other new approaches have recently been used to determine the availability of toxic metals for living organisms in aquatic environments. One of the most relevant contributions in this field is that provided by MT induction determination.

Due to the interest in MT application in environmental monitoring, different techniques and methodologies have been developed in the last few years for the quantification of total MTs, including chromatographic separation of soluble cytosolic MT-containing fraction associated with the evaluation of the metal concentration, HPLC-AAS (17,18), HPLC-ICP (19–21), metal substitution assays (22,23), radioimmunological techniques (24–26), electrochemical analysis (27,28) and a spectrophotometric method recently developed for routine application in biomonitoring (29). Molecular biology techniques are employed for the analysis of MT mRNA. However, it is important to point out that the expression of MT levels is not implicit from the presence of endogenous MT mRNA levels in tissues.

Thanks to the growing knowledge about MT quantification, induction of MT following heavy metal exposure has been reported in different aquatic species and tissues (30,31), especially in common bioindicator species such as *Mytilus galloprovincialis, Mytilus edulis, Littorina littorea, Ostrea edulis, Crassostrea virginica, Dreissena polymorpha*, and *Macoma balthica*.

Recent field studies using *Mytilus galloprovincialis* as a bioindicator have also demonstrated that measurements of MTs can provide an accurate indication of subtle environment increases in metal contamination (32,33), confirming their usefulness as a biomarker of trace metal exposure in aquatic environmental monitoring.

Therefore, these proteins have been proposed by the European Commission and other international scientific organizations to be included in environmental monitoring programs as a biomarker to assess metal pollution in aquatic environments.

It is known that biotic and abiotic factors such as seasonal variation, sex, age, size of the animal, and dietary factors affect MT levels in the aquatic organisms (30,34). Such factors are likely to interfere with MT synthesis in response to metal occurrence (toxic or excess of essential metals) in the aquatic ecosystems. Therefore, it is very important to know how these factors affect MT expression in the utilized bioindicator species before MT can be used in a monitoring program.

Ideally, information obtained by quantification of MT in indicator species could be used to assess adverse effects on the species themselves, on other components of the ecosystem and on humans. This information could be used in routine biomonitoring programs as early warning of potential human health effects, to make decisions about possible cleanup, and to evaluate the efficacy of past cleanup of hazardous waste.

BIBLIOGRAPHY

1. Margoshes, M. and Vallèe, B.L. (1957). *J. Am. Chem. Soc.* **79**: 4813–4814.
2. Nordberg, M. and Kojima, Y. (1979). *Experientia Suppl.* **34**: 48–55.
3. Fowler, B.A., Hildebrand, C.E., Kojima, Y., and Webb, M. (1987). *Experientia Suppl.* **52**: 19–22.
4. Binz, P.A. and Kägi, J.H.R. (1999). In: C. Klaassen (Ed.). Birkhäuser Verlag, Basel, pp. 7–13.
5. Kojima, Y., Binz, P.A., and Kägi, J.H.R. (1999). In: C. Klaassen (Ed.). Birkhäuser Verlag, Basel.
6. Maret, W. (1995). *Neurochem. Int.* **27**: 111–117.
7. Zeng, J., Vallee, B.L., and Kagi, J.H. (1991). *Proc. Natl. Acad. Sci. U.S.A.* **88**: 9984–9988.
8. Zeng, J., Heuchel, R., Schaffner, W., and Kagi, J.H. (1991). *FEBS Lett.* **279**: 310–312.
9. Jacob, C., Maret, W., and Vallee, B.L. (1998). *Proc. Natl. Acad. Sci. U.S.A.* **95**: 3489–3494.
10. Maret, W., Larsen, K.S., and Vallee, B.L. (1997). *Proc. Natl. Acad. Sci. U.S.A.* **94**: 2233–2237.

11. Stuart, G.W., Searle, P.F., and Palmiter, R.D. (1985). *Nature* **317**: 828–831.
12. Hamer, D.H. (1986). *Annu. Rev. Biochem.* **55**: 913–951.
13. NRC (National Research Council) (1987). *Environ. Health Perspect.* **74**: 3–9.
14. Adams, S.M. (1990). *Am. Fish. Soc. Symp.* **8**: 1–8.
15. McCarthy, F. and Shugart, L.R. (1990). Lewis Publishers, Chelsea, MI.
16. Rainbow, P.S. (1995). *Mar. Pollut. Bull.* **31**: 183–192.
17. Suzuki, K.T. (1980). *Anal. Biochem.* **160**: 160–168.
18. Lehman, L.D. and Klaassen, C.D. (1986). *Anal. Biochem.* **102**: 305–314.
19. Sunaga, H., Kobayashi, E., Shimojo, N., and Suzuki, K.T. (1987). *Anal. Biochem.* **160**: 160–168.
20. Mason, A.Z., Storms, S.D., and Jenkins, K.D. (1990). *Anal. Biochem.* **186**: 187–201.
21. Mazzucotelli, A., Viarengo, A., Canesi, L., Ponzano, E., and Rivaro, P. (1991). *Analyst* **116**: 605–608.
22. Scheuhammer, A.M. and Cherian, M.G. (1986). *Toxicol. Appl. Pharmacol.* **82**: 417–425.
23. Eaton, D.L. (1982). *Toxicol. Appl. Pharmacol.* **66**: 134–142.
24. Nolan, C.V. and Shaikh, Z.A. (1986). *Anal. Biochem.* **154**: 213–223.
25. Roesijadi, G., Morris, J.E., and Unger, M.E. (1988). *Can. J. Fish. Aquat. Sci.* **45**: 1257–1263.
26. Hogstrand, C. and Haux, C. (1990). *Toxicol. Appl. Pharmacol.* **103**: 56–65.
27. Olafson, R.W. and Sim, R.G. (1979). *Anal. Biochem.* **100**: 343–351.
28. Thompson, J.A.J. and Cosson, R.P. (1984). *Mar. Environ. Res.* **11**: 137–152.
29. Viarengo, A., Ponzano, E., Dondero, F., and Fabbri, R. (1997). *Mar. Environ. Res.* **44**: 69–84.
30. Bordin, G. (2000). *Cell. Mol. Biol.* **46**(2): special issue.
31. Langston, W.J., Bebianno, M.J., and Burt, G. (1998). In: W.J. Langston and M.J. Bebianno (Eds.). Chapman and Hall, London, pp. 219–284.
32. Bebianno, M.J. and Machado, L.M. (1997). *Mar. Pollut. Bull.* **34**: 666–671.
33. Lionetto, M.G. et al. (2001). *Aquat. Conserv. Mar. Freshwater Ecosys.* **11**: 305–310.
34. Serafim, M.A. and Bebianno, M.J. (2001). *Environ. Toxicol. Chem.* **20**: 552–554.

AMPHIPOD SEDIMENT TOXICITY TESTS

Augusto Cesar
Maria del Carmen Casado Martínez
Inmaculada Riba López
Tomás Ángel Del Valls Casillas
Universidad de Cadiz
Cadiz, Spain

INTRODUCTION

General

Amphipod sediment toxicity tests are technically developed and are widely accepted as useful tools for a wide variety of research and regulatory purposes (1–3). For example, they can be used to determine the sediment toxicity of single chemicals and chemical mixtures, the chemical bioavailability, the potential adverse effects of dredged material, and the magnitude and spatial and temporal distribution of pollution impacts in the field (4).

Various methods have been developed to evaluate sediment toxicity and these procedures range in complexity from lethal to sublethal tests that measure effects of chemical mixtures on the amphipod species. The evaluated sediment phase may include whole sediment, suspended sediment, elutriates, or sediment extracts (5–7).

The test organisms include amphipods, algae, macrophytes, fishes, and other benthic, epibenthic, and pelagic invertebrates (8). However, amphipod toxicity tests can provide rapid and effective information on the potential effects of contaminants in sediments.

Historical Background

Historically, the assessment of sediment quality has often been limited to chemical characterizations (9). However, quantifying contaminant concentrations alone cannot always provide enough information to adequately evaluate potential adverse effects that arise from interactions among chemicals, or that result from time-dependent availability of sediment-associated contaminants to aquatic organisms (7).

The evaluation of contaminant has primarily emphasized surface waters and effluents, not sediments, and the first incentive for sediment testing was dredged material (7). In 1977, the U.S. EPA and the U.S. Army Corps of Engineers recommended a series of 10-d toxicity and bioaccumulation tests with amphipods, clams, polychaetes, shrimps, and fishes to evaluate proposed discharge of dredged material into estuarine and marine waters.

Significance of Use

Sediment provides habitat for many benthic organisms and is a major repository for many of the more persistent chemicals that are introduced into surface waters. In the aquatic environment, most anthropogenic chemicals and waste materials including toxic organic and inorganic chemicals eventually accumulate in sediment (3).

The objective of an amphipod sediment toxicity test is to determine whether contaminants in sediment are harmful to amphipod species. The tests can be used to measure interactive toxic effects of complex contaminant mixtures in sediment.

The purpose of the amphipod solid-phase toxicity test is to determine if test sediment samples reduce survival (growth, reproduction, etc.) of exposed organisms relative to that of organisms exposed to control and reference sediment. Test results are reported as treatment (station) or combination of treatments (sites or chemicals) that produce statistically significant reduced survival (growth, reproduction, etc.) from control or reference sediments.

Sediment amphipod tests (3) can be used to (1) determine the relationship between toxic effects and bioavailability, (2) investigate interactions among contaminants, (3) compare the sensitivities of different organisms,

(4) determine spatial and temporal distribution of contamination, (5) evaluate hazards of dredged material, (6) measure toxicity as part of product licensing or safety testing or chemical approval, (7) rank areas for clean up, and (8) set cleanup goals and estimate the effectiveness of remediation or management practices.

Scope and Application

Procedures are described for testing amphipod crustaceans in the laboratory to evaluate the toxicity of contaminants associated with sediments. Sediments may be collected from the field or spiked with compounds in the laboratory. A toxicity method is outlined for diverse species of estuarine, marine, and freshwater sediment amphipods found within coastal and fresh waters. Generally, the methods described may be applied to all species, although acclimation procedures and some test conditions (temperature, salinity, etc.) will be species specific. Procedures described here are principally based on References 1–3. Although it is recognized that a variety of other nonstandardized toxicity test methods are used in ecotoxicologic research, emphasis is placed on standardized protocols provided by the U.S. EPA and ASTM, because these are the tests most commonly used in regulatory applications (10). Other countries such as The Netherlands have adapted and developed standardized protocols using local species for regulatory purposes (11).

Selection of Test Organisms

The choice of a test organism has a major influence on the relevance, success, and interpretation of a test. Test organism selection should be based on both environmental relevance and practical concerns (3). The species should be selected based on sensitivity to contaminant behavior in sediment and feeding habitat, ecological relevance, geographic distribution, taxonomic relation to indigenous organisms, acceptability for use in toxicity assessment (e.g., a standardized method), availability, and tolerance to natural geochemical sediment characteristics (7).

Amphipods have been used extensively to test the toxicity of marine, estuarine, and freshwater sediments (1–3). Ideally, a test organism should have a toxicological database demonstrating relative sensitivity to a range of contaminants of interest in sediment, have a database for interlaboratory comparisons of procedures, be in direct contact with sediment, be readily available year-round from culture or through field collection, be easily maintained in the laboratory, be easily identified, be ecologically or economically important, have a broad geographical distribution, be indigenous (either present or historical) to the site being evaluated or have a niche similar to organisms of concern (e.g., similar feeding guild or behavior to the indigenous organisms), be tolerant of a broad range of sediment physicochemical characteristics (e.g., grain size), and be compatible with selected exposure methods and endpoints.

The sensitivity of an organism is related to route of exposure and biochemical response to contaminants (3). Generally, benthic organisms can receive exposure via from three primary sources: interstitial water, whole sediment, and overlying water. Because benthic communities contain a diversity of organisms, many combinations of exposure routes may be important (3). Therefore, behavior and feeding habits of a test organism can influence its ability to accumulate contaminants from sediment and should be considered when selecting test organisms for sediment testing (3). Table 1 lists some commonly used amphipod species for sediment toxicity testing and some useful information when selecting the proper amphipod species for sediment toxicity assessment.

Amphipods

Amphipods are ecologically important members of benthic infaunal communities and are a primary food resource for a number of marine invertebrate, fish, and bird species worldwide. In general, crustacea are among the most sensitive members of benthic communities to anthropogenic disturbance, including pollution (10).

Table 1. Some Commonly Used Amphipod Species for Sediment Toxicity Testing

Amphipods	Test and Point(s)[a]	Test Period, d	Habitat[b]	References
Fresh water				
Diporeia sp.	S	10–28	B, I	1,12
Hyalella azteca	S, G, R	10–28	B, E	1,13
Salt water				
Ampelisca abdita	S, G, R	10	T, I	2,14
Ampelisca brevicornis	S, G, R	10–28	T, I	15,16
Corophium voluntator	S, I	10	T, I	17,18
Eohaustorius estaurius	S	10	B, I	2,14
Gammarus aequicauda	S	10	E	19,20
Gammarus locusta	S, R	10–28	E	21,22
Grandidierella japónica	S, G	10	T, I	2,14
Hyalella azteca	S, G, R	10–28	B, E	2,23
Leptocheirus plumulosus	S, G, R	10–28	B, I	2,24
Microdeutopus gryllotalpa	S	10	T, I	25,26
Repoxynius abronius	S	10	B, I	2,14
Tiburonella viscana	S	10	B, I	27,28

[a] S = survival, G = growth, I = immobilization, R = reproduction.
[b] B = burrow, E = epibenthic, I = infaunal, T = tube dweller.

Amphipods are a group of small crustaceans that can live in very different habitats. They are found in the sea, in estuaries, in continental waters, and even in certain terrestrial wetlands (29). Their biogeographical distribution embraces the polar waters all the way to the tropics. As for their bathymetric distribution, they extend from humid atmospheres of some forests all the way to abyssal depths. At the present time, more than 6000 species have been registered worldwide (29). Amphipods are widely distributed and common in unpolluted lotic and lentic systems; however, they are less common in hydrodynamics zones and are a primary food source for fish and voracious feeders of animal, plant, and detrital material (30).

Infaunal amphipods are excellent organisms for toxicity tests with sediment (Fig. 1) and are strongly recommended as appropriate test species for toxicity bioassays (14,30,31). Amphipods are often chosen for ecological and ecotoxicological studies due to their ecological relevance, sensitivity to environmental disturbance, and amenability for culture and experimentation (2,21,32). Overall, infaunal amphipods are excellent bioassay organisms for toxicity tests with whole sediment (31).

METHOD DESCRIPTION AND EXPERIMENTAL DESIGN

The test system described by Swartz et al. (14) for the amphipod *Rhepoxynius abronius* is recommended for bioassays with this and other amphipod species. This section describes a general laboratory method to determine the toxicity of contaminated sediments using marine, estuarine, and freshwater amphipod crustaceans, following compiled standards procedures (1–3).

Test sediments may be collected from marine, estuarine, and freshwater environments or spiked with compounds in the laboratory. The toxicity test usually is

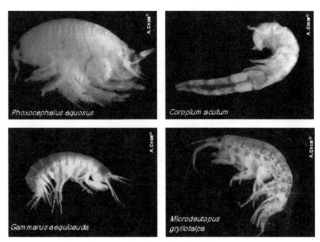

Figure 1. Some species used in amphipod sediment toxicity tests.

conducted in 1-L glass chambers containing 2 cm (175 mL) of sediment and 800 mL of overlying seawater (1:4 ratio of sediment to water). For 10-d acute tests the exposure is static, and the organisms are not fed during this period. Survival is most frequently used as the endpoint in studies, and reburial of surviving amphipods is an additional measurement that can be used. However, in chronic exposure, renewal systems can be static and organisms are fed over the 28-d exposure period. The endpoints commonly used are growth and reproduction.

A summary of general parameters and conditions to follow when developing an amphipod sediment toxicity test in the laboratory is included in Table 2.

Quality Control

If more than 10% mean mortality occurs in the control, the test must be repeated (3). However, in chronic tests 20% mean mortality in the control is acceptable. Unacceptably

Table 2. Summary of Some Conditions to Develop the Test Using Amphipods in the Laboratory

Parameter	Conditions
Test type	Whole sediment, suspended sediment, elutriates, or sediment extracts, static or renewal (depends on test type)
Temperature	Species dependent (10–25 °C)
Salinity	Species dependent (2–38 psu)
Light quality	Broad-spectrum fluorescent lights
Illuminance	500–1000 lux
Photoperiod	24 light: 0 dark
Test chambers	1-L glass beakers, recommended 10 cm Ø
Sediment volume	175–200 mL (1:4 sediment/water ratio)
Overlying water volume	600–800 mL (1:4 sediment/water ratio)
Water renewal	Not necessary or renewal (depends on test type)
Size and life stage of amphipods	Species dependent, 2–6 mm (no mature males or females)
Number of organisms per chamber	10–20
Number of replicates	At a minimum, 4 to 5 replicates must be used
Feeding regime	Species and test type dependent
Aeration	Trickle flow (<100 bubbles/min)
Overlying water	Clean natural or reconstituted water
Test exposure	10–28 d
Endpoints	Survival, growth, immobilization, reproduction, etc.
Test acceptability	Greater than or equal to 80–90% survival in controls

high control mortality indicates that the organisms are being affected by important stress factors other than contamination in the material being tested, and the test has to be reevaluated. These stresses may be due to injury or disease, unfavorable physical or chemical conditions, improper handling or acclimation, or possibly unsuitable sediment grain size. Species selection and the potential effects of these and other variables should be carefully reexamined in an attempt to reduce unacceptably high mortality.

Reference toxicant tests should be performed routinely in parallel on all populations of organisms in sample testing. When data for a particular reference toxicant have been generated on at least five populations of organisms of a species, two standard deviations above and below the mean are established as the bounds of acceptability. Reference toxicant tests should be conducted at least monthly on each species cultured in house and should be performed on each lot purchased or field collected organisms (3,33). Reference toxicant tests are most often acute lethality tests performed in the absence of sediment (3,7).

Statistical Analysis

The statistical comparisons and interpretations of the results should be appropriate to the experimental design. Study design depends on its objective and each study design has specific statistical design and analytical considerations.

Most amphipod sediment toxicity tests are based on the calculation of a medium effect or lethal and effective concentration or dose (LD_{50}, EC_{50}, IC_{50}, etc.) when sediments are spiked with individual contaminants, or upon the magnitude of response for field-collected sediments. However, the most useful information for predictive purposes is an estimation of the dose–response curve and the subsequent ranking of toxicity.

A variety of techniques are used to calculate these responses or endpoints: probit, moving angle average, and Spearman–Karber are examples. Most of these methods allow calculation of a confidence interval. These NOEC calculations depend on the number of replicates used in the toxicity test and the power of the ANOVA and other multiple comparison techniques used. When sediment samples are independently replicated, results can be statistically compared with control or reference sediments by t-test, analysis of variance (ANOVA), or regression analysis. An ANOVA is used to determine differences among treatments. When the assumptions of normality or homogeneity of variance are not met, transformations of the data may remedy the problem, so that the data can be analyzed by parametric procedures, rather than by nonparametric methods. Detailed statistical hypothesis testing, for making decisions about the appropriate method used, is discussed in Reference 3.

Confounding Factors

Interferences are characteristics of a sediment or sediment test system that can potentially affect test organism survival aside from those related to sediment-associated contaminants (3). These interferences can potentially confound interpretation of test results in two ways: (1) toxicity is observed in the test when contamination is not present, or there is more toxicity than expected; and (2) no toxicity is observed when contaminants are present at elevated concentrations, or there is less toxicity than expected.

There are three categories of interfering factors: those characteristics of sediments affecting survival independent of chemical concentration (i.e., noncontaminant factors); changes in chemical bioavailability as a function of sediment manipulation or storage; and the presence of indigenous organisms. For example, survival of some marine amphipods can be significantly affected by the grain size of the sediment (34) or ammonia concentration (35). Other confounding factors within sediment habitats include both abiotic (i.e., salinity, temperature, pH, dissolved oxygen, hydrogen sulfide, alkalinity, hardness, light) and biotic parameters (i.e., food limitation, predators, competitors, disease, biogenic disturbance).

Because of the heterogeneity of natural sediments, extrapolation from laboratory studies to the field can sometimes be difficult. Sediment collection, handling, and storage may alter the bioavailability and concentration by changing the physical, chemical, or biological characteristics of the sediment. Maintaining the integrity of a field-collected sediment during removal, transport, mixing, storage, and testing is extremely difficult and may complicate the interpretation of effects.

FUTURE RESEARCH

Research is continuing in the areas of chronic sediment toxicity methods, field validation of laboratory toxicity and bioaccumulation tests, toxicity identification and evaluations (TIE), and the multispecies experiments. Additional research is also needed to evaluate factors controlling the partitioning or sorption of a compound between water, colloids, and sediment, including aqueous solubility, pH, redox, affinity for sediment organic carbon and dissolved organic carbon, grain size of the sediment, sediment mineral constituents (oxides of iron, manganese, and aluminum), and the quantity of acid volatile sulfides in sediment. Additional research is also needed on aspects of fluid mechanics to quantify the physical processes of colloidal growth and transport and physical mass-transfer processes over a wide spectrum of solute–solid interfaces. The goal of this research would be to assimilate the biological, hydrological, geological, and ecological information necessary for the assessment and management of contaminated sediments. The end result of this research would be an improved understanding of the importance of contaminants in sediment relative to other factors influencing ecosystems.

Acknowledgments

A. Cesar thanks CAPES/MEC-Brasil for a postdoctoral scholarship (2558/03-3). M.C. Casado-Martínez has been funded by the Spanish Ministry of Education and Science (REN2002-01699/TECNO). This work was conducted under grants funded by the Spanish Ministry of Development (BOE 13-12-02) and of Education and Science (VEM2003-20563/INTER).

BIBLIOGRAPHY

1. American Society for Testing and Materials (1995). *Standard Test Methods for Measuring the Toxicity of Sediment-Associated Contaminants with Freshwater Invertebrates.* ASTME 1706–95a.

2. American Society for Testing and Materials (2000). *Standard Guide for Conducting 10-Day Static Sediment Toxicity Tests with Marine and Estuarine Amphipods.* ASTME 1367–99.

3. U.S. Environmental Protection Agency (1994). *Methods for Assessing the Toxicity of Sediment-Associated Contaminants with Estuarine and Marine Amphipods.* EPA/600/R-94/025.

4. Ferraro, S.P., and Cole, F.A. (2002). A field validation of two sediment–amphipod toxicity tests. *Environ. Toxicol. Chem.* **21**: 1423–1437.

5. Chapman, P.M. (1986). Sediment quality criteria for the sediment quality triad—an example. *Environ. Toxicol. Chem.* **5**: 957–964.

6. Swartz, R.C. et al. (1994). Sediment toxicity, contamination, and amphipod abundance at a DDT and dieldrin-contaminated site in San Francisco Bay. *Environ. Toxicol. Chem.* **13**: 949–962.

7. Ingersoll, C.G. (1995). Sediment tests. In: *Fundamentals of Aquatic Toxicology.* G.M. Rand (Ed.). Taylor & Francis, London.

8. Burton, G.A. Jr., Nelson, M.K., and Ingersoll, C.G. (1992). Freshwater benthic toxicity test. In: *Sediment Toxicity Assessment.* G.A. Burton Jr. (Ed.). Lewis, Ann Arbor, MI, pp. 213–240.

9. McCauley, D.J., DeGraeve, G.M., and Linton, T.K. (2000). Sediment quality guidelines and assessment: overview and research needs. *Environ. Sci. Pollut.* **3**: 133–144.

10. Anderson, B., Nicely, P., Gilbert, K., Kosaka, R., Hunt, J., and Phillips, B. (2004). *Overview of Freshwater and Marine Toxicity Tests.* California Environmental Protection Agency.

11. RIKZ (1999). *Standard Operating Procedure Specie-01 Marine Amphipod Corophium volutator Mortality Sediment Toxicity Test.* Riijksinstituut voor Kust en Zee. RIKZ/AB-99.114x, The Netherlands.

12. Lamdrum, P.F. (1989). Bioavailability and toxicokinetics of polycyclic aromatic hydrocarbons sorbed to sediments for the amphipod, *Pantoporeia hoyi. Environ. Sci. Technol.* **23**: 588–595.

13. Ingersoll, C.G. and Nelson, M.K. (1990). Testing sediment toxicity with *Hyalella azteca* (Amphipoda) and *Chironomus riparius* (Diptera). In: *Aquatic Toxicology and Risk Assessment.* W.G. Landis and W.H. van der Schalie (Eds.). ASTM, Washington, DC, pp. 93–109.

14. Swartz, R.C., DeBen, W.A., Jones, J.K.P., Lamberson, J.O., and Cole, F.A. (1985). Phoxocephalid amphipod bioassay for marine sediment toxicity. In: *Aquatic Toxicology and Hazard Assessment.* R.D. Cardwell, R. Purdy, and R.C. Bahner (Eds.). ASTM, Washington, DC, pp. 284–307.

15. Riba, I., DelValls, T.A., Forja, J.M., and Gómez-Parra, A. (2003). Comparative toxicity of contaminated sediments from a mining spill using two amphipod species: *Corophium volutator* and *Ampelisca brevicornis. Bull. Environ. Toxicol. Chem.* **71**: 1061–1068.

16. Riba, I., Zitko, V., Forja, J.M., and DelValls, T.A. (2003). Deriving sediment quality guidelines in the Guadalquivir estuary associated with the Aznalcóllar mining spill: a comparison of different approaches. *Cien. Mar.* **29**: 261–264.

17. van den Hurk, P., Chapman, P.M., Roddie, B., and Swartz, R.C. (1992). A comparison of North American and West European infaunal amphipod species in a toxicity test on North Sea sediments. *Mar. Ecol. Prog. Ser.* **91**: 237–243.

18. Brils, J.M. et al. (2002). Oil effect in freshly spiked marine sediment on *Vibrio fischeri, Corophium volutator* and *Echinocardium cordatum. Environ. Toxicol. Chem.* **21**: 2242–2251.

19. Cesar, A., Marín-Guirao, L.R., Vita, R., and Marín, A. (2002). Sensitivity of Mediterranean amphipods and sea urchins to reference toxicants. *Cien. Mar.* **28**: 407–417.

20. Cesar, A., Marín-Guirao, L.R., Vita, R., and Marín, A. (2004). Amphipod and sea urchin tests to assess the toxicity of Mediterranean sediments: the case of Portmán Bay. *Sci. Mar.* **68**: 205–213.

21. Neuparth, T., Costa, F.O., and Costa, M.H. (2002). Effects of temperature and salinity on life history of the marine amphipod *Gammarus locusta.* Implications for ecotoxicological testing. *Ecotoxicology* **11**: 61–73.

22. Costa, F.O., Correia, A.D., and Costa, M.H. (1996). Sensitivity of a marine amphipod to non-contaminant variables and to copper in the sediment. *Écologie* **27**: 269–276.

23. Nebeker, A.V. and Miller, C.E. (1988). Use of the amphipod crustacean *Hyalella azteca* in freshwater and esturine sediment toxicity tests. *Environ. Toxicol. Chem.* **7**: 1027–1033.

24. Schlekat, C.E., McGee, B.L., and Reinharz, E. (1991). Testing sediment toxicity in Chesapeake Bay using the amphipod *Leptocheirus plumulosus*: an evaluation. *Environ. Toxicol. Chem.* **11**: 225–236.

25. DelValls, T.A. and Chapman, P.M. (1998). The use of multivariate analysis to link the sediment quality triad components to site-specific sediment quality values in the Gulf of Cádiz (Spain) and in San Francisco Bay (USA). *Cien. Mar.* **24**: 313–336.

26. DelValls, T.A., Forja, J.M., and Parra, A.G. (1998). Integrated assessment of sediment quality in two littoral ecosystems from the gulf of Cádiz, Spain. *Environ. Toxicol. Chem.* **17**: 1073–1084.

27. Nipper, M. (2000). Current approaches and future directions for contaminant-related impact assessments in coastal environments: Brazilian perspective. *Aquat. Ecosys. Health Manage.* **3**(4): 433–447.

28. Abessa, D.M.S., Sousa, E.C.P.M., Rachid, B.R.F., and Mastroti, R.R. (1998). Use of the burrowing amphipod *Tiburonella viscana* as a tool in marine sediments contamination assessment. *Braz. Arch. Biol. Technol.* **41**: 225–230.

29. Ortiz, M. and Jimeno, A. (2001). Guía ilustrada para la identificación de las familias y los géneros de los anfípodos del suborden Gammaridea de la península Ibérica. *Graellsia* **57**: 3–93.

30. Burton, G.A. Jr. (1991). Assessing the toxicity of freshwater sediments. *Environ. Toxicol. Chem.* **10**: 1585–1627.

31. Swartz, R.C., DeBen, W.A., and Cole, E.A. (1979). A bioassay for the toxicity of sediment to marine macrobenthos. *Water Pollut. Control Fed.* **51**: 944–950.

32. DeWitt, T.H., Redmond, M.S., Sewall, J.E., and Swartz, R.C. (1992). *Development of A Chronic Sediment Toxicity Test for Marine Benthic Amphipods.* U.S. EPA, Chesapeake Bay Program.

33. Lee, D.R. (1980). Reference toxicants in quality control of aquatic bioassays. In: *Aquatic Invertebrate Bioassays.* A.L. Burkema Jr. and J. Cairms Jr. (Eds.). ASTM, Washington, DC, pp. 188–199.

34. De Witt, T.H., Ditsworth, G.R., and Swartz, R.C. (1988). Effects of natural sediments features on the phoxocephalid

amphipod, *Rheposynius abronius*: implications for sediment toxicity bioassays. *Mar. Environ. Res.* **25**: 99–124.

35. Ankley, G.T., Katko, A., and Arthur, J.W. (1990). Identification of ammonia as an important sediment-associated toxicant in the lower Fox River and Green Bay, Wisconsin. *Environ. Toxicol. Chem.* **9**: 313–322.

CILIATED PROTISTS AS TEST ORGANISMS IN TOXICITY ASSESSMENT

ANA NICOLAU
NICOLINA DIAS
NELSON LIMA
Centro de Engenharia Biológica
Braga, Portugal

TOXICITY ASSESSMENT, CYTOTOXICITY, AND TEST ORGANISMS

Increasing environmental pollution and the continuous development of new drugs have led to an ever growing concern about the potential effects of these compounds on human and/or environmental health, directly or indirectly. Toxicity data collected in the laboratory are the primary source of information that can be used for regulatory or risk assessment proposes.

In vitro cytotoxicity techniques are being increasingly developed and evaluated as alternatives to the use of vertebrates in testing environmental pollutants. Besides the reduction of laboratory animals, these lower cost tests allow for the use of specific endpoints to determine the targets of toxic effects with great precision and reproducibility (1). Because of the long-standing interest in aquatic toxicity testing, a myriad of test species, types of tests, and endpoints have been developed and used.

In recent years, much research has been done to assess the toxicity of various compounds in a series of biotests using several organisms. The appeal of these tests lies in their simplicity and high degree of reproducibility (2). On the other hand, test organisms for assessing environmental risk and impact must possess a lot of desirable features: they must be eukaryotic, their biology and general responses must be well known, the laboratory handling must be relatively easy, and a short generation time is desirable whenever studies of long-term effects are necessary (3).

Ciliated protists fulfill all those requirements. Furthermore, as they have a wide distribution and ecological significance, performing key functions in energy flow and elementary cycling in aquatic foodwebs, they can be ideal early warning indicators of aquatic ecosystem deterioration. Recently, the use of ciliates in toxicological tests has been investigated and their potential in standard bioassays has been demonstrated in aquatic environments and even elsewhere (4–6). Nevertheless, there are only a few works suggesting new test organisms in this group (7–9).

Toxicokinetics and toxicodynamics govern the response of exposed organisms (10). Single-celled organisms are useful test subjects because toxicokinetics is, in this case, very simplified and one can consider that it is proportional to water–lipid partitioning (11). Among protists, the ciliate *Tetrahymena*, and among the several species, *Tetrahymena pyriformis*, is the most commonly ciliated model used for laboratory research: it has been, for more than five decades, the organism of choice in analyses, evaluation of protein quality, and determination of effects of several toxic substances. It was the first protist to be cultivated axenically, that is, in a standard medium, free from bacteria or other organisms (12). This fact could explain the extensive use of this ciliate as a model cell system since the addition of a test compound can be, in principle, the only change in culture conditions (3). Therefore, under axenic culture conditions, standardization and guarantee of reproducibility of assays are obtained (9).

Beyond the rare reviews on this subject (3,13–16), some works demonstrate the potential of ciliates as test organisms, especially *Tetrahymena pyriformis* and *T. termophila*, but also *Colpidium campylum* and, more recently, *Spirostomum teres* and *S. ambiguum*.

Silverman (17) used the movement of *T. termophila*, in axenic cultures, to assess the toxicity of several compounds using the Draise test, concluding its validity. Later, Noever et al. (18) suggested another test to substitute for the Draise test, this time with *T. pyriformis*, using the mobility of the ciliate in only 30 s.

Yoshioka et al. (19) criticized the lack of rigorous establishment of experimental conditions in assays with *T. pyriformis* and validated a growth impairment assay of this ciliate in axenic culture, to assess the toxicity of a series of compounds, comparing the results with the calculation of LD_{50} of *Orizias latipe* and concluding for the validity of *T. pyriformis* in the assessment of toxicity in aquatic environments.

Dive et al. (7) presented, for the first time in a standardized form, a bioassay method using the ciliate *Colpidium campylum* as a test organism. Le Dû et al. (20) compared this bioassay to tests with *Daphnia* or the MICROTOX® test (Microbics) in the detection and modeling of the aquatic toxicity of soluble residuals, concluding for its validity as a complementary test or even a substitute for the other tests.

Nalecz-Jawecki et al. (8) developed a biotest using *S. ambiguum*, one of the biggest protists, 2–3 mm long, to assess water quality. Later, Nalecz-Jawecki and Sawicki (21) implemented a miniaturized and standardized bioassay based on the former and called it the SPIROTOX test. The development of the SPIROTOX test for the toxicity of volatile compounds enabled the creation of a database of toxicity of organic pollutants to *S. ambiguum* (22). Futhermore, comparison of the sensitivity of this test to other bioassay systems used worldwide indicated that the SPIROTOX test can be an important element of an assessment battery as the representative ciliated protists component (23,24).

Sauvant et al. (25) evaluated the toxicity of bottled water using two models in a series of *in vitro* assays: the mammalian fibroblasts L-929 and *T. pyriformis*. They concluded that both models are adequate and that *T. pyriformis* was more sensitive than the fibroblasts in samples bottled for longer periods.

Larsen et al. (26) proposed an international pilot ring test in order to standardize a growth inhibition test protocol with *T. pyriformis*. They concluded that the *Tetrahymena* test could be recommended for ecotoxicological screening purposes, being an ecologically relevant supplement to aquatic toxicity testing. In recent years, the *T. pyriformis* multigeneration growth impairment assay or TETRATOX (26,27) has been used quite often. It uses data from 40-h growth of a population of *T. pyriformis*.

Twagilimana et al. (9) suggested a new indicator, *Spirostomum teres*, a ciliated protist tolerant to low levels of dissolved oxygen; the assay was based on the calculation of the LD_{50} in nonaxenic cultures exposed to the toxic compounds for 24 h. The authors assured the high sensitivity of the assay, especially to heavy metals, in relation to the MICROTOX (Microbics) and to other conventional bioassays.

Schlimme et al. (28) suggested an assay to determine the toxicity of several bacteria strains: in the BACTOX test, these bacteria were given to *T. pyriformis* as prey. The authors highlighted the lack of toxicological assays with whole bacteria and, based on the results of their work, which seem to indicate that toxicity is specific to strains and not to species, they recommend that every bacteria strain should pass the BACTOX test before being produced in large scale.

TOXICOLOGICAL ASSESSMENT WITH PROTISTS

Growth rate and morphological changes have been used in the evaluation of toxicity for some decades. Population growth impairment is an often-used sublethal toxic endpoint for organic and inorganic compounds and does not require special technical expertise. It is considered the most sensitive sublethal parameter, reflecting the global state of a series of parcel effects. Other parameters such as cell motility, swimming patterns, cytoskeleton analysis, phagocytosis rate, and biochemical parameters can be assessed and have been proposed to determine the physiological and energetic state of protists when in contact with toxic compounds.

Tetrahymena pyriformis has been used extensively in studies on effects of heavy metals. Although the mechanisms by which heavy metals affect the microorganisms is not clear, authors have described copper as an essential metal that accumulated in the mitochondria (29) with its distribution observable as small electron-dense dots (3,30). Alteration of *Tetrahymena* morphology after aluminum exposure was reported by Sauvant et al. (31), who observed the presence of large and irregular-shaped monster cells in treated cultures. These morphological abnormalities were detected in zinc-treated cells but not in copper cultures. Most of the works focus on one metal only.

There are also several studies that use protists as test organisms to assess the toxicity of several compounds, both inorganic and organic. Some of these works are collected in Table 1.

Beyond these, Rogerson et al. (70) determined the effects of 17 hydrocarbons on the mortality of *Colpidium colpoda* and *T. ellioti* and Otsuka et al. (71) assessed the toxicity of diphenyl, *o*-phenylphenol, and 2-(4-thiazol)benzimidazol on the growth of *T. pyriformis*. Dias et al. (72) and Nicolau et al. (73) studied the toxicity of

Table 1. Toxicity Assessment of Different Compounds with Protists (*T—Tetrahymena* sp.; *C—Colpidium* sp.; *S—Spirostomum*)

Compound	Organism	References
Metals		
Aluminum	T	31
Cadmium	T; T; T; T; T; C	32; 33; 34; 35; 36; 37
Lead	T; T	38; 39
Copper	T; T; T; C	30; 29; 35; 37
Iron	T	40
Mercury	T; T	41; 42
Vanadate	T	43
Zinc	T; T	32; 33
Other		
Ceramide analogues	T	44
Cycloheximide	T	45
Chloramphenicol	T	46
Chloroquinone	T	47
Concanavalin A	T	48
Cocaine and derivates	T; T	49; 50
Dichloroisoproterenol	T	51
2,4-Dinitrophenol	T	52
DMSO (dimethylsulfoxide)	T; T; T	53; 54; 55
Ethidium bromide	T	56
Histamine	T	57
Histidinol	T	58
Insulin	T; T	59; 60
Lofepramine	T	61
Methotrexate	T	62
Octanol	T	63
Opioids	T	64
Penicillin	T	65
Phenols	T	66
Surfactants	T; T; S; T	24; 67; 68; 69

metals, Triton X-100, and cycloheximide on biochemical parameters of *T. pyriformis*. Nicolau et al. (74) had already assessed the toxicity of these compounds on the growth, mortality, and grazing of *T. pyriformis*.

NEW APPROACHES IN THE ASSESSMENT OF TOXICITY WITH PROTISTS

Population growth impairment is, as mentioned above, the most frequently used sublethal toxic endpoint. A growth/mortality bioassay using microscopic observation of morphological changes at low magnification is a simple reproducible technique that does not require technical expertise or operational expenses. Nevertheless, some limitations detected in such an assay can make it unsuitable with several toxicants. Recognizing dead cells was sometimes ambiguous, since nonmotile cells as well as altered shape cells were counted as dead cells. It was found that light microscopy observations underestimated the true number of viable cells (i.e., false-negative counting), since several abnormal cells were not in fact dead. Indeed, they recovered their normal shape immediately after toxicant removal and began growing in the recovery assay (75). A novel cellular cytotoxicity assay using two fluorescent dyes was developed by Dias and Lima (75) as an alternative method to standard direct counting of viable protozoa under light microscopy. The compound calcein AM is a nonfluorescent substance that diffuses passively into cells and is converted by intracellular esterases to green fluorescent calcein in viable cells. EthD-1, which binds to DNA-stained red dead cells, was also added. The live/dead assay was found to be a very sensitive method for toxicants that damaged the cell membrane, given that all green cells were established as viable cells even with an altered shape. Recently, an improvement of this approach was developed by Dayeh et al. (69) using *T. termophila* and similar sensitivity was found between the ciliate and piscine and mammalian cells.

Another problem that the toxicologist must deal with is that conventional toxicological assays are often slow and labor-intensive and become impractical when many compounds and/or concentrations are being tested rapidly. This has led to a greater interest in colorimetric and fluorimetric assays that can be miniaturized in 96-well microtiter plates and assessed using an ELISA spectrophotometric microtiter plate reader.

Assays of ATP content have been widely used to characterize biomass viability and to detect potential spoilage microorganisms in the beer and food industry (76,77) assuming that it is possible to use the concentration of ATP to measure the viable cells of a certain species. In cultures of *T. pyriformis*, ATP content was determined to give information about the general energetic state of the culture when submitted to the toxicants (73).

Previous works have been devoted to studying the acid phosphatase (ACP) activity of other hydrolases to detect digestive activity in protozoa (78–81), since there is an intimate relationship between lysosomal function and intracellular digestion in *Tetrahymena* and other ciliates. The ACP activity was used also as an indicator of the metabolic state of the cultures, namely, of the intracellular digestive function, when the ciliates were exposed to toxicants (73,82).

Other colorimetric assays are the assays based on tetrazolium dye reduction, like MTT, which have been extensively used in the *in vitro* evaluation of cellular proliferation and cytotoxicity as well as screening for additives, mycotoxins, or anticancer drugs. The MTT assay is a quantitative method that consists of metabolic reduction of the tetrazolium dye, a group of water-soluble quaternary ammonium compounds, by dehydrogenases of viable immobilized or suspended cells. Production of intensely colored formazan water-insoluble crystals is the result of the tetrazolium reduction assay. Formazan crystals can be either observed microscopically in the cell cytoplasm or extracted and dissolved with organic solvents, such as DMSO, enabling spectrophotometric quantification in cultures of *T. pyriformis* (72).

As mentioned previously, bioactivity of various chemicals can be assessed by means of *Tetrahymena*'s grazing rate determination. For this purpose, fluorescent-labeled latex beads (FLLB) can be used instead of labeled bacteria, assuming that ciliates do not discriminate between different particles on the basis of properties other than size and shape. Therefore, direct counting of fluorescent beads ingested through a period less than 30 min allows an estimate of the xenobiotic's influence on the grazing activity (68,73).

Counting beads one by one inside all digestive vacuoles of a protist is not only tedious but also time consuming. Recently, an automated method of FLLB counting was proposed by Dias et al. (68) to improve the FLLB ingestion toxicity test. It was found that the image analysis is superior to direct FLLB counting, when speed of analysis is the concern.

A different toxicological test approach using *T. pyriformis* was based on calorimetric measurements assuming that protists produce heat by metabolism and movement (83). The addition of toxic substances should result in a reduction of heat production. It was found that the calorimetric measurement showed some advantages compared to the turbimetry.

QSARS: PREDICTING TOXICITY

With the increase of chemical usage, more and more newly manufactured chemicals are created by industry. Today, it is impossible to test all these new products and the already existing ones. Consequently, gaps in toxicity data exist and predictive models have become a means of filling these gaps. Relating chemical structure and toxic potency is a way to predict toxicity. The result of such an approach is known as quantitative structure–activity relationships, that is, QSARs.

There are several ways to do it: the traditional one relies on the use of a set of homologous compounds, each having a functional group common to them all and an alkyl substituent of varied length. In such a set, chemical reactivity, due to the functional group, is ignored because it is the same for all compounds; the hydrocarbon moiety alters hydrophobicity and therefore toxic potency. The models obtained in this way are usually simple regressions of toxicity versus hydrophobicity.

A more complex but innovative approach uses analogous groups of compounds with multiple function groups. The models obtained in this case are often multiple regressions of toxicity versus hydrophobic and stereo-electronic properties (84). It must be assumed, however, that one model predicts only one mechanism of toxic action (85).

Acute toxicity can fall into one of two categories: reversible nonspecific toxicity or narcosis and irreversible specific toxicity (11). Over 70% of organic industrial chemicals cause narcosis (86), which is reversible, physical alterations in the membrane, because they are nonbiding to the macromolecules (10). The remainder of industrial organic chemicals elicited toxicity that is nonreversible (87). These cause reactions that are typically covalent, resulting in changes in the biological systems, and are known as electro(nucleo)philic. The electro(nucleo)philic mechanism of toxic action seems to be dependent on the chemical class of the compound (84).

CONCLUSIONS

Conventional toxicological assays are often labor intensive and become impractical when many compounds and/or concentrations are being tested rapidly. This has led to a greater interest in colorimetric and fluorimetric assays that can be miniaturized and, on the other hand, do not require technical expertise or operational expenses that are needed in bioassays carried out by regulatory agencies at higher trophic levels such as in fish. The use of protists in miniaturized assays seems an excellent option.

All that was referred emphasizes the importance of protists as test organisms and encourages further work in this area, namely, in the establishment of new response patterns to toxicants and in the research of new species to be validated as test organisms, especially species with high indicator value or significant species in certain specific environments (activated sludge, eutrophic waters, amended soils, etc.).

BIBLIOGRAPHY

1. Olabarrieta, I., L'Azou, B., Yuric, J., Cambar, J., and Cajaraville, M.P. (2001). *In vitro* effects of cadmium on two different animal cell models. *Toxicol. In Vitro* **15**: 511–517.
2. Cairns, J. Jr. and Pratt, J.R. (1987). Ecotoxicological effect indices. *Water Sci. Technol.* **19**: 1–2.
3. Nilsson, J.R. (1989). *Tetrahymena* in cytotoxicology: with special reference to effects of heavy metals and selected drugs. *Eur. J. Protozool.* **25**: 2–25.
4. Benitez, L. et al. (1994). The ciliated protozoa *Tetrahymena termophila* as a biosensor to detect micotoxins. *Lett. Appl. Microbiol.* **19**: 489–491.
5. Eckelund, F., Ronn, R.E., and Christensen, S. (1994). The effect of three different pesticides on soil protozoan activity. *Pestic. Sci.* **42**: 71–78.
6. Campbell, C.D., Warren, A., Cameron, C.M., and Hope, S.J. (1997). Direct toxicity assessment of two soils amended with sewage sludge contaminated with heavy metals using a protozoa (*Colpoda steinii*) bioassay. *Chemosphere* **34**: 501–514.
7. Dive, D. et al. (1989). A bioassay using the measurement of the growth inhibition of a ciliate protozoan: *Colpidium campylum* Stokes. *Hydrobiologia* **188/189**: 181–188.
8. Nalecz-Jawecki, G., Demkowicz-Dobrzanski, K., and Sawicki, J. (1993). Protozoan *Spirostomum ambiguum* as a highly sensitive bioindicator for rapid and easy determination of water quality. *Sci. Total Environ. Suppl.* 1227–1234.
9. Twagilimana, L., Bohatier, J., Groliere, C.-A., Bonnemoy, F., and Sargos, D. (1998). A new low-cost microbiotest with the protozoan *Spirostomum teres*: culture conditions and assessment of sensitivity of the ciliate to 14 pure chemicals. *Ecotoxicol. Environ. Safety* **41**: 231–244.
10. Seward, J.R., Sinks, G.D., and Schultz, T.W. (2001). Reproducibility of toxicity across mode of toxic action in the *Tetrahymena pyriformis* population growth impairment assay. *Aquat. Toxicol.* **53**: 33–47.
11. Schultz, T.W., Sinks, G.D., and Cronin, M.T.D. (1997). Quinone-induced toxicity to *Tetrahymena*: structure–activity relationships. *Aquat. Toxicol.* **39**: 267–278.
12. Lwoff, M.A. (1923). Sur la nutrition des infusoires. *C.R. Acad. Sci. Paris* **176**: 928–930.
13. Persoone, G. and Dive, D. (1988). Toxicity tests on ciliates: a short review. *Ecotoxicol. Environ. Safety* **2**: 105–114.
14. Lynn, D.H. and Gilron, G.L. (1992). A brief review of approaches using ciliated protozoa to assess aquatic ecosystem health. *J. Aquat. Ecosys. Health* **1**: 263–270.
15. Sauvant, M.P., Pepin, D., and Piccinni, E. (1999). *Tetrahymena pyriformis*: a tool for toxicological studies. A review. *Chemosphere* **38**: 1631–1669.
16. Nicolau, A., Dias, N., Mota, M., and Lima, N. (2001). Trends in the use of protozoa in the assessment of wastewater treatment. *Res. Microbiol.* **152**: 621–630.
17. Silverman, J. (1983). Preliminary findings on the use of protozoa (*Tetrahymena termophila*) as models for ocular irritation testing in rabbits. *Lab. Anim. Sci.* **33**: 56–59.
18. Noever, D.A., Matsos, H.C., Cronise, R.J., Looger, L.L., and Relwani, R.A. (1994). Computerized *in vitro* test for chemical toxicity on *Tetrahymena* swimming patterns. *Chemosphere* **29**: 1373–1384.
19. Yoshioka, Y., Ose, Y., and Sato, T. (1985). Testing for the toxicity of chemicals with *Tetrahymena pyriformis*. *Sci. Total Environ.* **43**: 149–157.
20. Le Dû, A., Dive, D., Guebert, M., and Jouany, J.M. (1993). The protozoan biotest *Colpidium campylum*, a tool for toxicity detection and toxic interaction modelling. *Sci. Total Environ. Suppl.* 809–815.
21. Nalecz-Jawecki, G. and Sawicki, J. (1999). Spirotox—a new tool for testing the toxicity of volatile compounds. *Chemosphere* **38**: 3211–3218.
22. Nalecz-Jawecki, G. and Sawicki, J. (2002). The toxicity of trisubstituted benzenes to the protozoan ciliate *Spirostomum ambiguum*. *Chemosphere* **46**: 333–337.
23. Nalecz-Jawecki, G. and Sawicki, J. (2002). A comparision of sensitivity of Spirotox biotest with standard toxicity tests. *Arch. Environ. Contam. Toxicol.* **42**: 389–395.
24. Nalecz-Jawecki, G., Grabinska-Sota, E., and Narkiewicz, P. (2003). The toxicity of cationic surfactants in four bioassays. *Ecotoxicol. Environ. Safety* **54**: 87–91.
25. Sauvant, M.P., Pepin, D., Bohatier, J., Groliere, C.A., and Veyre, A. (1994). Comparative study of two *in vitro* models (L-929 fibroblasts and *Tetrahymena pyriformis* GL) for the cytotoxicological evaluation of packaged water. *Sci. Total Environ.* **156**: 159–167.
26. Larsen, J., Schultz, T.W., Rassmussen, L., Hoofman, R., and Pauli, W. (1997). Progress in an ecotoxicological standard protocol with protozoa: results from a pilot ring test with *Tetrahymena pyriformis*. *Chemosphere* **35**: 1023–1041.
27. Schultz, T.W. (1997). TETRATOX: the *Tetrahymena pyriformis* population growth impairment endpoint—a surrogate for fish lethality. *Toxicol. Methods* **7**: 280–309.
28. Schlimme, W., Marchiani, M., Hansekmann, K., and Jenni, B. (1999). BACTOX, a rapid bioassay that uses protozoa to assess the toxicity of bacteria. *Appl. Environ. Microbiol.* **65**: 2754–2757.
29. Wakatsuki, T., Tazaki, Y., and Imahara, H. (1986). Respiratory inhibition by copper in *Tetrahymena pyriformis* GL. *J. Fermentol. Toxicol.* **64**: 119–127.
30. Nilsson, J.R. (1981). Effects of copper on phagocytosis in *Tetrahymena*. *Protoplasma* **109**: 359–370.
31. Sauvant, M.P., Pepin, D., Bohatier, J., and Groliere, C.A. (2000). Effects of chelators on the acute toxicity and bioavailability of aluminum to *Tetrahymena pyriformis*. *Aquat. Toxicol.* **47**: 259–275.
32. Chapman, G. and Dunlop, S. (1981). Detoxification of zinc and cadmium by the freshwater protozoan *Tetrahymena pyriformis*. I. The effect of water hardness. *Environ. Res.* **26**: 81–86.
33. Chapman, G. and Dunlop, S. (1981). Detoxification of zinc and cadmium by the freswater protozoan *Tetrahymena pyriformis*.

II. Growth experiments and ultrastructural studies on sequestration of heavy metals. *Environ. Res.* **24**: 264–274.

34. Houba, C., Remacle, J., and De Parmentier, F. (1981). Influence of cadmium on *Tetrahymena pyriformis* in axenic culture. *Eur. J. Appl. Microbiol. Biotechnol.* **11**: 179–182.

35. Piccinni, E., Irato, P., Coppellotti, O., and Guidolin, L. (1987). Biochemical and ultrastructural data on *Tetrahymena pyriformis* treated with copper and cadmium. *J. Cell Sci.* **88**: 283–293.

36. Lawrence, S.G., Holoka, M.H., and Hamilton, R.D. (1989). Effects of cadmium on a microbial food chain, *Chlamydomonas reinhardii* and *Tetrahymena vorax*. *Sci. Total Environ.* **87/88**: 381–395.

37. Sekkat, N., Le Dû, A., Jouany, J.M., and Guebert, M. (1992). Study on the interaction between copper, cadmium, and ferbam using the protozoan *Colpidium campylum* bioassay. *Ecotoxicol. Environ. Safety* **24**: 294–300.

38. Nilsson, J.R. (1978). Retention of lead within the digestive vacuole in *Tetrahymena*. *Protoplasma* **95**: 163–173.

39. Nilsson, J.R. (1979). Intracellular distribution of lead in *Tetrahymena* during continuous exposure to the metal. *J. Cell Sci.* **39**: 383–396.

40. Rasmussen, L., Toftlund, H., and Suhr-Jessen, P. (1984). *Tetrahymena*. Adaptation to high iron. *Exp. Cell Res.* **153**: 236–239.

41. Thrasher, J.D. and Adams, J.F. (1972). The effects of four mercury compounds on the generation time and cell division in the *Tetrahymena pyriformis*, WH14. *Environ. Res.* **5**: 443–450.

42. Tingle, L.E., Pavlat, W.A., and Cameron, I.L. (1973). Sublethal cytotoxic effects of mercuric chloride on the ciliate *Tetrahymena pyriformis*. *J. Protozool.* **20**: 301–304.

43. Nilsson, J.R. (1999). Vanadate affects nuclear division and induces aberrantly-shaped cells during subsequent cytokinesis in *Tetrahymena*. *J. Eukaryot. Microbiol.* **46**: 24–33.

44. Kovács, P. and Csaba, G. (1998). Effect of ceramide-analogues on the actin cytoskeleton of *Tetrahymena pyriformis* GL. A confocal microscopic analysis. *Acta Protozool.* **37**: 201–208.

45. Ricketts, T.R. and Rappit, A.F. (1975). The effect of puromycin and cycloheximide on vacuole formation and exocytosis in *Tetrahymena pyriformis* GL-9. *Arch. Microbiol.* **102**: 1–8.

46. Nilsson, J.R. (1986). Effects of chloramphenicol on the physiology and fine structure of *Tetrahymena pyriformis* GL: correlation between diminishing inner mitochondrial membrane and cell doubling. *Protoplasma* **135**: 1–11.

47. Nilsson, J.R. (1992). Does chloroquine, an antimalarial drug, affect autophagy in *Tetrahymena pyriformis*? *J. Protozool.* **39**: 9–16.

48. Scott, S.M. and Hufnagel, L.A. (1983). The effect of concanavalin A on egestion of food vacuoles in *Tetrahymena*. *Exp. Cell Res.* **144**: 429–441.

49. Stefanidou, M.E., Georgiou, M., Maravelias, C., and Koutselinis, A. (1990). The effects of morphine, cocaine, amphetamine and hashish on the phagocytosis of the protozoon *Tetrahymena pyriformis* strain W. *Toxicol. In Vitro* **4**: 779–781.

50. Stefanidou, M.E., Alevisopoulos, G., Maravelias, C., Loutsidis, C., and Koutselinis, A. (1999). Phagocytosis of the protozoon *Tetrahymena pyriformis* as an endpoint in the estimation of cocaine salt and cocaine freebase toxicity. *Addict. Biol.* **4**: 449–452.

51. Fok, A.F. and Shockley, B.U. (1985). Processing of digestive vacuoles in *Tetrahymena* and the effects of dichloroisoproterenol. *J. Protozool.* **32**: 6–9.

52. Nilsson, J.R. (1995). pH-dependent effects of 2,4-dinitrophenol (DNP) on proliferation, endocytosis, fine structure and DNP resistance in *Tetrahymena*. *J. Eukaryot. Microbiol.* **42**: 248–255.

53. Lee, D. (1971). The effect of dimethylsulphoxide on the permeability of lysosomal membrane. *Biochim. Biophys. Acta* **233**: 619–623.

54. Nilsson, J.R. (1974). Effects of DMSO on vacuole formation, contractile vacuole function, and nuclear division in *Tetrahymena pyriformis* GL. *J. Cell Sci.* **16**: 39–47.

55. Nilsson, J.R. (1980). Effects of dimethyl sulphoxide on ATP content and protein synthesis in *Tetrahymena*. *Protoplasma* **103**: 189–200.

56. Meyer, R.R., Boyd, C.R., Rein, D.C., and Keller, S.J. (1971). Effects of ethidium bromide on growth and morphology of *Tetrahymena pyriformis*. *Exp. Cell Res.* **70**: 233–237.

57. Darvas, Z., Madarász, B., and Lászlo, V. (1999). Study of histamine effects on phagocytosis and enzyme secretion of *Tetrahymena pyriformis*. *Acta Biol. Hung.* **50**: 325–334.

58. Ron, A. and Wheatley, D.N. (1984). Stress-proteins are induced in *Tetrahymena pyriformis* by histidinol but not in mammalian (L-929) cells. *Exp. Cell Res.* **153**: 158–166.

59. Fülöp, A.K. and Csaba, G. (1994). Insulin pretreatment (imprinting) produces elevated capacity in the insulin binding of *Tetrahymena*. Different binding by the cilia and the oral field. *Biosci. Rep.* **14**: 301–308.

60. Fülöp, A.K. and Csaba, G. (1997). Accumulation of insulin-gold particles in the oral apparatus of *Tetrahymena* after insulin pretreatment (imprinting). *Microbios* **90**: 123–128.

61. Darcy, P., Kelly, J.P., Leonard, B.E., and Henry, J.A. (2002). The effect of lofepramine and other related agents on the motility of *Tetrahymena pyriformis*. *Toxicol. Lett.* **128**: 207–214.

62. Nilsson, J.R. (1983). Methotrexate permits a limited number of cell doublings and affects mitochondrial substructure in *Tetrahymena*. *Protoplasma* **117**: 53–61.

63. Bearden, A.P. and Schultz, T.W. (1998). Comparision of *Tetrahymena* and *Pimephales* toxicity based on mechanism of action. *SAR QSAR Environ. Res.* **9**: 127–153.

64. Wu, C., Fry, C.H., and Henry, J.A. (2002). Membrane toxicity of opioids measured by protozoan motility. *Toxicology* **117**: 35–44.

65. Szablewski, L. (1981). The effect of selected antibiotics on *Tetrahynmena pyriformis*. *Acta Protozool.* **20**: 309–322.

66. Jaworska, J.S. and Schultz, T.W. (1991). Comparative toxicity and structure–activity relationships in *Chlorella* and *Tetrahymena*: monosubstituted phenols. *Bull. Environ. Contam. Toxicol.* **47**: 57–62.

67. Dias, N., Mortara, R.A., and Lima, N. (2003). Morphological and physiological changes in *Tetrahymena pyriformis* for the *in vitro* cytotoxicity assessment of Triton X-100. *Toxicol. In Vitro* **17**: 357–366.

68. Dias, N., Amaral, A.L., Ferreira, E.C., and Lima, N. (2003). Automated image analysis to improve bead ingestion toxicity test counts in the protozoan *Tetrahymena pyriformis*. *Lett. Appl. Microbiol.* **37**: 1–4.

69. Dayeh, V.R., Chow, S.L., Schirmer, K., Lynn, D.H., and Bols, N.C. (2004). Evaluating the toxicity of Triton X-100 to protozoan, fish and mammalian cells using fluorescent dyes as indicators of cell viability. *Ecotoxicol. Environ. Safety* **57**: 375–382.

70. Rogerson, A., Shiu, W.Y., Huang, G.L., Mackay, D., and Berger, J. (1983). Determination and interpretation of hydrocarbon toxicity to ciliate protozoa. *Aquat. Toxicol.* **3**: 215–228.

71. Otsuka, K., Yoshikawa, H., Sugitani, A., and Kawai, M. (1988). Effect of diphenyl, o-phenylphenol and 2-(4-thiazol)benzimidazole on growth of *Tetrahymena pyriformis*. *Bull. Environ. Contam. Toxicol.* **41**: 282–285.

72. Dias, N., Nicolau, A., Carvalho, G.S., Mota, M., and Lima, N. (1999). Miniaturization and application of the MTT assay to evaluate metabolic activity of protozoa in the presence of toxicants. *J. Basic Microbiol.* **39**: 103–108.

73. Nicolau, A., Mota, M., and Lima, N. (2004). Effect of different toxic compounds on ATP content and acid phosphatase activity in axenic cultures of *Tetrahymena pyriformis*. *Ecotoxicol. Environ. Safety* **57**: 129–135.

74. Nicolau, A., Mota, M., and Lima, N. (1999). Physiological responses of *Tetrahymena pyriformis* to copper, zinc, cycloheximide and Triton X-100. *FEMS Microbiol. Ecol.* **30**: 209–216.

75. Dias, N. and Lima, N. (2002). A comparative study using a fluorescence-based and a direct-count assay to determine cytotoxicity in *Tetrahymena pyriformis*. *Res. Microbiol.* **153**: 313–322.

76. Dowhanick, T.M. and Sobczak, J. (1994). ATP bioluminescent procedure for viability testing of potential beer spoilage microorganisms. *J. Am. Soc. Brew. Chem.* **52**: 19–23.

77. Gamborg, G. and Hansen, E.H. (1994). Flow-injection bioluminescent determination of ATP on the use of the luciferin–luciferase system. *Anal. Chem. Acta* **285**: 321–328.

78. Rothstein, R.L. and Blum, J.J. (1974). Lysosomal physiology in *Tetrahymena*. III. Pharmacological studies on acid hydrolase. *J. Cell Biol.* **62**: 844–859.

79. Rothstein, R.L. and Blum, J.J. (1975). Lysosomal physiology in *Tetrahymena*. II. Effect of culture age and temperature on the extracellular release of 3 acid hydrolases. *J. Protozool.* **21**: 163–168.

80. Ricketts, T.R. and Rappit, A.F. (1975). The adaptative acid phosphatase of *Tetrahymena pyriformis* GL-9. Effect of endocytosis of various nutrients. *Protoplasma* **85**: 119–125.

81. Williams, J.T. and Juo, P.S. (1976). Release and activation of a particulate bound acid phosphatase from *Tetrahymena pyriformis*. *Biochim. Biophys. Acta* **422**: 120–126.

82. Martin, A. and Clydes, M. (1991). Acid phosphatase endpoint for *in vitro* toxicity tests. *In Vitro Cell. Dev. Biol.* **27A**: 183–184.

83. Beermann, K., Buschmann, H.-J., and Schollmeyer, E. (1999). A calorimetric method for the rapid evaluation of toxic substances using *Tetrahymena pyriformis*. *Termochim. Acta* **337**: 65–69.

84. Akers, K.S., Sinks, G.D., and Schultz, T.W. (1999). Structure–toxicity relationships for selected halogenated aliphatic chemicals. *Environ. Toxicol. Pharmacol.* **7**: 33–39.

85. Cronin, M.T.D. and Dearden, J.C. (1995). QSAR in toxicology. 1. Prediction of aquatic toxicity. *Quant. Struct. Act. Relat.* **14**: 1–7.

86. Bradbury, S.P. and Lipnick, R.L. (1990). Introduction: structural properties for determining mechanisms of toxic action. *Environ. Health Perspect.* **87**: 181–182.

87. Mekenyan, O.G. and Veith, G.D. (1994). The electronic factor in QSAR: MO-parameters, competing interactions, reactivity and toxicity. *SAR QSAR Environ. Res.* **2**: 129–141.

SOFIE: AN OPTIMIZED APPROACH FOR EXPOSURE TESTS AND SEDIMENT ASSAYS

Jos P.M. Vink
Serge Rotteveel
Cornelis J.H. Miermans
Institute for Inland Water
Management and Waste Water
Treatment–RIZA
Lelystad, The Netherlands

INTRODUCTION

Regulatory quality standards are primarily developed to indicate, scale, or rank the environmental risks of contaminants. For soils and sediments, most countries legally prescribe the chemical extraction of total amounts of contaminants as a first step for risk assessment. Contaminants have mostly been sequestered over a long period of time and may only become available to organisms to a limited extent. The introduction of the bioassay, as a second step, generally increased the understanding of the concept of chemical versus biological availability. This is illustrated in Fig. 1. Aquatic oligochaetes that have been exposed to a number of sediments with a varying degree of cadmium content appear to be particularly susceptible to dissolved fractions. Although location L1 is ranked as seriously contaminated, based on total extracted amounts of cadmium from sediment, *Limnodrilus* spp. accumulated the most cadmium at the "clean" location L2.

The thought that chemical speciation generally determines the adverse toxic effects in biota is well accepted [reviewed by, among others, Campbell (1) in 1995]. In recent years, some techniques have become available to actually measure the availability (i.e., labile metal species) in "natural" solutions. These techniques make use of the (combined) electrochemical, diffusive, or competitive properties of metal ions. Table 1 summarizes some of the most promising techniques. However, only a few of these may find their use in exposure tests because of their disturbance (or destruction) of the sample, or the inability to measure speciation under anoxic conditions.

The purpose of bioassays is typically to determine the toxicity of ambient waters and sediments. Some assays focus on the determination of the exact concentration at which a chemical becomes toxic to an organism, so a biological response is used as a sensor. In order to be useful, this response should be repeatable. Ideally, bioassays are used to make predictions of environmental toxicity required for contaminant management (9,10) and should therefore at all times aim at extrapolating results to the field. Sediment bioassays are a relatively new approach to determine the environmental effects of sequestered metals in sediments. This type of assay has been discussed widely (11,12) in terms of limitations and advantages. Sediments typically consist of oxidized layers, overlying reduced ones. It is well known and documented how redox processes affect metal chemistry (3) and consequently the outcome of the test. Nevertheless, sediments are generally collected as bulk samples that contain this initial redox

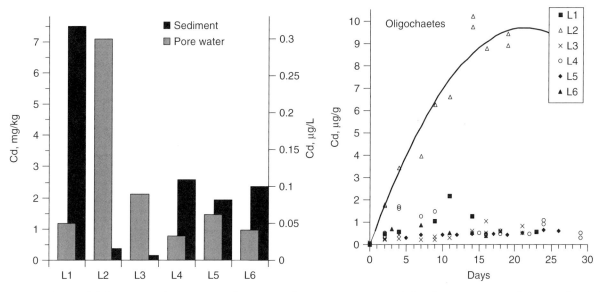

Figure 1. The availability concept in practice: cadmium in sediment, pore water, and oligochaetes at various locations (L1–L6).

zonation. Mostly, these samples are further manipulated by sieving, mixing, and oxygenation in order to increase homogeneity and repeatability. Indeed, variability is likely to be increased if sediments are not homogenized after collection, but these types of handling may destroy important reaction zones that are typical for the sediment in question and its environmental effect.

Data from standardized bioassays (i.e., protocols) generally provide the clear-cut answers that managers find useful as the basis for decisions, regardless of the accuracy to natural conditions. Therefore, these data continue to be the principal tool in the development of regulatory quality standards (12). If a bioassay is conducted in order to meet the objective of field-extrapolation, that is, to assess the environmental risks under natural conditions, a method of minimal sample manipulation should be considered.

Recently, a new approach was introduced that allows for a nondestructive handling and true-to-nature assessment of water/sediment systems. This approach is the "sediment or fauna incubation experiment," or SOFIE, and was developed to measure chemical speciation of metals over redox zones in undisturbed systems, while simultaneously

Table 1. Operational Metal Speciation Sensors

Method	Chemical Separation	Measured Species	Response Time	Remark	Reference
AdSV	Adsorptive stripping voltammetry	Electrochemical	Free metal + labile species	Minutes	Complex detection	2
CLE	Competing ligand exchange	Exchange reaction with ligands	Free metal (+ very labile species?)	Seconds	Critical modification and handling	3
DET	Diffusive equilibrium in thin film	Porous gel	Free metal + penetrable complexes	Days	Ionic strength limitations	4
DGT	Diffusive gradient in thin film	Porous gel	Free metal + labile penetrable complexes	Days	Critical gel composition	5
DMT	Donnan membrane technique	Charged pore membrane	Free metal + part of cationic penetrable complexes	Days	Critical pore size	
GIME	Gel impregnated microelectrode	Porous gel + electrochemical	Free metal + labile penetrable complexes	Minutes	Surface/lability interaction	6
ISE	Ion selective electrode	Liquid partition without countercharge transport	Free metal	Seconds	Interference from other metals and organic matter	7
PLM	Permeation liquid membrane	Liquid partition with countercharge transport	Free metal + labile complexes	Minutes	Diffusion-controlling step	8

conducting a single or multiple species bioassay. Here, we discuss the performance of this approach and compare it with the outcome of a standardized protocol for sediment or surface water bioassays. Tests were conducted with the sediment dweller *Chironomus riparius* (mosquito larva) and *Daphnia magna* (water flea) for a range of contaminants. Since this study aims at the straightforward comparison between the outcome of both methods, the speciation measurements conducted with SOFIE are not reported here.

METHODS OF EXPOSURE

The Standard Bioassay

The widely applied protocol for water and sediment testing for single species is what we call here the "standard bioassay." This protocol is based on the TRIAD guidelines of 1993 and describes the general procedure of the test, including handling of samples and test species. From this protocol, two test procedures were followed.

Standard Test with Sediment Dwellers. For tests using *Chironomus riparius* as a test organism, use is made of a water/sediment system. A bulk sample of the top 10 cm of a sediment is collected from the field and mechanically homogenized. The sample is then mixed 1:4 v/v with DSW ("standard water"; demineralized water with mineral additives). The water/sediment mixture is shaken for 24 h at ambient atmosphere. Next, 100 mL of the suspension is decanted into dishes to settle, after which test species are introduced. After 28 d of exposure, the organisms are collected by sieving, allowed to void their gut content for 24 h, and freeze dried.

Standard Test with Water Dwellers. For tests using *Daphnia magna*, the protocol prescribes the use of pore water, which is separated from the homogenized sediment by centrifugation (2500 g, 30 min) and stored in bottles. Exposure is carried out in 100-mL glass beakers and pore water is refreshed twice a week. Individuals are periodically collected, rinsed with Elendt medium, and analyzed for metals (not PAHs because of mass and detection limitations).

The Sediment or Fauna Incubation Experiment (SOFIE)

This study was performed with a novel experimental technique (EU patent 1018200/02077121.8, October 2001, J. Vink, Rijkswaterstaat), which was introduced as sediment or fauna incubation experiment, in short SOFIE. This device or "cell" is shown in Fig. 2.

SOFIE is based on the competing ligand exchange technique, which is integrated in a (pore) water probe and combined with a bioassay setting. Full and detailed experimental possibilities have been described earlier (3). The cell consists of a circular core, which is used as a sampling device to obtain undisturbed water/sediment systems. Field samples are taken including the overlying surface water in such a way that the physical and geochemical integrity of the sample (e.g., bulk density, redox status) is guaranteed. After sampling, the core is

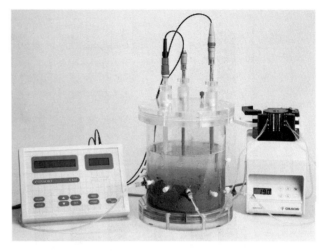

Figure 2. Sediment or fauna incubation experiment (SOFIE).

closed at the bottom with a shutter plate and is mounted on a base socket. The sample does not leave the body core but is now part of the cell. A top socket is attached, ensuring a gas-tight fitting with the body core with an internal silicone seal. The top plate of this socket has four gas-tight connectors for any commercial electrode to probe sediment or overlying water. For experimental scenario purposes, aerobic or anaerobic conditions in the water–sediment system may be manipulated by flushing the atmospheric head space with any desirable gas. The cell wall contains fifteen gas-tight connectors for probes. These probes (diameter 1 mm, length 50 mm) consist of a semipermeable polyethersulfon, which acts as a membrane to discriminate colloidal fractions greater than $0.1\ \mu m$. This mesh size is impermeable to bacteria, so pore water that passes the polymer is sterile. The probes can be directly coupled onto micro ion-exchange columns (MICs), as was described in detail by Vink (3), to separate free metal ions from other forms of organic and inorganic metal ligands in solution at the reigning geochemical status of the probed sediment layer.

For a simple, straightforward comparison between the standard bioassay and SOFIE, sediment samples were collected from the river Meuse, The Netherlands, and analyzed for a large amount of properties. An undisturbed water–sediment interface was taken with the cell's body core at 0.4 m water depth. After steady-state metal concentrations were determined, 125 individuals of *Chironomus riparius* were introduced into the SOFIE cell and sampled at the same time intervals as in the standard bioassay. Tests with *Daphnia magna* were performed in the same way.

The Presence of Food During Exposure

In an additional test with *Daphnia magna*, the effect of feeding during exposure was investigated. This test was performed in a two-compartment SOFIE cell, which has two separate chambers that divide the sample while it is cored. In both compartments, daphnids were introduced in comparable amounts as in the standard bioassay. Daphnids in compartment A were fed with the alga *Chlorella,* their natural food source, while those in compartment B were not fed. Organisms and water samples were collected from the cell compartments on several occasions.

Metal concentrations in chironomids and daphnids were determined by digestion in 500 μL of 14.9 M HNO_3 (Ultrex) at 180 °C in a microwave. The digest was analyzed using inductively coupled plasma mass spectrometry (ICP-MS, PerkinElmer Elan 6000). Polycyclic aromatic hydrocarbons (PAHs) in chironomids were determined by refluxing for 6 h with hexane. Analyses were done with an Agilent HP liquid chromatograph using a fluorescence detector and a Vydac 201TP54 reverse phase C18 column. Dolt-2 (certified by the Community Bureau of Reference, BCR, Brussels, Belgium) was used as biological reference material.

RESULTS AND DISCUSSION

The sediment was composed of 14% < 2 μm, 2.3% organic C, and 0.55 inorganic C, at pH = 7.2, and was moderately contaminated with metals (As, 8.5 mg/kg dw; Cd, 1.57; Cr, 27.7; Cu, 25.7; Hg, 0.21; Ni, 20.8; Pb, 79; Zn, 321) and various PAHs (ranging from 100 to 800 ng/g dw).

Collection of pore water from the sediment by homogenization and centrifugation, according to the protocol, generally increased metal concentrations, most significantly for Pb and Cd. Table 2 shows the concentrations that were measured at the start of the exposure test in the standard bioassay and in SOFIE's surface water.

The collected time data for chironomids are, for reasons of survey ability, summarized in Fig. 3, which shows body concentrations of polycyclic aromatic hydrocarbons (PAH) and heavy metals after four weeks of exposure. The data quite clearly show that, in the standard bioassay, chironomids take up PAHs in significantly larger amounts than in the SOFIE environment. Although we did not perform or aim at a mechanistic study here, it is likely that sample handling (and in particular stirring and homogenization) has liberated PAHs from the organic sediment matrix by unlocking and exposing organic surface areas. Many authors have described this phenomenon.

Bioaccumulated metals are shown in Figs. 3 and 4. Accumulated amounts for both test species are lower in the standard bioassay than those collected from SOFIE (note the log scale, differences are significant). This is valid for all metals. Although this observation may appear somewhat counterintuitive [oxidation leads to liberation

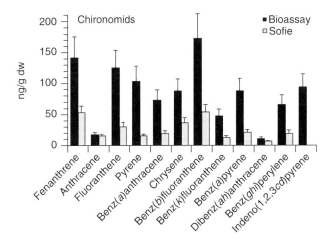
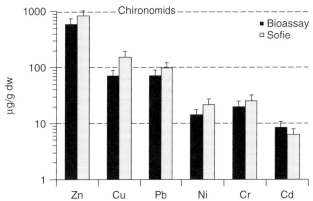

Figure 3. PAHs (top) and heavy metals (bottom) in chironomids after 28 d. Error bars denote the variation in analytical recovery.

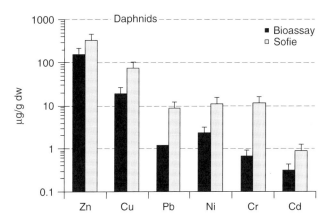

Figure 4. Metals in daphnids after 16 d of exposure.

of metals from sulfides and should therefore increase the (bio)available fraction of metals], the explanation is quite straightforward. Indeed, oxidation leads to enhanced metal concentrations, as is shown in Table 3. Following chemical thermodynamics, this primarily yields large free ion activities. Simultaneously, iron and manganese are oxidized from the reduced state into highly reactive oxyhydroxides (either precipitated or as colloids). Free metal ions have a high affinity for these newly formed sorption sites and are quickly adsorbed and immobilized. In the case of the standard *Daphnia* bioassay, where

Table 2. Pore Water Concentrations (μg/L) After Sample Handling (Standard Bioassay) and in Sediment of SOFIE

	Standard Bioassay[a]		SOFIE[a]	
Zn	23.3	(4.7)	28.1	(5.6)
Cu	6.5	(0.7)	2.7	(0.2)
Pb	17.8	(3.6)	1.3	(0.3)
Ni	6.4	(0.7)	2.0	(0.2)
Cr	4.2	(0.3)	2.4	(0.2)
Cd	0.26	(0.01)	0.03	(0.002)

[a] Values in parentheses are analytical standard deviations.

Table 3. Initial Concentrations C_i (μg/L) in the Exposure Medium, Model Parameters k_0, k_1, k_2 (1/d), and Correlation Coefficient

	Standard Bioassay						SOFIE					
	Zn	Cu	Pb	Ni	Cr	Cd	Zn	Cu	Pb	Ni	Cr	Cd
Chironomids												
C_i	9.2	3.7	3.8	2.8	2.0	0.08	45.8	2.7	2.5	4.9	1.0	0.39
k_0	−0.040	−0.11	−0.120	−0.030	−0.050	−0.100	−0.007	−0.015	−0.018	0.018	0.045	0.001
k_1	44.05	7.41	5.96	18.67	9.56	10.19	1.55	4.53	47.53	7.52	2.04	3.39
k_2	1.15	6.51	9.71	8.84	3.72	1.07	0.09	0.11	1.98	1.61	0.04	0.08
r^2	0.80	0.38	0.54	0.68	0.44	0.14	0.90	0.79	0.96	0.55	0.67	0.53
Daphnids												
C_i	23.3	6.5	17.8	6.4	4.2	0.26	45.8	2.7	2.5	4.9	1.0	0.39
k_0	−0.114	−0.258	−0.350	−0.215	−0.151	−0.272	−0.040	0.047	0.101	0.023	−0.062	0.106
k_1	10.54	0.94	0.005	0.03	0.26	0.31	0.66	1.44	0.01	0.01	0.11	0.07
k_2	8.31	12.94	12.72	2.24	12.80	12.67	0.11	−0.06	−0.35	−0.32	−0.17	0.01
r^2	0.92	0.13	0.71	0.89	0.58	0.47	0.97	0.99	1.00	1.00	0.99	0.85

sediment is absent, there is obviously no delivery from the sediment (e.g., by diffusion and desorption) to the aqueous phase. Daphnids take up metals from solution and eventually face declining exposure concentrations to a point where concentration levels may become inhibiting to uptake. Figure 5 shows copper for an example. This trend was observed for all metals in the standard bioassays, but not in the SOFIE cells where concentrations remained relatively constant during the exposure period or showed some release from the sediment.

Figure 6 shows the effect of feeding daphnids during exposure. The effect of adding the alga *Chlorella* is profound (note the log scale). In most cases, bioaccumulated amounts of metals are reduced some orders of magnitude compared to the nonfeeding test. Since alga are always present in surface waters in varying amounts, it is obvious that this test is closer to exposure conditions as they occur in the field.

Results show that dissolved metal concentrations change during exposure. These time-varying changes may follow first-order reaction kinetics; that is, $dC/dt = kC$, so the external concentration, or exposure concentration, at a given time $C(t)$, is written as

$$C(t) = C_i e^{k_0 t}$$

where C_i is the initial exposure concentration, and k_0 is a rate term describing the increase or decrease of the initial concentration during the test. A time/concentration-dynamic, two-compartment model approximated metal bioaccumulation patterns of both test species. Body concentrations vary in time and relate to uptake from the water phase at a certain rate. At the same time, uptake is accompanied by elimination, which is kinetically directed by the organism itself. Hence, one may write

$$dQ/dt = k_1 C(t) - k_2 Q(t)$$

where Q is the internal body concentration (μg·g^{-1}), k_1 and k_2 are uptake and elimination rate constants, respectively (d^{-1}), and t is time. This yields, for $Q(0) = 0$,

$$Q(t) = \frac{k_1 C_0}{k_2 - k_0}(e^{-k_0 t} - e^{-k_2 t})$$

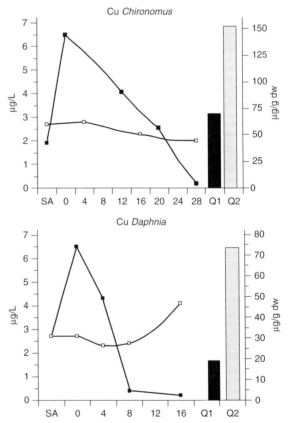

Figure 5. Copper concentrations in overlying water at time of sampling (SA) and during exposure (days) in the standard bioassay (■) and SOFIE (□) for chironomids (top) and daphnids (bottom). Q1 and Q2 are body concentrations from the standard bioassay and SOFIE, respectively. While concentrations in the undisturbed water–sediment system remain relatively constant or increase slightly, those in the standard bioassay show large variation due to the handling and manipulation of the sample, followed by exhaustion from solution.

With the introduction of k_0, time-dependent concentrations determine the overall exposure of organisms, and thus $Q(t)$.

Table 3 summarizes the results of the uptake model. In the standard bioassay, initial concentrations differ

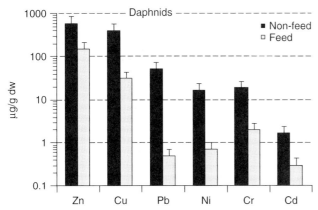

Figure 6. Heavy metals in *Daphnia magna* in two feeding scenarios.

between chironomids and daphnids. This is the result of sample handling and the absence of sediment in the daphnid test. From the viewpoint that bioassays should deliver generic, true-to-nature outcomes, this appears an undesirable starting point. Although *Chironomus* is a sediment dweller and *Daphnia* is not, both species encounter contaminant exposure in an environment that consists of both sediment and surface water, where bioavailability is a combined action of the two compartments.

The rate by which exposure concentrations change during the tests is represented by k_0. Positive values denote an increase, negative values a decrease in concentrations. The outcome clearly shows that in the standard bioassay, k_0 values are in all cases negative and larger than those observed in the assays that were performed with SOFIE. The effect is most profound in the daphnid tests, where k_0 values diverge both in direction and in magnitude. For all metals, and both test species, the two-compartment model performed better with the data that were acquired from the SOFIE setup.

CONCLUSIONS

Based on the outcome of water–sediment testing of two aquatic species with the standard protocol and with the sediment or fauna incubation experiment, SOFIE, the following conclusions are drawn:

- Sample handling such as homogenization and oxygenation affects the chemical speciation of the exposure medium for both organic and inorganic contaminants. Disadvantages of this manipulation (disrupting chemical speciation and bioavailability) outweigh the advantages (homogenization to increase reproducibility) by far and should therefore seriously be reconsidered or applied with great caution.
- Tests conducted with the standard protocol tend to *overestimate* uptake of PAHs by chironomids and *underestimate* uptake of heavy metals by chironomids and daphnids. This may generate both false-positive and false-negative results, respectively.
- Aquatic bioassays should be conducted in sediment/water systems, regardless of whether the tested species are sediment dwellers or not. The physical, chemical, and biological interaction between sediment and the water phase is a site-specific characteristic that determines the ultimate risk of contaminants.
- SOFIE provides the necessary tool to conduct risk assessments in a close-to-nature setting. It may therefore contribute to establish better-founded quality criteria for natural waters.

BIBLIOGRAPHY

1. Campbell, P.C.G. (1995). *Metal Speciation and Bioavailability in Aquatic Systems*. John Wiley & Sons, Hoboken, NJ.
2. Kaldova, R.A. and Kopanica, M. (1989). *Pure Appl. Chem.* **61**: 97–112.
3. Vink, J.P.M. (2002). *Environ. Sci. Technol.* **36**: 5130–5138.
4. Davison, W., Grime, G.W., Morgan, J.W., and Clarke, K. (1991). *Nature* **352**: 545.
5. Zhang, H. and Davison, W. (1995). *Anal. Chem.* **67**: 3391.
6. Tercier, M.L. and Buffle, J. (1996). *Anal. Chem.* **68**: 3670–3678.
7. Berner, R.A. (1963). *Geochim. Cosmochim. Acta* **27**: 563–575.
8. Buffle, J., Parthasarathy, N., Djane, N.K., and Matthiasson, L. (2000). *In Situ Monitoring of Aquatic Systems*. John Wiley & Sons, Hoboken, NJ.
9. Chapman, P.M. (1989). *Environ. Sci. Technol.* **8**: 589–601.
10. Blum, D.J.W. and Speece, R.E. (1990). *Environ. Sci. Technol.* **24**: 284b.
11. Schwartz, R.C. (1987). *Fate and Effects of Sediment Bound Chemicals in Aquatic Systems*. Pergamon Press, New York.
12. Luoma, S.N. (1995). *Metal Speciation and Bioavailability in Aquatic Systems*. John Wiley & Sons, Hoboken, NJ.

PASSIVE TREATMENT OF ACID MINE DRAINAGE (WETLANDS)

ARTHUR W. ROSE
Pennsylvania State University
University Park, Pennsylvania

Starting in about 1985, passive treatment has been increasingly used as a method of remediating acid mine drainage (AMD). A passive treatment facility uses natural reactions such as microbial oxidation or reaction with a bed of crushed limestone to improve water quality, with only minimal human attention after construction. The initial passive method was a simple wetland, which was found to improve water quality for some types of AMD. However, for other waters, wetlands were not effective, but a variety of other methods have been developed, as indicated in Fig. 1. Passive treatment contrasts with active treatment in which a chemical such as sodium hydroxide, lime, or similar alkaline agents is added to the water. Active methods require frequent human attention and continued replenishment of the reagents.

Passive methods applicable for a specific site are selected using a flowsheet, as in Fig. 2. The first step is to determine the flow and water chemistry. Key chemical

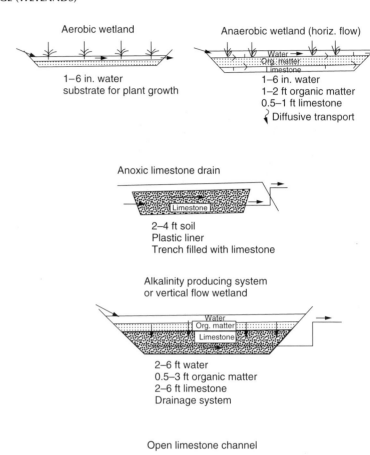

Figure 1. Schematic diagrams for passive treatment systems (after Ref. 1).

variables are the acidity, alkalinity, Fe, Al, Mn, and SO_4. The acidity should be a hot peroxide acidity such as presented in References 2 and 3. In these methods, the Fe and Mn are oxidized so that the total acidity to pH 8.3 is measured and the alkalinity is subtracted during the procedure so the result is a net acidity (4).

If the water is net alkaline (negative acidity), the water can easily be treated by a wetland. Sizing of wetlands is discussed by Hedin et al. (5). In the wetland, Fe and Mn are oxidized in the shallow water, precipitate as hydroxides, and settle in the slowly flowing water. The alkalinity is adequate to maintain a pH above 4.5, usually higher.

If the water is net acid (positive acidity), then the dissolved O_2, ferric iron, and aluminum must be measured. If these are negligible (less than about 1 mg/L each), then an anoxic limestone drain (ALD) can be successful. An ALD is a trench filled with crushed limestone ($CaCO_3$) and covered to eliminate access of air. Acidic water from a subsurface source flows into one end and passes along the limestone bed. The water acquires alkalinity and increases in pH by reaction with the limestone. On emerging from the limestone bed, Fe oxidizes and precipitates in an oxidation-settling pond.

If the water is net acid and contains appreciable dissolved O_2, ferric iron, or aluminum, then a common passive method is a vertical flow pond (also known as a SAPS, successive alkalinity producing system) (6,7). In this method, the water first flows into a settling pond to remove sediment and possibly some Fe. It then passes into the vertical flow pond that drains through the bottom. The top layer in the bottom of the pond is compost that consumes dissolved O_2 and reduces ferric iron to ferrous iron. Some Fe and acidity may be removed by microbial reduction of SO_4 in the compost. The water then enters the lower layer of crushed limestone, where acidity is consumed, alkalinity is added, and pH is increased. The water leaves the limestone layer through perforated pipes and flows to an oxidation-settling pond for Fe precipitation and removal. A wetland may be used to further clean up the water.

A problem in some vertical flow systems is dissolved Al, which can precipitate in the limestone bed and plug the system, or coat limestone leading to inhibited reaction. Some units are constructed to be flushed, so that precipitates are flushed out of the limestone bed. Recent measurements indicate that this method is only partially effective. Trials are underway with back-flushing systems, upflow limestone ponds, mixing of fine limestone into the compost, and other variations.

On relatively steep slopes, an open limestone channel can be successful in neutralizing acidity. To be successful, the flow must be vigorous enough to erode the Fe and Al coatings off the limestone.

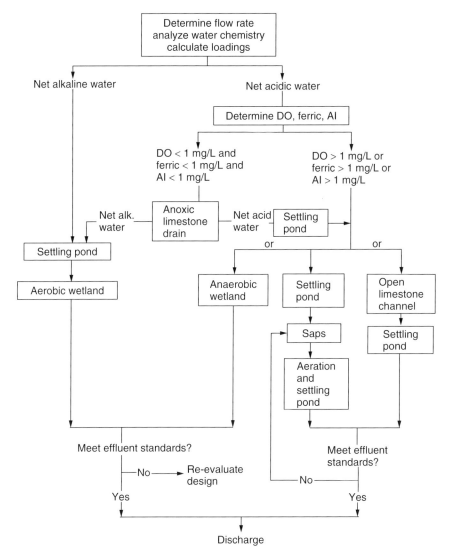

Figure 2. Flowsheet for selection of passive treatment method (after Ref. 1).

Anaerobic wetlands, in which the substrate for the wetland is compost and/or limestone, can be used to treat net acid water. However, the area of such wetlands must be much larger than aerobic wetlands or vertical flow ponds, so these are not common.

Discharge standards for mine waters specify removal of most manganese, which can be removed passively in aerobic limestone beds (8). The influent waters must be oxidizing and have negligible concentrations of Al and Fe. In these beds, Mn-oxidizing bacteria oxidize Mn and precipitate Mn oxides and hydroxides at pH 6.5 and higher. In some such facilities, the limestone bed is inoculated with specific Mn-oxidizing bacteria (9).

Recent publications indicate that periodic inspection, maintenance, and occasional rebuilding are necessary for most passive systems, especially those treating net acid water. Problems include washout during high flow, muskrat penetration of the dams, plugging of pipes by precipitates or vegetation, plugging of limestone by Al or Fe precipitates, accumulation of Fe oxides on top of compost, and channeling of flow through limited portions of the system.

BIBLIOGRAPHY

1. Skousen, J., Rose, A., Geidel, G., Foreman, J., Evans, R., and Hellier, W. (1998). *A Handbook of Technologies for Avoidance and Remediation of Acid Mine Drainage*. National Mine Land Reclamation Center, West Virginia University.

2. American Public Health Association (1998). Acidity (2310)/titration method. In: *Standard Methods for the Examination of Water and Wastewater*, 20th Edn. L.S. Clesceri et al. (Eds.). American Public Health Association, Washington, DC, pp. 2.24–2.26.

3. U.S. Environmental Protection Agency (1979). Method 305.1, acidity (titrimetric). In: *Methods for Chemical Analysis of Water and Wastes*, U.S. Environmental Protection Agency Report EPA/600/4-79-020 Available at (www.nemi.gov).

4. Cravotta, C.A. and Kirby, X. (2004). Acidity and alkalinity in mine drainage: practical considerations. In: *Proceedings, American Society of Mining and Reclamation*, April 18–22, 2004, Morgantown, WV, pp. 334–365.

5. Hedin, R.S., Watzlaf, G.R., and Kleinmann, R.L.P. (1994). *Passive Treatment of Coal Mine Drainage: U.S. Bureau of Mines*, Information Circular IC 9389.

6. Kepler, D.A. and McCleary, E.C. (1994). Successive alkalinity producing systems (SAPS) for the treatment of acidic mine

drainage. In: *Proceedings, International Mine Drainage and Land Reclamation Conference*, April 24–29, 1994, Pittsburgh, PA, U.S. Bureau of Mines Special Publication SP06A-94, pp. 195–204.

7. Rose, A.W. (2004). Vertical flow systems—effects of time and acidity relations. In: *Proceedings, American Society of Mining and Reclamation*, April 18–22, 2004, Morgantown, WV, pp. 1595–1616.

8. Rose, A.W., Means, B., and Shah, P.J. (2003). Methods for passive removal of manganese. In: *Proceedings, West Virginia Surface Mine Drainage Task Force Symposium*, Morgantown, WV, April 15–16, 2003, pp. 71–82.

9. Vail, W.J. and Riley, R.K. (2000). *The Pyrolusite Process, A Bioremediation Process for the Abatement of Acid Mine Drainage*, Vol. 30, Greenlands, pp. 40–46.

BIOMARKERS AND BIOACCUMULATION: TWO LINES OF EVIDENCE TO ASSESS SEDIMENT QUALITY

M.L. Martín-Díaz
Instituto de Ciencias Marinas de Andalucía
Cádiz, Spain

M.C. Morales-Caselles
N. Jiménez-Tenorio
I. Riba
T.A. Delvalls
Facultad de Ciencias del Mar y Ambientales
Cádiz, Spain

INTRODUCTION

Contaminated sediments pose a risk to aquatic life, human health, and wildlife throughout the world. There is an overwhelming amount of evidence that chemicals in sediments are responsible for toxicological (1) and adverse ecological effects (2). Frequently, the chemicals causing these effects are present in the sediment as mixtures of organic, metal, and other types of contaminants.

Identification of toxicants in sediments is useful in a variety of contexts. Adverse environmental effects from contaminated sediment resulted in international treaties and protocols for the environmental management of these dredged sediments. In this sense the OSPAR and Helsinki Conventions (North Sea, northeast Atlantic, Baltic Sea) proposed guidelines to control the disposal of sediment.

Once a toxicant is identified in the sediment, steps can also be taken to link a toxicant to a discharger and prevent further discharge. In addition, identification of major causes of toxicity in sediments may guide programs such as the development of environmental sediment guidelines and, retrospectively, aid regulators in determining the type of pesticide or manufactured chemical that may cause toxicity in the field. Here, we review the use of biomarkers and bioaccumulation in sediment quality assessment. These different tools provide a more sensitive measure of bioavailability and the effects of the different contaminants in the sediment.

SEDIMENT QUALITY ASSESSMENT

Sediments can act as deposits of contaminants entering the environment and, as a consequence, constitute a source of contamination. Today, remediation and management of contaminated sediments is becoming more and more economically and technologically demanding. Because of this, the scientific community has been developing science-based tools to identify sediments that are impaired and, ultimately, to support effective management decisions and priorities for dealing with contaminated sediments (3). When discussing the adverse effects of pollutants, some consider the accumulation of pollutant residues in the tissues of organisms to be adverse. Others consider an effect injurious only if changes occur in physiological processes in organisms, such as alterations in cellular morphology, metabolic activity, or physiological rates. Ecologists might restrict this definition still further to only those pollutant-induced effects that give rise to ecologically significant changes, that is, those at the population level (4). It is important to take into account the bioindicator species, the toxicity assay, and the different biological measurements that could show the exposure to a contaminant and the effect.

Different sensitive toxicity tests have been developed to assess sediment toxicity. The different toxicity tests can group pore water and whole sediment exposure tests. These tests are conducted using benthic organisms such as amphipods, polychaetes, algae, and juvenile fish (5). Use of these organisms allows one to estimate the effect of sediment ingestion and measure bioaccumulation and different biomarkers of exposure and effect. For example, the role of sediment ingestion, which should not be underestimated, has been identified as an important route of uptake of polycyclic aromatic hydrocarbons (PAHs) by deposit-feeding benthic organisms, as shown in experiments with polychaete worms (6,7).

More appropriately, three ecological life styles (filter feeding, deposit feeding, and burrowing) can be represented by the species used to assess sediment effects (8). Table 1 shows that the most appropriate phyla for potential impact testing in sediment is crustaceans, followed by annelids, mollusks, insects, fish, and echinoderms. Some of the species belonging to the taxonomic groups of crustaceans, mollusks, and fish are estuarine individuals that tolerate salinity fluctuation, which could affect the toxicity evaluation response in an environment affected by variation of physicochemical variables (9).

Bioaccumulation

Bioaccumulation values from different contaminant studies can help identify the bioavailability of chemicals in marine sediments and waters since, unlike chemical analysis, bioaccumulation provides a measurement of bioavailable contaminants (15).

Bioaccumulation is the uptake and retention of a bioavailable chemical from any possible external source (water, food, substrate, air). It is the net result of the uptake, distribution, and elimination of a substance in an organism due to exposure in water, food, sediment, and air. For bioaccumulation to occur, the rate of uptake from

Table 1. Summary of Species Appropriate for Testing Potential Impacts of Contaminated Sediment (10–14)

Taxonomic Group	Name	Test Species	Salinity Tolerance
Crustaceans	Mysid shrimp	*Americamysis* sp.	M
		Neomysis	M
		Americana	M
		Holmesimysis costata	
	Grass shrimp	*Palaemonetes* sp.	M, E
	Sand shrimp	*Crangon* sp.	M, E
	Shrimp	*Farfante penocus*	M
		Pandalus sp.	M
		Sicyonia ingentis	M
	Crab	*Callicnetes sapidus*	M, E
		Cancer sp.	M, E
		Carcinus maenas	M, E
	Amphipod	*Ampelisca* sp.	M
		Rhepoxynius sp.	M
		Eohaustarius sp.	M, E
		Grandiderella japonica	M, E
		Corophium insidiosum	M, E, F
		Leptocheirus plumulosus	M, E
		Hyalella azteca	M, E
		Corophium sp.	M
		Ampelisca brevicornis	M
		Mycrodentopus	M
		Gryllotalpa	M
	Cladoceran	*Daphnia magna*	F
		Ceriodaphnia dubia	F
Insects	Midge	*Chironomus tentans*	F
		C. riparius	F
	Mayfly	*Hexagenia limbata*	F
Molluscs	Mussel	*Anodonta imbecillis*	F
	Clam	*Yoldia limatula*	M
		Protothaca staminea	M
		Tapes japonica	M, E
		Ruditapes philippinarum	M
	Burrowing polychaete	*Nereis* sp.	M
		Neanthes arenaceodentata	M
		Nephthys sp.	M
		Glycera sp.	M
		Arenicola sp.	M
		Abarenicola sp.	M
Annelids	Oligochaete	*Pristina leidyi*	F
		Tubifex tubifex	F
		Lumbriculus variegates	F
Echinoderms	Sea urchin	*Strongylocentratus purparatus*	M
		Lytechinus pietus	M
		Echinocardium cordatum	M
		Paracentrotus lividus	M
Fish	Sheepshead minnow	*Cyprinodon variegates*	M, E
	Arrow gobi	*Clevelandia ios*	M
	Rainbow trout	*Oncorhynchus mykiss*	F, E
	Flat fish	*Solea senegalensis*	M
	Sea bream	*Sparus aurata*	M
	Turbot	*Scophthalmus maximus*	M

all sources must be greater than the rate of loss of the chemical from the tissues of the organism (16) (Fig. 1).

Toxic effects of anthropogenic compounds in biota and ecosystems are regarded in relation to their chemistry and fate in the environment. The bioavailability of chemicals, which is dependent on biogeochemical and physiological processes, is an important factor, often neglected in ecotoxicological evaluation and hazard assessment. The bioavailable fraction is critical for uptake and, ultimately, for the concentration at the target site in the organisms. The bioavailability of contaminants in sediments depends on several factors: physical (grain size of the sediment and suspended particulate materials), chemical (solubility, reactivity of compounds, complexing agents), and biological (benthic or pelagic organisms, mode of exposure) (17). Nevertheless, not all bioavailable

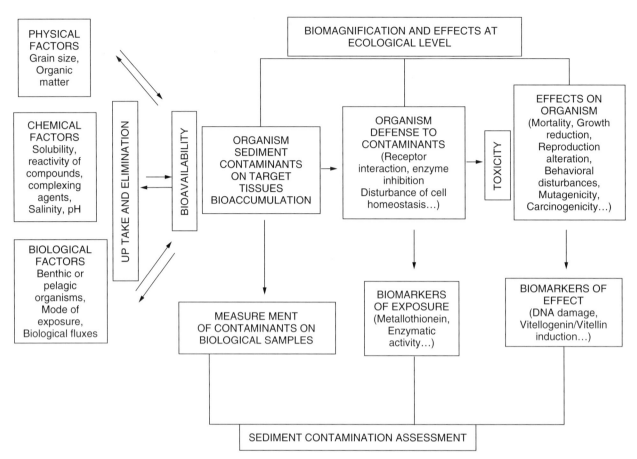

Figure 1. Ecotoxicological effects and assessment of the bioavailable fraction of pollutants in sediment.

chemicals can be bioaccumulated by the organisms in their tissues. For example, highly soluble chemicals, such as ammonia and some inorganic ions, are bioavailable and rapidly penetrate permeable tissues of marine organisms. However, they are not retained and are lost just as rapidly from the tissues by diffusion, metabolic transformation, or active transport. Other bioavailable chemicals are also taken up rapidly but are transformed and/or excreted rapidly by metabolic processes of the organism and are not bioaccumulated (16). In this sense, it is important to take into account the bioaccumulation factor (BAF), which is the ratio of the sum of the uptake rate constants of the chemical from all environmental compartments accessible to the organism to the sum of the release rate constants by active and passive mechanisms from the organism (18). The bioaccumulation factor allows one to predict if a bioavailable chemical is being bioaccumulated.

Once a chemical is bioaccumulated, the concentration of the contaminant at the target tissue in the organism induces molecular manifestations of defense and the effects can be determined (Fig. 1).

Persistent chemicals may accumulate in aquatic organisms through different mechanisms: via the direct uptake from water by gills or skin (bioconcentration), via uptake of suspended particles (ingestion), and via the consumption of contaminated food (biomagnification). Even without detectable acute or chronic effects in standard ecotoxicity tests, bioaccumulation should be regarded as a hazard criterion in itself, since some effects may only be recognized in a later phase of life, are multigeneration effects, or manifest only in higher members of a food web (e.g., impact of PCBs on the hatching success of eggs) (19).

Bioaccumulation of certain persistent environmental contaminants in animal tissues may be considered a biomarker of exposure (20,21). However, based on the definitions provided by Van Gestel and Van Brummelen (22), body burdens are not considered to be biomarkers or bioindicators since they do not provide information on deviations related to "health." To avoid confusion, Van der Oost et al. (23) proposed that analytical–chemical indicators (body burdens) be referred to as bioaccumulation markers, while all biological (biochemical, physiological, histological and morphological) indicators measured inside an organism or its products be referred to as biomarkers.

The relationship between concentration of contaminants in tissues and toxic effects measured in organisms has received increased attention during the last few decades (1). These relationships should allow one to derive tissue quality guidelines (TQGs), defined as the concentrations of the chemicals measured in the different tissues that are associated or not associated with the biological effects measured using the sediment quality guidelines (SQGs) widely used around the world (24). To identify possible toxic agent(s), one must have body burden data collected from the same organisms exhibiting toxicity.

Biomarkers

Biomarkers are specific biological responses related to metabolism, detoxification, or toxicity induced by pollutants and associated with disposed materials. In this respect, an important advantage of biomarkers in assessing the impact of contaminated sediment is the inherent capability to detect early occurrence of various stress conditions within the organism and monitor the temporal progression (or regression) of the disturbance at various levels of biological organization (25).

Laboratory studies have documented the use of biomarkers to provide rapid quantitative predictions of toxicity on individual organisms. At this time, the use of biomarkers is not a replacement for traditional monitoring techniques, but it can be a useful supplementary approach to demonstrate links between sublethal biochemical exposure and decreases noted in field population studies (26).

A biomarker is defined as a change in a biological response (ranging from molecular though cellular and physiological responses to behavioral changes) which can be related to exposure to or toxic effects of environmental chemicals (27). Van Gastel and Van Brummelen (22) redefined the terms "biomarker," "bioindicator," and "ecological indicator," linking them to different levels of biological organization. They considered a biomarker as any biological response to an environmental chemical at the subindividual level, measured inside an organism or its products (urine, feces, hair, feathers, etc.), indicting a deviation from the normal status that cannot be detected in the intact organism. A bioindicator is defined as an organism giving information on the environmental conditions of its habitat by its presence or absence or by its behavior. An ecological indicator is an ecosystem parameter, describing the structure and functioning of ecosystems.

One of the most important features of biomarkers is that they have the potential to anticipate changes at higher levels of the biological organization (i.e., population, community, or ecosystem). Thus, these "early warning" biomarkers can be used in a predictive way, allowing the initiation of bioremediation strategies before irreversible environmental damage of ecological consequences occurs. Biomarkers are then defined as short-term indicators of long-term biological effects.

According to the National Research Council (20) and the World Health Organization (21), biomarkers can be subdivided into three classes: biomarkers of exposure, biomarkers of effects, and biomarkers of susceptibility. The biomarkers of exposure cover the detection and measurement of an exogenous substance or its metabolite or the product of an interaction between a xenobiotic agent and some target molecule or cell that is measured in an organism. On the other hand, the biomarkers of effect show an established or possible health impairment or disease through a measurable biochemical, physiological, or other alteration in the organism. Finally, biomarkers of susceptibility indicate the inherent or acquired ability of an organism to respond to the challenge of exposure to a specific xenobiotic substance, including genetic factors and changes in receptors which alter the susceptibility of an organism to that exposure.

New techniques allow one to detect the effects of complex mixtures of contaminants. Many are diagnostic of causes, provide information on the bioavailability of contaminants, and allow more accurate assessments of potential ecological damage. Cellular and molecular indicators provide the greatest potential for identifying individuals and populations for which conditions have exceeded compensatory mechanisms and which are experiencing chronic stress, which, if unmitigated, may progress to severe effects at the ecosystem level.

The classification of biomarkers in the literature is very diffuse since biomarkers of exposure and those of effect are distinguished by the way they are used, not by an inherent dichotomy (28).

Good biomarkers are sensitive indices of both pollutant bioavailability and early biological responses (23,29–31). Recently, the biomarker approach has been incorporated into several pollution monitoring programs in Europe and the United States. Likewise, different methods for biological effect measurement have been evaluated in a series of practical workshops organized by the International Council for the Exploration of the Sea (ICES) and the Intergovernmental Oceanographic Commission (IOC), such as those in the North Sea (32). The United Nations Environment Programme has founded a biomonitoring program in the Mediterranean Sea including a variety of biomarkers (10). Biomarkers have also been included in the Joint Monitoring Programme of the OSPAR convention, where Portugal and Spain are members. Nevertheless, the biomarkers approach has not been included in the guidelines for the management and monitoring of dredging and disposal activities. The current guidelines for the control of these activities are based on several approaches, which take into account chemical measurements, analysis of benthic communities, and toxicity tests. Very few studies have been done on the utility of biomarkers.

In this sense, different studies on the use of biomarkers to assess the impact of contaminants on sediment (Table 2) are being carried out. Most of the studies focus on the determination of the activity of biotransformation enzymes, antioxidant enzymes, and biochemical indices of oxidative damage, DNA damage, and metallothioneins. Today, promising biomarker tools are the "genomics" and "proteomics." In some cases genomic is used as a broad term—including proteomic—and it includes genomic sequencing, functions of specific genes, genome architecture, gene expression at transcriptome level, and protein expression at proteome level and metabolite flux (metabolomics).

USING BIOMARKERS AND BIOACCUMULATION FOR REGULATION PURPOSES

Frequently, the chemicals causing these effects are present in the sediment as mixtures of organic, metal, and other types of contaminants. The foundation for sediment quality assessment in the context of environmental risk assessment (ERA) is the weight-of-evidence (WOE), whose objective is to integrate multiple lines-of-evidence (LOE), both chemical and biological.

Table 2. Biomarkers Used for Sediment Toxicity Assessment in the Laboratory, *In Situ* (Caged Individuals), and in the Field (11–14,26,33–36)

Parameter	Description	Toxicity Assay	Pollutant Response
Metallothionein	Induction of this protein indicates exposure to metals	Field, caged individuals, laboratory	Cd, Cu, Zn, Hg, Co, Ni, Bi, Ag
GSH	Assay that determines the total glutathione content, a natural antioxidant	Field	PAHs, PCBs
DNA damage	Assay that detects single-strand breaks in DNA, a measure of DNA damage	Laboratory, caged individuals	PAHs, PCBs
EROD	Assay for ethoxyresorufin-*O*-deethylase, a phase I detoxification enzyme	Laboratory, caged individuals, field	PAHs, PCBs
CAT	Assay for catalase, an antioxidant enzyme	Caged individuals, field	PAHs, PCBs
SOD	Assay for superoxide dismutase, an antioxidant enzyme	Field	PAHs, PCBs
GR	Assay for glutathione reductase, an antioxidant enzyme	Laboratory, field	PAHs, PCBs
GPX	Assay for glutathione peroxidase, an antioxidant enzyme	Field	PAHs, PCBs
GST	Assay for glutathione-*S*-transferase, a phase II detoxification enzyme	Laboratory, field, caged individuals	PAH, PCB, BNF
LPO	Assay to determine the level of thiobarbituric reactive substances from lipid peroxide breakdown	Laboratory, caged individuals	Cd, PCB
Vitellogenin/vitellin	Induction of this protein indicates the exposure to substances that could perturb endocrine function	Laboratory, caged individuals	Cd, Zn, PAHs, PCBs
TOSC	Total oxyradical scavenging capacity	Field	PAHs, PCBs
Imposex, intersex	Sex change	Field, laboratory	TBT

Different tools are proposed in order to obtain multiple LOE in sediment quality assessment: (1) sediment chemistry including numeric sediment quality guidelines (SQGs), (2) toxicity tests, (3) bioaccumulation tests, (4) biomarkers, and (5) resident aquatic community structure. These tools should provide adequate estimation of the influence of the physical, chemical, and biological factors in the level of exposure and bioavailability of the different xenobiotics in the sediment (37).

These tools, expressing different lines of evidence, are integrated in environmental risk assessment methodologies and utilized in monitoring and assessment sediment programs. In this regard, a tiered approach to testing is recommended. At each tier, it will be necessary to determine whether sufficient information exists to allow a management decision to be taken or whether further testing is required (8).

The toxicological significance and complexity increase with each tier (Fig. 2). On Tier 0, a recompilation of information that already exists is developed. Tier I testing involves physical–chemical studies and ecotoxicological screening of the dredged material. If toxic effects are observed at Tier I (major hazard concern), the material may be evaluated on this basis. If no toxic effects are observed at this stage, but chemical analyses indicate that there may be reasons for concern, Tier II is entered and further biotests are requested. Tier II testing covers a wide range of effects parameters, including long-term and sublethal toxicity. In most cases, the Tier II results will allow a comprehensive evaluation of the (dredged) sediment. Only if the results are still inconclusive will it be necessary to add further tests on specific effects depending on evidence for particular contaminants, assessment of biomarkers of exposure and effect, and/or verification of laboratory measurements in the field (Tier III). Tier III testing may be used to monitor possible impacts of dredging operations in the field. The available techniques comprise nonspecific monitoring methods that respond to a wide range of environmental contaminants, biomarkers (fish, shellfish) of specific exposures depending on evidence for particular (bioaccumulation) contaminants, and assessment of long-term effects on benthos community structure and function (33).

FUTURE RESEARCH NEEDS

The evaluation of sediment quality in the environmental risk assessment context requires the use of more sensitive tools for sediment evaluation and the extrapolation to the field. The use of bioaccumulation and biomarkers in different bioindicator species is being researched for the assessment and management of sediment. There is an increasing consensus and interest in the valuable information that they could provide. Nevertheless, although they are more sensitive tools, they show higher variability with respect to the physical–chemical characteristics of the sediment, their extrapolation from laboratory to the field, and their intercalibration.

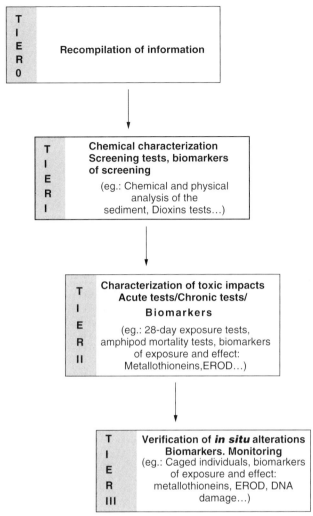

Figure 2. Schematic representation of the tier testing steps including biomarkers and bioaccumulation measurements.

In summary, some recommendations are proposed for the incorporation of biomarkers and bioaccumulation in research programs for the evaluation of dredged material quality using a weight of evidence approach:

1. Determination of biomarker screening responses to contaminated and control sediments for different sites and organisms in order to determine the advantages and disadvantages of the use of different biomarkers for the identification of different sources of contamination.
2. Study of the bioavailability of contaminants in sediment and biomarker responses through the exposure–dose–response triad methodology (exposure–chemical/biomarkers of exposure and bioaccumulation; dose/response—biomarkers of effect). It could allow the analysis of the sensitivity of different biomarkers to different availabilities of contaminants in sediment and to different contamination.
3. Measurement of biomarkers and bioaccumulation response to natural environmental conditions (pH, temperature, salinity). It could allow the application of biomarkers to environments that support high variability of environmental conditions, avoiding the "noise" that could produce this variability in a biomarker's response.
4. Determination of biomarkers and bioaccumulation response over time through a toxicokinetic study of these responses. It could provide knowledge of biomarker and bioaccumulation behavior in the different organisms.
5. Intercalibration of bioaccumulation and biomarker responses; determination of biomarkers and bioaccumulation of the same contaminated sites in different laboratories. It could provide knowledge of the biomarkers and bioaccumulation response variation due to handling of organisms and methodology of measurement.
6. Biomarkers and bioaccumulation extrapolation from laboratory to field conditions through the determination of biomarkers and bioaccumulation response to laboratory and field contaminated toxicity exposure. These measurements could allow the analysis of responses due to realistic environmental conditions. It should be developed for three species: filter feeding, deposit feeding, and burrowing. The link between both sets of data will allow derivation of tissue quality values that need to be validated for human health and regulatory purposes.

Acknowledgments

This research has been partially supported by Plan Nacional I + D + I, REN2002-01699/TECNO, Ministerio de Fomento and the Spanish Ministry of Development (BOE 13-12-02) and of Education and Science (VEM2003-20563/INTER).

BIBLIOGRAPHY

1. Chapman, P.M. (1998). Death by mud: amphipod sediment toxicity test. In: *Microscale Testing in Aquatic Toxicology*. P.G. Wells, K. Lee, and C. Blaise (Eds.). CRC Press, Boca Raton, FL.
2. Andersen, B. et al. (1998). Comparison of marine sediment toxicity test protocols for the amphipod *Rhepoxonius abronius* and the polychaete worm *Nereis (Neanthes) arenaceodentata*. *Environ. Toxicol. Chem.* **17**: 859–866.
3. Environment Canada (2004). *Quality Guidelines*. Environmental Protection Publications.
4. Depledge, M.H. and Hopkin, S.P. (1995). In: R.A. Linthurst, P. Bourdeau, and R.J. Tardiff (Eds.). John Wiley & Sons, Chichester, UK.
5. Carr, R.S. (1998). Marine and estuarine porewater toxicity testing. In: *Microscale Testing in Aquatic Toxicology*. P.G. Wells, K. Lee, and C. Blaise (Eds.). CRC Press, Boca Raton, FL, pp. 523–538.
6. Forbes, T.L., Forbes, V.E., Giessing, A., Hansen, R., and Kure, L.K. (1998). Relative role of pore water versus ingested sediment in bioavailability of organic contaminants in marine sediments. *Environ. Toxicol. Chem.* **17**: 2453–2462.
7. Weston, D.P. and Mayer, L.M. (1998). *In vitro* digestive fluid extraction as a measure of the bioavailability of sediment-associated polycyclic aromatic hydrocarbons: sources of variation and implications for partitioning models. *Environ. Toxicol. Chem.* **17**: 820–829.

8. U.S. Environmental Protection Agency and U.S. Army Corps of Engineers (U.S. EPA and U.S. ACE) (1998). *Evaluation of Dredged Material Proposed for Discharge in Waters of the US: Testing Manual*, EPA/823-B-94/002. EPA, Washington, DC.
9. DelValls, T.A., Casado-Martínez, M.C., Martín-Díaz, M.L., and Riba, I. IV (2004). Workshop SedNet.
10. UNEP (1997). *Report of the Meeting of Experts to Review the MEDPOL Biomonitoring Programme*, Athens, Greece. UNEP- (OCA)/MED WG, pp. 132–137.
11. CEDEX (2003). Recomendaciones para la gestión del material de dragado en los puertos Españoles. Centro de Estudios y Experimentación de Obras Públicas, Puertos del Estado, Madrid.
12. Meussy, J.J. and Payen, G.G. (1988). Female reproduction in malacostracan crustacea. *Zool. Sci.* **5**: 217.
13. Lafontaine, Y. et al. (2000). Biomarkers in zebra mussels (*Dreissena polymorpha*) for the assessment and monitoring of water quality of St. Lawrence River (Canada). *Aquat. Toxicol.* **50**: 51–71.
14. Regoli, F. et al. (2002). Application of biomarkers for assessing the biological impact of dredged materials in the Mediterranean: the relationship between antioxidant responses and susceptibility to oxidative stress in the red mullet (*Mullus barbatus*). *Mar. Pollut. Bull.* **44**: 912–922.
15. Furley, T.H. and Niencheski, L.F. (1993). Temporal and spatial variability of Cd, Pb, Zn, Cu and Mn concentration in mussels (*Perna perna*, Linne, 1758) from the coast of Rio Grande do Sul, Brazil. In: *Proceedings of the International Symposium on Perspectives for Environmental Geochemistry in Tropical Countries*. J.J. Abrao, J.C. Wasserman, and E.V.F. Silva (Eds.). Universidade Federal Fluminense, Niteroi, R.J., Brazil, pp. 39–40.
16. Neff, M. (2004). *Bioaccumulation in Marine Organisms*. Elsevier Ltd., Oxford, UK.
17. Borgmann, U. (2000). Methods for assessing the toxicological significance of metals in aquatic ecosystems: bioaccumulation–toxicity relationships, water concentrations and sediment spiking approaches. *Aquat. Ecosyst. Health Manage.* **3**: 277–289.
18. Farrington, J.W. and Westal, J. (1986). Organic chemical pollutants in the oceans and groundwater: A review of fundamental chemical properties and biogeochemistry. In: *The Role of the Oceans as a Waste Disposal Option*. G. Kullenberg (Ed.). Reidel Publishing, New York, pp. 361–425.
19. Tillit, D.E. et al. (1992). Polychlorinated biphenyl residues and egg mortality in double-crested cormorants from the Great Lakes. *Environ. Toxicol. Chem.* **11**: 1281–1288.
20. NRC: Committee on Biological Markers of the National Research Council (1987). Biological markers in environmental health research. *Environ. Health Perspect.* **74**: 3–9.
21. WHO: International Programme on Chemical Safety (IPCS) (1993). *Biomarkers and Risk Assessment: Concepts and Principles*, Environmental Health Criteria 155. World Health Organization, Geneva.
22. Van Gestel, C.A.M. and Van Brummelen, T.C. (1996). Incorporation of biomarker concept in ecotoxicology calls for a redefinition of terms. *Ecotoxicology* **5**: 217–225.
23. Van der Oost, R., Beyer, J., and Vermeulen, N.P.E. (2003). Fish bioaccumulation and biomarkers in environmental risk assessment. *Environ. Toxicol. Pharmacol.* **13**: 57–147.
24. Riba, I., Forja, J.M., Gómez-Parra, A., and Del Valls, T.A. (2004). Sediment quality in littoral regions of the Gulf of Cádiz: a triad approach to address the influence of mining activities. *Environ. Pollut.* 341–353.
25. Maycock, D.S. et al. (2003). Incorporation of *in situ* and biomarker assays in higher tier assessment of the aquatic toxicity of pesticides. *Water Res.* **37**: 4180–4190.
26. Hyne, R.V. and Maher, W.A. (2003). Invertebrate biomarkers: links to toxicosis that predict population decline. *Ecotoxicol. Environ. Safety* **54**: 366–374.
27. Peakall, D.W. (1994). Biomarkers: the way forward in environmental assessment. *Toxicol. Ecotoxicol. News* **1**: 55–60.
28. Suter, G.W. (1993). *Ecological Risk Assessment*. Lewis Publishers, Boca Raton, FL.
29. Stegeman, J.J., Brouwer, M., Richard, T.D.G., Förlin, L., Fowler, B.A., Sanders, B.M., van Veld, P.A. (1992). In: *Biomarkers: Biochemical, Physiological and Histological Markers of Anthropogenic Stress*. R.J. Huggett, R.A. Kimerly, P.M. Mehrle Jr., and H.L. Bergman (Eds.). Lewis Publishers, Chelsea, MI, p. 235.
30. Pedersen, S.N., Lundebye, A.K., and Depledge, M.H. (1997). Field application of metallothionein and stress protein biomarkers in the shore crab (*Carcinus maenas*) exposed to trace metals. *Aquat. Toxicol.* **37**: 183–200.
31. Fossi, M.C., Savelli, C., and Casini, S. (1998). Mixed function oxidase induction in *Carcinus aestuarii*. Field and experimental studies for the evaluation of toxicological risk due to Mediterranean contaminants. *Comp. Biochem. Physiol.* **121C**: 321–331.
32. Stebbing, A.R.D. and Dethlefsen, V. (1992). Introduction to the Bremerhaven workshop on biological effects of contaminants. *Mar. Ecol. Prog. Ser.* **91**: 1–8.
33. OSPAR (1995). *Report of the workshop on biological effects monitoring techniques*. Abeerden, Scotland, 2–6 October.
34. Fredrickson, H.L. et al. (2001). *Sci. Total Environ.* **274**: 137.
35. Martín Díaz, M.L., Sales, D., and Del Valls, T.A. (2004). Influence of salinity in hemolymph vitellogenin of the shore crab *Carcinus maenas* to be used as biomarker of contamination. *Bull. Environ. Contam. Toxicol.*, in press.
36. Martín Díaz, M.L., Blasco, J., Sales, D., and DelValls, T.A. (2004). *International Conference on Remediation of Contaminated Sediments*. Batelle Press, Columbus, OH.
37. Wenning, R.J. and Ingersoll, C.G. (2002). Summary booklet of a SETAC Pellston workshop on use of sediment quality guidelines and related tools for the assessment of contaminated sediments; Fairmont, Montana. Society of Environmental Toxicology and Chemistry (SETAC), Pensacola, FL.

LEAD AND ITS HEALTH EFFECTS

SHAHIDA QUAZI
DIBYENDU SARKAR
RUPALI DATTA
University of Texas
San Antonio, Texas

INTRODUCTION

Lead (Pb) is a bluish white, lustrous, naturally occurring element that people have used almost since the beginning of civilization. Because of its malleability, Pb was one of the first metals that humans began to use for industrial purposes. Plumbism, or Pb poisoning, has been recognized for many centuries and is one of the oldest ailments afflicting humans. Nikander (200 B.C.) was first to describe the symptoms of Pb poisoning, which included colic and pallor. The English aristocracy of the seventeenth and eighteenth centuries suffered from widespread Pb poisoning from consumption of Portuguese wine, transported with submerged Pb bars to enhance flavor and to prevent spoilage. Lancereaux provided the first description of kidney disease and interstitial nephritis by postmortem examination of a Pb poisoned artist.

Anthropogenic activities during the past century have resulted in increased accumulation of toxic metals in soils and natural waters (1). Lead contamination is widespread in urban areas due to industrial emissions, extensive use of alkyl-lead compounds as antiknock additives in gasoline in the past, and lead-based paints and pipes (2), resulting in soil contamination on a global scale with adverse effects on human and environmental health (3).

Lead occurs naturally in low concentrations in all rocks, soils, and dust, usually ranging from 2 to 200 ppm (4). The periodic position of Pb favors the formation of covalent rather than ionic bonds in Pb^{4+} compounds. Lead is similar to certain alkaline earth metals, particularly calcium. Both calcium and Pb form insoluble carbonates and phosphates. However, Pb carbonates and phosphates are less soluble compared to those of calcium. This insolubility of Pb together with its high affinity to biological donors (enriched in oxygen and possibly nitrogen) accounts for its toxicological properties (5).

Here, we provide an overview of the various sources of Pb, its impact on the ecosystem and its effects on human health. Future directions in research related to the effects of Pb on human health are also discussed.

SOURCES OF LEAD

Natural Sources

The average concentration of Pb in the earth's crust is estimated to be 16 mg/kg (4). Lead is mainly found in the form of its sulfide, galena. Some other ores of Pb include cerussite and anglesite. In general, granitic rocks that have high acidic content tend to have more Pb than basaltic rocks. Lead concentrations in soils (farming and city) range from 2 to 200 mg/kg, with most soils having Pb in the range of 5–25 mg/kg (6). Moreover, soils that are underlain by granitic rocks with high acid content have higher Pb concentrations than those overlying basaltic and other alkaline rocks (6). It has been estimated that global natural emissions of Pb to the atmosphere are in the range of $18.6-29.5 \times 10^6$ kg/yr.

Anthropogenic Sources

The major anthropogenic sources of Pb are industrial activities such as mining operations (nonferrous metal, secondary Pb smelters), vehicle emissions, coal burning, refuse incineration, and industrial outdoor paints applied to structural surfaces. However, with the phasing out of Pb in gasoline (which began in the 1970s), Pb in soils and house dusts originating from lead-based paints have become the principal sources of exposure in the United States (7).

In rivers or water bodies, the concentration of Pb mainly depends on local inputs. Since most Pb compounds have low solubility, high concentrations are very rare. However, they are found in acid mine drainage in high concentrations mostly in areas of Pb mineralization (8). Lead shots used for hunting and fishing can also result in significant Pb input to both soil and water and contribute to inputs to surface waters (9), ultimately adding to the Pb burden in the sediments.

MOVEMENT OF LEAD IN THE ENVIRONMENT

The passage of metals through the atmosphere is integral to their biogeochemical cycling. Because of the dynamic nature of the atmosphere, metals can be deposited in areas remote from their initial source. Air is the primary medium for the transport of Pb. Fine particles of Pb generated anthropogenically are transported over long distances until they settle through wet, dry, or cloud deposition (5). Lead transport in the environment depends on the size of the particles. While larger particles (comprising 95% of the emission) settle down after sometime, the smaller Pb particles can travel hundreds of kilometers before they are deposited (4). Lead in the soil and dust is relatively unavailable biologically. It normally reacts with the soil to form insoluble salts. Moreover, Pb also complexes with components of organic matter such as humic acid and fulvic acid, thus making Pb more unavailable (6). The downward movement of Pb from soil by leaching is very slow under most natural conditions (10). Mobility of Pb in the soil is dictated by a number of geochemical process, including oxidation/reduction, precipitation/dissolution, adsorption/desorption, and complexation/chelation. Metallic Pb can be oxidized to more labile forms, the rate of oxidation and the oxidation products being dependent on the site. Once Pb is oxidized, it can be precipitated in the form of hydroxides, sulfates, sulfides, carbonates, and phosphates. The solubility of these precipitates is dependent on factors such as pH and oxidation–reduction conditions. In general, Pb is more soluble under acidic conditions compared to neutral and alkaline conditions. Most of the Pb accumulated in the soil may be directly taken up by grazing animals and microorganisms or may be ingested by children through hand-to-mouth activity. From the grazing animals Pb may enter the food chain (5).

Lead in rivers comes from runoff, erosion, and direct deposition from air. Fresh water generally contains more inorganic and organic suspended material than marine water, and this suspended material has a tendency to adsorb any dissolved Pb. In the case of the oceans, most of the Pb comes from atmospheric deposition rather than from rivers (4).

ECOTOXICOLGICAL EFFECTS OF LEAD

The uptake of Pb in both aquatic and terrestrial ecosystems is determined by its bioavailability. The term "bioavailability" reflects the premise that, for some heavy metals, organisms may be exposed to less than the total amount present in their habitat (11). The extent to which heavy metals react with and affect biological components is immensely influenced by the physicochemical factors of the specific environment (12). In general, the systemic availability of Pb is matrix and chemical species dependent (13–15).

Toxicity to Aquatic Organisms

Dose-related increases in lead concentrations of submerged and floating plants have been observed upon their exposure to dissolved Pb, accumulation being the greatest in the roots, followed by stems and leaves (16). In a study conducted by Kay et al. (17) water hyacinth (*Eichhornia crassipes*) was exposed to solutions containing lead nitrate concentrations of 0–5 mg/L for 6 weeks. The accumulation of Pb was dose-dependent and in the order of roots > stems > leaves. Uptake of Pb from shot-contaminated sediments has been observed in aquatic plants. In an investigation of a skeet target shooting range located on the shores of Lake Merced, California, sediments contained Pb levels up to 1200 µg/g in the shotfall zone, and tule seedheads and coontails growing within these sediments exhibited Pb concentrations averaging 10.3 and 69.2 µg/g dry weight, respectively, compared to concentrations of 2.3 and 11.9 µg/g dry weight, respectively, at control sites (18). However, Van der Werff and Pruyt (19) reported that application of lead nitrate with a Pb concentration of up to 10^{-5} mol/L had no observable toxicity on four aquatic plants, *Elodea gibba*, *Callitiche platycarpa*, *Spirodela polyrhiza*, and *Leman gibba*. In the case of fishes, it has been found that *in vivo* sublethal dietary doses of Pb (II) and tributyltin (TBT) induced toxic effects in *H. malabaricus* (20). Davis et al. (21) conducted long-term bioassays with rainbow trout to establish a maximum acceptable toxicant concentration (MATC) limit for inorganic Pb. Fingerling trout were exposed to nominal total Pb concentrations of 0, 40, 120, 360, 1080, or 3240 mg/L. Actual dissolved Pb was measured and results were expressed in terms of dissolved salt. They reported that MATC values between 0.018 and 0.032 mg/L caused the "black tail effect." A similar bioassay with soft water in fishes hatched from exposed eggs suggested MATC values between 0.041 mg/L (where no black tails occurred) and 0.076 mg/L (where 4.7% of fish showed the black tail effect). When fingerlings from nonexposed eggs were used in a soft water bioassay, the MATC for the black tail effect was between 0.072 mg/L (where no black tails occurred) and 0.146 mg/L (where 41.3% of fish had black tails). There were no significant differences between the measured dissolved Pb concentrations in the two tests, indicating that fish from exposed eggs were more sensitive to the effects of Pb than those from nonexposed eggs.

Maddock and Taylor (22) investigated the uptake of organolead by shrimp, mussel, and dab (a flat fish) in short-term experiments and in mussel and dab in long-term experiments. Bioconcentration factors were higher for tetraalkyl Pb than for trialkyl Pb in the short-term exposure. In the long-term experiment conducted on mussels, the greatest tissue concentration of Pb occurred in the gills; with the digestive glands, gonads, and feet containing progressively less Pb (17).

Toxicity to Terrestrial Organisms

Toxicity to Plants. Although Pb is considered to be toxic to plants just like other life forms, sensitivity of plants to Pb and responses of different plant species to Pb vary according to their genetic and physiological makeup (23). The metal remains largely as a superficial deposit or topical aerosol coating on plant surfaces (24,25). Toxicity of Pb to a large extent depends on its absorption, transport, and cellular localization (23). Plants of *Cassia tora* and *C. occidentalis* growing by the roadside accumulated up to 300 mg/g dry weight of Pb (26). Furthermore, accumulated Pb content generally increases with the increase in Pb in the environment, as has been reported for maize and pea leaves and *Vigna* and sesame roots (23) and leaves (27, 28). Photosynthesis is considered as one of the most sensitive metabolic processes to Pb toxicity (23). Moreover, nitrate reduction is inhibited drastically in roots by the metal, but in the leaves a differential effect is observed in various cultivars (23). Nitrate assimilation is inhibited by 10–100 mM Pb^{2+} in sorghum leaves (29). Lead also inhibits nodulation, N fixation, and ammonium assimilation in the root nodules (23).

Toxicity to Invertebrates. Long-term exposure of terrestrial invertebrates to lead-contaminated soil can result in high tissue concentrations, but one of the detoxification mechanisms within their cells is the subcellular compartmentalization of Pb. This results in only a fraction of the total body Pb burden actually contributing to toxicity (30). Doelman et al. (31) incubated a mixed culture of bacteria in lead nitrate solutions and also grew the fungus *Alternaria solani* on malt agar to which lead nitrate had been added. The cultures were used as food for the nematodes *Mesorhabditus monohystera* and *Aphelenchus avenae,* which were reared for up to 22 days on bacteria and fungus, respectively. Lead was taken up by bacteria to give a range of doses to the nematodes of between 7.6 and 110 µg/g of food. All these exposures had a significant inhibitory effect on the reproduction of *Mesorhabditus monohystera*. However, when woodlice *(Porcellio scaber)* were exposed to treated soil litter containing between 100 and 12,800 mg/kg dry weight of Pb as lead oxide over 64 weeks, no significant effect was found on adult survival, number of young produced, or the survival of the young at exposures up to 6400 mg lead/kg (32).

Toxicity to Birds. In the case of birds, metallic Pb is highly toxic when it is given in the form of Pb shots. Lead poisoning through the ingestion of spent gunshots is a widely recognized waterfowl mortality factor (33). Water fowl ingest spent gunshot when searching for grit (and possibly for food) particles. Once in the bloodstream, Pb is rapidly (but reversibly) deposited in soft tissues, mainly liver and kidney, and (relatively irreversibly) into bone (34). If a large quantity of Pb (e.g., >10 shots) is ingested within a short time, acute Pb poisoning may occur, resulting in death within a few days. However, most of the mortality results from the ingestion of a smaller number of shots, with birds dying of chronic poisoning 2–3 weeks following ingestion (34). The U.S. Fish and Wildlife Service (35) reported that Pb poisoning from ingested sinker and jigs accounts for 10–15% mortality of common loons (*Gavia immer*) with a study revealing that 27% of adult loons had fishing tackle in their stomachs and high Pb levels in their blood. Signs of chronic Pb poisoning include green and watery feces, dropping wings, anemia, weight loss, and atypical behavior; the affected birds die approximately 2–3 weeks after ingesting the shot, often in a very weak condition (36). Secondary Pb shot poisoning can occur when a predator or scavenger consumes animals that have been shot with Pb shot shell ammunition or consumes the gizzard of a bird that has ingested Pb shot (36). Lead also causes a decrease in clutch and egg size, mortality of embryos, depression of growth, and deficits in behavior, thus affecting the survival of birds (37).

EFFECTS ON HUMANS

Lead is present in various forms in the environment, such as free hydrated ions, ion-pair salts/complexes, and organic complexes/chelates (38). Humans are most likely to be exposed to the inorganic forms of Pb such as halides, oxides, sulfides, carbonates, and chromates. Since Pb was phased out as a gasoline additive (tetraethyl Pb) in the 1970s and its use in paint and food containers (e.g., ceramic ware and tin cans) was curtailed, blood Pb concentrations (BLLs) have decreased significantly. However, other sources of Pb and its unknown threshold of subclinical toxicity continue to make Pb an issue of public health concern (39). Lead exposure in the general population (including children) occurs primarily through ingestion, although inhalation also contributes to Pb body burden and may be the major contributor for workers in Pb-related occupations (40). Children's hand-to-mouth activity, increased respiratory rates, and increased intestinal absorption of Pb make them more susceptible to Pb exposure than adults (41,42). Exposure to Pb can result in significant adverse health effects to multiple organ systems. Once Pb is absorbed into the blood plasma, it rapidly equilibrates with the extracellular fluid. More slowly, but within minutes, Pb is transferred from plasma into blood cells (43,44). From the blood plasma, the absorbed Pb is transferred to the liver and the kidneys. Lead tends to accumulate in areas with high levels of calcium; hence, the highest body burden of Pb is found in the bones. In a typical adult, 95% of the Pb is stored in the bones, whereas in children, it amounts to 70% (45,46). During childhood, Pb accumulates predominantly in the trabecular bones. But in the case of adults most of the Pb is found in both cortical and trabecular bones. In both the trabecular and cortical bones there are two physiological compartments, one labile and the other inert for Pb (7). While Pb may be exchanged between the blood and the bone through the labile component, Pb may be stored for years in the inert component (7).

The distribution of Pb in tissues reflects a state of constant, dynamic equilibrium. Any situation that mobilizes the very large, relatively stable pools of Pb within the body, particularly those in the bones, will lead to the redistribution of Pb to a variety of tissues. Redistribution is known to occur during pregnancy, which results in increased risk to the fetus, particularly in women with prior Pb poisoning (5).

The various detrimental human health effects of Pb include:

- Neurological effects including effects on the central and peripheral nervous systems.
- Effects on both male and female reproduction.
- Renal effects.
- Effects on vitamin D metabolism.
- Cardiovascular effects.
- Hematopoietic effects.

Neurological Effects

Lead affects both the central and peripheral nervous systems. An extensive database has provided a direct link between low level Pb exposure and deficits in the neurobehavioral–cognitive performance evidenced in childhood through adolescence (47). Many of the biological aberrations produced by Pb appear to be related to the ability of this heavy metal to either inhibit or mimic the action of calcium. Calcium ions play a special role in the release of neurotransmitters from presynaptic nerve endings (48). A large number of experimental studies have investigated the effects of prenatal and neonatal Pb exposure on central nervous system development and behavior. These studies have shown variable results, forming a spectrum that includes delayed nervous system development, deficits in visual motor function, abnormal social and aggressive behavior, hyperactivity and hypoactivity, as well as no changes in activity (49). The brain and the nervous systems of children and fetuses are generally vulnerable to Pb intoxication because of the incomplete blood–brain barrier and also because both the brain and the nervous system are still developing (40).

Effects on Central Nervous System

In the central nervous system, Pb toxicity is more common in children than adults and may produce either overt symptoms of acute encephalopathy, such as ataxia, headache, convulsions, and coma, or lesser deficits including learning disorders and hyperactive behavior (50,51). Lead disrupts the main structural components of the blood–brain barrier through primary injury to astrocytes with a secondary damage to the endothelial microvasculature. Within the brain, lead-induced damage occurs preferentially in the prefrontal cerebral cortex, hippocampus, and cerebellum (47) Lead enhances the spontaneous or basal release of neurotransmitters from presynaptic nerve endings at low concentrations (48). In addition to enhancing the spontaneous release of neurotransmitters, Pb blocks the release of neurotransmitters normally produced by depolarization of nerve endings. It modulates neurotransmitter release by altering calcium metabolism either by competing with calcium for entry into the cell or by increasing intracellular calcium levels (48). Moreover, high dose exposure to Pb (i.e., BLLs in excess of

4 μM) disrupts the blood–brain barrier (48). In general, blood–brain barrier excludes plasma proteins and most organic molecules and limits even the passage of ions such as sodium and potassium (52). However, at high BLLs large molecules such as albumin, which are normally excluded, freely enter the brain of immature animals exposed to high concentrations of Pb (53,54). Ions and water follow and edema is produced, resulting in the rise of intercranial pressure because of the physical restraint of the skull. When the intercranial pressure approaches the systemic blood pressure, cerebral perfusion decreases and brain ischemia occurs (48).

Children exposed to high levels of Pb suffer from encephalopathy and hyperirritability, ataxia, convulsions, stupor, and coma or death. In general blood, BLLs of about 70–80 μg/dL or greater in children pose a threat (7). However BLLs as low as 10 μg/dL or less can adversely affect the developing nervous system of a child (40). Studies have shown that for every 10 μg/dL increase in the BLL, there is a decrease of children's IQ by about four to seven points (55–60). In the case of adults, Pb encephalopathy may occur at BLLs of about 460 μg/dL (61).

Effects on Peripheral Nervous System

Peripheral neuropathy is a manifestation of Pb toxicity in adults with excessive occupational exposure. Children exposed to high levels of Pb may also suffer from peripheral neuropathy. While in adults the most common manifestation of peripheral neuropathy is wrist drop, in the case of children, general weakness and foot drop are more common (62).

Effects of Lead on Reproduction

Male Reproductive System. An adverse effect of Pb on the male reproductive system has been reported in several epidemiologic studies (63). It has been reported that BLLs below the currently accepted working protection criteria adversely affect spermatogenesis (64). Earlier studies on Pb workers had shown androgenic dysfunction including asthenospermia, hypospermia, and teratospermia that could be produced by direct toxicity to the testis (65). It has been shown that significant levels of asthenospermia and teratospermia were found in workers having BLLs over 500 μg/L (66). The effect of Pb poisoning on the male reproductive system can be seen at BLLs of about 40 μg/dL, with long-term Pb exposure resulting in diminishing sperm counts, sperm concentrations, and total sperm motility (7). Moreover, it was found that the risk of stillbirth or birth defects was elevated for preconception employment in a high Pb exposure environment compared with low Pb exposure jobs (67).

Female Reproductive System. There have been few well-documented reports on reproductive effects of Pb in humans (68). Exposure may result in increased preterm delivery. It was found that preterm delivery was statistically significantly correlated with maternal BLLs in a dose-responsive manner (69). Moreover, due to changes in the bone physiology and mineral metabolism during pregnancy, Pb may move from the bone into the maternal circulation. Lead does cross the human placenta as early as the 12th week of gestation (70). Cord-blood content of Pb shows a high correlation with maternal-blood samples, suggesting that infants are exposed in uterus to BLLs equivalent to those of the mother (71). Recent isotopic speciation studies have demonstrated that the skeletal contribution to BLLs increases from 9% to 65% during pregnancy (72). The maternal Pb burden has been reported to be negatively associated with the birth weight of infants (73). It has been found that prenatal exposure to low Pb levels (e.g., maternal BLLs of 14 μg/dL) may increase the risk of reduced birth weight and premature birth (7).

Effects on the Renal System

Lead affects the renal system in three stages. In the first stage, there is proximal tubular dysfunction (Fanconi's syndrome) manifested by aminoaciduria, glycosuria, and phosphaturia (74). Then chronic exposure results in the second stage that is characterized by gradual tubular atrophy and interstitial fibrosis. There is reduced incidence of inclusion bodies, and glomerular filtration is impaired (75). At the third stage, renal failure occurs and is characterized by renal tubular neoplasia or adenocarcinoma (76). In general, individuals with Pb levels exceeding 60 μg/dL are at a risk of developing renal failure (77,78).

It has been reported that Pb inhibits rBAT-induced amino acid transport in a noncompetitive, allosteric fashion. This blockade of rBAT-induced amino acid transport may be involved in aminoaciduria following Pb intoxication (79). Lead exposure also results in the onset of "saturnine gout" as a result of lead-induced hyperuricemia due to decreased renal excretion of uric acid. Most documented renal effects for occupational workers are observed in acute and high to moderate chronic exposures with BLLs of more than 60 μg dL (40).

Endocrine Effects/Effects on Vitamin D Metabolism

From the studies of children exposed to high Pb levels, it has been found that a correlation exists between BLLs and the level of vitamin D. Lead interferes with the conversion of vitamin D to its hormonal form, 1,25-dihydroxyvitamin D, which is responsible for the maintenance of extracellular and intracellular calcium homeostasis. Thus, impairment by Pb results in impaired cell growth, maturation, and tooth and bone development (40).

Cardiovascular Effects

Lead exposure is one of the factors that may contribute to the onset of hypertension. Other factors contributing to the development of hypertension include old age, increased weight, poor diet, and excess alcohol intake (40). Studies have shown that greater exposure to Pb in occupational environments may increase the risk of hypertensive heart disease and cardiovascular disease. There are several reports that have associated Pb exposure to elevation in blood pressure (80–83).

Hematopoietic Effects

Lead inhibits the body's ability to make hemoglobin by interfering with several enzymatic steps in the heme pathway (40). Heme is the prosthetic group of hemoglobin, myoglobin, and cytochromes. In mammals, heme is synthesized from succinyl-CoA and glycine in eight enzyme-mediated steps (Fig. 1) (84). Enzyme studies of the heme biosynthetic pathway have shown that Pb is an inhibitor of δ-aminolevulinate dehydratase (ALAD), corporphyrinogen oxidase, and ferrochelatase (85). However, the metal has the greatest influence on ALAD, and measurement of ALAD activity can be used as an indicator of Pb levels in the blood (84). Heme synthesis occurs partly in mitochondria and partly in cytoplasm. The three important steps in the synthesis of heme that are influenced by Pb intoxication are:

1. The condensation reaction of two molecules of δ-aminolevulinic acid that is catalyzed by PBG synthase (porphobilinogen synthase) or ALA dehydratase to form porphobilinogen (PBG). ALA-D is the enzyme that is most sensitive to Pb. Inhibition of the enzyme results in the prevention of the utilization of ALA and subsequently a decline in heme synthesis (68).
2. The rate-limiting step in heme biosynthesis is the ALA synthase catalyzed step. The oxidation product of heme is hemin, which acts as a negative feedback inhibitor of ALA synthase and inhibits the transport of ALA synthase from the cytosol to the mitochondria and also represses the synthesis of the enzyme. Lead interferes with the negative feedback control of heme synthesis; ALA synthetase activity is depressed, resulting in increased activity of the enzyme and increased synthesis of ALA (68).
3. The third reaction that is influenced by the intoxication of Pb is the incorporation of the ferrous iron into the porphyrin ring structure that is catalyzed by the enzyme ferrochelatase. Lead inhibits the enzyme ferrochelatase, thus preventing the introduction of iron into the protoporphyrin IX to form heme. The porphyrin chelates with Zn nonenzymatically to form ZnPP, which in turn is incorporated into hemoglobin. It has been found that ZnPP containing hemoglobin has a much lower oxygen capacity than Fe-containing hemoglobin (68). Moreover, iron in the form of ferritin and ferruginous micelles also accumulate in the mitochondria of bone marrow reticulocytes (68) (Fig. 1). In general, Pb induces two types of anemia. While acute, a high Pb level results in hemolytic anemia; chronic Pb exposures result in anemia that is induced by the interference of Pb with heme biosynthesis and also by diminished red blood cell survival. According to the EPA, in the case of occupationally exposed adults, the threshold BLL for a decrease in hemoglobin is 50 μg/dL, whereas in children, the threshold BLL has been observed to be 40 μg/dL (7).

FUTURE DIRECTION

Lead is ubiquitous in nature, but anthropogenic activities have further exacerbated Pb levels in the environment. Although lead-based petroleum has been phased out, Pb is still being used in the industry (such as lead-acid battery manufacturing industry) due to the absence of viable alternatives. Blood lead levels of 10 μg/dL have been associated with adverse health effects in children (86). The treatment of Pb poisoning involves chelation, which accelerates the process of reducing the Pb levels in the circulating blood. In children, chelation therapy is recommended only when the BLLs exceed 45 μg/dL (86). There is an abundance of evidence demonstrating that dietary calcium decreases gastrointestinal Pb absorption,

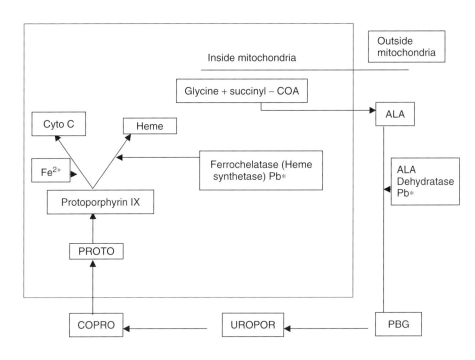

Figure 1. Scheme of heme synthesis showing sites where lead has an effect. PGB, phorbobilinogen; UROPOR, uroporphyrinogen; COPRO, coproporphrinogen; PROTO, protophorphyrinogen; Co A, coenzyme A; ALA, δ—aminolevulinic acid; Pb* site for Pb effect (from Ref. 68).

thereby reducing Pb toxicity (87). Experiments carried out on rats suggest that bone Pb concentrations increased by about fourfold in rats fed with a low calcium diet, compared to rats on a normal calcium diet, although the amounts of Pb ingested were equal (88). Short-term intake of calcium does not alter BLLs, but reports suggest that under conditions of physiological stress, skeletal minerals can be mobilized for calcium and that Pb will also be released along with calcium during this mineral mobilization (72,89,90). Ongoing research is now addressing whether calcium intake will reduce this mobilization.

Epidemiological studies have shown correlations between low levels of Pb contamination in the environment and subtle neurobehavioral effects in children. However, the biochemical mechanisms forming the basis of these subtle neurotoxic effects of low concentrations of Pb in the environment have not been clearly established (5). This is another area where research needs to be focused, to find out the biochemical interactions that may be primarily responsible for the neurotoxic effects of Pb.

BIBLIOGRAPHY

1. Pinheiro, J.P., Mota, A.M., and Benedetti, M.F. (1999). Lead and calcium binding to fulvic acids: salt effect and competition. *Environ. Sci. Technol.* **33**: 3398–3404.
2. Callender, E. and Rice, K.C. (2000). The urban environmental gradient: anthropogenic influences on the spatial and temporal distribution of lead and zinc in sediments. *Environ. Sci. Technol.* **34**(2): 232–238.
3. Markus, J. and McBratney, A.B. (2001). A review of the contamination of soils with lead II. Spatial distribution and risk assessment of soil lead. *Environ. Int.* **27**: 399–411.
4. Thorton, I., Rautia, R., and Brush, S. (2001). Lead the facts. Prepared by IC Consultants Ltd., London, UK. Available at http://www.ldaint.org/factbook/.
5. Mahaffey, K.R., McKinney, J., and Reigart, J.R. (2000). Lead and compounds. In: *Environmental Toxicants: Human Exposures and Their Health Effects*. M. Lippmann (Ed.). John Wiley & Sons, Hoboken, NJ, pp. 481–553.
6. Lansdown, R. and Yule, W. (1986). *Lead Toxicity, History of Environmental Impacts*, The John Hopkins University Press, Baltimore.
7. Agency for Toxic Substances and Disease Registry (ATSDR) (1999). *Toxicological Profile for Lead*. U.S. Department of Health and Human Services, Atlanta.
8. Alloway, B.J. and Ayres, D.C. (1997). *Chemical Principles of Environmental Pollution*. Blackie Academic and Professional, London.
9. Roorda, A.A.H. and Van Der Ven, B.L. (1999). *Lead sheet and the environment*, TNO Institute of Environmental Sciences, Energy Research and Process Innovation, Amsterdam, The Netherlands, Report no. TNOMEP-R98/503.
10. National Science Foundation (NSF) (1977). *Lead in the Environment*. W.R. Bogess (Ed.). NSF/RA-770214. NSF, Washington, DC.
11. Schmitt, C.S., Finger, S.E., May, T.W., and Kaiser, M.S. (1987). Bioavailability of lead and cadmium from mine tailings to the pocket book mussel (*Lampilis ventriosa*) In: *Proceedings of the Die-offs of Freshwater Mussel in the United States*. R. Neves (Ed.). U.S. Fish and Wildlife Service and the Upper Mississippi River Conservation Committee, pp. 115–122.
12. Erten-Unal, M., Wixson, B.G., Gale, N., and Pitt, J.L. (1998). Evaluation of toxicity, bioavailability and speciation of lead, zinc and cadmium in mine/mill wastewaters. *Chem. Speciation Bioavail.* **10**(2): 37–46.
13. LaVelle, J.M. et al. (1991). Bioavailability of lead in mining wastes: an oral incubation study of young swine. *Chem. Spec. Bioavail.* **3**: 105–111.
14. Davis, A., Ruby, M.V., and Bergstrom, P.D. (1992). Bioavailability of arsenic and lead in soils from the Butte, Montana, mining district. *Environ. Sci. Technol.* **26**: 461–468.
15. Gulson, B.L. et al. (1994). Lead bioavailability in the environment of children; blood lead levels in children can be elevated in a mining community. *Arch. Environ. Health* **49**: 326–331.
16. Fleming, S. (1994). *Scientific Criteria Document for Multimedia Environmental Standards Development—Lead*. PIBS 2832, Ontario Ministry of Environment and Energy, March.
17. Kay, S.H., Haller, W.T., and Garrard, L.A. (1984). Effects of heavy metals on water hyacinths (*Eichhornia crassipes* (Mart.) Solms). *Aquat. Toxicol.* **5**: 117–128.
18. Peterson, S., Kim, R., and Moy, C. (1993). *Ecological risks of lead contamination at a gun club: waterfowl exposure via multiple dietary pathways*. Prepared for Society of Environmental Toxicology and Chemistry, by Ecology and Environment Inc., San Francisco.
19. Van Der Werff, M. and Pruyt, M.J. (1982). Long-term effects of heavy metals on aquatic plants. *Chemosphere* **11**: 727–739.
20. Rabitto, I.S. et al. (2004). Effects of dietary Pb(II) and tributyltin on neotropical fish *Hoplias malabaricus*: histopathological and biochemical findings. *Ecotoxicol. Environ. Safety* **60**: 147–156.
21. Davis, P.H., Goettl, J.P., Sinley, J.R., and Smith, N.F. (1976). Acute and chronic toxicity of lead to rainbow trout *Salmo gairdneri*, in hard and soft water. *Water Res.* **10**: 199–206.
22. Maddock, B.G. and Taylor, D. (1980). The acute toxicity and bioaccumulation of some lead alkyl compounds in marine animals. In: *Lead in the Marine Environment*. M. Branica and Z. Konrad (Eds.). Pergamon Press, Oxford, pp. 233–261.
23. Singh, R.P. et al. (1997). Response of higher plants to lead contaminated environment. *Chemosphere* **34** (11): 2467–2493.
24. Schuck, E.A. and Locke, J.K. (1970). Relationship of automotive lead particulates to certain consumer crops. *Environ. Sci. Technol.* **4**: 324–330.
25. Zimdahl, R.L. and Faster, J.M. (1976). The influence of applied phosphorus, manure or lime or uptake of lead from soil. *J. Environ. Qual.* **5**: 31–34.
26. Krishnayappa, N.S.R. and Bedi, S.J. (1986). Effect of automobile lead pollution of *Cassia tora* and *Cassia occidentalis* l. *Environ. Pollut.* **40**: 221–226.
27. Kumar, G., Singh, R.P., and Sushila, X. (1993). Nitrate assimilation and biomass production in *Sesamum indicum* L. Seedlings in a lead enriched environment. *Water, Air Soil Pollut.* **66**: 163–171.
28. Bharti, M.B. and Singh, R.P. (1993). Growth and nitrate reduction by *Sesamum indicum* cv PB-I respond differentially to lead. *Phytochemistry* **33**: 531–534.
29. Venkataraman, S., Veeranianeyulu, K., and Ramdas, V.S. (1978). Heavy metal inhibition of nitrate reductase. *Ind. J. Exp. Biol.* **16**: 615–616.

30. Jones, R., Inouye, L., Bednar, A., and Boyd, R. (2004). Lead toxicity and subcellular compartmentalization in earthworms exposed to contaminated soil. The Society of Environmental Toxicology and Chemistry, 25th Annual Meeting, North America. Available at http://abstracts.co.allenpress.com/pweb/setac2004/document/?ID=42021.
31. Doelman, P., Nieboer, G., Scrooten, J., and Visser, M. (1984). Antagonistic and synergistic toxic effects of Pb and Cd in a simple food chain: nematodes feeding on bacteria or fungi. *Bull. Environ. Contam. Toxicol.* **32**: 717–723.
32. Beyer, W.N. and Anderson, A. (1985). Toxicity to woodlice of zinc and lead oxides added to soil litter. *Ambio* **14**: 173–174.
33. Sanderson, G.C. and Bellrose, F.C. (1986). *A Review of the Problem of Lead Poisoning in Waterfowl*. Illinois Natural History Survey, Champaign, Illinois, Special Publication 4. Northern Prairie Wildlife Research Center Jamestown, ND. Available at http://www.npwrc.usgs.gov/resource/othrdata/pbpoison/pbpoison.htm.
34. Pain, D.J. (1990). Lead shot ingestion by water birds in the Camargue, France: an investigation of levels and interspecific differences. *Environ. Pollut.* **66**: 273–285.
35. United States Fish and Wildlife Service (USFWS) (1999). Establishing "lead free fishing areas" and the prohibition of the use of certain fishing sinkers and jigs made with lead on specific units of the National Wildlife Refuge System. *Fed. Reg.* **64**(70).
36. Lead poisoning report (2000). 3rd International Update Report on Lead Poisoning in water birds. Available at http://www.wetlands.org/pubs&/Lead_P_Report.htm.
37. Burger, J. (1995). A risk assessment for lead in birds. *Toxicol. Environ. Health* **45**(4): 369–396.
38. Boline, D.R. (1981). Some speciation and mechanistic aspects of trace metals in biological systems. In: *Environmental Health Chemistry*. J.D. McKinney (Ed.). Science, Ann Arbor.
39. Kelada, S.N., Shelton, E., Kaufmann, R.B., and Khoury, M.J. (2001). δ-Aminolevulenic acid dehydratase genotype and lead toxicity: a huge review. *Am. J. Epidemiol.* **154**(1): 1–13.
40. Agency for Toxic Substances and Disease Registry (ATSDR) (2003). Case studies in environmental medicine. Lead toxicity. Available at http://www.atsdr.cdc.gov/HEC/CSEM/lead/physiologic_effects.html#Pregnancy%20Outcomes.
41. Lin-Fu, J.S. (1973). Vulnerability of children to lead exposure and toxicity: part one. *N. Engl. J. Med.* **289**: 1229–1233.
42. Ziegler, E.E. et al. (1978). Absorption and retention of lead by infants. *Pediatric Res.* **12**: 29–34.
43. Chamberlain, A.C. (1983). Effect of airborne lead on blood lead. *Atmos. Environ.* **17**: 677–692.
44. Simons, T.J.B. (1986). Passive transport and binding of lead by human red blood cells. *J. Physiol.* **378**: 267–286.
45. Barry, P.S. and Mossman, D.B. (1970). Lead concentrations in human tissues. *Br. J. Ind. Med.* **27**: 339–351.
46. Schroeder, H.A. and Tipton, I.H. (1968). The human body burden of lead. *Arch. Environ. Health* **17**: 965–978.
47. Finkelstein, Y., Markowitz, M.E., Rosen, J.F. (1998). Low-level lead-induced neurotoxicity in children: an update on central nervous system effects. *Brain Res. Rev.* **27**: 168–176.
48. Bressler, J.P. and Goldstein, G.W. (1991). Mechanisms of lead neurotoxicity. *Biochem. Pharmacol.* **41**(4): 479–484.
49. Damstra, T. (1977). Toxicological properties of lead. *Environ. Health Perspect.* **19**: 297–307.
50. Blackman, S.A. (1937). The lesions of lead encephalitis in children. *Bull. Johns Hopkins Hosp.* **61**: 1–61.
51. Needleman, H.L., Schell, A., Bellinger, D., Leviton, A., and Allred, E.N. (1990). The long-term effects of exposure to low doses of lead in childhood. *N. Engl. J. Med.* **322**: 83–88.
52. Bradbury, M.W.B. (1984). The structure and function of blood–brain barrier. *Fed. Proc.* **43**: 186–190.
53. Clasen, R.A. et al. (1973). Electron microscopic and chemical studies of the vascular changes and edema of lead encephalopathy. *Am. J. Pathol.* **74**: 215–240.
54. Goldstein, G.W., Asbury, A.K., and Diamond, I. (1974). Pathogenesis of lead encephalopathy. Uptake of lead and reaction of brain capillaries. *Arch. Neurol.* **31**: 382–389.
55. Yule, X., Lansdown, R., Miller, I., and Urbanowicz, M. (1981). The relationship between blood lead concentrations, intelligence, and attainment in a school population: a pilot study. *Dev. Med. Child Neurol.* **23**: 567–576.
56. Schroeder, Sr., Hawk, B., Otto, D.A., et al. (1985). Separating the effects of lead and social factors on IQ. *Environ. Res.* **38**: 144–154.
57. Fulton M., Raab, G., Thomson, G., et al. (1987). Influence of blood lead on the ability and attainment of children in Edinburgh. *Lancet* **1**: 1221–1226.
58. Lansdown R., Yule, W., Urbanowicz, M.A., and Hunter, J. (1986). The relationship between blood–lead concentrations, intelligence, attainment and behavior in a school population: the second London study. *Int. Arch. Occup. Environ. Health* **57**: 225–235.
59. Hawk, B.A., Schroeder, S.R., Robinson, G., et al. (1986). Relation of lead and social factors to IQ of low SES children: a partial replication. *Am. J. Ment. Defic.* **91**: 178–183.
60. Winneke, G., Brockhaus, A., Ewers, U., et al. (1990). Results from the European multicenter study on lead neurotoxicity in children: implications for risk assessment. *Neurotoxicol. Teratol.* **12**: 553–559.
61. Kehoe, R.A. (1961). The metabolism of lead in man in health and disease: present hygienic problems relating to the absorption of lead: the Harben 1960. *J. Inst. Public Health. Hyg.* **24**: 177–203.
62. Seto, D. and Freeman, J.M. (1964). Lead neuropathy in childhood. *Am. J. Dis. Child.* **107**: 337–342.
63. Xu, Bo., Chia, S.E., Tsakok, M., and Ong, C. (1993). Trace elements in blood and seminal plasma and their relationship to sperm quality. *Reprod. Toxicol.* **7**: 613–618.
64. Alexander B.H. et al. (1996). Paternal occupational lead exposure and pregnancy outcome. *Int. J. Occup. Environ. Health* **2** (4): 280–285.
65. Lancranjan, I. et al. (1985). Reproductive ability of workmen occupationally exposed to lead. *Arch. Environ. Health* **30**: 396–401.
66. Lerda, D. (1992). Study of sperm characteristics in persons occupationally exposed to lead. *Am. J. Ind. Med.* **22**: 567–571.
67. Alexander, B.H. et al. (1996). Semen quality of men employed at a lead smelter. *Occup. Environ. Med.* **53**: 411–416.
68. Chisolm, J.J., Jr. and O'Hara, D.M. (1982). *Pb Absorption in Children: Management, Clinical and Environmental Aspects*. Urban & Schwarzenberg, Baltimore.
69. McMichael, A.J. et al. (1986). The Port Pirie Cohort study: maternal blood lead and pregnancy outcome. *J. Epidemiol. Community Health* **40**: 18–25.
70. Barltrop, D. (1969). Transfer of lead to the human foetus. In: *Mineral Metabolism in Paediatrics*. D. Barltrop and W.L. Burland (Eds.). Blackwell, Oxford.

71. Gershanik, J.J., Brooks, G.G., and Litle, J.A. (1974). Blood lead values in pregnant women and their offspring. *Am. J. Obstet. Gynecol.* **119**: 508–511.

72. Gulson, B.L. et al. (1997). Dietary lead intakes for mother–child pairs and relevance to pharmacokinetic models. *Environ. Health Perspect.* **105**: 1334–1342.

73. Sanin, L.H. et al. (2001). Effect of maternal lead burden on infant and weight gain at one month of age among breastfed infants. *Pediatrics* **107**(5): 1016–1023.

74. Nolan, C.V. and Shaikh, Z.A. (1992). Lead nephrotoxicity and associated disorders: biochemical mechanisms. *Toxicology* **73**: 127–146.

75. Cramer, K., Goyer, R.A., Jagenburg, R. and Wilson, M.H. (1974). Renal ultra structure, renal function and parameters of lead toxicity in workers with different periods of lead exposure. *Br. J. Ind. Med.* **31**: 113–119.

76. Dobryszycka, W. et al. (1984). Morphological and biochemical effects of chronic low level cadmium and lead feeding rats. *Acta Pol. Pharm.* **41**: 111.

77. Lilis, R. et al. (1980). Kidney function and lead: relationships in several occupational groups with different levels of exposure. *Am. J. Ind. Med.* **1**: 405–412.

78. Nuyts, G.D. et al. (1991). Does lead play a role in the development of chronic renal disease? *Nephrol. Dial. Transplant.* **6**: 307–315.

79. Waldegger S. et al. (1995). Heavy metal inhibition of rBAT—induced amino acid transport. *Kidney Int.* **47**(6): 1667–1681.

80. Victery W., Throler, H.A., Volpe, R., et al. (1988). Summary of discussion sessions: symposium on lead blood pressure relationships. *Environ. Health Perspect.* **78**: 139–155.

81. Schwartz, J. (1995). Lead, blood pressure, and cardiovascular disease in men. *Arch. Environ. Health* **50**: 31–37.

82. Korrick, S.A., Hunter, D.J., Rotnitzky, A., et al. (1999). Lead and hypertension in a sample of middle-aged women. *Am. J. Public Health* **89**(3): 330–335.

83. Hu, H. et al. (1996). The relationship of bone and blood lead to hypertension. The normative aging study. *JAMA* **275**: 1171–1176.

84. Warren, M.J., Cooper, J.B., Wood, S.P., and Shoolongin-Jordan, P.M. (1998). Lead poisoning, haem synthesis and 5-aminolaevulinic acid dehydratase. *TIBS* **23**: 217–221.

85. Kappas, A., Sassa, S., Galbraith, R.A., and Nordmann, Y. (1995). In: *The Metabolic Basis of Inherited Diseases*, 7th Edn. Scriver, C.R., A.L. Beaudet, W.S. Sly, and D. Valle (Eds.). McGraw Hill, New York, pp. 2103–2159.

86. Centers for Disease Control (1991). *Preventing Lead Poisoning in Young Children. A Statement by the Centers for Disease Control*, United States Department of Health and Human Service, United States Public Health Service, Atlanta.

87. Bruening, K. et al. (1999). Dietary calcium intake of urban children at risk of lead poisoning. *Env. Health Perspect.* **107**: 431–435.

88. Mahaffey, K.R. (1995). Environmental health issues—commentary on nutrition and lead: strategies for public health. *Environ. Health Perspect.* **103**(S6): 191.

89. Gulson, B.L. et al. (1997). Pregnancy increases mobilization of lead from maternal skeleton. *J. Lab. Clin. Med.* **130**: 51–62.

90. Gulson, B.L. et al. (1998). Mobilization of lead from the skeleton during the postnatal period is even larger than during pregnancy. *J. Lab. Clin. Med.* **131**: 324–329.

MICROBIAL DETECTION OF VARIOUS POLLUTANTS AS AN EARLY WARNING SYSTEM FOR MONITORING OF WATER QUALITY AND ECOLOGICAL INTEGRITY OF NATURAL RESOURCES, IN RUSSIA

GALINA DIMITRIEVA-MOATS
University of Idaho
Moscow, Idaho

GENERAL INTRODUCTION

General

The U.S. Environmental Protection Agency (U.S. EPA) persistently supports programs that monitor water resources using bioassessments and that develop biocriteria for different water bodies throughout the country in accordance with the Clear Water Act, the Watershed Protection Approach, and the National Water-Quality Assessment (NAWQA) Program. Currently, assessments of biological integrity and water quality are based on habitat measurements and also on biosurveys of the status of indicator aquatic communities. In order to evaluate an integrative ecological situation in water resources, the U.S. EPA has established protocols for bioassessments and for developing biocriteria applicable to selected biological assemblages, such as phytoplankton, zooplankton, periphyton, macrophytes, macroinvertebrates, and fish. As the U.S. EPA literature mentions, at the present time, no established protocols exist for other groups of organisms, including bacteria, because the data about those organisms are not sufficiently numerous. However, bioassessments based on the use of currently selected indicator groups of hydrobionts do not indicate the cause of water resource impairment. Also, usually habitat measurements monitor already known impairing factors; therefore, if a new pollutant has appeared, it may remain undetected.

The author of this paper had long-term extensive experience working with water quality control and ecological integrity monitoring using microbial indices. This practice includes eight years of microbial bioassessments and the development of biocriteria implications for the detection of various organic pollutants, a number of heavy metals, and biological contamination in seawaters of Russia, which provided impressive results, demonstrating the advantages of such an approach for monitoring ecological integrity and water quality of estuaries and shallow and pelagic regions of marine coastal zones. The composition alteration of aquatic sediment or plankton microbial assemblages subjected to pollution resulted in an increase of pollutant-resistant (indicator) groups that detected expected and unexpected pollutants and their levels. The analysis of the proportion between all indicator groups assisted us in drawing conclusions about the natural or anthropogenic process or factors causing impairment of water resources.

As monitoring of water quality using microbial indices and habitat metrics can detect many different pollutants at a number of monitoring stations, we gathered sample point data via a Global Positioning System (GPS) and displayed the distribution of each pollutant over the water body on maps employing a Geographic Information System (GIS). Several pollutants have been monitored using microbial assessments and habitat metrics, including measurements of toxic compounds at the reference stations and at the monitoring stations. Microbiological assessments for the evaluation of the biological contamination included controlling fecal and pathogenic bacteria, such as *E. coli, Salmonella, Shigella,* and *Listeria*, detecting phenol-resistant bacteria as indicators of fecal phenols, and also evaluation of the proportion of proteolytics and lipolytics bacteria in microbial assemblages. In order to monitor the pollution of oil and oil by-products, we studied the percent of microorganisms resistant to those pollutants as well as the concentration of those compounds. The portion of lipolytic bacteria that might contribute to oil biodegradation was also detected and evaluated. We also applied microbiological assessments and habitat measurements in order to detect the presence and the origin of phenols that can appear in the water from point and nonpoint sources. The choice of monitored heavy metal pollution using microbial assessments, and also habitat metrics, was based on the environmental role of heavy metals and their impact. As a primer pool of elements, we controlled Cu, Pb, and Ni as indicators of industrial impact; Cd and Zn as indicators of anthropogenic impact; Co and Cs as detectors of radioactive contamination; and Fe as a marker of terrestrial influence.

Historical Background

Application of Microbial Assessments for Detection of Various Pollutants in the Environment

Response of Indicator Groups of Microorganisms to Pollution in the World Ocean. The concept of microbial indication of organic pollutants in the world ocean has been developed most actively during the last two decades. Progressively increasing anthropogenic load on the marine environment creates tense ecological situations, primarily in coastal areas. As had been established earlier, an increase of anthropogenic pressure on the marine environment influences structural and functional properties of hydrobiont communities (1). Pollutants cause various changes in marine organisms. Organisms that have adapted to new chemical compounds, or increased concentrations of pollutants in the environment, begin to dominate in the structure of biocenosis. The ability to quickly adapt to changing environmental conditions because of high rates of reproduction and growth and versatile enzymatic activity is a unique feature of microorganisms. For this reason, microorganisms are unique tools that scientists have started using to control marine pollution because they have the most direct connections to the surrounding environment, they react quickly by altering population size and composition, and they reflect any environmental changes. On this basis, microbial assemblages are considered indicators of physical, chemical, and biological processes in the world ocean (2–7). As is known, microorganisms are indicators of contamination of various organic substances, because they are destroyers and consumers of those compounds at the same time. The existence and sensitivity of the individual response of the plankton community to variations in the concentrations of metals in contaminated waters, with a complex of metals, are confirmed on the basis of microbiological data and the results of their comparison with the data of chemical analysis (8–11). A number of microorganisms, the composition of main physiological groups, and distinct genera in seawaters and sediments are characteristics of microbial assemblages that researchers used for the purposes of biomonitoring of pollutants in the marine environment (12–15). Biological contamination criteria were developed earlier, historically for fresh waters more than for seawaters. Microbial assessments of other kinds of pollution have longer records in the marine environment, particularly, in Russia.

Detection of Pathogenic Bacteria and Biological Contamination in Surface Waters. Indication of pathogenic microorganisms in surface waters is under the control of environmental services and laboratories. Observed indicator bacteria of biological contamination exist, as well as saprobe indices of waters in freshwater and seawater environments. Freshwater saprobe indices are established on the basis of the ratio between the number of bacteria grown on reach nutrient agar and the number of bacteria grown on pour agar at starving conditions. For barely contaminated sites, the ratio is 1.2–2.8, and for references, the ratio appears as 0.1–0.7 (16). Also, most of the microbial indicators of biological contamination have been studied in fresh waters. In the 1960s, seawaters in Russia were investigated and classified according to their saprobes (17). In the 1970s, indicator microorganisms for biological contamination were determined, and sanitary criteria were established according to international agreement (18,19). According to that classification, in secure waters, the number of *Escherichia coli* cells must not exceed 10 per mL. In 1 mL of dangerous waters that can cause infectious diseases, more than 24 cells were found. The control of biological contamination is also a very highly developed field at the present time.

As seawaters accept an increasing amount of domestic and industrial wastewaters, the biological contamination of seawaters is a subject of the continuing investigation. Distribution of fecal and pathogenic bacteria has been studied in the waters of the Atlantic Ocean (20) and in the benthos coastal zone of the Antarctic Ocean (21). Other regions of the world ocean have been studied in estuarial (22), coastal (13), and pelagic (23) zones. Among the indicators of fecal contamination indicated are such microorganisms as *E. coli, Nitrobacteria* spp., *Clostridium perfringens, Pseudomonas aeruginosa, Vibrio* spp., *Salmonella* spp., *Candida albicans, Klevsiella* spp., *Bacterioides* spp., enterococci, bifid-bacteria, and fecal streptococci (22). Intestinal viruses are also the subject of study in the water environments.

Use of Microorganisms as Indicators of Oil Pollution in Seawaters. Microorganisms were beginning to be used many years ago as detectors of oil pollution and of water quality in the marine environment. One of the first marine microbiologists studying this subject was ZoBell and his coauthors (24). These scientists have found more than 100 species of marine microbes representing 30 genera, which have been able to use hydrocarbons. Those microorganisms have been considered indicators of oil pollution in seawaters. By the present time, researchers have isolated more than 1000 strains of microorganism-oxidizing hydrocarbons (25–28). The most active hydrocarbon-oxidizing marine bacteria were determined to be *Mycobacterium* and *Arthobacter* (29).

In Russia, the first publications devoted to indication of oil contamination in seawaters were published in the 1950s (30,31). The evaluation criteria for oil pollution were based on the ratio of oil-oxidizing heterotrophic bacteria and saprotrophic bacteria or on the ratio of oil-oxidizing bacteria and the total bacterial quantity (29,32). Bacterial criteria for determining acceptable standards for clean water were also established. More recently, data appeared about microbial degradation of not only aliphatic hydrocarbons, but also of polycyclic hydrocarbons (33–35). Their results have demonstrated that indicator groups of microorganisms are developed weakly in noncontaminated pelagic seawaters. In oil-contaminated areas of the world ocean, those groups were very numerous. For this reason, researchers use the activity and a number of hydrocarbon-oxidizing microorganisms as an index of oil pollution in seawaters. Now this is the most heavily investigated field in the study of marine microbiology in Russia and in the world at large.

Microbial Detection of Phenols in Seawaters. Contamination of surface waters by phenols can have different origins. Fecal sterols, polyaromatic hydrocarbons, biphenyls, and Cl-phenols enter water environments from domestic wastewaters, from industrial oil pollution, from oil spills, and from cellulose factories. Contrary to microbial indication of oil pollution in seawaters, indication of phenols does not have long-term records. Over more than twenty years, some articles devoted to microbial detection of phenols in seawaters were published (7,36,37). The microbial detection of fecal sterols (38–40), Cl-phenols, and biphenyls has been studied in more detail (1,15). Microorganisms and degrading aromatic phenols are considered as indicators of phenol pollution in deep-water regions of the ocean (7,36,41). Remarkably, the composition of microbial assemblages from phenol-contaminated seawaters and marine sediments has been found to be close to the composition of the microbial community from active silts of wastewater treatment facilities, which were adapted to phenol by-products (42). Representatives from genera *Bacillus* (25%), *Pseudomonas* (12.5%), *Mycobacterium* (24%), *Micrococcus* (25%), and *Actinomyces* (12.5%) were found there. However, no indices were established to confirm the phenol contamination in the marine environment.

Bacteria-Degrading Lipids, Proteins, and Starch Are Additional Indicators of Organic Contamination in Coastal Waters. Seawater and marine neuston organisms contain considerable amounts of various lipids. The concentration of lipids in seawater varies from 0.01 to 0.12 mg·mL^{-1}, and concentration of lipids in marine plankton appears as 2–25%. Also, three gycerides, fatty acids, and phospholipids are oil by-products during the oil transformation process in the marine environment and bacteria synthesize them as well. As lipids are much lighter than water, they are concentrated on the margin of the surface of seawater and interfere with the gas and temperature exchange process between the ocean and the atmosphere, causing different negative consequences. Many marine bacteria can use lipids as a source of carbon. They possess lipolytic activity and provide self-clearance of lipids from seawaters. Bacterial neuston is particularly important for the biodegradation of lipids in the sea (43). For this reason, some authors suggest considering the appearance of lipolytic activity in the dominant part of neuston microbial assemblages, as confirmation of the presence of oil contamination in seawater.

Determination of a number of lipolytic microflora and measurement of lipolytic activity in marine bacteria serves as a test, characterizing the capability of the marine coastal environment for self-clearance (43,44). Nitcowski et al. (12) presented data in evaluating a portion of the physiological groups of lipolytic microorganisms in aquatic microbial association, as well as a portion of proteolytics, as an indicator of biological contamination. The authors have found that a number of bacteria from those groups were four times as numerous at contaminated sites. Among those physiological groups of microbes, 80% of the population has been identified as pseudomonades and vibrions.

Some marine bacteria, yeasts, and fungi are capable of degrading starch. Such microorganisms were detected among microbial assemblages living on the surface of marine algae and plants (45–47), their remains (44,46), and in the water (45). In our opinion, such microorganisms may also indicate contamination of seawaters by carbohydrates.

Application of Microbial Biosurveys for Detection of Heavy Metal Contamination in the Environment

Practical use of the microbial approach for detection of the presence of metals started in the 1950s in geological survey practice for the metal ore search in Russia. The method was based on the ability of certain bacteria to stimulate or to inhibit their growth in response to the presence of metals of interest in the ground and in the surface waters (48,49). In order to prove that the method of microbial detection can be implicated for metal deposit searches, scientists used the station grid approach, where habitat metrics included measurements of the concentrations of metals. Such field investigations covered extensive areas in different states and republics of the former Soviet Union in Eastern Siberia, Kazakhstan, Ural, and the Caucasus regions. Researchers have found a direct connection between the distribution of metals (Mo, Zn, Cu) and a number of cultured, metal-resistant, or metal-sensitive bacteria. All other similar applications in geological microbiology were reviewed later (50).

In the 1970s, the geochemical ecology of microorganisms was established as a science in Russia. The subject of geochemical ecology of microorganisms is the study of interactions between natural and industrial factors, including metals, in the environment (51,52). At the present time, environmental microbiology studies mechanisms of the interaction between microorganisms and heavy metals. Microorganisms absorb metals for their physiological needs, as do many other organisms, because metals are an important part of biological molecules, including enzymes, hormones, vitamins, pigments, and lipids.

Generated data demonstrate that microorganisms also have a unique system of cell organization and function, responsible for selective binding and absorption of metals by bacteria. The major element of this system is the cell wall. It is remarkable that the adsorptive capability of bacterial cell walls is so high and that it is comparable with the most efficient resins used for ion-exchange chromatography (53). As now established, the selective adsorption of distinct metals by cell walls varies significantly in different bacteria (52–54). In firmicutes, an affinity of cell walls in binding different metals is presented as follows: $La^{3+} > Cd^{2+} > Sr^{2+} > Ca^{2+} > Mg^{2+} > K^+ > Na^+ > Li^+$ (53). As a result of such high capability of the selective binding of metals, bacteria accumulate metals that greatly exceed the natural concentrations of metals in the environment.

The absorption process involves active and passive transport of metals through the plasma membrane. Specificity of active transport mechanisms is based on ligands selective to individual metals. Those minor membrane protein structures are a second element of the system responsible for selective binding and absorption of metals by bacteria. For example, it has been demonstrated that E. coli has three specific systems for transport of iron (55).

Bacteria have a high resistance to heavy metals because of different mechanisms of their detoxification. For this reason, microorganisms are the only live organisms on the Earth that are able to live in environments containing very high concentrations of heavy metals. The detoxification of heavy metals is the system that provides bacterial survival in heavy metal environments. In general, those mechanisms include: limitation or complete blocking of metal internalization into bacterial cells, active removal of absorbed metals from cells, and detoxification of metal ions by binding and transforming them into nontoxic forms or by the resulting intracellular accumulation (56,57). Thus, bacteria possess various mechanisms of self-protection against toxic effects of heavy metals in the environment.

On the basis of presented data, we may conclude that the variety of systems of selective binding and transport of heavy metals, and also the system of detoxification of heavy metals, are real cell functions that provide survival and the specific response of the microbial community to the presence of toxic metals in the aquatic and terrestrial environments. Of course, passive transport and nonspecific transport of metals also take place in microbial cells. Other mechanisms of interaction of bacteria with metals exist as well, but it is not an aim of this entry to discuss all aspects of the interaction of bacteria with metals, including details of detoxification of metals by microbial cells. Those aspects are presented in other papers (56–61). Also,

many environmental factors influence the efficiency of the interaction of metal ions with microbial cells, including the pH of various environments, the state of metal ions, and the presence of other ions (56). The information presented above aims to demonstrate a general knowledge that may help in the understanding of the true foundation for the prospective use of microbial assessments for monitoring heavy metals in water environments. However, this powerful tool for environmental monitoring is not yet widely used. Also, since recent times, the technique of microbial survey, employed to indicate different pollutants, their concentrations, and also the ecological state of marine or freshwater environments, was not sufficiently used in Russia, or in the remainder of the world for that matter.

Monitoring Marine Environmental Quality in Russia

The anthropogenic load and its influence on the condition of the world ocean, particularly in the marine coastal zones, continues to grow as a result of population growth and increasing human activity. Such pressure influences qualitative development of sea life and the structural and functional characteristics of hydrobiont assemblages in the ecological, morphological, physiological, biochemical, and genetic levels (1). One of the most important consequences of the pollution of the marine environment is an appearance of indicator forms that can possess harmful characteristics, which is one of the reasons why the methods for biological control of marine environmental quality were developed in the 1960s. Two major different strategic approaches have been applied for biomonitoring of pollution in the marine environment in Russia in the past. These approaches are bioassay and bioindication. Microbial detection of marine pollutants combines the advantages of both methods and gives new possibilities for efficient tracking of pollutants issued from point and nonpoint sources in surface waters.

Bioassay Approach. The bioassay approach is based on the ability of marine organisms to react through their longevity, by the reduction of some life functions, and by alteration of cell structure at the appearance and concentration of toxic compounds in the seawater. These variations are immediately observable and provide information about the degree of toxicity of the environmental samples containing the complex of pollutants for certain kinds of organisms. However, this method does not provide information about the concentration of pollutants or an indication of which pollutant causes this negative effect. The most suitable subjects used for such control are microalgae, larvae of marine invertebrates, young generations of fishes (62,63), sea urchins and molluscs, and also sexual cells of marine organisms and marine polychaete (64). Usually, those subjects are single cell or simply organized forms, which are very sensitive to unfavorable conditions and have short lifespans.

Bioindication Approach. Another technique widely used for the marine environmental quality control is bioindication, which pertains to the study of mature, long-term life forms and requires that organism monitors must be immobile or attached to the marine substrate, and that they must be resistant to pollution and also able to accumulate toxic compounds at a concentration exceeding environmentally acceptable levels to several significant digits. The organisms most often used for bioindication in the marine environment are macroalgae and molluscs (65). These organisms are used particularly to monitor heavy metals in seawaters. In such cases, chemical analysis of the organism's tissues displays every distinct pollutant and establishes the dynamic of its behavior in the marine environment over time. The disadvantage of this approach is that no information about how toxic a certain compound may be and what the concentration of that chemical was during a certain period may be determined.

Use of a Combination of the Bioassay and Bioindication Methods in Russia. The use of both the bioassay and bioindication methods in Russia is mostly applicable to scientific research at the present time. Currently, no state regulation of these methods exists in environmental policy. However, because such approaches are numerous and multidisciplinary, Russian Scientific Institutes always are afforded opportunities to submit results of expertise and to make strong recommendations to any federal government or local state and city organizations and agencies to influence their economical or political solutions. Such scientific investigations are always a subject of consideration for the Environmental Protection Committees and Ministry, and also for the Health Protection Ministry and the local Sanitary Committees of Russia.

It is fortunate to note that most of the time their final decisions are taken with consideration of the recommendations given by Scientific Organizations monitoring the marine environmental quality and condition of marine habitats. For example, several scientific marine institutions conducted long-term and multidisciplinary investigations of the ecological state of the marine environment in the important international economical zone near Vladivostok, Russia during 1996–1998. Named The Tumen River Economic Development Area Project (TREDA), it was planned by the United Nations Organization to help economical and political development of that Pacific Region, involving China, North Korea, Japan, and Russia. The first initiative was considered the only economic side of that project. No detailed information existed about the current complex ecological situation of the area and how the project application could change the surrounding marine environment. The existence of the unique Marine National Reservation in Russia was not considered at all. Subsequently, scientific investigations showed that nobody controlled the flow of wastewater into the Tumen River from China and North Korea; consequently, the pollution at this site has increased tremendously during recent years. The increased flow of domestic, agricultural, and industrial wastewaters into the Sea of Japan caused the appearance of many unhealthy forms among the marine habitats in the Marine State Reservation area. The expertise conducted made responsible persons and organizations plan additional efforts and funding to protect not only the Russian National Marine Reservation but also the marine environment of the Sea of Japan. I am proud to announce that microbial detection results of those dramatic changes were among the first data obtained through these efforts.

MICROBIAL BIOSURVEYS IN FAR EASTERN SEAS OF RUSSIA

Areas of Field Investigations

We began our investigation in this field 12 years ago. The areas that have been investigated in the Russian Pacific region have all involved international, economically, and politically important areas (Fig. 1), such as the Southeast Pacific Ocean coast of the Kamchatka peninsula (1995, 1997, 1999), the southwest (1995, 1997) and the northeast (1996, 1998) of Sakhalin island, and the North (1992, 1996–2000) and South parts of Primorie (1995–2001) (8–11,37,47,66–68). As a matter of background, the Far Eastern seas are influenced by global natural factors such as the Pacific ore "belt," volcanic activity, grand streams, upwelling, and by the action of bringing chemical elements to the surface water in active geochemical zones. As a result, the Far Eastern seas are distinguished as containing the highest level of toxic material, primarily heavy metals, which can provoke ecological stress and cause pollution.

Every water body chosen for our investigation had specific characteristics and was distinguished in the degree of anthropogenic impact, geochemical, geographic, and ecological conditions. In the past, Avachinskaya Guba Bay, Krasheninnikov Bight (Kamchatka peninsula) was known as the biggest Russian Navy base, and because of this status, the area became contaminated with radionuclides. Seawaters surrounding Sakhalin Island are

Figure 1. Areas of the studies on microbial indication of pollution of the near-shore waters by heavy metals: I—Kamchatka coast, area of Avachinskaya Guba; II—northeastern coast of Sakhalin Island, Nyiskii, Nabil', and Chaivo Bays; III—southwestern coast of Sakhalin Island, off Kholmsk; IV—Northern Primorie, Rudnaya Bight, and Lidovka Bights; V—Peter the Great Bay, the area of the mouth of the Tumen River Amur and Ussuri Bays, and Nakhodka Bight.

the most biologically productive in the Russian Far East and are used as one of the major international fishery regions. However, this region is affected by a strong anthropogenic impact. The waters near the Kholmsk area (South-West Sakhalin) is under the influence of the Tsushima Current, which carries oil, phenol, and heavy metal pollution from the coast of Japan. The shallow waters of bays of Northeast Sakhalin (Nyiski, Chaivo, and Nabil Bays) undergo oil extraction.

The Russian coast of the Sea of Japan (the Primorie region) is also an important local fishery and an international transportation and trade zone. Rudnaya Bay (Northern Primorie) lies within the region of production and processing of polymetal ores rich in lead, zinc, and also containing a great admixture of copper, cadmium, or rare-earth metals. Traditionally, that aquatic area has been investigated because of its particular characteristic of extensive heavy metal contamination. Kit Bay (Northern Primorie), one of the cleanest parts of the Primorie coast, was used as a reference site for the investigation.

Peter the Great Bay (Southern Primorie) presented itself as a rare type of habitat for aquatic life possessing unique geographic conditions, where the cold water of the Primorie stream mixes with the warm waters of the South Eastern Korean stream (a branch of the Tsushima Current) (69). The bay includes a highly productive shelf with a unique cenosis of aquatics. This area is composed of the only State Marine Reservation of Russia, the mouth of the Tumen River, Amur Bay, Ussuri Bay, Nakhodka Bight, and Golden Horn Bight (1995–2001). At the northwest end of Peter the Great Bay lie Golden Horn Bight, Amur Bay, Ussury Bay, and Nakhodka Bight, which are affected by the severe influence of domestic and industrial wastewaters from Vladivostok, Nakhodka, and other settlements. The ecological situation in the southwest part of Peter the Great Bay, during the longer warm period of the year, is determined by the influence of Tumen River running through industrial zones of China and North Korea. The State Marine Reservation is severely impacted by the pollution carried by that big river.

Parameters Analyzed

In selected aquatoria, a seasonal and monthly microbiological and chemical monitoring of the water quality was conducted. The following parameters were analyzed: (1) the number of colony-forming heterotrophic microorganisms and the proportion of metal-resistant groups of bacteria in the assemblages of colony-forming heterotrophic microorganisms in the near-surface and near-bottom water layers; (2) long-term variations in the number of metal-resistant colony-forming hetertrophic microorganisms; (3) seasonal dynamics in the number of metal-resistant colony-forming hetertrophic microorganisms; and (4) morphological, physiological, biochemichal, and molecular genetic peculiarities of metal-resistant heterotrophic microorganisms. All listed parameters have been investigated at reference sites and at impact sites.

Microbiological monitoring and chemical analyses were performed simultaneously on water samples from areas that had been subjected to a pronounced anthropogenic impact. Special selective media were developed in laboratory experiments and corrected during field observations for the selection of heterotrophic microorganisms that were highly resistant to contaminants. The composition of the pollutant-resistant groups of marine microbial communities was determined under conditions differing in the type of the anthropogenic impact and geochemical features.

Monitoring of Heavy Metal Contamination in Seawaters Using Microbial Biosurveys

The composition of microbial assemblages and also of physiological, biochemical, morphological, and molecular genetic characteristics of bacterial representatives were distinct in the contaminated and reference sites. However, because coastal seawaters usually contained a complex of pollutants, it was difficult to conclude which compound or factor caused that alteration. Only a number of bacteria in the planktonic microbial communities were resistant to each pollutant and seemed to be a clear indicator to characterize the presence of the distinct polluting compound and the degree of contamination. On the basis of the comparisons between the results from the synchronously conducted microbiological and chemical monitoring of the pollution of the near-shore seawaters, a positive correlation was established between the microbial indices of the abundance of resistant forms and the concentrations of each of the metal-treated and organic compounds. The microbial monitoring of heavy metal pollution was the least developed method. Therefore, as an example, I chose to demonstrate the approach of developing a microbial assessment for heavy metal monitoring in seawaters.

Design of Selective Media

The first step needed for the implication of microbial biosurveys was to design selective media that would allow calculation of a number of pollutant-resistant microorganisms in the water. A number of representative bacteria that commonly occur in coastal seawaters have been chosen in order to define the sensitivity of microorganisms to heavy metals. Bacteria of genera *Escherichia, Pseudomonas*, and *Bacillus* from laboratory collections that did not demonstrate a special resistance to metals were grown in nutrient media, containing metal salt at different concentrations. Every species has an individual sensitivity to the toxic metals. The concentration of a compound that completely inhibited any bacterial growth was considered as a selective factor and then used in selective media in order to detect the appearance and the proportion of metal-resistant microorganisms.

As Fig. 2 demonstrates, the reverse correlation took place between the content of metal in the media and the growth of bacteria, if microorganisms did not have the resistance to the introduced toxic compound. However, when metal contamination is present in water, an abundance of microorganisms resistant to contaminating factors appear among the aquatic microbial communities. The illustration below demonstrates a negative correlation between the concentration of heavy metals and number

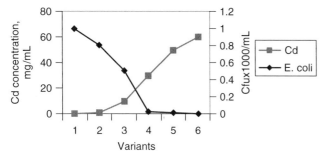

Figure 2. Dependence of bacterial growth on metal concentration in the growth medium.

of surviving microorganisms in the same samples of seawaters.

Chemical and Microbiological Monitoring of Heavy Metals in Seawaters of Peter the Great Bay

The synchronous sampling of heavy metal contents in the northern part of Peter the Great Bay allowed us to compare the results obtained. As graphs demonstrate (Fig. 3a–c), the alteration of a number of the bacteria resistant to heavy metals is an agreement with the concentration of dissolved forms of corresponding metal in the water. We can observe individual response of microbial community to the content of metals in different concentrations in the water under complex pollution conditions in Peter the Great Bay. It points to the existence and sensitivity of the individual response of microbial populations to the content of metals in different concentrations in the water under complex pollution. Moreover, in the case of each heavy metal, the dynamic of that alteration occurs differently within certain intervals of metal concentration. The boundaries between gradual and drastic variations of microbial indices were very clear, which allowed us to formulate microbial criteria that might adequately detect the concentration range of heavy metals in the water.

Average value of the percentage of the metal-resistant forms of bacteria in planktonic and benthal microbial assemblages allows comparison of the water quality from different parts of the large body of water. Moreover, microbial indices indicate from which part of the aquatorium the pollutant enters, from the surface or from the bottom. The next illustration demonstrates such a comparison made on the basis of microbial assessments conducted in several bights and secondary bays of Peter the Great Bay during the summer months of June–August of 1999 (Fig. 4a–d).

Morphological Features of Marine Bacteria Inhabiting Contaminated Waters of Peter the Great Bay

Golden Horn Bight is one of the most impaired water bodies among other bights of Peter the Great Bay. Fishery fleets, trade fleets, the military ship yard, and the city ferry transportation system are among industrial polluters of that aquatorium. Analysis of the morphological characteristics of bacteria in the planktonic assemblages can serve as an indirect characteristic of water pollution. It is a well-known fact that in heavily contaminated marine areas, rod-shaped (over 87%) and gram-negative (over

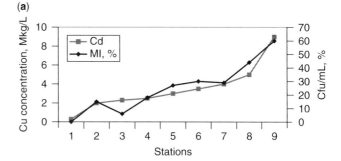

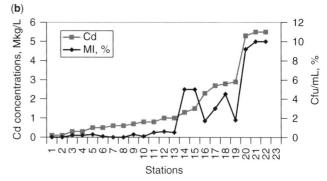

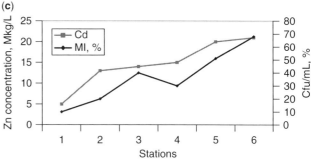

Figure 3. Changes in the values of microbial indices (line MI) and concentration of dissolved forms of metals in the water samples taken from Ussuri Bay: (**a**) copper, (**b**) cadmium, and (**c**) zinc. Along the horizontal axis, the station numbers were sorted with respect to increasing concentrations of heavy metal in the water.

60%) microflora prevail (70). We conducted microscopic examinations of 117 strains isolated from the near-surface water layer in the central part of Golden Horn Bight, Vladivostok, Russia. Among all microorganisms cultivated, rod-shaped bacteria made up 92%, and 72% of the bacteria were gram-negative.

The isolated microorganisms were resistant to several heavy metals at a time, but the combination of metals and the range of bacterial resistance were different, which once again confirms the individual character of microorganism response to the pollutant concentrations in the environment. The use of a transmission electron microscope demonstrated typical symptoms of absorbing specific substances from the environment in both bacilli and cocci and their accumulation both inside the cells and in the cell wall (11). Although these substances were not identified, but based on the electron density of the substances and the detection of multiple metal resistances

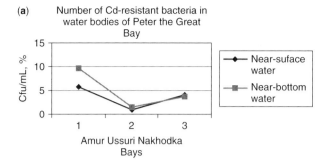
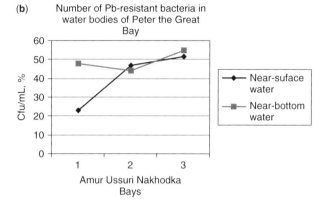
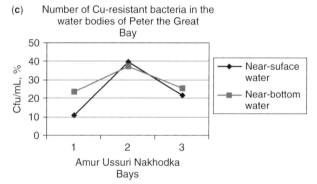
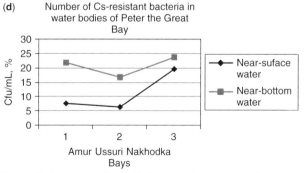

Figure 4. Average value of the number of metal-resistant bacteria in near-surface and near-bottom water samples taken from different water bodies of Peter the Great Bay in June–August of 1999.

in the test bacteria, we assumed that we had observed accumulation of heavy metals by the marine bacteria.

Study of the molecular genetics of Cd-resistant bacteria was also carried out. It was established that all Cd-resistant strains also displayed resistance to antibiotic ampicillin in the concentration of 250 mg/mL (71). Electrophoresis analysis of plasmid DNA from eight strains detected the presence of a small plasmid with a molecular mass of 4 MD in the form of a monomer, dimer, or trimer that were identical to marker plasmid p2728, containing transposon Tn 501 and designed from pUC19 and encoding resistance to Hg and ampicillin (72). Thus, we have demonstrated that marine bacteria, as well as terrestrial bacteria, possess cell mechanisms providing resistance to heavy metals. We assume that numerous bacteria with such morphological and molecular genetic features could serve as indicators of environmental disturbance and contamination in the investigated areas. The presence of such bacteria also suggests the active participation of microbial communities in the process of transformation of substances finding their way into the environment. Depending on the phenomena in question, marine microbial cenoses can act both as indicators of physicochemical and biological processes and as a powerful biotic factor promoting the elimination of pollutants from the marine environment.

Microbial Assessments for Evaluation of Heavy Metal Pollution Level in the Water of Far East Seas

Comparison of Environmental Conditions in Different Regions of Far Eastern Seas Using Microbial Biosurveys. Investigations carried out in a number of regions of Far Eastern seas, having distinguished impacts and ecological situations, resulted in data that confirmed the informative efficiency of the use of microbial assessments for the evaluation of water quality and ecological integrity in the marine environments over geographic and environmental variations taking place in different regions (Table 1).

For example, the data from Table 1 on the microbial detection of heavy metal pollution of Avachinskaya Guba show the presence of a significant portion of cobalt (Co)- and cesium (Cs)-tolerant microorganisms in the microbial associations in the areas where an increased content of radionuclides, such as Co-60 and Cs-137, had formerly been found (73). Krasheninnikov Bay (Fleet harbor), which has been exposed to a pronounced technogenic and anthropogenic impact, is the most contaminated by a complex of heavy metals in this aquatic area. Near the southwestern seaboard of Sakhalin Island (off Kholmsk), the microbial indication confirms the known intense influence of the Tsushima Current on the content of heavy metals in the water (65). In northern Primorie, in Rudnaya Bight, the microbial characteristics reflect the features in the composition of the lead ores mined and the intensity of the technogenic load. The distribution of the metal-resistant groups of microorganisms is the least uniform in Peter the Great Bay, which is caused by the active hydrodynamics and the presence of a large number of sources of heavy metal pollution of both natural and anthropogenic character. Nakhodka Bight, where major commercial, oil, and fishery ports, together with several dockyards are concentrated, is the region most contaminated by complex heavy metals.

Application of Microbial Assessments for Long-Term Monitoring and Evaluation of Water Quality in the Marine Environment. Long-term observation can give very important

Table 1. Relative Abundance of Metal-Resistant Forms (%) with Respect to the Total Amount of the Colony-Forming Heterotrophic Microorganisms in the Near-Shore Waters for Selected Aquatic Areas of the Far Eastern Seas

Bacteria, Resistant to Metal	Kamchatka Peninsula, Avachinskaya Guba July 1999		Sakhalin Island, Southwest, May 1997		Northern Primorie, Polymetal Ore Mining, July 1999		Peter The Great Bay, July 1999	
	Fleet Harbor	Malaya Sarannaya Bay, Reference	Kholmsk	Ore River Mouth	8 km to the South, Reference	Nakhodka Bight, Next to the Shipyard	Big Peles Island, Marine Reservation
Co	13	0	3	6.4	0	2.5	0
Cs	25	2.3	16	19	9.8	10.6	4.7
Cd	0	0	7	0.5	0	4.4	0
Zn	57.3	0.6	1.6	14.3	0.6	62.6	2.3
Ni	4.6	0	15	25.3	5.8	3.9	0.02
Cu	3	0	15.9	7.4	0	40.1	0
Pb	24.5	0.08	10	0.12	0	87.2	0
Fe	0	0	16.1	0	0	41.2	0

environmental information about the intensity of anthropogenic and industrial load on water bodies. Table 2 demonstrates the dynamic of the presence of metal-resistant forms of bacteria in seawater of the port, where Pb-Zn concentrate is transported by sea along the coast, which is located next to the lead smelter in Rudnaya Bight (see Fig. 5). Once again, when microbial indices are compared with the data of chemical analysis, they were in agreement. Moreover, both data of chemical and microbiological monitoring reflected the intensity of the ore processing and the production of the lead concentrate over time.

Years 1997 and 2000 are known as a period of intensive ore processing and the operation of the lead smelter at full production. On the contrary, the situation during 1998–1999 resulted in a drop in the volume of extracted and processed ore because of the general economic recession and long-term forced inactivity of the plant. The dynamics of the microbial assessments remarkably accurately reflect the environmental situation in Rudnaya Bight during that period.

Use of GIS Technology to Indicate Concentrations of Heavy Metals Found in Marine Sediments and to Show the Distribution of Metal-Resistant Bacteria in the Water. In the southwestern part of the bay, the environmental situation is mainly determined by the effect of the Tumen River flowing through the industrialized regions of China and North Korea. From the data of the microbiological analysis of this area, we found that stations located in the zone of riverine runoff waters were characterized in a state where the entire association of microorganisms cultivated showed an increased resistance to the presence of Ni, Zn, and Fe in the environment. In the communities mentioned, there exists a considerable portion of bacteria with an increased resistance to the presence of Cd, Pb, Cu, Co, and Cs in the environment. The data from the chemical analysis and the determination of the content of heavy metals in the tissues of hydrobionts confirm the contamination of the aquatic area by the pollutants listed in the scientific report (74). With a sufficient density in the network of monitoring stations, the data from microbial indication may be applied to preliminary biogeochemical

Table 2. A Number of the Highly Phenol-Resistant and Oil-Resistant Colony-Forming Heterotrophic Microorganisms (10^3 per mL) in the Near-Shore Waters for Selected Aquatic Areas of the Far Eastern Seas

Region	Kamchatka Peninsula, Avachinskaya Guba Bay, April 1996		Sakhalin Island, Southwest, Kholmsk, May 1997		Northwest Sakhalin, Nyi Bay, August 1996		Peter the Great Bay, July 1999	
Microbial Groups	Fleet Harbor	Ocean Waters, 5 km from Coast, Reference	Port	Destroyers Pulp and Paper Plant	Exit into Sea	Oil Drill	Amur Bay, Next to Domestic Wastewater	Big Peles Island, Reservation
Heterotrophic bacteria	800 ± 130	500 ± 110	600 ± 3	400 ± 3	3000 ± 300	1000 ± 80	24000 ± 307	60 ± 5
Destroyers of:								
Oil	40 ± 8	2 ± 0.3	100 ± 20	20 ± 0.1	1 ± 0.1	40 ± 4	130 ± 15	3 ± 0
Fuel oil	90 ± 7	3 ± 0.3	4 ± 0.7	30 ± 0.2	0.1 ± 0.01	40 ± 1	100 ± 16	0.8 ± 0.001
Engine fuel	7 ± 1	10 ± 2	2 ± 0.1	0	10 ± 0.8	0.7 ± 0.08	20 ± 4	0
Phenol	0	0	0.05 ± 0	20 ± 0	0.003 ± 0	0	3 ± 0.25	0

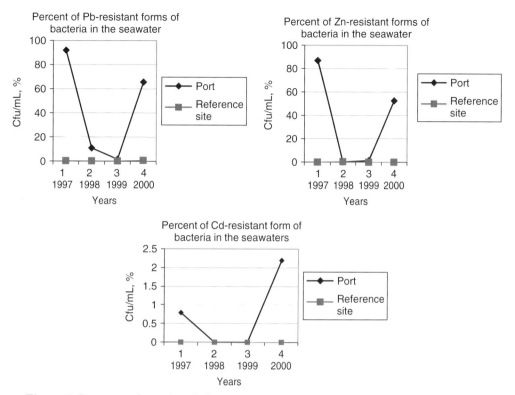

Figure 5. Long-term dynamics of the percent of metal-resistant forms of bacteria in the near-surface water layer of the Rudnaya Bight. Reference station was 2 miles away from the lead smelter.

mapping of the near-shore aquatic areas, which was demonstrated by the example in the area off the mouth of the Tumen River for two of the eight heavy metals analyzed (Fig. 6).

Use of Microbial Biosurveys for Assessments of Organic Pollution in Seawater. Microbial assessments proved to be even more efficient for monitoring organic pollution. This statement is based on the ability of microorganisms to use organic compounds as a source of carbon and energy. As it was reviewed above, the presence of such chemicals results in an increase of bacteria using hydrocarbons for their living needs.

As Table 3 demonstrates, microbial criteria at reference sites are significantly less than at impact areas. Also, microbial assessments give information about sources of the contamination. For example, phenols can be a part of fecal contamination or they can be the result of the waste from cellulose processing. Seawater samples contained phenol destroyers taken either from places where a number of ships are present (Kholmsk, Port; Nyi Bay, exit to the sea), near the domestic wastewater dumping site (in Amur Bay), or next to plants manufacturing products from cellulose (Kholmsk, paper plant). Aside from the case of the paper plant, the number of phenol destroyers was ten times as much in comparison to fecal contaminated sites and correlated with concentrations of phenol in contaminated waters.

A number of microorganisms destroying oil and oil by-products also had a connection with the content of those chemicals in the water. The presence of point or nonpoint contaminating sources close to the beach explains the reason for distribution of those specific microorganisms in the coastal waters. Moreover, the proportion between bacteria resistant to different oil by-products reflected the presence of those particular chemicals in seawaters, which may even explain the origin of the contaminant detected far away from land. For example, it is understandable why the water samples taken from port aquatoria have a significant number of oil and fuel oil microbial destroyers (Kamchatka, Fleet harbor). However, isolation of engine fuel microbial users from the surface of the open ocean water far away from Fleet harbor may cause us to think that bacteria trace the passage of ships that use that kind of fuel.

Example of the Implications of Microbial Criteria for Water Quality Control of Seawater in the Seas of Far East Russia. After years of monitoring the water quality in a large number of water bodies using microbial assessments and chemical habitat metrics, we worked out criteria that can be used for monitoring ecological situations and the quality of seawaters (Table 4). Those criteria can be used during the first evaluation of water samples in order to receive general information about the ecological status of the water body. Then, criteria should be adjusted with respect to all other metrics measured in the water.

Use of criteria, covering a wide number of polluting factors, gives information about the incredible value of

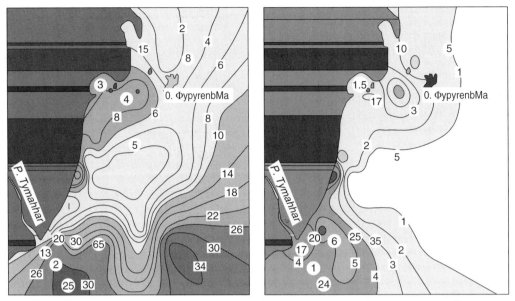

Figure 6. Distribution of the metal-resistant forms of bacteria over the aquatic area adjacent to the mouth of the Tumen River: (left) Cu- and (right) Co-resistant forms. Concentrations of metals in the bottom sediments (in μg/kg of dry mass) are given within the squares; microbial indices (%) are given in the circles.

this global approach, because we can present data on the ecological integrity of a larger system and complete an environmental situation report about the marine or freshwater water body.

Table 4 demonstrates the use and advantage of the microbial assessments for the evaluation of the environmental water quality in the example of data, presented above. The use of symbols better assists in presenting the data on ecological integrity of the marine environment at the moment.

On the basis of data received by a number of scientists over a long-term period, we can conclude that application of microbial criteria and microbial assessments is shown to serve as an operative method of monitoring and short-term forecasting of changes in the environmental conditions of the coastal waters of the sea. This method allows us to

Table 3. Criteria of Water Area Contamination on the Basis of Microbial Characteristics in Absolute Figures or in Percentage to the Maximum Number of Colony-Forming Microorganisms per 1 mL of Seawater

Pollutant	Level of Contamination				
	Background (Below the MAC)[a]	Insignificant (Around 1 MAC)[a]	Perceptible (1–3 MAC)	Considerable (Over 3 MAC)	Strong (Over 10 MAC)
Heterotrophic bacteria	$<10^3$	$10^3 < 10^4$	$10^4 < 10^5$	$10^5 – 10^6$	$>10^7$
Oil by-products	$10–10^3$	$>10^3 < 10^4$	$10^4 – 10^5$	$10^4 – 10^5$	$>10^5$
Phenol	0	1–9	$10 – <10^2$	$10^2 – <10^3$	$>10^3$
Pathogenic bacteria	0	—	—	various	various
Intestinal bacteria	0	0	1–9	10–24	>24
Proteins	<10%	—	10–46%	>46%	—
Polysaccharides	0	1–9	10–40%	>40%	—
Lipids	<10%	—	10–46%	>46%	—
Metals					
Cadmium	<0.01%	<0.5%	0.5–6%	>6%	—
Cobalt	<10%	10–46%	>46%	—	—
Cesium	<10%	10–46%	>46%	—	—
Lead	<0.1%	<10%	10–46%	>46%	—
Copper	<0.1%	<10%	10–46%	>46%	—
Zinc	<1%	<20%	20–46%	>46%	—
Nickel	<1%	<20%	20–46%	>46%	—
Iron	<20	<50%	>50%	—	—
Symbol	O	✽	□	▲	■

[a] MAC is a maximum admissible concentration of chemical in the water (for fishery water bodies).

Table 4. Evaluation of Environmental Quality of Krasheninnikov Bay (Kamchatka Peninsula, Avacha Bay, 1996)

Pollution \ Stations	Military Depot	Fleet Harbor	Rybachi, Navy Base Residences	Ferry-Boat Station	Ocean Waters, 3 miles Away from Coast
Heterotrophic bacteria	▲	▲	▲	▲	□
Intestinal bacteria	▲	O	■	O	■
Pathogenic bacteria	O	O	■	O	O
Oil	□	□	□	□	O
Fuel oil	□	□	□	□	O
Engine oil	*	□	□	□	□
Lipids	O	O	O	□	O
Peptides	O	O	O	O	O
Polysaccharides	O	O	O	O	O
Cd	O	O	O	O	□
Cu	*	□	*	*	*
Ni	*	*	*	*	*
Zn	O	*	O	O	
Co	O	*	O	O	O
Cs	*	*	O	O	O

O—Background (below the MAC); *—insignificant (around 1 MAC); □—perceptible (1–3 MAC); ▲—considerable (over 3 MAC); ■—strong (over 10 MAC); MAC is maximum admissible concentration of heavy metals in the water (for fishery water bodies in Russia).

obtain preliminary information, to create biogeochemical maps, preceding labor-intensive and expensive chemical analysis, and, e.g., at the initial stage when we select environmental quality monitoring stations. The microbial biosurvey does indeed permit control of the spread of pollutants in the near-shore waters.

BIBLIOGRAPHY

1. Izrael, Yu.A. and Tsiban, A.V. (1985). *Anthropogenic Ecology of the Ocean*. Hydrometeoizdat, Leningrad, p. 528.
2. Korneeva, G.A. (1996). Utilizing of enzyme test-systems for monitoring of sea waters condition of Black Sea. *Izvestiya RAN, Ser. Biol.* **5**: 589–597.
3. Novozhilova, M.I. and Popova, L.E. (1984). Bacterioplankton of meddle and south Kaspi. In: *Ecological Aspects of Aquatic Microbiology*. Nauka, Novosibirsk, pp. 35–43.
4. Sorokin, Yu.I. (1990). *Ecosystems of Coral Rife*. Nauka, Moskva, p. 503.
5. Teplinskaya, N.G. (1990). *Processes of Bacterial Production and Destruction of Organic Compounds in North Seas*. KNTS AN SSSR, Apatiti, p. 105.
6. Tsiban, A.V. (1970). *Bacterioneiston and Bacterioplankton of Shelf Zone of Black Sea*. Naukova Dumka, Kiev, p. 224.
7. Tsiban, A.V., Panov, G.V., and Barinova, S.P. (1990). Indicator microflora in Baltic Sea. In: *Investigation of Ecosystem of Baltic Sea*. Hydrometeoizdat, Leningrad, Issue 3, pp. 69–83.
8. Dimitrieva, G.Yu., Bezverbnaya, I.P., and Cemikina, G.I. (2000). Microbiological monitoring heavy metal pollution in coastal waters of Peter the Great Bay. *Proc. Pacific Institute Fishery and Oceanology (TINRO)*. Vol. 127, Vladivostok, Russia, pp. 657–676.
9. Dimitrieva, G.Yu., Bezverbnaya, I.P., and Khristoforova, N.K. (2001). Microbial indication is a possible approach for monitoring of heavy metal in Far Eastern Seas. *Proc. Pacific Institute Fishery and Oceanology (TINRO)*. Vol. 128, Vladivostok, Russia, pp. 719–736.
10. Dimitrieva, G.Yu. and Bezverbnaya, I.P. (2002). Microbial indication is the efficient tool for monitoring of heavy metal pollution in coastal sea waters. *Oceanology* **42**: 1–8.
11. Bezverbnaya, I.P., Dimitrieva, G.Yu., Tazaki, K., and Vatanabe, H. (2003). Evaluation of sea water quality in the coastal zone of Primirie using the method of microbial indication. *Water Res.* **30**: 199–208.
12. Nitcowski, F., Dudley, Sh., and Graikowski, J.T. (1977). *Mar. Poll. Bull.* **8**: 276–279.
13. Zoffman, M., Rodrigues-Valeria, F., and Peres-Filloi, M. (1989). *Mar. Poll. Bull.* **20**: 74–81.
14. Aubert, M. (1990). *Mar. Poll. Bull.* **21**: 24–29.
15. Tsiban, A.V., Panov, G.V., and Barinova, S.P. (1992). Heterotrophic saprotrophic microflora. Distribution, ecological and physiological properties. In: *Investigations of Ecosystems of Bering and Chukotsk Seas*. Hudrometeoizdat, St. Petersburg, Issue 3, pp. 143–165.
16. Gavrishina, N.A. (1980). O komplekse mikrobiologicheskih pokazatelei pri harakteristike kachestva void. In: *Smoochischenie I Bioindikatsiya Zagryaznennih Vod*. Nauka, Moskva, pp. 74–78.
17. Telitchenko, M.M. and Kokin, K.A. (1968). *Sanitary Hydrobiology*. MGU, Moskva, p. 102.
18. Kriss, A.E. (1976). *Microbiology of the Ocean*. Nauka, Moskva, p. 286.
19. Loransaki, D.N., Raskin, B.M., and Alhimov, N.N. (1975). *Sanitary Protection of the Sea*. Meditsina, Moskva, p. 168.
20. Fisher, B. (1984). Untersuchungen. *Ergebnisse der Plankton-Expedition der Humboldt-Stiftung* **4**: 1–83.
21. Edwards, D.D., McFeters, G.A., and Venkatesan, M.L. (1998). *Appl. Environ. Microbiol.* **64**: 2596–2600.
22. Eliot, E.L. and Colwell, R.R. (1985). *FEMS Microbiol. Rev.* **32**: 61–79.
23. Tabor, P.S. and Colwell, R.R. (1976). Initial investigations with a deep ocean *in situ* samples. In: *Proc. Oceanology, Marine Technol. Soc.*, Washington, DC, pp. 13D-1–13D-4.
24. ZoBell, C.T., Granth, C., and Haas, H. (1943). *Bull. Am. Assoc. Petrol. Geol.* **24**: 75.
25. Atlas, R.M. and Bartha, R. (1973). *Environ. Sci. Technol.* **7**: 538–541.
26. Guilott, M.L. (1983). *Sci. Techn.* **94**: 35–36.
27. Guerin, W.F. and Boyd, S.A. (1995). *Appl. Environ. Microbiol.* **61**: 4061–4068.
28. Iriberri, J. and Hernbel, G.J. (1995). *Micribiologia (Madrid)* **11**: 309–322.
29. Gusev, M.V., Koronelli, T.V., and Sentsova, O.Yu. (1985). Using of microorganisms as bioindicators to study ecological consequences of water pollution. In: *Ecological Consequences of the Ocean Pollution*. Hydrometeoizdat, Leningrad, pp. 113–127.
30. Izyurova, A.I. (1950). The rate of oil by-products destruction in water and soil. *Hygiene and Sanitary* **5**: 23–28.
31. Voroshilova, A.A. and Dianova, E.V. (1952). Oil-oxidizing bacteria as detectors of intensity of biochemical oxidizing of oil at natural conditions. *Microbiology* **24**: 64–73.
32. Mironov, O.G. (1980). Biological indication of carbohydrates in the sea. In: *Self-Purification and Bioindication of Contaminated Waters*. Nauka, Moskva, pp. 129–134.

33. Kong, H.-L. and Saylor, G.S. (1983). *Appl. Environ. Microbiol.* **46**: 666–672.
34. Hartmans, S., Bont, J.A., and Harder, W. (1989). *FEMS Microbiol. Rev.* **63**: 235–264.
35. Sigoillot, J.-C. and Nguen, M.-H. (1992). *Appl. Environ. Microbiol.* **58**: 1308–1312.
36. Divavin, I.A. and Kopilov, Y.P. (1977). Investigation of some biochemical aspects for biodegradation of carbohydrates at the condition of combined pollution. Abstracts of Workshop *Problems of the Marine Environment Protection*. Kaliningrad, Russia, pp. 31–33.
37. Dimitrieva, G.Yu. (1995). Microorganisms as bioindicators of phenol pollution of the marine coastal environment. *Russian J. Mar. Biol.* **21**: 407–411.
38. Dutka, B.J., Chay, A.S., and Corburn, J. (1974). *Water Res.* **8**: 1047.
39. Ayres, P.A. (1977). *Mar. Poll. Bull.* **8**: 283–285.
40. Goodfelow, R.M., Cardoso, J., and Eglinton, G. (1977). *Mar. Pol. Bull.* **8**: 272–276.
41. Janase, H., Zuzan, K., and Kita, K. (1992). *J. Ferm. Biol.* **74**: 297–300.
42. Kalmikova, G.Ya., Lazareva, M.F., Rogovskaya, Ts.I., Skirdov, I.V., and Tihaya, N.B. (1980). Metabolizm aromaticheskih soedinenii adaptirovannoi I neadaptirovannoi mikrofloroi aktivnih ilov. In: *Samoochischenie I Bioindikatsiya Zagryznennih Vod*. Nauka, Moskva, pp. 171–175.
43. Tsiban, A.V. (Ed.). *Rukovodstvo po Metodam Biologicheskogo Analiza Morskoi Void I Donnih Otlozhenii*. Hydrometeoizdat, Leningrad, p. 193.
44. Kondratyeva, L.M. (1996). Ekologicheskii potentsial morskih mikroorganizmov. Ph.D. (Biol.) Dissertation, HKI, Habarovsk, p. 412.
45. ZoBell, C.T. and Hittle, L.L. (1969). *J. Oceanogr. Soc. Japan*. **25**: 36–47.
46. Uchida, M. (1995). *Mar. Biol.* **123**: 639–644.
47. Dimitrieva, G.Yu. and Dimitriev, S.M. (1996). Symbiotic microflora of brown algae of genus *Laminaria* as a bioindicator of ecological condition of coastal laminarian biocenoses. *Russian J. Marine Biol.* **22**: 300–305.
48. Slavina, G.P. (1957). O vozmozhnosti ispol'zovaniya mikroorganizmov pri poiske rudnih mestorozhdenii. *Trudi 1 Vsesoyuznogo Coveschaniya po Geohimicheskim Metodam Poiska Rudnih Mestorozhdenii*. Gosgeoltehizdat, Moskva, pp. 302–305.
49. Belyakova, E.E., Reznikov, A.A., Kramarenko, L.E., Nechayeva, A.A., and Kronidova, T.F. (1962). *Gidrohimicheskii Method Poiskov Rudnih Mestorozhdenii v Aridnih I Poluaridnih Oblastyah*. Gosgeoltehizdat, Moskva, pp. 173–228.
50. Glazovskaya, M.A. and Dobrovolskaya, N.G. (1984). *Geochemical Functions of Microorganisms*. MGU, Moskva, p. 152.
51. Vernadski, V.I. (1967). *Biosphere*. Nauka, Moskva, p. 100.
52. Letunova, S.V. and Koval'ski, V.V. (1978). *Geohimicheskaya Ecologiya Mikroorganismov*. p. 147.
53. Gromov, B.V. (1985). *Stroenie Bacterii*. LGU, Leningrad, p. 190.
54. Mullen, M.D. et al. (1989). *Appl. Environ. Microbiol.* **55**: 3143–3149.
55. Grinyus, L.L. (1986). *Transport Molekul u Bacterii*. Nauka, Moskva, pp. 6–13.
56. Gadd, G.M. (1990). In: *Metal Tolerance*. C. Edwards (Ed.). Alden Press, Oxford, UK, pp. 178–211.
57. Gadd, G.M. (1993). *Phytology* **124**: 25–60.
58. Hingman, J.P., Sadler, P.L., and Scawen, M.D. (1985). *J. Gen. Microbiol.* **131**: 2539–2544.
59. Petterson, A., Kint, D., Bergman, B., and Roomans, Y.M. (1985). *J. Gen. Microbiol.* **131**: 2545–2548.
60. Schutt, C. (1989). *Microbial. Ecol.* **1**: 59–62.
61. Silver, S. (1990). *Forum. Microbiol.* **13**: 36.
62. Galaktionov, S.G. and Yurin, V.M. (1984). *Alga Danger Signals*. Nauka, Moskva, p. 356.
63. Kosolapov, V.N. and Sinyova, L.N. (1982). The action of phenol and waste waters from wood factory on motility of hydrobionts. In: *Toxicological and Ecological Aspects of the Environmental Pollution*. Irkutskii State University, Irkutsk, pp. 91–95.
64. Flerov, B.A. (1974). About using of animal behavior investigations in aquatic toxicology. *Hydrobiological J.* **5**: 114–118.
65. Khristoforova, N.K. (1985). *Bioindication of Seawater Contamination With Heavy Metals*. Ph.D. (Biol.) Dissertation, the Pacific Institute of Geography, Vladivostok, p. 394.
66. Dimitrieva, G.Yu. (1999). Plankton and epiphyte microorganisms: detection and stabilization of the marine coastal environment. Ph.D. (Biol.) Dissertation, Far Eastern State University, Vladivostok, Russia, p. 408.
67. Dimitrieva, G.Yu. (1999). The role of microorganisms in control and reservation of marine coastal environment. *Proc. Intern. Symp. "Earth—Water—Humans.* May 30–June 1, 1999, Kanazawa University, Kanazawa, Japan, pp. 22–35.
68. Dimitrieva, G.Yu., Dimitriev, S.M., Bezverbnaya, I.P., and Khristoforova, N.K. (2000). The role of microorganisms for control and restoration of the marine coastal environment. In: *Waste Recycling and Resource Management in the Developing Word*. B.B. Jana et al. (Ed.). Univ. of Kalyani, India and Intern. Ecol. Eng. Soc. Switzerland, pp. 437–447.
69. Bogdanov, D.V. (1981). *Short Physical and Geographic Characterization of Main Seas. The Pacific Ocean*. Nauka, Leningrad, pp. 129–143.
70. Izrael, Yu.A. and Tsyban, A.V. (1992). *Studies of the Bering and Chukotka Seas*. Gidrometeoizdat, Petersburg, Issue 3.
71. Dimitrieva, G.Yu. (2003). Microbial Detection as an effective technique for monitoring of heavy metal and organic pollution of coastal seawater. *Proc. Int. Workshop on Bioassay of the Marine Environment in the Northwest Pacific Region, Toyama, Japan*. March 15–16, 2003, Northwest Pacific Region Environmental Cooperation Center, Toyama, Japan, pp. 24–31.
72. Yanish-Perron, C., Viera, T., and Nessing, J. (1985). Improved M13 phage cloning vectors and in host strains: nucleotide sequence of the M13, mp 18, and pUC19 vectors. *Gen.* **33**: 103–119.
73. *Estimation of the Pollution Condition of the Area of Avachinskaya Inlet*. City of Petropavlovsk-Kamchatskii, and Primorskii and Rybachii Settlements. (1993). Report, Karpinskii Russian Geol. Inst., p. 346.
74. *Multidisciplinary Ecological Estimate of the Influence of the Tumen River Runoff on the Coastal Waters of Russia*. (1998). Report, Ross. Akad. Nauk, Vladivostok: IBM DVO, Part I.

READING LIST

Dimitrieva, G.Yu. et al. (1999). Phenol detoxification by microorganisms of the sea coastal zone. *Microbiology* **68**: 107–113.

LUMINESCENT BACTERIAL BIOSENSORS FOR THE RAPID DETECTION OF TOXICANTS

Maria B. Davoren
Dublin Institute of Technology
Dublin, Ireland

INTRODUCTION

The need to rapidly screen and assess the potential toxicity of aquatic contaminants has led to the development of several short-term bioassays utilizing bacteria, algae, protozoa, invertebrates, and cell cultures. Bioassays with microorganisms are frequently employed as ecotoxicological screening tools to monitor the hazards of chemical contaminants in the environment. In addition to their ecological importance as decomposers, microbial test systems have several advantages over traditional bioassays, including rapid response times, reproducibility of test conditions, amenability to genetic manipulations, increased sensitivity, and cost saving. Increasingly, luminescent microbial biosensors are being developed and employed for environmental monitoring (1–4). A biosensor integrates a biological sensing component (e.g., bacteria) with a transducer or detection system, which produces a measurable signal (e.g., luminescence) upon exposure of the sensing component to a specific or nonspecific analyte (5). Here the focus is on the use of both natural and genetically engineered luminescent bacterial biosensors for the assessment of toxicity and bioavailability.

BIOLUMINESCENT REPORTER GENES

Bioluminescence, which has been observed in a variety of organisms, is the production of light from living cells by the oxidation of a luciferin (a pigment) in the presence of the enzyme luciferase (6,7). Two frequently exploited bioluminescent reporter systems are the bacterial (the marine *Vibrio fischeri* and terrestrial *Photorhabdus luminescens*) and insect (the firefly, *Photinus pyralis*, and the click beetle, *Pyrophorus plagiophthalamus*) luciferase reactions.

Bacterial bioluminescence is encoded for by *lux* genes and light is produced through the oxidation of fatty aldehydes (8–10). The complete *lux*-gene cassette consists of five structural genes, *lux*CDABE. The *lux*AB genes encode the luciferase enzyme and the *lux*CDE genes encode an enzyme complex that synthesizes the aldehyde substrate for luciferase (3). Bacterial biosensors have been created using either the *lux*AB or the entire *lux*CDABE reporter system, but where the shorter gene construct is employed, an aldehyde substrate (usually *n*-decanal) must be added exogenously before bioluminescence can be measured (11,12).

The *luc*FF and *luc*GR genes encode for firefly and click beetle bioluminescence, respectively. In these eukaryotic systems, light is produced when the luciferase enzyme catalyzes the reaction between the substrate luciferin (which must be added exogenously), adenosine triphosphate (ATP), and oxygen (1,11). Rapid and sensitive detection and quantification of bioluminescence can be achieved using luminometers, liquid scintillation counters, or photographic films (4,13).

BACTERIAL BIOSENSORS BASED ON NATURAL BIOLUMINESCENCE

The naturally luminescent marine bacterium *V. fischeri* (formerly *Photobacterium phosphoreum*) has been extensively employed to generate acute toxicity data. Kaiser and Palabrica (14) published *P. phosphoreum* toxicity data for more than 1300 chemicals. A rapid screening test based on bioluminescence inhibition in *V. fischeri* (NRRL B 11177) has been commercialized as the Microtox test system (Azur Environmental, CA, USA) and is now an internationally adopted bioassay (15,16). The test system measures the light output of the luminescent bacteria after they have been challenged by a sample and compares it to the light output of a control (reagent blank). As bacterial bioluminescence is directly linked to cellular respiration, a difference in light output (between the sample and control) can be attributed to the toxic effect of the sample. The Microtox test has also been widely employed to generate acute toxicity data for environmental samples such as wastewaters (17), industrial effluents (18), and sediment samples (19). For a comprehensive review of the use of Microtox for assessing the toxicity of sediments and soils, the reader is referred to Doherty (20).

In addition to the well established Microtox test system, other toxicity bioassays have been developed which utilize *V. fischeri*. These include BioTox (Aboatox Oy, Finland), LUMIStox (Dr. Lange GmbH, Düsseldorf, Germany), and two portable systems ToxAlert (Merck Ltd., UK) and Deltatox (Azur Environmental, CA, USA). These test systems have been used either individually or in combination with each other in toxicity assessments (21–28). The ToxScreen-II assay (Checklight Ltd., Israel) employs another naturally luminescent marine bacterium, *Photobacterium leiognathii*, as a test organism (29). This test enables discrimination between cationic heavy metals and organic toxicants and has the added advantage of being able to be run at ambient temperatures.

ECOLOGICALLY RELEVANT BACTERIAL BIOSENSORS CONSTRUCTED USING LUMINESCENT-MARKED GENES

Despite the widespread use of Microtox and other test systems utilizing *Vibrio* sp., one frequently cited shortcoming of these assays is the lack of ecological relevance for the testing of freshwater and terrestrial environments (30,31). *Vibrio fischeri* is a marine bacterium and as such is sensitive to pH and osmotic conditions. The genes for bacterial bioluminescence (*lux*AB and *lux*CDABE) have therefore been cloned and inserted using plasmids into environmentally relevant terrestrial and freshwater bacterial test strains (32–35). As with the *V. fischeri* based biosensors, luminescence is dependent on metabolic activity; therefore, any chemical causing metabolic stress will result in a reduction in bioluminescence. Luminescent marking of terrestrial bacteria offers scope for environmentally relevant, whole-cell biosensors that can either be toxin specific or general indicators of pollutant toxicity. These genetically modified bacteria have the added advantage of not being limited by a narrow pH range.

RECOMBINANT LUMINESCENT BACTERIAL BIOSENSORS

Recent advances in reporter gene technology have led to the development of many luminescent marked bacterial biosensors, which have been employed as rapid bioassays to assess the toxicity of a wide range of pollutants. Some of the bacterial biosensors that have been constructed are used for general toxicity testing as they constitutively express luminescence and respond to toxicants in a "lights off" manner, with a decrease in light output related to exposure of the cells to pollutants (Table 1). These bacteria are not toxicant specific and cannot be used to determine the type or the actual concentration of a given toxicant, but rather are general indicators of adverse conditions. As no single biosensor will have a universal sensitivity to all toxicants, multiple luminescent bacteria are increasingly being employed, in a battery style approach to better characterize the toxicity associated with a chemical (50–52). Constitutively expressing biosensors have been used to assess the toxicity of a wide range of pollutants including heavy metals (32), chlorobenzenes (31), chlorophenols (37), industrial effluents (43), and wastewater (53).

In addition to constitutive expression, other biosensors have been developed that react to toxicants in a "lights on" manner, where light production is induced by bioavailable concentrations of toxicants. These recombinant bacteria have been constructed by the molecular fusion of two elements inside a bacterial host, a promoter gene regulated by specific chemical agents or their pathway intermediates, and a promoterless reporter gene (e.g., lucFF or luxCDABE) inserted as a plasmid. In these toxicant-specific bacterial sensors, a genetic regulatory unit controls the expression of the reporter gene. In the presence of a toxicant this regulatory unit is induced and generates a measurable signal relative to toxicant level, that is, receptor–reporter concept (54). These microbial biosensors offer a powerful new approach to environmental monitoring, providing an indication of the bioavailabilty of specific pollutants in complex environments, rather than the total concentrations obtained by traditional analytical techniques. Biosensors constructed for the detection of specific pollutants in the environment can complement analytical methods by distinguishing bioavailable from immobilized, unavailable forms of contaminants, and frequently an integrated approach using both methods is employed (36,55). In order to assess the overall toxicity of the sample, constitutively expressing reporter biosensors (non-toxicant-specific bioassays) are often run in parallel with toxicant-specific inducible biosensors (36,48,56). Many luminescent bacterial biosensors have been developed for assessing pollutant bioavailability in environmental samples, selected examples of which are presented in Table 2.

Table 1. Selected Examples of Non-Toxicant-Specific Luminescent Biosensor Strains

Bacterial Biosensor Strain	Plasmid	Luminescent Reporter Gene	Tested Compounds[a]	Reference
Bacillus subtilis BR151	pCSS962/pBL1	lucFF	Metal-contaminated soils	36
Escherichia coli HB101	pUCD607	luxCDABE	2,4-DCP; atrazine, simazine, propazine, mecoprop, MCPA, diuron, paraquat; As, CuSO$_4$, 2,4-DCP, 3,5-DCP, bronopol, TTAB, ZnSO$_4$	37–39
Escherichia coli MC1061	pCSS810	lucFF	Metal-spiked water samples	40
Pseudomonas putida F1	pUCD607	luxCDABE	Atrazine, simazine, propazine, mecoprop, MCPA, diuron, paraquat; oil-polluted soil	38,41
Pseudomonas fluorescens 10586r	pUCD607	luxCDABE	Atrazine, simazine, propazine, mecoprop, MCPA, diuron, paraquat	38
Pseudomonas fluorescens 10586s	pUCD607	luxCDABE	Copper in whiskey distillery effluent; papermill effluent; contaminated groundwater; benzene, catechol, phenol; chlorobenzenes; 2,4-DCP; metal-contaminated soils	31,37,42–46
Pseudomonas fluorescens OS8	pNEP01	lucGR	Soil-associated arsenite and mercury	47
Ralstonia eutropha ENV307	pUTK2	luxCDABE	Polychlorinated biphenyls	48
Staphylococcus aureus RN4220	pCSS810	lucFF	Metal solutions	49
Staphylococcus aureus RN4220	pTOO02	lucFF	Metal-contaminated soils	36

[a]As = sodium arsenite; bronopol = 2-brom-2-nitro-1, 3-propandiol; 2,4-DCP = 2,4-dichlorophenol; 3,5-DCP = 3,5-dichlorophenol; MCPA = 2-methyl-4-chlorophenoxyacetic acid; TTAB = tetradecyltrimethylammonium bromide.

Table 2. Selected Examples of Toxicant-Specific Luminescent Biosensor Strains

Bacterial Host Strain	Sensor Plasmid	Luminescent Reporter Gene	Inducing Compound(s)	Reference
Bacillus subtilis BR151	pTOO24	*luc*FF	Cadmium, antimony, zinc, tin; metal-contaminated soils	36,49
Escherichia coli AW3110	pTOO31	*luc*FF	Arsenic	40
Escherichia coli MC1061	pTOO31	*luc*FF	Arsenic; arsenic-contaminated sediments; arsenic-contaminated soils	40,47,57
Escherichia coli MC1061	pTOO11	*luc*FF	Mercury	40,58
Escherichia coli MC1061	pzntRluc	*luc*FF	Zinc, cadmium, mercury in spiked soil	59
Escherichia coli MC1061	pmerBR$_{BS}$luc	*luc*FF	Mercury; mercury, cadmium, zinc in spiked soil	59,60
Escherichia coli MT102-PIR	pUT-mer-lux	*lux*CDABE	Mercury	61
Staphylococcus aureus RN4220	pTOO24	*luc*FF	Cadmium, lead, antimony, tin; metal-contaminated soils	36,49
Staphylococcus aureus RN4220	pTOO21	*luc*FF	Arsenic	62
Pseudomonas fluorescens OS8	pTPT31	*luc*GR	Arsenite; arsenite-spiked soil; arsenic-contaminated sediments	47,56,63,64
Pseudomonas fluorescens OS8	pTPT11	*luc*GR	Mercury; mercury-spiked soil	56,63,64
Ralstonia eutropha AE104	pchrBluc	*luc*FF	Chromate	59
Ralstonia eutropha AE2515	pMOL1550	*lux*CDABE	Nickel, cobalt, bioavailable nickel in soils	65
Ralstonia eutropha ENV307	pUTK60	*lux*CDABE	Polychlorinated biphenyls	48

LUMINESCENT BACTERIAL BIOSENSORS FOR ASSESSING POTENTIAL GENOTOXICITY

Bioluminescent assay systems have also been employed to assess the potential genotoxicity of both chemicals and environmental samples. The Mutatox test (Azur Environmental, CA, USA) employs a dark mutant of *V. fischeri* (strain M169) and works in an opposite manner to the Microtox test, in that this strain exhibits increased light production (restoration of bioluminescence) when grown in the presence of sublethal concentrations of genotoxic agents. The Mutatox test is considered a promising tool for detection of genotoxic compounds in environmental samples (66–68). Both the Microtox and Mutatox test have been used in tandem to assess the potential acute toxicity and genotoxicity, respectively, of lipophilic contaminants in aquatic ecosystems (69).

Another test that utilizes a luciferase reporter gene system is the Vitotox assay (Thermo Electron Corporation, USA). This assay, which measures both genotoxicity and cytotoxicity, is based on recombinant *Salmonella typhimurium* TA104 bacteria containing the complete *V. fischeri lux* gene cassette under the control of the *rec*N promoter (70).

Vollmer et al. (71) have also described the construction and initial characterization of bacterial biosensors for the detection of DNA damage. In their research they constructed plasmids in which the DNA damage-inducible promoters *rec*A, *uvr*A, and *alk*A from *Escherichia coli* were fused to the *V. fischeri luxCDABE* operon. They reported that the fusion of the *rec*A promoter to *lux*CDABE yielded the most sensitive response. This recombinant strain (*E.coli* DPD2794) was later employed by Min et al. (72) to detect the genotoxicity of various chemicals and was demonstrated to be capable of distinguishing distinct bacterial responses for direct and indirect DNA damaging agents.

FUTURE DEVELOPMENT OF BIOSENSORS AS RAPID AND ROBUST TOOLS FOR ENVIRONMENTAL MONITORING

There is a growing demand for sensitive and easy to use biological assays for the rapid and cost effective detection of environmentally relevant concentrations of pollutants in soil, sediment, and water ecosystems. Recombinant luminescent bacterial biosensors have recently emerged as promising tools in environmental monitoring for the sensitive and specific quantification of both toxicity and bioavailability. These biosensors possess several of the characteristics of an ideal bioassay system, in that they are relatively inexpensive to conduct, easy to use, ecologically relevant, and amenable for on-line or *in situ* monitoring. In addition, previous research has demonstrated that freeze-drying of luminescent bacterial biosensors has only moderate effects on the performance with respect to sensitivity or induction coefficients (49,59). The availability of bacterial biosensors in a freeze-dried state would negate the need and cost of continuous culturing while ensuring a clonal strain, making reagent-like usage a viable option. Integration of the biosensor constructs into the host chromosome and the creation of stable constructs, which can be maintained even under nonselective conditions, have also been demonstrated (61,73). Optimization of immobilization techniques will advance the exploitation of the commercial potential of these biosensors, by providing a stable and practical system for use in real-time pollutant detection (74–77). A novel automated continuous toxicity test system using a recombinant bioluminescent freshwater bacterium has recently been described (78), giving further promise to the routine utilization of these biosensor systems for real-time biomonitoring of water toxicity in the near future.

BIBLIOGRAPHY

1. Steinberg, S.M., Poziomek, E.J., Engelmann, W.H., and Rogers, K.R. (1995). A review of environmental applications of bioluminescence measurements. *Chemosphere* **30**(11): 2155–2197.
2. Ramanathan, S., Ensor, M., and Daunert, S. (1997). Bacterial biosensors for monitoring toxic metals. *Trends Biotechnol.* **15**: 500–506.
3. D'Souza, S.F. (2001). Microbial biosensors. *Biosensors Bioelectron.* **16**: 337–353.
4. Keane, A., Phoenix, P., Ghoshal, S., and Lau, P.C.K. (2002). Exposing culprit organic pollutants: a review. *J. Microbiol. Methods* **49**: 103–119.
5. Simpson, M.L. et al. (1998). Bioluminescent-bioreporter integrated circuits form novel whole-cell biosensors. *Trends Biotechnol.* **16**: 332–338.
6. Grayeski, M.L. (1987). Chemiluminescence analysis. *Anal. Chem.* **59**: 1243A–1256A.
7. Wilson, T. and Hastings, J.W. (1998). Bioluminescence. *Annu. Rev. Cell Dev. Biol.* **14**: 197–230.
8. Hastings, J.W. et al. (1985). Biochemistry and physiology of bioluminescent bacteria. *Adv. Microbial Physiol.* **26**: 235–291.
9. Stewart, G.S.A.B. and Williams, P. (1992). Lux genes and the applications of bacterial luminescence. *J. Gen. Microbiol.* **138**: 1289–1300.
10. Meighen, E.M. and Dunlap, P.V. (1993). Physiological, biochemical and genetic control of bacterial luminescence. *Adv. Microbiol. Physiol.* **34**: 1–67.
11. Lampinen, J., Virta, M., and Karp, M. (1995). Use of controlled luciferase expression to monitor chemicals affecting protein synthesis. *Appl. Environ. Microbiol.* **61**: 2981–2989.
12. Tom-Petersen, A., Hosbond, C., and Nybroe, O. (2001). Identification of copper-induced genes in *Pseudomonas fluorescens* and use of a reporter strain to monitor bioavailable copper in soil. *FEMS Microbiol. Ecol.* **38**: 59–67.
13. Burlage, R.S. (1997). Emerging technologies: bioreporters, biosensors, and microprobes. In: *Manual of Environmental Microbiology*. C.J. Hurst, G.R. Knudsen, M.J. McInerney, L.D. Stetzenbach, and M.V. Walter (Eds.). ASM Press, Washington, DC, pp. 115–123.
14. Kaiser, K.L.E. and Palabrica, V.S. (1991). *Photobacterium phosphoreum* toxicity data index. *Water Pollut. Res. J. Can.* **26**: 361–431.
15. Bulich, A.A., Greene, M.W., and Isenberg, D.L. (1981). Reliability of the bacterial luminescence assay for the determination of the toxicity of pure compounds and complex effluents. In: *Aquatic Toxicology and Hazard Assessment: Fourth Conference*. D.R. Branson and K.L. Dickson (Eds.). STP 737, American Society for Testing and Materials, Philadelphia, pp. 338–347.
16. Vasseur, P., Ferrard, J.F., Rast, C., and Larbaight, D. (1984). Luminescent marine bacteria in ecotoxicity screening tests of complex effluents. In: *Toxicity Screening Procedures Using Bacterial Systems*. D.L. Liu and B.J. Dutka (Eds.). Marcel Dekker, New York, pp. 22–36.
17. Pardos, M. et al. (1999). Acute toxicity assessment of Polish (waste) water with a microplate-based *Hydra attenuata* assay: a comparison with the Microtox® test. *Sci. Total Environ.* **243/244**: 141–148.
18. Jung, K. and Bitton, G. (1997). Use of Ceriofast™ for monitoring the toxicity of industrial effluents: comparison with the 48-h acute *Ceriodaphnia* toxicity test and Microtox®. *Environ. Toxicol. Chem.* **16**: 2264–2267.
19. Cheung, Y.H. et al. (1997). Assessment of sediment toxicity using different trophic organisms. *Arch. Environ. Contam. Toxicol.* **32**: 260–267.
20. Doherty, F.G. (2001). A review of the Microtox® toxicity test system for assessing the toxicity of sediments and soils. *Water Qual. Res. J. Can.* **36**: 475–518.
21. Guzzella, L. (1998). Comparison of test procedures for sediment toxicity evaluation with *Vibrio fischeri* bacteria. *Chemosphere* **37**: 2895–2909.
22. Lappalainen, J., Juvonen, R., Vaajasaari, K., and Karp, M. (1999). A new flash method for measuring the toxicity of solid and colored samples. *Chemosphere* **38**: 1069–1083.
23. Lappalainen, J.O. et al.(2000). Comparison of the total mercury content in sediment samples with a mercury sensor bacteria test and *Vibrio fischeri* toxicity test. *Environ. Toxicol. Chem.* **15**: 443–448.
24. la Farre, M. et al. (2001). Wastewater toxicity screening of non-ionic surfactants by Toxalert® and Microtox® bioluminescence inhibition assays. *Anal. Chim. Acta* **427**: 181–189.
25. Jennings, V.L.K., Rayner-Brandes, M.H., and Bird, D.J. (2001). Assessing chemical toxicity with the bioluminescent photobacterium (*Vibrio fischeri*): a comparison of three commercial systems. *Water Res.* **35**(14): 3448–3456.
26. Mowat, F.S. and Bundy, K.J. (2001). Correlation of field-measured toxicity with chemical concentration and pollutant availability. *Environ. Int.* **27**: 479–489.
27. Fernández-Alba, A.R., Guil, M.D.H., Lopez, G.D., and Chisti, Y. (2002). Comparative evaluation of the effects of pesticides in acute toxicity luminescence bioassays. *Anal. Chim. Acta* **451**: 195–202.
28. Wang, C. et al. (2003). Ecotoxicological and chemical characterisation of selected treatment process effluents of municipal sewage treatment plant. *Ecotoxicol. Environ. Safety* **56**: 211–217.
29. Ulitzur, S., Lahav, T., and Ulitzur, N. (2002). A novel and sensitive test for rapid determination of water toxicity. *Environ. Toxicol.* **17**(3): 291–296.
30. Paton, G.I., Rattray, E.A.S., Campbell, C.D., Cresser, M.S., Glover, L.A., Meussen, J.C.L., and Killham, K. (1997). Use of genetically modified microbial biosensors for soil ecotoxicity testing. In: *Biological Indicators of Soil Health*. C.E. Pankhurst, B.M. Doube, and V.V.S.R. Gupta (Eds.). CAB International, Wallingford, UK, pp. 397–418.
31. Boyd, E.M., Meharg, A.A., Wright, J., and Killham, K. (1998). Toxicity of chlorobenzenes to a *lux*-marked terrestrial bacterium, *Pseudomonas fluorescens*. *Environ. Toxicol. Chem.* **17**: 2134–2140.
32. Paton, G.I., Campbell, C.D., Glover, L.A., and Killham, K. (1995). Assessment of bioavailability of heavy metals using lux modified constructs of *Pseudomonas fluorescens*. *Lett. Appl. Microbiol.* **20**: 52–56.
33. Layton, A.C., Gregory, B., Schultz, T.W., and Sayler, G.S. (1999). Validation of genetically engineered bioluminescent surfactant resistant bacteria as toxicity assessment tools. *Ecotoxicol. Environ. Safety* **43**: 222–228.
34. Chaudri, A.M. et al. (2000). Response of a *Rhizobium*-based luminescence biosensor to Zn and Cu in soil solutions from sewage sludge treated soils. *Soil Biol. Biochem.* **32**: 383–388.
35. Bundy J.G., Campbell, C.D., and Paton, G.I. (2001). Comparison of response of six different luminescent bioassays to

bioremediation of five contrasting oils. *J. Environ. Monitor.* **3**: 404–410.

36. Ivask, A. et al. (2004). Recombinant luminescent bacterial sensors for the measurement of bioavailability of cadmium and lead in soils polluted by metal smelters. *Chemosphere* **55**: 147–156.

37. Sinclair, G.M., Paton, G.I., Meharg, A.A., and Killham, K. (1999). *Lux*-biosensor assessment of pH effects on microbial sorption and toxicity of chlorophenols. *FEMS Microbiol. Lett.* **174**: 273–278.

38. Strachan, G. et al. (2001). Use of bacterial biosensors to interpret the toxicity and mixture toxicity of herbicides in freshwater. *Water Res.* **35**: 3490–3495.

39. Turner, N.L. et al. (2001). A novel toxicity fingerprinting method for pollutant identification with *lux*-marked biosensors. *Environ. Toxicol. Chem.* **20**: 2456–2461.

40. Tauriainen, S., Virta, M., Chang, W., and Karp, M. (1999). Measurement of firefly luciferase reporter gene activity from cells and lysates using *Escherichia coli* arsenite and mercury sensors. *Anal. Biochem.* **272**: 191–198.

41. Bundy, J.G., Paton, G.I., and Campbell, C.D. (2004). Combined microbial community level and single species biosensor responses to monitor recovery of oil polluted soil. *Soil Biol. Biochem.* **36**: 1149–1159.

42. Paton, G.I. et al. (1995). Use of luminescence-marked bacteria to assess copper bioavailability in malt whiskey distillery effluent. *Chemosphere* **31**: 3217–3224.

43. Brown, J.S. et al. (1996). Comparative assessment of the toxicity of a papermill effluent by respirometry and a luminescence-based bacterial assay. *Chemosphere* **32**: 1553–1561.

44. Boyd, E.M. et al. (1997). Toxicity assessment of xenobiotic contaminated groundwater using *lux* modified *Pseudomonas fluorescens*. *Chemosphere* **35**: 1967–1985.

45. Boyd, E.M., Meharg, A.A., Wright, J., and Killham, K. (1997). Assessment of toxicological interactions of benzene and its primary degradation products (catechol and phenol) using a *lux*-modified bacterial bioassay. *Environ. Toxicol. Chem.* **16**: 849–856.

46. McGrath, S.P. et al. (1999). Assessment of the toxicity of metals in soils amended with sewage sludge using a chemical speciation technique and a *lux*-based biosensor. *Environ. Toxicol. Chem.* **18**: 659–663.

47. Petänen, T. et al. (2003). Assessing sediment toxicity and arsenite concentration with bacterial and traditional methods. *Environ. Pollut.* **122**: 407–415.

48. Layton, A.C., Muccini, M., Ghosh, M.M., and Sayler, G.S. (1998). Construction of a bioluminescent reporter strain to detect polychlorinated biphenyls. *Appl. Environ. Microbiol.* **64**: 5023–5026.

49. Tauriainen, S., Karp, M., Chang, W., and Virta, M. (1998). Luminescent bacterial sensor for cadmium and lead. *Biosensors Bioelectron.* **13**: 931–938.

50. Gu, M.B. and Gil, G.C. (2001). A multi-channel continuous toxicity monitoring system using recombinant bioluminescent bacteria for classification of toxicity. *Biosensors Bioelectron.* **16**: 661–666.

51. Wiles, S., Whiteley, A.S., Philp, J.C., and Bailey, M.J. (2003). Development of bespoke bioluminescent reporters with the potential for *in situ* deployment within a phenolic-remediating wastewater treatment system. *J. Microbiol. Methods* **55**: 667–677.

52. Ren, S. and Frymier, P.D. (2005). Toxicity of metals and organic chemicals evaluated with bioluminescence assays. *Chemosphere* (in press).

53. Ren, S. (2004). Assessing wastewater toxicity to activated sludge: recent research and developments. *Environ. Int.* **30**: 1151–1164.

54. Lewis, J.C. et al. (1998). Applications of reporter genes. *Anal. Chem.* **70**: 579A–585A.

55. Bontidean, I. et al. (2004). Biosensors for detection of mercury in contaminated soils. *Environ. Pollut.* **131**: 255–262.

56. Petänen, T. and Romantschuk, M. (2003). Toxicity and bioavailability to bacteria of particle-associated arsenite and mercury. *Chemosphere* **50**: 409–413.

57. Turpeinen, R., Virta, M., and Häggblom, M.M. (2003). Analysis of arsenic bioavailability in contaminated soils. *Environ. Toxicol. Chem.* **22**: 1–6.

58. Virta, M., Lampinen, J., and Karp, M. (1995). A luminescence-based mercury biosensor. *Anal. Chem.* **67**: 667–669.

59. Ivask, A., Virta, M., and Kahru, A. (2002). Construction and use of specific luminescent recombinant bacterial sensors for the assessment of bioavailable fraction of cadmium, zinc, mercury and chromium in the soil. *Soil Biol. Biochem.* **34**: 1439–1447.

60. Ivask, A., Hakkila, K., and Virta, M. (2001). Detection of organomercurials with whole-cell bacterial sensors. *Anal. Chem.* **73**: 5168–5171.

61. Hansen, L.H. and Sørensen, S.J. (2000). Versatile biosensor vectors for detection and quantification of mercury. *FEMS Microbiol. Lett.* **193**: 123–127.

62. Tauriainen, S., Karp, M., Chang, W., and Virta, M. (1997). Recombinant luminescent bacteria for measuring bioavailable arsenite and antimonite. *Appl. Environ. Microbiol.* **63**: 4456–4461.

63. Petänen, T., Virta, M., Karp, M., and Romantschuk, M. (2001). Construction and use of broad host range mercury and arsenite sensor plasmids in the soil bacterium *Pseudomonas fluorescens* OS8. *Microbial Ecol.* **40**: 360–368.

64. Petänen, T. and Romantschuk, M. (2002). Use of bacterial sensors as an alternative method for measuring heavy metals in soil extracts. *Anal. Chim. Acta* **456**: 55–61.

65. Tibazarwa, C. et al. (2001). A microbial biosensor to predict bioavailable nickel in soil and its transfer to plants. *Environ. Pollut.* **113**: 19–26.

66. Thomas, K.V. et al. (2002). Characterisation of potentially genotoxic compounds in sediments collected from United Kingdom estuaries. *Chemosphere* **49**: 247–258.

67. Isidori, M. et al. (2003). *In situ* monitoring of urban air in Southern Italy with the tradescantia micronucleus bioassay and semipermeable membrane devices (SPMDs). *Chemosphere* **52**: 121–126.

68. Isidori, M., Lavorgna, M., Nardelli, A., and Parrella, A. (2004). Integrated environmental assessment of Volturno river in South Italy. *Sci. Total Environ.* **327**: 123–124.

69. Johnson, B.T. and Long, E.R. (1998). Rapid toxicity assessment of sediments from estuarine ecosystems: a new tandem *in vitro* testing approach. *Environ. Toxicol. Chem.* **17**: 1099–1106.

70. van der Lelie, D. et al. (1997). The Vitotox® test, an SOS bioluminescence *Salmonella typhimurium* test to measure genotoxicity kinetics. *Mutat. Res.* **389**: 279–290.

71. Vollmer, A.C. et al. (1997). Detection of DNA damage by use of *Escherichia coli* carrying *rec*A′::*lux*, *uvr*A′::*lux*, or *alk*A′::*lux* reporter plasmids. *Appl. Environ. Microbiol.* **63**: 2566–2571.

72. Min, J., Kim, E.J, LaRossa, R.A., and Gu, M.B. (1999). Distinct responses of a recA::luxCDABE *Escherichia coli* strain to direct and indirect DNA damaging agents. *Mutat. Res.* **442**: 61–68.
73. Hay, A.G., Rice, J.F., Applegate, B.M., Bright, N.G., and Sayler, G.S. (2000). A bioluminescent whole-cell reporter for detection of 2,4-dichlorophenoxyacetic acid and 2,4-dichlorophenol in soil. *Appl. Environ. Microbiol.* **66**: 4589–4594.
74. Corbisier, P. et al. (1999). Whole cell- and protein-based biosensors for the detection of bioavailable heavy metals in environmental samples. *Anal. Chim. Acta* **387**: 235–244.
75. Tauber, M., Rosen, R., and Belkin, S. (2001). Whole cell detection of halogenated organic acids. *Talanta* **55**: 959–964.
76. Sagi, E. et al. (2003). Fluorescence and bioluminescence reporter functions in genetically modified bacterial sensor strains. *Sensors Actuators* **B 90**: 2–8.
77. Timur, S., Pazarlioğlu, N., Pilloton, R., and Telefoncu, A. (2003). Detection of phenolic compounds by thick film sensors based on *Pseudomonas putida*. *Talanta* **61**: 87–93.
78. Cho, J.-C. et al. (2004). A novel continuous toxicity test system using a luminously modified freshwater bacterium. *Biosensors Bioelectron.* **20**: 338–344.

DEVELOPMENT AND APPLICATION OF SEDIMENT TOXICITY TEST FOR REGULATORY PURPOSES

Thomas Ángel del Valls Casillas
University of Cadiz
Puerto Real, Spain

INTRODUCTION

Since 1950 there is a growing concern about sediment contamination, and international conventions have been established for the protection of the marine environment (i.e., GIPME, 2000; London Convention, www.Londonconvention.org). Each country has developed particular guidelines for the protection of the environment but even these recommendations have no legislative force; they have been routinely used for characterizing contaminated sediments and for practical purposes such as dredging.

The first approaches to this issue were centered on sediment chemistry and thus the classification was made by comparing the concentration of determined contaminants of concern and sediment quality values established to effectively protect the aquatic environment (1). All these approaches follow the same schema and the same principles starting with an evaluation of the existing information. Contaminants of concern should be evaluated according to the information regarding physical, chemical, and biological properties of the sediments. The principal advantage of using this information is the cost effectiveness of the data collection and, if the data are consistent, decisions can be made to protect the environment without any further assessment.

For those sediments where existing information is not enough for decision-making, further assessments are needed. Traditionally, further assessment included the analysis of a list of contaminants, provided by regulators and assessors, that are to be considered in the evaluation. Concentrations are corrected for the particular characteristics and compared to sediment quality values defined as limit values to classify the sediments. These types of chemical guidelines are used worldwide and can be divided into two different categories depending on how they are derived: the empirical approach is based on statistical analyses of co-occurring chemical and biological effect data (2,3), while the mechanistic approach is based on calculation of equilibrium partitioning between a bound phase and a dissociated phase within pore water (4). Other guidelines use only a measure of the natural background condition or the contaminant concentration that satisfies the different concerns. The use of each one is under debate and new approaches are under development to meet existing regulatory requirements (5,6). One general limitation that applies for all matrices using numerical sediment quality guidelines is that they only address concerns regarding those chemicals for which guidelines have been developed. In addition, the actual availability and thus the effects depend on other factors such as the grain size distribution, composition of the organic matter, or concentration of sulfides in the sediment.

Since the mid-1970s research has focused on the development of sediment toxicity bioassays that could effectively predict potential biological adverse effects and could be used as an effective tool to characterize sediment toxicity. It is widely accepted that there are no chemical measurements that reliably predict sediment toxicity. In addition, sediments are a complex matrix that require complex tests for estimating the actual bioavailability of contaminants (7). Here we review the evaluation process that has been applied in the different initiatives carried out to introduce this new approach in a regulatory context. The design of toxicity tests for the characterization of dredged material in Spain is used as an example.

TEST DEVELOPMENT FOR USE IN REGULATORY PURPOSES

Once the new approach was established and widely accepted, much research was done regarding the selection of the proper battery of tests to include for regulatory purposes. The suitability of a toxicity test for decision-making is determined by the development degree of these tests through research (under academic or private initiatives) and the quantity and quality of the available scientific information. Involvement of the regulatory agencies is fundamental to evaluate the usefulness of the resulting information; they are the entities that should address the uncertainties associated with the different analytical tools used for the decision-making process. Before a bioassay can be included for regulatory purposes, some basic characteristics should be considered. Here we include some of the factors that should be taken into account when evaluating a bioassay for regulatory purposes:

Selection of Organisms

Test species should be autochthon and ecologically relevant in the areas of concern. This extends to those species widely distributed and adapted to the regional aquatic ecosystems naturally or by means of cultures. Their importance in the trophic chain should be established in relation to the geographical zone where the bioassay is going to be applied. It is preferable that they are going to be directly affected by contaminated sediments. Test species are more valuable if they are tolerant over a wide range of environmental variables (i.e., pH, salinity, grain size distribution, or volatile sulfides) and easy to maintain in the laboratory or purchased from commercial cultures. Information regarding the life cycle of the test species, information about its growth and reproduction, and information about the test species' sensitivity for the contaminants of concern are valuable tools for the final selection of the proper test organism. Easy taxonomical identification of the species must be taken into account. For example, the manila clam can be selected as a test organism for sediment toxicity assessment instead of *Macoma nasuta* because although it is not an autochthon species, it is widely cultured and meets all the criteria for its selection.

Selection of Endpoints

Criteria for selecting endpoints in environmental risk assessment processes (ERAs) include susceptibility to the stressor, ease of measurement, unambiguous definitions, and societal and ecological relevance (8). To link sediment quality with ERAs, similar criteria should be employed to select endpoints to assess sediment quality (9). The selection of the endpoint measured in the test must be related to a receptor, process, or effect of concern. The design of the test should pose and answer some questions of the decision-making process and its use should be justified in the followed assessment framework. As stated by Chapman et al. (10), the primary response is mortality, followed by reproductive or growth effects, and finally other sublethal endpoints such as behavior or biomarkers. However, these last ones only provide an indication of exposure and have not yet been linked directly to impacts at the organism level except using histopathology (11). The recommended tests measure mortality and reproductive or growth effects and use ecologically significant taxa similar to or related to resident taxa, which are likely to be exposed and which are appropriately sensitive. Even taking into account all these factors for the selection of the best measurable endpoint, sediment toxicity tests for predicting the effects of contaminants on natural populations, communities, and ecosystems are criticized because they do not take indirect effects into account.

Experimental Design

Two main conditions must be met to properly develop a laboratory toxicity test for regulatory purposes: the test should be designed with a specific goal for the decision-making process and the test must be consistently applied, with its experimental conditions perfectly defined, and should account for the influence of the experimental conditions on test organisms. Standard protocols, including technical reports from government or private organizations, or those endorsed by a standardization group [e.g., American Society of Testing and Materials (ASTM) or Environment Canada (EC)], or published papers in peer-reviewed journals, including interpretive guidance, are available for different species [e.g., amphipods (12,13) or polychaetes (14)]. Another factor affecting the experimental design and application of a test is adherence to standards to ensure the good health of test organisms and to limit the total costs of conducting the test. For example, the commercial bioassay Microtox® has been widely recommended and applied. As a commercial test it is highly standardized after a long period of research for its characterization. Although modifications to the initial protocols have been developed, the oldest design is recommended for use until the new one has been standardized and characterized.

Ruggedness

Test ruggedness is related to the sensitivity of test methods to variations in laboratory conditions such as the feeding regime, light and photoperiod, or grain size distribution. It is important to know the precision in defining features and to establish good quality assurance/quality control guidance.

Ring Tests

Standard methods are required in regulatory programs. In addition to standardization, other important prerequisites are good reproducibility, acceptable interlaboratory variability (defined as multilaboratory precision), and repeatability and intralaboratory precision. While intralaboratory precision reflects the ability of trained laboratory personnel to obtain consistent results repeatedly when performing the same test on the same organisms using the same toxicant (repeatability can include variability related to different operators), interlaboratory precision (also referred to as round-robin or ring tests) is a measure of how reproducible the method is when conducted by a large number of laboratories using the same method, organism, and samples (13). This information is a measure of confidence in the consistency of test performance and application.

Different studies have been successfully completed to evaluate sediment toxicity bioassays (including whole sediment and pore water or elutriate), for example, the whole sediment toxicity test using amphipods (15,16), the bioassay of inhibition of bioluminescence (17,18), and the liquid phase bioassay using sea urchin embryos (17,19).

Validation of the Tests

The validation process of a test can be accomplished through studies to find relationships between sediment toxicity and contaminant concentrations, using a different range of contamination, a different range of sediment conditions, or by means of comparison with other standardized tests. Test validation focuses on comparing data of field impacts to ensure the predictive value of such tools.

Data Management Procedures

When test results are accepted, they must be easily interpreted. The role of the test in the regulatory process is to answer the assessment questions formulated during the sample evaluation. Different approximations have been made to classify sediments according to the biological effects produced. In some countries and for some perspectives, strict guidelines have been established, although it seems more desirable to integrate multiple lines of evidence (terms that define categories of information) and to individuate in this way the process for each particular situation. This approach integrates different categories of information, for example, chemical, toxicological, and benthic community structure. The way in which this information is integrated is going to depend on the project itself and on the information available for each line of evidence.

APPLICATION OF SEDIMENT TOXICITY TEST FOR REGULATORY PURPOSES

In Fig. 1 a general tiered testing scheme to be followed when characterizing contaminated sediments and dredged materials is included as an example of the multiple approaches available. The first tier is accomplished if the available information, including geographical, geological, and pollutant sources, ascertains that the sediment contamination has no effects. If we find fine sediments and important sources of pollutants that result in high concentrations of contaminants, the characterization proceeds to the next tier and information regarding sediment chemistry and the characterization of other parameters, such as microbiology, should be accomplished. Some variables involved in the process are sampling, handling, and storage of the sediment, which influence the bioavailability of contaminants. Each regulatory agency should establish standard procedures that can guarantee the accuracy of the results. The regulatory agencies should also establish the parameters to be characterized, which normally include metals and metalloids, PCBs, PAHs, pesticides, and other compounds such as TBTs. Then the results are compared to the selected sediment quality guidelines for a single species. Such sediment chemical benchmarks have been used as rough limit values or regulatory levels. Depending on how the guidelines were derived (if they were intended to predict toxicity or the absence of toxicity) and what effects were to be assessed, one can judge if the regulatory requirements are met.

For the next tier, a secondary assessment regarding biological tests is used for the characterization. Three different categories can be distinguished according to the exposure routes: tests addressed to measure direct water column effects, those addressed to measure direct effects on benthic organisms, and those that measure the bioaccumulation effect. Because potential effects should include all possible exposure routes, water column effects are characterized by means of liquid phase bioassays and benthic effects by using benthic organisms exposed to the whole sediment. Differences are found when bioaccumulation tests are used. Some countries have included these last bioassays (i.e., Canada) as an important part of the battery of tests. Other countries only recommend them when contaminants accumulated by organisms are present and thus apply these more cost-effective tools to the particular cases where high concentrations of determined substances are present.

Some regulatory agencies have included a list of suitable or recommended bioassays but the final set of tools used for the characterization process is open. Others just include a list of recommendations and factors in selecting the proper battery of tests. The wise use of sediment bioassays is to design a battery of tests that will give a wide range of effects. A good example is the management of dredged materials in Canada (20). In this country, test methods currently used include an acute toxicity test using amphipods, a fertilization assay using echinoids, a solid phase toxicity test using the photoluminescent bacteria, and a bedded sediment bioaccumulation test using bivalves.

During the last decades different sediment toxicity tests have been carried out in Spain as part of a weight of evidence approach using the sediment quality triad (21). These studies used basically acute sediment toxicity tests that covered organisms from a bacterial population to juveniles of commercial species of fish (Table 1). Today, some of these acute toxicity tests shown in the table have been adapted for use in characterizing dredged material quality assessment in Spanish ports and are being considered for incorporation in the new recommendations for the management of dredged material in ports of Spain (RMDM). These tests include the next types and bioassays: screening, the Microtox STP, whole sediment, the amphipod, the polychaeta, the irregular sea urchin tests, and, using elutriates, the sea urchin larvae test (Table 1).

The design and selection of the different tests was carried out under a joint contract between CEDEX and the University of Cádiz and after a practical phase to validate the potential use of standardized tests by other countries but using autochthonous species of organisms in Spain. During this practical phase, different ports were sampled and analyzed using chemical and ecotoxicological tests. Results showed that the mentioned bioassays can be used to characterize dredged material in Spain and some biological guidelines or criteria were developed. These guidelines are similar to others designed and applied for the same purposes by different regulatory agencies, such as the U.S Environmental Protection Agency, Environment Canada, the Dutch Ministry for Commerce and Navigation, and Environment Australia (Table 2). Biological guidelines derived for Spain are shown in Table 3. These values are similar to those proposed by other agencies, although it should be noted that there is a double criteria for classification of toxicity in one sample based on an absolute value together with the demonstration of significant differences by means of statistical tools using a control or reference sediment.

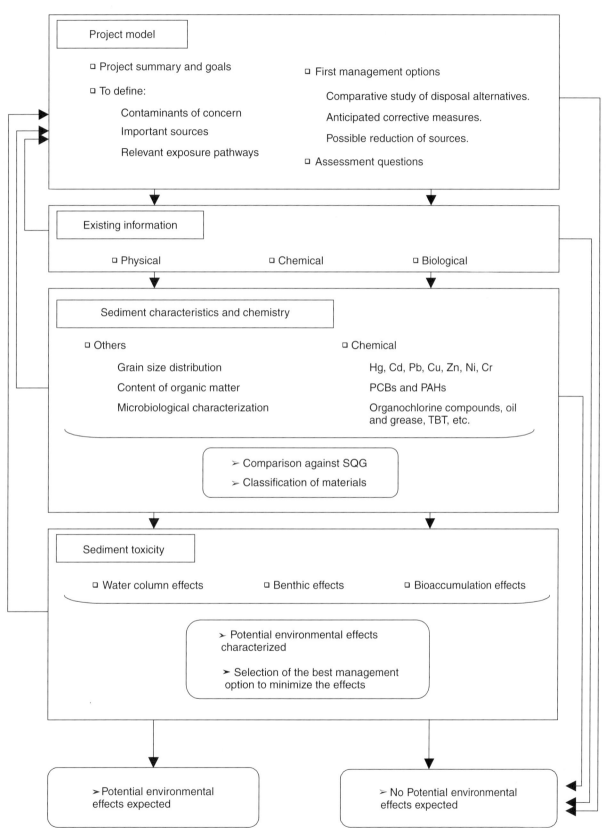

Figure 1. Tiered testing scheme to characterize contaminated sediments and dredged material. It has been adapted from different examples (Basque country, The Netherlands, Canada, or the United States).

Table 1. List of Different Bioassays Checked for Potential Uses in the Battery of Sediment Toxicity Tests Routinely Used to Characterize Dredged Material Disposal Options in Spain

Organisms	Exposure Route	Duration	Temperature, °C	Endpoints	Checked[a]	Acceptable[b]	References
Microtox® (SPT) (*Vibrio fischeri*)	Screening	15 min to 1 h	15	Luminescence inhibition	Yes	Yes	23
Echinoderms (*P. lividus*)	Elutriates	48 h	20	Survival (%)	Yes	Yes	23,24
Rotifer population (*B. plicatilis*)	Elutriates	7d	25	Survival (%)	Yes	?	25
Amphipod (*Corophium volutator, A. brevicornis*)	Whole sediment	10 d	20	Survival (%)	Yes	Yes	21,23
Benthic microalgae (*Cylindrotheca closterium*)	Whole sediment	24–72 h	20	Growth inhibition	0	?	26
Echinoderm (*Echinocardium cordatum*)	Whole sediment	14 d	15	Survival (%)	Yes	Yes	23,24
Polychaetae (*Arenicola marina*)	Whole sediment	10 d	15	Survival (%), bioaccumulation	Yes Yes	Yes ?	14,23
Bivalve (*Ruditapes philippinarum*)	Whole sediment	48 h to 15 d	20	Survival (%); burrowing (TB_{50}) bioaccumulation; histological damage; biomarkers; vitellogenin	Yes	?	27
Crab (*Carcinus maenas*)	Whole sediment	21 d to 3 m	15	Survival (%); burrowing (TB_{50}) Bioaccumulation; histological damage; biomarkers; vitellogenin	Yes/?	Yes/?	28
Fish: benthic and pelagic (*Solea senegalensis; Sparus aurata*)	Whole sediment	1–3 m	20	Survival; bioaccumulation; histological damage; biomarkers	Yes	Yes/?	11

[a] Yes = standardized; ? = in development; 0 = not checked.
[b] Yes means useful for dredged material characterization.
Source: Adapted from Reference 22.

Another important aspect is the use of autochthonous species from Spanish coasts.

FINAL REMARKS AND FUTURE RESEARCH

Research and development are focused on reducing the uncertainties and costs associated with assessing dredged sediment material. Both the number and complexity of environmental questions being posed to port and dredged material managers have steadily increased in recent years, and this trend will likely continue in the future. Ongoing research is being used to provide answers to these questions and to address sources of uncertainty in the assessment and decision-making process. Other confounding factors already discussed must be taken into account when selecting the battery of tests to be applied for characterizing toxicity. Although the effect of some of these factors can be estimated and corrected using a proper reference sediment, others are difficult and the lack of control on some of these variables can have serious implications if test results are used for regulatory purposes.

Acknowledgments

This work has been partially funded by grants from the Spanish Ministry of Development (BOE 13-12-02) and of Education and Science (REN2002_01699/TECNO and VEM2003_-20563/INTER). The design of the acute toxicity tests for dredged material characterization was conducted under joint contract between CEDEX and the University of Cádiz (2001 and 2003).

BIBLIOGRAPHY

1. CEDEX (1994). *Recommendations for the Management of Dredged Material in the Ports of Spain*. Puertos del Estado, Madrid.
2. Long, E.R., Field, L.J., and MacDonald, D.D. (1998). *Environ. Toxicol. Chem.* **17**: 714–727.
3. Riba, I., Casado-Martínez, M.C., Forja, J.M., and DelValls, T.A. (2004). *Environ. Toxicol. Chem.* **23**: 271.

Table 2. Summarized Description of Different Bioassays Selected for Dredged Material Characterization in Spain[a]

Type	Test[b]	Country	Criteria	References
Screening	Microtox	The Netherlands	100 toxic units	17
		Canada	1000 mg · L^{-1} dw	29,30
		Australia	—	30
		Italy	Toxicity assigned according to STI units after correction for fine fraction	
Solid phase	Amphipods	The Netherlands	35%	17
		Canada	+20% mortality compared to reference sediment or +30% compared to control sediment and statistical difference	20
		EEUU	+20% mortality compared to reference sediment or +30% compared to control sediment and statistical difference	31
		UK	40% mortality	32
		Italy	—	—
		Australia	—	30
	Benthic algae viability	Australia	—	30
Liquid phase	Sea urchin embryo development	Canada	+25% abnormality compared to control seawater and statistical difference	20
		Italy	—	—
	Sea urchin fertilization and larval development	Australia	—	30
	Bivalve larval development	Australia	—	30
	Tiger prawn survival (postlarvae)	Australia	—	30
	Algal growth inhibition test	Australia	—	30
Bioaccumulation	Bivalves	Canada	Statistical significant difference in tissue concentrations between test sediment and reference or control sediment	29

[a]The biological guidelines proposed for each test in different countries are also included and are used in the management of disposal options of these materials in each country.
[b]At least two different acute toxicity tests required, preferably three.

Table 3. Biological Guidelines for the Sediment Toxicity Tests Recommended in Spain to Be Part of the New Recommendations for the Management of Dredged Material (RMDM) in Ports of Spain that Are Under Discussion for Final Adoption by the Ministry of Development

Type	Test	Criteria	Comments
Screening	Microtox	1000 mg · L^{-1} dw	SPT protocol; correction for the fine fraction
Liquid phase	Sea urchin embryo development	+25% abnormality compared to control seawater and statistical difference	Test species: *Paracentrotus lividus*
Solid phase	Amphipods	+20% mortality compared to reference sediment or +30% compared to control sediment and statistical difference	*Corophium* sp. *Ampelisca* sp.

4. DiToro, D.M. et al. (1991). *Environ. Toxicol. Chem.* **10**: 1541.
5. Fairey, R. et al. (2001). *Environ. Toxicol. Chem.* **20**: 2276–2286.
6. McCauley, D.J., DeGraeve, G.M., and Linton, T.K. (2000). *Environ. Sci. Pollut.* **3**: 133–144.
7. O'Connor, T.P. and Paul, J.F. (1999). *Mar. Pollut. Bull.* **40**: 59–64.
8. Suter, G.W. (1993). *Ecological Risk Assessment*. Lewis, Chelsea, MI.
9. Clements, W.H. (1995). Ecological significance of endpoints used to assess sediment quality. In: *Ecological Risk Assessment of Contaminated Sediments*. C.G. Ingersoll, T. Dillon, and G.R. Biddinger (Eds.). SETAC Press, Pensacola, FL.
10. Chapman, P.M., Ho, K.T., Munns, W.R. Jr., Solomon, K., and Weinstein, M.P. (2002). *Mar. Pollut. Bull.* **44**: 271.
11. DelValls, T.A. et al. (1998). *Ecotoxicol. Environ. Safety* **41**: 157.
12. ASTM (1991). *Standard guide for conducting 10 day static sediment toxicity test with marine and estuarine amphipods*. In: *Annual Book of ASTM Standards*, Vol. 11.04, E 1367-90. American Society of Testing and Materials, Philadelphia, pp. 1052–1076.

13. U.S. EPA (1994). *Methods for assessing the toxicity of sediment-associated contaminants with estuarine and marine amphipods*. United States Environmental Protection Agency, EPA/600/R-94/025.
14. Thain, J.E. and Bifield, S.A. (1993). *A sediment bioassay using the polychaete Arenicola marina*. Test guideline for the PARCOM sediment reworker ring test. Unpublished report, MAFF Fisheries Laboratory, Burham-on-Crouch, UK.
15. Schlekat, C.E. et al. (1995). *Environ. Toxicol. Chem.* **14**: 2163.
16. Casado-Martínez, M.C. et al. Inter laboratory precision study of a whole sediment toxicity test using crustacean amphipods. *Cien. Mar.* (submitted).
17. Stronkhorst, J. (2003). *Ecotoxicological effects of Dutch harbour sediments*. The development of an effects-based assessment framework to regulate the disposal of dredged material in coastal waters of The Netherlands [Ph.D. thesis]. Vrije UniversiteitSuter.
18. Casado-Martínez, M.C. et al. Inter laboratory study of precision: the bioluminescence inhibition test for rapid sediment toxicity assessment. *Cien. Mar.* (submitted).
19. Casado-Martínez, M.C. et al. *Cien. Mar.* (submitted).
20. Environment Canada (2000). *Report On Biological Toxicity Tests Using Pollution Gradient Studies: Sydney Harbour*. Minister of Public Works and Government Service Report, Canada, EPS 3/AT/2.
21. DelValls, T.A., Forja, J.M., and Gómez-Parra, A. (1998). *Environ. Toxicol. Chem.* **17**: 1073.
22. DelValls, T.A., Casado.Martínez, M.C., and Buceta, J.L. (2003). Proposal of a tiered testing schema to characterize dredged material from Spanish ports. In: *Remediation of Contaminated Sediments—2003. Proceedings of the Second International Conference on Remediation of Contaminated Sediments*. M. Pellei and A. Porta (Eds.). Battelle Press, Columbus, OH.
23. DelValls, T.A. et al. (2001). *Investigación conjunta sobre la viabilidad de utilizar ensayos ecotoxicológicos para la evaluación de la calidad ambiental del material de dragado*, Technical Report for CEDEX, Puerto Real. Cádiz.
24. Casado-Martínez, M.C., DelValls, T.A., Fernández, N., Riba, I., and Buceta, J.L. (2004). *The use of acute sediment toxicity test to characterize dredged material disposal in Spanish Ports*. 4th SedNet Workshop Harmonization of Impact Assessment Tools for Sediment and Dredged Materials, San Sebastian, Spain 10–11 June.
25. DelValls, T.A., Lubián, L.M., Forja, J.M., and Gómez-Parra, A. (1997). *Environ. Toxicol. Chem.* **16**: 2323.
26. Moreno-Garrido, I., Hampel, M., Lubián, L.M., and Blasco, J. (2003). *Ecotoxicol. Environ. Safety* **54**: 290.
27. DelValls, T.A., Forja, J.M., and Gómez-Parra, A. (2002). *Chemosphere* **46**: 1033.
28. Martín-Díaz, M.L. (2004). *Determinación de la calidad ambiental de sistemas litorales y de estuario de la Península Ibérica utilizando ensayos de campo y laboratorio [Tesis doctoral]*, Universidad de Cádiz.
29. Environment Canada (2002). *Biological Test Method: Reference Method for Determining the Toxicity of Sediment Using Luminescent Bacteria in a Solid-Phase Test*. Report EPS 1/RM/42.
30. Environment Australia (2002). *National Ocean Disposal Guidelines for Dredged Material*. May.
31. U.S. EPA/U.S. ACE (1998). *Evaluation of Dredged Material Proposed for Discharge in Waters of the US. Testing manual* (The Inland Testing Manual). EPA-823-F-98-005.
32. ICES (2000). *Biological assessment of toxicity of marine dredged materials*. In: *Report of the ICES Working Group on Biological Effects of Contaminants*, Nantes, France, 27–31 March, Chap. 7.

E_h

SUDUAN GAO
USDA–ARS
Parlier, California

KENNETH K. TANJI
University of California
Davis, California

A reduction–oxidation (redox) reaction involves the transfer of electrons. Thus redox reactions change the oxidation state and affect the speciation, solubility, mobility, and toxicity of elements in the aquatic environment. Major elements involved in redox reactions are O, C, N, S, Mn, and Fe, and the major sensitive redox components include O_2/H_2O, $NO_3^-/(N_2, NH_4^+)$, Mn(III, IV)/(II), Fe(III)/Fe(II), SO_4^{2-}/H_2S, and CO_2/CH_4. In polluted waters, trace elements could include As(V)/(III), Cr(VI)/(III), Hg (II)/(0, −II), Se(VI)/(IV, 0, −II), and U(VI)/(IV). Strong oxidants such as chlorine and ozone are used in water treatment processes too. The theoretical aspects of redox potential (E_h) are based on chemical equilibrium. Redox reactions in nature are driven by chemical and, more importantly, biological processes toward equilibrium, but it is seldom achieved. The traditional method of E_h measurement in water is simple but it is difficult to interpret because redox systems are seldom at equilibrium and multiple redox components coexist. Here we review and summarize theoretical predictions, measurement, and the applications of E_h in aqueous chemistry.

WHAT IS E_h?

E_h is an intensity measure of the reducing or oxidizing conditions in a system, that is, the tendency of a solution to donate to or accept electrons from a chemical species or electrode into the solution (1). E_h is conventionally defined in terms of the potential of a cell composed of two half-reactions: the half-cell of particular interest and the standard hydrogen half-cell. The potential of the hydrogen half-cell [$H^+ + e^- = \frac{1}{2} H_2(g)$] at standard-state conditions, that is, (H^+) = 1.0 M and P_{H_2} = 1.0 atm, is set to 0.00 V.

Any redox reaction in nature is completed by two simultaneous half-reactions involving reduction and oxidation of two different chemical species and electrons are transferred from one species (reductant) to another (oxidant). The most important redox reaction, for example, is the microbially mediated oxidation of organic matter (represented by CH_2O) by oxygen (O_2) in an oxic environment (where oxygen is present) (2):

$$CH_2O + O_2 = CO_2 + H_2O \qquad (1)$$

where CH_2O loses and O_2 accepts the electrons, although the number of electrons transferred do not show up in this overall reaction.

Any of the half-reactions in a redox reaction can be generalized as a reduction reaction:

$$Ox + n\,e^- + m\,H^+ = Red + H_2O \qquad (2)$$

where Ox and Red represent the oxidized and the reduced species, respectively. Most reduction reactions consume protons while involved in transferring electrons (Table 1).

The reduction potential of Equation 2 is expressed by the Nernst equation,

$$E_h = E_h^0 - \frac{2.303RT}{nF} \log \frac{(Red)}{(Ox)(H^+)^m} \qquad (3)$$

where E_h = reduction potential for the half-reaction (V), E_h^0 = standard-state reduction potential for the half-reaction (V), n = moles of electrons transferred in the reaction, m = moles of protons involved in the reaction, R = natural gas constant (8.314 J · K^{-1} · mol^{-1}), T = absolute temperature, F = Faraday constant (96484.56 C · mol^{-1}), and (P) = activity. In dilute solutions, one may assume that activity approximately equals concentration.

The standard state is defined when all activities of redox species are at unity, that is, when $E_h = E_h^0$. At 25 °C, Equation 3 becomes

$$E_h = E_h^0 - \frac{0.05916}{n} \log \frac{(Red)}{(Ox)(H^+)^m} \qquad (4)$$

or

$$E_h = E_h^0 - \frac{0.05916}{n} \left(m\text{pH} + \log \frac{(Red)}{(Ox)} \right) \qquad (5)$$

Thus, the reduction potential is determined by the ratio of the redox couple and the pH. A high ratio of the reducing species to the oxidizing species yields a low E_h value, indicating the abundance of reducing species (an electron-rich system) and a high tendency of the solution to donate electrons. Thus, E_h is an intensity parameter representing the electron activity of a system.

The standard-state reduction potentials of half-reactions (E_h^0) encountered commonly in aquatic environments are listed in Table 1. The free energy change related to the reduction potential of the reaction is

$$\Delta G_r = -nFE_h \qquad (6)$$

or

$$\Delta G_r^0 = -nFE_h^0 \qquad (7)$$

The more positive E_h values indicate stronger tendency for the reaction to proceed for the reaction to occur from left to right as it relates to more free energy release (Eqs. 6 and 7). The redox couples with higher positive E_h^0 indicate stronger oxidants (e.g., O_2) and those with lower E_h^0 indicate stronger reductants (e.g., CH_4, H_2S). However, the standard-state conditions do not normally occur in natural systems, where activities of most dissolved redox species including protons are much below unity. The reduction potentials of the important reactions at pH = 7 and at 25 °C, $E_h^0(w)$, are calculated using Equation 5, and are given in Table 1. These values indicate redox reactions may proceed sequentially in a closed system when electron acceptors are consumed from the strongest to the weakest. It implies that in natural aquatic systems, where organic matter serves as the most abundant reductant, oxidation of organic matter may be accompanied by the consumption of electron acceptors in the sequence of oxygen, nitrate and Mn oxides, Fe oxides, sulfate and carbonate reductions. In reality, these reductions often overlap because microbially mediated reactions can be slow.

In aquatic systems, water reduction to H_2 or oxidation to O_2 sets the boundaries and limits the range of E_h^0. E_h can exceed these boundaries if stronger oxidants than oxygen are added to water, such as chlorine and ozone that are used in drinking water and wastewater treatment. The most important redox couples in natural waters are O_2/H_2O, NO_3^-/N_2, Mn (III/IV)/Mn(II), Fe

Table 1. Selected Standard-State Reduction Potentials of Half-Reactions Involving Important Elements in Water, Sediments, and Soils

Reaction	E_h^0 (V)[a]	$E_h^0(w)$ at pH = 7.0[b]	pe^0(w) at pH = 7.0[b]
$\frac{1}{4}O_2(g) + H^+ + e^- = \frac{1}{2}H_2O$	1.230	0.816	13.79
$\frac{1}{5}NO_3^- + 6/5\,H^+ + e^- = \frac{1}{10}N_2(g) + \frac{3}{5}H_2O$	1.248	0.751	12.70
$\frac{1}{8}NO_3^- + 5/4\,H^+ + e^- = \frac{1}{8}NH_4^+ + \frac{3}{8}H_2O$	0.881	0.363	6.14
$\frac{1}{2}MnO_2(s) + 2\,H^+ + e^- = \frac{1}{2}Mn^{2+} + H_2O$	1.230	0.579	9.79
γ-MnOOH(s) + 3\,H$^+$ + e$^-$ = Mn^{2+} + 2\,H$_2$O	1.503	0.616	10.41
Fe(OH)$_3$(s) + 3\,H$^+$ + e$^-$ = Fe^{2+} + 3\,H$_2$O	0.935	0.048	0.81
FeOOH(s) + 3\,H$^+$ + e$^-$ = Fe^{2+} + 2\,H$_2$O	0.769	−0.118	−1.99
$\frac{1}{8}SO_4^{2-} + \frac{5}{4}H^+ + e^- = \frac{1}{8}H_2S + \frac{1}{2}H_2O$	0.308	−0.210	−3.54
$\frac{1}{8}SO_4^{2-} + \frac{9}{8}H^+ + e^- = \frac{1}{8}HS^- + \frac{1}{2}H_2O$	0.254	−0.212	−3.58
$\frac{1}{8}CO_2(g) + H^+ + e^- = \frac{1}{8}CH_4(g) + \frac{1}{4}H_2O$	0.172	−0.242	−4.09
$H^+ + e^- = \frac{1}{2}H_2(g)$	0.000	−0.414	−7.00

[a] Data are from Reference 3.
[b] Calculated based on the E_h^0 and agreeable with data in Stumm and Morgan (6). $E_h^0(w)$ and pe^0(w) apply to the electron activity for unit activities of oxidant and reductant in neutral water (pH = 7.0) at 25 °C; except for the Fe and Mn couples that are based on Mn^{2+} or Fe^{2+} activity of 10^{-6} M because these values are more realistic than the unit activities of standard-state conditions.

(III)/Fe(II), SO_4^{2-}/HS^-, and CO_2/CH_4. Highly reducing conditions result in methanogenesis and fermentation. Multiple reaction paths result in an array of E_h values for some elements that depend on a particular biological or chemical reaction path (e.g., N, Mn, S, and C). For example, nitrate reduction to N_2 (denitrification) has a higher E_h^0 than reduction to ammonium (NH_4^+). Various forms of solid phase Mn and Fe oxyhydroxides result in varying reduction potentials.

USE OF pe

In lieu of E_h, electron activity of a system is defined as a pe analogous to pH: $pe = -\log(e^-)$.

From Equation 2, at equilibrium, we have

$$K = \frac{(\text{Red})}{(\text{Ox})(e^-)^n(H^+)^m} \quad (8)$$

$$e^- = \frac{(\text{Red})}{n(\text{Ox})(H^+)^m K} \quad (9)$$

Take the logarithmic form on both sides and rearrange:

$$pe = pe^0 - \frac{1}{n}\log\left[\frac{(\text{Red})}{(\text{Ox})(H^+)^m}\right] \quad (10)$$

or

$$pe = pe^0 - \frac{1}{n}\left(m\,pH + \log\left[\frac{(\text{Red})}{(\text{Ox})}\right]\right) \quad (11)$$

where

$$pe^0 = \frac{1}{n}\log K^0 \quad (12)$$

Both Equations 10 and 4 indicate that the activity of electrons can be quantitatively expressed by and directly related to the distribution of redox species and pH. Use of E_h or pe is a matter of preference. There are some advantages in using pe to examine redox phenomenon in some aspects. For example, pe + pH is considered a true variable in representing redox status as it can be used to differentiate the regions of redox species stability (3). When dealing with measured potential, however, it is more straightforward to use E_h than pe; although the conversion shown below is straightforward.

Conversion between E_h and pe at 25 °C is simplified as

$$E_h = 0.05916\,pe, \quad (13)$$

where E_h is in volts. E_h and pe values change in the same direction and higher values indicate electron-poor systems and lower values indicate electron-rich systems. Table 1 lists the correspondence between $pe^0(w)$ and $E_h^0(w)$.

E_h MEASUREMENT

Redox potential (E_h) of a solution is the difference in electrical potential measured between a noble metal (e.g., Pt or Au) electrode (also called indicator electrode) that is sensitive to redox couples in solution and a reference electrode that has a fixed potential in reference to the standard hydrogen electrode (SHE). Both electrodes are immersed in the solution and the potential is read on a potentiometer (e.g., pH meter). The noble electrodes are inert and resistant to chemical reactions; that is, they are not involved in the redox reactions and function only to conduct current. Because a standard hydrogen reference electrode ($E_h^0 = 0$) is not convenient for practical use, reference electrodes such as Ag/AgCl or calomel (Hg/Hg_2Cl_2) are commonly used in E_h measurement. Combination ORP (oxidation–reduction potential) electrodes consisting of both the indicator and reference electrodes are commercially available. These electrodes employ a platinum element to measure the potential due to the ratio of two oxidation states or species of an element in solution and each contains its own reference electrode. A commonly used Pt electrode system (electrochemical cell) is illustrated (3) as follows:

Pt electrode	**Solution**
Pt	Ox (aq) → Red (aq)
	Reference electrode
KCl (aq, saturated)	Hg(l) : Hg_2Cl_2(s)

In this system, the Pt electrode conducts the electrons from the reduction reaction. The calomel reference electrode consists of a gel composed of liquid Hg^0 and solid Hg_2Cl_2 bathed in a saturated KCl solution. An asbestos fiber connects the saturated KCl solution to the sample by allowing K^+ and Cl^- ions to diffuse from the reference electrode into the sample solution to maintain electrical neutrality. This connection is called a salt bridge or liquid junction and electrically connects the reference electrode to the sample solution. A small and unpredictable liquid junction potential is encountered due to the differences between K and Cl ion diffusion rates, which are affected by the types of ions and the ion concentrations. The junction potential can be minimized by using a concentrated salt bridge and can be corrected to some degree with calibration against a solution with a known redox potential (e.g., ZoBell solution).

ZoBell solution, which contains 0.1 M KCl and an equal amount (e.g., 3.33×10^{-3} M) of $K_4Fe(CN)_6$ and $K_3Fe(CN)_6$, is typically used for testing redox instruments. It is more stable than quinhydrone solution above 30 °C. Quinhydrone solution, however, is much easier to prepare and use. The potential of the ZoBell solution measured by a Pt electrode versus a saturated KCl, Ag/AgCl reference electrode is a linear function of temperature, up to 55 °C (4):

$$E_{h,\,\text{ZoBell}} = 0.428 - 0.0022\,(T-25), \quad (14)$$

where T is the solution temperature (in °C).

A solution E_h can be obtained from the measurements of the solution and the ZoBell solution by an ORP combination electrode (5):

$$E_{h,\,\text{system}} = E_{\text{observed}} + E_{\text{reference}}$$

where E_{observed} is the sample potential relative to the reference electrode used and

$$E_{\text{reference}} = E_{h,\,\text{ZoBell/reference}} - E_{h,\,\text{ZoBell/observed}}$$

Table 2. Potential of Pt Electrode Versus Selected Reference Electrodes at 25 °C in Standard Zobell Solution[a] (mV)

Calomel Hg/Hg$_2$Cl$_2$	Silver/Silver Chloride (Ag/AgCl)			Standard Hydrogen
Saturated KCl	1.00 M KCl	4.00 M KCl	Saturated KCl	
+183	+192	+228	+229	+428

[a] ZoBell solution contains 1.4080 g potassium ferrocyanide (K$_4$Fe(CN)$_6$ · 3H$_2$O), 1.0975 g potassium ferricyanide (K$_3$Fe(CN)$_6$), and 7.4555 g potassium chloride (KCl) in 1 L. The solution should be stored in a dark plastic bottle in a refrigerator.
Source: Reference 5.

where $E_{h, ZoBell/reference}$ is the theoretical E_h of the reference electrode and ZoBell's solution, relative to SHE (Eq. 14 for a Pt electrode and a saturated KCl, Ag/AgCl system; others are listed in Table 2) and $E_{h, ZoBell/observed}$ is the observed potential of ZoBell's solution relative to the reference electrode.

Care must be taken in measurement of water sample E_h. Proper instrument calibration and electrode testing must be performed. Freshwater samples must be used. Minimizing exposure to the atmosphere during E_h reading is necessary for low E_h samples. Flow-through cells are recommended for use for groundwater or a system with dynamic flow. Detailed procedures for measurement of E_h are available in Clesceri (5) and Nordstrom and Wilde (7).

INTERPRETATION

The concept of E_h measurement is derived based on chemical equilibrium, while the E_h measured in water does not reflect this aspect because redox reactions are seldom at equilibrium in natural waters. Thermodynamics only determines if a reaction is possible. Most redox reactions are mediated by microorganisms with slow kinetics. Lack of equilibrium is particularly prevalent among the various redox couples. Furthermore, many redox species (e.g., O_2, N_2, NO_3^-, SO_4^{2-}, HCO_3^-, and CH_4) are not electroactive; that is, they do not readily take up or give off electrons at the surface of the Pt electrode to contribute to the potential meant to be measured. The only exception is the couple Fe(III)/Fe(II), which are electroactive at low pH and where dissolved Fe is in certain concentrations (e.g., 10^{-5} M). Therefore, the measured E_h using a Pt electrode does not necessarily reflect the distribution of redox couples or species. Because it is a measurement of potential, the Pt electrode responds to changes in pH and other potentials. Comparison among E_h measurements should be based on the same pH and other conditions. Due to the presence of multiredox couples and nonequilibrium status in water, the measured potential is a mixed potential from only those electroactive redox species. Thus, E_h normally does not provide a quantitative measurement of the electron activity of the system measured; that is, it is normally only qualitative. Some authors believe that the measured E_h has no valuable theoretical meaning.

Other factors contributing to the difficulty in using E_h measurement to reflect electron activity of a solution include low concentrations of redox species, which cannot provide enough electron transfers that are detectable by the Pt electrode. The value of junction potential may differ greatly from that in the redox buffer solutions used to calibrate the cell. The Pt electrode can become contaminated by oxides or other coatings; thus, proper cleaning and polishing of the probe are needed before use. Permanently installed probes can be fouled either in high dissolved O_2 solution or strongly reducing conditions. Oxygen can react with Pt to form Pt(OH)$_2$ on the electrode surface. Pt(OH)$_2$ is electroactive and develops a potential with Pt ($\frac{1}{2}$Pt(OH)$_2$ + e$^-$ + H$^+$ → $\frac{1}{2}$Pt + H$_2$O; E_h^0 = 0.982 V) and masks the electrode response from electron transport between redox-sensitive solution species (3). Poising of permanent electrodes was also observed in reducing soils (8). The surface of the electrodes became dull after immersing into soil reducing solution for a few days and unable to respond to increased E_h values when nitrate was introduced into the system, while temporary electrodes detected the increase. Thus, a cleaning procedure was necessary to recover the function of the electrodes.

As a result of the discussion above, a large discrepancy is often observed between measured E_h values and predicted E_h values based on the redox couples present in the solution. Figure 1 shows an example of the theoretical E_h values and those observed in soil solutions at neutral pH for some important redox couples (compiled from Refs. 1 and 3). The large discrepancy for the O_2–H_2O, NO_3^-–N_2, and Mn oxides–Mn^{2+} are partially attributed to the inert behavior of O_2, NO_3^-, and N_2 at the Pt electrode and inappropriate choice of MnO_2 as electron acceptor as various forms of Mn oxides are present in a number of oxidation states (II, III, and IV) in nature. Similar reasons can apply for the SO_4^{2-}–H_2S and CO_2–CH_4 couples. The predicted and measured E_h values are closer only for the Fe(OH)$_3$–Fe^{2+} couple. In fact, wider E_h ranges for the same redox reactions than those shown in Fig. 1 were reported under various conditions. This should not be surprising when considering all the factors influencing the E_h measurement as discussed above.

APPLICATION

Although there are practical problems and difficulties in interpretation, E_h is a parameter that can be easily obtained and can be useful for better understanding of water chemistry. A decrease of E_h showed correspondence to sequential reduction of nitrate, Mn, and Fe in soil incubation experiments (9). The critical E_h at which all of the NO_3^- was reduced and Mn^{2+} appeared in solution was about 200 mV and the critical E_h at which Fe^{2+} appeared was 100 mV. The qualitative E_h measurement has been used to classify redox regions or zones into different categories in relation to specific redox reactions. For examples, the following zones in soils at pH 7 were defined: oxidized (E_h > 414 mV, O_2 present), moderately reduced (120–414 mV, nitrate reduction and Mn reduction), reduced (−120 to 120 mV,

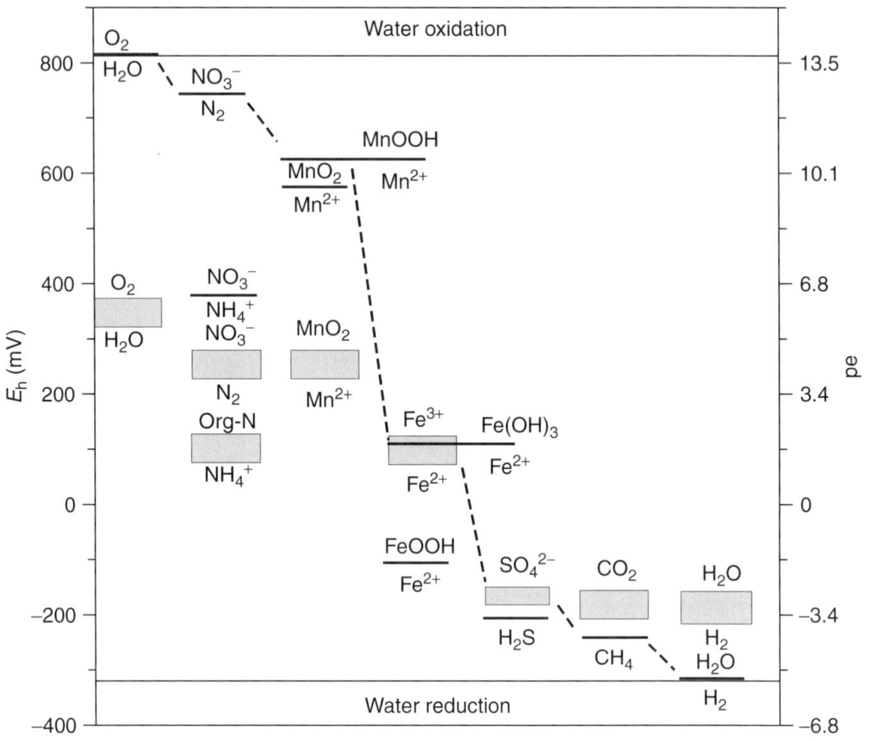

Figure 1. Theoretical and observed E_h for reduction sequences under neutral soil pH solutions. (Modified from Refs. 1 and 3.) The theoretical E_h values for the redox transformations to occur at pH 7.0 are indicated by the horizontal solid line. These values are based on equal activities of reduced and oxidized species unless noted. The activities of dissolved Fe^{2+} and Mn^{2+} are assumed at levels of 10^{-6} M and the pressure of H_2 is arbitrarily set at 10^{-3} atm. The observed E_h ranges for transformation from oxidized to reduced species are indicated by the shaded area. Corresponding pe values are shown on the right y axis.

Fe reduction), and highly reduced ($E_h < -120$ mV, sulfate reduction) (10). The following zones are further proposed at pH 7: oxic ($E_h > 414$ mV, oxygen and nitrate reduction), suboxic (120–414 mV, Mn and Fe oxide reductions), and anoxic ($E_h < 120$ mV, sulfate reduction) (11). Ranges of E_h for redox reactions may vary greatly from one observation to another. In fact, a wide range of E_h for the same redox couple has been reported from the literature (12). These variations originate from the differences among systems observed. Overlapping of redox processes is common especially in an organic matter-rich environment for a complex system such as paddy soil with straw amendment. E_h decrease corresponded to the progressing reducing environment developed upon flooding (13).

Redox buffering capacity (poising) in analogy to pH buffering capacity controls the E_h changes of a system. It is the capability of a system to maintain a redox potential upon addition of oxidizable or reducible materials. In a closed system with abundant organic matter or water near the organic matter-rich sediment (assuming originally equilibrated with oxygen in the atmosphere), oxygen will be quickly depleted due to oxidation of organic matter resulting in an abrupt drop in E_h (13). The E_h level may quickly decrease to and maintain the level indicative of sulfate reduction in most natural waters because nitrate concentrations are usually low and denitrification adds little buffering capacity to the water. In soils or groundwater in contact with oxidizing sediments, however, Mn and Fe oxides in the solid phase can add significant amount of buffering capacity to maintain a relatively higher E_h level for a period of time. The E_h ranges of these sequential redox reactions can be inferred from (Fig. 1).

In conclusion, E_h measurement in water is only a qualitative indication of the electron richness or poorness of a system. E_h measured in a solution is a mixed potential and usually does not quantitatively relate to specific redox reactions or redox couples. The easy measurement of E_h can have some practical use if its determination if handled carefully by following proper procedures (5,7). It can indicate the progressing of a system toward reduction or oxidation and can be used to predict potential redox reactions when comparing similar type systems. This qualitative measurement can assist in further quantitative determinations of redox reactions and chemical species. Nordstrom and Wilde (7) concluded that E_h data can be useful for gaining insights on the evolution of water chemistry and for estimating the equilibrium behavior of multivalent elements relative to pH for an aqueous system. E_h can delineate qualitatively strong redox gradients, such as in stratified lakes and rivers, in an oxidized surface flow that becomes anaerobic after passing through stagnant organic-rich systems, and in mine-drainage discharges.

OTHER APPROACHES

Many other alternative methods in describing and examining redox phenomenon of soil and water have been examined and proposed. A capacity parameter, oxidative capacity (OXC), was defined by Scott and Morgan (14). It integrates all the oxidized and reduced species into a single descriptive parameter. By setting an electron reference level, comparisons among oxidative capacities of various components results in identification of redox classes such as oxic (oxygen present), postoxic (nitrate,

Mn, and Fe reduction), sulfidic (sulfate reduction), and methanogenesis. Challenges remain in quantifying the capacity of individual components especially those involving solid phase and organic matter. A nonequilibrium approach to defining the dominant terminal electron-accepting processes (TEAPs) has been applied to groundwater systems (15). This method considers simultaneously the consumption of electron acceptors [dissolved O_2, $NO_3^- - N$, Fe(III), $SO_4^{2-} - S$, and CO_2], final products [Mn(II), Fe(II), H_2S, and CH_4], and the intermediate product (dissolved hydrogen gas, H_2). A high degree of confidence can be achieved if a combination of all three indicators yields a positive identification of the predominant TEAPs. This is difficult to achieve sometimes because of the incorporation with the solid phase for some electron acceptors and final products. Dissolved H_2 is considered as a master variable in characterizing organic matter degradation or redox chemistry in anoxic conditions (16,17). Molecular hydrogen can be produced by many fermentative microorganisms during metabolism of organic compounds and at the same time consumed by respiratory microorganisms that use different electron acceptors. Molecular hydrogen is a result of the energy threshold for the organisms catalyzing the terminal electron-accepting process. The more electrochemically positive electron acceptors [e.g., nitrate and Mn(IV)] give a greater potential energy yield and reducers of these compounds can obtain their threshold energy yield at lower H_2 concentrations. Thus, a correlation between H_2 concentrations and predominant TEAPs was applied (12,15).

BIBLIOGRAPHY

1. McBride M.B. (1994). *Environmental Chemistry of Soils*. Oxford University Press, New York.
2. Manahan, S.E. (1994). *Environmental Chemistry*, 6th Edn. CRC Press, Boca Raton, FL.
3. Essington, M.E. (2004). *Soil and Water Chemistry—An Interactive Approach*. CRC Press, Boca Raton, FL.
4. Wood, W.W. (1976). Guidelines for collection and field analysis of ground water samples for selected unstable constituents. In: *Techniques of Water-Resources Investigations of the United States Gelogical Survey, Book 1*. U.S. Geological Survey, Washington, DC, Chap. D2.
5. Clesceri, L.S., Greenberg, A.E., and Eaton, A.D. (1998). *Standard Methods for Examination of Water and Waste Water: Including Bottom Sediments and Sludges*, 20th Edn. American Public Health Association, Washington, DC.
6. Stumm, W. and Morgan, J.J. (1996). *Aquatic Chemistry*. John Wiley & sons, Hoboken, NJ.
7. Nordstrom, D.K. and Wilde, F.D. (1998). Reduction–oxidation potential (electrode method). In: Wilde, E.D. and Radtke, D.B. (Eds.), *National Field Manual for the Collection of Water-Quality Data*, U.S. Geological Survey Techniques of Water-Resources Investigations, Book 9, Chap. A6, Sec. 6.5. Available at http://water.usgs.gov/owq/FieldManual/Chapter6/6.5_contents.html.
8. Bailey, L.D. and Beauchamp, E.G. (1971). Nitrate reduction and redox potential with permanently and temporarily placed platinum electrodes in saturated soils. *Can J. Soil Sci.* **51**: 51–58.
9. Patrick, W.H. Jr. and Jugsujinda, A. (1992). Sequential reduction and oxidation of inorganic nitrogen, manganese, and iron in flooded soil. *Soil Sci. Soc. Am. J.* **56**: 1071–1073.
10. Sposito, G. (1981). *The Thermodynamics of Soil Solution*. Oxford University Press, New York.
11. Sposito, G. (1989). *The Chemistry of Soils*. Oxford University Press, New York.
12. Lovley, D.R. and Goodwin, S. (1988). Hydrogen concentrations as an indicator of the predominant terminal electron-accepting reactions in aquatic sediments. *Geochim. Cosmochim. Acta* **52**: 2993–3003.
13. Gao, S., Tanji, K.K., Scardaci, S.C., and Chow, A.T. (2002). Comparison of redox indicators in a paddy soil during rice-growing season. *Soil Sci. Soc. Am. J.* **66**: 805–817.
14. Scott, M.J. and Morgan, J.J. (1990). Energetics and conservative properties of redox systems. In: *Chemical Modeling of Aqueous Systems II*. X. Melchior and R.L. Barrett (Eds.). *CACS Symposium Series* 416. American Chemical Society, Washington, DC, Chap. 29, pp. 368–378.
15. Chapelle, F.H. et al. (1995). Deducing the distribution of terminal electron-accepting processes in hydrogeologically diverse groundwater systems. *Water Resour. Res.* **31**: 359–371.
16. Hoehler, T.M., Alperin, M.J., Albert, D.B., and Martens, C.S. (1998). Thermodynamic control on hydrogen concentrations in anoxic sediments. *Geochim. Cosmochim. Acta* **62**: 1745–1756.
17. Jakobsen, R. et al. (1998). H_2 concentrations in a landfill leachate plume (Grindsted, Denmark): *in situ* energetics of terminal electron acceptor processes. *Environ. Sci. Technol.* **32**: 2142–2148.

WATER RESOURCE DEVELOPMENT AND MANAGEMENT

WATER RESOURCES CHALLENGES IN THE ARAB WORLD

M.E. YOUNG
Conwy, United Kingdom

THE PROBLEMS

The water shortage in the Arab World (Fig. 1) is severe and worsening. The region receives only 1% of the world's renewable water resources yet contains more than 5% of the world's population. In most Arab states at the start of the twenty-first century, the gross volumetric human and environmental demand for water exceeds natural replenishment. Per capita water availability is worsened by a high regional population growth; a 3% annual population growth rate halves the per capita availability every 24 years.

The paramount need, common to all states, therefore, is to find a sustainable balance between use and replenishment by a combination of measures to increase supply and to reduce demand. Both approaches involve technical (engineering) developments, legal and financial measures, improved water resources management through institutional interventions, and adjustments to national policies of water allocation.

All states need, as a priority, to develop coherent national water resources policies and strategies for the integrated management of water in all sectors—agricultural, municipal, industrial and environmental—in the context of the level of supply that is sustainable and according to sectoral priority. Institutional capacities to assess and manage water resources require development in many countries.

Many states of the region share resources from river flows or aquifers. Protocols that provide for equitable, sustainable and secure sharing must be agreed upon.

Specific water resources problems vary from state to state and include

- low levels of access to safe water supplies;
- inadequate or obsolete supply systems in growing cities;
- waterlogging and salinization of agricultural areas by irrigation (e.g., along the Nile, Tigris, and Euphrates);
- unsustainable groundwater abstraction
- pollution of aquifers and watercourses by human, industrial, and agricultural contamination;
- pollution of overpumped coastal aquifers by seawater intrusion (e.g., Oman, the Gulf States, and Libya); and
- drought and seasonal variability of water supplies.

Most freshwater in the region is used for irrigated agriculture. However, the economic value of water used for irrigation, in terms of the value of food produced, is lower than its value for municipal or industrial use. This situation has been exacerbated in some states by now outdated national policies restricting production to low-value staple food crops that have high water requirements.

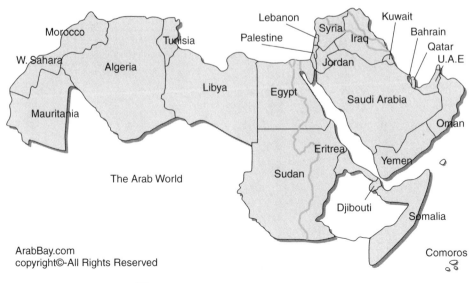

Figure 1. The Arab World.

Reducing water use in irrigation would be the most effective means of reducing the water deficit in most countries. At the same time, reallocating water to higher value sectors would enhance the GDP of all countries of the region. How to effect this transformation with the minimum social disruption in the agricultural sector, while maintaining food supplies, is the central issue of water resources management in the region.

SUPPLY AND DEMAND AT THE START OF THE TWENTY-FIRST CENTURY

Data on water supply and demand from Arab states vary in extent, reliability, and detail. Inventories of water resources are still in progress in most countries, and data on consumption, particularly, are generally inferred rather than measured.

Table 1 shows the estimated volumes of naturally renewable water for the region by state and per capita. Renewable resources comprise internally renewable resources from groundwater recharge and drainage to rivers and the net flows (inflows less outflows) of surface and groundwater from neighboring states. Inflows and outflows agreed to by treaty, or established by political circumstances, are implicit in these data; in the eastern Mediterranean states, particularly, such arrangements are complex and politically uncertain (3). The proportion of the actual renewable resource that comes from outside a state's borders is shown by the dependency ratio.

By 1995, resources per capita had fallen below the level of 1000 m^3 pcpa (per capita per annum) in four states and below 500 m^3 pcpa in 10 more. (An average level of 1000 m^3 is regarded as the level above which a state can be self-sufficient in food; less than 500 m^3 pcpa indicates severe water stress.) At the current rate of population growth, there would be four and 14 states, respectively, in these categories by 2020.

Figure 2 shows how much of this renewable resource is provided by recharge to groundwater, for those states where data exist. Those states that have the highest levels of water stress rely heavily on groundwater.

In many states, groundwater is abstracted faster than it is replenished. Figure 3 shows the proportion of the total annual water supply that is drawn unsustainably ("mined") from nonrenewable ("fossil") groundwater. Figure 4 shows groundwater abstraction as a percentage of renewal; in some states, groundwater is being mined several times faster than it is replenished.

This continued depletion of freshwater resources results from the increasing food demands of growing populations and from agricultural policies that seek national or local food self-sufficiency through irrigation, regardless of the unsustainability of the water supply. Figure 5 shows the proportion of national water supplies allocated to irrigation.

In many states, private well owners hold unrestricted traditional rights to groundwater for irrigation, regardless of the unsustainability of the resource (e.g., Oman, Yemen). Some governments have sought to exploit fossil resources to supply large-scale agriculture (e.g., the Great

Table 1. Renewable Water Resources[a]

	Internal, Million m^3	Actual Total, Million m^3	Actual, pcpa (1995), m^3/a	Dependency Ratio, %
Mauritania	400	11,400	5,013	96
Iraq	35,200	75,420	3,688	53
Sudan	35,000	88,500	3,150	77
Eritrea	2,800	8,800	2,480	68
Comoros	1,020	1,020	2,100	0
Syria	7,000	26,260	1,791	80
Somalia	6,000	15,740	1,702	62
Lebanon	4,800	4,407	1,465	1
Morocco	30,000	30,000	1,110	0
Egypt	1,800	58,300	926	97
Oman	1,650	1,650	775	0
Djibouti	300	300	520	0
Algeria	13,900	14,300	512	3
Tunisia	3,520	4,120	463	15
Yemen	4,100	4,100	283	0
Bahrain	4	116	206	97
Jordan	780	880	161	23
Saudi Arabia	2,400	2,400	134	0
Libya	600	600	111	0
Palestinian Terr.	600	225	100	0
Qatar	51	53	96	4
UAE	150	150	79	0
Kuwait	0	20	13	100

[a] References 1 and 2.

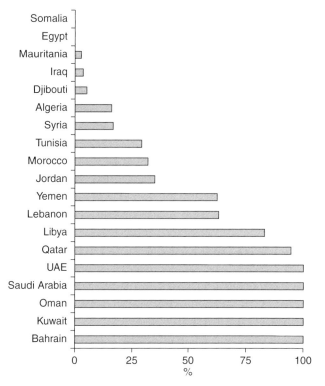

Figure 2. Groundwater recharge as a % of total renewable water resources (1,4).

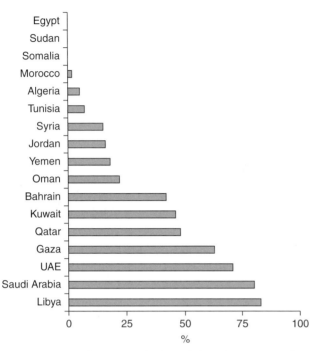

Figure 3. Groundwater overdraft (withdrawal less renewal) as a percentage of total supply (4).

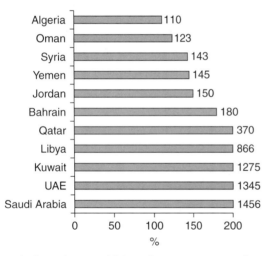

Figure 4. Groundwater withdrawal as a percentage of renewal (1,4).

Man-made River Project in Libya or for grain production in Saudi Arabia). Access to groundwater has been easier since about 1970, following the introduction of engine-driven groundwater pumps and the availability of deep drilling technology.

Figure 6 shows the volumes of "virtual" water already imported either as processed food, grain, or as live food animals, as a percentage of the total renewable water resource. The high volumes of virtual water show clearly that food self-sufficiency is an unrealistic target for many states.

These data show that the reason underlying the growing shortage of water in the Arab region is the increasing use of water, mainly for irrigation, at rates that exceed rates of replenishment. Historically, governments in the region have sought to address this problem by augmenting water supplies through engineering methods, rather than by encouraging conservation. Augmentation measures considered or implemented include interstate and interbasin water transfer, aquifer development, desalination, and minor processes such as cloud seeding and fog collection. However, the full production and delivery cost of water by all these methods typically exceeds the economic value of the food produced, and most water supplied by these means is now used only to supply higher value urban or industrial demands. The scope for further cost-effective augmentation of supply is limited by technology.

CONSERVATION AND DEMAND MANAGEMENT

Conservation methods used in the region include rainwater harvesting using small structures, recharging flood flows to aquifers, recycling of wastewater to irrigation, and leak reduction in urban networks. The storage of flood flows and treated wastewater in aquifer zones for later recovery and use needs to be further researched and promoted in the region. There is also substantial potential for increasing the conservation and re-use of treated wastewater. However, water can be most effectively conserved by improving farming and irrigation practices, in two ways: by improving food productivity per unit of water used and by reducing water losses in irrigation systems. The former approach includes the introduction of higher yielding and more water efficient (including salt tolerant) crop species and the use of scientific irrigation strategies specific to the crop type and conditions. Water reduction methods include the introduction of drip and sprinkler technology to replace spate irrigation and reduction in transmission and evaporation losses. In most countries of the region, significant savings can still be made by these technical means.

Despite these helpful measures, water resource deficits can be reversed only by fundamental changes in national water and food policies that reduce the water demands of the agricultural sector and gradually divert water to more economically productive sectors, while, at the same time, supporting rural communities in this transition.

POSSIBLE MANAGEMENT SOLUTIONS

It is clearly necessary for each country to develop and implement policies and strategies to produce a new and sustainable balance between supply and demand in irrigation. Most countries continue to pursue supply augmentation by various methods and to promote and subsidize conservation through engineering methods. Nonengineering interventions to promote conservation of water fall into four classes:

Political
    Policy revision
    Water resources assessment, monitoring, and control
    Pursuit of favorable international agreements for food importation
    Promoting efficient management of centralized water supplies

Legal
    Quotas (limits) on abstraction
    Water rights—possibly transferable or tradable within a basin
    Restrictions on water use

Fiscal
    Tariffs on water use
    Taxes on pump fuel
    Removal of tariffs on cheap agricultural imports
    Subsidy programs for irrigation and agricultural modernization

Institutional
    Expert consultancy
    Social pressure and public awareness
    Development of integrated management on a river basin or aquifer scale
    Outreach and participation, including water users' associations

The combination of approaches that is appropriate for each country depends on the extent and effectiveness of national and local skills and institutions.

Political

Revision of government policy and coherency of policies between ministries and internationally, is paramount. A lack of coordination may result in development plans that cause unsustainable depletion of resources. Some states in the region still actively pursue policies of food self-sufficiency despite the lack of sustainable water resources.

The task of government is first to establish an enabling environment in which government, industry, and non-governmental entities work coherently toward water resource conservation, informed by a good understanding of the extent and quality of the resources. Second, governments and the international community must seek favorable international trade agreements for importing food.

Practical political interventions, whereby centralized governments operate at least primary systems of water-lifting and distribution, are common throughout the region. Great investment is required to upgrade infrastructure and improve efficiency, particularly for the delivery of municipal and rural drinking water. The capital needed is beyond the capacity of some countries but private sector financial participation through public–private partnerships is an alternative. Full privatization of water supply is widely opposed in the region and governments need to develop partnership models that are locally acceptable. To reduce risk and encourage investment, governments need to establish clear water law and robust fiscal regulations, and improve transparency.

International treaties to define shared resources and protocols for sharing information are essential for securing transboundary supplies. The Neil Basin Initiative and the Nubian Sandstone Aquifer System are two vehicles through which several countries consult and collaborate on managing the surface and underground resources of the major resources of northeast Africa. Similar processes are needed for the shared resources of the northern Arabian Peninsula.

International cooperation at a wider level to promote the development and sharing of water technology and information is essential. Recent valuable initiatives and foundations include the Dialogue on Water and Climate in the Mediterranean Region, organized by the World Conservation Union; the Middle East Desalination Research Center, in Oman; and the International Center for Biosaline Agriculture, in Dubai.

Legal and Fiscal

Economists believe that a sustainable water demand will be achieved only by charging the full cost of water to all users. The principle of charging tariffs for water delivery is widely accepted in urban areas of the region, but costs are only partially recovered. Stepped tariff structures that subsidize low-volume users are common. Uncharged-for water losses are abnormally high in some cities of the region.

Other legal and fiscal measures, such as irrigation quotas, water rights, and full-cost pricing are applicable only where water law is clearly defined and where established economic and financial frameworks facilitate the operation of market-based systems. This is not yet the case in most of the Arab region, where volume restrictions on irrigators or fiscal restrictions have been and are likely to be opposed. A cultural perspective that water is a free natural resource is a major obstacle to the introduction of innovative legal and fiscal instruments.

Institutional

Decentralization of water resources management to regional and local management systems is widely regarded as the most effective means of improving water supply services at the level of use, coupled with a well defined water law and integrated water resources management policy at the national level (5).

Institutional approaches are most appropriate where organizations and networks of civil society are amenable to development through training, leadership, and incentives. For example, Egypt, Morocco, and Tunisia have developed water users' associations and basin management organizations that seek to bring all stakeholders together in cooperative and participative operations. They assess the availability and sustainability of the resources; allocate water-use rights based on need, environmental demand, social contingencies, and economic value; control and monitor water distribution and quality; provide training; and collect fees to finance administration and system maintenance.

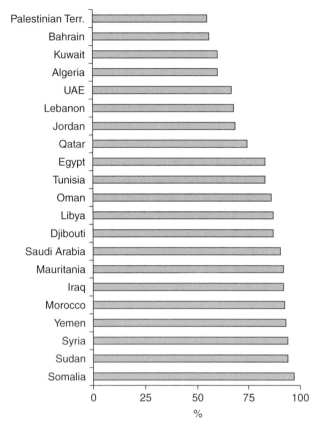

Figure 5. Share of water allocated to agriculture (1).

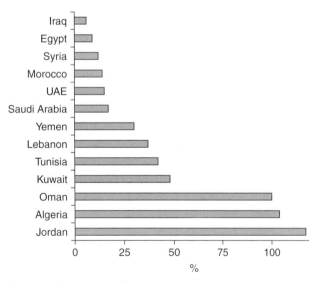

Figure 6. Water embedded in net food imports as a percentage of total renewable resource (4).

The Influence of Islam

Recent international prescriptions for water management, such as the Dublin Principles of 1992 and subsequent pronouncements, have emphasized economic value as a driving criterion for allocating water among sectors. This approach tends to ignore social priorities in rural Arab states and underestimates the practicalities of implementing water management changes in the contexts of Islamic law and, in some states, weak or inflexible political systems.

The holy writings from which Islamic law is derived contain many prescriptions relevant to water management and concur with the Dublin principles regarding community participation. Water is seen as a social good that must be managed sustainably and for the good of all; concepts of payment for delivery, the use of treated wastewater, and privatization of delivery are regarded as permissible (6).

Islam plays a prominent, central, and universally respected role throughout the region and, as an institution distinct from but working with both government and people, wields great influence. Islam has the potential to educate and lead consumers and drive the necessary changes in social attitudes toward water management. Islamic institutions can, therefore, provide leadership in promoting water resource conservation where political and fiscal interventions may fail.

BIBLIOGRAPHY

1. Food and Agriculture Organization (FAO), Water Resources Development and Management Service. (2001). AQUASTAT Information System on Water in Agriculture. World Wide Web address: http://www.fao.org/.
2. Inter-Islamic Network on Water Resources Development and Management (INWRDAM). (1999). Table of water scarcity in OIC states. INWRDAM, Amman, Jordan. World Wide Web address: http://www.nic.gov.jo/inwrdam/.
3. Brooks, D.B. and Mehmet, O. (Eds.). (2000). *Water Balances in the Eastern Mediterranean*. International Development Research Centre, Ottawa, Canada.
4. Lajaunie, M.-L. (1999). *MENA/MED regional water data report*. In: *Proceedings, Second Regional Seminar of the MENA/MED Initiative on Water Policy Reform, Amman, Jordan, May 1999*. World Bank, Washington, DC.
5. 3rd World Water Forum, March (2003), Kyoto, Japan. *Thematic and Regional Statements*. World Wide Web address: http://www.world.water-forum3.com/.
6. Faruqui, N.I., Biswas, A.K., and Bino, M.J. (Eds.). (2001). *Water Management in Islam*. United Nations University Press.

READING LIST

Guerquin, F., Ahmed, T., Hua, M., Ikeda, T., Ozbilen, V., and Schuttelar, M. (2003). *World Water Actions: Making Water Flow for All*. World Water Council. Earthscan Publications, UK.

Mahdi, K. (Ed.). (1998). *Water in the Arabian Peninsula: Problems & Policies*. Exeter University, UK.

Rogers, P. and Lydon, P. (Eds.). (1994). *Water in the Arab World*. Harvard University Press, Cambridge, MA.

Schiffler, M. (1998). *The Economics of Groundwater Management in Arid Countries*. Frank Cass, London, UK.

Shady, A. (2000). *Summary: Arab countries session*. In: *Conference on Water Security in the 21st Century*. The Hague, Netherlands, March 2000. World Wide Web address: www.worldwaterforum.net/index2.html

World Bank. (1998). *From Scarcity To Security—Averting a Water Crisis in the Middle East*. World Bank, Washington, DC.

EFFLUENT WATER REGULATIONS IN ARID LANDS

Richard Meyerhoff
CDM
Denver, Colorado

Robert Gensemer
Parametrix
Corvallis, Oregon

James T. Markweise
Neptune and Company, Inc.
Los Alamos, New Mexico

This article addresses the regulation of water resources in arid environments and focuses on the special considerations of effluent-driven ecosystems. Effluent serves as an increasingly important aquatic resource in the arid West, and the application of existing regulations for natural systems to effluent-based bodies of water is receiving greater scrutiny. Issues affecting regulation of effluent include defining appropriate attainable beneficial uses for these created ecosystems, net ecological benefits of effluent discharge, and impacts from the modification of extant flow regimes resulting from effluent addition. The practical application of regulations in effluent-driven ecosystems depends on recognizing these key issues.

EFFLUENT WATERS IN ARID ENVIRONMENTS

Effluent-based waterbodies are an important aquatic resource in arid environments because of the limited availability of alternate sources of water. In addition to general scarcity, water availability is sporadic and unpredictable in dry lands; consequently, effluents from residential, commercial, or agricultural activities are frequently one of the more stable sources of water in the arid West (1).

Effluent-driven ecosystems have proliferated, as dams and diversions, groundwater pumping, and the need to dispose of treated effluent in metropolitan areas have modified natural flow regimes.

Three basic natural flow regime types are perennial, intermittent, and ephemeral. Perennial waters flow continuously, intermittent waters flow continuously only at certain times of the year (e.g., because of spring snowmelt), and ephemeral waters flow only in direct response to precipitation. Effluent-based ecosystems are either "dependent" on (effluent is discharged to an ephemeral stream channel) or "dominated" by (effluent is discharged to an intermittent stream channel) treated wastewater effluent discharge.

The discharge of effluent to intermittent or ephemeral streams represents a discontinuity resulting in a disruption to the natural equilibrium as it exists at the time the discharge begins. The natural tendency for the created stream ecosystem is to restructure itself so that a new equilibrium is achieved. This restructuring will take some time; the length will depend on local factors and whether additional stressors are placed on the system (e.g., construction of physical structures in, across, or along the stream, or an increase in effluent flow) (2).

Riparian systems that develop as a result of wastewater discharged into normally dry channels may stand in stark contrast to the adjacent upland vegetation that is not influenced by discharge. In addition, the terrestrial community downstream of the discharge point can be distinctly different from the terrestrial community upstream of the discharge. The expected characteristics of an effluent-based waterway downstream of the effluent discharge can be modeled conceptually (Fig. 1). This conceptual model, which represents a created stream ecosystem where the effluent discharge is the only discontinuity, is based on the physical, chemical, and biological characteristics of 10 effluent-dependent and dominant ecosystems in the arid West (2). From a regulatory standpoint, created or modified stream ecosystems are a class of aquatic environments that are

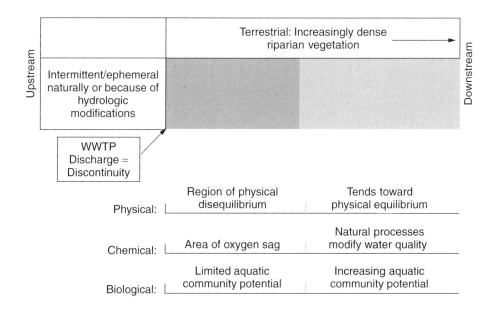

Figure 1. Conceptual model of an effluent-driven stream ecosystem.

typically not distinguished from natural ecosystems. Yet, these ecosystems may have unique attributes that result in different expectations for a "functioning" ecosystem.

Given the unique nature of effluent-based waterbodies, the challenge is to apply the existing regulatory framework in a way that is protective of this resource and still biologically meaningful. Below we address key regulatory issues for protecting effluent-derived ecosystems in arid lands. These considerations include defining "attainable uses" in effluent-driven ecosystems, using net ecological benefit as an alternative end point to measure use attainability, and implications of modifying the flow regime—is wetter better?

ATTAINABLE USES IN EFFLUENT-DRIVEN ECOSYSTEMS

States and tribes are required to establish beneficial uses for surface waters under their jurisdiction. At a minimum, existing uses must be protected (uses actually attained in the waterbody on or after November 28, 1975). States and tribes may go beyond the protection of existing uses and protect designated uses, which are expressed as goals for the waterbody. Designated uses must be protected only if the uses are actually attainable. The EPA defines attainable uses as those beneficial uses that can be achieved (1) when effluent limits under the Clean Water Act are imposed on point source dischargers and (2) when cost-effective and reasonable best management practices (BMPs) are imposed on nonpoint source dischargers. If all appropriate water quality control activities are implemented, then the pertinent question is, "what is the best beneficial use that can be achieved?" Determining what is attainable in a natural ecosystem can itself be difficult; however, when effluent creates the aquatic ecosystem, determining what is attainable is an even greater challenge.

The attainable use may be equal to the designated use or it may be something less. Consider a dry stream bed characterized as an ephemeral flow system. At the point of effluent input (Fig. 1), the water becomes perennial. The addition of water changes the potential of the system and thus what is attainable, especially with regard to the aquatic community. The effects of water addition, physical habitat, water quality, and other stressors define the potentially attainable aquatic community. The result of the interaction of these variables differs with time and also with distance downstream from the point of discharge (Fig. 1). Often, the water quality changes over time as a new technology is installed, new regulations are implemented, or the stream channel is modified. Each change modifies the potential outcome or potential aquatic community. In other words, defining what is attainable is a moving target.

Determining potential or what is attainable is typically defined in the context of the effluent quality, which is related to the degree/type of wastewater treatment. A general belief exists that improved treatment will result in improved aquatic community characteristics. This is likely to be true at low levels of treatment. For example, with the change from primary to secondary treatment, major improvements in the aquatic community are expected and do occur. It is not necessarily realistic, however, to expect the same degree of improvement with more advanced treatment changes, such as with the addition of nutrient removal or filtration, if the limiting factor is habitat-related.

Failing to establish appropriate end points for beneficial uses has significant regulatory and cost implications because the Clean Water Act mandates that states assess water quality in surface waters every 2 years and determine if any of these waters is "impaired." A waterbody is considered impaired if a determination is made that, even after implementing all required effluent limitations (technology and water-quality-based) and BMPs, the water quality still does not meet applicable water quality standards. This analysis presumes that beneficial uses and criteria are appropriately established. However, if the uses and criteria are inappropriate, the basis for the assessment is inaccurate, and the waterbody may be incorrectly assessed as impaired.

Impaired waters must have a total maximum daily load (TMDL) established under the Clean Water Act. A TMDL is the sum of the wasteload allocations for point sources and load allocations for nonpoint sources (including natural background), plus a margin of safety (uncertainty factor). TMDLs are implemented through permits, BMPs, or other water quality management methods. Pollutant allocations are based on meeting the applicable end point or water quality standard. TMDLs can require substantial resources to develop and implement; thus, it is very important to use appropriate end points as the basis for assessing waters correctly.

Under the current method for regulating created or effluent-based ecosystems, states typically establish the same water quality standards or end points for effluent-based waters as established for natural perennial waters. This approach is based on the assumption that the aquatic community in effluent-based waters should be the same as the aquatic community in natural perennial waters; that is, the attainable end point in a natural ecosystem should be the same as the attainable end point in a created ecosystem. When this assumption forms the basis for implementing water quality control programs, the same requirements for effluent limitations are applied to both types of waters.

From a regulatory perspective, establishing appropriate end points in effluent-based ecosystems is especially problematic. The approach outlined above can result in applying substantially more stringent effluent limitations on facilities discharging to effluent-based ecosystems than facilities that discharge to a perennial water. The reason for this difference is that when a facility discharges to a perennial water, credit is given to allow for mixing the effluent with in-stream water. This credit results in less stringent effluent limitations. No such credit is available when the receiving waterbody is dry. This approach results in regulation that is strictly discharge or end-of-pipe based. No consideration (positive or negative) is given to the ecosystem of the receiving water.

The discharge of effluent to dry riverbeds creates an aquatic ecosystem. This ecosystem may be "new," or it may replace an aquatic ecosystem that has been lost because of dams, diversions, groundwater withdrawals, etc. If effluent-based ecosystems are created and if they function differently from natural perennial waters such that what is attainable in the created ecosystem differs from what is attainable in the natural ecosystem, then the use of an alternative regulatory approach to implement a water quality control program might be appropriate.

NET ECOLOGICAL BENEFIT

A concept, called "net ecological benefit," addresses some of the concerns associated with establishing attainable end points for effluent-based waters (3,4). The net ecological benefit methodology is a tool for evaluating whether an alternative end point could be established if it can be demonstrated that the ecological benefits created by the discharge outweigh any identified detriments. A natural progression of this thought process is to consider the use of a "performance-based" approach that considers the functionality of the ecosystem. This approach could lead to establishing alternative success criteria or end points for evaluating ecosystem protection. Examples of net ecological benefit include enhancement of habitat/food for terrestrial native or threatened and endangered (T&E) species, enhancement or restoration of riparian values, preservation of existing habitat that could not be supported without effluent flow, restoration of aquatic and riparian values lost due to human activities, enhancement of water quality resulting in conditions conducive to ecosystem restoration and/or preservation, improvement or creation of habitat capable of supporting fish or allowing migration of anadramous species, and restoration of species diversity in aquatic ecosystems.

EPA (3) guidance notes key issues that should be considered when evaluating whether or not a net ecological benefit exists as a result of an effluent discharge. Foremost among these is that effluent discharges should not produce or contribute to concentrations of pollutants in tissues of aquatic organisms or wildlife that are likely to be harmful to humans or wildlife through food chain concentration. This requirement addresses EPA and USFWS policies to minimize persistent, bioaccumulative chemicals in the environment. Also, a continued discharge to the waterbody should not create or be likely to cause or contribute to violations of other end points, for example, downstream water quality standards or groundwater quality. Finally, all practicable pollution prevention programs, such as pretreatment and source reduction, are operational and the discharger has to be appropriately responsive to previous and ongoing compliance actions. These considerations are designed to ensure that (1) a minimum level of wastewater management and treatment is implemented and (2) pollutants are not exported from the water column into the ecosystem through the food chain.

Use of a net ecological benefit approach as an alternative for setting end points would change the regulatory focus in effluent-driven ecosystems from end-of-pipe based to ecosystem-based. Establishing this nontraditional approach would allow recognizing the environmental benefits of water flow in an otherwise dry channel. Moreover, alternative end points that include terrestrial elements would support efforts to implement permitting on a watershed basis, support ecological restoration efforts, address concerns about the increasing loss of riparian habitat in the arid West, support increased interest in habitat restoration in urban rivers, and potentially benefit threatened and endangered species by maintaining wildlife migration corridors.

MODIFYING THE FLOW REGIME—IS WETTER BETTER?

The addition of effluent to a riverbed has been portrayed as a benefit in the context that any water in a riverbed is better than no water in a riverbed. For sites where the effluent replaces historically natural flows, this thought process makes sense. However, when the effluent discharge creates a flowing river where none previously existed, the question can be asked, "what has been lost or changed by the addition of effluent?" This question is relevant because naturally ephemeral streams have important biological attributes that are as distinct as the biological attributes of a natural perennial river.

The addition of an artificial perennial flow makes the system clearly different from historical conditions (Fig. 1). Biological attributes such as aquatic community richness and diversity are likely to be greater. The increased biological productivity of the aquatic community will provide additional food resources for terrestrial organisms. In addition to these changes in the aquatic community, the terrestrial community will be substantially different, especially in terms of the types of organisms supported. Are these biological changes good? Is having a wetter channel better biologically? These questions have no simple answer. In fact, the answer will depend on public values and local needs. One can easily argue that the number of ephemeral channels, especially in arid regions, far exceeds the number of naturally perennial channels, and thus the creation of a perennial stream in a previously ephemeral stream is a positive benefit. However, in some areas, especially in rapidly developing urban environments, the number of lost ephemeral channels can be significant, and the loss of habitats as a result of effluent discharge can be an important issue for the public to consider.

The creation of effluent-driven ecosystems creates a relatively stable aquatic environment, at least from the standpoint of temperature and flow. Certainly, stormwater runoff will still occasionally cause spikes in flow, but the norm is a relatively constant flow with reduced temperature variation (compared to what might be expected in a natural environment). Native fish species are adapted to the relatively episodic flow regime of natural arid waters. These species have evolved where abiotic factors are the dominant driver in establishing biological communities. Accordingly, these species are better adapted to this environment than nonnative species. Modification of the flow regime removes this driver and if nonnative fish become established (as they have throughout the arid West), they may outcompete the native species.

Similar to fish, the richness and diversity of native riparian vegetation is influenced by the flow regime. For example, the selection of native cottonwoods over nonnative salt cedar can be driven by the distribution and frequency of flows across the floodplain during floods. Under a stable flow regime, native species may be outcompeted. In the end, ecological benefits may result from the discharge of effluent to an otherwise dry or intermittent riverbed, but the modified flow regime is not necessarily a positive benefit for native species.

FINDING A BALANCE IN REGULATING ARID LAND WATERS

Among the general public, a common desire exists to protect beneficial uses, establish water quality criteria to protect uses, and protect riverine habitat and associated wildlife. However, basic agreement on these fundamentals exists, but substantial disagreement may exist over how attainable end points should be defined and how uses should be protected. This disagreement often stems from differences of opinion over how end points should be calculated, what are appropriate assumptions, and how much conservatism or how much of a safety factor should be built into the regulations.

The EPA (4), states, and tribes often take the simplest, most conservative approach when establishing attainable end points because this approach requires the least amount of scientific data. Regulatory institutions cannot be faulted for this. Given the number of different types of aquatic systems (natural or created) and the various and often competing beneficial uses of water (from protection of native species to use of water for industry), the number of factors that could be considered in establishing end points for each waterbody is substantial. Resources are simply not available to establish site-specific uses and criteria for each waterbody. Although regulators are limited in what they can reasonably consider when they establish criteria for an entire state or area, options are available for modifying uses and criteria and establishing appropriate site-specific end points. These tools should be used, when appropriate, because the importance of establishing correct end points cannot be understated. If water quality control efforts are based on unattainable uses or criteria, substantial costs may be incurred with little or no improvement in water quality.

BIBLIOGRAPHY

1. Ryti, R.T. and Markwiese, J.T. (in press). Assessment of ecological effects in water-limited environments (WR-3). *Encyclopedia of Water*, J. Lehr (Ed.). John Wiley & Sons, New York.
2. Pima County Wastewater Management Department. (2002). *Habitat Characterization Study*. Prepared by URS Corporation and CDM for Arid West Water Quality Research Project, PCWMD, Tucson, AZ.
3. EPA. (1992). *Guidance for Modifying Water Quality Standards and Protecting Effluent-Dependent Ecosystems*. Interim Final. EPA Region 9.
4. EPA. (1993). *Supplementary Guidance on Conducting Use Attainability Analyses on Effluent Dependent Ecosystems*. EPA Region 9.

CALIFORNIA—CONTINUALLY THE NATION'S LEADER IN WATER USE

WILLIAM E. TEMPLIN
U.S. Geological Survey
Sacramento, California

In 1950, the U.S. Geological Survey began publishing a series of water-use circulars entitled "Estimated Use of Water in the United States." Every 5 years since then the report has been updated and now provides a valuable, long-term data set of national water-use estimates. Since the inception of this series, California has always reported the largest total fresh and saline withdrawals of all states in the Nation (Fig. 1). The most recent update in this series (1) reports 1995 conditions. California again accounts for the largest withdrawal of water for off-stream uses of all states, which is about 45.9 billion gallons per day, followed by Texas (29.6 Bgal/d), Illinois (19.9 Bgal/d), and Florida (18.2 Bgal/d, Figs. 2 and 3).

The poster demonstrates WHY and HOW California continues to lead the Nation in many withdrawal categories (Figs. 4 and 5). Withdrawals include water removed from the ground or diverted from a surface-water source for use.

WHY? California continues to be the most populous state in the United States, accounting for 12% of its people (more than 32 million in 1995) followed by Texas and New York with about 7% each (Fig. 6). In 1995, California accounted for 5.62 Bgal/d, 14% of the Nation's public supply freshwater use (Fig. 7). Public supply withdrawals are directly related to population supplied, but they also are influenced by withdrawals to supply industrial and commercial water users.

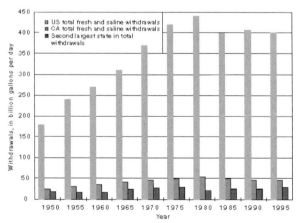

Figure 1. Trends in total fresh and saline withdrawals, 1950–1995.

This article is a US Government work and, as such, is in the public domain in the United States of America.

CALIFORNIA—CONTINUALLY THE NATION'S LEADER IN WATER USE

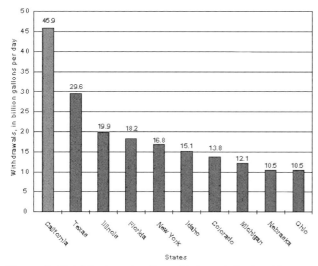

Figure 2. Total fresh and saline withdrawals by top 10 states, 1995.

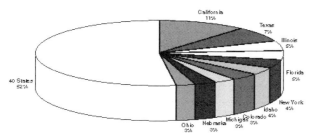

Top 10 states and balance of United States

Figure 3. Total fresh and saline withdrawals in the United States during 1995, 402 Bgal/d.

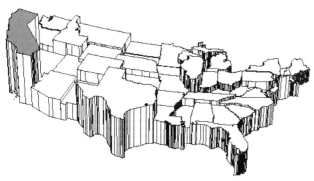

Figure 4. California led all states in freshwater withdrawals followed by Texas, Illinois, and Idaho, 1995.

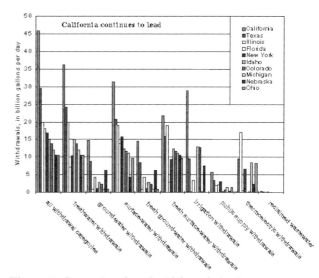

Figure 5. Categories of total withdrawals in the top 10 states, 1995.

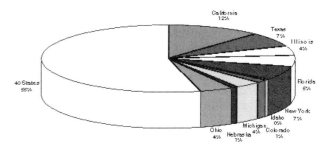

Figure 6. 1995 U.S. population—267 million.

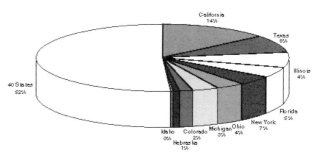

Figure 7. U.S. public supply withdrawals, 40.2 billion gallons per day, 1995.

- GOT MILK? California is the Nation's largest dairy state, producing 26 billion pounds of milk and cheese (2).
- GOT COTTON CLOTHING? California is the Nation's second largest cotton producer.
- HUNGRY? More than half of the Nation's fruits, vegetables, and nuts are grown in California.

HOW? The total withdrawals in the Tulare-Buena Vista Lake Hydrologic Cataloging Unit accounted for more than 8330 Mgal/d in withdrawals for irrigation uses in 1995 (Fig. 10) and 8800 Mgal/d of total freshwater. Only 13 states had more total freshwater withdrawals. This area

HOW? Los Angeles County's population exceeded 9.3 million in 1995 and contributed to the county's large public supply withdrawals. Only eight states have more people than Los Angeles County, which accounts for 29% of California's population (Fig. 8).

WHY? California accounted for 28.9 Bgal/d in total irrigation withdrawals in 1995, with 22% of the Nation's total irrigation withdrawals (Fig. 9).

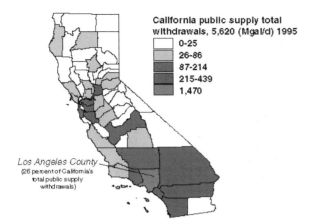

Figure 8. Public supply total withdrawals by county, in California, 1995.

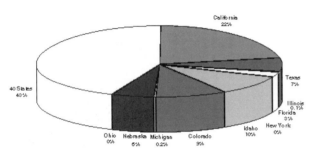

Figure 9. Irrigation withdrawals in the United States in 1995 totals 134 billion gallons per day.

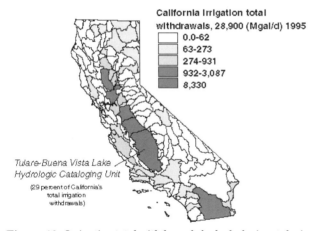

Figure 10. Irrigation total withdrawals by hydrologic cataloging unit, in California, 1995.

contains many of California's dairies and much of our irrigated acreage.

Even though California leads all states in total withdrawals, California's per-capita use rate for total off-stream use (1130 gal/d) ranks below the national average (1280 gal/d, Fig. 11).

Acknowledgments

The author acknowledges the assistance of Michael B. Roque, Illustrator, in the layout and preparation of the original poster presented at the American Water Resources Association Annual

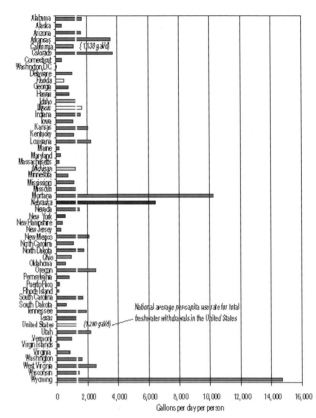

Figure 11. Per-capita use rates for total freshwater withdrawals for the United States and each state, 1995.

Conference on Water Resources, November 16–19 1998, Point Clear, Alabama.

The author also acknowledges the assistance of Tammy Shelton, Management Analyst, in the development of the web page version of this information.

BIBLIOGRAPHY

1. Solley, W.B., Pierce, R.R., and Perlman, H.A., 1998, *Estimated use of water in the United States in 1995: U.S. Geological Survey Circular 1200*, p. 71. http://water.usgs.gov/public/watuse/
2. California Department of Food and Agriculture. Home Page. October 1998. http://www.cdfa.ca.gov/

LESSONS FROM THE RISING CASPIAN*

Igor S. Zonn

INTRODUCTION

The Caspian Sea is the biggest inland body of water in the world. Its surface area is roughly equivalent to the combined area of The Netherlands and Germany (about 400,000 km^2, or 144,000 mi^2). The surface water inflow into the sea is formed by the flow of the Volga, Ural, Terek,

*From *Handbook of Weather, Climate, and Water: Atmospheric Chemistry, Hydrology, and Societal Impacts*, T.D. Potter and B.R. Colman (Eds.). John Wiley & Sons, Hoboken, NJ. pp. 885–892.

Sulak, Samur, and Kura Rivers, and small Caucasian and Iranian rivers. The watershed area of the Caspian Sea is 3.5 million square kilometers. The basin of the Volga River makes up nearly 40% of the territory of the catchment of the Caspian Sea, and it supplies about 80% of the total volume of annual water flow into the sea. All components of the Caspian ecosystem, directly or indirectly, to a greater or lesser extent, are influenced by river flow.

The Caspian Sea basin falls into three morphologically different parts: (1) the northern (25% of the sea area), a shallow area (less than 10 m deep; about 20% with depths less than 1 m) extending to a conventional line passing from the Terek River to the Mangyshlak Peninsula; (2) the medium (35%), with an average depth of 170 m (the maximum being 790 m); and (3) the southern (39%), the deepest area, with a maximum depth of 1025 m and an average depth of 325 m. Deep depressions in the northern and southern parts of the sea are divided by an underwater threshold running from the Apsheron Peninsula to Turkmenbashi (formerly Krasnovodsk) (1).

Before the breakup of the Soviet Union in December 1991, the USSR and Iran were the only two independent nations occupying the shores of the Caspian. With the breakup, three additional newly independent nations emerged along the coast: Azerbaijan, Kazakstan, and Turkmenistan. The Russian Federation's Caspian coastline is shared by three of its political units: Astrakhan Oblast, the Republic of Kalmykia, and the Republic of Dagestan.

NATURE OF SEA-LEVEL CHANGES IN CASPIAN SEA

The Caspian Sea is a closed basin in the inland part of Eurasia and this sea's water level is below that of the world ocean. The sea basin stretches almost 1200 km from north to south and its width varies between 200 and 450 km. The total length of the coastline is about 7000 km. Its water surface area is about 390,600 km^2 (as of January 1993). Water salinity in the northern part is 3–6‰ and reaches 12‰ in the middle and southern parts.

Fluctuations in sea level for various lengths of time can be found in the data of geomorphological and historical studies of the record of the Caspian Sea (Fig. 1). Within the last 10,000 years, the amplitude of fluctuations of Caspian Sea level has been 15 m (varying from −20 to −35 m). During the period of instrumental observations (from 1830 onward), this value was only about 4 m, varying from −25.3 m during the 1880s to −29 m in 1977. Annual increases in the level during this period met or exceeded 30 cm on three occasions (in 1867, 30 cm; in 1979, 32 cm; and in 1991, 39 cm). The mean annual increment in the level in the 1978–1991 period was 14.3 cm.

Natural factors are the primary cause of recent Caspian Sea level fluctuations (but not the only cause). Scientists have identified three distinct periods of level changes: 1830–1930, 1931–1977, and 1978 to the present. The first period of 100 years saw sea-level fluctuations not exceeding 1.5 m (5 ft). Researchers considered this period to have been relatively stable. The second period, from 1931 to 1977, is identified by a constant decline in level by 2.8 m (9.1 ft), and in 1977 the Caspian Sea reached its lowest level since the beginning of instrumental record-keeping in the 1830s.

As the sea level declined throughout the 1950s, 1960s, and early 1970s, Soviet scientists forecast that the decline would continue for at least a few decades into the future. Scientists have linked the reason for the decline to the regulation of Volga River flow. During these decades, major engineering activities were undertaken along the Volga, such as the construction of water diversion canals, reservoirs, and dams. The construction of such engineering facilities diverted water away from the Caspian.

In response to this major drop in sea level, human settlements bordering the sea coast began to move toward the receding coastline. Fields and pasturelands were prepared for use, roads and rail lines were constructed, and housing and factories were built on the newly exposed seabed. During the Soviet era, many people emigrated from other parts of the region to settle along the border of the sea. Development of infrastructure along the coast took place to support the increasing population.

In an attempt to save the Caspian from drying out, Soviet scientists and engineers proposed the construction of a dam to block the flow of Caspian water to Kara-Bogaz-Gol Bay, a large desert depression in Turkmenistan adjacent to the sea's eastern shore. Political decisions made in the mid-1970s ordered the construction of the dam, but due primarily to bureaucratic inertia, the dam was not completed until the early 1980s. This was a few years after the Caspian's sea-level change had reversed direction. Before the dam was constructed, the bay took in 40 km^3 (8.6 mi^3) of Caspian water annually. It served as a huge evaporation pond, as well as a natural location for the accumulation of commercially useful mineral salts.

Another Soviet government response to the decline in the Caspian's sea level was a diversion of water into the Volga River from other Soviet rivers that flowed northward into the Arctic Ocean. River water flowing into the Arctic was viewed as wasted and without value to the Soviet Union because it was unused by human activity.

THE CASPIAN RISES

To the surprise of Soviet scientists, the level of the sea began to rise suddenly in 1978, the beginning of its third period of level changes. Since then, the Caspian has risen

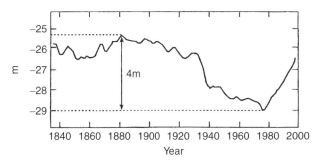

Figure 1. Observed Caspian Sea level, 1835–1999.

steadily by more than 2.5 m. One of the first actions the newly independent government of Turkmenistan took in 1992 was to tear down the dam in order to allow great amounts of water to flow into Kara-Bogaz-Gol Bay again and to replenish the supply of salts.

Scientists have proposed a variety of hypotheses about why the Caspian Sea level had increased so rapidly. These can be clustered into the following categories: tectonic plate movement on the seabed, climate fluctuations and change, and hydraulic construction along the Volga River, or some combination of these factors.

Tectonic Plate Movement Hypothesis

Tectonic movements over periods such as centuries and millennia have been the cause of many geologic changes in the Caspian basin. The region has been subjected to uplift, subsidence, overthrust of landforms, seabed mud-volcanic activity, and landslides, in addition to erosion processes and the accumulation on the Caspian seabed of river-transported sediments. However, it is difficult to see how tectonic movements could cause such sharp fluctuations in the Caspian's sea level over relatively short periods. Thus, it appears that such movements have had an insignificant impact on recent sea-level fluctuations.

Climate Change Hypothesis

Today, most Russian scientists believe that climatic factors are the real cause of the Caspian Sea level rise. Studies by Golitsyn (2) and Golitsyn and McBean (3) indicate that recent changes of the Caspian Sea level are 90% associated with corresponding changes in the water balance components of the sea, as opposed to possible tectonic activity. The volume of inflow from rivers to the sea increased sharply after 1978. During certain years (e.g., 1979, 1985, and 1990), more than 350 km^3 of river water entered the sea. From 1978 until 1990, Volga River flow exceeded 260 km^3/yr. At present, no arguments have challenged the view that the main contribution to seasonal and annual level fluctuations of the Caspian is accounted for by surface inflow and evaporation levels. Within recent decades, the sea's fluctuations have been subjected to anthropogenic impacts as well.

In this regard, climate has two dimensions: climate fluctuations and climate change. Climate fluctuations occur on various time scales, with those of interest to present-day society being on the order of decades and perhaps centuries. Climate-related fluctuation refers to the increase and decrease of sea level over the course of decades. During the past two centuries, the sea has undergone several fluctuations. Those of the twentieth century have adversely affected socioeconomic activities and infrastructure along the sea's coastline.

The view that climatic processes in the Volga basin are the dominant cause of sea-level fluctuations has been recently reinforced. Droughts in this basin and sharply reduced Volga flow into the Caspian from mid-1995 until early 1997 have been associated with a 25-cm (10-inch) drop in Caspian level. Nevertheless, Russian scientists still suggest that the sea level will continue to rise into the first decades of the twenty-first century.

Climate change associated with global warming induced by human activities has also been proposed as the forcing factor behind the Caspian's rise since 1978. Those who see global warming as the forcing factor suggest that the most recent sea-level rise can be associated with intensification of the hydrologic cycle (i.e., more active precipitation-producing processes), an intensification that some scientists have linked to the human-induced global warming of the atmosphere. An increase in precipitation within the Volga River basin would translate into increased sea level.

Hydraulic Construction Hypothesis

Some observers have argued that the recent fall and rise in sea level were the result of human activities. They suggest that the widespread development of hydraulic structures (e.g., dams, reservoirs, irrigation systems) in the Volga River basin, beginning in the 1950s, led to a sharp decline in Volga flow. The filling of many reservoirs built along the rivers flowing into the Caspian, the increase in industrial and municipal water use by several times, and changes in the water regime of the floodplains led to a decrease of streamflow into the sea. Such a hypothesis could be tested by constructing a water budget model for the Caspian. Such a model would need to identify all the inflows into the Caspian Sea (such as from rivers and groundwater) and all outflow from the sea (such as evaporation and water diversions). While it is a seemingly straightforward task, identifying all the sources and sinks of Caspian water is not easy.

There is also a hypothesis about an Aral Sea connection. Yet another suggestion that seems to be made at just about every Aral or Caspian Sea conference is that the decline in the level of the Aral Sea is linked to the rise in level of the Caspian. The reasoning is that water diverted from the Aral basin to the Caspian basin to irrigate the desert sands for cotton production in Turkmenistan ends up either being evaporated into the air or seeping into the groundwater, which eventually makes its way into the Caspian. However, it is important to point out that *both* the recent fall and rise in the Caspian Sea level occurred during three and a half decades of a constant decline in the Aral's level.

SOCIETAL IMPACTS OF SEA-LEVEL RISE

According to a UN Environment Programme estimate, the cost of the impact of the sea-level rise of the Caspian, as of 1994, was $30 to $50 billion (U.S.). Coastal ecosystems have been destroyed, villages inundated and populations evacuated, sea banks eroded, and buildings destroyed. Coastal plains have been invaded by subsurface seawater or have become waterlogged. Fauna have changed, and pasturelands and sturgeon spawning grounds have been destroyed.

Each of the five countries sharing the coasts of the Caspian Sea has suffered losses, and those losses increased until the mid-1990s. They suffer from the different impacts of sea-level rise because the territory along its coastline is neither uniformly settled nor uniformly developed economically. Economic losses in the big cities and villages have been higher than in the rural areas. More specifically, in Astrakhan Oblast (equivalent to an American state), about 10% of its agricultural land was out of production by 1995 because of sea-level rise. The coastline of the Republic of Dagestan (also part of Russia) was affected by the flooding of at least 40 factories in its cities of Makhachkala, Kaspiysk, Derbent, and Sulak. Nearly 150,000 hectares (370,000 acres) of land have been inundated, with a loss of livestock production and breeding facilities. Much of the 650-km (390-mile) Caspian coastline of Turkmenistan is made up of low-lying sandy beaches and dunes that are vulnerable to coastal flooding and erosion. In fact, some Turkmen villages that were once several kilometers from the sea are now coastal communities. Similar adverse impacts of sea-level rise on human settlements and ecosystems are found in Kazakstan, Azerbaijan, and Iran.

The Caspian has been referred to as a "hard currency sea" because of its large oil and natural gas reserves and because of its highly valued caviar-producing sturgeon. Regional reserves contain upward of 18 billion metric tons of oil and 6 billion cubic meters (215 billion cubic feet) of natural gas. Experts suggest that the Caspian is second only to the Persian Gulf with respect to the size of its oil and gas reserves, and that Turkmenistan is a "second Kuwait." If the sea level were to continue to rise, a large part of the oil and gas mains along the Turkmen coast would become submerged and would also be subjected to corrosion by seawater. Coastal settlements, which include the greater part of Turkmenistan's oil, gas, and chemical enterprises, would also be threatened. Similar environmental problems would certainly affect other Caspian coastal countries as well (4).

The Caspian Sea is unique in yet another respect: it contains about 90% of the sturgeon that produce the lucrative prized black caviar for export to foreign markets. Sturgeon roe is often referred to as "black gold." Today, however, Caspian sturgeon is at risk of extinction from overexploitation by illegal poachers and by destitute fishermen desperately seeking funds to buy food for their families. The sea-level rise, with its destruction of sturgeon spawning grounds, adds yet another threat to the endangered Caspian sturgeon.

Poachers hunt sturgeon only for its caviar. Today, they catch sturgeon directly in the open sea. However, in the early 1960s, prohibition was introduced by the former USSR against catching sturgeon in the open sea. Since that time, catching sturgeon has been carried out in the river deltas. Sturgeon reproduce very slowly: the fish do not spawn for the first time until they reach the age of 20–25 years. In 1990, the permissible catch of sturgeon in the USSR was set at 13,500 metric tons. In 1996, permissible (legal) catch was only 1200 metric tons (5).

SEA-LEVEL CHANGE AS A GLOBAL PROBLEM

Given the growing concern about, and possible evidence of, global warming, there has been considerable speculation about the potential impacts on coastal areas of a sea-level rise related to global warming. Scientists who participated in the 1995 Intergovernmental Panel on Climate Change (IPCC) Report (6) suggested that global sea level may well increase by an additional 15–70 cm (6–27 inches) by the end of the twenty-first century. The exact amount of rise would depend on the actual increase in global temperatures. Clearly, any additional increase in sea level could have devastating consequences for coastal communities.

All states that border bodies of water, whether along the global oceans or inland seas, should pay attention to fluctuations in sea level as well as to the rise in sea level linked to global warming. Inland seas, for example, can be viewed as living bodies in the sense that they can expand and can shrink. These changes can occur on different time scales: from daily to seasonally, from a year to a decade, or a century, or a millennium. In fact, they fluctuate and change on all these scales. The same can be said of the open oceans, but they tend to fluctuate on much longer time scales than do the inland seas, over periods of many decades and centuries. Such time scales are difficult to factor into the thinking of economic development planners, whose time frames are on the order of years to a few decades at most.

In essence, one can consider the Caspian as a laboratory of sea-level change and its potential societal and environmental consequences. For the Caspian to serve as a true "laboratory," its environmental-monitoring network, which collapsed with the breakup of the Soviet Union, must be restored and maintained by regional cooperation among the Caspian states. Impacts on ecosystems that are managed (farms and pastures) and unmanaged (wetlands, forests, deserts) can be identified. Effective human responses to changes in the coastal zone (both land and sea) can also be identified and assessed; environmental engineering proposals to deal with sea-level changes (such as seawall construction, higher oil platforms in the sea, diversion of water from the Caspian to the drying Aral Sea) can be evaluated for effectiveness, taking into consideration the scientific uncertainties surrounding sea-level fluctuations.

Whether the global climate gets warmer, cooler, or stays as it has been for the last several decades, the level of inland seas will likely continue to fluctuate (the mean ocean level has already gone up by 5–6 inches in the twentieth century alone). Societies must learn to cope with both short- and long-term fluctuations. In the Middle Ages, people in the Caspian region were not allowed to settle too close to the sea's shore, under the threat of death. Apparently, leaders were then aware of the dangers that the Caspian's fluctuating levels posed to their citizens. Today's leaders would be well advised to pay attention to traditional wisdom.

BIBLIOGRAPHY

1. Kosarev, A.N., and Yablonskaya, E.A. (1994). *The Caspian Sea*. SPB Academic Pub., The Hague.
2. Golitsyn, G.S. (1989). Once more about the changes of the Caspian Sea level. *Vestnik AN SSR* **9**: 59–63.
3. Golitsyn G.S., and McBean, G.A. (1992). Changes of the atmosphere and climate. *Proc. Russian Acad. Sci. Geogr. Ser.* **2**: 33–43.

4. Ragozin, A.L. (1995). Synergistic effects and consequences of the Caspian Sea level rise. In: *Proceedings of the International Scientific Conference: Caspian Region: Economics, Ecology, Mineral Resources*. Geocenter-Moscow, pp. 120–121 (in Russian).
5. Rosenberg, I. (1996). Catching sturgeon. *Itogi Magazine* 4 June (in Russian), pp. 48–50.
6. IPCC. (1996). *Climate Change 1995: The Science of Climate Change*. Cambridge University Press, Cambridge, UK.

INSTITUTIONAL ASPECTS OF WATER MANAGEMENT IN CHINA

Rongchao Li
Delft University of Technology
Delft, The Netherlands

INTRODUCTION

Located in East Asia, on the western shore of the Pacific Ocean, the People's Republic of China (PRC) has a land area of about 9.6 million km^2. China, the most populous country in the world, had 1.25909 billion people at the end of 1999, about 22% of the world's total. The population density in China is 130 people per sq km. A map of China is shown in Fig. 1.

China has a marked continental monsoonal climate characterized by great variety. The summer monsoons last from April to September. The warm and moist summer monsoons last from April to September, bringing abundant rainfall and high temperatures from the oceans, and little difference in temperature between the south and the north. China's complex and varied climate results in a great variety of temperature belts and dry and moist zones.

China has 50,000 rivers that cover catchment areas of at least 100 square kilometers, and 1500 of them cover catchment areas of more than 1000 square kilometers. Most of the rivers flow from west to east to empty into the Pacific Ocean. The main rivers include the Yangzi (Changjiang), the Yellow River (Huanghe), Heilongjiang, the Pearl River, Liaohe, Haihe, Qiantangjiang, and Lancang Rivers. At 6300 kilometers long, the Yangzi is the longest river in China. The second longest is the Yellow River at 5464 kilometers. The Grand Canal from Hangzhou to Beijing is a great water project of ancient China, 1794 kilometers in length. It is the longest canal in the world (*Source*: http://www.chinatour.com/countryinfo/countryinfo.htm#RIVER).

INSTITUTIONAL ASPECTS OF WATER MANAGEMENT IN CHINA

Legislative and Executive Structure

The agencies involved in WRM can be divided into legislative and executive organizations. The legislative structure is organized, in descending order, as follows:

- state level: state congress
- provincial level: provincial congress

Figure 1. Map of China (*Source*: http://www.maps-of-china.com/china-country.shtml).

- city level: municipal congress
- county level: county congress

The legislative structure in China is composed of four levels, shown in Fig. 2.

The executive structure closely relevant to water management is shown in Fig. 3.

Large water projects are managed by Basin Administrative Institutions that are subordinations of the MWR, such as the Yellow River Water Conservation Commission (YRCC). Medium and small water projects are managed by provincial, municipal, and county water administrative institutions, respectively, on the basis of ownership.

FUNCTIONS OF ADMINISTRATIVE INSTITUTIONS

In China, the water resource management organization has traditionally been organized along narrow, subsectoral

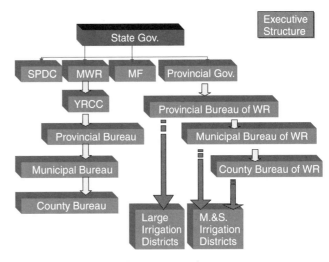

Figure 3. Executive structure of water management.

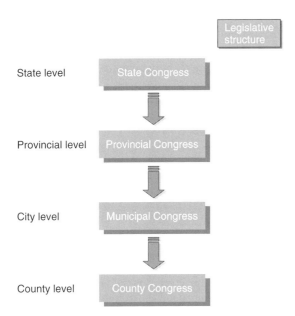

Figure 2. Legislative structure in China.

lines. The national government determines the general legislative and policy framework for water management. The Ministry of Water Resources (MWR) is responsible for the state water management and issues the national water legislation and policies. Other ministries, such as land and agriculture and electronic, are also involved in water management.

The following competent departments under the State Council and local government and their respective functions based on the duty division of the State Council compose the administrative structure of water resource management in China (Table 1).

Ministry of Water Resources

The Ministry of water resources (MWR) is the main department of water resources management under the state council. Its functions are:

- to establish respective regulations of water resources management on water law

Table 1. The 14 Sectors Involved in Water Management in China

Index	Sector	Task
1	Ministry of water resources	Surface water resource management
2	National environmental protection bureau	Water environmental protection
3	Ministry of geomineral	Underground water management
4	Ministry of construction	Construction of city water resource development and protection
5	Ministry of agriculture	Agricultural water management
6	Ministry of energy	Water and electronic construction
7	National planning and development commission	Permits for conservancy projects
8	Ministry of sanitation	Monitor and protect drinking water
9	Ministry of finances	Permits for flood control finance capital
10	National scientific commission	Management of water resource research
11	National meteorological bureau	Prediction and management of precipitation
12	Ministry of transportation	Management of inland navigation
13	Bureau of national land management	Project management of protected basins
14	Ministry of forestry	Protection of river basin forestry

- to carry out a national comprehensive scientific survey and investigation as well as an assessment of water resources
- to draw up the overall planning for the main river basins in China
- to formulate a long-term water supply plan on national and regional scales
- to issue permits for water use
- to construct and maintain water projects such as reservoirs, dikes, and weirs
- to manage the waterways, including rivers, lakes, canals, and flood discharge areas

Generally, there is a department or bureau of water resources under each level of local government which is responsible for local water resources development and management.

To strengthen the development and conservation of water resources in the large rivers in China, the Yangtze, Yellow, Pearl, Huai, Hai, Songhua, and Liao, there are seven river basin commissions established under the leadership of the MWR. These river basin commissions as agencies of the MWR are responsible for the tasks mentioned above in the basin concerned.

National Environmental Protection Agency

The National Environmental Protection Agency, the environmental protection department of the State Council, exercises unified supervision and management for the prevention and control of water pollution. It is in charge of the following responsibilities of water quality management:

- to establish the implementation rules and regulations of water quality management on the basis of prevention and control in the water pollution law
- to establish national water environmental quality standards
- to establish national pollutants discharge standards in accordance with national water environmental quality standards and the country's economic and technological conditions
- to amend national water environment quality standards and national pollutants discharge standards in due time
- to examine and approve the environmental impact statement and inspect water pollution prevention and control facilities of new projects, extensions, or reconstruction projects which discharge pollutants into waterbodies directly or indirectly
- to carry out the task of registration and application for permits for water pollution discharge
- to carry out the task of payment for water pollutant discharge
- to implement elimination and control of water pollution within a certain period
- to exercise the punishment authority of warning and fines on violators of prevention and control in the water pollution law
- to be in charge of pesticide registration

Ministry of Geology and Mineral Resources

The functions of the Ministry of Geology and Mineral Resources are:

- general investigation and exploitation of groundwater
- assessment and unified planning for the development and use of groundwater
- supervision and management of groundwater

Ministry of Agriculture

This ministry is in charge of the management and conservation of farmland irrigation water and fishery waterbodies.

Ministry of Public Health

- supervision and management of drinking water quality in urban and rural areas
- formulation of sanitary standards of drinking water
- monitoring drinking water quality
- definition of sources of drinking water in accordance with other competent departments
- participation in the management of protection zones for drinking water

Ministry of Construction

Construction and management of the water supply in urban, and rural areas and sewerage engineering as well as domestic wastewater treatment plants in cities.

Ministry of Transportation

Supervision and management of water quality for navigation and navigation ways in inland rivers.

In addition to the responsibility of the various Ministries under the State Councils as mentioned above, the responsibility for flood control belongs directly to the State Councils and local governments at various levels as well as the seven large river basin commissions.

LEGAL SYSTEM

In 1988, the Water Law of China was enacted and issued, representing the legal basis for water resources development and management. At present, other relevant laws or regulations have been enacted for effective water pollution control, water and soil conservation, the water-drawing permit system, water conservation in urban areas, flood control, safe management of dams, tariff collection, riverbank management, navigation management, land use for water works, and resident resettlement. A series of laws and regulations related to water resources development and management is also being drafted and processed for enactment, concerning integrated use of water resources, watershed management, flood control, water resources fee collection and management, water pricing of water supply projects, totaling more than 20 (see below).

List of Chinese Laws Related to WRM

Laws

- Constitution of the People's Republic of China (DOE*: 1982, RA**: National People's Congress)
- Forest Law of the People's Republic of China (DOE: 1984, RA: National People's Congress)
- Fishery Law of the People's Republic of China (DOE: 1987, RA: National People's Congress)
- Water Law of the People's Republic of China (DOE: 1988, RA: National People's Congress)
- Environmental Protection Law of the People's Republic of China (DOE: 1989, RA: National People's Congress)
- Water and Soil Conservation Law of the People's Republic of China (DOE: 1991, RA: National People's Congress)
- Mineral Resources Law of the People's Republic of China (DOE: 1994, RA: National People's Congress)
- Water Pollution Control Law of the People's Republic of China (DOE: 1996, RA: National People's Congress)
- Flood Control Law of the People's Republic of China (DOE: 1997, RA: National People's Congress)
- Land Administration Law of the People's Republic of China (DOE: 1998, RA: National People's Congress)

Regulations and Policies

- Working Regulation of Soil Conservancy (DOE: 1982, RA: State Council)
- Price Management Regulation (DOE: 1987, RA: State Council)
- Law on Land Administration (DOE: 1987, RA: State Council)
- Water Extraction Permit System Regulation (DOE: 1993, RA: State Council)
- Urban Water Supply Regulation (DOE: 1994, RA: State Council)
- Method for Management of Irrigation Districts (DOE: 1981, RA: Ministry of Water Resources)
- Method for Construction and Management of Irrigation and Drainage Projects (DOE: 1981, RA: Ministry of Water Resources)
- Technical and Economical Index of Irrigation Management (DOE: 1996, RA: Ministry of Water Resources)
- Industry Policies for the Water Sector (DOE: 1997, RA: Ministry of Water Resources)
- Urban Water Conservation Management Regulation (DOE: 1989, RA: Ministry of Construction)

*DOE: date of enactment;
**RA: responsible agency.
(*Source*: The government of the PRC and the Asian Development Bank, 1999)

FINANCIAL SOURCES FOR WATER MANAGEMENT IN CHINA

The main financial source for water management is composed of three parts: the infrastructure fund of the water conservancy construction fund, maintenance of projects, and institutional management fees. The sources of the infrastructure fund for water conservancy are:

- the infrastructure fund of the general budget
- the water conservancy construction fund
- loans from domestic banks and other financial organizations
- funds from the sale of bonds
- loans from foreign financial organizations
- other sources approved by state governments

Both the sources for maintenance of projects and institutional management fees are the general budget and the water tariff. Spending of the infrastructure fund for water conservancy is on flood control, drainage, irrigation, water supply, hydropower, navigation, and others. The Financing chart for water management of the YRCC is shown in Fig. 4.

POLICY, LAWS AND REGULATIONS

The Central Government lays the utmost stress on the Yellow River. A series of principles, policies, general water laws, water administration regulations, and basin-oriented water laws have been developed. They involve the water administration of the Yellow River, the management of water resources, management of a water conservancy project and engineering affairs of the Yellow River, management of flood control, water and soil conservation, and protection of water resources. Examples are given below.

Water Laws such as the "Water Laws of the People's Republic of China," "Laws of Water Pollution Prevention of the People's Republic of China," "Laws of Water and Soil Conservation of the People's Republic of China," and "Laws of Flood Prevention of the People's Republic of China"

Laws and Regulations of Water Administration: "Working Regulations of Water and Soil Conservation," "Management Regulations of River Channels of the People's Republic of China," "Management Means of Ratification and Collecting of Water Fees for a Water Conservancy Project," "Regulations of Water and Soil Conservation in Developing the Contiguous Areas of Shanxi, Shaanxi, and Inner Mongolia," "Collecting and Management Means of Water Fees for the Intake of the Irrigation Canal," "Means of Fee Collecting and Management of Sand Mining in River Channels," "Implementation Means of Permit Regulation of Abstracting Water," and "Means of Control and Management of Water Quantity"

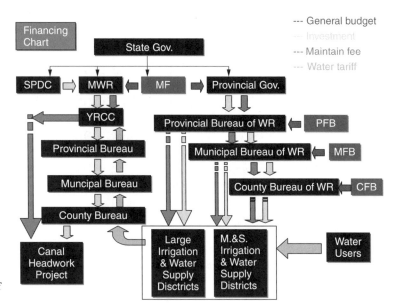

Figure 4. Financial chart for water management of the YRCC.

In addition, there are many local laws and regulations relevant to the management and development of the Yellow River. For example, "Management Means of Water Conservancy Project of the Yellow River of Shandong Province," and "Regulations for Certain Problems in River Channels and Flood Detention Areas in Henan Province."

READING LIST

General Information of the People's Republic of China: http://www.chinatoday.com/general/a.htm#POPU

WILL WATER SCARCITY LIMIT CHINA'S AGRICULTURAL POTENTIAL?

Bryan Lohmar
Economic Research Service,
U.S. Department of Agriculture

This article is a US Government work and, as such, is in the public domain in the United States of America.

Water shortages in important grain-producing regions of China may seriously compromise China's agricultural production potential. Rapidly increasing industrial and domestic water consumption and expanding irrigation have drawn down groundwater tables and disrupted surface-water deliveries. The problem is most severe in north-central China, where most of China's wheat and cotton is produced and irrigation is essential to maintaining high yields. The situation may worsen unless effective water conservation policies can be put into place rapidly.

China is responding to these concerns on several levels. At the national level, the Ministry of Water Resources began promoting water conservation through various measures in the late 1990s, such as strengthening the authority of National River Basin Commissions to enforce water withdrawal limits and promoting irrigation management reforms. Provincial and other local officials are mediating conflicts between users to improve overall water management. In villages, local water managers and farmers are adopting water management reforms and water-saving techniques, such as forming water user associations and alternating wet-dry irrigation for rice. In addition, reforms in the pricing and fee collection system may provide farmers with better incentives to conserve water. Pricing water deliveries to farms based on volume could improve efficiency, but would be costly to monitor since China has over 200 million farm households, each tending several tiny plots of land.

As water for agriculture becomes more scarce, changes in the pattern of crops are more likely than a reduction in cultivated acreage. Wheat is most likely to suffer declines, since wheat is irrigated in much of north China and brings low returns to water. Production of a variety of crops—corn, cotton, and high-value fruits and vegetables—may increase as farmers switch from irrigated wheat. High-value fruits and vegetables are often more water intensive, but are also more suited to water-saving irrigation technologies, such as drip irrigation and greenhouse production.

The success of current efforts to encourage water conservation in China will depend on a variety of factors. Policy reforms will depend critically on the enforcement of withdrawal limits both from surface-water systems and from ground water. Also important is the extent to which policies and local management practices motivate water users and water managers to conserve water resources.

WATER AND COASTAL RESOURCES

U.S. Agency for International
Development (USAID)

Every ecosystem, society, and individual on Earth depends on water. Food security and human health, energy supplies, and industrial production are all dependent on water to a large degree. Water plays an important role in regulating global climate. It is essential for plants and wildlife and ecosystems they inhabit. Water shortages and poor water management can lead to loss of biodiversity and agricultural production, increase in malnutrition and disease, reduced economic growth, social instability, and conflict.

Reliable water supply has been a fundamental component of the increased agricultural productivity achieved by the green revolution. Irrigated agriculture currently uses 70% of the world's developed water supplies. With increasing competition for water to meet domestic and industrial needs, and for servicing essential ecosystems, agriculture is faced with producing more food with less water.

Water resources are not distributed uniformly across the globe—nor are they necessarily located where the largest concentrations of people reside. Demand for water outstrips supply in a growing number of countries, and the quality of that supply is rapidly declining: 450 million people in 31 countries already face serious shortages of water. These shortages occur almost exclusively in developing countries, which are ill-equipped to adopt the policy and technology measures needed to address the crisis. By the year 2025, one-third of the world's population is expected to face severe and chronic water shortages.

Exacerbating the problem, human activities often contaminate the world's limited freshwater resources, making them unavailable for further human use and threatening the health of the lake, river, and wetland ecosystems they support. Likewise, coastal and ocean systems are under threat from the impact of a broad range of human activities. Coastal systems are particularly vulnerable to degradation from land-based activities, climate change, over-fishing, and damage to coral reefs, and they require active intervention to ensure their continued survival.

The U.S. Agency for International Development has made the preservation and environmentally sound development of the world's water resources a top priority. The

This article is a US Government work and, as such, is in the public domain in the United States of America.

Figure 1. USAID worked with local partners to improve water resources management in Morocco's Nakhla region of the Rif Mountains.

Agency's investments have helped improve access to safe and adequate water supply and sanitation, improve irrigation technology, enhance natural environments, and develop better institutional capacity for water resources management in countries around the world (Fig. 1). This has supported the Agency's underlying goals of reducing conflict and improving the welfare of people across the globe.

WATER USE CONSERVATION AND EFFICIENCY

Mario O. Buenfil-Rodriguez
National University of Mexico
Cuernavaca, Morelos,
Mexico

WATER USE AND USERS ON GLOBAL AND REGIONAL SCALES

Water has many uses and users, ranging from fluvial navigation to employing water as a component in a chemical experiment. On a global level, the four most water demanding human activities are energy—hydroelectricity—generation, irrigation for agriculture, urban supply, and nonurban industrial production. Each of them requires huge quantities of water; and in its usage at least some water characteristic is affected, which limits the options of the remaining users to profit from that same water. This causes a competition in which each user would like to have exclusive rights to manage and exploit the resource. Hydroelectricity affects the potential energy of water (topographic elevation), and the other three basically affect water quality. There is a fifth user, frequently neglected by humans, which has the most ancient right to benefit from water, nature. The natural world needs water in lakes, forest, woods, floods, and climate regulation and has been greatly affected by the human desire to conquer and transform and to grow and expand.

Worldwide, the agricultural sector is the biggest water consumer, although this balance is not equal in every continent or country. For instance, nations that have benign and frequent rains that allow good direct agricultural production may have no need for irrigation; others that are very dry or have an uneven distribution of water and rain require huge irrigation works to get reasonable crop yields. Figure 1 depicts the distribution of volumes among the three human uses that most affect water quality.

WATER SCARCITY, CONFLICTS, TECHNOLOGY, AND CONSCIOUSNESS

Water scarcity, which is rapidly growing in most countries, creates serious conflicts, competition for water, and evidently great unfairness in many cases. Figure 2 is a clear and simple illustration of why there is growing scarcity in most nations. Of course, its severity differs widely, depending on regional natural resources and demographic pressure. In general, there are growing crises in the whole world.

Legal rules establishing some priorities for water usage, or simply the legal support to historical privileges, sometimes help in attenuating real or apparent conflicts among users. Other regulatory mechanisms are water markets, where users can trade their water rights to other users. This last has advantages, but also great disadvantages when the market and the world economy (usually a totally artificial issue subject to political fancies and fashions) send erroneous signals about the value of things.

Competition among users is not absolute; water used upstream can later be freely used downstream, provided the residual water discharges from the first consumers are not dangerous enough to inhibit their later usage. Anyway, there always is some technological possibility of treating and reusing water. If there were enough money to apply technology, conflicts would be eliminated, but technology is costly, and most of the world is poor. This is worsened by the low value given to water by traditional economies and by the unequal competition for globalized trade in agricultural products. So unfairness and crises still have a long way to go.

An often misconceived economics paradigm suggests that water should be used where it renders the greater rate of return. This is a kind of taboo to be seriously questioned and reviewed because it leads us to accept the

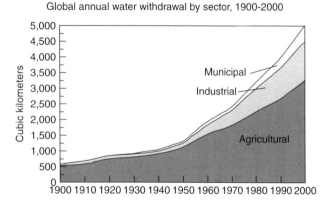

Figure 1. Rising water use. Global annual water withdrawal by sector, 1900–2000 (from Ref. 1).

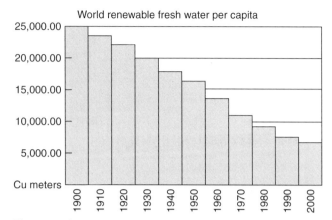

Figure 2. A century of changes in water availability: clear signs of scarcity and conflicts.

idea that uses in services such as tourism, electronics, and industry in general are more important than use in agriculture and food. A better approach is to accept that the latter use is greatly undervalued by the market and the others are exaggeratedly overvalued. The same situation occurs in ecological water use.

The mid-twentieth century vision of dealing with water needs and scarcity problems was to increase the supply by means of big dams and similar huge engineering works. They have proved insufficient and merely made the demand disproportionately grow, so the vision for this century must be radically changed. The approach now must be "water demand management" respecting natural limits and boundaries. Scarcity must be directly dealt with, by conservation and "efficient water use." These options also have a technological component but rather rest mainly on human habits and consciousness. They are radically different from options that rely on huge engineering works for water transfers and treatment.

WATER USERS AND DEMAND ON AN URBAN SCALE

Water users in cities are mostly the people who live and work there; they can be represented by existing houses, buildings, shops, offices, parks, hotels, schools, and factories. There are many types of cities and ample variations in these components. Tourist and recreational cities have lots of hotels and almost no factories; in industrial cities, it is the opposite. Housing for families usually is the greater component in any city and evidently the sector demanding more water.

The water demand in a city can be expressed in liters/person/day; it is comprised of the water used directly or indirectly by an average citizen and the leaks in the piping network. People use water directly in their houses, but also in their jobs or at school, and indirectly by others serving them, as in restaurants where they eat, by shops or services visited, and in general by every industry in the city. Leaks unavoidably occur so the water may arrive in the consumption area. The previous reasons explain why if the average in-house *consumption* is 300 L/capita/day, the full *demand* may be around 500 L/capita/day. Figure 3

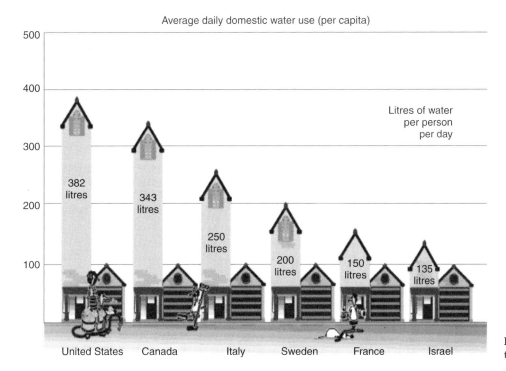

Figure 3. House water consumption in some countries (from Ref. 2).

contrasts some international statistics about in-house consumption. There are big variations, of course, even within a nation and even among different zones and persons in the same city. Figure 4 is a sample of in-house water usage distribution.

CONSERVATION OF WATER AND OTHER RESOURCES

Water emphasizes many conflicts and scarcities, although water may not be the real issue considering that there is plenty of water on our planet. Water quantities may not be the problem, but its quality, or its location in time or in space. Infrastructure, investments, and technology are always essential for actions such as water treatment, storage for other seasons, or transport to other places. Therefore, the real restrictions may not be water volumes or qualities, but those regarding energy resources, financial sources, and managerial capabilities to transport it or to preserve its purity. Energy is one important inhibitor; if it were freely available, water could be pumped from far away flood zones to deserts (this does not consider possible negative ecological impacts).

When institutions request people to save water, they do not merely think of saving water volumes, but rather of saving energy expenditures by eliminating the need for larger pipes and, mainly, by diminishing pollution caused by the water usage and discharge. Many propose conservation plans, so the volume saved is used to support demographic and demand growth. In this case, there are no real savings and compensations for previous damage and overexploitation, but merely variants of the twentieth century backing of human expansionism.

The need for conservation and for "efficient water use" has a much broader scope than merely using less water. The real purposes and issues may be protecting the whole environment, our economy, and the quality of life of future generations. They can have a rather dark side when used as a political and demagogic argument simply to maintain inertias for growth and expansionism, which require less money investments but possibly greater human and environmental sacrifices and discomforts that are not taken into account.

MEASURING THE EFFICIENCY OF URBAN SERVICES

Even in something as specific as public water supply, efficiency has many angles and goals. Simultaneously, it is important to save water volumes, to avoid pollution, to diminish unnecessary expenditures, and to distribute water more fairly among different users. For instance, if there were an optimal amount of water to be given to

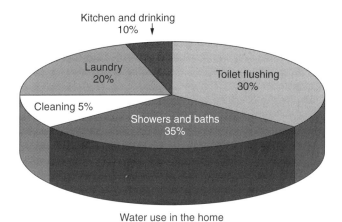

Figure 4. Typical household water use in Canada (from Ref. 2).

people, considering health, welfare, and other aspects (for example, 150 L/person/day), anything above or below such a quantity could be considered inefficient.

One common type of inefficiency occurs in water transportation because there always is some degree of leakage caused by breaks in the piping system. This efficiency is computed as the ratio of the volume delivered to clients, measured by micrometers, divided by the total volume extracted from the water source. The operating or "revenue efficiency" is the total money collected through bills paid by clients, compared to the total amount in issued bills. The "unaccounted for water" (UFW) indicator is a measure of the overall inefficiency of a water supplier, considering physical and commercial losses. Another strategic indicator is represented by the ratio of total expenditures to total income during a certain period.

Yet another important way to assess performance is by comparing the initial water quality at the source against the quality returned after discharge. An increasing degree of recycled or reused water within a city evidently is also a desirable way to demonstrate improvements.

It would be extremely expensive and totally impractical to have a system without inefficiencies, that is, 100% efficiency in all aspects. What is done by developed water utilities is to work simultaneously using an ample set of "performance indicators" and to apply several benchmarking rules and comparisons with other utilities, as suggested by the International Water Association (3).

Just as any water utility may have various types of efficiencies to be confronted routinely and evaluated, the private home owner, or any factory or hotel manager, may have many parameters by which to assess performance in water usage and its impact on the environment. Water audits (Fig. 5) are appropriate as systematic procedures for assessing performance and to detecting failures as well as opportunities for improvement.

CONSERVATION AND EFFICIENCY, THE NEED AND THE HOW

The need for conservation and efficient water use comes from scarcity and competition for the resource, each time more evident in many places. In this regard, many national and international institutions advise how to save water. The most classical are to take shorter showers, not to leave the faucet open while shaving, or changing the old toilet for one of newer design (maximum 6 liters/flush, see Fig. 6). There are lots of these recommendations, which are not bad at all, but maybe at sites of severe crises and scarcities, recommendations should be even more daring and definitive, to attack the roots of problems, not their symptoms. Some of these newer counsels are: use dry latrines instead of toilets, shower just once a week, and use perfumes; be a vegetarian and not a meat eater (meat production demands much more water than vegetables); family planning; bicycles instead of cars; reuse, recycle, rethink; charge "green taxes" for many environmentally damaging products or packaging.

The importance of systematically assessing and measuring efficiency lies in finding improvement opportunities. Performance or efficiency indicators should be compared against reference values to detect failures, changes, and trends. The reference standards should be carefully discussed and selected, because if not, some incongruence and evident mistakes may occur. One pitiable flaw would be to report efficiencies greater than 100% when abstaining from using water in certain industrial process or eliminating a superfluous water need. For instance, when comparing the change to dry latrines after using traditional toilets, it evidently does not mean infinite efficiency, although doubtlessly it is great environmental progress.

Excessive, wasteful, and polluting water uses should be restricted. However, how these restrictions should be

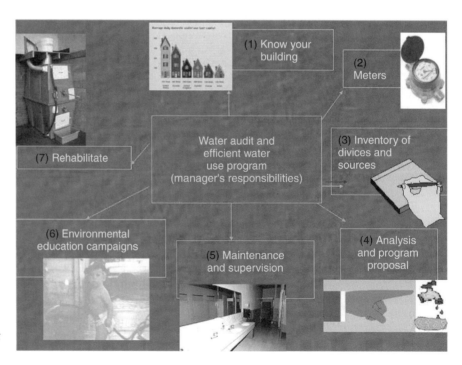

Figure 5. Seven stages for water-saving programs in buildings.

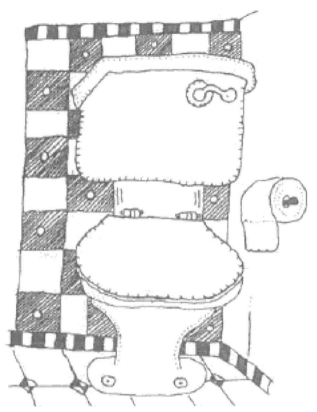

Figure 6. Toilet bowl: friend and foe. The most frequent cause of water consumption, leakages, and pollution.

attained is an interesting issue. The least desirable options would be laws for direct prohibitions. There are better choices, for instance, measuring water and contaminant loads in discharges and applying tariffs high enough to discourage those uses and encourage the search for better technologies and processes.

Besides metering and tariffs, there are many programs and actions that any water supplier must consider, which contribute to efficient water use in a city. Among them are user education and water culture, macrometering, information systems, client census, demand forecasting, automation of operations, leak detection and control, materials testing and quality control, employee training, improved maintenance procedures, bylaws, aquifer recharge, rain harvesting, water audits, improved domestic water-saving devices, water treatment, water reuse and recycling, and desalination. Each of these programs and tasks has different degrees of importance depending on the local conditions, and together make up a desirable set of simultaneous actions to be considered and used.

This article supplies only a list of these programs without further details. Each one deserves a specific and detailed explanation, which hopefully may appear in other sections of this Encyclopedia of Water. All of them are important and strategic for efficient water use and resource conservation.

LIMITS TO GROWTH AND EFFICIENCY

The options for solving water scarcity, along with the crises and conflicts it causes, rest mainly in the social, moral, and economic realm, rather than in the technical one, although at first glance, it may appear the other way.

Most likely, technology can always find a way to improve efficiency, but also it implies that costs sometimes are not apparent. Solutions often have clear and immediate economic components but often also may have vague, but possibly important, environmental, quality of life, and long-term consequences. Unfortunately, traditional economic and technical models merely deprecate and do not consider such impacts because they do not know how to deal with them.

Technological options for improving water efficiency sometimes are clear and simple, and can be adopted without much questioning and doubt; however, in many other cases, it is important to have broader vision. Here is a brief list of arguments to reflect on and consider:

— Efficient water use is not simply "using less water." There are many other options for managing demands. Reuse is an important one.
— Efficient water use depends more on the efficient use of resources other than merely on water saved. Energy, money, technology, people's willingness, forests, and climate are some of these other resources.
— Urban water scarcity is not a water availability problem. It may be a symptom and reflection of other problems so that, when confronted, the scarcity issue will be immediately relieved.
— Scarcity problems could be "easily" solved by stopping consumption and population growth.
— Urban water scarcity is a "moral and ecological problem," not a technical one.
— More severe and urgent laws are required for sites suffering extreme scarcity, particularly, where aquifers present continued water table drawdown.
— Scarcity is referred to as a problem of shortages but is a situation of "longages" (excesses) of bad habits and population or industrial growth in inappropriate zones.

Technology can succeed in allocating population increases by using and reusing the fixed available water volume more intensely. However, this style of technology application merely justifies expanding the coverage, without a clear improvement in the quality of life. Reuse, treatment, and desalination, are desirable options up to certain limits, but should never be used indiscriminately to justify further growth. On the contrary, they must promote diminishing human impacts to make more natural space available. Theoretically, it is possible to have houses or factories with zero water discharge, and almost zero water demand by intensive treatment and reuse.

Water desalination is an interesting option for increasing the supply of water. It has a history as ancient as the hydrologic cycle, its basis. Humans have used desalination techniques for survival on long ocean journeys and for supplying isolated islands for centuries. Nowadays, desalination processes use reverse osmosis that demands huge energy input, which mainly comes from thermoelectric power plants; so the process is equivalent to exchanging petrol for water. In that sense, besides being a rather expensive choice not for every one, it is not a sustainable option because fossil energy is a nonrenewable resource. It would be more desirable and sound to research and promote technology using solar energy for these processes.

Now that the failures and flaws of traditional economic models that cheer growth and expansion are evident, there is need for a new paradigm. Efficient water use and conservation should be seen as a route to reach this new paradigm, not as a means for reinforcing the old one.

When continuous, long lasting, and growing sacrifices or "improved efficiencies" are the alternative to make available water to increasing populations and when there is not a clear end to these situations, serious questioning must be done about the risk of failure of this route (plagued by decreasing safety levels) and of the worth of such sacrifices. It could be that opting to limit growth and technology would be more appropriate, sound, and simple.

BIBLIOGRAPHY

1. Hinrichsen, D., Robey, B., and Upadhyay, U.D. (1997). Solutions for a Water-Short World. *Populations Reports, Series M*, No. 14 Baltimore, Johns Hopkins School of Public Health, Population Information Program, http://www.jhuccp.org/pr/m14edsum.stm.
2. Environment Canada (1998). *The Management of Water*. National Water Issues Branch, Environment Canada, http://www.ec.gc.ca/water/en/manage/e_manag.htm.
3. Alegre, H., Baptista, J. M., Hirner, W., and Cabrera, E. (2000). *The IWA system of performance indicators for water supply services* International Water Association IWA, manual of best practices (performance indicators). http://www.iwapublishing.com/template.cfm?name=isbn1900 222272.

READING LIST

Arreguín, C. F. and Buenfil, R. M. (1997). *64 Recomendaciones Para Ahorrar Agua*, IMTA- Ministerio de Desarrollo Económico de Colombia. http://www.cepis.ops-oms.org/bvsadiaa/e/noticias/colombia/agua.pdf.

Asian Development Bank (McIntosh A., Yñiguez, C.) (1997). *Second Water Utilities Data Book. Asian and Pacific Region*. Asian Development Bank.

Barker, P. (1977). *Cost-Based Water Prices*, 23 WEDC conference, Durban South Africa, http://www.lboro.ac.uk/departments/cv/wedc/papers/23/groupf/barker.pdf.

Brown, L. (2001). *Eco-Economy: Building an Economy for the Earth*, Earth Policy Institute, http://www.earth-policy.org./Books/Eco_contents.htm.

Brown, L. (2001). *Worsening Water Shortages Threaten China's Food Security*. Earth Policy Institute, http://www.earth-policy.org/.

Buenfil, M. (2001). *Selfishness and Efficiency (in Water Use)*, 4th Information Water Summit, Panama. http://www.waterweb.org/wis/wis4/cd/Buenfil-S&E%20water%20use.pdf.

Environment Canada (1993). *Manual for Conducting Water Audits and Developing Water Efficiency Programs at Federal Facilities*. Environment Canada, water issues branch, http://www.ec.gc.ca/water/en/info/pubs/manual/e_contnt.htm.

EPA. (2001). *Residential Water Conservation Techniques*. Environmental Protection Agency, USA, http://www.epa.gov/seahome/watcon.html.

Gleick, P.H. (2000). *The Changing Water Paradigm. A Look at Twenty-First Century Water Resources Development*. International Water Resources Association. http://www.iwra.siu.edu/win/win2000/win03-00/gleick.pdf.

Gleick, P.H. (2001). Making every drop count. *Sci. Am.* http://www.scientificamerican.com/2001/0201issue/0201gleick.htm.

Hardin, G. (1968). Tragedy of the commons. *Science* http://www.sciencemag.org/cgi/content/full/162/3859/1243.

Hawken, P., Lovins, A., and Lovins, L. H. *Aqueous Solutions, Chapter 11 of Book "Natural Capitalism: Creating the Next Industrial Revolution*. Rocky Mountain Institute http://www.natcap.org, http://www.natcap.org/images/other/NCchapter11.pdf, http://www.rmi.org/sitepages/pid172.php, http://www.rmi.org/images/other/W-AqueousSol.pdf.

Hinrichsen, D., Robey, B., and Upadhyay, U.D. (1997). *Solutions for a Water-Short World*. Population Reports, Series M, No. 14. Johns Hopkins School of Public Health, Baltimore, Population Information Program, http://www.jhuccp.org/pr/m14edsum.stm.

James, W. (1999). *Urban Water Systems*. Guelph University, Canada. http://www.eos.uoguelph.ca/webfiles/james/homepage/Teaching/661/wj661Modules.html.

Mello, E.J. (2001). *Conservação e Medição da Água* various links and articles related to metering and water conservation, http://www.geocities.com/hidrometro/.

Postel, S. (1984). *Water: Rethinking Management in an Age of Scarcity*, Worldwatch Institute paper 62, Worldwatch Institute, Washington USA, http://isbn.nu/0916468623.

Postel, S. (1999). *Pillar of Sand. Can the Irrigation Miracle Last?* WorldWatch book.

Reimold, R.L. and Bloetscher, F. (1996). An innovative opportunity for water reuse. *Fla. Water Resour. J.* http://www.fwrj.com/articles/9607.pdf.

Vickers, A. (2000). *Handbook of Water Use and Conservation*. California Urban Water Conservation Council, http://www.cuwcc.com/publications/action.lasso?-database=cuwcc_catalog&-layout=CDML&-response=detailed_results.html&-recordID=33056&-search.

WaterMagazine (Cayford, Joel). (2002). Water services—regulation & pricing—article links. *Water Mag.* New Zealand, http://www.watermagazine.com/secure/pricereg.htm.

World Bank (Yepes, G. and Dianderas, A.). (1996). *Water and Waste Water Utilities Indicators 2nd Edition*, Water and Sanitation Division, TWUWS, The World Bank, http://www.worldbank.org/html/fpd/water/pdf/indicators.pdf.

CONSERVATION OF WATER

MUKAND SINGH BABEL
Asian Institute of Technology
Pathumthani, Thailand

Water is the main component of the natural ecosystem and an essential element for human life. It is also fundamental for the social and economic development of a country. Water is needed for drinking, washing, and sanitation at personal, domestic, and municipal levels; for agricultural purposes such as irrigation, livestock, and fisheries; for industrial development, navigation, hydropower production, recreation, and so on. In addition, there is increased realization that water is needed for maintaining ecosystems.

The higher rate of population growth associated with several anthropogenic activities leading to increased pollution of water resources has augmented critical situations worldwide related to scarcity of water resources either in quantity or quality to meet the vital necessities of humankind.

When the ratio of water availability to demand exceeds certain limits, the stress on resources becomes noticeable, and the conflicts among users intensify and increase in frequency. A region that have water availability of about 1700 m^3 per capita/year faces periodic water scarcity, and this is considered a warning if the population continues to grow. Below this limit, "water stress" is periodic. If this limit is even lower, about 1000 m^3 per capita/year, the region is under "chronic water scarcity." Economic development, well-being, and health are affected at these levels. The situation is "extremely scarce" if the per capita availability is below 500 m^3 per year. Many countries that used to have plentiful water resources are facing water scarcity, that is likely to be aggravated in coming years.

On one hand, there is the problem of scarce availability in relation to increasing demands; on the other hand, there is poor resources management and wasteful and inefficient use of freshwater. To address these water problems, it is necessary to control the overuse and pollution of water resources by improved management of water resources through conservation and water saving concepts and practices.

The main water-use sectors are agricultural, industrial, and domestic. At a global level, agricultural water use accounts for 70%, industrial use 21%, and 9% for domestic consumption. Though industrial water demand is much smaller than agricultural demand in low- and middle-income countries, it exceeds domestic and agricultural demands in developed and industrialized countries, as shown in Table 1.

Water is wasted nearly everywhere; there is great potential for better conservation and management, no matter how water is used. The term water conservation is associated mainly with the policy, managerial measure, or user practice aiming at conserving or preserving water resources and taking care of the degradation of water resources; water saving aims to limit or control water demand and avoid wastage and misuse. Both perspectives are complementary and interrelated in managing scarce water resources.

Table 1. Distribution of Water Use[a]

	Agricultural Use	Industrial Use	Domestic Use
World	70%	22%	8%
High-income countries	30%	59%	11%
Low- and middle-income countries	82%	10%	4%

[a] Reference 1.

AGRICULTURAL SECTOR

Most water is used for agriculture (Table 1). Efficient agricultural water management, therefore, is a major conservation option. It is common to find that more than half of the water drawn from a source does not even reach the fields being irrigated. Generally, less than 40% system efficiency prevails in many irrigation schemes. Such an appalling situation should be tackled by pursuing better systems of farm water management—improving irrigation efficiency by reducing water distribution losses, changing cropping patterns, improving irrigation scheduling, and adopting irrigation efficient technologies. Deterioration of irrigation network systems due to lack of proper and adequate maintenance are the main causes of increased losses through leakage and percolation. Rehabilitation of existing irrigation schemes is, therefore, required to alleviate the situation. It is also equally important, when implementing new irrigation schemes, to give sufficient attention to adequate operation and maintenance. Active participation by the users is vital for sustainable management of irrigation systems. Several governments are in the process of transferring management functions to farmers or developing joint (government agency and water users) management systems.

In gravity irrigation methods such as basin, furrow, and flooding, a large quantity of water is lost by leaching down the root zone and as runoff at the end of the field. Land leveling, along with other technical measures, provides conditions for reducing the advance time and water volumes required to complete the advance, for better water distribution as well as application efficiencies. For moderate to high infiltration soils, where system automation is possible, water can also be saved by adopting surge irrigation which consists of intermittent cycling of water applied to furrows and borders, thereby producing changes in the soil surface conditions which favor a faster advance of water.

Sprinkler irrigation methods have higher system efficiencies than gravity irrigation. Water in sprinkler irrigation can be conserved by improving the distribution uniformity—optimizing the overlapping of sprinkler jets, minimizing discharge variations, minimizing wind drift and evaporation losses, maximizing infiltration of applied water, and avoiding surface runoff losses. Microirrigation has the least loss of water among irrigation methods. It can be classified in two general categories: drip or trickle irrigation, where water is slowly applied through small emitter openings from plastic tubing; and microspray irrigation or microsprinkling, in which water is sprayed over the soil surface. Water conservation in microirrigation can be improved by using a single drip line for a double crop row, using microsprayers in high infiltration soils, adjusting the duration and timing to soil and crop characteristics, controlling pressure and discharge variations, adopting automation and fertigation, and chemigation (application of fertilizers and herbicides or other chemicals with irrigation water).

Water conservation measures are also very important in dry-land agriculture because crop growth depends solely on the moisture retained by the soil from rain. Conservation measures in rain-fed agriculture are carried out mainly by appropriate crop and soil management practices. Crop management techniques consist of managing crop risks, such as selecting the cropping pattern to suit the available seasonal rainfall; controlling water stress, such as adaptation of crop rotation to environmental constraints, including covered fallow for grazing; and water conservation cropping techniques, such as by employing conservation tillage, windbreaks to decrease wind impacts on evaporation, etc. Soil management practices are broadly categorized as runoff control, such as contour tillage, using mulches, furrow diking, etc; improvement of soil infiltration rates, such as application of organic matters and mulching; increasing soil water storage holding capacity, such as deep tillage; control of evaporation from soil by crop residue and mulching; runoff control in sloping areas by terracing, contour ridges, and strip cropping; and water harvesting.

Reuse of water, such as using wastewater (graywater) for agriculture, is considered a conservation option to increase the availability of water either currently or in the future. More than 500,000 ha of farmland are currently being irrigated with treated wastewater in 15 countries around the world. However, the use of such wastewater, if not treated well, may threaten the public health by spreading bacteria, viruses, and parasites. For example, the outbreak of typhoid fever in the mid-1980s in Santiago, Chile, was attributed to the use of wastewater, with unsatisfactory sanitary quality, to irrigate 16,000 ha of horticultural farms (2).

The major challenge to water planners and managers is that physical availability of water is fixed, but demand will increase to cater to the needs of an ever increasing population. The tendency today is to shift from water policies based on water supply management to those based on demand management. Agricultural water demand management consists of reducing crop irrigation requirements, adopting irrigation practices that lead to higher irrigation performance and water saving, controlling system water losses, and increasing yields and income per unit of water used. It works with agronomic, economic, and technical management decisions for decreasing irrigation water demand and recovering the costs of irrigation water supply. An agricultural water demand management program should consider the interaction between the quantity, quality, and biological aspects of both groundwater and surface waters; the sustainability of irrigation and drainage schemes; and environmental sustainability. Water demand management necessitates establishing structural incentives, regulations, and restrictions that help to bring about efficient water use in irrigation by farmers while encouraging innovations and saving technology.

INDUSTRIAL SECTOR

Water is used in industries in power generation, temperature control, manufacturing processes, and washing. The principal water users in industry are thermal and atomic power generation, chemistry and petroleum chemistry, ferrous and nonferrous metallurgy, wood pulp and paper industry, and machine building.

The quantity and quality of water required by industry vary with the type of industry and the industrial processes. The effluents produced by the industry are also of different qualities, which may require different treatment processes. Water recycling and reuse are the main water saving strategies in industry. Industrial water recycling, which began on a large scale in the 1970s to help cope with antipollution regulations, has also become an effective conservation measure. American steelmakers, which once consumed 280 tons of water for every ton of steel made, later needed to use only 14 tons of new water; the rest was recycled. Similarly, in Japan and western Germany, for example, a large increase in the number of factories has been accommodated without a considerable increase in total industrial water use (2).

The most common conservation measures in industrial water use are water-use monitoring, recycling, reuse, cooling tower use, equipment modification, and employee education. Of all industrial water use, cooling water accounts for a very large share. Replacement of once-through cooling processes with recirculating systems can provide some of the largest water savings. Often, cooling water can be reused for landscape irrigation and cleanup. The physical and chemical properties of water may change after use, requiring some simple treatment before being used again. Treated water can be reused for heating processes; water from washing raw food material can be reused for cooling, heating, or indoor washing.

DOMESTIC SECTOR

In urban areas, water supplies are used in households and for services such as city washing, fire fighting, and maintenance of swimming pools or recreational parks and lakes. A considerable amount of water is lost by leakage in the water distribution network. The unaccounted for water (UFW) comprises system leakages, illegal tapping, theft, incorrect meter reading and billing, and overflowing reservoirs. System monitoring and metering should be done in appropriate intervals to obtain information on the system state variables to prepare plans for water conservation measures as well as for estimating water losses, their location, and causes so as to prepare for appropriate corrective measures. Typical water losses as a percent of water supplied are 8–24% for developed countries, 15–24% for newly industrialized countries; for developing countries, it is 25–45%, as surveyed in 1991 by the International Water Services Association (IWSA). Examples of typical water losses for developing countries are well portrayed by Arequipa, Peru, with a 45% loss, and Hanoi, Vietnam, with a 68% loss (43% nonbilled, 20% leakage, 5% company use) (3). Between 20 to 40% of the supply was UFW in Indian cities and towns (4). On the municipal level, cities have saved from 10 to 25% of their water by repairing leaky pipes (2). Preventive as well as corrective maintenance can play a major role in water conservation.

Water pricing and billing can induce water saving habits in users if price increases with the volume used, particularly when the differential increases in price are large enough to encourage water savings. The reclamation of wastewater for nonpotable purposes has become a proven measure to help meet increasing demand. Where extreme water scarcity exists, a feasible but expensive solution is to have a dual distribution network for high quality water which is needed for uses such as drinking, food preparation, and bathing, and for treated reusable water which can be used for landscape and recreational ground irrigation, toilet flushing, and floor washing. Such reusable water can be available from, for example, treated urban drainage stormwater. Such examples of dual systems can be found in Australia, Hong Kong (China), Japan and the United States, to name a few. In Japanese cities, for example, such reclaimed water finds greatest use in toilet flushing (40%), compared with only 15% for irrigation in urban landscaping; in Hong Kong, seawater has been used extensively for decades for toilet flushing in commercial and multifamily residential buildings. However, a dual distribution system is difficult to implement because it requires very large investments from the public sector as well on the part of users in their homes. Financial incentives such as subsidies may be required for such investments in additional infrastructures and to implement other high-cost water conservation technologies.

In the domestic sector, most water is used for gardens and for bathrooms and toilets. In a typical household that has outdoor facilities, nearly 50% of the water is used for gardening; of the total indoor water use, 33% is spent on showers and bath, 24% for toilet flushing, and 22% in laundry (5). Water saving plumbing fixtures and appliances are cost-effective and provide permanent long-term economic advantages. Replacing traditional toilets which use 16 to 20 liters per flush with low-flow toilets (such as 6 liters per flush) or placing containers or bags filled with water in the tanks of large volume toilets can save a lot of water by reducing their capacity. A minor repair to control toilet leaks can be of immense help in conserving water. These leaks can be detected by adding a few drops of food coloring to water in the toilet tank and observing if the colored water appears in the bowl. Showers take second place in domestic water demand. Reducing the time for showering and preferring showers to immersion bathing can be very effective in conserving water. Low-flow showerheads and faucet aerators save water as well as energy used to heat water (where warm water is needed).

When washing clothes and dishes by machine, operating clothes washers and dishwashers with full loads can economize on water used. Outside the home, water can be saved by avoiding unnecessary spillage of water. Simple practices like using a broom, not a hose, to clean driveways, steps, and sidewalks; washing the car with water from a bucket; and irrigating plants either in the early morning or the late evening can save a lot of water. In landscaping, use of native plants that require less care and water than ornamental varieties can save a large quantity of water. While mowing the lawn, the lawn mower should be adjusted to a higher setting to provide natural ground shade and to promote water retention by the soil. In swimming pools, water can be continuously recycled, clarified, and purified by means of portable equipment and appropriate chemicals. Evaporation losses can be minimized by covering the pool when not in use.

INSTITUTIONAL MEASURES

Educational programs are very important in instilling water conservation awareness among users. Conservation literature can be distributed along with regular bills to water users. Local service organizations can also play a big role in disseminating water conservation information. An excellent time to provide this information to customers is just prior to the summer season when demand normally peaks. Local newspaper articles, radio and television public service announcements, information centers at local fairs and shopping centers, etc. can also be used to promote conservation awareness and education. Local schools can be a very good place to offer conservation knowledge to young people by means of water education activities such as water conservation posters, slogans, essays, or exhibit contests for children. High-use facilities such as schools and colleges, hospitals and institutions, country clubs, and health clubs are appropriate places to initiate water conservation programs.

Water laws provide a general framework for water use and conservation and are complemented by regulations. They can help water conservation by restricting excessive water use, for example, establishing standards for indoor plumbing fixtures, maximum volumes per flush in toilets, or making it compulsory for filtering, treating, and recycling the water in swimming pools, etc. The provision of incentive and penalties helps in enforcement of regulations as well as in the adoption of water saving technology by customers.

BIBLIOGRAPHY

1. World Bank. (2001). World Development Indicators (WDI). Washington, DC, Available in CD-Rom.
2. Beekman, G.B. (1998). Water conservation, recycling and reuse. *Water Resour. Dev.* **14**(3): 353–364.
3. WHO. (2001). Leakage Management and Control—A Best Practice Training Manual. http://www.who.int/docstore/water_sanitation_health/leakage/begin.html.

4. Venugopalan, V. (1997). Water conservation methods. *Water Supply* **15**(1): 51–54.
5. Stanger, G. (2000). Water conservation: what can we do? In: *Water Management, Purification & Conservation in Arid Climates*. M.F.A. Goosen and W.H. Shayya (Eds.). Technomics, Lancaster, PA.

THE DEVELOPMENT OF AMERICAN WATER RESOURCES: PLANNERS, POLITICIANS, AND CONSTITUTIONAL INTERPRETATION

Martin Reuss
Office of History Headquarters
U.S. Army Corps of Engineers

To understand the development of American water resources, one must first look at American political and social values and at American governmental institutions. Even a cursory examination shows the lasting influence of decisions and attitudes molded as the country took its first hesitant steps as a republic. Historian Joyce Appleby has argued that the first generation of Americans bequeathed "open opportunity, an unfettered spirit of inquiry, [and] personal liberty" to future generations—qualities, we might note, that often introduce an element of uncertainty into public administration. But if we extend the analysis a bit, we might not only gain an appreciation of the many challenges facing water resource developers, but also illuminate a fundamental question facing democratic nations: To what extent should human liberty be constrained in order to provide and manage a human necessity—water?

Beyond Appleby's observations, one notes at least pervasive elements woven into American political behavior. The first, the inescapable, element is distrust of powerful governments. Power corrupts, the first Americans agreed without much hesitation, and the challenge was how to minimize that corruption, how to ensure that good men will not be enticed to do evil, and how to disperse power to minimize oppression. Loudly over the years, Americans continue to proclaim their distrust of big government; even popular presidents generate skepticism when they appear to reach for increased power and authority.

Only as a last resort, and then with resignation, not enthusiasm, as during the Great Depression, do Americans turn to the national government to solve their problems. The result can be truly impressive: Grand Coulee and Bonneville dams, locks and dams on the Upper Mississippi, the California Central Valley Project, and the Los Angeles flood control system all came out of depression-era politics, but arguably all are aberrations in the story of American water resources.

The second element, almost as pervasive as the first, is that power and liberty are fundamental antagonists. The dispersion of power among the three branches of government purposely sets power at war with itself rather than with "life, liberty, and the pursuit of happiness." Each branch would be allowed only sufficient power to discharge official duties, and a system of checks and balances would guard against abuse. Recoiling from British monarchism, the constitutional drafters took special care to try to prevent executive branch intrusions into the duties of the other two branches. This was a system that, regardless of its merits, made implementation of rational planning enormously difficult, as water developers soon appreciated.

Political attitudes one thing; government structure was another. And here the Founding Fathers developed a system that guaranteed further complications. They fashioned a republican form of governments within the government. A century later, young political scientist Woodrow Wilson thought that this structure posed the principal challenge to American administration. Few water resource planners would disagree. Republican government, it must be remembered, began in the states, not in the new national capital; delegates to the Continental Congress delayed business so they could go home and participate in state constitutional conventions. The formation of these state governments may have excited Americans more than the latter formation of the union itself, and the American Constitution explicitly guaranteed to each state a republican form of government (Article IV, Section 4). Once the United States achieved its independence, many Americans pondered how citizens could owe allegiance to two governments, two legislatures, simultaneously. Were the states and national government partners or were the states meekly to accept national supremacy? No one at the Constitutional Convention quite knew what to expect from this layer-of-powers (or was it a marble-cake, twentieth century political scientists later debated), and numerous, contrary explanations emerged of what the delegates had actually achieved. In no area did the confusion become more manifest or disruptive than in internal improvements, especially in water projects that crossed state lines.

The term "internal improvements" came to mean many things to the citizens of the young republic. It included roads, canals, schools, lighthouses, fortifications, and even technological innovations-most anything that seemed to provide security and promote the economy. Gradually, it came to mean something a bit more specific, though still covering (pardon the pun) a large amount of ground: it applied to what we now call "infrastructure," and water transportation was a central concern. Benjamin Franklin had proposed at the Constitutional Convention that Congress have the power to construct canals, but opponents won the day, fearing that Congress would become too powerful. In fact, the term "internal improvements" cannot be found in the American Constitution, an obstacle for those seeking affirmative authority for federal involvement in public works. But neither did the Constitution proscribe the activity, which meant to internal improvement advocates that the function lay legitimately within federal authority. This ambiguity not only produced a constitutional quagmire for internal improvements, but it provided a platform upon which larger issues of the role of government and the nature of liberty could be debated. In short, the internal

---
This article is a US Government work and, as such, is in the public domain in the United States of America.

improvements issue amplified and sharpened the debates about the very nature of American republicanism. By any other name, it continues to serve that function to the present day.

Given Americans' distrust of government and emphasis on personal liberty, America's first politicians, and all the generations following, confronted the difficulty of promoting economic growth without expanding governmental authority. One answer was the corporation, a device that actually predated the Constitution but in the age of internal improvements became much favored. As presumed promoters of the public good, they effectively became agencies of government. In this way, legislatures could support economic and political development without necessarily involving tax monies. The fact that individual incorporators might thereby profit aroused little concern. The more important point was that corporations brought together sufficient capital to launch an enterprise, whether a canal or a municipal water system. Even if a number of these ventures brought forth charges of corruption, internal improvement advocates ceaselessly trumpeted the moral and intellectual gifts stemming from public works, as though canals were spiritual as well as economic enterprises. To complaints that corporations disenfranchised people and led to the inequitable distribution of wealth, champions argued-somewhat quaintly in light of what subsequently emerged corporations were nothing more than little republics eminently suited for the United States. For better or worse, the victory of the corporation in American life was almost as revolutionary as the victory of republicanism itself, and the alliance between government and corporations became a hallmark of American economic development. Government was not to replace business, but was to support and, within certain limits, protect it.

George Washington and other Federalists had ardently hoped that corporations might provide the capital and means to build internal improvements to bind the nation together and transcend local interests, perhaps leaving overall planning to the national government. But the chance slipped through their hands. The structure of Congress assured that state interests in internal improvements would prevail over national interest. There would be no national board, no national planning. Rather, Congress would periodically pass rivers and harbors acts that generally reflected parochial politics. To stimulate states and the private sector, Congress also provided a percentage of funds obtained from the sale of public lands in new states to finance roads and canals (the three and five percent funds dating back to 1802) and voted to turn over certain lands to states for reclamation (Swampland acts of 1849 and 1850). In a few cases, too, Congress might vote to subscribe to canal stock or even grant land to a company-a practice that presaged the enormous land grants given to railroad companies as they extended their presaged the enormous land grants given to railroad companies as they extended their lines across the continent later in the century.

Caught in a congressional quagmire that appeared to offer no rational plan for the development of the country's infrastructure, succeeding presidents attempted to develop some orderly process, but at the same time they worried over possibly unconstitutional intrusions into areas beyond federal authority. We turn our attention to the Executive Branch in part two of this series.

This article is based on Martin Reuss, "The Development of American Water Resources, Planners, Politicians, and Constitutional Interpretation," in Paul Slack and Julie Trottier (Eds.), *Managing Water Resources, Past and Present: The Twelfth Annual Linacre Lectures*. Oxford University Press, 2004. Used by permission of Oxford University Press.

The Development of American Water Resources: Planners, Politicians, and Constitutional Interpretation.

The Constitution and Early Attempts at Rational Water Planning.

The Expansion of Federal Water Projects.

WATER MARKETS: TRANSACTION COSTS AND INSTITUTIONAL OPTIONS

Terence R. Lee
Santiago, Chile

The costs involved in the transfer of property, or transaction costs, can significantly affect the capacity of any market to operate efficiently. If water marketing is to achieve its full potential, markets must be designed to minimize transaction costs. Water marketing may lead, however, to efficiency gains, even if transaction costs are high. Moreover, increasing water scarcity will raise the welfare gains from trade relative to these costs, as the value of water rights increases. Transaction costs can be lowered by technological advance and by institutional design. It can also be expected that transactions costs will fall as the market matures, as learning by doing has its effect.

TRANSACTION COSTS

Transaction costs can be defined as those resources dedicated to establish, operate, and enforce a market system. They may take one of two forms, the services which buyers or sellers must provide from their own resources; the differences (margins) between the buying price and the selling price of a water right, for example, due to the direct financial cost of any brokerage services.

In a water market, there are three potential sources of transaction costs:

- the costs of finding trading partners, of verifying ownership of water rights, and of describing the right for purposes of the proposed transfer;
- the costs involved in reaching an agreement to trade, such as negotiating the price; arranging financing and other terms of transfer; drawing up contracts; consulting with lawyers and other experts; paying fees for brokerage, legal and insurance services; and transferring legal titles;
- the costs of setting up a legal, regulatory and institutional framework, mitigating possible third-party

effects, and ensuring compliance with applicable laws.

Water markets tend to be highly regulated, so many transaction costs arise from public policies governing water transfers. The burden imposed by policy-induced transaction costs comes from the direct costs of a transfer and also from the delays due to waiting for regulatory approval. These costs can be high. For example, in the western United States, waiting periods for state agency decisions, vary from 4–6 months in New Mexico, to 5–9 months in Utah, and to an astonishing 20–29 months in Colorado (1,2). It is possible to place institutional restraints on such regulatory induced delays. In Chile, the Water Code regulates the maximum time periods permitted for each stage of any recourse to regulatory intervention (3). This type of institutional requirement, at least, reduces the arbitrariness inherent in regulatory delays.

Transaction costs prevent markets from operating efficiently and reduce the overall economic benefits of water marketing. They increase the cost of water rights and decrease the incentive to trade. Transaction costs introduce inefficiencies, which block equalization of marginal values among different uses, users, and locations. In extreme cases, transaction costs can prevent markets from forming altogether.

These considerations imply that transaction costs usually reduce welfare and that public policies to reduce costs will be welfare improving. This does not mean, however, that all regulation of water markets is bad. Some costs arise from justifiable efforts to protect third-party interests. Such policy-induced transaction costs are not necessarily wasteful or inefficient if they provide real protection for third parties that may be affected by water transfers.

Transaction costs can be independent of the quantity of water rights transferred. The title search, filing fees, and other similar costs are often fixed. This penalizes smaller trades and favors larger trades. Other transaction costs, such as brokers' commissions, may be proportional to the amounts or values traded. More extensive water infrastructure and highly developed user institutions can lower costs. A recent empirical study indicates that, in Chile, transaction costs are particularly low in the areas that have more modern infrastructure and well-developed water users associations (4).

Variations in transaction costs can affect the spatial concentration of market activity. In Texas, for example, the major market activity is concentrated in the Lower Rio Grande Valley where transaction costs are low (5). Transaction costs can also affect the choice of the type of transaction. For example, usually the transaction costs for a permanent transfer exceed those for a lease.

MARKET-RELATED TRANSACTION COSTS

The efficiency of any competitive markets rests on the assumption that good, reliable, and easily accessible information is available. In practice, such information is not always available for water markets, partly because producing and disseminating information is often costly and difficult and, partly, because the means to redistribute risk are incompletely developed due to uncertainty about the nature of the risk and the asymmetry of information between market participants. Altogether, however, the information requirements for efficient water marketing are no greater than those needed for effective administrative allocation of water.

A study of water markets in the western United States concluded that all markets studied were characterized by varying degrees of uncertainty and incomplete access to market information on water commodities, prices, and market opportunities (6). Where there is no ready means for buyers and sellers to obtain information, they face legal, hydrologic, and economic uncertainties. These constitute a substantial cost disincentive to engage in water marketing.

INSTITUTIONAL OPTIONS

Efficient construction of any market requires the existence of the necessary conditions for trading to occur: (1) well-defined property rights, (2) public information on the supply of and the demand for water rights, and (3) the physical and legal possibility for trading to take place. Of these three necessary conditions, by far the most important is the existence of well-defined property rights. In the case of water, property rights define and limit the rights and duties of their holders relative to one another and to the rest of society to the use of a certain amount of water, which may be defined either volumetrically or in terms of shares of a stream or canal flow. If rights are poorly defined, market processes cannot be relied upon to allocate water resources efficiently. It is a basic responsibility of governments, as far as markets are concerned, to define, allocate, and enforce property rights in water. Government policies play a critical role in defining the institutional setting for market operation and provide the basis for market activity by defining, allocating, and enforcing water rights.

The way property rights are defined will structure the incentives and disincentives that members of society face in their decisions regarding water ownership, use, and transfer. For market participants to estimate the value of a water right, they must be able to form secure expectations about the benefits and costs of owning and transferring it and the degree to which it is protected from impairment by others (7).

So, there is a need for a reliable and trusted means of registering rights. Hydrologic information is also required to permit the right to be defined. Various types of information are essential for rational decision-making by water rights holders: the legal and hydrologic characteristics of water rights and the costs of alternative means of obtaining water. This implies the existence of good data and monitoring systems, which can be provided only by public agencies.

The public availability of information on the supply of and the demand for water rights must also include the means to identify willing buyers, sellers and intermediaries or brokers and the means for entering into enforceable contracts. Such information is better provided by the market rather than by public agencies.

In addition, a clearly defined set of transfer rules is necessary to permit market transactions to take place when buyers and sellers determine. Transactions should be contingent only upon compliance with a known set of trading rules or transfer criteria. This is an essential prerequisite for a continuous water market. There must also be information ensuring that the physical possibilities for the transfers produced through trade can actually happen. This may require the legal possibility of easements or the purchase of rights of way across the property of third parties.

BIBLIOGRAPHY

1. Colby, B.G., McGinnis, M.A., and Rait, K.A. (1989). Procedural aspects of state water law: transferring water rights in the western states. *Ariz. Law Rev.* **31**.
2. MacDonnell, L. (1990). *The Water Transfer Process as a Management Option for Meeting Changing Water Demands.* Natural Resources Law Center, University of Colorado, Boulder.
3. Lee, T.R. (2000). *Water Management in the 21st Century: the Allocation Imperative.* Elgar, Cheltenham.
4. Hearne, R.R. and William, E.K. (1995). *Water allocation and water markets: an analysis of gains-from-trade in Chile*, Technical Paper Number 315, The World Bank, Washington, DC.
5. Chang, C. and Griffin, R.C. (1992). Water marketing as a reallocative institution in Texas. *Water Resour. Res.* March, No. 3.
6. Saliba and Colby B. (1987). Do water markets 'work'? Market transfers and trade-offs in the Southwestern states. *Water Resour. Res.* **7**.
7. Colby, B.G. (1988). Economic impacts of water law—state law and water market development in the Southwest. *Nat. Resour. J.* October, No. 4.

AVERTING WATER DISPUTES

JODY W. LIPFORD
PERC
Bozeman, Montana, and
Presbyterian College
Clinton, South Carolina

INTRODUCTION

At midnight on August 31, 2003, time ran out on a proposed agreement among the states of Alabama, Florida, and Georgia to allocate water in the Apalachicola-Chattahoochee-Flint (ACF) River Basin. The deal had been 13 years in the making, but it ended in failure. "It's a true shame that we were as close as we were and couldn't get an agreement," said Alabama's chief negotiator (1).

It was a shame. The collapse of these lengthy negotiations sends the matter to the courts, and the Supreme Court may ultimately decide how the disputed water will be divided. More broadly, the failure of the state governments to reach agreement reveals that water, long considered plentiful in the southeastern United States, is in danger of becoming a subject of intractable conflict. The failure signals that a water crisis may well emerge in the region unless new approaches to allocating water are adopted.

As the population of the Southeast increases, competing demands for water—for municipal use, for recreation, and for hydropower, to name just a few—are growing. Today the problem surfaces in the form of occasional interstate disputes such as this one, but the failure to resolve them casts an ever-longer shadow over the future of water resources in the region. When demands of competing users outstrip supply, there must be ways to ensure that water goes to the users who value it most and that the waterways of the Southeast are not roiled by unending conflict.

This article explains the reasons behind the conflict in the Apalachicola-Chattahoochee-Flint River Basin, why attempts at resolving it failed, and what alternatives should be considered. It explains how to allocate water to its most productive uses, restore peace to the areas around these waterways, and avert other conflicts that are emerging, not only in these states but elsewhere in the South.

THE BACKGROUND

As shown on the map in Fig. 1, the ACF Basin drains an area of 19,800 square miles in the states of Georgia, Alabama, and Florida. The Basin starts in the headwaters of the Chattahoochee River in northern Georgia, above Atlanta. The Chattahoochee flows through Georgia's Piedmont before turning sharply south, forming the southern half of Georgia's border with Alabama and a notch in Georgia's border with Florida. At the border, it meets the Flint River to form the Apalachicola River, which flows through the Florida panhandle into the Gulf of Mexico (2).

Historically, in the ACF basin, as in most of the southeastern United States, water has been abundant and has met the many demands for it. The demands include water for domestic, commercial, industrial, hydroelectric, navigational, and recreational uses.

Under riparian water rights—the system of water rights in the eastern part of the United States—landowners can use water that flows adjacent to their property as long as they do not appreciably diminish the quantity or quality of water available to downstream users. But riparian rights were effectively overridden in 1946, when Congress authorized the U.S. Army Corps of Engineers to construct dams for flood control, navigation, and hydroelectricity along the Chattahoochee River.[1] Later,

---

[1]The Army Corps of Engineers has long played a role in the ACF River Basin. To facilitate commercial traffic, the Corps began dredging the Chattahoochee River in the 1880s. At present, five Army Corps dams dot the ACF River Basin: the Buford Dam, which forms Lake Sidney Lanier; the West Point Dam, which forms Lake West Point; the Walter F. George Dam, which forms Lake Walter F. George; the George W. Andrews Dam; and the Jim Woodruff Dam, which forms Lake Seminole. Authorization and construction of these dams began in the 1940s, 1950s, and 1960s, and by the 1970s they were all operational (3, pp. 151, 155).

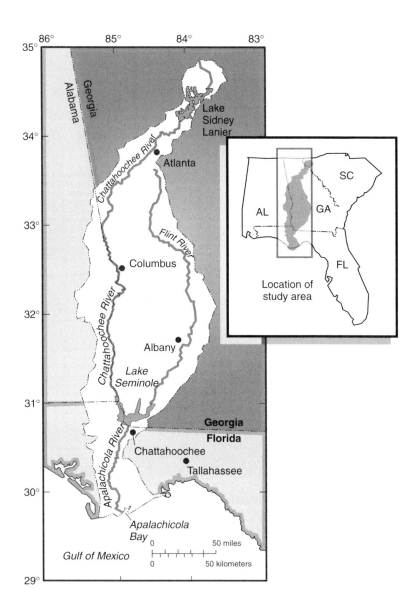

Figure 1. Apalachicola-Chattahoochee-Flint (ACF) River Basin. Courtesy of Mandy-Scott Bachelier.

the Corps added water supply and recreation as purposes of the dams and the reservoirs they created (4,p.2; 5). These projects transformed the waters of the basin from private property governed by the riparian doctrine to public property.

Today, the waters of the ACF River Basin continue to be owned and managed by the federal government through the Army Corps of Engineers. The Corps' managers meet weekly to consider various water needs, such as hydroelectric production, recreation, navigation, and environmental quality (6). In addition, all requests to increase water withdrawals must be approved by the Corps (7, p. 993).

Except for flood control, each of these purposes requires a minimal lake level or river flow rate. Electricity cannot be produced, nor can barges navigate, without sufficient water. Similarly, fish populations require stable lake levels during spawning season. Sufficient flow also dilutes pollution, helping to ensure water quality (6). The Corps also commonly provides recreational facilities, such as parking areas, boat ramps, and public restrooms. Response to the recreational amenities of the lakes has been heavy; millions of user-days are tallied each year (3, p. 158). The Basin also supplies water for public use. All these competing demands limit the amount of withdrawals that can be made.

Conflicts Over Increasingly Scarce Water

For a long time, the system of riparian doctrine and public management through the Army Corps of Engineers worked well. In the 1980s and 1990s, however, rapid population growth, particularly in metropolitan Atlanta, combined with recurrent drought led to increased pressure on the ACF River Basin's resources.

Atlanta's population grew from 2.2 million in 1980 to 3.0 million in 1990, and then to 4.1 million in 2000. Of 126 metropolitan statistical areas listed by the Census Bureau, only 17 had higher population growth rates from 1980 to 1990, and only eight had higher population growth rates from 1990 to 2000.[2]

---
[2] See Table 30 of U.S. Census Bureau (8, pp. 32–34).

Demand for water to satisfy this growing population increased dramatically. Metropolitan Atlanta's water use increased from 289 million gallons per day in 1980 to 459 million gallons per day in 1990, and then to 606 million gallons per day in 2000.[3] Metro Atlanta relies almost exclusively on surface water; over 70% comes from the Chattahoochee River and Lake Sidney Lanier, the lake formed north of Atlanta by the dam at the headwaters of the Chattahoochee.

Yet, the Chattahoochee River and Lake Lanier are ill suited to supplying Atlanta's water needs. The Chattahoochee is the smallest watershed in the country to supply a metropolitan area with the majority of its water (9). The largest share of Atlanta's water use—53.8%—is for residential use; commercial, government, and industrial users take 22.8%, 5.6%, and 4.2%, respectively (10).

A preliminary study by the Army Corps of Engineers indicates that Atlanta is already approaching, and at times exceeding, water use levels that were not expected until 2030. Whether these findings are accurate or not—the Georgia Environmental Protection Division says that the Chattahoochee River and Lake Lanier are sufficient to supply Atlanta through 2030—there is no doubt that future demands on the Chattahoochee River and Lake Lanier will be heavy and growing (11).

Drought has worsened this pressure on the River Basin's waters. During the 1980s and 1990s, the Southeast experienced recurrent and severe droughts. In the most severe drought, rainfall in Atlanta fell by as much as 25%, and annual average streamflows along the Apalachicola River fell to less than half their historical norms.[4]

The Crisis Begins

In 1989, recognizing that Atlanta's "finite supply of clean water is looming as a barrier to growth" (12, p. 68), Atlanta and the Army Corps of Engineers proposed to approximately double the water Atlanta drew from Lake Lanier, to 529 million gallons per day. Without sufficient water, Georgia officials feared the loss of 680,000 jobs and $127 billion in wages through 2010 (13, p. 26).

The proposal by Georgia and the Army Corps alarmed the citizens of Alabama. Increased withdrawals from Lake Lanier would reduce flows along the segment of the Chattahoochee River that forms the southern half of Alabama's border with Georgia and would stunt economic development there. So, in 1990, Alabama sued the Army Corps to keep it from allocating more of the ACF River Basin's waters to Atlanta.

The state of Florida quickly joined the lawsuit on the side of Alabama, fearing that reduced water flows would harm the oyster-rich Apalachicola Bay farther downstream and lead to deterioration of the Apalachicola's water quality. The state of Georgia then joined the lawsuit on the side of the Army Corps of Engineers to defend its withdrawal. The stage was set for 13 years of studies, proposals, counterproposals, and extended deadlines.

[3] Data supplied by Julia Fanning, U.S. Geological Survey, Atlanta, e-mail correspondence, November 7, 2003.
[4] For data on streamflow of the Apalachicola River at Chattahoochee, FL, see waterdata.usgs.gov/fl/nwis/annual/calendar_year/?site_no=02358000.

These actions reflected unique concerns in each state. Alabama officials worried that Atlanta's withdrawals would stifle Alabama's economic development by limiting water needed for domestic, industrial, and commercial use. Water quality would also suffer because reduced downstream flows would mean less dilution of polluted upstream water. Atlanta not only uses a large amount of water, but also discharges heavily polluted water back into the Chattahoochee (7, p. 996; 12, p. 68–69).[5] If dirtier water came from Atlanta, Alabama might have to raise water cleanup standards for industrial and municipal users, which would be costly and would put Alabama at a competitive disadvantage in attracting economic development (14, p. 3).

From Florida's perspective, the chief problem was oyster beds. Reduced flows, especially at critical times, and heavier pollution could threaten the Apalachicola Bay's oyster industry, which supplies approximately 10% of the country's oysters and employs over 1000 people. The river is also a commercial source of shrimp, blue crab, and finfish, as well as the home of an important sport fishery and the only commercial source of Tupelo honey. It has been recognized as an Outstanding Florida Water (15, p. 34; 16).

Other factors exacerbated the conflict. For example, recreational users want lakes kept full or nearly full; however, the competing objectives of hydropower and flood control require that lakes be drawn down, typically during summer and winter. Similarly, navigation requires minimal flows that reduce the water in lakes when river levels become too low for barge traffic. Finally, ecosystem preservation requires a pattern of flows that mimics nature's seasonal cycle and may conflict with other demands (4, p. 2; 14, p. 3).

Despite years of negotiation, the states never agreed how to allocate the basin's waters. Because these negotiations failed, the matter is now likely to be settled by the Supreme Court.

Constitutional Paths to Solution

Once Alabama took the Corps to court in 1989, the states had several constitutional options to choose from to settle the conflict over water allocation. They could go to Congress for a decision, their suits could reach the Supreme Court, or they could negotiate their own agreement or compact. Neither Congress nor the Supreme Court likes to get involved in interstate water disputes. Moreover, suits reaching the Supreme Court are costly, usually require lengthy negotiations, and yield uncertain outcomes.

For these reasons, the states opted for an interstate water compact. Alabama agreed to deactivate its lawsuit during the negotiation period, the Army Corps of Engineers agreed not to allocate additional water from Lake Lanier to Atlanta, and all parties agreed to a comprehensive study of the water resources in the basin (4,

[5] In response to federal consent decrees to stop spills of untreated wastewater into the Chattahoochee and to comply with the Clean Water Act, Atlanta is currently trying to raise over $3 billion to renovate its antiquated sewage treatment system.

pp. 3–4).[6] In early 1997, all three state legislatures ratified legislation authorizing the negotiation of an ACF River Basin Compact to allocate the basin's waters. These acts were subsequently signed by the three state governors. The U.S. Constitution requires congressional approval of interstate compacts, so in November 1997, Congress approved and President Clinton signed federal authorizing legislation. The goal of the compact was to assign property rights to water fairly and avoid future conflicts. This goal proved elusive, however.

The Failure of the Compact

The federal legislation set an initial deadline of December 31, 1998 for compact negotiations, unless the states agreed unanimously to extend that deadline.[7] This deadline proved much too optimistic, as each state presented proposals reflecting its parochial interests. To begin with, Alabama and Florida wanted consumptive uses of water defined and limited.[8] Georgia resisted this approach in favor of one that focused on reservoir levels. Specifically, Georgia wanted the ACF River Basin's reservoirs to be kept full or nearly full (19, p. 8), so that Georgia would have enough water to supply Atlanta (4, p. 14). In return, Georgia offered Alabama and Florida minimum flow guarantees. But Alabama and Florida rejected Georgia's proposal, fearing that the minimum flows might become the norm, in essence reducing the water flowing downstream.

To ensure adequate flows along its border, Alabama argued that the ACF River Basin's water should be allocated to meet the original objectives of dam construction. These included navigation (i.e., barge traffic), flood prevention, and hydroelectric production—but not water supply to municipalities or recreation (4, p. 2; 19, p. 9). Florida agreed with Alabama in opposing minimum flows, but also wanted downstream flows to be adjusted to mimic natural flow cycles. Additional problems plagued the negotiations, from definitional questions (e.g., how to define "severe drought") to the choice of the computer model for forecasting river flows and lake levels (19, pp. 9–10).

No agreement was forthcoming, so the states agreed to extend the deadline till January 1, 2000. Yet, the passing of another year did not appreciably advance the negotiations. Once again, the states set a 1-year deadline, establishing a pattern of deadline extensions that continued until July 22, 2003. At that point, progress seemed to have been made. The three governors signed a Memorandum of Understanding that set a blueprint for water allocation

---

[6]Until the compact was completed, the states agreed to "freeze" water at current use levels. Should increased withdrawals be needed, the states agreed to notify other states in advance (17, p. 202). For further details, see Public Law 105-104, Article VII (c).
[7][Public Law 105-104, Article VIII (3)].
[8]Consumptive use, also known as water consumed or water depleted, may be defined as the "part of water withdrawn that is evaporated, transpired, incorporated into products or crops, consumed by humans or livestock, or otherwise removed from the immediate water environment." The definition also includes "any water withdrawn in the basin and transferred out of the basin for use" (18).

in the ACF River Basin. The memorandum authorized water supply for Atlanta from Lake Lanier at 705 million gallons per day and left open the possibility of greater future withdrawals. The memo also established minimum flow requirements downstream from Atlanta; the most important was a flow of 5,000 cubic feet per second on the Apalachicola River at Chattahoochee, Florida. The deadline for final agreement on the memorandum was August 31, 2003 (20).

Although Florida Governor Jeb Bush signed the memorandum, he and other Florida officials had reservations, which they expressed in an accompanying statement (21). They insisted that minimum flows must not become targets, that Lake Lanier must be managed to deal effectively with drought, and that the governing ACF Commission must have authority to approve any withdrawals from Lake Lanier that exceeded the amount specified in the memorandum.[9]

Ultimately, the states could not agree. Florida feared that minimum flows, which had been less than 5000 cubic feet per second only twice during the recent droughts (on a mean monthly basis), might become the norm. Thus, Florida officials again raised the issue of Georgia's withdrawals from Lake Lanier. Georgia officials responded by agreeing either to limit Atlanta's withdrawals or to promise minimum flows through the basin, but not both, and accused Florida of trying to micromanage its waters (23,24). As a result, Florida officials refused to accept the Memorandum of Understanding, and the deal collapsed. The states have now reactivated their original lawsuits (1,5,24).[10]

THE ALLOCATION OF WATER

Economists recognize the scarcity of valuable resources. Without prices on these resources, there is not enough to satisfy all who want to use them. In most settings, however, market prices allocate resources and allow them to move to those users who value them the most. This market process allocates resources to their most productive uses and maximizes society's wealth. In the case of the ACF River Basin water, however, market prices do not currently allocate water; it is allocated politically.

Economists also recognize that resources have multiple uses. Water, for example, may be used to generate

---

[9]Sixty-four percent of the ACF River Basin's reservoir capacity is held in Lake Lanier (22).
[10]Complicating the legal proceedings is a deal struck by Georgia and the Army Corps of Engineers in January 2003 in which metropolitan Atlanta governments agreed to pay the Corps $2.5 million per year toward the operation of Lake Lanier's Buford Dam in exchange for greater withdrawals from the lake. With this deal, metropolitan Atlanta sought not only to obtain additional water, but also to mollify hydropower customers who pay for the dam and who had filed suit against the Corps in 2000 because the Corps had already allocated water from hydropower to supply metropolitan Atlanta. Georgia and the Corps negotiated this deal without informing Alabama or Florida and set it aside only when Alabama and Florida found out about it and threatened to withdraw from the compact negotiations. This deal, like the initial lawsuits between the states and the Corps, has been reactivated and will have to be settled by the courts (5,25).

electricity; aid oyster production; provide wildlife habitat; produce industrial products; provide channel depth for barges; provide recreational opportunities for boaters, skiers, and fishers; or supply households with water for drinking, watering lawns, or filling swimming pools. In the ACF Basin, some water is used in each of these ways.

Whether allocation occurs through market prices or other methods, it is rarely all-or-nothing. In the Southeast (and the United States generally), there is ample water to supply basic human needs, such as drinking water. Where conflicts occur, as in the ACF Basin, it is over shifting some water, not all water, from one use to another. Simply put, the ACF Basin issue is whether more water should be allocated to Atlanta, presenting Alabama and Florida with the prospect of less water but not complete deprivation.

Although compacts have some advantages over congressional or judicial apportionment, they are poorly suited to allocate water in ways that maximize water's productive value to society. Compacts are highly political and confront intractable information problems, and such was the case with the negotiations over the ACF Basin's waters.

Interest Groups

Groups with a vested interest in the outcome of the compact influenced the ACF River Basin negotiations. Each tried to get more water allocated in its favor, irrespective of water's most productive uses. The influence of these groups introduced conflict, making a workable agreement difficult to achieve. Industrial, environmental, municipal, and political interests all made their voices heard (19, p. 8). The *Atlanta Journal-Constitution* identified political and business leaders of metropolitan Atlanta, environmentalists, and Florida's shellfish and fishing industries as uncompromising interest groups who refused to yield to the demands of other users (1).

The Apalachicola Bay's oyster industry serves as an example of a small, well-organized interest group with strong influence; its employment of approximately 1000 people is minimal in a state with total employment of approximately 7.2 million.[11] Note, however, that the industry had support throughout the state of Florida from citizens who wanted the environmental amenities of their state's river preserved.

After the agreement failed, some interests, such as the Atlanta Regional Commission and homeowners and businesses on Lakes Lanier and West Point, seemed just as happy. They feared that Georgia had compromised too much already.

Informational Problems

Even if negotiators could be insulated from interest group influence, they would still face important informational questions. If their goal is to allocate the water to its most productive uses, negotiators must first know how much total water can be allocated and how that will vary over years of normal rainfall and drought. Perhaps most important, they need to determine whether society will benefit more from allocating water to Atlanta's developers, say, or to Florida's oyster producers. And, if they can decide objectively to allocate more water to Florida's oyster producers, they would still have to decide whether the extra water should come from Lake Lanier (thereby maintaining levels at downstream reservoirs) or from downstream reservoirs (thereby maintaining levels at Lake Lanier).

By making these decisions, policy makers are implicitly choosing who will benefit and who will be harmed. A decision to allocate more water to Atlanta lessens development in eastern Alabama and reduces Florida's seafood production. A decision to allocate more water to Alabama and Florida benefits the economies of these states, but curbs Atlanta's economic development. Similarly, a decision to supply downstream users from Lake Lanier diminishes recreational opportunities for users of that lake and maintains those opportunities for users of downstream reservoirs; the opposite decision would benefit Lake Lanier's recreational users but harm those who use downstream reservoirs. And even if policy makers could determine water's most productive uses, their decisions would soon be rendered obsolete by changes in the total supply of water, changes in the total demand for water, and marginal changes in allocation necessary to maximize the total productive value of the basin's water resources.

Negotiators did try to obtain answers to some of the technical questions through the use of computer software that forecast future river flows and reservoir levels based on consideration of "historic rainfall patterns over the last fifty-five years" and "anticipated water uses within the basin in a future year, typically 2030 or 2050" (19, p. 8). But repetition of historical rainfall patterns is not guaranteed. Nor are anticipated water uses easily forecast. Atlanta's rapid population growth and commensurate water use have been dramatically underpredicted by the experts.[12] To compound matters further, different software programs give different estimates, and, not surprisingly, the states have used different modeling programs (19, p. 10).

Practical Problems

Negotiators confronted two other factors that made agreement less likely: (1) The ACF River Basin's waters were already fully allocated, and (2) the drought was expected to end soon. That the water wealth of the ACF River Basin is already fully allocated made bargaining more contentious because changes will force redistribution of existing allocations. In contrast, for example, the country's first interstate water compact, the Colorado River Compact of 1922, was negotiated in the arid West, and more water was anticipated from the Boulder Canyon Project Act, which authorized the Hoover Dam and created Lake Mead. Moreover, negotiators knew that the 1998–2002 drought was unprecedented and would likely come to an end, reducing pressures on the ACF River Basin's waters. These expectations were borne out;

---

[11]The employment figure is taken from table 602 of U.S. Census Bureau (8, p. 393).

[12]The *State and Metropolitan Area Data Book, 1997–98* predicted that metropolitan Atlanta's 2000 population would be 3.682 million (26). The actual figure was 4.112 million, an error of 430,000 for a prediction published only 2 years in advance.

abundant rain fell during the latter half of 2002 and during 2003.[13]

PROPOSALS FOR MARKETING WATER

With compact negotiations now in disarray, policy makers must look to other alternatives. An obvious proposal is for Atlanta (and other municipalities in the basin) to charge a price for its water that at least approximates its market value. Ample evidence shows that higher water prices reduce consumption (27, p. 12–13). At present, water is underpriced in Atlanta, leading to overuse. Even in the most recent drought, the city of Atlanta raised its price to residential users by only 3%. For the average residential user, the monthly water bill rose from $16.55 to $17.05. The city plans additional rate hikes through 2004, but these will raise average residential bills by a mere $1.00 per month (28). During the drought, rather than raising the price of water further, officials imposed restrictions on outdoor water use that continue in effect (29,30).

Raising water prices to market levels is apparently not politically feasible. This means that the states of Alabama, Florida, and Georgia and the federal government should consider basinwide proposals to create water markets. Experience with markets in water has shown that they can overcome some of the most difficult challenges of water allocation. They can ensure that water is allocated to its most productive uses and can prevent conflicts among users.

To some, marketing water is still a strange idea. Long accustomed to the notion that water is a commonly owned resource, many readers may doubt that it is feasible to "trade" water and thereby satisfy various interests more readily than through political negotiations. Yet, there is a strong precedent for marketing of water. Much of the American West is arid; it receives less than 17 inches of precipitation a year. With water always precious, the West evolved a system of private property rights to water, and with it, water markets. This system, known as the prior appropriation doctrine, resulted from the need to divert water for mining and agriculture. In simplest terms, it allowed a person to divert water from a river or stream on the basis of seniority (or "first in time, first in right"); the right remained as long as the individual continued to use the water ("use it or lose it"). Water users could transfer their rights to others. The private provision of water flourished (31, p. 31–45), and continues to do so, although laws that guide the transfers of water are somewhat antiquated, and restrictions on transfers that made sense in the past do not necessarily encourage efficient use today.

In spite of these restrictions, water trades occur among agricultural users, between agricultural users and cities, and between agricultural users and environmentalists. Even interbasin and interstate trades are common (27,

[13]From July 2002 to August 2003, rainfall exceeded normal levels in Atlanta, Columbus, and Albany, Georgia, by 9.64 inches, 12.21 inches, and 6.57 inches, respectively (data supplied by Pam Knox, assistant state climatologist, Georgia State Climatology Office, University of Georgia, Athens, e-mail correspondence, October 15, 2003).

p. 14–21). In Texas, where both riparian and prior appropriation doctrines are recognized, a system of marketable permits similar to that described below allocates water along the Rio Grande River (32).

In a fully functioning water market, users pay a market price for water consumed, and that price serves as a rationing mechanism. Those who can put the water to the most productive use and demonstrate this by willingness to pay, will purchase the water, be they residential developers or oyster producers. Market prices motivate those with relatively less productive opportunities to sell the water to more productive users. Through markets, groups can work out slight or marginal changes that maximize the total value from all uses.

Markets also yield peace among transacting parties. In markets, only the parties considering buying or selling a resource take part in the negotiations. Outside influences from politicians, bureaucrats, or interest groups do not guide the negotiations, as they do in the political process. The terms of exchange, such as the price, must be voluntarily agreed upon for exchange to occur. Any would-be buyer or seller who does not like the price does not have to buy or sell.

In contrast, in the political sphere, resource users often do not pay a price for a resource they consume, or they pay less than the resource's market value. This encourages them always to want more and leads to conflicts among users and a state of perpetual unhappiness for all.

As economists often point out, the foundation for markets is private property rights that are defined—that is, rights with a clearly specified ownership claim; enforced—that is, rights with a claim that is secure; and transferable—that is, rights that may be sold to others. Clearly defined, enforced, and transferable property rights are necessary for exchange. Buyers will not purchase resources if the rights to those resources are uncertain or insecure, but when rights to property are certain, secure, and transferable, markets flourish. Market-based allocation of the ACF River Basin's waters would encourage allocation of the basin's waters to their most productive uses and foster peace among users.

The Army Corps' Role

To understand how markets might work, it is appropriate to begin with the Army Corps of Engineers, which is the effective owner of the water in the ACF Basin. At present, the Corps almost entirely depends on congressional appropriations. In the Mobile District in which the ACF River Basin is located, the Corps receives some fees for its services, but they represent a small part of the Corps' budget. The fees it receives are either insufficient to cover the costs of its services (as in recreation fees), the Corps does not retain the revenues (as in the case of revenues from hydropower), or it simply does not charge for the services it provides (as for navigational services, although commercial vessels do pay a fuel tax that is used to fund inland waterway projects).

The Corps is supported by taxpayer dollars and cannot receive financial benefit from the services it provides, so it has no incentive to determine which competing uses are most productive and thus to adopt market exchange as the way to allocate water in the Basin. In an ideal world, the Corps' financing and function would be changed to give it an incentive to allocate scarce water resources to their most productive uses, thereby raising the total wealth generated from the Basin's waters. This would happen if the Corps were to retain property rights and management authority over the Basin's waters, but taxpayer support of the Corps and its projects were reduced. In exchange, the Corps would be given the authority to charge whatever fees it deemed appropriate for the services it provided and to retain the revenues. For example, the Corps could implement or change fees it charges for hydropower, dredging, water supply, and recreational services. If drought or increased demand raised the relative scarcity of water, the Corps would have the authority to raise fees. Some taxpayer support is justified because the Corps also provides flood control that benefits all users.

Although the Corps is extremely powerful in Congress and therefore such a change in the financing of the Corps is unlikely, there is some precedent for this kind of institutional reform of a public agency. In 1996, the Fee Demonstration Program allowed the National Park Service, the Forest Service, the Bureau of Land Management, and the Fish and Wildlife Service each to choose 100 sites that could raise or implement new fees and retain 80% of the revenues. Although the Fee Demonstration Program does not intend self-sufficiency for the participating agencies or individual sites, the results from changed incentives are evident, as these agencies have improved services to visitors of public lands by allocating more funds to badly needed repair and maintenance of some of the country's most-valued natural and recreational resources (33). Like the Fee Demonstration program, reform of the Army Corps of Engineers could begin on a short-term, experimental basis.

With a mandate to balance its budget and the authority to set fees and retain revenues, the Corps would have an incentive to allocate water resources to their most productive uses. If Atlanta developers wanted more water, they would have to pay a price that reflected the value of the water to other users. If it did not, those other users would outbid it. The Corps would also have to take into account the costs of its services. If barge traffic was insufficient to generate revenues to cover the costs of dredging, the Corps would cease to dredge the basin. Through this system, those with the most productive opportunities for the water would be the ones to obtain it. Such allocation would also maximize the Corps' net revenues. Unfortunately, this outcome is not very likely to occur in a political setting.

A System of Marketable Permits

Under current political arrangements, marketable permits seem to be the most promising approach to creating a water market. Marketable permits depend on the assignment of property rights to water. To implement them in the ACF River Basin, the Army Corps of Engineers could first establish a daily "water budget," consisting of the total net withdrawals allowed from the basin, based on average daily withdrawals from some past period of consumption.

After this global budget is established, the Army Corps of Engineers could grant water allocations to each user based on average daily use, again from some period of past consumption. Even though hydropower producers, barges, and oyster producers do not strictly divert water, the water they use is "diverted" from the Basin into the Gulf of Mexico and therefore should be measured for the allocation. By making the allocations daily, the Corps would allow for seasonal variations in demand and flood control. Permit allocations would be divisible and transferable. And, of course, under no condition could water be allocated in a way that violates federal water use laws.

When the supply of rainfall was abundant, so that water in the Basin exceeded the global daily budget, all users could be satisfied without the need to transfer water among users. However, in the case of drought, the Army Corps of Engineers could cut daily permit allowances by an equal percentage for all users. The Corps could then serve as a water broker, facilitating transactions among users by matching buyers and sellers and helping to negotiate terms of exchange, while charging a fee to cover administrative costs. Similarly, if the demand for water rises to the point that it exceeds the global daily budget, users who want more water would have to purchase that water from other users.

To see how this might work, consider a simplified example with two users, a lake, and a river running downstream from the lake. Suppose the two users are Atlanta developers and Florida oyster producers, the lake named Lake Lanier, and the river the Chattahoochee. Suppose that for a given day, the water budget for this river basin is 1000 gallons, allocated between 800 gallons for Atlanta developers and 200 gallons for Florida oyster producers. (Actual quantities would, of course, be in the millions of gallons per day.)

If rainfall allows greater net withdrawals, say to 1100 gallons, each user's allotment can rise by 10%. A drought, however, might reduce net withdrawals to 900 gallons, forcing cuts in permitted allotments to 720 gallons for Atlanta developers and 180 gallons for Florida oyster producers. This is where trading comes in. If Atlanta's developers want to restore their allocation, they must offer to purchase an additional 80 gallons from Florida's oyster producers. If the contracting parties agree, the Army Corps will release less water from Lake Lanier, increasing the amount available to Atlanta's developers and reducing the downstream flow for Florida's oyster producers.

If, in contrast, Florida's oyster producers want to retain a flow of 200 gallons, they will offer to purchase an additional 20 gallons from Atlanta developers. If the parties agree, this time the Army Corps of Engineers will release more water from Lake Lanier, reducing the amount available for Atlanta's developers but increasing the flow for Florida's oyster producers.

One can envision associations of users with similar wants, such as an upstream association of developers and recreational users and a downstream association of hydroelectric utilities, barges, environmentalists, and oyster producers. At times, association members would benefit by combining funds and sharing costs.

Purchasing water allotments to retire (i.e., not use) them should also be allowed. For example, if electric utilities want to increase downstream flows to generate electricity at the same time that recreational users want lake levels held high, as on a summer holiday weekend, the recreational users could purchase water rights from the electric utilities, if the utilities agreed, and retire those rights. Instead of producing revenues through hydropower, the electric utilities would receive payments from recreational users. Similarly, environmentalists might want to purchase and use or retire rights during seasons when fish spawn.[14]

In each of the exchanges described, the amount of water traded would be a small portion, not all of the total allowances. For example, recreational users would be likely to purchase some, but not all, of the electric utilities' water. Lake levels would fall enough to generate some electricity, but not as much as they would fall if recreational users did not purchase some of the water rights.[15]

To be effective, these marketable permits must have the key characteristics of property rights: They must be clearly defined, so that each user knows its allocation for each day; enforced, with the Army Corps of Engineers serving as enforcer of the permit allowances through its monitoring of lakes and dams; and transferable, with transfer facilitated by the Army Corps of Engineers serving as broker. With defined, enforced, and transferable property rights, a water market could develop that would ensure an allocation of water to its most productive uses and peace among contracting parties.

As an alternative to water transfers among users at mutually agreed upon prices, the Corps could advance market allocation by establishing a water bank. The Corps could serve as an underwriter that buys and sells water at specified prices, with the spread between these prices used to cover the costs of administering the bank. Such banks have been used in times of drought. For example, water banks were used successfully in 1977 and 1991 in California to cope with drought. In 1991, California offered to purchase water at a price of $125 per acre-foot and to sell water at a price of $175 per acre-foot. The state purchased and sold 400,000 acre-feet of water, mostly to municipal and agricultural users (31, pp. 11–12, 102–103).

In the ACF River Basin, the Army Corps of Engineers could assign to users daily property rights to flows of water, based on historic use patterns, and then serve as a water banker, standing ready to buy and sell water at specified prices. Depositors could leave water in the Basin, and withdrawers could buy it. With price playing an allocative role and with voluntary transactions, the Basin's waters would be allocated more efficiently, and relations among the ACF River Basin's users would become more harmonious.

Implications for the Southeast

Making these kinds of changes in the ACF River Basin is critical because water conflicts are brewing throughout the Southeast. Along Georgia's border with South Carolina, a request by Habersham County, Georgia, to withdraw 12.5 million gallons per day from the Savannah River Basin provoked the South Carolina state legislature to introduce resolutions calling on Congress to stop the Army Corps of Engineers from granting the request, which would have transferred water from the Savannah River Basin to the ACF Basin.[16] In addition, Georgia is involved in a dispute with Alabama over water in the Alabama-Coosa-Tallapoosa River Basin. And North Carolina and South Carolina have disputed the flow of water in the Yadkin-Great Pee Dee River Basin (35–37). By establishing water markets in the ACF River Basin, the states of Alabama, Florida, and Georgia could serve as an example to other southeastern states to help them avoid the conflicts that have for so long plagued the attempts to allocate that Basin's waters.

WHAT CHANCE FOR CHANGE?

Market reforms offer great potential, but when government is in control, change typically comes about only with crisis. Higher rainfall since the summer of 2002 has diminished the sense of crisis. Does this mean that all hope is lost for market allocation of water in the ACF River Basin or elsewhere in the Southeast? Not at all.

By failing to achieve compact resolution, the states of Alabama, Florida, and Georgia have embarked down the risky path of judicial apportionment. The risk is that the Supreme Court could allocate the ACF River Basin's waters in a way that is unsatisfactory to each or all of the states (17, p. 226). Because these allocations are not transferable, states with an unsatisfactory allocation would have no way, short of further litigation, to change the allocation. As the states contemplate this possibility, they may find it prudent to drop their lawsuits and pursue a means of allocating water that relies on markets, such as one of the proposals offered in this article.

Once demand permanently outstrips supply under current arrangements, water in the Southeast will be rationed. The question will be how. Will water be rationed by markets, which promote productive use and harmony among users? Or will it be rationed by political processes that are likely to result in misallocation and conflict? As economic development continues, perhaps plagued by drought, the citizens of the Southeast may choose the efficiency and harmony of markets over the misallocation and contention of politics.

BIBLIOGRAPHY

1. Shelton, S. (2003). Water talks a washout—states point fingers. *Atlanta Journal-Constitution* September 6, G1.

---

[14] Retiring rights requires some ranking among users. If hydropower users have the higher ranking, recreational users would have to purchase rights from them to keep lake levels up. On the other hand, if recreational users had the higher ranking, hydropower users would have to purchase rights from them to drop lake levels. Coase (34) argues that clearly defined property rights and sufficiently low transaction costs will lead to allocating resources to their most productive uses.

[15] With less water for hydroelectric production, utilities might have to raise prices to their customers.

[16] Before this conflict could escalate, Habersham County withdrew its permit request.

2. U.S. Geological Survey. (2000). Apalachicola-Chattahoochee-Flint River Basin NAWQA study: Description of the ACF River Basin study area, 2000 (last modified August 23). Online: ga.water.usgs.gov/nawqa/main.description.html (cited Sept. 24, 2003).

3. Jeane, D.G. (2002). *A History of the Mobile District Corps of Engineers, 1815–1985.* Online: www.sam.usace.army.mil/MobileDistrictHistory (cited December 10, 2003).

4. Carriker, R.R. (2000). *Water Wars: Water Allocation Law and the Apalachicola-Chattahoochee-Flint River Basin.* University of Florida, Cooperative Extension Service, Institute of Food and Agricultural Services. Online: edis.ifas.ufl.edu/BODY_FE208 (cited Sept. 24, 2003).

5. Shelton, S. (2003). Water deal may buy Atlanta some time. *Atlanta Journal-Constitution* September 12, C3.

6. U.S. Army Corps of Engineers, Mobile District. (2001). How the river systems are managed (Dec. 2, 2001). Online: water.sam.usace.army.mil/narrativ.htm (cited Sept. 30, 2003).

7. Beaverstock, J.U. (1998). Learning to get along: Alabama, Georgia, Florida and the Chattahoochee River Compact. *Ala. Law Rev.* **49** (Spring): 993–1007.

8. Census Bureau, U.S. (2002). *Statistical Abstract of the United States: 2002.* Government Printing Office, Washington, DC.

9. Metro Atlanta Chamber of Commerce and the Regional Business Coalition. (2003). *A Primer on Water Resources in the Metro Atlanta Region.* Online: www.cleanwaterinitiative.com/background/primer.htm (cited October 28, 2003).

10. Atlanta Regional Commission. (2003). *Regional Water Supply Plan.* Online: atlreg.com/water/supplyplan.html (cited October 28, 2003).

11. Seabrook, C. (2002). Atlanta guzzling water; metro thirst exceeds projections. *Atlanta Journal-Constitution* May 15.

12. Walker, B.P. (2001). Using geographic information system mapping and education for watershed protection through better-defined property rights. In: *The Technology of Property Rights*, T.L. Anderson and P.J. Hill. (Eds.). Rowman & Littlefield, Lanham, MD, pp. 57–78.

13. *Economist.* (1991). River rivalry. March 30.

14. Hull, J.W. (2000). The War Over Water. *Regional Resource.* Council of State Governments, Atlanta.

15. American Rivers. (2002). *America's Most Endangered Rivers of 2002.* Online: www.amrivers.org/mostendangered/2002report.htm (cited December 10, 2003).

16. Apalachicola Bay Chamber of Commerce. (2003). *Apalachicola.* Online: www.apalachicolabay.org (cited Sept. 25, 2003).

17. Erhardt, C. (1992). The battle over "The Hooch": The federal-interstate water compact and the resolution of rights in the Chattahoochee river. *Stanford Environ. Law J.* **11**: 200–228.

18. Marella, R.L., Fanning, J.L., and Mooty, W.S. (1993). *Estimated Use of Water in the Apalachicola-Chattahoochee-Flint River Basin During 1990 with State Summaries from 1970 to 1990.* Water-Resources Investigation Report, 93–4084. U.S. Geological Survey, Tallahassee, FL.

19. Moore, C.G. (1999). Water wars: Interstate water allocation in the Southeast. *Nat. Resour. Environ.* **14**(Summer): 5–10; 66–67.

20. Alabama, Florida, and Georgia. (2003). *Memorandum of Understanding Regarding Initial Allocation Formula for the ACF River Basin (July 22).* Online: www.dep.state.fl.us/secretary/comm/2003/july/0722_acf.htm (cited Sept. 29, 2003).

21. Struhs, D.B. (2003). *Statement of Intent to Accompany the Memorandum of Understanding Regarding Initial Allocation Formula for the ACF River Basin, July 22.* Online: www.dep.state.fl.us/secretary/comm/2003/july/0722_acf.htm (cited Sept. 29, 2003).

22. Ritchie, B. (2003). River pact moves closer. *Tallahassee Democrat* July 23, A1.

23. Ritchie, B. (2003). Florida willing to take battle to court. *Tallahassee Democrat* August 27, B3.

24. Ritchie, B. (2003). High court may hear water fight. *Tallahassee Democrat* September 2, A1.

25. Seabrook, C. (2003). Water Costs Likely to Rise. *Atlanta Journal-Constitution* Sept. 8, F1.

26. U.S. Census Bureau. 1998. *State and Metropolitan Area Data Book, 1997–98.* 5th Edn. Washington, DC.

27. Anderson, T.L. and Snyder, P.S. (1997). Priming the invisible pump. *PERC Policy Series*, PS-9. PERC, Bozeman, MT.

28. Atlanta Bureau of Water. (2003). *Rates, Fees & Meter Price.* Online: www.atlantaga.gov/citydir/water/index.htm (cited Oct. 28, 2003).

29. Atlanta Regional Commission (2003). *Current Water Restrictions (May 9).* Online: atlreg.com/water/waterrestrictions.html (cited November 6, 2003).

30. Judd, A. (2000). State slaps water limits on 15 metro counties: outdoor water use banned from 10 a.m. to 10 p.m. *Atlanta Journal-Constitution* June 2.

31. Anderson, T.L. and Snyder, P.S. (1997). *Water Markets: Priming the Invisible Pump.* Cato Institute, Washington, DC.

32. Yoskowitz, D.W. (2001). Markets, mechanism, institutions, and the future of water. *Environ. Law Rep.* **31**(February).

33. Fretwell, H.L. (1999). Paying to play: The fee demonstration program. *PERC Policy Series*, PS-17. PERC, Bozeman, MT.

34. Coase, R. (1960). The problem of social cost. *J. Law Econ.* **3**(October): 1–44.

35. Henderson, Bruce. (2002). Who gets the water? *Charlotte Observer* Dec. 29.

36. Libaw, O.Y. (2000). Water wars: Drought-ridden Southeast battles over use of rivers (August 14). ABC News. Online: more.abcnews.go.com/sections/us/dailynews/water000811.html (cited Sept. 16, 2003).

37. Pompe, J.J. and Franck, D.P. (2003). *The Economic Impact of Water Transfer: Options for Policy Reform.* Working paper, Department of Economics, Francis Marion University, Florence, SC.

WATER SUPPLY AND WATER RESOURCES: DISTRIBUTION SYSTEM RESEARCH

MARK C. MECKES
ROY C. HAUGHT
U.S. Environmental
Protection Agency

Two water distribution system simulators (DSS) are in operation at the U.S. EPA Test and Evaluation (T&E) Facility in Cincinnati, Ohio. The T&E Facility is a multifaceted research resource in which a wide variety

---

This article is a US Government work and, as such, is in the public domain in the United States of America.

of water treatment and other environmental protection technologies are conceived, designed, and evaluated in the laboratory, bench, and pilot plant scale. The EPA designed and fabricated the DSS systems to evaluate and understand the dynamics which influence water quality within water distribution infrastructure systems in the United States and abroad. The first distribution system simulator (DSS-1) has continuous flow conditions. There are six individual 75 ft (25 meter) lengths of 6 in (15 cm) diameter ductile iron pipe arranged into "pipe loop" configurations to simulate a distribution system. This pipeloop system can be configured to operate as; independent/individual loops, collectively as one unit, or in various experimental configurations to complement EPA's experimental research. DSS-1 is equipped with two 1500 gallon reservoir tanks to simulate a comprehensive distribution infrastructure system. This unique engineering design permits operating any combination and configuration of the six loops under various experimental operating parameters (Table 1). Each loop is insulated and fitted with a heat exchanger to maintain constant temperature conditions during operation. The DSS is interfaced with a Supervisory Control and Data Acquisition (SCADA) system, which is used to monitor, control, and archive operating conditions and collected data continuously. Biofilm samples are collected on coupons [View1–View2] which are set flush with the interior surface of the pipe wall. The coupons may be removed for sampling and analysis without disrupting water flow.

The second distribution system (DSS-2) is over 300 feet long and is a once through system composed of six inch diameter PVC pipe. This unit is being utilized to evaluate water quality in a dead end branch of a distribution system and to develop design models for water distribution. Both DSS units are located above ground to permit easy access to the entire pipe network.

Table 1. Operating Parameters Of The Distribution System Simulator

Parameter	Normal Operation	Experimental Test Conditions
Distribution system simulator	Parallel (6 individual distribution system simulators)	Parallel or series in groups of 6, 3, 2, or 1
Housing	Ductile iron (Non-lined)	Ductile iron (Non-lined)
Flow	88 gpm or 1 ft/s	No Flow to 140 gpm or 0 to 1.6 ft/s
Temperature	60 °F (15.5 °C)	35 °F (1.6 °C) to Ambient
Chemical control	Free chlorine 1.0 ppm	Chemical control as needed
pH	7.0 to 7.5	Control/monitor as needed
Turbidity	< 0.5 NTU	Control/monitor as needed
Water supply	Cincinnati—tap water (Chlorinated)	Dechlorinated, deionized, tanked, surface water (River)

Experimental studies (Table 2) are ongoing to understand the physical, chemical, and biological activities that occur in drinking water distribution systems.

OBJECTIVES

The DSS provides researchers with a mechanism to study how water quality is affected during distribution. Results from research studies (Table 2) will be used to provide guidance on how to maintain a high level of water quality during distribution. A secondary objective to this project is to develop, evaluate, and demonstrate real time monitoring of water quality parameters within distribution systems using remote telemetry. Results from research will be used to provide guidance on how to utilize remote monitoring of water quality to detect changes in water quality within distribution systems.

RELEVANCE

Throughout the world there are millions of miles of water distribution pipe lines which provide drinking water for use by individuals and industry. Although these distribution systems provide drinking water to the world, very little is known about the physical, chemical, and biological activities that occur within them. Some of these water distribution systems have been in service well over one hundred years. DSS-1 has been in operation for over three years, as water moves through distribution systems, it comes into contact with a wide range of material some of which can cause significant changes to the quality of the finished water supply. Suspended solids in finished water can settle out under low flow conditions and can be suspended during high flow. Various disinfection agents and water additives react with organic and inorganic materials within a distribution system generating by-products which may be undesirable in the water supply. Oxidant resistant microorganisms colonize pipe surfaces producing a complex micro environment known as "biofilm". Biofilms are highly resistant to many disinfection methods and techniques. Our research looks at the various experimental test parameters which influence biofilm growth. This work will also help EPA

Table 2. Proposed DSS Studies

No	Title of Proposed Study
1	Preliminary Studies of Biofilm Formation in Pilot-Scale Distribution Systems
2	Opportunistic Pathogens in Biofilms
3	Effect of a Pollution Event on a Simulated Water Distribution System
4	Impact of Nutrient Removal on Growth Potential for Bacteria
5	Impact of Alternative Treatment on Biofilm Growth
6	Real-Time Monitoring and Control of Distribution Systems
7	Effects of pH Changes on Biofilm Growth in a Distribution System
8	Bacterial Growth in Distribution Systems

develop a better understanding of the dynamics inside distribution systems.

RESEARCH GOALS

Fabricate aboveground water distribution system simulator's which permit easy access and can be operated under controlled conditions. Conduct studies to develop a better understanding of the dynamics that occur in drinking water distribution systems. Determine what physical, chemical, and biological factors influence biofilm growth within such systems. Develop and test mechanisms for the enhancement and control of biofilm growth within a simulated distribution system. Additionally, develop and evaluate real time monitoring, data collection, and archiving of water quality parameters within water distribution systems using remote telemetry. Results will be used to develop and provide guidance on ways to maintain high levels of water quality through distribution systems.

DROUGHT IN THE DUST BOWL YEARS

National Drought Mitigation Center

INTRODUCTION

In the 1930s, drought covered virtually the entire Plains for almost a decade (1). The drought's direct effect is most often remembered as agricultural. Many crops were damaged by deficient rainfall, high temperatures, and high winds, as well as insect infestations and dust storms that accompanied these conditions. The resulting agricultural depression contributed to the Great Depression's bank closures, business losses, increased unemployment, and other physical and emotional hardships. Although records focus on other problems, the lack of precipitation would also have affected wildlife and plant life and would have created water shortages for domestic needs.

Effects of the Plains drought sent economic and social ripples throughout the country. For example, millions of people migrated from the drought areas, often heading west, in search of work. These newcomers were often in direct competition for jobs with longer-established residents, which created conflict between the groups. In addition, because of poverty and high unemployment, migrants added to local relief efforts, sometimes overburdening relief and health agencies.

Many circumstances exacerbated the effects of the drought, among them the Great Depression and economic overexpansion before the drought, poor land management practices, and the areal extent and duration of the drought. [Warrick et al. (2) and Hurt (3) discuss these issues in greater detail.] The peculiar combination of these circumstances and the severity and areal coverage of the event played a part in making the 1930s drought the widely accepted drought of record for the United States. To cope with and recover from the drought, people relied on ingenuity and resilience, as well as relief programs from state and federal governments. Despite all efforts, many people were not able to make a living in drought-stricken regions and were forced to migrate to other areas in search of a new livelihood. It is not possible to count all the costs associated with the drought of the 1930s drought, but one estimate by Warrick (1) claims that financial assistance from the government may have been as high as $1 billion (in 1930s dollars) by the end of the drought. Fortunately, several lessons were learned that were used in reducing the vulnerability of the regions to future droughts.

THE GREAT DEPRESSION AND OVERCAPITALIZATION

In the early 1920s, farmers saw several opportunities for increasing their production. New technology and crop varieties were reducing the time and costs-per-acre of farming, which provided a great incentive for agricultural expansion. This expansion was also necessary to pay for expensive, newly developed equipment (such as listers and plows) that was often purchased on credit, and to offset low crop prices after World War I.

When the national economy went into decline in the late 1920s because of the Great Depression, agriculture was even more adversely affected. In addition, a record wheat crop in 1931 sent crop prices even lower. These lower prices meant that farmers needed to cultivate more acreage, including poorer farmlands, or change crop varieties to produce enough grain to meet their required equipment and farm payments.

When drought began in the early 1930s, it worsened these poor economic conditions. The depression and drought hit farmers on the Great Plains the hardest. Many of these farmers were forced to seek government assistance. A 1937 bulletin by the Works Progress Administration reported that 21% of all rural families in the Great Plains were receiving federal emergency relief (4). However, even with government help, many farmers could not maintain their operations and were forced to leave their land. Some voluntarily deeded their farms to creditors, others faced foreclosure by banks, and still others had to leave temporarily to search for work to provide for their families. In fact, at the peak of farm transfers in 1933–34, nearly 1 in 10 farms changed possession, with half of those being involuntary (from a combination of the depression and drought).

POOR LAND MANAGEMENT PRACTICES

A number of poor land management practices in the Great Plains region increased the vulnerability of the area before the 1930s drought. Some of the land use patterns and cultivation methods in the region can be traced back to the settlement of the Great Plains nearly 100 years earlier. At that time, little was known of the region's

---

This article is a US Government work and, as such, is in the public domain in the United States of America.

climate. Several expeditions had explored the region, but they were not studying the region for its agricultural potential, and, furthermore, their findings went into government reports that were not readily available to the general public (5). Misleading information, however, was plentiful. "Boosters" of the region, hoping to promote settlement, put forth glowing but inaccurate accounts of the Great Plains' agricultural potential. In addition to this inaccurate information, most settlers had little money or other assets, and their farming experience was based on conditions in the more humid eastern United States, so the crops and cultivation practices they chose often were not suitable for the Great Plains. But the earliest settlements occurred during a wet cycle, and the first crops flourished, so settlers were encouraged to continue practices that would later have to be abandoned. When droughts and harsh winters inevitably occurred, there was widespread economic hardship and human suffering, but the early settlers put these episodes behind them once the rains returned. Although adverse conditions forced many settlers to return to the eastern United States, even more continued to come west. The idea that the climate of the Great Plains was changing, particularly in response to human settlement, was popularly accepted in the last half of the 19th century. It was reflected in legislative acts such as the Timber Culture Act of 1873, which was based on the belief that if settlers planted trees they would be encouraging rainfall, and it was not until the 1890s that this idea was finally abandoned (6). Although repeated droughts tested settlers and local/state governments, the recurrence of periods of plentiful rainfall seemed to delay recognition of the need for changes in cultivation and land use practices.

Several actions in the 1920s also increased the region's vulnerability to drought. Low crop prices and high machinery costs (discussed above) meant that farmers needed to cultivate more land to produce enough to meet their required payments. Since most of the best farming areas were already being used, poorer farmlands were increasingly used. Farming submarginal lands often had negative results, such as soil erosion and nutrient leaching. By using these areas, farmers were increasing the likelihood of crop failures, which increased their vulnerability to drought.

These economic conditions also created pressure on farmers to abandon soil conservation practices to reduce expenditures. Furthermore, during the 1920s, many farmers switched from the lister to the more efficient one-way disc plow, which also greatly increased the risk of blowing soil. Basically, reductions in soil conservation measures and the encroachment onto poorer lands made the farming community more vulnerable to wind erosion, soil moisture depletion, depleted soil nutrients, and drought.

DROUGHT DURATION AND EXTENT

Although the 1930s drought is often referred to as if it were one episode, there were at least four distinct drought events: 1930–31, 1934, 1936, and 1939–40 (7). These events occurred in such rapid succession that affected regions were not able to recover adequately before another drought began. Historical maps of U.S. climate divisions and graphs of U.S. river basins reflect this situation.

COPING WITH AND RECOVERING FROM DROUGHT

During the 1930s, many measures were undertaken to relieve the direct impacts of droughts and to reduce the region's vulnerability to the dry conditions. Many of these measures were initiated by the federal government, a relatively new practice. Before the 1930s drought, federal aid had generally been withheld in emergency situations in favor of individual and self-reliant approaches. This began to change with the development of the Great Depression in the late 1920s and the 1933 inauguration of President Franklin Delano Roosevelt. The depression helped "soften deep-rooted, hard-line attitudes of free enterprise, individualism, and the passive role of the government", thus paving the way for Roosevelt's New Deal programs, which in turn provided a framework for drought relief programs for the Great Plains (1).

Warrick et al. (2) describe these drought relief programs, which are credited with saving many livelihoods throughout the drought periods. The programs had a variety of goals, all of which were aimed at the reduction of drought impacts and vulnerability:

- Providing emergency supplies, cash, and livestock feed and transport to maintain the basic functioning of livelihoods and farms/ranches.
- Establishing health care facilities and supplies to meet emergency medical needs.
- Establishing government-based markets for farm goods, higher tariffs, and loan funds for farm market maintenance and business rehabilitation.
- Providing the supplies, technology, and technical advice necessary to research, implement, and promote appropriate land management strategies.

As important as these programs may have been, the survival of a majority of the families and enterprises undoubtedly rested solely with their perseverance and integrity. Whether they stayed or moved into the drought regions or migrated to other areas in hopes of a better life, families encountered new hardships and obstacles that would require ingenuity, resilience, and humility. Those who remained in the drought regions were forced to endure severe dust storms and their health effects, diminished incomes, animal infestations, and the physical and emotional stress over their uncertain futures. Humor helped; tales about birds flying backward to keep from getting sand in their eyes, housewives scouring pots and pans by holding them up to keyholes for a sandblasting, and children who had never seen rain were among the favorite stories of Dust Bowl inhabitants. In the end, it was a combination of willpower, stamina, humor, pride, and, above all, optimism that enabled many to survive the Dust Bowl. These qualities are succinctly expressed in the comments of one contemporary Kansan: "We have faith in the future. We are here to stay" (quoted in Ref. 3).

The 1930s drought and its associated impacts finally began to abate during spring 1938. By 1941, most areas of the country were receiving near-normal rainfalls. These rains, along with the outbreak of World War II, alleviated many of the domestic economic problems associated with the 1930s. In fact, the new production demands and positive climatic conditions brought the United States into a rapid economic boom.

Even though short-term conditions seemed to be relatively stable, there were some drawbacks to this production growth. One drawback (described in Ref. 3) was that the start of World War II shifted remaining funds and priorities away from drought-related programs. Men were taken off work programs to enter the armed forces and to produce for the war effort. Moreover, items such as gasoline and replacement parts were redirected from federal drought and conservation programs to the war efforts. This meant that conservation programs and research were significantly reduced during this period. Another drawback was that with the return of the rains, many people soon forgot about conservation programs and measures implemented during the 1930s droughts. This led to a return to some of the inappropriate farming and grazing practices that made many regions so vulnerable to drought in the 1930s.

1930s DROUGHT COSTS

Although the 1988–89 drought was the most economically devastating natural disaster in the history of the United States (7), a close second is undoubtedly the series of droughts that affected large portions of the United States in the 1930s. Determining the direct and indirect costs associated with this period of droughts is a difficult task because of the broad impacts of drought, the event's close association with the Great Depression, the fast revival of the economy with the start of World War II, and the lack of adequate economic models for evaluating losses at that time. However, broad calculations and estimates can provide valuable generalizations of the economic impact of the 1930s drought.

Overall Drought Costs

In 1937, the Works Progress Administration (WPA) reported that drought was the principal reason for economic relief assistance in the Great Plains region during the 1930s (4). Federal aid to the drought-affected states was first given in 1932, but the first funds marked specifically for drought relief were not released until the fall of 1933. In all, assistance may have reached $1 billion (in 1930s dollars) by the end of the drought (1).

According to the WPA, three-fifths of all first-time rural relief cases in the Great Plains area were directly related to drought, with a disproportionate amount of cases being farmers (68%) and especially tenant farmers (70% of the 68%). However, it is not known how many of the remaining cases (32%) were indirectly affected by drought. The WPA report also noted that 21% of all rural families in the Great Plains area were receiving federal emergency relief by 1936 (4); the number was as high as 90% in hard-hit counties (1). Thus, even though the exact economic losses are not known for this time period, they were substantial enough to cause widespread economic disruption that affected the entire nation.

LESSONS LEARNED: THE LEGACY OF THE 1930s DROUGHT

The magnitude of the droughts of the 1930s, combined with the Great Depression, led to unprecedented government relief efforts. Congressional actions in 1934 alone accounted for relief expenditures of $525 million (8); the total cost (social, economic, and environmental) would be impossible to determine.

If the Roosevelt era marked the beginning of large-scale aid, it also ushered in some of the first long-term, proactive programs to reduce future vulnerability to drought. It was in these years, for example, that the Soil Conservation Service (SCS)—now the Natural Resources Conservation Service—began to stress soil conservation measures. Through their efforts, the first soil conservation districts came into being, and demonstration projects were carried out to show the benefits of practices such as terracing and contouring (for a discussion of the activities of the SCS during this period, see Ref. 3).

Warrick et al. (2) note that the proactive measures continued in the years following the drought: conservation practices and irrigation increased, farm sizes grew larger, crop diversity increased, federal crop insurance was established, and the regional economy was diversified. Many other proactive measures taken after the 1930s drought also reduced rural and urban vulnerability to drought, including new or enlarged reservoirs, improved domestic water systems, changes in farm policies, new insurance and aid programs, and removal of some of the most sensitive agricultural lands from production (7).

Problems remained, but these programs and activities would play a fundamental role in reducing the vulnerability of the nation to the forthcoming 1950s drought. Although a larger area was affected during the 1950s drought, the conservation techniques that many farmers implemented in the intervening years helped prevent conditions from reaching the severity of the 1930s drought.

BIBLIOGRAPHY

1. Warrick, R.A. 1980. Drought in the Great Plains: A Case Study of Research on Climate and Society in the USA. In J. Ausubel and A.K. Biswas (eds.). *Climatic Constraints and Human Activities*, pp. 93–123. IIASA Proceedings Series, Vol. 10. Pergamon Press, New York.
2. Warrick, R.A.; P.B. Trainer; E.J. Baker; and W. Brinkman. 1975. *Drought Hazard in the United States: A Research Assessment*. Program on Technology, Environment and Man Monograph #NSF-RA-E-75-004, Institute of Behavioral Science, University of Colorado, Boulder.

3. Hurt, D.R. 1981. *An Agricultural and Social History of the Dust Bowl*. Nelson Hall, Chicago.
4. Link, I.; T.J. Woofter, Jr., and C.C. Taylor. 1937. *Research Bulletin: Relief and Rehabilitation in the Drought Area*. Works Progress Administration, Washington, DC.
5. Fite, G.C. 1966. *The Farmer's Frontier, 1865–1900*. Holt, Rinehart and Winston, New York.
6. White, R. 1991. *"It's Your Misfortune and None of My Own": A History of the American West*. University of Oklahoma Press, Norman.
7. Riebsame, W.E.; S.A. Changnon, Jr.; and T.R. Carl. 1991. *Drought and Natural Resources Management in the United States: Impacts and Implications of the 1987–89 Drought*. Westview Press, Boulder, Colorado.
8. United States House of Representatives. 1934. *Relief of the drought area*. Communication from the President of the United States, 73d Congress, 2d Session, Document No. 398, Washington, DC.

OTHER RESOURCES

Dust Storms and Their Damage. Photos from the collection of the Wind Erosion Research Unit of USDA's Agricultural Research Service at Kansas State University.

The Dust Bowl (photos). A collection of 20 photos for sale from The Social Studies Web.

American Memory from the Library of Congress. A multi-format look at daily life in Farm Security Administration migrant work camps in central California, 1940 and 1941.

Weedpatch Camp. The story of the Arvin Federal Government Camp, immortalized as "Weedpatch Camp" in Steinbeck's Grapes of Wrath.

Dust storm footage. A short film clip of a dust storm during the 1930s. From the website of the USDA/ARS Wind Erosion Research Unit at Kansas State University.

Surviving the Dust Bowl. The website of the 1998 Public Broadcasting Service (PBS) film "Surviving the Dust Bowl". Provides a good background on the people and events of the 1930s Dust Bowl.

1930s Dust Bowl. Excerpts from "The Dust Bowl, Men, Dirt and Depression" by Paul Bonnifield. On the Cimarron Heritage Center (Boise City, Oklahoma) website.

Woody Guthrie Dust Bowl Ballads. Discussion and analysis of Guthrie's "Dust Bowl Ballads" recording.

America from the Great Depression to World War II. Photos from the Farm Security Administration—Office of War Information Collection, 1935–1945.

The Plow that Broke the Plains. The story behind the 1936 documentary made for the Farm Security Administration on the destruction and suffering associated with drought in the early 1930s.

DROUGHT MANAGEMENT PLANNING

NEIL S. GRIGG
Colorado State University
Fort Collins, Colorado

It is impossible to "manage" a naturally occurring drought, so "drought management" refers to managing resources, including water, to mitigate the adverse effects of droughts, which are certain to occur. Other resources that can be managed are food supplies, including animal feed, public facilities of various kinds, and economic resources to aid businesses, farmers, and citizens.

As "creeping disasters," droughts are easy to ignore until too late, thus presenting a significant risk to water resources managers. The dilemma is that to prepare for them might be expensive and require seldom-used facilities, but without preparation, water supplies might run out on rare occasions. For this reason, drought management planning presents a challenge to water agencies, as well as to other governmental and private sector organizations.

Drought has two forms: a period without enough rain and a period of shortage: each requires drought management planning. A period without enough rain is a "meteorologic drought," and leads to shortages, which are called by names such as "hydrologic," "agricultural," and "socioeconomic" drought.

Drought management planning is a form of contingency planning that answers the question "what do we do if a drought occurs?" As in other forms of contingency planning, the process is to assess the threat, identify the vulnerable parts of systems, and take measures to prepare, mitigate, respond, and recover from the impact of a drought, should one occur.

Failures in water supply can have serious consequences for cities, industries, and other water users such as irrigation, hydropower, recreation, and wildlife. The primary impact of drought is due to real or feared interruption of supplies because water supply is critical to the economy and the natural environment. This risk increases with the interdependence and vulnerability of water systems, environmental stakes are also high. Drought is a serious threat to food supplies and farm income, especially in nations where food supplies are marginal.

Assessing the threat of drought requires knowledge of the security of supplies or the probability that a raw water supply system will run out of water. This is usually estimated for individual water agencies in terms of the return period of the drought planned for or the annual probability of running short. The concepts of return period and failure probability are in wide use, but for droughts they are complex because drought duration, system yield, and return period must be considered.

The concept of "safe yield" is used to describe the reliability of a water system, another measure of security. To estimate safe yield requires analysis of components and the systems that deliver water to users. Vulnerability analysis requires assessing of all possible modes of failure, not only hydrologic failure. For example, if a dam requires emptying for repair, supplies may run short from lack of storage, even when they are available from precipitation.

Examples of measures to prepare, mitigate, respond, and recover from drought impacts are shown in Table 1 (1).

Drought response plans are usually prepared by water management organizations to anticipate drought and plan activities to take place after a drought occurs. These are custom-tailored to the needs of each

Table 1. Drought Planning

Supply Augmentation	Demand Reduction	Impact Minimization
New sources	Legal restrictions	Forecasting
Water storage	Pricing	Mutual aid
Reuse of water	Devices to limit use	Insurance

organization but generally include the usual contents of emergency response plans, such as assessment of the hazard, identification of vulnerable components, mitigation measures, plans to meet critical water needs, arrangements for mutual aid, team organization and roles, and special conditions.

BIBLIOGRAPHY

1. Grigg, N.S. and Vlachos, E.C. (1993). Drought and water-supply management: roles and responsibilities. *J. Water Resour. Plann. Manage. American Society of Civil Engineers.* September/October.

READING LIST

Wilhite, D.A., Easterling, W.E. and Wood, D.A. (Eds.) (1987). *Planning for Drought: Toward a Reduction in Societal Vulnerability*, Westview Press, Boulder, CO.

DROUGHT AND WATER SUPPLY MANAGEMENT

NEIL S. GRIGG
Colorado State University
Fort Collins, Colorado

Drought is a normal, albeit extreme, aspect of hydrologic variation, so it should be considered when making plans for water. Planning for water supply management involves both supply augmentation and demand management. Supply augmentation can involve new supplies or extending use of existing supplies.

During drought, making water available to meet requirements of people, business, and the environment is critical. As explained in the article on "Drought Management Planning," preparing for drought requires contingency planning to assure adequacy of water supplies, and management of available water supplies during drought is the key element of drought response. Thus, effective water resources management requires actions that assure water availability during drought as well as normal times.

Technologies and methods for providing water supplies go back to early Rome, which brought in water supplies by aqueducts. Water supply systems in the United States date from early settlements, which used spring water pumped through bored logs. Some early pumps were horse-driven, followed later by steam power.

Water supply systems have three subsystems: source of supply, treatment and distribution, and delivery of enough water of high quality at sufficient pressure for intended uses. A percentage of the water is normally unaccounted for, mainly through leakage and other losses. Water for fire fighting must also, be provided. Providing enough water requires volumes of delivery to meet needs reliably during peak demand periods and during drought.

The required quantities of water can be estimated from statistics of water use for domestic, commercial, and industrial uses. These vary widely, depending on many socioeconomic factors. Average per capita use of water in the United States, including all uses, varies from about 100–200 gallons per capita per day (gpcd) (378–756 liters per capita per day). The highest rates are normally in dry regions, where water supplies are used for lawn irrigation. In highly concentrated urban areas, water uses may be mostly inside apartment buildings and in commercial facilities and may be close to 100 gpcd (378 liters per capita per day), without any requirement for lawn irrigation. In other countries, water demands may be less, depending on socioeconomic variables. In a developing country city, use might be much less than levels in the United States, and water may not be available to all residents.

Sources of freshwater supply include surface water, groundwater, and reclaimed waters. Surface water can include stored water in reservoirs or direct diversions from streams. Typical infrastructure components include dams, tunnels, outlet tubes, canals, gates and controls, spillways, and support structures. Groundwater sources include springs, wells, infiltration galleries, and aquifers that store recharged water. Infrastructure components include wells, casings, pumping systems, piping, housing, and other support facilities. Given the increasing restrictions on the development of surface water, groundwater development is receiving more attention, including the implementation of aquifer storage and recovery (ASR) systems.

In special cases, rainwater can be caught on roofs and stored in cisterns. This type of local supply system is found mostly in developing countries and in special, remote locations where centralized systems are not possible. Bottled water is an increasingly popular source, and point-of-use treatment systems are becoming more popular.

Demand reduction can be legal, physical, or voluntary. Drought plans normally contain provisions to restrict use during times of shortage. Physical controls can include reduced pressures and requirements for devices to impose water conservation. Voluntary systems rely primarily on education and calls for citizenship.

As population increases and environmental water needs are more recognized, it becomes more difficult to find new sources of supply. Therefore, innovative approaches being studied include

- dual use of water where reclaimed and impaired waters are used for nonpotable applications
- conservation systems, where "new" sources are created by saving water
- innovative storage systems, such as aquifer–storage–recovery (ASR) systems

- conjunctive use, where water from different sources, such as surface and groundwater, are managed jointly
- reuse, where wastewater is treated and used again in one form or another
- point-of-use treatment systems
- bottled water

ASSESSMENT OF ECOLOGICAL EFFECTS IN WATER-LIMITED ENVIRONMENTS

RANDALL T. RYTI
JAMES T. MARKWEISE
Neptune and Company, Inc.
Los Alamos, New Mexico

We consider that water-limited environments include arid lands that have less than 250 millimeters (10 inches) of annual precipitation and semiarid lands that have annual precipitation between 250 and 500 millimeters (1). This paper discusses the characteristics of arid and semiarid environments, approaches for evaluating the potential ecological impacts from anthropogenic contaminant sources, and the issues that make such assessments complex in arid and semiarid environments.

Precipitation defines water-limited environments, and plants and animals have developed specialized adaptations to the paucity/lack of water. For example, avian adaptation to desert environments is characterized by decreased metabolic rates and minimized water loss (2). Organisms have adapted to more efficient use of limited water quantities by behaviorally and/or physiologically optimizing their ability to acquire and retain water. For example, some beetles meet all their water needs by drinking the dew that collects on their bodies. Another strategy for surviving in dry environments includes adaptations to exploit periodic monsoons that are characteristic of arid and semiarid regions and involves rapid population growth to exploit periods of abundant resources (3).

Precipitation can be large, but it is infrequent. The spatial scale for precipitation also varies from thunderstorms that may impact a portion of a watershed to larger storm systems that impact an entire region. The predictability of storms varies widely between seasons or years for water-limited environments. Terms like intermittent or interrupted have been used to describe flow in water-limited environments. The pattern and intensity of flow change as a result of prolonged periods of wet or dry conditions in these environments.

Effluent and surface water runoff from residential, commercial, or agricultural activities are frequently more predictable sources of water in dry environments. Effluent/runoff is typically laden with metals and organic chemicals, and, it has been demonstrated, affects the composition and abundance of aquatic organisms in receiving streams (4). Wastewater effluent from anthropogenic sources may be inserted as a point source (i.e., an outfall pipe), or runoff may originate from nonpoint sources (e.g., distributed spatially from agricultural application). These types of contaminant sources frequently overlap and confound the interpretation and mitigation of ecological effects.

Directly related to effluent/runoff is the source of water used by people in water-limited environments. Two common water sources are impoundments along major river systems or pumping of groundwater. When dams are introduced, flow regimes in these water-limited environments change and lead to dramatic changes in the species in areas of modified flow regimes (5).

In addition to anthropogenic influences on water availability and water quality, the physical environment of arid/semiarid land potentially affects water resources. In the arid southwest United States, closed hydrologic basins exist; salt flats are one resultant landscape feature of closed basins. Surface depressions may also accumulate water during intense precipitation, and evaporation concentrates chemicals in these ponds during dry periods. This is in contrast to moist environments, where water transport of contaminants generally leads to diluted concentrations. Wind-driven contaminant transport is also more important in dry environments where the soil and rock lie exposed to wind erosion because of limited plant coverage.

Dry environments are also more susceptible to disturbance due to several factors. One is decreased resilience to stressors because of the extreme nature of the environment. Aridity in itself can be a stressor for plants and animals, and, it has been shown, influences species composition (6) and population characteristics (7). Adding anthropogenic stressors to this environment (e.g., through pollution) may decrease a population's viability if it is already near the individual tolerance limits for surviving in the dry environment. For example, Sjursen and others (8) showed that exposure to organic chemicals could reduce drought tolerance in soil invertebrates. In addition, populations depending on rainfall for reproduction could be locally exterminated if their source of water (e.g., an ephemeral rain pool) were impacted with chemical toxicants. Soils in arid environments also develop more slowly than in moist areas, recover from stress slowly (9), and are more susceptible to erosion after physical disturbance.

Adverse effects contributing to impacts can be defined as "Changes that alter valued structural or functional attributes of ecological entities ..." (10). Effects can be assessed for ecological entities on any scale of biological organization, including genetic, individual, population, community, and ecosystem levels. "An evaluation of adversity may include a consideration of the type, intensity, and scale of the effect as well as the potential for recovery" (10). These evaluations are typically categorized as either stressor-based or effects-based assessments. Stressor-based assessments are initiated to evaluate physical, chemical, or biological entities that can induce an adverse response (11). Effects-based assessments are initiated to determine the cause(s), once an adverse impact is observed.

Tools to evaluate and manage environmental effects come from a number of disciplines. Resource management fields (e.g., forestry, wildlife management) develop plans to manage or mitigate effects of logging, hunting, and fishing. Conservation biology develops plans to mitigate more general human impacts on the environment. For regulatory agencies, ecological risk assessment has been the tool used to evaluate ecological effects of planned or historic actions (10).

The U.S. Environmental Protection Agency (EPA) Office of Solid Waste and Emergency Response (a.k.a. Superfund) started developing guidance on ecological risk assessment in the late 1980s. This process culminated in the Risk Assessment Forum process for ecological risk assessment (12). Parallel to these efforts was the Ecological Risk Assessment Guidance for Superfund or ERAGS (11).

ERAGS starts by identifying the ecological entity that needs to be protected (the assessment endpoint). By defining an adverse effect, the measures used to evaluate effects will be identified. ERAGS encourages using multiple lines of evidence, including field information, laboratory studies, and literature studies for each measure. Effects are characterized through lines of evidence (13,14), and the weight of the evidence dictates how one manages adverse effects.

The complications from assessments in water-limited environments are often related to the spatial and temporal scales of the assessment. For example, one may be concerned about potential adverse ecological impacts on aquatic and terrestrial receptors of effluent releases to a stream from an industrial facility. One set of considerations is the physical environment and other anthropogenic impacts on the stream. As discussed before, flow in the stream may be expected to change dramatically as seasonal and annual changes in precipitation occur. Altered flow regimes and other anthropogenic impacts on this stream are also expected. These considerations may help select appropriate assessment endpoints.

The abundance of bats in roosts near a stream near an industrial facility could be selected as an assessment endpoint. However, bats have large foraging ranges and are likely to be impacted by contaminants from multiple sources. In addition, bats are migratory and could be exposed to contaminants from distant locations. Another assessment end point for this example could be abundance of benthic macroinvertebrates. This choice has the advantage of greater site fidelity and thus minimizes confounding effects from neighboring anthropogenic releases. Benthic macroinvertebrates also lend themselves to direct experimentation as a line of evidence without some of the ethical and logistical problems of vertebrates.

Another complication of adverse effects assessments in water-limited environments is the availability of appropriate toxicity bioassays for taxa commonly found in such environments (15). This problem is more obvious for terrestrial receptors. For example, earthworms are the most commonly used soil animals in soil toxicity bioassays. However, earthworms are not representative of the more important detritivores in arid/semiarid environments (mites). This is an area of active research for more biologically relevant toxicity testing organisms for water-limited environments.

Last, variation in timing and amount of precipitation complicates some empirical evaluation of ecological effects. For example, the phenology of most organisms in arid and semiarid environments is tied to precipitation, and most studies of adverse effects are tied to government funding cycles. This is why some empirical studies of biota abundance and diversity are dismissed for practical reasons and laboratory studies of toxicity are emphasized. In some cases, a field experiment where precipitation is artificially simulated is another option, if time represents a logistical constraint.

Acknowledgments

We thank two peer reviewers, R. Mirenda and M. Tardiff, for their constructive comments on an earlier draft of this paper.

BIBLIOGRAPHY

1. Walker, A.S. (1997). *Deserts: Geology and Resources*. Online Edition. United States Geological Survey, Washington, DC.
2. Tieleman, B.I. and Williams, J.B. (2000). The adjustment of avian metabolic rates and water fluxes to desert environments. *Physiol. Biochem. Zool.* **73**: 461–479.
3. van der Valk, H. (1997). Community structure and dynamics in desert ecosystems: potential implications for insecticide risk assessment. *Arch. Environ. Contam. Toxicol.* **32**: 11–21.
4. Ganasan, V. and Hughes, R.M. (1998). Application of an index of biological integrity (IBI) to fish assemblages of the rivers Khan and Kshipra (Madhya Pradesh), India. *Freshwater Biol.* **40**: 367–383.
5. Saunders, D.L., Meeuwig, J.J., and Vincent, A.C.J. (2002). Freshwater protected areas: strategies for conservation. *Conserv. Biol.* **16**: 30–41.
6. Tsialtas, J.T., Handley, L.L., Kassioumi, M.T., Veresoglou, D.S., and Gagianas, A.A. (2001). Interspecific variation in potential water-use efficiency and its relation to plant species abundance in a water-limited grassland. *Functional Ecol.* **15**: 605–614.
7. Spinks, A.C., Bennett, N.C., and Jarvis, J.U.M. (2000). A comparison of the ecology of two populations of the common mole-rat, *Cryptomys hottentotus hottentotus*: the effect of aridity on food, foraging and body mass. *Oecologia* **125**: 341–349.
8. Sjursen, H., Sverdrup, L.E., and Krogh, P.H. (2001). Effects of polycyclic aromatic compounds on the drought tolerance of *Folsomia fimerata* (Collembola, Isotomidae). *Environ. Toxicol. Chem.* **20**: 2899–2902.
9. Belnap, J. (1993). Recovery rates of cryptobiotic crusts: inoculant use and assessment methods. *Great Basin Naturalist* **53**: 89–95.
10. Committee on Environment and Natural Resources. (1999). *Ecological Risk in the Federal Government*. National Science and Technology Council, Washington, DC, CENR/5-99/001.
11. United States Environmental Protection Agency. (1997). *Ecological Risk Assessment Guidance for Superfund: Process for Designing and Conducting Ecological Risk Assessments*. U.S. Environmental Protection Agency Interim Final Report. Emergency Response Team, Edison, NJ.
12. United States Environmental Protection Agency. (1998). *Guidelines for Ecological Risk Assessment*. Risk Assessment Forum, EPA/630/R-95/002F, Final. U.S. Environmental Protection Agency, Washington, DC.

13. Menzie, C. et al. (1996). Report of the Massachusetts weight-of-evidence workgroup: a weight-of-evidence approach for evaluating ecological risks. *Hum. Ecol. Risk Assessment* **2**: 277–304.
14. United States Environmental Protection Agency. (2000). *Stressor Identification Guidance Document*. U.S. Environmental Protection Agency, Office of Water and Office of Research and Development, Washington, DC.
15. Markwiese, J.T., Ryti, R.T., Hooten, M.M., Michael, D.I., and Hlohowskyj, I. (2001). Toxicity bioassays for ecological risk assessment in arid and semiarid ecosystems. *Rev. Environ. Contam. Toxicol.* **168**: 43–98.

REACHING OUT: PUBLIC EDUCATION AND COMMUNITY INVOLVEMENT IN GROUNDWATER PROTECTION

JENNIFER NELSON
The Groundwater Foundation
Lincoln, Nebraska

INTRODUCTION

Throughout its existence, The Groundwater Foundation (TGF) has identified public education and community involvement as vital ingredients to successful groundwater protection activities. TGF is an international nonprofit organization based in Lincoln, Nebraska, with a mission to educate and motivate the public to care about and for groundwater. Groundwater provides drinking water to nearly half of the U.S. population (1) and is critical for other economic uses such as irrigation, agriculture, and industry. Unfortunately, because it is a hidden resource, the general public knows comparatively little about groundwater or how to protect it. Providing education to stakeholders about the value of groundwater is an excellent way to illustrate how individual and group decisions and activities can have a direct impact on the quality and quantity of local groundwater resources.

TGF was founded in 1985 and since has held that people could learn about groundwater in ways that were both scientifically accurate and user friendly. TGF works to make groundwater science accessible to and understandable by people everywhere, allowing these educated people to become active partners in protecting the environmental and economic vitality of their communities by protecting local and regional groundwater resources.

Groundwater Guardian is a program of TGF that provides a framework for community action and groundwater protection by providing recognition, support, and lessons learned. Over 300 communities from the United States and Canada have participated in Groundwater Guardian since the program began in 1994. Groundwater Guardian relies on voluntary efforts spearheaded at the local level to effectively address local groundwater-related concerns and issues through the creation of a diverse local team, which focuses the interests and resources of a community on the importance of long-term groundwater protection.

Groundwater Guardian serves as both an organizing and planning tool for communities interested in groundwater protection. A community within the program is broadly defined—it can be a city, county, watershed area, school, or other area where there are people committed to learning about and protecting groundwater. To enter the program, communities first form a team of diverse stakeholders, including citizens, local government, educators, business, and/or agriculture. The team is the heart of the program and is designed to be broad-based to draw on the different resources and expertise of various sectors of the community. These stakeholders who represent key local groups are in an excellent position to identify a community's groundwater problems, develop an education/action plan to address these issues, and document their progress and success.

After a team is formed, local issues of concern are identified and then addressed through activities addressing public education and awareness, pollution prevention, public policy, conservation, and best management practices. Teams submit plans for these activities in the beginning of the year and report back to TGF at the end of the year on progress made. Groundwater Guardian designation is awarded based on the information provided in these plans and reports.

USING EDUCATION AS A STARTING POINT FOR GROUNDWATER PROTECTION

Groundwater Guardian teams generally start with some sort of public education campaign to build awareness for future protection activities. This education may take the form of a youth festival, newspaper column, water bill insert, public service announcement, website, brochure, seminar, public meeting, or other means. Education is a popular starting point for most Groundwater Guardian teams because:

- Education is viewed as an accessible, common activity that invites broad community participation.
- Goals and objectives can easily be set and achieved and give community efforts valuable success to build upon.
- Awareness and understanding are the basis for future activity, for adults and children alike.
- Children are likely to communicate with and involve parents, relatives, neighbors, and friends with successful and memorable education experiences (2).

Public education allows a progressionary approach to groundwater resources protection. It first engages community residents to know that their individual actions and practices, like disposing of motor oil properly, using only appropriate amounts of fertilizer, and taking shorter showers, can have an impact on local groundwater quality and quantity. It then also builds a base of knowledge for future support of protection activities. Residents who know that their drinking water is supplied by groundwater, pumped from wells in a specific area, are more likely to support efforts such as wellhead protection and the implementation of best management practices that protect that source than a resident who simply thinks that drinking water comes from the tap. Education creates a foundation for action.

VALUE OF COMMUNITY INVOLVEMENT IN GROUNDWATER PROTECTION

In TGF's experiences, community involvement is necessary to successful, meaningful, and long-lasting groundwater protection, which provides numerous benefits to all community residents, including:

- *Protecting Natural Resources.* Groundwater is sometimes referred to as the environmental "bottom line" as an indicator of how successful all environmental protection activities are. By monitoring groundwater quality and quantity, communities can see the impacts of their groundwater and other natural resource protection activities. By protecting groundwater resources, communities are often also protecting a variety of other natural resources, such as wildlife habitat and wetlands.
- *Safeguarding Public Health.* Safe groundwater is directly related to good health. Contaminated drinking water can cause illness, disease, or even death. By protecting groundwater supplies that are used for drinking water, communities are protecting their own health.
- *Developing Economic Vitality.* Communities depend on local groundwater supplies not only as a source of drinking water but to provide businesses and agricultural producers with an input to create a market good. Without clean, safe and plentiful groundwater supplies, communities across the country would face serious challenges to their long-term sustainability.
- *Building Community Capacity.* Communities that make the commitment to groundwater are the very same communities who are working to be innovative and strive for long-term growth and sustainability of their community. Because of the variety of effective groundwater protection activities and options available, communities working to protect groundwater are also building their capacity to address other community issues, such as growth, strategic planning, solid waste, transportation, and economics.
- *Sustaining Supplies.* As populations continue to grow, increased demand is placed on groundwater supplies. Communities that are educated about the value of local groundwater resources are much more likely to manage those resources in a sustainable way.
- *Sharing Responsibility.* Whether it is for drinking or as an economic input, all those within a community use and benefit from groundwater. Consequently, protecting groundwater supplies is best done by a diverse team of community representatives who share a variety of viewpoints but are working toward a common goal. By sharing the responsibility for groundwater protection, strong communities will protect groundwater supplies for generations to come.

STRATEGIES FOR PUBLIC EDUCATION AND COMMUNITY INVOLVEMENT IN GROUNDWATER PROTECTION

There are numerous options for public education and community involvement for local groundwater protection. Protection efforts implemented by Groundwater Guardian communities often begin with public education activities, including:

- Organizing a community or school water festival
- Launching a public awareness campaign
- Distributing educational materials about local groundwater issues
- Holding public meetings
- Planning field trips

Communities often find success with these relatively easy to implement, yet high impact, public education activities and can then move on to more challenging efforts, such as:

Conservation
Endorse and encourage the use of water-saving devices
    Encourage sustainable lawn care and gardening practices
    Work with the local water department to foster conservation awareness

Pollution Prevention
Inventory local pollution sources
    Properly close abandoned wells
    Recognize local pollution sources and identify possible solutions

Public Policy
Establish and manage wellhead protection areas
    Support compliance with the Safe Drinking Water Act
    Encourage federal, state, and local interagency coordination
    Develop a community comprehensive plan

Best Management Practices (BMPs)
Encourage the adoption of BMPs in rural, urban, and commercial areas
    Adopt land use protection measures
    Provide voluntary management options

PUBLIC EDUCATION + COMMUNITY INVOLVEMENT = GROUNDWATER PROTECTION

Groundwater protection in small, groundwater-dependent communities is reliant on a high level of citizen involvement to be successful. These communities may lack the financial resources and/or professional expertise to ensure the protection of groundwater resources that are vital to the human and economic health of the community. Instead, they rely on an educated and empowered citizenry acting responsibly on behalf of the community. When citizens understand the fundamental importance of groundwater to their environmental and economic future, they are motivated and able to develop innovative and cost-effective strategies for its protection.

TGF has found that no matter how diverse communities are, those that are successful begin the groundwater protection process in generally the same way: by involving local stakeholders. The involvement of a diverse group of stakeholders is the most important part of the Groundwater Guardian program because a credible, broad-based team effectively represents community interests and therefore shares the responsibility for groundwater protection among community groups.

As is the case for any issue that depends on community involvement for its success, groundwater protection needs a long-term commitment that is continuously rejuvenated by creativity, new ideas, and leadership. The same ingredients that contribute to sustaining meaningful groundwater education and protection activities also apply to other beneficial community activities. By starting with public education and stressing diverse community involvement, communities are on the path to sustainable groundwater protection. As TGF has learned, "it is because of people that groundwater must be protected, but it is only through people that we can do so."

BIBLIOGRAPHY

1. Alley, W.M., Reilly, T.E., and Franke, O.L. (1999). *Sustainability of Ground-Water Resources*. U.S. Geological Survey Circular 1186.
2. Herpel, R., Jackson, J., Kuzelka, R., Rodenberg, S., and Seacrest, S. (1998). *Let's Make a Difference: Mobilizing for Community Action*. The Groundwater Foundation, Lincoln, NE.

INTEGRATION OF ENVIRONMENTAL IMPACTS INTO WATER RESOURCES PLANNING

MEHMET ALI YURDUSEV
Celal Bayar University
Manisa, Turkey

Nature is so complicated that we need to idealize or simplify to understand, interpret, and benefit from it. Simplifications are carried out by modeling natural events. Models give us the opportunity to input values representing the observed values of natural phenomena to see what the possible outcome would be. Water is one of the few natural substances without which human beings cannot survive. Water use, therefore, is much more complicated because it has to reconcile an irregular or random natural phenomenon and humans' attitudes toward it, which is not less complicated than water in nature itself. Therefore, it has to be managed by employing complicated computer-based mathematical models. The spatial and temporal dimensions of this complicated issue can then be handled. These modeling studies are generally referred to as water resources planning (WRP). WRP can be defined as matching future demands to potential resources, satisfying some preset objectives such as cost-effectiveness or environmental quality. The end product of water resources planning is a development plan for some future period, normally 20 to 30 years. As such, WRP mainly encompasses three types of activities:

- assessment of available water resources,
- assessment of future water requirements, and
- matching between available resources and forecast demands.

In the third phase of water resources planning, what is primarily involved is identifying the most appropriate development strategy for meeting projected demands at minimum cost, over the planning horizon. On a national scale, the planning process is characterized by the need to screen a large number of potential options (1) to determine

- which resources should be developed,
- the timing and order of that development, and
- the areas of demand to which each new resource should be assigned.

Typically, this has required information relating to future demand, the performance of different size sources at various locations, the possible links between sources and demand centers, together with all associated construction and operating costs. Faced with an enormous choice of possibilities for matching resources to needs, a systematic search procedure is normally required to achieve an objective assessment. The objective function of the procedure would be minimization of total discounted costs. To that end, various forms of mathematical programming have been used (2) to identify the minimum net present value of a development program. Alternatively, heuristic search techniques can be used which, although not as rigorous, provide a practical solution (3).

In the past, scant attention would have been given to environmental considerations within the water resources planning process until a scheme had already been selected. Then, detailed environmental assessment would be undertaken. At that stage, some attempts would be made to identify and ameliorate damaging environmental impacts. In contrast, the environmental considerations can be incorporated from the outset and included in selecting the scheme to be promoted. This is achieved by weighting the costs of the various options (both construction and operating) to reflect their environmental impacts, before including them in the economic evaluation. The objective function of the combined methodology is to minimize the total monetary and environmental discounted cost of meeting the projected demands. Quantification of such considerations is achieved by running an appropriate model with and without the environmental factors. Such a model should integrate an environmental impact assessment (EIA) and economic planning (EP) models. The EIA model develops what might be called environmental impact factors (EIF) to weigh the costs of schemes, and the EP model is to undertake the scheme selection and timing process using environmentally adjusted costs. The EIF is a weighting function whose role is to raise the cost of a water resource option if its environmental impact assessment indicates that it is environmentally damaging and to decrease it otherwise. Using such a factor in an economic planning model has the effect of influencing the solution toward selecting environmentally friendly schemes at the expense of environmentally damaging ones. In mathematical terms, an EIF is

$$0 < EIF < 1 \text{ and } 1 < EIF < 2$$

where 0 is the best possible outcome and 2 is the worst. Using such factors, environmental gain can be expressed by the former and the environmental loss by the latter: a value of 1 indicates neutrality.

A multicriterion decision-making technique referred to as composite programming (CP) can be used to derive the EIFs. The output from CP is a measure of the composite distance resulting from the aggregation of a series of basic indicators (4). For a given L, the composite distance from a so-called ideal point, where 0 represents the best possibility and 1 represents the worst possibility, the main features of CP are (1) $0 < L < 1$ and (2) the larger L, the worse the associated scheme. The real values of basic indicators are weighed to obtain a further indicator value. The procedure continues until it reaches a final indicator value that will represent the performance of the option according to the indicators considered. A simple mathematical function which deliberately exaggerates the extreme values can be used to transform L into EIF as follows:

$$\text{EIF} = 1.4907\sqrt{L} \quad \text{if } 0 < L < 0.45 \quad (1)$$

$$\text{EIF} = 2 - 1.4907\sqrt{1-L} \quad \text{if } 0.45 < L < 1 \quad (2)$$

In this way, it is possible to convert a series of basic indicators covering both detrimental and beneficial impacts to a corresponding EIF within the range of 0 to 2. When evaluating an impact, assessment values range from negative significant to positive significant with a central point at neutral. The impacts of water projects can be grouped according to their resource use, quality, ecosystem, and social implications. For example, the aggregation structure given in Table 1 can be considered as assessing the overall construction impact of a groundwater development scheme. The assessment values in Table 1 are assigned within a range between 1 and 8; 1 is the worst situation, 8 is the best, and 4 is neutral. The values in parenthesis show the normalized ones.

The values of second level indicators are calculated from the weighted sum of the associated basic impacts. The value of resource implications is directly taken as neutral because there are no associated basic impacts from which its value can be obtained. The overall value is, then, calculated as 0.41, and the EIF as 95% from Equation 1. This value indicates that the project is somewhat environmentally friendly and, therefore, the associated cost will be reduced accordingly during the optimization process.

Clearly, different types of resources/links will affect the environment in different ways, and therefore, each has its own specific set of considerations. Similarly, there are different considerations for each type of resource/link during the construction and operational phases. All these different impacts can be accommodated within the EIA model (5).

The economic planning model is intended to undertake the scheduling process to determine when the resources are to be built. Therefore, an existing model should be adopted for this purpose. Such a model would perform the planning process using the costs of water schemes contained in the case study under consideration. The objective is to find the least-cost solution for the whole exercise. As previously mentioned, the idea is to use the outputs of an EIA model (EIFs) in the economic planning model selected to reflect the environmental performance of each scheme. The two-way usage of an economic planning model would give the user the opportunity to compare results with and without environmental impacts. This may even be used to quantify the cost of environmental impacts in the form of additional investment required to apply environmentally friendly solutions (with environmental impact solutions). There are several economic water resources planning models that can be used for this purpose. The models can be based on unit costs (6), integer linear programming (2,7), heuristic programming (8), and genetic algorithms (9). The selection of an economic planning model depends on the availability of models, the amount to be invested, the nature of the exercise, and the capability of the models. Unfortunately, most of the models mentioned may not be commercially available. However, the owners may lend the models on request.

Combining economic planning and EIA models produces an overall environmentally-influenced economic planning system in which EIFs are used to weigh the real construction and operating costs for various development options to reflect their positive or negative environmental

Table 1. Constructional Environmental Impact Assessment for a Groundwater Scheme

Basic Impacts	Assessment Values	Weights	Second-Level Indicators	Indicator Values	Weights
			Resource use implications	0.50	0.18
Lack of visual intrusion	6(0.18)	1.00	Quality implications for physical system	0.18	0.29
Effect of construction on surface ecosystem	4(0.50)	0.44			
Pipeline laying impact on vegetation and wildlife	3(0.66)	0.56	Ecosystem implications	0.59	0.25
Impact of infrastructure construction; noise, mud, etc.	4(0.50)	0.27			
Landscape/land use impact of pipeline/structures	4(0.50)	0.27	Social implications	0.35	0.29
Implementation benefits	6(0.18)	0.45			

impacts. The coupling can be achieved simply by an economic model using the cost figures modified by the EIFs produced by the EIA model.

Such a methodology was applied to the whole of England and Wales using the National Rivers Authority's (NRA) data (3). The exercise was undertaken using the same economic planning model used by NRA coupled with an independent code for the EIA model (8). The environmentally sensitive solution was repriced in real terms by the model and therefore, the additional cost of preferring such a plan was estimated (5).

BIBLIOGRAPHY

1. Jamieson, D.G. (1986). An integrated multi-functional approach to water-resources management. *J. Hydrol. Sci.* **31**(3): 14.
2. O'Neill, P.G. (1972). A mathematical programming model for planning a regional water-resource system. *J. Inst. Water Engs.* **26**: 47–61.
3. National Rivers Authority. (1994). *Water-Nature's Precious Resource: An Environmentally Sustainable Water Resources Development Strategy for England and Wales*. HMSO, London, England.
4. United Nations Educational Social and Cultural Organisation (UNESCO). (1987). *Methodological Guidelines for the Integrated Environmental Evaluation of Water Resources Development*. United Nations Environmental Programme(UNEP), Paris, France.
5. Yurdusev, M.A. (2002). *Environmental Impacts for Water Resources Planning*. WIT Press, Southampton, UK.
6. Armstrong, R.B. and Clarke, K.F. (1972). Water resource planning in South East England. *J. Inst. Water Eng.* **26**: 11–46.
7. Schrage, L. and Cunningham, K. (1991). *LINGO Optimisation Modelling Language*. Lindo Systems Inc., Chicago.
8. Anglian Water Services. (1993). *Economic Planning Model of Water Resources: RESPLAN User Guide*. Anglian Water Services Ltd., Cambridge, England.
9. Dandy, G. and Connarty, M. (1995). Use of generic algorithms for project sequencing. In: *Integrated Water Resources Planning for the 21st Century*. M.F. Domenica (Ed.). Proc. 22nd Annu. Conf., ASCE, Cambridge, Massachusetts, USA, pp. 540–543, May 7–11.

THE EXPANSION OF FEDERAL WATER PROJECTS

MARTIN REUSS
Office of History Headquarters
U.S. Army Corps of Engineers

Instead of national planning, Congress settled on a piecemeal approach to public works development in the early nineteenth century and, with rare exceptions, has clung to this approach ever since. In May, 1824, President Monroe signed legislation appropriating $75,000 to improve navigation on the Ohio and Mississippi rivers—major routes to the western part of the country. The act empowered him to employ "any of the engineers in the public service which he may deem proper" and to purchase the "requisite water craft, machinery, implements, and force" to eliminate various obstructions. Along with the General Survey Act, signed a month before, the Mississippi-Ohio rivers legislation initiated the permanent involvement of the Army Corps of Engineers in rivers and harbors work. However, each act focused on one activity: the General Survey Act on planning; the other on construction. Two years later, Congress combined surveys and projects in one act, thus establishing a pattern that lasts until the present. The 1826 act, therefore, can be called the first true rivers and harbors legislation.

By the time of the Civil War, the federal contribution to river, harbor, and canal improvements amounted to about $17 million in appropriated monies. Some 4.6 million acres of public lands were given for canal improvements and another 1.7 million acres for river improvements. Land grants under the 1849 and 1850 Swamp Land acts and the 1841 land grant act totaled some 73 million acres. While these grants and appropriations were significant, they represented a modest amount of aid compared with state and private sector contributions, which by 1860 totaled well over $185 million for canals alone. Corporations and public agencies spent many millions more on the construction of urban water systems.

Many of the canal companies incorporated by the states ran into trouble. The 1837 depression had driven a number into bankruptcy; others survived, but only with a healthy influx of state money, state guaranteed bonds, and occasional federal and state land grants. Often, too, the national story was repeated at the state level, with rationally planned canal routes sacrificed to local political pressures to extend canals to uneconomical out-of-the-way villages. The one major exception to this sad story was the Erie Canal, whose success had spurred the canal boom that increasingly appeared more like a dismal bust, especially with new competition from the railroads. From Pennsylvania to Ohio to Indiana to Illinois and on into the states of the old Northwest, canal fever turned to canal panic, and the public lost faith in both the companies and the politicians who had supported the enterprises.

The American Civil War (1861–1865) also affected the development of water projects. Military action and wartime budgetary constraints took their toll on many of the nation's ports and navigable waterways, and after the war commercial development accelerated demands for waterway improvements. A business-oriented Republican Congress responded by authorizing a great deal more money for rivers and harbors. The federal government also took over many of the bankrupt canal companies, and the Corps of Engineers became the custodian of former private or state waterways. This, as one author put it, was the "Golden Age of the Pork Barrel". Between 1866 and 1882, the Presidents signed 16 rivers and harbors acts. The 1866 act appropriated $3.7 million for 49 projects and has been described as the first omnibus bill, so called because like a horse-drawn omnibus of the time,

---

This article is a US Government work and, as such, is in the public domain in the United States of America.

the legislation provided room for a great many people boosting various projects. Sixteen years later, though, the 1882 act appropriated five times more money. By that year, the federal government had spent over $111 million on rivers and harbors projects. "Willingness to pay"—the primary test of project implementation before the Civil War—now included unprecedented federal largess. In the so-called "Gilded Age," lack of federal or non-federal funds was about the only thing that prevented construction.

By the 1880s, the basic working relationship between Congress and the Army Corps of Engineers was set. Congress directed the Corps to survey potential projects, make recommendations, and provide cost estimates. Rivers and Harbors acts funded both the surveys and the projects that Congress chose to authorize. Also in the early 1880s, Congress mandated that the Corps of Engineers use more contractors and less hired labor. By the end of the century, contractors did nine-tenths of all waterways construction, and no Corps officer could use hired labor without the express authority of the Chief of Engineers. Increasingly, then, the Corps became a funding conduit to the private sector. This pattern did not stop private sector engineers from calling for the complete elimination of the Corps from public works, but Congress rejected all bills that leaned in that direction.

Fear of railroad competition and questions about federal aid to projects of apparently local benefit moved the Senate in 1872 to create a Select Committee on Transportation Routes to the Seaboard. Composed of nine senators, the committee was headed by Senator William Windom of Minnesota and known popularly as the Windom Committee. Its 1873 report promoted waterway over railway transportation wherever waterways were properly located. Of more relevance here is the committee's conclusion (on a five to four vote) that the sum of local rivers and harbors projects contributed to the national interest. Generally accepted by Congress, this conclusion justified federal contributions for waterway improvements. The result was the authorization of dozens of dubious projects. By 1907, the cumulative total for rivers and harbors appropriations was more than four times the 1882 figure; the federal role in navigation improvements continued to grow.

Meanwhile, the issue of constitutional authority had somewhat changed focus. In 1870, the Supreme Court ruled in The Daniel Ball case that the common law doctrine that navigability depended on tidal influence, a doctrine accepted in British courts, did not fit the American situation. However, the definition the Court substituted was extraordinary. The test of navigation was to be the river's "navigable capacity." That meant, the Court went on:

> Those waters must be regarded as public navigable rivers in law which are navigable in fact. And they are navigable in fact when they are used, or are susceptible of being used, in their ordinary condition, as highways for commerce over which trade and travel are or may be conducted in the customary modes of trade and travel on water.

In short, American rivers were navigable if they were, are, or could be navigable. This decision, in combination with the earlier 1824 Gibbons v. Ogden Court ruling, made the federal government the clear guardian and ultimate decision-maker on tens of thousands of miles of waterways in the United States. In practice it sufficed to show that a stream had the capacity to float logs to declare it navigable. However, with this issue more or less settled, another appeared: flood control.

Rivers always flood, but the floods do not always damage life and property. In the United States, we can trace floods as far back as 1543, when Mississippi River floods stopped Hernando De Soto's expedition. Naturally, as settlers moved into the floodplains and built villages, then cities, the damages increased. By the mid-nineteenth century, the problem was becoming critical along the lower Mississippi. Most people put their faith in technology to protect them. Indeed, the then popular term "flood prevention" testifies to an extraordinarily unrealistic idea when one thinks about it a bit. In the twentieth century, the term became "flood control," a somewhat more modest formulation. Nowadays we speak of "flood damage reduction," which probably comes closest to the mark. In any case, in the 1870s calls came for repairing and raising the levees on the Mississippi River. In 1879, Congress created a joint military—civilian Mississippi River Commission to develop and implement plans to improve navigation and flood control on the lower Mississippi. However, once again some Congressmen raised constitutional objections, expressing doubts that flood control was an appropriate federal activity. Until 1890, no appropriation could be used for repairing or constructing any levee in order to prevent damage to lands from overflow, or for any purpose other than deepening and improving the navigation channel. The 1890 floods along the lower Mississippi resulted in the removal of this restriction, which, in any event, had had little practical effect other than satisfying congressional scruples.

Floods in 1912, 1913, and 1916 along the Ohio and Mississippi rivers eventually led to passage of the 1917 Flood Control Act, the nation's first act dedicated solely to flood control. It provided funds on a cost-shared basis for levee construction along the lower Mississippi and another appropriation to improve the Sacramento River in California. While an important step towards federal involvement in flood control, it was comparatively modest compared to what followed in the coming decades, when flood control became intertwined with multipurpose development—the subject of the next essay.

This article is based on Martin Reuss, "The Development of American Water Resources, Planners, Politicians, and Constitutional Interpretation," in Paul Slack and Julie Trottier (Eds.), *Managing Water Resources, Past and Present: The Twelfth Annual Linacre Lectures*. Oxford University Press, 2004. Forthcoming. Used by permission of Oxford University Press.

The Development of American Water Resources: Planners, Politicians, and Constitutional Interpretation.

The Constitution and Early Attempts at Rational Water Planning.

The Expansion of Federal Water Projects.

FLOOD CONTROL HISTORY IN THE NETHERLANDS

Rongchao Li
Delft University of Technology
Delft, The Netherlands

The traditional flood control approach in The Netherlands is fighting against natural dynamics or, restricting the natural dynamics of a river system by canalization and embankments. However, such an approach results in destruction of river scenery, damage to nature and cultural values, as well as continuous dike raising and strengthening. Changing societal views in The Netherlands have aroused discussion of alternative flood management strategies. Around the beginning of the for twenty-first century, new visions appeared such as "living with floods" and the "room for the river" policy.

FLOOD CONTROL HISTORY IN THE NETHERLANDS

The Netherlands is a small country in Western Europe. The delta of the Rhine, Meuse, and Schelde Rivers lie within the country. One-third of The Netherlands needs and has artificial protection against floods from the sea or major rivers (see Fig. 1). Millions of people live, and large industries have settled in this area. In the western coastal area, the major cities such as Amsterdam, The Hague, and Rotterdam are situated. Large parts of the western coastal area lie below sea level. Also in the area along the rivers, cities and industries have settled. Therefore, flood risk management is an important issue for The Netherlands.

The Rhine and Meuse Rivers are the largest and most important rivers in The Netherlands. The Rhine River has a large catchment stretching from the high Alps in Switzerland, Germany, and France. The Meuse River flows through France and Belgium. Only the most downstream parts of these rivers are situated in the Netherlands. Therefore, flood risk management focuses on the discharge of water. Nowadays, after a long history of coping with floods, a high safety standard has been reached.

Before 1000 A.D., floods occurred very frequently. Mainly the higher parts in The Netherlands were inhabited, and in the lower parts, people lived on mounds, so no real disasters occurred in that period. Because people had no technological options to prevent flood then, they had no other option but to live with the floods and to avoid the floodwaters. They adapted their lives and land use to the rivers and sea.

About 1000 A.D., the inhabitants of The Netherlands started to build the first dikes around relatively small polders. The flood risk management strategy changed from living with floods to a strategy where the river and its floods were more controlled. This trend of increasing the control of the river continued, and by 1400 A.D., an almost completely closed dike system already existed along the rivers (2). Furthermore, huge changes in the course of the river and the riverbed in the form of regulation and canalization took place (3). The river was harnessed into a small area and now the people and land use did not adapt to the river any longer, but the river was adapted according to the wishes of the inhabitants. The main flood management strategy consisted of preventing floods. This strategy continued unchanged until 1953. After each dike breach, the broken dikes were reconstructed, raised, and improved resulting in increasingly higher and stronger dikes. The last great flood of the Rhine River occurred in 1926, after which the dikes were improved again (2). Since then, the rivers seemed controlled, and people feel safe.

In 1953, a major flood from the North Sea took place that killed more than 1800 people. This flood resulted in new safety regulations for the coast and also for the river area. Previously, dike heights were based on the maximal recorded water level, but after 1953, a more scientific base has been used. The optimal level of safety was defined as the accepted probability of flooding for the different areas in The Netherlands. To be able to use this new norm, it was simplified to the demand that dike levels should exceed water levels related to a discharge with a chosen return time. After several years of discussion, one safety level was chosen for the whole area threatened by river floods: a discharge with a probability of once in 1250 years (the design discharge). Flooding by causes other than overtopping of dikes and uncertainties in nature and in the calculations were considered by adding 0.5 m to the required height and some regulations for the design of the dikes. Every 5 years, the design discharge is calculated anew based on the recorded discharges.

Nowadays, the Dutch rivers are thus strongly human-influenced. Figure 2 shows a cross section of a typical

Figure 1. Almost 75% of The Netherlands is threatened by floods (1).

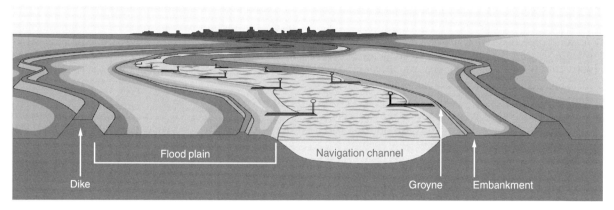

Figure 2. A cross section of a typical river in The Netherlands (4).

Dutch part of a river. The rivers consist of a small mainstream that has to enable navigation all year round. Next to this mainstream, are floodplains which carry water during peak flows, about twice each year in wintertime. These floodplains are used for agriculture and recreation in the summertime during lower discharges. The area outside these floodplains is protected from floods by huge embankments. The unprotected floodplains sometimes lie higher than the surrounding area due to sedimentation in the floodplains and subsidence of the surrounding area. To prevent the area along the rivers from becoming flooded at once, this area is subdivided into dike circles, which are areas surrounded by embankments or natural higher areas.

In 1993 and 1995, extremely high river discharges occurred in the Meuse and Rhine Rivers. Because of these extreme peak discharges, the design discharge with a probability of once in 1250 years increased from 15.000 m^3/s to 16000 m^3/s at the Dutch–German border. Using the traditional approach of flood risk management, this would lead to a further increase of dike heights. However, new solutions have developed nowadays creating room for the rivers (5).

NEW VISIONS AND IDEAS IN FLOOD RISK MANAGEMENT

Due to changes in the values of society and available technology in The Netherlands, flood risk management is changing. Some of these changes and the resulting ideas on flood risk management are mentioned here (1):

- The current strategy based on *one design discharge for the whole* area threatened by river floods leads to the same level of protection for different areas with different potential flood damage. Cities, agricultural areas, and nature reserves all have the same probability of flooding in The Netherlands. The question arises whether this is economically sensible. It also means that it is unknown which area will become flooded first when an extreme discharge occurs. Because all areas theoretically have the same probability of flooding, a large area has to be evacuated.

- Due to the fact that this strategy *focuses on preventing floods*, little attention has been paid to the consequences of possible floods. Relatively little attention is paid to plans for emergency situations, evacuation plans, flood mitigation measures, etc. which also might lower flood risks. New technology has increased the possibilities for good flood forecasts, of communication technology, and flood warnings. Also the possibility to anticipate peak flows by lowering a structure and using a detention pond has grown. These technologies make it easier to allow controlled floods and therefore reduce the need to prevent floods in all circumstances.

- By looking only at the probability of a design discharge, risks are not clearly visualized. The uncertainties in, for example, the design discharge, the translation of the discharge to water levels, the diversion of the peak flow through the different river branches, the strength of dikes and structures, and human behavior are not clearly considered. This results in a false sense of safety in the area and thus in rapid economic development and increasing potential flood damage.

- Another disadvantage of the strategy is that it includes an endless need for raising and improving the water defense structures. *Restricting* the natural dynamics of a river system by canalization and embankments requires continuous maintenance and improvements, otherwise the river dynamics will damage these works, and the river will try to return to its own natural behavior.

- The current strategy results in destruction of the river scenery and damages nature and cultural values. Recently, these values have become more important. A new strategy should consider these values and not focus only on reaching a "safe" situation by technical solutions.

- In these modern times, disasters are not accepted any longer. The government and water managers would be held responsible for a disaster, even if the event that caused the disaster is extreme and rare.

In The Netherlands, consciousness has increased that absolute safety cannot be guaranteed and that floods will always occur. However, disasters are not accepted any longer. Therefore, and because of the disadvantages of the current strategy, as mentioned above, new ways to deal with uncertainties and flood risks have to be found.

In 1990, the policy report "Room for Rivers" was published. This policy wants to create room for rivers to stop the continuous cycle of dike raising, increased feelings of safety, increased investments, increased flood risks, and needs for further dike raising. By moving the dikes away from the river, digging out the floodplains, removing obstacles from the floodplains and sometimes constructing bypasses for peak flows, room for the river is created. The purpose of this policy is to interrupt the continuous dike raising and lower the water levels in the river by giving the river more space. In reality, this policy has been applied at different locations. Numerous obstacles, such as roads to bridges, roads to ferries, and brick factories in the floodplains have been removed or have changed to permit water flow. At Nijmegen near the Dutch–German border, a dike has been moved to increase room for the river. The inhabitants of the area normally opposed these projects because the room for the river is taken from other functions. However, they are mostly also against dike raising because it deteriorates their scenery and sometimes it destroys houses and restaurants built on or near the dike.

Another new policy is the indication of so-called "emergency areas." These are areas that must be flooded first to protect other, more vulnerable areas. Officially these areas should be used only for discharges that exceed the design discharge, thus with a frequency of less than once in 1250 years. Official research has been published, but until now, no governmental decision has been made about the indication of such areas, the location of these areas, or the regulations for the inhabitants of these areas. It is unclear whether the villages in these areas will be protected by dikes and what damage compensation they will receive. The purpose of this policy is to limit the emergency measures to smaller areas and to know what area will become flooded and not leave the choice to the river.

Other new concepts are introduced in the National Policy on Water Management. According to this policy we should "aim at resilient and healthy water systems." Resilient water systems are systems that may be disturbed or show temporary change but will easily recover. To discuss resilience and resistance, a systems approach to flood risk management has to be adopted. The system on which flood risk management focuses can be defined as the whole area threatened by floods including the society, ecosystems, and the river itself in this area. Optimal flood risk management depends on the society and culture of the inhabitants, not only on the discharge regime of the river. Furthermore, the whole system should be studied, not just a small river stretch, because measures as well as floods downstream and upstream interact with each other. Measures may transport the flood problems downstream, and upstream floods may prevent downstream inundations. In resilience strategies for flood risk management, floods are not necessarily prevented, but the flood impacts are limited, and recovery after the flood is enhanced. Resilience strategies focus on the whole discharge regime, not on a certain design discharge or design event. These strategies are still only a subject of research and an option for the long term. For more information, see De Bruijn and Klijn (1) or Vis et al. (6).

CONCLUSIONS

One-third of The Netherlands needs and has artificial protection against floods from the sea or the major rivers. In this area, millions of people live and large industries have settled. This proves the importance of flood management for The Netherlands. After centuries of coping with floods, research and reconsideration of flood management strategies are still needed because society is changing and thus the optimal flood management strategy must also change.

BIBLIOGRAPHY

1. De Bruijn, K.M., Klijn, F. (2001). *Resilient Flood Risk Management Strategies*. Proc. IAHR Congress September 16–21, Beijing, Tsinghua University Press, Beijing, China, pp. 450–457.
2. Commissie Rivierdijken (1977). *Rapport commissie rivierdijken*. (In Dutch) Hoofddirectie van de Waterstraat, s-Gravenhage.
3. Janssen, J.P.F.M. and Jorissen, R.E. (1997). Flood management in The Netherlands; recent development and research needs. In: *Ribamod, River Basin Modelling, Management And Flood Mitigation Concerted Action*. R. Casale, K. Havnø, and P. Samuels (Eds.). pp. 89–104.
4. Silva, W., Klijn F., and Dijkman, J. (2000). *Ruimte Voor Rijntakken. What The Research Taught Us. Delft*, WL|Delft Hydraulics, The Netherlands.
5. Min.VROM En V&W (Ministeries VROM En V&W) (1997). *Beleidslijn Ruimte Voor De Rivier*. Den Haag (In Dutch).
6. Vis, M., Klijn, F., and Van Buuren, M. (Eds.). (2001). *Living with Floods, Resilience Strategies for Flood Risk Management and Multiple Land Use in the Lower Rhine River Basin*. Summary report. NCR, Delft, The Netherlands, de Bruijn, 2001.

FOOD AND WATER IN AN EMERGENCY

Federal Emergency
Management Agency

INTRODUCTION

If an earthquake, hurricane, winter storm or other disaster strikes your community, you might not have access to food, water and electricity for days, or even weeks. By taking some time now to store emergency food and water supplies, you can provide for your entire family. This brochure was developed by the Federal Emergency Management Agency in cooperation with the American Red Cross and the U.S. Department of Agriculture.

This article is a US Government work and, as such, is in the public domain in the United States of America.

WATER

Having an ample supply of clean water is a top priority in an emergency. A normally active person needs to drink at least two quarts of water each day. Hot environments can double that amount. Children, nursing mothers and ill people will need even more.

You will also need water for food preparation and hygiene. Store a total of at least one gallon per person, per day. You should store at least a two - week supply of water for each member of your family.

If supplies run low, never ration water. Drink the amount you need today, and try to find more for tomorrow. You can minimize the amount of water your body needs by reducing activity and staying cool.

HOW TO STORE WATER

Store your water in thoroughly washed plastic, glass, fiberglass or enamel-lined metal containers. Never use a container that has held toxic substances.

Plastic containers, such as soft drink bottles, are best. You can also purchase food-grade plastic buckets or drums. Seal water containers tightly, label them and store in a cool, dark place. Rotate water every six months.

EMERGENCY OUTDOOR WATER SOURCES

If you need to find water outside your home, you can use these sources. Be sure to purify the water according to the instructions listed below before drinking it.

- Rainwater
- Streams, rivers and other moving bodies of water
- Ponds and lakes
- Natural springs

Avoid water with floating material, an odor or dark color. Use saltwater only if you distill it first. You should not drink floodwater.

THREE WAYS TO PURIFY WATER

In addition to having a bad odor and taste, contaminated water can contain microorganisms that cause diseases such as dysentery, typhoid and hepatitis. You should purify all water of uncertain purity before using it for drinking, food preparation or hygiene.

There are many ways to purify water. None is perfect. Often the best solution is a combination of methods. Two easy purification methods are outlined below. These measures will kill most microbes but will not remove other contaminants such as heavy metals, salts and most other chemicals. Before purifying, let any suspended particles settle to the bottom, or strain them through layers of paper towel or clean cloth.

1. *Boiling*
   Boiling is the safest method of purifying water. Bring water to a rolling boil for 1 minute, keeping in mind that some water will evaporate. Let the water cool before drinking.
   Boiled water will taste better if you put oxygen back into it by pouring the water back and forth between two clean containers. This will also improve the taste of stored water.

2. *Disinfection*
   You can use household liquid bleach to kill microorganisms. Use only regular household liquid bleach that contains 5.25 percent sodium hypochlorite. Do not use scented bleaches, color-safe bleaches or bleaches with added cleaners.
   Add 16 drops of bleach per gallon of water, stir and let stand for 30 minutes. If the water does not have a slight bleach odor, repeat the dosage and let stand another 15 minutes.
   The only agent used to purify water should be household liquid bleach. Other chemicals, such as iodine or water treatment products sold in camping or surplus stores that do not contain 5.25 percent sodium hypochlorite as the only active ingredient, are not recommended and should not be used.
   While the two methods described above will kill most microbes in water, distillation will remove microbes that resist these methods, and heavy metals, salts and most other chemicals.

3. *Distillation*
   Distillation involves boiling water and then collecting the vapor that condenses back to water. The condensed vapor will not include salt and other impurities. To distill, fill a pot halfway with water. Tie a cup to the handle on the pot's lid so that the cup will hang right-side-up when the lid is upside-down (make sure the cup is not dangling into the water) and boil the water for 20 minutes. The water that drips from the lid into the cup is distilled.

HIDDEN WATER SOURCES IN YOUR HOME

If a disaster catches you without a stored supply of clean water, you can use the water in your hot-water tank, pipes and ice cubes. As a last resort, you can use water in the reservoir tank of your toilet (not the bowl).

Do you know the location of your incoming water valve? You'll need to shut it off to stop contaminated water from entering your home if you hear reports of broken water or sewage lines.

To use the water in your pipes, let air into the plumbing by turning on the faucet in your house at the highest level. A small amount of water will trickle out. Then obtain water from the lowest faucet in the house.

To use the water in your hot-water tank, be sure the electricity or gas is off, and open the drain at the bottom of the tank. Start the water flowing by turning off the water intake valve and turning on a hot-water faucet. Do not turn on the gas or electricity when the tank is empty.

Food

Short—Term Supplies. Even though it is unlikely that an emergency would cut off your food supply for two weeks, you should prepare a supply that will last that long.

The easiest way to develop a two-week stockpile is to increase the amount of basic foods you normally keep on your shelves.

Storage Tips

- Keep food in a dry, cool spot-a dark area if possible.
- Keep food covered at all times.
- Open food boxes or cans carefully so that you can close them tightly after each use.
- Wrap cookies and crackers in plastic bags, and keep them in tight containers.
- Empty opened packages of sugar, dried fruits and nuts into screw-top jars or air-tight cans to protect them from pests.
- Inspect all food for signs of spoilage before use.
- Use foods before they go bad, and replace them with fresh supplies, dated with ink or marker. Place new items at the back of the storage area and older ones in front.

Nutrition Tips

During and right after a disaster, it will be vital that you maintain your strength. So remember:

- Eat at least one well-balanced meal each day.
- Drink enough liquid to enable your body to function properly (two quarts a day).
- Take in enough calories to enable you to do any necessary work.
- Include vitamin, mineral and protein supplements in your stockpile to assure adequate nutrition.

Food Supplies

When Food Supplies Are Low. If activity is reduced, healthy people can survive on half their usual food intake for an extended period and without any food for many days. Food, unlike water, may be rationed safely, except for children and pregnant women.

If your water supply is limited, try to avoid foods that are high in fat and protein, and don't stock salty foods, since they will make you thirsty. Try to eat salt-free crackers, whole grain cereals and canned foods with high liquid content.

You don't need to go out and buy unfamiliar foods to prepare an emergency food supply. You can use the canned foods, dry mixes and other staples on your cupboard shelves. In fact, familiar foods are important. They can lift morale and give a feeling of security in time of stress. Also, canned foods won't require cooking, water or special preparation. Following are recommended short-term food storage plans.

SPECIAL CONSIDERATIONS

As you stock food, take into account your family's unique needs and tastes. Try to include foods that they will enjoy and that are also high in calories and nutrition. Foods that require no refrigeration, preparation or cooking are best.

Individuals with special diets and allergies will need particular attention, as will babies, toddlers and elderly people. Nursing mothers may need liquid formula, in case they are unable to nurse. Canned dietetic foods, juices and soups may be helpful for ill or elderly people.

Make sure you have a manual can opener and disposable utensils. And don't forget non-perishable foods for your pets.

SHELF-LIFE OF FOODS FOR STORAGE

Here are some general guidelines for rotating common emergency foods.

- Use within six months:
  - Powdered milk (boxed)
  - Dried fruit (in metal container)
  - Dry, crisp crackers (in metal container)
  - Potatoes
- Use within one year:
  - Canned condensed meat and vegetable soups
  - Canned fruits, fruit juices and vegetables
  - Ready-to-eat cereals and uncooked instant cereals (in metal containers)
  - Peanut butter
  - Jelly
  - Hard candy and canned nuts
  - Vitamin C
- May be stored indefinitely (in proper containers and conditions):
  - Wheat
  - Vegetable oils
  - Dried corn
  - Baking powder
  - Soybeans
  - Instant coffee, tea and cocoa
  - Salt
  - Non-carbonated soft drinks
  - White rice
  - Bouillon products
  - Dry pasta
  - Powdered milk (in nitrogen-packed cans)

IF THE ELECTRICITY GOES OFF

FIRST, use perishable food and foods from the refrigerator.

THEN, use the foods from the freezer. To minimize the number of times you open the freezer door, post a list of freezer contents on it. In a well-filled, well-insulated freezer, foods will usually still have ice crystals in their centers (meaning foods are safe to eat) for at least three days.

FINALLY, begin to use non-perishable foods and staples.

HOW TO COOK IF THE POWER GOES OUT

For emergency cooking you can use a fireplace, or a charcoal grill or camp stove can be used outdoors. You can also heat food with candle warmers, chafing dishes and fondue pots. Canned food can be eaten right out of the can. If you heat it in the can, be sure to open the can and remove the label first.

Disaster Supplies

Emergency Supplies. It's 2:00 a.m. and a flash flood forces you to evacuate your home-fast. There's no time to gather food from the kitchen, fill bottles with water, grab a first-aid kit from the closet and snatch a flashlight and a portable radio from the bedroom. You need to have these items packed and ready in one place before disaster strikes.

Pack at least a three-day supply of food and water, and store it in a handy place. Choose foods that are easy to carry, nutritious and ready-to-eat. In addition, pack these emergency items:

- Medical supplies and first aid manual n Money and matches in a waterproof
- Hygiene supplies container
- Portable radio, flashlights and n Fire extinguisher extra batteries n Blanket and extra clothing
- Shovel and other useful tools
- Infant and small children's needs (if appropriate)
- Household liquid bleach to purify drinking water.
- Manual can opener

Learn More

The Federal Emergency Management Agency's Community and Family Preparedness Program and the American Red Cross Community Disaster Education Program are nationwide efforts to help people prepare for disasters of all types. For more information, please contact your local emergency management office and American Red Cross chapter.

WATER DEMAND FORECASTING

STEVEN J. RENZETTI
Brock University
St. Catharines, Ontario, Canada

INTRODUCTION

People who are involved with planning, constructing, and operating water supply facilities frequently must have an idea of the current state of water use and also what future water demands are likely to be. This need is made all the more important when water use is rising rapidly and when the cost of major water infrastructure projects is considered. These factors imply that there is a need for accurate forecasts of water demands so that infrastructure projects are correctly sized and scheduled. The purpose of this article is to examine briefly the concepts and methods for forecasting water demands (for a more detailed discussion, see Refs. 3–5). We begin by considering the general nature of water demand forecasts.

ISSUES IN FORECASTING WATER DEMANDS

There are several issues surrounding the general practice of water demand forecasting. Boland (1) points out that any forecast has two essential components: explanation and prediction: "Explanation of water use usually takes the form of a model that relates the past observed level of water use to various variables. Replacing past values of the explanatory variables with those expected in the future produces a prediction of future water use" (pp. 162–163). Historically, the first of these two tasks—explaining water use—has been done by using 'fixed-coefficient' models. In this approach, it is assumed that water use is related to a single explanatory factor such as population and, further, that the relationship between water use and population is a fixed, proportional one. Dziegielewski (2), for example, provides a brief review of the history of urban water demand forecasting and demonstrates that, in the 'traditional' method of forecasting, total future demand is predicted as the product of expected population growth and a fixed per capita water use coefficient. This method was subsequently refined by disaggregating total water use by user classes, area, and time period. The fundamental forecasting method remained the same, however; 'unit water use coefficients' is multiplied by the projected growth in a particular user group in a specific location. The fundamental shortcoming of the fixed coefficient approach is that it fails to anticipate changes in the relationship between water use and the dominant explanatory variable that may arise from changes in other, neglected variables. For example, forecasts of residential water use that are based on the number of households in an area may overstate future water demands if they neglect the impacts of rising water prices. More recently, more complex models have sought to estimate statistically the relationship between water use and a set of explanatory variables, including water prices.

Estimated relationships form the basis for demand forecasts. Dziegielewski (2) describes how a computer model commonly used to forecast urban water demands (IWR-MAIN) is evolving to incorporate more complex models of water use.

Another feature of water demand forecasts is that they have different time horizons and that these different time horizons reflect, in part, the different uses to which water demand forecasts are put. These may be roughly categorized into short-, medium-, and long-term forecasts (3,6). Short-run forecasts are conducted by water utility managers and water supply agencies to anticipate daily peak loads and other seasonal features of water use. Medium-term forecasts often examine user group investment decisions (such as the locations of new housing developments and manufacturing facilities) to establish anticipated changes in water demands during the next 5–10 years. These forecasts are then used to guide water agency investment decisions. Finally, long-term forecasts can investigate the impacts of structural and technological changes in the economy (such as the shift in employment from manufacturing to service-based industries) as well as the impacts of major policy changes (such as water quality legislation).

CASES OF WATER DEMAND FORECASTING

As mentioned earlier, water demand forecasts can be conducted on a variety of geographic scales. This section draws on chapter 11 of Renzetti (5) and considers examples of water demand forecasts conducted on municipal, regional, and national scales. Boland (1) summarizes a forecasting exercise where the IWR-MAIN water demand model was combined with predictions of climate change to predict water use in Washington, DC. Boland combines the predictions of five climate change models with forecasts of other explanatory variables such as prices, incomes, and housing stock. These are then fed into IWR-MAIN, and the model provides predicted water use levels for 2030 under different climate and policy scenarios. In the absence of any change in climate patterns, summer water use in the region is expected to grow 100% by 2030. Once the impacts of climate change are incorporated, the predicted growth ranges from 74% to 138%, assuming no change in policies. However, implementing relatively modest nonprice conservation efforts (such as adopting a revised plumbing code) reduces the range of forecasted growth in summer water use to between 40 and 92%. Adding a 50% increase in real water prices reduces expected water use growth even further: predicted growth in summer water use between 1985 and 2030 in this case ranges from 26.6% to 73.6%.

Planners often also forecast the water demands of major water-using sectors within a region. Representative of this type of exercise is the California Department of Water's recent effort to forecast urban, agricultural, and in-stream water use in California to the year 2020 (7). Urban water use is assumed to be the product of population growth and changes in per capita water use the latter is modeled as determined by a number of factors such as income, water prices, and conservation measures. Statewide urban per capita water use in 1995 was 229 U.S. gallons/day; the forecasted level of per capita water use in 2020 is 243 gallons/day without new conservation measures and 215 gallons/day using new conservation measures. The California Department of Water adopts a sophisticated and multifaceted approach to forecasting irrigation water demands. Each crop's aggregate irrigation water use is calculated as the product of the crop's water "requirements" and the statewide irrigated acreage for the crop. Individual crop's water requirements are estimated using agronomic and climate data. Expected statewide acreage for each crop is a function of forecasted market conditions and expected government policies. The final type of water use to be factored into the Californian forecasts is in-stream needs. It turns out that these are largely determined by legislation and future climate conditions. After summing across sectors, total Californian water use in 1995 was 79,490 thousand acre-feet and, assuming a continuation of past climatic conditions, total Californian water use in 2020 is forecast at 80,500 thousand acre-feet. The prediction of constant total water use masks the fact that several sectors and regions are expected to exhibit different rates of growth. For example, total projected urban water use will rise from 8770 thousand acre-feet in 1995 to 12,020 thousand acre-feet in 2020, and agricultural water use is expected to decline from 33,780 thousand acre-feet to 31,500 thousand acre-feet for the same period.

Finally, there have been considerable efforts expended to forecast water demand growth on a national and even global scale. Gleick (8) critically surveys researchers' efforts to project global water use into the twenty-first century. Gleick points out that early projections were essentially extrapolations of current water use patterns under 'business as usual' assumptions regarding technology, water pricing, and water use efficiency. The extrapolations are then based on projections of expected growth in population, agricultural, industrial, and energy output. Gleick contends, however, that almost all of the projections based on fixed-coefficient models of water use have significantly overstated actual water use in the past because they cannot anticipate the extent to which water's growing scarcity induces technological and institutional innovation, changes in sectoral composition, and slowing rates of population growth. For example, Gleick compares a number of projections of global water use in 2000 with actual recorded water use and finds that actual withdrawals in the mid-1990s were only half of what most forecasters in the 1960s and 1970s projected.

A representative example of this type of effort is Seckler et al. (4). The authors project per capita and total withdrawals for 118 countries for the period 1990–2025. Projections of per capita residential withdrawals are based on population or GDP growth, and per capita irrigation water use is assumed constant ("business as usual") or declines due to improved technology ("high efficiency"). The impact of improved efficiency is dramatic. In the first scenario, global irrigation water use grows 60% for the period; in the second scenario, irrigation water use grows by only 13%. Similarly, global water withdrawals increase by 56% in the "business as usual" scenario and by 26% in the 'high efficiency' scenario.

CONCLUSIONS

The rapid growth in water demands that has been observed in many regions and sectors and the costs of servicing these demands highlight the need for accurate water demand forecasts. This article has demonstrated that researchers and utility operators have developed techniques to conduct forecasts for different time periods and different geographic scales. Historically, the fixed-coefficient method of explaining and predicting water demands has dominated, but experience with this model has demonstrated that it neglects many factors (such as prices, incomes and, water conservation programs) that influence demands. This shortcoming has typically led fixed-coefficient models to overstate expected water demand growth. It is expected that more sophisticated models that account for a wider range of factors that influence water use and that allow for feedback between water use and economic conditions will provide more accurate forecasts of water demands.

BIBLIOGRAPHY

1. Boland, J. (1997). Assessing urban water use and the role of water conservation measures under climate uncertainty. In: *Climate Change and Water Resources Planning Criteria*. K. Frederick, D. Major, and E. Stakhiv (Eds.). Kluwer Academic, Dordrecht, pp. 157–171.
2. Dziegielewski, B. (1996). Long-term forecasting of urban water demands. In: *Marginal Cost Rate Design and Wholesale Water Markets*. D. Hall (Ed.). *Advances in the Economics of Environmental Resources* vol. 1, JAI Press, Greenwich, Connecticut, pp. 33–47.
3. Billings, R.B. and Jones, C.V. (1996). *Forecasting Urban Water*. American Water Works Association, Demand Denver.
4. Seckler, D., Barker, R., and Amarasinghe, U. (1999). Water scarcity in the twenty-first century. *Int. J. Water Resour. Dev.* **15**(1/2): 29–42.
5. Renzetti, S. (2002). *The Economics of Water Demands*. Kluwer Academic, Norwell, Massachusetts.
6. Baumann, D. (1998). Forecasting urban water use: theory and principles. In: *Urban Water Demand Management and Planning*. D. Baumann, J. Boland and W.M. Hanemann (Eds.). McGraw-Hill, New York, pp. 77–94.
7. California Department of Water Resources (1998). *The California Water Plan Update*. Bulletin 160–98, Sacramento, California.
8. Gleick, P. (2000). *The World's Water 2000–2001*. Island Press, Washington, DC.

REMOTE SENSING AND GIS APPLICATION IN WATER RESOURCES

SANJAY KUMAR JAIN
National Institute of Hydrology
Roorkee, India

Water resources are essential for human life, agriculture, and hydroelectric power generation. Remote sensing and the Geographic Information System (GIS) help in creating an appropriate information base for efficiently managing these resources. The synoptic view provided by satellite remote sensing and the analytical capability provided by the GIS offer a technologically appropriate method for studying various features related to land and water resources.

Remote sensing is a tool that permits accurate and real-time evaluation, continuous monitoring or surveillance, and forecasts of inland water resources. Remote sensing systems are used to observe the earth's surface from different levels of platforms, such as satellites and aircraft, and make it possible to collect and analyze information about resources and environment across large areas. Remote sensors record electromagnetic energy reflected or emitted from the earth's surface. Different kinds of objects or features such as soils, vegetation, and water reflect and emit energy differently. This characteristic makes it possible to measure, map, and monitor these objects and features using satellite or aircraft-borne remote sensing systems. Satellite imagery offers a number of advantages over conventional survey techniques:

- areal synoptic coverage (gives areal information as against point information through conventional techniques)
- repetitive global coverage (for monitoring change)
- real-time processing
- sensing of surrogates rather than the desired specific observation
- multispectral coverage
- more automation, less human error

For many water related studies, remote sensing data alone are not sufficient; they have to be merged with data from other sources. Hence a multitude of spatially related (i.e., climatic and geographic) data concerning rainfall, evaporation, vegetation, geomorphology, soils, and rocks have to be considered. In addition, information is also required such as locations and type of tube wells and rain and river gauges. Thus, the fast storage, retrieval, display, and updating of map contents are important functions. A system that can store data, select and classify stations, and perform mathematical and sorting operations is called a database, and information can be extracted from it for a given purpose. If this information can also be displayed in the form of maps, we can speak of geographic information. So this complete set of information forms the Geographic Information System (GIS). The GIS is an effective tool for storing, managing, and displaying spatial data (Fig. 1) often encountered in hydrology and water resources management studies. The GIS technology integrates common database operations, such as queries and statistical analysis, with the unique visualization and benefits of geographic analysis that maps and spatial databases offer. The use of remote sensing data in the GIS is shown in Fig. 2.

One of the capabilities of the GIS most important to water resources management studies is describing the topography of a region. A digital elevation model (DEM) provides a digital representation of a portion of the earth's terrain across a two-dimensional surface. DEMs have proved to be a valuable tool for topographic parameterization of hydrologic models, especially for drainage analysis, hillslope hydrology, watersheds, groundwater flow, and contaminant transport.

Problems of water resources development and efficient use such as frequent floods and droughts, waterlogging and salinity in command areas, an alarming rate of reservoir sedimentation due to deforestation, and deteriorating water quality and environment are varied and numerous in nature. All must be tackled by systematic approaches involving a judicious mix of conventional methods and remotely sensed data in the GIS environment. The availability of remotely sensed data and the use of the GIS has provided significant impetus to hydrologic analysis and design and their use in water resources

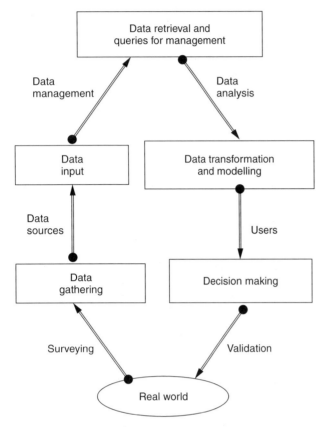

Figure 1. The GIS, a management tool.

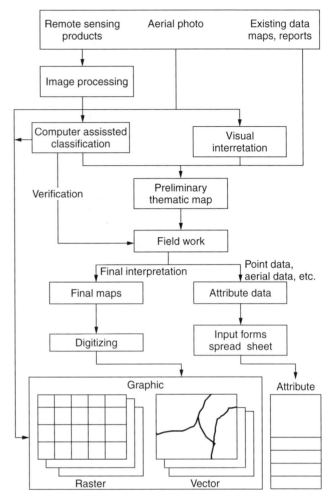

Figure 2. Remote sensing products in a GIS.

planning and development. Remote sensing and GIS techniques have been extensively used in various areas of water resources development and management, such as land use/cover classification, precipitation, snow cover, floodplain management, command area studies, waterlogging and soil salinity, sedimentation in reservoirs, and hydrologic modeling, which are defined below in brief.

LAND USE/LAND COVER CLASSIFICATION

Multispectral classification of land cover types was one of the first well-established remote sensing applications for water resources. Numerous investigators have used a classification of land cover from satellite data as input to various water resources studies. Land use features can be identified, mapped, and studied on the basis of their spectral characteristics. Healthy green vegetation has considerably different characteristics in the visible and near-infrared regions of the spectrum, whereas dry bare soil has a relatively stable reflectance in both regions of the spectrum. Water shows very low reflectance in the visible part of the spectrum and almost no reflectance in the infrared part of the spectrum. Thus, by using multispectral data suitably, different ground features can be differentiated from each other, and thematic maps depicting land use can be prepared from satellite data. Land use data devised from remote sensing are an ideal database for distributed hydrologic modeling changes. In many hydrologic models, the parameters of vegetation related model components (e.g., for interception and evapotranspiration) are chosen, dependent on land use.

PRECIPITATION STUDIES

Precipitation, one of the driving forces in the hydrologic cycle, is a prime candidate variable for remote sensing measurement. There are many remote sensing possibilities, including ground based radar, visible and thermal infrared satellite imagery, and microwave satellite data. The use of ground based radar has been successful especially when used with an integrated rain-gauge network and in areas of low relief. Visible and thermal infrared techniques using measurements of cloud top reflectance and temperature have been used in a variety of ways by meteorologists and other scientists to estimate monthly, daily, and storm precipitation totals. Visible and infrared measurements represent observations only of the upper surface of clouds. In contrast, it is often believed that microwave radiation is not affected by the presence of clouds. Measuring rain from a spaceborne radar is very attractive because it can discriminate in range (altitude when operated from space), and radar reflectivity is directly related to rainfall rate.

SNOW COVER STUDIES

Measurement of the area of a basin covered by snow is similar to land cover measurement. Conventional methods have limitations in monitoring the snow covered area in mountainous basins because of inaccessibility.

Very little information on snow is collected regularly in mountainous areas, so remote sensing remains the only practical way of obtaining some relevant information on the snow cover in the large number of mountain basins. At present the visible, near IR, and thermal IR data from various satellite (Landsat, IRS, NOAA) are being used operationally for mapping the areal extent of snow cover in mountainous basins. Visible and near-infrared wavelengths do not penetrate far into the snowpack, so they provide information mainly about the surface of the snowpack. Microwave remote sensing is promising because of its ability to penetrate the dry snowpack and its capability of acquiring data in cloudy or nighttime conditions. Snow cover data extracted by satellite remote sensing is immediately usable in snowmelt runoff models. In addition to the extent of snow cover, satellite data are useful in computing snow water equivalent from microwave data.

SOIL MOISTURE ASSESSMENT

Remotely sensed data have great potential for providing areal estimates of soil moisture rather than point measurements. Quantitative assessment of soil moisture regime is essential for water balance models, irrigation scheduling, crop management, and surface and subsurface flow predictions. Remote sensing of soil moisture can be accomplished to some degree in all regions of the electromagnetic spectrum. The use and application of remotely sensed data for soil moisture depends on measuring the electromagnetic energy that is reflected or emitted from the surface. Thus, either direct observations of the soil by remote sensors or indirect estimates using plant condition are useful in soil moisture observations. The soil moisture beneath microwave remote sensing can directly measure the dielectric properties of the earth's surface, which in turn is strongly dependent on moisture content.

FLOODPLAIN MANAGEMENT

Reliable data on river morphology, river meandering, the extent of flooding, and duration are required for proper planning of flood control projects. In conventional methods of flood risk zoning, the flood discharge is routed through a river reach to estimate the likely inundation due to spilling over banks/embankments based on topographic contour maps available and the configuration of the river geometry obtained from land surveys. Continuous availability of satellite based remote sensing data has made understanding the dynamics of floods much easier. Satellite remote sensing techniques provide wide area coverage, repetitiveness, and consistency, which enables collecting reliable information on all major floods.

Remote sensing can provide information on flood inundated areas for different magnitudes of floods, so that the extent of flooding in floodplains can be related to flood magnitude. The duration of flooding can be estimated with the help of multiple coverage satellite imagery of the same area within 2–3 days by satellites. High-resolution satellite data provide information on the floodplain and effectiveness of flood control works. The extent of inundation for specific flood return periods can be estimated. Using close contour information, the extent of inundation for a given elevation can be estimated, this is a vital input for risk zone mapping.

The GIS provides a broad range of tools for determining the area affected by floods and for forecasting areas that are likely to be flooded due to the high water level in a river. Spatial data stored in the digital database of the GIS, such as a digital elevation model (DEM), can be used to predict future floods. The GIS database may also contain agricultural, socioeconomic, communication, population, and infrastructural data. This can be used in conjunction with flooding data to adopt an evacuation strategy, rehabilitation planning, and damage assessment in a critical flood situation.

MAPPING AND MONITORING OF WATERSHEDS

Proper planning of watersheds is essential for conservation of water and land resources and their productivity. Characterization and analysis of watersheds are a prerequisite for this. Watershed characterization involves measuring parameters of geologic, hydrogeologic, geomorphological, hydrologic, soil, and land cover/land use. Remote sensing via aerial and spaceborne sensors can be used for watershed characterization, assessing management requirements, and periodic monitoring. The various physiographic measurements that can be obtained from remotely sensed data are size, shape, topography, drainage pattern, and landforms. Quantitative analysis of drainage networks enables relationships among different aspects of the drainage pattern of a basin to be formulated as general laws and to define certain useful properties/indexes of the drainage basin in numerical terms. The laws of stream numbers, stream length, and stream slopes can be derived from measurements made in the drainage basin. Remote sensing along with ground based information in the GIS mode can used for broad and reconnaissance level interpretations of land capability classes, irrigation suitability classes, potential land users, responsive water harvesting areas, monitoring the effects of watershed conservation measures, correlation for runoff and sediment yields from different watersheds, and monitoring land use changes and land degradation.

COMMAND AREA STUDIES

Water management in command areas requires serious attention in view of the disappointing performance of our irrigation projects, despite huge investments. The command area is the total area lying between drainage boundaries, which can be irrigated by a canal system.

Remote sensing can play a useful complementary role in managing the land and water resources of command areas to maximize production. Management of water supplies for irrigation in command areas is a critical problem to tackle with limited quantities. This requires information on total demand and the distribution of demand for irrigation in command areas. Moreover, the vastness of the areas involved, time constraints, and yearly changes demand fast inventories of conditions. As more area is brought under irrigation, crop monitoring also becomes essential for estimating agricultural production and efficient planning of water management. In all these, remote sensing can be looked upon as an aid in planning and decision making. The usefulness of remote sensing techniques in inventorying irrigated lands; identifying crop types, their extent, condition; and production estimation has been demonstrated in various investigations. Conjunctive use planning of surface and groundwaters can be done using remotely sensed information on surface water in conjunction with ground based data on groundwater availability. This would permit development of conjunctive use models for water allocations in a GIS environment.

WATERLOGGING AND SOIL SALINITY

Waterlogging and soil salinity are some of the major land degradation processes that restrict the economic and efficient use of soil and land resources in command areas. To assess waterlogging in command areas, multispectral and multitemporal remote sensing data are very useful. The satellite data thus provide a quick and more reliable delineation of waterlogged areas and standing water. The spatial distribution of soil affected by "positional waterlogging" (i.e., that due to its location in the landscape) can be modeled from digital topographic data using the concept of contributing area. This waterlogging depends on two topographic factors: (1) the local slope angle and (2) the drainage area. The probability of waterlogging increases with the contributing drainage area and decreases with increasing local slope angle. The waterlogging phenomenon is related to topography, so digital terrain modelling (DTM) can aid in detecting waterlogged areas. DTM provides information regarding slope and aspect, which in turn provide information about the areas that are susceptible to waterlogging.

One of the common practices for observing waterlogged areas in command areas is to take observations in existing open wells at regular intervals, twice a year in the pre- and postmonsoon seasons. Data are also collected on water quality. The information thus collected is used to draw hydrographs and the depth of the water table to prepare maps subsequently. So preparation of these maps can help a lot in identifying waterlogged zones in a command area. Using field data, which is available as point data, groundwater depth distributional maps can be prepared in the GIS. With the help of these maps, shallow GW areas, areas susceptible to waterlogging, can be identified. The areas falling within the 0–1.5 m range generally indicate waterlogged or salt-affected patches depending primarily on the soil characteristics, particularly texture.

RESERVOIR SEDIMENTATION

Most common conventional techniques for sedimentation quantification are (1) direct measurement of sediment deposition by hydrographic surveys and (2) indirect measurement of sediment concentration by the inflow–outflow method. Both methods are laborious, time-consuming, and costly and have their own limitations. Sampling and measurement of suspended sediments is a tedious and expensive program for either *in situ* or laboratory work.

The introduction of remote sensing techniques in the recent past has made it very cheap and convenient to quantify sedimentation in a reservoir and to assess its distribution and deposition pattern. Remote sensing techniques, offering data acquisition over a long time period and broad spectral range, are superior to conventional methods for data acquisition. The advantage of satellite data over conventional sampling procedures include repetitive coverage of a given area every 16–22 days, a synoptic view which is unobtainable by conventional methods, and almost instantaneous spatial data across the areas of interest. Remote sensing techniques provide a synoptic view of a reservoir in a form very different from that obtained by surface data collection and sampling.

WATER QUALITY STUDIES

Recently, the alarming proportions of water quality deterioration have necessitated rapid monitoring for efficient checks to prevent further deterioration and to cleanse our polluted water resources. Moreover, surveillance of water quality is an important activity for multiple uses such as irrigation and water supply. Water quality is a general term used to describe the physical, chemical, thermal, and/or biological properties of water.

A combination of ground (water) and remote sensing measurements are required to collect the data necessary to develop and calibrate empirical and semiempirical models and validate the more physically based models. Water samples analyzed for substances of interest (i.e., suspended sediment, chlorophyll) should be collected at the same time (or on the same day) that the remote sensing data are collected. Location of sample sites should be determined by GPS (or other available technique) so that the correct data (pixel information) can be extracted from remote sensing for comparison. Remote sensing applications to water quality are limited to measuring those substances or conditions that influence and change the optical and/or thermal characteristics of the apparent surface water properties. Suspended sediment, chlorophyll (algae), oil, and temperature are water quality indicators that can change the spectral and thermal properties of surface waters and are most readily measured by remote sensing techniques.

GROUNDWATER STUDIES

A remote sensing system is quite helpful in groundwater exploration because remotely sensed data provide a synoptic view of high observational density. The common current remote sensing platforms record features on the surface. Most of the information for groundwater, as yet, has to be obtained by qualitative reasoning and semiquantitative approaches. Remotely sensed information is often surrogate and has to be merged with geohydrologic data to become meaningful. Vegetation can be used as an indicator if local knowledge is available and the types can be identified on the satellite data. Apart from the contribution that remote sensing can make to understanding regional hydrogeology—necessary for managing groundwater resources—perhaps the strongest application for management is the evaluation of recharge, groundwater drafts for irrigation, and identification of flow systems in areas where there is a paucity of geohydrologic data. Surface conditions, soils, weathered zones, geomorphology, and vegetation determine recharge, suitability for artificial recharge, and soil and water conservation measures that can affect recharge.

Groundwater vulnerability to pollution is also directly related to surface conditions. Indexing methods for group depth to water table, net recharge, topography, impact of vadose zone media, and hydraulic conductivity of the aquifer (leading to the acronym DRASTIC) into a relative ranking scheme use a combination of weights and ratings to produce numerical values.

HYDROLOGIC MODELING

In the early days, the GIS was used mainly as a hydrologic mapping tool. Nowadays, it plays a more important role in hydrologic model studies. Its applications span a wide range from sophisticated analyses and modeling of spatial data to simple inventories and management tools. The GIS has evolved as a highly sophisticated database management system to put together and store the voluminous data typically required in hydrologic modeling. The application of the GIS has enhanced the capacity of models in data management, parameter estimation, and presentation of model results, but the GIS cannot replace hydrologic models in solving hydrologic problems.

Due to its data handling and manipulation capabilities, the GIS is increasingly being used as an interface and data manager for hydrologic models. There are four levels of linkage of hydrologic model with the GIS. These levels vary from considering the GIS and the model as separate systems to fully integrating the model and the GIS.

One typical application of a GIS in watershed analysis is predicting the spatial variability of surface erosion from spatial data sets obtained from maps of the vegetative cover, soils, and slope of the area. Solutions of a surface erosion (soil loss) prediction model, for example, the universal soil loss equation (USLE) or its modifications, combine spatial data sets, their derivatives, and other information necessary to predict the spatial variability of surface erosion on a watershed. This analysis can determine areas of potentially severe surface erosion and provide an initial step in the appraisal of surface erosion problems.

GIS FOR DECISION MAKING

The GIS is derived from multiple sources of data of different levels of accuracy. Though a single piece of data can be assigned an accuracy value, information derived from multiple sources of inaccurate data can also be assigned a level of accuracy. In any pictorial representation of data, uncertainty can be brought in as one of the dimensions to guide final decision making. Any decision today has to depend on a variety of factors, which are available in an information system like the GIS. However, weighting as well as the proper use of such data is still problematic.

The decision support system (DSS) can be designed as an interactive, flexible, and adaptable computer based information system (CBIS), especially developed to support the solution of a management problem for decision making. It uses data, provides an easy interface, and allows for decision-makers' own insights to illustrate the objectives and information characteristics of the various levels of decision making. DSS provides a framework for incorporating analytical modeling capabilities with a database to improve the decision-making process. Currently available DSS provides primarily for well-structured problems. It are also being used for decision making in water resources planning and management.

CONCLUSIONS AND FUTURE NEEDS

Many water resources applications can profit from the use of remote sensing and the GIS. The reason for adopting GIS technology is that it allows displaying spatial information in integrative ways that are readily comprehensible and visual. Remote sensing is now being widely regarded as a layer in the GIS. Although remote sensing is a specialized technique, it is now being accepted as a basic survey methodology and as a means of providing data for a resource database. The GIS provides a methodology by which data layers can be interrelated to arrive at wider decisions (1).

The research needs in the area of water resources are as follows:

- A first and very important aspect is data availability and compatibility in any GIS related study. Spatial information required for water resources studies should be readily available for timely execution. The data banks should provide digitized maps and their spatial data compatible with various systems. Such data availability could significantly speed up the analysis.
- One difficult task in incorporating the GIS in water resources modeling is interfacing water system models with the GIS. Automation of interfacing tasks is one of the areas to be researched in incorporating the GIS and available models.
- The recent development of the decision support system (DSS) to assist in water resources decision making holds the key for integrating the GIS and water resources models.

- Another area of potential research to enhance the modeling process further is the integration of expert systems and the GIS. Expert GIS systems can be used to provide regulatory information by linking regulatory facts stored in a database to sites located in a GIS through an expert system query interface.
- Further research is needed for comparing the GIS packages available on the market and their positive and negative aspects, providing check lists for GIS users.

BIBLIOGRAPHY

1. Tsihrintzix, V.A., Hamid, R., and Fuentes, H.R. (1996). Use of GIS in water resources: a review. *Water Resour. Manage.* **10**: 251–277.

READING LIST

Schultz, G.A. and Engment, E.T. (Eds.). (2000). *Remote Sensing in Hydrology and Water Management*. Springer Verlag, New York.

Rango, A. (1990). Remote sensing of water resources: accomplishments, challenges and relevance to global monitoring. *Proc. Int. Symp. Remote Sensing Water Resour.*, Enschede, The Netherlands, August 20–24.

GLOBALIZATION OF WATER

A.Y. Hoekstra
UNESCO–IHE Institute for
Water Education
Delft, The Netherlands

WHY WATER IS A GLOBAL RESOURCE

People in Japan can affect the hydrological system in the United States. People in Europe can affect regional water systems in Thailand or Brazil. There are basically two mechanisms that make global connections between seemingly local water systems. First, the climate system connects different places on earth, because evaporation in one place results in precipitation in another place. The climate system is inherently global: local emissions of greenhouse gases contribute to a changing global climate, thus affecting temperature, evaporation, and precipitation patterns elsewhere. In this way, human activities in the economic centers of the world affect base and peak flows in rivers throughout the world. There is, however, a second mechanism through which people can affect water systems in other parts of the world. A European consumer of Thai rice raises rice demand in Thailand and subsequently the use of water for rice irrigation in Thailand. Globally, roughly one-fifth of the water used in agriculture is applied in areas used for producing export commodities. This fraction is increasing in line with the increase of global trade. Scientists have known for several decades that human activities within a river basin strongly impact on the water flows and water quality in such a river basin. It is more recent that scientists have started to realize that local water problems are often, in their roots, problems that cannot be solved locally or regionally, because the driving forces lay outside the region. The gradual disappearance of the Aral Lake in Uzbekistan–Kazakhstan, for instance, is directly linked to the global demand for cotton.

EFFICIENT USE OF THE GLOBAL WATER RESOURCES

According to international trade theory, countries with abundant water resources might have a comparative advantage in producing water-intensive goods if compared to water-poor countries. Since water resources are quite unevenly distributed over the world, one should not be surprised that under conditions of free trade water-intensive production processes get concentrated in the water-rich parts of the world. A water-scarce country can thus aim at importing products that require a lot of water in their production (water-intensive products) and exporting products or services that require less water (water-extensive products). This is called *import of virtual water* (as opposed to import of real water, which is generally too expensive) and will relieve the pressure on the nation's own water resources. For water-abundant countries an argument can be made for *export of virtual water*. Import of water-intensive products by some nations and export of these products by others results in international "virtual water flows."

The idea of achieving "global water use efficiency" through virtual water trade between countries is relatively new. Traditionally, water managers have focused on achieving "local water use efficiency" at user level and "water allocation efficiency" at river basin level. "Local water use efficiency" can be increased by creating awareness among the water users, by charging prices based on full marginal cost, and by stimulating water-saving technology. "Water allocation efficiency" can be improved by allocating the water within a catchment or river basin to those types of use where water creates the highest added value. A key question today is which level—the user, river basin, or global level—will be most relevant in increasing overall water use efficiency and reducing overall impact on the globe's water systems.

THE CONCEPT OF "VIRTUAL WATER"

The volume of virtual water "hidden" or "embodied" in a particular product is defined as the volume of water used in the production process of that product (1,2). Not only agricultural products contain virtual water: most studies to date have been limited to the study of virtual water in crops. Industrial products and services also contain virtual water. As an example of virtual water content, one often refers to the virtual water content of grains. It is estimated that for producing 1 kg of grain, grown under rain-fed and favorable climatic conditions, we need about 1–2 m^3 of water, which is 1000–2000 kg of water. For the same amount of grain, but growing in an arid country, where the climatic conditions are not favorable (high temperature, high evapotranspiration), we need up to 3000–5000 kg of water.

If one country exports a water-intensive product to another country, it exports water in virtual form. In this way, some countries support other countries in their water needs. Worldwide both politicians and the general public increasingly show interest in the pros and cons of "globalization" of trade. The tension in the debate relates to the fact that the game of global competition is played with rules that many see as unfair. Knowing that economically sound water pricing is poorly developed in many regions of the world, this means that many products are put on the world market at a price that does not properly include the cost of the water contained in the product. This leads to situations in which some regions in fact subsidize export of scarce water.

METHOD TO CALCULATE VIRTUAL WATER FLOWS

Calculation of the Virtual Water Content of a Product

The virtual water content of a product is a function of the water volumes used in the different stages of the production process. The virtual water content of a particular crop in a particular country can, for instance, be calculated on the basis of country-specific crop water requirements and crop yields (3):

$$\text{VWC}[n,c] = \text{CWR}[n,c]/Y[n,c] \qquad (1)$$

Here, VWC denotes the virtual water content (m^3/ton) of crop c in country n, CWR the crop water requirement (m^3/ha), and Y the crop yield (ton/ha). The crop water requirement can be calculated from the accumulated crop evapotranspiration ET_c (in mm/d) over the complete growing period. The crop evapotranspiration ET_c follows from multiplying the "reference crop evapotranspiration" ET_0 with the crop coefficient K_c:

$$ET_c = K_c \times ET_0 \qquad (2)$$

The crop coefficient accounts for the actual crop canopy and aerodynamic resistance relative to the hypothetical reference crop. The crop coefficient serves as an aggregation of the physical and physiological differences between a certain crop and the reference crop. The concept of "reference crop evapotranspiration" was introduced by FAO to study the evaporative demand of the atmosphere independently of crop type, crop development, and management practices. The only factors affecting ET_0 are climatic parameters. The reference crop evapotranspiration ET_0 is defined as the rate of evapotranspiration from a hypothetical reference crop with an assumed crop height of 12 cm, a fixed crop surface resistance of 70 s/m, and an albedo of 0.23. This reference crop evapotranspiration closely resembles the evapotranspiration from an extensive surface of green grass cover of uniform height, actively growing, completely shading the ground, and with adequate water (4). Reference crop evapotranspiration can be calculated on the basis of the FAO Penman-Monteith equation (4,5).

The calculation of the virtual water content of livestock products is a bit more complex. First, the virtual water content (m^3/ton) of live animals is calculated, based on the virtual water content of their feed and the volumes of drinking and service water consumed during their lifetime (6). Second, the virtual water content is calculated for each livestock product, taking into account the product fraction (ton of product obtained per ton of live animal) and the value fraction (ratio of value of one product from an animal to the sum of the market values of all products from the animal).

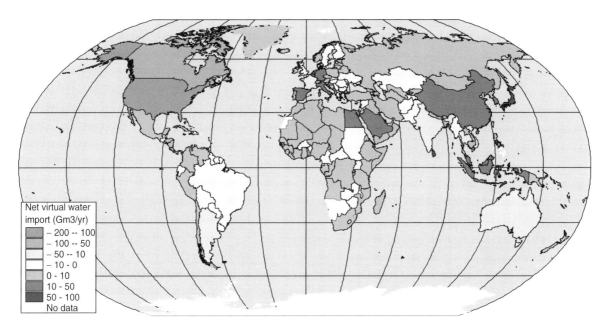

Figure 1. National virtual water balances over the period 1995–1999.

Table 1. Water Withdrawals, Virtual Water Import and Export, Water Scarcity, Water Self-sufficiency, and Water Dependency of Nations (1995–1999)

Country	Population	Water Availability, 10^6 m^3/yr	Water Withdrawal, 10^6 m^3/yr	Gross Virtual Water Import, 10^6 m^3/yr	Gross Virtual Water Export, 10^6 m^3/yr	Net Virtual Water Import, 10^6 m^3/yr	Water Scarcity, %	Water Self-sufficiency, %	Water Dependency, %
Afghanistan	25765766	65000	23261	58.5	287.5	−229	36	100	0
Albania	3387574	41700	1714	277.7	14.4	263.2	4	87	13
Algeria	29959010	14320	6074	9810.6	6.9	9803.7	42	38	62
Angola	12771448	184000	343	173.9	5.5	168.4	0	67	33
Argentina	36577450	814000	29072	1486.9	46755.4	−45268.4	4	100	0
Armenia	3798845	10529	2954	316.2	5.5	310.6	28	90	10
Australia	18963804	492000	23932	1011	30130.3	−29119.3	5	100	0
Austria	8095446	77700	2112	1281	976.1	304.9	3	87	13
Azerbaijan	7979460	30275	17247	1004.4	30	974.4	57	95	5
Bahrain	666956	116	299	137.4	0.2	137.1	258	69	31
Bangladesh	128837760	1210644	79394	8304.6	2562.6	5742	7	93	7
Belarus	10039496	58000	2789	1255.9	35	1220.9	5	70	30
Belgium-Lux	10227060	21400	8980	14412.4	2497	11915.4	42	43	57
Benin	6112575	24800	250	472.6	1077.8	−605.3	1	100	0
Bhutan	782229	95000	420	26.5	0	26.5	0	94	6
Bolivia	8139894	622531	1387	674.8	1732.1	−1057.3	0	100	0
Bosnia Herzegovina	3865576	37500	1354	238	63.8	174.2	4	89	11
Brazil	168220660	8233000	59298	23161.6	32161.8	−9000.2	1	100	0
Bulgaria	8213543	21300	10498	288.4	759.7	−471.4	49	100	0
Burkina Faso	11005226	12500	780	68.6	973.4	−904.7	6	100	0
Burundi	6677800	3600	234	3.7	0.3	3.4	6	99	1
Cambodia	11755836	476110	4091	130.1	27.7	102.4	1	98	2
Cameroon	14557762	285500	985	175.3	187.9	−12.6	0	100	0
Canada	30498614	2902000	45974	4814.4	59308.4	−54494	2	100	0
Central African Republic	3657263	144400	22	1.8	2.9	−1.1	0	100	0
Chad	7492965	43000	234	1.4	0	1.4	1	99	1
Chile	15013962	922000	12539	3262.6	1211.2	2051.4	1	86	14
China	1252042000	2896569	630289	30550.4	10114.9	20435.6	22	97	3
Colombia	41543956	2132000	10711	7535.4	865.2	6670.2	1	62	38
Comoros	544534	1200	13	39.3	0	39.3	1	25	75
Congo	2764600	832000	39	93.4	6.8	86.6	0	31	69
Costa Rica	3731672	112400	2677	1947.7	690	1257.7	2	68	32
Cote d'Ivoire	15580058	81000	931	773.5	83.5	690	1	57	43
Croatia	4395695	105500	1760	569.9	346	223.9	2	89	11
Cuba	11150144	38120	8204	1082.5	1304	−221.5	22	100	0
Czech Republic	10283004	13150	2566	1245.7	761.2	484.5	20	84	16
Denmark	5318089	6000	1267	1382.6	1843.6	−461	21	100	0
Djibouti	620352	300	8	109.5	0.2	109.3	3	7	93
Dominican Republic	8237523	20995	3386	731.8	2663.9	−1932.1	16	100	0
Ecuador	12409904	432000	16980	1594.2	2184.4	−590.2	4	100	0
Egypt	62782964	86800	68653	16937.1	901.6	16035.5	79	81	19
El Salvador	6155042	25230	1273	1142.3	94.6	1047.7	5	55	45
Eritrea	3988805	6300	304	74.8	0.3	74.6	5	80	20
Estonia	1388705	12808	163	631	100.4	530.6	1	24	76
Ethiopia	62782412	110000	2648	349.1	22.7	326.4	2	89	11
Fiji	802087	28550	69	174.6	0	174.6	0	28	72
Finland	5164368	110000	2478	918.9	1091.8	−172.9	2	100	0
France	58656600	203700	39959	9376.3	27051.4	−17675.1	20	100	0
Gabon	1198661	164000	128	100.7	0.7	100	0	56	44
Gambia	1263370	8000	32	319.3	164.2	155.1	0	17	83
Georgia	5188007	63330	3607	308.4	103	205.4	6	95	5
Germany	82109980	154000	47052	23260.4	9671.3	13589.1	31	78	22
Ghana	18875980	53200	520	671.2	217.5	453.8	1	53	47
Greece	10537058	74250	7759	3121.4	5088	−1966.6	10	100	0
Guatemala	11095762	111270	2005	1195.1	15536.6	−14341.5	2	100	0
Guinea Bissau	1174665	31000	110	8.3	5.4	2.9	0	97	3
Guyana	757015	241000	1642	67.9	226.6	−158.7	1	100	0
Haiti	7803032	14025	984	389	0	389	7	72	28
Honduras	6257825	95929	860	799.1	331.7	467.4	1	65	35
Hungary	10221682	104000	7641	635.6	4589.6	−3954	7	100	0
Iceland	277700	170000	153	64.8	1.6	63.2	0	71	29
India	997775760	1907760	645820	2413	34612.3	−32199.3	34	100	0
Indonesia	207029780	2838000	82773	21366.2	1139.2	20227	3	80	20
Iran	62762116	137510	72877	6623.1	803.4	5819.7	53	93	7
Iraq	22797032	96420	42702	1100.7	3.3	1097.4	44	97	3

Table 1. (Continued)

Country	Population	Water Availability, 10^6 m^3/yr	Water Withdrawal, 10^6 m^3/yr	Gross Virtual Water Import, 10^6 m^3/yr	Gross Virtual Water Export, 10^6 m^3/yr	Net Virtual Water Import, 10^6 m^3/yr	Water Scarcity, %	Water Self-sufficiency, %	Water Dependency, %
Ireland	3752276	52000	1129	945.8	201.9	743.9	2	60	40
Israel	6100032	1670	2041	5188.1	589.9	4598.2	122	31	69
Italy	57627528	191300	44372	19625.8	6762.1	12863.7	23	78	22
Jamaica	2604246	9404	409	392.8	137.3	255.5	4	62	38
Japan	126624200	430000	88432	59632	188.4	59443.6	21	60	40
Jordan	4742815	880	1016	4536	55	4481	115	18	82
Kazakhstan		109610	35008	41.8	7876	−7834.2	32	100	0
Kenya	29402552	30200	1576	970.2	169.7	800.5	5	66	34
Korea, Democratic People's Republic	22141004	77135	9024	643	2.2	640.8	12	93	7
Korea, Republic	46839720	69700	18590	22582.6	69	22513.6	27	45	55
Kuwait	1925635	20	445	497.8	0.1	497.7	2227	47	53
Kyrgyzstan	4844973	46450	10080	192.5	145.2	47.3	22	100	0
Lao	5159165	333550	2993	94.2	1.7	92.5	1	97	3
Latvia	2408205	35449	293	301.4	53.4	248	1	54	46
Lebanon	4267969	4837	1372	776.2	29.4	746.8	28	65	35
Liberia	3046804	232000	107	67.5	1.7	65.7	0	62	38
Libya	5176657	600	4811	789.1	45.4	743.7	802	87	13
Lithuania	3531820	24900	267	500.4	383.9	116.5	1	70	30
Macedonia	2020714	6400	847	149	97.9	51.1	13	94	6
Madagascar	15057966	337000	14970	320	131.7	188.3	4	99	1
Malawi	10096722	17280	1005	25.9	786.6	−760.8	6	100	0
Malaysia	22724518	580000	9016	11508.3	1255.9	10252.4	2	47	53
Mali	10588286	100000	6930	79.8	14.9	65	7	99	1
Mauritania	2579964	11400	1698	375.7	0.6	375.1	15	82	18
Mauritius	1173176	2210	612	564.7	274.9	289.7	28	68	32
Mexico	96615488	457222	78219	24361.3	15374.7	8986.7	17	90	10
Moldova Republic	4291104	11650	2308	82.8	455.5	−372.7	20	100	0
Mongolia	2377183	34800	444	24.6	10.1	14.5	1	97	3
Morocco	28240226	29000	12758	5617.8	87.4	5530.4	44	70	30
Mozambique	17331232	216110	635	337	85.1	251.9	0	72	28
Myanmar	47134402	1045601	33224	21.1	3501.3	−3480.2	3	100	0
Nepal	22507210	210200	10177	47.6	19	28.6	5	100	0
Netherlands	15812200	91000	7944	35002.3	5462.6	29539.7	9	21	79
New Zealand	3808760	327000	2111	1000.6	113.1	887.5	1	70	30
Nicaragua	4940828	196690	1300	583.6	333.1	250.5	1	84	16
Niger	10478080	33650	2187	309.5	107.9	201.6	6	92	8
Nigeria	123837060	286200	8004	5796.4	934.4	4862	3	62	38
Norway	4461300	382000	2185	2214.7	11.1	2203.6	1	50	50
Oman	2350640	985	1350	1228.1	119.6	1108.5	137	55	45
Pakistan	134871900	233770	169384	2547.1	2556.8	−9.8	72	100	0
Panama	2810118	147980	824	539.8	331.1	208.7	1	80	20
Papua New Guinea	5006703	801000	75	48.9	20.4	28.5	0	72	28
Paraguay	5358929	336000	489	343	8768.1	−8425.1	0	100	0
Peru	25230198	1913000	20132	5566.3	143.5	5422.8	1	79	21
Philippines	74178100	479000	28520	8206.7	7242	964.7	6	97	3
Poland	38654642	61600	16201	4210.1	452.4	3757.7	26	81	19
Portugal	10028200	77400	11263	6758	529.9	6228.1	15	64	36
Qatar	563710	53	294	59.3	0	59.3	554	83	17
Romania	22469358	211930	23176	877.7	2701.2	−1823.5	11	100	0
Russian Federation	146180880	4507250	76686	14534.5	12079.6	2454.9	2	97	3
Rwanda	8304804	5200	76	93	0.2	92.9	1	45	55
South Africa	42043988	50000	15306	6927.6	2558.3	4369.3	31	78	22
Saudi Arabia	20239432	2400	17320	11313.3	435.2	10878.1	722	61	39
Senegal	9279048	39400	1591	2680.4	43.6	2636.8	4	38	62
Sierra Leone	4932139	160000	380	83.2	0.6	82.6	0	82	18
Singapore	3957913	600	211	3839.2	435.2	3404.1	35	6	94
Slovakia	5395677	50100	1818	386.6	977.4	−590.8	4	100	0
Slovenia	1986239	31870	762	1062.9	21.8	1041.1	2	42	58
Somalia	8480576	13500	3298	299.7	22.7	277.1	24	92	8
Spain	39415552	111500	35635	22124.6	5621	16503.6	32	68	32
Sudan	30534126	149000	37314	561.3	1712.3	−1151.1	25	100	0
Suriname	415105	122000	665	27.5	114.5	−86.9	1	100	0
Sweden	8864128	174000	2965	737.3	1577.2	−839.9	2	100	0
Switzerland	7145332	53500	2571	2098.9	162.5	1936.5	5	57	43
Syria	15798242	46080	19947	884.5	5263.2	−4378.6	43	100	0
Tajikistan	6138744	99730	11962	46.7	83.7	−37.1	12	100	0
Tanzania	32902714	91000	1996	1211.5	283.5	928.1	2	68	32

(*continued overleaf*)

Table 1. (*Continued*)

Country	Population	Water Availability, 10^6 m³/yr	Water Withdrawal, 10^6 m³/yr	Gross Virtual Water Import, 10^6 m³/yr	Gross Virtual Water Export, 10^6 m³/yr	Net Virtual Water Import, 10^6 m³/yr	Water Scarcity, %	Water Self-sufficiency, %	Water Dependency, %
Thailand	60275202	409944	87065	4098	50763.3	−46665.4	21	100	0
Togo	4392474	14700	166	851.6	214.8	636.8	1	21	79
Trinidad Tobago	1293248	3840	305	679.6	90.6	589	8	34	66
Tunisia	9448461	4560	2726	3925.1	57.8	3867.4	60	41	59
Turkey	64341266	231700	37519	10297.6	8244.3	2053.1	16	95	5
Turkmenistan	5057637	60860	24645	57.4	0.6	56.9	40	100	0
Uganda	21616208	66000	295	208.8	294.8	−86	0	100	0
Ukraine	49904874	139550	37523	468.9	6832.6	−6363.8	27	100	0
United Arab Emirates	2800073	150	2306	2109.4	418.1	1691.2	1538	58	42
United Kingdom	59481556	147000	9541	14204.1	15174.5	−970.4	6	100	0
Uruguay	3312629	139000	3146	821.8	3223.3	−2401.5	2	100	0
United States	278035840	3069400	479293	29264.3	180924.3	−151660	16	100	0
Uzbekistan	24394002	72210	58334	532.7	123.7	409	81	99	1
Venezuela	23705676	1233170	8368	6250.8	1325.2	4925.6	1	63	37
Viet Nam	77508750	891210	71392	153.8	18185.7	−18031.9	8	100	0
Yemen	17056736	4100	6631	1448.9	11.5	1437.4	162	82	18
Zambia	9872326	105200	1737	34.8	132.9	−98.1	2	100	0
Zimbabwe	12382668	20000	2612	115.8	633.3	−517.5	13	100	0

Source: Reference 3.

Calculation of International Virtual Water Flows and Drawing a National Virtual Water Balance

Virtual water flows between nations can be calculated by multiplying international product trade flows by their associated virtual water content. The latter depends on the water needs in the exporting country where the product is produced. Virtual water trade is thus calculated as

$$\text{VWT}[n_e, n_i, p, t] = \text{PT}[n_e, n_i, p, t] \times \text{VWC}[n_e, p, t] \quad (3)$$

in which VWT denotes the virtual water trade (m³/yr) from exporting country n_e to importing country n_i in year t as a result of trade in product p. PT represents the product trade (ton/yr) from exporting country n_e to importing country n_i in year t for product p. VWC represents the virtual water content (m³/ton) of product p in the exporting country.

The gross virtual water import (GVWI) to a country is the sum of all product-related virtual water imports. The gross virtual water export (GVWE) from a country is the sum of all product-related virtual water exports. The virtual water balance of country x for year t can be written as

$$\text{NVWI}[x, t] = \text{GVWI}[x, t] - \text{GVWE}[x, t], \quad (4)$$

where NVWI stands for the net virtual water import (m³/yr) to the country. Net virtual water import to a country has either a positive or a negative sign. The latter indicates that there is net virtual water *export* from the country.

GLOBAL VIRTUAL WATER FLOWS

The global volume of international virtual water flows is roughly 1000 billion m³/yr (7). For comparison, the total water use by crops in the world has been estimated at 5400 Gm³/yr. These figures show that a substantial volume of water serves the global market. Countries with a large virtual water export are the United States, Canada, Thailand, Argentina, and India. Large importers of virtual water are Europe (except France), Japan, China, and Indonesia.

National virtual water balances over the period 1995–1999 are shown in Fig. 1. Full data on virtual water imports and exports are provided in Table 1. Some countries, such as Brazil, Syria, Pakistan, Tajikistan, and Uganda, have net export of virtual water over the period 1995–1999, but net import of virtual water in one or more particular years in this period. There are also countries that show the reverse, such as the Philippines, the Russian Federation, Uzbekistan, Kyrgyztan, Mongolia, Nicaragua, and Mexico. Developed countries generally have a more stable virtual water balance than the developing countries.

Countries that are relatively close to each other in terms of geography and development level can have a rather different virtual water balance. While European countries such as The Netherlands, Belgium, Germany, Spain, and Italy import virtual water in the form of crops, France exports a large amount of virtual water. In the Middle East we see that Syria has net export of virtual water related to crop trade, but Jordan and Israel have net import. In Southern Africa, Zimbabwe and Zambia had net export in the period 1995–1999, but South Africa had net import. (It should be noted that the balance of Zimbabwe has recently turned due to the recent political and economic developments.) In regions of the former Soviet Union, countries such as Kazakhstan and the Ukraine have net export of virtual water, but the Russian Federation has net import.

Net virtual water flows between thirteen world regions in the period 1995–1999 are shown in Fig. 2. The largest virtual water flows have been indicated with arrows. Regions with a significant net virtual water import are

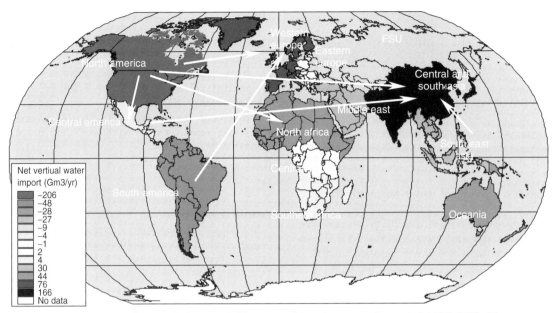

Figure 2. Virtual water balances of thirteen world regions over the period 1995–1999. The arrows show the largest net virtual water flows between regions (>20 Gm^3/yr).

Central and South Asia, Western Europe, North Africa, and the Middle East. Two other regions with net virtual water import, but less substantial, are southern Africa and Central Africa. Regions with substantial net virtual water export are North America, South America, Oceania, and Southeast Asia. Three other regions with net virtual water export, but less substantial, are the former Soviet Union, Central America, and Eastern Europe. North America is by far the biggest virtual water exporter in the world, while Central and South Asia is by far the biggest virtual water importer. Central and South Asia is the largest region in terms of population, so food demand is higher than in the other regions. This explains why the region is the biggest virtual water importer. The virtual water flows between countries within the region are also high, thus the countries within the region highly depend on both countries outside and countries within the region. Western Europe is the region with the biggest volume of internal virtual water flows. Besides, the volume appears to be rather stable over the years.

BIBLIOGRAPHY

1. Allan, J.A. (1997). *Virtual Water: A Long Term Solution for Water Short Middle Eastern Economies?* Occasional Paper 3, School of Oriental and African Studies (SOAS), University of London.
2. Hoekstra, A.Y. (1998). *Perspectives on Water: A Model-Based Exploration of the Future*. International Books, Utrecht, The Netherlands.
3. Hoekstra, A.Y. and Hung, P.Q. (2002). *Virtual Water Trade: a Quantification of Virtual Water Flows Between Nations in Relation to International Crop Trade*. Value of Water Research Report Ser. No. 11, UNESCO-IHE, Delft, The Netherlands.
4. Smith, M., Allen, R.G., Monteith, J.L., Perrier, A., Pereira, L.S., and Segeren, A. (1992). *Report on the Expert Consultation on Revision of FAO Methodologies for Crop Water Requirements*. FAO, Rome, Italy, 28–31 May 1990.
5. Allen, R.G., Pereira, L.S., Raes, D., and Smith, M. (1998). *Crop Evapotranspiration: Guidelines for Computing Crop Water Requirements*. FAO Irrigation and Drainage Paper 56, FAO, Rome, Italy.
6. Chapagain, A.K. and Hoekstra, A.Y. (2003). *Virtual Water Flows Between Nations in Relation to Trade in Livestock and Livestock Products*. Value of Water Research Report Ser. No. 13, UNESCO-IHE, Delft, The Netherlands.
7. Hoekstra, A.Y. (Ed.) (2003). Virtual water trade. In: *Proceedings of the International Expert Meeting on Virtual Water Trade*. Value of Water Research Report Ser. No. 12, UNESCO-IHE, Delft, The Netherlands.

WATER SCIENCE GLOSSARY OF TERMS

U.S. Geological Survey

Here's a list of water-related terms that might help you understand our site better. It is compiled from a number of sources and should not be considered an "official" U.S. Geological Survey water glossary. A detailed water glossary is kept by the Water Quality Association, and an **extremely detailed** water dictionary is offered by the Nevada Division of Water Planning.

A

acequia–acequias were important forms of irrigation in the development of agriculture in the American Southwest. The proliferation of cotton, pecans and green

---

This article is a US Government work and, as such, is in the public domain in the United States of America.

chile as major agricultural staples owe their progress to the acequia system.

acid—a substance that has a pH of less than 7, which is neutral. Specifically, an acid has more free hydrogen ions (H^+) than hydroxyl ions (OH^-).

acre-foot (acre-ft)—the volume of water required to cover 1 acre of land (43,560 square feet) to a depth of 1 foot. Equal to 325,851 gallons or 1,233 cubic meters.

alkaline—sometimes water or soils contain an amount of alkali (strongly basic) substances sufficient to raise the pH value above 7.0 and be harmful to the growth of crops.

alkalinity—the capacity of water for neutralizing an acid solution.

alluvium—deposits of clay, silt, sand, gravel, or other particulate material that has been deposited by a stream or other body of running water in a streambed, on a flood plain, on a delta, or at the base of a mountain.

appropriation doctrine—the system for allocating water to private individuals used in most Western states. The doctrine of Prior Appropriation was in common use throughout the arid west as early settlers and miners began to develop the land. The prior appropriation doctrine is based on the concept of "First in Time, First in Right." The first person to take a quantity of water and put it to Beneficial Use has a higher priority of right than a subsequent user. Under drought conditions, higher priority users are satisfied before junior users receive water. Appropriative rights can be lost through nonuse; they can also be sold or transferred apart from the land. Contrasts with Riparian Water Rights.

aquaculture—farming of plants and animals that live in water, such as fish, shellfish, and algae.

aqueduct—a pipe, conduit, or channel designed to transport water from a remote source, usually by gravity.

aquifer—a geologic formation(s) that is water bearing. A geological formation or structure that stores and/or transmits water, such as to wells and springs. Use of the term is usually restricted to those water-bearing formations capable of yielding water in sufficient quantity to constitute a usable supply for people's uses.

aquifer (confined)—soil or rock below the land surface that is saturated with water. There are layers of impermeable material both above and below it and it is under pressure so that when the aquifer is penetrated by a well, the water will rise above the top of the aquifer.

aquifer (unconfined)—an aquifer whose upper water surface (water table) is at atmospheric pressure, and thus is able to rise and fall.

artesian water—ground water that is under pressure when tapped by a well and is able to rise above the level at which it is first encountered. It may or may not flow out at ground level. The pressure in such an aquifer commonly is called artesian pressure, and the formation containing artesian water is an artesian aquifer or confined aquifer. See *flowing well*

artificial recharge—an process where water is put back into ground-water storage from surface-water supplies such as irrigation, or induced infiltration from streams or wells.

B

base flow—streamflow coming from ground-water seepage into a stream.

base—a substance that has a pH of more than 7, which is neutral. A base has less free hydrogen ions (H^+) than hydroxyl ions (OH^-).

bedrock—the solid rock beneath the soil and superficial rock. A general term for solid rock that lies beneath soil, loose sediments, or other unconsolidated material.

C

capillary action—the means by which liquid moves through the porous spaces in a solid, such as soil, plant roots, and the capillary blood vessels in our bodies due to the forces of adhesion, cohesion, and surface tension. Capillary action is essential in carrying substances and nutrients from one place to another in plants and animals.

commercial water use—water used for motels, hotels, restaurants, office buildings, other commercial facilities, and institutions. Water for commercial uses comes both from public-supplied sources, such as a county water department, and self-supplied sources, such as local wells.

condensation—the process of water vapor in the air turning into liquid water. Water drops on the outside of a cold glass of water are condensed water. Condensation is the opposite process of evaporation.

consumptive use—that part of water withdrawn that is evaporated, transpired by plants, incorporated into products or crops, consumed by humans or livestock, or otherwise removed from the immediate water environment. Also referred to as water consumed.

conveyance loss—water that is lost in transit from a pipe, canal, or ditch by leakage or evaporation. Generally, the water is not available for further use; however, leakage from an irrigation ditch, for example, may percolate to a ground-water source and be available for further use.

cubic feet per second (cfs)—a rate of the flow, in streams and rivers, for example. It is equal to a volume of water one foot high and one foot wide flowing a distance of one foot in one second. One "cfs" is equal to 7.48 gallons of water flowing each second. As an example, if your car's gas tank is 2 feet by 1 foot by 1 foot (2 cubic feet), then gas flowing at a rate of 1 cubic foot/second would fill the tank in two seconds.

D

desalinization—the removal of salts from saline water to provide freshwater. This method is becoming a more popular way of providing freshwater to populations.

discharge—the volume of water that passes a given location within a given period of time. Usually expressed in cubic feet per second.

domestic water use—water used for household purposes, such as drinking, food preparation, bathing, washing clothes, dishes, and dogs, flushing toilets, and watering lawns and gardens. About 85% of domestic water is delivered to homes by a public-supply facility, such as a county water department. About 15% of the Nation's population supply their own water, mainly from wells.

drainage basin–land area where precipitation runs off into streams, rivers, lakes, and reservoirs. It is a land feature that can be identified by tracing a line along the highest elevations between two areas on a map, often a ridge. Large drainage basins, like the area that drains into the Mississippi River contain thousands of smaller drainage basins. Also called a "watershed."

drip irrigation–a common irrigation method where pipes or tubes filled with water slowly drip onto crops. Drip irrigation is a low-pressure method of irrigation and less water is lost to evaporation than high-pressure spray irrigation.

drawdown–a lowering of the ground-water surface caused by pumping.

E

effluent–water that flows from a sewage treatment plant after it has been treated.

erosion–the process in which a material is worn away by a stream of liquid (water) or air, often due to the presence of abrasive particles in the stream.

estuary–a place where fresh and salt water mix, such as a bay, salt marsh, or where a river enters an ocean.

evaporation–the process of liquid water becoming water vapor, including vaporization from water surfaces, land surfaces, and snow fields, but not from leaf surfaces. See transpiration

evapotranspiration–the sum of evaporation and transpiration.

F

flood–An overflow of water onto lands that are used or usable by man and not normally covered by water. Floods have two essential characteristics: The inundation of land is temporary; and the land is adjacent to and inundated by overflow from a river, stream, lake, or ocean.

flood, 100-year–A 100-year flood does not refer to a flood that occurs once every 100 years, but to a flood level with a 1 percent chance of being equaled or exceeded in any given year.

flood plain–a strip of relatively flat and normally dry land alongside a stream, river, or lake that is covered by water during a flood.

flood stage–The elevation at which overflow of the natural banks of a stream or body of water begins in the reach or area in which the elevation is measured.

flowing well/spring–a well or spring that taps ground water under pressure so that water rises without pumping. If the water rises above the surface, it is known as a flowing well.

freshwater, fresh water–water that contains less than 1,000 milligrams per liter (mg/L) of dissolved solids; generally, more than 500 mg/L of dissolved solids is undesirable for drinking and many industrial uses.

G

gage height–the height of the water surface above the gage datum (zero point). Gage height is often used interchangeably with the more general term, stage, although gage height is more appropriate when used with a gage reading.

gaging station–a site on a stream, lake, reservoir or other body of water where observations and hydrologic data are obtained. The U.S. Geological Survey measures stream discharge at gaging stations.

geyser–a geothermal feature of the Earth where there is an opening in the surface that contains superheated water that periodically erupts in a shower of water and steam.

giardiasis–a disease that results from an infection by the protozoan parasite Giardia Intestinalis, caused by drinking water that is either not filtered or not chlorinated. The disorder is more prevalent in children than in adults and is characterized by abdominal discomfort, nausea, and alternating constipation and diarrhea.

glacier–a huge mass of ice, formed on land by the compaction and recrystallization of snow, that moves very slowly downslope or outward due to its own weight.

greywater–wastewater from clothes washing machines, showers, bathtubs, hand washing, lavatories and sinks.

ground water–(1) water that flows or seeps downward and saturates soil or rock, supplying springs and wells. The upper surface of the saturate zone is called the water table. (2) Water stored underground in rock crevices and in the pores of geologic materials that make up the Earth's crust.

ground water, confined–ground water under pressure significantly greater than atmospheric, with its upper limit the bottom of a bed with hydraulic conductivity distinctly lower than that of the material in which the confined water occurs.

ground-water recharge–inflow of water to a ground-water reservoir from the surface. Infiltration of precipitation and its movement to the water table is one form of natural recharge. Also, the volume of water added by this process.

ground water, unconfined–water in an aquifer that has a water table that is exposed to the atmosphere.

H

hardness–a water-quality indication of the concentration of alkaline salts in water, mainly calcium and magnesium. If the water you use is "hard" then more soap, detergent or shampoo is necessary to raise a lather.

headwater(s)–(1) the source and upper reaches of a stream; also the upper reaches of a reservoir; (2) the water upstream from a structure or point on a stream; (3) the small streams that come together to form a river. Also may be thought of as any and all parts of a river basin except the mainstream river and main tributaries.

hydroelectric power water use–the use of water in the generation of electricity at plants where the turbine generators are driven by falling water.

hydrologic cycle—the cyclic transfer of water vapor from the Earth's surface via evapotranspiration into the atmosphere, from the atmosphere via precipitation back to earth, and through runoff into streams, rivers, and lakes, and ultimately into the oceans.

I

impermeable layer—a layer of solid material, such as rock or clay, which does not allow water to pass through.

industrial water use—water used for industrial purposes in such industries as steel, chemical, paper, and petroleum refining. Nationally, water for industrial uses comes mainly (80%) from self-supplied sources, such as a local wells or withdrawal points in a river, but some water comes from public-supplied sources, such as the county/city water department.

infiltration—flow of water from the land surface into the subsurface.

injection well—refers to a well constructed for the purpose of injecting treated wastewater directly into the ground. Wastewater is generally forced (pumped) into the well for dispersal or storage into a designated aquifer. Injection wells are generally drilled into aquifers that don't deliver drinking water, unused aquifers, or below freshwater levels.

irrigation—the controlled application of water for agricultural purposes through manmade systems to supply water requirements not satisfied by rainfall. Here's a quick look at some types of irrigation systems.

irrigation water use—water application on lands to assist in the growing of crops and pastures or to maintain vegetative growth in recreational lands, such as parks and golf courses.

K

kilogram—one thousand grams.

kilowatt-hour (KWH)—a power demand of 1,000 watts for one hour. Power company utility rates are typically expressed in cents per kilowatt-hour.

L

leaching—the process by which soluble materials in the soil, such as salts, nutrients, pesticide chemicals or contaminants, are washed into a lower layer of soil or are dissolved and carried away by water.

lentic waters—ponds or lakes (standing water).

levee—a natural or manmade earthen barrier along the edge of a stream, lake, or river. Land alongside rivers can be protected from flooding by levees.

livestock water use—water used for livestock watering, feed lots, dairy operations, fish farming, and other on-farm needs.

lotic waters—flowing waters, as in streams and rivers.

M

maximum contaminant level (MCL)—the designation given by the U.S. Environmental Protection Agency (EPA) to water-quality standards promulgated under the Safe Drinking Water Act. The MCL is the greatest amount of a contaminant that can be present in drinking water without causing a risk to human health.

milligram (mg)—One-thousandth of a gram.

milligrams per liter (mg/l)—a unit of the concentration of a constituent in water or wastewater. It represents 0.001 gram of a constituent in 1 liter of water. It is approximately equal to one part per million (PPM).

million gallons per day (Mgd)—a rate of flow of water equal to 133,680.56 cubic feet per day, or 1.5472 cubic feet per second, or 3.0689 acre-feet per day. A flow of one million gallons per day for one year equals 1,120 acre-feet (365 million gallons).

mining water use—water use during quarrying rocks and extracting minerals from the land.

municipal water system—a water system that has at least five service connections or which regularly serves 25 individuals for 60 days; also called a public water system

N

nephelometric turbidity unit (NTU)—unit of measure for the turbidity of water. Essentially, a measure of the cloudiness of water as measured by a nephelometer. Turbidity is based on the amount of light that is reflected off particles in the water.

non-point source (NPS) pollution—pollution discharged over a wide land area, not from one specific location. These are forms of diffuse pollution caused by sediment, nutrients, organic and toxic substances originating from land-use activities, which are carried to lakes and streams by surface runoff. Non-point source pollution is contamination that occurs when rainwater, snowmelt, or irrigation washes off plowed fields, city streets, or suburban backyards. As this runoff moves across the land surface, it picks up soil particles and pollutants, such as nutrients and pesticides.

O

organic matter—plant and animal residues, or substances made by living organisms. All are based upon carbon compounds.

osmosis—the movement of water molecules through a thin membrane. The osmosis process occurs in our bodies and is also one method of desalinizing saline water.

outfall—the place where a sewer, drain, or stream discharges; the outlet or structure through which reclaimed water or treated effluent is finally discharged to a receiving water body.

oxygen demand—the need for molecular oxygen to meet the needs of biological and chemical processes in water. Even though very little oxygen will dissolve in water, it is extremely important in biological and chemical processes.

P

pH—a measure of the relative acidity or alkalinity of water. Water with a pH of 7 is neutral; lower pH levels indicate increasing acidity, while pH levels higher than 7 indicate increasingly basic solutions.

View a diagram about pH.

particle size–the diameter, in millimeters, of suspended sediment or bed material. Particle-size classifications are:

[1] *Clay* 0.00024–0.004 millimeters (mm);
[2] *Silt* 0.004–0.062 mm;
[3] *Sand* 0.062–2.0 mm; and
[4] *Gravel* 2.0–64.0 mm.

parts per billion–the number of "parts" by weight of a substance per billion parts of water. Used to measure extremely small concentrations.

parts per million–the number of "parts" by weight of a substance per million parts of water. This unit is commonly used to represent pollutant concentrations.

pathogen–a disease-producing agent; usually applied to a living organism. Generally, any viruses, bacteria, or fungi that cause disease.

peak flow–the maximum instantaneous discharge of a stream or river at a given location. It usually occurs at or near the time of maximum stage.

per capita use–the average amount of water used per person during a standard time period, generally per day.

percolation–(1) The movement of water through the openings in rock or soil. (2) the entrance of a portion of the streamflow into the channel materials to contribute to ground water replenishment.

permeability–the ability of a material to allow the passage of a liquid, such as water through rocks. Permeable materials, such as gravel and sand, allow water to move quickly through them, whereas unpermeable material, such as clay, don't allow water to flow freely.

point-source pollution–water pollution coming from a single point, such as a sewage-outflow pipe.

polychlorinated biphenyls (PCBs)–a group of synthetic, toxic industrial chemical compounds once used in making paint and electrical transformers, which are chemically inert and not biodegradable. PCBs were frequently found in industrial wastes, and subsequently found their way into surface and ground waters. As a result of their persistence, they tend to accumulate in the environment. In terms of streams and rivers, PCBs are drawn to sediment, to which they attach and can remain virtually indefinitely. Although virtually banned in 1979 with the passage of the Toxic Substances Control Act, they continue to appear in the flesh of fish and other animals.

porosity–a measure of the water-bearing capacity of subsurface rock. With respect to water movement, it is not just the total magnitude of porosity that is important, but the size of the voids and the extent to which they are interconnected, as the pores in a formation may be open, or interconnected, or closed and isolated. For example, clay may have a very high porosity with respect to potential water content, but it constitutes a poor medium as an aquifer because the pores are usually so small.

potable water–water of a quality suitable for drinking.

precipitation–rain, snow, hail, sleet, dew, and frost.

primary wastewater treatment–the first stage of the wastewater-treatment process where mechanical methods, such as filters and scrapers, are used to remove pollutants. Solid material in sewage also settles out in this process.

prior appropriation doctrine–the system for allocating water to private individuals used in most Western states. The doctrine of Prior Appropriation was in common use throughout the arid West as early settlers and miners began to develop the land. The prior appropriation doctrine is based on the concept of "First in Time, First in Right." The first person to take a quantity of water and put it to beneficial use has a higher priority of right than a subsequent user. The rights can be lost through nonuse; they can also be sold or transferred apart from the land. Contrasts with riparian water rights.

public supply–water withdrawn by public governments and agencies, such as a county water department, and by private companies that is then delivered to users. Public suppliers provide water for domestic, commercial, thermoelectric power, industrial, and public water users. Most people's household water is delivered by a public water supplier. The systems have at least 15 service connections (such as households, businesses, or schools) or regularly serve at least 25 individuals daily for at least 60 days out of the year.

public water use–water supplied from a public-water supply and used for such purposes as firefighting, street washing, and municipal parks and swimming pools.

R

rating curve–A drawn curve showing the relation between gage height and discharge of a stream at a given gaging station.

recharge–water added to an aquifer. For instance, rainfall that seeps into the ground.

reclaimed wastewater–treated wastewater that can be used for beneficial purposes, such as irrigating certain plants.

recycled water–water that is used more than one time before it passes back into the natural hydrologic system.

reservoir–a pond, lake, or basin, either natural or artificial, for the storage, regulation, and control of water.

return flow–(1) That part of a diverted flow that is not consumptively used and returned to its original source or another body of water. (2) (Irrigation) Drainage water from irrigated farmlands that re-enters the water system to be used further downstream.

returnflow (irrigation)–irrigation water that is applied to an area and which is not consumed in evaporation or transpiration and returns to a surface stream or aquifer.

reverse osmosis–(1) (Desalination) The process of removing salts from water using a membrane. With reverse osmosis, the product water passes through a fine membrane that the salts are unable to pass through, while the salt waste (brine) is removed and disposed. This process differs from electrodialysis, where the salts are extracted from the feedwater by using a membrane with an electrical current to separate the ions. The positive ions go through one membrane, while the negative ions flow through a different membrane, leaving the end product of freshwater. (2) (Water Quality) An advanced method of water or wastewater treatment that relies on a semipermeable membrane to separate waters from pollutants. An external force is used to reverse the normal osmotic process resulting in the solvent moving from a solution of higher concentration to one of lower concentration.

riparian water rights–the rights of an owner whose land abuts water. They differ from state to state and often depend on whether the water is a river, lake, or ocean. The doctrine of riparian rights is an old one, having its origins in English common law. Specifically, persons who own land adjacent to a stream have the right to make reasonable use of the stream. Riparian users of a stream share the streamflow among themselves, and the concept of priority of use (Prior Appropriation Doctrine) is not applicable. Riparian rights cannot be sold or transferred for use on nonriparian land.

river–A natural stream of water of considerable volume, larger than a brook or creek.

runoff–(1) That part of the precipitation, snow melt, or irrigation water that appears in uncontrolled surface streams, rivers, drains or sewers. Runoff may be classified according to speed of appearance after rainfall or melting snow as direct runoff or base runoff, and according to source as surface runoff, storm interflow, or ground-water runoff. (2) The total discharge described in (1), above, during a specified period of time. (3) Also defined as the depth to which a drainage area would be covered if all of the runoff for a given period of time were uniformly distributed over it.

S

saline water–water that contains significant amounts of dissolved solids.

Here are our parameters for saline water:

Fresh water. Less than 1,000 parts per million (ppm)
Slightly saline water. From 1,000 ppm to 3,000 ppm
Moderately saline water. From 3,000 ppm to 10,000 ppm
Highly saline water. From 10,000 ppm to 35,000 ppm

secondary wastewater treatment–treatment (following primary wastewater treatment) involving the biological process of reducing suspended, colloidal, and dissolved organic matter in effluent from primary treatment systems and which generally removes 80 to 95 percent of the Biochemical Oxygen Demand (BOD) and suspended matter. Secondary wastewater treatment may be accomplished by biological or chemical-physical methods. Activated sludge and trickling filters are two of the most common means of secondary treatment. It is accomplished by bringing together waste, bacteria, and oxygen in trickling filters or in the activated sludge process. This treatment removes floating and settleable solids and about 90 percent of the oxygen-demanding substances and suspended solids. Disinfection is the final stage of secondary treatment.

sediment–usually applied to material in suspension in water or recently deposited from suspension. In the plural the word is applied to all kinds of deposits from the waters of streams, lakes, or seas.

sedimentary rock–rock formed of sediment, and specifically: (1) sandstone and shale, formed of fragments of other rock transported from their sources and deposited in water; and (2) rocks formed by or from secretions of organisms, such as most limestone. Many sedimentary rocks show distinct layering, which is the result of different types of sediment being deposited in succession.

sedimentation tanks–wastewater tanks in which floating wastes are skimmed off and settled solids are removed for disposal.

self-supplied water–water withdrawn from a surface- or ground-water source by a user rather than being obtained from a public supply. An example would be homeowners getting their water from their own well.

seepage–(1) The slow movement of water through small cracks, pores, Interstices, etc., of a material into or out of a body of surface or subsurface water. (2) The loss of water by infiltration into the soil from a canal, ditches, laterals, watercourse, reservoir, storage facilities, or other body of water, or from a field.

septic tank–a tank used to detain domestic wastes to allow the settling of solids prior to distribution to a leach field for soil absorption. Septic tanks are used when a sewer line is not available to carry them to a treatment plant. A settling tank in which settled sludge is in immediate contact with sewage flowing through the tank, and wherein solids are decomposed by anaerobic bacterial action.

settling pond (water quality)–an open lagoon into which wastewater contaminated with solid pollutants is placed and allowed to stand. The solid pollutants suspended in the water sink to the bottom of the lagoon and the liquid is allowed to overflow out of the enclosure.

sewage treatment plant–a facility designed to receive the wastewater from domestic sources and to remove materials that damage water quality and threaten public health and safety when discharged into receiving streams or bodies of water. The substances removed are classified into four basic areas:

[1] greases and fats;
[2] solids from human waste and other sources;
[3] dissolved pollutants from human waste and decomposition products; and
[4] dangerous microorganisms.

Most facilities employ a combination of mechanical removal steps and bacterial decomposition to achieve the desired results. Chlorine is often added to discharges from the plants to reduce the danger of spreading disease by the release of pathogenic bacteria.

sewer–a system of underground pipes that collect and deliver wastewater to treatment facilities or streams.

sinkhole–a depression in the Earth's surface caused by dissolving of underlying limestone, salt, or gypsum.

Drainage is provided through underground channels that may be enlarged by the collapse of a cavern roof.

solute–a substance that is dissolved in another substance, thus forming a solution.

solution–a mixture of a solvent and a solute. In some solutions, such as sugar water, the substances mix so thoroughly that the solute cannot be seen. But in other solutions, such as water mixed with dye, the solution is visibly changed.

solvent–a substance that dissolves other substances, thus forming a solution. Water dissolves more substances than any other, and is known as the "universal solvent."

specific conductance–a measure of the ability of water to conduct an electrical current as measured using a 1-cm cell and expressed in units of electrical conductance, that is, Siemens per centimeter at 25 degrees Celsius. Specific conductance can be used for approximating the total dissolved solids content of water by testing its capacity to carry an electrical current. In water quality, specific conductance is used in ground water monitoring as an indication of the presence of ions of chemical substances that may have been released by a leaking landfill or other waste storage or disposal facility. A higher specific conductance in water drawn from downgradient wells when compared to upgradient wells indicates possible contamination from the facility.

spray irrigation–a common irrigation method where water is shot from high-pressure sprayers onto crops. Because water is shot high into the air onto crops, some water is lost to evaporation.

storm sewer–a sewer that carries only surface runoff, street wash, and snow melt from the land. In a separate sewer system, storm sewers are completely separate from those that carry domestic and commercial wastewater (sanitary sewers).

stream–a general term for a body of flowing water; natural water course containing water at least part of the year. In hydrology, it is generally applied to the water flowing in a natural channel as distinct from a canal.

streamflow–the water discharge that occurs in a natural channel. A more general term than runoff, streamflow may be applied to discharge whether or not it is affected by diversion or regulation.

subsidence–a dropping of the land surface as a result of ground water being pumped. Cracks and fissures can appear in the land. Subsidence is virtually an irreversible process.

surface tension–the attraction of molecules to each other on a liquid's surface. Thus, a barrier is created between the air and the liquid.

surface water–water that is on the Earth's surface, such as in a stream, river, lake, or reservoir.

suspended sediment–very fine soil particles that remain in suspension in water for a considerable period of time without contact with the bottom. Such material remains in suspension due to the upward components of turbulence and currents and/or by suspension.

suspended-sediment concentration–the ratio of the mass of dry sediment in a water-sediment mixture to the mass of the water-sediment mixture. Typically expressed in milligrams of dry sediment per liter of water-sediment mixture.

suspended-sediment discharge–the quantity of suspended sediment passing a point in a stream over a specified period of time. When expressed in tons per day, it is computed by multiplying water discharge (in cubic feet per second) by the suspended-sediment concentration (in milligrams per liter) and by the factor 0.0027.

suspended solids–solids that are not in true solution and that can be removed by filtration. Such suspended solids usually contribute directly to turbidity. Defined in waste management, these are small particles of solid pollutants that resist separation by conventional methods.

T

tertiary wastewater treatment–selected biological, physical, and chemical separation processes to remove organic and inorganic substances that resist conventional treatment practices; the additional treatment of effluent beyond that of primary and secondary treatment methods to obtain a very high quality of effluent. The complete wastewater treatment process typically involves a three-phase process: (1) First, in the primary wastewater treatment process, which incorporates physical aspects, untreated water is passed through a series of screens to remove solid wastes; (2) Second, in the secondary wastewater treatment process, typically involving biological and chemical processes, screened wastewater is then passed a series of holding and aeration tanks and ponds; and (3) Third, the tertiary wastewater treatment process consists of flocculation basins, clarifiers, filters, and chlorine basins or ozone or ultraviolet radiation processes.

thermal pollution–a reduction in water quality caused by increasing its temperature, often due to disposal of waste heat from industrial or power generation processes. Thermally polluted water can harm the environment because plants and animals can have a hard time adapting to it.

thermoelectric power water use–water used in the process of the generation of thermoelectric power. Power plants that burn coal and oil are examples of thermoelectric-power facilities.

transmissibility (ground water)–the capacity of a rock to transmit water under pressure. The coefficient of transmissibility is the rate of flow of water, at the prevailing water temperature, in gallons per day, through a vertical strip of the aquifer one foot wide, extending the full saturated height of the aquifer under a hydraulic gradient of 100-percent. A hydraulic gradient of 100-percent means a one foot drop in head in one foot of flow distance.

transpiration–process by which water that is absorbed by plants, usually through the roots, is evaporated into the atmosphere from the plant surface, such as leaf pores. See evapotranspiration.

Tributary–a smaller river or stream that flows into a larger river or stream. Usually, a number of smaller tributaries merge to form a river.

turbidity–the amount of solid particles that are suspended in water and that cause light rays shining through the water to scatter. Thus, turbidity makes the water cloudy or even opaque in extreme cases. Turbidity is measured in nephelometric turbidity units (NTU).

U

unsaturated zone–the zone immediately below the land surface where the pores contain both water and air, but are not totally saturated with water. These zones differ from an aquifer, where the pores are saturated with water.

W

wastewater–water that has been used in homes, industries, and businesses that is not for reuse unless it is treated.

wastewater-treatment return flow–water returned to the environment by wastewater-treatment facilities.

water cycle–the circuit of water movement from the oceans to the atmosphere and to the Earth and return to the atmosphere through various stages or processes such as precipitation, interception, runoff, infiltration, percolation, storage, evaporation, and transportation.

water quality–a term used to describe the chemical, physical, and biological characteristics of water, usually in respect to its suitability for a particular purpose.

water table–the top of the water surface in the saturated part of an aquifer.

water use–water that is used for a specific purpose, such as for domestic use, irrigation, or industrial processing. Water use pertains to human's interaction with and influence on the hydrologic cycle, and includes elements, such as water withdrawal from surface- and ground-water sources, water delivery to homes and businesses, consumptive use of water, water released from wastewater-treatment plants, water returned to the environment, and instream uses, such as using water to produce hydroelectric power.

watershed–the land area that drains water to a particular stream, river, or lake. It is a land feature that can be identified by tracing a line along the highest elevations between two areas on a map, often a ridge. Large watersheds, like the Mississippi River basin contain thousands of smaller watersheds.

watt-hour (Wh)–an electrical energy unit of measure equal to one watt of power supplied to, or taken from, an electrical circuit steadily for one hour.

well (water)–an artificial excavation put down by any method for the purposes of withdrawing water from the underground aquifers. A bored, drilled, or driven shaft, or a dug hole whose depth is greater than the largest surface dimension and whose purpose is to reach underground water supplies or oil, or to store or bury fluids below ground.

withdrawal–water removed from a ground- or surface-water source for use.

X

xeriscaping–a method of landscaping that uses plants that are well adapted to the local area and are drought-resistant. Xeriscaping is becoming more popular as a way of saving water at home.

More on xeriscaping: Texas Agricultural Extension Service

More on xeriscaping: Texas Natural Resource Center

Y

yield–mass per unit time per unit area

Some of this information is courtesy of the Nevada Division of Water Planning.

HARVESTING RAINWATER

Ganesh B. Keremane
University of South Australia
Adelaide, Australia

INTRODUCTION

Water is the genesis and continuing source of all life on the earth, and we all completely depend on this renewable source. According to the United Nation's World Water Development Report (1), "Mankind's most serious challenge in the 21st century might not be war or hunger or disease or even the collapse of civic order; it may be the lack of fresh water," which justifies the present scenario that freshwater is a scarce resource. Although it is the commonest and the most critical stuff on the earth, water has been always an undervalued resource. There are more than 1.4 billion cubic kilometers (km^3) of water—enough to give every man, woman, and child more than 230 million cubic meters (m^3) each if we were to divide it evenly (2). However, more than 98% of the world's water is saltwater, and for our basic vital needs, we have to depend on freshwater, most of which is locked in the polar ice caps. This leaves less than 1% of the earth's freshwater to be accessed in lakes, rivers, and groundwater aquifers. The major portion of the water diverted for human needs is drawn from this renewable, readily accessible part of the world's freshwater resources, thus making it a scarce resource.

NEED FOR WATER HARVESTING

The prevailing water scarcity experienced by most of the water-scarce countries (arid and semiarid) of the world can be attributed to high water stress and unsustainable rates of withdrawal. Dupont (3) attributes this widening gap between freshwater demand and supply to three main factors: a limited global supply of freshwater, increased water consumption, and poor water management. The global supply of freshwater is limited and unevenly distributed as against the demand for freshwater, which increased sixfold between 1900 and 1995 (4,5). Moreover, the exponential increase in the consumption of water from population growth, rapid industrial development, and the expansion of irrigated farmland (6) contributes to added pressure on the water supply. Of these, the trend toward irrigating farmland, which provides one-third of the world's harvest from just 17% of its cropland, is responsible for much of the pressure on the water supply and has risen from 50 million hectares to 250 million hectares since 1990 (7). Poor water management and conservation practices have contributed to the decline in the quality and availability of water in addition to overlogging and deforestation that are destroying water tables and causing siltation and salinization. Wasteful and environmentally destructive farming techniques, overuse of pesticides and chemical fertilizers, and expansion of urban areas are other factors that have further reduced freshwater supplies. The statistics provided in the United Nations World Water Development Report (1) support this:

- World population is likely to increase by half as much again, to about 9.3 billion by 2050.
- One liter of wastewater is sufficient to pollute about eight liters of freshwater, and every day about 2 million tons of wastes are dumped into rivers, lakes, and streams.
- As of today, it is estimated that across the world there are about 12,000 cubic kilometers of wastewater, which is more than the total amount in the world's 10 largest river basins at any given moment.
- At present, 48% of the earth's population lives in towns and cities; this will increase to 60% by 2030.

For these reasons, water has become a scarce resource in many parts of the world. About 80 countries containing 40% of the world's population suffer from serious shortages, mostly in the developing world (4). Despite widely available evidence of the crisis, ways to help ensure adequate water supplies for household, agricultural, and other uses are available to farms and communities. Rudimentary conservation measures, application of new technology, and sensible conservation practices could do much to stabilize water withdrawals (3). Water harvesting is one such technology.

WATER HARVESTING—TO MAKE EVERY DROP COUNT

Providing irrigation to crops in hot, dry countries (arid and semiarid) accounts for 70% of the water use in the world (8). Further, the rise in land pressure has resulted in using more and more marginal areas for agriculture. Much of this land is located in the arid and semiarid belts characterized by scanty, uneven distribution of rainfall, where much of the precious water is soon lost as surface runoff. Recent droughts have highlighted the risks from the failure of monsoons to all life on the earth. Irrigation may be the most obvious response to drought but has proved costly and can benefit only a fortunate few (9). Lately this has gained popularity for an old technology in a new way, and that technology is water harvesting.

Water harvesting is no new concept for humankind. Though its importance in mitigating the water scarcity problem has been highlighted much in the recent past, it traces its history to biblical times. Extensive water harvesting apparatus existed 4000 years ago in Palestine and Greece (10). Critchley and Chapman (9), while discussing the history of water harvesting systems, mentioned that the earliest evidence of the use of water harvesting is the well-publicized systems used by the people of the Negev Desert, where the hillsides were cleared of vegetation and smoothed to provide as much runoff as possible; the water was then channeled in contour ditches to agricultural fields and/or to cisterns. In the New World, about 400–700 years ago, people living in Colorado (North America) and Peru (South America) also employed relatively simple methods of water harvesting for irrigation (11). The practice of collecting water from rooftops is also a very ancient type of water harvesting, practiced since the earliest times up to the present. However, renewed interest in the technology of water harvesting arose in the 1950s in Israel, Australia, and the United States. Since then, this technology has developed manifold through experimentation and demonstration projects.

WHAT IS WATER HARVESTING?

Prinz (12) opines, "There is no generally accepted definition of water harvesting." But the term water harvesting in its crudest form can be understood as, *catching rain as it falls*. However, many experts have defined water harvesting and some of them are mentioned for reference: Critchley and Chapman (9), in their manual on water harvesting, define water harvesting in its broadest sense as the collection of runoff for its productive use. Thames (11) defines water harvesting systems as "artificial methods whereby precipitation can be collected and stored until it is beneficially used. The system includes a catchment area and a storage facility for the harvested water, unless the water is to be immediately concentrated in the soil profile." "Water harvesting is collecting and using precipitation from a catchment surface" (10).

Though defined differently, it is clear from the definitions that water harvesting is the practice of collecting and storing water from various sources for beneficial use; the catchment area, runoff, and the storage area are the central elements or components of any water harvesting system. These elements are briefly explained in Box 1.

Box 1: Components of Water Harvesting[a]

Catchment Area. An area that is reasonably impermeable to water which can be used to produce runoff. It may include natural surfaces such as rock outcrops, surfaces developed for other purposes such as paved highways or roof tops, and/or surfaces prepared with minimal cost and effort such as those cleared of vegetation.

> **Runoff.** That portion of precipitation which is not intercepted by vegetation, absorbed by the land surface, or evaporated, and thus flows overland into a depression, stream, lake, or ocean. Runoff may be classified as direct runoff or base runoff according to the speed of appearance after rainfall or melting snow, and as surface runoff, storm interflow, or groundwater runoff, according to its source.
>
> **Storage Area.** Based on where the water is stored, the storage area can be the soil profile, excavated ponds, or any underground structure.
>
> 1. *The Soil Profile*: Simple arrangements are made to direct water from hillsides onto cultivated areas to store the water immediately in the soil for plant use. Whether sufficient water can be stored to offset a prolonged drought is a disadvantage.
> 2. *Excavated Ponds*: Ponds excavated in the ground surface are used for storing large quantities of water. These are often economical means of storage, but evaporation and seepage are serious drawbacks.
> 3. *Underground Structures*: Can be in the form of aquifers (an underground layer of porous rock, sand, or gravel containing large amounts of water) or cisterns (man-made underground construction to store water). In the light of the disadvantages/drawbacks of the other two types, it is an interesting alternative.
>
> [a]*Source*: Reference 13.

FORMS OF WATER HARVESTING

As mentioned before, water harvesting has been practiced for ages and is still relevant worldwide. A great number of forms exist with various names. Prinz and Singh (14) group water harvesting techniques into three broad groups: rainwater harvesting, floodwater harvesting, and groundwater harvesting. A brief description of these techniques along with subtypes is given below.

Rainwater Harvesting

Defined as a method for inducing, collecting, storing, and conserving local surface runoff for agriculture or other beneficial use (15). This can be further grouped into three subtypes:

- Water collected from roof tops, courtyards, and similar compacted or treated surfaces that is used for domestic purposes or garden crops.
- Microcatchment water harvesting is a method of collecting surface runoff from a small catchment area (catchment length usually between 1 and 30 meters) and storing it in the root zone of an adjacent infiltration basin. The basin is planted with a tree, a bush, or with annual crops.
- Macrocatchment water harvesting, also called harvesting from external catchments, is done where runoff from a hillslope catchment (catchment length usually between 30 and 200 meters) is conveyed to a cropping area located at the hill foot on flat terrain.

Floodwater Harvesting

Defined as the collection and storage of creek flow for irrigation use. According to Critchley and Chapman (9), floodwater harvesting is also known as "large catchment water harvesting" or "spate irrigation" and may be further classified as

- *Floodwater harvesting within a steambed* in which case the water flow is dammed and, as a result, inundates the valley bottom of a floodplain. The water is forced to infiltrate and the wetted area can be used for agriculture or pasture improvement.
- *Floodwater diversion* wherein the wadi water is forced to leave its natural course and is conveyed to nearby fields.

Groundwater Harvesting

A rather new term employed to cover traditional as well as unconventional ways of groundwater extraction. Subsurface dams and sand storage dams are some fine examples of groundwater harvesting. They obstruct the flow of temporary or short-lived streams in a riverbed, and store the water in the sediment below the ground surface, which is useful for aquifer recharge.

WATER HARVESTING STRUCTURES

All the previously mentioned forms of water harvesting are done with the aid of different permanent or temporary structures. There are several structures by which water can be harvested for aquifer recharge. A few most widely used and cheapest structures are described below. Barton (16) ranks them among the most popular methods for two main reasons: low construction cost and ease of operation and maintenance.

- *Contour trenches* are excavations parallel to contours on slopes to conserve water and prevent soil erosion and thus help water to infiltrate.
- *Gully plugs* are stone barriers built across gullies and deep rills that trap sediments eroded from higher up a slope and impound runoff, encouraging infiltration.
- *Check dams* are temporary structures constructed from locally available material such as brushwood, loose rock, or woven wire to impede the soil and water removed from the catchment. The impeded water collects behind the dam and infiltrates the soil, recharging the aquifer. The cost of construction ranges from US $200–400 (for temporary structures) to US $1000–2000 (for permanent structures), depending on the material used, the size of the gully, and the height of the dam (16).

- *A percolation tank* is a dam built on permeable ground so that floodwater is held back long enough to percolate into the ground, hence, the name "percolation tank." This is constructed by excavating a depression to form a small reservoir or by constructing an embankment in a natural ravine or gully. The construction cost is estimated at approximately US $5,000–10,000.
- *Groundwater dams* are structures that intercept or obstruct the natural flow of groundwater. These are often built within riverbeds to obstruct and detain groundwater flow so as to sustain the storage capacity of the aquifer and meet demands during high periods. These may be either a subsurface dam that is constructed within the aquifer itself or a sand storage/silt trapping dam that is constructed above ground to trap sand transported with floodwater.
- *Rooftop harvesting* is a method that can be adopted by individuals. Several techniques to harvest rainwater from the roof tops are in use: abandoned dug well, abandoned/running hand pump, gravity head recharge, and recharge pit (16). Though in use mainly to harvest water for household and domestic use, there are evidences of groundwater recharge using this technique. CGWB (10) has observed additional recharge to groundwater by dug wells to the extent of 6.6 TCM with a benefited area of 1.3 hectare in the Jalgaon district of Maharashtra, and, also, a rise in water level from 1.43 to 2.15 m has been recorded in Delhi from the adoption of rooftop harvesting techniques.

CHOICE OF WATER HARVESTING AREA AND TECHNIQUES

Planning is an important process involved in implementing any water harvesting project by an individual or community. It starts with the identification of the area for water harvesting, followed by selection of the techniques/forms/structures. While doing so, it is of utmost importance to consider the following parameters (11,12).

- *Amount and Seasonal Distribution of Rainfall*: Knowledge of the intensity and distribution of rainfall for a given area is necessary for designing a water harvesting system.
- *Topography*: Knowledge of the landform along with the slope is the next important parameter to determine the system for harvesting water and also the structures to be included.
- *Soil Type and Depth*: Important for designing a water harvesting system, the characteristics of soils determine the water movement and the quantity of water that can be stored and also influence the rainfall runoff process.
- *Provisions for Maintenance*: A maintenance program, even when the water collected is not being used (off season), is a must. Failure to provide for maintenance will result in early failure of the system.
- *Acceptance by the Local Community*: Last but not the least, to be successful water harvesting requires local capacity building, cooperation, and extensive participation. The users must believe that the system proposed is the best for their needs. Hence, before implementing any system, acceptance by the end or final user is a must to ensure the sustainability of the system. Only then can it be a success.

CONCLUSIONS

The water harvesting concept used to be more applicable to arid and semiarid regions where rainfall is either not sufficient or precipitation is erratic. But, lately, due to ever increasing population, growing urbanization, unsustainable withdrawal of groundwater accompanied by erratic monsoons, other regions of the world are also facing the problem of water scarcity. Hence, water harvesting is being looked into as an effective method to mitigate this problem of water scarcity all over the world because it offers a method of effectively developing scarce water resources by concentrating the rainfall/runoff in parts of the total area. These systems have the potential to increase the productivity of arable and grazing land by increasing the yields and by reducing the risk of crop failure. As studied by Kakade et al. (17), they also facilitate afforestation, fruit tree planting, and agroforestry. Using harvested rainwater helps decrease the use of other valuable water sources such as groundwater, and helps in groundwater recharge (18). It is also a relatively inexpensive method of water supply that can be adapted to the resources and needs of the rural poor. As such, there is no "the best" system of water harvesting. However, the system can be designed to fit best within the constraints of a given location taking into account technical, social, physical, and economic factors. Thus, learning from the past experiences and adopting new technological developments that take place, any water harvesting system can be considered a successful or an ideal system if it is

- technically sound, properly designed and maintained,
- economically feasible for the users, and
- capable of being integrated into the social traditions and abilities of the users.

BIBLIOGRAPHY

1. United Nations World Water Development Report. (2003). Accessed at http://www.unesco.org/water/wwap/wwdr/table_contents.shtml.
2. Asian Development Bank. (1999). *Annual Report—Special theme-Water in the 21st Century*. Philippines. [http://www.adb.org/].
3. Dupont, A. (2003). Will there be water wars? *Dev. Bull.* **63**: November 16–20.
4. United Nations Environmental Programme. (1991). *Freshwater Pollution*. UNEP, Nairobi.
5. Gleick, P.H. (1998). *The Worlds Water 1998–1999: The Biennial Report on Freshwater Resources*. Island Press, Washington, DC, pp. 40–47.

6. Wallensteen, P. and Swain, A. (1997). *International Freshwater Resources Conflict or Cooperation?* Stockholm Environment Institute, Stockholm, Sweden.
7. Gleick, P.H. (1993). An introduction to global fresh water issues. In: *Water in Crisis: A Guide to the World's Fresh Water Resources*. P.H. Gleick (Ed.). Oxford University Press, New York, p. 6.
8. McCarthy, M. (2003). *Water scarcity could affect billions: Is this the biggest crisis of all?* Accessed at http://www.commondreams.org/headlines03/0305-05.htm.
9. Critchley, W. and Chapman, C. (1991). *Water Harvesting—A Manual for the Design and Construction of Water Harvesting Schemes for Plant Production*. Food and Agriculture Organization of the United Nations, Rome.
10. Central Ground Water Board. (2000). *Rainwater Harvesting and Artificial Recharge to Groundwater-A Guide to Follow*. CGWB and United Nations Educational, Scientific and Cultural Organisation, New Delhi, India.
11. Thames, J.L. (1985). *Water Harvesting*. Proceedings of the FAO Expert Consultation on the Role of Forestry in Combating Desertification, Saltillo, Mexico June 24–28, 1985, Thematic paper 3.11.
12. Prinz, D. (1994). Water harvesting: past and future. *Sustainability of Irrigated Agriculture*. Proc., NATO Advanced Research Workshop, Vimerio, March 21–26, 1994, Balkema, Rotterdam, pp. 135–144.
13. Nevada Division of Water Resources. *Nevada Water Words Dictionary*. Carson City, NV [http://www.water.nv.gov/].
14. Prinz, D. and Singh, A. (2000). *Technological Potential for Improvements of Water Harvesting*. Contributing paper to the World Commission on Dams final report Dams and Development: A new Framework for Decision Making, Earthscan, London, UK.
15. Boers, T.M. and Ben-Asher, J. (1982). A review of rainwater harvesting. *Agric. Water Manage.* **5**: 145–158.
16. Barton, J. (2002). *Water Harvesting for Groundwater Recharge—Is It Effective?* Seminar presented at the student seminars on water harvesting in drought prone areas, University of Wales, Bangor, April 2002.
17. Kakade, B.K., Neelam, G.S., Petare, K.J., and Doreswamy, C. (2001). *Revival of Rivulets Through Farm Pond Based Watershed Development*. BAIF Development Research Foundation, Pune, India.
18. Chandrakanth, M.G. and Diwakara, H. (2001). *Synergistic Effects of Watershed Treatments on Farm Economy Through Groundwater Recharge—A Resource Economic Analysis*. Department of Agricultural Sciences, Bangalore, India.

URBAN WATER RESOURCE AND MANAGEMENT IN ASIA: HO CHI MINH CITY

HIEP N. DUC
Environment Protection
Authority, NSW
Bankstown, New South Wales
Australia

INTRODUCTION

Ho Chi Minh City, the largest city in Vietnam, has a current population of about 5 million that is increasing rapidly due to internal migration and natural growth. The water demand and the stress on the environment due to increasing waste and pollution pose a serious challenge to the people and the authority administering the city.

The Ho Chi Minh City Service of Communication and Public Works (SCPW) admitted in September 1998 that the city's decades-old water supply system cannot meet the growing demand. According to the SCPW, the water consumption is estimated at 800,000 m^3 per day, but the combined daily capacity of its two water plants at Thu Duc and Hoc Mon and underground wells stops short at 750,000 m^3. But just only 500,000 m^3 is available to users each day after water loss in delivery is taken into account (the proportion of water loss is as high as 32% in some areas). However, in 2000, the daily demand jumped to 1,250,000 m^3/day but the Water Supply Company can provide only 850,000 m^3/day.

Ho Chi Minh City is situated next to the Saigon River, which joins the Dong Nai River to form the northern and eastern boundaries of the city. This means that the waterways can be both potential water sources and receivers of polluted wastes.

The Dong Nai River has a minimum flow of approximately 100 m^3/s. It originates in the Central Highlands of Vietnam and flows through Dong Nai and Ho Chi Minh City with tributaries from other provinces. The total catchment area is 42,665 km^2, and the total flow volume is 30.6 km^3/year (1). Forests cover approximately 30% of the basin. The upstream area in the lowland part of the river is the present Cat Tien National Park. The Tri An reservoir and hydropower plant are located downstream nearby.

WATER ENVIRONMENT AND POLLUTION

One of the important aspects of water resource usage management is the prevention and mitigation of sources of pollution going into the watercourses.

Sources of Pollution

The majority of industries are located in industrial zones around Ho Chi Minh City. In addition to old industrial zones such as Bien Hoa zone 1 built in 1963, a large number of new industrial zones were set up recently in the neighboring provinces of Dong Nai, Song Be, and Ba Ria-Vung Tau. They were established to attract foreign investment to set up factories to produce goods for domestic and export markets. The industrial zone in Bien Hoa (Dong Nai) is upstream from Ho Chi Minh City. The foremost concern is that the wastewater discharge from this zone to the Dong Nai River is near the Hoa An pumping station that supplies water to the residents of Ho Chi Minh City (2).

Altogether, it was estimated that about 200,000 m^3/day of wastewater are discharged into the Dong Nai–Saigon River from the industrial zones. Industrial source discharges are located mostly in the Thu Duc and Bien Hoa areas, but they can also be found scattered along many canals in and around Ho Chi Minh City. There are five main canal systems that receive direct wastes from various sources: Nhieu Loc-Thi Nghe, Tau Hu-Ben Nghe, Doi-Te, Tan Hoa-Lo Gom, and Tham Luong-Vam Thuat. The total length of these canal systems is about 56 km with about

100 km of smaller tributaries. According to statistical data from the Water Supply Company, there is 461.291 m^3 of wastewater discharge into the canal systems daily (2).

Water pollution is more visible during the rainy season when the inadequate drainage system cannot cope with the water volume. In many areas of the city, traffic cannot move due to extensive flooding after a rain.

Pollution Management

Wastewater treatment facilities are required for new and established industrial zones. In Binh Duong province, a wastewater treatment plant was built in 1998 for the Vietnam–Singapore Industrial Park and adjacent areas. The system includes an 8-km canal to drain rain and treated wastewater from the zone into the Saigon River. In Bien Hoa II Industrial Park, a plant for treating wastewater from factories was put into operation in 1999. The plant has an initial daily capacity for treating 4,000 m^3, and 20 of the 76 factories had been connected to the plant since it started operating. The 2-hectare plant has four reservoirs for biological, chemical, and physical treatment of liquid waste before it is conveyed to another reservoir for final treatment.

Another issue regarding the management of water pollution is the storm water system. In 2001, the municipal Communication and Public Works will spend VND101 billions to upgrade and improve the storm water drainage system. It is part of a 5-year plan to reduce flooding during the rainy season. To partly finance the maintenance and upgrading of the drainage system, the city authority decided in early 2001 to allow the Water Supply Company to collect a water drainage levy from water users starting in July 2001. The rest of the capital is in the form of loans from international bodies such as foreign governments or the World Bank.

The most significant plan for improving water quality and the environment is the Urban Drainage and Sewer Master Plan. This plan was prepared and developed in 1999 with assistance from the Japanese Government. The Nhieu Loc–Thi Nghe canal basin has been identified and given the highest priority to improve the water environment and associated infrastructure (3). The shoreline improvement via house clearance, relocation, and parkland development as well as canal dredging has been done since 1996 but without capital investment in the drainage and sewer infrastructure. Street flooding and traffic jams during the rainy season are a frequent phenomenon. The Nhieu Loc–Thi Nghe Project was started in early 2001; part of the financing came from a World Bank loan. The Project is managed and implemented by the Urban Drainage Company.

The city Urban Drainage and Sewer Master Plan conforms to the national Urban Wastewater Collection and Sanitation Strategy Study and the Government Decision No. 35/1999/QD-TTg of March 5, 1999 (3).

WATER SUPPLY AND DISTRIBUTION

Government and City Organizations

In 1977, the city People's Committee established the Environmental Protection Council to oversee the environment. In 1992, the Environment Committee was formed. In 1994, the Department of Science, Technology and Environment (DOSTE) was established as part of national body run by the Ministry of Science, Technology and Environment (MOSTE) when the Environment Law was passed by the National Assembly.

The Department of Transport Communication and Public Works is the city government body in charge of planning the water supply network and other infrastructures. HCM Water Supply Company is the city operating enterprise in charge of supplying water to the city. The Urban Water Drainage Company is another city authority that oversees the operation of the wastewater and storm water network. The Waste Treatment Company of Ho Chi Minh City is concerned with the disposal of solid wastes, including household and hospital wastes, produced in the city. It owns several landfill waste sites in and around the city. These three public service enterprises report to the city's Director of Transport Communication and Public Works.

All the above companies are city government enterprises. The first private water treatment plant in Vietnam was built on a build-operate-transfer (BOT) basis under a 20-year contract with the city People's Committee and the Water Supply Company of Ho Chi Minh City. The plant was built and run by Binh An Water Corporation, a consortium of Malaysian companies that has a loan from the International Finance Corporation (IFC) (4). The Binh An treatment plant will provide 100,000 m^3/day (about 10% of the current demand) to the Ho Chi Minh City and Bien Hoa industrial zone (5).

Water Supply Network

The main water supply plant for Ho Chi Minh City is the Thu Duc Water Station. It currently supplies 650,000 m^3/day to the city using water from the Dong Nai River. The Dong Nai Water Supply Project will increase the plant capacity to 750,000 m^3/day (6). Part of the project cost is funded by a loan from the Asian Development Bank (ADB). Lyonaise des Eaux is building another water plant at the existing Thu Duc Water station to supply 300,000 m^3 each day to districts 2, 7, 9, Thu Duc, and Nha Be. This is a 25-year joint BOT venture with the Water Supply Company.

In 1991, there was a project planned to build a new water supply plant using water from the Saigon River with funding aid from the Italian government. The planned supply plant has a capacity of 300,000 m^3/day and would provide water to District 12, 6, Tan Binh, and Binh Chanh. But as the project was underway in 1995, the aid was stopped by the Italian government due to faults in timing management. Since then, the HCM People Committee has tried to use various schemes to get it restarted, first as a joint venture with foreign companies, then as a BOT project, and finally in April 2000 as a total sale package including equipment and project works to private investors. But so far none was successful.

In surrounding areas of the city, the water demand was alleviated by some new projects. In 1997, the Hoc Mon Underground Water plant was upgraded to a capacity of 50,000 m^3/daily using a loan from the Asian Development

Bank (ADB). In Dong Nai province, construction of a new water supply plant is being carried out at Thien Tan. The US $25 million Thien Tan water project is funded by aid from the government of the Republic of Korea to provide clean water to more residents of the province. The plant will have a capacity of 100,000 m^3/day.

BIBLIOGRAPHY

1. Talaue-McManus, L. (2000). *Transboundary Diagnostic Analysis for the South China Sea*. EAS/RCU Technical Report Series No. 14, UNEP, Bangkok, Thailand.
2. Nguyen Thi Lan (6/3/1997). Bao Dong o nhiem moi truong nuoc Tp HCM, Tuoi Tre.
3. World Bank—Vietnam-Ho Chi Minh Environmental Sanitation Project. http://www.worldbank.org/pics/pid/vn52037.txt
4. IFC Press Release (1/7/1998). *IFC invests US $25 million in first private water treatment facility in Vietnam*.
5. World Bank—Vietnam Binh An Water Corp. Project, http://www.worldbank.org/pics/ifcers/vne08948.txt.
6. *Saigon Times Daily* (March 22, 2001). City seeks capital scaledown for water supply project.

HYDROPOWER—ENERGY FROM MOVING WATER

Energy Information
Administration—Department of
Energy

Of the renewable energy sources that generate electricity, hydropower is the most often used. It accounted for 6 percent of U.S. generation and 42 percent of renewable generation in 2001. It is one of the oldest sources of energy and was used thousands of years ago to turn a paddle wheel for purposes such as grinding grain. Our nation's first industrial use of hydropower to generate electricity occurred in 1880, when 16 brush-arc lamps were powered using a water turbine at the Wolverine Chair Factory in Grand Rapids, Michigan. The first U.S. hydroelectric power plant opened on the Fox River near Appleton, Wisconsin, on September 30, 1882. Until that time, coal was the only fuel used to produce electricity. Because the source of hydropower is water, hydroelectric power plants must be located on a water source. Therefore, it wasn't until the technology to transmit electricity over long distances was developed that hydropower became widely used.

Mechanical energy is derived by directing, harnessing, or channeling moving water. The amount of available energy in moving water is determined by its *flow* or *fall*. Swiftly flowing water in a big river, like the Columbia River along the border between Oregon and Washington, carries a great deal of energy in its flow. So, too, with water descending rapidly from a very high point, like

---

This article is a US Government work and, as such, is in the public domain in the United States of America.

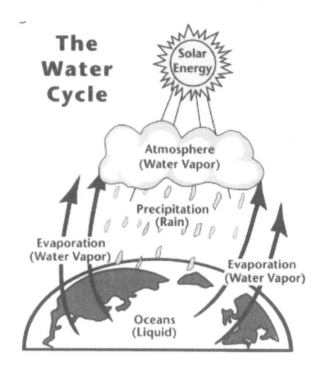

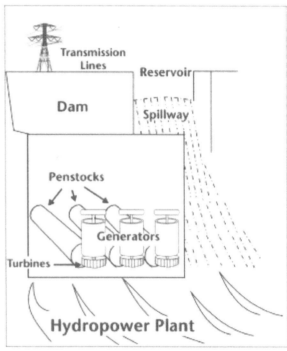

Niagara Falls in New York. In either instance, the water flows through a pipe, or *penstock*, then pushes against and turns blades in a turbine to spin a generator to produce electricity. In a *run-of-the-river system*, the force of the current applies the needed pressure, while in a *storage system*, water is accumulated in reservoirs created by dams, then released when the demand for electricity is high.

Meanwhile, the reservoirs or lakes are used for boating and fishing, and often the rivers beyond the dams provide opportunities for whitewater rafting and kayaking. Hoover Dam, a hydroelectric facility completed in 1936 on the

TOP HYDROPOWER PRODUCING STATES 1998

Colorado River between Arizona and Nevada, created Lake Mead, a 110-mile-long national recreational area that offers water sports and fishing in a desert setting.

Over one-half (52 percent) of the total U.S. hydroelectric capacity for electricity generation is concentrated in three States (Washington, California and Oregon) with approximately 27 percent in Washington, the location of the Nation's largest hydroelectric facility—the Grand Coulee Dam.

It is important to note that only a small percentage of all dams in the United States produce electricity. Most dams were constructed solely to provide irrigation and flood control.

Some people regard hydropower as the ideal fuel for electricity generation because, unlike the nonrenewable fuels used to generate electricity, it is almost free, there are no waste products, and hydropower does not pollute the water or the air. However, it is criticized because it does change the environment by affecting natural habitats. For instance, in the Columbia River, salmon must swim upstream to their spawning grounds to reproduce, but the series of dams gets in their way. Different approaches to fixing this problem have been used, including the construction of "fish ladders" which help the salmon "step up" the dam to the spawning grounds upstream.

WATER MARKETS IN INDIA: ECONOMIC AND INSTITUTIONAL ASPECTS

Halanaik Diwakara
University of South Australia
Adelaide, Australia

BACKGROUND

Historically, India's water resources have been perceived as an important input to production in agriculture for regional and economic development. India is an agrarian economy. Therefore, water is the elixir of life for millions of farmers in the country. Water markets are not akin to commodity markets. Because of ambiguous property rights to both surface and groundwater, informal selling and buying are popping up in different parts of India. These are not legal; surface water structure is built by government funding, so farmers have no right to sell it.

In addition, electricity pumping costs are not borne by farmers and a major portion of groundwater pumping is through electricity; so farmers have no right to sell groundwater also.

The water markets are prominent because of the profit from the sale of water for nonagricultural purposes. Electricity is virtually free, so selling groundwater for agriculture is not as profitable as it is for nonagricultural purposes. Even though nonagricultural purposes such as domestic water needs are not as large as for agriculture, consumer demand for domestic water is apparent, and consumers are willing to pay more than farmers who buy water for agriculture. Groundwater is reckoned as a poverty reduction tool in rural India, and its contribution to rural wealth is more than that of surface water (1).

Increasing demand for groundwater in India on account of the Green Revolution and commercial farming has led to enormous pressure on groundwater resources. Because of the growing population and expanding economy, the demand for water is surpassing the supply. Further, development of water resources is constrained by investment bottlenecks, environmental concerns, and political and legal snags inherent in interregional water transfers (3). Though the onus of achieving effective use of water falls on water markets, unless the true cost (value) of water is reflected in terms of actual electricity cost, any attempt by the government would be futile.

Reallocation of water through marketing is a response to growing demands on limited supplies of water (4). Neoclassical economic theory tells us that a market increases efficiency by allocating resources to their most valuable use. However, in the case of natural resources, the economic theory of markets may not always be true, for example, in central California, where the water market is immature (5). This is precisely true for India as well.

This article explores the nature and extent of water markets in India and focuses on groundwater markets and relevant economic and institutional (legislative) aspects. In the next section, the nature and extent of water markets in India are provided, followed by some economic aspects of water markets. In the following section, institutional aspects focusing on regulation of groundwater resource are discussed in the context of water markets. Finally, some water policy (institutional) issues under which water markets can be improved by a water reform process in India are discussed.

NATURE OF WATER MARKETS

Recent literature on water markets in India indicates that markets for groundwater are one of the ways of accessing irrigation (6). These water markets are niche markets, highly location-specific and endemic, at least in south India. In India, water markets are informal; the transfer of water takes place among farmers (between irrigators). In informal water markets, enforcement of contracts does not occur by recourse to legal and institutional measures but rather on personal trust (4). Some markets for water resemble close to perfect competition. Each seller is offering an identical product, so there is free entry and exit of farmers, no single buyer and seller can influence the price, and there is enough information as well (for instance, water markets in Gujarat). These water markets are seasonal and not permanent. Though the water transfer takes place mainly between irrigators, in some pockets like in periurban areas of Bangalore, a southern city of India, water is traded between irrigator and nonirrigators, water markets are localized, and the nature of competition resembles an oligopoly (7). Similar findings are reported by Palanisami (8) and Janakarajan (9). Nevertheless, the demand for groundwater remains large for agriculture.

EXTENT OF WATER MARKETS

The extent of the area irrigated through water markets, which is often considered a proxy of water trading, varies across regions as well as over time, depending on a number of factors such as rainfall, groundwater supply, cropping patterns, and the cost and availability of electricity (3).

It is estimated that 20% of the owners of the 14.2 million pump sets in India are likely to be involved in water trading (10); this indicates that water markets are providing water for about 6 million hectares, or 15% of the total area irrigated by groundwater (4). This is on the high side and therefore it cannot be a reliable estimate. Still, there is a lack of systematic estimates of the magnitude of water trading on the macrolevel. Using a methodology based on pump set rental data, Saleth (10) estimated that 6 million hectares or 15% of the total area under groundwater irrigation is on account of water trading. Further, by a net addition to output of $230/ha/year (based on the difference between the average irrigated and rain-fed yields, as reported by Government of India), the total value of output due to water sales is estimated at $1.38 billion per year (11). According to the irrigation and electricity acts, water selling is illegal, but it persists in niches.

ECONOMIC ASPECTS

The thesis behind the economics of groundwater use is that *water is used most efficiently when it is extracted at rates that maximize net benefits over time*. Hence, the benefit–cost calculus matters. The cost components include the cost of extracting and delivering the groundwater, and the opportunity or the uses to which the water is put determines user cost and benefits. For example, Diwakara and Nagaraj (7) indicated that the benefit–cost ratio for groundwater sales to an urban area in an informal setting is 2.67:1 for the construction and hotel industries compared to 1.35:1 for paddy crops. This signals that water transfer to nonagriculture is a lucrative business. The costs of extraction are primarily a function of pumping technology, the depth from which the groundwater can be pumped, and the costs of energy (mainly electricity in India). These costs increase with pumping depth and costs of energy and decrease as pump efficiency is improved. Thus the water market for agriculture is different from the water market for nonagriculture. Farmers selling water for nonagricultural purposes are relatively more efficient than farmers selling for agriculture because the marginal value product (MVP) of the former is higher than that of the latter.

In India, the cost of energy used to pump the water is charged on a flat rate (fixed) basis, $30.55 per horsepower (in Gujarat), indicating that the marginal extraction cost of pumping groundwater is zero; this encourages farmers to use water to the point where the marginal value of production is close to zero.

The appropriators and policy makers largely ignore the user pays principle, so water markets are inefficient and illegal because user costs are not borne either by the buyer or seller. This has encouraged water sellers to use electric motors to their full capacity at no marginal cost. Further, the cost of extraction also includes the value of the opportunity foregone by extracting and using the water immediately rather than at sometime in the future (opportunity cost). The user cost is a measure of the economic consequences of pumping now and thereby lowering the water table and increasing the cost of extraction for all future periods, that is, increasing negative economic effects that are not internalized. Therefore, sellers are more into a neoclassical return through water rates and interlocked markets.

INSTITUTIONAL (LEGAL) ASPECTS

So far in India, there has been little or no effort to devise appropriate water institutions to cope with the increasing demand for water in a dynamic agrarian economy. Solutions to the technical problems of groundwater resources are well known and are enforceable. However, efficient allocation and use of water resources depend on institutions, which have received lip service. An influential article, "The Tragedy of the Commons" by Garrett Hardin (12) encouraged scholars to hold the view that lack of private property rights to the resource has led to degradation and overexploitation of common pool resources (13). The solution is to promote self-governing institutions (14,15).

Albeit there has been an attempt to regulate groundwater resources through the "Model Bill to Regulate and Control the Development of Groundwater" in 1992, but it failed to address the groundwater extraction and use issues; rather, more emphasis was given to administrative control mechanisms. This misses the potential management opportunities represented by water markets (16). The National Water Policies of 1987 and 2002 have also failed to address regulatory issues to govern the private extraction of groundwater resources.

The issue of property rights has fueled the recent debate on groundwater ownership. It is believed that assigning property rights to groundwater and permitting competitive markets to function can finesse most of the inefficiencies of the common pool problem (17), but this neglects the fact that the appropriative activities by which individuals create effective property rights are themselves an alternative use of scarce resources (18). The priority for exclusive private ownership may be an obstacle to incompletely defining resources situations where some collective ownership may be more appropriate (19). The main function of property rights is to encourage greater internalization of effects (20,21). The rights to extract groundwater are by and large in the hands of private individuals who own land. So, the landlords are also water lords in India, provided the groundwater is available from the vulnerable aquifers. Groundwater is a common pool resource unit, so its yield is subtractable and its exclusion is not an issue of rivalry (22). Hence, defining the rights to use is a challenging task.

Current energy pricing (flat rate) is not desirable for India's groundwater sustainability. Even though some big farmers use high-energy diesel pumps to lift groundwater, the water extraction will not be and cannot be as much as that in the use of electrical pumpsets because farmers have to pay for diesel and not for electricity. Hence, groundwater mining and overuse of energy are interlinked and necessitate a high opportunity cost in water-scarce regions. In Gujarat State, energy pricing is US $37 per year per pump horsepower (23). Therefore, marginal cost pro-rata pricing for electricity could be one of the welfare-improving institutional measures that could be considered. However, as Zilberman (24) notes, these welfare measures would be felt only when the efficiency benefits from reform are more significant and apparent.

On the one hand, groundwater is treated as a "free good," and on the other, it is exploited competitively. This underscores the importance of well-defined, clearly enforceable rights to extract groundwater. Groundwater resources are largely extracted by individual well owners in India. In the absence of well-defined and enforceable rights to extract water, the availability of groundwater is determined and subject to the law of capture, that is, whoever taps the groundwater first gets to use it. In such a situation, pumpers have an incentive to extract as much water as possible, subject to the constraints imposed by pumping costs (25).

Access to groundwater, a fugitive (dynamic) resource, is largely a function of rights to extract. Though a legal framework that ties rights to groundwater to land ownership governs groundwater use in India, there is no legal limit on the amount of water a landowner can draw. Hence, markets for groundwater in India are informal and illegal and fluctuate in pricing as well as in pricing methods. For example, if a farmer is selling groundwater to his neighbor, it is also in return for some labor shortage situations or may be some interest payments and so on. These, too, are subsumed in the water charge paid by the buyer.

In the absence of a limit on the volume of groundwater extracted, farmers are tapping the resource with myopic behavior ignoring the fact that one's extraction is a function of the neighboring well's extraction at a time and over time leading to cumulative interference of wells that affects the lives of the wells (26). The lack of quantity regulation of groundwater extraction might lead to the *tragedy of the water law*. Hence, quantity regulations are common practices around the world and are preferred to taxes (27).

CONCLUDING REMARKS

Though water policy has generally been considered an important issue, its rational formulation and implementation have received lip service, and the water policies of the twenty-first century have to be significantly different from the past policies because of several changes that have already occurred in the water sector (28). India is gearing up to meet the challenges of the millennium with a fast developing economy. Throughout history, institutions have been a panacea for managing natural resources. Without suitable water institutions, the dream of achieving sustainable use of water resources will be in vain. Recognizing water markets as informal institutions that tackle the problem (access) and efficiency (high value use), appropriate measures through regulation of groundwater are the need of the hour for India. It is indicative that markets and states are now seen not as competing but as complementary institutions in the quest to "*get the rules right*," and many formulations see a broader range of institutions of economic governance as essential in this task, including small scale communities-neighborhoods, nongovernmental associations, and the like (29). In this regard, a blend of regulatory and economic measures applicable to Indian situation is suggested herewith.

There is a lack of explicit policy on water markets in India. Keeping in view the reckoned benefits of formal water markets, an attempt could be made to devise a set of rules focusing on property rights (tradable), quantity regulation, energy pricing, well-spacing norms, water laws to promote self-governing institutions that manage groundwater, and establishing an oversight commission to settle disputes between users and water authorities.

PROPERTY RIGHTS

Groundwater in India is largely developed by private investment, as in many parts of the world. However, the rights to extract water are still chattel to the land. The potential benefit of well-defined property rights is that it reduces transaction costs, and assuming access to credit, small farmers can have water rights. Considerable consensus exists that, for markets to be effective, legally defining property rights to water as private and transferable should be in place (4,30,31) along with a declaration of rights for efficient use of the resource (32). As the sovereign power and guardian of the nation's resources, as well as the protector of the law and rights, the government has the power to define necessary property rights and then dispose them to individuals (33). However, the third party effects of water markets need attention because the effects of water transfers and the costs of administering efficient and fair solutions to these third party effects could lead to restricting trade (34). In this regard, quantity regulation would be a suitable strategy.

QUANTITY REGULATION

The groundwater resource is extracted in unlimited quantities in India without regard to its impact on neighboring wells and environmental consequences. Evidence indicates that, in a variety of situations, governments regulate undesirable activities by controlling quantity rather than charging taxes (27). So, simply defining property rights is not a sufficient condition. It should be blended with quantity regulation. If quantity regulation is opted for and implemented, still, the access to groundwater resources is a function of number of users across time and space. Hence, to facilitate efficiency (physical and economic), the resource has to be priced to reflect its true value. This could be done through marginal cost pricing of electricity and groundwater.

MARGINAL COST PRICING

The energy used to lift the groundwater from the deeper layers of the aquifer in many parts of India is electricity and diesel to some extent. Electricity is priced on a fixed (flat rate) basis. Hence, the marginal cost of extraction is zero, and there is no incentive for users to conserve the resource. In a true economic sense, users extract groundwater until the marginal value product becomes zero. The value of groundwater needs to be reflected through marginal cost pricing of electricity and groundwater as well. This could be achieved by pricing electricity on a pro-rata basis.

WELL-SPACING NORMS

In the absence of well spacing norms, the well interference problem poses serious threats to sustainability and equity in well irrigation, especially in water-scarce regions in India. The withdrawal of groundwater without regard to recharge efforts goes unabated, and the resource itself becomes unsustainable by reducing the lives of wells. So, to avoid the well interference problem, which underscores sustainability, there is a need for spacing norms between wells.

PROMOTING SELF-GOVERNING INSTITUTIONS

The community should accept a law to allocate water rights, regulate quantities of water, implement marginal cost pricing, and impose well-spacing norms. Evidence indicates that, if the economic rents from water trading are concentrated in the hands of a few individuals or the negative effect on third-party users is large and unmitigated, the community is not likely to obey the law (4). Therefore, self-governing institutions, such as water users' cooperatives, associations, or small private well organizations need to be promoted by enacting a suitable water law. In any society, there is no guarantee that all stakeholders obey the law. To settle disputes between violators of the rules, there is a need for an oversight commission with representatives from users, local, state, and federal governments.

In conclusion, emphasis should be on collective action through user groups and cooperatively managed irrigation systems for sustainable use of water resources. Therefore, a combination of local water rights and user organizations should be encouraged. Nevertheless, the institutional linkage between water markets and rental markets for irrigation assets needs attention because the rental market allows farmers to irrigate their farms by renting irrigation assets from neighbors, and they contribute both to equity in water use and better use of irrigation assets, especially when groundwater rights are not defined (35).

These are some of the measures that are believed to be possible in the Indian scenario and could be incorporated into legislation, provided that the sociopolitical economy realizes the need and accepts them. Further, there shall not be any underenforced water law, especially in the context of the economic and environmental damage that has been created and unaccounted for largely due to policy failure.

BIBLIOGRAPHY

1. International Water Management Institute. (2002). The socio-ecology of groundwater in India. *Water Policy Briefing* **4**: 1–6.
2. Bhatia, B. (1992). Lush fields and parched throats: Political economy of groundwater in Gujarat. *Economic and Political Weekly* **XXVII**(51–52): A12–170.
3. Saleth, R.M. (1994). Groundwater markets in India: A legal and institutional perspective. *Indian Econ. Rev.* **29**(2): 157–176.
4. Easter, K.W., Rosegrant, W.M., and Dinar, A. (1999). Formal and informal markets for water: Institutions, performance, and constraints. *World Bank Res. Observer* **14**(1): 99–116.
5. Carey, J., Sunding, L.D., and Zilberman, D. (2002). Transaction costs and trading behaviour in an immature water market. *Environ. Dev. Econ.* **7**: 733–750.
6. Dubash, K.N. (2000). Ecologically and socially embedded exchange: Gujarat model of water markets. *Economic and Political Weekly* **April 15**: 1376–1385.
7. Diwakara, H. and Nagaraj, N. (2003). Negative impacts of emerging informal groundwater markets in peninsular India: Reduced local food security and unemployment. *J. Soc. Econ. Dev.* **5**(1): 90–105.
8. Palanisami, K. (1994). M.W. Rosegrant and R.G. Schleyer (Eds.). Tradable Water Rights Experiences in Reforming Water Allocation Policy, USAID, Arlington, VA.
9. Janakarajan, S. (1994). M. Moench (Ed.), *Selling Water: Conceptual and Policy Debates over Groundwater Markets in India*, VIKSAT, Ahmedabad.
10. Saleth, R.M. (1998). Water markets in India: Economic and institutional aspects. In: *Markets for Water: Political and performance*, K.E. William, M.W. Rosegrant, and A. Dinar (Eds.). Kluwer Academic, Boston, MA.
11. Mohanty, N. and Gupta, S. (2002). Breaking the gridlock in water reforms through water markets, Working Paper Series, Julian L. Simon Centre for Policy Research.
12. Hardin, G. (1968). The tragedy of the commons. *Science* **162**: 1243–1248.
13. Sethi, R. and Somanathan, E. (1996). The evolution of social norms in common property resource use. *Am. Econ. Rev.* **86**(4): 766–788.
14. Ostrom, E. and Gardner, R. (1993). Coping with asymmetries in the commons: Self-governing irrigation systems can work. *J. Political Econ.* **7**(4): 93–112.
15. Bardhan, P. (1993). Analytics of the institutions of informal cooperation in rural development. *World Dev.* **21**(4): 633–639.
16. Moench, M. (1998). M. Moench (Ed.), *Groundwater Law: The growing debate*, VIKSAT, Ahmedabad pp. 35–68.
17. Gisser, M. (1983). Groundwater: Focusing on the real issue. *J. Political Econ.* **91**(6): 1001–1027.
18. Grossman, I.H. (2001). The creation of effective property rights. *Am. Econ. Rev.* **91**(2): 347–352.
19. Dragun, K.A. (1987). Property rights in economic theory. *J. Econ. Issues* **XXI**(2): 859–868.

20. Demsetz, H. (1967). Toward a theory of property rights. *Am. Econ. Rev.* **57**(2): 347–359.
21. Dragun, K.A. (1983). Externalities, property rights, and power. *J. Econ. Issues* **XVII**(3): 667–680.
22. Walker, J.M. and Gardner, R. (1992). Probabilistic destruction of common pool resources: Experimental evidence. *Econ. J.* **102**: 1149–1161.
23. Gupta, K.R. (2002). Water and energy linkages for groundwater exploitation: A case study of Gujarat State, India. *Water Resour. Dev.* **18**(1): 25–45.
24. Zilberman, D. (2003). Water marketing in California and the West. *Int. J. Public Adm.* **26**(3): 291–315.
25. Committee on Valuing Groundwater. (1997). *Valuing Groundwater: Economic Concepts and Approaches*. National Academy Press, Washington, DC.
26. Chandrakanth, M.G. and Arun, V. (1997). Externalities in groundwater irrigation in hard rock areas. *Indian J. Agric. Econ.* **52**(4): 761–771.
27. Glaeser, L.E. and Shleifer, A. (2001). A reason for quantity regulation. *Am. Econ. Rev.* **91**(2): 431–435.
28. Biswas, K.A. (2001). Water policies in the developing world. *Water Resour. Dev.* **17**(4): 489–499.
29. Bowles, S. and Gintis, H. (2000). Walrasian economics in retrospect. *Q. J. Econ.* **November**: 1411–1439.
30. Bauer, J.C. (1997). Bringing water markets down to earth: The political economy of water rights in Chile, 1976–95. *World Dev.* **25**(5): 639–656.
31. Hearne, R.R. and Easter, K.W. (1995). Water Allocation and Water Markets: An Analysis of Gains-from-Trade in Chile, World Bank Technical Paper Number 315. The World Bank, Washington, DC.
32. Schlager, E. and Ostrom, E. (1992). Property rights regimes and natural resources: A conceptual analysis. *Land Econ.* **68**(3): 249–262.
33. Dragun, K.A. (1985). Property rights and Pigovian taxes. *J. Econ. Issues* **XIX**(1): 111–122.

READING LIST

Brennan, D. and Scoccimarro, M. (1999). Issues in defining property rights to improve Australian water markets. *Aust. J. Agric. Resour. Econ.* **43**(1): 69–89.

Richards, A. and Singh, N. (2001). No easy exit: property rights, markets, and negotiations over water. *Water Resour. Dev.* **17**(3): 409–425.

Saleth, R.M. (2004). Strategic analysis of water institutions in India: applications of a new research paradigm, Research Report, 79. International Water Management Institute, Colombo, Sri Lanka.

Shah, T. (1993). *Groundwater Markets and Irrigation Development: Political Economy and Practical Policy*. Oxford University Press, Bombay.

WATER RESOURCES OF INDIA

Sharad K. Jain
National Institute of Hydrology
Roorkee, Uttranchal, India

India occupies an important place in Southeast Asia. Its geographic area of 3.29 million square kilometers extends from 8° 4′ N and 37° 6′ N latitude and between 68° 7′ E and 97° 25′ E longitude. The country is bounded by the Himalayas in the north and has a large peninsular region tapering toward the Indian Ocean. The coastline of India extends for more than 6000 km. This unique combination of location and topography, a large peninsula, very high mountains followed by wide river plains, and the ocean in the south, create a complex climate and hydrology in the country.

From the physiographical point of view, India can be divided into seven broad regions: (1) northern mountains; (2) Plains of Indus and Ganga Basin; (3) Central Highlands; (4) peninsular plateau, the east coast region; and (5) the west coast region. Two chains of islands, the Andaman and Nicobar in the Bay of Bengal and Lakshadweep in the Arabian Sea, are also part of India.

The Himalayas in the north are one of the major mountain ranges of the world. In India, they extend over 2500 km from east to west, and the width varies from 250 to 400 km. Topographically, the Himalayas can be subdivided into three major ranges, the Greater Himalayas, the Middle Himalayas, and the Shivaliks whose average heights are 6000, 4000, and 1000 meters, respectively. The Himalayas are home to many of the highest peaks of the world and have large permanently snow-covered areas. Geologically, the Himalayas are comparatively young mountain chains that are formed by sedimentary rocks. This coupled with steep slopes and intense monsoon rainfall yields high sediment loads in the rivers. The other prominent mountains include the Aravalis, the Vindhyachals, the Satpuras, the Eastern Ghats, and the Western Ghats. The mountains are the primary source of rivers that derive their flow from rainfall and snow and glacier melt (Himalayan rivers). Plateaux, another striking feature of the topography in India, range in elevation from 300 to 900 meters. The major plateaux are the Malwa, the Vindhyas, the Chota Nagpur, the Satpura, and the Deccan.

India is traversed by a number of mighty rivers that can be broadly divided into two groups: the rivers of the Himalayan region and the peninsular rivers. The snow-fed Himalayan rivers are mostly perennial, whereas many small rivers of the peninsular region remain dry during the summer months.

CLIMATE OF INDIA

The mighty mountains, extensive plateaux, and the ocean in the south have an important bearing on the climate of India that can be described as a tropical monsoon climate. India is a country with extremes of climate. The climate varies from extremely hot to extremely cold, from extremely arid to extremely humid, and from continental to oceanic. Here, temperature varies from more than 47° C at some places in summers to below −40° C at many places in the Himalayas; rainfall varies from almost negligible to about 1100 cm. The eastern state of Meghalaya is the home to Cheerapunji, which is famous for the most rainfall in the world—on the order of 1142 cm in a year. Rainfall amounting to nearly 104 cm has been recorded at this place in a day. The potential evapotranspiration over the country varies from 150 to 350 cm. This wide range of climatic conditions working in conjunction with a range of topographic and soil/rock properties produces a complex and interesting pattern of water resource distribution across the country.

In India, rainfall is received through southwest and northeast monsoons, cyclonic depressions, and western disturbances. Most of the rainfall in India takes place under the influence of the southwest monsoon between June and September, except in Tamil Nadu where it occurs under the influence of the northeast monsoon from October–November. The rainfall in India shows great variations, unequal seasonal distribution, unequal geographic distribution, and frequent departure from the normal. The flow of monsoon winds is significantly influenced by high mountains in the north which are an effective barrier. The peninsular shape of the southern part that is close to the ocean provides a big source of moisture.

Monsoons

The southwest monsoon starts from the equatorial belt and hits the Indian subcontinent in two distinct currents. The Bay of Bengal branch sets in the northeastern part of the country, and the Arabian sea branch hits at the southern part of the peninsula. The first branch moves westward, and the second northward; together they cover the whole country. Normally, the monsoon sets in over the entire country by the beginning of July. The monsoon withdraws in September.

As the southwest monsoon withdraws, a northeasterly flow of air begins. This air picks up moisture from the low-pressure areas in the Bay of Bengal, hits the coastal areas of Orissa and Tamil Nadu, and causes rainfall. During this period, severe tropical cyclones are also formed in the Bay of Bengal and the Arabian Sea. These cyclones are responsible for intense rainfall in the coastal areas.

The water vapor carried by the monsoon from June–September amounts to about 11,100 km^3. About 3000 km^3 ($=$ 27% of the total) of this moisture precipitates as rainfall. During the remaining 8 months of the year, precipitation is of the order of 1000 km^3. In South Indian states such as Tamil Nadu, Andhra Pradesh, and Kerala, the northeastern monsoon also contributes significantly to precipitation.

At the beginning of December, weather disturbances originating in extratropical regions enter India from Afghanistan and Pakistan. These are known as western disturbances and cause moderate to heavy rain and snowfall in the northern mountainous region. Light to moderate rainfall is also experienced in the northern plains.

The annual precipitation over the country, including snowfall, is about 4000 km^3 which is equivalent to about 120 cm of depth. The variation of average annual rainfall over the country is shown in Fig. 1. As seen from the figure, there is a considerable amount of areal variation in the annual rainfall in India. The annual rainfall varies from about 10 cm in western deserts to about 1100 cm in northeastern parts of the country. More than half of the precipitation takes place in about 15 days and that too in about 100 hours in a year. The number of rainy days varies from about five in western deserts to 150 in the northeast. Such a wide variation makes the task of water resource engineers very challenging.

River Basins of India

A river basin, the natural context in which water occurs, is the most appropriate unit for planning, development, and management of water resources. The Ganga–Brahmaputra–Meghna basin, the largest basin in India, receives waters from an area that comprises about one third of the total area of the country. The second largest basin, that of Godavari, covers about 10% of the total area of India. The rivers are well spread over the entire country except for the Thar desert in Rajasthan. The catchment areas of the major river basins and the states through which they flow are given in Table 1. A map of India showing important rivers is given in Fig. 2.

Based on topography, the river systems of India can be classified into four groups: (1) Himalayan rivers, (2) Deccan rivers, (3) coastal rivers, and (4) rivers of the inland drainage basin. The Himalayan rivers receive input from rain as well as snow and glaciers and, therefore, are perennial. During the monsoon months, the Himalayas receive very heavy rainfall, and these are the periods when the rivers carry about 80% or more of the annual flow. This is also the period when floods are commonly experienced. The Deccan rivers are rain-fed and therefore have little or no flow during the nonmonsoon season. The coastal streams, especially on the west coast, have limited catchment areas and are not lengthy. Most of them are nonperennial. The streams of the inland drainage basin of Western Rajasthan are few and far between. They flow for a time only during the monsoon.

The main river systems in the Himalayas are those of the Indus and the Ganga–Brahmputra–Meghna. The River Indus rises near Mansarovar Lake in Tibet. Flowing through Kashmir, it enters Pakistan and empties into the Arabian Sea near Karachi. A number of important tributaries of the Indus, the Sutlej, the Beas, the Ravi, the Chenab, and the Jhelum, flow through India. The Bhagirathi and Alakhnanda are two important rivers that originate in the Garhwal Himalayas. These join at Dev Prayag to form the Ganga, the most sacred river of India. It traverses Uttranchal, Uttar Pradesh, Bihar, and West Bengal and thereafter enters Bangladesh. Some important tributaries of the Ganga are the Yamuna, the Ramganga, the Ghaghra, the Gandak, the Kosi, and the Sone. Many of these tributaries are mighty rivers themselves. For example, River Yamuna has its own mighty tributaries, Chambal and Betwa. The Brahmaputra rises in Tibet (where it is known by the name Tsangpo). It enters India in Arunachal Pradesh and after traversing Assam, enters Bangladesh. Its important tributaries are the Dibang, Luhit, Subansiri, Manas, and Tista. The Ganga and Brahmputra Rivers meet at Goalundo in Bangladesh. The Barak river, the head waters of the Meghna rises in the hills in Manipur. The Meghna is part of the Ganga–Brahmaputra–Meghna system. The combined Ganga–Brahmaputra River meets the Meghna in Bangladesh, and this combined system flows into the Bay of Bengal.

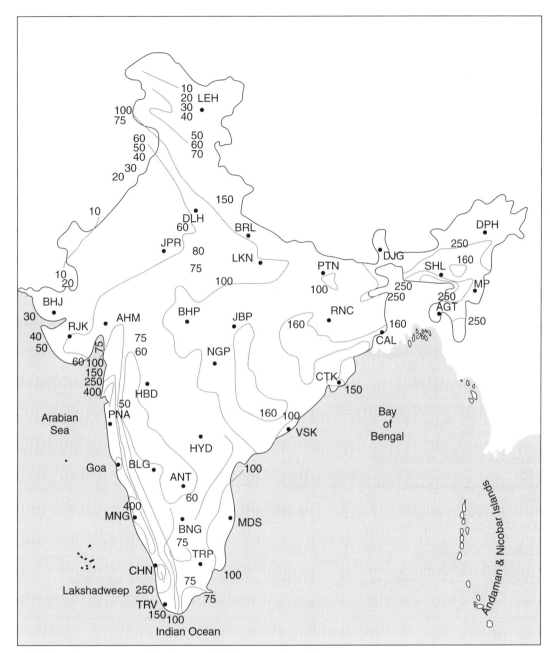

(i) Based upon survey of India map with the permission of the surveyor general of India
(ii) © Government of India copyright 1990.
(iii) The territorial waters of India extend into the sea to a distance of twelve nautical miles measured from the appropriate base line.

Figure 1. Annual rainfall across India (1).

The Deccan rivers can be further classified in two groups: west flowing rivers and east flowing rivers. The Narmada and the Tapi Rivers flow westward into the Arabian Sea. The important east flowing rivers are the Brahmani, the Mahanadi, the Godavari, the Krishna, the Pennar, and the Cauvery. These rivers flow into the Bay of Bengal. The rivers on the west coast are important because they contain as much as 14% of the country's water resources and drain only 3% of the land.

SURFACE WATER RESOURCES OF INDIA

The water resources of a basin are based on stream flow measured at a terminal site on the river. The average annual flow at the terminal point of a river is normally denoted as the water resource of the basin. Note that this refers to availability of water with a probability of 50%. To plan a water resource project, dependability at other levels such as 75% and 90% is needed. Most Indian rivers

Table 1. Major River Basins of India[a]

Sl. No.	River Basin	Catchment Area, km²	States in the Catchment
1	Ganga–Brahmaputra–Meghna Basin		
1a	Ganga Subbasin	862,769	Uttranchal, Uttar Pradesh, Himachal Pradesh, Haryana, Rajasthan, Madhya Pradesh, Bihar, West Bengal, and Delhi UT.
1b	Brahmaputra subbasin	197,316	Arunachal Pradesh, Assam, Meghalaya, Nagaland, Sikkim, and West Bengal
1c	Meghna (Barak) subbasin	41,157	Assam, Meghalaya, Nagaland, Manipur, Mizoram, and Tripura
2	Indus	321,289	J&K, Punjab, Himachal Pradesh, Rajasthan, and Chandigarh
3	Subernarekha	29,196	Bihar, West Bengal, and Orissa
4	Brahmani–Baitarani	51,822	Madhya Pradesh, Bihar, and Orissa
5	Mahanadi	141,589	Madhya Pradesh, Maharashtra, Bihar, and Orissa
6	Godavari	312,812	Maharashtra, Andhra Pradesh, Madhya Pradesh, Orissa, and Pondicherry
7	Krishna	258,948	Maharashtra, Andhra Pradesh, and Karnataka
8	Pennar	55,213	Andhra Pradesh and Karnataka
9	Cauvery	87,900	Tamil Nadu, Karnataka, Kerala, and Pondicherry
10	Tapi	65,145	Madhya Pradesh, Maharashtra, and Gujarat
11	Narmada	98,796	Madhya Pradesh, Maharashtra, and Gujarat
12	Mahi	34,842	Rajasthan, Gujarat and Madhya Pradesh
13	Sabarmati	21,674	Rajasthan and Gujarat
14	West flowing rivers of Kachchh, Saurashtra, and Luni	334,390	Rajasthan, Gujarat, Daman, and Diu
15	West flowing rivers south of Tapi	113,057	Karnataka, Kerala, Goa, Tamil Nadu, Maharashtra, Gujarat, Daman, Diu, and Nagar Haveli
16	East flowing rivers between Mahanadi and Godavari	49,570	Andhra Pradesh and Orissa
17	East flowing rivers between Godavari and Krishna	12,289	Andhra Pradesh
18	East flowing rivers between Krishna and Pennar	24,649	Andhra Pradesh
19	East flowing rivers between Pennar and Cauvery	64,751	Andhra Pradesh, Karnataka, and Tamil Nadu
20	East flowing rivers south of Cauvery	35,026	Tamil Nadu and Pondicherry Union Territory
21	Area of North Ladakh not draining into the Indus	28,478	Jammu and Kashmir
22	Rivers draining into Bangladesh	10,031	Mizoram and Tripura
23	Rivers draining into Myanmar	26,271	Manipur, Mizoram, and Nagaland
24	Drainage areas of Andaman, Nicobar and Lakshadweep Islands	8,280	Andaman, Nicobar, and Lakshadweep
	Total	3,287,260	

[a] Reference 1.

carry large flows during the monsoon season (June to September). The average annual flow in important Indian rivers is given in Table 2.

Usable Surface Water Resources

Though the estimated surface water availability is of the order of 1869 km³, the amount of water that can be put to beneficial uses is much less because India experiences monsoon climate, implying that nearly 80–90% of the annual runoff occurs during monsoon months. As enough storage capacity is not available, a large part of this flow goes to waste, and some of it also causes flood damage.

As is clear from Table 2, the storage space available from the projects completed by 1995 is about 174 km³; the projects under construction would provide 76 km³, a total of about 250 km³ of storage space, ignoring small structures. This is about 36% of the possible storage space of 690 km³. To harness the usable surface water, about 400 km³ of additional live storage needs to be created. Thus, the development of surface water has a long way to go to achieve full potential.

GROUNDWATER RESOURCES ASSESSMENT

Groundwater resources can be classified as static and dynamic. The static resource can be defined as the amount of groundwater available in the permeable portion of the aquifer below the zone of water level fluctuation. The dynamic resource is the amount of groundwater available in the zone of water level fluctuation. The usable groundwater resource is essentially a dynamic resource which is recharged annually or periodically

Figure 2. Important rivers of India and flood-prone areas (1).

by rainfall, irrigation return flows, canal seepage, and influent seepage.

Due to diversified meteorologic, topographic, and geologic features, the groundwater distribution is quite divergent across the country. Rock formations vary from Archean to Recent, and the topography ranges from the mountain terrain of Himalayas to the flat alluvial plains and coastal tracts. The northern plains in the Indus–Ganga–Brahmputra Basins that extend over a distance of 2400 km from east to west constitute one of the largest groundwater reservoirs in the world. Almost the entire peninsular part is covered by a variety of hard and fissured rock formations. The coastal and deltaic tracts form a narrow strip around the peninsula; some of them contain highly productive aquifers. The basin-wise groundwater potential of the country is given in Table 3.

The groundwater resources available for irrigation are about 360.3 km^3; out of this, the usable quantity (90% of potential) is about 324.27 km^3. Thus the total usable groundwater resource is $324.27 + 71.08$(domestic and other uses) $= 395.35 \text{ km}^3$. The present level of development is about 32%.

The water that is present in the aquifer below the zone of water level fluctuation is known as static groundwater. The static groundwater resource of the country is estimated at $10,812 \text{ km}^3$.

To summarize, the usable volume of water from surface resources is 690 km^3, and from groundwater sources 396 km^3, giving total annual usable water resources of

Table 2. Surface Water Resource Potential of the River Basins of India, km³

Sl. No.	Name of the River Basin	Average Annual Potential in the River	Estimated Usable Flow, Excluding Groundwater	Live Storage		
				Completed Projects	Ongoing Projects	Proposed Projects
1	Indus (area in Indian territory)	73.31	46.00	13.83	2.45	0.27
2	(a) Ganga	525.02	250.00	36.84	17.12	29.56
	(b) Brahmaputra, Barak, and others	585.60	24.00	1.10	2.40	63.35
3	Godavari	110.54	76.30	12.51	10.65	8.28
4	Krishna	78.12	58.00	34.48	7.78	0.13
5	Cauvery	21.36	19.00	7.43	0.39	0.34
6	Pennar	6.32	6.86	0.38	2.13	—
7	East flowing and rivers from Mahanadi to Godavari and Krishna to Pennar	22.52	13.11	1.63	1.45	0.86
8	East flowing rivers between Pennar and Kanyakumari	16.46	16.73	1.42	0.02	—
9	Mahanadi	66.88	49.99	8.49	5.39	10.96
10	Brahmani and Baitarani	28.48	18.30	4.76	0.24	8.72
11	Subarnarekha	12.37	6.81	0.66	1.65	1.59
12	Sabarmati	3.81	1.93	1.35	0.12	0.09
13	Mahi	11.02	3.10	4.75	0.36	0.02
14	West flowing rivers of Kutch and Saurastra, including Luni	15.10	14.98	4.31	0.58	3.14
15	Narmada	45.64	34.50	6.60	16.72	0.46
16	Tapi	14.88	14.50	8.53	1.00	1.99
17	West flowing rivers from Tapi to Tadri	87.41	11.94	7.10	2.66	0.84
18	West flowing rivers from Tadri to Kanyakumari	113.53	24.27	10.25	2.31	1.70
19	Area of inland drainage in Rajasthan desert	Negligible		—	—	—
20	Minor rivers draining to Myanmar (Burma) and Bangladesh	31.00		0.31	—	—
	Total	1869.00	690.00	173.73	75.42	132.30

1086 km³. In addition, about 10–15% of this quantity will be available as return flow from irrigation and domestic uses. As against this, the total withdrawal for all uses in the year 1990 was 552 km³. The irrigation sector accounted for about 83% of this use.

Most of the groundwater development in India is up to a depth of about 50 meters and is financed mainly by institutional and private sources. In the shallow region, the wells include dug wells, dug and bore wells, and shallow tube wells. The exploitation in the deeper zone (50 m and more) is usually through the public sector for community irrigation, water supply, or industrial purposes. Groundwater development has been intensive in the plains of the Indus–Ganga basin of Punjab, Haryana, and Uttar Pradesh.

WATER RESOURCES REQUIREMENT

The increase in population and expansion of economic activities inevitably lead to rising demands for water for diverse purposes. Surface water is either used in-stream for hydropower and navigation or is diverted for off-stream use. A small part of diverted water returns to streams or aquifers, and the rest is consumed. Groundwater is used mostly for irrigation or for domestic requirements.

The Indian economy has traditionally been agriculturally based. At independence, it was of crucial importance to develop irrigation to increase agricultural production to make the country self-sustaining and to alleviate poverty. Accordingly, the irrigation sector was assigned a very high priority in the 5-year plans. Giant schemes such as the Bhakra Nangal, Hirakud, Damodar Valley, Nagarjunasagar, and Rajasthan Canal project were undertaken to increase irrigation potential and maximize agricultural production.

Long-term planning has to take into account the growth of population. A number of individuals and agencies have estimated the likely population of India by the year 2025 and 2050. According to the estimates adopted by NCIWRD(1), by the year 2025, the population is expected to be 1,333 million in the high growth scenario and 1,286 million in the low growth scenario. In the year 2050, the high rate of population growth is likely to result in about 1,581 million people, whereas the low growth projections place the number at nearly 1,346 million. Keeping in view the level of consumption, the estimated food grain requirements per capita per year would be 218 kg for the year 2025 and 284 kg for the year 2050. After considering factors such as feed requirement, losses in storage and transport, seed requirements, and buffer stock, the projected food-grain and feed demand for 2025 would be 320 million tonnes (high demand scenario) and 308 million tonnes (low demand scenario). The requirements for the year 2050 would be 494 million

Table 3. Groundwater Potential in the River Basins of India (Pro Rata Basis) in km^3/yr^a

Sl. No.	Name of the Basin	Total Replenishable Groundwater Resources	Provision for Domestic, Industrial, and Other Uses	Available Groundwater for Irrigation	Net Draft	Balance of Groundwater Potential	Level of Groundwater Development, %
1	Brahmani with Baitarni	4.05	0.61	3.44	0.29	3.16	8.45
2	Brahmaputra	26.55	3.98	22.56	0.76	21.80	3.37
3	Chambal Composite	7.19	1.08	6.11	2.45	3.66	40.09
4	Cauvery	12.30	1.84	10.45	5.78	4.67	55.33
5	Ganga	170.99	26.03	144.96	48.59	96.37	33.52
6	Godavari	40.65	9.66	30.99	6.05	24.94	19.53
7	Indus	26.49	3.05	23.43	18.21	5.22	77.71
8	Krishna	26.41	5.58	20.83	6.33	14.50	30.39
9	Kutch & Saurashtra Composite	11.23	1.74	9.49	4.85	4.64	51.14
10	Chennai and South Tamil Nadu	18.22	2.73	15.48	8.93	6.55	57.68
11	Mahanadi	16.46	2.47	13.99	0.97	13.02	6.95
12	Meghna	8.52	1.28	7.24	0.29	6.95	3.94
13	Narmada	10.83	1.65	9.17	1.99	7.18	21.74
14	Northeast Composite	18.84	2.83	16.02	2.76	13.26	17.20
15	Pennar	4.93	0.74	4.19	1.53	2.66	36.60
16	Subarnarekha	1.82	0.27	1.55	0.15	1.40	9.57
17	Tapi	8.27	2.34	5.93	1.96	3.97	33.05
18	Western Ghat	17.69	3.19	14.50	3.32	11.18	22.88
	Total	431.43	71.08	360.35	115.21	245.13	31.97

^{a}Reference 2.

tonnes (high demand scenario) and 420 million tonnes (low demand scenario).

Irrigation

The irrigated area in the country was only 22.6 million hectare (Mha) in 1950–1951. Food production was much below the requirement of the country, so due attention was paid to expansion of irrigation through surface and groundwater projects. The ultimate irrigation potential of India has been estimated at 140 Mha. Out of this, 76 Mha would come from surface water and 64 Mha from groundwater sources.

The quantity of water used for irrigation in the last century was of the order of 300 km^3 of surface water and 128 km^3 of groundwater. The estimates indicate that by the year 2025, the water requirement for irrigation would be 561 km^3 for a low demand scenario and 611 km^3 for a high demand scenario. These requirements are likely to increase to 628 km^3 for a low demand scenario and 807 km^3 for a high demand scenario by the year 2050.

Domestic Use

Water for the community water supply is the most important requirement, and it is about 5% of total water use. It is estimated that about 7 km^3 of surface water and 18 km^3 of groundwater are being used for community water supply in urban and rural areas. Along with the increase in population, another important change in water supply is a higher rate of urbanization. Per the projections, the higher the expected growth, the higher would be urbanization. It is expected that nearly 61% of the population will be living in urban areas by the year 2050 in a high growth scenario as against 48% in a low growth scenario.

Different organizations and individuals have given different norms for water supply in cities and rural areas. The figure adopted by NCIWRD (1) was 220 liters per capita per day (lpcd) for class I cities. For cities other than class I, the norms are 165 for the year 2025 and 220 lpcd for the year 2050. For rural areas, 70 lpcd and 150 lpcd have been recommended for the years 2025 and 2050. Based on these norms and the projection of population, it is estimated that by the year 2050, the water requirements per year for domestic use will be 90 km^3 for a low demand scenario and 111 km^3 for a high demand scenario. It is expected that about 70% of the urban water requirement and 30% percent of the rural water requirement will be met by surface water sources and the remaining from groundwater.

Hydroelectric Power

The hydropower potential of India has been estimated at 84,044 MW at a 60% load factor. At independence (1947), the installed capacity of hydropower projects was 508 MW which was about 37% of the total installed capacity. By the end of 1998, the installed hydropower capacity was about 22,000 MW which was 24.85% of the total installed capacity of 88,543 MW. The status of hydropower development in major basins is quite uneven. For example, the hydropower potential of Brahmputra Basin is about 34,920 MW which is 41.5% of the total potential of the country. Only 1.3% of this was developed by the year 1998. The percentage development in Indus was about 14.7% and 17.26% in Ganga. The major hydroelectric projects (above 750 MW) in India are Bhakra, Dehar, Koyna, Nagarjunasagar, Srisailam, Sharavarthy, Kalinadi, and Idukki.

Table 4. Annual Water Requirement for Different Uses in km³

Uses	Year 1997–1998	Year 2010			Year 2025			Year 2050		
		Low	High	%	Low	High	%	Low	High	%
Surface Water										
Irrigation	318	330	339	48	325	366	43	375	463	39
Domestic	17	23	24	3	30	36	5	48	65	6
Industries	21	26	26	4	47	47	6	57	57	5
Power	7	14	15	2	25	26	3	50	56	5
Inland navigation		7	7	1	10	10	1	15	15	1
Flood control		—	—	0	—	—	0	—	—	0
Environment (1) afforestation		—	—	0	—	—	0	—	—	0
Environment (2) ecology		5	5	1	10	10	1	20	20	2
Evaporation losses	36	42	42	6	50	50	6	76	76	6
Total	399	447	458	65	497	545	65	641	752	64
Groundwater										
Irrigation	206	213	218	31	236	245	29	253	344	29
Domestic	13	19	19	2	25	26	3	42	46	4
Industries	9	11	11	1	20	20	2	24	24	2
Power	2	4	4	1	6	7	1	13	14	1
Total	230	247	252	35	287	298	35	332	428	36
Total Water Use										
Irrigation	524	543	557	78	561	611	72	628	807	68
Domestic	30	42	43	6	55	62	7	90	111	9
Industries	30	37	37	5	67	67	8	81	81	7
Power	9	18	19	3	31	33	4	63	70	6
Inland navigation	0	7	7	1	10	10	1	15	15	1
Flood control	0	0	0	0	0	0	0	0	0	0
Environment (1) afforestation	0	0	0	0	0	0	0	0	0	0
Environment (2) ecology	0	5	5	1	10	10	1	20	20	2
Evaporation losses	36	42	42	6	50	50	6	76	76	7
Total	629	694	710	100	784	843	100	973	1180	100

Industrial Water Requirement

Rough estimates indicate that present water use in the industrial sector is of the order of 15 km³. The water use by thermal and nuclear power plants with installed capacities of 40,000 MW and 1,500 MW (1990 figures), respectively, is estimated at about 19 km³. In view of the shortage of water, the industries are expected to switch over to water efficient technologies. If the present rate of water use is continued, the water requirement for industries in the year 2050 would be 103 km³, and if water saving technologies are adopted on a large scale, the requirement is likely to be nearly 81 km³.

Water Quality

Water quality issues are gaining recognition as river waters are getting heavily polluted in many places and groundwater quality at many places is beginning to deteriorate. Among the main concerns for water quality are (1) arsenic in drinking water in West Bengal (affecting about 5 million persons); (2) fluoride levels are high in Andhra Pradesh, Gujarat, Haryana, Karnataka, Punjab, Rajasthan, Tamil Nadu, and Madhya Pradesh (affecting 14 million persons); (3) iron levels are high in the northeastern and eastern parts of the country (affecting 29 million persons); and (4) salinity is high in Gujarat, Haryana, Karnataka, Punjab, Rajasthan, and Tamil Nadu.

Total Water Requirements

Based on the various studies, the total annual requirement for freshwater from various sectors is estimated at about 1050 billion m³ by 2025 A.D. The breakdown is shown in Table 4. Estimates show that this demand will be met by harnessing 700 billion m³ of surface water and 350 billion m³ of groundwater.

FLOODS AND DROUGHTS

Floods and drought are two water-related natural calamities that are recurrent phenomena in India.

Floods

Floods are the most frequent natural calamities faced by India in different magnitudes, year after year. About 80–90% of the annual precipitation in India takes place during 4 months of monsoon; this is also the season when floods are mostly experienced. The main causes of floods in India are inadequate capacity of river sections to contain high flows, silting of river beds, and drainage congestion. Floods are also caused by cyclones, cloud bursts, and snow/glacier melt. Based on the causes of floods, the country can be divided into four basins/regions. In the Brahmaputra and Barak Basin, the main problems in the state of Assam are inundation caused by overflow of the Brahmaputra and its tributaries and erosion along the river bank. In the Ganga Basin, the flooding problem is confined to small portions in the middle and terminal reaches. In general, the severity of the problem increases from west to east and from south to north. The worst flood affected states in the Ganga Basin are Uttar Pradesh, Bihar, and West Bengal. In Uttar Pradesh, flooding is confined to the eastern districts: the rivers that cause flooding include the Sarada, the Ghagra, the Rapti, and the Gandak. The major causes here are drainage congestion and bank erosion. North Bihar is in the grip of floods almost every year due to spillage of rivers. In West Bengal, floods are caused by drainage problems as well as tidal effects. Flood problems are not very severe in the rivers in the northwest part of India, and here the cause of floods is inadequate drainage capacity. Flash floods are also experienced here. Flash floods are characterized by a sharp rise and recession of river flow, and the damage is mainly due to the sudden rise of the river, leaving a very short time for evacuation. In the central and southern part of the country, floods occur in the Narmada, Tapi, Godavari, Mahanadi, and Krishna Rivers. The coastal regions of Orissa, Andhra Pradesh, and Tamil Nadu also face problems due to cyclones.

According to the estimates, the average area annually affected by floods is 7.52 Mha of which the agricultural area is 3.52 Mha. According to NCIWRD (1), during the second half of the twenty-first century, on average, 1515 lives and 95,285 head of cattle were lost every year. The area liable to floods was assessed at 40 Mha which is about one-eighth of the geographical area of the country.

Drought

Normally, the term drought is used for a shortage of precipitation or water at a place when it is expected. Meteorologic drought is said to occur at a place when there is more than a 25% decline in rainfall from the normal value. Meteorologic drought over an extensive area triggers hydrologic drought which is the condition of significant depletion of water resources in rivers, reservoirs, lakes, and springs. A marked depletion of soil moisture and precipitation leads to agricultural drought. Based on detailed drought studies for the country, 99 districts have been identified as drought prone. The total geographic area of these districts is about 1,081,131 sq. km.

BIBLIOGRAPHY

1. NCIWRD. (1999). *Integrated Water Resources Development—A Plan for Action*. Report of *The National Commission for Integrated Water Resources Development*. Govt. of India, Ministry of Water Resources, New Delhi.
2. IWRS. (1998). *Theme Paper "Five decades of water resources development in India"*. Indian Water Resources Society, Roorkee.

READING LIST

CGWB. (1995). *Ground Water Resources of India, Central Ground Water Board*, Govt. of India, New Delhi.
IWRS. (1999). *Theme Paper "Water: Vision 2050"*. Indian Water Resources Society, Roorkee.
Rao, K.L. (1973). *India's Water Wealth*. Orient Longman Limited, New Delhi.

WATER INFRASTRUCTURE AND SYSTEMS

Neil S. Grigg
Colorado State University
Fort Collins, Colorado

Water infrastructure is the set of constructed elements that operate within water resources systems and interact with environmental elements, or natural systems, such as watersheds, streams, lakes, and aquifers (see article on Water Resources Management). Table 1 gives examples of structures used in constructed systems for different purposes of water management.

In a watershed, sets of these structures may control water for different purposes. Most apparent might be a multiple-purpose reservoir, that has a dam and hydroelectric plant connected to electric transmission lines. Below might be a diversion dam to enable a canal to take irrigation water from the stream, if it is located in a dry region. There might be a re-regulating reservoir with locks for navigation. Dams are a special class of hydraulic structure, and they provide barriers to the flow to enable reservoirs to store water.

Dams have a number of special components, including service and emergency spillways, outlet structures,

Table 1. Structures for Water Resources Management

Purpose	Conveyance	Storage	Treatment[a]	Pump or Generate	Control
Water supply	Water pipes	Reservoir	WTP	Pumps	Valves
Water quality	Sewer pipes	Tanks, ponds	WWTP	Sewage pumps	Gates
Flood control	Channels	Detention ponds	SWTP	Flood pumps	Inlets
Hydropower	Penstocks	Reservoir	–	Turbines	Gates
Navigation	Locks	Reservoir	–	–	Valves
Environment	Rivers	Natural storage	Wetland	–	–

[a]WTP = water treatment plant; WWTP = wastewater treatment plant; SWTP = stormwater treatment plant.

and drains. Outlet works might discharge to a municipal water supply treatment plant, and further downstream a wastewater treatment plant might discharge to the stream.

A reservoir is a lake where water is stored, either natural or dammed artificially. Reservoirs require special management attention because they control flow throughout river systems. As a consequence, they tend to be both strategically important and controversial. Great dams such as Hoover Dam on the Colorado River, Aswan Dam on the Nile River, and Three Gorges Dam on the Yangtze River in China influence whole regions and even nations.

To control flows, there will be valves, gates and spillways, but they are mostly invisible. Wells might provide water for users such as farming operations or irrigation of golf courses.

Diversion structures on a stream change its direction or flow patterns. They might include intakes, boat chutes, or river training structures. A fish ladder is a special kind of diversion structure meant to provide migrating fish species with methods for swimming upstream, usually around a dam.

The structures and components that are associated with conveyance systems include natural open channels and canals, pipelines, pipe networks and sewers, bridges, and levees. Open channels are either natural or man-made. A river is an example of a natural open channel. Levees are part of conveyance systems that form the banks of channels to protect land areas from flooding. A canal or lined drainage ditch would be an example of a man-made channel. A lock raises or lowers boats and ships up or down a river.

Pipes, or closed conduits, can be classified as tunnels, transmission pipelines, pressure pipe networks, or sewer networks.

A bridge is part of a conveyance system in the sense that it provides a method for separating a stream bed from a road, a rail line, or some other structure crossing a stream. A culvert has a function somewhat like a bridge that enables a separation of grades between a small stream and a roadway or other embankment above.

Hydroelectric plants are facilities that generate electric energy from water discharge. They include turbines, generators, and associated hydraulic components.

Pumps impart energy to water and raise it in elevation or pressure. There are several different types of pumps, built mainly around the centrifugal or turbine pump format, and they have numerous applications.

Valves, gates and spillways are control devices for conveyance systems or dams. Spillways are used as emergency overflow devices to protect dams. There are a number of types, normally classified as the service spillway or emergency spillway. Water flowing over a spillway normally flows into some type of energy dissipation device to avoid erosion downstream.

Coordinating the operation of these components places the focus on the water resources *system*, as opposed to individual pieces. A water resources system is a combination of constructed water control facilities and environmental elements that work together to achieve water management purposes (see article on WATER RESOURCES MANAGEMENT). A constructed system consists of structural facilities that control water flow and quality, and a natural water resources system comprises environmental elements such as watersheds, stream channels, and groundwater systems.

OVERVIEW AND TRENDS IN THE INTERNATIONAL WATER MARKET

RANDY T. PIPER
Dillon, Montana

OVERVIEW AND TRENDS

The term "water market" is a slippery concept. Like water, the term water market is free flowing and can take a variety of forms, depending on the cultural environment and who invokes the term. In "The Importance of Getting Names Right: The Myth of Markets for Water," Joseph Dellapenna argues for greater precision when using the language of water markets (1). Consequently, when we speak of water markets, we need to identify specifically the rules that governments employ and the roles that governments play in defining, divesting, and defending water markets. In most geographic settings, water markets do not exist without some formal role for government administration.

In Brazil, local water markets have evolved with minimal government oversight. Dating from 1854 in the Cariri region of northeast Brazil, sugar cane farmers established a water market. The farmers themselves defined the amount of water for each farm. Moreover, farmers developed water measuring devices and monitoring rules independent of formal government administration. However, this water market limits who can own the water, that is, water is neither auctioned to the highest bidder nor sold to nonfarmers who would divert the water for fisheries or for nonagricultural urban consumption (2).

Technology can and does play an important role in water markets. For example, technology can make bottled water more affordable to consumers and more aesthetically appealing to consumers. In 2001, Norwegian water in American markets gained fame and market share largely because of the package design. Technology can improve the quality of water, from wastewater treatment plants to desalination. For example, 120 countries now employ some form of desalting equipment (3). At the 2001 Bahrain meeting of the World Congress of International Desalination Association, the primary focus was on the technologies of water markets, not the institutions of water markets (4). Technology can improve the institutions of water markets and make it more cost-effective for governments and nongovernments to define and enforce the boundaries of water markets and to monitor the consumption and transactions of water markets (5).

Technology in the narrow sense can and does improve the product and package of water and water markets. Technology in the broad sense can and does improve the institutions of water markets. However, the technology and science of water markets per se are not the paramount challenges in many international markets. In 2000, the 10th Stockholm Water Symposium reached this conclusion: "Long-term challenges in water management are not so much linked to classical scientific and technical aspects as to institutional innovations" (6).

In "Water Challenge and Institutional Response," R. Maria Saleth and Ariel Dinar reviewed the institutional innovations of 11 countries: Australia, Brazil, Chile, China, India, Israel, Mexico, Morocco, South Africa, Spain, and Sri Lanka. These World Bank researchers found several commonalities among the 11 countries. For example, countries have reached a supply limit of water resource development and are now focusing on the rules of allocation of their supplies. Countries have also begun to experiment with decentralization and privatization. Among the countries (Australia, Israel, Mexico, and Spain) that have national water plans, Israel and Mexico delegate much of the allocation administration to subnational levels that are organized primarily along hydrogeologic lines in the form of basins (7). These river basin organizations then may develop rules of participation and the roles of market-oriented institutions for allocation. According to R. Maria Saleth and Ariel Dinar, "Countries have begun to recognize the functional distinction between centralized mechanisms needed for coordination and enforcement and decentralized arrangements needed for user participation and local level solutions" (8).

In the urban water subsector, another institutional innovation is privatization or corporatization of water and water infrastructures. Privatization can take a number of forms, from management, service, and lease contracts to build, own, operate, and transfer contracts to full private ownership under a regulatory regime. In the 1990s, several transnational corporations acquired water firms and/or formed subsidiaries to compete in international markets. This trend is illustrated by noting the home country's parent company followed by its water subsidiary: England's Kelda and Alcontrol, France's Vivendi and Vivendi Environment, Germany's RWE and Bovis Thames, and America's Enron and Azurix. In comparison to the 1990–1994 period, during the 1995–1999 period, the monetary value of water and sanitation privatization projects increased by 100% to approximately $9 billion in Latin America and approximately $9 billion in East Asia and Pacific regions, by more than 1000% to approximately $1 billion in sub-Saharan Africa and approximately $2 billion in Europe and Central Asia regions, and by more than 2000% to approximately $5 billion in the Middle East and North Africa regions (9). The ownership and names of the transnational corporations will change based on the ebb and flow of supply–demand interactions. However, the role of any newly named corporation will remain the same: to maximize its returns on equity for various privatization projects.

Some institutional analysts consider the two trends of decentralization and privatization innovative and preferable to centralized national ownership and management. However, other institutional analysts view these trends as less than innovative and undesirable. A statement from the conclusions of the 10th Stockholm Water Symposium captures this sentiment: "The economically based approach that dominates at present, and has the market as a recommended mode of interpretation, was seen as having severe limitations when it comes to acceptable ways of coping with the emerging water crisis" (10).

Even when countries can overcome the cultural and linguistic barriers to crafting water market-oriented institutions, these countries still encounter other barriers, primarily cost-effective technology. Cost-effective technology exists for the defining, defending, and divesting of water *quantity*. For example, affordable geographic information systems (GIS) allow governments and nongovernments to monitor whether water users are "rustling water" and consuming illegal quantities. However, when the focus turns to water *quality*, the costs of technology for measuring and monitoring pollution trespassers increase dramatically (11). The wealthier countries of the European Union, South America, Asia, and Middle East can afford some of these emerging technologies—both for water quantities and qualities—but less affluent countries cannot allocate sufficient resources to create water markets, whether for quantities or qualities.

SUMMARY

Generalizations remain precarious. Institutional successes at one national level may not transfer to other national levels within the same region. Institutional successes at one national level may not transfer to subnational levels within the same country. The barriers to diffusion of institutional innovations are numerous and varied, flowing from economic, political, and social to technological and linguistic barriers.

Governments will continue with cross-country agreements that more clearly define their respective rights, rules, and roles. National governments may play a more active role in achieving hydrosolidarity among themselves, but this external hydrosolidarity does not preclude internal hydrosolidarity in the form of denationalization and water-market-oriented institutions. In a few countries, governments may well continue to finance, manage, and own water resources and view water as a nationalized "means of production." However, as the World Bank's R. Maria Saleth and Ariel Dinar observe, "But nowhere, even in China, is the state's absolute ownership of water established to exclude private use rights" (8).

Many governments will still continue to denationalize and privatize water resources. In urban areas, private corporations will continue to build and manage water systems, while the government finances the construction and capital costs, and consumers finance the operating costs via user fees. When urban consumers cannot finance the full capital and operating costs, governments will experiment with full-cost pricing mechanisms, transparency of subsidies, and "water vouchers." In rural areas, national governments may allocate some of the financing, managing, and owning to local governments. In turn, localized governments may experiment with a variety of market-oriented institutions. The structure of these institutions could include allowing current owners and users to sell their water for nonagricultural uses.

During the next few decades, it is unlikely that many governments will permit the following scenario for international water markets to develop. Both national and local governments give full private property rights to agriculture and/or urban "owners." With full rights, these owners could then sell their water to whomever they wanted. Willing higher and highest bidding buyers and consumers would include rural and urban residents, transnational corporations, environmental groups, and even water speculators (12). Once governments and nongovernments have experimented with internal hydrosolidarity and learned the lessons from the multiple forms of water-market-oriented institutions such as water banks and water vouchers, governments and nongovernments may well transfer this knowledge to the realm of external hydrosolidarity.

Finally, when we interpret "international water markets" and we more precisely speak in terms of water-market-oriented institutions, our language and logic must accurately reflect the multiple roles that governments play in the creation and evolution of international water markets.

BIBLIOGRAPHY

1. Dellapenna, J.W. (2000). The importance of getting names right: the myth of markets for water. *William and Mary Environ Law Policy Law Rev.* **25**(Winter): 317–378.
2. Kemper, K.E. (1999). Jose Yarley de Brito Goncalves, and Francisco William Brito Bezerra, Water allocation and trading in the Cariri region—Ceara, Brazil. In: *Institutional Frameworks in Successful Water Markets: Brazil, Spain, and Colorado, USA.* M. Marino and K.E. Kemper (Eds.). World Bank Technical Paper 427, February, Washington, DC, pp. 1–9.
3. Gleick, P.H. (2000). *The World's Water, 2000-2001: The Biennial Report on Freshwater Resources.* Island Press, Covelo, CA, p. 96.
4. International Desalination Association. Accessed at http://www.ida.bm on August 24, (2001).
5. Landry, C.J. (2001). The role of geographic information systems in water rights management. In: *The Technology of Property Rights.* L.A. Terry and P.J. Hill (Eds.). Rowman & Littlefield Publishers, Lanham, MD, p. 23.
6. Stockholm International Water Institute. *The Final Overall Conclusions: The 10th Stockholm Water Symposium.* August 14–17, 2000, p. 1. Accessed at http://www.siwi.org/sws2000/eng/sws2000conclusions.html on September 24, 2001.
7. Maria, R.S. and Ariel, D. (1999). *Water Challenge and Institutional Response: A Cross-Country Perspective.* World Bank Policy Research Working Paper 2045, January, Washington, DC, p. 36.
8. Saleth and Dinar, p. 37.
9. The Water Page. Accessed at http://www.thewaterpage.com/ppp_new_main.htm on October 22, 2001.
10. Stockholm International Water Institute. *The Final Overall Conclusions: The 10th Stockholm Water Symposium.* August 14–17, 2000, p. 3. Accessed at http://www.siwi.org/sws2000/eng/sws2000conclusions.html on September 24, 2001.
11. Yandle, B. (2001). Legal foundations for evolving property rights technologies. In: *The Technology of Property Rights.*

T.L. Anderson and P.J. Hill (Eds.). Rowman & Littlefield Publishers, Lanham, MD, pp. 12–18.
12. For a review of the dynamics of domestic and international water markets, see *Water Marketing—The Next Generation.* T.L. Anderson and P.J. Hill (Eds.). 2000. *The Political Economy of Water Pricing Reforms* Ariel Dinar (Ed.). 2000, and Water Bank at http://www.WaterBank.com.

BEST MANAGEMENT PRACTICES FOR WATER RESOURCES

MOHAMMAD N. ALMASRI
An-Najah National University
Nablus, Palestine

JAGATH J. KALUARACHCHI
Utah State University
Logan, Utah

INTRODUCTION

Best management practices (BMPs) for water resources are affordable, practical, and effective methods for eliminating or reducing the movement of pollutants from the ground surface to groundwater or surface waterbodies. Such pollutants include nutrients and pesticides, among others, that may have adverse effects on human health. In addition, BMPs can be viewed as methods to ensure a sustainable and safe yield of water resources through optimal conjunctive management of these resources. The main premise in developing BMPs is to achieve a balance between the goal of protecting water resource quality, for instance, and the economic and social ramifications of adopting these BMPs. Before adopting BMPs for a specific site, it is essential to understand the BMPs thoroughly and to ensure their technical feasibility through research, experience, field pilot studies, and surveys. In addition to technical feasibility, BMPs should be economically well founded through cost-effectiveness analysis and acceptable via adopting and maintaining them. BMPs may consist of single practices or combinations of them based on the problem at hand and the social and economic ramifications. BMPs for groundwater resources are important because they preserve and protect groundwater quality and quantity. Preservation of groundwater resources minimizes public health problems and maintains the social and economic values of groundwater. For instance, the cost of mitigating polluted groundwater can be extremely high and by adopting effective BMPs, such mitigation costs can be reduced dramatically.

This document focuses on BMPs to protect and preserve groundwater resources from nitrate contamination.

NITRATE CONTAMINATION OF GROUNDWATER

Groundwater is the primary source of drinking water in many parts of the world and the sole supply of potable water in many rural communities (1). There is increasing awareness that groundwater is vulnerable to contamination from domestic, industrial, and agricultural wastes (2). Sources of groundwater contamination are widespread and include accidental spills, landfills, storage tanks, pipelines, agricultural activities, and many other sources. Of these sources, agriculture activities produce nonpoint source pollution in small to large watersheds especially due to nitrogen in fertilizers and various carcinogenic substances found in pesticides. Due to its high mobility, nitrate is the primary nitrogen species lost from soils by leaching (3). Groundwater contamination by nitrate has been confirmed throughout the United States (4,5) and worldwide (6). Elevated nitrate concentration in drinking water can cause *methemoglobinemia* in infants and stomach cancer in adults (7–9). Therefore, the U.S. Environmental Protection Agency (U.S. EPA) has established a maximum contaminant level (MCL) of 10 mg/L NO_3-N (10). Nitrate may indicate the presence of bacteria, viruses, and protozoa in groundwater if the source of nitrate is animal waste or effluent from septic tanks. Likewise, nitrate contamination of surface water has health and environmental effects. Transport of nitrate to surface water occurs mainly via discharge of groundwater during baseflow conditions (11). Hence, prevention of groundwater contamination protects surface water quality as well.

Agricultural activities are probably the most significant anthropogenic source of nitrate contamination in groundwater. Nitrogen is a vital nutrient for plant growth. Nevertheless, when nitrogen-rich fertilizer application exceeds the plant demand and the denitrification capacity of the soil, nitrogen can leach to groundwater usually in the form of nitrate (12). Many studies have shown high correlation between agricultural land use and nitrate concentration in groundwater (6,13,14). Agricultural practices result in nonpoint source pollution of groundwater and the effects of these practices accumulate over time (7). Nonpoint sources of nitrogen from agricultural practices include fertilizer, dairy farms, manure application, and leguminous crops (15). Point sources of nitrogen such as septic tanks and dairy lagoons contribute to nitrate pollution of groundwater.

BMPs TO MINIMIZE NITRATE CONTAMINATION OF GROUNDWATER

Identification of areas with heavy nitrogen loadings from point and nonpoint sources is important for land use planners and environmental regulators. Once such high-risk areas have been identified, BMPs can be implemented to minimize the risk of nitrate leaching to groundwater. The knowledge of the spatial distribution of nitrogen loading can identify the areas where groundwater needs to be protected. This assessment is of great importance in designating areas that can benefit from pollution prevention and monitoring programs. In large areas such as large watersheds or basins, proper understanding of on-ground nitrogen loading from different sources and corresponding transformations in the soil are needed before areas can be designated for future protective measures (16).

BMPs developed to protect groundwater quality are improvements in agricultural practices and land use patterns (17). Improving agricultural practices focuses on tillage practices, crop rotation, rational fertilizer and manure applications, vegetation cover, and irrigation management (18). For instance, on-ground application of manure should be in accordance with agronomic requirements and the nitrogen mass in the soil (19). Land use changes include reevaluation of land use distribution and changes in the land use in areas that are vulnerable to contamination. However, this option may not be feasible in many instances due to competing stakeholder concerns arising from economic constraints and these constraints may, sometimes, override environmental concerns. BMPs imply conflicting objectives. On the one hand, the main aim of a BMP is to reduce nitrate concentrations in groundwater below the MCL. On the other hand, a protection alternative should consider the minimization of economic losses incurred from implementing the protection alternatives. Therefore, a multicriteria decision analysis might be developed to determine the most viable BMP that accounts for the different decision criteria.

Fertilizer and manure applications are required to replace crop land nutrients that have been consumed by previous plant growth. They are essential for economical yields. However, excess fertilizer and manure use and poor application methods can lead to nitrate pollution of groundwater and surface waterbodies. Therefore, it is important to match nitrogen applications to crop uptake to minimize nitrate leaching. This is the core of the BMPs that address the problem of nitrate contamination of groundwater resources.

Effective BMPs limit movement by minimizing the quantity of nutrients available for leaching below the root zone. This is achieved by developing an inclusive nutrient management plan that uses field studies, soil and groundwater models, uses only the types and amounts of nutrients necessary to produce the crop, applies nutrients at the proper times and by proper methods, implements additional farming practices to reduce nutrient losses, and follows proper procedures for fertilizer storage and handling. In the following, the main BMPs pertaining to fertilizer and manure applications for reducing nitrate occurrences in the groundwater are summarized based on the work provided by Waskom (20), the EPA (21), Almasri (22), and Almasri and Kaluarachchi (23), among other studies listed in the references.

Application Rates and Fertilizer Types

One component of a comprehensive nutrient management plan is determining proper fertilizer application rates (24). The objective is to limit fertilizer application to an amount sufficient to achieve a realistic crop yield. Allowing for other nitrogen sources in the soil is also part of the concept. Previous legume crops, irrigation water, manure, and soil organic matter all contribute nitrogen to the soil. Along with soil samples and fertilizer credits from other sources, nitrogen fertilizer recommendations are based on yield goals established by crop producers. Yield expectations are established for each crop and field based on soil properties, available moisture, yield history, and management level.

Applying the appropriate form of nitrogen fertilizer can reduce leaching.

Nitrate forms of nitrogen fertilizer are readily available to crops but are subject to leaching losses. Nitrate fertilizer use should be limited when the leaching potential is moderate to high. In these situations, ammonium nitrogen fertilizers should be used because they are not subject to immediate leaching. However, ammonium nitrogen transforms rapidly into nitrate via the nitrification process, especially when soils are warm and moist. More slowly available nitrogen fertilizers should be used in these conditions. Nitrification inhibitors can also delay the conversion of ammonium to nitrate under certain conditions (23).

Fertilizer Application Timing

Nitrogen fertilizer applications should be timed in periods of maximum crop uptake. Fertilizer applied in the fall causes groundwater quality degradation. Partial application of fertilizer in the spring, followed by small additional applications as needed, can improve nitrogen uptake and reduce leaching risk.

Fertilizer Application Methods

Fertilizer application equipment should be checked at least once a year and should be properly calibrated to ensure that the recommended amount of fertilizer is applied. Fertilizer placement in the root zone can greatly enhance plant nutrient uptake and minimize leaching losses. Subsurface applied or incorporated fertilizer should be used instead of surface broadcast fertilizer. An efficient application method for fertilizer is to place dry fertilizer into the ground in bands closer to the seed, so that it can be recovered by the crop very efficiently. To reduce losses through surface runoff and volatilization, all fertilizers should be mechanically incorporated into the soil. Fertilizer application to frozen ground should be avoided and limited on slopes and areas with high runoff or overland flow.

Irrigation Water Management

Irrigation water should be managed to maximize efficiency and minimize runoff or leaching. Irrigated crop production has the greatest potential for source water contamination because of the large amount of water applied and the movement of nutrients via irrigation flux. Nitrate can leach into groundwater when excess water is applied to fields. Irrigation systems, such as sprinklers, low-energy precision applications, surges, and drips, allow producers to apply water uniformly and with great efficiency. Efficiency can also be improved by using delivery systems such as lined ditches and gated pipe, as well as reuse systems such as field drainage recovery ponds that efficiently capture sediment and nutrients. Gravity-controlled irrigation or furrow runs should be shortened to prevent overwatering at the top of the furrow before the lower end is adequately watered.

Manure Application Reduction

Manure is an excellent source of plant nutrients, especially nitrogen. Nevertheless, excessive applications of manure lead to nitrogen buildup in the soil that eventually leads to groundwater pollution. Prior to application, manure should be analyzed to determine its nutrient content. The application of increasing quantities of manure to the same land area may result in groundwater quality problems and adverse environmental consequences. If it turns out that the land base is insufficient for manure application, then BMPs should be introduced. In the following subsections, BMPs pertaining to dairy manure such as reducing herd size, composting/exporting manure, or implementing feeding strategies to reduce nutrient content in the excrement are summarized (25).

Dairy Herd Size Reduction. Manure production and loading are functions of dairy herd size, so downsizing the dairy herd is apparently the most straightforward and effectual BMP that minimizes manure loading (25). Nonetheless, such an option has serious economic ramifications that may prohibit the adoption of this alternative.

Manure Composting/Exporting. Manure exporting is a viable alternative because it does not involve herd size reduction. For manure to be exported, it should be composted. Composting is a biological process that converts organic manure to a more stable material such as humus. Composting is the aerobic decomposition of organic matter by certain microorganisms such as bacteria that consume oxygen and use nutrients such as carbon, nitrogen, phosphorus, and potassium as they feed on the organic waste. They produce heat, carbon dioxide, and water vapor. Heat destroys pathogens and weed seeds. The resulting composted manure is humus-like organic material, fine-textured, low in moisture, and has a nonoffensive earthy odor. Because of the carbon dioxide and water vapor that escape during the process, the resulting compost can be approximately half the volume and weight of the original material. Efficient composting requires maintaining proper temperatures and oxygen levels in the composting material (26). Warm temperatures help microorganisms to grow best. Elevated temperatures are needed to destroy pathogens and weed seeds. Bulking materials can be mixed with manure to provide structural support when manure solids are too wet to maintain air spaces within the composting pile. Composting is best accomplished by placing the manure in windrows. The windrows can be turned and mixed periodically to maintain oxygen levels for proper composting. The windrows may be 3 to 6 feet high and 10 to 20 feet wide, depending mostly on the type of machinery used in turning them. The width of the windrow must allow air movement into the manure to introduce oxygen. Composting reduces spreading costs, facilitates land application, eliminates manure odors, and promotes manure transporting and marketing. Composted manure can be used mainly on lawns and gardens to keep and add organic matter in the soil. Organic matter provides good aeration porosity, holds more water, and releases important nutrients.

Improve Dairy Cow Diets. The future challenge for dairy producers and nutritionists will be to formulate rations properly for high milk production levels while simultaneously minimizing the environmental impact of excessive nitrogen content in the excrement. A properly formulated ration that precisely meets the cow's requirements for milk production, maintenance, and growth minimizes excessive nitrogen in the manure. Many dairy producers overfeed crude protein to support high levels of milk production. However, this results in excessively high nitrogen in the excrement. The protein that is not used for milk production or maintenance and growth is excreted as urea or organic-N. This practice has an adverse effect on the environment. Feeding strategies can be adopted to minimize manure nitrogen content while maintaining the same milk production levels. Van Horn (27) showed that when cows are precisely fed to meet the National Research Council recommendations (28), the nitrogen content in manure dropped by 14%.

BIBLIOGRAPHY

1. Solley, W.B., Pierce, R.R., and Perlman, H.A. (1993). Estimated use of water in the United States in 1990. U.S. Geological Survey Circular 1081.
2. Almasri, M.N. and Kaluarachchi, J.J. (2004). Assessment and management of long-term nitrate pollution of ground water in agriculture-dominated watersheds. *J. Hydrol.* **295**: 225–245. doi:10.1016/j.jhydrol.2004.03.013.
3. Tesoriero, A., Liecscher, H., and Cox, S. (2000). Mechanism and rate of denitrification in an agricultural watershed: Electron and mass balance along ground water flow paths. *Water Resour. Res.* **36**(6): 1545–1559.
4. Spalding, R.F. and Exner, M.E. (1993). Occurrence of nitrate in groundwater—a review. *J. Environ. Qual.* **22**: 392–402.
5. Nolan, B.T. (2001). Relating nitrogen sources and aquifer susceptibility to nitrate in shallow ground waters of the United States. *Ground Water* **39**(2): 290–299.
6. Hudak, P.F. (2000). Regional trends in nitrate content of Texas groundwater. *J. Hydrol.* **228**: 37–47.
7. Addiscott, T.M., Whitmore, A.P., and Powlson, D.S. (1991). Farming, fertilizers and the nitrate problem. CAB International, Wallingford, United Kingdom.
8. Lee, Y.W. (1992). Risk assessment and risk management for nitrate-contaminated groundwater supplies. Unpublished Ph.D. Dissertation. University of Nebraska, Lincoln, NE.
9. Wolfe, A.H. and Patz, J.A. (2002). Reactive nitrogen and human health: Acute and long-term implications. *Ambio* **31**(2): 120–125.
10. U.S. Environmental Protection Agency. (1995). Drinking water regulations and health advisories. Office of Water, Washington, DC.
11. Bachman L.J., Krantz, D.E., and Böhlke, J, (2002). Hydrogeologic framework, ground-water geochemistry, and assessment of N yield from base flow in two agricultural watersheds, Kent County, Maryland. U. S. Environmental Protection Agency, EPA/600/R-02/008.
12. Meisinger, J.J. and Randall, G.W. (1991). Estimating N budgets for soil-crop systems. In: *Managing N for groundwater quality and farm profitability*. R.F. Follet, D.R. Keeney, and R.M. Cruse (Eds.). Soil Science Society of America, Madison, WI, pp. 85–124.
13. Ator, S.W. and Ferrari, M.J. (1997). Nitrate and selected pesticides in ground water of the Mid-Atlantic Region. United States Geological Survey. Water-Resources Investigation Report 97–4139.
14. Harter, T., Davis, H., Mathews, M., and Meyer, R. (2002). Shallow groundwater quality on dairy farms with irrigated forage crops. *J. Contam. Hydrol.* **55**: 287–315.
15. Cox, S.E., and Kahle, S.C. (1999). Hydrogeology, groundwater quality, and sources of nitrate in lowland glacial aquifer of Whatcom County, Washington, and British Columbia, Canada. USGS Water Resources Investigation Report 98–4195, Tacoma, Washington.
16. Schilling, K.E. and Wolter, C.F. (2001). Contribution of base flow to nonpoint source pollution loads in an agricultural watershed. *Ground Water* **39**(1): 49–58.
17. Latinopoulos, P. (2000). Nitrate contamination of groundwater: Modeling as a tool for risk assessment, management and control. In: *Groundwater Pollution Control*. K.L. Katsifarakis (Ed.). WIT Press, Southampton, United Kingdom, pp. 7–48.
18. Goss, M.J. and Goorahoo, D. (1995). Nitrate contamination of groundwater: Measurement and prediction. *Fert. Res.* **42**: 331–338.
19. Livingston, M.L. and Cory, D.C. (1998). Agricultural nitrate contamination of ground water: An evaluation of environmental policy. *J. Am. Water Resour. Assoc.* **34**(6): 1311–1317.
20. Waskom, R.M, (1994). Best Management Practices for Nitrogen Fertilization. Colorado State University Cooperative Extension, (XCM-172).
21. U.S. Environmental Protection Agency. (2001). Source water protection practices bulletin: Managing agricultural fertilizer application to prevent contamination of drinking water. Office of Water, 4606, EPA 916-F-01-028, Washington, DC.
22. Almasri, M.N. (2003). Optimal management of nitrate contamination of ground water. Ph.D. Dissertation, Utah State University, Logan, UT.
23. Almasri, M.N. and Kaluarachchi, J.J. (2004). Implications of on-ground nitrogen loading and soil transformations on ground water quality management. *J. Am. Water Resour. Assoc.* **40**(1): 165–186.
24. Yadav, S.N. and Wall, D.B. (1998). Benefit-cost analysis of best management practices implemented to control nitrate contamination of groundwater. *Water Resour. Res.* **34**(3): 497–504.
25. Davis, J., Koenig, R., and Flynn, R. (1999). Manure best management practices: A practical guide for dairies in Colorado, Utah, and New Mexico. Utah State University Extension, AG-WM-04.
26. Pace, M.G., Miller, B.E., and Farrell-Poe, K.L. (1995). The composting process. Utah State University Extension, AG-WM 01.
27. Van Horn, H.H. (1992). Recycling manure nutrients to avoid environmental pollution. In: *Large Dairy Herd Management*. H.H. Van Horn and C.J. Wilcox (Eds.). American Dairy Science Association, Champaign, IL, pp. 640–654.
28. National Research Council. (1989). *Nutrient Requirements of Dairy Cattle*, 6th Revised Edn. National Academic Sciences, Washington, DC.

INTEGRATED WATER RESOURCES MANAGEMENT (IWRM)

MUKAND SINGH BABEL
Asian Institute of Technology
Pathumthani, Thailand

TODAY'S WATER PROBLEMS

Population and economic growth has led to rising demand for water for human consumption, food production, industrial uses, and other development activities, and opportunities for increasing the supply are becoming prohibitively expensive. The growing concern for maintaining natural ecosystems is exerting further pressure on water resources. The threatening environmental problems originating from unsustainable use of water resources are of great concern. Pollution of water is connected with human activities. Proliferation of development activities and modern living styles has also contributed to the worsening of water quality. Adverse conditions such as social inequity, economic marginalization, and lack of poverty alleviation programs in many countries also force people living in extreme poverty to overexploit soil and forestry resources, which often results in negative impacts on water resources. Deteriorating water quality caused by pollution makes safe use difficult for downstream users and threatens human health and the functioning of aquatic ecosystems.

In most countries, water policies have been largely dominated by a supply-oriented and compartmentalized sector mentality. Top-down approaches have prevailed in development and management of water resources without participation by stakeholders in decision making. In these traditional approaches, the important distinction between managing water as a commonly owned natural resource and providing services for water users has not been fully recognized, leading to a confusion of roles and responsibilities of different organizations dealing with water. Thus, inefficient governance and increased competition for finite water resources are the root cause of water problems.

Balancing and compromising the need for water by people, different economic sectors, and nature is the main challenge faced by the water sector. About 1 billion people in the world are without access to safe drinking water, and about 3 billion people do not have access to adequate sanitation. Securing water for people is a prime concern in many countries. According to the UN medium projections, the population on earth in 2025 will reach about 7.8 billion, a 38% increase over present levels. Providing water to meet the needs of food and fiber for the world population in years ahead is another main challenge. Even with improved irrigation management and water productivity, it is estimated that 17% more water will be required for this population (1). Ecosystems are sources of various benefits to society such as providing timber, fuel, wildlife habitats, aquatic resources, etc. Ecosystem security through maintenance of the natural environment and proper management of land and water resources is therefore essential.

Variation of precipitation over time and space leads to the classical problem of too much or too little water. The variations in water flow and groundwater recharge, whether of climatic origin or due to land mismanagement, can add to drought and floods, which can have catastrophic effects in terms of large-scale loss of human life and damage to economic, social, and environmental systems. Managing the variability of water in time and space and the associated risks are other challenges.

CONCEPTS OF IWRM

Compared to traditional approaches to water resources problems, integrated water resources management (IWRM) takes a broader holistic view and examines a more complete range of solutions. It looks outside the narrow water sector for policies and activities to achieve sustainable water resources development. It also considers how different actions affect and can reinforce each other. IWRM has attracted particular attention since the International Conferences on Water and Environmental in Dublin in 1992. IWRM can be defined as follows:

"IWRM is a process which promotes the coordinated development and management of water, land and related resources in order to maximize the resultant economic and social welfare in an equitable manner without compromising the sustainability of vital ecosystems" (2).

IWRM principles place its overriding importance over economic, social, and natural facets. It draws its inspiration from the Dublin principles, culminating from the International Conference on Water and the Environment in Dublin, 1992. These are the four Dublin principles:

- Freshwater is a finite and vulnerable resource, essential to sustain life, development and the environment.
- Water development and management should be based on a participatory approach, involving users, planners, and policy makers at all levels.
- Water has an economic value in all its competing uses and should be recognized as an economic good.
- Women must play a central part in the provision, management and safeguarding of water.

In the traditional approach, the planning and operation of water systems is usually fragmented, causing a lack of coordination, wastage of resources, and conflict among stakeholders. Water issues are also generally neglected when decisions are made about crop patterns, trade and energy policies, and urban design and planning, all of which are critical determinants of water demand. For the sustainable use of water resources, there is a need to create institutions and frameworks that can transcend these traditional boundaries and involve a variety of users and stakeholders. IWRM promotes a holistic view and looks at the entire hydrologic cycle and the interaction of water with other natural and socioeconomic systems. IWRM focuses on integration, participation, consultation, gender awareness, and consensus.

Integration among different stakeholders and different domains of water resources is one of the highlights of IWRM. At its most fundamental level, IWRM is as concerned with the management of water demand as with its supply. Thus, integration can be considered in two basic categories—the natural system and the human system. The natural system puts critical importance on resource availability and quality; the human system fundamentally determines the resource use, waste production, and pollution of the resource. Integration of the natural system implies a concern with upstream–downstream water-related interests, integration of land and water management, integration of freshwater management and coastal zone management, a unified management of surface water and groundwater, a shift to management at a river basin level, integration of quantity and quality, and matching water management with other sectoral policies with a collateral impact (trade, housing, energy, agriculture, etc.). Human system integration involves mainstreaming of water resources, cross-sectoral integration in national policy development, analyzing macroeconomic effects of water development on overall economic development, influencing economic sector decisions, and integrating all stakeholders in the planning and decision-making process.

As one of the principles of the IWRM, the concept of water as an economic good introduces a wholly new model to water resources management. Therefore, it differs from the traditional approach in that it invariably associates economic aspects of water with water resources management issues. It distinguishes between the value of water and its pricing. The concept of the value of water is used in assisting allocation processes; the pricing concept is used as part of the cost recovery issue. The economic value of water is highlighted through greater stress on demand management rather than supply-side management, a recognition (and estimation, where possible) of the economic value of water in different uses, acceptance of the notion of opportunity cost (what is lost to other uses from taking it for a particular purpose), and attention to cost recovery, though with concern for affordability and securing access for the poor.

Water conservation is another critical aspect of IWRM. Freshwater ecosystems face enormous threats, directly and indirectly, from human activities. Conserving freshwater systems is essential for the future survival of all living species on the earth. Water conservation and management call for the protection, improvement, and use of water according to principles that will assure their highest economic and social benefits. Maintaining the delicate and sustainable balance of demand and supply of water is a critical element of any sound water strategy.

To fulfill desirable social, economic, and natural conditions, IWRM places its overriding criteria on economic efficiency in water use, equity and access for all, and sustainability of vital ecosystems. The increasing scarcity of water as well as financial resources, under increasing pressure from the excessive demands upon it, demands that water be used with maximum possible efficiency. The "social equity and access for all" criteria recognize the basic human right to access to water of adequate quantity and quality. The third criterion (ecological sustainability) is about ensuring that the present use of the resource does not undermine ecological stability and that future generations are not adversely affected.

Three basic elements highlight the IWRM framework: the enabling environment, the institutional roles, and the management instruments, as shown in Fig. 1.

The enabling environment refers to the general framework of national policies, legislation, and regulations. The

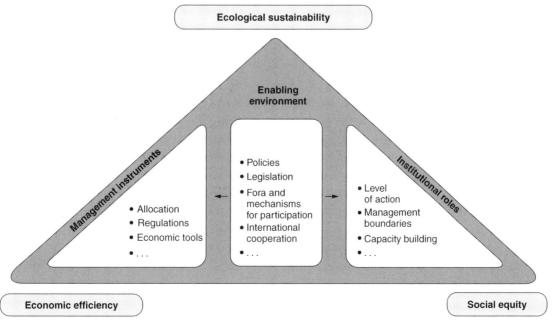

Figure 1. General framework for IWRM (2).

institutional framework defines the roles and responsibilities of government agencies at different levels and other stakeholders. The management instruments include water resources assessment, communication and information systems, allocation, economic tools, technology, and so on.

In spite of considerable advantages in IWRM, its implementation requires a much greater commitment from all stakeholders. There are many challenges, which have to be overcome for successful implementation. The short-term costs and disadvantages may seem more apparent than tangible benefits in the pursuit of longer term advantages. Integration is difficult to implement at first and may not guarantee equity for all; simply involving a wider range of stakeholders in decision making may not automatically ensure fair treatment for them.

BIBLIOGRAPHY

1. Seckler, D., Amarasinghe, U., Molden, D., de Silva, R., and Barker, R. (1998). *World Water Demand and Supply, 1990 to 2025: Scenarios and Issues, Research Report 19*. International Water Management Institute, Colombo, Sri Lanka.
2. GWP. (2000). *Integrated Water Resource Management, TAC Background Paper No. 4*. Global Water Partnership Technical Advisory Committee, Stockholm, Sweden.

MANAGEMENT OF WATER RESOURCES FOR DROUGHT CONDITIONS*

WILLIAM R. WALKER
Virginia Water Resources
Research Center

MARGARET S. HREZO
Radford University
Virginia

CAROL J. HALEY
Virginia Water Resources
Research Center

INTRODUCTION

Droughts have been a part of our environment since the beginning of recorded history, and humanity's survival may be testimony only to its capacity to endure this climatic phenomenon. According to Yevjevich and others (1, p. 41), one of the earliest records of efforts to plan for droughts is found in the biblical story of Joseph.

This article is a US Government work and, as such, is in the public domain in the United States of America.
*This article is reproduced in its entirety from Walker, W.R., Hrezo, M.S., and Haley, C.J. (1991). Management of water resources for drought conditions. In: *National Water Summary 1988-89—Hydrologic Events and Floods and Droughts*. Paulson, R.W., Chase, E.B., Roberts, R.S., and Moody, D.W., (Compilers). U.S. Geological Survey Water-Supply Paper 2375, pp. 147–156.

The pharaoh of Egypt gave Joseph the authority to plan and implement the world's first recorded national drought strategy. One can ask if improvements in society's planning capability over the past several millenniums have increased our ability to deal with droughts in other than catastrophic terms. Certainly, droughts will continue to be a randomly occurring climatic phenomenon. Have we built on any past successes pertaining to the planning for droughts and managing water resources during droughts; have we learned from past actions or inactions; or are we doomed to repeat our mistakes? The evidence indicates that society tends to be unwilling to plan for droughts.

To better understand why society is still reacting to droughts rather than planning for their eventuality, the activities of various levels of government in planning for droughts must be examined. Until very recently, the policies of most State governments for dealing with drought have been to "wait 'til it rains" and in the meantime provide some form of emergency assistance to localities and hope a catastrophe can be avoided. Many States also, as a matter of implicit policy, place primary responsibility for drought action with the Federal Government and local governments who, throughout the Nation's history, have been the primary levels of government involved when droughts occur. The policies of both the Federal Government and the local governments also have been to provide emergency relief and to try to reduce water demand to match the available supply. This governmental approach, however, has not reduced the economic losses or the level of inconvenience and suffering of the Nation's citizens. With each succeeding drought, the cycle repeats itself. As a result, the following questions arise: Do the States have a responsibility for planning for droughts? If the States do have such a responsibility, what type of planning should take place? Is there a possibility that drought-related activities of the Federal Government and the local governments are likely to change in the future, and if so, how and why?

In strictly climatic terms, a drought can be defined as an interval of time, generally months or years in duration, during which the actual moisture supply at a given place consistently is less than the climatically expected or climatically appropriate moisture supply (2, p. 3). Drought generally is defined as being meteorological, hydrological, or agricultural. However, the ultimate consequences of droughts have to be placed in the context of the effects on the social and economic activities of a given region (Evan C. Vlachos, Colorado State University, written commun., January 1988). Thus, the climatic attributes of drought also need to be defined in social and economic terms because it is in these contexts that water management becomes important (Evan C. Vlachos, Colorado State University, written commun., January 1988). Yevjevich and others (1, p. 32) have suggested that we think in terms of "sociological drought," which is defined as the meteorological and hydrological conditions under which less water is available than is anticipated and relied on for the normal level of social and economic activity of the region.

Due in part to our agrarian heritage and because water deficiencies can develop rather quickly in the root zone, most droughts are thought of as agricultural droughts. This characterization can be confusing, however, because an agricultural drought can occur in the midst of a hydrological wet period. The distribution of precipitation during a year can be such that there is a shortage of moisture (drought) during a critical growing period for a variety of crops, yet the total precipitation for the year can be greater than the historical yearly average. In this discussion of institutional and management aspects, droughts are considered as moisture shortages that seriously affect the established economy and the quality of life of a region.

The following analysis of the roles of government in planning for droughts begins with an examination of the traditional activities of government during a drought. If the traditional approach by government seems less than desirable, then the obstacles to greater involvement need to be identified and evaluated to determine if they can be removed or circumvented. The suggestions by King and others (3) in their "Model Water Use Act" for managing water resources during droughts are examined in some detail to determine if they provide a minimum approach for States to consider in developing a plan for managing water resources for drought conditions. To better assess the status of State involvement, the drought programs and drought-related actions of the 48 conterminous States are reviewed in terms of the minimum criteria recommended by King and others (3). Lastly, the expanded role that the Federal Government might have in planning for and mitigating the effects of droughts pursuant to existing Federal water-quality legislation is considered.

TRADITIONAL GOVERNMENT APPROACH TO DROUGHTS

Droughts have two components—climatic (decrease in precipitation) and demand (use of water). In responding to droughts, governments tend to concentrate most of their activity on reducing the demand for water, although they might have limited options for controlling the climatic component. Cloud seeding, for example, might make increasing the snowpack possible and, in some instances, obtaining a limited summer rainfall. Because climatic control, however, generally is less reliable, most measures focus on management, reallocation, and distribution of existing water sources and on establishing priorities accordingly for different uses (4, p. 41).

LOCAL GOVERNMENT

Traditionally, managing water resources during droughts has been based on immediate reactions to a current crisis. The focus of most action is to reduce the daily demand for water, and local governments usually are responsible for reducing water demand within their jurisdictions. Broader issues that impinge on their programs to reduce water demands are not within their purview; for example, local governments are not permitted to allocate surface water and groundwater among competing users. Local governments, however, do not covet the existing authority for reducing water demands. Even in times of impending crisis, there is a great reluctance to impose water-conservation measures, if there is any hope that rain will fall in time to save officials from having to do so. Decision makers are not popular when they must halt or reduce industrial activity, curtail domestic use, or prohibit "nonessential" services. As a result, timely action rarely is achieved (Evan C. Vlachos, Colorado State University, written commun., January 1988).

The local approach to the management of water resources during droughts is not responsive to other drought-induced issues such as minimum instream flows because these issues are not within the purview of local government. Yet water for wildlife, increased contamination due to low flows, and decreased navigation and hydroelectric-power generation are issues equally as important as the ones being addressed by local government. Water shortages cause low streamflows that have an adverse effect on fish and wildlife habitat. If there are no limits as to how much flows can decrease or for how long (the longer the period of low flow, the greater the stress), the recovery time for a habitat can be extremely long, or a habitat can be lost permanently. Low flows can increase salt-water intrusion, increase health hazards because of increased concentration of toxic substances and pathogens, decrease hydroelectric-power generation, and curtail recreational opportunities. Decreased precipitation also increases the potential for brush and forest fires and wind erosion of topsoil.

With a few exceptions, the response of government at any level to the shortages caused by decreased precipitation has been to react rather than to adopt a proactive approach to minimize the effect of droughts. Drought planning at the local level in many areas appears to be given a low priority because of the randomness of droughts, the limited resources for planning, the limited jurisdiction (local government might not be able to control streamflow levels), and the programs of the Federal Government to provide disaster relief in time of crisis. As a result, local governments are encouraged to accept an implicit policy of doing only what can be done after a crisis has occurred.

A notable exception to inaction at the local level to plan for drought conditions occurred in the Washington, DC, area. The leadership role in this case was taken by the Interstate Commission on the Potomac River Basin (ICPRB), but the implementation of the water resources plan was by local government. The plan finally adopted saved between $200 million and $1 billion compared to longer scale structural solutions previously proposed (5, p. 106). Implementation of the plan was through eight separate but interlocking contracts executed in 1982 (5, p. 106).

STATE GOVERNMENT

With the exception of eight States—Arkansas, California, Connecticut, Delaware, Florida, Minnesota, New Jersey, and South Carolina—the activities of the State governments in managing water resources during droughts have been minimal. Most State governments have not passed legislation providing for additional drought planning beyond slight modifications in their water laws. Governors, on occasion, will declare counties or designated areas as disaster areas in order to make individuals eligible for Federal relief.

FEDERAL GOVERNMENT

The Federal Government generally has limited its activities to providing direct relief to drought victims and to farmers in general. For example, the Federal response to the 1976-77 drought, which affected about two-thirds of the country, was to enact the Emergency Drought Act of 1977 (Public Law 95-18), the Community Emergency Relief Act of 1977 (Public Law 95-31), and certain provisions of the Supplemental Appropriations Act of 1977 (Public Law 95-26) to bolster existing emergency-assistance programs. As the result of these laws, 40 Federal programs, administered by 16 agencies, offered drought relief in the form of loans, grants, indemnity payments, and other forms of assistance to State and local governments, households, farms, and private businesses (6, p. 1–2). The overriding objective was to reduce impending damage by implementing short-term actions to augment existing water supplies. Other Federal programs available to drought victims were designed to provide assistance after damage had occurred (6, p. 2). Typical of these were the disaster loan programs of the Farmers Home Administration. Four Federal agencies—the Departments of Agriculture, Commerce, and Interior and the Small Business Administration—were responsible for implementing emergency drought programs at a cost of $5 billion, which included an additional $1 billion for short-term emergency actions to augment existing water supplies (6, p. 11). The Comptroller General of the United States (6, p. 11) concluded that:

- Some drought programs were enacted or implemented too late to have much effect in augmenting water supplies.
- Inadequate standards for determining the worthiness of projects meant that many projects were funded that had little, if any, effect in mitigating the effects of the drought.
- Drought victims were treated in an inconsistent, inequitable, and confusing manner.
- Inadequate coordination among the agencies resulted in inefficient and inequitable distribution of funds.

The report (6, p. 21) also recommended that a national plan be developed for providing assistance in a more timely, consistent, and equitable manner. Issues to be considered in the development of such a plan are:

- Identification of respective roles of agencies involved to avoid overlap and duplication of activities.
- Need for legislation to more clearly define agency roles and activities.
- Need for standby legislation to permit more timely response to drought-related problems.

In 1988, in response to another severe drought, Congress passed the Disaster Assistance Act of 1988 (Public Law 100–387). This legislation is less comprehensive than the 1977 legislation in that its primary purposes are to protect farm income in an efficient and equitable manner, protect the economic health of rural communities affected by the drought, and help assure a continued adequate supply of food for American consumers. Little in this legislation reflects the major recommendations of the Comptroller General of the United States (6) regarding the development of a national plan for providing future assistance in a more timely, consistent, and equitable manner. Although the Secretary of the U.S. Department of Agriculture is authorized to make grants and provide other assistance to combat water shortages, the Disaster Assistance Act of 1988 is primarily an agricultural-relief act. Thus, to date, the Federal Government has limited its involvement in droughts to the provision of water-resources information, technical assistance, and financial relief to mitigate the costs incurred once a drought has occurred.

OBSTACLES TO EFFECTIVE PLANNING FOR DROUGHTS

Although governments can plan effectively for droughts, fundamental problems that deter action need to be examined and understood before drought planning can become a reality. Five obstacles to planning for droughts—specificity, randomness, drought phenomenon, cost of droughts, and political considerations—are discussed here.

SPECIFICITY

Planning for and management of hazardous events presuppose that those events are well defined and discernible to all. The planning necessary to reduce the effects of most natural hazards is difficult because the intensity and frequency of the events are unknown, although there is never any question as to their eventual occurrence. Although no technical expertise is required to determine when floods, volcanic eruptions, or earthquakes have been experienced, considerable uncertainty exists as to when droughts start and end. A drought is almost a "non-event" (Evan C. Vlachos, Colorado State University, written commun., January 1988). Any discussion about planning for drought conditions and management of water resources, therefore, requires some definition as to what constitutes a drought. This lack of specificity can be a major contributing factor, although unstated, as to why the planning for droughts and the managing of water resources during droughts have received less attention than they deserve.

RANDOMNESS

Droughts, when placed in historical perspective, have not received much attention from governments at any level until severe water shortages occur. What most inhibits the planning for water shortages associated with droughts is their random nature. It is this inherent variability that makes "reacting" to a water-shortage crisis and instituting relief efforts when the shortages occur appear to be more rational than does "planning" for droughts. People tend to overlook that droughts are a normal part of the climatic regime and that they will recur. Droughts remain a certainty; only their frequency and severity are unknown. Thus, it makes good sense to plan to reduce both the costs that result from droughts and the associated personal hardships.

DROUGHT PHENOMENON

Another obstacle to planning for droughts is inherent in the drought phenomenon. Although droughts do affect individuals, in reality droughts are a community problem having characteristics of Hardin's famous "Tragedy of the Commons" (1, p. 34). The self-interest of each individual using communal property is to maximize it for immediate gain. The net result can well be the destruction or deterioration of the communal property. In Hardin's example, the overgrazing and destruction of the common pastures occurred because each person sought to graze all the animals possible. The sum of the actions by each individual was not "best" for the sum of the individuals. This phenomenon makes the best policy infeasible unless the individuals reach a consensus themselves or are compelled to do so by a government.

Droughts produce the same type of situation. For individuals experiencing a drought, options to deal effectively with water shortages are limited. Collective action by all the individuals provides the best solution. The use of groundwater during times of drought in many Eastern States provides a comparable situation to the one described by Hardin in the case of common pastures. The common law in many Eastern States deals with groundwater as a common resource that is appropriated under the Rule of Capture (7, p. 115). What you capture is what you get, and those having the deepest wells and the largest pumps get the most water. Given these circumstances, the solution for an individual during drought conditions might be to drill a well. If all individuals act in the same manner, a variety of consequences can occur to the detriment of each. A shallow aquifer eventually can be depleted, the individuals can be competitors for water and cause larger and larger cones of depression as deeper wells and larger pumps are utilized, or the increased pumping can cause saltwater intrusion, which will destroy the quality of the water in the aquifer for all. The best solution for all parties might be an agreement between the individual well owners, or restriction by the local government, to curtail the time and rate of pumping. Under these circumstances, an individual who pumps water from a common aquifer cannot plan effectively for droughts. If an individual well owner seeks to conserve groundwater or plan for a water shortage, he or she needs to be aware that the water he or she does not pump will probably be pumped by others.

Because droughts affect larger geographic areas than those occupied by single communities, the ability of an individual community to respond effectively is affected by the actions of similar communities in the drought area. The position of each community in this larger arena can be analogous to that of the individual in the community. Individual actions by each community can be counterproductive to the policy best for the region as a whole. For one community, the solution may be the building of a reservoir on a stream that is the water source for other communities downstream. As each community opts to resolve its water needs without regard to its neighbors, the stream can become an inadequate water source for all. The development of a regional water supply for all communities might be the best solution, but this will require the consensus of all the communities.

COST OF DROUGHTS

The lack of information about the cost of droughts is another reason why only marginal interest exists in planning for droughts, especially at the State level. The magnitude of drought costs is assumed to be less than that of other natural hazards because the losses associated with other natural hazards are more evident and generally are incurred during short periods of time. In contrast, drought losses generally are distributed over longer time periods. When the true costs of drought are known, drought losses can dwarf the losses from other natural hazards. For example, Australia determined that, for the period 1945–1975, the costs of droughts were four times the costs of other natural hazards (8, p. 226). In addition, all the costs associated with droughts are not clearly defined. The social effects of droughts and the associated costs, how the effects propagate throughout society, and who is ultimately affected need to be better understood. Human suffering is less likely to be factored into the cost assessment even though these costs are real and can continue for years, long after other costs have been absorbed. The aggregated indirect costs probably are far greater than the direct costs, but because of their diffused nature they are difficult to identify and quantify and, thus, generally go unrecognized (1, p. 32). These indirect costs, which are disbursed among large groups and throughout large geographic areas, nevertheless constitute a major proportion of the total costs resulting from droughts (1, p. 32). Again, the random nature of droughts, coupled with the rapid decrease of public interest in droughts after normal precipitation resumes and the limited resources available for planning, make the determination of in-direct costs associated with the droughts less urgent. As long as these indirect, diffused costs remain undisclosed, decision makers will have incomplete knowledge of the costs of drought. If the past is any guide, the total costs of droughts probably never will be reliably assessed.

The length of a drought also has a significant effect on the total costs; long droughts are more costly than shorter droughts. A sustained drought, such as the one in the 1930s, can have economic and social costs that are never quantified. During the 1930s drought, for example, agriculture was abandoned in some sections of the Nation; this abandonment, in turn, caused dislocation of people and severe impairment of the economic substructure that supported agriculture. The 1930s drought had an effect on a whole generation of Americans, wherever they lived and however they made their living (4, p. 34). Even if the value of human life is ignored, the total economic losses from droughts can be staggering.

POLITICAL CONSIDERATIONS

Lastly, political considerations affect action that might lessen the effects of droughts through better planning and management. The randomness of droughts induces the public to believe that little can be done to reduce the costs of droughts before they occur. In addition, the public's memory of past tragedies usually is short, and political attention shifts quickly to new political problems. The public, lacking an analysis of the total costs associated with a drought, has the illusion that droughts are affordable, although inconvenient. Thus, decision makers lack the public support needed to take aggressive action in planning for droughts and managing water resources during droughts. This is not strictly an American phenomenon, as witnessed by the inaction of the British Parliament during the mid-1970s when the country experienced its worst drought in 500 years (9, p. 51). Early in the drought, efforts were made to have Parliament enact legislation to extend the responsibilities of the river authorities to mitigate the effects of the drought. Before responding, Parliament waited until there was widespread public awareness of the need for the legislation; thus, the damage factor was increased substantially compared to what it would have been had Parliament acted earlier.

In contrast to a lack of public support, special-interest groups at the State level might oppose activities that are essential for an effective water-management plan applicable to droughts. For example, farm groups in Virginia strongly oppose any Federal, State, or regional water-management plan because they believe there should be no regulation of water apart from the Riparian Doctrine (Mark Tubbs, Virginia Farm Bureau, commun., 1981). Water management, of necessity, must be at the core of any program to mediate the effects of water shortages that occur during droughts, but political factors can substantially dampen the interest in managing water resources even during droughts.

FRAMEWORK FOR STATE ACTION

The "Tragedy of the Commons" phenomenon, which characterizes the problems associated with any management plan to mitigate the costs associated with droughts, illustrates that, in the absence of agreement among all the parties affected by the drought, the management responsibility needs to be at the lowest possible level of government that will permit the attainment of management's objectives (1, p. 34). The State, in most situations, represents the unit of government that has the authority to allocate water, to set policy objectives that are concerned with water-use efficiency and equity, to consider interboundary issues and externalities associated with matters such as minimum instream flows, and to coordinate the activities of local governments in meeting water-supply needs during times of severe water shortages.

The responsibility for managing water resources during droughts, once assumed by the State, needs to be vested in such a manner as to require timely action and not be vulnerable to legal challenges by groups who do not favor an approach taken by the State. Although expanding the Governor's powers to deal with disasters by including droughts might be expeditious, the action taken by most States generally is to group management activities with the authority primarily designed to respond to disasters after they have occurred rather than to undertake planning activities to reduce the cost of droughts in advance of their occurrence. Colorado, Delaware, New Jersey, and North Carolina are examples of States that have used executive power to develop statutory guidelines that define droughts and delineate interaction among State agencies responsible for water resources (10, p. 162–163).

PLANNING TOOLS

The authors of the "Model Water Use Act" (3, 1958) developed five planning tools—identification of drought indicators, designation of government authority, notification of the public, curtailment of water use and maintenance of revenues, and monitoring of water-user compliance—to cope with planning for water shortages associated with droughts. These tools addressed the following fundamental questions:

- How does a State know when there is a drought?
- If there is a drought, who is in charge?
- How is the public informed?
- How are current allocations and uses of water to be modified?
- How is compliance assured?

The answers or responses to these fundamental questions need not be identical for each State having a functioning water-management plan applicable to droughts. Each State, however, needs to address each question in terms of its own circumstances. Failure to address each of the questions will detract seriously from the effectiveness of a water-management plan. Each of these planning tools is now examined in detail.

IDENTIFICATION OF DROUGHT INDICATORS

Because of the difficulty of deciding when droughts start and end, specific drought indicators must be used to decide when to implement a water-management plan.

When such indicators have been identified, water users can formulate contingency plans and make decisions on future economic investments (11, p. 47). The drought indicators must be precise and susceptible to little, if any, subjective decision making. The latter makes the indicators vulnerable to court action by those who oppose advanced planning.

A variety of drought indicators can be used, including the Palmer Index (a drought-severity index), in stream flows, historical data on the present and anticipated needs for water, the degree of subsidence or saltwater intrusion, the potential for irreversible adverse effects on fish and wildlife, and reservoir or groundwater conditions relative to the number of days of water supply remaining (11, p. 47). Usually it is desirable to select a number of drought indicators to reflect the seasonal relation of supply versus demand. The Delaware River Basin Commission (11, p. 51) relies on five drought indicators—precipitation, groundwater levels, reservoir storage, streamflow, and the Palmer Index. Ranges of values for each of these indicators are assigned to one of four drought stages—normal, drought watch, drought warning, and drought emergency (11, p. 51). To activate any one of the drought stages, three of the five drought indicators must indicate a given drought stage (12, p. 34). The drought indicators should not be so complex as to cause uncertainty about whether some stage of the water-management plan should be activated; for example, if precipitation and reservoir storage are two drought indicators, the decision to activate the plan will be unclear if the precipitation is less than normal while reservoir storage is normal.

When there is only one source of water supply, one drought indicator may be sufficient; for example, when a city's only source of water is a reservoir, the water-management plan can be activated if reservoir storage, expressed as a percentage of normal seasonal capacity, decreases below a specified percentage. The phasing criteria used by Manchester, Connecticut, are an example (Table 1): a drought watch goes into effect when reservoir storage is at 70 percent of normal seasonal capacity, and stage I of the water-management plan becomes operational when reservoir storage is 57 percent of normal seasonal capacity.

A sliding scale for drought indicators also can be shown graphically in terms of the storage in a reservoir, as shown by the operation curves for three reservoirs in New York that also are part of the Delaware River basin (Fig. 1). When the actual reservoir level drops below the drought-warning zone or drought zone, schedule of reduced diversions from the basin to the various localities takes effect. The Pennsylvania Drought Contingency Plan for the Delaware River Basin is based on these criteria. Where groundwater is one of the main sources of water supply, drought indicators based on groundwater levels can be used; the Alameda County Water District in California has such a plan (12, p. 38).

Most water-management plans have correlated successive stages of a drought strategy to certain deficit-reduction goals (12, p. 39). Fewer than three stages in a plan can result in marked differences in the actions to be implemented between the first and the second stages. More than five stages in a plan, however, can cause frequent transitions between stages, which can decrease the effectiveness of the plan. An example of a workable plan is the Seattle Water Department five-stage plan for reducing water use (Table 1). Agencies having water-management

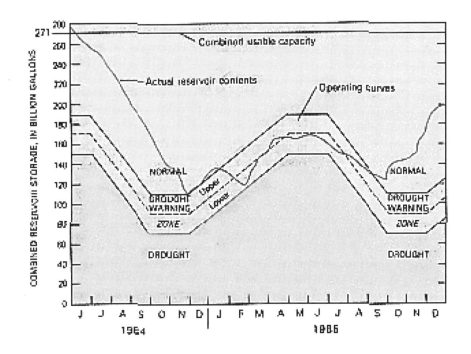

Figure 1. Operation curves for three New York City reservoirs (Cannonville, Pepacton, and Neversink) in the Delaware River Basin (from Ref. 13, p. 31).

Table 1. Drought-Contingency-Plan Phasing Criteria Used By Two Cities in the United States

MANCHESTER, CONNECTICUT
(South System)

Drought Stage	Drought Stage Initiating Conditions	Demand-Reduction Objective
Drought watch	Reservoir water levels at 70 percent of normal seasonal capacity.	Informational only, raise public awareness
Stage 1	Reservoir water levels at 57 percent of normal seasonal capacity.	Cut back withdrawals from reservoirs by 5 percent or reduce total system use by 3.8 percent.
Stage 2	Reservoir water levels at 40 percent of normal seasonal capacity.	Cut back withdrawals from reservoirs by 30 percent or reduce total system use by 20 percent.
Stage 3	Reservoir water levels at 0 percent of normal seasonal capacity	Eliminate withdrawals from reservoirs and reduce total system use by 70 percent.

Source: Data from Reference 12, pp. 33, 35, and 36.

SEATTLE, WASHINGTON, WATER DEPARTMENT
(Deficit reduction objective is based on demand levels in a 1-in-50-year drought, which are expected to be above normal averages due to warm, dry conditions)

Summer-Shortage Response Plan

		Demand-Reduction Amount, in Million Gallons Per Day	
Stage 1 Minor shortage potential	Total system storage is not filled to capacity as of June 1. Streamflow and snowmelt forecasts indicate that inflows will be inadequate to fill storage facilities before the beginning of the peak-use season.	Water-system management Customer Total	3.0 0.0 3.0
Stage 2 Moderate shortage potential	Total system storage is predicted to fall below the level required to meet expected demands during a 1-in-50-year drought. System inflows continue to be low. Weather forecasts predict a continuing trend of warmer, drier than normal conditions.	Water-system management Customer Total	3.0 4.7 7.7
Stage 3 Serious shortage	Total system storage drops below the level required to meet expected demands during a 1-in-50-year drought. System inflows continue to be low. Weather forecasts predict a continuing trend of warmer, drier than normal conditions.	Water-system management Customer Total	5.0 15.0 20.0
Stage 4 Severe shortage	Conditions described for stage 3 occur near the end of the peak-use season.	Water-system management Customer Total	5.0 16.1 21.1
Stage 5 Critical emergency	Customer demands and system pressure requirements cannot be met.	Not applicable.	

Fall-Shortage Response Plan

		Demand-Reduction Amount, in Million Gallons Per Day	
Stage 1 Minor shortage potential	Total system-storage levels are dropping due to the increased use associated with a warm, dry summer. Weather forecasts predict a continuing trend of warmer, drier than normal conditions.	Water-system management Customer Total	3.0 0.0 3.0
Stage 2 Moderate shortage potential	Total system storage is expected to fall below the level required to meet expected demands during a 1-in-50-year drought.	Water-system management Customer Total	5.0 2.4 7.4
Stage 3 Serious shortage	System inflows continue to be low.	Water-system management Customer Total	5.0 6.2 .11.2
Stage 4 Severe shortage	Weather forecasts predict a continuing trend of warmer, drier than normal conditions.	Water-system management Customer Total	5.0 19.6 24.6
Stage 5 Critical emergency	Customer demands and system pressure requirements cannot be met.	Water-system management Customer Total	5.0 52.4 57.4

plans have determined that fall droughts have a lesser probability of occurrence, but if they do occur, they are likely to develop more quickly and be more severe.

DESIGNATION OF GOVERNMENT AUTHORITY

The designation of a governmental unit or agency having specific drought-planning authority in advance of a drought is one of the critical aspects of providing for managing water supplies during droughts (10, p. 162). For a water-management plan to be effective, the designated agency needs to have authority to declare that a drought exists and to alter water-use patterns. The location of a unit or agency that has this authority within the State's administrative structure will vary. Statutory authority needs to be detailed and specific. If discretionary authority is given, well-defined guidelines for its use need to be given. Failure to adequately delineate the limits wherein action needs to be taken can create situations where administrators postpone action in order to avoid conflict with user groups. This postponement lessens the protection afforded to both water quantity and quality (10, p. 164).

NOTIFICATION OF THE PUBLIC

When drought conditions activate the implementation of the water-management plan, the public needs to be notified. The notice needs to contain information about the provisions to curtail use, when conservation measures become effective, the availability of variances, and the procedures for obtaining a variance. The exact notification procedure can be developed to reflect local conditions. Examples of States having well-defined notification procedures include Florida, Georgia, North Carolina, and South Carolina (10, p. 165).

CURTAILMENT OF WATER USE AND MAINTENANCE OF REVENUES

A system of priorities of water-use categories needs to be in place before droughts occur, so that each user knows, in advance of a drought, in what order water restrictions will be applied. If industries and commercial establishments know, in advance of a drought, the procedures to be used in reducing water availability, they can establish their own contingency plans for those reductions; for example, they can arrange for alternative water supplies or plan reductions in production schedules. Unlike most Western States, the majority of Eastern States, which allocate water according to the riparian doctrine, do not have a water-use-priority system to deal with water shortages caused by droughts. Some attempt, however, has been made in most States to show a preference for certain uses (10, p. 162).

A reduction in water use will cause a reduction in the revenues of the water suppliers. The reduced revenues come at a time when costs are greater because of expenditures made to deal with the drought. In the absence of a water-revenue reserve or a drought-emergency account, water suppliers need to either increase water rates or impose a drought surcharge (12, p. 49). The use of a drought surcharge has several advantages compared to a simple increase in water rates. For example, the drought surcharge is easier to administer, and the amount of revenue to be generated is more predictable. This surcharge probably is more acceptable to the customer because it is a one-time charge that is understandable and allays the fear that a water-rate increase to make up revenues lost during droughts will continue when the drought is over.

MONITORING OF WATER-USER COMPLIANCE

Experience has indicated that reductions in water use greater than 20–25 percent cannot be obtained with a request for voluntary conservation (12, p. 29). There does not seem to be general agreement as to whether mandatory conservation regulations or water-rate increases and drought surcharges are the most effective means of reducing water use to a volume less than that obtained by voluntary conservation. Utah determined that price increases to reduce water use were not as effective as mandatory conservation regulations for a short drought (12, p. 28). New Jersey, in contrast, recently enacted a "water emergency price schedule" in preference to mandatory restrictions (New Jersey Administrative Code, title 7, section 19B-1.5(a)). Some utilities managers argue that the availability of enforcement mechanisms is the important feature of the plan and that their application is rare (12, p. 29). Much of the monitoring of customers for compliance comes from peer-group pressure, but governmental employees, such as supervisors of streets and wastewater departments and inspectors for buildings, plumbing, electric, construction, and health services, can be empowered to issue citations (12, p. 29). This is an effective method of monitoring a service area with a minimum of expense and with minimal disruption of employees' regular work schedules.

SYNOPSIS OF PLANNING AND MANAGEMENT ACTIVITIES OF THE STATES

The planning tools suggested by King and others (3) in their "Model Water Use Act" provide a minimum set of criteria for evaluating the activities of States in the area of water management during droughts. A survey in 1983 and its update in 1986 by the Virginia Water Resources Research Center (10) determined that, of the 48 conterminous States, only 8 States have comprehensive water-shortage plans, 27 States have emergency drought provisions within their water-rights system, and 13 States do not appear to have plans in place for managing water resources during droughts (Fig. 2).

The eight States that have comprehensive water-management plans incorporate all of the basic concepts suggested in the "Model Water Use Act." However, the approach by each State has been different and reflects individual State needs and existing water-allocation systems. The drought indicators that are used

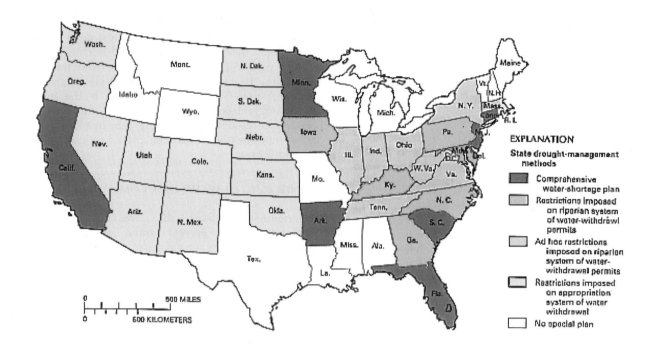

Figure 2. State-level methods for managing water resources during drought conditions (From Ref. 10, p. 146).

to determine when the plans are to begin and to end probably were the least precise component in the plan of each of the eight States. All the water-management plans are activated at the State level except in Florida, where activation is the responsibility of existing water-management districts, and in California, where activation is in the charge of the intrastate districts mandated by statute.

Water laws of 27 States have been modified to establish systems of water-use categories (Fig. 2). Although having these systems in place eases the management of water supplies once a drought occurs, most of the changes are designed to meet emergency water shortages rather than to provide a strategy to mitigate the effects of droughts before an emergency develops (10, p. 149). These 27 States can be subdivided into those that have modified the riparian doctrine with some form of a permit system; that use a modified appropriation system; or that rely on civil-defense, disaster, or emergency legislation that incorporates droughts within its definitions (10, p. 148). A description of these subdivisions follows.

Six States (Fig. 2) in the East have modified then common-law riparian doctrine by requiring water-use permits for some uses. Many of the permits are subject to water-use restrictions during times of severe water shortages. Georgia has water-use restrictions that are implemented during water emergencies. Iowa has made all use of surface and groundwater subject to a statewide permit system; permit users are denied water whenever the minimum instream-flow standard is reached. In Kentucky, permits give no guarantee of a water right during droughts. Conservation programs are required under all permits in Maryland when the safety is threatened by existing or projected water demand. North Carolina's permit system is limited to designated capacity-use areas; other laws in North Carolina grant water-emergency powers that affect the capacity-use areas and the rest of the State. Pennsylvania requires permits only of public water suppliers using surface water; permits are conditioned on there being an emergency water plan in place. All these States have taken some initial steps to manage their water resources during droughts, but they utilize few of the planning tools suggested by King and others (3) in their "Model Water Use Act" for water management (10, p. 155).

Twelve States (Fig. 2) in the West have attempted to deal with severe water shortages by modifying the appropriation doctrine for allocating water. Under this doctrine only the most senior appropriation in time will receive water during water shortages. All these States have modified the appropriation doctrine to some degree to accommodate water shortages caused by droughts. New Mexico, for example, provides for changing the place of diversion, storage, or use of water if an emergency exists (10, p. 152). In Oregon, the Director of Water Resources can order State agencies and political subdivisions to develop water-conservation and water-use-curtailment plans that encourage conservation, reduce nonessential water use, prevent waste, provide for reuse of water, and allocate or rotate the supply to domestic, municipal, and industrial uses (10, p. 152). Utah allows its State Engineer to use regulatory authority to prevent waste in order to mandate rotation of irrigation water where no use will benefit from a diversion of the water supply (10, p. 152). In these 12 States, legislation provides some management tools to assist in decreasing the effects of a water-shortage crisis. Little of the modifying legislation includes the planning tools suggested by King and others (3) in their "Model Water Use Act" to manage water shortages due to droughts in a comprehensive way.

Nine States (Fig. 2) in the Midwest and Mid-Atlantic region have left the management of droughts, from a State perspective, almost entirely to the Governor. These States perceive water shortages as primarily the responsibility of local government. The State has a role when the shortage becomes extensive in terms of time and scope and affects such a large area of the State that the water shortage is perceived as a severe crisis approaching a disaster. These States choose to deal with water shortages due to droughts on an ad hoc basis. New York passed a law that took effect in 1990 and requires a conservation plan for surface withdrawals.

INTRASTATE REGIONAL AUTHORITIES AND INTERSTATE COMMISSION COMPACTS

Intrastate regional authorities can perform the water-management function during droughts if droughts are less than statewide in scope. Politically, such authorities are difficult to create because of the rivalry that exists among local units of government. If intrastate regional authorities are established, such regional grouping must be done carefully to avoid possible constitutional challenge as unlawful delegation of legislative authority (10, p. 162). The water-management districts in Florida are examples of this approach.

When droughts affect more than one State but are not national in scope, interstate commissions and compacts can provide the management function to mitigate drought effects. Their effectiveness is predicated on having well-publicized plans and specific rules for planning purposes so that all users know how they will fare when the river flows cannot accommodate all the withdrawal demands (10, p. 164). The Delaware River Basin Compact (Public Law 87–328, 75 Stat. 688, 1961) and the 1978 Potomac River Low Flow Agreement are examples of interstate compacts that have been used to plan for the problems associated with droughts. Such commissions or compacts require, however, the approval of all of the involved State legislatures, which could be a difficult task politically. The fact that the number of such entities is small is ample evidence of the difficulty and time required to establish them.

FEDERAL RESPONSIBILITIES FOR WATER QUALITY AS RELATED TO DROUGHTS

Since 1972, with the passage of the Water Pollution Control Act Amendments (Public Law 92–500), the Federal Government has assumed a more dominant role on water-quality issues related to surface water. Because droughts and the resulting low flows have a substantial effect on water quality, it is possible, on the basis of existing legislation and regulations, that the Federal Government might become more interested in drought management. For example, Section 208(b)(2)(I) of the Clean Water Act (Public Law 92–500) provides:

> Any plan prepared under such process shall include, but not be limited to, (1) a process to (i) identify, if appropriate, salt water intrusion into rivers, lakes, and estuaries resulting from reduction of fresh water flow from any cause, including irrigation, obstruction, groundwater extraction, and diversion, and (ii) set forth procedures and methods to control such intrusion to the extent feasible where such procedures and methods are otherwise a part of the waste treatment management plan.

The legislative history amplifies on this point (1972 U.S. Code Congressional and Administrative News, p. 3706) (emphasis added):

> Salt water intrusion no less than point sources of discharge, alters significantly the character of the water and the life system it supports. Salt water intrusion often devastates the commercial shellfish industry. It must be accounted for and controlled in any pollution control program. It makes no sense to control salts associated with industrial or municipal waste point sources and allow, at the same time, similar effects to enter the fresh water as a result of intrusion of salt water. *Fresh water flows can be reduced from any number of causes. The bill requires identification of those causes and establishment of methods to control them so as to minimize the impact of salt water intrusion.*

Droughts can be one of the major causes for reduced freshwater flows, and the law requires that methods be established to control or to minimize the causes of reduced freshwater flows that allow saltwater intrusion.

The Siting Requirement under the Safe Drinking Water Act (Public Law 93–523) contains language (emphasizes added) that can be construed to include drought conditions and, thus, impose on States the requirement to adapt siting criteria that include drought conditions:

> Before a person may enter into a financial commitment for or initiate construction of a new public water system or increase the capacity of an existing public water system, he shall notify the State and, to the extent practicable, avoid locating part or all of the new or expanded facility at a site which: (a) Is subject to a significant risk from earth-quakes, floods, fires *or other disasters which could cause a breakdown of the public water system or a portion thereof.*

Droughts could conceivably be "other disasters" provided for in the Safe Drinking Water Act. A combination of these water-quality considerations might, in the future, cause the Federal Government to give greater consideration to managing the effects of droughts, at least with respect to those effects that impinge on water-quality issues.

Lastly, the quality of surface-water bodies is affected markedly by the runoff that occurs when precipitation increases after a drought. During the drought, pollutants accumulate on the land surface and on other surfaces, such as pavement and structures. It is not uncommon for droughts to be followed by a period of abnormally high precipitation that tends to aggravate the already existing water-quality problems by rapidly flushing large loads of pollutants into surface-water bodies. After the drought in England in the 1970s, the nitrate concentration in the Thames River increased to the point where the public-supply intakes had to be closed (9, p. 54). This kind of post drought problem may be reflected in what the Federal Government requires States to do to meet water-quality standards. Some drought planning may occur at the State level as a byproduct of the action taken to address this water-quality problem.

CONCLUSIONS

The planning for and the management of the effects of droughts appear to have a low priority in all but a few States, although all have experienced severe water shortages. For the most part, accommodating the inconvenience caused by droughts is considered a local-government responsibility. The Federal Government's role has been to provide financial assistance to citizens after the droughts have occurred. Water-quality legislation may cause the Federal Government to take a more proactive approach to managing the effects of droughts. Several factors will have to coexist before many States will undertake development of plans to mitigate the effects of droughts. Such factors may be the occurrence of a drought that is long and extensive, thereby increasing demands on a fixed water supply, and a public awareness of the economic costs of droughts. As water demands continue to grow, even minor droughts will become more serious, and States will be compelled to become leaders in developing water-management plans.

BIBLIOGRAPHY

1. Yevjevich, V., Hall, W.A., and Salas, J.D. (Eds.). (1978). *Drought Research Needs-Conference on Drought Research Needs, Colorado State University, Fort Collins, Colorado,* December 12–15, 1977, Proceedings: Fort Collins, CO, Water Resources Publications, p. 288.
2. Palmer, W.C. (1965). *Meteorological Drought.* U.S. Weather Bureau Research Paper 45, p. 64.
3. King, D.B., Lauer, T.E., and Zieglar, W.L. (1958). Model water use act with comments. In: *Water Resources and the Law.* University of Michigan, Ann Arbor, pp. 533–614.
4. Harrison, R. (1977). Response to Droughts. *Water Spectrum* **9**(3): 34–41.
5. Sheer, D.P. (1986). *Managing Water Supplies to Increase Water Availability, in National Water Summary 1985—Hydrologic Events and Surface-Water Resources.* U.S. Geological Survey Water-Supply Paper 2300, pp. 101–112.
6. Comptroller General of the United States. (1979). *Federal Response to the 1976-77 Drought-What Should Be Done Next.* Government Printing Office, Washington, DC, p. 25.
7. Cox, W.E. (1982). Water law primer. *American Society of Civil Engineers, Water Resources Planning and Management Division, Proceedings,* v. 18 (WRI), pp. 107–122.
8. Heathcode, R.L. (1986). Drought mitigation in Australia. *Great Plains Quarterly* **6**: 225–237.
9. Blackburn, A.M. (1978). Management strategies—dealing with drought. *Am. Water Works Assoc. J.* **1978**: 51–59.
10. Hrezo, M.S., Bridgeman, P.G., and Walker, W.R. (1986). Integrating drought planning into water resources management. *Nat. Resour. J.* **26**: 141–167.
11. Hrezo, M.S., Bridgeman, P.G., and Walker, W.R. (1986). Managing droughts through triggering mechanisms. *Am. Water Works Assoc. J.* **1986**: 46–51.
12. California Department of Water Resources. (1988). *Urban Drought Guidebook.* California Department of Natural Resources, Department of Water Resources, Office of Water Conservation, Water Conservation Guidebook 7, p. 144.
13. U.S. Geological Survey. (1986). *National Water Summary 1985-Hydrologic Events and Surface-Water Resources.* U.S. Geological Survey Water-Supply Paper 2300, p. 506.

WATER RESOURCES MANAGEMENT

Neil S. Grigg
Colorado State University
Fort Collins, Colorado

Water is an essential asset for all life and economic activity, and the term "water resources management" describes activities necessary to balance supplies and demands of water. It involves applying structural and nonstructural measures to control natural and man-made water resources systems to achieve beneficial human and environmental purposes, which may be listed as water supply, wastewater service, flood control, hydropower, recreation, navigation, and environmental protection (1). Along with these, water resources management deals with mutual impacts between water and other systems, such as land, wildlife, and economic systems.

Water resources management controls *water resources systems*, that are combinations of constructed water control facilities and natural, or environmental elements that work together to achieve water management. A constructed water resources system, consisting of structural facilities, provides control of water flow and quality and includes facilities for water supply and wastewater management; for drainage of land and control of floods; and for water control in rivers, reservoirs, and aquifers. Examples include conveyance systems (channels, canals, and pipes), diversion structures, dams and storage facilities, treatment plants, pumping stations and hydroelectric plants, wells, and all appurtenances.

Natural water resource systems comprise sets of environmental or hydrologic elements in nature that include the atmosphere, watersheds, stream channels, wetlands, floodplains, aquifers and groundwater systems, lakes, estuaries, seas, and the ocean.

In addition to structural measures to control water flow and quality, nonstructural measures are used for programs or activities that do not require constructed facilities. Examples of nonstructural measures include pricing schemes, zoning, incentives, public relations, regulatory programs, and insurance.

Water resources management takes place within a "water industry" that consists of water service organizations, regulators, coordinators, and support organizations. This industry places great value on public involvement in decisions because water affects all people and activity in interdependent relationships.

Water supply serves four categories of water users: people, industries, farms, and the general environment. Thus, we speak of water supply for people (domestic water supply), for cities (urban water supply), for farms (irrigation), for industries (industrial water supply), and for cities and industries (municipal and industrial or M&I water), or we can speak of water for the environment (water for natural systems and habitat). Wastewater management serves the same categories, as urban wastewater, industrial wastewater, and drainage for farms.

In-stream uses of water include hydropower generation, navigation, recreation, and sustenance of fisheries and ecological systems.

Stormwater and flood control are different types of water management activities that handle excessive water. As "protective" services, they do not provide water, but remove or store the excess water.

Given its many facets, water resources management should be comprehensive, coordinated, and integrated. Comprehensive water resources management includes all purposes and stakeholders in its activities. When water management is coordinated, there should be linkages between activities so that they occur with due consideration of each other. Integrated water resources management is a term that includes aspects of comprehensive and coordinated water management, and it has been explained in several ways; usually the goal is describing a holistic management process.

For example, a working group of the Awwa Research Foundation (2) used the term "Total Water Management" to explain the integrated nature of water resources management:

> "Total Water Management is the exercise of stewardship of water resources for the greatest good of society and the environment. A basic principle of Total Water Management is that the supply is renewable, but limited, and should be managed on a sustainable use basis. Taking into consideration local and regional variations, Total Water Management
>
> - encourages planning and management of natural water systems through a dynamic process that adapts to changing conditions;
> - balances competing uses of water through efficient allocation that addresses social values, cost-effectiveness, and environmental benefits and costs;
> - requires the participation of all units of government and stakeholders in decision-making through a process of coordination and conflict resolution;
> - promotes water conservation, reuse, source protection, and supply development to enhance water quality and quantity; and
> - fosters public health, safety, and community good will."

In the past, water resources management was primarily an engineering arena for building dams, laying pipelines, installing pumps, and operating systems. Now, managing water resources requires skills and approaches that go beyond pure engineering, science, management, and law. Water resources managers must deal with complexity and conflict to unravel interdependency of systems and resolve political and legal dilemmas. Skills from several disciplines are required, including law, finance, and public administration, along with engineering and science.

BIBLIOGRAPHY

1. Grigg, N.S. (1996). *Water Resources Management: Principles, Cases, and Regulations*. McGrawHill, New York.
2. Awwa Research Foundation. (1996). *Minutes of Workshop on Total Water Management*. August 1996, Seattle, Denver.

NASA HELPING TO UNDERSTAND WATER FLOW IN THE WEST

LINDSAY RENICK MAYER
KRISHNA RAMANUJAN
Goddard Space Flight Center
Greenbelt, Maryland

To do their jobs, water resource managers in the Columbia River Basin have mostly relied on data from sparsely located ground stations among the Cascade Mountains in the Pacific Northwest. But now, NASA and partnering

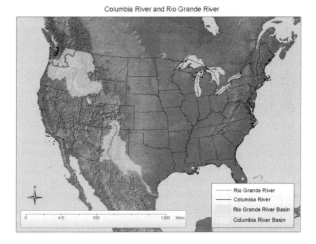

Figure 1. Map of Rio Grande and Columbia River Basins. *Credit*: Image by Robert Simmon, NASA GSFC Earth Observatory, Michael Tischler, NASA/GSFC.

This article is a US Government work and, as such, is in the public domain in the United States of America.

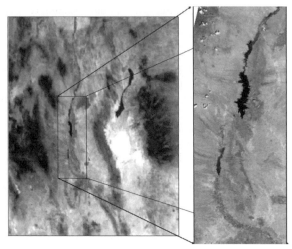

Figure 2. Rio Grande River Basin, SeaWiFS Image with MODIS higher resolution inset. This graphic shows a larger view (1 kilometer resolution) of the Rio Grande River Basin taken by the Sea-viewing Wide Field-of-view Sensor (SeaWiFS) on the OrbView-2 satellite, with an inset of the same area (250 meter resolution) taken by the Moderate Resolution Imaging Spectroradiometer (MODIS) instrument aboard the Terra satellite. In the inset, the largest dark area indicates Elephant Butte Reservoir. *Credit*: Image by Andrew French of NASA GSFC/NRC, Kristi Arsenault, NASA GSFC and Univ. of Maryland-Baltimore County.

demands range from hydropower, to farming, fishing, boating and protecting endangered species (Fig. 1). Water resource managers in these areas grapple with the big money stakes of distributing a finite amount of water to many groups. NASA satellite data offer to fill the data gaps in mountainous and drought-ridden terrain, and new computer models let users quickly process that data.

Land Surface Models (LSMs) from NASA, other agencies and universities, and NASA satellite data can be used to determine snowpack, amounts of soil moisture, and the loss of water into the atmosphere from plants and the soil, a process known as evapotranspiration. Understanding these variables in the water cycle is a key to managing water in such resource-limited areas.

"The latest satellites provide so much up-to-date and wide-ranging data, which we can use in the models to monitor and better understand what is happening with the water cycle in these areas," said Kristi Arsenault, research associate for the Land Data Assimilation System (LDAS) team at NASA's Goddard Space Flight Center, and Research Associate at University of Maryland, Baltimore County.

"These efforts are designed to improve the efficiency of the analysis and prediction of water supply and demand using the emerging technologies of the Land

Figure 3. MODIS image of Columbia River Basin snowcover, February 24, 2003. This image from the Moderate Resolution Imaging Spectroradiometer (MODIS) shows snowcover for the Columbia River Basin in the Cascade Mountains of Washington State, taken on February 24, 2003 (250 meter resolution). *Credit*: Jeff Schmaltz MODIS Land Rapid Response Team, NASA/GSFC.

agencies are going to provide United States Bureau of Reclamation water resource managers with high resolution satellite data, allowing them to analyze up-to-date water-related information over large areas all at once.

The pilot program is now underway with the Rio Grande and Columbia River basins where water is scarce while

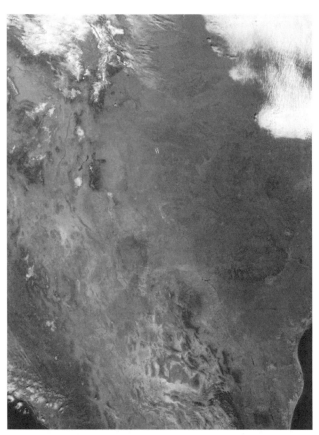

Figure 4. MODIS image of Rio Grande River Basin, February 22, 2003. This image from the Moderate Resolution Imaging Spectroradiometer (MODIS) shows the Rio Grande River Basin taken on February 22, 2003 (250 meter resolution). *Credit*: Jeff Schmaltz MODIS Land Rapid Response Team, NASA/GSFC.

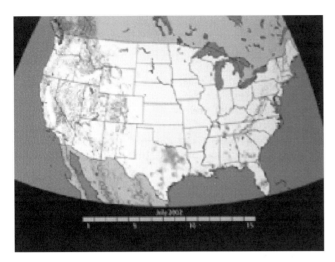

Figure 5. U.S. rainfall (July 1 to 15, 2002). This animation shows 3-hourly rainfall from July 1 through 15, 2002. Rainfall is one of the main variables that drive land surface models used by water resource managers and scientists. *Credit*: Images by Robert Simmon, NASA GSFC Earth Observatory, based on data provided by Kristi Arsenault, NASA GSFC and Univ. of Maryland-Baltimore County.

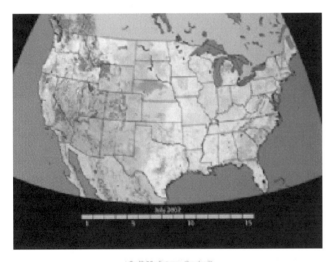

Figure 6. U.S. soil moisture (July 1 to 15, 2002). This animation shows changes in soil moisture from July 1 through 15, 2002, derived from a land surface model. Soil moisture is important for knowing how much water is contained in the soil, which is vital for crop production, flood prediction, and evapotranspiration estimates. *Credit*: Images by Robert Simmon, NASA GSFC Earth Observatory, based on data provided by Kristi Arsenault, NASA GSFC and Univ. of Maryland-Baltimore County.

Data Assimilation System," said Dr. Dave Matthews, manager of the River Systems and Meteorology Group of the Technical Services Center, U.S. Bureau of Reclamation (Reclamation). Computer models, known as decision support systems, that factor in ecological, human, and legal restrictions are vital to managing and allocating water, Matthews added. These systems will incorporate NASA satellite and model data.

NASA's tools may be of vital use in the Rio Grande and Columbia River basins where the disparate and numerous water demands have enormous economic implications. In the Rio Grande Basin, for example, water managers dole out water to farmers so they can irrigate their land. At the same time, under the Endangered Species Act, states are required by law to maintain river water levels to protect the habitat of the endangered silvery minnow. A recent seven-year drought has exacerbated these demands.

Similarly, the Columbia River Basin provides water for the Coulee Dam, the largest concrete dam in North America, and a means for controlling floods. This hydroelectric dam is the third largest producer of electricity in the world. At the same time, the basin is a source of water for a billion dollar agricultural area.

To help make big decisions of allocating water, NASA's special technologies can provide a unique perspective from space. For example, satellites can classify vegetation, a task that is essential to calculating evapotranspiration, which accounts for up to 60 percent of water loss into the air in a region like the Rio Grande Basin. Some managers have been relying on vegetation maps that dated back to 1993, in areas where wild-lands, crops and farming practices are subject to change.

Landsat data can provide highly detailed spatial information, but these images may only be available once a month, and are very expensive. The newer technologies of the Moderate Resolution Imaging Spectroradiometer (MODIS) instrument on the Terra and Aqua satellites provides more frequent passes and day-to-day and week-to-week changes in vegetation production (Fig. 2). In addition, other variables of interest, like snow cover and land surface temperatures, are updated more regularly by MODIS, which can aid in identifying areas with potential flooding and help with the daily management of the water resources (Figs. 3 and 4).

LDAS has also begun to evaluate soil moisture data from NASA's Advanced Microwave Scanning Radiometer (AMSR-E) aboard the Aqua satellite and 3-hour rainfall estimates from NASA's Tropical Rainfall Measuring Mission (Figs. 5 and 6). All this data helps determine how much water is being absorbed into the ground, versus how much is evaporating into the atmosphere. These observations will then be assimilated into Land Surface Models so that water managers can assess flood risks and other factors and act accordingly in a timely manner.

Reclamation brings water to more than 31 million people and provides one out of five Western farmers with irrigation water for 10 million acres of farmland.

One mission of NASA's Earth Science Enterprise is to expand and accelerate the realization of economic and societal benefits from Earth science information and technology.

TRANSBOUNDARY WATER CONFLICTS IN THE NILE BASIN

Rongchao Li
Delft University of Technology
Delft, The Netherlands

WATER SCARCITY IN THE NILE

The Nile, the longest river in the world, has nourished livelihoods, supported a vast array of ecosystems, and played a central role in a rich diversity of cultures. As shown in Fig. 1, from its major source, Lake Victoria in east central Africa, the White Nile flows generally north through Uganda and into Sudan where it meets the Blue Nile at Khartoum, which rises in the Ethiopian highlands. From the confluence of the White and Blue Nile, the river continues to flow northward into Egypt and on to the Mediterranean Sea. From Lake Victoria to the Mediterranean Sea, the length of the Nile is 5584 km (3470 mi). From its remotest headstream, the Ruvyironza River in Burundi, the river is 6671 km (4145 mi) long. The river basin has an area of more than 3,349,000 sq km (1,293,049 sq mi) (source: http://www.nilebasin.org/IntroNR.htm).

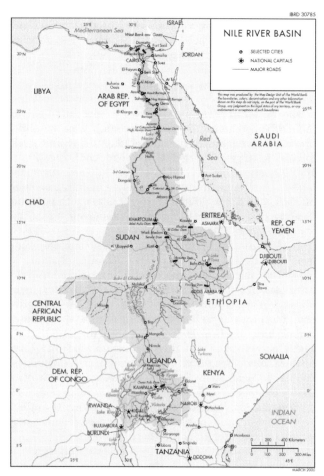

Figure 1. Nile basin map. (*Source*: http://www.nilebasin.org/nilemap.htm.)

As a typical transboundary case, the Nile River Basin is shared by ten countries: Burundi, Democratic Republic of Congo, Egypt, Eritrea, Ethiopia, Kenya, Rwanda, Sudan, Tanzania, and Uganda.

Since the early 1970s, water scarcity has become a matter of concern for the downstream states of the Nile. The engagement of the two downstream states (Egypt and the Sudan) in effective development of irrigated agriculture marked the beginning of real scarcity. The 1960s, in particular, marked the beginning when the two countries engaged in constructing large-scale water projects; by the end of the decade, three dams (Roseires and Khashm Al-Girba in Sudan and the High Dam in Egypt) were constructed. Large-scale agricultural schemes have become dependent on these dams.

In addition to damming the river and increasing demand for irrigation, population growth is considered a major cause of water scarcity in the Nile Basin. Since the late 1940s, probably all Nile Basin countries started to witness steady population increases. In the 1990s, the Nile Basin region experienced one of the highest population increments among the world's regions. Striking evidence of this is that 6 out of the top 20 countries in the world in population increments in 1991 were Nile riparians. The population of Ethiopia, Sudan, and Egypt will double from 157 million to 388 million by the year 2050. The total population of the Nile Basin will more than double when it increases from its 1992 figure of 259 million (1, p. 82) to 607 million by the year 2025. The Nile riparians will remain, well into the first half of this century, among those with the highest growth rate in the world; their rates range from 2.4% to 3.6% (1, p. 82). The most important conclusion to be derived is that more significant increases are taking place in upstream than in downstream countries (2, p. 2). Moreover, it is observed that the countries that have large populations are increasing their population faster than those that have smaller populations. Thus, except for the Democratic Republic of Congo, all large-populated countries will be in dire need of Nile waters as irrigated agriculture increases.

The 1980s witnessed yet another driving force in the region that contributed to water scarcity, the recurrent droughts and their consequential population concentration along the banks of the Nile and its tributaries. The 1980s and 1990s seem to provide new causes that have turned migration into "displacement." Referring to the Nile, Mageed (3, p. 159) stated that "The arid and semi-arid regions of the Basin are experiencing a serious breakdown of the environmental fabric and the spread of desertification along with the collapse of socio-economic systems." The part of the Nile Basin that is crossed by the Sudano–Sahelian zone has scored the highest rate (88%) of desertification in the early 1980s compared to the African average (84%) and the world average (61%). The consequences of drought and land degradation surfaced soon after the 1980s decade started and reached their peaks in famines in Ethiopia and the Sudan simultaneously in 1984. The effect of these, however, has hit all the Nile riparians in different degrees. These consequences have led/will lead to radical changes in the population map in Ethiopia and the Sudan. These famines have pushed large groups of population to the Nile River zone and thus contributed to real population concentration along the banks of the Nile and cultivable irrigable lands in its vicinity. Computations from population census data reveal that the river zone in central Sudan has increased its relative population size by 5.61% between 1983 and 2000, and similarly, in Ethiopia, the relative size of the population inside the Nile Basin increased by 6.5% in just a decade between 1984 and 1994. The same census data show that the population of urban areas in both countries is increasing along the banks and tributaries of the Nile/inside the Nile Basin. The population of Egypt has always been concentrated along the borders of the Nile except for some small oases in the deserts. Although the government of Egypt is developing urban areas outside the Nile valley, most of these developments still depend on Nile water for their water supply.

Unlike population concentration on a regional scale, population concentration in urban areas generates scarcity of a much higher magnitude (4, p. 90). This concentration generates water scarcity in several ways, including high demand for safe water for drinking and sanitation and high pollution rates, which render some water useless, or even harmful. However, as no further facilities have been installed, the concentrating of population has negatively affected the provision of safe water for urban dwellers; in Sudan, the rates decreased from 100% in 1980 to 86% in 2001. In all Nile Basin countries, except for Egypt, provision of safe water has been very limited. Water scarcity caused by population concentration is increased by the in-stream demand for hydropower generation, by the evaporation of reservoirs developed for hydropower generation, by increasing demand due to agricultural development, and by climate change due to global warming.

WATER ALLOCATION AND DISPUTES IN THE NILE BASIN

In reality, water scarcity already causes violence and conflict within and between nations, which is a threat to social and political stability. Peter Gleick, who makes an ongoing effort to understand the connections between water resources and international conflict and security issues, points out that "disputes over control of water resources may reflect either political power disputes, disagreements over approaches to economic development, or both. It is evolving as international and regional politics evolves..." (5). Water disputes typically are erupting within countries in the downstream regions of stressed river basins. Even if water disputes between countries do not lead to war directly, they have fueled decades of regional tensions and thwarted economic development before eventually giving way to cooperation. The Nile is an example in this respect.

Legal arrangements for the Nile in the nineteenth century were much more comprehensive than today in terms of legal obligations abided by all governments ruling over almost all the territory of the Nile Basin. Britain had been the main party to all these agreements; it signed agreements with Italy, Congo Free State, Ethiopia and Egypt. These agreements remained intact until independence. The decolonization process left the Nile with colonial agreements that were not respected by all newly independent riparians other than Egypt.

Immediately after acquiring its independence, the Sudan sought a new deal with Egypt over the Nile water to replace the 1929 agreement signed between the Sudan government (British administration) and Egypt. Conflicts arising at the postindependence stage, unlike the previous conflict, which was basically for securing the flow of rivers, were about the allocation of specified quotas of the now known as large quantities of water, yet scarce to meet the heightening demand. Hence, the Sudan started to confront the unfair division of water; it considered that Egypt's share was more than it actually should be, given that all the Nile water flows into Egypt, save 4 billion m^3 allocated to the Sudan by the 1929 agreement. For the Sudan, a new deal was thus seen as necessary. The challenge of the Sudan did not stem from preserving rights to the future realization of its agricultural potential; rather it was the obstruction of its already developed policies for bringing on board this competent potential. "The Sudan, to prove its case, embarked on drawing up the 'Nile Valley Plan'. The main object of the plan was to provide for the irrigation needs of Egypt and the Sudan and the assumed needs of the other riparian states for irrigation, together with the full development of hydro-electric potential in Uganda, Ethiopia, the Sudan and Egypt" (6, p. 73). The whole exercise was challenging Egypt's monopoly of the Nile by adding an element of egalitarian multilateralism. What made this grow to a challenge was that Sudan was seeking to mobilize the other riparians behind its plan. Egypt, however, was striving to start the implementation of the High Dam at Aswan. The Sudan, while opposing the Aswan Dam project in principle, had put forward three principles for resettling the water question. First, that Sudan's share should be determined before work began on the High Dam. Second, that Egypt must provide for the resettlement and "adequate alternative livelihood" of the people of the Wadi Halfa town and district who would be removed by the High Dam construction. Finally, that the Sudan should have the right in the future to build whatever works were deemed necessary for using her share of the Nile waters (7, pp. 173–174).

Upon accepting these conditions, the conflict between Egypt and the Sudan was bilaterally settled by the Full Utilisation of the Nile Waters Agreement in (1959). The 1959 agreement "established a joint committee (PJTC) and specifies quantitative allocations for the two countries: assuming an inflow of 84 billion m^3 from the Blue Nile and the White Nile sources, the agreement allocates 55.5 billion m^3 to Egypt and 18.5 billion m^3 to Sudan" (8, p. 17; 9).

The deliberations leading to the signing of the 1959 agreement, however, had created a serious conflict with Ethiopia. In 1957, Ethiopia sent a note to Egypt and Sudan in which it asserted its "natural rights" to use the water originating in her territory and in which it also referred to Ethiopia's projected water needs and requirements. These projected needs would certainly decrease the discharge of the Nile (10). The 1959 agreement was considered to favor the Sudan (11,12), dismember Ethiopia (12), and neglect the other upstream countries. The outcome of the 1959 Agreement had effectively put Egypt and the Sudan in one camp against the camp of all other riparians. The newly independent upstream states—individually, but sharing the same view—started to criticize Egypt and the Sudan's monopoly of the Nile waters. Egypt seemed to have accepted the idea that the upstream states are bound by agreements signed during the colonial time, but Sudan's position was different, namely, recognizing the rights of upstream states (10, p. 12).

In the eyes of the Egyptians, Ethiopia has always represented a potential threat to the Nile water. According to Hultin (13, p. 37): "It is not so much what Ethiopian governments—or other riparian governments for that matter—have done with regard to the waters of the Nile, but rather what they might be doing, that is the cause of anxiety in Cairo." Repeatedly, the Ethiopian governments asserted that they are not bound by any agreements on the Nile and that those agreements that were signed during colonial rule were either not signed by them or they were forced to sign by a colonial power. An important element in the dispute is the difference between the states in their starting argument. Egypt and the Sudan, bound by an agreement, based their contest for the Nile waters on real needs, but Ethiopia is driven by the awareness of its future needs and, therefore, is not ready to risk her unused resources (12). Egypt and the Sudan see that their 1959 Full Utilisation of the Nile Waters agreement can be the foundation for any future agreement among all riparians, but the other riparians, especially Ethiopia, want new arrangements.

FORMULATION OF A BASIN-WIDE WATER SHARING FRAMEWORK

The current environmental condition in general and the water scarcity issue in particular require new regimes of thinking and social organization or reorganization. As stated above, there are a variety of driving forces in water scarcity. For instance, until recently, the quarrel over the Nile waters was driven by the states' ambitions to develop economically. In the last two decades, this necessity to develop, in the Nile riparian, is being replaced by a much more urgent need—food security. Two conditions in this respect can be illustrated: one is a condition of abundance that allowed states to expand their irrigated large-scale agriculture endlessly to engage more and more in the market and the second is the condition of scarcity which brought new pressures that affect political stability and necessitate responses from states to groups demands. The condition of scarcity makes it a necessity that states adopt certain regulations to cater to the pressures arising.

The first attempt at cooperation among the Nile countries took place in 1960 through what is referred to as the Hydromet Survey Project, basically aiming to build a database for the Basin. In 1992, Hydromet was replaced by TECCONILE (Technical Cooperation Committee for the Promotion of the Development and Environmental Protection of the Nile Basin). TECCONILE continued until 1999 when it was replaced by the Nile Basin Initiative (NBI) as a wider framework with different operating organs to pave the way for final legal and institutional arrangements.

When competition reaches the crisis stage that the likely win–lose balance turns into a lose–lose balance, the contestants are likely to cooperate to survive the ordeal (14). Developments in Nile Basin politics in the last half of 1999 indicate that the co-Basin states have already reached the situation indicated above. Unlike the first half of the 1990s of national assertiveness and unilateral approaches to use of the Nile waters, the co-Basins now have (in the second half of the 1990) a new initiative, using the Technical Advisory Committee as its technical arm. The objective of the initiative is to reach a solution through the equitable use of the Nile waters. The conference of the Nile riparians in Addis Ababa in May 1999 moved further in this direction, and the Nile co-riparians have emphasized the sustainable development of the river. Thus, already the issue of equitable water use has been expressed together with the issue of sustainability.

Out of these initiatives, the shared vision of the NBI arose that is meant "to achieve sustainable socio-economic development through the equitable utilization of, and benefit from, the common Nile Basin water resources." The initiative as a comprehensive cooperative framework provides for significant changes in the old formula that would protect the co-riparians from inflicting harm on each other. Significant harm is always the consequence of competition, whereas equity and reasonableness can be attained through cooperation. The first is often defined by ideological and political imperatives, whereas the latter is for calculating mutual benefits or avoiding mutual losses—the economic imperatives. In this regard, the request of the Nile-COM for the World Bank and its partners to host a consultative group (ICCON) as a forum for seeking funds is a significant move toward emphasizing the economic imperatives. The Ethiopia entry into the debate on legal arrangements is likely to cool down the previous tension and create a condition conducive to cooperation.

The emphasis on sustainability is what makes the NBI different from the previous initiatives. Moreover, unlike the previous ones, it is a vision shared by all riparians. It means, in short, that awareness of the current environmental problems that affect all riparians is finally on its way toward institutionalization. It is the emphasis on sustainability that gives way to the newest part in the Initiative: the subsidiary action subprogram. The principle of subsidiary is considered "an important approach to cooperative action within a Basin-wide framework" and is meant "to take decisions at the lowest appropriate level, to facilitate the development of real action on the ground." It is this part which gives place to alternative developments by decentering the definition-making and decision-making about the Nile waters, which has been always the monopoly of central governments or bodies representing them. Building the pillars of the temple of cooperative framework (confidence building and stakeholder involvement; socioeconomic, environment and sectoral analysis; development and investment planning; and applied training) in itself involves a necessary learning process in which different levels will contribute. It, thus, informs the new priorities to be set for the Nile Basin countries as it translates to "capacity building and human resource development." As well, it makes place for priorities depending on the needs and specific problems facing each country or a sub-Basin that combines more that one country. It caters to the communities dwelling in the Basin that would be affected by the grand policies to be launched.

Another important aspect of NBI is that it will also consider measures outside the basin when this will ease the problems in the basin, in particular, measures with respect to population and agriculture. Their approach will be based on equitable regional economic development to ensure the balance between upstream and downstream *and* between inside and outside the basin.

What complicates the issues of water scarcity in the future is that the Nile Basin countries are far from providing water efficient alternatives. Thus, the development of the river, necessarily sustainable, is an urgent necessity.

PLAUSIBLE SOLUTIONS

Despite the fact that the Nile River is viewed as the archetype for any future conflict over international water, recent developments show that through cooperation, the Nile riparians can overcome many of the hurdles and establish a peaceful and sustainable solution to the problems of water scarcity in the Nile. What is needed, therefore, is:

1. Principle of equity. Egypt and the Sudan must accommodate the position and demands of Ethiopia and other upstream countries in an institutional manner (15). Ethiopia which views itself as being dismembered by the 1959 Nile water agreement will likely offer greater cooperation should the downstream riparians recognize its right to development of its part of the Basin and facilitate it. If Ethiopia is allowed to go ahead with its Blue Nile basin plan, Egypt and the Sudan would benefit from it (see also Infrastructure Solutions). Governments and international organizations must act early and constructively. Tensions among the Nile basin countries are finally easing, thanks in part to unofficial dialogue among scientists and technical specialists that have been held since the early 1990s and more recently at a ministerial level in the NBI. The key is establishing a process of early cooperation before serious hostilities erupt that make it difficult for nations to sit around a negotiating table together.

2. Strong institutions. Treaties that provide for effective monitoring and enforcement are often remarkably resilient, holding even when the signatories are engaged in hostilities over nonwater issues. Long-term programs of joint fact-finding, technical cooperation, and other initiatives that establish a climate of cooperation among countries can pave the way for resolving disputes when they do arise.

3. Effective rural development. To our understanding, the major problems that the Nile Basin countries are facing now are not merely the need for sharing water; rather they have to do with what makes the quest for sharing more feverish. In other words, environmental degradation and population concentration makes the quest for more water endless. So dealing with these problems and similar ones is what is urgent for all Nile riparians. The NBI can be seen as a framework which allows overcoming some of the factors contributing to water scarcity, including rural development. The multiple billions of dollars that are expected for the development of the Nile basin should be spent on large-scale hydraulic constructions and should also partly be spent on rural development in the territory of the co-riparians—not necessarily inside the Basin.

4. The import of "virtual water" represents one important element in how states could solve their water scarcity problems. In the Nile, this virtual water importation can be thought of in a more innovative way. The Nile Basin countries differ in climate; therefore, crops produced as well as the natural hazards, solutions, and therefore cooperation should consider these differences. Ethiopia faces severe soil erosion due to deforestation, and this very soil erosion, among others, is what pushes Ethiopia toward using Nile water. A fund for forestation and then benefiting from forest products (service for the ecosystem, fruits and other products which become part of the livelihood of rural population, and products for exportation) would thus produce a double advantage. Downstream as well as upstream Nile riparians can import virtual water from Ethiopia necessary for the rehabilitation of the latter. The Sudan, viewed as the breadbasket of Africa and the Middle East, could import virtual water from Ethiopia and export virtual water too. The difference, however, is that virtual water from Ethiopia, in the form of afforestation products, would help rehabilitate this country's highlands, therefore, decrease its demand for Nile water. Similarly Egypt can import virtual water from Ethiopia and the Sudan, through a formula of less virtual water from downstream and more virtual water from upstream—less water-consuming products versus more water-consuming products (rice).

5. Infrastructure solutions. Establishment of reservoirs on the Blue Nile in Ethiopia is suggested as "the greatest opportunity over the long term for dramatic improvement in the overall management of Nile resources" (16, p. 152). What is interesting in this regard is that the water savings so made, which could be of the order of 12–21.4 billion m^3 per year (17), would quadruple Ethiopia's irrigated area without reducing supplies to Egypt and the Sudan. Some other solutions are transfer of storage from reservoirs in Egypt and the Sudan to Lake Tana and the Blue Nile basin combined with water conservation in the Ethiopian highlands (by way of reducing evaporation and annual flooding losses) and will provide regulated flow to downstream countries. Similar schemes could be implemented in the equatorial lake regions of the upper White Nile riparian countries,

resulting in additional water saving construction of a bypass canal (the Jonglei diversion canal) in southern Sudan reducing the enormous evaporation losses and increasing the contribution of the White Nile to the main Nile appreciably. The Machar Marsh and Bahr El Ghazal in southern Sudan can contribute 11 billion m^3 (18, p. 139; also see 1, p. 92).

6. Water saving technologies. In all cases, maximum use should be made of water saving technologies (more crop per drop). This includes water saving options in the surface water system (reduction of losses of seepage and evaporation), more efficient irrigation techniques (land leveling, drip irrigation, etc.), and other management techniques (reuse, nightirrigation, etc.).

BIBLIOGRAPHY

1. Shapland, G. (1997). *Water in The Middle East and North Africa: The Political Economy and Ethics of Scarce Resource Use*, compiled and Edited by Timothy C. Weiskel, For the Harvard Seminar on Environmental Values (1997–98) "Water—Symbol and Substance of Life".
2. Omar, A. and Takele, T. (1998). *Egypt and the Horn of Africa: That Serpent of Old Nile*, Ethiopia Tribune (26.06.1998).
3. Mageed, Y.Abdel (1994). The Nile Basin: Lessons from the Past. In: *International Waters of the Middle East: From Euphrates-Tigris to Nile*. A.K. Biswas (Ed.). Oxford University Press, Bombay.
4. Falkenmark, M. Lindh (1993). *Global Changes and Water Resources*, www.water.hut.fi/wr/research/glob/egloshow/egloshow2.pdf.
5. Gleick, P.H. (2000). The world's water 2000–2001. *The Biennal Report on Freshwater Resources*. Island Press (Ed.). Island Press, Washington, DC, p. 315.
6. Mageed, Y.A. (1981). *Integrated River Basin Development: A Challenge to the Nile Basin Countries*, in Sudan Notes and Records No. 62, Vol. LXII.
7. Holt, P.M. and Daly M.W. (1979). *The History of the Sudan. From the Coming of Islam to the Present Day*, 3rd Edn. Weidenfeld and Nicolson, London.
8. Tadesse, H. (1998). The Nile river basin. *Ethioscope* **4**(1).
9. Waterbury, J. (1979). *Hydropolitics of the Nile Valley*, Syracuse, Syracuse University Press.
10. Khalid, M. (1984). The Nile waters: the case for an integrated approach. In: *The Nile Valley Countries: Continuity and Change* (Vol. 1), M.O. Beshir (Ed.). Institute of African and Asian Studies—University of Khartoum—Sudanese Library Series (12), Khartoum.
11. Davies, H.R.J. (1984). "Continuity and change in the nile valley" a geographical viewpoint. In: *The Nile Valley Countries: Continuity and Change*, Vol. 2, M.O. Beshir (Ed.). Institute of African and Asian Studies—University of Khartoum—Ethiopiaese Library Series (12), Khartoum.
12. Tilahun, W. (1979). *Egypt's Imperial Aspirations over Lake Tana and the Blue Nile*. United Printers Ltd, Addis Ababa.
13. Hultin, J. (1995). The Nile: source of life, source of conflict. In: *Hydropolitics: Conflicts over Water as Development Constraint*, Leif Ohlsson (Ed.). University Press, Dhaka and Zed Books, London.
14. Wolf, A.T. (1997). *Water Wars' and Water Reality: Conflict and Cooperation Along International Waterways*. NATO Advanced Research Workshop on Environmental Change, Adaptation, and Human Security. Budapest, Hungary, 9–12 October.
15. Tafesse, T. (2001). The Nile question: hydropolitics, legal wrangling, modus vivendi and perspectives. Münster, Lit. pp. 131–145.
16. Whittington, D. and McClelland, E. (1992). Opportunities for regional and international cooperation in the Nile Basin. *Water Int.* **17**: 144–154.
17. Kliot, N. (1994). *Water Resources and Conflict in the Middle East*. Routledge, London.
18. Naff, T. and Matson, R. (Eds.) (1984). *Water in the Middle East: Conflict or Cooperation?* Westview Press, Boulder, CO.

PLANNING AND MANAGING WATER INFRASTRUCTURE

NEIL S. GRIGG
Colorado State University
Fort Collins, Colorado

Water infrastructure consists of the constructed elements that control water for water supply, wastewater service, flood control, hydropower, recreation, navigation, and environmental protection (see articles on WATER RESOURCES MANAGEMENT and WATER INFRASTRUCTURE AND SYSTEMS). Generally, these constructed elements involve conveyance systems, including channels, canals, pipes, and bridges; diversion structures; dams; reservoirs; locks; treatment plants for water supply and wastewater management; pumping stations; hydroelectric plants; spillways, valves, and gates; and wells. Planning and managing water system capital assets require provisions for capital investment, operations, and maintenance.

Capital investment in water facilities requires a careful process that begins with planning them, preparing the designs, and constructing the systems. The goal of capital investment is to obtain a maximum return over the project's "life cycle." This means considering costs and management needs over the full life of the facility and requires that operations and maintenance, including corrective maintenance, be carefully conducted to gain maximum usage and life from the investment.

The planning process involves a series of steps that include identifying the need; searching out alternative ways to meet the need; evaluating economic, environmental, and social costs and benefits; making decisions about the alternatives to pursue; and implementing the plan. The steps in planning drive the process, but they involve many complex issues and substeps. Each project has unique features and stakeholders, political issues in gaining approval for infrastructure often control the pace and agenda. Traditional planning processes now incorporate sophisticated environmental, social, and economic investigations and are now much more complex than in the past. For example, the United States' experience with environmental impact statements required by the National Environmental Policy Act shows that it costs more to add environmental impact analysis to planning, and the decision process becomes more complex.

In an attempt to rationalize complex criteria, the concept of "multiobjective evaluation" was developed to consider economic development, social benefits, and environmental enhancement when planning for water purposes such as water supply, power, and flood control. This concept was included in the "Principles and Standards" that emerged from the Water Resources Planning Act. These later were changed to "Principles and Guidelines."

For a minor component, such as a new pump station, the process may be straightforward and short term. For major facilities such as dams, the process may take many years and, in some cases, ultimately not succeed at all. In addition to the planning steps cited, practical assessments such as how to pay for facilities and identifying legal obstacles are required. Also, stakeholders must be involved at each step of the process through public involvement and consultation.

The design process involves creative decision-making about the configuration and details of projects. Design includes drawings, documents, and plans necessary to initiate construction. Construction begins with preparing contract documents and involves bidding, review, award, organization, construction itself, inspection, and acceptance. Keeping all of the phases going in the proper direction and order is the task of the "project management process." Designing and constructing major water resources projects require high levels of skill and organization involving risk to the public and environment. For example, dam construction may require control of an entire river over several seasons when floods might occur. A wastewater treatment plant might involve large-scale and sensitive biological processes.

After construction is completed, the project enters the "operations and maintenance" or "O&M" phase. Large water systems such as dams and reservoirs are subject to special operating rules. Water treatment plants require trained operators who can monitor chemical and biological parameters while adjusting operating controls. Far-flung systems extending over several watersheds may have sophisticated supervisory control and data acquisition (SCADA) systems, consisting of instruments, wireless and wired communications, computers, and control systems.

Decisions about water system operation are controlled by complex sets of rules and guidelines. In some cases, regulatory controls are very stringent, and violations result in heavy penalties. Operations may be based on computer simulations and multiobjective criteria, as in following rule curves for reservoir operation. A science of *water resources systems analysis* has emerged to provide tools and techniques for operating complex systems.

Maintenance management is also required in a capital management system. Without adequate maintenance, water resources infrastructure systems will usually not perform well and will wear out sooner than planned. The maintenance management system brings activities together for investment, organization, scheduling, and monitoring. Its general functions are inventory, condition assessment, preventive maintenance, and corrective maintenance.

An inventory of facilities, components, and equipment is the cornerstone of maintenance management. Inventory methods differ for real property, fixed assets, and equipment. An inventory can be as simple as, for example, a set of drawings indicating where sewer pipes are located in a section of a city. These drawings, with annotations, are used by maintenance forces to locate and service pipes. On a more sophisticated level, the drawings could indicate other, nearby facilities, such as water and electricity lines, and can also be used to coordinate between services. Even more sophisticated would be coordinated data in a GIS and database format, available on a common basis to different sections of the organization whose work involves shared data, processes, facilities, and staff.

Condition assessment requires inspection and analysis of condition so that maintenance activities can be planned and scheduled. It is used to plan and schedule capital management activities, such as maintenance, needs assessments, budgeting, and capital improvement programs.

Preventive maintenance (PM) is the ongoing program of care given to equipment or components. In general, PM requires consistent, timely completion of tasks prescribed by documented procedures according to set schedules that include regular follow-up. Information sources for PM are O&M manuals, product information, and experience of workers. PM records include equipment data, the preventive maintenance record, the repair record, and a spare parts stock card.

Generally speaking, corrective maintenance means to repair equipment or components that have failed or deteriorated. It can range from minor to major repair and drives the "3 R's" of infrastructure—repair, rehabilitation, and replacement. Corrective maintenance requires a decision if the deficiency is minor or major enough to require capital budgeting. If the problem is major, the capital plan and budget incorporate information about new standards and growth forecasts to lead to decisions about rehabilitation and/or replacement.

When a water resources facility has reached the end of its useful life, it must be removed, rehabilitated, or replaced. Many dams around the world are now 50 or more years old and will face these decisions in the future. Wastewater treatment plants built during the 1970s in the United States are now three decades old. In addition to new investment to handle population growth, a discussion will occur in the future about strategies for reinvestment and modernization of these key infrastructure components.

APPLICATION OF THE PRECAUTIONARY PRINCIPLE TO WATER SCIENCE

Leonard I. Sweet
Engineering Labs Inc.
Canton, Michigan

Discussions of applying the precautionary principle have received new momentum from several highly contentious policy issues, including the bovine spongiform encephalopathy crisis, blood donor screening, genetically modified foods, antibiotic feed additives, and phthalates in baby toys. The level of precautionary interpretation and its application to "decision-making under ambiguity" has wider ramifications for waste policy, ecological toxicology, and especially water science. A balanced debate is warranted over the merits of the two seemingly rival environmental risk management paradigms—risk assessment and the precautionary principle.

The basic definition of the precautionary principle has three elements: threat of harm, scientific uncertainty, and the precautionary action. As such, the principle questions whether scientific knowledge is strong enough to permit reasonably confident decision-making, and it favors preventive action in the face of uncertainty as well as a participatory public in decision-making. The precautionary principle can be applied in general on two fronts, the *ex ante* side that reflects a desire to avoid or prevent undesirable adverse effects in policy frameworks and legislation, and on the *ex post* side, reflecting the fact of having to live with consequences of past decisions, side effects, or accidents (e.g., historical liability or natural resource damage assessments).

The relative constraints and uncertainties of risk assessment suggest to some that risk managers should abandon the risk framework in favor of the precautionary approach. In water science, it is not prudent to propose one as an alternative to the other. An immediate concern is that the precautionary-based approach fails to account for risk transfers, priority substances, and costs versus benefits. It is proposed that a variety of precautionary initiatives and decision analysis tools in risk management (e.g., life cycle assessment, weight of evidence, cost–benefit), as well as the open exchange of information and rational thinking, will aid in bridging the dichotomy between risk assessment and the precautionary principle.

INTRODUCTION

The momentum toward a paradigm shift to precautionary approaches will be faced globally by government and industry, and those in water science will have to rethink radically their current safety, risk management, and regulatory control schemes. The premise of precautionary decision-making is that environmental or human health threats are challenging to comprehend and resolve and that it is necessary to establish an environmental management paradigm to regulate suspect activities before they cause harm. Although originally intended for environmental management issues, the precautionary principle has now expanded to address human health and safety issues (i.e., toxicology, epidemiology, industrial hygiene, medicine). For example, it is likely that a precautionary approach was used in addressing the health and safety risks of using thalidomide because the precise mechanism by which thalidomide impacts the fetus was not characterized until the late 1980s.

In general, precautionary approaches instruct environmentally protective decisions before any potentially harmful effects of a given action or technology are fully proven. There are varying degrees of interpretation of the precautionary principle, including the most stringent or maximally precautionary versus the most permissive or minimally precautionary: (1) strong precaution (where the scientific evidence is uncertain but the public is informed and the evidence indicates that the worst case is the most probable outcome), (2) strong precaution (where the evidence is too weak to justify abrupt changes, given costs, but policymakers take a holding pattern and halt further intervention until more risk evidence is available), and (3) weak precaution (where the evidence is overly weak to justify change in public policy, given costs and other factors, and the public is informed of possible risk).

It can be seen that a regime based strictly on the precautionary principle places the burden of proving the harmlessness of a given action on the party who would engage in that action. This version of the burden of proof policy is a departure from the traditional toxic tort approach in which no harm is presumed to result from the activity of another until a party can demonstrate damage and causation. In the environmental arena, the tort-oriented nonprecautionary approach postpones the regulation of activities until harm occurs and compels a response to potential degradation. Consistent with this continuum of approaches is the presumption that precautionary action is acceptable as an interim measure because, as the scientific evidence accumulates, it is only a matter of time until an adequate level of certainty is reached.

It is apparent that in the governance of risk, the risk-based and precautionary approaches are divergent with regard to scope, risk, and how best to handle uncertainties and values (Table 1).

Rather than seeing that precaution is in conflict with science-based regulations, some suggest that the precautionary approach is consistent with sound scientific practice: (1) in acknowledging and responding to difficult problems in risk assessment such as ignorance (we don't know what we don't know) and incommensurability (we have to compare apples and pears) and (2) in critically thinking about the "use" of scientific knowledge.

GLOBAL SCOPE

Reflecting the divergent values of risk management, the precautionary principle has been defined, and the relative importance weighted differently by various nations (e.g., lack of firmly established international or legal meanings of the precautionary principle). The inconsistency in the precautionary approach is further complicated because actions within one state may cause identifiable harm to another state's environment, which in turn may create international responsibilities.

In a global economy, there are certainly different country-specific circumstances for interpreting the precautionary principle, including the level of industrialization, governance, economic and technical contexts, and human resources available for monitoring and evaluating problems connected with chemical risks. The current perception is that Europe believes that the earlier one exercises precaution, especially when cumulative pollution and diverse chemical risks do not respect boundaries, the more effectively it can be done.

Table 1. Highlights of Risk-Based and Precautionary Frameworks

Risk Framework	Precautionary Framework
Scope:	Scope:
• origins in quantification and hypothesis testing	• origins in subjectivity and emotion
• premise of threshold and assimilative capacity	• does not support assimilative capacity or threshold in environmental/human health
• focus on individual and population risk	• focus on values and a unifying perspective of precaution
• methods (i.e., hazard identification and characterization, appraisal of exposure, risk characterization)	• methods not well defined; no cause–consequence chain
• adheres to tenets of toxicology (i.e., Paracelsus, Pare's, Haber's, Arndt–Schulz)[a]	• adheres to caveat of *primum non nocere* or "first do no harm"
Chemical innocent until proved guilty:	Chemical guilty until proved innocent:
• "prove harm" regulatory approach	• "prove safety" regulatory approach
• burden of proof on regulatory agency, customer, or public	• burden of proof on innovator or manufacturer
• pollution prevention/minimization opportunity if established threat	• ensure safety or ban agent/activity before established threat
• focus on preventing false positives	• focus on avoiding false negatives
Optimistic view of	Pessimistic view of
• best available data	• data (inaccuracies, gaps, unsuitability)
• default values	• modeling (overly complex, nonrandom)
• analytical techniques	• analytical (detection limits, interference)
• modeling methods	• timescales (inadequate for decision-making and legislation)
• margins of error	• uncertainty (e.g., smokescreen vs. scientific indeterminacy)
• statutory time constraints	
• reasonable estimates	
Risk recommendations viewed as	Risk recommendations viewed as
• entrenched in model and method uncertainty	• proactive and adaptive in the face of new information or the unforeseen
• carried out on product versus process	
• subject to cost/benefit analysis	• not subject to cost/benefit analysis

[a] Paracelsus: dose–response contains the notion of threshold of effect; Arndt/Arndt–Schulz: predicts continuum of physiological mode of action dependent on strength of stimuli, that is, from hormesis/stimulation at low levels to retardation at high levels; Haber's: includes the variables of time and dose as a function of exposure to compare hazards better and formulate thresholds of toxicity; Pare's: the specificity and toxicity of chemicals are due to their unique structure and the laws of biology that govern the response interaction.

Regardless of the geographical complexities, however, there appear to be two consistent global elements of the precautionary principle: first, insufficient scientific information and second, significant but uncertain risks.

Consider the European policy on chemicals and the environment. Europe regards the precautionary principle highly, and erring on the side of caution when knowledge is limited and prediction speculative is viewed as a sensible and sound approach to good governance. The European community has also adopted a variety of principles that are directly or indirectly also designed to improve implementation of the precautionary principle, including the following: (1) polluter pays, (2) preventive principle, (3) sustainable development, and (4) inter-/intragenerational equity. Taken together, the core direction of these principles suggests that liability, at a minimum, should enhance incentives for more conservative, proactive, preventive, future-oriented, and responsible actions.

The European community's approach to the "polluter pays" principle could lead to a structure of environmental liability based on increased levels of precaution and prevention. Such an approach may also introduce new damage assessments, such as environmental damages to nature or biological diversity (e.g., due to the discharge of hazardous substances into waterways) when polluters are identifiable, where the damages are quantifiable and concrete, and where a causal link is established between polluters and damage.

Although there is no fully agreed upon version of the precautionary principle, it is being embraced extensively in international articles, treaties (International Joint Commission on Great Lakes; EU Institutional Treaty), summits (Rio Earth Summit), laws (e.g., German Vorsorgeprinzip legislation; Dutch wildlife laws), conventions (e.g., United Nations' Convention on Law of the Sea, Climate Change, and Biodiversity; UN General Assembly World Charter for Nature), conferences (e.g., 1998 Wingspread Conference Center), and other policy-making formats (e.g., Communication from the European Commission about guidelines for EU research activities; Expert Advisory Board on Children's Health and the Environment). The precautionary principle is clearly viewed as a logical corollary of the established international law norm that no state has the right to engage in activities within its borders that cause harm to other states.

Although the various interpretations of the role and application of the precautionary principle revolve around science and policy, the issue is steadily subjected to political motivations in administration and legislation. In particular, precautionary action has elevated such issues as genetically modified organisms, acid rain, radio-frequency fields, toxic waste disposal, and climate change. However, the precautionary principle is currently lacking in providing evidentiary standards for safety, (e.g., with medicines it revolves around quality, efficacy, and dose–response constancy), as well as criteria for obtaining regulatory approval. This is a serious shortcoming because it is conceivable that practitioners of precaution could be overcautious or unscientific in considering whether new technologies pose unique, extreme, or unmanageable risks, without properly characterizing the many new aspects that may confer net benefits and total hazard reduction. Such a discriminatory policy could inflate the costs of research or remedies, result in high taxation of products, waste resources, inhibit the development of new technologies, restrict consumer choice, and result in zero hazard reduction, all potentially for theoretical risks.

DRIVERS TO ADOPT

Proponents of the precautionary principle believe that existing environmental regulations and other decisions based on risk appraisal techniques have failed to protect environmental and industrial health. Recognized drivers for adopting the precautionary principle include the following:

- risk aversion
- reduced reliance on the quantitative risk in favor of qualitative techniques and participative deliberation
- strong commonsense appeal to general public
- working to safe minimum standards
- adopting sustainability efforts
- following a win–win/least regrets regime with regard to costs and benefits
- primary prevention
- data deficiency or uncertainty challenges, modeling difficulties, analytical difficulties, and insufficient time for adequate research
- commonsense appeal of having the polluter pay
- those in risk governance are frequently asked to provide the final word of advice on safety or risk before they have sufficient proof
- concern that too much focus on risk assessment could delay needed protections
- provides additional rationale for environmental and public health surveillance efforts, to use indicators, and detect the probability of adverse consequences as early as possible.

But in practice or in policy implementation, the precautionary approach is not very popular among experts in the risk arena because the potential benefits from the drivers may not warrant the cost and it is a challenge to make the principle a unified perspective that addresses many of the concerns.

ANTIDRIVERS

Clearly, however, there are antidrivers to adopting the precautionary principle. The main ones include:

1. the reverse onus that shifts the duty of care to be placed on the developer's actions to "prove safety," which may ignore the tenets of toxicology that demonstrate a need to regulate chemicals based on both the beneficial as well as adverse; the other concern is that this is potentially counter to notions of liberty and freedom by shifting the burden of proof from government onto innovators
2. the apparent advocating of controversial measures, including bans or phaseouts, clean production, alternatives assessments, organic agriculture, ecosystem management, health based occupational exposure limits, and a requirement for premarket testing
3. the apparent lack of cost–benefit analyses, in that for many cases that call for precautionary action, the marketplace does not necessarily play a critical role and good cost–benefit information is not available
4. the belief that the precautionary principle allows controlling important resource allocation issues by fear, emotion, and politics and that becoming more "safe" in some areas may involve spending resources that cannot be used doing more 'good' in other areas
5. the continuing challenge for proponents to frame an adequate and harmonized version of the principle. This is a reflection that the principle is still evolving—to some, it is already part of the risk appraisal arsenal, to others it is not yet on the radar screen, and to still others it is the antithesis of risk assessment.

Other concerns in adopting precautionary approaches revolve around the implications for the relationship between environmental protection and economic development, the economic consequences. Technologists and scientists are concerned that the precautionary principle is irrational and potentially based on nonobjective science and that it is anti innovation/antibusiness. The concern is that the precautionary principle is biased toward noncorporate and nonindustrial opinions, is susceptible to manipulation with regard to trade and equality, and may become an excuse for arbitrary restrictions that may stifle progress, prevent the use of newer and safer technologies, and possibly create more risk or uncertainty. Not every action or technology that has associated uncertainty should trigger precautionary action, and furthermore, every chemical should not be a presumed or preestablished carcinogen or endocrine disruptor.

The reticence to implement precautionary approaches fully is likely to remain until a more wise and balanced approach to chemical risk management is established. Most would agree that uncertainties in major decisions need to be weighed, analyzed, and communicated to impacted parties, but a combination of options in addition to the precautionary principle may prove most effective in managing risk (e.g., perhaps prioritized precautionary thinking in tandem with risk characterization and the use of current and better/best control technologies). It is also clear that the growing interest in precautionary approaches is more likely a reflection of shifting values on risk tolerance and governance versus a preference for subjective versus objective problem-solving techniques. Herein lies much of the controversy around its implementation—what constitutes sufficient proof, proof of safety, or proof of negligence?

RISK TOLERANCE

One certainly must consider that the tolerability of individual risk and the regulation of societal risk are integral to our government and culture. Yet the concepts of safety and risk are relative and subjective. In some areas of regulation, an explicit treatment of risk is warranted, whereas in other areas, risk appraisal is more implicit and reliant on best professional judgment. But how safe is safe enough with regard to personal, national, or international risk criteria in water science?

To address this question practically, one must characterize the time frame and magnitude of the hazard and decide on a common language—the language of risk analysis or the precautionary principle.

- *Risk Analysis*. Risk-based target levels, health-based criteria, standards as low as reasonably practicable, reasonable probability of no harm, an adequate margin of safety, proportionality, weighing costs and benefits, virtually safe doses, allowable daily intakes, benchmarks, maximum allowable tissue concentrations
- *Precautionary Principle*. Common sense, an ounce of prevention is worth a pound of cure; better safe than sorry; a stitch in time saves nine; if in doubt, don't pump it out; sentence first, verdict afterward

Apparently lacking in the language of the precautionary principle are a cost–benefit analysis, risk communication issues (e.g., regarding whether the hazard is imposed upon, entertained voluntarily, controllable, inequitable, and to what level the public is averse to the issue), and a good characterization of the boundaries and dimensions of uncertainty (i.e., ignorance, systemic, indeterminacy).

Risk informed decision-making must involve some level of cost–benefit analysis to focus resources and set priorities so that the greatest social benefits are achieved at the lowest cost. Some practitioners may strive to make a formal cost–benefit analysis the determining factor for risk tolerability. However, a formal cost–benefit analysis seemingly contradicts the precautionary principle, as it assumes certainty where, by definition, certainty does not exist. Furthermore, there are some concerns that requiring a cost–benefit analysis would render the precautionary principle less effective, cumbersome, and bureaucratic. Cost–benefit analysis may not be an appropriate methodology for managing uncertainty, as it can be applied only if there is full knowledge of the cost of that damage—and this is challenging because precautionary actions are not based on demonstrated risk, but rather on anticipated risks that are considered plausible.

UNCERTAINTY

Uncertainty refers to the failure to address fully a standard of proof/safety required by decision-makers and to the recognition that one cannot fully characterize the extent and potential seriousness of the consequences of not meeting that standard. In the face of inherent and irreducible uncertainty, it is clear that risk management and precautionary principle paradigms need to be adaptive and consider statistical significance (i.e., power, probability of type I or type II error), as well as environmental significance.

What level of causation, threshold of confidence, or lack of proof is required to trigger use of the precautionary principle to link an exposure to some possible harm? Various trigger criteria have emerged, including reasonable suspicion, potential damages are serious or irreversible, scientific certainty, cost-effectiveness, a single complaint, a single case, a single animal study, a combination, only where the probability and cost of impacts are completely unknown, or some gradient or level lower than the balance of probabilities or prevailing probability. Whatever the trigger, the precautionary principle should be subject to review and assigning responsibility, as well as subject to the European Commission criteria of proportionality, nondiscrimination, consistency, and cost–benefit analysis. The precautionary measures must also be proportionate to the benefits to be achieved. Hence, there is a need for an open and transparent procedure to identify the best options for harm avoidance.

Uncertainty in detecting and evaluating chemical hazards leads to challenges when conducting risk analyses (e.g., robust estimates of exposure and risk for pharmaceuticals in drinking water are not available). In the absence of a scientifically rigorous approach to evaluating the risk potential, actions may be guided by societal perceptions of risk or safety (e.g., refractory pharmaceuticals in wastewater) and may lead to malignment, restriction, or banning of chemical agents. One has to balance carefully the perceived reduction in health risks to society from precautionary action and the denial of public access to materials or technologies that, under appropriate conditions of safe handling and use, would not result in notable risks and may have substantial net benefits.

In the end, one must recognize that the scientific basis of modern water policy involves varying degrees of uncertainty. But it is more certain that risk analysis and precaution converge at the level where they are issue driven and that, regardless of the paradigm or language agreed upon, risk informed decision-making in water science must be efficient and rational, even when data are uncertain and the lines of evidence are imperfect.

POTENTIAL RELEVANCE TO WATER AND ENVIRONMENTAL SCIENCE ISSUES

The precautionary principle can be applied to emerging public concerns, such as large-scale environmental and health issues, and to the associated changes in regulatory frameworks. Most notably, issues include major industrial accidents, agricultural and food sector (e.g., mad cow disease), dioxin exposure and long-range transport, the irreversibility of global issues (e.g., ozone layer depletion, enhanced greenhouse effect), and biotechnologies (e.g., cloning and genetic modification). For watersheds, management decisions based on the relationship between stressors and resources may include act now, protect, restore, or watch.

These are current examples of the precautionary principle in action in the regulatory and environmental arenas:

- *DDT Use to Combat Malaria.* The precautionary principle demands no risk taking when there are unpredictable consequences—the potential environmental and health risks posed by future use of DDT—and hence there is a strong push not to allow it to be used or manufactured. The millions of people suffering from malaria that could have benefited from using DDT simply do not understand the argument of focusing only on potential future risks versus present and clear benefits: the economic benefits of the chemical if the threat of malaria were reduced, health care costs reduced, needless morbidity and mortality reduced, that is, the need to balance the potential risks against the certainties and benefits.
- *Global Climate Change.* With climate change, proof will be validated only after many years of data collection. Given the scope, potential injury, and irreversible consequences, the available information should be given great deference.
- *Biodiversity and Wetlands Loss.* Consider unforeseen circumstances with regard to the Endangered Species Act in the United States; it does not require landowners to protect unlisted but declining species on their lands.
- *Fisheries Sustainability (Disease, Transgenics, Overfishing, etc.).* The use of transgenic salmonid fish that are reproductively sterile. Aquatic organism risk analysis is proposed to evaluate the environmental safety of transgenic aquatic organisms, but its primary weakness is that the probabilities and consequences are based on subjective evaluations and opinions. There is no requirement for scientific analysis or collection of experimental data in support of the risk analysis process. The role of risk analysis as a basis for fish health management is questioned, given the uncertainties, including unrealistic expectations of the ability to identify the right fisheries management decision, stakeholder participation, inadequate predictive ability of models, lack of integration of management and science, and a tendency to overestimate possible versus likely hazards. There are also considerations of failing to account for the unknown, unexpected, or synergistic impacts.
- *Substances of Concern.* California's Proposition 65 assumes that a listed chemical of concern is released to waterways and that it is "guilty before proven innocent."
Banning persistent, bioaccumulative, toxic substances. Sweden is considering banning substances that persist in the environment and accumulate in living tissue, regardless of whether they are proven significantly toxic such as lead in leaded glass. Under the current system, regulations are aimed at controlling rather than preventing end of pipe versus upstream pollution.
Restricting pesticide use. Canada is moving closer to adopting the precautionary principle based on a year 2001 Supreme Court ruling that upholds the right of local governments to restrict pesticide use to essential uses to protect human health and the environment.

Northern Ireland is banning many pesticides used for food production amid rising consumer concerns about impacts on human health and the environment due to chemical residues. This, of course, involves a ban on any pesticide where there is doubt about its safety, even if the weight of scientific evidence is insufficient or inconclusive.
Potential bans on antibiotics, given the uncertainties and risk of microbial resistance in the environment.
- *Ocean Shipments or Dumping.* Ocean shipments of radioactive materials and the reprocessing of nuclear wastes and acquiring stockpiles of plutonium. Proponents may want to cite explicitly the precautionary principle as a framework to regulate such shipments. The principle might require users of the ocean to exercise precaution, alertness, and effort by undertaking relevant research, developing nonpolluting technologies, and avoiding activities that present uncertain risks to the marine ecosystem—and further it would reject the current notion of a measurable ability of the environment to assimilate wastes.
Proposals to dump sewage sludge on the seemingly barren areas of the deep seabed. There is little known about the ecology of this region, the harm to the seabed and interconnected ecosystems. A traditional tort approach would allow dumping until harm is recognized, but a precautionary regime would require the dumping state to prove that the dumping is harmless before engaging in that activity.
- *Water Reuse, Reclamation, and Safety.* A precautionary approach might stipulate that climate change and increased urbanization may threaten to dry up wetlands, pushing species of those habitats to extinction. The fix may be to increase the efficiency of water use and stop wasteful irrigation practices where water losses are highest. The principle might also stipulate that all types of water abstraction and water transfers should be controlled. Certainly economic instruments and regulation can provide a flexible means of improving water resource issues.
The events of 9/11 may warrant conducting a security sweep of the nation's drinking water systems or for facilities to develop specific security plans based on vulnerability assessments.
- *Sewage Sludge.* In the United States, the biosolids rule imposes limits on molybdenum and alkylphenol. Some critics claim that the risk assessment process is inadequate and does not address all potential or foreseeable risks. Environmentalists, farmers, and scientists have encouraged the U.S. EPA to adopt a more protective program that uses the precautionary principle to determine the safety of applying water to land.
- *Habitat Conservation.* Involves dealing with data deficiencies in species status and magnitude of impacts in a manner that guards target species from irreversible habitat loss yet does not preclude development.
- *Biotechnology.* Consider genetically modified organisms or foods. The benefits of genetically modified

plants include reduced use of fertilizer and pesticides, increased yields, better adaptation to extreme environmental conditions, and reduced allergenicity versus the cost of impacts that may include allergenicity and exotic species that outcompete native and nonmodified organisms.

- *Antidegradation.* Antidegradation and antibacksliding rules are likely to become increasingly important in water policy because the intention of the Clean Water Act is to prevent perceptible downgrading of designated water uses. The protection of high-quality waters under the antidegradation rule might be interpreted differently by risk analysis versus the precautionary principle: For example, the precautionary principle might interpret the rule as an absolute prohibition on lowering water quality. The risk-based approach would consider that avenue a no-growth interpretation and propose a structure that provides systematic evaluation of all available information regarding the social, environmental, and economic impacts of lowering water quality.
- *Sustainable Development.* Some would argue that to achieve sustainable development, policies must be based on the precautionary principle. Environmental measures must anticipate, prevent, and attack the causes of environmental degradation. Science alone cannot determine the appropriate focus of the management inquiry or how to define a just or politically acceptable solution. Setting standards is the confluence of science and policy determination.

Consider the example of applying the precautionary approach to the risks of contaminants in surface and groundwater, including biological, chemical, and radiological agents. Certainly, on a global scale, pathogenic contamination of drinking water can be considered a significant health risk due to disease outbreaks, but there are also exposures to nonpathogenic toxicants. Water pollution comes from many routes, sources, and in various forms; in each case, there are incomplete or lacking data, certainly in the exposure, probability, and effects stages.

In contrast to the precautionary principle, risk-based objectives in water science are typically technology-based (e.g., best available technology, or best available technology economically achievable). One clear advantage of an approach based on technology includes national consistency in treatment level, predictable economics, and relative ease of enforcement—although it does not guarantee improved environmental performance because some areas may be under or overmanaged.

For some environmental practitioners, the precautionary principle appears to focus only on the risk from the agent or technology versus the clear benefits that may result from its use or from the risk trade-offs or transfers that may result from an alternative. Because the risks of a new technology are often not apparent until the technology is in use, some suggest it is then too late to undo the harm (e.g., various case studies in the Great Lakes have suggested to some investigators that the length of time between the introduction of a new technology and the discovery of its deleterious effects and the regulatory action to reduce exposure is typically greater than 25 years). Certainly, unpredictable consequences are not new to water science, considering the unanticipated challenges caused by the use of asbestos and PCBs. The concepts of risks and benefits in water science are not as clearly etched as some may think.

RECOMMENDATIONS FOR BALANCING PROOF VERSUS PRECAUTION

Proponents should proceed cautiously with the precautionary principle. Potential problems in abandoning the risk assessment framework in favor of following the precautionary principle are highlighted in Table 2:

There are certainly some refractory questions to consider, such as:

1. What grounds justify applying the precautionary principle, and what will be the trade-offs? Stifle new technology and associated benefits (e.g., genomic understanding of microbes, plants, animals) from which there is little experience and lots of uncertainty; increase the costs of developing and marketing; threaten freedom and trust of basic research.
2. Who will trigger use of the precautionary principle, and how will they interpret the volume and breadth of data generated from such areas as genomics and proteomics in a positive manner? Risk related data are increasingly in the form of biological markers, gene expression, and protein level change, for which there is an abundance of data generated; yet these are possibly laden with uncertainty and lack of validation. How will this data be processed by the precautionary principle in characterizing new products and assessing efficacy, safety, and so on?
3. What is "sustainable" about the precautionary principle? If we apply the recognized working definition that sustainability is "a challenge that economic growth and development must take place and be maintained over time within the limits set by the physical, chemical, biological, social, and political environments," it follows that environmental protection, product stewardship, and economic development are complementary rather than antagonistic processes, as the precautionary principle would suggest.
4. What is the role of scientific evidence, and to what extent can the precautionary regulatory approach be thought of as without factual or scientific justification, arbitrary and irrational, defensible or enforceable in court, and vulnerable to scientific, legal, and constitutional challenges?
5. To what extent can the precautionary regulatory approach for setting risk standards be flexible enough to consider site-specific conditions in which it is applied (i.e., natural and anthropogenic factors, chemical agent, receptor type and size, exposure conditions, bioavailability, and cost)?

In the end, the relative strength and quality of the precautionary policy will depend on legitimacy, degree of

Table 2. Issues and Risks

Issue	Recommendation
• Process and structure not costless and likely has high expenditures associated given uncertainties	• Utilize cost–benefit and regulatory flexibility analyses • Rational decisions should account for the impact of benefit reduction
• Chemical use does not equal chemical risk	• Account for the way chemical is used, materials handling, exposure sources, potency to receptors • Adopt guidelines to ensure chemical risk and substitution are balanced and not harmful
• Potential risk transfers	• Do not trade a known manageable risk for an uncertain alternative • Actions to remove hazard may remove benefits; loss of benefits introduces new risks • Brainstorm unforeseen risk potentials, such as danger of responding to wrong threat • Utilize a "total hazard" approach
• Undefined and arbitrary methods or interpretation	• Rigorous, prioritized; avoid overregulation that may lead to loss of potential benefits
• Technological stagnation	• Cessation or phaseout of regulatory prescription threatens to kill the patient while inventing a cure for a disease that does not necessarily exist; it may lead to arbitrary unscientific rejection of beneficial technologies
• Lack of prioritization	• Final assessment should involve prioritization of substances and relative risk; should consider impacts of cessation or phaseout on international legislation, production and use, economics, and alternatives
• No zero risk in a technology-based society	• Neither risk assessment nor the precautionary principle can eliminate all risks; these paradigms can only be refined to reduce the relevant risks to a perceived acceptable level; proving no risk is a difficult if not impossible task
• Risks of no action	• The precautionary principle must include a full range of options, including no action
• Tools for precautionary thinking	• Life cycle analysis, waste-free product goals, cleaner production, green building, environmental justice, ecoefficiency, benchmarking, best practices
• Tolerability and meaningful restraints	• Need to define and quantify risk acceptability criteria • The margins of safety for descriptions of low risk must be defined and should consider factors of uncertainty, cost–benefits, and lack of knowledge • Goal is to introduce reasonable/plausible conservatism
• Politicization • Codes of ethics	• Balance technical, scientific, economic, and political • Practitioners should adhere to a code of ethics that re-iterates the need for validity of data and sound scientific approaches that are transparent, systematic, subject to peer review, accountable, and independent

public trust, and actual quality of products and decisions. Whether by traditional risk analysis or by precautionary thinking, risk management should be determined case by case and population by population, with a full and transparent accounting of the bigger picture, the consequences of different actions including no action, and the public values surrounding those actions; even if the details are obscure.

READING LIST

Burke, M. (2000). Promoting a greener European Union. *Environ. Sci. Technol.* **17**: 388A–393A.

Chapman, P.M. (1999). Does the precautionary principle have a role in ecological risk assessment? *Hum. Ecol. Risk Assessment* **5**: 885–888.

Charnley, G. and Donald Elliott, E. (2002). Risk versus precaution: environmental law and public health protection. *The Environ. Law Report* **32**: 10363–10366.

European Commission. (2000). *Communication from the Commission on the Precautionary Principle*. COM 02.02.2000, Brussels.

Fairbrother, A. and Bennett, R.S. (1999). Ecological risk assessment and the precautionary principle. *Hum. Ecol. Risk Assessment* **5**: 943–949.

Foster, K.R., Vecchia, P., and Repacholi, M.H. (2000). Science and the precautionary principle. *Science* **288**: 979–981.

Gilbertson, M. (2002). The precautionary principle and early warnings of chemical contamination of the Great Lakes. In: *EU Communication "Late Lessons from Early Warnings: The Precautionary Principle 1896–2000*.

Graham, J.D. (1999). Making sense of the precautionary principle. *Risk Perspect.* **7**: 1–6.

Graham, J.D. (2000). Decision-analytic refinements of the precautionary principle. *J. Risk Res.* in press.

Hansson, S.O. (1999). Adjusting scientific practices to the precautionary principle. *Hum. Ecol. Risk Assessment* **5**: 909–921.

Mckinney, W.J. and Hammer Hill, H. (2000). Of sustainability and precaution: The logical, epistemological, and moral problems of the precautionary principle and their implications for sustainable development. *Ethics Environ.* **5**: 77–87.

Pittinger, C.A. and Bishop, W.E. (1999). Unraveling the chimera: a corporate view of the precautionary principle. *Hum. Ecol. Risk Assessment* **5**: 951–962.

Raffensperger, C. and Tickner, J. (Eds.). (1999). *Protecting Public Health and the Environment: Implementing the Precautionary Principle*, Island Press.

Raffensperger, C. and deFur, P.L. (1999). Implementing the precautionary principle: rigorous science and solid ethics. *Hum. Ecol. Risk Assessment* **5**: 933–941.

Rasmussen, F.B. (1999). Precautionary principle and/or risk assessment. *Environ. Sci. Pollut. Res.* **6**: 188–192.

Rogers, M.D. (2001). Scientific and technological uncertainty, the precautionary principle, scenarios and risk management. *J. Risk Res.* **4**: 1–15.

Rogers, M.F., Sinden, J.A., and De Lacy, T. (1997). The precautionary principle for environmental management: a defensive-expenditure application. *J. Environ. Manage.* **51**: 343–360.

Sandin, P. (1999). Dimensions of the precautionary principle. *Hum. Ecol. Risk Assessment* **5**: 889–907.

Santillo, D. and Johnston, P. (1999). Is there a role for risk assessment within precautionary legislation? *Hum. Ecol. Risk Assessment* **5**: 923–932.

Santillo, D., Stringer, R.L., Johnston, P.A., and Tickner, J. (1998). The precautionary principle: protecting against failures of scientific method and risk assessment. *Mar. Pollut. Bull.* **36**: 939–950.

Shipworth, D. and Kenley, R. (1999). Fitness landscapes and the precautionary principle: the geometry of environmental risk. *Environ. Manage.* **24**: 121–131.

Stijkel, A. and Reijnders, L. (1995). Implementation of the precautionary principle in standards for the workplace. *Occup. Environ. Med.* **52**: 304–312.

Tucker, A. (2002). Life-cycle assessment and the precautionary principle. *Environ. Sci. Technol.* **36**(3): 70A–75A.

Underwood, A.J. (1997). Environmental decision-making and the precautionary principle: what does this principle mean in environmental sampling practice? *Landscape Urban Plann.* **37**: 137146.

Wiener, J.B. and Rogers, M.D. (2001). *Comparing Precaution in the United States and Europe*. Working Paper August 2001-01.

WATER PRICING

STEVEN J. RENZETTI
Brock University
St. Catharines, Ontario, Canada

INTRODUCTION

The price of water is measured by the value of the economic resources that must be sacrificed to acquire potable water. In many cases, this price is easily understood as the number of dollars and cents that must be paid to obtain a specific volume of water. In other cases, often in low-income countries, the price of water is best understood as the combination of out-of-pocket expenses plus the time required for travel to and from a water source.

The purpose of this article is twofold. First, it describes the nature and range of water prices around the world. Second, it explains the significance of the price of water for decisions regarding the supply of water and the demand for water. The article concludes that, despite the potentially important role of water prices, a lack of attention to water pricing has been a major contributing factor to worldwide problems of water allocation and water quality.

WATER PRICES AROUND THE WORLD

The level and form of water prices vary widely around the world. Figure 1 gives an indication of the range of residential water prices across a number of countries. The form or structure of water prices also varies across countries and across sectors (agricultural, residential, commercial, and industrial). On the one hand, there are many examples where users face a price of zero for their water. This occurs in municipalities where household water use is not metered and in rural areas where households rely on, but do not pay for, groundwater. There is also another way in which the price of using water can be said to be zero in many cases. If a factory or sewage treatment plant discharges wastes into a waterbody without paying for that right, then these facilities are using water free. It is important to remember, however, that a zero price for water does not mean that there is no cost for using it. This cost may come in the form of the infrastructure needed to purify and transport the water to users, in the form of reduced water quality, or it may come in the form of foregone benefits that would have been enjoyed by other water users had they access to another's water.

On the other hand, water prices can be quite complex and have several components: a one-time connection charge, a recurring fixed charge, a volumetric charge, and supplemental charges for sewage treatment and environmental protection. Further, the volumetric component of the price of water can, itself, vary in structure. It can be a constant amount (such as $0.76 U.S. per cubic meter in Toronto, Canada in 2002) or it can be variable in the sense that the price charged on the last unit of consumption either rises or falls with consumption (these are termed increasing and decreasing block rate structures, respectively). The City of Los Angeles, for example, has a very complex rate structure for residential consumers; the volumetric component varies by season, the household's location, and the quantity of water consumed (2).

Table 1 provides a snapshot of water prices in different sectors and countries. There are a number of noteworthy features of the information presented. First, the information is somewhat fragmentary because information for a number of sectors is not included. Most national governments do not collect data on water prices in their countries. Second, in many sectors and countries, the price of water is very low. The implications of this observation for the viability of water supply systems and for the level of water use are discussed in the next

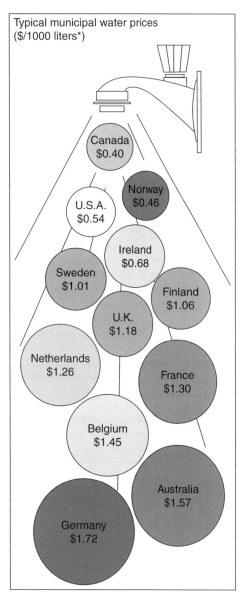

Figure 1. International comparison of water prices. (*Source*: Reference 1. Note that 1 Canadian dollar in 1992 is the approximate equivalent of $0.75 U.S. in 2001).

section. As Dinar and Subramanian (3) point out, within any sector, prices may vary by the quantity consumed, by the type of water use (e.g., what type of crop is grown), by the time of consumption, and by location. Prices also vary significantly across sectors; industry typically faces the highest prices. Interestingly, prices appear to differ less across countries than they do across sectors within a given country. This may be due to some sectors' political power that has allowed them to avoid facing the full costs of their water use. Finally, there does not appear to be a strong connection between a country's water prices and the scarcity of water in that country. Canada's low water prices might be explained, in part, by its abundant water supplies, but low water prices are also observed in water-scarce countries such as Australia and Spain.

THE FUNCTIONS OF WATER PRICES

In principle, the price of water can play several important roles. From the viewpoint of society as a whole, the price of water informs users and suppliers of the value and scarcity of water. From the viewpoint of the agency supplying water, water prices generate the revenues needed to ensure the financial viability of the agency. From the viewpoint of an individual household, farm, or business, water prices inform decision-makers what they must give up in exchange for getting water. Thus, it can be seen that water prices play many roles and that they have the potential for influencing the demand for water, the supply of water, and the allocation of water among competing users (4,5).

To see how the price of water plays these multiple roles, consider the impact of rising water prices (for

Table 1. Water Prices[a] for Various Sectors and Countries (1996 U.S.$/m^3)[b]

Country	Agriculture[c]	Domestic[c]	Industry[c]
Algeria	0.019–0.22	0.057–2.7	4.64
Australia	0.0195	0.23–0.54	7.82
Botswana	NA	0.28–1.48	NA
Brazil	0.0042–0.032	0.4	NA
Canada	0.0017–0.0019	0.34–1.36	0.17–1.52
France	0.11–0.39	0.36–2.58	0.36–2.16
India	NA	0.0095–0.082	0.136–0.290
Israel	0.16–0.26	0.36	0.26
Italy	NA	0.14–0.82	NA
Madagascar	NA	0.325–1.75	NA
Namibia	0.0038–0.028	0.22–0.45	NA
New Zealand	NA	0.31–0.69	NA
Pakistan	NA	0.06–0.10	0.38–0.97
Portugal	0.0095–0.0193	0.153–0.529	1.19
Spain	0.0001–0.028	0.0004–0.0046	0.0004–0.0046
Sudan	NA	0.08–0.10	0.08–0.10
Tanzania	0.260–0.398	0.062–0.241	0.261–0.398
Tunisia	0.02–0.078	0.096–0.529	0.583
Uganda	NA	0.38–0.59	0.72–1.35
United Kingdom	NA	0.0095–0.0248	NA
United States	0.0124–0.438	NA	NA

[a] Prices refer only to the volumetric component. They do not include any connection fees or other charges not related to the volume of water consumed.
[b] A cubic meter is approximately 220 Imperial gallons or 264 U.S. gallons.
[c] "NA" means that no number was reported in the original source.
Source: Adapted from Table 1.2 in Reference 3.

example, that might be brought on by the increased energy needed to pump water from shrinking aquifers). Water users see increased prices as a signal of the increased costs that result from using an increasingly scarce (and, thus, valuable) resource. Users can then be expected to reconsider their planned water uses with an eye to conservation and/or substitution toward less expensive alternatives. These types of responses can include changing landscaping practices, retrofitting plumbing fixtures, introducing water recycling, and investing in more efficient water-using equipment. It is important to notice that changing water prices, then, can have an influence on water-use practices and also on the use of other goods and services.

From the perspective of water suppliers, rising prices are a signal that alternative sources of supply are now more attractive options than previously thought. Securing additional water can take the form of reducing system losses, investing in new technologies (perhaps more efficient pumps or water meters), extending the water collection network, or purchasing or leasing water rights.

Rising water prices also provide valuable signals to other decision-makers in society. Rising water prices (and their signal of growing water scarcity) indicate to government that there is an increasing possibility of heightened conflicts among competing water users. As a result, that there may be a need to review the adequacy of water rights laws and water allocation regulations. In addition, rising water prices are a signal to those firms who design and build equipment that delivers or uses water. These firms will see that there is a stronger incentive to innovate and design improved forms of water-related technologies and equipment.

ASSESSING WATER PRICES

The preceding section indicates that water prices, in principle, play a variety of roles, including raising revenue for water utilities and signaling relative scarcity to both suppliers and demanders. Water prices are best able to play these roles when they are free to rise and fall according to changes in supply and demand. Furthermore, the signal they send will be most informative when prices accurately reflect the full costs to society of water use (including any environmental damage). This section briefly examines how well prices perform these roles.

Many researchers and analysts (6,7) are critical of the history and current state of water pricing and argue that, in most cases throughout the world, water prices fail either to reflect accurately the full social costs arising from water use or to fulfill their other functions. Renzetti (8) demonstrates that municipal water prices in Ontario, Canada, often understate the incremental cost of supply by a factor of 2. Munasinghe (9) examines water pricing in Manila and finds that it underestimates the cost of supply by failing to reflect declining groundwater levels. Underpricing water is hardly limited to municipal water supply systems. Research indicates that low prices for irrigation water have historically provided significant subsidies to the agricultural sector. Roodman (7) asserts that "The U.S. government spent an estimated $45–93 billion more than it earned on public irrigation projects between 1902 and 1986 (in 1995 dollars)" (p. 37). These researchers believe that inadequate water prices are a major factor in many of the problems relating to water management (6,10). These problems include overuse of water, poor system reliability, reduced water quality, and the diversion of productive capital to subsidized water projects.

It is sometimes thought that low water prices benefit low-income farmers and households. Unfortunately, this is unlikely to be the case. First, low water prices are an inefficient way to assist the poor because water generally makes up a relatively small share of total household expenditures. In any case, a better option is to design water prices so that families who use relatively little water can be given a free allowance of a specific quantity of water. Second, low water prices often translate into water agencies that do not have adequate financial resources to maintain water quality and system reliability. Middle and high-income households may be able to shield themselves by finding alternative sources of water supply, but the poor can rarely do this. Thus, the general principle is to set prices so that they may reflect, as fully as is politically and administratively feasible, the forces of supply and demand.

CONCLUSIONS

This article has examined the nature of water prices and has demonstrated that the price of water can play an important role in many decisions related to the supply and demand of water and to the allocation of other scarce resources such as investment capital. Despite this potential, water prices have not played an important part in the history of water management. For example, many people remember the building of the Grand Coulee and High Aswan dams, but how many know the prices charged for water in these projects? This lack of attention to water pricing has had important consequences. These include prices that are far below the cost of supply, overuse of water, reduced water quality, and the delayed development of water-conserving technologies. Perhaps because of these conditions, a growing amount of attention is being directed toward water pricing.

BIBLIOGRAPHY

1. Environment Canada (2002). Freshwater website http://www.ec.gc.ca/water/e_main.html. Accessed February 27, 2002.
2. City of Los Angeles Department of Water Services (2002). Water Rates For Service In The City And County Of Los Angeles http://www.ladwp.com/water/rates/sch_b.htm. Accessed February 27, 2002.
3. Dinar, A. and Subramanian, A. (Eds.). (1997). *Water Pricing Experiences: An International Perspective*. World Bank Technical Paper 386, The World Bank, Washington, DC.
4. Hall, D. (Ed.). (1996). *Marginal Cost Rate Design and Wholesale Water Markets*, Advances in the Economics of Environmental Resources. Vol. 1. JAI Press, Greenwich, CT.
5. Renzetti, S. (2002). *The Economics of Water Demands*. Kluwer Academic, Boston.
6. Postel, S. (1996). Forging a sustainable water strategy. In: *State of the World 1996*. L. Brown (Ed.). Worldwatch Institute, Washington, DC, pp. 40–59.
7. Roodman, D.M. (1996). *Paying the Piper: Subsidies, Politics and the Environment*. Research paper 133, WorldWatch Institute, Washington, DC.
8. Renzetti, S. (1999). Municipal water supply and sewage treatment: costs, prices and distortions. *Can J. Econ.* **32**(2): 688–704.
9. Munasinghe, M. (1992). *Water Supply and Environmental Management: Developing World Applications*. Studies in Water Policy and Management. Westview Press, Boulder, CO.
10. Easter, K.W., Feder, G., Le Moigne, G., and Duda, A. (1993). *Water Resources Management: A World Bank Policy Paper*. The World Bank, Washington, DC.

SPOT PRICES, OPTION PRICES, AND WATER MARKETS

Terence R. Lee
Santiago, Chile

Markets in water rights do not operate only for buying and selling permanent water rights. Using a market makes available a variety of other tools for increasing the efficiency of water use. Among the most interesting of these tools is the use of spot markets and water use options or future markets. Both alternatives lead to a broadening of the market and to an increase in the number of participants on both the buyer and the seller sides of the market. Their use also results in more stable demand and supply conditions as the market has a larger number of buyers and sellers. They also permit water users to hedge against changes in the price and availability of water rights and thus eliminate the need to hoard as a hedge against this risk. On the other hand, the use of spot and option or future markets requires developing storage facilities and interconnected conveyance systems if their full benefits are to be obtained.

Spot markets involve the leasing, not the sale of water rights. The sale of rights is replaced by the sale of water. In a spot market, the water right remains with the owner. The temporary sale of water in a spot market is a preferred market response to short-term changes in demand and supply.

Water users may find it advantageous to engage to engage in a spot market for a variety of reasons. The seller has an opportunity to earn revenue in the temporary trade of surplus water but does not give up any water rights. This is particularly useful for accommodating a short-term demand for additional water for any use that has a predictable and fixed life span or for a use of uncertain duration. For example, a farmer facing an unexpected variation in normal supplies may buy on a spot market when it is not economical to transport water in periods of sufficient water supply. There are, however, examples of the longer term use of spot markets when a water user is unable or unwilling to commit the resources necessary to buy a water right.

One example of a permanent spot market, where farmers always pay spot prices for water, is found in the community of Huerta in Alicante, Spain. Here the ownership of water is separate from the ownership of land. Water is distributed by rotation at a fixed rate, approximately the same quantity of water in each successive rotation, and the proportion of water available to any water right holder varies for each rotation, depending on the water rights acquired. Before each rotation, a notice is posted that announces the date on which the rotation will commence and informs water rights holders that, within a prescribed period, they should claim their "*albalaes*" or tickets for this rotation. Once allocated, tickets, available in twelve denominations for a constant supply of water from 1 hour to 1/3 minute, are freely tradable in a public auction and an informal market. The community makes a genuine effort to provide information to farmers so they can buy and sell water intelligently; there are brokers who facilitate trading. A simulation model comparison of this system with those found elsewhere in Spain, where trading is not permitted, indicated that the spot market approach adopted in Alicante is the most efficient in terms of net increases in regional income. The differences are not great for moderate water shortages but are significant in conditions of severe water shortage (1).

In a typical spot market contract, the buyer pays the owner of the water right, generally in periodic installments, but there can also be an up-front payment to initiate the sale. Such contracts are often renewable. The length of a contract in irrigation districts is typically a single season. Contracts can be, however, very short, sometimes for as little as a few hours. In the short run, spot prices usually provide a source of water cheaper than the purchase of a permanent water right, but water on a spot market can and will fluctuate in price. The supply is secure only for the period of the contract. There is also an expense in the constant renewal costs for those who depend on short-term purchases in a spot market.

By providing immediate access to water, spot markets can, however, accommodate the most varying needs. The flexibility of spot markets makes them an attractive option for many users, and they are often very active, particularly among neighboring water rights holders in irrigation districts, even where markets in permanent rights have not been established. Spot purchases, often informal, are usually the predominant form of market transactions, whether or not a formal market exists.

In Chile, seasonal spot purchases of water have been a much more active form of water reallocation than water rights sales (2). The most common transaction is an arrangement between neighboring farmers whose water requirements differ through the cropping cycle (3,4). Spot trading between neighbors is easier because it does not require investment in storage facilities or elaborate systems of interconnection between canal systems. In California, as well, water marketing is characterized by an emphasis on seasonal spot markets, although there are good facilities for transferring water over very large distances (5).

Option prices are based on longer term agreements to lease, or sometimes, but less commonly, to sell a water right. The option is taken up when a given event occurs, typically a drought. There are examples of the use of options in both the United States and in Chile. A typical arrangement in Chile is the payment by a fruit farmer of a prenegotiated fee to a farmer growing annual crops for an option on water supply in the case of drought (6).

Options can be and are also used to transfer water from agricultural to nonagricultural users; this has been common practice in the western United States. Dry year option contracts are an attractive alternative when supplies are normally adequate. Options provide supplies during droughts at a cost lower than purchases or leases of water rights.

Options are particularly attractive because the buyer secures long-term water supplies and the seller receives compensation for the option, including the income lost when the option is exercised. For the buyer of an option, the contract also provides a means of obtaining additional water supply, under predetermined conditions, and at a specified price.

Option contracts are, however, complex. The contract must address the risk to the buyer that the water right will not be available when the time comes to exercise the option. The establishment of an option contract limits the rights of the owner of the water right to sell. This issue can be addressed by including in the contract the "right of first refusal" which allows the seller to retain the option of selling the right, but gives the option holder the right to match the offered price (7). Moreover, to produce the greatest benefits, options require long-term contractual commitments, up to 20 or more years, which can introduce many uncertainties into the arrangement.

Option contracts can include payments adjusted over time to allow for changes in water use, production costs, technology, and other conditions. Alternative methods of charging can be used; a lump sum, annual payments, or a combination of annual payments with a lump sum when the option is exercised. The latter is a particularly attractive alternative, because neither party needs to anticipate fully the number of times the option will be exercised over the contract period (8). The holders of water rights on which option contracts have been agreed to can also be compensated in kind, for example, by lower rates for the option holder's production; as in irrigation to hydroelectricity generation transfers, farmers can be compensated by lower electricity rates.

BIBLIOGRAPHY

1. Maass, A.R. and Anderson, R.L. (1978). *. . . and the Desert Shall Rejoice: Conflict, Growth, and Justice in Arid Environments*. MIT Press, Cambridge, MA.
2. Chile, *Trabajo de Asesoría Económica al Congreso Nacional, 'Nuevo proyecto de modificación al Código de Aguas'*, (1996). Trabajo de Asesoría Económica al Congreso Nacional, No. 74, December, Programa de Postgrado en Economía, ILADES/Georgetown University.
3. Gazmuri, R. (1994). Chile's market-oriented water policy: institutional aspects and achievements. In: *Water Policy and Water Markets*. Selected papers and proceedings from the World Bank's Ninth Annual Irrigation and Drainage Seminar, Annapolis, Maryland, December 8–10, 1992. G. Le Moigne, K.W. Easter, W.J. Ochs, and S. Giltner (Eds.). Technical Paper Number 249, The World Bank, Washington, DC.
4. Gazmuri, R. and Rosegrant, M.W. (1994). *Chilean Water Policy: The Role of Water Rights, Institutions, and Markets*.

Tradable water rights: experiences in reforming water allocation policy by M.W. Rosegrant and R. Gazmuri, Irrigation Support Project for Asia and the Near East.
5. Howitt, R.E. (1997). *Initiating Option and Spot Price Water Markets: Some Examples from California*, Seminar on Economic Instruments for Integrated Water Resources Management: Privatization, Water Markets and Tradable Water Rights. Proceeding, Inter-American Development Bank, Washington, DC.
6. Thobani, M. (1997). *Formal Water Markets: Why, When and How to Introduce Tradable Rights in Developing Countries*, Seminar on Economic Instruments for Integrated Water Resources Management: Privatization, Water Markets and Tradable Water Rights. Proceedings, Inter-American Development Bank, Washington, DC.
7. Michelsen, A.M. and Young, R.A. (1993). Optioning agricultural water rights for urban water supplies during droughts. *Am. J. Agric. Econ.* **73**: 4.
8. Hamilton, J.R., Whittlesey, N.K., and Halverson, P. (1989). Interruptible water markets in the Pacific Northwest. *Am. J. Agric. Econ.* **71**: 1.

WATER MANAGED IN THE PUBLIC TRUST

Laurel Phoenix
Green Bay, Wisconsin

Managing water in the public trust requires considering the water demand of current and future citizens as well as that of ecosystems, fish, and wildlife. This presumes that water will be managed sustainably for human and ecosystem benefit, despite the fact that ecosystems have no voice in the political process. To understand how complicated it is to accomplish this, the definition and history of the public trust is examined. Next, water management is briefly described to contrast current water management methods with the finesse that managing water in the public trust would require. Finally, some of the difficulties of achieving this in a world of rising demand and finite resources are highlighted, followed by a set of principles describing government obligations to uphold human rights.

WHAT IS THE PUBLIC TRUST?

The public trust is an enduring doctrine obligating government to manage natural resources in the best interests of its citizens. Because some vital natural resources like oceans, air, and wildlife cannot be owned in the same way a person can hold a property interest in land, it is the responsibility of government to hold ownership over these common resources in trust for its citizens, so that their use may be regulated to ensure universal access and prevent their overuse or degradation.

HISTORY OF THE PUBLIC TRUST

The common law doctrine in European history is first attributed to the Roman Emperor Justinian, who declared in 533–534 B.C.E. that running water, seas, and wildlife were part of a "negative community" of things that could not be owned but could be used under regulation to ensure vital common resources were not exploited (1). A later example is the Magna Carta signed in 1215, granting guaranteed liberties to the people that not even King John or future kinds could overrule. Because of its importance to fisheries, commerce, and navigation, the public trust doctrine in America was initially used at federal and state levels only to apply to waterways, but court rulings since then have expanded the public trust to include many other environmental issues, such as protection of ecosystems, recreation like swimming and boating, preservation of views (scenic values), habitat (wildlife values), open space, and beaches (2). Some states have taken a step further to apply the public trust to nonenvironmental issues, such as historical sites; civic sites or economically important areas, such as aging downtown districts; or economic development such as tearing down old, low value single homes to be replaced with many more valuable units, such as expensive condos.

Consequently, the public trust has evolved from rules for equitably sharing the natural "commons" of the land to case law expanding the multiple uses of water resources, as well as rules looking at any social or economic good that could be said to advance the public welfare.

HOW IS WATER "MANAGED"?

Water resources management has been defined as "the application of structural and nonstructural measures to control natural and man-made water resources systems for beneficial human and environmental purposes" (3). Structural facilities include wells, dams, diversion structures, pipes, pumps, treatment plants, and hydroelectric plants. Nonstructural measures are economic incentives, regulations, public education, and planning. Numerous individuals, agencies, and jurisdictions are involved in these structural and nonstructural components. Beyond this complexity, water managers must concomitantly address needs for navigation, municipal water supply, irrigation, fishing, industrial use, hydropower, recreation, public access, and traditionally sacred sites, among others. Water must also be managed to protect ecosystems, fish, and wildlife and should not be unsustainably used to the detriment of future generations.

In America, water resources management is a fragmented collection of federal, state, and local laws managed by numerous and sometimes competing agencies, water districts, ditch companies, acequias, and so on. There are a haphazard mix of treatment and pollution laws at the federal level, water allocation laws at the state level, and sporadic conservation laws at some local levels when supplies run low. It is complicated further by outdated international treaties and interstate compacts. Some states permit, measure, and track individual water use, and others do not.

Two reasons for this lack of a more organized and efficient water management system are (1) a long history, until recently, of having enough water to meet the needs of most users and (2) a variety of political influences on water management decisions. First, competing human uses will have great influence in politics, whereas ecosystems or future generations have few advocates. Second, of the current competing uses, some interest groups have more political power than others, so water may be inequitably or inefficiently managed, allocated, and used. Third, of the current competing uses, some are mutually exclusive in that they cannot happen at the same time in the same location, causing more conflict and subsequent pressure on water managers. For example, building a dam on a prime salmon run damages their ability to spawn, reducing their populations by 80–90%. Another incompatible use of a stream or lake is jet skiing and fishing. Fourth, political influences can cause the best science to be ignored, so, for example, irrigators will get the water they want even after biologists predict large fish kills will result (e.g., Klamath River fish kills of 2002–2003). Fifth, politicians and other decision-makers will be pressured to make decisions with short-term benefits, ignoring long-term solutions. Finally, land use decisions, usually made at a local level, rarely are made with the goal of preserving/protecting water resources. Thus, there are small but cumulative negative effects on surface and ground water everywhere.

MANAGING WATER IN THE PUBLIC TRUST

Managing water in the public interest, therefore, would be a complex and deliberate allocation of water resources to meet the requirements of as many uses as possible, consistent with protecting the resources from pollution and allocating some water for users with no political voice (e.g., ecosystems and future generations)—a Sisyphean task. (Punished for his misdeeds, the Greek Sisyphus was condemned by the gods to rolling a large boulder up a hill for eternity, only to have it roll back down again and force him to start over.) This is complicated enough, but how is this to be done in a world with a fixed amount of water but rapidly growing and often impoverished populations? Already there is evidence of a current global crisis of not enough water to meet human demand. If over 1.1 billion people have no access to safe drinking water today (4), and water conflicts have already been mapped over a large part of the world (5), how can water be wisely managed to provide for the larger populations predicted for the future? There are 6.4 billion people on the Earth now, and by 2050 there will be 9.3 billion (6). Can technological and regulatory improvements produce enough water for these future generations? Besides the inevitable greater demand from additional people, water managers will also have to prepare for climate changes that could alter the historic precipitation and evaporation rates of different regions. Set with such a task, where do we start?

GOVERNMENTAL OBLIGATIONS TO MANAGE WATER IN THE PUBLIC INTEREST

The United Nations has set some guidelines for how governments can fulfill their obligations to their people to support their human rights. These directives of "respect, protect, fulfill" were applied to water to give examples of these governmental obligations (7):

1. The obligation to respect means no interference with anyone's current access to water. Nations could no longer be:

    "engaging in any practice or activity that denies or limits equal access to adequate water;

    artibrarily interfering with traditional arrangements for water allocation;

    unlawfully polluting water;

    limiting access to, or destroying, water services and infrastructure."

2. The obligation to protect means protecting citizens' access to water from third parties, such as warring clans or multinational water companies. Nations would need to:

    "adopt the necessary and effective legislative and other measures to restrain third parties from denying access to adequate water and from polluting and inequitably extracting from water resources;

    prevent third parties from compromising equal, affordable, and physical access to sufficient and safe water where water services are operated or controlled by third parties."

3. The obligation to fulfill means government would need to institute laws and policies to ensure that *every* citizen finally had equal access to water. This could be, for example, "legislative implementation, adoption of a national water strategy and plan of action to realize this right while ensuring that water is affordable and available for everyone."

This framework of respect, protect, and fulfill would use structural and nonstructural measures, but the goals and responsibilities would be more clearly defined than they are now, driving the laws, policies, and choices for governmental expenditures.

FUTURE CHALLENGES

A world where water is managed in the public trust will require major changes. First, it will need governments responsive to and under the control of their citizens, where there is full public participation and transparency in decision-making regarding equitable allocation of water and effective pollution control. Prioritization will then need to be established in advance to deal with a triaged allocation strategy in times of scarcity. Nonpoint and point sources of pollution will need to be far more rigorously monitored to prevent drinking water and habitat degradation. We will no longer be able to pollute with impunity. And finally, inefficient consumption across all human uses, domestic, municipal, industrial, and agricultural will need to be replaced with the most careful consumption, as water becomes increasingly precious.

BIBLIOGRAPHY

1. Getches, D.H. (1997). *Water Law in a Nutshell*, 3rd Edn. West Publishing, St. Paul.
2. Bray, P. (2004). *An Introduction to the Public Trust Doctrine*. Government Law Center, Albany Law School, 80 New Scotland Ave., Albany, NY 12208. Available at http://www.responsiblewildlifemanagement.org/an_introduction_to_public_trust_doctrine.htm.
3. Grigg, N.S. (1996). *Water Resources Management: Principles, Regulations, and Cases*. McGraw-Hill, New York.
4. Intergovernmental Council of the IHP. *IHP contribution to the World Water Assessment Programme (WWAP)*. Document # SC.2004/CONF.203/CLD.32; IHP/IC-XVI/INF.19.
5. Basins at Risk Project. A study by Dr. Aaron Wolf with maps. Available at http://www.transboundarywaters.orst.edu/projects/bar/.
6. The world population highlights (2004). Available at http://www.prb.org/.
7. Scanlon, J., Cassar, A., and Nemes, N. (2004). *Water as a Human Right?* IUCN Environmental Policy and Law Paper # 51. International Union for Conservation of Nature and Natural Resources (IUCN), p. 22.

WATER RECYCLING AND REUSE: THE ENVIRONMENTAL BENEFITS

Environmental Protection Agency

The Experience at Koele Golf Course, on the Island of Lanai, has used recycled water for irrigation since 1994. The pond shown is recycled water, as is all the water used to irrigate this world-class golf course in the state of Hawaii.

> "Water recycling is a critical element for managing our water resources. Through water conservation and water recycling, we can meet environmental needs and still have sustainable development and a viable economy."
> —*Felicia Marcus, Regional Administrator*

This article is a U.S. Government work and, as such, is in the public domain in the United States of America.

WHAT IS WATER RECYCLING?

Recycle: verb 1.a. To recover useful materials from garbage or waste, b. To extract and reuse.

While recycling is a term generally applied to aluminum cans, glass bottles, and newspapers, water can be recycled as well. Water recycling is reusing treated wastewater for beneficial purposes such as agricultural and landscape irrigation, industrial processes, toilet flushing, and replenishing a groundwater basin (referred to as groundwater recharge). Water is sometimes recycled and reused onsite; for example, when an industrial facility recycles water used for cooling processes. A common type of recycled water is water that has been reclaimed from municipal wastewater, or sewage. The term water recycling is generally used synonymously with water reclamation and water reuse.

Through the natural water cycle, the earth has recycled and reused water for millions of years. Water recycling, though, generally refers to projects that use technology to speed up these natural processes. Water recycling is often characterized as "unplanned" or "planned." A common example of unplanned water recycling occurs when cities draw their water supplies from rivers, such as the Colorado River and the Mississippi River, that receive wastewater discharges upstream from those cities. Water from these rivers has been reused, treated, and piped into the water supply a number of times before the last downstream user withdraws the water. Planned projects are those that are developed with the goal of beneficially reusing a recycled water supply.

HOW CAN RECYCLED WATER BENEFIT US?

Recycled water can satisfy most water demands, as long as it is adequately treated to ensure water quality appropriate for the use. Figure 1 shows types of treatment processes and suggested uses at each level of treatment. In uses where there is a greater chance of human exposure to the water, more treatment is required. As for any water source that is not properly treated, health problems could arise from drinking or being exposed to recycled

The Palo Verde Nuclear Generating Station, located near Phoenix, Arizona, uses recycled water for cooling purposes.

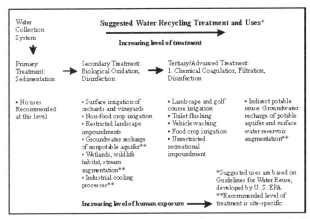

Figure 1. While there are some exceptions, wastewater in the United States is generally required to be treated to the secondary level. Some uses are recommended at this level, but many common uses of recycled water such as landscape irrigation generally require further treatment.

The Irvine Ranch Water District provides recycled water for toilet flushing in high rise buildings in Irvine, California. For new buildings over seven stories, the additional cost of providing a dual system added only 9% to the cost of plumbing.

water if it contains disease-causing organisms or other contaminants.

The U.S. Environmental Protection Agency regulates many aspects of wastewater treatment and drinking water quality, and the majority of states in the United States have established criteria or guidelines for the beneficial use of recycled water. In addition, in 1992, EPA developed a technical document entitled "Guidelines for Water Reuse," which contains such information as a summary of state requirements, and guidelines for the treatment and uses of recycled water. State and federal regulatory oversight has successfully provided a framework to ensure the safety of the many water recycling projects that have been developed in the United States.

Recycled water is most commonly used for nonpotable (not for drinking) purposes, such as agriculture, landscape, public parks, and golf course irrigation. Other nonpotable applications include cooling water for power plants and oil refineries, industrial process water for such facilities as paper mills and carpet dyers, toilet flushing, dust control, construction activities, concrete mixing, and artificial lakes.

Although most water recycling projects have been developed to meet nonpotable water demands, a number of projects use recycled water indirectly for potable purposes. (Indirect potable reuse refers to projects that discharge recycled water to a water body before reuse. Direct potable reuse is the use of recycled water for drinking purposes directly after treatment. While direct potable reuse has been safely used in Namibia (Africa), it is not a generally accepted practice in the United States.) These projects include recharging groundwater aquifers and augmenting surface water reservoirs with recycled water. In groundwater recharge projects, recycled water can be spread or injected into groundwater aquifers to augment groundwater supplies, and to prevent salt water intrusion in coastal areas. For example, since 1976, the Water Factory 21 Direct Injection Project, located in Orange County, California, has been injecting highly treated recycled water into the aquifer to prevent salt water intrusion, while augmenting the potable groundwater supply.

While numerous successful groundwater recharge projects have been operated for many years, planned augmentation of surface water reservoirs has been less common. However, there are some existing projects and others in the planning stages. For example, since 1978, the upper Occoquan Sewage Authority has been discharging recycled water into a stream above Occoquan Reservoir, a potable water supply source for Fairfax County, Virginia. In San Diego, California, the Water Repurification Project is currently being planned to augment a drinking water reservoir with 20,000 acre-feet per year of advanced treated recycled water.

WHAT ARE THE ENVIRONMENTAL BENEFITS OF WATER RECYCLING?

In addition to providing a dependable, locally controlled water supply, water recycling provides tremendous environmental benefits. By providing an additional source

For over 35 years, in the Montebello Forebay Groundwater Recharge Project, recycled water has been applied to the Rio Hondo spreading grounds to recharge a potable groundwater aquifer in south-central Los Angeles County.

In California, Mono Lake's water quality and natural resources were progressively declining from lack of stream flow. In 1994, the Los Angeles Department of Water and Power was required to stop diverting one-fifth of the water it historically exported from the basin. The development of water recycling projects in Los Angeles has provided a way to partially offset the loss of Mono Basin water, and to allow the restoration of Mono Lake to move ahead. Copyright 1994, Mono Lake Committee.

of water, water recycling can help us find ways to decrease the diversion of water from sensitive ecosystems. Other benefits include decreasing wastewater discharges and reducing and preventing pollution. Recycled water can also be used to create or enhance wetlands and riparian habitats.

WATER RECYCLING CAN DECREASE DIVERSION OF FRESHWATER FROM SENSITIVE ECOSYSTEMS

Plants, wildlife, and fish depend on sufficient water flows to their habitats to live and reproduce. The lack of adequate flow, as a result of diversion for agricultural, urban, and industrial purposes, can cause deterioration of water quality and ecosystem health. Water users can supplement their demands by using recycled water, which can free considerable amounts of water for the environment and increase flows to vital ecosystems.

Incline Village, Nevada, uses a constructed wetland to dispose of wastewater effluent, expand the existing wetland habitat for wildlife, and provide an educational experience for visitors.

Recycled water has been used for a number of years to irrigate vineyards at California wineries, and this use is growing. Recently, Gallo Wineries and the City of Santa Rosa completed facilities for the irrigation of 350 acres of vineyards with recycled water from the Santa Rosa Subregional Water Reclamation System.

WATER RECYCLING DECREASES DISCHARGE TO SENSITIVE WATER BODIES

In some cases, the impetus for water recycling comes not from a water supply need, but from a need to eliminate or decrease wastewater discharge to the ocean, an estuary, or a stream. For example, high volumes of treated wastewater discharged from the San Jose/Santa Clara Water Pollution Control Plant into the south San Francisco Bay threatened the area's natural salt water marsh. In response, a $140 million recycling project was completed in 1997. The South Bay Water Recycling Program has the capacity to provide 21 million gallons per day of recycled water for use in irrigation and industry. By avoiding the conversion of salt water marsh to brackish marsh, the habitat for two endangered species can be protected.

RECYCLED WATER MAY BE USED TO CREATE OR ENHANCE WETLANDS AND RIPARIAN (STREAM) HABITATS

Wetlands provide many benefits, which include wildlife and wildfowl habitat, water quality improvement, flood diminishment, and fisheries breeding grounds. For streams that have been impaired or dried from water diversion, water flow can be augmented with recycled water to sustain and improve the aquatic and wildlife habitat.

WATER RECYCLING CAN REDUCE AND PREVENT POLLUTION

When pollutant discharges to oceans, rivers, and other water bodies are curtailed, the pollutant loadings to these bodies are decreased. Moreover, in some cases, substances that can be pollutants when discharged to a body of water can be beneficially reused for irrigation. For example, recycled water may contain higher levels of nutrients, such as nitrogen, than potable water. Application of recycled water for agricultural and landscape irrigation can provide an additional source of nutrients and lessen the need to apply synthetic fertilizers.

At West Basin Wastewater Treatment Plant in California, reverse osmosis, an advanced treatment process, is used to physically and electrostatically remove impurities from the wastewater.

WHAT IS THE FUTURE OF WATER RECYCLING?

Water recycling has proved to be effective and successful in creating a new and reliable water supply, while not compromising public health. Nonpotable reuse is a widely accepted practice that will continue to grow. However, in many parts of the United States, the uses of recycled water are expanding in order to accommodate the needs of the environment and growing water supply demands. Advances in wastewater treatment technology and health studies of indirect potable reuse have led many to predict that planned indirect potable reuse will soon become more common.

While water recycling is a sustainable approach and can be cost-effective in the long term, the treatment of wastewater for reuse and the installation of distribution systems can be initially expensive compared to such water supply alternatives as imported water or groundwater. Institutional barriers, as well as varying agency priorities, can make it difficult to implement water recycling projects. Finally, early in the planning process, agencies must implement public outreach to address any concerns and to keep the public involved in the planning process.

As water demands and environmental needs grow, water recycling will play a greater role in our overall water supply. By working together to overcome obstacles, water recycling, along with water conservation, can help us to conserve and sustainably manage our vital water resources.

STATE AND REGIONAL WATER SUPPLY

TIMOTHY J. DOWNS
Clark University
Worcester, Massachusetts

INTRODUCTION

Integrated watershed management (IWM) seeks to combine interests, priorities, and disciplines as a multistakeholder planning and management process for natural resources within a watershed ecosystem, centered on water quantity and water quality. Driven bottom-up by local needs and priorities and top-down by regulatory responsibilities, it must be adaptive, evolving dynamically as conditions change (1). Water resource spatial and political scales range from local/municipal, subregional, state, regional (multistate), national, multinational ("regional" in a global sense) to continental and global. Large watersheds occupy state and regional spatial scales, presenting considerable challenges for IWM and water supply, in particular, because of the number and diversity of human consumers and biodiverse habitat needs.

A watershed spatial scale is often used by ecologists to analyze species distribution and biogeochemical cycling. Watersheds can be characterized by the proportions of land area taken up by five main land use/land cover categories (2): urban land, agricultural land, rangeland, forestland, and water. It is arguable whether a watershed or ecoregion scale (region of relative homogeneity based on soil, climate, and vegetative criteria) is most appropriate for managing water quality and aquatic habitats (2). Even in terms of water stocks and flows, the differences between precipitation and evapotranspiration in one watershed that determine the runoff and infiltration that can be used for water supply are influenced by regional climatic and hydrological processes.

Water has many users and uses: drinking, industrial processing, irrigation, hydroelectricity, cooling water for power plants, and recreation (swimming and fishing). It also satisfies the needs of nonhuman species and participates in hydrogeochemical cycles and climate. Competing uses can be sources of conflict, especially when supplies are scarce or degraded by pollution. For example, using a river for industrial processing upstream may conflict with its use as a drinking water supply downstream if toxic discharges are significant. On a state and regional scale, the number of users and competing uses increases, and successful water supply depends on effective communication and negotiation among constituencies. Just as in small watersheds, conflicts in one portion can affect other portions.

The state and regional hydrographic area becomes an operational unit of analysis when water supply and wastewater sanitation are administered on this scale; this means that policies and practices are carried out on a geopolitical scale larger than a municipality or a subregional watershed. Such is the case in many countries, like France, Mexico, parts of the United States, and the United Kingdom. Examples of regional approaches are (1) in France, six water agencies administer the six large hydrographic regions of the country; (2) in the United States, examples of regional management include the Great Lakes Commission and the Ohio River Valley Water Sanitation Commission; and (3) in Mexico, Watershed Councils administer 26 hydrographic regions; several have portions of two or three states.

Much of the emphasis on regional watershed protection in the United States focuses on pollution and water quality, which directly impacts water supply sources and calls for an approach based on risk assessment. For example, in the Mid-Atlantic region (consisting of 18 U.S.

Geological Survey hydrologic units), the Environmental Protection Agency (EPA) is coordinating efforts to identify watersheds that make the most significant contribution to pollution (2). The GIS is employed to model the relationships between water quality indicators and land use/land cover. Water resources management on a state or regional scale is sensitive to the resolution of data gathered and the spatial scale on which predictive models are run and boundaries for analysis set. The pollution of lakes and rivers by nutrients and sediment from nonpoint sources (NPS), it has been shown, is well predicted by land use/land cover patterns (2). Two management strategies are employed to manage NPS pollution: (1) planning and source reduction, for example, limiting impervious surfaces, keeping grazing animals out of streams; and (2) structural and best management practices (BMPs), for example, vegetative borders, conservation tillage.

Water supply on a state and regional hydrographic scale requires drawing up plans that address the following (3): inter- and intra-agency coordination, technical assistance to local agencies and the public, public participation, and program boundaries (e.g., coastal zones must include watersheds contributing NPS pollution). And specifically NPS pollution control programs must address (3) urban sources of pollution, marinas and boats, agricultural sources of pollution, forestry, hydromodification, and wetlands.

INSTITUTIONAL STRUCTURE

Howe (4) has identified key water management requirements for a watershed/multiwatershed scale: (1) coordinated management of surface water and groundwater resources, (2) coordinated management of both water quantity and water quality, (3) provision of incentives for greater economic and technical efficiencies in water use, and (4) protection of public values associated with water service (e.g., reliable, safe, clean, affordable supply). To achieve this, the water institutions must develop the following characteristics (after Ref. 4):

- the capacity to coordinate water plans with other agencies (e.g., urban planning, agriculture, public health, environment, industrial);
- the capacity to solve water problems creatively using a variety of options and approaches (e.g., laws, pricing, taxes, tradeable supply and/or pollution permits, subsidies);
- the foresight to separate roles and responsibilities for water resource planning and management activities from construction activities (i.e., avoid conflicts of interest);
- the multidisciplinary capacity to undertake multicriteria/multiobjective planning and evaluation of alternatives;
- devolve decision-making power to the lowest level—national, state/regional (provincial), local or municipal—consistent with the scale of the water issue;
- capacity and willingness to use appropriate participatory methods involving different stakeholders at different stages of a project (preplanning/conceptual, planning and design, implementation, maintenance, monitoring, and evaluation);
- ability to reward innovation and adapt to changing conditions and priorities.

Federal, state, and regional laws and regulations are crucial instruments that help us respond to the challenge of water sustainability. Such responsive legal instruments should do the following (after Ref. 5):

- Encourage administration at the appropriate hydrological scale: watershed, multiwatershed or aquifer system.
- Foster internalization of the values and ethics of sustainable resource development.
- Encourage integrated approaches to water supply and sanitation, ecology, and public health.
- Prevent water allocation and usage policy making from fragmenting among agencies.
- Promote integrated appraisals, notably environmental, economic, and sociopolitical impacts assessment of alternative actions.
- Encourage integrated capacity strengthening of governmental institutions, NGOs, community associations, and businesses to transition into more sustainable policies and practices and work collaboratively.
- Enforce reward and penalty incentives that encourage sustainability.

WATER SUPPLY PLANNING AND MANAGEMENT

These are the five categories of human consumption (6):

1. domestic (in-house and out-of-house);
2. industrial (e.g., factories, power stations), commercial (e.g., shops and hotels), and institutional (e.g., hospitals, schools);
3. agricultural (e.g., farms, crops, livestock);
4. public (e.g., parks, fire fighting, sewer flushing;
5. Losses—distribution losses, consumer wastage, metering errors, and/or unrecorded consumption.

Table 1 shows the water supply categories used by the U.K. National Rivers Authority for watershed and regional multiwatershed planning and the principal demand considerations.

For state and regional water supply that may include several watersheds and aquifers, coordinating the forecasting of demand is essential for sustainable supply. Forecasting beyond 15 years is highly uncertain; so for longer time frames, we employ scenario projections that ask the question: If these conditions or assumptions apply (e.g., a certain population growth rate, unit demand, supply losses), then what will the future water demand be? The estimation of demand involves (6):

Table 1. Regional Water Use Categories Employed by the U.K. National Rivers Authority and Demand Considerations

Use Category	Typical Uses	Demand Considerations
Drinking water supply	Municipal water supply (surface and/or groundwater; private wells)	Population levels, growth rates and consumption patterns, pricing and wastage
Industrial water supply	Process water supply; cooling waters	Production process technologies and efficiencies
Agricultural water supply	Irrigation waters; livestock watering; milkhouse wash water; livestock housing wash water	Cropland area and crop types and cultivation methods; livestock populations; dairy production
Flood control	Impoundment of high flows for controlled release; construction of dams, reservoirs, levees, channel protection	Flood risks, especially in the floodplain; seasonality and intensity of rainfall and snowmelt; vegetative buffers
Thermal power generation	Cooling waters; settling pond waters; water for pipe flushing and maintenance	Thermal cycle efficiency; power demand and scheduling; capacity
Hydroelectric power generation	Impoundment of water for power generation; construction of dams and reservoirs; pumping and drawdown of water levels	Minimum head required for expected power generation
Navigation	Recreational boating; commercial transport shipping; commercial tourist shipping	Minimum depths and flows required for safe passage
Water-based recreation	Recreational fishing; boating; swimming; hiking, picnicking; nature enjoyment (e.g., bird watching); aesthetic enjoyment	Cultural values, including water odor and color, nature appreciation; seasonality of recreation; numbers of recreators; laws
Fish and wildlife habitat	Aquatic and riparian habitats; protection of community structure; protection of rare and endangered species	Sensitivity of species to disturbance; ecosystem type; types and intensities of stressors (e.g., water loss, pollution)
Water quality management	Protection of minimal flows for water quality preservation; low-flow augmentation from reservoirs; assimilation of waste discharges from municipalities and industries; assimilation of storm and combined sewer discharges	Magnitude, duration and frequency of waste flows; point or nonpoint sources; types and toxicities of contaminants; health risks to humans and other species

Source: After References 7 and 8.

1. plotting the population trend for the past 10–20 years and estimating the proportion likely to be due to immigration and that due to natural increase (births minus deaths);
2. dividing the supply area into different socioeconomic classes of domestic use;
3. estimating the typical domestic consumption per capita in each class;
4. seeking values of future immigration and natural increase for the different classes of housing;
5. Estimating distribution pipe loss rates, consumer wastage, and unsatisfied demand;
6. Estimating the growth in industrial, commercial, and service demands (a function of population growth);
7. Estimating growth in agricultural demand, hydroelectric demand, and public and ecosystem maintenance demand.

Equation 1 is the general supply equation for any bounded area (6).

$$\begin{aligned}\text{Total supply} = &\ \text{total legitimate potential demand} \\ &+ \text{consumer wastage} + \text{distribution} \\ &\ \text{losses} - \text{unsatisfied demand} \\ =&\ \text{water supply unreturned} \\ &+ \text{wastewater return} \end{aligned} \quad (1)$$

Water can be supplied, and demand can be met by one or a combination of the following options: (1) surface freshwater withdrawal (lakes, reservoirs, rivers and streams), (2) saltwater desalination, (3) groundwater withdrawal (springs and aquifers), (4) rainfall harvesting, (5) wastewater reuse, and (6) demand management (efficiency measures and absolute reduction). For the latter, pricing water equitably is an important consideration, and it has been postulated that there is a inverse power relationship between demand (Q) and price (P), such that (6)

$$Q = k\,P^{-x} \quad (2)$$

where k is a constant and x is the elasticity between Q and P.

In addition, in developing countries, the price of water must be tiered according to ability to pay, such that the poorest people (those at a subsistence level without surplus income) are fully subsidized, whereas high-end consumers (more than 300–400 liters/person/day) pay the most and have a financial incentive to reduce their consumption.

REGIONAL WATER TREATMENT AND DISTRIBUTION

Water treatment technologies seek to produce adequate and continuous supplies of water that are of sufficient quality—chemically, microbiologically, and aesthetically—for their intended uses. Treatment consists of a series of physical and chemical unit processes that are designed as a function of the source and quality of the raw water, the required water quality produced, and the throughput (or flow rate). A typical full water treatment system consists of eight stages (9): (1) intake, (2) pretreatment (e.g., screening, neutralization, aeration); (3) primary treatment (e.g., coagulation, flocculation, sedimentation); (4) secondary treatment (rapid and slow sand filtration); (5) disinfection (e.g., chlorination or ozonation); (6) advanced treatment (activated carbon, membrane diffusion); (7) fluoridation; and (8) distribution.

On a state and regional scale, sufficient water treatment capacity must be built and planned for using regional demand forecasting. The location of water treatment plants depends on the location of the source of raw water, the location of the users, and whether or not it is more cost-effective to build a few large plants to serve the region or a network of smaller local plants. This treatment plant spatial configuration will also determine, and be determined by, the distribution system design. The distribution system consists of a network of pipes (9): (1) trunk mains that bring large flows from the source to the treatment plant and then to storage reservoirs or towers; and (2) distribution mains that deliver water from the service storage to the users, a highly branched network. Service reservoirs and water towers cope with the diurnal variation in demand by satisfying peak demand, supplementing the average daily flows. Service reservoirs must be able to satisfy the population for at least 24–36 hours and provide enough pressure to reach the storage tanks of individual users, perhaps more than 50,000 households (9). The hydraulic head for fire fighting is at least 30m, and maximum head is about 70m (above this, leaks and bursts are common). In flat areas, water is fed by gravity from water towers, which can incur high pumping costs. High-rise buildings require their own pumping. Typically, each user has a supply pipe that connects to the communication pipe at the property boundary, which in turn connects to the service main. Water distribution networks serve different zones, each one supplied by a service reservoir and/or tower (9). On a regional and state scale, the planning and management of such an infrastructure must be carefully coordinated to meet the growing demands of a diverse set of users. In times of drought, regional water supply is severely tested, and the risk of drought must be addressed with a careful regional provision of adaptive reserve storage and aggressive water conservation strategies that penalize nonessential use.

WATER SUPPLY IN DEVELOPING COUNTRIES

Table 2 summarizes the sources, decision criteria, delivery methods, and levels of service for water supply in developing countries. Three main factors affect the choice of water supply in developing countries (11):

- institutional—land tenure, administrative structure, finance level;
- population—demography, size, growth rate, socioeconomic levels, attitudes, and behaviors;
- technical—scale, geophysical conditions, urban layout, alternatives, resources.

For the rural subsistence poor, the main water supply problem is stable access to safe water. For the urban poor and periurban marginalized communities, the main problem of water supply is that the connection and supply costs are too high.

On a state or regional scale, these water supply issues become magnified, and it is no surprise that the most controversial aspect of water supply in developing countries is selling of water administration concessions to private water companies by a central government. Recognizing water as an economic good is essential for efficient management, but it should be priced equitably according to users' ability to pay. For subsistence communities, a full

Table 2. Key Aspects of Water Supply in Developing Countries

Sources	Selection Factors	Delivery Methods	Levels of Service
• Springs • Rivers • Lakes • Aquifers • Rainfall • Snowmelt • Recycled wastewater • Desalinated saltwater	• Distance from source to user • Water quality at source • Cost of raw water and delivery • Scale of supply • Alternatives available • Sanitation method • Return drainage method • Preferences of users • Land and water rights • Risk of shortfall, outage, pollution	• Hand transport in containers • Vehicular transport • Pipelines (gravity, pumped) • Pipelines to troughs, reservoirs, standpipes, communal ablutions, households, businesses, public places, fields of crops • Channels	1. Simple source with disinfection 2. Pump and supply line to reservoir or standpipe 3. Household standpipes or connections 4. Individual household supplies

Source: After Reference 10.

public subsidy is appropriate for basic needs (1): overprivatization risks occur when water services are run purely for profit and marginalized communities unable to pay incur crippling water debts (12). This has occurred in South Africa and touches a nerve in many other places where water is viewed as both a public good and a human right that should not be controlled by commercial interests. We should also recognize that the problem with user willingness to pay is cultural (habits of free water) and also one of poor service quality: Why should a person pay for water that is not clean and comes only intermittently?

WATER POLLUTION AND HEALTH RISKS

Pollution produced in one part of a watershed or regional hydrographic area poses potential health risks to other parts. Microbiological water pollution causes many persistent water-related diseases worldwide, whereas chemical water pollution adversely affects humans and other species, especially sensitive aquatic organisms. Much of the serious chemical pollution stems from runoff of agricultural chemicals (fertilizers containing nutrients and pesticides such as DDT) and industrial wastes such as solvents and heavy metals. For example, many tributaries of the Amazon Basin have been contaminated by mercury from mining waste, resulting in methylmercury levels in fish well above the World Health Organization (WHO) standard (13). Another dramatic example of regional mercury contamination occurred in Canada in 1984, where 64% of all Cree Indian residents of Chiasabi had levels above the WHO standard (13). Nash (14) claims that the main obstacles to effective water quality management are cost and a lack of information. Downs argues that the problems go much deeper to cultural obstacles and weak societal capacity (15).

AQUATIC ECOSYSTEMS

Freshwater lakes, ponds, rivers, streams, wetlands, and aquifers are inextricably linked by hydrologic cycles acting on watershed, multiwatershed, regional hydrographic, and continental scales. A mere 1% of land surface is covered by freshwater, but it contains 12% of the animal species (16). Conservationists claim that freshwater biota are being destroyed by anthropogenic impacts faster than they can be studied or protected: overconsumption by human populations, cultural eutrophication, acidification, dam construction and hydromodification, the introduction of invasive species, habitat loss from development, and climate change. Examples are all too plentiful (16): (1) the invasion by lampreys (an ocean parasite) of the Laurentian Great Lakes after the construction of the Erie Canal, resulting in the decimation of salmonid species; (2) the invasion of the same system by the zebra mussel and spiny water flea, upsetting native invertebrates and zooplankton; and (3) the massive diversion of river water to the Aral Sea on the border of Kazakhstan and Uzbekistan to irrigate cropland, leading to a 65% decrease in volume from 1926–1990, destruction of the fisheries, economic collapse, and health problems.

The Aral Sea is a dramatic example of the need to take a regional approach to water supply and wastewater sanitation and to account for the coupled needs of humans and other species. Covitch (16) argues that technology alone—such as GIS monitoring and remote sensing—cannot hope to address the pace or complexity of freshwater ecosystem degradation. Rather, institutional reforms must provide water to aquatic habitats and minimize adverse impacts, while enabling multidisciplinary research is done on regionally representative biotas and hydrologic regimes (16).

WATER SUPPLY FOR AGRICULTURAL PRODUCTION

The twentieth century's Green Revolution boosted agricultural production and was made possible in large part by doubling the amount of irrigated land from 1950–1980 (17). In 1989, a third of the world's food harvested came from the 17% of the cropland that was irrigated, and two-thirds of the freshwater supply is used for irrigation (17). Clearly, on a state and regional scale, for example, California, agricultural demands are often the largest of all demands, and so savings derived from more efficient irrigation technology and/or diversions to other uses can have a huge impact on water supply sustainability. There is concern that growing populations, especially in water-scarce developing countries, will overwhelm the capacity of irrigated cropland to provide sufficient food and that water stress will become even more acute. The cost of adding to land currently irrigated is high in places such as Africa because of weak capacity and the need to build supporting infrastructure. In addition, many rivers are seasonal.

There is great scope for improving the efficiency and management of irrigation water: (1) using much less water per hectare [often less than half the water applied benefits the plant (17)] and (2) ensuring that water is delivered at the right time in sufficient quantities. Many schemes that involve massive irrigation cause fields to become waterlogged and soils to become salinized, leach nutrients, and lose fertility. Grossly inefficient schemes also contribute directly to the overexploitation of aquifers and surface waters and the degradation of aquatic ecosystems. Postel (17) has estimated that about 20% of the irrigated land in the United States relies on overexploited groundwater and causes major state and regional water supply impacts. Similar problems exist in many other countries, such as China, India, and the Middle East. Already, in the Great Plains of the United States, farms have been abandoned because of falling water tables; in the Texas High Plains Region, irrigated land area shrunk by 34% from 1974–1989 because the cost of overpumping the Ogallala Aquifer System exceeds the value of crops grown there (17).

Diversions of irrigation water to other uses, especially urban areas, is becoming a more common practice; some farmers sell water for more than they would earn from using it to grow crops. Los Angeles, Mexico City, and Beijing are just three examples of megacities that rely on importing water from rural farming zones. Unless laws and regulations adequately compensate the regions that lose the water to thirsty cities, serious issues of environmental justice exacerbate rural poverty.

WATER SUPPLY FOR ENERGY PRODUCTION

Water supply and energy supply are closely related, and conflicts over access, costs, pricing, and the overarching sustainability challenges for both resources invariably run parallel (18). Energy production relies on reliable water supplies: fossil fuel, geothermal, and nuclear power plants draw vast amounts of water from lakes, rivers, or the ocean for cooling. In water-scarce regions, either withdrawals alter hydrologic regimes and impact ecosystems, water scarcity places absolute limits on energy production, or scarcity forces interbasin water transfers. Droughts in Africa and Asia and even California have significantly impacted hydroelectric production, causing shortfalls that either go unmet or must be met by scaling up other energy sources (18). For example, a decade of drought in Egypt severely reduced the capacity of the Aswan Dam that supplied half of Egypt's power. The massive investment worldwide in dams, aqueducts, and reservoirs was made to overcome both the uneven distribution of freshwater for water supply and to provide hydroelectricity for development. Cheap energy has allowed water resources to be superexploited, that is, pumping of deep fossil groundwater and pumping of interbasin water transfers over mountains in water-stressed regions. In this way, just as coal and oil allowed industry to escape the solar budget, cheap energy also allows settlements to escape the immediate constraints of local hydrology. But on a regional scale, such constraints are not escaped, merely reconfigured temporarily. Sustainable regional water and energy supply must be considered as coupled resources and managed accordingly.

WATER SUPPLY POLITICS AND LAW

All regional issues impacting water supply—urban demand, public health, ecosystem integrity, agricultural demand, energy needs—can be effectively addressed only by state and regional institutions that design and apply the aforementioned responsive regulatory instruments and allow the participation of regional stakeholders in water supply planning and management. McCaffrey (19) states that there are two major principles that underpin most international water treaties: (1) the principle of equitable utilization—the apportionment of uses and benefits of a shared watercourse should be made in an equitable way; and (2) that a country, through actions that affect an international watercourse, may not significantly harm other countries. The key to successful comanagement is sharing of information and communication, especially about development plans. These same ideas should be applied to state and regional water supply within a country.

CLOSING REMARKS

Water supply on a state and regional scale poses special challenges for planning and management because of the large number and wide diversity of users and habitats to support; the related uncertain forecasting of demand; the need to provide a reliable, affordable service; and the number of potential multisectoral priority issues (energy, agriculture, health, ecology) to address. In practice, the success of this depends on communication and coordination on three levels: (1) between each village, town, city, farm and industry of the region; (2) among the water, industrial, energy and agricultural development sectors of the state or region; and (3) among the water utility companies, their customers, and local and state government regulators. Responsive institutions and regulations are vitally important to achieve this, to mitigate the risks of drought and pollution, and to meet the growing demand, especially in water-scarce/water-stressed regions where the competition for safe water is increasingly intense.

BIBLIOGRAPHY

1. Downs, T.J. (2002). A participatory integrated capacity building approach to the theory and practice of sustainability—Mexico and New England Watershed Case Studies. In: *International Experiences on Sustainability*. W.L. Filho (Ed.). Peter Lang, Frankfurt am Main, pp. 179–205.

2. Hunsaker, C.T., Jackson, B.L., and Simcock, A. (1998). Regional assessment for watershed management in the Mid-Atlantic States. In: *Watershed Management—Practice, Policies and Coordination*. R.J. Reimold (Ed.). McGraw-Hill, New York, pp. 11–29.

3. Smith, J.P., Carlisle, B.K., and Donovan, A.M. (1998). Watershed protection and coastal zone management. In: *Watershed Management—Practice, Policies and Coordination*. R.J. Reimold (Ed.). McGraw-Hill, New York, p. 99.

4. Howe, C. (1995). *Guidelines for the Design of Effective Water Management Institutions Utilizing Economic Instruments*. Report presented at Workshop on the use of economic principles for the integrated management of freshwater resources, United Nations Environment Programme (UNEP), Nairobi, Kenya.

5. Smith, D., and Rast, W. (1998). Environmentally sustainable management and use of internationally shared freshwater resources. In: *Watershed Management—Practice, Policies and Coordination*. R.J. Reimold (Ed.). McGraw-Hill, New York, p. 294.

6. Twort, A.C., Ratnayaka, D.D., and Brandt, M.J. (2000). *Water Supply*, 5th Edn. Arnold and IWA, London, pp. 1–35.

7. NRA (National Rivers Authority) (1993). *National Rivers Authority Strategy*. NRA Corporate Planning Branch, Bristol, UK.

8. Heathcote, I.W. (1998). *Integrated Watershed Management, Principles and Practice*. John Wiley & Sons, Hoboken, NJ, p. 74.

9. Gray, N.F. (1999). *Water Technology, An Introduction for Water Scientists and Engineers*. Arnold, London, pp. 205–226.

10. Stephensen, D. (2001). Problems of developing countries. In: *Frontiers in Urban Water Management*. C. Maksimovic and J.A. Tejada-Guibert (Eds.). UNESCO and IWA, London, p. 284.

11. Lyonnaise de Eaux. (1998). *Alternative Solutions for Water Supply and Sanitation in Areas with Limited Financial Resources*. Paris.

12. Gleik, P., Wolff, G., Chalecki, E.L., and Reyes, R. (2002). *The New Economy of Water: The Risks and Benefits of*

Globalization and Privatization of Fresh Water. Pacific Institute, Oakland, CA.

13. Gleik, P. (1993). An introduction to global fresh water issues. In: *Water in Crisis, a Guide to the World's Fresh Water Resources.* P. Gleik (Ed.). Oxford University Press, New York, pp. 1–12.
14. Nash, L. Water quality and health. In: *Water in Crisis, a Guide to the World's Fresh Water Resources.* P. Gleik (Ed.). Oxford University Press, New York, pp. 26–37.
15. Downs, T.J. (2001). Changing the culture of underdevelopment and unsustainability. *J. Environ. Plann. Manage.* **43**(5): 601–621.
16. Covitch, A.P. (1993). Water and ecosystems. In: *Water in Crisis, a Guide to the World's Fresh Water Resources.* P. Gleik (Ed.). Oxford University Press, New York, pp. 40–55.
17. Postel, S. (1993). Water and agriculture. In: *Water in Crisis, a Guide to the World's Fresh Water Resources.* P. Gleik (Ed.). Oxford University Press, New York, pp. 56–66.
18. Gleik, P. (1993). Water and energy. In: *Water in Crisis, a Guide to the World's Fresh Water Resources.* P. Gleik (Ed.). Oxford University Press, New York, pp. 67–79.
19. McCaffrey, S.C. (1993). Water, politics and international law. In: *Water in Crisis, a Guide to the World's Fresh Water Resources.* P. Gleik (Ed.). Oxford University Press, New York, pp. 92–104.

RIVER BASIN DECISIONS SUPPORT SYSTEMS

CHARLES D.D. HOWARD
Water Resources
Victoria, British Columbia,
Canada

The interacting components of water resource systems incorporate physical, biological, chemical, and institutional constraints and objectives (Fig. 1). Water projects are difficult to manage efficiently within potentially conflicting objectives and constraints on water quality, river flows, and lake levels. There are many types of decisions, and these must be made on a timely and reliable basis. Decisions vary from timescales of minutes for assigning loads to hydroelectric generating units to weeks and months for managing water quality, reservoir levels, and pumping from groundwater. On shorter timescales, decisions are based on deterministic analyses. On longer timescales, decisions are dominated by consideration of probabilities.

Planning can benefit from decision support systems developed for actual operations. Sound plans require realistic analysis that these systems provide for changes

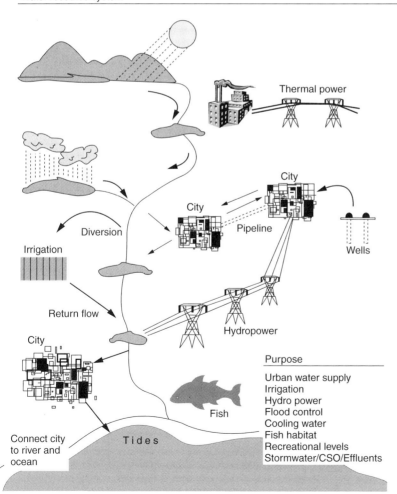

Figure 1. Water resource systems are complex.

in operating parameters, capacity expansion, and the benefits and costs of potential new environmental objectives. Figure 2 illustrates how many of the same program modules and data are used for planning studies and for actual operations.

Until recently, the data and computer simulation analysis behind water management decisions simply supported experience and judgments. In the latter half of the twentieth century, *decision support systems* based on optimization techniques were developed to recommend the best possible decisions. These software systems develop recommendations by continuous analysis of constantly changing data and evolving priorities. The key components of decision support systems are data, computer models of processes, and trained people.

The purpose of a decision support system is to acquire data and to increase its informational content. To be effective, a decision support system should execute rapidly in response to changing data. Computer input and output should be in a form that is most convenient for the person making the decisions. Modern computer systems minimize the manual input required and automatically suggest optimal operating decisions that recognize all current factors.

COMPUTER MODELS FOR WATER MANAGEMENT

There are two related, but fundamentally different types of water resource models—scientific models and management models. Scientific models are used primarily to investigate physical relationships in some detail. They provide a surrogate for extensive field monitoring by using available field data and scientific principles to interpolate and to extrapolate. They in-fill missing information on how the physical system is operating. A secondary purpose for scientific models is to test how different assumptions about future or past physical conditions would affect the system.

Management models are used primarily to suggest water management decisions—they are the main tools of decision support systems. They include a description of the scientific, institutional, and economic factors that bear on the decision-making process. To accomplish this, they must be comprehensive in modeling the interactions among components of these systems. For some types of decisions, an approximate description of the physical processes is adequate; for most types of decisions, the institutional and economic factors must be comprehensively described because they make up the key objectives in water management decisions.

Scientists should not expect management models to be absolutely accurate in scientific matters. Managers should be prepared to have recommendations from management models reality-checked by scientists.

SPREADSHEETS IN DECISION SUPPORT SYSTEMS

Spreadsheets can be the engine of a decision support system for planning studies but are unsuitable for supporting short-term operating decisions for data-intensive water resource systems. They are used routinely for hydrologic studies and for economic analysis of water resource systems. A spreadsheet can model operation of reservoirs, aqueducts, treatment plants, and water quality. The sensitivity of decisions to data

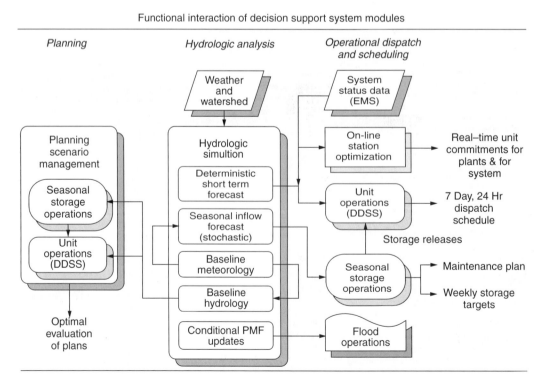

Figure 2. A decision support system for hydroelectric planning and operations.

can be examined, and specific recommendations for improvements in other models can be developed with the aid of spreadsheets.

Macro capabilities in spreadsheets are used for managing data files, displaying and printing graphs and tables, and for eliminating the need for others to be experts in programming. Macros can readily merge spreadsheets and acquire data from servers without user interaction. The detailed logic for water planning can be examined to determine how it might be structured for optimization routines in a more complex model.

Spreadsheet add-ons provide routines that can be very useful for optimizing water management decisions. A spreadsheet SOLVER function deals with nonlinear constraints and automatically drives the spreadsheet to maximize objectives within specified constraints. This eliminates trial and error in using the spreadsheet, reduces the number of explicit logical operators, and simplifies technology transfer and user training. Aftermarket optimization routines from third-party vendors extend this capability to larger more complex management issues.

A spreadsheet offers the following advantages: It can be a platform for quickly trying out new ideas, other computer models and their output can be conveniently evaluated, spreadsheets are widely used and upgradeable as computer technology advances, and output tables and graphics can be readily tailored to different needs. A spreadsheet is an easy link with past methods of operating a water resource system, and it can be carried forward into the future with minimal risk of obsolescence.

Spreadsheets are not a good choice for large complex systems that require comprehensive data management and quality control, a high degree of automation, and rapid execution. Special purpose water management computer programs are more effective for large water resource systems if they are structured conveniently for the specific application. But, as part of an overall decision support system, the spreadsheet format offers advantages in rapid implementation, capabilities for data manipulation, flexibility in presentation of results, and ease of understanding.

DEALING WITH UNCERTAINTY IN HYDROLOGIC DATA

On longer timescales, allowances for uncertainty may be intuitive—primarily colored by recent experience or based on hydrologic ensembles. A hydrologic ensemble is a set of inflow records used in a computer model to test river and reservoir operations across a range of possible inflow conditions. The ensemble members are input to simulation or optimization routines that model the operating constraints and objectives and provide information that supports actual decisions.

In some applications, hydrologic ensembles are developed with a calibrated hydrologic forecasting model from initial hydrologic conditions and historical weather data. Meteorologic ensembles generated by medium-range weather forecasting models offer the potential for conditioning future weather on current atmospheric conditions, and they eventually may replace historical weather data for forecasting hydrologic ensembles.

Hydrologic ensembles also may consist of the records of river discharge or synthetic streamflow sequences based on the autocorrelation and cross correlation characteristics of basin discharge records. Ensembles based on historical discharges are missing the atmospheric and basin moisture conditioning information that limits the range of likely future discharges. It can be observed that historical stream flow records contain a wider range of inflow possibilities than it is realistic to expect for the current conditions that are faced during operations. Heuristic methods may consider that each member of an ensemble is equally likely or some members of the ensemble may be given more weight, for example, if an El Niño condition is relevant or there is a heavy snow cover. Uncertainty of hydrology is reduced to some extent in forecasts that realistically deal with the initial watershed state and with improved future medium-range weather forecasts.

The realities of water project operations are often avoided in project planning. Postconstruction project evaluations and benefits seldom, if ever, relate back to the projections made during planning. Part of this difficulty can be avoided if the planning methodology simulates how actual operating decisions will be made. It is unrealistic to plan a system without considering the forecasted inflows during actual operations and how the operators will respond. Deterministic optimization does not consider uncertainty at all, and simulation studies based on rule curves and historical flow data lack both the finesse of optimization and a realistic expectation of near-term inflows. Water resource planning studies should include some measure to simulate the forecasts of water supply and demand that are updated at each decision point during actual operations.

EXAMPLE OF A HYDROELECTRIC DECISION SUPPORT SYSTEM

With the advent of market competition, there are clear advantages to being in position to provide quickly an optimum generation schedule based on projections of market prices and the availability and value of water. For a cascade of dams, this is a complex decision process. When water is in short supply, maximizing revenue or ensuring that all generating stations are loaded so that overall system generation is as efficient as possible will optimize the 168-hour generating schedule. This requires allocating load so that plants operate at the relatively few discrete points of maximum efficiency determined by optimal use of the generating units within each plant while meeting operational constraints (Fig. 3). Water movement between plants is considered in the optimization so that head is maximized, spill is minimized, and generation takes advantage of time of day variations in loads and energy prices.

Additional considerations include very short-term adjustments to generation and longer term plans for the most effective use of reservoir storage. Planning for maintenance of the machinery considers availability of crews, seasonal variability of hydrology, environmental constraints, and energy prices. The decision support

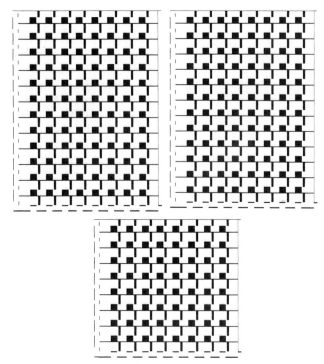

Figure 3. Hydroelectric plants have discrete points of maximum efficiency.

system recommends the optimum decisions by considering hydrologic probabilities and risks.

Each decision support system is slightly different, depending on the requirements for water management and the date when the software was developed. For a hydroelectric decision support system, the following functions are now supported by software (1):

- water management through reservoir release scheduling,
- systemwide maintenance planning for generators,
- forecasting hourly and seasonal local unregulated inflows,
- near-real-time unit loading optimization,
- hourly schedules for all generators in the system, and
- convenient data acquisition, quality control, archiving, and retrieval

Each of these functions requires access to a common pool of consistently reliable data. Static data describe the physical facilities and operational constraints such as minimum river flows and allowable ramping of river and reservoir levels. Dynamic data include hydrometerologic data, weather forecasts, discharge throughout the river system, projected hourly value of energy, the state of generating machinery, reservoir and tailwater levels, spill, current system electrical load, and special operating goals and constraints.

Some of these data are current, some are not—some have been previously checked, and some are in a raw form and may contain errors. The software provides quality control that takes data from several sources, on different time steps, of different quality, and concatenates it to provide currently useful information.

Recent installations are PC-based; local workstations use the Windows operating system. Data are stored on dedicated central servers with an SQL or Oracle database. The database sends and receives data from various servers and workstations around the system via an internal network. An archive of historical operating data supports projections and comparisons with past conditions.

REQUIREMENTS FOR A WATER RESOURCES MANAGEMENT DECISION SUPPORT SYSTEM

A customized modeling and database system will provide convenient graphical current and historical information and operating recommendations that are automatically optimized by the system. Data handling must be convenient with minimal opportunities for introducing human error and maximum flexibility for editing storing, retrieving, and displaying data and model results. There should be an emphasis on realistically designed graphical user interfaces and interactive graphical editing that reflect the preferences of the intended operators.

The data management model consists of a database system and site-specific capability for editing and displaying data in a manner that is keyed to the basin mapping. The model will be informative and practical for routine quality control, storage, retrieval, and graphic display. The database will provide a home for the monitoring data and an integrated backbone for all of the modeling software. The data will probably come from more than one source, so the database must be compatible with the communications facilities.

The hydrologic forecasting model establishes local inflow forecasts for all of the reservoirs. Two types of forecasts are necessary: deterministic hourly forecasts for one week that support daily operating decisions and weekly conditional probability forecasts for 1 year that support longer term management strategies. Both types are driven by weather data and weather forecasts. Initial conditions for each forecast reflect the computed or measured current moisture state of the basin, including soil moisture and water in storage as snow. Conditional probabilities are derived from historical weather data and the current initial moisture conditions of the basin. Deterministic forecasts include hydrodynamic routing in river systems.

An effective implementation often uses a combination of a calibrated conceptual model of the hydrologic cycle and field monitoring of river discharges, soil moisture, and snow cover. River discharge measurements provide forecast verification and information for model readjustment during operation. The local inflow forecasts are routinely updated daily and more frequently when there is concern for floods.

The water demand forecasting model considers the planning objectives, past demands, current demands and forecasts, and hydrologic forecasts. The model includes individual subarea requirements that place significant demands on the system. This is an important component for ongoing water management planning in response to changing circumstances, including water and energy demand management during drought.

The reservoir operations models must be practical, fast, and provide comprehensive optimizations that reflect complex site-specific physical and institutional details and prioritized or multiple objectives. River basin operation may be simulated by suitably constraining the optimization model(s) or by a special simulation model. Generic river and reservoir models used for planning are unsuitable for operational scheduling within a decision support system if they lack features that minimize manual data input and have excessive execution time during optimization.

The river and reservoir operations optimization model(s) use forecasts of local unregulated inflows to recommend the best long-term and short-term strategies. The objective function may include water supply for irrigation and cities, hydroelectric production, and flood management within environmental and facilities management constraints, specific discharge and water level targets, and physical limitations of the system of reservoirs and river channels. All components and operating constraints of important facilities are represented in the optimization modeling system. A simulation capability can be used to test the viability of operating plans developed by judgments that do not follow the recommendations from deterministic and probabilistic optimization models.

When there are several operating objectives, optimization can be accomplished in the software by using a set of prioritized objectives. For example, in flood management, the first priority might be to protect downstream property from flooding. With this as the objective, the software determines an operating sequence for spillway gate operations. Then, with the resulting minimized downstream flood damage as a constraint, a new objective function is used to determine the gate operations that will minimize upstream flooding. Finally, with both the upstream and downstream previously optimized flood damages as constraints, a third optimization minimizes the number of nighttime gate changes. This last step determines the sequence of gate operations actually recommended for managing the flood with available reservoir storage. With a suitable optimization model, the method is quick because each successive optimization uses the feasible solution from the previous analysis as a starting point.

The method of prioritized objectives also may be useful as a practical procedure for defining trade-offs for other types of water management decisions that require considering multiple objectives. Revising the priorities and reoptimizing can examine trade-offs. This application of a decision support system has the advantage of avoiding the difficult and often confusing task of assigning relative weights to objectives. The significance of priorities may be more readily grasped than the relative value of objectives that may be incommensurate.

BASIC REQUIREMENTS

The entire system of software must run on an operating system that is compatible with standard word processing and spreadsheet software. Maintenance for the entire system should be minimal. The software and data should be capable of being maintained and upgraded by computer specialists readily available to the owner of the decision support system. The water management models, operating system software, and computer hardware should have sufficient redundancy to provide a high level of reliability. The entire modeling system must run on the owner's computer network and must provide remote access to the database by users who are not connected to the network. The system must have levels of password security to control access to various portions of the database and modeling modules.

WHY HAVE A DECISION SUPPORT SYSTEM?

A common bottom line question among water resource system managers is: What is the value of a decision support system? The benefits are not uniform for all systems because of the relative importance of uncertainties and the multitude of institutional and physical constraints on operation. The value of a specific decision support system can be estimated by reconstructing the actual operating history and optimizing at each step as each operating decision was made. The decision support software can be designed to automate such an audit as a routine business practice aimed at raising overall efficiency within an enterprise.

Since 1987, a decision support system has been used to guide weekly reservoir release decisions at two hydroelectric plants in the coastal mountains of British Columbia, Canada. Studies of 1970–1974 operations (a period before the decision support system became operational) showed that, compared to operation with perfect foresight, as determined by a deterministic optimization model, the rule curve based operation had produced 83.4% of the maximum attainable energy compared to 95.1% with the full decision support system. Without the hydrologic forecast component, the optimization component would have produced 92.8% by simply using long-term average monthly inflows in place of the forecast. The actual energy produced by operating with the DSS in each year between 1989 and 1993 was 100, 93, 98, 94, and 96% compared to the maximum possible. The decision support system provides accessible data and a consistent framework for improving operating decisions.

Decision support systems can be expensive, but they bring improvements through more effective management to achieve objectives, reduced personnel requirements, fewer violations of environmental constraints, more effectively trained personnel, and objective technical performance evaluations. For a complex river system with many control points and reservoirs, a team of operators is trained to use the software modules and to interpret their output. For systems with only two or three major facilities, one or two trained operators can manage a highly automated, complex decision support system.

BIBLIOGRAPHY

1. Livingstone, A., Smith, D., Van Do, T., and Howard, C.D.D. (1999). Optimizing the hydro system. *Hydro Rev.* **18** (5).

READING LIST

Van Do, T., and Howard, C.D.D. Hydro-power stochastic forecasting and optimization. *Proc. ASCE 3rd Water Resour. Operations Manage. Workshop*, Ft. Collins, CO, June 1988.

Howard, C.D.D. (1999). Death to rule curves. *26th ASCE Conf., WRPM*, Tempe, AZ, June.

Stedinger, J.R. and Howard, C.D.D. (1996). Control room III: finding work for humans. *Hydro Rev.* August.

Stedinger, J.R., and Howard, C.D.D. (1995). The control room of the future II. *Hydro Rev.* July.

Howard, C.D.D., Shamir, U., and Crawford, N. (1995). Hydrologic forecasting for reservoir operations. *ASCE WRPM Div. Conf.* Boston, May.

Howard, C.D.D. (1994). Optimal integrated scheduling of reservoirs and generating units. *Leading Edge Technol. Hydro-Vision Conf.*, Phoenix, AZ, August.

Howard, C.D.D. (1993). Sewer and treatment plant operations control for receiving water protection. *Proc. 6th Int. Conf.* Urban Drainage Syst., Niagara Falls, Ont., Sept.

Stedinger, J.R. and Howard, C.D.D. (1993). A look at the control room of the not-too-distant future. *Hydro Rev.* August.

Howard, C.D.D. (1992). Experience with probabilistic forecasts for cumulative stochastic optimization. *AWRA Annu. Conf.*, Reno, NV., Nov.

Howard, C.D.D. (1992). Optimal operation of reservoirs and hydro plants during floods. *4th Water Resour. Operations Manage. Workshop, ASCE Water Resour. Plann. Manage. Div.*, Mobile, AL, Mar.

Howard, C.D.D. (1992). Optimization methodology for hydropower and instream flow operations. *Annu. Conf. Northwest Hydropower Assoc.*, Bellevue, WA., Jan.

Howard, C.D.D. (1990). Achieving effective technology transfer. In: *Decision Support Systems in Water Resources Planning*, Loucks et al. (Eds.). NATO sponsored workshop in Ericiera, Portugal, Springer-Verlag, Berlin.

Howard, C.D.D. (1988). Expert systems analysis of hydropower scheduling with forecasting uncertainty. *15th Specialty Conf., ASCE Water Resour. Plann. Manage. Div.*, Norfolk, VA, June.

Howard, C.D.D. (1984). Very short term forecasts for optimal flood control and hydroelectric peaking operations. *AWRA Symp.* Forecasting Water Manage., Seattle, WA, June.

Howard, C.D.D. and Flatt, P.E. (1976). Multiple objective optimization model for real-time operation of a water supply system. *Int. Symp. Large Scale Syst.*, Univ. of Manitoba, Winnipeg, Aug.

WATER RESOURCE SUSTAINABILITY: CONCEPTS AND PRACTICES

TIMOTHY J. DOWNS
Clark University
Worcester, Massachusetts

INTRODUCTION

The most famous definition of sustainable development is "development that meets the needs of the present without compromising the ability of future generations to meet their own needs" (1). This unifying concept was the theme of the 1992 Earth Summit in Rio de Janeiro that produced the so-called Blueprint for Sustainable Development, *Agenda 21* (2). At the core of the sustainability paradigm is the desire to reframe the perceived competition between development and environment (economy vs. ecology) into new forms of development ethics, thought, practice, and policy, reconciling enhancement of the quality of (human) life and social well-being with the conservation of ecological integrity. Underpinning this lofty goal was the ethical pursuit of a more equitable society both today (intragenerational equity) *and* in the future (intergenerational equity). The term *sustainable development* is somewhat of an oxymoron because it implies stability, yet change, and begs the coupled questions: What to sustain, for how long, and for whom? This has led to confusion and a plethora of debate over the conceptual validity of the model. A useful working definition of a sustainable project, program, practice, or policy is one that yields a steady stream of benefits (or positive impacts) that exceed costs (negative impacts) over intergenerational timescales (25 years and beyond), while conserving ecological integrity.

Too much talk of "achieving" sustainable development has led to some disenchantment with the concept when it is not delivered, or even approximated. The real power of the sustainable development idea lies in its ability to change the way we think about intra- and intergenerational equity and our responsibility as custodians of biodiversity (3). It is much more helpful to view sustainability as a *relative*, dynamic state of development practice, one that can be degraded or improved substantially, but never reached absolutely. Sustainability becomes very useful when we work with stakeholders to explore development alternatives that are compared against the status quo, business-as-usual baseline state. So, one key criterion for comparison is the relative sustainability of an option compared to others, for example, one water supply alternative versus others. In practice, this criterion is broken down into social, economic, and ecological subcriteria, the comparative evaluation of alternatives is done by a multicriteria method such as environmental and social impacts assessment (ESIA).

Viewing development in terms of helping people help themselves, finding alternative policies, plans, programs, and actions, and comparing them for positive and negative impacts is a useful simplifying approach that cuts through skepticism about sustainability and its inherent complexity. This is exactly what lawmakers had in mind when they enacted the landmark U.S. National Environmental Policy Act (NEPA) in 1969, the law that gave birth to environmental impact assessment (EIA). But in few cases has compliance with this law led to more sustainable solutions, and the public is skeptical that it is merely a technical process that endorses preconceived actions (3).

The fact that Earth Summit 2002 in Johannesburg was oriented strongly toward understanding why the plans and lofty goals of Earth Summit 1992 have not been reached, even approximated,—prospects are worsening for sustainability according to global trends (4–6)—adds considerable impetus to our quest to understand what makes the sustainability concept operational. What are the underlying concepts and philosophies that describe

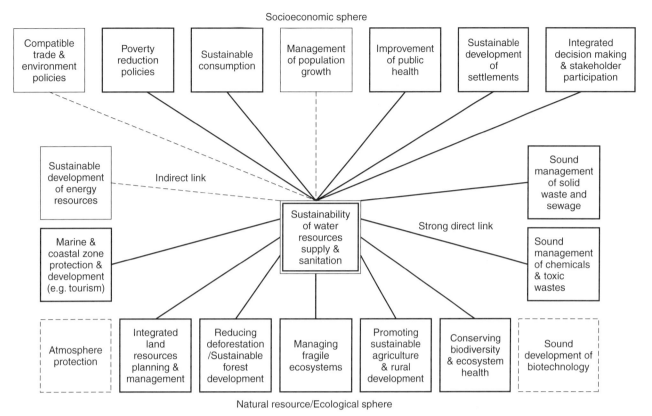

Figure 1. Agenda 21 components of sustainable development. Water resource sustainability influences many sectors and is a strategic operational focal point (7) (permission of University of Newcastle–upon-Tyne).

sustainability, and how can these be put into practice? Notably, freshwater resource sustainability is of great strategic importance to sustainable development as a whole because it has by far the highest degree of structural and functional interdependency with the other sectors/topics considered in *Agenda 21* (Fig. 1). This means, in theory, if we can solve water problems, many others will follow.

SUSTAINABILITY CONCEPTS

Human activities—agriculture, industry, urban development, and the like—stress environment and ecological health (including human health) in three main ways that often overlap and magnify each others' effects (3):

- through overexploitation of natural capital/resources (soils, water, plants, animals),
- through overproduction of wastes beyond the assimilative/attenuative capacity of ecosystems, and
- through land use/land cover changes that alter ecological structure and function.

The literature of the late 1980s and early 1990s is replete with discussions of sustainability as a concept, model, idea, paradigm, and philosophy. Hardin's seminal article called "The Tragedy of the Commons" in 1968 (8) was an important work that provoked discussion and raised awareness of the risk of overexploiting common natural resources without property rights. Hardin's theory held that individuals who have access to a common resource would seek to overexploit the resource to maximize their benefits and that the aggregate cost of overexploitation borne by the population as a whole would exceed the benefits. Such a scheme would be neither optimal for the society nor sustainable. The field of Common-Pool Resources (CPR) seeks to find governance solutions to sustainable natural resource management (9,10), investigating institutional predisposing conditions for successful governance of common natural resources, a partial refutation of Hardin's theory. Critics of CPR state that it is predicated on a single resource concept isolated from the ecosystem context of many resources, that studies are about self-governance that excludes the state, and that it is based on assumptions that resources and people do not change (11).

Another seminal publication was *The Limits to Growth* (12), which predicted the collapse of natural resources such as fossil fuel, metals, timber, and fish from overexploitation. This "crisis" scenario proved largely erroneous because of a gross underestimation of available reserves. Initially, it heightened concern over the sustainability of development, but then it undermined these same concerns by exaggerating the risks of collapse and losing legitimacy.

In 1987, the United Nations Commission on Environment and Development produced a report called *Our*

Common Future (1), widely known as the Bruntland Report after the chairperson, Norwegian Prime Minister Gro Harlem Bruntland. Focusing on satisfying basic human needs—water, food, shelter, energy, health—the Commission promoted seven strategic imperatives for sustainability and seven preconditions (Table 1). The report went on to suggest that the transition to sustainability will be driven by industrial wealth and that wealth creation be made more equitable and environmentally responsible. Its value has been to stimulate activity, though it has been criticized for failing to fully consider the dependence of industrial economic growth on natural resource exploitation (13).

The physical thermodynamic aspects of sustainability have been addressed by the Natural Step philosophy, set up in Sweden, based on a set of four systemic principles (14):

1. Substances from the earth's crust may not be extracted at a rate faster than their slow redeposit into the earth's crust.
2. Substances must not be produced by society faster than they can be broken down in nature or deposited into the earth's crust.
3. The physical basis for nature's productivity and diversity must not be allowed to deteriorate.
4. There must be fair and efficient use of energy and other resources to meet human needs.

Despite the obvious difficulty of meeting the first three principles, the World Business Council for Sustainable Development (WBCSD), a coalition of 125 companies worldwide, endorses *Natural Step*, as do conservation groups such as the International Union for the Conservation of Nature (IUCN) (14). The interests of the business sector are addressed by the concept of *triple bottom line*, the three dimensions being economic, social, and environmental. The main issues among these three are as follows (after Ref. 14):

- Economy–environment issues: How to promote ecoefficiency, clean production, and waste prevention and minimization? How to employ ecotaxation and create a market for tradable environmental permits? How to account for nonmarket costs and benefits?
- Society–environment issues: How to raise awareness and environmental literacy? How to counter environmental injustice and foster generational equity?
- Economy–society issues: How companies consider social costs and benefits of their investments? How wealth generation can address human rights and business adhere to an ethical code of practice.

Pearce and Turner (15) have defined sustainable development as "maximizing the net benefits of economic development subject to maintaining the services and quality of natural resources over time." They state several sustainability rules: (1) use renewable resources at rates less than or equal to the natural rate at which they can regenerate; (2) always keep waste flows to the environment at or below the assimilative capacity of the environment; and (3) optimize the efficiency with which nonrenewable resources are used, subject to substitutability between resources and technological progress. The first two are absolute physical thermodynamic sustainability criteria readily applicable to water quantity and water quality, whereas the third is an economic criterion. Examples of violations of rule 1 and 2 are the depletion of aquifers by withdrawing more water than is being recharged and the degradation of water quality, respectively.

Sustainability fundamentally challenges cultural norms, values, and behaviors. Using a historical cultural evolution argument, Downs (16) argues strongly that a combination of ethics, productive social interaction, and knowledge integration are cultural prerequisites for improving sustainability. Professionals and academics in the environmental field are now (and for the foreseeable future) centrally concerned with the pursuit of 'sustainable solutions' to priority problems of natural resource degradation. Reflecting the triple bottom line, sustainable solutions are those that are at the same time ecologically viable, economically feasible, and socially desirable (Fig. 2).

The International Union for the Conservation of Nature (IUCN) has actively promoted *green accounting*,

Table 1. Sustainability Imperatives and Preconditions

Strategic Imperatives	Preconditions[a] (Capital Needs)
1. Economic growth must be revived in developing nations to alleviate poverty and reduce pressure on the environment.	1. A responsive political decision-making process (S).
2. Equity and nonmaterial value must be included in the consideration of growth.	2. More efficient economic systems that use less natural resources (N, S, H, F, P)
3. Basic human needs for food, water, shelter, energy must be met, accepting that changing (more equitable) patterns of consumption are needed.	3. Responsive social systems that redistribute the costs and benefits of development (S, H, F).
4. Reduce population growth by reducing economic pressures to have children.	4. Production processes that operate within ecological limits (N, S, H).
5. Conserve and enhance the natural resource base.	5. Technology development that supports efficient energy and resource solutions (N, S, H, F, P).
6. Environmental risk management technology must be developed and made available to the developing world.	6. International order that maintains cohesion globally (S).
7. Decision making should consider ecological and economic criteria.	7. Responsive, flexible, self-correcting government institutions (S, H).

[a] S: social capital; H: human capital; F: financial capital; P: physical capital; N: natural capital.
Source: After Reference 1.

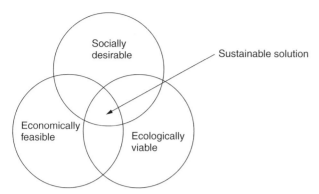

Figure 2. Sustainable solutions must satisfy three sets of criteria: social, ecological, and economic (17).

a process by which the economic costs and benefits of natural environment exploitation are measured (14). Norway has institutionalized this process to track the state of natural resource stocks and pollution rates. The United States has developed a supplement to normal economic accounting called the Integrated Economic and Environmental Satellite Account (IEESA) that seeks to account for the economic impact of environmental changes in water quantity, water quality, mineral stocks, forest stocks, and fish stocks (14). To do this will require new concepts, methods, and data, and impetus that can counter lobbying against such policies by special interests.

Spearheading what has been called the *Blue Revolution* for water sustainability (14) is the method of integrated watershed management (IWM)/integrated water resources management (IWRM). IWM/IWRM seeks to combine interests, priorities, and disciplines as a multistakeholder planning and management process for natural resources within the watershed ecosystem, centered on water quantity and water quality. Driven bottom-up by local needs and priorities and top-down by regulatory responsibilities, it must be adaptive, evolving dynamically as conditions change (3). Two global institutions play a key role in strengthening the social capital to sustain water resources: (1) the Global Water Partnership (GWP, founded 1996) that seeks to translate principles into practice through partnerships based on integrating knowledge and experience and (2) the World Water Council (WWC, founded 1996) that seeks to raise awareness of water problems and promote solutions (14).

IWM/IWRM is new and still evolving: "IWRM has neither been unambiguously defined nor has the question of how it is to be implemented been fully addressed. What has to be integrated and how is it best done? Can the broad principles of IWRM be operationalized in practice—and, if so, how?" (18). There are four broadly accepted IWRM principles, as set down by the 1992 *International Conference on Water and the Environment* in Dublin (19):

- Fresh water is a finite and vulnerable resource, essential to sustain life, development, and environment.
- Water development and management should be based on a participatory approach, involving users, planners, and policy-makers at all levels.
- Women play a central part in the provision, management, and safeguarding of water.
- Water has an economic value in all its competing uses and should be recognized as an economic good.

Building on these principles, water resource analysis, planning and management must be (after Ref. 3)

- holistic—incorporate the governing aspects of the "whole ecosystem" (watershed links to other watersheds, aquifer systems, bays, and estuaries), including interrelated natural resources (air, water, soil, biota), and humans as an integral, supermodifier of the ecosystem;
- integrated and participatory—incorporate relevant knowledge from the natural sciences, engineering, social sciences, and humanities, and accommodate the diverse set of interests, concerns, roles, and responsibilities of different stakeholders (including marginalized groups like periurban poor and rural subsistence farmers);
- strategic—identify and focus primarily on governing dynamics (the few drivers and responses that govern the way the system behaves), gather priority data, identify and address priority problems most cost-effectively;
- adaptive—adapt to changing geophysical and sociopolitical conditions; and
- sustainable—provide a steady stream of use benefits (or positive impacts) to present and future populations of humans and other species that exceed costs (negative impacts), while conserving ecosystem integrity and environmental quality.

Laws and regulations are crucial instruments that help us respond to the challenge of water sustainability. Such responsive legal instruments should do the following (after Ref. 20):

- Encourage administration at the appropriate hydrological scale: watershed, multiwatershed, or aquifer system.
- Foster internalization of the values and ethics of sustainable resource development.
- Encourage integrated approaches to water supply and sanitation, ecology, and public health.
- Prevent water allocation and usage policy making from fragmenting among agencies.
- Promote integrated appraisals, notably environmental, economic, and sociopolitical impacts assessment of alternative actions.
- Encourage integrated capacity strengthening of governmental institutions, NGOs, community associations, and businesses to transit into more sustainable policies and practices, and work collaboratively.
- Enforce reward and penalty incentives that encourage sustainability.

SUSTAINABILITY PRACTICES

The numerous rules and principles that populate the concepts of sustainability can be translated into indicators that can be measured quantitatively (e.g., waste flows are being assimilated/waste accumulation is zero) or qualitatively (e.g., the quality of stakeholder collaboration). These ecological, economic, and social sustainability indicators should be used to assess existing conditions and identify priorities for action, comparing action alternatives (impact assessment), and monitoring/evaluating the performance of the chosen alternative.

Earth Summit 2002 in Johannesburg tried to come to grips with ways to make the concept operational, how to put into practice the aspirations of Rio 1992. Howe (21) identified key water management requirements for a watershed/multiwatershed scale: (1) coordinated management of surface water and groundwater resources, (2) coordinated management of both water quantity and water quality, (3) provision of incentives for greater economic and technical efficiencies in water use, and (4) protection of public values associated with water service (e.g., reliable, safe, clean, affordable supply). And to achieve this, water institutions must develop the following characteristics (21):

- the capacity to coordinate water plans with other agencies (e.g., urban planning, agriculture, public health, environment, industrial);
- the capacity to solve water problems creatively using a variety of options and approaches (e.g., laws, pricing, taxes, tradable supply and/or pollution permits, subsidies);
- the foresight to separate roles and responsibilities for water resource planning and management activities from construction activities (i.e., avoid conflicts of interest);
- the multidisciplinary capacity to undertake multicriteria/multiobjective planning and evaluation of alternatives;
- devolve decision-making power to the lowest level—national, state/regional (provincial), local, or municipal—consistent with the scale of the water issue;
- the capacity and willingness to use appropriate participatory methods involving different stakeholders at different stages of a project (preplanning/conceptual, planning and design, implementation, maintenance, monitoring, and evaluation).
- the ability to reward innovation and adapt to changing conditions and priorities.

It is clear that to do most of what sustainability challenges society to do requires us to strengthen our *capacity to respond* to those challenges and opportunities. "Capacity building is the sum of efforts needed to develop, enhance and utilize the skills of people and institutions to follow a path of sustainable development" (22). A UNDP program *Capacity 21* (22), seeks to build capacity to implement *Agenda 21*.

Considerable community-based water management experience worldwide provides empirical evidence of what works on the community scale and why (see 23–26 among many others). IRC (23) found that the following are major factors for success in community water projects: baseline community capacity, community demand, donor and government support (financial and as policy), sufficient water resources, and the capacity of implementation agencies. But work has tended to focus on one population context alone (e.g., rural subsistence), and on one aspect (water supply and/or sanitation, or irrigation, or land degradation), instead of the more integrated watershed management approach advocated for sustainability. A considerable body of knowledge exists on participatory methods, methods include participatory rural appraisal (PRA), participatory action research (PAR), rapid rural appraisal (RRA), and participatory action development (PAD); all have been applied to strengthening community water management. But calls have been made to move beyond traditional participatory methods to more integrated analysis and planning methods with community and government stakeholders (27).

How do we make sustainability progress in a field that builds on successes, learns from failures? From 1998–2000, local working groups in collaboration with the Mexican National Water Commission (CNA), coordinated by Downs (7), developed a participatory integrated capacity-building (PICB) approach to water sector sustainability. Following an analysis comparing relatively sustainable development projects worldwide during the past 10–15 years (those yielding a steady stream of benefits *after* external support was removed) with a much larger number of *un*sustainable projects, six broad synergistic levels of capacity building emerged as critical components for success: (1) strengthening political and financial commitment; (2) strengthening human resources, including education, training, and awareness-raising; (3) strengthening information resources for policymaking (e.g., monitoring and GIS tools for data integration); (4) strengthening policies, regulations, enforcement, and verification; (5) applying appropriate technology and basic infrastructure (e.g., for water and wastewater treatment); and (6) stimulating local enterprise development (i.e., support products and services providing socioeconomic sustainability). Each one builds on those before it with positive feedback. Crucially, we see that operational sustainability is a function of participation and integrated capacity building (7).

Figure 3 shows a desirable participatory project sequence. On the analytical (science) side, we need strategic information on watershed conditions, especially water quantity, water quality, and use requirements (present and future users are humans and other species). On the policy and management sides, we need strategic planning methods that accommodate stakeholder inputs, evaluate options, choose a preferred option (by multiple criteria), and support the implementation and sustainable operation of this option. PICB is used to make the selected preferred management option (PMO) sustainable (for more

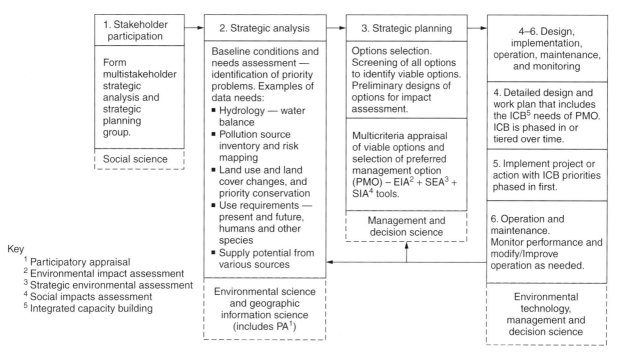

Figure 3. Six stages of operational water sustainability. Strategic participatory planning selects an alternative using sustainability criteria (impact assessment), and integrated capacity building sustains it (after Ref. 3).

information, see the article INTEGRATED CAPACITY BUILDING NEEDS FOR WATER SUPPLY AND WASTEWATER SANITATION).

An integrated, sustainable watershed management plan must be based on reliable information (PICB Level 3) about baseline geophysical, hydrologic, water quality, pollution source, ecological and socioeconomic conditions. Identifying key indicators, integrating data on them, interpreting and disseminating information to watershed–bay stakeholders informs policy and raises awareness of priority issues. It also facilitates collaborative actions to mitigate those issues. This strategic data can be fed into a GIS database that can be manipulated for decision-making. The GIS format can be used to engage community, NGO and government stakeholders in the *strategic analysis* process that must underpin the watershed–bay planning, design, implementation, and maintenance/monitoring stages (Fig. 3). A web-accessible GIS can also provide actors with a powerful means to understand how their interests fit into the wider multi-interest context of a watershed ecosystem. Such visualization and information is conducive to multistakeholder participation and dispute mediation: Actors internalize the stakes of others and can better negotiate compromises to meet common goals (3). Such information resource/planning tools do not yet exist, data are scattered, of varied formats, and have significant gaps that flag primary data gathering needs. Close collaboration with watershed NGOs engaging in community capacity building and state and local environmental agencies will ensure that results have the greatest impact and utility (3).

The term *strategic* is stressed: We must choose to collect data of the highest information value for our objectives and only this data (3). In this way, what appears to be a huge project scope is reduced to its critical components, is manageable, and most importantly, effective. This also keeps volunteer data gatherers engaged because they see the impacts of their efforts on policy. We already know what most of these key data are (priority pollutants, use data, hydrologic balance, etc.); often some exists, though dispersed and must be integrated. Other priority data needs emerge as part of the analysis.

Empirical evidence points strongly to community-based natural resource management as a key to social sustainability. But how can this be done on the watershed scale required for hydrologic sustainability? The answer lies in strengthening the social capital of the watershed by forming collaborative networks of communities. Often in developing countries, marginalized rural agroforesters occupy the upper reaches of the watershed and become, *de facto*, producers of water resources for downstream users. Evidently, both producers and users share the common interest of watershed sustainability, and both must be involved in strategic planning. By networking and through capacity building coordinated by NGOs, marginalized communities can gain the power they need to negotiate with influential user groups and government agencies.

CLOSING REMARKS

Integrated watershed management and integrated capacity building seem to confirm that the challenges of sustainability are primarily cultural and sociopolitical, not economic, scientific, or technical. Echoing the cultural prerequisites of Downs (16), any sustainability process, method, or approach—certainly IWM, PICB, EIA—will ultimately depend for its success on two cultural determinants:

- *an ethical core*—values and attitudes respectful of intra- and intergenerational equity and natural resource conservation; and
- *a participatory culture*—swapping "win–lose" for "win–win"; choosing the philosophy of collaboration and mutual gains over conflict negotiation and trade-offs.

Providentially, international case studies reveal that water sustainability challenges and opportunities are similar across cultures and geographical contexts and yield significant economies of scale in the development of theory and practice.

BIBLIOGRAPHY

1. WCED (World Commission on Environment and Development). (1987). *Our Common Future*, Oxford University Press, New York.
2. UNCED (United Nations Conference on Environment and Development). (1992). *Earth Summit 1992—the United Nations Conference on Environment and Development*, The Regency Press, London.
3. Downs, T.J. (2002). A participatory integrated capacity building approach to the theory and practice of sustainability—Mexico and New England Watershed Case Studies. In: *International Experiences on Sustainability*. W.L. Filho (Ed.). Peter Lang, Frankfurt am Main, pp. 179–205.
4. UN. Earth Summit 2002 Briefing Documents, available at: http://www.earthsummit2002.org/, consulted 2002.
5. UN. Earth Summit +5: Programme of action adopted by the Assembly, Special Session of the Assembly to review and appraise the implementation of Agenda 21, UN Department of Economic and Social Affairs, New York, available at: http://www.un.org/esa/earsummit/, authored 1997, consulted 2000.
6. World Watch. (2002). *State of the World 2002*. WW Norton, New York.
7. Downs, T.J. (2001). Making sustainable development operational: Capacity building for the water supply and sanitation sector in Mexico. *J. Environ. Plann. Manage.* **44**(4): 525–544.
8. Hardin, G. (1968). The tragedy of the commons. *Science* **162**: 1243–1248.
9. Ostrom, E. (1999). Coping with the tragedy of the commons. *Annu. Rev. Political Sci.* **2**: 493–535.
10. Dietz, T. et al. (2002). The drama of the commons. In: *The Drama of the Commons*. E. Ostrum (Ed.). National Academy Press, Washington, DC.
11. Steins, N., Roling, N., and Edwards, V. (2000). Redesigning the principles: an interactive perspective to CPR theory. *Proc., 8th Conf. Int. Assoc. Study Common Property*, Bloomington, IN, June 1–4.
12. Meadows, D.H., Meadow, D.L., Randers, J., and Behrens, W.W. (1972). *The Limits to Growth: A Report for the Club of Rome's Project on the Predicament of Mankind*. Universe Books, New York.
13. Clayton, M.H. and Radcliffe, J. (1996). *Sustainability, a Systems Approach*. Earthscan, London, p. 65.
14. Calder, I.R. (1999). *The Blue Revolution, Land Use and Integrated Water Resources Management*. Earthscan, London, pp. 62–71.
15. Pearce, D. and Turner, R.K. (1990). *Economics of Natural Resources and the Environment*, Johns Hopkins University Press, Baltimore, pp. 24–44.
16. Downs, T.J. (2001). Changing the culture of underdevelopment and unsustainability. *J. Environ. Plann. Manage.* **43**(5): 601–621.
17. Doyle, K. (2002). *Environmental Career Trends*. Environmental Careers Organization, Boston.
18. GWP (Global Water Partnership). (2000). *Framework for Action: Responding to the Forum*. Report reflecting on the Framework for Action presented at the Second World Water Forum, Stockholm, Sweden, March.
19. GWP/TAC (Global Water Partnership/Technical Advisory Committee). (1999). *The Dublin Principles for Water*. TAC Background Papers, No. 3, Global Water Partnership, Stockholm, Sweden.
20. Smith, D. and Rast, W. (1998). Environmentally sustainable management and use of internationally shared freshwater resources. In: *Watershed Management—Practice, Policies and Coordination*. R.J. Reimold (Ed.). McGraw-Hill, New York, p. 294.
21. Howe, C. (1995). *Guidelines for the Design of Effective Water Management Institutions Utilizing Economic Instruments*. Report presented at Workshop on the use of economic principles for the integrated management of freshwater resources, United Nations Environment Programme (UNEP), Nairobi, Kenya.
22. UNDP (United Nations Development Programme), About Capacity 21, Available at: www.sdnp.undp.org/c21, consulted 2001.
23. IRC. (2001). *Community Management: The Way Forward*. Report of the workshop, November 19–27, Rockanje, Netherlands.
24. IRC. (2001). *From System to Service—Scaling-up Community Management*. Report of the conference, The Hague, Netherlands, December 12–13.
25. World Bank. (1994). *World Development Report: Infrastructure for Development*, Oxford University Press, Oxford, UK.
26. WSSCC. (2000). *Vision 21: Water for People—a Shared Vision for Hygiene, Sanitation and Water Supply*. Water Supply and Sanitation Collaborative Council, Geneva, Switzerland.
27. Hailey, J. *Beyond the PRA Formula*, Paper presented at the *Symposium* on *Participation: the New Tyranny?* Institute for Development Policy and Management, Manchester, UK, November 3, 1998.

THE PROVISION OF DRINKING WATER AND SANITATION IN DEVELOPING COUNTRIES

Terence R. Lee
Santiago, Chile

Few die from a lack of water, but many die from water. Despite all the resolutions made at international congresses and all the pledges from governments and international organizations, more than 3 million people die each year from water-related diseases (Fig. 1). These deaths occur in circumstances where safe water and sanitation would have cut by at least one-third the number of diarrhea cases (currently at 4 billion worldwide every year) which result in 2.2 million of those deaths (1).

The U.S. National Academy of Engineering includes drinking water supply as one of the great engineering achievements of the twentieth century. For those fortunate to live in a developed country, it was. However, the failure to provide universal safe drinking water and adequate sanitation is amongst the gravest contradictions of our world at the beginning of the twenty-first century. The necessary technology has existed for at least two millennia, but there remain serious obstacles to its application. The result is that at least 1 billion people remain without access to a reliable and safe source of water, and more than 2.5 billion lack safe sanitation. This population is concentrated in countries at the lowest levels of development.

The deficits in services are greatest in Africa where 38% of its population remains without safe water and 40% without sanitation; Asia has 19% of its population without access to safe water, 52% without adequate sanitation; and in Latin America and the Caribbean, 15% of the population is without a potable water supply and 22% without adequate sanitation. Although, it has been estimated that a large number of additional people obtained access to services in the 1990s (800 million to water and 750 million to sanitation), population migrations and growth have meant that the proportion of the urban population with access to a safe water supply actually decreased, whereas the absolute number of people without access to water and sanitation remained the same (3).

An example of what these statistics can mean in reality for the population of a city can be illustrated by the situation in Dakar, the capital of Senegal in West Africa. In Dakar, the urban area can be subdivided into three zones:

- The first zone is completely laid out in lots and endowed with water supply and sanitation. It corresponds to the area of the colonial city founded in the second half of the nineteenth century. Today, it is the commercial and administrative center.

- The second zone is the formal suburban area, which is partly parceled in legally constituted lots. However, only part of this area has public water supply and sanitation. In the rest, where the majority of houses have been built on public land without permission, the inhabitants eliminate domestic wastes *in situ*. Of the inhabitants, 40% use latrines, 40% use abandoned wells, and 20% use dug holes.

- The third zone is entirely without any public water supply or sanitation (4).

In the suburbs of Dakar, only 40% of the population receives drinking water from a protected source, usually public taps; the remainder of the population relies on water from shallow wells, running the risk of contracting diseases as the groundwater quality is menaced through the existence of poor quality latrines and of excreta deposited directly on the soil.

The menace of well contamination is present even in rural areas that have much lower population densities than those of the suburbs of Dakar. The problem is the reliance on shallow wells when excreta are also disposed of into the shallow aquifer. Problems due to contamination are compounded by the restrictions on water consumption imposed by the need to carry water over long distances (Table 1). People will go a very long distance to get water, even more than 2000 meters, but consumption drops drastically when water has to be carried so far. It has been estimated by the World Health Organisation that something of the order of 50 liters per person a day is needed to live safe from disease, which, the data in the table suggest, means a tap in the dwelling.

For the rural population, water may be free, even if not close by. In urban areas, even for low levels of supply, water has to be paid for. It has long been known that the poor pay more when the source is not a public utility (Table 2).

Rural water supply service can be extended through the successful change to a demand-led provision of both technologies for the individual household (hand dug wells, hand pumps, roof catchments) and community technologies (gravity flow and simple pumped systems manageable by community members). The main issue is to overcome the operations and maintenance problems, which, in the past, have made up to one-third of even such

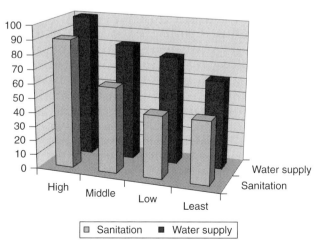

Figure 1. Percentage of population with access to improved water source and sanitation, by level of development. (*Source*: Reference 2.)

Table 1. Water Consumption in Rural Areas

Type of Water Supply	Average Daily Consumption, Liters per Person
Communal source	
Well or tap farther than 1000 meters	7
Well or tap between 250 and 1000 meters	12
Well less than 250 meters	20
Tap less than 250 meters	30
Tap in yard	40
Tap in house	
Single tap	50
Multiple taps	150

Source: Reference 5.

Table 2. Ratio Between the Price Charged by Water Vendor and by Public Utilities, Selected Cities

City	Ratio
Dacca, Bangladesh	12–25
Cali, Colombia	10
Guayaquil, Ecuador	20
Port-au-Prince, Haiti	17–100
Tegulcigalpa, Honduras	16–34
Jakarta, Indonesia	4–60
Adidjan, Ivory Coast	5
Nairobi, Kenya	7–11
Lagos, Nigeria	4–10
Karachi, Pakistan	28–83
Lima, Peru	17
Istanbul, Turkey	10
Kampala, Uganda	4–9

Source: Reference 6.

relatively low technology systems out of order soon after installation (7). Any improvement will require continuing support once the systems are installed. It is now accepted that increased private sector involvement and the transfer of many water points from community to private household management can improve performance. For example, a World Bank survey in the 1980s showed that 60% of the public hand pumps in rural India were out of order, compared to only 10% of privately owned pumps (8).

Rural sanitation normally takes the form of on-plot sanitation, which does not impose high external investment costs or significant operation and maintenance responsibilities outside the household. Public support is required for the critical tasks of health and sanitation education and promotion, and there is a need for effective provision of small-scale credits to allow the purchase of components, such as latrine slabs. Some of these activities could be successfully delegated to the private sector, as has been done with both rural water supply and sanitation installation and maintenance support in Chile (9).

In urban areas, self-help solutions have limited application. Urban water supply systems, even where they exist, commonly do not cover the entire population, have not expanded to keep pace with population growth, and fail to provide either adequate pressure or a continuous supply. The shortfall in coverage always means the absence of service for lower income groups. In addition to the failure of systems to expand or provide continuous supply, demand for water grows as industrial and commercial demands rise with economic growth. At the same time, increased household demand for drinking water is in part driven by higher living standards resulting from economic growth and by the use of conventional waterborne sewerage, which significantly increases per capita water use.

Because of economies of scale and the low value of the product supplied, urban water supply has been considered a 'natural' monopoly. It is certainly not practicable or economical to have direct competition in the provision of network services at the household level. Traditionally, providing water supply and sewerage services has been a local responsibility where services are overwhelmingly run through municipal governments.

One consequence of local municipal ownership and management of services has been the creation of serious difficulties for the transfer of successful experience. In the nineteenth and early twentieth centuries, technological and administrative transfers, even if restricted, did occur through colonial governments. Now, transfer has been largely left to international aid organizations which, whether governmental or nongovernmental, have failed to provide continuity in commitment over the long term.

Even where historically services have been provided by private companies, these too have tended to be small and local. Water supply and sewerage services have largely resisted globalization and even today foreign operation and investment in services is very limited. There has been, therefore, little transfer of the ways and means of providing service from the successful to the rest of the world.

Where transfers have occurred, advances and innovation in water supply and sewerage have been related mainly to the technical aspects of the operation and administration of large organized urban systems. There has been little or no innovation for the provision of services to the informal suburbs of the large cities in developing countries, and this is why the use of the shallow well for water supply and the latrine for excreta disposal remain the methods of choice.

Moreover, it has been the conventional wisdom that a large and relatively sophisticated institution is required to plan, organize, and control this monopoly provision. However, there is considerable potential for private sector involvement in managing, or supporting the management of, all or parts of a system, as well as in contributing to the capital requirements in middle-income metropolitan areas.

Dar-es-Salaam, Tanzania, presents an example of one possible alternative approach. Residents of Dar-es-Salaam face chronic water shortages and often have to combine several sources of water at different costs and quality to fill their needs. The Dar-es-Salaam Water and Sewerage Authority provides most of the bulk water supplies to the city. The Authority is relatively efficient; some 80% of the water produced is billed, although only 50% of the bills is collected. However, only one-third of households receive their water directly from the utility, creating a large niche for the private sector in water distribution. Large parts of the distribution system—both the piped system and nonpiped supply—are in private hands (10). This "privatization by default" is manifested in two ways:

- through the private redistribution of water, that is, water vending, ranging from home-based, quasi-legal resellers and informal low-income pushcart vendors to more formalized and regulated distributors delivering water to people's homes by tank truck; and
- by the "spaghettization" of the piped water network, where privately financed individual service lines are, in practice, directing investments and expansion of the distribution system.

Urban sanitation needs may be met through on-plot systems, communal facilities, or traditional sewerage systems. Privatization by default of network sewerage services is rarely, if ever, found. On-plot systems involve households in providing their own facilities through different types of latrines, but contamination of the groundwater results in densely populated communities. Communal facilities economize on investment and water but may not be acceptable in all cultures, however. Alternative systems to traditional sewerage systems can reduce costs and could be provided communally. The provision of simplified sewers is one alternative. Simplified sewerage is essentially conventional sewerage without any of its conservative design requirements. It can be considered as the latter stripped down to its hydraulic basics. It is also called shallow sewerage, and its in-block variant is often called backyard or condominial sewerage. Under this system, the sewers are laid on private rather than public land, household labor is accepted as partial payment for the connection, and smaller bore sewers are used. It is estimated that the costs of service delivery can be reduced by 30% using this system.

The provision of sewerage through conventional waterborne sewers has considerable economies of scale leading to monopolistic provision and needs complex institutional support. However, smaller scale provision is possible with simplified sewerage systems. This means that the private sector and local communities could become more involved in supplying and servicing such systems as well as in providing on-plot sanitation, where this is possible, and also in providing and managing communal sanitation facilities.

The provision of services to the rural population, or even to less dense urban areas, should not and does not require such high levels of investment, and the same technology can be applied that has existed for milennia. However, a serious obstacle to improving water supply and sanitation in large third-world cities is the cost of modern centralized water supply and waterborne sewage systems. These require a capital investment of, at least, US$ 1000 per connection even without sewage treatment. It is in the cities that innovation is required and alternative systems must be sought.

BIBLIOGRAPHY

1. World Health Organization (WHO), UNICEF and the Water Supply and Sanitation Collaborative Council. (2001). Global Water Supply and Sanitation Assessment 2000 Report. Available at: www.who.int/water-sanitation-health/Globassesment/GlasspdfToC.htm.
2. United Nations Development Programme. (2003). *Human Development Report*, New York.
3. Report of the World Panel on Financing Water Infrastructure. (2003). *Financing Water for All*. Presentation to the 3rd World Water Forum, Kyoto, Japan.
4. Tandia, A.A., Gaye, C.B., and Faye, A. (1998). *Origin, Process and Migration of Nitrate Compounds in the Aquifer of Dakar Region, Senegal. IAEA-TECDOC-1046*, Application of isotope techniques to investigate groundwater pollution, Joanneum Research, Vienna, pp. 67–80.
5. International Referral Centre. (1981). *Small Community Water Supplies*. International Reference Centre for Community Water Supply and Sanitation, Technical Paper No.18, The Hague.
6. Bhatia, R. and Falkenmark, M. (1992). *Water Resource Policies and the Urban Poor: Innovative Approaches and Policy Imperatives*. Background paper ICWE, Dublin.
7. Franceys, R. (1997). *Private Waters?—A Bias Towards the Poor*. Department for International Development, Occasional paper No. 3, London.
8. Cairncross, S. (1989). Water supply and sanitation: An agenda for research. *J. Trop. Med. Hyg.* **92**: 571–577.
9. Chile, Ministerio de Obras Públicas. (1999). *Memoria*, Santiago, Chile.
10. Kjellén, M. (2001). *Water Provisioning in Dar-es-Salaam, Tanzania: The Public—Private Interface*. Paper presented to the UNESCO symposium: Frontiers in Urban Water Management, Marseilles, France.

SUSTAINABLE MANAGEMENT OF NATURAL RESOURCES

LUCAS REIJNDERS
University of Amsterdam
Amsterdam, The Netherlands

INTRODUCTION

It is easier to agree on what is *un*sustainable than on what is sustainable. The once great Newfoundland cod fishery collapsed in 1992, causing 20,000 people to lose their jobs (1). It is uncontroversial to call the fishing practices that led to this collapse unsustainable. Similarly, it is uncontroversial to call mining of fossil water for corn production in Saudi Arabia unsustainable. However, by now there are probably more than 100 definitions of sustainability. Such definitions include environmental and often also social and economic aspects of human activities. They may either stress matters that can in principle be objectively established (such as fish stocks) or be social constructivist, emphasizing divergent views of stakeholders, different world views, or procedures to come to agreement about what sustainability means in a specific social context.

Here sustainability will be defined according to the World Conservation Strategy (2) that introduced the concept among policy makers. This definition fits a steady-state economy (3)—an equilibrium relation between human activities and the physical environment. The definition is also in line with long-standing approaches to safeguard the long-term productivity of forestry and fisheries (4). The main underlying principle of so defined sustainability is that of justice between the generations. This principle comes back in the most famous of later definitions of sustainability, given in the report *Our Common Future* (5): "meeting the needs of the present without compromising the ability of future generations to meet their needs."

"Natural resources" is a concept less controversial than sustainability. It includes all that we use or take from nature to provide for our existence. Mineral oil, river water, North Atlantic cod, and iron ore all belong to our natural resources. Efforts are needed to obtain these resources. They can be considered natural capital and are not technologically replaceable by other forms of capital, such as social or financial capital (6). There is also a category of services of nature that we use, whether or not we are aware of it. These include degradation of toxic substances, maintenance of atmospheric composition, sequestration and cycling of minerals, and structuring services. Many of these services are ecosystem services. There is no reason to believe that financial or social capital can be an effective substitute for nature in producing ecosystem services (7).

In this article, sustainable management of natural resources comes down to maintaining natural capital so that it can be used by all future generations of humans. This is based on a production level that can be sustained indefinitely without threatening the living conditions of future generations (8).

First, sustainability is discussed in the context of four different types of natural resources. Thereafter, three aspects of sustainability are discussed in more detail:

- sustainable management of fresh water resources,
- sustainable management of fisheries, and
- sustainable management of ecosystems.

TYPES OF NATURAL RESOURCES AND THEIR SUSTAINABLE USE

In the context of resource management, several important types of natural resources can be distinguished. First, there are flux-type natural resources. These can be used without any impact on future availability. Examples of such flux-type natural resources are wind and solar radiation. Second, there are renewable natural resources. In this case, there are currently substantial additions to the stock. Such resources include freshwater in rivers and fish in the seas. Aquifers that are replenished, mostly at a rate of 0.1–0.3%/year (9), also belong to the category of renewable resources.

A third type of natural resource is the product of slow geological processes. This is the category of the (virtually) nonrenewables and includes metal ores, fossil water, and fossil carbon compounds. Compared with the total stock, there are no or relatively small additions to these resources. Total stocks may vary strongly. For instance, stocks of iron ore are large, but the stocks of, for instance, Ag (silver), Sn (tin), or Pt (platinum) ores and fossil water are relatively small. A fourth type of resource covers living nature, which provides for ecosystem services.

From the viewpoint of sustainable resource management, flux-type resources may be freely used. This is not so for the other three types of resources.

To conserve renewables over long periods of time, the usage of renewables should not exceed the addition to stock. Moreover, the quality of the stock should be maintained. For instance, as to groundwater, the condition should be met that water tables do not fall, and also pollution that negatively affects the future value of the resource should be prevented.

The concern for (virtually) nonrenewable natural resources is mainly with geochemically scarce natural resources that are currently being depleted rapidly. In this category belong fossil water resources in North Africa and Southwest Asia that are currently "mined" (10). Fossil carbon compounds and ores for phosphates and metals such as silver, platina, tin, and nickel are other examples. If the stocks that are formed by slow geological processes are to last during the normal lifetime of a primate species ($\sim 10^6$ years), losses from stocks (including stocks in the economy) due to dissipation should be kept small. When substitutes that are dependent on flux-type or renewable resources are made available in a sustainable way, relatively large reductions of the stock that was formed by geological processes are defensible (11).

In ecosystem functions that are important to us, both the area available to natural ecosystems and species composition ("biodiversity") of ecosystems matter. Moreover, for proper functioning, a number of conditions regarding the abiotic aspects of the ecosystem should be met. Requirements as to the quantity and quality of water available function, often prominently, among such conditions (12).

SUSTAINABLE MANAGEMENT OF FRESHWATER RESOURCES

Freshwater is a vital natural resource for humankind. Only 3% of the world's water is fresh, and of that over 98% occurs as groundwater and less than 2% in surface water such as rivers and lakes. Desalinized seawater is used to a very limited extent as a resource. This is closely related to its high price (9).

To meet the criterion of sustainability, as defined here, the usage of renewable freshwater resources should not exceed additions to stocks. Withdrawal of seawater is by itself insignificant in view of sustainability, but desalinization should be powered by energy from flux-type resources (wind solar) or by sustainable biomass chains. The use of fossil freshwater stocks may be considered to bridge gaps temporarily on the way to sustainability.

Traditionally, management of freshwater resources has been strongly focused on the supply side: trying to supply adequate amounts of good-quality water and limiting the negative impacts of excessive water. Interest in sustainable management has shifted the focus. Demand-side management, reuse of water, and water conservation have become important. Though there is substantial scope for improvement at current price levels, it has rightly been argued that sustainable management of water resources is dependent on major changes in economic arrangements for dealing with water resources. A number of proposals for such changes have focused on rationing, where the sum of rations reflects sustainable management. Most proposals, however, argue in favor of water pricing that reflects the true full cost, including external costs.

In demand-side management, providing for freshwater-based services with less water ("improving efficiency") matters. As agriculture is the most important consumer of freshwater, requiring an estimated 65–75% of all water use, agriculture is a useful primary focus for efforts to improve water efficiency and reuse (9,12,13). This is especially so in semiarid areas that are currently "water starved" or where current practices cannot be sustained. Moreover, the share of agriculture in water withdrawals in such areas is often >85% (14). Water losses from agricultural systems in semiarid areas due to leakage during storage and conveyance (in case of irrigation), and to runoff, drainage, and evaporation are usually of the order of 70–85%. In some parts of sub-Saharan Africa, the loss for rain-fed agriculture may be 95% (13). Though a substantial part of these losses is unavoidable, there is also scope for improvement. For instance, when current water efficiency in sub-Saharan Africa is 5%, an increase to 10% (doubling vegetation yield) can often be achieved by the introduction of dwarf shrubs and improved tillage and nutrient management (13).

In irrigated agriculture, now responsible for about one-third of worldwide food and fiber harvests, better efficiencies are also possible. Options for efficiency improvement include better provisions for storage and conveyance, better scheduling, furrow diking, direct seeding, land leveling, microirrigation systems (drip irrigation and microsprinklers), and better regulation of the groundwater table (12,13).

Reuse of water in agriculture is a way to limit water withdrawals. In practice, both water from wastewater treatment plants and drainage water are reused (10,13–16). In the latter case, such water is often used to irrigate relatively salt-resistant crops. In managing reuse, the quality of water is important. The salinity and sodicity of water may negatively affect the future productivity of agricultural land. Excessive levels of nitrate, pesticides, and the presence of agents causative of infectious diseases may pose a direct threat to animal and human health. Receiving soils may, furthermore, act as a "sink" for hazardous compounds that may subsequently enter human and animal food chains. When persistent hazardous pollutants enter deep groundwater, they can result in long-term contamination of this important freshwater stock. Negative aspects of the secondary water supply may be influenced by pollution prevention ("source reduction") and treatment. Often there are possibilities for source reduction that are profitable at current prices (18).

Though both improvements in agricultural water efficiency and reuse offer scope for a more sustainable management of resources, it should be realized that in an increasing number of countries, especially due to increasing population pressure, such improvements will not be sufficient to feed the population in a sustainable way. This leads to the need for such countries to import "water-intensive" food from countries that have no water shortages.

Improvements in efficiency and reuse are also possible outside agriculture. An obvious object for improved efficiency is excessive loss and illegal withdrawal from distribution systems. In industrialized countries, losses of the order of 6% to more than 25% are common (17,19). Higher percentages have been noted elsewhere. In Jordanian cities, for instance, 55% and more of the input in distribution systems is "unaccounted for water" (10).

Water-efficiency improvement in industry and households has advanced. In industry, much of the improvement has come from internal reuse and process optimization, and there is still scope for further reductions in industrialized countries. For instance, a detailed study in the United Kingdom (20) suggested that industrial water consumption can be cut by 30% in a profitable way at current price levels. When water prices for industry increase to reflect true costs, profitable efficiency gains may increase considerably (17,21). Studies pertinent to developing and emerging industrial countries also tend to show a large potential for profitable water efficiency improvements (22–24).

Substantial improvements are underway in household water efficiency. Water use by water closets has been reduced by 75% or more, and major improvements in water efficiency have also been achieved in showers and appliances such as (dish and clothes) washing machines. Such efficiency improvements are usually profitable if current lifetime costs of appliances are considered (12,25). Internal reuse of household water, for instance, by using gray water for flushing water closets is practiced to a limited extent.

Finally, stocks of water may be improved by better water conservation. An eye-catching example of water conservation is in Sun Valley in the Los Angeles basin (USA). This valley was beset by floods from periodic winter rains. This led to regular press coverage showing vehicles driving in heavy (rain) water. To solve this problem, the County's Public Works Department initially opted for a storm drain to carry surplus water to the ocean. However, public discussion led to another solution. Water from winter rain is now increasingly collected in Sun Valley to recharge groundwater. In Europe, new urban developments increasingly rely on keeping rainwater out of the sewer systems and allowing it to infiltrate to add to groundwater stocks.

SUSTAINABLE MANAGEMENT OF FISHERIES

Fisheries catch renewable natural resources. When catches systematically exceed additions to stock, such resources can be depleted. If catches that exceed additions to stock are stopped, in a number of cases, the population rebounds. However, there are also cases where stocks do not rebound and remain at low equilibrium populations.

To be sustainable, catches should not exceed additions to stock in quality and quantity. Regarding cod, for instance, it is not only important to maintain a constant number of fish but also to safeguard that such fish are in the right age and size classes to maintain predation and reproduction (8). Moreover, for economic, biological, and social reasons, fisheries should manage stocks toward the higher end of abundance (1).

Sustainable management of fisheries is a long-standing issue. At least since the thirteenth century, European kings and other people in high places are on record opposing unsustainable fishing in their realms. The earliest recorded government-sponsored efforts to manage fisheries in a sustainable way in Europe date from the same time (26).

There is evidence that a number of traditional management systems have operated sustainably while managing stocks toward the higher end of abundance. Such systems have usually been characterized by ownership and controlled access; closed seasons during which fishing was prohibited; a taboo on fishing in refuge areas; and gear, size, and species restrictions (27,28). By present standards, fishing effort has been low under such traditional arrangements. When efforts increase (usually linked to financial pressures, population pressure, and technological development), stocks tend to decrease—especially when there is open access. The unregulated dynamic of fisheries is to deplete a stock as far as markets and technology will allow and then move on to the next stock (1,29).

A variety of approaches have been tried to counter the tendency of fisheries to deplete stocks. In many cases, such efforts have failed. Consequently, prices paid for the produce of fisheries are now often much higher than they would have been under conditions of sustainable management (8). Moreover, it has been estimated that the revenues of the worlds fishing fleets are approximately US$20–40 $\times 10^9$ below their operating costs (30,31).

Against this sobering background, there is no shortage of suggestions for improvement, all focusing on the supply side. Some have argued in favor of a revival of traditional arrangements that have worked well, such as the community-based fisheries management systems in the Pacific Islands and Japanese inshore fisherman's traditional user rights (28,32). Whether such a revival can be successful is dependent on meeting a number of conditions (28). Current developments are not necessarily conducive to this (27).

Recently, based on an initiative of the Worldwide Fund for Nature (WWF) and the company Unilever, a nongovernmental initiative emerged to certify sustainable fisheries. This is the Marine Stewardship Council that, by March 2004, had certified eight fisheries as sustainable. There is considerable favorable consumer interest in this approach (33), but it is too early to offer a verdict on the success of this scheme.

Most proposals for improving management practices have focused on government intervention. The use of incentive-based price instruments, such as landings and vessel taxes, has been advocated (31). Though, in the case of freshwater, adapting prices is increasingly popular as a means to attain sustainable management, this is different for fisheries. To the extent that efforts have been undertaken by governments, the focus has been firmly on trying to manage "physical" aspects of fisheries by measures such as limitation of effort, minimum allowable sizes for individual species, maximum allowable catches, and protected areas.

There have been both failures and cases of successful governmental management of fisheries focusing on physical aspects of fisheries. From the successes and failures so far, a number of lessons can be drawn (1,32,34–37).

- The problem of illegal catches exists, especially if stocks are close to the market. When the policing effort is not up to the ingenuity of illegal fishing, management efforts may well become a failure.

- With few exceptions ("dolphin safe" tuna fishing, the use of turtle excluders, and the Norwegian no-discard policy), management systems have been poor at limiting the negative impacts on stocks of by-catch and discards. Globally, discarding may be about 30% of landed tonnage.

- Many management strategies have opted for limitation of effort, for instance, by regulating the efficiency of fishing gear; engine size; fishing season; and/or limiting the number, size, and storage capacity of fishing boats. Only in a limited number of cases (e.g., the rock lobster fishery in West Australia) have such approaches been successful.

- A number of efforts have opted for allocating total allowable catches to political entities such as countries. In some cases, these efforts have been successes and in other cases, they have been failures. An important reason for failure, apart from illegal catches, has been that conservation goals were not separated from allocation, which resulted in losing conservation goals in allocation battles. Electoral considerations favoring perceived (short-term) interests of the fishing industry have also contributed to failure.

- More recently, there has been the tendency to base management on the allocation of long-term or even perpetual individual rights or individual transferable rights for their fishermen. In most cases, these rights have been based on past catch histories ("grandfather rights"). In some cases, such individual rights have been auctioned. There have been failures of such systems, such as the abalone fishery of British Columbia and the Southern school sharks fishery of Australia. But in other cases, this approach has worked relatively well. Success may be partly linked to the fact that individual rights that are allocated for a long time become more valuable if fishing practices are sustainable.

- The establishment of marine protected areas that are protected from fishing (which are aimed at conservation of fish stock but may also conserve or restore ecosystems) has so far not been practiced to an extent that allows a verdict on its effectiveness. The same holds for the allocation of fishing rights to cooperatives.

SUSTAINABLE MANAGEMENT OF ECOSYSTEMS

Natural ecosystems provide for a wide variety of services. Part of the services of natural ecosystems relate to water. Natural ecosystems are important in determining levels of greenhouse gases, and thereby climate, that in turn is a determinant of the global water cycle. Such ecosystems are also a major determinant of worldwide evapotranspiration. There are, furthermore, local quantitative and qualitative links between ecosystems and water. For instance, deep-rooted natural vegetation along watercourses catches minerals and returns these to topsoils. Thereby, it reduces the mineral content of surface waters, while increasing primary productivity on topsoils (38,39). Tropical rain forests are characterized by high levels of water recycling. When there is clear felling of such forests at low levels of externally added precipitation, the area may be converted into a savanna (40).

Marine, freshwater, and wetland ecosystems generate extensive services. It is estimated that the value of a number of ecosystem services in the entire biosphere is in the range of US$16–54 × 10^{12}; the average (US $33 × 10^{12}) is in excess of the gross world product (41). Of the average value of US $33 × 10^{12}, somewhat over US $27 × 10^{12} comes from marine, freshwater, and wetland ecosystem services. There is even a case to attribute infinite monetary value to ecosystem services provided by living nature because human life would be impossible without them.

For sustainability, as defined here, ecosystem services should be maintained at their present level. To attain this, the area allocated to nature and the abiotic conditions under which nature functions should be such that ecosystem services do not change.

A debate has developed regarding the question, how much biodiversity can be reduced without reducing ecosystem services. Loss of a species should not lead to loss of function when there is functional redundancy. However, experimental evidence suggests that the statistical relation between species diversity and ecosystem functions important to humans may well be strong, though it is likely that actual contributions may show large differences between species (42–45). Even when a species seems functionally redundant, it has been found that the actual loss of apparent functional redundancy may come at great cost (46). Also a "redundant species" may still have an insurance function in case of ecosystem perturbation (47). So sustainable management is presumably strongly dependent on maintaining species diversity in ecosystems as that has developed.

A main problem in attaining criteria for sustainability is that many ecosystem services in practice do not have a price. There are economic incentives that may be conducive to the sustainable management of ecosystems. These include user fees for national parks, royalties from bioprospecting, conservation easements, and mitigation banking. So far, however, the application and effect of economic incentives are very limited (35). Legal systems for protecting nature are, with a few exceptions, rather focused on preventing the extinction of specific species than on ecosystem services. A major effort is needed to base nature conservation on the true value of ecosystem services.

BIBLIOGRAPHY

1. Hilborn, R. et al. (2003). State of the world's fisheries. *Annu. Rev. Environ. Resour.* **28**: 359–399.
2. IUCN, UNEP, WWF. (1980). *The World Conservation Strategy*. WWF, Gland, Switzerland.
3. Daly, H. (1973). *Toward a Steady State Economy*. Freeman, San Francisco.
4. Becker, B. (1998). *Sustainability Assessment: A Review of Values, Concepts and Methodological Approaches*. World Bank, Washington, DC.
5. World Commission on Environment and Development. (1987). *Our Common Future*. Oxford University Press, Oxford, UK.
6. Ayres, R.U. (1996). Statistical measures of unsustainability. *Ecol. Econ.* **16**: 239–255.
7. Batabyal, A.A., Kahn, J.R., and O' Neill, R.V. (2004). On the scarcity value of ecosystem services. *J. Environ. Econ. Manage.* in press.
8. Hueting, R. and Reijnders, L. (2004). Broad sustainability contra sustainability: the proper construction of sustainability indicators. *Ecol. Econ.*, in press.
9. Pimentel, D. and Houser, J. (1997). Water resources, agriculture, the environment and society. *Bioscience* **47**: 97–107.
10. Haddadin, M.J. (2002). Water issues in the Middle East. Challenges and opportunities. *Water Policy* **4**: 205–222.
11. Reijnders, L. (2000). A normative strategy for sustainable resource choice and recycling. *Resour. Conserv. Recycling* **28**: 121–133.
12. Gleick, P.H. (2003). Water use. *Annu. Rev. Environ. Resour.* **28**: 275–314.
13. Quadir, M. et al. (2003). Agricultural water management in water-starved countries. Challenges and opportunities. *Agric. Water Manage.* **62**: 165–185.
14. Araus, J.L. (2004). The problems of sustainable water use in the Mediterranean and research requirements for agriculture. *Ann. Appl. Biol.* **144**: 259–272.
15. Crites, R., Reed, S., and Bastian, R. (2001). Applying treated wastewater to land. *Biocycle* **April**: 32–36.
16. Tsagarakis, K.P., Dialynas, G.E., and Angelakis, A.N. (2004). Water resources management in Crete (Greece) including water recycling and reuse and proposed quality criteria. *Agric. Water Manage.* **66**: 35–47.
17. Levin, R.B. et al. (2002). US drinking water challenges in the twenty-first century. *Environ. Health Perspect.* **110**: 43–52.
18. Reijnders, L. (1996). *Environmentally Improved Production Process and Products*. Kluwer, Dordrecht, The Netherlands.
19. Antolioni, B. and Filipppini, M. (2001). The use of a variable cost function in the regulation of the Italian water industry. *Utilities Policy* **10**: 181–187.
20. Holt, C.P., Philips, P.S., and Bates, M.P. (2002). Analysis of the role of waste minimization clubs in reducing industrial water demand in the UK. *Resour. Conserv. Recycling* **30**: 315–331.
21. Braeken, L., van der Bruggen, B., and Vandecasteele, C. (2004). Regeneration of brewery waste water using nanofiltration. *Water Res.* **38**: 3073–3082.
22. Van Berkel, C.W.M. (1996). *Cleaner Production in Practice*. IVAM Environmental Research, Amsterdam, The Netherlands.
23. Rao, J.R. et al. (2003). Recouping the wastewater: A way forward for cleaner leather processing. *J. Cleaner Prod.* **11**: 591–599.
24. Puplampu, E. and Siebel, M. (2005). Minimisation of water use in a Ghanaian brewery: Effects of personnel practices. *J. Cleaner Prod.* in press.
25. Allen, J. (2004). Water efficiency. *Environ. Design Constr.*, May, pp. 65–66.
26. Hoffmann, R.C. (1996). Economic development and aquatic ecosystems in medieval Europe. *Am. Hist. Rev.* **101**: 631–699.
27. Evans, S.M. et al. (1997). Traditional management practices and the conservation of the gastropod (*Trochus nilitocus*) and

fish stocks in the Maluku Province (eastern Indonesia). *Fish. Res.* **31**: 83–91.

28. Ruddle, K. (1998). The context of policy design for existing community-bases fisheries management systems in the Pacific Islands. *Ocean Coastal Manage.* **40**: 105–126.

29. Kirby, M.X. (2004). Fishing down the coast: historical expansion and collapse of oyster fisheries along continental margins. *Proc. Natl. Acad. Sci. USA* **101**: 13096–13099.

30. Cochrane, K.L. (2000). Reconciling sustainability, economic efficiency and equity in fisheries: the one that got away. *Fish Fish.* **1**: 2–21.

31. Sanchirico, J.N. (2003). Managing marine capture fisheries with incentive based price instruments. *Public Finance Manage.* **3**: 67–93.

32. Caddy, J.F. (1999). Fisheries management in the twenty-first century: Will new paradigms apply? *Rev. Fish Biol. Fish.* **9**: 1–49.

33. Jaffry, S. et al. (2004). Consumer choices for quality and sustainability labeled seafood products in the UK. *Food Policy* **29**: 215–228.

34. National Research Council. (1999). *Sharing the Fish, Toward a National Policy on Individual Fishing Quota*. National Academy Press, Washington, DC.

35. Leva, C.E.di (2001). The conservation of nature and natural resources through legal and market-based instruments. *Reciel* **11**: 8495.

36. Baskaran, R. and Anderson, J.L. (2004). Atlantic sea scallop management: an alternative rights-based cooperative approach to resources sustainability. *Mar. Policy* in press.

37. Daw, T. and Gray, T. (2004). Fisheries science and sustainability in international policy: a study of failure in the European Union's Common Fisheries Policy. *Mar. Policy*, in press.

38. Schlesinger, W.H. (1991). *Biogeochemistry*. Academic Press, San Diego.

39. Christinaty, L., Kimmins, J.P., and Maily, S. (1997). Without bamboo, the land dies. *Forest Ecol. Manage.* **91**: 83–91.

40. Silveira, L.da and Sternberg, L. (2001). Savanna-forest hysteresis in the tropics. *Global Ecol. Biogeography* **10**: 369–378.

41. Constanza, R. et al. (1997). The value of the world's ecosystem services and natural capital. *Nature* **387**: 253–260.

42. Tilman, D., Wedin, D., and Knops, J. (1996). Productivity and sustainability influenced by biodiversity in grassland ecosystems. *Nature* **379**: 718–720.

43. Duarte, C.M. (2000). Marine biodiversity and ecosystem services: an elusive link. *J. Exp. Mar. Biol. Ecol.* **250**: 117–131.

44. Hector, A. et al. (2001). Community diversity and invasion resistance. *Ecol. Res.* **16**: 819–831.

45. Symstad, A.J. et al. (2003). Long term and large scale perspectives on the relationship of biodiversity and ecosystem functioning. *Bioscience* **53**: 89–98.

46. Bellwood, D.R., Hughes, T.P., Folke, C., and Nystrom, M. (2004). Confronting the coral reef crisis. *Nature* **429**: 827–833.

47. Walker, B. (1995). Conserving biological diversity through ecosystems resilience. *Conserv. Biol.* **9**: 747–752.

SUSTAINABLE WATER MANAGEMENT ON MEDITERRANEAN ISLANDS: RESEARCH AND EDUCATION

MANFRED A. LANGE
DÖRTE POSZIG
ANTONIA A. DONTA
University of Münster
Centre for Environmental Research
Münster, Germany

INTRODUCTION

The availability of water in adequate quality and sufficient quantity is indispensable for human beings and the functioning of the biosphere. Shortcomings in integrated water management practices aimed at linking natural, social, and economic needs in a given region, wasteful and inadequate use of water, and inadequate administrative practices have often resulted in depleted water resources. Pollution has exacerbated these problems.

Addressing issues of water quality and water supply, the European Union-Water Framework Directive (WFD) came into force on December 22, 2000 (1). Two basic goals are to be achieved by the parties within the following 10 to 15 years: (1) users will have to bear the full cost of the water supplied (1, Annex III, WFD), and (2) water will have to be provided based on "good ecological status" by adhering to practices that do not harm the environment. Moreover, the WFD strives toward implementing the "polluter pays principle."

Considering the long-term perspectives for sustainable water use in the European Union, the Mediterranean in general and its islands in particular deserve special attention for a number of reasons:

- Island water resources are largely influenced by the (non)existence, timing, and amount of precipitation. In some parts of the Mediterranean, the amount of (winter) precipitation has decreased (2), and the main precipitation period has shortened during the last few decades. This trend may continue as a result of regional climate change, thus exacerbating the problem (2,3).

- Overabstraction of aquifers, in many instances, has led to lowering of the water table and the deterioration of water quality, primarily through saltwater intrusion in coastal areas. In some cases, this has resulted in exploitation of lower lying aquifers, thereby further reducing the remaining reserves of the islands.

- The demand for water is expected to increase on Mediterranean islands because of, among others, an expected increase in the population and the number of tourists. Conflicts between competing users of water will inevitably intensify.

- Agriculture is the predominant consumer of water on most Mediterranean islands (Fig. 1). Agricultural activities threaten the availability (quantity) but also the quality of water due to the use of fertilizers, pesticides, and the release of olive-oil-mill wastes. During the summer months, tourism also becomes a major water consumer, and competition between agriculture and tourism can lead to serious stakeholder conflicts.

- The rational distribution and use of water requires resolving conflicts among the different end users of water. Such conflicting interests have contributed to the unsustainable use of water in the past.
- Given the need for the protection and the sustainable use of island water resources, coupled with the need to satisfy the increasing water demand, the islands have to formulate strategic policies based on integrated water management and should pay proper attention to the objectives of the WFD.

The WFD may introduce an effective avenue to address these problems, but at the same time its implementation presents a challenge. The problems that govern the availability of water in the Mediterranean have specific manifestations on each of the islands, but the islands also share a number of general characteristics and patterns. Thus, learning how to implement the WFD best on selected Mediterranean islands through comparisons, mutual learning, and experience sharing offers the prospects of deriving a more generic set of recommendations.

The overriding characteristic that governs most of the problems mentioned lies in their complex and multifaceted nature, which requires a holistic approach involving the integration of natural and social sciences. Thus, in considering water availability and water use and deriving strategies for sustainable and equitable distribution of water, interdisciplinarity is a necessary condition.

Moreover, lasting solutions for sustainable water management in the Mediterranean will be found only through recommendations and/or regulations that are based on mutually agreed principles among the stakeholders. This requires a stakeholder-based, participatory process that builds on the results of scientific investigations, on the one hand, and on the consent of major stakeholders, on the other (5).

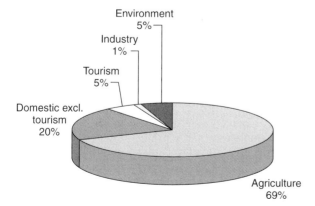

Figure 1. Distribution of water consumers on Cyprus in 2000 (after Reference 4). The predominance of agriculture as the major user is typical of most Mediterranean islands.

TOWARD SUSTAINABLE, STAKEHOLDER-BASED WATER MANAGEMENT PRACTICES: THE MEDIS PROJECT

To address the problems of water management on Mediterranean islands in accordance with the criteria just introduced and by observing the need to implement the WFD swiftly and effectively on these islands, an EU-funded research project, MEDIS (toward sustainable water use on Mediterranean islands: addressing conflicting demands and varying hydrological, social, and economic conditions), has been implemented. MEDIS is carried out by a consortium of 12 institutions from seven countries.

Goals and Objectives

The overall goal of MEDIS is to contribute to sustainable and equitable use of water on Mediterranean islands. This is to be achieved through a set of recommendations to implement sustainable, equitable water management regimes on each island that reflect the current and possible future (climatic) conditions on each island. These recommendations will comply with the WFD and will build on the conclusions reached in consultation with stakeholders from all five islands involved. More specifically, work in MEDIS

- concentrates on a synthesis of results from previous studies and integrating information from various disciplines;
- is carried out on a catchment scale on major islands of the Mediterranean stretching from the west to the east: Majorca, Corsica, Sicily, Crete, and Cyprus;
- seeks to develop an infrastructure for participatory stakeholder involvement that will enable establishing sustainable and equitable water management schemes;
- considers (possibly drastically) altered conditions that may arise due to changed climatic conditions through a set of what-if scenarios;
- concentrates on agriculture as the predominant consumer of water on Mediterranean islands; and
- undertakes comparative analyses between the islands to derive generic conclusions/recommendations and to use possible common solutions to water management problems in compliance with the WFD.

Methodology

The major stages of the research strategy applied in MEDIS, which are carried out during a 4-year period, are depicted in Fig. 2. The initial phase of the project comprises a thorough assessment of the major characteristics that determine the availability and the demand of water on each island. This includes (current and future) climatic conditions; (geo)physical, pedological, and hydrologic characteristics as well as the vegetative cover on each island; major sources and consumers of water; agricultural practices in irrigation and agrochemical use; conditions governing water management and administration; basic demographic and social conditions; and major economic indicators, particularly related to water-dependent economic sectors. This assessment leads to a typology of water management regimes for each island, based on a number

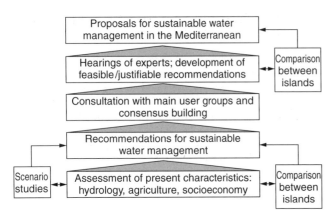

Figure 2. Schematic of major stages of the MEDIS research strategy (for details, see text).

of well-defined parameters, that share similar quantitative or qualitative ranges for each of the types specified. The final result of this project stage is a first set of recommendations for improved water management practices geared to each of the management types identified.

An integral part of the work during this stage of the project is involving stakeholders and gaining their help in specifying some of the previously mentioned parameters. They are either part of the supply network for water and/or consumers of water, so their experience and expertise are indispensable for the initial assessment. Therefore, an important activity during the first project stage is identifying relevant stakeholders and involving them in Intra-Island Workshops. In addition, Inter-Island Workshops among the islands under consideration are conducted.

Current physical/climatic conditions and possible future variations (i.e., scenario studies) are also considered. For climate, the results of the Intergovernmental Panel on Climate Change (IPCC) scenarios (6) are used in a downscaled format (or through a dedicated regional climate model). When considering prospective situations on each of the islands, future scenarios of other major determinants of water supply and demand are taken into account accordingly.

The spatial scales of the investigations depend on the specific questions asked. MEDIS is being carried out on a local (hundreds of m to km) and/or catchment scale, and inferences for all islands are drawn by appropriate upscaling. The Mediterranean scale is being addressed by comparing islands. As for temporal scales, MEDIS focuses on two time slices of 20 years centered at 2020 and 2050, corresponding to the IPCC assessments.

The second stage in the project comprises intensive consultation with the stakeholders on each island. Typical stakeholders include regional or national water administrative bodies, representatives of institutions or enterprises representing major water users, agricultural or industrial enterprises that develop and/or maintain water management schemes, nongovernmental organizations concerned with water issues, farmers and farming cooperatives, tourist enterprises and operators, and communities/residents in need of water.

The major goal comprises a consensus among the different stakeholder groups concerning their major objectives for an improved water management regime. At the same time, the stakeholders should bear in mind the recommendations from the first stage of MEDIS and specify possible alterations and amendments to these recommendations. Ultimately, a new set of recommendations will be obtained that represents both the factual basis and the mutually agreed upon principles for an improved water management regime on each island.

During the third stage of the project the second set of recommendations will be subject to scrutiny by legal and institutional experts. The objective here is to ensure that the recommendations comply with regional and national laws and regulations, and agree with the goals and principles of the WFD. This may result in a third set of recommendations that reflect the factual basis, the stakeholder consensus, and the legal and institutional requirements on national and European scales.

In the final stage of the project, island-specific conclusions will be discussed in the broader context of the Mediterranean Basin through Inter-Island Workshops and attempts to synthesize conclusions. The common solutions and possible deviations in the conclusions among the islands will be discussed and resolution attempted. Ultimately, this results in desired recommendations for improved, sustainable and equitable water management in compliance with the WFD and national rules and regulations. It is expected that these recommendations will significantly enhance the implementation of the WFD in the Mediterranean Basin.

First Results

Essential information was compiled in a series of Island Reports. In summary, the islands

- comprise areas between 3600 km^2 (Majorca) and 28,000 km^2 (Sicily),
- have permanent populations between 0.26 million (Corsica) and 5.6 million (Sicily),
- receive mean annual precipitation between 500 (Cyprus) and 1000 mm (Corsica, Sicily),
- attain a total annual water supply (through precipitation) from 2400×10^6 m^3 (Majorca) to $14,000 \times 10^6$ m^3 (Sicily; these numbers have partly been derived from the mean annual precipitation and the total area of the islands under consideration and are thus only rough estimates),
- have increasing water demand trends in the foreseeable future, and
- a water consumption structure clearly dominated by agricultural needs.

Though climatically similar, the amount of water supplied to various users differs greatly among the islands. This will have obvious consequences for the measures to be taken to implement the WFD.

The Island Reports and their synthesis led to recommendations aimed at reducing the vulnerability to water scarcity and implementing a sustainable and stakeholder-driven water management regime on the islands. In considering vulnerabilities, three different dimensions of water management have to be taken into account: a physical/environmental dimension, an economic/regulatory dimension, and a social/institutional/political dimension. Details of current and future vulnerabilities to water scarcity are described elsewhere (7), but some of the recommendations are presented in the following. These recommendations are interrelated and overlapping in many ways. They require a holistic, interdisciplinary approach to implement them.

Physical/Environmental Dimensions. The following recommendations have been formulated with regard to the physical and environmental dimensions of water scarcity:

- *Reduce Water Consumption*: The most promising strategies embrace (1) water pricing, an increase in consumer prices or alternatively/additionally quotas on water extraction may be imposed (possibly differentiated between different user groups); and (2) establishment of incentives for reduced water consumption, for example, subsidies for water saving and not—as has often been the case—for water consumption.
- *Change in Water Allocation*: Support of economic sectors accounting for maximum gross-domestic-product generation and employment through government programs. However, care must be taken to avoid possible one-sided economic advantages for specific sectors. Incentives for saving water in water-intensive sectors should be introduced in parallel.
- *Reduce Losses*: (1) Loss of water to the sea either as riverine discharge or through subsea groundwater discharge should be reduced (8), though care must be taken to avoid adverse impacts on biogeochemical cycles and on marine ecosystems in near-coastal waters; (2) reduce losses and contamination of water in distribution networks. On Cyprus, annual losses of drinking water in the distribution network account for 40×10^6 m^3, corresponding to 15% of the total demand and 23% of the total domestic demand (G. Constantinou, Geological Survey of Cyprus, pers. comm., March 2003). Improvement and/or renewal of distribution networks should be pursued where appropriate.
- *Increase Use of Additional Water Resources*: Though largely neglected for a long time, increasing attention is paid to this possible remedy. In particular, potable or irrigation water can be obtained through wastewater recycling, use of brackish water, and rainwater harvesting.

Economic/Regulatory Dimensions. The following actions addressing vulnerability related to economic/regulatory dimensions are recommended:

- Support sectors with high economic potential and small water needs.
- Change agricultural practices: (1) Consult and inform stakeholders about possible alternatives to current cropping patterns with regard to water requirements of plants, watering schedules, and the amount of irrigation water applied per plant (for a summary of possible alternatives, see Reference 9), optimize tillage systems as well as weeding and harvest controls to minimize irrigation; (2) optimize the application of agrochemicals to avoid adverse effects on groundwater quality. The use of agrochemicals and changes in agricultural techniques should be considered holistically to arrive at best practices that maximize yield and also minimize water use and the possible impact of agrochemicals on groundwater quality.
- Eliminate/reduce subsidies for water prices (in the past, subsidies have often contributed either to excessive water use and/or—in the case of agriculture—to cropping patterns that are economically and environmentally unsustainable).
- Promote cultivation of crops that have a high potential on the domestic and the foreign market (eliminate wasting products and water).
- Provide assistance in capacity building of farmers and for investments in modern irrigation technology.
- Provide economic incentives for rational water use in all sectors (e.g., specific water tariffs or relaxation of quotas and limitations in water consumption to those sectors that strive for a more rational use of potable water).

Social/Institutional/Political Dimensions. Reducing vulnerabilities related to social, institutional, and political aspects may be pursued through the following actions:

- Increase public awareness (water use, ownership, conflicts) to enhance the capacity to deal with water scarcity problems.
- Implement comprehensive monitoring of water extraction.
- Improve enforcement of existing rules and regulations.
- Simplify administrative mechanisms to enhance the efficiency of water administration.
- Transfer power from central government institutions to regional and local decision makers that seem better suited to become partners in the EU-WFD than individual users or single municipalities (e.g., TOEBs, Farmers Irrigation Organizations, in Greece).
- Encourage stakeholder-controlled water management structures.
- Ensure/improve an adequate factual basis for political decision making through input from science and stakeholders.

THE NEED FOR INTERDISCIPLINARY WATER STUDIES: THE ADVANCED STUDY COURSE SUSTAINIS

In addressing water management through research and technology projects, one is often faced with a problem different from those outlined. This is the lack of well-trained and motivated young professionals able to face the challenge of interdisciplinary water studies and the implementation of management options. The underlying reason for this problem is in the prevalent focus of traditional universities on specialized and discipline-focused curricula. Though quite adequate and successful in many instances, this kind of training leaves graduates ill-prepared for the type of work required in cross-disciplinary investigations. To resolve this problem at least partly, an Advanced Study Course, SUSTAINIS (Sustainable Use of Water on Mediterranean Islands: Conditions, Obstacles and Perspectives), was devised and implemented. SUSTAINIS was funded by the European Commission and was closely linked to the MEDIS project and its objectives.

Goals and Rationale

The course aimed at introducing 24 graduate students from various disciplinary and national backgrounds to the patterns of water supply and demand, to obstacles preventing rational use of water resources, and to perspectives for sustainable, equitable, and stakeholder-based use of water on Mediterranean islands. To provide them with an opportunity to learn about the specific environmental and socioeconomic features of Mediterranean islands, the course was carried out in Cyprus.

Curriculum and Structure

The emphasis in the teaching lay on demonstrating the complex interrelationships between environmental and societal systems that govern water regimes on Mediterranean islands and the importance of stakeholder-based strategies.

The course comprised lectures by 23 researchers from a variety of professional and national backgrounds, field trips, and discussion forums in the form of role games. The discussion forums were prepared by small, interdisciplinary, international groups of students and exposed the participants to the challenges but also to the advantages of interdisciplinary work.

The curriculum included eight modules:

Module 1: Introduction, Historical and Cultural Background of Water Use in the Mediterranean and Basic Concepts of Environmental Ethics. Cultural practices of management of freshwater resources from antiquity to the present, the recent history of water management and its importance for health, well-being, and prosperity of inhabitants of Mediterranean islands, and concepts of environmental ethics addressing normative issues in water management.

Module 2: Perspectives on Sustainable, Equitable, and Stakeholder-Based Water Management on Mediterranean Islands. Principles and tools for supporting participative water resource planning and management, design and implementation of an integrated water management approach and portrayal of research projects on integrated and sustainable water resources management in the Mediterranean.

Module 3: Essential Geologic and Hydrologic Characteristics of Mediterranean Islands. Characteristics and properties of the geology and hydrology of Mediterranean islands, description of subsurface aquifer conditions and problems related to overabstraction of groundwater resources (saltwater intrusion), and application of models in water management.

Module 4: Current and Possible Future Climatic Conditions and Their Impacts on the Hydrologic Regime. Climatic conditions in the Mediterranean and their effects on the timing and magnitude of precipitation, global and regional climate models and their strengths and weaknesses, review of expected future trends of climate development and their repercussions for water availability in the Mediterranean, and basic concepts and methodologies for integrated climate impact assessments.

Module 5: Current Patterns of Water Consumption, Major Stakeholders, Existing Conflicts, and Resource Policies. Conditions of water resources and common trends of water use with case studies from Cyprus, Malta, and Majorca; irrigation scheduling and water quality for agriculture; threat of deteriorating water quality through the application of agrochemicals and wastewater practices; water and tourism; examples of conflicts between users and possible solutions; and basic resource policies and patterns of political decision making.

Module 6: A Review of Nonconventional Technologies Aimed at Rational Water Management. Optimization of water consumption by employing appropriate and innovative technology like novel irrigation practices in agriculture, water recycling, and reuse for domestic and agricultural purposes, desalinization.

Module 7: Socioeconomic and Administrative Conditions of Water Use on Mediterranean Islands. Estimation of the value of water in its different uses, cost–benefit analysis of water projects, water consumption, public awareness, and alternative models of water management.

Module 8: Political Options, Initiatives, and Conditions for Sustainable Water Management, Specifically in the Context of the EU-WFD. Description of the EU Water Framework Directive, water pricing policies and the implications of the WFD, economic approaches in Integrated River Basin Management, and examples of integrated water management from European countries.

Results of the discussion forums as well as the lectures are published as course proceedings. In addition, a documentary film about the course and the issue of sustainable water use on Mediterranean islands for universities and passing on to stakeholders has been produced.

Student Feedback and Perspective

The student feedback showed that exposure to a wide variety of disciplines and research fields, and interacting with a group of individuals from a wide spectrum of disciplinary and cultural backgrounds were the most important contributing factors for the success of SUSTAINIS. It is expected that the SUSTAINIS course succeeded in teaching students about sustainable water use and management in the Mediterranean and in motivating them to pursue integrated regional studies in their careers.

Acknowledgments

MEDIS and SUSTAINIS were funded by the European Commission (contract numbers EVK1-CT-2001-00092 and EVK1-CT-2002-60001, respectively). We would like to thank Dr. Panagiotis Balabanis, the responsible officer at the Commission, for his support. Thanks are also extended to all partners in MEDIS, all colleagues who served as lecturers in SUSTAINIS, as well as the participants who contributed substantially to the success of the course.

BIBLIOGRAPHY

1. Commission of the European Communities. (2000). *Directive 2000/60/EC of the European Parliament and of the Council of 23 October 2000 establishing a framework for community action in the field of water policy*, Off. J. Eur. Communities **L 327**: L 327/1–L 327/72.

2. Xoplaki, E. (2002). *Climate variability over the Mediterranean*, Ph.D. thesis, University of Bern, Switzerland.

3. Bolle, H.-J. (Ed.). (2003). *Mediterranean Climate—Variability and Trends*, Vol. 1 of Series in Regional Climate Studies, H.-J. Bolle, M. Menenti, and I. Rasool (Eds.), Springer-Verlag, New York.

4. Savvides, L., Dörflinger, G., and Alexandrou, K. (2001). *The assessment of water demand of Cyprus*. Ministry of Agriculture, Natural Resources and Environment, and Food and Agriculture Organization of the United Nations report TCP/CYP/8921.

5. Lange, M.A., Cohen, S.J., and Kuhry, P. (1999). Integrated global change impacts studies in the Arctic: the role of the stakeholders. *Polar Res.* **18** (2): 389–396.

6. Houghton, J.T., Ding, Y., Griggs, D.J., Noguer, M., van der Linden, P.J., Dai, X., Maskell, K., and Johnson, C.A. (Eds.). (2001). *Climate Change 2001: The Scientific Basis. Contribution of Working Group I to the Third Assessment Report of the Intergovernmental Panel on Climate Change*, Cambridge University Press, Cambridge, UK.

7. Lange, M.A. and MEDIS Consortium. (2004). Water management on Mediterranean islands: current issues and perspectives. In *Proc. IDS Water 2004*, http://www.idswater.com/Admin/Images/Paper/Paper_47/Manfred%20Lange-Paper.pdf, 15 pp., May 10–28 2004.

8. Burnett, W.C., Bokuniewicz, H., Huettel, M., Moore, W.S., and Taniguchi, M. (2003). Groundwater and pore water inputs to the coastal zone. *Biogeochemistry* **66**: 3–33.

9. Olesen, J.E. and Bindi, M. (2002). Consequences of climate change for European agricultural productivity, land use and policy, *Eur. J. Agron.* **16**: 239–262.

MEETING WATER NEEDS IN DEVELOPING COUNTRIES WITH TRADABLE RIGHTS

Terence R. Lee
Santiago, Chile

The most serious current issue in water management is not water scarcity, but the allocation of water among competing uses and users. The issue of allocation overshadows all other aspects of water management. Unfortunately, almost everywhere, the methods used for water allocation are woefully inadequate to achieve an effective, efficient, and timely distribution of water among uses and users. This failure to allocate water efficiently is the root cause of the widespread perception that water is becoming so scarce as to amount to a crisis. Almost everywhere, water is allocated either according to tradition or by bureaucratic processes subject to political pressures. Economists have long argued that water should be treated as an economic good. If it is treated as an economic good, then its use can be allocated by creating water rights through a market.

In most developing countries, where agriculture accounts for around 80% of water use, the reallocation of the use of water is crucial if water is not to be a limiting resource for future economic growth. It has been convincingly argued that permitting trade in water rights will increase available supplies because consumption will be reduced as water is used more efficiently by all users. This will not only alleviate, but probably even reverse, what many perceive as the coming water crisis.

In any country, however, the decision to introduce a system of tradable water rights requires considering of many issues so that the proposed water market can function smoothly and equitably to meet water needs:

1. In the initial distribution of rights, consideration must be given to both the rights of existing water users and the need to limit any windfall gains that they may receive as the owners of rights.

2. Once rights are allocated, the holder must be assured that the right is clearly and securely defined and appropriately registered.

3. The establishment of individual rights must not give the holders benefits at the expense of society as a whole.

4. Whatever decisions are taken and policies adopted, the system must be as simple as possible.

5. Once a system of water rights is clearly defined and the market is used as the mechanism for water allocation, then the intervention of government should be kept to a minimum.

The approach taken in the initial distribution of water rights is very important if trading in water rights is, in the future, to be the main means for water allocation. The size of any political opposition to the introduction of a water rights market is usually proportionate to the fairness seen in the initial distribution of water rights.

PRINCIPAL ISSUES

Under free market conditions, the trading of rights would result ultimately in the same allocation of water rights regardless of their initial distribution. However, in the expected presence of significant costs in any exchange of rights, the nature of the initial distribution of rights can affect the future efficiency of the market. Information, bargaining, contracting, and enforcement are required for a market to function, and they are not without cost. The initial distribution of water rights can also affect the quantity of transactions, the equilibrium allocation of rights, and the aggregate benefits of water marketing. If the costs of trading in rights are large enough to deter trading, then exchanges will not occur. There will always be a trade-off between promoting efficiency and equity considerations in the initial distribution of water rights.

Alternative initial assignments of water rights among users, whether individuals, ethnic groups, local governments, or environmental protection agencies, will also result in entirely different future sets of bargaining relationships and potentially different patterns of transfers.

Basing the initial distribution of water rights on historic water use is the easiest and the most commonly used system. The rule that has been usually adopted is the historic record of possession of licences or permits for water use under the previous allocation system, or, where such evidence is lacking, according to other benchmarks, such as land holdings. This is usually called "grandfathering." However, it does represent a transfer of wealth, the new water rights, to current water users.

Once a market is established, water rights will be reallocated through voluntary transfers between willing buyers and sellers in the same way as other goods are traded. It must be emphasized that the social impact of market transactions will depend on the initial distribution of water rights and bargaining power. The initial distribution will determine how the benefits from owning water rights are distributed, who has the protection of the state to use water, and who must pay to obtain water rights in the future.

An issue often raised in opposition to establishing markets in water rights is the possibility of windfall gains if existing water users are favored in the initial distribution of rights. Proposed solutions to this issue include taxing transfers, prohibiting transfers, and other restrictions on trading. Such restrictions, however, defeat the purpose of assigning rights and establishing a market. It might be simpler to ignore the problem, which will facilitate the reallocation of water to higher value uses but provide windfall gains to original users. One mechanism that has been used for the initial distribution of rights to avoid windfall gains is auctions, but these can raise the problem of rights going only to those who can pay.

In the initial distribution, care needs to be taken to respect the rights of the disadvantaged, such as poor farmers and indigenous peoples. In Chile, in introducing the market allocation of water, the government put in place and maintains a program to facilitate the legalization of property titles to water rights by poorer farmers; it has been spending more than US$320,000 annually for this purpose during the last 20 years.

In the initial distribution of water rights, consideration should also be given to establishing minimum flows to protect aquatic and riparian habitats and other uses which, because of strong public goods characteristics from which everyone benefits, cannot compete in the market for water. Where historic uses have preempted the total supply, an argument can be made for a one-time reallocation of water. In Chile, although no provision of this kind is included in the law, the interpretation of the protection of the rights of third parties has been expanded to include environmental protection and ecological flows.

The main advantage of adopting "grandfathering" is that it avoids conflicts and reduces the opposition of existing water users, typically farmers, to the introduction of a market. Irrigation farmers usually argue that they are entitled to receive water rights without charge because they have already paid for the rights implicitly in the purchase price of the land. Under the auction approach, although payments for rights do not represent real economic costs to society as a whole but merely transfers from one group to another, to the users, the payments constitute a financial burden.

THE MARKET ALLOCATION OF WATER

A water market is a water management tool. It is, moreover, a tool that spreads the burden and difficulties of water management among a larger population; it permits greater popular participation in management decisions and can introduce greater flexibility into management systems. The creation of a water market will also demand new skills and attitudes from public administration, judicial systems, and water users, as well as investment in the registration of rights, monitoring and measurement systems and, possibly, in improving water distribution and transportation systems.

A water market will allocate water rights at a price set by the free exchange of the right either for a limited period of time through a lease or in perpetuity by a sale. For a market to work, it is necessary only that there be a tradable margin in water rights, even if the number of rights traded is marginal to the total supply. Water markets can normally be expected to be relatively small or "thin," with few transactions.

Any water allocation system should be both flexible and secure. At the same time, all costs and benefits should be reflected in the decisions that participants make; otherwise, these decisions will be inefficient from an overall social perspective. The system must also be predictable, equable, and fair. This is particularly important in developing societies where economic growth and water use efficiency require achieving a balance in the allocation of water between flexibility and security. Water markets are flexible because markets are by their very nature a decentralized and incentive-oriented institution, rather than centralized and regulatory.

SECURITY OF TENURE

Markets require security of ownership to function. Therefore, it is important that rights be clearly defined and publicly registered. Security of ownership of the right, in turn, helps encourage efficient use, resource conservation, and capital investment. It also helps strengthen and consolidate the autonomy of water user organizations. Security of ownership and the possibility of acquiring water rights in the market, it has been shown, encourage investment and growth in activities that require secure water supplies. Moreover, the reallocation of rights by voluntary exchanges allows market systems to defuse potential political conflicts over water allocation.

The political dimension of a system of secure and transferable water rights arises from the definition and clarification of property rights, which the introduction of such a system requires. The market transfer of water rights reduces political conflicts, as these transfers are always voluntary transactions in which owners of rights will participate only if they believe that it is in their best interest, given the alternative opportunities available. Administrative allocation, in contrast, often generates intense conflicts because granting a water right to one user necessarily precludes another.

Markets can, therefore, reduce conflicts among environmental interests, water suppliers, and polluters by providing natural economic incentives for water conservation and wastewater treatment. Environmental economists often prefer property-right systems to pricing systems because property-right systems can be ecologically more dependable than pricing systems.

WATER MARKETS IN CHILE

Chile is the only country where tradable rights are universal and all water allocation is the result of market transactions. The introduction of water markets in Chile coincided with a major increase in agricultural production and productivity. It is reasonable to conclude that the introduction of tradable, and particularly secure, property rights in water made a noticeable contribution to this overall growth in the value of Chile's agricultural production. This increase occurred within an agricultural sector largely dependent on irrigation, without a significant increase in either the supply of water or the area under irrigation. There has been, however, considerable private investment in improving the existing irrigation infrastructure both on and off the farm. The influence of water markets, however, cannot be fully separated from the effects of other economic factors, especially stable economic policies, trade liberalization, and secure land rights. This notwithstanding, it is recognized that trading in water rights reduced the need for new hydraulic infrastructure, improved overall irrigation efficiency, and has reduced the number of conflicts over water allocation. It also appears to have facilitated the shift from low-value, water-intensive crops to higher value, less-water-intensive crops.

In addition, market transfers of water rights have produced substantial economic gains from trade in some river basins. These gains occur both in trades between farmers and in trades between farmers and other sectors. The economic gains from trade tend to be large between farmers but relatively modest in intersector trade because water has been transferred from profitable farmers to urban drinking water supply; even though the financial gain to the farmer who sells is large, the overall economic gains of the reallocation are relatively small, because both uses have similar social benefits. In Chile, as in other areas of irrigation agriculture, if, prior to the sale, water was not being taken from river or canal by the owner of the right, other farmers would have used it downstream. Transfers among farmers, however, lead to using the water for higher value crops.

READING LIST

Anderson, T.L. and Snyder, P. (1997). *Water Markets*. Cato Institute, Washington, DC.

Holden, P. and Thobani, M. (1995). Tradable water rights: a property rights approach to improving water use and promoting investment. *Cuadernos de Economía* **32**(97): 263–316.

Frederick, K.D. (1993). *Balancing Water Demands with Supplies. The Role of Management in a World of Increasing Scarcity*. Technical Paper No. 189, The World Bank, Washington, DC.

Lee, T. and Jouravlev, A. (1998). *Prices, Property and Markets in Water Allocation*. United Nations, Economic Commission for Latin America and the Caribbean, Santiago.

Lee, T. (2000). *Water Management in the 21st Century: The Allocation Imperative*. Elgar, Cheltenham.

Peña, H. (1996). Water markets in Chile: what they are, how they have worked and what needs to be done to strengthen them. *Paper Presented at the Fourth Annual World Bank Conference on Environmentally Sustainable Development*. Washington, DC, pp. 25–27.

Rosegrant, M.W. and Binswanger, H.P. (1994). Markets in tradable water rights: potential for efficiency gains in developing country water resource allocation. *World Dev.* **22**(11): 1613–1625.

WATER USE IN THE UNITED STATES

Frédéric Lasserre
Université Laval
Ste-Foy, Québec, Canada

A STABILIZATION IN TOTAL WITHDRAWALS?

Total water withdrawals in the United States are estimated at 402 billion gallons per day (Bgal/day) in 1995. This is 1.5% less than withdrawals for 1990 and confirms an overall stabilization of withdrawals throughout the country since 1975. That was the first time that withdrawals growth stopped outpacing population increases, implying an intensification of the use of water withdrawn. All major users, thermoelectric, irrigation, and industrial uses, saw their withdrawals stabilize; rural (domestic and livestock) and urban use, on the contrary, witnessed an ongoing trend of withdrawal increase.

The decline in overall water use does not translate into the same pattern for all water sources (Table 1). Saline surface water use grew rapidly after 1965, to reach 71 Bgal/d in 1980, but has declined since then to 59.7 Bgal/d; groundwater use also declined after the 1980 peak. After a similar trend from 1980 to 1990, surface freshwater use, however, started growing again by 1.9% between 1990 and 1995.

On the other hand, after a decrease between 1980 and 1985 and a small increase from 1985 to 1990, consumptive use was back to growing faster than that of population. This trend could be linked to increased urban water use in suburbs. Particularly as urban sprawl takes place in the West, that implies a large share of consumption, such as lawn watering and pools, as will be seen below.

Table 1. Evolution of Water Use in the United States, 1950–1995[a]

	1950	1955	1960	1965	1970	1975	1980	1985	1990	1995
Population, in millions	150.7	164	179.3	193.8	205.9	216.4	229.6	242.4	252.3	267.1
Variation, %		8.8	9.3	8.1	6.2	5.1	6.1	5.6	4.1	5.9
Total withdrawals, Bgal/d	180	240	270	310	370	420	440	399	408	402
Variation, %		33.3	12.5	14.8	19.4	13.5	4.8	−9.3	2.3	−1.5
Of which:										
Thermoelectric	40	72	100	130	170	200	210	187	195	190
Other industrial	37	39	38	46	47	45	45	30.5	29.9	29.1
Irrigation	89	110	110	120	130	140	150	137	137	134
Rural domestic and livestock	3.6	3.6	3.6	4	4.5	4.9	5.6	7.8	7.9	8.9
Public supply	14	17	21	24	27	29	34	36.5	38.5	40.2
Source of water										
Ground freshwater	34	47	50	60	68	82	83	73.2	79.4	76.4
Surface, fresh	140	180	190	210	250	260	290	265	259	264
Surface, saline	10	18	31	43	53	69	71	59.6	68.2	59.7
Consumptive[b] use	NA	NA	61	77	87	96	100	92.3	94	100
Variation, %				26.2	13	10.3	4.2	−7.7	1.8	6.4

[a] Water-related data in billion gallons per day, Bgal/d.
[b] Consumptive use as of 1970: freshwater only.
Source: Adapted from Reference 1.

Table 2. Total Water Withdrawals by Water-Use Category Among Selected States, 1995, in Million Gallons per Day

State	Public, Domestic, and Commercial	Agriculture	Industrial, Incl. Mining	Thermoelectric
Alabama	880	268	753	5200
Arizona	867	5702	197	62
Arkansas	519	6294	187	1770
California	6125	29359	801	9655
Colorado	741	12759	192	114
Connecticut	475	29	12	3940
Delaware	104	52	64	1326
Florida	2417	3526	649	11636
Georgia	1295	770	677	3073
Idaho	560	14460	76	0
Illinois	2053	236	2407	17100
Kansas	399	3489	77	1270
Maine	146	29	16	135
Maryland	931	97	331	6360
Michigan	1535	241	1913	8370
Montana	161	8602	80	22
New Jersey	1144	127	486	4360
New York	3344	64	320	13060
Ohio	1628	54	650	8190
Texas	3464	9765	2916	13460
Utah	511	3638	252	55
Vermont	92	9	12	453

Source: Adapted from Reference 1.

THE DRY WEST STILL THE LARGEST USER AND CONSUMER

Regional Water Use

Wide discrepancies show up immediately when the structure of water use in the United States is examined.

Patterns do appear: they highlight very industrialized and urbanized states such as New Jersey, New York, and Michigan, where water used is concentrated in domestic, industrial, and power generating uses (Table 2). States that have rain-fed agriculture and a small urban population, such as Maine and Vermont, use water mainly for their power and their cities. In highly urbanized and industrialized Texas and California, all uses are strong, except industrial use in California because of an industrial base that does not use a large amount of water; other Western States do show a large dominance of agricultural use, such as in Arizona, Colorado, Idaho, and Montana.

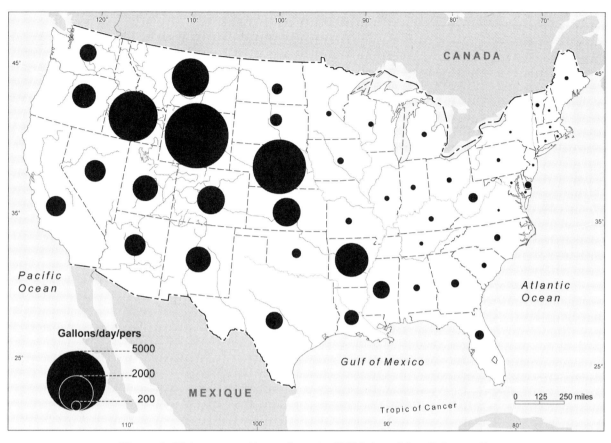

Figure 1. Water consumptive use by state, 1995 (adapted from Reference 2).

Consumptive Use

Regionally, states in the West do show much greater per capita consumptive use, up to 4288 gallons per day per person (gal/d/p) for Nebraska, but it is not the urban use that is responsible for their high consumption (Table 3). Highly urbanized California displays a more moderate figure of 795 gal/d/p (Fig. 1). Large withdrawals for irrigation are required in a dry climate.

AMERICAN WATER USE STRUCTURE COMPARED TO OTHER COUNTRIES

Agriculture remains the main user and consumer of water in the United States, as it is, by a large margin, among developing countries (Table 4). The picture is somewhat subtler among developed countries. Drier countries such as Spain and Greece, unsurprisingly, use a large share of water for agriculture, but water can also be largely used in more humid places that have traditionally rain-fed agriculture, such as Belgium, Germany, and Denmark (Table 5). This trend is explained, as observed in the United States, by farmers turning to irrigation to avoid all risk of yield decrease.

WATER USE MANAGEMENT : RECONCILING DIFFERENT USE STRUCTURES ACROSS THE COUNTRY

The geography of withdrawals and consumption, illustrated by the consumptive use map shown above, is all the harder to manage globally as Western states consume more and more (states west of the 100th meridian accounted for 75.1% of total consumption in 1995), but draw on a smaller share of the resource; water is far more plentiful in the eastern part of the country. From the end of the Civil War, this relative scarcity of water never prevented development schemes and led to large engineering designs that were built to harness the West's major rivers and make up for nature's pattern of groundwater, surface water, and precipitation distribution.

On the other hand, urbanized states in the eastern United States display small consumptive use structures, as low as 19 gal/d/p for Rhode Island and 26 for New York. It is the sheer size of the urban area that is responsible for New York's freshwater supply, not the profligacy of its inhabitants. However, New York and

Table 3. Consumptive Use in Selected States, in gal/d/p, 1995

Arizona	California	Colorado	Florida	Idaho
908	795	1396	196	3732
Illinois	Maine	New York	Ohio	Rhode Island
72	39	26	71	19

Source: Adapted from Reference 1.

Table 4. Share of Withdrawals in Selected Countries, by Economic Sector

Country	Year	Withdrawals, Share of Total, %		
		Agriculture	Industry	Domestic and Commercial
Developing countries				
India	1992	91	4	5
Uzbekistan	1994	94	2	4
Thailand	1992	90	6	4
Tanzania	1994	89	2	9
Vietnam	1992	78	9	13
China	1993	77	18	5
Developed countries				
Greece	1990	63	29	8
Spain	1994	62	26	12
Italy	1990	53	33	14
Portugal	1990	48	37	15
Denmark	1990	43	27	30
United States	1995	35	54	11
Germany	1990	18	68	14
France	1994	15	69	16
Ireland	1990	10	74	16
Norway	1985	8	72	20
Austria	1991	8	73	19
Belgium	1990	4	85	11
United Kingdom	1994	3	77	20

Source: Reference 3, pp. 208–211.

Table 5. Comparison of Withdrawals and Consumption by Main Sector Between France and the United States, 1995

In % of Total	Domestic/Commercial	Industrial	Thermoelectric	Agriculture
United States:				
Withdrawals	12.2	8.2	38.7	40.9
Consumption	8	4.1	3.3	84.6
France:				
Withdrawals	15	10	62.5	12.5
Consumption	24	5	3	68

Source: Reference 4, p. 102.

Los Angeles authorities, for instance, have aggressively promoted water-saving programs with effective results.

Local authorities are acutely conscious of serious water problems in the country. These are twofold. First, although withdrawals are on a stabilizing trend, and consumption was up until a few years ago, withdrawal levels are often locally unsustainable in the long term, especially in the West. Aquifers are being overpumped. The now famous Ogallala aquifer, in the Midwest, is expected to run unusable—not totally dry—in 50 to 70 years at present withdrawal rates. California saw its water consumption increase by 41.5% between 1980 and 2000, still untapped available resources are dwindling rapidly. As other upstream states withdrew small amounts of water for small populations, even Colorado water that California long took for granted is now being reallocated to the "New West" states under the Colorado Compact provisions because Nevada, Arizona, and New Mexico boast extremely fast population growth (Table 6). Between 1980 and 2000, consumptive uses also grew by 42% in Texas, 58.2% in Florida, and 70% in Arizona. Groundwater withdrawals in California exceed its aquifer renewal capacity by 15% (5).

Second, the American society changes, and its geographical setting and consumption structures evolve. The

Table 6. Population of Selected States, 1970–2000, in Millions

	1970	1980	1990	2000	Average Annual Growth, 1970–2000, %
Florida	6.79	9.75	12.94	16	+2.9
California	19.95	23.67	29.76	33.9	+1.8
Arizona	1.77	2.72	3.67	5.13	+3.6
Nevada	0.49	0.8	1.2	2	+4.8
Colorado	2.21	2.89	3.3	4.3	+2.2
New Mexico	1.02	1.3	1.52	1.82	+1.9
Utah	1.06	1.46	1.72	2.23	+2.5

Source: Adapted from Reference 6. Average population growth in the United States 1970–2000: +1.1%.

Table 7. Reclaimed Wastewater Use, 1995, in Gallons/Days/Person, Top 10 States

State	Arizona	California	Colorado	Florida	Maryland
Reclaimed water	42.7	10.4	2.9	16.7	13.9
State	Missouri	Nevada	Texas	Utah	Wyoming
Reclaimed water	2.1	15.7	5.8	7.2	19

Source: Reference 8.

Table 8. Desalination Capacity by Country, January 1999

Country	Total Capacity, Mgal/day	Total Capacity,[a] Mm^3/day	Country	Total Capacity, Mgal/day	Total Capacity,[a] Mm^3/day
Saudi Arabia	1.344	5.1	Japan	204.7	0.78
United States	851	3.2	Libya	185	0.7
United Arab Emirates	574.7	2.18	Qatar	149	0.57
Kuwait	338.4	1.29	Italy	137.1	0.52
Spain	210	0.8	Iran	115.3	0.44

[a] 1 m^3 = 263.16 gal.
Source: Reference 3, p. 288.

past 20 years have witnessed several phenomena that will question the water use structure: the acceleration of suburban growth, especially in the West; and the persistence of population shifts toward the Sun Belt in the West, Texas, and Florida; and more particularly toward what some geographers call the "New West," interior states such as Nevada, Arizona, Colorado, and New Mexico, where water resources are even more delicate to manage than in California.

This fast increasing population fuels the suburban sprawl that poses a great challenge to water management and distribution authorities; the newcomers are quick to adopt a new lifestyle and behavior that embodies, in their eyes, economic and societal achievement—green lawns, pools, frequent car washes. As a whole, Americans consume about 40 gal/d/p; in the Southwest, per capita daily consumption averages 120 gallons. In cities such as Phoenix and Las Vegas, residents consume more than 300 gal/d/p. In Denver, more than half of the water consumed is attributable to outdoor landscaping (7).

CONSUMING LESS

Water saving measures have thus been enforced by local authorities in various ways. Higher water pricing, for instance, led to a sharp rationalization of industrial water use, water had become a substantial cost for large water users, although trends in industrial restructuring led to a decline in water-intense processes. The same basic economic mechanism is also expected to prevent citizens from using too much water.

Total irrigation withdrawals decreased since 1980 to stabilize around 136 Bgal/d, although risk-averse farmers are more prone to irrigating their fields for fear of a decrease in their yield, even in more humid areas (4, p. 86); irrigated acreage remained stable at 58 million acres since 1980. This stabilization does not mean that there is no room for improvement, especially in the West, where extensive irrigation still often uses flood irrigation. In Idaho, porous soils require constant watering, up to 12 feet of water an acre per season.

This decrease in withdrawals stems from pricing, though it remains very low for farmers compared to urban and industrial users. This discrepancy fuels "water wars" in the West, as urban communities are confronting the first appropriation rights of rural irrigation districts to increase their share of a relatively scarce resource. It also stems from the idea the main resource would become too scarce. These two factors drove farmers to adopt improved irrigation systems and techniques.

DEVELOPING NEW SOURCES

Using and consuming less is an option more and more favored in the United States due to the transition from water-supply management to water-demand management. This shift in focus is attested to by changes in pricing policies as well as in recurring awareness campaigns targeted at citizens, industry managers, and farmers. But increasing supply is not a forgotten option. Grand schemes of damming Canada's untamed rivers and diverting them toward the thirsty Midwest were part of some engineering firms' and government agendas, notably in Québec, up until 1986; large-scale water exports were again contemplated in Canada from 1997 to 2001 but are now banned by a federal law. These projects no longer have the favor of an environmentally aware public, especially in a decade that saw large dams become the focus of environmentalists' anger and campaigning in the United States and throughout the Western world and even against some large projects in developing countries, such as the Narmada Dam project in India.

Reclaimed Wastewater

Reclaimed wastewater, for instance, was long thought of as irrelevant because of cost and safety issues. In 1995, total wastewater use amounted to 1.02 Bgal/d, a 36% increase since 1990 (Table 7). Water recycling is mainly a feature of the dry West. Florida, confronted with acute freshwater supply problems, has invested in extensive reuse programs. Arizona, Wyoming, and Florida appear to be the most effective states in recycling wastewater.

Desalting

Desalting also used to be considered an irrelevant solution because it produced very expensive water. During the last 15 years, improved technological processes have led to a fast decrease in the cost per gallon of water produced, although only urban or industrial users can afford it. Local governments are keenly interested in this technology, as attested to by recent major contracts in Florida, where a plant of 95,000 m^3/ day (25 million gal/d) approved by regional water officials in March 1999, is to be located near Tampa Bay. The water produced should cost between 45 and 55 ¢/m^3, or $1.71 to $2.08/1000 gal.

Desalination is proving very popular in the United States, where built capacity in early 1999 ranked second in the world (Table 8).

BIBLIOGRAPHY

1. USGS. (1995). *Water Use in the United States*.
2. USGS data. (2001).
3. Gleick, P. (Ed.). (2000). *The World's Water 2000–2001*. Island Press, Washington, DC, pp. 208–211, 288.
4. Lasserre, F. (2002). L'eau rare? Des solutions pour assurer l'approvisionnement. In: Lasserre and Descroix, *Eaux et territoires: tensions, coopérations et géopolitique de l'eau*. Presses Universitaires du Québec, Québec, p. 102.
5. Mayrand, K. (1999). *Les marchés internationaux de l'eau: exportations d'eau douce et marché des infrastructures et des services urbains*. Ministry of International Relations, Québec, pp. 8–9.
6. US Census Bureau. (2001).
7. Riebsame, W.E. (Ed.). (1997). Water for the west. *Atlas of the New West*. WW Norton & Company, New York, p. 81.
8. USGS. (1995). *Offstream Use*.

HOW WE USE WATER IN THESE UNITED STATES

U.S. Environmental Protection Agency

Water use is usually defined and measured in terms of withdrawal or consumption—that which is taken and that which is used up. Withdrawal refers to water extracted from surface or ground water sources, with consumption being that part of a withdrawal that is ultimately used and removed from the immediate water environment whether

This article is a US Government work and, as such, is in the public domain in the United States of America.

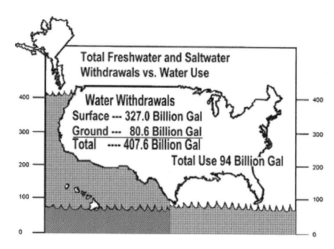

Figure 1. Total daily offstream water use in the United States.

by evaporation, transpiration, incorporation into crops or a product, or other consumption. Conversely, return flow is the portion of a withdrawal that is actually not consumed but is instead returned to a surface or ground water source from a point of use and becomes available for further use.

Water use can also be divided into offstream and instream uses. Offstream water use (see Fig. 1) involves the withdrawal or diversion of water from a surface or ground water source for

- Domestic and residential uses
- Industrial uses
- Agricultural uses
- Energy development uses

Instream water uses are those that do not require a diversion or withdrawal from the surface or ground water sources, such as:

- Water quality and habitat improvement
- Recreation
- Navigation
- Fish propagation
- Hydroelectric power production

NATIONAL TRENDS IN WATER USE

National patterns of water use indicate that the largest demand for water withdrawals (fresh and saline) is for thermoelectric generation (47 percent), followed by irrigation (34 percent), public supply (9 percent), industrial (6 percent), mining (1 percent), livestock (1 percent), domestic (1 percent), and commercial uses (1 percent). While thermoelectric generation represents the largest demand for fresh and saline withdrawals, irrigation represents the largest demand for freshwater withdrawal

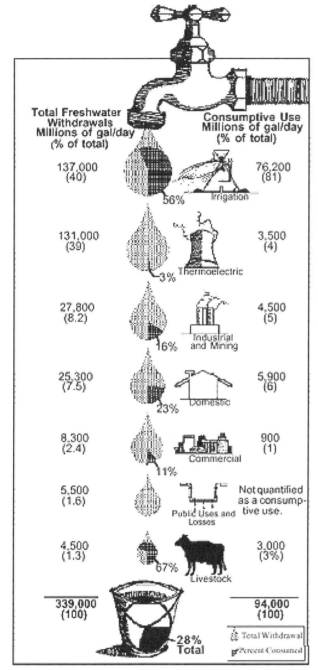

Figure 2. Comparison of freshwater consumptive use in the United States for 1990 by category.

alone (see Fig. 1). Activities that reduce the need to withdraw surface and ground water will lead to many of the beneficial effects of conserving water.

NATIONAL CONSUMPTION PATTERNS

Water consumption varies by water use category, with irrigation consuming the highest percent (81 percent) and commercial the lowest (1 percent) (see Fig. 2). The difference between the volume of water withdrawn and that consumed is the return flow. As more good-quality water is available in return flows, more water is available for other beneficial uses.

Some categories of water use, such as irrigation and livestock watering, consume a high percentage of water that is withdrawn from surface and ground water sources. Thus, less water is available for return flows from these high-consumption activities. Other categories of use like thermoelectric power consume only a small fraction of the water they withdraw.

CATEGORIES OF WATER USE

With several different ways to categorize water use in the United States, this article separates offstream uses into

- Municipal/public supply
- Domestic and commercial
- Industrial and mining
- Agricultural
- Thermoelectric power

MUNICIPAL/PUBLIC WATER SUPPLY

While water withdrawals for public use can be applied to street cleaning, fire fighting, municipal parks, and public swimming pools, keep in mind that municipalities and private suppliers might also provide water for other purposes (domestic/commercial, agricultural, thermoelectric power).

Per capita (per person) use of public water supplies in the United States (1990) averaged 183 gallons per day (gal/d). The average per capita use can vary greatly between communities for any number of reasons, including, but not limited to:

- Climate differences
- The mix of domestic, commercial, and industrial uses
- Household sizes
- Lot sizes
- Public uses
- Income brackets
- Age and condition of distribution system

For instance, per capita use of public water is about 50 percent higher in the West than the East mostly due to the amount of landscape irrigation in the West (see Fig. 3). However, per capita use can also vary greatly within a single state. For example, in 1985 the demand for municipal water in Ancho, New Mexico, totaled 54 gallons per capita per day (gal/cap/day) while in Tyrone, New Mexico, municipal demand topped off at 423 gal/cap/day. Rural areas typically consume less water for domestic purposes than larger towns.

In 1990, water withdrawn nationwide for public supplies totaled 38,530 million gallons per day (Mgal/d) (see Table 1). Although this withdrawal rate represents a 5 percent increase over 1985 amounts, the number of people supplied with water distributed through public systems also increased 5 percent during that same 5-year period. Again in 1990, surface water supplied about

Figure 3. Average use per person (gal/day) of public water in the United States by USGS water region.

Table 1. Fate of Water in Public Water Supplies of the United States, 1990

Receiving Category	Volume, Mgal/day	Percentage of Total
Domestic	21,900	57
Commercial	5,900	15
Public Use Losses	5,460	14
Industrial	5,190	13
ThermoelectricPower	80	<1
Total	**38,530**	**100**

61 percent of the public water supply, with ground water supplying the other 39 percent.

Of the total water withdrawn in 1990 for public supplies representing 11 percent of total U.S. offstream freshwater withdrawals 72 percent went to domestic and commercial uses, 13 percent to industrial uses, and 0.2 percent to thermoelectric power. The remaining 14 percent went to public uses such as fire protection or was lost during distribution (usually due to leaks).

DOMESTIC/COMMERCIAL

Domestic water use includes everyday uses that take place in residential homes, whereas commercial water uses are those which take place in office buildings, hotels, restaurants, civilian and military institutions, public and private golf courses, and other nonindustrial commercial facilities. Combined freshwater withdrawals for domestic and commercial use in 1990 totaled 33,600 Mgal/d, or 10 percent of total freshwater withdrawals for all offstream categories (see Fig. 2).

Typical categories of residential water use include normal household uses such as

- Drinking and cooking
- Bathing
- Toilet flushing
- Washing clothes and dishes
- Watering lawns and gardens
- Maintaining swimming pools
- Washing cars

When divided into indoor uses and outdoor uses, the amount of indoor water use remains fairly constant throughout the year, with the breakdown of typical indoor water uses depicted in Fig. 4. By far the largest percentage of indoor water use occurs in the bathroom, with 41 percent used for toilet flushing and 33 percent for bathing.

Outdoor residential water use, however, varies greatly depending on geographic location and season. On an annual average basis, outdoor water use in the arid West and Southwest is much greater than that in the East or Midwest. Figure 5 compares the national average for residential outdoor water use with that of Pennsylvania and California, with landscape irrigation the primary application. While average outdoor water use in Pennsylvania represents only approximately 7 percent of the total residential demand, in California average outdoor use climbs to about 44 percent of the demand.

INDUSTRIAL AND MINING

Industrial water uses, estimated to be 8 percent of total freshwater use for all offstream categories, include cooling in factories and washing and rinsing in manufacturing processes. Some of the major water-use industries include mining, steel, paper and associated products, and chemicals and associated products.

Water for both industrial and mining uses comes from public supplies, surface sources, and ground water. During the 5-year span from 1985 to 1990, industrial water use in the United States decreased approximately 13 percent. In the same period, mining water use increased about 24 percent.

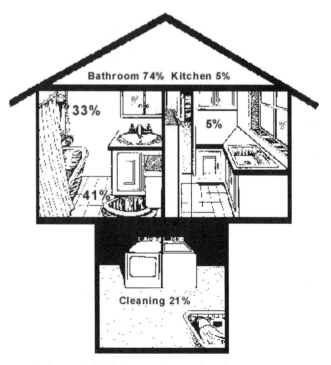

Figure 4. Typical breakdown of interior water use.

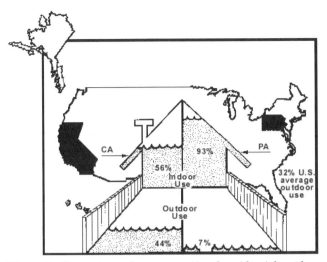

Figure 5. Comparison of average national residential outdoor water use with that of Pennsylvania and California.

AGRICULTURAL

Agricultural water use can be divided between irrigation and livestock. Irrigation includes all water applied to farm or horticultural crops; livestock incorporates water used for livestock, dairies, feedlots, fish farms, and other farm needs.

Estimated annual water use for irrigation remained at about the same level between 1985 and 1990, with approximately 63 percent of the water used for irrigation in 1990 coming from surface water. Approximately 60 percent of the water used for livestock came from ground water sources and the remaining 40 percent from surface water sources. Combined water use for irrigation and livestock represents about 41 percent of total offstream freshwater use for 1990, (see previous Fig. 1) with 40 percent going to irrigation and the lone 1 percent to livestock uses.

Not only can the loss of water from irrigation conveyance systems be significant, but the percentage of consumptive water use for agriculture is high as well an estimated 54 percent consumption in 1985. By 1990 this had climbed to an estimated 56 percent consumption for irrigation uses and 67 percent for livestock uses (see Fig. 2).

THERMOELECTRIC POWER GENERATION

This final category includes water used for the production of energy from fossil fuels, nuclear energy, or geothermal energy. Most water withdrawn for thermoelectric power production is used for condenser and reactor cooling. While 1990 estimates of freshwater withdrawals remained constant from 1985, nearly half again as much saline water was also used.

More than 99 percent of the water used for thermoelectric power production comes from self-supplied surface water, less than 0.2 percent from public supplies. In 1990, water used for thermoelectric power production represented close to 39 percent of total offstream freshwater use in the United States, but only about 3 percent was consumed.

The Mid-Atlantic, South Atlantic Gulf, Ohio, and Great Lakes water resource regions use the largest amounts of water for thermoelectric production. The eastern United States uses about five times more water than the West to produce about twice as much thermoelectric power.

VALUING WATER RESOURCES

Diane Dupont
Steven J. Renzetti
Brock University
St. Catharines, Ontario,
Canada

INTRODUCTION

Suppose that consideration is being given to dam a river. The dam will provide benefits in the form of hydroelectric power, water for irrigation, and protection from floods. Conversely, building the dam will also destroy wetlands, fish spawning grounds, and impair recreational opportunities. In deciding whether to build the dam, one important type of information is the value to society of the benefits and damages that will arise. Obtaining estimates of these values requires an understanding of the services that society derives from water, how those services will change with the building of the dam, and the value that society attaches to each service. It is the last of these tasks that this article addresses. Specifically, the purpose of this article is to review briefly how economists estimate the value of water. Because of the importance of water and the increasing costs of maintaining clean water, there has been a large amount of research into this topic (1–5). This article reviews the methods developed by economists to estimate water's value in the next section. Empirical estimates of the various uses of water are discussed later.

THE ECONOMIC PERSPECTIVE ON VALUATION

The starting point for most economic methods for measuring the value of water is an understanding of the structure of household tastes or preferences and firms' productive technologies. This understanding can then be used to develop measures of the benefits of water use and the change in household welfare or firm profits that arise from changes in the price, quantity, or quality of water. These measures, in turn, allow the analyst to determine the maximum willingness to pay (in the case of a change leading to an improvement in welfare) for water. This idea of willingness to pay forms the cornerstone of economic measures of value. (This approach to understanding and measuring value is not without its critics; see the discussion in Reference 1.)

These measures provide estimates of the value of water to the individual user (i.e., the "private" value). This may differ from the social value of water use (the value to all of society) by an individual household, firm, or farm.

There are a number of reasons that social and private values of water use may differ. First, an individual's water use may result in lowered water quality from pollution, and this, in turn, would lower the value of that water for other users and, thus, to society. Second, a farm may use water to increase production of crops that are supported by government subsidies. In this case, the farmer's valuation of the increased output will be higher than society's because of the cost of the subsidies to society.

Another issue complicating the measurement of the value of water is that, like many environmental goods, the value of some of the uses to which water is put are reflected in market transactions and others are not. An example of the former is a farmer purchasing water to irrigate crops—an analyst may be able to infer the value of water to the farmer by using market information on the increase in crop yield brought on by irrigation. An example of the latter is a family enjoying a day of swimming at a municipal beach. In this case, analysts must use "nonmarket" methods to estimate the family's valuation of its recreation.

As Table 1 indicates, economic valuation methods may be categorized into indirect and direct methods. Indirect valuation methods are those that infer a household or firm's valuation of water from its observed market behavior. There are a number of indirect valuation techniques. The derived demand method applies statistical or mathematical programming techniques to observations on market transactions in water to estimate a household's or firm's demand for water. The estimated demand relationship can, in turn, be used to calculate the user's willingness to pay for water. The avoided cost method assesses the value of water by examining the costs that are avoided because water is available. (A related method is the averting behavior method. If households purchase water filters to remove waterborne impurities, then the value to the household of removing the impurities can be estimated by the cost of water filtration.) For example, if a firm is able to release effluents into a lake rather than treating them, then the avoided cost of treatment provides an estimate of the value of the lake's assimilative capacity to the firm. The residual imputation method calculates the value of water to a firm or farm as the difference between the operation's revenues and the cost of all of its nonwater inputs. The hedonic method uses statistical techniques to determine the contribution of water to the total value of marketed commodities. For example, a house of a given size is usually more valuable, the closer it is to a body of water, and hedonic methods can estimate the increase in value that is related to the proximity to water. Finally, the travel cost method is used to estimate households' valuation of water-related recreation experiences by examining the time traveled to a recreational site and out-of-pocket expenses incurred by a family. A greater time traveled implicitly raises the cost of recreation because it usually means more time away from work.

Direct valuation methods are those that survey households to obtain their valuations. The most important direct valuation technique is contingent valuation. In this approach, individuals are presented with information concerning a hypothetical or constructed market and asked to indicate their willingness to pay to achieve a desired good or service. For example, a respondent might be asked to value a potential improvement in water treatment that reduces the risk of illness. This method is valuable for assessing projects before they are undertaken and for assessing households' "nonuse" values of water. For example, some households may be willing to pay a positive amount to ensure access to water some time in the future (6).

EMPIRICAL ESTIMATES OF THE VALUE OF WATER

The preceding section demonstrates that economic researchers have developed the conceptual framework and measurement methods for estimating the value of water. This section briefly points to a number of examples of the results of estimation studies. (The interested reader

Table 1. Economic Valuation Methods

Method	Applications	Comments
Indirect		
Derived demand	Residential, industrial	Estimated demand curve yields user's valuation of water
Avoided cost	Residential, industrial	Access to clean water allows firms and homes to avoid costs such as water filters
Residual imputation	Agricultural, industrial	Requires information on revenues and costs of nonwater inputs
Hedonics	Residential, recreation	Good for assessing the impacts of changes in water quality
Travel cost	Recreation	One of the earliest methods—in use since the 1960s
Direct		
Contingent valuation	Recreation, changes in water quality, ecosystems	Only method that can estimate nonuse values (values that arise from an aesthetic appreciation of water or a desire to hold an option to use water in the future)

Table 2. Representative Valuation Studies

Author	Sector/Use	Method	Estimate/Year
Indirect			
Moore (8)	Agriculture	Derived demand	$42–70/acre-foot (1989 US$)
Young and Gray (9)	Industrial waste disposal	Avoided cost	$0.07–1.28/acre-foot (1972 US$)
Gibbons (2)	Industrial water intake	Avoided cost	$6–78/acre-foot (1980 US$).
Abdalla (10)	Residential	Avoided cost	$125–330/household (1987 US$)
Smith, Desvousges and McGivney (11)	Recreation	Travel cost	$10–20/household (1977 US$)
Direct			
Whitehead and Blomquist (12)	Wetland preservation	Contingent valuation	$6–21/household (1992 CAN$).
Carson and Mitchell (13)	Recreation	Contingent valuation	$70–90/household (1990 US$)

should consult the valuable compilation of empirical estimates of water's value in Reference 7.) The results of a number of representative studies are reported in Table 2 and are discussed briefly below.

Indirect Valuation Studies

Moore (8) compares estimates of the value of irrigation water using the derived demand and residual imputation methods. The author finds that the two methods yield quite different results (the derived demand method provides higher estimated values). This is an important finding as the price paid by some American farmers for irrigation water is based, in part, on their estimated valuation of that water.

Young and Gray (9) and Gibbons (2) employ the avoided cost approach to value industrial water use. In both cases, the authors calculate the costs that firms are able to avoid by being able to dispose of effluents in waterbodies and by using increased water intake rather than in-plant water recirculation. Young and Gray find that the value of water used in industrial waste disposal is relatively low as few costs are actually avoided. Conversely, Gibbons finds a wider range of values for increased intake water substituting for water recirculation because, in some applications, in-plant recirculation must also address reduced water quality.

Abdalla (10) examines household valuation of improvements in the quality of their drinking water supplies by measuring the cost savings of those improvements. These cost savings arise from decreased purchases of water filters and bottled water and from less time needed to boil water. The authors also find that households with children place a higher value on water quality improvements than those without.

Smith et al. (11) address the change in the valuation of sport fishing from changes in water quality at specific sites. The authors do this by investigating the role of water quality in approximately 20 published studies of water-based recreation that use the travel cost method. The principal finding of the Smith et al. study is that increases in water quality increase the demand for, and valuation of, water-based recreation. For example, when applied to data derived from users of the Monongahela River, the average consumer surplus (a measure of willingness to pay net of fishing expenses) for water quality improvements from boatable to game fishing conditions was $9.96 per household per season (1977 US$).

Direct Valuation Studies

One of the most important advantages of direct valuation methods is that they can be employed to consider household 'nonuse' values of water. These values arise not from household use of water but rather because households may have an interest in knowing that a particular waterbody has been preserved (possibly for their use in the future). For example, Whitehead and Blomquist (12) employ a contingent valuation survey to examine household willingness to pay for wetlands preservation and the role that information provision plays in determining that value. The authors find that, while many households are willing to contribute toward the preservation of wetlands, as more information is provided about the availability of substitutes for an endangered wetland, respondents' willingness to pay for preservation falls off.

Finally, Carson and Mitchell (13) report on an influential study of American household valuation of enhanced water-based recreational opportunities that could arise from improved water quality. A particularly important feature of this study's methodology is the contingent valuation survey that employs the water quality "ladder" to set out for respondents the various levels of water quality that they are asked to consider. In doing so, the authors frame or present differ levels of water quality ("boatable, fishable, and swimmable", p. 2447) in a way that is comprehensible for most respondents. The results indicate that the sample of households places significant value on improving water quality.

CONCLUSION

The growing scarcity of water creates the need to develop and implement rational methods for allocating water across competing needs. Part of this process of decision-making requires understanding the value that individual households, firms, and farms obtain from their various uses of water. This article has briefly outlined the methods developed by economists to estimate those values.

BIBLIOGRAPHY

1. Donahue, J. and Johnston, B. (Eds.). (1998). *Water, Culture and Power: Local Struggles in a Global Context*. Island Press, Washington, DC.
2. Gibbons, D. (1986). *The Economic Value of Water*. Resources for the Future, Baltimore.
3. Young, R. (1996). *Measuring Economic Benefits for Water Investments and Policies*. Technical Paper No. 338. The World Bank, Washington, DC.
4. Bergstrom, J., Boyle, K., and Poe, G. (2001). Economic value of water quality: introduction and conceptual background. In: *The Economic Value of Water Quality*. J. Bergstrom, K. Boyle, and G. Poe (Eds.). Edward Elgar, Cheltenham, UK, pp. 1–17.
5. Renzetti, S. (2002). *The Economics of Water Demands*. Kluwer Academic Press, Norwell, MA.
6. Desvousges, W., Smith, V.K., and Fisher, A. (1987). Option price for water quality improvements: a contingent valuation study for the monongahela river. *J. Environ. Econ. Manage.* **14**: 248–267.
7. Frederick, K., VandenBerg, T., and Hanson, J. (1997). *Economic Values of Freshwater in the United States*. Discussion Paper 97-03, Resources for the Future, Washington, DC.
8. Moore, M. (1999). Estimating irrigators' ability to pay for reclamation water. *Land Econ.* **75**(4): 562–578.
9. Young, R., Gray, S., Held, R., and Mack, R. (1972). *Economic Value of Water: Concepts and Empirical Estimates Report to the National Water Commission*, NTIS no. PB210356, National Technical Information Service, Springfield, MA.
10. Abdalla, C. (1994). Groundwater values from avoidance cost studies: implications for policy and future research. *Am. J. Agric. Econ.* **76**: 1062–1067.
11. Smith, V.K., Desvousges, W., and McGivney, M. (1983). Estimating water quality benefits: an econometric analysis. *South Econ. J.* **50**: 422–437.
12. Whitehead, J. and Blomquist, G. (1991). Measuring contingent valuation for wetlands: effects of information about related environmental goods. *Water Resour. Res.* **27**(10): 2523–2531.
13. Carson, R.T. and Mitchell, R.C. (1993). The value of clean water: the public's willingness to pay for boatable, fishable, and swimmable quality water. *Water Resour. Res.* **29**(7): 2445–2454.

WATER—HERE, THERE, AND EVERYWHERE IN CANADA

Environment Canada

How much water is there in the world? Scientists estimate over one billion cubic kilometers (one cubic kilometer of water would fill 300 Olympic-sized swimming pools). Water covers nearly three quarters of the earth's surface in oceans as well as rivers, lakes, snow and glaciers. There is water in the atmosphere and water underground. Water evaporates and returns to the land surface in what is known as the *hydrologic cycle*.

In the hydrologic cycle, water evaporates from the ocean into the atmosphere, from there it can precipitate back into the ocean, or onto the land surface. From the land, it can evaporate or transpire back into the atmosphere, or flow overland or percolate underground before flowing back into the ocean. The distribution of the water around the globe varies from season to season and year to year, but the total quantity of water on the earth's surface remains essentially constant. The hydrologic cycle is discussed in detail in Freshwater Series No. A-1, "Water—Nature's Magician."

Although water exists in other forms in the hydrologic cycle, this issue in the Freshwater Series focuses on surface water as it is this water which we see in our everyday lives. Most of the earth's water is salty or permanently frozen. Figure 1 illustrates the proportion of fresh water that is available to us from the world's water supply.

This article is produced by Environment Canada and is reproduced with permission.

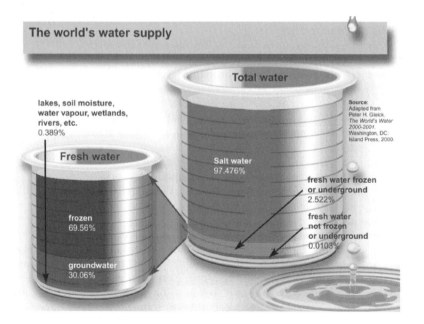

Figure 1.

Major stocks of water on earth
(thousand cubic kilometres)

	Volume (1000 km³)	Percentage of total water	Percentage of total fresh water
Saltwater stocks			
Oceans	1 338 000	96.54	
Saline/brackish groundwater	12 870	0.93	
Saltwater lakes	85	0.006	
Total salt water stocks	1 350 955	97.476	
Freshwater stocks			
Glaciers, permanent snowcover	24 064	1.74	68.7
Fresh groundwater	10 530	0.76	30.06
Ground ice, permafrost	300	0.022	0.86
Total frozen and underground freshwater stocks	34 894	2.522	99.62
Freshwater lakes	91	0.007	0.26
Soil moisture	16.5	0.001	0.05
Atmospheric water vapour	12.9	0.001	0.04
Marches, wetlands*	11.5	0.001	0.03
Rivers	2.12	0.0002	0.006
Incorporated in biota*	1.12	0.0001	0.003
Total not frozen or underground freshwater stocks	135	0.0103	0.389
Total freshwater stocks	35 029	2.5323	100
TOTAL WATER ON EARTH (1000 km³)	1 385 984	100	

Source: Adapted from Peter H. Gleick. *The World's Water 2000-2001*. Washington, D.C. Island Press, 2000.

Note: Totals may not add due to rounding

*Marshes, wetlands and water incorporated in biota are often mixed salt and fresh water.

Figure 2.

World's largest rivers

Rank	Based on drainage area		Based on length		Based on average annual total discharge	
	Name	Drainage area (1000 km²)	Name	Length (km)	Name	Average annual total discharge (km²/yr)
1	Amazon	6 915	Nile	6 670	Amazon	6 923
2	Congo	3 680	Mississippi*	6 420	Ganges	1 386
3	Murray	3 520	Amazon	6 280	Congo	1 320
4	La Plata	3 100	Yangtze	5 520	Orinoco	1 007
5	Ob	2 990	Mackenzie*	5 472	Yangtze	1 006
6	Mississippi*	2 980	La Plata	4 700	La Plata	811
7	Nile	2 870	Hwang Ho	4 670	Yenisei	618
8	Yenisei	2 580	Mekong	4 500	Lena	539
9	Lena	2 490	Lena	4 400	Mississippi*	510
10	Niger	2 090	Congo	4 370	Mekong	505
11	Amur	1 855	Niger	4 160	Chutsyan	430
12	Yangtze	1 800	Ob	3 650	Ob	404
13	Mackenzie*	1 790	Yenisei	3 490	Amur	360
			Murray	3 490		
14	Ganges	1 730	Volga	3 350	Mackenzie*	325
15	Volga	1 380	Indus	3 180	St. Lawrence*	318
16	Zambezi	1 330	St. Lawrence*	3 060	Niger	302
17	St. Lawrence*	1 030	Ganges	3 000	Volga	255
			Yukon*	3 000		

Source: Adapted from World Water Resources and Their Uses. Joint SHI/UNESCO Product, prepared by Prof. Igor A. Shiklomanov. 1999.

* Partly or entirely in Canada

Figure 3.

Figure 2 lists quantities that scientists have estimated for the various types of water that make up the world's supply. These amounts should be regarded as indicators of the relative quantities of water on earth. Owing to the difficulties in estimating volumes of water on a global scale, especially water underground, estimates can vary considerably. What is important is the overall picture that these estimates give.

RIVERS

A river's watershed or drainage basin—the area supplying it with water—is separated from the watersheds of neighboring rivers by higher lands called *drainage divides*. The map shows Canada's continental watersheds, one to each surrounding ocean: the Pacific, the Arctic, and the Atlantic as well as to Hudson Bay and to the Gulf of Mexico. Small watersheds combine to make up regional watersheds, which in turn join others to form continental watersheds.

The world's largest rivers are shown in Fig. 3.

Sculpting the Earth

As a swiftly flowing river, water can erode the underlying terrain. Where the river slope is flatter, the river slows down and deposits materials. This usually occurs in the lower reaches and especially near the mouth of the river, either at a lake or an ocean. A river can carve steep valleys, especially in higher parts of the drainage basin.

In the lower parts of the basin, deposits may create deltas at the river's mouth.

The volume of water flowing in a river, together with the speed and timing of the flows, determines how a river shapes the surrounding landscape and how people can use its waters. Rainfall, snowmelt, and groundwater all contribute to the volume of flow, producing variations from season to season and year to year.

In Canada, most high flows are caused by spring snowmelt. This is the season when floods are most likely to occur. Rainstorms can also cause high flows and floods, especially on small streams. The effects of floods and storms can be much less severe on rivers with large drainage basins. The lowest flows for rivers in Canada generally occur in late summer, when precipitation is low and evaporation along with consumption by plants is high, and in late winter, when rivers are ice covered and the precipitation is stored until spring in the form of ice and snow.

LAKES

Canada has more lake area than any other country in the world (Fig. 4), with 563 lakes larger than 100 square kilometers. The Great Lakes, straddling the Canada–U.S. boundary, contain 18% of the world's fresh lake water.

How is Water Measured?

The Water Survey of Canada, Environment Canada, along with many contributing agencies, measure the rate of flow (discharge) in rivers and record the levels of lakes and rivers at more than 2,600 locations in Canada. Typical river flows are listed in Fig. 5.

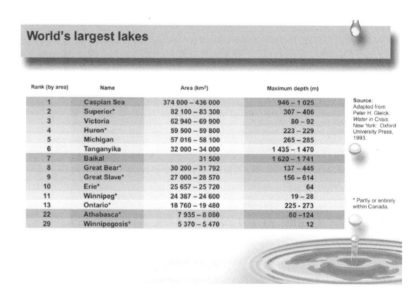

Figure 4.

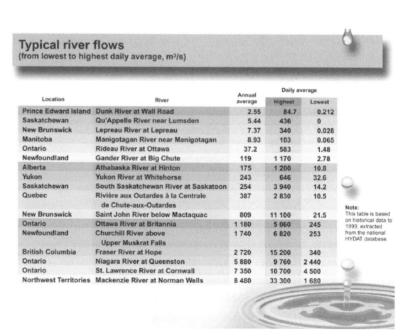

Figure 5.

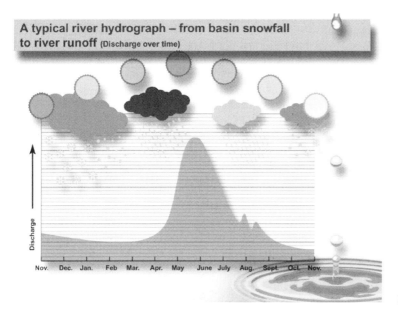

Figure 6.

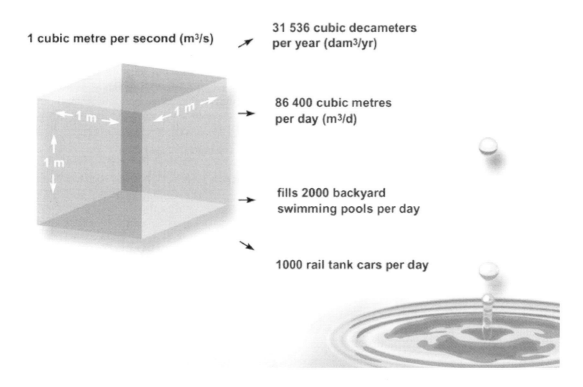

- Water levels are read manually by gauge readers or continuously recorded either digitally or on graph paper.
- Rate of flow (or discharge) requires multiple measurements of channel depth, width, and flow velocity to yield the average discharge in the stream crossing for a given water level. Measurements can be made from a bridge, by wading in a stream, by boat or from a cableway strung across the river. In winter, the measurements are made through the ice.

- With sufficient measurements of flow over a variety of water levels (including extreme lows and highs), a water level-discharge relationship is established at each location. The discharge rate can then be computed from measured water levels.
- Historical records from 5,000 active and discontinued sites permit the estimation of streamflow at ungauged locations.

Creating a Balance—Naturally

The importance of lakes lies in their ability to store water during times of plenty and release it gradually. Thus lakes perform an extremely valuable task in balancing the flow of the rivers on which they are located. For example, the Saskatchewan River, with few lakes, has a maximum recorded flow of 59 times its minimum flow. On the other hand, the St. Lawrence River, which drains the Great Lakes, has a maximum flow of only twice its minimum flow. The difference in flow patterns in these two rivers is partly due to precipitation differences, but results mainly from the vast storage provided by the Great Lakes for the St. Lawrence River compared with the negligible lake storage on the Saskatchewan River.

Creating a Balance—Artificially

Since ancient times, people have built dams to control the outflow from existing lakes or to create new lakes. Dams and their reservoirs have provided:

- A stable source of inexpensive energy
- A more dependable water supply throughout the year
- Flood control downstream
- Recreation

THE UNDERGROUND RESERVOIR

Beneath the surface of the earth is a huge reservoir of fresh water. Groundwater does not rest; it moves continuously, but at a snail's pace, from its point of entry to areas of natural discharge. Groundwater moves so slowly that its speed is measured in meters per day, and even per year. (Surface water velocities are described in meters per second.) Wells intercept some groundwater but most of it continues until it reappears naturally in a spring or a seepage area and joins a watercourse.

Groundwater contributes to Canada's water supply by:

- Feeding streams, producing the entire flow of some streams during dry periods
- Replenishing wells—a valuable source of supply for individuals, communities, industries, and irrigated farms
- Supporting important ecological systems such as wetlands
- Moderating the adverse impacts of acid rain on surface water systems

Additional information on groundwater can be found in: Freshwater Series No. A-5, *Groundwater—Nature's Hidden Treasure*.

NATURE'S FROZEN RIVERS

A huge quantity of fresh water is frozen in polar ice caps and in high mountain glaciers. Snow that is packed down over many years at high elevations becomes glacial ice, which slowly proceeds downslope like a frozen river, under the pull of gravity, and eventually melts to become part of streamflow at lower elevations. If the rate of melting is greater than the rate of accumulation, the glacier recedes; if it is less, the glacier advances.

Glaciers exert a direct influence on the hydrologic cycle by slowing the passage of water through the cycle. Like lakes and groundwater reservoirs, glaciers are excellent natural storehouses, releasing water when it is needed most. Glaciers, however, can release water when you need it least. Glacier-outburst floods, called *jökulhlaups*, can be devastating. Glacier-fed rivers reach their peak during hot summer weather.

SNOWFALL

Much of Canada's annual precipitation comes as snow: in the North, 50%; in the Prairies, 25%; and on both coasts and in southern Ontario, as little as 5%. Snow exerts a marked effect on the distribution of streamflow throughout the year. Instead of immediately infiltrating the soil or running off into stream channels as rainfall does, this water is first stored in the snowpack for several months.

The relatively quick melting of snow in spring causes peak flows, sometimes resulting in floods. Some of the worst and most unpredictable flooding occurs when ice that has not yet melted is carried along in the swollen rivers until it jams, blocking the flow of water and creating a lake behind the jam with attendant flooding. When the ice jam breaks, a tremendous amount of water is suddenly released downstream, and more flooding may result.

FRESHWATER SERIES A-2

Note: A resource guide, entitled *Let's Not Take Water For Granted*, is available to help classroom teachers of grades 5–7 use the information from the Water Fact Sheets.

WATER CONSERVATION—EVERY DROP COUNTS IN CANADA

Environment Canada

The importance of protecting our water resource cannot be overstated. In economic terms, the measurable contribution of water to the Canadian economy is difficult to estimate. In environmental terms, water is the lifeblood of the planet. Without a steady supply of clean, fresh water, all life, including human, would cease to exist.

This article is produced by Environment Canada and is reproduced with permission.

The perception that Canada is blessed with an abundance of fresh water has led to misuse and abuse of the resource: from household toilets that use 18 liters per flush where 6 liters would do, to industrial plants—and some municipalities—that use water bodies as convenient sewers.

In 1999, the average Canadian daily domestic use of fresh water per capita was 343 liters. The quantity, quality and economic problems we face as a result of our use of water are complex but, at least *one* of the causes of these problems is easy to manage—the way we waste water. And, the solution is straight forward—*water conservation*. Simply stated, water conservation means doing the same with less, by using water more efficiently or reducing where appropriate, in order to protect the resource now, and for the future. Using water wisely will reduce pollution and health risks, lower water costs, and extend the useful life of existing supply and waste treatment facilities.

And it's easy. With little change to the way we do things now, or the equipment we use, we can reduce water consumption in the home, and in business, by 40% or more. These pages outline the role of water conservation in addressing problems related to water use and water quality. It also shows us what part we can play as residential consumers in finding solutions.

SUSTAINING OUR WATER SUPPLY

Water is considered a renewable resource: "renewable" referring to that portion which circulates back and forth in the hydrological cycle (described more fully in Freshwater Series A-1, *Water—Nature's Magician*). However, pressures on the resource are growing. For example, between 1972 and 1996, Canada's rate of water withdrawals increased by almost 90%, from 24 billion m^3/yr (cubic meters per year) to 45 billion m^3/yr. But, our population increased by only 33.6% over the same period, illustrating the growth in our thirsty lifestyles. As the readily available supplies of fresh water are being used up, we begin to see that there are real limits to how much water we can count on.

LIMITING FACTORS

- Although Canada has a significant amount of fresh water, we possess only 7% of the world's *renewable* freshwater supply.
- In Canada, 84% of the population lives in a narrow southern band, while 60% of our water supply flows north to the Arctic Circle.
- Our growing population, and our growing thirst for water, are being concentrated in expanding metropolitan areas, and are forcing water regulators and policy makers to find ways to stretch available supplies even further.
- Increasing pollution of surface and groundwater is further reducing the supplies of readily available, clean water.
- Because our water use almost always leads to some degree of deterioration in water quality, the less water we withdraw, the less we upset the natural balance of our aquatic ecosystems. And, the less we upset the ecosystem, the less we have to spend to restore the water quality to an acceptable standard for public use.
- Finally, financing by municipal governments for the treatment of water supplies and wastewater is becoming increasingly constrained.

We can, however, make a significant contribution to solving these problems by reducing unnecessary levels of water use. To do so requires that we identify the areas within our homes, businesses, buildings and processes where we waste water and then make appropriate changes, either in our fixtures, or in our water-using habits.

HOW WATER IS USED

We use water in many ways, and assign different values to those uses. *Instream* uses (e.g., for transportation and recreation) are valued highly, but it has proven difficult to give them a dollar value that has any real meaning. For example, just what would the average consumer be willing to pay to swim in a clean lake or for a chance to catch fish in a clean, unpolluted river?

By far the greater number and variety of water uses occur on land. These are called *withdrawal* uses and, although important to our daily lives, they have tended to be assigned a low value. Water is withdrawn, used and then discharged. Most withdrawal uses "consume" some of the water, which means less is returned to the source than was taken out. And, after it has been used, the quality of the water that is returned is often diminished, which has a negative impact on both the environment and recreational instream uses.

In 1996, five main withdrawal uses accounted for a total annual water intake (extractive uses) of 44.6 billion m^3. These uses are described more fully in Freshwater Series A-4, *Water Works!*.

Thermal power generation includes both conventional and nuclear power generating plants, which withdrew slightly more than 64% of the total water intake in 1996.

Manufacturing accounted for 14% of water withdrawals in 1996. Paper and allied products, primary metals and chemicals were the main industrial users.

Agriculture accounted for nearly 9% of total withdrawals, with the semi-arid Prairie region of Canada accounting for 75% of this total. Agriculture consumes a large portion of what it uses, returning less than 30% to its source where it can be used again. Irrigation is the largest agricultural consumer of water.

Municipal use accounted for 10% of all water withdrawals in 1996, or 12% when similar rural uses were included (excludes industrial uses and large-scale agriculture). In the municipal sector, more than half of the water demand is a result of residential use.

Mining use, including metal mining, non-metal mining, and the extraction of coal, accounted for 1% of all water withdrawals in 1996. Water is used by the mining industry to separate ore from rock, to cool drills, to wash the ore during production, and to carry away unwanted material.

Solutions—the Municipal Challenge

Municipal governments across Canada are beginning to take action to manage the demand for water, instead of seeking new sources of supply. Demand management, incorporating water efficient applications, is rapidly gaining popularity as a low cost, effective way to get more service out of existing systems, thus delaying or deferring the need for constructing new works. The benefits of water efficient techniques apply equally well to rural, private wells and septic disposal systems, as they do to central water and sewer systems in the city.

The wide range of water efficiency initiatives currently being undertaken, can be grouped under four principal categories:

1. Structural,
2. Operational,
3. Economic, and
4. Socio-political.

Most of these water conservation activities fall within the jurisdiction of municipal governments and/or public utilities.

Structural

- metering
- water recycling systems
- wastewater re-use
- flow control devices
- distribution system pressure reduction
- water saving devices (efficient fixtures, appliances and retrofits)
- drought resistant landscaping (xeriscaping)
- efficient sprinkling/irrigation technology
- new process technologies
- plant improvements

Operational

- leak detection and repair
- water use restrictions
- elimination of combined sanitary/storm sewers to reduce loadings on sewage treatment plants
- plant improvements

Economic

- rate structures
- pricing policies
- incentives through rebates and tax credits
- other sanctions (fines)

Socio-political

- public education
- information transfer and training
- regulatory (legislation, codes, standards and by-laws)

SUSTAINING OUR INFRASTRUCTURE

While many communities have access to an abundant water supply, the costs of the infrastructure that provides homes and industry with water and sewer services are straining the available municipal financial resources.

By *infrastructure*, we mean the *water treatment plants* that purify our water, the *water mains* in the ground that transport water, and the *towers* and *reservoirs* that store water. The term includes the *sewer pipes* that carry away wastewater and the *sewage treatment plants* that treat wastewater before returning it to the environment where it often becomes the source of water for communities downstream.

Experts are predicting a growing problem involving municipal water and sewer infrastructure in Canada. In 1991, the value of this investment was estimated to be worth over 90 billion dollars, of which a significant amount is deteriorating with age.

An increasing number of Canadian municipalities are considering water conservation as the key to keeping expansion needs to a minimum. Water conservation also optimizes plant efficiency, while assisting municipalities in financing the replacement of infrastructure that may be over 50 years old in some communities and up to 100 years old in several others.

Communities with older systems in need of extensive repairs or replacement face the most difficult problems.

With all levels of government adopting policies of realistic water pricing and user pay principles, many municipalities have instituted *full cost pricing* to recover the total cost of providing both water and sewer services—including the costs of financing the replacement of older systems and the upgrading of overloaded treatment plants. Higher municipal costs, in turn, mean higher water—and sewer—bills.

The problem of stressed treatment systems is not restricted to communities with piped water and sewer systems. Over the past 25 years, there has been a substantial migration of urban dwellers to the countryside. City-bred water using habits and attitudes are, in many instances, lowering the water table. And, the flood of wastewater produced is stressing the soil's ability to treat septic effluent adequately.

For both urban and rural communities, water conservation can extend the life of this over-stressed infrastructure.

Metering

Tied to price increases, metered households generally show reductions in water use, with the greatest savings occurring during the summer months, when water use is usually much higher due to frequency of lawn watering, car washing and other outdoor uses. In 1999, water use was 70% higher when consumers faced flat rates rather than volume-based rates. And yet, only about 56% of Canada's *urban* population was metered in 1999.

Metering of industry has been common for some time. What's new is the metering of the return flow to the sewer system, particularly as it relates to the industrial sector. Case studies show that including sewage treatment in rate calculations generates greater water savings. An increasing number of municipalities are applying sewer surcharges to residential water bills.

Water Efficient Residential Technology

More than half of municipal water is used by the residential sector. As a consequence, the residential sector represents a logical target for demand management activities. Depending on the nature of the water efficiency program developed, each household can reduce water use by 40% or more.

Leak Detection and Repair

Up to 30% of the total water entering supply-line systems is lost to leaking pipes. In most cases, if unaccounted for water in a municipal system exceeds 10 to 15%, a leak detection and repair program is cost-effective. For example, studies have shown that for every $1.00 spent in communities with leak detection programs, up to $3.00 can be saved.

Rates, Pricing and Public Education

About 55% of Canadians served municipal water pay in ways that do not promote conservation. A 2001 study of rate structures by Environment Canada showed that in 1999, 43% of the population was under a *flat rate* structure (where the charge or assessment is fixed, regardless of the amount of water used). Another 12% were under a *declining block rate* structure (where the consumer's bill rises at a slower rate as higher volumes of water are used); i.e., the more you use, the less you pay per unit.

Only about 45% of the population served was found to be under a rate structure that provided a definite incentive to conserve water: 36% were under a *constant rate* structure (where the bill to the consumer climbs uniformly with the volume used); and 9% were under an *increasing block* rate structure (where a successively higher price is changed as larger volumes of water are used).

Introducing conservation-oriented pricing or raising the price has reduced water use in some jurisdictions, but it must be accompanied by a well articulated public education program that informs the consumer what to expect.

SUSTAINING OUR WATER QUALITY

In addition to water supply and infrastructure issues, water quality is a problem in many Canadian communities. Generally speaking, the decline in water quality is a function of the way we use water. Even something as simple as rinsing dishes in the kitchen creates wastewater that is contaminated to some degree. Once this water enters the sewer system, it must be treated in a sewage treatment plant. These facilities are never 100% effective, which means that some water quality deterioration remains after the treatment process.

Specific causes of impaired water quality are numerous, including: agricultural runoff containing the residues of fertilizers, pesticides and other chemicals, industrial pollution, either directly from the facility, or indirectly from the leaching of chemicals from landfills, or pollution from average households in the form of improperly treated municipal sewage (refer to Freshwater Series A-3, *Clean Water—Life Depends on It*). Nearly 75% of Canadians are serviced by municipal sewer systems. In 1999, 97% of the Canadian population on sewers received some form of wastewater treatment. The remaining 3% of Canadians served by sewage collection systems were not connected to wastewater treatment facilities in 1999 and discharged their untreated sewage directly into receiving water bodies.

For the roughly 25% of the Canadian population served by private wells and septic disposal systems, the news is not much better. These systems were originally designed for houses that were widely separated from their nearest neighbor, such as farmhouses and the occasional rural residence. Yet, today, in many parts of the country, individual private wells are being installed in subdivisions at suburban densities. The primary danger here is that too many wells may pump too much water for the aquifer to sustain itself.

Septic treatment systems associated with these developments can stress the environment in a number of other ways. They are often allowed in less than satisfactory soil conditions and are seldom maintained properly. They are also unable to treat many household cleaners and chemicals which, when flushed down the drain or toilet, often impair or kill the bacterium needed to make the system work (The same applies in urban systems). The end results are improper treatment of wastewater—if not outright failure of the system—and the contamination of adjacent wells with septic effluent containing bacterium, nitrates and other pollutants.

Once these contaminants are in the groundwater, they eventually reach rivers and lakes. In other words, once we have a *pollution* problem, we may be only a step away from a *water supply* problem.

Solutions—in the City and in the Countryside

The irony in all of this is that water quality impacts from overloaded or poorly maintained and operated municipal and private sewage disposal systems are the number one preventable type of pollution in Canada. The answer lies in better, more thorough treatment. And, one of the ways to enhance the treatment process is to limit the amounts of wastewater entering the wastewater stream. Again, water conservation is one of the easiest and cheapest ways to reduce the volume of wastewater flows and improve water quality.

Following a few common sense rules, it should be possible to safeguard your water supply while extending the life of your sewage disposal system, regardless of whether you live in the country or in the city.

Think carefully about the quantities of wastewater your household or business produces, as well as the quality of the wastewater. Do you make it a habit of discarding solvents, cleaners and related chemicals down your drains? If you do, you may be introducing substances that are toxic to the bacterium and other organisms that play a vital role in the treatment of sewage. This statement applies equally well to urban and rural households and businesses.

INDIVIDUAL ACTION—CONSERVING WATER IN THE HOME, COMMUNITY AND AT WORK

As we have seen, water quality and quantity are two sides of the same coin. How does saving water help water quality? Because water saved is water that does not end up in the wastewater stream requiring treatment. This, in turn, reduces municipal pumping and treatment costs and frees up monies that can be used for infrastructure renewal and replacement and protection of supply sources. Less wastewater in the sewage treatment plant also means that the plant has a better chance of doing the job it was intended to do.

So where do we start? The first step is to identify where we use water in the home. Then we need to decide on what to do to reduce the amount of water we use, either by eliminating wasteful practices and habits, or by improving the efficiency of our water using fixtures and devices. Since we waste so much, this should be a relatively easy and painless process. The prime area to target is the bathroom, where nearly 65% of all indoor water use occurs.

What follows are some suggestions for how to get your house or business in order. Based on the three rules of water conservation—*reduce*, *repair* and *retrofit*—a typical household can reduce water consumption by 40% or more, with or no effect on lifestyle.

REDUCE

Much of the water "consumed" in our daily activities is simply wasted. Taps are left running while we brush our teeth. Dishwashers and laundry machines are operated without full loads. Really, everywhere we use water there is room for improvement. Here are just a few examples for both indoor and outdoor water use.

- Don't use the toilet as a wastebasket or flush it unnecessarily.
- A quick shower uses less hot water than a full tub (and saves energy too).
- Keep a bottle of drinking water in the refrigerator rather than letting your tap run to get cold water when you want a drink. (Rinse the bottle every few days.)
- More than 50% of the water applied to lawns and gardens is lost due to evaporation, or run-off because of overwatering. Find out how much water your lawn really needs. As a general rule, most lawns and gardens require little more than 2 to 3 cm (1 inch) of water per week.
- To reduce loses due to evaporation, water early in the morning (after the dew has dried).
- When washing a car, fill a bucket with water and use a sponge. This can save about 300 liters of water.

REPAIR

Leaks can be costly. A leak of only one drop per second wastes about 10,000 liters of water per year. Most leaks are easy to find and to fix, at very little cost.

- Leaking faucets are often caused by a worn out washer that costs pennies to replace. Most hardware stores will have faucet repair kits with illustrations showing how to replace a washer.
- A toilet that continues to run after flushing, if the leak is large enough, can waste up to 200,000 liters of water in a single year! To find out if your toilet is leaking, put two or three drops of food coloring in the tank at the back of the toilet. Wait a few minutes. If the colour shows up in the bowl, there's a leak.
- Toilet leaks are often due to a flush valve or flapper valve that isn't sitting properly in the valve seat, bent or misaligned flush valve lift wires, or a corroded valve seat. All of these can be fixed easily and inexpensively. To get at the valve seat, which surrounds the outlet hole at the bottom of the tank,

you must first empty the tank. This is accomplished by turning off the inlet tap under the tank and flushing the toilet, making sure to keep depressing the flush lever until no more water drains out of the tank. Then, holding the valve out of the way, sand the corroded or warped valve seat smooth with a piece of emery cloth, if, however, the leak is around the base of the toilet where it sits on the floor, call a professional.

RETROFIT

Retrofit means *adapting* or *replacing* an older water-using fixture or appliance with one of the many water-efficient devices now on the market. While these solutions cost more, they also save the most water and money. Retrofitting offers considerable water saving potential in the home and business, so this issue in the Freshwater Series is devoting considerable space to it.

Toilet Retrofits

When it comes to retrofitting, the prime fixture to target is the toilet. You can: (i) *adapt* your existing toilet in a number of ways, by installing certain water-saving devices inside the tank at the back of the toilet; or, (ii) if the toilet is more than fifteen years old—which means it probably uses about 18 or more liters of water per flush—you can *replace* it with one of the growing number of ultra-low-volume (ULV) toilets, that can be ordered from most plumbing outlets, and use only 6 liters or less per flush.

There are many *toilet adaptations* you can install in the tank of an existing toilet to reduce the amount of water used in a flush cycle. These devices fall into three generic categories:

- water retention devices;
- water displacement devices; and,
- alternate flushing devices.

The most common water retention device available is the toilet dam. A set will save about 5 liters per flush when installed properly. Their main attraction is their low cost (under $10.00 per set) and the fact that they are easy to distribute and install for example, as part of a wider municipally-sponsored retrofit program. Their main disadvantage is that they tend to leak over time by slipping out of adjustment and can slip free and interfere with the moving parts inside the toilet tank, if not routinely checked.

Toilet dam and displacement bag:
The water displacement devices familiar to most people are the *plastic bags* or *bottles* filled with water which are suspended inside the toilet tank. As the name implies, these devices displace several liters of water, saving an equivalent amount during each flush. Like the toilet dam, most displacement devices are inexpensive and easy to install. Their chief disadvantage is that they don't save as much water as other devices and, if they are not installed carefully, they can interfere with the proper operation of the toilet.

One displacement device to stay away from is the *brick*! It can disintegrate inside the toilet tank, leading to excessive leakage at the flapper valve and may even be heavy enough to actually crack the tank.

There are essentially two types of alternative flush devices: *early-closure* and *dual-flush*. They are usually attached to the overflow tube inside the toilet tank. In both cases, they close the flush valve or flapper after the tank is only partially emptied. In theory, this interruption in the flush cycle occurs after the bowl has been cleared. In the case of the dual-flush mechanism, the amount of water saved is dependent upon how long the flush lever is activated—a partial flush for light duty or full flush or heavy duty.

While all of the above toilet adaptations appear to work as intended when first installed, their performance may vary considerably, depending on the toilet design. The best advice is to monitor the performance of the devices periodically. If you discover that it becomes necessary to double flush the toilet, something is in need of adjustment or replacement. Remember that double flushing defeats the purpose of your water conservation efforts and is costing you money.

If you decide that it is time for a *toilet replacement* in your home or business, you are well on your way to significant water savings that you can bank on over the life of the toilet. Replacing a 18 liter per flush toilet with an ultra-low-volume (ULV) 6 liter flush model represents a 70% savings in water flushed and will cut indoor water use by about 30%. Keep in mind that 18 liters per flush, assuming 4.5 flushes per person per day, translates into nearly 30,000 liters of clean, fresh water per year just to get rid of 650 liters of body waste. A 6 liter flush toilet only uses about 10,000 liters to do the same task. Low flush toilets are available for less than $150.00 at most plumbing and supply stores.

Remember, the ULV toilet not only uses less water, *it produces less wastewater*. If your municipality applies a *sewer surcharge* on your water bill, the investment in the better toilet could translate into a 50% reduction in your combined water/sewer bill. If you are on a private well and septic system, you are significantly reducing the loading on your tile field while extending its useful life. To a lesser degree, the same applies to the other water-saving devices described here.

Showerheads and Faucets

After the toilet, the shower and bath consume the most water inside the home. Conventional showerheads have flow rates up to 15 to 20 liters per minute. A properly designed *low-flow showerhead* can reduce that flow by half and still provide proper shower performance. Low-flow showerheads can be purchased in most plumbing supply outlets.

Depending on your preference for finish and appearance, you can select a serviceable low-flow showerhead starting at around ten dollars. Consider one with a shut-off button. The advantage of the shut-off button is that it allows you to be really water efficient if you so choose, by being able to interrupt the flow, while you lather up or shampoo, and then resume at the same flow rate and temperature.

Beware of the type of showerheads that produce such a fine mist that the water is quite cool by the time it reaches your feet. And, stay away from so-called *flow restrictors* that are inserted inside your existing showerhead. They look like a small plastic washer and can produce a fierce, stinging spray pattern which may significantly reduce the enjoyment of taking a shower.

Conventional faucets have an average flow rate of 13.5 liters of water per minute. Install *low flow aerators* to reduce this flow. In the bathroom, a flow rate of 2 liters per minute should do the trick, and in the kitchen a flow rate of 6 to 9 liters per minute is sufficient. Don't bother retrofitting the tap in the utility sink; it is intended to provide large volumes of water quickly, for example, for cleaning or washing, such that low flows will only inconvenience the user.

Outdoors

During the growing season water use can increase by as much as 50%. While lawns require a lot of water, much of this water is wasted—lost due to overwatering and evaporation (see Reduce).

Watering equipment also plays a part in how much water is saved and lost. Ideally, sprinklers should be suited to the size and shape of the lawn. That way, you avoid watering driveways and sidewalks. Installing timers on outdoor taps can be a wise investment.

Sprinklers that lay water down in a flat pattern are better than oscillating sprinklers which lose as much as 50% of what they disperse through evaporation. Drip irrigation systems which apply water only to the roots zone are the most efficient—and the most expensive—alternative.

The water you use to water your lawn doesn't have to come out of a tap. A cistern, which captures and stores rainwater, can be used as a source of irrigation water. A rain barrel can adequately fulfil this function.

Finally, consider a low-maintenance landscape—one which requires little more water than nature provides. Often called xeriscaping, the principles of a low-maintenance landscape are as follows:

- a reduced amount of lawn;
- proper plan selection making use of native grasses, shrubs and trees;
- mulching to reduce evaporative losses around shrubs and trees;
- the use of rain barrels and roof drainage;
- improvements to soils;
- a proper irrigation system; and
- planned maintenance.

The most significant savings of course, come from a reduction in lawn area and switching from exotic plant forms to native species which require less water. In general, lawn areas should not exceed what is useful for play and social activities, and should be limited to the backyard where the family spends the majority of its time.

Solutions—At Work and in the Community

Many of the suggestions made for reducing water use in the home have wider application, both in the workplace, and in the community at large. Low-flow equipment are available for most commercial and toilet applications, instituting them may mean taking a leading role yourself, for example, forming and leading a committee that would address the following questions:

- do your workplace bathrooms, kitchens, etc. have water-efficient toilets, faucets, etc. similar to those discussed for the household?
- if your workplace uses water in its production process or for washing goods or equipment, is this being done efficiently?
- does your community have a water-efficiency assistance program that helps households and business improve their water-use efficiency?
- is the water distribution system properly maintained so that no pollution leaks into it and so that no water is wasted through leaky mains?

THE BOTTOM LINE

Water conservation. The message is clear. If we each save a little, it can add up to major savings in water, energy and money. For the average household, reductions in water use as high as 40% or more are feasible, just by following the steps outlined in this issue in the Freshwater Series.

The benefits don't stop at the household or business. The municipal water and sewer department gets a break on the amount of water it has to pump to our homes and businesses *and* on the amount of wastewater it has to treat in sewage treatment plants. Water conservation can extend the useful life of municipal water supply and treatment plants, and will benefit the operating efficiency—and life expectancy—of private septic disposal systems.

And, finally, water conservation can generate *significant* environmental benefits. It can reduce water diverted and the pollution loadings on our lakes and rivers by reducing the volumes of wastewater which we have to treat. This can help to protect our drinking water and the ecological balance in sensitive aquatic ecosystems.

If we all practice water conservation, everyone—and everything—benefits.

FRESHWATER SERIES A-6

Note: A resource guide, entitled *Let's Not Take Water For Granted*, is available to help classroom teachers of grades 5–7 use the information from the Water Fact Sheets.

ECOREGIONS: A SPATIAL FRAMEWORK FOR ENVIRONMENTAL MANAGEMENT

Md. Salequzzaman
Khulna University
Khulna, Bangladesh

DEFINITION

An ecoregion may be defined as a scale for planning and analysis in the hierarchy of the earth's ecosystem. It has broad applicability for modeling and sampling, strategic planning and assessment, and international planning of domain, division, and provincial ecological units.

Ecoregionization is a process of delineating and classifying ecologically distinctive areas of ecological land. Each area can be viewed as a discrete system that has resulted from the mesh and interplay of the geologic, landform, soil, vegetative, climatic, wildlife, water, and human factors where ecological functions and processes are continuing. The dominance of any one or more of these factors varies with the given ecological land unit. This holistic approach to land classification can be applied incrementally on a scale-related basis from site-specific ecosystems to very broad ecosystems (1). Ecological processes, evolutionary mechanisms, and geological forces are continually reshaping landscapes across various scales of time and space and result in distinctive but dynamic ecoregions. All of the world's food and most medicines and raw materials are derived from these processes and associated biodiversity. Thus ecoregions gain their identity through spatial differences in a combination of landscape characteristics. Several factors such as topography, hydrology, and nutrients are important to identify these characteristics that may vary from one place to another in an ecoregion.

OBJECTIVE OF ECOREGIONIZATION

The main objective of using the ecoregion as a spatial framework for environmental management is to maximize ecoefficiency by minimizing resource-use conflicts and harmonizing sociocultural needs in the ecoregion. Ecoregions are also used to isolate areas for the interpretation of environmental quality (2). Countries such as Canada and the United States are now using the ecoregion as the coarsest scale for evaluating representative areas (3). This process needs lots of information. The baseline information on ecosystem properties and changing conditions of ecosystems in any specific ecoregion may be collected through a series of monitoring, which affects the management practices in that ecoregion. Knowledge of the ecoregion facilitates; the allocation or linking of sites to a standard ecological hierarchy minimizes the sampling variance and increases the ability to extrapolate results for areas with similar properties.

IMPORTANCE OF THE STUDY OF AN ECOREGION

Ecological integrity, natural capital, and biodiversity are three ecological requirements for environmental management in any ecosystem (4). These three factors should be present as a prerequisite for a healthy ecoregion because an ecoregion itself is a big ecosystem. Among these factors, biodiversity has global importance and combines wildlife value with a rich cultural heritage. Thus each ecoregion has an immense natural value with a colorful and rich cultural heritage bound up with the unique natural heritage of age-old cultural traditions. As each ecoregion is ecologically distinct and strikingly beautiful, an ecoregion attracts the ecotourism that has a great potential to improve the local economy as well as develop a global network for ecobusiness and globalization.

SOME IMPORTANT ISSUES FOR ENVIRONMENTAL MANAGEMENT IN THE ECOREGIONS AROUND THE WORLD

Ensure a Balanced Ecosystem

Every organism, including animals and plants, no matter how small they are, plays a vital role in keeping the systems of the earth as healthy and functioning ecoregions. Removing any species from the complex web of life disturbs the balance of the entire system, resulting in the impoverished life zones left today. Scientific studies have made it clear that a balanced ecosystem is teeming with abundant life in quality and quantity, such as a wetland ecosystem or the ocean. Historical evidence clearly suggests that once the oceans were filled with whales, sea turtles, fish, and other forms of life that seem more identical to a scientific fantasy than reality. Many of them are now extinct. The reality is that the actual populations were exponentially greater than they are today (before human predation began). Once humans began hunting, their appetites were insatiable, and a pattern of imbalance started that may culminate in the complete collapse of some ecosystems of the ecoregion. Thus one of the important aspects of ecoregionization is to keep the ecosystem balance in any ecoregion.

Ensure the Protection of the Natural Heritage and the Environment

The natural heritage and its environment are the identity of any ecoregion, which are also the natural wealth that sustained human communities in comfort and plenty for millennia, such as the rain forest in many countries of the world, for example, Alaska. Recently, the Alaskan rain forest is being destroyed in such a way that the region faces a graver, more permanent threat: the same razed-earth logging that has already devastated the Pacific Northwest. Rain forest logging is perfectly in step with the boom-and-bust rhythm of the Alaskan economy due to poorly planned development spread out over wildlands, slashing of shorelines for beach homes and resorts, limiting public access, and destroying fragile barrier beaches and wetlands. Besides this destruction, it is also destroying a lot of wildlife habitats as well.

Ensure the Protection of the Ethnic Community of the World

Each ecoregion has a special characteristic of the specific group of ethnic community people. They preserve a sustainable livelihood from generation to generation. Now, this sustainability livelihood is challenged for survival. The ethnic community people attempt overexploitation of natural resources such as fish, oil, gas, forest products, and minerals to satisfy their mounting needs induced by globalization and modernization. Recently, some nations have agreed to monitor environmental damage in their ecoregion, establish an emergency response program, protect the sensitive environment, and conserve endemic flora and fauna. However, huge chunks of the ecoregion still lack essential wilderness protections. Oil and gas development threatens the coastal plain and millions of acres at the heart of biological refuges. Recently, a large portion of coral bleaching (hot spot) in coral reefs was discovered throughout the world (5). Coral bleaching is the loss of color from the living coral animal and can cause their death. The color of coral comes from microalgae living in the coral's tissue; when the water surrounding the coral becomes too warm, corals become stressed and expel the microalgae. These microalgae also provide the coral with food, and as the algae fade away, corals starve to death.

Therefore, environmental management and conservation should be adopted first to quickly protect the ecoregion, including its organisms, people, and its total effective surroundings from misuse or damage by human activities; second, to ensure that life support systems and renewable resources are used in an ecologically sustainable manner while supplying the conditions for a satisfactory human existence. Community-based projects will demonstrate the benefits for local people and the environment combining sustainable development and conservation and building up coordination among various activities in the ecoregion. In addition, ecologically sustainable management should be adopted in every ecoregion that will be used to assess the ecological sustainability of the biological resources in the ecoregion.

Environmental justice often plays itself out as a disparity between the rich and the poor, between the "haves" and the "have-nots." Just as environmental justice speaks to issues of poor people and people of color who are disproportionately subject to environmental pollution and public health risks, it also should speak of issues where poor people and people of color do not equally derive opportunities to obtain benefits and values associated with environmental resources.

CONCLUSION

Sustaining ecological processes, the landscape function, and the biological diversity of any ecoregion are important globally, nationally, and locally. Pressure to accommodate a rapidly increasing human population together with increased provision of goods and services has been growing for decades. This has impacted each ecoregion in its sociocultural, economic, and environmental characteristics. Therefore, a multiple scale, transdisciplinary integrated approach, and strategic framework for sustainable environmental planning and management of an ecoregion are needed. In addition, pressure to take action to halt land degradation, breakdown in production systems, decreasing water quality, loss of biodiversity, and global climate change is also needed for any government to improve the way it conserves, protects, and manages ecological processes, habitats, and species in the ecoregion. Furthermore, it must integrate planning and management across multiple scales, nesting the functional requirements of ecological systems and social systems to uphold and restore resilience in the ecological and social systems of the ecoregion for long-term sustainability.

BIBLIOGRAPHY

1. Wiken, E.B. (1986). Terrestrial ecozones of Canada. Ecological Land Classification Series No. 19. Environment Canada, p. 26.
2. Jarvis, I.E., MacDonald, K.B., and Betz, T. (1995). Development and application of a Canadian ecological framework. In: *Proc. GIS 95, 9th Ann. Symp. Geogr. Inf. Syst.*, Vancouver, British Columbia. GIS World Inc., Ft. Collins, CO, pp. 605–612.
3. Gauthier, D. (1992). Framework for developing a nationwide system of ecological areas: Part 1-A Strategy. Occasional Paper No. 12, Canadian Council on Ecological Areas. Secretariat, Canadian Wildlife Service, Environment Canada, p. 47.
4. Brunckhorst, D.J. (2000). *Bioregional Planning: Resource Management- Beyond the New Millennium.* Harwood Academic, Australia, p. 5.
5. Greg, R. (2002) Coral Threat Renews Push on Protocol, Wednesday 9 January 2002, http://www.theage.com.au/news/national/2002/01/09/FFX715DK6WC.html (access on 12 September 2002).

FLOOD OF PORTALS ON WATER

Dick de Jong
IRC International Water and
Sanitation Centre
Delft, The Netherlands

Marta Bryce
CEPIS/PAHO
Delft, The Netherlands

INTRODUCTION

The Internet has been inundated with new gateways and portals on water over the last couple of years, as witnessed by information specialists at the IRC International Water and Sanitation Centre. The New on the Net section in Source Water and Sanitation News (http://www.irc.nl/source/section.php/12) carries 54 news items since July 2002, of which seven refer to "gateway" or portal. Unfortunately, many of them only serve to confuse instead of help people who search the web for water-related information. Part of the problem resides in defining what constitutes a "portal" and what doesn't. Another issue is that most portals do not seem to have a clear notion of their intended visitors—and their motives for visiting. As a result, the portals lack much-needed focus.

Here, we give some recent examples, try to create more clarity by providing definitions and types, discuss key lessons learned, and raise key issues for further discussion and research. This is based on various Internet searches and an annotated selection of various gateways/portals by IRC staff.

Here are five examples of sites, none of which have all the functionality of a portal yet.

1. The closest example of a portal is the BBC Water Portal—The Water Debate (http://news.bbc.co.uk/2/hi/in_depth/world/2003/world_forum/water/default.stm) that was launched early June 2003. This is not only an impressive interactive site for the general public with video, audio, expert views, fact files, and water stories; it also provides fun options such as online opinion polls and a water quiz. However, it does not contain an option for online collaboration.
2. The IRC International Water and Sanitation Centre (established in 1968) family of websites (http://www.irc.nl) provides news and information (including advocacy/communication, publications, and bibliographic database) on low cost water supply and sanitation in developing countries. It has a number of topic sites. It also provides a starting point for the Source Water and Sanitation News service and the InterWATER guide to more than 650 organizations, in collaboration with the Water Supply and Sanitation Collaborative Council. The search engine (Atomz) provides access to all these related sites. The IRC DOC online database contains 15,000 bibliographical entries on developing country water, sanitation and hygiene documents, books, periodicals, videos, and so on. The (www.worldwaterday.org) site maintained by IRC contains information on one topic per year that is selected by the UN. Visitors can add comments and events.
3. Sanitation Connection (http://www.sanicon.net/) is an Internet-based resource that gives access to accurate, reliable, and up-to-date information on technologies, institutions, and financing of sanitation systems around the world. Institutions of international standing contribute to the information base by providing and maintaining a topic of their specialization.
4. Water Portal of the Americas (http://www.WaterPortal-Americas.org) is a combination of a gateway (entryway or portal) and a community of practice. It is basically a search engine, a structure and system for other organizations to copy or fit in. It is still a prototype that may grow into a collaborative portal. "The goal is to provide both an entryway (portal) to water information and to create a water information network, community, and resource that will provide qualified, trusted, and verifiable information and contacts."

   At the March 2003 Third World Water Forum the prototype of a "Water Portal for the America's" was shown by UNESCO/World Water Assessment Programme and the Waterweb Consortium, a year and a half after it was first announced at the Fifth Water Information Summit. The prototype gave an impression of what will be on the site, with only limited content. This is the first regional water portal that is scheduled to become part of the UNESCO WWAP/Waterweb Global Water Portal, as was announced in Japan. Funding for this started in April 2002.
5. The U.S.-based WaterTechOnline (http://www.watertechonline.com/news.asp?mode=4&N_ID=31857) is one of the commercial portals addressing business needs. It contains breaking daily news stories about the water and wastewater treatment industry, as well as industry bulletin boards, searchable supplier directories and article archives, industry opinion polls, new product announcements, and other industry-specific information.

These examples illustrate the proliferation of water portals, gateways, and websites. It also raises questions about the definitions, users, and relevance of these internet-based initiatives.

DEFINITIONS AND TYPOLOGIES

General information on portals can be found on the Portals Community site. This advertises on Google.com as "The Definite Enterprise Portal Resource" and offers Portals information, White papers, networking, and research (http://www.portalscommunity.com/registration/member.cfm?code=TDXTL05).

Defining Water Portals

From About.com and our own sources, a portal is a kind of website and provides a single point of access to aggregated information—a virtual front door. The term originated with large, well-known Internet search engine sites that expanded their features to include e-mail, news, stock quotes, and an array of other functionality. Some corporations took a similar approach in implementing their intranet sites, which then became known as enterprise information or corporate portals.

There are general portals and specialized or niche portals. Some major general portals include Yahoo, CNET, AOL, and MSN. Examples of niche portals that are accessible to the public include Garden.com (for gardeners), Fool.com (for investors), and DPReview.com (for photographers). These are sometimes also called vortals—vertical portals. Private niche portals are those that are used by employees of a company.

Typical services offered by public portal sites include a start page with rich navigation, a directory of websites, a facility to search for other sites, news, a collection of loosely integrated features (some of which may be provided by partners or other third parties) like weather information, stock quotes, phone, and map information, e-mail, and sometimes a community forum. Private portals often include access to payroll information, internal phone directories, company news, and employee documentation.

Here, we define a portal as "a one-stop, client-oriented website that offers visitors a broad array of interactive resources such as news, databases, discussion forums, search options, space to collaborate online, and links on water-related topics."

Searches

IRC staff did various online literature searches:

- On "Water Portals" on all the web search engines in April, May, and June 2003. Of the 45 results shown, 32 were relevant. On Google.com, 22 out of 42 were relevant. The other results were mainly related to game sites.
- On the specialized ICT site Whatis.com there is a lot of information on corporate portals through Best web links, news headlines expert technical advice, and web results.

CONCLUSIONS

Our main conclusions are:

1. There is a wide variety of "portals" on water, definition wise, topic wise, and language wise.
2. Even in the world of corporate portals the concept is subject to confusion in the marketplace (Peter A. Buxbaum, Nov 2001, searchEBusiness). Buxbaum quotes:

   There are several different types of corporate portals, each sporting different capabilities and each requiring different technologies and technology providers. Experts segment the enterprise portal arena into at least three components. *Information portals* emphasize data and document retrieval and management. *Knowledge portals* enable collaboration and information sharing as well as expertise and knowledge capture. *Application portals* provide end-users with interfaces to enterprise systems. What do all these have in common? As IDC defines it, they all manage end-user access to multiple applications and information sources on the corporate intranet.

   Looking at the software giant SAP's recent steps in its portal strategy, Buxbaum leaves us with a nagging question: If a software giant like SAP can't explain portals clearly, how are its potential customers supposed to understand them?

3. Most of the sites that claim to provide a water "portal" can be more correctly described as providing a start page or gateway to web information. There is in fact a flood of gateways, but too many providers are drowning in ambition.
4. An interesting paper that describes pros and cons of a portal from a university point of view in easily understandable language is Kevin Lowey's University Portals FAQ, *University Web Developers' Mailing List 2002* (http://www.usask.ca/web_project/uwebd/portals_faq.html).
5. Some interesting findings emerge from a user survey of electronic information services (EIS) among 25 U.K. institutions of higher education. The effect of the Internet on information seeking by staff and students is hugely significant; search engines and known sites are the first resort for most academic queries, as well as for many personal domestic queries. There was a wide range of engines used and indications of haphazard searching.
6. Another major conclusion from the same survey is that subject gateways are notable only for their lack of mention among students and academic staff, although there is some use among library and information staff.

LESSONS LEARNED

The majority of the gateways/portals described here focus on the broadest possible water topic: water resources and management.

The claims for "The Water Portal" cannot be justified. At best we can help clarify what the sites we have selected are focusing on, what they are, what they deliver to visitors (who?), and how interactive they are.

Two good examples of the different approaches are The Netherlands-based start page on water (www.water-pagina.nl) that provides four pages of clickable names divided over 32 categories, compared to the recently established www.h2o-scanner.com. This has a start page from where searches can be done combining countries, segments, issues, and content.

With a portal the emphasis is shifting from a public website showing online pamphlets to a user-oriented website that provides tools, reports, and services specifically designed for that individual.

WHAT NEXT?

By offering its new website architecture to interested partners, IRC started since mid-2003 to develop a family of related websites on water sanitation and hygiene, using the same technical foundation: *A Water Portal Is Born*. The STREAMS coalition of resource centers is going to be the first partner using this opportunity to contribute content through its own segment of this shared platform. More partners will be added.

In 2004 the Portal functionality will be expanded to also contain login functionality to provide Intra- and Extranet opportunities, moderated lists to enable E-conferences, workflow management, and others.

INTERNET ACCESS GAP

A few data on Internet connectivity according to CIA's World Factbook (http://cyberatlas.internet.com/big_picture/geographics/article/0,,5911_151151,00.html) show the enormous gap between Internet use in the developed and developing world.

The percentage of the population of Internet users in the world in 2002 was around 9%. In The Netherlands and Denmark it was around 61%, in the United States some 59%, in Canada 50%, and in Belgium 36%. In Colombia it was about 2.8%, in Bolivia 0.9%, in India 0.7%, in Nepal 0.23%, in Mozambique and Bangladesh 0.1%, and in Ethiopia 0.03%.

Kamel from Nepal Telcom indicates that there are 8 phone lines per 1000 persons in Nepal. E-mail was introduced in 1993 and Internet in 1994. They have 50 cyber cafes in Kathmandu and 20 in Pokhara. Government official Vandya indicates that IT development is important and Nepal has to continue to develop it. This needs to be done in parallel with other media such as wall newspapers, taking into account that basic literacy is 55% and English literacy 2%.

In Burkina Faso the literacy is only 20% and they have 60 spoken local languages in a population of 2.6 million. A local priest is making a very interesting effort by publishing a quarterly farmers magazine in some of the local languages. He indicates that Internet is useful to find information, but a considerable part of the information in the magazine comes from local farmers.

In terms of difficulties it is clear that the number of users is very low in the South and even more so in rural areas. Another difficulty is that Internet performance gives problems. Downloading in the South, for example, is slow and the cost of access is high. It is often a struggle to access a telephone line let alone high bandwidth facilities. In the preparation for a knowledge management workshop at the Sixth Water Information Summit in September 2003, it was already very clear that files over 1 MB gave a lot of problems. Only after simplifying and zipping them did they become more manageable.

LATIN AMERICA: VIRTUAL LIBRARY ON ENVIRONMENTAL HEALTH

A good example from the achievements and challenges of web use in the developing world is the Virtual Library on Environmental Health (VLEH) of the Pan American Information Network on Environmental Health, REPIDISCA, which has been operating since 1982 in 21 countries of Latin America and the Caribbean with 370 Cooperating Centres. The Pan American Center for Sanitary Engineering and Environmental Sciences, CEPIS/PAHO, of the Pan American Health Organization, PAHO/WHO, is in charge of the regional coordination.

The VLEH disseminates information on water supply, sanitation, and related topics (http://www.cepis.ops-oms.org/indexeng.html). It uses the same methodology as the Latin American and Caribbean Health Sciences Information Centre, BIREME, another information system sponsored by PAHO/WHO. Its architecture is based on six types of information sources and services: electronic publications in full text, secondary sources, teaching material, SDI service, news and discussion lists, and integrating components. The bibliographic database contains more than 134,000 records with abstracts, of which 7553 are full text.

The principles of the VLEH are:

- Equitable access to environmental health information.
- Alliances and consortia for maximizing resource sharing.
- Cooperative work and exchange of experiences.
- Decentralized development and operation at all levels.
- Development based on local conditions.
- Integrated mechanisms for evaluation and quality control.

The VLEH architecture (Fig. 1) is based on the characteristics of its information sources. An information source is any resource that responds to information needs, including information products and services, persons or networks of

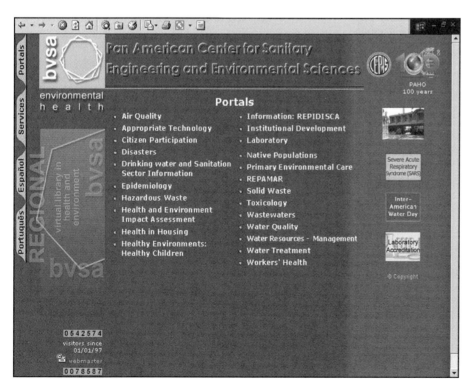

Figure 1. The VLEH main page.

Table 1. Number of Records in Each Information Source of the VLEH

Number	Source	1981–1997	1998	1999	2000	2001	2002	5 June 2003
	Secondary							
1	Bibliographic	96,467	106,577	117,925	119,321	120,784	130,087	134,972
2	Institutions	1,953	1,986	2,043	1,914	1,779	2,106	2,283
3	Specialists					103	103	136
4	Events (current)		186	475		1,065	1,005	1,100
5	Academic programs					129	284	298
6	Journals			451	475	483	505	516
7	Videos			398	404	475	494	500
	Primary							
8	Full texts		404	927	1,568	3,159	6,071	7,553
9	Legislation			165	547	1,079	1,235	1,004
10	HDT	70	73	76	79	82	88	88
11	Repindex	63	66	69	72	75	78	78
12	Indicators					ALC	ALC	ALC
	Teaching Material							
13	Teaching material					607	829	879
14	Self-instruction courses						9	9
	SDI Service							
15	SDI data base		452	519	654	765	2,020	2,114
	News/Discussion Lists							
16	News			119	263	322	973	1356
17	Open discussion lists					7	11	11
18	Restricted discussion lists					12	11	13
	Integrating Components							
19	LISA (www links)			175	314	1,448	1,682	1,754
20	Thesaurus in five languages					3,000	3,143	3,259

persons, and computer programs. The VLEH architecture is organized as follows (Table 1):

1. *Secondary Sources.* Composed of the following databases: bibliographic (more than 134,000 records); institutions (includes the address and names of authorities); events (includes seminars, congresses, and short courses); academic programs (includes diplomas, master, and doctoral programs); journals (includes the titles, electronic availability, and access to full texts); videos (includes material available at CEPIS and Cooperating Centres).

2. *Primary Sources.* Composed of the following databases: full text documents (more than 7500 documents in Acrobat); legislation (more than 1000 laws and regulations in full text related to water quality); HDT (includes 78 issues of the series Technical Dissemination Sheets, published by REPIDISCA); Repindex (includes 88 issues of this series also published by REPIDISCA); indicators (includes data from 45 countries of the Latin American and Caribbean region based on the global assessment of water and sanitation services, 2000).

3. *Teaching Material.* The material training database contains electronic and multimedia sources with added value for teaching purposes; the database on self-instruction courses contains nine courses prepared by PAHO/CEPIS which are also available on CD-ROM.

4. *SDI Service.* Selective dissemination of information (SDI) to update users based on their specific interest profiles. It is also a mechanism for providing information through e-mail to users that lack or have communication constraints via the Internet.

5. *News and Discussion Lists.* Fosters communication among persons, including discussion lists, forums, and virtual communities.

6. *Integrating Components.* Ensure the integration of decentralized information sources, like the REPIDISCA Thesaurus (more than 3000 terms in five languages: Spanish, English, Portuguese, French, and German, with its corresponding hierarchical structure and synonyms) and the Health and Environment Information Locator (gathers other Internet sources on environmental health, based on the Global Information Locator Service, GILS, adopted by the Global Program of the Information Society and the Dublin Core).

Building the virtual library implies the development, adoption, and adaptation of tools to operate information sources according to its architecture.

PAHO's Headquarters, in Washington, DC, hosts the virtual library because its server has higher connection speed. Statistics show that the number of visitors increases every year: 549,469 users have visited our website up to June 18, 2003 (Fig. 2).

Figure 3 shows the number of users per information source from January to June 2003. The source of full-text documents had the highest number of users, indicating that efforts should be concentrated on this information resource.

FLOOD OF PORTALS ON WATER   673

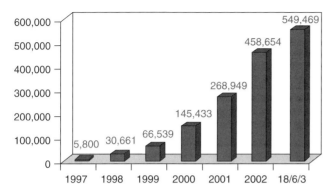

Figure 2. VLEH visitors, 1997–2003 (cumulative).

The VLEH includes 17 additional topics related to CEPIS/PAHO working scope and that of the Sustainable Development and Environmental Health Area of PAHO/WHO. In 2002, the topic with the highest number of visitors was Workers' Health, in 2001 it was the Assessment in Drinking Water and Sanitation, and in 2002 it was the Solid Wastes Assessment.

Throughout its 20 years, REPIDISCA has carried out several user surveys. The last one in 2000 focused on web users and 424 persons replied to the questionnaire. Figure 5 shows the percentage of users per sector.

Regarding the type of user, 35% of them were professionals and technicians, as shown in Fig. 6. It is worthwhile to note that the use of the VLEH increases during weekends.

Every year, comparisons of statistics are done to analyze the trend of the VLEH use, as shown in Fig. 4 regarding the number of visitors per water-related subject.

DIFFICULTIES

The difficulties that CEPIS/PAHO faces now with VLEH are:

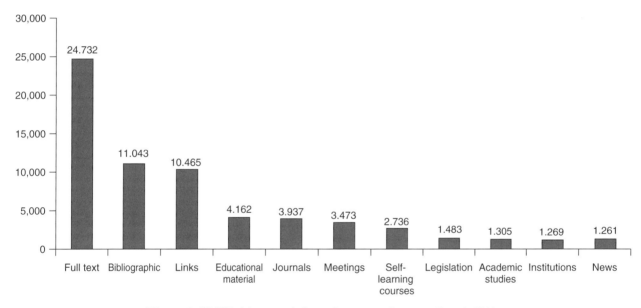

Figure 3. VLEH visitors per information source, January–June 6, 2003.

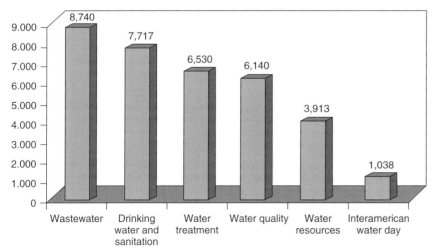

Figure 4. Visitors per water-related subject, January–June 2003.

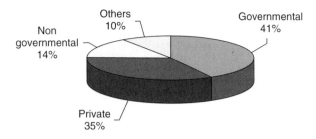

Figure 5. Users by sectors.

- Weakness of national institutions; in many countries water and sanitation utilities have undergone a privatization process.
- Staff reduction in public and private organizations; lack of investment in technical information and lack of resource sharing.
- Poor budgets for acquisition of information.
- Use of commercial software and methodologies in large universities that would not be willing to change their information system to follow the VLEH procedures; thus they need to duplicate the recording of the information they want to record into the VLEH.
- Inadequate updating of the VLEH at the regional and national levels.

Now CEPIS/PAHO has a great deal of information, but the VLEH cannot be just a space to make links with other websites. Users need a place where information is gathered, organized, and contained. Toward this end, CEPIS/PAHO and BIREME are implementing *Shared*, a Dutch technology that allows the production of "fingerprints" of every piece of information that is physically located in another web. These "fingerprints" are kept in a centralized file and weights may also be assigned to the topics of every information item. Institutions using the VLEH methodology will do this work more easily. In the future, it will be necessary to sign agreements to extract information automatically using *Shared*.

If *Shared* is implemented successfully, CEPIS would be able to enter into the Water Web Portal and other relevant sites updating their server of "fingerprints" and repeating the operation every month. The same procedure may be applied by each Cooperating Centre; thus national VLEH would be updated automatically. This will also avoid their regular export and sending of new information to CEPIS and the subsequent follow-up and control of duplicates that is done currently. Finally, regular training should be considered as a must to enable national institutions the use of the VLEH methodology. Workshops have been carried out in eleven countries and as a result they are already operating their own national VLEH.

FUZZY CRITERIA FOR WATER RESOURCES SYSTEMS PERFORMANCE EVALUATION

SLOBODAN P. SIMONOVIC
The University of Western Ontario
London, Ontario, Canada

INTRODUCTION

One of the main goals of engineering design is to ensure that a system performs satisfactorily under a wide range of possible future conditions. This premise is particularly true of large and complex water resources systems. Water resources systems usually include conveyance facilities such as pipes and pumps, treatment facilities such as sedimentation tanks and filters, and storage facilities such as reservoirs and tanks. These elements are interconnected in complicated networks serving broad geographical regions. Each element is vulnerable to temporary disruption in service due to natural hazards or human error whether unintentional, as in the case of operational errors and mistakes, or due to intentional causes such as a terrorist act. Most of the hazards cannot be controlled or predicted with an acceptable degree of accuracy. Uncontrollable external factors also affect the capacity and the performance of water resource systems. The determination of the load pattern presents unique challenges. Ang and Tang (1) point out that there is uncertainty in all engineering-based systems because these systems rely on the modeling of physical phenomena that are either inherently random or difficult to model with a high degree of accuracy.

The sources of uncertainty are many and diverse and, as a result, impose a great challenge to water resources systems design, planning, and management. The goal to ensure failsafe system performance may be unattainable. Adopting high safety factors is one means to avoid the uncertainty of potential failures. However, making

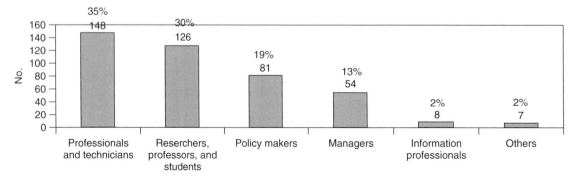

Figure 6. Types of user.

safety the first priority may render the system solution infeasible. Therefore, known uncertainty sources must be quantified. Engineering risk and reliability analysis is a general methodology for the quantification of uncertainty and the evaluation of its consequences for the safety of engineering systems (2). The first step in any risk analysis is to identify the risk, clearly detailing all sources of uncertainty that may contribute to the risk of failure. The quantification of risk is the second step, where the effects of the uncertainties are measured using different system performance indices and figures of merit. The early works of Hashimoto et al. (3,4) are the basis for the use of performance indices to evaluate the risk and reliability of water resources systems. They suggest *reliability, resiliency, vulnerability*, and *robustness* indices as criteria for evaluating the performance of water resources systems. Prior knowledge of the probability density functions of both resistance and load and/or their joint probability distribution function is a prerequisite to the probabilistic approach. In practice, data on previous failure experience is usually insufficient to provide such information. The subjective judgment of the decision-maker to estimate the probability distribution of a random event [subjective probability approach of Vick (5)] is another approach to deal with data insufficiency. The third approach is Bayes's theory, where engineering judgment is integrated with observed information. Until now the probabilistic approach was the only approach for water resources systems reliability analyses. But it fails to address the problem of uncertainty that goes along with human input, subjectivity, and lack of history and records. There is a practical and urgent need to investigate new approaches that can compensate for the ambiguity or uncertainty of human perception.

This contribution investigates the different approaches used to handle the problem of system reliability. Three new fuzzy reliability measures are presented (6): (1) the combined reliability–vulnerability index, (2) the robustness index, and (3) the resiliency index. They are developed to evaluate the operational performance of water resources systems. These measures could be useful decision-making aids in a fuzzy environment where subjectivity, human input, and lack of previous records impede the decision-making process.

FUZZY SETS

Zadeh (7), the founder of the theory of fuzzy sets, defines it as a formal attempt to capture, represent, and work with objects with unclear or ambiguous boundaries. This concept, although relatively new, has its origins in the early application of a multivalued logic notion to overcome the difficulties faced by the dual-logic representation in set theory. Water resources systems are potentially vulnerable to a wide variety of hazards that could limit their satisfactory performance. As a result, risks of future systems failure are often unavoidable (1). The engineering risk index characterizes the safety of water resources systems. The uncertainties associated with various sources of risk undermine the efficiency of this index. Because of these inevitable inefficiencies, all properties of water resources systems are subject to unavoidable and uncertain risk conditions. Different performance measures have been used in water resources systems reliability analyses to describe the system performance under extreme loading conditions (2). The majority of engineering reliability analyses rely on the use of a probabilistic approach. Both resistance and load are considered random variables. However, the characteristics of resistance and/or load cannot always be measured precisely or treated as random variables. Therefore, the fuzzy representation of either must be examined. The case of both fuzzy resistance and fuzzy load is rarely addressed in studies on the subject (6,8).

FUZZY CRITERIA FOR SYSTEM PERFORMANCE EVALUATION

Key Definitions

Failure. The calculation of performance indices depends on the exact definition of unsatisfactory system performance. It is difficult to arrive at a precise definition of failure because of the uncertainty in determining system resistance, load, and the accepted unsatisfactory performance threshold. Figure 1 depicts a typical system performance (resistance time series), with a constant load during the operation horizon. According to the classical definition, the failure state is the state when resistance falls below the load, margin of safety $M < 0.0$ or safety factor $\Theta < 1.0$, which is represented by the ratio between the system's resistance and load, shown in Fig. 1 by the dashed horizontal line.

Sometimes water resources systems fail to perform their intended function. For example, the available resistance from different sources in the case of water resources systems is highly variable. The actual load may also fluctuate significantly. Consequently, in the design of water resources systems, certain periods of water shortage may be a given. The precise identification of failure is neither realistic nor practical. It is more realistic to build in the inevitability of partial failure. A degree of acceptable system failure was introduced using the solid horizontal line, as shown in Fig. 1.

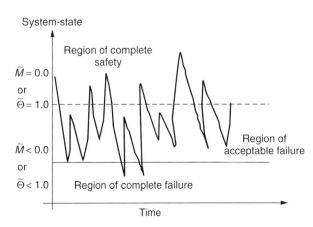

Figure 1. Variable system performance.

The boundary of the acceptable failure region is ambiguous and varies from one decision-maker to the other depending on personal perception of risk. Boundaries cannot be determined precisely. Fuzzy sets, on the other hand, are capable of representing the notion of imprecision better than ordinary sets. The acceptable level of performance can be represented as a fuzzy membership function in the following form:

$$\tilde{M}(m) = \begin{cases} 0, & \text{if } m \leq m_1 \\ \phi(m), & \text{if } m \in [m_1, m_2] \\ 1, & \text{if } m \geq m_2 \end{cases}$$

or

$$\tilde{\Theta}(\theta) = \begin{cases} 0, & \text{if } \theta \leq \theta_1 \\ \phi(\theta), & \text{if } \theta \in [\theta_1, \theta_2] \\ 1, & \text{if } \theta \geq \theta_2 \end{cases} \quad (1)$$

where $\tilde{M}$ is the fuzzy membership function of the margin of safety; $\phi(m)$ and $\phi(\theta)$ are functional relationships representing the subjective view of the acceptable risk; m_1 and m_2 are the lower and upper bounds of the acceptable failure region, respectively; $\tilde{\Theta}$ is the fuzzy membership function of the factor of safety; and θ_1 and θ_2 are the lower and upper bounds of the acceptable failure region, respectively.

Figure 2 is a graphical representation of the definition presented in Equation 1. The lower and upper bounds of the acceptable failure region are introduced in Equation 1 by m_1 (or θ_1) and m_2 (or θ_2). The value of the margin of safety (or factor of safety) below m_1 (or θ_1) is definitely unacceptable. Therefore, the membership function value is zero. The value of the margin of safety (or factor of safety) above m_2 (or θ_2) is definitely acceptable and therefore belongs in the acceptable failure region. Consequently, the membership value is one. The membership of the in-between values varies with the subjective assessment of a decision-maker. Different functional forms may be used for $\phi(m)$ [or $\phi(\theta)$] to reflect the subjectivity of different decision-makers' assessments.

High system reliability is reflected through the use of high values of the margin of safety (or factor of safety),

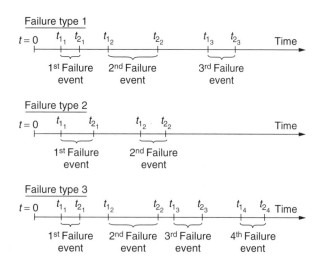

Figure 2. Fuzzy representation of acceptable failure region.

that is, high values for both m_1 and m_2 (or θ_1 and θ_2). The difference between m_1 and m_2 (or θ_1 and θ_2) inversely affects the system reliability: that is, the higher the difference, the lower the reliability. Therefore, the reliability reflected by the definition of an acceptable level of performance can be quantified in the following way:

$$LR = \frac{m_1 \times m_2}{m_2 - m_1} \quad \text{or} \quad LR = \frac{\theta_1 \times \theta_2}{\theta_2 - \theta_1} \quad (2)$$

where LR is the reliability measure of the acceptable level of performance.

The subjectivity of decision-makers will always result in a degree of ambiguity of risk perception. This alternate definition of failure allows for a choice among the lower bound, upper bound, and the function $\phi(m)$ [or $\phi(\theta)$]. This approach also provides an easy and comprehensive tool for risk communication.

Fuzzy System-State. System resistance and load can be represented in a fuzzy form to capture the uncertainty inherent in the system performance. The fuzzy form allows for the determination of the membership function of the resistance and load in a straightforward way even when there is limited available data. Fuzzy arithmetic can be used to calculate the resulting margin of safety (or factor of safety) membership function as a representation of the system-state at any time:

$$\tilde{M} = \tilde{X}(-)\tilde{Y} \quad \text{and} \quad \tilde{\Theta} = \tilde{X}(/)\tilde{Y} \quad (3)$$

where $\tilde{M}$ is the fuzzy margin of safety; $\tilde{X}$ is the fuzzy resistance capacity; $\tilde{Y}$ is the fuzzy load requirement; $(-)$ is the fuzzy subtraction operator; $(/)$ is the fuzzy division operator; and $\tilde{\Theta}$ is the fuzzy factor of safety.

Compatibility. The purpose of comparing two fuzzy membership functions is to illustrate the extent to which the two fuzzy sets match. Several classes of methods are available, none of which can be described as the best method. The reliability assessment, presented here, involves a comparative analysis of the system-state membership function and the predefined acceptable level of the performance membership function. Therefore, the compliance of two fuzzy membership functions can be quantified using the fuzzy compatibility measure.

Possibility and necessity lead to the quantification of the compatibility of two fuzzy sets. The possibility measure quantifies the overlap between two fuzzy sets, while the necessity measure describes the degree of inclusion of one fuzzy set into another fuzzy set (9). However, in some cases, high possibility and necessity values do not reflect clearly the compliance between the system-state membership function and the acceptable level of performance membership function. For example, let's consider the two system-state functions A and B with the same possibility and necessity values. However, assume that system-state A has a larger overlap with the performance membership function than the system-state B. The overlap area between the two membership functions, as a fraction of the total area of the system-state, illustrates compliance more clearly than the possibility and necessity measures; that is,

$$C_{S,L} = \frac{OA_{S,L}}{A_S} \quad (4)$$

where $C_{S,L}$ is the compliance between the system-state membership function (S) and the acceptable level of performance membership function (L); $OA_{S,L}$ is the overlap area between the system-state membership function (S) and the acceptable level of performance membership function (L); and A_S is the area of the system-state membership function (S). An overlap in a high significance area (area with high membership values) is preferable to an overlap in a low significance area. The compatibility measure can be calculated using

$$CM_{S,L} = \frac{WOA_{S,L}}{WA_S} \quad (5)$$

where $CM_{S,L}$ is the compatibility measure between the system-state membership function (S) and the acceptable level of performance membership function (L); $WOA_{S,L}$ is the weighted overlap area between the system-state membership function (S) and the acceptable level of performance membership function (L); and WA_S is the weighted area of the system-state membership function (S).

Combined Fuzzy Reliability–Vulnerability Criteria

Reliability and vulnerability were used to provide a complete description of system performance in case of failure and to determine the magnitude of the failure event. Once an acceptable level of performance is determined in a fuzzy form, the anticipated performance in the event of failure as well as the expected severity of failure can be determined.

When certain values are specified for the lower and upper bounds [m_1 and m_2 (or θ_1 and θ_2) in Equation 1], thus establishing a predefined acceptable level of performance, the anticipated system failure is limited to a specified range. Systems that are highly compatible with the predefined acceptable level of performance will yield a similar performance; that is, the expected system failure will be within the specified range ([m_1, m_2] or [θ_1, θ_2]). In order to calculate system reliability, several acceptable levels of performance must be defined to reflect the different perceptions of the decision-makers. A comparison between the fuzzy system-state membership function and the predefined fuzzy acceptable level of performance membership function provides information about both system reliability and system vulnerability at the same time (Fig. 3).

The system reliability is based on the proximity of the system-state to the predefined acceptable level of performance. The measure of proximity is expressed by the compatibility measure suggested in Equation 5. The new combined fuzzy reliability–vulnerability index is formulated as follows:

$$RE_f = \frac{\max_{i \in K}\{CM_1, CM_2, \ldots, CM_i\} \times LR_{\max}}{\max_{i \in K}\{LR_1, LR_2, \ldots, LR_i\}} \quad (6)$$

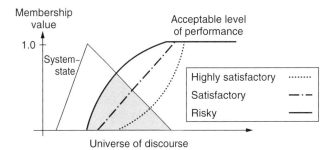

Figure 3. Compatibility with different level of performance membership functions.

where RE_f is the new combined fuzzy reliability–vulnerability index; $LR_{\max}$ is the reliability measure of acceptable level of performance corresponding to the system-state with maximum compatibility value; LR_i is the reliability measure of the ith acceptable level of performance; CM_i is the compatibility measure for the system-state with the ith acceptable level of performance; and K is the total number of the defined acceptable levels of performance. The reliability–vulnerability index is normalized to attain a maximum value of 1.0, by the introduction of the $\max_{i \in K}\{LR_1, LR_2, \ldots, LR_i\}$ value as the maximum achievable reliability.

Fuzzy Robustness

Robustness measures the system's ability to adapt to a wide range of possible future load conditions, at little additional cost (4). The fuzzy form of change in future conditions can be obtained through a redefinition of the acceptable level of performance and a change in the system-state membership function. As a result, the system's robustness is defined as the change in the compatibility measure:

$$RO_f = \frac{1}{CM_1 - CM_2} \quad (7)$$

where RO_f is the new fuzzy robustness index; CM_1 is the compatibility measure before the change in conditions; and CM_2 is the compatibility after the change in conditions.

Equation 7 reveals that the higher the change in compatibility, the lower the value of the fuzzy robustness. Therefore, high robustness values allow the system to better adapt to new conditions.

Fuzzy Resiliency

The time required to recover from the failure state can be represented as a fuzzy set. Because the reasons for a failure may be different, system recovery times will vary depending on the type of failure, as shown in Fig. 4. A series of fuzzy membership functions can be developed to allow for various types of failure. The maximum recovery time is used to represent the system recovery time (10):

$$\tilde{T}(\alpha) = (\max_{j \in J}[t_{1_1}(\alpha), t_{1_2}(\alpha), \ldots, t_{1_J}(\alpha)]$$
$$\times \max_{j \in J}[t_{2_1}(\alpha), t_{2_2}(\alpha), \ldots, t_{2_J}(\alpha)]) \quad (8)$$

where $\tilde{T}(\alpha)$ is the system fuzzy maximum recovery time at α level; $t_{1,j}(\alpha)$ is the lower bound of the jth recovery time at α level; $t_{2,j}(\alpha)$ is the upper bound of the jth recovery time at α level; and J is total number of failure events.

The center of gravity of the maximum fuzzy recovery time can be used as a real number representation of the system recovery time. Therefore, system resilience is determined to be the inverse value of the center of gravity:

$$RS_f = \left[\frac{\int_{t_1}^{t_2} t\, \tilde{T}(t)\, dt}{\int_{t_1}^{t_2} \tilde{T}(t)\, dt} \right]^{-1} \quad (9)$$

where RS_f is the new fuzzy resiliency index; $\tilde{T}(t)$ is the system fuzzy maximum recovery time; t_1 is the lower bound of the support of the system recovery time; and t_2 is the upper bound of the support of the system recovery time.

The inverse operation can be used to illustrate the relationship between the value of the recovery time and the resilience. The longer the recovery time, the lower the system's ability to recover from the failure, and the lower the resilience.

CONCLUSIONS

Water resources systems are vulnerable to a wide variety of hazards that could potentially limit their ability to perform satisfactorily. The diversity of uncertainty sources presents a great challenge to water resources systems design, planning, and management. The probabilistic approach usually fails to address the problems of human error, subjectivity, and the lack of system performance history and records. The fuzzy set approach addresses those issues. A fuzzy system reliability analysis is ideally based on the comparison between the fuzzy sets representing both the system's state of safety and the potential for system failure.

A fuzzy reliability–vulnerability measure clearly quantifies the reliability and the vulnerability of multicomponent systems. The quantification is based on the use of appropriate fuzzy compatibility measures to illustrate the relationship between the fuzzy system's state of safety and the fuzzy failure events. The contribution also shows a fuzzy measure, capable of determining the system's ability to adapt to changing conditions. This measure of fuzzy robustness provides a vital tool to assess the system's behavior through the introduction of a wide variety of uncertain conditions. A fuzzy resiliency measure was also presented to capture the system's response to uncertain future failure events. This measure is able to incorporate all types of system responses to potential failure events throughout the life of the design.

The presented fuzzy performance criteria were evaluated using two simple hypothetical cases by El-Baroudi and Simonovic (6).

BIBLIOGRAPHY

1. Ang, H.-S. and Tang, H. (1984). *Probability Concepts in Engineering Planning and Design*. John Wiley & Sons, Hoboken, NJ.
2. Ganoulis, J.G. (1994). *Engineering Risk Analysis of Water Pollution: Probabilities*. VCH, Weinheim, The Netherlands.
3. Hashimoto, T., Stedinger, J.R., and Loucks, D.P. (1982). Reliability, resiliency, and vulnerability criteria for water resources system performance evaluation. *Water Resour. Res.* **18**(1): 14–20.
4. Hashimoto, T., Loucks, D.P., and Stedinger, J.R. (1982). Robustness of water resources systems. *Water Resour. Res.* **18**(1): 21–26.
5. Vick, S. (2002). *Degrees of Belief: Subjective Probability and Engineering Judgment*. ASCE Press, Reston, VA.
6. El-Baroudy, I. and Simonovic, S.P. (2004). Fuzzy criteria for the evaluation of water resources systems performance. *Water Resour. Res.* **40**(10): W10503.
7. Zadeh, L.A. (1965). Fuzzy sets. *Inf. Control* **8**: 338–353.
8. Shersrtha, B. and Duckstein, L. (1998). A fuzzy reliability measure for engineering applications. In: *Uncertainty Modelling and Analysis in Civil Engineering*. CRC Press, Boca Raton, FL, pp. 120–135.
9. Pedrycz, W. and Gomide, F. (1998). *An Introduction to Fuzzy Sets*. MIT Press, Cambridge.
10. Kaufmann, A. and Gupta, M. (1985). *Introduction to Fuzzy Arithmetic: Theory and Applications*. Van Nostrand Reinhold, New York.

PARTICIPATORY MULTICRITERIA FLOOD MANAGEMENT

SLOBODAN P. SIMONOVIC
The University of Western Ontario
London, Ontario, Canada

INTRODUCTION

Flood management in general comprises different water resources activities aimed at reducing the potentially harmful impact of floods on people, the environment, and the economy of a region. Sustainable flood management decision-making requires integrated consideration of economic, ecological, and social consequences of disastrous

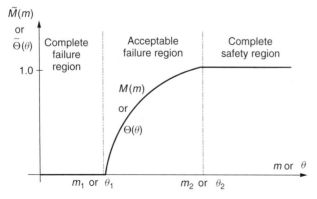

Figure 4. Recovery times for different types of failure.

flood. While economic consideration gets priority in traditional approaches to decision-making, empowerment of stakeholders is an issue that is demanding increasing attention today in many decision-making processes. Flood management activities (i.e., disaster mitigation, preparedness, and emergency management) may be designed and achieved without the direct participation of stakeholders; however, it cannot be implemented without them (1). In order to decide what flood control measures to adopt in a floodplain, the decision-making process should include different stakeholders.

Government policy-makers and professional planners are the first to consider. However, others such as the general public, communities affected by the decision outcomes, nongovernmental organizations, and different interest groups should be included as well. This contribution describes a multicriteria decision-making methodology for participatory flood management (2,3). The methodology should (1) evaluate potential alternatives based on multiple criteria under uncertainty; (2) accommodate the high diversity and uncertainty inherent in human preferences; and (3) handle a large amount of data collected from stakeholders.

METHODOLOGY

Flood management decision-making problems are complex due to their multicriteria nature. For a given goal, many alternative solutions may exist that provide different levels of satisfaction for different issues, such as environmental, social, institutional, and political. These concerns naturally lead to the use of multicriteria decision-making techniques, in which trade-offs among the single objectives can lead to the most desirable solution. Multicriteria decision-making becomes more complicated with the increase in the number of individuals/groups involved in the decision-making process. In reality, the decision-making process often involves multiple stakeholders/decision-makers. Multiple stakeholders' participation introduces a great deal of complexity into the analysis. The decision problem is no longer limited to the selection of the most preferred alternative among the possible solutions according to a single set of preferences. The analysis must also be extended to account for the conflicts among different stakeholders with different objectives. Most of the existing approaches in multicriteria decision-making with a single stakeholder/decision-maker consist of two phases (4): (1) the aggregation of the judgments with respect to all criteria and per decision alternatives; and (2) the ranking of the decision alternatives according to the aggregated judgment. In the case of multiple stakeholders, an additional aggregation is necessary with respect to the judgments of all the stakeholders.

Consider a multicriteria multiparticipant decision-making problem where m alternatives are to be evaluated by n decision-makers, who are using p objectives. The general conceptual decision matrix for the discrete multicriteria multiparticipant problem is shown in Table 1.

In Table 1, **A** denotes the alternative, **O** is the criterion, and **DM** is the decision-maker/stakeholder. The preference

Table 1. Conceptual Decision Matrix for a Discrete Multicriteria Multiparticipant Decision Problem

	O_1	...	O_p
A_1	a_{11}	...	A_{1p}
...	...	...	...
A_m	a_{m1}	...	a_{mp}
DM_1	w_{11}	...	w_{1p}
...	...	...	...
DM_n	w_{n1}	...	w_{np}

of the decision-maker k ($k = 1, \ldots, n$) for the objective j ($j = 1, \ldots, p$) is expressed by w_{jk}, and a_{ij} is the performance evaluation of the alternative i ($i = 1, \ldots, m$) for each objective j.

The classical outcome of the decision matrix is the ranking of the alternatives. To obtain that, a number of steps are necessary, such as establishing the preference structure, the weights, and also the performance evaluations. All these can be termed as the inputs for the decision matrix. These inputs come from the stakeholder/decision-maker. The decision matrix shows that the inputs can be for the preference of criteria as well as for the performance evaluations. The decision-maker might also have a preference structure for the alternatives. In the case of a multiparticipant decision-making problem, these inputs are to be collected from all the stakeholders.

Equation 1 is a general mathematical formulation of this multicriteria, multiparticipant problem. A payoff matrix can be obtained for the problem where m alternatives are to be evaluated by n stakeholders/decision-makers, who are using p criteria:

$$A^k = (a_{ij})^k = \begin{bmatrix} a_{11} & \ldots & \ldots & a_{1p} \\ a_{21} & \ldots & \ldots & a_{2p} \\ \ldots & \ldots & \ldots & \ldots \\ a_{m1} & \ldots & \ldots & a_{mp} \end{bmatrix}, \quad (k = 1, \ldots, n) \quad (1)$$

Here $A_i^k = (a_{i1}, \ldots, a_{ip})^k$ means that alternatives i are being evaluated by criteria from 1 to p by decision-maker k. The symbol $A_j^k = (a_{1j}, \ldots, a_{mj})^k$ means that the objective j is being used by decision-maker k to evaluate all alternatives from 1 to m.

The solution to this problem is to have each alternative evaluated by all the decision-makers using all criteria. The process can be summarized as the following mapping function:

$$\Psi : \{A^k | k = 1, \ldots, n\} \to \{G\} \quad (2)$$

where G is a collective weighted agreement matrix. It is crucial that this mapping function represents all criteria that the decision-makers use in judging all the alternatives.

Flood management decision-making is always associated with some degree of uncertainty. This uncertainty could be categorized into two basic types: uncertainty caused by inherent hydrologic variability and uncertainty due to a lack of knowledge (5). Uncertainty of the first type is associated with the spatial and temporal changes of hydrologic variables such as flow, precipitation, and water quality. The second type of uncertainty occurs when the particular value of interest cannot be assessed exactly because of the limitation in the available knowledge. The second type of decision uncertainty is more profound in the area of public decision-making such as in the case of flood management. Capturing the views of individuals presents the problem of uncertainty. The major challenge while collecting these views is to find out the technique that will capture those uncertainties and also will be usable in a multicriteria tool.

Participation of Multiple Stakeholders

An aggregation procedure is one of the ways to include information from the participating decision-makers into the decision matrix. The available methods do not seem to be appropriate for flood management for two reasons. The methodology (3) includes representation of inputs from a large number of participants and the analysis of inputs to make them usable for application to various multicriteria decision-making methods. Fuzzy set theory and fuzzy logic are used to represent the uncertainties in stakeholders' opinions. Three possible types of fuzzy input have been considered to capture the subjectivity of the responses from stakeholders. When a stakeholder is asked to evaluate an alternative against a particular criterion, the answer may take one of the following forms: (1) a numeric scale response; (2) a linguistic answer (e.g., poor, fair, good, very good); or (3) an argument (e.g., "if some other condition is satisfied then it is good"). For the first type, the input is quite straightforward. For a type (2) answer, it will be necessary to develop the membership functions for the linguistic terms. Type (3) input can be described by using a fuzzy inference system, which includes membership functions, fuzzy logic operators, and the if–then rule. For this, the membership functions for the input arguments need to be developed first. Then fuzzy operator and fuzzy logic are applied to obtain the output. It should be noted that the interpretation of type (2) and type (3) input values is highly dependent on the shape of the membership functions and the degree of severity chosen by the expert for a particular application.

After receiving the inputs from all stakeholders, the next step is to aggregate those inputs to find a representative value. It is obvious that for all input types considered above, the responses are sure to be influenced by a number of repetitions. This means many respondents can provide the same response. This implies that the general methodologies of fuzzy aggregation cannot be applied for deriving the resultant input from a large number of decision-makers. The fuzzy expected value (FEV) method can be used instead to get the resulting opinion of the stakeholders. Following is the definition of the fuzzy expected value: Let χ_A be a B-measurable function such that $\chi_A \in [0, 1]$. The FEV of χ_A over the set A, with respect to the fuzzy measure μ, is defined as

$$\text{FEV}(\chi_A) = \sup_{T \in [0,1]} \{\min[T, \mu(\xi_T)]\} \quad (3)$$

where

$$\xi_T = \{x | \chi_A(x) \geq T\} \quad (4)$$

and

$$\mu\{x | \chi_A(x) \geq T\} = f_A(T) \quad \text{is a function of the threshold } T \quad (5)$$

Figure 1 provides a geometric interpretation of the FEV. Performing the minimum operator, the two curves create the boundaries for the remaining triangular curve. The supremum operator returns the highest value of $f_A(T)$, which graphically represents the highest point of the triangular curve. This corresponds to the intersection of the two curves where $T = H$.

The FEV can be computed for all three types of inputs mentioned earlier in this section. For type (1) input, the resultant FEV should be a numeric value between 0 and 1. For both type (2) and type (3) inputs, the resultant FEVs are membership functions. The crisp numeric equivalents of these membership functions can be obtained by applying a defuzzification method and can then be compared with type (1) answers. The centroid of area defuzzification method has been used to return a value obtained by averaging the moment area of a given fuzzy set. Mathematically, the centroid, $\bar{x}$, of a fuzzy set, A, is defined as

$$\bar{x} = \frac{\int_0^1 x \cdot \mu_A(x) dx}{\int_0^1 \mu_A(x) dx} \quad (6)$$

where $\mu_A(x)$ is the membership function of the fuzzy set A.

The resultant FEVs are now the aggregated evaluation of the alternatives from all the stakeholders. They can now be used as the input value in the decision matrix for the multicriteria analysis.

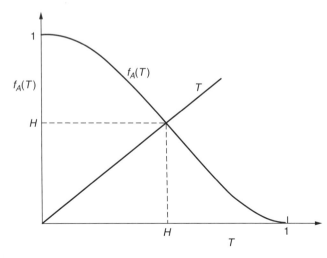

Figure 1. Participatory multicriteria flood management.

Participatory Multicriteria Decision-making Under Uncertainty

Here an innovative modification has been made to the compromise programming multicriteria decision-making technique to accommodate participatory flood decision-making under uncertainty. Bender and Simonovic (6) fuzzified compromise programming entirely and thus formulated fuzzy compromise programming (FCP). The driving force for the transformation from a classical to a fuzzy environment is that there is a need for accurate representation of subjective data in the flood decision-making. It is the theory of fuzzy sets that can represent the subjective data well. Thus, instead of using crisp numbers in the compromise programming distance metric equation, fuzzy numbers are used; instead of using classical arithmetic, fuzzy arithmetic is applied; instead of simply sorting distance metrics, fuzzy set ranking methods must be applied to sort the fuzzy distance metrics. In other words, the fuzzy transformation complicates the interpretation of the results but, on the other hand, models the decision-making process more realistically. Mathematically, the compromise programming distance metric in its discrete form can be presented as

$$L_j = \left[\sum_{z=1}^{t}\left\{w_z^p \left(\frac{f_z^* - f_z}{f_z^* - f_z^-}\right)^p\right\}\right]^{1/p} \quad (7)$$

where $z = 1, 2, 3, \ldots, t$ and represents t criteria; $j = 1, 2, 3, \ldots, n$ and represents n alternatives; L_j is the distance metric of alternative j; w_z corresponds to a weight of a particular criterion; p is a parameter ($p = 1, 2, \infty$); f_z^* and f_z^- are the best and the worst value for criterion z, respectively (also referred to as positive and negative ideals); and f_z is the actual value of criterion z. The parameter p is used to represent the importance of the maximal deviation from the ideal point. Varying the parameter p from 1 to infinity allows one to move from minimizing the sum of individual regrets (i.e., having a perfect compensation among the criteria) to minimizing the maximum regret (i.e., having no compensation among the criteria) in the decision-making process. The choice of a particular value of this compensation parameter p depends on the type of problem and desired solution. The weight parameter, w_z, characterizes decision-makers' preference concerning the relative importance of criteria. Simply stated, the parameter places emphasis on the criteria that the decision-maker deems important. The parameter is needed because different participants in the decision-making process have different viewpoints concerning the importance of a criterion.

Bender and Simonovic (6) fuzzified compromise programming and thus formulated fuzzy compromise programming (FCP). The driving force for the transformation from a classical to a fuzzy environment is that there is a need for accurate representation of subjective data in the flood decision-making. It is the theory of fuzzy sets that can represent the subjective data well. Thus, instead of using crisp numbers in the distance metric, Equation 7, fuzzy numbers are used; instead of using classical arithmetic, fuzzy arithmetic is applied; instead of simply sorting distance metrics, fuzzy set ranking methods must be applied to sort the fuzzy distance metrics. In other words, the fuzzy transformation complicates the interpretation of the results but, on the other hand, models the flood decision-making process more realistically.

In fuzzy compromise programming, obtaining the smallest distance metric values is not easy, because the distance metrics are also fuzzy. To pick out a smallest fuzzy distance metric, from a group of distance metrics, fuzzy set ranking methods have to be used. A study by Prodanovic and Simonovic (7) compared fuzzy set ranking methods for use in fuzzy compromise programming and recommended using the method of Chang and Lee (8). This recommendation was founded on the fact that Chang and Lee's (8) method gave the most control in the ranking process—with degree of membership weighting and the weighting of the subjective type. The overall existence ranking index (OERI) suggested by Chang and Lee (8) has the following mathematical form:

$$\text{OERI}(j) = \int_0^1 \omega(\alpha)[\chi_1 \mu_{jL}^{-1}(\alpha) + \chi_2 \mu_{jR}^{-1}(\alpha)]d\alpha \quad (8)$$

where the subscript j stands for alternative j, while α represents the degree of membership; χ_1 and χ_2 are the subjective type weighting indicating neutral, optimistic, or pessimistic preferences of the decision-maker, with the restriction that $\chi_1 + \chi_2 = 1$; parameter $\omega(\alpha)$ is used to specify weights, which are to be given to certain degrees of distance metric membership (if any); and $\mu_{jL}^{-1}(\alpha)$ represents an inverse of the left part, and $\mu_{jR}^{-1}(\alpha)$ the inverse of the right part of the distance metric membership function.

For χ_1 values greater than 0.5, the left side of the membership function is weighted more than the right side, which in turn makes the decision-maker more optimistic. Of course, if the right side is weighted more, the decision-maker is more of a pessimist (this is because he/she prefers larger distance metric values, which means the solution farther from the ideal solution). In summary, the risk preferences are: if $\chi_1 < 0.5$, the user is a pessimist (risk averse); if $\chi_1 = 0.5$, the user is neutral; and if $\chi_1 > 0.5$, the user is an optimist (risk taker). Simply stated, Chang and Lee's (8) overall existence ranking index is a sum of the weighted areas between the distance metric membership axis and the left and right inverses of a fuzzy number.

FLOOD MANAGEMENT IN THE RED RIVER BASIN, MANITOBA, CANADA

The proposed methodology is applied to flood management in the Red River Basin (9–11). One of the flood management problems at the planning stage in the Red River Basin is the complex, large-scale problem of ranking potential flood management alternatives. During the evaluation of alternatives, it is necessary to consider multiple criteria that may be quantitative and qualitative. The flood management process in the basin also involves numerous stakeholders. They include different levels of government, different agencies, private organizations, interest groups, and the general public. They all have different and specific

Table 2. Resultant FEVs

Alternative Type	Structural			Nonstructural			Combination		
	A	B	C	A	B	C	A	B	C
Question Number	FEV	FEV	FEV	FEV	FEV	FEV	FEV	FEV	FEV
Community Involvement									
1	0.600	0.650	0.544	0.647	0.650	0.544	0.600	0.625	0.544
2	0.529	0.517	0.500	0.500	0.517	0.491	0.500	0.570	0.544
3	0.618	0.700	0.529	0.559	0.625	0.529	0.600	0.625	0.544
4	0.600	0.650	0.544	0.657	0.650	0.559	0.686	0.650	0.544
5	0.700	0.700	0.559	0.629	0.650	0.544	0.700	0.650	0.544
6a	0.800	0.825	0.677	0.704	0.770	0.588	0.800	0.825	0.647
6b	0.771	0.770	0.588	0.714	0.717	0.574	0.743	0.770	0.574
6c	0.700	0.700	0.574	0.629	0.650	0.574	0.686	0.700	0.574
7	0.800	0.825	0.735	0.829	0.850	0.718	0.857	0.825	0.718
8	0.700	0.717	0.574	0.700	0.650	0.574	0.700	0.700	0.574
Personal Loss									
1	0.800	0.770	0.718	0.700	0.700	0.574	0.700	0.717	0.671
2	0.588	0.570	0.544	0.600	0.650	0.544	0.600	0.625	0.574
3a	0.500	0.570	0.574	0.559	0.625	0.574	0.559	0.570	0.574
3b	0.700	0.717	0.625	0.700	0.717	0.588	0.706	0.717	0.588
4	0.771	0.770	0.574	0.700	0.650	0.574	0.700	0.717	0.544
5	0.500	0.570	0.529	0.700	0.570	0.544	0.571	0.570	0.544

needs and responsibilities during all stages of flood management—planning, emergency management, and flood recovery. There has been increasing concern by the general public about the decisions to be taken on the selection of flood control measures. During the 1997 flood, it was indicated that certain stakeholders in the basin, particularly the floodplain residents, did not have adequate involvement in flood management decision-making. Dissatisfaction has been observed among the stakeholders about evacuation decisions during the emergency management and about compensation decisions during the postflood recovery (10).

The methodology presented in the previous section has been used to collect information from the stakeholders across the Canadian portion of the Red River Basin. In order to evaluate the utility of the methodology, a generic experiment was considered for the study to evaluate three alternative options for improved flood management. Three generic options considered are (1) structural alternatives, (2) nonstructural alternatives, and (3) a combination of both. The selection of criteria against which the alternatives are ranked is one of the most difficult but important tasks of any multicriteria decision analysis. The following two social objectives have been considered in our case study: level of community involvement and amount of personal losses (include financial, health, and psychological losses). A detailed survey has been conducted in the basin to collect the information on the two selected social criteria (12). All three types of inputs obtained from all the stakeholders were processed using the fuzzy expected value method. Table 2 summarizes the results of all three types of inputs (scale, linguistic, and conditional types, which are termed A, B, and C, respectively, in the table) as the evaluation of three alternatives (structural, nonstructural, combination) against two criteria (community development, personal loss). Obtained results show good correlation between the numeric scale type and linguistic type of inputs with an average difference of only 0.029.

The FEVs obtained in Table 2 are used further to rank the three generic alternatives. All questions are considered to carry the same weight. A set of ranking experiments has been conducted to evaluate the impact of different stakeholder groups on the final rank of alternatives: (a) experiment 1—all stakeholders interviewed; (b) experiment 2—stakeholders from the city of Winnipeg; (c) experiment 3—stakeholders from the Morris area (south of Winnipeg); and (d) experiment 4—stakeholders from the Selkirk area (north from Winnipeg).

The final results of four ranking experiments with three generic alternatives and two social criteria are shown in the Table 3 (defuzzified distance metric value and the rank in parentheses). It is obvious that the final rank varies with the experiment, therefore confirming that preferences of different stakeholders are being captured by the developed methodology.

CONCLUSIONS

The analyses of flood management options in the Red River Basin, Manitoba, Canada (3) show the applicability of the methodology for a real flood management decision-making problem. The stakeholders can now express their concerns regarding flood hazard in an informal way, and

Table 3. Final Rank of Flood Management Alternatives

Participants	Alternative 1	Alternative 2	Alternative 3
All stakeholders	13.22 (1)	13.72 (3)	13.29 (2)
Morris	15.43 (2)	16.09 (3)	13.63 (1)
Selkirk	14.63 (3)	14.42 (1)	14.58 (2)
Winnipeg	13.74 (1)	15.25 (3)	13.92 (2)

that can be incorporated into the multicriteria decision-making model. The application of methodology helps in solving the problem of incorporating a large number of stakeholders in the flood decision-making process.

BIBLIOGRAPHY

1. Affeltranger, B. (2001). *Public Participation in the Design of Local Strategies for Flood Mitigation and Control*, UNESCO IHP-V, Technical Documents in Hydrology, Vol. 48.
2. Simonovic, S.P. (2004). Sustainable floodplain management—participatory planning in the Red River Basin, Canada. In: *Proceedings of the Workshop on Modelling and Control for Participatory Planning and Managing Water Systems*, IFAC, CD-ROM.
3. Akter, T. and Simonovic, S.P. (2004). Aggregation of fuzzy views of a large number of stakeholders for multiobjective flood management decision making. *J. Environ. Manage.* **22**(1–2): 57–72.
4. Zimmerman, H.J. (2001). *Fuzzy Set Theory—and Its Application*, Academic Publishers, Boston.
5. Simonovic, S.P. (2000). Tools for water management: one view of the future. *Water Int. IWRA* **25**(1): 1–8.
6. Bender, M.J. and Simonovic, S.P. (2000). A fuzzy compromise approach to water resources planning under uncertainty. *Fuzzy Sets Syst.* **115**(1): 35–44.
7. Prodanovic, P.P. and Simonovic, S.P. (2002). Comparison of fuzzy set ranking methods for implementation in water resources decision-making. *Can. J. Civil Eng.* **29**: 692–701.
8. Chang, P.T. and Lee, E.S. (1994). Ranking of fuzzy sets based on the concept of existence. *Comput. Math. Appl.* **27**: 1–21.
9. Simonovic, S.P. (1999). Decision support system for flood management in the Red River Basin. *Can. Water Res. J.* **24**(3): 203–223.
10. International Joint Commission (2000). *Living with the Red*. Ottawa, Washington. Available at http://www.ijc.org/php/publications/html/living.html.
11. Simonovic, S.P. and Carson, R.W. (2003). Flooding in the Red River Basin—lessons from post flood activities. *Nat. Hazards* **28**: 345–365.
12. Salonga, J. (2004). Aggregation methods for multi-objective flood management decision making with multiple stakeholders: Red River Basin case study [Civil Engineering thesis]. The University of Western Ontario, London.

WATER RESOURCES SYSTEMS ANALYSIS

SLOBODAN P. SIMONOVIC
The University of Western Ontario
London, Ontario, Canada

INTRODUCTION

Water resources planning and management process is a search for the solution of how to meet the needs of a population with the available resources (1). Water resources planning and management is as old as humanity. However, with knowledge and technology development, a change in the living standard of people, and further economic development, the analysis procedure changes.

Principal Objectives for Industrialized Countries

An example of the principal water resources planning and management objectives for industrialized countries is based on the *Principles and Standards for Planning Water and Related Land Resources* used in the United States and introduced by the U.S. Water Resources Council in 1973 and modified in 1979 and 1980 (2). According to them the overall purpose of water resources planning and management is improvement of the quality of life through contributions to national economic development, environmental quality, regional economic development, and other social effects. Under these "principles and standards," the water resources planning process consists of six major steps: (1) specification of the water and related land resources problems and opportunities; (2) inventory, forecast, and analysis of water and related land resource conditions within the planning area relevant to the identified problems and opportunities; (3) formulation of alternative plans; (4) evaluation of the effects of the alternative plans; (5) comparison of alternative plans; and (6) selection of a recommended plan based on the comparison of alternative plans.

Principal Objectives for Developing Countries

Water resources planning and management in developing countries should correspond to a set of criteria distinct from those used in industrialized countries. These criteria should reflect prevailing constraints on physical, financial, and human resources and the need to allocate critically sparse resources to programs that correspond to short- and long-term sociopolitical objectives and promise to be the most cost effective.

The United Nations Industrial Development Organization (UNIDO) (3) recommended that planning analysis consider the following objectives: (1) aggregate consumption, (2) income redistribution, (3) growth rate of national income, (4) employment level, (5) self-reliance, and (6) merit wants. The UN publication treats the problems of evaluating the extent to which projects advance each of the objectives and presents their combination as a measure of "aggregate economic profitability." The UNIDO guidelines, in developing a system of objectives, do not lay any stress on the quality of the environment or other intangible descriptors applied to the quality of human life.

SYSTEMS ANALYSIS

Systems analysis is the use of rigorous methods to help determine preferred plans and designs for complex, often large-scale systems. It combines knowledge of the available analytic tools, understanding of when each is more appropriate, and skill in applying them to practical problems. It is both mathematical and intuitive, as is all planning and design.

Systems analysis is a relatively new field. Its development parallels that of the computer, the computational power of which enables us to analyze complex relationships, involving many variables, at reasonable cost. Most of its techniques depend on the use of the computer for practical applications. Systems analysis may be thought of as the set of computer-based methods essential for the planning of major water resources projects. It is thus central to a modern water resources engineering curriculum.

Systems analysis covers much of the same material as operations research, in particular, linear and dynamic programming and decision analysis. The two fields differ substantially in direction, however. Operations research tends to be interested in specific techniques and their mathematical properties. Systems analysis focuses on the use of the methods.

Systems analysis emphasizes the kinds of real problems to be solved; considers the relevant range of useful techniques, including many besides those of operations research; and concentrates on the guidance they can provide toward improving plans and designs. Use of systems analysis instead of the more traditional set of tools generally leads to substantial improvements in design and reductions in cost. Gains of 30% are not uncommon. These translate into an enormous advantage when one is considering projects worth tens and hundreds of millions of dollars.

DEFINITIONS

There are many variations in the definition of what a system is, but all of the definitions share many common traits. Some kind of system is inherent in all but the most trivial water resources engineering planning and design problems. To understand a problem, the engineer must be able to recognize and understand the system that surrounds and includes it. Some of the reasons for poor system definition in former projects include poor communications, lack of knowledge of interrelationships, politics, limited objectives, and transportation difficulties.

What, then, is a system? The dictionary definition of the term "system" is a mass of verbiage providing no less than 15 ways to define the word. In the most general sense, a system may be defined as a collection of various structural and nonstructural elements that are connected and organized in such a way as to achieve some specific objective by the control and distribution of material resources, energy, and information.

A more formal definition of a system can be stated as

$$S : \mathbf{X} \to \mathbf{Y} \qquad (1)$$

where $\mathbf{X}$ is an input vector and $\mathbf{Y}$ is an output vector. So, a system is a set of operations that transforms input vector $\mathbf{X}$ into output vector $\mathbf{Y}$.

The usual representation of the system definition is presented in Fig. 1. The system's objects are input, output, process, feedback, and a restriction. Input energizes the operation of a given process. The final state of the process is known as the output. Feedback performs a number of operations to compare the actual output with an objective and identifies the discrepancies that exist between them.

To avoid any misunderstanding, let's define some terms often in use: *mathematical model*—a set of equations that describes and represents the real system; *decision variables*—the controllable and partially controllable inputs; *policy*—resulting set of decision variables, when each decision variable is assigned a particular value; *objective function*—quantity used to measure the effectiveness of a particular policy, expressed as a function of the decision variables; *constraints*—physical, economic, or any other restrictions applied to the model; *feasible policy*—a policy that does not violate any constraints; and *policy space*—the subset consisting of all possible feasible policies.

WATER RESOURCES SYSTEM

General formulation of water resources system (4) may be presented as the transformation of available water resources,

$$W_A = \{Q_A, K_A, L_A\} \qquad (2)$$

into required water resources,

$$W_D = \{Q_D, K_D, L_D\} \qquad (3)$$

taking into consideration water quality protection, flood control, and regional and national development plans. Notation used in Equations 2 and 3 includes: Q_A—available quantity of water; Q_D—required quantity of water; K_A—quality of available water; K_D—required water quality; L_A—location of available resources; and L_D—location of demand points.

Mathematical expression of the transformation problem can be given as

$$W_D = T \times W_A \qquad (4)$$

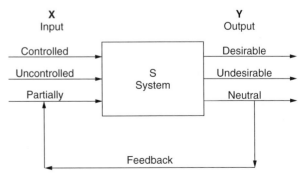

Figure 1. System presentation.

This transformation may be performed through the following three processes:

1. *Water Resources Planning*—formulation of goals and objectives that are consistent with political, social, environmental, economic, technological, and aesthetic constraints; and the general definition of procedures designed to meet those goals and objectives.
2. *Water Resources Design*—selection of a particular combination of resources and a way to use them.
3. *Water Resources Operations*—implementation of procedures in order to achieve preselected goals and objectives.

SYSTEMS APPROACH

The systems approach is a general problem-solving technique that brings more objectivity to the water resources planning/design processes. It is, in essence, good design; a logical and systematic approach to problem solution in which assumptions, goals, objectives, and criteria are clearly defined and specified. Emphasis is placed on relating system performance to these goals. A hierarchy of systems, which allows handling of a complex system by looking at its component parts or subsystems, is identified. Quantifiable and nonquantifiable aspects of the problem are identified, and immediate and long-range implications of suggested alternatives are evaluated.

The systems approach establishes the proper order of inquiry and helps in the selection of the best course of action that will accomplish a prescribed goal by broadening the information base of the decision-maker; by providing a better understanding of the system and the interrelatedness of the system and its component subsystems; and by facilitating the prediction of the consequences of several alternative courses of action.

The systems approach is a framework for analysis and decision-making. It does not solve problems, but it does allow the decision-maker to undertake resolution of a problem in a logical, rational manner. While there is some art involved in the efficient application of the systems approach, other factors play equally important roles. The magnitude and complexity of decision processes require the most effective use possible of the scientific (quantitative) methods of systems analysis. However, one has to be careful not to rely too heavily on the methods of systems analysis. Outputs from simplified analyses have a tendency to take on a false validity because of their complexity and technical elegance.

The steps in the systems approach include (5): definition of the problem; gathering of data; development of criteria for evaluating alternatives; formulation of alternatives; evaluation of alternatives; choosing the best alternative; and final design/plan implementation. Often several steps in the systems approach are considered simultaneously, facilitating feedback and allowing a natural progression in the problem-solving process.

The systems approach has several defining characteristics. It is a repetitive process, with feedback allowed from any step to any previous step. Frequently, because systems analysis takes such a broad approach to problem-solving, interdisciplinary teams must be called in. Coordination and commonality of technique among the disciplines is sometimes hard to achieve. Close communication among the parties involved in applying the systems approach is essential if this understanding is to be achieved.

Definition of the Problem

Problem definition may require iteration and careful investigation, because problem symptoms may mask the true cause of the problem. A key step of problem definition requires the identification of any systems and subsystems that are part of the problem, or related in some way to it. This set of systems and interrelationships is called the environment of the problem. This environment sets the limit on factors that will be considered when analyzing the problem. Any factors that cannot be included in the problem environment must be included as inputs to, or outputs from, the problem environment.

Gathering of Data

Gathering of data to assist in planning and design decision-making through the systems approach will generally be done in conjunction with several steps. Some background data will have to be gathered at the problem definition stage and data gathering and analysis will continue through the final plan/design and implementation stage. Data that are gathered as the approach continues will help to identify when feedback to a previous step is required.

Data will be required at the problem definition stage to evaluate if a problem really exists; to establish what components, subsystems, and elements can reasonably be included in the delineation of the problem environment; and to define interactions between components and subsystems. Data will be needed during later steps to establish constraints on the problem and systems involved in it, to increase the set of quantifiable variables and parameters (constants) through statistical observation or development of measuring techniques, to suggest what mathematical models might contribute effectively to the analysis, to estimate values for coefficients and parameters used in any mathematical models of the system, and to check the validity of any estimated system outputs. When feedback is required, the data previously acquired can assist in redefining the problem, systems, or system models.

Development of Evaluative Criteria

Evaluative criteria must be developed to measure the degree of attainment of system objectives. These evaluative criteria will facilitate a rational choice of a particular set of actions (from among a large number of feasible alternatives), which will best accomplish the established objectives. Some evaluative criteria will provide an absolute value of how good the solution is, such as the cost of producing one unit of some product. Other evaluative criteria will only produce relative values that can be compared among the alternatives to rank them in order of preference, as in economic comparisons such as benefit/cost analysis.

In most complex real-world water problems, more than one objective can be identified. A quantitative or qualitative analysis of the trade-offs between the objectives must be made. For many problems, cost effectiveness would be the primary objective. Cost effectiveness can be defined as the lowest possible cost for a set level of control of a system, or the highest level of system control for a set cost.

Formulation of Alternatives

Formulation of alternatives is essentially the development of system models that, in conjunction with evaluative criteria, will be used in later analysis and decision-making. If at all possible, these models should be mathematical in nature. However, it should not be assumed that mathematical model building and optimization techniques are either required or sufficient for application of the systems approach. Many problems contain unquantifiable variables and parameters that would render results generated by even the most elegant mathematical model meaningless. If it is not practical to develop mathematical models, subjective models that describe the problem environment and systems included can be constructed. Models allow a more explicit description of the problem and its systems and facilitate the rapid examination of alternatives. Effective model building is a combination of art and science. The science includes the technical principles of mathematics, physics, and engineering science. The art is the creative application of these principles to describe physical or social phenomena.

Evaluation of Alternatives

To evaluate the alternatives that have been developed, some form of analysis procedure must be used. Numerous mathematical techniques are available, including the simplex method for linear programming models, the various methods for solving ordinary and partial differential equations or systems of differential equations, matrix algebra, various economic analyses, and deterministic or stochastic computer simulation. Subjective analysis techniques may be used for multiobjective analysis, or subjective analysis of intangibles. The appropriate analysis procedures for a particular problem will generate a set of solutions for the alternatives that can be tested according to the established evaluative criteria. In addition, these solution procedures should allow efficient utilization of personnel and computational resources.

Choosing the Best Alternative

Choice of the best alternative from among those analyzed must be made in the context of the objectives and evaluative criteria previously established, but also must take into account nonquantifiable aspects of the problem such as aesthetic and political considerations. The chosen alternative will greatly influence the development of the final plan/design and will determine in large part the implementability of the suggested solution. Preferably, the best alternative can be chosen from the mathematical optimization within feasibility constraints. Frequently, however, a system cannot be completely optimized. Near optimum solutions can still be useful, especially if sensitivity analysis has shown that the solution (and thus the objective function) is not sensitive to changes in the decision variables near the optimum point.

Final Plan/Design Implementation

Actual final planning/design is primarily a technical matter that is conducted within the constraints and specifications developed in earlier stages of the systems approach. One of the end products of final planning or design is a report that describes the recommendations made.

MATHEMATICAL MODELING

In general, to obtain a way to control or manage a water resources system we use a mathematical model that closely represents the physical system. Then the mathematical model is solved and its solution is applied to the physical system. Models, or idealized representations, are an integral part of everyday life. Mathematical models are also idealized representations, but they are expressed in terms of mathematical symbols and expressions. Such laws of physics as $F = ma$ and $E = mc^2$ are familiar examples. Similarly, the mathematical model of a business problem is the system of equations and related mathematical expressions that describe the essence of the problem. Thus, if there are n related quantifiable decisions to be made, they are represented as decision variables (say, $x_1, x_2, \ldots, x_n$) whose respective values are to be determined. The appropriate measure of performance (e.g., profit) is then expressed as a mathematical function of these decision variables (e.g., $P = 3x_1 + 2x_2 + \cdots + 5x_n$). This function is called the objective function. Any restrictions on the values that can be assigned to these decision variables are also expressed mathematically, typically by means of inequalities or equations (e.g., $x_1 + 3x_1x_2 + 2x_2 \leq 10$). Such mathematical expressions for the restrictions often are called constraints. The constants (coefficients or right-hand sides) in the constraints and the objective function are called the parameters of the model. The mathematical model might then say that the problem is to choose the values of the decision variables so as to maximize the objective function, subject to the specified constraints.

Mathematical models have many advantages over a verbal description of the problem. One obvious advantage is that a mathematical model describes a problem much more concisely. This tends to make the overall structure of the problem more comprehensible, and it helps to reveal important cause-and-effect relationships. In this way, it indicates more clearly what additional data are relevant to the analysis. It also facilitates dealing with the problem in its entirety and considering all its interrelationships simultaneously. Finally, a mathematical model forms a bridge to the use of high-powered mathematical techniques and computers to analyze the problem. Indeed, packaged software for both microcomputers and mainframe computers is becoming widely available for many mathematical models.

The procedure of selecting the set of decision variables that maximizes/minimizes the objective function, subject to the systems constraints, is called the optimization procedure. The following is a general optimization problem. Select the set of decision variables $x_1^*, x_2^*, \ldots, x_n^*$ such that

$$\text{Min or Max } f(x_1, x_2, \ldots, x_n)$$

subject to

$$\begin{aligned} g_1(x_1, x_2, \ldots, x_n) &\leq b_1 \\ g_2(x_1, x_2, \ldots, x_n) &\leq b_2 \\ g_m(x_1, x_2, \ldots, x_n) &\leq b_m \end{aligned} \quad (5)$$

where $b_1, b_2, \ldots, b_m$ are known values.

If we use matrix notation, Equation 5 can be rewritten as

$$\text{Min or Max } f(x) \quad (6)$$

subject to

$$g_j(\mathbf{x}) \leq b_j, \quad j = 1, 2, \ldots, m$$

When optimization fails, due to system complexity or computational difficulty, a reasonable attempt at a solution may often be obtained by simulation. Apart from facilitating trial and error design, simulation is a valuable technique for studying the sensitivity of system performance to changes in design parameters or operating procedure.

According to Equations 5 and 6, our main goal is the search for an optimal or best solution. However, it needs to be recognized that these solutions are optimal only with respect to the model being used. Since the model necessarily is an idealized rather than an exact representation of the real problem, there cannot be any utopian guarantee that the optimal solution for the model will prove to be the best possible solution that could have been implemented for the real problem. There just are too many imponderables and uncertainties associated with real problems. However, if the model is well formulated and tested, the resulting solution should tend to be a good approximation to the ideal course of action for the real problem. Therefore, rather than be deluded into demanding the impossible, the test of the practical success of an operations research study should be whether it provides a better guide for action than can be obtained by other means.

The eminent management scientist and Nobel Laureate in Economics, Herbert Simon, points out that satisficing is much more prevalent than optimizing in actual practice. In coining the term *satisficing* as a combination of the words *satisfactory* and *optimizing*, Simon is describing the tendency of water resources managers to seek a solution that is "good enough" for the problem at hand. Rather than trying to develop various desirable objectives (including well-established criteria for judging the performance of different segments of the organization), a more pragmatic approach may be used. Goals may be set to establish minimum satisfactory levels of performance in various areas, based perhaps on past levels of performance or on what the competition is achieving. If a solution is found that enables all of these goals to be met, it is likely to be adopted without further ado. Such is the nature of satisficing. The distinction between optimizing and satisficing reflects the difference between theory and the realities frequently faced in trying to implement that theory in practice.

BIBLIOGRAPHY

1. Goodman, A.S. (1984). *Principles of Water Resources Planning*. Prentice-Hall, Englewood Cliffs, NJ.
2. U.S. Water Resources Council (1973). *Principles and Standards for Planning Water and Related Land Resources*, September 10, 1973; Rev. December 14, 1979; Rev. September 29, 1980.
3. Dasgupta, P., Sen, A., and Marglin, S. (1972). *Guidelines for Project Evaluation*. United National Industrial Development Organization, Vienna.
4. Simonovic, S.P. (1989). Application of water resources systems concept to the formulation of a water master plan. *Water Int.* **14**: 37–50.
5. Jewell, T.K. (1986). *A Systems Approach to Civil Engineering Planning and Design*. Harper and Row, New York.

INDEX

NOTE: Page numbers followed by f refer to figures, page numbers followed by t refer to tables.

Acequia, defined, 2:541–542
Acetic acid, 2:295
Acidity, negative and positive, 2:424
Acid mine drainage (AMD), 2:1–2
  iron-oxidizing bacteria in, 2:150–151
  passive treatment of, 2:423–426
Acidophilic iron bacteria, 2:149
Acid oxidation, 2:85
Acid phosphatase (ACP) activity, 2:415
Acids
  defined, 2:542
  strength of, 2:294
Acinetobacter, 2:241
Acre-foot, defined, 2:542
Activated carbons, 2:79–86. *See also* Powdered activated carbon (PAC)
  acid surface functional groups on, 2:81
  background of, 2:80
  physical and chemical properties of, 2:81
  precursors of, 2:80
  preparation and properties of, 2:82–83
  sorption of trace metals onto, 2:83–85
  surface functional groups on, 2:82
Activation process, 2:80–81
Active biomonitoring (ABM), 2:33–37
  advantages and disadvantages of, 2:33–34
  system for, 2:28
Acute sediment toxicity tests, 2:351, 385t
Acute toxicity, categories of, 2:415
Adenosine triphosphate (ATP), 2:42
Adsorbents, micropores and mesopores in, 2:79
Adsorption, 2:374
Adsorptive columns, sizing, 2:363
Advanced Microwave Scanning Radiometer (AMSR-E), 2:589
Advanced nucleic acid analyzer (ANNA), 2:339
Advanced oxidation processes (AOP), 2:318
Adverse effects assessments, 2:516–517
Aeration, 2:374
Aerobic bacteria, 2:20
  in cooling systems, 2:241
Aerobic plate count (APC), in dishwater, 2:113, 114

Africa, unsafe water in, 2:631
Agricultural activities, as a nitrate contamination source, 2:571
Agricultural Nonpoint Source Pollution Software (AGNPS), 2:325
  Model 2001 (AGNPS 2001), 2:251
Agricultural practices
  algal control via, 2:3–4
  improving, 2:571
Agricultural water
  Arab World share of, 2:474f
  United States use of, 2:653
Agriculture
  in the Arab World, 2:470–471
  in China, 2:488–489
  groundwater role in, 2:392
  irrigated, 2:489
  nonpoint source pollution from, 2:186
  water conservation in, 2:495–496
  as a water consumer, 2:489–490, 662
  water reuse in, 2:635
  water supply for, 2:617–618
Air oxidation, 2:85
Air sampling techniques, 2:307
Alaskan rain forest, 2:667
Algae, 2:269
Algal bioremediation, 2:44–45
Algal growth, 2:241–242
Algal populations
  biomanipulation of, 2:6
  chemical control of, 2:5–6
  environmental control of, 2:3–5
Algal toxins, 2:387–392
Alkaline phosphatase (APase), 2:231
Alkalinity
  defined, 2:542
  in natural waters, 2:394
Alluvium, defined, 2:542
Alternative formulation, in a systems approach, 2:686
Ambient water quality, optimal level of, 2:129
American Society of Testing and Materials (ASTM), 2:459
American water resource development, 2:498–499
American Water Works Association (AWWA) survey, 2:91
Americium (Am). *See* Am-humic colloids
Am-humic colloids, dissociation kinetics of, 2:105–106

Aminolevulinate dehydratase (ALAD)
  activity, 2:236
  protein, 2:236–237, 238f
Aminolevulinic acid (ALA), 2:236
Ammonia
  impact on aquatic organisms, 2:123
  in inhibition of biological iron removal, 2:153–154
Ammonium, in source waters, 2:226
Amoebas, 2:313
Amoebic dysentery, 2:313
Amphipods
  characteristics of, 2:409–410
  commonly used, 2:409t
Amphipod sediment toxicity tests, 2:408–413
  application of, 2:409
  experimental design in, 2:410–411
  future research related to, 2:411
  organism selection in, 2:409
  significance of, 2:408–409
Anaerobic bacteria, 2:20–21
  iron-reducing, 2:151
  oil-field brine and, 2:288
Anaerobic wetlands, 2:425
Analysis, systems approach to, 2:685
Analysis of variance (ANOVA) comparison technique, 2:411
Anatoxin-a, 2:390
Anatoxins, 2:388–389
Animal facilities, managing, 2:186
Annual water quality report (AWQR), 2:346
Anodic stripping voltammetry (ASV), 2:207
Anoxic limestone drain (ALD), 2:424
Anthropogenic activities, toxic metals accumulation from, 2:433
Anthropogenic compounds, toxic effects of, 2:427
Anthropogenic load, 2:443
Anthropogenic sources of lead, 2:433
Antidegradation, precautionary principle regarding, 2:601
Antimony (Sb), maximum contaminant level of, 2:234t
Apalachicola-Chattahoochee-Flint (ACF) River Basin water allocation dispute, 2:501–502
  Army Corps of Engineers role in, 2:506–507
  crisis related to, 2:503
  role of interest groups in, 2:505
  water allocation and, 2:504–505

689

Apalachicola-Chattahoochee-Flint (ACF) River Basin water allocation dispute, (*continued*)
   water marketing proposals related to, 2:506–508
Aplysiatoxin, 2:389
Appliances, water-saving, 2:497
Application portals, 2:670
Appropriation doctrine, 2:542
Aquaculture, defined, 2:542
Aquatic ecosystems, 2:617
Aquatic environments
   classification and environmental quality assessment in, 2:94–98
   impact of road salt on, 2:320–321
   lipophilic contaminant monitoring in, 2:170–172
   trace metal pollution in, 2:64–65
Aquatic Life Criterion, 2:355–357
Aquatic macroinvertebrates, effect of road salt on, 2:321–323
Aquatic macrophytes. *See also* Macrophytes
   as biomonitors, 2:65–66
   hyperaccumulation potential of, 2:66
   research opportunities related to, 2:66–67
Aquatic organisms, toxicity of lead to, 2:434
Aquatic sediments, 2:384
Aquatic systems
   assessing the quality of, 2:25
   impact on, 2:24
   water reduction in, 2:465–466
Aquatic toxicity
   measures of, 2:381
   standardized test protocols for evaluating, 2:278
   testing, 2:380–381
Aqueduct, defined, 2:542
Aquifers
   defined, 2:542
   overpumped, 2:648
Aquifer–storage–recovery (ASR) systems, 2:515
Arab World
   conservation and demand management in, 2:472
   renewable water resources by state, 2:471t
   supply and demand in, 2:471–472
   water management solutions in, 2:472–474
   water resources in, 2:470–474
Arid countries/lands. *See also* Dry environments; Dry-land agriculture
   crop irrigation in, 2:549

effluent water regulations in, 2:475–478
water regulation in, 2:478
Army Corps of Engineers
   role in Apalachicola-Chattahoochee-Flint River Basin water allocation dispute, 2:501–502
   water marketing and, 2:506–507
   working relationship with Congress, 2:523
Aromatic hydrocarbon, 2:110
Arsenates, 2:16, 17
Arsenic (As)
   health effects of, 2:15–18
   inorganic, 2:8–11
   maximum contaminant level of, 2:234t
   metabolism and disposition of, 2:16–17
   particulate, 2:11
   toxicity mechanisms of, 2:17
   uncharacterized, 2:11
   uptake of, 2:217
Arsenic compounds, 2:7–15
   methylated, 2:11
Arsenic exposure, acute and chronic, 2:17–18
Arsenic species, 2:7t
   organic, 2:12–13t
Arsenite, 2:17
Arsenothiols, 2:11–13
Artesian water, defined, 2:542
"Artificial" communities, 2:379
Artificial recharge, defined, 2:542
Asbestos
   analysis of, 2:310
   maximum contaminant level of, 2:234t, 235t
Asbestos-containing material (ACM), 2:310
Asia, urban water resource management in, 2:552–554
Assays, standardization and reproducibility of, 2:378
Assimilable organic carbon (AOC), 2:175, 224–225, 244–245
Atomic absorption (AA) spectroscopy, 2:308–309
ATP content assays, 2:415
Atrazine, 2:312
Attainable uses, for surface waters, 2:476–477
Automatic model calibration, 2:332
Avoided cost water valuation approach, 2:655

Background pollutant concentration, 2:18–20
   inorganic pollutants, 2:19–20

   organic pollutants, 2:20
   water chemistry and, 2:19
Backyard sewerage, 2:633
Bacteria. *See also* Marine bacteria
   algal control using, 2:6
   identification of, 2:21
   indicator, 2:293
   iron-oxidizing, 2:149–152
   luminescent, 2:172–176
   metal-resistant, 2:446, 448–449
   molecular and biochemical techniques related to, 2:23–24
   resistance to heavy metals, 2:443
   waterborne, 2:20–24
   as water quality indicators, 2:23t
Bacterial bioassay, 2:389
Bacterial bioluminescence, 2:453
   toxicity and, 2:47
Bacterial biosensors
   availability of, 2:455
   constructed, 2:453–454
   luminescent, 2:453–458
Bacterial/fecal contamination, indicators of, 2:21
Bacterial growth, dependence on metal concentration, 2:446f
Bacterial infections, 2:371
Bacterial neuston, 2:442
Bacterial source tracking (BST), 2:21–22
Bacteriological monitoring, 2:267
Bacteriological parameters, of River Yamuna groundwater, 2:396
BACTOX test, 2:414
Balanced ecological community, 2:193
Balantidium coli, 2:313–314
Barium (Ba), maximum contaminant level of, 2:234t
Barley straw, in algal control, 2:5–6
Base flow, defined, 2:542
Baseline stations, 2:178
Bases
   defined, 2:542
   strength of, 2:294
BASINS. *See* Better Assessment Science Integrating Point and Nonpoint Sources (BASINS)
Bautzen Reservoir, biomanipulation trials in, 2:57t
BAX® system, 2:140
BBC Water Portal, 2:669
Bed depth-service time model (BDST), 2:363
Bedrock, defined, 2:542
Beneficial use classification, 2:191
Beneficial uses
   protection of, 2:478
   for surface waters, 2:476

Benthic chambers, 2:352
Benthic organisms/species
   exposure tests with, 2:212
   sediment as habitat for, 2:408
   surface water metals and, 2:217
Beryllium (Be), maximum
   contaminant level of, 2:234t
Best management practices (BMPs),
   2:192–193
   to minimize groundwater nitrate
     contamination, 2:571–573
   for nonpoint source control,
     2:187–188
   for water resources, 2:570–573
Better Assessment Science
   Integrating Point and Nonpoint
   Sources (BASINS), 2:328,
   249–250
BGP carbon, 2:83–85
Bioaccumulated metals, 2:421f
Bioaccumulation, 2:44, 69
   in assessing sediment quality,
     2:426–428
   bioassays, 2:351
   characteristics of, 2:70t
   of chemical contaminants, 2:125
   recommendations concerning,
     2:431
   of selenium, 2:357t
   use in regulation, 2:429–430
Bioassay batteries, 2:383–384
Bioassays
   in marine environmental
     monitoring, 2:443, 444
   purpose of, 2:418
   sediment toxicity, 2:458–464
   subcellular, 2:278–279
Bioassessment procedures
   current, 2:25–26
   development of, 2:24–25
   preserving biotic integrity and,
     2:26
Bioavailability
   in aquatic ecosystems, 2:170
   metal, 2:235
Biochemical oxygen demand (BOD),
   2:37–41, 175, 332. *See also* BOD_5
   test; Rapid BOD technologies
   historical context of, 2:37–38
   measurements in, 2:37
   operation of, 2:38
   uses for, 2:38
Biocides, types of, 2:242
Biodegradable dissolved organic
   carbon (BDOC), 2:224–225, 244,
   245
Biodegradable organic carbon,
   control of, 2:224
Biodegradation, 2:41–45
   acclimation phase of, 2:43
   algal bioremediation and, 2:44–45

   bioavailability and, 2:43–44
   enrichment cultures and, 2:42
   enzymatic mechanisms in, 2:42–43
   factors limiting, 2:42
   groundwater bioremediation and,
     2:44
   marine oil spills and, 2:44
   microbial diversity and, 2:42
   nonaqueous phase liquids and, 2:44
   recalcitrant molecules and, 2:43
Biodiversity
   precautionary principle regarding,
     2:600
   reducing, 2:637
Biofilms
   channel structures in, 2:230
   detachment of, 2:232
   in distribution systems, 2:244
   formation of, 2:228–229
   metabolic heterogeneity of, 2:230
   microbial dynamics of, 2:228–233
   microbial structure and dynamic
     heterogeneity in, 2:230–232
   nitrification/denitrification, 2:231
   oxygen limitation in, 2:231–232
   structural heterogeneity of,
     2:229–230
   sulfate-reducing, 2:231
Biofouling
   caused by iron bacteria, 2:151–152
   in a cooling tower, 2:241–242
   in a membrane water treatment
     system, 2:242
   methods of controlling, 2:242
   microbial forms in, 2:239–243
   nutrient adsorption in, 2:239
Biofouling layers, relationship of
   microorganisms inside,
   2:240–241
"Biogeochemical reflux" scheme,
   2:359f
Bioindication approach, to marine
   environmental monitoring,
   2:443–444
Bioindicators, 2:29–30
   species used as, 2:34
   versus biomonitors, 2:64
Biological contaminants, 2:343–344
   detection in surface waters, 2:441
Biological integrity, 2:24
Biologically stable water, 2:224
   parameter values that define,
     2:224t
Biological monitoring, 2:26, 378
   with bivalve molluscs, 2:35–36
   submitochondrial particle assay as
     a tool for, 2:376–379
Bioluminescence tests, 2:47–48
Bioluminescence toxicity tests,
   marine, 2:48
Bioluminescent assay systems, 2:455

Bioluminescent biosensors. *See also*
   Recombinant luminescent
   bacterial biosensors
   in direct toxicity assessment,
     2:47–48
   for toxicity testing, 2:45–50
   in WWTP monitoring, 2:48
Bioluminescent recombinant
   bacteria, 2:174, 454
Bioluminescent reporter genes, 2:453
Biomanipulation, 2:50–58
   lake restoration methods and, 2:50
   principles of, 2:50–51
Biomanipulation trials, review of,
   2:51–58
Biomarkers, 2:28–33
   in assessing sediment quality,
     2:429
   classes of, 2:429
   metallothioneins as, 2:407
   recommendations concerning,
     2:431
   relevance of, 2:32
   use in regulation, 2:429–430
   use in sediment toxicity
     assessment, 2:430t
Biomass immobilization, 2:72
Biomonitoring. *See also* Biomonitors
   developing tools for, 2:62–63
   genomic technologies/tools in,
     2:58–64
   high throughput technologies and,
     2:59–62
   importance of, 2:67
Biomonitors. *See also* Biomonitoring
   trace metal, 2:64–68
   versus bioindicators, 2:64
Bioremediation
   iron-reducing bacteria in, 2:151
   of selenium-contaminated waters,
     2:355–360
Biosensors
   future development of, 2:455
   microbial aminolevulinate
     dehydratase, 2:236–237
   non-toxicant-specific, 2:454t
   toxicant-specific, 2:455t
Biosorbent materials
   broad-range, 2:69
   comparison of, 2:72t
   immobilization of, 2:72
Biosorption, 2:69
   future of, 2:73
   of heavy metal anions, 2:72
   by living microorganisms, 2:71
   mathematical modeling of, 2:71–72
   technologies, 2:73
   of toxic metals, 2:68–74
Biosorption treatment plants, 2:73
Biosurveys, microbial, 2:442–443,
   444–451

Biotechnology, precautionary principle regarding, 2:600–601
Bioterrorism Act, 2:347, 348
Biovolatilization, 2:359
Birds, toxicity of lead to, 2:434–435
Bivalve molluscs
　active biomonitoring of, 2:33–37
　biomonitoring with, 2:34–36
　chemical monitoring with, 2:35
　illnesses associated with, 2:360–361
Blackstone River model, 2:256–260
Blue baby syndrome, 2:219
Blue Book, 2:315
Blueprint for Sustainable Development, 2:624
Blue Revolution, 2:627
BOD_5 test. See also Biochemical oxygen demand (BOD)
　components of, 2:39
　measurement protocol of, 2:38
　problems associated with, 2:40
　quality controls for, 2:39–40
　quantity measured by, 2:39–40
　standards for, 2:39
Body concentrations, time-dynamic speciation and, 2:215–219
Boron (B), in irrigation water, 2:159
Brain, lead-induced damage to, 2:435–436
Brazil, water markets in, 2:568
Br^-/DOC ratio, 2:75, 76–78
Bridges, 2:568
Brine, oil-field, 2:284–290
Brine injection technology, 2:285
Bromide
　health effects and regulations related to, 2:74
　influence on trihalomethane and haloacetic acid formation, 2:74–79
Buffer land, 2:168
Buffer solutions, 2:295
Buffer strips, 2:3–4
Business sector, interests of, 2:626

Cadmium (Cd). See also Cd-resistant bacteria
　in groundwater, 2:148
　maximum contaminant level of, 2:234t
Caged animals, use at dredged sites, 2:351–352
Calderon-Sher Safe Drinking Water Act (California), 2:345
California
　source-water protection in, 2:312
　water use in, 2:478–480
California Department of Health Services (CDHS), 2:345

California Department of Water, demand forecasting by, 2:530
California Toxics Rule, 2:356
Canada. See also Environment Canada (EC)
　annual precipitation in, 2:660
　continental watersheds in, 2:657
　flood management in, 2:681–682
　infrastructure in, 2:662–663
　water conservation in, 2:660–666
　water misuse in, 2:661
　water prices in, 2:604
　water quality in, 2:663–664
　water resources in, 2:656–660
　water supply in, 2:661
　water use in, 2:661–662
Canadian Environmental Protection Act, 2:319
Candida albicans, 2:240–241
Capacity building, 2:628
"Cap and trade" program, 2:133, 134
Capillary action, 2:542
Capillary lysimeter, 2:340–341
Carbon (C). See also Organic entries
　control of, 2:224–225
　uptake capacity of, 2:83t
Carbonization process, 2:80
Carcinogenic agents
　arsenic and arsenic compounds as, 2:7, 17–18
　luminescent bacteria for determining, 2:174–175
Carcinogenic byproducts, 2:116
Cardiovascular effects, of lead, 2:436
Cartridge filters
　advantages and disadvantages of, 2:155
　for iron removal, 2:152–155
Caspian Sea level rise, 2:480–484
　climate change and, 2:482
　hydraulic construction and, 2:482
　nature of, 2:481
　societal impacts of, 2:482–483
　tectonic plate movement and, 2:482
Catchment water quality modeling software, 2:325–326
cDNA libraries, 2:62. See also Complementary DNA (cDNA)
cDNA microarrays, 2:63
Cd-resistant bacteria, 2:447. See also Cadmium (Cd)
Cell membrane, metal transport across, 2:70
Cellular biomarkers, 2:31
Centers for Disease Control (CDC), 2:343
Central Ground Water Authority (CGWA), 2:181
Central nervous system, effect of lead on, 2:435–436

Central Pollution Control Board (CPCB), "designated best use" yardstick of, 2:183, 184
CE-QUAL-RIV1 software, 2:328
CE-QUAL-W2 software, 2:328
Chadha's diagram, 2:396, 397f
Channel structures, in biofilm, 2:230
Check dams, 2:550
Chemical coagulation process, 2:374
Chemical concentration, measurement of, 2:59. See also Concentration techniques
Chemical contaminants, 2:344–345. See also Contaminants
　bioaccumulation of, 2:125
Chemical monitoring, 2:267
　with bivalve molluscs, 2:35
Chemical oxidation, 2:300. See also Oxidation entries
Chemical oxygen demand (COD), 2:40
　of dishwater, 2:113, 114
Chemicals. See also Compounds
　bioavailability of, 2:427–428
　impact of, 2:28
Chemical sensors, impairment of, 2:283
Chemical standards, U.S., 2:268
Chemical treatment, for iron bacteria, 2:151–152
Chemosynthetic bacteria, 2:21
Chile
　tradable rights in, 2:645
　water markets in, 2:645
　water reallocation in, 2:607
China
　legal system in, 2:486–487
　policy, laws, and regulations in, 2:487–488
　water management financial sources in, 2:487
　water management in, 2:484–488
　water management institutions in, 2:484–486
　water scarcity in, 2:488–489
Chironomids, metal concentrations in, 2:421
Chironomus, heavy metal uptake by, 2:215
Chironomus riparius, as a test organism, 2:420
Chloraminated systems, nitrification in, 2:226
Chloramines, 2:89, 118–119
　advantages of, 2:244
Chloride. See also Chlorine (Cl)
　in drinking water, 2:395
　levels, 2:321
Chlorinated dibenzofurans (CDFs), 2:107
　chemical properties of, 2:109t

Chlorinated dibenzo-*p*-dioxins (CDDs), 2:107
　chemical properties of, 2:109t
Chlorinated disinfection byproducts (DBPs), 2:115
Chlorinated hydrocarbons, 2:109, 266
Chlorinated solvents, 2:317
Chlorinated trihalomethanes, 2:77–78
Chlorination, 2:88–90, 374
　advantages and disadvantages of, 2:398
　application and design of, 2:90
　carcinogenic byproducts of, 2:398
　chemistry of, 2:88–89
　Escherichia coli O157:H7 susceptibility to, 2:138–139
　factors affecting, 2:89
　for hepatotoxins, 2:390
　products of, 2:89
Chlorination byproducts, 2:91–94
　health effects of, 2:92
Chlorine (Cl). *See also* Chloride; Cl-phenols
　compounds of, 2:88
　disinfection using, 2:244
　dose response of mussels to, 2:401–406
　efficacy as an antifoulant, 2:402
　health effects of, 2:92
Chlorine (gas), safety of, 2:90. *See also* Free chlorine
Chlorine dioxide, 2:119
Chlorine residual, 2:398–399
Chlorine toxicity, factors influencing, 2:402–405
Chloroform, 2:92
Chlorophyll-α concentrations (CHL), in water quality models, 2:262
Chromatographic columns, standard packed, 2:304–305
Chromium (Cr). *See also* Hexavalent chromium
　in groundwater, 2:147–148
　maximum contaminant level of, 2:234t
Chronic bioassays, 2:351
"Chronic water scarcity," 2:495
Ciliated protists, as toxicity assessment test organisms, 2:413–418
Ciliates, 2:313–314
Ciliophora, 2:313–314
Cities, water demand in, 2:490–491
Civil War, effect on water projects, 2:522–523
Clean Lakes Program, 2:315
Clean Water Act (CWA), 2:96, 194–195, 292, 315
　goals of, 2:127
　improvements under, 2:191
　nonpoint source control regulations under, 2:185
Climate change, 2:600
　Caspian Sea level rise and, 2:482
Climatic conditions, in the Mediterranean islands, 2:642
Closure, of municipal solid waste landfills, 2:168
Cl-phenols, microbial detection of, 2:442
Coagulation, 2:98
　disinfection byproduct precursor removal by, 2:116–117
　enhanced, 2:120
　in precursor minimization, 2:116–117
Coastal resources, 2:489
Coastal waters
　arsenic in, 2:11
　indicators of organic contamination in, 2:442
Coastal Zone Act Reauthorization Amendments, 2:185
Coliform group, as indicator organisms, 2:293
Coliforms, in groundwater, 2:396. *See also Escherichia coli;* Fecal coliforms (FC)
Colloids, 2:99–100
　biophysicochemical interactions with filters, 2:100–102
Colorado River, 2:555
Colorimetric assays, 2:415
Colorimetry, in pH measurement, 2:295
Colpidium campylum, as a test organism, 2:413
Columbia River, 2:554, 555
Columbia River Basin, 2:589
Column experiments
　categories of, 2:103
　column preparation in, 2:103–104
　in contaminant transport, 2:103–106
　examples of, 2:105–106
　procedure for, 2:104–105
Columns, dimensions of and materials for, 2:104
Combined chlorine residual, 2:398
Cometabolic oxidation, 2:300
Cometabolism, 2:42
"Command and control" regulation, 2:132
Command area studies, remote sensing and GIS application in, 2:533–534
Commercial water use, 2:542
　in the United States, 2:652
Common-Pool Resources (CPR), 2:625
Community, water conservation in, 2:664–666
Community-based fisheries management, 2:636
Community-based resource management, 2:629
Community-based water management, 2:628
Community involvement
　in groundwater protection, 2:518–520
　strategies for, 2:519
Competing ligand exchange technique, 2:420
Complementary DNA (cDNA), 2:59. *See also* cDNA entries
Complexation constants, 2:206
Composite materials, sorption characteristics of, 2:364
Composite programming (CP), 2:521
Composting manure, 2:572
Compounds, in groundwater, 2:316–319. *See also* Chemicals
Comprehensive Environmental Response, Compensation and Liability Act (CERCLA), 2:96
Comprehensive models, 2:272
Computer models, water-management, 2:620
Concentration measurement, 2:59
Concentration techniques, 2:102. *See also* Chemical concentration
Condensation, defined, 2:542
Condominial sewerage, 2:633
"Confined" disposal of dredged sediments, 2:124–125
Conservation
　in the Arab World, 2:472
　industrial, 2:496
Constructed wetlands, algal control via, 2:3
Consumer Confidence Report (CCR), 2:346
Consumptive use, defined, 2:542
Consumptive water use, in the United States, 2:647
Containment, as a remediation strategy, 2:357–358
Contaminant mixtures, effects of, 2:429
Contaminant release, from redeposited sediment, 2:125
Contaminants. *See also* Chemical contaminants; Contamination
　field sampling and monitoring of, 2:263–269
　groundwater, 2:183
　lipophilic, 2:170–172
　natural, 2:181
　regulatory standards for, 2:346t
　sensitivity to, 2:409

Contaminants. *See also* Chemical contaminants; Contamination *(continued)*
  use of biodegradation to remove, 2:42
Contaminant transport, column experiments in, 2:103–106
Contaminated sediments, characterizing, 2:460
Contaminated sediment testing, species appropriate for, 2:427t
Contamination. *See also* Contaminants
  groundwater, 2:182
  vulnerability to, 2:313
Contingent valuation (CV) method, 2:130
Continuous chlorination, 2:398
Contour trenches, 2:550
Control costs, 2:131
Controlled-tension lysimeter, 2:341–342
Control water, selecting, 2:382–383
Control-water exposures, 2:382
Conveyance loss, defined, 2:542
Conveyance system structures, 2:568
Cooking, without power, 2:529
Cooling tower, biofouling in, 2:241–242
Cooperative agreements, 2:401
Copper (Cu)
  concentrations, 2:422t
  in groundwater, 2:147
  maximum contaminant level of, 2:234t
  removal by sorption, 2:362
  sorption capacity, 2:83
  toxicity to aquatic life, 2:147
Corporations, American, 2:499
Corporatization, of water, 2:569
Corrective maintenance, 2:595
Cost–benefit analysis, 2:129–130
Cost-effectiveness analysis, 2:131
Cost-effective technology, 2:569
Costs. *See also* Transaction costs; Pricing; Water pricing
  of droughts, 2:579–580
  of the 1930s drought, 2:513
  of water quality management, 2:369
CREAMS model, 2:250–251
Crop coefficient, 2:537
Crop management techniques, 2:496
Crustaceans, in sediment impact testing, 2:426. *See also* Amphipod entries
Cryptosporidium parvum, 2:343–344
Crystal Lake, biomanipulation trials in, 2:57t
Cubic feet per second, 2:542

Cumulative absolute relative error (CARE), 2:335
Cumulative probability plots, 2:20
Cyanide
  analysis, 2:309–310
  maximum contaminant level of, 2:234t, 235t
Cyanobacteria, 2:21
  blooms of, 2:390
  toxic, 2:388
Cyanosis, 2:219
Cyanotoxins, 2:387–388
  detection of, 2:389–390
  health problems related to, 2:388–389
  removal of, 2:390
  stability of, 2:390
Cylindrospermopsins, 2:388, 390
CYP1A induction, 2:109–110
Cytochrome P450 enzymes, 2:107–108
Cytochrome P450 monooxygenase
  as a bioindicator, 2:110
  as an indicator of pollutants in fish, 2:110
  as a PCB/dioxin-like compound indicator, 2:106–111
Cytotoxicity, 2:413–414

Dairy cow diets, improving, 2:573. *See also* Livestock entries
Dairy herd size reduction, 2:572
Dams, 2:567–568
  in the United States, 2:555
Dangerous Substances Directive (EU), 2:267
Daphnia, heavy metals in, 2:423f
Daphnia magna, as a test organism, 2:420–421
Daphnids, metal concentrations in, 2:421
Data gathering, in a systems approach, 2:685
Data management, following test results, 2:460
Data management model, 2:622
DDT use, 2:600
Decay rate estimation, 2:334
Dechlorination, 2:399
  reductive, 2:299–300
Decision making
  flood-management, 2:679–680
  participatory multicriteria, 2:681
  precautionary, 2:596
  risk-informed, 2:599
  systems approach to, 2:685
Decision matrix, 2:679
Decision Support System for Evaluating River Basin Strategies (DESERT), 2:328

Decision support systems (DSSs), 2:328, 535, 589
  hydroelectric, 2:621–622
  river basin, 2:619–624
  spreadsheets in, 2:620–621
  value of, 2:623
  for water resources management, 2:622–623
Decision variables, in a mathematical model, 2:686–687
Definite Enterprise Portal Resource, 2:669
Degradative enzymes, 2:43
Deicers, chemical, 2:319. *See also* Ice control
Delaware River Basin Commission drought indicators, 2:581
Delaware River Basin Compact, 2:585
Demand management
  in the Arab World, 2:472
  in Canada, 2:662
Demand reduction systems, 2:515
Depth filters, 2:99
Depth filtration, role of colloids and dissolved organics in, 2:99–103
Dermatotoxins, 2:389
Desalination, 2:374, 493–494
  defined, 2:542
  in the United States, 2:650
Design, municipal solid waste landfill, 2:168
Desorption, 2:70
Desorptive agents, 2:70
Detailed water quality models, 2:249
Detectors, gas chromatography, 2:307–308
Deterministic techniques, 2:332
Developing countries
  drinking water and sanitation in, 2:630–633
  meeting water needs in, 2:643–645
  water resources planning and management objectives for, 2:683
  water supply in, 2:616–617
Development
  managing urban runoff from, 2:187
  patterns of, 2:191
Diarrhea illnesses, 2:371
Dichloromethane (DCM), 2:317
Diffuse double layer (DDL) model, 2:363–364
Digital elevation model (DEM), 2:531
Digital terrain modelling (DTM), 2:534
Dilution, in algal control, 2:4
Dilution water, 2:382
  characteristics of, 2:383
1,4-Dioxane, 2:317–318

Dioxin-like compounds, as environmental toxicants, 2:106–107
Direct electron transfer, 2:41
Direct potable reuse, 2:611
Direct search method, 2:332
Direct toxicity assessment (DTA), 2:46–48
Direct water valuation studies, 2:655
Disaster Assistance Act of 1988, 2:578
Disaster supplies, 2:529
Discharge, defined, 2:542
Discriminant function analysis (DFA), 2:282
Disease outbreaks
 Cryptosporidium, 2:343–344
 Escherichia coli, 2:344
Disease pathogens, waterborne, 2:343
Diseases, water-related, 2:630–631
Dishwashing detergent, antibacteriological properties of, 2:114
Dishwater
 microbiological content of, 2:114
 temperature of, 2:113–114, 115
Dishwater quality, 2:112–115
 environmental aspects of, 2:114
Disinfectant application, changing the order of, 2:120
Disinfectant residual concentrations, 2:226t
Disinfection. See also Chlorination entries
 in distribution systems, 2:244
 regulations related to, 2:92
Disinfection byproduct precursors
 removal by coagulation, 2:116–117
 removal from natural waters, 2:115–117
Disinfection byproducts (DBPs), 2:398
 aspects of, 2:77f
 formation and occurrence of, 2:91–92
 minimization of, 2:119–120
 regulations related to, 2:92
 routes of exposure for, 2:92–93
Disinfection Byproducts Rule (DBPR), 2:92
Disinfection practices, alternative, 2:118–119
Dissociation kinetics, of Am-humic colloids, 2:105–106
Dissolved matter, transport phenomena of, 2:105
Dissolved organic carbon (DOC), 2:224
Dissolved organic matter (DOM), 2:212, 213

Dissolved organics, 2:100
 biophysicochemical interactions with filters, 2:100–102
Dissolved oxygen (DO), 2:142
 production of, 2:275
 as a water quality parameter, 2:270
Distillation, 2:527
Distribution, regional, 2:615–616
Distribution systems, 2:616
 control of nitrification in, 2:226
 disinfection in, 2:244
 microbial nutrients in, 2:244–245
 microbiological quality control in, 2:243–247
 research on, 2:511
Diversion structures, 2:568
DNA, shotgun sequencing of, 2:63
DNA arrays, 2:60
DNA damage, bacterial biosensors for detecting, 2:455
DNA probes, in Salmonella, 2:337–338
"Dolphin safe" tuna fishing, 2:636
Domestic sector, water conservation in, 2:496–497
Domestic wastewater, disposal of, 2:182
Domestic water, quality of, 2:371
Domestic water usage, 2:371, 542
 in India, 2:565
 in the United States, 2:652
Dong Nai–Saigon Rivers, wastewater discharge into, 2:552–553
Dose-response data, 2:381
Dose response of mussels to chlorine, 2:401–406
Double-composite landfill liner, 2:165
Drainage basins, 2:543
 bioremediation in, 2:359
Drainage divides, 2:657
Drawdown, defined, 2:543
Dredged material
 acute sediment toxicity tests for, 2:351
 characterization of, 2:351–352, 354t
 uncertainties associated with testing, 2:353t
Dredged sediment
 regulating disposal of, 2:125–126
 water quality aspects of managing, 2:122–127
Dredged sites, use of caged animals at, 2:351–352
Dredging, hydraulic, 2:122
Dredging/barge disposal, mechanical, 2:122
Dredging/disposal operations, impact of, 2:122–125

Dresden, Germany, biomanipulation trials in, 2:56t
Drinking water
 aquatic macrophytes as biomonitors of, 2:66
 in developing countries, 2:630–633
 disinfection byproduct standards for, 2:92t
 microbes in, 2:243
 nitrates in, 2:219–220
 nutrient limitations of bacterial growth in, 2:245
 organic compounds in, 2:372–373
 Salmonella detection and monitoring in, 2:337–340
 toxic chemicals in, 2:371–372
 types of organic carbon in, 2:224–225
Drinking water purification, Br^-/DOC changes in, 2:75–76
Drinking water regulations, history of, 2:194–198
Drinking water standards, 2:345–347
 national primary, 2:373t
Drip irrigation, 2:543
Drought(s)
 collective action against, 2:579
 costs of, 2:513, 579–580
 in India, 2:567
 legacy of, 2:513
 in the Nile Basin, 2:590–591
 in the 1930s, 2:511–514
 obstacles to planning for, 2:578–579
 planning tools for, 2:580
 political considerations related to, 2:580
 public notification of, 2:583
 recovering from, 2:512–513
 traditional government approach to, 2:577
 water resources management for, 2:576–586
 water use curtailment during, 2:583
 water-user compliance monitoring during, 2:583
Drought-contingency-plan phasing criteria, 2:582t
Drought indicators, identifying, 2:580–583
Drought management, 2:515–516, 585–586
 planning in, 2:514–515
Drought planning, 2:577
 specificity of, 2:578
Drought-planning authority, designating, 2:583
Drought programs, federal, 2:578

Drought-related planning and management activities, state, 2:583–585
Drought relief programs, 2:512
Drought-severity index, 2:581
Drought surcharge, 2:583
Dry electrolytic conductivity detectors (DELCD), 2:308
Dry environments, disturbance of, 2:516. *See also* Arid countries/lands
Dry-land agriculture, water conservation in, 2:496
Dry tomb landfilling, 2:163–164
Dual-flush device, 2:665
Dublin principles, 2:574
Duplex PCR, 2:338
Dynamic data, 2:622
Dynamic models, 2:272

Early-closure flush device, 2:665
"Early warning" biomarkers, 2:429
Earth Summits, 2:624, 628
Ecohydrology, 2:368
Ecological benefit. *See* Net ecological benefit concept
Ecological effects, assessment for water-limited environments, 2:516–518
Ecological health, 2:182
  human activities as stressors on, 2:625
Ecological Risk Assessment Guidance for Superfund (ERAGS), 2:517
Ecological surveys, 2:30
Economic activity, in India, 2:564
Economic growth, American, 2:499
Economic planning (EP) model, 2:520, 521
Economics. *See also* Costs; Global economy; Pricing; Water economic analysis
  during the Great Depression, 2:511
  of water quality, 2:127–135
  of water resource valuation, 2:653–654
Ecoregions, environmental management of, 2:667–668
Ecosystems. *See also* Sensitive ecosystems
  balanced, 2:667
  effluent-driven, 2:476–478
  sustainable management of, 2:636–637
Ecotoxicogenomic research, developing tools for, 2:62–63
Ecotoxicological effects, of lead, 2:433–435
Ecotoxicological sediment assessment, 2:383
Ecotoxicology, purpose of, 2:379
Education. *See* Public education
Educational programs, 2:497
"Efficient water use," 2:491
Effluent(s). *See also* Whole effluent toxicity controls
  defined, 2:543
  diversion of, 2:3
  industrial, 2:496
  runoff, 2:516
  trading, 2:134
Effluent discharges, 2:373
  to dry riverbeds, 2:476–477
  national primary drinking water standards, 2:373t
Effluent-driven ecosystems
  attainable uses in, 2:476–477
  flow regime modification in, 2:477–478
  net ecological benefit in, 2:477
  water regulation in, 2:478
Effluent treatment, of algal populations, 2:3
Effluent water regulations, in arid lands, 2:475–478
Egypt-Sudan water dispute, 2:591–592
E_h, 2:464–469
  application of, 2:467–468
  difficulty in using, 2:467
  interpreting, 2:467
  measurement of, 2:466–467
Electron activity. *See* pe
Electron capture detector (ECD), 2:307–308
Electronic nose
  environmental applications for, 2:283
  future of, 2:283–284
  for odor monitoring, 2:281–284
  user needs for, 2:281–283
Electron transfer (ETr) assay, 2:376
Electron transfer reaction (ETR) assay, 2:279, 280
Electrospray ionization MS quadrupole (ESI-MS), 2:61
Emergencies, food and water in, 2:526–529
"Emergency areas" policy, 2:526
Emergency outdoor water sources, 2:527
Emergency Response Plan (ERP), 2:348
Emergency supplies, 2:529
Emergency water purification, 2:527
Enabling environment, in the integrated water resources management framework, 2:575–576
Endocrine system, effects of lead on, 2:436
Endpoint selection
  in environmental risk assessment, 2:459
  for sediment toxicity tests, 2:459
Energy pricing, in India, 2:557
Energy production
  iron-reducing bacteria in, 2:151
  water supply for, 2:618
Engineered carbons, 2:82–83, 85
Engineering risk index, 2:675
Engineering works, increasing water supply via, 2:490
Enhanced Stream Water Quality Model (QUAL2E), 2:326–327, 329–330, 334
Enterococcus group, as indicator organisms, 2:293
Enterohemorrhagic *Escherichia coli* (EHEC), 2:138
Environment(s)
  benefits of water recycling to, 2:610–613
  lead movement in, 2:433
  water-limited, 2:516–518
Environmental contaminants, bioaccumulation of, 2:428
Environmental economics, 2:127
Environmental Fluid Dynamics Code (EFDC) model, 2:256
Environmental impact assessment (EIA) model, 2:520
Environmental impact factors (EIF), 2:520–521
Environmental impacts
  integrating into water resources planning, 2:520–522
  tools to evaluate, 2:516–517
Environmental justice, 2:668
Environmental management, of ecoregions, 2:667–668
Environmental monitoring
  biosensors as tools for, 2:455
  considerations for, 2:263
  using submitochondrial particle assay, 2:378
Environmental Protection Agency (EPA). *See also* EPA entries; United States Environmental Protection Agency (USEPA); U.S. EPA Subtitle D dry tomb landfilling
  nonpoint source control and, 2:185
  primary and secondary water quality standards of, 2:373
  priority pollutants list, 2:19
  state reporting to, 2:315
Environmental Protection Council, Ho Chi Minh City, 2:553

Environmental quality
  aquatic-environment assessment of, 2:94–98
  standards for, 2:46
Environmental risk assessment processes (ERAs), 2:459
Environmental science, precautionary principle regarding, 2:599–601
Environmental toxicants, 2:106–107
Environmental water quality, evaluation of, 2:450
Environment Canada (EC), 2:321, 459
Enzymatic biodegradation mechanisms, 2:42–43
Enzyme assays, 2:110
Enzyme-linked immunosorbent assay (ELISA), 2:139, 337, 390
EPA priority pollutants, 2:305t. See also Environmental Protection Agency (EPA)
EPA Superfund, 2:517. See also Superfund sites
EPA toxic constituent manuals, 2:381
Ephemeral waters, 2:475
Epidemiological studies, lead-related, 2:438
Epidemiology, of *Escherichia coli* O157:H7, 2:137–138
Equations, water quality model, 2:273–278
Equilibrium analysis, 2:71
Equity, water quality and, 2:131–132
Erosion, 2:543
  river, 2:657–658
Erosion–Productivity Impact Calculator (EPIC), 2:251
Escherichia coli, 2:21, 22
  lux-gene-carrying, 2:174
Escherichia coli O157:H7, 2:136–142, 344
  clinical features and treatment for, 2:136–137
  diagnostic procedures for, 2:139–140
  epidemiology of, 2:137–138
  history of, 2:136
  monitoring in water, 2:140
  pathogenesis of, 2:137
  susceptibility to chlorination, 2:138–139
  transmission routes of, 2:138
  waterborne transmission of, 2:138
Estuary, 2:543
Ethnic community, protecting, 2:667–668
EU Freshwater Fish Directive, 2:267
Eukarya, 2:313
European policy, precautionary principle and, 2:597

European Union (EU), water quality standards, 2:267–268. See also EU Freshwater Fish Directive
European Union-Water Framework Directive (WFD), 2:638, 639, 640, 641
Eutrophication, 2:2, 181, 269
  relation to organic loading, 2:142–143
Eutrophic water bodies, algal control in, 2:2–7
Evaluative criteria, in a systems approach, 2:685–686
Evaporation, 2:543
Evapotranspiration, 2:543
Evolutionary algorithms (EAs), 2:333
EXAFS spectroscopy method, 2:207, 208
"Explosimeters," 2:308
Export of virtual water, 2:536
Exposure, presence of food during, 2:420–421
Exposure Analysis Modelling System (EXAMSII), 2:327
Exposure biomarkers, 2:28–29, 31
Exposure concentrations, 2:422, 423
  time-varying, 2:213–215
Exposure methods, 2:420
Exposure tests, sediment or fauna incubation experiment for, 2:418–423
Expressed sequence tag (EST) databases, 2:62
Extracellular biodegradation, 2:43
Extracellular polymeric substances (EPSs), 2:229–230
Exxon Valdez oil spill, 2:292

Failure, definition of, 2:675
Failure region boundary, 2:676
Far Eastern Russia seas
  biosurveys in, 2:444–451
  evaluation of heavy metal pollution in, 2:447
  seawater quality control in, 2:449–451
Farming. See Agricultural entries
Farmland, irrigated, 2:549
Farm runoff, 2:400
Fast atom bombardment mass spectroscopy (FABMS), 2:389
Fatty acids, 2:442
Faucets, low-flow, 2:665–666
Fecal bacteria, as water quality indicators, 2:23t
Fecal coliforms (FC), 2:293, 344. See also Coliforms
  in groundwater, 2:396
  as indicator bacteria, 2:361–362

Federal Emergency Management Agency, 2:526, 529
Federal government. See also Federal regulations; United States; U.S. entries
  approach to droughts, 2:578
  drought-related water quality and, 2:585–586
  as waterway decision maker, 2:523
Federal Highway Administration (FHWA), 2:186
  model, 2:253
Federal regulations. See also Federal government
  history of, 2:194
  water-quality-related, 2:190
Federal Water Pollution Control Act (FWPCA), 2:125, 194
  amendments to, 2:195, 314–315
Federal water projects, expansion of, 2:522–523
Feeding status, chlorine toxicity and, 2:403–404
Female reproductive system, effects of lead or, 2:436
Fern Lake, biomanipulation trials in, 2:57t
Ferric hydroxides, 2:363. See also Iron entries
Fertilizer application
  methods of, 2:572
  rates of, 2:571–572
Field sampling
  of contaminants, 2:263–269
  when and how often to sample, 2:264–266
  where to sample, 2:263–264
Filamentous bacteria, 2:150
Filters. See also Filtration
  biofouling of, 2:99–100
  biophysicochemical interactions with colloids and dissolved organics, 2:100–102
  iron-removal, 2:152–155
  large-pore, 2:101
  types of, 2:100
Filtration, 2:374
  sorptive, 2:362–366
Fingerprint biomarkers, 2:29, 31
"Fingerprinting" analysis, 2:307
Fire management, 2:201
Fiscal solutions, in the Arab World, 2:473
Fish
  biomanipulation of, 2:51–58
  effect of road salt on, 2:323
  impacts of PCBs and dioxin-like compounds on, 2:108–109
Fisheries, sustainable management of, 2:635–636

Fishery sustainability, precautionary principle regarding, 2:600
Fish removal, algal control and, 2:6
Fixed-bed sorption, 2:363–364
Flagellates, 2:313
Flame ionization detector (FID), 2:307
Flame photometric detectors, 2:308
Flat rate structure, 2:663
Flocculation, 2:98–99, 374
Flood Control Act, 2:523
Flood management
  participatory multicriteria, 2:678–683
  Red River Basin, Manitoba, Canada, 2:681–682
  stakeholder participation in, 2:680
Flood plain, 2:543
  remote sensing and GIS application in managing, 2:533
Flood prevention, 2:525
Flood risk management, 2:525–526
  optimal, 2:526
Floods, 2:543
  in India, 2:566–567
  in the United States, 2:523
Flood stage, defined, 2:543
Floodwater harvesting, 2:550
Flowing well/spring, 2:543
Flow regime modification, 2:477–478
Flow restrictor showerheads, 2:666
Fluorescent-labeled latex beads (FLLB), 2:415
Fluorescent labeling, 2:59
Fluoride
  in groundwater, 2:396
  maximum contaminant level of, 2:234t
Flushing, in algal control, 2:4
Flux-type natural resources, 2:634
Food(s)
  emergency, 2:526–529
  low supplies of, 2:528
  shelf-life of, 2:528
  short-term supplies of, 2:527–528
  storage tips for, 2:528
Forest
  cutting, 2:200
  regeneration of, 2:200
Forested landscape, water quality management in, 2:199–202
Forest fire, impact on water quality, 2:201
Forest hydrology research, 2:199
Forest management, 2:201
Forestry, nonpoint source pollution from, 2:187
Forest soil, disturbance of, 2:200
FORTRAN Hydrological Simulation Program (HSPF), 2:249–250, 326
Fouling, mussel size and, 2:402

Fredericksborg Castle Lake, biomanipulation trials in, 2:56t
Free chlorine. *See also* Chlorine entries
  bromide oxidation by, 2:74–75
  residual, 2:89
Free chlorine residual, 2:398
Free ion activity model (FIAM), 2:212
Free water knock out (FWKO), 2:285–286
Freshwater, 2:543
Freshwater resources, sustainable management of, 2:634–635
Freshwater supply
  global, 2:548–549
  sources of, 2:515
Freundlich equation, 2:71, 87
Full cost pricing, 2:663
Funding, of municipal solid waste landfills, 2:168. *See also* Costs
Fungal biofouling, 2:240
Fuzzy compromise programming (FCP), 2:681
Fuzzy criteria, for system performance evaluation, 2:674–678
Fuzzy expected value (FEV), 2:680, 682
Fuzzy inference system, 2:680
Fuzzy membership function, 2:676
Fuzzy reliability–vulnerability criteria, 2:677
Fuzzy resiliency, 2:677–678
Fuzzy robustness, 2:677
Fuzzy sets, 2:675

Gage height, 2:543
Gaging station, 2:543
Galge site, biomanipulation trials in, 2:55t
Ganga–Brahmaputra–Meghna Basin, 2:560
Gas chromatography (GC)
  air sampling techniques in, 2:307
  detectors in, 2:307–308
  headspace technique in, 2:306
  for odor analysis, 2:281
  purge and trap procedure in, 2:306
  sample introduction in, 2:305–306
  solvent extraction in, 2:307
  in water analysis, 2:304–308
Gastroenteritis, 2:138
GEMS/WATER program, objectives of, 2:177–178
Gene Chips®, 2:60
Gene-probe technology, 2:352
Generalized Watershed Loading Function (GWLF), 2:252–253
Genes, luminescent-marked, 2:453–454

GENESIS genetic algorithm optimization software, 2:327, 334
Genetic algorithms (GAs), 2:333–335
Genetic engineering, 2:48
Genomics, 2:59–61, 429
Genomic technologies
  in biomonitoring, 2:58–64
  high-throughput, 2:59–62
Genomic tools, 2:339
  in biomonitoring, 2:63
Genotoxicity assessment, luminescent bacterial biosensors for, 2:455
Genotoxic response, 2:171
Geobacter metallireducens, 2:151
Geographic Information System (GIS). *See also* GIS technology
  application in water resources, 2:531
  in command area studies, 2:533–534
  in floodplain management, 2:533
  future need for, 2:535–536
  in groundwater studies, 2:534–535
  in hydrologic modeling, 2:535
  in land use classification, 2:532
  in precipitation studies, 2:532
  in reservoir sedimentation, 2:534
  in snow cover studies, 2:533
  in waterlogging and soil salinity, 2:534
  in water quality studies, 2:534
  in watershed mapping and monitoring, 2:533
Geyser, 2:543
Giardiasis, 2:543
GIS technology, 2:252
  to indicate heavy metal concentrations, 2:448–449
Glaciers, 2:543, 660
  Canadian, 2:660
Glass pH sensor, 2:296
GLEAMS model, 2:251
Gliding cell formation, 2:240
Global economy, precautionary principle and, 2:596–597. *See also* Economics
Global impacts, of sea-level change, 2:483
Globalization, of water, 2:536–541
Global search methods, 2:333
Global water resources, efficient use of, 2:536
Global water use, 2:489–490
Glucose–glutamic acid (GGA) solution, in BOD_5 tests, 2:39
Glycerides, 2:442
Gold Book, 2:315
Golden Horn Bight, pollution in, 2:446

Government. *See also* Federal entries; State entries; United States; U.S. entries
  drought-planning authority of, 2:583
  drought policies of, 2:576, 577–578, 580
  market operation and policies of, 2:500–501
  public interest water management and, 2:609
  structure of American, 2:498
Granular activated carbon, 2:80
  effects of treatment with, 2:76–77
Granular biological activated carbon (GBAC), oil- field brine and, 2:288–289
Gravity-flow units
  advantages and disadvantages of, 2:155
  for iron removal, 2:154
Gravity irrigation methods, 2:495
Great Britain. *See* U.K. National Rivers Authority
Great Depression, 2:511
  development of, 2:512
Great Lakes, 2:658
Green accounting, 2:626–627
Green Book, 2:315
Green Revolution, 2:617
Greywater, defined, 2:543
Groundwater, 2:95. *See also* Groundwater assessment; Groundwater quality
  in Arab states, 2:471
  bioremediation of, 2:44
  Canadian, 2:660
  compounds in, 2:316–319
  contaminants in, 2:183
  defined, 2:543
  EU monitoring of, 2:268
  geochemistry of, 2:19
  harvesting of, 2:550
  hydrochemical data on, 2:158t
  impact of river water quality on, 2:396
  in India, 2:555
  investigating humic matter in, 2:209–210
  iron-oxidizing bacteria in, 2:150
  nitrate contamination/content in, 2:156, 570–571
  particulate arsenic in, 2:11
  pollution of, 2:400
  role in agriculture, 2:156
  trace element contamination in, 2:143–148
Groundwater assessment
  experimental methodology for, 2:144
  recommendations following, 2:148
  results of, 2:144–148
Groundwater dams, 2:551
Groundwater Foundation, The (TGF), 2:518, 519
Groundwater Guardian, 2:518, 520
Groundwater monitoring, for municipal solid waste landfills, 2:165
Groundwater overdraft, in the Arab World, 2:472f
Groundwater protection, public education and community involvement in, 2:518–520
Groundwater quality, 2:160, 182–183
  land-use threats to, 2:183t
  near River Yamuna, India, 2:392–398
Groundwater recharge, 2:393
  projects, 2:611
Groundwater resources, in India, 2:562–564
Ground Water Rule (GWR), 2:195, 345
Groundwater scheme, environmental impact assessment for, 2:521t
Groundwater sources, high control of, 2:400
Groundwater studies, remote sensing and GIS application in, 2:534–535
Growing areas, shellfish, 2:361–362
Growth-linked biodegradation, 2:42
Guided random search method, 2:333
Gully plugs, 2:550

Habitat conservation, precautionary principle regarding, 2:600. *See also* Environmental entries
Haloacetic acids (HAAs), 2:91–92
  bromide influence on formation of, 2:74–79
  brominated, 2:75
Hardness
  defined, 2:543
  of drinking water, 2:394
Headspace technique, 2:306
Headwater, 2:543
Health, ecological, 2:182
Health Assessment on Perchlorate, 2:345
Health effects
  of arsenic, 2:15–18
  of bromide, 2:74
  of chlorine and chlorination byproducts, 2:92
  of lead, 2:432–440
  of polychlorinated biphenyls, 2:107
  of water pollution, 2:617
Heavy metal anions, biosorption of, 2:72
Heavy metal pollution, 2:407
  in Far Eastern Russia seas, 2:447
Heavy metal removal technologies, performance characteristics of, 2:69t
Heavy metals
  microbial biosurvey detection of, 2:442–443, 445, 446
  uptake rates of, 2:211–219
  as water quality indicators, 2:266
Heavy metal toxicity detection, microbial enzyme assays for, 2:233–238
Heavy oil sorption, 2:363
Hematopoietic effects of lead, 2:437
Hemolytic uremic syndrome (HUS), 2:136–137
Hemorrhagic colitis (HC), 2:136
Hepatotoxins, 2:388
Herbicides, in algal control, 2:5
Heterotrophic bacteria, 2:21, 175
Heterotrophic microorganisms, 2:38, 448t
Hexavalent chromium, samples, 2:309
Hidden water sources, in the home, 2:527–528
High-density polyethylene (HDPE) landfill liners, 2:164–165
High-performance liquid chromatography (HPLC), 2:308, 389
High-throughput technologies, biomonitoring and, 2:59–62
Himalayas, 2:559
  river systems in, 2:560
Hin primers, 2:338
H-1i primers, 2:338
Ho Chi Minh City
  water environment and pollution in, 2:552–553
  water resource management in, 2:552–554
  water supply and distribution in, 2:553–554
Home
  hidden water sources in, 2:527–528
  water conservation in, 2:664–666
Homeland Security Act, 2:347
Hoover Dam, 2:554–555
Hopper-dredge disposal, 2:123–124
Household water efficiency, 2:635
Hubenov Reservoir, biomanipulation trials in, 2:53t
Human activities
  contamination from, 2:489
  most water demanding, 2:489

Human infections, *Escherichia coli* shedding patterns in, 2:137–138
Humans, toxicity of lead to, 2:435–437
Human water consumption, categories of, 2:370, 614
Humic acid, properties of, 2:205–206
Humic colloids, 2:105–106
Humic/fulvic acid
  complexation to, 2:207–208
  inorganic colloids stabilized by, 2:208–209
Hydraulic construction, Caspian Sea level rise and, 2:482
Hydraulic dredging, 2:122
Hydrochemical classification, 2:396, 397f
Hydrodynamic and Water Quality Model for Streams (CE-QUAL-RIV1) software, 2:328
Hydrodynamic simulation, 2:256–259
Hydroelectric decision support system, 2:621–622
Hydroelectric plants, 2:568
Hydroelectric power
  in India, 2:565
  water use for, 2:543
Hydrogen ion concentration, 2:294–295
Hydrographic area, as an operational unit of analysis, 2:613
Hydrological Simulation Program–FORTRAN (HSPF), 2:249–250, 326
Hydrologic cycle, 2:95, 656
  defined, 2:544
Hydrologic data, uncertainty in, 2:621
Hydrologic ensembles, 2:621
Hydrologic forecasting model, 2:622
Hydrologic information, availability of, 2:500–501
Hydrologic modeling, remote sensing and GIS application in, 2:535
Hydrology research, forest, 2:199
Hydropower, 2:554–555
Hyperaccumulation potential, of aquatic macrophytes, 2:66
Hypochlorites, 2:88

Ice control, alternatives to NaCl for, 2:320t. *See also* Deicers
Idaho, source-water protection in, 2:312
Immunodetection methods, 2:109–110
Immunohistochemistry, 2:109
Immunologic testing, for *Escherichia coli* O157:H7, 2:139
Impermeable layer, 2:544
Import of virtual water, 2:536
Index of Biotic Integrity (IBI), 2:25
India. *See also* Indian water markets
  climate of, 2:559–561
  floods and droughts in, 2:566–567
  groundwater resources in, 2:562–564
  irrigation water quality in, 2:155–161
  marginal cost pricing in, 2:558
  promoting self-governing institutions in, 2:558
  property rights in, 2:557
  regions of, 2:559
  surface water resources of, 2:561–562
  trace element groundwater contamination in, 2:143–148
  water quantity regulation in, 2:557–558
  water resource requirements for, 2:564–566
  water resources of, 2:559–567
  well-spacing norms in, 2:558
Indian water markets, 2:555–559
  economic aspects of, 2:556
  extent of, 2:556
  legal aspects of, 2:556–557
  nature of, 2:555–556
Indicator forms, harmful characteristics in, 2:443
Indicator groups, response to pollution, 2:441–442
Indicator monitoring, 2:267
Indicator organisms, 2:25, 292–294
  coliform group, 2:293
  enterococcus group, 2:293
  future of, 2:293–294
Indirect potable reuse, 2:611
Indirect search method, 2:332
Indirect water valuation studies, 2:655
Indoor residential water use, 2:652
Induction, of cytochrome P450 monooxygenase, 2:108
Inductively coupled plasma (ICP), 2:308, 309
Industrialized countries, water resources planning and management objectives for, 2:683
Industrial pollutants, macrophytes as biomonitors of, 2:66
Industrial wastewater, 2:375
Industrial water requirement, in India, 2:566
Industrial water use, 2:544
  in the United States, 2:652
Industry
  water conservation in, 2:496
  water consumption in, 2:370–371
  water-efficiency improvement in, 2:635
Infaunal amphipods, 2:409, 410
Infection, fecal-oral cycle of, 2:112
Infiltration, defined, 2:544
Infiltration rates, on forest soils, 2:199
Information portals, 2:670
Infrastructure, in the Nile Basin, 2:593–594
Injection wells, 2:544
  drinking water contamination by, 2:286–287
Inorganic chemical monitoring, 2:267
Inorganic colloids, stabilized by humic/fulvic acid, 2:208–209
Inorganic–humic agglomerates, metal ion binding to, 2:208
Inorganic pollutants, background concentration of, 2:19–20
In situ biodegradation, 2:44
In situ dredged material characterization, 2:351–352
In situ monitoring, of odorous emissions, 2:282–283
In situ sediment toxicity assessments, 2:384
Institutional solutions, in the Arab World, 2:473–474
Institutional water conservation measures, 2:497
Instream water use, 2:650, 661
Integrated Economic and Environmental Satellite Account (IEESA), 2:627
Integrated Pollution Prevention and Control (IPPC) Directive, 2:45–46
Integrated water quality modeling software tools, 2:328
Integrated water resources management (IWRM), 2:263, 574–576, 627
Integrated watershed management (IWM), 2:613, 627
Interdisciplinary water studies. *See* SUSTAINIS advanced study course
Intergovernmental Panel on Climate Change (IPCC), 2:640
  Report by, 2:483
Interim Enhanced Surface Water Treatment Rule (IESWTR), 2:195, 344
Interlaboratory precision tests, 2:459
Intermittent waters, 2:475
"Internal improvements," American, 2:498–499
International treaties, 2:473, 597

International Union for the Conservation of Nature (IUCN), 2:626–627
International virtual water flows, calculating, 2:540
International water market, trends in, 2:568–570
International Water Services Association (IWSA), 2:497
International water treaties, 2:618
Internet access gap, 2:670–671
Internet portals, water-related, 2:668–674
Interstate commission compacts, drought management by, 2:585
Interstate Commission on the Potomac River Basin (ICPRB), 2:577
Interstate water compact, 2:503–504
Interstitial water, 2:384
Intracellular biodegradation, 2:43
Intrastate regional authorities, drought management by, 2:585
Invertebrates, toxicity of lead to, 2:434
In vitro bioassays, 2:280
In vitro cytotoxicity techniques, 2:413
In vivo bioassays, 2:280
Ion-selective field effect transistor (ISFET), 2:298–299
IRC International Water and Sanitation Centre websites, 2:669
Iron. *See* Ferric hydroxides
Iron bacteria. *See also* Iron-oxidizing bacteria
 biofouling and water color caused by, 2:151–152
 metabolic activities of, 2:153
Iron oxidation rates, 2:152
Iron-oxidizing bacteria, 2:149–152. *See also* Iron- reducing bacteria
 in acid mining drainage, 2:150–151
 in groundwater and surface water, 2:150
Iron-reducing bacteria, 2:151
Iron removal
 ammonia and manganese inhibition of, 2:153–154
 cartridge filters for, 2:152–155
Iron removal systems, biological, 2:152–153
Irreversible specific toxicity, 2:415
Irrigated agriculture, 2:635
Irrigation. *See also* Irrigation water
 defined, 2:544
 diversions of, 2:617–618
 efficiency of, 2:495
 groundwater for, 2:156
 in India, 2:565

managing, 2:186
methods of, 2:375
Irrigation water, 2:370
 annual use of, 2:653
 efficiency and management of, 2:572, 617
 electrical conductivity for, 2:159t
 quality of, 2:155–161, 371
 use of, 2:544
Islamic law, water management prescriptions in, 2:474
Island Reports (Mediterranean), 2:640–641
IWR-MAIN water demand forecast model, 2:529, 530

Kansas, source-water protection in, 2:312
Kinetic analysis, 2:71
Kinetic processes, in water quality models, 2:271
Kinetics, of metal ion–humic colloid interaction, 2:209–210
Knowledge portals, 2:670
Krasheninnikov Bay, environmental quality of, 2:451t

Lability, of metal species, 2:212
Laboratory safety, 2:161–163
Lake and pond water, 2:95
Lake Bleiswijkse Zoom, biomanipulation trials in, 2:55t
Lake Bysjön, biomanipulation trials in, 2:54t
Lake Cristina, biomanipulation trials in, 2:54t
Lake Duiningermeer, biomanipulation trials in, 2:55t
Lake Feldberger, biomanipulation trials in, 2:57t
Lake Froylandsvatn, biomanipulation trials in, 2:56t
Lake Gjersjøen, biomanipulation trials in, 2:54t
Lake Haugatjern, biomanipulation trials in, 2:54t
Lake Lilla Stocke-lidsvatten, biomanipulation trials in, 2:53t
Lake Mead, 2:555
Lake Michigan, biomanipulation trials in, 2:53t, 56t
Lake Mosvatn, biomanipulation trials in, 2:54t
Lake purification, 2:50
Lake restoration methods, classification of, 2:50
Lake Rusutjärvi, biomanipulation trials in, 2:56t
Lakes, Canadian, 2:658–660

Lake Severson, biomanipulation trials in, 2:53t
Lake Søbygård, biomanipulation trials in, 2:55t
Lake Tenkiller model, 2:260–262
Lake Trummen, biomanipulation trials in, 2:53t
Lake Væng, biomanipulation trials in, 2:54t
Lake Washington, biomanipulation trials in, 2:53t
Lake Wolderwijd, biomanipulation trials in, 2:55t
Lake Zwemlust, biomanipulation trials in, 2:55t
Land Data Assimilation System (LDAS), 2:588–589
Land degradation processes, remote sensing and GIS application in, 2:534
Landfill buffer land, 2:166
Landfill gas
 collecting, 2:168
 from municipal solid waste landfills, 2:165–166
Landfill regulations, 2:166–167
Landfills, municipal solid waste, 2:163–169
Land management practices, in the 1930s, 2:511–512
Landscape irrigation, 2:651
Land Surface Models (LSMs), 2:588
Land use, effect on water quality, 2:169
Land-use activities, as threats to groundwater quality, 2:183t
Land use classification, remote sensing and GIS application in, 2:532
Langmuir isotherm equation, 2:87
Latin America, Web use in, 2:671–674
Laws, water sustainability, 2:627. *See also* Legal entries; Legislation; Regulations
Leachate collection and removal, from municipal solid waste landfills, 2:164–165
Leachate toxicity evaluation, 2:280
Leaching, 2:544
Lead (Pb). *See also* Pb bioavailability
 cardiovascular effects of, 2:436
 ecotoxicological effects of, 2:433–435
 effects on central nervous system, 2:435–436
 effects on peripheral nervous system, 2:436
 effects on renal system, 2:436
 effects on reproduction, 2:436
 endocrine effects of, 2:436

Lead (Pb). *See also* Pb bioavailability (*continued*)
  future research on, 2:437–438
  in groundwater, 2:148
  health effects of, 2:432–440
  hematopoietic effects of, 2:437
  intoxication by, 2:435
  maximum contaminant level of, 2:234t
  movement in the environment, 2:433
  neurological effects of, 2:435
  poisoning by, 2:437
  sources of, 2:433
Lead concentration, along the Blackstone River, 2:259
Lead uptake, from sediment pore water, 2:217
Leak detection and repair, 2:663, 664–665
Least squares method, 2:332
Legal considerations, in water quality management, 2:183
Legal solutions, in the Arab World, 2:473
Legal system, Chinese, 2:486–487
Legislation. *See also* Laws; Regulations
  drought-related, 2:578
  water, 2:96
  water-quality, 2:586
"Legislative" toxicity, 2:47
Lentic waters, 2:544
Leptothrix ferrooxidans, 2:149
Levee, 2:544
Ligand chemicals, 2:236–237
Ligand groups, 2:204
Light production, measuring, 2:48–49
Limits to Growth, The, 2:625
Limnodrilus, heavy metal uptake by, 2:215
Lines of evidence (LOEs), 2:353, 429–430
  biomarkers and bioaccumulation as, 2:426–432
Lipids, as indicators of organic contamination, 2:442
Lipophilic contaminants, SPMD-TOX paradigm monitoring of, 2:170–172
Liquid chromatography, high performance, 2:308
Liquid junction, 2:466
Liquid/liquid solvent extraction, 2:307
Liquid-phase carbon adsorption, 2:364
Liquid traps, 2:307
Livestock grazing, managing, 2:186. *See also* Dairy entries

Livestock water use, 2:544
Living filters, 2:102
Local government, approach to droughts, 2:577
Local management agencies, 2:193, 195
Local search methods, 2:332
"Local water use efficiency," 2:536
Locus of enterocyte effacement (LEE), 2:137
Logging, rain forest, 2:667
Los Angeles County, water withdrawals by, 2:479
Lotic waters, 2:544
Low flow aerators, 2:666
Low impact development (LID), 2:188
Low-water-solubility compounds, 2:44
Luminescent bacteria
  for determining mutagenic and carcinogenic agents, 2:174–175
  for determining nutrients in water, 2:175
  in online water toxicity monitoring, 2:174
  in water quality determination, 2:172–176
Luminescent bacterial biosensors
  for genotoxicity assessment, 2:455
  recombinant, 2:454
  for toxicant detection, 2:453–458
Luminescent-marked genes, 2:453–454
Luminometers, 2:49f
Luminometry, 2:48
*lux*CDABE reporter system, 2:453
lux genes
  for determining stress-inducing agents or toxic chemicals, 2:174
  in water quality determination, 2:172–176
Lyngbyatoxin A, 2:389
Lysimeter soil water sampling, 2:340–343
Lysosomes, 2:31

Macroarrays, 2:60
Macroinvertebrates, aquatic, 2:321–323
Macrophytes. *See also* Aquatic macrophytes
  as biomonitors of industrial pollutants, 2:66
  as trace metal biomonitors, 2:64–68
Maintenance, of municipal solid waste landfills, 2:168

Malaria, DDT use for, 2:600
Male reproductive system, effects of lead on, 2:436
Management
  as a remediation strategy, 2:358
  of water infrastructure, 2:594–595
Management activities, impacts of, 2:248
Management models, 2:369, 620
Manchester, Connecticut, drought-contingency-plan phasing criteria, 2:581, 582t
Manganese (Mn)
  in groundwater, 2:147
  in inhibition of biological iron removal, 2:153–154
  microbial oxidation of, 2:226
  removal of, 2:425
Manmade reservoirs, 2:367
Manufacturing, water use for, 2:662
Manure, composting/exporting, 2:572
Manure application, nitrate contamination and, 2:572
"Marble-in-a-cup" theory, 2:50, 52f
Marginal cost pricing, in India, 2:558
Margin of safety, 2:676
Marine bacteria, morphological features of, 2:446–447
Marine bioluminescence toxicity tests, 2:48
Marine environment
  monitoring quality of, 2:443–444
  monitoring water quality in, 2:447–448
Marine oil spills, 2:44, 291–292
Marine Protection, Research, and Sanctuaries Act, 2:125
Marine sediments, heavy metal concentrations in, 2:448–449
Marine Stewardship Council, 2:636
Marketable permits system, 2:507–508
Market-related transaction costs, 2:500
Mass conservation equations, 2:271
Massively parallel signature sequencing (MPSS) technology, 2:60–61
Mass spectroscopic methods, advanced, 2:206
Mass spectroscopy (MS), 2:308, 389
Mastigophora, 2:313
Mathematical models, 2:333
  role in water storage, 2:368
  in water resources systems analysis, 2:686–687
Matrix-assisted laser desorption ionization time of flight (MALDI-TOF), 2:61, 206

Maximum acceptable toxicant concentration (MATC), 2:434
Maximum concentration limit (MCL), of arsenic, 2:15
Maximum contaminant level goals (MCLGs), 2:195, 268
Maximum contaminant levels (MCLs), 2:233, 268, 373
  for arsenic, 2:8
  defined, 2:544
  for perchlorate, 2:345
  water contaminants with, 2:234–235t
Mean liquor suspended solids (MLSS), 2:48
Measurement programs, optimizing, 2:177
Mechanical dredging/barge disposal, 2:122
Mechanical treatment, for iron bacteria, 2:152
Median effective concentration (EC_{50}), 2:377–378
Mediated sensors, 2:41
Medications, nitrate-containing, 2:220
MEDIS project, 2:639–641
Mediterranean islands, sustainable water management on, 2:638–643. *See also* SUSTAINIS advanced study course
Membrane filtration, role of colloids and dissolved organics in, 2:99–103
Membrane fouling, limiting, 2:364
Membrane water treatment system, biofouling in, 2:242
Mercury (Hg), maximum contaminant level of, 2:234t
Metabolic heterogeneity, in biofilms, 2:230
Metabolic potential, 2:42
Metabolomics, 2:62
Metal biosorption. *See also* Metal sorption
  characteristics of, 2:70t
  mechanisms of, 2:70–71
Metal complex formation, 2:204
Metal concentration(s)
  dissolved, 2:422
  impact on plants, 2:65
  pervasiveness of, 2:237
Metal discharges, regulations concerning, 2:68
Metal-induced humic agglomeration, 2:209
Metal ion binding, to inorganic–humic agglomerates, 2:208
Metal ion concentrations, in groundwater, 2:144

Metal ion–humic colloid interaction, 2:205
  kinetics of, 2:209–210
  mechanisms of, 2:207–209
  modeling approaches to, 2:206
  speciation methods related to, 2:206–207
Metal ion speciation, estimating, 2:206. *See also* Metal speciation
Metalloenzymes, microbial, 2:233–235
Metallothioneins (MTs)
  biological function of, 2:406–407
  as biomarkers, 2:407
  classes of, 2:406
  factors affecting levels of, 2:407
  as indicators of trace metal pollution, 2:406–408
  quantification of, 2:407
Metal-oxide semiconductor field effect transistor (MOSFET) technology, 2:298
Metal-resistant bacteria
  distribution of, 2:448–449
  in Peter the Great Bay, 2:447f
  in Rudnaya Bight, 2:449f
Metal response elements (MREs), 2:407
Metals. *See also* Heavy metal entries
  bioavailability of, 2:211–212
  in key enzyme functions, 2:237t
  permissible concentrations of, 2:235t
  precipitation of, 2:70
  reduction of, 2:300
  sample preparation of, 2:309
  simulation, 2:256–259
Metal sorption, 2:363. *See also* Metal biosorption
Metal speciation
  chemical reactions affecting, 2:203–204
  measurement of, 2:204
  modeling, 2:204
  sensors, 2:419t
Metering, 2:496
  in Canada, 2:663
Methemoglobinemia, 2:219–223
  diagnosis and treatment of, 2:221
  hereditary causes of, 2:220–221
  susceptibility to, 2:219
  water-related and non-water-related causes of, 2:220
Methylated arsenic compounds, 2:11
Methyl-*tertiary*-butyl ether (MTBE), 2:312, 317, 344–345
Metropolitan Washington Council of Governments (MWCOG) "simple method," 2:253
Microarray analysis, 2:60

Microarray experiments, 2:62–63
Microarray Gene Expression Data (MGED) Society, 2:60
Microbes, in drinking water, 2:243. *See also* Bacteria; Marine bacteria; Microorganisms
Microbial activities management, 2:223–228
  biofilm control, 2:223–226
  control of biodegradable organic carbon, 2:224–225
Microbial activity
  in "dirty water" formation, 2:226
  disinfection to control, 2:225–226
  effect of temperature on, 2:226
Microbial aminolevulinate dehydratase (ALAD), 2:236–237
Microbial assessments, 2:450–451
  historical background of, 2:441–442
  for monitoring marine environment water quality, 2:447–448
Microbial biomass, as a sorbent, 2:365–366
Microbial biosurveys
  in Far Eastern Russia seas, 2:444–451
  for heavy metal detection, 2:442–443, 445, 446
  for organic pollution assessments, 2:449
Microbial BOD biosensors, 2:40. *See also* Biochemical oxygen demand (BOD)
Microbial detection, of phenols, 2:442
Microbial diversity, 2:42
Microbial dynamics, of biofilms, 2:228–233
Microbial enzyme assays, for heavy metal toxicity detection, 2:233–238
Microbial forms, in biofouling events, 2:239–243
Microbially available phosphorus (MAP), 2:245
Microbial metabolism, 2:299
Microbial metalloenzymes, 2:233–235
Microbial motility, 2:228–229
Microbial nutrients, in distribution systems, 2:244–245
Microbial pathogens, 2:371
  in aquatic systems, 2:352
Microbial pollutant detection, water quality monitoring and, 2:440–452
Microbial products, algal control and, 2:6
Microbial structure, in biofilms, 2:230–232

Microbiological monitoring, 2:267
Microbiological quality control, in distribution systems, 2:243–247
Microbiological standards, U.S., 2:268
Microcolumns (MIC), 2:212
Microcoulometric detectors, 2:308
Microcystins, 2:388, 390
Microflora, lipolytic, 2:442
MicroMac-ToxScreen, 2:174
Microorganisms. *See also* Bacteria; Microbes; Single-celled organisms
  adhesion of, 2:239
  geochemical ecology of, 2:442–443
  as indicators of oil pollution, 2:441–442
  inside biofouling layers, 2:240–241
  metabolic diversity of, 2:44–45
  oil-destroying, 2:449
  role in biodegradation, 2:42
  role in water vectored diseases, 2:112
  selective binding and absorption of metals by, 2:443
Microscale toxicity testing, in natural systems, 2:379–380
MICROTOX® bioassay, 2:171, 172, 173, 278, 351, 413, 453
Microtox-OS Test System, 2:174
Midrange water quality models, 2:248
Mineralization, 2:42
Mine waters, discharge standards for, 2:425
Minimum Information About a Microarray Experiment (MIAME), 2:60
Minimum National Standards (MINAS), 2:183
Mining water use, 2:544, 662
  in the United States, 2:652
Ministry of Agriculture (China), 2:486
Ministry of Construction (China), 2:486
Ministry of Geology and Mineral Resources (China), 2:486
Ministry of Public Health (China), 2:486
Ministry of Transportation (China), 2:486
Ministry of Water Resources (China), 2:485–486, 488
Models. *See also* Mathematical models
  calibration methods for, 2:332–333
  comprehensive, 2:255–256
"Model Water Use Act," 2:580, 583
Moderate Resolution Imaging Spectroradiometer (MODIS), 2:589
MODSIM water quantity network flow allocation model, 2:327
Molecular assays, for *Escherichia coli* O157:H7, 2:139–140
Molecular methods, 2:352
Molluscs. *See* Bivalve molluscs; Mussel entries; Shellfish growing
Monitoring
  of contaminants, 2:263–269
  of municipal solid waste landfills, 2:168
Monitoring parameters, criteria for, 2:266–267
Monohalogenated haloacetic acids, 2:75
Monsoons, Indian, 2:560
Morphological changes, studies of, 2:65–66
Most probable number (MPN) technique, 2:337
"Most sensitive species" assay, 2:376
Mother–daughter (shedding) cell formation, 2:239
Mouse bioassay, 2:389
Moving water, flow (fall) of, 2:554
MTT assay, 2:415
Multiconcentration tests, 2:382
"Multiobjective evaluation" concept, 2:595
Multispecies Freshwater Biomonitor (MFB), 2:31
Municipal solid waste landfills, 2:163–169. *See also* Landfill entries
  cover for, 2:164
  dry tomb, 2:163–164
  groundwater monitoring for, 2:165
  improving, 2:167–168
  landfill gas from, 2:165–166
  leachate collection and removal from, 2:164–165
  postclosure monitoring and maintenance of, 2:166–167
  reuse, recycling, and reduction related to, 2:167
  siting issues related to, 2:167
Municipal water supply, in the United States, 2:651–652
Municipal water supply system, 2:371, 544
Municipal water use, 2:662
Mussels, dose-response to chlorine, 2:401–406
"Mussel Watch Program," 2:35
Mutagenic agents, luminescent bacteria for determining, 2:174–175
Mutagenicity assay, 2:175
Mutatox assay, 2:171, 172, 174, 455
Mytilus galloprovincialis, as a bioindicator, 2:407
NADH, monitoring, 2:376–377. *See also* Nicotinamide-adenine dinucleotide (NAD$^+$)
Nanofiltration, effect of, 2:77–78
National Aeronautics and Space Administration (NASA), water-flow satellite data from, 2:587–589
National Environmental Policy Act, 2:594
National Environmental Protection Agency (China), 2:486
National Estuary Program (NEP), 2:185
National Oceanic and Atmospheric Administration (NOAA), 2:185
National Policy on Water Management, 2:526
National Pollutant Discharge Elimination System (NPDES), 2:96, 315
  permit program, 2:132, 188, 373
  pollution control policy of, 2:127, 128
National primary drinking water standards, 2:373t
National Shellfish Sanitation Program (NSSP), 2:361
National water policies, India, 2:556
National Water-Quality Assessment (NAWQA), 2:311–312
Natural bioremediation, of selenium, 2:358–359. *See also* Bioremediation
Natural contaminants, 2:181
Natural heritage, protecting, 2:667
Naturally occurring radioactive materials (NORM), 2:287
Natural organic matter (NOM), 2:74, 75
  removal of, 2:116
Natural resources. *See also* Natural water resource systems
  monitoring integrity of, 2:440–452
  sustainable management of, 2:633–638
  types of, 2:634
Natural Resources Conservation Service, 2:513
Natural systems, microscale toxicity testing in, 2:379–380
Natural water resource systems, 2:586
Natural waters
  cause of alkalinity in, 2:394
  disinfection byproduct precursor removal from, 2:115–117

trace metals in, 2:202
Nephelometric turbidity unit (NTU), 2:544
Net ecological benefit concept, 2:477
Netherlands, flood control history in, 2:524–526
Neurological effects, of lead, 2:435
Neurotoxins, 2:388–389
New Jersey, source water protection in, 2:312
Nhieu Loc–Thi Nghe Project, 2:553
Niagara Falls, 2:554
NICA–Donnan model, 2:206
Niche portals, 2:669
Nickel (Ni), in groundwater, 2:147
Nicotinamide-adenine dinucleotide (NAD^+), 2:42. See also NADH
Nile Basin
  solutions to water conflicts in, 2:593–594
  transboundary water conflicts in, 2:590–594
  water allocation in, 2:591–592
  water scarcity in, 2:590–591
  water-sharing framework for, 2:592–593
Nile Basin Initiative (NBI), 2:592
Nile Valley Plan, 2:591
Nile Waters Agreement, 2:591–592
Nitrate, in drinking water, 2:219–220, 395–396
Nitrate contamination
  of groundwater, 2:570–571
  maximum level of, 2:234t, 235t
Nitrate leaching, 2:200
Nitrate Leaching and Economic Analysis Package (NLEAP), 2:252
Nitrifer studies, 2:231
Nitrification, in distribution systems, 2:226
Nitrification/denitrification biofilm, 2:231
Nitrite, maximum contaminant level of, 2:235t
Nitrite-oxidizing bacteria (NOB), 2:231
Nitrogen (N). See also N/P (nitrogen/phosphorus) ratio
  forms of, 2:269
  maximum contaminant level of, 2:234t, 235t
  as a nutrient of concern, 2:269
  retention of, 2:200
Nitrogen availability reporter system, 2:175
Nitrogen fertilizer, 2:572
Nitrogen loading, identifying areas with, 2:571
Nodularins, 2:388, 390

NOEC calculations, 2:411. See also "No observed effect concentration" (NOEC)
Nonaqueous phase liquids (NAPLs), 2:44
Nonengineering water conservation interventions, 2:472–473
"Nonhazardous" waste management, 2:163
Nonoxidizing biocides, 2:242
Nonoxygen-based sensors, 2:41
Nonpathogenic microbes, 2:243
Nonpoint source control, 2:184–189
  federal programs for, 2:185–186
  regulations related to, 2:185
  by source categories, 2:186–187
  total maximum daily load and, 2:188–189
Nonpoint source (NPS) pollution, 2:181, 184, 544, 614
  management of, 2:311
Nonpoint source pollutants, 2:373–374
Nonpoint sources, 2:169, 190
Nonpoint source water quality models, 2:248–249
Nonpriority pollutant petroleum hydrocarbon analyses, 2:310–311
Nonrenewable natural resources, 2:634
Non-toxicant-specific luminescent biosensors, 2:454t
Nontoxic chemicals, 2:278
"No observed effect concentration" (NOEC), 2:381. See also NOEC calculations
"No observed effect level" (NOEL), 2:381
"Not in my backyard" (NIMBY), 2:167
N/P (nitrogen/phosphorus) ratio, 2:143
Nucleic add sequence-based amplification (NASBA), 2:337–338
Numerical models, 2:272
Nutrient adsorption, in biofouling, 2:239
Nutrient loading, 2:3
Nutrient management, 2:571
Nutrients. See also Microbial nutrients
  for algal growth, 2:269–270
  luminescent bacteria for determining, 2:175
  managing, 2:186
  in water bodies, 2:3
  as water quality indicators, 2:266

Nutrient transformation processes, 2:270
Nutrition, emergency, 2:528
Occupational Safety and Health Administration (OSHA) water handling standards, 2:162
Ocean pollution, response of indicator groups to, 2:441–442
Ocean shipments/dumping, precautionary principle regarding, 2:600
Odor analysis, state-of-the-art developments in, 2:281
Odor measurement methods, basis of, 2:281
Odor monitoring, electronic nose for, 2:281–284
Odors, landfill, 2:166
Office of Solid Waste and Emergency Response (EPA), 2:517
Offstream water use, 2:650
Oil-destroying microorganisms, 2:449
Oil-field brine (OFB), 2:284–290
  controlling by wastewater treatment methods, 2:288–289
  historical control practice for, 2:285
  impact on water quality, 2:287t
  pollution aspects of, 2:286–288
  use in oil production, 2:285–286
Oil pollution, 2:290–292
  microorganisms as indicators of, 2:441–442
Oil-resistant microorganisms, 2:448t
Oil spills
  biggest, 2:291t
  case study of, 2:172
Olfactometry, 2:281
Oligochaetes, accumulation patterns for, 2:215
Online monitors, bioluminescence-based, 2:174
Online water toxicity, monitoring luminescent bacteria in, 2:174
On-site disposal systems, managing runoff for, 2:187
On-site water quality determination, luminescent bacteria for, 2:173
Open limestone channel, 2:424
Optical pH sensors, 2:299
Option contracts, 2:607
Option prices, 2:607
Organic arsenic species, 2:12–13t
Organic carbon. See also Carbon (C)
  control of, 2:224–225
  in distribution systems, 2:244–245
  types of, 2:224–225
Organic chemical monitoring, 2:267
Organic compounds, in drinking water, 2:372–373

Organic contamination, indicators of, 2:442
Organic loading, relation to eutrophication, 2:142–143
Organic matter, 2:544
    microbially mediated oxidation of, 2:464–465
Organic pollutants, microbial indication of, 2:441
Organic pollution assessments, microbial biosurveys for, 2:449
Organic pollution measures, alternative, 2:40
Organisms, selection for sediment toxicity tests, 2:459. See also Indicator organisms; Microorganisms; Sediment dwelling organisms
Organization for Economic Cooperation and Development (OECD) BOD_5 standard, 2:39
Organolead, uptake of, 2:434. See also Lead (Pb)
Osmosis, 2:544
Our Common Future, 2:625–626, 633
Outdoor residential water use, 2:652
Outdoor watering equipment, 2:666
Outdoor water sources, emergency, 2:527
Outfall, defined, 2:544
Overall existence ranking index (OERI), 2:681
Overcapitalization, during the Great Depression, 2:511
Oxidation
    chemical, 2:300
    cometabolic, 2:300
Oxidation–reduction potential (ORP), 2:466. See also E_h; Oxygen redox potential (ORP); Redox entries
Oxidative capacity (OXC), 2:468
Oxidative precursor removal, 2:120
Oxidized carbons, sorptive capacity and selectivity of, 2:84
Oxidizing biocides, 2:242
Oxygen (O). See Dissolved oxygen (DO)
Oxygen-based sensors, 2:40–41
Oxygen complexes, in activated carbons, 2:81
Oxygen demand, 2:544
Oxygen-demanding wastes (ODWs), as water quality indicators, 2:266
Oxygen depletion, in waterbodies, 2:181
Oxygen limitation, in biofilms, 2:231–232

Oxygen redox potential (ORP), 2:231. See also Oxidation–reduction potential (ORP); Redox entries
Ozonation, 2:119, 244, 374
Packing cell formation, 2:240
PACT process, 2:87–88
Palmer Index, 2:581
Pan American Center for Sanitary Engineering and Environmental Sciences (CEPIS/PAHO), 2:671, 673, 674
Pan American Health Organization (PAHO/WHO), 2:671, 672, 673
Paralytic shellfish poisons (PSPs), 2:389
Parameter optimization, 2:332
Parasites, waterborne, 2:344t
Participatory integrated capacity-building (PICB), 2:628–629
Participatory multicriteria flood management, 2:678–683
Particle size, 2:545
Particulate arsenic, 2:11
Passive biomonitoring (PBM), 2:33
Passive treatment
    of acid mine drainage, 2:423–426
    selecting, 2:425f
    systems, 2:424f
Pathogenic bacteria, detection in surface waters, 2:441
Pathogenic pollution, 2:182
Pathogens, 2:545
    waterborne, 2:344t
    as water quality indicators, 2:266
Paul Lake, biomanipulation trials in, 2:56t
Pb bioavailability, microbial aminolevulinate dehydratase as a biosensor for, 2:236–237. See also Lead (Pb)
PCB-contaminated sediments, Hudson River, 2:126
PCB/dioxin-like compounds, cytochrome P450 monooxygenase as an indicator of, 2:106–111
PCR-based monitoring, of *Salmonella*, 2:338–339
pe, 2:466. See also E_h; pH
Peak flow, 2:545
Pennsylvania, source-water protection in, 2:312
Pennsylvania Drought Contingency Plan, 2:581
Penstock, 2:554
Pentavalent arsenic compounds, 2:11, 16
Per capita use, 2:545
Perchlorate, 2:317, 345
Perchlorate Community Right to Know Act of 2003, 2:345

Perchloroethylene (PCE), 2:317
    removal of, 2:299–301
Percolation, 2:545
Percolation tank, 2:551
Perennial flow, artificial, 2:477
Perennial waters, 2:475
Performance evaluation, for water resource systems, 2:674–678
Peripheral nervous system, effects of lead on, 2:436
Peripheral neuropathy, 2:436
Permanent spot market, 2:606–607
Permeability, 2:545
Pesticide Root Zone Model (PRZM), 2:252
Pesticides
    managing, 2:186
    water quality and, 2:266, 303
Peter the Great Bay, 2:445
    features of marine bacteria in, 2:446–447
    heavy metal monitoring in, 2:446
Petroleum pollution, 2:290–292. See also Oil entries
Petroleum products, 2:304
pH, 2:294–299, 544
    of acid mine drainage, 2:1
    dishwater, 2:113
    impact on DBP concentrations, 2:119–120
    measurement of, 2:295–299
    research on, 2:352–353
    theory of, 2:294–295
pH buffers, 2:295
Phenol-resistant microorganisms, 2:448t
Phenols, microbial detection of, 2:442
pH glass element, 2:296
PhoP primers, 2:338
Phospholipids, 2:442
Phosphorus (P). See also N/P (nitrogen/phosphorus) ratio
    in distribution systems, 2:245
    forms of, 2:269
    as a nutrient of concern, 2:269–270
Phosphorus precipitation, in algal control, 2:4
Photocatalytic oxidation, 2:300
Photoionization detector (PID), 2:307
Photosynthesis, sensitivity to Pb toxicity, 2:434
Photosynthetic bacteria, 2:21
PHREEQC program, 2:363–364
pH reference element, 2:296–297
Physicochemical monitoring, 2:267
Phytochelatin, 2:406
Phytoplankton, 2:142–143, 269
Pipeline sediment disposal, 2:124
Piper trilinear diagram, 2:396
Pipe wall biofilm communities, 2:223

Planned water recycling, 2:610
Planning
  decision support systems and, 2:619–620
  for droughts, 2:578–579
  water-infrastructure, 2:594–595
Planning tools, drought-related, 2:580
Pleasant Pond, biomanipulation trials in, 2:53t, 56t
PLOAD model, 2:253
Plumbing, water-saving, 2:497
Plumbism, 2:432–433
Point estimation techniques, 2:381
Point-of-use treatment methods, 2:343
Point-of-zero charge (PZC), 2:98
Point-pollution sources, 2:265
Point-source pollution, 2:545
Point sources, 2:169, 190
Political solutions, in the Arab World, 2:473
Pollutants. See also Pollution
  affinity to suspended particulate matter, 2:384
  background concentration of, 2:18–20
  inorganic, 2:19–20
  microbial detection of, 2:440–452
  organic, 2:20
  versus pollution, 2:190
"Polluter pays" principle, 2:597, 638
Pollution
  groundwater, 2:182–183
  indicator group response to, 2:441–442
  nonpoint source, 2:181, 184, 186–187
  oil, 2:290–292
  from oil-field brine, 2:286–288
  pathogenic, 2:182
  trace metal, 2:64–65
  urbanization-related, 2:181
  water recycling and, 2:612
Pollution control, 2:97
  affordable, 2:132
  in stored water, 2:367
Pollution loading, variations in, 2:264
Pollution management
  in Ho Chi Minh City, 2:553
  in urban landscapes, 2:190
Pollution monitoring programs, biomarker approach in, 2:429
"Pollution offsets," 2:133
Pollution taxes, 2:132–133
Poltruba backwater, biomanipulation trials in, 2:56t
Polychlorinated biphenyls (PCBs), 2:545
  chemical properties of, 2:109t
  as environmental toxicants, 2:106

Polycyclic aromatic hydrocarbons (PAHs), in chironomids, 2:421
Polymerase chain reaction (PCR), 2:139
  relative quantitation of products of, 2:59
Polymeric flocculants, 2:98–99
Polynuclear aromatic compounds, 2:310
Population growth, 2:574
  efficiency and, 2:493–494
  in the Nile Basin, 2:590
  water scarcity and, 2:495
Population growth impairment, as a sublethal toxic endpoint, 2:414–415
Pore water, 2:384
  concentrations, 2:421
Porosity, 2:545
Porous cup samplers, 2:341
Portals
  general, 2:669
  water-related, 2:668–674
Porter-Cologne Water Quality Act, 2:358
Potable water, 2:545
  regulatory and security requirements for, 2:343–350
  security requirements for, 2:347–348
Potassium (K), in groundwater, 2:395
Potassium permanganate, 2:300
Potomac River Low Flow Agreement, 2:585
Powdered activated carbon (PAC), 2:86–88. See also Activated carbons
  advantages of, 2:88
  dosages of, 2:87, 88
  oil-field brine and, 2:288
  in water treatment, 2:86–87
Powdered activated carbon contactor, analysis and design of, 2:87
Precautionary principle
  application to water science, 2:595–603
  concerns related to, 2:598
  current examples of, 2:599–601
  drivers and antidrivers related to, 2:598
  global elements of, 2:596–597
  global scope of, 2:596–598
  interpretation of, 2:596
  recommendations for, 2:601–602
  relevance to water and environmental science issues, 2:599–601
  risk and, 2:601
  risk tolerance and, 2:599
Precipitation, 2:516, 545
  in India, 2:560, 561f

variation in, 2:517, 574
Precipitation studies, remote sensing and GIS application in, 2:532
Precursors, physical removal of, 2:120
Predictive models, 2:26
Pressurized filters
  advantages and disadvantages of, 2:155
  for iron removal, 2:154
Preventive maintenance (PM), 2:595
Pricing, tiered, 2:615. See also Water pricing
Primary wastewater treatment, 2:545
Primary water quality standard, 2:373
Prior appropriation doctrine, 2:545
Prioritized objectives method, 2:623
Priority pollutants
  EPA, 2:305t
  list of, 2:304
Private value of water, 2:653–654
Private well owners, in Arab states, 2:471
Privatization, of water, 2:569
"Privatization by default," 2:632–633
Probabilistic approach, 2:675
Problem definition, in a systems approach, 2:685
Programs, conservation, 2:493
Property rights, 2:500, 506
  in India, 2:556–557
Protein array technologies, 2:61–62
Protein expression analysis, high throughput, 2:61
Protein phosphatase inhibition assay, 2:389–390
Proteins, as indicators of organic contamination, 2:442
Proteomics, 2:61–62, 429
Protists, as toxicity assessment test organisms, 2:413–418
Protozoa, 2:313
Pseudomonas, 2:241
Pseudosolubilization, 2:44
Public concerns, precautionary principle and, 2:599
Public education
  in Canada, 2:663
  in groundwater protection, 2:518
  strategies for, 2:519
Public health goal (PHG), 2:345
Public Health Security and Bioterrorism Preparedness and Response Act, 2:347
Public Health Service (PHS), 2:194
Publicly owned water treatment work (POTW), 2:96
Public Notification Rule, 2:347
Public supply, 2:545

Public trust, water managed in, 2:608–610
Public water supplies
 disinfection of, 2:118
 in the United States, 2:651–652
Public water systems (PWSs), 2:96, 371
Public water use, 2:545
Purgeable halocarbons, gas chromatogram of, 2:306f
Purge and trap procedure, 2:306
Pyrite, weathering of, 2:1

Quality, of irrigation water, 2:155–161. See also Water quality
Quality assessment. See Environmental quality assessment
Quality control, in amphipod sediment toxicity tests, 2:410–411
Quantitative reverse transcription polymerase chain reaction (qRT-PCR), 2:59
Quantitative structure–activity relationships (QSARs), 2:415

Radially entrained vapor extraction (REVEX), 2:286
Radionuclide monitoring, 2:267
Radionuclide standards, U.S., 2:268
Rainfall, in India, 2:559–560, 561f
Rainwater, 2:95
Rainwater harvesting, 2:548–552
 components of, 2:549–550
 defined, 2:549
 need for, 2:548–549
 structures related to, 2:550–551
 techniques for, 2:551
Random search method, 2:333
Rapid Bioassessment Protocols (RBPs), 2:25–26
Rapid BOD technologies, 2:40–41. See also Biochemical oxygen demand (BOD)
Rating curve, 2:545
rBAT-induced amino acid transport, 2:436
Recharge, defined, 2:545
Reciprocal SSH libraries, 2:62
Reclaimed wastewater, 2:545, 649–650
Recombinant bacteria, 2:454, 455
 bioluminescent, 2:174
Recombinant luminescent bacterial biosensors, 2:454
Recycled water, 2:545
 in the United States, 2:613

Red Book, 2:315
Redeposited sediment, contaminant release from, 2:125
Redox buffering capacity, 2:468
Redox gradients, 2:213
Redox phenomena, alternative methods of examining, 2:468–469. See also Oxygen redox potential (ORP); Reduction–oxidation (redox) reaction
Redox potential. See E_h
Redox reactions, ranges of E_h for, 2:468
"Red-rot" disease of water, 2:147
Reduction–oxidation (redox) reaction, 2:464. See also Redox entries
Reductive dechlorination, 2:299–300
Reference crop evapotranspiration, 2:537
Reference site criteria, 2:26
Reference toxicant tests, 2:411
Regional watershed protection, 2:613–614
Regional water supplies, 2:613–619
Regional water treatment, 2:615–616
Regional water use, 2:489–490
 in the United States, 2:646–647
Regression methods, 2:253
 in odor analysis, 2:282
Regulation. See also Regulations; Water regulation
 using biomarkers and bioaccumulation for, 2:429–430
 of water markets, 2:500
Regulations. See also Effluent water regulations; Federal regulations; Laws; Regulation
 bromide-related, 2:74
 Chinese, 2:487–488
 disinfection and disinfection byproduct, 2:92
 drinking water, 2:194–198, 345–347
 metal-discharge-related, 2:68
 nonpoint source control, 2:185
 Subtitle D, 2:166, 167
 water quality 314, 2:315
 water quantity, 2:557–558
 water supply degradation and, 2:401
 water sustainability and, 2:614, 627
Regulatory purposes
 sediment toxicity tests for, 2:383–387
 test development for, 2:458–464
Remedial investigation/feasibility study (RIFS), 2:318

Remediation, of selenium-contaminated waters, 2:355–360
Remediation target setting, 2:355–357
Remote sensing
 application in water resources, 2:531–536
 in command area studies, 2:533–534
 in floodplain management, 2:533
 future need for, 2:535–536
 in groundwater studies, 2:534–535
 in hydrologic modeling, 2:535
 in land use classification, 2:532
 in precipitation studies, 2:532
 in reservoir sedimentation, 2:534
 in snow cover studies, 2:533
 in soil moisture assessment, 2:533
 in waterlogging and soil salinity, 2:534
 in water quality studies, 2:534
 in watershed mapping and monitoring, 2:533
Removal, as a remediation strategy, 2:358
Renal system, effects of lead on, 2:436
Renewable freshwater resources, 2:634
Renewable natural resources, 2:634
Renewable water resources, in Arab states, 2:471t
REPIDISCA Technical Dissemination Sheets, 2:672
Reporter genes, bioluminescent, 2:453
Reproduction, effects of lead on, 2:436
Research
 on aquatic macrophytes, 2:66–67
 biodegradation, 2:42
Research biomarker, 2:30
Reservoir operations models, 2:623
Reservoirs, 2:545, 367, 568
Reservoir sedimentation, remote sensing and GIS application in, 2:534
Reservoir system, components of, 2:367
Residential water use, categories of, 2:652
Resource Conservation and Recovery Act (RCRA), 2:163
Resources. See American water resource development; Coastal resources; Community-based resource management; Flux-type natural resources; Freshwater resources; Global water resources; Groundwater resources; Integrated water resources management (IWRM);

Natural resources; Renewable entries; Water resource entries
Respirometers, 2:40–41
Retrofitting, 2:665–666
Return flow, 2:545
Reverse electron transfer (RET) assay, 2:279, 280, 376, 377–378
Reverse osmosis (RO), 2:545–546
  oil-field brine and, 2:289
Reversible nonspecific toxicity, 2:415
Reversibly adhering (rolling) cell formation, 2:240
Revised Universal Soil Loss Equation (RUSLE), 2:251
Rhepoxynius abronius, sediment toxicity tests using, 2:410–411
Ribosome content, cellular, 2:230
Ring tests, 2:459
Rio Grande Basin, 2:589
Riparian habitats. See also River entries
  wastewater discharge and, 2:475–476
  water recycling and, 2:612
Riparian water rights, 2:501, 546
Risk assessment, toxicological methods for, 2:278
Risk-informed decision making, 2:599
Risk management. See Flood risk management
Risk tolerance, 2:599
River and reservoir operations optimization models, 2:623
River basin decision support systems, 2:619–624
River basins, Indian, 2:560–561, 562t, 564t
River chemistry, global, 2:19
River discharge measurements, 2:622
River model development, 2:333–334
Rivers
  Canadian, 2:657–658
  Indian, 2:560–561, 563f
River (stream) water, 2:95
River water quality, impact on groundwater, 2:396–397
River water quality model calibration, 2:331–333
  genetic algorithms for, 2:333–335
  procedure for, 2:334
  single optimized parameter set selection in, 2:335
River water quality modeling software tools, 2:325–331
  evaluation and selection of, 2:328–330
River Yamuna, India, ground water quality near, 2:392–398
Road construction, managing, 2:187
Road salt, 2:319–325
  alternatives to, 2:320

effects on aquatic macroinvertebrates, 2:321–323
effects on zooplankton and fish, 2:323
field and laboratory studies of, 2:322t
impacts on aquatic environments, 2:320–321
indirect effects of, 2:321
loading, 2:321
properties of, 2:320
runoff, 2:321
Rooftop harvesting, 2:551
"Room for Rivers" policy report, 2:526
Root Zone Water Quality Model (RZWQM), 2:252
Round Lake, biomanipulation trials in, 2:57t
Runoff, 2:546
  effluent, 2:516
  farm, 2:400
  road salt, 2:321
  stormwater, 2:190
  surface water, 2:516
  urban, 2:186–187, 190, 192
Run-of-the-river system, 2:554
Rural development, in the Nile Basin, 2:593
Rural sanitation, in developing countries, 2:632
Rural water supply, in developing countries, 2:631–632
Russia, marine environmental monitoring in, 2:443–444. See also Far Eastern Russia seas

Safe Drinking Water Act (SDWA), 2:195, 286, 345, 585
  Amendments of 1986, 2:8, 299
Safety, of chlorine gas, 2:90
"Safe yield" concept, 2:514
Saline water, 2:546
Salinity. See also Desalination; Soil salinity
  of irrigation water, 2:157
  research on, 2:352–353
  in water bodies, 2:181
Salmonella, 2:337–340
  detecting using molecular tools, 2:337–339
  direct detection of, 2:338
  DNA probes in, 2:337–338
  future options in detecting, 2:339
  PCR based monitoring of, 2:338–339
Salt(s), as water quality indicators, 2:266. See also Road salt; Saline water; Salinity
Salt mines, oil-field brine and, 2:288
Sample contamination, 2:342

Sampling. See also Water sampling
  aquatic environmental, 2:170–171
  design, 2:265
  frequency and time of, 2:179–180
  locations and points related to, 2:178
  timing of, 2:180
Sampling parameters, criteria for, 2:266–267
Sampling site, documenting, 2:162
Sandy media, in column experiments, 2:104
Sanitary landfills, 2:163
Sanitation, in developing countries, 2:630–633
Sanitation Connection Internet resource, 2:669
Sanitation surveys, 2:361
Sarcodina, 2:313
Satellite data, water-flow-related, 2:587–589
Saturating recycle equation, 2:277
"Saturnine gout," 2:436
Saxitoxins, 2:389
Scanning electron microscopy (SEM), 2:310
Scientific water management models, 2:620
Sea-level rise
  as a global problem, 2:483
  societal impacts of, 2:482–483
Seattle Water Department, water use reduction plan, 2:581, 582t
Seawater, 2:95
  contamination criteria for, 2:450t
  microbial detection of oil pollution in, 2:441–442
  microbial detection of phenols in, 2:442
  quality control of, 2:449–451
Secondary physicochemical monitoring, 2:267
Secondary wastewater systems, powdered activated carbon in, 2:88
Secondary wastewater treatment, 2:546
Secondary water quality standard, 2:373
Security requirements, for potable water, 2:347–348
Sediment(s)
  aquatic, 2:384
  contaminated, 2:426
  defined, 2:546
  from forest water, 2:199
  heterogeneity of, 2:411
  properties of, 2:213t
  sealing and removal of, 2:5
Sedimentary rock, 2:546

Sediment assays, sediment or fauna incubation experiment for, 2:418–423. *See also* Sediment quality assessment/characterization; Sediment toxicity assessment/tests
Sedimentation, 2:374
  managing, 2:186
Sedimentation tanks, 2:546
Sediment cleanup, dredging for, 2:126
Sediment-dwelling organisms
  body concentrations in, 2:215–219
  exposure tests with, 2:212, 420
  heavy metal uptake rates among, 2:211–219
Sediment Or Fauna Incubation Experiment (SOFIE®), 2:212, 418–423
  exposure method for, 2:420
  results of, 2:421–423
Sediment oxidation, algal control via, 2:4–5
Sediment pollutants, ecotoxicological effects of, 2:428t
Sediment quality
  criteria for, 2:126
  future research on, 2:430–431
Sediment quality assessment/characterization
  bioaccumulation in, 2:426–428
  biomarkers in, 2:429
  chronic, bioaccumulation, and sublethal bioassays for, 2:351
  future developments in, 2:352–353
  historical background of, 2:408
  weight of evidence approach to, 2:350–355
Sediment quality guidelines (SQGs), 2:428, 430
Sediment Quality Triad (SQT), 2:350, 351
Sediment simulation, 2:256–259
Sediment testing, incentive for, 2:408
Sediment toxicity assessment/tests
  amphipod, 2:408–413
  application of, 2:460–462
  bioassays checked for use in, 2:462t
  biological guidelines for, 2:463t
  biomarkers used for, 2:430t
  design of, 2:459
  development and application of, 2:383–387
  development for regulatory purposes, 2:458–464
  systems used for, 2:384–385
Seed microorganisms, 2:39
Seepage, 2:546
Select Committee on Transportation Routes, 2:523

Selective dissemination of information (SDI), 2:672
Selenium (Se)
  Aquatic Life Criterion for, 2:355–357
  bioaccumulation of, 2:357t
  biogeochemistry of, 2:355
  biogeochemical cycling of, 2:356f
  maximum contaminant level of, 2:235t
  volatilization of, 2:359
Selenium-contaminated waters, remediation and bioremediation of, 2:355–360
Self-governing institutions, promoting, 2:558
Self-supplied water, 2:546
Semipermeable membrane devices (SPMDs), 2:170–172
Senec gravel pit, biomanipulation trials in, 2:55t
Sensitive ecosystems, water recycling and, 2:612
Sensor-based pH measurement, 2:295–299
Sensor drift, 2:283
Septic tank, 2:546
Septic treatment systems, 2:663–664
Serial analysis of gene expression (SAGE), 2:60
Settling pond, 2:546
Sewage
  contaminants from, 2:182
  dioxin-contaminated, 2:110
Sewage liquid, electronic nose analysis of, 2:282
Sewage sludge, precautionary principle regarding, 2:600
Sewage treatment plant, 2:546
Sewerage, simplified, 2:633
Sewer overflows, 2:263
Sewer systems, urban, 2:96
Shallow sewerage, 2:633
Shellfish growing, water classification for, 2:360–362
Shiga-like toxins, 2:136
Shiga toxin-producing *Escherichia coli* (STEC), 2:136, 137
Shiga toxins, 2:137
Showerheads, low-flow, 2:665–666
Simple water quality models, 2:248
Simulated oil spill case study, 2:172
Simulator for Water Resources in Rural Basins (SWRRB) model, 2:250
Single-celled organisms, as test subjects, 2:413
Sinkhole, 2:546–547
Site-specific sampling, 2:162
Siting issues, for municipal solid waste landfills, 2:167

Size fractionation procedures, 2:102
Skin cancer, arsenic-related, 2:17
Slow migratory (spreading) cell formation, 2:240
SMP reagents, 2:376
Snow control, alternatives to NaCl for, 2:320t
Snow cover studies, remote sensing and GIS application in, 2:533
Snowfall, in Canada, 2:660
Social value of water, 2:653–654
"Sociological drought," 2:576
Sodium (Na), in groundwater, 2:394
Sodium adsorption ratio (SAR)
  for irrigation water, 2:157–159
  values of, 2:160t
Sodium carbonate, in irrigation water, 2:159
Sodium chloride, properties of, 2:320t. *See also* Road salt
SOFIE. *See* Sediment Or Fauna Incubation Experiment (SOFIE®)
Software tools
  catchment water quality modeling, 2:325–326
  integrated water quality modeling, 2:328
  river water quality modeling, 2:325–331
  steady-state, 2:326–327
  unsteady-state, 2:327–328
Soil and Water Assessment Tool (SWAT), 2:250
Soil Conservation Service (SCS), 2:513
Soil management practices, water quality models for developing, 2:248–255
Soil moisture assessment, remote sensing and GIS application in, 2:533
Soil salinity, remote sensing and GIS application in, 2:534
Soil water
  lysimeter sampling of, 2:340–343
  solute concentration of, 2:340
Solid/liquid solvent extraction, 2:307
Solid state pH sensors, 2:298–299
Solid trap materials, 2:307
Solute, 2:547
Solution
  defined, 2:547
  redox potential of, 2:466
Solvents, 2:317t, 547
  extraction of, 2:306–307
  stabilizers for, 2:317
SOLVER function, 2:621
Sorbents, chemically modified, 2:72
Sorbitol-containing media, 2:139
Sorption, 2:362–364

fixed-bed, 2:363–364
kinetics, 2:72
Sorption coefficient, of cadmium onto sea sand, 2:105
Sorptive filtration, 2:362–366
  combined process in, 2:364–366
Source Loading and Management Model (SLAMM), 2:253
Source-water protection, 2:120, 311–313
Source water quality, degraded, 2:400
Source water quality management, 2:399–401
  future of, 2:401
Southeast, water demands in, 2:501
Soxhlet extractor, 2:307
Spain
  dredged material characterization in, 2:351–352
  permanent spot market in, 2:606
  sediment toxicity tests in, 2:460–462
Spawning season, chlorine sensitivity and, 2:402–403
Speciation, effects of treatment processes on, 2:75–78
Species, taxonomical identification of, 2:459
Specific conductance, 2:547
Spectral libraries, 2:308
Spectroscopic methods, 2:207
Spectroscopy, atomic absorption, 2:308–309
Spillways, 2:568
Spirostomum ambiguum, biotest using, 2:413
Spirostomum teres, as a test organism, 2:414
SPIROTOX test, 2:413
SPMD-TOX paradigm, monitoring lipophilic contaminants using, 2:170–172
Sporozoans, 2:314
Spot markets, 2:606–607
Spray irrigation, 2:547
Spreadsheet models, 2:253
Spreadsheets, decision support system, 2:620–621
Sprinkler irrigation methods, 2:495–496
Sprinklers, 2:666
Stakeholders
  integration among, 2:575
  participation in flood management, 2:680
Standard bioassay, as an exposure method, 2:420
Standardized bioassays, 2:419, 420
Standards
  disinfection byproduct, 2:92t

water-handling, 2:162–163
water-quality, 2:302
Standard-state reduction potentials, 2:465t
Starch, as an indicator of organic contamination, 2:442
State government, approach to droughts, 2:577–578, 580
State planning and management activities, drought-related, 2:583–585
States, drought policies of, 2:576
State treatment capacity, 2:616
State variable interactions, 2:274f
State water laws, 2:584–585
State water supplies, 2:613–619
Static data, 2:622
Statistical analysis, in amphipod sediment toxicity tests, 2:411
Steady-state models, 2:272
Steady-state software tools, 2:326–327
Steam flooding, of oil-field brine, 2:286
Stochastic methods, 2:333
Stockholm Convention on Persistent Organic Pollutants (SCOPOP), 2:106
Storage system, 2:554
Storage volume, water quality and, 2:367–368
Storage water quality, 2:367–370
  management of, 2:368–369
STORET system, 2:315
Storm sewer, 2:547
Stormwater management, 2:192
Storm Water Management Model (SWMM), 2:325–326
Stormwater runoff, 2:190
Strategic analysis, 2:629
Straw, in algal control, 2:5–6
Streamflow, 2:547
StreamPlan (Spreadsheet Tool for River Environmental Assessment Management and Planning), 2:328
Streamside Management Areas (SMAs), 2:187
STREAMS resource centers, 2:670
Stress-inducing agents, *lux* genes for determining, 2:174
Stressor-based assessments, 2:516
Stx-producing *Escherichia coli*, 2:137. See also *Escherichia coli* O157:H7
Subchronic toxicity tests, sediment, 2:385t
Sublethal bioassays, 2:351
Submitochondrial particle (SMP) assay, 2:278–280

as a biological monitoring tool, 2:376–379
environmental monitoring with, 2:378
literature review of, 2:377–378
sensitivity of, 2:377
variants of, 2:378
Subsidence, 2:547
Subtitle D landfill regulations, 2:166, 167
Successive alkalinity producing system (SAPS), 2:424
Suction cups, 2:341
Sudan-Egypt water dispute, 2:591–592
Sulfate, in groundwater, 2:395
Sulfate-reducing biofilm, 2:231
Sulfur-oxidizing bacteria, 2:241
Sun Valley, water conservation in, 2:635
Superfund. See EPA Superfund
Superfund sites, adsorptive filtration at, 2:365
Supervisory Control and Data Acquisition (SCADA) system, 2:510
Supplies, disaster, 2:529
Supply equation, 2:615
Suppression subtractive hybridization (SSH) libraries, 2:62
Surface complexation models, 2:363
Surface functional groups, on activated carbons, 2:82
Surfaces, biofilm formation on, 2:228
Surface supplies, variability in, 2:400–401
Surface tension, 2:547
Surface water(s)
  defined, 2:547
  detection of pathogenic bacteria and biological contamination in, 2:441
  EU monitoring of, 2:267
  iron-oxidizing bacteria in, 2:150
  relationship to groundwater, 2:396–397
  runoff, 2:516
  quality of, 2:585–586
  use-based classification of, 2:184
Surface water resources, of India, 2:561–562
Surface water sources, high control of, 2:399–400
Surface Water Treatment Rule, 2:140, 195
Surface water use, in arid lands, 2:476
Survey sampling, 2:161
Suspended sediment, 2:547
  concentration of, 2:547

Suspended sediment, (*continued*)
  discharge, 2:547
Suspended solids, 2:547
Sustainability
  concepts related to, 2:625–627
  defined, 2:633
  imperatives and preconditions related to, 2:626t
  in the Nile Basin Initiative, 2:592
  practices in, 2:628–629
  thermodynamic aspects of, 2:626
  water resource, 2:624–630
Sustainable development, 2:624
  precautionary principle regarding, 2:601
Sustainable flood-management decision making, 2:678–679
Sustainable management. *See also* Sustainable water management; Water resource sustainability
  of ecosystems, 2:636–637
  of fisheries, 2:635–636
  of freshwater resources, 2:634–635
  of natural resources, 2:633–638
Sustainable water management. *See also* MEDIS project
  on Mediterranean islands, 2:638–643
  practices in, 2:639–641
SUSTAINIS advanced study course, 2:641–642
SYMTOX4 (Simplified Method Program–Variable Complexity Stream Toxics Model) software, 2:327
Synthetic dilution water, 2:383
Synthetic organic chemicals (SOCs), 2:268
System performance evaluation, fuzzy criteria for, 2:674–678
Systems analysis, 2:683–687
Systems approach, 2:685–686

Tanzania, "privatization by default" in, 2:632–633
"Target concentration," 2:357
Tariffs, water delivery, 2:473
TCA, 2:317, 318t
Technical Cooperation Committee for the Promotion of the Development and Environmental Protection of the Nile Basin (TECCONILE), 2:592
Technologies
  high-throughput, 2:59–62
  role in water markets, 2:568–569
Tectonic plate movement, in the Caspian basin, 2:482
Temperature, chlorine sensitivity and, 2:404–405

Tensiometers (TE), 2:341
Tension-free lysimeter, 2:340
Tension lysimeter, 2:341
Tentatively identified compounds (TICs), 2:308
Terminal electron-accepting processes (TEAPs), 2:469
Ternary complexes, 2:208
Terrestrial bacteria, luminescent marking of, 2:453–454
Terrestrial organisms, toxicity of lead to, 2:434
Tertiary wastewater treatment, 2:547
Test development, for regulatory purposes, 2:458–464
Test kits, commercial, 2:384
Test organisms, ciliated protists as, 2:413–418
Test ruggedness, 2:459
Test systems, for sediment toxicity assessment, 2:384–385
Test validation, 2:459–460
Tetrachlorodibenzo-*p*-dioxin (TCDD), 2:108–109
Tetrachloroethylene. *See* Perchloroethylene (PCE)
Tetrahymena, in laboratory research, 2:413
Tetrahymena pyriformis, as a test organism, 2:413, 414
Tetrahymena termophila, in toxicity assessment, 2:413
TETRATOX test, 2:414
Thallium (Tl), maximum contaminant level of, 2:235t
The Groundwater Foundation (TGF), 2:518, 519
Thermal conductivity detectors, 2:308
Thermal pollution, 2:547
  as a water quality indicator, 2:266
Thermal power generation, water use for, 2:662
Thermoelectric power water use, 2:547
  in the United States, 2:653
Thiobacillus ferrooxidans, 2:149, 150
THM formation potential (THMFP), 2:117. *See also* Trihalomethanes (THMs)
THM precursor removal, 2:116–117
Threat contaminant (TC), 2:348
Thrombotic thrombocytopenic purpura (TTP), 2:137
Thu Duc Water Station, Ho Chi Minh City, 2:553
Tiered testing, 2:430, 460, 461f
Timber Culture Act of 1873, 2:512
Timber harvesting, managing, 2:187

Time-dependent models, 2:272
Time-dynamic speciation, body concentrations and, 2:215–219
Time-resolved laser fluorescence spectroscopy (TRLFS), 2:207–208
Time-varying exposure concentrations, 2:213–215
Tissue analysis, for trace metal concentration, 2:65
Tissue quality guidelines (TQGs), 2:428
Tissue quality values (TQVs), 2:351
TMDL plan, 2:188. *See also* Total maximum daily load (TMDL)
Toilet dam, 2:665
Toilet leaks, 2:664–665
Toilet retrofits, 2:665
Tool development, for ecotoxicogenomic research and biomonitoring, 2:62–63. *See also* Software tools
Tordon 75D investigation, 2:280
Tory Lake, biomanipulation trials in, 2:57t
Total chlorine residual, 2:399
Total Coliform Rule, 2:140
Total dissolved solids (TDSs), 2:157
  in groundwater, 2:394
Total maximum daily load (TMDL), 2:21, 185, 188, 233
  cost-effectiveness analysis for, 2:131
  of impaired waters, 2:476
  programs, 2:127, 128, 188–189
Total oil and grease analysis, 2:311
Total organic carbon (TOC), 2:40
Total petroleum hydrocarbon (TPH) oil and grease, 2:311
Total phenolics analysis, 2:309–310
Total suspended solids (TSS), along the Blackstone River, 2:258
"Total Water Management," 2:587
TOXcontrol-BioMonitor, 2:174
Toxic agents/chemicals/substances, 2:29
  in drinking water, 2:371–372
  lux genes for determining, 2:174
  precautionary principle regarding, 2:600
Toxicant detection, luminescent bacterial biosensors for, 2:453–458
Toxicants
  defined, 2:170
  in sediment, 2:426
Toxicant-specific luminescent biosensors, 2:455t
Toxic constituent manuals (EPA), 2:381

Toxic cyanobacteria, ecology of, 2:388
Toxic metal biosorption, 2:68–74
  mechanisms of, 2:70–71
Toxic metals, persistence in the water environment, 2:233
Toxicity. *See also* Aquatic toxicity
  of arsenic, 2:15
  bacterial bioluminescence and, 2:47
  chlorine, 2:402
  numbers associated with, 2:47
  persistence of, 2:381
  predicting, 2:415
Toxicity assays, of SPMD dialysates, 2:171. *See also* Toxicity assessment; Toxicity bioassays; Toxicological assays
Toxicity assessment
  ciliated protists in, 2:413–418
  direct, 2:46–47
  new approaches in, 2:414–415
  quantitative structure–activity relationships in, 2:415
Toxicity bioassays, 2:453. *See also* Toxicity assays
Toxicity biomarkers, 2:29, 31
Toxicity consents, 2:46
Toxicity controls, whole effluent, 2:382–383
Toxicity identification and evaluations (TIEs), 2:380–382, 383, 411
"Toxicity reduction evaluation" (TRE), 2:381
Toxicity testing
  bioluminescent biosensors for, 2:45–50
  environmental quality standards and, 2:46
  legislative drivers for, 2:45–46
  of sediment, 2:351, 383–387
Toxicogenomic experiments, 2:62–63
Toxicological assays, problems with, 2:415. *See also* Toxicity assays
Toxicological database, 2:409
Toxicological methods, 2:278
Toxins, algal, 2:387–392
ToxScreen test, 2:173–174
Trace element contamination, in groundwater, 2:143–148
Trace elements, in River Yamuna groundwater, 2:396
Trace metal biomonitors, macrophytes as, 2:64–68
Trace metal pollution
  in the aquatic environment, 2:64–65
  metallothioneins as indicators of, 2:406

Trace metals, sorption onto activated carbon, 2:83–85
Trace metal speciation, 2:202–205
  chemical reactions affecting, 2:203–204
  measurement of, 2:204
  modeling, 2:204
Tradable permits system, 2:133–134
Tradable rights
  issues related to, 2:643–644
  market allocation of water and, 2:644
  meeting water needs with, 2:643–645
  security of tenure and, 2:644–645
"Tragedy of the Commons, The," 2:556,580,625
Transaction costs, water market, 2:499–500
Transboundary water conflicts, Nile Basin, 2:590–594
Transfer rules, 2:501
Transmissibility, ground water, 2:547
Transmission electron microscopy (TEM), 2:310
Transpiration, 2:547
Transport phenomena, of dissolved matter, 2:105
Travel cost models, 2:130
Treatment
  effect on assimilable organic carbon and biodegradable dissolved organic carbon, 2:225
  effect on speciation, 2:75–78
  loss of, 2:48
  as a remediation strategy, 2:358
Treatment systems, stressed, 2:663
Trend stations, 2:178
Tributary, 2:547
Trichloroethylene (TCE), 2:317
Trickling filters
  advantages and disadvantages of, 2:155
  for iron removal, 2:154–155
Trihalomethanes (THMs), 2:91–92, 115, 116, 268. *See also* THM entries
  bromide influence on formation of, 2:74–79
  brominated, 2:74–75
Trondheim Biomonitoring System (TBS), 2:28, 30–33
Tumen River Economic Development Area Project (TREDA), 2:444
Tumorigenicity, of arsenic, 2:8
Turbidity
  defined, 2:547
  measurements of, 2:344
Typhoid fever, 2:337

U.K. National Rivers Authority, water supply categories used by, 2:614, 615t
Ultrafiltration, pore sizes for, 2:101–102
Ultra-low-volume (ULV) toilets, 2:665
Ultraviolet (UV) radiation, 2:119, 244, 374
Unaccounted for water (UFW) indicator, 2:492, 496
Uncertainty, 2:599
  in hydrologic data, 2:621
  participatory multicriteria decision-making under, 2:681
  sources of, 2:674–675
  types of, 2:680
United Nations, public trust guidelines of, 2:609
United Nations Industrial Development Organization (UNIDO), planning analysis objectives, 2:683
United Nations World Water Development Report, 2:548, 549
United States. *See also* Federal entries; U.S. entries; Western United States
  aquatic environment clean up by, 2:184
  drinking water quality in, 2:373
  evolution of water use in, 2:646t
  new water sources in, 2:649–650
  selenium containment strategy in, 2:358
  source-water protection in, 2:312–313
  total water withdrawals in, 2:646t
  water consumption patterns in, 2:651
  water markets in, 2:500
  water quality management in, 2:193–198
  water quality policy in, 2:127–128
  water quality standards in, 2:268
  water saving measures in, 2:649
  water use in, 2:645–650, 650–653
United States Environmental Protection Agency (USEPA). *See also* USEPA Test and Evaluation (T&E) Facility
  maximum contaminant levels, 2:233
  water quality standards, 2:373
  water reuse guidelines, 2:611
Universal soil loss equation (USLE), 2:535
Unplanned water recycling, 2:610
Unregulated Contaminant Monitoring Rule (UCMR), 2:345
Unsaturated zone, 2:548

Unsteady-state software tools, 2:327–328
Upflow filtration, 2:366
Upland-"confined" dredged sediment disposal, 2:124–125
Upper Mississippi River Source Water Protection Initiative, 2:312
Urban Drainage and Sewer Master Plan, Ho Chi Minh City, 2:553
Urbanization, pollution due to, 2:181
Urban landscape, water quality management in, 2:189–193
Urban river sites case study, 2:172
Urban runoff, 2:190, 192
　indicator parameters of, 2:192t
　managing, 2:187
　nonpoint source pollution from, 2:186–187
Urban services, efficiency of, 2:491–492
Urban Wastewater Treatment Directive, 2:45
Urban water demand, 2:490–491
Urban water resource management, in Asia, 2:552–554
Urban watershed protection approach, 2:191
Urban water supply, in developing countries, 2:632
U.S. Agency for International Development, 2:489
U.S. Army Corps of Engineers. *See* Army Corps of Engineers
U.S. coastal waters, oil spills in, 2:291t
U.S. Department of Agriculture (USDA), 2:186
U.S. Department of the Interior (DOI), 2:186
U.S. Department of Transportation (DOT), 2:186
U.S. EPA Subtitle D dry tomb landfilling, 2:163–164
USEPA Test and Evaluation (T&E) Facility, 2:509–510
U.S. Geological Survey (USGS), 2:311, 312
U.S. National Environmental Policy Act (NEPA), 2:624
U.S. Salinity Laboratory classification, 2:159–160, 161t
U.S. waterways, contamination of, 2:122

VALIMAR monitoring project, 2:31, 32
Vertical flow pond, 2:424
Vibrio fischeri luminescent bacterium, 2:453
Vietnam, water treatment in, 2:553
Vinyl chloride (VC), 2:266–267
Viral disease, 2:371
Virtual Library on Environmental Health (VLEH), 2:671–674
　difficulties with, 2:673–674
Virtual water, 2:536–537
　import and export of, 2:538–540t
　importation in the Nile Basin, 2:593
Virtual water balance, 2:537f
　national, 2:540
Virtual water content, calculating, 2:537
Virtual water flows
　calculating, 2:537–540
　global, 2:540–541
Vitamin D metabolism, effects of lead on, 2:436
Vitotox® assay, 2:175, 455
Volatile aromatic compounds, 2:310
Volatile organic chemicals (VOCs), 2:268, 343. *See also* Perchloroethylene (PCE)
　detection of, 2:283
　as water quality indicators, 2:266–267
Vulnerability assessment (VA), 2:347–348

Washington State, source-water protection in, 2:311, 312
Wasteload allocations (WLA), 2:188
Waste management practices, 2:374–375
Wastewater(s)
　defined, 2:548
　EU monitoring of, 2:267–268
　from oil production, 2:285–286
　reclamation of, 2:497
Wastewater biofilms, 2:230
Wastewater treatment
　biofilm in, 2:230–232
　powdered activated carbon in, 2:87–88
Wastewater treatment facilities, Ho Chi Minh City, 2:553
Wastewater treatment methods, oil-field brine control by, 2:288–289
Wastewater-treatment return flow, 2:548
Water. *See also* Aquatic entries; Water quality; Water resource entries
　algal toxins in, 2:387–392
　bacteria, viruses, and pathogens in, 2:303
　balancing the need for, 2:574
　classification of, 2:95
　competing uses for, 2:613
　emergency, 2:526–529
　globalization of, 2:536–541
　market allocation of, 2:644
　marketing proposals for, 2:506–508
　measurement of, 2:658–660
　protozoa in, 2:313–314
　as a renewable resource, 2:661
　storing, 2:527
Water Act of 1974, 2:183
Water allocation, 2:504–506. *See also* Water distribution
　efficient, 2:643
Water allocation law, 2:558
Water allocation system, 2:644
Water allotments, purchasing, 2:508
Water analysis, tests for, 2:178, 180t
Water analysis methods, 2:304–311
　gas chromatography, 2:304–308
Water assessment and criteria, 2:24–28
　biotic integrity and, 2:26
　current, 2:25–26
　development of, 2:24–25
　reference sites and biocriteria, 2:26
Water audits, 2:492
Waterbodies
　effluent-based, 2:475
　"impaired," 2:476
Waterborne bacteria, 2:20–24
　identification of, 2:21
Waterborne disease outbreaks, 2:293
Waterborne transmission, of *Escherichia coli* O157:H7, 2:138
"Water budget," 2:507
Water chemistry, factors determining, 2:19
Water color, caused by iron bacteria, 2:151–152
Water conflicts. *See also* Water disputes
　in the Nile Basin, 2:590–594
　over scarce water, 2:502–503
Water conservation, 2:489–494, 495–498. *See also* Water saving technologies; Water use
　in Canada, 2:660–666
　efficiency and, 2:492–493
　in the home, community, and workplace, 2:664–666
　in integrated water resources management, 2:575
Water conservation interventions, nonengineering, 2:472–473
Water consumption, 2:370–371
　reducing, 2:664
　rise in, 2:375
Water cycle, 2:548
Water debts, 2:617
Water demand, urban, 2:490–491
Water demand forecasting, 2:529–531

cases of, 2:530
model, 2:622–623
Water dependency, global, 2:538–540t
Water disputes. *See also* Water conflicts
  averting, 2:501–510
  constitutional solutions to, 2:503–504
  water allocation and, 2:504–506
Water distribution, fair, 2:491–492. *See also* Water allocation entries
Water distribution system simulators (DSS), 2:509–510
Water dwellers, exposure method using, 2:420
Water economic analysis, components of, 2:128–129
Water efficiency, technological options for, 2:493
Water efficient residential technology, 2:663
"Water emergency price schedule," 2:583
Water environment, persistence of toxic metals in, 2:233
Water Erosion Prediction Project (WEPP), 2:251–252
Water facilities, capital investment in, 2:594
Water-flow satellite data, from NASA, 2:587–589
Waterfowl, lead toxicity to, 2:434–435
Water Framework Directive, 2:45
Water harvesting
  choosing techniques for, 2:551
  components of, 2:549–550
  defined, 2:549
  forms of, 2:550
  need for, 2:548–549
  structures related to, 2:550–551
Water industry, 2:587
Water infrastructure, 2:567–568
  planning and managing, 2:594–595
Watering equipment, outdoor, 2:666
Water Law of China, 2:486
Water laws/legislation, 2:96
  state, 2:584–585
Water-limited environments, ecological effects assessment for, 2:516–518
Waterlogging, remote sensing and GIS application in, 2:534
Water management
  in the Arab World, 2:472–474
  in China, 2:484–488
  computer models for, 2:620
  institutional structure of, 2:614–615

in the Mediterranean islands, 2:642
in the public trust, 2:608–610
Water-management plans, drought strategy in, 2:581
Water markets, 2:499–501. *See also* International water market; Spot markets
  Chilean, 2:645
  in India, 2:555–559
  institutional options related to, 2:500–501
  reform of, 2:508
  regulation of, 2:500
  transaction costs of, 2:499–500
Water needs, tradable rights and, 2:643–645
Water policies, supply-oriented, 2:574
Water pollution, 2:95–96
  health risks of, 2:617
  tests used for measuring, 2:180t
Water Portal of the Americas, 2:669
Water portals, 2:668–674
  types of, 2:669
Water pricing, 2:497, 603–606. *See also* Costs; Option prices
  assessment of, 2:605–606
  in Canada, 2:663
  components of, 2:603
  functions of, 2:604–605
  global, 2:603–604
  increase in, 2:604–605
Water problems, current, 2:574
Water project operations, 2:621
Water projects, federal, 2:522–523
Water purification, emergency, 2:527
Water quality, 2:96–97, 301–304, 314–316. *See also* Storage water quality; Water quality management; Water quality modeling; Water supply quality
  in Canada, 2:663–664
  chronology of events related to, 2:314
  costs and benefits of, 2:129
  criteria for, 2:315
  data on, 2:192
  defined, 2:302, 548
  in dredged sediment management, 2:122–127
  early pollutant detection and, 2:440–452
  economics of, 2:127–135
  effect of human activities on, 2:303
  effect of natural processes on, 2:302–303
  effect of scale on, 2:190
  electronic nose applications related to, 2:282, 283
  ethic for, 2:315
  evaluating, 2:368

federal regulations regarding, 2:190
in forested land, 2:199
guidelines for, 2:337
impaired, 2:663
incentive-based mechanisms for improving, 2:132–134
in India, 2:566
information about, 2:303–304
issues related to, 2:181–182
land use effects on, 2:169
microbial criteria for controlling, 2:449–451
in municipal solid waste landfills, 2:163–169
oil-field brine and, 2:287
parameters for characterizing, 2:180t
protection measures for, 2:120
public awareness of, 2:194
regulations related to, 2:314–315
uncertainties in analyzing, 2:369
using luminescent bacteria and *lux* genes for determining, 2:172–176
variations in, 2:179–180
Water Quality Act of 1965, 2:194, 314
Water Quality Analysis Simulation Program (WASP5), 2:327, 329–330
Water quality impairment, urban-landscape-related, 2:191t
Water Quality Index (EPA), 2:368
Water quality indicators, 2:266–267
Water quality indices, 2:368
Water quality "ladder," 2:655
Water quality management, 2:176–184, 188
  approach to, 2:183–184
  determinants related to, 2:178
  in a forested landscape, 2:199–202
  legal considerations in, 2:183
  nonpoint source control and, 2:184–189
  in the United States, 2:193–198
  in an urban landscape, 2:189–193
Water quality modeling. *See also* Water quality models
  case studies in, 2:255–263
  of Lake Tenkiller, Oklahoma, 2:260–262
  software tools for, 2:325–328
Water quality models, 2:248–255. *See also* Water quality modeling; Water quality simulations
  Agricultural Nonpoint Source Pollution Model 2001, 2:251
  chemical principles related to, 2:269–273
  CREAMS and GLEAMS, 2:250–251

Water quality models, 2:248–255.
    See also Water quality
    modeling; Water quality
    simulations (*continued*)
  detailed, 2:249
  equations for, 2:273–278
  Erosion–Productivity Impact
    Calculator, 2:251
  Generalized Watershed Loading
    Function, 2:252–253
  hydrologic simulation program,
    2:249–250
  kinetic parameters used in, 2:276t
  mathematical framework of,
    2:273–278
  midrange, 2:248
  Nitrate Leaching and Economic
    Analysis Package, 2:252
  Pesticide Root Zone Model, 2:252
  regression methods and
    spreadsheet models, 2:253
  review of, 2:249–253
  Revised Universal Soil Loss
    Equation, 2:251
  Root Zone Water Quality Model,
    2:252
  selecting, 2:255–256
  simple, 2:248
  Soil and Water Assessment Tool,
    2:250
  Source Loading and Management
    Model, 2:253
  variables in, 2:270–272
  Water Erosion Prediction Project,
    2:251–252
  Watershed Modeling System, 2:250
Water quality monitoring, 2:177–180
Water quality policy, 2:127–128
  economic instruments for
    informing, 2:129–131
  instruments of, 2:132–134
Water quality programs, 2:315
Water quality simulations,
  2:261–262
Water quality standards, 2:267–268,
  302
Water quality studies, remote
  sensing and GIS application in,
  2:534
Water quality variables, notations
  and source terms for, 2:275t
Water quantity regulation, in India,
  2:557–558
Water recycling
  defined, 2:610
  environmental benefits of,
    2:610–613
  future of, 2:613
  pollution reduction and, 2:612
  for sensitive ecosystems, 2:612
  for wetlands and riparian habitats,
    2:612
Water reduction methods, in Arab
  states, 2:472
Water regulation, history of,
  2:193–198
Water related diseases, 2:111–112
Water resource development,
  American, 2:498–499
Water resource requirements, for
  India, 2:564–566
Water resources, 2:370
  in the Arab World, 2:470–474
  best management practices for,
    2:570–573
  Canadian, 2:656–660
  Geographic Information System
    and remote sensing in,
    2:531–536
  in India, 2:559–567
  protection of, 2:374–375
  stress on, 2:495
Water resources management,
  2:586–587
  in America, 2:608
  Chinese laws related to,
    2:488
  decentralization of, 2:473
  decision support system for,
    2:622–623
  for drought conditions, 2:576–586
  structures for, 2:567t
Water resources planning,
  integrating environmental
  impacts into, 2:520–522
Water resources systems, 2:586
  components of, 2:619
  performance evaluation of,
    2:674–678
Water resources systems analysis,
  2:683–687
  defined, 2:684
  mathematical modeling in,
    2:686–687
Water resource sustainability,
  2:624–630
  concepts related to, 2:625–627
  practices in, 2:628–629
  stages of, 2:629f
Water resource valuation, 2:653–656
  economic perspective on,
    2:653–654
  empirical estimates of, 2:654–655
Water reuse
  in agriculture, 2:496
  in industry, 2:496
Water rights, 2:500–501
  distribution of, 2:643–644
  security of, 2:644–645
Water safety, precautionary principle
  regarding, 2:600
Water sampling, 2:161–163
  security and integrity of, 2:162
Water-saving techniques, domestic,
  2:497
Water-saving technologies, in the
  Nile Basin, 2:594
Water scarcity, 2:181, 490
  conflicts over, 2:502–503
  global, 2:538–540t
  in the Nile Basin, 2:590–591
Water science. *See also* Water studies
  issues and risks related to, 2:602t
  precautionary principle regarding,
    2:595–603
  terms related to, 2:541
Water self-sufficiency, global,
  2:538–540t
Watershed, 2:548
  assessments of, 2:264–265
  water control in, 2:567
Watershed approach, 2:189
Watershed management, 2:192, 248
  plan for, 2:629
Watershed mapping/monitoring,
  remote sensing and GIS
  application in, 2:533
Watershed Modeling System (WMS),
  2:250
Watershed spatial scale, 2:613
Water shortage planning tools,
  Palmer Index, 2:581
Water-shortage plans, state, 2:583
Water, Soil and
  Hydro-Environmental Decision
  Support System
  (WATERSHEDSS), 2:328
Water source emergencies, planning
  for, 2:401
Water sources
  groundwater supplies variability
    in, 2:400–401
  little control of, 2:400–401
Water stocks, improving, 2:635
Water studies, interdisciplinary,
  2:641–642
Water supplies
  for agricultural production,
    2:617–618
  in developing countries, 2:616–617
  for energy production, 2:618
  failures in, 2:514
  political and legal impacts on,
    2:618
  state and regional, 2:613–619
  treatment of, 2:374
  types of, 2:371
  uncertainty of, 2:400–401
Water supply allocation, 2:569
Water supply management,
  2:515–516

Water supply planning/
    management, 2:614–615
Water supply quality, 2:370–376
    factors affecting, 2:373–374
Water supply subsystems, 2:515
Water Survey of Canada, 2:658
Water system operation, decisions
    about, 2:595
Water table, 2:548
WaterTechOnline portal, 2:669
Water transportation, inefficiency in,
    2:492
Water treatment
    powdered activated carbon for,
        2:86–87
    regional, 2:615–616
Water treatment system, stages in,
    2:616
Water
    use
        in California, 2:478–480
        categories of, 2:651–653
        defined, 2:548
        economic efficiency in, 2:575
        excessive, 2:492–493
        global and regional, 2:489–490
        national trends in, 2:650–651
        regulatory mechanisms regarding,
            2:490
        in the United States, 2:645–650
Water use categories, regional, 2:615t
Water use curtailment, during
    droughts, 2:583
Water use management, in the
    United States, 2:647–649
Water-use-priority system, 2:583
Water-user compliance, monitoring
    during droughts, 2:583
Water users, categories of, 2:587
Water-use sectors
    main, 2:495–497

water demand forecasting for,
    2:530
Water valuation studies, 2:655
Water vectored disease, 2:112
"Water wars," 2:649
Water washed diseases, 2:112
Waterways, gross pollution of, 2:314
Water withdrawal(s), 2:478. *See also*
    Withdrawal
    in Canada, 2:661
    by country, 2:648t
    global, 2:538–540t
    for public supplies, 2:651–652
    in the United States, 2:645–650
Weight of evidence (WOE) approach,
    2:381, 429
    to sediment quality
        characterization, 2:350–355
Well, defined, 2:548
Well contamination, in developing
    countries, 2:631
Well-spacing norms, in India, 2:558
Well water, blue baby syndrome and,
    2:219
Western blotting, 2:109
Western United States
    water markets in, 2:500
    water use in, 2:646–647
Wetlands
    passive treatment of, 2:423–426
    water recycling for, 2:612
Wetlands loss, precautionary
    principle regarding, 2:600
Whole effluent toxicity controls,
    2:382–383
Whole effluent toxicity testing
    protocols (EPA), 2:382
Wick lysimeter, 2:340
Windermere humic aqueous model
    (WHAM), 2:206

Withdrawal. *See also* Water
    withdrawal(s)
    defined, 2:548
    uses for, 2:661–662
Workplace, water conservation in,
    2:664–666
Works Progress Administration
    (WPA), 2:511, 513
World Health Organization (WHO)
    arsenic-related guidelines, 2:8
    water quality standards,
        2:267
WWTP monitoring, 2:48

XANES spectroscopy method,
    2:207
Xenobiotics, 2:42
Xeriscaping, 2:548

Yamuna. *See* River Yamuna, India
Yarra River, Australia, water quality
    software for, 2:328–330
Yarra River Water Quality Model
    (YRWQM), 2:334, 335
    linking with genetic algorithm
        software, 2:334
Yield, defined, 2:548

Zero-tension lysimeter, 2:340
Zero valent metals (ZVMs), 2:300
Zinc (Zn)
    in groundwater, 2:148
    uptake of, 2:217
ZoBell solution, 2:466–467
Zooplanktivorous fish, algal control
    and, 2:6
Zooplankton, effect of road salt on,
    2:323